MODERN ECONOMICS

An Analytical Study

Twentieth Edition

ABOUT THE AUTHOR

Dr H L Ahuja, a distinguished academician, is a distinction holder in MA Economics. He was the recipient of McConnel Levin Prize from Punjab University in 1961 for his paper "*Investment Criteria in Development Planning*." He continued his academic pursuit and received his doctorate from Delhi University in 1982.

An eminent professor, Dr Ahuja has taught Advanced Economic Theory (Micro and Macro), Development Economics and Problems of Indian Economy to postgraduate students at the Kurukshetra University. He subsequently taught at Zakir Husain Delhi College, Delhi University, where he taught Micro and Macro Economic Theories, Development Policy and Indian Economy. He was a member of the Academic Council of Delhi University for a period of five years (1987-92). A prolific writer, Dr Ahuja has more than a dozen books to his credit.

MODERN ECONOMICS

An Analytical Study

Twentieth Edition

H L AHUJA
MA, PhD (DSE)
Former Associate Professor
Department of Economics
Zakir Husain Delhi College
University of Delhi
Delhi

S Chand And Company Limited
(ISO 9001 Certified Company)

S Chand And Company Limited

(ISO 9001 Certified Company)

Head Office: D-92, Sector–2, Noida – 201301, U.P. (India), Ph. 91-120-4682700

Registered Office: A-27, 2nd Floor, Mohan Co-operative Industrial Estate, New Delhi – 110 044, Phone: 011-49731800

www.**schandpublishing.com;** e-mail: **info@schandpublishing.com**

Marketing Offices:

Chennai : Ph: 23632120; chennai@schandpublishing.com

Guwahati : Ph: 2738811, 2735640; guwahati@schandpublishing.com

Hyderabad : Ph: 40186018; hyderabad@schandpublishing.com

Jalandhar : Ph: 4645630; jalandhar@schandpublishing.com

Kolkata : Ph: 23357458, 23353914; kolkata@schandpublishing.com

Lucknow : Ph: 4003633; lucknow@schandpublishing.com

Mumbai : Ph: 25000297; mumbai@schandpublishing.com

Patna : Ph: 2260011; patna@schandpublishing.com

First Edition 1980
Subsequent Editions and Reprints 1982, 84, 86, 88, 90, 92, 96, 97, 98, 99, 2002, 2004, 2006, 2008, 2009, 2010, 2012, 2013, 2014, 2015
Twentieth Revised Edition 2016, Reprints 2018, 2020, 2021, 2022, 2023

Reprint 2025

ISBN: 978-93-525-3146-2 **Product Code:** H5ECT41ECON10ENAT16O

PRINTED IN INDIA

By Vikas Publishing House Private Limited, Plot 20/4, Site-IV, Industrial Area Sahibabad, Ghaziabad – 201 010 and Published by S Chand And Company Limited, A-27, 2nd Floor, Mohan Co-operative Industrial Estate, New Delhi – 110 044.

PREFACE TO THE TWENTIETH EDITION

It is a matter of great pleasure for me to bring out the revised twentieth edition of this comprehensive book which covers several subjects of modern economics, namely, microeconomics, macroeconomics, monetary theory and policy, international economics, public finance and fiscal policy and, above all, economics of growth and development. In the present edition, I have made exhaustive revision of various parts of the book. Among the changes made, mention may be made of the following:

VOLUME–1

1. In Chapter 3 of the introductory part, the scientific nature of economic models has been explained. In Chapter 5 of Part I, the problem of inefficiency in production and resource allocation, the concept of opportunity cost and social choice have been explained with the help of the concept of production possibility curve.

2. In Part II, concerning theory of demand, the relation between demand and utility and the issue of intertemporal choice between present consumption and future consumption and determination of saving by it with the help of indifference curves has been discussed.

3. A very important new Chapter 14A deals with the choice by an individual under conditions of risk and uncertainty and how this choice depends on his/her attitude towards risk has been clearly brought out and the tendency to get insurance by risk-averters and gambling by risk-lovers has been explained and illustrated.

VOLUME–2

4. Chapter 14, regarding unemployment and wage-price flexibility to restore full employment equilibrium, has been substantially revised and the impact of real balance effect on employment has been examined.

5. A new Chapter 12 has been devoted to the study of nature of unemployment in developing countries. In this connection, Amartya Sen's analysis of disguised unemployment in a labour-surplus economy has been explained and graphically illustrated.

6. A new Chapter 16 deals with the objectives and instruments of monetary policy.

7. In Chapter 17, RBI's approach to measure money supply, the relation between fiscal deficit and money supply, and the relation of balance of payments in the open economy with the money supply have been explained.

8. In Chapter 37, role of multinational corporations (MNCs) in economic growth of developing countries, the policy of 'Make in India' and of big push to FDI by NDA government have been discussed.

9. In Chapter 37A, trade facilitation agreement at Bali in 2013 and Doha Round declaration of WTO at Nairobi in December 2015 have been critically examined.

10. Part XIV concerning economics of growth and development has been thoroughly revised and updated. In this connection, mention may be made of (a) Sen's Capabilities Approach to Development (b) Core Values of Development (c) Human Development Index (HDI) and its Critique in Chapter 38 and (d) Myrdal's Theory of Backwash Effect on Development in Chapter 39.

11. Besides, the significant additions in the part of economics of growth and development include the critical study of the following subjects:

(a) Technological Progress and Economic Development

(b) Human Resource Development: Education and Health

(c) Income Inequality and Economic Growth incorporating Kuznets inverted-U hypothesis.

(d) Chapter 54 on Poverty and Economic Growth has been thoroughly revised. It explains the concept of poverty, its measurement, international comparison of the magnitude of poverty, causes of poverty and relation between poverty and economic growth.

(e) Chapters 57 and 58 have been substantially revised where I have discussed the important issue of environment and economic development. The concept of sustainable development and climate change has been explained at length and the recent agreement at Paris to take measures to reduce rise in global temperature to 1.5° by the end of 21st century has also been discussed.

I hope with the above revision and changes, the book will be more useful for the students of both the undergraduate and postgraduate classes of economics and commerce. Since its coverage is very comprehensive and the treatment of the subject matter is highly analytical, candidates preparing for IAS, IES and Allied Public Services examinations will find it greatly helpful for them. It incorporates the study of almost all the topics prescribed in their new syllabi.

Suggestions for further improvement of the book from students and fellow teachers are most welcome. The author can be approached through E-mail: drhlahuja@gmail.com

69, Vaishali
Pitampura
Delhi—110 034

H L AHUJA

PREFACE

Since I brought out my Advanced Economic Theory for postgraduate students of economics, there were requests and suggestions from various parts of India for writing a comprehensive text-book on Modern Economics which should cover all the three main branches of economics, namely, Microeconomics, Macroeconomics and Development Economics. Accordingly, to meet the needs of undergraduate and postgraduate students, I have endeavoured to write this book on Modern Economics incorporating new trends and tendencies in economic theory. I have taken great pains to make it an authoritative textbook on Modern Economics, both from the point of view of the subject-matter and exposition. I am quite aware that this is a tall claim. But how far I have succeeded in achieving my objective, only the students and fellow teachers can judge and assess.

I hope the book caters to the needs of not only the students of Indian Universities but also of candidates preparing for competitive and professional examinations such as IAS, IES, and other Allied Services.

Suggestions for further improvement of the book from fellow teachers and students are most welcome.

January 26, 1980 **H L AHUJA**
New Delhi

CONTENTS

VOLUME–1

MICROECONOMICS: THEORY AND APPLICATIONS

PART–I

SCOPE AND METHODOLOGY OF ECONOMICS

1. What Economics is About 3—14

Economics is a Science of Wealth – Critical Evaluation of 'Wealth Definition' of Economics – Marshall's Definition: Economics is a Science of Material Welfare – Critical Evaluation of Marshall's Definition and other Welfare Definitions – Robbins' Definition: Economics is a Science of Scarcity and Choice, Critical Evaluation of Robbins' Definition – Positive Economics, Normative Economics and Welfare Economics – Normative Economics and Value Judgements.

2. Micro and Macro Economics 15—25

Microeconomics – Importance and Uses of Microeconomics – Macroeconomics – Macroeconomics as a Study of Determination of Income and Employment – Macroeconomics and General Level of Prices – Macroeconomics and Theory of Economic Growth – Interdependence between Micro and Macro Economics.

3. The Methodology of Economics 26—41

Nature of Scientific Theory – Methods of Economic Analysis – Deductive Method – Testing of Economic Hypotheses through Statistical Methods – Inductive Method – Conclusion: Integration of Two Methods – Economic Models – Nature of Economic Laws and Generalisations – Economic Laws as Statements of Tendencies – Scientific Nature of Economic Laws – Economic Laws are less Exact and Definite than Laws of Physical Sciences – Role of Assumptions in Economics: Friedman's View.

4. Economic Statics and Dynamics 42—50

Stationary and Changing Phenomena – Economic Statics – Importance of Economic Statics – Economic Dynamics – Frisch's Time Period Analysis – Harrod's Concept of Economic Dynamics: Rates of Change Analysis – Expectations and Dynamics: Need and Significance of Economic Dynamics – Comparative Statics.

5. Central Problems of an Economy 51—65

The Economic Problem – Central Problems of an Economy – What to Produce: How to Produce – For whom to Produce – What provision should be made for Economic Growth – What, How and For Whom: Subject Matter of Microeconomics – Macroeconomics: Whether Resources are Fully Utilized? – How are the Central Problems Solved? – Production Possibility Curve – The Problem of Unemployment and Production Possibility Curve – Inefficiency in Production – Economic Growth and Shift in Production Possibility Curve and its Uses.

6. Market Economy and Role of Price Mechanism 66—85

Chief Features of a Free Market (Capitalist) Economy – Functioning of the Capitalist Economic System – Deciding What to Produce; Deciding How to Produce; Deciding for Whom to Produce: Distribution of National Product – Deciding about Rate of Economic Growth – Solving all the Problems Simultaneously – Critical Evaluation of the efficiency of Market Economy and Price Mechanism – Alternative Economic Systems – Mixed Economic System – Two Forms of Mixed Economic Systems – Indian Economy as a Mixed Economic System – Functioning of a Mixed Economy – Role of Government in a Mixed Economic System.

PART–II
THEORY OF DEMAND

7. Demand and Law of Demand **89—100**

The Meaning of Demand – Demand and Utility – The Law of Demand – Demand Schedule and Demand Curve – Market Demand Curve – Reasons for the Law of Demand: Why does Demand Curve Slope Downward? – Exceptions to the Law of Demand – Determinants of Demand – Extension and Contraction in Demand – Changes in Demand: Increase and Decrease in Demand – Demand Function and Demand Curve: Market Demand Function – Bandwagon Effect and Snob Effect.

8. Demand: Marshall's Cardinal Utility Analysis **101—115**

Introduction – Assumptions of Cardinal Utility Analysis – Law of Diminishing Marginal Utility – Applications and Uses of Diminishing Marginal Utility – Principle of Equi-marginal Utility: Consumer's Equilibrium – Limitations of the Law of Equi-marginal Utility – Derivation of the Demand Curve and the Law of Demand – Critical Evaluation of Marshall's Cardinal Utility Analysis.

9. Indifference Curve Analysis of Demand **116—164**

Indifference Curve Approach – What are Indifference Curves? Indifference Map – Marginal Rate of Substitution – Principle of Diminishing Marginal Rate of Substitution – Assumptions of Indifference Curves – Properties of Indifference Curves – Indifference Curves of Perfect Substitutes and Perfect Complements – Budget Line – Budget Space – Consumer's Equilibrium: Maximising Satisfaction – Second Order Condition for Consumer's Equilibrium – Income Effect: Income Consumption Curve – Income Consumption Curve and Engel Curve – Substitution Effect – Price Effect: Price Consumption Curve – Breaking up Price Effect into Income and Substitution Effects – Price-Demand Relationship: Deriving Law of Demand – Price-Demand Relationship: Inferior Goods – Price Demand Relationship: Giffen Goods or Giffen Paradox – Derivation of Individual's Demand Curve from Indifference Curve Analysis – Consumer's Equilibrium and Corner Solution – Complements and Substitutes.

Appendix B to Chapter 9: Slutsky Substitution Effect **165—171**

Slutsky Substitution Effect for a Fall in Price – Slutsky Substitution Effect for a Rise in Price – Merits and Demerits of Hicksian and Slutsky Methods – Price Effect Broken up into Income and Substitution Effects: Slutsky Method – Slutsky Equation.

10. Marshall's Cardinal Utility Analysis vs. Indifference Curve Analysis **172—179**

Superiority of Indifference Curve Analysis – Is Indifference Curve Analysis "Old Wine in a New Bottle"? – Critique of Indifference Curve Analysis.

11. Applications and Uses of Indifference Curves **180—198**

Effect of Subsidies to Consumers: Price Subsidy vs Cash Subsidy – Food-Stamp Programme: In–kind Food Subsidy – Direct Tax Versus Indirect Tax – Effect of Rationing on Consumer's Welfare – Income-leisure Choice – Need for Higher Overtime Wage Rate – Wage Offer Curve and the Supply of Labour – Income Effect and Substitution Effect of the Change in Wage Rate – Backward-Bending Supply Curve of Labour – Intertemporal Choice between Present Consumption and Future Consumption: Indifference Curve Analysis.

12. Revealed Preference Theory of Demand **199—209**

Preference Hypothesis and Strong Ordering – Deriving Demand Theorem from Revealed Preference Hypothesis – Breaking up of Price Effect into Substitution Effect and Income Effect – Critical Appraisal of Revealed Preference Theory.

13. Elasticity of Demand **210—246**

Various Concepts of Demand Elasticity – Price Elasticity of Demand – Measurement of Price Elasticity – Total Revenue, Total Expenditure and Price Elasticity of Demand – Measurement of Price Elasticity of Demand at a Point on the Demand Curve – Comparing Price Elasticity of two Demand Curves with Different Slopes – Determinants of Price Elasticity of Demand

– Importance of Price Elasticity of Demand – Cross Elasticity of Demand – Income Elasticity of Demand – The Elasticity of Substitution – Relationship between Price Elasticity, Income Elasticity and Substitution Elasticity – Price Consumption Curve and Price Elasticity of Demand.

14. Consumer's Surplus **247—259**

Meaning of Consumer Surplus – Marshall's Measure of Consumer Surplus – Measurement of Consumer's Surplus through Indifference Curve Analysis – Critical Evaluation of the Concept of Consumer's Surplus – Uses and Applications of Consumer Surplus: Water-Diamond Paradox; Evaluating Loss of Benefit from Tax; Evaluating Gain from a Subsidy; Use of Consumer Surplus in Cost-benefit Analysis.

14A. Choice Under Risk and Uncertainty **260—270**

Introduction – The Concept of Risk – Measuring Risk: Probability of an Outcome – Expected Value and Expected Utility – Attitude Toward Risk: Risk Averter; Risk Lover and Risk Neutral – Insurance and Risk Premium – Risk Lover and Gambling – Management of Risk: Reducing Risk.

PART–III
THEORY OF PRODUCTION AND COST

15. Factors of Production **273—284**

Factors of Production: Meaning – Classification of Factors of Production – Land: Meaning and Significance – Land as Renewable and Non-renewable Resource – Inelastic Supply of Land – Capital: Physical and Human – Meaning of Capital – Fixed Capital and Working Capital – Human Capital – Role of Capital – Human Capital, Education and Economic Growth – Supply Curve of Saving – Factors Affecting Saving – Labour – Division of Labour – Entrepreneur – Functions of Entrepreneur.

16. The Theory of Production: Returns to a Factor **285—297**

Importance of the Theory of Production – Production Function: Transforming Inputs into Output – Concepts of Product – Law of Variable Proportions – Three Stages of the Law of Variable Proportions – Causes of Diminishing Marginal Returns to a Factor – Applicability of the Law of Diminishing Returns – Technological Progress and Diminishing Returns.

17. Production Functions with Two Variable Factors **298—323**

Isoquants – Marginal Rate of Technical Substitution – General Properties of Isoquants – Isoquants of Perfect Substitutes and Complements – Linear Homogeneous Production Function – Cobb-Douglas Production Function – Elasticity of Technical Substitution (Between Factors) – Returns to Scale – Constant Returns to Scale – Divisibility of Factors, Proportionality and Constant Returns to Scale – Increasing Returns to Scale – Decreasing Returns to Scale – Varying Returns to Scale in a Single Production Process – Cobb-Douglas Production Function and Returns to Scale – Returns to Scale and Marginal Returns to a Variable Factor.

18. Optimum Factor Combination **324—337**

The Economic Region of Production and Ridge Lines – Iso-cost Line – Optimum or Least-cost Combination of Factors – Output Maximisation for a Given Level of Outlay – Expansion Path – Expansion Path of a Linear Homogeneous Production Function – Factor Substitution and Changes in Factor Prices.

19. Cost of Production and Cost Curves **338—373**

The Concepts of Cost: Accounting Costs and Economic Costs; Opportunity Cost – Private Cost versus Social Cost – Short-Run Costs: – Total Fixed and Variable Costs – The Short-run Average Cost Curve – Marginal Cost – The Relationship between the Average and Marginal Cost Curves – Derivation of Long-run Average Cost Curve – Why Long-run Average Cost Curve is of U-shape? – The Concept of Optimum Cost – Derivation of Long-run Marginal Cost Curve, External Economies, External Diseconomies and Cost Curves – Learning Curve

– L-shaped Long-run Average Cost Curve: Modern Cost Theory: A Critical Evaluation of L-shaped Long-run Average Cost Curve Concept – Numerical Problems.

20. **Supply and its Elasticity** 374—386
The Meaning of Supply – Supply Function – Law of Supply – Explanation of the Law of Supply: Why does Supply Curve Generally Slope Upward? – Assumptions of Law of Supply – The Concept of Supply Curve and Perfect Competition – Changes in Supply: Increase and decrease in Supply – Factors Determining Supply – Elasticity of Supply – Measurement of point Elasticity of Supply – Factors Determining Elasticity of Supply.

PART–IV

THEORY OF FIRM AND PRICING IN DIFFERENT MARKET STRUCTURES

21. **Main Market Forms and Concepts of Revenue** 389—408
Meaning of Market – Classification of Market Forms: Perfect Competition; Monopolistic Competition; Oligopoly; Monopoly – Market Classification and Cross Elasticity of Demand – Concepts of Average Revenue and Marginal Revenue – Relationship between AR and MR Curves – Average Revenue, Marginal Revenue and Price Elasticity of Demand – Three Types of Revenue (AR, MR, TR) and Price Elasticity (E).

22. **Objectives and Equilibrium of the Firm: A General Analysis** 409—427
Meaning of Firm's Equilibrium – Objectives of the Firm: Profit Maximisation Objective – A Critique of Profit Maximisation Objective – Achieving a Steady flow of Secure Profits in the Long run – Hall and Hitch's Mark-up Pricing Approach: Seeking Normal Profits – Sales Maximisation – Utility Maximization – Growth Maximisation – Satisficing Behaviour – Case for the Objective of Maximising Profits – Equilibrium of the Firm: Maximising Profits – Firm's Equilibrium with Total Revenue – Total Cost Approach; Marginal Revenue-Marginal Cost Approach – Second Order Condition for Equilibrium of the Firm – Numerical Problems.

23. **Equilibrium of the Firm and Industry under Perfect Competition** 428—451
Meaning and Conditions of Perfect Competition – The Demand Curve of a Product Facing a Perfectly Competitive Firm – Profit Maximisation by a Competitive Firm: TR-TC Approach – Equilibrium of the Firm under Perfect Competition – Long-run Equilibrium of the Firm under Perfect Competition – Equilibrium of the Industry – The Concept of Supply Curve – Short-run Supply Curve of the Perfectly Competitive Firm – Short-run Supply Curve of the Competitive Industry – Long-run Supply Curve of the Industry under Perfect Competition – Numerical Problems.

24. **Price Determination under Perfect Competition** 452—476
Two Approaches in Price Theory: Partial and General Equilibrium Analyses – Price Determination: Equilibrium between Demand and Supply – Changes in Market Equilibrium – Demand-Supply Model of Pricing: Mathematical Analysis – Importance of Time Element in the Theory of Price – Determination of Market Price – Determination of Short-run Price – Determination of Normal Price and Derivation of Long-run Supply Curve of the Industry – Producer Surplus and Efficiency of Competitive Market – Economic Efficiency and Equity.

24A. **Stability of Equilibrium and Cobweb Model** 477—484
Stable and Unstable Equilibrium – Walras Price Adjustment Approach – Stability of Equilibrium: Marshall's Quantity Adjustment Approach – Static Stability vs Dynamic Stability – Dynamic Stability with Lagged Adjustment: Cobweb Model.

25. **Applications of Demand and Supply Analysis** 485—499
Price Control and Rationing – Rent Control – Minimum Support Price – Incidence of Indirect Taxes – Price and Income Stabilisation in Agriculture – Effect of Unplanned Fluctuations in Agricultural Supply on Prices; Effect of Unplanned Fluctuations in Production on Incomes of the Farmers; Stabilisation of Agricultural Prices and Incomes – Buffer Stock Operations by the Government.

26. Price and Output under Monopoly 500— 530

Monopoly: Its Meaning and Conditions – The Nature of Demand and Marginal Revenue Curves under Monopoly – Price-output Equilibrium under Monopoly – Long-run Equilibrium Under Monopoly – Sources or Reasons of Monopoly Power – Monopoly Equilibrium and Perfectly Competitive Equilibrium Compared – Monopoly and Inefficiency: Social Cost of Monopoly – Dead-weight Loss (Social Cost) under Monopoly in Case of Increasing Marginal Cost – Absence of Supply curve under Monopoly – Measurement of Monopoly Power – The Rule of Thumb for Pricing – Regulation of Monopoly – Numerical Problems.

27. Price Discrimination 531—544

Meaning of Price Discrimination – When is Price Discrimination Possible? – When is Price Discrimination Profitable? – Equilibrium under Price Discrimination – Equilibrium under Price Discrimination in the Dumping Case – Comparing between Simple Monopoly Output and Discriminating Monopoly Output – Price Discrimination and Social Welfare – Numerical Problems.

28. Price under Monopolistic Competition 545—575

Imperfect Competition: Monopolistic Competition and Oligopoly – Product Differentiation and Monopolistc Competition – Important Features of Monopolistic Competition – The Nature of Demand and Marginal Revenue Curves under Monopolistic Competition – Price-Output Equilibrium under Monopolistic Competition: Individual Firm's Equilibrium under Monopolistic Competition – Selling Costs and Advertising – Optimum Level of Advertising Outlay (Selling Cost): With Price and Product Design as Constants – Optimum Level of Advertising Expenditure with both Price and Output as Variable – Effect of Advertising (Selling Costs) on Elasticity of Demand – Effect of Advertising (Selling Costs) on Price and Output – Excess Capacity under Monopolistic or Imperfect Competition – Causes of Excess Capacity-benefits of Excess Capacity- equilibrium under Monopolistic Competition: Chamberlin's Alternative Approach Price-output Equilibrium under Monopolistic Competition Compared with that under Perfect Competition – Monopolistic Competition and Economic efficiency.

29. Price and Output under Oligopoly 576—598

Characteristics of Oligopoly – Are Price and Output under Oligopoly Indeterminate? – Various Approaches to Determination of Price and Output under Oligopoly – Collusive Oligopoly: Cartels – Price Leadership – Price-output Determination under Low-cost Price Leadership – Price-output Determination under Price Leadership by the Dominant Firm – Difficulties of Price Leadership – The Kinked Demand Curve Theory of Oligopoly – Why Price Rigidity under Oligopoly? Critical Appraisal of Kinked Demand Curve Theory – Prisoner's Dilemma and Oligopoly Theory: Prisoner's Dilemma and Oligopolistic Behaviour: Instability of a Cartel – The Theory of Contestable Markets.

Appendix to Chapter 29: Classical Models of Oligopoly 599—613

Cournot's Duopoly Model – Cournot's Approach to Equilibrium of the Duopolists – Cournot's Duopoly Equilibrium – A Critique of Cournot's Oligopoly Model – Numerical Problems – Bertrand's Duopoly Model – Edgeworth's Duopoly Model – Comments over Classical Models of Duopoly – Chamberlin's Oligopoly Model.

30. Sales Maximisation Model of Oligopoly 614—619

Sales Maximisation vs. Profit Maximisation – Sales Maximisation: Price-Output Determination – Emphasis on Non-Price Competition – Sales Maximisation Model: Optimal Advertising Expenditure – Critical Appraisal of Sales Maximisation Theory.

31. Theory of Games and Competitive Strategy 620—630

Cooperative and Non-cooperative Games – Dominant Strategy – Choice of an Optimal Strategy in the Absence of a Dominant Strategy – The Nash Equilibrium – Prisoner's Dilemma and Oligopoly Theory – Repeated Games and TIT-for-TAT Strategy – Strategic Moves: Threats, Commitment and Credibility – Entry Deterrence.

PART–V
FACTOR PRICING AND INCOME DISTRIBUTION

32. Factor Pricing and Income Distribution: A General View **633—667**

Functional vs Personal Distribution – Theory of Distribution as a Special Case of the Theory of Price – Micro and Macro Theories of Distribution – Marginal Productivity Theory of Distribution – Critical Evaluation of Marginal Productivity Theory – Concepts of Factor Productivity-Marginal Revenue Product (MRP) – Value of Marginal Product (VMP) – A Competitive Firm's Equilibrium regarding Factor Employment: General Conditions – Demand Curve for a Factor – Competitive Industry's Demand Curve for a Factor – The Supply of Factors – Competitive Firm's Demand Curve for a Single Variable Factor – Firm's Demand Curve for a Factor: With all Factors (inputs) Variable – Determination of Factor Prices under Perfect Competition – Determinants of Demand for Factors – Euler's Theorem and Product Exhaustion Problem – Hicks-Samuelson's Solution to the Product Exhaustion Problem – Pricing of Factors in Imperfectly Competitive Markets – Determination of Factor Prices under Monopsony.

33. Wage Determination in Competitive and Imperfectly Competitive Markets **668 —691**

Supply Curve of Labour – Wage Determination under Perfect Competition in the Labour Market – Wage Determination under Monopsony – Exploitation of Labour – Role of Trade Unions and Collective Bargaining in Raising Wages–Wage Determination under Bilateral Monopoly when Monopsonist Buyer Faces a Monopolist Seller – Wage Differentials – Numerical Problems.

Appendix to Chapter 33: An Application of the Theory of Wages **692—696**

Fixation of Minimum Wages – Effect on Income of Workers – Case For Minimum Wage.

34. The Theory of Rent **697—715**

Concepts of Rent – Ricardian Theory of Rent: Scarcity Rent and Differential Rent – Critical Evaluation of Ricardian Theory of Rent – Explaining Determination of Land Rent through Demand and Supply – Land Rent, Cost and Price – Modern Theory of Rent: Economic Rent as a Surplus over Transfer Earnings – Economic Rent Arises when the Supply of a Factor is less than Perfectly Elastic – Quasi Rent.

35. Alternative Theories of Interest **716—733**

Classical Theory of Interest – Critical Appraisal of the Classical Theory of Interest – Loanable Funds Theory of Interest – Critical Evaluation of Loanable Funds Theory – Keynes's Liquidity Preference Theory of Interest – Critical Appraisal of Keynes's Liquidity Preference Theory of Interest.

36. The Theory of Profits **734—742**

Profits as a Dynamics Surplus: Clark's Dynamic Theory of Profits – Schumpeter's Innovations Theory of Profits – Risk, Uncertainty and Profits: Knight's Theory of Profits – Role and Functions of Profits – Social Responsibility of Business and Profit Motive.

PART–VI
GENERAL EQUILIBRIUM ANALYSIS AND WELFARE ECONOMICS

37. General Equilibrium Analysis **745—760**

Partial Equilibrium Analysis and General Equilibrium Analysis – General Equilibrium of Exchange and Consumption: Distribution of Goods between Individuals – General Equilibrium of Production – Edgeworth Production Box and General Equilibrium of Production – Transformation Curve and General Equilibrium of Production – General Equilibrium of Production and Exchange (Consumption) – General Equilibrium and

Perfect Competition – Perfect Competition and General Equilibrium of Exchange – Perfect Competition and General Equilibrium of Production – Perfect Competition and General Equilibrium of Production and Exchange – General equilibrium determines only relative prices.

38. Welfare Economics: An Introduction **761—765**

What Welfare Economics is about – Three Concepts of Social Welfare – Role of Value Judgements in Welfare Economics.

39. Economic Efficiency and Pareto Optimality **766—780**

Notion of Pareto Optimality and Economic Efficiency – Pareto Criterion of Social Welfare – Marginal Conditions of Pareto Optimality – The Optimum Distribution of Products among the Consumers: Efficiency in Exchange – Optimum Allocation of Factors: Pareto Efficiency in Production – Optimum Direction of Production: Efficiency in Product Mix – The Second Order and Total Conditions of Pareto Optimality – A Critical Evaluation of Pareto Criterion and Pareto Optimality – Prof. Amartya Sen's Critique of Pareto Optimality – Perfectly Competitive Equilibrium and Pareto Optimality – Fundamental Theorem of Welfare Economics and its Critique.

40. Market Failures, Externalities and Public Goods **781—792**

Monopoly as an Obstacle to the Attainment of Pareto Optimality – Externalities and Market Failure – How Externalities Cause Market Failure – Public Goods and Market Failure – Free-rider's Problem and Public Goods.

41. New Welfare Economics: Compensation Principle **793—799**

Kaldor-Hicks Welfare Criterion: Compensation Principle – Scitovsky Paradox – Scitovsky's Double Criterion of Welfare – A Critique of the Compensation Principle.

42. Social Welfare Function **800—805**

Bergson-Samuelson Social Welfare Function – Social Welfare Function and Value Judgements – Grand Utility Possibility Frontier and Position of Constrained Bliss – A Critical Evaluation of Bergson -Samuelson Social Welfare Function.

PART–VII
THEORY OF INTERNATIONAL TRADE

43. International Trade: Comparative Cost Theory **809—819**

Theory of Absolute Advantage – The Theory of Comparative Cost – Comparative Cost Theory: Opportunity Cost Approach – Critical Appraisal of Comparative Cost Theory – Conclusion.

44. Heckscher-Ohlin Theory of International Trade **820—831**

Heckscher-Ohlin Theory – Equalisation of Factor Prices – Critical Evaluation of Heckscher-Ohlin Theory of International Trade – Gains from Trade: Static Gains from Trade – Dynamic Gains from Trade: International Trade and Economic Growth.

45. Terms of Trade **832—840**

Concepts of Terms of Trade: Net Barter Terms of Trade; Gross Barter Terms of Trade; Income Terms of Trade – Determination of Terms of Trade: Theory of Reciprocal Demand – Critical Evaluation of the Reciprocal Demand Theory – Determination of Terms of Trade and Offer Curves – Effect of Tariff on Terms of Trade.

VOLUME–2

MACROECONOMICS AND ECONOMICS OF GROWTH AND DEVELOPMENT

PART–VIII

MACROECONOMICS: THEORY OF INCOME AND EMPLOYMENT

1. Nature and Scope of Macroeconomics 3—13

What is Macroeconomics – Say's Law and Keynesian Revolution – Issues in Macroeconomics: Determination of National Income and Employment; General Price Level and Inflation; Business Cycles; Stagflation; Economic Growth – Balance of Payments and Exchange Rate – Post-Keynesian Schools of Macroeconomic Thought: Monetarism; Supply-side Economics; New Classical Macroeconomics: Rational Expectations Theory; New Keynesian Economics – Why a Separate Study of Macroeconomics – Importance of Macroeconomics.

2. Circular Flow of Income and National Income Accounting 14—44

Meaning of National Income – Circular Flow of Income – Money Flows in the Four Sector Open Economy Adding foreign Sector – National Income and National Product – Concepts of National Income – Measurement of National Income: Value Added Method; Income Method; Expenditure Method – Nominal GNP and Real GNP – Limitations of GDP Concept as a Measure of Social Welfare – The Concept of Green GDP.

3. Classical Theory of Income and Employment 45—62

Economy in the Long Run: The Classical Full Employment Model – Classical Theory of Employment: An Introductory Analysis: Say's Law and Classical Theory; Wage-Price Flexibility and Full-employment – Classical Theory of Employment and Output (Income): A Formal Model: Determination of Income and Employment in the Short Run without Saving and Investment; Say's Law and No Deficiency of Demand – Classical Model: Determination of Income and Employment with Saving and Investment; Classical Theory of Income and Employment; Aggregate Demand, Money and Prices – Classical Aggregate Supply Curve – Classical Theory of Output and Employment : Complete Classical Model – Neutrality of Money and Classical Dichotomy – Keynes's Critique of Say's Law and Classical Theory.

4. Keynes's Theory of Employment and Unemployment 63—73

Keynesian Revolution – Keynesian Theory of Employment: Principle of Effective Demand: Aggregates Supply Function; Aggregate Demand Function; Determination of the Equilibrium Level of Employment by Effective Demand; Effective Demand and the Determination of Employment; Underemployment Equilibrium – The Problem of Demand Deficiency – Summary of Keynes's Theory of Employment – Involuntary Unemployment – Keynes' Money-Wage Rigidity Model.

5. Determination of National Income: Keynes's Basic Two-sector Model 74—96

Keynes's Income Expenditure Approach – Aggregate expenditure (with a fixed price level) – Aggregate Output – Equilibrium Level of National Income – Principle of Effective Demand – Equilibrium is not Necessarily established at full-Employment Level – Determination of Equilibrium Level of National Income – Algebraic Analysis – How to Overcome Recession – Shifting Aggregate Expenditure Curve Upward – Determination of National Income: Saving-Investment Approach – What Causes Depression or Recession – Keynes's View – Policy Implications of Keynes's Theory of Income and Employment – Concepts of Inflationary and Deflationary Gaps – Some Numerical Problems.

Appendix to Chapter 5: Keynes's Complete Macroeconomic Model: Integrating Goods Market with Money Market 97—99

Linking Goods Market with Money Market – Keynes's Complete Macroeconomic Model.

5A Determination of National Income in an Open Economy with Government and Foreign Trade: Three Sector and Four Sector Models* **100—120**

Determination of National Income with Government: Three Sector Model – The Lump-tax Multiplier – Balanced Budget Multiplier – Determination of National Income in an Open Economy: The four Sector Model – Trade Balance (Net Exports) and Foreign Capital Flows.

6. Consumption Function **121—140**

Concept of Consumption Function – Average and Marginal Propensity to consume – Saving Function – Keynes's Theory of Consumption – Determinants of Propensity to Consume – Life Cycle and Permanent Consumption Theories of Consumption – Consumption Function Puzzle: Keynes's Consumption Function and Kuznets' Findings – Importance of Consumption Function – Important Features of Keynes' Consumption Function – Consumption Function Puzzle: Keynes' Consumption Function and Kuznets' Findings.

7. Post-Keynesian Theories of Consumption* **141—152**

Relative Income Theory of Consumption: Demonstration Effect, Ratchet Effect –Life Cycle Theory of Consumption – Permanent Income Theory of Consumption – Conclusion.

8. Investment Demand **153—166**

Meaning of Investment – Autonomous Investment and Induced Investment –Determinants of Investment: Marginal Efficiency of Capital, Rate of Interest and Investment Demand Curve – Business Expectations and Investment demand Curve – Factors Causing Shift in Investment Demand Curve – Accelerator Theory of Investment – Relationship Between Saving and Investment: Ex-post Saving and Ex-post Investment are always equal; Ex-ante Saving and Ex-ante Investment are equal only in equilibrium.

9. Theory of Multiplier **167—181**

The Concept of Multiplier – Derivation of Investment Multiplier: Algebraic Derivation of Multiplier, Calculating the size or value of multiplier, Assumptions of multiplier theory – Diagrammatic Representation of Multiplier – Leakages in the Multiplier Process – Importance of the Concept of Multiplier – The Paradox of Thrift – The Keynesian Explanation of the Great Depression: The Impact of Multiplier – Limitations of the working of Keynesian Multiplier in the Developing Countries.

10. Aggregate Demand and Aggregate Supply Model (With Price Flexibility) **182—201**

AD-AS Model with Flexible Prices – Aggregate Demand – Derivation of Aggregate Demand Curve from Income − expenditure Analysis – Shift in Aggregate Demand Curve – Aggregate Supply Curve: Shifts in Aggregate Supply Curve, Changes in Long-run Aggregate Supply Curve – Determination of GDP and Price Level: AS-AD Model – Multiplier with Changes in Price Level: AS-MD Model – Shift in Aggregate Supply and Stagflation.

11. Unemployment, Full Employment and Price-wage Flexibility **202—215**

Meaning of Unemployment – Types of Unemployment: Frictional Unemployment; Structural Unemployment; Cyclical Unemployment – The Concept of Full Employment – Wage Cut and Employment: Classical View – Keynes' Critique of Classical View – A General Equilibrium Analysis of Wage Cut and Employment, Real Balance Effect or Pigou Effect – Insider-Outsider model – Anti-recessionary Policy to remove Cyclical Unemployment.

12. Unemployment in Developing Countries with Special Reference to India **216—228**

Unemployment in Developing Countries: Nature and Causes – Wage-goods Constraint – Causes of Unemployment in the Indian Economy – Measurement of Unemployment in the Indian Economy: Measurement of Unemployment in India: Dantwala Committee Approach – Disguised Unemployment: Prof. Amartya Sen's Analysis – Employment Strategy to Reduce Unemployment.

PART–IX
MONETARY DEMAND AND SUPPLY

13. Money: Nature, Functions and Role 231—242

Difficulties of Barter System and Invention of Money – Evolution of Money – Definition of Money – Functions of Money – Forms of Money – Modern Paper Currency System or Managed Currency Standard – Importance of Money – Role of Money in Economic Development of the Developing Countries.

14. Credit and Commercial Banking 243—256

Credit Creation by Banks: Meaning – Functions of Bank Credit – Purposes or End-uses of Credit – Origin and Evolution of Commercial Banking – Balance Sheet of a Bank: Liabilities and Assets Structure – Functions of Commercial Banks – Role of Commercial Banks in Economic Development – Credit Creation by Banks: Balance Sheet Approach – Limitations on the Credit-Creating Power of the Banks.

15. Central Banking 257—266

Principles of Central Banking – Functions of Central Bank – Methods of Credit Control – Selective Credit Controls.

16. Objectives and Instruments of Monetary Policy 267—276

Objectives of Monetary Policy: Price Stability; Economic Growth – Instruments of Monetary Policy: Bank Rate Policy, Repo rate in India – Instruments of Bank Rate and Repo rate Policy, Limitations of Open Market Operations, Changing the Cash Reserve Ratio – Selective Credit Controls.

16A Role of Monetary Policy in Economic Growth 277—280

Monetary Policy and Savings – Monetary Policy and Investment – Allocation of Funds – Monetary Policy to Check Inflation.

17. Supply of Money and its Determinants 281—295

Importance of Money Supply – The Concept of Money Supply and its Measurement – Four Measures of Money Supply – Determinants of Money Supply – Factors Determining Money Supply: RBIs Approach – Fiscal Deficits and Money Supply – Money Supply and the Open Economy – Sterilization by the Central Bank.

18. Demand for Money and Keynes's Liquidity Preference Theory of Interest 296—307

Fisher's Transactions Approach to Demand for Money – Cambridge Cash-Balance Theory of Demand for Money – Keynes's Theory of Demand for Money – Keynes's Liquidity Preference theory of Rate of Interest.

19. Post-Keynesian Theories of Demand for Money 308—316

Tobin's Portfolio Approach to Demand for Money – Tobin's Liquidity Preference Function – Baumol's Inventory Approach to Transactions Demand for Money: Baumol's Analysis of Transactions Demand for Money – Friedman's Theory of Demand for Money.

PART–X
MONEY, PRICES AND INFLATION

20. Money and Prices: Quantity Theory of Money 319—328

Value of Money and Price Level – Quantity Theory of Money: Fisher's Transactions Approach – Quantity Theory of Money: Income version – Quantity Theory of Money: Cambridge Cash-Balance Approach – Keynes's Critique of Quantity Theory of Money.

21. Keynesian's Monetary Theory: Money, Income and Prices 329—337

Keynes's Monetary Theory: Integrating Money Market with Goods Market – Keynes' Theory of Money and Prices – Money Supply, Aggregate Demand and Price Level.

22. Monetarism and Friedman's Modern Quantity Theory of Money 338—359

Monetarism: An Introduction – Demand for Money and Friedman's Restatement Quantity Theory of Money – Determination of Nominal Income: Friedman's Approach: Increase in Money Supply and Changes in National Income Price Level – Long-run Impact of Expansion in Money Supply: Views of Friedman and other Monetarists – Key propositions of Monetarism – A critique of Monetarism –Monetarism and Keynesianism Compared.

23. Inflation and Hyperinflation: Causes, Social Costs, Effects and Cure **360—390**
Inflation: Meaning and Causes – Demand-Pull Inflation – Cost-Push Inflation – Monetarist Theory of Inflation – Structuralist Theory of Inflation – Inflation in Developing Countries: Demand-Pull or Cost Push Inflation – Social Costs of Inflation – Effects of Inflation – Effect of inflation on Long-Run Economic Growth – Hyperinflation and Economic Crisis – Measures to Control Demand-Pull Inflation.

PART–XI
IS-LM CURVES MODEL, PHILLIPS CURVE CONCEPT AND BUSINESS CYCLES

24. The IS-LM Curve Model **393—409**
The Goods Market and Money Market: Links between them – Goods Market Equilibrium: The Derivation of the IS Curve – Shift in IS Curve – Money Market Equilibrium: Derivation of LM Curve – Shift in the LM Curve – Simultaneous Equilibrium of the Goods Market and Money Market – Critique of the IS-LM Curve Model – IS-LM Curve Model: Explaining Roles of Government's Fiscal and Monetary Policies – Effect of Fiscal Policy – Effectiveness of Fiscal Policy: Crowding out Effect – Monetary Accommodation – Impact of Monetary Policy – Effectiveness of Monetary Policy.

25. Inflation and Unemployment: Phillips Curve and Rational Expectations Theory **410—418**
Inflation-unemployment Trade-off: Philips Curve – Explanation of Phillips Curve – Collapse of Phillips Curve (1971-91) – Causes of Shift in Phillips Curve – Natural Rate Hypothesis and Adaptive Expectations: Friedman's Views Regarding Phillips Curve – Long-run Phillips Curve: Natural Rate Hypothesis – Rational Expectations Theory.

26. Stagflation and Supply-side Economics **419—430**
Stagflation–Causes of Stagflation: Adverse Supply Shocks – Inflationary Expectations – End of Stagflation in the USA: 1982-88 – Supply-side Economics – Basic Propositions of Supply-side Economics: Taxation and Labour Supply; Incentives to Save and Invest; Cost-push Effect of the Tax Wedge–Tax Revenue and Laffer Curve – Reaganomics and Supply-side Economics: Reduction of Taxes; Reducing the Burden of Government Regulations – A Critical Appraisal of Supply-Side Economics; The Threat of Inflation: Demand-side Effects of Tax Cuts; Increase in Budget Deficits – Effect on the Distribution of Income.

27. Business Cycles **431—451**
Phases of Business Cycles – Features of Business Cycles – Theories of Business Cycles: Sun-spot Theory; Hawtrey's Monetary Theory of Business Cycles; Under-consumption Theory – Keynes's Theory of Business Cycles – Samuelson's Model of Business Cycles: Interaction Between Multiplier and Accelerator – Hicks' Theory of Business Cycles.

27A. Friedman's Monetarist and New Classical (Rational Expectations) Theories of Business Cycles **452—459**
Friedman's Monetarist Theory of Business Cycles: Explaining Expansion; Importance of Lags; Critical Evaluation – Lucas' New Classical (Rational Expectations) Theory of Business Cycles: Introduction; Rational Expectations: Explaining Recession – Rational Expectations: Explaining Expansion – Critical Evaluation.

28. Economic Stabilisation: Fiscal Policy **460—470**
Macroeconomic Policy and Stabilisation – Goals of Macroeconomic Policy – Discretionary Fiscal Policy for Stabilisation – Financing Increase in Government Expenditures and Budget Deficit – Reduction in Taxes to Overcome Recession – Fiscal Policy Options: Increase in Government Expenditure or Reduction in Taxes – Fiscal Policy to Control Inflation – Disposing of Budget Surplus – Non-discretionary Fiscal Policy: Automatic Stabilisers – Crowding Out and Effectiveness of Fiscal Policy.

29. Economic Stabilisation: Monetary Policy 471—481

Tools of Monetary Policy – Expansionary Monetary Policy to Cure Recession or Depression – How Expansionary Monetary Policy Works: Keynesian Transmission Mechanism – Tight Monetary Policy to Control Inflation – How the Tight Monetary Policy Works: Keynesian Transmission Mechanism – Liquidity Trap and Ineffectiveness of Monetary Policy – Monetary Policy: Monetarist View – Monetary Rule: Monetary Policy Prescription – The Policy of Quantitative Easing (QE).

PART–XII
GOVERNMENT AND THE MACROECONOMY: PUBLIC FINANCE AND FISCAL POLICY

30. Public Finance: Public Expenditure 485—494

What is Public Finance? – Public Expenditure – Importance of Public Expenditure – Classification of Public Expenditure – Growth of Public Expenditure – Effects of Public Expenditure on Production and Distribution: Public Expenditure and Production; Public Expenditure on Distribution.

31. Public Revenue: Taxation and Budget 495—510

Government Budget Constraint – What is Tax? – Classification of Taxes – Specific and Ad Valorem Taxes – Progressive, Proportional and Regressive Taxes – Principles or Canons of a Good Tax System – Principle of Equity in Taxation – Benefits Received Theory – Ability to Pay Theory – Direct Taxes vs Indirect Taxes.

32. Incidence of Taxation 511—517

Shifting the Burden of a Tax – Incidence of Indirect Taxation and Elasticity of Demand – Elasticity of Supply and Incidence of a Tax – Incidence of an Indirect Tax and Cost Conditions of the Industry.

33. Fiscal Policy and Economic Growth in a Developing Country 518—542

Objectives of Fiscal Policy in Developing Countries – Role of Fiscal Policy for Mobilisation of Resources in Developing Countries – Taxation and Mobilisation of Resources for Economic Growth – Role of Taxation in Promoting Private Savings and Investment – Government Borrowing and Resource Mobilisation for Economic Development – Role of debt-financing (*i.e.*, Borrowing) in Resources Mobilisation: The Case of India – Money Financing and Resource Mobilisation for Economic Growth – Fiscal Policy, Equity and Price Stability; Fiscal Policy and Equitable Income Distribution – Fiscal Policy for Achieving Price Stability – Fiscal Policy and Fiscal deficit in India – Revenue Deficit – Fiscal Deficit – Implications of large Fiscal Deficit – Primary Deficit – Monetisation of Fiscal Deficit – Measures for Fiscal Consolidation – Fiscal Deficit and Economic Growth.

PART–XIII
OPEN ECONOMY MACROECONOMICS

34. Balance of Payments 545—561

Balance of Trade and Balance of Payments – Distinction between Current Account and Capital Account – Determinants of Balance of Payment– Balance of Payments on Current Account – Balance of Payments on Capital Account – Does Balance of Payments always balance? – Equilibrium and Disequilibrium in the Balance of Payments – Causes of Disequilibrium in the Balance of Payments: Cyclical Disequilibrium; Secular or Long-run Disequilibrium; Technological Disequilibrium; Structural Disequilibrium – How Disequilibrium can be Corrected? – Globalisation, Capital Flows and Balance of Payments.

35. Foreign Exchange Rate 562—588

Foreign Exchange Rate and Foreign Exchange Market – Floating (Flexible) and Fixed Exchange Rate System – Appreciation and Depreciation of Currencies – Determination of Exchange Rate – Changes in Exchange Rate – Factors Affecting Exchange Rate – Purchasing Power Parity Theory – Foreign Exchange Rate and Balance of Payments – Effects of Changes in Exchange Rate (Depreciation or Devaluation) on the Economy – Devaluation or Depreciation and the Balance of Trade: The J-curve – Fixed Exchange Rate and the Bretton Woods System – Collapse of the Bretton Woods System – Fixed and Flexible Exchange Rates – Currency Convertibility – Meaning of Currency Convertibility – Current Account and Capital Account Convertibility of Currency; Convertibility of Indian Rupee – Advantages of Currency Convertibility – Capital Account Convertibility of Rupee.

36. Free Trade versus Protection **589—599**

Case for Free Trade: Gain in Output and Well-being from Specialisation; Gains from Economies of Scale; Long-Run Dynamic Gains; Promotes Competition and Prevents Monopoly; Political Gains from Free Trade – Case for Protection: Nationalism; Employment Argument; Anti-dumping Argument; Correcting Balance of Payments Deficit – Trade Barriers: Tariffs and Import Quotas – Effects and Efficiency Cost of a Tariff – Effects and Dead-weight Loss of Quotas.

37. Foreign Investment and Role of Multinational Corporations **600—610**

Foreign Portfolio Investment – Benefits of Portfolio Capital Flows – Foreign Direct Investment – Merits of Foreign Direct Investment (FDI) – Role of Multinational Corporations (MNCs).

Appendix to Chapter 37: Foreign Direct Investment (FDI) Policy in India **611—614**

Successful Foreign Firms in India – Liberalisation of Foreign Direct Investment (FDI) – Big Push to Foreign Investment by NDA Government.

37A. Globalisation, Commercial Policy and WTO **615—638**

Case for Globalisation of Indian Economy – Measures Adopted in India to Promote Globalisation – Dangers and Risks of Globalisation – Global Commercial Policy – Effects of Regional Trade Agreements: Trade Creation and Trade Diversion – GATT (General Agreement on Tariffs and Trade) – World Trade Organisation (WTO) – Object of WTO – WTO: Status, Administration and Functions – The Guiding Fundamental Principles and Features of WTO – WTO Agreements and Liberalisation; GATS, TRIPS and TRIMS – India's Role in WTO Talks – India's Gain from WTO – Failure of WTO Talks at Cancun – WTO's Hong Kong Declaration – Trade Facilitation Agreement at Bali.

PART–XIV
ECONOMICS OF GROWTH AND DEVELOPMENT

38. Economic Growth and Development **641—656**

Meaning of Economic Growth – Meaning of Economic Development: Traditional View – Concept of Economic Development: Modern View – Sen's 'Capabilities' Approach to Development: Freedoms and Capabilities; Three Core Values of Development – The Objectives of Development – Human Development Index (HDI) – Capital-Output Ratio.

39. Underdevelopment and Common Characteristics of Developing Countries **657—663**

Meaning of Underdevelopment – Common Characteristics of Developing(Underdeveloped) Economies: Low Per Capita Income or Poverty; Excessive Dependence on Agriculture; Lack of Capital and Low Rate of Capital Formation; Rapid Population Growth and Disguised Unemployment; Dualistic Structure of the Underdeveloped Economies.

40. Obstacles to Economic Development **664—676**

Imbalance between Capital and Population – Vicious Circle of Poverty (Supply-Side) – Vicious Circle of Poverty (Demand side) – Breaking the Vicious Circle of Poverty – Deficiency of Dynamic Entrepreneurship and Weak Propensity to Accumulate – Inadequate Infrastructure – Myrdal's Backwash Effect and Economic Growth – Demonstration Effect and Economic Growth – How to Overcome Demonstration Effect.

Appendix to Chapter 40: Rostow's Stages of Growth 677—679

41. Capital Formation: Determinants and Acceleration 680—698

Meaning of Capital Formation – Widening of Capital and Deepening of Capital – Human Capital – Role of Capital Formation and Economic Growth – Role of Capital: Diagrammatic Illustration – Role of Capital: A Dissenting View – Determinants of Capital Formation – Public Savings – Investment – Capital Formation and Economic Growth in India – Measures to Accelerate Capital Formation.

42. Technological Progress and Economic Development 699—706

Meaning of Technological Change – Technological Change and Economic Growth – Technological Change and Capital Formation – Technological Progress to overcome Stationary State – Technological Change: Embodied and Disembodied.

43. Human Resource Development: Education and Health 707—722

The Concept of Human Capital – Role of Education or Human Capital in Economic Development – Cost of Human Capital or of Acquiring Skills – Return on Investment in Education – Measuring Contribution of Education to Economic Growth – Health – Importance of Healthcare as a Factor in Economic Development – Selected Health Indicators: A Comparative Study – Evaluation of Views of Amartya Sen and Jean Dreze – Undernutrition and Disease Prevention: Energy Intake and Level of Productivity.

44. Population Growth and Economic Development 723—730

How Population Growth Adversely Affects Economic Development – Population Control Policy.

45. Harrod–Domar Model of Growth 731—741

Domar's Growth Model – Demand or Income Effect of Investment – Condition for Equilibrium Growth – Harrod's Growth Model – Relevance of Harrod-Domar Growth Model for Developing Countries.

46. Neoclassical Theory of Growth 742—754

Neoclassical Growth Theory: Production Function and Saving; Fundamental Growth Equation – Growth Process and Steady Growth Rate – Impact of Increase in the Saving Rate – Effect of Increase in Population Growth Rate – Long-run Growth and Technological Change – Conclusion: Key Results of Solow's Neoclassical Model – Sources of Economic Growth: Production Function Approach – Limitations of Neoclassical Growth Model – Knowledge or Education is the Missing Factor – Ignores Economies of Scale as a Factor Contributing to Economic Growth.

47. New Theory of Economic Growth (Endogenous Growth Model) 755—761

Endogenous Growth Models – Investment in Human Capital and Learning by Doing – Policy Implications of New Growth Theory.

48. Mahalanobis Growth Model and Heavy-industry Strategy of Development 762—770

Mahalanobis Model of Growth – Mahalanobis Growth Model and Development Strategy in India's Five Year Plans – Critical Evaluation of Mahalanobis Heavy-Industry Strategy of Development.

49. Wage-goods Model and Strategy of Development 771—777

Wage-goods Model – Arguments in Favour of Wage-goods Strategy – The Critical Evaluation of Wage-goods Strategy.

50. Strategies of Balanced and Unbalanced Growth 778—787

Strategy of Balanced Growth – Size of Market and Inducement to Invest – Nurksian Strategy of Balanced Growth – A Critique of Balanced Growth Doctrine – Hirschman's Strategy of Unbalanced Growth – A Critique of Unbalanced Growth Strategy.

51. Lewis Model of Development with Unlimited Supply of Labour 788—794

Lewis Model of Development with Surplus Labour – Reinvestment of Capitalist Surplus (Profits) – Profit as the Main Source of Capital Formation – Credit-financed Money:

Another Source of Capital Formation – Tapering off of the Expansionary Process – A Critical Appraisal of Lewis Model.

52. **Choice of Techniques in Developing Countries** **795—801**
Factor-price Ratio and Choice of Technique – Choice of Techniques: Maximum Reinvestible Surplus Criterion – Amartya Sen's Time Series Criterion and Choice of Technique – Choice between Maximising Employment and Maximising Production – Conflict between Maximising Current Employment and Maximising Rate of Growth of Production and Employment – Appropriate or Intermediate Technology.

53. **Employment and Economic Growth in Developing Countries** **802—816**
Factors Determining Growth of Employment – Various Development Strategies for Employment Generation in Developing Countries – Industrialisation-Led Strategy of Development – Heavy-industry Strategy and Mahalanobis Growth Model – Wage-goods Strategy of Development and Employment in the Developing Countries – Development Strategy using Labour-intensive Technology – Strategy of Rural Public Works for Employment Generation.

54. **Poverty and Economic Growth** **817— 850**
Poverty and Economic Growth – Concepts of Poverty: Absolute Poverty; Head Count Ratio; Poverty Gap; Relative Poverty – Why is Relative Poverty Bad? Regional Poverty Estimates of the World – The New Index of Poverty: Multidimensional Poverty Index (MPI) by the United Nations – World Bank's Notion of Shared Prosperity – Latest Estimates of Global Poverty at Updated Poverty Line of 1.90 PPP $ a Day by World Bank – Problem of Poverty in India – Dandekar-Rath Measure of Poverty – Issues in Estimating Poverty – Trends in Poverty in India Since 1973-74 – Who are the Poor in India? – Causes of Poverty in India – Poverty and Economic Growth – Strategy to Reduce Poverty.

55. **Income Inequalities and Economic Growth** **851—866**
Inequality: Meaning and Measurement – Measurement of Income Inequality – Income Inequality: A Global Comparison – Economic Growth and Inequality: Kuznets' Inverted-U Hypothesis – Income Inequality, Economic Efficiency and Economic Growth: Inequality, Incentives and Efficiency – Inequality, Saving and Capital Formation – Society's Tolerance of Inequalities.

56. **Economic Reforms, Structural Adjustment and Growth** **867—880**
Rationale for Economic Reforms – Washington Consensus, Development Policy and Structural Adjustment Reforms – Structural Adjustment Reforms in India – Structural Economic Reforms: A Critical Evaluation – Adverse Effects of Globalisation – Market Failures and Role of Government.

57. **Environment and Economic Development** **881—890**
Population and Environment – Poverty and Environment – Sources of Pollution and Environmental Degradation – Industrialisation, Urbanisation and Environment – Rural and Agricultural Development and Environment – Negative Externalities and Environment Resources – Common Property Resources.

58. **Sustainable Development and Climate Change** **891—912**
Sustainable Development: Meaning and Conditions – The Size of World's Greenhouse Gas Emission Problem – Allocation of Sparse Carbon Budget – Sustainable Development and Climate Change in the Indian Context – Sustainable Development and Poverty Eradication – The United Nations Conference on Sustainable Development, RIO + 20, 2012. Climate Change: Issues, Causes and Effects – What is Climate Change? – Why should India be Concerned about Climate Change? – Adverse Effects of Climate Change – Mitigating Climate Change to Ensure Sustainable Development – India's Policies to Tackle Climate Change – Climate Change Negotiations: Berlin Meeting and Kyoto Protocol; Doha Climate Change Conference 2012; – Paris Treaty on Climate Change, 2015.

VOLUME–1

MICROECONOMICS: THEORY AND APPLICATIONS

PART I

SCOPE AND METHODOLOGY OF ECONOMICS

- What Economics is about
- Micro and Macro Economics
- The Methodology of Economics
- Economic Statics and Dynamics
- Central Problems of an Economy
- Market Economy and Role of Price Mechanism

CHAPTER 1
WHAT ECONOMICS IS ABOUT

Knowledge has many branches and economics is an important and useful branch of knowledge. In recent years, the science of economics has assumed greater significance in view of the fact that knowledge of economics is being used for initiating and accelerating growth in the economies of the world and thus for eradicating want, poverty and unemployment from the human race. Besides, the nature of so many other problems such as inflation, food, stagnation and recession, population explosion, adverse balance of payments and so on that confront the economies of today cannot be understood and solution for them cannot be provided without the adequate knowledge of the science of economics.

The science of economics in the form in which we have it today is just two hundred years old.[1] It has made a remarkable progress during this period. In this book we shall explain the principles of modern economics, that is, economics in the form in which we have it today. But before we start the study of the principles of modern economics, it will be desirable to explain to the students what economics is about or what we study in economics, or, in other words, what is the subject matter of economics. Every science or a branch of knowledge is concerned with a particular subject. Thus, the science of physics deals with the properties of matter and energy, the science of chemistry deals with the constitution or composition of matter. Political science studies the nature of State and Government. In biology, we study the constituion and evolution of living organisms. Therefore, the question which a beginner is likely to ask is: "What is the subject matter of economics ?"Therefore, economists have also tried to define the subject matter of economics. By defining the subject matter of economics they have tried to delimit the scope of economics and to distinguish it from other branches of knowledge.

Perhaps there is no other science or a body of knowledge in regard to which there has been such a great controversy about its subject matter or its proper definition as about economics. J.N. Keynes is right when he says that "Political Economy is said to have strangled itself with definitions."[2] Economics has been variously defined by different economists from time to time. This is partly because "*economics is an unfinished science*."[3] With the passage of time there have been significant developments in economic theory and new subjects have been included in it. It is still in the process of growth and development. Therefore, the old definitions of economics have become irrelevant. The definition of a science or a body of knowledge delimits its subject matter that is already in existence. But in a science like economics that is growing and developing, its correct and satisfactory definition can be given only after it has sufficiently developed and grown. Now that the science of economics has sufficiently developed, we can provide an adequate and satisfactory definition of economics.

After discussing some old famous definitions of economics, we shall give some modern

---

1. Adam Smith who is generally known as father of economics brought out his famous book "*The Nature and Causes of Wealth of Nations*" in the year 1776. Before Adam Smith, there were other writers who expressed important economic ideas but economics as a distinct branch of knowledge started with Adam Smith's *Wealth of Nations*. It is worth noting that in the beginning the name of economics was "*Political Economy*."
2. J. N. Keynes *Scope and Method of Political Economy* p. 153.
3. F. Zuethen, *Economic Theory and Method* (1953), p. 3.

definitions of economics. Various old definitions of economics can be grouped together under three heads:

(*i*) Wealth Definitions, (*ii*) Welfare Definitions and (*iii*) Scarcity Definitions. We shall critically examine below these definitions.

ECONOMICS IS A SCIENCE OF WEALTH

Pioneers of the science of economics defined it as a science of wealth. Adam Smith, who is known as father of economics, named his famous book on economics as "*An Enquiry into the Nature and Causes of the Wealth of Nations*." Thus, according to Adam Smith, economics enquires into the factors that determine wealth of the country and its growth. In this book Adam Smith analyses the factors that determine the growth of the volume of production. That the emphasis of Adam Smith is on the wealth and riches of a nation is clear from the following quotation from his book. "The great object of Political Economy of every country is to increase the riches and power of that country." [4] As the wealth and riches of a country cannot grow without the proper utilization of its resources and this is what forms the subject matter of his book "*The Wealth of Nations*". Thus, Adam Smith empahsised the production and expansion of wealth as the subject matter of economics. However, Ricardo shifted the emphasis from the *production of wealth* to the *distribution of wealth*. Ricardo writes: "The produce of the earth—all that is derived from its surface by the united application of labour, machinery and capital is divided among three classes of the community, namely, the proprietor of the land, the owner of the stock of capital necessary for its cultivation, and the labourers by whose industry it is cultivated."[5] He further writes, "*To determine the laws which regulate this distribution, is the principal problem in Political Economy*."

Besides Adam Smith and Ricardo, other classical economists too regarded economics as study of wealth. Thus, according to J.B. Say, a French classical economist, "economics is the science which treats of wealth." Likewise, F.A. Walker writes, "Political economy or Economics is the name of that part of knowledge which relates to wealth."[6]

Critical Evaluation of 'Wealth Definition' of Economics

That economics is a 'science of wealth' has been severely criticised. In the seventeenth and eighteenth centuries when religion and ethics had a strong hold on minds of men, anything which was connected with wealth and riches was looked upon as sordid and mean. Since economics was defined as study of wealth, it was dubbed by men of letters, especially Carlyle and Ruskin, as 'Gospel of Mammon•, a 'pig science' and a 'dismal science'. It was alleged that economists have ignored the higher values of life and are hankering after formulation of laws which seek "to enrich both the people and the sovereign." Economics, it was said, has elevated wealth into an 'object of scientific study.' However, this is improper, unwarranted and biased attitude towards economics. Classical economists like Adam Smith, Ricardo and Malthus did not teach people to worship mammon or wealth. They sought to expound the principles of wealth because it is wealth understood in the sense of goods or commodities, which is needed to provide physical subsistence to men and to raise their standards of living. How a nation wrests wealth from the niggardly nature, how this wealth is distributed and exchanged within the nation, these enquiries with which Adam Smith and Ricardo were concerned still remain even today an important *object* of economic science. By studying the laws of production, exchange and distribution of wealth—useful goods or commodities, economics makes an important contribution to the promotion of social welfare. Therefore, it is unjest to dub economics as a sordid and mean science.

Besides, well-known topics of discussion in modern economics such as determination of income, employment and economic growth are intimately connected with the production and distribution of

4. Adam Smith. *Wealth of Nations* (Modern Library Edition), p. 352.
5. David Ricardo, Preface to his *Principles of Political Economy*.
6. F.A. Walker, *Political Economy*, New York, 1883, p.3.

wealth with which Adam Smith, Ricardo and other classical economists were concerned. As a matter of fact, it goes to the credit of Adam Smith and Ricardo that they addressed themselves to the vital problem of economic growth that brings about increase in the wealth or the production of goods. As is well-known, the main problem which confronts the developing countries like India is how to initiate and accelerate economic growth in their economies. The abject poverty, huge magnitude of unemployment and under-employment prevailing today in developing countries like India cannot be removed without expanding the production of wealth and distributing it equitably.

It is worth noting that the term 'wealth' has been differently interpreted by classical economists, that is, a variety of meanings has been attached to the term 'wealth'. However, most of the classical economists including Adam Smith, Ricardo and Malthus restricted its meaning exclusively to the *material* goods. That is, by wealth they generally meant *material* wealth. It must be remembered , however, that issue was not between materialistic economics preaching the accumulation of material goods and spiritual economics advocating higher values of life and spiritual desires of man. In other words, the classical economics was concerned with material wealth because it was easier to measure it accurately.[7]

Malthus, a renowned classical economist, makes it clear as to why only material goods are regarded as wealth. He writes: "If we wish to attain anything like precision in our inquiries, when we treat of wealth we must narrow the field of inquiry, and draw some line, which will leave us only those objects, the increase or decrease of which is capable of being estimated with more accuracy."[8] But in our view this does not truly reflect the views of Adam Smith. Adam Smith drew distinction between productive labour and unproductive labour .[9] The labour which produced material goods was called "productive" and that which produced immaterial services such as the services of teachers, actors, dancers, musicians, *etc*., were regarded by him as "unproductive". Thus Adam Smith seems to think that economics was concerned with material wealth because it was the result of productive labour, the study of immaterial wealth (*i.e.* services just noted above) which are the result of unproductive labour lies outside the scope of economics. It is this viewpoint of not recognising immaterial services as the subject of study in economics that has been rightly criticised by L.Robbins and others. According to Robbins, things which satisfy the wants of the people and are scarce in relation to the wants and therefore involve the problem of choice are worth studying in economics irrespective of whether they are material or immaterial. In fact, abilities and services such as teaching, singing, acting, providing music, curing diseases and bad health (*i.e.* the services of doctors) satisfy the important wants of the people and are also very scarce. The *economic aspects* of these services, that is, how to use them so as to attain maximum social welfare and how their prices are determined come under the purview of economics. Thus, in our view, classical economists by confining themselves to studying the laws of 'material wealth' and the neglect of immaterial services by them unduly restricted the scope of economics.

Further, keeping such immaterial things, as health, education, good administration outside the definition of wealth and therefore beyond the bounds of economic science shows that the classical economists did not realise and recognise the importance of these immaterial things for the economic growth of the nation, or to put in their own words, for the expansion of material wealth. Several modern economists, especially Prof. Amartya Sen, a Nobel Prize winner in economics, have empahsised the role of education, health and good administration and have shown that they greatly raise the productivity of man and promote economic growth of the nation. In view of their vital importance in raising production and productivity, good health, education and skills have in fact been called *human capital* by modern economists . That is why investment in promoting education, health, etc., has been called by modern economists as *investment in human capital or investment in man*.

---

7. M. Kirzner, *The Economic Point of View*, D. Van Nostrand Company, 1960, pp. 29-30.
8. Thomas R. Malthus, *Principles of Political Economy* (1820), p. 27.
9. Adam Smith, *op. cit*.

Another drawback in making economics a "science of wealth" by classical economists is that in doing so they laid conspicuous emphasis on wealth and put man to the secondary place in economic studies. They did not lay adequate stress on *man's behaviour* in relation to wealth. Nor did they emphasise the end or ultimate objective of economics which is the promotion of human and social welfare. As a matter of fact, wealth is only a means to an end, the end being the welfare of man and society. To regard wealth as the be-all and the end-all of economic science is to make the means as the end. The credit goes to Alfred Marshall, a prominent English economist, for shifting the emphasis from wealth to man and also from wealth to welfare. According to him, economics is "*on the one side a study of wealth; and on the other, and more important side, a part of the study of man*". [10] He further writes, "Economics examines that part of individual and social action which is most closely connected with the attainment and with the use of material requisites of well-being".[11]

Another feature of definition of economics in terms of the study of wealth which is open to question is that it involves the acceptance and justification of property rights in various forms of wealth. "Despite the variety of meanings that have been attached to the term "wealth" by classical economists, almost all these meanings find some common ground with a definition of wealth as *consisting in the objects of ownership*."[12] Classical economics with its emphasis on wealth regards the institution of private property as legally, morally, and naturally just. Gunnar Myrdal has shown that ideas of political and social philosophers who consider property rights as natural rights *account for* the classical, especially the Ricardian, theories of value and distribution.[13] Read, a classical economist, has described economics as "an investigation concerning the right to wealth" and explaining what the rights and duties of men in society are with regard to property. [14] "Commenting on this aspect of classical economics Kirzner writes, "According to Ricardo, economics shows how wealth is distributed among the factors of production ; according to Reed, economics in so doing, at the same time lay down the law of *natural rights* of the factors of production on their several shares."[15]

But that the property rights have natural and moral basis is not accepted by many modern economists, especially those who believe in socialist philosophy. Rights to private wealth or property are conferred by the society and if the national interests so require they can be vested in the State or the rights of private ownership of property can be greatly restricted. In India, various land reforms which involve restriction of private property in the ownership of land are advocated by the economists because they will help in increasing both output and employment. Thus the rights of private property or wealth are no longer considered as natural and moral rights.

Conclusion

By considering the problems of production, distribution and exchange of wealth, classical economists focussed attention on the important issues with which economics is concerned. However, by restricting the definition of wealth to *material* wealth and the neglect of immaterial services in their economic studies they narrowed down the scope of economics. Further, they also showed a biased attitude by regarding rights to private property or wealth as natural and moral rights.

MARSHALL'S DEFINITION: ECONOMICS IS A SCIENCE OF MATERIAL WELFARE

Marshall is one of the economists who contributed a good deal to economic theory. Even his definition of economics occupies an important place in the literature of economics. Marshall was the first economist who lifted the science of economics from the disrepute it had fallen into due to its

10. Alfred Marshall, *Principles of Economics*, 8th edition, p.1.
11. *Ibid*, p.1.
12. I.M. Kirzner, *The Economic Point of View*, p. 25.
13. Gunnar Myrdal, *The Political Element in the Development of Economic Theory*, Harvard, 1954, p. 696.
14. For the views of Reed, see I.M. Kirzner, *The Economic Point of View*, p. 32.
15. *Ibid*, p.32.

being associated with the study of wealth. Marshall pointed out that, for economics, wealth is not an end in itself but it is only a means to an end; the end being the promotion of human welfare. Thus, according to Marshall, wealth is only a secondary thing, it is man and his ordinary business of life which is the primary object of economic study. In fact, Marshall tried to make the study of economics an engine of social betterment. With this end in view, Marshall gave the following definition of economics; "*Political Economy or economics is the study of mankind in the ordinary business of life; it examines that part of individual and social action which is most closely connected with the attainment and with the use of the material requisites of well-being.*" [16]

Three things are worth noting in the above definition provided by Marshall. First, it is a study of man as such and not of wealth. No doubt, according to this definition, economics is concerned with wealth but it is concerned with wealth in the sense that it studies man's action regarding how he earns wealth and how he spends it. It is thus clear that it is the study of man which occupies the prominent place in the economic study. Thus Marshall writes "Economics on the one side is study of wealth and, on the other and more important side, a part of the study of man." Secondly, Marshall's definition implies that economics is concerned with a particular aspect of man's life. There are many aspects of man's life—social, religious, political, etc. Economics studies man's life in the ordinary business of life. The ordinary business of life means how a man gets his living and how he spends it. Thus Marshall says "Economics is a study of mankind in the ordinary business of life." At another place he says, "Economics is study of man's action in the ordinary business of life. It enquires how he gets his income and how he uses it." Thirdly, according to the above Marshall's definition, the primary object and end of economics is the promotion of material welfare. It should be noted that economics is not concerned with the totality of human welfare but only a part of it. Economists are concerned with only one aspect of human welfare which is connected with the achievement and the use of the material means of welfare. Thus Marshall incorporated in his definition that economics "examines that part of individual and social action which is most closely connected with attainment and with the use of the material requisites of well-being." It is the phrase material requisites of well-being that stands for material welfare. It is thus clear that Marshall laid emphasis on material welfare as the primary concern of the science of economics.

There are other economists who have also defined economics in terms of economic welfare. Thus Cannan defines economics in the following words: "The aim of Political Economy is the explanation of the general causes on which the material welfare of human beings depends."[17] Thus in this definition also emphasis has been laid on material welfare of man and, according to Cannan, economics enquires into the factors which determine the material welfare of human beings. Likewise, Pigou describes that "*the range of our enquiry becomes restricted to that part of social welfare which can be brought directly or indirectly into relationship with the measuring rod of money.*"[18]

It should be noted that in the above Pigou's definition the part of social welfare which can be related directly or indirectly with the measuring rod of money means the material welfare of human beings because it is material welfare which can be measured by the measuring rod of money. Non-material welfare cannot be measured with the aid of money. The total human welfare also cannot be measured with the help of money. It is only a part of the human welfare, that is, material welfare which is measurable in terms of money.

Critical Evaluation of Marshall's Definition and other Welfare Definitions

After Marshall had given the definition of economics it began to be thought that the problem of defining economics had ended because it was considered that Marshall's definition provided a correct scope and objective of economics. Many economists accepted it and even now many

16. Alfred Marshall, *Principles of Economics* (8th edition), p.1.
17. E. Cannan, *Wealth*, p. 17.
18. A.C. Pigou, *Economics of Welfare*, 4th ed. p. 11.

modern economists would express agreement with Marshall. But Marshall has also not been without its critics. His definition and other welfare definitions have been severely criticised by Lionel Robbins, a prominent English economist. Marshall's definition has been criticised by Robbins on the following grounds:

1. Firstly, Robbins is of the view that economics should not have any connection with material welfare. Robbins points out that in economics we study not only material things but also immaterial things. Therefore, according to him, it is incorrect to say that economics is concerned with material things alone. He points out that in economics we also enquire how the prices of immaterial services such as professional singers, actors, and actresses, dancers, *etc.*, are determined and they are important topics of price theory. He thus says: "A theory of wages which ignored all those sums which were paid for immaterial services or were spent on immaterial ends would be intolerable."[19] Moreover, it is very difficult to separate the material welfare from the other types of welfare. The welfare is entity as such and we cannot divide it into various parts. Even with the measuring rod of money we cannot precisely and exactly segregate the material welfare from the total welfare.

2. Robbins has also objected to the word 'welfare' in the above welfare definitions. According to Robbins, the concept of welfare is not fixed and definite one; it differs in different countries and at different times. Welfare is a subjective thing and it varies from person to person. Therefore, according to Robbins, it cannot be said in objective terms which things would promote welfare and which will not. Moreover, according to Robbins, economics is concerned with many goods and activities which are generally thought to be harmful to human welfare but they are studied in economics. Goods like liquor, cigarettes, opium are hardly conducive to human welfare but their pricing problem is studied in economics. In fact, according to him, economics is not concerned with welfare at all. He explains that economics studies the problems which have arisen because of the scarcity of resources. Goods and services which are scarce in relation to demand for them would carry price in the market. Therefore, we should study all those goods whether they promote welfare or not.

Goods like liquor, cigarettes and opium though harmful to human welfare, are studied in economics because some people in the society want them and they are scarce in relation to demand for them. Therefore, economists have to study the pricing problem and other aspects of such goods whether they promote welfare or not. Robbins remarks, "Why talk of welfare at all ? Why not throw away the mask altogether." Thus , according to Robbins, if economics is thought to be concerned with the causes of material welfare, then we have to give decision regarding what will promote welfare and what will reduce welfare. Therefore, we would be entering into the question of ethics, namely, what ought to be and what ought not to be. But Robbins is of the view that economics is neutral between ends and it cannot prescribe what should be done and what should not be done, what promotes welfare and what not, what is good and what is bad. Therefore, Robbins remarks whatever economics is concerned with it is not concerned with causes of material welfare as such.

ROBBINS' DEFINITION: ECONOMICS IS A SCIENCE OF SCARCITY

Robbins not only criticised Marshall's definition and other welfare definitions of economics but also provided a new definition which he considered to be more scientific and correct. He has given this definition in his famous book, "*An Essay on the Nature and Significance of Economic Science*, " which he brought out in 1931. According to Robbins, economics studies the problems which have arisen because of the scarcity of resources. Nature has not provided mankind sufficient resources to satisfy all its wants. Therefore, the people have to choose for which ends or for which wants the resources are to be utilised. Thus, according to Robbins, economics is the science of scarcity and it studied how the scarce resources are allocated among their different uses. Thus he has

19. Lionel Robbins, *An Essay on The Nature and Significance of Economic Science*, 1935.

given the following definition: "*Economics is the science which studied human behaviour as a relationship between ends and scarce means which have alternative uses*." This definition is based upon the following three facts :

1. **Unlimited wants.** The first fact on which Robbins' definition is based is that man's wants are unlimited. In his definition 'ends' implies wants and for statisfying wants the man uses resources. That man's wants are unlimited is a very important and fundamental fact of economic life of the people. If man's wants were limited, then no economic problem would have arisen. But in the real life of the people there is no limit to their wants; when one want is satisfied, another want crops up. An important things to know about wants is that they are not of equal intensity; some are more intense than others. It is because of the different intensities of the wants that people are able to allocate the resources to satisfy their different wants.

2. **Scarce means.** The second element which gives rise to economic problem is that resources are scarce in relation to wants. If the resources like wants were unlimited, no economic problem would have arisen because in that case all wants could have been satisfied and there would have been no problem of choosing between the wants and allocating the resources between them. Because the resources are scarce, all wants cannot be satisfied. Therefore, human beings have to decide for the satisfaction of which wants the resources should be used and which wants should be left unfulfilled. It should be noted that means or resources here refer to natural productive resources, man-made capital goods, consumer goods, money and time available with men, *etc*. If the means or resources were unlimited, then we would have obtained goods in the desired quantities because in that state of affairs goods would have been free goods. But in the real life we cannot obtain goods free or without price; we have to pay price for them and make efforts to obtain them.

3. **Alternative uses of means.** The third fact on which Robbins' definition is based is that resources or means have various alternative uses. In other words, the resources can be put to various uses. For instance, coal can be used as a fuel for the production of industrial goods, it can be used for running trains, it can be used for domestic cooking purposes and for so many other purposes. Likewise, monetary resources can be utilised for the production of essential consumer goods, for the production of capital goods and so many other goods. It has to be decided how the resources have to be allocated among different uses. The man or society has, therefore, to choose the uses for which resources have to be employed. If the resources would have a single use only, the question of choice would not have arisen at all. In the case of single use of resources, they would have been employed for the uses which they have. It is because of the various alternative uses of the resources that we have to decide about the best allocation of resources.

We thus see that Robbins' definition stands on the above-mentioned three facts, namely, unlimited wants, scarce resources and alternative uses of the resources. According to him, economics studies human behaviour regarding how he satisfies his wants with the scarce resources. According to him, economics is a human science and not a mere social science. It studies man in society or without society when he is confronted with the problem of allocating scarce resources to satisfy his unlimited wants. An important thing to note about Robbins' definition is that Robbins does not distinguish between material and non-material, between welfare and non-welfare. According to him, economics studies man's activities in regard to all goods and services, whether they are material or immaterial, provided they satisfy the wants of the people.

Economics is neutral between ends

Besides, whether the goods and services are conducive to human welfare or not, economics should study them if they satisfy the wants of some men. It is also worth noting that in view of Robbins, economics does not deal with the question as to what ends should be achieved, that is, what wants should be satisfied and what not, because in this regard man himself has to decide. Economics itself does not make a choice. Economist only tells in what ways the given ends or wants can be achieved with the minimum possible resources. What ends or wants should be selected for satisfaction is not the concern of economists. Whether the ends chosen by man are good or bad, noble or ignoble, economics should study them, because the task of economist is not to praise or condemn but only to analyse and explain. To decide about the desirability or otherwise of a thing is beyond the scope of economics. Therefore, *according to Robbins, economics is neutral between ends.*

Economics as a Science of Choice

It follows from the definition of Robbins that economics is a *science of choice*. It deals with how the resources of society should be allocated to the satisfaction of different wants. Whenever the resources are scarce and the wants are many, the question of choice arises. The man has to choose between the wants to which resources are to be allocated. Thus, Professor Robbins remarks, "When time and means for achieving ends are limited and capable of alternative application and the ends are capable of being distinguished in order of importance, then behaviour necessarily assumes the form of choice."[20] It is thus clear that economics is the science of choice.

Like Robbins many other economists have defined economics in terms of scarcity of resources and choice. Thus, Professor Wicksteed says that economics is a "study of those principles on which the resources of a community should be so regulated and administered as to secure social ends without waste."[21] Likewise, Professor Stigler defines economics in the following words: "Economics is the study of the principles of governing the allocation of scarce means among competing ends when the objective of allocation is to maximise the attainment of the ends."[22] Similarly, Scitovosky says, "Economics is a social science concerned with the administration of scarce resources."[23] Professor Erich Roll has also defined economics in terms of scarcity of resources and the choice. Thus, according to him, "The economic problem is essentially a problem arising from the necessity of choice; choice of the manner in which limited resources with the alternative uses are disposed of. It is the problem of the husbandry of resources.... economics studies the activity of husbandry."[24]

It is thus clear that after Robbins economics has assumed the character of science which is concerned with the scarcity of resources and the problem of choice which arises because of the scarcity. Robbins' definition has been claimed to be more scientific. It has broadened the scope of economics whereas the material welfare definitions had narrowed down the scope of economic study.

With this definition of economics which lays a stress on scarcity of resources and the problem of choice, economics can no longer be called dismal science. Economics has no responsibility about the choice of ends; ends may be good or bad, economics has no concern with it. When the ends are many and resources scarce, the science of economics is required to study this problem.

Critical Evaluation of Robbins' Definition

No doubt, Robbins has made economics a scientific study and his definition has become very popular among the economists. This definition brings to light the basic economic problem which confronts the society. But Robbins' definition has also been criticised on several grounds. The main charge against Robbins is that he has made economics quite impersonal, colourless and devoid of any normative element. He says equilibrium is just an equilibrium. He does not seek to make economics as study of human welfare. Therefore, many economists like Durbin, Fraser, Beveridge and Wootton have tried to defend Marshall's idea about the true scope of economics and its objective of the promotion of social welfare. Thus Wootton has said that "It is very difficult for economists to divest their discussions completely of all normative significance." We give below the various criticisms which have been levelled against Robbins' definition.

---

20. Lionel Robbins, *Nature and Significance of Economic Science,* p.14.
21. Wicksteed, *Commonsense of Political Economy*.
22. G.J. Stigler, *The Theory of Price*, 1949, p.12.
23. Tibor Scitovosky, *Welfare and Competition*, 1952, p. 3.
24. Erich Roll, *The Elements of Economic Theory*, 1950, p. 16.

1. Firstly, Robbins' definition has been criticised on the ground that it is not justified on the part of Robbins to oppose making economics as an engine of social welfare. In fact it has been contended that even in Robbins' definition the idea of welfare is present. If we closely analyse Robbins' definition we would come to the conclusion that it says that economics is concerned with how a man and society uses its scarce resources so as to achieve maximum possible satisfaction of its wants. But this maximum satisfaction of wants is nothing else but maximum welfare. Robbins' definition implies that allocation of resources has to be made in such a way that maximum satisfaction is achieved. Thus, without the satisfaction or welfare being brought into consideration the question of allocation of scarce resources does not arise.

2. Secondly, Robbins has been criticised on the ground that he has made economics neutral between ends. Many economists are of the view that if economics has to play an important role in promoting social welfare and economic growth, it has to give its decision regarding what is good and what is bad to achieve these ends. That is, if economics is to serve as an engine of social betterment, it would have to abandon the neutrally between ends or objectives. Economist would have to tell what is good or bad for welfare and progress and whether efforts should be made to attain some given ends or not. Professor Thomas rightly remarks, "the *function of the economist is not only to explain and explore but also to praise and condemn*."[25]

3. A serious objection against Robbins' definition is that it has reduced economics to a mere value theory, i.e., the theory of product and factor pricing. In other words, according to him, economics should study only the allocation of resources among the production of various goods and consequently how the prices of goods and factors are determined. But the scope of economics is wider than the allocation of resources and the price theory. These days the importance of macro-economics has increased in which we study how the national income of a country and its total employment are determined. But the determination of national income and employment does not fall within the purview of Robbins' definition which only lays stress on the allocation of resources. The history of Western economies has revealed that there has been great instability in them. There has been periodic occurrence of mass unemployment and depression. On the other hand, at other times, these economies have witnessed periods of boom and inflation. It is the basic task of economics to explain this instability and to prescribe policies for removing fluctuations in the levels of income, output and employment. But Robbins' definition leaves this issue untouched. It is thus clear that macro-economics which has become more important these days does not come within the purview of economics as defined by Robbins.

4. Robbins' definition is also criticised on the ground that it does not cover the theory of economic growth and development. The theory of economic growth and development studies how the national income and per capita income of a country increase over a long period of time and what factors cause such increases. With economic growth productive capacity of the country expands which brings about increases in national income, per capita income and the level of employment. While Robbins takes the resources as given and talks about their allocation, the theory of economic growth is concerned with *reducing the scarcity of resources,* through raising the level of national income and accumulating more capital and wealth. In developing countries the question of economic growth is more important because these countries are making efforts to remove poverty of their people and to raise their living standards through economic growth. In the recent years, many theories regarding how to initiate and accelerate economic growth in developing countries have been propounded. The theory of economic growth has now become the core of the science of economics both in developed and developing countries. Robbins' definition is defective because it does not cover an important subject like economic growth.

---

25. S.E. Thomas, *Elements of Economics.*

5. Even the problem of unemployment which is being faced by both the developing and developed countries of the world is not covered by Robbins' definition. As has been said above, Robbins' definition deals with the problem of scarcity, however the problem of unemployment is not the problem of scarcity but one of abundance of manpower in relation to demand for it. As is well known, labour or manpower is an important factor of production and its unemployment implies that it is abundant and not scarce. It is the task of economists to study the causes of the problem of unemployment and surplus labour and to suggest means to solve it.

6. Robbins' definition has also been attacked for its making our economics a human science instead of social science. In one sense the scope of economics has been unnecessarily widened by Robbins. In accordance with Robbins' views, economist would also study a *sadhu,* who lives in a cave of Himalaya, because that *sadhu* would also be faced with the problem of how to distribute his time between various ends. That is, a *sadhu* has also to face the problem of choice and therefore comes within the purview of Robbins' definition. But many economists are of the view that economics is a social science and it should study the problem of choice when it has a social aspect, that is, when a man's choice affects other members of the society. Therefore, many economists though laying stress on the scarcity of resources and the problem of choice consider economics as a social science and not a human science as Robbins does. Thus Scitovsky defines economics as "*a social science concerned with the administration of scarce resources." According to Cairncross, "Economics is a social science studying how people attempt to accommodate scarcity to their wants and how these attempts interact through exchange.*" It is thus clear that, contrary to the views of Robbins, economics has been regarded by many economists as a social science; economics studies the problem of choice if an individual's choice affects other members of the society.

In modern economics the subject matter of economics is divided into two main parts: (1) *micro-economics* and (2) *macro-economics*. The distinction between micro-economics and macro-economics and the subjects which fall under them will be discussed in the next chapter.

POSITIVE ECONOMICS, NORMATIVE ECONOMICS AND WELFARE ECONOMICS

It is important to know the difference between positive economics and normative economics. Positive economics is concerned with explaining '*what it is,*' that is, it describes theories and laws to explain observed economic phenomena, whereas normative economics is concerned with '*what should be*' or '*what ought to be*' the things. J.N. Keynes drew the distinction between the two types of economics in the following manner: "*A positive science may be defined as a body of systematised knowledge concerning what it is; normative science or a regulative science as a body of systematised knowledge relating to criteria of what ought to be, and concerned with the ideal as distinguished from the actual....* The objective of a positive science is the establishment of uniformities (*that is, scientific laws*); of a normative science, the determination of the ideals."[26] Thus, in positive economics we derive propositions, theories and laws following certain rules of logic. These theories, laws and propositions explain the cause and effect relationship between economic variables. In positive micro-economics, we are broadly concerned with explaining the determination of relative prices and the allocation of resources between different commodities. In positive macro-economics, we are broadly concerned with how the level of national income and employment, aggregate consumption and investment and the general level of prices are determined. In these parts of positive economics, what should be the prices, what should be the saving rate, what should be the allocation of resources, and what should be the distribution of income are not discussed. These questions of what should be and what ought to be, fall within the purview of normative economics. Thus, given the profit maximization assumption, positive economics states that monopolist will fix a price which will equate marginal cost with marginal revenue. The question what price should or ought to be fixed so that maximum social welfare is achieved lies outside the purview of positive economics. Similarly, given the

26. J.N. Keynes, *Scope and Method of Political Economy,* MacMillan, London, 1930, p. 46., (italics added)

monopsony in the labour market, positive economics explains how actual wage rate is determined. It does not go into the question what wage rate should be paid to the workers so that they should not be exploited. Likewise, how national income between different individuals is distributed falls within the domain of positive economics. But positive economics is not concerned with the question how income should be distributed. On the other hand, normative economics is concerned with describing what should be the things. It is, therefore, also called *prescriptive economics*. Thus, wehat price for a product should be fixed, what wage rate should be paid, how income should be distributed, *etc.*, fall within the purview of normative economics.

Normative Economics and Value Judgements

It should be noted that, normative economics involves *value judgements* or what are simply known as *values*. By value judgements or values, we mean the conceptions of the people about what is good or bad. These conceptions regarding values of the people are based on the ethical, political, philosophical and religious beliefs of the people and are not based upon any scientific logic or law. Because normative economics involves value judgements eminent economist Professor Robbins contended that economics should not become normative in character.[27] He opined that it was unscientific to include the value judgements in the economic analysis. To quote him, "the role of the economist is more and more conceived of as that of the expert, who can say what consequences are likely to follow certain actions, but who cannot judge as an economist the desirability of ends."[28] While drawing difference between economics and ethics he further writes, "economics deals with ascertainable facts, ethics with valuations and obligations is outside its purview. The two fields of inquiry are not on the same plane of discourse. Between the generalisations of positive and normative economics, there is a logical gulf fixed which no ingenuity can disguise and no juxtaposition in space or time bridge over. Propositions involving the verb 'ought' are different in kind from propositions involving the verb 'is'.[29]

Value judgements of various individuals differ and their rightness or wrongness cannot be decided on the basis of scientific logic or laws. Therefore, in our view, positive economics should be kept separate and distinct from normative economics. However, because normative economics involves value judgements, it does not mean that it should be considered as useless or not meaningful and should not be the concern of economics. As a matter of fact, many vital issues concerning economic welfare of the society necessarily involve some value judgements. If economics is to become an 'engine for social betterment', it has to adopt certain norms, ideals or criteria with which to evaluate economic issues and pass judgements on what is good and what is bad from the viewpoint of social welfare. We agree with Professor A.C. Pigou, "*Our impulse is not the philosopher's impulse, knowledge for the sake of knowledge but rather the physiologist's knowledge for the healing that knowledge may help to bring*."[30] The economist should not refrain from making value judgements if there is a wide consensus about them among the community. Using his knowledge of economics and these value judgements he should comment upon the desirability or otherwise of certain policies and issues. Professor Paul Streeten rightly says, "Economists cannot and should not refrain from making value judgements if their studies are to be more than a purely formal technique of reasoning, an algebra of choice. The technique, the algebra is important and ought to be as scientific as possible, but it is significant only as a means to study of wealth and welfare and of the ways to improve them."[31]

As is clear from above, normative ecnomics is concerned with welfare propositions, since what is good or what is bad ultimately depends upon its effect on the welfare of the individual and the

---

27. L. Robbins, *"Nature and Significance of Economic Science."* MacMillan, London, 2nd Edition, 1938.
28. *Ibid.*
29. *Ibid.*
30. A.C. Pigou, *Economics of Welfare*, 4th Edition, MacMillan, London, 1932, p.5.
31. Paul Streeten, Economics and Value Judgements, *Quarterly Journal of Economics*, 1950, p. 595.

society. In recent years, a branch of economics, known as *welfare economics*, has been developed. This welfare economics seeks to evaluate the social desirability of alternative social states or economic policies. Thus, Professor Scitovsky writes, " welfare economics is that branch of economic analysis which is concerned primarily with establishment of criteria that can provide a positive basis for adopting policies which are likely to maximise social welfare."[32]

QUESTIONS FOR REVIEW

1. "Economics is a science of wealth which enquires into the nature and causes of wealth of nations". Who gave this definition of economics and make your comments on it?
2. According to Marshall, economics is a science of material welfare of man. Critically examine this. How is it different from Robbin's definition?
3. According to Marshall, "economics is a study of mankind in the ordinary business of life". Explain and critically examine this.
4. Explain Robbin's definition of economics. How is it based on unlimited wants and scarce means? Critically examine it. Are subjects of unemployment and economic growth covered in it ?
5. What are the fundamental propositions of Robbin's definition of economics? How is it superior to Marshall's definition?
6. "Economics is a science which studies human behaviour as a relationship between ends and scarce means which have alternative uses". Discuss.
7. "Economics is a science of choice". Discuss.

   OR

   In Economics, the fundamental problem is the problem of choice. Elucidate this statement.
8. What is meant by scarcity in economics? Scarcity is mother of all economic problem. Discuss.
9. Distinguish between positive economics and normative economics. Explain some of the questions which are explained in each of them.
10. Economics is neutral between ends. Explain and critically examine this view point.
11. The task of an economist is not to praise and condemn but only to analyse and explain.
12. Which of the following statements are true and which false? Give reasons in brief for your answer.
    (*a*) Robbin's definition treats economics as a purely positive science.
    (*b*) Price theory is an important constituent of economic theory.
    (*c*) The concept of scarcity is valid only in poor economies.
    (*d*) According to Robbins, economics is neutral between ends.
13. Some important definitions of economics are given below. Name the economist who gave each definition and explain them.
    (*a*) Economics is a science which studies human behaviour as a relationship between ends and scarce means which have alternative uses.
    (*b*) Economics is an enquiry into the nature and causes of wealth of nations.
    (*c*) Economics is a study of mankind in the ordinary business of life.
    (*d*) Economics is a study which enquires into the material welfare of man.
    (*e*) The produce of the earth — all that is derived from its surface by the united application of labour, machinery and capital is divided among three classes of the community. To determine the laws which govern this distribution is the principal problem in political economy.
14. What are value judgements? Critically examine their role in economics.

---

32. Tibor Scitovsky, *Papers on Welfare and Growth*, 1962, p. 174.

CHAPTER 2
MICRO AND MACRO ECONOMICS

The subject-matter of economics has been divided into two parts : Microeconomics and Macroeconomics. These terms were first coined and used by Ragnar Frisch and have now been adopted by economists all the world over. *Nowadays one can hardly come across a text-book on modern economic analysis which does not divide its analysis into two parts, one dealing with microeconomics and the other with macroeconomics*. The term microeconomics is derived from the Greek word *mikros*, meaning "small" and the term macroeconomics is derived from the Greek word *makros*, meaning "large." Thus microeconomics deals with the analysis of small individual units of the economy such as individual consumers, individual firms and small aggregates or groups of individual units such as various industries and markets. On the other hand, macro-economics concerns itself with the analysis of the economy as a whole and its large aggregates such as total national output and income, total employment, total consumption, aggregate investment. Thus, according to K. E. Boulding, "Microeconomics is the study of particular firms, particular households, individual prices, wages, incomes, individual industries, particular commodities."[1] About macroeconomics he writes, "Macroeconomics deals not with individual quantities as such but with aggregates of these quantities; not with individual incomes but with national income; not with individual prices but with price levels; not with individual outputs but with the national output."[2]

MICROECONOMICS

As stated above, microeconomics studies the economic actions and behaviour of *individual units* and *small groups* of individual units. In microeconomic theory we discuss how the various cells of economic organism, that is, the various units of the economy such as thousands of consumers, thousands of producers or firms, thousands of workers and resource suppliers in the economy do their economic activities and reach their equilibrium states. In other words, in microeconomics we make a *microscopic study* of the economy. But it should be remembered that microeconomics does not study the economy in *its totality*. Instead, in microeconomics we discuss equilibrium of innumerable units of the economy *piece meal* and their inter-relationship to each other. Thus, micro-economics consists of looking at the economy through a microscope, as it were, to see how the millions of cells in the economy —the individuals or households as consumers, and the individuals firms as producers— play their part in the working of the whole economic organisation. For instance, in microeconomic analysis we study the demand of an individual consumer for a good and from there go on to derive the *market demand* for the good (that is, demand of a group of individuals consuming a particular good). Likewise, microeconomic theory studies the behaviour of the individual firms in regard to the fixation of price and output and their reactions to the changes in the demand and supply conditions. From there we go on to establish price-output fixation by an industry (Industry means a group of firms producing the same product).

Thus, microeconomic theory seeks to determine the mechanism by which the different economic units attain the position of equilibrium, proceeding from the individual units to a *narrowly*

1. K. E. Boulding, *A Reconstruction of Economics*, (1950), p. 3.
2. K. E. Boulding, *Economic Analysis*, p. 25.

defined group. Microeconomic analysis concerns itself with narrowly defined groups since it does not study the *totality of behaviour* of all units in the economy. In other words, the study of economic system or economy as a whole lies outside the domain of microeconomic analysis.

Microeconomics and Allocation of Resources

Microeconomic theory takes the total quantity of resources as given and *seeks to explain how they are allocated to the production of various goods*. It is the allocation of resources that determines what goods shall be produced and how they shall be produced. The allocation of resources to the production of various goods in a free-market economy depends upon prices of the various goods and prices of the various resources or factors of production. Therefore, to explain how the allocation of resources is determined, microeconomics proceeds to analyse how relative prices of goods and factors are determined. Thus the theory of product pricing and the theory of factor pricing (or the theory of distribution) fall within the domain of microeconomics. The theory of product pricing explains how relative prices of cotton cloth, foodgrains, jute, kerosene oil, *Vanaspti ghee* and thousands of other goods are determined. The theory of distribution explains how *wages* (price for the use of labour), *rent* (payment for the use of land), *interest* (price for the use of capital) and *profits* (the reward for the entrepreneur) are determined. Thus, the theory of product pricing and the theory of factor pricing are the branches of microeconomic theory.

Prices of products depend upon the forces of demand and supply. The demand for goods depends upon the consumers' behaviour pattern, and the supply of goods depends upon the conditions of production and cost and the behaviour pattern of the firms or entrepreneurs. Thus the demand and supply sides have to be analysed in order to explain the determination of prices of goods and factors. Thus the theory of demand and the theory of production are two subdivisions of the theory of pricing.

Microeconomics and Economic Efficiency

Besides analysing the pricing of products and factors and the allocation of resources based upon the price mechanism, microeconomics also seeks to explain whether the allocation of resources determined is *efficient*. Efficiency in the allocation of resources is attained when the resources are so allocated that maximises the satisfaction of the people. Economic efficiency involves three efficiencies; *efficiency in production, efficiency in distribution* of goods among the people (This is also called *efficiency in consumption*) and *allocative economic efficiency*, that is, efficiency in the *direction of production*. Microeconomic theory shows under what conditions these efficiencies are achieved. Microeconomics also shows what factors cause departure from these efficiencies and result in the decline of satisfaction from the maximum possible level.

Efficiency in production involve producing the maximum possible amount of various goods from the given available amount of resources. When such productive efficiency is attained, then it is no longer possible by any reallocation of the productive resources or factors among the production of various goods and services to increase the output of any good without a reduction in the output of some other good. Efficiency in consumption consists of distributing the given amount of produced goods and services among millions of the people for consumption in such a way as to maximise the total satisfaction of the society. When such an efficiency is achieved it is no longer possible — by any redistribution of goods among the people — to make some people better off[3] without making some other ones worse off.[4] Allocative economic efficiency or optimum direction of production consists of producing those goods which are most desired by the people, that is, when the amounts of different goods produced is such that maximises the satisfaction of the people.

In other words, allocative economic efficiency implies that pattern of production (*i.e.* amounts

3. Making some people *better off* means increasing their satisfaction.
4. Making some people *worse off* means reducing their satisfaction.

of the various goods and services produced) should correspond to the desired pattern of consumption of the people. Even if efficiencies in consumption and production of goods are present, it may be that the goods which are produced and distributed for consumption may not be those preferred by the people. There may be some goods which are more preferred by the people but which have not been produced and vice versa. To sum up, *allocative efficiency (optimum direction of production) is achieved when the resources are so allocated to the production of various goods that the maximum possible satisfaction of the people is obtained.* Once this is achieved, then by producing some goods more and others less by any rearrangement of the resources will mean loss of satisfaction or efficiency. The question of economic efficiency is the subject matter of theoretical *welfare economics* which is an important branch of microeconomic theory.

That microeconomic theory is intimately concerned with the question of efficiency and welfare is evident from the following remarks of A. P. Lerner, a noted American economist. "In micro-economics we are more concerned with the avoidance or elimination of waste, or with inefficiency arising from the fact that production is not organised in the most efficient possible manner. Such inefficiency means that it is possible, by rearranging the different ways in which products are being produced and consumed, to get more of something that is scarce without giving up any part of any other scarce item, or to replace something by something else that is preferred. Microeconomic theory spells out the conditions of efficiency (*i.e.*, for the elimination of all kinds of inefficiency) and suggests how they might be achieved. These conditions (called Pareto-optimal conditions) can be of the greatest help in raising the standard of living of the population."[5]

The four basic economic questions namely, (1) what goods shall be produced and in what quantities, (2) how they shall be produced, (3) how the goods and services produced shall be distributed among the people and (4) whether the production of goods and their distribution for consumption is efficient fall within the domain of micro-economics. The whole content of microeconomic theory is presented in the following chart :

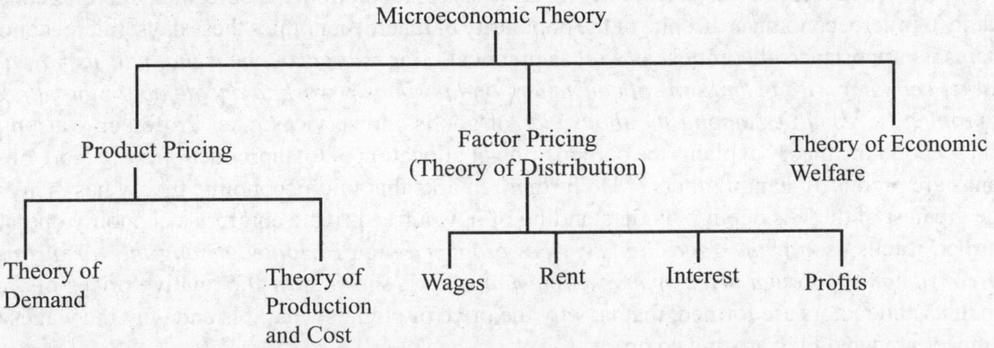

Microeconomics as a Microscopic Study of the Economy

It is generally understood that microeconomics does not concern itself with the economy as a whole and an impression is created that microeconomics differs from macroeconomics in that whereas the latter examines the economy as a whole, the former is not concerned with it. But this is not correct. That microeconomics is concerned with the economy as a whole is quite evident from its discussing the problem of allocation of resources in the society and judging the efficiency of the same. Both microeconomics and macroeconomics analyse the economy as a whole but with two different ways or approaches. Microeconomics examines the economy as a whole, so to say *microscopically*, that is, it analyses the behaviour of individual economic units of the economy, their inter-relationships

5. Abba P. Lerner, Microeconomic Theory, printed in *Perspectives in Economics*, edited by Brown, Neuberger and Palmatier (Preliminary edition, 1968), p. 50.

and equilibrium adjustment to each other which determine the allocation of resources in the society. This is known as *general equilibrium analysis*.

No doubt, microeconomic theory mainly makes particular or partial equilibrium analysis, that is the analysis of the equilibrium of the individual economic units, taking other things remaining the same. But microeconomic theory, as stated above, also concerns itself with general equilibrium analysis of the economy wherein it is explained how all the economic units, various product markets, various factor markets, money and capital markets are inter-related and interdependent to each other and how through various adjustments and readjustments to the changes in them, they reach a general equilibrium, that is, equilibrium of each of them individually as well as collectively to each other. Professor A. P. Lerner rightly points out, "Actually microeconomics is much more intimately concerned with the economy as a whole than is macroeconomics, and can even be said *to examine the whole economy microscopically*. We have seen how economic efficiency is obtained when the "cells" of the economic organism, the households and firms, have adjusted their behaviour to the prices of what they buy and sell. Each cell is then said to be 'in equilibrium.' But these adjustments in turn affect the quantities supplied and demanded and therefore also their prices. This means that the adjusted cells then have to readjust themselves. This in turn upsets the adjustment of others again and so on. An important part of microeconomics is to examine whether and how *all* the different cells get adjusted at the same time. This is called *general equilibrium analysis* in contrast with *particular equilibrium* or *partial equilibrium analysis*. General equilibrium analysis is the microscopic examination of the inter-relationships of parts within the economy as a whole. Overall economic efficiency is only a special aspect of this analysis."[6]

Importance and Uses of Microeconomics

Microeconomics occupies a vital place in economics and it has both theoretical and practical importance. It is highly helpful in the formulation of economic policies that will promote the welfare of the masses. Till recently, especially before Keynesian Revolution, the body of economics consisted mainly of microeconomics. In spite of the popularity of macroeconomics these days, microeconomics retains its importance, theoretical as well as practical. It is *microeconomics that tells us how a free-market economy with its millions of consumers and producers work to decide about the allocation of productive resources among the thousands* of goods and services. As Professor Watson says, "microeconomic theory explains the composition or allocation of total production, why more of some things are produced than of others."[7] He further remarks that microeconomic theory has many uses. The greatest of these is depth in understanding of how a free private enterprise economy operates."[8] Further, it tells us *how the goods and services produced are distributed among the various people for consumption through price or market mechanism*. It shows how the relative prices of various products and factors are formed, that is, why the price of cloth is what it is and why the wages of an engineer are what they are and so on.

Moreover, as described above, microeconomic theory *explains the conditions of efficiency in consumption and production* and highlights the factors which are responsible for the departure from the efficiency or economic optimum. On the basis of this microeconomic theory *suggests suitable policies* to promote economic efficiency and welfare of the people. Thus, not only does microeconomic theory describe the actual operation of the economy, it has also a normative role in that it suggests policies to eradicate "inefficiency" from the economic system so as to maximise the satisfaction or welfare of the people. The usefulness and importance of microeconomics has been nicely stated by Professor Lerner. He writes, "Microeconomic theory facilitates the understanding of what would be a hopelessly complicated confusion of billions of facts by constructing simplified models of behaviour

6. Abba P. Lerner, *op. cit.*, pp. 36-37.
7. D. S. Watson, *Price Theory and its Uses* (1963), p. 5.
8. *Ibid.*, p. 11.

which are sufficiently similar to the actual phenomena to be of help in understanding them. These models at the same time enable economists to explain the degree to which the actual phenomena depart from certain ideal constructions that would most completely achieve individual and social objectives. They thus help not only *to describe* the actual economic situation but to *suggest policies* that would most successfully and most efficiently bring about desired results and to predict the outcomes of such policies and other events. Economics thus has descriptive, normative and predictive aspects."[9]

We have noted above that microeconomics reveals how a decentralised system of a free private enterprise economy functions without any central control. It also brings to light the fact that the functioning of a completely centrally directed economy with efficiency is impossible. Modern economy is so complex that a central planning authority will find it too difficult to get all the information required for the optimum allocation of resources and to give directions to thousands of production units with various peculiar problems of their own so as to ensure efficiency in the use of resources. To quote Professor Lerner again, "Microeconomics teaches us that completely 'direct' running of the economy is impossible — that a modern economy is so complex that no central planning body can obtain all the information and give out all the directives necessary for its efficient operation. These would have to include directives for adjusting to continual changes in the availabilities of millions of productive resources and intermediate products, in the known methods of producing every thing everywhere, and in the quantities and qualities of the many items to be consumed or to be added to society's productive equipment. The vast task can be achieved, and in the past has been achieved, only by the development of a decentralised system whereby the millions of producers and consumers are induced to act in the general interest without the intervention of anybody at the centre with instructions as to what one should make and how and what one should consume."[10]

Microeconomic theory shows that welfare optimum or economic efficiency is achieved when there prevails perfect competition in the product and factor markets. Perfect competition is said to exist when there are so many sellers and buyers in the market that no individual seller or buyer is in a position to influence the price of a product or factor. Departure from perfect competition leads to a lower level of welfare, that is, involves loss of economic efficiency. It is in this context that a large part of microeconomic theory is concerned with showing the nature of departures from perfect competition and therefore from welfare optimum (economic efficiency). The power of giant firms or a combination of firms over the output and price of a product constitutes the problem of monopoly. Microeconomics shows how monopoly leads to misallocation of resources and therefore involves loss of economic efficiency or welfare. It also makes important and useful policy recommendations to regulate monopoly so as to attain economic efficiency or maximum welfare. Like monopoly, *monopsony* (that is, when a large buyer or a combination of buyers exercises control over the price) also leads to the loss of welfare and therefore needs to be controlled. Similarly, micro- economics brings out the welfare implications of oligopoly (or oligopsony) whose main characteristic is that individual sellers (or buyers) have to take into account, while deciding upon their course of action, how their rivals react to their moves regarding changes in price, product and advertising policy.

Another class of departure from welfare optimum is the problem of *externalities*. Externalities are said to exist when the production or consumption of a commodity *affects other people* than those who produce, sell or buy it. These externalities may be in the form of either *external economies* or *external diseconomies*. External economies prevail when the production or consumption of a commodity by an individual benefits other individuals and external diseconomies prevail when the production or consumption of a commodity by him *harms other individuals*. Microeconomic theory reveals that

9. A. P. Lerner, *op cit.*, p. 29.
10. A. P. Lerner, *op. cit.*, p. 33.

when the externalities exist, free working of the price mechanism fails to achieve economic efficiency since it does not take into account the benefits or harms made to those external to the individual producers and the consumers. The existence of these externalities requires government intervention for correcting imperfections in the price mechanism in order to achieve maximum social welfare.

Microeconomic analysis is also usefully applied to the various applied branches of economics such as Public Finance, International Economics. It is the microeconomic analysis which is used to explain the factors which determine the distribution of the *incidence or burden* of a commodity tax between producers or sellers on the one hand and the consumers on the other. Further, microeconomic analysis is applied to show the *damage* done to the social welfare or economic efficiency by the imposition of a tax. If it is assumed that resources are optimally allocated or maximum social welfare prevails before the imposition of a tax, it can be demonstrated by microeconomic analysis that what amount of the damage will be caused to the social welfare. The imposition of a tax on a commodity (*i.e.*, indirect tax) will lead to the loss of social welfare by causing deviation from the optimum allocation of resources, the imposition of a direct tax (for example, income tax) will not disturb the optimum resource allocation and therefore will not result in loss of social welfare. Further, microeconomic analysis is applied to show the *gain from international trade* and to explain the factors which determine the distribution of this gain among the participant countries. Besides, microeconomics finds application in the various problems of international economics. Whether devaluation will succeed in correcting the disequilibrium in the balance of payments depends upon the elasticities of demand and supply of exports and imports. Furthermore, the determination of the foreign exchange rate of a currency, if it is free to vary, depends upon the demand for and supply of that currency.

We thus see that microeconomic analysis is a very useful and important branch of modern economic theory.

MACROECONOMICS

We now turn to explain the approach and content of macroeconomics. As said above, word macro is derived from the Greek word '*makros*' meaning '*large*' and therefore macroeconomics is concerned with the economic activity *in the large*. Macroeconomics analyses the behaviour of the whole economic system in totality or entirety. In other words, macroeconomics studies the behaviour of the large aggregates such as total employment, the national product or income, the general price level of the economy. Therefore, macroeconomics is also known as *aggregative economics*. Macroeconomics analyses and establishes the functional relationship between these large aggregates. Thus Professor Boulding says, "Macroeconomics deals not with individual quantities as such but with the aggregates of these quantities; not with individual incomes but with the national income; not with individual prices but with the price level; not with individual output but with the national output.[11]

In his other famous work, "*Economic Analysis*", he similarly remarks, "Macroeconomics, then, is that part of the subject which deals with large aggregates and averages of the system rather than with particular items in it and attempts to define these aggregates in a useful manner and to examine their relationships."[12] Professor Gardner Ackley makes the distinction between the two types more clear and specific when he says, "macroeconomics concerns itself with such variables as the aggregate volume of output in an economy, with the extent to which its resources are employed, with the size of the national income, with the "general price level". Microeconomics, on the other hand, deals with the *division* of total output among industries, products and firms and the allocation of

11. K. E. Boulding, "*A Reconstruction of Economics*" (1950), p. 3.
12. K. E. Boulding, "*Economic Analysis*". p. 259.

resources among competing uses. It considers problems of income distribution. Its interest is in relative prices of particular goods and services.[13]

Macroeconomics should be carefully distinguished from microeconomics. It should be noted that microeconomics also does deals with some "aggregates" but not of the type with which macroeconomics is concerned. Microeconomics examines the behaviour of the industry in regard to the determination of its product price, output and employment, and the industry is an *aggregate of the various firms* producing the same or similar product. Likewise, micro-economic theory seeks to explain the determination of price of a product through the interaction of market demand and market supply for a product. Market demand for a product is the aggregate of individual demands of all consumers wishing to buy the product and the market supply of a product is the aggregate of the productions of many firms producing that product. Similarly, demand for and supply of labour in an industry of a city through which microeconomics explains wage determination are aggregative concepts.

But the aggregates with which macroeconomics is concerned are of somewhat different variety. Macroeconomics concerns itself with these *aggregates which relate to the whole economy*. Macroeconomics also discusses the sub-aggregates of the large aggregates relating to the whole economy, but these sub-aggregates of macroeconomics, unlike the aggregates of microeconomics which examines aggregates relating to a particular product, a particular industry or a particular market, cut across various products and industries. For example, the total production of consumer goods (*i.e.*, total consumption) and the total production of capital goods (*i.e.*, total investment) are two important sub-aggregates dealt with in macroeconomics but these aggregates are not confined to a single product or a single industry but instead they refer to all industries producing consumer goods and all industries producing capital goods. Moreover, the sub-aggregates, discussed in macroeconomics, add up to an aggregate for the whole economy. For instance, total consumption and total investment, two important sub-aggregates in macroeconomics, together constitute the total national product. Likewise, the total wage income *(i.e.*, total share of labour) and total profits (defined as total property income) add up to the national income. Professor Ackley thus says, "Macroeconomics also *uses* aggregates smaller than for the whole economy but only in a context which makes them sub-divisions of an economy wide total. Microeconomics also uses aggregates, but not in a context which relates them to *an economy-wide* total."[14]

Macroeconomics as a Study of Determination of Income and Employment

As we said above, the subject matter of microeconomics consists in explaining the determination of relative prices of products and factors and the allocation of resources based upon them. On the other hand, the subject matter of macroeconomic analysis is to explain what determines the *level of national income and employment*, and what causes *fluctuations* in the level of national income, output and employment. Further, it also explains the *growth* of national income over a long period of time. In other words, macroeconomics examines the determination of the level, fluctuations (cycles) and trends (growth) in the overall economic activity (*i.e.*, national income, output and employment).

There were no doubt pre-Keynesian theories of business cycles and the general price level which were "macro" in nature but it was late Lord J. M. Keynes who laid great stress on macro-economic analysis and put forward a general theory of income and employment in his revolutionary book, "*A General Theory of Employment, Interest and Money*" published in 1936. Keynes's theory made a genuine break from the classical economics and produced such a fundamental and drastic change in economic thinking that his macro-economic analysis has earned the names "*Keynesian Revolution*" and "*New Economics*". Keynes in his analysis made a frontal attack on the classical

---

13. Gardner Ackley, *Macroeconomic Theory*, 1961, p. 4.
14. *Op. cit.*, p. 5.

"*Say's Law of Markets*" which was the basis of full-employment assumption of classical economics and challenged the classical dictum that involuntary unemployment could not prevail in a free private enterprise economy. He showed how the equilibrium level of national income and employment was determined by aggregate demand and aggregate supply and further that this equilibrium level of income and employment may well be established at a far less than full employment level in a free private enterprise economy and thereby causing involuntary unemployment of labour on the one hand and excess productive capacity (*i.e.*, under-utilization of the existing capital) on the other. His macro-economic model reveales how consumption function, investment function, liquidity preference function, conceived in aggregative terms, interact to determine national income, employment, interest and the general price level.

Therefore, before showing how a level of income and employment is determined, we have to study the determinants of consumption function and investment function. The analysis of consumption function and investment function are the important subjects of macroeconomic theory. It is the total consumption demand and total investment demand taken together that constitute the level of aggregate demand which is the crucial determinant of the level of income and employment in the country.

Macroeconomics and General Level of Prices

Besides studying how a level of income and employment is determined in the economy, macroeconomics also concerns itself with showing how the general level of prices is determined. Keynes made a significant improvement over the Quantity Theory of Money by showing that the increase in the supply of money does not always bring about rise in prices. Important topic in this field is to explain the causes of inflation. Keynes, who before the Second World War showed that involuntary unemployment and depression were due to deficiency of aggregate demand, during the war period when prices rose very high he explained in a booklet entitled "*How to Pay for War*" that just as unemployment and depression were caused by deficiency of aggregate demand, inflation was due to excessive aggregate demand. Since Keynes's theory of inflation has been further developed and many types of inflation depending upon various causes have been pointed out. The problem of inflation is a serious problem faced these days, both by the developed and developing countries of the world. Theory of inflation is an important subject of macroeconomics.

Macroeconomics and Theory of Economic Growth

Another distinct and more important branch of macroeconomics that has been developed recently is the *theory of economic growth*, or what is briefly called *growth economics*. The problem of growth is a long-run problem and Keynes did not deal with it. It was Harrod [15] and Domar [16] who extended the Keynesian analysis to the long-run problem of growth with stability. They pointed out the dual role of investment— one of income generating, which Keynes considered, and the second of increasing capacity which Keynes ignored because of his pre-occupation with the short run. In view of the fact that an investment adds to the productive capacity (*i.e.*, capital stock), then if growth with stability (*i.e.* without secular stagnation or secular inflation) is to be achieved, income or demand must be increasing at a rate sufficient enough to ensure the full utilisation of the increasing capacity. Thus, macroeconomic models of Harrod and Domar have revealed the rate of growth of income that must take place if steady growth of the economy is to be achieved. These days growth economics has been further developed and extended a good deal. Though a general growth theory applies to both the developed and developing economies, special theories which explain the causes of under-development and poverty in developing countries and which also suggest strategies for initiating and accelerating growth in them have been propounded. These special growth theories relating to developing countries are generally known as *Economics of Development*.

---

15. R. F. Harrod, *Towards a Dynamic Economics* (1948).
16. E. D. Domar, Expansion and Employment, *American Economic Review* (March 1947).

INTERDEPENDENCE BETWEEN MICRO AND MACRO ECONOMICS

Actually micro and macro-economics are interdependent. The theories regarding the behaviour of some macroeconomic aggregates (but not all) are derived from theories of individual behaviour. For instance, the theory of investment, which is a part and parcel of the microeconomic theory, is derived from the behaviour of individual entrepreneur. According to this theory, an individual entrepreneur in his investment activity is governed by the expected rate of profit on the one hand and rate of interest on the other. And so is the aggregate investment function. Similarly, the theory of aggregate consumption function is based upon the behaviour patterns of individual consumers. It should be noted that we are able to derive aggregate investment function and aggregate consumption function from individual functions because in this respect the behaviour of the aggregate is in no way different from the behaviour patterns of individual components. Moreover, we can derive the behaviour of these aggregates only if either the composition of aggregates is constant or the composition changes in some regular way as the size of aggregates changes. From this it should not be understood that behaviour of all macroeconomic relationships is in conformity with behaviour patterns of individuals composing them. As we saw above, saving-investment relationship and wage-employment relationship for the economic system as a whole are quite different from the corresponding relationships found in case of individual parts.

Microeconomic theory contributes to macroeconomic theory in another way also. The theory of relative prices of products and factors is essential in the explanation of the determination of general price level. Even Keynes used microeconomic theory to explain rise in the price level as a result of increase in the money supply in the economy. According to Keynes, when as a result of the increase in money supply, aggregate demand in the economy increases and more output is produced, the cost of production rises. With the rise in the cost of production, the price level rises. According to Keynes, cost of production rises because of (1) the law of diminishing returns operates and (2) wages and prices of raw materials may rise as the economy approaches full-employment level. Now, the influence of cost of production, diminishing returns, etc., on the determination of prices are parts of microeconomics.

Not only does macroeconomics depend upon to some extent on microeconomics, micro-economics also depends upon to some extent on macroeconomics. The determination of rate of profit and rate of interest are well-known microeconomic topics, but they greatly depend upon the macroeconomic aggregates. In microeconomic theory, the profits are regarded as reward for uncertainty bearing but microeconomic theory fails to show the economic forces which determine the magnitude of profits earned by the entrepreneur and why there are fluctuations in them. The magnitude of profits depends upon the level of aggregate demand, national income, and the general price level in the country. We know that at times of depression when the levels of aggregate demand, national income and price level are low, entrepreneurs in the various fields of the economy suffer losses. On the other hand, when aggregate demand, incomes of the people, the general price level go up and conditions of boom prevail, entrepreneurs earn huge profits.

Now, take the case of the rate of interest. Strictly speaking the theory of the rate of interest has now become a subject of macroeconomic theory. Partial equilibrium theory of interest which belongs to microeconomic theory does not reveal all the forces which take part in the determination of the rate of interest. Keynes showed that the rate of interest was determined by the liquidity preference function and the stock (or supply) of money in the economy. The liquidity preference function and the stock of money in the economy are macroeconomic concepts. No doubt, the Keynesian theory of interest has also been shown to be indeterminate, but in the modern theory of interest Keynesian aggregative concepts of liquidity preference and stock of money play an important role in the determination of rate of interest. Moreover, in the modern interest theory (intersection of

LM and IS curves), along with liquidity preference and supply of money, the other two forces which are used to explain the determination of interest are saving and investment functions which are also conceived in aggregative or macro terms.

It is thus clear from above that the determination of the profits and rate of interest cannot be explained without the tools and concepts of macroeconomics. It follows that though microeconomics and macroeconomics deal with different subjects, there is great interdependence between them. In the explanation of many economic phenomena, both micro and macro economic tools and concepts have to be applied. About interdependence between microeconomics and macroeconomics, Professor Ackley's remarks are worth quoting. He says, "The relationship between macroeconomics and theory of individual behaviour is a two-way street. On the one hand, microeconomic theory should provide the building blocks for our aggregate theories. But macroeconomics may also contribute to microeconomic understanding. If we discover, for example, empirically stable macroeconomic generalisations which appear inconsistent with microeconomic theories, or which relate to aspects of behaviour which microeconomics has neglected, macroeconomics may permit us to improve our understanding of individual behaviour."[17]

QUESTIONS FOR REVIEW

1. Distinguish between microeconomics and macroeconomics.

2. What is microeconomics? Explain the subjects which are studied in microeconomics.

3. What is meant by economic efficiency? In which branch of economics, the subject of economic efficiency is studied? Explain the following concepts of economic efficiency:
   (*i*) Production efficiency
   (*ii*) Efficiency in distribution
   (*iii*) Efficiency in resource allocation.

4. Microeconomic theory is intimately concerned with the question of efficiency and welfare. Explain.

5. Microeconomics is a microscopic study of the economy. Explain.

6. Explain the importance of microeconomic theory and describe the fields of economics in which it has been usefully applied.

7. What is macroeconomics? Explain important issues which form the subject matter of macroeconomics .

8. Macroeconomics is a study of economic aggregates. Discuss.

9. Why is there a need for a separate study of macroeconomics? Explain giving some examples of *macroeconomic paradoxes*.

10. Distinguish between Micro and Macroeconomics. Show how the two are interdependent.

11. Which of the following subjects fall under microeconomics and which ones under macroeconomics?
   (1) Determination of wage rate
   (2) Determination of the level of employment in the economy
   (3) How is general price level determined?
   (4) Determination of price of wheat
   (5) How is national income determined?
   (6) What determines rate of economic growth of a country?

---

17. Gardner Ackley, *op. cit.*

12. Tick the correct answer. Macroeconomics is concerned with:

(*a*) The level of output of goods and sources.
(*b*) The determination of the level of employment and unemployment.
(*c*) The general level of prices and inflation.
(*d*) The growth of real output.
(*e*) All the above.

13. The study of which of the following problems fall within the purview of microeconomics:

(1) Allocation of resources
(2) How to produce
(3) For whom to produce
(4) Unemployment of labour
(5) Economic efficiency in resource allocation
(6) Economic growth of a country.
(7) The level of national income.

14. Tick which of the following topics fall within the purview of macroeconomics:

(1) Inflation in the economy
(2) Price of steel
(3) Level of national income
(4) Output of cotton fabrics
(5) Income from film industry
(6) Unemployment in the economy
(7) Growth of the economy
(8) Per capita income of a country
(9) Cement production in a year.

15. What is meant by economic efficiency? How do monopoly and externalities prevent the achievement of economic efficiency?

16. Explain how the relationship between macroeconomics and microeconomics is a two-way street?

CHAPTER 3
THE METHODOLOGY OF ECONOMICS

Every science derives hypotheses, generalisations, principles, laws and theories which explain the behaviour of phenomena it studies. For deriving generalisations and theories it has to adopt a certain methodology. One of the controversies in economics relates to the kind of methods to be adopted to discover generalisations and theories about the relationship between economic facts. Economists have in fact adopted both deductive and inductive methods of reasoning. Recently econometrics, that is, application of statistical methods to test economic hypotheses have gained much popularity among economists. These different methods will be explained below.

The other aspect of the method or technique of economic analysis is whether it should be of the nature of statics, comparative statics or dynamics. These methods along with the concept of equilibrium shall be explained in the next chapter.

Nature of a Scientific Theory

A scientific theory sets up a relationship between facts or, in other words, it describes cause and effect relationship between various variables. The variables with which economists are concerned are prices, quantities demanded and supplied, the money supply, national income, employment, wages, profits, etc. Every theory is based upon a set of assumptions, often called premises or postulates. It is worth mentioning that some assumptions are taken merely to simplify the analysis, though they may not be entirely realistic. In economics, these assumptions may be behavioural, that is, relating to the behaviour of economic variables or they may be technological pertaining to the production technology and the availability of productive factors. From the assumptions or postulates taken implications or conclusions are deduced through logical process of reasoning. The process of logical deduction to discover relevant conclusions from a set of definitions and assumptions is carried out either in words or in the language of symbolic logic or it may be done with the aid of geometry or more formal mathematics. It is these conclusions drawn from the assumptions through deductive logic which are called hypotheses.

It is worth noting that a *scientific hypothesis or a theory states the proposition about relationship between facts or variables in a form that is testable or falsifiable, that is, proposition which is capable of being refuted.* If the predictions based on a hypothesis are refuted by the direct observation of actual facts or through the statistical methods of interpreting actual facts, a hypothesis stands rejected. If on testing the predictions based on a given hypothesis are proved correct, it stands established as a scientific theory. For instance, the quantity demanded varies inversely with price is one of the important economic hypotheses established in economics. If a sales tax is imposed on a commodity and as a result the price of the commodity rises, the prediction will be that the quantity demanded will decline, other things remaining constant. This has not been falsified and in fact has been corroborated by the facts of the real world. So the law of demand stating that there is inverse relationship between price and quantity demanded is a scientific economic law. Likewise, the generalisation regarding the direct relationship between price and quantity supplied, that is, the higher the price, the greater will be the quantity supplied, other factors held constant, has also been found to be consistent with facts in several cases.

Further, the Keynesian theory that under conditions of less than full-employment of resources, levels of national income and employment are determined by the magnitude of aggregate effective demand is also a well established economic hypothesis regarding the developed capitalist economies. The predictions based on the above Keynesian hypothesis that the increase in aggregate demand through deficit budgeting by the Government under conditions of less than full-employment will lead to the rise in national income and employment has been found to be consistent with facts. Therefore, Keynesian principle of effective demand is a very important hypothesis about the advanced capitalist economies which has been proved to be valid by empirical evidence.

If the predictions based on a hypothesis are falsified by the facts of the real world, either some error would have been committed during the process of logical deduction or the assumptions made would have been too unrealistic, wrong or irrelevant to the subject of economic enquiry. Thus, in order to establish scientific economic hypotheses, error in logic and mistake of making unrealistic assumptions be avoided. It is worth emphasising that *every hypothesis or theory is based upon some simplifying assumptions which are not quite realistic*, that is, it is an abstraction from reality. But good hypotheses and theories abstract from reality in a useful and significant way. Indeed, if we do not abstract from reality we would merely duplicate the real world in a camera like manner and will not gain any understanding of it. The crucial test of a theory is that whether predictions which follow from it are falsified or not by the empirical evidence, that is, by the facts in the real world. If the predictions of a hypothesis or theory are found to be consistent with the facts, its assumptions would be justified even if they are unrealistic. Thus, an economic hypothesis or theory should not be criticised simply because assumptions it makes are unrealistic, if its predictions are found to be consistent with the facts.

METHODS OF ECONOMIC ANALYSIS

After having explained the scientific nature of economic theory we are now in a position to explain in detail how the generalisations in economics are derived and to clearly bring out the nature of economic reasoning. Economic generalisations describe the laws or *statements of tendencies* in various branches of economics such as production, consumption, exchange and distribution of income. In the view of Robbins, economic generalisations or laws are statements of uniformities which describe human behaviour in the allocation of scarce resources between alternative ends. The generalisations of economics, like the laws of other sciences, state cause and effect relationships between variables and describe those economic hypotheses which have been found consistent with facts or, in other words, have been found to be true by empirical evidence. But a distinction may be drawn between a generalisation (law) and a theory. A law or generalisation *just describes* the relationship between variables; it does not provide any explanation of the described relation. On the other hand, a theory provides an explanation of the stated relation between the variables, that is, it brings out the logical basis of the generalisation. An economic theory or a model derives a generalisation through process of logical reasoning and explains the conditions under which the stated generalisation will hold true.

Deductive Method

Generalisations in economics have been derived in two ways: (1) *Deductive Method*, (2) *Inductive Method*. We shall first explain the deductive method of deriving economic generalisations. The deductive method is also called *abstract, analytical* and *a priori* method and represents an abstract approach to the derivation of economic generalisations and theories. The principal steps in the process of deriving economic generalisations through deductive logic are: (*a*) perception of the problem to be enquired into; (*b*) defining precisely the technical terms and making appropriate assumptions, often called postulates or premises; (*c*) deducing hypotheses, that is, deriving conclusions from the premises through the process of logical reasoning; and (*d*) testing of hypothesis deduced.

(*a*) **Perception of the Problem.** In any scientific enquiry, the analyst or theorist must have a clear idea of the problem to be enquired into. He must know the significant variables regarding whose behaviour and interrelationship he wants to derive generalisations. The perception of the problem is by no means an easy task.

(*b*) **Definition of Technical Terms and Making of Assumptions.** The next step in the process of deriving economic generalisations is to define precisely and unambiguously the various technical terms to be used in the analysis as well as to state clearly the assumptions he makes to derive generalisations. As mentioned above, assumptions may be behavioural pertaining to the behaviour of the economic variables or they may be technological relating to the state of technology and the factor endowments. The crucial assumptions are made on the basis of observations or introspection. A crucial assumption that has been taken in economics is that consumers try to maximise their satisfaction and producers try to maximise their profits. Likewise, it is assumed that investors try to minimise their risk and maximise the expected rate of their profits. Some of the assumptions are made merely to simplify the analysis and may not be quite realistic. The actual economic world is quite complex and full of details in which numerous factors play a part and act and interact on each other. The introduction of simplifying assumptions is quite necessary in order to bring out the importance of really significant factors having a bearing on the problem under investigation. According to Prof. Boulding, economic theory represents just a 'map' of real world phenomenon and not a perfect picture of it. To quote him, "Just as we do not expect a map to show every tree, every blade of grass in a landscape, so we should not expect economic analysis to take into account every detail and quirk of real economic behaviour."[1] It, therefore, follows that each and every assumption made by a theory may not be realistic. The crucial factor in building up a valid theory is whether its predictions are corroborated by the facts in the world. A correct scientific theory or generalisation must be expressed in the form of a hypothesis that is conceivably refutable. As mentioned above, Professor Friedman in his now well-known article, "*The Methodology of Positive Economics*" has expressed the view that undue importance should not be given to the 'realism' of assumptions. What matters most from the viewpoint of scientific theory, according to him, is whether it enables us to predict things accurately.

(*c*) **Deducing Hypotheses through Logical Deduction.** The next step in deriving generalisations through deductive logic is deducing hypotheses from the assumptions or premises taken. A hypothesis describes relationship between factors affecting a phenomenon; it establishes the cause and effect relationship between the variables having a bearing on the phenomenon. Then, through logical process, hypothesis is deduced from the assumptions made. This logical reasoning may be carried out verbally or it may be conducted in symbolic terms using the language of what is known as symbolic logic. The geometric or graphic technique is also usually employed to deduce the hypotheses about the relationship between factors. Besides, the process of logical deduction may be done with the help of more formal mathematics. These days in almost all branches of modern economics, mathematics as tool of analysis for deriving economic theories and generalisations is being increasingly used. The use of mathematics in economic analysis proves extremely useful where geometrical methods make the analysis more complicated to comprehend. Besides, the use of mathematical method makes the derivation of economic hypotheses more rigorous and exact.

It is worthwhile to note that in deriving analytically sound hypotheses, one should guard against committing *logical fallacy* in the process of logical deduction. For instance, it is inappropriate to conclude that A must be the cause of B, if A happens to precede B. Further, it is logically fallacious to argue that since there exists a high degree of correlation between the two factors, say between the supply of money and the general price level, the former must be the cause of the latter, unless the causation must be logically developed.

(*d*) **Testing or Verification of Hypotheses.** Hypotheses obtained above have to be verified before they are established as generalisations or principles of economics. For the verification of hypoth-

1. K. E. Boulding, *Economic Analysis*, Hamish Hamilton, New York, 1956, p. 11.

eses, economists cannot make controlled experiments, because they have to discover uniformities in behaviour patterns of man. We cannot make experiments with man under controlled conditions, such as in laboratories as physical scientists make experiments with inanimate objects of nature and biologists make these with animals and plants. Therefore, economists have to rely on uncontrolled experience and observations. The informations regarding uncontrolled experience about the behaviour patterns concerning variables about man and the economy are quite amply available. The reliance by economists on uncontrolled experiences, however, does increase the number of observations required to verify the hypotheses or to establish the generalisations. Besides, the need to rely on uncontrolled experiences complicates the analysis and requires that facts must be carefully interpreted to discover successfully the significant relationship between relevant economic variables. Prof. Baur rightly remarks, "The need to rely on uncontrolled experiences does, however, increase the number of observations required, and also complicates their successful analysis and interpretation, before we can discern successfully the significant uniformities and ascertain their limits."[2]

It may, however, be pointed out that in spite of the complexities and difficulties involved in verifying economic hypotheses through successful analysis and proper interpretation of uncontrolled experiences and observations, several useful and significant generalisations have been established in economics. In the field of microeconomics, the well-established generalisations relate to the inverse relationship between price and quantity demanded, the direct relation between price and quantity supplied, the tendency of the price of the product to be equal to the marginal cost under conditions of perfect competition, and the tendency for the wages to be equal to the value of marginal product under conditions of perfect competition and several others. In the field of macro-economics, established generalisations relate to the determination of the level of national income by aggregate demand and aggregate supply in a capitalist economy, the multiple increase in income and employment as a result of a given initial increase in investment depending upon the size of marginal propensity to consume, the dependence of the amount of investment on the marginal efficiency of capital and the rate of interest and several others.

It is worth noting that the absence of controlled experiments in economics affects the exactness of economic laws and generalisations.[3] This means that the generalisations in economics are not as exact as those of physical sciences and they are therefore not universally applicable under all circumstances. Because of the absence of controlled experiments economic generalisations lack in firmness, they are not easily accepted by all and even generalisations that are refuted by empirical evidence are not abandoned for good by all. Prof. Baur rightly points out, the absence of the vivid and dramatic evidence provided by the controlled experiments adds greatly to the difficulty of securing acceptance for generalisations which are amply justified by the analysis of the available evidence."[4] Likewise, absence of controlled experiments, according to Friedman, renders the weeding out of unsuccessful hypotheses slow and difficult. They are seldom downed for good and are always cropping up again.[5]

In regard to framing and testing of economic generalisations, two related distinctions must be borne in mind. First, functional relationship between economic variables and a historically sequence of events must be distinguished. For instance, the law of demand stating inverse relationship between price and quantity demanded does not become invalid in view of the fact that both prices and quantities sold of many commodities increase during boom periods. This is because certain other forces such as a rise in aggregate investment demand operate which causes increase in both

---

2. P. T. Baur, *Economic Analysis and Policy in Underdeveloped Countries*, Routledge and Kegan Paul Ltd., 1957, p. 8.
3. P. T. Baur, *op. cit.*, p. 9.
4. *Ibid*, p. 9.
5. Milton Friedman, *Essays in Positive Economics*, Chicago, 1953, p. 11.

the price and quantity sold during a boom period. Second, prediction of a generalisation to show its validity must be carefully differentiated from the forecasting of future events; actual events may not exactly come about as predicted by a generalisation and yet that generalisation may be correct. This is because, as mentioned above, the actual course of events is governed by several other factors assumed by a generalisation which remains constant under the qualification "other things remaining the same". Thus, "even if the prediction that producers of a particular crop respond to a higher price by producing more is correct, this prediction does not enable us to forecast accurately next year's output (still less the harvest in the more distant future) which in the event will be affected by many factors besides changes in price."

In the absence of controlled experiments, for the verification of their theories economists have to rely on the direct observations of events in the real world. By direct observations we mean "gathering of information personally or reliance on comparatively unprocessed material such as files of business firms and government departments, locally published reports, proceedings of representative assemblies, newspapers, advertisements, market reports, auction notices and the like." In order to prove the validity of hypotheses and therefore to establish laws or generalisations, importance of direct observations cannot be underrated. Thus Prof. Baur correctly assets, "The depth and significance of economic generalisations depend on the quality of the underlying observations and analysis."[6]

Testing of Economic Hypotheses through Statistical Methods

In recent years a very useful method to test economic hypothesis has been developed. This is the *statistical method* or what is now popularly called *econometric method.* The statistical[7] or econometric method to verify and establish the theoretical generalisations occupies an important place because of the limited applicability of controlled experimentation in economics. The various statistical methods such as regression analysis have been developed to empirically test the economic hypotheses on the basis of collected economic data. The merit of econometrics is that the degree of functional relationship between relevant economic variables in precise quantitative terms is obtained by it and also the level of significance of the results can also be estimated. Recently, econometric method has been used to establish the precise relationships between money supply and the price level, quantity of money and the national income, consumption and income, capital accumulation and rate of economic growth and so forth.

It may, however, be pointed out that statistical analysis or econometrics alone cannot be used to derive and establish economic principles and theories. Economic hypotheses or theories must be developed logically before we can meaningfully use statistical analysis to test and verify them. Indeed, a theory or hypothesis is needed beforehand for selecting the relevant facts and data regarding relevant variables which can be subjected to empirical testing through the methods of econometrics. Prof. Myrdal is quite right, when he says, "theory, therefore, must always be *a prior* to the empirical observation of facts. Facts come to mean something only as ascertained and organised in the frame of a theory. Indeed, facts as part of scientific knowledge have no existence outside such a frame. Questions must be asked before answers can be obtained and, in order to make sense, the questions must be part of a logical co-ordinated attempt to understand social reality as a whole. A non-theoretical approach is, in strict logic, unthinkable."[8]

Principal steps followed in formulation of economic theories and generalisations through deductive method can be summarised as given below.

6. P. T. Baur, *Economic Analysis and Policy in Underdeveloped Countries*, 1957, pp. 10-11.
7. *Ibid*, p. 9.
8. Gunnar Myrdal, *Value in Social Theory*, Routledge and Kegan Paul Ltd., London, 1958, p. 233.

Various Steps in Deductive Method

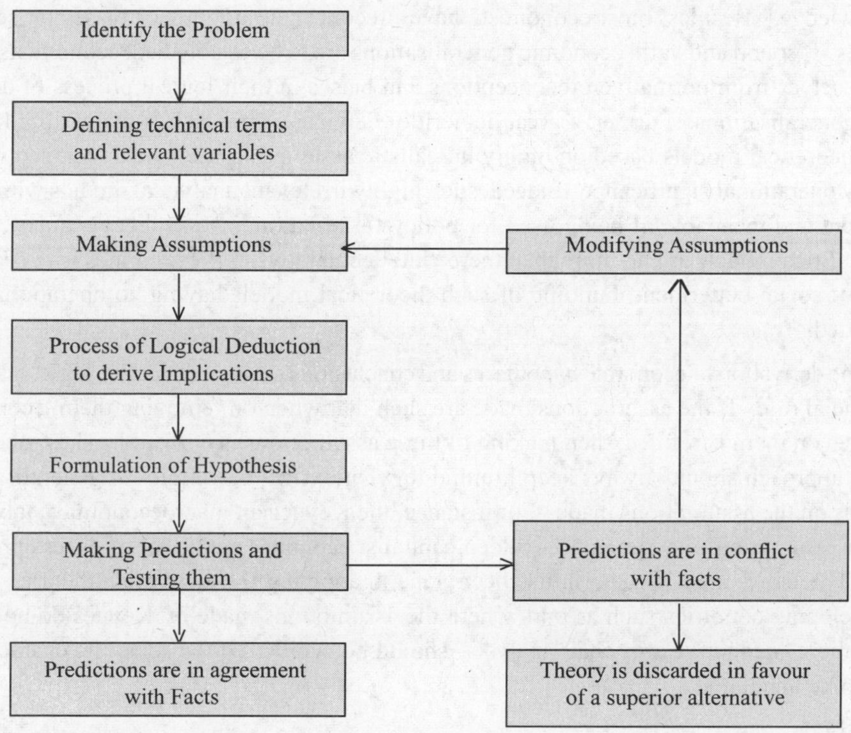

Fig. 3.1

Merits and Demerits of Deductive Method

The deductive approach to establish economic generalisations was extensively used by Classical and Neo-Classical economists such as Ricardo, Malthus, Senior, J. S. Mill, Marx, Marshall and Pigou. It still remains popular with modern economists as it has several merits. First, useful mathematical techniques can be employed to derive laws and theories of economics. With the aid of rigorous mathematical logic, economic theories can be developed through the process of deduction which can successfully explain economic phenomena. Secondly, through deductive logic useful economic theories can be derived without the tenuous and detailed collection and analysis of data which are required under the inductive method. Thus, as compared to inductive method, deduction is less time-consuming and less expensive. Thirdly, in view of the limited scope for controlled experimentation in economics, the method of deduction is extremely useful method of constructing economic theories. This is because several forces act simultaneously on an economic phenomenon and it is not possible to eliminate some of these by means of a controlled experiment. This indicates the crucial importance of deductive logic for building up economic principles or theories. Fourthly, the use of sophisticated mathematical methods in the deductive approach enables the economists to introduce accuracy and exactness in economic principles and theories.

In spite of the above-mentioned merits, shortcomings of the deductive approach should not be overlooked. The use of deductive method in deriving economic generalisations requires the use of a high-level competence in logic and theoretical abstraction. A good deal of care and objectivity is needed to avoid bad logic or faulty economic reasoning. Prof. Blaug rightly opines, "It is perfectly true that economists have often deceived themselves and their readers by engaging in what Leontief once called "implicit theorising" presenting tautologies in the guise of substantive contributions to economic knowledge." Besides, most economists have preconceived notions or biases on several economic issues. If sound and valid economic generalisations are to be established, economists must dissociate themselves from normative preconceptions and biases in their logical process of deducing economic generalisations. Further, a great demerit of deductive approach is that with it highly sophisticated theoretical models based on highly unrealistic assumptions may be developed which do not have any operational significance. Indeed, such highly irrelevant analytical models with little empirical content and incapable of being used for policy formulation have in fact been developed by economists. Such models are no more than mere "intellectual toys". If economics is to serve as an instrument of social betterment, building of such theoretical models having no operational use should be avoided.

Lastly, in the derivation of economic hypotheses and conclusions through deductive logic, assumptions play a crucial role. If the assumptions made are such that when on removing them, economic hypothesis based on them is refuted, then making of these assumptions is not valid. Thus, one who uses deductive approach should always keep in mind to what extent the validity of generalisations derived depends on the assumptions made. For instance, the Keynesian macroeconomic analysis is based upon the assumption of a depression-ridden capitalist economy with a lot of excess productive capacity. Therefore, a positive harm has been done in applying the Keynesian theories in the context of developing countries such as ours where the assumptions made by Keynes do not hold good. Hence, mere "deductive arm-chair analysis" should be avoided, if the scientific character of economics is to be maintained.

Inductive Method

The inductive method which is also called empirical method derives economic generalisations on the basis of experience and observations. In this method detailed data are collected with regard to a certain economic phenomenon and effort is then made to arrive at certain generalisations which follow from the observations collected. But, it is worth mentioning that the number of observations has to be large if it can yield a valid economic generalisation. One should not generalise on the basis of a very few observations. There are three ways which can be used for deriving economic principles and theories. They are: (*a*) experimentation, (*b*) observations, (*c*) statistical or econometric method. As has been mentioned above, the experimentation, that is, the use of contrived experiments is of limited applicability in economics. First, unlike natural sciences which are concerned with analysing the behaviour of either inanimate objects or obedient animals such as rats and rabbits under the influence of chloroform, economics deals with the behaviour of man who is quite fickle, wayward and unmanageable. Besides, man cannot tolerate the idea of being experimented upon, either individually or collectively. Secondly, an economic phenomenon is the result of multiplicity of factors and causes acting and interacting upon each other. Therefore, economic phenomenon does not repeat itself in the same uniform pattern. Numerous factors acting on an economic phenomenon 'disturb' it and make its exact repetition unlikely. Thus, as compared with the natural phenomena, economic phenomena are of less uniform pattern, less repetitive and more variable.

Thirdly, economists study the economic phenomena in which pressure groups such as employers' associations, trade unions, farming lobby, political parties with their different ideologies play a crucial part and their activities render it difficult to make controlled experiments in the economic world. However, in spite of these difficulties, experimental method can be used in some fields. For instance, experiments have been conducted to find out which law of production is valid, that is, whether law of diminishing returns, law of constant returns or law of increasing returns operates in the real world. Besides, public undertakings or big industrial firms often try to assess the effect of the changes in the prices of their products on the demand for it and thus find out the demand elasticity of their products.

Various Steps in Inductive Method

Various steps are gone through in developing economic theories through inductive method. The first step, as in the deductive approach, is to *identify the problem.* The second step is *defining technical terms and variables related to the problem.* It is the next step which is peculiar to the inductive method, namely, the *collection of data* about the variables related to the problem and doing some preliminary thinking about the possible functional relationships between the relevant variables.

The next important step in the construction of economic theories in this method is the *processing of data* collected and finding out what relations between the variables actually hold good. From this, a theory is developed which can be further refined and tested through statistical methods. Once the theory has been developed one can make predictions on its basis, as is done in the deductive approach. If predictions of theory are in agreement with the facts and actual behaviour of the economy, then a new reliable theory has been developed. If a *new theory explains "how things work" better than the existing ones, it replaces them.*

However, if predictions are in conflict with actual facts and behaviour of the economy, either the theory is discarded or fresh efforts are made to *modify and refine it by collecting more data and processing them.* The various steps in the construction and development of economic theories through inductive method are illustrated in Fig. 3.2.

Various Steps in Inductive Method

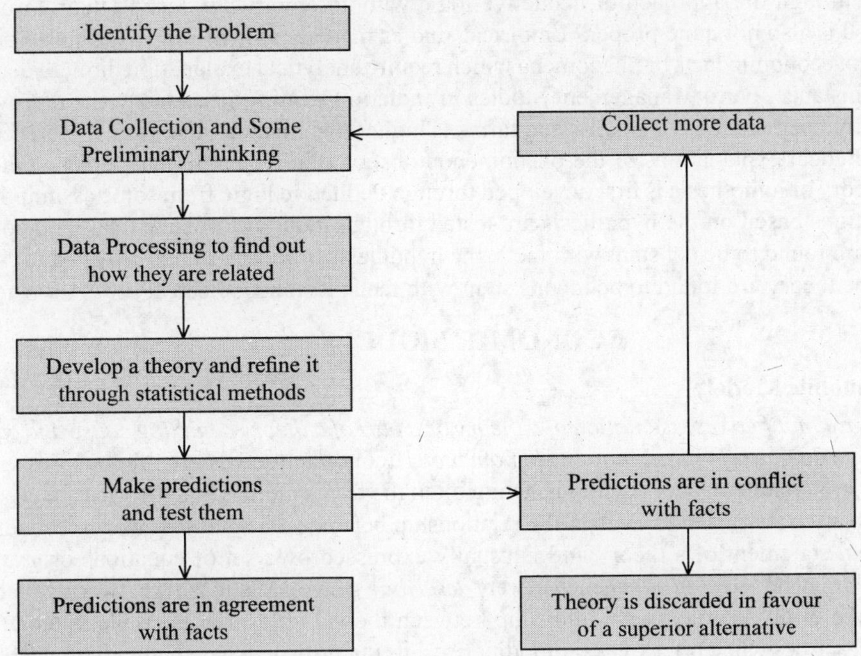

Fig. 3.2

Evaluation of Inductive Method

As has been explained above, observations of facts through collection of detailed data and the use of statistical methods to arrive at economic generalisations establishing relationship between facts are being increasingly made. Some of the recent researches in the field of macroeconomics, such as the nature of consumption function describing the relation between income and consumption, the principle of acceleration describing the factors which determine investment in the economy have been obtained through the use of mainly inductive method.

However, it needs to be emphasised again that the use of induction or empirical method is not of much value if it is not supported by the economic hypothesis or theory developed by deductive logic.

The inductive method can at best be used to empirically test the theory or hypothesis as to whether it is consistent with or refuted by facts. The inductive method has another limitation in that there is a great risk of conclusions being drawn from insufficient data. To obtain generalisations through inductive method, one should take care that sufficient number of observations or data has been taken into account. Besides, the collection of data itself is also not an easy task. And a researcher who wants to employ the inductive method to arrive at generalisations must have good knowledge of statistical methods, that is, he must know the art of collecting, processing and interpreting data. It is obvious that as compared with the deductive method, the inductive method is time-consuming and expensive.

Conclusion: Integration of Two Methods

Now, the controversy which existed among the earlier economists as to whether deductive or inductive approach is more appropriate in developing economic theories and principles has been resolved. The modern viewpoint in this regard is that both are needed for the proper development of scientific economic theories. Indeed, the two are complementary rather than competitive. The modern economists first derive economic hypotheses through the process of logical deduction and then empirically test them through statistical or econometric methods. Marshall rightly pointed out, "*induction and deduction are both needed for scientific thought as the right and left foot are both needed for walking*."[9] Empirical studies made through statistical or inductive method without a theoretical hypothesis to serve as a guide for the selection of data are quite useless. The derivation of economic generalisations through the approach of deductive logic without empirically testing them through inductive method is also not quite proper. Empirical studies made in inductive approach also bring to light significant economic facts or phenomena which require analytical explanation through deductive logic. For instance, Farm Management Studies in India in the mid- fifties led to the discovery of a fact that output per acre on the small-sized farms is higher than that on large farms. This led to the various theoretical explanations of the phenomenon observed in the empirical studies. On the other hand, a theory or hypothesis is first developed through deductive logic from some assumptions and then predictions based on the hypothesis are tested through inductive or statistical method. If the predictions are found to be constant with facts, the hypothesis or theory stands proved and if the predictions of the theory are found to be inconsistent with facts, it stands rejected.

ECONOMIC MODELS

What is an Economic Model?

A model is a simplified representation of a theory or a part of a theory depicting cause and effect relationships regarding a certain economic phenomenon. Economic models are often expressed in words or graphs or in mathematical equations. Quite often the terms models and theories are used as synonym as both are constructed to explain the relationship between variables. However, economic model is a *formal* statement of a theory and is usually expressed by a set of equations or graphs. Economic models may be '*general*' when they merely *describe* a general relationship between variables without saying the exact quantitative relationship between the variables. For example, a model of demand which describes when price of a commodity rises, its quantity demanded will fall is a general model. Models which describe a *quantitative relationship* between the variables are called *quantitative models*. For example, if a model describes that if price of cars rises by 10 per cent, the people will buy 15 per cent less of cars is a quantitative model. In the modern economics econometric model which are quantitative in nature have been built to describe relationship between economic variables in both the microeconomics and macroeconomics. Like other scientists, economists also build models *to analyse and predict economic events*. The economists have constructed *analytical models* to analyse and predict the behaviour of an individual consumer, a producer or the economy as a whole. A model is composed of a number of assumptions from which conclusions or implications have been derived. Two things must be noted with regard to the analytical models built by the economists. First, a model

9. Alfred Marshall, *Principles of Economics*, 8th Edition, p. 29.

must always abstract from reality. That is, in order to build a model one has to make some unrealistic assumptions to simplify it. Secondly, an analytical economic model consists of a number of equations or graphs that represent cause and effect relationship. Let us explain below these two attributes of analytical economic models in some detail.

Assumptions of a Model

An economic model is always based in some assumptions which do not represent the real economic situation in its totality. Although the assumptions made in building a model must have some relationship to the type of situation it is to analyse and explain, it need not be replica of the real world situation. A model represents only main significant features of real economic world. In fact, the real world is so complex that to understand it, it is necessary to simplify and abstract. One has to abstract from reality to some extent so that some useful and meaningful features of reality are brought out. However, a model is not a complete abstraction from reality, it abstracts from reality in some ways in order to focus those features of reality which are significant and useful for explaining the behaviour of a consumer, producer or the economy as a whole.

However, a care should be taken. We should not over-simplify so that a model that we build does not yield a distorted picture of the real world phenomenon. An over-simplified and distorted model is useless. Therefore, we should construct economic models in such a way that only irrelevant and unimportant considerations and factors are overlooked. The important factors that have a bearing on the real-world phenomena must be included in the construction of models, if they are to be useful in analysing and predicting things.

Equations or Graphs and Economic Models

Secondly, an economic model is usually composed of a set of equations or graphs expressing relationship between variables that are relevant to the problem to be investigated. Each equation attempts to explain the behaviour of one variable, that is, it seeks to establish cause and effect relationship in respect of an individual variable. It is worth mentioning that causation does not always run in one direction. There is a mutual relationship among various variables, that is, a variable influences the other variables and in turn is influenced by them. For instance, consumption depends upon income and also consumption being an important constituent of aggregate demand influences income. Therefore, in such a system values of various variables are to be determined simultaneously. Therefore, models which involve more than one equation attempt to solve these equations simultaneously.

Purposes or Uses of Economic Models

Now, an important question is why economists are interested in building models. Economic models are built for purposes of (*a*) *analysis* and (*b*) *prediction.* By analysis we mean how adequately we can explain the behaviour of an economic agent, that is, consumer, producer or the economic system. From a set of assumptions we derive through deductive logic certain laws which describe the behaviour of economic agents (consumer, producer or the whole economy) and which have a quite general application. On the other hand, prediction implies the ability of a model to forecast the effects of changes in some variables or magnitudes in the economy. For instance, a model of price determination through demand and supply is generally used to predict the effect of imposition of an excise duty or sales tax on the price of a commodity.

Usefulness of a model may be judged on the basis of either its explanatory or predictive power, or the realism of its assumptions, or the extent of its applicability. Economists differ as to what is more important attribute of a valid model. According to Milton Friedman, *the most important attribute of a model is its predictive power*, that is, to what extent it can correctly predict the behaviour of an economic unit. If the model has a good predictive power, then, according to his view, it is immaterial whether its assumptions are realistic or not. On the contrary, Paul Samuelson is of the view that *realism of assumptions and the analytical power of the model to explain the behaviour of consumers, producers or the economic system, are the essential attributes of a valid and satisfactory model.*

It may be noted that the general view among economists is that the most important attribute of a model depends on its purpose, that is, whether the model builder wants to use it for predicting the effect of a change in some variable or for analysing and explaining the particular behaviour of an economic agent (consumer, producer or economic system). Realism of assumptions and explanatory power are important features of a good model if the purpose of the model is the explanation of why an agent behaves as it does. On the other hand, if the purpose of building a model is to predict some event, then the power of the model to predict correctly will be its important feature whether or not its assumptions are realistic or not.

It will be useful for understanding economic models if we give an example of a economic models. We give below a widely used microeconomic model to illustrate how economists use economic model to get insight into the mechanism behind an economic phenomenon.

An Example of a Model: Demand-Supply Model

This is a simple microeconomic model which relates to the determination of price and output of a product through demand and supply. In this model, it is assumed that demand of consumers for a product is function of price, other factors influencing demand such as income, tastes, prices of other goods remaining constant. Further, for the sake of simplicity, it is assumed that this demand function is of linear type. Similarly, in this model, the quantity supplied of a commodity in the market is assumed to be a function of price alone, other factors affecting supply such as prices of inputs, the state of technology, prices of other products are assumed to remain constant. Besides, for the sake of simplification it is assumed that the quantity supplied is linearly related to price.

Further, in this demand-supply model it is assumed that perfect competition prevails in the market for the product. This implies that there are a large number of sellers and buyers of an identical product so that no single seller or buyer has any influence over the price of the product. This demand-supply model consists of the following three equations:

$$Q^d = a - bp \qquad \dots (i)$$

$$Q^s = c + dp \qquad \dots (ii)$$

$$Q^d = Q^s \qquad \dots (iii)$$

Where,

Q^d stands for quantity demanded

Q^s stands for quantity supplied

p stands for price of a commodity

a, b, c and d are constants

The above set of three equations constitutes a model. Here, in this model, the variables considered are price, quantity demanded, quantity supplied. The objective is to obtain the price and quantities demanded and sold when the system attains the equilibrium position. To find the equilibrium values of these variables, the above three equations are required to be solved simultaneously. The equation (*i*) describes that the quantity demanded is a decreasing function of price, that is, as price falls quantity demanded increases and vice versa. Equation (*ii*) describes that quantity supplied is increasing function of price, that is, as price rises quantity supplied increases and vice versa. Equation (*iii*) describes the equilibrium condition meaning thereby that the price that equates quantity demanded and quantity supplied will clear the market. Equilibrium price implies that both the buyers and sellers are satisfied with their purchases and sales and hence there is no tendency on their part to raise or lower the price. It is in this way that Marshall explained the determination of price of a commodity.

Equations of the model can be solved mathematically to obtain the equilibrium values of the variables. The equations of the model can also be represented by graphs to determine geometrically the equilibrium values of the variables. Let us illustrate this famous demand-supply micro-model of price determination of a commodity in a free market by taking specific demand and supply functions. Suppose the following demand and supply functions of a commodity are given:

$$Q^d = 100 - 20\,P$$
$$Q^s = -\,5 + 15\,P$$

Where Q^d stands for quantity demanded, Q^s for quantity supplied of a commodity and P for the price of the commodity.

Since for an equilibrium price the quantity demanded is equal to the quantity supplied, we put the two demand and supply equations as equal to each other. Thus, we have

$$Q^d = Q^s$$
$$100 - 20\,P = -\,5 + 15\,P$$
$$15\,P + 20\,P = 100 + 5$$
$$35\,P = 105$$
$$P = \frac{105}{35} = 3$$

Thus we get ₹ 3 as price per unit of the commodity. Putting the value of P in either the demand or supply equation we can obtain the equilibrium quantities Q^d and Q^s. Thus

$$Q^d = 100 - 20 \times 3 = 40$$

Or, $$Q^s = -5 + 15 \times 3 = 40$$

Thus, equilibrium price of the commodity equals ₹ 3 per unit and equilibrium quantity produced and purchased equal 40 units per period of time. We can now use the model to predict how price and quantity bought and sold will be affected if government imposes an excise duty on a commodity.

Keynes's Macroeconomic Model

Let us take Keynes's simple two sector model of determination of natural income. According to this model, national income is determined by effective aggregate demand or expenditure which consists of consumption and investment expenditure (I). This macromodel can be written as

$$AE = C + I \qquad \ldots (i)$$
$$C = a + bY \qquad \ldots (ii)$$
$$I = \bar{I} \qquad \ldots (iii)$$

Where C is consumption, a is autonomous consumption, b is marginal propensity to consume, $\bar{I}$ is autonomous investment and Y is national income.

In equilibrium, $Y = AE$

$$= C + I \qquad \ldots (iv)$$

Substituting $a + bY$ for C and $\bar{I}$ for I we have

$$= a + bY + \bar{I}$$
$$Y - bY = a + \bar{I}$$
$$Y(1 - b) = a + \bar{I}$$
$$Y = \frac{1}{1-b}(a + \bar{I})$$

Thus, national income is the sum of autonomous factors $(a + \bar{I})$ multiplied by the multiplier whose value is given by $\frac{1}{1-b}$.

NATURE OF ECONOMIC LAWS AND GENERALISATIONS

We have discussed the subject matter of economics and have seen that it is very wide and comprehensive. Economics is a science and like other sciences it also has its laws. Economic laws are also known as *generalisations, principles and uniformities*. The economic laws describe how a man behaves as a producer and a consumer. The economic laws are also concerned with how the economic system works and operates. Man in his economic life produces wealth, consumes wealth, exchanges it with others. Therefore, economic laws have been framed which govern production, consumption and exchange of wealth by men. Besides, economic laws are also concerned with how the national product produced is distributed and how the level of income and employment are determined. Lastly, economic laws also describe growth of the economy as well as international trade between the various countries of the world. In fact, economic laws have been framed in all fields of the subject matter of economics, namely, consumption, production, price determination, determination of income and employment, growth of the economy, foreign trade, etc. Among the important laws of economics, mention may be made of Law of Demand, Law of Diminishing Marginal Utility, Law of Variable Proportions or Diminishing Returns, Keynesian Psychological Law of Consumption, the Principles of Multiplier and Accelerator, Malthusian Law of Population, Law of comparative Advantage.

According to Marshall, "economic laws are these social laws which relate to branches of conduct in which the strength of the motives chiefly concerned can be measured by money price."[10] By this Marshall means that economists have framed laws and theories which explain the conduct or behaviour of man who tries to maximise some things or seek to fulfil his objectives, and these things and objectives must be measurable in terms of money. What is not measurable in terms of money does not come within the purview of economics. However, this is a very narrow view about economic laws. Therefore, Robbins broadened the scope of economic laws and brought out that whether any objective or conduct of man is concerned with money or not, it can come within purview of laws of economics if it is concerned with the problem of choice, that is, allocation of scarce resources among unlimited wants. According to Robbins, economic laws are statements of tendencies which govern human behaviour concerning the utilisation of scarce resources for the achievement of unlimited wants. In other words, when resources are scarce and wants are unlimited and therefore the problem of choice arises, economists have to frame the laws of choice, whether or not variables and objectives involved in it are measurable with money.

Economic Laws as Statements of Tendencies

The nature of economic laws has been a subject of controversy. Marshall thought that the laws of economics are not exact and definite; they are mere *statements of tendencies*. According to him, this is unlike the laws of physical sciences which are quite exact, precise and definite. Because of the exactness and definiteness, the laws of physical sciences can predict the course of events. But laws of economics lack this predictive value. Laws of economics are conditional and are associated with a number of qualifications and assumptions and these assumptions and qualifications are generally contained in the phrase "other things remaining the same, "or *Ceteris Paribus* which is attached to every law and theory of economics. But, in the real world, these other things generally do not remain the same because the economic world is dynamic and ever-changing. For example, according to the law of demand when price of a commodity rises, its quantity demanded by the consumer will fall. But if along with the rise in price of the commodity, income of the consumer increases, then the consumer may demand more of the commodity even at the higher price. This seems to be contrary to the law of demand but in fact this is not so because the law of demand assumes that other things, such as income, tastes and prices of the related goods remain the same, and in our case this qualification has not been fulfilled because income of the consumer has increased. Law of demand will hold good only if other things such as income, tastes and preferences, the prices of related goods remain constant and unchanged.

Likewise, according to the law of diminishing returns, when amount of labour to a given piece of land is increased, marginal product of labourer would diminish beyond a certain stage. But again in actual practice this may not happened so. It may happen that when the use of labour is increased on a given piece of land, improved and more productive technology is employed, then the marginal product of labour may increase rather than decline. Actually this has happened in the present-day developed countries where along with the growth of population and labour force on land, marginal product of labour has increased and this is due to the fact that there has been rapid progress in agricultural technology in these countries due to which marginal productivity has increased. But this also does not prove that law of diminishing returns is invalid. This is because law of diminishing returns also assumes that other things such as technology amount of capital *etc.* remain unchanged.

10. Alfred Marshall, *Principles of Economics*, 8th edition.

It is worth mentioning that laws of economics are not like those of the legal laws passed by the Government or Parliament. These legal laws are made by the Government or the Parliament in order to maintain law and order in the country, and it is compulsory for the citizens to obey those laws. If any man violates these legal laws he invites punishment from the government. On the other hand, economic laws tell us how a rational man behaves in his economic life.

Scientific Nature of Economic Laws

Laws of economics are of scientific nature. All scientific laws establish relationship between cause and effect. Economic laws also establish relationship between cause and effect about economic behaviour of man and economic phenomena. If we observe man in using his scarce resources to satisfy his unlimited wants, then we will see that he behaves in a particular manner. By observing the behaviour of several people, economists have established certain generalisations or general principles which are called economic laws. Therefore, these economic laws are general tendencies of man's behaviour in his economic life. Therefore economic laws are related to economic life of man. In his economic life, man produces wealth and consumes it. Besides, the distribution and exchange of wealth also affect the economic life of man. Economists have made several laws regarding production, consumption, distribution and exchange of wealth. Like other scientific laws, economic laws also establish relationship between cause and effect. For example, according to the law of demand, when price of a commodity falls, its quantity demanded increases, other things remaining the same. Here the fall in the price is the cause, and the rise in quantity demanded is the effect. Law of Diminishing Marginal Utility describes that as a man has more units of a commodity, its marginal utility goes on diminishing. Here, the increase in the quantity of the commodity is the cause and the fall in marginal utility is the effect. This holds goods in case of other economic laws also.

It is thus clear that *economic laws are hypothetical and conditional*. But this does not mean that economic laws are not scientific or that they are useless. As a matter of fact, all scientific laws are conditional. The famous scientific law of gravitation is also conditional. According to the law of gravitation, when any commodity is thrown in the air, then it falls down to the ground. This is because the earth has the power to attract and pull other things to itself. But this famous scientific law also depends upon the fulfilment of certain conditions. The condition for the law of gravitation to apply is that no opposing forces obstruct the commodity from falling to the ground. We often see that aeroplanes, birds, balloons, etc. fly in the air and don't fall to the ground. This is because some opposing forces are at work which prevent these things from falling to the ground. Due to the working of these opposing forces, law of gravitation does not apply in these cases. Therefore, law of gravitation also applies when the opposing forces do not operate. Take another example of a scientific law from the science of chemistry. According to a well-known law of chemistry, water is formed if two atoms of hydrogen are mixed with one atom of oxygen: but this law will also apply under specific conditions. Water will be formed with two atoms of hydrogen and one atom of oxygen under certain conditions of temperature and pressure. Thus, all scientific laws hold good under certain conditions. Therefore, the fact that economic laws are conditional and hypothetical is not a unique thing. Nor does this conditional and hypothetical nature destroy the scientific nature of economic laws. Economic laws are, therefore, as important and useful as the laws of physical sciences.

Economic Laws are Less Exact and Definite than Laws of Physical Sciences

But it is worth mentioning that laws of economics are less exact and definite than the laws of physical sciences. Law of gravitation which we have mentioned above is so exact and definite that we can calculate and measure exactly the movements of the solar system and we can predict their exact position at a particular time. But this is not true in case of economic laws. As said above, economic laws are not so exact and definite; we cannot say with certainty how a man would behave under certain conditions and, therefore, cannot exactly predict his behaviour. We can only say that a man would tend to behave in a particular manner. But because of ignorance, habit, blind faith, emotion, it is quite possible that he behaves contrary to a relevant economic law. That is why, Marshall said that "*the laws of economics are to be compared with the laws of tides, rather than with the simple and exact law of gravitation*." [11] Laws of tides tell us that how tides rise and fall twice in a day. Besides, they tell us that the height of the tides is maximum on a full-moon day. In other words, on a full-moon day tides are most furious. We can know from these laws of tides that what would be the length of these tides at a certain place on a certain day. But all this we cannot say with definiteness and certainty because tides are affected by external conditions such as air, weather, rain and storm. It cannot be said that what these external conditions will be at a particular place and time. Therefore, how much tides rise at a particular date and time cannot be predicted with certainty. We can say only this much that if these other conditions remain the same, what will be the height of these tides at a certain place and time. Thus, like economic laws, laws of tides are also statements of tendencies.

Now the question arises why economic laws are less definite and exact than those of natural sciences. This is because the object of economic study is such that economic laws cannot be completely definite and exact. The object of economic study is man with a free will and with certain wants and resources. The behaviour of man is governed by several external forces which are beyond his control. Economic behaviour of man is affected by his tastes, fashions, social conditions and customs, family conditions, etc. and all these go on changing. Therefore, economic laws framed by establishing relation between cause and effect in regard to human behaviour cannot be exact and definite. Man's tastes and other external conditions often change and, therefore, man may behave contrary to the established laws of economics. On the other hand, the objects of physical sciences are inert things which lack free will and whose nature always remain the same. Therefore, the laws of physical sciences are more exact and definite. Besides the laws of physical sciences are formulated by making controlled experiments.

By controlling many factors the functional relationship between physical variables are established, but it is not possible to make such controlled experiments regarding human behaviour. We cannot shut a man in a laboratory and keep many things constant and study his attitude and reactions to a particular change and thus establish relationship between relevant economic variables. In order to establish economic laws regarding human behaviour, we have to observe man's actions and reactions in the actual world. Empirical observations are an important way of establishing laws of economics. Another way to formulate laws of economics is the *use of introspective or psychological method*. Economists by knowing their reactions to a certain economic phenomenon also think that others too would behave in the same way, human nature being the same. Thus, by their own psychological reactions they derive generalisations about economic behaviour of the people.

Although laws of economics are less exact and definite than the laws of physical sciences, yet they are more exact and definite than the laws of other social sciences. Economics is fortunate enough to have a measure of money with which we can assess the motives or objectives as well as the results. Such a measure is not available to other social sciences.

Role of Assumptions in Economics: Friedman's View

As has been pointed out above, every law and generalisation of economics is based upon some assumptions. Now, the question is whether for the formulation of proper economic laws these assumptions should be realistic or not. One view is that laws of economics if they are to be valid and unseful must be based upon assumptions which are realistic. Thus, according to this view, making unrealistic assumptions and establishing laws on their basis will make these laws invalid. However, a contrary view has been put forward by Prof. Milton Friedman of Chicago University in his now well-known article, "*The Methodology of Positive Economics*."[12] In this context Prof. Friedman draws

11. Alfred Marshall, *Principles of Economics*, 8th edition, p. 26.
12. See his book "*Essays on Positive Economics*."

a distinction between positive economics and normative economics. According to Prof. Friedman, positive economics explains "a system of generalisations that can be used to make correct predictions about the consequences of any change in circumstances."[13] Because the predictions of this positive economics have to be tested with the empirical evidence it is as much a science as any other physical science even though the assumptions made may be unrealistic. The crucial point is whether the predictions based on economic laws of positive economics are confirmed by the facts and empirical evidence. According to Friedman, assumptions cannot be realistic; since they are made merely to simplify the analysis. However, it may be pointed out that while drawing conclusions from economic theories and laws regarding economic policies it must be known whether the assumptions made do not make the policy conclusions invalid if these assumptions are removed. Dr. K. N. Raj has rightly said, "some of the differences between economists regarding policy questions can be traced to the assumptions they choose to make when faced with the problems of this kind."[14] He further adds, "it is, however, essential in the interest of clarity and intellectual honesty that economists state clearly the assumptions on which one set of policies and programmes is advanced in preference to another and the reasons for making these assumptions"[15]

QUESTIONS FOR REVIEW

1. Explain the nature of a scientific theory? Is economics a science?

2. Differentiate between deductive and inductive methods of constructing economic theories. Explain their merits and demrits.

3. What is deductive method? How economic theories are established through deductive method?

4. How economists test or verify their theories? What happens if predictions of a theory are found to be in conflict with facts.

5. Explain the various steps in formulating economic theories through deductive method.

6. What is inductive method? How is inductive method used for constructing an economic theory?

7. Explain the various steps of constructing economic theories through inductive method.

8. "Induction and deduction are both needed for scientific thought as the right and left foot are both needed for walking." Do you agree? Discuss.

9. Explain why there is limited scope for making controlled experiments in economics.

10. Explain and illustrate the role of deductive and inductive methods in economics.

11. Explain Deductive and Inductive methods of economic analysis.

12. Explain the nature of economic laws and compare them with laws of physical sciences.

13. Laws of economics have been compared to the complex laws of tides rather than to simple laws of gravitation. Why?

14. Economic laws are mere statements of tendencies. Do you agree? Discuss.

15. Are economic laws universal ? Give some examples of economic laws which though valid for developed countries are not applicable to the developing countries.

16. Economic laws are unscientific because they are based on unrealistic assumptions. Do you agree? Discuss referring to Prof. Milton Friedman's views in this regard.

17. If economics a science, then why do economists disagree?

---

13. Milton Friedman, *op. cit.*
14. See his Presidential Address to the 55th Annual Conference of the Indian Economic Association, held at Bodh Gaya, Magadh University, Dec. 1972, reprinted in *The Indian Economic Journal*. Vol. XX, No. 3.
15. *Ibid.*

CHAPTER 4
ECONOMIC STATICS AND DYNAMICS

In the methodology of economics, techniques of economic statics and dynamics occupy an important place. A greater part of economic theory has been formulated with the aid of the technique of economic statics. However, during the last eighty years (since 1925) dynamic technique has been increasingly applied to the various fields of economic theory. Prior to 1925, dynamic analysis was mainly confined, with some exceptions, to the explanation of business cycles. After 1925, dynamic analysis has been used extensively not only for the explanation of business fluctuations but also for income determination, growth and price determination. More recently, economists like Samuelson, Goodwin, Smithies, Domar, Metzler, Haavelmo, Klein, Hicks, Lange, Koopmans and Tinter have further extended and developed dynamic models concerning the stability and fluctuations around any equilibrium point or path which cover the four important fields of economic theory, namely, business cycles, income determination, economic growth and price theory.

We shall explain below the meaning and nature of economic statics, dynamics and comparative statics and shall bring out the distinction between them. There has been a lot of controversy about their true meaning and nature, especially about economic dynamics.

Stationary and Changing Phenomena

In order to make the difference between the natures of economic statics and dynamics quite clear, it is essential to bring out the distinction between two sorts of phenomena, *stationary and changing*[1]. An economic variable is said to be stationary, if value of the variable does not change over time, that is, if its value is constant over time. For instance, if price of a good does not change as time passes, price will be called stationary. Likewise, national income is stationary if its magnitude does not change through time. On the other hand, the variable is said to be changing (non-stationary) if its value does not remain constant through time. Thus, the whole economy can be said to be *stationary* (*changing*), if value of all important variables are constant through time (are subject to changes). It may be noted that the various economic variables whose behaviour over time is studied are prices of goods, quantity supplied, quantity demanded, national income, level of employment, size of the population, level of investment, etc.

It is worth mentioning that it is quite possible that whereas a variable may be changing from the micro point of view, but stationary from the macro point of view. Thus, the prices of individual goods may be changing, of which some may be rising and some falling, but the general price level may remain constant over time. Likewise, the national income of a country may be stationary while the incomes generated by various industries may be changing. On the other hand, the particular variables may be stationary, while the economy as a whole may be changing. For example, even if the level of net investment in the economy is stationary, the economy as a whole may not be stationary. When there is a constant amount of net positive investment, the economy will be growing (changing) since addition to its stock of capital will be occurring.

---

1. *See* J. R. Hicks, *Capital and Growth* (Oxford Clarendon Press), 1965.

It should be carefully noted that there is no necessary relationship between stationary phenomenon and economic statics, and the changing phenomenon and dynamics. Although economic dynamics is inherently connected with only a changing phenomenon but the static analysis has been extensively applied to explain the changing phenomena. The distinction between statics and dynamics is the difference between the two different techniques of analysis and not the two sorts of phenomena. Prof. Tinbergen rightly remarks, "The distinction between Statics and Dynamics is not a distinction between two sorts of phenomena but a distinction between two sorts of theories, *i.e.*, between two ways of thinking. The phenomena may be stationary or changing, the theory (the analysis) may be Static or Dynamic".[2]

ECONOMIC STATICS

The task of economic theory is to explain the functional relationships between a system of economic variables. These relationships can be studied in two different ways. *If a functional relationship is established between two variables whose values relate to the same point of time or to the same period of time, the analysis is said to be static.* In other words, the static analysis or static theory is the study of static relationship between relevant variables. A functional relationship between variables is said to be static if values of the economic variables relate to the same point of time or to the same period of time. Numerous examples of static relationships between economic variables and the theories or laws based upon them can be given. Thus, in economics the quantity demanded of a good at a time is generally thought to be related to the price of the good at the same time. Accordingly, the law of demand has been formulated to establish the functional relationship between the quantity demanded of a good and price of that good at a *given moment or period of time.* This law states that, *other things remaining the same*, the quantity demanded varies inversely with price at a given point or period of time.

Similarly, the static relationship has been established between quantity supplied and price of goods, both variables relating to the same point of time. Therefore, the analysis of this relationship is a static analysis.

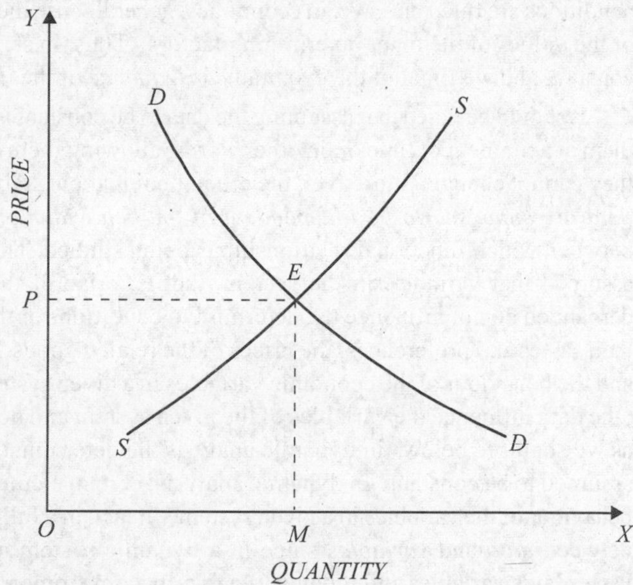

Fig. 4.1. *Static Equilibrium*

Generally, economists are interested in the equilibrium values of the variables which are attained as a result of the adjustment of the given variables to each other. That is why economic theory has sometimes been called *equilibrium analysis*. Till recently, the whole price theory in which we explain the determination of equilibrium prices of products and factors in different market categories were mainly static analysis, for the values of the various variables, such as demand, supply, price were taken to be relating to the same point or period of time. Thus, according to the price theory, equilibrium at a given moment of time under perfect competition is determined by the intersection of given demand function and the supply functions (which relate the values of variables at the same point of time). Thus in Fig. 4.1 given a demand function as demand curve *DD*

2. Tinbergen, Significant Developments in General Economic Theory, *Econometrica*, 1934.

and a supply function *SS*, the equilibrium price *OP* is determined. The equilibrium amount supplied and demand so determined is *OM*. This is a static analysis of price determination, for all the variables such as, quantity supplied, quantity demanded and the price refer to the same point or period of time. Moreover, the equilibrium price and quantity determined by their interaction also relate to the same time as the determining variables.

Professor Schumpeter describes the meaning of static analysis as follows: "By static analysis we mean method of dealing with economic phenomena that tries to establish relations between elements of the economic system – prices and quantities of commodities all of which have the same time subscript, that is to say, refer to the same point of time. The ordinary theory of demand and supply in the market of an individual commodity as taught in every text book will illustrate this case: it relates demand, supply and price as they are supposed to be at any moment of observation."[3]

A point worth mentioning about static analysis is that in it certain determining conditions and factors are assumed to remain *constant* at the point of time for which relationship between the relevant economic variables and outcome of their mutual adjustment is being explained. Thus, in the analysis of price determination under perfect competition described above, the factors such as incomes of the people, their tastes and preferences, prices of the related goods which affect demand for a given commodity are assumed to remain constant. Similarly, prices of the productive resources and production techniques which affect cost of production and thereby the supply function are assumed to remain constant. These factors or variables do change with time and their changes bring about shift in the demand and supply functions and therefore affect prices. But because in static analysis we are concerned with establishing the relationship between certain given variables and their adjustment to each other at a given point of time, changes in the other determining factors and conditions are ruled out. We, in economics, generally use the term *data* for the determining conditions or the values of the other determining factors. Thus, in static analysis, data are assumed to remain constant and we find out the eventual *consequence* of the mutual adjustment of the given variables.

It should be noted that assuming the data to be constant is very much the same thing as considering them at a moment of time or, in other words, allowing them a very short period of time[4] within which they cannot change. Moreover, the crucial point about static analysis is that the *given conditions or data are supposed to be independent of the behaviour of variables or units in the given system* between which functional relationship is being studied. Thus, in the above static price analysis it is assumed that variables in the system, that is, price of the good, quantity supplied and quantity demanded do not influence the determining conditions or the data regarding incomes of the people, their tastes and preferences, the prices of the related goods, etc. Thus, relationship between the data and the behaviour of the economic variables in a given system is assumed to be one-way relationship ; the data influence the variables of the given system and not the other way around. On the contrary, as we shall see below, in dynamic analysis the determinant data or determining conditions are not assumed to be constant. In dynamic analysis, certain elements in the data are not independent of the behaviour of the variables in a given system. In fact, in a fully dynamic system, it is hard to distinguish between *data* and *variables* since in a dynamic system over time "today's determinant data are yesterday's variables and today's variables become tomorrow's data. The successive situations are interconnected like the links of a chain."[5]

Since in static analysis, we study the behaviour of a system at a particular time, or in other

3. Joseph A. Schumpeter, *History of Economic Analysis*.
4. Time does nto wait or remain constant, it is always passing. Therefore, by a 'moment' or point of time, we actually mean a very short period of time. A point or moment of time may be a day, a week or a month.
5. Haberler, *Prosperity* and *Depression*, p. 249.

words, in *economic statics, we do not study the behaviour of a system over time*, therefore how the system has proceeded from a previous position of equilibrium to the one under consideration is not studied in economic statics. Prof. Stanley Bober rightly remarks, "A static analysis concerns itself with the understanding of what determines an equilibrium position at any moment in time. It focuses attention on the outcome of economic adjustments and is *not concerned with the path by which the system*, be it the economy in the aggregate or a particular commodity market, has proceeded from a previous condition of equilibrium to the one under consideration."[6]

To sum up, in static analysis we ignore the passage of time and seek to establish the causal relationship between certain variables relating to the same point of time, assuming some determining factors as remaining constant.

Importance of Economic Statics

The method of economic statics is very important and a large part of economic theory has been developed using the technique of economic statics. Now, the question arises as to why the technique of static analysis is used which appears to be unrealistic in view of the fact that determining conditions or factors are never constant. Static techniques are used because *it makes the otherwise complex phenomena simple and easier to handle*. To establish an important causal relationship between certain variables, it becomes easier if we assume other forces and factors constant, not that they are inert but for the time it is helpful to ignore their activity. According to Prof. Robert Dorfman, "statics is much more important than dynamics, partly because it is the ultimate destination that counts in most human affairs, and partly because the ultimate equilibrium strongly influences the time paths that are taken to reach it, whereas the reverse influence is much weaker".[7]

To sum up, in static analysis we ignore the passage of time and seek to establish the causal relationship between certain variables relating to the same point of time, assuming some determining factors as remaining constant.

ECONOMIC DYNAMICS

Now, we turn to the method of Economic Dynamics which has become very popular in contemporary economics. Economic dynamics is a more realistic method of analysing the behaviour of the economy or certain economic variables through time. The definition of economic dynamics has been a controversial question and it has been interpreted in various different ways. We shall try to explain the standard definitions of economic dynamics.

Frisch's Time Period Analysis

The course thorough time of a system of economic variables can be explained in two ways. One is the method of economic statics described above, in which the relations between the relevant variables in a given system refer to the same point or period of time. On the other hand, if the analysis considers the relationship between relevant variables whose values belong to *different points of time* is known as Dynamic Analysis or Economic Dynamics. The relations between certain variables, the values of which refer to the different points or different periods of time are known as *dynamic relationships*. Thus, Professor Schumpeter says, "We call a relation dynamic if it connects economic quantities that refer to different points of time. Thus, if the quantity of a commodity that is offered at a point of time (t) is considered as dependent upon the price that prevailed at the point of time ($t-1$), this is a dynamic relation."[8] In a word, economic dynamics is the analysis of dynamic relationships.

---

6. Stanley Bober, *The Economics of Cycles and Growth*, First Wiley Eastern Reprint Edition, 1971, p.-2.
7. Robert Dorfman, *Prices and Markets*, Prentice Hall of India, Private Limitd, 2nd edition, New Delhi, 1972.
8. Joseph A. Schumpeter, *History of Economic Analysis*, Oxford, 1954.

We thus see that in economic dynamics we duly recognise the element of time in the adjustment of the given variables to each other and accordingly analyse the relationships between given variables relating to different points of time. Professor Ragnar Frisch who is one of the pioneers in the use of the technique of dynamic analysis in economics defines economic dynamics as follows: "A system is dynamical if its behaviour over time is determined by functional equations in *which variables at different points of time are involved in an essential way*."[9, 10] In dynamic analysis, he further elaborates, "We consider not only a set of magnitudes in a given point of time and study the interrelations between them, but *we consider the magnitudes of certain variables in different points of time, and we introduce certain equations which embrace at the same time several of those magnitudes belonging to different instants*. This is the essential characteristic of a dynamic theory. Only by a theory of this type we can explain *how one situation grows out of the foregoing*."

Many examples of dynamic relationships from both micro and macro economic fields can be given. If one assumes that, the supply (S) for a good in the market in the given time (t) depends upon the price that prevails in the preceding period (that is, $t - 1$) the relationship between supply and price is said to be dynamic. This dynamic functional relation can be written as :

$$S_t = f(P_{t-1})$$

where S_t stands for the supply of a good offered in a given period t and P_{t-1} for the price in the preceding period. Likewise, if we grant that the quantity demanded (D) of a good in a period t is a function of the expected price in the succeeding period ($t + 1$), the relation between demand and price will be said to be dynamic and the analysis of such relation would be called dynamic theory or economic dynamics.

Similarly, examples of dynamic relationship can be given from the macro field. If it is assumed that the consumption of the economy in a given period depends upon the income in the preceding period ($t - 1$) we shall be conceiving a dynamic relation. This can be written as:

$$C_t = f\left(Y_{t-1}\right)$$

When macroeconomic theory (theory of income, employment and growth) is treated dynamically, that is, when macroeconomic dynamic relationships are analysed, the theory is known as "*Macro dynamics*". Paul Samuelson, J.R. Kalecki, Post-Keynesians like R.F. Harrod, J.R. Hicks have greatly dynamised the macroeconomic theory of Keynes.

It should be noted that the change or movement in a dynamic system is *endogenous*, that is, it goes on independently of the external changes in it; one change grows out of the other. There may be some *initial* external shock or change but in response to that initial external change, the dynamic system goes on moving independently of any fresh external changes, successive changes growing out of the previous situations. In other words, the development of a dynamic process is self-generating. Thus, according to Paul Samuelson, "It is important to note that each dynamic system generates its own behaviour over time either as an autonomous response to a set of "initial conditions" or as a response to some changing external conditions. *This feature of self-generating development over time is the crux of every dynamic process*.'[11] Likewise, Professor J. K. Mehta remarks, "In simple words, we can say that an economy can be said to be in a dynamical system when the various variables in it such as output, demand, prices have values at any time dependent on their values at some other time. If you know their values at one moment of time, you should be able to know their

9. & 10. Ragnar Frisch, Propagation Problems and Impulse Problems in Dynamic Economics, *Economic Essays in Honour of Gustav Cassel*, George Allen and Unwin, Ltd., London, 1933, pp. 171-172.

11. Paul A. Samuelson, Dynamic Process Analysis, printed in *The Collected Scientific Papers of Paul A. Samuelson*, Vol. 1, edited by Joseph E. Stiglitz (MIT Press 1966), p. 590.

values at subsequent points of time. Prices of goods in a causal dynamic system do not depend on any outside, exogenous forces. A dynamic system is self-contained and self-sustained."[12]

The concept or technique of economic dynamics which we have presented above was first of all put forward by Ragnar Frisch in 1929. According to his view, like static analysis, *economic dynamics is a particular method of explanation of economic phenomenon*, economic phenomena themselves may be stationary or changing. Although technique of dynamic analysis has great scope in a changing and a growing system but it may also be applied even to stationary phenomena. A system or phenomenon may be stationary in the sense that, the values of relevant economic variables in it may remain constant through time, but if the values of the variables at a time are dependent upon the values at another time, then dynamic analysis can be applied. But, as stated above, the greater scope of economic dynamics lies in the field of changing and growing phenomena.

Harrod's Concept of Economic Dynamics: Rates of Change Analysis

We have explained above the concept of economic dynamics which has been associated with the name of Ragnar Frisch, though it has been propounded by others also. Prof. R. F. Harrod, an eminent Cambridge economist in his famous book "*Towards a Dynamic Economics*", has given a different concept of economic dynamics. According to Harrod, economic dynamics deals with *rates of change*. An analysis or a theory is dynamic if the rates of change of certain variables are considered to depend on the rates of change of other variables. In his view dynamics studies an "*economy in which rates of output are changing*."[13] He defines economic dynamics as the study of "*the necessary relations between the rates of growth of the different elements in a growing economy*".[14]

Harrod's concept of dynamics covers both the technique (method) as well as the scope of economic dynamics. According to him, economic dynamics as a technique has to consider *rates of change* of certain variables and how they are related to the rate of change of some other variables. Since only in a growing and changing economy, magnitudes of variables undergo a change, it is the growing and changing economy with which, according to Harrod, dynamics deals with.

In modern economic theory both Frischian and Harrodian concepts of economic dynamics have been adopted. Thus, in modern economics, dynamic analysis concerns itself with either establishing functional relationships between economic variables relating to different points of time, or considering rates of change of variables in a growing economy and how they are related to each other. While the former involves *period analysis*, the latter *rates of change analysis*.

Expectations and Dynamics

We have described above that economic dynamics is concerned with explaining dynamic relationships, that is, the relationships among variables relating to different points of time. The variables at the present moment may depend upon the variables at other times, past and future. Thus, when the relationship between the economic variables belonging to different points of time is considered, or when rates of change of certain variables in a growing economy are under discussion, the question of future creeps into the theoretical picture. The economic units (such as consumers, producers and entrepreneurs) have to take decisions about their behaviour at the present moment. The consumers have to decide what goods they should buy and what quantities of them. Similarly, producers have to decide what goods they should produce, what factors they should use and what techniques they should adopt. These economic units decide about their present course of action on the basis of their *expected values* of the economic variables in the future. *When their expectations are realised*, they continue behaving in the same way and the dynamic system is in equilibrium. In other words, when the expectations of the economic units are fulfilled, they repeat the present pattern

12. J. K. Mehta, *Lectures on Modern Economic Theory*, 3rd edition 1967, p. 212
13. R. F. Harrod, *Towards a Dynamic Economics*, (1948). p. 4.
14. Ibid., p.19

of behaviour and there exists what has been called *dynamic equilibrium*, unless some external shock or distubring force disturbs the dynamic system.

The expectations or anticipations of the future held by the economic units play a vital role in economic dynamics. In a purely static theory expectations about the future have practically no part to play since static theory is mainly concerned with explaining the conditions of equilibrium positions *at a point of time* as well as under the assumptions of *constant tastes, techniques and resources*. Thus, in static analysis expectations about the future play a little part since under it no *processes over time* are considered. On the other hand, since dynamic analysis is concerned with dynamic processes over time, that is, changing variables over time and their action and interaction upon each other through time, expectations or anticipations held by the economic units about the future have an important place.

Need and Significance of Economic Dynamics

The use of dynamic analysis is essential if we want to make our theory realistic. In the real world, various key variables such as prices of goods, output of goods, income of the people, investment and consumption, *etc.* are changing over time. Both Frischian and Harrodian dynamic analyses are required to explain these changing variables and to show how they act and react upon each other and what results flow from their action and interaction. Many economic variables take time to make adjustment to the changes in other variables. In other words, there is a *lag* in the response of some variables to the changes in the other variables, which make it necessary that dynamic treatment be given to them. We have seen that changes in income in one period produce influence on consumption in a later period. Many other similar examples can be given from micro and macro fields.

Besides, it is known from the real world that the values of certain variables depend upon *the rate of growth* of other variables. For example, we have seen in Harrod's dynamic model of a growing economy that investment depends upon expected rate of growth in output. Similarly, the demand for a good may depend upon the rate of change of prices. Similar other examples can be given. In such cases where certain variables depend upon the rate of change in other variables, application of both the period analysis and the rate of change analysis of dynamic economics become essential, if we want to understand their true behaviour.

Till recently, dynamic analysis was mainly concerned with explaining business cycles, or economic fluctuations . But, after Harrod's[15] and Domar's[16] path-breaking contributions, the interest in the problems of growth has been revived among economists. It is in the study of growth that dynamic analysis becomes more necessary. These days economists are engaged in building dynamic models of optimum growth both for developed and developing countries of the world. Thus, in recent years, the stress on dynamic analysis is more on explaining growth rather than cycles or oscillations. Prof. Hansen is right when he says, "In my own view mere oscillation represents a relatively unimportant part of economic dynamics. Growth, not oscillation, is the primary subject-matter for study in economic dynamics. Growth involves changes in technique and increases in population. Indeed that part of cycle literature (and cycle theories are a highly significant branch of dynamic economics) which is concerned merely with oscillation is rather sterile."[17]

---

15. R. F. Harrod, *Towards a Dynamic Economics*, Macmillan & Co. Ltd. (London), 1948.
16. E. Domar, Capital Expansion, Rate of Growth and Employment "*Econometrica*", Vol. 14, 1946, pp. 137-47
17. Erich Schneider, *Pricing and Equilibrium*, George Allen and Unwin, Ltd., 1962, pp. 235-36.

COMPARATIVE STATICS

We have studied above static and dynamic analysis of the equilibrium position. To repeat, static analysis is concerned with explaining the determination of equilibrium values with a given set of data and the dynamic analysis explains how with a change in the data the system gradually grows out from one equilibrium position to another. Midway between the static and dynamic analyses is the comparative static analysis. Comparative static analysis compares one equilibrium position with another when the data have changed and the system has finally reached another equilibrium position. It does not analyse the whole path as to how the system grows out from one equilibrium position to another when the data have changed; *it merely explains and compares the initial equilibrium position with the final one reached after the system has adjusted to a change in data. Thus, in comparative static analysis, equilibrium positions corresponding to different sets of data are compared.*

It should be noted that for better understanding of the changing system, comparative statics studies the effect on the equilibrium position of a change in *only a single datum at a time* rather than the effects of changes in the many or all variables constituting the data. By confining ourselves to the adjustment in the equilibrium position as a result of alteration in a single datum at a time, we keep our analysis simple, manageable and at the same time useful, instructive as well as adequate enough to understand the crucial aspects of the changing phenomena. To quote Professor Schneider:

"The set of data undergoes changes in the course of time, and each new set of data has a new equilibrium position corresponding to it. It is therefore of great interest to compare *the different equilibrium positions corresponding to different sets of data*. In order to understand the effect of a change in the set of data on the corresponding position of equilibrium, we must only alter a *single datum at a time*. Only in this way it is possible to understand fully the effects of alterations in the individual data. In this way we compare the equilibrium values for the system corresponding to the two equilibrium positions with one another. This sort of comparative analysis of two equilibrium positions may be described as comparative-static analysis, since it studies the alteration in the equilibrium position corresponding to an alteration in a single datum".[18]

Let us give some examples of comparative static analysis from the microeconomic theory. We know that given the data regarding consumer's tastes, incomes, prices of other goods on the one hand and the technological conditions, costs of machines and materials, and wages of labour we have given demand and supply functions which by their interaction determine the price of a good. Now suppose that, other things remaining constant, incomes of consumers increase. With the increase in incomes, the demand function would shift upward. With the change in the demand as a result of the change in the income, the supply would adjust

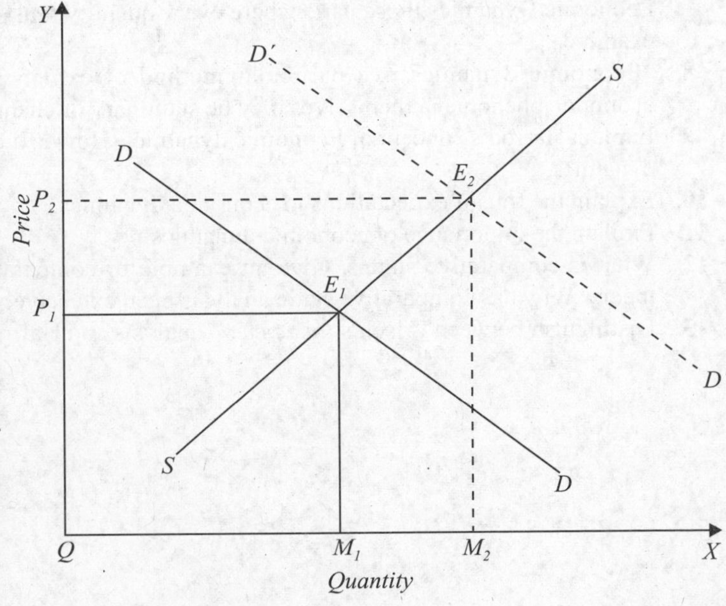

Fig. 4.2 *Comparative Statics*

18. Erich Schneider, op. cit., pp. 235-36

itself and final new equilibrium position would be determined. To explain the determination of new equilibrium price and how it differs from the initial one is the task of comparative statics. In Fig. 4.2, initially the demand and supply functions are *DD* and *SS* and with their interaction price OP_1 is determined. When the demand function changes to $D'D'$ as a result of changes in consumer's income, it intersects the given supply function *SS* at E_2 and the new equilibrium price OP_2 is determined. In comparative-static analysis, we are only concerned with explaining the new equilibrium position E_2 and comparing it with E_1 and not with the whole path the system has traversed when it gradually grows out from E_1 to E_2. As we shall study in the part of price theory, comparative static analysis was extensively used by Alfred Marshall in his time-period analysis of pricing under perfect competition.

No doubt, more realistic, complete and true analysis of the changing phenomena of the real world would be the dynamic analysis, nevertheless comparative statics is a very useful technique of explaining the changing phenomena and its crucial aspects without complicating the analysis.

QUESTIONS FOR REVIEW

1. What is meant by Economic Statics? Illustrate it with the demand-supply model of determination of price.
2. Distinguish between economic statics and dynamics. Does economic statics analyse stationary phenomenon only?
3. What is meant by '*data*' in economics? In static analysis the data are assumed to be constant and we seek to find out the eventual consequence of the mutual adjustment of the given variables. Explain.
4. Explain why static analysis has been extensively used in economics though it appears to be unrealistic in view of the fact that the determining conditions never remain constant.
5. What is meant by *economic dynamics*? Give an examples of dynamic relationships from micro-economic theory and macro-economic theory.
6. A distinctive feature of dynamic analysis is to show how a dynamic process is self-generating.
7. "I call Economic Statics those parts of economic theory where we do not trouble about dating, Economic Dynamics those parts where every quantity must be dated" (J.R. Hicks). Critically examine.
8. "Economic dynamics is a particular method of explanation of economic phenomenon, economic phenomena themselves may be stationary or changing" (Ragnar Frisch). Explain.
9. Explain Harrod's concept of Economic dynamics. How is it different from Frisch's concept of dynamics.
10. Explain the role of expectations in economic dynamics.
11. Explain the importance of economic dynamics.
12. What is comparative statics. Give an example of comparative statics from microeconomic theory. Why has comparative static analysis been extensively used in economics?
13. Distinguish between Frisch's *time period* analysis and Harrod's *rate of change analysis*.

CHAPTER 5
CENTRAL PROBLEMS OF AN ECONOMY

THE ECONOMIC PROBLEM

Economics is mainly concerned with the achievement and use of material requirements to satisfy human wants. Human wants are unlimited and productive resources such as land and other natural resources, raw materials, capital equipment with which to produce goods and services to satisfy those wants are scarce and limited. Thus goods and services which satisfy human wants are scarce because productive resources with which to produce goods and services are scarce. The problem of scarcity of resources is felt not only by individuals but also by the society as a whole. With wants being unlimited and resources scarce, we individually as well as collectively cannot satisfy all our wants. This gives rise to the problem of how to use scarce resources to attain maximum satisfaction. This is generally called '*the economic problem*' as it lies at the root of all economic problems faced by the society. Every economic system, be it capitalist, socialist or mixed, has to contend with this central problem of scarcity of resources relative to wants for them.

Thus the economic problem derives from the scarcity of resources relative to human wants. This gives rise to the struggle of man for existence and efforts by him to promote his well-being. That the scarcity of resources in relation to human wants is the fundamental economic problem can be easily understood in the context of poor and developing countries like India where quite a large number of population lives at a bare subsistence level. The struggle for existence due to the scarcity of resources is too obvious in them to need any elaborate explanation. However, to say that the developed countries, such as U.S.A., where affluence and prosperity have been brought about also confront the scarcity problem raises some doubts. But the fact is that, despite of their affluence and riches, developed societies too face the problem of scarcity. Of course, their possession of goods and services has enormously increased, but so has their wants. Indeed, their wants for goods and services has been multiplying during the course of economic growth so that their present wants still remain ahead of their resources and capacity to produce.

As has been said above, the problem of scarcity of resources arises not merely due to the limited resources and capacity to produce alone but also due to immense human wants. So long as human wants for goods and services remain ahead of the resources, both natural and acquired, the economic problem of scarcity would be there. If Americans today, for example, were to content to live at the level of the Indian middle class people, all their wants would be fully satisifed with their available resources and capacity to produce. In that situation they would face little or no scarcity and economic problem for them would disappear. To conclude, the affluent and developed countries of U.S.A. and Western Europe face the problem of scarcity even today as their present wants remain ahead of their increased resources and capability to produce.

Scarcity of resources requires that efficient and optimum use of resources be made so that we should get most out of them and thus maximum possible satisfaction of the people is achieved. Further, since all wants cannot be satisfied due to scarcity of resources we face *the problem of choice* — choice among multiple wants which are to be satisfied. If it is decided to use more resources in one line of production, then some resources must be withdrawn from another commodity. Thus, the problem of choice from the viewpoint of the society as a whole refers to which goods and in what quantities are to be produced and how productive resources allocated for their production accordingly so as to achieve the greatest possible satisfaction of the people. We have seen in the first chapter that an eminent English Economist Lord Robbins defines economics in terms of this basic economic problem. According to him, "*Economic is a science which studies human behaviour as a relationship between ends and scarce resources which have alternative uses*." Ends refer to wants which are considered to be unlimited. The use and allocation of scarce resources to produce goods and services has to be such as would maximise satisfaction. This applies both to the behaviour of the individual as well as to that of the society as a whole.

The scarcity of resources also compels us to decide how the different goods should be produced, that is, what production methods should be employed for the production of goods so as to make best possible use of the available resources. If the resources were unlimited, the problem of how goods should be produced would not have arisen. This is because with unlimited resources it would not matter whichever method, efficient or inefficient, was employed for the production of goods.

Further, due to scarcity of resources goods cannot be produced in sufficient quantities to satisfy all wants of all the people of a society. This raises the question who should get how much from the national output. That is, how the national output is distributed among various members of a society.

Lastly, people are not content with the meagre resources they have. They want to improve their material well-being by increasing their resources. Thus human society wants to have economic growth and development so as to overcome or reduce the problem of scarcity. The promotion of economic growth and development requires accumulation of capital and improvement in technology.

CENTRAL PROBLEMS OF AN ECONOMY

As stated above, the scarcity of resources gives rise to various *basic economic problems* which have to be solved by an economy if it is to fulfil its purpose. These basic economic problems are also called central problems of the economy. They are:

1. *What* to produce
2. *How* to produce
3. *For Whom* to produce
4. *What Provision* (if any) be made for economic growth.

We explain below these problems in some detail.

What to produce

This implies that society has to decide which goods and in what quantities are to be produced, "Guns or butter" has been the popular way of describing this dilemma of choice posed by the scarcity of resources. But this choice between war goods and goods for civilian consumption is not the only problem of choice faced by the society. The society has to choose among hundreds of consumer goods themselves and decide about allocation of resources between them. Of special mention in this regard is the choice between necessities and luxuries. Further, an important choice about the production of goods and resource allocation is to decide about what amounts of consumer goods and capital goods are to be produced. As shall be made clear later, this decision about the allocation of resources between consumer goods and capital goods is of crucial importance from the point of view of economic growth.

Society has to decide not only the relative amounts of consumer and capital goods to be produced but it has also to determine the specific quantity of each type of capital good and each type of consumer good. More generally, in answering the question what goods shall be produced, society would have to choose somehow or other among scores of goods such as cars, hospitals, schools, houses, radios, televisions, nuclear bombs, wheat , rice, cloth, machinery, steel, soap, lipsticks, terylene, nylon, etc. But this decision is only half the battle. Once society has decided which goods shall be produced, it must then give proper weights to each of the good it selects to produce. Suppose society has decided to produce wheat, hospitals, schools and cloth from the above list. The fact that resources are scarce means that society cannot produce unlimited amounts of even these selected goods. Therefore, society must decide how much wheat, how many hospitals, how many schools and how many metres of cloth are to be produced. In fact, most of the goods in the above list will be decided to be produced. Only the question of what quantity of each of the goods is to be produced would have to be decided.

How to Produce

This means with what combination of resources a society decides to produce goods. A combination of resources (or factors) implies a technique of production. Usually, there are various alternative techniques of producing a commodity and the society has to choose among them, each technique using a different combination of resources like labour and capital. For instance, cotton cloth can be produced with either handlooms, or powerlooms or automatic looms (as used in modern textile mills). Production with technique of handlooms involves the use of more labour relatively to capital and therefore represents a labour-intensive technique. On the other hand, the production of cloth with automatic looms of the modern textile mills involves use of more capital relative to labour and hence represents a capital-intensive technique of production. Likewise, alternative techniques involving different degrees of capital and labour intensity are available in case of other commodities. Obviously, the choice between different techniques would depend on the available supplies of different factors of production and their relative prices.

Scarcity of resources demands that goods should be produced with the most efficient method. If the economy uses its resources inefficiently, the output would be smaller and there would be unnecessarily sacrifice of goods that otherwise would have been available. Therefore, it is in society's interest that those techniques of production are used that 'make greater use of relatively abundant factors and economises as much as possible those factors which are relatively scarce.

For whom to produce

For whom to produce implies how the national product is to be distributed among the members of the society. Productive resources and the resulting output being scarce, we cannot satisfy all wants of all the people. Therefore, a society has to decide who should get how much from the total output of goods and services. This is, as it were, sharing of national cake among the people constituting a society. Obviously, in a free market economy who would get how much of national output depends on his money income. The greater the money income a person enjoys, the greater the amount of goods he would be able to buy from the market. Therefore, the greater the inequalities in the distribution of money incomes, the greater the inequalities in the distribution of national output. Therefore, this raises the question as to how a free market economy decides about the distribution of money incomes.

Money income can be obtained in two ways. First, it can be obtained through work, that is, selling one's labour service. Wages (and salaries) represent the incomes from work. Differences in wages earned by various people arise due to differences in production activities, skill, education as well as due to the bargaining powers of various social classes and a host of social and institutional factors. Second, income can be made from property such as land, factories and other forms of capital. Rent, interest and profits are the examples of income from property. Differences in the ownership of property in a free market economy cause differences in incomes from property. Thus, money incomes of the people made in these ways go to determine the distribution of national output. Therefore, in order to expelain the determination of output, we have to explain how distribution of income, that is, wages of labour, rent of land, interest on capital and profits of enterprise are determined.

How the national income is to be distributed has been a burning topic not only in the field of economics but also in politics. There is perhaps no topic in the whole of economics on which discussion has been so hot and furious as the distribution of national product and income. Some have argued with a good deal of justification that all people should get equal incomes and hence equal shares from the national product. According to Karl Marx, the distribution of national income should be on the basis of *"from each according to his ability, to each according to his needs"*. Another important view has been that each individual should get income equal to the contribution he makes to the national production.

What Provision Should be Made for Economic Growth

Both an individual and a society would not like to use all its scarce resources for current consumption only. This is because if all resources are used for producing consumer goods only and no provision in terms of allocating some resources for investment, that is, for production of capital goods, is made, the future resources or productive capacity would not increase. This implies that incomes or standards of living of the people would remain stagnant. Indeed, if no such provision is made for future, the future productive capacity and hence the levels of living may decline. For, in the absence of this provision, the stock of capital would fall due to depreciation as a result of its use for producing consumer goods in a period. This is better expressed by quoting an old Chinese saying which states. "*He who cannot see beyond the dawn will have much good wine to drink at noon, much green wine to cure his headache at dusk and only rain water to drink for the rest of days*." Therefore, a wise individual or a society likes to provide for its growth of productive capacity. This requires that a part of its resources should be devoted to the production of capital goods and for promotion of research and development activity. The capital accumulation and technological progress achieved in this way would enable the society to produce more of consumer goods in the future and ensure higher living standards for its people. It follows that provision for capital accumulation and technological progress implies sacrifice of some current consumption. Therefore, a society has to decide how much saving and investment (that is, how much sacrifice of current consumption) should be made for future economic progress. The pangs of the sacrifice of some current consumption for promoting economic growth are greatly felt in less developed countries like India where not only the present *consumption* levels are very low (and therefore any sacrifice of current consumption is hardwritting) but also the need for economic growth is very pressing.

What, How and For Whom: Subject Matter of Microeconomics

We have explained above the four basic problems which have been the concern of economists. The function and purpose of all economic systems is to solve these problems. It is noteworthy that the analysis of first three problems, namely, what, how and for whom to produce are discussed in what is now called microeconomics. Microeconomics studies behaviour of individual consumers as to what they will consume and demand, of producers or firms as to what and how they will produce, of individual industries as to how demand for their products and the supply by them will determine their prices. It is their demand for and prices of various products that determine the allocation of resources between them.

Besides, microeconomics studies how prices of factors such as wages of labour, rent of land, interest on capital and profits of enterprise are determined. It is these factor prices that, on the one hand, determine the incomes of various individuals and hence the demand for various products and, on the other, determine the cost of the products It is the money incomes of the various people that determine who would get how much from national output.

It may be noted that not only does microeconomics explain the different mechanisms through which the above four problems are solved but also analyses the implications of alternative choices made in this regard for welfare of the society.

Macroeconomics: Whether Resources are Fully Utilised?

Now, the pertinent question is which of the above problems are studied by what is now called macroeconomics. This requires some elaboration. Despite the fact that not enough resources are available to satisfy all human wants, in a free market economic system it so happens that they are not fully utilised. This goes contrary to the view held by the Classical and Neo-Classical economists who believed in the existence of full employment of resources. A noted English economist challenged this viewpoint during nineteen thirties during which a severe depression overtook the capitalist economies. At this time, strange phenomenon of a large number of workers (human resources) were rendered unemployed and a good deal of the factories and capital equipment installed in then remained idle or under-utilised. Keynes in his well known work "*General Theory of Interest, Money and Employment*" explained that despite the scarcity of resources, the unemployment may be caused due to the lack of effective demand. Thus the level at which employment of labour and (non-human resources will be used and hence the level of national income will be determined depends on effective demand. Keynesian analysis greatly improved our understanding of the working of the capitalist system which cause involuntary unemployment of human and capital resources due to lack of effective demand. Macroeconomics is concerned with explaining how the employment of resources and hence the level of national income is determined and what causes fluctuations and growth in them.

HOW ARE THE CENTRAL PROBLEMS SOLVED?

We have explained above the four central problems of an economy which must be solved somehow or other by an economy. Now, the important thing to know is how an economy solves these central problems which are essential to its functioning. There are two main methods to solve these central problems. The first method is to solve these problems through *market or price mechanism*. That is, what goods are to be produced and what quantities, which methods for production are to be employed for the production of goods and how the output is to be distributed, should be decided by the free play of the forces of demand and supply. In such an economic system, capital and other means of production are the private property of the individuals and private entrepreneurs who undertake the work of production. Consumers have freedom to buy the goods they want. In such a system, those goods are produced more for which there is greater demand and those goods are produced less for which there is less demand. The demand for and supply of various goods determine the prices and quantities produced of various goods. Besides, the prices of factors such as labour capital, land *i.e.*, wages, interest, rent also depend upon their demand and supply and these factor prices determine the incomes of the owners of these factors and it is these incomes which determine the distribution of goods among the various individuals in a society.

Given the prices of the factors, entrepreneur decides which technique of production is to be employed (that is, which combination of factors or resources is to be used) for production of goods. This method of solving these central problems through market mechanism, that is, through forces of demand and supply is used by a free-enterprise capitalist economy.

The second method which can be employed to solve the central problems is the adoption of *economic planning*. In this method, the solution of the various basic problems is not achieved through the free working of demand for and supply of goods and factors. But to solve these problems, Government sets up a central planning authority which has been called by several names, such as planning commission, planning ministry or planning board. What goods should be produced and in what quantities, bow should they be produced, how should they be distributed among the population and how much should be invested to bring about economic growth are all decided by the central planning authority. This method of solving the central problems through the use of economic planning had been adopted in the former communist countries such as Soviet Russia, Poland, etc. In such an economic system, the capital and property are collectively owned by the society and work of production is organised by the Government. In this system, private entrepreneurs have no right to undertake the work of production and consumers lose their freedom to choose and consume the goods of their choice.

It is worth noting that there is no any country in the world today, which allows the solution of these central problems through unfettered market mechanism: In capitalist countries such as England, the U.S.A., France, Governments intervene in the economy and regulate the market mechanism, directly or indirectly. In this way, in these capitalist countries also, Government plays an active role in the solution of the various central problems, regarding production, distribution and investment. In the so-called capitalist countries, Government interferes in the economy and affects the solution of these problems through the adoption of proper monetary policy (in which the policies regarding supply of money, bank credit and rate of interest are included), fiscal policy (which is concerned with the Government expenditure and taxation policies) and direct controls, such as industrial licencing policies, control on prices, imports and exports.

The Government interferes in the working of the capitalist economy through these various policy instruments because it has now been realised that the unfettered functioning of the market mechanism or the forces of demand and supply, results in wide economic fluctuations, price instability and unemployment and lack of economic growth. In other words, with the free working of the market mechanism, the achievement of the objectives of economic stability, full employment and rapid economic , growth is not possible. During 1929-33, when in all the capitalist countries, there occurred a great depression which resulted in widespread unemployment among the population, then famous economist J.M. Keynes laid stress on the adoption of appropriate fiscal and monetary policies so as to achieve the objective of full employment. After Keynes main problem of the capitalist countries has been of inflation *i.e.* a severe rise in general level of prices. Besides inflation, these countries also desire more rapid economic growth and the problem is how to ensure steady rate of economic growth. Therefore, after Keynes, economists have laid stress on the intervention by the Government to tackle the problem of inflation and also to ensure steady rate of economic growth. As a matter of fact, the Governments of these countries have taken active steps to hold the price line and to promote steady growth. Therefore, economists such as Paul Samuelson, Hansen, Galbraith, call these so-called capitalist countries as *mixed capitalist enterprise system* or simply *mixed economies*. This is because in these countries, now-a-days, market or price mechanism does not work freely to solve the central problems and instead the Government plays an active role through the adoption of appropriate policy instruments to regulate these economies.

It is worth noting here that the nature of mixed economy of India is different from that of the mixed economies of America and England. In India, Government not only regulates and controls the private enterprise and market mechanism through monetary and fiscal measures and direct controls but also itself undertakes the work of production. Various important industrial and power projects have been set up in the public sector, the working of which does not depend upon market and price mechanism. Besides, in India public enterprises and other economic activities of the Government are conducted according to the policies laid down in Five-Year Plans. Thus, in the mixed economy of India, direct participation of the Government *i.e.*, public enterprise and economic planning have an important place in the working of Indian economy. So far as private sector is concerned, its working primarily depends upon the market or price mechanism. But Government tries to regulate and control the activities of the private sector through appropriate direct controls, monetary and fiscal policies. As regards public sector, the decisions regarding what to produce, how to produce and how much to invest are taken by the Government and Planning Commission.

We have only indicated here that there are two main methods of solving the central problems. How these central problems are solved by a capitalist economy and a centrally planned socialist economy will be explained in detail in a later chapter.

PRODUCTION POSSIBILITY CURVE

The nature of basic economic problems explained above can be better understood and distinguished from each other with the aid of an important tool of modem economics known as production possibility curve. Production possibility curve is also called *production possibility frontier*. We shall explain below the concept of production possibility curve and bring out its relation with the basic economic problems stated above.

Production possibility curve (frontier) represents graphically alternative production possibilities facing an economy. As the total productive resources of the economy are limited, the economy has to choose between different goods. The productive resources can be employed for the production of various alternative goods. It has, therefore, to be decided which goods to be produced more and which ones less. In deciding what amounts of various goods are to be produced, society would in fact be deciding about the allocation of resources among different possible goods. How much labour should go into raising wheat on the farms and how much should be employed for manufacturing cloth. How many factories should produce arms for the army and how many should produce consumer goods for the civilians. In order to simplify our analysis we shall assume that two types of goods–wheat and cloth–are to be produced. We shall explain the production possibilities with these two goods but the analysis made will equally apply to the choice between any other two goods.

Let us assume that there is a given amount of productive resources and they remain fixed. Although resources are fixed in quantity, yet they can be shifted from the production of one good to another. Further, we assume that the given resources are being used fully and with utmost efficiency. In other words, we assume that resources are neither unemployed nor underemployed. This means that economy is working at the level of full-employment and using its productive capacity fully. We also presume that technology does not undergo any change. In other words, we rule out any progress in technology. In short, we assume fixed resources, full employment, full use of production capacity and a given technology. All these assumptions imply that we are looking at our economy at some particular point in time or over a very short period of time. This is because it will be very unrealistic to rule out progress in technology and growth in the supply of resources over a long period of time.

With the given amount of resources and a given technology, we have constructed the following table showing various production possibilities between wheat and cloth. If all the given resources

Table 5.1

Alternative Production Possibilities

Production Possibilities	*Cloth (in thousand metres)*	*Wheat (in thousand quintals)*
A	0	15
B	1	14
C	2	12
D	3	9
E	4	5
F	5	0

are employed for the production of wheat, it is supposed that 15 thousand quintals of wheat are produced. On the other hand, if all the resources are devoted to the production of cloth, 5 thousand metres of cloth is made. But these are the two extreme production possibilities. In between these two, there will be many other production possibilities such as *B*, *C*, *D* and *E*.

With production possibility *B*, the economy can produce with given resources 14 thousand quintals of wheat and one thousand metres of cloth, and with production possibility *C*, the economy, can have 12 thousand quintals of wheat, and 2 thousand metres of cloth and so on. As we move from possibility *A* towards *F*, we draw away some resources from the production of wheat and devote them to the production of cloth. In other words, we give up some units of wheat in order to have some more units of cloth. As we move from alternative *A* to *B*, we sacrifice one thousand quintals of wheat for the sake of one thousand metres of cloth. Again, our movement from alternative B to C involves the sacrifice of two thousand quintals of wheat for the sake of one thousand more metres of cloth. A look at the table will show that our sacrifice of wheat goes on increasing as we move further from *C* towards *F*. It is, therefore, clear that in a full- employment economy and with a full-use of production capacity, more of one good can be obtained only by cutting down the production of another good. Thus, we conclude that with a full-employment and a full use of *production capacity, an economy must always give up something of one good to obtain* some more of another. The basic fact that resources are limited prevents an economy from having more of both the goods.

The alternative production possibilities can be illustrated graphically by plotting the data of the Table 5.1 on a graph paper. The curve *AF* in Figure 5.1 is obtained when the data of the table are plotted, This curve *AF* is called the *production possibility curve which shows the various combinations of two goods or two classes of goods which the economy can produce with a given amount of resources and a given technology*. This production possibility curve *AF* like the Table 5.1 illustrates that in a full-employment economy and with a given technology, an increase in the amount of cloth necessitates a decrease in the amount of wheat. As we move from *A* towards *F* on the production possibility curve we sacrifice some units of wheat for having more of cloth. On the other hand, if we move up from *F* towards *A*, we will be giving up some of cloth for the sake of more of wheat.

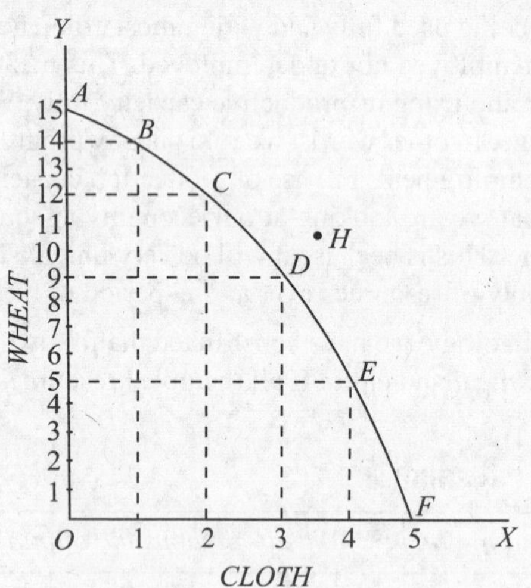

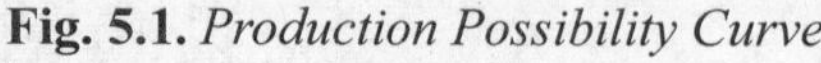
Fig. 5.1. *Production Possibility Curve*

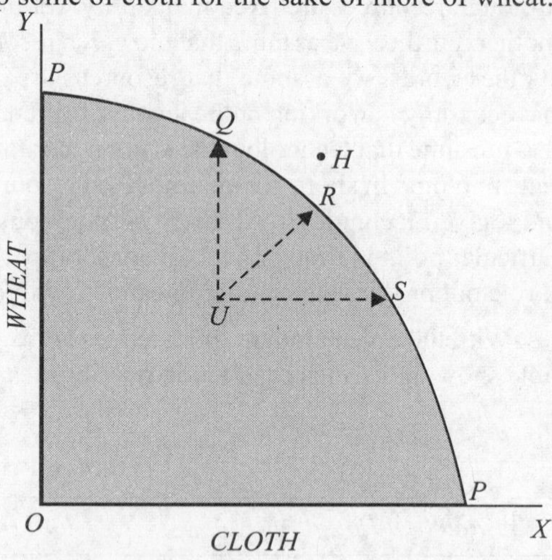

Fig. 5.2. *The Problem of Underutilisation of Resources*

The production possibility curve is also called *transformation curve* because in moving from one point to another on it, one good is 'transformed' into another, not physically but by transferring resources from one use to the other. With the given resources being fully employed and utilised, the combination of two goods produced can lie anywhere on the production possibility curve *AF* but not inside or outside it. For example, the combined output of two goods can neither lie at *U* nor at *H* (see Fig. 5.2). This is so because at point *U* the economy would not be utilising its resources fully, and the output of two goods represented by point *H*, given the productive resources, would lie beyond the capacity of the economy to produce.

The Problem of Unemployment and the Production Possibility Curve

However, during those periods when the economy is not fully utilising its resources, or not using them most efficiently, that is, when there is either unemployment, under-employment or inefficiency in the use of resources, output combination of two products can lie below the economy's production possibility frontier, such as at a point like *U* in Fig. 5.2, where the economy can produce more of both the goods or more of either of the two goods (as indicated by arrows) by putting idle resources to work. As shown in Fig. 5.2 if the economy is working at *U*, *then by using its idle resources fully and most efficiently, it can move from U* to *Q*, or to *R*, or to *S* on the production possibility curve *PP*.

Inefficiency in Production

Another important reason for the economy to find itself operating inside the production possibility curve is when it is using its resources inefficiently. We achieve production efficiency in the use of resources if with the given resources and technology, more of one good cannot be produced without a reduction in the output of the other. If the resources are not being used efficiently, then the production of both the goods can be increased or the production of one good can be increased without producing less of the other. For example, if land of a village which is better suited for the production of wheat is wrongly employed for the production of cotton, the economy will not be using its resources efficiently which will result in loss of output. This will make the economy work below or inside the production possibility curve. Similarly, if we employ textile engineers who have skills to perform a specialised job in the production of cotton cloth for the production of wheat, we will not be maximising output from the given resources. We could get more wheat and more cloth if we employ the workers and engineers in the tasks that closely match their skills.

Take another example, if managers of firms lack incentives to maximise output from the given resources (or to minimise cost to produce a given output), the economy will not be making optimum use of its resources resulting in loss of output. As a consequence, the economy will be operating below the production possibility curve. This applies both to the managers of public sector and private corporate sector. It is claimed by some that it is the lack of incentives on the part of managers in the former Communist Soviet Russia and in the various public enterprises in India which are incurring heavy losses that is responsible for low productivity and poor performance of public enterprises.

Economic Growth and Shift in Production Possibility Curve

Let us turn to the question of economic growth and see what happens to the production possibility curve when the economy's productive capacity increases over time. As already pointed out, the production possibility curve is drawn with a given amount of productive resources like land, labour and capital equipment. Now, if these productive resources increase, the production possibility curve will shift outward and to the right showing that more of both goods can be produced than before.

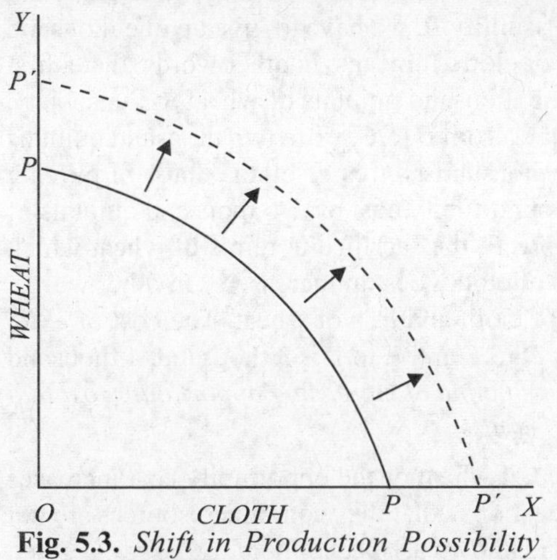

Fig. 5.3. *Shift in Production Possibility Curve due to Economic Growth*

Further, when the economy makes progress in technology, that is, when the scientists and engineers discover new and better ways of doing things, the production possibility curve will shift to the right and will indicate the possibility of producing more of both the goods. Technological progress by improving productive efficiency allows the society to produce more of both the goods with a given and fixed amount of resources.

From above it follows that when the supplies of resources increase or an improvement in technology occurs, the production possibility curve shifts outward such as from *PP* to *P′P′*, in Fig. 5.3. On production possibility curve *P′P′*, the economy can produce more goods than on curve *PP* . The increase in the amount of capital, natural and human resources and the progress in technology are determinants of economic growth. Thus with the growth of the economy production possibility curve shifts outward.

It is very important to understand the distinction between (*i*) the movement of the economy from a point inside the production possibility curve to a point on it, such as from point *U* to point *Q* in Fig. 5.2 and (*ii*) the movement of the economy from one production possibility curve to another. In both these cases national product or output of goods and services will increase. But the former involves fuller employment of the given resources while the latter involves the increase in resources or productive capacity. While the first type of movement is dealt with by the short–run macroeconomic theory, the latter is dealt with by the theory of economic growth. The fact that in both these movements of the economy the national product or income increases they are likely to be confused with each other. But the two movements are of quite different nature and different types of measures are required to bring them about. When the economy is working at a point below its production possibility curve due to the lack of aggregate demand as it happens in times of depression in free-market economies then those policy measures should be adopted which raise the level of aggregate demand. The increase in aggregate demand under such circumstances will bring about a shift of the economy from a point below the production possibility curve to a point on it. This will mean full employment or full utilisation of available labour and capital resources and as a result the levels of national income, output and employment will rise and the existing unemployment and under-utilisation of productive capacity will be removed.

On the other hand, when the economy is fully utilising its given resources and is, therefore, working at a point on the production possibility curve, the increase in national output and employment cannot be achieved by simply raising aggregate demand. Under such circumstances, national income and employment can be increased by adopting measures which generate economic growth. These measures aimed at generating economic growth involve stepping up of rate of capital accumulation and making a progress in technology.

Production Possibility Frontier and the Law of Increasing Opportunity Cost

The production possibility frontier *AF* in Figure 5.1 shows an important principle of economics. That principle is the law of increasing opportunity cost. From looking at the Table 5.1 it will be clear that as we move from possibility A to possibility *B*, we have to give up one thousand quintals of wheat in order to have one thousand metres of cloth. It means, in other words, that a first thousand metres of cloth has the opportunity cost of one thousand quintals of wheat to the society. But as we step up the production of cloth and move further from *B* to *C*, extra two thousand quintals of wheat have to be forgone for producing extra one thousand metres of cloth. Thus, in moving from *B* to *C* one thousand metres of cloth involves the opportunity cost of two thousand quintals of wheat. As we move further from *C* to *D*, *D* to *E* and *E* to *F*, the sacrifice in terms of wheat which we have to make for having extra one thousand metres of cloth goes on increasing. In other words, opportunity cost goes on increasing as we have more of cloth and less of wheat. The cost of extra one thousand metres of cloth as we move from *C* to *D*, *D* to *E* and *E* to F is 3 thousand, 4 thousand and 5 thousand quintals of wheat respectively. It is this *principle of increasing opportunity cost that makes the production possibility curve concave to the origin.*

But the question now arises: Why does the sacrifice of wheat or the opportunity cost increases as we produce more of cloth. A simple answer to this question is that the economic resources are not completely suited or adaptable to alternative uses. This is known as *specificity of resources*: a given resource is more suited to the production of one good than another. Thus, land is more suited to the production of wheat than cloth. The production of wheat requires relatively larger use of land than cloth. As we increase the production of cloth, resources which are less and less adaptable or productive in the production of cloth would have to be pushed in that line of production. As we move from *A* towards *F*, we will first transfer those resources which are more productive in making cloth. As we move further from to *C*, *C* to *D* and so forth, we will have to transfer those resources to the production of cloth which are more productive for producing wheat but less productive for making cloth.

It is, therefore, obvious that as the resources which are more suited to the production of wheat are withdrawn, extra loss of wheat for the sake of producing extra one thousand metres of cloth will go on increasing. This law equally holds good if we move from *F* to towards *A*; more and more amount of cloth would have to be given up for the sake of a given extra increase in the amount of wheat.

USES OF PRODUCTION POSSIBILITY CURVE

Scarcity and Resource Allocation

Production possibility frontier or curve is an important concept of modern economics. This concept is used to explain the important economic problems and theories. The basic economic problem of scarcity on which Robbins' definition of economics is based, can be explained with the aid of production possibility curve. According to the problem of scarcity, because of the limited availabilities of the resources, all wants of the society for goods cannot be satisfied; if a society decides to allocate more resources to the production of one good, it has to withdraw resources from the production of another good, as has been seen above. Given the amount of resources and the state of technology, the economy has to operate on the given production possibility curve. As has been brought out above, 'when we increase the production of one commodity moving along the production possibility curve, we have to reduce the production of some other commodity. If the given resources are being fully used and technology remains constant, an economy cannot increase the production of both the goods represented on the two axes. This illustrates the basic economic problem. Thus, the basic economic problem is that, in view of the scarcity of resources, at what point of the production possibility curve, the economy should produce so as to maximise social welfare.

Choice and Opportunity Cost

A production possibility curve illustrates three basic economic concepts, namely, (1) scarcity of resources (2) choice by a society from the **opportunity set** and (3) opportunity cost. Scarcity is depicited by the boundary of the production possibility curve beyond which the economy cannot produce goods and services. That is why the production possibility curve is also called **production possibility frontier**. Given the resources and the state of technology, the economy has to operate either on the production possibility curve or inside it.

The issue of choice arises because when the resources are fully employed and efficiently used and allocated and also the technology remains the same, we cannot have more of both the goods represented by a production possibility curve. *Given these constraints of resources and technology, a society has to make a choice at what point on the production possibility curve it should operate to produce goods and services. Thus, a society has to make a constrained choice.*

The concept of opportunity cost can be illustrated with the use of a production possibility curve. As we move along a production possibility curve we give up or sacrifice some units of a commodity to obtain a unit of the other commodity. That is, there is trade-off between the two commodities; more of one commodity is obtained at the cost of less of the other commodity. *The amount of one commodity which is forgone to obtain a unit of the other commodity is the opportunity cost of the latter.* Thus, when we move from point *B* to point *C* in Fig. 5.1, we forgo 2 thousand quintals of wheat to obtain one thousand metres of cloth. Therefore, the opportunity cost of one thousand metres of cloth is here equal to 2 thousand quintals of wheat.

Production Possibility Curve and Central Economic Problems

Another use of production possibility frontier is that with its aid we can explain the central problems of *what, how and for whom to produce*. Which goods should be produced and in what quantities, implies that on what point of the production possibility curve the economy should operate. A glance at Figure 5.1 will reveal that if the economy is operating at point *B* on the production possibility curve *AF*, then one thousand metres of cloth and fourteen thousand quintals of wheat are being produced. If the economy operates at point *E* on this curve, four thousand metres of cloth and five thousand quintals of wheat are being produced. Thus, *operating at different points of the production possibility curve implies different allocation of resources between the production of two goods*. At which point of the production possibility curve, a free market economy will operate depends upon the consumers' demand for different goods. In other words, in a free market economy, how the resources would be allocated between the two goods on a given production possibility curve is determined by the demand of the consumers.

How the goods are to be produced implies which methods or techniques should be employed for the production of various goods. In. other words, what resource combination should be used for the production of goods so as to maximise the output or to minimise the cost. A factor would be used for the production of a product for which it is more efficient. It is obvious that this is the problem of production efficiency. If for producing goods such resource combinations as will minimise cost of production are not employed, the economy will be operating at a point below the given production possibility curve. Thus, *if in the production of various goods, efficient methods are not used or if the resources are not employed in their efficient uses, the economy will not be operating at a point on the production possibility curve, instead it will be operating at a point below the production possibility curve such as U in Figure* 5.2. The working of the economy below the production possibility curve indicates that less than maximum possible production is being done which will lower the welfare and standard of living of the people. The loss of production is the result of inefficient use of the resources.

For whom to produce or how the national product is being distributed is not directly revealed by the production possibility curve. However, we can obtain some knowledge of the distribution of goods from the production possibility curve. If a production possibility curve is constructed in which necessaries are represented on the one axis and luxuries on the other, we can know from the actual position of the economy on this curve that how the national output is being distributed. Consider Figure 5.4 in which on the X -axis necessary goods and on the Y–axis luxury goods have been measured. If the economy is working at point R on the production possibility curve PP in this figure, the economy would be producing relatively more of luxury goods such as refrigerators, televisions, motor cars, air conditioners and would be producing relatively less quantities of essential consumer goods, such as foodgrains, cloth, edible oil, which indicates that distribution of national income is very much uneven and the richer sections of the society will be getting relatively more of luxury goods, whereas the poorer sections would be deprived of even the necessaries of life. On the contrary, if the economy is operating at point S on the production possibility curve PP, then it implies that essential consumer goods will be produced relatively more and luxury goods will be produced relatively less by the economy. This indicates that the distribution of income and output in the society in this case will be relatively more equal.

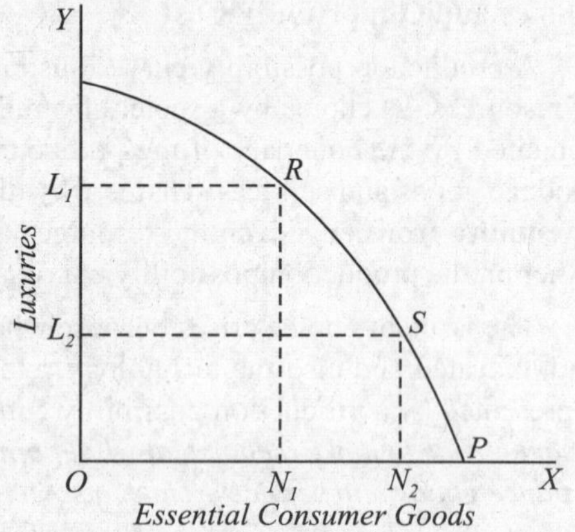

Fig. 5.4. *Resource Allocation between Basic Consumer Goods and Luxuries*

What quantities of various goods will be produced in a free market economy i.e. how much of luxury goods and how much of necessaries would be produced, depends upon the pattern of demand of the consumers. In other words, pattern of production will correspond to the pattern of demand. But it should be remembered that the pattern of demand depends upon the distribution of income in a society. The more unequal is the distribution of income in the society, the greater the amount of luxury goods produced in it.

The Problem of Unemployment and Underemployment of Resources. As we have studied above, the problem of unemployment and underemployment of resources can be illustrated and understood with the aid of the production possibility curve. When all resources are being fully used the economy will operate at a point on the production possibility curve. But the economy will operate at a point on the production possibility curve if aggregate demand is large enough to buy the total output produced by the full employment of resources. If aggregate demand is somehow smaller, the economy will not be able to use its productive capacity fully, that is, it will not be able to utilise its resources fully, which will result in unemployment and underemployment of resources. In case of unemployment and underemployment of resources, the economy will be working at a point below the production possibility curve (such as point *U* in Figure 5.2). In such a situation if aggregate demand for goods increases, the demand for resources and, therefore, their employment will rise and as a result unemployment and underemployment will disappear and national income will increase. Thus, it follows that as a result of increase in aggregate demand the economy moves from a point below the production possibility curve to a point on the production possibility curve. Renowned economist J.M. Keynes, who attributed unemployment and underemployment to the lack of aggregate demand recommended construction of public works on a large scale by the Government financed by deficit financing so as to raise the aggregate demand which will help in utilisation of resources fully and therefore in solving the problem of unemployment and underemployment.

The Problem of Capital Formation and Economic Growth

Another important use of the production possibility curve is that with it we can explain with it the problem of capital formation and economic growth. In order to explain the problem of capital formation we have to construct such a production possibility curve in which on one axis capital goods and on the other axis consumer goods are measured. This has been done in Figure 5.5 in which along the *X* -axis consumer goods and along the Y-axis, capital goods are measured. If the economy is allocating the available resources between capital and consumer goods in such a way that it operates at point A on the production possibility curve *PP*, it will be producing OC_1 of consumer goods and OK_1 of capital goods. Now suppose that the society decides to produce more of capital goods. To implement this decision, society will have to withdraw some resources from the production of consumer goods and use them for the production of capital goods. As a result, the production of consumer goods will decline. It is clear from Figure 5.5, that if the economy reallocates its resources between consumer and capital goods and shifts from point *A* to point *B* on the production possibility curve *PP*, it will now produce OK_2 of capital goods and OC_2 of consumer goods. That is, K_1K_2 amount of capital goods will be produced more and C_1C_2 amount of consumer goods will be produced less than before. We, therefore, conclude that in order to step up the rate of capital formation the production of consumer goods and therefore consumption has to be reduced.

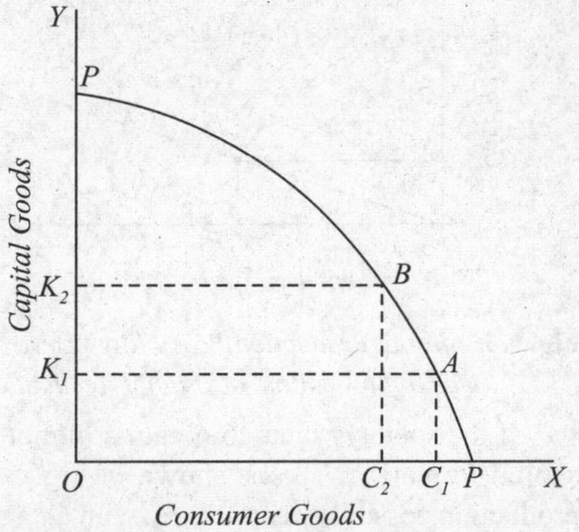

Fig. 5.5. *Resource Allocation between Consumer and Capital Goods*

But the above conclusion is based on the assumption that the economy is using its resources fully and most efficiently and is operating at a point on the production possibility curve. However, if some available resources are lying unemployed and idle or the economy is not using them more efficiently, the economy will be working below the production possibility curve. When the economy is working at a point below the production possibility curve, then more capital can be created without a reduction in the production of consumer goods because by employing idle and unemployed resources, economy can produce more of capital goods. But, as has been explained above, if the economy is utilising its resources fully then the rate of capital formation cannot be increased without the reduction in consumption. But it is worth noting that when the rate of capital formation is raised, this does not mean that amount of consumption is reduced for ever. The accumulation of more capital enables economy to increase its production of consumer goods in the future. That is, the accumulation of capital raises the productive capacity of the economy. Thus, capital accumulation implies that "*less jam today for more jam tomorrow*."

Since the accumulation of capital raises the productive capacity, national production will increase, that is, economic growth will take place. As a result, the economy will not remain on the same production possibility curve and its production possibility curve will shift outward which indicates that the economy will be able to produce more than before. The greater the rate of capital formation, the greater the extent of shift in the production possibility curve, and the greater the rate of economic growth. Consider Figure 5.6 in which in the beginning the economy is producing OC_1 of consumer goods and OK_1 of capital goods on the production possibility curve P_1P_1. If the economy maintains this rate of capital formation, then the production possibility curve will go on shifting and the economy will be growing annually at a certain fixed rate. It should be noted that in Figure 5.6, as a result of low rate of capital formation, production possibility curve shifts outward at a relatively slow speed. Thus growth path *OR* in Fig. 5.6 represents a lower rate of economic growth.

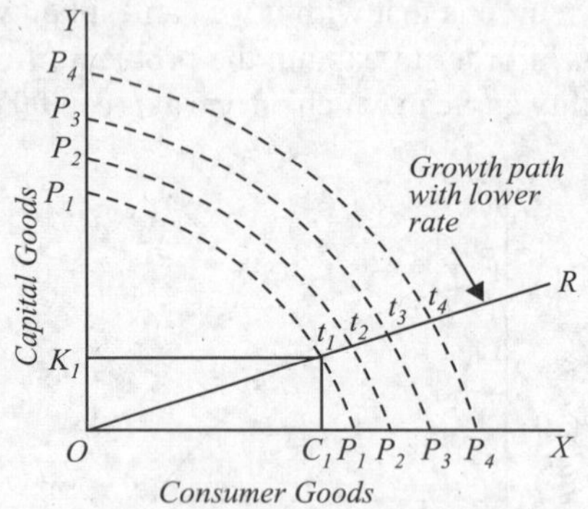

Fig. 5.6. *Shift in Production Possibility Curve due to Lower Rate of Capital Accumulation*

Fig. 5.7. *Shift in Production Possibility Curve due to Higher Rate of Capital Accumulation*

If the society wants to obtain a higher rate of economic growth, it will have to raise its rate of capital formation. This is shown in Fig. 5.7 in which the economy is producing at point t_1 on the production possibility curve P_1P_1, with OK_2 of capital goods at OC_2 of consumer goods. If the economy maintains this rate of capital formation, production possibility curve will go on shifting outward to a greater extent than in Figure 5.6. This means that the rate of economic growth will now be relatively greater than in Figure 5.6. In the two Figures 5.6 and 5.7, it will be noticed that, in the beginning in Fig. 5.7; the production of consumer goods is less than in Figure 5.6, but when as a result of higher rate of economic growth, production possibility curves reach their position P_4 P_4 at time t_4, it will be producing more consumer goods in Fig. 5.7 exhibiting higher rate of economic growth than in Figure 5.6, where the rate of capital formation and therefore the rate of economic growth is relatively less.

We have explained above economic growth which has been brought about by capital formation. Besides capital formation, there are other factors which determine rate of economic growth. Progress in technology and expansion in education also favorably affect rate of economic growth' and cause production possibility curve to shift outward.

We have explained above only some important uses of production possibility curve. There are several other uses of production possibility curve. We can understand better the concept of opportunity cost with the aid of production possibility curve. The concept of production possibility curve has also been extensively used in welfare economics and in the theory of international trade. In the modern economic theory gains from international trade have also been explained with the aid of production possibility curve.

QUESTIONS FOR REVIEW

1. Define '*The Economic Problem*'. How does it give rise to basic economic problems faced by an economy.

2. Scarcity is the mother of all economic problems. Do you agree? If so, how?

3. What are the central problems of an economy? Explain them briefly.

4. (*a*) Explain the following basic problems of the economy.
1. What to produce?
2. How to produce?
3. For whom to produce?
(*b*) How do they arise from the problem of scarcity?

5. What are the basic problems of an economic system? Which of these constitute the subject-matter of micro-economic analysis? Discuss briefly. (*DU B.A.* (*Hons*) 1988)

6. An eminent British economist J.M. Keynes explained that at times of depression, there was unemployment and surplus of both labour and capital resources. Is it not inconsistent with the problem of scarcity?

7. Explain the problem of economic growth. How is it related to the problem of scarcity? On what factors does economic growth depend?

8. How would you explain the following problems with the help of production possibility curve?
(*i*) The choice between the production of basic consumer goods and luxuries.
(*ii*) The problem of unemployment
(*iii*) The problem of economic growth

9. Explain how the central problems of an economy are solved in a (*i*) Free market economy and, (*ii*) Centralised planned economy.

10. What is production possibility curve? What assumptions are made white drawing it? Why is it generally concave towards the origin?

11. What is opportunity cost of a commodity? Why does it increase when more of it is produced?

12. What is meant by economic growth? On what factors does it depend? How does economic growth affect the production possibility curve of the economy?

CHAPTER 6
MARKET ECONOMY AND ROLE OF PRICE MECHANISM

Introduction

We have explained in the earlier chapters the basic issues, problems which the economists study. What goods and services are produced, how they are produced, how the output of goods and services produced are distributed among the people of a country and what provision is made for promoting economic growth are the main problems analysed in economics. We have also explained that these basic problems arise due to the scarcity of resources in relation to human wants. Now, the important questions is who make these choices of what, how and for whom to produce and how much resources be allocated to capital accumulation and bringing about technological progress for promoting economic growth. The present chapter is devoted to answer who and how the above mentioned social choices are made. It is important to note that economics has been evolved and developed in framework of a free market economy in which the resources of a society are owned by individuals and firms. Thus, a market economy recognises and protects *property rights*, that is, the rights that govern the ownership, use and disposal of resources and the goods and services produced with their help. Historically, such a social arrangement in which resources are privately owned and consumers (i.e. individuals or households) and producing firms interact to determine the choices regarding production and distribution of goods and services has been called capitalism. Thus, according to W.N. Loucks, "*Capitalism is a system of economic organisation featured by the private ownership of and the use of man-made and nature-made capital*"[1].

Since the term capitalism has been associated with exploitation of man by man, we will avoid using the term capitalism or capitalist economy and prefer to use the term '*market economy*'. In the basic model of a market economy we also assume that competitive markets prevail in a market economy in which no individual consumer and firm exercises any control over the prices of product and resources. Hence we call this model of a market economy as *basic competitive model*. In what follows we will first explain the essential features of a competitive market economy and the basic competitive model and then analyse how the basic units of the economy — individual households and firms — make choices when they are faced with scarcity of resources in relation to their needs or wants. Besides, we will study how the interaction of these units determine social choices regarding how the scarce resources are used and allocated. It is important to mention here that for determining the social choices of what to produce, how to produce and for whom prices play a crucial role. Therefore, in the discussion of basic competitive model we will study the role of price mechanism in making the above social choices.

CHIEF FEATURES OF A FREE MARKET (CAPITALIST) ECONOMY

The efficient working of a free market economy requires that the producer firms must have incentives to work hard and produce goods and services at the lowest possible cost per unit of output Market economies provide incentives to the firms and individuals by recognising and enforcing the property rights of the individuals and firms to own the resources and goods and services produced by using them. It also allows the individuals and firms to use the resources owned by them

I. W. N. Loucks, *Comparative Economic Systems*.

for making profits. *Earning of profits from using resources owned by them also provides incentives to the firms and individuals to produce goods and services efficiently*. Profits depend on cost per unit incurred and price charged of goods and services produced. The prices of goods and resources provide information to the individuals and firms about the relative scarcity of different goods and services. Thus property rights, profits, prices and incentives play a crucial role in the functioning of a market economy. We briefly explain them below.

Property Rights

Property rights are social institutions that govern the ownership, use and disposal of resources, goods and services. There are different types of property which individuals and firms can privately own. (1) *Real property* which includes land, buildings, durable goods such as plant, capital equipment etc. (2) *Financial property* which includes shares and bonds, bank deposits, money kept at home. (3) *Intellectual property*, which is the latest addition to the list of property, represents the products of creative effort and includes books written, audio and video material, computer programmes. It is interesting to note here that Bill Gates have become the richest man of the world by possessing intellectual property in the form of computer programmes such as windows, Miscrosoft Office.

The right of private property is a fundamental feature of a capitalist economy. The right of private property means that productive resources such as land, factories, machinery, mines, etc. are under the private ownership. In other words, an individual has a right to acquire and to use the means of production. Besides, the owner is free to sell or dispose of his property in any way he likes. Because of this right of private property, goods like land, factories, mines, machinery, houses, and other producer and consumer goods are privately owned by the people. They use these means of production for their individual benefit. Moreover, the right of private property includes in itself the **right of inheritance**. The right of inheritance means that on the death of a person, his sons or daughters or some relatives become the owners of his property. No doubt, there are always some restrictions on the right to private property imposed by the Government for social harmony and benefit. But, except for some restrictions capitalist economic system tries to protect and enforce the right of private property.

The two attributes of property rights, namely, *the right of the owner to use the property as he likes and the right to sell it provide incentives to the owners to use their property efficiently*. For example, the owner of a factory building will try to make most profitable use of it by producing a commodity which yields him maximum profits. If he makes a wrong decision and produces a commodity for which there is insufficient demand he will incur losses. This gives him incentive to think carefully and try to obtain adequate information before making a decision about what to produce and in what quantity. The owners of the property would also try to *maintain it* properly so that when in future he wants to sell it he can do so at a good price.

It is important that government should enforce the property rights. If property rights are not enforced, then the incentives to use the property efficiently is weakened and the potential gain from its efficient use will be lost. *Enforcing intellectual property rights is proving to be a big challenge these days* in view of the modern technologies that make it relatively easy to copy books, audio and video material and computer programmes.

Freedom of Private Enterprise

Allied to the property rights is freedom of enterprise which is another basic feature of a market economy. Freedom of enterprise means that everybody is free to engage in whatever economic activity he pleases. In other words, he is free to choose to work in any industry he likes or adopts any occupation or trade he desires. More specifically, freedom of enterprise means that an entrepreneur is free to set up any firm or business unit to produce goods or make investment in shares or bonds of corporate sector. True, he may not have enough capital to invest in business or a productive unit like factory or he may not have enough ability or training to follow certain occupations.

But subject to these limitations and to the laws passed by the Government in public interest, an individual or firm is free to engage in any economic activity he feels most desirable or profitable. In India before 1991, there was a *system of industrial licensing* which restricted the freedom of enterprise. For setting up industrial units in various industries, the private sector firms had to obtain licenses which were not easily available. Similarly, for procuring necessary raw materials and inputs such as steel, cement, businessmen had to get *permits* from the government to purchase these products and to get permits businessmen had to bribe the government officials. This license-permit system greatly restricted the freedom of enterprise of the private sector firms. This not only caused allocative inefficiency but also obstructed economic growth of the country. Under the structural economic reforms initiated in 1991 by Dr. Manmohan Singh who was then Finance Minister, industrial licensing has been largely abolished and the system of permits has been done away with. Grant of licenses to import goods, machines and raw materials has been greatly liberalised and custom duties have been drastically reduced. As a result, private sector now enjoys greater freedom to invest in industries they think it profitable and diversify their productive capacity by producing multiple products.

Abolition of industrial licensing system and liberalisation of imports and ending of permit system has tended to promote competition in the Indian economy. The greater competition has helped to keep inflation under check and to achieve higher industrial growth rate.

Profits and Prices: Incentives and Information

No individuals would work and save if adequate incentives are not provided to them in the form of wages and interest respectively. Similarly, the firms will not produce goods and services and bear risk of losing money if sufficient incentives are not given to them. Profits are earned from undertaking the task of producing goods and services and introducing new products and new techniques of production. Profits earned by the firm depend on prices of goods and services produced and cost incurred. As seen above, in a perfectly competitive market firms are price takers, that is they take price of a product or service as given. With a given price, firm's profits will be larger, if cost per unit of output is smaller. Therefore, in order to maximise profits, firms try to minimise cost for producing a given level of output. *Thus profits serve as incentive for the firm to produce efficiently*.

The working of price system ensures that those individuals and firms will get goods who are willing and able to pay for them. Prices of goods and services indicate how much money individuals are prepared to pay for them. In other words, *prices convey information to the firms about how individuals value different goods and services*. The goods and resources which are relatively more scarce, will have higher prices in the market. On the other hands, the goods and resources which are relatively less scarce will have low market prices. The profit motive induces firms to respond to the prices of different goods. The firms will earn more profits if they produce goods which people want most efficiently, that is, economise the use of scarce resources and use relatively more those resources which are relatively less scarce in supply. Thus, by producing efficiently the firms are able to increase profits. It is important to note that for profit motive to be effective the firms must be allowed to keep a good part of the profits earned by them and not taxed away by government. As seen above, this is implied by property rights enjoyed by the firms. Similarly, if incentives to individuals or households have to be provided to work hard and acquire skills and training, they should be permitted to retain a good amount of what they earn from their work or what they receive as a return on their investments.

Importance of Information and Incentives. It follows from above that *for market economies to solve its basic economic problems and work efficiently individuals and firms must have adequate information and have incentives to act on the available information*. In market economies it is prices through which individuals and firms get information about the relative scarcity of goods and resources. Relatively more scarce goods and resources tend to have higher prices than the relatively less scarce resources. Further, in market economies incentives play a crucial role for the working of individuals and firms. Professor Stiglitz, a Nobel laureate in economics writes, "*incentives can be viewed as being at the heart of economics. Without incentives, why would individuals go to work in the morning? Who would undertake the risks of bringing out new products ? Who would put aside savings for a rainy day*?"[2] Emphasising the importance of incentives he further writes, "Providing appropriate incentives is a fundamental economic problem. In modern market economies, profits provide incentives for firms to produce the goods individuals want and wages provide incentives for individuals to work, property rights also provide people with important incentives not only to invest and to save but also to put their assets to the best possible use."[3]

It follows from above that property rights are important features of market economies and play a significant role in their efficient working.

Incentives vs. Inequality. While incentives play a significant role in market economies for their working efficiently to solve the basic problems of what to produce, how to produce, for whom to produce and how much to invest for bringing about economic growth, but they involve a cost which must be taken into account. *The cost of providing incentives is the inequality in the economy they cause*. Any system of incentives must provide compensation to individuals and firms in terms of higher wages, larger profits, higher return on investment for better performance or higher efficiency. If pay for work, return on investment is tied to performance etc. then since performances differ, the compensation for them too will differ and this will bring about inequality in incomes. This raises the question how much incentives should be given for better performance or higher efficiency so that inequality of incomes is not much higher. Since the greater the incentives given, the greater will be the inequality of incomes, there is *incentive -inequality trade-off*. If a society provides greater incentives to promote efficiency and growth, there is likely to be more inequality of incomes.

It may be noted that one basic problem facing governments of market economies is to devise a tax system (*i.e.* kinds of taxes and rates of taxes) that tends to reduce incentives on the one hand but provide revenue to finance welfare programmes for those who have low incomes. Thus, providing incentives to earn higher wages or return and profits encounters serious problem of increasing inequalities of income in a society.

Competitive Markets

In ordinary speech the word *market* means a place where buyers and sellers meet to buy and sell goods. In economics, market has a wider meaning and is interpreted to mean any arrangement that enables buyers and sellers to exchange things; they may contact each other through telephone, fax or direct computer link to negotiate prices of goods they buy and sell. In our above example, rational and self-interested consumers and profit-maximising firms contact each other through any media and buy and sell things on the negotiated price.

The self-interested consumers are interested in low prices of goods they buy and, on the other hand, the profit-maximising firms will be interested in charging a higher price of the goods they produce and offer for sale in the market. However, through their interaction they may agree to buy and sell goods at the agreed price.

In the basic competitive model of a market economy, it is assumed that neither the consumers, nor the firms have any market power to influence the prices of the goods and services they want to buy and sell. In fact, the economists generally assume that *perfect competition* prevails in the market. Under perfect competition, there are a large number of firms producing a good and since no one is in a position to influence the price, they take prices of goods as given and constant, that is, each firm is a *price taker*. Each firm must accept the price prevailing in the market. This prevailing price is

2. Joseph Stiglitz. *Principles of Economics*, p.25.
3. *Op. cit.*, p. 26.

determined by the interaction of *all* firms and buyers. We will see in a later chapter how demand for a good by all the consumers who want to buy the goods and supply by all the firms who produce and supply a good together determine the market price of the goods. Suffice it to say here, if a firm in this perfectly competitive market charges a higher than the prevailing price, it will not be able to sell any quantity of the goods because his customers will switch to other firms producing and supplying the same good.

Similarly, in such a perfectly competitive market since each consumer cannot influence the market price, it must accept it as given and make a choice about how much to buy the various goods and services, given his budget constraint and preferences for different goods. Thus, in the basic competitive model of a market economy, rational and self-interested consumers and profit-maximising firms interact to determine prices of goods and resources. As will be explained later, these prices play a crucial role in determining social choices regarding what, how and for whom to produce.

Consumer Sovereignty

In a free market economy there is a freedom of choice for the consumers to buy goods and services which suit their tastes and preferences. In a market economy only those goods and services are produced if the firms producing and supplying them are able to sell them at a profit. Now, profits are made if goods or services produced are sold for more than what it costs to produce them. Therefore, firms cannot expect to make profits, if they do not make goods or services which are not in accordance with the preferences and demands of the consumers and also for which consumers are not willing to pay adequate price. Since firms are guided by profit motive, they will produce those goods which ensure greater profits to them. Therefore, in what goods and in what quantities will be produced by firms is dictated ultimately by the tastes and preferences of consumers who in a way '*vote*' for the goods they buy. This is generally called the principle of *consumer sovereignty*. This means in a market economy the consumers are just like a king or sovereign who dictate what goods and services and what quantities of them are produced.

FUNCTIONING OF THE FREE MARKET (CAPITALIST) ECONOMIC SYSTEM

A market economy has no central planning commission or bureau to decide what and how goods are to be produced and how the total national product is to be distributed among the consumers. In a free market or capitalist economy there is a freedom of enterprise and choice. Consumers are free to buy whatever they like. Businessmen are free to produce and sell what they choose and suppliers of factors of production are free to supply their material and human resources to whatever occupations they prefer. Now, in the absence of any central co-ordinating body it looks like a miracle as to how such an economy functions and is able to take decisions regarding what, how and for whom to produce? If consumers want cloth, producers choose to make motor cars and resource suppliers wish to give their services for making bread, there will be complete confusion and chaos because of the inconsistency of such choices. So there must be some co-ordinating and organising mechanism to prevent such confusion and chaos posed by the freedom of enterprise and choice. But if there is no central planning authority to make the fundamental economic decisions and thus to allocate productive resources among various competing uses, how can then a free-enterprise economy solve its central problems ? The answer is that a free market economy uses the impersonal forces of market to solve its central problems and thus determine the allocation of its resources. Market forces are reflected in price changes

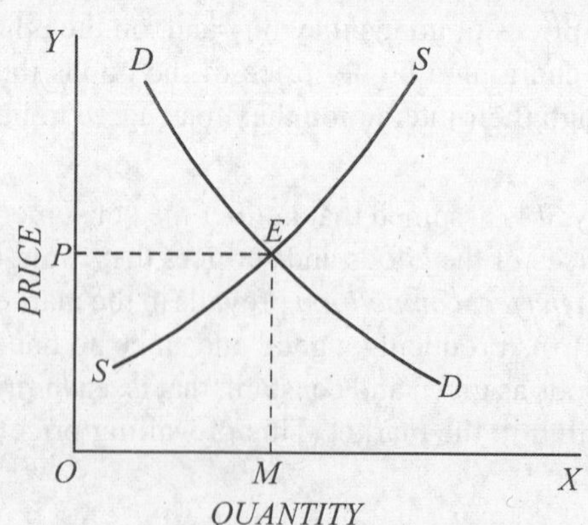

Fig. 6.1. *Market mechanism determining price*

and prices determine what goods are to be produced, how they are to be produced, and for whom they are to be produced. Thus price mechanism plays a vital role in the working of a free-enterprise market economy.

Now the question is what does market mechanism or price mechanism mean? Market stands for the contact between buyers and sellers of goods through which the buyers get goods in exchange for the price they pay for them. The buyers bring the demand for the goods and the sellers offer supply of the goods to meet the wants of the buyers. The interaction of demand for and supply of goods determine their prices. Thus the market stands for the forces of demand and supply of goods. How price is determined by market mechanism, that is, through demand and supply is shown in Fig. 6.1. The price at which the demand and supply curves intersect each other will come to stay in the market. In Fig. 6.1, *OP* price is determined at which *OM* quantity of the good is demanded and sold. The productive resources will be allocated to the production of the good to the extent that *OM* quantity of the good is produced.

How Does Price Mechanism Allocate Resources?

Now, how does price mechanism allocate resources among goods in a market economy. It is worth noting that competition among the sellers or producers and among the buyers of goods is essential for the working of a market economy. Production of goods requires resources such as labour, capital (mechanies, etc), raw materials, fuel or power for their production. In any economic system, scarce resources are allocated among various goods and services. A market economy relies on demand for and supply of goods and services for allocating resources among goods and services. Demand for and supply of goods and services determine their prices. *Prices are signals that direct the allocation of resources among production of goods and services*. Since a firm is driven by profit motive, it will produce, what is profitable to produce. If a firm can sell a product at a price which is more than the cost per unit incurred on it, the firm is likely to produce it. In a free market system resources flow towards the production of those goods which yield profits to the firms. Higher prices of certain goods mean higher profit opportunities in the production those of goods. As a result, firms will devote more resources to the production of products whose prices are relatively higher and therefore prospects of making profits in the production of these products is greater.

It is worth mentioning that *demand curve of a product represents what people are willing to pay*. If people's preferences or income change, the demand for different goods will change causing changes in their prices and therefore change in profit opportunities in the production of different goods. As a result, there will be reallocation of resources among different goods and services. To illustrate how resource are allocated consider Figure 6.2 where as a result of change in preferences of individuals or increase in their incomes, demand curve for cars increases and as a result demand curve for cars shifts to the right from D_1D_1 to D_2D_2. It will be seen from Fig. 6.2 that shift in demand curve of cars from D_1D_1 to D_2D_2 causes price of cars to go up from P_1 to P_2. Rise in price of cars relative to any increase in average cost of producing cars makes it profitable to produce more cars. Towards that end, the existing firms will expand their capacity by buying more capital (*i.e.* installing new machines), procure new buildings to install these machines and employ more labour

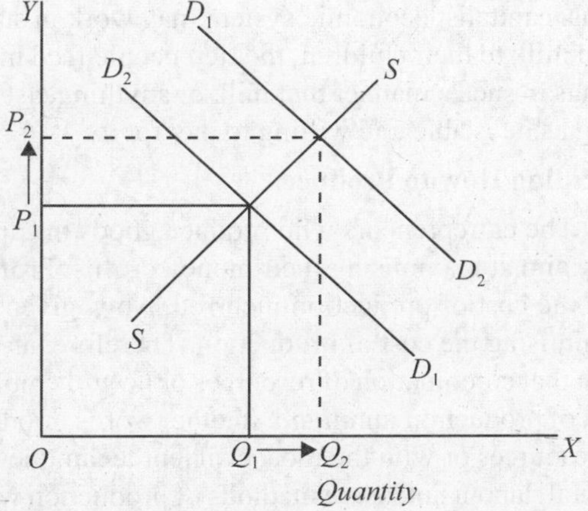

Fig. 6.2. *Allocation of Resources to the Production of Cars*

and raw materials to produce more cars. Besides, to avail of the profit opportunities in car production new firms will enter the business of car manufacturing. As a result of all these, more resources will be allocated to the production of cars.

We shall now explain in detail how the central problems of an economy are solved through price or market mechanism in a free-enterprise capitalist economy. In other words, we shall describe below at length what role price mechanism plays in the functioning of the capitalist economy.

Deciding What to Produce

In a free market economy, there are thousands of entrepreneurs or businessmen to produce goods for the society. The aim of entrepreneurs or businessmen is to earn as much money profits as possible. The desire for profits causes businessmen to compete with one another to produce those goods which consumers wish to buy. For instance, when the consumers raise their demand for a particular good, price of that good will rise. A rise in price of that good, cost remaining the same, will lead to more profits. The increase in profits will attract more productive resources into the production of that good. In order to earn more profits businessmen will produce that good in a larger quantity. On the other hand, if consumers' demand for a good decreases, price of the good will fall. The fall in the price will reduce profits and may cause losses. The businessmen will, therefore, reduce the output of that good and as a result some resources will be withdrawn from the production of that good and employed somewhere else. Thus in a free market economy, the question regarding what to produce and in what amount is decided ultimately by the preferences of the consumers as shown by their demand for goods and services in the market. The consumer is, so to say, a king, who decides what is to be produced and in what quantity by making his wishes known by the types and quantities of the goods he buys in the market. Thus, in a free enterprise economic system consumers determine what is to be produced and producers only act and produce such goods and in such quantities that are wanted by the consumers. This is known as the *principle of consumer sovereignty*.

Determination of 'what to produce' can be better understood by comparing it to an election. When a consumer spends a rupee on a particular good, it is as if he was casting a vote for the production of that good. Various goods will be produced in proportion to the number of 'rupee votes' cast for them. That good will be produced in a larger quantity for which the 'rupee votes' cast are greater. But the economic election is not democratic because every consumer does not have equal number of votes to cast. The greater the income of a consumer, the greater the number of his money-votes. Therefore, in a capitalist economy while luxuries are produced for the rich people who have got enough purchasing power or ability to pay for such goods, the necessities are not produced in sufficient quantities for the poor people because they do not have much ability to pay. Thus, capitalist economic system may work in such a way that while the poor people are unable to feed milk to their children, the rich people feed milk to their dogs. The market mechanism arranges things in such a manner that milk or anything else goes to anyone in whatever quantity he wishes so long as he is able and willing to pay for it.

Deciding How to Produce

The entrepreneurs who produce goods in capitalist economies are in search of profits. In fact, they aim at maximising their money profits. Profits are the difference between the revenue earned and the cost of production incurred. Now, given the level of output, profits can be maximised by minimising the cost of production. Therefore, an entrepreneur will produce a given level of output with that combination of resources or he will employ that technique of production which renders his cost of production minimum. In other words, producers will produce with the least-cost combination of resources or with the most efficient technique of production. If labour is relatively cheaper than capital, labour-intensive methods of production will be used. On the other hand, if capital is cheaper than labour, capital-intensive technique will be employed for production. Which combination of resources is the least cost one depends upon the prices of the various factors of production. In the least-cost combination of resources the quantity of that productive factor will be larger whose price is relatively lower. In order to minimise cost, the producers will use various factors of production in such a proportion that marginal products of factors are proportional to their prices.

In a free market economy the factor of competition and the motive of profit force all businessmen to produce goods with least-cost combination of factors and with most efficient technique. The inefficient firms who fail to employ the least-cost method of production will have to go out of the field because of competition. The maximum efficiency in production or least-cost production is very desirable from the viewpoint of the society because it ensures that the scarce resources of the society are being used to produce largest possible output of goods desired by the consuming public.

Deciding for Whom to Produce: Distribution of National Product

Who gets the goods and how much depends upon the money income one has to spend on goods? Distribution of total output of goods and services depends upon the distribution of national income. More unequal the distribution of national income, more uneven will be the distribution of national product. There are thus two steps in the process of distribution.

The first is the distribution of money income. Incomes can be earned either by doing work or by lending the services of one's property such as land, capital and houses. Thus, there are two sources of getting money income: first, income from work; and second, income from property. The forces of competition in a free-enterprise economy force the entrepreneurs to pay the workers according to the contribution made to production by them. Therefore, a worker gets income or wage equal to the value of their marginal products in the production of the goods he helps to produce. Similarly, in a capitalist economy, the owners of the land, that is landlords, and the owners of capital, that is, capitalists, get rent and interest respectively according to the value of the marginal products of the services of their respective resources. And the entrepreneurs or businessmen earn profits by setting up business and bearing the risks of production. Wages, rent, interest and profits are respectively the prices of factors of production, labour, land, capital and enterprise. Just as there are markets for goods, similarly there are markets for the factors of production. Demand and supply for the different factors determine their market prices. These factor prices constitute the incomes of the owners of the factors. It is thus clear that price mechanism (*i.e.*, demand and supply of the factors and their prices) work to determine the incomes of the various individuals and hence their shares of output.

How much income an individual will be able to make depends not only on the amount of work he does and the prices of the factors he owns but also on how much property in the form of land and capital he owns. A person who owns a large amount of property will be enjoying a higher income. In a capitalist economy, because of the large inequalities in the ownership of property, there are glaring inequalities of incomes.

Next step in the distribution process is the sharing of goods and services between different individuals or consumers. As in a free-enterprise capitalist economy, incomes of different individuals as determined by price mechanism and ownership of property are very much unequal, the distribution of total output of goods and services is also very much unequal. People with low incomes .receive small shares from the national product. Poor people cannot obtain high-priced goods because incomes being small their ability to pay is limited. They may not even get necessities in sufficient quantities. Whom some goods are in short supply, their prices are bid up and goods are delivered to those who are able and willing to pay the most for them. Not only the prices of factors but also the prices of goods also determine the share of each individual in the national output. Suppose an individual has earned income of ₹ 4,000 as determined by the price and the amount of factors he owns; Now, how much amount of the goods he will be able to get with ₹ 4,000 from the market depends upon the prices of the goods as determined by their demand and supply. The higher the prices of the goods, the smaller the amount of goods he will be able to obtain.

We thus see that just as prices determine what goods are produced, in what quantities, and by what methods, so, under free private enterprise, these also determine the .share of each individual in the national output.

Deciding about Rate of Economic Growth

Rate of economic growth depends on rate of saving and investment. In solving this problem also a free market or capitalist economy depends largely on the freedom of choice between consumption and saving by the consumers and free decisions regarding investment by private entrepreneurs. Money and capital market play an important role in carrying out the decisions by the consumers and entrepreneurs. Each individual decides how much of his income he wants to save. His saving is governed by the rate of interest prevailing in the money market. Rate of interest is the price for the use of money. Thus, we see that the price governs the savings made by the individuals. Investment decisions made by entrepreneurs depend upon the profit opportunities in making various types of capital goods. In other words, investment depends upon the rate of return on capital. The greater the profit expectations, that is, the larger the rate of return on capital, the greater will be the amount of investment made by the entrepreneur. Thus we see that it is the price system – the rate of interest on savings and rate of return on capital – which regulates the amount of savings and investment in a free-market economy.

It may, however, be pointed out that this is the problem where a free market economy has often run into trouble. If the aggregate investment planned by entrepreneurs is equal to the aggregate savings made by the households and individuals, then the free market economy will run smoothly. But there is no guarantee that planned savings and planned investment will be equal. This is because savings and investments are made by two different groups of people. Whereas savings are largely made by households, the investment is made by entrepreneurs. When planned investment exceeds planned saving, income and employment will tend to rise and there may arise conditions of boom and inflation. And when planned investment falls short of planned savings, income and employment will tend to fall resulting in depression and unemployment. Thus, the fluctuations in economic activity, or what are called business cycles, are inherent in the working of a free market economy.

Solving All the Problems Simultaneously

We have discussed above the solutions of basic problems by a free market economy separately. But it is worth mentioning that in the actual working of the free market economy, these decisions are not made separately. All of the four problems are solved simultaneously. Solutions of all the four problems are completely interdependent. Economic system is in fact a huge inter-connected set of markets each with buyers and sellers. The problems are simultaneously solved by the millions of decisions made by consumers, workers, entrepreneurs, bankers, etc., who constitute buyers and sellers in various markets. In fact, the solution of the problem in the free market economy is like that of solving a set of simultaneous equations in mathematics. The various interdependent decisions and markets represent a sort of simultaneous equations.

The working of free-market capitalist economy through price and market mechanism can be more easily understood with the help of a flow diagram depicted in Fig. 6.3. In the upper part of this diagram a rectangle is constructed to show the working of product markets in which demand and supply of various goods and services determine their prices. On the right hand side of this diagram, a rectangle representing the households has been made. These households which represent the whole population of the society serve two functions in an economy. Firstly, these households repre sent the consumers of the society who provide demand for goods and services. Secondly,

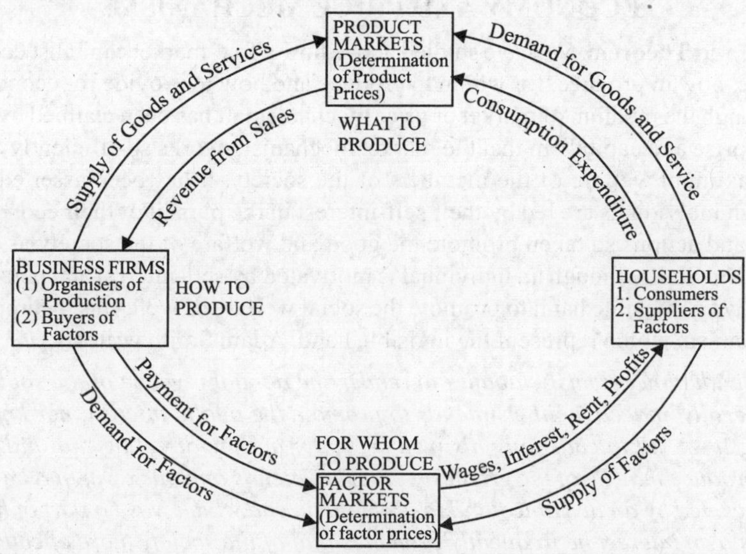

Fig. 6.3. *Functioning of a Free-Market Capitalist Economy*

these households are suppliers of the various resources or factors such as labour, land, capital and entrepreneurial ability. On the left hand side is shown a rectangle representing the businesses or productive firms which are organisers of production for the society as well as the sellers who supply goods and services in the product markets. For their function of organising production, business firms have to obtain the factors of production from the households through factor markets shown at the bottom of the diagram. In the factor markets demand for factors by the business firms and the supply of these factors by the households interact to determine their prices, such as wages of labour, rent of land, interest on capital, etc.

It is shown by arrow marks in the diagram who supply the goods and services and who demand them in the product markets. Likewise in the factor markets also, it is shown by arrow marks who supply and who demand the factors in the factor markets. Along with these supply and demand for goods and factors, money flows are also shown. Thus, with the determination of prices of goods in the product markets, the money will be spent by the households, which will become the revenue receipts of the business firms. It is transactions in these product markets where demand and supply interact to determine prices of goods and services that decide what goods shall be produced and in what quantities. In other words, it is the forces of demand and supply working in the product markets that decide the allocation of resources between the production of various goods.

Equally important in the working of a free market economy is factor markets through which the flow of money expenditure returns to the households in the form of wages, interest, rent and profits. It is here in the factor markets that the problem for whom to produce, that is, the distribution of income (and therefore the shares in the. total output) is determined. Lastly, there are the business firms themselves who organise the production work of the society by combining the factors. It is here that the problem of 'how to produce', that is, the methods or techniques of production to produce goods and services are decided.

It is thus evident from this flow diagram as to how a free-market capitalist economy works through market mechanism.

CRITICAL EVALUATION OF THE EFFICIENCY OF MARKET ECONOMY AND PRICE MECHANISM

Invisible Hand Theorem. We have studied above how a free-market capitalist economy decides what to produce, how to produce, for whom to produce and how to provide for economic growth of the country through the medium of market or price mechanism. It has been claimed by the supporters of private enterprise and capitalism that the market mechanism works so efficiently and effectively as to ensure maximum welfare of the members of the society. It has been asserted that in such a system, although individuals are led by their self-interest in the pursuit of their economic activities, their decisions and actions so taken promote the good and welfare of the society as a whole. Thus, according to Adam Smith, though an individual is motivated by self-interest and personal gain, he is led, as it were, by an invisible hand to promote the social welfare as well. It is the impersonal forces of market mechanism which represent the invisible hand. Adam Smith writes.

"*Every individual necessarily labours to render the annual revenue of the society as great as he can. He generally, indeed, neither intends to promote the public interest, nor knows how much he is promoting it ..... he intends only his own security; and by directing that industry in such a manner as its produce may be of the greatest value, he intends only his own, and he is in this, as in many other cases, led by an invisible hand to promote an end which was no part of his intention..... By pursuing his own interest he frequently promotes that of the society more effectively than when he really intends to promote it*"[4]

It has been claimed on behalf of the freedom of enterprise and the free and unfettered working of price mechanism that it is the one and only way in which people can achieve the maximum satisfaction of their wants with a minimum sacrifice. It is contended that we can always trust the free private enterprise and market mechanism to decide about what goods shall be produced, how they shall be produced, and how they shall be distributed among the members of the society, better than any other method. The case for freedom of enterprise and free functioning of market mechanism that they ensure maximum satisfaction of the people with minimum cost has been ably summed up by Professor Slicher in the following words:

"The reasoning in support of the belief that freedom of enterprise ensures the maximum of satisfaction at the minimum of cost is very simple. Each individual, it is said, is better able than anyone else to judge his own interests. If men are at liberty to spend their money as they choose, they will naturally purchase those things that will yield them most satisfaction. Consequently, the very commodities which give consumers the greatest pleasure are the most profitable for business enterprises to produce. Likewise, if men are free to use such methods of production as they wish, they will select those which involve the least cost per unit of output. With the goods which give the greatest gratification being made by the methods which are least costly, it follows, according to the theory, that there will be the maximum surplus of satisfaction over sacrifice."[5]

But in the modern times it is generally believed that free enterprise and unfettered working of price mechanism does not work so effectively and efficiently as to ensure maximum satisfaction of the society at the minimum social cost. The free and unfettered working of market system has led to so many evils that even in capitalist countries the governments regulate and control it through various measures such as physical controls, monetary and fiscal measures. We discuss below why and in

4. Adam Smith, *The Wealth of Nations* Book IV.
5. Sumner H. Slicher, Free Private Enterprise, published in *Readings in Economics*, edited by Paul A. Samuelson, 6th edition, 1970.

what ways market mechanism fails to function effectively and efficiently so as to achieve maximum satisfaction of the people at the least social cost.

1. Free market system does not ensure maximum social satisfaction at minimum social cost. We have described above how a free private -enterprise economy decides about what and for whom goods are to be produced. Maximum social satisfaction can result only if goods produced go to those consumers who will derive the greatest satisfaction from them. We have further seen that in such a system goods go into the hands of those people who can offer best prices for them. In other words, goods are distributed according to the ability to pay of the people rather than their needs and desires.

Now, it has been rightly pointed out by critics that the distribution of goods in accordance with the ability to pay of the people does not ensure maximum satisfaction of the members of the society. The people who are willing and able to pay most for goods are not generally those who will derive greatest satisfaction from consuming them. In that case, it is possible to "increase the satisfaction of the society as a whole by getting goods distributed according to the needs or wants rather than ability to pay for them. Professor Slicher rightly remarks: "*We have no way of comparing the amount of pleasure which two persons derive from consuming an article. And yet it seems ridiculous to assert that ability to derive satisfaction from goods is proportionate to ability to pay for them.* Assume that *A* and *B* each wish a pair of shoes, A, who is well-to-do, is willing to pay $ 12; B, who is poor, will offer only $ 7. Obviously A will get the shoes. But because he is rich and well supplied with shoes, an additional pair is only a slight convenience to him. *B*, poor and scantily supplied, has urgent need for another pair. It seems clear that the sum total of satisfaction would be greater if *B* obtained the shoes and yet it seems equally clear that under freedom of enterprise they will go to A."[6]

Thus, the market mechanism distributes goods in such a manner that anything goes to anyone, in whatever quantity he wants, who is able and willing to pay for it. As a result, in a market system while luxuries are produced for the rich people who have got enough purchasing power or ability to pay for such goods, the necessities or wage-goods as they are called these days which are consumed most by the poor people may not be produced in adequate quantities for the poor because they have not much ability to pay. Total satisfaction of the community can be increased by withdrawing productive resources from the production of luxuries and devoting them to the production of necessities. But this is not possible under free market mechanism where production is governed by market prices rather than basic human needs and wants. Market prices of goods do not accurately measure human needs and wants; they merely indicate the ability to pay for them.

However, distributing goods according to the needs rather than ability to pay and assigning various jobs to those who can do them with the least sacrifice might increase the satisfaction of the society and reduce its total sacrifice in the short run, but any such measure will be reducing the total satisfaction in the long run by adversely affecting the total production. People have incentive to produce more and work more for the purpose of earning more so that they have larger ability to pay and hence a larger share from the national output. If the national output is distributed according to the needs and not according to the ability to pay, then people will lose incentive to produce more which will result in reduction in total production. Fall in total production will mean fall in social welfare. Any attempt to distribute the 'Cake' according to the needs of the people will lead to the reduction in its size. They assert that the Marxian principle, "*from each according to his capacity, to each according to his needs*", will not work.

The argument regarding the role of incentives in promoting production' has a some force in it, but large inequalities in incomes and shares in national output as exist in the real-world market

6. J.K. Galbraith, op. cit.

economies cannot be justified on the basis of incentives. No doubt some incentives are necessary to promote production, hard work and efficiency, but the prevailing inequalities in incomes or shares in national output are much greater than those required on the basis of incentives. It is also wrong on the part of advocates of market system to argue that capacity to derive satisfaction from the goods is proportionate to the ability to pay, *i.e.* incomes of the people.

2. Economic Inefficiency in Free Market Economies. A free market economy fails to achieve economic efficiency when externalities (especially, external diseconomies) are present. The business firms while pursuing their productive activities pollute the air or dump toxic wastes into streams or rivers which prove greatly harmful and costly to the people in general. For these harms or costs inflicted on the public, the private firms do not pay any price to the people. This causes inefficiency in resource allocation and lowers the welfare of the people. These are often called *market failures*. In recent years Government of free market economies have intervened to check these bad external effects by imposing taxes on the private firms and adopting other measures

3. Economic Instability and Unemployment. Another serious shortcoming of market mechanism is that it does not ensure a high and stable level of employment. Nor does it guarantee stability of the price level. As a result of the free working of market mechanism, there take place cyclical fluctuations in levels of economic activity with ups and downs in the levels of employment and prices. In these cyclical fluctuations, business booms are followed by business depressions. These alternating periods of booms and depressions are generally called business or trade cycles in economics. Both depressions and (inflation) are bad and cause great sufferings to the people. The depression causes huge unemployment and lot of human sufferings. The depressions of nineteen thirties in the Western capitalist countries caused so much unemployment and human suffering that the revolutions were feared to take place there. Similarly, the period of inflation is also quite harmful. The consumers, especially the people having fixed incomes, are hit hard by inflation.

During the early thirties (1929-33) there was severe economic depression in the free-market capitalist economies which caused huge losses in output and employment. Such was severity of this depression that rate of unemployment in the US shot up to 25% of labour force causing a lot of human suffering. This led to the loss of faith in the smooth and efficient working of free-market capitalist system and many economists turned to socialism as preached by Karl Marx. However, British economist J.M. Keynes rescued capitalism by advocating intervention by the government through adoption of expansionary fiscal policy to get the economy out of depression.

That the free market system does not guarantee economic stability and full employment has been proved by the **recent global financial crisis (2007-09)** which started in the US and spread to other countries (including India) through global linkages of free trade and capital flows and caused global slowdown. The financial crisis started with the burst of sub-prime housing bubble causing widespread mortgage defaults and large losses to the banks and other financial institutions. Due to heavy losses causing lack of liquidity and risk of default banks stopped giving loans to the corporate sector for investment. The decline in investment resulted in recession in the American economy which through global linkages affected other countries causing world-wide economic slowdown.

But the important point to note is that market system, which was believed by the classical economists, monetarists led by Milton Friedman and new classical economists led by Robert Lucas to correct itself automatically and recover from the crisis, failed to deliver. As a result, there were loud protests about the failure of free market capitalist system to ensure economic stability and full employment. Again it was Keynesian thought which came to the rescue of the US and other countries which adopted Keynesian expansionary fiscal and monetary policies to give stimulus to their economies so as to get out of recessionary conditions. These stimulus packages seem to have worked and in economies of the US, European economies, Japan, China and India recovery has started and it is expected that these economies would return to the growth path in 2010.

4. Decline of Competition and its Adverse Effect on the Working of the Price System. We saw above in the discussion of the working of private-enterprise market system that the existence of perfect competition among the producers or sellers of the products was essential for the achievement of production and sale of goods at minimum possible prices and the attainment of efficient allocation of resources. Thus competition is thought to be the controlling mechanism through which a free market system realises its ideal of maximum satisfaction of the people with the most efficient use of scarce resources. But in the real world, competition has greatly weakened. What we actually find is the world of oligopolies and monopolies so that industries are dominated by a small number of firms. Two factors have been responsible for the decline of competition. Firstly, producers in order to get rid of the irksome competition try to restrict competition through merger and by elimination of rivals through cut-throat competition and higher productive efficiency. Secondly, technological progress that has taken place has contributed a good deal to the decline of competition. Modern technology requires (l) the use of extremely large amount of capital equipment, (2) large markets, (3) complex and huge managerial set-up and (4) large and reliable sources of raw materials.

This requires operation of firms on a very large scale not only in the absolute sense but also in relation to the size of the market. Thus, the achievement of maximum productive efficiency through the employment of modern technology often requires the existence of a few large firms rather than a large number of small firms.

As a result of the decline of competition and the emergence of oligopolies and monopolies, prices charged from the consumers are considerably higher than they would be under competitive conditions. The pricing power or market power enjoyed by the oligopolistic or monopolistic business firms goes against the public interest and violates the principle of consumer sovereignty. "*To the degree that competition declines, the price system will be weakened as a mechanism for efficiently allocating resources. Producers and resource suppliers will be less subject to the will of consumers, the sovereignty of producers and resource suppliers will then challenge and weaken the sovereignty of consumers*."[7]

We thus see that with the decline of competition and the predominance of oligopolies and monopolies in the real-world price system does not provide the goods most wanted by the people, that is, price system has become less efficient in allocating resources in accordance with the wishes of consumers. Further, according to Prof. J.K. Galbraith, the emergence of multinational corporations or big businesses in the present-day market economies has significantly reduced consumer soverignty. These big corporations do not merely *satisfy or fill* the wants of the people, they themselves create these wants by massive expenditure on advertisements which influence the public to buy their products.

6. Concentration of Wealth and Income as a Result of the Free Working of the Price System. Another defect of the free working of the private-enterprise market system has been that the more efficient, more cunning entrepreneurs as well as those who enjoy large monopolistic power have been able to accumulate large amounts of wealth and property. These inequalities of wealth and property have further increased through time by the right of inheritance. The concentration of wealth and property in the hands of a few has led to the extremely large inequalities in money incomes in free market economies. Owing to the large inequalities of money incomes, the people differ in their abilities to express their wants in the market; the rich have more "rupee votes" than poor and are, accordingly, in a better position to get goods produced for them. We have already noted above the effect of the unequal distribution of income on the production of luxuries and necessities in a free market economy. In order to reduce inequity in income distribution the Governments today.

Summing up. We have discussed above the various defects that result from the free working of market mechanism. There have been two forces at work to correct these defects of the market

7. Robert, L. Heilbroner, *op. cit*., p.126.

system. First, there has emerged what Professor Galbraith has called *"countervailing power"* to check the monopolistic power of the giant firms and its evils. By countervailing power, Galbraith means that the growth of giant and monopolistic firms on one side of the market has also led to the growth of powerful firms on the opposite side of the market. The decline of competition and emergence of strong economic power on each side of the market has, therefore, been neutralised to some extent by the growth of equally strong power on the opposite side of the market. This has prevented to some extent the evils of decline of competition and the growth of monopoly power in the free-market economy. "*To-day powerful corporate sellers often face equally powerful corporate buyers. The giant raw materials producer who faces little or no competition within his industry must sell to the giant chemical or other processing plant; the giant steel mill to the giant auto firm; the giant canner to the giant supermarket chain. Not least important, the large firm no longer bargains with the individual employee, but with the large and powerful unions*.,,[8] Of course, the growth of such countervailing power is not fully effective in every market and in every situation but it is the contention of Professor Galbraith that it is effective in sufficient markets and sufficient situations so as to work as powerful restraining force on the unhindered exercise of monopoly power. A second force that has tended to maintain competition is the existence of competition among different products. "Even if all steel prices are kept at "administered" levels, steel as a whole must compete with aluminium. Nor does the competition ends here. What we find, indeed, is an immense chain of interproduct competition steel against aluminium, aluminium against glass, glass against plastics, plastics against wood, wood against concrete, concrete against steel. And this competition is without doubt effective." [9]

ALTERNATIVE ECONOMIC SYSTEMS

In the foregoing analysis we have explained how the free market economic system which is historically called *capitalism* works and what are its merits and drawbacks. We have also studied how at present in the market economies government plays a role in improving the functioning of market economic system. Now, an important question is if there is any altenative economic system that can solve the basic economic problem faced by the nations. In fact an alternative economic system known as authoritarian socialist system or *command economy* was adopted in erstwhile Soviet Russia, China, East European countries such as Poland, Czechoslovakia. In this authoritarian socialist system or command economy basic economic problem of what, how and for whom to produce etc. are resolved through a combination of state ownership of productive enterprises and centralised planning. It is government or any central authority set up by it that decides output targets for each public enterprise, allocates resoures to them and fixes prices of various goods and services and incomes (wages, salaries) of various types of workers.

In a centrally planned socialist economy, there exists a central planning authority, which organises and regulates the various economic activities. Thus it is the central planning authority which determines what the various production units shall produce, what the various individuals shall obtain for consumption and as a result what shall be left for investment. Further, it decides what type of investment shall be made.

It should be realised that the organization of the various economic activities by one planning authority is full of administrative and economic difficulties. The authority is faced with the political and administrative problems of getting their decisions implemented. In other words, there are problems of putting the various decisions into effect. Moreover, the planning authority is faced with an overall economic problem, namely, how to use the limited productive resources for achieving the best possible results. The particulars purposes for which resources are to be used have to be decided. In other words, allocation of resources as between various kinds of

8. J.K. Galbraith, *op, cit*.
9. *Ibid*.

consumer goods, and as between consumer goods and capital goods, and as between civilian goods and war goods, has to be decided. Taking of all these decisions is indeed a very difficult task. Besies, decisions have to be taken about the particular techniques with which various products are to be produced and about distributing goods among the people.

It may be noted that planning of actual production of various products is greatly difficult due to the fact that technical relations between different kinds of products are very much complicated. For example, to build ships it apparently seems that only steel and other materials which make up ships are required. But little more thinking will reveal that arrangements for the complicated process of production of steel and other materials have to be made before steel and other materials become available for use in ship building. Moreover, various types of machinery would have to be made for converting iron ore into steel as well as converting steel into ships. Further, the arrangements for coal, electricity and transport have to be made if the production of steel and ships is to become possible. Thus, technical relations between different kinds of output are so complicated that for carrying out the decision to produce more of one kind of product, adjustments on several other lines have to be made.

Collapse of the Command Economy or Authoritarian Socialist System

It is obvious from what has been said above that the planning of productive activity by one central authority is a very difficult task. The planning authority in the centrally planned economies like those of erstwhile Soviet Russia and East European countries which used to plan production with the aid of several technical experts before the system broke down in the late eighties. In the late eighties the socialist countries, Soviet Russia and countries of Eastern Europe such as Poland, Czechoslovakia had to face severe economic problems which led to the collapse of Soviet-type authoritarian socialism or what is called command economy. *In its place the market system has been introduced and this is proclaimed as victory of capitalism over socialism.*

There is more than one reason for the collapse of the command economy in Soviet Russian and Eastern Europe. The important reason for the collapse of socialism and centralised planning *was lack of efficiency in reasource allocation*. It was found that for a central planning authority consisting of bureaucrats it was difficult to make right decisions to allocate scarce resources according to the wants and perferences of the people. As result, shortage in some goods and surplus in others emerged. These could not be easily corrected due to corrupt, inefficient and unresponsive government officials. This caused a lot of wastage of national resources. The most glaring example of inefficiency in allocation of resources in Soviety Russia and othe command economies was the emergence of *acute scarcity of essential consumer goods*. This led to *long-queues* of the people at the consumer stores to get these goods.

This acute scarcity of consumer goods led to the lack of faith in socialism to deliver the goods. Indeed, it was said that Soviet-type authoritarian socialism promised bread at the expense of liberty of the people. But in actual practice people got neither bread nor liberty. Since by the late nineteen eighties the then Soviet President Mickhail Gorbachev had introduced some freedom for the people, the wide expression of resentment about the scarcity of consumer goods could not be suppressed. This led to the collapse of authoritarian socialism and introduction of reforms aimed at introduction of private enterprise and the market system.

The other important reason for the failure of centralised planning was that the managers of public enterprises *lacked incentives to improve productivity and efficiency*. This was due to the absence of profit motive on the part of managers of public enterprises. This caused inefficiency in the use and allocatiion of scarce resources.

The most important economic problem faced by these socialist countries was *slowdown in economic growth* in these countries. In Soviet Russia the growth rate which in the post-war period averaged about 7 per cent in 1950s and about 5 per cent in 1960s slipped into 3 per cent range during the 1970s and was nearly zero in the 1980s. Similar experience of slowdown in economic growth rate was observed in other socialist countries of Eastern Europe. Now, when growth rate in them was higher, more employment opportunities were increasing and more goods were forthcoming which enabled people to improve their levels of living. However, when there was sharp slowdown in economic growth, to satisfy the ever-rising wants of the people for goods became difficult and a strong need was felt to make a switch over to the market system to speed up the rate of economic growth.

Lastly, *problems of centralised planning with its inflexibility and inefficiency* have been present throughout in the post revolution period in Soviet Russia. Lacking profit-motive the managers of productive enterprises do not have sufficient incentives to produce the right type of goods in the most efficient way. This causes inefficiency in the use and allocation of resources which accounted for *low productivity and efficiency* in the socialist countries. Owing to lack of incentives managers of productive enterprises were more concerned with fulfilling the quotas assigned to them (which they succeeded in achieving them by deliberately setting them at lower levels) rather than improving efficiency. But still deeper problem posed by the absence of profit motive was *managers of state enterprises lacked incentives to introduce innovations and undertake risks* which caused inefficiency and slow economic growth and ultimately stagnation.

To conclude, the above problems led to the downfall of authoritarian socialism or command economy system in Soviet Russia and other socialist countries. This led to the process of liberalisation of Soviet Russia and other socialist countries. It is because of these problems of economic inefficiency and slowdown in economic growth that even China and India introduced economic reforms which lay stress on privatisation and greater role of market mechanism for solving the economic problems. It may be noted here tha India has adopted the path of a mixed economy. In India's mixed economy both private sector and economic planning by the Government play roles in the working of the Indian eocnomy.

MIXED ECONOMIC SYSTEM

In the real word today a pure free market economy and pure command economy or a centrally planned economic system are not found. Almost all economies have now become mixed economic system in which government plays an important role in allocation of resources and distribution of income. Thus mixed economy or mixed economic system refers to an economic system in which both free market and Government have significant roles in the working of the economy. In a mixed economy elements of both free market system and control or regulation by the Government are present. The proportion of free market working and Government regulation varies from country to country.

Many economies of the present-day world are examples of mixed economy. Mixed economy functions through private enterprise as well as Government. The Government interferes with or regulates the private enterprise in various ways. It has now been realised that the free and unfettered functioning of the private enterprise gives rise to many evils. As a result of free working of the private enterprise, there are violent fluctuations in economic activities; sometimes the conditions of depression and unemployment are created and sometimes the conditions of boom and inflation emerge. Thus a free-market economy suffers from what are called business cycles, with all their consequent evils. Besides, as a result of free functioning of the private enterprise and price mechanism, extreme inequalities of income and wealth are produced. The following of '*Laissez Faire*' policy by the Government, weaker sections of the society are not protected. On the other hand, the command economic system in which resource allocation and income distribution is decided through centralised planning collapsed due to various deficiencies and failures as explained above. Therefore, in both Russia and China free-market economic system along with some government control has been adopted.

In has also been realised that in a developing country like India, desired rate of economic growth cannot be achieved under free private enterprise. Therefore, in order to avoid the above-mentioned evils of the free private enterprise and free functioning of the market mechanism and to achieve the desired rate of economic growth the Government takes an active part in the functioning of the economic system in most economies of the world. Today all the capitalist economies have become mixed form of economies, because in all of them economic role of the Government has greatly increased. '*Laissez Faire*' policy advocated by Adam Smith and other classical economists has now been given up because of its several shortcomings and evils. Therefore, the economies of the United States of America and Britain have also become mixed systems. Paul Samuelson and A.H. Hansen, eminent American economists, have called the economies of America, Britain and France as '*Mixed Capitalist System*' or '*Mixed Enterprise System*', because in them the Government now interferes in the economic activities and takes active part in the functioning of the economic system and regulates and controls the private enterprise by various methods.

Two Forms of Mixed Economic Systems

It is worth noting that mixed economies can be classified into two forms. One form of mixed economy is that in which means of production are in the ownership of private sector and the Government regulates and controls the activities of private enterprise through direct controls (such as price control, licensing system, control over imports, etc.) as well as through monetary and fiscal policies. In such a type of economy, the government does not take over the means of production, and if it does, it does so relatiely on a small scale. That is why such a type of 'mixed system' has been called *Mixed Capitalist System* because such an economic system is basically capitalist and the Government regulates and controls the economic activities through various types of controls and various measures of monetary and fiscal policies so that the various evils of the free working of the private enterprise and the price mechanism are avoided and the economic system is directed towards the desired goals. In such a type of 'mixed economy,' Government itself does not undertake the work of production on a large scale. Government's production work is merely confined to the production of equipments and materials for the army and the working of the public utility services. Such a type of mixed economic system is also called *controlled capitalism*.

Indian Economy as a Mixed Economic System

The second form of a mixed economy is that in which the Government not only regulates and controls the private enterprise through various types of direct controls and appropriate monetary and fiscal policies, *it also directly participates in the production of various goods and services*. In such a type of mixed economy, various basic industries and infrastructure industries are generally in the ownership of the public sector and it is the government which organises and runs them. The remaining industries are in the ownership of private enterprise and it is the private enterprise which is assigned the task of production in them. But it is worth remembering that the government regulates and controls the private enterprise in such industries also through direct control and appropriate monetary and fiscal measures.

The economies of the United States of America and Britain have also become mixed economies. But the nature of mixed economy of India is quite different from them, because in the mixed economy of India the public sector takes more active, more important and more extensive part in the working and growth of the economy. Whereas the mixed economies of the USA and England are biased towards capitalism, the mixed economy of India is biased towards socialism.

Functioning of a Mixed Economy

A mixed economy functions through both price mechanism and planning by the Government. So far as the industries in the public sector are concerned the price, output and investment decisions are taken by the government or the authorities appointed by the government according to the strategy and policy frame of the development plans. But the private sector of the mixed economy is governed and regulated by the price and market mechanism and therefore in regard to the industries in the private sectors, the decisions regarding price, output and investment are taken by the private entrepreneurs or the industrialists with a view to making profits and these decisions are mainly based on price mechanism. The private sector in the mixed economy is, however, influenced, regulated and controlled by the government through monetary and fiscal measures as well as through direct or physical controls.

Prior to 1991 in India, there was a 'licensing system' according to which to start and establish factories in certain industries, the government's permission or licence was necessary. However, under the economic reforms initiated since 1991 in India, licensing and permit system has been done away with and private sector has been given a lot of freedom for deciding about production and expansion in their enterprises. Even foreign investment by multinational companies has been allowed to make investment in India and repatriate the profits to their home countries. However the government also regulates and controls the investment and production in the private sector through appropriate monetary and fiscal measures. By providing concessions in taxation, and by making available cheap credit facilities, the government provides incentives to private entrepreneurs for investment in the desired lines of production decided in development plans.

Role of Government in a Mixed Economic Systems

As noted above, in the mixed economic system the Government plays a significant role in allocation of scarce resources and distribution of income. However, despite the important role of Government, the people and private enterprises are free to make economic transactions. They voluntarily decide whether or not to transact, have the right to buy and sell what they want. Workers are free to accept or refuse any work they do not want to do, free to move to different places for employment of their choice. **The right to own property and freedom of contract are maintained by the Government.** The Government makes laws which provide right to ownership of property and freedom of contract and courts are set up by it to enforce these laws. Besides, the Government maintains law and order to protect life and liberty of the people and to enforce property rights of the people. It also makes arrangements for defence of the country against foreign aggression.

Besides the above basic functions, Government intervenes in the economy to correct what are called ***market failures.*** These are the situations in which free market does not work efficiently. For example, natural resources such as forests, mines, common pastures tend to be overexploited resulting in their destruction under free market conditions. Further, some goods, called *public goods* such as defence, law and order are not produced at all by private enterprises which are driven by profit motive. This is because once produced the people who do not even pay for them cannot be prevented from using them.

Moreover, private enterprises while producing goods often impose costs that are called ***detrimental externalities*** on others by their economic activities. For example, the production of goods in factories *pollute air and water* surrounding them. This *environment pollution* causes a lot of harm to the people but the private enterprises do not have to pay for the harms they do to others. Nor do they take into account these externalities in their cost calculations while making their investment decisions. As a result, there is overproduction of such goods whose production pollutes environment and impose costs on others and this causes misallocation of resources. The Government can prevent them by imposing taxes on private enterprises which pollute environment.

Above all, the Government in modern mixed economic systems takes fiscal and monetary policies to ***stabilise the economy*** against fluctuations in national income, employment and prices. Recession causes huge unemployment which deprives people of their livelihood. On the other hand, when there is inflation, cost of living of the people rises and as result there is a lot of human suffering. Inflation is a tax that hurts the poor people most. Following J.M. Keynes the Government adopts discretionary fiscal and monetary policies to lift the economy out of recession. For example, recently in 2007-09 when due to global financial crisis originated in the US due to bursting of sub-prime housing loan bubble which caused global recession and meltdown, the Governments in the US, Britain, Japan, France and India too increased public expenditure and cut taxes to revive their economies. In the US and other free market economies the Government even helped private banks with funds to bail them out of the turmoil. In India the Government borrowed heavily in 2008-09 and 2009-10 to increase its expenditure and also to cut taxes to prevent slowdown of the Indian economy due to the impact of global financial crisis. Similarly, when there is inflation the Central Bank of a country adopts tight monetary policy and the Government adopts contractionary fiscal policy to control rise in prices.

Last but not the least, it is a vital function of the modern Government to ***promote economic growth.*** Economic growth requires capital accumulation, progress in technology and investment in social sectors such as education and health. Besides, in developing countries like India the Government has to adopt measures ***to eradicate poverty*** that prevails on a large scale in these countries. For example, the Indian Government has started Mahatma Gandhi National Rural Employment Guarantee Scheme (MNREGS) under which it provides guaranteed employment to the poor households. Further, to help the poor the Government in India has also begun Public Distribution System under which it provides subsidised foodgrains to the poor households. Just as there are ***market failures***, there are ***Government failures too.*** Mixed economy represents a middle path between a purely free-market economy and purely command economy.

QUESTIONS FOR REVIEW

1. Explain the essential elements of basic competitive model.
2. Why might government policy to make the distribution of income more equitable lead to less efficiency.
3. Explain how information and incentives play a crucial role for the efficient working of market economies.
4. Examine the role of prices, property rights and profits for the efficient working of the market economy.
5. What is a free market economy? Explain its important features.
6. Explain the role of price mechanism in solving the basic economic problems in a free market economy.
7. Explain the main characteristics of a free market economy ? How does it solve its central economic problems ?
8. What is meant by '*invisible hand*' ? How does a free market economy function through it ?
9. What are the functions performed by price mechanism ? How does a free market system determine the following :
   1. Rationing or distribution of goods 2. Allocation of resources
   3. Choice of production techniques
10. Who controls the functioning of a free market economy ? Consumers or producers.
11. It is said in a free-market (capitalist) economy, consumers is a king who directs its functioning? Is this valid in free-market economies of the world today ? Explain.
12. A free market economic system ensures maximum social welfare at minimum social cost. Do you agree ? Discuss
13. What is meant by '*Countervailing power*'? How has it affected the working of a capitalist economy?
14. Explain the factors which led to the collapse of command economy system in Soviet Russia and other socialist countries.
15. Profit motive and private property rights provide incentives for rational individuals and firms to work hard and efficiently. Analyse. *D.U., BA(Hons.) Economics 2009*
16. Explain Indian economy as a mixed economy. Explain the role which the Government plays in India's mixed economy.

PART II

THE THEORY OF DEMAND

- Demand and Law of Demand
- Demand: Marshall's Cardinal Utility Analysis
- Indifference Curve Analysis of Demand
- Marshall's Cardinal Utility Analysis

  vs.

  Indifference Curve Analysis
- Applications and Uses of Indifference Curves
- Revealed Preference Theory of Demand
- Elasticity of Demand
- Consumer's Surplus
- Choice Under Risk and Uncertainty

CHAPTER 7
DEMAND AND LAW OF DEMAND

People demand goods because they satisfy the wants of the people. The utility means want-satisfying power of a commodity. *It is also defined as the amount of satisfaction which a person derives from consuming a commodity*. Utility is a subjective entity and resides in the minds of men. Being subjective it varies with different persons, that is, different persons derive different amounts of utility from a given good. People know utility of goods by means of introspection. The desire for a commodity by a person depends upon the utility he expects to obtain from it. The greater the utility he expects from a commodity, the greater his desire for that commodity. It should be noted that no question of ethics or morality is involved in the use of the word 'utility' in economics. The commodity may not be useful in the ordinary sense of the term, even then it may provide utility to some people. For instance, alcohol may actually harm a person but it possesses utility for a person whose want it satisfies. Thus, the desire for alcohol may be considered immoral by some people but no such meaning is conveyed in the economic sense of the term. Thus, in economics, the concept of utility is ethically neutral.

The Meaning of Demand

It is useful to know what economists mean by the demand of the goods by consumers. The demand for a commodity is essentially consumers' attitude and reaction towards that commodity. Demand for a good is in fact a photograph or a panoramic picture of consumers' attitude towards a commodity. This consumers' attitude gives rise to actions in purchasing units of a commodity at various given prices. *Precisely stated, the demand for a commodity is the amount of it that a consumer will purchase or will be ready to take off from the market at various given prices in a period of time*. Thus, demand in economics implies both the desire to purchase and the ability to pay for a good. It is noteworthy that mere desire for a commodity does not constitute demand for it, if it is not backed by the ability to pay. For example, if a poor man who hardly makes both ends meet wishes to have a car, his wish or desire for a car will not constitute the demand for the car because he cannot afford to pay for it, that is, he has no purchasing power to make his wish or desire effective in the market. Thus, in economics unless demand is backed by purchasing power or ability to pay it does not constitute demand. Demand for a good is determined by several factors, such as, the tastes and desires of the consumer for a commodity, income of the consumer, the prices of related goods, substitutes or complements. When there is a change in any of these factors, demand of the consumer for a good changes. Individual consumer's demand and market demand for a good may be distinguished. Market demand for a good is the total sum of the demands the individual consumers who purchase the commodity in the market. We shall discuss in detail later in this chapter the various factors which determine demand for a commodity and also how a demand curve for a commodity is derived.

The Law of Demand

An important generalisation about demand is described by the law of demand. This law of demand expresses the functional relationship between price and quantity demanded. The law of demand or functional relationship between price and quantity demanded is one of the best known and most important laws of economic theory. According to the law of demand, other things being equal, *if price of a commodity falls, the quantity demanded of it will rise, and if price of the commodity rises, its quantity demanded will decline.* Thus, according to the law of demand, there is inverse relationship between price and quantity demanded, other things remaining the same. These other things which are assumed to be constant are the tastes or preferences of the consumer, the income of the consumer, and the prices of related goods. If these other factors which determine demand also undergo a change, then the inverse price-demand relationship may not hold good. Thus, the constancy of these other things is an important qualification or assumption of the law of demand.

Demand Schedule and Demand Curve

The law of demand can be illustrated through a demand schedule and a demand curve. A demand schedule is presented in Table 7.1. It will be seen from this demand schedule that when price of a commodity is ₹ 12 per unit, consumer purchases 10 units of the commodity. When price of the commodity falls to ₹ 10, he purchases 20 units of the commodity. Similarly, when price further falls, quantity demanded by him goes on rising until at price ₹ 2, the quantity demanded by him rises to 60 units. We can convert this demand schedule into a demand curve by graphically plotting the various price-quantity combinations, and this has been done in Fig. 7.1, where along the *X*-axis, quantity demanded is measured and along the *Y*-axis, price of the commodity is measured. By plotting 10 units of the commodity at price ₹ 12, we get point *Q* in Fig. 7.1. Likewise, by plotting 20 units of the commodity demanded at price ₹ 10, we plot point *S* in Fig. 7.1. Similarly, points *S, T, U* and *V* are plotted, representing other combinations of price and quantity demanded presented in Table 7.1. By joining these various points, *Q, R, S, T, U,* and *V,* we get a curve *DD*, which is known as the demand curve. Thus, the *demand curve is a graphic statement or presentation of quantities of a good which will be demanded by the consumer at various possible prices at a given moment of time.*

Table 7.1
Demand Schedule of an Individual Consumer

Price (₹)	*Quantity Demanded*
12	10
10	20
8	30
6	40
4	50
2	60

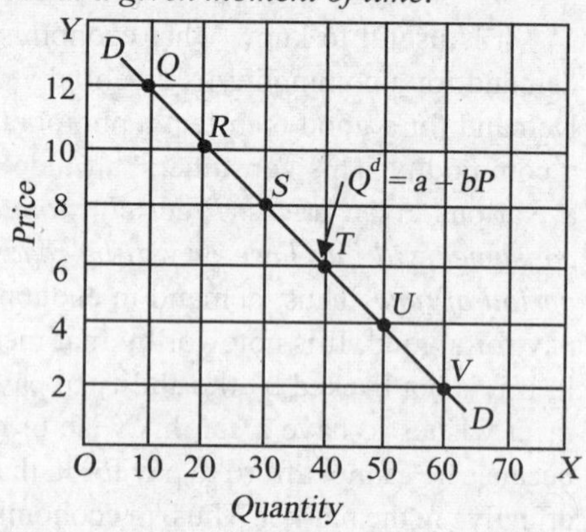

Fig. 7.1. *The Demand Curve of a Consumer*

It should be noted that a demand schedule or a demand curve does not tell us what the price is; it only tells us how much quantity of the good would be purchased by the consumer at various possible prices. Further, it will be seen both from the demand schedule and the demand curve that as price of a commodity falls, more quantity of it is purchased or demanded. Since more is demanded at a lower price and less is demanded at a higher price, the demand curve slopes downward to the right. Thus, the downward-sloping demand curve is in accordance with the law of demand which, as stated above, describes inverse price-demand relationship. It is important to note here that behind this demand curve or price-demand relationship always lie the tastes and preferences of the consumer, his income, the prices of substitutes and complementary goods, all of which are assumed to be constant in describing price-demand relationship. If any change occurs in any of these factors, the whole demand schedule or demand curve will change and a new demand schedule or demand curve will have to be drawn. In drawing a demand curve, we assume that the buyer or consumer does not exercise any influence over the price of a commodity, that is, he takes the price of the commodity as given and constant for him.

Market Demand Curve

We can add or sum up the various quantities demanded by the number of consumers in the market and by doing so we can obtain the *market demand curve* for a commodity which like the individual consumer's demand curve will slope downward to the right. How this summation is done is illustrated in Fig. 7.2. Suppose there are three individual buyers of the good in the market. In Fig. 7.2 diagrams (*a*), (*b*) and (*c*) show the demand curves of the three independent individual buyers. Now, the market demand curve can be obtained by adding together the amounts of the good which individuals wish to buy at each price. Thus, at price P_1 the individual *A* wishes to buy Oa_1 of the good; individual *B*

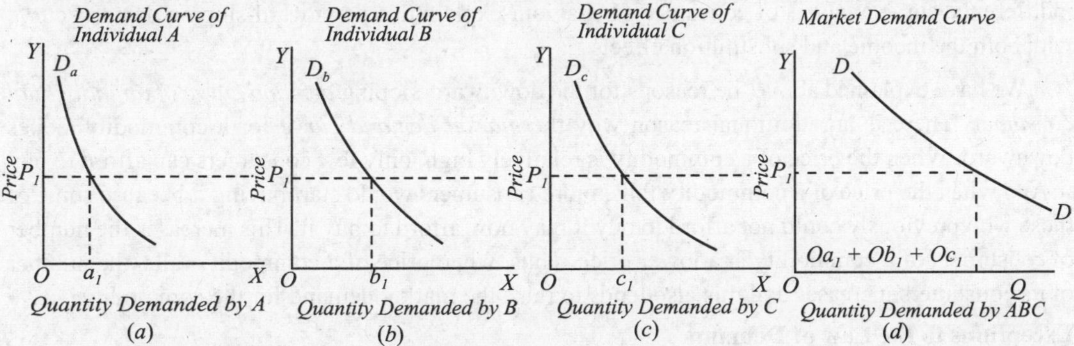

Fig. 7.2. *Derivation of Market Demand Curve*

wishes to buy Ob_1 of the good and individual *C* wishes to buy Oc_1. The total quantity of the good that all the three individuals plan to buy at price P_1 is therefore $Oa_1 + Ob_1 + Oc_1$ which is equal to *OQ* in Fig. 7.2 (*d*). Similarly, we can plot the quantity of the good that will be demanded by all the three individuals at every other price of the good. When all the points showing the amounts demanded of the good at various prices are joined together we get a market demand curve for the good. For the sake of convenience we have supposed that there are three individuals or buyers in the market for a good. Whatever be the number of individuals in the market, their demand curves can be added together, as illustrated above, to get a market demand curve for the good.

The market demand curve will slope downward to the right since the individual demand curves, whose lateral summation gives us the market demand curve, normally slope downward to the right. Besides, as price of the good falls, it is very likely that the new buyers will enter the market and will further raise the quantity demanded of the good. This will be another reason why the market demand curve should slope downward to the right.

Reasons for the Law of Demand: Why does Demand Curve Slope Downward?

We have explained above that when price falls the quantity demanded of a commodity rises and *vice versa*, other things remaining the same. It is due to this law of demand that demand curve slopes downward to the right. Now, the important question is why the demand curve slopes downward, or in other words, why the law of demand describing inverse price-demand relationship is valid. We can explain this with marginal utility analysis and also with the indifference curve analysis which we will do in the next chapters.

When price of a commodity falls, the consumer can buy more quantity of the commodity with his given income. Or, if he chooses to buy the same amount of the commodity as before, some money will be left with him because he has to spend less on the commodity due to its lower price. In other words, as a result of fall in price of the commodity, consumer's real income or purchasing power increases. This increase in real income induces the consumer to buy more of that commodity.

This is called *income effect of the change in price* of the commodity. This is one reason why a consumer buys more of a commodity whose price falls. The other important reason why the quantity demanded of a commodity rises as its price falls, is the *substitution effect*. When the price of a commodity falls, it becomes relatively cheaper than other commodities. This induces the consumer to substitute the commodity whose price has fallen for other commodities which have now become relatively dearer. As a result of this substitution effect, the quantity demanded of the commodity, whose price has fallen, rises. This substitution effect is more important than the income effect. Marshall explained the downward-sloping demand curve with the aid of this substitution effect alone, since he ignored the income effect of the price change. But in some cases even the income effect is very significant and cannot be ignored. As will be discussed in a later chapter, Hicks and Allen who put forward indifference curve analysis of consumer's behaviour explain this downward-sloping demand curve with both the income and substitution effects.

We have explained above the reasons for the downward-sloping *demand curve of an individual consumer*. There is an additional reason why the *market demand curve* for a commodity slopes downward. When the price of a commodity is relatively high, only few consumers can afford to buy it. And when the price of a commodity falls, more consumers would start buying it because some of those who previously could not afford to buy it may now afford to buy it. This increases the number of consumers of a commodity at a lower price. Thus, when price of a commodity falls, the number of its consumers increases and this also tends to raise the market demand for the commodity.

Exceptions to the Law of Demand

Law of demand is generally believed to be valid in most of the situations. However, some exceptions to the law of demand have been pointed out.

Goods having Prestige Value: Veblin Effect. One exception to the law of demand is associated with the name of the economist Thorstein Veblin who propounded the doctrine of conspicuous consumption. According to Veblin, some consumers measure the utility of a commodity entirely by its price, *i.e.,* for them, the greater the price of a commodity, the greater its utility. For example, diamonds are considered as prestige good in the society and for the upper strata of the society the higher the price of diamonds, the higher the prestige value of them and therefore the greater utility or desirability of them. In this case, the consumer will buy less of the diamonds at a lower price because with the fall in price its prestige value will go down. On the other hand, when price of diamonds goes up, their prestige value will go up and therefore their utility and desirability. As a result, at a higher price the quantity demanded of diamonds by a consumer will rise. This is called *Veblin effect*.[1]

Giffin Goods. Another exception to the law of demand was pointed out by Sir Robert Giffin who observed that when the price of bread increased, the low-paid British workers in the early 19th century purchased more bread and not less of it, and this is contrary to the law of demand described above. The reason given for this is that these British workers consumed a diet of mainly bread and when the price of bread went up they were compelled to spend more on given quantity of bread. Therefore, they could not afford to purchase as much meat as before. Thus, they substituted even bread for meat in order to maintain their intake of food. After the name of Robert Giffin, such goods in whose case there is a direct price-demand relationship are called *Giffin goods.* It is *important to note that when with the rise in the price of a commodity*, its quantity demand increases and with the fall in its price its quantity demanded decreases, the demand curve will slope upward to the right and not downward.

---

1. See his book, *The Economics of Leisure Class.*

Some False Exceptions

There are some other exceptions to the law of demand which are only apparent and not genuine In other words, these exceptions to the law of demand are false. One of the false exceptions relates to the changes in the expectations of the people regarding prices of the commodities in the future. Suppose, the rainfall in India in any year does not occur in adequate quantity and there is widespread drought, the expectations of the people will be that the prices would rise in the future. Therefore, even if the prices of foodgrains are higher at the present, they would demand greater quantities since they will be expecting that the prices in the future would be still higher. This is contrary to the law of demand, but in this case applying the law of demand is not valid. This is because in this case the increase in the quantity demanded is not due to the rise in price but due to the rightward shift in the demand curve as a result of changes in the price expectations of the people.

Furthermore, statistical data regarding the quantities purchased of various goods reveal that larger amounts of them have been purchased at high prices and smaller quantities have been purchased at lower prices over the course of the business cycle. This is also apparently the violation of the law of demand but properly interpreted it does not contradict the law of demand, for this only shows that demands for many commodities increase in times of prosperity periods of the business cycles due to the increase in the incomes of the people, and the demands for the commodities decrease in times of the depression periods of the business cycles due to the fall in incomes of the people.

Another false exception to the law of demand is found when a commodity is sold under two brand names whose prices greatly differ. It is often found that some people demand high-priced brand more than the low-priced one, although they are almost identical. This author has seen that some of his friends buy more of 'Supreme Lux' having a much higher price than 'Lux' having a low price even though the two are of almost the same quality. However, this is not a real exception to the law of demand. This is because those who buy more of high-priced brand *think* that the high-priced brand is different from and superior to the low-priced brand. Therefore, for analysing the nature of demand for these brands, they should be regarded as two different commodities.

DETERMINANTS OF DEMAND

We have explained above demand changes as a result of changes in price. Demand schedule and law of demand state the relationship between price and quantity demanded by assuming *"other things remaining the same"*. When there is a change in these other things, the whole demand schedule or demand curve undergoes a change. In other words, these other things determine the *position and level* of the demand curve. If these other things or the determinants of demand change, the whole demand schedule or the demand curve will change. As a result of the changes in these factors or determinants, a demand curve will shift above or below as the case may be. The following are the factors which determine demand for goods.

1. **Tastes and Preferences of the Consumers.** An important factor which determines demand for a good is the tastes and preferences of the consumers for it. A good for which consumers' tastes and preferences are greater, its demand would be large and its demand curve will lie at a higher level. People's tastes and preferences for various goods often change and as a result there is change in demand for them. The changes in demand for various goods occur due to the changes in fashion and also due to the pressure of advertisements by the manufacturers and sellers of different products. For example, a few years back when Coca-Cola plant was established in New Delhi demand for it was very small. But now people's taste for Coca-Cola has undergone a change and become favourable to it because of large advertisement and publicity done for it. The result of this is that the demand for Coca-Cola has increased very much. In economics we would say that the demand curve for Coca-Cola has shifted upward. On the contrary when any good goes out of fashion or people's tastes and

preferences no longer remain favourable to it the demand for it decreases. In economics we say that the demand curve for these goods will shift downward.

2. **Incomes of the People.** The demand for goods also depends upon incomes of the people. The greater the incomes of the people, the greater will be their demand for goods. In drawing a demand schedule or a demand curve for a good we take incomes of the people as given and constant. When as a result of the rise in incomes of the people the demand increases, the whole of the demand curve shifts upward and *vice versa.* The greater income means the greater purchasing power. Therefore, when incomes of the people increase, they can afford to buy more. It is because of this reason that the increase in income has a positive effect on the demand for a good. When the incomes of the people fall they would demand less of the goods and as a result the demand curve will shift below. For instance, during the planning period in India the incomes of the people have greatly increased owing to the large investment expenditure on the development schemes by the Government and the private sector. As a result of this increase in incomes, demand for foodgrains has greatly increased which has resulted in rightward shift in the demand curve for them. Likewise, when because of drought in a year the agricultural production greatly falls, incomes of the farmers decline. As a result of the decline in incomes of the farmers, they demand less of cotton cloth and other manufactured products.

3. **Changes in the Prices of the Related Goods.** The demand for a good is also affected by the prices of other goods, especially those which are related to it as substitutes or complements. When we draw a demand schedule or a demand curve for a good we take the prices of the related goods as remaining constant. Therefore, when the prices of the related goods, substitutes or complements change the whole demand curve would change its position; it will shift upward or downward as the case may be. When price of a substitute for a good falls, the demand for that good will decline and when the price of the substitute rises, the demand for that good will increase. For example, when price of the tea as well as the incomes of the people remain the same but price of the coffee falls, the consumers would demand less of tea than before. Tea and coffee are very close substitutes, therefore when coffee becomes cheaper, the consumers substitute coffee for tea and as a result the demand for tea declines. The goods which are complementary to each other, the change in the price of any of them would affect the demand of the other. For instance, if price of the milk falls, the demand for sugar would also be affected. When people would take more milk or would prepare more *khoya, burfi, rasgullas* with milk, the demand for sugar will also increase. Likewise, when price of cars falls, the demand for them will increase which in turn will increase the demand for petrol. Cars and petrol are complementary to each other.

4. **The Number of Consumers in the Market.** We have already explained that the market demand for a good is obtained by adding up the individual demands of the present as well as prospective consumers or buyers of a good at various possible prices. The greater the number of consumers of a good, the greater the market demand for it. Now, the question arises on what factors the number of consumers of a good depends. If the consumers substitute one good for another, then the number of consumers of that good which has been substituted by the other will decline and for the good which has been used in its place, the number of consumers will increase. Besides, when the seller of a good succeeds in finding out new markets for his good and as a result the market for his good expands, the number of consumers of that good will increase. Another important cause for the increase in the number of consumers is the growth in population. For instance, in India the demand for many essential goods, especially foodgrains, has increased because of the increase in the population of the country and the resultant increase in the number of consumers for them.

5. **Changes in Propensity to Consume.** People's propensity to consume also affects the demand for them. The income of the people remaining constant, if their propensity to consume rises, then out of the given income they would spend a greater part of it with the result that the demand for goods will increase. On the other hand, if propensity to save of the people increases, that is, if propensity to

consume declines, then the consumers would spend a smaller part of their income on goods with the result that the demand for goods will decrease. It is thus clear that with income remaining constant, change in propensity to consume of the people will bring about a change in the demand for goods.

6. **Consumers' Expectations with regard to Future Prices.** Another factor which influences the demand for goods is consumers' expectations with regard to future prices of the goods. If due to some reason, consumers expect that in the near future prices of the goods would rise, then in the present they would demand greater quantities of the goods so that in the future they should not have to pay higher prices. Similarly, when the consumers hope that in the future they will have good income, then in the present they will spend greater part of their incomes with the result that their present demand for goods will increase.

7. **Income Distribution.** Distribution of income in a society also affects the demand for goods. If distribution of income is more equal, then the propensity to consume of the society as a whole will be relatively high which means greater demand for goods. On the other hand, if distribution of income is more unequal, then propensity to consume of the society will be relatively less, for the propensity to consume of the rich people is less than that of the poor people. Consequently with more unequal distribution of income, the demand for consumer goods will be comparatively less. This is the effect of the income distribution on the propensity to consume and demand for goods. But the change in the distribution of income in the society would affect the demand for various goods differently. If progressive taxes are levied on the rich people and the money so collected is spent on providing employment to the poor people, the distribution of income would become more equal and with this there would be a transfer of purchasing power from the rich to the poor. As a result of this, the demand for those goods will increase which are generally purchased by the poor because the purchasing power of the poor people has increased and, on the other hand, the demand for those goods will decline which are usually consumed by the rich on whom progressive taxes have been levied.

Extension and Contraction in Demand

We have studied above the demand schedule, demand curve and law of demand. All these show that when price of a good falls, quantity demanded of it rises, and when its price rises, its quantity demanded falls, other things remaining the same. *When as a result of changes in price, the quantity demanded rises or falls, extension or contraction in demand is said to have taken place. Therefore, in economics, the extension and contraction in demand are used when the quantity demanded rises or falls as a result of changes in price and we move along a given demand curve.* When the quantity demanded of a good rises due to the fall in price, it is called *extension of demand* and when the quantity demanded falls due to the rise in price, it is called *contraction of demand.* For instance, suppose the price of bananas in the market at any given time is ₹12 per dozen and a consumer buys one dozen of them at that price. Now, if other things such as tastes of the consumer, his income, prices of other goods remain the same and price of bananas falls to ₹ 8 per dozen and the consumer now buys 2 dozen bananas, then extension in demand is said to have occurred. On the contrary, if the price of bananas rises to ₹ 15 per dozen and consequently the consumer now buys half a dozen of the bananas, then contraction in demand is said to have occurred.

It should be remembered that *extension and contraction in the demand takes place as a result of changes in the price alone when other determinants of demand such as tastes, income, propensity to consume and prices of the related goods remain constant.* These other factors remaining constant means that the demand curve remains the same, that is, it does not change its position; only the consumer moves downward or upward on it.

The extension and contraction in demand is illustrated in Fig. 7.3. Assuming other things such as income, tastes and fashion, prices of related goods remaining constant, a demand curve *DD* goods remaining constant, a demand curve *DD* has been drawn. It will be seen in this figure that when the price of the good is *OP*, then the quantity demanded of the good is *OM*. Now, if the price of the good falls to *OP′* the quantity demanded of the good rises to *ON*. Thus, there is extension in demand by the amount *MN*. On the other hand, if price of the good rises from *OP* to *OP″* the quantity demanded of the good falls to *OL*. Thus, there is contraction in demand by *ML*. We thus see that *as a result of changes in price of a good the consumers move along the given demand curve; the demand curve remains the same and does not change its position.*

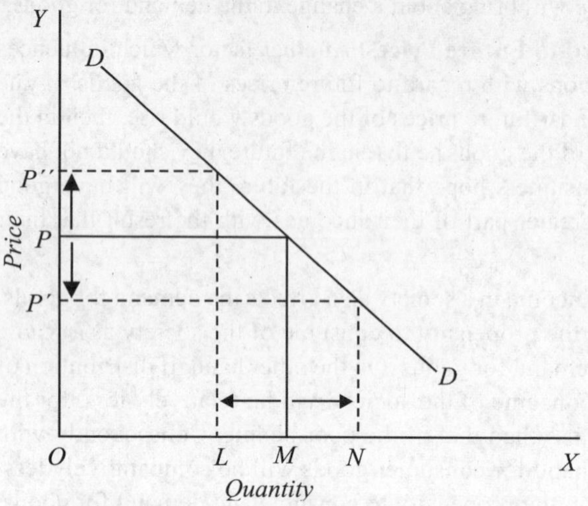

Fig. 7.3. *Extension and Contraction in Demand*

Demand and Quantity Demanded. It is important to understand the distinction between the concepts of demand and quantity demanded as they are often confused with each other. Demand represents the whole demand schedule or demand curve and shows how price of a good is related to quantity which the consumers are willing and able to buy, other fators which determine demand being held constant. On the other hand, quantity demanded refers to the quantity which the consumers buy at a particular price. The quantity demanded of a good varies with changes in its price; it increases when price falls and decreases when price rises. The changes in demand for a commodity occur when there is a change in the factors other than price, namely, tastes and preferences of the people, incomes of the consumers, and prices of related goods.

Changes in Demand: Increase and Decrease in Demand

When demand changes due to the factors other than price, there is shift in the whole demand curve. As mentioned above, apart from price, demand for a commodity is determined by incomes of the consumers, their tastes and preferences, prices of related goods. Thus, when there is any change in these factors, it will cause a shift in demand curve. For example, if incomes of the consumers increase, say due to the hike in their wages and salaries or due to the grant of dearness allowance, they will demand more of a good, say cloth, at each price. This will cause a shift in the demand curve to the right as is shown in Fig. 7.4. To begin with, *DD* is the demand curve. With the increase in income, the demand curve shifts to the right to *D′D′* which implies that at each price such as P_1, P_2, P_3 the consumers demand more of the commodity than before. Similarly, if preferences of the people for a commodity, say colour TV, become greater, their demand for colour TVs will increase, that is, the demand curve will shift to the right and, therefore, at each price demand for colour TVs will increase.

The other important factor which can cause an increase in demand for a commodity is the *expectations about future prices*. If people expect that price of a commodity is likely to go up in future, they will try to purchase the commodity, especially a durable one, in the current period which will boost the current demand for the good and cause a shift in the demand curve to the right.

The *prices of related commodities such as substitutes and complements* can also change the demand for a commodity. For example, if price of coffee rises, other factors remaining constant, this will cause the demand for tea, a substitute for coffee, to increase and its demand curve to shift to the right.

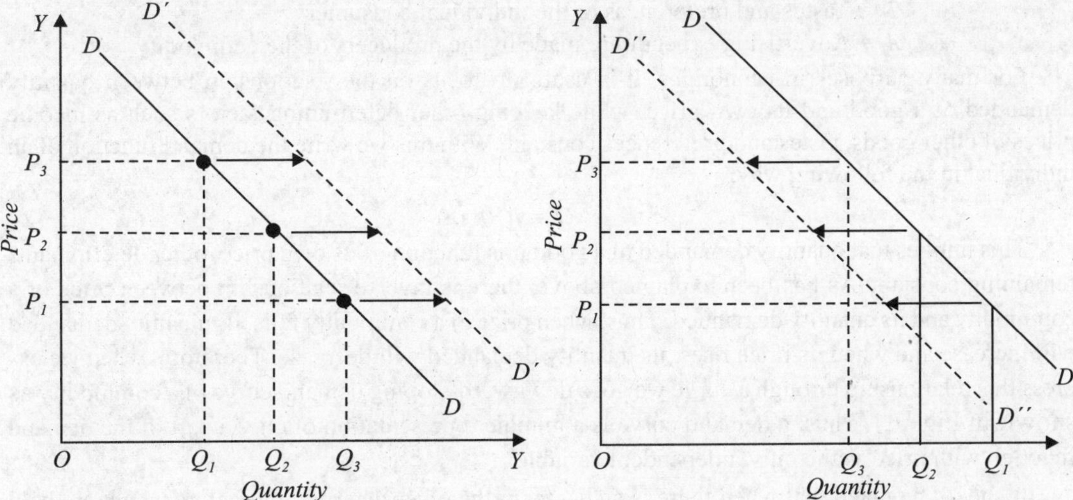

Fig. 7.4. *Increase in demand leads to the shift in the demand curve to the right*

Fig. 7.5. *Decrease in demand leading to the shift in the demand curve to the left*

Decrease in Demand and Shift in the Demand Curve. If there are adverse changes in the factors influencing demand, it will lead to the decrease in demand causing a shift in the demand curve to the left as shown in Fig. 7.5. For example, if due to inadequate rainfall agricultural production in a year declines, this will cause a fall in the incomes of the farmers. This fall in incomes of the farmers will cause a decrease in the demand for industrial products, say cloth, and will result in a shift in the demand curve to the left as shown in Fig. 7.5. It will be seen from Fig. 7.5 that as a result of decrease in demand, demand curve shifts to the left to *D''D''* and at each price such as P_1, P_2, P_3 the farmers demand less of cloth than before. Similarly, change in preferences for commodities can also affect the demand. For example, when colour TVs came to India people's greater preference for them led to the decrease in demand for black and white TVs causing leftward shift in demand curve for these black and white TVs.

Conclusion

From the above analysis it is clear that in case of extension and contraction in demand, a consumer moves along the same demand curve. He goes up and down on it due to the changes in price of the good alone. On the other hand, in the case of increase and decrease in demand the whole demand curve changes or shifts; in case of increase in demand it shifts to the right and in case of decrease in demand it shifts to the left.

DEMAND FUNCTION AND DEMAND CURVE

Individual's demand for a commodity depends on the own price of a commodity, his income, prices of related commodities (which may be either substitutes or complements), his tastes and preferences, and advertising expenditure made by the producers for the commodity in question. Individual demand for a commodity can be expressed mathematically in the following *general functional* form:

$$Q_d = f(P_x, I, P_r, T, A) \quad \ldots (1)$$

where P_x = Own price of the commodity X

I = Income of the individual

P_r = Prices of related commodities

T = Tastes and preferences of the individual consumer
A = Advertising expenditure made by the producers of the commodity

For many purposes in economics, it is useful to focus on the relationship between quantity demanded of a good and its own price, while keeping other determining factors such as income prices of other goods, tastes and preferences constant. With this we write the demand function of an individual in the following way:

$$Q_d = j(P) \quad \text{... (2)}$$

This implies that quantity demanded of a good X is function of its own price, other determinants remaining constant. As has been explained above, there is inverse relationship between price of a commodity and its quantity demanded. Thus, when price of a commodity falls, its quantity demanded will increase and when its price rises, its quantity demanded will decrease. Therefore, when we express this relationship through a curve we get a downward-sloping demand curve of a commodity as shown in Fig. 7.1. Thus, a demand curve is a graphic representation of only a part of the demand function with price as the only independent variable.

It should be noted that when there is a change in the other determining factors which are held constant such as income, tastes, prices of related commodities, the whole demand curve will shift. For example, if income increases, the whole demand curve will shift to the right and, on the contrary, if income decreases, then the whole demand curve shifts to the left. Similarly, changes in other determining factors such as tastes, prices of related commodities, advertising cause shift in the demand curve and are therefore called *shift factors.*

The individual's demand function in (2) above is a *general* functional form and does not show how much quantity demanded of a consumer will change following a unit change in price (P_x). For the purpose of actually estimating demand for a commodity we need a specific form of the demand. function. Generally, demand function is considered to be of a *linear* form. The specific demand function of a linear form is written as

$$Q_d = a - bP_x \quad \text{... (3)}$$

where a is a constant intercept term on the X–axis and b is the coefficient showing the slope of the demand curve. If on estimating the demand function (3) from the information about monthly quantities demanded of sugar at its various prices by an individual consumer, we find the constant a to be equal to 12 and the constant b to be equal to 2. We can write individuals demand function as

$$Q_d = 12 - 2P_x$$

This is interpreted as *one rupee fall* in price of sugar will cause its *quantity demanded to increase by 2 units* of sugar.

Market Demand Function

A market consists of several individuals. Market demand function is obtained by summing up the demand functions of the individuals constituting the market.

Example 1

A market for a commodity consists of three individuals A, B and C whose demand functions for the commodity are given below. Find out the market demand function.

$$Q_A = 40 - 2P$$
$$Q_B = 25.5 - 0.75P$$
$$Q_C = 36.5 - 1.25P$$

When individual demand functions are expressed as *'quantity as a function of price'* as is the case in our problem stated above, market demand function can be obtained by summing up the individual demand functions. Thus, market demand function is

$$Q_M = Q_A + Q_B + Q_C$$
$$= (40 - 2P) + (25.5 - 0.75P) + (36.5 - 1.25P)$$
$$= (40 + 25.5 + 36.5) - (2 + 0.75 + 1.25)P$$
$$= 102 - 4P$$

However, note that when individual demand functions are expressed as *"price as function of quantity"*, then in order to obtain the market demand, they have to be first converted into 'quantity as function of price'.

Example 2

Suppose a market consists of three consumers, A, B and *C, whose inverse demand functions are given below:*

(A): $P = 35 - 0.5Q_A$

(B): $P = 50 - 0.25Q_B$

(C): $P = 40 - 2.00Q_C$

(*i*) *Find out the market demand function for the commodity.*

(*ii*) *If the market supply function* is *given by* $Q_S = 40 + 3.5P$, *determine the equilibrium price and quantity*

Since the individual demand functions are expressed as 'price as function of quantity, that is, we are given *"inverse demand functions"* we have first to transform them into 'quantity demanded as function of price'. Transforming them yields the following demand functions:

$$Q_A = 70 - 2P$$
$$Q_B = 200 - 4P$$
$$Q_C = 20 - 0.5P$$

Market demand function:

$$Q_D = (70 - 2P) + (200 - 4P) + (20 - 0.5P)$$
$$Q_D = 70 + 200 + 20 - (2 + 4 + 0.5P)$$
$$= 290 - 6.5P$$

Market supply function $(Q_S) = 40 + 3.5P$

In equilibrium, $Q_D = Q_S$

$$290 - 6.5P = 40 + 3.5P$$
$$10P = 250$$
$$P = \frac{250}{10} = 25$$

Substituting the equilibrium value of price in the demand function equation, we have

$$Q = 290 - 6.5\,(25)$$
$$= 290 - 162.5$$
$$= 127.5$$

Thus, the equilibrium price is ₹ 25 and equilibrium quantity is 127.5 units.

Bandwagon Effect and Snob Effect

It is worth noting that making horizontal summation of individual demand curves to obtain market demand curve (function) we have made an important assumption *that demand functions of individual consumers are independent.* This is however not always the case. Most important factor causing interdependence of demands of individual consumers is the **bandwagon effect.** Bandwagon effect arises because individuals demand some commodities because others are doing so, that is, those commodities happen to be in fashion. The desire to look fashionable and to keep up with Joneses makes demands of individuals interdependent. For example, desire or demand for wearing jeans by girls has been influenced by the number of other girls purchasing and demanding jeans. The bandwagon effect tends to make the market demand curve flatter or more elastic than given by the simple horizontal summation of the individual demand curves.

The second factor which makes the demands of different individuals interdependent is snob effect. The *snob effect*, which is opposite of bandwagon effect, refers to the desire to purchase a commodity having prestige value so as to look different or exclusive than others. Therefore, individuals demand less of a commodity having a prestige value as more individuals buy it. This is because when more people have a commodity, it loses its prestige value. The existence of snob effect tends to make the demand steeper or less elastic than indicated by the horizontal summation of individual demand curves.

QUESTIONS FOR REVIEW

1. Define demand for a commodity. Explain the various factors which determine demand for a commodity.
2. Demand for a commodity refers to:
   (*a*) desire for a commodity (*b*) need for a commodity
   (*c*) desire for a commodity backed by ability to pay for it.
   (*d*) ability to pay for a commodity.
3. What is meant by *Ceteris Paribus*? What factors are covered under *Ceteris Paribus* condition in relation to demand for a commodity?
4. You are given the following demand function for a commodity:
   $$Q_d = 20 - 2P$$
   where Q_d stands for quantity demanded of a commodity and P for its price.
   Draw the demand curve representing the above demand function.
   [**Hints.** Take some prices such as ₹ 3, 5, 7 etc. and substitute them in the given demand function to find the quantities demanded at various prices. Then plot these various price-quantity combinations.
5. Distinguish between (1) extension and increase in demand, (2) contraction and decrease in demand. Illustrate with diagrams.
6. Distinguish between the movement along a demand curve and a shift in the demand curve.
7. State the law of demand. How would you explain it with substitution effect and income effect? Are there any exceptions to this law?
8. Why does demand curve usually slope downward to the right? Are there any exceptions to this?
9. Distinguish between Giffin paradox and Veblin effect.
10. Distinguish between a demand function and demand curve. What are the factors that cause a shift in the demand curve?

CHAPTER 8

DEMAND: MARSHALL'S CARDINAL UTILITY ANALYSIS

Introduction

The price of a product depends upon the demand for and the syupply of it. In this part of the book we are concerned with the theory of demand, which explains the demand, for a good and the factors determining it. The factors governing the supply of a good will be discussed in the next part of the book. Individual's demand for a product depends upon price of the product, income of the individual, the prices of related goods. It can be put in the following functional form:

$$D_x = f(P_x, I, P_y, P_z, T\ etc.)$$

where D_x stands for the demand of good *X* P_x for price of good *X*, *I* for individual's income, P_y, P_z, for the prices of related goods and T for tastes and preferences of the individual. But among these determinants of demand, economists single out price of the good in question as the most important factor governing the demand for it. Indeed, the function of a theory of demand is to establish a relationship between quantity demanded of a good and its price and to provide an explanation for it. From time to time, different theories have been advanced to explain consumer's demand for a good and to derive a valid demand theorem. Cardinal utility analysis is the oldest theory of demand which provides an expalantion of consumer's demand for a product and derives the law of demand which establishes an inverse relationship between price and quantity demanded of a product. Recently, cardinal utility approach to the theory of demand has been subjected to severe criticisms and as a result some alternative theories, namely, *Indifference Curve Analysis, Samuelson's Revealed Preference Theory, Hicks' Logical Weak Ordering Theory* have been propounded. We will discuss indifference curve analysis of demand in the next chapter, while in the present chapter we shall be concerned with the cardinal utility analysis of demand. Though cardinal utility approach to the theory of demand is very old, its final shape emerged at the hands of Marshall. Therefore, it is Marshallian cardinal utility analysis of demand which has been discussed in this chapter.

Assumptions of Cardinal Utility Analysis

Cardinal utility analysis of demand is based upon certain important assumptions. Before explaining how cardinal utility analysis explains consumer's equilibrium in regard to the demand for good, it is essential to describe those basic assumptions on which the whole utility analysis rests. As we shall see later, cardinal utility analysis has been criticised because of its unrealistic assumptions. the basic assumptions or premises of utility analysis are as follows:

The Cardinal Measurability of Utility. The exponents of cardinal utility analysis regard utility to be a cardinal concept. In other words, they hold that utility is a measurable and quantifiable entity. According to them, a person can express utility or satisfaction he derives from the goods in the quantitative cardinal terms. Thus, a person can say that he derives utility equal to 10 units from the consumption of a unit of good *A*, and 20 units from the consumption of a unit of good *B*. Moreover, the cardinal measurement of utility involves that a person can compare in respect of size, that is, how much one level of utility is greater than another. A person can say that the utility he gets from the consumption of one unit of good *B* is double the utility he obtains from the consumption of one unit of good *A*.

According to Marshall, marginal utility is *actually measurable in terms of money*.[1] Money represents the general purchaing power and it can therefore be regarded as a command over alternative utility-yielding goods. Marshall argues that the amount of money which a person is prepared to pay for a unit of a good rather than he without it is a measure of the utility he derives from that good. Thus, according to him money is the measuring rod of utility. Some economists belonging to the cardinalist school measure utility in imaginary units called "utils". They assume that a consumer is capable of saying that one apple provides him utility equal to 4 utils. Further, on this ground, he can say that he gets twice as much utility from an apple as from an orange.

The Hypothesis of Independent Utilities. The second important tenet of the cardinal utility analysis is the hypothesis of *independent utilities*. On this hypothesis, the utility which a consumer derives from a good is the function of the quantity of that good and of that good only. In other words, the utility which a consumer obtains from a good does not depend upon the quantity consumed of other goods; it depends upon the quantity purchased of that good alone. On this assumption, then the total utility which a person gets from the whole collection of goods purchased by him is simply the total sum of the separate utilities of the goods. Thus, the cardinalist school regards utility as '*additive*', that is, separate utilities of different goods can be added to obtain the total sum of the utilities of all goods purchased.

Constancy of the Marginal Utility of Money. Another important assumption of the cardinal utility analysis is the constancy of the marginal utility of money. Thus, while the marginal utility analysis assumes that marginal utilities of commodities diminish as more of them are purchased or consumed, but the marginal utility of money remains constant throughout when the individual is spending money on a good and due to which the amount of money with him varies. Daniel Bernoulli first of all introduced this assumption but later Marshall adopted this in his famous book "*Principles of Economics*'. As stated above, Marshall measured marginal utilities in terms of money. But measurement of marginal utility of goods in terms of money is only possible if the marginal utility of money itself remains constant. It should be noted that the assumption of constant marginal utility of money is very crucial to the Marshallian analysis, because otherwise Marshall could not measure the marginal utilities of goods in terms of money. If the money which is the unit of measurement itself varies as one is measuring with it, it cannot then yield correct measurement of the marginal utility of the good.

When the price of a good falls and the real income of the consumer rises, the marginal utility of money to him will fall but Marshall ignored this and assumed that marginal utility of money did not change as a result of the change in price. Likewise, when the price of a good rises, the real income of the consumer will fall and his marginal utility of money will rise. But Marshall ignored this and assumed that marginal utility of money remains the same. Marshall defended this assumption on the ground that "his (the individual consumer's) expenditure on any one thing.....is only a small part of his whole expenditure."

Introspective Method. Another important hypothesis of the marginal utility analysis is the use of introspective method in judging the behaviour of marginal utility. "Introspection is the ability of the observer to reconstruct events which go on in the mind of another person with the help of self-observation. This form of comprehension may be just guesswork or intution or the result of long lasting experience."[1] Thus, the economists construct with the help of their own experience the trend of feeling which goes on in other men's mind. From his own response to certain forces and by experience and observation one gains understanding of the way other people's minds would work in similar situations. To sum up, in introspective method we attribute to another person what we know

1. Emil Kauder, *A History of Marginal Utility Theory*, (Princeton, New Jersy, 1965). p. 120.

of our own mind. That is, by looking into ourselves we see inside the heads of other individuals. So the law of diminishing marginal utility is based upon introspection. We know from our own mind that as we have more of a thing, the less utility we derive from an additional unit of it. We conclude from it that other individuals' mind will work in a similar fashion, that is, marginal utility to them of a good will diminish as they have more units of it.

With the above basic premises, the founders of marginal utility analysis have developed tow laws which occupy an important place in economic theory and have several applications and uses. These two laws are : (1) Law of Diminishing Marginal Utility and (2) Law of Equi-Marginal Utility. It is with the help of these two laws about consumers' behaviour that the exponents of utility analysis have derived the law of demand. We explain below these two laws in detail.

LAW OF DIMINISHING MARGINAL UTILITY

An important tenet of cardinal utility analysis relates to the behaviour of marginal utility. This familiar behaviour of marginal utility has been stated in the Law of Diminishing Marginal Utility according to which marginal utility of a good diminishes as an individual consumes more units of a good. In other words, as a consumer takes more units of a good, the extra utility or satisfaction that he derives from an extra unit of the good goes on falling. It should be carefully noted that it is the marginal utility and not the total utility that declines with the increase in the consumption of a good. The law of diminishing marginal utility means that the total utility increases at a decreasing rate.

Marshall who was the famous exponent of the cardinal utility analysis has stated the law of diminishing marginal utility as follows:

"*The additional benefit which a person derives from a given increase of his stock of a thing diminishes with every increase in the stock that he already has.*"

This law is based upon two important facts. First, while the total wants of a man are virtually unlimited, each single want is satiable. Therefore, as an individual consumes more and more units of a good, intensity of his want for the good goes on falling and a point is reached where the individual no longer wants any more units of the good. That is, when saturation point is reached, marginal utility of a good becomes zero. Zero marginal utility of a good implies that the individual has all that he wants of the good in question. The second fact on which the law of diminishing marginal utility is based is that the different goods are not perfect substitutes for each other in the satisfaction of various particular wants. When an individual consumes more and more units of a good, the intensity of his particular want for the good diminishes but if the units of that good could be devoted to the satisfaction of other wants and yielded as much satisfaction as they did initially in the satisfaction of the first want, marginal utility of the good would not have diminished.

It is obvious from above that the law of diminishing marginal utility describes a familiar and fundamental tendency of human nature. This law has been arrived at by introspection and by observing how people behave.

Illustration of the Law of Diminishing Marginal Utility

Consider Table 8.1. where we have presented the total and marginal utilities derived by a person from cups of tea consumed per day. When one cup of tea is taken per day, the total utility derived by the person is 12 utils. And because this is the first cup its marginal utility is also 12. With the consumption of 2nd cup per day, the total utility rises to 22 but marginal utility falls to 10. It will be seen from the table that as the consumption of tea increases to six cups per day, marginal utility from the additional cups goes on diminishing (*i.e.* the total utility goes on increasing at a diminishing rate). However, when the cups of tea consumed per day increases to seven, then instead of giving *positive* marginal utility, the seventh cup gives *negative* marginal utility equal to – 2. This is because too many cups of tea consumed per day (say more that six for a particular individual) may cause acidity and gas trouble. Thus, the extra cups of tea beyond six to the individual in question gives him disutility rather than positive satisfaction.

Table 8.1
Diminishing Marginal Utility

Cups of Tea consumed per day	*Total Utility (utils)*	*Marginal Utility (utils)*
1	12	12
2	22	10
3	30	8
4	36	6
5	40	4
6	41	1
7	39	– 2
8	34	– 5

We have graphically represented the data of the above table in Fig. 8.1. We have constructed rectangles representing the toal utility obtained from various numbers of cups of tea consumed per day. As will be seen from the Fig. 8.1 the length of the rectangle goes on increasing up to the sixth cup of tea and beyond that length of the rectangle declines, indicating thereby that up to the sixth cup of tea total utility obtained from the increasing cups of tea goes on increasing whereas beyond the 6th cup, total utility declines. In other words, marginal utility of the additional cups up to the 6th cup is positive whereas beyond the sixth cup marginal utility is negative. The marginal utility obtained by the consumer from additional cups of tea as he increases the consumption of tea has been shaded. A glance at the Fig. 8.1 will show that this shaded area goes on declining which shows that marginal utility from the additional cups of tea is diminishing.

We have joined the various rectangles by a smooth curve which is the curve of total utility which rises up to a point and then declines due to negative marginal utility. Moreover, the shaded areas of the rectangle representing marginal utility of the various cups of tea have also been shown separately in the figure given at the bottom. We have joined the shaded rectangles by a smooth curve which is the curve of marginal utility. As will be seen, this marginal utility curve goes on declining throughout and even falls below the X-axis. Portion below the X-axis indicates the negative marginal utility. This downward-sloping marginal utility curve has an important implication for consumer's behaviour regarding demand for goods. We shall explain below how the demand curve is derived from marginal utility curve.

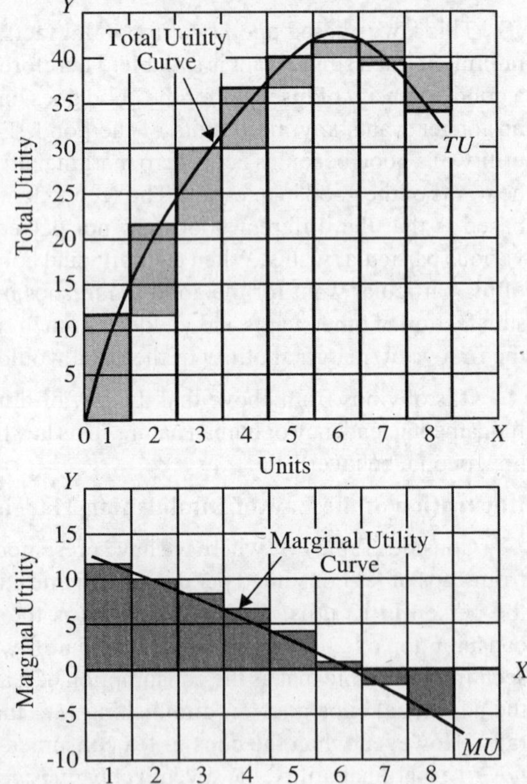

Fig. 8.1. *Diminishing Marginal Utility*

The significance of the diminishing marginal utility of a good for the theory of demand is that the quantity demanded of a good rises as the price falls and *vice versa*. Thus, it is because of the diminishing marginal utility that the demand curve slopes downward. This will be explained in detail later in this chapter. If properly understood the law of diminishing marginal utility applies to all objects of desire including money. But it is worth mentioning that marginal utility of money is generally never zero or negative. Money represents purchasing power over all other goods, that is, a man can satisfy all his material wants if he possesses enough money. Since man's total wants are practically unlimited, therefore the marginal utility of money to him never falls to zero.

Applications and Uses of Diminishing Marginal Utility

The marginal utility analysis has a good number of uses and applications in both economic theory and policy. The concept of marginal utility is of crucial significance in explaining determination of the prices of commodities. The discovery of the concept of marginal utility has helped to explain the *paradox of value* which troubled Adam Smith in *The Wealth of Nations*. Adam Smith was greatly surprised to know why water which is so very essential and useful to life has such a low price (indeed no price), which diamonds which are quite unnecessary, have such a high price. He could not resolve this water-diamond paradox. But modern economists can solve it with the aid of concept of marginal utility. According to the modern economists, the total utility of a commodity does not determine the price of a commodity and it is the marginal utility which is crucially important determinant of price. Now, the water is available in abundant quantities so that its relative marginal utility is very low or even zero. Therefore, its price is low or zero. On the other hand, the diamonds are scarce and therefore their relative marginal utility is quite high and this is the reason why their prices are high. Prof. Samuelson explains this paradox of value in the following words:—"The more there is of a commodity, the less the relative desirability of its last little unit becomes, even though its total usefulness grows as we get more of the commodity. So, it is obvious why a large amount of water has a low price. Or why air is actually a free good despite its vast usefulness. The many later units pull down the market value of all units."[2]

Further, as shall be seen below, with the aid of the law of diminishing marginal utility, we are able to derive the law of demand and to show why the demand curve slopes downward. Besides, the Marshallian concept of consumer's surplus is based upon the principle of diminishing marginal utility.

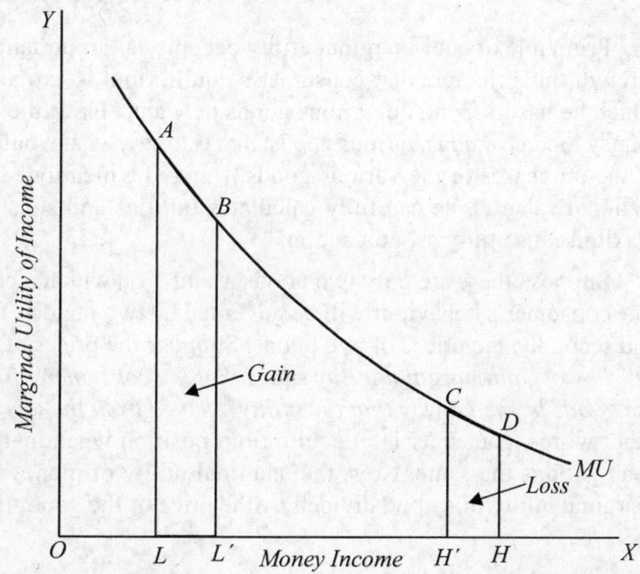

Fig. 8.2. *Redistribution of Income to Increase Social Welfare*

Another important use of marginal utility is in the field of fiscal policy. In the modern Welfare State, the governments redistribute income so as to increase the welfare of the people. This redistribution of income through imposing progressive income taxes on the rich sections of the society and spending the tax proceeds on social services for the poor people is based upon the diminishing marginal utility. The

2. Paul Samuelson, *Economics,* McGraw Hill, 8th edition, p. 417.

concept of diminishing marginal utility demonstrates that transfer of income from the rich to the poor will increase the economic welfare of the community. As has been pointed out above, law of diminishing marginal utility also applies to the money; as the money income of a consumer increases, the marginal utility of money to him falls. How the redistribution of income will increase welfare of the community, is illustrated in Fig. 8.2. In this figure, money income is measured along *X*-axis and marginal utility of income is measured along *Y*-axis. *MU* is the marginal utility curve of money which is sloping downward. Suppose *OL* is the income of a poor person and *OH* is the income of a rich person. If rich person is subjected to the income tax and amount of money equal to *HH′* is taken from him and the same amount of money *LL′* (equal to *HH′*) is given to the poor man, it can be shown that the welfare of the community will increase. As a result of this transfer of income, the income of the rich man falls by *HH′* and the income of the poor person rises by *LL′* (*HH′* = *LL′*). Now it will be seen from Fig. 8.2 that the loss of satisfaction or utility of the rich man as a result of decline in his income by *HH′* is equal to the area *HDCH′*. Further, it will bee seen that the gain in satisfaction or utility by the increase of an equivalent amount of income *LL′* of the poor man, is equal to *LABL′*.

It is thus obvious from the figure that the gain in utility of the poor person is greater than the loss of utility of the rich man. Therefore, the total utility or satisfaction of the two persons taken together will increase. Thus, on the basis of the diminishing marginal utility of money many economists and political scientists have advocated that Government must redistribute income in order to raise economic welfare of the society. However, it may be pointed out that some economists challenge the validity of such redistribution of income to promote social welfare. They point out that the above analysis of marginal utility is based upon interpersonal comparison of utility which is quite invalid and unscientific. They argue that people differ greatly in their preferences and capacity to enjoy goods and, therefore, it is difficult to know the exact shapes of the marginal utility curves of the different persons. Therefore they assert that the losses and gains of utility of the poor and the rich cannot be measured and compared.

PRINCIPLE OF EQUI-MARGINAL UTILITY: CONSUMER'S EQUILIBRIUM

Priniciple of equi-marginal utility occupies an important place in cardinal utility analysis. It is through this principle that consumer's equilibrium is explained. A consumer has a given income which he has to spend on various goods he wants. Now, the question is how he would allocate his money income among various goods, that is to say, what would be his equilibrium position in respect of the purchases of the various goods. It may be mentioned here that consumer is assumed to be 'rational', that is, he carefully calculates utilities and substitutes one good for another so as to maximise his utility or satisfaction.

Suppose there are only two goods *X* and *Y* on which a consumer has to spend a given income. The consumer's behaviour will be governed by two factors: first, the marginal utilities of the goods and secondly, the prices of two goods. Suppose the prices of the goods are given for the consumer. *The law of equi-marginal utility states that the consumer will distribute his money income between the goods in such a way that the utility derived from the last rupee spent on each good is equal.* In other words, consumer is in equilibrium position when marginal utility of money expenditure on each good is the same. Now, the marginal utility of money expenditure on a good is equal to the marginal utility of a good divided by the price of the good. In symbols,

$$MU_e = \frac{MU_x}{P_x}$$

where MU_e is marginal utility of money expenditure and MU_x is the marginal utility of *X* and P_x is the price of *X*. The law of equi-marginal utility can therefore be stated thus: the consumer will spend his money income on different goods in such a way that marginal utility of money expenditure each good is equal. That is, consumer is in equilibrium in respect of the purchases of two goods *X* and *Y* when

$$\frac{MU_x}{P_x} = \frac{MU_y}{P_y}$$

Now, if $\frac{MU_x}{P_x}$ and $\frac{MU_y}{P_y}$ are not equal and $\frac{MU_x}{P_x}$ is greater than $\frac{MU_y}{P_y}$, then the consumer will substitute good *X* for good *Y*. As a result of this substitution, the marginal utility of good *X* will fall and marginal utility of good *Y* will rise. The consumer will continue substituting good *X* for good *Y* till $\frac{MU_x}{P_x}$ becomes equal to $\frac{MU_y}{P_y}$. When $\frac{MU_x}{P_x}$ becomes equal to $\frac{MU_y}{P_y}$ the consumer will be in equilibrium.

But the equality of $\frac{MU_x}{P_x}$ with $\frac{MU_y}{P_y}$ can be achieved not only at one level but at different levels of expenditure. The question is how far does a consumer go in purchasing the goods he wants. This is determined by the size of his money income. With a given income and money expenditure a rupee has a certain utility for him: this utility is the marginal utility of money to him. Since the law of diminishing marginal utility applies to money income also, the greater the size of his money income the smaller the marginal utility of money to him. Now, the consumer will go on purchasing goods till the marginal utility of money expenditure on each good becomes equal to the marginal utility of money to him. Thus, the consumer will be in equilibrium when the following equation holds good:

$$\frac{MU_x}{P_x} = \frac{MU_y}{P_y} = MU_m$$

Where MU_m is marginal utility of money expenditure (that is, the utility of the last spent on eachy good).

If there are more than two goods on which the consumer is spending his income, the above equation must hold good for all of them.

Let us illustrate the law of equi-marginal utility with the aid of an arithmetical table given below:

Table 8.2
Marginal Utility of Goods X and Y

Units	MU_x *(Utils)*	MU_y *(Utils)*
1	20	24
2	18	21
3	16	18
4	14	15
5	12	9
6	10	3

Let the prices of goods *X* and *Y* be ₹ 2 and ₹ 3 respectively. Reconstructing the above table by dividing marginal utilities of *X* (MU_x) by ₹ 2 and marginal utilities of *Y*(MU_y) by ₹ 3 we get the Table 8.3.

Table 8.3
Marginal Utility of Money Expenditure

Units	$\frac{MU_x}{P_x}$	$\frac{MU_y}{P_y}$
1	10	8
2	9	7
3	8	6
4	7	5
5	6	3
6	5	1

Suppose a consumer has money income of ₹ 24 to spend on the two goods. It is worth noting that in order to maximise his utility the consumer will not equate *marginal utilities of the goods* because prices of the two goods are different. He will equate the marginal utility of the last rupee (*i.e.* marginal utility of money expenditure) spent on these two goods. In other words, he will equate $\frac{MU_x}{P_x}$ with $\frac{MU_y}{P_y}$ while spending his give money income on the two goods. By looking at the Table 8.3 it will become clear that $\frac{MU_x}{P_x}$ is equal to 5 utils when the consumer purchases 6 units of good *X* and $\frac{MU_y}{P_y}$ is equal to 5 utils when he buys 4 units of good *Y*. Therefore, consumer will be in equilibrium when he is buying 6 units of good *X* and 4 units of good *Y* and will be spending (₹ 2 × 6 + ₹ 3 × 4) = ₹ 24 on them. Thus, in the equilibrium position where he maximises his utility,

$$\frac{MU_x}{P_x} = \frac{MU_y}{P_y} = MU_m$$

$$\frac{10}{2} = \frac{15}{3} = 5$$

Thus, marginal utility of the last rupee spent on each of the two goods he purchases is the same, that is, 5 utils.

Consumers' equilibrium is graphically protrayed in Fig. 8.3. Since marginal utility curves of goods slope downward, curves depicting $\frac{MU_x}{P_x}$ and $\frac{MU_y}{P_y}$ also slope downward. Thus, when the consumer is buying *OH* of *X* and *OK* of *Y*, then

$$\frac{MU_x}{P_x} = \frac{MU_y}{P_y} = MU_m$$

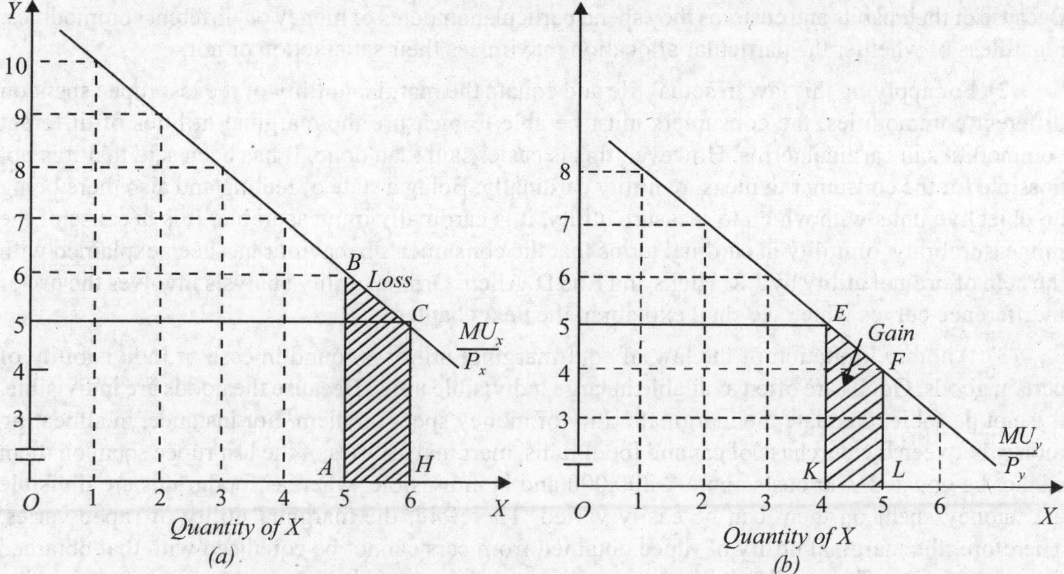

Fig. 8.3. *Equi-marginal Principle and Consumer's Equilibrium*

Therefore, the consumer is in equilibrium when he is buying 6 units of *X* and 4 units of *Y*. No other allocation of money expenditure will yield greater utility than when he is buying 6 units of commodity X and 4 units of commodity *Y*. Suppose if the consumer buys one unit less of good *X* and one unit more of good *Y*. This will lead to the decrease in his total utility. It will be observed from Fig. 8.3 (a) that the consumption of 5 units instead of 6 units of commodity *X* means a loss in satisfaction equal to the shaded area *ABCH* and from Fig. 8.3(b) it will be seen that consumption of 5 units of commodity *Y* instead of 4 units will mean gain in utility by the shaded area *KEFL*. It will be noticed that with this rearrangement of pucrchases of the two goods, the loss in utility *ABCH* exceeds gain in utility *KEFL*. Thus, his total satisfaction will fall as a result of this rearrangement of purchases. Thus when the consumer is making purchases by spending his given income in such a way that $\frac{MU_x}{P_x} = \frac{MU_y}{P_y}$, he will not like to make any further changes in the basket of goods and will therefore be in equilibrium situation by maximizing his utility.

The above equi-marginal condition for the equilibrium of the consumer will hold even when a consumer spends his money income on several goods. Thus

$$\frac{MU_x}{P_x} = \frac{MU_y}{P_y} = \ldots\ldots\ldots\ldots = \frac{MU_n}{P_n} = MU_m$$

Limitations of the Law of Equi-marginal Utility

Like other laws of economics, law of equimarginal utility is also subject to various limitations. This law, like other laws of economics, brings out an important tendency among the people. This is not necessary that all people exactly follow this law in the allocation of their money income and therefore all may not obtain maximum satisfaction. This is due to the following reasons:–

(1) For applying this law of equi-marginal utility in the real life, consumer must weigh in his mind the marginal utilities of different commodities. For this he has to calculate and compare the marginal utilities obtained from different commodities. But it has been pointed out that the ordinary consum-

ers are not so rational and calculating. Consumers are generally governed by habits and customs. Because of their habits and customs they spend particular amounts of money on different commodities, regardless of whether the particular allocation maximises their satisfaction or not.

(2) For applying this law in actual life and equate the marginal utility of the last rupee spent on different commodities, the consumers must be able to measure the marginal utilities of different commodities in cardinal terms. However, this is easier said than done. It has been said that it is not possible for the consumer to measure utility cardinally. Being a state of feeling and also there being no objective units with which to measure utility, it is cardinally immeasurable. It is because of the immeasurability of utility in cardinal terms that the consumer's behaviour has been explained with the help of ordinal utility by J.R. Hicks and R.G.D. Allen. Ordinal utility analysis involves the use of indifference curves which we shall explain in the next chapter.

(3) Another limitation of the law of equi-marginal utility is found in case of indivisibility of certain goods. Goods are often available in large indivisible units. Because the goods are indivisible, it is not possible to equate the marginal utility of money spent on them. For instance, in allocating money between the purchase of car and foodgrains, marginal utilities of the last rupee spent on them cannot be equated. Car costs about ₹ 200,000 and is indivisible, whereas foodgrains are divisible and money spent on them can be easily varied. Therefore, the marginal utility of rupee varies. Therefore, the marginal utility of rupee obtained from cars cannot be equalised with that obtained from foodgrains. Thus, indivisibility of certain goods is a great obstacle in the way of equalisation of marginal utility of a rupee from different commodities.

DERIVATION OF THE DEMAND CURVE AND THE LAW OF DEMAND

Marshall derived the demand curves for goods from their utility functions. It should be further noted that in his utility analysis of demand Marshall assumed the utility functions of different goods to be independent of each other. In other words, Marshallian technique of deriving demand curves for the goods from their utility functions rests on the hypothesis of *additive utility functions*, that is, utility function of each good consumed by the consumer does not depend on the quantity consumed of any other good. As has already been noted, in case of independent utilities or additive utility functions, the relations of substitution and complementarity between goods are ruled out. Further, in deriving demand curve or law of demand Marshall assumes the marginal utility of money expenditure (MU_m) to remain constant.

We now proceed to derive demand curve from the cardinal utility analysis. Consider the case of a consumer who has a certain given income to spend on a number of goods. According to the law of equi-marginal utility, the consumer is in equilibrium in regard to his purchases of various goods when marginal utilities of the goods are proportional to their prices. Thus, the consumer is in equilibrium when he is buying the quantities of the two goods in such a way that satisfies the following proportionality rule:

$$\frac{MU_x}{P_x} = \frac{MU_y}{P_y} = MU_m$$

where MU_m stands for marginal utility of money income.

With a certain given income for money expenditure the consumer would have a certain *marginal utility of money* (MU_m) *in general*. In order to attain the equilibrium position, according to the above proportionality rule, the consumer will equalise his marginal utility of money (expenditure) with the ratio of the marginal utility and the price of each commodity he buys. It follows therefore that a rational consumer will equalise the marginal utility of money (MU_m) with $\frac{MU_x}{P_x}$ of good X, with

$\frac{MU_y}{P_y}$ of good *Y* and so on. Given *Ceteris Paribus* assumption, suppose the price of good X falls. With the fall in the price of good *X*, the price of good *Y*, consumer's income and tastes remaining unchanged, the equality of the $\frac{MU_x}{P_x}$ with $\frac{MU_y}{P_y}$ and MU_m in general would be disturbed. With the lower price than before $\frac{MU_x}{P_x}$ will be *greater than* $\frac{MU_y}{P_y}$ or MU_m (It is assumed of course that the marginal utility of money does not change as a result of the change in the price of one good). Then in order to restore the equality, marginal utility of *X* or MU_x must be reduced. And the marginal utility of *X* or MU_x can be reduced only by the consumer buying more of the good *X*. It is thus clear from the proportionality rule that as the price of a good falls, its quantity demanded will rise, other things remaining the same. This will make the demand curve downward sloping. How the quantity purchased of a good increases with the fall in its price and also how the demand curve is derived is illustrated in Fig. 8.4.

In the upper portion of the Fig. 8.4, on the *Y*-axis $\frac{MU_x}{P_x}$ is shown and on the *X*-axis is the quantity demanded of good *X* is shown. Given a certain income of the consumer, the marginal utility of money is equal to *OH*. The consumer is buying Oq_1 of good *X* when the price is Px_1 since at the quantity Oq_1 of *X*, the marginal utility of money *OH* is equal to $\frac{MU_x}{P_{x1}}$. Now, when price of good *X* falls, to Px_2 the curve will shift upward to the new position $\frac{MU_x}{P_{x2}}$. In order to equate marginal utility of money (*OH*) with the new $\frac{MU_x}{P_{x2}}$ the consumer increases the quantity demanded to Oq_2. Thus, with the fall in price of good *X* to P_{x2}, the consumer buys more of it. It should be noted that no account is taken of the increase in real income of the consumer as a result of fall in price of good *X*. This is because if the change in real income is taken into account then the marginal utility of money will also change and this would have an effect on

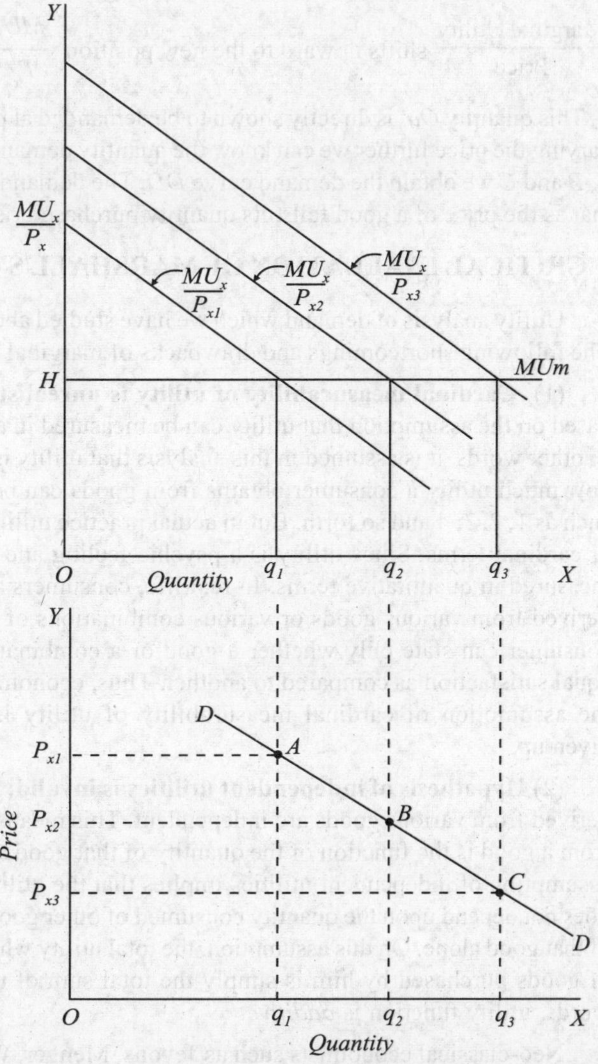

Fig. 8.4. *Derivation of Demand Curve*

the purchases of goods. Marginal utility of money can remain constant in two cases. Firstly, when the elasticity of marginal utility curve (price elasticity of demand) is unity so that even with increase in the purchase of a commodity following the fall in price, the money expenditure made on it remains the same. Second, marginal utility of money will remain approximately constant for small changes in price of unimportant goods, that is, goods which account for negligible part of consumer's budget. In case of these unimportant goods increase in real income following the fall in price is negligible and therefore can be ignored.

At the bottom of Fig. 8.5 the demand curve for *X* is derived. In this lower diagram, price is measured the on *Y*-axis. As in the upper portion, the *X*-axis represents quantity. When the price of good *X* is Px_1, the relevant curve of $\frac{\text{Marginal Utility}}{\text{Price}}$ is $\frac{MU_x}{P_{x_1}}$ which is shown in the upper portion. With $\frac{MU_x}{P_{x1}}$, as explained earlier, he boys Oq_1 of good *X*. Now, in the lower portion this quantity Oq_1 is directly shown to be demanded at the price Px_1. When price of *X* falls to Px_2, the curve of $\frac{\text{Marginal Utility}}{\text{Price}}$ shifts upward to the new position $\frac{MU_x}{P_{x_2}}$. With $\frac{MU_x}{P_{x_2}}$ the consumer buys Oq_2 of *X*. This quantity Oq_2 is directly shown to be demanded at price Px_2 in the lower portion. Similarly, by varying the price further we can know the quantity demanded at other prices. Thus, by joining points *A*, *B* and *C* we obtain the demand curve *DD*. The demand curve *DD* slopes downward which shows that as the price of a good falls, its quantity purchased rises.

CRITICAL EVALUATION OF MARSHALL'S CARDINAL UTILITY ANALYSIS

Utility analysis of demand which we have studied above has been criticised on various grounds. The following shortcomings and drawbacks of marginal utility analysis have been pointed out:

(1) **Cardinal measurability of utility is unrealistic:** Cardinal utility analysis of demand is based on the assumption that utility can be measured in absolute, objective and quantitative terms. In other words, it is assumed in this analysis that utility is cardinally measurable. According to this how much utility a consumer obtains from goods can be expressed or stated in cardinal numbers such as 1, 2, 3, 4 and so forth. But in actual practice utility cannot be measured in such quantitative or cardinal terms. Since utility is a psychic feeling and a subjective thing, it cannot therefore be measured in quantitative terms. In real life, consumers are only able to *compare* the satisfactions derived from various goods or various combinations of the goods. In other words, in the real life consumer can state only whether a good or a combination of goods gives him more, or less, or equal satisfaction as compared to another. Thus, economists like J.R. Hicks are of the opinion that the assumption of cardinal measurability of utility is unrealistic and therefore it should be given up.

(2) **Hypothesis of independent utilities is invalid:** Utility analysis also assumes that utilities derived from various goods are independent. This means that the utility which a consumer derives from a good is the function of the quantity of that good and of that good alone. In other words, the assumption of independent utilities implies that the utility which a consumer obtains from a good does not depend upon the quantity consumed of other goods; it depends upon the quantity purchased of that good alone. On this assumption, the total utility which a person gets from the whole collection of goods purchased by him is simply the total sum of the separate utilities of the good. In other words, utility function is *additive*.

Neo-classical economists such as Jevons, Menger, Walras and Marshall considered that utility functions were additive. But in the real life this is not so. In actual life the utility or satisfaction derived from a good depends upon the availability of some other goods which may be either substitutes for or complementary with each other. For example, the utility derived from a pen depends upon whether ink is available or not. On the contrary, if you have *only tea*, then the utility derived from it would be greater but if along with tea you also have the coffee then the utility of tea to you would be comparatively less. Whereas pen and ink are *complements* with each other, tea and coffee are *substitutes for each other*. It is thus clear that various goods are related to each other in the sense that some are complements with each other and some are substitutes for each other. As a result of this, the utilities derived from various goods are independent, that is, they depend upon each other. Therefore, the utility obtained from a good is not the function of its quantityt alone but also depends upon the existence or consumption of other related goods (complements or substitutes). It is thus evident that the assumption of the independence of utilities by Marshall and other supporters of marginal utility analysis is a great defect and shortcoming of their analysis. As we shall see below, the hypothesis of independent utilities along with the assumption of constant marginal utility of money reduces the validity of Marshallian demand theorem to the one-commodity model only.

(3) **Assumption of constant marginal utility of money is not valid.** An important assumption of cardinal utility analysis is that when a consumer spends varying amount on a good or various goods or when the price of a good changes, the marginal utility of money remains unchanged. But in acutal practice this is not correct. As a consumer spends his money income on the goods, money income left with him declines. With the decline in money income of the consumer as a result of increase in his expenditure on goods, the marginal utility of money to him rises. Further, when the price of a commodity changes, the real income of the consumer also changes. With this change in real income, marginal utility of money will change and this would have an effect on the demand for the good in question, even though the total money income available with the consumer remains the same. But utility analysis ignores all this and does not take cognizance of the changes in real income and its effect on demand for goods following the change in the price of a good.

As has been explained earlier, according to Marshall, utility from a good can be measured in terms of money (that is, how much money a consumer is prepared to sacrifice for a good). But, to be able to measure utility in terms of money marginal utility of money itself should remain constant. Therefore, assumption of constant marginal utility of money is very crucial in Marshallian demand analysis. But in case the consumer has to spread his money income on a number of goods, there is a necessity for revision of marginal utility of money with every change in the price of a good. In other words, in a multi-commodity model marginal utility of money does not remain invariant or constant.

(4) **Marshallian demand theorem cannot genuinely be derived except in a one commodity case.** J.R. Hicks and Tapas Majumdar have further criticised the Marshallian utility analysis on the ground that "Marshallian demand theorem cannot genuinely be derived from the marginal utility hypothesis except in a one-commodity model without contradicting the assumption of constant marginal utility of money".[3] In other words, Marshall's demand theorem and constant marginal utility of money are incompatible except in a one commodity case. As a result, Marshall's demand theorem cannot be validity derived in the case when a consumer spends his money on more than one good. In order to know the truth of this assertion consider a consumer who has a given amount of money income to spend on some goods with given prices. According to utility analysis, consumer will be in equilibrium when he is spending money on goods in such a way that ratios of marginal utilities of various goods to their prices are equal.

Let us assume that in this equilibrium position, consumer is buying q_1 quantity of a good X at a price p_1. Since the consumer is buying q_1 quantity of good X at price p_1, he will be spending $p_1 q_1$ amount of money on it. Suppose that the price of good X rises from p_1 to p_2 and as a result quantity demanded falls from q_1 to q_2 so that the new expenditure will be euqal to $p_2 q_2$. Now, the important

3. Tapas Majumdar, *Measurment of Utility*, p. 55.

thing to see is that whether his new expenditure $p_2 q_2$ on good *X* is equal to, smaller or greater than $p_1 q_1$. This depends upon the elasticity of marginal utility curve (*i.e.* price elasticity of demand). If elasticity of marginal utility curve of good *X* is unity, then the new expenditure on *X* (that is, $p_2 q_2$) after the rise in price of *X* from p_1 to p_2, will be equal to the initial expenditure $p_1 q_1$. When the monetary expenditure made on the good remains constant as a result of change in price, then the Marshallian demand theory is valid. But constant monetary expenditure on a commodity following a price change is only a rare phenomenon.

Thus, in case of more than one good, Marshallian demand theorem cannot be genuinely derived while keeping the marginal utility of money constant. If, in Marshallian demand analysis, this difficulty is avoided "by giving up the assumption of constant marginal utility of money, then money can no longer provide the measuring rod, and we can no longer express the marginal utility of a commodity in units of money[4].

(5) **Cardinal utility analysis does not split up the price effect into substitution and income effects.** The third shortcoming of the cardinal utility analysis is that *it does not distinguish between the income effect and the substitution effect of the price change*. We know that when price of a good falls, the consumer becomes better off than before, that is, a fall in price of a good brings about an increase in real income of the consumer. In other words, if with the fall in price the consumer purchases the same quantity of the good as before, then he would be left with some income. With this extra income he would be in a position to purchase more of this good as well as other goods. This is the income effect of the fall in price on the quantity demanded of the good. Besides, when price of a good falls, it becomes relatively cheaper than other goods and as a result the consumer is induced to substitute that good for others. This results in the increase in the quantity demanded of that good. This is the substitution effect of the price change on the quantity demanded of the good.

Thus, with the fall in price of a good, the quantity demanded of it rises because of income effect and substitution effect. But cardinal utility analysis does not make clear the distinction between the income and the substitution effects of the price hange by assuming the constancy of marginal utility of money. Thus, according to Tapas Majumdar, "the assumption of constant marginal utility of money obscured Marshall's insight into the truly *composite character* of the price-demand relationship."[5] Marshall explained the changes in demand as a result of change in the price of a good on the basis of substitution effect on it. Thus, marginal utility analysis does not tell us about how much quantity demanded increases due to income effect and how much due to substitution effect as a result of the fall in price of a good. J.R. Hicks rightly remarks, "*the distinction between income effect and substitution effect of a price change is accordingly left by the cardinal theory as an empty box which is crying out ot be filled*". [6]

(6) **Marshall could not explain Giffen Paradox.** By not visualising the price effect as a combination of substitution and income effects and ignoring the income effect of the price change, Marshall could not explain Giffen Paradox. He treated it merely as an exception to his law of demand. In contrast to it, indifference curve analysis has been able to explain satisfactorily the Giffen good case. According to indifference curve analysis, in case of Giffen Paradox or Giffen good negative income effect of the price change is more powerful than the substitution effect so that when the price of a of a Giffen good falls the negative income effect outweighs the substitution effect with the result that quantity demanded of it falls. Thus, in case of a *Giffen good quantity demanded varies directly with the price and the Marshall's law of demand does not hold good*. It is because of the constant

4. Tapas Majumdar, *ot.cit. p.*56.
5. Ibid, p. 57.
6. J.R. Hicks, *A Revision of Demand Theory*, Oxford University Press, 1956.

marginal utility of money and therefore the neglect of income effect of the price change that Marshall could not explain why the quantity demanded of the Giffen good falls when its price falls and rises when its price rises. This is a serious lacuna in Marshallian's utility analysis of demand.

Because of the above drawbacks, cardinal utility analysis has been given up in modern economic theory and demand is analysed with indifference curves which we shall explain in the next chapter.

QUESTIONS FOR REVIEW

1. State and explain the law of diminishing marginal utility? How is law of demand related to it?
2. What is meant by consumer equilibrium ? How does a consumer maximise his satisfaction in cardinal utility analysis ?
3. Explain the Law of Equi-marginal utility ? How does it explain consumer equilibrium ?
4. You are given the following marginal utilities of goods *X* and *Y* obtained by a consumer. Given that price of *X* = ₹ 5, price of *Y* = ₹ 2 and income = ₹ 22, find out the optimal combination of goods.

Number of units consumed of a commodity	*MUx (utils)*	*MUy (utils)*
1	30	20
2	25	18
3	20	16
4	15	14
5	10	12
6	5	10
7	1	8

5. On what grounds Marshall's cardinal utility analysis has been criticised ?
6. Distinguish between cardinal utility and ordinal utility. Which is more realistic ?
7. Why assumption of constant marginal utility of money is essential in Marshall's cardinal utility analysis of demand. What argument Marshall gave in defence of this assumption.
8. What is Giffen Paradox ? Why Marshall could not explain it ?
9. Derive the demand curve for a commodity from Marshall's cardinal utility analysis. Explain in terms of cardinal utility theory why demand curve for a commodity slopes downward.
10. Deduce the inverse relationship between the quantity demanded of a commodity and its price in terms of Marshallian marginal utility analysis, indicating the underlying assumptions. How would you explain in terms of this analysis, the phenomenon that a fall in price of salt does not make a consumer buy more of it ? (*D.U.B.A.* (Hons.) 2001)
11. Suppose the *MU* of good *X* is 20, its price is ₹ 4 and the *MU* of good *Y* is 50 and its price is ₹ 5. The individual to whom this information applies is spending ₹ 20 on each good. Is he or she maximising satisfaction ? Why or why not ? [*D.U. B.A. (Hons.) 1997*)]
12. What is value paradox ? How does diminishing marginal utility explain this paradox?

CHAPTER 9
INDIFFERENCE CURVE ANALYSIS OF DEMAND

In the last chapter we explained and critically evaluated Marshall's cardinal utility analysis of demand. A popular alternative theory of consumer's demand is the Indifference Curve Analysis which forms the subject-matter of the present chapter. The technique of indifference curves was first of all invented by a classical economist Edgeworth but he used it only to show the possibilities of exchange between two persons and not to explain consumer's demand.

Two English economists, J.R. Hicks and R.G.D. Allen in their now well-known paper '*A Reconsideration of the Theory of Value*' severely criticized Marshall's cardinal utility analysis based upon cardinal measurement of utility and put forward the indifference curve approach based on the notion of ordinal utility to explain consumer's behaviour. In 1939 Hicks reproduced the indifference curve theory of consumer's demand in his book '*Value and Capital*' modifying somewhat the version of the original paper.

INDIFFERENCE CURVE APPROACH

Indifference curve method has been evolved to supersede the marginal utility analysis of demand which was discussed in the last chapter. The indifference curve method seeks to derive all rules and laws about consumer's demand that are derivable from the cardinal utility analysis. At the same time the inventors and supporters of new method contend that their analysis is based on fewer and more reasonable assumptions. The indifference curve analysis has, however, retained some of the assumptions of Marshall's cardinal utility analysis. Thus, the indifference curve approach, like the old cardinal utility approach, assumes that the *consumer possesses* '*complete information*' about all the relevant aspects of economic environment in which he finds himself. For example, the prices of goods, the markets in which they are available, the satisfaction to be obtained from them etc. are all known to the consumer. Further, it is assumed that the *consumer acts rationally* in the sense that, given the prices of goods and the money income, he will choose the combination from among the various possible combinations that gives him maximum satisfaction. Moreover, the *assumption of* '*continuity*' has also been retained by Hicks-Allen indifference curve method. Continuity assumption means that the consumers are capable of ordering or ranking all *conceivable combinations of goods* according to the satisfaction they yield.

The fundamental approach of indifference curve analysis is that it has abandoned the concept of cardinal utility and instead has *adopted the concept of ordinal utility*. According to the supporters of the indifference curves theory, utility is a psychic entity and it cannot therefore be measured in quantitative cardinal terms. In other words, utility being a psychological feeling is not quantifiable. The concept of cardinal utility, according to the exponents of the indifference curve theory, is therefore untenable. On the other hand, the assumption of ordinal utility, according to them, is quite reasonable and realistic. *The ordinal utility implies that the consumer is capable of simply 'comparing the different levels of satisfaction'*. In other words, according to the ordinal utility hypothesis, while the consumer may not be able to indicate the exact amounts of utilities that he derives from commodities or any combination of them, but he is capable of judging whether the satisfaction obtained from a good or a combination of goods is equal to, lower than, or higher than another.

Notions of Indifference and Preference

For deriving the theory of consumer's behaviour, it is sufficient to assume that the consumer is able to rank his preferences consistently. Thus, the basis of indifference curve analysis of demand is the *preference-indifference hypothesis*. This means that if the consumer is presented with a number of various combinations of goods, he can order or rank them in '*scale of preferences*'. If the various combinations are marked *A*, *B*, *C*, *D*, *E*, etc. the consumer can tell whether he prefers *A* to *B*, or *B* to *A*, or is indifferent between them. Similarly, he can indicate his preference or indifference between any other pair of combinations. *The concept of ordinal utility implies that the consumer cannot go beyond stating his preference or indifference*. In other words, if a consumer happens to prefer *A* to *B*, he cannot tell by how much he prefers *A* to *B*. Thus, under ordinal utility hypothesis, the consumer cannot tell the '*quantitative differences*' between various levels of satisfaction; he can simply compare them 'qualitatively', that is, he can merely judge whether one level of satisfaction is higher than, lower than or equal to another. Further, according to the supporters of indifference-curve method, by 'how much' one combination of goods is preferred to another is not even needed for deriving laws concerning consumer's behaviour. It is sufficient to assume that the consumer is able to tell whether one combination of goods gives him greater, equal, or less satisfaction than another.

It may be noted that the *consumer formulates his scale of preferences independently of the market prices of goods keeping in view only the satisfaction which he hopes to get from various combinations of goods*. In consumer's scale of preferences some combinations will occupy the same place, *i.e.*, the consumer will be indifferent among them. Combinations occupying a higher place in the scale will be preferred to the combinations occupying lower places in the scale. Moreover, the indifference curve analysis assumes that the preference and indifference relations are '*transitive*'. The transitivity of preferences or indifference relations means that if a consumer prefers *A* to *B*, and *B* to *C*, then he will also prefer *A* to *C* and likewise, if he is indifferent between *A* and *B*, and between *B* and *C*, then he will also be indifferent between *A* and *C*.

WHAT ARE INDIFFERENCE CURVES?

The basic tool of Hicks-Allen ordinal analysis of demand is the indifference curve which represents all those combinations of goods which give same satisfaction to the consumer. Since all the combinations on an indifference curve give equal satisfaction to the consumer, he will be indifferent between them, that is, it will not matter to him which one he gets. In other words, all combinations of the goods lying on a consumer's indifference curve are equally desirable to or equally preferred by him. To understand indifference curves, it is better to start with indifference schedules. In Table 9.1, two indifference schedules are given. In each schedule the amounts of goods *X* and *Y* in each combination are so much that the consumer is indifferent among the combinations in each schedule. In schedule 1, the consumer has to start with 1 unit of *X* and 12 units of *Y*. Now, the consumer is asked to tell how much of good *Y* he will be willing to give up for the gain of an additional unit of *X* so that his level of satisfaction remains the same. If the gain of one unit of *X* compensates him fully for the loss of 4 units of *Y*, then the next combination of 2 units of *X* and 8 units of *Y* ($2X + 8Y$) will give him as much satisfaction as the initial combination ($1X + 12Y$). Similarly, by asking the consumer further how much of *Y* he will be prepared to forgo for successive increments in his stock of *X* so that his level of satisfaction remains unaltered, we get combinations $3X + 5Y$, $4X + 3Y$, and $5X + 2Y$, each of which provides same satisfaction as combination $1X + 12Y$ or $2X + 8Y$. Since his satisfaction is the same whichever combination of goods in the schedule is offered to him, he will be indifferent among the combinations of two goods included in the schedule.

Table 9.1. Two Indifference Schedules

I		*II*	
Good *X*	Good *Y*	Good *X*	Good *Y*
1	12	2	14
2	8	3	10
3	5	4	7
4	3	5	5
5	2	6	4

In schedule II, the consumer has initially 2 units of *X* and 14 units of *Y*. By asking the consumer how much of *Y* he will be prepared to abandon for the successive additions of *X* in his stock so that his satisfaction remains equal to what he derives from the initial combination ($2X + 14Y$), we get combinations $3X + 10Y$, $4X + 7Y$, $5X + 5Y$ and $6X + 4Y$. Thus, each of the combinations in schedule II will be equally desirable to the consumer and he will be indifferent among them. But it should be borne in mind that the consumer will prefer any combination in schedule II to any combination in schedule I. That is, any combination in schedule II will give him more satisfaction than any combination in shcedule I. This is because it is assumed that more of a commodity is preferable to less of it (in other words, the greater quantity of a good gives an individual more satisfaction than the smaller quantity of it), the quantities of other goods with him remaining the same. Initial combination in schedule II contains more of both the goods than the initial combination in schedule I, therefore the former will give greater satisfaction to the consumer than the latter. Now, since each of the other combinations in indifference schedule II provides the same satisfaction as the initial combination ($2X + 14Y$) of the schedule and also each of other combinations in indifference schedule I gives the same satisfaction as the initial combination ($1X + 12Y$), any combination of the schedule II will be preferred to (will yield greater satisfaction than) any combination of schedule I.

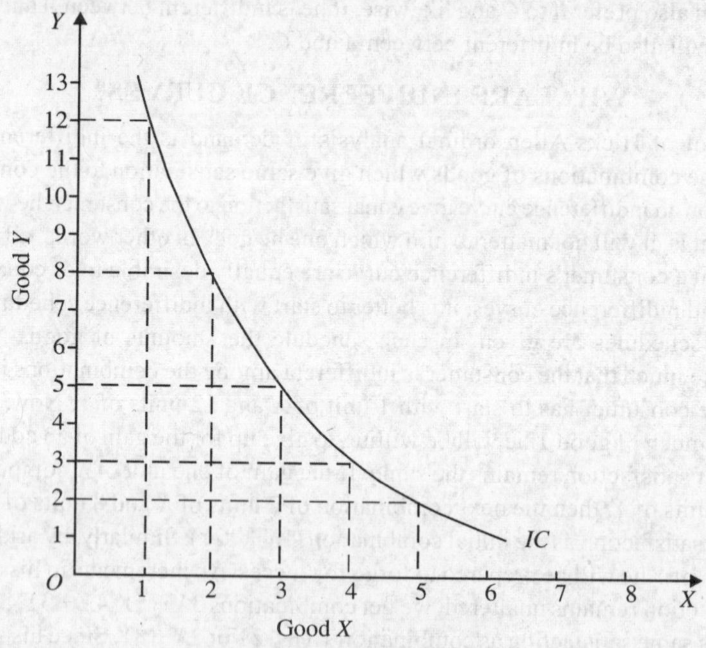

Fig. 9.1. *An Indifference Curve.*

Now, we can convert the indifference schedules into indifference curves by plotting the various combinations on a graph paper. In Fig. 9.1 an indifference curve *IC* is drawn by plotting the various combinations of the indifference schedule I. The quantity of good X is measured on the horizontal axis, and the quantity of the good *Y* is measured on the vertical axis. As in an indifference schedule, combinations lying on an indifference curve will also be equally desirable to the consumer, that is, will give him the same satisfaction. The smoothness and continuity of an indifference curve means that goods in question are assumed to be perfectly divisible. If the indifference schedule *II* is also converted into indifference curve, this will lie above the indifference curve *IC*.

Any combination on a higher indifference curve will be preferred to any combination on a lower indifference curve. It is thus clear that the indifference curve lying above and to the right of an indifference curve will indicate higher level of satisfaction than the latter. It may be noted that while an indifference curve shows all those combinations of two goods which provide equal satisfaction to the consumer, it does not indicate *exactly how much* satisfaction is derived by the consumer from those combinations. This is because the concept of ordinal utility does not involve the quantitative measurability of utility. Therefore, no attempt is made to label an indifference curve by the quantity or amount of satisfaction it represents.

Indifference Map

A complete description of consumer's tastes and preferences can be represented by an *indifference map* which consists of a set of indifference curves. Because the field in a two-dimensional diagram contains an infinite number of points each representing a combination of goods *X* and *Y*, there will be an infinite number of the indifference curves each passing through combinations of goods that are equally desirable to the consumer. In Fig. 9.2 an indifference map of a consumer is shown which consists of five indifference curves. The consumer regards all combinations on the indifference curve *I* as giving him equal satisfactions. Similarly, all the combinations lying on indifference curve *II* provide the same satisfaction but the level of satisfaction on indifference curve *II*, for the reason already explained, will be greater than the level of satisfaction on indifference curve I. Likewise, all the higher indifference curves, *III*, *IV* and *V* represent progressively higher and higher levels of satisfaction. It is importnat, to remember that while the consumer will prefer any combination on a higher indifference curve to any combination on a lower indifference curve, but by *how much he prefers* one combination to another cannot be said. In other words, a higher indifference curve represents a higher level of satisfaction than a lower indifference curve but "*how much higher*" cannot be indicated. This is because the indifference curve system is based upon the concept of ordinal utility according to which the consumer is able to sate only the '*qualitative*' differences in his various levels of satisfaction. It is not possible for the consumer to specify 'quantitative' differences in his various levels of satisfaction (*i.e.*, by how much more or by how much less cannot be stated by him). Therefore, in an indifference map successively higher indifference curves can be denoted by any ascending series, 1, 3, 7, 9...; or 1, 4, 6, 8, 13...; or 1, 2, 5, 8, 10...; etc., the magnitude of these various numbers and the quantitative differences among them having no relevance. It is more usual to label the indifference curves by ordinal numbers as *I*, *II*, *III*, *IV*, *V* as is done in Fig. 9.2.

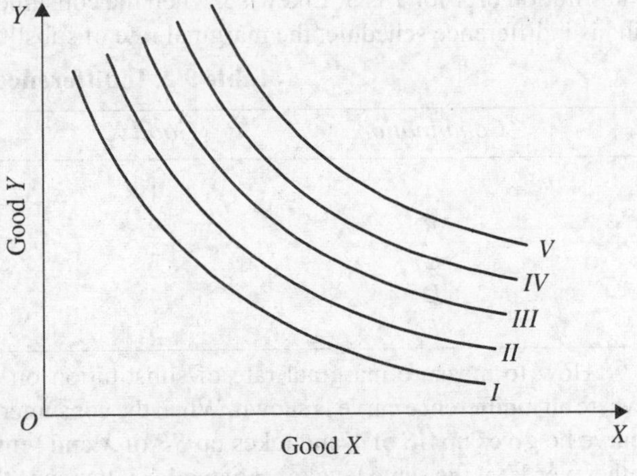

Fig. 9.2. *Indifference Map*

An indifference map of a consumer represents, as said earlier, his tastes and preferences for the two goods and his preferences between different combinations of them. In other words, *an indifference map portrays consumer's scale of preferences*. Scale of preferences of indifference curve analysis replaces Marshall's utility schedule. So long as consumer's tastes and preferences remain unchanged, the whole indifference map will remain the same. If the consumer's tastes and preferences undergo a change, then a new indifference map corresponding to new tastes and preferences will have to be drawn. If, for instance, good *Y* is eggs and good *X* is bread, and if the doctor advises our consumer to take more of eggs to overcome some diseases, the shapes of all his indifference curves will change and his indifference map will have to be redrawn. Since the doctor's advice will intensify our consumer's desire for eggs, now a smaller quantity of eggs than before will be given up by him for a given increment in bread.

MARGINAL RATE OF SUBSTITUTION

The concept of marginal rate of substitution is an important tool of indifference curve analysis of demand. The rate at which the consumer is prepared to exchange goods *X* and *Y* is known as marginal rate of substitution. In our indifference schedule I above, which is reproduced in Table 9.2, in the beginning the consumer gives up 4 units of *Y* for the gain of one additional unit of *X* and in this process his level of satisfaction remains the same. It follows that one unit gain in *X* fully compensates him for the loss of 4 units of *Y*. It means that at this stage he is prepared to exchange 4 units of *Y* for one unit of *X*. Therefore, at this stage consumer's marginal rate of substitution of *X* for *Y* is 4. Thus, *we may define the marginal rate of substitution of X for Y as the amount of Y whose loss can just be compensated by one unit gain in X*. In other words, marginal rate of substitution of *X* for *Y* represents the amount of *Y* which the consumer has to give up for the gain of one additional unit of *X so that his level of satisfaction remains the same*.

In Table 9.2, when the consumer moves from combination *B* to combination *C* on his indifference schedule he forgoes 3 units of *Y* for additional one unit gain in *X*. Hence, the marginal rate of substitution of *X* for *Y* is 3. Likewise, when the consumer moves from *C* to *D*, and then from *D* to *E* in his indifference schedule, the marginal rate of substitution of *X* for *Y* is 2 and 1 respectively.

Table 9.2. Indifference Schedule

Combination	*Good X*	*Good Y*	MRS_{xy}
A	1	12	
			4
B	2	8	
			3
C	3	5	
			2
D	4	3	
			1
E	5	2	

How to measure marginal rate of substitution on an indifference curve ? Consider Fig. 9.3 where an indifference curve is shown. When the consumer moves from point *A* to *B* on this indifference curve he gives up *AS* of *Y* and takes up *SB* of *X* and remains on the same indifference curve (or, in other words, at the same level of satisfaction). It means that the loss of satisfaction caused by giving up *AS* of *Y* equals the gain in satisfaction due to the increase in good *X* by *SB*. It follows that the consumer is prepared to exchange *AS* of *Y* for *SB* increase in *X*. In other words, marginal rate of substitution of *X* for *Y* (MRS_{xy}) is equal to $\frac{AS}{SB}$. Now, a small change in the amount *Y* such as *AS*, along an indifference curve can be written as ΔY and the change in the amount of *X* as ΔX. Thus, ΔY shows the amount which the consumer has to give up for the ΔX increase in *X* if he is to remain on the same indifference curve. Therefore, it follows that:

$$\text{Marginal rate of substitution of } X \text{ for } Y\ (MRS_{xy}) = \frac{AS}{SB} = \frac{\Delta Y}{\Delta X}$$

Now, suppose that points *A* and *B* are very close to each other so that it can be assumed that both of them lie on the same tangent $t\,T$. Now, in a right-angled traingle ASB, $\frac{AS}{SB}$ is equal to the tangent of the angle ABS. It therefore follows that:

$$MRS_{xy} = \frac{AS}{SB} = \frac{\Delta Y}{\Delta X} = \text{ tangent of } \angle ABS$$

But in Fig. 9.3 $\angle ABS = tTO$

Hence MRS_{xy} = tangent of $\angle tTO$

But the tangent of $\angle tTO$ is equal to $\frac{Ot}{OT}$

The tangent of $\angle tTO$ indicates the slope of the tangent line tT drawn at point A or B on the indifference curve. In other words, the slope of the indifference curve at point A or B is equal to the tangent of $\angle tTO$. It therefore follows:

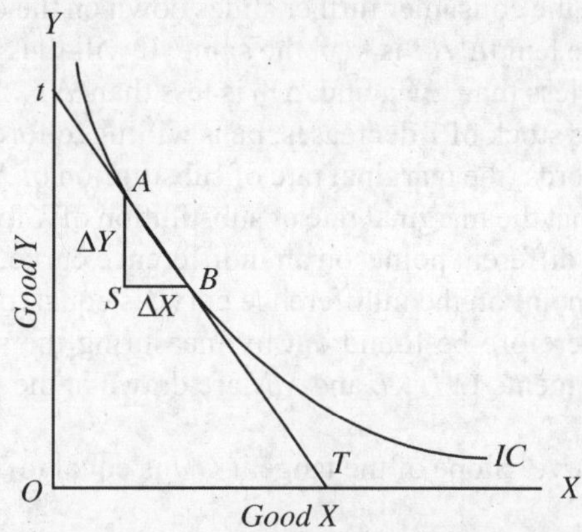

Fig. 9.3. *Graphic Illustration of the Concept of Marginal Rate of Substitution*

$$MRS_{xy} = \text{tangent of } \angle tTO = \text{slope of the indifference curve on } A \text{ or } B = \frac{Ot}{OT}$$

It is thus clear from above that if we have to find out the MRS_{xy} at a point on the indifference curve we can do so by drawing tangent at the point on the indifference curve and then measuring the slope by estimating the value of the tangent of the angle which the tangent line makes with the X-axis.

Principle of Diminishing Marginal Rate of Substitution

An important principle of economic theory is that marginal rate of substitution of X for Y diminishes as more and more of good X is substituted for good Y. In other words, as the consumer has more and more of good X, he is prepared to forego less and less of good Y. The principle of diminishing marginal rate of substitution is illustrated in Fig. 9.4. In Fig. 9.4 (a) when the consumer slides down from A to B on the indifference curve he gives up ΔY_1 of good Y for the compensating gain of ΔX of good X. Therefore, the marginal rate of substitution (MRS_{xy}) is here equal to $\frac{\Delta Y_1}{\Delta X}$. But

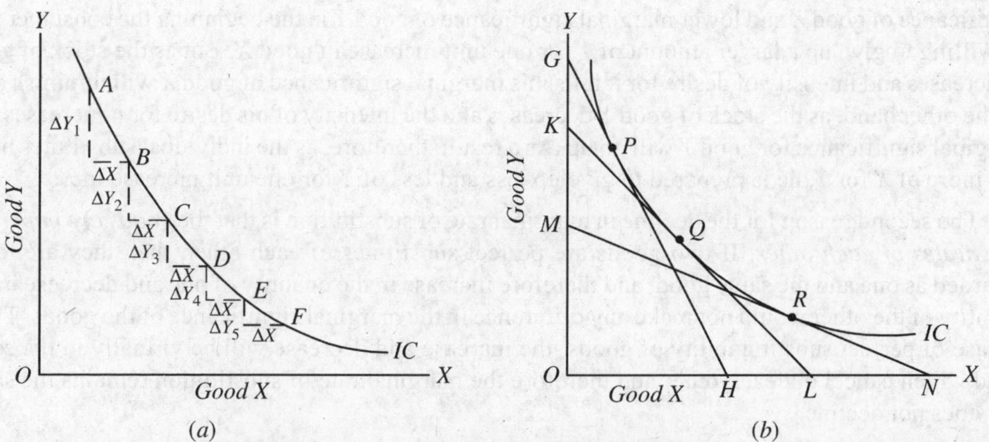

Fig. 9.4. *Diminishing Marginal Rate of Substitution*

as the consumer further slides down on the curve, the length ΔY becomes shorter and shorter, while the length ΔX is kept the same. It will thus be seen from Fig. 9.4 (*a*) that ΔY_2 is less than ΔY_1; ΔY_3 is less than ΔY_2; and ΔY_4 is less than ΔY_3. It means that as the consumer's stock of *X* increases and his stock of *Y* decreases, he is willing to forego less and less of *Y* for a given increment in *X*. In other words, the marginal rate of substitution of *X* for *Y* falls as the consumer has more of *X* and less of *Y*. That the marginal rate of substitution of *X* for *Y* diminishes can also be known from drawing tangents at different points on an indifference curve. As explained above, the marginal rate of substitution at a point on the indifference curve is equal to the slope of the indifference curve at that point and can therefore be found out by measuring the slope of tangent drawn at a point. In Fig. 9.4 (*b*) three tangents *GH*, *KL* and *MN* are drawn at the points *P*, *Q* and *R* respectively to the given indifference curve. Slope of the tangent *GH* is equal to $\frac{OG}{OH}$. Hence, the marginal rate of substitution of *X* for *Y* at point *P* is equal to $\frac{OG}{OH}$. Likewise, the marginal rate of substitution at point *Q* is equal to $\frac{OK}{OL}$ and at point *R* it is equal to $\frac{OM}{ON}$. It will be noticed that $\frac{OK}{OL}$ is smaller than $\frac{OG}{OH}$ and $\frac{OM}{ON}$ is smaller than $\frac{OK}{OL}$. It follows that MRS_{xy} diminishes as the consumer slides down on his indifference curve.

That the marginal rate of substitution falls is also evident from the Table 9.2. In the beginning the marginal rate of substitution of *X* for *Y* is 4 and as more and more of *X* is obtained and less and less of *Y* is left, the MRS_{xy} keeps on falling. Between *B* and *C* it is 3; between *C* and *D*, it is 2; and finally between *D* and *E*, it is 1.

Now, the question is what accounts for the diminishing marginal rate of substitution. In other words, why is it that the consumer is willing to give up less and less of *Y* for a given increment in *X* as he slides down on the curve? The following three factors are responsible for diminishing marginal rate of substitution.

First, *the want for a particular good is satiable* so that as the consumer has more and more of a good the intensity of his want for that good goes on declining. It is because of this fall in the intensity of want for a good, say *X*, that when its stock increases with the consumer, he is prepared to forego less and less of good *Y* for every increment in *X*. In the beginning, when the consumer's stock of good *Y* is relatively large and his stock of good *X* is relatively small, consumer's marginal significance for good *Y* is low, while his marginal significance for good *X* is high. Owing to higher marginal significance of good *X* and lower marginal significance of good *Y* in the beginning the consumer will be willing to give up a larger amount of *Y* for one unit increase in good *X*. But as the stock of good *X* increases and intensity of desire for it falls, his marginal significance of good *X* will diminish and, on the other hand, as the stock of good *Y* decreases and the intensity of his desire for it increases, his marginal significance for good *Y* will go up. As a result, therefore, as the individual substitutes more and more of *X* for *Y*, he is prepared to give up less and less of *Y* for one unit increase in *X*.

The second reason for the decline in marginal rate of substitution is that the *goods are imperfect substitutes of each other*. If two goods are perfect substitutes of each other, then they are to be regarded as one and the same good, and therefore increase in the quantity of one and decrease in the quantity of the other would not make any difference in the marginal significance of the goods. Thus, in case of perfect substitutability of goods, the increase and decrease will be virtually in the same good which cancel out each other and therefore the marginal rate of substitution remains the same and does not decline.

Assumptions of Indifference Curves

We now turn to discuss the properties or attributes which the indifference curves normally possess. It will be userful if we first mention the assumptions about the behaviour of the consumer, which are generally made in indifference curve analysis.

1. More of a commodity is better than less. It is assumed that the consumer will always prefer a larger amount of a good to a smaller amount of that good, provided that the other goods at his disposal remains unchanged. This is a very reasonable and realistic assumption. This assumption implies that the consumer is not over-supplied with any good. When a consumer is over-supplied or over-satiated with one good, he will prefer a smaller quantity of that good to its larger quantity. It is thus assumed that the consumer has not yet reached the point of satiety in the consumption of any good. This assumption is therefore known as *non-satiety assumption.*

2. Preferences or indifferences of a consumer are transitive. Suppose there are three combinations of two goods : *A*, *B* and *C*. If the consumer is indifferent between *A* and *B* and also between *B* and *C*, it is then assumed that he will be indifferent between *A* and *C* too. This condition implies that consumer's tastes are quite consistent. This assumption is known as *assumption of transitivity*.

3. Diminishing marginal rate of substitution. In indifference curve analysis the principle of diminishing marginal rate of substitution is assumed. In other words, it is assumed that as more and more units of *X* are substituted for *Y*, the consumer will be willing to give up fewer and fewer units of *Y* for each additional unit of *X*, or when more and more of *Y* is substituted for *X*, he will be willing to give up successively fewer and fewer units of *X* for each additional unti of *Y*. This rule about consumer's behaviour is described as the principle of diminishing, marginal rate of substitution. As seen above, this principle follows as a matter of logical necessity from the assumption that particular wants are satiable and that various goods are not perfect substitutes for one another.

PROPERTIES OF INDIFFERENCE CURVES

We now proceed to deduce from the above-mentioned assumptions the important properties of typical indifference curves.

Property I. Indifference curves slope downward to the right.

This property implies that an indifference curve has a negative slope. This property follows from assumption I. Indifference curve being downward sloping means that when the amount of one good in the combination is increased, the amount of the other good is reduced. This must be so if the level of satisfaction is to remain the same on an indifference curve. If, for instance, the amount of good *X* is increased in the combination, while the amount of good *Y* remains unchanged, the new combination will be preferable to the original one and the two combinations will not therefore lie on the same indifference curve.

A little reflection will make it clear that an indifference curve on which those combinations of two goods lie that yield the same satisfaction to the consumer cannot take a shape other than downward-sloping to the right. If the indifference curve had the shape of a horizontal straight line (parallel to the *X*-axis), as in Fig. 9.5, that would mean as the amount of the good *X* was increased, while the amount of good *Y* remained the same, the consumer would remain indifferent as between various combinations. But this

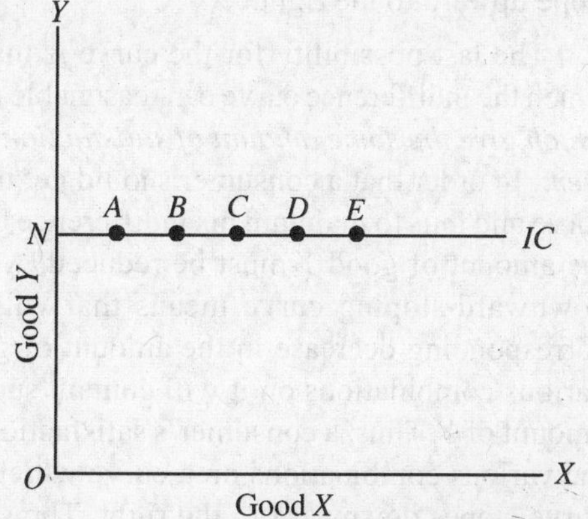

Fig. 9.5. *Indifference curve cannot be a horizontal straight line.*

cannot be so if our assumption I is to hold good. According to assumption I, the consumer always prefers a larger amount of a commodity to the smaller amount of it, other things being given. In Fig. 9.5 in various combinations such as *A*, *B*, *C* and *D* on indifference curve *IC* while the amount of good *X* is successively larger, the amount of good *Y* remains unchanged (= *ON*). If the consumer is to prefer a larger amount of good *X* to a smaller amount, then how he can be indifferent between combinations *A*, *B*, *C*, and *D* etc. It is thus concluded that indifference curve cannot be a horizontal straight line.

Likewise, indifference curve cannot be a vertical straight line becaue a vertical straight line would mean that while the amount of good *Y* in the combination increases, the amount of good *X* remains the same. Thus, in Fig. 9.6, a vertical straight line *IC* is drawn on which are shown combinations A, *B*, *C*, *D* and *E*. While all these combinations contain the same amount of *X*, the amount of *Y* is successively larger. Thus, combinations *A*, *B*, *C*, *D*, etc. would not yield the same amount of satisfaction to the consumer and therefore cannot be the points of an indifference curve. We therefore conclude that indifference curve cannot be a vertical straight line too.

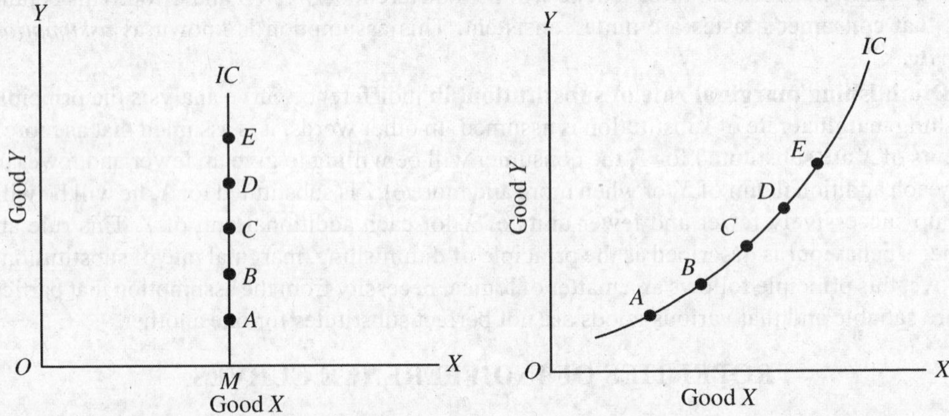

Fig. 9.6. *Indifference curve cannot be a vertical line.* **Fig. 9.7.** *Indifference curve cannot slope upward.*

A third possibility for a curve is to slope upward to the right as in Fig. 9.7. But indifference curve cannot be of this shape too. Upward-sloping curve means that the amounts of both the goods increase as one moves to the right along the curve. If the indifference curve were upward sloping to the right, it would mean that a combination which contains more of both the goods gave the same satisfaction to the consumer as the combination which had smaller amounts of both the goods. This is clearly invalid in view of our assumption I. It follows therefore that indifference curve cannot slope upward to the right.

The last possibility for the curve is that it slopes downward to the right and this is the shape which the indifference curve can reasonable take. *An indifference curve represents those combinations which give the same amount of satisfaction to the consumer and he is therefore indifferent between them.* In order that a consumer should get the same satisfaction from the various combinations of a curve and thus to maintain his indifference between them, then as the amount of good *X* is increased, the amount of good *Y* must be reduced. And this is what a downward-sloping curve indicates. A downward-sloping curve means that with every increase in the amount of good *X*, there is corresponding decrease in the amount of good *Y*. As one moves to the right on such a curve, the various combinations on it will contain successively larger amount of *X*, but successively smaller amount of *Y*. Thus, a consumer's satisfaction can remain the same and he can be indifferent between the various combinations on a curve which slopes downward. We thus conclude that indifference curve slopes downward to the right. The slope of the indifference curve at its various points will depend upon how much of good *Y* the consumer is willing to give up for an additional unit of good *X*.

Property II: Indifference curves are convex to the origin.

Another important property of indifference curves is that they are usually convex to the origin. In other words, the indifference curve is relatively flatter in its right-hand portion and relatively steeper in its left-hand portion. This property of indifference curves follows from assumption 3, which is that the marginal rate of substitution of *X* for *Y* (MRS_{xy}) diminishes as more and more of *X* is substituted for *Y*. Only a convex indifference curve can mean a diminishing marginal rate of substitution of *X* for *Y*. If indifference curve was concave to the origin it would imply that the marginal rate of substitution of *X* for *Y* increased as more and more of *X* was substituted for *Y*, as shown in Fig. 9.8 (*a*). It will be clear from Fig. 9.8 (*a*) that as more and more of *X* is acquired for each extra unit of *X*, the consumer is willing to part with more and more of *Y*, that is, MRS_{xy} increases as more and more of *X* is substituted for *Y*. That the concave indifference curve shows increasing MRS_{xy} is also evident from Fig. 9.8 (*b*). We know the slope at a point on an indifference curve shows the marginal rate of substitution of *X* for *Y* (MRS_{xy}) at that point. It will be seen that the slope at the point *E* on the indifference curve *IC* is greater than at point *A* (the tangent at *E* is steeper than the tangent at *A*). It therefore follows that MRS_{xy} is greater at *E* than at *A* on the indifference curve *IC*. In other words, as more of *X* has been substituted for *Y*, MRS_{xy} has increased.

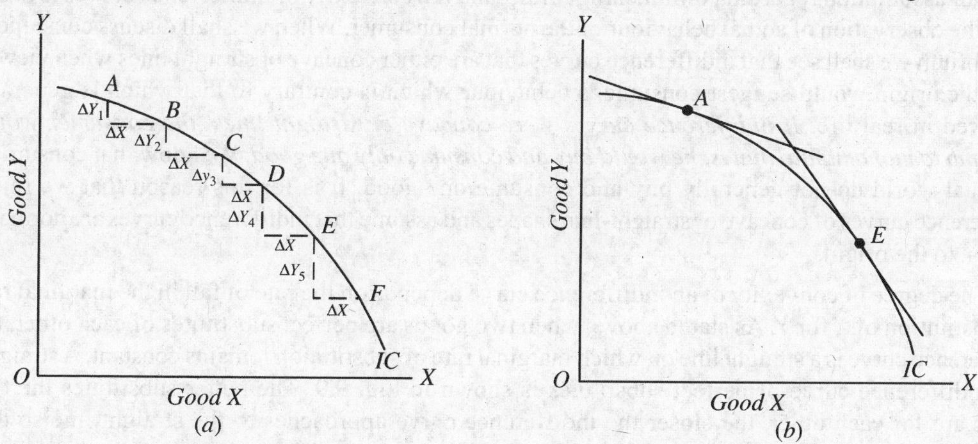

Fig. 9.8. *Indifference curve generally cannot be concave to the origin.*

The above Fig. 9.8 (*a*) and Fig. 9.8 (*b*) have been drawn to show that if indifference curves were concave to the origin, it would mean that the marginal rate of substitution of *X* for *Y increased* as the consumer obtained more and more of *X* in place of *Y*. But this clearly violates our fundamental assumption about the consumer's behaviour (assumption 3) which states that MRS_{xy} declines as consumer substitutes more and more of *X* for *Y*. If the principle of diminishing marginal rate of substitution is valid, then the indifference curve cannot be concave to the origin.

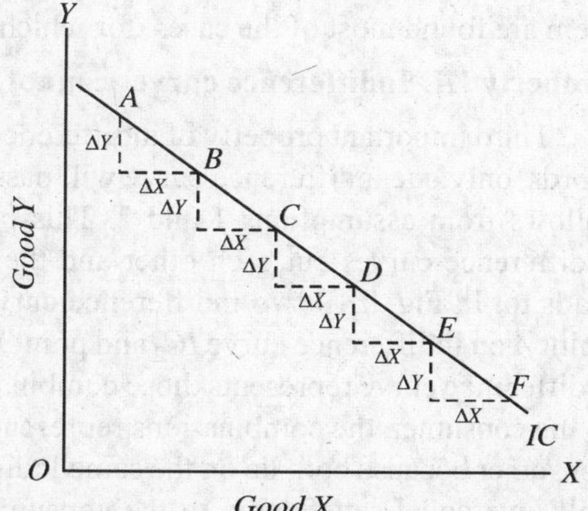

Fig. 9.9. *Except in case of perfect substitutes an indifference curve cannot be a straight line.*

Likewise, indifference curve cannot be a straight-line, except when goods are perfect substitutes. A straight-line indifference curve would mean that MRS_{xy} remains constant as more units of *X* are acquired in place of *Y*. As shown in Fig. 9.9 that on a straight-line indifference curve the amount of *Y* which the consumer is willing to give up for each additional unit of *X* remains the same as more and more of X is substituted for Y, that is, MRS_{xy} remains constant. Since MRS_{xy} is equal to the slope of indifference curve at a point on it, and because a straight-line has the same slope throughout, therefore the straight-line indifference curve will mean the same MRS_{xy} throughout. But, as said above, MRS_{xy} cannot remain constant except when goods happen to be perfect substitutes. Normal consumer's behaviour reveals that when goods are less than perfect substitutes MRS_{xy} usually falls as more of good X is substituted for Y (Assumption 3). It is therefore concluded that indifference curve cannot normally be a straight line.

The third possibility for indifference curve in this regard is that it may be convex to the origin and this is the shape which indifference curves normally possess. This is so because indifference curves being convex to the origin are consistent with the principle of diminishing marginal rate of substitution of X for Y. As shown in Fig. 9.4, when the indifference curve is convex to the origin, MRS_{xy} diminishes as more of X is substituted for Y. We therefore conclude that indifference curves are generally convex to the origin.

Our assumption regarding diminishing MRS_{xy} and the convexity of indifference curves is based upon the observation of actual behaviour of the normal consumer. When we shall discuss consumer's equilibrium we shall see that indifference curves that are either concave or straight lines when viewed from the origin would suggest consumer's behaviour which is contrary to that which is generally observed in real life. *If indifference curves were concave or straight lines, the consumer would succumb to monomania, that is, he would buy and consume only one good.* We know that consumers in actual world do not generally buy and consume one good. It is for this reason that we reject indifference curves of concave or straight-line shapes and assume that indifference curves are normally convex to the origin.

The degree of convexity of an indifference curve depends on the rate of fall in the marginal rate of substitution of X for Y. As stated above, when two goods are perfect substitutes of each other, the indifference curve is a straight line on which marginal rate of substitution remains constant. A straight-line indifference curve of perfect substitutes is shown in Fig. 9.9. The better substitutes the two goods are for each other, the closer the indifference curve approaches to the straight line so that when the two goods are perfect substitutes, the indifference curve is a straight line.

The greater the fall in marginal rate of substitution, the greater the convexity of the indifference curve. The less the ease with which two goods can be substituted for each other, the greater will be the fall in the marginal rate of substitution. In the extreme case when two goods cannot be substituted for each other and are used in a given fixed ratio the indifference curve will be of right-angle shape. Perfect substitutes and perfect complements stand at opposite ends of the substitution scale. Between them are found most of the cases, for which indifference curves are convex to the origin.

Property III: Indifference curves cannot intersect each other.

Third important property of indifference curves is that they cannot intersect each other. In other words, only one indifference curve will pass through a point in the indifference map. This property follows from assumptions 1 and 2. This property can be easily proved by first making the two indifference curves cut each other and then showing the absurdity or self-contradictory result it leads to. In Fig. 9.10 two indifference curves are shown cutting each other at point C. Now take point A on indifference curve IC_2 and point B on indifference curve IC_1 vertically below A. Since an indifference curve represents those combinations of two commodities which give equal satisfaction to the consumer, the combinations represented by points A and C will give equal satisfaction to the consumer because both lie on the same indifference curve IC_2. Likewise, the combinations B and C will give equal satisfaction to the consumer; both being on the same indifference curve IC_1. If combination A is equal to combination C in terms of satisfaction, and combination B is equal to combination C, it follows that the combination A will be equivalent to B in terms of satisfaction. But a glance at Fig. 9.10 will show that this is absurd conclusion since combination A contains more of good *Y* than combination *B*, while the amount of good *X* is the same in both the combinations. Thus, the consumer will definitely prefer *A* to *B*, that is, *A* will give more satisfaction to the consumer than *B* (Assumption 1). But the two indiffference curves cutting each other lead us to an absurd conclusion of *A* being equal to *B* in terms of satisfaction. We therefore conclude that indifference curves cannot cut each other.

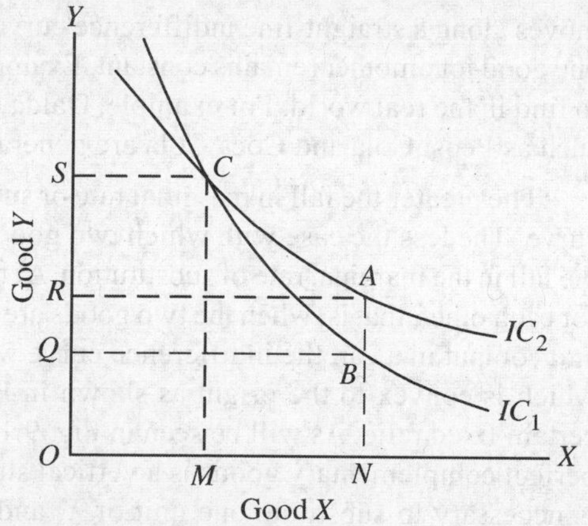

Fig. 9.10. *No two indifference curves can cut each other.*

Another point which is worth mentioning in this regard is that indifference curves cannot even meet or touch each other or be tangent to each other at a point. The meeting of two indifference curves at a point, will also lead us to an absurd conclusion. The same argument holds good as developed above in the case of intersection of indifference curves.

Property IV: A higher indifference curve represents a higher level of satisfaction than a lower indifference curve.

The last property of indifference curve is that a higher indifference curve will represent a higher level of satisfaction than a lower indifference curve. In other words, the combinations which lie on a higher indifference curve will be preferred to the combinations which lie on a lower indifference curve. Consider indifference curves IC_1 and IC_2 in Fig. 9.11. IC_2 is a higher indifference curve than IC_1. Combination *Q* has been taken on a higher indifference curve IC_2 and combination S on a lower indifference curve IC_1. Combination *Q* on the higher indifference curve IC_2 will give a consumer more satisfaction than combination *S* on the lower indifference curve IC_1 because the combination *Q* contains more of both goods *X* and *Y* than the combination *S* (Assumption *I*). Hence by assumption I, the consumer must prefer *Q* to *S*. And by transitivity assumption *II*, he will prefer any other combination on IC_2 (all of which are indifferent with *Q*) to any combination on IC_1 (all of which are indifferent with *S*). We, therefore, conclude that a higher indifference curve represents a higher level of satisfaction and combinations on it will be preferred to the combinations on a lower indifference curve.

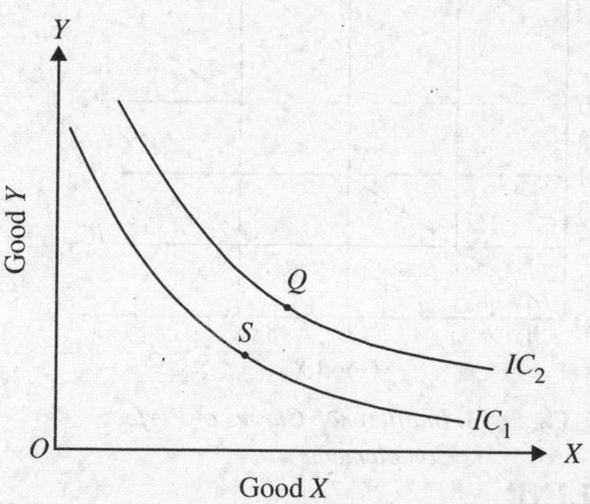

Fig. 9.11. *A higher indifference curve shows a higher level of satisfaction.*

Indifference Curves of Perfect Substitutes and Perfect Complements

The degree of convexity of an indifference curve depends upon the rate of fall in the marginal rate of substitution of *X* for *Y*. As stated above, when two goods are perfect substitutes of each other, the indifference curve is a straight line on which marginal rate of substitution remains constant. Straight-line indifference curves of perfect substitutes are shown in Fig. 9.12. The better substitutes the two goods are for each other, the closer the indifference curve approaches to the straight-line so that when the two goods are perfect substitutes, the indifference curve is a straight line. In case of perfect substitutes, the indifference curves are parallel straight lines because the consumer equally prefers the two goods and is welling to exchange one good for the other at a constant rate. As one moves along a straight-line indifference curve of perfect substitutes, marginal rate of substitution of one good for amother remains constant. Examples of goods that are perfect substitutes are not difficult to find in the real world. For example, Dalda and Rath Vanaspati, two different brands of cold drink such as Pepsi Cola and Coca Cola are generally considered to be perfect substitutes of each other.

The greater the fall in marginal rate or substitution, the greater the convexity of the indifference curve. The less the ease with which two goods can be substituted for each other, the greater will be the fall in the marginal rate of substitution. At the extreme, when two goods cannot at all be substituted for each other, that is, when the two goods are perfect complementary goods, as for example gasoline and coolant in a car, the indifference curve will consist of two straight lines with a right angle bent which is convex to the origin as shown in Fig. 9.13. perfect complementary goods are used in a certain fixed ratio. As will be seen in Fig. 9.13, the left-hand portion of an indifference curve of the perfect complementary goods is a vertical straight line which indicates that an infinite amount of Y is necessary to substitute one unit of X, and the right-hand portion of the indifference curve is a horizontal straight line which means that an infinite amount of X is necessary to substitute one unit of Y. All this means that the two perfect complements are used in a certain fixed ratio and cannot be substituted for each other. In Fig. 9.13 two perfect complements are consumed in the ratio, $3X : 2Y$. *Complements are thus those goods which are used jointly in consumption so that their consumption increases or decreases simultaneously*. Pen and ink, right shoe and left shoe, automobile and petrol, sauce and hamburger, type writer and typists are some examples of perfect complements.

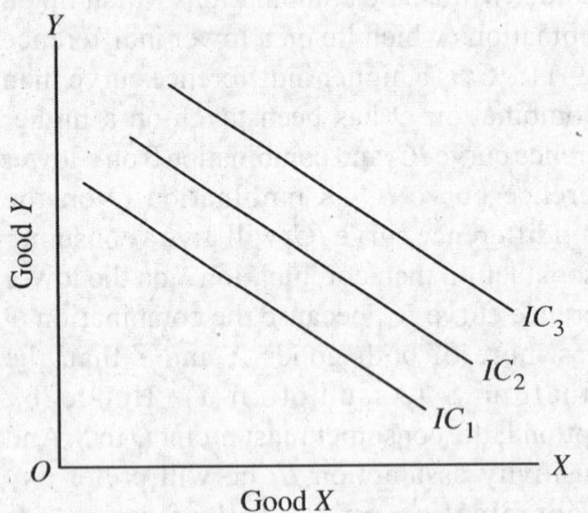

Fig. 9.12. *Indifference Curves of Perfect Substitutes*

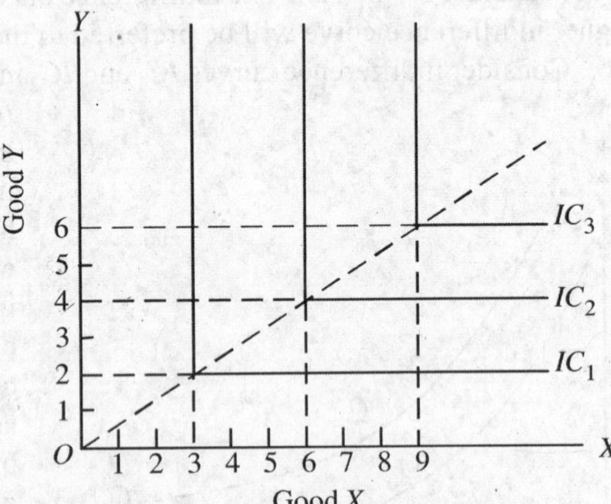

Fig. 9.13. *Indifference Curves of Perfect Complements*

BUDGET LINE

The knowledge of the concept of budget line[1] is essential for understanding the theory of consumer's equilibrium. As explained above, a higher indifference curve shows a higher level of satisfaction than a lower one. Therefore, a consumer in his attempt to maximise his satisfaction will try to reach the highest possible indifference curve. But in his pursuit of buying more and more goods and thus obtaining more and more satisfaction he has to work under two constraints: first, he has to pay the prices for the goods and, secondly, he has a limited money income with which to purchase the goods. Thus, how far he would go in for his purchases depends upon the prices of the goods and the money income which he has to spend on the goods. As explained above, indifference

1. The budget line has been variously named by different authors. Other names in the literature for the budget line are price line, price opportunity line, price-income line, outlay line, budget constraint, expenditure line and consumption possibility line.

map represents consumer's scale of preferences between two goods. Now, in order to explain consumer's equilibrium there is also the need for introducing into the indifference curve analysis the budget line which represents the prices of the goods and consumer's money income.

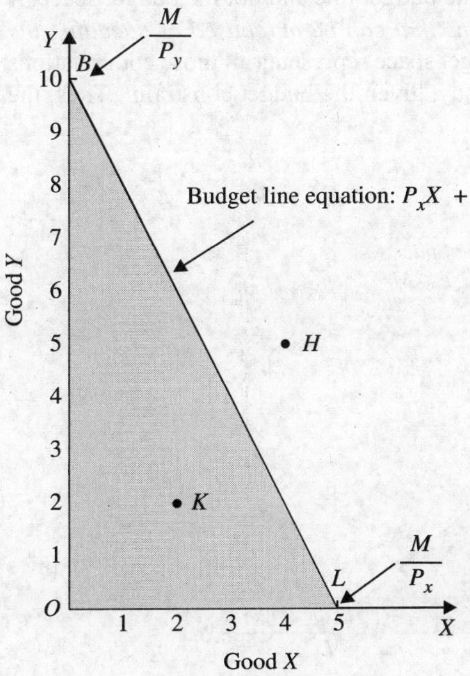

Fig. 9.14. *Budget Line*

Suppose our consumer has got income of ₹ 50 to spend on two goods X and Y. Let price of the good X in the market be ₹ 10 per unit and that of Y ₹ 5 per unit. If the consumer spends his whole income of ₹ 50 on good X, he would buy 5 units of X; if he spends his whole income of ₹ 50 on good Y he would buy 10 units of Y. If a straight line joining $5X$ and $10Y$ is drawn, we will get what is called the price line or the budget line. *Thus budget line shows all those combinations of two goods which the consumer can buy by spending his given money income on the two goods at their given prices*. A look at Fig. 9.14 shows that with ₹ 90 and the prices of X and Y being ₹ 10 and ₹ 5 respectively the consumer can buy $10Y$ and $0X$, or 8Y and $1X$; or $6Y$ and $2X$, or $4Y$ and $3X$ etc. In other words, he can buy any combination that lies on the budget line with his given money income and given prices of the goods. It should be carefully noted that any combination of the two goods such such as H ($5Y$ and $4X$) which lies above and outside the given budget line will be beyond the reach of the consumer. But any combination lying within the budget line such as K ($2X$ and $2Y$) will be well within the reach of the consumer, but if he buys any such combination he will not be spending all his income of ₹ 50. Thus, with the assumption that whole of the given income is spent on the given goods and at given prices of them, the consumer has to choose from all those combinations which lie on the budget line.

It is clear from above that budget line graphically shows the *budget constraint*. The combinations of commodities lying to the right of the budget line are *unattainable* because income of the consumer is not sufficient to be able to buy those combinations. Given his income and prices of the two goods, the combinations of goods lying to the left of the budget line are *attainable*, that is, the consumer can buy any one of them. It is also important to remember that the intercept OB on the Y-axis in Fig. 9.14 equals the amount of his entire income divided by the price of commodity Y. That is, $OB = M/Py$. Likewise, the intercept OL on the X-axis measures the total income divided by the price of commodity X. Thus $OL = M/P_x$.

The budget line can be written algebraically as follows:

$$P_xX + P_yY = M \qquad \ldots (9.1)$$

where P_x and P_y denote prices of goods X and Y respectively and M stands for money income:

The above budget-line equation (9.1) implies that, given the money income of the consumer and prices of the two goods, every combination lying on the budget line will cost the same amount of money and can therefore be purchased with the given income. The budget line can be defined as *a set of combinations of two commodities that can be purchased if whole of a given income is spent on them* and its slope is equal to the negative of the price ratio.

Budget Space

It should be carefully understood that the budget equation $P_xX + P_yY = M$ or $Y = M/P_y - Px/P_y.X$ depicted by the budget line in Fig. 9.14 only describes the budget line and not the budget space. *A budget space shows a set of all commodity combinations that can be purchased by spending the whole or a part of the given income.* In other words, budget space represents all those combinations of the commodities which the consumer can afford to buy, given the budget constraint. Thus, the

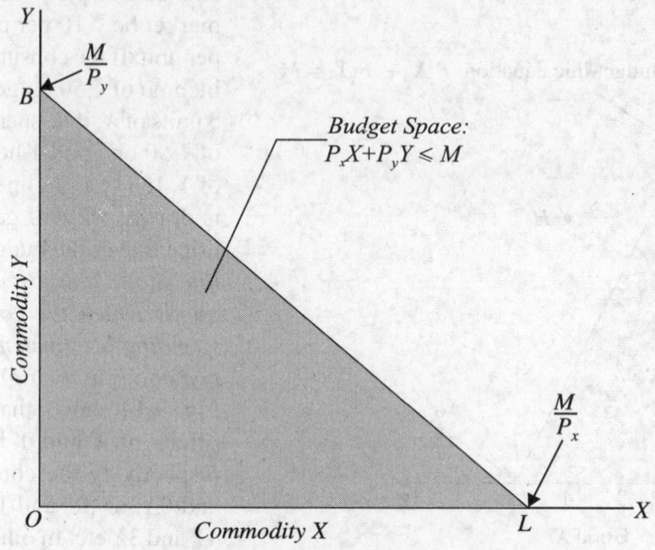

Fig. 9.15. *Budget Space*

budget space implies the set of all combinations of two goods for which income spent on good X (*i.e.*, P_xX) and income spent on good Y (*i.e.*, $P_y Y$) *must not exceed* the given money income. Therefore, we can algebraically express the budget space in the following form of inequality:

$$P_xX + P_yY \leq M, \text{ or } M \geq P_xX + P_yY$$

The budget space has been graphically shown in Fig. 9.15 as the shaded area. The budget space is the entire area enclosed by the budget line BL and the two axes.

Changes in Price and Shift in Budget Line

Now, what happens to the price line if either the prices of goods change or the income changes. Let us first take the case of the changes in prices of the goods. This is illustrated in Fig. 9.16. Suppose the budget line in the beginning is BL, given certain prices of the goods X and Y and a certain income. Suppose price of X falls, the price of Y and income remaining unchanged. Now, with a lower price of X the consumer will be able to purchase more quantity of X than before with his given income. Let at the lower price of X, the given income purchases OL' of X which is greater than OL. Since the price of Y remains the same, there can be no change in the quantity purchased of good Y with the same given income and as a result there will be no shift in the point B. Thus, with the fall in the price of good X, the consumer's money income and the price of Y remaining constant, the price line will take the new position BL'.

Now, what will happen to the budget line (initial budget line BL) if the price of good X rises, the price of good Y and income remaining unaltered. With higher price of good X, the consumer can purchase smaller quantity of *X*, say *OL′′* than before. Thus, with the rise in price of *X* the price line will assume the new position *BL′′*.

Fig. 9.17 shows the changes in the price line when price of good Y falls or rises, with the price of *X* and income remaining the same. In this the initial budget line is *BL*. With the fall in price of good *Y*, other things remaining unchanged, the consumer could buy more of *Y* with the given money income and therefore budget line will shift to *LB′*. Similarly, with the rise in price *Y*, other things being constant, the budget line will shift to *LB′′*.

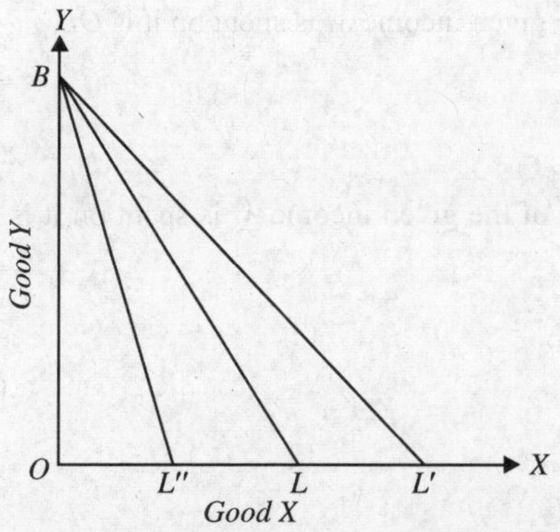

Fig. 9.16. *Changes in Budget Line as a Result of Changes in Price Good X*

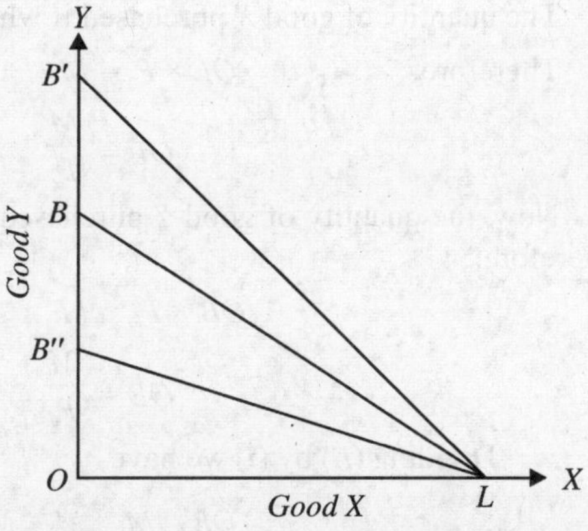

Fig. 9.17. *Changes in Price Line as a Result of Changes in Price of Good Y*

Changes in Income and Shifts in Budget line

Now, the question is what happens to the budget line if income changes, while the prices of goods remain the same. The effect of changes in income on the budget line is shown in Fig. 9.18. Let *BL* be the initial budget line, given certain prices of goods and income. If the consumer's income increases while the prices of both goods *X* and *Y* remain unaltered, the price line shifts upward (say, to *B′L′*) and is parallel to the original budget line *BL*. This is because with the increased income the consumer is able to purchase proportionately larger quantity of good *X* than before if whole of the income is spent on *X*, and proportionately greater quantity of good *Y* than before if whole of the income is spent on *Y*. On the other hand, if income of the consumer decreases, the prices of both goods *X* and *Y* remaining unchanged, the budget line shifts downward (say, to *B′L′*) but remains parallel to the original price line *BL*. This is because a lower income will purchase a proportionately smaller quantity of good *X* if the whole of the income is spent on *X* and proportionately smaller quantity of good *Y* if the whole of the income is spent on *Y*.

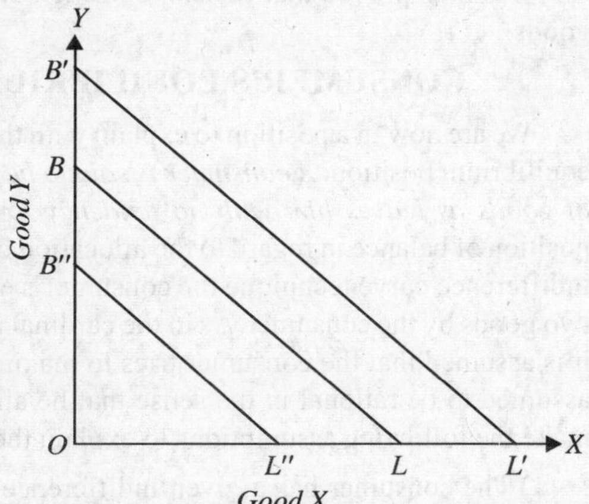

Fig. 9.18. *Shifts in Budget Line as a Result of Changes in Income*

It is clear from above that the budget line will change if either the prices of goods change or the income of the consumer changes. Thus, the two determinants of the budget line are : (*a*) *the prices of goods, and (b) the consumer′s income to be spent on the goods.*

Slope of the Budget Line and Prices of two Goods

It is also important to remember that the *slope* of the budget line is equal to the ratio of the prices of two goods. This can be proved with the aid of Fig. 9.14. Suppose the given income of the consumer is M and the given prices of goods X and Y are P_x and Py respectively. The slope of the budget line BL is $\frac{OB}{OL}$. We intend to prove that slope is equal to the ratio of the prices of goods X and Y.

The quantity of good X purchased if whole of the given income M is spent on it is OL.

Therefore, $OL \times P_x = M$

$$OL = \frac{M}{P_x} \qquad ...(i)$$

Now, the quantity of good Y purchased if whole of the given income M is spent on it is OB. Therefore,

$$OB \times P_y = M$$

$$OB = \frac{M}{P_y} \qquad ...(ii)$$

Dividing (*ii*) by (*i*) we have

$$\frac{OB}{OL} = \frac{M}{P_y} \div \frac{M}{P_x} = \frac{M}{P_y} \times \frac{P_x}{M} = \frac{P_x}{P_y}$$

$$\text{Thus, slope of budget line } = \frac{OB}{OL} = \frac{P_x}{P_y}$$

It is thus proved that the slope of the budget line BL represents the ratio of the prices of two goods.

CONSUMER'S EQUILIBRIUM: MAXIMISING SATISFACTION

We are now in a position to explain with the help of indifference curves how a consumer reaches equilibrium position. *A consumer is said to be in equilibrium when he is buying such a combination of goods as leaves him with no tendency to rearrange his purchases of goods*. He is then in a position of balance in regard to the allocation of his money expenditure among various goods. In the indifference curve technique the consumer's equilibrium is discussed in respect of the purchases of two goods by the consumer. As in the cardinal utility analysis, in the indifference curve analysis also it is assumed that the consumer tries to maximise his satisfaction. In other words, the consumer is assumed to be rational in the sense that he aims at maximising his satisfaction. Besides, we shall make the following assumptions to explain the equilibrium of the consumer:

(1) The consumer has a given indifference map exhibiting his scale of preferences for various combinations of two goods, X and Y.

(2) He has a fixed amount of money to spend on the two goods. He has to spend whole of his given money on the two goods.

(3) Prices of the goods are given and constant for him. He cannot influence the prices of the goods by buying more or less of them.

(4) Goods are homogeneous and divisible.

To show which combination of two goods, X and Y, the consumer will decide to buy and will be in equilibrium position, his indifference map and budget line are brought together. As seen above, while indifference map portrays consumer's scale of preferences between various possible combinations of two goods, the budget line shows the various combinations which he can afford to buy with his given money income and given prices of the two goods. Consider Fig. 9.19 in which we depict consumer's indifference map together with the budget line *BL*. Good *X* is measured on the *X*-axis and good *Y* is measured on the *Y*-axis. With a given money to be spent and given prices of the two goods, the consumer can buy any combination of the goods which lies on the budget line *BL*. Every combination on the budget line *BL* costs him the same amount of money. In order to maximise his satisfaction the consumer will try to reach the highest possible indifference curve which he could with a given expenditure of money and given prices of the two goods. Budget constraint forces the consumer to remain on the given budget line, that is, to choose a combination from among only those which lie on the given budget line.

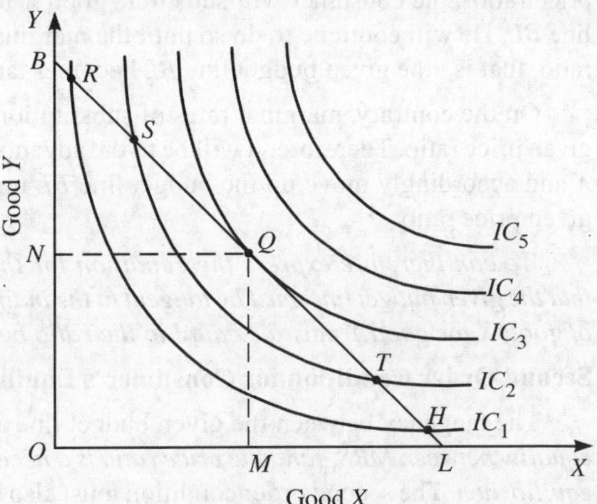

Fig. 9.19. *Consumer's Equilibrium*

It will be seen from Fig. 9.19 that the various combinations of the two goods lying on the budget line *BL* and which therefore the consumer can afford to buy do not lie on the same indifference curve; they lie on different indifference curves. The consumer will choose that combination on the budget line *BL* which lies on the highest possible indifference curve. The highest indifference curve to which the consumer can reach is the indifference curve to which the budget line *BL* is tangent. Any other possible combination of the two goods either would lie on a lower indifference curve and thus yield less satisfaction or would be unattainable.

In Fig. 9.19 budget line *BL* is tangent to indifference curve IC_3 at point *Q*. Since indifference curves are convex to the origin, all other points on the budget line *BL*, above or below the point *Q*, would lie on lower indifference curves. Take point *R* which also lies on the budget line *BL* and which the consumer can afford to buy. Combination of goods represented by *R* costs him the same as the combination *Q*. But, as is evident, *R* lies on the lower indifference curve IC_1 and will therefore yield less satisfaction than *Q*. Likewise, point *S* also lies on the budget line *BL* but will be rejected in favour of *Q* since S lies on the indifference curve IC_2 which is also lower than IC_3 on which *Q* lies. Similarly, *Q* will be preferred to all other points on the budget line *BL* which lies to the right of *Q* on the budget line, such as *T* and *H*. It is thus clear that of all possible combinations lying on budget line *BL*, combination *Q* lies on the highest possible indifference curve IC_3 which yields maximum possible satisfaction. Of course, combinations lying on indifference curves IC_4 and IC_5 will give greater satisfaction to the consumer than *Q*, but they are unattainable with the given money income and the given prices of the goods as represented by the budget line *BL*. It is therefore concluded that with the given money expenditure and the given prices of the goods as shown by *BL* the consumer will obtain maximum possible satisfaction and will therefore be in equilibrium position at point *Q* at which the budget line *BL* is tangent to the indifference curve IC_3. In this equilibrium position at *Q* the consumer will buy *OM* amount of good *X* and *ON* amount of good *Y*.

At the tangency point *Q*, the slopes of the budget line *BL* and indifference curve IC_3 are equal. Slope of the indifference curve shows the marginal rate of substitution of *X* for *Y* (MRS_{xy}), while the slope of the budget line indicates the ratio between the prices of two goods P_x/P_y. Thus, at the equilibrium point *Q*.

$$MRS_{xy} = \frac{\text{Price of good } X}{\text{Price of good } Y} = \frac{P_x}{P_y}$$

When the marginal rate of substitution of X for Y (MRS_{xy}) is greater or less than the price ratio between the two goods, it is advantageous for the consumer to substitute one good for the other. Thus, at points R and S in Fig. 9.19, marginal rates of substitution (MRS_{xy}) are greater than the given price ratio[2], the consumer will substitute good X for good Y and will come down along the budget line BL. He will continue to do so until the marginal rate of substitution becomes equal to the price ratio, that is, the given budget line BL becomes tangent to an indifference curve.

On the contrary, marginal rates of substitution at points H and T in Fig. 9.19 are less than the given price ratio. Therefore, it will be to the advantage of the consumer to substitute good Y for good X and accordingly move up the budget line BL until the MRS_{xy} rises so as to become equal to the given price ratio.

We can therefore express the condition for the equilibrium of the consumer by either saying that the given budget line must be tangent to the indifference curve, or the marginal rate of substitution of good X for good Y must be equal to the ratio between the prices of the two goods.

Second Order Condition for Consumer's Equilibrium

The tangency between the given budget line and an indifference curve or, in other words, *the equality between MRS_{xy} and the price ratio is a necessary but not a sufficient condition of consumer's equilibrium*. The second order condition must also be fulfilled. *The second order condition is that at the point of equilibrium indifference curve must be convex to the origin, or to put it in another way, the marginal rate of substitution of X for Y must be falling at the point of equilibrium.* It will be noticed from Fig. 9.19 above that the indifference curve IC_3 is convex to the origin at Q. Thus at point Q both conditions of equilibrium are satisfied. Point Q in Fig. 9.19 is the optimum or best choice for the consumer and he will therefore be in stable equilibrium at Q.

But it may happen that while budget line is tangent to an indifference curve at a point but the indifference curve may be concave at that point. Take for instance, Fig. 9.20 where indifference curve IC_1 is concave to the origin around the point J. Budget line BL is tangent to the indifference curve IC_1 at point J and MRS_{xy} is equal to the price ratio, P_x/P_y. But J cannot be a position of equilibrium because consumer's satisfaction would not be maximum there. Indifference curve IC_1 being concave at the tangency point J, there may be some points on the given budget line BL such as U and T, which will lie on an indifference curve higher than IC_1. Thus the consumer by moving along the given budget line BL can go to points such as U and T and obtain greater satisfaction than at J. We therefore conclude that for the consumer to be in equilibrium, the following two conditions are required:

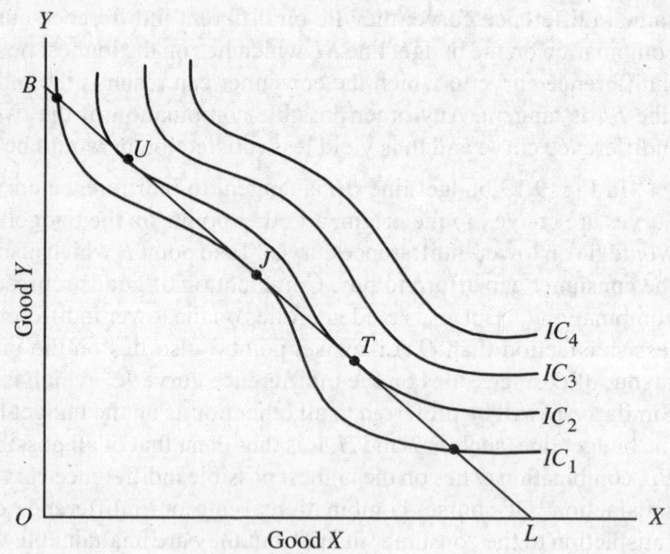

Fig. 9.20. *Second Order Condition for Consumer's Equilibrium*

1. *A given budget line must be tangent to an indifference curve, or marginal rate of substitution*

2. Tangents drawn at point R and S on indifference curves IC_1 and IC_2 respectively have greater slopes than the given slope of the budget line BL.

of X for Y (MRS_{xy}) *must be equal to the price ratio of the two goods* $\frac{P_x}{P_y}$.

2. *Indifference curve must be convex to the origin at the point of tangency.*

The above explanation of consumer's equilibrium in regard to the allocation of his money expenditure on the purchases of two goods has been made *entirely in terms of the consumer's relative preferences of the various combinations of two goods*. In this indifference curve analysis of consumer's equilibrium no use of cardinal utility concept has been made which implies that satisfaction or utility obtained from the goods is measurable in the quantitative sense.

INCOME EFFECT: INCOME CONSUMPTION CURVE

With a given money income to spend on goods, given prices of the two goods and given an indifference map (which portrays given tastes and preferences of the consumers), the consumer will be in equilibrium at a point in an indifference map. We are now interested in knowing how the consumer will react in regard to his purchases of the goods when his money income changes, prices of the goods and his tastes and preferences remaining unchanged. Income effect shows this reaction of the consumer. Thus, *the income effect means the change in consumer's purchases of the goods as a result of a change in his money income.* Income effect is illustrated in Fig. 9.21.

With given prices and a given money income as indicated by the budget line P_1L_1, the consumer is initially in equilibrium at point Q_1 on the indifference curve IC_1 and is having OM_1 of X and ON_1 of Y. Now suppose that income of the consumer increases. With his increased income, he would be able to purchase larger quantities of both the goods. As a result, budget line will shift upward and will be parallel to the original budget line P_1L_1. Let us assume that the consumer's money income increases by such an amount that the new budget line is P_2L_2 (consumer's income has increased by L_1L_2 in terms of X or P_1P_2 in terms of Y). With budget line P_2L_2, the consumer is in equilibrium at point Q_2 on indifference curves IC_2 and is buying OM_2 of X and ON_2 of Y. Thus, as a result of the increase in his income the consumer buys more quantity of both the goods. Since he is on the higher indifference curve IC_2 he will be better off than before *i.e.*, his satisfaction will increase. If his income increases further so that the budget line shifts to P_3L_3, the consumer is in equilibrium at point Q_3 on indifference curve IC_3 and is having greater quantity of both the goods than at Q_2. Consequently, his satisfaction further increases. In Fig. 9.21 the consumer's equilibrium is shown at a still further higher level of income and it will be seen that the consumer is in equilibrium at Q_4 on indifference curves IC_4 when the budget line shifts to P_4L_4. As the consumer's income increases, he switches to higher indifference curves and as a consequence enjoys higher levels of satisfaction.

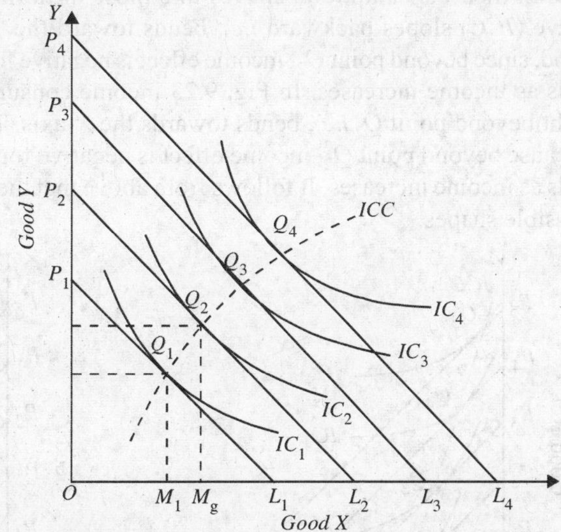

Fig. 9.21. *Income Consumption Curve: Income Effect*

If now various points Q_1, Q_2, Q_3 and Q_4 showing consumer's equilibrium at various levels of income are joined together, we will get what is called *Income Consumption Curve* (*ICC*). Income consumption curve is thus the locus of equilibrium points at various levels of consumer's income. *Income consumption curve traces out the income effect on the quantity consumed of the goods.*

Income effect can either be *positive or negative. Income effect for a good is said to be positive when with the increase in income of the consumer, his consumption of the good also increases*. This is the normal case. When the income effect of both the goods represented on the two axes of the figure is positive, the income consumption curve (*ICC*) will slope upward to the right as in Fig. 9.21. Only the upward-sloping income consumption curve can show rising consumption of the two goods as income increases.

However, for some goods, income effect is negative. *Income effect for a good is said to be negative when with the increases in his income, the consumer reduces his consumption of the good.* Such goods for which income effect is negative are called *Inferior Goods*. This is because the goods whose consumption falls as income of the consumer rises are considered to be some way 'inferior' by the consumer and therefore he substitutes superior goods for them when his income rises. When with the increase in his income, the consumer begins to consume superior goods, the consumption or quantity purchased by him of the inferior goods falls. When the people are poor, they cannot afford to buy the superior goods which are often more expensive. Hence as they become richer and can afford to buy more expensive goods they switch to the consumption of superior and better quality goods. For instance, most of the people in India consider cheaper common foodgrains such as maize, jawar, bajra as inferior goods and therefore when their income rises, they shift to the consumption of superior varieties of foodgrains like wheat and rice. Similarly, most of the Indian people regard Vanaspati Ghee to be inferior and therefore as they become richer, they reduce its consumption and use '*Desi Ghee*' instead.

In case of inferior goods, indifference map would be such as to yield income consumption curve which either slopes backward (*i.e.*, toward the left) as in Fig. 9.22, or downward to the right as in Fig. 9.23. It would be noticed from the two figures that income effect becomes negative only after a point. It signifies that only at higher ranges of income, some goods become inferior goods and up to a point their consumption behaves like those of normal goods. In Fig. 9.22 income consumption curve (*ICC*) slopes backward *i.e.*, bends toward the *Y*-axis. This shows good *X* to be an inferior good, since beyond point Q_2, income effect is negative for good *X* and as a result its quantity demanded falls as income increases. In Fig. 9.23 income consumption curve (*ICC*) slopes downward to the right beyond point Q_2 *i.e.*, bends towards the *X*-axis. This signifies that good *Y* is an inferior good because beyond point Q_2, income effect is negative for good *Y* and as a result its quantity demanded falls as income increases. It follows from above that the income consumption curve can have various possible shapes.

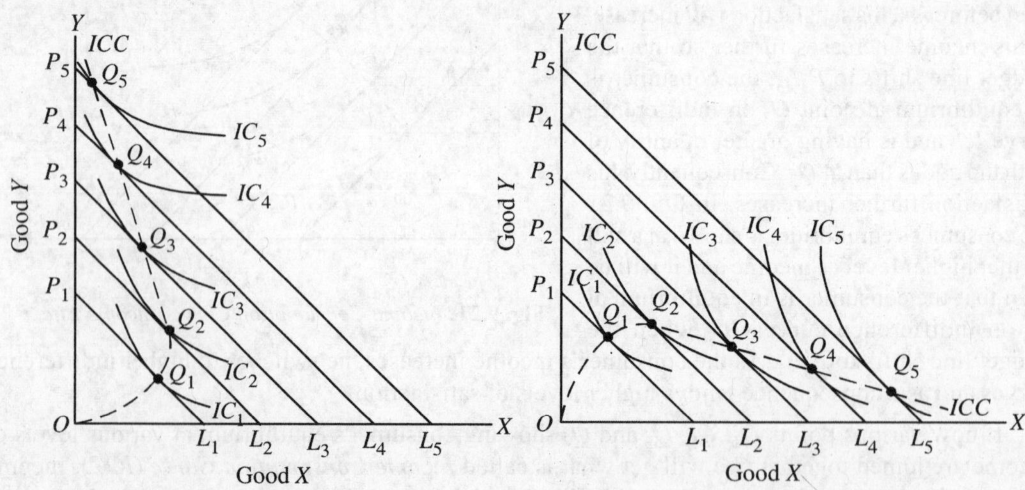

Fig. 9.22. *Income Consumption Curve in Case of Good X being Inferior Good.*

Fig. 9.23. *Income Consumption Curve in Case of Good Y being Inferior Good.*

But normal goods can be either necessities or luxuries depending upon whether the quantities purchased of the goods by the consumers increase less than or more than proportionately to the increases in his income. If the quantity purchased of a commodity rises less than proportionately to the increases in his income, the commodity is known as a *necessity*. On the other hand, if the quantity purchased of a commodity increases more than proportionately to the increases in his income, it is called a *luxury*. In Fig. 9.24, the slope of income consumption curve ICC_1 is increasing which implies that the quantity purchased of the commodity X increases less than proportionately to the increases in income. Therefore, in this case of ICC_1, good X is a necessity and good Y is luxury. On the other hand, the slope of income consumption curve ICC_3 is decreasing which implies that the quantity purchased of good X increases more than proportionately to increases in income and therefore in this case good X is luxury and good Y is necessity. It will be seen from Fig. 9.24 that the income consumption curve ICC_2 is a *linear* curve passing through the origin which implies that the increases in the quantities purchased of both the goods are rising in proportion to the increase in income and therefore neither good is a luxury or a necessity.

If income effect is positive for both the goods X and Y, the income consumption curve will slope upward to the right as in Fig. 9.21. But upward-sloping income consumption curves to the right for various goods may be of different slopes as shown in Fig. 9.24 in which income consumption curves, with varying slopes, are all sloping upward and therefore indicate both goods to be normal goods having positive income effect. *If income effect for good X is negative, income consumption curve will slope backward to the left as ICC′ in Fig. 9.25. If good Y happens to be an inferior good and income consumption curve will bend towards X-axis as shown by ICC″ in Fig. 9.25.* In Figs. 9.24 and 9.25, various possible shapes which income consumption curve can take are shown bereft of indifference curves and budget lines which yield them. It may however be pointed out that given an indifference map and a set of budget lines there will be one income consumption curve.

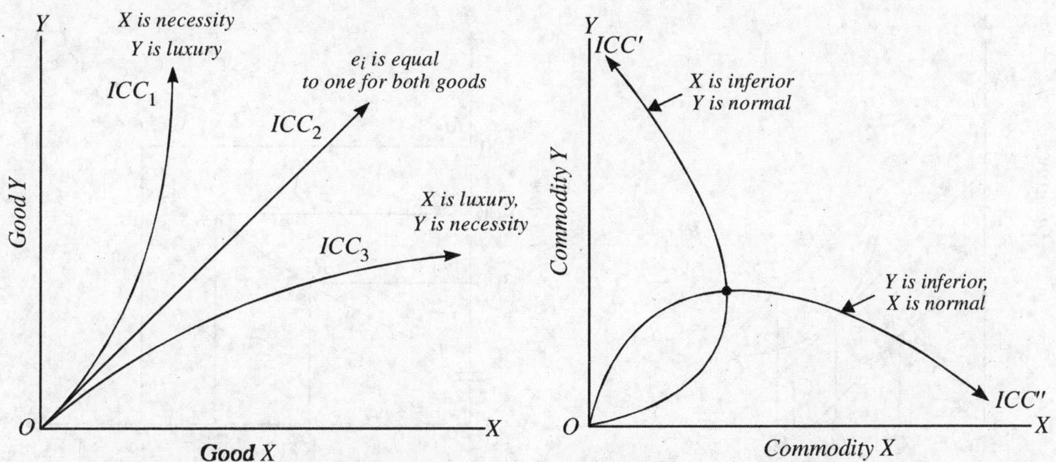

Fig. 9.24. *Income Consumption Curves of Normal Goods.* **Fig. 9.25.** *Income Consumption Curves of Inferior Goods.*

A noteworthy point is that it is not the indifference curves which explain why a good happens to be an inferior good. In other words, indifference curves do not explain why income effect for a good is negative. Indifference curves can only illustrate the inferior-good phenomenon.

INCOME CONSUMPTION CURVE AND ENGEL CURVE

As seen above, income consumption curve is the locus, in indifference curve map, of the equilibrium quantities consumed by an individual at different levels of his income. Thus, the income consumption curve (*ICC*) can be used to derive the relationship between the level of consumer's income and the quantity purchased of a commodity by him. A nineteenth century German statistician Ernet Engel (1821–1896) made an empirical study of family budgets to draw conclusions about the pattern of consumption expenditure, that is, expenditure on different goods and services by the households at different levels of income. The conslusions he arrived at are still believed to be generally valid. According to Engel's studies, as the income of a family increases, the proportion of its income spent on necessities such as food falls and that spent on luxuries (consisting of industrial goods and services) increases. In other words, the poor families spend relatively large proportion of their income on necessities, whereas rich families spend a relatively a large part of their income on luxuries. *This change in the pattern of consumption expenditure (that is, decline in the proportion of income spent on food and other necessities and increase in the proportion of income spent on luxuries) with the rise in income of the families has been called Engel's law.*

Though Engel dealt with the relationship between *income and expenditure* on different goods, in order to keep our analysis simple we will describe and explain the *relationship between income and quantities* purchased of goods. However, both types of relations will convey the same information about individual's consumption behaviour as in our analysis of Engel's curve, the prices of goods are held constant. *The curve showing the relationship between the levels of income and quantity purchased of particular commodities has therefore been called Engel curve.* In what follows we explain how an Engel curve is derived from income consumption curve. In our analysis of Engel curve we relate *quantity purchased* of a commodity, rather than *expenditure* on it, to the level of consumer's income.[3]

It is worth noting that like the demand curve depicting relationship between price and quantity purchased, other factors remaining the same, Engel curve shows relationship between income and quantity demanded, other influences on quantity purchased such as prices of goods, consumer preferences are assumed to be held constant.

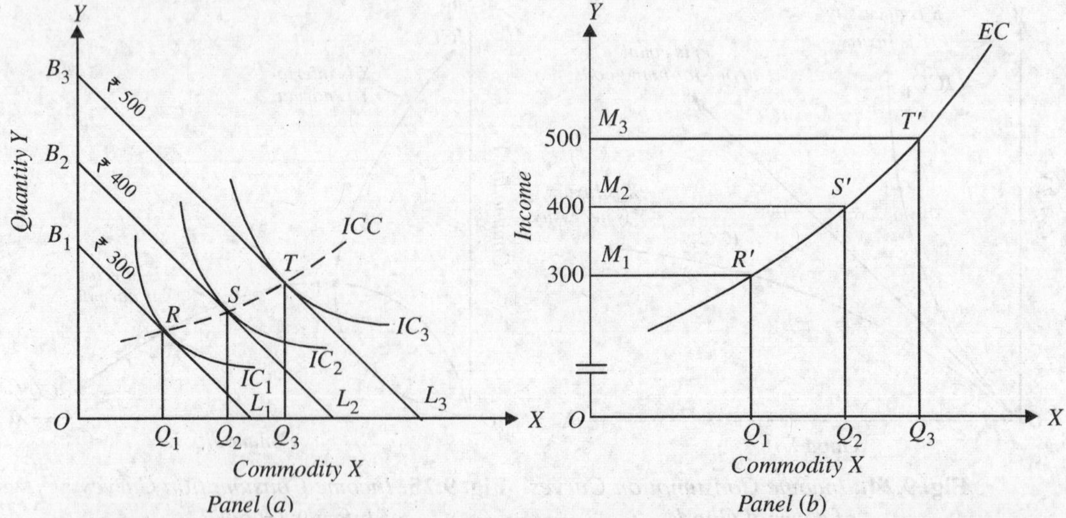

Fig. 9.26. *Deriving Engel Curve from Income Consumption Curve in Case of Necessities*

For deriving Engel curve from income consumption curve we plot level of income on the *Y*-axis and quantity purchased of a commodity on the *X*-axis. Consider panel (*a*) in Fig. 9.26. Given the

3. It may be noted that modern economists also study the relationship between expenditure and level of income and represent this by a curve which they call Engel Expenditure Curve. When they study the relationship between income and quantity purchased and derive a curve of this relationship, they call it simply Engel Curve as we are doing in the present chapter.

indifference map representing the preferences of a consumer and the prices of two goods X and Y, ICC is the income consumption curve showing the equilibrium quantities purchased of a commodity by the consumer as his income increases from ₹ 300 to ₹ 400 and to ₹ 500 per day. It will be seen from panel (*a*) of Fig. 9.26 that when income is ₹ 300, given prices of goods X and Y, the consumer is buying OQ_1 quantity of the commodity.

In panel (*b*) of Fig. 9.26 in which level of income is represented on the vertical axis and quantity purchased of commodity X on the horizontal axis we directly plot quantity OQ_1 against income level of ₹ 300. As income increases to ₹ 400, prices of goods remaining constant, the budget line in panel (*a*) shifts outward to the left to the new position B_2L_2 with which consumer is in equilibrium at point S and the consumer buys OQ_2 quantity of good X. Thus, in panel (*b*) of Fig. 9.26 we plot quantity purchased OQ_2 of commodity X against income level of ₹ 400. Likewise, as income further rises to ₹ 500, budget line in panel (*a*) shifts to B_3L_3 and the consumer buys OQ_3 quantity of X in his new equilibrium position at T. Therefore, in panel (*b*) of Fig. 9.26, we plot OQ_3 against income of ₹ 500. Thus equilibrium points constituting the income consumption curve in consumer's indifference map have been transformed into Engel curve depicting quantity-income relationship. Each point of an Engel curve corresponds to a relevant point of income consumption curve. Thus R' of the Engel curve EC corresponds to point R on the ICC curve. As seen from panel (*b*), *Engel curve for normal goods is upward sloping which shows that as income increases, consumer buys more of a commodity.*

The slope of Engel curve EC drawn in panel (*b*) of Figure 9.26 equals $\Delta M/\Delta Q$ where ΔM stands for income and ΔQ a for change in quantity demanded of good X and has a positive sign. It is important to note that *the slope of the Engel curve in Fig. 9.26 (panel (b)) increases as income increases*. This indicates that with every equal increase in income, expansion in quantity purchased of the good successively declines. This upward-sloping Engel curve with increasing slope as income rises depicts the case of necessities, consumption of which increases relatively less as income rises. For instance, in Fig. 9.26 when income is initially ₹ 300 (= M_1) per week, the quantity purchased of the good X equals OQ_1 and when income rises by ₹ 100 to ₹ 400 (= M_2) per week he increases his consumption to OQ_2, that is, by quantity Q_1Q_2. Now, when his income per week further increases by ₹ 100 to ₹ 500 per week, the quantity consumed increases to OQ_3, that is, by Q_2Q_3 which is less than Q_1Q_2. Thus, in Engel curve drawn in panel (*b*) of Fig. 9.26 the quantity purchased of the commodity increases with the increase in income but *at a decreasing rate*. This shape of the Engel curve is obtained for *necessities*.

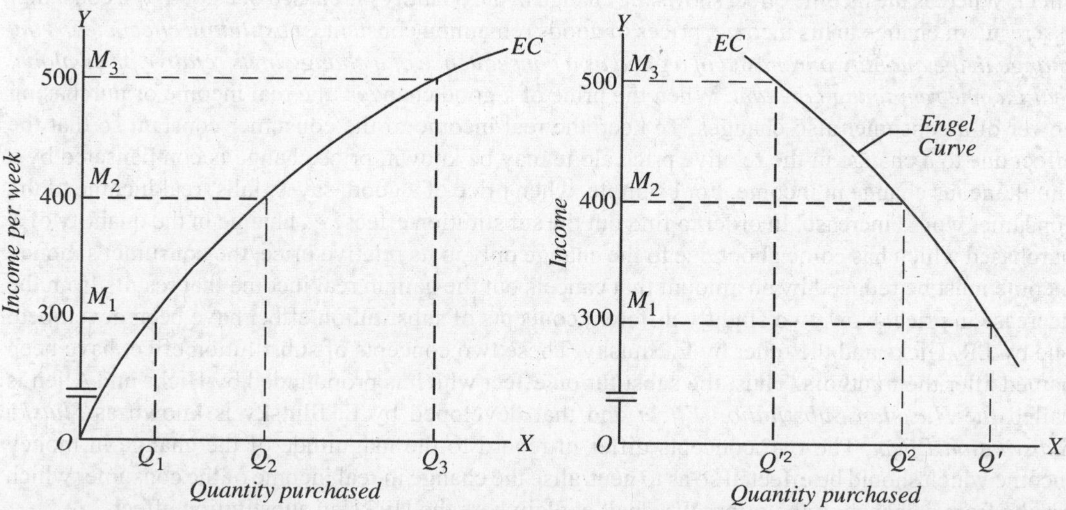

Fig. 9.27. *Engel Curve of a Luxury.*

Fig. 9.28. *Backward Bending Engel Curve of an Inferior Good.*

The Engel curve drawn in Fig. 9.27 is upward sloping but is concave. This implies that slope of the Engel curve ($\Delta M/\Delta Q$) is declining with the increase in income. That is, in the Engel curve of a commodity depicted in Fig. 9.27 the equal increments in income result in successively larger increases in the quantity purchased of the commodity. Thus, in Fig. 9.27 at income of ₹ 300 the consumer purchases OQ_1 quantity of a commodity. The increase in income by ₹ 100 to ₹ 400 results in increase in quantity purchased of the commodity equal to Q_1Q_2. With the further increase in income by the same amount of ₹ 100 to ₹ 500, the quantity purchased increases by Q_2Q_3 which is larger than Q_1Q_2. This implies that as a consumer becomes richer he purchases relatively more of the commodity. Such commodities are called luxuries. Example of luxuries are air travel, butter, costly woollen suits, air conditioners, costly fruits, etc.

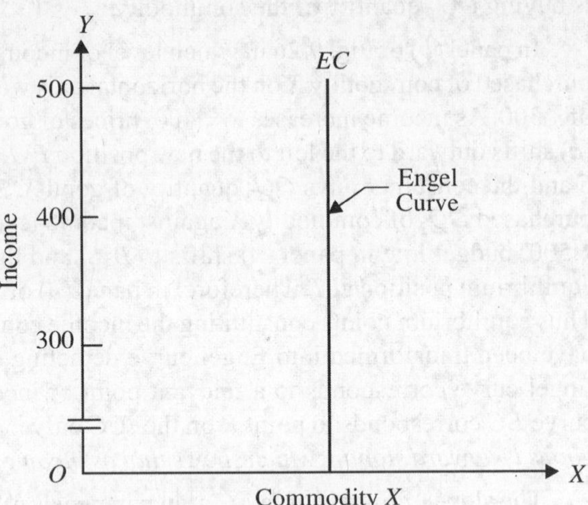

Fig. 9.29. *Engel Curve of a Neutral Good*

In case of *inferior goods*, consumption of the commodity declines as income increases. Engel curve of an inferior good is drawn in Fig. 9.28 which is backward bending indicating a fall in the quantity purchased of the good as income increases.

An extreme case of Engel curve is a vertical straight line as drawn in Fig. 9.29. This represents the case of a *neutral* commodity which is quite unresponsive to the increase in income. The Engel curve of the shape of a vertical straight line shows that a person goes on consuming the same amount of a commodity whatever the level of his income. For example, the quantity of common salt purchased by a family remains the same, determined as it is by food habits, with the increase in their income.

SUBSTITUTION EFFECT

We have explained above the effect of changes in income on purchases or consumption of a good. Another important factor responsible for the changes in consumption of a good is the substitution effect. Whereas the income effect shows the change in the quantity purchased of a good by a consumer as a result of change in his income, prices of goods remaining constant, *substitution effect means the change in the quantity purchased of a good as a consequence of a change in its relative price alone, real income remaining constant*. When the price of a good changes, the real income or purchasing power of a consumer also changes. To keep the real income of the consumer constant so that the effect due to a change in the relative price alone may be known, price change is compensated by a simultaneous change in income. For example, when price of a good, say *X*, falls, real income of the consumer would increase. In order to find out the substitution effect *i.e.*, change in the quantity of *X* purchased which has come about due to the change only in its relative price, the consumer's money income must be reduced by an amount that cancels out the gain in real income that results from the decrease in price. Now, two slightly different concepts of substitution effect have been developed; one by J.R. Hicks and the other by E. Slutsky. These two concepts of substitution effect have been named after their authors. Thus, the substitution effect which is propounded by Hicks and Allen is called the *Hicksian Substitution Effect* and that developed by E. Slutsky is known as *Slutsky Substitution Effect*. The two concepts differ in regard to the magnitude of the change in money income which should be effected so as to neutralise the change in real income of the consumer which results from a change in the price. We shall explain here the Hicksian substitution effect.

In the Hicksian substitution effect price change is accompained by a so much change in money income that the consumer is neither better off nor worse off than before. In other words, money income of the consumer is changed by an amount which keeps the consumer on the same indifference curve on which he was before the change in the price. Thus the Hicksian substitution effect takes place on the same indifference curve. *The amount by which the money income of the consumer is changed so that the consumer is neither better off nor worse off than before is called* Compensating *Variation in Income*. In other words, *compensating variation in income is a change in the income of the consumer which is just sufficient to compensate the consumer for a change in the price of a good.*

Thus, in the Hicksian type of substitution effect, income is changed by the magnitude of the *compensating variation in income*. Hicksian substitution effect is illustrated in Fig. 9.30. With a given money income and given prices of the two goods as represented by the budget line *PL*, the consumer is in equilibrium at point *Q* on the indifference curve *IC* and is purchasing *OM* of good *X* and *ON* of good *Y*. Suppose that the price of good *X* falls (price of *Y* remaining unchanged) so that the budget line now shifts to *PL′*. With this fall in price of *X*, the consumer's real income or purchasing power would increase. In order to find out the substitution effect, this gain in real income should be wiped out by reducing the money income of the consumer by such an amount that forces him to remain on the same indifference curve *IC* on which he was before the change in price of the good *X*. When some money is taken away from the consumer to cancel out the gain in real income, then the budget line which shifted to position *PL′* will now shift downward but will be parallel to *PL′*. In Fig. 9.30, a budget line *AB* parallel to *PL′* has been drawn at such a distance from *PL′* that it touches the indifference curve *IC*. It means that reduction of consumer's income by the amount *PA* (in terms of *Y*) or *L′B* (in terms of *X*) has been made so as to keep him on the same indifference curve. *PA* or *L′B* is thus just sufficient to cancel out the gain in the real income which occurred due to the fall in the price of *X*. *PA* or *L′B* is therefore compensating variation in income.

Now, budget line *AB* represents the new relative prices of goods *X* and *Y* since it is parallel to the budget line *PL′* which was obtained when the price of good *X* had fallen. In comparison to the budget line *PL*, *X* is now relatively cheaper. The consumer would therefore rearrange his purchases of *X* and *Y* and will substitute *X* for *Y*. That is, since *X* is now relatively cheaper and *Y* is now relatively dearer than before, he will buy more of *X* and less of *Y*. It will be seen from Fig. 9.30 that budget line *AB* represents the changed relative prices but a lower money income than that of *PL*, since consumer's income has been reduced by compensating variation in income.

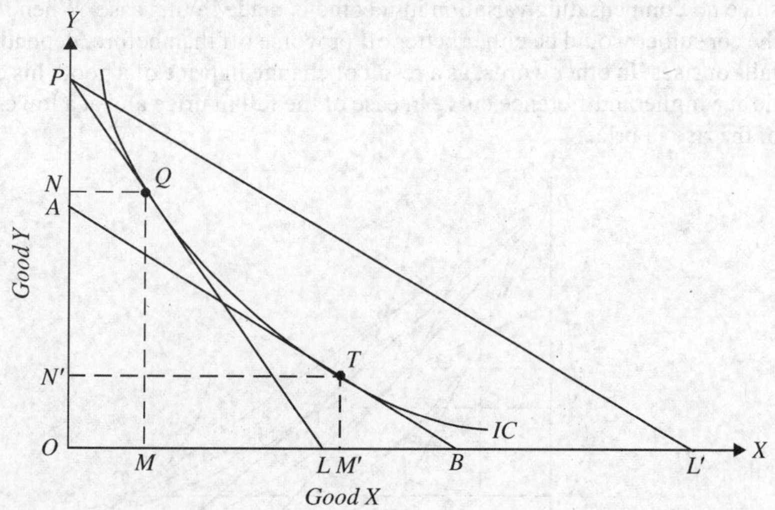

Fig. 9.30. *Hicksian Substitution Effect*

It will seen from Fig. 9.30 that with budget line *AB* the consumer is in equilibrium at point *T* and is now buying *OM′* of *X* and *ON′* of *Y*. Thus in order to buy *X* more he moves on the same indifference curve *IC* from point *Q* to point *T*. This increase in the quantity purchased of good *X* by *MM′* and the decrease in the quantity purchased of good *Y* by *NN′* is due to the change only in the relative prices of goods *X* and *Y*, since effect due to the gain in real income has been wiped out by making a simultaneous reduction in consumer's income. Therefore, movement from *Q* to *T* represents the substitution effect. Substitution effect on good *X* is the increase in its quantity purchased by *MM′* and substitution effect on *Y* is the fall in its quantity purchased by *NN′*. It is thus clear that as a result of the Hicksian substitution effect the consumer remains on the same indifference curve; he is however in equilibrium at a different point from that at which he was before the change in price of good *X*. The less the convexity of the indifference curve, the greater will be the substitution effect. As is known, the convexity of indifference curve is less in the case of those goods which are good substitutes. It is thus clear that the substitution effect in case of good substitutes will be large.

It is thus clear that, a fall in relative price of a commodity always leads to the increase in its quantity demanded due to the substitution effect, the consumer's satisfaction or indifference curve remaining the same. Thus the *substitution effect is always negative*. The negative substitution effect implies that the relative price of a commodity and its quantity demanded change in opposite direction, that is, the *decline* in relative price of a commodity always causes *increase* in its quantity demanded. This inverse relationship between relative price and quantity demanded holds good in *case the indifference curves are convex to the origin*. Given that indifference curves are convex to the origin, a fall in the relative price of a commodity causing an increase in its quantity demanded is known as *Slutsky theorem* as this proposition was originally put forward by Slutsky. It is this negative substitution effect which lies at the root of the famous law of demand stating inverse relationship between price and quantity demanded.

PRICE EFFECT: PRICE CONSUMPTION CURVE

We will now explain how the consumer reacts to charges in the price of a good, his money income, tastes and prices of other goods remaining the same. Price effect shows this reaction of the consumer and measures the full effect of the change in the price of a good on the quantity purchased since no compensating variation in income is made in this case. When the price of a good charges, the consumer would be either better off or worse off than before, depending upon whether the price falls or rises. In other words, as a result of change in price of a good, his equilibrium position would lie at a higher indifference curve in case of the fall in price and at a lower indifference curve in case of the rise in price.

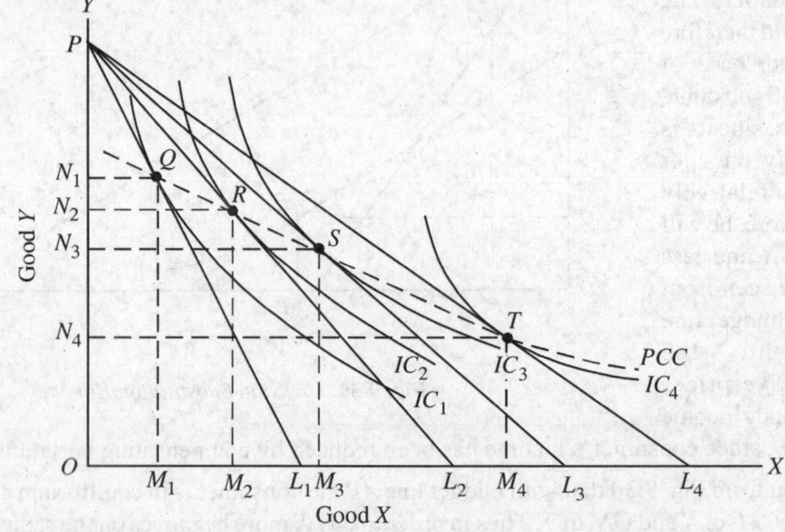

Fig. 9.31. *Downward-Sloping Price Consumption Curve*

Price effect is shown in Fig. 9.31. With given prices of goods *X* and *Y*, and a given money income as represented by the budget line PL_1, the consumer is in equilibrium at *Q* on indifference curve IC_1. In this equilibrium position at *Q*, he is buying OM_1 of *X* and ON_1 of *Y*. Let price of good *X* fall, price of *Y* and his money income remaining unchanged. As a result of this price change, budget line shifts to the position PL_2. The consumer is now in equilibrium at *R* on a higher indifference curve IC_2 and is buying OM_2 of *X* and ON_2 of *Y*. He has thus become better off, that is, his level of satisfaction has increased as a consequence of the fall in the price of good *X*. Suppose that price of *X* further falls so that PL_3 is now the relevant price line. With budget line PL_3 the consumer is in equilibrium at *S* on indifference curve IC_3 where he has OM_3 of *X* and ON_3 of *Y*. If price of good *X* falls still further so that budget line now takes the position of PL_4, the consumer now attains equilibrium at *T* on indifference curve IC_4 and has OM_4 of X and ON_4 of *Y*. When all the equilibrium points such as *Q*, *R*, *S*, and *T* are joined together, we get what is called *Price Consumption Curve* (*PCC*). *Price consumption curve traces out the price effect. It shows how the changes in price of good X will affect the consumer's purchases of X, price of Y, his tastes and money income remaining unaltered.*

In Fig. 9.31 price consumption curve (*PCC*) is sloping downward. Downward sloping price consumption curve for good *X* means that as price of good *X* falls, the consumer purchases a larger quantity of good *X* and a smaller quantity of good *Y*. This is quite evident from Fig. 9.31. As we shall discuss in detail in the chapter concerning elasticity of demand, we obtain downward-sloping price consumption curve for good *X* when demand for it is elastic (*i.e.*, price elasticity is greater than one). But downward sloping is one possible shape of price consumption curve. Price consumption curve can have other shapes also. In Fig. 9.32 upward-sloping price consumption curve is shown. *Upward-sloping price consumption curve for X means that when the price of good X falls, the quantity demanded of both goods X and Y rises*. We obtain the upward-sloping price consumption curve for good *X* when the demand for good is inelastic, (*i.e.*, price elasticity is less than one).

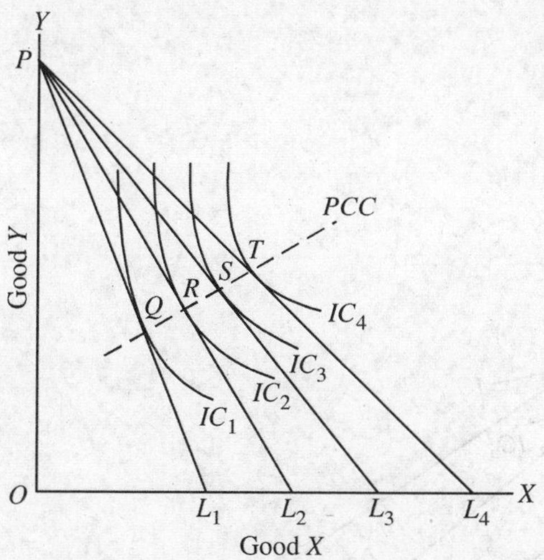

Fig. 9.32. *Upward-Sloping Price Consumption Curve*

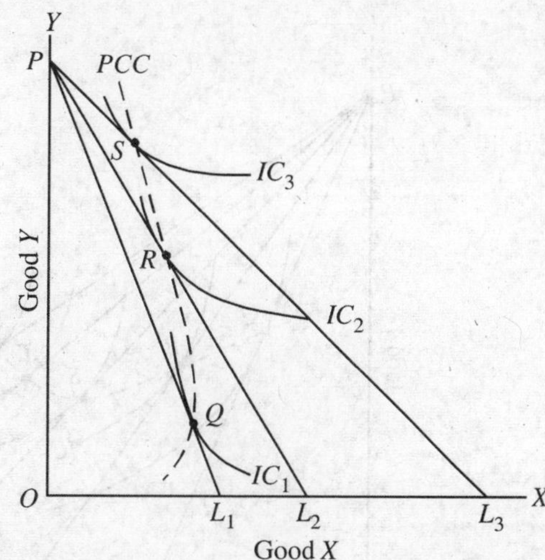

Fig. 9.33. *Backward-Sloping Price Consumption Curve in Case of Giffen Goods*

Price consumption curve can also have a backward-sloping shape, which is depicted in Fig. 9.33. *Backward-sloping price consumption curve for good X indicates that when price of X falls, after a point smaller quantity of it is demanded or purchased.* We shall see later in this chapter that this is true in case of exceptional type of goods called Giffen Goods.

Price consumption curve for a good can take horizontal shape too. It means that when the price of the good X declines, its quantity purchased rises proportionately but quantity purchased of Y remains the same. Horizontal price consumption curve is shown in Fig. 9.34. We obtain horizontal price consumption curve of good X when the price elasticity of demand for good X is equal to unity.

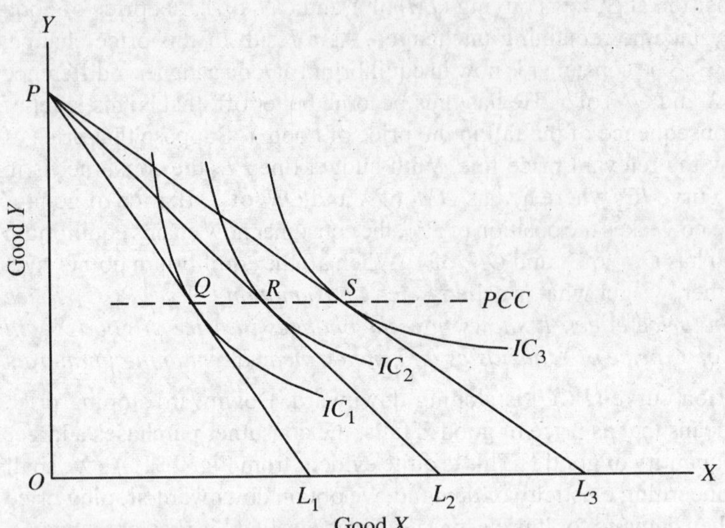

Fig. 9.34. *Horizontal Price Consumption Curve.*

But it is rarely found that price consumption curve slopes downward throughout or slopes upward throughout or slopes backward throughout. More generally, price consumption curve has different slopes at different price ranges. At higher price levels it generally slopes downward, and it may then have a horizontal shape for some price ranges but ultimately it will be sloping upward. For some price ranges it can be backward sloping as in case of Giffen goods. A price consumption curve which has different shapes or slopes at different price ranges is drawn in Fig. 9.35. We shall study in a subsequent chapter that such a type of price consumption curve means that price elasticity of demand varies at different price ranges.

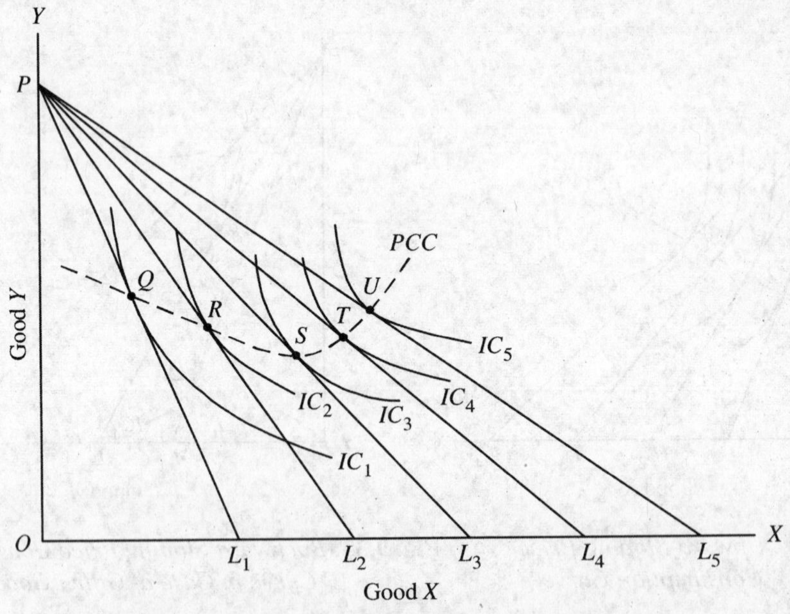

Fig. 9.35. *Price Consumption Curve with Varying Slopes*

BREAKING UP PRICE EFFECT INTO INCOME AND SUBSTITUTION EFFECTS

It has been explained above that as price of a good X falls, other things remaining the same, consumer would move to a new equilibrium position at a higher indifference curve and would buy more of good X at the lower price unless it is a Giffen good. Thus, in the Fig. 9.36 the consumer who is initially in equilibrium at Q on indifference curve IC_1 moves to the point R on indifference curve IC_2 when the price of good X falls and the budget line twists from PL_1 to PL_2. The movement from Q to R represents the price effect. It is now highly important to understand that this price effect is the net result of two distinct forces, namely, substitution effect and income effect. In other words, price effect can be split up into two different parts, one being the substitution effect and the other income effect.

There are two approaches for decomposing price effect into its two parts, substitution effect and income effect. They are the Hicksian approach and Slutsky approach. Further, Hicksian approach uses two methods of splitting the price effect, namely *(i) Compensating variation in income (ii) Equivalent variation in income*. Slutsky uses *cost-difference method* to decompose price effect into its two component parts. How the price effect can be decomposed into income effect and substitution effect by the Hicksian methods is explained below, whereas Slutsky's cost-difference method will be explained in an appendix to this chapter.

1. Decomposing Price Effect: Compensating Variation in Income

In the method of decomposing price effect by compensating variation we adjust the income of the consumer so as to offset the change in satisfaction resulting from the change in price of a good and bring the consumer back to his original indifference curve, that is, his initial level of satisfaction which he was obtaining before the change in price occurred. For instance, when the price of a commodity falls and consumer moves to a new equilibrium position at a higher indifference curve his satisfaction increases. To offset this gain in satisfaction resulting from a fall in price of the good we must take away from the consumer enough income to force him to come back to his original indifference curve. *This required reduction in income (say, through levying a lump sum tax) to cancel out the gain in satisfaction or welfare occurred by reduction in price of a good is called compensating variation in income*. This is so called because it compensates (in a negative way) for the gain in satisfaction resulting from a price reduction of the commodity. How the price effect is decomposed into substitution effect and income effect is through the method of compensating variation in income illustrated in Fig. 9.36.

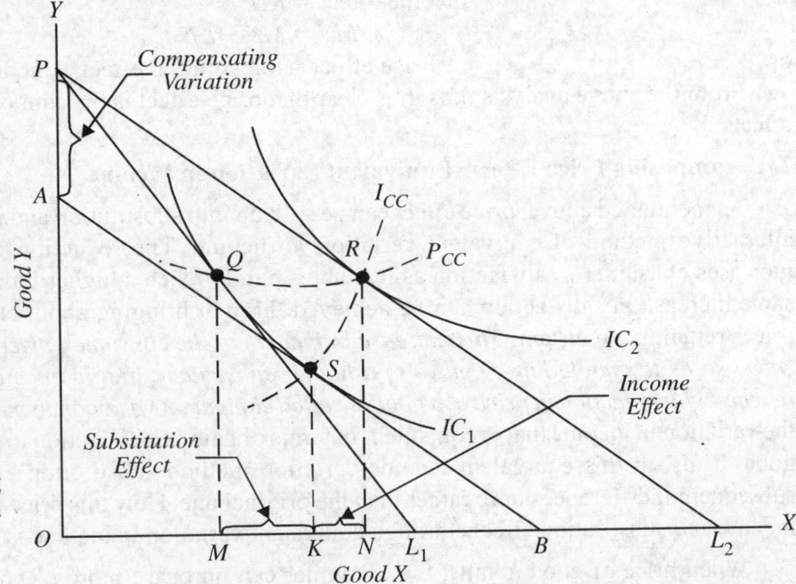

Fig. 9.36. *Price Effect Split up into Substitution and Income Effects through Compensating Variation Method*

When the price of good X falls and as a result budget line shifts to PL_2, the real income of the consumer rises, *i.e.*, he can buy more of both the goods with his given money income. That is, price reduction enlarges consumer's opportunity set of the two goods. With the new budget line PL_2 he is in equilibrium at point R on a higher indifference curve IC_2 and thus gains in satisfaction as a result of fall in price of good X. Now, if his money income is reduced by the compensating variation in income so that he is forced to come back to the indifference curve IC_1 as before, he would buy more of X since X has now become relatively cheaper than before. In Fig. 9.36 as result of the fall in price of X, price line switches to PL_2. Now, with the reduction in income by compensating variation, budget line shifts to AB which has been drawn parallel to PL_2 so that it just touches the indifference curve IC_1 where he was before the fall in price of X. Since the price line AB has got the same slope as PL_2, it represents the changed relative prices with X being relatively cheaper than before. Now, X being relatively cheaper than before, the consumer in order to maximise his satisfaction in the new price-income situation substitutes X for Y. Thus, when the consumer's money income is reduced by the compensating variation in income (which is equal to PA in terms of Y or L_2B in terms of X), the consumer moves along the same indifference curve IC_1 and substitutes X for Y. With price line AB, he is in equilibrium at S on indifference curve IC_1 and is buying MK more of X in place of Y. This movement from Q to S on the same indifference curve IC_1 represents the substitution effect since it occurs due to the change in relative prices alone, real income remaining constant. If the amount of money income which was taken away from him is now given back to him, he would move from S on indifference curve IC_1 to R on a higher indifference curve IC_2. The movement from S on a lower in difference curve to R on a higher in difference curve is the result of income effect. Thus the movement form Q to R due to price effect can be regarded as having been taken place into two steps : first from Q to S as a result of substitution effect and second from S to R as a result of income effect. In is thus mainfest that price effect is the combined result of a substitution effect and an income effect.

In Fig. 9.36 the various effects on the purchases of good X are:

$$\text{Price effect} = MN$$
$$\text{Substitution effect} = MK$$
$$\text{Income effect} = KN$$
$$MN = MK + KN$$

or

$$\text{Price effect} = \text{Substitution effect} + \text{Income effect}$$

From the above analysis, it is thus clear that price effect is the sum of income and substitution effects.

2. Decomposing Price Effect: Equivalent Variation in Income

As mentioned above, price effect can be split up into substitution and income effects through an alternative method of equivalent variation in income. The reduction in price of a commodity increases consumer's satisfaction as it enables him to reach a higher indifference curve. Now, the same increase in satisfaction can be achieved through bringing about an increase in his income, prices remaining constant. *This increase in income of the consumer, prices of goods remaining the same, so as to enable him to move to a higher subsequent indifference curve at which he in fact reaches with reduction in price of a good is called equivalent variation in income* because it represents the variation in income that is equivalent in terms of gain in satisfaction to a reduction in price of the good. Thus, in this equivalent income-variation method substitution effect is shown along the subsequent indifference curve rather than the original one. How this price effect is decomposed into income and substitution effects through equivalent variation in income is shown in Fig. 9.37.

When price of good X falls, the consumer can purchase more of both the goods, that is, the purchasing power of his given money income rises. It means that after the fall in price of X, if the consumer buys the same quantities of goods as before, then some amount of money will be left over.

In other words, the fall in price of good X will release some amount of money. Money thus released can be spent on purchasing more of both the goods. It therefore follows that a change in price of the good produces an income effect. When the power to purchase goods rises due to the income effect of the price change, or in other words, when some amount of money is released as a result of the fall in price, the consumer has to decide how this increase in his purchasing power is to be spread over the two goods he is buying. How he will spread the released purchasing power over the two goods depends upon the nature of his income consumption curve which in turn is determined by his preferences about the two goods.

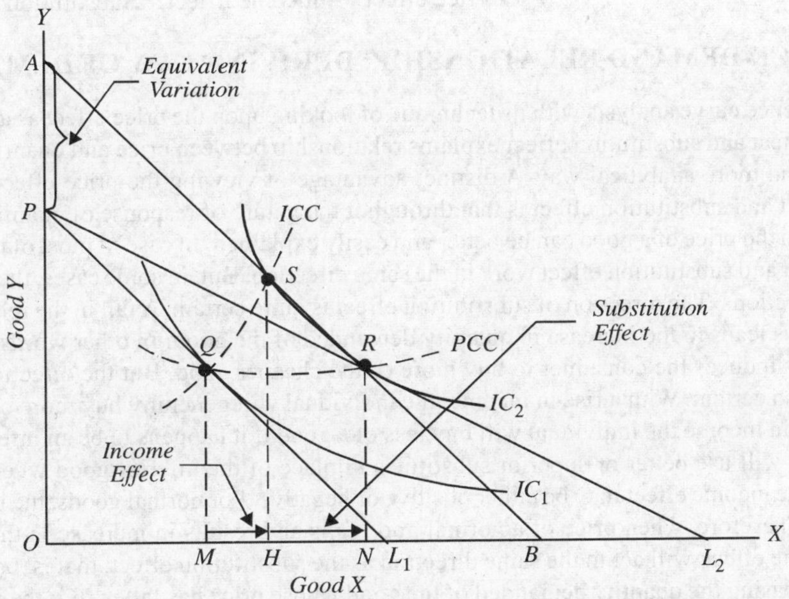

Fig. 9.37. *Price Effect Split Up into Income and Substitution Effects through Equivalent Variation Method*

From above it follows, that, as a result of the increase in his purchasing power (or real income) due to the fall in price, the consumer will move to a higher indifference curve and will become better off than before. It is as if the price had remained the same but his money income was increased. In other words, a fall in price of good X does to the consumer what an equivalent rise in money income would have done to him. As a result of fall in price of X, the consumer can therefore be imagined as moving up to a higher indifference curve along the income consumption curve as if his money income had been increased, prices of X and Y remaining unchanged. Thus, a given change in price can be thought of as an equivalent to an appropriate change in income.

It will be seen from Fig. 9.37 that with price line PL_1, the consumer is in equilibrium at Q on indifference curve IC_1. Suppose price of good X falls, price of Y and his money income remaining unaltered, so that budget line is now PL_2. With budget line PL_2, he is in equilibrium at R on indifference curve IC_2. Now, a line AB is drawn parallel to PL_1 so that it touches the indifference curve IC_2 at S. It means that the increase in real income or purchasing power of the consumer as a result of the fall in price of X is equal to PA in terms of Y or L_1B in terms of X. Movement of the consumer from Q on indifference curve IC_1 to S on the higher indifference curve IC_2 along the income consumption curve is the result of income effect of the price change. But the consumer will not be finally in equilibrium at S. This is because now that X is relatively cheaper than Y, he will substitute X, which has become relatively cheaper, for good Y, which has become relatively dearer. It will be a profitable thing for the consumer to do so. Thus the consumer will move along the indifference curve IC_2 from S to R. This movement from S to R has taken place because of the change in relative prices alone and therefore represents substitution effect. Thus the price effect can be broken up into income and substitution effects, showing in this case substitution along the subsequent indifference curve. In Fig 9.37 the magnitudes of the various effects are :

Price effect = MN

Income effect = MH

Substitution effect = HN

In Fig. 9.37, $MN = MH + HN$

or Price effect = Income Effect + Substitution Effect

PRICE-DEMAND RELATIONSHIP: DERIVING LAW OF DEMAND

Indifference curve analysis with its technique of looking upon the price effect as a combination of income effect and substitution effect explains relationship between price and quantity demanded in a better and more analytical way. A distinct advantage of viewing the price effect as a sum of income effect and substitution effect is that through it the nature of response of quantity purchased to a change in the price of a good can be better and easily explained. In case of most of the goods, the income effect and substitution effect work in the same direction. But, in some cases, they may pull in different directions. The direction of substitution effect is quite certain. A fall in the relative price of a good always leads to the increase in quantity demanded of the good. In other words, substitution effect always induces the consumer to buy more of the cheaper good. But the direction of income effect is not so certain. With a rise in income, the individual will generally buy more of a good. But with the rise in income the individual will buy less of a good if it happens to be an inferior good for him since he will use better or superior substitutes in place of the inferior good when his income rises. Thus the income effect may be either positive or negative. For normal goods, the income effect is positive. Therefore, when price of a normal good falls and results in increase in the purchasing power, income effect will act in the same direction as the substitution effect, that is, both will work towards increasing the quantity demanded of the good whose price has fallen. For the inferior good in which case income effect is negative, income effect of the price change will work in opposite direction to the substitution effect. The net effect of the price change will then depend upon the relative strengths of the two effects. To sum up, price effect is composed of income effect and substitution effect and further that the direction in which quantity demanded will change as a result of the fall in price will depend upon the direction and strength of the income effect on the one hand and strength of the substitution effect on the other.

Price Demand Relationship: Normal Goods

In order to understand the way in which price-demand relationship is established in indifference curve analysis, consider Fig. 9.36. Given the price of two goods and his income represented by the budget line PL_1, the consumer will be in equilibrium at Q on indifference curve IC_1. Let us suppose that, price of X falls, price of Y and his money income remaining unchanged so that budget line now shifts to PL_2. The consumer will now be in equilibrium at a point on the new budget line BL_2. If the equilibrium position on BL_2 lies to the right of Q such as at R in Fig. 9.36, it will mean that the consumer buys more quantity of good X than at Q. Now, it can be proved that in case of normal goods the *new equilibrium point on* BL_2 *will lie to the right of* Q*, meaning thereby that the quantity demanded of the good* X *will increase as its price falls.*

As seen above, the direction and magnitude of the change in quantity demanded as a result of the fall in price of a good depend upon the direction and strength of income effect on the one hand and substitution effect on the other. As for normal goods, the income effect is positive, it will work towards increasing the quantity demanded of good X when its price falls. The substitution effect which is always negative and operates so as to raise the quantity demanded of the good if its price falls and reduces the quantity demanded of the good if its price rises. Thus, both the income effect (when positive) and negative substitution effect works in the same direction and cause increase in the quantity purchased of good *X* whose price has fallen with the result that the new equilibrium point will lie to the right of the original equilibrium point *Q* such as point *R* in Fig. 9.36 above. Substitution effect causes *MK* increase in quantity demanded. Income effect which is positive here also leads to the increase in quantity demand by *KN*. Each effect therefore reinforces the other. As a result, the total effect of a fall in price of *X* from the level indicated by PL_1 to the level indicated by PL_2 is the rise in quantity demanded of good *X* from *OM* to *ON*, that is, quantity demanded increases by *MN* which is equal to *MK* + *KN*. To sum up, *the income effect and substitution effect in case of normal goods work in the same direction and will lead to the increase in quantity demanded of the good whose price has fallen.* In other words, quantiy purchased of a normal good will vary inversely with its price as in its case income effect is positive.

Price-Demand Relationship: Inferior Goods. In case of inferior goods the income effect will work in opposite direction to the substitution effect. When price of an inferior good falls, its negative income effect will tend to reduce the quantity purchased, while the substitution effect will tend to increase the quantity purchased. But normally it happens that negative income effect of the change in price is not large enough to outweigh the substitution effect. This is so because a consumer spends a very small proportion of his income on a single commodity and when price of the commodity falls, a very little income is released. In other words, income effect even when negative is generally too weak to outweigh the substitution effect. It follows therefore that as a result of the fall in price of a good the substitution effect which always induces the consumer to buy more of the good whose price has fallen will usually outweigh the negative income effect. Thus even in most cases of inferior goods the net result of the fall in price will be increase in its quantity demanded. It is thus clear that

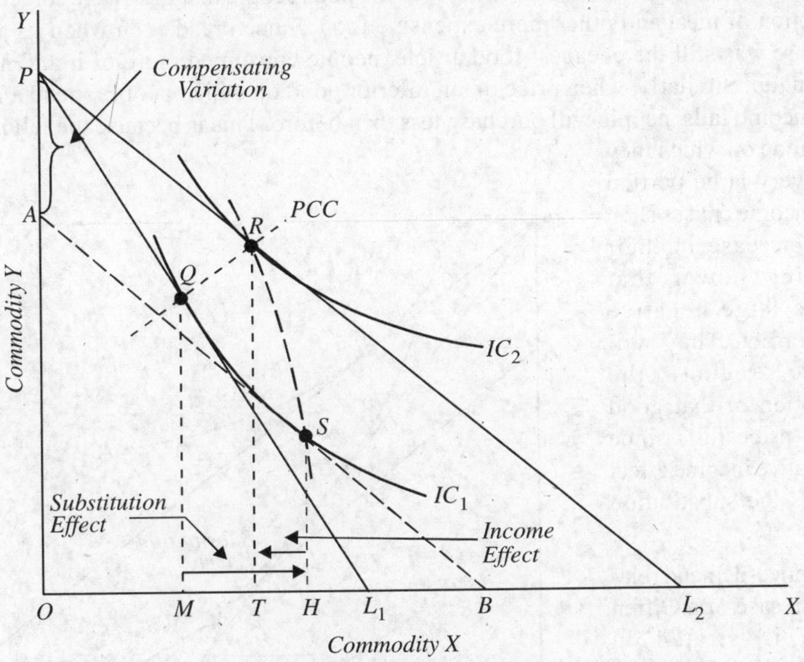

Fig. 9.38. *Price-Demand Relationship in Case of an Inferior Good.*

in a majority of inferior goods quantity demanded of the good will vary inversely with price and the Marshallian Law of Demand will hold good. The price-demand relationship, in case of inferior goods having weaker income effect is illustrated in Fig. 9.38. It will be seen from Fig. 9.38 that the fall in price of good *X* makes the consumer to shift from equilibrium *Q* to a new equilibrium at *R*. As a result, quantity purchased rises from *OM* to *OT*. But the income effect is negative and is equal to *HT*. If income effect alone was working, it would have caused the consumer to buy *HT* less of good *X*. But substituion effect is universally present and always induces the consumer to buy more of the relatively cheaper good. In Fig. 9.38 substitution effect is equal to *MH* and is greater than negative income effect *HT*. Therefore, the net effect of the fall in price of good *X* is the rise in quantity demanded by *MT*. Hence we conclude that *in case of inferior goods, quantity demanded varies inversely with price when negative income effect is weaker than the substitution effect.* In other words, even in case of inferior goods having weaker income effect, the demand curve will be downward sloping.

Price-Demand Relationship: Giffen Goods or Giffen Paradox. There is a third possibility. This is that there may be some inferior goods for which the negative income effect is strong or large enough to outweigh the substitution effect. In this case, quantity purchased of the good will fall as its price falls and quantity purchased of the good will rise as its price rises. In other words, quantity purchased or demanded will vary directly with price. Now, the income effect can be substantial only when the consumer is spending a very large proportion of his income on the good in question so that when price of the good falls, a good amount of income is released. If that good happens to be inferior good, the income effect will be negative as well as strong and may outweigh the substitution effect so that with the fall in price, the consumer will buy less of the good. *Such an inferior good in which case the consumer reduces its consumption when its price falls and increases its consumption when its price rises is called a Giffen good* named after the British statistician, Sir Robert Giffen, who in the mid-nineteenth century is said to have claimed that when price of cheap common foodstuff like bread went up the people bought and consumed more bread. A rise in the price of bread caused such a large decline in the purchasing power of the poor people that they were forced to cut down the consumption of meat and other more expensive food. Since bread even when its price was higher than before was still the cheapest food article, people consumed more of it and not less when its price went up. Similarly, when price of an inferior good on which people spend a large proportion of their income falls, people will purchase less than before. This is because the fall in the price of an inferior good on which they spend a very large portion of their income causes such a large increase in their purchasing power that creates a large negative income effect. They will therefore reduce the consumption of that good when its price falls since large negative income effect outweighs the substitution effect.

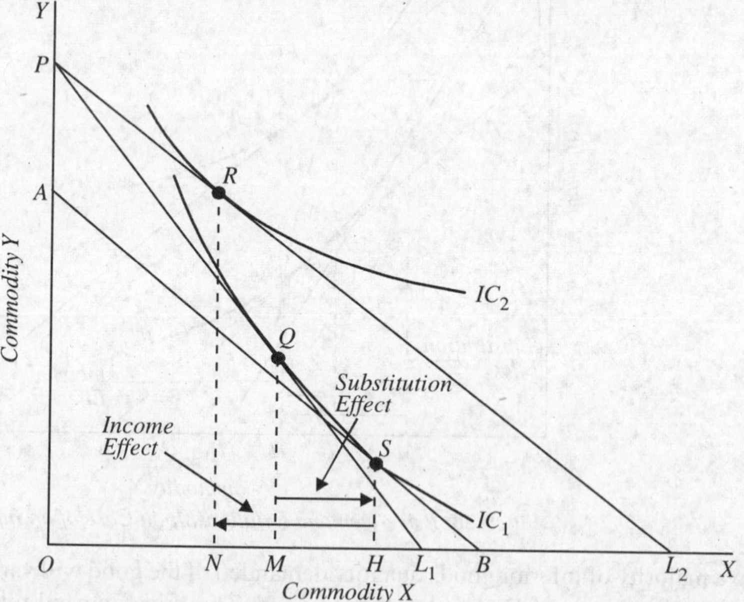

Fig. 9.39. *Price-Demand Relationship in case of a Giffen Good*

The price-demand relationship in case of a Giffen good is illustrated in Fig. 9.39. With a certain given price-income situation depicted by the budget line *PL*, the consumer is initially in equilibrium at *Q* on indifference curve IC_1. With a fall in the price of the good, the consumer shifts to point *R* on indifference Curve IC_2. It will be seen From Fig. 9.39 that with the fall in price and, as a result, the shift of the price line from PL_1 to PL_2 the consumer reduces his consumption of the good *X* from *OM* to *ON*. This is the net effect of the negative income effect which is here equal to *HN* which induces the consumer to buy less of good *X* and the substitution effect which is equal *MH* which induces the consumer to buy more of the good. Since the negative income effect *HN* is greater than the substitution effect *HM*, the net effect is the fall in quantity purchased of good *X* by *MN* with the fall in its price. Thus, *the quantity demanded of a Giffen good varies directly with price.* Therefore, if a demand curve showing price demand relationship, is drawn it will slope upward.

Thus, *the quantity demanded of a Giffen good* varies directly with price. For a good to be a Giffen good, the following three conditions are necessary:[4]

(1) The good must be inferior good with a large negative income effect;
(2) The substitution effect must be small; and
(3) The proportion of income spent upon the inferior good must be very large.

Three Demand Theorems Based on Indifference Curve Analysis

It follows from above that, indifference curve analysis enables us to derive a more general law of demand in the following composite form, consisting of three demand theorems to which the Marshallian law of demand constitutes a special case.[5]

(*a*) The quantity demanded of a good varies inversely with price when the income effect is positive or nil.

(*b*) The quantity demanded of a good varies inversely with price when the income effect for the good is negative but is weaker than the substitution effect.

(*c*) The quantity demanded of a good *varies directly with price* when the income effect for the good is negative and this negative income effect of a change in price is larger than the substitution effect.

In the case, (*a*) and (*b*) the Marshallian law of demand holds good and we get a downward sloping demand curve. The case (*a*) applies to normal goods in which income effect and substitution effect work *in the same direction*. The case (*b*) applies to inferior goods which are not Giffen goods. When the third case occurs, we get a Giffen good of positively sloping demand curve. Marshallian law of demand does not hold true in the third case. Marshall mentioned a Giffen good case as an exception to his law of demand. Thus the indifference curve analysis is superior to Marshallian analysis in that it yields a more general law of demand which covers the Giffen-good case. The explanation for the occurrence of a Giffen good is that in its case the negative income effect outweighs the substitution effect. Since Marshall ignored the income effect of the change in price, he could not provide a satisfactory explanation for the reaction of the consumer to a change in price of a Giffen good.

However, it may be pointed out that it is very hard to satisfy the above mentioned third conditions for the occurrence of the Giffen good, namely, the consumer must be spending a very large proportion of his income on an inferior good. Therefore, although Giffen case is theoretically possible the chance of its occurrence in the actual world is almost negligible.[6] This is because consumption of the people is generally diversified so that people spend a small proportion of their income on a single commodity with the result that price-induced income effect even when negative is generally small and cannot therefore outweigh the substitution effect. As mentioned earlier, Marshall believed that quantity demanded in actual practice could vary directly with price, and, as mentioned

4. J.R. Hicks, *A Revision of Demand Theory*, p. 66.
5. Tapas Majumdar, *Measurement of Utility*, pp. 74-75.
6. J.R. Hicks, *op.cit*, p. 67.

above, Sir Robert Giffen is said to have actually observed this phenomenon. But there is a controversy about the intrepretation of this so-called Giffen good. But from our analysis it is clear that Giffen good case can occur in theory. As explained above, when negative income effect of the fall in the price of an inferior good is larger than substitution effect we get a positively-sloping demand curve of Giffen good. Thus Giffen good is theoretically quite possible. But, since income effect of the change in price of a single commodity in the real world is small, the negative income effect of the change in price of an inferior good is too weak to outweigh the substitution effect and therefore a Giffen good, although theoretically conceivable, rarely occurs in practice.

DERIVATION OF INDIVIDUAL'S DEMAND CURVE FROM INDIFFERENCE CURVE ANALYSIS

Price consumption curve traces the effect of a change in price on the quantity demanded of a good. But price consumption curve does not directly relate price with quantity demanded. In indifference curve diagram price is not explicitly shown on the *Y*-axis. On the other hand, demand curve directly relates price with quantity demanded, price being shown on the *Y*-axis and quantity demanded on the *X*-axis. A demand curve shows how much quantity of a good will be purchased or demanded at various prices, assuming that tastes and preferences of a consumer, his income, prices of all related goods remain constant. This demand curve showing explicit relationship between price and quantity demanded can be derived from price consumption curve of indifference curve analysis.

In Marshallian utility analysis, demand curve was derived on the assumptions that utility was cardinally measurable and marginal utility of money remained constant with the change in price of the good. In the indifference curve analysis, demand curve is derived without making these dubious assumptions.

Let us suppose that a consumer has got income of ₹ 300 to spend on goods. In Fig. 9.40 money is measured on the *Y*-axis, while the quantity of the good *X* whose demand curve is to be derived is measured on the *X*-axis. An indifference map of a consumer is drawn along with the various budget lines showing different prices of the good *X*. Budget line PL_1 shows that price of the good *X* is ₹ 15 per unit. As price of good *X* falls from ₹ 15 to ₹ 10, the budget line shifts to PL_2. Budget line PL_2 shows that price of good *X* is ₹ 10. With a further fall in price to ₹ 7.5 the budget line takes the position PL_3. Thus PL_3 shows that price of good *X* is ₹ 7.5. When price of good *X* falls to ₹ 6, PL_4 is the relevant budget line. The various budget lines obtained are shown in the column 2 of the following table. Tangency points between the various budget lines and indifference curves, which when joined together by a line constitute the price consumption curve which shows the amounts of good *X* purchased or demanded at various prices. With the budget line PL_1 the consumer is in equilibrium at point Q_1 on the price consumption curve at which the budget line PL_1 is tangent to indifference curve IC_1. In his equilibrium position at Q_1 the consumer is buying *OA* units of the good *X*. In other words, it means that the consumer demands *OA* units of good *X* at price ₹ 15. When price falls to ₹ 10 and thereby the budget line shifts to PL_2, the consumer comes to be in equilibrium at point Q_2 of the price-consumption curve where the budget line PL_2 is tangent to indifference curve IC_2. At Q_2, the consumer is buying *OB* units of good *X*. In other words, the consumer demands *OB* units of the good *X* at price ₹ 10. Likewise, with budget lines PL_3 and PL_4, the consumer is in equilibrium at points Q_3 and Q_4 of price consumption curve and is demanding *OC* units and *OD* units of good *X* at price ₹ 7.5 and ₹ 6 respectively. It is thus clear that from the price consumption curve we can get information which is required to draw the demand curve showing directly the amounts demanded of the good *X* against various prices. With the above information we draw up the demand schedule in the table given as follows:

Demand Schedule

Price of good X ₹	*Budget Line*	*Quantity Demanded*
15	PL_1	OA
10	PL_2	OB
7.5	PL_3	OC
6	PL_4	OD

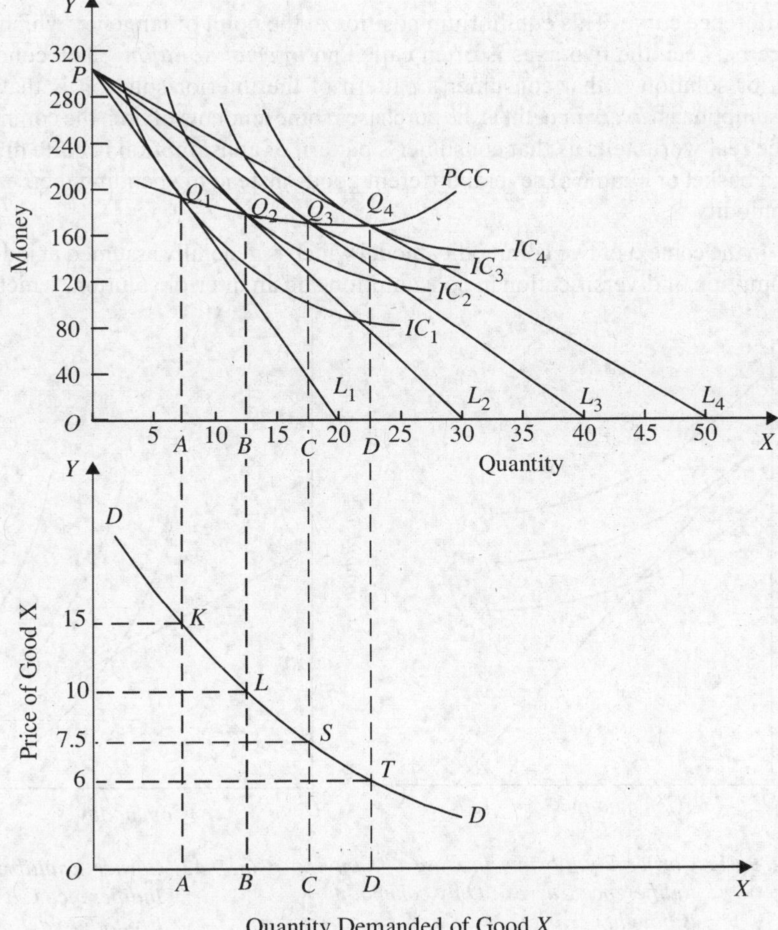

Fig. 9.40. *Derivation of Individual's Demand Curve.*

The adjoining demand schedule which has been derived from the indifference curve diagram can be easily converted into a demand curve with price shown on the *Y*-axis and quantity demanded on the *X*-axis. It is easier to understand if this demand curve is drawn rightly below the indifference curve diagram. This has been done so in Fig. 9.40. In the diagram at the bottom, on the *X*-axis the quantity demanded is shown as in indifference curves diagram in top pannel, but on the *Y*-axis in the diagram at the bottom *price per unit* of the good *X* is shown instead of total money. In order to obtain the demand curve, various points *K*, *L*, S and *T* representing the demand schedule of the above table are plotted. By joining the points *K*, *L*, *S* and *T* we get the required demand curve *DD*. In most cases the demand curve of individuals will slope downward to the right, because as the price of a good falls both the substitution effect and income effect pull together in extending the demand of the good. Even when the income effect is negative, the demand curve will slope downward to the right if the substitution effect is strong enough to overwhelm the negative income effect. Only when the negative income effect is powerful enough to outweigh the substitution effect can the demand curve slope upward to the right instead of sloping downward to the left.

Consumer's Equilibrium: Corner Solution

When a consumer's preferences are such that he likes to consume some amount of both the goods, he reaches an equilibrium position at the point of tangency between the budget line and his indifference curve. This equilibrium position at the point of tangency which lies within commodity space between the two axes is often called an *interior solution.* The economic implication of the interior solution is that consumer's pattern of the interior solution is that consumer's pattern of consumption is *diversified*, tht is, he purchases some amounts of both the commodities. Our knowledge of the real world tells us that consumer's pattern of consumption is quite diversified and they often buy a basket or bundle of several different goods instead of spending their entire income on a single commodity.

In the context of two commodity model which is generally assumed in indifference curve analysis, assumption of diversification in consumption and an interior solution, which implies that consumer

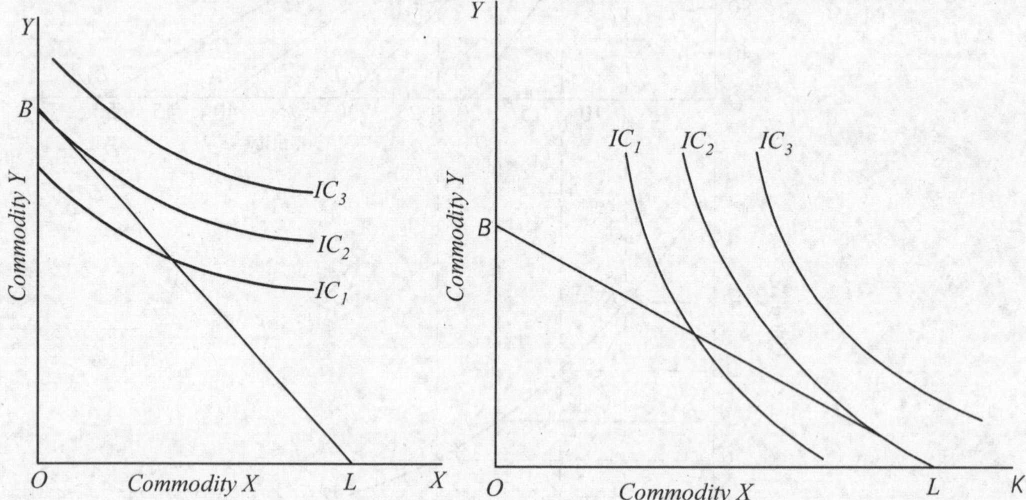

Fig. 9.41. *Corner Equilibrium in Case of Convex Indifference Curves : Only commodity Y is bought.*

Fig. 9.42. *Corner Equilibrium in Case of Convex Indifference Curves : Only commodity X is purchased.*

purchases some amount of both goods, is correct. However, in the real world of many commodities we often find that a typical consumer does not buy positive amounts of all the goods and services available in the market. In fact, a typical consumer buys only a small number of goods available in the market. How to explain this real world phenomenon?

Convex Indifference Curves and Corner Equilibrium

The reason for not purchasing a commodity by a consumer may be that the price or opportunity cost of that particular commodity may be too high for him. One may like to have a Maruti car, air conditioner or a colour TV, but may not actually have them on account of their prices being too high. The indifference curve analysis enables us to explain even this phenomenon. Consider Fig. 9.41, where indifference map between goods *X* and *Y* and budget line *BL* are such that the interior solution is not possible and consumer in its equilibrium position at point *B* will not consume any quantity of commodity *X*. This is because as seen in the Fig. 9.41 the price of commodity *X* is so high that budget line is steeper than the indifference curves between the two commodities. In economic terms it means that the price or opportunity cost of commodity *X* in the market is greater than the marginal rate of substitution of *X* for *Y* which indicates willingness to pay for the commodity *X*. The price of goods *X* is so high as compared to marginal rate of substitution (willingness to pay for *X* or the marginal valuation of the first unit of the commodity *X* that the consumer does not purchase even one unit of the commodity *X* ($P_x/P_y > MRS_{xy}$). Thus the consumer maximises his satisfaction or is in equilibrium at the corner point *B* where he buys only commodity *Y* and none of commodity *X*. Thus we have a corner solution for consumer's equilibrium.

On the other hand, in Fig. 9.42 the indifference map between the two goods is such that the budget line *BL* is less steep than the indiffrence curves between the two goods so that the $MRS_{xy} > P_x/P_y$ for all levels of consumption along the budget line *BL*. Therefore, he maximizes his satisfaction at the corner point *L* where the buys only commodity *X* and none of *Y*. In this case price of commodity *Y* and willingness to pay (*i.e. MRS*) for it are such that he does not consider it worth while to purchase even one unit of it.

Corner Equilibrium and Concave Indifference Curves

As said above, indifference curves are usually convex to the origin. Convexity of indifference curves implies that the marginal rate of substitution of *X* for *Y* falls as more of *X* is substituted for *Y*. Thus, indifference curves are convex to the origin when principle of diminishing marginal rate of substitution holds good which is generally the case. But the possibility of indifference curves being concave to the origin cannot be ruled out in some exceptional cases. Concavity of the indifference curves implies that the marginal rate of substitution of *X* for *Y* increases when more of *X* is substituted for *Y*. It will be clear from the analysis made below that in the case of indifference curves being concave to the origin, the consumer will choose or buy only one good. In other words, *concavity of indifference curves implies that the consumer has a distaste for variety, that is, does not like diversification in consumption*. However, distaste for variety cannot be considered a normal or model behaviour, so we regard convexity to be the general case. But when consumer have a distaste for variety and diversification the case of concave indifference curves will occur.

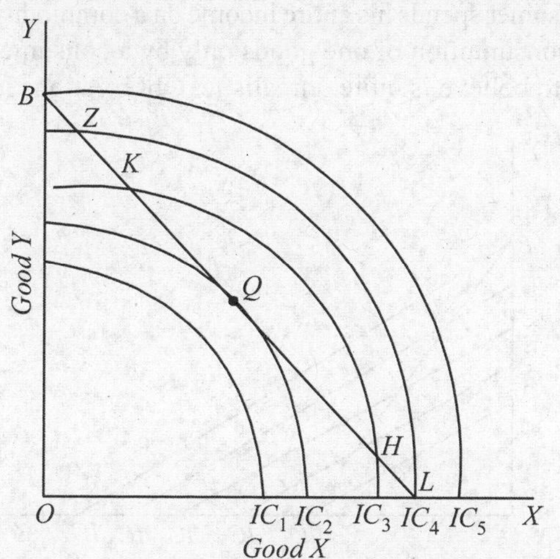

Fig. 9.43. *Consumer Equilibrium in Case of Concave Indifference Curves*

In case of concave indifference curves, the consumer will not be in equilibrium at the point of tangency between budge line and indifference curve, that is, in this case interior solution will not exist. Instead, we would have *corner solution* for consumer's equilibrium. Let us take Fig. 9.43 where indifference curves are shown to be concave. The given budget line *BL* is tangent to the indifference curve IC_2 at point *Q*. But the consumer cannot be in equilibrium at *Q* since by moving along the given budget line *BL* he can get on to higher indifference curves and obtain greater satisfaction than at *Q*. Thus, by moving to *K* on the given budget line *BL*, he will get more satisfaction than at *Q* since *K* lies on a higher indifference curve than *Q*. He can increase his satisfaction still more by moving to point *Z* on the budget line *BL*.

Thus, as he moves upward from tangency point Q on the budget line his satisfaction will go on increasing until he reaches the extremity point B. Likewise, if from Q he moves downward on the budget line, he will get on to higher indifference curves and his satisfaction will go on increasing till he reaches the other extremity point L. In these circumstances the consumer will choose only one of two goods: he will buy either X or Y depending upon whether L or B lies on the higher indifference curve. In the situation depicted in Fig. 9.43 point B lies on a higher indifference curve than point L. Therefore, the consumer will choose only Y and will buy OB of Y. It should be carefully noted that at B the budget *line is not tangent to the indifference curve I_5 even though the consumer is here in equilibrium*. Thus, in case of concave indifference curve, we have a corner solution to consumer's equilibrium. It is clear that when a consumer has concave indifference curves, he will succumb to monomania, that is, he will consume only one good.

Conclusion

In our analysis above, we have shown that corner solution of consumer's equilibrium is possible even when his indifference curves between goods are convex. It is worth noting that *in case of convex indifference curves, corner equilibrium is however not inevitable; it occurs only when price of a commodity is too high as compared to the marginal rate of substitution of even the first unit of the commodity. However, when the indifference curves are concave, consumer's equilibrium will inevitably be a corner solution.* This implies that more of commodity X a consumer has the more useful or significant in terms of satisfaction an extra unit of it becomes. Therefore, the concave indifference curves do not seem to be plausible or realistic. Now, as seen above, the concavity of indifference curves for a consumer implies that the consumer spends his entire income on a commodity and therefore buys only one commodity. However, consumption of one goods only by a consumer which the concavity of indifference curves leads us to believe is quite unrealistic. Observations in the real world reveal that consumers do not spend their entire income on a single commodity and in fact purchase a multitude of different goods and services. This rejects the existence of concave indifference curves.

Our anyalsis of inevitability of corner equilibrium in case of concave indifference curves provides us an important *economic rationale for indifference curves being convex rather than concave*. If indifference curves were predominantly concave the consumers would spend their entire income on a single commodity alone and thus consume only one commodity. This is quite inconsistent with the observed behaviour of consumers. This reinforces our belief that indifference curves of consumers are generally convex.

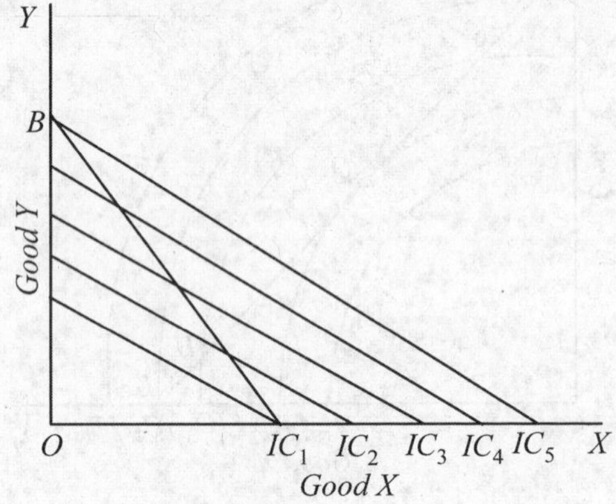

Fig. 9.44. *Corner Equilibrium in Case of Perfect Substitutes*

Corner Solution in Case of Perfect Substitutes and Perfect Complements

Another case of corner solution to the consumer's equilibrium occurs in case of perfect substitutes. As seen above, indifference curves for perfect substitute are linear. In their case too, *tangency or interior solution for consumer's equilibrium is not possible since the budget line cannot be tangent to a point of the straight-line indifference curve of substitutes.* In this case budget line would cut the straight-line indifference curves. Two possibilities can be visualised: either the slope of the budget line BL can be greater than the slope of indifference curves, as in Fig. 9.44 or the slope of the budget line can be less than the slope of indifference curves, as in Fig. 9.45. If the slope of the budget line is greater than the slope of indifference curves, *B* would lie on a higher indifference curve than *L* and therefore the consumer will buy only *Y*. If the slope of the budget line is less than the slope of indifference curves as in Fig. 9.45, *L* would lie on a higher indifference curve than *B* and the consumer will buy only *X*. It should be noted that in these cases too, the consumer will not be in equilibrium at any point between *B* and *L* on the budget line since in case of Fig. 9.44 of all the points on the given budget line extremity point *B* would lie on the highest possible indifference curve and in case of Fig. 9.45 of all the points of the budget line extremity point *L* would lie on the highest possible indifference curve. It is thus manifest that *even in case of perfect substitutes, the consumer will succumb to monomania* and equilibrium will be achieved at a corner point.

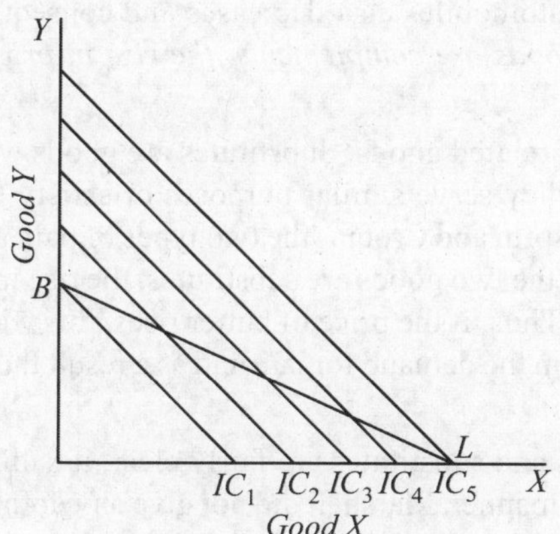

Fig. 9.45. *Corner Equilibrium in Case of Perfect Substitutes*

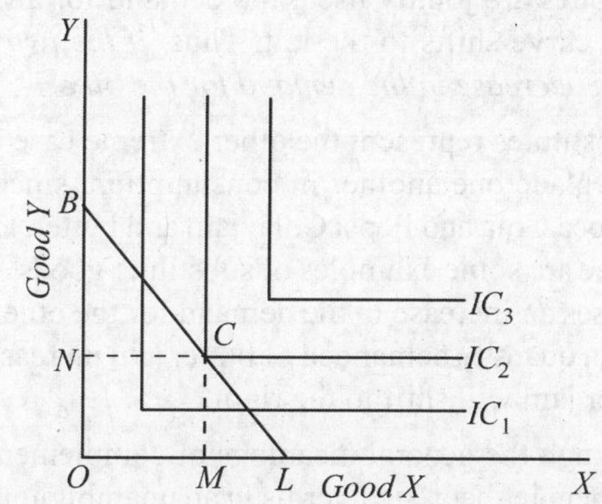

Fig. 9.46. *In case of perfect complementary equilibrium exists at the corner of an indifference curve.*

Another non-normal case is of *perfect complementary goods*, which is depicted in Fig. 9.46. Indifference curves of perfect complementary goods have a right-angle shape. In such a case the equilibrium of the consumer will be determined at the corner point of the indifference curve through which the budget line passes. At no other point of the indifference curve of perfect complementary goods, equilibrium is possible. In Fig. 9.46 given the budget line *BL* the consumer will be in equilibrium at the corner point *C* on indifference curve IC_2 and consumer will buy *OM* of *X* and *ON* of *Y*. It should be noted, in strict mathematical sense, point *C* in Fig. 9.46 is not a tangency point as slope of IC_2 at point *C* is not well defined. Thus, even in case of perfectly complementary goods tangency solution to consumer's equilibrium is not achieved.

EXERCISE: *Is tangency between an indifference curve and the budget line a necessary condition for a bundle to be an optimal. Substantiate your answer through the following situations*:

(1) *kinked indifference curves, and*

(2) *strictly convex preferences leading to a corner optimum.*

[**Hints.** Tangency between an indifference curve and budget line is not a necessary condition for consumer's equilibrium. This is firstly because even in case of convex indifference curves a consumer *may be in equilibrium at the corner* where he buys only one commodity as we have seen in Figure 9.44 and 9.45 where the shopes of budget line are no where equal to the slopes of indifference curves in the interior combination of goods. Secondly, we have seen in case of perfect complements in whose case indifference curves are kinked, the consumer is in equilibrium at the corner or kink of the indifference curve and consumer's equilibrium is not a tangency solution as slope of a corner or kink point is not well-defined.

COMPLEMENTS AND SUBSTITUTES

An important use of indifference curves is to explain complements and substitutes in a better way. It will be recalled that complements and substitutes represent the case of related goods. Whenever, a demand curve for a good is drawn it is assumed that the other things such as income, consumer preferences, the prices of other goods remain constant because they also affect the demand for a particular good. Complements are those goods which are used jointly in consumption so that their consumption increases or decreases simultaneously. Pen and ink, right shoe and left shoe, automobile and petrol, typewriter and typists, sauce and hamburger are some examples of complements. If the price of petrol rises, it affects the demand for automobiles such as car, scooters etc. This is because the rise in price of petrol leads to the fall in its quantity demanded and due to the fact that petrol and automobiles are jointly used, the demand for an automobiles also decreases and consequently its demand curve shifts to the left. Thus, *if the two goods are complements, the rise in price of one causes a decrease in the demand for the other*.

Substitutes represent the other extreme case of related goods. Substitutes are goods which can easily replace one another in consumption, since they serve similar purposes or satisfy the same want. Coca Cola and Pepsi Cola, jam and butter. Disprin and Crocin (the two types of pain pills), tea and cofee are some examples of substitute goods. If the two goods are substitutes, the rise in price of one causes an increase in the demand for the other. Thus, if the price of butter rises, this will lead to the fall in quantity demanded of butter and increase in the demand for jam and as a result the demand curve for jam will shift to the right.

Though the general meanings of complements and substitutes are fairly clear, it is difficult to define them in theoretical terms in an unambiguous manner. Marshall did not give any definitions of substitute and complementary goods. However, before Marshall, Edgeworth and Pareto had provided the definitions of substitute and complementary goods in terms of marginal utility. According to Edgeworth-Pareto definitions, "*Y is a complementary with X in the consumer's budget if an* increase in the supply of *X* (*Y* constant) raises the marginal utility of *Y*; *Y* is competitive with *X* (or a substitute for *X*) if an increase in the supply of *X* (*Y* constant) lowers the marginal utility of *Y*"[7]

From above Edgeworth-Pareto definition, it follows that assuming marginal utility of money remains constant if the price of good *X* falls and consequently the quantity demanded of good *X* increases, this will bring about an increase in the marginal utility of good *Y* if goods *X* and *Y* are complementary, and will therefore bring about increase the demand for *Y*. On the contrary, if goods *X* and *Y* are substitutes, according to Edgeworth-Pareto definition, the fall in the price of good *X* and consequently the increase in the quantity demanded of *X* will lower the marginal utility of *Y* and thereby bring about a decline in the demand for *X*. Thus, it is in this way that Edgeworth and Pareto explained the demanded for inter-related goods–complementary and substitute goods.

However, J.R. Hicks, the pioneer of indifference curve analysis of demand, criticized Edgeworth-Pareto definition of complements and substitutes. According to him, Edgeworth-Pareto definition of complementary and substitute goods is based on the assumption that utility is cardinally measurable. But Pareto regarded the utility to be immeasurable in cardinal or quantitative sense. Therefore, Pareto contradicted himself by defining complementary and substitute goods, in terms of measurable utility. Thus, according to Hicks, "Edgeworth-Pareto definition sins against Pareto's own principle of the immeasurability of utility. The distinction between complemetary and competitive goods will differ according to the arbitrary measure of utility which is adopted."[8]

Hicksian Explanation of Complementary and Substitute Goods

With indifference curves analysis of demand in which price effect was split up into substitution

7. J.R. Hicks, *Value and Capital*, p. 42
8. *Op., cit.*, p.43

effect and income effect, Hicks was able to explain in a satisfactory way the cases of substitute and complementary goods. As seen above, before Hicks, substitutes and complementary goods were generally explained in terms of *total price effect*. According to Hicks, if income effect is taken into account, then even if with the fall in price of *X*, the quantity demanded of good *Y* may also increase even though the good *Y* may be substitute or competitive good. Suppose with the fall in price of good *X*, there is a large income effect which more than offsets the substitution effect. Income effect of the fall in price of good *X* tends to increase the quantity demanded of good *Y* (as also of the good *X*) and the substitution effect of the fall in price of *X* works in favour of *X* (that is, tends to increase its quantity demanded) and against good *Y* (that is, tends to reduce its quantity demanded). When this income effect for *Y* is stronger than substitution effect, then the quantity demanded of *Y* increases as a result of the fall in price of *X*, even though the two may be substitute goods. Therefore, effect for the commodity *Y* which has become relatively dearer due to the fall in price of good *X*, the purchases of both goods *X* and *Y* increase as a result of the fall in price of good *X*. Thus, on the basis of total price effect, the goods would be described as complements, even though they are in fact substitute goods. Therefore, according to Hicks, goods can be classified as substitutes or complements more accurately by reference to the substitution effect or preference function alone. Hence, *in the opinion of Hicks, we can define substitute and complementary goods correctly and precisely only in a situation when we have eliminated the income effect of the price change by making a compensating variation in income*. When with a change in price compensating variation in income is also made, the effect which remains is the substitution effect.

Since indifference curve analysis splits up the price effect into income and substitution effects, it is greatly helpful in analysing the relations of substitution and complementarily. Take two goods *X* and *Y*. If the price of good *X* falls, price of *Y* remaining constant, the quantity demanded of good *X* will increase due to the substitution effect and income effect (we suppose that good *X* is not an inferior good). Now, if after the income of the consumer is reduced by compensating variation in income so that with reduced price of good *X* he is no better off than before, the quantity demanded of *X* increases and the quantity demanded of *Y* declines, then good *Y* is a substitute for *X*. In this case due to the relative fall in its price, good *X* has been substituted for good *Y* and because of compensating variation in income, consumer is no better off than before.

Now, if the price of good *X* falls and after making compensation variation in income, the quantity demanded of *X* increases due to the substitution effect and if with it the quantity demanded of *Y* also increases, then *Y* is complement of *X*. Thus, in this case of complements, the quantity purchased of both the goods increases and *both substitute some other good*. Consumer is no better off than before, since compensating variation in income having been made, the quantities purchased of two complementary goods increase due to the substitution effect alone. In view of the above analysis, Professor Hicks defines the substitutes and complements in the following way :

"*I shall say Y is a substitute of X if a fall in the price of X leads to a fall in the consumption of Y; Y is a complement of X if a fall in the price of X leads to a rise in the consumption of Y; a compensating variation in income being made of course in each case*. Thus, a fall in the price of X, combined with a compensated variation in income, which must tend to increase the consumption of *X* itself (by the first substitution theorem), will increase the consumption of complements, but diminish the consumption of substitutes."[9]

We have been above that the type of relation of substitutability or complementarity depends on the substitution effect. To determine the substitution effect is quite simple if there are only two commodities on which the consumer has to spend his money income. We know that a fall in the price of good *X* always leads to the substitution of *X* for the other goods; and if *Y* was the only other goods available to the consumer, then the substitution effect of the fall in price of good *X* must necessarily

9. J.R. Hicks, *A Revision of Demand Theory*, Oxford University Press, 1956, p. 128

reduce the quantity demanded of *Y*. However, when there are more than two goods, a fall in the price of good *X* may not reduce the quantity demanded of *Y*; it may in fact increase the quantity purchased of good *Y*, if the two goods *X* and *Y* happen to be complements. Here the two goods *X* and *Y* are substituted for *some other things*.

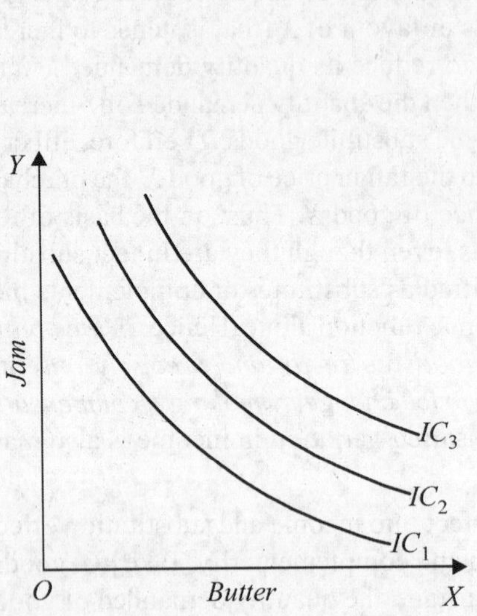

Fig. 9.47. *Close Substitutes*

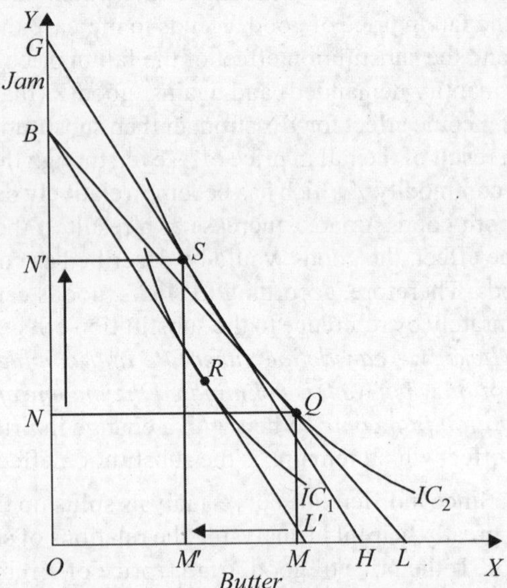

Fig. 9.48. *Close Substitutes : A large replacement of one good by another resulting from a small change in relative price.*

From the above description, it is clear that the definition and proper analysis of substitutes and complementary goods require at least three goods. That is why J.R. Hicks in his *value and Capital* defined them by taking three commodities, *X*, *Y* and money and in terms of the concept of marginal rate of substitution. It should be remembered that money stands for all other goods lumped together and is known as a *composite commodity*.

However, in order to keep our analysis simple we confine ourselves to depict and explain substitutes and complements taking two goods case which tables us to explain them with two dimensional indifference curves. We first take up substitutes for our analysis.

Analysis of Substitutes with Indifference Curve

As mentioned above, the indifference curves of substitute goods are nearly linear, that is, they have a slight convexity. Jam and butter, *HCL* computers and Wipro computers, wheat and rice, Kashmiri apples and H.P. apples are some examples of good substitutes. The indifference curves between two substituttes, jam and butter, and drawn in Fig. 9.47, which are nearly linear with a very small degree of convex curvature. In this case of chose substitutes a very small change in the relative price of one good leads to a relatively large change in quantity demanded of the other. Consider Fig. 9.48. Suppose to begin with, budget constraint and price ratio of two goods, jam and butter, are such that we have a budget line *BL*. As will be seen from Fig. 9.48, the budget line is tangent to the indifference curve IC_2 at point *Q* and accordingly demand *OM* of butter and *ON* of jam. Now suppose that price of butter rises slightly so that the budget line rotates and attains the new position *BL'*. In order to find out the true degree of substitutability, we need to increase the income of the individual to compensate him for the rise in price (that is, to increase his income equal to compensating variation, as has been suggested by J.R. Hicks). Thus, if following the rise in price of butter, if his income is raised by *BG* in terms of jam (see Fig. 9.48) so that the budget line representing new price ratio shifts from *BL′* to *GH*. With budget line *GH* he reaches in equilibrium position at point *S* indicating that with the slight rise in price of butter and after the compensating variation in income has been made the individual substitutes a large quantity *NN′* of jam for *MM′* of butter. This shows that the two goods, jam and butter, are close substitutes.

In case of perfect substitutes, the indifference curves are parallel straight lines because the consumers equally prefer the two goods and is willing to exchange one good for the other at a constant rate. As one moves along the straight-line indifference curve of perfect substitutes, marginal rate of substitution of one good for the other remains constant.

Examples of goods that are perfect substitutes are not difficult to find in the real world. For example, Dalda and Rath Vanaspati, two different brands of cold drink such as Pepsi Cola and Coca Cola, two different brands of tooth paste (such as Colgate and Binaca), two different brands of butter with equal quality and fat content. It is however worth noting that if the price ratio of the two perfect substitute goods differs from the slope of their straight-line indifference curves, there will be a *corner solution* with respect to consumer's equilibrium. That is, the consumer will consume only one good in this case, and not any combination of the two. This is illustrated in Fig. 9.45 and where five parallel straight-line indifference curves between two perfect substitute goods have been drawn. If the price-ratio of the two goods *X* and *Y* is represented by the slope of the budget line *BL*, he will be in equilibrium at the corner point *L* which lies on the higher indifference curve IC_5 and will consume commodity *X* alone. On the other hand, if the price ratio of the two commodities is such that the slope of the budget line *BL* is less than the slope of the indifference curve the consumer will be in equilibrium at the corner point *B* which lies on the higher indifference curve IC_3 (as compared to point *L* which lies on lower indifference curve IC_1) and will consume commodity *Y* alone (see Fig. 9.44).

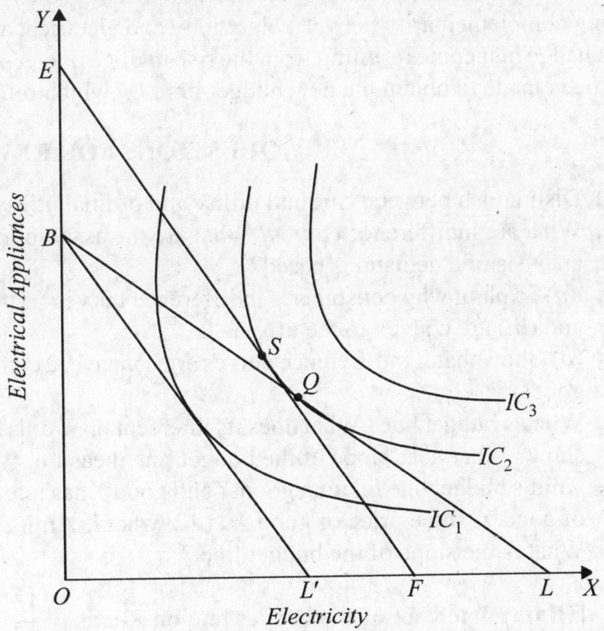

Fig. 9.49. *In case of complementary goods, a large change in price ratio causes only a small change in their quantities demanded.*

Analysis of Complements with Indifference Curves

In indifference curve analysis, the case of *perfect complements* is depicted by L-shaped or right angled indifference curves as shown in Fig. 9.49. The two perfect complementary goods are used not only jointly but they are also used in a fixed ratio, that is; one is useless without the required quantity of the other. In other words, perfect complements are those jointly used goods which cannot be substituted for each other at all; the substitution effect of the change in price of one good is zero in case of perfect complements. Right and left shoes, right nd left gloves are important examples of perfect complements. However, it may be noted that as in case of perfect substitutes the perfect complements also do not have any practical significance. This is because consumers never consider right and left shoes, right and left gloves as *separate* goods. They lend to consider *pairs* of shoes or *pairs* of gloves in their demand for them.

But, of practical significances are the cases of two goods which are *strong* (but not perfect) complements. Their indifference curves have appearance as shown in Figure 9.49, where the indifference curves between electricity and electrical appliances with a very high degree of convexity have been drawn. It is worth mentioning that in case of these strong complements relatively large changes in their price ratio produces only a small changes in their quantities used or demanded. A glance at Fig. 9.49, reveals that with a given price ratio and the budget constraint, consumer is in equilibrium at point Q. With a relatively large change in price ratio from the slope of budget line BL to the slope of the budget line EF (that is, price of electricity has risen very high), *only a small decline in quantity demanded of electrical appliances has taken place*. This shows the presence of strong complementarity between electricity and electrical appliances. It may be noted that to offset the change in income resulting from the rise in price of electricity, compensating variation in income has been made to obtain the new budget lime EF which reflects the new price ratio.

QUESTIONS FOR REVIEW

1. Distinguish between cardinal utility and ordinal utility? Which is more realistic?
2. What are indifference curves? What are the assumptions on which indifference curve analysis of a demand is based?
3. (*a*) Explain why consumer's indifference curves (*i*) have negative slope, (*ii*) do not intersect and (*iii*) are convex to the origin.
   (*b*) Show that if indifference curves are concave, a consumer will consume only one of the two goods.
4. What is budget line? What does its intercept on X-axis show? What does its intercept on Y-axis show? What does slope of the budget line measure?
5. Amit's budget line relating good X and good Y has intercepts of 50 units of good X and 20 units of good Y. If the price of good X is 12, what is Amit's income? What is the price of good Y? What is the slope of the budget line ?

[**Hints** : Intercept of the budget line on X-axis $= \dfrac{M}{P_x}$ (where M = income and P_x = Price of good X)

$$50 = \frac{M}{12}, \text{ or } M = 600$$

$$\text{Intercept on } Y-\text{axis} = \frac{M}{P_y}$$

$$\text{or} \quad 20 = \frac{600}{P_y}$$

$$P_y = \frac{600}{20} = 30$$

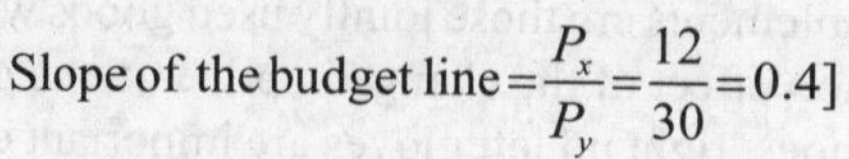

$$\text{Slope of the budget line} = \frac{P_x}{P_y} = \frac{12}{30} = 0.4]$$

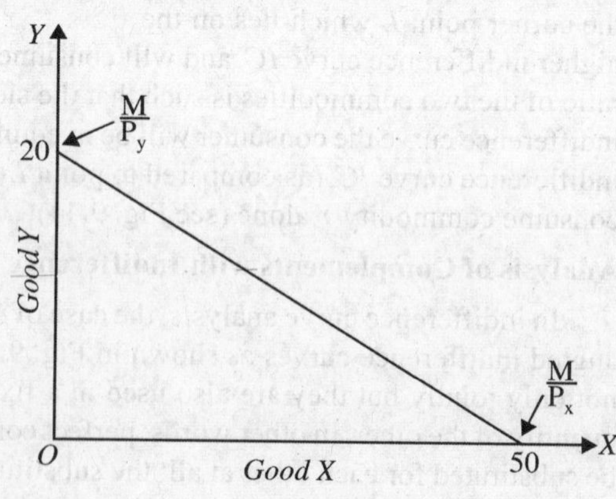

Fig. 9.50

6. What is marginal rate of substitution? An indifference curve of Sonia contains the following market baskets of apples and bananas. Each of these baskets gives her equal satisfaction.

Market basket	*Apples*	*Bananas*
1	2	16
2	3	11
3	4	7
4	5	4
5	6	2
6	7	1

Find out marginal rate of substitution of Sonia. How does marginal rate of substitution vary as she consumes more of apples and less of bananas? Give reasons.

7. Explain consumer's equilibrium condition with the help of indifference curve approach. How will a change in consumer's income affect his equilibrium ? *(DU,BA(H), B.Com (H).*

8. Explain why a consumer will choose a market basket so that marginal rate of substitution (*MRS*) equals price ratio.

9. A consumer spends all her income on food and clothing. At the current prices of P_f= ₹ 10 and P_c= ₹ 5, she maximises her utility by purchasing 20 units of food and 50 units of clothing.

(*a*) What is the consumer's income?

(*b*) What is the consumer's marginal rate of substitution of food for clothing at the equilibrium position ?

[**Hints:** (*a*) Income (*M*) can be obtained from budget equation :

$$M = P_f.Q_f + P_c.Q_c$$
$$= 10 \times 20 + 5 \times 50$$
$$= 200 + 250 = ₹\ 450$$

(*b*) Consumer's marginal rate of substitution at the equilibrium point is equal to the ratio of prices of goods. Thus, $MRS_{fc} = \frac{P_f}{P_c} = \frac{10}{5} = 2$

10. What is the inferior good? How is it different from normal goods? Show them with the help of indifference curves.

11. Use indifference curves to explain what happends to the demand for an inferior good as consumers's real income increases at constant relative prices. *[C.U.B. Com., (H)1999]*

12. What is price consumption curve? What is the relationship between price consumption curve and price elasticity of demand?

(*b*) Derive demand curve from Price Consumption Curve *[D.U. B.Com., (H)]*

13. What is income consumption curve. Draw indifference curve diagrams showing the income consumption curve in the following case:

(*a*) both *X* and *Y* are normal goods

(*b*) good *X* is normal good and good *Y* is inferior ,

(*c*) good *X* is inferior good and good *Y* is normal good.

14. (*a*) What is an Engel curve? How is Engel curve derived from income consumption curve?

(*b*) Draw an Engel curve for (*i*) a necessary good, (*ii*) luxury good (*iii*) neutral good.

15. Using indifference curve analysis, show how price effect of a commodity is decomposed into income effect and substitution effect. *[D.U.B. Com.(H) 2000; C.U.B. Com., 1998]*

16. With the help of indifference curve analysis, derive demand curve for a normal commodity? Explain why it slopes downward to the right.

17. What is Giffen good? How does indifference curve analysis explain Giffen Paradoix? What is the shape of price consumption curve for a Giffen good? Illustrate it with an indifference curve diagram.

18. Derive demand curve for a Giffen good. Why has it a positive slope? *(D.U.B.com (H) 1996)*

19. What is an inferior good? What is the shape of a demand curve for an inferior good which is not a Giffen good?

20. .'All Giffen' goods are inferior goods but all inferior goods are not Giffen goods." Do you agree? Give reasons. *(D.U. B.Com.(H)2001)*

21. Why is an indifference curve generally convex to the point of orgin ? When will it be : (*i*) a straight line (*ii*) right-angled *(D.U. B.Com. (H) 2000)*

22. What is compensating variation in income ? Using the concept of compensating variation in income distinguish between income and substitution effects of a fall in price of a commodity with the help of indifference curve technique in case of (*a*) normal good, and (*b*) a Giffen good.

23. (*a*) How does Hicks define substitute and complementary goods ? How do his definitions differ from those given by Edgeworth and Pareto

(*b*) Analyse substitute and complementary goods with Hicksian indifference curve analysis of demand.

24. Differentiate between inferior goods and Giffen goods with the help of indifference curve analysis of demand.

25. An individual's marginal utilities for commodities are given by the following relations : $MU_x = 40 - 5X$ and $MU_y = 20 - 3Y$.

What is his marginal rate of substitution in consumption at the consumption basket $X=3, Y=5$? If $P_x =$ ₹5 and $P_y =$ Re. 1, is the basket a consumption equilibrium ? *(DU. BA (H))*

[**Hints :** $MRS_{ny} = \frac{MU_x}{MU_y} = \frac{40-5X}{20-3Y} = \frac{40-(5\times3)}{20-(3\times5)} = \frac{25}{5} = 5$ and $\frac{P_x}{P_y} = \frac{5}{1}$

Thus $MRS_{xy} = \frac{P_x}{P_Y}$

Thus, the given basket of consumption ($X = 3$ & $Y = 5$) is an equilibrium basket].

26. If we observe an individual to demand less of a good as its price falls, we may conclude that it is an inferior good for him. Do you agree ? Explain *(D.U. BA. (Hons) 1993)*

(**Hint :** When an individual demands less of a good when it price falls, this is the case of a Giffen good. All Giffen goods are inferior goods. Therefore, we can conclude that the good will be inferior good).

27. A consumer is observed to consume only one of two goods available. Illustrate and explain in terms of the indifference curve model when this can happen. Mention in particular how the usual first order condition has to be interpreted in this case. *(D.U. B.A. (Hons.) 1986)*

28. If a consumer's indifference curves are concave, will he consume both the goods in equilibrium ? Why or why not ?

29. If a consumer spends all his income on two goods, can both the goods be inferior ? Explain *D.U. B.A. (Hons) 1997*

30. Assume that consumer's income consumption curve for good *X* is a vertical line. Show that his demand curve for *X* must be downward sloping. *D.U. B.A. (Hons) 2001*

[**Hints.** Vertical income consumption curve indicates that income effect for goods *X* is zero. Since substitution effect is always negative and operates to *increase the quantity* demanded of a good whose *price falls*, the demand curve for good *X* will be downward sloping]

APPENDIX B TO CHAPTER 9*

SLUTSKY SUBSTITUTION EFFECT

In Chapter 9, we have explained the concept of substitution effect put forward by J.R. Hicks. There is another important version of substitution effect put forward by E. Slutsky. The treatment of the substitution effect in these two versions has a significant difference. As explained earlier, while Hicksian substitution effect occurs, when with the change in price of a good money income is adjusted by such an amount that leaves *consumer's satisfaction* unchanged, Slutsky substitution effect comes about when money income is adjusted by such an amount that leaves *consumer's purchasing power* unchanged. Since Slutsky substitution effect has an important empirical and practical use, we explain below Slutsky's version of substitution effect in some detail.

In Slutsky's version of substitution effect when the price of a good changes and consumer's real income or purchasing power increases, the income of the consumer is changed by the amount equal to the change in its purchasing power which occurs as a result of the price change. His purchasing power changes by the amount equal to the change in the price multiplied by the :number of units of the good which the individual used to buy at the old price. In other words, in Slutsk's approach, income is reduced or increased (as the case may be), by the amount which leaves the consumer to be just able to *purchase the same combination of goods*, if he so desires, which he was having at the old price. That is, the income is changed by the difference between the cost of the amount of good *X* purchased at the old price and the cost of purchasing the same quantity if *X* at the new price. Income is then said to be changed by the *cost difference*. Thus, in Slutsky substitution effect, income is reduced or increased not by compensating variation as in case of the Hicksian substitution effect but by the cost difference.

Slutsky Substitution Effect For a Fall in Price

Slutsky substitution effect is illustrated in Fig. 9B.l With a given money income and the given prices of two goods as represented by the price line *PL*, the consumer is in equilibrium at *Q* on the indifference curve IC_1 buying *OM* of *X* and *ON* of *Y*. Now suppose that price of X falls, price of *Y* and money income of the consumer remaining unchanged. As a result of this fall in price of *X*, the price line will shift to *PL′* and the real income or the purchasing power of the consumer will increase. Now, in order to find out the Slutsky substitution effect, consumer's money income must be reduced by the cost-difference or, in other words, by the amount which will leave him to be just able to purchase the old combination *Q*, if he so

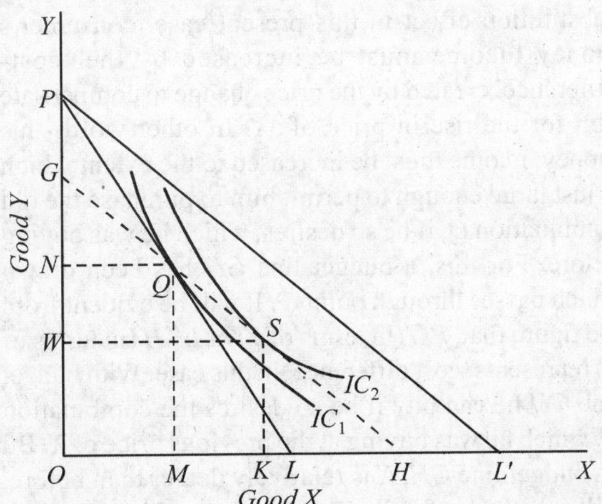

Fig. 9B.1. *Slutsky Substitution Effect (For a Fall in Price)*

---

* This appendix is meant for B.A (Hons.) and B.Com. (Honours) classes and should therefore be omitted by B.A. (Pass) Course students.

desires. For this, a price line *GH* parallel to *PL′* has been drawn which passes through the point *Q*. It means that income equal to *PG* in terms of *Y* or *L′H* in terms of *X* has been taken away from the consumer and as a result he can buy the combination *Q*, if he so desires, since *Q* also lies on the price line *GH*. Actually, he will not now buy the combination *Q* since *X* has now become relatively cheaper and *Y* has become relatively dearer than before. The change in relative prices will induce the consumer to rearrange his purchases of *X* and *Y*. He will substitute *X* for *Y*. But in this Slutsky substitution effect, he will not move along the same indifference curve IC_1, since the price line *GH*, on which the consumer has to remain due to the new price-income circumstances is nowhere tangent to the indifference curve IC_1. The price line *GH* is tangent to the indifference curve IC_2 at point *S*. Therefore, the consumer will now be in equilibrium at a point *S* on a higher indifference curve IC_2. This movement from *Q* to *S* represents Slutsky substitution effect according to which the consumer moves not on the same indifference curve, but from one indifference curve to another. A noteworthy point is that movement from *Q* to *S* as a result of Slutsky substitution effect is due to the change in relative prices alone, since the effect due to the gain in the purchasing power has been eliminated by making a reduction in money income equal to the cost-difference. At *S*, the consumer is buying *OK* of *X* and *OW* of *Y*; *MK* of *X* has been substituted for *NW* of *Y*. Therefore, Slutsky substitution effect on *X* is the increase in its quantity purchased by *MK* and Slutsky substitution *effect on* *Y* is the decrease in its quantity purchased by *NW*.

Slutsky Substitution Effect for a Rise in Price

We have graphically explained above Slutsky substitution effect for a fall in price of good *X*. It will be instructive to explain it also for a rise in price of *X*. This is demonstrated in Fig. 9B.2. Initially, the consumer is in equilibrium at point *Q* on the indifference curve IC_1, prices of the two goods and his money income being given. Now suppose that price of good *X* rises, price of *Y* remaining unchanged. As a result of the rise in price of *X*, budget line will shift downward to *PL″* and consumer's real income or purchasing power of his given money income will fall. Further, with this price change, good *X* has become relatively dearer and good *Y* relatively cheaper than before. In order to find out Slutsky substitution effect in this present case, consumer's money income must be increased by the 'cost-difference' created by the price change to compensate him for the rise in price of *X*. In other words, his money income must be increased to the extent which is just large enough to permit him to purchase the old combination *Q*, if he so desires, which he was buying before. For this, a budget line *GH* has been drawn which passes through point *Q*. It will be evident from the figure that, *PG* (in terms of *Y*) or *L″H* (in terms of *X*) represents 'cost difference' in this case. With budget line *GH* he can buy if he so desires the combination *Q*, which he was buying at the previous price of *X*. But actually he will not buy combination *Q*, since on budget line *GH*, *X* is relatively dearer than before, he will therefore replace some *X* by *Y* (*i.e.*, he will substitute of *Y* for *X*). As is shown Fig. 9B.2, with budget line *GH* he is in equilibrium position at *S* on a higher indifference curve of IC_2 and is buying *OK* of *X* and *OW* of *Y*. *MK* of *X* has been replaced by *NW* of *Y*. Movement from point *Q* to *S* is the result of Slutsky substitution effect; the effect due to the fall in purchasing power has been cancelled by giving him money equal to *PG* of *Y* or *L″ H* of *X*. In this present case of stipulated rise in price of *X*, Slutsky substitution effect on *X* is

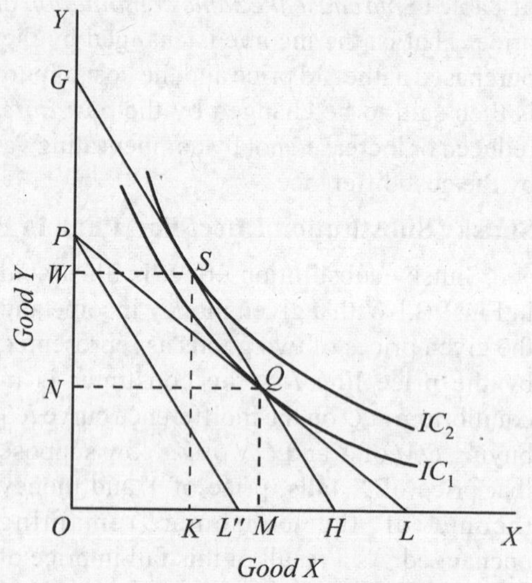

Fig. 9B.2. *Slutsky Substitution Effect (for a rise in price)*

the fall in its quantity brought by *MK* and Slutsky substitution effect an *Y* is the increase in its quantity brought by *NW*.

From the above analysis it is clear that whereas Hicks-Allen substitution effect takes place on the same indifference curve, Slutsky substitution effect involves the movement from one indifference curve to another curve, a higher one. The difference between the two versions of the substitution effect arises solely due to the magnitude of money income by which income is reduced or increased to compensate for the change in income. The Hicksian approach just restores to the consumer his initial level of satisfaction, whereas the Slutsky approach "over-compensates" the consumer by putting him on a higher indifference curve.

Merits and Demerits of Hicksian and Slutsky Methods

Prof. J.R. Hicks points out that the method of adjusting the level of money income by the compensating variation "has the merit that on this interpretation, the substitution effect measures the effect of change in relative price, *with real income constant*, the income effect measures the, effect of the change in real income. Thus the analysis which is based upon the compensating variation is a resolution of the price change into two fundamental economic 'directions', we shall not encounter a more fundamental distinction upon any other route,.[1]

But Slutsky method has a distinct advantage in that it is easier to find out the amount of income equal to the 'cost difference' by which income of the consumer is to be adjusted. On the other hand, it is not so easy to know the compensating variation in income. Thus, the cost-difference method has the advantage of being dependent on observable market data, while for knowing the amount of compensating variation in income, knowledge of indifference curves (that is, tastes and preferences) of the consumer between various combinations of goods is required.

It follows from what has been said above that both the cost-difference and compensating variation methods have their own merits. While the law of demand can be easily and adequately established by the method of cost-difference, method of compensating variation is very useful for the analysis of consumer's surplus and welfare economics. With the help of the cost-difference, the income effect can be easily separated from the substitution effect but the substitution effect so found out involves some gain in real income (since it causes movement from a lower indifference curve to a higher indifference curve). It is because of this that, on cost-difference method, substitution effect is not a theoretically distinct concept.

A. Numerical Examples. Let us explain the concept of cost-difference and Slutsky substitution effect with a numerical example stated below:

When the price of petrol is ₹ 20.00 per litre, Amit consumes 1,000 litres per year. The price of petrol rises to ₹ 25.00 per litre.. Calculate the cost difference equal to which the Government should give him extra money income per year to compensate him for the rise in price of petrol. Will Amit be better off or worse off after the price rise plus the cash compensation equal to the cost difference than he was before? What will happen to petrol consumption?

As explained above, the cost difference is equal to $\Delta P.Q$ where ΔP stands for the change in price of a good and Q stands for the quantity of commodity he was consuming prior to the change in price. Thus, in our above example.

$$\Delta P = ₹25 - 20 = ₹5$$

$$Q = 1{,}000 \text{ litres per year}$$

$$\text{Cost-difference} = \Delta P.Q$$

$$= ₹5 \times 1{,}000 = ₹\,5{,}000.$$

Now, with higher price of petrol of ₹25.00 per litre and cash compensation of ₹5,000 equal to the cost difference he can buy, if he so desires, the original quantity of 1,000 litres of petrol per year.

1. J.R. Hicks. *A Revision of Demand Theory*; 1956 p. 62.

However, he may not buy this original quantity of petrol in the new price-income situation if his satisfaction is maximum at some other point. Consider Fig. 9B.3. where we measure petrol on the *X*-axis and money income representing other goods on the *Y*-axis. Suppose, BL_1 is the initial budget line when price of petrol is ₹20.00 per litre and consumer is in equilibrium at point *Q* on the indifference curve IC_1 where he is consuming 1,000 litres of petrol per year. Now, with the rise in price of petrol to ₹25.00 per litre, suppose the budget line shifts to BL_2. Now, if to compensate for the rise in price, his, money income is raised by ₹5,000, that is, equal to the cost difference, the budget line shifts in a parallel manner to the left so that it reaches the position *GH* which passes through the original point of consumption *Q*. A glance at Fig. 9B.3, will reveal that the consumer with higher price of petrol and having received monetary compensation equal to the cost difference of ₹5,000 will not be in equilibrium at the original point *Q* and instead he will maximising his satisfaction in the new situation at point *S* on a higher indifference curve IC_2 where his consumption of petrol has decreased to *ON* litres (that is, the decrease in consumption of petrol by *MN* is the Slutsky substitution effect.) Since, with the rise in price and simultaneous increase in his income equal to the cost-difference has enabled him to attain a higher indifference curve, he has become better off than before the rise in price.

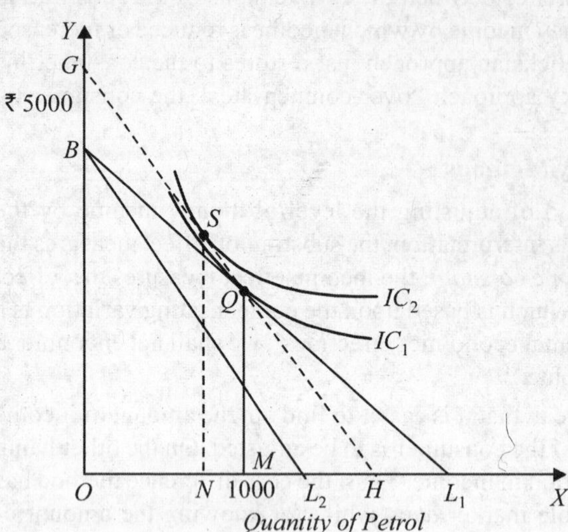

Fig.9B.3. *Cash Compensation for a Rise in Price of Petrol.*

Price Effect Broken Up into Income and Substitution Effects: Slutsky Method

In our discussion of substitution effect we explained that Slutsky presented a slightly different version of the substitution and income effects of a price change from the Hicksian one. In fact it was Slutsky who first of all divided the price effect into income and substitution effects. His way of breaking up the price effect is shown in Fig. 9B.4. With a certain price-income situation, the consumer is in equilibrium at *Q* on indifference curve IC_1 With a fall in price of *X*, other things remaining the same, budget line shifts to PL_2. With budget line PL_2, the consumer would now be in equilibrium at *R* on the indifference curve IC_3. This movement from *Q* to *R* represents the price effect. As a result of this he buys *MN* quantity of good *X* more than before. Now, in order to find out the substitution effect his money income be reduced by such an amount that he can buy, if he so desires, the old combination *Q*. Thus, a line *AB*, which is parallel to PL_2, has been so drawn that it passes through point *Q*. Thus *PA* in terms of good *Y*

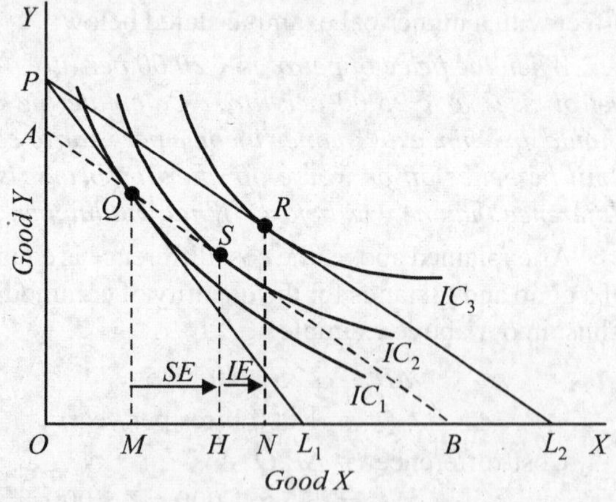

Fig. 9B.4. *Price Effect is decomposed into Substitution and Income Effects with Slutsky's Cost-Difference Method.*

represents the cost difference. With budget line AB, the consumer can have combination Q if he so desires, but actually he will not buy combination Q because X is now relatively cheaper than before. It will pay him to substitute X for Y. With budget line AB he is in equilibrium at S on indifference curve IC_2. The movement from Q to S represents Slutsky substitution effect which induces the consumer to buy MH quantity more of good X. If now the money taken away from him is restored to him, he will move from S on indifference curve IC_2 to R on indifference curve IC_3. This movement from S to R represents income effect. Thus, movement from Q to R as a result of price effect can be divided into two steps. First, movement from Q to S as a result of substitution effect and secondly, movement from S to R as a result of income effect. It may be pointed out here again that, unlike the Hicksian method, Slutsky substitution effect causes movement from a lower indifference curve to a higher one. While separately discussing substitution effect above, we pointed out the merits and demerits of the Hicksian and Slutskian methods of breaking up the price effect.

SLUTSKY EQUATION

We have graphically shown above how the effect of change in price of a good can be broken up into its two component parts, namely, substitution effect and income effect. The decomposition of price effect into its two components can be derived and expressed mathematically. Suppose price of good X falls, its substitution effect on quantity demanded of the good arises due to substitution of the relatively cheaper good X for the now relatively dearer good Y and as a result in the Hicksian method the consumer moves along the same indifference curve so that his level of utility remains constant. The overall effect of change in its own price on the quantity demanded can be expressed as $\frac{dq_x}{dp_x}$ and the substitution effect can be expressed $\left.\frac{\partial q_x}{\partial p_x}\right|_{U=\bar{U}}$. The term $\left.\frac{\partial q_x}{\partial p_x}\right|_{U=\bar{U}}$ shows change in quantity demanded resulting from a relative change in price of X while utility or satisfaction of the consumer remains constant.

However, expressing income effect of the price change mathematically is rather a ticklish affair. Suppose a unit change in income (∂I) causes a (∂q_x) change in quantity demanded of the good. This can be written as $\frac{\partial q_x}{\partial I}$. But how much income changes due to a change in price of the good is determined by how much quantity of the good (q_x) the consumer was purchasing on the one hand and change in price of the good (∂p_x) that has taken place on the other. The change in income due to a change in price can be measured by q_x (∂p_x). How much this change in income will affect the quantity demanded of the good X is determined by $\frac{\partial q_x}{\partial I}$ which shows the effect of a unit change in income on the quantity demanded of the good X. Thus the overall effect of change in price of the good X on its quantity demanded can be expressed by the following equation which is generally called *Slutsky equation* because it was Russian economist E. Slutsky who first of all divided the price effect into substitution effect and income effect.

$$\frac{\partial q_x}{\partial P_x} = \left.\frac{\partial q_x}{\partial p_x}\right|_{U=\bar{U}} - q_x . \partial p_x \frac{\partial q_x}{\partial I} \qquad \ldots (i)$$

The first term on the right hand side of the equation represents the substitution affect obtained after income of the consumer has been adjusted to keep his level of utility constant. The second term on the right hand side of the equation shows the income effect of the fall in price of the good. The term $q_x . \partial P_x$ measures the increase in income or purchasing power caused by the fall in price and

$\frac{\partial q_x}{\partial I}$ measures the change in quantity demanded resulting from a unit increase in income (I). Therefore, income affect of the price change is given by $q_x . \partial p_x \frac{\partial q_x}{\partial I}$.

Since the *fall* in price *increases* income or purchasing power of the consume which in case of normal goods leads to the *increase in quantity demanded* of the good, sign of the income affect has been taken to be negative (–).

It may be further noted that if *price falls by Re.* 1 *per unit of commodity X*, then the change in $q_x . \partial P_x$, will be equal to q_x *in money terms* and therefore the expression for total price effect can be written as

$$\frac{dq_x}{dp_x} = \frac{\partial q_x}{\partial p_x}\bigg|_{U=\bar{U}} - q_x \frac{\partial q_x}{\partial I} \qquad \ldots (ii)$$

Further, a point needs to be clarified. In the above analysis of *Slutsky equation*, we have considered substitution effect when with a change in price, consumer is so compensated as to keep his utility constant. However, strictly speaking in obtaining Slutsky substitution effect income of the consumer is adjusted to *keep his purchasing power constant* so that he could buy the original combination of goods if he so desires. This difference was later emphasised by J.R. Hicks, but since it was Slutsky who first of all split up the price effect into substitution effect and income effect the above equation is popularly known as *Slutsky equation*. It is proper to call it *Slutsky-Hicks equations*.

An important result follows from the Slutsky equation. If the commodity is a normal good, then $\frac{\partial q_x}{\partial I}$ is positive by definition. It follows that a *fall* in price will lead to the *increase* in income causing *increase in quantity demanded* of the good and therefore the expression for income effect of the price change $q_x \times \left(\frac{\partial q_x}{\partial I}\right)$ is taken to be negative in the Slutsky equation. Coupled with the fact that substitution effect is always negative, the Slutsky equation tells us that when commodity X is normal, the price effect $\frac{dq_x}{dp_x}$ is *necessarily negative* implying that *fall* in price will cause quantity demanded of the good to *rise* because of both the substitution effect and income effect. It should be carefully noted that since in case of normal goods a fall in price of a commodity leads to rise in quantity demanded due to both the substitution effect and income effect both have to be added to obtain the price effect.

The second important conclusion which follows from Slutsky equation is that as the quantity of commodity (q_x) becomes smaller and smaller, the income effect of the price change will become smaller and smaller. Thus if the quantity consumed of a commodity is very small, then the income effect is not very significant.

QUESTIONS FOR REVIEW

1. What is Slutsky Substitution effect? Explain Slutsky substitution effect for a rise in price of good X with indifference curves analysis.
2. Distinguish between Hicks' and Slutsky approach to the measurement of income and substitution effects. (*D.U.B.A (Hons)* 1990)

3. Show graphically the substitution and the real income effect by Hicks and Slutsky. What is the advantage of Slutskys measure?

4. A consumer spends all his income on two goods *X* and *Y*. When the price of goods *X* is ₹ 2. he consumes 100 units of goods *X*. If the price of goods *X* increases to ₹3 and income is increased by ₹100 to enable him to purchase the original bundle, price of *Y* remaining constant, (*i*) would the level of satisfaction of the consumer change; and
(*ii*) what would happen to the quantity consumed of *X*? *D.U. B.Com (Hons)*

5. State the Slutsky equation algebraically and explain each term in it. Using the equation explain. (*i*) what must be the sign and magnitude of the income effect for a downward sloping demand curve ? (*ii*)Under what circumstances we will get a Giffen good case? *D.U. B.A. (Hons.)*

CHAPTER 10
MARSHALL'S CARDINAL UTILITY ANALYSIS
VS.
INDIFFERENCE CURVE ANALYSIS

We are now in a position to compare the indifference curve analysis with Marshallian cardinal utility analysis. Barring the views of some economists like Dennis Robertson, W. E. Armstrong, F. H. Knight, it is now widely believed that indifference curve analysis makes a definite improvement upon the Marshallian cardinal utility analysis. It has been asserted that whereas Marshallian cardinal utility analysis assumes 'too much' it explains 'too little'. On the other hand, the indifference curve analysis explains more by taking fewer as well as less restrictive assumptions.

It may be noted that there are some similarities in these two theories of demand. First, both these theories assume that consumers are rational and therefore try to maximize utility or satisfaction. Second, both these theories use psychological or introspective method to explain consumer's behaviour. Thirdly, both these theories of demand assume in some form diminishing marginal utility or desire for a commodity as a consumer takes more units of the commodity. In this connection it is generally believed that the principle of diminishing marginal rate of substitution is similar to the law of diminishing marginal utility of Marshallian cardinal utility analysis.

SUPERIORITY OF INDIFFERENCE CURVE ANALYSIS

After having pointed out the similarities between the two types of analysis, we now turn to study the difference between the two and to show how far indifference curve analysis is superior to the Marshallian cardinal utility analysis.

1. **Ordinal vs. Cardinal Measurability of Utility.** In the first place, Marshall assumes utility to be *cardinally measurable*. In other words, he believes that utility is *quantifiable*, both in principle and in actual practice. According to this, the consumer is able to assign a specific amount to the utility obtained by him from the consumption of a certain amount of a good or a combination of goods. Further, the amounts of utility can be manipulated in the same manner as weights, lengths, heights, etc. In other words, the utilities can be compared and added. Suppose, for instance, utility which a consumer gets from a unit of good *A* is equal to 15, and from a unit of good *B* equal to 45. We can then say that the consumer prefers *B three times as strongly as A* and the utility obtained by the consumer from the combination containing one unit of each good is equal to 60. Likewise, even the differences between the utilities obtained from the various goods can be so compared as to enable us to say *A* is preferred to *B* twice as much as *C* is preferred to *D*.

According to the critics, the Marshallian assumption of cardinal measurement of utility is very restrictive; it demands too much from the human mind. They assert that utility is a psychological feeling and the precision in measurement of utility assumed by Marshall and others is therefore unrealistic. Critics hold that the utility possesses only ordinal magnitude and cannot be expressed in cordial terms.

According to the sponsors of the indifference curve analysis, utility is merely *orderable* and not quantitative. In other words, indifference curve technique assumes what is called '*ordinal measurement of utility*'. According to this, the consumer need not be able to assign specific amounts to the utility he derives from the consumption of a good or a combination of goods but it is assumed that he is capable of *comparing* the different utilities or satisfactions in the sense whether one level of satisfaction is equal to, lower than, or higher than another. He cannot say by *how much* one level of satisfaction is higher or lower than another. That is why the indifference curves are generally labelled by the ordinal numbers such as I, II, III, IV, etc., showing successively higher levels of satisfaction. The advocates of indifference curve technique assert that for the purpose of explaining consumer's behaviour and deriving the theorem of demand, it is quite sufficient to assume that the consumer is able to *rank his preferences consistently*.

It is obvious that the ordinal measurement of utility is a less restrictive assumption and sounds more realistic than cardinal measurement of utility. This shows that the indifference curve analysis of demand which is based upon the ordinal utility hypothesis is superior to Marshall's cardinal utility analysis.

2. **Analysis of Demand Without Assuming Constant Marginal Utility of Money.** Another distinct improvement made by indifference curve technique is that unlike Marshall it explains consumer's behaviour and derives demand theorem without the assumption of constant marginal utility of money. In indifference curve analysis, it is not necessary to assume constant marginal utility of money. As has already been seen, Marshall assumed that the marginal utility of money remained constant when there occured a change in the price of a good. It has been shown in the last chapter that the Marshallian demand analysis based upon constancy of marginal utility of money is not self-consistent. In other words, "the Marshallian demand theorem cannot genuinely be derived from the marginal utility hypothesis except in one commodity model, without contradicting the assumption of constant marginal utility of money." [1] It means that "the constancy of marginal utility of money is incompatible with a proof of the demand theorem in a situation where the consumer has more than a single good to spread his expenditure on."[2] To overcome this difficulty in Marshallian cardinal utility analysis, if the assumption of constant marginal utility of money is abandoned, then money can no longer serve as a measuring rod of utility and we can no longer measure marginal utility of a commodity in units of money.

On the other hand, indifference curve technique using ordinal utility hypothesis can validly derive the demand theorem without the assumption of constant marginal utility of money. In fact, as we shall see below, the abandonment of the assumption of constant marginal utility of money enables the indifference curve analysis to derive a *more general* demand theorem.

3. **Greater Insight into Price Effect.** The superiority of indifference curve analysis further lies in the fact that it makes greater insight into the effect of price change on the demand for a good by distinguishing between income and substitution effects. The indifference technique splits up the price effect analytically into its two component parts substitution effect and income effect. The distinction between the income effect and the substitution effect of a price change enables us to gain a better understanding of the effect of a price change on the demand for a good. The amount demanded of a good generally rises as a result of the fall in its price due to two reasons. First, real income rises as a result of the fall in price (income effect) and, secondly, because the good whose price falls becomes relatively cheaper than others and therefore the consumer substitutes it for others (substitution effect). In indifference curve technique, income effect is separated from the substitution effect of the price change by the methods of 'compensating variation in income' and 'equivalent variation in income'.

---

1. Tapas Majumdar, *op. cit.*, p. 55.
2. *Ibid*, p. 55–56.

But, by assuming constant marginal utility of money Marshall ignored the income effect of a price change. He failed to understand the composite character of the effect of a price change. Prof. Tapas Majumdar rightly remarks, "the assumption of constant marginal utility of money obscured Marshall's insight into the truly composite character of the unduly simplified price-demand relationship"[3]. In this context, remarks made by J. R. Hicks are worth noting, "The distinction between direct and indirect effects of a price change is accordingly left by the cardinal theory as an empty box, which is crying out to be filled."[4] Commenting on the improvement made by Hicks-Allen indifference curve approach over the Marshallian utility analysis, Prof. Tapas Majumdar says : "The efficiency and precision with which the Hicks-Allen approach can distinguish between the 'income' and 'substitution' effects of a price change really leaves the cardinalist argument in a very poor state indeed."[5]

4. **Deriving a more general and adequate 'Demand Theorem'.** A distinct advantage of the technique of dividing the effect of a price change into income and the substitution effects employed by the indifference curve analysis is that it enables us to enunciate a *more general* and a more *inclusive* theorem of demand than the Marshallian law of demand. In the case of most of the normal goods in this world, both the income effect and the substitution effect work in the same direction, that is to say, they tend to increase the amount demanded of a good when its price falls. The income effect ensures that when the price of a good falls the consumer buys more of it because he can now afford to buy more; the substitution effect ensures that he buys more of it because it has now become relatively cheaper and is, therefore, profitable for him to substitute it for others. This thus accounts for the inverse price-demand relationship (Marshallian law of demand) in the case of normal goods.

When a certain good is regarded by the consumer to be an inferior good, he will tend to reduce its consumption as a result of the increase in his income. Therefore, when price of an inferior good falls, the income effect so produced would work in the opposite direction to that of the substitution effect. But so long as the inferior good in question does not claim a very large proportion of consumer's total income, the income effect will not be strong enough to outweigh the substitution effect. In such a case therefore the net effect of the fall in price will be to raise the amount demanded of the good. It, therefore, follows that even for most of the inferior goods, the Marshallian law of demand holds good as much as for normal goods.

But it is possible that there may be inferior goods for which the income effect of a change in price is larger in magnitude than the substitution effect. This is the case of Giffen goods for which the Marshallian law of demand does not hold good. In such cases the negative income effect outweighs the substitution effect so that the net effect of the fall in price of the good is the reduction in demand for it. Thus, amount demanded of a Giffen good varies directly with price.

It is clear from above that by breaking the price effect into income effect and substitution effect, the indifference curve analysis enables us to arrive at a general and a more inclusive theorem of demand in the following composite form.[6]

(*a*) The demand for a commodity varies *inversely* with price when the income effect for that commodity is nil or positive.

(*b*) The demand for a commodity varies inversely with price when the income elasticity is negative but the income effect of the price change is smaller than the substitution effect.

(*c*) The demand for a commodity varies *directly* with price when the income effect is negative and this income effect of the price change is larger than the substitution effect.

In the case of (*a*) and (*b*) the Marshallian law of demand holds while in (*c*) we have a Giffen good

---

3. Tapas Majumdar, *op cit*., p. 76.
4. J. R. Hicks, *A Revision of Demand Theory*, p. 14.
5. Tapas Majumdar, *op. cit*., p. 57.
6. Tapas Majumdar, *op. cit, pp*., 74–75.

case which is exception to the Marshallian law of demand. Marshall could not account for 'Giffen Paradox', because by assuming constant marginal utility of money, he ignored the income effect of the price change. The indifference curve technique by distinguishing between the income and substitution effects of the price change can explain the Giffen good case. According to this, the Giffen good case occurs in the case of an inferior good for which the negative income effect of the price change is so powerful that it outweighs the substitution effect, and hence when the price of a Giffen good falls, its quantity demanded also falls instead of rising. Thus, a great merit of Hicks-Allen indifference curve analysis is that it offers an explanation for the Giffen-good case, while Marshall failed to do so.

5. **Significance of a Price Change in terms of Income and Welfare Increments.** Another distinct improvement of Hicks-Allen ordinal theory is that through it the welfare consequences of a change in price can be translated into those of a change in income. As seen above, a fall in the price of a good enables the consumer to shift from a lower to a higher level of welfare (or satisfaction). Likewise, a rise in the price of the good would cause the consumer to shift down to a lower indifference curve and therefore to a lower level of welfare. This means that a fall in price of a good causes a change in consumer's welfare exactly as the rise in income would do. In other words, the consumer can be thought of reaching a higher level of welfare through an equivalent rise in income rather than the fall in price of a good. In Fig. 10.1 with the fall in price of good X from PL_1 to PL_2 the consumer shifts from indifference curve IC_1 to the indifference curve IC_2 showing an increase in the level of welfare. Now, if instead of the fall in price from PL_1 to PL_2 the consumer's income is increased by the amount equal to PA or L_1B, he will reach the indifference curve IC_2. Thus the increase in consumer's welfare due to the rise in income by PA or, L_1B is equal-to that of the change in price of X from PL_1 to PL_2. "*The equivalence of a given change in price to a suitable change in income is a major discovery of ordinal utility analysis*."[7] This fundamental relation necessarily remains obscure in cardinal utility analysis with its single good model and the assumption of constant marginal utility of money.

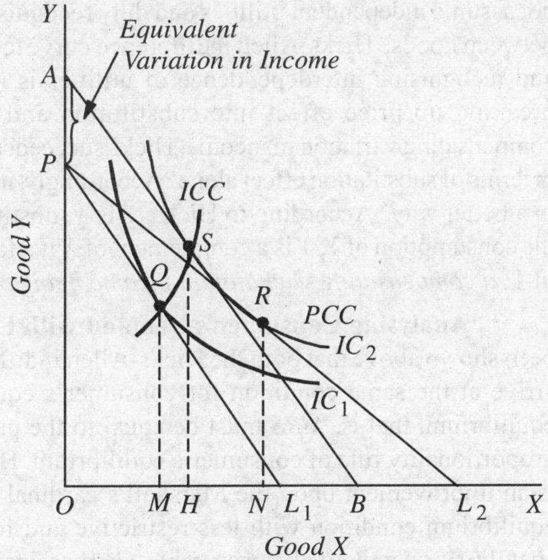

Fig. 10.1. *Income Equivalent of a Price Change*

The discovery of a suitable change in income equivalent in terms of welfare to a given change in price has enabled Hicks to extend Marshall's concept of consumer's surplus. Marshall's concept of consumer's surplus was based upon the assumption that utility was cardinally measurable and also that the marginal utility of money remained constant when the price of a good changed. Hicks has freed the concept of consumer's surplus from these dubious assumptions and by using ordinal utility hypothesis along with the discovery that the welfare effect of a price change can be translated into a suitable change in income he has not only been able to rehabilitate the concept of consumer's surplus but also to extend it further.

6. **Hypothesis of Independent Utilities Given Up.** Marshall's cardinal utility analysis is based upon the *hypothesis of independent utilities*. This means that the utility which the consumer derives from any commodity is a function of the quantity of that commodity and of that commodity alone. In other words, the utility obtained by the consumer from a commodity is independent of that derived

7. Tapas Majumdar, op. cit., p. 72.

from any other. *By assuming independent utilities Marshall completely bypassed the relation of substitution and complementarity between commodities.*

Demand analysis based upon he hypothesis of independent utilities, as shown in the last chapter, leads us to the conclusion "that in all cases a reduction in the price of one commodity only will either result in an expansion in the demand for all other commodities or in a contraction in the demands for all other commodities." But this is quite contrary to the common cases found in the real world. In the real world, it is found that as a result of the fall in price of a commodity the *demand for some commodities expands while the demand for others contracts.* We thus see that Marshall's analysis based upon `independent utilities' does not take into account the complementary and substitution relations between goods. This is a great flaw in Marshall's cardinal utility analysis.

On the other hand, this flaw is not present in Hicks-Allen indifference curve analysis which does not assume independent utilities and duly recognises the relation of substitution and complementarity between goods. Hicks-Allen indifference curve technique by taking more than one commodity model and recognising interdependence of utilities is in a better position to explain related goods. By breaking up price effect into substitution and income effects by employing the technique of compensating variation in income, Hicks succeeded in explaining complementary and substitute goods in terms of substitution effect alone. Accordingly, it can define and explain substitutes and complements in a better way. According to Hicks, Y is a substitute for X if a fall in the price of X leads to a fall in the consumption of Y; Y is a complement of X if a fall in the price of X leads to a rise in the consumption of Y, *a compensating variation in income being made in each case so as to maintain indifference.*

7. **Analysing Consumer's Demand with Less Restrictive and Fewer Assumptions.** It has been shown above that both the Hicks-Allen indifference curve theory and Marshall's cardinal theory arrive at the same condition for consumer's equilibrium. Hicks-Allen condition for consumer's equilibrium, that is, *MRS* must be equal to the price ratio amounts to the same thing as Marshall's proportionality rule of consumer's equilibrium. But even here indifference curve's ordinal approach is an improvement upon the Marshall's cardinal theory in so far as the former arrives at the same equilibrium condition with less restrictive and fewer assumptions. Dubious assumptions such as (*i*) utility is quantitatively measurable, (*ii*) marginal utility of money remains constant, and (*iii*) utilities of different goods are independent of each other, on which Marshall's cardinal theory is based, are not made in indifference-curve's ordinal utility theory.

IS INDIFFERENCE CURVE ANALYSIS "OLD WINE IN A NEW BOTTLE"?

But superiority of indifference curve theory has been denied by some economists, foremost among them are Prof. D. H. Robertson, F. H. Knight, W. E. Armstrong. Prof. Knight remarks, "indifference curve analysis of demand is not a step forward; it is in fact a step backward." Prof. D. H. Robertson is of the view that the indifference curve technique is merely "*the old wine in a new bottle.*" The indifference curve analysis, according to him, has simply substituted new concepts and equations in place of the old ones, while the essential approach of the two types of analyses is the same. Instead of the concept of `utility', the indifference curve technique has introduced the term 'preference' and scale of preferences. In place of cardinal number system of one, two, three, etc., which is supposed to measure the amount of utility derived by the consumer, the indifference curves have the ordinal number system of first, second, third, etc., to indicate the order of consumer's preferences. The concept of marginal utility has been substituted by the concept of marginal rate of substitution. And against the Marshallian 'proportionality rule' as a condition for consumer's equilibrium, indifference curve approach has advanced the condition of equality between the marginal rate of substitution and the price ratio.

Prof. Robertson's view that the concept of marginal rate of substitution of indifference curve analysis represents the reintroduction of the concept of marginal utility in demand analysis requires further consideration. Prof. Robertson says : "In his earlier book *Value and Capital* Hicks's treatment involved making an assumption about the convexity of those 'indifference curves' which appeared to some of us to involve reintroduction of marginal utility in disguise."[8] It has thus been held that the *use of marginal rate of substitution implies the presence of cardinal element in indifference curve technique.* In going from one combination to another on an indifference curve, the consumer is assumed to be able to tell what constitutes his compensation in terms of a good for the loss of a marginal unit of another good. In other words, the consumer is able to tell his marginal rate of substitution of one good for another. Now, the marginal rate of substitution has been described by Hicks and others as the ratio of the marginal utilities of two goods.

$$\left(MRS_{xy} = \frac{MU \text{ of } X}{MU \text{ of } Y} \right)$$

But ratio cannot be measured unless the two marginal utilities in question are at least measurable in principle. It has, therefore, been held that the concept of marginal rate of substitution and the idea of indifference based upon it essentially involves an admission that utility is quantifiable in principle.

Against this, Hicks contends that we need not assume measurability of marginal utilities in principle in order to know the marginal rate of substitution. He says, "All that we shall be able to measure is what the ordinal theory grants to be measurable – namely the ratio of the marginal utility of one commodity to the marginal utility of another." This means that *MRS* can be obtained without actually measuring marginal utilities. If a consumer, when asked, is prepared to give up 4 units of good *Y* for the gain of one marginal unit of *X*, *MRS* of *X* for *Y* is 4 : 1. We can thus directly derive the ratio indicating *MRS* by asking him how much quantity of good *Y* the consumer would forego for the gain of a marginal unit of *X*. The contention that the concept of marginal rate of substitution is a mere reintroduction of the marginal utility (a cardinal concept) in disguise is therefore not valid.

It has been further contended by Robertson and Armstrong that it is not possible to arrive at the Hicksian *principle of diminishing marginal rate of substitution* without making use of the concept of marginal utility and the principle of diminishing marginal utility. It is asked why *MRS* of *X* for *Y* diminishes as more and more of *X* is substituted for *Y*? The critics point out that the marginal rate of substitution (MRS_{xy}) diminishes and the indifference curve becomes convex to the origin, because as the consumer's stock of *X* increases, the marginal utility of *X* falls and that of *Y* increases. They thus hold that Hicks and Allen have not been able to derive the basic principle of diminishing marginal rate of substitution independently of the law of diminishing marginal utility. They contend that by a stroke of terminological manipulation, the concept of marginal utility has been relegated to the background, but it is there all the same. They, therefore, assert that "the principle of diminishing marginal rate of substitution is as much determinate or indeterminate as the poor law of diminishing marginal utility".

However, even this criticism of indifference curve approach advanced by the defenders of the Marshallian cardinal utility analysis is not valid. As shown above, the derivation of marginal rate of substitution does not depend upon the actual measurement of marginal utilities. While the law of diminishing marginal utility is based upon the cardinal utility hypothesis (*i.e.*, utility is quantifiable and actually measurable), the principle of marginal rate of substitution is based upon the ordinal utility hypothesis (*i.e.*, utility is mere orderable). As a consumer gets more and more units of good *X*, his strength of desire for it (though we cannot measure it in itself) will decline and therefore he will be prepared to forego less and less of *Y* for the gain of a marginal unit of *X*. It is thus clear that the principle of diminishing marginal rate of substitution is based upon purely ordinal hypothesis and is derived independently of the cardinal concept of marginal utility, though both laws reveal essentially the same phenomenon. We therefore agree with Prof. Hicks who claims that "*the replacement of the principle of diminishing marginal utility by the principle of diminishing marginal rate of substitution is not a mere translation. It is a positive change in the theory of consumer demand*".

8. Sir Dennis Robertson, *Lectures on Economic Principles*, The Fontana Library Edition, 1963, p. 85.

It follows from what has been said above that indifference curve analysis of demand is an improvement upon the Marshallian utility analysis and the objections that the former too involves cardinal elements are groundless. It is of course true that the indifference curve analysis suffers from some drawbacks and has been criticized on various grounds, as explained below, but as far as the question of indifference curve technique versus Marshallian utility analysis is concerned, the former is deci-dedly better.

CRITIQUE OF INDIFFERENCE CURVE ANALYSIS

Indifference curve analysis has come in for criticism on several grounds. In the first place, it is argued that the indifference curve approach for avoiding the difficulty of measuring utility quantitatively is forced to *make unrealistic assumption that the consumer possesses complete knowledge of all his scale of preferences or indifference map*. The indifference curve approach, so to say, falls from the frying pan into the fire. The indifference curve analysis envisages a consumer who carries in his head innumerable possible combinations of goods and relative preferences in respect of them. It is argued: Is not this carrying into his head all his scales of preferences too formidable a task for a real human being ? Hicks himself admits this drawback. When revising his demand theory based on indifference curves, he says that "one of the most awkward of the assumptions into which the older theory appeared to be impelled by its geometrical analogy was the notion that the consumer is capable of ordering all conceivable alternatives that might possibly be presented to him – all the positions which might be represented by points on his indifference map. This assumption is so unrealistic that it was bound to be a stumbling block."[9] This is one of the reasons that Hicks gave up indifference curves in his *Revision of Demand Theory*.

Further, there is another unrealistic element present in indifference curve analysis. It is pointed out that *such curves include even the most ridiculous combinations which may be far removed from his habitual combinations*. For example, while it may be perfectly sensible to compare whether three pairs of shoes and six shirts would give a consumer as much satisfaction as two pairs of shoes and seven shirts, the consumer will be at a loss to know and compare the desirability of an absurd combination such as eight pairs of shoes and one shirt. The way the indifference curves are constructed, they include absurd combinations like the one just indicated.

A further shortcoming of the indifference curve technique is that *it can demonstrate and analyse consumer's behaviour effectively only in simple cases, especially those in which the choice is between the quantities of two goods only*. In order to demonstrate the case of three goods, three-dimensional diagrams are needed which are difficult to understand and handle. When more than three goods are involved geometry altogether fails and recourse has to be taken to the complicated mathematics which often tends to conceal the economic point of what is being done. Prof. Hicks also admits this shortcoming of indifference curve technique.

Another demerit of indifference curve analysis because of its geometrical nature is that it involves the *assumption of continuity* "a property which the geometrical field does have, but which the economic world in general does not". The real economic world exhibits discontinuity and it is quite unrealistic and analytically bad if we do not recognise it. This is why Hicks also abandoned the assumption of continuity in his later work *A Revision of Demand Theory*.

Prof. Armstrong has criticized the relation of transitivity involved in indifference curve technique. He is of the view that in most cases the consumer's indifference is due to his imperfect ability to perceive difference between alternative combinations of goods. In other words, the consumer indicates his indifference between the combinations which differ very slightly from each other not because they

9. J.R. Hicks, *A Revision of Demand Theory*, p.20.

give him equal satisfaction but because the difference between the combinations is so small that he is unable to perceive the difference between them. *If this concept of indifference is admitted, then the relation of indifference becomes non-transitive. Now, with non-transitivity of indifference-relation; the whole system of indifference curves and the demand analysis based upon it breaks down.*

It may, however, be pointed out that Prof. Armstrong's interpretation of indifference is not correct. Actually, the relation of indifference in the ordinal theory is the exact equivalent of the relation of 'equality' in the cardinal sense. In other words, the consumer is said to be indifferent between *A* and *B*, for instance, because he derives equal utility from the two combinations and not because the difference between the utilities from *A* and *B* is imperceptible. If such is the case then "the axiom of transitivity of ordinal indifference emerges automatically and is no more subject to dispute than is the axiom of transitivity of numerical equality".

Further, *indifference curve analysis has been criticised for its limited empirical nature.* Indifference curve analysis is neither based upon purely imaginary and subjective utility functions, nor is it based upon purely empirically derived indifference functions. It is because of this fact that Professor Schumpeter dubbed indifference curve analysis as 'a midway house'. It would have been quite valid if indifference curve analysis was based upon experimentally obtained quantitative data in regard to the observed market behaviour of the consumer. But, in Hicks-Allen theory, indifference curves are based upon *hypothetical experimentation*. The indifference curve theory of demand is, therefore, based upon imaginary drawn indifference curves. Commenting on Hicks-Allen theory of demand Prof. Schumpeter remarked, "If they use nothing that is not observable in principle they do use "potential" observations which so far no body has been able to make in fact : from a practical standpoint we are not much better off when drawing purely imaginary indifference curves than we are when we speak of purely imaginary utility functions."[10]

It may, however, be pointed out that attempts have recently made by some economists and psychologists to derive or measure indifference curves experimentally. But a limited success has been achieved in this regard. This is because such experiments have been made under controlled conditions which render these experiments quite unfit for drawing conclusions regarding real consumer's behaviour in 'free circumstances'. So, for all intents and purposes, indifference curves still remain imaginary.

An important criticism against Hicks-Allen ordinal theory of demand is that *it cannot formalise consumer's behaviour when uncertainty or risk is present.* In other words, consumer's behaviour cannot be explained by ordinal theory when he has to choose among alternatives involving risk or 'uncertainty of expectation'. Von Neumann and Morgenstern have asserted that while cardinal utility theory can, the ordinal utility theory cannot formalise consumer's behaviour when we introduce "uncertainty of expectations with regard to the consequences of choice."

We thus find that Hicks-Allen ordinal utility system cannot formalise consumer's behaviour when there exists uncertainty of expectations with regard to the consequences of choice. On the other hand, cardinal utility theory can formalise consumer's behaviour in the presence of uncertainty of expectations since it involves quantitative estimates of utilities or preference intensities. Commenting on indifference preference hypothesis, Neumann and Morgenstern remark : "If the preferences are not all comparable, then the indifference curves do not exist. If the individual preferences are all comparable, then we can even obtain a (uniquely defined) numerical utility which renders the indifference curves superfluous."[11]

---

10. J. A. Schumpeter, *History of Economic Analysis*, P. 1067
11. See their work "*The Theory of Games and Economic Behaviour*"

CHAPTER 11
APPLICATIONS AND USES OF INDIFFERENCE CURVES

We have studied the indifference curve analysis of demand. But the technique of indifference curves has been used not only to explain consumer's behaviour and demand but also to analyse and explain several other economic problems. In other words, besides analysing consumer's demand, indifference curves have several other applications. Thus, indifference curves have been used to explain the concept of consumer's surplus, substitutability and complementarity of goods, supply curve of labour of an individual, several principles of welfare economics, burden of different forms of taxation, gain from foreign trade, welfare implications of subsidy granted by the Government, index number problem, mutual advantage of exchange of goods between two individuals and several other things. We shall discuss applications of indifference curves in some of the above stated fields in their relevant chapters. We shall explain below only few applications.

EFFECT OF SUBSIDIES TO CONSUMERS: PRICE SUBSIDY vs. CASH SUBSIDY

An important application of indifference curves is to analyse with its aid the effect of subsidies to the consumers. Several kinds of subsidies are paid to the individuals these days by the Government for promoting welfare of the people. We will explain and compare the effects of two types of subsidies, price subsidy and lump sum cash grant, on consumer's welfare. It is worth noting that *price subsidy on a commodity is also generally called excise subsidy. Under price or excise subsidy the Government pays a part of the price of a good and allows the consumer to buy as many units of the good as he desires at the subsidised price.* On the other hand, in case of cash subsidy the government provides a lump sum cash income to the consumer. Let us take the case of food subsidy which is given by the Government to help the needy families. Suppose that under food-subsidy programme, the needed families are entitled to purchase food at half the market price, the other half of the market price is paid by the Government as subsidy. The effect of this subsidy on consumer's welfare and money value of this subsidy to the consumer is illustrated in Fig. 11.1 where the quantity of food is measured on the *X*-axis and money on the *Y*-axis. Let us suppose that the individual has *OP* money income. Given

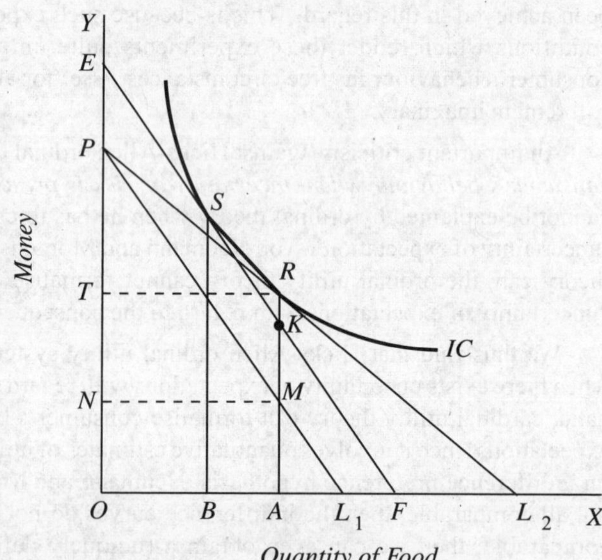

Fig. 11.1. *Effect of Subsidies to Consumers*

this money income and given the market price of food, the price line is PL_1. Since we are assuming that subsidy paid by the Government is half the market price of food, the consumer would pay half the market price. Therefore, with subsidy the individual will face the price line PL_2 where $OL_1 = L_1 L_2$. With price line PL_2, the individual is in equilibrium at point R on the indifference curve IC at which he is purchasing OA quantity of food. By purchasing OA quantity of food, the individual is spending PT amount of money. Now, if no food subsidy was given and therefore the price line was PL_1, then for buying OA quantity of food, the individual would have spent PN amount of money. In other words, PN is the market price of OA quantity of food. Since PT amount of money is paid by the individual himself, the remaining amount TN or RM (the vertical distance between the price lines PL_1 and PL_2 at OA amount of food) is paid by the Government as food subsidy for the individual.

Now, the important question is what is the *money value* of this price subsidy (RM) on food to the individual. When no price subsidy is paid, the individual faces the price line PL_1. In order to find the money value of the subsidy to the individual, draw a line EF parallel to PL_1 so that it touches the same indifference curve IC where the individual comes to be in equilibrium when subsidy is paid. It will be seen from Fig. 11.1 that budget line EF touches the indifference curve IC at a point S and is buying OB quantity of food. This means that if individual is paid PE amount of money (say as a cash grant), he reaches the same indifference curve IC (same level of welfare) at which he is when price subsidy is paid by the Government on food. Thus PE is money value of the subsidy to the individual. It will be seen from Fig. 11.1 that PE is less than RM which is the amount of money paid by the Government as subsidy. In our figure $PE = MK$ (the vertical distance between two parallel lines) and RM is greater than MK. Therefore, RM is also greater than PE. It follows that PE is less than RM. If instead of giving RM as price subsidy on food, Government pays the individual cash money equal to PE, the individual will reach the same level of welfare as he does with RM subsidy.

Thus, the cash money equivalent of the price subsidy to the individual is less than the cost of the subsidy to the Government. "In fact, it would always be so whatever the subsidy and whatever the preferences of consumers so long as only the indifference curves remain convex and smooth. *Thus the cost of giving subsidies to consumers is always greater than the money equivalent of the subjective gain to the consumers*". Here, of course, is a special case of general principle that, apart from considerations of etiquette and sentiments, *you can make someone happier if you give him cash instead of a commodity even if the commodity is something he wants.*[1]

Lump Sum Cash Subsidy

Now, if instead of providing price subsidy on food, the Government gives lump sum cash grant to the consumer *equivalent to the cost of price subsidy on food*, what will be its impact on the individual's welfare and consumption of food by him? As explained above in Fig. 11.1, cost of price subsisdy on food to the Government equals RM amount of money. If the Government provides the consumer lump sum cash grant of RM instead of price subsidy on food, this will amount to increasing the money income of the consumer by RM amount. With this extra cash transfer equal to RM (=PC), the budget line will shift to the right to the position CD in Fig. 11.2 which passes through point R. It will be seen from Fig. 11.2 that with the budget line CD though the individual *can buy the same market basket R, if he so desires*, which he was purchasing with price subsidy on food, he is actually in equilibrium at point H on higher indifference curve IC_2. Thus, the cash transfer equivalent to the cost of price subsidy has led to the greater increase in welfare or satisfaction of the individual as compared to the price subsidy. Further, as will be seen from Fig. 11.2, *with a cash grant the individual buys less food and more of other goods* relative to the situation under price subsidy with the equivalent monetary cost.

1. D. S. Watson, *Price Theory and Its Uses*, 1963, p. 94.

That the individual with cash transfer must be better off and his food consumption must be less as compared with price subsidy on food is due to the fact that indifference curves being convex, the budget line *CD* obtained with cash transfer *must intersect* the indifference curve IC_1 at point *R* reached with the equivalent price subsidy. Therefore, given that the consumer is free to spend money as he likes, with cash grant his new equilibrium position must be to the left of point *R on the budget line CD* where it will be tangent to the higher indifference curve than IC_2. This implies that in *case of lump sum cash subsidy, the consumer will be better off and consume less food* relative to the equilibrium position under price subsidy on food.

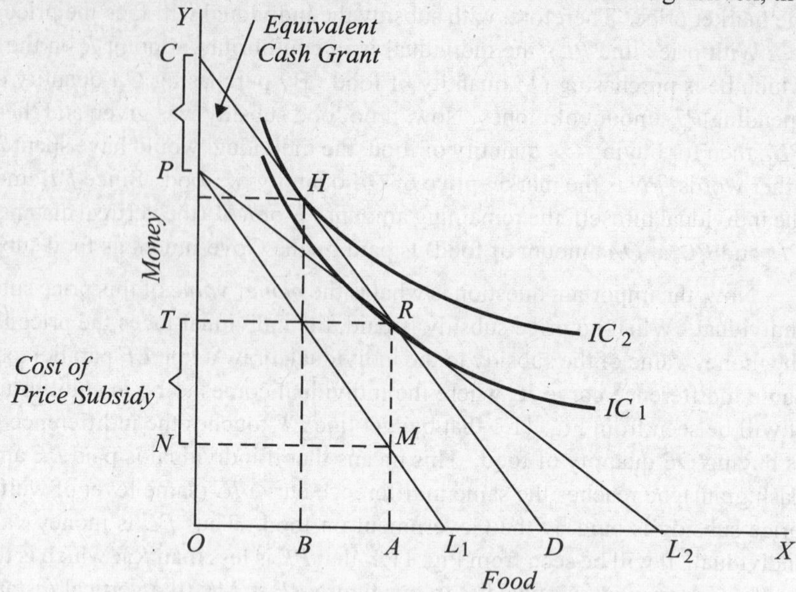

Fig. 11.2. *Lump Sum Cash Grant is Better than Price Subsidy*

The superiority of cash grant in terms of its impact on the welfare of the individuals can be explained in a slightly different way. Though both the lump sum cash transfer and price subsidy on a commodity produce income effect making the individual better off, under *cash grant the individual is free to buy different goods according to his own tastes and preferences which ensures a higher level of welfare as compared to the policy of price subsidy on food which imposes a certain pattern of consumption favouring food.* Besides, a lower price of food due to price subsidy on it induces the consumer to substitute food for other goods causing greater consumption of food as compared to the scheme of lump sum cash grant which have no such substitution effect and permits free choice of goods to the individual according to his own preference. Thus, to quote Prof. Watson again, "*You can make someone happier if you give him cash instead of a commodity, even if the commodity is something he wants*."[2] Likewise, Professor Scitovsky remarks, "*One can make a man happier by giving him cash and letting him spend it as he thinks best than by forcing him to take all his relief in the form of one commodity. Hence, relief payments in cash are preferable to a food subsidy because they are economically more efficient, giving the relief receipts either a greater gain at the same cost to the Government or the same gain at a lower cost.*"[3]

But the above principle regarding the programme of subsidised food, subsidised housing etc. cannot always be validly applied to the Government subsidy programme since the above principle is based upon the *subjective benefits to the individuals* which is not always the correct criteria to judge the desirability of Government subsidy programme. For instance, the aim of Government's food subsidy programme may be that the needy families should consume more food so that their health and efficiency may be improved. It will be seen from Fig. 11.2 that with food subsidy *RM*, the

2. D. S. Watson, *Price Theory and Its Uses*, 1963, p. 94.

3. Tibor Scitovsky, *Welfare and Competition*, revised edition, 1971, p. 70.

individual is having OA amount of food, whereas with equivalent cash payment of PC the individual purchases OB amount of food which is less than OA. Thus the food subsidy has induced the individual to consume more food than in case of cash payment. Similarly, if a country has food surpluses and wants to dispose them of, then the food subsidy to the needy families will be the ideal measure to increase the consumption of foodgrains and thereby to dispose of the food surpluses.

Food Stamp Programme: In-kind Food Subsidy

Food stamp programme is a type of food subsidy to provide poor people with adequate quantity of food. It is a form of *in-kind food subsidy* in contrast to the subsidy provided in the form of cash income, often called *cash subsidy*. In the United States it was introduced in 1964 and was amended in 1979 and since then it continues there in the amended form. In India also food stamp programme has been suggested in recent years as an anti-poverty measure. Under food stamp programme, some stamps or coupons are given to the eligible persons or households. *With these stamps, the recipient can buy food and only food.* That is, these food stamps cannot be used to buy non-food goods. Further, these stamps cannot be traded or transferred to the other people.

Let us explain how a receipt of food stamps affects the budget line, consumption of food and welfare of the individual. We will also demonstrate how the effect of food stamp subsidy differs from cash subsidy. Consider Fig. 11.3 where along the X-axis we measure quantity of food and along the Y-axis we measure money which represents all other goods, (*i.e.* goods other than food). With a given money income OB_1 of an individual and given market price of food, B_1L_1 is the budget line whose slope represents the price of food (Note that price of money represented on the Y-axis is Re. 1, that is, price of rupee one is ₹ 1.). Before the receipt of food stamps the individual is in equilibrium at point E_1 on indifference curve IC_1 and is consuming OF_1 quantity of food and ON_1 quantity of other goods per week.

Now, suppose the individual is given food stamps of ₹ 200 per week which he can spend on food alone. Suppose further that price of food is ₹ 10 per kg. With stamps of ₹ 200 he can therefore buy 20 kg of food which is equal to B_1C at the given market price. Since the consumer cannot use food stamps to buy non-food items (other goods) he cannot spend more than his initial income OB_1 on other goods. Thus above the horizontal line B_1C, the combinations of other goods and food are not attainable when he is given the food stamps of ₹ 200. Since at the given market price of food, he can buy B_1C amount of food with the food stamps of ₹ 200 provided to him, while spending his entire income OB on other goods. *For instance, if price of food is ₹ 10 per kg., then with ₹ 200 he can buy 20 kg of food.* In this case, therefore, B_1C will be equal to 20 kg. If the individual wants to buy more foodgrains than B_1C, then he will spend some part of his initial income to purchase additional food. Since the food stamps are in addition to his initial income OB_1 his budget line with food stamps becomes a kinked line B_1CL_2.

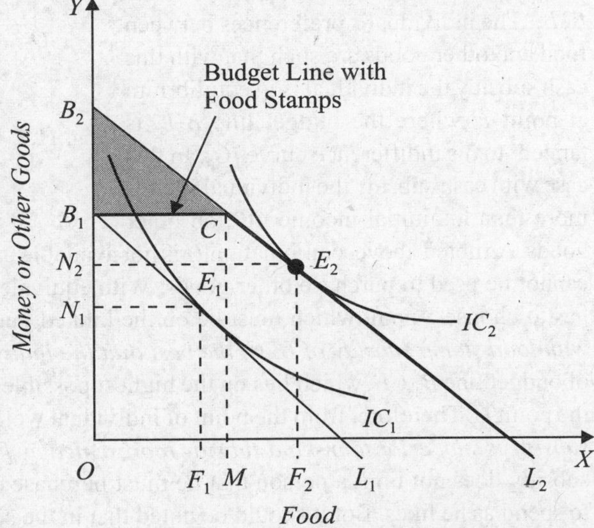

Fig. 11.3. *Effects of Food Stamp Programme on Consumption*

The food stamp programme can affect the recipient in two ways. One possibility is that with the food stamp subsidy and resultant kinked budget line B_1CL_2 in Fig. 11.3, the individual maximises his satisfaction at point E_2 where his budget line is tangent to indifference curve IC_2. At this new equilibrium point E_2 he is purchasing OF_2 quantity of food and ON_2 of other goods. Thus, as compared to the situation prior to food-stamp subsidy, he is on higher indifference curve showing a greater level of satisfaction or welfare and consuming greater quantities of food and other goods. Thus, *food stamp subsidy has led him to buy not only more food but also more of other goods. This means that a part of food stamp subsidy has been indirectly used for financing the purchases of non-food commodities.*

It is important to note that in this possible case, the effect of food stamp subsidy is exactly the same as would be the case if cash subsidy is granted to the individual. Thus, if instead of the food stamps the individual is given equivalent cash income of $B_1 B_2$ (*Note that with given market price of food, cash income of B_1B_2 can buy B_1C quantity of food and thus the two are equivalent*), the budget line will shift from $B_1 L_2$ to $B_2 L_2$. But given the preferences of the individual between food and other goods, he is in equilibrium at the same point E_2 at which his budget line B_2L_2 is tanget to the indifference curve IC_2. Thus, in this possibility, the effect of equivalent cash subsidy is exactly the same as the effect of food stamp subsidy. This happens because the preferences of the individual between food and other commodities are such that he wants to have more than B_1C quantity of food which is the quantity of food provided under the food stamp subsidy.

Another important conclusion from this possible case is that with either food stamp subsidy or cash subsidy the individual buys more of both food and other goods than he buys before the grant of subsidy. This is because food and other goods are here considered as normal goods whose quantity demanded increases with the increase in income.

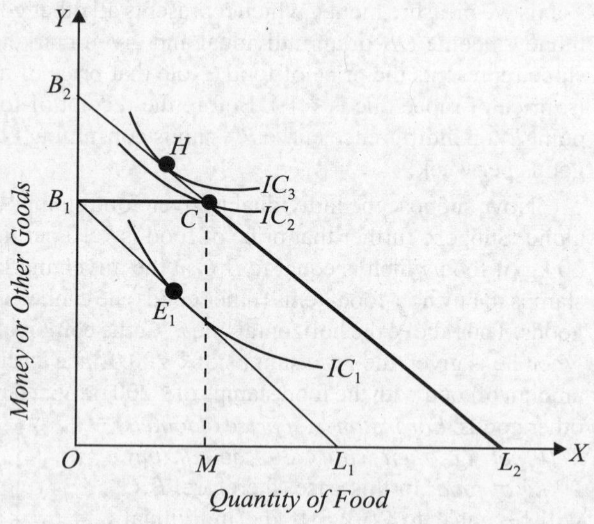

Fig. 11.4. *Effects of Food Stamp Subsidy and Cash Subsidy on Consumption and Welfare*

The second possibility of the effect of food stamp subsidy is illustrated in Fig. 11.4. Prior to grant of any subsidy, and given his budget line $B_1 L_1$, the individual is in equilibrium at point E_1 on indifference curve IC_1. Now let us assume that he is given the cash subsidy of $B_1 B_2$ so that with the given market price of food, budget line shifts to B_2L_2. The individual's preferences between food and other goods are such that with this cash subsidy the individual is in equilibrium at point H where the budget line B_2L_2 is tangent to his indifference curve IC_3. In this case with cash subsidy the individual spends more than his initial income OB_1 on other goods. As noted above, combination H is not available under food stamp subsidy because food stamps cannot be used to purchase other goods. With equivalent food stamp subsidy of B_1C the individual has to choose a point which must be on the kinked budget line B_1CL_2. With the budget line B_1CL_2 *with food stamp subsidy of B_1 C, the best that the individual can do is to choose the corner point C* of budget line B_1 CL_2 which lies on the highest possible indifference curve IC_2 which passes through the point C. Therefore, from the point of individual welfare we reach our earlier conclusion that *cash subsidy is supperior to in-kind subsidy represented by food stamps programme*. This is because cash subsidy does not limit a person that he must purchase certain amount of food and is therefore free to spend as he likes. But it should be noted that in the second possible case the consumption of food is more under the food stamp subsidy programme than in case of cash subsidy. If the purpose is to increase the consumption of food and therefore provide adequate diet to the people, then food stamp subsidy is better than cash subsidy as under the former, the individual is constrained to buy at least a given quantity of food.

Another important result obtained from our above analysis is that *even with food stamps programme the individual increases the consumption of all other goods* (*i.e.*, non-food items) too. This shows that a part of food stamp subsidy is *indirectly* used to finance the increased consumption of other goods. This is because some part of the income which the individual was spending on food prior to food stamp subsidy get released because of the food stamps being used for its purchase and this released income is spent on non-food items. This increases the consumption of non-food items also. This result is of special importance because supporters of food stamp subsidy have been emphasising that food subsidy should not be used to finance any part of non-food unnecessary items such as liquor. However, as seen above, in practice it is difficult to make a plan that will increase the consumption of subsidised food and will not affect the consumption of other goods.

Lastly, out of the two possibilities of the effects of food stamp subsidy and cash subsidy which is the most common result, that is, the most common outcome of the two possible cases presented in Fig. 11.3 and Fig. 11.4 . However, the final result of the two types of subsidies depends on the value of food stamps relative to preferences and incomes of the individuals to whom subsidies are granted. We cannot predict the specific result purely on theoretical grounds. Empirical research conducted in the USA however reveals that most recipients of food stamp programme represent situation depicted in Fig. 11.3. This means for most of the recipients, food stamp programme has the same effect as a cash subsidy.

DIRECT TAX VERSUS INDIRECT TAX

An important application of indifference curves is to judge the welfare effects of direct and indirect taxes on the individuals. In other words, if the Government wants to raise a given amount of revenue whether it will be better to do so by levying a direct tax or an indirect tax from the view-point of welfare of the individuals.[4] We shall study below that indirect tax such as excise duty income causes excess burden on the individuals, that is, indirect tax reduces welfare more than the direct tax, say income tax when an equal amount of revenue is raised through them. Consider Fig. 11.5 where on the *X*-axis, good *X* and on the *Y*-axis money is measured. With a given income of the individual and the given price of good *X*, the price line is PL_1 which is tangent to indifference curve IC_3 at point Q_3 where the individual is in equilibrium position.

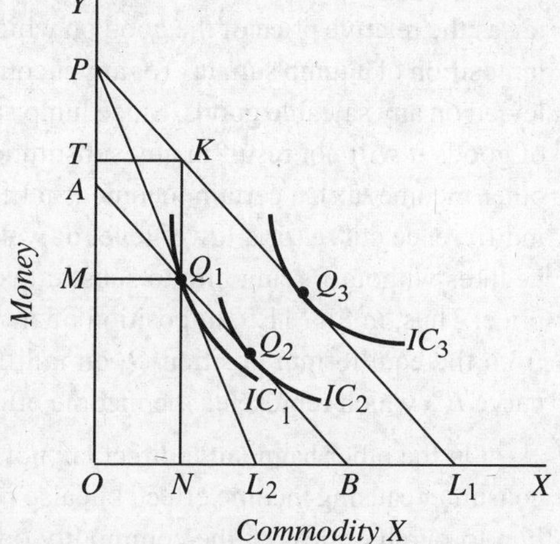

Fig. 11.5. *Indirect tax causes excess burden.*

Suppose now that Government levies an excise duty (an indirect tax) on good *X*. With the imposition of excise duty, the price of good *X* will rise. As a result of the rise in price of good *X*, the price line rotates to a new position PL_2 which is tangent to indifference curve IC_1 at point Q_1. It is thus clear that as a result of the imposition of excise duty, the individual has shifted from a higher indifference curve IC_3 to a lower one IC_1, that is, his level of satisfaction or welfare has declined. It is worth noting that the movement from Q_3 on indifference

4. Direct taxes are those taxes whose incidence cannot be shifted to others. Income tax, wealth tax, death duty are the examples of direct tax. On the other hand, an indirect tax is one which can be passed on or shifted to others by raising the prices of the goods. The excise duty, sales tax are the examples of indirect tax.

curve IC_3 to Q_1 on indifference curve IC_1 is the combined result of the income effect and substitution effect caused by the excise duty.

It should be further noted that at point Q_1 (that is, after the imposition of excise duty) the individual is purchasing *ON* amount of good *X* and has paid *PM* amount of money for it. At the old price (before the excise duty was imposed), he could purchase *ON* quantity of good *X* for *PT* amount of money. Thus, the difference *TM* (or KQ_1) between the two is the amount of money which the individual is paying as the excise duty. Direct taxes are those taxes whose incidence cannot be shifted to others. Lump sum tax, proportionate and progressive income taxes, wealth tax, death duty are the examples of direct tax. On the other hand, an indirect tax is one which can be passed on or shifted to others by raising the prices of the goods. The excise duty, sales tax are the examples of indirect tax.

Now, suppose that instead of excise duty, Government levies a direct tax of the type of lump sum tax on the individual when the individual is initially at point Q_3 on indifference curve IC_3. With the imposition of lump sum tax, the price line will shift below but will be parallel to the original price line PL_1. Further, if the same amount of revenue is to be raised through lump sum tax as with the excise duty, then the new price line *AB* should be drawn at such a distance from the original price line PL_1 that it passes through the point Q_1. So, it will be seen from Fig. 11.5 that with the imposition of lump sum tax equivalent in terms of revenue raising to the excise duty, we have drawn the budget line *AB* which is passing through the point Q_1. However, with *AB* as the price line, individual is in equilibrium at point Q_2 on indifference curve IC_2 which lies at a higher level than IC_1. In other words, at point Q_2 individual's level of welfare is higher than at Q_1. Lump sum tax has reduced the individual's welfare less than that by the excise duty. Thus, *indirect tax (excise duty) causes an excess burden on the individual.*[5]

Now, the important question is why an indirect tax (an excise duty or a sales tax on a commodity) causes excess burden on the consumer in terms of loss of welfare or satisfaction. The basic reason for this is that whereas both the lump sum tax (or any other general income tax) and an indirect tax reduce consumer's income and produce income effect, the indirect tax in addition to the income effect, also raises the relative price of the good on which it is levied and therefore causes substitution effect. The imposition of a lump sum tax (or any income tax) does not affect the prices of goods because it is not levied on any saleable goods. Since lump sum tax or any income tax does not alter the relative prices of goods it will not result in any substitution effect. With the imposition of a lump sum tax (or any other income tax), a certain income is taken away from the consumer and he is pushed to the lower indifference curve (or a lower level of welfare) but he is free to spend the income he is left with as he likes without forcing him to substitute one commodity for another due to any change in relative price. Thus, in Fig. 11.5, imposition of an equivalent lump sum or income tax, the consumer moves from the equilibrium position Q_3 on indifference curve IC_3 to the new position Q_2 on indifference curve IC_2 which represents the income effect.

On the other hand, an indirect tax not only reduces the purchasing power or real income of the consumer causing income effect, but also produces price-induced substitution effect and thus forcing him to purchase less of the commodity on which indirect tax has been levied and buy more of the

5. This conclusion regarding excess burden of indirect tax compared to the direct tax is however based on several assumptions about initial conditions. Whether or not indirect tax reduces welfare more than direct tax depends upon so many initial conditions before the taxes are introduced. For knowing the analysis regarding welfare implications of direct and indirect taxes, see (1) I.M.D. Little, "Direct versus Indirect Taxes," *Economic Journal*, Sept. 1951. (2) Milton Friedman, "The Welfare Effects of Taxes," *Essays in Positive Economics*, University of Chicago Press, 1953 and (3) R.A. Musgrave, *Theory of Public Finance*.

non-taxed commodity. And this later substitution effect caused due to the price distortion by the indirect tax further reduces his welfare. As will be seen from Fig. 11.5, as a result of income effect of the indirect tax the consumer moves from point Q_3 on indifference curve IC_3 to point Q_2 on lower indifference curve IC_2 and as a result of substitution effect he is further pushed to point Q_1 on still lower indifference curve IC_1.

EFFECT OF RATIONING ON CONSUMER'S WELFARE

Indifference curve analysis can be used to explain under what conditions rationing of goods by the Government can act as a binding or a constraint on consumer's choices and further how it affects his welfare. It may be noted that income of a consumer along with the prices of goods serves as a constraint on his choices and is often called a *budget constraint*. This budget constraint can be written as follows :

$$P_x \cdot X + P_y \cdot Y \leq M$$

The above inequality implies that consumer can choose a combination of goods from within or on the *market opportunity set*. With given income of the consumer and prices of the two goods we draw a budget line *BL* in Fig. 11.6. The shaded region bounded by the budget line *BL* and the coordinate axes represent market opportunity set from which the consumer can make a choice of the two commodities. If now the Government introduces a rationing for commodity *X* and fixes a ration of *X* equal to OR_x (At point R_x we have shown a vertical line showing the constraint or ration limit imposed by the rationing fixed at OR_x). It will be seen that with ration limit fixed at OR_x rationing does not act as a binding at all and proves to be quite ineffective in restricting the consumption of good *X* which is the objective of the policy. However, such a situation is relevant in case of a poor family whose income is so small that it cannot buy even the rationed quantity. It is the income that serves as a binding on his consumption choice and not the ration limit.

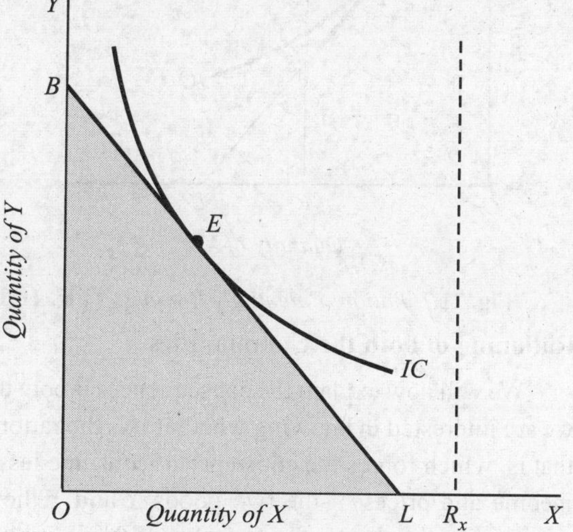

Fig. 11.6. *Ration: Not Binding*

Now consider Fig. 11.7 where ration limit is fixed at R_x which lies to the left of *L*. This ration limit reduces or truncates his market opportunity set (*i.e.* set of attainable combinations of two goods *X* and *Y*) as shown by the reduced shaded area in Fig. 11.7 and therefore in this case the *ration limit is only potentially binding* on him. Though the consumer can buy the ration amount, that is, ration limit is attainable but he is not willing to consume good *X* as much as the ration limit permits him. He is in equilibrium at point *E* where he consumes quantity of good *X* which is smaller than the ration amount R_x. Thus, his preferences are such that rationing is not actually binding for him. Ration limit is potentially binding here because due to ration limit the consumer *cannot* buy any set of two commodities lying in the region SLR_x, if he so desires though his price-income situation permits him to do so.

However, the more important and relevant case of rationing is depicted in Fig. 11.8. In this case ration limit fixed is R_x which lies to the left of his equilibrium position *E*. Without the restriction of rationing he will consume a larger amount of commodity *X* and will be at indifference curve IC_1. With the budget line *BL* and the rationed quantity R_x he will be at point *E′* which lies at lower indifference curve IC_0. Thus ration limit serves as a binding for him and forces him to consume less of good *X* and more of good *Y* than he prefers. That is why at point *K* in Fig. 11.8 he is at lower indifference curve IC_0 indicating his lower level of welfare. Thus, in this case, rationing is actually binding for the consumer and reduces his welfare.

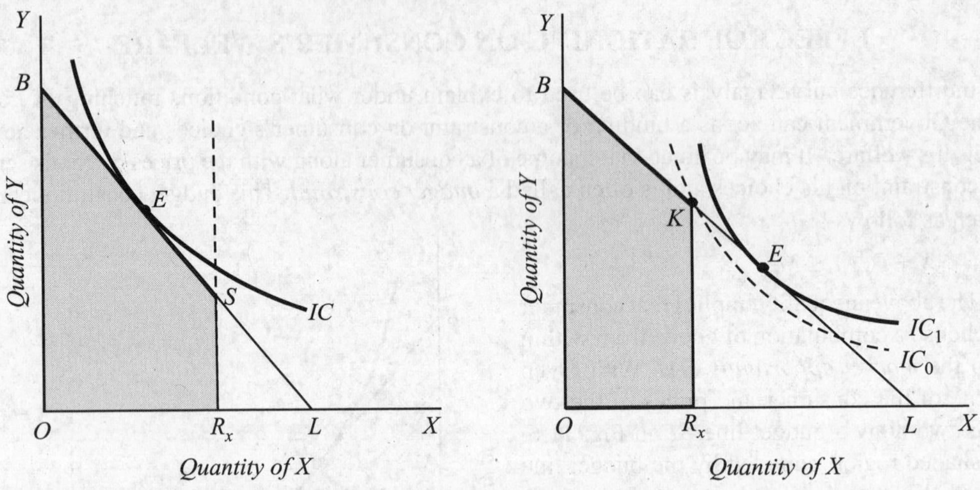

Fig. 11.7. *Ration: Potentially Binding* **Fig. 11.8.** *Ration is binding and reduces consumer's welfare.*

Rationing of both the Commodities

We will now explain the consequences if both the commodities *X* and *Y* are rationed. In particular we are interested in knowing whether it is the ration limits or income of the consumer that is binding, that is, which forces the consumer to consume less amounts of the goods. In Fig. 11.9 with a given income and prices of the two goods *X* and *Y* the consumer is in equilibrium at *E* buying *OM* of commodity *X* and *ON* of commodity *Y*. Now, suppose with the introduction of rationing, ration limit *R* is fixed for good *X* and R_y for good *Y*. It will be seen from Fig. 11.9 that ration amounts of R_x and R_y of goods *X* and *Y* respectively are greater than *OM* and *ON* which the consumer is buying with his price-income situation. Therefore, the ration limits in this case are *not actually binding* since these do not any more constrain his consumption. Of course, these ration limits narrow down or truncate his market opportunity set at both ends on the *X*-and *Y*-axes, and in this way they are *potentially binding* but they are *not effective* in restraining his consumption. Therefore, this can be interpreted to be the case of a poor family whose optimum consumption basket of the two goods is small because of its low income and therefore remains unaffected if ration limits are set at higher levels.

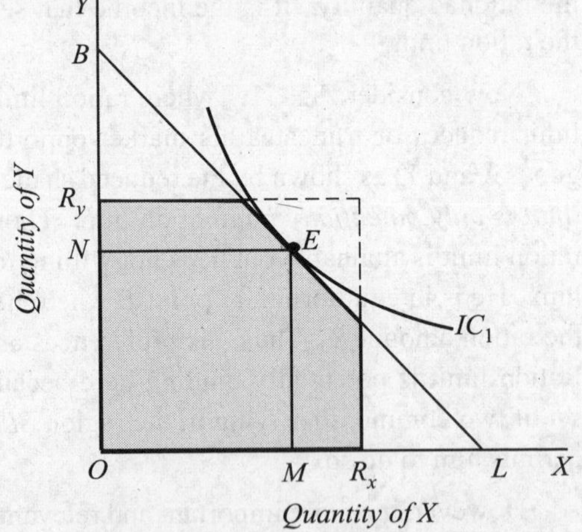

Fig. 11.9. *Ration is potentially binding*

Now, consider Fig. 11.10 where it will be seen that ration limit R_y is smaller than consumer's optimum consumption of *Y* as indicated by combination *E* of the two goods on the budget line

BL which has been drawn with his given income and prices of two goods. But ration limit R_x for good X is larger than his optimum or equilibrium consumption quantity of good X. It follows therefore that for good Y ration limit is actually binding as it forces the consumer to point K on a lower indifference curve IC_0 than optimum point E on indifference curve IC_1 at which he would have been without the restriction of rationing. On the other hand, in case of good X in Fig. 11.10 the ration limit R_x does not seem to be effective, though it is potentially binding as it truncates his market opportunity set. It will be observed that binding by rationing lowers his level of welfare as he is forced to come to the point L on a lower indifference curve IC_0 where he consumes less of good Y and more of good X than he prefers.

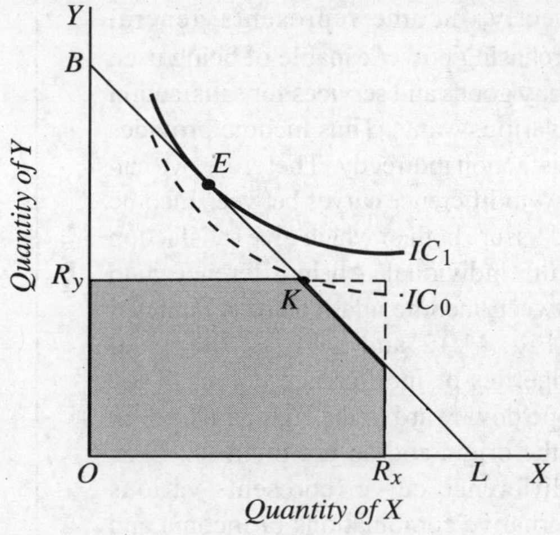

Fig. 11.10. *Y–Ration is Binding, X–Ration is Not Binding.*

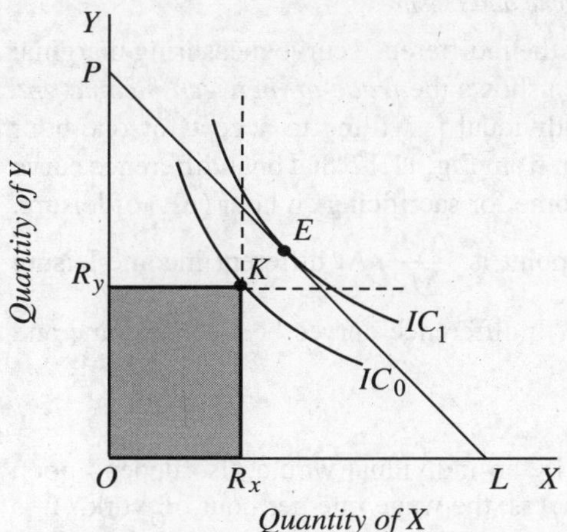

Fig. 11.11. *Both Ration Limits are Binding and Reduce Consumer's Welfare.*

In Fig. 11.11 both ration limits R_x and R_y are actually binding and force the individual to consume less of both the commodities than he consumes at his equilibrium position E in the absence of rationing of the two commodities. Fig. 11.11 depicts the case of a relatively rich person who without the binding of rationing is at point E on indifference curve IC_1 and is consuming greater quantities of the two goods than the ration amounts. Introduction of rationing limits forces him to come to point K on the lower indifference curve IC_0 and consume less amounts of both the goods than he would do without the restriction of rationing. Thus in this case rationing is actually binding on him and reduces his welfare.

INCOME-LEISURE CHOICE

Indifference curve analysis can be used to explain an individual's choice between income and leisure and to show why a higher overtime wage rate must be paid if more hours of work are to be obtained from the workers. It is important to note that income is earned by devoting some of the leisure time to do some work. That is, income is earned by sacrificing some leisure. The greater the amount of this sacrifice of leisure, that is, the greater the amount of work done, the greater income an individual earns. Further, income is used to purchase goods, other than leisure for consumption. Leisure time can be used for resting, sleeping, playing, listening to music on radios and television etc. all of which provide satisfaction to the individual. Therefore, *in economics leisure is regarded as a normal commodity the enjoyment of which yields satisfaction to the individual*. While leisure yields satisfaction to the individual directly, income represents general purchasing power capable of being used to buy goods and services for satisfaction of various wants. Thus income provides satisfaction indirectly. Therefore, we can draw indifference curves between income and leisure both of which give satisfaction to the individual. An indifference map between income and leisure is depicted in Fig. 11.12 and have all the usual properties of indifference curves. They slope downwàrd to the right, are convex to the origin and do not intersect. Each indifference curve represents various alternative combinations of income and leisure which provide equal level of satisfaction to the individual and the farther away an indifference curve is from the origin, the higher the level of satisfaction it represents for the individual. The slope of the indifference curve measuring marginal rate of substitution between leisure and income (MRS_{LM}) shows the *trade-off between income and leisure*. This trade-off means how much income the individual is willing to accept for one hour sacrifice of leisure time. In geometric terms, it will be seen from Fig. 11.12 that on indifference curve IC_1 at point A the individual is willing to accept ΔM income for sacrificing an hour (ΔL) of leisure. Thus the trade-off between income and leisure at this point is $\frac{\Delta M}{\Delta L}$. At different income-leisure levels, the trade-off between leisure and income varies. Indifference curves between income and leisure are therefore also called *trade-off curves*.

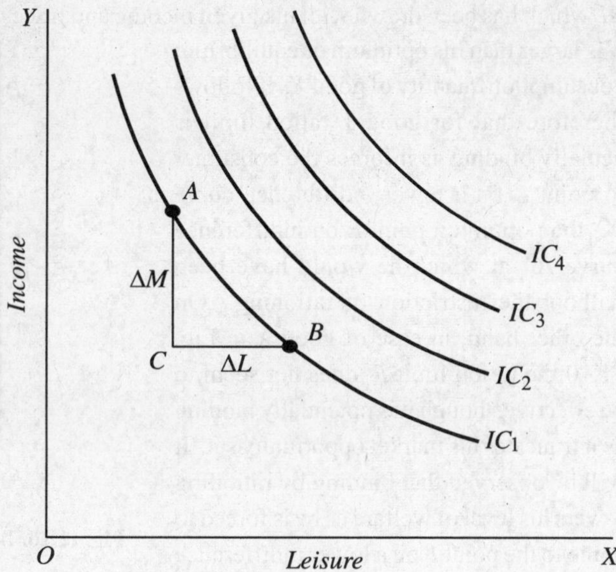

Fig. 11.12. *Trade-off Curves: Inidfference Curves between Income and Leisure*

Income-Leisure Constraint

However, the actual choice of income and leisure by an individual would also depend upon what is the market rate of exchange between the two, that is, the wage rate per hour of work. It is worth noting that wage rate is the opportunity cost of leisure. In other words, to increase leisure by one hour, an individual has to forego the opportunity of earning income (equal to wage per hour) which he can earn by doing work for an hour. This leads us to income-leisure constraint which together with the indifference map between income and leisure would determine the actual choice by the individual. The maximum amount of time available per day for the individual is 24 hours. Thus, the maximum amount of leisure time that an individual can enjoy per day equals 24 hours. In order to earn income for satisfying his wants for goods and services, he will devote some of his time to do work. Consider Fig. 11.13 where leisure is measured in the rightward direction along the horizontal axis and the maximum leisure time is OT (equal to 24 hours). If the individual can work for all the 24 hours in a day, he would earn income equal to OM. Income OM equals OT multiplied by the hourly wage rate ($OM = OT.w$) where w represents the wage rate. The straight line MT is the budget constraint, which in the present context is generally referred to as income-leisure constraint and shows the various combinations of income and leisure among which the individual will have to make a choice. Thus, if a person chooses combination C, this means that he has OL_1 amount of leisure time and OM_1 amount of income. He has earned OM_1 amount of income by working TL_1 hours of work. Choice of other points on income-leisure line MT will show different amounts of leisure, income and work.

Income $OM = OT.w \quad \frac{OM}{OT} = w$

Thus, the slope of the income-leisure curve OM/OT equals the wage rate.

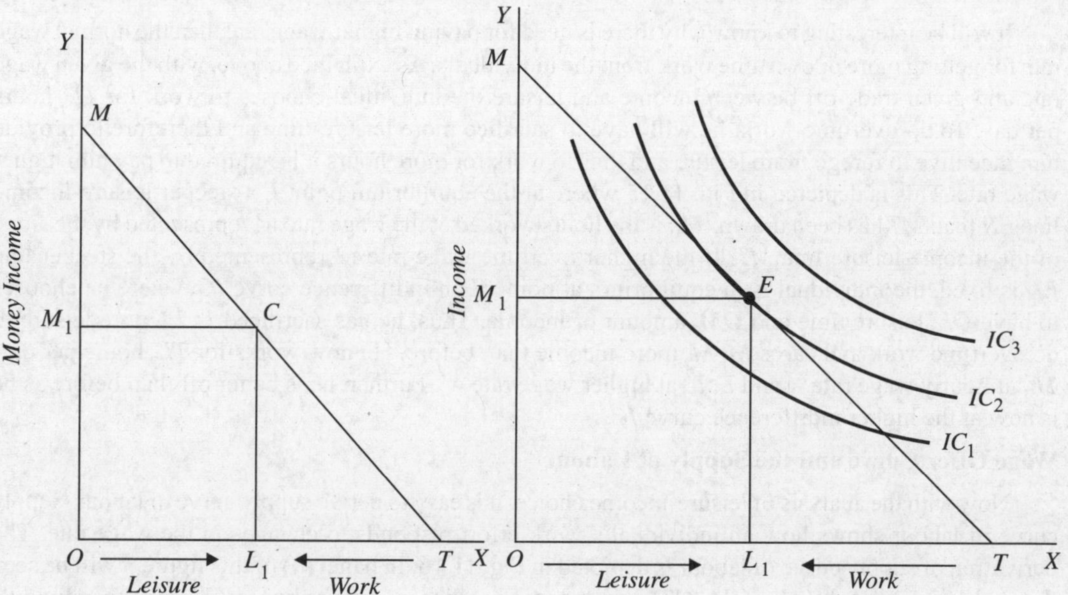

Fig. 11.13. *Income-Leisure Constraint* **Fig. 11.14.** *Income-Leisure Equilibrium*

Income-Leisure Equilibrium

Now, we can bring together the indifference map showing ranking of preferences of the individual between income and leisure and the income-leisure line to show the actual choice of leisure and income by the individual in his equilibrium position. We will further show how much work effort (*i.e.* supply of labour in terms of hours worked) he would put in this optimal situation. Our analysis is based on two assumptions. First, he is *free to work as many hours per day as he likes.* Second, wage rate is the same irrespective of the number of hours he chooses to work.

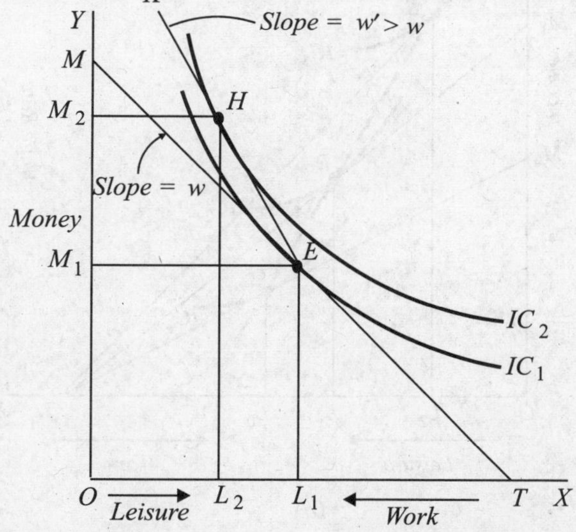

Fig. 11.15. *Need for higher overtime wage rate.*

Fig. 11.14 displays income-leisure equilibrium of the individual. With the given wage rate, the individual will choose a combination of income and leisure lying on the income-leisure line *MT* that maximises his satisfaction. It will be seen from Fig. 11.15 that the given income-leisure line *MT* is tangent to the indifference curve IC_2 at point *E* showing choice of OL_1 of leisure and OM_1 of income. In this optimal situation, income-leisure trade-off (*i.e. MRS* between income and leisure equals the wage rate (w), that is, the market exchange rate between the two. In this equilibrium position the individual works for TL_1 hours per day ($TL_1 = OT - OL_1$). Thus, he has worked for TL_1 hours to earn OM_1 amount of income.

Need for Higher Overtime Wage Rate

It will be interesting to know why there is need for paying higher wage rate than the normal wage rate for getting more or overtime work from the individuals. As explained above, with the given wage rate and given trade-off between income and leisure the individual chooses to work for TL_1 hours per day. To do overtime work, he will have to sacrifice more leisure-time and therefore to provide him incentive to forego more leisure and thus to work for more hours it is required to pay him higher wage rate. This is depicted in Fig. 11.15 where at the equilibrium point E a steeper leisure-income line EK than MT has been drawn. TL_1 is the hours worked at the wage rate w' represented by the slope of the income-leisure line MT. If the higher overtime wage rate w' represented by the steeper line EK is fixed, the individual is in equilibrium at point H on indifference curve IC_2 where he chooses to have OL_2 leisure time and OM_2 amount of income. Thus, he has sacrificed $L_1 L_2$ more leisure to do overtime work and earns $M_1 M_2$ more income than before. He now works for TL_2 hours per day, TL_1 at hourly wage rate w and $L_1 L_2$ at higher wage rate w'. Further, he is better off than before as he is now at the higher indifference curve IC_2.

Wage Offer Curve and the Supply of Labour

Now with the analysis of leisure-income choice, it is easy to derive supply curve of labour. Supply curve of labour shows how an individual's work effort responds to changes in the wage rate. The derivation of supply curve of labour is depicted in Fig. 11.16. In panel (*a*) of this figure it will be seen that at the wage rate w_0 ($w_0 = OM_0/OT$), the wage line or income-leisure line is TM_0 and the individual is in equilibrium at point Q where he chooses OL_0 leisure time and works for TL_0 hours. That is, at wage rate w_0 he supplies TL_0 amount of labour. This supply of labour is directly shown against wage rate w_0 in panel (*b*) of Fig. 11.16. Now, when the wage rate rises to w_1, wage line or income-leisure

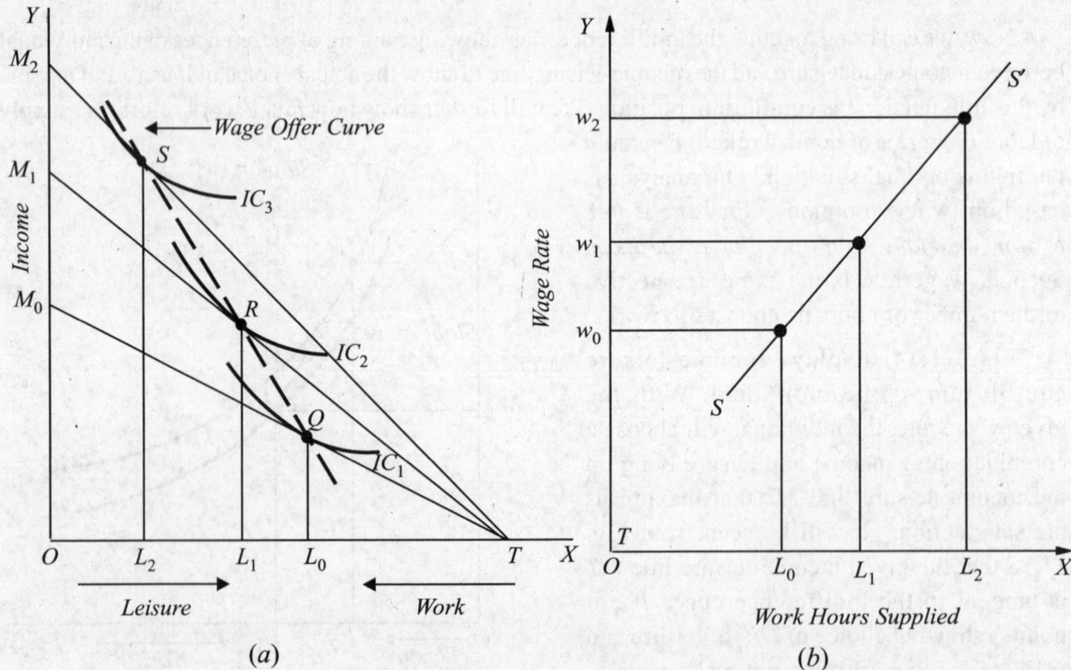

Fig. 11.16. (*a*) *Wage Offer Curve,* (*b*) *Supply Curve of Labour*

line shifts to TM_1 ($w_1 = OM_1/OT$), the individual reduces his leisure to OL_1 and supplies TL_1 hours of work; $L_1 L_0$ more than before (see panel (*a*) in Fig. 11.16). Thus, TL_1 number of work-hours supplied is shown against w_1 in panel (*b*) of Fig. 11.16. Likewise, when the wage rate rises to w_2 (w_2, $= OM_2/OT$), income-leisure line shifts to TM_2 the individual chooses to have leisure time OL_2 and supplies TL_2 work-hours. In panel (*a*) on joining points *Q*, *R* and *S* we get what is often called wage-offer curve which is similar to price-consumption curve. In panel (*b*), the information supplied by the wage-offer curve, that is, the supply of labour (work-hours) by the individual at different wage rates is shown directly as, in this panel, supply of labour (hours worked) is measured along the *X*-axis and wage rate along the *Y*-axis. A glance at panel (*b*) of Fig. 11.16 will reveal that supply curve of labour is upward sloping indicating positive response of the individual to the rise in wage rate.

Income Effect and Substitution Effect of the Change in Wage Rate

Now the supply curve of labour does not always slope upward as is shown in Fig. 11.16(*b*). It can slope or bend backward too which implies that at a higher wage rate, the individual will supply less labour (*i.e.* will work for less hours). Under what conditions supply curve of labour (*i.e.* work-hours) slopes upward and under what circumstances it bends backward can be explained in terms of income effect and substitution effect of a change in wage rate. As in case of change in price, rise in wage rate has both the substitution effect and income effect. The net combined effect on the supply of labour (hours worked) depends on the magnitude of the substitution effect and income effect of the rise in wage rate. It is important to note that *leisure is a normal commodity which means that increase in income leads to the increase in leisure enjoyed (i.e. less work-hours supplied)*. That is, income effect of the rise in wage rate on leisure is positive, that is, leads to the *increase in the hours of leisure enjoyed* (that is, tends to *decrease labour supply*).

On the other hand, the rise in wage rate increases the opportunity cost or price of leisure, that is, it makes enjoyment of leisure relatively more expensive. Therefore, as a result of rise in wage rate individual substitutes work (and therefore income) for leisure which leads to the increase in supply of labour. That is, a substitution effect of the rise in wage rate tends to reduce leisure and *increase labour supply* (*i.e.* number of hours worked). It is thus clear that *for an individual supplier of labour income effect and substitution effect work in opposite directions*[6]. Whereas income effect of the rise in wage rate tends to reduce supply of labour, substitution effect tends to increase it. If the income effect is stronger than the substitution effect, the net combined effect of rise in wage rate will be to reduce labour supply. On the other hand, if substitution effect is relatively larger than the income effect, the rise in wage rate will increase labour supply.

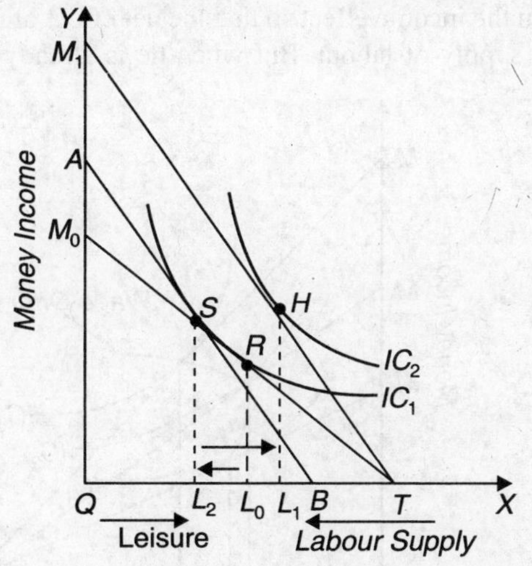

Fig. 11.17. *Decomposing the Effect of Rise in Wage Rate into Income Effect and Substitution Effect*

How the effect of rise in wage rate is split up into income effect and substitution effect is shown in Fig. 11.17. In this figure we measure money income on the *Y*-axis and leisure (reading from left to right) and labour supply (reading from right to left) on the *X*-axis. Suppose to begin with the wage rate is w_0 and if all the available hours *OT* are used

6. Note that this is unlike the case of *price effect of normal goods* in which case substitution effect and income effect work in the *same* direction.

to do work, OM_0 money income is earned. This gives us TM_0 as the budget constraint or which in the present context is also called leisure-income constraint. It will be seen from Fig. 11.17 that TM_0 is tangent to indifference curve IC_1 between leisure and income at point R. Thus, with wage rate w_0 the individual is in equilibrium when he enjoys OL_0 leisure and therefore he is supplying TL_0 work-hours of labour. Now suppose that wage rate rises to w_1 with the result that income-leisure constraint line rotates to TM_1. Now, with TM_1 as new income-leisure constraint line, the individual is in equilibrium at point H at which he supplies TL_1 work-hours of labour which are less than TL_0. Thus, with the rise in wage rate, supply of labour has decreased by L_0L_1. To break up this wage effect on labour supply, we reduce his money income by compensating variation in income.

To do so we take away so much income from the individual that he comes back to the original indifference curve IC_1. AB is such line obtained after reducing his money income by compensating variation. AB is tangent to indifference curve IC_1 at point S at which he supplies TL_2 hours for work. This shows with rise in wage rate from w_0 to w_1 resulting in leisure becoming relatively expensive, he substitutes work (*i.e.* labour supply) L_0L_2 for leisure. This is substitution effect of rise in wage rate which tends to increase labour supply by L_0L_2. Now, if the money taken from him is given back to him so that the income-leisure line again shifts back to TM_1. With TM_1 he reaches his old equilibrium position at point H where he supplies TL_1 work-hours. Thus, movement from point S to H and as a result the decrease in labour supply by L_2L_1 represents the income effect of the rise in wage rate. Thus, while income effect of the increase in wage rate causes decrease in labour supply by L_2L_1, its substitution effect causes increase in labour supply by L_0L_2. It will be seen from Fig. 11.17 that income effect is stronger than substitution effect so that the net result is reduction in labour supply by L_0L_1 work-hours. Now, if substitution effect had been larger than income effect, work-hours supplied would have increased as a result of rise in wage rate.

Backward Bending Supply Curve of Labour

It may, however, be noted that on theoretical grounds it cannot be predicted which effect will be stronger. It has, however, been empirically observed that when the wage rate is small so that the demand for more income to purchase goods and services is very strong, substitution effect is larger than the income effect so that the net effect of rise in wage rate will be to reduce leisure and increase the supply of labour. But when he is already supplying a large amount of labour and is earning

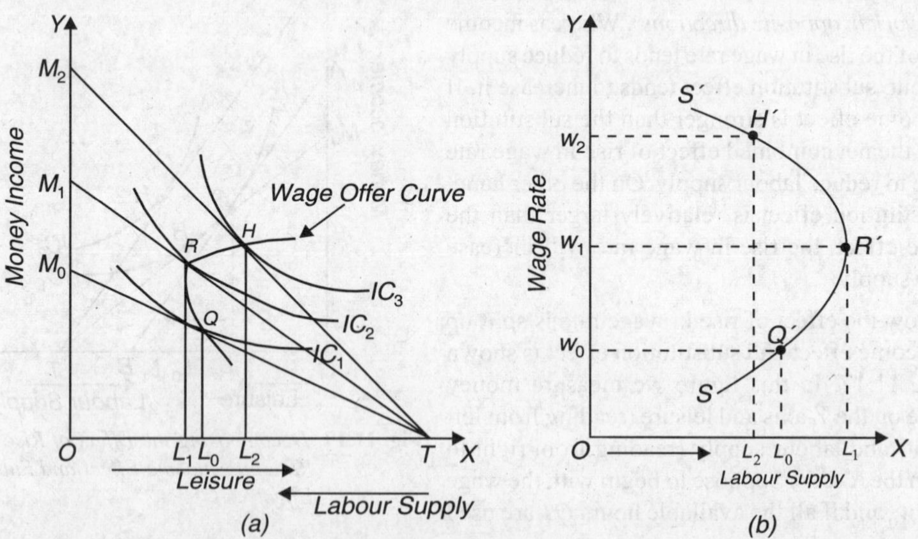

Fig. 11.18. *Backward Bending Supply Curve of Labour*

sufficient income, further increases in wage rate may induce the individual to demand more leisure so that income effect may outweigh the substitution effect at higher wage rates. This implies that at higher wage rates, labour supply may be reduced in response to further rise in wage rates. This means that up to a point substitution effect is stronger than income effect so that labour supply curve slopes upward, but beyond that at higher wage rates supply curve of labour bends backward. This is illustrated in Fig. 11.18 where in panel (*a*) wage offer curve is shown, and in panel (*b*) supply curve of labour is drawn corresponding to leisure-work equilibrium points in panel (*a*). Thus, to start with at wage rate w_0 (*i.e.* TM_0 as budget constraint) L_0, amount of work-hours (labour) are supplied. This is directly plotted against the wage rate w_0 in panel (*b*) of Fig. 11.18. When the wage rate rises to w_1 (budget constraint becomes TM_1 in panel (*a*) of Fig. 11.18), the greater amount of labour L_1 is supplied, which is greater than L_0. Amount of labour L_1 is directly plotted against higher wage rate w_1 in panel (*b*) of Fig. 11.18. With the further increase in wage rate to w_2, the income-leisure constraint rotates to TM_2 and the individual is in equilibrium when he supplies L_2 work-hours which are smaller than L_1. Thus, with the rise in wage rate above w_1 labour supply decreases. In other words, up to wage rate w_1, labour supply curve slopes upward and beyond that it starts bending backward. This is quite evident from panel (*b*) of Fig. 11.18.

INTERTEMPORAL CHOICE BETWEEN PRESENT CONSUMPTION AND FUTURE CONSUMPTION: INDIFFERENCE CURVE ANALYSIS

Capital formation in a year depends on savings of the people. Now the savings of the people are determined by choice between present consumption and future consumption. The indifference curve analysis can be used to show how an individual will choose between present and future consumption so as to maximise his total satisfaction over time. To make our analysis simple (we consider only two time periods, working period and the retirement period). In real life there are more than two time periods over which consumption and saving decisions are made by an individual but the two-period model provides us useful insights into the basic issues involved. As mentioned above, the exchange of present consumption for future consumption is done through saving and lending these savings to others or keeping them in bank deposits which yield interest.

In our analysis we denote present period income by Y_0, future period income by Y_1, current consumption by C_0, future consumption by C_1 and rate of interest by i. Given his present income Y_0 and rate of interest (r), an individual has to decide how much he will consume in the present (*i.e.* C_0) and how much he will save for consumption in the future (*i.e.* C_1). The consumer's preferences between the present consumption and future consumption are represented by a set of indifference curves such as I_1, I_2 which are downward-sloping and convex to the origin as shown in Fig. 11.19. The downward sloping feature of indifference curves implies that the individual is willing to substitute some amount of present consumption for future consumption and the rate at which he is willing to substitute present consumption for future consumption depends on the particular pattern of consumption he likes to have.

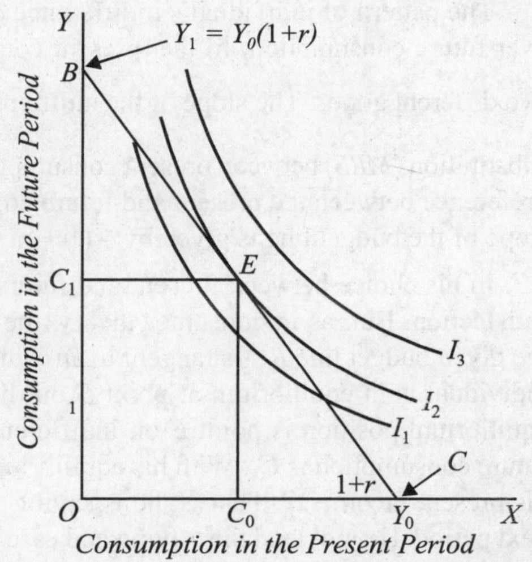

Fig. 11.19. *Choice between Present and Future Consumption: Intertemporal Choice*

The convexity of indifference curves implies that the individual would like to have some '*average*' *amount of consumption* in each period rather than extremely large consumption in the present and no or little consumption in the future or *vice versa.*

The Budget Constraint. To understand the choice, it is important to know the nature of budget constraint an individual faces in this regard. There are two possible kinds of consumption choices. First, he consumes his entire income in present period, that is, he consumes Y_0 in the present period and saves nothing for the future *(i.e.* next period). The second possibility is that he chooses to consume less than his present income and saves some for future consumption. That is, some present consumption is exchanged for more consumption in the future. How much more future consumption he would like to get for sacrifice of some consumption in the present depends on the rate of interest. Thus, if market real rate of interest is 10 per cent, then if he saves ₹ 100 in the present year, he will have ₹ 110 in the next year (110 = 100 + 10% of 100). In other words, he has exchanged ₹ 100 of the present consumption for consumption of ₹ 110 in the next year.

We now derive the budget constraint or what is called ***intertemporal budget line*** which has been shown by *BC* in Fig. 11.21. We denote the individual's income in the present period by Y_0 and the next period's income by Y_1. In order to draw the intertemporal budget constraint we have to determine what would be the future income if the individual saves all his present income equal to Y_0. Given the interest rate equal to *r*, the next period's income will be $Y_1 = Y_0(1 + r)$. We represent it by point *B*.

Thus, if the individual decides to save his entire present income Y_0 for future consumption, then his next period's consumption which we denote by Y_1 will be

$$Y_1 = Y_0(1 + r)$$

We represent Y_1 by point *B* on the vertical axis in Fig. 11.21. If we join *B* with *C* we get the intertemporal budget line *BC* representing the budget constraint.

The slope* of the budget constraint line $BC = -(1+r)$. Thus the slope of the budget line depends on the rate of interest which is equal to $-(1 + r)$ where *r* is the market rate of interest. It may be noted that rate of interest is in fact the return on saving.

Intertemporal Choice: Individual's Equilibrium

The pattern of individual's indifference curves depicts his preferences of present consumption over future consumption. In fact present consumption and future consumption can be regarded as two different goods. The slope of the indifference curve, that is, $\frac{\Delta C_1}{\Delta C_0}$ measures the marginal rate of substitution (*MRS*) between present consumption and future consumption and reflects his rate of time preference between the present and future consumption. On the other hand, as explained above, the slope of the budget line is given by $-(1 + r)$ where *r* is rate of interest.

In his choice between present and future consumption, the consumer will try to maximise his satisfaction. But, as in consumer theory, the individual will be maximising his satisfaction where the given budget line *BC* is tangent to an indifference curve. It will be seen from Fig. 11.21 that the individual is in equilibrium at point *E* on the indifference curve I_2. Thus the consumer's optimal equilibrium position is point *E* on indifference curve I_2 where his present consumption is C_0 and future consumption is C_1. With his equilibrium position at *E* present his consumption C_0 is less than his present income Y_0. That is, he is saving $Y_0 - C_0$ of his present income for consumption in the next period. He will lend this saving and earn interest on it so that his consumption in the next period is C_1. Thus, the individual saves a part of his income in the present period and lends it to others to earn interest.

It may be noted that at the equilibrium point *E*, the slope of indifference curve of the individual depicting his marginal rate of substitution between present consumption and future consumption

($MRS_{C_0C_1}$) is equal to the slope of the intertemporal budget line BC which in absolute terms is equal to $(1 + r)$. Thus, at equilibrium or optimal point E,

$$MRS_{C_0C_1} = (1 + r)$$

Saving and Interest Rate

It is useful to know what will be the impact of change in the rate of interest on individual's intertemporal choice or on his savings. This is depicted in Fig. 11.20. With the equilibrium position at E_0, suppose rate of interest rises. The rise in rate of interest will change the budget constraint. With a higher interest rate than the previous one, the new budget line $B'C$ will be steeper than the previous budget line BC. But. It will be seen from Fig. 11.20 the new budget line $B'C$ representing higher rate of interest is steeper than the previous budget line BC. With this new budget line $B'C$ the individual is in equilibrium at point E_1 on his higher intertemporal indifference curve I_2 and is therefore better off than before. Besides, it will be noticed that with the new position E_1 of intertemporal choice on indifference curve I_2, the individual's present consumption has decreased from C_0 to C_0', that is, he has saved more at the higher rate of interest to increase his consumption in the future period by C_1C_1'.

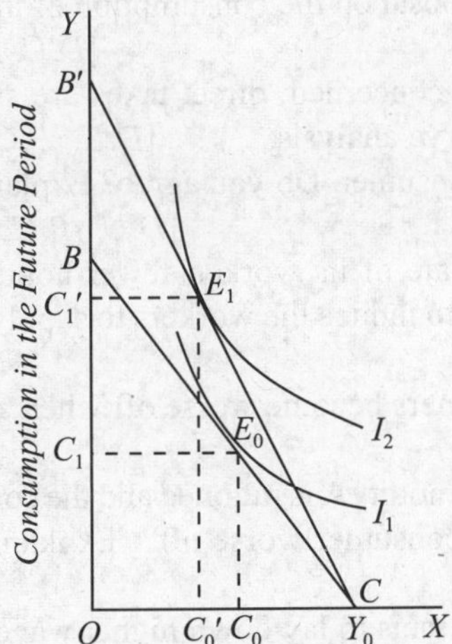

Fig. 11.20. *Impact of Rise in Interest Rate on Saving: Substitution Effect Dominates.*

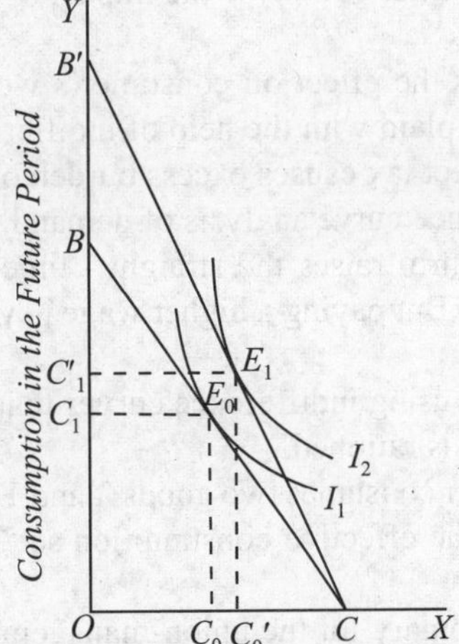

Fig. 11.21. *Impact of Rise in Interest Rate on Saving: Income Effect Dominates.*

The effect of rise in interest rate has income effect as well as substitution effect. A rise in rate of interest increases income of the individual and therefore makes him better off. This induces him to consume more in the present. This is the income effect which tends to reduce savings. But the rise in interest rate also increases return on savings. This induces him to postpone consumption because every sacrifice of consumption (*i.e.*, savings) will mean more consumption in the next period due to higher interest earned. This is substitution effect which tends to increase savings. Thus the substitution effect and income effect of rise in interest rate work in opposite direction. Therefore, net effect of rise in interest rate on saving is quite uncertain. Either of the two effects may dominate. As a result, higher interest rate may cause more or less savings. In Fig. 11.20 we have shown the case where substitution effect dominates over income effect and therefore rise in interest rate causes savings to increase. In Fig. 11.21 we have shown the case when income effect dominates and therefore higher interest results in lower savings. However, *empirical evidence in the US shows that substitution effect slightly outweighs the income effect. As a result, there is a small net positive effect of higher interest rate on savings.*[7]

7. See J. Stiglitz, *Principles of Economics*, p. 166.

QUESTIONS FOR REVIEW

1. Will an increase in the wage rate always lead to an increase in the number of hours worked? Why or why not? [*D.U. B.A. (Hons.) 2002*]
2. Compare the effects of the following policies on consumer welfare:
   (*i*) Lump sum cash income.
   (*ii*) Price subsidy on food [*D.U.B. Com.(Hons.) 2000*]
3. Explain the effect of a food subsidy on the welfare of consumers and compare it with an *outright cash transfer of the same cost*. What considerations determine government's policy in this regard? [*D.U. B.A. (Hons.) 1990*]
4. One can make a man happier by giving him cash and letting him spend as he thinks best than by forcing to take all his relief in the form of one commodity (Scitovsky). Do you agree? Discuss with reference to the effect of excise subsidy on sugar.
5. Explain the effect on consumption and welfare of food subsidiy in the form of food stamps.. Compare it with the effects of cash subsidy.
6. During the energy crisis of 1990-91 a proposal was made to increase the tax on petrol by ₹ 2.0 per litre. The revenue collected from the increased tax was to be given as a cash subsidy to motor vehicle owners. Analyse the impact of the proposal on the consumption of the petrol. [*D.U. B.A. (Hons.)1996*]
7. As far as the effect on consumer's welfare is concerned, direct taxes are superior to indirect taxes. Explain with the help of indifference curve analysis. [*D.U. B.Com.(Hons.) 1998*]
8. An indirect tax causes excess burden on the consumer. Do you agree? Explain with the help of indifference curve analysis of demand. [*D.U. (B.Com. (Hons.) 1995*]
9. When a firm raises the straight - time wage rate of its worker, fewer hours of work may be supplied. But paying a higher wage never fails to induce the workers to supply additional hours. Explain.
10. Illustrate using indifference curves that consumers become worse off when a product that they consume is rationed.
11. Consumer consumes two goods *X* and *Y*. If commodity *X* is rationed and the constraint is binding, what is the effective consumption set? Is the consumer worse off? Explain with indifference curves.
12. It is customary for the union-management contracts to lay down higher wage rate for overtime work. Does such an arrangement always lead to more work on the part of the workers?
13. If a leisure is a normal good, can a rise in wage rate lead to fewer hours worked?
14. Explain the intertemporal choice between present consumption and future consumption. Distinguish between income effect and substitution effect of rise in interest rate.
15. In the context of intertemporal choice explain under what conditions rise in interest rate will cause savings to increase.

CHAPTER 12

REVEALED PREFERENCE THEORY OF DEMAND

In both the Marshallian cardinal utility theory of demand and Hicks-Allen indifference curve theory of demand introspective method has been applied to explain the consumer's behaviour. In other words, both these theories provide psychological explanation of consumer's demand; they derive laws about consumer's demand from how he would react psychologically to certain hypothetical changes in price and income. But the Revealed Preference Theory which has been put forward by Prof. Samuelson seeks to explain consumer's demand from his actual behaviour in the market in various price-income situations. Thus, in sharp contrast to psychological or introspective explanation *Prof. Samuelson's revealed preference theory provides behaviouristic explanation of consumer's demand.* Besides, revealed preference theory is based upon the concept of ordinal utility. In other words, revealed preference theory regards utilities to be merely comparable and not quantifiable. Prof. Tapas Majumdar has described Samuelson's revealed preference theory as "*Behaviourist Ordinalist*."[1] The description "Behaviourist Ordinalist" highlights the two basic features of the revealed preference theory : first, it applies behaviouristic method, and secondly it uses the concept of ordinal utility.

PREFERENCE HYPOTHESIS AND STRONG ORDERING

Prof. Samuelson's revealed preference theory has *Preference Hypothesis* as a basis of his theory of demand. According to this hypothesis when a consumer is observed to choose a combination *A* out of various alternative combinations open to him, then he 'reveals', his preference for *A* over all other alternative combinations which he could have purchased. In other words, when a consumer chooses a combination *A*,it means he considers all other alternative combinations which he could have purchased to be inferior to *A*. Still in other words, it means he rejects all other alternative combinations open to him in favour of the chosen combination *A*. Thus, according to Prof. Samuelson, *choice reveals preference*. Choice of a combination *A* reveals his definite preference for *A* over all other rejected combinations. From the hypothesis of 'choice reveals preference' we can obtain definite information about the preferences of a consumer from observing his behaviour in the market. By comparing preferences of a consumer revealed in different price-income situations we can obtain certain information about his preference scale.

Let us graphically explain the preference hypothesis. Given the prices of two commodities *X* and *Y* and the income of the consumer, budget line *PL* is drawn in Fig. 12.1. The budget line *PL,* represents a given price-income situation. Given the price-income situation as represented by *PL*, the consumer can buy or choose any combination lying within or on the triangle *OPL*.

It may be noted that the combinations of two goods lying on line *AH* and line *AS* would give higher satisfaction to the consumer as they contain one good more with the quantity of the other good remaining the same as compared to the combination *A but they are not available* to the consumer as they lie beyond his budget line. Similarly, the combinations of two goods lying in the area *in between the lines AH and AS* (which has been shaded) as they contain more of the two goods and therefore will be preferred by the consumer, but they also lie beyond his budget constraint.

1. Tapas Majumdar, *Measurement of Utility*, Chapter VII

Thus, all combinations lying on the line *PL* such as *A*, *B*,*C* and lying below the line *PL* such as *D*, *E*, *F* and *G* are alternative combinations open to him, from among which he has to choose any combination. If our consumer chooses combination *A* out of all those open to him in the given price-income situation, it means he reveals his preference for *A* over all other combinations such as *B*, *C*, *D*, *E* and *F* which are rejected by him. Therefore all these other combination lying on the price line *PL* or below it are *inferior* to the chosen combination *A*. As is evident from Fig. 12.1, in his observed chosen combination *A*, the consumer is buying *OM* quantity of commodity *X* and *ON* quantity of commodity *Y*.

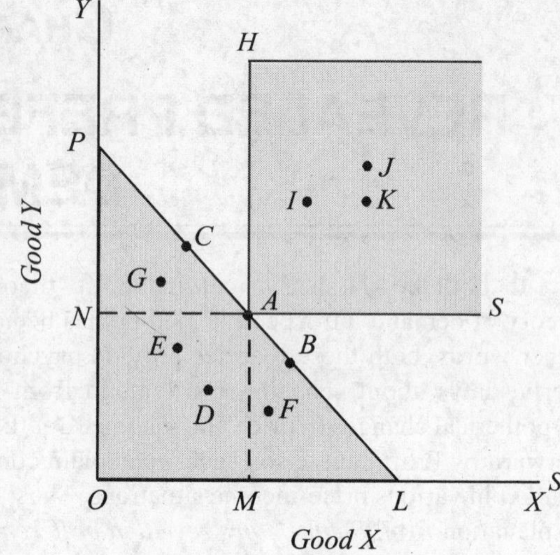

Fig. 12.1. *Choice reveals preference*

Strong Form of Preference Hypothesis

It should be carefully noted that Prof. Samuelson's revealed preference theory is based upon the strong form of preference hypothesis. In other words, in revealed preference theory, strong-ordering preference hypothesis has been applied. Strong ordering implies that there is *definite ordering* of various combinations in consumer's scale of preferences and therefore the choice of a combination by a consumer reveals his definite preference for that over all other alternatives open to him. Thus, *under strong ordering, relation of indifference between various alternative combinations is ruled out*. When in Fig. 12.1 a consumer chooses a combination *A* out of various alternative combinations open to him, it means he has a definite preference for *A* over all others, the possibility of the chosen combination *A* being indifferent to any other possible combination is ruled out by strong ordering hypothesis.

Consistency Postulate or Weak Axiom of Revealed Preference (WARP)

The revealed preference theory rests upon another basic assumption which has been called the '*consistency postulate*'. In fact, the consistency postulate is implied in strong ordering preference hypothesis. The consistency postulate can be stated thus :'no two observations of choice behaviour are made which provide conflicting evidence to the individual's preference." In other words, *consistency postulate asserts that if an individual chooses A rather than B in one particular instance, then he cannot choose B rather than A in any other instance*. If he chooses *A* rather than *B* in one instance and chooses *B* rather than *A* in another when *A* and *B* are present in both the instances, then he is not behaving consistently. Thus, consistency postulate requires that if once *A* is revealed to be preferred to *B* by an individual, then *B* cannot be revealed to be preferred to *A* by him at any other time when *A* and *B* are present in both the cases. Since comparison here is between two situations, consistency involved in this case has been called '*two term consistency*' by J. R. Hicks.

If a person chooses a combination *A* rather than combination *B* which he could purchase with the given budget constraint, then it cannot happen that he would choose (*i.e., prefer*) *B* over *A* in some other situation in which he could have bought *A* if he so wished. This means his choices or preferences must be consistent. This is called *revealed preference axiom*. We illustrate, revealed preference axiom in Fig. 12.2. Suppose with the given prices of two goods *X* and *Y* and given his money income to spend on two goods, *PL* is the budget line facing a consumer. In this budgetary situation *PL*, the consumer chooses *A* when he could have purchased *B* (note that combination *B* would have even cost him less than *A*). Thus, his choice of *A* over *B* means he prefers the combination *A* to the combination *B* of the two goods.

Now suppose that price of good *X* falls, and with some income adjustment, budget line changes to *P′L′*. Budget line *P′L′* is flatter than *PL* reflecting relatively lower price of *X* as compared to the budget line *PL*. With this new budget line *P′L′*, if the consumer chooses combination *B* when he can purchase combination *A* (as *A* lies below the budget line *P′L′* in Fig. 12.2), then the consumer will

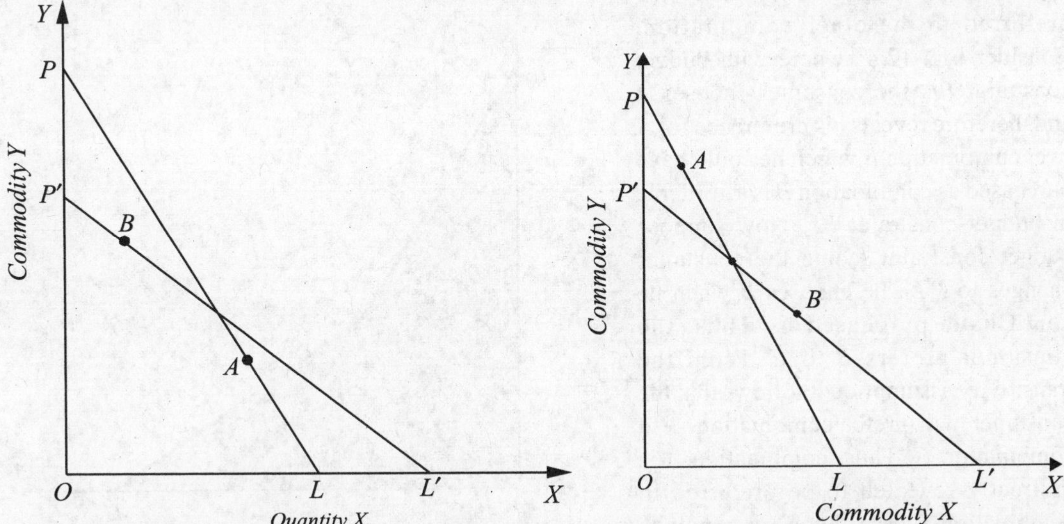

Fig. 12.2. *Consumer's Preferences are inconsistent.*

Fig. 12.3. *Consumer's choices satisfy axiom of revealed preference.*

be inconsistent in his preferences that is, he will be violating the axiom of revealed preference. Such inconsistent consumer's behaviour is ruled out in revealed preference theory based on strong ordering. This axiom of revealed preference according to which consumer's choices are consistent is also called '*Weak Axiom of revealed Preference* or simply *WARP*. To sum, up according to the axiom of revealed preference , *if combination A is directly revealed preferred to another combination B, then in any other situation, the combination B cannot be revealed to be preferred to A by the consumer when combination A is also affordable.*[2]

Now consider Fig. 12.3 where to start with a consumer is facing budget line *PL* where he chooses combination *A* of two goods *X* and *Y*. Thus, consumer prefers combination *A* to all other combinations within and on the triangle *OPL*. Now suppose that budget constraint changes to *P′L′* and consumer purchases combination *B* on it. as combination *B* lies outside the original budget line *PL* it was not affordable when combination *A* was chosen. Therefore, choice of combination *B* with the budget line *P′L′* is consistent with his earlier choice *A* with the budget constraint *PL* and is in accordance with the axiom of revealed preference .

Transitivity Assumption of Revealed Preference Theory

The axiom of revealed preference described above provides us a consistency condition that must be satisfied by a rational consumer who makes an optimum choice. Apart from the axiom of revealed preference, revealed preference theory also assumes that *revealed preferences are transitive.* According to this, if an optimizing consumer prefers combination *A* to combination *B* of the goods and prefers combination *B* to combination *C* of the goods, then he will also prefer combination *A* to combination *C* of the goods. To put briefly, assumption of transitivity of preferences requires that if $A > B$ and $B > C$, then $A > C$.

2. See Hal R. Varian, *Intermediate Microeconomics*, 1990, p. 124.

In this way we say that combination *A* is *indirectly* revealed preferred to combination *C*. Thus, *if a combination A is either directly or indirectly revealed to be preferred* to another combination we say that combination *A* is revealed to be preferred to the other combination. Consider Fig. 12.4 where with budget constraint *PL*, the consumer chooses *A* and therefore reveals his preference for *A* over combination *B* which he could have purchased as combination *B* is affordable in budget constraint *PL*. Now suppose budget constraint facing the consumer changes to *P′L′*, he chooses *B* when he could have purchased *C*. Thus, the consumer prefers *B* to *C*. From the transitivity assumption it follows that the consumer will prefer combination *A* to combination *C*. Thus, combination *A* is indirectly revealed to be preferred to combination *C*. We therefore conclude that the consumer prefers *A* either directly or indirectly to all those combination of the two goods lying in the shaded region in Fig. 12.4.

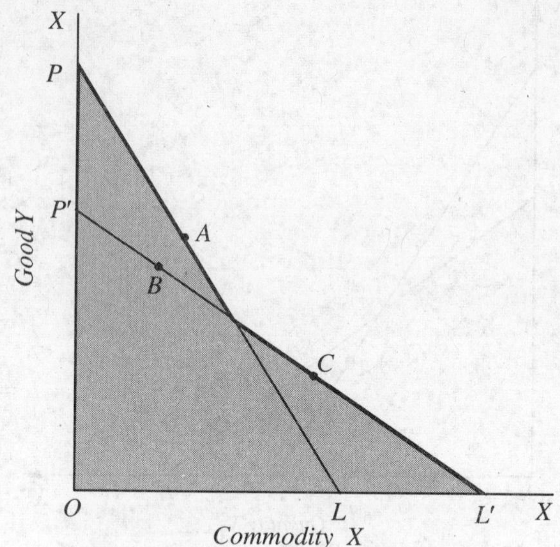

Fig. 12.4. *Revealed preferences are transitive.*

It is thus evident from above that concept of revealed preference is a very significant and powerful tool which provides a lot of information about preferences of a consumer who behaves in an optimising and consistent manner. By merely booking at the consumer's choices in different price-income situations we can get a lot of information about the underlying consumer's preferences.

DERIVING DEMAND THEOREM FROM REVEALED PREFERENCE HYPOTHESIS

Revealed preference hypothesis can be utilised to establish the demand theorem. Prof. Samuelson has derived the Marshallian law of demand from his revealed preference hypothesis. Marshallian law of demand, as is well known, states that a rise in the price of a good must, if income and other prices are held constant, results in the reduction of amount demanded of the good and *vice versa.* In other words, according to Marshall's law of demand, there is inverse relation between price and amount demanded of a good. Samuelson proceeds to establish relationship between price and demand by *assuming that income elasticity of demand is positive*. From positive income elasticity, he deduces the Marshallian inverse price-demand relationship. He states the demand theorem what he calls the *Fundamental Theorem of Consumption Theory* as under:

"*Any good (simple or composite) that is known always to increase in demand when money income alone rises must definitely shrink in demand when its price alone rises*".

It is clear from the above statement of Fundamental Theorem of Consumption that positive income elasticity of demand has been made a necessary qualification to the inverse price-demand relationship. The geometrical proof of the Fundamental Theorem is illustrated in Fig. 12.5. Let us suppose that the consumer spends his entire income on two goods *X* and *Y*. Further suppose that his income in terms of good *X* is *OB*, and in terms of *Y* is *OA*. Now the budget line *AB* represents the price-income situation confronting the consumer. All the combinations of goods *X* and *Y* lying within or on the triangle *OAB* are available to the consumer, from which he can buy any combination. Suppose that the consumer is observed to choose the combination *Q*. This means that *Q* is revealed to be preferred to all other combinations that lie in or on the triangle *OAB*.

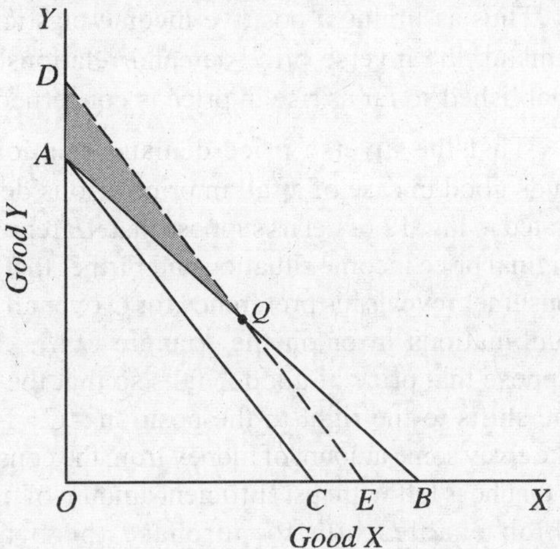

Fig. 12.5. *Proving Fundamental Demand Theorem in Case of Rise in Price*

Now, suppose that price of good *X* rises, price of *Y* remaining unchanged. With the rise in price of *X* the budget line shifts to the new position *AC*. The budget line *AC* represents new price-income situation. We now want to know what is the effect of this rise in price of good *X* on its quantity demanded, assuming that demand varies directly with income (*i.e.*, income elasticity of demand is positive). It is evident from Fig.12.5 that combination *Q* is not available to the consumer in price-income situation *AC*. Let us compensate the consumer for the higher price of *X* by granting him extra money so that he can buy the same combination *Q* even at the higher price of *X*. The amount of money which is required to be granted to the consumer so that he could buy the original combination *Q* at the higher price of *X* has been called *Cost-difference* by Prof. J. R. Hicks. In Fig. 12.5, a line *DE* parallel to *AC* has been drawn so that it passes through *Q*. *DE* represents the higher price of *X* and the money income after it has been increased by cost difference.

Now, the question is which combination will be chosen by the consumer in price-income situation *DE*. The original combination *Q* is available in price-income situation *DE*. It is evident from Fig. 12.5 that he will not choose any combination lying below *Q* on the line *DE*. This is because if he chooses any combination below *Q* on the line *DE*, his choice would be inconsistent. All combinations below *Q* on *DE*, that is, all combinations on *QE* could have been bought by the consumer but had been rejected by him in price-income situation *AB* in favour of *Q*. (All points on *QE* were contained in the original choice triangle *OAB*.). Since we are assuming consistency of choice behaviour on the part of the consumer he will not choose in price-income situation *DE*, any combination below *Q* on *QE* in preference to *Q* when *Q* is available in the new situation. It follows, therefore, that in the price-income situation *DE* the consumer will either choose the original combination *Q* or any other combination on *QD* segment of *DE* or within shaded area *QAD*. It should be noted that choice of any other combination on *QD* or within the shaded area *QAD* in preference to *Q* by the consumer will not be inconsistent since combinations lying above *Q* on *QD* or within shaded region *QAD* were not available in price-income situation *AB*. In price-income situation *DE* if the consumer chooses the original combination *Q*, it means he will be buying the same amount of goods *X* and *Y* as before, and if he chooses any combination above *Q* on *QD* or within the shaded area *QAD*, it means that he will be buying less amount of commodity *X* and greater amount of *Y* than before.

Thus, even after sufficient extra income has been granted to the consumer to compensate him for the rise in price of good *X*, he purchases either the *same* or the *smaller* quantity of *X* at the higher price. Now, if the extra money granted to him is *withdrawn*, he will definitely buy the smaller amount of *X* at the higher price, if the demand for good *X* is known always to fall with the decrease in income (that is, if income elasticity of demand fox *X* is positive). In other words, when the price of good *X* rises and no extra money is granted to the consumer so that he faces price-income situation *AC*, he will purchase less amount of good *X* than at *Q*. Thus assuming a positive income-elasticity of demand, the inverse price-demand relationship is established so far as rise in price is concerned.

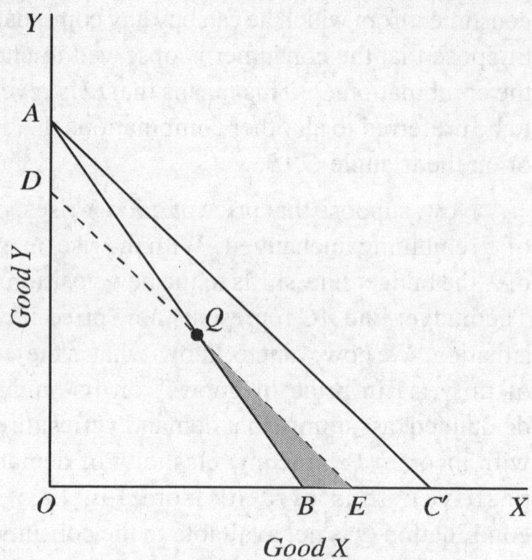

Fig. 12.6. *Proving Fundamental Demand Theorem in Case of Fall in Price*

That the inverse price-demand relationship holds good in case of a fall in price also is demonstrated in Fig. 12.6. Let us suppose that *AB* represents original price income situation and further that other consumer reveals his preference for *Q* over all other combinations in or on the triangle *OAB*. Now, suppose that price of good *X* falls so that the price line shifts to the right to the position *AC*. Let us take away some amount of money from the consumer so that he is left with just sufficient amount of money which enables him to purchase the original combination *Q* at the lower price of good *X*. Thus, in Fig. 12.6, a line *DE* is drawn parallel to *AC* so that it passes through *Q*. Price line *DE* represents lower price of *X* as given by *AC* and the money income after it has been reduced by the cost difference.

It is obvious that in price-income situation *DE*, the consumer cannot choose any combination above *Q* on *QD*, since all such combinations were available to him in the original price-income situation *AB* and were rejected by him in favour of *Q*. The consumer will, therefore, choose either *Q* or any other combination on *QE* or from within the shaded region *QEB*. In price-income situation *DE*, his choice of *Q* means that he buys the same quantity of goods *X* and *Y* as in original price-income situation *AB*, and his choice of any other combination on *QE* or from *within the shaded region QEB* means that he buys a larger amount of good *X* and a smaller amount of good *Y* than in the original price-income situation *AB*. Thus, even after consumer's income has been reduced, he buys either the same quantity of *X* or more at the lower price. And if we give him back the amount of money taken away from him so that he confronts again price-income situation *AC* he will definitely buy more of *X* at the lower price, provided that his demand for *X* rises with the rise in income (*i.e.*, his income elasticity of demand for good *X* is positive).

The two demonstrations given above together prove the fundamental theorem of consumption theory, according to which any good whose demand varies directly with income must definitely shrink in demand when its price rises and expands in demand when its price falls. It may be noted that Samuelson's theory involves two implicit assumptions which have not been explicitly stated. In the first place the consumer is always shown to choose a combination *on* the price line. In other words, he is never shown to choose a combination from within the triangle. This is based upon the assumption that a consumer always prefers a larger collection of goods to a smaller one. Secondly, another implicit assumption involved in Samuelson's theory is that the consumer is shown to choose only one combination of goods in every price-income situation. With these two implicit assumptions the inverse price-demand relationship is deduced by Samuelson by making explicit assumptions of consistency of choice and a positive income elasticity of demand.

Breaking up of Price Effect into Substitution Effect and Income Effect

Having now explained the derivation of law of demand from revealed preference approach we are now in a position to show how in the revealed preference approach price effect can be broken up into substitution and income effects. We will explain this by considering the case of fall in price of a commodity. Now consider Fig.12.7 where, to begin with, price-income situtation faced by a consumer is given by the budget the *AB*. With price-income situation represented by the budget line, *AB* suppose the consumer chooses combination *Q* and buys *OM* quantity of commodity *X*.

Now, suppose price of commodity *X* falls and as a result budget line shifts to the new position *AC*. Now, income of the consumer is reduced so much that the new budget line *DE* passes through the original chosen combination *Q*. That is, income is reduced equal to the cost difference so that gain in real income caused by the fall in price of commodity *X* is cancelled out . As seen above, with the new budget line *DE*, to be consistent in his behaviour the consumer can either choose the original combination *Q* or any combination lying on the segment *QE* of the budget line *DE*. If he chooses again the original combination *Q*, the Slutsky substitution effect well be zero. However, suppose that the consumer actually chooses combination *S* on the segment *QE* of the new budget line *DE*. Now, choice of the combination *S* shows that there will be substitution effect due to which the consumer will buy *MN* more of good *X*.

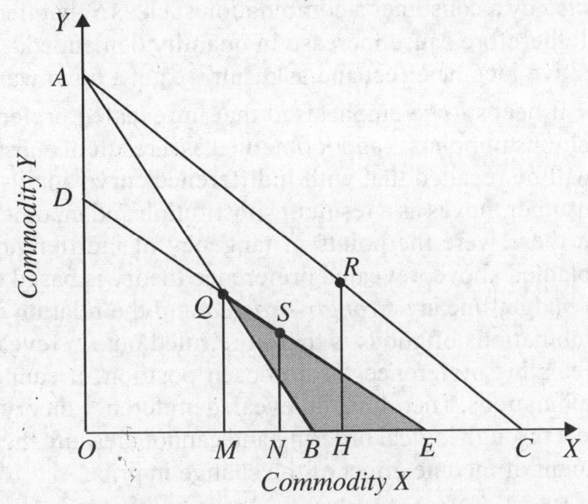

Fig. 12.7. *Breaking up Price Effect into Substitution and Income Effects.*

Note that subsitution effect is negative in the sense that the relative *fall* in price has led to the *increase in quantity demanded* of X, that is, change in quantity demanded *is in opposite direction* to the change in price. It should be noted that choice of combination *S* on segment *QE* in preference to combination *Q* of the budget line *DE* is not inconsistent because combinations on *QE* segment and within the shaded area were not available before when combination *Q* was earlier chosen in price-income situation *AB*. Thus, with the new budget line *DE* after consumer's income has been adjusted to cancel out the gain in real income resulting from a relative fall in price of *X*, the consumer chooses either *Q* (when substitution effect is, zero or a combination such as *S* on segment *QE* when substitution effect leads to the increase in quantity demanded of good *X* by *MN*. This is generally known as **Slutsky theorem** *which states that if income effect is ignored substitution effect will lead to the increase in quantity demanded of the good whose price has fallen and therefore the Marshallian law of demand describing inverse relationship between quantity demanded and price of a good will hold good,* that is, due to substitution effect alone demand curve slopes downward.

Now, if the consumer choose the combination *S* on the line segment *QE* of budget line *DE* it means that he buys *MN* more due to the substitution effect. Thus he prefers combination *S* to combination *Q*. In other words, his choice of *S* instead of *Q* reveals that he will be *better off* at *S* as compared to *Q*. Now, if money income withdrawn from him is restored to him so that he is faced with the budget line *AC*′. If income effect is positive, he will choose a combination, say *R*) on the budget line *AC*′ to the right of point *S* indicating that as a result of income effect he buys *NH* more of the commodity *X*.

Thus quantity demanded of commodity *X* increases by *MN* as a result of substitution effect and by *NH* as a result of income effect. This proves the law of demand stating inverse relationship between price and quantity demanded.

On budget line *DE*, if the consumer chooses combination *Q* and consequently substitution effect is zero, the whole increase in quantity demanded *MH* as a result of decline in price of good *X* will be due to positive income effect. However, which is more likely the substitution effect will lead to the choice by a consumer a combination such as *S* that lies to the right of *Q* on the line segment *QE* and will therefore cause increase in quantity demanded.[3] This substitution effect, is reinforced by the positive income effect and as result we get a downward sloping demand curve.

It needs to be emphasised that in revealed preference theory it is not possible to locate exact positions of points *S* and *R* obtained as a result of substitution effect and income effect respectively. It will be recalled that with indifference curve analysis we could obtain precise points to which a consumer moves as a result of substitution and income effects and as we saw in the previous chapter that these were the points of tangency of indifference curves with the relevant budget lines. As explained above, revealed preference theory is based on the assumption that all points on or below the budget line are *strongly ordered* and the relation of indifference of a consumer between some combinations of goods is therefore ruled out. In revealed preference theory, choice of a consumer reveals his preference for a chosen position; it cannot reveal indifference of consumer between combinations. Therefore, in revealed preference theory we can say about the *direction* of substitution effect through logical ordering and cannot measure the exact size of it, nor we can measure the exact amount of income effect of the change in price.

Besides, the substitution effect obtained through variation in income through cost difference method, does not represent *pure* substitution effect in the Hicksian sense in which the consumer's satisfaction remains constant. In substitution effect obtained by the revealed preference theory through Slutskian method of cost difference, the consumer moves from point *Q* to point *S* on budget line *DE*. His choice of *S* on budget line *DE* instead of *Q* under the influence of substitution effect shows that he prefers *S* to *Q*. That is, he is *better off* in position *S* as compared to the position *Q*. Therefore, it is maintained by some economists that the substitution effect obtained in revealed preference theory is not a pure one and contains also some income effect.[4]

However, the present author is of the view that the two types of substitution effect (Hicksian and the one obtained in revealed preference theory) differ with regard to the *concept of real income* used by them. In indifference curve analysis the term real income is used in the sense of level of satisfaction obtained by the consumer, whereas in revealed preference theory real income is used in the sense of purchasing power. Thus, Hicksian substitution effect involves the change in the quantity demanded of a good when its *relative price alone changes, level of his satisfaction remaining the same*. On the other hand, *revealed preference* theory considers substitution effect as the result of change in relative price of a good on its quantity demanded, *purchasing power remaining the same*. In Fig. 12.7 we obtain budget line *DE* after variation in income by cost difference so that it passes through the original combination *Q* chosen by the consumer before the fall in price. This implies that with budget line *DE*, he can buy, if he so desires the original combination *Q*, that is, the gain in purchasing power or real income caused by fall in price of *X* has been cancelled out by the reduction in his money income.

CRITICAL APPRAISAL OF REVEALED PREFERENCE THEORY

Samuelson's revealed preference theory has gained some advantages over the Marshallian

---

3. However, it may be noted that substitution effect, if not equal to zero, is *always negative*. A bit confusion has been created by Professors Stonier and Hague in their well-known book "*A Textbook of Economic Theory*" wherein they write substitution effect is *always positive* (see, op. cit. second edition pp.96-97). What Professors Stonier and Hague want to convey is that "Change in quantity demanded of good *X* whose price has fallen is *more that zero*" (that is, $\Delta X > 0$). This is of course correct. However, when examining substitution effect we refer to the '*relation* between *change in quantity demanded and a change in relative price of a good,* it is this relationship which is always negative and not positive. This is because change in quantity demanded (ΔX) is *in opposite direction* to the relative change in price (ΔP).
4. See Stonier and Hague, *op. cit.*, p. 96.

cardinal utility theory and Hicks-Allen indifference curve theory of demand. It is the first to apply behaviouristic method to derive demand theorem from observed consumer's behaviour. In contrast, both the earlier theories, namely, Marshallian utility analysis and Hicks-Allen indifference curve theory were psychological or introspective explanations of consumer's behaviour.

Now, the question is whether it is the behaviouristic approach or the psychological approach which is more correct to explain consumer's demand. We are of the opinion that no prior ground for choosing between behaviourist and introspective methods can be offered which would be acceptable irrespective of personal inclinations. Commenting on the behaviourist-ordinalist controversy, Professor Tapas Majumdar says, "*Behaviourism certainly has great advantages of treading only on observed ground; it cannot go wrong.*" But whether it goes far enough is the question. It may also be claimed for the method of introspection that operationally it can get all the results which are obtained by the alternative method, and it presumes to go further, it not only states, but also expalins its theorems"[5]. We may conclude that which of the two methods is better and more satisfactory depends upon one's personal philosophical inclinations. However, behaviourist method has recently gained wide support from the economists and has become very popular.

The concept of reveal preference is a powerful tool which can provide a significance information about consumer's preferences from which we can derive law of demand or downward sloping demand curve. Revealed preference theory does this without assuming that a consumer possesses complete information about his preferences and indifferences. In indifference curve analysis it was supposed that consumers had complete and consistent scale of preferences reflected in a set of indifference curves. His purchases of goods were in accordance with the scale of preferences. It is as if consumers were carrying complete indifference maps in their mind and purchasing goods accordingly. Therefore, it was considered better to derive demand theorem by observing consumer's behaviour in making actual choices. Most economists now-a-days believe that it is unrealistic to assume that a consumers have complete knowledge of their scale of preferences depicted in a set of indifference curves. The merit of revealed preference theory is that it has made possible to derive law of demand (*i.e.* downward sloping demand curve) on the basis of revealed preference without using indifference curves and associated restrictive assumptions. Further, it has enabled us to divide the price effect into its two component parts, namely, substitution and income effects through cost difference method and axiom of revealed preference. Cost difference method requires only market data regarding purchases of goods in different market situations. Cost difference (ΔC) can be simply measured by change in price (ΔP) multiplied by the quantity initially purchased by him. Thus,

$$\Delta C = \Delta P_x . Q_x$$

where ΔC stands for the cost difference, ΔP_x stands for the change in price of good *X*. Q_x is the quantity purchased by the consumer before the change in price of the good *X*.

Further, with revealed preference theory we can even establish the *existence* of indifference curves and their important property of *convexity*. However, it is notworthy that indifference curves are not required for deriving law of demand or downward sloping demand curve. Indifference curve analysis requires less information than Marshall's cardinal utility theory. But it still requires a lot of information on the part of a consumer since indifference curve analysis requires him to be able to rank consistently all possible combinations of goods. On the other hand, in Samuelson's revealed preference theory of demand the consumer does not require to rank his preferences on the basis of his introspection. It is based on the preferences revealed by his purchases or choices in the different market situations and on the axiom of revealed preference. If consumer's preferences and tastes do not change, revealed preference theory enables us to derive demand theorem just from observation of his market behaviour, that is, what purchases or choices he makes in different market situations.

5. *Op. cit.*

It is however assumed that his preference pattern or tastes do not change. As said above, we can even construct indifference curves from consumers' revealed preferences even though they are not required for establishing law of demand.

A Critique of Revealed Preference Theory

Although Samuelson's revealed preference approach has made some important improvements upon the earlier theories of demand but it is not free from all flaws. Various criticisms have been levelled against it.

First, Samuelson does not admit the possibility of indifference in consumer's behaviour. As has been explained above, the rejection of indifference by Samuelson follows from his strong ordering preference hypothesis. Prof. J.R. Hicks in his later work "*A Revision of Demand Theory*" does not consider the assumtion of strong ordering as satisfactory and instead employs weak ordering from of preference hypothesis. Whereas under strong ordering, the chosen combination is shown to be preferred to all other combinations *in* and *on* the triangle, under weak ordering the chosen combination is preferred to all positions *within* the triangle but *may be either preferred to or indifferent to other combinations on* the same triangle (*i.e. on* the budget line).

Further, in Samuelson's theory, preference is considered to be revealed for a single act of choice. It has been pointed out that if preference is to be judged from a large number of observations, then the possibility of indifference also emerges. Thus, an individual reveals preference for *A* over *B* if he chooses *A* rather than *B more frequently* than he chooses *B* rather than *A* over a given number of observations. Now, we can say that an individual is indifferent between the two situations *A* and *B* if a definite preference for either does not emerge from a sufficiently large number of observation. Thus only because Samuelson regards preference to be revealed from a single act of choice that indifference relation is methodologically inadmissible to his theory. The possibility of indifference relation clearly emerges if the existence of preference or otherwise is to be judged from a sufficiently large number of observations.

Furthermore, if we assume that an *individual is able to compare his ends*, which is a very valid assumption to be made about the individual's behaviour. Then the possibility of indifference or in other words, remaining at the same level of satisfaction by sacrificing some amount of one good for a certain amount of another good will emerge clearly. Thus, commenting on the Samuelson's revealed preference theory from 'welfare' point of view Prof. Tapas Majumdar remarks : "It may be remembered that in all forms of welfare theory, indeed in any integral view of human activity, we have to assume that the individual can always compare his ends. If this axiom is not granted, the whole of welfare economics falls to the ground. And if this axiom is granted, then the idea of remaining on the same level of welfare while sacrificing something of one commodity for something else of another will emerge automatically."[6]

Since Samuelson proves his demand theorem on the basis of positive income elasticity of demand, it cannot derive the demand theorem when income effect or income elasticity is negative. Thus, Samuelson is able to establish the demand theorem in case in which, in terms of Hicksain indifference curve theory, substitution effect has been reinforced by positive income effect of the price change. When the income elasticity is negative, Samuelson's revealed preference theory is unable to establish the demand theorem. In other words, given negative income elasticity of demand, we cannot know on the basis of revealed preference theory as to what will be the direction of change in demand as a result of change in price. Thus Samuelson's revealed preference theory cannot derive the demand theorem when (*i*) the income elasticity is negative and the negative income effect is smaller than the substitution effect; and (*ii*) the income elasticity is negative and the negative income effect is greater than the substitution effect.

---

6. Tapas Majumdar, *Measurement of Utility,* p. 90.

From above it follows that Samuelson's theory cannot account for Giffen's Paradox. The case of Giffen goods occurs when the income effect is negative and this negative income effect is so powerful that it outweighs the substitution effect. In case of Giffen goods, demand varies directly with price. Since he assumes income elasticity to be positive in his establishment of demand theorem, his theory cannot explain Giffen good. We thus conclude that though Samuelson makes improvement over Hicks-Allen indifference curves theory of demand in respect of methodoology adopted (that is, its behaviourist method is superior to Hicks-Allen's introspective method) but in respect of the content of the demand theorem established by it , it is a few steps backward as compared to Hicks-Allen indifference curve theory of demand.

Lastly, Samuelson's fundamental axiom 'choice reveals preference' has been criticised. Under conditions of perfect competition choice of a collection by a consumer may well reveal his preference for it but "this axiom is invalid for situations where the individual choosers are known to be capable of employing strategies of a game theory type" [7] It should, however, be noted that even indifference curve theory does not apply to the situation where strategies of a game theory type are to be employed.

In the end, we may emphasise the point that superiority of Samuelson's theory lies in his applying scientific or behaviouristic method to the consumer's demand and his description of preference hypothesis.

QUESTIONS FOR REVIEW

1. What is meant by 'revealed preference hypothesis'? Explain Samuelson revealed preference theory of demand based on it.

2. Carefully state the assumptions of the revealed preference theory of demand. Explain how law of demand is derived from these assumptions. *(D.U.B. A.(Hons.)1998*

3. Define substitution effect. *Substitution effect is alwasy negative.* How can you prove it with revealed preference approach to demand theory. *(D.U. 1990)*

4. Distinguish between strong ordering and weak ordering forms of preference hypothesis.

5. What is Slutsky equation? How will you establish it with the help of revealed preference approach to demand analysis?

6. Explain how price effect is broken up into income effect and substitution effect with the aid of revealed preference theory of demand. *(B.A. (Hons.) D.U. 1996)*

7. The price of a good increases. We know that the substitution effect is always negative. Does it follow that the income and substitution effects of the price rise always work in the same direction if the income elasticity of demand for the good is negative? Will the demand curve be downward sloping? Explain. *(D.U. B.A.(Hons.)1998)*

8. Show with the help of a diagram that the substitution effect of a fall in price of a commodity will make a consumer either consume the same quantity or more but not less of it. *(D.U. B.A. (Hons.) 1987)*

9. (*a*) State Samuelson's "*Weak Axiom of Revealed Preference*" (WARP).

(*b*) Consider two commodities, apples and oranges initially when both apples and oranges cost ₹ 1 per unit each, a consumer buys 4 apples and 6 oranges. When the price of apples rises to ₹ 2 per unit and that of oranges falls to ₹ 0.50 per unit, she buys 8 apples and 3 oranges. Does she violate WARP. *(D.U. B.A.(Hons.) 1998)*

10. Revealed preference theory makes a major advancement in the theory of demand. Discuss.

---

7. Tapas Majumdar, *op. cit*.

CHAPTER 13
ELASTICITY OF DEMAND

Various Concepts of Demand Elasticity

We have discussed in the preceding chapter that when the price of a good falls, its quantity demanded rises and when the price of it rises, its quantity demanded falls. This is generally known as *law of demand*. This law of demand indicates only the *direction* of change in quantity demanded in response to a change in price. This does not tell us by *how much* or to what extent the quantity demanded of a good will change in response to a change in its price. This information as to how much or to what extent the quantity demanded of a good will change as a result of a change in its price is provided by the concept of elasticity of demand.

It is price elasticity of demand which is usually referred to as elasticity of demand. But, besides price elasticity of demand, there are various other concepts of demand elasticity. As we have seen in the previous chapter, demand for a good is determined by its price, incomes of the people, prices of related goods, etc. Quantity demanded of a good will change as a result of change in any of these determinants of demand. The concept of elasticity of demand therefore refers to the degree of responsiveness of quantity demanded of a good to a change in its price, consumers' income and prices of related goods. Accordingly, there are three concepts of demand elasticity : price elasticity, income elasticity, and cross elasticity. Price elasticity of demand relates to the degree of responsiveness of quantity demanded of a good to the change in its price. Income elasticity of demand refers to the sensitiveness of quantity demanded to a change in income. Cross elasticity of demand means the degree of responsiveness of demand of a good to a change in the price of a related good, which may be either a substitute for it or a complementary with it.

The concept of elasticity of demand forms the subject-matter of the present chapter. The concept of elasticity has a very great importance in economic theory as well as for formulation of suitable economic policies.

PRICE ELASTICITY OF DEMAND

As mentioned above, price elasticity of demand indicates the degree of responsiveness of quantity demanded of a good to the change in its price, other factors such as income, prices of related commodities that determine demand are held constant. Precisely, *price elasticity of demand is defined as the ratio of the percentage change in quantity demanded of a commodity to a percentage change in price. Thus*

$$e_P = \frac{\text{Percentage change in quantity demanded}}{\text{Percentage change in price}}$$

It follows from the above definition of price elasticity of demand that when the percentage change in quantity demanded a commodity is greater than the percentage change in price that brought it about, price elasticity of demand (e_p) will be greater than one and in this case demand is said to be *elastic*. On the other hand, when a given percentage change in price of a commodity leads to a smaller percentage change in quantity demanded, elasticity will be less than one and demand in this case is said to be *inelastic*. Further, when the percentage change in quantity demanded of a commodity is equal to the percentage change in price that caused it, price elasticity is equal to one. Thus in case of elastic demand, a given change in price causes quite a large change in quantity demanded. And in case of inelastic demand, a given change in price brings about a very small change in quantity demanded of a commodity.

It is a matter of common knowledge and observation that there is a considerable difference between different goods in regard to the magnitude of response of demand to the changes in price. The demand for some goods is more responsive to the changes in price than those for others. In terminology of economics, we would say that demand for some goods is *more elastic* than those for the others or *the price elasticity of demand of some* goods is greater than those of the others. Marshall who introduced the concept of elasticity into economic theory remarks that the *elasticity or responsiveness of demand in a market is great or small according as the amount demanded increases much or little for a given fall in price, and diminishes much or little for a given rise in price.*[1] This will be clear from Fig. 13.1 and 13.2 which respresent two demand curves. For a *given fall* in price,

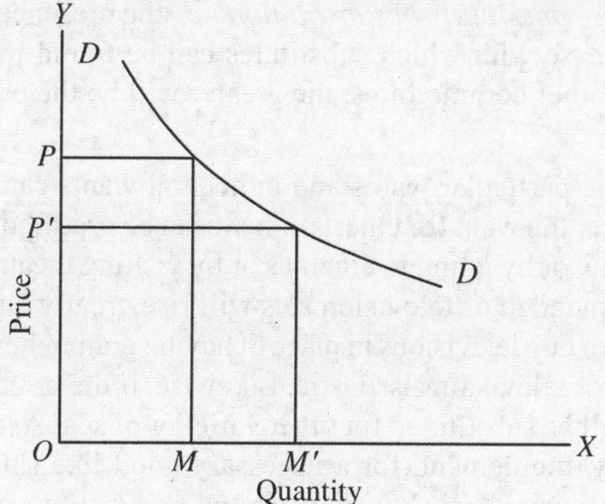

Fig. 13.1. *Elastic Demand*

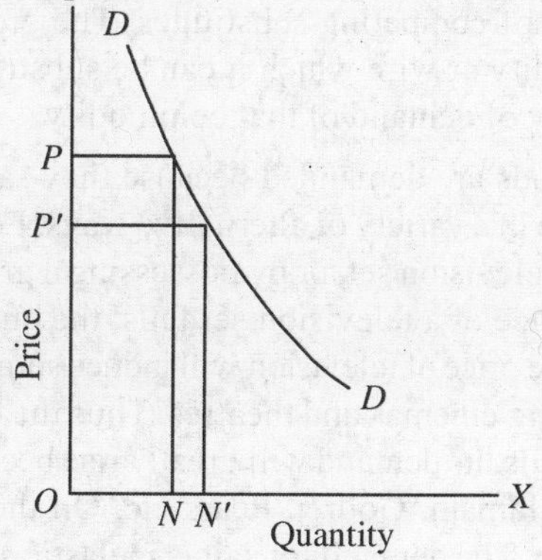

Fig. 13.2. *Inelastic Demand*

from *OP* to *OP*′, increase in quantity demanded is much greater in Fig.13.1 than in Fig. 13.2. Therefore, demand in Fig. 13.1 is more elastic than the demand in Fig. in 13.2 for a given fall in price for the portion of demand curves considered. Demand for the good represented in Fig. 13.1 is generally said to be elastic and the demand for the good in Fig. 13.2 to be inelastic.

It should, however, be noted that *terms elastic and inelastic demand are used in the relative sense*. In other words, elasticity is a matter of degree only. Demand for some goods is only more or less elastic than others. Thus, when we say that demand for a good is elastic, we mean only that the demand for it is relatively more elastic. Likewise, when we say that demand for a good is inelastic, we do not mean that its demand is absolutely inelastic but only that it is relatively less leastic. In economic theory elastic and inelastic demands have come to acquire precise meanings. Demand for a good is said to be *elastic* if price elasticity of demand for it is *greater than one*. Similarly, the demand for a good is called *inelastic* if price elasticity of demand for it is *less than one*. Price elasticity of demand equal to one, or in other words, *unit elasticity* of demand therefore represents the dividing line between elastic and inelastic demand. It will now be clear that by inelastic demand we do not mean perfectly inelastic but only that price elasticity of demand is less than unity, and by elastic demand we do not mean absolutely elastic but that price elasticity of demand is greater than one.

---

1. Alfred Marshall, *Principles of Economics* (8th Edition), Vol. 2.

Thus,

Elastic demand	:	$e_p > 1$
Inelastic demand	:	$e_p < 1$
Unitary elastic demand	:	$e_p = 1$

Price Elasticity of Demand for Different Goods Varies a Good Deal

As said above, goods show great variation in respect of elasticity of demand *i.e.*, their responsiveness to changes in price. Some goods like common salt, wheat and rice are very unresponsive to changes in their prices. The demand for salt remains practically the same for a small rise or fall in its price. Therefore, demand for common salt is said to be 'inelastic'. Demand for goods like televisions, refrigerators etc., is elastic, since changes in their prices bring about large changes in their quantity demanded. We shall explain later at length those factors which are responsible for the differences in elasticity of demand of various goods. It will suffice here to say that the main reason for differences in elasticity of demand is the *possibility of substitution i.e.*, the presence or absence of competing substitutes. The greater the ease with which substitutes can be found for a commodity or with which it can be substituted for other commodities, the greater will be the price elasticity of demand of that commodity.

Goods are demanded because they satisfy some particular wants and in general wants can be satisfied in a variety of alternative ways. For instance, the want for entertainment can be gratified by having television set, or by possessing a gramophone, or by going to cinemas or by visiting theatres. If the price of a television set falls, the quantity demanded of television sets will rise greatly since fall in the price of television will induce some people to buy televisions in place of having gramophones or visiting cinemas and theatres. Thus the demand for televisions is elastic. Likewise, if the price of `Lux' falls, its demand will greatly rise because it will be substituted for other varieties of soapÅsuch as Jai. Hamam, Godrej, Pears etc. On the contrary, the demand for a necessary good like salt is inelastic. The demand for salt is inelastic since it satisfies a basic human want and no substitutes for it are available. People would consume almost the same quantity of salt whether it becomes slightly cheaper or dearer than before.

Perfectly Inelastic and Perfectly Elastic Demand

We will now explain the two extreme cases of price elasticity of demand. First extreme situation is of perfectly inelastic demand which is depicted in Fig. 13.3. In this case changes in price of a

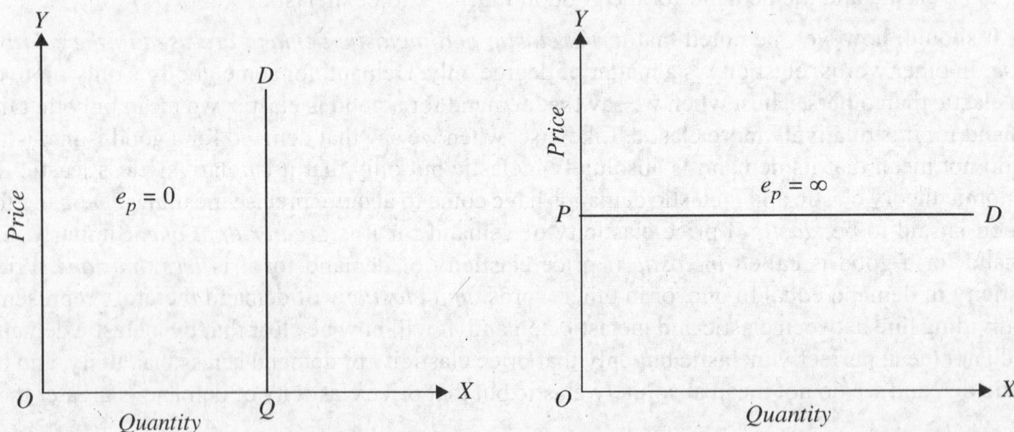

Fig. 13.3. *Perfectly Inelastic Demand ($e_p = 0$)* **Fig. 13.4.** *Perfectly Elastic Demand $e_p = \infty$*

commodity does not affect the quantity demand of the commodity at all. In this perfectly inelastic demand, demand curve is a vertical straight line as shown in Fig. 13.3. As will be seen from this figure, whatever the price quantity demanded of the commodity remains unchanged at *OQ*. An approximate example of perfectly inelastic demand is the demand of acute diabetic patient for insulin. He has to get the prescribed doze of insulin per week whatever its price.

The second extreme situation is of perfectly elastic demand in which case demand curve is a horizontal straight line as shown in Fig. 13.4. This horizontal demand curve for a product implies that a small reduction in price would cause the buyers to increase the quantity demanded from zero to all they could obtain. On the other hand, a small rise in price of the product will cause the buyers to switch completely away from the product so that its quantity demanded falls to zero. We will see in later chapters that perfectly elastic demand curve is found for the product of an individual firm working under perfect competition. Products of different firms working under perfect competition are completely identical. If any perfectly competitive firm raises the price of its product, it would lose all its customers who would switch over to other firms and if it reduces its price somewhat it would get all the customers to buy the product from it.

MEASUREMENT OF PRICE ELASTICITY

As said above, price elasticity of demand expresses the response of quantity demanded of a good to changes in its price, given the consumer's income, his tastes and prices of all other goods. Thus price elasticity means the degree of responsiveness or sensitiveness of quantity demanded of a good to a change in its price. An important method to measure price elasticity of demand is the percentage method which we explain below.

Percentage Method. Price elasticity can be precisely measured by dividing the the percentage change in quantity demanded in response to a small change in price, divided by the percentage change in price. Thus we can measure price elasticity by using the following formula:

$$\text{Price Elasticity} = \frac{\text{Percentage change in quantity demanded}}{\text{Percentage change in price}}$$

$$e_p = \frac{\dfrac{\Delta q \times 100}{q}}{\dfrac{\Delta p}{p}} = \frac{\Delta q}{q} \div \frac{\Delta p}{p}$$

$$= \frac{\Delta q}{q} \times \frac{p}{\Delta p}$$

$$= \frac{\Delta q}{q} \times \frac{p}{q}$$

where e_p stands for price elasticity
q stands for the original quantity
p stands for the original price
Δ stands for a small change

Mathematically speaking, price elasticity of demand has a *negative sign* since the change in quantity demanded of a good is in opposite direction to the change in its price. When price falls, quantity demanded rises and vice versa. But for the sake of convenience in understanding the *magnitude of response* of quantity demanded of a good to a change in its price we ignore the negative sign and take into account only the numerical value of the elasticity. Thus, if 2% change in price leads to a 4% change in quantity demanded of good *A* and 8% change in that of *B*, then the above formula of elasticity will give the value of price elasticity of good *A* equal to 2 and that of good *B* equal to 4. It indicates that the quantity demanded of good B changes relatively much more than that of good A in response to a given change in price. But if we had written minus signs before the numerical values of elasticities of the two goods, that is, if we had written the elasticity as – 2 and –4 respectively as strict mathematics would require us to do, then since – 4 is smaller than – 2, we would have been misled in concluding that price elasticity of demand of B is less than that of good A. But, as we have noted above, response of demand for good B to the change in its price is greater than that of A, it is better to ignore minus sign and draw conclusion from the absolute values of elasticities. Hence, *by convention minus sign before the value of price elasticity of demand is generally ignored in economics* unless otherwise mentioned.

Midpoint Method: Calculating Percentage Changes

When we calculate the price elasticity of demand we face a problem whether to use the *initial price* as the base for calculating percent change in price and *initial quantity* as the base for calculating the percent change in quantity demanded in response to a given percent change in price. For example, suppose price of a commodity rises from ₹ 4 to ₹ 6 per unit and as a result, the quantity demanded falls from 120 units to 80 units. If we take initial price ₹ 4 as the base for change in price, then change in price by ₹ 2 amounts to 50 percent change in price $\left(\frac{6-4}{4}\times100=\frac{1}{2}\times100=50\right)$ And taking initial quantity 120 units as the base for calculating percent change in quantity demanded, then there is 33.3 percent change in quantity demanded $\left(\frac{120-80}{120}\times100=\frac{1}{3}\times100=33.3\right)$. Thus we get price elasticity of demand as

$$e_p=\frac{33}{50}=0.66$$

Let us now reverse the direction. Suppose the price of the commodity falls from ₹ 6 to ₹ 4 per unit, and as a result quantity demanded increases from 80 units 120 units, then now taking initial ₹ 6 as the base for calculating percentage change in price, then there is 33 per cent change in price $\left(\frac{6-4}{6}\times100=\frac{2}{6}\times100=33\right)$ and taking 80 units as the base for calculating percentage change in quantity, then the quantity demanded rises by 50 percent $\left(\frac{120-80}{80}\times100=\frac{40}{80}\times100=50\right)$. Thus, we will now get $\frac{50}{33}=1.5$ as the price elasticity of demand. It therefore follows that for the same absolute change in price and absolute change in quantity demanded we get different values of price elasticity of demand if we use ₹ 4 or ₹ 6 as the base for calculating percentage change in price and 120 units or 80 units as the base for calculating percentage change in quantity demanded. To avoid this problem we use *midpoint method* for calculating the percentage changes in price and quantity demand. In mid-point method we calculate the percentage change in price or quantity demanded by *taking midpoint of the initial and final values of price and quantity demanded* respectively as the base. Thus, in our above example, midpoint (or, in other words, average) of prices of ₹ 4 and ₹ 6 is $\frac{4+6}{2}=5$ and midpoint (or average) of quantities demanded is $\frac{80+120}{2}=100$. Using this midpoint method the percentage change in price is $\frac{6-4}{5}\times100=40$ and percentage change in quantity demanded is $\frac{120-80}{100} \times 100 = 40$. With these percentage changes in price and quantity demanded price of elasticity of demand will be

$$e_p = \frac{40}{40} = 1$$

It should be carefully noted that for *large changes in price, we must use midpoint method of calculating rice elasticity of demand.* If change in price is *very small,* then we can use initial price and initial quantity demanded.

If p_1 stands for initial price and p_2 for the new price and q_1 for the initial quantity and q_2 for the new quantity, then midpoint formula for calculating price elasticity of demand (e_p) can be written as

$$e_p = \frac{(q_2 - q_1)}{\frac{q_1 + q_2}{2}} \div \frac{p_2 - p_1}{\frac{p_1 + p_2}{2}}$$

$$= \frac{\Delta q}{\frac{q_1 + q_2}{2}} \times \frac{\frac{p_1 + p_2}{2}}{\Delta p}$$

$$= \frac{\Delta q}{\Delta p} \times \frac{p_1 + p_2}{q_1 + q_2}$$

Arc Elasticity of Demand

The concept of arc elasticity of demand should be distinguished from the *point elasticity of demand.* Point elasticity of demand refers to the price elasticity at a point on the demand curve or, in other words, it refers to the price elasticity when the changes in the price and the resultant changes in quantity demanded are infinitesimally small. In this case if we take the initial or the original price and original quantity or the subsequent price and quantity after the change in price as the basis of measurement, there will not be any significant difference in the coefficient of elasticity. However, as explained above, when the price change is quite large or we have to measure elasticity over an *arc of the demand curve* rather than at a *specific point* on it, the measure of point elasticity, namely, $\frac{\Delta q}{\Delta p} \cdot \frac{p}{q}$ does not provide us the true and correct measure of price elasticity of demand and therefore we use the midpoint method of measuring price elasticity of demand. Therefore, in cases when large changes in price and quantity demanded are involved or in other words when we have to find the price elasticity of the arc of a demand curve as between points *A* and *B* on demand curve *DD* in Fig. 13.4(a). We use *midpoint formula* (or what is also known as arc elasticity formula) for correctly calculating price elasticity of demand. As explained above, in such cases, the coefficient of price elasticity would be different depending upon whether we choose original price and

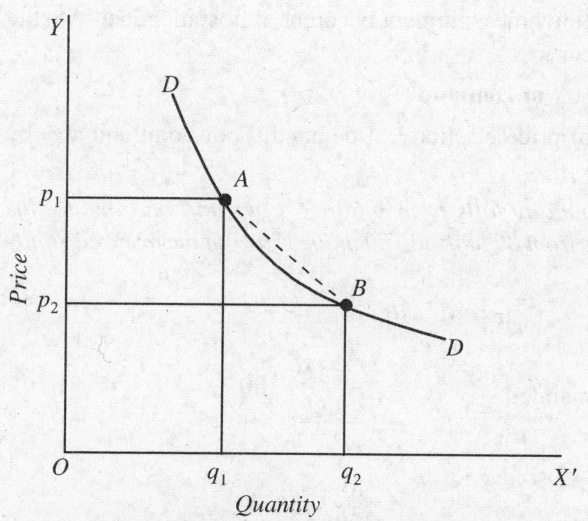

Fig. 13.4(a). *Arc Elasticity of Demand*

quantity demanded or the subsequent price and quantity demanded as the base for measurement of price elasticity and therefore there will be significant difference in the two coefficients of elasticity obtained from using the two bases.

Therefore, when the change in price is quite large, say more than 5 percent, then accurate measure of price elasticity of demand can be obtained by taking the *average of original price and subsequent price as well as average of the original quantity and subsequent quantity* as the basis of measurement of percentage changes in price and quantity. Thus, in Fig. 13.4(a) if price of a good falls from p_1 to p_2 and as a result its quantity demanded increases from q_1 to q_2, the average of the two prices is given by $\frac{p_1+p_2}{2}$ and average of the two quantities (original and subsequent) is given by $\frac{q_1+q_2}{2}$. Thus, in Fig. 13.4(a) the formula for measuring arc elasticity, we use midpoint method of measuring price elasticity which is given by :

$$e_q = \frac{\Delta q}{\frac{q_1+q_2}{2}} \div \frac{\Delta p}{\frac{p_1+p_2}{2}}$$

$$= \frac{\Delta q}{q_1+q_2} \cdot \frac{p_1+p_2}{\Delta p} = \frac{\Delta q}{\Delta p} \cdot \frac{p_1+p_2}{q_1+q_2}$$

It should be carefully ntoed that, the greater the convexity of the demand curve between *A* and *B*, the greater the divergence between the dashed line *AB* and the true demand curve and therefore the poorer the approximation of arc elasticity measure (of the dashed line *AB*) for the true curve between *A* and *B*. Moreover, the larger the distance between *A* and *B* on the demand curve, the greater will be the divergence between the dashed straight line *AB* and the true curve from *A* to *B* and consequently the greater will be the discrepancy between the elasticity on the true curve from *A* to *B* and consequently the greater will be the discrepancy between the elasticity on the true curve and the elasticity on the dashed line *AB* measured by the arc elasticity formula given above. That is why even the concept of arc elasticity is relevant in case when the arc involved in small, that is, two points *A* and *B* lie close together. Therefore, the arc elasticity formula or midpoint method of measuring price elasticity should be used when the change in price is quite *large*. On the other hand, when the two points on the demand curve are *very close* together, arc (dashed straight line) becomes almost indentical with the true curve and the arc elasticity measurement becomes almost identical with the point elasticity measurement on the demand curve.

Some Numerical Problems on Price Elasticity of Demand

Let us solve some numerical problems of price elasticity of demand (both point and arc) by percentage method.

Problem 1. *Suppose the price of a commodity falls from ₹ 6 to ₹ 4 per unit and due to this quantity demanded of the commodity increases from 80 units to 120 units. Find out the price elasticity of demand.*

Solution: Change in quantity demand $(Q_2 - Q_1) = 120 - 80$

Percentage chagne in quantity demanded $= \frac{Q_2 - Q_1}{\frac{Q_2 + Q_1}{2}} \times 100$

$$= \frac{40}{\frac{200}{2}} \times 100$$

$= 40$

Change in price $= P_2 - P_1 = 4 - 6 = -2$

% Change in price $= \dfrac{\dfrac{P_2 - P_1}{\dfrac{P_2 + P_1}{2}}}{} \times 100 = \dfrac{-2}{\dfrac{10}{2}} \times 100$

$= -40$

Price elasticity of demand $= \dfrac{\%\text{ change in quantity demanded}}{\%\text{ Change in price}}$

$= \dfrac{40}{-40} = -1$

We ignore the minus sign. Therefore, price elasticity of demand is equal to one.

Problem 2. *A consumer purchases 80 units of a commodity when its price is ₹ 1 per unit and purchases 48 units when its price rises to ₹ 2 per unit. What is the price elasticity of demand for the commodity ?*

Solution: It should be noted that the change in price from ₹ 1 to ₹ 2 in this case is very large (*i.e.*, 100%). Therefore, to calculate the elasticity coefficient in this case midpoint elasticity formula should be used.

Change is price (Δp) = ₹ 2 – 1 = 1

Average of the original and subsequent prices $= \dfrac{p_1 + p_2}{2}$

$= \dfrac{1+2}{2} = \dfrac{3}{2} = 1.5$

Change in quantity demanded (Δq) $= 80 - 48 = 32$

Average of the original and subsequent quantities $= \dfrac{q_1 + q_2}{2}$

$$e_p = \frac{\Delta q}{\frac{q_1 + q_2}{2}} \div \frac{\Delta p}{\frac{p_1 + p_2}{2}} = \frac{80+48}{2} = \frac{128}{2} = 64$$

$$= \frac{32}{64} \div \frac{1}{1.5}$$

$$= \frac{32}{64} \times \frac{1.5}{1} = \frac{1}{2} \times \frac{15}{10} = \frac{3}{4} = 0.75$$

Thus, the price elasticity of demand obtained is equal to 0.75.

Problem 3. *Suppose a seller of a textile cloth wants to lower the price of its cloth from ₹ 150 per metre to ₹ 142.5 per metre. If its present sales are 2000 metres per month and further it is estimated that its elasticity of demand for the product equals – 0.7. Show*

(a) Whether or not his total revenue will increase as a result of his decision to lower the price; and

(b) Calculate the exact magnitude of its new total revenue.

Solution. (*a*) Price elasticity $= \dfrac{\Delta q}{\Delta p} \cdot \dfrac{p}{q}$

p = ₹ 150

$$q = 2000 \text{ metres}$$
$$\Delta p = 150 - 142.5 = 7.5$$
$$e_p = 0.7$$
$$\Delta q = ?$$

Substituting the values of p, q, Δp and e_p in the price elasticity formula we have

$$0.7 = \frac{\Delta q}{7.5} \times \frac{150}{2000}$$

$$\Delta q = \frac{0.7 \times 7.5 \times 2000}{150} = 70$$

Since the price has fallen the quantity demanded will increase by 70 metres. So the new quantity demanded will be 2000 + 70 = 2070.

(*b*) Total Revenue before reduction in price = 2000 × 150 = ₹ 3,00,000

Total revenue after price reduction = 2070 × 142.5 = ₹ 2,94,975

Thus with reduction in price his total revenue has decreased.

TOTAL REVENUE, TOTAL EXPENDITURE AND PRICE ELASTICITY OF DEMAND

Total revenue is the amount received by the seller from the sale of the quantity of the good sold in the market. It is worthnoting noting that total revenue received by the seller from sale of the quantity of a good is the expenditure made by the buyers. Total revenue (or total expenditure) from the sale of a good is intimately related to the price elasticity of demand. Total revenue is the product of price (*P*) and the quantity (*Q*) of the commodity sold (*i.e.*, Total revenue or total expenditure = *P* × *Q*) . Consider Fig. 13.4(b) where *DD* is the demand curve. At price *OP* or ₹ 5, the quantity demanded and sold of the commodity is *OQ* (or 100 units). Thus the area of the rectangle *OPAQ* (or 5× 100 = 500) represents the total revenue made by the sellers or the expenditure made by the buyers on the commodity :

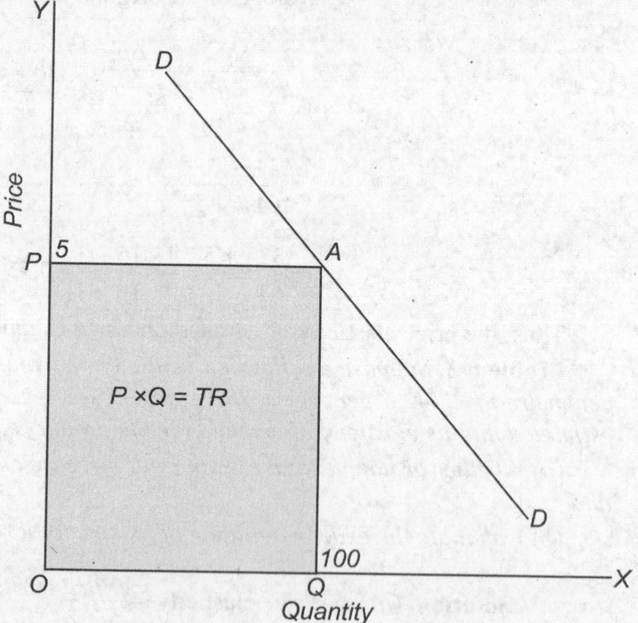

Fig. 13.4(b). *Total Revenue = P × Q*

Change in total revenue following a change in price of the good depends on price elasticity of demand. We can classify the commodities into the following three categories.

1. When demand is elastic (e_p>1). When demand for a commodity is elastic (e_p> 1), the percentage increase in quantity demanded of the commodity will be greater than the percentage fall in price that caused the former. As a result, total revenue will increase following the reduction in price of the commodity. Conversely, this also implies that when demand is elastic (e_p >1), rise in price will cause the total revenue to decrease.

2. **When Demand is Inelastic (e_p <1).** When demand for a commodity is inelastic, with a fall in price the quantity demanded of a good increases

proportionately less then the fall in price. Therefore, in this case of inelastic demand, with a fall in price, total revenue (or expenditure) decreases. This also implies that when demand is inelastic, the rise in price will lead to a *percentage decline in quantity demanded that is proportionately less than the rise in price*. As a result, if demand is inelastic, rise in price will cause total revenue to increase.

3. **When demand is unitary lastic (e_p = 1).** When the demand for a commodity is unitary elastic, with the fall or rise in price, the quantity demanded rises or falls in the same proportion as the change in price . As a result, in case of unitary elastic demand, total revenue ($P \times Q$) remains the same when price changes.

Thus we arrive at the following conclusions about the relationship between changes in price of a good and total revenue.

1. *When demand for good is elastic ($e_p > 1$), a fall in price raises total revenue and a rise in price reduces total revenue.*

2. *When demand for a good is inelastic ($e_p < 1$) , a fall in price reduces total revenue and a rise in price raises total revenue.*

3. *When demand for a commodity is unitary elastic ($e_p = 1$), changes in its price do not affect total revenue.*

Estimating Price Elasticity of Demand With Expenditure Method

We can now use the relationship between changes in price and total revenue explained above for estimating price elasticity of demand. It may be noted that total revenue received by sellers is the expenditure made by the buyers on a commodity. Therefore, estimating price elasticity from changes in total revenue is also called *expenditure method* of measuring price elasticity of demand. Moreover, from changes in total revenue or expenditure caused by changes in price we can only estimate whether price elasticity is greater than one, equal to one or less than one, we cannot measure exact coefficient of price elasticity of demand. Thus, for estimating price elasticity from changes in total expenditure or revenue, we use the following lists.

1. *If fall in price of a commodity causes total expenditure (or total revenue) to increase, demand for the good is elastic ($e_p > 1$).*

2. *If fall in price of a commodity causes total expenditure (or total revenue) on a commodity to decrease, demand for the commodity is inelastic ($e_p < 1$).*

3. *If changes in price of a good does not bring about any change in total expenditure, (i.e. total revenue) demand for it is unitary elastic.*

Illustration of Expenditure Method

Let us illustrate how we judge the price elasticity of demand as to whether it is greater than one, equal to one or less than one. Consider Table 13.1, which gives quantity demanded of pens at various prices. It will be seen from Table 13.1 that quantity demanded increases from 30 pens at price ₹ 5 per pen to 87 pens at price ₹ 3.25. We have calculated the total outlay by multiplying the quantity demanded with the corresponding price of the pen. It will be observed from the table that when price of the pen falls from ₹ 5 to ₹ 4.75, from ₹ 4.75 to ₹ 4.50, from ₹ 4.50 to ₹ 4.25 and from ₹ 4.25 to ₹ 4 the quantity demanded increases so much that the total outlay on pens increases indicating thereby that elasticity of demand is greater than one at these prices.

When price falls from ₹ 4.00 to ₹ 3.75, the quantity demanded increases from 75 pens to 80 pens so that total outlay remains the same at ₹ 300. This shows that price elasticity of demand is unity. When price of the pen further falls from ₹ 3.75 to ₹ 3.50 and then to ₹ 3.25 the total outlay spent on pens decreases. Thus, the price elasticity of demand for pens at these prices is less than unity.

Table 13.1. Price Elasticity of Demand and Total Expenditure

Price of Pen (₹) P	*Quantity Demand Q*	*Total Expenditure or Revenue (₹) P × Q*	*Price Elasticity of Demand (e_p)*
5.00	30	150	
4.75	40	190	e > 1
4.50	50	225	e > 1
4.25	60	255	e > 1
4.00	75	300	e > 1
3.75	80	300	e = 1
3.50	84	294	e < 1
3.25	87	282.75	e < 1

Graphic Illustration of Expenditure Method

This relationship between price elasticity of demand and total outlay can also be graphically illustrated with the aid of the demand curve. This relationship between price elasticity of demand and total expenditure made on a good is illustrated with the aid of Fig. 13.5 where demand curve *DD* is given. In the Fig. 13.5 when the price is *OP*, the total expenditure made on the good is equal to *OP* × *OQ* *i.e.*, area *OPRQ* and when the price falls to *OP′*, the total expenditure is equal to *OP′R′Q′*. It will be seen from he Fig. 13.5 that the area *OP′ HQ* is common in both the rectangles *OPRQ* and *OP′R′Q′*. Now, by comparing the remaining areas *PRHP′* and *QHR′Q′* we find that two are equal to each other. That is, the total outlay *OP′R′Q′* is equal to the original outlay *OPRQ*. This means that with the fall in price the total outlay (expenditure) made on the good has remained the same. Hence the price elasticity of demand is here equal to unity.

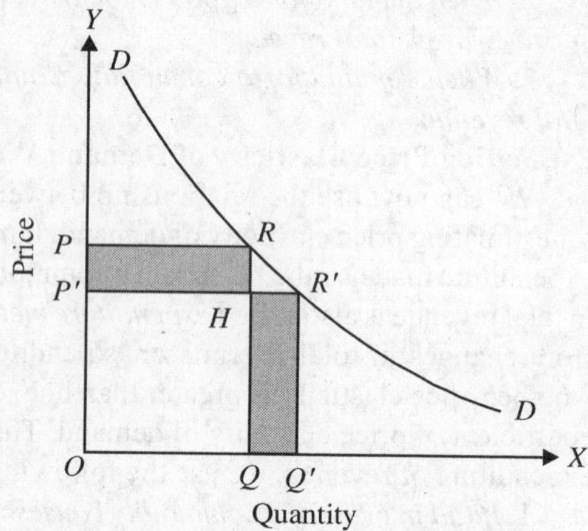

Fig. 13.5. *Price Elasticity is equal to one*

That when the price elasticity of demand is greater than one, the total expenditure made on the good increases with the fall in price of the good is illustrated in Fig. 13.6 in which a demand curve *DD* is given. When price of the good is *OP*, *OQ* quantity of the good is demanded. At the price of *OP*, the total expenditure made on the good is equal to the area *OPRQ*. Now, if price of the good falls to *OP′*, the quantity demanded of the good rises to *OQ′*. Therefore, now at price *OP′* total expenditure on the good is equal to the area *OP′R′Q*. It will be seen in Fig. 13.6 that the area *OP′HQ* is present in both the rectangles *OPRQ* and *OP′R′Q′*. The remaining areas in the two rectangles are *PRHP′* and *QHR′Q′*. Now, a galance at Fig. 13.6 will show that the area *QHR′Q* is greater than the area *PRHP′*. It is thus clear that the expenditure (*OP′R′Q*) on the good at price *OP′* is greater than the expenditure (*OPRQ*) at price *OP*. That is, with the fall in price the total expenditure on the commodity has increased. Hence, the price elasticity of demand is here greater than unity.

Now, consider Fig. 13.7. In this figure demand for the good is such that with the fall in price the total expenditure made on the good declines. At price *OP* the total expenditure is *OPRQ* and when the price falls to *OP′* the total expenditure made on it is equal to *OP′R′Q′*. Now, by comparing the two total expenditure it is evident that the expenditure *OP′R′Q′* is less than the expenditure *OPRQ* in Fig. 13.7. Therefore, the price elasticity of demand is here less than unity.

From the above analysis it is clear that from the changes in the total expenditure as a result of the changes in price we can know the price elasticity of demand for a good. We repeat once again that with the total expenditure method we cannot know the exact and precise measure of the price elasticity; with this we can only know whether price elasticity is equal to one, greater than one or less than one.

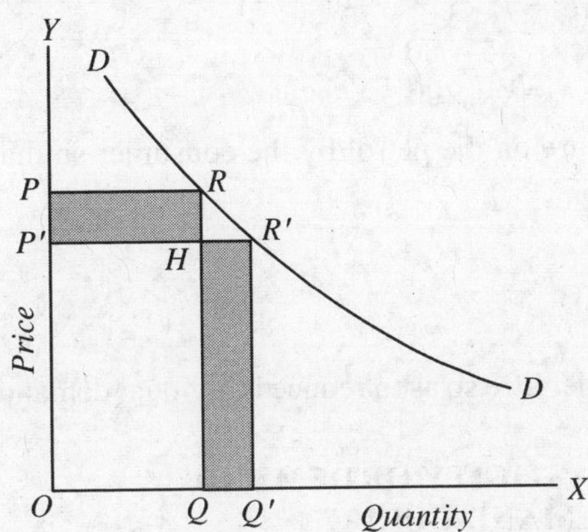

Fig. 13.6. *Price Elasticity is greater than one*

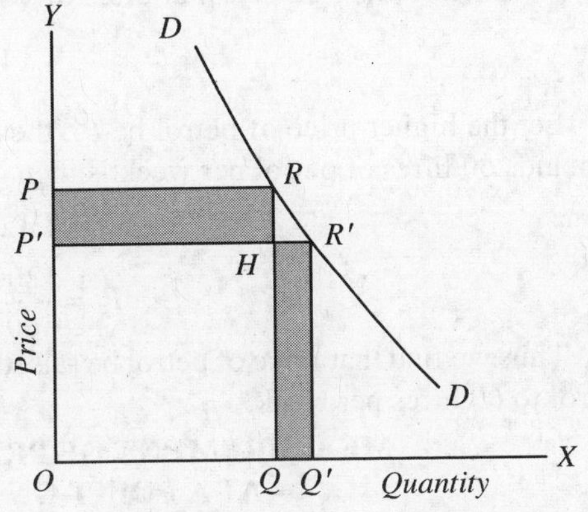

Fig. 13.7. *Price Elasticity is less than*

Problem 4. *Suppose price of a good falls from ₹ 10 to ₹ 8 per unit. As a result, its quantity demanded increases from 80 units to 100 units. What can we say about price elasticity of demand by total outlay method.*

Solution.

At price ₹ 10, quantity demand is 100 units

Therefore, total outlay at price ₹ 10 = $P \times Q$

= 10 × 80

= ₹ 800

At the lower price ₹ 8, quantity demand increases to 100 units

Therefore, total expenditure at price ₹ 8 = $P \times Q$

= 8 × 100

= ₹ 800

We thus find that with the change in price of the good, total outlay on the commodity remains constant. Hence price elasticity of demand is equal to one.

Problem 5. *Suppose price of a commodity rises from ₹ 15 to ₹ 16 per unit. As a result, its quantity demanded falls from ₹ 100 units to 80 units. Find out the price elasticity of demand by expenditure method.*

Solution.

Expenditure on the commodity at price ₹15 = P × Q

= 15 × 100

= ₹ 1500

Expenditure on the commodity at price ₹ 16 = 16 × 80

= ₹ 1,280

Thus we find with the rise in price, expenditure on the commodity decreases. This means the price elasticity of demand is greater than one ($ep > 1$).

Problem 6. *Suppose that price elasticity of demand for petrol is equal to unity and at ₹ 15 per litre an individual consumes (i.e., demands) 80 litres of petrol in a week. How much price of petrol should be fixed so that he demands 60 litres of petrol ?*

Solution.

Since the given price elasticity of demand for petrol is equal to unity, the expenditure on the petrol by the individual will remain constant at different prices.

Expenditure on petrol at price of ₹ 15 per litre $= P \times Q$
$= 15 \times 80$
$=$ ₹ 1,200

Let the higher price of petrol be P', the expenditure on the petrol by the consumer so that he demands 60 litres of petrol per week is then given by

$$P' \times 60 = ₹\ 1200$$

$$P = \frac{1200}{60} = ₹\ 20$$

Thus we find that price of petrol be raised to ₹ 20 per litre so as to reduce individual demand for petrol to 60 litres per week.

MEASUREMENT OF PRICE ELASTICITY OF DEMAND AT A POINT ON THE DEMAND CURVE

Let a straight line demand curve DD' is given and it is required to measure price elasticity at a point R on this demand curve. It will be seen from Fig. 13.8 that corresponding to point R on the demand curve DD', price is OP and quantity demand at it is OQ.

The measure of price elasticity of demand is given by :

$$e_p = \frac{\Delta q}{\Delta p} \cdot \frac{p}{q}$$

The first term in this formula, namely, $\frac{\Delta q}{\Delta p}$ is the reciprocal of the slope of the demand curve DD' (Note that the slope of the demand curve DD' is equal to $\frac{\Delta p}{\Delta q}$ which remains constant all along the straight line demand curve). The second term in the above point elasticity formula is the original price (p) divided by the original quantity (q). Thus

$$e_p = \frac{1}{slope} \cdot \frac{p}{q}$$

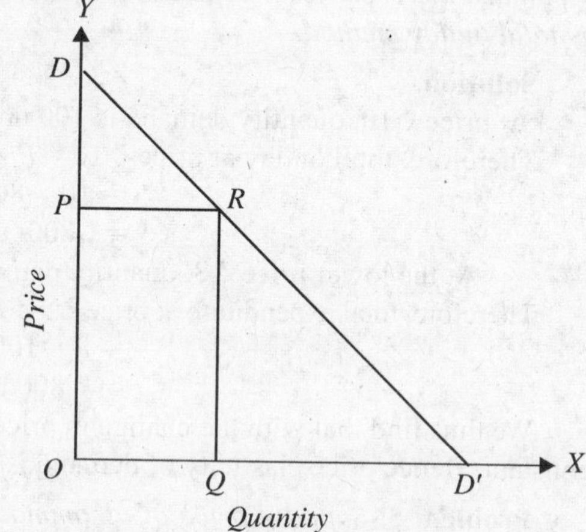

Fig. 13.8. *Measuring Price Elasticity at a Point on a Straight-Line Demand Curve*

It will be seen from Fig. 13.8 that at point R, original price $p = OP$ and original quantity $q = OQ$.

Further, slope of the demand curve DD' is $\frac{\Delta P}{\Delta Q} = \frac{PD}{PR}$

Substituting these values in the above formula we have

$$e_p = \frac{1}{\frac{PD}{PR}} \times \frac{OP}{OQ}$$

$$= \frac{PR}{PD} \times \frac{OP}{OQ}$$

A glance at Fig. 13.8 reveals that $PR = OQ$ and they will cancel out in the above expression.

Therefore, $$e_p = \frac{OP}{PD} \qquad ...(1)$$

Measuring price elasticity by taking the ratio of these distances on the vertical axis, that is, $\frac{OP}{PD}$ is called *vertical axis formula*.

In a right-angled triangle *ODD′*, *PR* is paralled to *OD′*. Therefore,

$$e_p = \frac{OP}{PD} = \frac{RD'}{RD}$$

RD′ is the lower segment of the demand curve *DD′* at point *R* and *RD* is its upper segment. Therefore,

$$e_p = \frac{RD'}{RD} = \frac{\text{lower segement}}{\text{upper segment}} \qquad ...(2)$$

Measuring price elasticity at a point on the demand curve by measuring the ratio of the distances of lower segment and upper segment is another popular method of measuring point price elasticity on a demand curve.

Measuring price elasticity on a non-linear demand curve. If the demand curve is not a straight line like *DD′* in Fig. 13.8 but is, as usual, a non-linear curve, then how to measure price elasticity at a given point on it. For instance, how price elasticity at point *R* on the demand curve *DD* in Fig. 13.9 is to be found. In order to measure elasticity in this case, we have to draw a tangent *TT′* at the given point *R* on the demand curve *DD′* and then measure price elasticity by finding out the value of $\frac{RT'}{RT}$.

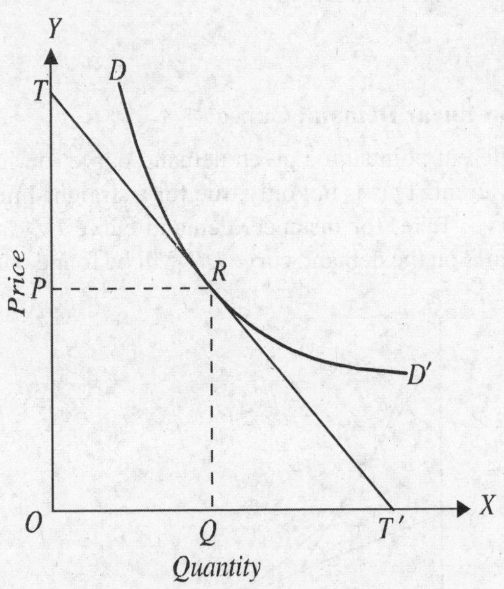

Fig. 13.9. *Measuring Price Elasticity at a Point on a Non-Linear Demand Curve*

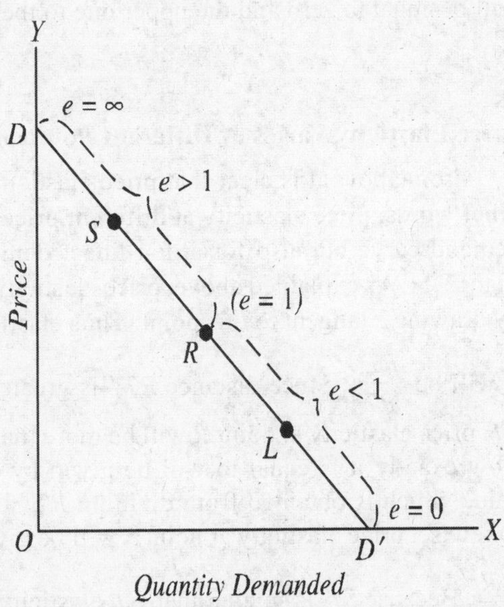

Fig. 13.10. *On a linear demand curve price elasticity varies from infinity to zero.*

On a linear demand curve price elasticity varies from zero to infinity. Now again, take the straight-line demand curve *DD′* (Fig. 13.10). If point *R* lies exactly at the middle of this straight-line demand curve *DD′*, then the distance *RD* will be equal to the distance *RD′*. Therefore, elasticity which is equal to $\frac{RD'}{RD}$ will be equal to one at the middle point of the straight-line demand curve. Suppose a point S lies above the middle point on the straight-line demand curve DD'. It is obvious that the distance SD' is greater than the distance SD and price elasticity which is equal to $\frac{SD'}{SD}$ at point S will be more than one. Similarly, at any other point which lies above the middle point on the straight-line demand curve, price elasticity will be greater than unity. Moreover, price elasticity will go on increasing as we move further towards point D and at point D price elasticity will be equal to infinity. This is because price elasticity is equal $\frac{RD'}{RD}$ *i.e.*, $\frac{\text{lower segment}}{\text{upper segment}}$ and as we move towards D the lower segment will go on increasing while the upper segment will become smaller. Therefore, as we move towards D on the demand curve, the price elasticity will be increasing. At point D, the lower segment will be equal to the whole DD', and the upper segment will be zero. Therefore,

$$\text{Price elasticity at point } D \text{ on the demand curve } DD' = \frac{DD'}{0} = \text{infinity.}$$

Now, suppose a point L lies below the middle point on the linear demand curve DD'. In this case, the lower segment LD' will be smaller than the upper segment LD and therefore price elasticity at L which is equal to $\frac{LD'}{LD}$ will be less than one.

Moreover, price elasticity will go on decreasing as we move towards point D'. This is because whereas lower segment will become smaller and smaller, the upper one will be increasing as we move towards point D'. At point D′ the price elasticity will be zero, since at D' the lower segment will be equal to zero and the upper one to the whole DD'. At point D',

$$e_p = \frac{0}{DD'} = 0$$

Price Elasticity Varies at Different Points on a Non-linear Demand Curve

From above it is clear that price elasticity at different points on a given demand curve (or, in other words, price elasticity at different prices) is different. This is not only true for a straight-line demand curve but also for a non-linear demand curve. Take, for instance, demand curve DD in Fig. 13.11. As explained above, price elasticity at point R on the demand curve DD will be found out by drawing a tangent to this point. Thus elasticity at R will be $\frac{RT'}{RT}$. Since distance RT' is greater than RT, price elasticity at point R will be more than one. How exactly it is equal to will be given by actual value which is obtained from dividing RT' by RT. Likewise, price elasticity at point S will be given by $\frac{SJ'}{SJ}$. Because SJ' is smaller than SJ, elasticity point at S will be less than one. Again, how exactly it is, will be found from actually dividing SJ' by SJ. It is thus evident that elasticity at point S is less than that at point R on he demand curve DD. Similarly, price elasticity at other points of the demand curve DD will be found to be different.

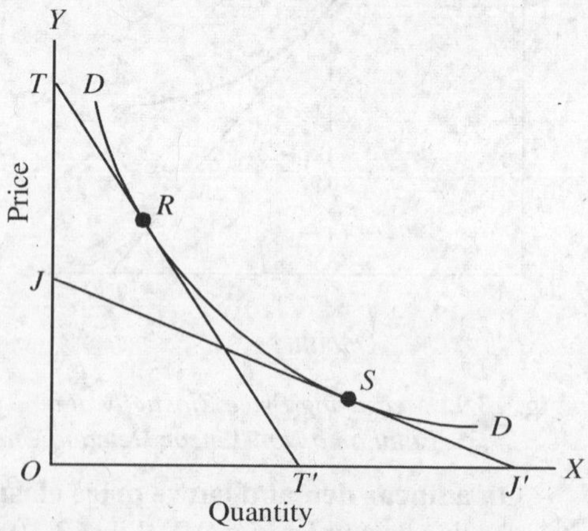

Fig. 13.11. *Price elasticity declines as we move down on a demand curve.*

Comparing Price Elasticity of Two Demand Curves With Different Slopes

Having explained the concept of price elasticity of demand we will now explain how to compare price elasticity on two demand curves. First, we take up the case of two *demand curves with different slopes* starting from a given point on the *Y*-axis. This case is illustrated in Fig. 13.12 where two demand curves *DA* and *DB* which have different slopes but are starting from the same point *D* on the *Y*-axis. Slope of demand curve *DB* is less than that of *DA*. Now, it can be proved that at any given price the price elasticity on these two demand curves would be the same. If price is *OP*, then according to demand curve *DA*, *OL* quantity of the good is demanded and according to demand curve *DB*, *OH* quantity of the good is demanded. Thus, at price *OP* the corresponding points on the two demand curves are *E* and *F* respectively. We know that price elasticity at a point on the demand curve is equal to $\frac{\text{lower segment}}{\text{upper segment}}$. Therefore, the price elasticity of demand at point *E* on the demand curve *DA* is equal to $\frac{EA}{ED}$ and the price elasticity of demand at point *F* on the demand curve *DB* is equal to $\frac{FB}{FD}$.

Now, take triangle *ODA* which is a right-angled triangle in which *PE* is parallel to *OA*.

It follows that in it, $\frac{EA}{ED}$ is equal to $\frac{OP}{PD}$. Thus, the price elasticity at point *E* on the demand curve *DA* is equal to $\frac{OP}{PD}$.

Now, in the right-angled triangle *ODB*, *PF* is parallel to *OB*. Therefore, in it $\frac{FB}{FD}$ is equal to $\frac{OP}{PD}$. Thus, price elasticity of demand at point *F* on the demand curve *DB* is also equal to $\frac{OP}{PD}$. From above it is clear that price elasticity of demand on points *E* and *F* on the two demand curves respectively is equal to $\frac{OP}{PD}$, that is, elasticities of demand at points *E* and *F* are equal though the slopes of these two demand curves are different. *It follows therefore that price elasticity is not the same thing as slope. Therefore, price elasticity on two demand curves should not be compcred by considering their slopes alone.*

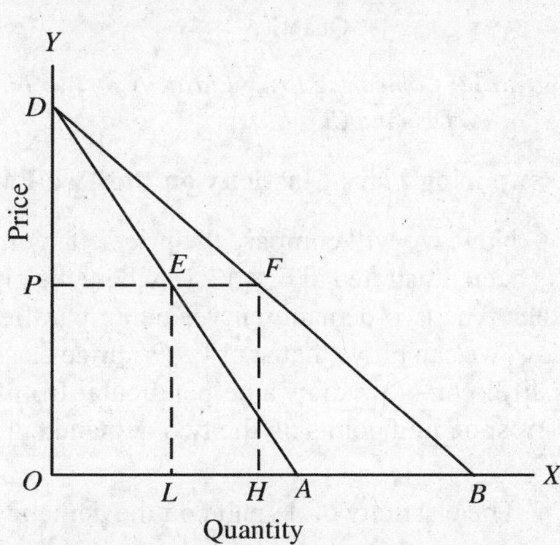

Fig. 13.12. *Comparing Price Elasticity on the Two Demand Curves with Different Slopes*

Comparing Price Elasticity on Two Intersecting Demand Curves

We now take up the case of *comparing price elasticity at a given price where the two demand curves intersect*. In Fig. 13.13 we have drawn two *demand curves AB* and *CD* which intersect at point *E*. It will be noticed from the figure that demand curve *CD* is flatter than the demand curve *AB*. Now, it can be easily proved that at every price on the flatter demand curve *CD*, price elasticity will be greater than that on the relatively steeper demand curve *AB*. For example at price *OP*, corresponding to the intersecting point *E*, using the vertical axis formula, elasticity at point *E* on demand curve

$CD = \frac{OP}{PC}$. Similarly, elasticity at point *E* on the demand curve $AB = \frac{OP}{PA}$. It will be seen from Fig. 13.13 that $\frac{OP}{PC} > \frac{OP}{PA}$ because distance *PC* is less than the distance *PA*. Hence at the price *OP*, elasticity is greater on the flatter demand curve *CD*, as compared to the steeper demand curve *AB*. Likewise, it can be shown at any other given price, price elasticity of demand will be greater on the flatter demand curve *CD* as compared to the steeper demand curve *AB*.

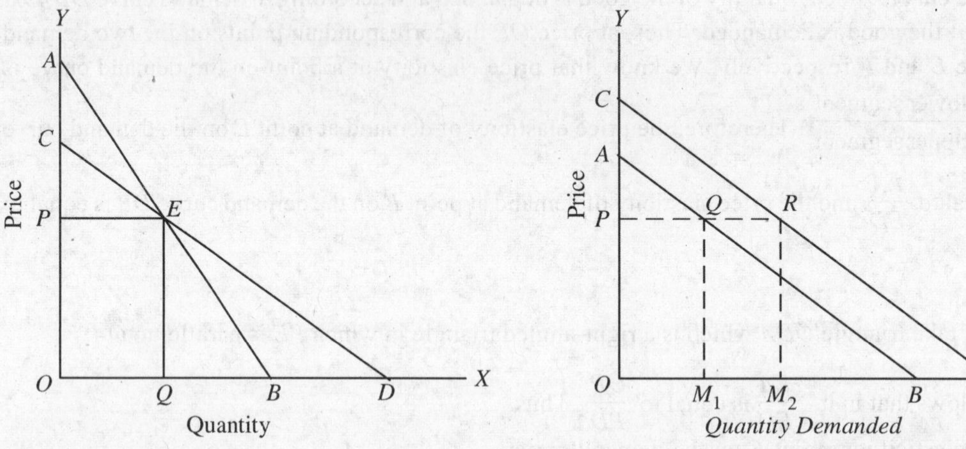

Fig. 13.13. *Comparing Price Elasticity of Two Intersecting Demand Curves*

Fig. 13.14. *Comparing Price Elasticity of Two Parallel Demand Curves*

Comparing Price Elasticity on the Two Parallel Demand Curves

Now, we will compare the price elasticity at two parallel demand curves at a given price. This has been illustrated in Fig. 13.14 where the given two demand curves *AB* and *CD* are parallel to each other. The two demand curver being parallel to each other imlies that they have the same slope. Now, we can prove that at price *OP* price elasticity of demand on the two demand curves *AB* and *CD* is different. Now, draw a perpendicular from point *R* to the point *P* on *Y*-axis. Thus, at price *OP* the corresponding points on the two demand curves are *Q* and *R* respectively.

The elasticity of demand on the demand curve *AB* at point *Q* will be equal to $\frac{QB}{QA}$ and at point *R* on the demand curve *CD* it is equal to $\frac{RD}{RC}$.

Because in a right-angled triangle *OAB*, *PQ* is parallel to *OB* :

Therefore, $$\frac{QB}{QA} = \frac{OP}{PA}$$

Hence, price elasticity at point *Q* on the demand curve $AB = \frac{OP}{PA}$

At point *R* on the demand curve *CD*, price elasticity is equal to $\frac{RD}{RC}$. Because in the right-angled triangle *OCD*, *PR* is parallel to *OD*.

Therefore, $\frac{RD}{RC}=\frac{OP}{PC}$. Hence, on point R on the demand curve CD, price elasticity = $\frac{OP}{PC}$

On seeing Fig. 13.14 it will be clear that at point Q on the demand curve AB, the price elasticity = $\frac{OP}{PA}$ and at point R on the demand curve CD, price elasticity = $\frac{OP}{PA}$ which are not equal to each other. Because PC is greater than PA,

Therefore $$\frac{OP}{PC}<\frac{OP}{PA}$$

It is, therefore, clear that at point R on the demand curve CD the price elasticity is less than that at point Q on the demand curve AB, when the two demand curves being parallel to each other have the same slope. It also follows that *as the demand curve shifts to the right the price elasticity of demand at a given price goes on declining*. Thus, as has been just seen, price elasticity at price OP on the demand curve CD is less than that on the demand curve AB.

DETERMINANTS OF PRICE ELASTICITY OF DEMAND

The following are the main factors which determine price elasticity of demand for a commodity.

The Availability of Substitutes. Of all the factors determining price elasticity of demand the availability of the number and kinds of substitutes for a commodity is the most important factor. If for a commodity close substitutes are available, its demand tends to be elastic. If price of such a commodity goes up, the people will shift to its close substitutes and as a result the demand for that commodity will greatly decline. The greater the possibility of substitution, the greater the price elasticity of demand for it. If for a commodity good substitutes are not available, people will have to buy it even when its price rises, and therefore its demand would tend to be inelastic.

For instance, if the price of Coca Cola were to increase sharply, many consumers would turn to other kinds of cold drinks, and as a result, the quantity demanded of Coca Cola will decline very much. On the other hand, if the price of Coca Cola falls, many consumers will change from other cold drinks to Coca Cola. Thus, the demand for Coca Cola is elastic. It is the availability of close substitutes that makes the consumers sensitive to the changes in the price of Coca Cola and this makes the demand for Coca Cola elastic. Likewise, demand for common salt is inelastic because good substitutes for common salt are not available. If the price of common salt rises slightly, the people would consume almost the same quantity of salt as before since good substitutes are not available. The demand for common salt is inelastic also because people spend a very little part of their income on it and even if its price rises it makes only negligible difference in their budget allocation for the salt.

The Proportion of Consumer's Income Spent On a commodity. Another important determinant of the elasticity of demand is how much it accounts for in consumer's budget. In other words, the proportion of consumer's income spent on a particular commodity also influences the price elasticity of demand for it. The greater the proportion of income spent on a commodity, the greater will be generally its elasticity of demand, and vice versa. The demand for common salt, soap, matches and such other goods tends to be highly inelastic because the households spend only a fraction of their income on each of them. When price of such a commodity rises, it will not make much difference in consumers' budget and therefore they will continue to buy almost the same quantity of that commodity and, therefore, the demand for them will be inelastic. On the other hand, demand for cloth in a country like India tends to be elastic since households spend a good part of their income on clothing. If price of cloth falls, it will mean great saving in the budget of many households and therefore they will tend to increase the quantity demanded of the cloth. On the other hand, if price of cloth rises many households will not afford to buy as much quantity of cloth as before, and therefore, the quantity demanded of cloth will fall.

The Number of Uses of a Commodity. The greater the number of uses to which a commodity can be put, the greater will be its price elasticity of demand. If price of a commodity having several uses is very high, its demand will be small and it will be put to the most important uses and if price of such a commodity falls it will be put to less important uses also and consequently its quantity demanded will rise significantly. To illustrate, milk has several uses. If its price rises to a very high level, it will be used only for essential purposes such as feeding the children and sick persons. If price of milk falls, it would be devoted to other uses such as preparation of curd, cream, ghee and sweets. Therefore, the demand for milk tends to be elastic.

Complementarity Between Goods. Complementarity between goods or joint demand for goods also affects the price elasticity of demand. Households are generally less sensitive to the changes in prices of goods that are complementary with each other or which are jointly used as compared to those goods which have independent demand or used alone. For example, for the running of automobiles, besides petrol, lubricating oil is also used. Now, if the price of lubricating oil goes up, it will mean a very small increase in the total cost of running the automobile, since the use of oil is much less as compared to other things such as petrol. Thus, the demand for lubricating oil tends to be inelastic. Similarly, the demand for common salt is inelastic, partly because consumers do not use it alone but along with other things.

It is worth mentioning here that for assessing the elasticity of demand for a commodity all the above three factors must be taken into account. The three factors mentioned above may reinforce each other in determining the elasticity of demand for a commodity or they may operate against each other. The elasticity of demand for a commodity will be the net result of all the forces working on it.

Time and Elasticity. The element of time also influences the elasticity of demand for a commodity. Demand tends to be more elastic if the time involved is long. This is because consumers can substitute goods in the long run. In the short run, substitution of one commodity by another is not so easy. The longer the period of time, the greater is the ease with which both consumers and businessmen can substitute one commodity for another. For instance, if the price of fuel oil rises, it may be difficult to substitute fuel oil by other types of fuels such as coal or cooking gas. But, given sufficient time, people will make adjustments and use coal or cooking gas instead of the fuel oil whose price has risen. Likewise, when the business firms find that the price of a certain material has risen, then it may not be possible for them to substitute that material by some other relatively cheaper one. But with the passage of time they can undertake research to find substitute material and can redesign the product or modify the machinery employed in the production of a commodity so as to economise in the use of the dearer material. Therefore, given the time, they can substitute the material whose price has risen. We thus see that demand is generally more elastic in the long run than in the short run.

IMPORTANCE OF PRICE ELASTICITY OF DEMAND

The concept of elasticity of demand plays a crucial role in the pricing decisions of the business firms and the Government when it regulates prices. The concept of price elasticity is also important in judging the effect of devaluation or depreciation of a currency on its export earnings. It has also a great use in fiscal policy because the Finance Minister has to keep in view the price elasticity of demand when it considers to impose taxes on various commodities. We shall explain below the various uses, applications and importance of the elasticity of demand.

Pricing Decisions by Business Firms. *The business firms take into account the price elasticity of demand when they take decisions regarding pricing of the goods*. This is because change in the price of a product will bring about a change in the quantity demanded depending upon the coefficient of price elasticity. This change in quantity demanded as a result of, say a rise in price by a firm, will affect the total consumer's expenditure and will therefore, affect the revenue of the firm. If the demand for a product of the firm happens to be elastic, then any attempt on the part of the firm to raise the price of its product will bring about a fall in its total revenue. Thus, instead of gaining from the increase in price, it will lose if the demand for its product happens to be elastic. On the other hand, if the demand for the product of a firm happens to be inelastic, then the increase in price by it will raise its total revenue. Therefore, for fixing a profit-maximising price, the firm cannot ignore the price elasticity of demand for its product.

Price elasticity of demand can be used to answer the following types of questions:

1. What will be the effect on sales if a firm decides to raise the price of its product, say by 5 per cent.

2. How large a reduction in price of a product is required to increase sales, say by 25 percent.

It has been found by some empirical studies that business firms often fail to take elasticity into account while taking decisions regarding prices, or they give insufficient attention to the coefficient of price elasticity. No doubt, the main reason for this is that they don't have the means to calculate price elasticity for their product, since sufficient data regarding past prices and quantity demanded at those prices are not available. Even if such data are available, there are difficulties of interpretation of it because it is not clear whether the changes in quantity demanded were the result of changes in price alone or changes in some other factors determining the demand. However, recently big corporate business firms have established their research departments which estimate the coefficient of price elasticity from the data concerning past prices and quantities demanded. Further, they are also using statistical techniques to isolate the price effect on the quantity demanded from the effects of other factors.

Uses in Economic Policy Regarding Price Regulation and Crop Restriction of Farm Products

Governments of many countries, especially the United States of America provide subsidies per unit of the products produced to the farmers to give them incentives to produce more as due to inelastic nature of demand for farm products more production causes such a steep fall in the prices of farm products that leads to the decrease in incomes of farmers.

However, during the last over three decades, government in the United States helps farmers by adopting an unusual policy of requiring them to restrict production. To induce them to restrict output government provides subsidy to them for not planting crops on all their land (that is, for keeping some land uncultivated). The purpose of restricting production in this way is to reduce their supply in the market so that price of the agricultural product in the market rises. In view of the fact that demand for agricultural product is inelastic a fall in production will cause their revenue or income to rise and will thus make them better off. Consider Fig. 13.14(a) where without intervention by government demand curve *DD* and supply curve S_1 determine equilibrium price P_1 and the farmer's

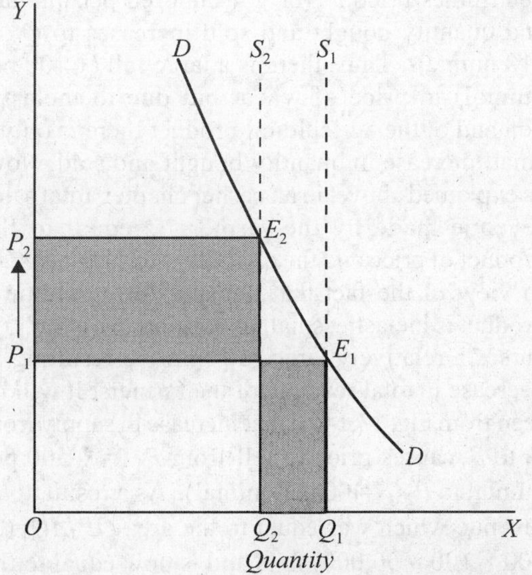

Fig. 13.14(a). *Crop restriction programme raises both price and income of the farmers.*

sales revenue is equal to the area $OP_1E_1Q_1$. As a result of crop restriction programme of government supply curve shifts to the left to S_2 (For sake of simplicity we have assumed perfectly inelastic supply curves). With the intersection of new supply curve S_2 with the given inelastic demand curve *DD*, price of agricultural product rises to P_2 and the quantity sold has fallen to Q_2 so that the new total revenue is $OP_2E_2Q_2$ which is greater than the initial revenue $OP_1E_1Q_1$ *before crop restriction.* Thus, crop restriction programme of government has led to the increase in total revenue and hence it will raise the incomes of the farmers.

It should be noted that whereas crop restriction has raised the incomes of the farmers, it has hurt the consumers as they have to pay higher prices for foodgrains as they would if flood or drought conditions had created shortage of foodgrains. Thus, interests of farmers and consumers clash with each other.

Explanation of the Paradox of Poverty Amidst Plenty

The concept of price elasticity of demand also helps us to explain the so-called 'paradox of plenty' in agriculture, namely, that a bumper crop reaped by the farmers brings a smaller total income to them. The fall in the income or revenue of the farmers as a result of the bumper crop is due to the fact that with greater supply the prices of the crops decline drastically and in the context of inelastic demand for them, the total expenditure on the crop output declines, bringing about fall in the incomes of the farmers. Thus, bumper crop instead of raising their incomes, reduces them. Therefore, in order to ensure that the farmers do not lose incentive to raise their production, they need to be assured certain minimum price by the Government. At that minimum price the Government should be prepared to buy the crop from the farmers.

Similarly, if due to research there is improvement in agricultural technology which leads to the substantial increase in agricultural output instead of raising farmers' income may actually brings about reduction in it. This strange phenomenon is due to the *inelastic nature of demand for agricultural products.* Consider figure 13.14(*b*), Suppose initially the demand curve *DD* and supply curve S_1 determine price of an agricultural crop, say wheat, equal to P_1 (or ₹ 5 hundred per quintal) at which the quantity Q_1 (or 100 quintals, are bought and sold. Suppose due to good weather (or in case of India due to good monsoon), there is bumper crop of wheat and as result supply curve of wheat shifts from S_1 to S_2. The new supply curve S_2 intersects the inelastic demand curve *DD* at point *B* and determines price P_2 (or ₹ 4 hundred per quintal) and quantity bought and sold increases to Q_2 or 110 quintals. Thus, there is a large fall (₹ 100 per quintal) in price of wheat but due to inelastic demand of the agricultural product there is only a small increase in quantity bought and sold. Now, as explained above, in an earlier chapter, total sales revenue made by the farmers is equal to the product of price and the quantity sold ($TR = P \times Q$). In view of the fact that demand for agricultural product is inelastic, small expansion in output. (Q) curses a relatively large full in price resulting in decrease in total revenue of the farmers. It will be seen from Fig. 13.14(b) that increase in supply from S_1 to S_2 causes price to fall from P_1 (= ₹ 500 per quintal to P_2 (₹ 400 per quintal). As a result, total revenue which was equal to the area OP_1AQ_1 (or 500 × 100 = 50,000) falls and is now equal to the area OP_2BQ_2 (or 400 × 110 = 44,000). Thus, bumber crop has led to the fall in total rvenue or

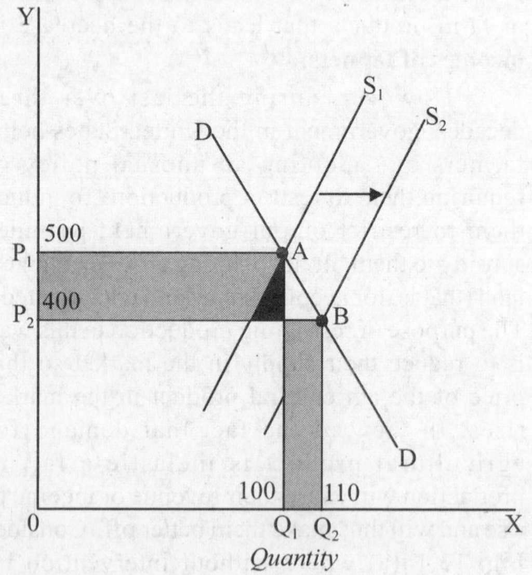

Fig. 13.14(b). *Explaining Paradox of 'Poverty Amidst Plenty*

income of the farmers. This is usually described as farmers face a paradox of 'poverty amongst plenty' or a good news (in the form of good weather or good monsoon) turns out to be actually a bad news for the farmers as it leads to the decline in their incomes and they become worse off.

Use in International Trade

The of price elasticity of demand is also crucially important in the field of international economics. The Governments of the various countries have to decide about whether to devalue their currencies or not when their exports are stagnant and imports are mounting and as a result their balance of payments position is worsening. The effect of the devaluation is to raise the price of the imported goods and to lower the prices of the exports. *If the demand for a country's exports is inelastic, the fall in the prices of exports as a result of depreciation will lower their foreign exchange earnings rather than increasing them.* This is because, demand being inelastic, as a result of the fall in prices quantity demanded of the exported products will increase very little and the country would suffer because of the lower prices. On the other hand, if the demand for a country's exports is elastic, then the fall in the prices of these exports due to depreciation will bring about a large increase in their quantity demanded which will increase the foreign exchange earnings of the country and will thus help in solving the balance of payments problem. Thus, the decision to devalue or not, depends upon the coefficient of the demand elasticity of exports.

Likewise, if the objective of depreciation or devaluation is to reduce the imports of a country, then this will be realised only when the demand for the imports is elastic. With elastic demand for imports, the imports will decline very much as a result of rise in their prices brought about by devaluation and the country will save a good amount of foreign exchange. On the other hand, if the demand for imports is inelastic, the increase in prices as a result of devaluation will adversely affect the balance of payments, because at higher prices of the imports and almost the same quantity of imports, the country would have to spend more on the imports than before.

Importance in Fiscal Policy

The elasticity of demand is also of great significance in the field of fiscal policy. The Finance Minister has to take into account price elasticity of damand of the product on which he proposes to impose the excise duty or sales tax if the revenue for the Government is to be increased. The imposition of an indirect tax, such as excise duty or sales tax, raises the price of a commodity. Now, if the demand for the commodity is elastic, the rise in price caused by the tax, will bring about a large decline in the quantity demanded and as a result the Government revenue will decline rather than increase. *The Government can succeed in increasing its revenue by the imposition a of commodity tax only if the demand for the commodity is inelastic.*

The elasticity of demand also determines to what extent a tax on a commodity can be shifted to the consumer. Thus, the incidence of a commodity tax on the consumers depends on their price elasticity of demand for the commodity. A commodity tax, excise duty or sales tax drives a wedge between the price paid by the buyers and the price received by the sellers. Consider Figure 13.14(*c*) where the demand curve *DD* is inelastic and supply curve *SS* is elastic. Suppose now sales tax equal to *CB* per unit is imposed. The producers or sellers will be willing to sell a *given quantity of a commodity*, if they receive the same *net price* as before. That is, the producers or sellers will treat the sales tax *CB* per unit as an extra cost of production and, therefore, they would add it to the cost per unit. As a result of the imposition of a sales tax per unit of the commodity, supply curve will shift upward to S_2 and will be parallel to the supply curve S_1 without a tax. Demand curve *DD* will remain unaffected as a result of the imposition of sales tax. It will be seen from Fig. 13.14(c) that the new supply curve S_2 intersects the demand curve *DD* at point *B* and determines higher equilibrium price OP_2 and the quantity sold falls to OQ_2. It will be seen from Fig. 13.14(*c*) that in this present case when demand is inelastic and supply elastic, the burden of the tax falls more on the buyers and less on the sellers. The buyers have to pay *EB* more price than before and sellers receive *EC* less price than before. And $EB > EC$.

Now consider Fig. 13.14(d) where demand curve *DD* is elastic and supply curve S_1S_1 is relatively inelatic. Before the imposition of tax P_1 is the price at which quantity Q_1 is being bought and sold consequent to imposition of tax equal to *CB*, supply curve shifts to S_2S_2. Price rises from P_1 to P_2 and equilibrium quantity sold and bought falls to Q_2. In this case when demand is elastic and supply relatively inelastic, burden of tax *EB* per unit borne by the buyers is much less than *CE* borne by the sellers.

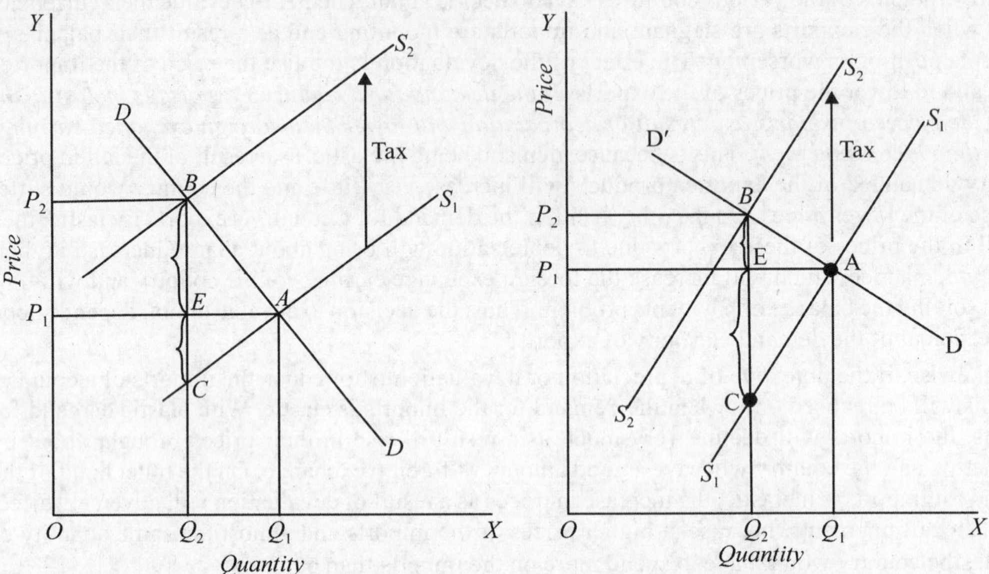

Fig. 13.14(c). *In case of inelastic demand the burden of a commodity tax falls more on the buyers.*

Fig. 13.14(d). *In case of elastic demand the burden of a commodity tax falls more on sellers.*

It follows from above that the burden or *the incidence of taxes borne by the producer and the consumer will depend upon the elasticity of demand as well as elasticity of supply. The lower the elasticity of demand, the greater will be the incidence of tax borne by the consumer.*

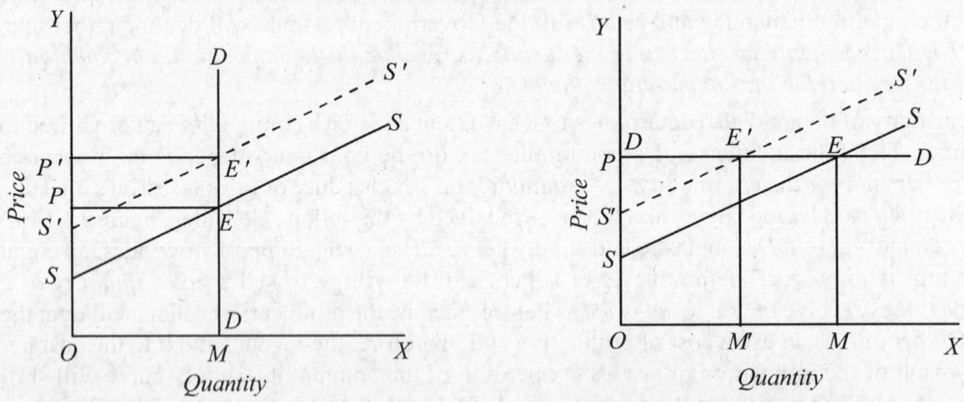

Fig. 13.14(e). *Incidence of Tax in Case of Perfectly Inelastic Demand.*

Fig. 13.14(f). *Incidence of Tax in Case of Perfelectly Elastic Demand*

If the demand for a commodity is perfectly inelastic, the whole of the burden of the commodity tax will fall on the consumers as is seen from Fig. 13.14(e). When a tax is imposed on a commodity, its price will rise. As in the case of perfectly inelastic demand, the quantity demanded for the commodity remains the same, whatever the price, the price will rise to the extent of the tax per unit.

Therefore, the consumers will bear the whole burden of the tax in the form of a higher price they pay for the same quantity demanded.

On the contrary, *if the demand for a commodity is perfectly elastic, the imposition of the tax on it will not cause any rise in price and, therefore, the whole burden of the tax will be borne by the manufacturers or sellers as will be seen from Fig. 13.14(f)*. When demand is neither perfectly inelastic, nor perfectly elastic, then, as seen above, respective burdens borne by the consumers and the producers will depend upon the elasticity of demand as well as on the elasticity of supply. We thus see that a Finance Minister cannot ignore price elasticity of demand for products while levying taxes.

CROSS ELASTICITY OF DEMAND

Very often demands for two goods are so related to each other that when the price of any of them changes, the demand for the other good also changes, its own price remaining the same. Therefore, *degree of responsiveness of demand for one good in response to the change in price of another good represents the cross elasticity of demand of one good for the other.*

The concept of cross elasticity of demand is illustrated by Fig. 13.15 and 13.16 where demand curves of two goods *X* and *Y* respectively are given. Initially, the price of good *Y* is OP_1, at which OQ_1 quantity of it is demanded and the price of good *X* is *OP* at which OM_1 quantity of it is demanded.

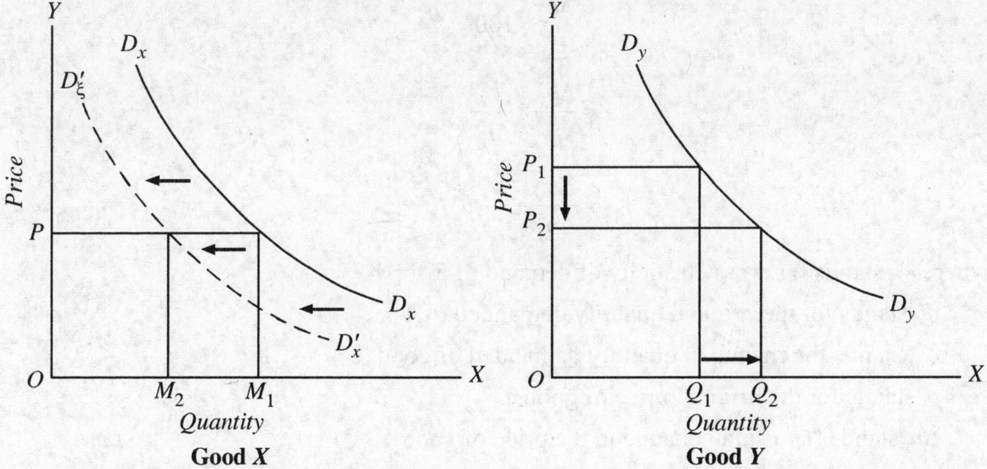

Fig. 13.15 & 13.16. *The Effect of Change in Price of Good Y on the Demand for Good X*

Now suppose and 13.16 the price of good *Y* falls from OP_1 to OP_2, while price of good *X* remains constant at *OP*. As a consequence of the fall in price of good *Y* from OP_1 to OP_2, its quantity demanded rises from OQ_1 to OQ_2. In drawing the demand curve, D_xD_x for good *X*, it is assumed that the prices of other goods (including good *Y*) remain the same. Now that the price of good *Y* has fallen and as a result its quantity demanded has increased, it will have an effect on the demand for good *X*. *If good Y is a substitute for good X, then as a result of the fall in price of good Y from OP_1 to OP_2, in Fig. 13.16 the demand curve of good X in Fig. 13.15 will shift to the left, that is, the demand for good X will decrease*. This is because *as the quantity of a good increases, the marginal utility of its substitute good declines and therefore the entire marginal utility curve of the substitute good shifts to the left*. As shall be seen from the Fig. 13.16 that with the fall in price of good *Y*, its quantity demanded increases for Q_1 to Q_2, and as a result the demand curve of good *X* in 13.15 shifts from D_xD_x to the dotted position $D'_xD'_x$ so that at price *OP* now less quantity OM_2 of *X* is demanded. M_1M_2 of good *X* has been substituted by Q_1Q_2 of good *Y*.

It should be noted that if good *X* instead of being substitute is complement of good *Y*, the resultant increase in its quantity demand of good *Y* due to fall in its price would have caused the increase in demand for good X and as a result the entire demand curve of good X, instead of shifting to the left, would shift to the right. This is because *when the price of a good falls and consequently its quantity demanded increases, the marginal utility of its complement would increase and therefore its entire demand curve would shift to the right.* With a rightward shift of the demand curve of good X, the greater quantity of it will be demanded at the given price OP. It should be noted again that in the concept of cross elasticity of demand, while the price of *one good* changes, there is a change in the quantity demanded of *another good.*

When the quantity demanded of good X falls as a result of the fall in the price of good Y, the coefficient of cross elasticity of demand of X for Y will be equal to the percentage change in the quantity demanded of good X in response to a given percentage change in the price of good Y. Therefore,

$$\text{Coefficient of cross elasticity} = \frac{\text{Percentage change in the quantity demanded of } X}{\text{Percentage change in the price of good } Y}$$

$$e_c = \frac{\frac{\Delta q_x}{q_x} \times 100}{\frac{\Delta p_y}{P_y} \times 100} = \frac{\Delta q_x}{q_x} \div \frac{\Delta p_y}{P_y}$$

$$= \frac{\Delta q_x}{q_x} \times \frac{P_y}{\Delta p_y}$$

$$= \frac{\Delta q_x}{\Delta p_y} \cdot \frac{P_y}{q_s}$$

where e_c, stands for cross elasticity of demand of X for Y

Q_q stands for the original quantity demanded of X

Δ_q, stands for change in quantity demanded of good X

p_y stands for the original price of good Y

Δp_y stands for a small change in the price of good Y

Mid-Point Method. When changes in price and quantities demanded are large, ***mid-point method*** of measuring elasticity of demand with average of initial and subsequent prices and quanties should be used as the basis for measuring percentage changes in them. Thus mid-point formula for measuring cross elasticity of demand is:

$$e_c = \frac{q_{x2} - q_{x1}}{q_{x2} + q_{x1}} \div \frac{p_{y2} - p_{y1}}{p_{y2} + p_{y1}} = \frac{q_{x2} - q_{x1}}{p_{y2} - p_{y1}} \times \frac{p_{y2} + p_{y1}}{q_{x2} + q_{x1}}$$

$$= \frac{\Delta q_x}{\Delta p_y} \times \frac{p_{y2} + p_{y1}}{q_{x2} + q_{x1}}$$

Numerical Problem

Let us take an example. If the *price of coffee* rises from ₹ 45 per 250 grams to ₹ 55 per 250 grams per pack and as a result the consumer's *demand for tea* increases from 600 to 800 packs then the cross elasticity of demand of tea for coffee can be found out as follows:

In the above example:

$$\Delta q_x = 800 - 600 = 200 \text{ packs}$$
$$\Delta p_y = ₹\ 55 - 45 = 10$$

$$\text{Cross elasticity of demand} = \frac{\Delta q_x}{\Delta p_y} \times \frac{p_{y1} + P_{y_2}}{q_s \times q_{x_2}} = \frac{200}{10} \times \frac{100}{1400}$$

$$= \frac{2000}{1400} = \frac{20}{14} = 1.43$$

Substitute and Complementary Goods

As we have seen in the example of tea and coffee above, when two goods are substitutes of each other, then as *a result of the rise in price of one good, the quantity demanded of the other good increases*. Therefore, the *cross elasticity of demand between the two substitute goods is positive,* that is, in response to the *rise in price of one good, the demand for the other good increases.* Substitute goods are also known as *competing goods*. On the other hand, when the two goods are complementary with each other just as bread and butter, tea and milk etc., the rise in price of one good brings about the *decrease in demand* for the other. Therefore, the *cross elasticity of demand between* the two complementary goods is negative. Therefore *according to the classification based on the concept of cross elasticity of demand, goods X and Y are substitutes or complements according as the cross elasticity of demand is positive or negative.*

The concept of cross elasticity of demand is very important in economic theory. The substitute and complementary goods, as we have seen above, are defined is terms of cross elasticity of demand. *The goods between which cross elasticity of demand is positive are known as substitute goods and the goods between which cross elasticity of demand is negative are complementary goods*.

Besides, classification of various types of market structures is made on the basis of cross elasticity of demand. Thus, Professor Triffen has employed the concept of cross elasticity of demand in distinguishing the various forms of markets. Perfect competition is defined as that in which the cross elasticity of demand between the products produced by many firms in it is infinite. Monoploy is said to exist when a producer produces a product the cross elasticity of demand for which with any other product is very low. In fact, the pure or absolute monopoly is sometimes defined as the production by a single producer of a product whose cross elasticity of demand with any other product is zero. Monopolistic competition is said to prevail in the market when a large number of firms produces those products between which cross elasticity of demand is large and positive, that is, they are close substitutes of each other.

Importance of Cross Elasticity of Demand

The concept of cross elasticity of demand is of great importance in managerial decision making for formulating proper price strategy. Multi-product firms often use this concept to measure the effect of change in price of one product on the demand for other products. For example, Maruti Udyog Ltd. produces Maruti Vans, Alto and Maruti SX-4. These products are good substitutes of each other and therefore cross elasticity of demand between them is very high. If Maruti Udyog decides to lower the price of Alto, it will significantly affect the demand for Maruti Vans and Maruti SX-4. So it will formulate a proper price strategy fixing appropriate price for its various products. Further, Gillete Company produces both razors and razor blades which are complements with high cross elasticity of demand. If it decides to lower the price of razors, it will greatly increase the demand for razor blades. Thus there is need for adopting a proper price strategy when a firm produces products with high positive or negative cross price-elasticity of demand.

Second, the concept of cross elasticity of demand is frequently used in defining the boundaries of an industry and in measuring interrelationship between industries. An industry is defined as a group of firms producing similar products (that is, products with a high positive cross elasticity of demand. For example cross elasticity of demand between Maruti SX-4, Hyndui's Verna, Tata's Indigo is positive and quite high. They therefore belong to the same industry (*i.e.*, automobiles). It should be noted that because of interrelationship of firms and industries between which cross price-elasticity of demand is positive and high, any one cannot raise the price of its product without losing sales to other firms.

Further, the concept of cross elasticity of demand is extremly used in the United States in deciding cases relating to Antitrust laws and monopolistic practices used by firms. It so happens that in order to reduce competition that one dominant firm producing a product with high cross elasticity of demand with the products of other firms tries to take over them and thereby establish a monopoly, or various firms producing close substitutes with high cross elasticity of demand try to merge with each other to form a cartel to enjoy monopolistic profits. These actions are held illegal by Antitrust or anti-monopoly laws. An interesting attempt was made in India by Caca-Cola in 1995 when it returned to India following the adoption of policy of liberalisation. In order to reduce competition, Coca-Cola company purchased the firm producing Thums Up, Gold Spot, Limca which have high positive cross elasticity of demand with Coca-Cola. Thus, it succceeded in significantly reducing competition. With this its competition has been mainly with other multinational rival firm Pepsi-Cola.

Numerical Problems on Cross Elasticity of Demand

1. *Suppose the following demand function for coffee in terms of price of tea is given. Find out the cross elasticity of demand when price of tea rises from ₹ 50 per 250 grams pack to ₹ 55 per 250 grams pack.*

$$Q_C = 100 + 2.5P_t$$

where Q_C is the quantity demanded of coffee in terms of packs of 250 grams and P_t is the price of tea.

Solution. The positive sign of the derivative of P_t shows that rise in price of tea will cause an increase in quantity demanded of coffee. This implies that tea and coffee are substitutes. In order to determine cross elasticity of demand between tea and coffee, we first find out quantity demanded of coffee when price of tea is ₹ 50 per 250 grams. Thus,

$$Q_C = 100 + 2.5 \times 50 = 225$$

$$\text{Cross elasticity}, e_c = \frac{dQ_c}{dP_t} \times \frac{P_t}{Q_c}$$

$$\frac{dQ_c}{dP_t} = 2.5$$

$$e_c = 2.5 \times \frac{50}{225} = \frac{125}{225} = 0.51$$

22. *Two goods have a cross-price elasticity of demand of +1.2 (a) Would you describe the goods as substitutes or complements ? (b) If the price of one of the goods rises by 5 per cent, what will happen to the demand for the other good, holding other factors constant ?*

Solution. (*a*) The goods with positive cross-price elasticity of demand are substitute goods. (*b*) If the price of one of the two goods increases by 5 per cent, it will be substituted by the other good so that the quantity demanded of this other good will rise. With positive cross-price elasticity being equal to 1.2, the quantity demanded of the other good will increase by 1.2 × 5 = 6 per cent.

INCOME ELASTICITY OF DEMAND

Another important concept of elasticity of demand is income elasticity of demand. *Income elasticity of demand shows the degree of responsiveness of quantity demanded of a good to a small change in the income of consumers*. The degree of response of quantity demanded to a change in income is measured by dividing the proportionate change in quantity demanded by the proportionate change in income. Thus, more precisely, the income elasticity of demand may be defined as the ratio of the percentage change in purchases of a good to a percentage change in income which induces the former.

$$\text{Income elasticity} = \frac{\text{Percentage change in purchases of a good}}{\text{Percentage change in income}}$$

Let Y stand for an initial income, ΔY for a small change in income, q for the initial quantity purchased, Δq for a change in quantity purchased as a result of a change in income and e_i for income elasticity of demand. Then,

$$e_i = \frac{\frac{\Delta q}{q} \times 100}{\frac{\Delta y}{y} \times 100} = \frac{\Delta q}{q} \div \frac{\Delta Y}{Y} = \frac{\Delta q}{\Delta Y} \times \frac{Y}{q}$$

If, for instance, consumer's daily income rises from ₹ 300 to ₹ 320, his purchase of the good X increases from 25 units per week to 30 units, then his income elasticity of demand for X is :

$$e_i = \frac{\frac{5}{25} \times 100}{\frac{20}{300} \times 100} = \frac{\frac{1}{5}}{\frac{1}{15}} = \frac{1}{5} \times 15 = 3$$

When changes in income and quantity demanded are large, **mid-point method** should be used to measure percentage changes in income and quantity demanded. The mid-point method of measuring income elasticity of demand can be stated as under :

$$e_i = \frac{q_2 - q_1}{\frac{q_1 + q_2}{2}} \div \frac{Y_2 - Y_1}{\frac{Y_1 + Y_2}{2}} = \frac{\Delta q}{q_1 + q_2} + \frac{Y_1 + Y_2}{\Delta Y}$$

$$= \frac{\Delta q}{\Delta Y} \times \frac{Y_1 + Y_2}{q_1 + q_2}$$

Income elasticity of demand being zero is of great significance. Zero income elasticity of demand for a good implies that a given increase in income does not at all lead to any increase in quantity demanded of a good or expenditure on it. In other words, zero income elasticity signifies that quantity demanded of the good is quite unresponsive to changes in income.

Income Elasticity, Normal Good and Inferior Goods

Besides, zero income elasticity is significant because it represents dividing line between positive income elasticity on the one side and negative income elasticity on the other. On the one side, when income elasticity is more than zero (that is, positive), then an increase in income leads to the increase in quantity demanded of the good. This happens in case of normal goods. On the other side of zero income elasticity are all those goods whose income elasticity is less than zero (that is, negative) and in such cases increases in income will lead to the fall in quantity demanded of the goods. *Goods having negative income elasticity are known as inferior goods*. Goods with positive income elasticity are called *normal goods*. We thus see that zero income elasticity is a significant value, for it helps us to distinguish normal goods from inferior goods.

Income Elasticity, Luxuries and Necessities

Another significant value of income elasticity is unity. This is because when income elasticity of demand for a good is equal to one, then proportion of income spent on the good remains the same as consumer's income increases. Income elasticity of unity also represents a useful dividing line. If the income elasticity for a good is greater than one, the proportion of consumer's income spent on the good rises as consumer's income increases, that is, that good bulks larger in consumer's expenditure as he becomes richer. On the other hand, if income elasticity for a good is less than one, the proportion of consumer's income spent on it falls as his income rises, that is, the good becomes relatively less important in consumer's expenditure as his income rises. *A good having income elasticity more than one and which therefore bulks larger in consumer's budget as he becomes richer is called a luxury. A good with an income elasticity less than one and which claims declining proportion of consumer's income as he becomes richer is called a necessity*. It should, however, be noted that the definitions of luxuries and necessities on the basis of income elasticity may not conform to their definitions in Engligh dictionary because the dictionary's luxuries may be necessities and its necessities may be luxuries according to the above definition. But in economic theory it is useful to call the goods with income elasticity greater than one as luxuries and goods with income elasticity less than one as necessities.

Importance of Income Elasticity

The concept of income elasticity is important for decision making both by business firms and industries. First, the firms producing products which have a high income elasticity have great potential for growth in an expending economy. For example, if for a firm's product income elasticity of demand is greater than one, it means that it will gain more than proportionately to the increase in national income. Thus firms which are producing products having high income elasticity are more interested in forecasting the level of aggregate economic activity (*i.e.*, level of national income) because the demand for their products will greatly depend on the level of overall econmic activity. Further, as seen above, the demand for luxuries is highly income elastic. Therefore, the demand for luxuries fluctuate very much during different phases of business cycles. During boom periods, demand for luxuries increase very much, and decline sharply during recessionary periods.

On the other hand, the demand for products with low income elasticity will not be greatly affected by the fluctuations in aggregate economic activity. During booms the demand for their products will not increase much and during recessions it will not decrease sharply. Therefore, the firms with low income elasticity for their products would not be much interested in forecasting future business activity. Remember it is generally necessities for which demand is not much income elastic. However, there is one good thing for the firms which face low income elasticity. They are to a good extent recession-proof. In the periods of recession, their incomes do not fall to the extent of decline in aggregate income. Of course, to share the benefits of increasing national income firms currently producing products with low income elasticity would try to enter the industries demand for whose products is highly income elastic as this would ensure better growth opportunities.

The knowledge of income elasticity of demand also plays a significant role in designing marketing strategies of the firms. If income of people is an important determinant of demand for a product, the firms producing product with high income elasticity of demand will be located in those areas or set up their sales outlets in those cities or regions where incomes are increasing rapidly. Besides, the firms will direct their advertising campaigns and other sales production activities to those segments of people whose income is high and also increasing rapidly. This is to ensure higher growth of sales of their products.

The concept of income elasticity of demand shows clearly why farmers income do not rise equal to that of urban people engaged in manufacturing industries. Income elasticity of demand for agricultured products such as foodgrains is less than one. This implies that it is defficult for the farmers' income from agriculture to increase in proportion to the expanding national income. Thus farmers cannot keep up with the urban people who derive their incomes from industries producing goods with high income elasticity of demand.

THE ELASTICITY OF SUBSTITUTION

The elasticity of substitution is another important concept of demand elasticity. The elasticity of substitution between the two goods is a relative measure of the degree of substitution possibility between the two goods for consumption by the consumers. As we move along an indifference curve, we substitute one commodity for another in our combination. Thus elasticity of substitution is measured as the proportionate change in the ratio of the quantities of the two goods consumed divided by the proportionate change in the marginal rate of substitution (MRS_{xy}) between the two goods along a given indifference curve.

$$\sigma = \frac{\text{Proportionate change in the ratio of the two goods consumed}}{\text{Proportionate change in } MRS_{xy}}$$

$$\sigma = \frac{\Delta\left(\frac{q_x}{q_y}\right) / \frac{q_x}{q_y}}{\Delta\, MRS_{xy} \,/\, MRS_{xy}}$$

where σ stands for elasticity of substitution q_x are q_y are the quantities of the two goods X and Y consumed.

As a consumer moves along an indifference curve downward to the right, he substitutes the

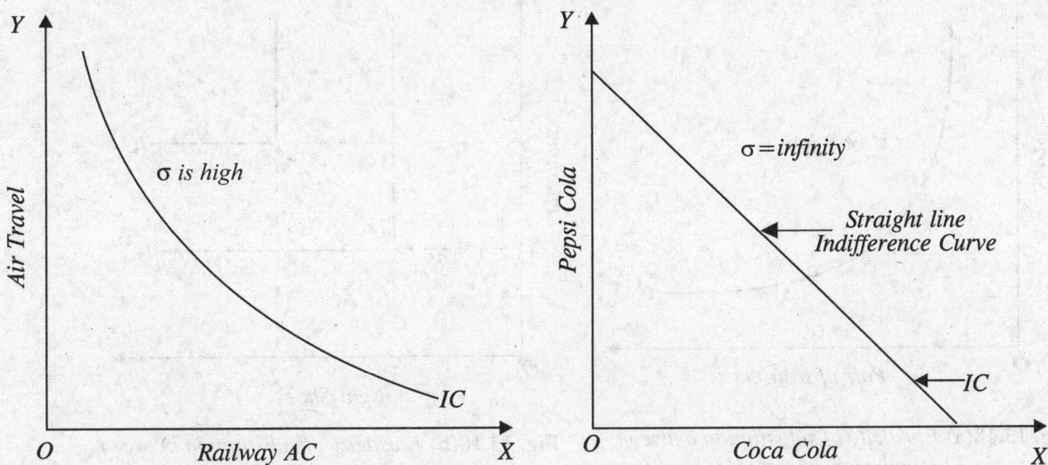

Fig. 13.17(a) *Elasticity of substitution (σ) is high.*

Fig. 13.17(b) *Elasticity of substitution (σ = ∞).*

commodity X for commodity Y in consumption and as a result the ratio of consumption of the two commodities changes, his level of statisfaction remaining the same. Besides, with the movement along an indifference curve marginal rate of substitution (MRS_{xy}) between the two commodities changes. It may be noted that change in the marginal rate of substitution between the two goods is measured by the change in absolute value of the slope of the indifference curve at relevant points on it. Further, as a consumer moves downward along an indifference curve he consumes more and more of X and less and less of Y, marginal rate of substitution of good X for good Y diminishes. Changes in the marginal rate of substitution indicate the extent to which one good can be substituted for another, his level of satisfaction remaining the same. When the substitution between the two goods is easy, then as a consumer moves along an indifference curve a proportionate change in the ratio of the two goods consumed (q_x/q_y) is large compared with the relative change in the marginal rate of substitution (MRS_{xy}) between the two goods. In such cases, elasticity substitution between the two goods is large. Therefore, the goods which are close substitutes of each other as, for example, railway AC coach and air travel, tea and coffee, the elasticity of substitution between them is high and the indifference curve is close to the straight line as shown in Fig. 13.17(a).

In the limiting case when the two goods are perfect substitutes, as perhaps Coca Cola and Pepsi Cola, the indifference curve between the two goods is a straight line as shown in Fig. 13.17(b) where there are infinitely large possibilities of substitution between the two goods as compared with the relative change in marginal rate of substitution which is zero (Note that the slope of a straight line is constant). Thus, in case of perfect substitutes.

$$\sigma = \frac{\Delta\left(\frac{q_x}{q_y}\right) / \frac{q_x}{q_y}}{\Delta MRS_{xy} \ / \ MRS_{xy}} = \frac{\Delta\left(\frac{q_x}{q_s}\right) / \frac{q_x}{q_y}}{0} = \infty$$

It may be noted that in real life it is very difficult to find the examples of perfect substitutes and if such goods do exist then from economic point of view they should be considered as the same goods.

On the other hand, when the two goods are bad substitutes and therefore it is difficult to substitute

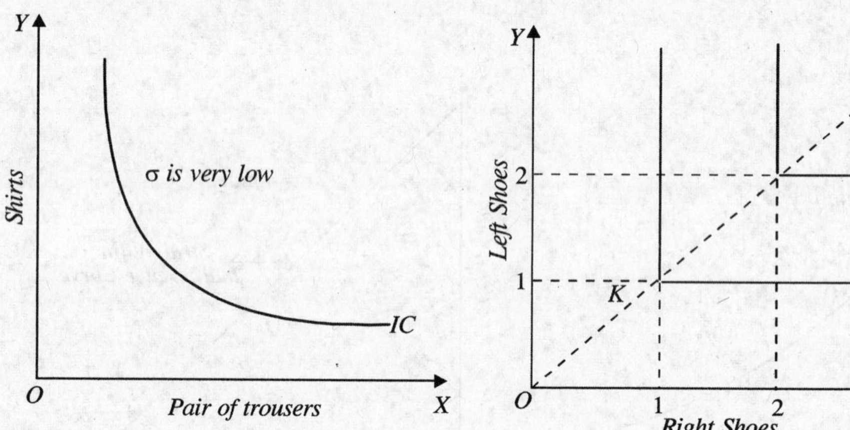

Fig. 13.18(a) *Elasticity of substitution between complementary goods is very low*

Fig. 13.18(b) *Elasticity of substitution between perfect complements is zero ($\sigma = 0$)*

one good for another, as for instance, between shirts and trousers, the indifference curves between them are highly convex, and in their case the ratio of the two goods consumed by an individual changes very little. For example, one would not like to have more or less than two or three shirts to each pair of trousers. The convexity of indifference curve between complementary goods is very high and elasticity of substitution between them is very small. In the extreme case when the goods are perfect complements to each other they cannot be substituted for each other and must be used in a fixed proportion as, for example, right shoes and left shoes. The elasticity of substitution between them is zero as depicted in Fig. 13.18(b). Therefore, indifference curves of perfect complements are L-shaped (i.e. right angled) and elasticity of substitution between them is zero. In case of perfect complements since $\Delta\left(\frac{q_x}{q_y}\right)$ is zero, we have

$$\sigma = \frac{0}{\text{Proportionate change in } MRS_{xy}} = 0$$

Since the cases of perfect substitutes and perfect complements are remote possibilities, the absolute value of elasticity of substitution between the two goods in the real world is greater than zero but less than infinity ($0 < \sigma < \infty$).

RELATIONSHIP BETWEEN PRICE ELASTICITY, INCOME ELASTICITY AND SUBSTITUTION ELASTICITY

As already explained, price effect, that is, the effect on the quantity demanded of a good due to a change in price, depends upon income effect on the one hand and substitution effect on the other. Similarly, price elasticity of demand which is the relative measure of the price effect depends upon the income elasticity on the one hand and substitution elasticity on the other. Thus, *price elasticity, in a way, is a compromise between income elasticity and substitution elasticity of demand.* The relationship between these three elasticities can be expressed in the form of *Slutsky equation*. Thus, the price elasticity of demand for good X is given by the following Slutsky equation :

$$e_p = KX.e_i + (1 - KX)e_s$$

where

e_p stands for price elasticity of demand.

e_i stands for income elasticity of demand.

e_s stands for substitution elasticity of demand.

K_t stands for the proportion of consumer's income spent on good X.

The first part of the equation, that is, $KX.e_i$ shows the influence of income effect on the price elasticity of demand. In other words, it shows that the change in quantity demanded of good *X* due to the fall in its price depends in part upon the magnitude of income effect of the change in price. This income effect of the change in price depends, on the one hand, on the proportion of income spent on good *X* (that is, *KX*) and income elasticity of demand for good *X* (that is, e_i) on the other. The proportion of income spent on good *X* determines what amount of the income spent on good *X* will be released as a result of fall in price of good *X*. The income thus released will be spent on increasing the purchases of good *X* as well as of other goods. Given the income elasticity of demand, the greater the proportion of income spent on good *X*, the greater the amount of income released for purchasing good *X* and other goods as a result of a fall in price of *X* and consequently greater the increase in amount demanded of good *X*. Besides the proportion of income spent on good *X*, the increase in amount demanded of good *X* due to income effect of the price fall depends on the income elasticity of demand for good *X*. This is because income elasticity of demand for good *X* determines how much of the income released by the fall in price of good *X* will be spent on good *X* whose price has fallen and how much will be spent on other goods. Proportion of income spent on good *X* being given, the greater the income elasticity of demand for good *X*, greater the part of income released will be spent on good *X* and consequently there will be a greater increase in quantity demanded of good *X*. We thus see that first part of the equation, that is *KX. e.i.*, shows the impact of income effect on the price elasticity of demand.

The quantity demanded of the good does not increase only because of the income effect of the fall in price, it also increases because of substitution effect. As the price of good *X* falls, more of it is purchased at the expense of others because *X* has now become relatively cheaper than others. The second part of the equation, that is, $(1 - KX)\ e_s$ shows the influence of the substitution effect on the price elasticity of demand for good *X*. The magnitude of substitution effect in turn depends in part on the elasticity of substitution (e_s), that is, the extent to which good *X* can be substituted for other goods now that it is relatively cheaper. The substitution effect also depends on the amount of other goods which were being purchased before the fall in price of good *X*. This is because only to the extent one is already purchasing other goods, the substitution of *X* for other goods is possible. Elasticity of substitution being given, the greater the amount of other goods being purchased by the consumer, the greater is the possibility of substitution of good *X* for other goods when *X* becomes cheaper. Since KX stands for the proportion of income spent on good *X*, (1–*KX*) will give the proportion of income not being spent on good *X*. In other words, (1–*KX*) stands for the proportion of income spent on goods other than *X*. Therefore, (1–*KX*) indicates the proportion of consumer's income within which substitution of *X* for other goods is possible.

From the above analysis it follows that the price elasticity of demand for a good is determined by the following four factors :

1. *Proportion of income spend on a good*
2. *Income elasticity of demand*
3. *Elasticity of substitution*
4. *Proportion of income spent of other goods*

Price elasticity can be known if the first three factors are known. Let us consider the following examples.

Suppose that a consumer is spending 1/5th of his income on any good *X* and the income elasticity of demand for the good *X* is 2 and elasticity of substitution between good *X* and all other good is 3. What will be the price elasticity of demand in this case ?

$$e_p = KX.e_i + (1 - KX)e_s$$
$$= 1/5 \times 2 + (1-1/5) \times 3$$
$$= 2/5 + 4/5 \times 3$$
$$= 2/5 + 12/5$$
$$= 14/5 = 2.8$$

Thus, price elasticity of demand for good *X* is equal to 2.8.

From the above formula of price elasticity of demand, it follows that whatever the proportion of income spent on a good, if income elasticity and substitution elasticity are equal to one, then price elasticity will also be equal to one. For instance, if a proportion of income spent on a good is 1/5 and income and substitution elasticities are equal to one, then price elasticity will be:

$$e_p = KX.e_i + (1-KX)e_x$$
$$= 1/5 \times 1 + 4/5 \times 1$$
$$= 1/5 + 4/5 = 1$$

Likewise, if proportion of income spent on a good is 1/3, and given that both the income and substitution elasticities are equal to one, price elasticity will be found to be equal to one.

PRICE CONSUMPTION CURVE AND PRICE ELASTICITY OF DEMAND

It is also possible to know with indifference curve analysis whether price elasticity is more than one, equal to one or less than one. It is from the slope of *price consumption curve* that we are able to judge the price elasticity of demand. Let us take Fig. 13.19 where on the *Y*-axis money income is measured and on *X*-axis the quantity of a commodity *X*. It is assumed that the consumer has *OA* amount of money to spend. Each of the indifference curves drawn between the two axes will show the various combinations of money and good *X* among which consumer is indifferent. To begin with, *AB* is the price line. The slope of the price line *AB*, *i.e.*, $\frac{OA}{OB}$ will give the price of good *X*. At this price (*i.e.*, with price line *AB*) the consumer is in equilibrium at point Q_1 on indifference curve IC_1 and is buying OX_1 of good *X*. Thus, in this equilibrium position, he is having combination of OX_1 of good *X* and OY_1 of money. It means that he has spent AY_1 of money on the good *X* and has obtained OX_1 of its quantity. Let price of good *X* falls, money income of the consumer remaining the same, so that we get a new price line *AC*. The new price of good *X* will be given by the slope of the new price line *AC*, *i.e.*, $\frac{OA}{OC}$. With this lower price or with price line *AC*, the consumer is in equilibrium at Q_2 on indifference curve IC_2. At this new equilibrium position Q_2 the consumer is getting OX_2 of good *X* and amount OY_2 of money is let with him. It means that at the lower price of good *X* he has spent AY_2 amount of money on it which is greater than the amount AY_1 of money which he spent at the original price. Thus, with the fall in price, his expenditure on the good *X* has increased. Similarly, when the price of good *X* falls further so that *AD* is now the relevant price line, consumer is in equilibrium at Q_3 where he is spending AY_3 amount of money and is having OX_3 quantity of the good *X*. Money expenditure AY_3 is greater than AY_2. It is thus clear that in the present case when the price *consumption curve is sloping downward* (*i.e*, *PCC* has a negative slope), with the reduction in price of the good *X* the consumer's money outlay on the good *X* increases. In Fig. 13.19 indifference map depicting preferences of the consumer

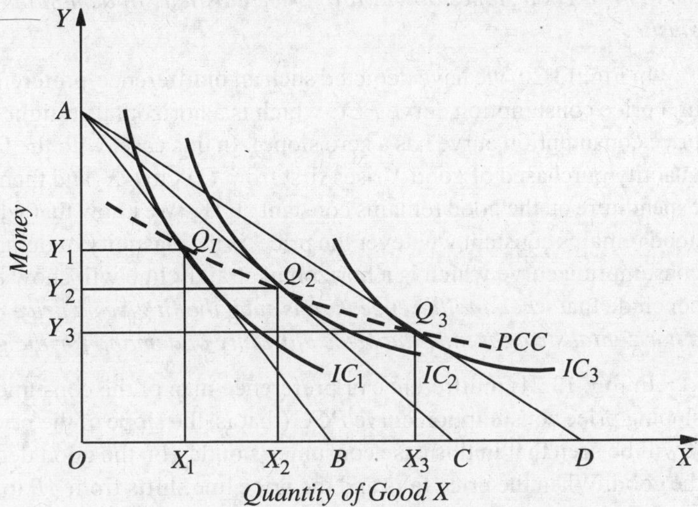

Fig. 13.19. *Elastic Demand*

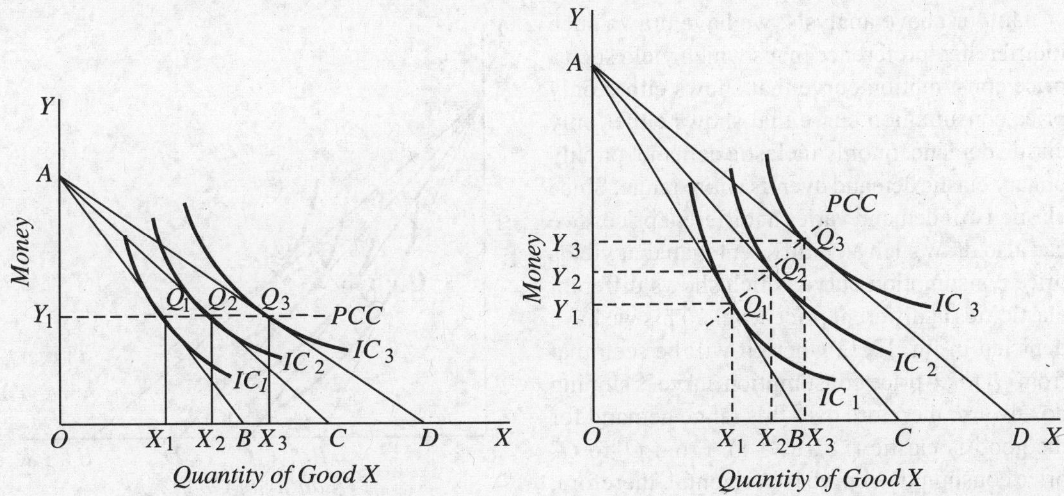

Fig. 13.20. *Unitary Elastic Demand*

Fig. 13.21. *Inelastic Demand*

is such that we get a downward sloping consumption curve which means, as explained above, that with the fall in the price of good *X*, consumer's expenditure on it rises. We know that when consumer's money expenditure on a good rises with the fall in price of the good, the demand for the good is elastic, i.e., elasticity of demand is more than one. We thus conclude that *when price consumption curve for a good slopes downward, price elasticity of demand is more than one, that is, demand is elastic*.

In Fig. 13.20, we have depicted such an indifference-preference map of the consumer that gives us a price consumption curve *PCC* which is a horizontal straight line, parallel to the *X*-axis (that is, price consumption curve has a zero slope). In this case, with the fall in the price of the good, though quantity purchased of good *X* rises first from OX_1 to OX_2 and then from OX_2 to OX_3, but consumer's expenditure on the good remains constant at AY_1. We know that when consumer's expenditure on the good remains constant whatever the price, price elasticity of demand is equal to one. Thus, the price consumption curve which is a horizontal straight line will show unit elasticity of demand. We thus conclude that *when indifference map is such that it gives a price consumption curve of the shape of a horizontal straight line, the price elasticity of demand for the good X is equal to unity*.

In Fig. 13.21, indifference or preference map of the consumer is such that it yields an upward-sloping price consumption curve *PCC* (that is, the slope of the price consumption curve is positive). It will be seen that in this case consumer's outlay on the good decreases with the fall in the price of the good. When the price falls and the price line shifts from *AB* to *AC*, the quantity demanded of the good rises from OX_1 to OX_2 but consumer's expenditure on the good *X* falls from AY_1 to AY_3.

Likewise, when price falls further and as a result price line shifts from *AC* to *AD*, though quantity demanded of the good *X* rises from OX_2, to OX_3, consumer's expenditure on the good *X* falls from AY_2 to AY_3. Thus upward sloping price consumption curve means a decline in consumer's expenditure as the price of the good *X* falls. Since a fall in consumer's expenditure as a result of the fall in price means that demand for the good is inelastic, *upward-sloping price consumption curve will therefore show inelastic demand, i.e., elasticity will be less than one.*

To sum up, downward-sloping price consumption curve for a good means that demand for the good is elastic, upward-sloping price consumption curve means that demand for the good is inelastic and horizontal straight-line price consumption curve means that demand for the good is unit elastic.

In our above analysis, we have drawn such indifference-preference maps which yield such a price consumption curve that shows either only price consumption curve that shows either only elastic demand, or only inelastic demand, or only unitary elastic demand over its entire range. Since elasticity of demand varies at different prices, we can also draw such an indifferent map that yields price consumption curve which shows different elasticities at different price levels. This we have depicted in fig. 13.22 where it will be seen that from Q_1 to Q_2 price consumption curve is sloping downward, therefore over this range demand for the good is elastic (*i.e.*, $e_p > 1$). From Q_2 to Q_3 price consumption curve is horizontal, therefore, the elasticity is here equal to unit (*i.e.* $e_p = 1$). From Q_3 to Q_4 and onward price consumption curve slopes upward, so over this range demand for the good is inelastic (*i.e.* $e_p < 1$).

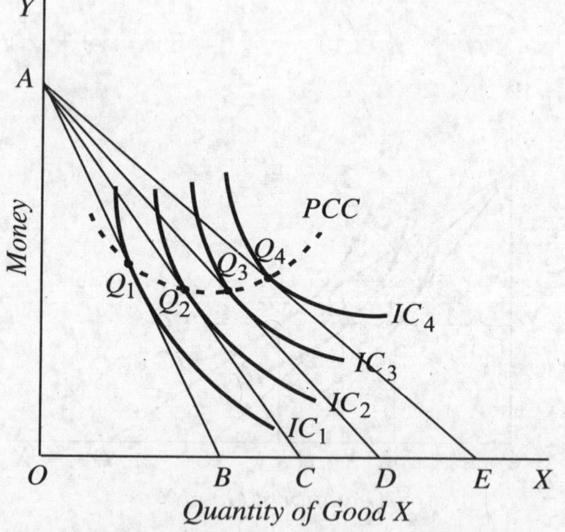

Fig. 13.22. *Varying Elasticity of Demand*

QUESTIONS FOR REVIEW

1. Explain the following concepts of elasticity of demand:
   (*a*) Price elasticity of demand
   (*b*) Income elasticity of demand
   (*c*) Cross elasticity of demand
   (*d*) Point price elasticity of demand
   (*e*) Arc price elasticity of demand
2. (*a*) How would you measure point price elasticity of demand at a point on the demand curve?
   (*b*) What happens to point price elasticity at a given price when the demand curve shifts to the right in a parallel manner?
3. What are the factors which determine price elasticity of demand? What role does price elasticity of demand play in decision-making by business firms?
4. What would you say about the price elasticity of demand for a good when
   (*a*) price consumption curve slopes downward
   (*b*) price consumption curve slopes upward
   (*c*) price consumption curve slopes backward
5. Suppose that in a year the excise duty on cigarettes is doubled and as a result the total revenue from the excise duty decreases. What conclusions about price elasticity of demand for cigarettes would you draw?
6. Do you think that price elasticity of demand would be greater for car industry as a whole or for Maruti 800 of the Maruti firm.
7. Demand for a firm's product has been estimated to be

   $$Q_d = 1000 - 200P$$

   If price of the product is ₹ 3 per unit, find out the price elasticity of demand at this price.

   [Hints : Price elasticity of demand is

   $$e_p = \frac{\Delta Q}{\Delta P} \cdot \frac{P}{Q}$$

   In the given demand function 200 is the coefficient of price which measures $\frac{\Delta Q}{\Delta P}$.

   In order to find out price elasticity of demand at price ₹ 3, we have first to find out the quantity demanded at this price by using the given demand equation. Thus,

   $$Q = 1000 - 200 \times 3 = 400$$

   Thus, at $P =$ ₹ 3, the quantity demanded at this price is 400 units. Substituting the values of $\frac{\Delta Q}{\Delta P}$, P and Q in the price elasticity formula, we have

   $$e_p = \frac{\Delta Q}{\Delta P} \times \frac{P}{Q} = 200 \times \frac{3}{400} = 1.5$$

8. Given : $P.Q = 500$, what happens to price elasticity of demand as price falls ($P =$ price, and Q = quantity purchased) ? *D.U. B.A. (H) Economics 2001*

   [**Hints :** $P.Q$ represents expenditure on a commodity which remains constant at 500 even when price changes. When expenditure on a commodity remains constant as price changes, its price elasticity of demand is equal to one]
9. An individual spends all his income on two goods, *X* and *Y*. if with the rise in price of good *X*, the quantity purchased of good *Y* remains unchanged, what is the price elasticity of demand for *X*? *D.U. B.A. (Hons.) 1996, 2001*

[**Hints :** Quantity purchased of good *Y* remaining the same even when price of good *X* rises, means that expenditure on good *X* ($P_x \cdot Q_x$) remains constant. This implies that price elasticity of demand for good *X* equals one.]

10. *The price elasticity of demand for colour TVs is estimated to be –2.5. If the price of colour TVs is reduced by 20 per cent, how much percentage increase in the quantity of colour TVs sold do you expect?*

[**Hints :** Price elasticity of demand being equal to –2.5 means that one per cent change in price causes 2.5 per cent change in quantity demanded or sold. Besides, minus sign only indicates there is *inverse* relationship between price and quantity sold. Thus, 20 per cent *reduction in price* of colour TVs will cause *increase in quantity sold* by 2.5 × 20 = 50 per cent.]

11. In an attempt to increase sales and profits, a firm is considering 5 per cent increase in price and 15 per cent increase in advertising expenditure. If the price elasticity of demand is –1.5 and advertising elasticity of demand is +0.7, would there be increase or decrease in total revenue?

[**Hints :** 5 per cent increase in price will cause 5 × 1.5 = 7.5 *per cent decrease* in quantity sold. However, 15 per cent increase in advertising will cause 0.7 × 15 = 10.5 per cent *increase* in quantity sold. Thus, there is net *increase in quantity sold* by 3 per cent. Since there is also 5 per cent *increase in price*, the total revenue (which is equal to *P.Q.*) will increase.]

12. Explain the relationship between the total revenue of a firm and the price elasticity of demand for price reduction.

(**Hints :** Note that expenditure on a product by the consumers is revenue for the firm. Therefore, the relationship between price elasticity and total expenditure on it applies to relation between total revenue and price elasticity of demand.

13. *(a) Explain why a firm facing a downward sloping demand curve would never produce at the inelastic ($e_p < 1$) portion of the demand curve.* *D.U.BA. (Hons) 2002*

14. Show that on a linear demand curve, price elasticity of demand decreases continuously from infinity at the price axis to zero at the quantity axis.

15. From the demand schedule given below calculate price elasticity of demand and total revenue.

Price (₹	*Quantity demanded*	*Price elasticity*	*Total revenue*
10	*20*	–	–
8	*30*	–	–
6	*35*	–	–
4	*40*	–	–

16. Explain the concept of income elasticity of demand. How would you define necessities and luxuries on the basis of income elasticity of demand?

17. Explain the importance of income elasticity of demand for business firms especially in designing marketing strategies.

18. Explain the concept of cross elasticity of demand. Using the concept of cross elasticity of demand, define substitutes and complements.

19. Explain the importance of the concept of cross elasticity of demand in (*a*) formulating proper price strategy by a firm (*b*) in analysing the degree of competition prevailing in an industry.

20. *Colgate sells its standard size toothpaste for ₹ 25. Its sales have been on an average 8000 units per month over the past year. Recently, its close competitor Binaca reduced the price of its same standard size toothpaste from ₹ 35 to ₹ 30. As a result, colgate sales declined by 1500 units per month.*

(i) Calculate the cross elasticity between the two products.

(ii) What does your estimate indicate about the relationship between the two?

CHAPTER 14
CONSUMER SURPLUS

Meaning of Consumer Surplus

The concept of consumer surplus was first formulated by Dupuit in 1844 to measure social benefits of public goods such as canals, bridges, national highways. Marshall further refined and popularised this in his '*Principles of Economics*' published in 1890. The concept of consumer surplus became the basis of old welfare economics. Marshall's concept of consumer's surplus was based on the cardinal measurability and interpersonal comparisons of utility. According to him, every increase in consumer's surplus is an indicator of the increase in social welfare. As we shall see below, *consumer's surplus is simply the difference between the price that 'one is willing to pay' and 'the price one actually pays' for a particular product.*

Concept of consumer's surplus is a very important concept in economic theory, especially in theory of demand and welfare economics. This concept is important not only in economic theory but also in formulation of economic policies such as taxation by the Government and price policy pursued by the monopolistic seller of a product. The essence of the concept of consumer's surplus is that a consumer derives extra satisfaction from the purchases he daily makes over the price he actually pays for them. In other words, people generally get more utility from the consumption of goods than the price they actually pay for them. It has been found that people are prepared to pay more price for the goods than they actually pay for them. This extra satisfaction which the consumers obtain from buying a good has been called consumer surplus. Thus, Marshall defines the consumer's sruplus in the following words: "*excess of the price which a consumer would be willing to pay rather than go without a thing over that which he actually does pay is the economic measure of this surplus satisfaction.... it may be called consumer's surplus.*"[1]

The amount of money which a person is willing to pay for a good indicates the amount of utility he derives from that good; the greater the amount of money he is willing to pay, the greater the utility he obtains from it. Therefore, the marginal utility of a unit of a good determines the price a consumer will be prepared to pay for that unit. The total utility which a person gets from a good is given by the sum of marginal utilities (ΣMU) of the units of a good purchased and the total price which he actually pays is equal to the price per unit of the good multiplied by the number of units of it purchased. Thus:

Consumer's surplus = What a consumer is willing to pay minus what he actually pays.

= Σ Marginal utility – (Price × Number of units of a commodity purchased)

The concept of consumer surplus is derived from the law of diminishing marginal utility. As we purchase more units of a good, its marginal utility goes on diminishing. It is because of the diminishing marginal utility that consumer's willingness to pay for additional units of a commodity declines as he has more units of the commodity. The consumer is in equilibrium when marginal utility from a commodity becomes equal to its given price. In other words, consumer purchases the number of

1. Alfred Marshall, *Principles of Economics*, 8th edition, p. 103.

units of a commodity at which marginal utility is equal to price. This means that at the margin what a consumer will be willing to pay (*i.e.*, marginal utility) is equal to the price he actually pays. *But for the previous units which he purchases, his willingness to pay (or the marginal utility he derives from the commodity) is greater than the price he actually pays for them. This is because the price of the commodity is given and constant for him and therefore price of all the units is the same.*

Marshall's Measure of Consumer Surplus

Consumer surplus measures extra utility or satisfaction which a consumer obtains from the consumption of a certain amount of a commodity over and above the utility of its market value. Thus the total utility obtained from consuming water is immense while its market value is negligible. It is due to the occurrence of diminishing marginal utility that a consumer gets total utility from the consumption of a commodity greater than its market value. Marshall tried to obtain the monetary measure of this surplus, that is, how many rupees this surplus of utility is worth to the consumer. It is the monetary value of this surplus that Marshall called consumer surplus. To determine this monetary measure of consumer surplus we are required to measure two things. First, the total utility in terms of money that a consumer expects to get from the consumption of a certain amount of a commodity. Second, the total market value of the amount of commodity consumed by him. It is quite easy to measure the total market value as it is equal to market price of a commodity multiplied by its quantity purchased (*i.e.*, *P.Q.*). An important contribution of Marshall has been the way he devised to determine the monetary measure of the total utility a consumer obtained from the commodity. Consider Table 14.1 which has been graphically shown in Fig. 14.1.

Table 14.1. Marginal Utility and Consumer Surplus

No. of Units	*Marginal Utility*	*Price*	*Net Marginal Benefit*
1	₹ 20	₹ 12	₹ 8
2	₹ 18	₹ 12	₹ 6
3	₹ 16	₹ 12	₹ 4
4	₹ 14	₹ 12	₹ 2
5	₹ 12	₹ 12	₹ 0
6	₹ 10	₹ 12	₹ (– 2)
	Total Consumer Surplus (from 5 units)		= 20

Suppose for the first unit of the commodity the consumer is prepared to pay ₹ 20. This means that the first unit of the commodity is at least worth ₹ 20 to him. In other words, he derives marginal utility equal to ₹ 20 from the first unit. For the second unit of the commodity, he is willing to pay ₹ 18, that is, the second unit is at least worth ₹ 18 to him. This is in accordance with the law of diminishing marginal utility. Similarly, the marginal utility of the third, fourth, fifth and sixth units of the commodity falls to ₹ 16, 14, 12 and 10 respectively. However, actually the consumer has not to pay the sum of money equal to the marginal utility or marginal valuation he places on them. For all the units of the commodity he has to pay the current market price of the commodity. Suppose the current market price of the commodity is ₹ 12. It will be seen from the Table 14.1 and Fig. 14.1 that the consumer will buy 5 units of the commodity at this price because his marginal utility of the fifth unit just equals the market price of ₹ 12. This shows that his marginal utility of the first four units is greater than the market price which he actually pays for them. He will therefore obtain surplus or net marginal benefit of ₹ 8 (₹ 20 – 12) from the first unit, ₹ 6 (= ₹ 18 – 12) from the second unit, ₹ 4 from the third unit and ₹ 2 from the fourth unit and zero from the fifth unit. He thus obtains total consumer surplus or *total net benefit* equal to ₹ 20.

Measurement of Consumer Surplus as an Area Under the Demand Curve

The analysis of consumer surplus made above is based on discrete units of the commodity. If we assume that the commodity is perfectly divisible, which is usually made in economic theory, the

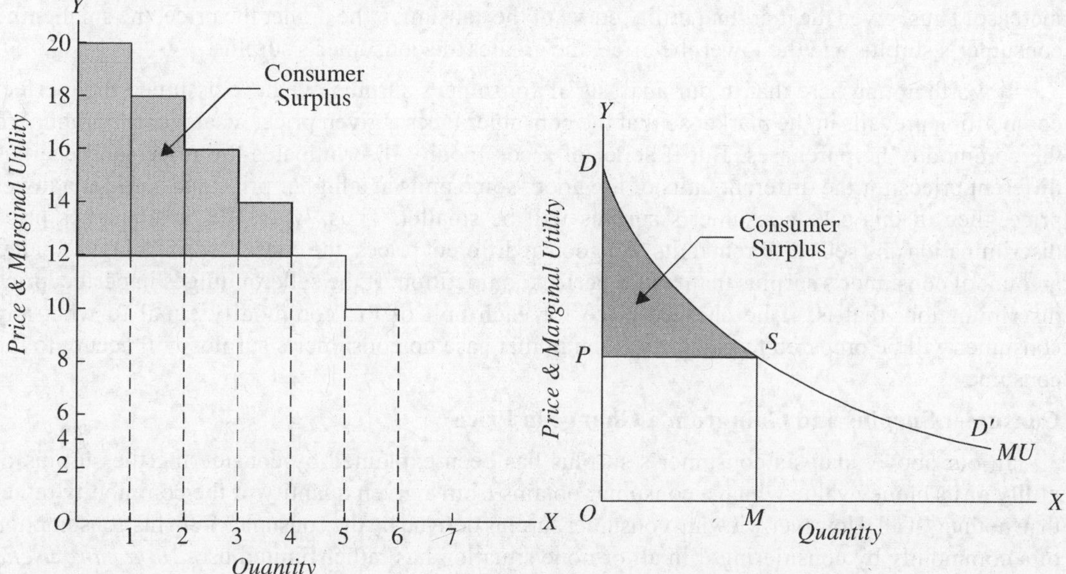

Fig. 14.1. *Consumer Surplus*

Fig. 14.2. *Marshall's Measure of Consumer's Surplus*

consumer surplus can be represented by an area under the demand curve. The measurement of consumer surplus from a commodity from the demand or marginal utility curve is illustrated in Fig. 14.2 in which along the *X*-axis the amount of the commodity has been measured and on the *Y*-axis the marginal utility (or willingness to pay for the commodity) and the price of the commodity are measured. *DD'* is the demand or marginal utility curve which is sloping downward, indicating that as the consumer buys more units of the commodity falls, marginal utility of the additional units of the commodity. As said above, marginal utility shows the price which a person is willing to pay for the different units rather than go without them. If *OP* is the price that prevails in the market, then the consumer will be in equilibrium when he buys *OM* units of the commodity, since at *OM* units, marginal utility from a unit of the commodity is equal to the given price *OP*. The *M*th unit of the commodity does not yield any consumer's surplus to the consumer since this is the last unit purchased and for this price paid is equal to the marginal utility which indicates the price that he is prepared to pay rather than go without it. But for the intra-marginal units *i.e.*, units before *M*th unit, marginal utility is greater than the price and, therefore, these units yield consumer's surplus to the consumer. The total utility of a certain quantity of a commodity to a consumer can be known by summing up the marginal utilities of the various units purchased.

In Fig. 14.2, the total utility derived by the consumer from *OM* units of the commodity will be equal to the area under the demand or marginal utility curve up to point *M*. That is, the total utility of *OM* units in Fig. 14.2 is equal to *ODSM*. In other words, for *OM* units of the good the consumer will be prepared to pay the sum equal to ₹ *ODSM*. But given the price equal to *OP*, the consumer will actually pay the sum equal to ₹ *OPSM* for *OM* units of the good. It is thus clear that the consumer derives extra utility equal to *ODSM* minus *OPSM* = *DPS*, which has been shaded in Fig. 14.2. To conclude when we draw a demand curve, the monetary measure of consumer surplus can be obtained by the area under the demand curve over and above the rectangular area representing the total market value (*i.e.*, *P.Q.* or the area *OPSM*) of the amount of the commodity purchased.

If market price of the commodity rises above *OP*, the consumer will buy fewer units of the commodity than *OM*. As a result, consumer's surplus obtained by him from his purchase will decline. On the other hand, if price falls below *OP*, the consumer will be in equilibrium when he is purchasing more units of the commodity than *OM*. As a result of this, the consumer's surplus will increase. Thus, given the marginal utility curve of the consumer, the higher the price, the smaller the consumer's surplus and the lower the price, the greater the consumer's surplus.

It worth noting here that in our analysis of consumer's surplus, we have assumed that perfect competition prevails in the market so that the consumer faces a given price, whatever the amount of the commodity he purchases. But if seller of a commodity discriminates the prices and charges different prices for the different units of the good, some units at a higher price and some at a lower price, then in this case consumer's surplus will be smaller. Thus, when the seller makes price discrimination and sells different units of a good at different prices, the consumer will obtain smaller amount of consumer's surplus than under perfect competition. If the seller indulges in perfect price discrimination, that is, if he charges price for each unit of the commodity equal to what any consumer will be prepared to pay for it, then in that case no consumer's surplus will accure to the consumer.

Consumer Surplus and Gain from a Change in Price

In our above analysis consumer's surplus has been explained by considering the surplus of utility or its money value which a consumer obtains from a given quantity of the commodity rather than nothing at all. However, viewing consumer surplus derived by the consumer from his consumption of a commodity by considering it in all or none situation has rather limited uses. *In a more useful way, consumer's surplus can be considered as net benefit or extra utility which a consumer obtains from the changes in price of a good or in the levels of its consumption.*

Consider Fig. 14.3 where *DD* shows the demand curve for food. At a market price *OP* of the food, the consumer buys *OQ* quantity of the food. The total market value which he pays for *OQ* food is equal to the area *OPEQ*, that is, price *OP* multiplied by quantity *OQ*. The total benefit, utility or use-value of *OQ* quantity of food is the area *ODEQ*. Thus, consumer's surplus obtained by the consumer would be equal to the area *PED*. Now, if price of food falls to *OP′*, the consumer will buy *OQ′* quantity of food and the consumer surplus will increase to *P′TD*. The net increase in the consumer's surplus as a result of fall in price is the shaded area *PETP′*, (*P′ TD* – *PED* = *PETP′*). This measures the net benefit or extra utility obtained by the consumer from the fall in price of food. This net benefit can be decomposed into two parts. First, the increase in consumer surplus arising on consuming previous *OQ* quantity of food due to fall in price.

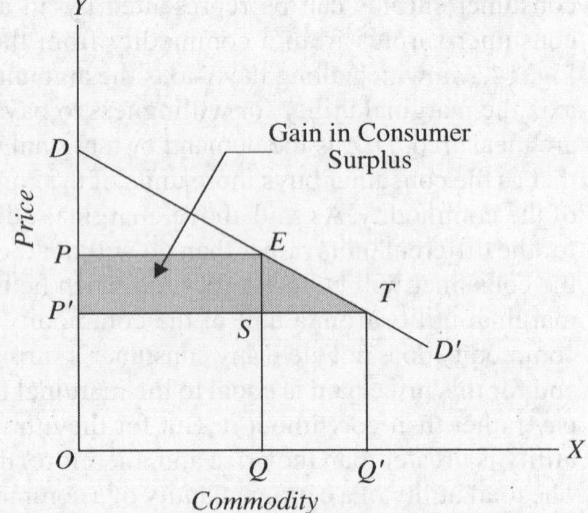

Fig. 14.3. *Gain in Consumer Surplus with a Fall in Price*

Second, the increase in consumer surplus equal to the small triangle *EST* arising due to the increase in consumption of the food following the lowering of its price (*PETP′* = *PESP′* + *EST*).

Measurement of Consumer's Surplus through Indifference Curve Analysis

We have explained above the Marshallian method of measuring consumer's surplus. Marshallian method has been criticised by the advocates of ordinal utility analysis. Two basic assumptions made by Marshall in his measurement of consumer's surplus are: (1) utility can be quantitatively or cardinally measured, and (2) when a person spends more money on a commodity, the marginal utility of money does not change or when the price of a commodity falls and as a result consumer becomes better off and his real income increases, the marginal utility of money remains constant. Economists like Hicks and Allen have expressed the view that utility is a subjective and psychic entity and, therefore, it cannot be cardinally measured. They further point out that marginal utility of money does not remain constant with the rise and fall in real income of the consumer following the changes in price of a commodity. The implication of Marshallian assumption of constant marginal utility of money is that he neglects the income effect of the price change. But in some cases income effect of the price change is very significant and cannot be ignored. Marshall defended his assumption of constancy of marginal utility of money on the ground that an individual spends a negligible part of his income on an individual commodity and, therefore, a change in its price does not make any significant change in the marginal utility of money. But this need not be so in case of all commodities.

Prof. J.R. Hicks rehabilitated the concept of consumer's surplus by measuring it with indifference curve technique of his ordinal utility analysis. Indifference curve technique does not make the assumption of cardinal measurability of utility, nor does it assume that marginal utility of money remains constant. However, without these invalid assumptions, Hicks was able to measure the consumer's surplus with his indifference curve technique. The concept of consumer's surplus was criticised mainly on the ground that it was difficult to measure it in cardinal utility terms. Therefore, Hicksian measurement of consumer's surplus in terms of ordinal utility went a long way in establishing the validity of the concept of consumer's surplus.

How consumer's surplus is measured with the aid of Hicksian indifference curve technique is illustrated in Fig. 14.4. In Fig. 14.4, we have measured the quantity of commodity *X* along the *X*-axis, and money along the Y-axis. It is worth noting that money represents *other goods* except the commodity *X*. We have also shown some indifference curves between the given commodity *X* and money for the consumer, the scale of his preference being given. We know that consumer's scale of preferences depends on his tastes and is quite independent of his income and market prices of the good. This will help us in understanding the concept of consumer's surplus with the aid of indifference curves.

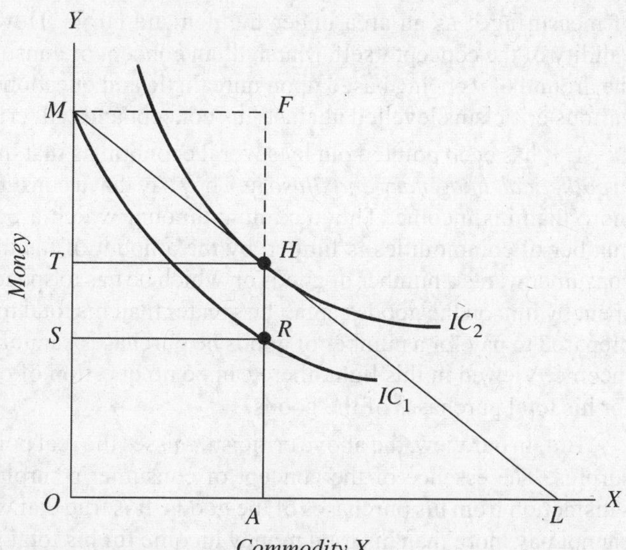

Fig. 14.4. *Measurement of Consumer Surplus with Indifference Curves*

Suppose, a consumer has *OM* amount of money which he can spend on the commodity *X* and the remaining amount on other goods. The indifference curve IC_1 touches the point *M* indicating thereby that all combinations of money and commodity *X* represented on the indifference curve IC_1 give the same satisfaction to the consumer as *OM* amount of money. For example, take combination *R* on an indifference curve IC_1. It follows that *OA* amount of commodity *X* and *OS* amount of money will give the same satisfaction to the consumer as *OM* amount of money because both *M* and *R* combinations lie on the same indifference curve IC_1. In other words, it means that the consumer is willing to pay *MS* amount of money for *OA* amount of the commodity *X*. It is thus clear that, given the scale of preferences of the consumer, he derives the same satisfaction from *OA* amount of the commodity *X* as from *MS* amount of money. In other words, he is prepared to give up *MS* (or *FR*) for *OA* amount of commodity *X*.

Now, suppose that the price of commodity *X* in the market is such that we get the budget line *ML* (price of *X* is equal to $\frac{OM}{OL}$). We know from out analysis of consumer's equilibrium that consumer is in equilibrium where the given budget line is tangent to an indifference curve. It will be seen from Fig. 14.4 that the budget line *ML* is tangent to the indifference curve IC_2 at point *H*, where the consumer is having *OA* amount of commodity *X* and *OT* amount of money. Thus, given the market price of the commodity *X*, the consumer has actually spent *MT* amount of money for acquiring *OA* amount of commodity *X*. But, as mentioned above, he was prepared to forego *MS* (or *FR*) amount of money for having *OA* amount of *X*. Therefore, the consumer actually pays *TS* or *HR* less amount of money than he is prepared to pay for *OA* amount of the commodity *X* rather than go without it. Thus, *TS* or *HR* is the amount of consumer's surplus which the consumer derives from purchasing *OA* amount of the commodity. In this way, Hicks explained consumer's surplus with his indifference curves technique without assuming cardinal measurability of utility and without assuming constancy of the marginal utility of money. Since Marshall made these dubious assumptions for measuring consumer surplus, his method of measurement is regarded as invalid and Hicksian method of measurement with the technique of indifference curves is regarded as superior to the Marshallian method.

Critical Evaluation of the Concept of Consumer's Surplus

The concept of consumer's surplus has been severely criticised ever since Marshall propounded and developed it in his *Principles of Economics*. Critics have described it as quite imaginary, unreal and useless. Most of the criticisms of the concept has been levelled against the Marshallian method of measuring it as an area under the demand curve. However, some critics have challenged the validity of the concept itself. Marshallian concept of consumer's surplus has also been criticised on the ground of its being based upon unrealistic and questionable assumptions. We explain below the various criticisms levelled against this concept and will critically appraise them.

1. It has been pointed out by several economists that the concept of *consumer's surplus is quite hypothetical, imaginary and illusory*. They say that a consumer cannot afford to pay for a commodity more than his income. The maximum amount which a person can pay for a commodity or for a number of commodities is limited by the amount of his money income. And, as is well-known, a consumer wants a number of goods on which he has to spend his money. Total sum of money actually spent by him on the goods cannot be greater than his total money income. Thus what a person can be prepared to pay for a number of goods he purchases cannot be greater than the amount of his money income. Viewed in this light, there can be no question of consumer getting any consumer's surplus for his total purchases of the goods.

But, in our view, the above criticism misses the real point involved in the concept of consumer's surplus. The essence of the concept of consumer's surplus is that consumer gets excess psychic satisfaction from his purchases of the goods. It is true that with his limited money income, consumer cannot pay more than his total money income for his total purchases than that he actually pays. But nothing prevents him from feeling and thinking that he derives more satisfaction from the goods than the price he actually pays for them and if he had the means he would have paid much more for the goods than he actually pays for them.

2. Another criticism against consumer's surplus is that it is *based upon the invalid assumption that different units of the goods give different amount of satisfaction to the consumer*. We have explained above how Marshall calculated consumer's surplus derived by the consumer from a good. Consumer purchases the amount of a good at which marginal utility is equal to its price. It is

assumed that marginal utility of a good diminishes as the consumer has more units of it. This means that while at the margin of the purchase, marginal utility of the good is equal to its price, for the previous intra-marginal units, marginal utility is higher than the price and, on these intra-marginal units, a consumer obtains consumer's surplus. Now, the critics point out that when a consumer takes more units of a commodity it is not only the utility of the marginal unit that declines but also all previous units of the commodity he has taken. Thus, as all units of a commodity are assumed alike, all would have the same utility. And when, at the margin, price is equal to the marginal utility of the last unit purchased, the price will also be equal to the utility of the previous units and consumer would, therefore, not get any consumer's surplus.

But this criticism is also not acceptable because even though all units of a commodity may be alike, they do not give same satisfaction to the consumer; as the consumer takes the first unit, he derives more satisfaction from it and when he takes up the second unit, it does not give him as much satisfaction as the first one, because while taking the second unit, a part of his want has already been satisfied. Similarly, when he takes the third unit, it will not give him as much satisfaction as the previous two units, because now a part of his want has been satisfied. Similarly, when he takes the third unit, it will not give him as much satisfaction as the previous two units. If we accept the above criticism, we then deny the law of diminishing marginal utility. But diminishing marginal utility from a good describes the fundamental human tendency and has also been confirmed by observation of actual consumer's behaviour. The concept of consumer's surplus is derived from the law of diminishing marginal utility. If law of diminishing marginal utility is valid, the validity of the Marshallian concept of consumer's surplus cannot be challenged.

3. The concept of consumer's surplus has also been criticised on the ground that it *ignores the interdependence between the goods*, that is, the relations of substitute and complementary goods. Thus, it is pointed out that if only tea were available and no other substitute drinks such as milk, coffee, etc., were there, then the consumer would have been prepared to pay much more price for tea than that in the presence of substitute drinks. Thus, the magnitude of consumer's surplus derived from a commodity depends upon the availability of substitutes. This is because if only tea were available, consumer will have no choice and would be afraid that if he does not get tea, he cannot satisfy his given want by consuming any other commodity. Therefore, he will be willing to pay more for a cup of tea rather than go without it. But if substitutes of tea are available, he would not be prepared to pay as much price since he will think that if he is deprived of tea, he will take other substitute drinks like milk and coffee. Thus, it is said that consumer's surplus is not a definite, precise and unambiguous concept; it depends upon the availability of substitutes. The degree of substitutability between different goods is different for different consumers, and this makes the concept of consumer's surplus a little vague and ambiguous. Marshall was aware of this difficulty and, to overcome this, he suggested that for the purpose of measuring consumer's surplus, substitute products like tea and coffee be clubbed together and considered as one single commodity.

4. *Prof. Nicholson described the concept of consumer's surplus as hypothetical and imaginary.* He writes, "of what avail is it to say that the utility of an income of (say) £100 a year is worth (say) £1000 a year". According to Prof. Nicholson and other critics, it is difficult to say how much price a consumer would be willing to pay for a good rather than go without it. This is because consumer does not face this question in the market when he buys goods; he has to pay and accept the price that prevails in the market. It is very difficult for him to say how much he would be prepared to pay rather than go without it. However, in our view, this criticism only indicates that it is difficult to measure consumer's surplus precisely. That a consumer gets extra satisfaction from a good than the price he pays for it is undeniable.

Moreover, as J.R. Hicks has pointed out "*the best way of looking at consumer's surplus is to regard it as a means of expressing it in terms of money income gain which accrues to the consumer as a result of a fall in price.*"[2] When the price of a commodity falls, the money income of the consumer being given, the budget line will switch to the right and the consumer will be in equilibrium at a higher indifference curve and as a result his satisfaction will increase. Thus, consumer derives more satisfaction at a lower price than that at the higher original price of the good. This implies that fall in the price of a commodity, and, therefore, the availability of the commodity at a cheaper price adds to the satisfaction of the consumer and this is in fact the change in consumer's surplus brought about by change in the price of the good. Prof. J.R. Hicks has further extended the concept of consumer's surplus considering it from the viewpoint of gain which a consumer gets from a fall in price of a good.[3] Moreover, the concept of consumer's surplus is useful and meaningful and not unreal because it indicates that he gets certain extra satisfaction and advantages from the use of amenities available in modern towns and cities.

5. The concept of consumer's surplus has also been criticised on the ground that *it is based upon questionable assumptions of cardinal measurability of utility and constancy of the marginal utility of money*. Critics point out that utility is a psychic entity and cannot be measured in quantitative cardinal terms. In view of this, they point out that consumer's surplus cannot be measured by the area under the demand curve, as Marshall did it. This is because Marshallian demand curve is based on the marginal utility curve in drawing which it is assumed that utility is cardinally measurable.

Further, as has been explained in earlier chapters, by assuming constant marginal utility of money, Marshall ignored income effect of the price change. Of course, income effect of the price change in case of most of the commodities is negligible and can be validly ignored. But in case of some important commodities such as foodgrains, income effect of the price change is quite significant and cannot be validly ignored. Therefore, the Marshallian method of measurement of consumer surplus as area under the demand curve, ignoring the income effect, is not perfectly correct. However, this does not invalidate the concept of consumer's surplus. As has been explained above, J.R. Hicks has been able to provide *a money measure of consumer's surplus* with his indifference curve technique of ordinal utility analysis which does not assume cardinal measurement of utility and constant marginal utility of money. Hicks has not only rehabilitated the concept of consumer's surplus but has also extended and developed it further.

Despite some of the shortcomings of the concept of consumer surplus, some of which are based on wrong interpretation of the concept of consumer surplus, it is of great significance not only in economic theory but also in the formulation of economic policies by the Government. The concept of consumer's surplus has a great practical importance in the formulation of economic policies by the Government. We explain below some important uses and applications of consumer surplus.

USES AND APPLICATIONS OF CONSUMER SURPLUS

The concept of consumer surplus has several applications both in economic theory and economic policy. This concept has been used to resolve water-diamond paradox of value theory, to explain the effects of taxes and subsidies on people's welfare, to make cost-benefit analysis of public projects, to show gains from trade etc. We will explain below some of the applications of the concept of consumer surplus.

Explaining Water-Diamond Paradox

One of the most famous puzzles in economic theory is why diamonds are more expensive than water. Water is essential for life; it is so useful that without its consumption one cannot live or survive. On the other hand, diamonds, though attractive and beautiful, satisfy less important human needs than water. Then, how it can be that in the market a less useful commodity like diamond is so

2. J.R. Hicks, *Value and Capital*, Oxford University Press, 2nd edition 1946, p. 40.
3. In the chapter on Indifference curve analysis we have explained the concept of Compensating Variation in Income and Equivalent Variation in Income. They are in fact measures of consumer surplus obtained by the consumer from a fall in price of a good.

expensive and a highly useful commodity as water is very cheap. Some thinkers in the past therefore complained that something was wrong with the market system which determines high price of commodities such as diamond, gold etc. which are least useful and low price of a commodity such as water which is necessary and highly useful. Therefore, this came to be known as water-diamond paradox. However, for modern economists there is no paradox about it as they are able to explain the large price differential between water and diamond.

The notion of marginal utility or marginal benefit of a commodity and the concepts of consumer surplus based on it, can be used to resolve the water-diamond paradox. The marginal benefit or marginal valuation per litre of water for the consumer is very low as the actual supply of water per period is very large. On the other hand, the marginal utility or marginal benefit of diamonds is very high because the amount of diamond actually available is very small. If, in fact, only a few litres of water were available marginal valuation of water would have been much greater than that of diamonds. Note that *marginal valuation of a commodity reflects how much amount of money consumer is prepared to pay for a commodity*. This indicates marginal utility or use-value of the commodity for the consumer. It is worth noting that downward-sloping demand curve for a commodity can be interpreted as showing the marginal valuation or marginal utility in terms of money to the consumer of various units of a commodity. If the quantity actually available of a commodity in the market is very large, its marginal valuation or marginal utility will be very small, though its total use-value or total benefit may be very large. On the other hand, as the actually available quantity of a commodity such as diamonds, gold etc. is very small, its marginal valuation or marginal utility is very high, though its total value-in-use or total utility is small.

Market price of a commodity is determined not by its total use-value but by its marginal valuation or marginal utility which in turn depends on the actually available quantity. The total use-value or total utility which a consumer gets from a quantity of a commodity equals the amount actually paid and the consumer surplus he obtains from it. In case of water market price as determined by its marginal utility is very low but consumer surplus from it is very large. On the other hand, in case of diamond due to their greater scarcity, marginal utility and hence its price is very high but consumer surplus from it is very small. Thus, the concept of consumer's surplus shows that price should not be confused with total use-value of a commodity and this helps us to resolve the water-diamond paradox.

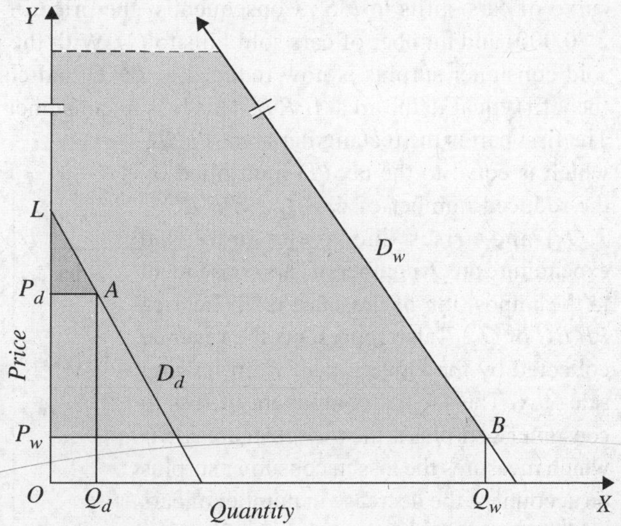

Fig. 14.5. *Water-Diamond Paradox*

This is illustrated in Fig. 14.5 where consumer's demand curve D_d depicts the marginal valuation curve for diamonds. On the *X*-axis quantity of diamonds in grams per time period and along the *Y*-axis marginal utility or valuation and price of diamonds are measured. Since the total use-value (*i.e.*, total utility) of diamonds is small, the demand curve is at a low level. Suppose the quantity of diamonds actually available is OQ_d and, as will be seen from the figure, price of diamond determined by demand and supply is P_d which is quite high, whereas the consumer surplus equal to LAP_d (shaded area) obtained by the consumers over and above what they actually pay is small.

Figure 14.5 also illustrates the price determination of water. Demand curve D_w representing marginal valuation or marginal utility of different quantities of water is at a higher level. If the quantity of water available is a very large quantity OQ_w, its marginal utility equals Q_wB and therefore price determined is OP_w. Though the market price of water is very low, consumer surplus obtained by the consumers will be the whole shaded area (not fully shown) above the price line P_wB which is very large compared with those of diamonds.

To sum up, the total utility or satisfaction derived from water consumed is much greater compared with diamonds but its marginal utility is low due to its abundant supply. The difference is large consumer surplus. On the other hand, whereas total utility (value in use) of diamonds is very small, due to its scarcity its marginal utility and therefore its price is relatively very high as compared with water. The difference is very small consumer surplus.

Evaluating Loss of Benefit from Tax

The notion of consumer surplus is applied for evaluating benefits and losses from certain economic policies. The losses and gains from taxes and subsidies to the consumers can be analysed using market demand curve and the concept of consumer's surplus. First, we explain the loss in consumer's surplus or welfare caused by the imposition of an indirect tax (say, sales tax) on cars. We assume that supply curve of cars is perfectly elastic, indicating constant-cost conditions under which the car industry is working. Under these conditions, imposition of a sales tax, say ₹ 10,000 per car sold would raise the price of car by exactly this amount, say from ₹ 2,60,000 to ₹ 2,70,000. The rise in price of car will result in fall in its quantity demanded and sold.

The loss in benefits incurred by the consumers as a result of the sales tax is illustrated in Fig. 14.6. The *DD* represents the demand curve for cars. This demand curve for cars can also be interpreted as marginal utility or marginal valuation curve of the cars for the consumers. Before the imposition of a sales tax *PS* is the supply curve of cars. The demand and supply for cars are in equilibrium at quantity Q_1 and price *OP*. Thus Q_1 quantity of cars is sold at price *OP*. In this situation consumers obtain *APD* amount of consumer surplus which measures net benefit to the consumers from the use of Q_1 number of cars. Now, with the imposition of sales tax of ₹ 10,000 (or *PP*′ = T) the supply curve of cars shifts to *P*′*S*′. Consequently, the price of car rises from *OP* (₹ 2,60,000) to *OP*′ (₹ 2,70,000) and number of cars sold falls to Q_2. With the rise in price and fall in the number of cars sold consumer surplus is now reduced to *BP*′*D* and consumers incur a loss in consumer surplus (benefit) equal to the area *APP*′*B*. This loss in consumer surplus can be decomposed into two parts. The first part is the rectangular area, *PP*′*BC* which is equal to the tax (*T*) multiplied by the reduced number of cars Q_2, ($PP'BC = T.Q_2$) and arises due to the increased expenditure on Q_2 number of cars consequent to the imposition of the sales tax. The area *PP*′*BC* or Q_2. *T* also represents the revenue collected by the Government from levying sales tax. The second component of loss in consumer's surplus is the triangular area *ABC* which measures the loss in consumer surplus on account of the decrease in number of cars sold, that is, ΔQ or Q_1Q_2 as a result of levying of sales tax. In this way, the total loss in consumer surplus is $PP'BA = PP'BC + ABC$. Where *PP*′*BC* is also equal to tax per car (*i.e.*, *T*) multiplied by the reduced numbers

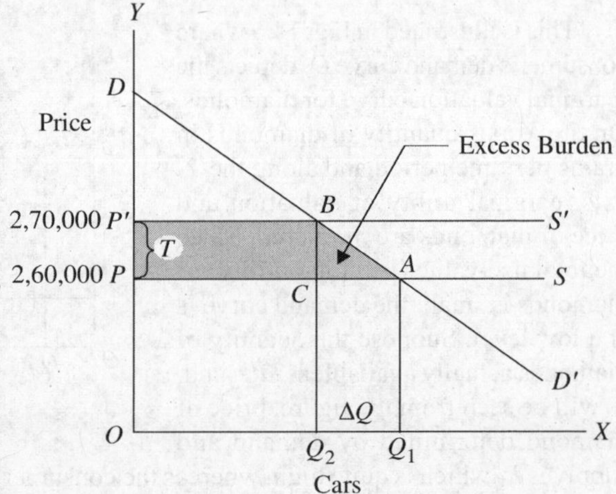

Fig. 14.6. *Evaluating Loss of Consumer Surplus from Sales Tax*

of cars sold (*i.e.*, Q_2), that is, $T.Q_2$ and the triangular area *ABC* which equals 1/2 $T.\Delta Q$. Thus, loss in consumer surplus ($PP'BA$) = $T.Q_2 + 1/2\ T.\Delta Q$.

It is worth noting that the loss in consumer surplus *PP′BA* is greater than the revenue collected by the Government which is equal to the area *PP′BC* and this extra loss in consumer surplus is equal to the triangular area *ABC*. This area *ABC* represents the *excess burden of sales taxation.* The net loss in welfare or consumer surplus in excess of tax revenue received by the Government *is also called dead weight loss*. If instead of sales tax, a lump-sum tax of an equal amount were levied, there would have been no excess burden. Thus, sales tax distorts the price of cars, reduces the number of cars sold and thereby results in loss in consumer welfare in excess of the amount of tax collected. Therefore, *economists dub indirect taxes such as sales tax or excise duty as economically inefficient.*

Thus, it follows that the burden of indirect tax (such as sales tax or excise duty) is greater than the direct tax such as lump-sum tax or income tax. Therefore, many economists hold that from the viewpoint of social welfare or optimum allocation of resources, the direct taxes such as a lump-sum tax or income tax is superior to an indirect tax such as sales tax or excise duty. Let us show and illustrate it with the help of Fig. 14.6 *which represents the case of constant cost industry*. With the given demand and supply curves *DD′* and *PS* respectively price *OP* and equilibrium quantity OQ_1 are determined. This is the position before the imposition of any tax and represents the maximum satisfaction or welfare of the people. Now, if the sales tax equal to *PP′* per unit of the commodity is imposed, the supply curve will shift upward to the position *P′S′*. As a result, the price will rise to *OP′* and the quantity demanded and sold will fall to Q_2. As explained above, tax collected by the Government in this case will be equal to *PP′BC*, but the loss in consumer's surplus suffered by the consumer will be equal to *PP′BA*. Now, if the Government takes away the sum equal to *PP′BC* through a direct tax, say lump sum tax or income tax, then the people would not have to suffer the loss equal to the area *ABC*. This is so because whereas *an indirect tax distorts the price of a commodity* (*i.e.* it raises the price), *a lump-sum or income does not affect the price*. Thus, it is clear that a direct tax causes less loss of welfare than a price-distorting indirect tax.

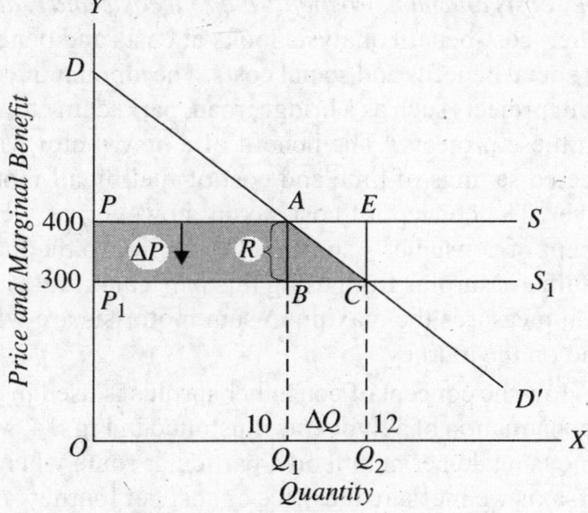

Fig. 14.7. *Gain from a Price Subsidy*

Evaluating Gain from a Subsidy

The concept of consumer surplus can be used to evaluate the gain from subsidies. The Government these days provide subsidies on many commodities such as foodgrains, fertilizers, power. Let us take the example of subsidy on foodgrains production being given by the Government. Suppose the subsidy reduces the price of foodgrains from ₹ 400 to ₹ 300 per quintal. As a result of the fall in price of foodgrains due to subsidy being provided for its production, the quantity demanded of foodgrains increases from 10 thousand quintals (Q_1) to 12 thousand quintals (Q_2). Now, the question to be answered is what will be net social benefit or gain from this subsidy. Consider Fig. 14.7 where *DD* is the demand curve for foodgrains which, as explained above, can also be interpreted as *marginal utility (or marginal valuation)* curve. To begin with, *PS* is the supply curve, assuming constant cost conditions. Price determined is *OP* or ₹ 400 per quintal. With the grant of subsidy equal to ₹ 100 per quintal, supply curve shifts below to P_1S_1 and as a result price falls to OP_1 or ₹ 300 per quintal. With the reduction in price to OP_1 (*i.e.*, ₹ 300 per quintal) quantity demanded increases to from OQ_1 to OQ_2. It will be seen from Fig. 14.7 that the total gain in consumer surplus is equal to the area $PACP_1$ which can be divided into two parts, namely, the area $PABP_1$ ($= R.Q_1$) where R is the subsidy per quintal of foodgrains plus triangle ABC, which equals 1/2 $R.\Delta Q$.

Thus, the gain in consumer surplus $= R.Q_1 + \frac{1}{2} R.\Delta Q$.

where $R.Q_1$ represents the reduction in expenditure on the quantity Q_1 that would have cost ₹ R (= ₹ 100) per quintal more without subsidy. Thus $R.Q_1$ represents the benefit or gain in consumer surplus to those who were purchasing foodgrains before the grant of subsidy but would now do so at a lower price. The amount 1/2 $R.\Delta Q$ represents the gain in consumer surplus due to the increase in quantity demanded at a lower price made possible by the grant of subsidy. Thus, the total gain in consumer surplus is the area $PACP_1$ which equals $R.Q_1 + 1/2\ R.\Delta Q$.

But the cost of subsidy to the Government is $R.Q_2$ or the area $P_1\,PEC$ which is greater than the gain in consumer surplus by the area of the triangle ACE. Thus, if the buyers would have been given the lump-sum grant equal to the area $PACP_1$ they would have been as well off as in case of subsidy which costs more to the Government. Thus, *subsidy causes excess burden equal to the area of triangle ACE as compared to the lump sum grant.*

Use of Consumer Surplus in Cost-benefit Analysis

An important application of consumer surplus is its use in cost-benefit analysis, especially of public investment projects. In fact, Dupuit, the originator of the idea of consumer surplus in his paper. "*On the Measurement of Public Works*" in 1844 used the concept of consumer surplus for describing the impact of public investment projects on social welfare. In recent years Prof. E.J. Mishan has based his cost-benefit analysis on consumer surplus approach. Consumer's surplus has been treated as benefits in various cost-benefits analysis of investment projects. The cost-benefit analysis has become very popular these days to judge the desirability of public investment in particular projects. It should be noted that *costs and benefits in cost-benefit analysis do not merely mean money costs and money benefits but real costs and real benefits in terms of satisfaction and resources*. Further, cost-benefit analysis looks at costs and benefits from social point of view; it is concerned with social benefits and social costs. The amount of consumer's surplus expected to be derived from certain projects such as a bridge, road, park, dam etc. are considered as an important benefit flowing from these projects.[4] The benefit of a new motor way or flyover is estimated by reference to the expected savings of time and cost of fuel by all motorists who will make use of the new road or flyover. The concept of cost-saving however, as we shall see below, is derived directly from the concept of consumer's surplus. Thus, prior to the introduction of the new flyover in question, the consumer's surplus from using this particular route is the triangle under the relevant demand curve which measures the maximum sum motorists are willing to pay above the amount they currently spend on the journey.[5]

How the concept of consumer surplus is used in cost-benefit analysis of public investment, say the construction of a flyover is illustrated in Fig. 14.8 where on the X-axis we measure the number of journeys made per month on a particular route where flyover is proposed to be undertaken and on the Y-axis we measure the price or cost per journey. DD is the demand curve for the journey on that route which, as explained above, shows the maximum price the motorists are willing to pay for making journeys on that route. If the current price or cost per journey, that is, prior to the construction of flyover is OP_1, the motorists make OQ_1 number of journeys on that route and pay $OP_1 . OQ_1$ as the total cost for the OQ_1 journeys made. It will be seen from the demand curve that the *total amount of money that the motorists will be willing to pay for OQ_1 journeys equals the area $ODAQ_1$* and thus the triangle P_1DA represents the consumer surplus they derive from making OQ_1 journeys. Now suppose the flyover is constructed which by reducing their fuel consumption reduces the cost per journey to OP_2. At the lower price or cost per journey, they will make OQ_2 number of journeys and their consumer surplus will increase by the shaded area P_1ACP_2. This is the benefit the motorists receive from the construction of flyover.

4. See E.J. Mishan, *Cost-Benefit Analysis: An Informal Introduction*, George Allen & Unwin Ltd.
5. E.J. Mishan, *op. cit.*, p. 33.

This increase in consumer surplus can be divided into two parts. First, we have the cost-saving component equal to the rectangular area P_1P_2TA which is calculated as the saving per journey multiplied by the original number of journeys OQ_1. The other part of the increase in consumer surplus is represented by the area of the triangle ATC which is the gain in consumer's surplus obtained from the additional journeys made by the same motorists or the new ones. It is worth noting that it is the cost-saving segment of the increment in consumer surplus that often enters into the cost-benefit calculations of the investment projects. But, as we have just seen, this cost-saving is the main component of the addition to the consumer surplus due to fall in the cost per journey brought about by the construction of the flyover.

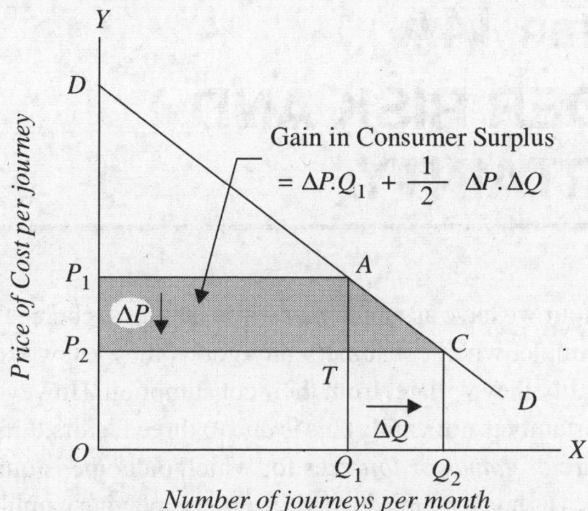

Fig. 14.8. *Use of Consumer Surplus in Cost-Benefit Analysis of the Construction of a Flyover*

We have seen above that the concept of consumer's surplus in the context of an individual and in the context of a particular good is a meaningful and useful concept. However, it is worth noting here that the use of the concept of consumer's surplus as a tool for formulation of policies such as choice of investment projects based on cost-benefit calculation, requires the summation of consumer's surpluses derived from a good or project by various consumers belonging to different income groups. Such summation and comparison of consumer's surplus of different individuals can be validly made if one rupee worth of consumer's surplus means the same thing to different individuals. However, this is based on the assumption that marginal utility of money is the same for all individuals regardless of the size of their income. This lands us into interpersonal comparison of the utility which is not regarded as scientific and justified by many economists.

QUESTIONS FOR REVIEW

1. Define consumer surplus show *h* one would you measure it on a consumer demand curve for a commodity.
2. What is value paradox ? Using demand curves for water and diamonds and their associated price explain how you will resolve the paradox using the concept of consumer's surplus. *D.U. B.A. (Hons) 1999.*
3. Explain the concept of consumer surplus. How did Marshall measure it ?
4. How is Consumer's surplus measured with the help of indifference curves ? Under what conditions consumer surplus as measured by the area under the demand curve is exactly equal to that measured by indifference curves. *I.A.S. (Main Examination 1996)*
5. Using the concept of consumer surplus show that indirect tax, that is, excise duty or sales tax, causes excess burden on the consumer as compared to a direct tax.
6. Using the concept of consumer surplus show that a consumer will prefer the subsidy in cash rather than a price subsidy on a commodity.
7. Explain the use of consumer surplus in cost-benefit analysis of projects.
8. With the help of consumer surplus show that social welfare will be increased by imposing a tax on the product of an increasing cost industry and revenue so collected be used for subsidising the product of a decreasing cost industry.
9. Consumer's surplus is hypothetical having no practical importance. Examine critically.

CHAPTER 14A
CHOICE UNDER RISK AND UNCERTAINTY

Introduction

So far in our analysis of consumer's behaviour we have assumed that while making a choice the consumers do not face any uncertainty. For example, when consumers buy goods they know with certainty what they are getting and how much utility they will get from their consumption. However, in the real world situations this assumption of certainty is not valid. This is due to three factors. First, some goods that individuals buy are in the nature of *games or lotterie*s for which outcome is quite uncertain. For example, in case of some games such as horse racing, buying insurance, playing gambles outcome is quite uncertain. You may view a gamble and get a certain amount of money which will increase your utility. On the other hand, you may lose the game which will reduce money with you and cause a loss of utility. Buying of such goods such as gambles does not guarantee an individual any particular outcome with certainty.

Secondly, individuals who make investment in assets face uncertainty as the outcome (*i.e.*, return) from the assets or from any combination (*i.e.*, portfolio) of assets cannot be known with certainty. This is because cost conditions and future prices can vary and cannot be known with certainty. Therefore, the individuals have to make a choice of assets or investment portfolio of assets in uncertain and therefore risky situation.

The third case in which the individuals face uncertainty is when there is lack of information about the quality of goods they buy for example, while buying cars, colour TVs, refrigerators, the individuals are not quite sure the carm refrigerator or colour TV they buy will offer best quality for money. The problem of uncertainty arising out of lack of information is more true when individuals buy used or second-hand cars or second-hand bikes.

Of the above three cases of uncertainty we will be concerned with individual choice in the first two cases of uncertainty of outcomes. Choice or decision making in case of uncertainty date to lack of information on the part of buyers will be analysed in the last part of the book.

In what follows we first explain the concept of risk and how to measure it. We will then explain the different preferences toward risk. Risk is considered to be undesirable. Many individuals are risk averse and like to avoid risks, but some others are willing to bear risk if it yields them sufficient return. We also explain why some people gamble. In the end we will explain the ways such as diversification, insurance, gathering of information by which risk can be reduced.

The Concept of Risk

The risk refers to a situation when the outcome of a decision is uncertain but when the probability of each possible outcome is known or can be estimated. The analysis of decision making and choice involving risk or uncertainty requires that the individual knows all the possible outcomes and also have some idea of the probability of occurrence of each possible outcome. For example, in tossing a coin there is equal or 50-50 chance of getting either a head or tail. Likewise, when an individual invests in the shares of a company, the probability of outcome (that is, how much dividend it will yield and how much its price will rise in a year) can be estimated from the past experience. The out- come can vary a good deal. The greater the variability of possible outcome, the greater the risk involved in making the investment decision. Therefore, in the theory of choice or decision making under uncertainty it is necessary to know the meaning of probability of outcome and its variability.

It is important to mention the distinction between risk and uncertainty. *The uncertainty refers to the situation when there is more than one possible outcome of a decision but where the probability of occurrence of each particular outcome is not known or even cannot be estimated.* This may be due to lack of sufficient past information or the great instability of the variables involved that determine the outcome. In some extreme cases even outcomes themselves are not known. For example, in case of drilling for exploration of oil in an *unproven field*, investor does not know either the probability of oil being struck or the possible output of oil from it. Though the distinction between risk and uncertainty is theoretically significant the two terms are generally used interchangeably in decision making analysis.

In decision making involving risk or uncertainty the three terms are quite often used. They are: (1) *strategy*, (2) *state of nature* and (3) *outcome*. A strategy refers to one of several alternative courses of action or plans that can be implemented to achieve the desired goal. For example, a manager may be considering the following three strategies aimed at increasing profits of the firm.

(*i*) Building a new technologically more efficient plant so as to reduce the cost of production.

(*ii*) Launching of a new advertising and marketing campaign to increase sales.

(*iii*) Redesigning of the product so as to lower the cost of production and to increase in demand for it by getting greater consumer acceptance.

A state of nature refers to the conditions that prevail in future and which have a significant effect on the success or failure of the strategy. For example, in case of building a technologically superior plant the different states of nature that may exist in some future years are (a) boom, (b) recession, or (c) normal conditions. It is worthwhile to note that the decision maker has little or no control over the state of nature. For example, whether economic conditions of boom or recession will exist in future is beyond the control of the decision maker.

Finally, ***outcome*** refers to the results which are usually in the form of profits that come about as a result of implementation of a strategy. A *payoff matrix* lists the outcomes associated with combinations of each strategy and state of nature. It is important to note that *risk refers to the amount of variability in the outcome as a result of the adoption of a particular strategy*.

Measuring Risk: Probability of an Outcome

We have seen above that risk refers to the situation when there is more than one possible outcome of a decision and the probability of each outcome is either known or can be estimated. Therefore, to measure the degree of risk we need to know the probability of each possible outcome of a decision. *The probability means the likelihood of occurring of an event*. Thus, if possibility of an outcome occurring is $\frac{1}{4}$ or 0.25, this means that there is 1 chance in 4 for the outcome to occur. For example, suppose a person is considering to invest in a company which is engaged in the new exploration of offshore oil. If the exploration is successful, the price of company's stock will rise to ₹ 50 per share and if its exploration meets with a failure, the company's stock will fall to ₹ 10 per share. Given these two possible outcomes, namely, ₹ 50 per share price and ₹ 10 per share price, if the past information reveals that the chance of oil exploration being successful is $\frac{1}{4}$ and chance of its failure is $\frac{3}{4}$, then we say the probability of success of oil exploration is $\frac{1}{4}$ and the probability of its failure is $\frac{3}{4}$ (The success and failure are the two possible outcomes). Thus, *probability is a number that indicates the likelihood of an event or outcome occurring.*

These are two concepts of probability depending on how it is measured. The first is the frequency concept of probability. If past information or data is available regarding occurrence of outcomes or events, the probability is defined as the proportion of times an outcome occurs if the situation is repeated in the long run over and over again. In general, if a situation is repeated over a large number of times, say M, and if outcome, say X, occurs m times, then

$$P(X)=\frac{m}{M}$$

Thus, in our example if we know from the past data of oil exploration that rate of success is 25 per cent, then the probability of getting success is $\frac{1}{4}$ or 0.25. The measurement of probability based on the past experience is generally known as the ***objective measure of probability.***

But in many cases there are no similar past situations which helps us in measuring probability. In that case the ***concept of subjective probability*** is used. The subjective probability is an individual's personal view about the chance of an outcome to occur and is based on his personal judgement, experience or knowledge about the subject and not on the frequency with which outcome actually took place in the past. Obviously, when probability is subjectively determined and not based on the past data, the different individuals will attach different probabilities to the occurrence of various outcomes and therefore they will make different choices. In whatever way the probability is arrived at, it helps us to measure two important concepts, namely, *expected value and variability of outcome.* With the use of these two concepts we compare the profitability of various strategies involving risk and uncertainty which helps us to make a choice among them.

Expected Value and Expected Utility. In the theory of making a choice under risk and uncertainty an individual will make a decision on the basis of **expected utility of outcome** rather than **expected money value.**[1] Let us explain these two concepts. Suppose a vegetable vendor who moves around in the streets of a town to sell his vegetables. How much vegetables he will be able to sell and earn income depends on weather, namely, whether it rains or not. If it does not rain he will be able to sell more and earn more income or profits, say equal to ₹ 700 and if it rains he will not be able to sell much and his earnings or profit will be ₹ 100. Weather forecast is that there is 50 per cent chance of raining. Now in this situation what is the expected money value, $E(V)$ or expected money earnings. Now, the expected money value or earnings is the sum of values of each possible outcome times the probabilities of those outcomes.

Thus,

$E(V) = P_r$ (Earnings if it rains) $+ P_r$ (Earnings if it does not rain)

where P_r stands for probability

1. Daniel Bernoulli provided a convincing explanation of behaviour of a rational individual in deciding on the basis of expected utility of outcome and is known as **Bernoulli hypothesis.**

or $E(V) = \frac{1}{2}(700) + \frac{1}{2}(100)$

$= 350 + 50 = 400$

Note that the expected value represents the amount of earnings on an average per day if the above sale by the vendor is repeatedly made for several days. Now, in order to find the expected utility of outcome of the above event of sale we have to know the utility function of the individual. With the information of utility function the individual's expected utility $E(U)$ is found as under:

$E(U) = P_r$ (no rain) $\times$ U (Earnings in case of no rain) + P_r (rain) $\times$ Earnings in case of rain)

Since probability (P_r) of both raining and not raining is $\frac{1}{2}$. $P_r = \frac{1}{2}$ in both cases. Suppose from the utility function of the individual we know that the utility of the earnings of ₹ 700 in case of no rain is 140 and expected utility of earnings of 100 (in case of no rain) is 70, then expected utility is given by

$$E(U) = \frac{1}{2} \times U(₹ 700) + \frac{1}{2} \times U(₹100)$$

$$= \frac{1}{2} \times 140 + \frac{1}{2} \times 70$$

$$= 70 + 35 = 105$$

Now, it is important to know that utility function of an individual depends on his attitude toward risk as to whether he is risk averter, risk lover or risk neutral. Now the choice in the risky and uncertain situation depends the individual's attitude towards risk.

It may be further noted that risk is mainly measured in two ways. First, by using the concept of variance or standard deviation. Second, by using utility function approach. In this chapter, we adopt utility function approach to measure risk.

ATTITUDES TOWARD RISK

We analyse the problem of investment choice to describe how people makes a choice among alternative investments involving different degree of risk involved. But the underlying principles of making a choice in risk and uncertain situation, namely, expected return and the degree of risk involved apply equally well to other choices. In this section, we focus on examining individual's choices in the face of risk. In the theory of consumer's behaviour we saw that in making choices among commodity bundles when there is no risk and uncertainty, the consumer maximises his utility. We will analyse below how an individual maximises his expected utility when risk or uncertainty is present.

People's preferences toward risk greatly differ. Most individuals generally prefer the less risky situation (that is, the situation with less variability in outcomes). In other words, most individuals seek to minimise risk and are called ***risk averter*** or ***risk averse.*** However, some individuals prefer risk and are therefore called *risk seekers* or *risk lovers.* Some other individuals are indifferent toward risk and are called ***risk neutral.*** But it is important to note that these different preferences toward risk depend on whether for an individual *marginal utility of money diminishes or increases or remains constant as his income or wealth increases.* As shall be explained below, *for a risk averse individual marginal utility of money diminishes* as he has more money, while *for a risk seeker marginal utility of money increases as money with him increases.* In case of *risk neutral individual marginal utility of money remains constant as he has more money.*

Risk Averter. To explain the preference toward risk we will consider a single composite commodity, namely, money income. An individual's money income represents the market basket of goods that he can buy. It is assumed that the individual knows the probabilities of making or gaining money income in different situations. But the outcomes or payoffs are measured in terms of utility[2] rather than rupees.

In Fig. 14A.1 we have drawn a curve *OU* showing utility function of money income of an individual who is risk-averse. It will be seen from this figure that the slope of total utility function *OU* decreases as the money income of the individual increases. Note that we measure money income on the *X*-axis and utility on the *Y*-axis. It will be seen from Fig. 14A.1 that as money income of the individual increases from 10 to 20 thousand rupees, his total utility increases from 45 to 65 units (that is by 20 units) and when his money increases from 20 thousand to 30 thousand rupees, his total utility increases from 65 to 75 units (that is, by 10 units). Thus in this **concave utility function** depicted in Fig. 14A.1 marginal utility of money of an individual decreases as his money income decreases and therefore it represents the case of risk-averse individual.

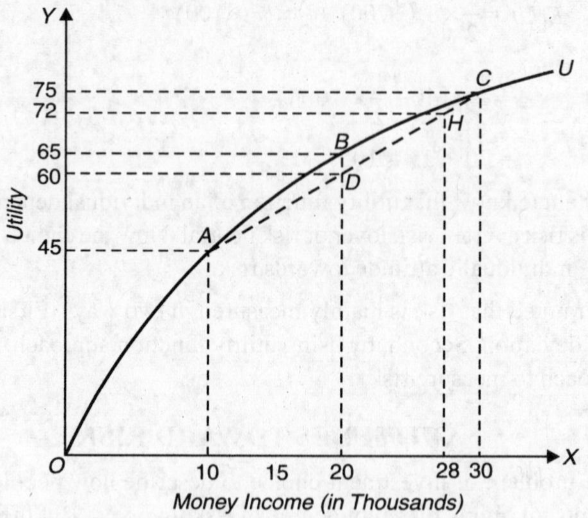

Fig. 14A.1. *Utility Function of a Risk Averter*

Suppose the individual is currently employed on a fixed monthly salary basis of ₹ 20,000. There is no uncertainty about the income from this present job on the fixed salary basis and hence no risk. Now, suppose that the individual is considering to join a new job of a salesman on a *commission basis*. This new job involves risk because his income in this case is not certain. This is because if he proves to be a successful salesman, his income may increase to ₹ 30 thousand per month but if he does not happen to be a good salesman his income may go down to ₹ 10 thousand per month. Suppose in this new job there is 50-50 chance of either earning ₹ 30 thousands or ₹ 10 thousands (that is, each has a probability of 0.5). When there is uncertainty, the individual does not know the actual utility from taking a particular action. He has to decide on the basis of expected utility. *Given the probabilities of alternative outcomes, we can calculate his expected utility*. Whether the individual will choose the

2. *Neumann and* M_O *rgenstern in their important contribution. "**The Theory of Games and Economic Behaviour**" suggested a way of measuring utility in quantitative terms in situations involving risk and uncertainty. They construct index numbers, termed as N-M Index Numbers, to measure utility. However, to simplify our analysis we just assume that utility can be measured in some units.*

new risky job or retain the present salaried job with *a certain* income can be known by comparing the expected utility from the new risky job with the utility of the current job. It will be seen from the utility function curve *OU* in Fig. 14A.1 that the utility of his current certain money income of ₹ 20,000 with *certainty* is ₹ 65. Further, in case of new risky job if he is proved to be a successful salesman and his income rises to ₹ 30 thousands, his utility from ₹ 30 thousands is 75, and if he fails as a good salesman, his income falls to ₹ 10 thousands which yields him utility of ₹ 45. (Note that in the new risky job, the *expected value of income* is 20,000 which is given by $E(V) = 0.5 \times 10{,}000 + 0.5 \times 30{,}000 =$ ₹ 20,000). It will be seen from the utility function in Fig. 14A.1 that utility of ₹ 10,000 is 45 and that of ₹ 30,000 is 75. Given that the probability of success or failure as a salesman is 0.5, the expected utility of the new job is given by

$$\begin{aligned} E(U) &= 0.5\ U(10{,}000) + 0.5\ U(30{,}000) \\ &= 0.5 \times 45 + 0.5 \times 75 \\ &= 22.5 + 37.5 \\ &= 60.0 \end{aligned}$$

Thus with the present job with a fixed salary of ₹ 20,000 with no uncertainty utility is 65, whereas the expected utility of the new job as salesman on commission basis is 60. Since, the individual is risk averse as revealed by the nature of his utility function of money income, and since the expected utility of the risky job is less than the utility of the present job with a *certain* income, he will reject new risky job.

Let us now change the data suppose the individual is more confident of being a successful in his new job so that his subjective probability of being a successful is 90 per cent (*i.e.*, 0.9). As a result, his probability of being non-successful is 10 per cent (*i.e.*, 0.1). With this the expected value of income in the new risky job is ₹ 28,000.

Thus,

$$\begin{aligned} E(X) &= 0.1(10{,}000) + 0.9(30{,}000) \\ &= 1000 + 27000 \\ &= 28000 \end{aligned}$$

As expected utility from ₹ 10,000 is 45 and expected utility from ₹ 30,000 is 75 (see Fig. 14A.1), the expected utility from the new risky job in this case when subjective probability of being successful salesman is 90 per cent is

$$\begin{aligned} \mathrm{E(U)} &= 0.1 \times 45 + 0.9(75) \\ &= 4.5 + 67.5 \\ &= 72 \end{aligned}$$

In Fig. 14A.1 this expected utility of 72 is depicted by point *H* on the chord *AC* against expected value of income equal to 28 thousands. Now, with this changed probabilities, the expected utility of 72 is greater than 65 with the income of ₹ 30,000 with certainty. Thus, in this case the individual who is risk averse as shown by concave utility curve will choose the new risky job. Thus, *a risk-averse individual will choose a riskier option only if it has a sufficiently higher expected value.*

We are now in a position to provide a precise definition of risk-averse individual. Precisely speaking, *a person who prefers a certain given income to a risky job with the same expected income is called risk averter or risk averse.* Risk aversion is the most common attitude toward risk.

Risk Seeker. On the other hand, *a person is risk seeker or risk lover* who *prefers a risky job with the same expected income to a certain income.* In case of a risk loving individual, marginal utility of income to the individual increases as his money income increases as shown by the **convex total**

utility function curve *OU* in Fig. 14A.2. Suppose this risk-seeking individual has a present job with a certain income of ₹ 20 thousands. It will be seen from Fig. 14A.2 that the utility of ₹ 20 thousands is 43 units to this individual. Now, if he is offered a risky job with his income of ₹ 30 thousands if he

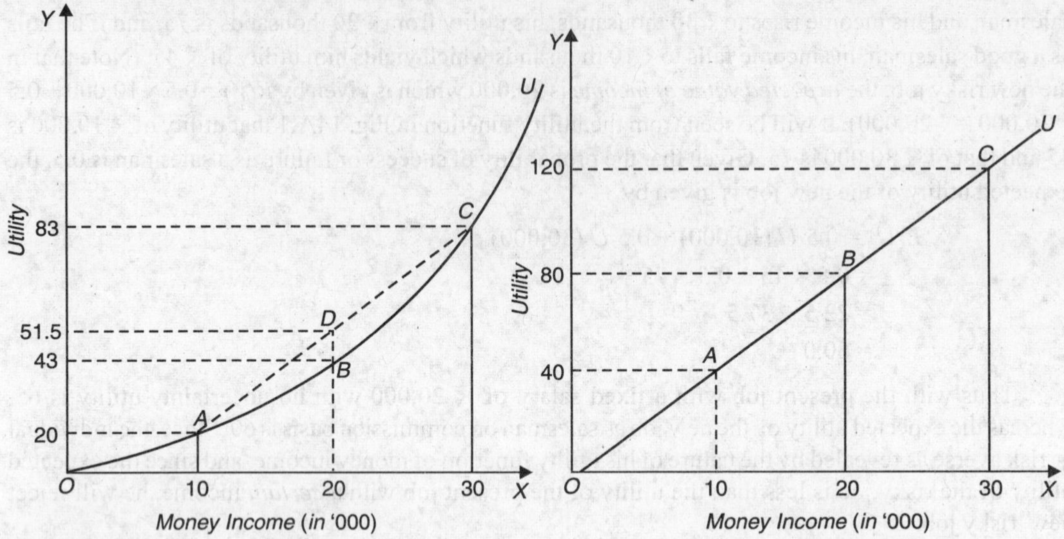

Fig. 14A.2. *Utility Function of a Risk Seeker or Risk lover*

Fig. 14A.3. *Utility Function of a Risk Neutral*

happens to be highly efficient and ₹ 10 thousands if he happens to be not so efficient in the new job with the equal probability of 0.5 in these two jobs, then the expected utility from the new is given by

$$E(U) = 0.5\ U(10{,}000) + 0.5\ U(30{,}000)$$

It will be seen from Fig. 14A.2 that the utility of ₹ 10 thousands to this individual is 20 while utility of ₹ 30 thousands to him is 83. Therefore,

$$\begin{aligned} E(U) &= 0.5\,(20) + 0.5\,(83) \\ &= 10 + 41.5 \\ &= 51.5 \end{aligned}$$

Since the expected utility from the new risky job is 51.5 which is greater than the utility of 43 from the present job with a certain income of ₹ 20 thousands, the *risk-loving individual will prefer the new risky job even though the expected income in the new risky job is also ₹ 20,000* as (0.5 × 10,000) + 0.5 (30,000) = ₹ 20,000). But expected utility of 51.5 in the new risky job is greater than 43 in the present job with certain income of ₹ 20,000.

As mentioned above, most of the individuals are risk averse but there is a good deal of evidence of people who are risk seekers or risk lovers. It is risk-loving individuals who indulge in gambling, buy lotteries, engage in criminal activities such as robberies, big frauds even at risk of getting heavy punishment if caught.

Risk Neutral. *A person is called risk neutral, if he is indifferent between a certain given income and an uncertain income with the same expected value. An individual will be risk neutral if his marginal utility of money income remains constant with the increase in his money.* The total utility function of a risk neutral person is shown in Fig. 14A.3. It will be seen from this figure that utility of a *certain* income of ₹ 20 thousands to him is 80. Now, in a risky job when income increases to ₹ 30 thousands if he proves to be a successful salesman, the utility of ₹ 30 thousands is 120 units. On the other hand, if in a new risky job, he proves to be a bad salesman, his income goes down to ₹ 10,000 whose utility to the individual is 40 units. We assume that there is equal probability of high and low income in the new risky job. Note that expected value of income in the new job with an uncertain income is 20,000 as $(0.5 \times 10{,}000 + 0.5\,(30{,}000) = 20{,}000)$. The expected utility of the new risky job is given by

$$E(U) = 0.5\,U(10{,}000) + 0.5\,U(30{,}000).$$
$$= 0.5\,(40) + 0.5\,(120)$$
$$= 20 + 60$$
$$= 80$$

It is seen from above that in *case of risk-neutral person expected utility of an uncertain income with the same expected value* (₹ 20,000 *in the present case*), is equal to utility of an assured or a certain income of the same amount. That is, risk-neutral person is indifferent between them.

INSURANCE AND RISK PREMIUM

Our foregoing analysis shows why most people buy insurance when they are faced with a risky and uncertain situation. As mentioned above, most people are risk averters and therefore they buy insurance to avoid risk. Now an important question is how much money or premium a risk-averse individual will pay to the insurance company to avoid risk and uncertainty facing him. Suppose the individual buys a house which yields him income of ₹ 30 thousands per month. But if the house catches fire and due to the damage caused, his income from it falls to ₹ 10 thousands per month and thus he suffers a loss of income. For the sake of simplifying analysis suppose there is 50 per cent chance of the house catching fire. Then the expect value of income in this risky and uncertain situation is

$$E(V) = 0.5 \times 30{,}000 + 0.5 \times 10{,}000$$
$$= 15{,}000 + 5{,}000$$
$$= 20{,}000$$

The utility function *OU* with a diminishing marginal utility of money income of a risk-averse individual is shown in Fig. 14A.4. With money income of ₹ 30 thousands, his utility is 75 and with his lower income of ₹ 10 thousands his utility is 45. Given that there is probability of 0.5 for each outcome, expected utility of the two outcomes is given by

$$E(U) = 0.5\,U(30{,}000) + 0.5\,U(10{,}000)$$
$$= 0.5 \times 75 + 0.5 \times 45$$
$$= 37.5 + 22.5$$
$$= 60$$

It will be seen from Fig. 14A.4 that we have drawn a straight line *AB* joining the utilities of 75 and 45. It is on this straight line *AB* that the amount of expected utility will be corresponding to the expected value of income in the present risky and uncertain situation. It will be seen from Fig.14A.4 that on this straight line *AB* and corresponding to the expected value of income of ₹ 20 thousands, the expected utility is 60 which corresponds to point *D* on the straight line *AB*. But it will be seen from the individual's utility function *OU*, that utility of 60 is associated with *an assured and certain income* of ₹16 thousands. Thus the individual with an uncertain income of ₹ 20 thousands will be willing to forego ₹ 4 thousands (or *DC*) to get a certain or guaranteed income of ₹ 16 thousands as the expected utility of *uncertain expected income* of ₹ 20 thousands is equal to the utility of a *certain income* of

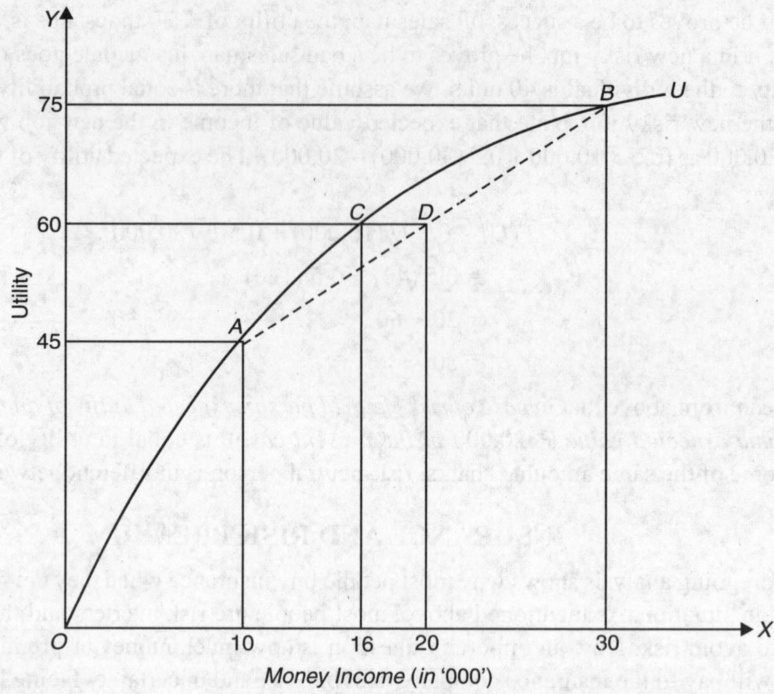

Fig. 14A.4. *Insurance and Risk Premium*

₹ 16 thousands. This means that if the individual gives up ₹ 4 thousands (20 – 16 = 4) from his uncertain expected income he will get the same utility of 60 as with a certain income of ₹ 16 thousands. ₹ 4 thousands equal to the distance *DC* is called the *risk premium.* Therefore, *the risk premium is the maximum amount of money that a risk-averse individual will be willing to pay to avoid the risk.* By paying the risk premium the individual can insure himself against a large loss from a fire and to get an assured or certain income.

It is clear from above way people buy insurance for fire, accident, ill health and even life.

RISK LOVER AND THE GAMBLING

Now, we explain how a risk lover responds to a fair gamble. It may be noted that a risk lover is one who prefers a gamble with the same expected value of income to the equivalent income with certainty. In Fig. 14A.5 the utility curve of a risk lover has been drawn and, as it will be seen, it is *convex.* The convexity of the utility curve implies that marginal utility of money income of the person increases as his income increases. Now, suppose that present income of the individual is ₹ 3,000 and is offered a gamble with 50:50 chance of winning or losing ₹ 1,000. Therefore, if he wins, his money income will rise to ₹ 4,000, and if he loses his income will fall to ₹ 2,000. The expected money value of the gamble will therefore be equal to 1/2(4000) + 1/2(2000) = ₹ 3,000. And the expected utility from the expected value of income of ₹ 3,000, as it will be seen from Fig. 14A.5 will be M_2D or OU_2 which is more than utility M_2C or OU_1 of income of ₹ 3,000 with certainty.

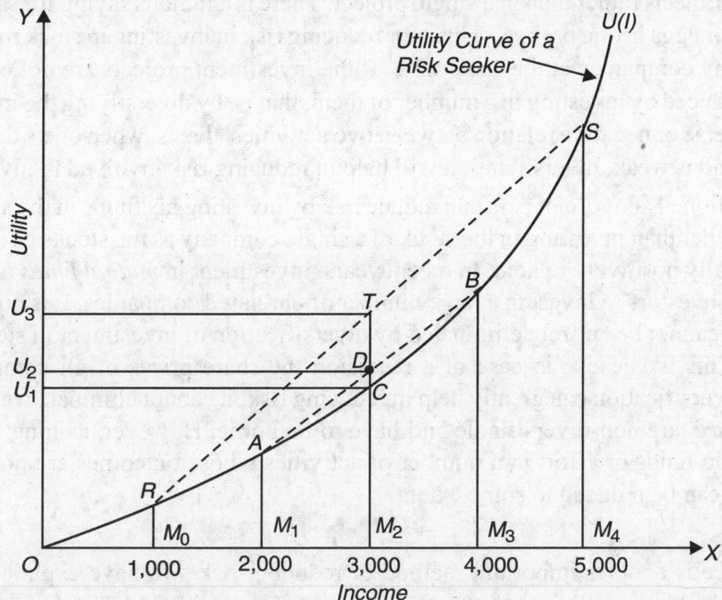

Fig. 14A.5. *The Utility Curve of a Risk-Lover and his Preference for Gambling*

Since his expected utility from the gamble is greater than that from his income with certainty, he will accept the gamble. Further, it can be shown that risk lover, a person whose marginal utility of money increases with increase in his income, will prefer a gamble having greater risk, that is, the gamble which has a larger variability of outcome to a gamble with less risk (that is, with smaller variability of outcome). Suppose a person with increasing marginal utility of income is offered two fair gambles: One with 2,000 chance of winning or losing ₹ 1,000 and the other with a 50:50 chance of winning or losing ₹ 2,000 as outcome of the gamble. Given his present income of ₹ 3,000, the expected value of the two gambles is the same.

$E(V)$ of the Gamble I = 1/2 (4000) + 1/2 (2000) = 3000

$E(V)$ of the Gamble II = 1/2 (5000) + 1/2 (1000) = 3000

However, it will be seen from Fig. 14A.5 that the expected utility of the two gambles differs; the expected utility of the second gamble, as will be seen from the straight-line segment *RS*, is equal to M_2T or OU_3 which is greater than M_2D or OU_2 of the first gamble with less variability (or risk) of outcome. It therefore follows that the risk lover will prefer the second gamble with greater risk and variability of outcome.

MANAGEMENT OF RISK: REDUCING RISK

We have seen above that though some individuals are risk seekers, most of the individuals are risk averse and try to reduce risk or uncertainty they face. There are three methods by which individuals or consumers can reduce risk. They are:

1. Diversification
2. Insurance
3. Gathering more Information.

We discuss below these methods.

Diversification

An important way of reducing risk is diversification, that is, spreading the risks. For example, an investor can reduce risk involved in investment by investing his investible funds in a number of independent projects rather than in a single project. There is a famous saying for safe investment, "Do not put all your eggs in one basket." Thus, for reducing risk in investing in stock market is to purchase shares of many companies at the same time. If the investment projects are not closely related, then risk can be reduced by investing in a number of them, that is, by diversifying the investment portfolio. If there is inverse or negative relation between two activities, that is, when one's demand is strong, the other's demand is weak, diversification will help in reducing risk involved in investment or activity.

As mentioned above, a person can reduce risk by investing his funds in the stock of a number of companies rather than investing in the stock of a single company as the stocks of different companies are not generally positively related. In recent years, investment in *mutual funds* are preferred as they enable small investors to invest in a large number of unrelated companies. It is however important to note that risk cannot be entirely eliminated by diversification of investment in stocks of a number of companies. This is because in case of a recession the share prices of all companies tend to fall. Therefore, diversification can greatly help in reducing risk; it cannot eliminate it completely. Risks of *systemic nature* are non-diversifiable and have to be borne. However, as long as you can allocate your investible funds or effort in a number of activities whose outcomes are not closely positively related, risks can be reduced to some extent.

Insurance

Insurance is another important method of reducing risk. We have explained above that risk-averse individuals are willing to give up some income to avoid risk. *The maximum price which an individual is willing to forego or pay for insurance is called risk premium.* Therefore, if the cost of insurance is equal to the amount of *risk premium* which they are prepared to give up to assure themselves a certain income in a risky and uncertain situation, they will buy the insurance policy. For example, if an individual owns a house worth ₹ 20 lakhs and faces a probability of 0.01 *i.e.*, one per cent chance of his house getting burnt down, then the expected value of loss is $\frac{1}{100} \times 20{,}00{,}000 = 20{,}000$. If a fire insurance policy is offered to the house owner for ₹ 20,000, he will definitely buy it. Such an insurance premium is fairly priced as it is equal to the expected loss that may assure to the individual in case his house is burnt down due to catching fire. In fact, a risk-averse individual may be willing to pay more than this depending on the degree of his risk aversion as measured by the risk premium he is prepared to pay.

The individuals generally buy insurance from the companies which specialize in the insurance business. These insurance companies work on the basis of *a law of large numbers.* The insurance companies not only insure against the loss suffered by the individuals but also earn profits for them. For example, from the past information, the insurance companies know that 1 per cent houses will catch fire. To provide insurance cover they pool a large number of policies. It insures against the expected loss to 100 policy holders, they will actually pay the loss to one policy holder if the probability of houses catching fire is one per cent. They fix such an amount of insurance premium that they are not only able to compensate the loss to the individuals but also earn profits for themselves.

Obtaining More Information

Collecting more accurate information is one important way in which individual can reduce risk while making a choice under conditions of uncertainty. More information will enable them to increase expected value and expected utility. With more information about the new good you want to purchase or about the new job you are offered will reduce the possibility of suffering a loss. For example, for buying a car or refrigerator of a company many individuals seek information from friends, read reviews of the car or refrigerator in the newspapers or on television to know how frequently a particular brand is likely to break down and requires repairs more often. By gathering such information you can reduce the possibility of making a wrong decision.

PART III

THEORY OF PRODUCTION AND COST

- Factors of Production
- The Theory of Production: Returns to a Factor
- Production Functions with Two Variable Factors
- Optimum Factor Combination
- Cost of Production and Cost Curves
- Supply and its Elasticity

CHAPTER 15
FACTORS OF PRODUCTION

Factors of Production: Meaning

Production of goods and services requires the use of factors of production which are also called agents of production. Volume of production and therefore national income depends upon the supply of various factors of production and their productivity. Since economic growth consists in expansion in total production or national income, the supply and productivity of the factors are vital to the process of economic growth.

But, what do we exactly mean by a factor of production. Fraser defines "factor of production as a group or class of productive sources".[1] More explicitly, the term 'factor' of production is used for a class of productive elements, the individual members of which are known as '*units*' of the factor and all units of a factor are homogeneous and interchangeable. Technically speaking, all units of a factor are perfect substitutes of one another. This also implies that two different factors of production cannot be perfect substitutes of each other.

It may also be noted that modern economists prefer to talk in terms of *productive services* or more recently of merely *inputs* rather than the conventional factors of production such as land, labour, capital and entreprepreneur. Thus, the goods or services which are used for the production of goods are called inputs and what they produce are called outputs. This *input-output analysis* has become an important tool of modern economic analysis.

Classification of Factors of Production

As mentioned above, factors of production have been classified as *land, labour, capital and organisation* (*or entrepreneurship*). Land in economics does not mean mere soil. Land in economics is synonymous with all natural resources available from air, water, from above the land surface and below it which can be used for production. Likewise, labour does not mean merely physical exertion, but all types of work, physical or mental, done by man for a monetary reward. By capital is meant the whole of the stock of wealth consisting of machines, tools, implements, raw materials etc. which is used for the production of further goods. Organisation or entrepreneurship consists in bringing the above three factors together, assigning work to each and bearing the risk and uncertainty of production. The entrepreneur who supplies this factor, initiates and directs production and bears risk and uncertainty.

Some economists have classified factors into two categories, land and labour (or nature and man) on the ground that they are the only *original or primary factors*. It is said that capital has no independent origin and is merely the outcome of combined efforts of land and labour, while organisation or enterprise is merely a special form of labour. But, whatever the origin, being man -made instrument of production, capital is capable of being expanded while the supply of land as a whole is fixed. This has made capital an important factor of production as a separate and distinct factor from land. Similarly, the entrepreneur's work is of a distinct type from that of labour and he is remunerated on a different basis. Thus, the four-fold classification of factors into land, labour, capital and enterprise is being followed even today.

---

1. L. M. Fraser, *Economic Thought and Language,* 1947, Ch.12.

LAND

Meaning and Significance

The term 'land' has been given a special meaning in economics. It does not mean soil, as in the ordinary speech, but it is used in a much wider sense. In the words of Marshall, land means "the materials and the forces which nature gives freely for man's aid, in land and water, in air and light and heat."[2] Land stands for all natural resources which yield an income or which have exchange value. "It represents those natural resources which are useful and scarce, actually or potentially."[3]

In every stage of economic evolution, nature has been man's most useful ally. In the hunting and the fishing stage, nature supplied food freely and sustained human life. In the pastoral stage, but for land surface and the pastures and meadows, herds of cattle and sheep could not have been reared and kept. The utility of land is obvious in the agricultural stage for how else could man grow his crops without soil, air and sunshine? When the agricultural stage has given place to the industrial stage, land still remains essential. Land is the chief agent in the production of wage goods, such as foodgrains, cloth and sugar. Every commodity that we use can, directly or indirectly, be traced ultimately to land. Look in whatever direction we may, our debt to nature is deep and obvious. Our very existence would be impossible without it. In Marshall's worlds. "Earth's surface is a primary condition of anything that a man can do, it gives him room for his action.[4]

The quantity and quality of natural resources (including agricultural land) plays a vital role in the economic development of a country. Important natural resources are those of agricultural land, minerals and oil resources, water, forests, climate, etc. The quantity of natural resources available in a country puts a limit on the level of output of goods which can be attained. Without a minimum of natural resources there is not much hope for economic development. It should, however, be noted that resource availability is not a sufficient condition for economic growth. For instance, India, though sufficiently rich in natural resources, has remained poor and under-developed. This is because resources have not been fully utilised for productive purposes. Thus, it is not only the availability of natural resources but also the ability to use them for production, which determines growth of an economy.

Supplies of natural resources can be increased as a result of new discoveries of resources in a country and technological changes which facilitate new discovery or transform certain previously useless materials into highly useful ones. It should also be noted that the scarcity of certain natural resources can be overcome by synthetic substitutes. For example, the synthetic rubber is being increasingly used in place of natural rubber in advanced countries. Further, nylon which is a synthetic commodity is being largely used in place of silk which is a natural commodity.

The use of natural resources and the role they play in the process of economic growth depends, among other things, on the type of technology. One does not have to go back very far in history to find that an item currently as valuable as petroleum was of little or no significance. In many developing economies there are deposits of many minerals which are not being used because of technological deficiencies, Thus, natural resources are in themselves passive in their influence on economic growth. They remain dormant and largely unknown till such time as the change in technology reveals their use-value. They acquire a dynamic character largely through changes in technology introduced by man.

Land as Renewable and Non-renewable Resource

Natural resources are often divided into two categories – (*a*) the exhaustible resources or nonrenewable resources, (*b*) the renewable resources. *The exhaustible resources are those natural resources which when used once cannot be renewed.* Natural resources such as mineral deposits of iron ore, copper, deposits of coal and petroleum get depleted as they are used by the economy. Their

---

2. Marshall,

3. L. M. Fraser, *Ibid* p. 222.

4. Alfred Marshall, *Principles of Economics,* Eighth Edition, p.138.

stocks are limited and cannot be renewed, that is, their stocks cannot be increased. According to the data available with the Planning Commission if we assume roughly an annual average growth rate in exploitation of say about 6% to 7%, cooking-coal reserves will be exhausted by about 2150 A.D., iron ore by' about 2050 A.D., manganese ore and chromite by about 2000 A.D. and bauxite by about 2033 A.D.[5] It may, however, be noted that all potential stock of these non-renewable resources may not be known at a given time and therefore their supply may increase through new discoveries as has happened in India where deposits of mineral oil and gas have been discovered at several places in the last few years.

On the other hand, *renewable resources are those which go on being used again and again and year after year for production*. Thus, agricultural land which is a very important natural resource can be cultivated again and again for production of foodgrains and other crops without much loss of productivity. Moreover, not only the productivity of land can be maintained but it can be significantly improved by human efforts. Thus, land in the sense of agricultural soil is renewable resource. Water resources, fisheries and forest resources are other examples of renewable resources: If some part of the forest resources are used, they can be replanted and their stock can be increased in the long run.

It is clear from above that land in its wider meaning which includes all natural resources is in part renewable and in part non-renewable. Agricultural land, water resources, forest resources, etc. are renewable but mineral and power resources such as deposits of iron ore, coal, petroleum, etc., are non-renewable resources.

Inelastic Supply of Land

An important characteristic of land is that it is fixed in supply. Land is a free gift from nature and its quantity is fixed by nature. Therefore, more land cannot be produced in response to greater demand for it. Whatever the rent, high or low, for the use of land its supply to the economy as a whole remains unchanged. In other words, the supply of land to the entire economy is not dependent on the price *i.e.*, rent for its use. Hence, from the standpoint of the whole economy, the supply of land (which includes natural resources) is *perfectly inelastic*. Since supply of land is a free gift from nature and not a produced factor, cost of production has no relevance for supply. For the society as a whole, land has got no cost of production, since society did not produce it; it got it free from nature.

But the supply of land to a *single use or a particular* industry is not perfectly inelastic. The supply of land to a particular use or industry can be increased by the shifting of land from other uses or industries. By offering attractive rents, the supply of land for a particular use can be taken away from other competitive uses. Therefore, *the supply curve of land to a particular use or industry is elastic and slopes upward from left to right.*

CAPITAL: PHYSICAL AND HUMAN

Meaning of Capital

The term capital is used in economics in various senses. In ordinary language and sometimes in economics also capital is used in the sense of money. But when we talk of capital as a factor of production, to confuse capital with money is quite wrong. Of course, money is used to purchase various factors such as raw materials, machinery, labour which help to produce goods, but money *itself* does not directly help in the production of goods. The money which is available for investment and productive purposes has been called *money capital or financial capital* by some economists. But money capital is not the real capital. The real capital consists of machinery, raw materials, factories; fertilizers, etc., which *directly* assist in the production of goods. Similarly, capital has been defined by some economists as *that part of man's wealth other than land which yields an income*. But this too, is not a satisfactory definition of capital. Securities, bonds, stocks, etc., possessed by a man yield

5. See Planning Commission, *Revised Fifth Plan*, Chapter 2, pp. 11-12.

income to him but they cannot be called real capital because they represent only titles of ownership rather than factors of production.

Capital has been rightly defined as "*produced means of production*". This definition distinguishes capital from both land and labour because both land and labour are not produced factors. Land and labour are often considered as *primary or original factors of production*. But capital is not a primary or original factor, it is a produced factor of production. Capital has been produced by man by working with nature. Therefore, capital may well be defined as *man-made instrument of production*. Capital thus consists of those physical goods which are produced for use in future production. Machines, tools and instruments, factories, canals, dams, transport equipment, stocks of raw materials are some of the examples of capital. All of them are produced by man to help in the production of further goods. According to Prof. Richard T. Gill, "*A country's capital is its stock of produced or man-made means of production, consisting of such items as buildings, factories, machinery, tools, equipment and inventories of goods in stock.*"[6]

Fixed Capital and Working Capital

Capital may be classified into fixed capital and working capital. *Fixed Capital* are durable-use producr goods which are used in production again and again till they wear out. Machinery, tools, railways tractors, factories *etc*., are all fixed capital. Fixed capital does not mean fixed in location. Capital like plant, tractors, and factories are called fixed because if money spent upon these durable goods, remains fixed for a long period in contrast with the money spent on purchasing raw materials which is released as soon as goods made with them are sold. *Working Capital*, on the other hand, are the single-use producers' goods like raw materials, goods in process and fuel. They are used up in a single act of production. Moreover, money spent on them is fully recovered when goods made with them are sold in the market.

Human Capital

We have explained above the concept of *physical capital*. For a long time in economics, it was thought that it was the physical capital which played a crucial role in expanding production. In recent years, a new concept of '*human capital*' has been evolved and emphasised. *By human capital is meant the stock of people equipped with education, skills, health*, etc. It has now been found that the rate of growth achieved in the developed countries cannot be wholly explained by the increases in physical capital and advances in technology. A good part of economic growth has occurred due to the accumulation of human capital. It has now been realised that human capital formation is as important in increasing production and productivity as the physical capital formation. An educated, trained and skilled man is much more productive than an uneducated, untrained, and unskilled. Likewise, a person with good health contributes to production to a greater degree than the person with frail and poor health. Since investment in education, skill and health adds greatly to the productivity of men, investment in human capital has also been called *investment in men*, or *investment in human beings*. One of the major tasks confronting the developing countries is the building of human capital. There is a growing realisation that a rapid rate of 'human capital formation' is as important a pre-condition of economic growth as the rapid rate of 'physical capital formation.'

ROLE OF CAPITAL

Capital plays a vital role in the modern productive system. Production without capital is hard for us even to imagine. Nature cannot furnish goods and materials to man unless he has the tools and machinery for mining, farming, forestry, fishing, etc. If man has to work with his hands on barren soil, productivity would be very low indeed. Even in the primitive stage, man used some tools and implements to assist him in the work of production. Primitive man made use of elementary tools like

6. Richard Gill "*Economic Development*: *Past and Present*," Third Edition; 1975, Prentice-Hall of India Private Limited, p. 14.

bow and arrow for hunting and fishing net for catching fishes. With the growth of technology and specialisation, capital has become more complex and is of superior and advanced type. More goods can be produced with the aid of capital. In fact, greater productivity of the modern economy like that of U.S.A. is mainly due to the extensive use of capital, *i.e.* machinery, tools or implements in the production process. Capital adds greatly to the productivity of workers and hence of the economy as a whole. Because of its strategic role in raising productivity, capital occupies a central position in the process of economic development. In fact, capital accumulation is the very core of economic development.

Capital goods are man-made instruments of production and increase the *productive capacity* of the economy. Therefore, accumulation of capital goods every year greatly increases the national product or income. Capital accumulation is necessary to provide people with tools and implements of production. If the population goes on increasing and no net capital accumulation takes place, then the growing population would not be able to get necessary tools, instruments, machines and other means of production with the result that their capacity to produce would be seriously affected.

Capital Formation and Economic Growth

Thus, capital accumulation makes possible the use of ***indirect or roundabout methods of production*** which greatly increase the productivity of the workers. Under these indirect or roundabout methods of production, workers instead of working with bare hands, work with the aid of more productive tools, instruments and machinery. Under these indirect or roundabout methods some labourers and other productive resources are first employed in producing capital goods and then with the help of these capital goods workers produce consumer goods. The greater the extent to which the methods, of production would be indirect or roundabout, the greater their productivity and efficiency. But, as we have seen above, for the use of indirect or roundabout methods of production capital has to be accumulated. Therefore, we see that capital accumulation makes the use of indirect or roundabout methods of production possible and thereby greatly increases the national product and is helpful in bringing about rapid economic growth. Moreover, productivity of the workers depends upon the amount of capital per worker. The greater the quantity of capital per worker, the greater the productivity and efficiency of the worker.

From the viewpoint of economic growth capital formation is important also because it make large scale production and greater degree of specialisation possible. Thus, with capital accumulation the *advantages of large-scale production and specialisation* are obtained. The advantage of large-scale production and specialisation is that they greatly increase output and productivity and thereby bring down the *cost of production per unit*. Without adequate capital accumulation neither the scale of production can be increased nor greater specialisation and division of labour in the production process is possible. Hence, capital accumulation by increasing the scale of production and specialisation increases the production and productivity in the economy and thereby promotes economic growth.

Capital Formation and Technological Progress

Another contribution of capital accumulation is that it makes the *technological progress of the economy* possible. Different technologies need different types of capital goods. Therefore, when new, superior and better technology is discovered, its use can be made for production only if that technology is embodied in new capital goods, that is, if capital goods according to that technology are made. Therefore, without capital accumulation, not much technical progress can be made. If there is no capital accumulation, then the various new inventions or discoveries will remain unused for production. It is therefore, clear that capital accumulation promotes technical progress and thereby accelerates the economic growth of the country.

Capital Formation and Employment Generation

Another important economic role of capital formation is the *creation of employment opportunities* in the country. Capital formation creates employment at two stages. First, when the capital is produced, some workers have to be employed to make capital like machinery, factories, dams, irrigation works, etc. Secondly, more men have to be employed when capital has to be used for producing further goods. In other words, many workers have to be engaged to produce goods with the help of machines, factories etc. Thus, one sees that employment will increase as capital formation is stepped up in the economy. Now. if the population grows faster than the increase in the stock of capital the entire addition to the labour force cannot be absorbed in productive employment because not enough instruments of production will be there to employ them. This results in unemployment. The rate of capital formation must be kept sufficiently high so that, employment opportunities are enlarged to absorb the additions to the working force of the country as a result of population growth. In India the stock of capital has not been growing at a fast enough rate so as to keep pace with the growth of population. That is why there is huge unemployment and underemployment in both the urban and rural areas. The fundamental solution to this problem of unemployment and under-employment is to speed up the rate of capital formation so as to enlarge employment opportunities.

HUMAN CAPITAL AND EDUCATION ECONOMIC GROWTH

Gross domestic product of a country depends on not only the amount of labour (i.e. work-hours) used for producing goods and services but also on its productivity. One of the important factors that determines productivity of worker is human capital, that is, education. As seen above, in the modern economics the concept of capital is not confined to physical capital such as machines, tractors, capital equipment etc. that raises productivity of labour but also includes what is called human capital. *By human capital we mean the skills and knowledge that workers acquire through education and training*. This human capital, that is, knowledge and skills, are accumulated by human beings through education during the time they spend in primary and secondary schools and upto graduation and post graduation in college or universities. Besides, specialised professional education such as engineering, computer training, management education and others raises greatly the productivity of workers. Therefore, accumulation of human capital is generally called *investment in people*.

Human capital though less tangible than physical capital, is similar to it in many ways. First, like physical capital, accumulation of human capital increases the ability of a country to produce more goods and services. Second, like physical capital, human capital is a produced factor of production. Producing human capital requires investment in inputs such as student time, teachers, books and libraries, college buildings. Thirdly, like the physical capital, the accumulation of human capital increases the productivity of workers and therefore causes their wages to rise. Fourthly, like the accumulation of physical capital, the accumulation of human capital requires the sacrifice of some present consumption so as to have more consumption in future.

Cost of Human Capital: Human capital is built through acquiring more education by spending more time in school, college or university. Thus, in spending more time in acquiring more education, one not only delays one's entry into labour force but also sacrifices income or wages which he could have earned by working during the time he spends for acquiring education. This is the *opportunity cost* of acquiring more education, skills and grades (*i.e.* accumulating more human capital). But they spend time in acquiring more human capital with the expectation that they would be able to earn higher income, or wages in the future as more education and higher skills raises the productivity of the workers and therefore their wages. Thus, the students, like the people who save, face a trade-off between less consumption today for more consumption in the future.

Thus, "spending more on education today (reducing consumption) raises future income but each additional investment in education provides a smaller and smaller return"[7]. It follows from above that, like physical capital, expenditure on education also represents investment in capital which raises productivity in the future. Besides, investment in education is tied to specific human being and

7. Joseph, E. Stiglitz. *Economics*, W.W. Norton & Co. New York 1907, p.170.

therefore it is called human capital. Thus, according to Mankiw, "*Like all forms of capital, education represents an expenditure of resources at one point in time to raise productivity in the future. But unlike an investment in other forms of capital investment in education is tied to a specific person and this linkage is what makes it human capital*"[8]. It is important to note that workers endowed with higher education (*i.e.* more human capital) earns more income or wages than those having less education. The differences in wages between those with more education and those with less education are quite large and has been increasing. For example "College graduates in the United States earn about twice as much as those workers who end their education with a high school diploma"[9] It may be noted that this difference in earnings between workers with more human capital and those with less human capital on an average tends to be even larger in developing countries where educated workers are in scarce supply.[10]

Human Capital and Sources of Economic Growth

The supply of human capital plays a crucial role in promoting economic growth. In an important study Edward F. Denison[11] has measured the relative importance of various sources of economic growth in the United States for the 50 years period, 1929-69.

Table 15.1: Sources of US Economic Growth

	Per cent of Growth Rate	
Source	*1929-1969*	*1948-1969*
Advances in knowledge	31.1	34.1
More work done	28.7	23.9
Capital accumulation	15.8	21.6
Increased education	14.1	11.9
Others	10.3	8.5
Total	100	100

Rate of economic growth in the US economy was on an average 3.3 percent during the 50 years period (1929-69) of the study. It will be seen from Table 15.1 that the most important source of economic growth (31.1 to 34 per cent) was *advances in knowledge* (which includes improvement in technological knowledge) that takes place as a result of increase in education and in research and development activity. The next important source of economic growth was increase in labour force (*i.e.* more work done) as measured by increase in hours worked by labour which contributed about 28 per cent during the 50 years period of study. Contrary to expectations of economists, accumulation of physical capital contributed only 15.8 per cent during the entire period (1929-69) and 21.6 per cent during 1948-69 to the growth in GDP. For the entire 50 years period of study (1929-69) increase in eduction (*i.e. number of years spent in* schools, colleges and universities) contributed almost the same as accumulation of physical capital (14.1 percent as against 15.8 per cent of physical capital. However, if we club the contribution of increased education and advances in knowledge their contribution is as high as about 45 percent to growth.

Other sources include the *efficient use of available resources, economies of scale,* a shift of labour force from low-productivity agricultural sector to higher productivity industrial sector which contributed about 10 per cent to growth in GDP. It may however be noted that Denison could not

8. N. Gregory Mankiw, *Principles of Economics*, Thomson, 2nd edition, 2001, p. 419
9. Mankiw op. cit., p.419
10. See, Stiglitz, op. cit. p. 171. and Mankiw, op. cit. pp-419-420.
11. Edward, E. Denison, *Accounting for US Economic Growth, 1929-1965* (Wshington D.C. Brookings Institution, 1974.

measure directly the contribution of advances in knowledge. He measured the contributions of other four sources of growth, namely, more labour work, accumulation of physical capital accumulation, increased education, and others (as mentioned above) and assumed that all economic growth that could not be explained by other sources was the contribution of advances in knowledge.

Supply Curve of Saving

As explained in Chapter 11 in the discussion of intertemporal choice, a rise in the rate of interest induces the individual to save more. In this way, we can determine the saving of an individual at different rates of interest. The higher the rate of interest, the greater is the supply of saving by him. By summing up the saving of various individuals at various rates of interest we can obtain market supply curve of savings, which will slope upward showing positive relationship between rate of interest and savings. This positively sloping curve of saving is shown in Fig. 15.1. It may be noted that in case of some individuals who have a target of having a certain fixed level of income or consumption in future, a higher rate of interest will enable them to earn that fixed income by saving and lending less in this year. That is, for these individuals supply curve of saving will be backward sloping. However, those individuals who save and lend more at higher rates of interest predominate and therefore the overall supply curve of savings slopes upward, though at very high rates of interest backward-bending shape of the supply curve of saving cannot be ruled out. However, empirical evidence shows that a rise in interest rate has a positive effect on the rate of saving. Therefore, the supply curve of saving as shown in Fig. 15.4 shows positive relationship between rate of interest and savings.

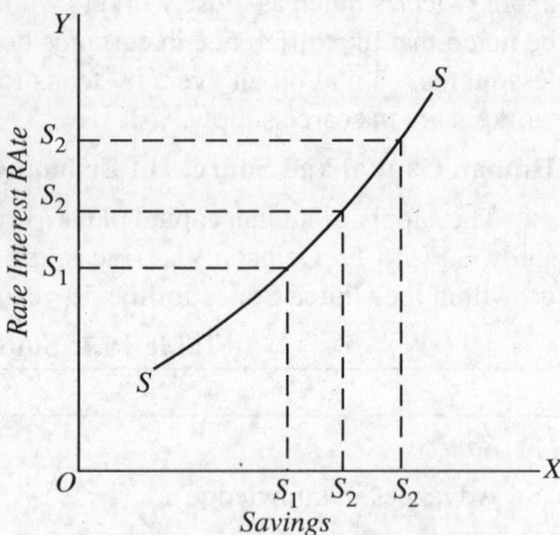

Fig. 15.1. *Positively-sloping Supply Curve of Saving*

Factors Affecting Savings

We have seen above that people save during working years so that they can consume enough in the years of retirement. And interest rate affects the saving for retirement. In addition to interest rate rate there are other factors that affect savings. We explain below all these factors.

1. Interest Rate. We have explained above that interest rate is one factor that determines supply of savings in the economy. The rise in interest rate generally brings about increase in supply of savings, though this effect is not significant. This is because rise in interest rate gives rise to two effects – income effect and substitution effect – which work in opposite direction. While income effect of rise in wage rate tends to reduce savings and its substitution effect tends to increase it. Since substitution effect generally dominates there is a net positive effect, though a small one, of rise in interest rate on supply of savings.

2. Income. Keynes emphasised that it is level of disposable income that determines savings. As the income rises, both consumption and saving increase. However, according to Keynes, average propensity to consume declines as income increases and therefore saving rate rises at higher levels of income.

3. Social Security Provisions. Social security measures adopted by the government such as state pension, free health care lower savings. Generous pension scheme reduces the need for saving for the retirement years. Similarly, the provision of free health care through the National Health Service also reduces the need for savings to meet the medical expenses. It has been pointed by some economists that low saving rate of the UK and USA is mainly due to the comprehensive social security system in these countries.

4. Taxation System. Taxes also affect savings by individuals. In India and other countries income tax is the main direct tax levied on individuals' incomes including interest and dividend income. It has been pointed that income tax discourages saving. For people who keep their savings in bank deposits or use them for buying bonds and shares, tax on interest and dividends reduce after-tax return from them. "*The tax on interest income substantially reduces the future payoff from current saving and as a result reduces the incentives for people to save*"[12]. According to N.G. Mankiw, "*Low rate of saving in the United States is at least partly attributable to tax laws that discourage savings*"[13].

Aggregate Savings

Savings of all individuals in a society constitute aggregate savings of the economy. At any time, some individuals who are in working life save and others who are old and leading a retired life dissave. The retired people withdraw money from their bank accounts and/or sell shares and bonds to supplement income from their pension and other social security schemes. In determination of aggregate savings, *demographic factor*s also play a role. A more slowly growing population has a proportionately larger population of old and retired people and this depresses aggregate saving rate of the economy.

Bequest Motive. An important factor that determines saving of a society is bequest motive. In order to leave behind a good amount of wealth for their children some rich people save during their life. If high inheritance taxes are levied, they would discourage saving on this account. Since bequest motive is an important reason for savings by individuals, especially the wealthy people, the lower inheritance taxes will have a large effect on increasing the saving rate of the economy.

Precautionary Motive. The people save for the unforeseen contingencies such as illness and periods of unemployment to protect themselves against these emergencies for which these is no insurance coverage.

Purchasing Power of Assets. The real value, that is, purchasing power of assets such as shares, bonds, houses possessed by the people also influence savings of the people. If the prices of these assets rise, people feel better off than before. This induces them to consume more and therefore save less. It has been found that when prices of assets rise, people even get more loans from the banks and spend them for consumption purposes.

We have explained above some important factors that affect savings of a society.

LABOUR

The contribution of labour to the national product and income depends not only on the size of labour force but also on its quality. By quality of labour we mean how much productive it is, that is, what is its level of productivity. A given labour force with higher productivity will yield a larger national product and income. Therefore, in this chapter we shall discuss the factors on which productivity or efficiency of labour depends. An important factor determining efficiency of labour force is the division of labour.

Whereas land and capital are material factors of production, labour is a human factor. Because of this, labour should not be treated as a saleable commodity like land and capital.

The contribution of labour to the national product and income depends not only on the size of labour force but also on its quality. By quality of labour we mean how much productive it is, that is, what is its level of productivity. A given labour force with higher productivity will yield a larger

12. See J. Stiglitz, op.cit p. 176.
13. N. Gregory Mankiw, *Principles of Economics*, Second Edition, 2001, P. 566

national product and income. Therefore, an important factor determining efficiency of labour force is the division of labour. We, therefore, start our study of the qualitative aspect of labour force with the discussion of division of labour.

Division of Labour

Division of labour is an important feature of modern industrial organisation. Division of labour occupies so important a place in the modern production process and adds so greatly to the total output and wealth of a country that Adam Smith, the father of economics, selected it as the first topic for discussion in his book. "*An Enquiry into the Nature and Causes of Wealth of Nations*."

Division of labour may be simple or complex. Simple division of labour refers to the *production of a single commodity by a person*. In a society where simple division of labour prevails every worker would be producing only one commodity. Some would be producing cloth, some will be making shoes, some others will be producing farm products but every body would be producing either farm products or cloth or shoes. In the old village societies there prevailed this simple division of labour and, therefore, the old village society composed of farmers who produced agricultural goods, weavers who made cloth, cobblers who made and repaired shoes, etc.

But, in the modern days, division of labour is of complex type. In fact, it is the complex division of labour which has increased so greatly the productivity of the modern productive system. *Complex division of labour means that the making of an article is split up into several processes and each process is carried out by a separate worker or a separate group of workers*. Thus, complex division of labour refers to the specialisation in one operation or process. Every group of workers specialises only in one process in the production of a commodity. The process may be split up into sub-processes and thus division of labour may become more complex. In the modern tailoring shops, making of a shirt is broken up into different processes. Some workers only do the job of "cutting", some others only do the work of "sewing", and still a separate group of workers puts buttons on it, etc. This is complex division of labour in the making of shirts.

Several advantages are claimed for the system of division of labour. The system of division of labour has proved to be of immense benefit to society.

Advantages of Division of Labour

1. Increase in Productivity. The greatest advantage of division of labour is that it increases immensely the productivity per worker. This point can be illustrated by the famous example of pin making given by Adam Smith. The process of pin making is divided into 18 distinct operations. Ten men make 48,000 pins in a day. One worker may, therefore, be considered to have made 4,800 pins in a day. In the absence of division of labour one man could have made only one pin in a day or at the most only 20 pins. Thus, with the division of labour the productive capacity of the individual and of the community has greatly increased.

2. The Right Man in the Right Place. Another great advantage is that the work under division of labour is allotted according to the ability and capacity of the individual worker. This ensures a high degree of efficiency as the right man is put in the right job. Thus, it eliminates the possibility of fitting a square peg in a round hole.

3. Dexterity and Skill. The worker becomes highly skilled and acquires high degree of dexterity because of the repeated performance of the same operation. As the age-old dictum goes, *practice makes a man perfect*. The worker acquires perfection in his skill because he has to carry out the same operation over and over again. This adds to his productivity.

4. Inventions are Facilitated. Another significant advantage is that it promotes the development of new ideas and better techniques of doing the work. It is due to the fact that when a worker is performing the same operation over and over again, he can think of doing that process in a better and improved manner. Even some mechanical device may occur to him to do that task easily and more efficiently. Thus, division of labour results in inventions of new machinery and better tools.

5. Saving in Time. Under division of labour, a worker performs a part of the whole process and therefore he needs to learn only that much. Long periods of training are, therefore, rendered unnecessary. This saves great time and money.

Further as the worker remains employed on the same process, he does not waste his time in moving from one process to another. He, therefore, goes on working without loss of time. This further results in saving in time.

6. Economy in the Use of Tools. A worker is not provided with a complete set of tools required for the whole process. He is provided with only those tools which are required by him for the performance of that part of the process which is allotted to him. Thus, one set of tools can be made use of by many workers. This is a great economic gain.

7. Use of Machinery Encouraged. By breaking up the production of a commodity into small and simple operations, division of labour encourages the use of machinery and its introduction. These simple operations easily can be carried out by suitable machines. Machines can be economically used only when they are fully used, that is, when their productive capacity is fully utilised. Division of labour involves production on a large scale and therefore permits the economical use of machinery.

8. Cheaper Goods. Another distinct advantage of division of labour is that the economies of large scale are reaped. The cost per unit tends to fall down when the commodity is produced on a large scale which ensures the production of cheaper goods. Even poor people can buy them. As a result of this, the standards of living of the people rise.

9. Rise of Entrepreneurs. Since the work is divided into various processes, someone is required to coordinate the work. This has resulted in the rise of entrepreneurs whose job is to specialise in the art of organisation. The rise of entrepreneurs has greatly contributed to the increase in efficiency and productivity of the modern economy.

ENTREPRENEUR

Having explained three factors–land, labour and capital–we now turn to the explanation of the fourth important factor, namely, the entrepreneur. It is not enough to say that production is a function of land, capital and labour. There must be some factor which combine these factors in the right proportion and initiates the process of production and also bear the risk involved in it. This factor is known as the entrepreneur. He has also been called the organiser, the manager and the risk taker. But in these days of specialisation, the task of manager and organiser has become different from that of the entrepreneur. While the organisation and management include decision-making of routine type, *the real task of the entrepreneur is to initiate the production work* and not to manage the business affairs. Indeed, the entrepreneur may employ on wage basis managers who manage day-to-day affairs of the organisation. Further, whereas the hired managers get fixed wages, *the entrepreneurs bear risk and uncertainty of the production work.*

Functions of the Entrepreneur

There is a controversy about the proper and true functions of the entrepreneur. Various economists have laid stress on the different functions of the entrepreneur. According to Schumpeter, the function of the entrepreneur is to introduce innovations. By innovations, he means new combinations of factors or new processes or the discovery of new products or the discovery of new markets or any change in production methods which reduce cost of production. In other words, anything which reduces cost of production or raises the demand for the product is called innovation by Schumpeter. F.H. Knight emphasised the uncertainty bearing function of the entrepreneur. Alfred Marshall assigns to the entrepreneur the function of risk bearing, provision of finance, introduction of innovations and management of a business enterprise. We shall explain below these various functions of the entrepreneur, but it should be carefully noted that different economists have emphasised different functions. Further, with the passage of time and change in the organisational structure of the business enterprises, there has been division of labour and specialisation in the various tasks which were previously assigned to entrepreneur. Thus, in a corporate form of business organisation co-ordination between the various factors, day-to-day management of the organisation and decisions regarding price, output, selling costs, investment are taken by specialised or professional managers. The important functions of entrepreneure are explained below:

(1) **Initiating a business enterprise and bringing about resource co-ordination**. The first and foremost function of an entrepreneur is to initiate a business enterprise. For this he has to collect different factors of production, such as labour, capital, land or factory building and bring about co-ordination among them. In other words, he does all what is required to set up a new firm or a productive enterprise. These various other factors of production are paid fixed contractual remunerations; the labourers are paid at the fixed rate of wages, the land or factory building gets fixed rate of rent for its use and on capital the fixed rate of interest is paid. But the reward for the entrepreneur, that is, profits is not fixed. If his planning and anticipation go very much wrong, he may get even negative profit, that is, losses. Thus, to start a business, the entrepreneur has to confront the possibility of earning profits or making losses. On the other hand, the other factors get their fixed contractual payments irrespective of whether the entrepreneur makes profits or losses.

(2) **Risk taking or uncertainty bearing.** Ultimate responsibility of the business lies with the entrepreneur. As pointed out above, what is planned and anticipated by the entrepreneur may not come true and the actual course of events may differ from what was anticipated and planned. The economy is dynamic and changes in it are occurring every day: the demand for commodities may change, the cost structure may change, fashions and tastes of the people may change. Changes in Government's policy regarding taxation, interest may also change. All these changes bring about changes in the cost or demand conditions of the business firms. It may happen that as a result of certain particular changes, which were not anticipated by the entrepreneur, the firm has to incur a heavy loss. To bear the financial losses arising out of the uncertainty is the task of the entrepreneur. Therefore, many economists have emphasised that true function of the enterprise is to bear risk and uncertainty. F. H. Knight distinguished between risk taking and uncertainty bearing. He divided the risk into two kinds. The first kind of risks are those which can be insured against and the entrepreneur can avoid them by getting them insured and pay premium for such insurance. This premium is included in cost of production. Therefore, such risks are not borne by the entrepreneur, Risks of fire, theft, burglary, etc. are this type of risks which can be insured. The second type of risks are those which cannot be insured. The risk of incurring losses as a result of changes in the demand or cost conditions cannot be insured. The bearing of this non-insurable risk has been called uncertainty bearing by Knight. According to Knight, the entrepreneur earns profits because he bears uncertainty in the economy where dynamic changes are occurring every day.

(3) **Innovations.** As said above, according to Schumpeter, the true function of the entrepreneur is to introduce innovations. Schumpeter uses innovations in a very broad sense which include introduction of new production methods, utilisation of new source of raw materials, adoption of new forms organisation, introduction of a new product including quality improvements, discovering new markets. The first three types of innovations are primary cost reducing and the last two are primarily of a demand creating nature. There is always uncertainty attached to the introduction of new innovations. They may succeed in the sense that they may bring profits to the entrepreneur or they may fail in the sense that entrepreneur has to suffer losses for introducing them. The tasks of the entrepreneur is to continuously introduce new innovations. When, a new innovation is introduced for the first time by the entrepreneur and it proves successful, it will be imitated by others and all will be adopting that innovation after some time lag. Therefore, the innovations may yield profits for the entrepreneur for some time but when it is widely adopted by others, the profits would disappear. But the entrepreneur may introduce further new innovations to make profits.

CHAPTER 16
THE THEORY OF PRODUCTION: RETURNS TO A FACTOR

So far we have been discussing the demand side of the pricing problem. In this chapter and the next few ones we shall be discussing the supply side of the pricing of products. The supply of a product, as we shall see in a later chapter, depends upon its cost of production, which in turn depends upon (a) the physical relationship between inputs and output, and (b) the prices of inputs. The physical relationship between inputs and output plays an important part in determining the cost of production. It is the general description of this physical relation between inputs and output which forms the subject-matter of the theory of production. In other words, the theory of production relates to the physical laws governing production of goods.

The act of production involves the transformation of inputs into outputs. The word production in economics is not merely confined to bringing about physical transformation in the matter, *it is creation or addition of value*. Therefore, production in economics also covers rendering of services such as transporting, financing, marketing. Laws of production, or in other words, the generalisations regarding relations between inputs and outputs developed in this chapter will apply to all these types of production.

The relation between inputs and output of a firm has been called the 'Production Function'. Thus, the theory of production is the study of production functions. The production function of a firm can be studied by holding the quantities of some factors fixed, while varying the amount of other factors. This is done when the *law of variable proportions* is derived. The production function of a firm can also be studied by varying the amounts of all factors. The behaviour of production when all factors are varied is the subject-matter of the laws of *returns to scale*. Thus, in the theory of production, the study of (a) the law of variable proportions and (b) the laws of returns to scale is included. Besides this, the theory of production is also concerned with explaining *which combination of inputs (or factors of production) a firm will choose* so as to minimise its costs of production for producing a given level of output or to maximise output for a given level of cost.

IMPORTANCE OF THE THEORY OF PRODUCTION

The theory of production plays a double role in the price theory. First, it provides a basis for the analysis of relation between costs and amount of output. Costs govern supply of a product which, together with demand, determines the price of a product. The prices of inputs (factors) of production influence the cost of production and hence play a part in determining the prices of product. Secondly, the theory of production provides a basis for the theory of firm's demand for factors (inputs) of production. Demand for factors of production or inputs, together with the supply of them, determines their prices.

The theory of production has great relevance for the theory of firm. The theory of firm is concerned with what level of output it will produce so as to maximise its profits. In order to fix this profit-maximizing output, besides the demand conditions (average and marginal revenues), the firm will be guided by the marginal and average costs of production. In addition to the prices of the factors of production, the changes in marginal and average costs of production as a result of increase in output are determined by the physical relationship between inputs and output.

Demand for factors of production depends upon the marginal revenue productivity of the factors and therefore demand curves for factors of production are derived, given the price for output, from their marginal productivity curves. And the theory of production explains the forces which determine the marginal productivity of the factors. As we shall study in a later part concerning the theory of distribution, the *relative prices of factors*, that is, wages of labour, rent of land, interest on capital etc. depend a good deal on the demand for them and hence on their marginal productivity. Thus, the theory of production has a great relevance to the theory of relative prices of factors, that is, to the microtheory of distribution.

The theory of production is also relevant to the macrotheory of distribution. The aggregate distributive shares of the various factors, for instance, aggregative shares of wages and profits in national income, depend upon the elasticity of substitution between factors which is an important concept of the theory of production. In fact, in the neo-classical macro-theory of distribution, elasticity of substitution is the crucial factor which determines the aggregative shares of the various factors.

PRODUCTION FUNCTION: TRANSFORMING INPUTS INTO OUTPUT

The act of production involves the transformation of inputs into output. Production is a transformation of physical inputs into physical output. The output is thus a function of factors which are also called *inputs*. The functional relationship between physical inputs and physical output of a firm is known as production function. Algebraically, production function can be written as

$$Q = f(L, K, M)$$

where Q stands for the quantity of output, L, K and M stand for the quantities of factors labour, capital and raw materials respectively.

The above equation shows that the quantity (Q) of output produced depends upon the quantities of the factors used. The production function expresses the relationship between the quantity of output and the quantities of the various inputs used for the production. More precisely, *the production function states the maximum quantity of output that can be produced with any given quantities of various inputs*. If a small firm produces wooden tables in a day, its production function will consist of the *maximum* number of tables that can be produced from a given quantities of various inputs such as wood, varnish, labour time, machine time, floor space.

Two things must be noted in respect of production function. First, production function, like the demand function, must be considered with reference to a *particular period of time*. Production function expresses *a flow of inputs* resulting in *a flow of output* in a specific period of time. Secondly, production function of a firm is determined by the state of technology. When the technology advances, the production function changes with the result that greater flow of output can be obtained from the given inputs, or smaller quantities of inputs can be used for producing a given quantity of output.

In economic theory we are interested in two types of production functions. First, we study the production function when the quantities of some inputs such as capital and land are kept constant and the quantity of one input such as labour (or quantities of few inputs) is varied. This kind of production function [$Q = f(L, \bar{K})$] is called *short-run production function*. The study of short-run production is the subject-matter of the law of diminishing returns which is also called the *law of variable proportions*. Secondly, we study production function (input-output relation) by varying all inputs and this is called *long-run production* function and can be expressed as $Q = f(L, K, M)$. This forms the subject-matter of the *law of returns to scale*. Generally, the terms constant and increasing returns are used with reference to constant and increasing returns to scale.

CONCEPTS OF PRODUCT

Regarding physical production by factors there are three concepts (1) *Total Product*, (2) *Average Product*, and (3) *Marginal Product*.

Total Product. Total product of a factor is the amount of total output produced by a given amount of the factor, other factors held constant. As the amount of a factor increases, the total output increases. It will be seen from Table 16.1 that when with a fixed quantity of capital (K), more units of labour are employed total product is increasing in the beginning. Thus, when one unit of labour is used with a given quantity of capital 80 units of output are produced. With two units of labour 170 units of output are produced, and with three units of labour total product of labour increases to 270 units and so on. After 8 units of employment of labour total output declines with further

Table 16.1. Returns to Labour

Units of Labour	*Total Product (Quintals)*	*Marginal Product (Quintals)*	*Average Product (Quintals)*
L	Q	$\frac{\Delta Q}{\Delta L}$	$\frac{Q}{L}$
1	80	80	80
2	170	90	85
3	270	100	90
4	368	98	92
5	430	62	86
6	480	50	80
7	504	24	72
8	504	0	63
9	495	–9	55
10	480	–15	48

increase in labour input. But the *rate* of increase in total product varies at different levels of employment of a factor. Graphically the total product curve is shown by *TP* curve in Fig. 16.1. It will be seen that in the beginning total product curve rises at an increasing rate, that is, the slope of the *TP* curve is rising in the beginning. After a point total product curve starts rising at a diminishing rate as the employment of the variable factor is increased. The fact that ultimately total product increasses at a diminishing rate has been proved by empirical evidence, as shall be seen later in our discussion of the law of diminishing returns.

Average Product. Average product of a factor is the total output produced per unit of the factor employed. Thus,

$$\text{Average Product} = \frac{\text{Total Product}}{\text{Number of units of a factor employed}}$$

If Q stands for total product, L for the number of a variable factor employed, then average product (AP) is given by:

$$AP = \frac{Q}{L}$$

We can measure the average product from the total product data given in Table 16.1. Thus when two units of labour are employed, the average product is $\frac{Q}{L} = \frac{170}{2} = 85$. Similarly, when three units of labour are employed, average product is $\frac{270}{3} = 90$ and so on. From a total product curve *TP* in

Fig. 16.1, we can measure the average product of labour. Thus, when OL_1 units of labour are employed, total product is equal to L_1A and therefore average product of labour equals $\frac{L_1A}{OL_1}$ which would be equal to the slope of the ray OA. Similarly, when OL_2 units of labour are employed, total product (TP) is L_2B which would give us average product to be equal to $\frac{L_2B}{OL_2}$ or the slope of the ray OB. Further, with the employment of labour equal to OL_3, the average product will be measured by the slope of the ray OC.

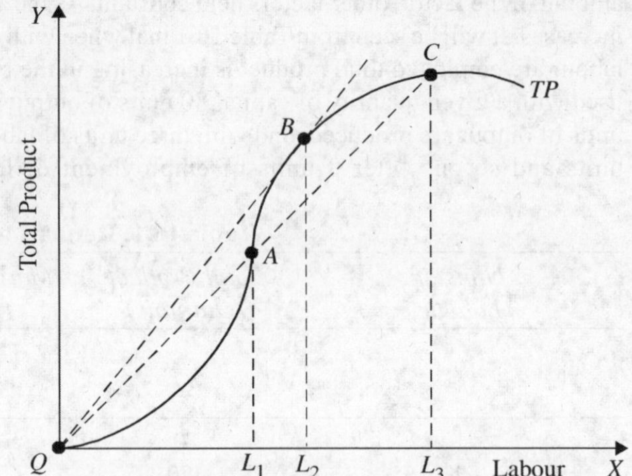

Fig. 16.1. *Measuring Average Product on a Total Production Function Curve*

It has been generally found that as more units of a factor are employed for producing a commodity, the average product first rises and then falls. As shall be seen from Table 16.1 and the Fig. 16.1, the average product curve of a variable factor first rises and then it declines. That is, the average product curve has an inverted U-shape as shown in the bottom of Fig. 16.3 where AP curve first rises and then slopes downward.

Marginal Product. *Marginal product of a factor is the addition to the total production by the employment of an extra unit of a factor*. Suppose when two workers are employed to produce wheat in an agricultural farm and they produce 170 quintals of wheat per year. Now, if instead of two workers, three workers are employed and as a result total product increases to 270 quintals, then the third worker has added 100 quintals of wheat to the total production. Thus 100 quintals is the marginal product of the third worker.

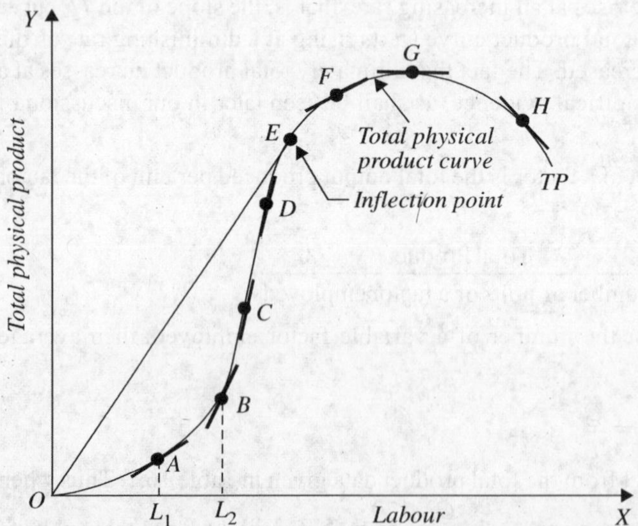

Fig. 16.2. *Measuring Marginal Physical Product of Labour*

It will be seen from Table 16.1 that marginal product of labour increases in the beginning and then diminishes. Marginal product of 8th unit of labour is zero and beyond that it becomes negative.

Mathematically, if employment of labour increases by ΔL units which yield an increase in total output by ΔQ units, the marginal physical product of labour is given by $\frac{\Delta Q}{\Delta L}$. That is,

$$MP_L = \frac{\Delta Q}{\Delta L}$$

The marginal physical product curve of a variable factor can also be derived from the total physical product curve of labour. At any given level of employment of labour, the marginal product of labour can be obtained by measuring the slope of the total product curve at a given level of labour employment. For example, in Fig. 16.2 when OL_1 units of labour are employed, the marginal physical product of labour is given by the slope of the tangent drawn at point A to the total product curve *TP*. Again, when OL_2 units of labour are employed, the marginal physical product of labour is obtained by measuring the slope of the tangent drawn to the total product curve *TP* at point *B* which corresponds to OL_2 level of labour employment and so on for further units of labour employed.

The marginal product of a factor will change at different levels of employment of the factor. It has been found that due to the law of variable proportions the marginal product of a factor rises in the beginning and then ultimately falls as more of it is used for production, *other factors remaining the same*. That is why in Fig. 16.2 marginal product (*MP*) of labour as measured by the slopes of the tangents drawn to the total product curve *TP* at various points has been shown to be rising in the beginning and then diminishing till it becomes zero at the maximum point *G* of the total product curve. Thereafter, the marginal product of labour becomes negative. As shown in the lower part of Fig. 16.3, the *MP* curve first rises and then slopes down. The relationship between average product and marginal product and how both of them are related to the total product will be explained in detail below in our analysis of the law of variable proportions.

LAW OF VARIABLE PROPORTIONS

Law of variable proportions occupies an important place in economic theory. This law examines the production function with one factor variable, keeping the quantities of other factors fixed. In other words, it refers to the input-output relation when output is increased by varying the quantity of one input. When the quantity of one factor is varied, keeping the quantity of other factors constant, the proportion between the variable factor and the fixed factor is altered; the ratio of employment of the variable factor to that of the fixed factor goes on increasing as the quantity of the variable factor is increased. *Since under this law we study the effects on output of variation in factor proportions, this is also known as the law of variable proportions*. Thus law of variable proportions is the new name for the famous "*Law of Diminishing Returns*" of classical economics. This law has played a vital role in the history of economic thought and occupies an equally important place in modern economic theory. This law has been supported by the empirical evidence about the real world. The law of variable proportions or diminishing returns has been stated by various economists in the following manner:

As equal increments of one input are added; the inputs of other productive services being held constant, beyond a certain point the resulting increments of product will decrease, i.e., the marginal products will diminish."[1] (G. Stigler)

"*As the proportion of one factor in a combination of factors is increased, after a point, first the marginal and then the average product of that factor will diminish*."[2] (F. Benham)

"*An increase in some inputs relative to other fixed inputs will, in a given state of technology, cause output to increase; but after a point the extra output resulting from the same addition of extra inputs will become less*."[3] (Paul A. Samuelson)

Marshall discussed the law of diminishing returns in relation to agriculture. He defines the law as follows: "*An increase in the capital and labour applied in the cultivation of land causes in general a less than proportionate increase in the amount of product raised unless it happens to coincide with an improvement in the arts of agriculture*."[4]

It is obvious from the above definitions of the law of variable proportions (or the law of diminishing returns) that it refers to the behaviour of output as the quantity of one factor is increased,

1. G.J. Stigler, *Theory of Price*, The MacMillan Co., 1953, p. 111.
2. F.Benham, *Economics*, 2nd edition, 1950 p. 110.
3. Paul A. Samuelson, *Economics*, 8th edition, p. 25.
4. Alfred Marshall, *Principles of Economics*, 6th edition, p. 150.

keeping the quantity of other factors fixed and further it states that the marginal product and average product will eventually decline.

Assumptions of the Law. The law of variable proportions or diminishing returns, as stated above, holds good under the following conditions:

1. First, the state of technology is assumed to be given and unchanged. If there is improvement in the technology, then marginal and average products may rise instead of diminishing.

2. Secondly, there must be some inputs whose quantity is kept fixed. This is one of the ways by which we can alter the factor proportions and know its effect on output. This law does not apply in case all factors are proportionately varied. Behaviour of output as a result of the variation in all inputs is discussed under "*returns to scale*".

3. Thirdly, the law is based upon the possibility of varying the proportions in which the various factors can be combined to produce a product. The law does not apply to those cases where the factors must be used in fixed proportions to yield a product. When the various factors are required to be used in rigidly fixed proportions, then the increase in one factor would not lead to any increase in output, that is, the marginal product of the factor will then be zero and not diminishing. It may, however, be pointed out that products requiring fixed proportions of factors are quiet uncommon. Thus, the law of variable proportion applies to most of the cases of production in the real world.

The law of variable proportions is illustrated in Table 16.1. and Fig. 16.3. We shall first explain it by considering Table 16.1. Assume that there is a given fixed amount of land, with which more units of the variable factor, labour, is used to produce agricultural output. With a given fixed quantity of land, as a farmer raises employment of labour from one unit to 7 units, the total product increases from 80 quintals to 504 quintals of wheat. Beyond the employment of 8 units of labour, total product diminishes. It is worth noting that upto the use of 3 units of labour, total product increases at an increasing rate. This fact is clearly revealed from column 3 which shows successive marginal products of labour as extra units of labour are used. *Marginal product of labour, it may be recalled, is the increment in total output due to the use of an extra unit of labour.*

It will be seen from Col. 3 of Table 16.1, that the marginal product of labour initially rises and beyond the use of three units of labour, it starts diminishing. Thus when 3 units of labour are employed, marginal product of labour is 100 and with the use of 4th and 5th units of labour marginal product of labour falls to 98 and 62 respectively. Beyond the use of eight units of labour, total product diminishes and therefore marginal product of labour becomes negative. As regards average product of labour, it rises upto the use of fourth unit of labour and beyond that it is falling throughout.

Three Stages of the Law of Variable Proportions

The behaviour of output when the varying quantity of one factor is combined with a fixed quantity of the other can be divided into three distinct stages. In order to understand these three stages it is better to graphically illustrate the production function with one factor variable. This has been done in Fig. 16.3. In this figure, on the *X*-axis the quantity of the variable factor is measured and on the *Y*-axis the total product, average product and marginal product are measured. How the total product, average product and marginal product a variable factor change as a result of the increase in its quantity, that is, by increasing the quantity of one factor to a fixed quantity of the others will be seen from Fig. 16.3. In the top panel of this figure, the total product curve *TP* of a variable factor goes on increasing to a point and after that it starts declining. In the bottom panel, average and marginal product curves of labour also rise and then decline; marginal product curve starts declining earlier than the average product curve. The behaviour of these total, average and marginal products of the variable factor as a result of the increase in its amount is generally divided into three stages which are explained below.

Stage 1. In this stage, total product curve *TP* increases at an increasing rate up to a point. In Fig. 16.3. from the origin to the point *F*, slope of the total product curve *TP* is increasing, that is, up to the point *F*, *the total product increases at an increasing rate* (the total product curve *TP* is concave upward upto the point *F*), which means that the marginal product *MP* of the variable factor is rising. From the point *F* onwards during the stage 1, the total product curve goes on rising but its slope is declining which means that from point *F* onwards *the total product increases at a diminishing rate* (total product curve *TP* is concave down-ward), *i.e.*, marginal product falls but is positive. The point *F* where the total product stops increasing at an increasing rate and starts increasing at the diminishing rate is called the *pont of inflection. Vertically corres-ponding to this point of inflection marginal product is maximum,after which it starts diminishing*. Thus, marginal product of the variable factor starts diminishing beyond *OL* amount of the variable factor. That is, law of diminishing returns starts operating in stage 1 from point *D* on the *MP* curve or from *OL* amount of the variable factor used.

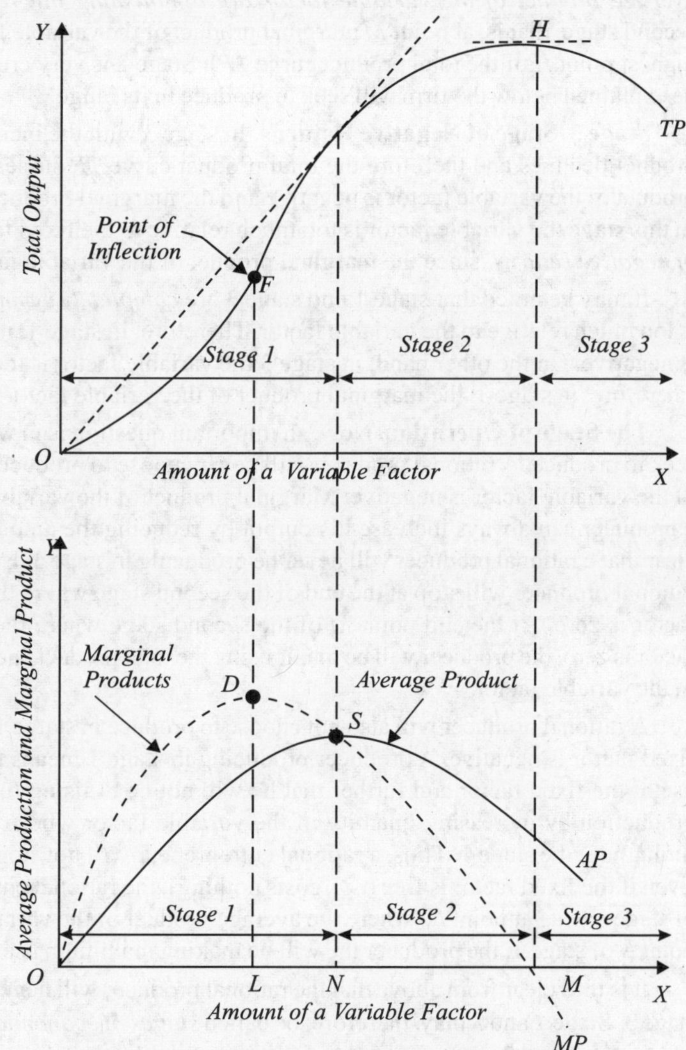

Fig. 16.3. *Three Stages of the Law of Variable Proportions*

This first stage ends where the average product curve *AP* reaches its highest point, that is, point *S* on *AP* curve or *ON* amount of the variable factor used. During stage 1, when marginal product of the variable factor is falling it still exceeds its average product and so continues to cause the average product curve to rise. Thus, during stage 1, whereas marginal product curve of a variable factor rises in a part and then falls, the average product curve rises throughout. In the first stage, the quantity of the fixed factor is too much relative to the quantity of the variable factor so that if some of the fixed factor is withdrawn, the total product will increase. Thus, in the first stage *marginal product of the fixed factor is negative*..

Stage 2. In stage 2, the total product continues to increase at a diminishing rate until it reaches its maximum point *H* where the second stage ends. In *this stage both the marginal product and the average product of the variable factor are diminishing but remain positive*. At the end of the second stage, that is, at point *M* marginal product of the variable factor is zero (corresponding to the highest point *H* of the total product curve *TP*). Stage 2 is very crucial and important because as will be explained below the firm will seek to produce in its range.

Stage 3. Stage of Negative Returns. In stage 3 with the increase in the variable factor the total product declines and therefore the total product curve *TP* slopes downward. As a result, marginal product of the variable factor is negative and the marginal product curve *MP* goes below the *X*-axis. In this stage the variable factor is too much relative to the fixed factor. This stage is called the *stage of negative returns*, since the marginal product of the variable factor is negative during this stage.

It may be noted that stage 1 and stage 3 are *completely symmetrical*. In stage 1 the fixed factor is too much relative to the variable factor. Therefore, in stage 1, marginal product of the fixed factor is negative. On the other hand, in stage 3 the variable factor is too much relative to the fixed factor. Therefore, in stage 3, the marginal product of the variable factor is negative.

The Stage of Operation. Now, an important question is in which stage a rational producer will seek to produce. A rational producer will never choose to produce in stage 3 where marginal product of the variable factor is negative. Marginal product of the variable factor being negative in stage 3, a producer can always increase his output by reducing the amount of the variable factor. It is thus clear that a rational producer will never be producing in stage 3. Even if the variable factor is free, the rational producer will stop at the end of the second stage where the marginal product of the variable factor is zero. At the end point *M* of the second stage where the marginal product of the variable factor is zero, the producer will be maximising the total product and will thus be making maximum use of the variable factor.

A rational producer will also not choose to produce in statge 1 where the marginal product of the fixed factor is negative. A producer producing in stage 1 means that he will not be making the best use of the fixed factor and further that he will not be utilising fully the opportunities of increasing production by increasing quantity of the variable factor whose average product continues to rise throughout the stage 1. Thus, a rational entrepreneur will not stop in stage 1 but will expand further. Even if the fixed factor is free (*i.e.*, costs nothing), the rational entrepreneur will stop only at the end of stage 1 (*i.e.*, at point *N*) where the average product of the variable factor is maximum. At the end point *N* of stage 1, the producer the will be making maximum use of the fixed factor.[5]

It is thus clear from above that the rational producer will never be found producing in stage 1 and stage 3. Stage 1 and 3 may, therefore, be called *stages of economic absurdity or economic non-sense*. The stages 1 and 3 represent non-economic regions in production function. A rational producer will always seek to produce in stage 2 where both the marginal product and average product of the variable factor are diminishing. At which particular point in this stage, the producer will decide to produce depends upon the prices of factors. The stage 2 represents the range of rational production decisions.

We have seen above *how output varies as the factor proportions are altered* at any given moment. We have also noticed that this input-output relation can be divided into three stages. Now, the question arises as to what causes increasing marginal returns to the variable factor in the beginning, diminishing marginal returns later and negative marginal returns to the variable factor ultimately.

---

5. The statement in respect of the non-operation by a rational producer in stage 1 applies to the producer who is working under *perfect competition* in both the product and factor markets where the price of the product sold by him and the prices of factors bought by him remain fixed. In monopoly and imperfect competition in product and factor markets where the prices of product and factors do not remain fixed with the changes in the level of output and employment, the producer might find his most profitable level of output in stage 1.

Causes of Initial Increasing Marginal Returns to a Factor

In the beginning, the quantity of the fixed factor is abundant relative to the quantity of the variable factor. Therefore, when more and more units of a variable factor are added to the constant quantity of the fixed factor, the *fixed factor is more intensively and effectively utilised*. This causes the production to increase at a rapid rate. When, in the beginning the variable factor is relatively smaller in quantity, some amount of the fixed factor may remain unutilised and therefore when the variable factor is increased *fuller utilisation of the fixed factor* becomes possible with the result that increasing returns are obtained. The question arises as to why the fixed factor is not initially taken in an appropriate quantity which suits the available quantity of the variable factor. Answer to this question is provided by the fact that generally those factors are taken as fixed which are *individisble*. Indivisibility of a factor means that due to technological requirements a minimum amount of that factor must be employed whatever the level of output. Thus, as more units of variable factor are employed to work with an indivisible fixed factor, output greatly increases in the beginning due to fuller and more effective utilisation of the latter. Thus, we see that *it is the indivisibility of some factors which causes increasing returns* to the variable factor in the beginning.

The second reason why we get increasing returns to the variable factor in the initial stage is that as more units of the variable factor are employed the efficiency of the variable factor itself increases. This is because *when there is a sufficient quantity of the variable factor, it becomes possible to introduce specialisation or division of labour which results in higher productivity*. The greater the quantity of the variable factor, the greater the scope of specialisation and hence the greater will be the level of its productivity or efficiency.

Causes of Diminishing Marginal Returns to a Factor

The stage of diminishing marginal returns in the production function with one factor variable is the most important. The question arises as to why we get diminishing marginal returns after a certain amount of the variable factor has been added to a fixed quantity of the other factor. As explained above, increasing returns to a variable factor occur initially primarily because of the more effective and fuller use of the fixed factor becomes possible as more units of the variable factor are employed to work with it. Once the point is reached at which the amount of the variable factor is sufficient to ensure the efficient utilisation of the fixed factor, then further increases in the variable factor will cause marginal and average products of a variable factor to decline because the fixed factor then becomes inadequate relative to the quantity of the variable factor. In other words, *the contributions to the production made by the variable factor after a point become less and less because the additional units of the variable factor have less and less of the fixed factor to work with*. The production is the result of the co-operation of various factors aiding each other. Now, how much aid one factor provides to the others depends upon how much there is of it. Eventally, the fixed factor is abundant relative to the number of the variable factor and the former provides much aid to the later. Evantally, the fixed factor becomes more and more scarce in relation to the variable factor so that as the units of the variable factor are increased they receive less and less aid from the fixed factor. As a result, the marginal and average products of the variable factor decline ultimately.

The phenomenon of diminishing marginal returns, like that of increasing marginal returns, rests upon the indivisibility of the fixed factor. As explained above, the important reason for increasing returns to a factor in the beginning is the fact that the fixed factor is indivisible which has to be employed whether the output to be produced is small or large. When the indivisible fixed factor is not being fully used, successive increases in a variable factor add more to output since fuller and more efficient use is made of the indivisible fixed factor. But there is generally a limit to the range of employment of the variable factor over which its marginal and average products will increase. There will usually be a level of employment of the variable factor at which indivisible fixed factor is being as fully and efficiently used as possible. It will happen when the variable factor has increased to such an amount that the fixed indivisible factor is being used in the '*best or optimum proportion*' with the variable factor.

Once the optimum proportion is disturbed by further increases in the variable factor, returns to a variable factor (*i.e., marginal product and average product*) will diminish primarily because the indivisible factor is being used too intensively, or in other words, the fixed factor is being used in non-optimal proportion with the variable factor. Just as the marginal product of the variable factor increases in the first stage when better and fuller use of the fixed indivisible factor is being made, so the marginal product of the variable factor diminishes when the fixed indivisible factor is being worked too hard. *If the fixed factor was perfectly divisible, neither the increasing nor the diminishing returns to a variable factor would have occurred.* If the factors were perfectly divisible, then there would not have been the necessity of taking a large quantity of the fixed factor in the beginning to combine with the varying quantities of the other factor. In the presence of pefect divisibility, the optimum proportion between the factors could have always been achieved. Perfect divisibility of the factors implies that a small firm with a small machine and one worker would be as efficient as a large firm with a large machine and many workers. The productivity of the factors would be the same in the two cases. Thus, we see that if the factors were perfectly divisible, then the question of varying factor proportions would not have arisen and hence the phenomena of increasing and diminishing marginal returns to a variable factor would not have occurred. Prof. Bober rightly remarks: "Let divisibility enter through the door, law of variable proportions rushes out through the window."[6]

Joan Robinson goes deeper into the causes of diminishing returns. She holds that *the diminishing marginal returns occur because the factors of production are imperfect substitutes for one another*. As seen above, diminishing returns occur during the second stage since the fixed factor is now inadequate relatively to the variable factor. Now, a factor which is scarce in supply is taken as fixed. When there is a scarce factor, quantity of that factor cannot be increased in accordance with the varying quantities of the other factors, which, after the optimum proportion of factors is achieved, results in diminishing returns. If now some factors were available which was perfect substitute of the scarce fixed factor, then the paucity of the scarce fixed factor during the second stage would have been made up by the increase in supply of its perfect substitute with the result that output could be expanded without diminishing returns. Thus, even if one of the variable factors which we add to the fixed factor were perfect substitute of the fixed factor, then when, in the second stage, the fixed factor becomes relatively deficient, its deficiency would have been made up the increase in the variable factor which is its perfect substitute.

Thus, Joan Robinson says, "*What the Law of Diminishing Returns really states is that there is a limit to the extent to which one factor of production can be substituted for another, or, in other words, that the elasticity of substitution between factor is not infinite*. If this were not true, it would be possible, when one factor of production is fixed in amount and the rest are in perfectly elastic supply, to produce part of the output with the aid of the fixed factor, and then, when the optimum propertion between this and other factors was attained, to substitute some other factor for it and to increase output at constant cost."[7] *We, therefore, see that diminishing returns operate because the elasticity of substitution between factors is not infinite.*

Explanation of Negative Marginal Returns to a Factor

As the amount of a variable factor continues to be increased to a fixed quantity of the other factor, a stage is reached when the total product declines and the marginal product of the variable factor becomes negative. *This phenomenon of negative marginal returns to the variable factor in stage 3 is due to the fact that the number of the variable factor becomes too excessive relative to the fixed factor* so that they obstruct each other with the result that the total output falls instead of

6. M.M. Bober, *Intermediate Price and Income Theory*.
7. Joan Robinson, *The Economics of Imperfect Competition*

rising. Besides, too large a number of the variable factor also impairs the efficiency of the fixed factor. The proverb "*too many cooks spoil the broth*" aptly applies to this situation. In such a situation, a reduction in the units of the variable factor will increase the total output.

APPLICABILITY OF THE LAW OF DIMINISHING RETURNS

We have discussed above the law of variable proportions which states marginal physical product of a variable factor eventually diminishes, even if it increases in the beginning. Uptill Marshall, it was thought that there were three laws of production–diminishing, constant and increasing returns – which were quiet distinct and separate. Now, modern economists have veered round to the view that diminishing, constant and increasing returns to a factor are not three separate laws but they are three phases of one general law of variable proportions. Moreover, uptill Marshall it was thought that law of diminishing returns applied to agriculture while the manufacturing industries were characterised by increasing or constant returns. But this is no longer believed to be correct, Law of diminishing returns has a vast general applicability. This law applies as much to industries as to agriculture. Whenever some factors are fixed and the amount of an other increases, then the technology remaining the same, diminishing returns to factor are bound to occur eventually both in agriculture and industries. We have given above the various definitions of the law of variable proportions which lay stress on its general and universal applicability.

We have given the theoretical reasons for the occurrence of diminishing returns. The occurrence of diminishing marginal physical returns after a point has been confirmed by the overwhelming empirical evidence. Indeed, *if the diminishing returns did not occur we could grow sufficient amount of foodgrains even in a flower pot by using more doses of labour and capital*. If the constant returns could be obtained by applying more labour on a given piece of land, then as the population increased we could use more labour on that land to get proportionate increase in agricultural output. In that case the world, especially the developing countries like India, would not have to face the problems of food shortage and over-population. Professor R.G. Lipsey is right when he says, "Indeed, were the hypothesis of diminishing returns incorrect, there would need to be no fear that the present population explosion will bring with it a food crises. If the marginal product of additional workers applied to a fixed quantity of land were constant, then world food production could be expanded in proportion to the increase in population merely by keeping the same proportion of the population on farms. As it is, diminishing returns means an inevitable decline in the marginal product of each additional labourer as an expanding population is applied with static techniques, to a fixed world supply of aggricultural land."[8]

But from above it should not be understood that because of diminishing returns there can be no hopes for raising the living standards of mankind, especially of the people in developing countries. To predict such a gloomy prospects for the future of mankind on the basis of the law of diminishing returns is wholly unwarranted.

Technological Progress and Diminishing Returns

Some people misunderstood the law and asserted that as the population increased, the quantity of land remaining unchanged, the productivity per person would necessarily decline. But this is quite wrong. The law of diminishing returns, as stated above, has a great proviso that technical knowledge, equipment, etc., remain the same. In the present-day developed countries, though population has increased, agricultural productivity has greatly gone up instead of diminishing. This is so because present-day developed countries have made an impressive progress in technical knowledge, resulting in new and superior machinery and other equipments and the use of fertilisers. Capital equipment per worker engaged in agriculture has greatly increased. As a result, agricultural productivity has registered a tremendous increase in the present-day advanced countries. On the other hand, developing countries have not made much progress in technical knowledge and in

8. Richard G.Lipsey, *Introduction to Positive Economics*, 3rd edition, p. 216.

accumulating and using sufficient capital equipment like machinery, tools, fertilisers, etc. It is no wonder, therefore, that agricultural productivity has not risen. In fact, marginal productivity of labour has gone down. The phenomenon of disguised unemplyment found in agriculture of developing countries reveals that marginal productivity of a worker is zero or nearly zero. It is thus clear that actual experience regarding the behaviour of agricultural productivity in both developed and developing countries is in no way a contradiction to the law of diminishing returns, the operation of which is subject to the condition that technical knowledge, capital equipment and other aids of production remain the same.

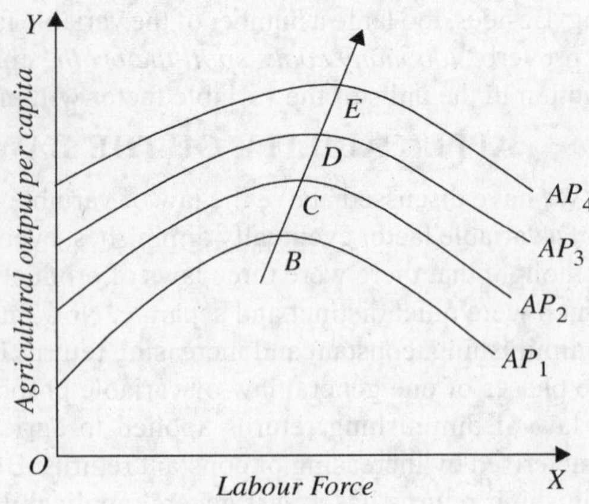

Fig. 16.4. *Rising Agricultural Productivity due to Technological Progress*

That the increase in agriculture productivity or output per capita in agriculture on account of the advances in technology is in no way contradiction to the principle of diminishing returns can be illustrated with the help of Figure 16.4. It will be seen from Fig. 16.4 that the curves of agricultural output per capita rise up to a certain point and then begin to fall due to the diminishing returns. However, in the developed countries agrigultural productivity curve has been shifting upward because of the progress in technology in agriculture over time. It is this progress in agricultural technology that has brought about shift in the productivity curve AP_1 to AP_2, from AP_2 to AP_3, and from AP_3 to AP_4. If there had been no progress in technology and increase in capital equipment, the agricultural productivity curve would have remained unchanged at AP_1 and therefore with increase in population and hence in labour force, the agricultural productivity would have diminished as is indicated by the falling part of AP_1 curve. But in actual practice along with the increase in the labour force, the average productivity of labour in agriculture has been rising due to the progress in technology, that is, due to the upward shift of *AP* curve. As a result of progress in technology, the economies of developed countries have been moving along the arrow mark line *BE* which indicates the rising average productivity despite the increase in the labour force. Thus, we see that diminishing returns with a proviso that 'other things remaining unchanged' is in no way contradiction to the actual experience of rise in average agricultural productivity over time in developed countries. It is the proviso 'other things remaining unchanged' that has not been fulfilled.

We thus see that despite the validity of the diminishing returns developing countries like India can rapidly increase their agricultural production by making advancement in agricultural technology. As demonstrated above, we can suspend the operation of diminishing returns by continually improving the techniques of production through progress in science and technology. Of course, if we fail to improve our technology sufficiently, diminishing returns would assert themselves and create the problems of food shortage and starvation. We, therefore, conclude "unless there is continual and rapidly accelerating improvements in the techniques of production, the population explosion must bring with it declining living standards over much of the world and eventual widespread famine."[9]

QUESTIONS FOR REVIEW

1. What is production function ? Distinguish between fixed inputs and variable inputs. Is the distinction between the two relevant in the long run ?

9. R.G. Lipsey, op. cit., p. 216

2. Explain the concept of production function ? Why is it useful in the analysis of firm's behaviour ?
3. What is the relationship between marginal product and average product of labour (or a variable input) ?
4. State the law of diminsihing returns. Why does diminsihing returns to a variable input occur eventually ? Can they become negative? If so, why ?
5. What are the three stages of short-run production function? Why does it not make any economic sense to produce in stage 1 or 3?

   [**Hints :** It is irrational for the firm to produce in either stage I or stage III. The operating in stage III will be irrational because the marginal product of the variable factor is negative in this stage. As a result, total production can be increased by using less of the variable factor. Therefore, even if the variable factor is absolutely free (that is, its price is zero), the firm will never use it beyond the end of stage II.

   The production in stage I is also irrational. In stage 1, the fixed factor, capital or land whichever be the fixed factor, is so abundant in relation to the variable factor that its marginal product is negative. Since a firm operating in stage I where the marginal product of the fixed factor is negative, can increase the level of output by reducing the amount of the fixed factor, it does not make any economic sense to produce in stage 1.]
6. How is the law of diminishing returns reflected in the shape of the total produce curve ? If the total product curve increases at a decreasing rate from the very beginning what would be the shapes of corresponding marginal and average product curves ?
7. Explain the law of diminishing returns. Mention on what assumptions it is based. How Malthus used the law to predict gloomy forecast for future mankind ? What mistake did he commit in making this gloomy forceast ?
8. Fill in the blanks in the following table :

Number of variable input	Total output (number of of units)	Marginal product of the variable input	Average product of the variable input
3	—	18	30
4	—	20	—
5	130	—	—
6	—	5	—
7	—	—	19.5

9. As the quantity of a variable input increases, explain why the point where marginal product begins to decline is reached before the point where average product begins to decline. Also explain why the point where average product begins to decline is reached before the point where total output begins to decline.

   [**Hints.** This is due to the usual average-marginal relationship in the production process. Due to the operation of diminishing returns, marginal returns begin to decline at some point but for some range though diminishing it remains greater than average product. Therefore, average product continues to increase. Only when in its diminishing phase marginal product becomes less than average product, average product starts declining. That is, why marginal product curve cuts the average product curve at the latter's highest point. Marginal product continues diminishing after it is equal to the maximum average product but remains positive for some range which causes the total output to continue increasing. When marginal product becomes zero, the total product reaches its maximum level. As a result, total output continues increasing even after the maximum average product and begins to decline only when marginal product becomes negative]

Chapter 17
PRODUCTION FUNCTIONS WITH TWO VARIABLE FACTORS

In the last chapter we explained the production function with a single variable factor, holding other factors constant. In the present chapter we are concerned with the analysis of production function when two factors are taken as variables in the production process. For the analysis of production function with two variable factors we make use of the concept called isoquants or iso-*product curves* which are similar to indifference curves of the theory of demand. Therefore, before we explain the production function with two variable factors and returns to scale, we shall explain the concept of isoquants (that is, equal product curves) and their properties.

ISOQUANTS

Isoquants, which are also called equal product curves, are similar to the indifference curves of the theory of consumer's behaviour. An isoquant represents all those factor combinations which are capable of producing the same level of output. The *isoquants are thus contour lines which trace the loci of equal outputs*. Since an isoquant represents those combinations of inputs which will be capable of producing an equal quantity of output, the producer would be indifferent between them. Therefore, isoquants are also often called equal product curves *production-indifference curves*.

Table 17.1

Factor Combinations to Produce a Given or Level of Output

Factor Combinations	*Labour*	*Capital*
A	1	12
B	2	8
C	3	5
D	4	3
E	5	2

The concept of isoquant can be easily understood from Table 17.1. It is presumed that two factors labour and captial are being employed to produce a product. Each of the factor combinations *A, B, C, D* and *E* produces the same level of output, say 100 units. To start with, factor combination *A* consisting of 1 unit of labour and 12 units of capital produces the given 100 units of output. Similarly, combination *B* consisting of 2 units of labour and 8 units of capital, combination *C* consisting of 3 units of labour and 5 units of capital, combination *D* consisting of 4 units of labour and and 3 units of capital, combination *E* consisting of 5 units of labour and 2 units of capital are capable of producing the same amount of output, *i.e.*, 100 units. In Fig. 17.1 we have plotted all these combinations and by joining them we obtain an isoquant showing that every combination represented on it can produce 100 units of output.

Though isoquants are similar to be indifference curves of the theory of consumer's behaviour, there is one important difference between the two. An indifference curve represents all those combinations of two goods which provide the same satisfaction or utility to a consumer but no attempt is made to specify the level of utility in exact quantitative terms it stands for. This is so

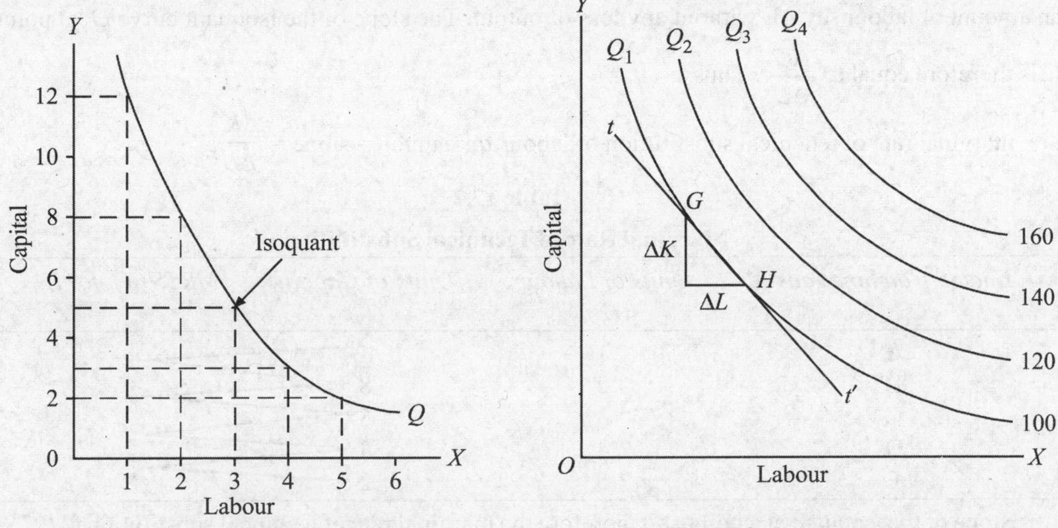

Fig. 17.1. *Isoquant* **Fig. 17.2.** *Isoquant map*

because the cardinal measurement of satisfaction or utility in unambiguos therms is not possible. That is why we usually label indifference curves by ordinal numbers as I, II, III etc. indicating that a higher indifference curve represents a higher level of satisfaction than a lower one, but the information as to how much one level of satisfaction is greater than another is not provided. On the other hand, we can label isoquants in the physical units of output without any difficulty. Production of a good being a physical phenomenon lends itself easily to absolute measurement in physical units. Since each isoquant represents a specified level of production, it is possible to say by how much one isoquant indicates greater or less production than another. In Fig. 17.2 we have drawn an *isoquant-map or equal-product map* with a set of four isoquants which represent 100 units, 120 units,140 units and 160 units of output respectively. Then, from this set of isoquants it is very easy to judge by how much production level on one isoquant curve is greater or less than on another.

MARGINAL RATE OF TECHNICAL SUBSTITUTION

Marginal rate of technical substitution in the theory of production is similar to the concept of marginal rate of substitution in the indifference curve analysis of consumer's demand. Marginal rate of technical substitution indicates the rate at which factors can be substituted at the margin without altering the level of output. More precisely, *marginal rate of technical substitution of labour for capital may be defined as the number of units of capital which can be replaced by one unit of labour, the level of output remaining unchanged.* The concept of marginal rate of technical substitution can be easily understood from Table 17.2.

Each of the factor combinations *A*, *B*, *C*, *D*, and *E* yields the same level of output. Moving down from combination *A* to combination *B*, 4 units of capital are substituted by 1 unit of labour in the production process without any change in the level of output. Therefore, marginal rate of technical substitution of labour for capital is 4 at this stage. Switching from input combination *B* to input combination *C* involves the replacement of 3 units of capital by an additional unit of labour, output remaining the same. Thus, the marginal rate of technical substitution is now 3. Likewise, marginal rate of technical substitution of labour for capital between factor combinations *C* and *D* is 2, and between factor combinations *D* and *E* it is 1.

The marginal rate of technical substitution at a point on an isoquant (an equal product curve) can be known from the slope of the isoquant at that point. Consider a small movement down the equal product curve from G to H in Fig. 17.2 where a small amount of capital, say ΔK is substituted by an amount of labour say ΔL without any loss of output. The slope of the isoquant curve Q_1 at point G is therefore equal to $\frac{\Delta K}{\Delta L}$. Thus,

$$\text{marginal rate of tehcnical substitution of labour for capital} = \text{slope} = \frac{\Delta K}{\Delta L}.$$

Table 17.2.

Marginal Rate of Technical Substitution

Factor Combinations	*Units of Labour (L)*	*Units of Capital (K)*	*MRTS of L for K*
A	1	12	
B	2	8	4
C	3	5	3
D	4	3	2
E	5	2	1

Slope of the isoquant at a point and therefore the marginal rate of technical substitution (*MRTS*) between factors can also be known by the slope of the tangent drawn on the isoquant at that point. In Fig. 17.3 the tangent TT' is drawn at point K on the given isoquant Q. The slope of the tangent TT' is equal to $\frac{OT}{OT'}$. Therefore, the marginal rate of substitution at point K on the isoquant Q is equal to $\frac{OT}{OT'}$. JJ' is the tangent at point L drawn to the isoquant Q. Therefore, the marginal rate of technical substitution of labour for capital at point L is equal to OJ/OJ'.

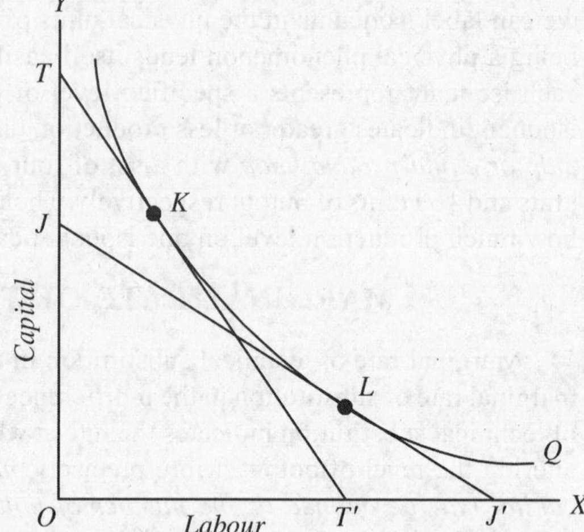

Fig. 17.3. *MRTS is given by the slope of an isoquant at a point.*

An important point to be noted about the marginal rate of technical substitution is that it is equal to the ratio of the marginal physical products of the two factors. Since, by definition, output remains constant on an isoquant the loss in physical output from a small reduction in capital will be equal to the gain in physical output from a small increment in labour. The loss in output is equal to the marginal physical product of capital (*MP*) multiplied by the amount of reduction in capital. The gain in output is equal to the marginal physical product of labour (*MP*) multiplied by the increment in labour.

Accordingly, along an isoquant

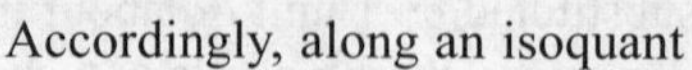

$$\Delta K.MP_K + \Delta L.MP_L = 0$$

$$\Delta K \times MP_K = \Delta L \times MP_L$$

$$\frac{\Delta K}{\Delta L} = \frac{MP_L}{MP_K}$$

But $\frac{\Delta K}{\Delta L}$, by definition, is the marginal rate of technical substitution of labour for capital

Therefore, $MRTS_{LK} = \frac{MP_L}{MP_K}$

We thus see that marginal rate of technical substitution of labour for capital is the ratio of marginal physical product of the two factors.

Diminishing Marginal Rate of Technical Substitution. An improtant characteristic of marginal rate of technical substitution is that it diminishes as more and more of labour is substituted for capital. In other words, as the quantity of labour used is increased and the quantity of capital employed reduced, the amount of capital that is required to be replaced by an additional unit of labour so as to keep the output constant will diminish. This is known as the principle of diminishing marginal rate of technical substitution. This principle of diminishing marginal rate of technical substitution is merely an extension of the law of diminishing returns to the relation between the marginal physical products of the two factors. Along an isoquant as the quantity of labour is increased and the quantity of capital is reduced, the marginal physical product of labour diminishes and the marginal physical product of capital increases. Therefore, less and less of capital is required to be substituted by an additional unit of labour so as to maintain the same level of output.

It may also be noted that the rate at which marginal rate of technical substitution diminishes is a measure of the extent to which the two factors can be substituted for each other. The smaller the rate at which the marginal rate of technical substitution diminishes, the greater the degree of substitutability between the two factors. If marginal rate of substitution between any two factors does not diminish and remains constant, the two factors are *perfect substitutes of each other*.

GENERAL PROPERTIES OF ISOQUANTS

The isoquants normally possess properties which are similar to those generally assumed for indifference curves of the theory of consumer's behaviour. Moreover, the properties of isoquants can be proved in the same manner as in the case of indifference curves. The following are the improtant properties of isoquants.

1. **Isoquants, like indifference curves, slope downward from left to right (i.e., they have a negative slope).** This is so because when the quantity of a factor, say labour, is increased, the qunatity of other capital *i.e.,* capital must be reduced so as to keep output constant on a given isoquant. This downward-sloping property of isoquants follows from a valid assumption that the marginal physical products of factors are positive, that is, the use of additional units of factors yield positive incre*ments* in output. In view of this when one factor is increased yielding positive marginal products, the other factor must be *reduced* to hold the level of output constant otherwise the output will increase and we will switch over to a higher isoquant.

The assumption that the marginal physical product of a factor is positive is quite reasonable. In the discussion of the law of variable proportions we saw that in the stage III, when the units of the variable factor, say labour, become excessive, it causes such an overcrowding on a fixed capital equipment (or on a given piece of land if land is the fixed factor) that they obstruct each other resulting in negative marginal products of labour, that is, the use of additional units of labour reduce total output. This could happen but no rational producer who aims to minimise cost or maximise profits will employ units of a factor to the point where its marginal product has become negative because positive prices have to be paid for them. Thus, in view of the positive prices that have to be paid for the units of a factor, we rule out the use of the units of a factor that have negative or zero marginal products.

Thus, with labour measured on the *X*-axis and capital on the *Y*-axis if the isoquant is a horizontal straight line, this would indicate that the marginal products of labour (MP_L) are zero. Likewise, vertical isoquant would indicate marginal products of capital (MP_K) are zero. Further, an upward sloping isoquant implies that either the marginal products of the two factors are zero or one of the two factors has negative marginal products and the other has positive marginal products. It is also worth noting that the *upward-sloping isoquant implies that the same output can be produced with the use of less of both the factors*, that is, marginal products of at least one factor is negative. In this situation when every reduction in both the factors used does not affect output, the producer will not reach an equilibrium position. It follows from above that over the economically relevant stage of production when the marginal products of the factors are positive we have *downward sloping isoquants*.

2. **No two isoquants can intersect each other.** If the two isoquants, one corresponding to 20 units of output and the other to 30 units of output intersect each other, there will then be a common factor combination corresponding to the point of intersection. It means that the same factor combination which can produce 20 units of output according to one isoquant can also produce 30 units of output according to the other isoquant. But this is quite absurd. How can the same factor combination produce two different levels of output, technique of production remaining unchanged.

3. **Isoquants, like indifference curves, are convex to the origin.** The convexity of isoquant curves means that as we move down the curve successively smaller units of capital are required to be substituted by a given increment of labour so as to keep the level of output unchanged. Thus, the convexity of equal product curves is due to the diminishing marginal rate of technical substitution of one factor for the other.

If the isoquants were concave to the origin, it would mean that the marginal rate of technical substitution increased as more and more units of labour were substituted for capital. This could be valid if the law of increasing returns applied. Since it is the law of diminishing returns which is more true of the real world, the principle of diminishing marginal rate of technical substitution generally holds good and it makes the isoquants convex to the origin. We have seen above that marginal rate of technical substitution diminishes because of diminishing marginal returns to a factor as we increase its quantity used. Therefore, the *convexity of isoquants implies the diminishing returns to a variable factor*. We have seen that there are diminishing returns to a factor because of the fact that different factors are imperfect substitutes of each other in the production of a good.

In general, convexity of isoquants implies that *it becomes progressively more difficult or harder to substitute one factor for another* as we move along an isoquant and increase the use of one factor substituting the other factor. Thus, if it is difficult to substitute a factor, say labour for capital, it will then require a relatively larger amount of labour to replace a unit of capital, the level of output being held constant. Thus, *the curvature or convexity of the isoquants indicates the ease with which one factor can be substituted for another. The more convex an isoquant, the harder it is to substitute one factor for another.*

Isoquants of Perfect Substitutes and Complements

There are two exceptions to this general property of the convexity of isoquants. One is the case of factors which are perfect substitutes of each other. When the two factors are perfect substitutes of each other, then each of them can be used equally well in place of the other. For all intents and purposes they can be regarded as the same factor. Therefore, the *marginal rate of technical substitution between two perfect substitute factors remains* constant. Since marginal rate of technical substitution remains the same throughout, the isoquants of perfect substitutes are straight lines, as shown in Fig. 17.4 instead of being convex to the origin.

Another exceptional case is of factors which are perfect complements and for which the isoquants are right-angled as shown in Fig. 17.5. The *perfect complementary factors are those which are jointly used for production in a fixed proportion*. Thus, in Fig. 17.5, *OA* of factor *X* and *OB* of factor Y are used to yield a level of output represented by isoquant Q_1. An increase in one factor without the required proportionate increase in the other factor will yield no additional output whatsoever. That is why the isoquant is right-angled (with two arms, one is a vertical straight line and the other is a horizontal straight line) at the combination consisting of a given proportion of the two factors.

Consider isoquant or equal product curve Q_1 in Fig. 17.5 where output Q_1 can be produced by the combination H consisting of OA amount of factor X and OB amount of factor Y. If now the amount

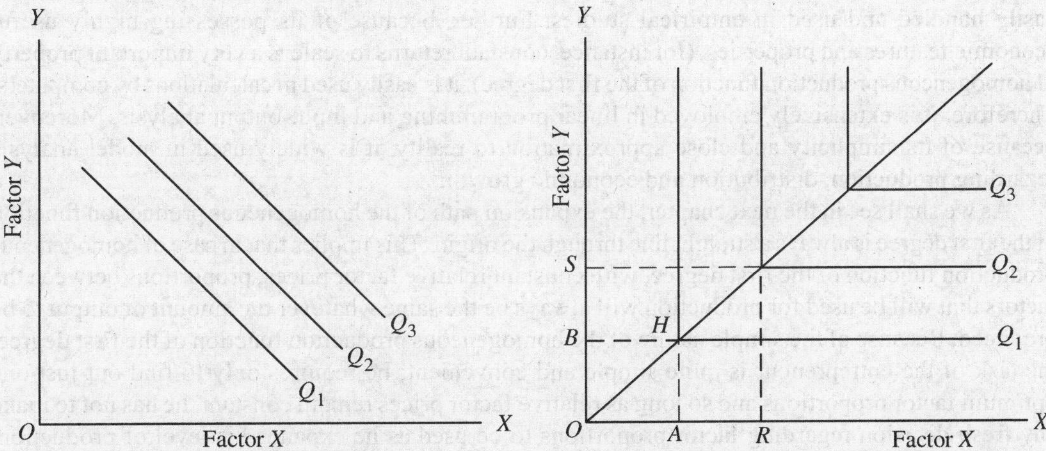

Fig. 17.4. *Perfect Substitutes* **Fig. 17.5.** *Perfect Complementary Factors*

of factor X is increased beyond OA without the increase in the factor Y, output will not rise and hence the lower portion of isoquant is a horizontal straight line. Likewise, if the amount of factor Y is increased beyond OB without the increase in factor X, the output will remain the same and hence the upper portion of the equal product curve is a vertical straight line. In case of perfect complementary factors, output can be increased only by increasing the amount of both the factors by the required given proportion. Thus, in Fig. 17.5, if the amount of factor X is increased to OR (which is twice OA), then the amount of factor Y will have to increase by OS (which is twice OB) so that we have the same factor proportion, output increases by the same proportion as the increase in factors and we have a new isoquant Q_2. It should be noted that no substitution of factors is possible in case of perfect complements.

LINEAR HOMOGENEOUS PRODUCTION FUNCTION

Production function can take several forms but a particular form of production function enjoys wide popularity among the economists. This is a linear homogeneous production function, that is, production function which is homogeneous of the first degree. *Homogeneous production function of the first degree implies that if all factors of production are increased in a given proportion, output also increases in the same proportion.* Hence linear homogeneous production function represents the case of constant returns to scale. If there are two factors X and Y, then homogeneous production function of the first degree can be mathematically expressed as:

$$mQ = f(mX, mY)$$

where Q stands for the total production and m is any real number.

The above function means that if factors X and Y are increased by m-times, total production Q also increases by m-times. It is because of this that homogeneous function of the first degree yields constant returns to scale.

More generally, a homogeneous production function can be expressed as

$$Qm^k = (mX, mY)$$

where m is any real number and k is constant. This function is homogeneous of the kth degree. If k

is equal to one, then the above homogeneous function becomes homogeneous of the first degree. If k is equal to two, the function becomes homogeneous of the 2nd degree. If k is greater than one, the production function will yield increasing returns to scale. If, on the other hand, k is less than 1, it will yield decreasing returns to scale.

Linear homogeneous production function is extensively used in empirical studies by economists. This is because in view of the limited analytical tools at the disposal of the economists, it can be easily handled and used in empirical studies. Further, because of its possessing highly useful economic features and properties, (for instance, constant returns to scale is a very important property of homogeneous production function of the first degree), it is easily used in calculations by computers. Therefore, it is extensively employed in linear programming and input-output analysis. Moreover, because of its simplicity and close approximation to reality, it is widely used in model analysis regarding production, distribution and economic growth.

As we shall see in the next chapter, the expansion path of the homogeneous production function of the first degree is always a straight line through the origin. This implies that in case of homogeneous production function of the first degree, with constant relative factor prices, proportions between the factors that will be used for production will always be the same whatever the amount of output to be produced. Because of the simple nature of the homogeneous production function of the first degree, the task of the entrepreneur is quite simple and convenient; he requires only to find out just one optimum factor proportions and so long as relative factor prices remain constant, he has not to make any fresh decision regarding factor proportions to be used as he expands his level of production. Moreover, the use of the same optimum factor proportions (with constant relative factor prices) at different levels of output in homogeneous production function of the first degree is also very useful in input-output analysis. Homogeneous production function of the first degree, which, as said above, implies constant returns to scale, has been actually found in agriculture as well as in many manufacturing industries. In India, farm management studies have been made for various states and data have been collected for agricultural inputs and outputs. Analysing the data collected in these farm management studies, Dr. A.M. Khusro has reached the conclusion that constant returns to scale prevail in Indian agriculture.[1] Likewise, empirical studies conducted in the United States and Britain have found that many manufacturing industries are characterised by a long phase of constant long-run average cost (*LAC*) curve which again implies constant returns to scale and homogeneous production function of the first degree.

Cobb-Douglas Production Function

Many economists have studied actual production functions and have used statistical methods to find out relations between changes in physical inputs and physical outputs. A most familiar empirical production function found out by statistical methods is the Cobb-Douglas production function. Originally Cobb-Douglas production function was applied not to the production process of an individual firm but to the whole of the manufacturing industry. Output in this function was thus manufacturing production. Two factor Cobb-Douglas production function takes the following mathematical form:

$$Q = AL^aK^b$$

where Q is the manufacturing output, L is the quantity of employed, K is the quantity of capital employed and A, a and b are parameters of the function. Economists use statistical methods to estimates A, a and b which determine the exact shape of the production function. It is important to know that the larger A is, the more output can be produced by a firm from *a given amounts* of labour and capital. We will show below that exponents a and b of labour and capital respectively in Cobb-Douglas production function measure the output elasticity of labour and capital respectively.

1. See his article "Returns to Scale in Indian Agriculture, *The Indian Journal of Agricultural Economics*, Vol. XIX-Dec. 1964, reprinted in "*Readings in Agricultural Development*", edited by A.M. Khusro, Allied Publishers, 1968.

Roughly speaking, Cobb-Douglas production function found that about 75% of the increase in manufacturing production was due to the labour input and the remaining 25% was due to the capital input. Cobb-Douglas production function can be estimated by regression analysis by first converting it into the following log form.

$$\log Q = \log A + a \log L + b \log K$$

Cobb-Douglas production function in log form is a linear function.

Cobb-Douglas production function is used in empirical studies to estimate returns to scale in various industries as to whether they are increasing, constant or decreasing. Further, Cobb-Douglas production function is also frequently used to estimate output elasticities of labour and capital. *Output elasticity of a factor shows the percentage change in output as result of a given percentage change in the quantity of a factor.*

Cobb-Douglas production has the following useful properties:

1. *The sum of the exponents of factors in Cobb-Douglas production function, ($Q = AL^aK^b$), that is, a+b measures returns to scale.*

   If $a + b = 1$, returns to scale are constant

   If $a + b > 1$, returns to scale are increasing

   If $a + b < 1$, returns to scale are decreasing

2. In Cobb-Douglas production function, there is important relationship between average product of labour (AP_L) and marginal product of labour (MP_L). Let us first know average product and marginal product of labour in Cobb-Douglas production ($Q = AL^aK^{1-a}$).

$$AP_L = \frac{AL^a K^{1-a}}{L} = \frac{Q}{L}$$

$$MP_L = \frac{\Delta Q}{\Delta L} = aAL^{a-1} K^{1-a} = a\frac{AL^a K^{1-a}}{L}$$

$$= a\frac{Q}{L}$$

   Thus, marginal product of labour is a times the average product of labour $\left(\frac{Q}{L}\right)$.

3. *In a linear homogeneous Cobb-Douglas production function, $Q = AL^aK^{1-a}$, average and marginal products of a factor depend on* ***ratio of factors*** *used in production and are independent of the absolute quantities of the factors used.* In the linear homogeneous Cobb-Douglas production function :

$$Q = A\,L^aK^{1-a} \qquad \text{where} \quad a + 1 - a = 1$$

Average product of labour can be obtained from dividing the production function by the amount of labour L

$$\text{Average product of labour} = \frac{AL^a K^{1-a}}{L} = \frac{AK^{1-a}}{L^{1-a}} = A\left(\frac{K}{L}\right)^{1-a}$$

Since A and a are *constants, average product of labour will depend on the ratio of the factors* $\left(\frac{K}{L}\right)$ used *and will not depend upon their absolute quantities.*

Like the average product of a factor, the marginal product of a factor of a *linear homogeneous* Cobb-Douglas production function also depends upon the *ratio* of the factors used and is *independent of the absolute quantities of the factors used.* Note that marginal product of a factor, say labour, is first derivative of the production function with respect to labour. The marginal product of labour of Cobb-Douglas production can be obtained as under:

$$Q = AL^a K^{1-a}$$

Marginal Product of labour, $$\frac{dQ}{dL} = AaL^{a-1} K^{1-a}$$

$$= \frac{AaL^a k^{1-a}}{L}$$

$$= \frac{AaL^{a-a} K^{1-a}}{L^{1-a}} = \frac{AaK^{1-a}}{L^{1-a}} = Aa\left(\frac{K}{L}\right)^{1-a}$$

Since A and a are constants, marginal product of labour will depend on capital-labour ratio $\left(\frac{K}{L}\right)$, that is, capital per worker and is independent of the *absolute quantities* of the factors employed.

4. *Linear homogenous Cobb-Douglas Production Function exhibits diminishing returns to a variable factor.*

   To prove that the linear cobb-Douglas production function exhibits diminishing marginal product of a variable factor we have to show that the second derivative of the function with respect to a variable factor (say labour) is negative. Linear Cobb-Douglas production function can be written as

   $Q = AL^{0.75}K^{0.25}$ where $0.75 + 0.25 = 1$

   $$MP \text{ of labour } \left(\frac{\partial Q}{\partial L}\right) = 0.75\, AL^{0.75-1}\, K^{0.25}$$

   $$= 0.75AL^{-0.25}\, K^{0.25}$$

   $$\text{Second derivative } \left(\frac{\partial^2 Q}{\partial L^2}\right) = -0.25 \times 0.75\, AL^{-1.25}\, K^{0.25}$$

   Thus, since second derivative of Cobb-Douglas production function is negative, it shows diminishing marginal returns to the variable factor i.e., labour.

5. Fourthly, the exponents of labour and capital in *Cobb-Douglas production function measure output elasticities of labour and capital respectively*. Remember that output elasticity of a factor shows the percentage change in output resulting from a given percentage change in the quantity of a factor, other factors remaining constant. Thus,

   $$\text{Output elasticity of labour} = \frac{\partial Q}{\partial L} \cdot \frac{L}{Q}$$

   where $\frac{\partial Q}{\partial L}$ = marginal productivity of labour. Substituting the value of marginal productivity of labour in Cobb-Douglas production function as obtained above (*i.e.*, $MP_L = a\frac{Q}{L}$) in the output-elasticity expression we have

   $$\text{Output elasticity of labour} = \frac{\partial Q}{\partial L} \cdot \frac{L}{Q}$$

   Substituting $a\frac{Q}{L}$ for $\frac{\partial Q}{\partial L}$ we have

Output eleasticity of labour $= a\frac{Q}{L}.\frac{L}{Q} = a$ . . . (*i*)

Thus exponent '*a*' of labour in the Cobb-Douglas production function is equal to the output elasticity of labour

Output elasticity of capital $= \frac{\partial Q}{\partial K}.\frac{K}{Q}$

Marginal productivity of capital, $\frac{\partial Q}{\partial K} = b.\frac{Q}{K}$

Therefore, output elasticity of capital $= b.\frac{Q}{K}.\frac{K}{Q} = b$ . . . (*ii*)

It follows from equation (*i*) above that 1% increase in labour in Cobb-Douglas production ($Q = AL^aK^b$), holding capital constant, causes *a* % increase in output. Similarly, from equation (*ii*), it follows that 1% increase in capital, holding labour constant, causes *b* % increase in output.

6. *When the sum of exponents (a + b) in the two factor Linear Homogenous Cobb-Douglas production function ($Q = AL^aK^b$) is equal to one, it would show constant returns to scale.* We can easily prove this. When the sum of exponents $a + b = 1$, we can write *b* as $1 - a$. Writing Cobb-Douglas production function in this way we have

$$Q = AL^a K^{1-a}$$

If the inputs of labour (*L*) and capital (*K*) are increased by a constant *g*, then the quantity of output will be increased to

$A(gL)^a (gK)^{1-a} = g^a g^{1-a} AL^aK^{1-a}$

Since $g^a g^{1-a} = g$,

Therefore, $A(gL)^a (gK)^{1-a} = g\ AL^a\ K^{1-a}$

$= gQ$

Thus, when the inputs of capital and labour are increased by a constant *g*, the output *Q* also increases by *g*.

7. *The elasticity of substitution between two factors, labour and capital, in Cobb-Douglas production function is equal to unity*. This unit elasticity of factor substitution in Cobb-Douglas production lies in between *infinite substitution elasticity* in case of perfect substitute factors and *zero substitution elasticity* between two perfect complementary factors. Due to this unit elasticity of substitution between two factors in this production function, isoquants curves are convex to the origin as shown in Fig. 17.6.

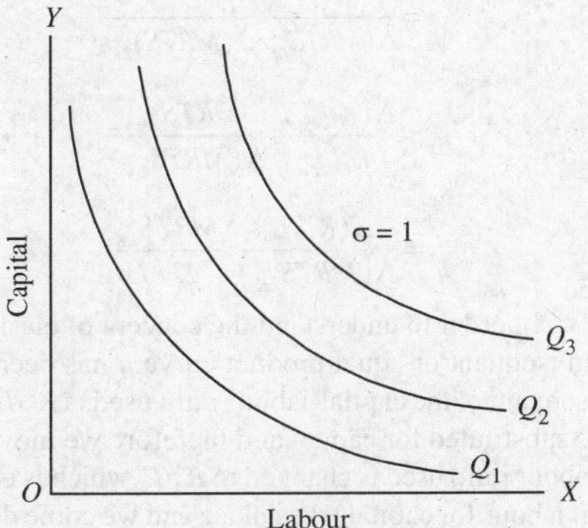

Fig.17.6. *Isoquants of Cobb-Douglas Production Function*

ELASTICITY OF TECHNICAL SUBSTITUTION (BETWEEN FACTORS)

In part 2 concerning the theory of demand, we explained the concept of elasticity of substitution between goods (products) in consumption. In the theroy of production we are concerned with the elasticity of substitution between factors (or inputs) in the production of goods. Thus, in the theory of production we are concerned with what may be called *elasticity of technical substitution*.

As seen above, marginal rate of technical substitution (*MRTS*) or factor *X* for factor *Y* declines as more of factor *X* is substituted for factor *Y* along an isoquant. In other words, marginal rate of technical substitution is different at different factor-proportions (*i.e.*, input ratios) used in the production of a good. This responsiveness of the proportions or ratios in which factors (or inputs) are used as there is a movement along an isoquant may be compared with the change in substitution possibilities in production as measured by the change in the marginal rate of technical substitution. *The proportionate change in the factor-proportions (or input ratios) divided by change in the marginal rate of technical substitution is known as elasticity of technical substitution between factors.* If the elasticity of substitution between the two factors is high, one factor can easily be substituted by another.

The concept of elasticity of technical substitution has been widely discussed in economic literature in connection witht the substitution between labour and capital. Therefore, we too shall explain the concept of elasticity of substitution with reference to capital and labour as factors of production. If *K* stands for the quantity of capital, *L* for the quantity of labour and σ for the elasticity of substitution, then in accordance with the above definition, elasticity of substitution between labour and capital can be expressed as follows:

$$\sigma = \frac{\text{proportionate change in the ratio of } (K/L) \text{ used}}{\text{proportionate change in the marginal rate of technical substitution of } L \text{ for } K}$$

$$= \frac{\text{proportionate change in } K/L}{\text{proportionate change in } MRTS_{LK}}$$

$$= \frac{\dfrac{\Delta(K/L)}{K/L}}{\Delta(MRTS_{LK})/(MRTS)_{LK}}$$

$$= \frac{\Delta(K/L)}{K/L} \cdot \frac{MRTS_{LK}}{\Delta(MRTS_{LK})}$$

$$= \frac{\Delta(K/L)}{\Delta(MRTS_{LK})} \cdot \frac{MRTS_{LK}}{K/L}$$

In order to understand the concept of elasticity factor of substitution, consider Fig. 17.7 where an isoquant or equal product curve *q* has been drawn. It will be seen that at point *A* on the given isoquant *q*, the captial-labour ratio used is K_1/L_1 which is equal to the slope of the ray *OA*. As labour is substituted for capital and therefore we move down on the isoquant *q* from point *A* to *B*, capital-labour ratio used is changed to K_2/L_2 which is equal to the slope of the ray *OB*. If further substitution of labour for capital takes place and we come down to point *C* on the isoquant *q*, capital-labour ratio further falls to K_3/L_3 which is equal to the slope of the ray *OC*. It is, therefore, clear that as we substitute more labour for capital along the isoquant, while labour-capital ratio is rising, capital-labour ratio is falling.

Now, as we have already seen, marginal rate of technical substitution at point *A* is given by the absolute value of the slope of the isoquant at that point (which is equal to the slope of the tangent t_1t_1 drawn to that point). At point *B*, the $MRTS_{LK}$ is equal to the absolute value of the slope of tangent t_2t_2 and at point *C*, the $MRTS_{LK}$ is equal to the absolute of value of the slope of tangent t_3t_3. It will be seen that the slope of tangent t_2t_2 is less than that of t_1t_1; slope of t_3t_3 is less than that of t_2t_2, that is, $MRTS_{LK}$ diminishes as more labour is substituted for capital.

The greater the ease with which one factor can be substituted for another to produce a given level of output, the higher will be the elasticity of technical substitution between them. Thus, in case of good substitutable factors, since the proportionate change in the rate of the quantities of the two factors used as one moves along an isoquant is relatively large as compared with the change in marginal rate of technical substitution (*MRTS*) between them, elasticity of technical substitution between them is very high. In the extreme case of factors which are perfect substitutes, there is infinite possibilities of substituting one factor for another to produce a given level of output without causing any change in *MRTS*, the elasticity of factor substitution is equal to infinity and the isoquants between them are straight lines as shown in Fig. 17.7. Thus, in case of perfect substitutes:

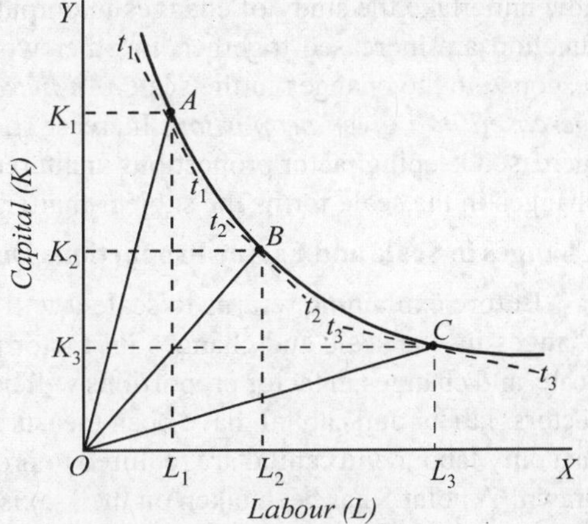

Fig. 17.7. *$MRTS_{LK}$ and Capital-Labour Ratios*

$$\sigma = \frac{\Delta\left(\frac{K}{L}\right) \Big/ \frac{K}{L}}{\Delta\, MRTS_{LK} \,/\, MRTS_{LK}} = \frac{\infty}{0} = \infty$$

On the other extreme case when the two factors are perfect complements which are used in a fixed ratio in the production of a commodity and hence there is no possibility of substitution between them at all (*i.e.*, $\Delta\left(\frac{K}{L}\right) \Big/ \frac{K}{L} = 0$), the substitution elasticity between them is zero. Therefore, in case of perfectly complementary factors, isoquants are L = shaped as depicted in Fig. 17.7.

Elasticity of Substitution and Factor Price Ratio

Since, in equilibrium position, the marginal rate of technical substitution is equal to the ratio of factor prices, the marginal rate of substitution in the formula for elasticity of substitution may therefore be replaced by the ratio of factor-prices. Thus,

$$\sigma = \frac{\text{proportionate change in the factor ratio}}{\text{proportionate change in the ratio of factor prices}}$$

$$= \frac{\Delta(K/L)}{K/L} \times \frac{P_L/P_K}{\Delta(P_L/P_K)} = \frac{\Delta(K/L)}{\Delta(P_L/P_K)} \times \frac{P_L/P_K}{K/L}$$

The replacement of factor-price ratio for marginal rate of technical substitution in the formula of elasticity of substitution is greatly helpful in practical applications of the concept of elasticity of substitution. This is because information regarding prices is more easily available than the information regarding marginal rate of technical substitution. Further, changes in factor-proportions in which factors are used are generally induced by the relative factor prices.

Elasticity of factor substitution occupies an important place in the theory of distribution. According to neo-classical theory of distribution, distributive shares of labour and capital in national income depends on the elasticity of factor substitution.

RETURNS TO SCALE

In the previous chapter we explained the behaviour of output when alteration in factor proportions is made. Factor proportions are altered by keeping the quantity of one or some factors fixed and varying the quantity of the other. The changes in output as a result of the variation in factor proportions, as seen before, forms the subject-matter of the "law of variable proportions." We shall now undertake the study of changes in output when all factors or inputs in a particular production function are increased together. In other words, we shall now study the behaviour of output in response to the changes in the scale. *An increase in the scale means that all inputs or factors are increased in a given proportion.* Increase in the scale thus occurs when all factors or inputs are increased keeping factor proportions unaltered. The study of changes in output as a consequence of changes in the scale forms the subject-matter of "returns to scale".

Changes in Scale and Factor Proportions Distinguished

Before explaining returns to scale it will be instructive to make clear the distinction between changes in the scale and changes in factor proportions. The difference between the changes in scale and changes in factor proportions will become clear from the study of Fig. 17.8 where the two factors, labour and capital, have been measured on the *X*-axis and *Y*-axis respectively. We suppose that only labour and capital are required to produce a particular product. An isoquant map has been drawn. A point *S* has been taken on the *Y*-axis and the horizontal line *ST* parallel to *X*-axis has been drawn. *OS* represents the amount of capital which remains fixed along the line *ST*. As we move towards right on the line *ST*, the amount of labour varies while the amount of capital remains fixed at *OS*. In other words, proportion between the two factors undergoes a change along the line *ST* ; the ratio of the variable factor 'labour' to the fixed factor 'capital' rises as we move towards right on the line *ST*. Thus, the movement along the line *ST* represents variation in factor proportions. Likewise, a vertical line *GH* parallel to the *Y*-axis has been drawn which also indicates changes in factor proportions. But in this case the quantity of labour remains fixed while the quantity of capital varies.

Now, draw a straight line *OP* passing through the origin. It will be seen that along the line *OP* the inputs of both the factors, labour and capital, vary. Moreover, because the line *OP* is a straight line through the origin, the ratio between the two factors along *OP* remains the same throughout. Thus, the upward movement along the line *OP* indicates the increase in the absolute amounts of two factors employed with the proportion between two factors remaining unchanged. Assuming that only labour and capital are needed to produce a product, then the increase in the two factors along the line *OP* represents the *increase in the scale* since along the line *OP* both the factors increase in the same proportion and therefore proportion between the two factors remains unaltered. If any other straight line through the origin such as *OQ* or *OR* is drawn, it will show, like the line *OP*, the changes in the scale but it will represent a different given proportion of factors which remains the same along the line. That is, the various straight lines through the origin will indicate different proportions between the two factors but on each line the proportion between the two factors remains the same throughout.

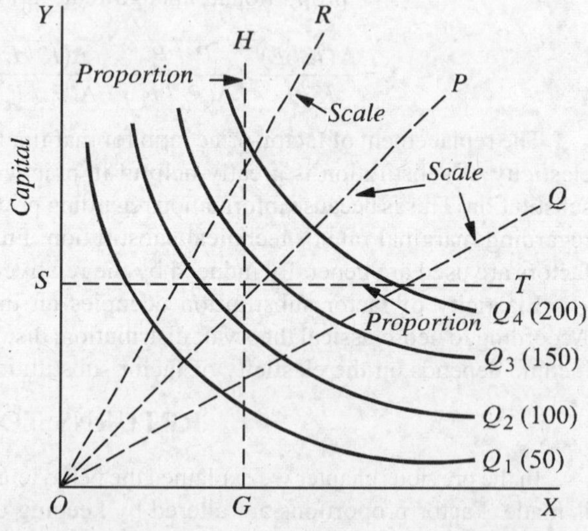

Fig. 17.8. *Changes in Scale and Factor Proportions Distinguished*

Validity of the Concept of Returns to Scale

We now proceed to discuss how the returns vary with the changes in scale, that is, when all *factors* are increased in the same proportion. But some economists have challenged the concept of returns to scale on the ground that all factors cannot be increased and therefore the proportions between factors cannot be kept constant. For instance, it has been pointed out that entrepreneurship is a factor of production which cannot be varied (in the single firm), though all other factors can be increased. The entrepreneur and his decision-making are indivisible and incapable of being increased. Thus, the entrepreneur is a fixed factor in all production functions. If labour and capital could produce a product with no one to supervise and take decisions, then the returns to scale in the sense of returns to all factors could be validly visualised. But the idea that labour and capital can produce goods without an entrepreneur is quite unrealistic. Thus, the concept of returns to scale involves a puzzle for economists which still remains unresolved. However, this puzzle can be resolved by assuming entrepreneurship to be variable in the sense that the greater the other inputs or factors, the greater the entrepreneurial work to be performed. This will be valid even if an entrepreneur is a single person because as he employs more units of other factors he uses his entrepreneurial ability to that extent greater. Thus, when labour and capital are increased in a proportion, the entrepreneurship can be assumed to be increased automatically in the same proportion. In this sense, returns to scale, that is, returns when *all factors* are varied, can be conceived.

But we shall explain below the concept of returns to scale by assuming that only two factors, labour and capital, are needed for production. This makes our analysis simple and also enables us to proceed our analysis in terms of isoquants or equal-product curves.

CONSTANT RETURNS TO SCALE

Returns to scale may be constant, increasing or decreasing. *If we increase all factors (i.e., scale) in a given proportion and the output increases in the same proportion, returns to scale are said to be constant*. Thus, if a doubling or trebling of all factors causes a doubling or trebling of outputs, returns to scale are constant. But, if a given percentage increase in all factors leads to a more than proportionate increase in output, returns to scale are said to be increasing. Thus, if all factors are doubled and output increases by more than a double, then the returns to scale are increasing. On the other hand, if the increase in all factors leads to a less than proportionate increase

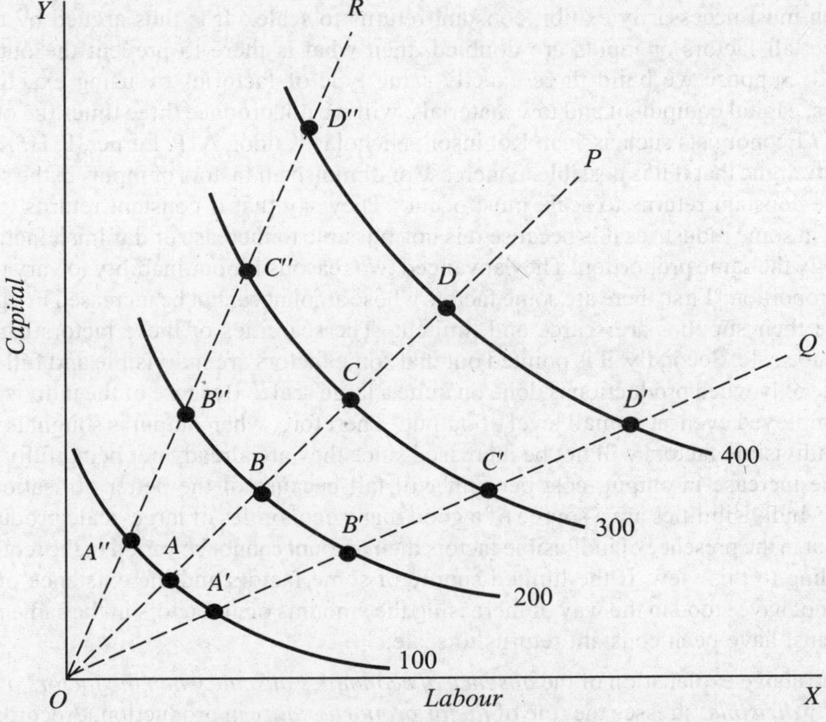

Fig. 17.9. *Constant Returns to Scale*

in output, returns to scale are decreasing. We shall explain below these various types of returns to scale.

As said above, the constant returns to scale means that with a given percentage increase in the scale or the amounts of all factors leads to the same percentage increase in output, that is, doubling of all inputs doubles the output. In mathematics the case of constant returns to scale is called *linear homogeneous production function or homogeneous production function of the first degree*. Production function exhibiting constant returns to scale possesses very important mathematical properties which make it very useful for theoretical analysis. There are a number of special theorems which apply when production function exhibits constant returns to scale. Empirical evidence suggests that production function for the economy as a whole is not too far from being homogeneous of the first degree. Empirical evidence also suggests that in the production function for an individual firm there is a long phase of constant returns to scale.

Let us illustrate diagrammatically the constant returns to scale with the help of equal product curves *i.e.* isoquants. Fig. 17.9 depicts an isoquant map. It is assumed that, in the production of the good, only two factors, labour and capital, are used. In order to judge whether or not returns to scale are constant, we draw some straight lines through the origin. As shown above, these straight lines passing through the origin indicate the increase in scale as we move upward. It will be seen from the figure that successive isoquants are equidistant from each other along each straight line drawn from the origin. Thus along the line OP, $AB = BC = CD$, and along the line OQ, $A'B' = B'C' = C'D'$ and along the line OR, $A''B'' = B''C'' = C''D''$. The distance between the successive equal product curves being the same along any straight line through the origin, means that if both labour and capital are increased in a given proportion, output expands by the same proportion. Therefore, Fig. 17.9 displays constant returns to scale.

Divisibility of Factors, Proportionality and Constant Returns to Scale

Some economists are of the view that if the factors of production are perfectly divisible production function must necessarily exhibit constant returns to scale. It is thus argued by them that if, for instance, all factors or inputs are doubled, then what is there to prevent the output from being doubled. Suppose we build three exactly same type of factories by using exactly same type of workers, capital equipment and raw materials, will we not produce three times the output of a single factory? Economists such as Joan Robinson, Nicholas Kaldor, A. P. Lerner, F. H. Knight who hold this view argue that if it is possible to increase or diminish all factors or inputs in the same proportion, then the constant returns to scale must occur. They say that if constant returns to scale does not prevail in some industries it is because it is not possible to increase or diminish factors used in them in exactly the same proportion. They advanced two reasons for our inability to vary the factors in the same proportion. First, there are some factors whose amount cannot be increased in a given proportion because their supplies are scarce and limited. The scarcities of these factors cause diminishing returns to scale. Secondly, it is pointed out that some factors are indivisible and full use of them can be made only when production is done on quite a large scale. Because of the indivisibility they have to be employed even at a small level of output. Therefore, when output is sought is to be expanded, these indivisible factors will not be increased since they are already not being fully utilised. Thus, with the increase in output, cost per unit will fall because of the better utilisation of indivisible factors. Indivisibilities are a source of a good many economies of large-scale production. It is thus clear that in the presence of indivisible factors their amount cannot be varied in the required proportion. According to this view, if the limited supply of some factors and the existence of indivisibilities would not have stood in the way of increasing the amounts of all factors in the same proportion, then there must have been constant returns to scale.

The above explanation of the *absence of economies of scale when the factors of production are perfectly divisible*, stresses the role of *factor proportionality* in production. According to this view, for achieving best results in production, there is a certain *optimum proportion* of factors. When the factors of production are perfectly divisible, they can be increased or decreased by suitable amounts so as to achieve always the optimum proportion between the factors. When factors are indivisible, that is, available in discrete units, some of them quite large or lumpy, production on a small scale would mean the use of non-optimum factor proportions and therefore the inefficiency of small-scale production. Thus, in case of perfect divisibility, factors could be divided and subdivided by appropriate amounts and any amount of output, no matter how small or large, can be produced with optimum factor proportions and as a result economies and diseconomies of scale would be non-existent and we would get constant returns to scale.

The above view has been criticised by Professor E. H. Chamberlin. Prof. Chamberlin and others of his view have argued that constant returns to scale cannot prevail. They say that even if all factors could be varied in required quantities and even if all factors were perfectly divisible there could be increasing returns to scale. In their view even in case of perfect divisibility and variation of the factors, increasing returns to scale can occur with the increase in the scale or size (*i.e.*, increase in all factors, because at a larger scale, (1) greater specialization of labour becomes possible and (2) introduction of specialized machinery or other inputs of a superior technology is made possible by a wise selection from among the greater range of technical possibilities opened up by greater resources. Thus, Professor Chamberlin lays stress on *size* (or *scale*), in causing economies of scale. According to him, when the size or scale of operations, or in other words when the *absolute amounts* of all factors increase, the efficiency of the factors is increased by the use of greater specialisation of labour and by the introduction of specialised and superior machinery. Thus, according to Chamberlin, the above view which stresses divisibility and proportionality neglects the effect of *scale* on the efficiency of factors.[2]

It has been further pointed out that one cannot meaningfully speak of doubling all the factors in a given situation. For instance, two factories existing nearby is simply not the same thing as doubling of one factory in isolation. The existence of another factory at a close distance affects labour discipline, air pollution, cost of labour training etc. It is thus argued that in practice it is not possible to vary all the factors in a given proportion and obtain increases in output in the same proportion.

More significantly, it is pointed out that if a large single factory is more efficient than two small factories (the two having total capacity equal to the large one), then there would be no incentive on the part of an entrepreneur to double or duplicate his factors in the sense of setting up another small factory, near to his previous one. In other words, when the entrepreneur sees the opportunity of getting increasing returns to scale by setting up a large factory, then he would not set up a duplicate factory of his previous size and obtain constant returns to scale. In this connection, it is pointed out that there are many types of economies of scale due to which there is a great possibility of getting, at least in the beginning, increasing returns to scale.

Cobb-Douglas Production Function and Constant Returns to Scale. As seen above, though some economists look at constant returns to scale with suspicion, empirical evidence shows that in the expansion of a single firm, after a phase of increasing returns to scale, there is a long phase of constant returns to scale covering a wide range of output. Cobb-Douglas production function found out empirically and which applies to the manufacturing industry as a whole and to several individual industries also exhibits constant returns to scale. As noted above, Cobb-Doughlas production function takes the following form: $Q = AL^aK^{1-a}$ where Q stands for quantity of output, L for the quantity of labour and K for the quantity of capital employed. A and a are positive constants and where $a < 1$. As shown above, according to the Cobb-Douglas production function returns to scale are constant. The production function of the type of constant returns to scale plays

2. For the views of Professor Chamberlin regarding the controversial question of Divisibility, Proportionality and Economies of Scale, see his article in *Quarterly Journal of Economics*. Vol. LXII, Feb. 1948.

an important role in the theory of distribution, input-output analysis and linear programming analysis of production.

INCREASING RETURNS TO SCALE

As stated above, increasing returns to scale means that output increases in a greater proportion than the increase in inputs. If, for instance, all inputs are increased by 25%, and output increases by 40% then the increasing returns to scale will be prevailing. When a firm expands, the increasing returns to scale are generally obtained in the beginning. Several factors account for increasing returns to scale, at least in the initial stages.

Indivisibility of the Factors. Many economists, such as Joan Robinson, Kaldor, Lerner and Knight ascribe increasing returns to scale to the indivisibility of factors. Some factors are available in large and lumpy units and can therefore be utilised with utmost efficiency at a large scale of output. Therefore, in the case of some indivisible and lumpy factors, when output is increased from a small level to a large one, indivisible factors are better utilised and therefore increasing returns are obtained. According to this view, as stated above, if all factors are perfectly divisible, increasing returns to scale would not occur.

Greater Possibilities of Specialisation of Labour and Machinery. As stated above, Chamberlin is of the view that returns to scale increase because of greater possibilities of specialisation of labour and machinery. According to him, even if the factors were perfectly divisible, with the increase in the scale, returns to scale can increase because the firm can introduce greater degree of specialization of labour and machinery (because now greater resources or amounts of factors become available) and also because it can install technologically more efficient machinery.

Integration of Processes. Integration of successive processes in the production of a commodity is another factor which gives rise to some economies of large-scale production. For example, a book publishing company may set up its own printing press, opens up a binding section with suitable binding machinery, and also starts varnishing the book covers itself. Likewise, a steel firm may start doing various processes such as melting of iron, making of steel and converting steel into different shapes and sizes. This integration of successive stages in a single firm increases the productivity and efficiency of the firm and proves to be highly economical.

Dimensional Economies. Another important cause of increasing returns to scale lies in dimensional relations, which have been emphasised by Professor Baumol.[3] A wooden box of 3 foot-cube contains 9 times greater wood than the wooden box of 1 foot-cube, that is, 3 foot-cube wooden box contains 9 times greater input. But the capacity of the 3 foot-cube wooden box is 27 times greater than that of 1 foot-cube. Another example is the construction of a warehouse. Suppose a rectangular whorehouse is proposed to be constructed. Most important input used in this construction-work is the number of bricks and other inputs which almost vary in proportion to the number of bricks used. The number of bricks used depends upon the wall area of the building. The elementary mathematics tells us that the wall area will increase equal to the square of the perimeter of the warehouse, while its volume, that is, its storage area will increase equal to the cube of

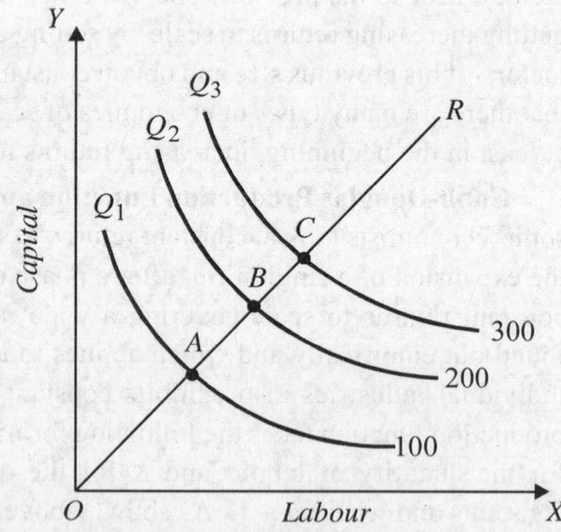

Fig. 17.10. *Increasing Returns to Scale*

3. W.J. Baumol, *Economic Theory and Operations Analysis;* Prentice Hall, 3rd edition, p. 382

the perimeter. In other words, double the number of bricks and other inputs that go with them, the storage capacity of the warehouse will be more than doubled. This is thus a case of increasing returns to scale. Similarly, if the diameter of a pipe is doubled, the flow through it is more than doubled.

Increasing returns to scale can be shown through isoquants. When increasing returns to scale occur, the successive isoquants will lie at decreasingly smaller distances along a straight line ray *OR* through the origin. In Fig. 17.10 the various isoquants Q_1, Q_2, Q_3 are drawn which successively represent 100, 200 and 300 units of output. It will be seen that distances between the successive isoquants decrease as we expand output by increasing the scale. Thus, increasing returns to scale occur since $AB < OA$ and $BC < AB$ which means that *equal increases in output are obtained by smaller and smaller increments in inputs*.

DECREASING RETURNS TO SCALE

As stated above, when output increases in a smaller proportion than the increase in all inputs, decreasing returns to scale are said to prevail. When a firm goes on expanding by increasing all his inputs, *eventually* diminishing returns to scale will occur. But among economists there is no agreement on a cause or causes of diminishing returns to scale. Some economists are of the view that the entrepreneur is a fixed factor of production ; while all other inputs may be increased, he cannot be. According to this view, decreasing returns to scale is therefore actually a special case of the law of variable proportions. Thus, they point out that we get diminishing returns to scale beyond a point because varying quantities of all other inputs are combined with a fixed entrepreneur. Thus, according to this view, decreasing returns to scale is a special case of the law of variable proportions with entrepreneur as the fixed factor. Other economists do not treat decreasing returns to scale as the special case of the law of variable proportions and argue that decreasing returns to scale eventually occur because of the *increasing difficulties of management, co-ordination and control with the expansion in scale and output*. When the firm has expanded to a too gigantic size, it is difficult to manage it with the same efficiency as previously.

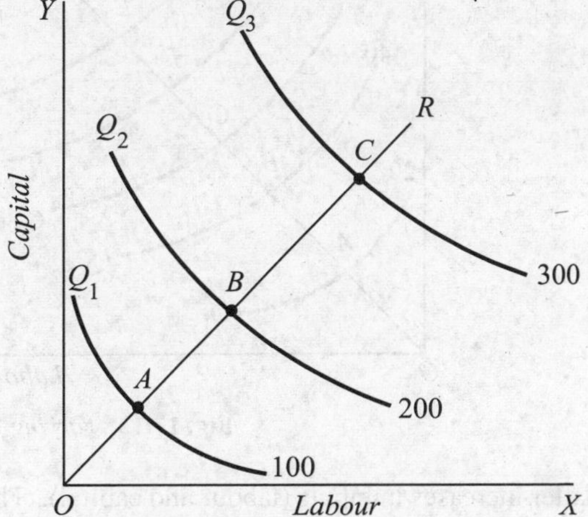

Fig. 17.11. *Decreasing Returns to Scale*

The case of decreasing returns to scale can-be shown with an isoquant map. When successive isoquants lie at progressively larger and larger distance on a ray through the origin, returns to scale will be decreasing. In Fig. 17.11 successively decreasing returns to scale occur since $AB > OA$, and $BC > AB$. It means that successively more and more of inputs (labour and capital) are required to obtain equal increments in output.

Varying Returns to Scale in a Single Production Process

It should be noted that it is not always the case that different production functions should exhibit different types of returns to scale. It generally happens that there are three phases of increasing constant and diminishing returns to scale in a single production function. In the beginning when the scale increases, increasing returns to scale are obtained because of greater possibilities of specialization of labour and machinery. After a point, there is a phase of constant returns to scale where output increases in the same proportions as inputs. Empirical evidence suggests that the phase of constant returns to scale is quite long. If the firm continues to expand, then eventually a point will be reached beyond which decreasing returns to scale will occur due to the mounting difficulties of co-ordination and control. These varying returns to scale have been shown in Fig. 17.12. It will be seen from Figure 17.12 that upto point C on a ray *OR* from the origin, the distance between the successive isoquants showing equal increments in output goes on decreasing. This implies that upto point *C* equal increments in output are obtained from the use of successively

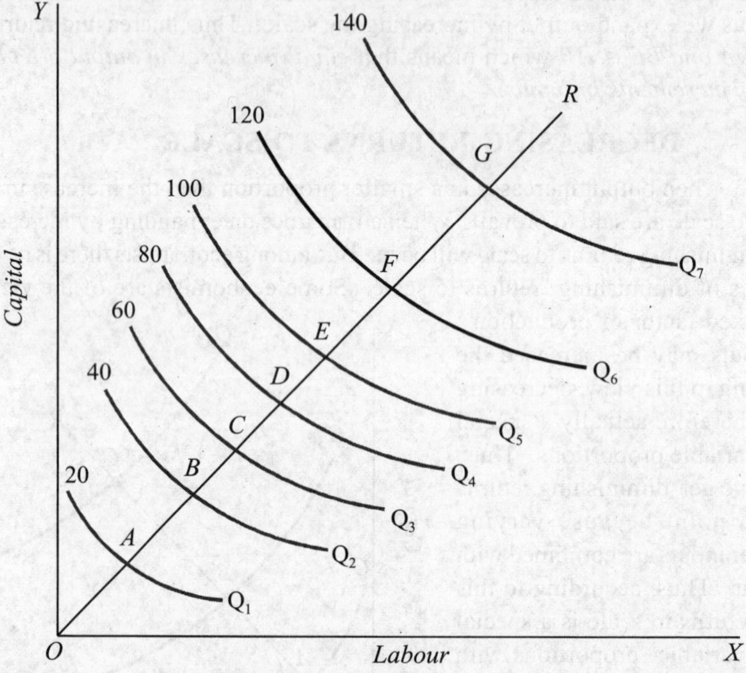

Fig. 17.12. *Varying Returns to Scale*

smaller increases in inputs (labour and capital). Thus, upto point *C* on ray *OR* increasing returns to scale occur. Further, it will be seen from Fig. 17.12 that from point *C* to point *E* constant returns to scale are obtained as the same proportionate increments in output are obtained from the proportional increase in inputs of labour and capital. Beyond point *E*, the distance between the successive isoquants representing equal increments in output is decreasing along the ray *OR* from the origin which implies that less than proportionate increase in output is obtained from a given proportionate increments in the use of the two factors, labour and capital on the ray *OR* from the origin, $EF > DE$ and $FG > EF$.

COBB-DOUGLAS PRODUCTION FUNCTION AND RETURNS TO SCALE

An important production function which has been found by empirical studies by eminent economists Cobb and Douglas and known after their name as Cobb-Douglas production function is given below:

$$Q = AL^{\alpha}K^{\beta}$$

In this production function the sum of exponents $(\alpha + \beta)$ measures returns to scale. Multiplying each input, labour (*L*) and capital (*K*), by a constant factor *g*, we have

$$Q' = A\,(gL)^{\alpha}\,(gK)^{\beta}$$

$$= g^{\alpha+\beta} (AL^{\alpha} K^{\beta})$$

Expression in bracket $AL^{\alpha} K^{\beta} = Q$. Therefore,

$$Q' = g^{\alpha+\beta} Q$$

This means that when each input is increased by a constant factor g, output Q increases by $g^{\alpha+\beta}$. Now, If $\alpha + \beta = 1$ then, in this production function

$$Q' = g^1 Q$$

or $$Q' = gQ$$

That is, when $\alpha + \beta = 1$, output (Q) increases by the same factor g by which both inputs are increased. This implies that production function is *homogenous of first degree* or, in other words, returns to scale are constant.

When $\alpha + \beta > 1$, say it is equal to 2, then, in this production function new output

$$Q' = g^{\alpha+\beta} AL^{\alpha} K^{\beta} = g^2 Q$$

In this case multiplying each input by constant g, then output (Q) increases by g^2, that is by more than g. Therefore, when $\alpha + \beta > 1$. *Cobb Douglas production exhibits increasing returns to scale.*

When $\alpha + \beta < 1$, say it is equal to 0.8, then in this production function new output

$$Q' = g^{\alpha+\beta} AL^{\alpha} K^{\beta} = g^{0.8} Q$$

That is, increasing each input by constant factor g will cause output to increase by $g^{0.8}$, that is, less than g. Returns to scale in this case are decreasing.

RETURNS TO SCALE AND MARGINAL RETURNS TO A VARIABLE FACTOR

Constant Returns to Scale and Returns to a Variable Factor

We now turn to study the relationship between returns to scale and marginal returns to a variable factor. In other words, we want to study whether marginal productivity of a factor diminishes or rises when the returns to scale are constant, diminishing and increasing. We first take the case of constant returns to scale. As seen above, in case of constant returns to scale, or what is called in mathematics, homogeneous production function of the first degree, the distance between the successive equal product curves along a straight line through the origin is always the same. Consider Fig. 17.13 in which here equal product curves showing 100, 200 and 300 units of output respectively are drawn. A straight line *OL* through the origin and cutting the equal product curves is drawn. Line *OL* displays returns to scale. Since returns to scale are constant, *PQ* is equal to *QR*. A horizontal line *ST* is also drawn, which shows the behaviour of output when the amount of the variable factor labour is increased to a fixed quantity *OS* of capital. Now, it is to be proved that when the returns to scale are constant, the marginal physical product of the variable factor diminishes. In terms of Fig. 17.13 where returns to scale are constant, we can prove that more units of the variable factor labour will be required to be added to a fixed quantity of capital to obtain equal increments in output, that is, *QN* is greater than *MQ*. Thus, given that *PQ* = *QR* (*i.e.*, returns to scale are constant), we have to prove that *QN* is greater than *MQ*. *QN* being greater than *MQ* means that marginal physical product of labour diminishes as varying amounts of it are added to the fixed quantity of *capital*. This can be proved in the following way:

In triangles *QRF* and *QPE* in Fig. 17.13

$QR = PQ$ (given, because of constant returns to scale)

$\angle QRF = \angle QPE$ (alternat]e angles)

$\angle RQF = \angle PQE$ (vertically opposite angles)

Therefore, triangles *QRF* and *QPE* are congruent, *i.e.*, equal in every respect.

Hence, $QF = EQ$

It will be seen from Fig. 17.13 that *MQ* is less than *EQ* and *QN* is greater than *QF*. Therefore, *QN* is greater than *MQ*.

It is thus proved that *when returns to scale are constant, (or when production function is homogeneous of the first degree) marginal physical product (i.e., marginal returns) of a variable factor diminishes. In other words, constant returns to scale imply diminishing marginal returns to a varaible factor.*

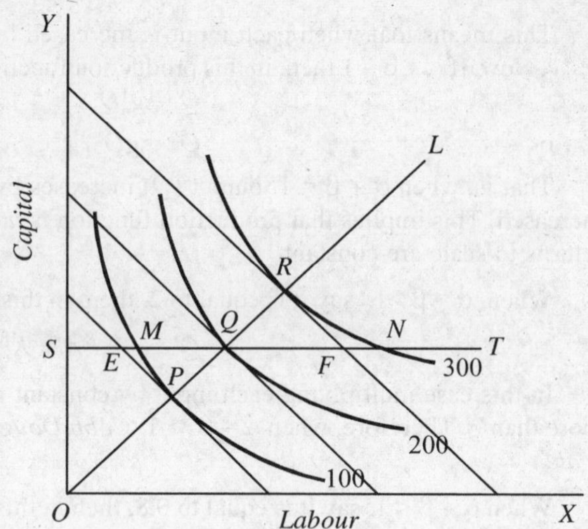

Fig. 17.13. *When returns to scale are constant, returns to a variable factor diminish*

Mathematical Proof[4]. That when returns to scale are constant, marginal product of a variable factor diminishes can be proved mathematically. Let us take Cobb-Douglas production function with sum of exponents equal to one which implies constant returns to scale. Such a Cobb-Douglas production function can be written as

$$Q = AL^{\frac{3}{4}} K^{\frac{1}{4}} \qquad \ldots (i)$$

Since in the above Cobb-Douglas production function sum of exponents $\left(\frac{3}{4}+\frac{1}{4}\right)$ equals one, this represents constant returns to scale.

Marginal product of labour $(MP_L) = \frac{\partial Q}{\partial L}$

Now, in order to obtain marginal product of labour we differentiate the given production function (*i*) with respect to labour. Thus,

$$MP_L = \frac{\partial Q}{\partial L} = \frac{3}{4} AL^{\frac{3}{4}-1} K^{\frac{1}{4}}$$

$$= \frac{3}{4} AL^{-\frac{1}{4}} K^{\frac{1}{4}} \qquad \ldots (ii)$$

For marginal product of labour to diminish as more labour is used, its second derivative must be negative. To obtain the second derivative we differentiate (*ii*) with respect to labour

Thus $$\frac{\partial^2 Q}{\partial L^2} = -\frac{1}{4} \cdot \frac{3}{4} AL^{-\frac{1}{4}-1} K^{\frac{1}{4}}$$

Thus marginal product of labour in Cobb-Doughlas production function is diminishing.

Likewise, it can be shown that with a given quantity of labour, if capital increases, marginal product of capital would diminish.

4. Those who are not interested in mathematical proof, they may omit this section

Decreasing Returns to Scale and the Marginal Physical Product of the Variable Factor

It also follows from above that when returns to scale are decreasing, the marginal physical In Fig. 17.14 returns to scale are diminishing, since *QR* is greater than *PQ*. It is quite obvious from Fig. 17.14 that along the horizontal line *ST* which represents the varying quantities of factor *X* with a

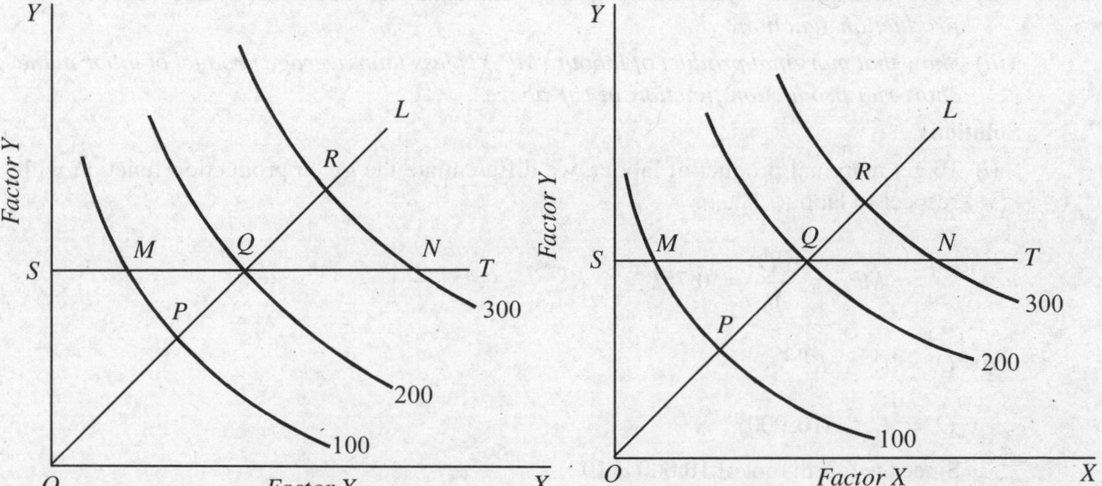

Fig. 17.14. *When returns to scale are decreasing returns to a variable factor diminish rapidly*

Fig.17.15. *Marginal physical product can diminish even when returns to scale are increasing.*

fixed quantity of factor *Y*, the intercept *QN* is much greater than the intercept *MQ*. It follows *therefore that when returns to scale are diminishing, marginal physical product of the variable factor falls rapidly*.

Increasing Returns to Scale and Marginal Physical Product of the Variable Factor

Finally, how the marginal physical product of the variable factor behaves when returns to scale are increasing. This case is illustrated in Fig. 17.15 where returns to scale are increasing *strongly*. *QR* is much less than *PQ*. It will be seen from Fig. 17.15 that along the horizontal line *ST*, *QN* is less than *MQ*. It means that as the varying quantities of factor *X* are successively added to the given quantity of the fixed factor *Y*, smaller quantity of factor *X* than before is required to be increased to get an equal increment in output. Thus, *QN* being less than *MQ* means that marginal physical product of the variable factor *X* increases. It is therefore concluded that when returns to scale are *strongly* increasing, the marginal returns to a variable factor (used with a fixed quantity of the other factor) increases. However, when the returns to scale are *slightly increasing*, then the marginal returns of the variable factor will diminish.

Thus, the law of diminishing marginal returns to a factor is even consistent with increasing returns to scale. As just seen above, increasing returns to scale can yield increasing returns as well as diminishing returns to a variable factor depending upon whether returns to scale are strongly increasing or slightly increasing.

From the foregoing analysis we conclude that if returns to scale are constant the marginal physical product of a variable factor used in combination with a fixed factor will always diminish as more of the variable factor is used. Similarly, when returns to scale are decreasing, marginal physical product of a variable factor always diminish, and when returns to scale are increasing marginal physical product of a factor will still diminish unless the returns to scale are increasing sufficiently strongly.

NUMERICAL PROBLEMS

Problem 1. *You are given the following production function:* $Q = L^{0.75} K^{0.25}$

(*i*) *Find the marginal product of labour*

(*ii*) *If the fixed quantity of capital in the short run equals 10,000 units, what is the short-run production function?*

(*iii*) *Show that marginal product of labour (MP_L) is less than average product of labor in the short-run production function in (ii) above*

Solution:

(*i*) To get marginal product of labour, we differentiate the given production function with respect to labour. Thus

$$MP_L = \frac{\partial Q}{\partial L} = 0.75L^{-0.25} \cdot K^{0.25} = 0.75\left(\frac{K}{L}\right)^{0.25}$$

(*ii*) $Q = L^{0.75} \cdot K^{0.25}$

$Q = L^{0.75} \cdot (10,000)^{0.25}$

Since the fourth root of 10,000 is 10

Thus short-run production function is

$Q = L^{0.75} \cdot 10 = 10L^{0.75}$

(*iii*) $MP_L = \frac{\partial Q}{\partial L} = 10 \times 0.75L^{-0.25} = \frac{7.5}{L^{0.25}}$

(*ii*) $AP_L = \frac{Q}{L} = \frac{10L^{0.75}}{L} = 10L^{0.75-1} = \frac{10}{L^{0.25}}$

Comparing the values of MP_L and AP_L we find that $MP_L < AP_L$

Problem 2. *You are given the following production functions. What types of production function they are? Draw isoquants for these production functions.*

(*i*) $Q = 2L + K$ (*ii*) $Q = \text{Min}(2L, K)$

Solution:

(*i*) Production function, $Q = 2L + K$ is a **linear production function.** This is proved as follows.

Let L and K are increased by a given number λ, thus

$$Q' = 2\lambda L + \lambda K$$

λ can be factored out. Thus

$$Q' = \lambda(2L + K) = \lambda Q$$

Thus increasing each inputs by λ, output also increases by λ. This shows this production function is linear homogeneous.

(*ii*) *Production function,* $Q = \text{Min}(2L, K)$ is a **Leontief production function.** This is so called because a noted American economist Wassely Leontief used this production function to describe the American economy. This is a fixed proportion production function in which labour and capital are combined in the ratio $2L$ and $1K$.

QUESTIONS FOR REVIEW

1. What are isoquants? Why does an isoquant slope downward? Why cannot isoquants cut each other? Why are they convex to the origin?
2. What is meant by marginal rate of technical substitution between factors ($MRTS_{LK}$)? How is it related to marginal products of factors? Why does marginal rate of technical substitution of labour for capital diminish as more labour is used by substituting capital?
3. "The slope of an isoquant is a measure of the relative marginal productivities of the factors." Explain.
4. Distinguish between fixed proportions and variable proportions production function. Draw the isoquants of fixed proportions production function.
5. (*a*) In case of perfect substitutes, what can you say about the equilibrium of the producer?
6. What is meant by linear homogenous production function. What are its important properties? [**Hint.** Cobb. Douglas production function is the chief example of linear homogeneous production function. Therefore, you can explain the properties of linear homogenous production by taking the example of Cobb-Douglas production function)
7. What is Cobb-Douglas production function? What are its important properties? Explain.
8. Show that in Cobb-Douglas production function $Q = AL^aK^b$ when

   a + b > 1, returns to scale are constant
   a + b > 1, returns to scale are increasing
   a + b < 1, returns to scale are decreasing
9. Explain the concept of elasticity of substitution between two factors of production. What shope does the isoquant take when the elasticity of substitution between factors is (*i*) zero, (*ii*) one, (*iii*) infinity. *D.U. B.A. (Hons) 1988*
10. Distinguish between returns to scale and returns to a variable factor with the help of isoquants.
11. What is meant by constant returns to scale? Represent it by an isoquant map. Show that empirically discovered Cobb-Dough las production function represents constant returns to scale.
12. If all factors were perfectly divisible, constant returns to scale would have occurred". Examine critically. On what grounds E.H. Chamberlin challenged this viewpoint.
13. What is meant by *increasing returns to scale*? Explain the factors that cause increasing returns to scale.
14. A production function is subject to constant returns to scale. What can you say about the returns to a variable factor. Explain diagrammatically. *D.U. B.A. (Hons) 1986*
15. If a production function reveals increasing returns to scale, what can you say about the returns to a variable factor? Explain diagrammatically. *D.U. B.A (Hons) 1998*
16. What are increasing returns to scale? Show them on an isoquant map. Explain the causes of increasing returns to scale.
17. Show the increasing cost and and decreasing returns to scale with the help of isoquants. What causes decreasing returns to scale beyond a point?
18. Is decreasing returns to a variable factor compatible with constant returns to scale? Explain *D.U. B.A. (Hons) 2001*
19. Show that when returns to scale are constant, marginal returns to a variable factor diminish (*a*) Prove geometrically (*b*) Give an algebraic proof of it using a linear homogeneous production function.

20. A firm's production function is given by $Q = K.L$, where K and L are the inputs used by the firm and Q represents output.

(*i*) Check the returns to scale for this production function.

(*ii*) Draw the isoquant map.

(*iii*) Calculate the marginal rate of technical substitution between L and K. What will the expansion path look like?

[**Hints.** This production function ($Q = K.L$) is a Cobb-Douglas type production function ($Q = K^a L^b$) with each of the two exponents, a and b, are equal to one (implied). Thus, the given production function can be written as

$$Q = K^1.L^1$$

(*i*) Returns to scale are measured by the sum of exponents, they are increasing in this production function as a + b is here greater then one (1 + 1 = 2)

(*ii*) As in Figure 17.10 in this chapter.

(*iii*) $MRTS_{LK} = \frac{MP_L}{MP_K}$ Now, in the production function, $Q = K.L$

$$MP_L = \frac{\partial Q}{\partial L} = K$$

$$MP_K = \frac{\partial Q}{\partial K} = L$$

Therefore, $MRTS_{LK} = \frac{K}{L}$

Since along an expansion path $MRTS_{LK}$ remains constant, the ratio of two factors, $\frac{K}{L}$ which is equal to $MRTS_{LK}$ in the given production function will also remain constant. Hence expansion path will be straight line through the origin.

21. When sum of exponents ($\alpha + \beta$) of Cobb-Douglas production function is equal to one, it shows constant returns to scale. Prove.

22. What is meant by constant returns to scale? Show them with an iso-product map. Is it correct to say that returns to scale would have been constant if the factors of production had been perfectly divisible?

23. Consider the following data on output and inputs. What type of returns to scale does it represent and why?

K	L	Q
5	8	3
10	16	6
20	32	12
40	64	24

Where K denotes units of capital, L denotes units of labour used and Q denotes output produced.

24. Explain the laws of returns to scale. Show the three kinds of returns to scale with the help of isoquants. Why do we get decreasing returns to scale? *D.U. B.Com (H) 2000*

25. You are given the following production functions. Which ones represent constant returns to scale, which one increasing returns to scale and which one decreasing returns to scale and why?

(*a*) $Q = AK^{0.5} L^{0.7}$ (*b*) $Q = AK^{0.25} L^{0.75}$

(*c*) $Q = AK^{0.3} L^{0.6}$ (*d*) $Q = 10\sqrt{L}\sqrt{K}$

(**Hints** : Sum of the exponents in Cobb-Douglas production function indicates returns to scale. In (*b*) and (*d*) sum of exponents is equal to one. Hence these show constant returns to scale. In (*a*) sum of exponents (0.5 + 0.7 = 1.2) exceeds one, it is therefore the case of increasing returns to scale type of production function. (*c*) depicts decreasing returns to scale as in it the sum of exponents is less than 1 (0.3 + 0.6 = 0.9)

26. A firm has the following production function:

$$Q = K^{\frac{1}{2}} L^{\frac{1}{2}}$$

Calculate marginal product function for labour and capital. Also calculate $MRTS_{LK}$.

What type of returns to scale does it represent?

27. You are given the following production function:

$Q = L^a K^b$

(*i*) Does this production function represent increasing, decreasing or constant returns to scale.

(*ii*) If $(a + b) > 1$, what would the isoquant map look like? *D.U. B.A. (Hons) 2001*

28. *Consider the production function represented by $Y = AL^{\alpha} K^{\beta}$ where Y is output and α and β are positive constants.*

(a) Show that this production function exhibits constant returns to scale when $\alpha + \beta = 1$.

(b) Show that marginal product of labour depends on the input ratio $\frac{K}{L}$ only.

(c) When is the expansion path linear?

(d) Define the range of values of α for which this function depicts diminishing MP of labour

[**Hints:** (*c*) Expansion path is linear when the sum of $\alpha + \beta = 1$ and production function becomes linear homogenous. For proof, see next chapter of this book]

(*d*) $$MP_L = \frac{\partial Y}{\partial L} = AaL^{a-1} K^{\beta} \quad ...(i)$$

$$\text{Slope of } MP_L = \frac{\partial^2 Y}{\partial L^2} = Aa(a-1)L^{\alpha-2} K^b \quad ...(ii)$$

For diminishing marginal product of labour to occur, $\frac{\partial^2 Y}{\partial L^2}$ must be negative.

From (*ii*) it follows that MP_L will be diminishing if $0 < a < 1$.

CHAPTER 18
OPTIMUM FACTOR COMBINATION

In the last three chapters we explained the law of variable proportions and returns to scale which underlie the process of production. An important problem facing an entrepreneur is to decide about a particular combination of factors which should be employed for producing a product. There are various technical possibilities open to a firm from which it has to choose, that is, there are various combinations of factors which can yield a given level of output and from among which a producer has to select one for production. As explained in an earlier chapter, various combinations of factors which produce equal level of output are represented by an equal product curve or what is also called insoquant. An isoquant or iso-product map represents various technical possibilities of producing different levels of output.

It is assumed that the entrepreneur aims at maximising his profits. A profit maximising entrepreneur will seek to minimise his cost for producing a given output, or to put it in another way, he will maximise his output for a given level of outlay. The choice of a particular combination of factors by an entrepreneur depends upon (*a*) technical possibilities of production, and (*b*) the prices of factors used for the production of a particular product. Technical possibilities of production are represented by the isoquant map. Before explaining how a producer will arrive at the optimum or least-cost combination of factors, we shall first explain how the prices of factors can be introduced into the study.

The Economic Region of Production and Ridge Lines

Before explaining which factor combination a firm will use for production, it will be useful to demonstraste the region in which the optimal factor combination will lie. The traditional economic theory focuses on only those combinations of factors which are technically efficient and the marginal products of factors are diminishing but positive. According to this isoquants are sloping downward (*i.e.* their slope is negative) and convex to the origin. however, there are regions in a production function, where isoquants may have positively sloped segments that is, bend backwards. In fig. 18.1 we represent a production function through isoquants and measure labour along the *X*-axis and capital along the Y-axis. It will be seen from this figure that above the line *OA* and below the line *OB* slope of the isoquants is *positive* which means that increases in both capital and labour are *required to produce a given fixed quantity of output*. Obviously, the production techniques (that is, factor combinations) lying on these positively sloping segments of the isoquants are technically inefficient. It may be recalled that a *technique or factor combination is technically inefficient if it requires more quantity of both the factors for producing a given level of output*. The positively sloping segments of isoquants implies that marginal product of one of the factors has become negative. Thus, above the line *OA*, marginal product of capital has become negative, which means output can be increased by using less capital, while the amount of labour is held constant. On the other hand, below the line *OB*, marginal product of labour becomes negative, which means output can be increased by using less labour, keeping capital constant. The lines *OA* and *OB* are called the **ridge lines** which bound a region in which marginal products of the two factors are positive. The ridge line

OA connects those points of the isoquants where marginal product of capital is zero ($MP_k = 0$). On the other hand, the ridge line *OB* connects those points of the isoquants where marginal product of labour is zero ($MP_L = 0$). Thus, the ridge lines are the locus of points of isoquants where marginal product of one of the factors is zero.

No rational enterpreneur will operate at a point outside the ridge lines since marginal product of one of the factors is negative and production is technically inefficient. In other words, production outside the ridge lines is inefficient, because *same output can be produced with less quantity of the factors which must be cheaper*. This can be better understood from fig. 18.1. Consider point *R* on isoquant Q_2, *R* is the point where the isoquant is positively sloping and therefore lies outside the ridge line. It will be seen from fig. 18.1 that production at point *R* to produce output Q_2 requires more of both capital and labour than some other points, such as point *H*, on the same isoquant. Since, both capital and labour have to paid positive prices, it will be cheaper to produce a given quantity of output at point *H* than at point *R*. Thus, since production outside the ridge lines are technically inefficient and marginal product of one or the other factor is negative, no rational entre-preneur will like to operate outside the ridge lines if he aims at minimising cost to produce a given output. Thus, *regions outside the ridge lines are called regions of economic nonsense*. A rational producer will produce in the region bounded by the two ridge lines *OA* and *OB* where the isoquants are negatively sloping, marginal products of factors are diminishing but positive. Therefore, the region bounded by the two ridge lines, *OA* and *OB* is called the *region of economic production* which has been shaded by us. Exactly at what point in the economic region, a firm will operate depends on the outlay it has to make on purchasing the factors and also on prices of the factors. In what follows we now turn to explain this choice by a firm. We will first explain the concept of iso-cost line which is used in the study of optimum factor combination.

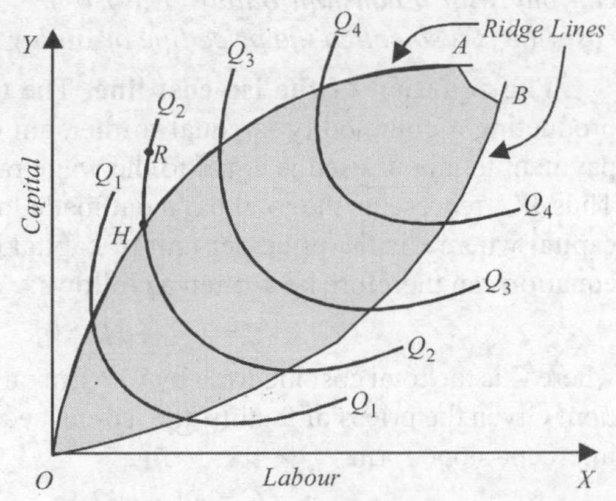

Fig. 18.1. *Economic Region of Production*

The above analysis also shows that there is a limit to which one factor can be substituted for another. As the substitution of one factor for another is carried out mor and more, it becomes progressively more difficult unti a point is reached beyond which substitution between factors becomes impossibe. As a result, the marginal product of the increasing factor first becomes zero and then it becomes negative so that isoquant becomes positively sloping.

ISO-COST LINE

The prices of factors are represented by the iso-cost line. The iso-cost line plays an important part in determining what combination of factors the firm will choose for production. An iso-cost line shows various combinations of two factors that the firm can buy with a given outlay. How the iso-cost line is drawn is shown in Fig. 18.2 where on the *X*-axis we measure units of labour and on the *Y*-axis we measure units of capital. We assume that prices of factors are given and constant for the factor. In other words, we are considering a firm which is working under perfect competition in the factor markets. Further suppose that a firm has ₹ 300 to spend on the factors, labour and capital and price of labour is ₹ 4 per labour hour and the price of capital is ₹ 5 per machine hour. With outlay of ₹ 300, he can buy 75 units of labour or 60 units of machine hours (*i.e.*, capital). Let *OB* in Fig. 18.2 represent 75 units of labour and *OA* represent 60 units of capital. In other words, if the firm spends its entire outlay of ₹ 300 on factor *X*, it buys 75 units or *OB* of labour hours and if it spends its entire outlay of ₹ 300 on capital it buys 60 units or *OA* of machine hours. The straight line *AB* which joins points *A* and *B* will pass through all combinations of labour and capital which the firm can buy with outlay of ₹ 300, if it spends the entire sum on them at the given prices. This line *AB* is called *iso-cost* line, for whichever combination lying on it the firm buys it has to incur the same cost or outlay at the given prices. *An iso-cost line is defined as the locus of various combinations of factors which a firm can buy with a constant outlay. The iso-cost line is also called the price line or outlay line.*

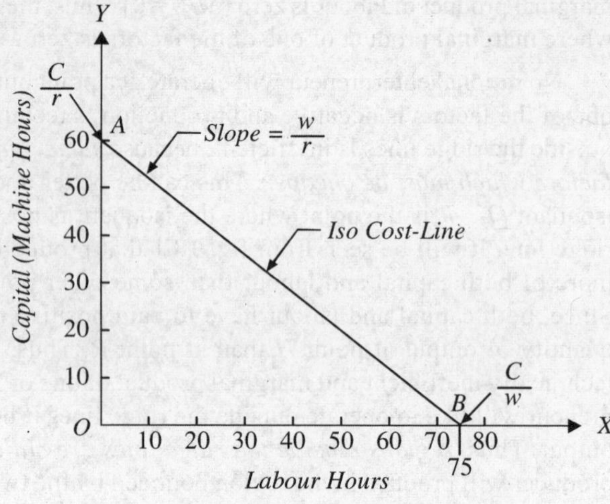

Fig. 18.2. *Iso-Cost Line*

The equation of the iso-cost line. The total cost incurred on the factors of production for producting a commodity is equal to the sum of the payments made to labour and capital. Now, payment to labour used is equal to the wage rate (*w*) multiplied by the amount of labour used (*L*). Thus *W.L* represents the total payment made to labour. Similarly, *rK* is the total payment made for capital where *r* is the price per unit of capital and *K* is the quantity of capital used. The total cost equation can therefore be written as follows:

$$C = w.L + rK$$

where C is the total cost incurred by the firm on purchasing the quantities of factors used for production. Given the prices of factors, the iso-cost equation can be rearranged as under to express it in the intercept-slope form:

$$C = wL + rK$$

$$rK = C - w.L$$

$$K = \frac{C}{r} - \frac{w}{r}.L \qquad ...(i)$$

where $\frac{C}{r}$ represents the intercept of the iso-cost line on the *Y*-axis and $\frac{w}{r}$ represents the factor price ratio and is equal to the slope of the iso-cost line.

Slope of the iso-cost line. The slope of the iso-cost line can be proved to be equal to the ratio of price of labour (*w*) and price of capital (*r*). Let, according to the iso-cost line *AB*, which given the factor prices, represents the total outlay or cost incurred on the two factors, labour and capital, the total cost equals *C*.

As explained above, the vertical intercept *OA* that represents the quantity of capital if entire cost-outlay is spent on it is equal to $\frac{C}{r}$. Similarly, the horizontal intercept *OB* representing the quantity of labour purchased if entire cost is incurred on purchasing it is equal to.

Now, the slope of the iso-cost line is:

$$\frac{OA}{OB} = \frac{C}{r} \div \frac{C}{w} = \frac{C}{r} \cdot \frac{w}{C} = \frac{w}{r}$$

Thus, the slope of the iso-cost line is equal to the ratio of factor-prices $\left(\frac{w}{r}\right)$.

Shifts in the Iso-cost Line

Now, the iso-cost line will shift if the total outlay which the firm wants to spend on the factors changes. Suppose if the total outlay to be made by the firm increases to ₹ 400, prices of factors remaining the same, then it can buy 100 units of labour hours (*i.e., OB′* of labour) or 80 units of machine hours (*i.e., OA′* of capital) if it spends the entire sum on either of them. Thus, the new iso-cost line will be *A″ B″* which will be parallel to the original iso-cost line *AB* (see Fig. 18.3). If the outlay which the firm intends to make further increases to ₹ 500, then iso-cost line will shift to the position *A′ B′*. Thus any number of iso-cost lines can be drawn, all parallel to one another, and each representing the various combinations of two factors that can be purchased for a particular outlay. The higher the outlay, the higher the corresponding iso-cost line.

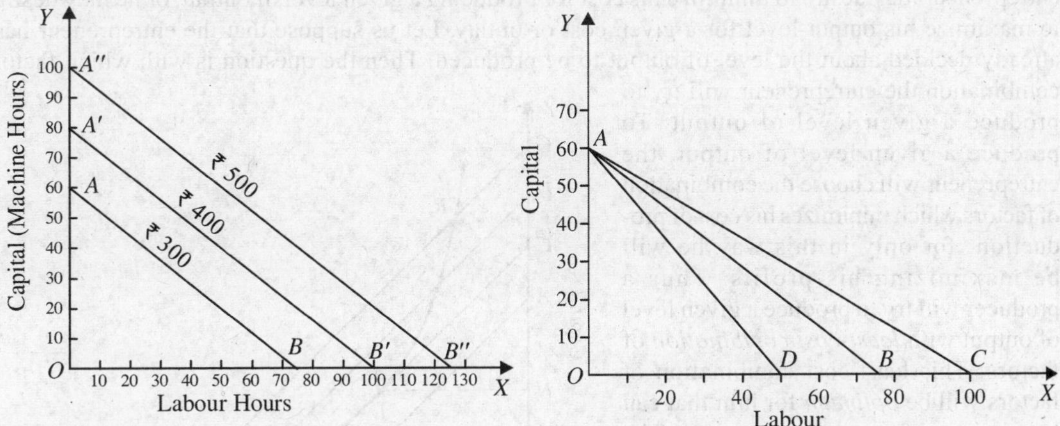

Fig. 18.3. *Iso-cost Line Resulting from increase in outlay or total cost.*

Fig. 18.4. *Changes in Iso-Cost Line as a Result of Changes in the Price of Labour X*

The iso-cost line will also change if the prices of factors change, outlay remaining the same. Suppose the firm's outlay is ₹ 300 and the prices of labour and capital are ₹ 4 and ₹ 5 respectively. Then the iso-cost line will be *AB* as shown in Fig. 18.4. If now the price of labour falls to ₹ 3, then with the outlay of ₹ 300 and ₹ 3 as the price of labour the firm can buy 100 units of labour if it spends the entire outlay on it. *OC* represents 100 units of labour. Therefore, as a result of the fall in price of labour from ₹ 4 to ₹ 3, the price line changes from *AB* to *AC*. If the price of labour rises from ₹ 4 to ₹ 6 per hour the iso-cost line will shift to *AD*. Likewise, if the price of capital changes, the outlay and the price of labour remaining the same, the iso-cost line will shift.

It is clear from above that the iso-cost line depends upon two things : (*i*) prices of the factors of production, and (*ii*) the total outlay which the firm has to make on the factors. Given these two things, an iso-cost line can be drawn. It should also be noted that the slope of the iso-cost line, like that of the price line in indifference curve analysis of demand, is equal to the ratio of the price of two factors. Thus, slope of the iso-cost line *AB*

$$= \frac{\text{Price of Labour}}{\text{Price of Capital}} = \frac{w}{r}$$

OPTIMUM OR LEAST-COST COMBINATION OF FACTORS

An equal product map or isoquant map represents the various factor combinations which can yield various levels of output, every equal product curve or isoquant showing those factor combinations each of which can produce a specified level of output. Thus, *an equal product map represents the production function of a product with two variable factors.* Therefore, an equal product map represents the technical conditions of production for a product. On the other hand, a family of iso-cost line represents the various levels of total cost or outlay, given the prices of two factors. The entrepreneur may desire to minimize his cost for producing a given level of output, or he may desire to maximize his output level for a given cost or outlay. Let us suppose that the entrepreneur has already decided about the level of output to be produced. Then the question is with which factor combination the entrepreneur will try to produce a given level of output. To produce a given level of output, the entrepreneur will choose the combination of factors which minimizes his cost of production, for only in this way he will be maximizing his profits. Thus a producer will try to produce a given level of output with *least-cost combination* of factors. This least-cost combination of factors will be *optimum* for him that can be purchased for a particular outlay. The higher the outlay, the higher the corresponding iso-cost line.

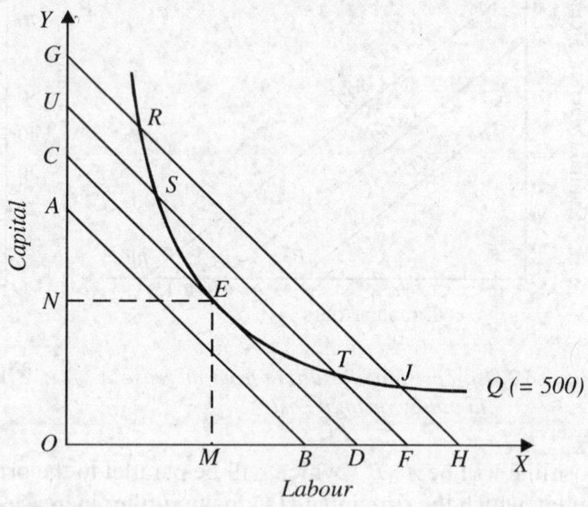

Fig. 18.5. *Minimising Cost for a Given Level of Output*

Which will be the least-cost combination of factors can be understood from considering Fig. 18.5. Suppose the entrepreneur has decided to produce 500 units of output which is represented by isoquant *Q*. The 500 units of output can be produced by any combination of labour and capital such as *R*, *S*, *E*, *T* and *J* lying on the isoquant. Now, a glance at the Fig. 18.5 will reveal that for producing the given level of output (500 units) the cost will be minimum at point *E* at which the iso-cost line *CD* is tangent to the given isoquant. At no other point such as *R*, *S*, *T* and *J*, lying on the isoquant *Q* the cost is minimum. It will be seen from Fig. 18.5 that all other points on isoquant *Q*, such as *R*, *S*, *T*, *J* lie on higher iso-cost lines than *CD* and which will therefore mean greater total cost or outlay for producing the given ouput. Therefore, the entrepreneur will not choose any of the combinations *R*, *S*, *T* and J. We thus see that factor combination *E* is the least-cost combination of labour and capital for producing a given output. Factor combination *E* is therefore an *optimum combination* for him under the given circumstances. Hence we conclude that the entrepreneur will choose factor combination *E* (that is, *OM* units of labour and *ON* units of capital) to produce 500 units of output. It is thus clear that the tangency point of the given isoquant with an iso-cost line represents the least-cost combination of factors for producing a given output.

How an entrepreneur arrives at the least-cost factor combination can also be explained with the help of the concept of marginal rate of technical substitution (*MRTS*) and the price ratio of the two factors. As has been shown earlier, the marginal rate of technical substitution (*MRTS*) is given by the slope of the isoquant at its various points. On the other hand, the price ratio of the factors is given by the slope of the iso-cost line. The entrepreneur will not choose to produce a given output at point *R* because at point *R* marginal rate of technical substitution of labour for capital is greater than the price ratio of the factors (at point *R* the slope of the isoquant *Q* is greater than the slope of the iso-cost line *GH*). Therefore, if he is at point *R* he will use more of labour in place of capital and go down on the isoquant. Likewise, he will not stop at point *S*, since the marginal rate of technical substitution of labour for capital is still greater than the price ratio of the factors; slope of the isoquant at point *S* being greater than the slope of the iso-cost line *UF*. Therefore, the entrepreneur will further substitute labour for capital and will go down further on the isoquant *Q*.

When the entrepreneur reaches point *E*, marginal rate of technical substitution of labour for capital is here equal to the price ratio of the factors, since the slopes of the isoquant and the iso-cost; line *CD* are equal to each other. The entrepreneur will have no incentive to go further down, for he will not be lowering his cost in this way, but in fact he will be reaching higher iso-cost lines. At points *J* and *T* on the isoquant *Q* marginal rate of technical substitution of labour for capital is smaller than the price ratio of the factors and the entrepreneur will try to substitute capital for labour and move upward on the isoquant *Q* until he reaches the point of tangency *E*, where marginal rate of technical substitution is equal to the price ratio of the factors. It is thus clear that the entrepreneur will be minimising his cost when the factor combination for which marginal rate of technical substitution is equal to the price ratio of the factors. Thus at his equilibrium point *E*.

$$MRTS_{LK} = \frac{w}{r}$$

where *w* stands for the wages rate of labour and *r* for the price of capital

But, as we saw in the last chapter, the marginal rate of technical substitution of labour for capital is equal to the ratio of the marginal physical products of the two factors.

Therefore,

$$MRTS_{LK} = \frac{MP_L}{MP_K} = \frac{w}{r}$$

$$\frac{MP_L}{MP_K} = \frac{w}{r}$$

We can rearrange the above equation to have

$$\frac{MP_L}{w} = \frac{MP_K}{r}$$

We therefore reach an important conclusion about the entrepreneur's choice of the quantities of the two factors. The entrepreneur will be in equilibrium in regard to his use and purchases of the two factors when he is using such quantities of the two factors that the marginal physical products of the two factors are proportional to the factor prices. If, for instance, the price of labour is twice as much as that of capital, then the entrepreneur will purchase and use such quantities of the two factors that the marginal physical product of labour is twice the marginal physical product of capital.

Output Maximisation for a Given Level of Outlay (I.E. cost)

We have explained earlier in this chapter the condition for minimisation of cost to produce a given level of output, namely, marginal rate of technical substitution between factors should be equal to the ratio of factor prices $\left(MRTS_{LK} = \frac{w}{r}\right)$. The dual of cost-minimization problem for a given level of output is of output maximization for a given level of cost or outlay. Suppose the firm has decided upon an outlay which it has to incur for the production of a commodity. With a given level of outlay, there will be a single iso-cost line that represents the outlay that firm has decided to spend. The firm will have to choose a factor combination lying on the given iso-cost line. Obviously, with a given cost or outlay, a rational producer will be interested in maximising output of the commodity. Consider Fig. 18.6. Suppose the firm has decided to incur an outlay of ₹ 200 on labour and capital which is represented by the iso-cost line *AB*. The firm has a choice to use any factor combination of labour and capital such as *R, S, E, T, J* etc. lying on the given iso-cost line *AB* to produce the product. An isoquant map showing a set of isoquants that represents various levels of output (200, 300, 400, 500 units) has been superimposed on the given iso-cost line *AB*. A glance at the Fig. 18.6 reveals that the firm will choose the factor combination *E* consisting of *ON* of labour and *OH* of capital. This is because of all the factor combinations that lie on the given iso-cost line *AB*, only the factor combination *E* enables the firm to reach the highest possible isoquant Q_3 and thus produce 400 units of output. All other combinations of labour and capital that lie on the given iso-cost line *AB* such as *R, S, T, J* etc., lie on lower isoquants showing lower levels of output than 400 units.

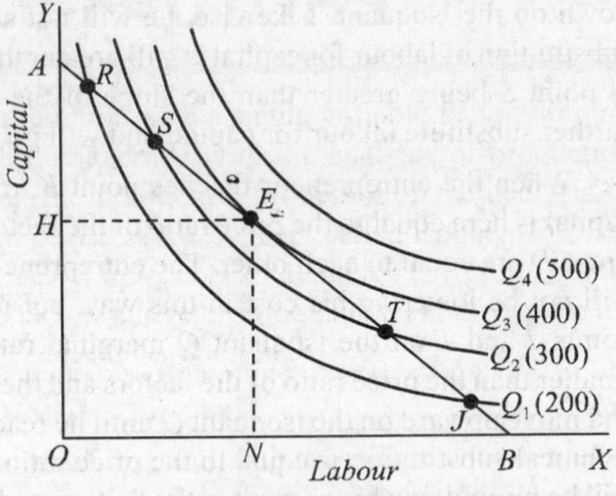

Fig. 18.6. *Output-maximisation for a given cost.*

$$\text{At point } E, \; MRTS_{LK} = \frac{w}{r}$$

Cost-Minimisation for a Given Output and Output-Maximization for a Given Cost-Outlay yield same results

This can be easily shown with the help of Fig. 18.7. Suppose *AB* is the given iso-cost line depicting the given cost constraint. Maximisation of output subject to this cost-constraint is achieved at point *E* which lies on the highest possible isoquant Q_3, while other points on the iso-cost line *AB* such as *R, S, T* or *J* lie on lower isoquants. Hence *E* represents maximum-output factor combination. Now, *given the output level* Q_3, point *E* also represents the least-cost factor combination as other points on isoquant Q_3 such as *G,H,D, M* lie on higher iso-cost curves.

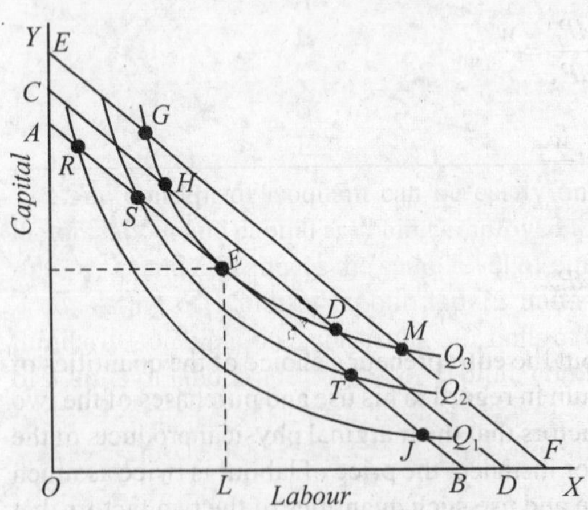

Fig. 18.7. *Cost Minimisation for a Given Output and Output Maximisation for a Given Cost.*

It is quite clear from above that the entrepreneur's behaviour in choosing the quantities of factors is exactly symmetrical with the behaviour of the consumer. Both the entrepreneur and the consumer purchase things in such quantities as to equate marginal rate of substitution with their price ratio. The consumer, to be in equilibrium, equates marginal rate of substitution (or the ratio of the marginal utilities of two goods) with the price-ratio of the goods. The entrepreneur equates marginal rate of technical substitution (or, the ratio of the marginal physical products of the two factors) with the price-ratio of the twofactors.

EXPANSION PATH

We explained above which factor combination a firm will choose to produce a specified level of output, given the prices of the two factors. We are now interested to study how the entrepreneur will change his factor combination as he expands his output, given the factor prices. To begin with, suppose the prices of the two factors *X* and *Y*, are such that are represented by the slope of the iso-cost line *AB*. In Fig. 18.8, four iso-cost lines, *AB, CD, UF*, and *GH* are drawn which show different levels of total cost or outlay. All iso-cost lines are parallel to one another indicating that prices of the two factors remain the same. If the firm wants to produce the output level denoted by Q_1 (= 100 units of output), it will choose the factor combination E_1 which minimisses cost of production; E_1 is the point of tangency between theequal product curve Q_1 and the iso-cost line *AB*. Now, if a firm wants to produce a higher level of output as denoted by the equal product curve Q_2, then it will choose the factor combination E_2 which is the least-cost combination for new output. Likewise, for still higher output levels as denoted by Q_3 and Q_4, the firm will respectively choose tangency combination E_3 and E_4 which minimise cost for the given outputs.

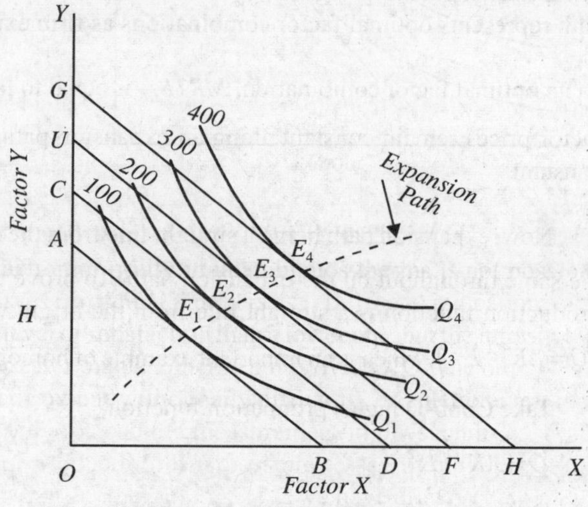

Fig. 18.8. *Expansion Path*

The line joining the minimum cost combinations such as E_1, E_2, E_3, E_4 is called the *expansion path* because it shows how the factor combination with which the firm produces will alter as the firm expands its level of output. *Thus the expansion path may be defined as the locus of the points of tangency between the isoquants and the iso-cost lines*. The expansion path is also known as *scale-line* because it shows how the entrepreneur will change the quantities of the two factors when it increases the scale of production. The expansion path can have different shapes and slopes depending upon the relative prices of the productive factors used and the shape of the isoquants (*i.e.* equal product curves). As we shall prove below, when production function exhibits constant returns to scale, the expansion path will be a straight line through the origin. Further, for a given equal product map there will be different expansion paths for different relative prices of the factors.

Since expansion path represents the minimum-cost combinations for various levels of output, it shows the cheapest way of producing each level of output, given the relative prices of the factors. When two factors are variable, the entrepreneur will choose to produce at some point on the expansion path. One cannot say exactly at which particular point on the expansion path the entrepreneur will in fact be producing unless one knows either the output which he wants to produce or the size of the cost or outlay it wants to incur. But this is certain that where both factors are variable and the prices of factors are given, a rational entrepreneur will seek to produce at one point or the other on the expansion path.

Expansion Path of a Linear Homogeneous Production Function

Whether expansion path is linear or non-linear depends on the nature of technology involved in the production function. An *important property of a linear homogeneous production function is that its expansion path is straight line from the origin* as shown in Fig. 18.9. As we saw above expansion path represents optimal factor combinations as firm expands its output, given the prices of factors. At an optimal factor combination, $MRTS_{LK}$ is equal to factor price ratio $\left(MRTS_{LK} = \frac{w}{r}\right)$. Since the factor prices remain constant along an expansion path, this implies that $MRTS_{LK}$ will also remain constant.

Now, expansion path being a straight line from the origin implies that factor ratio $\left(\frac{K}{L}\right)$ remains the same throughout on the expansion path. To prove that expansion path of a linear homogenous production function is a straight line from the origin we take Cobb–Douglas production function $\left(Q = AK^{1/2}L^{1/2}\right)$ which is an important example of homogenous production function of the first degree.

Take Cobb-Douglas production function,

$$Q = AK^{1/2}L^{1/2}$$

Differentiating it with respect to labour

$$\frac{\partial Q}{\partial L} = \frac{1}{2}AK^{1/2}L^{-1/2} \qquad \text{...(1)}$$

Differentiating it with respect to capital,

$$\frac{\partial Q}{\partial K} = \frac{1}{2}AK^{-1/2}L^{1/2} \qquad \text{...(2)}$$

Dividing (1) by (2) we have

$$\frac{MP_L}{MP_K} = \frac{\partial Q}{\partial L} \Big/ \frac{\partial Q}{\partial K} = \frac{\frac{1}{2}AK^{\frac{1}{2}}L^{\frac{-1}{2}}}{\frac{1}{2}AK^{\frac{-1}{2}}L^{\frac{1}{2}}} = \frac{K}{L}$$

or $$MRTS_{LK} = \frac{MP_L}{MP_K} = \frac{K}{L}$$

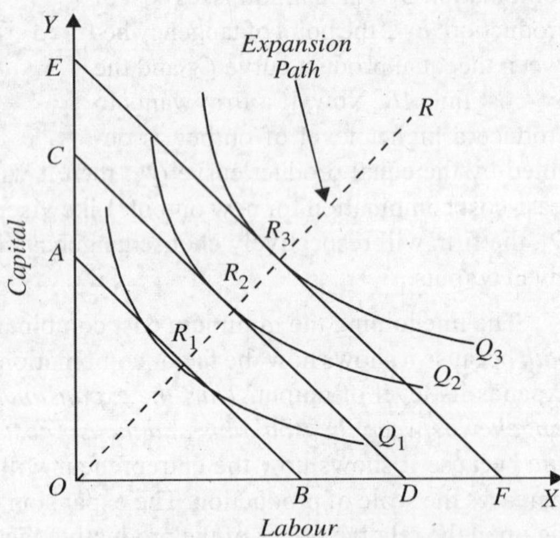

Fig. 18.9. *Expansion path of a linear homogeneous production function is a straight line from the origin.*

Thus, $MRTS_{LK}$ in the given linear homogenous Cobb-Douglas production function is equal to

$\frac{K}{L}$. As explained above, at optimal factor combinations on the expansion path, $MRTS_{LK}$ equals $\frac{w}{r}$ and, therefore, given the factor prices, $MRTS_{LK}$ remains constant along an expansion path. Thus $\frac{K}{L}$ which is equal $MRTS_{LK}$ in a linear homogenous Cobb-Douglas production will remain constant.

Constant factor ratio $\frac{K}{L}$ along the expansion path implies that it is a straight line from the origin.

FACTOR SUBSTITUTION AND CHANGES IN FACTOR PRICES

We have seen above that the cost-minimising factor combination depends on the *relative prices* of factors used. As shown above, given the prices of factors, the cost of producing a level of output is minimised by using a factor combination at which

$$MRTS_{LK} = \frac{w_0}{r_0}$$

or

$$\frac{MP_L}{w_0} = \frac{MP_k}{r_0}$$

where w_0 is the price of labour *i.e.* wage rate and r_0 is the price of capital.

Now, if either the price of labour (w) or the price of capital (r) changes, the producer will respond to this change in factor prices as their cost-minimisation state will be disturbed. For example, if the wage rate *rises* from w_0 to w_1, then at the initial equilibrium position,

$$\frac{MP_L}{w_1} < \frac{MP_k}{r_0} \text{ or, } \frac{MP_K}{r_0} > \frac{MP_L}{w_1}$$

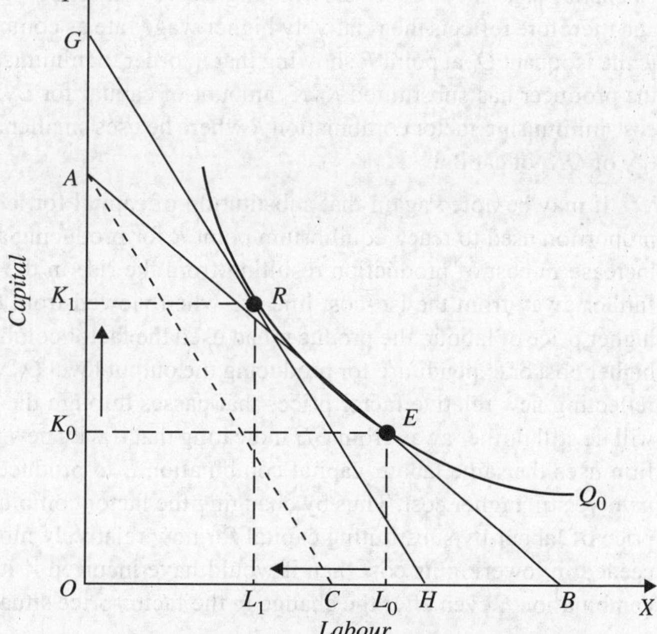

Fig. 18.10. *Rise in wage rate (price) of labour causes substitution of capital for labour.*

This will induce a rational producer to substitute capital for relatively more expensive labour. That is, he will try to use more capital and less labour and continue substituting capital for labour until

$$MRTS_{LK} = \frac{w_1}{r} \text{ or } \frac{MP_L}{w_1} = \frac{MP_K}{r}.$$

The substitution of one factor for another is graphically illustrated by using isoquants in Fig. 18.10, where with factor prices w_0 and r_0 respectively of labour and capital, AB, which is the iso-cost line for a given amount of outlay, is tangent to the isoquant Q_0 at point E. In this equilibrium situation, he is using OL_0 of labour and OK_0 of capital. Now suppose the price of labour (*i.e.*, the wage rate) rises so that the iso-cost line, price of capital (r) and outlay remaining constant, rotates to the new position AC. It will be seen from Fig. 18.10 that none of the factor combinations lying on the iso-cost line AC will be sufficient to produce the level of output Q_0 as the iso-cost line AC lies at a lower level than the isoquant Q_0. In other words, with higher wage rate w_1, the given amount of outlay is not enough to buy the required amounts of the two factors to produce the level of output Q_0. Thus, if the producer wants to produce the same level of output Q_0, it will have to increase its outlay. The increase in outlay on factors implies moving to a higher iso-cost line that will be parallel to the new iso-cost line AC. Now, with new relative prices of labour and capital, the iso-cost line GH is drawn parallel to AC so that it is tangent to the isoquant Q_0.

It will be observed from the Fig. 18.10 that the new iso-cost line GH will not be tangent at the initial equilibrium point E since its slope reflecting the new relative factor prices differs from the slope of the initial iso-cost line AB. Thus, *the initial point E no longer minimises cost in the context of new relative factor prices*.

Now that the wage rate is higher, that is, the labour is relatively more expensive, to produce the initial level of output, a producer will substitute capital for labour by moving upward along the isoquant Q_0. It will be observed from Fig. 18.10 that the new iso-cost line GH which is parallel to AC and therefore reflects the relatively higher wage rate as compared to the iso-cost line AB, is tangent to the isoquant Q_0 at point R showing that in order to minimise cost at the new relative factor prices, the producer has substituted $K_0 K_1$ amount of capital for $L_0 L_1$ amount of labour to reach the new cost-minimising factor combination R where he uses smaller amount OL_1 of labour and larger quantity of OK_1 of capital.

It may be noted again that substitution of capital for labour and thereby changing the factor-proportion used to reach equilibrium point R for producing a given level of output Q_0 involves the increase in cost of production resulting from the rise in the price of labour (iso-cost line GH lies further away from the iso-cost line AC when viewed from the origin). However, if with the new higher price of labour, the producer had used the factor combination E, he would have incurred still higher cost or expenditure for producing the output level Q_0. If iso-cost line is drawn parallel to AC reflecting new relative factor prices that passes through the original factor combination point E it will lie still further away from GH indicating that if with new relative prices of labour and capital the firm uses the same labour-capital combination E to produce that initial level of output Q_0, it will involve still higher cost. Thus by changing the factor combination from E to R following the rise in price of labour by substituting capital for now relatively more expensive labour, the firm has succeeded in lowering its cost than it would have incurred if it had continued to use the same factor combination E even after the change in the factor-price situation.

From the foregoing analysis we arrive at the conclusion that change in relative factor prices causes a substitution of a factor that has become relatively more expensive by a factor that has become relatively . In the real world there are several examples of factor substitution in response to changes in relative factor prices. When price of petroleum increased many countries tried to substitute other types of energy resources using inputs such as coal, electricity to reduce costs of production. Again, in the United States the firms use more machines (*i.e.* capital) and relatively less labour as labour there is very costly than is the case in some developing countries where wages are comparatively low. Further, recently when prices of computers have fallen, there has been substitution of manual labour for doing such work as book-keeping, making architectural maps, composing books and journals by computers (*i.e.* capital).

QUESTIONS FOR REVIEW

1. Explain the concept of production function. The information about production function facing a firm is inadequate for making decision regarding economically efficient use of factors or resources. Explain.

[**Hints:** Production function describes the technological aspect of production, that is, maximum possible output that can be produced by various combinations of factors. On the other hand, *economic efficiency in resource use implies least-cost combination of factors to produce a given output or alternatively it implies maximisation of output for a given cost outlay*. Thus, for deciding about optimal or economically efficient resource use we require not only the data about the production function but also about prices of factors.]

2. What is meant by efficient or optimum factor combination in production ? Explain with the help of isoquants (*i.e.*, equal product curves) and iso-cost lines how a producer achieves this combination of factors.

3. Show with the help of isoquants that a firm will be in equilibrium regarding use of a factor combination when marginal rate of technical substitution (MRTS) between factors is equal to the ratio of factor prices.

4. Given the prices of the two factors for the individual firm, explain the conditions for producing a given output at least cost. [*C.U.*, *B.Com.*, (*Hons.*), 1998]

5. Show with the isoquant-iso-cost apparatus, a firm is in equilibrium with regard to the use of factors when the ratios of marginal products of factors to their respective prices are equal. *D.U. B.Com (Hons.)*

6. Show that maximisation of output subject to a given cost constraint and minimisation of cost subject to a given output will yield identical results. [*C.U. B.Com., 1993*]

(*a*) Suppose a firm is using a combination of labour and capital to produce a certain level of output. Now suppose that wage rate of labour rises, price of capital remaining the same. Explain using diagram what will be its effect on the use of labour and capital, while output of the firm remains constant.

(*b*) (*i*) What are ridge lines? Explain the economic region of product using isoquant map .

(*ii*) Suppose labour is free, show with the help of an isoquant map how much labour will be employed, given a certain fixed quantity of capital.

D.U. B.Com (Hons.) and B.A. (Hons.)

7. Define substitute and complementary factors. Show that in case of perfectly complementary factors substitution effect is zero.

8. What is an expansion path? Show that expansion path of a linear homogeneous production function is a straight line through the origin. [*D.U. B.A.* (*Hons.*) *1996*]

[**Hint:** For the later part see mathematical portion in the text of the book.]

9. Explain using isoquants (*i.e.*, equal product curves) (*a*) the effect of rise in wage rate of labout, price of capital remaining, constant on the use of labour and capital, (*b*) the effect of fall in price of capital, wage rate of labour remaining the same.

10. Explain using isoquants-iso-cost apparatus how a change in price of a factor is split up into output effect and substitution effect.

11. What is meant by economic efficiency in the use of resources in production. A firm is producing output using labour and capital in such quantities that marginal product of labour is 15, and marginal product of capital is 8. The wage of labour is ₹ 3 and price of capital is ₹ 2. Is the firm using efficient factor combination for production ? If not, what it should do to achieve economic efficiency?

[**Hints:** Efficiency condition for factor use (that is, optimal factor combination) requires that the following condition should be fulfilled:

$$\frac{MP_L}{w} = \frac{MP_K}{r}$$

Now, $\frac{MP_L}{w} = \frac{15}{3}$ and $\frac{MP_K}{r} = \frac{8}{2}$

Thus, $\frac{15}{3} > \frac{8}{2}$

The given factor combination cannot therefore be efficient or optimal factor combination *because the firm is getting more output per unit of rupee spent on labour than on capital.* To achieve efficiency or maximum profits the firm should substitute labour for capital so that $\frac{MP_L}{w}$ becomes equal to $\frac{MP_K}{r}$.]

12. A firm reports that marginal product of labour is 5 and marginal rate of technical substitution of labour for capital is 2. What is the marginal product of capital ?

[**Hints :** $MRTS_{LK} = \frac{MP_L}{MP_K}$

$$2 = \frac{5}{MP_K}, \quad MP_K = \frac{5}{2} = 2.5$$]

13. A firm employs labour as the only variable factor along with a fixed quantity of capital. Wage rate of labour is ₹ 100 per day and its marginal product is 20 units of output. What is *marginal cost* of the product.

[**Hints:** Marginal cost is the additional cost incurred on producing an extra unit of output. Marginal cost can be obtained from dividing the wage that has to be paid to the additional unit of labour by the number of units of output produced by it (that is, its marginal product). Thus,

$$MC = \frac{W}{MP_L} = \frac{100}{20} = 5.$$

14. A biscuit producing company reports the following data about its production and factors used :

Q	K	L	Q	K	L
490	15	99	470	14	100
500	15	100	500	15	100

where Q = output, K = capital, L = labour.

If wage rate of labour is ₹ 5 and price of capital is ₹ 10. Does the input combination of 15K and 100L represent the least-cost factor combination. If not, should it use more labour and less capital or less labour and more capital?

[**Hints** : Least-cost factor combination requires that the following condition must be met:

$\frac{MP_L}{w} = \frac{MP_K}{r}$, where w is wage rate and r is price of capital.

Now, in the problem when factor combination is changed from $15K + 100L$ to $15K + 99L$, the output declines by 10 units (500 – 490 = 10). Thus, $MP_L = 10$. Similarly, when factor combination is changed from $15K + 100L$ to $14K + 100L$, output declines by 30 (500 – 470 = 30). Thus, $MP_K = 30$. Substituting these values of marginal products of labour and capital and the factor prices in the condition for least-cost factor combination, we have

$$\frac{MP_L}{w} = \frac{MP_K}{r}$$

$$\frac{10}{5} \neq \frac{30}{10} \text{ or } \frac{10}{5} < \frac{30}{10}$$

The company will use more capital and less labour as output per unit of a rupee (3) spent on capital is greater than the output per unit of rupee $\left(\frac{10}{5} \text{ or } 2\right)$ spent on labour.

15. Suppose that a product requires two inputs for its production, then is it correct to say that if the prices of inputs are equal, optimal behaviour on the part of producers will dictate that these inputs be used in equal amounts.

[**Hints:** The answer is 'NO'. This is because optimal use of factors depends not only on the prices of factors. but also on their marginal products which are governed by the nature of production function (that is, technology of production)]

CHAPTER 19
COST OF PRODUCTION AND COST CURVES

In the last chapter we studied the laws of returns underlying the production conditions of goods. These production conditions determine to an appreciable degree the supply of goods. In this chapter we carry further the analysis of the forces determining supply of goods. We shall examine here how the cost of production of a firm changes with the change in its output. In other words, cost-output relations form the subject matter of the present chapter. The relation between cost and output is called "*cost function*". The cost function of the firm depends upon the production conditions and the prices of the factors used for production. How much costs a firm will incur on production depends on the level of output. Moreover, the quantity of a product that will be offered by the firm for supply in the market depends to a great degree upon the cost of production incurred on the various possible levels of output. Cost of production is the most important force governing the supply of a product. It should be pointed out here that it is assumed that a firm chooses a combination of factors which minimises its cost of production for a given level of output. It is thus assumed that whatever the level of output a firm produces, it is produced at the minimum cost possible.

In microeconomic theory, economists are generally interested in two types of cost function : the short-run cost function and the long-run cost function. Accordingly, they derive the short-run and long-run cost curves. We first explain below the various concepts of cost which are used in modern economic theory and then turn to study the derivation of the short-run and the long-run cost curves.

THE CONCEPTS OF COST

Accounting Costs and Economic Costs

It is necessary for the proper understanding of the price theory to know the various concepts of cost that are often employed. When an entrepreneur undertakes production of a commodity he has to pay prices for the factors which he employs for production. He thus pays wages to the labourers employed, prices for the raw materials, fuel and power used, rent for the building he hires for the production work, and the rate of interest on the money borrowed for doing business. All these are included in his cost of production. An accountant will take into account only the payments and charges made by the entrepreneur to the suppliers of various productive factors.

But an economist's view of cost is somewhat different from this. It generally happens that the entrepreneur invests a certain amount of his own money capital in his productive business. If the money invested by the entrepreneur in his own business had been invested elsewhere, it would have earned a certain amount of interest or dividends. Moreover, an entrepreneur devotes time to his own work of production and contributes his entrepreneurial and managerial ability to it. If the entrepreneur had not set up his own business, he would have sold his services to others for some positive amount of money. Therefore, economists would also include in the cost of production (*i*) the *normal return on money capital* invested by the entrepreneur himself in his own business, which he could have earned if invested outside and (*ii*) *the wages or salary* he could have earned if he had sold his services to others. The accountant would not include these two items in a firm's cost of production but the economists consider them as *bonafide* costs and will accordingly include them in cost. Likewise, the money rewards for other factors owned by the entrepreneur himself and employed by him in his own business are also considered by the economists as parts of the cost of production.

It follows from above that the accountant considers those costs which involve cash payments to *others* by the entrepreneur of the firm. The economist takes into account all of *these accounting costs*, but in addition, he also takes into account the *amount of money the entrepreneur could have earned if he had invested his money and sold his own services and other factors in next best alternative uses*. The accounting costs are contractual cash payments which the firm makes to other factor owners for purchasing or hiring the various factors are also known as *explicit costs*. The normal return on money-capital invested by the entrepreneur and the wages or salary for his services and the money rewards for other factors which the entrepreneur *himself owns and employs* them in his own firm are known as *implicit costs or imputed costs*. The economists take into consideration both the explicit and implicit costs. Therefore,

Economic Costs = Accounting Costs + Implicit Costs

It may be pointed out that the firm will earn *economic profits* only if it is making revenue *in excess of the total of accounting and implicit costs*. Thus, when the firm is in no profit and no loss position, it means that the firm is making revenue equal to the total of accounting and implicit costs and no more. Therefore,

Economic Profits = Total Revenue – Total Economic Costs

Opportunity Cost

The concept of opportunity cost occupies a very important place in modern economic analysis. *The opportunity cost of any good is the next best alternative good that is sacrificed.* The factors which are used for the manufacture of a car may also be used for the production of an equipment for the army. Therefore, the opportunity cost of production of a car is the output of the army equipment foregone or sacrificed, which could have been produced with the same amount of factors that have gone into the making of a car. To take another example, a farmer who is producing wheat can also produce potatoes with the same factors. Therefore, the opportunity cost of a quintal of wheat is the amount of output of potatoes given up. Professor Benham defines the opportunity cost thus : "*the opportunity cost of anything is the next best alternative that could be produced instead by the same factors or by an equivalent group of factors, costing the same amount of money.*"[1]

Two points be noted in the above definition of opportunity cost. First, the opportunity cost of anything is only the *next-best alternative* foregone. That is to say, the opportunity cost of producing a good is not any other alternative good that could be produced with the same factors; it is only the *most valuable other good* which the same factors could produce. Second point worth noting in the above definition is the addition of the qualification "or by an equivalent group of factors costing the same amount of money". The need for the addition of this qualification arises because all the factors used in the production of one good may not be the same as are required for the production of the next best alternative good. For instance, the farmer who is employing land, workers, water, fertilisers, wheat seed, etc., for the production of wheat may use the same land, the same workers, the same water, the same fertilisers for the production of potatoes, but a different type of seed will be needed. Likewise, a manufacturing firm may shift from the production of one product to another without any changes in plant and equipment or its workers but it will require different types of raw materials. In such cases therefore the opportunity cost of a good should be viewed as the next best alternative good that could be produced with the *same value* of the factors which are more or less the same.

1. Frederic Benham, *Economics*, 6th edition, p. 195.

The concept of opportunity cost is very fundamental to economics. Robbins' famous definition of economics goes in terms of the scarcity of resources and their ability to be put into various uses. If the production of one good is increased, then the resources have to be withdrawn from the production of other goods. Thus, when the resources are fully employed, then more of one good could be produced at the cost of producing less of the others. If 100 units more of good *X* are produced by withdrawing resources from the industry producing good *Y*, then the opportunity costs of producing additional hundred units of *X* is the amount of good *Y* sacrificed.

The alternative or opportunity cost of a good can be given a money value. In order to produce a good the producer has to employ various factors of production and has to pay them sufficient prices to get their services. These factors have alternative uses. *The factors must be paid at least the price they are able to obtain in the next best alternative uses*. The total alternative earnings of the various factors employed in the production of a good will constitute the opportunity cost of the good.

A significant fact worth mentioning is that *relative prices of goods tend to reflect their opportunity costs*. The resources will remain employed in the production of a particular good when they are being paid at least the money rewards that are sufficient to induce them to stay in the industry, *i.e.*, equal to the value they are able to obtain and create elsewhere. In other words, a collection of factors employed in the production of a good must be paid equal to their opportunity cost. The greater the opportunity cost of a collection of factors used in the production of a good, the greater must be the price of the good. Thus, if the same collection of factors can produce either one tractor or 2 scooters, then the price of one tractor will be twice that of one scooter.

Private Cost Versus Social Cost

It is important to note the distinction made by the economists between the private cost and social cost. The sum of the explicit and implicit costs incurred by a firm to produce a product constitutes his private cost. It is these private costs that it takes into account while making decisions regarding price and output of the commodity it produces. However, in addition to the private costs, the *production of a commodity by a firm bestows some benefits or causes some damage or loss to others which he does not take into account while taking decisions regarding price and output of the commodity*. These external damages or benefits, if any, caused by the production of a firm to other firms or consumers also constitute a part of the social cost which are ignored by a firm but may be of great significance from the social point of view. For example, a firm producing pesticides pay prices for the purchase of raw materials, labour and capital used by it. These are the private costs of the firm. But in the production process of pesticides some by-products such as foul smell, polluted air and effluents that are emitted along the production of pesticides cause economic damage or costs to those people who live in the adjoining areas. These harmful external effects are also called *negative externalities*. The beneficial external effects created by firm are called *positive externalities. The social costs are the sum of the private costs and the net of negative externalities over positive externalities.* If there are no any negative externalities, social cost will be greater than private costs. If social welfare is to be maximised, it is the total social cost and not the private cost alone that has to be minimised. We shall discuss in detail the concept of social cost in the part relating to the welfare economics where we will explain when there occur market failures due to the presence of external damages and benefits of the production by firms, the intervention by the government becomes necessary to promote social welfare.

Short Run and Long Run Defined

There are some inputs or factors which can be readily adjusted in the short run with the changes in the output level. Thus, a firm can readily employ more workers, if it has to increase output. Likewise, it can secure and use more raw materials, more chemicals without much delay if it has to expand production in the short run. Thus, factors such as *labour, raw materials, chemicals etc., which can be readily varied with the change in output are called variable factors*. On the other hand, there are factors such as capital equipment, building, top management personnel which cannot be so readily varied. It requires a comparatively long time to make variations in them. It takes time to expand a factory building or to build a new factory building with a large area or capacity. Similarly, it also takes time to order and install new machinery. The *factors such as capital equipment, factory building which cannot be readily varied and require comparatively a long time to make adjustment in them are called fixed factors*.

Corresponding to the distinction between variable factors and fixed factors, economists distinguish between the short run and the long run. *The short run is a period of time in which output can be increased or decreased by changing only the amount of variable factors such as labour, raw materials, chemicals, etc*. On the other hand, in the long run, quantities of the fixed factors such as capital equipment, factory building, etc., can also be varied for making changes in output.

Thus, in the long run, the output can be increased not only by using more quantities of labour and raw materials but also by expanding the size of the existing plant or by building a new plant with a larger productive capacity.

Short-Run Costs: Total Fixed and Variable Costs

Fixed costs are those which are independent of output, that is, they do not change with changes in output. These costs are a fixed amount which must be incurred by a firm in the short run, whether output is small or large. Even if the firm closes down for some time in the short run but remains in business, these costs have to be borne by it. Fixed costs are also known as *overhead costs* and include charges such as contractual rent, insurance fee, maintenance costs, property taxes, interest on the capital invested, minimum administrative expenses such as manager's salary, watchman's wages etc. Thus *fixed costs are those which are incurred in hiring the fixed factors of production whose amount cannot be altered in the short run*.

Variable costs, on the other hand, are those costs which are incurred on the employment of variable factors of production whose amount can be altered in the short run. Thus the total variable costs change with changes in output in the short run, *i.e.*, they increase or decrease when the output rises or falls. These costs include payments such as wages of labour employed, prices of the raw materials, fuel and power used, the expenses incurred on transporting and the like. If a firm shuts down for some time in the short run, then it will not use the variable factors of production and will not therefore incur any variable costs. Variable costs are made only when some amount of output is produced and the total variable costs increase with the increase in the level of production. Variable costs are also called *prime costs* or *direct costs*. Total costs of a business is the sum of its total variable costs and total fixed costs. Thus:

$$TC = TFC + TVC$$

where *TC* stands for total cost, *TFC* for total fixed cost and *TVC* for total variable cost.

Because one component, *i.e.*, the total variable cost (*TVC*) varies with the change in output, the total cost of production (*TC*) will also change with the changes in the level of output. The total cost increases as the level of output rises. The concepts of total cost, total variable cost and total fixed cost in the short run can be easily understood with the help of the following Table 19.1.

It will be seen from the table that the total fixed costs are equal to ₹ 50 and remain constant when the output is increased from 1 to 8 units of output. Even if no output is produced, the firm has to bear the fixed costs of production. This is because, as said above, the firm cannot dispense with the fixed factors of production in the short run. It has therefore to keep the fixed factors idle in the short run and bear costs incurred on them if demand conditions are not favourable for production.

Table 19.1. Total Cost, Total Fixed Cost, and Total Variable Cost

No. of Units of Output	*Total Fixed Cost (TFC)*	*Total Variable Cost (TVC)*	*Total Cost (TC)*
0	50	0	50
1	50	20	70
2	50	35	85
3	50	60	110
4	50	100	150
5	50	145	195
6	50	190	240
7	50	237	287
8	50	284	334

As regards the variable costs, it will be seen from Table 19.1 that variable costs are equal to ₹ 20 when only one unit of output is produced and they rise to ₹ 284 when eight units are produced. Since variable costs are incurred on factors such as labour, raw materials, fuel etc., which vary with the change in the level of output, the total variable costs increase with the increases in output throughout.

As the total cost is the sum of fixed cost and the variable cost, it can be obtained by adding the figures of column 2 (fixed cost) and column 3 (variable cost). For example, when two units of output are produced, the total cost works out to be ₹ 70 (₹ 50 + ₹ 20 = 70). The total cost also varies directly with output because a significant part of it (*i.e.* variable cost) increases as output is increased.

Total fixed cost and total variable cost are portrayed in Fig. 19.1 where output is measured on the *X*-axis and cost on the *Y*-axis. Since the total fixed cost remains constant whatever the level of output, the total fixed cost curve (*TFC*) is a horizontal straight line. It will be seen from Fig. 19.1 that total fixed cost curve (*TFC*) starts from a point *B* on the *Y*-axis meaning thereby that the total fixed cost equal to *OB* will be incurred even if the output is zero. On the other hand, the total variable cost curve (*TVC*) rises upward showing thereby that as output is increased, the total variable cost also increases. The total variable cost curve *TVC* starts from the origin which shows that when output is zero the variable costs are nil.

It should be noted that total cost (*TC*) is a function of total output (*Q*); the greater the output, the greater will be the total cost. In symbols, we can write:

$$TC = f(Q)$$

where *Q* stands for output.

We can prove this as follows :

$$TC = TFC + TVC.$$

Suppose *TFC* is equal to *K* which is a constant amount whatever the level of output. *TVC* is equal to the amount used of the variable factor, say, *L*, multiplied by the given price of the variable, say, *w*

$$TVC = L \cdot w \qquad \ldots (i)$$

$$TC = \overline{K} + L \cdot w \qquad \ldots (ii)$$

Now, *L*·*w*, that is, *TVC* must rise with the increase in output, because only by increase in the amount of variable factor, that is, by increase in *L*, the output can be increased. From equation (*ii*) it

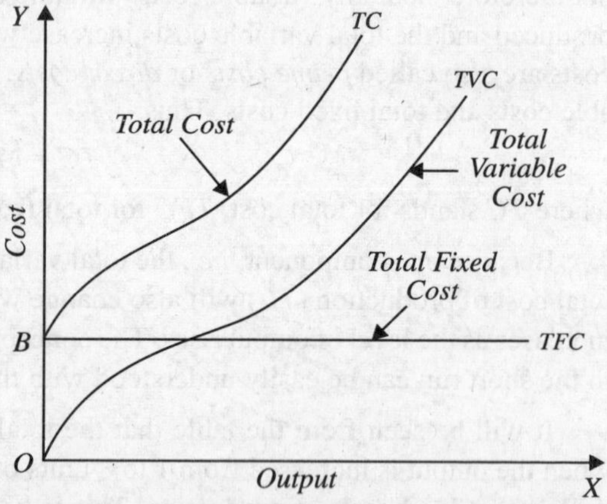

Fig. 19.1. *Total Fixed Cost, Total Variable Cost and Total Cost*

follows that with the increase in $L.w$ as output rises, TC must also rise. In other words, total cost (TC) is a function of total output (Q) and varies directly with it.

Total cost curve (TC) is obtained by adding up vertically total fixed cost and total variable cost curves because the total cost is sum of total fixed cost and total variable cost. It will be seen from Fig. 19.1 that the vertical distance between the TVC curve and TC curve is constant throughout. This is because the vertical distance between the TVC and TC curves represents the amount of total fixed cost which remains unchanged as output is increased in the short run. It should also be noted that the vertical distance between the total cost curve (TC) and the total fixed curve (TFC) represents the amount of total variable costs which increase with the increase in output. The shape of the total cost curve (TC) is exactly the same as that total of variable cost curve (TVC) because the same vertical distance always separates the two curves.

THE SHORT-RUN AVERAGE COST CURVES

We have explained above the short-run total cost curves. However, the cost concept is more frequently used both by businessmen and economists in the form of cost per unit, or average costs rather than as total costs. We, therefore, pass on to the study of short-run *average cost curves*.

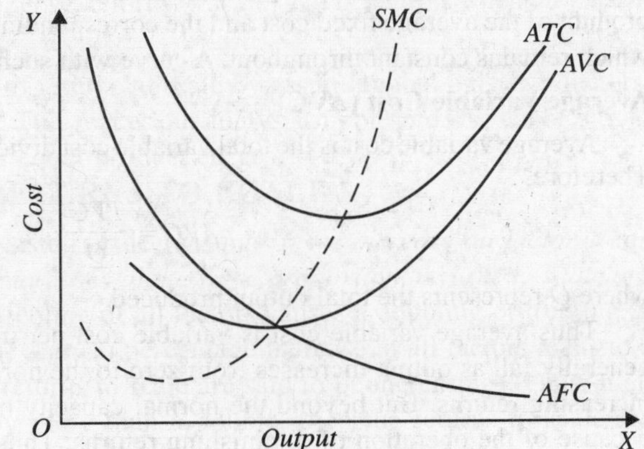

Fig. 19.2. *Short-Run Average and Marginal Cost Curves*

Average Fixed Cost (AFC)

Average fixed cost is the total fixed cost divided by the number of units of output produced. Therefore,

$$AFC = \frac{TFC}{Q}$$

where Q represents the number of units of output produced.

Thus average fixed cost is the fixed cost per unit of output. Suppose for a firm the total fixed cost is ₹ 2,000 when output is 100 units, average fixed cost (AFC) will be ₹ 2,000/100 = ₹ 20 and when output is expanded to 200 units, average fixed cost will be ₹ 2,000/200 = ₹ 10. Since total fixed cost is a constant quantity, average fixed cost will steadily fall as output increases. There

Table 19.2. Average Fixed Cost, Average Variable Cost and Average Total Cost

Units of Output (Q)	*Total Fixed Cost (TFC)*	*Total Variable Cost (TVC)*	*Total Cost (TC)*	*Average Fixed Cost (AFC)*	*Average Variable Cost (AVC) TVC/Q*	*Average Total Cost (ATC) TC/Q*
0	50	0	50	0	0	0
1	50	20	70	50.00	20.00	70.00
2	50	35	85	25.00	17.50	42.50
3	50	60	110	16.67	20.00	36.67
4	50	100	150	12.50	25.00	37.50
5	50	145	195	10.00	29.00	39.00
6	50	190	240	8.33	31.67	40.00
7	50	237	287	7.14	33.86	41.00
8	50	284	334	6.25	35.50	41.75

fore, average fixed cost curve slopes downward throughout its length. As output increases, the total fixed cost spreads over more and more units and therefore average fixed cost becomes less and less. When output becomes very large, average fixed cost approaches zero. Consider Table 19.2 where total cost is ₹ 50 when one unit of output is produced, the average fixed cost is obviously ₹ 50(50/1=50). On raising output to 2 units, average fixed cost will be ₹ 25. (*i.e.* 50/2 = 25). Further, if output is increased to 8 units, average fixed cost falls to ₹ 6.25 (*i.e.* 50/8 = 6.25). Average fixed cost curve (AFC) is shown in Fig. 19.2. It will be seen that average fixed cost curve continuously falls throughout. Mathematically speaking, average fixed cost curve approaches both axes asympotically. In other words, *AFC* curve gets very nearer to but never touches either axis.

The average fixed cost curve, *AFC,* possesses another important property. If we pick up any point on the average fixed cost curve and multiply the average fixed cost at that point with the corresponding quantity of output produced, then the product is always the same. This is because the product of the average fixed cost and the corresponding quantity of output will yield total fixed cost which remains constant throughout. A curve with such a property is called *rectangular hyperbola.*

Average Variable Cost (AVC)

Average variable cost is the total variable cost divided by the number of units of output produced. Therefore,

$$AVC = \frac{TVC}{Q}$$

where Q represents the total output produced.

Thus average variable cost is variable cost per unit of output. The average variable cost will generally fall as output increases from zero to the normal capacity output due to the occurrence of increasing returns. But beyond the normal capacity output average variable cost will rise steeply because of the operation of diminishing returns. Thus, in Table 19.2 average variable cost can be obtained from dividing total variable cost (*TVC*) by output. It will be seen from Table 19.2 that when two units of output are being produced, average variable cost can be found by dividing ₹ 35 by 2 which is equal to ₹ 17.50. Likewise, when five units of output are being produced, average variable cost becomes ₹ 29. The average variable cost curve is shown in Fig. 19.2 by the curve *AVC* which first falls, reaches a minimum and then rises.

Average total cost (*ATC*) is the sum of the average variable cost and average fixed cost. Therefore, as output increases and average fixed cost becomes smaller and smaller, the vertical distance between the average total cost curve (*ATC*) and average variable cost curve (*AVC*) goes on declining. When average fixed cost curve (*AFC*) approaches the *X*-axis, the average variable cost curve approaches the average total cost curve (*ATC*).

Relationship between AVC and Average Product

Average variable cost bears an important relationship with the average product per unit of the variable factor. Let Q stand for quantity of total product produced; L for the amount of the variable factor, say labour, used and w for the price per unit of the variable factor and AP for the average product of the variable factor. We assume that the price of the variable factor remains unaltered as more or fewer units of the variable factor are employed.

Total product (or output Q) $= AP \times L$

where AP stands for average product of labour, the variable factor and L for the amount of labour used.

$$\text{Average variable cost } (AVC) = \frac{TVC}{Q}$$

Since the total variable cost (*TVC*) is equal to the amount of the variable factor (L) employed multiplied by the price per unit (w) of the variable factor, ($TVC = L.w.$), therefore

$$AVC = \frac{L.w}{Q}$$

Since $$Q = AP \times L$$

$$\therefore \quad AVC = \frac{L.w}{AP \times L} = \frac{w}{AP} = w\left(\frac{1}{AP}\right)$$

Thus, given the price of the variable factor *w*, the average variable cost is equal to the reciprocal of the average product $\left(\frac{1}{AP} \text{ is the reciprocal of } AP\right)$ multiplied by a constant *w*. It follows that average variable cost and average product of the variable factor vary inversely with each other. Therefore, when average product of the variable factor rises in the beginning as more units of the variable factor are employed, the average variable cost must be falling. And when the average product of the variable factor falls, the average variable cost must be rising. At the level of output at which the average product of the variable factor is maximum, the average variable cost is minimum. Thus the average variable cost (*AVC*) curve looks like the average product (*AP*) curve turned upside down with minimum point of the *AVC* curve corresponding to the maximum point of the *AP* curve.

Average Total Cost (ATC)

The average total cost or what is simply called average cost is the total cost divided by the number of units of output produced.

$$\text{Average total cost} = \frac{\text{Total Cost}}{\text{Output}}$$

or $$ATC = \frac{TC}{Q}$$

Since the total cost is the sum of total variable cost and the total fixed cost, the average total cost is also the sum of average variable cost and average fixed cost. This can be proved as follows :

$$ATC = \frac{TC}{Q}$$

Since $$TC = TVC + TFC$$

Therefore, $$ATC = \frac{TVC + TFC}{Q}$$

$$= \frac{TVC}{Q} + \frac{TFC}{Q}$$

$$= AVC + AFC$$

Average total cost is also known as *unit cost*, since it is cost per unit of output produced. As the average total cost is the sum of average variable cost and average fixed cost, in Table 19.2 it can be obtained by summing up the figures of columns 5 and 6 corresponding to different levels of output. Thus, for example, with two units of output, average total cost is ₹ 25 + ₹ 17.50 = ₹ 42.50 and with three units of output it is equal to ₹ 16.67 + ₹ 20 = ₹ 36.67 and so on for other levels of output. Alternatively, the average total cost can be obtained directly from dividing the total cost by the number of units of output produced. Thus average total cost of 2 units of output is equal to ₹ 85/2 or ₹ 42.50. Likewise, when output is raised to 6 units, total cost rises to 240 and average total cost works out to be ₹ 240/6 = ₹ 40.

It follows from above that the behaviour of the average total cost curve will depend upon the behaviour of the average variable cost curve and average fixed cost curve. In the beginning, both

AVC and *AFC* curves fall, the *ATC* curve therefore falls sharply in the beginning. When *AVC* curve begins rising, but *AFC* curve is falling steeply, the *ATC* curve continues to fall. This is because during this stage the fall in *AFC* curve weighs more than the rise in the *AVC* curve. But as output increases further, there is a sharp rise in *AVC* which more than offsets the fall in *AFC*. Therefore the *ATC* curve rises after a point. Thus, the average total cost curve (*ATC*) like the *AVC* curve first falls, reaches its minimum value and then rises. The average total cost curve (*ATC*) is therefore almost of a 'U' shape.

MARGINAL COST (MC)

The concept of marginal cost occupies an important place in economic theory. Marginal cost is *addition* to the total cost caused by producing one more unit of output. In other words, marginal cost is the addition to the total cost of producing *n* units instead of *n* – 1 units (*i.e.* one less) where *n* is any given number. In symbols:

$$MC_n = TC_n - TC_{n-1}$$

Suppose the production of 5 units of a product involves the total cost of ₹ 206. If the increase in production to 6 units raises the total cost to ₹ 236, then marginal cost of the sixth unit of output is ₹ 30 (236 – 206 = 30). Let us illustrate the computation of marginal cost from a table of total cost and output. In Table 19.3 when output is zero in the short run, the producer is incurring total cost of ₹ 100 which represents the total fixed cost of the production. When one unit of output is produced, the total fixed cost rises to ₹ 125. The marginal cost of the first unit of output is therefore ₹ 25 (*i.e.* 125 – 100 = 25). When output is increased to 2 units, the total cost goes up to ₹ 145. Therefore, the marginal cost is now ₹ 20 (*i.e.* 145 – 125 = 20). In this way marginal cost can be found for further units of output.

$$MC = \frac{\Delta TC}{\Delta Q}$$

where ΔTC represents a change in total cost and ΔQ represents a unit change in output or total product. Since in the short run its is only the total variable cost (*TVC*) that varies with change in output, $MC = \dfrac{\Delta TVC}{\Delta Q}$

If we consider the total cost curve, $\dfrac{\Delta TC}{\Delta Q}$ represents the slope of it. Therefore, if we want to measure the marginal cost at a certain output level, we can do so by measuring the slope of the total cost curve corresponding to that output by drawing a tangent at it.

Table 19.3. Computation of Marginal Cost

Output Q	*Total Cost* TC	*Marginal Cost* $MC = \dfrac{\Delta TC}{\Delta Q}$
0	100	—
1	125	25
2	145	20
3	160	15
4	180	20
5	206	26
6	236	30
7	273	37

It is worth pointing out *that marginal cost is independent of the fixed cost and depends on the changes in the variable factors*. Since fixed costs do not change with output, there are no marginal fixed costs when output is increased in the short run. It is only the variable costs that vary with output in the short run. Therefore, the marginal costs are in fact due to the changes in variable costs, and whatever the amount of fixed cost, the marginal cost is unaffected by it.

The Relationship between Marginal Cost and Marginal Product of a Variable Factor

It should be noted that *marginal cost of production is intimately related to the marginal product of the variable factor*. Thus, if *MC* stands for marginal cost of output, *MP* for marginal product of the variable factor, *w* for the price of the variable factor, then

$$MC = \frac{\Delta TVC}{\Delta Q} \quad ...(i)$$

Since, given the price of the variable factor, change in total variable cost can occur by increasing the quantity of the variable factor (*e.g.,* labour), we have

$$\Delta TVC = w. \Delta L$$

$$MC = \frac{\Delta TVC}{\Delta Q} = \frac{w.\Delta L}{\Delta Q} \quad ...(ii)$$

where *w* is the given price of the variable factor 'labour', $\frac{\Delta L}{\Delta Q}$ is the reciprocal of marginal product of labour which we simply write as *MP*.

Thus, from (*ii*) we have,

$$MC = = w. \frac{1}{MP} = \frac{w}{MP} \quad ...(iii)$$

Thus, marginal cost of production is equal to the reciprocal of the marginal product of the variable factor multiplied by the price of the variable factor. In other words, *marginal cost is the price of the variable factor divided by its marginal product*. Therefore, marginal cost varies inversely with the marginal product of the variable factor. Now, if the price of the variable factor, *i.e.*, *w* is assumed to be constant, then from the relation between *MC* and *MP* represented in the above equation, we can ascertain the shape of the marginal cost curve. We know from the study of the law of variable proportions that as output increases in the beginning, marginal product of the variable factor rises. This means that constant *w* in equation (*iii*) is being divided by increasingly larger *MP*. This will cause the marginal cost (*MC*) to decline as output increases in the beginning. Further, according to the law of variable proportions, marginal product of a variable factor

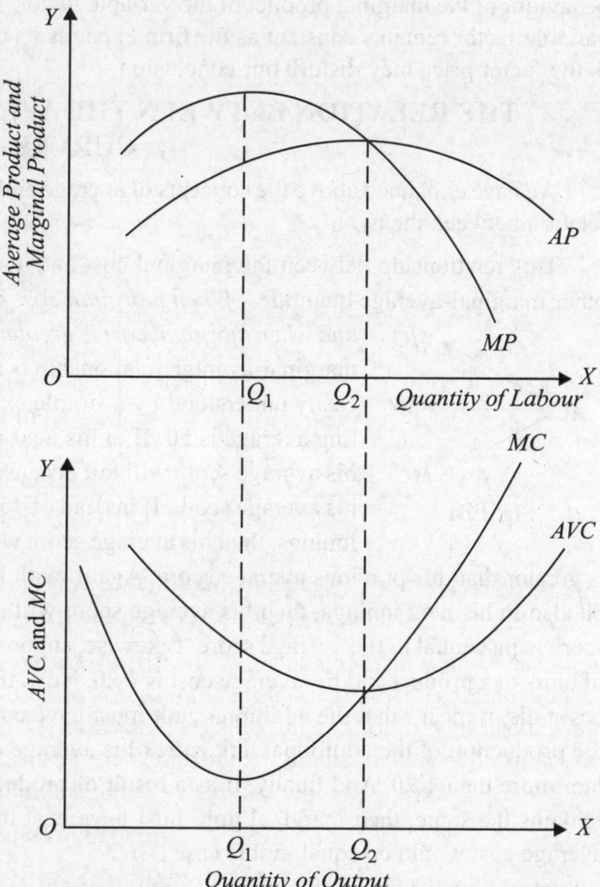

Fig. 19.3. *The Relationship between Product Curves and Cost Curve*

falls after a certain level of output, which means that now constant *w* in the above equation (*iii*) is being divided by increasingly smaller *MP*. This causes the marginal cost (*MC*) to rise after a certain level of output. Thus, the fact that marginal product first rises,reaches a maximum and then declines ensures that the marginal cost curve of a firm declines first, reaches a minimum and then rises. In other words, marginal cost curve of a firm has a U-shape. The relation between marginal product of labour and marginal cost curve is shown in Fig. 19.3, where marginal product curve (*MP*) of the variable factor labour is shown at the top panel, the marginal cost curve is shown in the panel at the bottom of the figure labelled as *MC*.

It is clear from above that the law of variable proportions, or in other words, the behaviour of marginal product (*MP*) curve determines the shape of marginal cost (*MC*) curve. Indeed, marginal cost (*MC*) curve is an inverse of the marginal product (*MP*) curve, with maximum of marginal product curve corresponding to the minimum of marginal cost curve. Marginal cost is simply the transformation of marginal product from physical terms into money terms. The relation between marginal product and marginal cost is quite similar to the relationship between average product and average cost. Three points are worth noting in regard to our above analysis of marginal cost. First, marginal cost is due to the changes in variable cost and is therefore independent of the fixed cost. Secondly, the shape of the marginal cost curve is determined by the law of variable proportions, that is, by the behaviour of the marginal product of the variable factor. Thirdly, the assumption that the price of the variable factor remains constant as the firm expands its output is greatly significant, since a change in the factor price may disturb our conclusion.

THE RELATION BETWEEN THE AVERAGE AND MARGINAL COST CURVES

We have explained above the concepts of average and marginal cost curves. There is an important relation between the two.

This relationship between the marginal cost and average cost is the same as that between any other marginal-average quantities. *When marginal cost is less than average cost, average cost falls and when marginal cost is greater than average cost, average cost rises*. This marginal-average relationship is a matter of mathematical truism and can be easily understood by a simple example. Suppose that a cricket player's batting average is 50. If in his next innings he scores less than 50, say 45, then his average score will fall because his marginal (additional) score is less than his average score. If instead of 45, he scores more than 50, say 55, in his next innings, then his average score will increase because now the marginal score is greater than his previous average score. Again, with his present average runs of 50, if he scores 50 also in his next innings, then his average score will remain the same because now the marginal score is just equal to the average score. Likewise, suppose a producer is producing a certain number of units of a product and his average cost is ₹ 20. Now, if he produces one unit more and his average cost falls, it means that the additional unit must have cost him less than ₹ 20. On the other hand, if the production of the additional unit *raises* his average cost, then the marginal unit must have cost him more than ₹ 20. And finally, if as a result of production of an additional unit the average cost remains the same, then marginal unit must have cost him exactly ₹ 20, that is, marginal cost and average cost would be equal in this case.

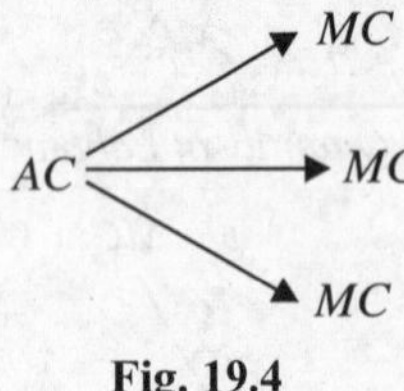

Fig. 19.4

The relationship between average and marginal cost can be easily remembered with the help of Fig. 19.4. It is illustrated in this figure that when marginal cost (*MC*) is above average cost (*AC*), the average cost rises, that is, the marginal cost (*MC*) pulls the average cost (*AC*) upwards. On the other hand, if the marginal cost (*MC*) is below the average cost (*AC*), average cost falls, that is, the marginal cost pulls the average cost downwards. When marginal cost (*MC*) stands equal to the average cost (*AC*), the average cost remains the same, that is, the marginal cost pulls the average cost horizontally.

Now, take Fig. 19.5 where short-run average cost curve *AC* and marginal cost curve *MC* are drawn. As long as short-run marginal cost curve *MC lies below* short-run average cost curve, the average cost curve *AC* is falling. When marginal cost curve *MC lies above* the average cost curve *AC*, the latter is *rising*. At the point of intersection *L* where *MC* is equal to *AC*, *AC* is neither falling nor rising, that is, at point *L*, *AC* has just ceased to fall but has not yet begun to rise. It follows that point *L*, at which the *MC* curve crosses the *AC* curve to lie above the *AC* curve, is the minimum point of the *AC* curve. Thus, *marginal cost curve cuts the average cost curve at the latter's minimum point*.

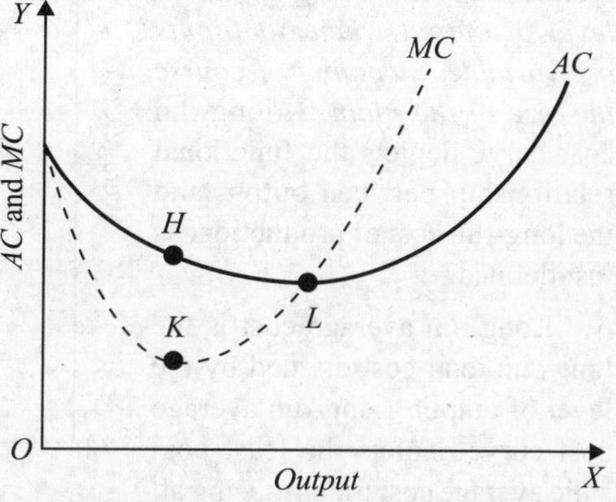

Fig. 19.5. *The Relation between AC and MC Curves*

It is important to note that *we cannot generalise about the direction in which marginal cost is moving from the way average cost is changing*, that is, when average cost is falling we cannot say that marginal cost will be falling too. When average cost is falling, what we can say definitely is only that the marginal cost will be below it but the marginal cost itself may be either rising or falling. Likewise, when average cost is rising, we cannot deduce that marginal cost will be rising too. When average cost is rising, the marginal cost must be above it but the marginal cost itself may be either rising or falling. Consider Fig. 19.5 where up to the point *K*, marginal cost is falling as well as below the average cost. As a result, the average cost is falling. But beyond point *K* and up to point *L* marginal cost curve lies below the average cost curve with the result that the average cost curve is falling. But it will be seen that between *K* and *L* where *the marginal cost is rising, the average cost is falling*. This is because though *MC* is rising between *K* and *L*, it is below *AC*. It is therefore clear that when the average cost is falling, marginal cost may be falling or rising. This can also be easily illustrated by the example of batting average. Suppose a cricket player's present batting average is 50. If in his next innings he scores less than 50, say 45, his batting average will fall. But his marginal score of 45, though less than the average score may itself have risen. For instance, he might have scored 40 in his previous innings so that his present marginal score of 45 is greater than his previous marginal score. Thus one cannot deduce about marginal cost as to whether it will be falling or rising when average cost is falling or rising.

DERIVATION OF LONG-RUN AVERAGE COST CURVE

We now turn to explain the cost curves in the long run. The long run, as noted above, is a period of time during which the firm can vary all its inputs. In the short run, some inputs are fixed and others are varied to increase the level of output. In the long run, none of the factors is fixed and all can be varied to expand output. The long-run production function has therefore no fixed factors and the firm has no fixed costs in the long run. It is conventional to regard the size or scale of plant as a typical fixed input. The term 'plant' is here to be understood as consisting of capital equipment, machinery, land etc. In the short run, the size of the plant is fixed and it cannot be increased or reduced. That is to say, one cannot change the amount of capital equipment in the short run, if one has to increase or decrease output. On the other hand, long run is a period of time sufficiently long to permit the changes in plant, that is, in capital equipment, machinery, land etc. in order to expand or contract output. Thus whereas in the short run the firm is tied with a given plant, in the long run the firm moves from one plant to another; the firm can make a larger plant if it has to increase its output and a smaller plant if it has to reduce its output. *The long-run average cost of production is the least possible average cost of production of producing any given level of output when all inputs are variable, including of course the size of the plant.* A long-run cost curve depicts the functional relationship between output and the long-run cost of production, as just defined.

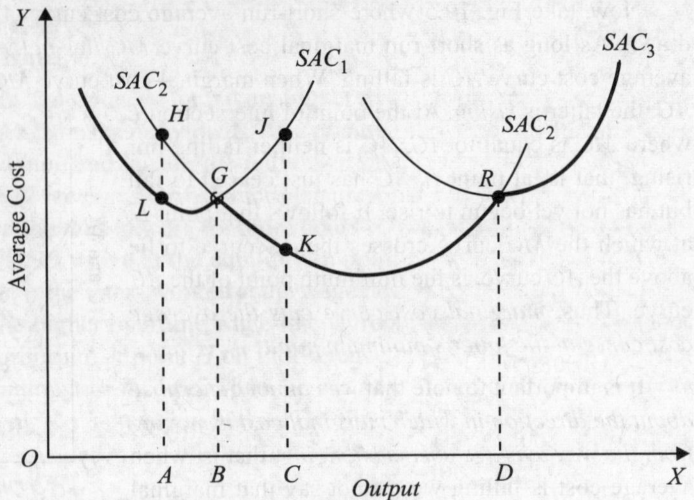

Fig. 19.6. *The Relation between AC and MC Curves*

Long-run average cost is the long-run total cost divided by the level of output. Long-run average cost curve depicts the least possible average cost for producing all possible levels of output. In order to understand how the long-run average cost curve is derived, consider the three short-run average cost curves as shown in Fig. 19.6. These short-run average cost curves are also called *plant curves*, since in the short run plant is fixed and each of the short-run average cost curves corresponds to a particular plant. In the short run, the firm can be operating on any short-run average cost curve, given the size of the plant. Suppose that there are only these three technically possible sizes of plant, and that no other size of the plant can be built. Given a size of the plant or a short-run average cost curve, the firm will increase or decrease its output by varying the amount of the variable inputs. But, in the long run, the firm can choose among the three possible sizes of plant as depicted by short-run average cost curves SAC_1, SAC_2 and SAC_3. In the long run the firm will decide about with which size of plant or on which short-run average cost curve it should operate to produce a given level of output at the minimum possible cost.

It will be seen from Fig. 19.6. that up to *OB* amount of output, the firm will operate on the short-run average cost curve SAC_1, though it could also produce with short-run average cost curve SAC_2, because up to *OB* amount of ouput, production on SAC_1 curve entails lower cost than on SAC_2. For instance, if the level of output *OA* is produced with SAC_1, it will cost *AL* per unit and if it is produced with SAC_2 it will cost *AH* per unit. It will be seen from Fig. 19.6 that *AL* is smaller than *AH*. Similarly, all other output levels up to *OB* can be produced more economically with the smaller plant SAC_1 than with the larger plant SAC_2. It is thus clear that in the long run the firm will produce any output up to *OB* on SAC_1. If the firm plans to produce an output which is larger than *OB* (but less than *OD*), then it will not be economical to produce on SAC_1. It will be seen from Fig. 19.6 that the output larger than *OB* but less then *OD* can be produced at a lower cost per unit on SAC_2 than on SAC_1. Thus, the output *OC* if produced on SAC_2 costs *CK* per unit which is lower than *CJ* which is the cost incurred when produced on SAC_1. Therefore, if the firm plans to produce between output *OB* and *OD*, it will employ the plant corresponding to short-run average cost curve SAC_2. If the firm has to produce an output which exceeds *OD*, then the cost per unit will be lower on SAC_3 than on SAC_2. Therefore, for output larger than *OD*, the firm will employ plant corresponding to the short-run average cost curve SAC_3.

It is thus clear that in the long run the firm has a choice in the use of a plant, and it will employ that plant which yields possible minimum unit cost for producing a given output. *The long-run average cost curve depicts the least possible average cost for producing various levels of output when all factors including the size of the plant have been adjusted.* Given that only three sizes of plants, as shown in Fig. 19.6, are technically possible, then the long-run average cost curve is the curve which has scallops in it. This heavily scalloped long-run average cost curve consists of some segments of all the short-run average cost curves as explained above.

Suppose now that the *size of the plant can be varied by infinitely small gradations* so that there are infinite number of plants corresponding to which there will be numerous short-run average cost curves. In that case, the long-run average cost curve will be a smooth and continuous curve without any scallops. Such a smooth long-run average cost curve has been shown in Fig. 19.7 and has been labelled as *LAC*. There will be infinite short-run average cost curves in such a case, though only seven have been shown in Fig. 19.7. This long-run average cost curve *LAC* is drawn so as to be tangent to each of the short-run average cost curves. Since an infinite number of short-run average cost curves is assumed, every point on the long-run average cost curve is a tangency point with some short-run average cost curve. In fact, *the long-run, average cost curve is nothing else but the locus of all these tangency points with short-run average cost curves*. It is again worth noting that the long-run average cost curve shows the least possible average cost of producing any output when all productive factors are variable. If a firm desires to produce particular output in the long run, it will pick a point on the long-run average cost curve corresponding to that output and it will then build a relevant plant and operate on the corresponding short-run average cost curve.

In the situation as depicted in Fig. 19.7, for producing output *OM*, the corresponding point on the long-run average cost curve *LAC* is *G* at which the short-run average cost curve SAC_2 is tangent to the long-run average cost curve *LAC*. Thus, if a firm desires to produce output *OM*, the firm will construct a plant corresponding to SAC_2 and will operate on this curve at point *G*. Similar would be the case for all other outputs in the long run. Further, consider that the firm plans to produce output *ON*, which corresponds to point *K* on the long-run average cost curve *LAC*. As already noted, every

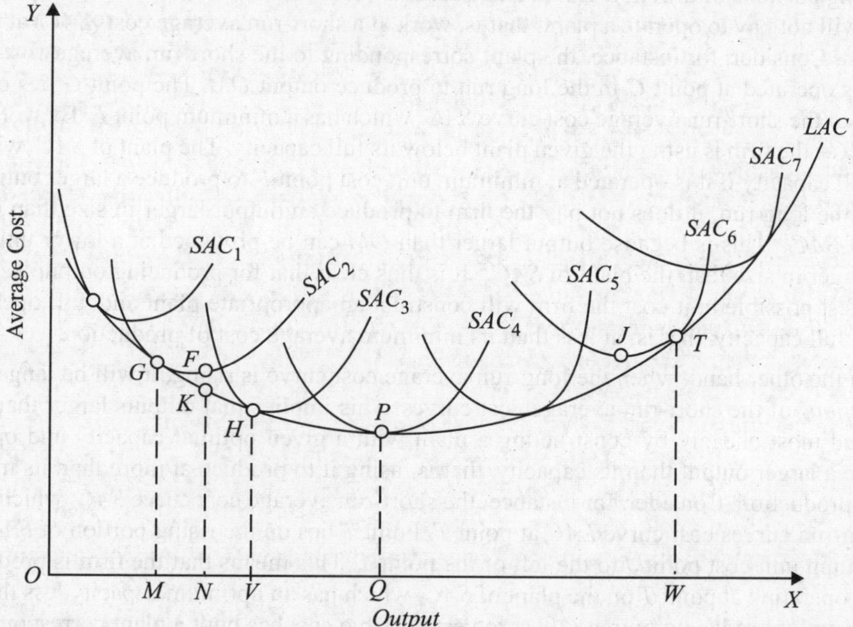

Fig. 19.7. *Deriving Long-Run Average Cost Curve from Short-Run Average Cost Curves*

point on the long-run average cost curve is a tangency point with some short-run average cost curve and that there are infinite number of short-run average cost curves, so there will be some short-run average cost curve (not shown in Fig. 19.7) which will be tangent to the long-run average cost curve

LAC at point *K* corresponding to *ON* output. Thus, for producing output *ON*, the firm will build a plant which will correspond to that short-run average cost curve which is tangent to the long-run average cost curve *LAC* at point *K* corresponding to *ON* output. The long-run average cost curve *LAC* is also called '*envelope*' since it envelops or supports a family of short-run average cost curves from below.

It is evident from Fig. 19.7 that larger outputs can be produced at the lowest cost with the larger plants, whereas smaller outputs can be produced at the lowest cost with smaller plants. Thus, output *OV* can be produced with the lowest possible cost with the plant represented by the SAC_3. To produce *OM* output with a larger plant corresponding to SAC_3, will entail higher unit cost than that on SAC_2. But a larger output *OV* can be produced most economically with a larger plant represented by SAC_3 while to produce *OV* with the smaller plant of SAC_2 will mean higher unit cost. This is as it should be expected. A larger plant which is more expensive when employed to produce a small output will not be fully utilised and its underutilisation will cause higher unit cost. In other words, using a larger plant and to operate it much below its capacity in order to produce a small output will naturally mean higher average cost. On the other hand, a large output with a small plant will also involve higher cost per unit because of its limited capacity.

It will be seen in Fig. 19.7 that the long-run average cost curve first falls and then beyond a certain point it rises, that is, the long-run average cost curve is U-shaped, though the U-shape of the long-run average curve is less pronounced than that of the short-run average cost curve. In Fig. 19.7 long-run average cost is minimum at output *OQ*. The long-run average cost falls up to the output *OQ* and it rises beyond output *OQ*. Why does the long-run average cost first decline and then after some point rises will be explained a little later.

An important fact about the long-run average cost curve is worth mentioning. It is that the *long-run average cost curve LAC is not tangent to the minimum points of the short-run average cost curves*. When the long-run average cost curve is declining, that is, for output less than *OQ*, it is tangent to the *falling portions* of the short-run average cost curves. This means that for any output smaller than *OQ*, it will not pay to operate a plant, that is, work at a short-run average cost (*SAC*) at its minimum unit cost. Consider, for instance, the plant corresponding to the short-run average cost curve SAC_2, which is operated at point *G* in the long run to produce output *OM*. The point *G* lies on the falling portion of the short-run average cost curve SAC_2 which has a minimum point *F*. By working at point *G* of SAC_2, the firm is using the given plant below its full capacity. The plant of SAC_2 will be utilised to its full capacity if it is operated at minimum unit cost point *F* to produce a larger output than *OM*. But, in the long run, it does not pay the firm to produce an output larger in size than *OM* with the plant of SAC_2. This is because output larger than *OM* can be produced at a lower unit cost with a plant larger in size than the plant of SAC_2. It is thus clear that for producing output less than *OQ* at the lowest possible unit cost the firm will construct an appropriate plant and will operate it at less than its full capacity, that is, at less than its minimum average cost of production.

On the other hand, when the long-run average cost curve is rising, it will be tangent to the *rising portions* of the short-run average cost curves. This implies that outputs larger than *OQ* will be produced most cheaply by constructing a plant with a given optimal capacity and operating it to produce a larger output than its capacity, that is, using it to produce at more than its minimum unit cost of production. Consider, for instance, the short-run average cost curve SAC_6 which is tangent to the long-run curves cost curve *LAC* at point *T*. Point *T* lies on the rising portion of SAC_6 which has a minimum unit cost point *J* to the left of the point *T*. This means that the firm is producing output *OW* by operating at point *T* on the plant of SAC_6 which has an optimum capacity less than *OW*. That is, the firm for producing output *OW* at lowest possible cost has built a plant corresponding to SAC_6 and works it at more than its capacity.

Long-run average cost curve is often called the '**planning curve**' of the firm by some economists, because a firm plans to produce any output in the long run by choosing a plant on the long-run average cost curve corresponding to the given output. The long-run average cost curve reveals to the firm that how large should be the plant for producing a certain output at the least possible cost. Thus while making decisions regarding the choice of a plant, the firm has to look at its long-run average cost curve enveloping a family of plant or short-run overage cost curves. What different sizes of plants are available at a time and what short-run average cost curves they will have for being used for production are known to the firm either from experience or from engineering studies.

WHY LONG-RUN AVERAGE COST CURVE IS OF U-SHAPE?

In Fig. 19.7, we have drawn the long-run average cost curve as having an approximately U-shape. It is generally believed by economists that the long-run average cost curve is normally U-shaped, that is, the long-run average cost curve first declines as output is increased and then beyond a certain point it rises. Now, what is the proper explanation of such a behaviour of the long-run average cost curve?

We saw above that the U-shape of the short-run average cost curve is explained with the law of variable proportions. *But the shape of the long-run average cost curve depends upon the returns to scale*. Since in the long run all inputs including the capital equipment can be altered, the relevant concept governing the shape of this long-run average cost curve is that of returns to scale. In a previous chapter we have explained that returns to scale increase with the initial increases in output and after remaining constant for a while, the returns to scale decrease. *It is because of the increasing returns to scale in the beginning that the long-run average cost of production falls as output is increased and, likewise, it is because of the decreasing returns to scale that the long-run average cost of production rises beyond a certain point.*

Why does LAC Fall in the Beginning? Economies of Scale

But the question is why we first get increasing returns to scale due to which long-run average cost falls and why after a certain point we get decreasing returns to scale due to which long-run average cost rises. In other words, what are the reasons that the firm first enjoys *internal economies of scale* and then beyond a certain point it has to suffer *internal diseconomies of scale*? Three main reasons have been given for the economies of scale which accrue to the firm and due to which cost per unit falls in the beginning.

First, as the firm increases its scale of operations, *it becomes possible to use more specialized and efficient form of all factors, especially capital equipment and machinery*. For producing higher levels of output, there is generally available a more efficient machinery which when employed to produce a large output yields a lower cost per unit of output.

Secondly, when the scale of operations is increased and the amount of labour and other factors becomes larger, *introduction of a great degree of division of labour or specialisation* becomes possible and as a result the long-run cost per unit declines. Thus, whereas the short-run decreases in cost (the downward sloping segment of the short-run average cost curve) occur due to the fact that the ratio of the variable input comes nearer to the optimum proportion, decreases in the long-run average cost (downward segment of the long-run average cost curve) take place due to the use of more efficient forms of machinery and other factors and to the introduction of a greater degree of division of labour in the productive process.

Indivisibility of Factors. Some economists explain economies of scale as arising from the *imperfect divisibility of factors*. In other words, they think that the economies of scale occur and therefore the long-run average cost falls because of the 'indivisibility' of factors'. They argue that most of the factors are 'lumpy', that is, they are available in *large indivisible units* and can therefore yield lower cost of production when they are used to produce a larger output. If a small output is produced with these costly indivisible units of the factors, the average cost of production will naturally be high. If the factors of production were perfectly divisible, then, according to them, suitable adjustment in the factors could be made so that the optimum proportions between the factors were maintained even for producing small amounts of output and hence the average cost of production would not have been higher. Thus, according to them, if the factors were perfectly divisible, the small-scale production would be as good and efficient as the large-scale production and the economies of scale would be non-existent. Thus, Joan Robinson remarks, "If all the factors were finely divisible, like sand, it would be possible to produce the smallest output of any commodity with all the advantages of large-scale industry."[2]

Why does LAC Rise Eventually? Diseconomies of Scale

So much for the downward sloping segment of the long-run average cost curve. As noted above, beyond a certain point the long-run average cost curve rises which means that the long-run average cost increases as output exceeds beyond a certain point. In other words, beyond a certain point a firm experiences net *diseconomies of scale*. There is also divergence of views about the proper explanation for this upward sloping of the long-run average cost curve. The first view as held by Chamberlin and his followers is that when the firm has reached a size large enough to allow the utilisation of almost all the possibilities of division of labour and the employment of more efficient machinery, further increases in the size of the plant will entail higher long-run unit cost because of the *difficulties of management*. When the scale of operations exceeds a certain limit, the management may not be as efficient as when the scale of operations is relatively small.

After a certain sufficiently large size these inefficiencies of management more than offset the economies of scale and thereby bring about an increase in the long-run average cost and make the *LAC* curve upward-sloping after a point. It should be noted that this view regards the entrepreneurial or managerial functions to be divisible and variable and explains the diseconomies of scale or the rising part of the long-run average cost curve as arising from the mounting difficulties of management (*i.e.* of supervision and coordination) beyond a certain sufficiently large-scale of operations.

The second view considers *the entrepreneur to be a fixed indivisible factor*. In this view, though all other factors can be increasesd, the entrepreneur cannot be. The entrepreneur and his functions of decision-making and ultimate control are indivisible and cannot be increased. Therefore, when a point is reached where the abilities of the fixed and indivisible entrepreneur are best utilised, further increases in the scale of operations by increasing other inputs cause the cost per unit of output to rise. In other words, *there is a certain optimum proportion between an entrepreneur and other inputs* and when that optimum proportion is reached, further increases in the other inputs to the fixed entrepreneur means the proportion between the inputs is moved away from the optimum and, therefore, this results in the rise in the long-run average cost. Thus, in this view, increases in the long-run average cost is explained by the law of variable proportions. Economists who hold this view think that the decreasing returns to scale or rising long-run average cost is actually a special case of variable proportions with entrepreneur as the fixed factor.

Long-Run Average Cost Curve in Case of Constant Returns to Scale

If the production function is linear and homogeneous (that is, homogeneous of the first degree) and also the prices of inputs remain constant, then the long-run average cost will remain constant at all levels of output. As explained in the previous chapter, linear homogeneous production function implies constant returns to scale which means that when all inputs are increased in a certain proportion, output increases in the same proportion. Therefore, with the given prices of inputs, when returns to scale are constant, the cost per unit of output remains the same. In this case, the long-run average cost curve will be a horizontal straight line as depicted in Fig. 19.8. Though there will be infinite

2. Joan Robinson, *The Economics of Imperfect Competition*, p. 334.

number of short-run average cost curves as we continue to assume that the size of the plant can be varied by infinitely small gradations, only *SAC* curves of three plants have been shown in Fig. 19.8.

It will be noticed from Fig. 19.8 that all short-run average cost curves such as SAC_1, SAC_2, SAC_3 have the same minimum average cost of production. This means whatever the size of the plant, the minimum average cost of production is the same. This implies that all factors can be adjusted in the long run in such a way that the proportions between them always remain optimum. In such a case, the optimum size of the firm is indeterminate, since all levels of output can be produced at the same long-run average cost which represents the same minimum short-run average cost throughout. It is useful to note that though all levels of output will be produced at the same minimum cost of production the different sizes of plants will be used for producing different levels of output. Thus, for producing output *OA*, the plant of SAC_1 will be employed; for output *OB*, the plant of SAC_2 will be employed; and for output *OC* the plant of SAC_3 will be employed and so on. This is because the production at the lowest possible cost for output *OA* is possible with plant SAC_1, for output *OB* with plant SAC_2 and for output *OC* with plant SAC_3.

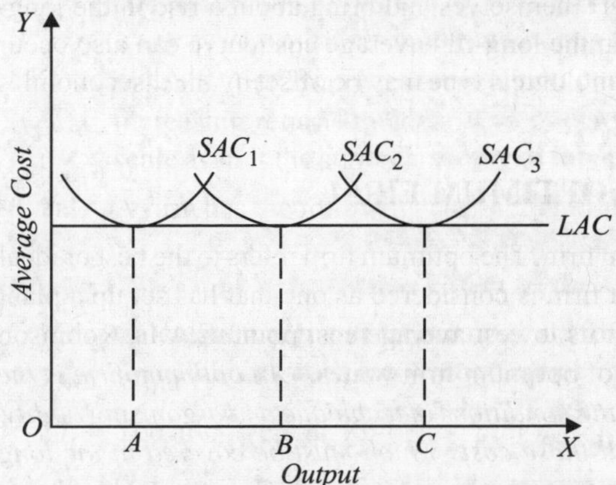

Fig. 19.8. *When returns to scale are constant, long-run average cost curve is a horizontal straight line.*

Some economists like Kaldor, Joan Robinson, Stigler are of the view that when all factors of production are "*perfectly divisible*" then there would be no internal economies of scale (and no internal diseconomies). Therefore, according to them, in case of 'perfect divisibility' of all factors, the long-run average cost curve will be a horizontal straight line showing that the long-run average cost is constant whatever the level of output. In their view, all internal economies of scale are due to the indivisibility of some factors. Therefore, they argue that if perfect divisibility of factors is assumed, then it implies the absence of internal economies of scale and therefore in such a case the long-run average cost curve will still be a horizontal straight line. But Prof. Chamberlin has challenged this viewpoint. According to him, perfect divisibility has nothing to do with efficiency, that is to say, perfect divisibility does not mean the absence of internal economies of scale. Thus, according to him, even if all factors were perfectly divisible, the economies of scale will still be reaped due to the use of more specialised machinery and a greater degree of division of labour at higher levels of output. Therefore, according to Prof. Chamberlin, constant returns to scale cannot exist and long-run average cost cannot remain constant.

Saucer-shaped Long-Run Average Cost Curve

However, many empirical studies have shown that U-shape of the long-run average cost curve is not smooth and regular as is shown in Fig. 19.7, but a wavy and irregular one. Further, a very important feature of long-run average cost curve revealed by empirical studies is that there is *a relatively very large flat portion* or, in other words, a large horizontal region in the centre of long-run average cost curve, as is depicted in Fig. 19.9. In such a real case, long-run average cost curve has a saucer-

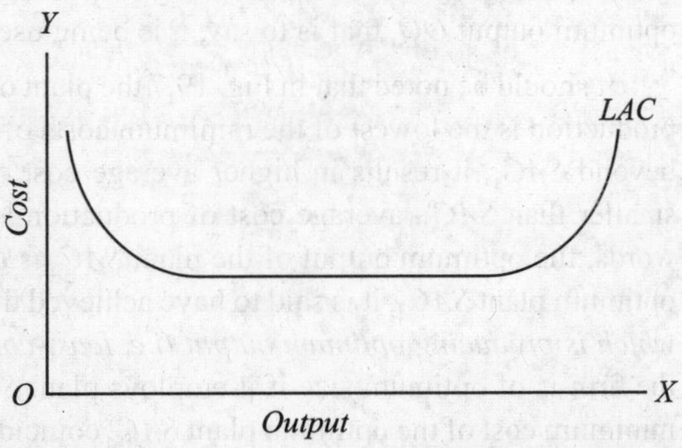

Fig. 19.9. *Saucer-shaped Long-Run Average Cost Curve*

shaped appearance. Such a long-run average cost curve with a very large flat portion in the centre can arise if the economies of scale are exhausted at a very modest scale of operation and then for a relatively large further expansion in output, diseconomies of scale do not occur. Only after a very large increase in output, diseconomies of scale exert themselves and bring about a rise in the long-run average cost. A long horizontal or flat section in the long-run average cost curve can also occur because economies of scale which are mostly of technological type may be offset by the diseconomies over a wide range of output.

THE CONCEPT OF OPTIMUM FIRM

It is important to explain the concept of optimum firm. The optimum firm refers to the best or ideal size of the firm. More specifically optimum or best firm is considered as one that has set up a plant with lowest possible cost and is also operating it at its lowest average cost point. E.A.G. Robinson who has done a good deal of research on the issue of optimum firm writes, "*An optimum firm is the one which operates at the scale at which, in the existing conditions of technique and organising ability, has the lowest average cost of production when all those costs which must be covered in the long run are included*". This means that optimum firm is one which operates at the lowest point of long-run average cost curve of production. Production at the minimum point of the long-run average cost curve is called optimum because at it resources of the society are most efficiently used. When a firm expands its size to the lowest point of the long-run average cost (*LAC*), it sets up a plant which, given the state of technology, has the lowest unit cost of production when operated at its full capacity. As has been explained above that long-run average cost of a firm is influenced by the various economies and diseconomies of scale. These economies and diseconomies are determined by various technical, managerial, financial and marketing factors. The optimum size is achieved when all the internal economies of scale such as division of labour, use of specialised machinery, managerial advantage of large scale production etc. are being fully enjoyed by the firm and the internal diseconomies of scale have not yet started accruing to it.

It is worth noting that in the determination of optimum size, the state of technology and methods of organising business remain unchanged. In Fig. 19.7 optimum firm is one that has set up a plant represented by short-run average cost curve SAC_4 and operating at the minimum point P on it and producing output OQ.

It is clear from Fig. 19.7 that in the continuous long-run average cost curve both for outputs less than OQ and more than OQ no plant is used at its point of minimum average cost. It is only the plant, the minimum point of whose short-run average cost curve coincides with the minimum point of the long-run average cost curve, which is operated at the point of minimum point of its short-run average cost curve. In Fig. 19.7 for producing output OQ, the plant of SAC_4 is being utilized to produce its optimum output OQ, that is to say, it is being used at its full capacity.

It should be noted that in Fig. 19.7 the plant of SAC_4 is optimum plant, since its minimum cost of production is the lowest of the minimum costs of all other plants. If the size of the plant is increased beyond SAC_4, it results in higher average cost of production. Similarly, if the size of the plant is smaller than SAC_4, average cost of production is higher. Further, the least-cost output, or in other words, the optimum output of the plant SAC_4 is OQ. Now, if the firm produces output OQ with the optimum plant SAC_4, it is said to have achieved the optimum size. Thus, *an optimum firm is that firm which is producing optimum output (i.e. least-cost output) with the optimum plant*. In our Fig. 19.7 the firm is of optimum size if it employs plant SAC_4 and uses it to produce OQ. Since the point of minimum cost of the optimum plant SAC_4 coincides with the minimum point of the long-run average cost curve, *the optimum firm can also be defined as one which produces at the minimum point of the long-run average cost curve* (*LAC*).

The optimum size of the firm varies a great deal in different industries. In agriculture, extractive industries, wholesale and retail trade, optimum size is relatively small, that is, the minimum point of the long-run average cost curve is reached at a comparatively small output. Fig. 19.10 shows a firm whose optimum size is relatively small. On the other hand, the optimum size of the firm in steel industry, automobile industry, other heavy industries and public utilities is relatively very large, that is, the minimum point of their long-run average cost curve is reached at a relatively

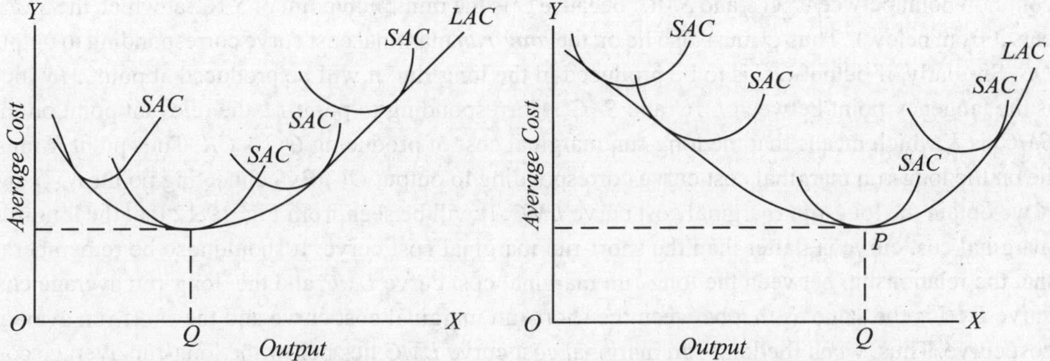

Fig. 19.10. *Small Optimum Size* **Fig. 19.11.** *Large Optimum Size*

very large output. Fig. 19.11 depicts a firm whose optimum size is very large. In the industry in which the optimum size of the firm is very large, there are generally a few number of firms, each with a large size.

DERIVATION OF LONG-RUN MARGINAL COST CURVE

In a previous section we explained what is the marginal cost and how the short-run marginal cost curve is obtained and what relation it bears to the short-run average cost curve. Since marginal cost curve is important both from the viewpoint of the short run and the long run, it will be useful to know how the long-run marginal cost curve is derived. The long-run marginal cost curve can be directly derived from the long-run total cost curve, since the long-run marginal cost at a level of output is given by the slope of the total cost curve at the point corresponding to that level of output. Besides, the long-run marginal cost curve can be derived from the long-run average cost curve, because the long-run marginal cost curve is related to the long-run average cost curve in the same way as the short-run marginal cost curve is related to short-run average cost curve. In Fig. 19.12, it is depicted that how the long-run marginal cost curve *LMC* is derived from a long-run average cost curve *LAC* enveloping a family of short-run average and marginal cost curves.

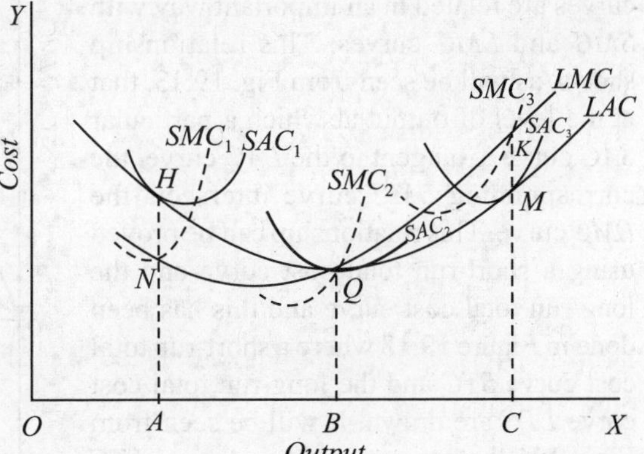

Fig.19.12 . *Derivation of Long-Run Marginal Cost Curve*

If the output *OA* is to be produced in the long run, then it must be produced on the long-run average cost curve *LAC* at point *H* which is a tangency point with the short-run average cost curve SAC_1. Thus, when output *OA* is to be produced in the long run, it will be produced with the plant corresponding to the short-run average cost curve SAC_1 and the short-run marginal cost curve SMC_1.

Corresponding to the tangency point *H* between the short-run average cost curve SAC_1 and the long-run average cost curve *LAC*, there is a point *N* on the short-run marginal cost curve *SMC*. This means that the production of output *OA* in the long run involves the marginal cost *AN*. Therefore, point *N* must lie on the long-run marginal cost curve corresponding to output *OA*. If output *OB* is to be produced in the long run, it will be produced at point *Q* which is the tangency point between *LAC* and SAC_2. *Q* is also the point on the short-run marginal cost curve SMC_2, corresponding to output *OB*. (*Q* is the common point between SAC_2 and SMC_2 because *Q* is the minimum point of SAC_2 at which the SMC_2 cuts it from below). Thus *Q* must also lie on the *long-run* marginal cost curve corresponding to output *OB*. Similarly, if output *OC* is to be produced in the long run, it will be produced at point *M* which is the tangency point between *LAC* and SAC_3. Corresponding to point *M*, the relevant point on the SMC_3 is *K* which means that the long-run marginal cost of producing *OC* is *CK*. Thus point *K* must lie on the long-run marginal cost curve corresponding to output *OC*. By connecting points *N*, *Q* and *K* we obtain the long-run marginal cost curve *LMC*. It will be seen from Fig. 19.12 that the long-run marginal cost curve is flatter than the short-run marginal cost curve. It should also be remembered that the relationship between the long-run marginal cost curve *LMC* and the long-run average cost curve *LAC* is the same as that between the short-run marginal cost curve and the short-run average cost curve. Thus, when the long-run marginal cost curve *LMC* lies below the long-run average cost curve, the latter will be falling, and when the long-run marginal cost curve lies above the long-run average cost, the latter will be rising. When the the long-run marginal cost is equal to the long-run average cost, the latter will be neither rising nor falling.

Relationship of LAC and LMC with SAC and SMC

It is important to note that *LAC* and *SAC* curves are related in an important way with *SMC* and *LMC* curves. This relationship shows, as will be seen from Fig. 19.13, that at the level of output at which a particular *SAC* curve is tangent to the *LAC* curve, the corresponding *SMC* curve intersects the *LMC* curve. This relationship can be proved using a short-run total cost curve and the long run total cost curve and this has been done in Figure 19.13 where a short-run total cost curve *STC* and the long-run total cost curve *LTC* are drawn. It will be seen from Fig. 19.13 that long-run total cost curve *LTC* lies below the short-run total cost curve *STC* at all levels of output except at output *OA* at which the two curves are tangent. This means that *LAC* is less than *SAC* at all levels of output other than *OA*. As will be seen from the bottom panel of Fig. 19.13 that long-run average cost *LAC* will be equal to the short-run average cost *SAC* at output *OA* at which *LTC* curve is tangent to the *STC* curve. But the long-run marginal cost *LMC* must also

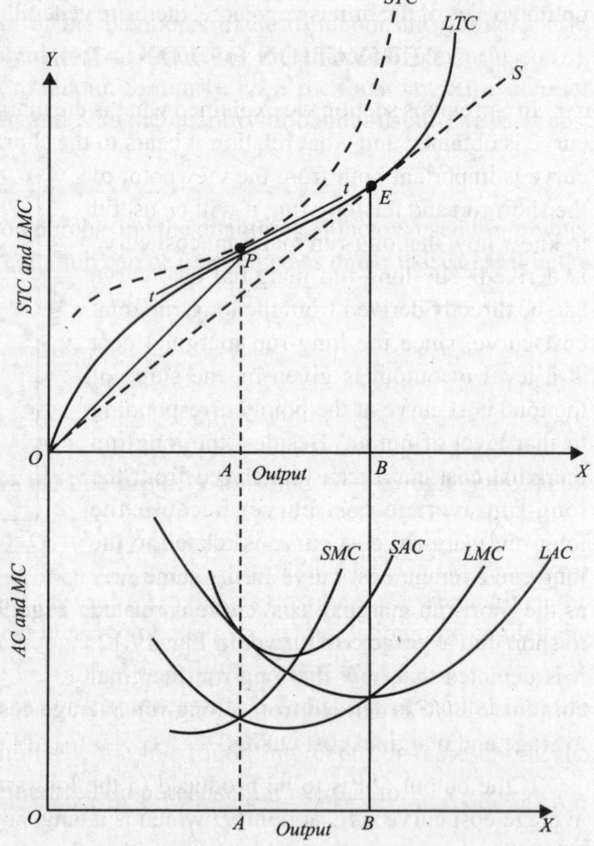

Fig. 19.13. *Relationship between SAC and LAC with SMC and LMC*

be equal to short-run marginal cost *SMC* at output *OA, i.e.,* the tangency point *P*. This is because marginal cost is given by the slope of the total cost curve at any point, and the *LTC* curve and *STC* curve have the same slope at the tangency point *P*. It will be seen from the bottom panel of Fig. 19.13 that at output level *OA*, where the *SAC* is equal to the *LAC, SMC* is also equal to the *LMC*.

EXTERNAL ECONOMIES AND DISECONOMIES AND COST CURVES

We have explained above that the long-run average cost curve falls downward in the beginning because of economies of scale, namely, the use of greater degree of division of labour and the specialised machinery at higher levels of output. The use of greater degree of division of labour and the

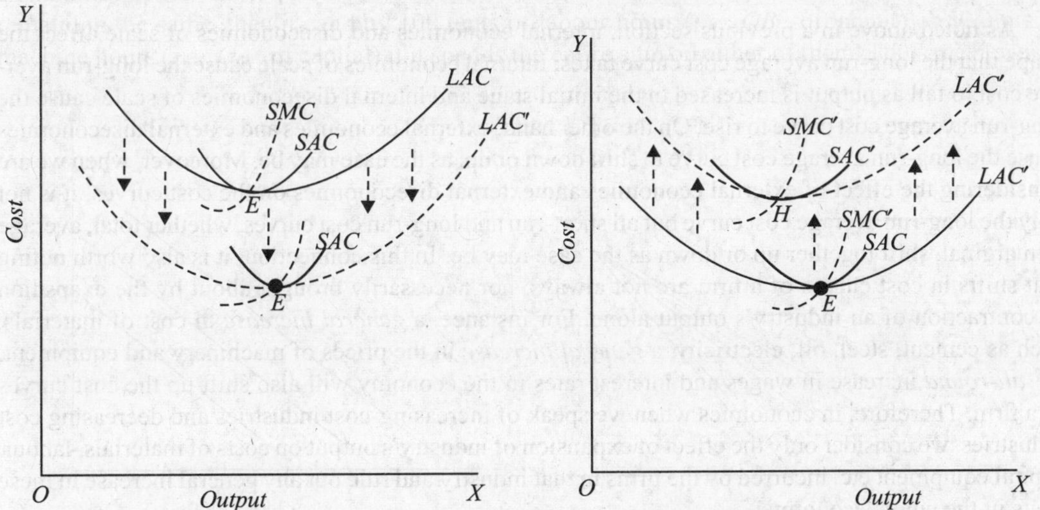

Fig. 19.14. *Downward Shift in the Cost Curves due to External Economies*

Fig. 19.15. *Downward Shift in the Cost Curves due to External Diseconomies*

specialised machinery at higher levels of output are called the *internal economies*. They are internal in the sense that they accrue to the firm when *its own output or scale* increases. Besides internal economies, Marshall introduced the concept of *external economies* which plays an important role in Marshall's partial equilibirum theory of value especially in his analysis of equilibrium problem under conditions of increasing returns or decreasing costs. Costs of a firm depend not only on its output level but also on the output level of the industry as a whole. *External economies and diseconomies are those which accrue to the firms as a result of the expansion in the output of the whole industry* and they are not dependent on the output level of individual firms. They are external in the sense that they accrue to the firms not out of its internal situation but from outside it, *i.e.*, from expansion of the industry. Marshall defined external economies as "those dependent on the general development of the industry."[3] In a more precise manner Jacob Viner has defined external economies as "those which accrue to particular concerns as the result of expansion of output by the industry as a whole and which are independent of their own individual output."[4]

External economies accrue to the individual firms, if the increase in the output of industry *lowers* the cost curves of each firm in the industry. On the other hand, external diseconomies accrue to the firms, when the expansion of the output of the industry *raises* the cost curves of each firm. Thus, when the industry expands and as a result of it certain external economies accrue to the firms, the cost

3. A. Marshall, *Principle of Economics*, 8th edition, p. 266.
4. Jaccob Viner, *Cost Curves and Supply Curves*: *Readings in Price Theory*, A.E.A., p. 217.

curves of a firm will shift down as is shown in Fig. 19.14. It should be noted that external economies will cause all types of firm's cost curves—long-run average and marginal cost curves, short-run average and marginal cost curves—to shift down. In Fig. 19.14, initially the long-run average cost curve is *LAC* (thick curve) and as a result of the expansion of the whole industry and the creation of external economies it shifts down to a new position *LAC′* (dotted).

On the other hand, when the external diseconomies accrue to the firms as a result of the expansion of the industry, cost curves of the firms will shift upward as is depicted in Fig. 19.15. In the beginning, the long-run average cost curve is *LAC* and with the expansion of the industry output and consequent emergence of external diseconomies causes the long-run average cost curve (along with its short-run average and marginal cost curves) to shift upward to a new position *LAC′* (dotted).

As noted above in a previous section, internal economies and diseconomies of scale affect the shape that the long-run average cost curve takes; internal economies of scale cause the long-run average cost to fall as output is increased in the initial stage and internal diseconomies of scale cause the long-run average cost curve to rise. On the other hand, external economies and external diseconomies cause the long-run average cost curve to shift down or up, as the case may be. Moreover, when we are considering the effect of external economies and external diseconomies on the cost curves, it is not only the long-run average cost curve but all short-run and long-run cost curves, whether total, average or marginal, shift together up or down as the case may be. In this connection, it is also worth noting that shifts in cost curves of a firm are not always, nor necessarily brought about by the exapsnion or contraction of an industry's output alone. For instance, a *general increase* in cost of materials, such as cement, steel, oil, electricity, a *general increase* in the prices of machinery and equipment, an *all-round* increase in wages and interest rates in the economy will also shift up the cost curves of a firm. Therefore, in economics when we speak of increasing-cost industries and decreasing cost industries, we consider only the effect of expansion of industry's output on costs of materials, labour, capital equipment etc. incurred by the firms in that industry and rule out any general increase in these costs in the whole economy.

Types of External Economies

Now, the question arises when an industry grows or expands its output, what types of external economies it generates which reduce the costs of the firms in it. The chief examples of external economies provided by Marshall are: (*i*) "improved methods of machinery which are accessible to the whole industry".[5] when it expands, (*ii*) economies which result from the growth of correlated branches of industry which mutually assist one another and "being concentrated in the same localities"[6] encourage the development of "*hereditary skill*", 'the growth of subsidiary trades supplying it with implements and machinery', (*iii*) economies which are "connected with the growth of knowledge and the progress of arts,[7] especially in matters of trade knolwedge: newspapers, trades, trade and technical publications."[8]

Like Marshall, Joan Robinson who analysed the phenomenon of increasing returns (*i.e.* decreasing costs) in the context of partial equilibrium analysis, provided the following main examples of external economies: (*i*) the cases "where the machinery can be bought more cheaply when the industry presents a large market to the machine-making industry"[9] and (*ii*) the cases "where a large labour force is accustomed to work in a certain trade" and develops "traditional skill"[10].

5. A. Marshall, *Principle of Economics*, 8th edition, p. 615.
6. *Ibid*, p. 317.
7. *Ibid*, p. 267.
8. *Ibid*, p. 284
9. Joan Robinson, *The Economics of Imperfect Competition*,. p. 340.
10. *Ibid*, p. 341.

From the above examples mentioned by Marshall and Joan Robinson we explain below some of important external economies which accrue to the firms and reduce their costs of production when the industry as a whole expands.

1. **Cheaper Materials and Capital Equipment.** First, the expansion of an industry may lead to new and cheaper raw materials, machinery and other types of capital equipment. The expansion of an industry means that demand for the various kinds of materials and capital equipment required by it increases. This makes it possible to produce them on a large scale by other industries. This large-scale production of materials and capital equipment lowers their costs of production and hence their prices. Thus the firms in the industry which use these materials and capital equipment will be able to get them at lower prices. This will favourably affect their costs of production. This, of course, will happen in cases where there are increasing returns (*i.e.* decreasing costs) in the industries supplying the materials and capital equipment.

2. **Technological External Economies.** Secondly, with the growth of an industry some external economies of technological type may accrue to the firms of that industry. In our discussion of returns to scale we mentioned that as an individual firm expands its scale it may become possible for it to use more specialised and productive machinery and to introduce greater degree of division of labour. These are *internal* technological economies, which change the technical coefficients of production and improve the firm's productivity. Similarly, when the whole industry expands, it may lead to the discovery of new technical knowledge and in accordance with that the use of improved and better machinery than before. This will change the technical coefficients of production and will enhance the productivity of the firms in the industry and will reduce their costs of production.

3. **Development of Skilled Labour.** Another example of external economies that has been suggested is the development of hereditary or traditional skills among labour. When an industry expands in an area, the labour in that area is well accustomed to do the various productive processes and learns a good deal from the experience. As a result, with the growth of an industry in an area a pool of trained labour equipped with the traditional skills is developed which has a favourable effect on the level of productivity and costs of the firms in the industry.

4. **The Growth of Subsidiary and Correlated Industries.** Another external economy accruing to the firms from the growth of an industry is the growth of subsidiary and correlated industries. These subsidiary and correlated industries may specialise in the production of raw materials, tools and machinery and therefore can provide them at lower prices to the main industry. Likewise, some specialised firms may come into existence, which process the '*waste product*' of the industry into some useful product, when the expansion of the industry makes the waste product large enough to make it worthwhile to set up separate plants for transforming the waste products into useful ones. When this happens, then the firms of the industry can sell their waste products at a good price. This will tend to reduce their cost of production.

5. **Improved Transportation and Marketing Facilities.** These external economies are greatly relevant when an infant industry grows up in a new territory. In the beginning, transportation and marketing facilities both for the purchase of materials and for the sale of its product may not be well-developed. However, the expansion of the industry by the entry of new firms in it may make possible the development of transportation and marketing facilities which will greatly reduce the costs of the firms.

6. **Development of Industry Information Services.** As an industry expands, the firms may form a trade association that distributes information regarding technical knowledge and market possibilities about the industry through publication of trade and technical journals. With the expansion of the industry the firms may jointly set up a central research institute which will be engaged in discovering new improved techniques for the firms in the industry. Thus, besides providing market information, the growth of the industry may help in discovering and spreading improved technical knowledge.

External Diseconomies

We have explained above the external economies which accrue to the firms as a result of the growth of the industry. But, as said above, the expansion of an industry is also likely to generate external diseconomies which raise the cost curves of the firms. The main example of external diseconomies is *the rise in some factor prices* when the industry expands and its demand for various factors needed by it increases. The expansion of an industry will definitely raise the prices of those raw materials and capital goods which are in short supply. Likewise, the expansion of the industry is likely to raise the wages of skilled labour, at least in the short run, since it always takes time for the labour to get training and acquire specialised skill needed in a particular industry.

Since the productive factors such as various types of raw materials, cement, steel, various kinds of machinery and tools and skilled labour are scarce, the increase in demand for them resulting from the expansion in the industry is likely to push up their industry will expand by snatching away the scarce resources from other industries, it will bid up their prices. Thus, in the real world of scarcity, an expanding industry will create more external diseconomies than external economies. Therefore, most industries in the real world encounter rising costs when they expand.

LEARNING CURVE

The learning curve is an important modern concept according to which cumulative experience in the production of a product over time increases efficiency in the use of inputs such as labour and raw materials and thereby lowers cost per unit of output. K.J. Arrow, one of the pioneers in putting forward this concepts calls it "*Learning by doing*"[11]. According to Arrow, as a firm or its manager produces successive lots of output over various periods of time, it learns to produce more with a given quantity of resources or it is capable of producing a given output by using lesser quantities of inputs or resources than before. Thus, either with the increase in efficiency of resources or with saving in resources such as labour and raw materials, cost per unit of output declines. This learning curve effect mostly occurs in the reduction of labour requirements per unit of output.

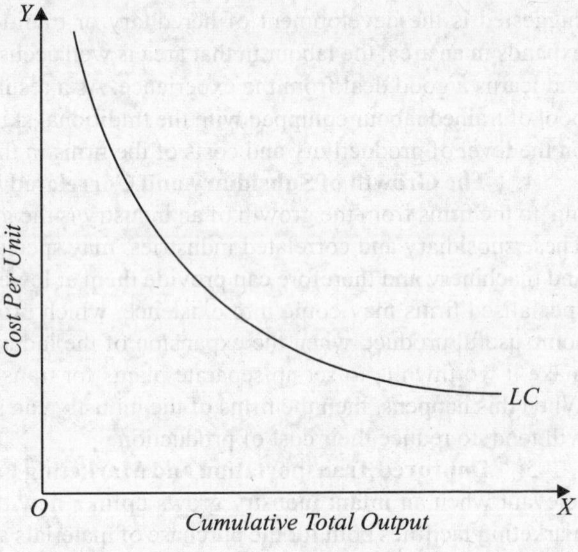

Fig. 19.16. *The Learning Curve*

A number of factors bring this learning curve effect. As cumulative volume of output over successive periods of time increases, labour and supervisors become more familiar with the work methods or the production process, which leads to the reduction in the amount of scrap and other types of wastes. Besides, raw materials cost per unit of output may also decline as cumulative volume of output in successive periods over time increases and as a result a firm gains more experience in doing a production process repeatedly over successive time periods. The learning curve is graphically shown in Fig. 19.16 where on the *X*-axis cumulative total output over successive periods of time and on the *Y*-axis cost per unit of output are measured. It will be

11. K.J. Arrow, "The Economic Implications of Learning by Doing", *Review of Economic Studies*, Vol. 29 (June 1962), pp. 154-179.

seen from Fig. 19.16 that the learning curve slopes downward which shows declining cost per unit of output as cumulative output increases over time and the firm learns from its work experience.

The learning curve effect is usually expressed as a constant percentage. This percentage represents the proportion by which cost per unit of output declines with the increase in cumulative output in each successive time period. For example, if in a production process labour-input cost experiences 80 per cent learning curve effect, this means that if in the first period production of a unit of output requires labour cost of ₹ 1000, in the next period labour cost per unit will decline to ₹ 800 and so forth. This learning curve relationship between cost output is expressed algebraically as follows:

$$C = aA^b$$

where C is the input cost of Qth unit of output, Q is successive unit of output produced, a is the input cost per unit of output in the first period and b is the rate of decline in cost per unit of output in the successive period. Since the learning curve is downward sloping, the value of b is negative.

It is important to note that reduction in cost per unit due to the learning curve effect is different from economies of scale. Whereas economies of scale refer to the decline in cost per unit of output as a *firm's output per time period increases, the learning curve describes the reduction in cost per unit of output as a firm's commulative output over successive time periods increases, while output per period may remain the same.*

L-SHAPED LONG-RUN AVERAGE COST CURVE: MODERN COST THEORY

A significant recent development in cost theory is that the long-run average cost curve is L-shaped rather than *U*-shaped. The L-shape of the long-run average cost curve implies that in the beginning when output is expanded through increase in plant size and associated variable factors, cost per unit falls rapidly due to economies of scale. Even after a sufficiently large scale of output, the long-run average cost does not rise; it may either remain constant or it may even go on falling slightly. At a very large scale of production, the managerial cost per unit of output may rise, but the technical or production economies more than offset the managerial diseconomies so that the total long-run average cost does not rise or may even fall continuously, though at a very small rate. Thus the empirical evidence gathered by economists in recent years does not indicate U-shaped long-run average cost curve.

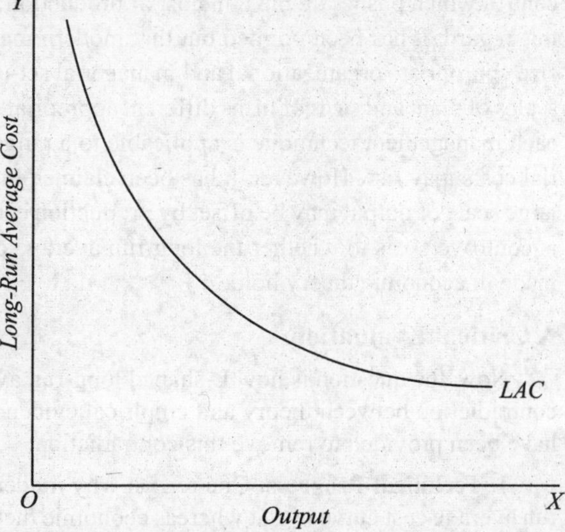

Fig. 19.17. L-*Shaped Long-Run Average Cost Curve*

Empirical evidence indicates that initially the long-run average cost rapidly falls but after a point it remains flat throughout or at its right-hand end it may even slope gently downward. For example, Joel Dean[12] in his cost function studies funds that long run average cost curve is L-shaped. Likewise, J. Johnston[13] in his empirical study of cost functions found strong

12. See Joel Dean, *Managerial Economics,* Prentice Hall, 1951, pp. 292-31.
13. J. Johnston, *Statistical Cost Analysis*, McGraw Hill, New York.

evidence for L-shaped long-run average cost curve. Besides, using data of Indian industries Vinod Gupta[14] who studied long-run average cost functions for 29 manufacturing Indian industries found that in 18 of them long-run cost curve was L-shaped. *L*-shaped long-run average cost curve is illustrated in Fig. 19.17. The difference between L-shaped *LAC* of Fig. 19.17 and the *U*-shaped *LAC* is that there is no rising portion in the former. Indeed, as stated just above, the empirical evidence shows that *LAC* may even slope gently downward at its right hand end. Thus, there is an apparent contradiction between traditional economic theory according to which *LAC* is U-shaped and the results of empirical cost-function studies which find *LAC* to be L-shaped. Two reasons have been given for the continuous occurrence of economies of scale which cause falling cost per unit of output:

1. Production or Technical Economies. First, it is pointed out that there are substantial technical economies of scale enjoyed by a firm when it expands its scale of output in the beginning. This causes the long-run average cost to fall steeply with the initial increases in scale of production. However, it has been asserted that even after most of the economies of scale have been achieved and the firm reaches a minimum *optimal scale*, given the technology of the industry, the *unit cost of production may fall due to some technical economies which it can continue to enjoy even after the minimum optimal scale*. First, the *new techniques* of production are adopted at a large scale of production due to which cost per unit falls. Besides, even with the existing known techniques some economies can always be obtained due to (1) decentralisation and improved skills and productivity of labour, (2) lower repair costs after a certain scale is achieved and (3) producing by a firm itself some of the materials and equipment it needs at a lower cost for its production process instead of buying them from other firms.

2. Development of Appropriate Managerial Techniques. In the traditional cost theory, the rising part of the *LAC* is explained by the difficulties in management, supervision, coordination and control which pushes up the unit cost of production after a certain scale of production is attained. In this regard, it has been pointed out that modern management science has developed for each plant size appropriate organizational and managerial set-up for efficient working of the firm. For different scales of plant and sizes of firms different appropriate management techniques have been evolved and each management technique is applicable to a range of output. Only at a very large scale, managerial costs may rise. However, it has been claimed that any rise in managerial costs even after a very large scale of output may be offset by production economies that continue to accrue. However, there is controversy as to whether the long-run average cost is really L-shaped when usual assumptions made in economic theory hold.

A Critical Evaluation

Now, the question is how L-shaped long-run average cost curve can be explained and apparent contradiction between theory and empirical evidence be removed. The following two explanations have been provided to remove this contradiction:

1. Technical Progress. One reason why modern empirical studies do not find U-shaped long-run average cost curve is that whereas economic theory assumes that technology remains unchanged or there is no technological progress, in the real world, technological progress does take place over time. As a result of technological progress in the real world, long-run average cost curve will shift downward over time. The empirical investigations which are based on times series data would not find the rising average cost in view of the existence of technological progress. Thus while in

14. Vinod K. Gupta, "Cost Function, Concentration and Barriers to Entry in Twenty Nine Manufacturing Industries in India," *Journal of Industrial Economics*, Vol. 17, Nov. 1948.

the case of unchanged technology, as is assumed in traditional cost theory, long-run average cost curve (*LAC*) is *U*-shaped, empirical studies conducted on the basis of data belonging to different points of time between which technological progress had taken place or inter-cross-section studies where both small and large firms using different technologies are included would find the average cost falling. In view of the technological improvement we cannot therefore find the average cost rising in empirical cost function studies. This is illustrated in Fig. 19.18.

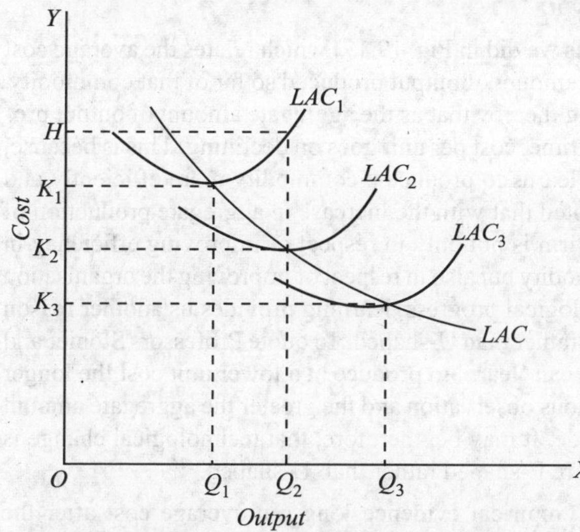

Fig. 19.18. *Shifting Down of LAC due to Technological Progress*

Suppose initially the firm is producing OQ_1 output at cost OK_1 by operating on LAC_1. If the demand for the firm's product increases to OQ_2, then in the context of unchanged technology the firm will expand production along LAC_1 and will produce OQ_2 at cost *OH*. Now, if the technological progress has taken place and in accordance with new technology the firm instals a new plant whose long-run average cost is LAC_2, the firm will produce on the new curve LAC_2 the output OQ_2 at cost per unit which is less than both OK_1 and *OH*. Likewise, when the firm at some later date has expanded output to OQ_3 in response to the increase in demand for its product, the technology might have advanced in the mean time so that the firm produces OQ_3 at OK_3 cost per unit which is less than OK_2. By joining the minimum points of long-run average cost curves we get a curve *LAC* which gently slopes downward due to technological improvement that has taken place over time, whereas with unchanged technology long-run average cost curves LAC_1, LAC_2, and LAC_3 each with a different but unchanged technology is *U*-shaped. Empirical cost-function studies made by economists at different points of time would estimate OK_1, OK_2 and OK_3 unit costs at output OQ_1, OQ_2 and OQ_3 respectively and would therefore suggest that long-run average cost curve was like the thick curve *LAC*.

We therefore conclude that while with a given and unchanged technology long-run average cost curves are *U*-shaped, the empirical evidence would find *L*-shaped long-run average cost curve due to technological progress that takes place over time. The fact that technology changes does not in itself contradict our convention that, if only because it is harder to manage a larger firm than a smaller one, long-run average cost curves will be *U*-shaped in a given state of technology. What the empirical evidence does suggest is that technological progress may often be rapid enough to reduce unit cost even in a situation where, with given technology, the problem of managing a bigger firm would increase unit costs.[15]

2. **Learning Curve Effect.** Learning curve effect or learning by doing is another factor which causes the long-run average cost curve to slope downward throughout. It is now common knowledge that a person learns while doing some productive work. The greater the amount of work he has done since the time he started doing a particular work, the greater the experience he attains and with the experience he learns to do things in a better way than before. This tends to reduce cost per unit. A firm learns to produce a commodity more efficiently as the aggregate amount of its output produced by it increases over time. A good deal of empirical evidence is available which goes to prove that

15. Stonier and Hague, *A Textbook of Economic Theory*, 4th edition, 1972, p.140.

firm's cost of production depends not only on the amount of output of a commodity it produces each month or year but also on the aggregate amount of that commodity produced since the time it started its production. This is because the aggregate output by a firm to date determines the degree of learning it has acquired and the efficiency gained by it.

Therefore, we can draw a learning curve (as we did in Fig. 19.17) which relates the average cost of production of a commodity to the aggregate amount of output produced so far of that commodity. This learning curve slopes downward indicating thereby that as the aggregate amount of output produced of a commodity by a firm increases over time, cost per unit goes on declining. This is because, as said above, with greater production, a firm learns to produce a commodity more efficiently and therefore cost per unit declines. It should be noted that with the increase in aggregate production of a commodity over time, learning gained by a firm is not only in respect of improving efficiency in physical operations in the production of a commodity but also in respect of improving the organisation of the plant. Thus, besides the factor of technological progress, learning provides us another reason why long-run average cost curves are L-shaped rather than U-shaped. To quote Professors Stonier and Hague, "Even, with a given technology, a firm can 'learn' to produce at a lower unit cost the longer the period of time that has elapsed since a previous observation and the greater the aggregate amount of that product that has consequently been made. It may be, therefore, that technological change is not the only reason why long-run cost curves are L-shaped rather than U-shaped."[16]

We have stated above that according to empirical evidence long-run average cost after the initial rapid fall, either remains constant or declines throughout; it does not rise. That is, long-run average cost does not turn up as is required if it is U-shaped. Professor G.A. Smith[17] who has examined empirical evidence in this connection has concluded that with a very large size of firm, labour costs, assembly costs and distribution costs increase very much and therefore large-sized plants *with increasing average cost are not set up in actual practice* and therefore empirical evidence cannot assess the cost situation in them. Therefore, according to Professor Smith, empirical evidence does not refute the U-shaped nature of long-run average cost curve. Walters[18] after examining the empirical studies by Dean and Johnston reached the conclusion that, "there is no large body of data which convincingly contradicts the hypothesis of a U-shaped long-run average cost curve and the fruitful results which depend on it".

NUMERICAL PROBLEMS

Problem 1. *Suppose a firm faces a cost function of* $C = 8 + 4q + q^2$

(i) What is the firm's fixed cost?

(ii) Derive an expression for the firm's variable cost and marginal cost.

[*DU. BA (Hons.) 2003*]

Solution :

(*i*) As fixed cost of the firm does not vary with output, the term in the given cost function which has no output (q) variable will be the fixed cost. From the given cost function it is evident that fixed cost is 8.

(*ii*) Total variable cost (TVC) = $TC - TFC$

$$= (8 + 4q + q^2) - 8 = 4q + q^2$$

and,
$$AVC = \frac{TVC}{Q} = \frac{4q + q^2}{q} = 4 + q$$

16. Stonier and Hague, *op. cit.*, p. 141.
17. C.A. Smith, "Empirical Evidence of Economies of Scale," printed in *Theory of Firm* (Penguin).
18. A. Walters, "Production and Cost Functions, An Econometric Survey," *Econometrica,* Vol. 31, 1963, p. 52.

Marginal cost is the first derivative of total cost function or total variable cost function. Thus, taking the first derivative of the total variable cost function we have

$$MC = \frac{d(TVC)}{dq} = 4 + 2q$$

Problem 2. *Given the total cost function* $TC = 1000 + 4Q + 0.5Q^2$
Determine the functions for TFC, TVC, MC, AVC and AC

Solution :

$$TC = 1000 + 4Q + 0.5Q^2$$

Total fixed cost is independent of level of output to be produced.

Therefore, $TFC = 1000$

$TVC = TC - TFC$

Therefore, $TVC = 4Q + 0.5Q^2$

$$AVC = \frac{TVC}{Q}$$

$$\therefore \quad AVC = \frac{4Q + 0.5Q^2}{Q} = 4 + 0.5Q$$

$$MC = \frac{\Delta TVC}{\Delta Q} = 4 + 1.0Q$$

or $MC = 4 + Q$

$$AC = AFC + AVC$$

$$AFC = \frac{TFC}{Q} = \frac{1000}{Q}$$

Therefore, $AC = \frac{1000}{Q} + 4 + 0.5Q$

Problem 3. *We give below short-run total cost function:*

$$TC = 100 + 50Q - 12Q^2 + Q^3$$

where TC is total cost and Q is level of output.

(*i*) *Determine* : (*a*) *total fixed cost function,* (*b*) *total variable cost function,* (*c*) *average variable cost function,* (*d*) *marginal cost function.*

(*ii*) *Calculate total cost, ATC, AVC and MC when the firm produces 10 units of output.*

(*iii*) *Calculate the level of output at which AVC is minimum.*

Solution. (*a*) From the total cost function it is evident that when level of output is zero, the firm's cost will be ₹ 100. Thus, *TFC* = 100

(*b*) If we deduct ₹ 100 from the total cost function we obtain the total variable cost function.

Thus, $TVC = TC - TFC$

$\therefore \quad TVC = 50Q - 12Q^2 + Q^3$

(*c*) If we divide the *TVC* function by output (*Q*), we get *AVC* function. Thus,

$$AVC = \frac{50Q - 12Q^2 + Q^3}{Q}$$

$$= 50 - 12Q + Q^2$$

(*d*) Marginal cost is the first derivative of total variable cost function. Thus,

$$MC = \frac{dTVC}{dQ} = 50 - 24Q + 3Q^2$$

(*ii*) Given total output (Q) equal to 10, we have

$$TC = 100 + 50(10) - 12(10)^2 + (0)^3$$
$$= 100 + 500 - 1200 + 1000$$
$$= 400$$
$$ATC = \frac{100}{Q} + 50 - 12Q + Q^2 = \frac{100}{10} + 50 - 12 \times 10 + (10)^2$$
$$= 10 + 50 - 120 + 100$$
$$= 40$$
$$AVC = 50 - 12Q + Q^2$$
$$= 50 - 12 \times 10 + 100$$
$$= 30$$
$$MC = 50 - 24Q + 3Q^2$$
$$= 50 - 240 + 300$$
$$= 110$$

(*iii*) The level of *AVC* is minimum at the level of output at which first derivative of *AVC* function is zero. Thus,

$$AVC = 50 - 12Q + Q^2$$
$$\frac{d(AVC)}{dQ} = -12 + 2Q = 0$$
$$2Q = 12$$
$$Q = 6$$

Thus at 6 units of output, *AVC* is minimum.

Problem 4. *Given the following total cost function* $TC = 1500 + 15Q - 6Q^2 + Q^3$

(i) Determine the total fixed cost for producing 1000 units of output and 500 units of output.

(ii) What is AFC at (a) 1000 units of output, and (b) 500 units of output

(iii) Determine TVC, AVC, MC and AC at 50 units of output.

Solution :

(*i*) *TFC* = 1500 irrespective the level of output

(*ii*) *AFC* at 1000 units of output :

$$AFC = \frac{TFC}{Q} = \frac{1500}{1000} = 1.5$$

AFC at 500 units of output

$$AFC = \frac{TFC}{Q} = \frac{1500}{500} = 3$$

(*iii*)

$$TVC = TC - TFC$$
$$\therefore \quad TVC = 15Q - 6Q^2 + Q^3$$

If output is 50 units

$$TVC = 15 \times 50 - 6 \times (50)^2 + (50)^3$$

$$= 750 - 15000 + 125000$$

$$= 1,10,750$$

$$AVC = \frac{TVC}{Q} = \frac{1,10,750}{50} = 2215$$

$$MC = \frac{dTVC}{dQ} = 15 - 12Q + 3Q^2$$

$$MC \text{ with } Q = 50, = 15 - 600 + 7500 = 6915$$

$$AC = AFC + AVC$$

$$AC \text{ with } Q = 50, = \frac{1500}{50} + 2215 = 30 + 2215 = 2245$$

Problem 5. *A firm producing hockey sticks has a production function given by $Q = 2\sqrt{KL}$. In the short run, the firm's amount of capital equipment is fixed at K = 100. The rental rate for K is ₹ 1 and the wage rate is ₹ 4.*

(i) Calculate the firm's short-run total and average costs.

(ii) What are STC, SAC and SMC for producing 25 sticks. [*D.U. BA (Hons.) 2003*]

Solution.

The given production function of firm is

$$Q = 2\sqrt{KL}$$

With $K = 100$ in the short run, the short-run production function is

$$Q = 2\sqrt{100L} = 2 \times 10\sqrt{L} = 20\sqrt{L}$$

Cost, $\quad C = wL + rK$

Given that $\quad w = 4$ and $r = 1$

$$C = 4L + 1K$$

With the given $\quad K = 100$

$$C = 4L + 100 \quad \text{... (1)}$$

The short-run production function is

$$Q = 20\sqrt{L}$$

Taking square of both sides, we have

$$Q^2 = 400L$$

or $$\frac{Q^2}{400} = L \quad \text{... (2)}$$

Substituting (2) in (1), we have

$$C = 100 + 4 \cdot \frac{Q^2}{400}$$

or $$C = 100 + \frac{Q^2}{100} \quad \text{... (3)}$$

The above equation (3) represents the short-run total cost function. To get the short-run average cost function, we divide the short-run total cost function in (3) by output (Q). Thus,

$$SAC = \frac{100 + \frac{Q^2}{100}}{Q} = \frac{100}{Q} + \frac{Q}{100}$$

Short-Run Marginal Cost Function:

Short-run marginal cost function can be obtained by taking the first derivative of short-run total cost function.

Short-run total cost function is

$$C = 100 + \frac{Q^2}{100}$$

$$SMC = \frac{dC}{dQ} = 2\frac{Q}{100} = \frac{Q}{50}$$

If output of hockey sticks = 25, then

$$STC = 100 + \frac{(25)^2}{100} = 100 + \frac{625}{100}$$

$$= 106.25$$

$$SAC = \frac{STC}{Q} = \frac{106.25}{25} = 4.25$$

$$SMC = \frac{Q}{50} = \frac{25}{50} = 0.5$$

Problem 6. *If $Q = A(KL)^{0.5}$, what is the short-run cost function when $K = 100$? What is MC function?*

Solution. With $K = 100$, short-run production function can be written as

$$Q = A(100L)^{0.5} = 10A(L)^{0.5}$$

Squaring both sides, we have

$$Q^2 = 400A^2L \qquad \text{... } (i)$$

Now, the short-run cost function is

$$C = TFC + TVC$$

$$= 100r + wL \qquad \text{... } (ii)$$

where r is the rental price of capital and w is the wage rate of labour and given $K = 100$

From equation (*i*), we have

$$L = \frac{Q^2}{100A^2}$$

Substituting this value of L in (*ii*), we get the following short-run cost function:

$$C = 100r + w.\frac{Q^2}{100A^2}$$

$$MC = \frac{TVC}{dQ} = \frac{2wQ}{100A^2}$$

Problem 7. *Given the production function, $Q = A(LK)^{0.5}$ derive the cost function.*

Solution. A firm minimises its cost production for a given level of output when

$$MRTS_{LK} = \frac{w}{r} \qquad \text{... } (i)$$

where w = wage rate of labour

r = rental price of use of capital

In the given Cobbe-Douglas production function,

$$Q = A(LK)^{0.5}, \qquad MRTS_{LK} = \frac{K}{L}$$

Therefore, at cost-minimisation

$$\frac{K}{L} = \frac{w}{r}$$

or $$wL = r.K$$

$$K = \frac{wL}{r} \qquad \text{... }(ii)$$

Iso-cost curve equation can be written as

$$C = wL + rK \qquad \text{... }(iii)$$

Substituting the optimum value of *K* as obtained in (*ii*) in the iso-cost equation (*iii*), we have

$$C = wL + r\cdot\frac{wL}{r}$$

$$C = 2wL$$

or $$L = \frac{C}{2w} \qquad \text{... }(iv)$$

Similarly, the optimum value of *K* can be obtained by eliminating *L*. Thus,

$$K = \frac{C}{2r} \qquad \text{... }(v)$$

Substituting the optimum values of *L* and *K* in the given production function, we have

$$Q = A\left(\frac{C}{2w}\right)^{0.5}\left(\frac{C}{2r}\right)^{0.5}$$

Solving for *C*, we have

$$C = \frac{2Q(w.r)^{0.5}}{A}$$

QUESTIONS FOR REVIEW

1. Distinguish between economic costs and accounting costs. Which should be taken into account for calculating the economic profits of the firm?

2. What is the difference between explicit costs and implicit costs? Should both be considered for optimal business decision-making by the firm?

3. Explain the concepts of total fixed costs, total variable costs and total costs. How are they related to each other ? Illustrate them through curves. Is the distinction between the fixed costs and variable costs relevant in the long run?

4. Explain the following concepts of cost:

(*a*) Average fixed cost (*AFC*)

(*b*) Average variable cost (*AVC*)

(*c*) Average total cost (*ATC*)

(*d*) Marginal cost (*MC*)

Draw these curves with usual shapes.

Why does *ATC* curve reach its lowest point after the *AVC* curve? Why does the *MC* curve intersect both the *AVC* and *ATC* curves at their lowest points?

5. Fill in the blanks in the table below:

Output	*Total Cost* (₹)	*Total Fixed Costs* (₹)	*Total Variable Costs* (₹)	*AFC* (₹)	*AVC* (₹)
0	50	—	—	—	—
1	70	—	—	—	—
2	100	—	—	—	—
3	120	—	—	—	—
4	135	—	—	—	—
5	150	—	—	—	—
6	160	—	—	—	—
7	165	—	—	—	—

[**Hint:** Since at zero output level variable costs are zero, the whole total cost of ₹ 50 at zero output level represents total fixed cost. Thus, ₹ 50 is the total fixed cost at all levels of output.]

6. From your knowledge of the relationships between various concepts of cost of production, fill in the blanks in the following table :

Q	*TC*	*TFC*	*TVC*	*ATC*	*AFC*	*AVC*	*MC*
0	125	—	—	—	—	—	—
10	—	—	—	—	—	—	5
20	—	—	—	10.5	—	—	—
30	—	—	110	—	—	—	—
40	255	—	—	—	—	—	—
50	—	—	—	—	—	3	—
60	—	—	—	—	—	—	3
70	—	—	—	5	—	—	—
80	—	—	295	—	—	—	—

7. What is the relationship between average cost and marginal cost? If the marginal cost is rising, does it mean that average cost must also be rising?
8. Derive long-run average cost curve from short-run average cost curves. How are they related to each other ? Does the curve joining the minimum points of the short-run average cost curves constitute long-run average cost curve?
9. Give reasons for the U-shape of long-run average cost curve. Why is long-run average cost curve usually called 'planning curve'?
10. Explain the various economies of scale and diseconomies of scale that accrue to the firm when it expands its scale of production.
11. Empirical studies have found that long-run average cost curve (*LAC*) is L-shaped. How would you explain it ? Does it not contradict the U-shaped long-run average cost curve?
12. What would be the shape of long-run average cost (*LAC*) curve when constant returns to scale occur?
13. What is Learning Curve? What are the factors that bring about learning curve effect? How is reduction in cost per unit due to the learning curve effect different from economies of scale?'
14. (*a*) Describe relationship between *AVC* and average product (*AP*), between marginal cost (*MC*) and marginal product (*MP*).

    (*b*) How is U-shape of the average variable cost explained by the law of variable proportions?
15. The production function for a firm exhibits constant returns to scale. Input prices paid by the firm are constant, regardless of the quantity purchased. Draw the LRAC curve for the firm. Draw a SRAC for the same firm at some plant size. [*D.U. B.A. (Hons.) 1987*]

16. Briefly discuss the factors affecting the shape of the long-run average cost curve.

17. Explain how the long-run envelope cost curve is derived from short-run average cost curves. Illustrate with (*i*) the case of a few plant sizes, and (*ii*) the case of many and continuously varied plant sizes. [*D.U. B.A. (Hons.)*]

18. Given U-shaped long-run average cost curve and the relevant short-run cost curves, derive long-run marginal cost curve. [*D.U. B.A.(Hons.)*]

19. *Distinguish between a firm's short-run cost of production from its long-run cost of production. Why is it that at the output where SAC = LAC short-run marginal cost also equals long-run marginal cost*? [*D.U. B.A. (Hons.)* 1989]

[Hints. Proof of the later part

$SAC = LAC$

Since $SAC.\ Q$ = short-run total cost and $LAC.Q$ is long-run total cost, we have $STC = LTC$

Differentiating both with respect to output (Q) we have

$$\frac{d(STC)}{dQ} = \frac{d(LTC)}{dQ}$$

Therefore, $SMC = LMC$]

20. Of the increasing segment of a firm's long-run marginal cost curve and short-run marginal cost curve which is steeper? Why?

[**Hints.** *SMC* is steeper. This is because in the short-run size of the plant is fixed and output is varied by increasing the amount of the variable factors. In the long run fixed factors represented by plant size can also be valued to minimise cost of production].

21. When the law of diminishing returns operates, the *TVC* rises at decreasing rate. True or false? Give reasons.

22. (*a*) A typical production function has three stages. Explain and illustrate. Relate these stages of production function with the shape of the total cost of production curve. Also explain why it implies a U-shaped marginal cost curve. [*D.U. BA(Hons.) 1991*]

23. What can you infer about the average cost curve from the information that marginal costs are less than average costs? Give reasons.

[**Hint:** When *MC* is less than *AC*, average cost curve slopes downward]

24. The long-run average cost curve is the locus of the minimum points of the short run average cost curves. True or false? Explain. [*D.U. B.A.(Hons.) 1999*]

[**Hint.** False; the long-run average cost curve is the locus of the *tangency points* of the short-run average cost curves with the long-run average cost curve.]

25. "If wage rate falls, marginal production cost must fall." True or False? Explain. [*D.U. B.A.(Hons.)*]

[**Hint:** $MC = \frac{W}{MP_L}$. Therefore, given marginal product of labour (MP_L) when, wage rate (w) falls, marginal cost (*MC*) must fall].

26. (*a*) What is the relationship between average cost and marginal cost as the average cost is falling, rising and at its minimum point?

(*b*) Why does the minimum of the average variable cost curve occurs before the minimum of the average cost curve? [*D.U. B.A.(Hons.) 1999*]

CHAPTER 20

SUPPLY AND ITS ELASTICITY

Price of a commodity is determined by the demand for and supply of a commodity. In some previous chapters we have studied the factors which determine the demand for a commodity. We have also studied the law of demand according to which the demand for a commodity is inversely related to its price. Supply of a commodity depends upon how the physical returns and costs change as more output of a commodity is produced. After having explained the behaviour of costs and returns in the previous chapters, we are now in a position to explain the law of supply and the concept of elasticity of supply.

The Meaning of Supply

As demand is defined as a schedule of the quantities of good that will be purchased at various prices, similarly the supply refers to the schedule of the quantities of a good that will be offered for sale at various prices. To be more correct, supply of a commodity is the schedule of the quantities of a commodity that would be offered for sale at all possible prices during a period of time, for example, a day, a week, a month and so on.

Supply should be carefully distinguished from stock. *Stock is the total volume of a commodity which can be brought into the market for sale at a short notice* and supply means the quantity which is *actually brought* in the market. For perishable commodities like fish and fruits, supply and stock are the same because whatever is in stock must be disposed of. The commodities, which are not perishable, can be held back if prices are not favourable. If the price is high, larger quantities of non-perishable commodities are offered by the sellers from their stock. And if the price is low, only small quantities are brought out for sale. In short, *stock is potential supply*.

Supply Function

The quantity of a commodity that firms will be able and willing to offer for sale in the market depends on several factors. The important factors determining supply of a commodity are :

1. The price of the commodity
2. The prices of inputs (i.e. resources) used for the production of the commodity
3. The state of technology
4. The number of firms producing and selling the commodity
5. The prices of related goods produced
6. Future expectations regarding prices.

We will explain these factors determining supply of a commodity in detail in a later section. However, it may be noted that out of the above determinants of supply the own price of the commodity, the prices of inputs (i.e. resources) used to produce the commodity, and the technology are three important factors and therefore the supply function of a commodity is often written taking these factors as independent variables. *Thus supply function of a commodity is written as*

$$Q^s_x = S(P_x, F_1, F_2 \ldots\ldots\ldots, F_m)$$

where Q^s_x is the quantity supplied of the commodity X, P_x is it own price, $F_1, F_2 \ldots\ldots F_m$ are the prices of inputs used to produce the commodity X and the state of technology determines the form of supply function S. It must be noted that the *form of the supply function refers to the precise quantitative relation between the independent variables such as the own price of the commodity X and prices of factors* such as F_1, F_2 etc.

If the factor prices (F_1, F_2, F_3 etc) and state of technology are taken as given and constant, then quantity supplied of a good is simply a function of its own price and is usually written as

$$Q_x^s = f(P_x)$$

Law of Supply

Supply of a commodity is functionally related to its price. The law of supply relates to this functional relationship between price of a commodity and its supply. In contrast to the inverse relationship between the quantity demanded and the changes in price, the quantity supplied *generally varies directly with price*. That is, the higher the price, the larger is the quantity supplied of a commodity.

The supply schedule and supply cruve reflect the law of supply. According to the law of supply, *when the price of a commodity rises, the quantity supplied of it in the market increases, and when the price of a commodity falls, its quantity demanded decreases, othe factors determining supply remaining the same*. Thus, according to the law of supply, the quantity supplied of a commodity is directly or positively related to price. It is due to this direct relationship between price of a commodity and its quantity supplied that the supply curve of a commodity slopes upward to right as seen from supply curve *SS* in Fingure. 20.1. When price of wheat rises from ₹ 520 to ₹ 530 per quintal the qunatity supplied of wheat in the market increases from 200 quintals to 225 quintals per period.

Table 20.1
Supply Schedule of Wheat

Price Per Quintal (₹)	*Quantity Supplied (in quintals)*
500	100
510	150
520	200
530	225
540	250
550	275

Corresponding to the demand schedule already explained we can construct an individual's supply schedule. Also by totalling up the amount supplied at various prices by all the sellers in a market, we can obtain the supply schedule of the market. Supply schedule represents the relation between prices and the quantities that the firms are willing to produce and sell. We have given in Table 20.1 a supply schedule of wheat per day in a market.

It will be seen from the above table that when price of wheat is ₹ 500 per quintal, the 100 quintals of wheat are supplied in the market. When price of wheat rises to ₹ 510 per quintal, 150 quintals of wheat are supplied. When price of wheat goes up to ₹ 550 per quintals, its quantity supplied in the market has risen to 275 quintals. By plotting the above supply schedule of wheat on a graph paper we have obtained supply curve *SS* in Fig. 20.1 In Fig. 20.1, the quantity supplied has been measured along the

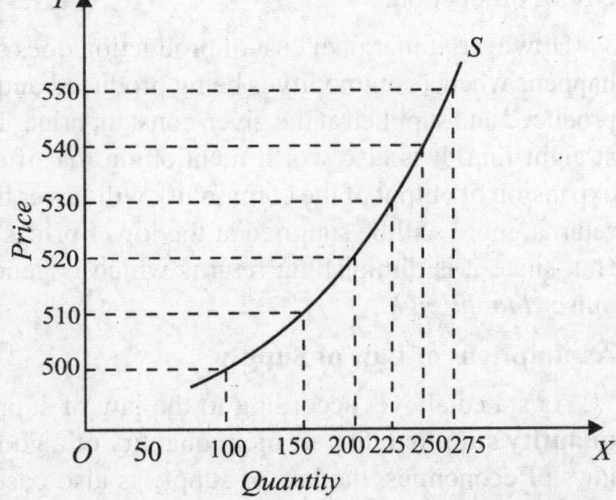

Fig. 20.1. *Supply Curve Showing Direct Relationship Between Price and Quantity Supplied*

X-axis and price of wheat has been measured along the Y-axis. It will be seen from this figure that supply curve slopes upward from left to right, which indicates that as the price of wheat rises, quantity supplied increases and *vice versa*. This is in a sharp contrast to the demand curve of a commodity which slopes downward from left to right.

Reserve Price. It may be noted that if price falls too much, supply may dry up altogether. The price below which the seller will refuse to sell any quantity of a commodity is called the *reserve price*. At this price, the seller buys his own stock as it were.

Explanation of the Law of Supply: Why does supply curve generally slope upward?

It has been observed from the experiences in the real world that price of a product and quantity supplied of it by firms producing it are positively related to each other, that is, at a higher price more is supplied and *vice versa,* other things remaining the same. This positive relationship between price and quantity supplied is an important law of economics. How do we explain it? In analysing the relation between price of a commodity and the quantity supplied, given *ceteris paribus* assumption we are in fact dealing with the supply function [$Q_x^s = S(P_x$]. It should be remembered that firms are driven by profit motive. The higher, price cost per unit of a product output makes it more profitable to expand more output and offer more quantity of the product for sale in the market. Thus, the high price of a product serves as an incentive for the producer to produce more of it. The higher the price, the greater the incentive for the firm to produce and supply more of a commodity in the market, other things remaining the same.

Further, the changes in quantity supplied of a product following the changes in its price depends on the *possibilities of substitution of one product for an other*. For example, if price of wheat in the market rises, the farmers will alter the cropping pattern so as to produce more of wheat by with-drawing land and other natural resources from the cultivation of gram and devoting them to the production of wheat. This is because high market price for wheat relative to gram induces farmers, who aim at maximising profits, to use more resources for production of wheat and fewer resources for production of gram.

The basic reason behind the law of supply (i.e., positive relation between price and quantity supplied) is the way cost changes as output is expanded to offer more for sale. To produce more of a product, firms have to devote more resources to its production. As seen above, when production of a product is expanded by using more resources, diminishing returns occur. Due to the diminishing returns average and marginal costs of production increase. This implies that more quantity of commodity would be produced and supplied in the market only at a higher price so as to cover higher cost of production.

However, if marginal cost of production does not rise with the increase in output as, for instance, happens when a commodity is being produced under conditions of *constant returns*, the more will be produced and supplied at the given constant price. That is, supply curve in this case will be a horizontal straight line. It is also worth mentioning that if a commodity is subject to increasing returns, the expansion of output of the commodity will lower the unit cost of production. As a result of increasing returns, more will be supplied at the lower prices and the supply curve will be sloping downward. But, since it is diminishing returns which is generally the rule, *the supply curve generally slopes upward to the right*.

Assumptions of Law of Supply

As stated above, according to the law of supply there is **positive relation between price and quantity supplied**, that is, more quantity of a good is supplied at a higher price. As in case of other laws of economics, the law of supply is also based on some assumptions. First, the law of supply assumes that firms operate in a competitive market structure and no individual firm has a monopoly over the production of the commodity. This law does not apply to the supply by the monopolist or any producer working under oligopoly and monopolistic competition as individual firms in these market structures exercise some control over the price of their products. Second, this law of supply applies when the short-run marginal cost is rising in the region where it is profitable to produce. It may be recalled that short-run marginal cost rises due to the diminishing marginal returns to the variable factors. In case of the rise in marginal cost the firms would supply more only at a higher price so as to cover the rise in marginal cost when more quantity of output is produced by hiring additional variable inputs.

Third, for the law of firms to apply the firms should aim at maximising profits or sales revenue. If they do not maximise profits or sales, they may not produce more to increase the quantity supplied. Fourth, for the law of supply to hold it is assumed that technology or input prices remain constant when price of a good rises. If there is improvement in technology or fall in input prices when firms expand their output, then more supply of the product may be made at a lower price.

Fifth, *in case of the competive industry the law of supply will hold if on the expansion of industry the external diseconomies exceed external economies so that we have an increasing-cost industry* and long-run supply curve of the industry slopes upward. Therefore, in this case too more is supplied at a higher price. In case of constant-cost industry external diseconomies, if any, cancel out external economies when the industry expands in the long run by the entry of new firms, the long-run supply curve of the industry is a horizontal straight line. Therefore, in constant cost industry more is supplied at the same price in the long run and the law of supply doe not apply in this case. On the other hand, if on the expansion of industry *external economies* outweigh the *external diseconomies*, the industry will be *decreasing-cost industry* and its long-run supply curve will be downward sloping. As a consequence, in case of decreasing-cost industry in the long-run *more will be supplied at a lower price*. This behaviour is quite opposite to the law of supply.

The Concept of Supply Curve and Perfect Competition

As 'demand' is defined as a schedule of the quantities of a good that will be purchased at different prices, the supply similarly, refers to a schedule of the quantities of a good that will be offered for sale at different prices. Supply curve is therefore, a graphic representation of what quantities of a good will be offered for sale at all possible prices. Supply curve depicts the sellers' quantity reactions to various prices. Thus the quantity supplied, like the quantity demanded, is a function of price. There is, however, an important difference between the reaction of quantity supplied and that of quantity demanded to changes in price of a good. Whereas the quantity demanded of a good generally increases with the fall in price of the good and decreases with the rise in its price, the quantity supplied decreases with the fall in price of a good and increases with the rise in its price. In other words, while the quantity demanded has a negative or inverse relation with the price, the quantity supplied bears generally a direct relation with the price. The positive relation between quantity supplied and price lies in the nature of the costs of production which generally rise as more quantity of a good is produced. This is because of positive relation between price and quantity supplied that supply curve slopes upward from left to right. *While the short-run supply curve always slopes upward to right, the long-run supply curve may slope either upward or downward, or it may be of a horizontal straight line depending upon whether the industry is working under increasing cost or decreasing cost or constant cost conditions.* However, the upward-sloping supply curve showing increasing cost is a more typical case even in the long run.

It is worth nothing here that the concept of supply curve, as it is used in economic theory, is relevant only for the case of perfect or pure competition and it is quite inapplicable to the cases of imperfect competition—monopolistic competition, monopoly and oligopoly. This is because the *notion of supply curve refers to the question as to how much quantity of a commodity a firm will supply at various given prices*. In other words, *notion of supply curve refers to the quantity reactions of a firm when the firm itself exercises no influence over the determination of price and takes price as given datum for it and adjusts its quantity produced or supplied.* Since only in perfect or pure competition a firm exercises no influence over the price which is determined by impersonal market mechanism of demand and supply and is beyond the control of individual firms, the concept of supply curve is relevant only for perfect or pure competition. So far as short-run supply curve of the industry under perfect competition is concerned, it is a mere lateral summation of the supply curves of the firms.

Under various forms of imperfect competition, an individual firm does not take the price as given and is not a mere quantity adjuster. In fact, under various forms of imperfect competition, a firm sets its own price. For a firm under imperfect competition it is not a question of adjusting output or supply at a given price but of choosing *price-output combination* which maximises its profits. Commenting on the relevance of supply curve, Prof. Baumol writes, "*The supply curve is, strictly speaking, a concept which is usually relevant only for the case of pure (or perfect) competition. The reason for this lies in its definition–the supply curve is designed to answer question of the form, 'How much will firm A supply if it encounters a price which is fixed at P dollars*'[1]. But such a question is most relevant to the behaviour of firms that actually face prices over whose determination they exercise no influence.

Changes in Supply: Increase and Decrease in Supply

As stated above, the supply of a commodity in economics means the entire schedule or curve depicting the relationship between price and quantity supplied of the commodity, given the other factors influencing supply. These other factors are the state of technology, prices of inputs (resources), prices of other related commodities, etc., which are assumed constant when the relationship between price and quantity supplied of a commodity is examined. It is the change in these factors that cause a shift in the supply curve due to the changes in factors other than price. For example, when prices of inputs such as labour and raw materials used for the production of a commodity decline, this will result in lowering the cost of production which will induce the producers to produce and make available a greater quantity of the commodity in the market at each price. This increases in supply of a commodity due to the reduction in prices of inputs will cause the entire supply curve to shift to the

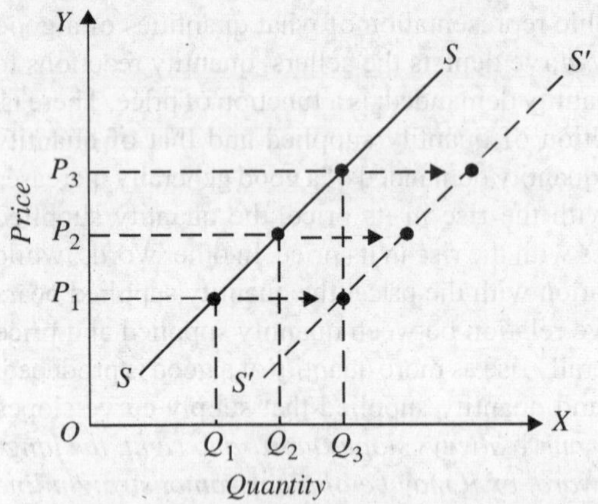

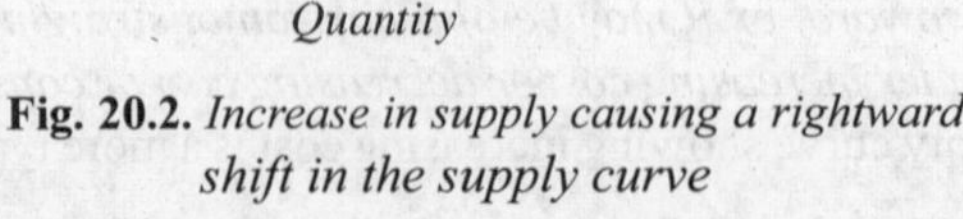
Fig. 20.2. *Increase in supply causing a rightward shift in the supply curve*

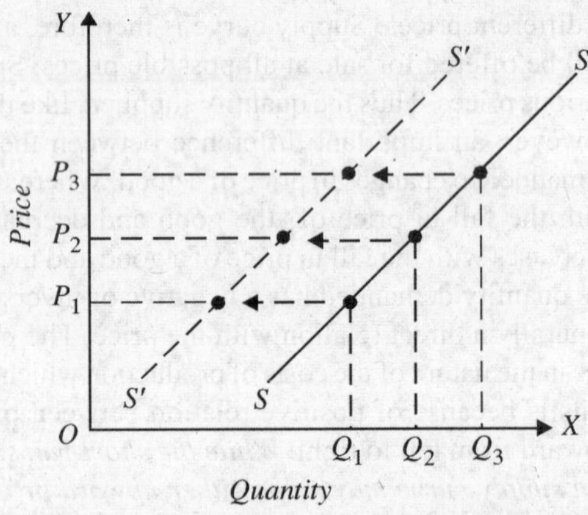

Fig. 20.3. *Decrease in supply causing a leftward shift in the supply curve*

right as shown in Fig. 20.2 where the supply curve shifts from *SS* to *S′S′*. As shown by arrow marks, at price P_1, P_2 and P_3 quantity supplied increases when supply increases causing a rightward shift in the supply curve. Similarly, progress in technology used for production of a commodity which increases productivity and reduces cost per unit will also cause the supply curve to shift to the right.

1. W. J. Baumol, *Economic Theory and Operations Analysis,* 2nd edition, p.342.

On the other hand, *decrease in supply means the reduction in quantity supplied at each price of the commodity* as shown in Fig. 20.3 where as a result of decrease in supply the supply curve shifts to the left from *SS* to *S'S'*. As shown by the arrow marks, at each price such as P_1, P_2, P_3, the quantity supplied on the supply curve *S''S''* has declined as compared to the supply curve *SS*. The decrease in supply occurs when the prices of factors (inputs) used for the production of a commodity is produced at a higher cost per unit which causes a reduction in quantity supplied at each price. Similarly, the imposition of an excise duty or sales tax on a commodity means that each quantity will now be supplied at a higher price than before so as to cover the excise duty or sales tax per unit. This implies that quantity supplied of the commodity at each price will decrease as shown by the shift of the supply curve to the left.

Another important factor causing a decrease in supply of a commodity is the *rise in price of other commodities* using the same factors. For example if the price of wheat resies sharply, it will become more profitable for the farmers to grow it. This will induce the farmers to reduce the cultivated area under other crops, say sugarcane, and devote it to the production of wheat. This will lead to the decrease in supply of sugarcane whose supply curve will shift to the left.

Further, agricultural production in India greatly depends on the rainfall due to Monsoon. If Monsoon comes in time and rainfall is adequate, there are bumper crops, the supply of agricultural products increases. However, in a year when Monsoon are untimely or highly inadequate, there is a sharp drop in agricultural output which cause shift in the supply curve of agricultural output to the left.

We thus see that there are several factors other than price which determine the supply of a commodity and any change in them will cause a shift in the entire supply curve.

Factors Determining Supply

It is clear from the supply schedule (Table 20.1) and the supply curve (Fig. 20.1) given above that the quantity supplied varies directly with price of the product. A supply schedule and supply curve shows that the supply of a product is function of its price. However, the supply depends not only on the price of a product but on several other factors too. It should be remembered that in economic theory whereas the effect of changes in price of a product on the quantity supplied of it is depicted and explained by movement along a given supply schedule or curve, the effect of other factors is represented by the changes or shifts of the entire supply schedule or supply curve. While making a supply schedule or drawing a supply curve we assume that these other factors remain the same. Thus when these other factors change, they cause a shift in the entire supply curve. The factors other than price which determine price are the following :–

(*a*) **Production Technology**–The change in technology affects the supply function by altering the cost of production. If there occurs an improvement in production technology used by the firm, the unit cost of production declines and consequently the firms would supply more than before at the given price. That is, the supply would increase implying thereby that the entire supply curve would shift to the right.

(*b*) **Price of Factors**– Changes in prices of factors or resources also cause a change in cost of production and consequently bring about a change in supply. For example, if either wages of labour increase or prices of raw materials and fuel go up, the unit cost of production will rise. With higher unit cost of production, less would be supplied than before at various given prices. This implies that supply curve would shift to the left.

(*c*) **Prices of other Products**–When we draw a supply curve we assume that the prices of other products remain unchanged. Now, any change in the prices of other products would influence the supply of a product by causing substitution of one product for another. For example, if the market price of wheat rises, if will lead to the reduction in the production and supply of gram by the farmers as they would withdraw land and other resources from the production of gram and devote them to the production of whcat. This will causc a lcftwards shift in the supply curve of gram.

(*d*) **Objective of the Firm**—The objective of a firm also determines the supply of a product produced by it. If the firms aim to maximise sales or revenue rather than profits, the production of the product produced by them and hence its supply in the market would be larger.

(*e*) **Number of Producers (or firms)**–If the number of firms producing a product increases, the market supply of the product will increase causing a rightward shift in the supply curve. When, in the short run, firms in an industry are making large profits, the new firms enter that industry in the long run and consequently the total production and supply of the product of the industry increases. On the other hand, due to losses if some firms leave the industry, the supply of its product will decline.

(*f*) **Future Price Expectations**–The supply of a commodity in the market at any time is also determined by sellers' expectations of future prices. If, as happens during inflationary periods, sellers expect the prices to rise in future, they would reduce supply of a product in the market at would instead hoard the commodity. The hoarding of huge quantities of goods by traders is an important factor in reducing their supplies in the market and thus causing further rise in their prices.

(*f*) **Taxes and Subsidies**– Taxes and subsidies also influence the supply of a product. If an excise duty or sales tax is levied on a product, the firms will supply the same amount of it at a higher price or less quantity of it at the same price. This implies that imposition of a sales tax or excise duty causes a leftward slift in the supply curve. The opposite happens in case of the supply of a commodity on which government provides subsidies.

If follows from above that technology, prices of factors and products, expectations regarding future prices and objective of the firms are the important determinants of supply which cause rightward or leftward slift in the whole supply curve.

ELASTICITY OF SUPPLY

When a small fall in price leads to a large contraction in supply, the supply is comparatively elastic. But when a big fall in price leads to a very small contraction in supply, the supply is said to be comparatively inelastic. Conversely, a small rise in price leading to a big extension in supply shows more elastic supply, and a big rise in price leading to a small extension in supply indicates inelastic supply. The elasticity of supply is really the measure of the extent to which the supply of output by an industry can be expanded and changes in of the marginal costs of production.

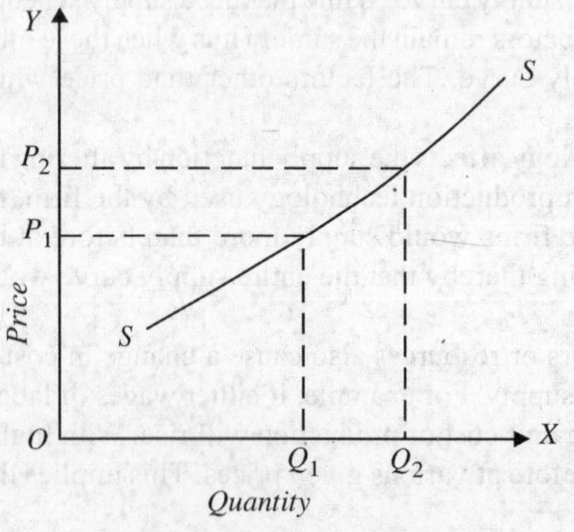

Fig. 20.4. *Elastic Supply*

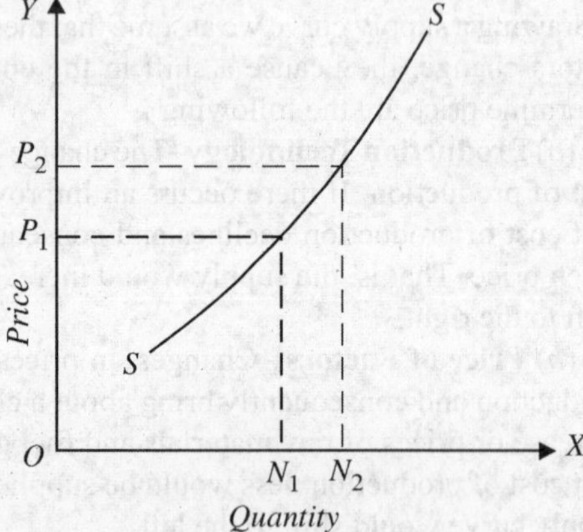

Fig. 20.5. *Inelastic Supply*

Consider Figs. 20.4 and 20.5 where two supply curves *SS* have been drawn. At price OP_1, the quantity supplied in Fig. 20.4 is OQ_1, and the quantity supplied in Fig. 20.5 is ON_1. With rise in price of the product, quantity supplied increases from OQ_1 to OQ_2 in Fig. 20.4 and from ON_1 to ON_2 in

Fig. 20.5. Whereas the relative change in price is the same in both the figures, the increase in quantity supplied Q_1Q_2 in Fig. 20.4 is much larger as compared to the increase in quantity supplied N_1N_2 in Fig. 20.5. Therefore, supply in Fig. 20.4 is said to be elastic, whereas that in Fig. 20.5 is inelastic. The elasticities of supply of various products differ very much from each other. What are the factors which determine elasticity of supply of products will be explained in some detail later.

The concept of elasticity of supply, like the elasticity of demand is a relative measure of the responsiveness of quantity supplied of a commodity to the changes in its price. The greater the responsiveness of quantity supplied of a commodity to the changes in its price, the greater its elasticity of supply. In precise terms, the elasticity of supply can be defined as under :–

$$\text{Elasticity of supply} = \frac{\text{Proportionate change in quantity supplied}}{\text{Proportionate change in price}}$$

$$e_s = \frac{\Delta q}{q} \div \frac{\Delta p}{p} = \frac{\Delta q}{q} \times \frac{p}{\Delta p} = \frac{\Delta q}{\Delta p} \times \frac{p}{q}$$

For an accurate measure of elasticity of supply ***midpoint method,*** as explained in case of elasticity of demand, should be used. Using midpoint formula, elasticity of supply can be measured as

$$e_s = \frac{q_2 - q_1}{\frac{q_1 + q_2}{2}} \div \frac{p_2 - p_1}{\frac{p_1 + p_2}{2}}$$

$$q_2 - q_1 = \Delta q \text{ and } p_2 - p_1 = \Delta p$$

Therefore,
$$e_s = \frac{\Delta q}{q_1 + q_2} \times \frac{p_1 + p_2}{\Delta P}$$

$$e_s = \frac{\Delta q}{\Delta p} \times \frac{p_1 + p_2}{q_1 + q_2}$$

If the price of a refrigerator rises from ₹ 2,000 per unit to ₹ 2500 per unit and in response to this rise in price the quantity supplied increases from 2500 units to 3500 units, the elasticity of supply by midpoint method will be:–

$$\Delta q = 3500 - 2500 = 1000$$

$$\frac{q_2 + q_1}{2} = \frac{3500 + 2500}{2} = 3000$$

$$\Delta p = 2500 - 2000 = 500$$

$$\frac{p_1 + p_2}{2} = \frac{2500 + 2000}{2} = 2250$$

$$e_s = \frac{1000}{3000} \div \frac{500}{2250}$$

$$= \frac{1000}{3000} \times \frac{2250}{500} = \frac{1}{3} \times \frac{4.5}{1} = \frac{4.5}{3} = 1.5$$

If the supply curve of a commodity is upward sloping as is generally the case (See Fig. 20.1) the coefficient of elasticity of supply will have a positive sign. When the supply curve is upward sloping, the elasticity of supply will be anything between zero and infinity.

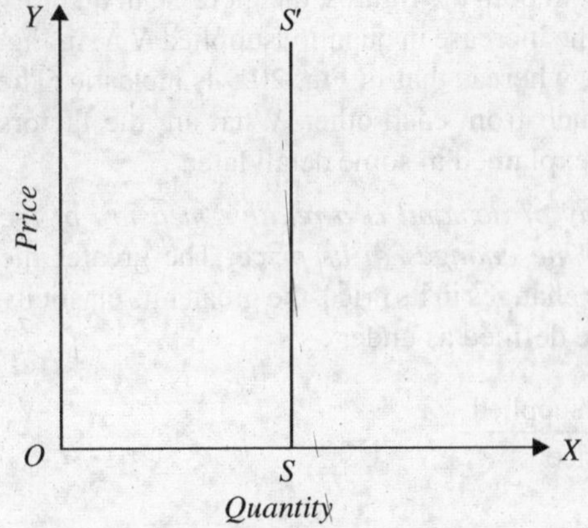

Fig. 20.6. *Perfectly Inelastic Supply*

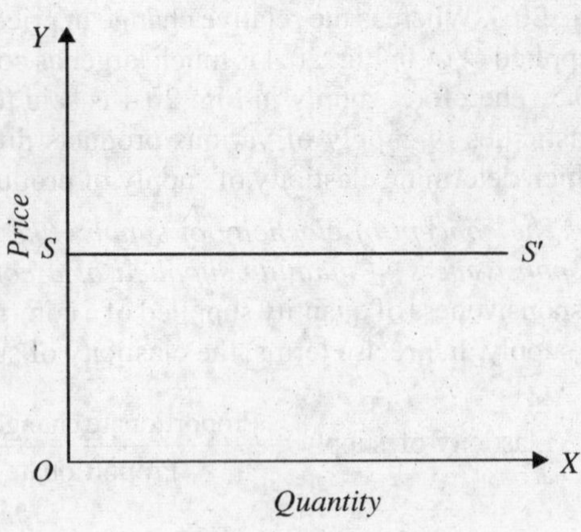

Fig. 20.7. *Perfecly Elastic Supply*

When the quantity supplied of a commodity does not change at all in response to the changes in response to the changes in its place, the elasticity of supply is zero. In the case of zero elasticity of supply, the supply curve will be vertical straight line parallel to the X-axis and is said to be *perfectly inelastic* (see Fig. 20.6). On the other hand, if at a price, any quantity of the good is supplied, its elasticity will be equal to infinity and its supply curve will be a horizontal straight line parallel to the quantity axis and is said to be perfectly elastic. Perfectly elastic supply curve is shown in Fig. 20.7.

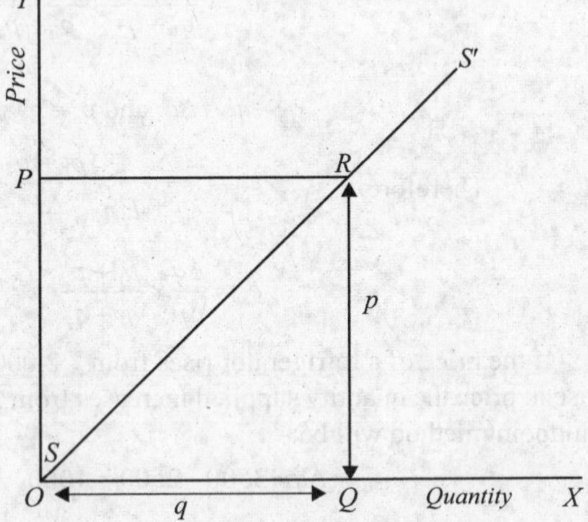

Fig. 20.7(a). *Unitary Elastic Supply Curve*

A third special case of supply curve is found when a straight line supply curve passes through the origin such as the one shown in Fig. 20.7(a). In this case elasticity of supply at any point such as R is unity. This follows from the elasticity formula; $\frac{\Delta q}{\Delta q}\cdot\frac{p}{q}\cdot$ It will be seen from Fig. 20.7(a) that in case of supply curve passing through the origin, the ratio $\frac{p}{q}$ is equal to the slope *i.e.* $\frac{\Delta p}{\Delta q}$ of the supply curve SS'. Thus substituting $\frac{\Delta p}{\Delta q}$ for $\frac{p}{q}$ in the elasticity formula we have $e_s = \frac{\Delta q}{\Delta p} \times \frac{\Delta p}{\Delta q} = 1$.

Measurement of Point Elasticity of Supply

The concept of elasticity of supply, like the elasticity of demand, occupies an important place in price theory. As explained above, *the elasticity of supply is the degree of responsiveness of supply to changes in the price of a good.* More precisely, the elasticity of supply can be defined as a proportionate change in quantity supplied of a good in response to a given proportionate change in price of the good. Therefore,

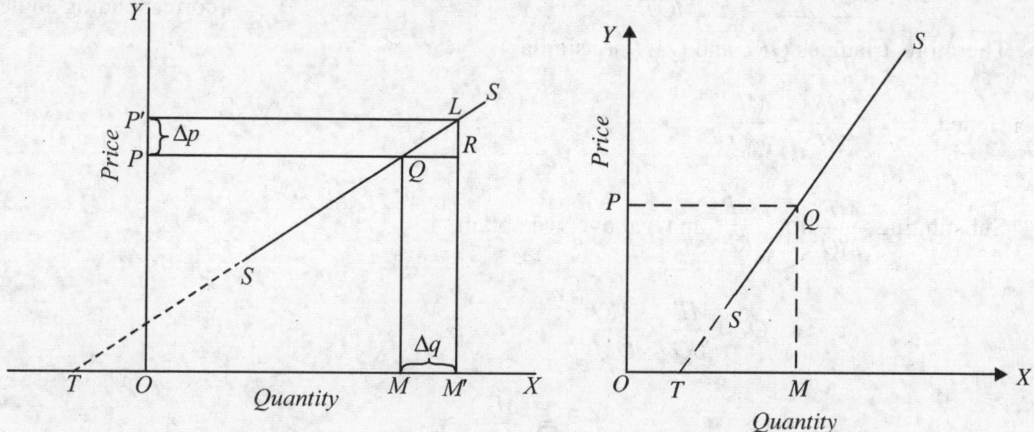

Fig. 20.8. *Measurement of Point Elasticity of Supply ($e_s > 1$)*

Fig. 20.9. *Measurement of Point Elasticity of Supply, $e_s < 1$*

$$e_s = \frac{\text{proportionate change in quantity supplied}}{\text{proportionate change in price}}$$

In terms of symbols, we can write elasticity of supply as :

$$e_s = \frac{\Delta q}{q} \div \frac{\Delta p}{p} = \frac{\Delta q}{q} \times \frac{p}{\Delta p}$$

$$= \frac{\Delta q}{\Delta p} \times \frac{p}{q}$$

The elasticity of supply at a point on the supply curve can be easily measured by a formula. We shall derive this formula below.

In Fig. 20.8 the supply curve *SS* is given and elasticity of supply at point *Q* is required to be measured. At price *OP* the quantity supplied is *OM*. With the rise in price from *OP* to *OP'* the quantity supplied increases from *OM* to *OM'*. Extend supply curve *SS* downward so that it meets the *X*-axis at *T*.

The elasticity of supply at point $Q = \frac{\Delta q}{q} \div \frac{\Delta p}{p}$

or, $$e_s = \frac{MM}{OM} \div \frac{PP}{OP} = \frac{MM}{OM} \times \frac{OP}{PP}$$

$$= \frac{MM}{PP} \times \frac{OP}{OM}$$

Substituting *QR* for *MM'*, *RL* for *PP'* and *MQ* for *OP*, we get

$$e_s \text{ at point } Q = \frac{QR}{RL} \times \frac{MQ}{OM} \qquad ...(i)$$

Now, in triangles *QRL* and *QMT*

$\angle LQR = \angle QTM$ (corresponding angles)

$\angle QRL = \angle QMT$ (right. angles)

$\angle QLR = \angle MQT$ (corresponding angles)

Therefore, triangles *QRL* and *QMT* are similar

Hence, $\dfrac{QR}{RL} = \dfrac{MT}{MQ}$

Substituting $\dfrac{MT}{MQ}$ for $\dfrac{QR}{RL}$ in (*i*) above, we obtain

$$e_s \text{ at } Q = \frac{MT}{MQ} \times \frac{MQ}{OM}$$

$$= \frac{MT}{OM}$$

Thus, we can get the value of elasticity of supply from dividing *MT* by *OM*. Since in Fig. 20.8, *MT* is greater than *OM*, supply elasticity $\dfrac{MT}{OM}$ will be greater than one.

In Fig. 20.9 supply curve when extended meets the *X*-axis to the right of the point of origin so that *MT* is smaller than *OM*. Therefore, in Fig. 20.6 the elasticity of supply $\dfrac{MT}{OM}$ is less than unity.

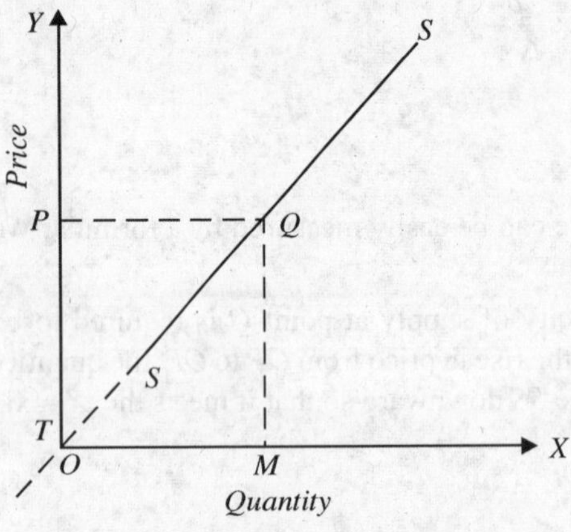

Fig. 20.10. $e_s = 1$

In Fig. 20.10 supply curve *SS* when extended meets the *X*-axis exactly at the point of origin so that the *MT* is equal to *OM*. Therefore, in Fig. 20.10 elasticity of supply will be equal to one.

In Fig. 20.8 elasticity of supply will be greater than one at every point of the curve, but it will differ from point to point. Similarly, in Fig. 20.9 supply elasticity is less than one at every point of the curve, but it will differ from point to point. However, in Fig. 20.10 elasticity of supply will be equal to one at every point of the curve.

We have studied above how the elasticity of supply is measured at a point on the straight-line supply curve. But now the question is how the point elasticity of supply can be measured on the real curve-type supply curve. Consider Fig. 20.11 where a supply curve of true curve type has been drawn and it is required to measure elasticity at point *A* on it. The general principle involved is the same as described above. In order to apply the above principle for estimating the point elasticity at point *A* on the supply curve *SS*, we have to draw a tangent at it.

Now, a tangent $t_1 t_1$ has been drawn at point *A*. On being extended, tangent t_1t_1 meets the *X*-axis at point T_1. Therefore, elasticity of supply at point *A* on the supply curve is $\dfrac{M_1T_1}{OM_1}$.

Likewise, we can find out the elasticity of supply at point *B* on the supply curve. For this, tangent t_2t_2 has been drawn at point *B* and has been extended to meet the *X*-axis at point T_2. Thus, the elasticity at point *B* on the supply curve *SS* is equal to $\dfrac{M_2T_2}{OM_2}$. It is also evident from Fig. 20.11 that elasticity

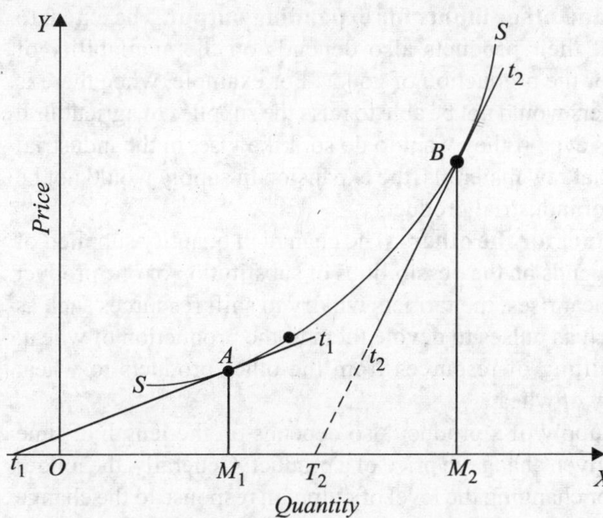

Fig. 20.11. *Varying Elasticity at Different Points on the Supply Curve*

of supply at point A and B is different. Since $\frac{M_2T_2}{OM_2}$ is less than $\frac{M_1T_1}{OM_1}$, the elasticity of supply at point B is less than that at A.

Factors Determining Elasticity of Supply

Elasticity of supply plays an important role in determining prices of products. To what extent price of a product will rise following the increase in demand for it depends on the elasticity of supply. The greater the elasticity of supply of a product, the less the rise in its price when demand for it increases. We explain below the factors which determine elasticity of supply of a product.

Changes in Marginal Cost of Production. Elasticity of supply of a commodity depends upon the ease with which increases in output can be obtained without bringing about rise in cost of production. If with the increase in production, the marginal cost of production goes up the elasticity of supply to that extent would be less. In the short run, with some factors of production being fixed, the increase in the amount of a variable factor eventually causes diminishing marginal returns and as a result with the expansion of output marginal cost of production rises. This causes supply of a commodity in the short run less elastic. However, in the long run, the firms can increase output by varying all factors and also the new firms can enter the industry and thereby add to the supply of a commodity. Therefore, the long-run supply curve of a commodity is more elastic than that of the short run.

In the increasing cost industry, that is, the industry which experiences increases in cost when industry expands through the entry of new firms, the long-run supply curve, like the short-run one, is upward sloping, but will be more elastic than in the short run. In the constant-cost industry, *i.e.*, the industry wherein costs do not change with the expansion of the industry as a whole, the long-run supply curve is perfectly elastic, because in this case increases in the industrial output can be obtained at the same cost of production. In the decreasing-cost industry, that is, industry which is subject to increasing returns, long-run supply curve is downward sloping and has therefore a negative elasticity of supply. This is because in the case of decreasing-cost industry expansion in the industry brings down the cost of production and therefore additional output is forthcoming at a lower supply price.

Response of the Producers. Besides the change in cost of production, the elasticity of supply for a product depends on the responsiveness of producers to changes in its price. If the producers do not respond positively to the increase in prices, the quantity supplied of a product would not increase as a result of rise in its price. A profit-maximising producer will increase the quantity supplied of a product following the rise in its price. However, producers who not always exhibit profit-maximising behaviour may not raise supply in response to the rise in price. For example, it has been argued by some with some empirical evidence that farmers in developing countries respond negatively to the rise in price of their agricultural products. They point out that with the higher agricultural prices, their need for fixed money income is met by selling smaller quantities of foodgrains and therefore at higher prices they may produce and sell smaller quantities rather than more.

Availability of infrastructure facilities and other inputs for expanding output. The extent to which the producers would raise supply of their products also depends on the availability of infrastructure facilities and inputs required for the production of goods. For example, when there is lack of fertilisers, irrigation facilities, the farmers would not be able to raise the supplies of agricultural products in response to the rise in their prices even if they want to do so. Likewise, in the industrial field if there is shortage of power, fuel, essential raw materials, the expansion in supply would not be forthcoming in response to the rise in prices of industrial products.

Possibilities of Substitution of one Product for the others. The change in quantity supplied of a product following the change in its price depends on the possibilities of substitution of one product for others. For example, if market price of wheat rises, the farmers will try to shift resources such as land, fertilisers away from other products such as pulses to devote them to the production of wheat. The greater the extent of possibilities of shifting of resources from the other products to wheat production, the greater the elasticity of supply of wheat.

The Length of Time. The elasticity of supply of a product also depends on the length of time during which producers have to respond to a given change in price of a product. Generally, the longer the time producers get to make adjustments for changing the level of output in response to the change in price, the greater the response of output, that is, the greater the elasticity of supply. From the viewpoint of the influence of the length of time on the elasticity of supply we distinguish between three time periods : (1) market period or very short run, (2) short run and (3) long run. The market period is a very short period and during this no more production is possible. Therefore, market-period supply curve is a vertical straight line (*i.e.*, perfectly inelastic). In the short run, firms can change output by changing the amounts of only variable factors, short-run supply curve is somewhat elastic. In the long run since firms can adjust all factors of production and also new firms can enter or leave the industry, long-run supply curve is more elastic.

QUESTIONS FOR REVIEW

1. Distinguish between supply and stock. Explain the factors that determine the elasticity of supply.
2. What is the law of supply? Are there any exceptions to this law?
3. Why does the supply curve generally slope upward to the right?
4. Distinguish between extension and contraction of supply on the one hand and increase and decrease in supply on the other. Illustrate the difference with a diagram.
5. Explain the factors which determine supply of a commodity.
6. What is meant by Elasticity of Supply? How is it measured?
7. What are the factors which determine the elasticity of supply of a commodity. Give some examples of elastic supply.
8. The coefficient of elasticity of supply of a commodity X is 2, what quantity of the commodity will the seller supply at a price of ₹ 5 per unit, if he supplies 80 units of it at ₹ 4 per unit.
9. A vertical supply curve means that elasticity of supply
   (*a*) infinity, (*b*) zero,
   (*c*) equal to one, (*d*) greater than one but less than infinity.
10. A horizontal supply curve means that elasticity of supply is
    (*a*) zero, (*b*) infinity,
    (*c*) equal to one, (*d*) greater than one but less than infinity.
11. A linear supply curve cuts the *Y*-axis (*i.e.*, price axis). What can you say about its elasticity?
    D.U. B.A. (Hons.)
12. Show that an upward sloping supply curve curve with its intercept on the X-axis has a price elasticity of supply less than one. *D.U. B.A. (Hons.)*

PART IV

THEORY OF FIRM AND PRICING IN DIFFERENT MARKET STRUCTURES

- Main Market Forms and Concepts of Revenue
- Objectives and Equilibrium of the Firm: A General Analysis
- Equilibrium of the Firm and Industry Under Perfect Competition
- Price Determination Under Perfect Competition
- Stability of Equilibrium and Cobweb Model
- Applications of Demand and Supply Analysis
- Price and Output under Monopoly
- Price Discrimination
- Price Under Monopolistic Competition
- Price and Output Under Oligopoly
- Sales Maximisation Model of Oligopoly
- Theory of Games and Competitive Strategy

CHAPTER 21

MAIN MARKET FORMS AND CONCEPTS OF REVENUE

The determination of prices and outputs of various products depends upon the type of market structure in which they are produced, sold and purchased. In this connection economists have classified the various markets prevailing in a capitalist economy into (*a*) perfect competition or pure competition, (*b*) monopolistic competition, (*c*) oligopoly and (*d*) monopoly. Three market forms, monopolistic competition, oligopoly and monopoly, are generally grouped under the general heading of imperfect competition, since these three forms of market differ with respect to the degree of imperfection in the market. Monopolistic competition is highly imperfect and monopoly is the most imperfect form of market structure.

But before explaining the salient features of various market forms, it will be useful to explain what is meant by market in economics.

Meaning of Market

Market is generally understood to mean a particular place or locality where goods are sold and purchased. However, in economics, by the term market we do not mean any particular place or locality in which goods are bought and sold. The idea of a particular locality or geographical place is not necessary to the concept of the market. What is required for the market to exist is the contact between the sellers and buyers so that transaction (*i.e.*, sale and purchase of a commodity) at an agreed price can take place between them. The buyers and sellers may be spread over a whole town, region or a country but if they are in close communication with each other either through personal contact, exchange of letters, telegrams, telephones, *etc*. so that they can sell and buy a good on agreed price, the market would be said to exist. Further, it is notworthy that because in a market, there is close and free communication between various buyers and sellers price of a homogeneous commodity settled between different sellers and buyers tends to be the same. Thus, in the words of Cournot, a French economist, *"Economists understand by the term market not any particular market place in which things are bought and sold but the whole of any region in which buyers and sellers are in such free intercourse with one another that the price of the same good tends to equality easily and quickly"*.

Thus, the essentials of a market are: (*a*) commodity which is dealt with; (*b*) the existence of buyers and sellers; (*c*) a place, be it a certain region, a country or the entire world; and (*d*) such communication between buyers and sellers that only one price should prevail for the same commodity at the same time.

CLASSIFICATION OF MARKET FORMS

The popular basis of classifying market structures rests on three crucial elements, (1) the number of firms producing a product, (2) the nature of product produced by the firms, that is, whether it is homogenous or differentiated, and (3) the ease with which new firms can enter the industry. The price elasticity of demand for a firm's product depends upon the number of competitive firms producing the same or similar product as well as on the degree of substitution which is possible between the product of a firm and other products produced by rival firms. Therefore, a distinguishing feature of different market categories is the degree of price elasticity of demand faced by an individual firm.

We present in the table given below the classification of market forms based on the number of firms, the nature of product produced by them and price elasticity of demand.

1. **Perfect Competition.** As is evident from Table 21.1 perfect competition is said to prevail where there is a large number of producers (firms) producing a homogeneous product. The maximum output which an individual firm can produce is very small relatively to the total demand of the industry product so that a firm cannot affect the price by varying its supply of output.

There are following four important features of perfect competition :

1. There is a large number of firms (producers and sellers) and buyers of a product,
2. Products of all firms are homogeneous,
3. There is freedom of new firms to enter the industry and old ones to to leave it,
4. All firms and buyers have perfect information about the prevailing market price of the product.

With many firms and homogenous product under perfect competition, no individual firm in it is in a position to influence the price of the product and therefore the demand curve facing it will be

Table 21.1

The Classification of Market Forms

Form of Market Sructure	*Number of Firms*	*Nature of Product*	*Price Elasticity of Demand for an Individual Firm*	*Degree of Control over Price*
(*a*) Perfect Competition	A large number of firms	Homogeneous product	Infinite	None
(*b*) **Imperfect Competition**				
(*i*) Monopolistic Competition	A large number of firms	Differentiated Products (but they are close substitues of each other)	Large	Some
(*ii*) Pure Oligopoly, (*i.e.*, Oligopoly without Product Differentiation)	Few firms	Homogeneous Product	Small	Some
(*iii*) Differentiated Oligopoly (*i.e.*, Oligopoly with Product Differentiation)	Few firms	Differentiated Products (which are close substitutes of each other)	Small	Large
(*c*) Monoply	One	Unique Product without Close Substitutes	Very small	Very large

horizontal straight line at the level of the prevailing price of the product in the market, that is, price elasticity of demand for a single firm will be infinite. Mrs. Joan Robinson writes: "Perfect competition prevails when the demand for the output of each producer is perfectly elastic. This entails, first, that the number of sellers is large so that the output of any one seller is negligibly small proportion of the total output of the commodity and the second that the products of various sellers are homogeneous from the viewpoint of consumers[1]. Many economists use perfect competition and pure competition interchangeably. But some such as E.H. Chamberlin, F.H. Knight draw a distinction between 'perfect' competition and 'pure' competition which we shall explain in a later chapter.

2. **Imperfect Competition**. Imperfect competition is an important market category wherein individual firms exercise control over the price to a smaller or larger degree depending upon the degree of imperfection present in a case. Control over price of a product by a firm and therefore the existence of imperfect competition can be caused either by the 'fewness' of the firms or by the product differentiation. Therefore, imperfect competition is divided into the following sub-categories.

(*a*) **Monopolistic Competition**: The first important sub-category of imperfect competition is monopolistic competition, on which Professor E.H. Chamberlin laid a great stress in his original thought-provoking work "*The Theory of Monopolistic Competition*". Monopolistic Competition, as is now understood, is characterised **by a large number of firms and product differentiation.**[2] That is, in *monopolistic competition* a large number of firms produce somewhat different products which are close substitutes of each other. As a result, demand curve facing a firm under monopolistic competition is highly elastic and this indicates that a firm working in it enjoys some control over the price. Besides, there is freedom of entry and exit under monopolistic competition.

(*b*) **Pure Oligopoly**: The second sub-category of imperfect competition is **oligopoly without product differentiation** which is also known as *pure oligopoly*. Under it there is competition among the few firms producing homogeneous or identical product. The fewness of the firms ensures that each of them will have some control over the price of the product and the demand curve facing each firm will be downward sloping which indicates that the price elasticity of demand for each firm will not be infinite.

(*c*) **Differentiated Oligopoly**: The third sub-category of imperfect competition is **oligopoly with product differentiation** which is also called *differentiated oligopoly*. As its name shows, it is characterised by competition among the few firms producing differentiated products which are close substitutes of each other. The demand curve facing individual firms under oligopoly with product differentiation is downward sloping[3] and the firms would have fairly large control over the price of their individual products.

---

1. Joan Robinson, *The Economics of Imperfect Competition*. Macmillan and Co., London, 1954, p. 18.
2. It should be carefully noted that Prof. E.H. Chamberlin called any market structure, which was a blend of monopoly and competition, as monopolistic competition. A blend of monopoly and competition could be there if there was either a few firms producing a identical or differentiated product or there was a large number of firms producing differentiated products. The former is the case of oligopoly with and without product differentiation and the latter is the monopolistic competition with many firms. Thus, in Chamberlin's terminology, the concept of monopolistic competition covers both oligopoly (with as well as without product differentiation) and monopolistic competition with a large number of firms producing differentiated products. However, in standard modern economic theory, the term monopolistic competition is used only for the latter.
3. The nature of demand curve under oligopoly is, however, a controversial matter. In some models of oligopoly without product differentiation, it is assumed that the demand curve for an oligopolist is perfectly elastic (*i.e.*, horizontal straight line). Further, another important model of oligopoly assumes a kinked demand curve (with a kink at the prevailing price) for an oligopoly firm.

3. Monopoly.[4] Monopoly, as is now generally understood, means the existence of *a single producer or seller which is producing or selling a product which has no close substitutes*. Since a monopoly firm has a sole control over the supply of a product, which can have only remote substitutes, it has a very large control over the price of its product. The expansion and contraction in output of the product by a monopolist will considerably affect the price of the product, contraction in output will raise its price and expansion in output will lower it. Therefore, the demand curve facing a monopolist is downward sloping and has a steep slope. Thus, according to Prof. F. Machlup, "Monopolistic Competition would then comprise the cases of closer substitutes and more elastic demand curves, while monopoly would comprise those of remote substitutes and steeper demand curves."[5] Besides, in monopoly there are strong barriers to the entry of new firms in the industry.

Market Classification and Cross Elasticity of Demand

The concept of cross elasticity of demand has been used by Robert Triffin[6] to measure the amount and kind of competition among firms and therefore for classifying the market structure. In perfect competition the cross elasticity of demand for the product of a single firm with respect to a change in the price of the rest of the industry is infinite. That is to say, the proportionate fall in the demand for the product of a single firm is infinitely large in response to any given proportionate fall in the price of the product of the whole industry. Similarly, in monopolistic competition the cross elasticity of demand for the product of a single firm with respect to a change in the price of other products made in the monopolistic 'group' is very high. The cross elasticity of demand for the product of a monopolist with respect to a fall in the price of the other products in the economy is very low. We present below the different market categories and the cross elasticity of demand (e_c) found in them in a tabular form.

Table 21.2
Market Classification on the Basis of Cross Elasticity of Demand

Market Form	*Cross Elasticity of Demand*
1. Perfect Competition	Cross elasticity is infinite ($e_c = \infty$)
2. Monopolistic Competition	Cross elasticity (e_c) is very high
3. Monopoly	Cross elasticity (e_c) is very low or zero

The concept of cross elasticity of demand as a measure of classifying market forms is very much inadequate and sometimes leads to wrong conclusions. "The basic objection to the concept of cross elasticity is that it causes neglect of the two basic determinants of market structure: the degree of closeness or remoteness of substitution among products and the number of firms in the relevant group or industry. The concept of cross elasticity leads to the neglect of the two elements which are fundamental to an understanding or market structure."[7]

---

4. For monopoly, sometimes the terms *absolute* monopoly, *pure* monopoly, *complete* monopoly are used. All these terms have the same meaning since all visualise a single producer or seller producing a product which has no close substitutes. However, Sraffa used the term 'Pure Monopoly' in a different sense. *By pure monopoly Sraffa means that a single seller of a product is so powerful that he is in a position to take the whole income of all the consumers whatever the level of his output. Obviously, such a pure or absolute monopoly cannot exist in the real world.*
5. Fritz Machlup, Monopoly and Competition: A classification of Market Positions, *American Economic Review*, September 1937, p.448.
6. *See* his book *"Monopolistic Competition and General Equilibrium Theory"*.
7. M. Olson and D. Mcfarland "The Restoration of Pure Monopoly and the Concept of Industry", *Quarterly Journal of Economics*. Vol. 76 (November 1962), pp. 613-31.

It has been pointed out by several economists[8] such as E.H. Chamberlin, W. Fellner, E.F. Beach and A.G. Papandreou that cross elasticity of demand of any perfectly competitive firm, instead of being infinity, is zero. Now, as said above, under pure monopoly, cross elasticity of demand is also zero. Thus, in two market situtaions of pure competition and pure monopoly which are two opposite extreme cases, cross elasticity is found to be the same. Therefore, cross elasticity of demand is a very unsatisfactory measure of a market structure. The cross elasticity of demand of a perfectly or purely competitive firm is zero because it produces a product which has so many homogeneous substitutes produced by other firms in the competitive industry. The number of firms producing homogeneous products, that is, perfect substitutes, is so great that *no one of them* will be noticeably affected when a single firm changes its price or output. If a purely competitive firm tries to reduce its price unilaterally, it will not be able to fully meet the demand for its product as the consumers of the homogeneous products of other firms will be attracted to it. But the crucial point is that no other single firm will perceive any noticeable change because the number of firms involved is very large.

We, therefore, conclude that cross elasticity of demand as a means of market classification is unsatisfactory. The classification of market forms made on the basis of (*i*) number of firms in the industry, (*ii*) the nature of product (*i.e.* the closeness or remoteness of substitution among products and the ease with which new firms can enter the industry or the old ones can leane it, which we have explained above, is quite adequate and satisfactory.

CONCEPTS OF AVERAGE REVENUE AND MARGINAL REVENUE

In the previous few chapters we discussed the nature of demand from the viewpoint of the consumer. But a producer or seller of good is also very much concerned with the demand for a good, because revenue obtained by him from selling the good depends mainly upon the demand for the good. He is, therefore, interested in knowing what sort of demand curve faces him. The demand curve of the consumers for a product is the average revenue curve from the standpoint of the sellers, since the price paid by the consumers is revenue of the sellers.

Average Revenue

Price paid by the consumer for the product forms the revenue or income of the seller. The whole income received by the seller from selling a given amount of the product is called *total revenue*. If a seller sells 15 units of a product at price ₹ 10 per unit and obtains ₹ 150 from this sale, then his total revenue is ₹ 150. Thus total revenue can be obtained from multiplying the quantity of output sold by the market price of the product ($P.Q$). On the other hand, average revenue is revenue earned *per unit of output*. Average revenue can be obtained by dividing the total revenue by the number of units sold. Thus,

$$\text{Average revenue} = \frac{\text{total revenue}}{\text{total output sold}}$$

$$AR = \frac{TR}{Q}$$

where AR stands for average revenue, TR for total revenue and Q for total output produced and sold.

In our above example, when total revenue Q equal to ₹ 150 is received from selling 15 units of the product, the average revenue will be equal to ₹ 150/15 = ₹ 10. ₹ 10 is here the revenue earned per unit of output. Now the question is whether average revenue is different from price or these two

8. E.H. Chamberlin, *Toward a More General Theory of Value*, pp. 79–81 : William Fellner. *Competition among the Few*, pp. 50–54; E.F. Beach, Triffin's Classification of Market Positions, *Canadian Journal of Economics and Politics*, Feb. 1943, pp. 69–74; A.G. Papandreou, Market Structure and Monopoly Power, *American Economic Review*, September 1949, pp. 883–97.

concepts mean the same thing. If a seller sells various units of a product at the same price, then average revenue would be the same thing as price. But when he sells different units of a given product at different prices, then the average revenue will not be equal to price. An example will clarify this point. Suppose a seller sells two units of a product, both at a price of ₹ 10 per unit. Total revenue of the seller will be ₹ 20 and the average revenue will be 20/2 = ₹ 10. Thus average revenue is here equal to the price of the product. Now suppose that the seller sells the two units of his product, one unit to the consumer *A* at price ₹ 12 and one unit to the consumer *B* at price ₹ 10. His total revenue from the sale of two units of the product will be ₹ 22. Average will be here equal to 22/2 = ₹ 11. Thus in this case when two units of the product are sold at different prices, average revenue is not equal to the prices charged for the product.

But in the actual life we find that different units of a product are sold by the seller at the same price in the market (except when he discriminates and charges different prices for different units of the good), average revenue equals price. Thus in economics we use average revenue and price as synonyms except when we are discussing price discrimination by the seller. Since the buyer's demand curve represents graphically the quantities demanded or purchased by the buyers at various prices of the good, it also, therefore, shows the average revenue at which the various amounts of the good are sold by the seller. This is because the price paid by the buyer is revenue from seller's point of view. Hence, average revenue curve of the firm is really the same thing as the demand curve of the consumers.

Marginal Revenue

On the other hand, marginal revenue is the *net revenue* earned by selling an additional unit of the product. In other words, *marginal revenue is the addition made to the total revenue by selling one more unit of a commodity*. Putting it in algebraic expression, marginal revenue is the addition made to total revenue by selling n units of a product instead of $n - 1$ where n is any given number. If a producer sells 10 units of a product at price ₹ 15 per unit, he will get ₹ 150 as the total revenue. If he now increases his sales of the product by one unit and sells 11 units, suppose the price falls to ₹ 14 per unit. He will, therefore, obtain total revenue of ₹ 154 from the sale of 11 units of the good. This means that 11th unit of output has added ₹ 4 to the total revenue. Hence ₹ 4 is here the marginal revenue.

Total revenue when 10 units are sold at price of ₹ 15 = 10 × 15 = ₹ 150

Total revenue when 11 units are sold at price of ₹ 14 = 11 × 14 = ₹ 154

Marginal revenue = 154 – 150 = ₹ 4

The word *net* in the first definition of marginal revenue given above is worth noting. The full understanding of the word 'net' in the definition will reveal why the marginal revenue is not equal to the price. The question is, taking our above numerical example, why the marginal revenue due to the 11th unit is not equal to the price of ₹ 14 at which the 11th unit is sold. The answer is that the 10 units which were sold at the price of ₹ 15 before will now all have to be sold at the reduced price of ₹ 14 per unit. This will mean the loss of one rupee on *each* of the previous 10 units and total loss on the previous 10 units due to price fall will be equal to ₹ 10. The loss in revenue incurred on the previous units occurs because the sale of additional 11th unit reduces the price to ₹ 14 *for all*. Thus in order to find out the net addition made to the total revenue by the 11th unit, the loss in revenue (₹ 10) on previous units should be deducted from the price of ₹ 14 at which the 11th unit is sold along with others. The marginal revenue in this case will, therefore, be equal to ₹ 14 – 10 = 4. Marginal revenue is thus less than the price at which the additional unit is sold.

It is clear from above that marginal revenue can either be found directly by taking out the difference between total revenue before and after selling the additional unit, or it can be obtained by subtracting the loss in revenue on previous units due to the fall in price from the price at which the additional unit is sold.

Therefore, marginal revenue = difference in total revenue in increasing sales from $n - 1$ units to n units.

= price of the additional unit minus loss in revenue on previous units resulting from price reduction.

It follows from above that when the price falls as additional unit is sold, marginal revenue is less than the price. But when the price remains the same as additional unit is sold, as under perfect competition, the marginal revenue will be equal to average revenue, since in this case there is no loss incurred on the previous units due to the fall in price. The relationship between average revenue and marginal revenue is the same as between any other average and marginal values. When average revenue falls marginal revenue is less than the average revenue. When average revenue remains the same, marginal revenue is equal to average revenue.

If *TR* stands for total revenue and *Q* stands for output, then marginal revenue (*MR*) can also be expressed as follows :

$$MR = \frac{\Delta TR}{\Delta Q}$$

$\frac{\Delta TR}{\Delta Q}$ indicates the slope of the total revenue curve.

Thus, if the total revenue curve is given to us, we can find out marginal revenue at various levels of output by measuring the slopes at the corresponding points on the total revenue curve.

Average and Marginal Revenue Under Imperfect Competition

The meaning of the concepts of total, average and marginal revenues under conditions of imperfect competition will become clear from Table 21.3. As has been stated above, when imperfect competition prevails in the market for a product, an individual firm producing that product faces a downward sloping demand curve. In other words, as a firm working under conditions of imperfect competition increases production and sale of its product its price falls. Now, when all units of a product are sold at the same price, the average revenue equals price. How marginal revenue can be obtained from the changes in total revenue and what relation it bears to average revenue will be easily grasped from looking at Table 21.3.

Table 21.3
Total, Average and Marginal Revenues

No. of Units Sold	*Total Revenue (Rupees)*	*Average Revenue or Price (₹)*	*Marginal Revenue (Addition made to Total Revenue) (Rupees)*
I	II	III	IV
1	16	16	16
2	30	15	14
3	42	14	12
4	52	13	10
5	60	12	8
6	66	11	6
7	70	10	4
8	72	9	2
9	72	8	0
10	70	7	–2

It will be seen from the Col. III of the table that price (or average revenue) is falling as additional units of the product are sold. Marginal revenue can be found out by taking out the difference between the two successive total revenues. Thus, when 1 unit is sold, total revenue is ₹ 16. When 2 units are sold, price (or *AR*) falls to ₹ 15 and total revenue increases to ₹ 30. Marginal revenue is therefore here equal to 30–16 = 14, which is recorded in Col. IV. When 3 units of the product are sold, price falls to ₹ 14 and total revenue increases to ₹ 42. Hence marginal revenue is now equal to ₹ 42–30 = ₹ 12 which is again recorded in Col. IV. Likewise, marginal revenue of further units can be obtained by taking out the difference between two successive total revenues. Marginal revenue is positive as long as total revenue is increasing. Marginal revenue becomes negative when total revenue declines. Thus when in our table 22.3 quantity sold is increased from 9 units to 10 units the total revenue declines from ₹ 72 to 70 and therefore the marginal revenue is negative and is equal to –2.

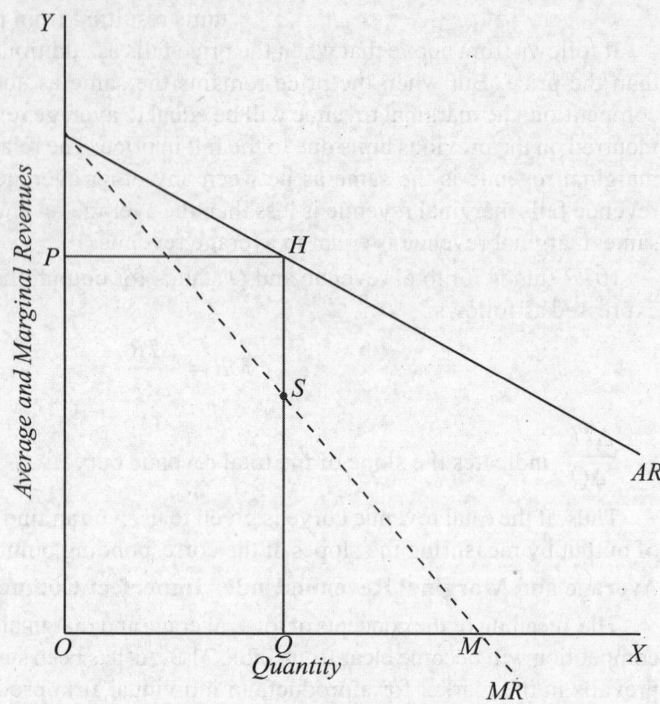

Fig. 21.1. *Average and Marginal Revenue Curves under Imperfect Competition*

It may be noted that in all forms of imperfect competition, that is, monopolistic competition, oligopoly and monopoly, average revenue curve facing an individual firm slopes downward as in all these market forms when a firm lowers the price of its product, its quantity demanded and sales would increase and vice versa.

The case, when average revenue (or price) falls when additional units of the product are sold in the market is graphically represented in Fig. 21.1. In Fig. 21.1 it will be observed that average revenue curve (*AR*) is falling downward and marginal revenue curve (*MR*) lies below it. The fact that *MR* curve is lying below *AR* curve indicates that marginal revenue declines more rapidly than average revenue. When *OQ* units of output are sold, *AR* is equal to *QH* or *OP* and *MR* is equal to *QS*. When *OM* units of the product are sold, marginal revenue is zero. If the quantity sold is increased beyond *OM*, marginal revenue becomes negative.

Average and Marginal Revenue uder Perfect Competition

When there prevails perfect competition in the market for a product, demand curve facing an individual firm is perfectly elastic and the price is beyond the control of a firm, average revenue remains constant. If the price or average revenue remains the same when more units of a product are sold, the marginal revenue will be equal to average revenue. This is so because if one more unit is sold and the price does not fall, the addition made to the total revenue by that unit will be equal to the price at which it is sold, since no loss in revenue is incurred on the previous units in this case. Consider the following table:

Table 21.4
Average and Marginal Revenues Under Perfect Competition

No. of Units Sold	*Price (or AR) (Rupees)*	*Total Revenue (Price × Output) (Rupees)*	*Marginal Revenue (MR) (Rupees)*
I	II	III	IV
1	16	16	16
2	16	32	16
3	16	48	16
4	16	64	16
5	16	80	16
6	16	96	16
7	16	112	16
8	16	128	16
9	16	144	16
10	16	160	16

In the above table, price remains constant at the level of ₹ 16 when more units of the product are sold. Col. III shows the total revenue when various quantities of the product are sold. Total revenue has been found out by multiplying the quantity sold by the price.

It will be found from taking out the difference between two successive total revenues that marginal revenue in this case is equal to the price *i.e.*, ₹ 16. Thus, when two units of the good are sold instead of one, the total revenue rises from ₹ 16 to ₹ 32, the addition made to the total revenue *i.e.* marginal revenue will be equal to ₹ 32 –16 = ₹ 16. Similarly, when three units of the product are sold, the total revenue increases to ₹ 48, and the marginal revenue will be equal to ₹ 48 –32 = ₹ 16 Likewise, it will be found for further units of the product sold that marginal revenue is equal to price. The case of perfect competition when for an individual firm average revenue (or price) remains constant and marginal revenue is equal to average revenue is graphically shown in Fig. 21.2 Average revenue curve in this case is a horizontal straight line (*i.e.*, parallel to the *X*-axis). Horizontal-straight-line average revenue curve (*AR*) indicates that price or average remains the same at *OP* level when quantity sold is increased. Marginal revenue (*MR*) curve coincides with average revenue (*AR*) curve since marginal revenue is equal to average revenue.

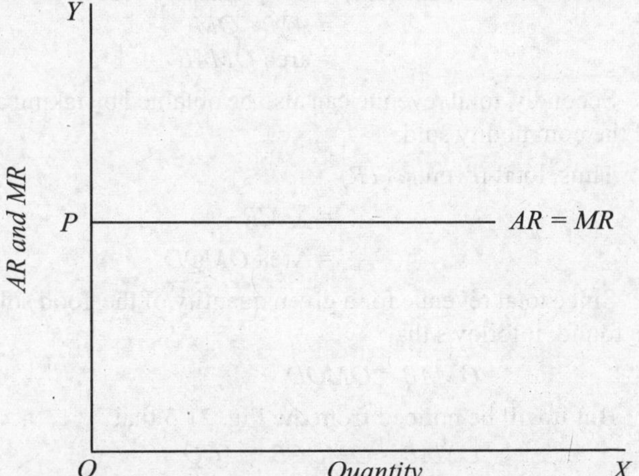

Fig. 21.2. *Average and Marginal Revenue Curves under Perfect Competition*

RELATIONSHIP BETWEEN AR AND MR CURVES

Given that both average revenue (*AR*) and Marginal revenue (*MR*) curves are of straight-line shape, it can be shown that (*MR*) curve will cut the distance between *AR* curve and the *Y*-axis in the middle. In other words, when both AR and MR curves are straight lines, then if a perpendicular is drawn from a point on the AR curve to the Y-axis, MR curve will cut this perpendicular at its middle point. Consider Fig. 21.3, where both AR and MR curves are straight lines. Point A is taken on the average revenue curve and a perpendicular AB is drawn to the Y-axis. MR curve cuts the perpendicular AB at point C. Now, if MR curve cuts halfway the distance between AR curve and the Y-axis, then AC must be equal to BC. So in order to show that MR cuts halfway the distance between AR and the Y-axis, we have to prove in Fig. 21.3 that $AC = BC$.

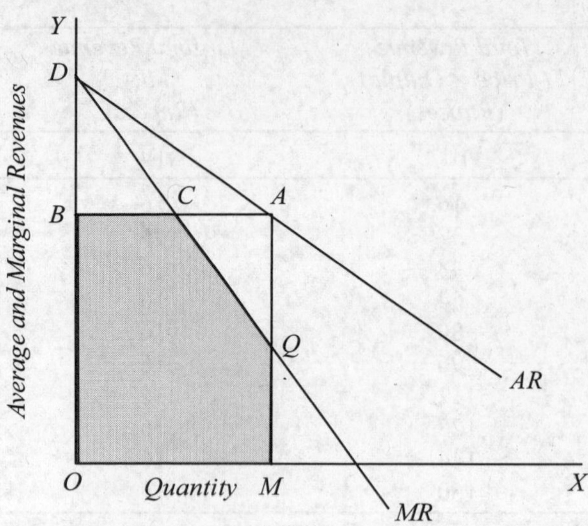

Fig. 21.3. *Relationship between AR and MR*

Draw a vertical straight line from A so as to meet the X-axis at M. It means that when OM quantity of the commodity is sold, average revenue is equal to AM. Now, there are two ways in which we can find out the total revenue earned by the sale of OM units of the commodity.

First, total revenue (TR) = AR × quantity sold
$= AM \times OM$
$=$ area $OMAB$ . . .(*i*)

Secondly, total revenue can also be obtained by taking a sum of marginal revenues of all the units of the commodity sold.

Thus, total revenue (TR)
$= \Sigma MR$
$=$ Area $OMQD$ . . .(*ii*)

Since total revenue for a given quantity of the good sold is to be the same whichever way it may be found, it follows that

$$OMAB = OMQD$$

But it will be noticed from the Fig. 21.3 that

$$OMAB = OMQCB + ACQ$$
$$OMQD = OMQCB + BDC$$

From above it follows that :

$$OMQCB + ACQ = OMQCB + BDC$$
$$ACQ = BDC$$

or

Thus triangles ACQ *and* BDC *are equal in area*

Now, in Δs ACQ and BDC

$QAC = \angle DBC$ (right angles)

$\angle ACO = \angle BCD$ (vertically opposite angles)

$\angle AQC = \angle BDC$ (alternate angles)

Therefore, ΔACQ and ΔBDC are similar.

We have proved above that triangles ACQ and BDC are equal in area as well as similar. Now, when the two triangles are both equal and similar, then they are congruent (*i.e.* equal in all respects).

Therefore Δs *ACQ* and *BDC* are congruent

Hence, $AC = BC$

It is thus proved that given the straight-line average and marginal revenue curves, the marginal revenue curve will lie half-way from the average revenue curve.

From above, we also learn the way of drawing *MR* curve corresponding to a given *AR* curve. If any *AR* curve is given to you, and you are asked to draw *MR* curve corresponding to it, then you should first extend the *AR* curve so that it meets the *Y*-axis (if it is not already so). After that you

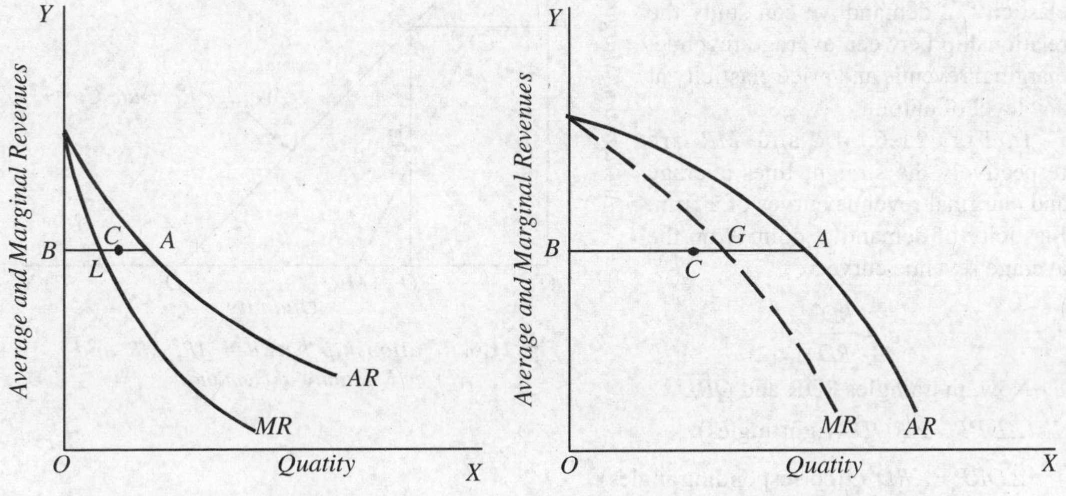

AR and MR curves when they are convex.

Fig. 21.4.

AR and MR curves when they are concave.

Fig. 21.5.

should draw *MR* curve starting from the *Y*-axis so that it *bisects* any perpendicular line drawn from a point on the *AR* curve to the *Y*-axis.

Marginal revenue curve corresponding to a convex or concave average revenue curve is not of straight-line shape but is either convex or concave to the origin. What relationship *MR* curve will bear to *AR* curve when average and marginal revenue curves are either convex or concave ? In either of these cases the marginal revenue curve will not lie halfway from the average revenue curve. If the average revenue curve is convex to the origin as in Fig. 21.4 the marginal revenue curve *MR* will also be convex to the origin and will cut any perpendicular drawn from *AR* curve to the *Y*-axis *more than halfway as* measured from the average revenue curve. On the other hand, if the average revenue curve is *concave* to the origin as in Fig. 21.5, the marginal revenue curve will also be concave and will cut any perpendicular line from the average revenue curve to the *Y*-axis *less than halfway* as measured from the average revenue curve.

In Figs. 21.4 and 21.5, *C* is the middle point on the perpendicular line *AB*.

AVERAGE REVENUE, MARGINAL REVENUE AND PRICE ELASTICITY OF DEMAND

There is a very useful relationship between elasticity of demand, average revenue and marg[illegible] revenue at any level of output. We will make use of this relation extensively when we come to the study of price determination under different market conditions. Let us study what this relation is.

We have stressed above that the average revenue curve of a firm is really the same thing as the demand curve of consumers for the firm's product. Therefore, elasticity of demand at any point on a consumer's demand curve is the same thing as the elasticity of demand on the given point on the firm's average revenue curve. We know that elasticity of demand at point R on the average revenue curve DD' in Fig. 21.6 $= \frac{RD'}{RD}$. With this measure of point elasticity of demand we can study the relationship between average revenue, marginal revenue and price elasticity at any level of output.

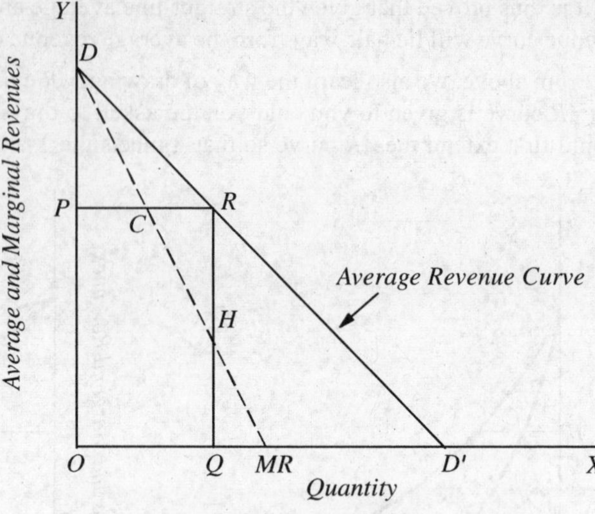

Fig. 21.6. *Relationship between AR, MR and Price Elasticity of Demand*

In Fig. 21.6, AR and MR are respectively the straight lines average and marginal revenue curves of a firm. Elasticity of demand at point R on the average revenue curve :

$$e_p = \frac{RD'}{RD}$$

Now, in traingles PDR and QRD'

$\angle DPR = \angle RQD'$ (right angles)

$\angle DRP = \angle RD'Q$ (corresponding angles)

Third $\angle PDR = \angle QRD'$

Therefore, triangles PDR and QRD' are equiangular

$$\text{Hence} \quad \frac{RD'}{RD} = \frac{RQ}{PD} \qquad \ldots (i)$$

In the triangles *PDC and CRH*

$PC = RC$

$\angle PCD = \angle RCH$ (vertically opposite angles)

$\angle DPC = \angle CRH$ (right angles)

Therefore, triangles PDC and CRH are congruent (*i.e.*, equal in all respects).

$$\text{Hence} \quad PD = RH \qquad \ldots (ii)$$

From (i) and (ii), we get

$$\text{Price elasticity at } R = \frac{RD'}{RD} = \frac{RQ}{PD} = \frac{RQ}{RH}$$

Now, it is seen from Fig. 21.6 that

$$\frac{RQ}{RH} = \frac{RQ}{RQ - HQ}$$

$$\text{Hence, price elasticity at point } R = \frac{RQ}{RQ - HQ}$$

It will be seen from Fig. 21.6 that RQ is the average revenue (AR) and HQ is the marginal revenue (MR) at the level of output OQ corresponding to point R on the demand or average revenue curve DD'. Therefore,

$$\text{Price Elasticity at } R = \frac{\text{Average Revenue}}{\text{Average Revenue} - \text{Marginal Revenue}}$$

If A stands for average revenue
M stands for marginal revenue
e stands for point elasticity on the average revenue curve,

then, $$e = \frac{A}{A-M}$$

$$eA - eM = A$$
$$eA - A = eM$$
$$A(e-1) = eM$$
$$A = \frac{eM}{e-1}$$

Hence, $$A = M\left(\frac{e}{e-1}\right)$$

And also, $$M = A\left(\frac{e-1}{e}\right)$$

Or, $$M = A\left(1-\frac{1}{e}\right)$$

Since price (P) equals average revenue (A) we have

$$M = P\left(1-\frac{1}{e}\right)$$

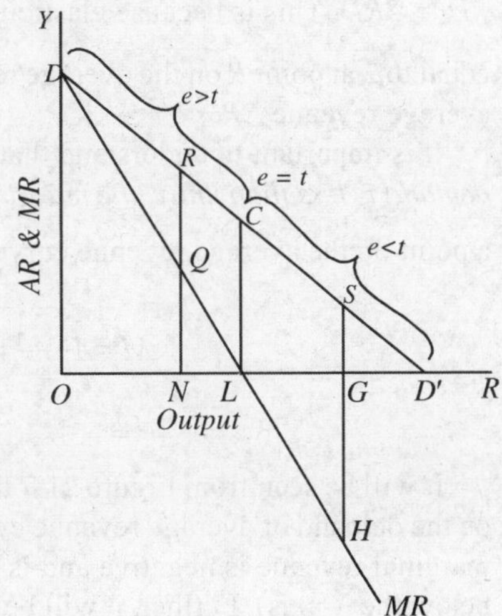

Fig. 21.7. *Average Revenue, Marginal Revenue and Price Elasticity of Demand*

where e stands for price elasticity of demand at a given point on the average revenue curve.

With the help of the above formulae we can find out marginal revenue at any level of output from average revenue at the same output provided we know the point price elasticity of demand on the average revenue curve. *If the price elasticity of a firm's average revenue curve at a given level of output is equal to one, marginal revenue equals zero.* This can be proved as under :

$$M = A\left(1-\frac{1}{e}\right)$$
$$= A\left(1-\frac{1}{1}\right)$$
$$= A \times 0$$
$$= 0$$

It will be seen from Fig. 21.7 that corresponding to the middle point C on the average revenue curve DD' where elasticity of demand equals unity, the marginal revenue is zero.

By applying the above formula it can be shown that at a point on the average revenue curve where elasticity of demand is greater than one, marginal revenue will be positive though less than the average revenue. *Thus when demand elasticity on a firm's average revenue curve is 2, the marginal revenue will be positive and will equal half the average revenue.* This is because

$$M = A\left(1-\frac{1}{e}\right)$$

$$= A\left(1-\frac{1}{2}\right)$$

$$= \frac{1}{2}A$$

If price elasticity of demand at a point on the average revenue curve equals 2, then the relevant point on the average revenue curve *DD′* in Fig. 21.7 is *R* corresponding to output *ON* such that *RD′* = 2*RD*. This is because elasticity at point $R = \frac{RD'}{RD} = 2$. Now, with elasticity of demand being equal to 2 at point *R* on the average revenue curve the marginal revenue *NQ* will be found half of the average revenue *NR*.

It is important to understand that *at a point on the average revenue curve at which elasticity of demand is less than unity, the marginal revenue will be negative.* For instance, suppose elasticity at a point on the average revenue curve is $\frac{1}{4}$. Then,

$$M = A\frac{\frac{1}{4}-1}{\frac{1}{4}} = A\left(-\frac{3}{4}\right)\times 4$$

$$= -3A$$

It will be seen from Figure 21.7 that corresponding to point *S* which lies below the middle point on the demand or average revenue curve *DD′* and therefore at which elasticity is less than unity, the marginal revenue is negative and is equal to *GH* (note that beyond output *OL*, the *MR* curve goes below the *X*-axis). Further, it will be noticed from Figure 21.7 that point *S* lies at such a position on the average revenue curve *DD′* so that $SD' = \frac{1}{4}SD$ and, therefore, $e_p = \frac{SD'}{SD} = \frac{1}{4}$.

Thus on measurement it will be found that corresponding to point *S* at which elasticity is equal to $\frac{1}{4}$ the marginal revenue *GH* is three times the average revenue *GS*.

To sum up, *marginal revenue is always positive at any point or output where the elasticity of the average revenue curve is greater than one and marginal revenue is always negative where the elasticity of average revenue curve is less than one and marginal revenue is zero corresponding to unit elasticity at the average revenue curve.*

THREE TYPES OF REVENUE (*AR, MR, TR*) AND PRICE ELASTICITY (*E*)

We are now in a position to describe the relationship between three types of revenue, namely, *AR*, *MR*, and *TR* on the one side and price elasticity of demand on the other. From the formula $MR = AR\left(\frac{e-1}{e}\right)$ we can know what would be the marginal revenue, if elasticity and *AR* are given to us. When *the elasticity is equal to one, it follows from the above formula that marginal revenue will be equal* to zero.

Thus
$$MR = AR\left(\frac{e-1}{e}\right)$$

$$MR = AR\left(\frac{1-1}{1}\right)$$

$$MR = AR \times 0 = 0$$

Likewise, it can be proved that,

If $e > 1$, MR is positive, and

if $e < 1$, MR is negative.

In a straight-line demand curve we know that the elasticity at the middle point is equal to one. It follows that marginal revenue corresponding to the middle point of the demand curve (or *AR* curve) will be equal to zero.

Consider Fig. 21.8. *C* is the middle point of the average revenue or demand curve *DD′*. At point *C* price elasticity is equal to one. Corresponding to *C* on the *AR* curve, marginal revenue will be zero. Thus *MR* curve is shown cutting the *X*-axis at point *N* which corresponds to point *C* on the *AR* curve. At a quantity greater than *ON* price elasticity on the demand curve, curve is less than one and the marginal revenue is negative. Marginal revenue being negative beyond *ON* means that total revenue will diminish if a quantity greater than *ON* is sold. Total revenue will be increasing upto *ON* output, since upto this marginal revenue remains positive. It follows therefore that total revenue will be maximum where elasticity is equal to one. Thus *TR* curve drawn in the bottom panel of Fig. 21.8 is shown to be at its highest level corresponding to the point *C* on *AR* curve or *ON* output where marginal revenue is zero and elasticity is equal to one.[9]

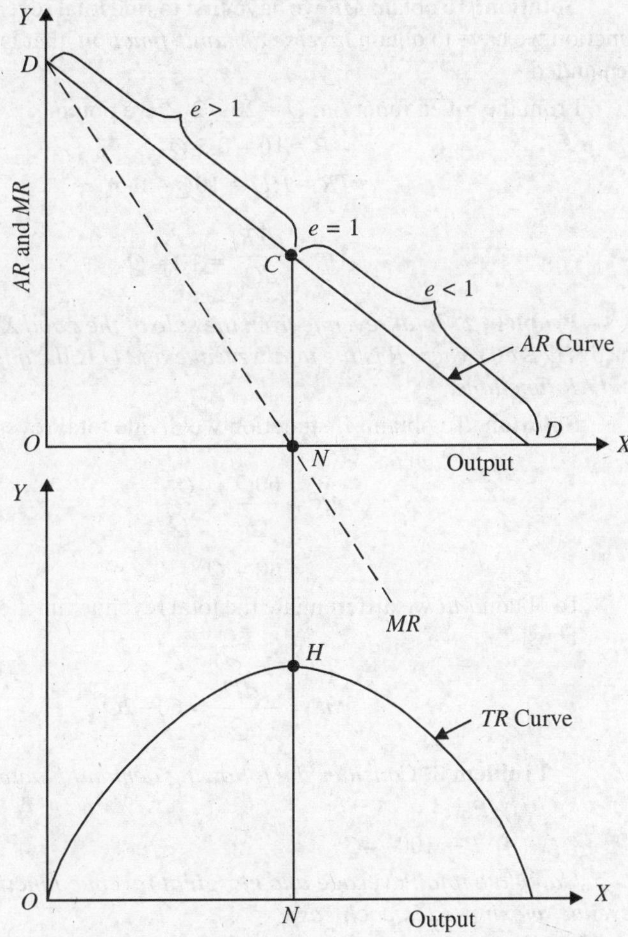

Fig. 21.8. *Relationship between AR, MR, TR and Price Elasticity*

We can also prove that total revenue is maximum corresponding to the unit elasticity point on the *AR* curve even without bringing in the marginal revenue. We know from the relationship between elasticity and total outlay (or total revenue) that the total revenue increases when elasticity is greater than one and total revenue diminishes when elasticity is less than one. Thus, in Fig. 21.8 beginning from point *D* on the average revenue or the demand curve *DD′* and coming down to the middle point *C*, elasticity remains greater than one and therefore the total revenue will go on increasing as we descend from point *D* to point *C* on the demand curve. Below point *C* on the demand or *AR* curve, elasticity is less than one, therefore the total revenue will start diminishing as we descend from point *C*

9. It should be noted that the *scale* on the *Y*-axis for *TR* curve is different from the scale used for *AR* and *MR* curves at the top panel.

downward. It, therefore, follows that corresponding to the middle point C on the demand curve where elasticity is equal to one, the total revenue will be maximum.

It will be seen from Fig. 21.8 that total revenue curve starts from O and goes on rising till it reaches its peak at point H. Then it starts declining till it meets point D' on the X-axis. It means that at output OD' total revenue is zero. This is because at output OD' average revenue or price is zero.

NUMERICAL PROBLEMS

Problem 1. *Suppose demand function facing a firm is $Q = 20 - 2P$ where Q is output demanded and P is price. Find the marginal revenue function.*

Solution: To obtain MR we have first to find total revenue function. For obtaining total revenue function we have to obtain *inverse demand function*, that is, price expressed as a function of output demanded.

From the given function, $Q = 20 - 2P$, we obtain

$$P = 10 - 0.5Q$$

$$TR = P.Q = 10Q - 0.5Q^2$$

$$MR = \frac{dTR}{dQ} = 10 - Q$$

Problem 2. *Total revenue from the sale of the good X is given by the equation $R = 60Q - Q^2$ for $0 \le Q \le 60$ (where R is the total revenue and Q is the quantity bought at price P). Calculate MR and AR functions.*

Solution: To obtain AR function we divide total revenue function by output. Thus

$$AR = \frac{60Q}{Q} - \frac{Q^2}{Q}$$

$$= 60 - Q$$

To obtain MR we differentiate the total revenue function ($R = 60Q - Q^2$) with respect to output. Thus,

$$MR = \frac{dTR}{dQ} = 60 - 2Q$$

Problem 3. *Consider the following demand functions :*

(*i*) $P = 4$

(*ii*) $P = 100 - 4Q$

Calculate total, average and marginal revenue functions for each one. At what point is total revenue maximised in each case. *(D.U.B.A (Hons.) 1997)*

Solution :

(*i*) $P = 4$

$$TR = P.Q = 4Q$$

$$AR = \frac{PQ}{Q} = P = 4$$

$$MR = \frac{dTR}{dQ} = \frac{d(4Q)}{dQ} = 4$$

Note that in this case MR equals AR.

(*ii*) $P = 100 - 4Q$

$TR = P.Q = 100Q - 4Q^2$

$$AR = \frac{P.Q}{Q} = 100 - 4Q$$

$$MR = \frac{dTR}{dQ} = 100 - 8Q$$

TR is maximum at output level where $MR = 0$

Thus, setting *MR* function equal to zero we have

$100 - 8Q = 0$

$8Q = 100$

$$Q = \frac{100}{8} = 12.5$$

Thus, at output level 12.5, *TR* will be maximum.

Problem 4. *The demand function equation faced by the firm HP for its personal computers is given by*

$$P = 50{,}000 - 4Q$$

1. *Write the total revenue function and marginal revenue equation.*
2. *At what price and quantity marginal revenue will be zero?*
3. *At what price and quantity total revenue will be maximized?*

Solution. (1) To obtain total revenue function, we multiply the given inverse demand equation by *Q*.

$$TR = PQ = 50{,}000Q - 4Q^2 \quad \text{..... (1)}$$

Taking the first derivative of the total revenue function would give us marginal revenue equation. Thus, seting *MR* function equal to zero we have :

$$MR = \frac{d(TR)}{dQ} = \frac{d(PQ)}{dQ} = 50{,}000 - 8Q \quad \text{..... (2)}$$

(2) Setting *MR* equation equal to zero would give us the quantity at which *MR* is zero. Thus, setting *MR* function equal to zero we have :

$$50{,}000 - 8Q = 0$$

$$8Q = 50{,}000$$

$$Q = 6250$$

Now, to obtain price associated with the quantity (6250), we substitute the value of *Q* in the inverse demand equation ($P = 50{,}000 - 4Q$). Thus.

$$P = 50{,}000 - (4 \times 6250)$$

$$= 50{,}000 - 25{,}000 = 25{,}000$$

It follows from above that at the quantity of 6250 units and price of ₹ 25,000 per computer, marginal revenue is zero.

(3) To obtain quantity at which *TR* is maximum, we set the *first derivative of TR function* (*i*), that is, *MR* equal to zero.

$$TR = 50.000Q - 4Q^2$$

$$\frac{d(TR)}{dQ} = MR = 50{,}000 - 8Q$$

Setting *MR* function equal to zero we have :

$$50{,}000 - 8Q = 0$$
$$8Q = 50{,}000$$
$$Q = 6250$$

Thus, at the quantity of 6250 units the total revenue is maximum.

Important Conclusion. It is clear from the answers to the questions (2) and (3) of this problem that the *quantity at which MR is equal to zero, total revenue is maximum*. This is because, as stated above, it follows from the very definition of marginal revenue. Marginal revenue is the increase in total revenue resulting from the sale of an additional unit of output. As long as marginal revenue is positive, total revenue is increasing and beyond the output level where total revenue is maximum, marginal revenue becomes negative and, therefore, total revenue declines. Thus, at the output level where marginal revenue is zero, total revenue is maximum.

Problem 5. *Outside an airport, a shopkeeper is running a coffee shop. The demand function for coffee cups is*

$$Q = 150 - 10P$$

where Q is the quantity demanded of coffee cups and P is price per coffee cup.

1. Write total revenue function and determine at what quantity of coffee cups sold and price fixed, total revenue is maximised.

2. Show that at the quantity sold where total revenue is maximised, price elasticity of demand is equal to unity.

Solution. 1. The given demand function is

$$Q = 150 - 10P \quad \text{...... }(i)$$

In order to determine the total revenue, we have to first obtain *inverse demand function*, that is, demand expressed in terms of price as a function of quantity demanded. Rearranging the demand equation (*i*) above we have

$$10P = 150 - Q$$

$$P = 15 - \frac{Q}{10} \quad \text{...... }(ii)$$

Now,

$$TR = PQ$$

$$= Q\left(15 - \frac{Q}{10}\right)$$

Thus total revenue function is : $TR = 15Q - \dfrac{Q^2}{10}$ ..... (*iii*)

Taking the first derivative of the *TR* function (Note that the first derivation of *TR* function measures *MR*). We have :

$$\frac{d(TR)}{dQ} = MR = 15 - \frac{2}{10}Q = 15 - \frac{1}{5}Q$$

In order to find out quantity sold at which *TR* is maximised, we set first derivative of *TR* i.e., *MR* function, equal to zero.Thus

$$15 - \frac{1}{5}Q = 0$$

$$\frac{1}{5}Q = 15$$

$$Q = 75$$

Substituting the value of Q into the demand equation (*ii*)

$$P = 15 - \frac{1}{10}.75 = 15 - 7.5$$

$$= ₹\,7.5$$

Thus, the quantity at which total revenue is maximized is 75 and associated price is ₹ 7.5 per coffee cup.

2. Price elasticity of demand $(e_p) = \frac{dQ}{dP} \cdot \frac{P}{Q}$. From the demand function equation (1) we find that the coefficient of P is 10 which measures $\frac{dQ}{dP} = 10$. Further, we have found that at 75 units of output, TR is maximum. Thus, price elasticity of demand at TR-maximising quantity is equal to one. Substituting these values into the price elasticity formula.

$$(e_p) = \frac{dQ}{dP} \cdot \frac{P}{Q} = 10 \times \frac{7.5}{75} = 1$$

QUESTIONS FOR REVIEW

1. Define market. What are the essentials of a market? Does market necessarily mean a particular price where goods and services are bought and sold?
2. Into what main categories market structures are classified? Explain the basis of such classification.
3. Differentiate between perfect competition, monopoly and monopolistic competition. Briefly explain their features with regard to
   (1) the number of firms
   (2) the nature of product being produced
   (3) freedom to enter the industry
   (4) control over the price of the product.
4. What is meant by cross elasticity of demand? Explain how it is used as a basis for classifying various market structures.
5. Explain the concepts of average revenue, marginal revenue and total revenue. Show how they are related to each other.
6. Show that in case of linear demand curve marginal revenue curve cuts *half-way the distance* between the average revenue curve and the Y-axis. Prove geometrically.
7. Explain how average revenue and marginal revenue are related to each other through price elasticity of demand. Give a geometric proof of this relation.
8. Show how marginal revenue and total revenue depend on the price elasticity of demand on a demand curve. Represent this relationship graphically.
9. Explain and illustrate with diagrams the nature of average and marginal revenue curves of a firm working under :
   (*a*) Perfect competition
   (*b*) Monopoly
   (*c*) Monopolistic Competition
10. Define marginal revenue. How is it related to (*a*) average revenue, and (*b*) total revenue?
11. Complete the following table.

Price (₹)	*Output* (*Units*)	*Total Revenue* ₹	*Marginal Revenue* ₹	*Price Elasticity Demand*
6	0			
5	1			
4	2			
2	4			
1	5			
0	6			

12. Why is marginal revenue curve of a firm under perfect competition identical with its average revenue curve? Explain.

13. When average revenue curve of a firm is falling, marginal revenue curve lies below it. Why?

14. When marginal revenue is zero, total revenue is maximum. Why ? Explain.

15. *If AR curve is rectangular hyperbola, draw the corresponding MR curve. Comment on the nature of price elasticity of demand on such a curve:* *D.U.B.Com. (H), 2000*

[**Hints:** *AR* curve being rectangular hyperbola shows that total area which represents total revenue or total expenditure (*P.Q*) remains constant as price of the commodity changes. Total revenue remaining constant with changes in price implies that price elasticity of demand is equal to one. Now,

$$MR = P\left(1 - \frac{1}{e}\right) = P\left(1 - \frac{1}{1}\right)$$

$$= P(0) = 0$$

Thus, *MR* curve will overlap the *X*-axis throughout].

CHAPTER 22

OBJECTIVES AND EQUILIBRIUM OF THE FIRM: A GENERAL ANALYSIS

In last chapter we have discussed the various forms of market in which a producer or firm may sell his product. Analysis of equilibrium of the firm and the industry under various market forms occupies an important place in economic theory. The theory of price with which we are concerned in this book is primarily an analysis of equilibrium of the firm and the industry under various market forms. When different firms are producing differentiated products, it is difficult to define an industry and the analysis of equilibrium of the industry under such conditions is full of conceptual difficulties. When different firms are producing differentiated products, each would have its separate demand and supply of its particular product. Therefore, in this case we cannot sum up the demand and supply of the various firms producing differentiated products to obtain the supply and the demand for the industry. *It was in connection with the industry composed of various firms producing homogeneous, undifferentiated products that the concepts of supply and demand were forged by Marshall.*

When various firms are producing same or homogeneous products, it is possible to identify an industry and the supply and the demand for the product of that industry can be ascertained. But when different firms are producing differentiated (but similar) products, it is not easy to identify an industry having its own supply and demand. Prof. Chamberlin in his concept of monopolistic competition where various firms produce differentiated but similar products has called the collection of these firms as *'group'* rather than industry. Moreover, it is because of this difficulty of viewing supply and demand for 'group' of firms producing differentiated products and also because of greater importance of behaviour of individual firms having control over their own products that in recent years emphasis has been shifted from the equilibrium of the industry to the equilibrium of the firm. However, equilibrium of the industry under conditions of perfect competition, where various firms produce homogeneous products, retains its importance and usefulness.

Meaning of Firm's Equilibrium

Word 'equilibrium' means a state of balance. When two opposing forces working on an object are in balance so that the object has no tendency to change, it is said to be in equilibrium. In other words, when the object under the pressure of forces working in opposite directions has no tendency to move in either direction, the object is in equilibrium. Thus, by consumer's equilibrium we mean that in regard to the allocation of money expenditure among various goods the consumer has reached the state where he has no tendency to reallocate his money expenditure. Similarly, *a firm is said to be* in *equilibrium when it has no tendency to change its level of output, that is, when it has no tendency either to increase or to contract its level of output.* The firm will produce the equilibrium level of output and will charge the price at which the equilibrium output can be sold in the market. In fact, in a previous chapter, we discussed the equilibrium of the firm in respect of the combination of factors it has to employ for producing a given level of output. But the problem for a producer is not simply what combination of factors it has to choose for producing a given level of output but it has to go further to decide what level of output it should produce. Thus, the equilibrium of the firm is generally conceived in respect of output it has to produce. In this regard, to repeat what has been just said above, the firm is in equilibrium, when, given certain demand and cost conditions, it has struck upon the level of output at which it will stick on and will have no tendency to change it.

OBJECTIVES OF THE FIRM

Before analysing conditions of equilibrium of the firm, we shall first explain the objectives of the firm because only when the firm is achieving its objective it will have no tendency to change the level of its output, that is, it will be in equilibrium position. Like that of a consumer the entrepreneur or firm is assumed to behave in a rational manner. The rationality on the part of a firm is generally understood to imply that a firm attempts to maximise its profits. Until a few years back, maximisation of profits was considered to be the valid objective of a firm. However, in recent years apart from profit maximisation, several other objectives of the firm have been pointed out by several economists. The various alternative objectives of the firm that have been stressed by different economists are:—

1. Profit Maximisation
2. Security Motive or Maximisation of Profits in the Long Run.
3. Profit Satisficing
4. Maximisation of Utility
5. Maximisation of Utility
6. Maximisation of Growth

We explain below the above objectives of the firm.

Profit Maximization Objective

Profit maximisation objective about the behaviour of the firm is one of the most fundamental assumptions of economic theory. The attempt of the entrepreneur to maximise his profits is regarded as a rational behaviour. It has been said that as the rationality on the part of the consumer means that he tries to maximize his satisfaction, the rationality on the part of the entrepreneur implies that he tries to maximise his profits.

It should be carefully noted as to what the entrepreneur is supposed to maximise under profit maximisation objective. An entrepreneur's income consists of two elements. Firstly, he gets wages for his work of routine management and supervision which he is supposed to pay to himself and include in his regular cost calculations. Thus, the total costs of output comprise not only the costs incurred on the hired factors by the entrepreneur but also the entrepreneur's own wages of routine management and supervision. When we say that the entrepreneur tries to maximise the difference between total revenue and total costs, these total costs include entrepreneur's wages of management and supervision. Thus the entrepreneur's wages of management do not form part of that income which the entrepreneur has to maximise. Secondly, the entrepreneur gets what is left after meeting all costs (including his own wages of routine management). This surplus of total revenue over total cost which is called his residual income. This residual income is the profits which is his true or net profits that the entrepreneur is assumed to maximise. Thus, we see that the entrepreneur's income comprises his own wages of routine management and the residual income which accrues to him.

Marshall called entrepreneur's wages of management and supervision as normal profits, and the residual income as super-normal profits.This dichotomy of entrepreneur's income is very fundamental to the theory of firm. *The normal profits are the minimum income which the entrepreneur must get in order to stay in a business or industry*. As has been just said above, the normal profits are included in costs and do not come under the maximising problem. It is the supernormal profits, *i.e*, true or pure profits which is the residual income which the entrepreneur aims at maximising. These *true or pure profits are of the nature of economic rent since they are over and above the normal profits which must be paid to the entrepreneur to make him to remain in the industry or business.* As has been just said above, *the normal profits are included* in costs and do not come under the maximising problem. It is the supernormal profits, i.e., true or pure profits which is the residual income which the entreperneur aims at maximising problem. These are called *true or pure profits as they over and above the normal profits and are reward for entrepreneurs for bearing risk and uncertainty* involved in production of goods and services. Another important thing to note is that in the traditional theory of the firm maximisation of pure profits is considered to be *maximisation of short-run profits*. This short run is generally considered to be a year or few years only.

A CRITIQUE OF PROFIT-MAXIMIZATION OBJECTIVE

The profit maximization objective has been subjected to severe criticisms in recent years. It has been pointed out that all firms do not maximise profits. Further, some economists have asserted that apart from making profits, firms in the real world also try to achieve other objectives. Some other alternative objectives have been pointed out by prominent economists. First, Prof. Rothschild[1] has asserted that entrepreneurs try to achieve a steady flow of profits in the long run, subject to a certain reasonable level of profits. Prof. Baumol[2] has put forward the view that firms try to *maximise sales i.e.*, the money value of the sales. Prof. Scitovsky, Reder and B. Higgins have asserted that the owner of the firms or entrepreneurs try to *maximise their utility or satisfaction*. Williamson[3] has argued that in case of corporate business firms, managers maximise their own utility function rather than maximising profits for shareholders who are the owners of the firm. On the other hand, H.A. Simon[4] and Cyert and March express the view that rather than maximise any thing the firms justs *satisfice*, that is, want satisfactory performance regarding profits made, market share and sales revenue. We explain below various criticisms levelled against profit maximisation and alternative objectives proposed.

Achieving a Steady Flow of Secure Profits in the Long Run

Prof. Rothschild maintains that maximization of profits may be the valid objective to be pursued by the firms working under conditions of perfect competition and monopolistic competition when a large number of firms are competing with each other to sell a product and under monopoly when a firm controls the supply of a product which has no close substitutes. Under these market conditions, firms do not feel insecure in making profits over a long period as they do not have to face effective competition and do not expect their profits will be competed away by the actions of their *existing* rivals. But under conditions of oligopoly where the firms are quite interdependent and face a lot of uncertainty regarding the activities of their few rivals, the firms desire to obtain steady flow of profits in the long run profits. This is because in deciding about his price and output policies the entrepreneur does not maximise his profits at a particular time or for a particular period of time but tries to have a *steady flow of profits* over a long period of time. Therefore, this has also been called the security objective.

It is generally conceded that under conditions of oligopoly the firms strive for maximising profits over a long period of time rather than maximum short-run profits.

Hall and Hitch's Mark-up Pricing Approach: Seeking Normal Profits

Further, reference may also be made to the empirical study made by Oxford economists, Hall and Hitch, who interviewed some thirty-eight entrepreneurs on price policy. From their empirical study Professors Hall and Hitch concluded that the businessmen did not try to maximise profits. They also concluded from their study that businessmen charged prices that cover their average cost of production and they add profit mark-up to this average cost to fix the price of their products. According to this

1. R.W. Rothschild, Price Theory and Oligopoly, *Economic Journal*, 1947 pp. 229-320.
2. W.J. Baumol: *Business Behaviour, Value and Growth,* pp. 47-48.
3. B.E. Williamson, *The Economics of Discretionary Behaviour. Managerial Objectives in a Theory of the Firm*, Marsham Publishing Co. Chicago, 1967.
4. Important works by H.A. Simon, in which he has developed his behavioural theory of the firm are : (1) A Behavioural Model of Rational Choice, *Quarterly Journal of Economics.* Feb. 1955.(2) Theories of Decision Making in Economics and Behavioural Sciences, *American Economic Review*. June 1959. (3) *Models of Men,* Wiley, New York, 1957.

principle, businessmen do not seek abnormal profits, that is, more than conventional profits which are considered as reasonable. Thus the mark-up practice of which Hall and Hitch found in their enquiry, is claimed to be opposed to the principle of profit maximisation. It may however be pointed out that the market situation in which businessmen of Hall and Hitch enquiry were placed was one of monopolistic competition with an admixture of oligopoly elements. In such a market situation, the desire to obtain secure profits in the long run greatly governs the businessmen in charging prices of their products. If they try to earn large economic profits by charging high prices, new firms will invade their field. Thus in a market situation where the obstacles for new comers to enter the field are very small and as a result the businessmen already in the field fear that new comers will enter the industry, they will not seek to maximise economic profits.

It is thus asserted that the practice of mark-up pricing contradicts the hypothesis of profit maximisation. However, it may be noted that a relevant question in this regard is what will determine the profit mark-up on the basis of which price will be fixed. On the face of it, it appears that since in this mark-up pricing principle demand is not taken into account in determining the profit mark-up it cannot possibly lead to the maximisation of profits. However, in our view this profit mark-up is not a fixed magnitude but varies depending on the price elasticity of demand or the intensity of competition in the market. In actual practice, businessmen determine this profit mark-up keeping in view the price elasticity of demand for their product or the intensity of competition from rival products. Empirical studies made in the USA confirm this varying profit mark-up in case of different products. For example, in an empirical study made for pricing by the U.S. Steel Corporation, it was found that profit mark-up or margins fixed in case of steel rails was relatively high because this was the product in which US steel faced little competition. On the other hand, profit mark-up fixed in case of stainless steel and tin plates was low because for them competition from aluminium and lumber products were quite strong. We thus see that mark-up principle of price fixation can be consistent with the objective of profit maximisation.

Sales Maximization

Prof. Baumol has also challenged the objective of profit maximization. He has argued that maximisation of sales rather than of profits is the ultimate objective of the firm. He says that the firm tries to promote sales not merely as a means to further its other objectives, namely, operational efficiency and profits, but for businessman "sales have become an end of themselves." He, therefore, thinks that sales maximization is the most valid assumption about the behaviour of the firm. By sales he means the total revenue earned by selling the product.

Prof. Baumol thinks that empirical evidence for his hypothesis that sales rank ahead of profits as the main object of the oligopolist's concern is quite strong. He says, "Surely it is common experience that when one asks an executive,. "How's business?' he will answer that his sales have been increasing (or decreasing), and talk about profits only as an after-thought, if at all."[5] Thus Prof. Baumol very strongly believes that sales maximisation has become the ultimate objective of the firms and therefore they direct their energies in promoting and maximising sales instead of profits.

But Prof. Baumol softens his sales maximisation hypothesis by pointing out that in their attempt to promote sales businessmen do not completely disregard cost incurred on output, and profits to be made. He also concedes that there is some conflict between the firm's sales goal and its profit objective. He points out that in the actual word, businessmen usually promote sales subject to the limitation that costs incurred are covered plus a usual rate of return on investment is earned. According to him, "management is not concerned to obtain profits higher than this. Once this minimum profit level is achieved, sales rather than profits become the overriding goal."[5] Thus Prof. Baumol asserts that "the typical oligopolist's objective can usefully be characterized approximately as sales maximization subject to minimum profit constraint. Doubtless this premise overspecifies a rather vague set of attitude but I believe it is not too far from truth. So long as profits are high enough to keep

5. *Ibid*., p. 49.

stockholders satisfied and contribute adequately to the financing of company growth, management will bend its efforts to the augmentation of sales revenue rather than to further increase its profits".[6]

It may be objected that maximisation of sales instead of profits means irrational behaviour of the entrepreneur. But Prof. Baumol rightly points out that his hypothesis in no way conflicts with the assumption of rationality. He presents a different conception of rationality which is more scientific . According to him, rationality does not consist in choosing the ends, it only means *pursuing the ends efficiently and consistently*. He says, "People's objectives are whatever they are. Irrationality surely must be defined to consist in decision patterns which make it more difficult to attain one's own ends that are for some reason considered to be right. Unless we are prepared to determine other people's values, or unless they pursue incompatible objectives, *we must class behaviour as rational if it efficiently pursues whatever goals happen to have been chosen*",[7] Thus he thinks that given the sales-maximisation as objective, the entrepreneur will be rational if he works most efficiently and consistently towards maximising his sales.

Utility Maximization

Since satisfaction or utility is the ultimate end which an individual aspires to get, some economists have pointed out that entrepreneur-owners and managers of joint stock companies try to maximise their utility rather than money profits. The objective of utility maximisation has been discussed in the context of two types of firms: First, in case of firms owned and managed by the entrepreneur himself, the utility maximization implies that in choosing an output level, *the entrepreneur-owner not only considers the money profits which he will make but also the sacrifice of leisure which he would have to make in doing the necessary activity for producing that level of output*. Second, in case of large firms owned and organised in the form of Joint Stock Companies where there exists separation of management from ownership. The utility function of managers or business executives of these Joint Stock Companies includes not only the profits which they earn for shareholders but also the promotion of sales, maintaining lavish offices, seeking to have a larger member of staff under their supervision *etc*. In this case manager will maximise his utility by attaining a best combination of profits and of the above-mentioned other objectives. We discuss below utility maximisation in both these cases.

Utility Maximization by Entrepreneur-Owner. It has been pointed out by some economists as Higgins, Reder and Scitovsky that profit maximisation does not necessarily mean utility or satisfaction maximization. If the entrepreneur is supposed to maximise his utility, then not only the satisfaction which he gets from material goods which are obtained with the money profits earned from putting in entrepreneurial activity or work, but also the satisfaction which he obtains from the leisure at his disposal. The leisure or what Hicks calls 'quiet life' is an essential ingredient of an individual's welfare. But more activity or work put in by the entrepreneur, the less leisure he will be able to enjoy. The preference for leisure must, therefore, be incorporated into the analysis of an entrepreneur who is supposed to maximise his satisfaction or utility.

We now proceed to show that as long as we do not make a special assumption about the entrepreneur's attitude or behaviour regarding work and leisure, maximisation of profits will not ensure maximum utility or satisfaction. Let us draw the entrepreneur's indifference curves between money profits and leisure. In Fig. 22.1 money profits are measured on *Y*-axis and leisure (from left to right) is measured on the *X*-axis. An indifference curve in such a diagram will represent the various combinations of money profits and leisure which will give the entrepreneur equal amount of satisfaction. The higher the level of such an indifference curve, the greater the level of entrepreneur's satisfaction or utility.

Further starting from point *W* and moving towards left entrepreneurial activity (that is, time spent to produce output) is measured. The greater the entreprenoural activity, the greater the output and larger the profits accrued to the owner-entrepreneur. However, after *WM* entrepreneurial activity has

6. *Ibid*., pp. 49-50 (italics added).
7. Baumol, *op.cit*.

been put in, the total profits start declining and therefore profit curve *PC* slopes downward. It will be seen from Fig. 22.1 that the entre-preneur's utility or satisfaction is maximum at point *S* where indifference curve IC_2 is tangent to the profit curve *PC*. At point *S* he is doing *WL* entrepreneurial activity and enjoying *OL* leisure. However, it will be seen from Fig. 22.1 that profits are maximised when he puts in larger entrepreneurial activity *WM* which results in lower amount of leisure *OM*.

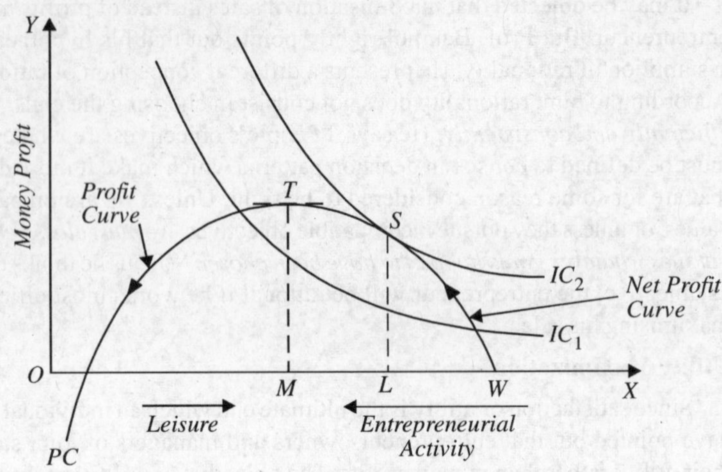

Fig. 22.1 *Utility Maximisation vs. Profit Maximisation*

Growth of Managerial Capitalist and Objectives of the Firm

With the growth of managerial capitalism, the two developments have taken place which have been the basis of criticism of profit maximisation objective. The traditional assumption of profit maximisation implies that a firm was owned by an entrepreneur himself and therefore it was perfectly rational for him to maximise profits. Now, in large corporate firms it is managers who make business decisions while the corporate firm is owned by shareholders. In other words, there is separation between **control** and **ownership**. It has been asserted that managers who make business decisions may not be interested in maximising profits; instead they may try to achieve their own goals or objectives which may not be in the interests of owners *i.e.* shareholders. This situation has been described as **managerial capitalism**. Prof. Nellis and Parker rightly write "With managers in control it is easy to question the validity of the profit maximisation assumption of traditional theory. Some managers may seek to keep shareholders happy by reporting a certain level of profit while leaving themselves the flexibility to achieve, perhaps personal, objectives (such as business growth, diversification, salary, etc)".[8]

The second development on the basis of which the validity of profit maximisation has been questioned is the emergence of oligopoly as the most common form of market structure. In the traditional theory the firms which were assumed to work under condition of perfect competition and monopoly, they were assumed to work *independently* and possess full and accurate information about demand and cost conditions. They could therefore easily determine profit maximisation by equating marginal cost with marginal revenue. However, the assumption of perfect maximisation fails to provide a satisfactory explanation of decision making under oligopoly because in this type of market structure (1) the firms are quite **interdependent** and (2) there exists a **lot of uncertainly** about demand and cost conditions. Mutual interdependence arises in oligopoly because there are a few firms in this market structure and each of them produces a sufficiently large proportion of the industry's output so that its price-output decisions affect the market share of its rival firms which are expected to retaliate. Uncertainty exists because under oligopoly form of market structure a firm's decision is influenced not just by what its competitors are doing but also by what it thinks its rivals might do in reponse to its initiative regarding changes in price, quantity of output, product variation and advertising.

8. Joseph G. Nellis and David Parker, *The Essence of Business Economics*, Prentice Hall of India 2002, p. 102

In the context of development of managerial capitalism and the emergence of oligopoly as the chief form of market structure, Baumol put forward the view that managers, of firms, maximise sales value (*i.e.* total revenue) rather than profits, O.E. Williamson[9] laid stress on the view that managers or business executives of large corporate firms are motivated by self-interest and therefore they **maximise their own utility function** rather than profits for the shareholders. Still another view has been presented by Marris[10] according to whom the managers try to **maximise growth of their companies** which raise their status, power and prestige rather than maximising profits which are mostly bagged by shareholders.

In addition to the above altrenative maximising objectives in place of profit maximisation, there are two other theories of the firm which emphasize that managers or firms **do not maximise any thing but pursue non-maximising goals.** There are mainly two non-maximising approaches to firm's behaviour. First on the basis of empirical study two Oxford economists professors Hall and Hitch expressed the view that businessmen do not maximise profits but charge price according to what is called **mark-up principle** to achieve normal profits. According to this principle, firms calculate average cost on the basis of expected output or sales of the product and add to it, (*i.e.* mark up) a normal profit margin. In this way they prevent the entry of new firms in the industry which enable them to earn a steady flow of profits over time.

The second approach to non-maximising behaviour is the behaviourist theory initially put forward by Professor H.A. Simon[11] but was further developed by R.M Cyert and J.G. March.[12] According to this behaviourist approach managers of the firms **do not try to maximise any thing,** whether it is profits, sales value, utility or growth. They just pursue the goal of **satisficing.** According to this, they try to achieve only a **satisfactory performance** with regard to profits, sales, market share. We shall explain these approaches in detail later. We explain below alternative objectives of firm which have been proposed by some economists.

Utility Maximisation by Managers of Corporate Business Firms

According to O.E. Williamson, managers or business executives of large firms are motivated by self-interest and they maximise their own utility function. Willianson argued that managers of large firms have enough discretion to pursue those polices which increase their personal utility. Utility function of managers include their salaries, the number of staff under their control, lavishly furnished office, *discretional* non-essential investment expenditure. However, the objective of utility maximisation by managers is subject to the constraint that after-tax profits are large enough to pay acceptable dividends to shareholders and also to pay for necessary investment expenditure. According to Willianson, utility maximisation by self-seeking managers depend on the following factors.

1. **Salaries.** The higher the salaries and other forms of monetary compensation which the managers receive from the business firms, the greater the utility they have. The high salaries ensure them high standard of living and high status.
2. **Staff under their control.** The utility desired by managers also depend upon the number of staff under their control. The greater the number of staff under the control of a manager, the higher his status, power and prestige.
3. **Managerial Slack.** Mananger utility also depends on what Williamson calls 'managerial slack' which consists of non-essentical eapenditure and includes such benefits such as lavishly furnished office, a luxurious company car, free air travel.

---

9. Oliver E. Williamson, *The Economies of Discretionary Behaviour*; *Managerial Objecties of the Firm*, Prentice-Hall, 1964
10. R. Marris, A Model of Managerial Enterprise, *Quarterly Tournal of Economics*, 1963.
11. H.A. Simon, A Behavioural Model of Rational Choice, *Quarterly Journal of Economics*, Feb-1955.
12..R.M. Cyert and J.C. March, *A Behavioural Theory of the Firm*, Prentice Hall, 1963.

4. **The Discretionary Investment Expenditure**- This includes expenditure which manager can spend according to his discrection. This discretionary expenditure is over and above the essential investment expenditure which is necessary for the growth of the firm.

The utility function of the managers and the factors on which it depends can be written as under:

$$U = f(S, N, M, I_d)$$

where, U = utility of a manager,

S = the salary and other forms of monetary compensation which a manager obtains from a business firm,

N = the number of staff under the control of a manager,

M = **management slack** which means the amount of non-essential expenditure by management such as lavishaly furnished offices, luxurious company car, large expense accounts, etc.

I_d = the amount of discretionary non-essential investment expendiutre by the manager.

Thus the managers maximise the above utility function, that is, the composite utility derived from the above mentioned four factors. However, as mentioned above, the objective of utility maximization by a manager is subject to the constraint that after-tax profits are large enough to pay acceptable dividends to the shareholders and also to pay for economically necessary investments (as discretionary investment expenditure by the managers).

Growth Maximisation

According to another important theory the managers of corporate firms try to **maximise the growth rate of their firms** rather than maximisation of profits. This theory was put forward by the Cambridge economist Robert Marris[13] in 1960s, Prof. Marris also considers the case of market structure in which competition is limited. Besides, he is concerned with the behaviour of the corporate firm where management is separated from ownership so that there is ample scope for managerial discretionary behaviour. Prof. Marris regards the corporator firm as a typically bureaucratic organisation where corporate growth and the security associated with it is a desirable goal. According to him, the goal of the corporate manager is to achieve balanced rate of growth of the firm which requires maximising rate of growth of demand for the products on the one hand and rate of growth of capital supply for increase in investment on the other.

Rationale for Maximising Growth of the Firm. Now an important question is why managers seek to maximise the balanced growth rate of the firm, that is, why do they jointly maximise the rate of growth of demand for firm's products and the growth rate of capital supply. This is because by doing so they maximise their own utility function and the utility function of their owners. Before Marris, it was generally argued by the management theorists that the goals of the manager and the owner often cash because the utility functions which they try to maximise greatly differ. The utility function which managers seek to maximise include variables such as salaries, status, job security. On the other hand, utility function which owners seek to maximise include variables such as profits, capital supply, size of output, market share and image or reputation in the public.

According to Marris, despite the difference in the variables in the utility functions of managers and owners, the most of the variables included in both of them are positively correlated with a single variable, namely, the growth rate of the firm. Further, according to him, the growth of the firm may be measured by the increase in the level of output, capital supply, sales revenue or market share. However, Marris regards steady balanced growth rate over time as the objective of the managers because most of the variables such as sales, output, capital supply, included in their utility function increase

13. R. Marris, A Model of Managerial Enterprise, *Quarterly Journal of Economics* (1963). Also R. Marris, *Theory of Managerial Capitalism*, Macmillan, 1964

simultaneously so that maximising long-run growth of any variable amounts to maximising long-run growth of others.

It is evident from above that Marris thinks that corporate managers would recognise the relationship between the growth of the company with profits being ploughed back into investment for expansion of production capacity on the hand and their own personal goals (such as increased status, power and salary) on the other. Besides, according to Marris, managers try to balance growth against the impact of their decision on profits and dividends. They are expected to be aware of the risk of low dividends lowering share prices which put the firms vulnerable to being taken over by the rival companies. Thus, according to Marris, minimisation of risk requires Prudent approach in deciding about investment and ralsing of capital. Prof. Nellis and Parker[14] rightly state that there may be trade off between securing profits to pay dividends and taking risk when investing to increase the growth of the firm at the same time while profits provide the retained earnings to help finance. New investment which leads to growth of excessive company liquidity may attract predators. Cash rich companies attract takeover bids In Marris's model this conflict is summarised as management seeking "**optimal dividend-to-profit retention ratio**".

Satisficing Behaviour

According to the satisficing hypothesis, corporate manageras aim at satisficing rather than maximising profits. We shall explain this approach in detail in chapter 32. The advocates of this hypothesis say that a corporate manager sets for himself a minimum standard for performance or what is called the aspiration level. Once this satisfactory performance according to this *aspiration level* is achieved, the firm will slack off. This approach is known as using behavioural theory and is concerned with how firms *actually behave*.

H.A. Simon[15], one of the pioneers of the behavioural approach to the theory of the firm, points out that most psychological theories assume that instead of maximising, rational men normally *satisfice.* Applying this to the business decisions of the firm, he suggests that instead of maximising *profits,* firms aim at *satisficing,,* that is, want to achieve *satisfactory* performance with regard to rate *of profit*, market share, sales revenue, grwoth, etc. Simon has further postulated that a firm has normally an *'aspiration level'*. An aspiration level of a firm is based on its goal as well as its past experience, and in fixing it uncertainties are duly taken into account. If the actual performance of the firm reveals that a given aspiration level can be easily achieved, it will be revised upward. On the other hand, if it is found that a given aspiration level is difficult to be achieved, it will be lowered. Simon points out that when the actual performance of a firm falls short of an aspiration level, 'search' activity is started to find out the ways of better performance in the future and therefore achieving the aspiration level. But, according to Simon, there is a limit to 'searching activities' which a firm will undertake because for searching activities such as obtaining of information firms have to incur cost. And therefore the gain from search activity must be balanced against its cost. That is why if searching activities relatively cost more, aspiration level is adjusted downward to a level which is more likely to be achieved. Since the firm limits its searching activity on account of its cost, it does not maximize profits. Therefore, the firms behave rationally when they aim at *'satisficing'* rather than 'maximising'.

According to an other prominent satisficing theory put forward by Cyert and March[16], in these days of large-scale corporate type of a business firm, we can no longer consider it as a *single major*

14. Joseph G. Nelli's and David Parker, The *Essence of Business Economics*, Prentice Hall, New Delhi 2001 p.112.
15. Important works by H.A. Simon, in which he has developed his behavioural theory of the firm are: (1) A Behavioural Model of Rational Choice, *Quarterly Journal of Economics*. Feb. 1955. (2) Theories of Decision Making in Economics and Behavioural Sciences, *American Economic Review.* June 1959. (3) *Models of Men,* Wiley, New York l957.
16. R.M. Cyert and J.C. March, A *Behavioural Theory of the Firm,* Prentice-Hall, 1963.

decision maker (i.e., the entrepreneur), but instead we should look at it as complex group or complex organisation composed of various individuals whose interests may conflict with each other. Cyert and March call this complex organisation or group as organisational *coalition* which may include managers, stockholders, workers, customers and so on. They assert that all of these different individuals participate in setting the goals of the organisation. Another argument for satisficing behaviour on the part of the corporate managers advanced is that top management serves as trustees of the organisation which has a responsibility not only to shareholders but also to employers, customers, creditors, suppliers etc. Thus, corporate managers pursuing a satisficing goal strike a statesman-like balance among the claims of shareholders for dividends and higher share price, the demands of employees for higher wages, the pressures from consumers for lower prices and better quality products.

Thus, the hypothesis of satisficing behaviour implies that instead of maximizing profits for owners, corporate managers strive for attaining satisfactory perormance with regard to profit, sales revenue market share, growth of profit. The satisficing model of behaviour of the corporate managers rightly stresses that the problem of decision making in large firms, especially in oligopolistic environment, is quite complex as they have to reconcile the interests of various pressure groups in the organisation. But the major problem with the satisficing hypothesis is that it does not provide a clear definition of *satisfactory rate of profits.* A number of standards of profits which may be considered as satisfactory may be mentioned. Thus, on the one hand, a satisfactory rate of profit may be that which is *high enough* to attract outside capital on a sustained basis. On the other hand, the firms may fix their satisfactory rate of profit at a *low level* so as to prevent the entry of new firms which may offer a strong competition and erode their profits. The satisfactory rate of profits may also be fixed at a low level in order to prevent the government control and regulation. Thus the standard of satisfactory profit may vary a good deal depending on the nature of competition and environment in which a particular firm may find itself. The satisficing model, therefore, does not provide us any general guideline for determination of satisfactory rate of profits for the fixation of output and price.

It may be noted that a noted American economist, Prof. J.K. Galbraith[17] who has made an indepth study of big modern corporations has found that managers, which he calls technostructure, *pursue multiple goals* in which along with sales maximisation and utility maximisation, the objective of achieving the highest possible growth of output is paramount. Further, Galbraith points out that managers of the big business corporations make every effort to increase their prestige, market power and technical superiority. In his view, the corporate technocrats who are highly skilled persons are able to pursue these multiple goals as they can greatly influence the consumers through effective advertising on a large scale. Prof. Galbraith further points out that the highly salaried managers or technocrats of the modern business corporations desire to have an easy life for themselves and try to avoid risk and for that purpose extensive business planning is done by them for taking appropriate decisions.

Case for the Objective of Maximising Profits

We have explained above the various alternatives to profit maximization objective. However, the various alternatives to profit maximisation are not free from drawbacks and no comprehensive theory of the firm has been developed on the basis of non-profit maximising assumption so that till now the theory of the firm based on profits-maximizing behaviour dominates the economic theory. Several reasons can be given in favour of the assumption of profit-maximizing behaviour of the firms.

In the first place there is a question of survivorship. The firm which is working in a very competitive environment if it does not maximise profits it will run the risk of not being able to survive in the long run. Thus, profit maximising is quite a rational behaviour in the fields where intense competition prevails. It may appear under certain circumstances that the firms are not maximising profits, but they

17. J.K. Galbraith, *The New Industrial* State

may be doing so only for the short run. For instance, the firms working in oligopolistic or monopolistic market structures do not maximise profits in the short run in order to prevent the potential competitors to enter the industry. Under these circumstances, the firms try to maximise profits in the long run. Similarly, some other goals such as maximising the growth of output, sales maximisation increasing the market share are only the means to achieve maximum profits in the long-run and therefore from the long-run point of view, they are not inconsistent with the goal of profit maximisation.

In defence of profit maximisation hypothesis, it may also be noted that the managers are not permanent in a firm and are likely to be changed by the owners (shareholders in the corporate firms) if they feel that managers are not providing them adequate return or profits on their investment. Thus, given the fact that managers are liable to be changed, if they deviate much from profit maximisation, they will not be allowed to continue for long in the firm. Of course, if control over management is absent, the managers may continue to behave in a non-profit maximising manner. However, if the managers of corporate firms are not maximising profits in the long-run, the prices of its shares will fall greatly and it may be taken over by others who will change the current management and install a new team of managers who are efficient and try to maximise profits in the long run.

Finally, it may be said that no model, nor its assumptions can be completely realistic. Models are built and assumptions are made so as to bring out the crucial aspects and relations of the economic phenomena. For this purpose we need not fully take into account the massive and confusing details of the real world. We must abstract from reality to draw purposeful conclusions which can adequately explain the economic phenomenon. This is also true of the profit maximisation assumption. The profit maximisation may not truly and exactly reflects the behaviour of the managers in the real world, but on the basis of the profit maximisation assumption, correct predictions regarding determination of prices and outputs of commodities have been made. In this regard we may refer again to the viewpoint of Friedman[16] who has argued that the *ultimate test of the validity of an assumption is its capability to predict correctly*; the assumption itself may be unrealistic. Defending the assumption of profit maximisation on these grounds he writes, "unless the behaviour of businessmen in some way or other is approximated behaviour consistent with the maximisation of returns, it seems unlikely that they would remain in business for long."[11] He points out that profit maximisation assumption is valid because predictions regarding changes in prices and output based on it have been shown to be correct. .

EQUILIBRIUM OF THE FIRM: MAXIMISING PROFITS

As stated above, *a firm is said to be in equilibrium when it has no tendency either to increase or to contract its output*. Since we are assuming that the firm aims at maximizing its profits, it will, therefore, be in equilibrium when it is making maximum money profits. In order to simplify our analysis we also assume that our firm produces a single product. It is true that a firm in the real world may produce more than one product and our assumption of single product firm may, therefore, be unrealistic. But the assumption of multi-product firm which seems to be more realistic will not, given the assumption of profit maximization, involve any significant modification in the method or results of our analysis. It is to make our analysis simple that we are making the assumption of a single product firm.

It should be noted that in the present chapter we are concerned with the analysis of equilibrium of the firm in general terms. We shall explain the equilibrium of the firm with reference to specific market forms, namely, perfect competition, monopoly, monopolistic competition when we take up their separate detailed study. Here we shall derive general conditions of equilibrium which are valid under all types of market.

There are two ways of explaining how a firm reaches its equilibrium position by maximising profits. In the first method we use the concepts of total cost and total revenue. In the second method, which is generally used in modern economic theory, we use the concepts of marginal revenue and marginal cost to explain firm's maximising behaviour. We explain below the equilibrium of the firm in both these ways.

Firm's Equilibrium: Total Revenue and Total Cost Approach

A firm will go on increasing its output if its profits are thereby increasing. It will fix its output at the level where it is making maximum money profits. The profits is the difference between total revenue (*TR*) and total cost (*TC*). Therefore, a firm will maximise its profits at a level of output where the difference between total revenue and total cost is the largest. Consider Table 22.1 which shows the changes in total revenue and total cost when it raises its output from one to 10 units.

It will be seen from the table that when firm is producing 2 units of output, its total revenue is ₹. 90 and total cost is ₹. 80. This yields profits of ₹. 10. Now when the firm raises its output level to 7 units, its profits go up to ₹. 108. Increase in output beyond 7 units will lower profits, and further if it raises output beyond 8 the losses will accrue to the firm. It is therefore clear that profits are maximum when the firm produces 7 units of output. Thus the firm will be in equilibrium by producing 7 units of output.

Table 22.1. Firm's Equilibrium: Maximising Profits

Output *Q*	*Total Revenue* *TR*	*Total Cost* *TC*	*Total Profits* *TR–TC*
1	45	50	–5
2	90	80	+10
3	135	90	+45
4	180	105	+75
5	225	130	+95
6	270	165	+105
7	315	207	+108
8	350	270	+80
9	375	385	–10
10	395	505	–110

Fig. 22.2 portrays what is called break-even chart by businessmen. In this are shown total revenue curve *TR* and total cost curve *TC*. Total revenue curve *TR* starts from the origin which means that when no output is produced the revenue is zero. Total revenue goes on increasing as more output is produced. However, it will be noticed that total cost curve *TC* starts from a point *F* which lies above the origin. In other words, it is assumed that even when there is no production the firm has to incur some costs equal to *OF*. For instance, when the firm has to stop production in the short run, it has to bear the fixed costs. Thus our Fig. 22.2 depicts short-run total revenue and total cost curves of the firm. As a firm starts from zero output and increases its production of the good, in the very initial stages total cost is greater than total revenue and the firm is not making any profits at all. When it is producing *OL* level of output, total revenue just equals total cost and the firm is therefore, making neither profits nor losses, that is, the firm is only breaking even. Thus the point *S* corresponding to *OL* output is called *Brak-Even Point.*

When the firm increases its output beyond *OL*, total revenue becomes larger than total cost and profits begin to accrue to the firm. It will be seen from the figure that profits are rising as the firm increases production to output *OM*, since the distance between the total revenue curve (*TR*) and total cost curve (*TC*) is widening. At *OM* level of output the distance between the *TR* curve and *TC* curve is the greatest and, therefore, the profits will be maximum. Thus the firm will be in equilibrium at the *OM* level of output. The firm will not produce any output larger than *OM* since after it the gap between *TR* and *TC* curves goes on narrowing down and therefore the total profits will be declining.

At *OH* level of output *TR* and *TC* curves again intersect each other, which means that total revenue is equal to total cost at output *OH*. Thus point *K* (corresponding to *OH* output) is again a break-even point. Beyond output *OH*, total revenue is less than total cost and the firm will make losses if it produces any output larger than *OH*.

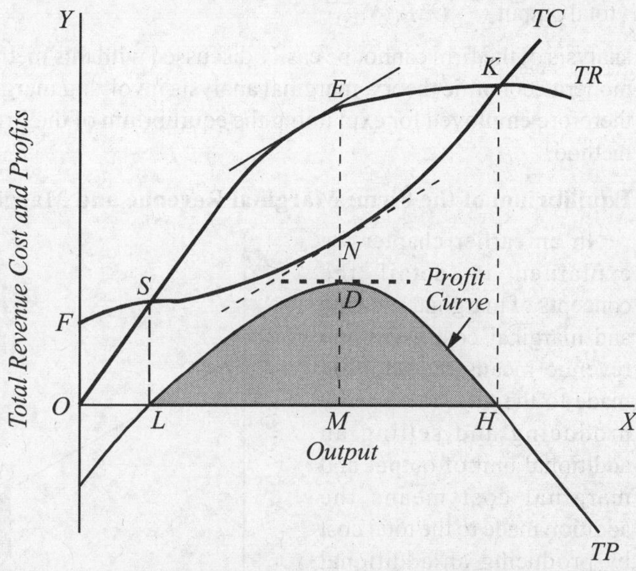

Fig. 22.2. *Profit Maximisation by a Firm*

From above it is clear that the firm will be in equilibrium at *OM* level of output where the total revenue exceeds total cost by the largest amount and hence profits are maximum. Now the question is how to locate this profit-maximising output level. By vision it is not easy to locate where exactly is the largest distance between *TR* and *TC* curves. In order to do so we have to draw tangents at the various points on the *TR* and *TC* curves. Where the tangents at the corresponding points on the *TR* and *TC* curves are parallel to each other, as is shown in Fig. 22.2 by tangents at points *E* and *N* on *TR* and *TC* curves respectively, the distance between *TR* and *TC* curves will be the largest and hence profits the maximum.

Another way to find out the profit-maximizing output is to draw directly total profit curve showing the difference between total revenue and total cost at various levels of output. In Fig. 22.2, *TP* is such a total profit curve which indicates the distance between revenue and total total cost at various levels of output. The level of output at which this profit curve stands highest from the *X*-axis will be the profit maximising level of output.

It will be noticed from Fig. 22.2 that total profit curve *TP* lies below the *X*-axis upto point *L* which shows that the firm is making negative profits (*i.e.*, losses) upto *OL* level of output. At *L*, the profit curve cuts the *X*-axis indicating that at output *OL*, the profits are equal to zero. As the firm increases its output beyond *OL*, profits curve is rising which indicates that total revenue and total profits are increasing. At output *OM*, profit curve stands highest from the *X*-axis and beyond *OM* the profit curve slopes downward indicating that total profit are declining when output is raised beyond *OM*. It may also be pointed out that tangent drawn to point *D* on the profit curve (corresponding to output *OM*) will be parallel to the *X*-axis which indicates the greatest distance between the profit curve and the *X*-axis at output *OM*. It follows that at output *OM* the firm will be making maximum profits and hence will be in equilibrium position. The profits earned at *OM* level of output are equal to *NE* or *MD*.

This method of finding out the profit-maximizing level of output by curves of total revenue and total cost seems to be quite reasonable and is also often employed by businessmen too, but it has some limitations. In the first place, the longest vertical distance between total revenue and total cost curves is difficult to find out at a glance. Many tangents have to be drawn before one finds the corresponding tangents to two curves to be parallel to each other indicating the level of output which yields maximum money profits. Of course, when profit curve is also drawn, it is relatively less difficult to locate the maximum profit point, because output corresponding to the highest point of profit curve is profit-maximising output. Secondly in this method, price per unit of output cannot be known at first sight from the diagram since price is not directly shown in the diagram. In order to know the price we have to divide the total revenue at the profit-maximising point by the total output. Thus in Fig. 22.2 at maximum profit output *OM* the total revenue is *ME*. The price charged by the firm will be equal to $\frac{\text{total revenue}}{\text{total output}} = \frac{ME}{OM}$ in Figure 22.2. With these limitations, complicated problems of the equilibrium analysis of the firm cannot be easily discussed with this method of showing equilibrium of the firm. In modern economic theory, marginal analysis involving marginal cost and marginal revenue curves are therefore employed for explaining the equilibrium of the firm. We now turn to explain this alternative method.

Equilibrium of the Firm: Marginal Revenue and Marginal Cost Approach

In an earlier chapter we explained in detail the concepts of marginal revenue and marginal cost. Marginal revenue means the addition made to the total revenue by producing and selling an additional unit of output and marginal cost means the addition made to the total cost by producing an additional unit of output. Now, a firm will go on expanding its level of output so long as an extra unit of output adds more to revenue than to cost, since it will be profitable to do so. The firm will not produce an additional unit of the product which adds more to cost than to revenue because to produce that unit will mean losses. In other words, it will pay the firm to go on producing additional units of output so long as the marginal revenue exceeds marginal cost. The firm will be increasing its total profits by increasing its output to the level at which marginal revenue just equals marginal cost. It will not be profitable for the firm to produce a unit of output for which marginal cost is greater than marginal revenue.

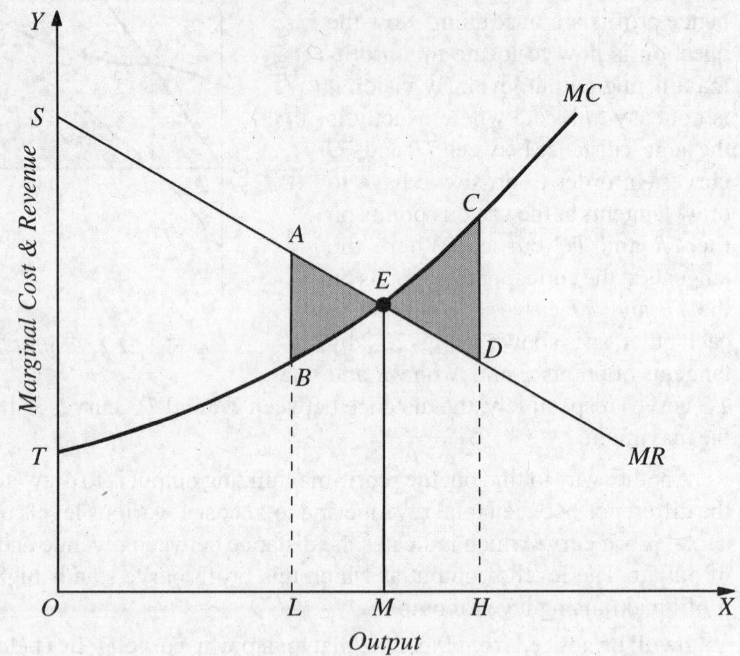

Fig. 22.3. *Equilibrium of the Firm: Profit-Maximising Output*

The firm will be making maximum profits by expanding output to the level where marginal revenue is equal to marginal cost. If it goes beyond the point of equality between marginal revenue and marginal cost, it will be incurring losses on the extra units of output and therefore will be reducing its total profits. Thus, *the firm will be in equilibrium position when it is producing the amount of output at which marginal revenue equals marginal cost*. It will be earning maximum profits at the point of equality between marginal revenue and marginal cost. Therefore, the condition for the equilibrium of the firm is that marginal revenue should be equal to marginal cost or *MR* = *MC*.

The whole argument can be better understood with the aid of Fig. 22.3 which depicts hypothetical marginal revenue and marginal cost curves of the firm. In this Fig. 22.3 firm's marginal revenue curve *MR* is sloping downward and firm's marginal cost curve *MC* is sloping upward and they intersect each other at point *E* which corresponds to output *OM*. Up to *OM* level of output, marginal revenue exceeds marginal cost and at *OM* the two are just equal to each other. The firm will be maximising its profits by producing *OM* output. The total profits will be less if it produces less than or more than *OM*. For instance, if the firm produces *OL* level of output (which is less than *OM*), its total profits will be less than at *OM*, because by producing *OL* it will be forgoing the opportunity to earn more profits which it can if it raises output to *OM*. This is so because additional units between *L* and *M* add more to revenue than to cost (*i.e.*, their *MR* is greater than *MC*) and it will therefore be profitable for the firm to produce them. The extra units between *L* and *M* can give to the firm extra profits equal to the area *ABE* which it would be forgoing if it produces *OL* output.

It is now clear that at any output level which is smaller than *OM*, the profits will be smaller than at *OM*. Likewise, profits will be smaller if the firm produces more than *OM*. Thus, at the greater output *OH*, profits will be less than at *OM*. This is so because extra units beyond *OM* add more to cost than to revenue (*i.e.*, their marginal cost is greater than their marginal revenue) and therefore the firm will be incurring a loss on these extra units with the result that its total profits will be reduced to that extent. On the units from *M*th to *H*th, the firm will be incurring a loss equal to the area *CDE* and total profits made by producing *OH* output will be equal to area *STE* minus area *CDE*. It is thus clear that total profits at *OH* output will be less than at *OM* output.

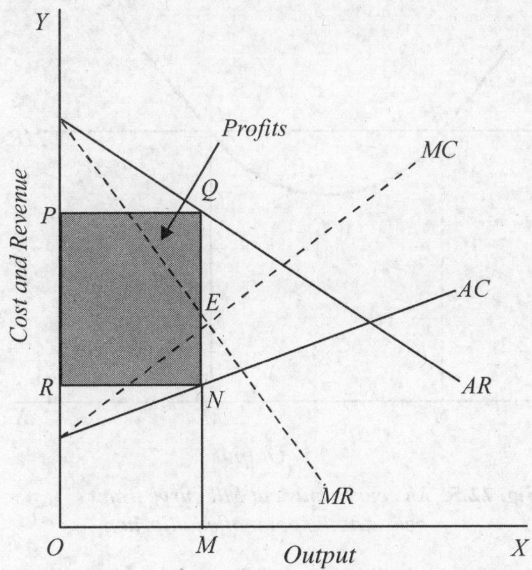

Fig. 22.4. *Equilibrium of the Firm: Equilibrium Ouput, Price and Profits*

To conclude, the firm will be making maximum profits and will therefore be in equilibrium at the level of output at which marginal revenue is equal to marginal cost, or where the marginal revenue and marginal cost curves intersect each other. The amount of total profits earned by the firm in its equilibrium position at *OM* output will be equal to the area *STE*. A point about profits is worth noting. It will be seen from Fig. 22.3 that as output is increased from zero tó point *M*, the gap between *MR* and *MC* curves is narrowing down till it disappears to *OM* output. Since gap or distance between *MR* and *MC* indicates the amount of profit earned on the additional unit of output, it means therefore the marginal profit earned on successive units of output goes on declining and is equal to zero at *OM*. Thus at *OM* output, marginal profit on the *M*th unit is zero but total proits will be maximum. It means that by expanding output to *OM* level, the firm has availed of the whole opportunity of making profits and therefore its total profits are maximum at *OM*.

In Fig. 22.3 only marginal revenue and marginal cost curves are shown and price cannot be directly read or known from this figure. In order to read price directly from the figure, it is customary to draw corresponding average revenue and average cost curves along with the marginal curves. This is exhibited in Fig. 22.4. The firm is in equilibrium at output *OM* at which marginal revenue and marginal cost curves cut each other. It will be seen from *AR* curve that *OM* output yields *MQ* (=*OP*) as average revenue or, in other words *OM* output can be sold at price *MQ*(=*OP*). Thus, we can directly read the price at which the firm sells its output from this figure. In Fig. 22.4 the total profits made by the firm can also be represented and known in a different way from that of Fig. 22.3.

Total profits = Total Revenue – Total Cost
= *AR* (or Price) × Output – *AC* × Output

In Fig. 22.4 *AR* or price = *MQ*, output = *OM*, average cost or *AC* = *MN*

Total profits $= MQ \times OM - MN \times OM$
= area $OMQP$ – area $OMNR$
= area $RNQP$

Second Order Condition for Equilibrium of the Firm

The first order condition for the equilibrium output of the firm is that its marginal revenue should equal marginal cost. There is a second order condition which must also be fulfilled if the firm is to be in a stable equilibrium position. Thus, *equality between marginal revenue and marginal cost is a necessary but not a sufficient condition of firm's equilibrium. The second order condition requires that for a firm to be in equilibrium marginal cost curve must cut marginal revenue curve from below at the point of equilibrium.* If at point of equality between *MR* and *MC*, the *MC* curve is cutting *MR* curve from above, then beyond this equality point, *MC* would be lower than *MR* and it will be profitable for the firm to expand output beyond this equality point. It is thus clear that the output at which marginal revenue is equal to marginal cost but marginal cost curve is cutting marginal revenue curve from above cannot be a position of equilibrium because the firm will have a tendency to increase its output further in spite of the equality between marginal revenue and marginal cost. For instance, consider Fig. 22.5. Here *MR* curve is a horizontal straight line and *MC* curve is U-shaped and is cutting *MR* curve at two points, *F*

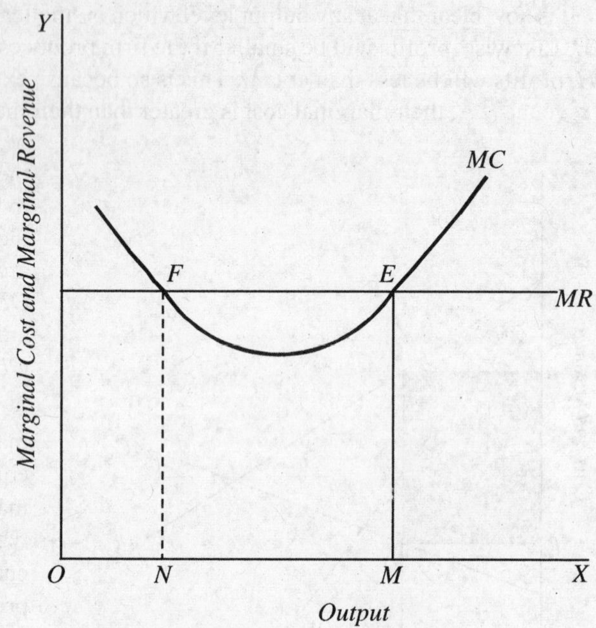

Fig. 22.5. *MC curve mus cut MR curve from below at the point of equilibrium*

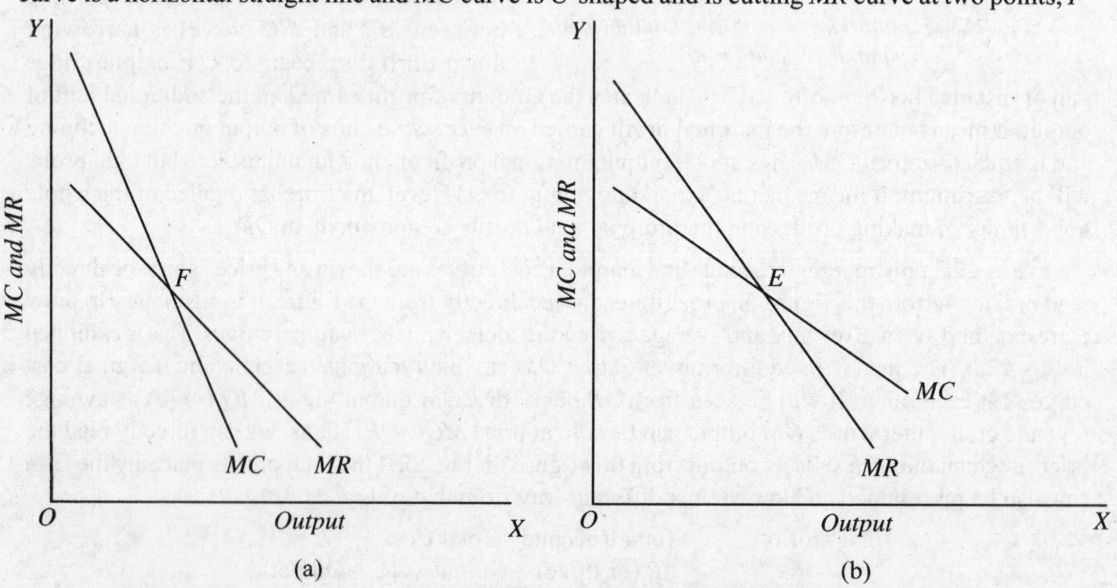

Fig. 22.6 (a) *Firm is not in equilibrium at F*

Fig. 22.6 (b) *Firm is in equilibrium at E*

and *E*. The firm cannot be in equilibrium at point *F* (or output *ON*) at which *MC* is equal to *MR*. This is because *MC* curve is cutting *MR* curve from above at point *F* corresponding to *ON* output with the result that beyond *ON* output *MC* is lower than *MR* and it is therefore profitable for the firm to expand output beyond *ON*. Actually, at *ON* output the firm is making losses equal to the area between *MC* curve and *MR* curve (*MC* curve lies above *MR* curve up to *F*). It is thus clear that despite the fact that marginal revenue and marginal cost are equal at *ON*, the firm is not in equilibrium since it is profitable for it to expand further.

At *OM* output (or point *E*) where marginal cost and marginal revenue are equal and also marginal cost curve is cutting marginal revenue curve *from below*, the firm will be in equilibrium position. This is so because beyond *OM*, marginal cost curve lies above marginal revenue curve and therefore it will not be worthwhile to produce more than *OM*. The firm will not stop short of *OM* output, for it can increase its profits by expanding output up to *OM*. To conclude, in Fig. 22.5 the firm is in equilibrium at *OM* output (point *E*) and not at output *ON* (point *F*).

Similarly, point *F* in Fig. 22.6(*a*) cannot be the position of equilibrium though *MC* equals *MR*. This is because at *F*, *MC* curve is cutting *MR* curve from above with the result that *MC* is lower than *MR* after *F*. The firm will go on expanding output beyond *F* since the additional units will add more the revenue than to cost. In the situation depicted in Fig. 22.6(*a*) there is no determinate position of firm's equilibrium. In this figure determinate position of equilibrium can only be established if marginal cost curve begins to rise so that it then cuts *MR* curve from below at some point. It should, however, be noted that point *E* in Fig. 22.6(*b*) denotes position of equilibrium of the firm. This is because *MC* curve is cutting *MR* curve from below at *E* with the result that *MC* is higher than *MR* after point *E*. It will not therefore be worthwhile to produce more output than that indicated by point *E*.

Another way in which we can state this second order condition of firm's equilibrium is:

For a firm to be in equilibrium, for the output greater than the equilibrium output, MC > MR, and for output less than the equilibrium output, MC < MR.

To sum up, for a firm to be in equilibrium, the following two conditions must be satisfied.

(1) *MR* = *MC*;

(2) *MC curve must cut MR curve from below at the point of equilibrium.*

NUMERICAL PROBLEMS

Problem 1. *Suppose a firm is facing the following demand function:*

$P = 100 - 2Q$

Firm's AC = MC = 20. Determine the equilibrium levels of firm's output, price and profits.

Solution:

We multiply the given inverse demand function by output (Q) to obtain total revenue (TR).

Thus $TR = P.Q = 100\,Q - 2Q^2$

$$MR = \frac{d(TR)}{dQ} = 100 - 4Q$$

Equating *MR* with *MC* we have

$$100 - 4Q = 20$$
$$4Q = 100 - 20 = 80$$
$$Q = {}^{80}\!/_{4} = 20 \quad \text{.... (i)}$$

To obtain equilibrium price we substitute the value of $Q = 20$ in the demand function. Thus

$$P = 100 - 2\,(20)$$
$$P = 60 \quad \text{...(ii)}$$

Now, profits, $\pi = TR - TC$

$$= (100Q - 2Q^2) - AC \times Q$$

Substituting the value of $Q = 20$ and the given $AC = 20$ we have

$$\pi = (2000 - 800) - (20 \times 20)$$
$$= 1200 - 400 = 800$$

Problem 2. *Given the demand curve $Q = 12 - 2P$, find the output and price that will maximise total revenue.* *D.U. B.A. (Hons.) 1994*

Solution: *TR is maximised at the level of output where $MR = 0$.* We first find the *inverse demand function* to obtain the total revenue from the given demand function.

$$Q = 12 - 2P$$
$$P = 6 - 0.5\,Q \text{ (inverse demand function)} \qquad ...(i)$$

Multiplying both sides of (*i*) by Q we have

$$TR = P.Q = 6Q - 0.5\,Q^2$$
$$MR = \frac{dTR}{dQ} = 6 - Q$$

Setting MR function (*i.e.* the first derivative of total revenue function) equal to zero we have

$$6 - Q = 0$$
$$Q = 6$$

At output level of 6 units, total revenue will be maximum

Substituting the value of $Q = 6$ in (*i*) we have

$$P = 6 - 0.5 \times 6$$
$$= 6 - 3 = 3$$

Thus, ₹. 3 per unit is the revenue-maximising price of the product.

Problem 3. *A firm faces demand curve given by $q = 100 - 2P$. Marginal and average costs for the firm are constant at ₹. 10 per unit. What output levels should the firm produce to (i) maximise profits and (ii) maximise slaes revenue? What are the respective profits at each output level.*

Solution: The given demand function is $q = 100 - 2P$

We first convert this demand function into inverse demand function. Thus

$$2P = 100 - q$$
$$P = 50 - \frac{1}{2}q \qquad ... (i)$$

$$\text{Total revenue is } P.q = 50q - \frac{1}{2}q^2 \qquad ... (ii)$$

$$MR = \frac{d\,(P.q)}{dq} = 50 - q \qquad ... (iii)$$

$$MC = AC = ₹.\ 10$$

Equating MR with MC to find out profit-maximising output level we have

$$50 - q = 10$$
$$q = 50 - 10 = 40$$

With $q = 40$, price (P) is

$$P = 50 - \frac{1}{2} \cdot 40 = 30$$

(*ii*) **Maximisation of Sales revenue**

Sales revenue is maximised at the output level whose total revenue (*TR*) is maximum. *TR* is maximum at the output level at which $MR = 0$.

Thus, setting MR function obtained in (*iii*) equal to zero we have

$$50 - q = 0$$
$$q = 50$$

(*iii*) Profits at profit - maximising output level 40 are :

$$\begin{aligned} \text{Profits} &= TR - TC \\ &= p.q - AC.q \\ &= (30 \times 40) - 10(40) \\ &= 1200 - 400 \\ &= ₹.\ 800 \end{aligned}$$

To obtain profits at sales revenue maximising output level 50 we substitute $q = 50$ in the inverse demand function we have is

$$\begin{aligned} \text{Profits} &= TR - TC \\ &= p.q - AC.q \\ &= 25 \times 50 - 10 \times 50 \\ &= 1250 - 500 = 750 \end{aligned}$$

QUESTIONS FOR REVIEW

1. Critically examine important objectives of a business firm. Are they consistent with each other?

2. What is meant by profit maximisation? On what grounds it has been criticised by some economists.

3. Critically evaluate profit-maximisation objective of business firms. What arguments would you give in defence of profit maximisation?

4. What is meant by economic profits? Is profit maximisation an appropriate goal for owners of a firm? For managers? What tends to happen if owners are not themselves managers?

D.U. B.A.(Hons.) 1990

5. Two Oxford economists, Hall and Hitch contended that businessmen do not maximise profits but follow "*mark-up pricing principle*" to obtain only normal profit. Is mark-up pricing principle consistent with profit-maximisation principle? Examine.

6. Given the separation of ownership from control in the modern corporate form of business organisation managers often strive to maximise their utility rather than maximising profits for the owners. Evaluate critically. Also mention the factors that enter into the utility functions of managers.

7. According to Prof. Baumol sales maximisation rather than profit maximisation is the valid objective of business executive today. Examine critically. Is it not irrational on the part of business managers to seek sales maximisation rather than profit maximisation?

8. H.A. Simon, one of the pioneers of the behavioural approach to the theory of the firm, suggests that instead of maximising profits, managers of business firms aim at '*satisficing*'. Discuss and explain what is meant by satisficing.

9. Write short critical notes on the following objectives of firm,

1. Security objective (*i.e.*, survival objective),
2. Maximisation of Growth
3. Staff maximisation

10. What is meant by equilibrium of the firm? Using total revenue total cost approach explain how firm maximises profits to attain equilibrium.

11. Define firm's equilibrium. Show that equality of marginal revenue with marginal cost is the necessary condition for firm's equilibrium regarding price and output. Illustrate with a diagram.

12. Equality of marginal cost with marginal revenue is necessary but not a sufficient condition of firm's equilibrium.

13. "Profit maximising output is smaller than the sales maximising output". Explain.

14. An author is to receive 10% of the sales revenue of a book as royalty. Would he favour a price which is lower than the one suggested by profit maximising publishers? Why?

D.U.B.A. (Hons.) 1994

[**Hints:** Author's royalty will be maximum when sales revenue is maximum. The price is lower at the output level where total sales revenue is maximum as compared to that at profit-maximising level of output. Therefore, the author will prefer a lower price].

CHAPTER 23

EQUILIBRIUM OF THE FIRM AND INDUSTRY UNDER PERFECT COMPETITION

MEANING AND CONDITIONS OF PERFECT COMPETITION

In an earlier chapter we made a general analysis of the equilibrium of the firm and its conditions. In this chapter we will study the equilibrium of the firm and industry under conditions of perfect competition. It would be in the fitness of things if we first describe what we mean by perfect competition. Perfect competition, as is generally understood, is said to prevail when the following conditions are found in the market :

1. There are a large number of firms producing and selling a product.
2. The product of all firms is homogeneous.
3. Both the sellers and buyers have perfect information about the prevailing price in the market.
4. Entry into and exit from the industry is free for the firms.

We shall discuss below in detail the above four conditions of perfect competition.

Large Number of Firms

The first condition of perfect competition is that there are a large number of firms in the industry. The position of a single firm in the industry containing numerous firms is just like a drop in the ocean. The existence of a large number of firms producing and selling the product ensures that an individual firm exercises no influence over the price of the product. The output of an individual firm constitutes a very small fraction of the total output of the whole industry so that any increase or decrease in output by an individual firm has a negligible effect on the total supply of product of the industry. As a result, a single firm is not in a position to influence the price of the product by increasing or reducing its output. The individual firm under perfect competition therefore takes the price of the product as a given datum and adjusts its output to earn maximum profits. In other words, *a firm under perfect competition is price-taker and output-adjuster*.

Homogeneous Products

The second condition of perfect competition is that the products produced by all firms in the industry are fully homogeneous and identical. It means that the products of various firms are indistinguishable from each other; they are perfect substitutes for one another. In other words, cross elasticity between the products of the firms is infinite. In case of homogeneous products, trade marks, patents, special brand labels etc. do not exist since these things make the products differentiated. It should be noted that if there are many firms, but they are producing differentiated products, each one of them will have influence over the price of his own variety of the product. The control over price is completely eliminated only when all *firms are producing homogeneous products*.

But whether or not products are homogeneous should be judged from the viewpoint of the buyers. Products would be homogeneous only when the buyers consider them to be so. Even if the buyers find some imagined differences between the products, the products would not be homogeneous, howsoever physically alike they may be. Anything which makes buyers prefer one seller to another, be it personality, reputation, convenient location, or the tone of his shop, differentiates the product to that degree, since what is bought is really a bundle of utilities of which these things are a part. Therefore, for the products to be homogeneous utilities offered by all sellers to buyers must be identical. If the bundle of utilities offered by all sellers is not the same, then the buyers would have a preference for some sellers who will have a degree of control over their individual prices. Thus the existence of homogeneous products signifies that the products of all sellers are completely identical in the eyes of consumers who therefore do not have any preference for one seller over another. Under such conditions it is evident that "buyers and sellers will be paired in random fashion in a large number of transactions. It will be entirely a matter of chance from which seller a particular buyer makes his purchases, and purchases over a period of time will be distributed among all sellers according to the law of probability. After all this is only another way of saying that the product is homogeneous."[1]

Perfect Information about the Prevailing Price

Another condition for perfect competition to prevail is that both the buyers and sellers are fully aware of the ruling price in the market. Because only when all buyers know fully the current price of the product in the market, sellers cannot charge more than the prevailing price. If any seller tries to charge a higher price than that ruling in the market, then the buyers will shift to some other sellers and buy the good at the ruling price since they know what the ruling price in the market is. Similarly, all sellers are also aware of the prevailing price in the market and no one will charge less price than this.

Free Entry and Exit

Lastly, perfect competition requires that there must be complete freedom for the entry of new firms or the exit of the existing firms from the industry in the long run. There must be no barriers to the entry of firms. Since, in the short run, firms can neither change the size of their plants, nor new firms can enter or old ones can leave the industry, the condition of free entry and free exit therefore applies only to the long-run equilibrium under perfect competition. If the existing firms are making super-normal profits in the short run, then this condition requires that in the long run new firms will enter the industry to compete away the profits. If, on the other hand, firms are making losses in the short run, some of the existing firms will leave the industry in the long run with the result that the price of the product will go up and the firms left in the industry will be earning at least normal profits.

The Demand Curve of a Product Facing a Perfectly Competitive Firm

The first three conditions ensure that a single price must prevail under perfect competition and the demand curve or average revenue curve faced by an individual firm under perfect competition is perfectly elastic at the ruling price in the market. Perfectly elastic demand curve signifies that the firm does not exercise any control over the price of the product but can sell any amount of the product as it likes at the ruling price. If the firm raises its price slightly above the ruling price, it will lose all its customers to its rivals. Because it can sell as much as it likes at the prevailing price it has no incentive to lower it. Without being able to raise the price and having no incentive to lower it, the firm is content to accept the ruling price in the market. Once the price in the market is established, a firm accepts as a given datum and adjusts its output at the level which gives it maximum profits. Consider Fig. 23.1. To begin with, demand curve *DD* and supply curve *SS* intersect at point *E* and determine price *OP*. Now, the firm, having no influence over the price, will take the price *OP* as given and therefore average-marginal revenue curve facing it will be a horizontal straight line at the level of *OP*. When the demand increases and as a result the price rises to *OP*′, the firm will

1. E.H. Chamberlin, *Theory of Monopolistic Competition,* 7th ed., p.8.

now confront average-marginal revenue curve at the level of OP'. And if the demand decreases and price falls to OP'' the firm's average-marginal revenue curve will shift below to the level of OP''.

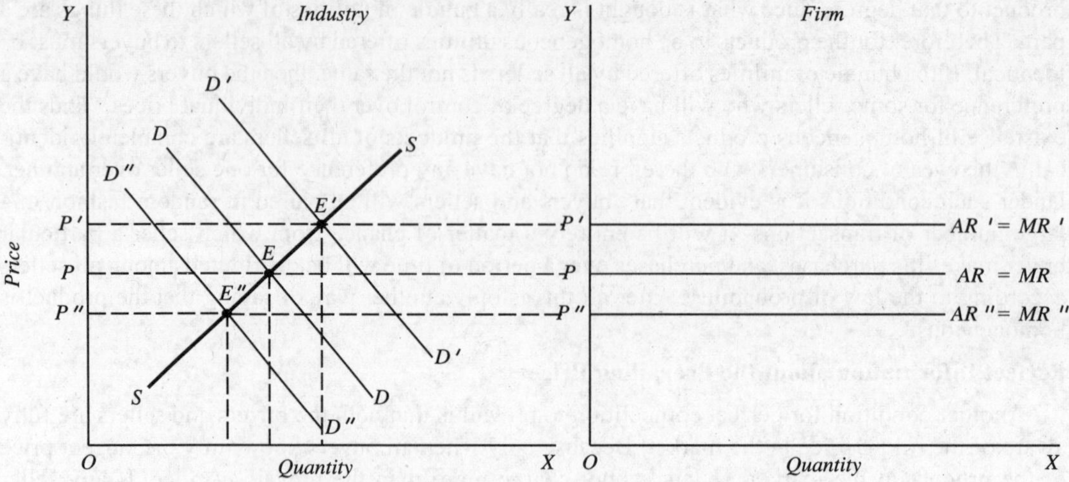

Fig. 23.1 *Demand Curve Facing an Individual Firm*

The fourth condition, namely, free entry and free exit, ensures that the firm will make only normal profits in the long run. On the one hand, super-normal profits will disappear by the entry of new firms in the industry and, on the other, losses will disappear as a result of some firms leaving the industry.

PROFIT MAXIMISATION BY A COMPETITIVE FIRM : TR-TC APPROACH

As seen in the previous chapter, there are two aproaches to explain the equilibrium or profit maximisation by the firm; (*a*) Total revenue and total cost approach, and (*b*) *MR* and *MC* approach. Though in modern economic analysis, the equilibrium of a firm is usually explained through *MR* and *MC* approach, for some purposes total revenue and total cost approach is highly useful (for example, to show break even level of output). Therefore, we first explain how a perfectly competitive firm maximises its economic profits using total revenue and total cost curves. Since under perfect competition marginal revenue is constant, total revenue curve is an upward-sloping straight line. On the other hand, due to varying returns to the factors used, total cost (*TC*) curve first increases at a decreasing rate and then after a point it increases at an increasing rate.

Equilibrium of a firm working under perfect competititon which aims at profit maximisation is graphically illustrated in Fig. 23.1 (*a*) where *TR* represents total revenue curve and *TC* represents total cost curve. Total revenue curve starts from the origin which means that when no output is produced, total revenue is zero. As output is increased total revenue goes on increasing at a constant rate. This is because price for a firm working under perfect competition remains constant whatever its level of output. Consequently, total revenue curve *TR* is a straight line from the origin. However, it will be noticed that the total cost curve *TC* starts from a point *F* which lies above the origin. It means that *OF* is the fixed cost which the firm has to incur even if it stops production in the short run. It will be seen that the short-run total cost curve *TC* initially increases at a decreasing rate and then after a point it increases at an increasing rate. This implies that average total cost curve is roughly of U-shape. Total profits can be measured as the vertical distance between the *TR* and *TC* curves. It will be observed from Fig. 23.1 (*a*) that up to the level of output OQ_B, *TC* curve lies above *TR* curve showing that as the firm raises its output in the initial stages total cost is greater than total revenue and the firm is incurring losses. When the firm produces OQ_B level of output, total revenue just equals total cost and the firm is therefore neither making profits nor losses. That is, the firm is only breaking even at output level OQ_B. Thus the point B or output level OQ_B is called *Break-Even Point*.

When the firm increases its output beyond OQ_B total revenue becomes larger than total cost and therefore profits begin to accrue to the firm. It will be noticed from Fig. 23.1 (*a*) that profits are increasing as the firm increases production to output OQ_M, since the distance between the total revenue curve (*TR*) and total cost curve (*TC*) is widening. At OQ_M level of output, the distance between the *TR* curve and *TC* curve is the greatest and therefore profits will be maximum. If the firm expands output beyond Q_M, the gap between *TR* and *TC* curves goes on narrowing down and therefore the total profits will be declining. It is therefore clear that the firm will be in equilibrium at O_M level of output where total revenue exceeds total cost by the largest amount and hence its profits are maximum.

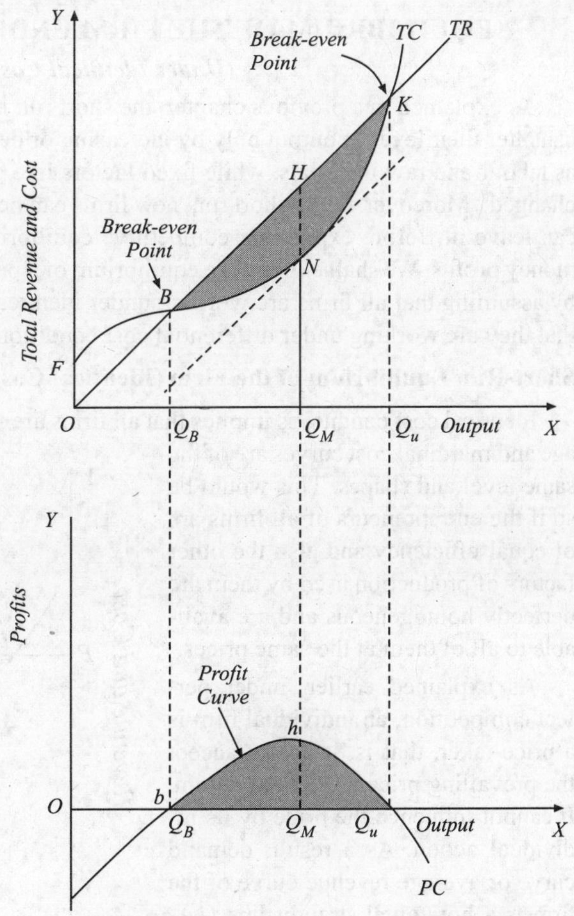

Fig. 23.1 (a). *Profit Maximisation under Perfect Competition: TC -TR Approach*

It will be observed from Fig. 23.1 (*a*) that at output level Q_U, total revenue is again equal to total cost (*TR* curve cuts *TC* curve at point *K* corresponding to output Q_U). Thus, point *K* is again a break-even point, usually called *upper break-even point*. It may however be noted that this upper break-even point *K* or output level Q_U is not of much relevance as it lies beyond firm's profit maximising level and may actually lie beyond firm's capacity to produce. It is the first break-even point *B* or output level Q_B which is highly significant as a firm will not plan to produce if it cannot sell output equal to at least Q_B at which total revenue just covers total cost of production so that its economic profits are zero.

For a more vivid representation of profit-maximising level of output we have drawn in the lower panel of Fig. 23.1 (*a*) the profit curve *PC* which measures the distance between *TR* and *TC* curves. It will be seen from this lower panel of Fig. 23.1 (*a*) that up to output level Q_B profit curve lies below the *X*-axis showing that the firm is making losses if it produces less than Q_B. At output level Q_B the firm's economic profits are zero because at this output level total revenue just covers total cost of production. Therefore, output Q_B *is break-even level of output*. As the firm expands its level of output beyond Q_B, profit curve is rising until it reaches its maximum point corresponding to level of output Q_M. Beyond level of output Q_M, profit curve slopes downward indicating that profits decline beyond output Q_M. Thus, *at output level Q_M, the firm maximises economic profits*. At a higher output level Q_U, profits are again zero indicating the upper break-even point.

There is a major shortcoming of the *TC* and *TR* approach of analysing firm's equilbrium. It is that in this approach what price will be charged by the firm in its equilibrium position is not directly known. Therefore, in what follows we explain the modern *MC* = *MR* approach to firm's equilibrium.

EQUILIBRIUM OF THE FIRM UNDER PERFECT COMPETITION

(Under Identical Cost Conditions)

As explained in a previous chapter, the short run means a period of time within which the firms can alter their level of output only by increasing or decreasing the amounts of variable factors such as labour and raw materials, while fixed factors like capital equipment, machinery etc. remain unchanged. Moreover, in the short run, new firms can neither enter the industry, nor the existing firms can leave it. Before explaining competitive equilibrium we assume that a firm tries to maximise money profits. We shall explain the equilibrium of a perfectly competitive firm in two stages : firstly, by assuming that all firms are working under identical cost conditions and, secondly, by assuming that they are working under differential cost conditions.

Short-Run Equilibrium of the Firm (Identical Cost Conditions)

Identical cost conditions implies that all firms are facing same cost conditions, that is, their average and marginal cost curves are of the same level and shapes. This would be so if the entrepreneurs of all firms are of equal efficiency and also the other factors of production used by them are perfectly homogeneous and are available to all of them at the same prices.

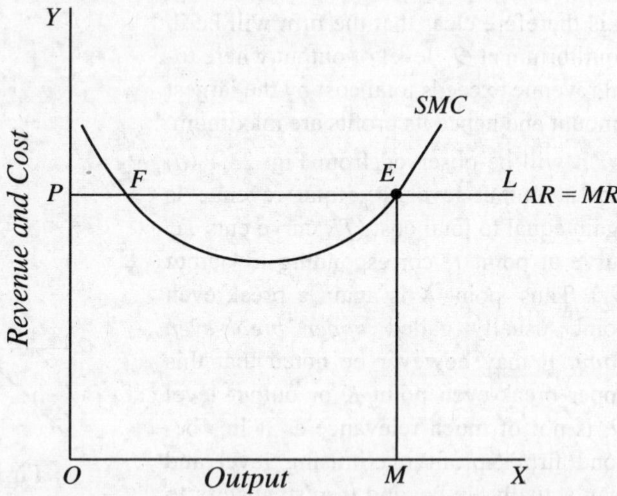

Fig. 23.2. *Firm's Equilibrium under Perfect Competition*

As explained earlier, under perfect competition, an individual firm is a price-taker, that is, it has to accept the prevailing price as a given datum. It cannot influence the price by its individual action. As a result, demand curve or average revenue curve of the firm is a horizontal straight line (*i.e.*, perfectly elastic) at the level of the prevailing price. Since perfectly competitive firm sells additional units of output at the same price, marginal revenue curve coincides with average revenue curve. Marginal cost curve, as usual, is U-shaped. Now, in order to decide about its equilibrium output, the firm will compare marginal cost with marginal revenue. It will be in equilibrium at the level of output at which marginal cost equals marginal revenue and marginal cost curve is cutting marginal revenue curve from below. At this level it will be maximising its profits. Since marginal revenue is the same as price (or average revenue) ***under perfect competition, the firm will equalise marginal cost with price to attain equilibrium output.***

Consider Fig. 23.2 in which price *OP* is prevailing in the market. *PL* would then be the demand curve or the average and marginal revenue curve of the firm. It will be seen from Fig. 23.2 that marginal cost curve cuts average and marginal revenue curve at two different points, *F* and *E*. *F* cannot be the position of equilibrium, since at *F* second order condition of firm's equilibrium, namely, that marginal cost curve must cut marginal revenue curve from below at the point of equilibrium, is not satisfied. The firm will be increasing its profits by increasing production beyond *F* because marginal revenue is greater than marginal cost. The firm will be in equilibrium at point *E* or output

OM since at *E* marginal cost equals marginal revenue (or price) as well as marginal cost curve is cutting marginal revenue curve from below. As under perfect competition marginal revenue curve is a horizontal straight line, the marginal cost curve must be rising so as to cut the marginal revenue curve from below. Therefore, in case of perfect competition the second order condition of firm's equilibrium requires that marginal cost curve must be rising at the point of equilibrium. Hence the twin conditions of firm's equilibrium under perfect competition are :

(1) $MC = MR =$ Price

(2) *MC curve must be rising at the point of equilibrium.*

But the fulfilment of the above two conditions does not guarantee that the profits will be earned by the firm. In order to know whether the firm is making profits or losses and how much of them, average cost curve must be introduced in the figure. This has been done in Fig. 23.3 where *SAC* and *SMC* curves are short-run average cost and short-run marginal cost curves respectively. Profit per unit of output is the difference between average revenue (price) and average cost. In Fig. 23.3, at the equilibrium output *OM*, average revenue is equal to *ME*, and average cost is equal to *MF*. Therefore, the profit per unit of output is *EF*, the difference between *ME* and *MF*. The total profits earned by the firm will be equal to *EF* (profit per unit) multiplied by *OM* or *HF* (total output). Thus, the total profits will be equal to the area *HFEP*. Because normal profits are included in average cost, the area *HFEP* indicates super-normal profits.

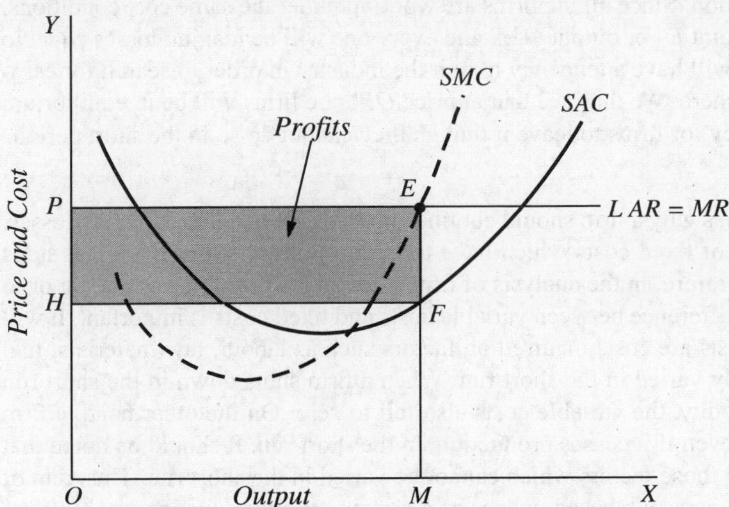

Fig. 23.3. *Short-Run Equilibrium with Profits*

Since we are assuming that all firms in the industry are working under same cost conditions and also for all of them price is *OP*, all will be earning super-normal profits equal to the area *HFEP*. Thus, while all firms in the industry will be in short-run equilibrium, but the industry will not be in equilibrium since there will be a tendency for the new firms to enter the industry to compete away the super-normal profits. But the short run is not a period long enough for the new firms to enter the industry. The exist-

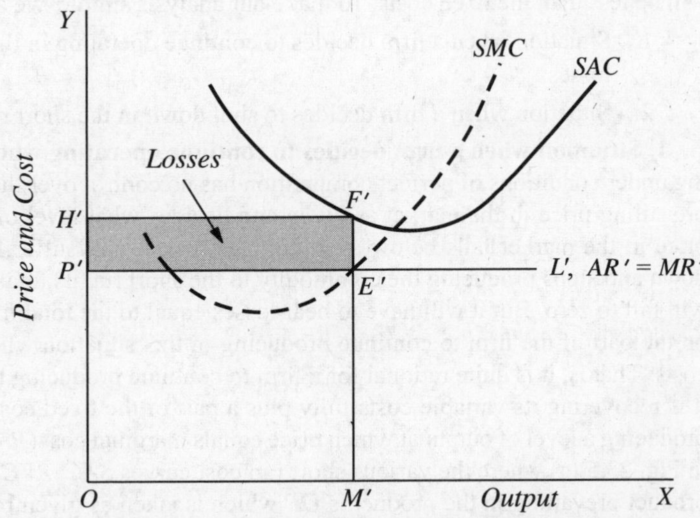

Fig. 23.4. *Short-Run Equilibrium with Losses*

ing firms will therefore continue earning super-normal profits equal to *HFEP* in the short period. It is evident that in the situation depicted in Fig. 23.3 all firms will be in equilibrium at *E* and each will be producing *OM* output, but the tendency for the new firms to enter the industry will be present, though they cannot enter during the short period.

Now suppose that the prevailing market price of the product is such that the price line or average and marginal revenue curve lies below average cost curve throughout. This case is illustrated in Fig. 23.4 where the ruling price is *OP′* which is taken as given by the firm. *P′ L′* is the price line which lies below *AC* curve at all levels of output. The firm will be in equilibrium at point *E′* at which marginal cost is equal to price (or marginal revenue) and marginal cost curve is rising. Firm would be producing *OM′* output but would be making losses, since average revenue (or price) which is equal to *M′E′* is less than average cost which is equal to *M′F′*. The loss per unit of output is equal to *E′F′* and total loss will be equal to *P′E′F′H′* which is the minimum loss that a firm can make under the given price-cost situation. Since all the firms are working under the same cost conditions, all would be in equilibrium at point *E′* or output *OM′* and every one will be making losses equal to *P′E′F′H′*. As a result, the firms will have a tendency to quit the industry in order to search for earning at least normal profits elsewhere. We thus see that at price *OP′* the firms will be in equilibrium at *E′* but there will be a tendency for firms to leave it though they cannot do so in the short period.

Deciding to Shut Down

Now, an important question is why a firm should continue operating when it is incurring losses. The answer lies in the concept of fixed costs which have to be borne by the firm even if it stops production in the short run. Therefore, in the analysis of firm's decision to continue operating or to shut down in the short run, the difference between variable costs and fixed costs is important. It will be remembered that variable costs are costs incurred on factors such as labour, raw materials, fuel or electricity which can be easily varied in the short run. When a firm shuts down in the short run and stops producing the commodity, the variable costs also fall to zero. On the other hand, a firm cannot escape from fixed costs even if it ceases production in the short run. It should be noted that fixed costs are costs incurred on those factors which cannot be varied in the short run. Thus rent of factory building, costs on machinery purchased, wages of a certain minimum managerial staff are some examples of fixed costs. *When a firm stops production, that is, shuts down in the short run, it will have to bear losses equal to the fixed costs*. Therefore, it will be wise to continue operating in the short run when firm's total revenue exceeds total fixed costs because in that case firm's losses will be less than the fixed costs. To make our analysis simple, we examine the question in two parts.

1. Situation when a firm decides to continue operating in the short run even when incurring losses.
2. Situation when a firm decides to shut down in the short run.

1. Situation when a firm decides to continue operating when incurring losses. A firm working under conditions of perfect competition has no control over the price of the product. It takes the prevailing price in the market as given and decides what level of output it should produce. When price in the market falls below average total cost, it will suffer losses. To avoid losses if it shuts down and stops producing the commodity in the short run its total revenue as well as variable costs will fall to zero. But it will have to bear losses equal to the total fixed costs. Therefore, it is prudent on the part of the firm to continue producing in this situation when losses are less than total fixed costs. That is, it is quite rational for a firm to continue producing the commodity in the short run, if it is recovering its variable costs fully plus a part of the fixed costs. But it will minimise losses by producing a level of output at which price equals marginal cost ($P = MC$). This situation is illustrated in Fig. 23.5 (a) where the various short run cost curves *SAC*, *AVC* and *SMC* are shown. Price of the product prevailing in the product is *OP* which is taken as given by the firm. The firm is in equilibrium at point *E* where it produces *OQ* output at which the given price *OP* is equal to marginal cost of production (*SMC*). It will be seen from Fig. 23.5 (a) that at the equilibrium output *OQ*, average variable cost is *QL*, which is less than the price *OP* (=*QE*) or Price > *AVC*. This means the firm is recovering variable costs plus a part of the fixed cost. Total revenue (*TR*) earned by producing output *OQ* is equal to the area *OPEQ*, while the total costs are equal to the area *ORTQ*.

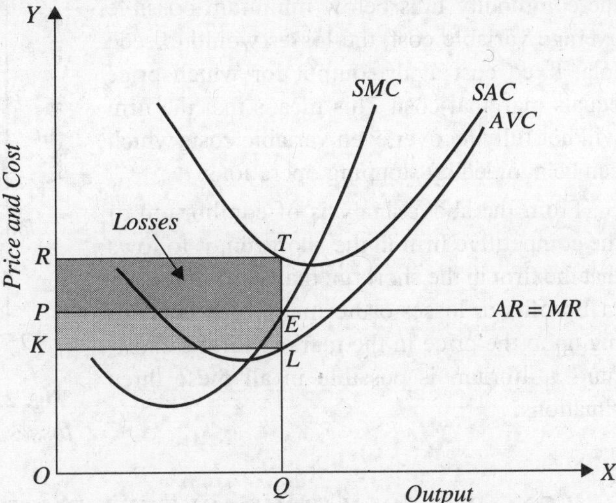

Fig. 23.5 (a). *It is rational to operate when losses < TFC*

It is evident from Fig. 23.5 (a) that when price is *OP* total revenue is less than the total costs and the firm is making losses equal to the area *RTEP*. It should be noted that average fixed cost at output level *OQ* is given by the vertical distance *TL* between short-run average total cost (*SAC*) and the average variable cost (*AVC*). Multiplying this average fixed cost by output *OQ* (= *KL*) we get the total fixed costs being equal to the area *RTLK*. It is thus clear by working at point *E* and producing output *OQ*, the firm is recovering the entire variable costs equal to the area *OQLK* and a part of the fixed cost equal to the area *KLEP*. Thus losses made equal to the area *RTEP* are less than the total fixed cost equal to the shaded area *RTLK*. If a firm shuts down in the short run and ceases to produce the product, its losses will be equal to the total fixed cost *RTLK*. It will therefore be a rational decision on the part of the firm to continue operating as shutting down in this situation will mean greater losses equal to the entire total fixed cost.

To conclude, *the firm will continue operating in the short run at a loss when total revenue exceeds total variable costs. This enables the firm to earn revenue to recover a part of the fixed costs.* We state below the condition when it is rational for the firm to continue production in the short run even when it is incurring losses:

$$TR > TVC$$

Since $TR = P.Q$, and $TVC = AVC.Q$

Therefore, $P.Q > AVC.Q$

$$P > AVC$$

2. Situation when a firm decides to shut down in the short run. This situation is depicted in Fig. 23.5 (b) where it will be seen that price has fallen to the level OP_1. With price OP_1, equilibrium is attained at point *D* corresponding to output OQ_1 at which price is equal to both marginal cost (*MC*) and minimum average variable cost. By producing OQ_1 output and selling it at price OP_1, the firm earns total revenue equal to the area OQ_1DP_1. The total cost of producing OQ_1 output is equal to the area OQ_1HB. Thus with price OP_1 the firm is incurring losses equal to the area P_1DHB. It should be noted that average fixed cost is *DH* at OQ_1 output, that is, the vertical distance between *SAC* and *AVC*. The total fixed cost is then given by the area P_1DHB. Thus when price falls to OP_1, firm's losses are equal to the total fixed cost. Even when the firm closes down, its losses will be equal to the total fixed cost. Therefore, if price falls below OP_1 which is equal to the minimum possible average variable cost (*AVC*), the losses will become greater than the fixed costs and the firm will shut down. Point *D* which indicates the minimum possible average variable cost represents the *shut-down point*. The situation when firm actually shuts down when price falls below average variable cost is explained below.

3. Situation when firm actually shuts down and does not operate. When price of the commodity falls below minimum possible average variable cost, the losses would exceed total fixed cost at the output for which price equals marginal cost. This means that the firm will not fully recover even variable costs which can be avoided by stopping operations.

From the above analysis of equilibrium of the competitive firm in the short run, it follows that the firm in the short run may earn supernormal profits or losses or normal profits depending upon the price in the market. Firm's short-run equilibrium is possible in all these three situations.

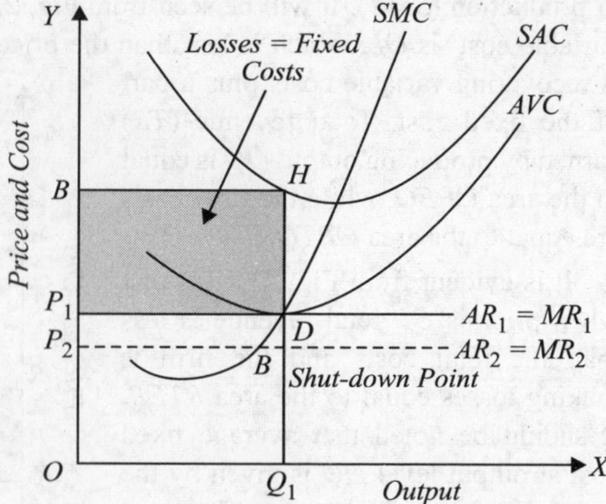

Fig. 23.5 (b). *When Price (P) = minimum AVC, losses are equal to total fixed costs. If price falls below it, the firm will shut down.*

LONG-RUN EQUILIBRIUM OF THE FIRM UNDER PERFECT COMPETITION

The long run is a period of time which is sufficiently long to allow the firms to make changes in all factors of production. In the long run, all factors are variable and none fixed. The firms, in the long run, can increase their output by changing their capital equipment; they may expand their old plants or replace the old lower-capacity plants by the new higher-capacity plants or add new plants. Besides, in the long run, new firms can enter the industry to compete the existing firms. On the contrary, in the long run, the firms can contract their output level by reducing their capital equipment; they may allow a part of the existing capital equipment to wear out without replacement or sell out a part of the capital equipment. Moreover, the firms can leave the industry in the long run. The long-run equilibrium then refers to the situation when free and full adjustment in the capital equipment as well as in the number of firms has been allowed to take place. It is therefore long-run average and marginal cost curve which are relevant for deciding about equilibrium output in the long run. Moreover, in the long run, it is the average total cost which is of determining importance, since all costs are variable and none fixed.

As explained above, a firm is in equilibrium under perfect competition when marginal cost is equal to price. But for the firm to be in long-run equilibrium, besides marginal cost being equal to price, the price must also be equal to average cost. For, if the price is greater or less than the average cost, there will be tendency for the firms to enter or leave the industry. If the price is greater than the average cost, the firms will earn more than normal profits. These supernormal profits will attract other firms into the industry. With the entry of new firms in the industry, the price of the product will go down as a result of the increase in supply of output and also the cost will go up as a result of more intensive competition for factors of production. The firms will continue entering the industry until the price is equal to average cost so that all firms are earning only normal profits.

On the contrary, if the price is lower than the average cost, the firms would make losses. These losses will induce some of the firms to quit the industry. As a result, the output of the industry will fall which will raise the price. On the other hand, with some firms going out of the industry, cost may go down as a result of fall in the demand for certain specialised factors of production. The firms will continue leaving the industry until the price is equal to average cost so that the firms remaining in the field are making only normal profits. It, therefore, follows that for a perfectly competitive firm to be in long-run equilibrium, the following two conditions must be fulfilled:

1. *Price = Marginal Cost*
2. *Price = Average Cost*

If price is equal to both marginal cost and average cost, then we have a double condition of long-run perfectly competitive equilibrium :

Price = Marginal Cost = Average Cost

But from the relationship between marginal cost and average cost we know that marginal cost is equal to average cost only at the minimum point of the average cost curve. Therefore, the condition for long-run equilibrium of the firm can be written as :

Price = Marginal Cost = Minimum Average Cost

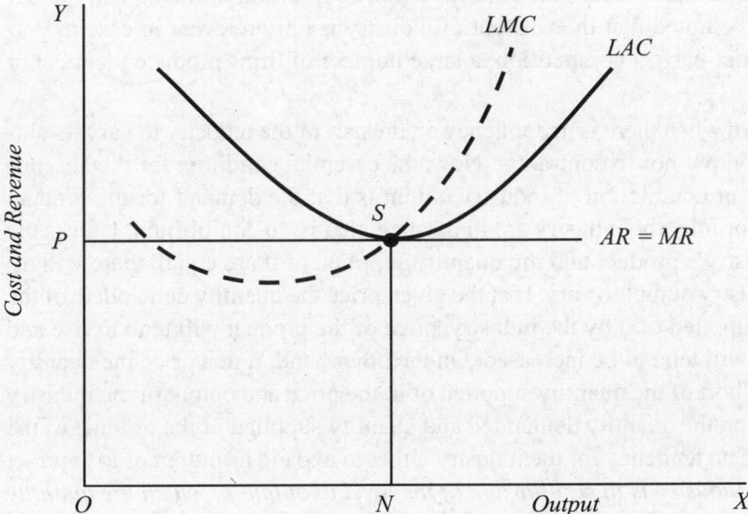

Fig. 23.6. *Long-Run Equilibrium of the Firm*

Fig. 23.6 represents long-run equilibrium of firm under perfect competition. The firm cannot be in the long-run equilibrium at a price greater than *OP* in Fig. 23.6. This is because if price is greater than *OP*, then the price line (demand curve) would lie somewhere above the minimum point of the average cost curve so that marginal cost and price will be equal where the firm is earning abnormal profits. Since there will be tendency for new firms to enter and compete away these abnormal profits, the firm cannot be in long-run equilibrium at any price higher than *OP*. Likewise, the firm cannot be in long-run equilibrium at a price lower than *OP* in Fig. 23.6 under perfect competition. If price is lower than *OP*, the average and marginal revenue curve will lie below the average cost curve so that the marginal cost and price will be equal at the point where the firm is making losses. Therefore, there will be tendency for some of the firms in the industry to go out with the result that price will rise and the firms left in the industry make normal profits. We therefore conclude that the firm can be in long-run equilibrium under perfect competition only when price is at such a level that the horizontal demand curve (that is, *AR* curve) is tangent to the average cost curve so that price equals average cost and firm makes only normal profits.

It should be noted that a horizontal demand curve can be tangent to a U-shaped average cost curve only at the latter's minimum point. Since at the minimum point of the average cost curve the marginal cost and average cost are equal, price in long-run equilibrium is equal to both marginal cost and average cost. In other words, double condition of long-run equilibrium is fulfilled at the minimum point of the average cost curve.

It is clear from above that *long-run equilibrium of the firm under perfect competition is established at the minimum point of the long-run average cost curve.* Working at the minimum point of the long-run average cost curve signifies that the *firm is of optimum size*, that is, it is producing output at the lowest possible cost. The fact that the firm, working under conditions of perfect competition, tends to be of optimum size in the long run is beneficial from the social point of view in two ways. Firstly, working at optimum size implies that the resources of the society are being utilised in the most efficient way. Secondly, it signifies that the consumers are getting the goods at the lowest possible price.

EQUILIBRIUM OF THE INDUSTRY

Since the price of a product under perfect competition is determined by the intersection of the demand and supply curves of the product of an industry, we need to know the nature and shape of the supply curve of a product under perfect competition. In our chapters on demand theory we have already explained how the market demand curve for a product is derived and what shape it normally takes. We shall now explain how the supply curve of a product under condition of perfect competition is derived and the shape it takes both in the short run and the long run. Before explaining the derivation of the supply curve, we shall discuss the concept of the equilibrium of the industry under perfect competition. It should be noted that the concept of industry is only relevant in case of perfect competition since only under perfect competition, a large number of firms produce identical or homogeneous product.

An industry is in equilibrium when there is no tendency on the part of the industry to vary its output, that is, neither to expand output nor to contract it. Now, the essential condition for the absence of any tendency for expansion or contraction of industry output is that the demand for the product of the industry and the supply of it by the industry are in balance, that is, in equilibrium. Unless the quantity demanded of the industry's product and the quantity supplied of it are equal, there will always be a tendency for the industry output to vary. If at the given price the quantity demanded of the product exceeds the quantity supplied of it by the industry, price of the product will tend to rise and also the output of the industry will tend to be increased. On the other hand, if at a price the quantity demanded of the product falls short of the quantity supplied of it, the price and output of the industry will tend to fall. Thus, only when the quantity demanded and quantity supplied of the product of the industry are equal, there will be no tendency for the industry either to expand its output or to contract it. We therefore conclude that *industry is in equilibrium at the level of output at which the quantity demanded and quantity supplied of its product are equal*, or in other words, at which the demand curve for the product of the industry and the supply curve of it by the industry intersect each other.

Now, the output of the product of an industry can vary in two ways. Firstly, the output of an industry can vary if the existing firms in it vary their output levels. And the firms will have no tendency to vary their output when they are individually in equilibrium by equating market price with marginal cost and are thus maximising their profits. Secondly, the output and therefore the supply of the product of an industry can vary by a change in the number of firms in it; the industry output will increase if new firms enter the industry and the industry output will decline if some of the existing firms leave it.

Thus, given the external conditions regarding demand for the product an industry would be in equilibrium when neither the individual firms have incentive to change their output nor there is any tendency for the new firms to enter the industry or for the existing firms to leave it. Therefore, in addition to the equality of demand and supply of the industry's product, two conditions which must be satisfied if there is to be equilibrium of the industry. First, each and every firm should be in equilibrium. It should be noted here that, according to Marshall, for the equilibrium of the industry each firm in it may not be in equilibrium. To Marshall, equilibrium of the industry meant the equality of demand and supply of the industry's product, nothing more. According to Marshall, given the equilibrium of the industry some firms in it may be growing (*i.e.* expanding their output), some may be declining (*i.e.*, contracting their output), and some others may be holding their outputs constant. It is in this connection that Marshall evolved the concept of representative firm, the firm which was in equilibrium (*i.e.*, holding its output steady) when the industry was in equilibrium. There was a serious shortcoming in Marshall's concept of the equilibrium of the industry in that he did not demonstrate that for the industry to be in equilibrium, the outputs of growing firms are equal to the outputs of the declining firms. In the modern economic theory, Marshall's concept of representative firm has not been adopted and therefore in modern microeconomics, for the industry to be in equilibrium, all firms must also be in equilibrium. This will happen at the output of a firm where marginal cost is equal to marginal revenue and marginal cost curve cuts the marginal revenue curve from below. Second, the number of firms should be in equilibrium, *i.e.* there should be no tendency for the firms either to move into or out of the industry. This will happen when all the entrepreneurs, *i.e.*, owners of the firms of the industry, are making zero economic profits which means they earn only '*normal profits*', *that is, profits which are just sufficient to induce them to stay in the industry, and when no entrepreneur outside the industry thinks that he could earn at least normal profits if he were to enter it.*

Normal Profits and Equilibrium of the Industry

Thus, the concept of normal profits is important in defining and describing equilibrium of the industry. If we assume that all entrepreneurs in a certain industry have the same opportunity costs, that is, the same transfer earnings if they leave the industry, then there would be a given fixed amount of normal profits for the whole industry. Every entrepreneur must earn at least this fixed amount of normal profits if he is to stay in the industry. If all the firms in the industry are earning profits above normal, there will be incentive for the firms outside the industry to enter it since there is every reason for the entrepreneurs outside the industry to expect that they would be able to earn at least normal profits if they enter. Thus, there will be a tendency for the firms in that industry to increase. If, on the other hand, the firms in the industry are earning profits below normal (*i.e.* when they are incurring losses), it implies that the firms cannot cover their opportunity costs. Therefore, some of them will leave the industry and search for profits elsewhere. Thus, the number of firms in that industry will tend to diminish. In conclusion, we can say that *equilibrium of the industry or full equilibrium, as it is sometimes called, would be attained when the number of firms in the industry is in equilibrium, (i.e., no movement into or out of the industry) and also all the individual firms in it are in equilibrium, that is, they are equating marginal cost with marginal revenue, and MC curve cuts MR curve from below.*

It should be noted that normal profits of entrepreneur are included in the average cost of production. Therefore, if the price is equal to the average cost of production, it means that the entrepreneur is earning only normal profits.

Short-Run Equilibrium of the Industry

We must distinguish between the *short-run and the long-run equilibrium of the industry.* In the short run only existing firms can make adjustment in their output while the number of firms remains

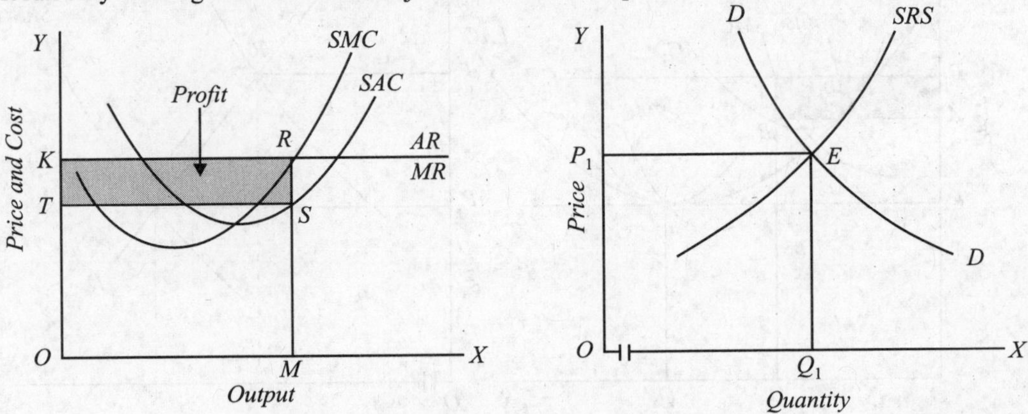

Fig. 23.7. *Short-Run Equilibrium of the Industry*

the same, that is, no new firms can enter the industry, nor any existing firms can leave it. Since, in the short run, by definition, the entry or exit of the firms is not permitted, for the short-run equilibrium of the industry, the condition of making only normal profits by the existing firms (or, in other words, the equality of average cost with average revenue) is not required. Thus, the industry is in short-run equilibrium when the short-run demand for and supply of the industry's product are equal and all the firms in it are in equilibrium. In the short-run equilibrium of the industry, though all firms must be in equilibrium, they all may be making supernormal profits or all may be having losses depending upon the demand conditions of the industry's product.

Short-run equilibrium of the industry is illustrated in Fig. 23.7 where in the right hand panel, industry's demand curve *DD* and short-run supply curve *SRS* respectively are shown. These curves intersect at point *E* and thereby determine the equilibrium price OP_1 and the equilibrium output OQ_1 of the industry. Firms will take price OP_1 as given and will adjust their output at the profit-maximising level. The left hand panel of Fig. 23.7 shows that a firm in the industry will be in equilibrium at *OM* output. (It should be noted that the scale on the *X*-axis of the diagrams in the right-hand and left-hand panels are different). With *OM* output, the firm is making economic profits equal to the area *KRST*. If it is assumed, as is being done here, that all firms in the industry are alike in respect of cost conditions, then all firms like the one shown in Fig. 23.7 (left-hand panel) will be making profits. Thus, while the industry is in short-run equilibrium, that is, the demand and supply of its product are equal and also all the firms in it are in equilibrium; and they are making economic profits.

If the demand conditions for the product of the industry are not so favourable, for instance, if the demand curve of the industry's product is at a much lower position than shown in Fig. 23.7, then the intersection of demand and supply curves may take place at the price at which the firms will be having losses in their equilibrium positions. In this case too the industry will be in short-run equilibrium. We, therefore, reach the following two conditions for the short-run equilibrium of the industry under perfect competition:

1. *The short-run demand for and supply of the product of the industry must be equal.*

2. *All the firms in the industry should be in equilibrium whether they are making economic profits or having losses.*

Long-Run Equilibrium of the Industry

In the long run, number of firms can vary. Lured by economic (super-normal) profits earned by

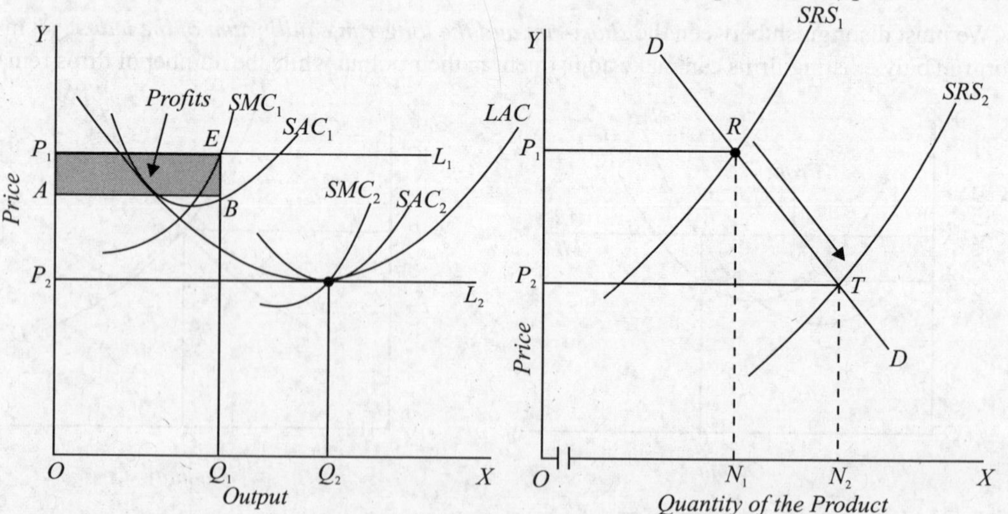

Fig. 23.8. *Long-Run Adjustment and Equilibrium of the Perfectly Competitive Industry*

the existing firms in the short run, the new firms will enter the industry and compete away these economic profits and thereby all the firms will be in long-run equilibrium at the minimum points of their long-run average cost curves and will therefore be making only normal profits, that is, zero economic profits. In the short run, if the existing firms make losses, some of the firms will leave the industry so that the output of the industry will fall and as a result the price will go up to the level of average cost. Thus, as a consequence of the exit of some firms, the remaining firms come to be in long-run equilibrium where they are earning only normal profits.

The industry is in long-run equilibrium when in addition to the equality of long-run supply of and demand for the industry's product, all firms are in equilibrium and also there is neither a tendency for the new firms to enter the industry, nor for the existing firms to leave it. The long-run equilibrium of the industry is depicted in Fig. 23.8 in which, in the right-hand panel, demand curve *DD* and short-run supply curve SRS_1 of the industry are shown which interesect at point *R* and thereby determine the price OP_1. It will be seen from left-side panel of Fig. 23.8 that with price OP_1, the firm is in equilibrium at point *E* and producing OQ_1 output and making supernormal profits. As explained above, with the expansion of scale of production by the existing firms and the entry of new firms in the long run, the short-run supply curve of the industry will continue shifting to the right until it interesects the demand curve *DD* at point *T* at which price OP_2 is determined. It will be seen from the left-hand panel of Fig. 23.8 that when short-run supply curve has shifted to the position SRS_2 and its intersection with demand curve *DD* determines price OP_2 which is equal to the minimum long-run average cost curve *LAC,* the firm is making only normal profits. Since all firms have identical cost conditions, all firms in the industry will be making only normal profits. Under these circumstances, there will be no tendency for the firms to enter or leave the industry. We therefore arrive at the following three conditions for the long-run equilibrium of the industry:

1. *The quantity supplied and demanded of the product of the industry should be equal.*
2. *All firms in the industry should be in equilibrium.*
3. *There should be no tendency for the new firms to enter the industry or for the existing firms to leave it. In other words, number of firms should be in equilibrium.* This implies that all firms in the indsutry will be earning only normal profits in the long run.

SHORT-RUN SUPPLY CURVE OF THE PERFECTLY COMPETITIVE FIRM

As is known, the short run is a period in which more quantity of the good is produced by working the given capital equipment or plant more intensively by employing more amounts of the variable factors. We have seen above that the firm under perfect competition produces that amount of the good at which marginal cost equals price. Since the price for a perfectly competitive firm is given and constant for it, price line will be a horizontal straight line. The horizontal coordinate of a point on the rising marginal cost curve measures the quantity of the good that the firm will produce at that price. The short-run marginal cost curve of the firm therefore indicates the quantities which the firm will produce in the short run at different prices.

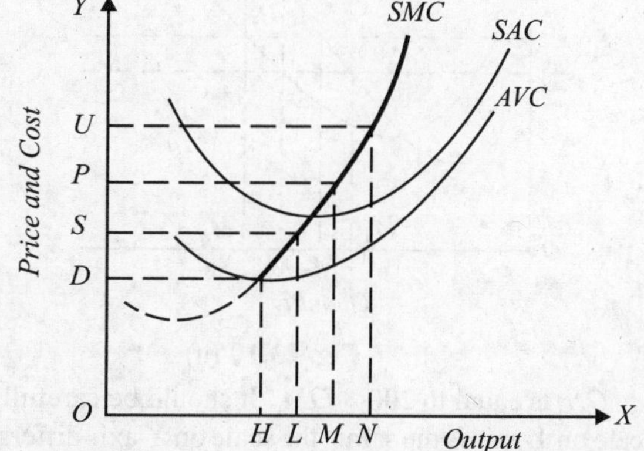

Fig. 23.9. *Derivation of Short-Run Supply Curve of the Firm*

Consider Fig. 23.9 at price *OP*, the firm will produce and offer for sale *OM* quantity of the good, because at *OM* quantity of the good, price *OP* equals marginal cost. Similarly, at price *OU* the quantity produced or supplied will be *ON*, since price *OU* equals marginal cost at output *ON*. Likewise, at price *OS*, the firm will produce and supply *OL* quantity of the product. It is thus clear that *short-run marginal cost curve of the firm is in fact the short-run supply curve of the firm*. The firm will not produce any output at a price below *OD*, since it will not be fully recovering its variable costs. Thus, *only the part of the short run marginal cost curve which lies above the average variable cost forms the short-run supply curve of the firm*. In Fig. 23.9 the *thick portion* of the short-run marginal cost curve *SMC* represents the short-run supply curve of the firm. Since under perfect competition marginal cost must be rising at the equilibrium output, *the short-run supply curve of the firm must always slope upward to the right*.

It should be noted that in our analysis of deriving short-run supply curve of the firm, we have assumed that following the rise in price when the firm expands its output or supply, prices of resources or inputs it uses for production do not go up. It is a valid assumption because an individual firm under perfect competition is only one among many and its demand for inputs or resources is insignificant part of the total market demand for them and therefore the increase in demand for these resources by the firm as it expands will have no effect on their prices.

SHORT-RUN SUPPLY CURVE OF THE COMPETITIVE INDUSTRY

We now proceed to derive the short-run supply curve of the competitive industry. As the market demand curve is found by the horizontal summation of demand curves of all individual consumers of a product, similarly *the supply curve of the industry is obtained by lateral summation (horizontal addition) of short-run supply curves of all individual firms in the industry*. How the short-run supply curves (short-run marginal cost curves) of the firms are added to obtain the short-run supply curve of the competitive industry is illustrated in Fig. 23.10. Suppose there are 200 firms in the competitive industry. We further assume that all firms are alike in respect of cost of production. In Fig. 23.10 (*a*) *SMC* represents short-run supply curve of an individual firm. At price OP_1, an individual firm will produce and supply OM_1 quantity of the product. Since there are 200 such firms in the industry, the whole industry will produce and supply 200 × OM_1 quantity of the product. Therefore, in Fig. 23.10 (*b*), 200 × OM_1 quantity is plotted against the price OP_1.

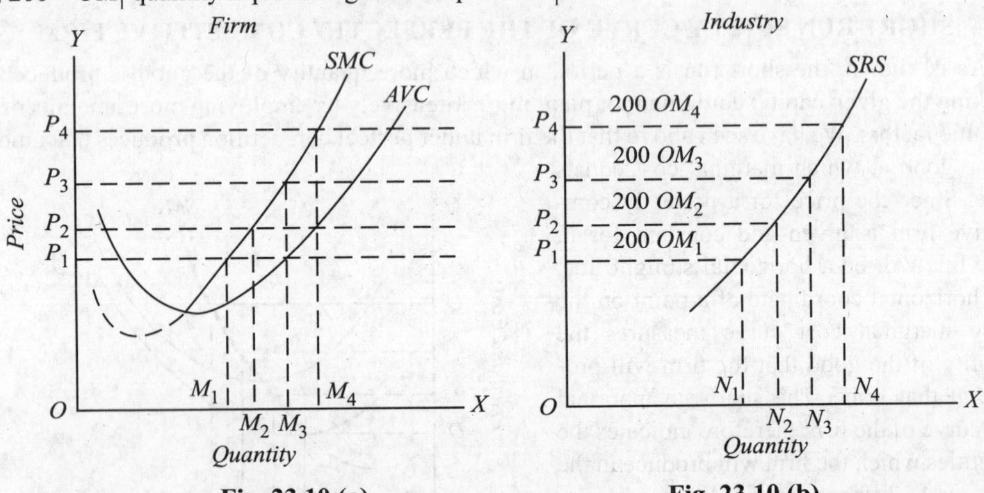

Fig. 23.10 (a) **Fig. 23.10 (b)**

ON_1 is equal to 200 × OM_1. It should be carefully noted that in Figs. 23.10 (*a*) and (*b*) while the scale on *Y*-axis is the same, the scale on *X*-axis differs very much. The scale on *X*-axis in Fig. 23.10 (*b*) has been compressed very much to accommodate large quantities. At price OP_2 the individual firm will produce and supply OM_2 amount of the good, while the whole industry will supply ON_2 which is equal to 200 × OM_2 amount of the product. In the same way, industry will produce and supply

$(200 \times OM_3) = ON_3$ output at price OP_3, and $(200 \times OM_4)$ or ON_4 at price OP_4. Likewise, the industry's supply can be determined for all other prices. *The short-run supply curve of the industry will always slope upward.* This is because the short-run marginal cost curves of the firms (*i.e.* their short-run supply curves) always slope upward above the minimum point of the average variable cost curves. The slope and the elasticity of short-run supply curve of the industry will obviously depend upon the slope and elasticity of the marginal cost curves of individual firms in the industry.

That the short-run supply curve of the industry under perfect competition is a lateral or horizontal summation of the short-run supply curves (*i.e. SMC*) of the firms in it is subject to an important qualification. This is that the simultaneous expansion of output by all the firms in it (*i.e.* the expansion of output by the industry) and therefore the increase in demand for the resources or inputs to be used for production will have no effect on the prices of these resources, that is, for the whole industry these resources or inputs are perfectly elastic. But whereas the expansion or contraction of output of the individual firm and therefore the changes in its demand for resources is not likely to affect their prices, the simultaneous expansion or contraction of all firms in the industry may mean a significant change in the demand for these resources and will therefore affect their prices. If expansion of industry output and therefore the increase in demand for resources raises the prices of these resources, then the cost curves of the individual firms will shift upward. On the other hand, if the expansion of the industry brings about a fall in the prices of resources, the cost curves of individual firms will shift downward. It may be that prices of some resources may rise and some others may fall with the expansion of the industry. In that case, the shift in the cost curves of the firms will depend upon whether the increase or decrease in resource prices is predominant.

When the cost curves of individual firms shift due to the change in resource prices, then the supply curve of industry cannot be obtained by summing up laterally the short-run supply curves of the firms, because then with every increase in the industry output, cost curves of firms change. In this case, therefore, when external effects are present, that is, when resource prices change with the expansion of industry, the short-run supply curve of the industry can be obtained by summing up the *equilibrium outputs* with different cost curves of all firms at each possible market price of the product.

LONG-RUN SUPPLY CURVE OF THE INDUSTRY UNDER PERFECT COMPETITION

In the long run the firms can change their capital equipment and the other fixed factors and also the number of firms can vary in response to changes in the demand for a commodity. In the long run, when new firms can enter and old ones can leave the industry the firm is in equilibrium at the minimum point of the long-run average cost curve, where the long-run marginal cost curve intersects it. Thus a firm under perfect competition in the long-run equilibrium is forced to produce only at one point of the long-run marginal cost curve at which it cuts the average cost curve. Price in the long run is equal to both long-run marginal cost and minimum average cost. Therefore, the firm in the long run will produce and supply an output indicated by the minimum point of its long-run average cost curve, that is, its optimum size. Of course, this optimum size of the firm may change with the change in the number of firms in the industry. The long-run supply in the industry will be determined by the changes in the optimum size of the firms (i.e. long-run supply of output by individual firms) but mainly by the variation in the number of firms in the industry.

It should be noted that *long-run supply curve is defined as the supply by the existing as well potential firms in the industry in the long run.* A little reflection will show that the long-run supply curve of the industry cannot be the lateral summation of the long-run marginal cost curves of a given number of firms. This is because of three reasons. Firstly, as explained above, *whole* of the long-run marginal cost *curve* does not constitute the long-run supply curve of an individual firm, *only one point* of the long-run marginal cost curve at which it cuts the average cost curve (that is, the minimum point of the average cost curve) constitutes long-run supply of the individual firm. Secondly, the number of firms varies at different prices or demand conditions in the long run.

Thirdly, we cannot sum up any existing long-run marginal cost curves of the firms to obtain the long-run supply curve of the industry because with the expansion of the industry in the long run cost curves of the firms shift due to the emergence of external economies and diseconomies. In order to know the output supplied by the firms of an industry in the long run, we need to know the position or level of the cost curves of the firms as well as number of firms in the industry at a given demand and price of the product.

As we explained in a previous chapter, *external economies and diseconomies are those which are realised by all firms in an industry as a result of the expansion of the industry as a whole.* The creation of external economies by an expanding industry will shift the cost curves of the firms downward. On the other hand, the creation of external diseconomies will shift the cost curves of the firms downward. Whether a given industry will experience upward or downward shift in the cost curves depends upon the *net or combined effect* of the external economies and diseconomies. When with the expansion of an industry the external economies outweigh the external diseconomies so that there are *net external economies,* cost curves of the firms will shift downward. On the other hand, if with the expansion of an industry external diseconomies are stronger than the external economies so that there are *net external diseconomies*, cost curves of the firms will shift upward.

In a previous chapter, we mentioned the important examples of external economies which may accrue to the firms of an expanding industry. To repeat in brief, main examples of external economies which an expanding industry may reap are : (*i*) the availability of tools, machinery, raw materials etc., at lower prices, (*ii*) discovery and diffusion of a superior technical knowledge, and (*iii*) economic use of the waste products.

With the growth of an industry some raw materials, tools, capital equipment etc. may become available because some specialised subsidiary and correlated firms may be set up which produce them on a large scale at lower cost per unit and are therefore in a position to supply them at lower prices. Thus, while explaining the reasons for falling costs in an industry due to external economies. Professor Robert Heilbroner writes: "The source of these changes in cost does not lie within the firm, in the relative efficiency of various factors mixed. Rather they are changes thrust upon the firm—for better or worse—by the interaction of the growing industry of which it is a part and the economy as a whole. A new industry, for example, by its very expansion may bring into being satellite firms that provide some of its necessary inputs; and as the main new industry grows, the *satellites also expand and thereby realise economies of scale* that will benefit the main industry itself. The automobile industry was surely an instance of such long-run falling costs (for a long period, at least) resulting from the economies of scale enjoyed by makers of tires, makers of batteries, and other equipment"[2]. Further, as the industry expands, trade journals may appear which help in discovering and spreading technical knowledge. Again, with the growth of an industry some specialised firms may come into existence which work up its waste products. The industry can then sell them at good price. There is every possibility of external economies to be reaped when a young industry grows in a new territory.

On the contrary, when a well-established and good-sized industry expands further, it may experience external diseconomies. *As more firms enter the industry, competition among them may push up the prices of scarce raw materials, skilled labour and other specialised inputs*. Moreover, the additional units of productive inputs being obtained by the industry may be less efficient than the previous ones. All these external diseconomies will raise the average and marginal cost curves of the firms. To quote Professor Heilbrone again, "Industries may also experience long-run rising costs if their expansion pushes them up against factor scarcity of a stubbornly inelastic kind. Extractive

2. Robert L. Heilbroner, *The Economic Problem*, Prentice Hall, 1970, p. 512 (italics added)

industries, for example, may be forced to use progressively less accessible mineral deposits; or agricultural industries may be forced to use progressively less fertile or less conveniently located land. Such industries would experience a gradual rise in unit costs as their output is increased".[3]

We thus see that external economies and diseconomies play a vital role in shaping the long-run supply of an industry. *Whether a particular industry on expansion will experience the phenomenon of rising costs or falling costs or constant costs will depend upon the net result of external economies and diseconomies.* The long-run supply curve of perfectly competitive industry will therefore have different shapes depending upon whether the industry in question is a (*i*) constant cost industry, (*ii*) increasing cost industry; (*iii*) decreasing cost industry.

It follows from above that at a given price the quantity supplied by an industry in the long run will be determined by the optimum output of a firm in the long run (*i.e.*, output corresponding to minimum long-run average cost) multiplied by the number of firms in the industry at that price. With the change in the price of the product following a change in demand conditions, the number of firms in the industry will change and also the cost curves of the firms will shift on account of the creation of external economies and diseconomies. As a result, the quantity supplied by the industry will change at a new price. *The long-run supply curve of the industry may either be sloping upward or be a horizontal straight line, or be sloping downward depending upon whether the industry in question is increasing cost, constant cost, or decreasing cost.* How the long-run supply curve of a perfectly competitive industry under these three types of cost conditions is obtained is explained in the next chapter.

NUMERICAL PROBLEMS

Problem 1. *Suppose a firm is operating under perfectly competitive conditions in the market in the short run. It faces the following revenue and cost conditions :*

$$TR = 12Q$$
$$TC = 2 + 4Q + Q^2$$

Determine the equilibrium level of output and total profits made.

Solution. Profits are maximised when the firm equates marginal cost (MC) with MR and marginal cost is rising. Thus, in order to obtain the equilibrium output we equate $MC = MR$.

$$TR = 12Q$$
$$MR = \frac{dTR}{dQ} = 12$$
$$MC = \frac{dTC}{dQ} = 4 + 2Q$$

In equilibrium,

$$MC = MR$$
$$4 + 2Q = 12$$
$$Q = 4$$
$$\text{Total profits } (\pi) = TR - TC = 12Q - (2 + 4Q + Q^2)$$

Substituting $Q = 4$, we have

$$\pi = (12 \times 4) - 2 - (4 \times 4) - 16$$
$$= 48 - 34 = 14$$

Note that in order to ensure the fulfilment of second order condition, we have to test whether MC is rising. For this, we take the derivative of MC, that is, the second derivative of TC function.

3. *Ibid*, p. 512.

$$MC = \frac{dTC}{dQ} = 4 + 2Q$$

The second derivative of *TC* function $\frac{d^2TC}{dQ^2} = +2$

The positive sign of the second derivative of *TC* function implies that *MC* is rising.

Problem 2. *Suppose that revenue and total cost of a firm are given by the equations $R = 60Q$ and $C = 10 + 5Q^2$, (Q = output).*

What will be the profit-maximising output and total profit of the firm ?

[*C.U.,B.Com., (H), 1998*]

Solution. We have solved the above example through $MC = MR$ approach. Let us solve this example through *TC* and *TR* approach.

$$\pi = TR - TC$$

$$= 60Q - 10 - 5Q^2 \quad ...(i)$$

Profits will be maximum at the level of output at which first derivative of total profit function = 0. Thus,

$$\frac{d\pi}{dQ} = 60 - 10Q = 0$$

$$10Q = 60$$

$$Q = 6$$

For the second order condition to be fulfilled, the second derivative of profit function should be negative. Taking the second derivative of the profit function, we have

$$\frac{d^2\pi}{dQ^2} = -10$$

Thus, the second order condition is satisfied. Substituting $Q = 6$ in the profit function (*i*), we have

$$\pi = 60 \times 6 - 10 - 5(6)^2$$

$$= 360 - 10 - 5 \times 36$$

$$= 360 - 10 - 180 = ₹\ 170$$

Problem 3. *A firm's total variable cost is given by the following :*

$$TVC = 75Q - 10Q^2 + Q^3$$

Will the firm produce the product if price of the product is ₹ 40?

Solution. A firm produces a product if price of the product exceeds its minimum average variable cost,

$$AVC = \frac{TVC}{Q} = \frac{75Q - 10Q^2 + Q^3}{Q}$$

$$= 75 - 10Q + Q^2 \quad ...(1)$$

AVC is minimised at the output level when

$$\frac{d(AVC)}{dQ} = 0$$

Taking a derivative of *AVC,* we have

$$\frac{d(AVC)}{dQ} = -10 + 2Q$$

Therefore, AVC will be minimum when

$$-10 + 2Q = 0$$

$$2Q = 10$$

$$Q = \frac{10}{2} = 5$$

Now substituting the value of Q in equation for AVC, we have

$$\text{Minimum} \quad AVC = 75 - 10 \times 5 + 25$$

$$= 100 - 50 = 50$$

Thus price of the product is less than the minimum average cost, the firm will therefore not produce the product at price equal to ₹ 40

Problem 4. *Given the following short-run cost function of a firm*

$$TC = 1000 + 10Q^2$$

What is the firm's shut-down point ? Derive the expression for firm's short-run supply curve.

Solution. A firm's short-run supply curve is firm's short-run marginal cost curve. To obtain marginal cost function we have to obtain the derivative of total cost function. Thus,

$$MC = \frac{dTC}{dQ} = 2 \times 10Q = 20Q.$$

To get the short-run supply curve of a firm we set price equal to marginal cost. Thus,

$$P = 2Q$$

or

$$Q = \frac{P}{2} \qquad \text{... (1)}$$

The supply curve of a firm is that portion of marginal cost curve that lies above the minimum point of of the average variable cost (AVC) curve. AVC is minimised at the output level where its first derivative equals zero.

$$\text{AV} = \frac{TVC}{Q} = \frac{10Q^2}{Q} = 10Q$$

Setting the first derivative of AVC equal to zero we have

$$\frac{d\,(10Q)}{dQ} = 0$$

or

$$Q = 0$$

Thus, AVC is minimised when output (Q) is equal to zero. Thus, the firm's shut-down point is $Q = 0$. It therefore follows that the entire marginal cost function found in (1) above, namely, $Q = \frac{P}{20}$ or $P = 20Q$ represents the short-run supply curve of the firm.

Problem 5. *The following table shows the total cost schedule of a competitive firm. What is total fixed cost of the firm? Using marginal analysis find the profit-maximising level of output if price of the product for the firm is ₹ 8. Calculate the profits at this output level.*

Quantity sold (units)	*TC* (₹)
0	5
1	7
2	10
3	12
4	15
5	23
6	33
7	40

Solution. Since the firm incurs a cost of ₹ 5 even at zero level of output, the firm's total fixed cost is ₹ 5. In order to determine the profit-maximising level of output (*i.e.* equilibrium output) we have to first calculate marginal costs at different levels of output. We do so in the following table.

Quantity sold (units)	*TC* (₹)	*MC* (₹) $(TC_n - TC_{n-1})$
0	5	–
1	7	2
2	10	3
3	12	2
4	15	3
5	23	8
6	33	10
7	45	12
8	58	13

To maximise profits the competitive firm will equate price with the marginal cost. It will be seen from the table that at output level of 5 units, price equals marginal cost (₹ 8). Thus the firm's maximising profit level of output is 5 units.

$$\text{Profits} = TR - TC = (P.Q) - TC$$
$$= 8 \times 5 - 23 = ₹\ 17$$

Problem 6. *A competitive firm has the following data:*

Output (Q)	*TFC* (₹)	*TVC* (₹)
0	100	0
1	100	50
2	100	90
3	100	140
4	100	200
5	100	280
6	100	380

(*i*) *If price = ₹ 60, how many units will the firm produce* ?

(*ii*) *What will be level of profits/losses at this level of production* ?

(*iii*) *Will the firm operate in the short run* ? (*iv*) *What happens in the long run* ?

Solution. (*i*) In order to obtain how many units of output the competitive firm will produce, given the price, depends on the marginal cost of production. Note that $MC = \dfrac{\Delta TVC}{\Delta Q}$ where $\Delta Q = 1$ or, in other words, $MC = TVC_n - TVC_{n-1}$. We calculate below *MC* at different levels of output.

Output (Q)	*TVC* (₹)	*MC* $TVC_n - TVC_{n-1}$
0	0	–
1	50	50
2	90	40
3	140	50
4	200	60
5	280	80
6	380	100

In order to be in equilibrium a perfectly competitive firm will produce the number of units of output at which price = *MC*. It will be seen from the above table that price of ₹ 60 equals marginal cost when it is producing 4 units of output.

(*ii*) With 4 units of output total cost = $TVC + TFC = 200 + 100 = 300$. Total revenue ($TR$) earned by producing 4 units of output will be $P \times Q = 60 \times 4 = 240$. Since total cost (₹ 300) exceeds total revenue (₹ 240), the firm will be incurring losses equal to $300 - 240 = 60$ rupees though it will be minimising losses.

(*iii*) Since at 4 units of output total revenue of ₹ 240 exceeds the total variable cost (200), the firm will continue to operate in the short run as it will be recovering a part of fixed costs in addition to total variable costs.

(*iv*) In the long run, some firms will leave the industry causing price of the product to rise. The firms that remain in the industry must earn zero economic profits, that is, for them price must be equal to the long-run average cost.

QUESTIONS FOR REVIEW

1. Explain the various features of a perfectly competitive market. How is price of a commodity determined under it?
2. What is meant by firm's equilibrium? Explain the conditions of short-run equilibrium of a firm under perfect competition. Is equality of marginal revenue with marginal cost sufficient condition for equilibrium of the firm?
3. Can a firm under perfect competition operate in the short run when it is making losses? If so, under what conditions?
4. When does a firm working under perfect competition decide to (*a*) shut down in the short run, (*b*) leave the industry in the long run?
5. Explain the conditions of long-run equilibrium of a firm operating under conditions of perfect competition. A firm operating under perfect competition tends to be of optimum size. Explain. What is significance of this for the consumer and the community as a whole?
6. "If there is free entry of new firms in the competitive industry, price must fall to the level of minimum long-run average cost." Explain.
7. "In the long-run equilibrium, every firm in a competitive industry earns zero profits. Thus, if the price falls, all of these firms will be unable to stay in business." Examine this statement.
8. *A perfectly competitive firm has the following total cost function* :

Total output (Units)	*Total cost* (₹)
0	20
1	30
2	42
3	55
4	69
5	84
6	100
7	117

How much the firm will produce if the price of the product in the market is ₹ 14 per unit? How will it change its output if price rises to ₹ 16 per unit?

[**Hints** : Under perfect competition MR = Price which is given to be equal to ₹ 14 or ₹ 16. For equating this MR or Price with MC, make the third column in the above table calculating MC at various levels of output and see at what output level $MC = MR$.]

9. (*a*) Explain the concept of supply curve. How is this relevant only under perfect competition?

(*b*) Derive a short-run supply curve of the firm operating under perfect competition. Explain that short-run supply curve of a firm *always* slopes upward.

10. What is the short-run supply curve of a competitive firm? What is its likely shape and why? *D.U. B.A (Hons) 1988*

11. What is meant by equilibrium of the industry? How would you derive short-run supply curve of the competitive industry? Does it always slope upward ?

12. What are external economies and external diseconomies? Give some examples. What role do they play in determining the long-run supply curve of a competitive industry?

13. Explain why a firm working under perfect competition is in equilibrium at the level of output at which marginal cost is equal to price and also at which marginal cost is rising.

14. Show for a perfectly competitive firm to be in long-run equilibrium price should not only be equal to marginal cost but must also be equal to long-run average cost of production.

15. Explain under perfect competition how the conditions of firm's long-run equilibrium differ from those of short-run equilibrium.

16. Firms working under perfect competition earn *zero economic profits* in their long-run equilibrium. Why then firms do not quit the industry in the long run?

17. Tick the right answer. The demand curve for a perfectly competitive firm :

(*a*) is a horizontal straight line (*b*) is downward sloping

(*c*) is perfectly inelastic (*d*) is positively sloping

18. Tick the right answer. Price in equilibrium of a perfectly competitive firm

(*a*) equals marginal cost (*MC*) of production.

(*b*) is greater than *MC* (*c*) is less than *MC*

19. Which ones of the following are correct? Perfectly competitive firms :

(*a*) are price makers (*b*) are price takers

(*c*) are quantity adjusters, but not price makers

(*d*) do not earn even normal profits in the long run.

[**Hint.** Both (*b*) and (*c*) are correct answers.]

20. Which of the following is *not correct* in case of long-run equilibrium of a firm under perfect competition?

(*a*) occurs where average revenue curve of a firm is tangent to its long-run average total cost curve (*LAC*).

(*b*) occurs when a firm earns zero economic profits.

(*c*) occurs when a firm makes only nromal profits in the long run.

(*d*) some firms may suffer losses.

[**Hint.** (*a*) is incorrect because firms leave the industry in the long run if they suffer losses.]

21. Short-run supply curve of a perfectly competitive firm :

(*a*) is a horizontal straight line

(*b*) always slopes upward

(*c*) can slope downward

Give reasons for the correct answer.

22. Which is correct in case of long-run supply curve of a perfectly competitive industry?

(*a*) is a lateral summation of the long-run marginal cost curves of the firms

(*b*) always slopes upward

(*c*) is always perfectly elastic

(*d*) none of the above

23. Explain why a perfectly competitive firm has a *horizontal demand curve*. Does it imply that a firm can sell as much as it likes at the given price? If so then why does not a perfectly competitive firm expand its output and sales without limit at the current given price?

24. Which of the following features of perfect competition is primarily responsible for a firm's zero economic profits in the long run?

(*a*) A large number of firms operating in it

(*b*) Identical costs

(*c*) Freedom of entry and exit

(*d*) Perfect information on the part of firms and buyers.

[**Hint.** (*c*) *is the correct answer*]

CHAPTER 24
PRICE DETERMINATION UNDER PERFECT COMPETITION

We are now in a position to explain how price is determind under perfect competition . In the preceding chapters we have explained in detail the forces of demand and supply which by their interaction determine the price of the product. In the discussion of the theory of demand we assumed that an individual buyer is unable to influence the price of the product and therefore takes the prevailing price of the product as a given for him and spends so much money on various goods as to obtain maximum satisfaction from his total outlay. On this assumption we derived the demand curve for a product with the aid of Marshallian cardinal utility analysis as well as with indifference curve technique. Likewise, we discussed the theory of supply by assuming that an individual entrepreneur or a firm cannot individually affect the price of the product and accepts the ruling price in the market as given and constant for him. With this assumption we have explained how much quantity of a product the firm and indsutry will supply at various prices.

Now, the intersection of demand and supply curves determines the price of the product. It is not the demand of a single buyer and the supply of a single seller that go to determine the price of the product, but the demand of all buyers of a product taken together (*i.e.*, market or industry's demand curve) and the supply of all firms selling the product taken together (*i.e.*, the supply curve of the industry) that determine the price of the product. The industry's demand curve shows the various quantities of the good which will be demanded by the consumers at different prices. On the other hand, the industry's supply curve shows the various quantities of the good that the industry will be ready to supply at different prices. That particular price at which quantity demanded is equal to quantity supplied will finally settle down in the market.

TWO APPROACHES IN PRICE THEORY

In regard to pricing under perfect competition, two main approaches have been adopted. One approach has been followed by famous English economist Alfred Marshall who adopted the *partial equilibrium approach* and the second approach has been adopted up by Walras and is called *general equilibrium approach*. We shall explain below both these approaches in price theory.

Marshall's Partial Equilibrium Analysis

In partial equilibrium approach to the pricing, we seek to explain the price determination of commodity, keeping the prices of other commodities constant and also assuming that the various commodities are not interdependent. In explaining partial equilibrium approach, Marshall writes: "The forces to be dealt with are, however, so numerous that it is best to analyse a few at a time and to work out a number of partial solutions as auxiliaries to our main study. Thus we begin by isolating the primary relations of supply, demand and price in regard to a particular commodity. We reduce to inaction all other forces by the phrase '*other things being equal*'. We do not suppose that they are inert, but for the time we ignore their activity. This scienctific device is a great deal older than science; it is the method by which consciously or unconsciously sensible men dealt from time immemorial with every difficult problem of everyday life."[1]

1. Alfred Marshall, *Principles of Economics*, 8th ed. Macmillan.

Thus in Marshallian explanation of pricing under perfect competition, demand function for a commodity is drawn with the assumption that prices of other commodites, tastes and incomes of the consumers remain constant. Similarly, supply curve of commodity is constructed by assuming that prices of other commodities, prices of resources or factors and production function remain the same. Then Marshall's partial equilbrium analysis seeks to explain the price determination of a single commodity through the intersection of demand and supply curves, with prices of other goods, resource prices etc., remaining the same. Prices of other goods, resource prices, incomes, *etc*., are the *data* of the system which are taken as given to explain the determination of price-output equilibrium of a single commodity. Given the assumption of *ceteris paribus* it explinas the determination of a price of a goods, say *X, independently of the prices of all other goods*. With the change in the data, new demand and supply curves will be formed and, corresponding to these, new price of the commodity will be determined. Thus partial equilibrium analysis of price determination also studies how the equilibrium price changes as a result of change in the data. But given the independent data the partial equilibrium analysis explains only the price determination of a commodity in isolation and does not analyse how the prices of various goods are interdependent and inter-related and how they are *simultaneously* determined.

It should be noted that *partial equilibrium analysis is based on the assumption that the changes in a single sector do not significantly affect the rest of the sectors*. Thus, in partial equilibrium analysis, if the price of a good changes, it will not affect the demand for other goods. Prof. Lipsey rightly writes: "All partial equilibrium analyses are based on the assumption of *ceteris paribus*. Strictly interpreted, the assumption is that all other things in the economy are unaffected by any changes in the sector under consideration (say sector *A*). This assumption is always violated to some extent, for anything that happens in one sector must cause changes in some other sectors. What matters is that the changes induced throughout the rest of the economy are sufficiently small and diffiuse so that the effect they in turn have on the sector A can be safely ignored".[2]

Walras General Equilibrium Analysis

In general equilibrium analysis, put forward by French Economist Walras the price of a good is not explained to be determined independently of the prices of other goods. Since the changes in price of good *X* affect the prices and quantities demanded of other goods and in turn the changes in prices and quantities of other goods will affect the quantity demanded of the good *X, the general equilibrium approach explains the simultaneous determination of prices of all goods and factors*.

As stated above, partial equilibrium approach assumes that the effect of the change in price of a good *X* will be so diffused in the rest of the economy (*i.e*., over all other goods) so as to have negligible effect on the prices and quantities of other individual goods. Therefore, where the effect of a change in the price of a good on the prices and quantities of some other goods is significant, as is there in the case of inter-related goods, that is, substitutes and complementary goods, the partial equilibrium approach cannot be validly applied in such cases and therefore there is need for applying general equilibrium analysis which should explain the mutual and simultaneous determination of their prices and quantities.

General equilibrium analysis deals with inter-relationship and inter-dependence between equilibrium ajustment with each other. General equilibrium exists when at the going prices, the quantities demanded of each product and each factor are equal to their respective quantities supplied. A change in the demand or supply of any good, or factor would cause changes in prices and quantities of all other goods and factors and there will begin the process of adjustment and readjustment in demand, supply and prices of other goods and factors till the new general equilibrium is established. Indeed, the general equilibrium analysis is solving a system of simultaneous equations. In a general

2. R.G. Lipsey, *An Introduction to Postive Economics*, 3rd edition 1971. p. 404

equilibrium system, the quantity demanded of *each good* is described by an equation in which its quantity demanded is a function of *prices of all goods*. Likewise, in general equilibrium analysis, quantity supplied of each good is considered to be the function of price of all factors of production.

In a general equilibrium system the prices of all *goods* affect the quantity demanded of *each* good. Further, the prices of the all factors affect the quantity supplied of each good. Besides these crucial equations, there will be equations determining *the price of each of the factors of production*. As noted above, a change in any of the demand or supply equations would cause changes in *all prices and quantities* and as a result the system will tend to move to the new general equilibrium.

To explain the inter-relationship and interdependence among the prices and quantities of goods and and factors and ultimately to explain the determination of the relative prices of all goods and factors, the proportion in which different goods are being produced and different factors are being used for the production of different goods is the essence of general equilibrium analysis. However, in this book we shall mainly confine ourselves to the partial equilibrium approach to the determination of relative prices. In a separate part of this book we shall explain the general equilibrium analysis using Edgeworth Box diagram.

PRICE DETERMINATION: EQUILIBRIUM BETWEEN DEMAND AND SUPPLY

There was a dispute among earlier economists who came befroe Marshall as to whether it is supply of a good or the demand for it that determines its price. Broadly speaking, there were two schools of thought in this regard. One school of thought believed that it is cost of production, that is, the force working on the supply side, which determiens the price of the product. The other school of thought held the view that it is the utility or more precisely the marginal utility that determines the demand and therefore the price of the product. But each school of thought took one-sided view of the pricing problem. The credit of finding the true answer to the pricing problem goes to Marshall who held that both demand and supply were equally important in determining the price of a product. In other words, Marshall said that both the marginal utility of a product and the cost of producing it took part in determining price. He likened the price determination to the cutting of a piece of a paper by a pair of scissors. This famous analogy is worth quoting here. "We might as reasonably dispute whether it is the upper or the under blade of a pair of scissors that cuts a piece of paper as whether value is governed by utility or cost of production. It is true that when one blade is held still and the cutting is effected by moving the other, we may say with careless brevity that the cutting is done by the second, but the statement is not strictly accurate and is to be excused only so long as it claims to be merely a popular and not a strictly scientific account of what happens."[3]

Neither the upper blade nor the lower one taken individually can cut the paper; both are required to do the work of cutting. The lower blade may be held stationary and only the upper one may be moved, yet both are indispensable for cutting the paper. Similarly, both demand and supply are essential forces for determining prices of goods. Professors Stonier and Hague rightly remark, "The only really accurate answer to the question whether it is supply or demand which determines price is that it is both. At times it will seem that one is more important than the other, for one will be active and the other passive. For example, if demand remains constant but supply conditions vary, it is demand which is passive and supply active. But neither is more or less important than the other in determining price."[4]

In the chapters on theory of demand we saw that the demand curve of a commodity normally slopes downward. In other words, with the fall in price, quantity demand rises and *vice versa*. In the theory of firm we studied that the supply curve of a commodity usually slopes upward. In other words, an industry will offer to sell more quantity of a good at a high price than at a lower one. The

3. Alfred Marshall, *Principles of Economics*, 8th edition, p. 344
4. Stonier and Hague: *A Textbook of Economic Theory*, 4th edition, 1972, p. 155

level of price at which demand and supply curves intersect each other will finally come to stay in the market. In other words, the price which will come to prevail in the market is one at which quantity demanded is equal to quantity supplied. *The price at which quantity demanded equals quantitty supplied is called equilibrium price*, for at this price the two forces of demand and supply exactly balance each other. *The quantity of the good which is purchased and sold at this equilibrium price is called equilibrium amount*. Thus, the intersection of demand and supply curves determines price-quantity equilibrium.

Only at the equilibrium price wishes of both the buyers and sellers are satisfied. If prices were greater or less than the equilibrium price, the buyers' and sellers' wishes would be inconsistent; either the buyers would demand more than the amount offered by sellers, or the sellers would be ready to supply more than the amount demanded by the buyers. If price was greater than the equilibrium price, quantity supplied would exceed quantity demanded. It means some of the sellers will not be able to sell the amount of the good they wanted to supply. These unsatisfied sellers would try to dispose of the unsold amount of the good by bidding down price. The price will go on declining until the quantity demanded equals quantity supplied.

On the other hand, if price was lower than the equilibrium price, the quantity demanded would exceed quantity supplied. Some buyers would not be able to obtain the amount of the good they wanted to pruchase at the prevailing price. They will therefore bid up price in their effort to get all that they desired to buy. The price will go on rising until the quantity demanded and quantity supplied are again equal. We thus see that the price which will settle down can be neither greater than, nor less than the equilibrium price. It is the equilibrium price which will finally come to stay in the market.

We illustrate below the process of price determination with the aid of the demand and supply schedules and the curves which will make the whole thing very clear. In the table below are given demand and supply schedules relating to a woollen cloth. It will be seen there that only at a price of ₹ 40, quantity demanded is equal to quantity supplied. Therefore, price of ₹ 40 will settle down in the market. Only at this price are all buyers and sellers satisfied. Thus, price of ₹ 40 is the equilibrium price and quantity exchanged at this price is 120 thousands metres. 120 thousand metres is the equilibrium amount. Once this equilibrium price comes to prevail in the market, there will be no tendency for it to change. If price is greater or lower than ₹ 40, then it will tend to change and come to the level of ₹ 40. For instance, if the price was ₹ 50, the buyers would demand 80

Table 24.1
Equilibrium Between Demand and Supply

Price ₹	*Quantity Demanded (in thousand-metres)*	*Quantity Supplied (in thousand metres)*	*Pressure on Price*
10	240	20	↓
20	200	40	↓
30	160	80	↓
40	120	120	Equilibrium
50	80	160	↓
60	40	200	↓
70	20	240	↓

thousand metres, while the sellers will be ready to supply 160 thouand metres. Thus, the sellers would not be able to sell the whole quantity of the cloth they want to supply and would therefore compete with each other to sell the extra quantity of the cloth by bidding down the price. In this way, the price will go on falling until the equilibrium level of ₹ 40 is reached where the quantity which the buyers would demand equals the quantity which the sellers would be prepared to sell. Further, suppose the price was below the equilibrium price, say ₹ 30. At price of ₹ 30, the quantity demanded by the buyers is 160 thousand metres while the sellers are willing to supply only 80 thousand metres. The unsatisfied buyers will then bid up the price. In this way, price will go on rising till it fianlly comes to settle down at ₹ 40 where all buyers and sellers are satisfied.

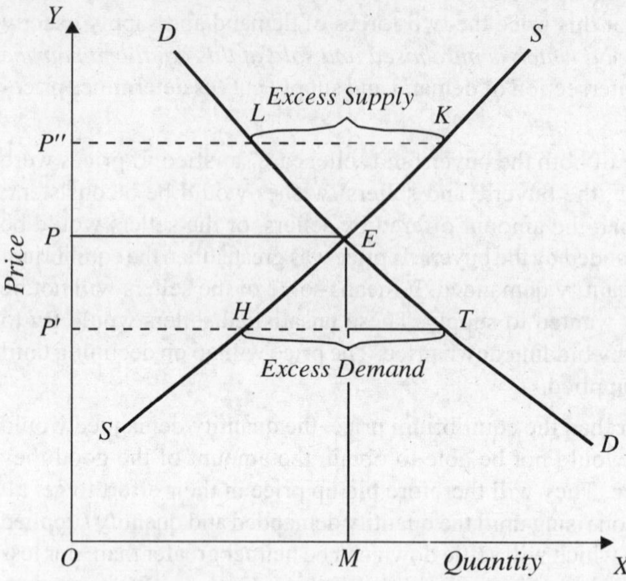

Fig. 24.1. *Determination of Price through Intersection of Demand and Supply Curves*

In graphical terms, the equilibrium between demand and supply is depicted in Fig. 24.1 where *DD* is the demand curve sloping downward and *SS* is the supply curve sloping upward. Demand and supply are in equilibrium at point *E* where two curves intersect each other. It means that only at price *OP* (corresponding to the intersection point), the quantity demanded is equal to quantity supplied. *OM* is the equilibrium quantity which is exchanged at price *OP*, if price is greater than the equilibrium price, say *OP″*, the quantity demanded by the buyers is *P″L*, while the quantity offered to supply by the sellers is *P″K*. Thus *LK* is the excess supply which the buyers will not take off from the market at price *OP″*. In order to dispose of this excess supply, the sellers will compete with each other and in doing so they will bring down the price. Thus there will be tendency for the price to fall to the level of equilibrium price *OP*.

At price *OP′*, which is less than the equilibrium price, the buyers demand *P′T*, the sellers are prepared to supply only *P′H*. *HT* represents excess demand. The unsatisfied buyers will compete with each other to obtain the limited supply of cloth and in their effort they will bid up the price. Thus there will be tendency for the price to rise to the level of equilibrium price *OP* where all wanting to buy and all wanting to sell will be satisifed.

From above it follows that if the price is above or below the equilibrium price, certain forces in the system will operate to bring the price to the level at which there is equilibrium between demand and supply or what is often called market equilibrium. The price at which demand and supply curves intersect each other will eventually come to prevail in the market.

CHANGES IN MARKET EQUILIBRIUM

Changes in either demand or supply cause changes in market equilibrium. Several forces bringing about changes in demand and supply are constantly working which cause changes in market equilibrium, that is, equilibrium prices and quantities. In the earlier chapters we have explained the factors or forces which cause shift in demand and supply curves. The demand may increase or decrease, the supply curve remaining unchanged. This would cause a change in equilibrium price and quantity. Similarly, the increase or decrease in supply, the demand curve remaining constant, would have an impact on equilibrium price and quantity. Both supply and demand for goods may change simultaneously causing a change in market equilibrium. Supply-demand analysis is an important tool of economics with which we can make forecasts about how prices and quantities will change in response to changes in demand and supply. We explain below the impact of changes in demand and supply on equilibrium price and quantity.

Impact of Increase in Demand on Market Equilibrium

Increase in demand affect prices and quantities. Suppose there is increase in income of the working class due to the enhancement of their salaries by the Pay Commission. As a result of this increase in income, their demand for cloth for shirting will increase causing a shift in the entire demand curve for cloth to the right. This will raise the equilibrium price and quantity of cloth, the supply curve of cloth remaining unchanged as is shown in Fig. 24.2. It is important to understand the chain of causation which leads to the increase in price and quantity as a result of increase in demand.

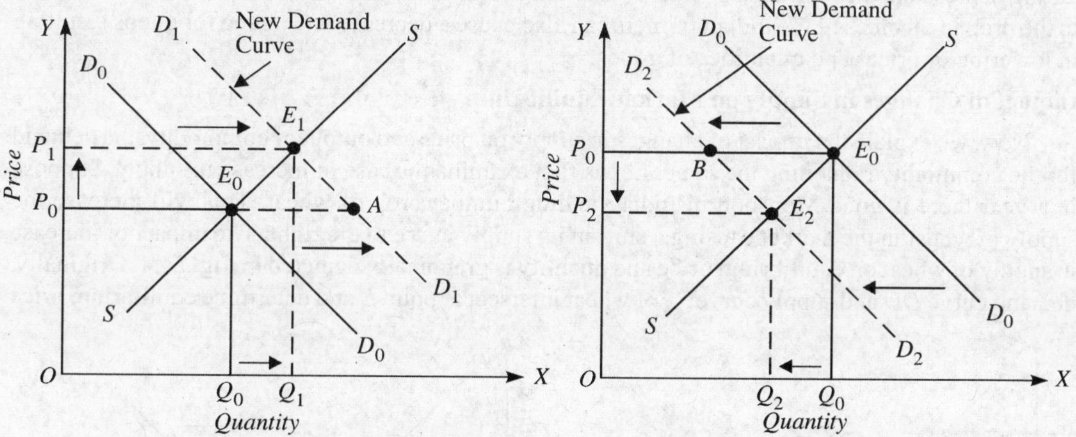

Fig. 24.2. *Impact of increase in Demand on Price and Quantity (P↑, Q↑).*

Fig. 24.3. *Impact of Decrease in Demand on Price and Quantity. (P↓, Q↓).*

Consider Fig. 24.2, in which D_0D_0 and *SS* are the initial demand and supply curves of cloth. The increase in income causes a shift in the entire demand curve to the right to the new position D_1D_1 while the supply curve *SS* remains constant. It will be observed from Fig. 24.2, that with the shift in demand curve to D_1D_1 at the old price OP_0 excess demand of cloth equal to E_0A has emerged. This excess demand of the good exerts upward pressure on price. This will result in rise in price to OP_1 where again quantity demanded equals quantity supplied and new market equilibrium is attained and excess demand is eliminated. It is worth noting that increase in demand is the most important factor causing inflation, that is, rise in prices and is generally described as *demand-pull* inflation. Though the term inflation is used in the context of *a rise in general price level*, but it has roots at the micro level (*i.e.*, in case of individual goods). Apart from increase in income, a favourable change in consumer's preferences for a particular good, rise in price of its substitutes will also cause an increase in demand for a good.

Impact of Decrease in Demand on Market Equilibrium

Now, take the opposite case of the impact of decrease in demand on market equilibrium, the supply curve remaining the same. As explained earlier, the decrease in demand causes a shift in the entire demand curve to the left. This is graphically shown in Fig. 24.3, where originally demand curve D_0D_0 intersects the supply curve *SS* of eggs at point E_0 and determines equilibrium price equal to OP_0 and equilibrium quantity OQ_0. Now, suppose that doctors advise the people to take less eggs as it contains greater quantity of cholestrol which increases the risk of heart disease. Consequently, demand for eggs decreases causing a shift in the demand curve to the left to the new position D_2D_2. The new equilibrium between demand and supply is attained at price P_2 and quantity Q_2, which are lower than the initial equilibrium price OP_0 and quantity OQ_0. Thus, *the decrease in demand leads to the fall in both price and quantity*. How does this come about? With the decrease in demand and consequently leftward shift in the demand curve to D_2D_2, supply curve remaining unchanged, at the original price OP_0, the surplus E_0B of the quantity supplied over the quantity demanded emerges which exerts a downward pressure on price. The sellers which cannot sell the quantity which they want to sell at the original price will make offers to sell eggs at a lower price. As a result, price will fall. As price falls, the quantity supplied of eggs is reduced. At the new price OP_2, the quantity supplied again equals quantity demand and surplus is eliminated.

Apart from the changes in preferences for a good as in case of eggs considered above, the decrease in incomes of the people such as when a large number of people are rendered unemployed during depression, the reduction of crop production in agriculture due to failure of Monsoon leading to the drop in incomes of the Indian farmers can also cause a decrease in demand for goods resulting in lowering of prices and quantities of goods.

Impact of Changes in Supply on Market Equilibrium

Now, we explain the impact of changes in supply on price and output of commodity, the demand for the commodity remaining the same. Let us first examine the case of increase in supply. Suppose in a year there is good Monsoon in India yielding bumper crop of wheat. This will increase the supply of wheat in the market causing a shift in its supply curve to the right. The impact of increase in supply of wheat on equilibrium price and quantity is graphically depicted in Fig. 24.4. Originally, demand curve *DD* and supply curve *SS* of wheat intersect at point *E* and determine equilibrium price

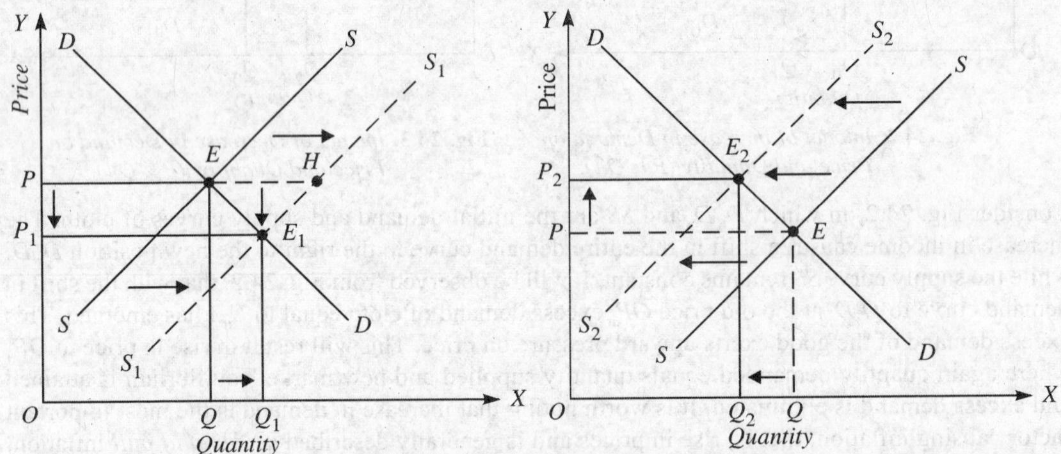

Fig. 24.4. *Increase in supply results in lowering of price and increase in quantity*

Fig. 24.5. *Decrease in supply cause rise in price and fall in quantity.*

equal to *OP* and equilibrium quantity *OQ* exchanged between the sellers and buyers. Now, due to good monsoon resulting in bumper crop of wheat the supply curve of wheat shifts to the right from *SS* to the new position S_1S_1. The new supply curve S_1S_1 intersects the given demand curve *DD* at point E_1 at which the new lower equilibrium price OP_1 and larger quantity OQ_1 are determined. Thus, the *increase in supply leads to the fall in price and increase in equilibrium quantity*.

Improvements in technology, reduction in the prices of factors and resources used in the production of a commodity or lowering of excise duty on a commodity also leads to the increase in supply of the commodity. For example, in recent years improvements in technology in the manufacture of personal computers have served to increase the supply of personal computers causing their supply curve to shift to the right. This has resulted in lowering the prices of personal computers. A personal computer which was available at price above ₹ 60,000 a few years ago are now available at about

₹ 20,000. Similarly, in the Central Budget for 1993-94, the Finance Minister Dr. Manmohan Singh reduced excise duties on several commodities with the hope that producers it would pass it on to the consumers and result in shifting their supply curve to the right and thereby causing the drop in their prices. At lower prices, he argued, more of these commodities would be demanded and therefore it would help the industries which were facing demand recession.

DEMAND-SUPPLY MODEL OF PRICING: MATHEMATICAL ANALYSIS

It is quite often more useful and illuminating to study economic problems through mathematical equations. Demand-supply model of pricing can be presented in mathematical equations and solved for equilibrium price and quantity. In this model it is assumed that both demand and supply functions are of linear type. Further, in this demand-supply model of pricing it is assumed that perfect competition forevails in the market for the product. This implies that there are a large number of sellers and buyers of a homogeneous product so that no single seller or buyer has any influence over the price of the product. This demand-supply model consists of the following three equations:

$$Q^d = a - bP \qquad \ldots (i)$$

$$Q^s = c + dP \qquad \ldots (ii)$$

$$Q^d = Q^s \qquad \ldots (iii)$$

Where,

Q^d stands for quantity demanded

Q^s stands for quantity supplied

P stands for price of a commodity

a, b, c and d are constants

The above set of three equations constitutes a model. Here, in this model, the variables considered are price, quantity demanded, quantity supplied. The objective is to obtain the price and quantities demanded and sold when the system attains the equilibrium position. To find the equilibrium values of these variables, the above three equations are required to be solved simultaneously. The equation (*i*) describes that the quantity demanded is a decreasing function of price, that is, as price falls quantity demanded increases and *vice versa.* Equation (*ii*) describes that quantity supplied is increasing function of price, that is, as price rises quantity supplied increases and *vice versa.* Equation (*iii*) describes the equilibrium condition meaning thereby that the price that equates quantity demanded and quantity supplied will clear the market. Equilibrium price implies that both the buyers and sellers are satisfied with their purchases and sales and hence there is no tendency on their part to raise or lower the price. It is in this way that Marshall explained the determination of price of a commodity.

Equations of the model can be solved mathematically to obtain the equilibrium values of the variables. The equations of the model can also be represented by graphs to determine geometrically the equilibrium values of the variables.

Numerical Example

Let us illustrate this famous demand-supply micro-model of price determination of a commodity in a free market by taking specific demand and supply functions. Suppose the following demand and supply functions of a commodity are given :

$$Q^d = 100 - 20\,P$$

$$Q^s = -5 + 15\,P$$

Where Q^d stands for quantity demanded, Q^s for quantity supplied of a commodity and P for the price of the commodity.

Since for the equilibrium price the quantity demanded is equal to the quantity supplied, we put the two demand and supply equations equal to each other. Thus, we have

$$Q^d = Q^s$$

$$100 - 20\ P = -5 + 15\ P$$

$$15\ P + 20\ P = 100 + 5$$

$$35\ P = 105$$

$$P = \frac{105}{35} = 3$$

Thus we get ₹ 3 as price per unit of the commodity. Putting the value of P in either the demand or supply equation we can obtain the equilibrium quantities Q^d and Q^s. Thus

$$Q^d = 100 - 20 \times 3 = 40$$

$$\text{Or, } Q^s = -5 + 15 \times 3 = 40$$

Thus equilibrium price of the commodity equals ₹ 3 and the equilibrium quantity sold and purchased equals 40 units in a period.

Are Demand and Supply Final Answers to the Pricing Problem?

From above it follows that price is determined by equilibrium between demand and supply. But it is worth mentioning that demand and supply do not provide final answer to the princing problem. Demand and supply which by their intersection determine price are themselves governed by several factors. There are many forces or factors which work behind demand and supply to determine price. 'Demand and supply' is only a superficial formula. Real factors that determine the prices of goods are those upon which demand and supply depend. But demand and supply are very useful concepts serving outer cover for two distinct categories of factors. Prof. Samuelson rightly remarks: "*Supply and demand are not ultimate explanations of price. They are simply useful catch-all categories of analysing and describing multitude of forces, causes and factors impinging on price.* Rather than final answers, supply and demand simply represent initial questions. Our work is not over but just begun."[5] For instance, market demand for a good depends upon the incomes of the people, their perferences for the goods, the total population, availability and prices of substitute goods. Any change in them will cause a change in demand for the good and hence in its equilibrium price. Similarly, supply depends upon the availability and price of labour, raw materials, machines, chemicals etc., and techniques of production, all of which govern cost of production. Any change in them will bring about change in cost of production and supply and will therefore alter the equilibrium price. It is thus clear that factors like availability of raw materials, cost of production, incomes of the people, the size of population, consumers' preferences for goods, etc., are ultimate determinants of the prices of various goods but all of them work through either demand or supply. For explaining any price behaviour of a good we have to go behind demand and supply and search for those factors which are responsible for the price behaviour for a commodity.

IMPORTANCE OF TIME ELEMENT IN THE THEORY OF PRICE

Marshall, who propounded the theory that price is determined by both demand and supply, also gave a great importance to the time element in the determination of price. Time element is of great relevance in the theory of value, since one of the two determinants of price, namely supply, depends on the time allowed to it for adjustment. It is worth mentioning that *Marshall divided time into different periods from the viewpoint of supply and not from the viewpoint of demand*. Time is short

5. Paul A. Samuelson, *Economics*, 8th edition, 1970, pp. 369-70

or long according to the extent to which supply can adjust itself. Marshall felt it necessary to divide time into different periods on the basis of response of supply because it always takes time for the supply to adjust fully to the changed conditions of demand. The reason why supply takes time to adjust itself to a change in the demand conditions is that nature of technical conditions of production is such as to prohibit instantaneous adjustment of supply to changed demand conditions. A period of time is required for changes to be made in the size, scale and organisation of firms as well as of the industry.

Another point is worth noting. When Marshall distinguished short and long periods he was not using clock or calendar time as his criterion, but '*operational*' time in terms of economic forces at work. In this regard, as said above, supply forces were given the major attention and a time was short or long according to the extent of adjustment in the forces of supply. The greater the adjustability of the supply forces, the greater the length of the time irrespective of the length in clock-time.

Marshall divided time into following three periods on the basis of response of supply to a given and permanent change in demand.

(1) **Market Period.** The market period is a very short period in which the supply is fixed, that is, no adjustment can take place in supply conditions. In other words, supply in the market period is limited by the existing stock of the good. The maximum that can be supplied in the market period is the stock of the good which has already been produced. In this period more good cannot be produced in response to an increase in demand. This market period may be a day or a few days or even a few weeks depending upon the nature of the good. For instance, in case of perishable goods, like fish, the market period may be a day and for a cotton cloth, it may be a few weeks.

(2) **Short Run.** Short run is a period in which supply can be adjusted to a limited extent. During the short period the firms can expand output with given equipment by changing the amounts of variable factors employed. Short periods is not long enough to allow the firm to change the plant or given capital equipment. The plant or a capital equipment remains fixed or unaltered in the short run. Output can be expanded by making intensive use of the given plant or capital equipment by varying the amounts of variable factors.

(3) **Long Run.** The long run is a period long enough to permit the firms to build new plants or abandon old ones. Further, in the long run, new firms can enter the industry and old ones can leave it. Since in the long run all factors are subject to variation, none is a fixed factor. During the long period forces of supply fully adjust themselves to a given change in demand; the size of individual firms as well as the size of the whole industry expands or contracts according to the requirements of demand.

From above, it is clear that because of the varying response of supply over a period of time to a sudden and once-for-all increase in demand Marshall found, it necessary and useful to study the pricing process in (*a*) the market period, (*b*) the short-run and (*c*) the long-run depending respectively upon whether the supply conditions have time to make (*i*) *no adjustment*, (*ii*) *some adjustment of labour and other variable factors, and* (*iii*) *full adjustment of all factors and all costs*. Therefore, Marshall explained how the equilibrium between demand and supply was established in three time periods and determined market price, short-run price and long-run price.

We thus see that the price that will prevail depends upon the period under consideration. If a sudden and a once-and-for all increase in demand takes place, the market price will register a sharp increase, since supply cannot increase in the market period. In this market period, firms can sell only the output that has already been produced. However, in the short run some limited adjustment in supply will take place as a result of the firms moving along their short run marginal cost curves by expanding output with the increase in the amount of variable factors. Consequently, the short run price will come down from the new high level of the market price. But this short-run price will stand above the level of original market price which prevailed before the increase in demand occurred. In the long run the firms would expand by building new plants, that is, by increasing the size of their capital equipment. In other words, firms would expand along the long-run marginal cost curves. Besides, the new firms will enter the industry in the long run and will add to the supply of output. As a result of this long-run adjustments in supply, the price will decline. Thus the long run price will be lower than the short-run price. But this long-run price will be higher than the original price which ruled before the increase in demand took place, if the industry happens to be increasing-cost industry.

The adjustment of supply over a period of time and consequent changes in price is illustrated in Fig. 24.6 where long-run supply curve *LRS* of an increasing-cost industry along with the market-period supply curve *MPS* and the short-run supply curve *SRS* have been drawn. Originally, demand curve *DD* and market-period supply curve *MPS* intersect at point *E* and price *OP* is determined. Suppose that there is a once-for-all increase in demand from *DD* to *D′ D′*. Supply cannot increase in the market period and remains the same at *OM*. Market-period supply curve *MPS* intersects the new demand curve *D′D′* at point *Q*. Thus, the market price sharply rises to *OP′* . Short-run supply curve *SRS* intersects the new demand curve *D′D′* at point *R*. The short-run price will therefore be *OP″* which is lower than the new market price *OP′*. As a result of the long-run adjustment the price will fall to *OP‴* at which the long-run supply curve *LRS* intersects the demand curve *D′ D′*. The new long-run price *OP‴* is lower than the new market price *OP′* and the short-run price *OP″*, but will be higher than the original price *OP* which prevailed before the increase in demand took place. This is so because we are assuming an increasing-cost industry. If the industry is subject to constant costs, the long-run price will be equal to the original price. Further, if the industry is subject to decreasing costs, the long-run price will be lower than the original price.

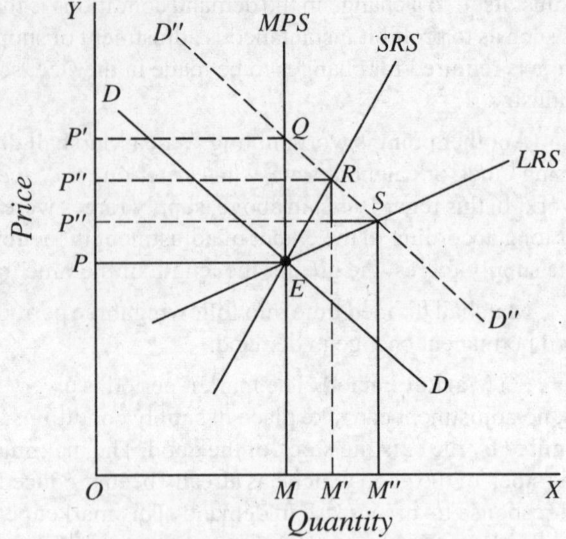

Fig. 24.6. *The Role of Time Element in the Determination of Price*

It follows from above that the price which prevails in the market depends upon the period under consideration. It is thus clear that the time plays an important role in the determination of price. Another significance of the time-period analysis of pricing is that it enabled Marshall to resolve the controversy current among economists whether it is demand or supply which determines price. Marshall propounded the view that both demand and supply took part in the determiantion of price. But, "as a general rule", said Marshall, "*the shorter the period which one considers the greater must be the share of our attention which is given to the influence of demand on value, and the longer the period more important will be the influence of cost of production on value*. Actual value at any time—the market value as it is often called—is often influenced by passing events and causes whose action is fitful and shortlived than by those which work peristently. But in the long run these fitful and irregular causes in a larger measure efface one another's influence so that in the long run persistent causes dominate value completely".[6]

From the above quotation from Marshall it follows that in the market period, demand exercises a predominant influence over price but in the long run it is the supply which is of overwhelming

6. Alfred Marshall, *Principles of Economics,* 8th edition,p.350.

importance as a determinant of price. Roughly speaking, we can say that in the market period it is the force of demand which determines price and in the long period it is the force of supply which governs price. Thus those economists who held that value was governed by demand were in a way right and so were those who contended that cost of production (*i.e.*, force working on the supply side) determines price. The difference in the two views was due to the fact that one gorup of economists was emphasising the determination of the *marekt price* over which demand has determining influence and over which cost of production does not exercise much influence, while the other group was stressing on the determination of *long-run price* over which cost of production has got paramount influence. It is thus clear that Marshall by putting forth the view that both demand and supply determine price by their interaction brought about synthesis between the views of earlier economists.

Both the two opposite views of earlier economists were in a way right but each was one-sided. Each view provided us with a force which governed price. The two forces of supply and demand furnished by the two opposing views were sufficient determining factors. Therefore, Marshall gave equal importance to both demand and supply as determinants of price, though the influence of the two varied in different time periods. Marshall introduced time period analysis into pricing process to bring out the varying influence of each of two forces over price of the product in different time periods.

It follows from what has been said above that Marshall and modern economists following him study the effect of the varying response in supply in different time periods on price to a sudden and permanent change in demand conditions. On the contrary, economists do not study the effect on price of the adjustment in demand over time in response to a change in supply conditions. The reason why we do not study adjustment in demand to a change in supply and consequent effect on price is better brought out in the worlds of Professors Stonier and Hague. "There is no reason why, if supply conditions change, demand conditions should change as well, or if they do, why they should change differently in the short run and the long run. Change in consumer's tastes are not dependent on technology in the way that supply conditions are. Admittedly, consumers' tastes may and probably will change as time goes on. But this will be a change of data and not a change induced by changed supply conditions. There is no necessary reason why the long-run demnad curve should differ from the short-run demand curve, however odd the behaviour of supply has been–we must expect that the longer is the period during which demand and supply are coming into equilibrium, the more changes will have time to take place. If we were to study the changes in demand and supply which would take place in respect to any change of data during many successive very short periods of time, we should find that we had introduced unnecessary and intolerable detail into the analysis."[7]

We shall explain below in detail the market-period equilibrium, short-run equilibrium, long-run equilibrium between demand and supply and thus the determination of market price, short-run price and long-run price under conditions of perfect competition.

DETERMINATION OF MARKET PRICE

Market price is the price of a good which prevails at any given time. Market price is determined by the momentary equilibrium between demand and supply at a time. In the market period, as has been already mentioned, the supply of the good is limited by the available stock of the good. But the quantity that may be supplied in the market period may not be equal to the given stock of the good. Whether the whole stock of the good is offered for sale or not depends upon whether the good in question is perishable or durable. The perishable goods like fish and milk cannot be stored or kept back; they will go waste if stored. Therefore, the whole of the given stock of a perishable good has

7. Stonier and Hague, *A Textbook of Economic Theory*. 4th edition, 1972. p.178

to be supplied in the market whatever the price of the good. As a result, the market-period supply curve of a perishable commodity is perfectly inelastic, or a vertical straight line. On the other hand, the sellers can hold back a durable good and wait for the time when the price of the good rises. Out of a given stock of the good, they will be prepared to sell a lesser amount at lower price, and a greater amount at a higher price. At some price, they will be willing to supply the whole stock of the good and beyond that price the supply of the good will be completely inelastic. Consequently, the supply curve of a durable commodity slopes upward to a point but becomes a vertical straight line after that.

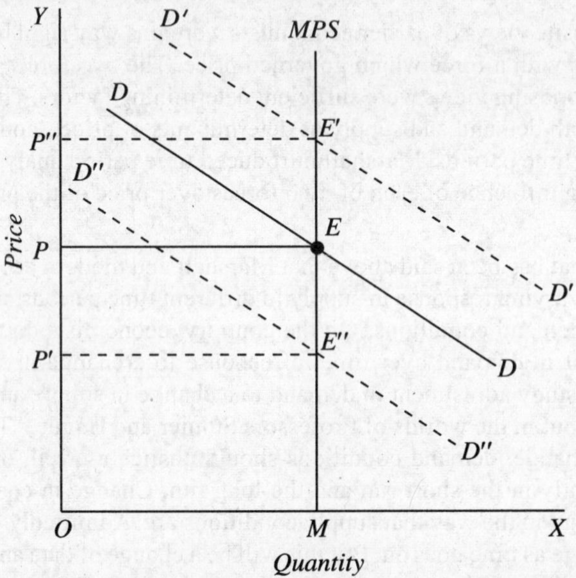

Fig. 24.7. *Determination of Market Price of a Perishable Commod-*

Fig. 24.7 illustrates the determination of market price of a perishable commodity. *OM* is the given stock of the good and *MPS* is the market-period supply curve. Suppose, to begin with, *DD* is the demand curve of the commodity. Demand and supply are in equilibrium at price *OP*. Therefore, *OP* is the equilibrium market price. Now if the demand incereases from *DD* to *D′D′*, the market price will sharply rise from *OP* to *OP′*, the supply of the good remaining fixed at *OM*. On the contrary, if the demand decreases from *DD* to *D″D″*, the market price falls from *OP* to *OP″* the supply of the good again remaining constant at *OM*. We thus see that changes in demand produce sharp changes in price in the market period, supply being constant during this period.

It is noteworthy that in the determination of market price, cost of production incurred in the past on the good does not have any influence on the market price of the good. Cost of production has a relevance for determining price during a period in which output can be varied. Since, in the market period, what has already been produced is to be supplied in the market, costs incurred in producing the good does not have much influence over the market price. However, cost of production exercises a determining influence over the short-run price and long-run price of the good. This is because in the short run and the long-run output can be increased or decreased and what amount of the good will be supplied in the marekt depends to a great extent upon the cost of producing it.

Since a product under perfect competition is homogeneous, it will have a single uniform price in the market. Once the market price is established an individual firm will take this price as given and constant for it and will so adjust its level of output as to get maximum profits.

DETERMINATION OF SHORT-RUN PRICE

Short-run price is determined by short-run equilibrium between demand and supply. Supply curve in the short run under perfect competition is a lateral summation of the short-run marginal cost curves of the firm. Also, the short-run supply curve of the industry always slopes upward, since the short-run marginal cost curves of individual firms slope upward. It was explained in a previous chapter that in the short run fixed costs are not taken into account while deciding about whether to produce or not.If the price does not cover fixed cost, the firms will continue producing provided the price stands above the average variable cost. It is, therefore, the average variable cost and not the average total cost (which includes average fixed cost) which determines whether to produce or not. If the price falls below average variable cost, the firms will close down (suspend production in order to avoid unnecessary losses). Thus, the average variable cost sets a minimum limit to the price in the short run, since at prices below it no amount of output will be produced and offered for sale.

Figure 24.8 illustrates the process of the price determination in the short run. In Fig. 24.8 (b) *DD* represents the demand curve for the product of the industry, and *SRS* is the short-run supply curve of the industry. To begin with, *OP* is the short-run price since the given demand curve *DD* intersects the short-run supply curve *SRS* at point *E*. The individual firm will take the price *OP* as a

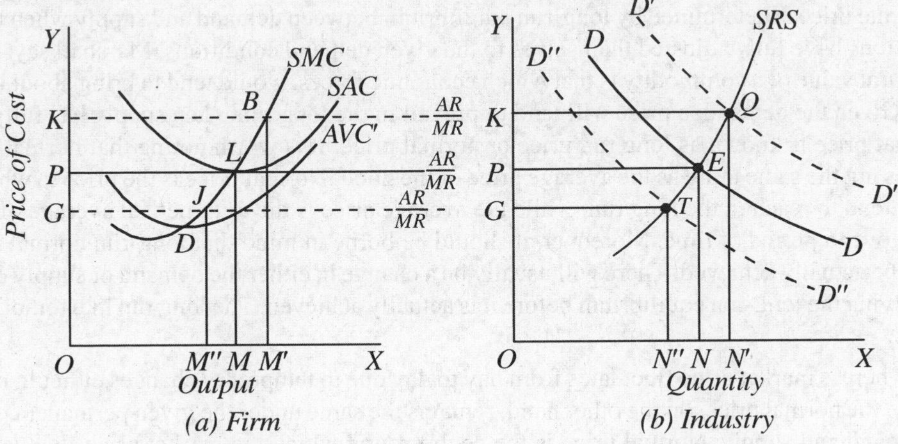

Fig. 24.8. *Determination of Short-Run Price under Perfect Competition*

given datum for it and will adjust its output at *OM* at which the price *OP* is equal to it short-run marginal cost

In order to further clarify the process of determination of short-run price, we suppose that in Fig. 24.8 (b) the demand increases from *DD* to *D′D′*. In the market period since the supply cannot be increased in response to this increase in demand and consequently with the increase in the the demand from *DD* to *D′D′* the market period price will sharply rise. But, in the short period, the supply of output will be increased by increasing the employment of variable factors to the given fixed capital equipment and therefore the short-run supply curve *SRS* slopes upward to the right. The new demand curve *D′D′* intersects the short-run supply curve *SRS* at point *Q* and therefore new short-run price *OK* is determined, which is higher than the original price *OP*. This is because the marginal cost rises as more is produced by the firms in the short run, short-run marginal cost curve being upward sloping. The individual firms will take this new short-run price *OK* as given and will produce *OM′* output at which price *OK* equals the marignal cost. It is clear from Fig. 24.8 (a) that at price *OK* the firm will be making supernormal profits, since price *OK* is greater than the average cost at equilibrium output *OM′*.

Now, suppose that instead of increase in demand, there is a decrease in demand for the product of the industry from *DD* to *D″D″*. In the market period, supply will remain fixed at *ON* and the new marekt price will fall sharply. But in the short period the firms will reduce output by reducing the employment of labour and other variable factors in response to the decrease in demand. New demand curve *D″D″* cuts the short-run supply curve *SRS* at point *T* and short-run price *OG* is determined, which is lower than the original short-run price *OP*. At price *OG*, the firm is producing *OM″* output and will be making losses, since the price *OG* is smaller than the average total cost at output *OM″*. But because the price *OG* is greater than the average variable cost, the firm will continue producing the product. A firm will not produce at a price which is lower than the average variable cost, since it will not be recovering variable costs fully and will be incurring losses more than the fixed costs. A firm will rather close down than to produce at a price lower than average variable cost. Thus we see that the short-run price cannot fall below the average variabel cost. Average variable cost sets a minimum limit to the price in the short run.

DETERMINATION OF NORMAL PRICE AND DERIVATION OF LONG-RUN SUPPLY CURVE OF THE INDUSTRY

Long-run price is also known as long-run normal price or simply normal price. Long-run price or normal price is determined by long-run equilibrium between demand and supply when the supply conditions have fully adjusted themselves to the given demand conditions. Marshall says, "Normal or natural value of a commodity is that which economic forces, would tend to bring about in the long run". Given the demand, a price will tend to prevail in the long run when supply has fully adjusted and that price is known as long run price or normal price. It is worth noting that normal price of a good is not the same thing as the average price of the good. Normal price is the price to which actual prices tend to reach in the long run, while the average price is the arithmetical average of all actual prices over a period of time. Moreover, it should be borne in mind that long-run normal price may never be actually achieved. There will usually be a change in either the demand or supply conditions underlying the long-run equilibrium before it is actually achieved. The long run like tomorrow never comes.

Whereas market price fluctuates from day to day due to temporary changes either in demand or supply, the normal price, on the other hand, remains the same under the given permanent conditions of demand and supply. Normal price is the centre round which the market price fluctuates due to temporary changes in demand or supply. However, it should be noted that normal price is not a permanently fixed level. If there is a permanent change in either demand or supply, the normal price will change.

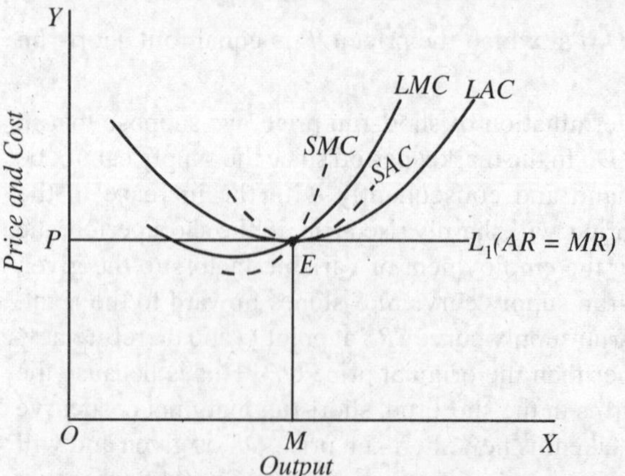

Fig. 24.9. *Long-run price is equal to minimum long-run average cost.*

As stated above, long-run normal price is determined by the long-run equilibrium between demand and supply. This long-run price under perfect competition cannot be above or below the long-run average cost. A firm under perfect competition is in long-run equilibrium at the output where price is equal to both marginal cost and average cost. If the price is greater than average cost, the firms will earn more than normal profits. Attracted by the supernormal profits new firms will enter the industry in the long run and will add to the supply of the product. As a result of this process, the price will fall to the level of minimum long-run average cost. On the other hand, if the price is below the average cost, the firms will be making losses and as a result some of the existing firms will quit the industry in the long run. With some of the firms going out of the industry, the supply of the product will fall and price will rise to the level of long-run average cost so that the remaining firms make only normal profits.

It follows from above that when price is either above or below the long-run average cost certain forces in the system so operate as to bring the long-run normal price to the level of minimum long-run average cost. It will be seen from Fig. 24.9 that long-run equilibrium of a firm is established at the minimum point of the long-run average cost and the long-run price *OP* is established which is equal to the minimum long-run average cost. It will be noticed from Fig. 24.9 that long-run normal price is equal to both the minimum short-run as well as long-run average cost. Thus

$$\text{Long-Run Price} = SMC = SAC = LAC = LMC$$

But to say that, in the long run, price will be equal to the minimum long-run average cost is not enough since what would be the level of minimum long-run average cost and hence of price depends upon what happens to the cost curves when industry expands or contracts in response to a given increase or decrease in demand. The long-run average cost shifts upward or downward or remains at the same level when the industry expands depending upon whether the industry in question is increasing-cost industry, decreasing-cost industry or constant-cost industry. We shall explain below what causes the industry to be increasing-cost, decreasing-cost or constant-cost and how long-run normal price is determined in each case. Besides, we shall also explain how long-run supply curve of the perfectly competitive industry is derived in these three types of industry.

Determination of Normal Price and Derivation of Long-Run Supply Curve in Increasing-Cost Industry

When an industry expands in response to an increase in demand it experiences some *external economies* as well as some *external diseconomies*. Whereas external economies tend to reduce the cost and thereby tend to shift the long-run average cost downward, the external diseconomies tend to raise the costs and thereby shift the long-run average cost curve upward. So we have an increasing-cost industry when external diseconomies outweigh the external economies, that is, when there are *net external diseconomies*.

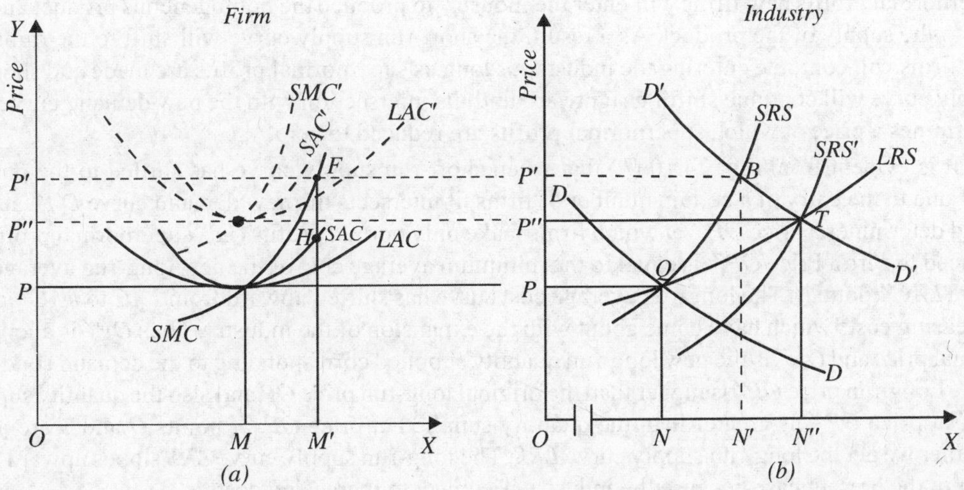

Fig. 24.10. *Long-Run Normal Price and Derivation of Long-Run Supply Curve of the Increasing Cost industry*

The external diseconomies which accrue to an expanding industry are generally the rise in prices of raw materials or factors used for production in the industry. As more firms enter the industry in response to the increase in demand for the product, the prices of scarce factors rise due to the increase in demand for them. In other words, more intensive bidding by the increased number of firms pushes up the prices of scarce raw materials. Wages of specialised labour, rent of land and the prices of scarce raw materials, capital equipment etc. are bound to rise as the demand for them increases as a result of the expansion of the industry. Generally, the same resources are used in different industries. As an industry expands, it has to take away the scarce resources from other industries by offering them higher prices. The rise in prices of the scarce factors raises the costs of production. Furthermore, the additional factors of production coming into the industry are generally less efficient or of inferior quality than the previous ones and this also brings about increase in cost of production as the industry expands. When these external diseconomies outweigh the external economies, the industry experiences rising costs and the average and marginal cost curves of the firms in the industry shift upward.

Figure 24.10 illustrates the determination of long-run normal price in an increasing-cost industry. The left-hand side of the diagram in Fig. 24.10 shows long-run equilibrium of the firm at two different long-run normal prices, while the right-hand side of diagram shows the demand and supply curves of the industry with the horizontal scale compressed. To begin with, a given demand curve intersects the short-run supply curve at *Q* therefore, the short-run price *OP* is determined. It will be seen from the left side of the diagram that price *OP* is equal to minimum long-run average cost of firms. It means that the sizes of firms as well as the number of firms have already been fully adjusted to the given demand conditions as represented by the curve *DD*. Therefore, price *OP* is also a long-run normal price corresponding to the demand curve *DD*.

Now, let us see what happens to the long-run normal price when a once-for-all increase in demand from *DD* to *D′D′* occurs. In the short-run price will rise to *OP′* and quantity exchanged to *ON′* as a result of the given increase in demand. From the left-hand diagram it will be seen that in the short run the individual firm will expand its output along the short-run marginal cost curve and will be in equilibrium at point *F* on *SMC*. The firms will therefore expand output from *OM* to *OM′* in response to the increase in demand from *DD* to *D′D′* and the rise in short-run price from *OP* to *OP′*. It will be further seen from the left-hand side of Fig. 24.10 that the individual firm will be making supermormal profits equal to *FH* per unit of output in the short run at price *OP′*. Lured by these supernormal profits new firms will enter the industry to produce the homogeneous product and will add to the supply of the product. As a result, the short-run supply curve will shift to the right. The new firms will continue entering the industry as long as supernormal profits are made and short-run supply curve will continue shifting rightward until its intersection with the new demand curve *D′D′* determines a price at which supernormal profits are reduced to zero.

It is evident from Fig. 24.10 (*b*) that when short-run supply curve has shifted to the position *SRS′* due to the entry of a certain number of firms, it intersects the new demand curve *D′D′* at point *T* and determines a price *OP″* at which firms make only normal profits (*i.e.*, supernormal profits are reduced to zero). Price *OP″* is equal to the minimum average cost of the new long-run average cost curve *LAC″*(dotted). The long-run average cost curve has shifted upward from *LAC* to *LAC″* due to increasing costs which have come about with the expansion of the industry. Thus *OP″* is a long-run normal price and *ON″* is the new long-run quantity supplied corresponding to the demand conditions *D′D′*. Long-run price *OP″* is higher than the original long-run price *OP* and also the quantity supplied *ON″* at price *OP″* is greater than the quantity supplied at price *OP*. If points *Q* and *T* are joined together we get the long-run supply curve *LRS*. The long-run supply curve *LRS* slopes upward to the right in the present case because the industry is subject to increasing costs.

It follows from above that the long-run price of a product rises as demand increases in the case of increasing-cost industry. In other words, in the case of increasing-cost industry, more quantity of the product can be obtained or supplied only at a higher price. The extent to which the new long-run price differs from the original long-run price depends upon the extent to which the increase in cost occurs following the expansion of the industry. It must be borne in mind *that every point of the long-run supply curve LRS represents a long-run equilibrium as the demand shifts to the right.* It is clear from the analysis made above that long-run supply curve of an increasing-cost industry slopes upward to the right and is more elastic than the short-run supply curve.

This case of increasing-cost industry or, in other words, *rising supply price* is believed to be the most typical of the competitive industries in the actual world. This is so because productive resources are scarce and are currently being used in various industries. Therefore, when an industry expands and requires more resources, it has to take away resources from others by paying higher prices for them. Thus, Professor Samuelson writes : The case of increasing cost is "the normal one to be met in most sizeable competitive industries. Why normal? Because when a large industry (which has already achieved the economies of large-scale production) expands, it must coax men and other productive factors away from the other industries by bidding up their prices and thus its own cost. *So the long-run supply curve will usually be sloping gently upward.*"

Normal Price and Derivation of Long-run Supply Curve in the Constant Cost Industry

If an industry on its expansion gives rise to some external economies and external diseconomies which cancel each other so that the constituent firms do not experience any shift in the cost curves, then that industry is a constant-cost industry. *In case of constant-cost industry, we have neither net external economies, nor net external diseconomies*. An industry can also be a constant-cost industry if its expansion breeds neither external economies, nor external diseconomies. As more firms enter an industry the demand for productive factors like raw materials, labour, chemicals and machinery will increase and if prices of these productive resources go up as a result of the increase in demand, then the cost is bound to rise. It is, therefore, evident that an industry can be constant-cost industry if it makes little impact on the market for these productive resources, that is, if its demand for productive resources is a negligible part of the total demand for them so that the increase in demand for them by the industry does not push up their prices. To quote Prof. Samuelson, "*only if the industry is small compared with the total of all other uses will Marshall's long-run supply curve be horizontal—which is called the case of constant-cost.*"[8] If the industry on its expansion does not create external diseconomies, it will be a constant-cost industry only when its expansion also does not give rise to any external economies.

Thus an industry can be constant-cost in two ways. First, when the expansion of an industry creates both external economies and diseconomies but they cancel each other in their effect on costs, and, secondly, when the expansion of an industry creates neither external economies nor external diseconomies. Fig. 24.11 depicts the determination of long-run normal price in the constant-cost industry. To begin with, a given demand curve *DD* intersects the short-run supply curve *SRS* at point *Q* and short-run price *OP* is determined. The price *OP* is equal to the minimum long-run average cost of the firm. This means that the number of firms has already been fully adjusted to the given demand conditions. Therefore, price *OP* is also the long-run price corresponding to the demand conditions represented by the curve *DD*. Now, suppose that the demand increases from *DD* to *D′D′*. As a result, price will rise to *OP′* in the short run at which the short-run supply curve *SRS* intersects the new demand curve *D′D′* . The quantity supplied by the industry will increase to *ON′* in the short run. At price *OP* a firm will be in equilibrium at *F* and will be producing *OM′* amount of the product. It will be seen from the figure that at price *OP′* the firm is making abnormal proits equal to *FH* per unit of output. This will attract other firms to the industry in question. As more firms enter the industry, the short-run supply curve will shift to the right. The new firms will go on entering the

8. Paul A. Samuelson, *Economics, 8th edition*, p.366

industry and short-run supply curve will go on shifting righward until its intersection with the demand curve $D'D'$ determines a price at which super-normal profits of the firms disappear completely.

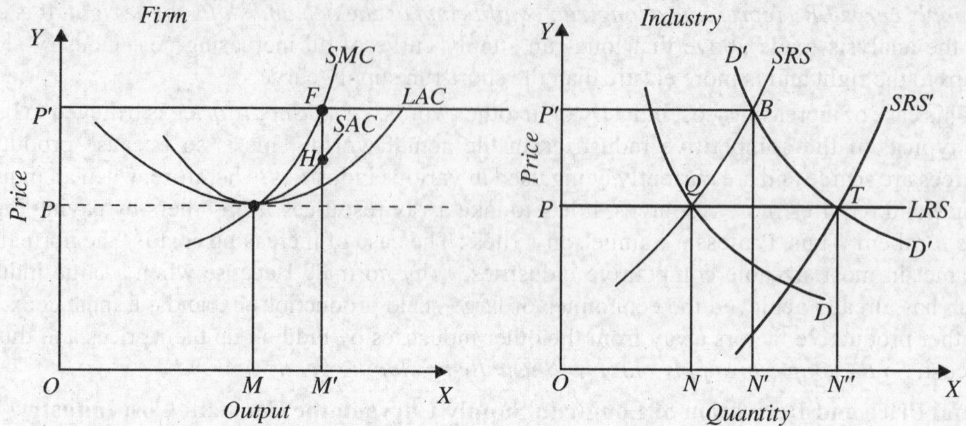

Fig. 24.11. *Long-Run Pricing in a Constant Cost Inustry*

It will be seen from Fig. 24.11 that when the short-run supply curve has shifted to *SRS'* it intersects the new demand curve $D'D'$ at point *T* and once again determines price *OP* at which firms will make only normal profits. Thus, price *OP* is also a long-run price corresponding to the new demand curve $D'D'$. The cost curves of the firms have not shifted because external economies and diseconomies have offset each other's effect. As a result of the increase in demand, the price went up in the short run but has returned to the original level of *OP* in the long run. *In a constant-cost industry, long-run normal price remains the same whatever the level of demand.* If points like *Q* and *T* are joined together we get a long-run supply curve *LRS* which is a horizontal straight line in the present case. It is thus clear that in the constant-cost industry new firms enter the industry without raising or lowering cost curves of firms and make available additional supplies of output at the same price which is equal to the minimum long-run average cost of the firms. *Since long-run supply curve of the constant-cost industry is a horizontal straight line, any increase or decrease in demand will not produce any effect on the long-run price.* In this case, increase or decrease in demand will in the long run only change the supply of output by causing a change in the number of firms without producing any effect on the long-run normal price.

Here, a point is worth noting. *While the long-run marginal cost curve of the firm will be sloping upward in its relevant part the long-run supply curve of the industry in this case is a horizontal straight line.* It is manifest from this that the long-run supply curve of an industry cannot be the lateral summation of the long-run marginal cost curves of the firms.

Determination of Normal Price and Derivation of Long-run Supply Curve in the Decreasing Cost Industry

When an industry grows in size by the increase in the number of firms, it may be that the external economies outweigh the external diseconomies so that there is a decline in the production costs of the firms. In other words, when an industry reaps *net external economies* as it expands, it is a decreasing-costs industry. There is every possibility of external economies to outstrip the external diseconomies when an industry grows in a new territory. In earlier stages of the growth of an industry in a new territory, not much diseconomies are created, while external economies are likely to accrue in greater amount. Costs may decline with the expansion of the industry due to the following external economies.

1. Some raw materials, tools and capital equipment may be made available to the industry at reduced prices since as the industry grows subsidiary and correlated firms may spring up in the vicinity of the industry which produce them on a large scale and therefore can provide the expanding industry with raw materials, tools and capital equipment at the reduced rates.

2. Cheaper and better trained labour may become available with the expansion of the industry.

3. With the growth of industry, certain specialised firms may come into existence which work up its waste products. The industry can then sell its waste products at good prices, while previously it may be throwing them away.

4. As the industry expands it may become worthwhile to publish trade journals which may help in discovering and spreading technical knowledge concerning the industry.

5. When there is enlargement of the industry, better information centres, research institutions etc. may emerge which greatly benefit the firms in the industry.

In case the above-mentioned external economies are more powerful than the external diseconomies that may arise, the cost curves of all firms in the industry will shift to a lower position as the industry expands.

Long-run pricing in a decreasing cost industry is illustrated in Fig. 24.12. To start with, demand curve *DD* and short-run supply curves *SRS* intersect at *Q* and determine short-run price *OP*. Price *OP* is also the initial long-run price since price *OP* is equal to the miminum long-run average cost of the firms which are therefore making only normal profits. With the increase in demand from *DD* to *D′ D′* short-run price rises to *OP′* and the quantity exchanged to *ON′*. As a result of the rise in price from *OP* to *OP′*, each firm of the industry will expand output to *OM′* in the short run. Price *OP′*

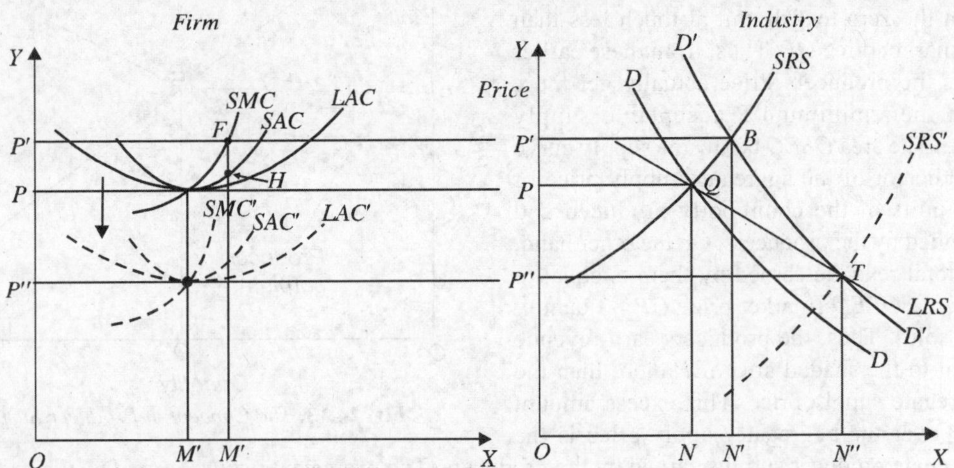

Fig. 24.12. *Long-Run Pricing in the Decreasing Cost Industry*

yields supernormal profits to the firms. These supernormal profits will attract other firms into the industry and as a result the industry will expand.

As the industry expands, its short-run supply curve will shift to the right and the industry being a decreasing cost industry, cost curves of the firms will shift downward. The short-run supply curve will continue to shift to the right and cost curves of the firms continue to shift downward until the price *OP″* is reached at which firms make only normal profits. Price *OP″* is therefore long-run price corresponding to the demand curve *D′D′* . As is evident from Fig. 24.12, new long-run price *OP′* is lower than the original long-run price *OP*. Connecting points *Q* and *T*, we get the downward sloping long-run supply curve *LRS*. We thus see that the *long-run supply curve of a decreasing-cost industry slopes downward from left to right*. In this case the greater supplies of the product will be forthcoming at reduced prices in the long run. Every increase in demand in this case will bring about a fall in the long-run price of the good. This is in sharp contrast to the increasing-cost industry in which case long-run price rises as the demand increases.

It follows from the analysis of long-run pricing made above that with the increases in demand, Ion run price rises, remain constant, or falls depending upon whether the industry is subject to increasing cost, constant cost or decreasing cost.

PRODUCER SURPLUS AND EFFICIENCY OF COMPETITIVE MARKET

Producer Surplus: The concept of producer surplus is similar to the concept of consumer surplus of the theory of demand. Recall that consumer surplus obtained by the consumers from buying a product i the price that they are willing to pay over and above the price which they actually pay for a commodity. The consumer surplus is welfare benefit which consumers obtain from buying the commodity. Similarly, producer surplus is the excess of market price at which producers sell the quantity of a commodity over and above the minimum price at which they would be willing to supply it. The minimum price which the producers would accept to supply a unit of the commodity is its marginal cost which reflects the opportunity cost of resources used for its production. Consider Fig. 24.13 where demand curve *DD* and supply curve *SS* of a commodity intersect at point *E* and determine *OP* as the market price and *OQ* as the quantity sold and bought. It will be seen from the supply curve SS in Fig. 24.13. that producers produce the last *Q*th unit of the commodity at the marginal cost which is just equal to the market price *OP*. However, as is indicated by the supply curve SS the producers will be ready to supply the earlier successive units from the zero to *Q*th unit at much less than the market price *OP*. Thus, from these earlier units the producers Price actually get more than their minimum P acceptable supply price. The area *OSEQ* below the supply curve is indicator of the aggregate supply price of *OQ* units of the commodity produced and supplied by the producers. On the other hand, the total revenue earned by them is equal to the area *OPEQ* (market price *OP* × quantity *OQ* sold). Thus, the producers earn revenue equal to the shaded area *SEP* more than the aggregate supply price. This excess amount *SEP* over the aggregate supply price is the aggregate producer surplus earned by the producers. The *producer surplus earned by the producers is the measure of benefits obtained by them for producing and exchanging the commodity.*

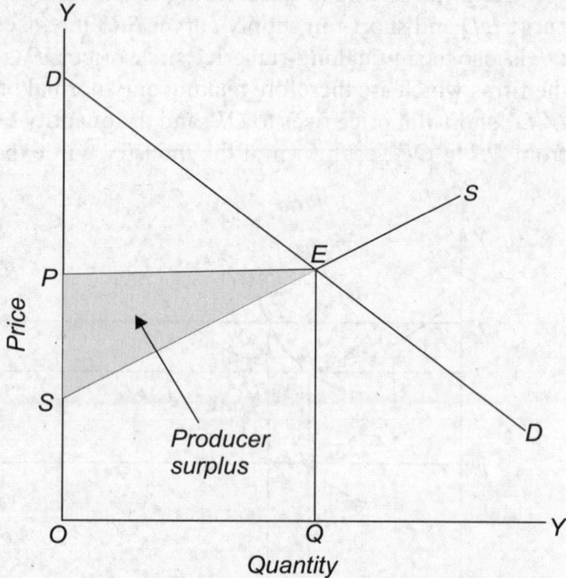

Fig. 24.13. *The Concept of Producer Surplus*

Efficiency of Perfectly Competitive Equilibrium

One of the most beneficial feature of perfectly competitive market is that it results in allocative efficiency which means that it enables the use of resources for production of goods that ensures maximum welfare·benefits to the persons in a society. Allocative efficiency of perfectly competitive markets also implies the mutually beneficial exchange between consumers and producers of goods. That is, both are better ofT in participating in the exchange. Under conditions of perfectly competitive markets optimum production and exchange of goods takes place which ensures the achievement of maximum social welfare. This can be shown by the concept of producer surplus explained above and of consumer surplus explained in an earlier chapter.

Recall that consumer surplus measures the welfare gain to the consumers from buying a commodity equal to the amount which they are willing to pay (which reflects their utility derived from the commodity) over and above the market price they actually pay for the good they buy from the producers. Consumer surplus is measured by the area under the 6: demand curve over the market price.

Thus, in Fig. 24.14 where demand curve *DD* of consumers and supply curve SS of consumers through their intersection at point E determine market price *OP* and quantity *OQ* of the good produced and exchanged between them. Consumer surplus obtained by the consumers is equal to the area *DPE*. On the other hand, as explained above, *PES* is producer surplus obtained by the producers which measures the excess revenue obtained by them over above the sum of marginal costs incurred by them for producing *OQ* amount of the good.

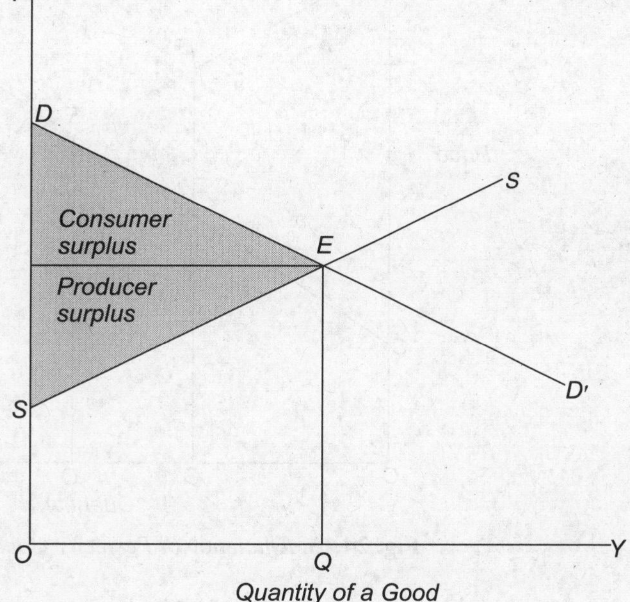

Fig. 24.14. *Maximisation of Total Economic Surplus*

It follows from above that both the consumers and producers gain from exchange and production of goods. *From the society's point of view, the total gain is the sum of consumer surplus and producer surplus. This sum of consumer surplus and producer surplus. That is, the sum of areas DPE and PES is known as the total economic surplus*. The allocative efficiency in consumption, production and exchange of goods and therefore maximum social benefit is attained when the total economic surplus (consumer surplus plus producer surplus) is maximised. And, as seen above, at price and output determined by demand and supply of goods in a perfectly competitive market, this maximum total economic surplus is actually attained.

An alternative way of viewing the total of consumer surplus and producer surplus is to consider it as the total benefits obtained from the consumption of two goods over and above the *total variable cost of producing a commodity*. Thus, the total benefit obtained by consuming *OQ* quantity of the good is the whole area *ODEQ* under the demand curve *DD′* which means the total utility or benefit attained by the consumers and the area *OSEQ* is the sum of total costs of producing *OQ* quantity of the good. Thus, the area *DES* measures the total surplus (*i.e.,* the sum of consumer surplus and producer surplus). *This sum of consumer and producer surplus therefore measures the net social benefit from producing and consuming OQ quantity of the good.*

The key question is whether the total surplus is maximised at the competitive equilibrium as determined by demand for and supply of the good. If some other production and consumption of the good (or in other words, some other resource allocation to the good in question) leads to a higher total surplus, the output *OQ* of the good determined by competitive equilibrium would not be efficient This is because it would then be possible to make consumers and producers better off collectively. On the other hand, if there is no alternative resource allocation to the good in question which generates a higher level of total surplus, the competitive equilibrium is economically efficient.

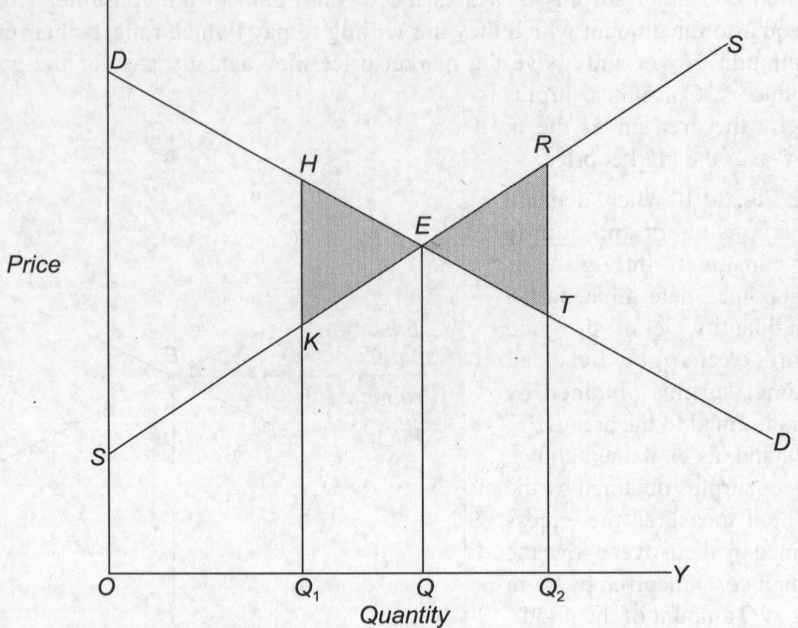

Fig. 24.15. *Efficiency of Perfectly Competitive Equilibrium*

To understand this consider Fig. 24.15 where competitive equilibrium is attained by producing *OQ* quantity of the good. Now, suppose a smaller quantity OQ_1 of the good is produced, then as will be seen from Figure 24.16 the consumers and producers will suffer a loss of total economic surplus equal to the area *HKE*. Likewise, any other quantity smaller than *OQ* would mean a lower total economic surplus than that obtained from producing and consuming *OQ* quantity of the good by allocating required resources to its production.

If a greater quantity than *OQ*, say OQ_2 is produced and consumed the consumers would get less utility (or welfare benefit) as indicated by the demand curve in the range of quantity QQ_2 than the cost of producing the good as indicated by the higher level of the supply curve in this quantity range (*i.e.*, *QQ*). Thus, the production of greater quantity OQ_2 of the good than the competitive output *OQ* leads to the loss of total economic surplus equal to the area *ERT*.

From above it follows that production of competitive equilibrium *OQ* quantity of the good ensures maximum total economic surplus and therefore leads to maximum social benefit. Thus is an important conclusion which tells us that competitive markets perform quite well in allocating scarce social resources.

But perfect competition rarely exists in the real world. It is market forms of monopolies, oligopolies and monoplistic competition that exists in the real world and not of perfect competition. We shall study in the next few chapters how market categories of monopolies, oligopolies and monopolistic competition create economic inefficiencies in resource allocation and hence loss of social welfare.

Further, as shall be studied in chapters on welfare economics that *economic efficiency is achieved under conditions of even perfect competition only when there do not exist any externalities, negative or positive*. The existence of externalities causes divergence between private benefit and social benefit as well as between private cost and social cost. Thus, perfectly competitive equilibrium based on private cost and private benefit would not lead to the maximum social welfare when externalities are present.

Economic Efficiency and Equity

That under conditions of perfect competition, the conclusion t t welfare is maximum is baied on the thinking that the total economic surplus is the true measure of so' w . To consider so is to ignore the effect of distribution of output and income on social welfare. The equity, that is, fairness in distribution is as important as maximisation of total economic surplus. To consider the size of total economic surplus as the sole criterion of assessment of social welfare is to implicitly make the value judgement that a rupee to each person is given the same weight irrespective of whether the person is consumer, or producer, rich or poor. To clarify this let us consider the price of a product is raised. As a result, the consumers would lose some consumer surplus and producers would gain some extra producer surplus. The criterion of total economic surplus is base on the net change in surplus in rupee terms. Suppose as a result of rise in price of a good consumers suffer a loss of consumer surplus of ₹ 1000 and producers gain extra producer surplus of ₹ 1000. Thus, the total economic surplus remains unchanged. But, if consumers are poor households whereas producers are rich individuals, the redistribution of income from the poor consumers to the rich producers, the total economic surplus 'remaining the same, will lead to the loss of social welfare. To conclude that in this case social welfare remains unchanged is based on the implicit assumption that the poor derive utility from Rs. 1000 equal to that of the rich. That is, marginal utility of one rupee is the same for the rich and the poor. This is a questionable assumption and involves value judgement which is not justified on any scientific grounds. In other words, maximising total economic surplus leads to economic efficiency, but it will not be necessarily 'fair' or 'equitable'. The maximisation of total economic surplus does not capture the equity aspect of social welfare. In the light of this fact, the total economic surplus (*i.e.*, the sum of producer surplus and economic surplus) is not a very good measure of social well-being. Thus, maximisation of total surplus leads to the outcome which is economically efficient but it may not be necessarily fair or equitable.

It may be noted that some economists consider maximisation of total surplus as a valid criterion . of social welfare as they think that once total surplus is maximised, it can be redistributed in accordance with society's notion of equity or fairness. It is argued that "make the pie as big as possible and then distribute according to society's notion of equity. However, in our view, it is difficult to redistribute output and income so as to ensure equity and thereby to increase social well-being. Besides, in the redistribution, demand and supply of curves of a good which generated the maximum total surplus, are likely to change which may result in deviation of the outcome from the maximum total economic surplus (or economic efficiency). We will discuss in detail this issue of economic efficiency vs equity in chapters on welfare economics where we will critically examamine various welfare criteria put forward by economists to evaluate alternative resource allocations.

QUESTIONS FOR REVIEW

1. Distinguish between partial equilibrium analysis and general equilibrium analysis of pricing. When are they used in explaining prices of commodities?
2. What is meant by partial equilibrium analysis of pricing? On what basic assumption is it based? Explain how Marshall defended the use of partial equilibrium analysis of pricing of commodities.
3. "General equilibrium analysis explaines simultaneous determination of prices of all goods and factors". Explain when it is more useful to adopt general equilibrium approach to explain the determination of prices of goods.
4. The credit of finding the true answer to the pricing problem goes to Marshall who held that both demand and supply were equally important in determining prices of commodities". Disscuss.

5. "Supply and demand are not ultimate explanations of price. They are simply useful catch-all concepts of describing many factors and forces that go to determine price" Discuss
6. Explain the importance of time element in the determination of prices in perfectly competitive markets.
7. Distinguish between a perishable commodity and durable commodity. How are their market prices determined in perfectly competitive markets.
8. Suppose there is once-and-for. all increase in demand for a commodity. How will it affect its (a) market price, (b) short-run price and (c) long-run price? Illustrate diagramatically.
9. How is short-run price of a commodity determined in a competitive market. Does increase in demand for a commodity always lead to rise in its price in the short run ? Explain what is meant by long-run normal price? Show that long-run normal price under perfect competition is equal to the minimum long-run average cost.
10. What are external economies and external diseconomies? When do we get increasing cost conditions in a competitive industry?
11. Derive a long-run supply curve of increasing-cost industry working under conditions of perfect competition. *D.U B.A.(Hons), 1997, 1999*
12. If all firms in a perfectly competitive industry have U-shaped cost curve, can the supply curve of the industry be a horizontal straight line? Explain and derive a long-run supply curve of a constant-cost industry.
13. Suppose there is a permanent increase in demand for a product produced under conditions of perfect competition by an increasing-cost industry. What would be its effect on price and output in the short run and long run.
14. A perfectly competitive industry is in long-run equilibrium. What are the long-run effects of an increase in demand for its product given that it is an increasing-cost industry. *D. UB.A. (Hons) 2002*
15. Under what conditions long-run supply curve of a competitive industry can slope downward? If in such an industry demand for the product increases, how will its price change in (a) the short run, (b) the long run.
16. If all the firms in a perfectly competitive industry have V-shaped cost curve, can the supply curve industry be downward sloping? Give reasons? *D. U B.A. (Hons) 1998*
17. Is is possible for an industry to be constant-cost industry though each firm in the industry has increasing marginal costs. Explain.

CHAPTER 24 A
STABILITY OF EQUILIBRIUM AND COBWEB MODEL

Stable and Unstable Equilibrium: Walras Price Adjustment Approach

We now turn to the problem of stability of market equilibrium. The equilibrium between demand and supply is attained at a price-quantity combination at which both buyers and sellers are satisfied about what they are buying and selling respectively and therefore they do not have any incentive to change their behaviour. Furthermore, changes in demand due to changes in preferences of the consumers and changes in supply due to technological changes or variation in factor prices often occur which disturb the equilibrium. These changes either in demand or supply will determine new equilibrium price but there is no guarantee that the new equilibrium price will be restored. Thus, when actual price differs from the equilibrium price due to certain disturbances, it is very relevant to ask whether some forces will come into operation which would cause the system to return to the equilibrium level. *The equilibrium is stable, if following a disturbance in it, the equilibrium is established again, and unstable if the system tends to move away from the equilibrium situation.* Whether or not equilibrium is restored when it is disturbed depends on the behaviour of agents, that is, buyers and sellers. Further when there is disequilibrium whether it is changes in price or changes in quantity through which adjustment takes place to restore equilibrium. *Walras thought it is through changes in price that adjustment takes place.* According to him, behaviour of buyers is such that they tend to bid up the price when quantity demanded exceeds the quantity supplied at a given price and the behaviour of sellers is such that they tend to lower it when the quantity supplied of the commodity is greater than the quantity demanded of it. With such changes in price equilibrium will be stable.

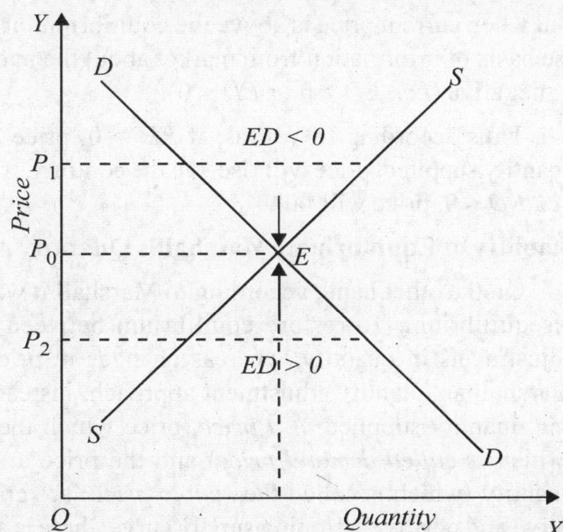

Fig. 24A.1. *Stable Equilibrium Walras Approach*

If demand curve is downward sloping (*i.e.*, has a negative slope) and supply curve is upward sloping (*i.e.*, has a positive slope), as is normally the case, the equilibrium between demand and supply; if it exists, will be stable. This is illustrated in Fig. 24A.1. At price, P_1, the quantity supplied exceeds quantity demanded (that is, *at price P_1, excess demand is negative* ($ED < 0$). This means at price P_1, the sellers would not be able to sell the quantity they would like, that is, disequilibrium exists. According to the profit-maximising behaviour, in these circumstances the sellers would bid down price until price P_0 is reached at which demand and supply are again in equilibrium. On the other land, if price is P_2, in Fig. 24A.1, the quantity demanded exceeds quantity supplied (that is, excess demand ED is positive), buyers would push up the price until P_0, is reached. Thus, *with negatively sloping demand curve and positively sloping supply curve equilibrium when exists is stable as market price when disturbed moves toward equilibrium due to the assumed behaviour of buyers and sellers.*

In Fig. 24A.2 both demand and supply curves are negatively sloping but the supply curve is less steep as compared to the demand curve. In this situation quantity demanded equals quantity supplied at price P_0. If the market price is initially at this equilibrium level P_0, it will stay there unless there is some external disturbance. However, if due to some disturbance, price happens to be P_1 which is above the equilibrium price P_0, quantity demanded exceeds quantity supplied (that is, $ED > 0$ where ED stands for excess demand), Now, given the behaviour pattern of buyers, they would tend to bid up the price and therefore the market price instead of returning to the equilibrium level P_0, will tend to move away from it. This shows *the equilibrium at price P_0 in Fig. 24A.2 is unstable*. On the other hand, in Fig. 24A.2 if due to some disturbance, price happens to be P_2, which is below the equilibrium price, the quantity supplied exceeds the quantity demanded (that is, $ED < 0$), the sellers will tend to lower the price and price will not go up to the equilibrium level but will move away from it. This again shows the *equilibrium at P_0 in Fig. 24A.2 represents unstable equilibrium in Walrasian approach*.

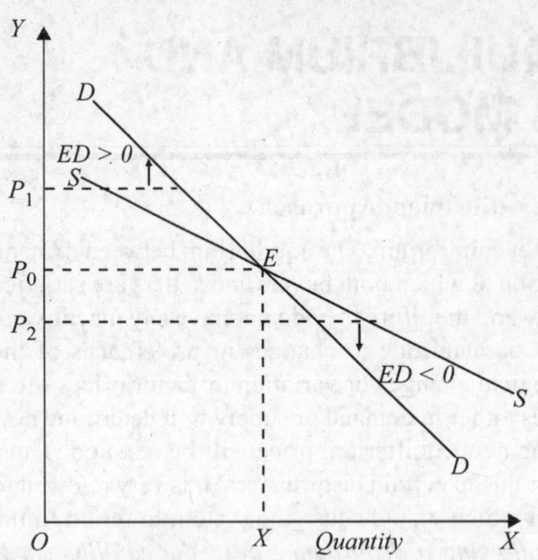

Fig. 24A.2. *Unstable Equilibrium under Walrasian Price Adjustment Approach*

Walrasian Stability of Equilibrium and Excess Demand. We have seen above that due to a certain external disturbance market may go out of equilibrium. This creates an adjustment in the market. But whether adjustment is made in price or quantity has been a controversial issue. Walras thought that it is price which is adjusted. When current price is below the equilibrium price, price is bid up and when current price is above the equilibrium, it is bid down. These changes in price are made on the basis of information from market about the nature of excess demand as to whether it is positive or negative (*i.e.*, $ED > 0$ or $ED < 0$).

Thus according to Walras, if $ED > 0$, price goes up, that is, if quantity demanded exceeds quantity supplied, price will rise. On the contrary, if quantity demanded is less than quantity supplied, *i.e.*, $ED < 0$, price will fall.

Stability of Equilibrium: Marshall's Quantity Adjustment Approach

On the other hand, according to Marshall it was quantity which was adjusted in case of market disequilibrium. To restore equilibrium between demand and supply, buyers and sellers make adjustments in quantity, whereas changes in prices follow from the adjustments in quantity. In Marshallian quantity adjustment approach, instead of viewing the situation as quantity demanded and quantity supplied *at a price*, price which the buyers are willing to pay for a *given quantity* (which is *called demand price*) and the price at which the producers are willing to supply that quantity (which is called the *supply price*) are considered. In case of negatively sloping demand curve and positively sloping supply curve, there is a *stable equilibrium* at the quantity where demand and supply curves intersect. This is illustrated in Fig. 24A.3. At the quantity Q_0 at which demand and supply curves intersect (that is, demand price equals supply price at the quantity Q_0), market equilibrium is reached and this equilibrium is stable even according to Marshallian approach, as was the case in Walrasian price adjustment approach. For example, suppose the quantity Q_1 is being produced and sold. It will be observed from Fig. 24A.3 that for quantity Q_1, demand price which is equal to Q_1H exceeds the supply price which is equal to Q_1G (the price at which producers are willing to sell quantity Q_1 and which depends upon the marginal cost of production). Therefore, it is profitable for the producers to increase the quantity produced, and this adjustment in quantity will continue until the quantity Q_0 is reached where demand price becomes equal to supply price. Further, it will be observed from Fig. 24A.3 that if quantity will happens to be Q_2, demand price at it is equal to Q_2L which is less than the supply price Q_2K at it. As a result, quantity will be reduced to the equilibrium quantity Q_0. This shows, according to Marshallian quantity adjustment approach, equilibrium at quantity Q_0 in Fig. 24A.3 is *stable*, as was so under Walrasian price adjustment approach in this case of negatively sloping demand curve and positively sloping supply curve.

However, in case of both demand and supply curves sloping downward with supply curve being less steep, as shown in Fig. 24A.4, the two approaches lead to different results. In this case whereas according to *Walrasian price adjustment, equilibrium is found to be unstable, under Marshallian quantity adjustment approach*, the equilibrium at Q_0in Fig. 24A.4 is *stable*. It will be seen from Fig. 24A.4 that if current quantity of commodity produced is Q_1 demand price exceeds supply price at it with the result that this would tend to increase the quantity until Q_0 is reached. On the contrary, if at

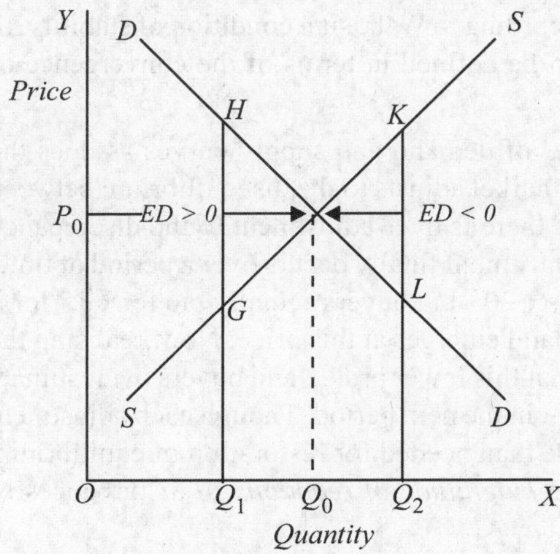

Fig. 24A.3. *Stability of Equilibrium: Marshallian Quantity Adjustment*

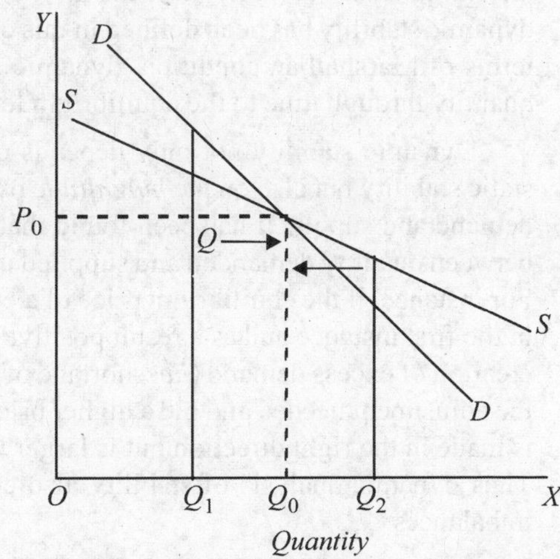

Fig. 24A.4. *Under Marshallian Quantity Adjustment Approach, Equilibrium at Q_0 is Stable.*

any time output happens to be Q_2, demand price is less than the supply price. Consequently, producers would be induced to reduce the quantity to Q_0 where demand price equals supply price and therefore quantity adjustment ceases.

STATIC STABILITY VS DYNAMIC STABILITY

Two types of stability of equilibrium may be distinguished. As mentioned above, when an equilibrium is disturbed, it creates an adjustment process in the market. For example, if price is below the equilibrium level, some buyers bid up the price leading to the restoration of equilibrium. *Static analysis of stability does not consider how price and quantity change or adjust through time when there is disequilibrium between demand and supply*; it merely considers the nature of changes or tendencies that exist as to whether the market will move towards or away from equilibrium. It is important to note that stability conditions are derived from assuming a certain behaviour pattern on the part of buyers and sellers. In the foregoing analysis we have made static analysis of stability of equilibrium and distinguished between Walrasian and Marshallian conditions of static stability. We now turn to explain dynamic stability of equilibrium.

Dynamic stability considers how price and quantity change through time when the system is thrown into disequilibrium state. In other words, starting from a disequilibrium state, the dynamic analysis of stability investigates *the time path of the adjustment process* of the movement of price and quantity towards the equilibrium levels. It always takes time for the adjustment to be made in price and quantity and therefore instantaneous adjustment to the new equilibrium is not feasible. The process of dynamic adjustment is often explained through what is called *recontracting model.* According to this, buyers and sellers enter into series of provisional contracts about price and quantity until they reach a final equilibrium situation when recontracting ceases. Let us explain this recontracting in some greater detail.

Suppose that at an initial price, the quantity demanded by buyers is not equal to the quantity supplied, that is, there is, disequilibrium. Adjustment process starts and recontracting takes place. If at the new price there is still a discrepancy between demand and supply, the price will be further changed by what is called an auctioneer — a middle man through whom contracts are being made. This process of recontracting continues and price bids go on changing from period to period over the course of time until equilibrium price is reached at which quantity demanded equals quantity supplied. Thus, "*equilibrium is stable in the dynamic sense if the price converages to or approximates the equilibrium over time* ; *it is unstable if the price change is away from equilibrium*".[1] Note that dynamic stability has been defined in this quotation according to Walrasian condition of stability. In terms of Marshallian condition, dynamic stability can be defined in terms of the convergence of quantity through time to the equilibrium level.

Dynamic stability not only depends on the *slopes* of demand and supply curves as does the static stability but also on the *magnitude* by which the market adjusts to the disequilibrium between demand and supply. It has been found that quite often there is over-adjustment to the discrepancy between quantity demanded and supplied until right adjustment finally occurs over a period of time. For instance, if the equilibrium price of a commodity is ₹ 20/- but buyers actually bid for ₹ 15/for it in the first instance and as a result positive excess demand emerges at this price. Now, realising the creation of excess demand (*i.e.* shortage of commodity at this lower price), and buyers overestimate the adjustment needed and bid a higher price, say ₹ 22/-, in the next period. Though each adjustment is made in the right direction but is larger in magnitude than needed for restoration of equilibrium. Thus dynamic analysis of stability also considers the *magnitude of reactions* to disturbances or imbalances.

Dynamic Stability with Lagged Adjustment: Cobweb Model

We shall now explain the notion of dynamic stability with a well known theorem often referred to as cobweb model. Cobweb theorem is the simplest model of economic dynamics when equilibrium reached over time between demand, supply and price is investigated. Producers' supply function (curve) shows how producers adjust their output to changes in price. At a higher price, they respond to produce more and at a lower price they cut down their production. But this adjustment in production in response to changes in price does not occur instantaneously but takes a good deal of time. Thus, there is a *time lag* between a change in price and appropriate adjustment in supply in response to it. The time gap between the decision to change the quantity supplied in response to a given price and its actually being supplied is known as *supply lag*. The supply lag is often found in case of agricultural commodities, animal rearing etc. For instance, it takes four-five months for wheat crop to be grown and raised.

It should be noted that supply lag often results in *cyclical movements or oscillations* in price and quantity over time. The dynamic analysis of stability investigates whether these oscillations coverage to the equilibrium values or move away from them.

Cobweb Model. Let us explain the dynamics of cobweb model. We consider the market for wheat as an example of a market where the supply of wheat is lagged function of price. In order to

1. J.M. Henderson, R.E. Quandt, *Microeconomic Theory*, 2nd ed. 1971. p. 136 (Italics added).

keep our analysis simple we assume that there is a one-year lag in the response of quantity supplied of wheat to a given market price of it. Thus

$$S_t = f(P_{t-1})$$

which means that the quantity of wheat supplied in a year t depends on the price prevailing in the previous period (P_{t-1}). Further, in accordance with this the present year price (i.e. P_t) will determine the quantity supplied in the next period (*i.e.* S_{t+1}).

On the side of demand function there is no lag, that is, the quantity demanded of this year depends on price of this year. The equilibrium price in any year is determined at the level at which the quantity demanded of any year t equals the quantity supplied in that year t. This means the price in a period adjusts to bring about equality in the quantity demanded in the year (D_t) and the quantity supplied in that year (S_t) which of course in this model is determined by the price prevailing in the previous year, *i.e.* $t - 1$. It should be noted that there is perfectly inelastic supply in any year as in case of agricultural commodities like wheat no more quantity can be grown until next year to augment supply. Thus in any year t

$$D_t = S_t \quad or \quad D_t - S_t = 0$$

This is the equilibrium situation brought about by adjustment in price in a year. In any year there is clearing of the market, that is, no producer is left with unsold stocks and no consumer with unsatisfied demand. Possible time paths of change in price and consequently the nature of oscillations depends upon the particular demand and supply functions. The demand and supply functions according to cobweb model can be written as :

$$D_t = bP_t + a \qquad \ldots (i)$$

$$S_t = gP_t + C \qquad \ldots (ii)$$

where b and a in the demand function are constants ; whereas b is a slope coefficient showing relation between price of a commodity and demand for it, a is the constant intercept term. Similarly, g and C in the supply functions are constants; whereas g denotes the slope coefficient showing the relation between price of a commodity and the quantity supplied which is forthcoming at it and C is the intercept term of this supply function.

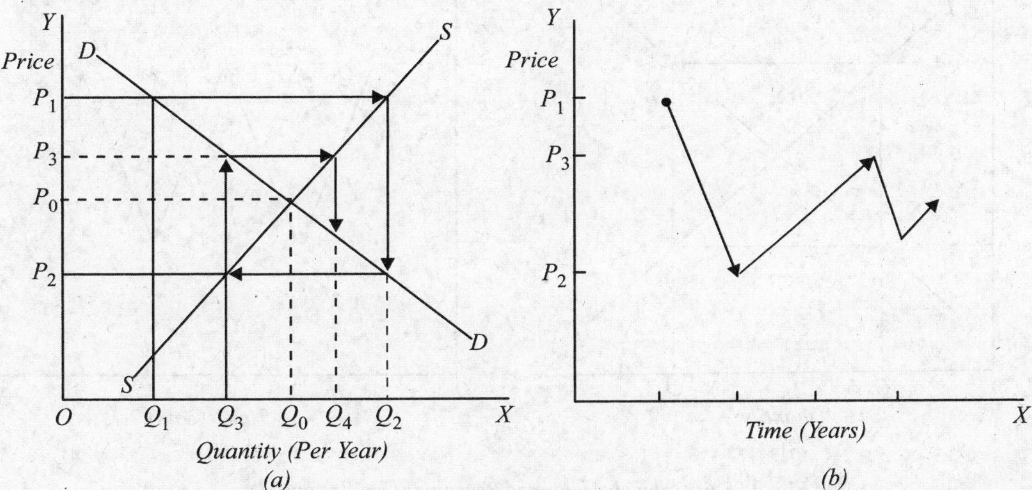

Fig. 24A.5. *Damped Oscillations: Stable Dynamic Equilibrium*

Now, the path of price as a function of time depends on the absolute values of the slope coefficient of the demand function [*i.e.* b in (i) above] and the slope coefficient of supply function [*i.e.* g in (ii) above]. If the slope of the supply curve i.e. $\frac{1}{|g|}$ is greater (in absolute value) than the slope of the demand curve i.e. $\frac{1}{|g|}$ (that is, supply curve is *steeper* or less elastic than the demand curve), we get the case of *damped oscillations*, which as will be seen from panel (a) of Fig. 24A.5 appears like a cobweb. Fig. 24A.5 describes the time path of price and quantity as we start from price-quantity combination (P_1, Q_1). Let us explain this path of price in greater detail. It will be seen from panel (a) of Fig. 24A. 5 that demand and supply curves intersect at price P_0 and Q_0 which are therefore equilibrium price and quantity.

Suppose there is a disturbance such as occurrence of a drought which causes the output (quantity) of wheat to fall to Q_1. Now, with Q_1 as the quantity supplied and given the demand curve *DD*, the price P_1 is determined. From the supply curve *S* in panel (a) of Fig. 24A.5 it will be observed that at the price P_1, the quantity supplied Q_2 will be forthcoming in the next year t_2. With Q_2 as the supply and the demand curve *DD*, price P_2 is determined. Thus price falls from P_1 in year 1 to price P_2 in year 2. Now, as will be observed form Fig. 24A.5 [panel (a)] price P_2 will induce the quantity supplied Q_3 in the next year 3. With Q_3 as the quantity supplied and demand curve *DD*, the price P_3 is the equilibrium price in period 3. This will further cause a change in supply in the next period and bring about further change in price. But, as will be seen from Fig. 24A.5 price is oscillating but it is damping, that is, tending towards the equilibrium price P_0. It is thus clear that in the present case *when the absolute slope of the supply curve exceeds that of the demand curve we have damped oscillations in price*, that is, price fluctuates but over a time it tends to move towards the equilibrium level. Therefore, the cobweb model of damped oscillations represents the case *of stable dynamic equilibrium.* It will be seen from panel (*b*) of Fig. 24A.5 which directly depicts the time path of price over the years and shows that over time the price tends to converge to the equilibrium level.

Perpetual Oscillations. If the absolute slopes of the supply and demand curves are equal, we will get a cobweb model of perpetual oscillations as shown in Fig. 24A.6. The equal slopes of supply and demand curves in economic sense imply that the response of quantity demanded and supplied to the changes in price is the same. Consider panel (a) of Fig. 24A.6. To start with, in year

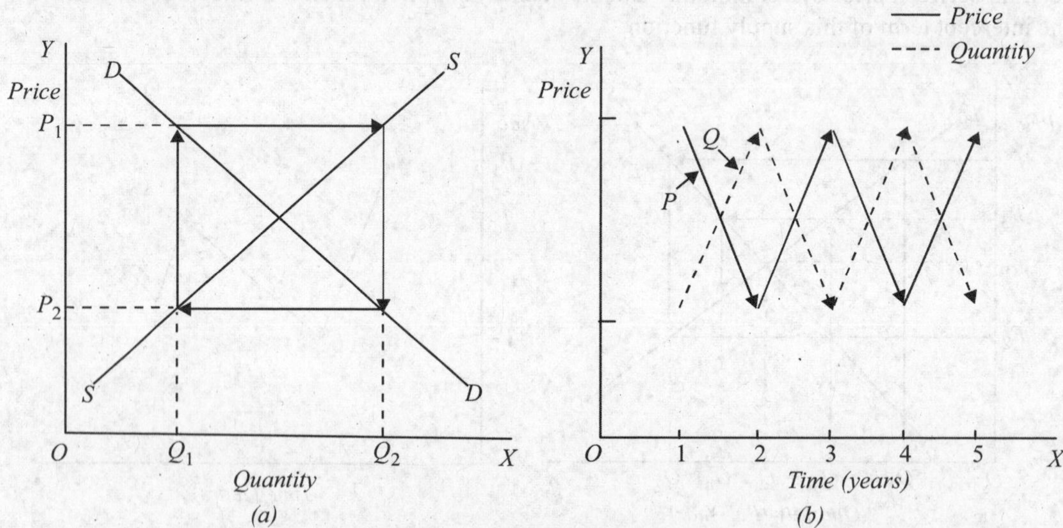

Fig. 24A.6. *Cobweb Theorem: Perpetual Oscillations.*

1 the quantity of the good produced is Q_1 and, given the demand curve *DD*, the price P_1 is determined in year 1. Price P_1 in year 1 calls forth quantity produced and supplied Q_2 in year 2. With Q_2 as the quantity supplied and the demand curve *DD*, a lower price P_2 is determined in year 2. It will be seen

from supply curve *SS* that at the lower price P_2, the producers reduce the quantity produced and supplied again to Q_1. With Q_1 as the quantity supplied and the demand curve *DD*, price again rises to P_1 in year 3 and so on. The price goes on fluctuating between P_1 and P_2 perpetually and quantity goes on oscillating ceaselessly between Q_1 and Q_2. It should be noted that these perpetual oscillations remain constant in amplitude. As will be seen from Fig. 24A.6 (*b*) that *in this case of perpetual oscillations, price and quantity do not move towards equilibrium price and quantity*. Therefore, equilibrium once disturbed is never restored. In fact equilibrium is never attained in this case. It should be noted that when quantity is low, price is high, and when quantity is high, price is low.

Explosive Oscillations. Finally, we have cobweb model in which prices and quantities move further away from equilibrium instead of moving towards equilibrium as indicated by the intersection of demand and supply curves. This occurs when the *absolute value* of the slope of the supply curve is less relative to that of the demand curve. In economic terms this implies that when the supply responds to changes in prices to a greater degree than the demand, (that is, supply is more elastic than the demand) the amplitude of oscillations of price and quantity go on increasing over time and become too much excessive. Therefore, these ever-widening oscillations of price and quantity are

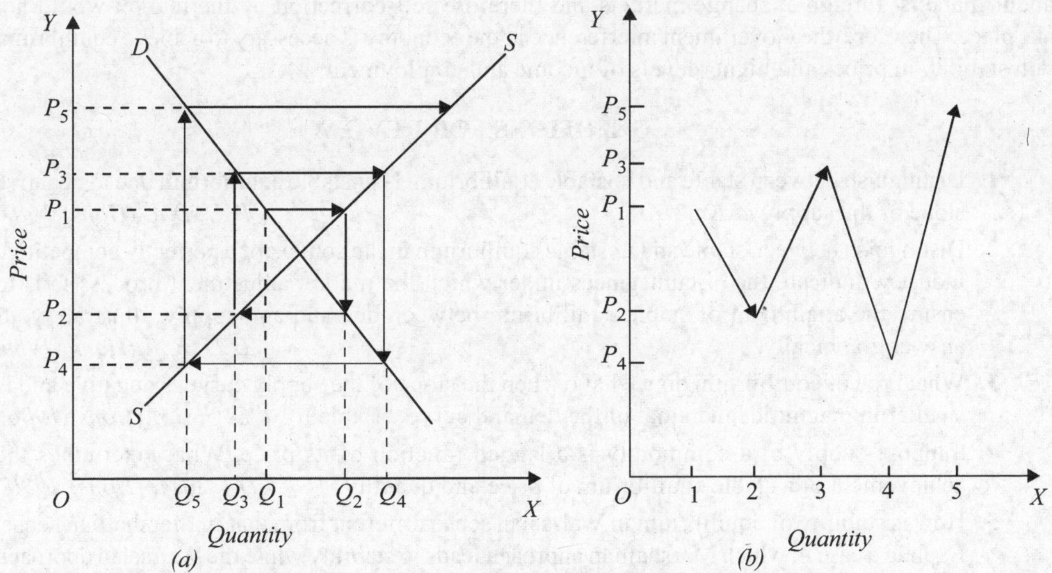

Fig. 24A.7. *Cobweb Model: Explosive Oscillations.*

called *explosive oscillations* and are depicted in Fig. 24A.7. Starting from price P_1 in year 1 it falls to P_2 in year 2 and then it goes up sharply to P_3 in year 3 and then have a drastic decline to P_4 in year 4. The price again rises sharply to P_5 in year 5. The price movements are also shown in panel (*b*) of Fig. 24A.7 where as will be seen price tends to move further away from the equilibrium price indicated by the intersection of demand and supply curves. Thus, *the explosive oscillations model represents and unstable dynamic equilibrium*.

Stable and Unstable Equilibrium: Policy Implications

The students are likely to gather impression that the discussion of stability of equilibrium is a futile theoretical exercise without any economic significance. But this is far from truth. The stability of equilibrium of a market has important implications for economic policy. If a competitive market is in stable equilibrium, it implies that it can survive any external disturbances, however large and drastic they may be. That is, if any disturbance causes disequilibrium, the stable nature of market equilibrium ensures that certain forces will automatically come into play to bring it back into equilibrium. For instance, if due to certain factors, given the demand and supply curves, if price rises above the equilibrium level, that is, there is inflation in the economy, then if market equilibrium is stable, rise in price will be self-corrected without any interference from the Government. Similarly, if due to some reasons equilibrium is disturbed and consequently there prevail deflation and unemployment, the stability of equilibrium, if it is stable, ensures that the adjustment in wages, prices and labour employment would occur which will bring the market back into equilibrium without unemployment. This is a very important conclusion because if market equilibrium is stable and automatic correction of disequilibrium is possible, theoretically there is no need for Government interference in a free market economy in times of inflation, recession or unemployment. On the contrary if market equilibrium is unstable the Government has to play an important role by intervening in the economy to restore equilibrium. The economists who are in favour of market-friendly approach tend to argue that less the interference by the Government in the working of a free market private enterprise system, the better it is. However, this argument in favour of free market and private sector is theoretically justified only if the market equilibrium is stable. In case of unstable market disequilibrium free competitive market could collapse under the pressure of either high inflation or recession. In fact, in the real world we witness unstable eqilibrium in some commodity markets, labour markets, foreign exchange markets and therefore self-correction by the market would not take place. Therefore, the Government interference in the economy is necessary to achieve equilibrium with stability in prices and higher levels of income and employment.

QUESTIONS FOR REVIEW

1. Distinguish between stable and unstable equilibrium. Is unstable equilibrium due to negative slope of the supply curve ? *D.U. B.A. (Hons) 1987*
2. Distinguish between stable and unstable equilibrium in the context of a perfectly competitive industry. Indicate the circumstances under which the market adjustment process fails to ensure the attainment of stable equilibrium between demand and supply. Illustrate your answer graphically. *D.U. B.A. (Hons) 1989*
3. What type of equilibrium do we have when the slope of the supply curve is negative and is smaller in magnitude than that of the demand curve ? Explain *D.U. B.A.(Hons) 1986.*
4. Suppose supply of a commodity is a lagged function of its price. What determines the achievement of a stable equilibrium of price and quantity *D.U. B.A. (Hons) 1990.*
5. How is stability of equilibrium in Walrasian sense different from that in Marshallian sense. Explain a case in which Marshallian approach leads to stability white the Walrasian approach does not. *D.U. B.A. (Hons) 1996*
6. Explain Cobweb model. Show how it explains cyclical movements or oscillations in price and quantity of agricultural products.
7. When there is lagged response of supply to changes in price of a commodity, under what conditions we get

    1. damped oscillations
    2. perpetual oscillations
    3. explosive oscillations.

CHAPTER 25
APPLICATIONS OF DEMAND AND SUPPLY ANALYSIS

Introduction

In the last chapter we have explained how prices of goods are determined by demand and supply. The analysis of price determination in terms of demand and supply is not merely of great theoretical significance but it has important several practical applications in economic life of a country. This analysis of demand and supply has been used to explain the implications of price control and rationing, minimum price fixation, incidence of taxes, several other economic problems and policies. In the present chapter we shall explain some of these applications of demand and supply analysis.

In the analysis of the previous chapter we have assumed that market mechanism is allowed to function without interference by the government. But government in the modern mixed economies interferes with the functioning of the market system to influence prices so as to promote social welfare when it is felt that free working of market will not produce desirable results. The government can interfere with the working of the economy in two main ways. The first government fixes the *maximum price* (often called *price ceiling*) or fix the minimum price often called floor price). *Price control* of foodgrains, rent controls are the examples of fixation of maximum price or price ceiling above which the sellers cannot charge the price. *Agricultural price support programme* is the example of fixation of minimum price to assure minimum remunerative prices to farmers so as to protect their interests.

The second way in which government interferers with the price or market system is *working through the market*. In the second way government can impose taxes on the commodities or provide subsidies. These taxes and subsidies affect the market supply or demand curves which determine prices of goods and services. Imposition of heavy excise duties on cigarettes or other drugs and providing subsidies on agricultural products are examples of interference by government through market. In what follows we will explain both types of intervention with the functioning of markets by government we begin our analysis with the imposition of price control and rationing by government.

PRICE CONTROL AND RATIONING

In times of war imposition of price control is quite common and was introduced by several countries during the Second World War. Even in peace time, price controls on essential commodities have been introduced in several countries to help the poor against inflation. *Under price control the maximum price of a good is fixed above which the sellers cannot charge from the consumers*. Price control is imposed or price ceiling is set below the equilibrium price. This is because if the price ceiling is set above the equilibrium price that balances supply and demand, it will have no effect or in other words, *it will not be binding*. Consider Fig. 25.1 where demand and supply balance each other at price P_1. At this equilibrium price both buyers and sellers are satisfied, buyers are getting the quantity of the good they want to buy at this equilibrium price and the sellers are selling what they want to sell at this price. Therefore, the higher price P_2 than the price P_1 fixed by government will have no effect.

When it is realised that the equilibrium price of a commodity is too high and consequently some buyers go unsatisfied, for they lack the means to pay for it, the Government may pass a law through which *it fixes the maximum price of the commodity* at a level below the equilibrium price. Now, at a price lower than the equilibrium price, quantity demanded will be larger than the quantity supplied and thus *shortage* of the commodity will emerge; some consumers who are willing and able to buy at that price will go unsatisfied. Buyers would, if permitted, bid up the price to the equilibrium level.

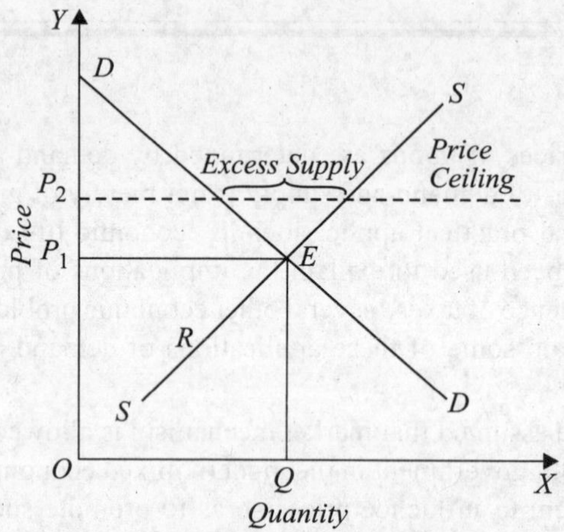

Fig. 25.1. *Price Celing above the Equilbrium Price is Not Binding and is Ineffective*

Fig. 25.2. *Effect of Price Control (i.e., Fixing Price Ceiling below the Equilbrium Level)*

But under price control by the Government, price is not free to move to equate quantity demanded with the quantity supplied. Thus, when the Government intervenes to fix the maximum price for a commodity, price loses its important function of a rationing device.

Price control and problems raised by it are graphically illustrated in Fig. 25.2 where demand and supply curves, *DD* and *SS* of sugar are given. As will be seen from this figure that demand and supply curves intersect at point *E* and accordingly OP_1 is the equilibrium price of sugar. Suppose that this equilibriumm price OP_1 of sugar is very high so that many poor people are not able to obtain any quantity of it. Therefore, the Government intervenes and fixes the maximum price of sugar at the level OP_0 which is below the equilibrium price OP_1. As will be seen from Fig. 25.2 at the controlled price OP_0 quantity demanded exceeds the quantity supplied. At price OP_0, whereas the producers offer to supply P_0R quantity of sugar, the consumers are prepared to buy P_0T quantity of it. As a result, shortage of sugar equal to the amount *RT* has emerged and some consumers will go unsatisfied.

In the absence of Government intervention fixing the maximum price at OP_0 level, the excess demand equal to *RT* would have led to rise in the price to the equilibrium level OP_1 where quantity demanded is equal to quantity supplied. *But, under price control by the government, to charge a price higher than the legally fixed maximum price* OP_0 is punishable under the law. Therefore, the available supply *OM* at the fixed price OP_0 has to be somehow allocated or rationed between the consumers. The rationing can take many forms. This task of rationing the available supply *OM* may be done by the producers or sellers themselves. The sellers may adopt the principle of "*first come first served*" and distribute the available supply of sugar among those who are *first in the queue* before their shops. This system of rationing is therefore called *queue rationing*.The second method of rationing or allocating the scarce supply of the good is to distribute it on the basis of what has been called, "*allocation by sellers' preferences*". Under this, the available supply of good is sold by the sellers to certain preferred consumers at the controlled price. The sellers may sell the good at the controlled price to their regular customers. They may also adopt the policy of selling the available supply to the buyers belonging to certain caste, religion, colour etc. and not to others.

If the Government does not like the rationing of a commodity among the population on the basis of either "first come first served" or arbitrary allocation by sellers' preferences, it may introduce *coupon rationing* of the commodity. Under the coupon rationing system consumers are given ration coupons just sufficient to buy the available quantity of the commodity. The number of ration coupons issued to a family may depend on the age of its members, sex, and the number of family members or on any other criterion considered desirable.

Black Market. A point worth noting is that price control with or without rationing is likely to give rise to the *black market* in the commodity. *By black market we mean the sale of a commodity by the producers or sellers at a price higher than the controlled price.* As mentioned above, at the controlled maximum price *fixed below the equilibrium price*, the quantity demanded would exceed the quantity supplied and consequently shortage of the commodity would develop. It is thus clear that some buyers of the commodity will not be fully satisfied as they will not be able to get the quantity of the good they wish to buy at the controlled price. Therefore, they will be prepared to pay a higher price for getting more quantity of the good, but they can do so only in the black market. Sellers will also be interested in selling the commodity, at least some quantity of it, in the black market at a higher price as it will fetch them larger profits.

Even when coupon rationing is introduced there will be pressure for the black market to develop. This is because the consumers are willing to buy more quantity of the commodity than is available at the controlled price, whereas rationing only distributes the available quantity of the commodity. Therefore, the consumers who want to procure larger quantity than the rationing amount will be prepared to pay a higher price to get some quantity in the black market.

There is sufficient evidence in India and abroad to confirm the predictions based on demand and supply analysis. *When price control and rationing system for some commodities which were in shortage were introduced during the Second World War and after, black markets developed in spite of punitive measures taken by the authorities.*

RENT CONTROL

Rent controls is an other example of maximum price that Government fixes on the rental price of housing units. Under rent control, the Government fixes the rent per month per housing unit of a standard size which is *below the equibrium rent* that would otherwise prevail in the marekt. The maximum rent fixed by the Government helps the tenants, who generally belong to lower and middle income groups and intend to prevent their exploitation by rich landlords who would charge a very high market determined rates of rent. Market determined equilibrium rent rate happens to be high because demand for rental housing tend to be relatively greater than supply of it. In several important cities such a New York, London, Mumbai, Delhi. Government has imposed rent control to help the lower and middle income people by ensuring rental houses at fair rents. In Delhil under Delhi Rent Control Act, 1958 which has now been amended by the recently passed New Rent Control Act 1995. This law specifies some monthly rental rates of housing units of some standard sizes above which the landlord cannot charge rent. Besides, the landlord cannot evict the tenants easily except under some conditions laid down in the law. It is however important to understand both the short-run and long-run effects of rent control.

Economists often point out the adverse effects of rent control and hold the view that it is highly inefficient way of helping the poor and lower middle class people. The adverse effects of rent control are evident only in the long run because it always takes time to construct new housing units/ apartments and also for the tenants to adjust to the rent and housing accommodation available on rent. So the long-run effect of rent control is different from the short run. In the short run the landlords have almost a fixed number of housing units/appartments to give on rent. Therefore, the supply curve of rental units is inelastic in the short-run. On the other hand, people searching for rental-housing units are also not very responsive in the short run as it always take time for them to adjust their housing arrangements. Thus, even demand for rental housing in relatively inelastic in the short run.

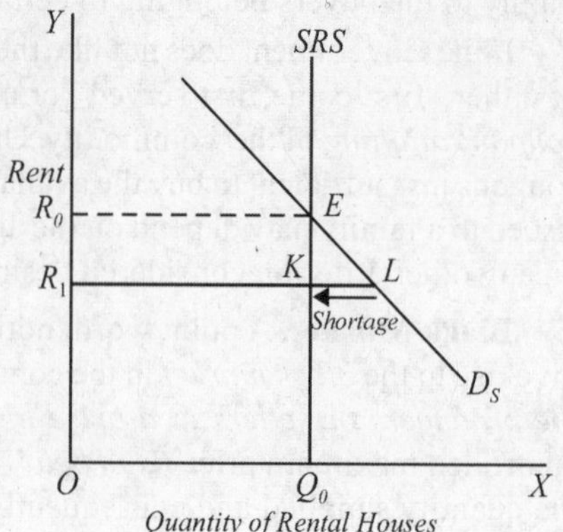

Fig. 25.3. *Effect of Rent Control in the Short Run*

Therefore, short-run supply curve of housing units is perfectly inelastic at Q_0 number of housing units available for renting. D_S is the short-run demand curve which is also relatively inelastic. If left free to the market force, rent equal to R_0 will be determined at which there is equilibrium between demand and supply. Suppose R_0 is too high for the poor and middle class people to pay. To help them, government fixes ceiling on rent at R_1. It will seen from Fig. 25.3 that at R_1, people demand R_1L housing units whereas supply of them remains at R_1K or OQ_0. Thus KL shortage of housing units has emerged since the demand and supply of housing units in the short run is inelastic, shortage caused by rent control is small. The main effect of rent control in the short-run is to reduce rents.

Although in the short run, landlords cannot do much to the lowering of rent through control, *further investmetn in constructing houses and apartments* by them will be reduced causing reduction in the supply of rental houses in the long run. In addition to this, the landlords will not spend any money *onrepairs and maintenance* of rental houses when rents are lowered. These steps will ultimately lead to the poor quality of rental houses and apartments. Thus, *in the long run, rent control has an important effect on the availability or supply of rental houses and their quantity*. It will be seen from Fig. 25.4 that at the lower controlled rent OR_1 the quantity demanded of rental housing increases to OQ_2 and the quantity supplied of rental housing units falls to OQ_1. Thus, fixation of lower controlled rent OR_1 results in increase in the quantity demanded and decrease in the quantity supplied of rental houses and thereby leads to the emergene of the large shortage of rental houses equal to Q_1Q_2 or KL as will be see from Fig. 25.4. *The greater the elasticity of supply and demand for rental housing in the long run, the greater will be the shortage of rental housing units as a result of imposition of rent control act.*

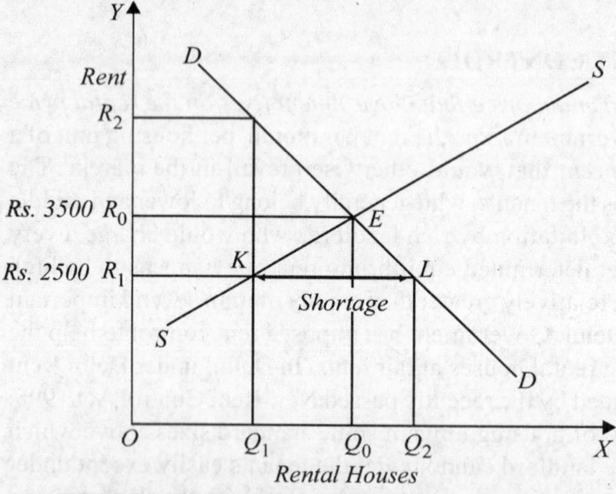

Fig. 25.4. *Effect of Rent Control in the Long Run*

It many be noted that this shortage of rental housing represents the conditions of excess demand for rental housing. An important question is whether fixation of maximum rent which is lower than the equilibrium rent can be effectively enforced when conditions of excess demand or shortage of rental housing units emerges. Of course, no one can openly or explicitly charge a rent higher than the controlled rate. However, the emergence of the conditions of excess demand or shortage of rental housing will tend to put upward pressure on the actual rents received. Due to the excess demand conditions landlords have devised various ways to circumvent the rent control act and charge higher actual rents. Consider Fig. 25.4 where it will be seen that at the controlled rent OR_1 of rental housing the quantity supplied of rental housing is OQ_1. Further, for OQ_1 rental housing units tenants are willing to pay rent equal to OR_2. *Under these conditions of excess demand and shortage of rental housing units, landlords tend to extract side payments from tenants, though explicitly they charge controlled rent*. For example, in Delhi and New York, two of the important world cities where rent control law operates, landlords require tenants to make a large non-refundable pay deposits or a large advance payments adjustable against monthly rents. Further, landlords may also require from tenants to make costly fittings or get expensive wood work done in the rental houses as a condition for rent and further also requiring them to pay for the repairs and maintenance of the rental housing units. All these ways of evading the rent control act have been observed. Unless the law explicity prohibits such practices, they will be operating and will have the effect of nullifying the rent control policy. That is, tenants will pay controlled rent OR_1 explicitly but extra expenditure and payments they are required to incur may add up to R_1R_0 per month so that the actual effective monthly rent may amount to the equilibrium rent OR_0.

It is evident from above that the consequence of rent control, like that of any other price control, is the emergence of shortage. However, in case of shortage of rental housing units those who are unable to get them, will make efforts to make another living arrangements. They may decide to live in other cities or satellite towns which are not covered by rent controls. Further, the disappointed seekers of rental housing may turn to the construction of their own self-occupied houses. But this requires a lot of finance which have to be arranged by them.

MINIMUM SUPPORT PRICE

In the price control we examined the case when the government fixed a *price ceiling* (that is, *maximum price*) to prevent it from rising to the equilibrium level. For many agricultural products the Government policy has been to fix a *price floor*, that is, the *minimum support price* above the equilibrium level which is considered to be low and unremunerative to the farmers. While in case of price control or fixation of price ceiling the Government simply announces the maximum price above which price cannot be charged by the producers or sellers of a product, in case of minimum support price, the Government becomes an active buyer of the product in the market. It is not only in India but also in the developed countries such as the USA that price support policy for agricultural products is adopted to provide reasonable prices to the farmers and increase their income. The effects of imposition of minimum support price for wheat, an important agricultural product, in India is illustrated in Fig. 25.5 where demand curve *DD* and supply curve *SS* of wheat intersect at point *E*. Thus if price of wheat is allowed to be determined by the free working of demand for and supply of wheat, equilibrium price is *OP* and equilibrium quantity determined is *OQ*.

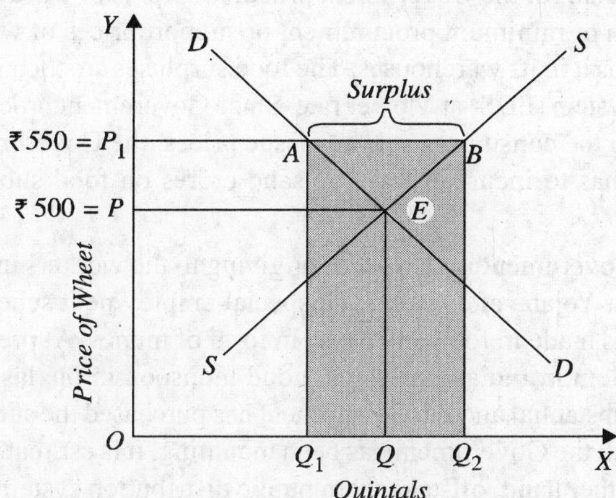

Fig. 25.5. *Minimum Support Price for Agriculture*

Now suppose this free market determined equilibrium price *OP* (= ₹ 500 per quintal) is considered to be unremunerative which does not provide incentives to the farmers to produce wheat or expand its production. Therefore, to promote the interests of the farmers, the Government intervenes and fixes a higher minimum support price OP_1 (₹ 550 per quintal) for wheat. It will be seen from Fig. 25.5 that at price OP_1 of wheat, the quantity demanded of wheat decreases to OQ_1 (– P_1A). On the other hand, at higher price OP_1 farmers expand their output and supply a greater quantity OQ_2 (= P_1B) of wheat. Thus at minimum support price OP_1 the quantity of wheat supplied by the farmers exceeds the quantity demanded of it by the consumers in the market. This means that the imposition of minimum support price of wheat higher than the equilibrium price *OP* leads to the emergence of *surplus of wheat* equal to *AB* or Q_1Q_2. If the Government does not purchase this surplus this will tend to depress the price of wheat. Therefore, in order to ensure this minimum price of wheat OP_1 (= ₹ 550 per quintal) to the farmers the Government will have to purchase the entire surplus *AB* or Q_1Q_2 from the farmers. It should be noted that to purchase the surplus Q_1Q_2 from the farmers, the Government will have to make expenditure equal to $OP_1 \times Q_1Q_2$, that is, equal to the area Q_1ABQ_2. This expenditure on purchase of wheat surplus may be financed by taxation of the people.

It follows from above that under minimum support price OP_1 the farmers sell OQ_1 quantity of wheat in the free market and quantity Q_1Q_2 to the Government. At the free market determined equilibrium price *OP* and quantity *OQ*, the total income of the farmers will be equal to the area *OPEQ*. Now, with minimum support price equal to OP_1 and the total quantity sold equal to OQ_2, the income of the farmers has increased to OP_1BQ_2. Thus minimum support price policy has greatly benefited the farmers both in terms of price they receive for their product and the income they are able to earn.

A major problem facing the Government is how to dispose of the surplus it purchases from the farmer at the higher minimum support price. If the Government sells it in the market, the price of wheat in the market will fall which will defeat the purpose of support price policy. Alternatively, the Government may store the surplus and in this case the Government will incur storage costs. Besides, the wheat and any other foodgrains get rottoned if kept for longer time in storage bins. Thus while to produce surplus requires valuable resources such as labour, fertilizers, irrigation and other inputs, yet it is quite often left to decay in government warehouses. In America, one important way of disposing of surplus is to give them to the developing countries as food aid. But this food aid is not without problems. The American food aid to developing countries has tended to depress prices of foodgrains in these countries and therefore has harmed the interests of farmers of these developing countries.

In India Food Corporation of India on behalf of the Government procures surplus of wheat and rice production created as a result of fixation of minimum procurement or support prices of wheat and rice. Food Corporation of India then keeps it in its warehouses. The food surpluses are then used for distribution through Public Distribution System (PDS) at a lower rate. Since Government procures these food grains at a higher rate and sells to the consumers at lower issue prices, the Government subsidises the foodgrain consumption and has to incur several thousand crores on food subsidy annually.

Besides, food surplus procured by the Government is also used for giving to the workers under '*food for work*' programme, Jawahar Rozgar Yojana and other such special employment schemes launched in India. A part payment of wages is made in food and a part in form of money. At present food surplus of wheat is also posing a problem in India. Due to the good monsoon in the last six successive years, food production has been substantial and the Government has purchased the surplus at higher procurement prices. The surplus with the Government has been mounting. It is estimated to about 50 million tons in June 2003. On the other hand, off-take from public distribution system has fallen. There is a real danger of these food surpluses getting rottoned in warehouses of the Food Corporation of India. Therefore, Government of India deciding to export of wheat.

It should be noted that in Indian Government has been raising year after year the procurement or support prices for wheat and rice. This raises the food cost in every sector of the economy which must lead to higher prices all round. Thus, increase in procurement prices of wheat and rice have been an important factor that has created inflationary pressures in the Indian economy.

We summarise below the important results of price support policy:–

1. Price paid by the consumers who buy from the open market increases when the minimum support price of the agricultural product is fixed at a higher level than the equilibrium price. This is because supply of the agricultural product in the open market decreases as a result of Government purchases of it from the farmers.
2. Fixation of minimum support price (*i.e.,* price floor) leads to the emergence of wheat surplus which the Government has to purchase from the farmers. This is quite obvious from the Indian experience where fixation of higher minimum support price (*MSP*) has resulted in mountain of foodgrains with Food Corporation of India.
3. Taxpayers pay more tax money to finance the Government's wheat purchases as well as storage costs.
4. How to dispose of the surplus purchased from the farmers poses a big problem. There are several ways to dispose of the surplus procured. One way is to sell it at subsidised rate to the persons below the poverty line through public distribution system. Second, the surplus can be used to make a part payment of wages in terms of foodgrains under '*food for work*' programme. Third, food surplus can be given to other countries as foreign aid or it can be exported.
5. Incomes of the farmers increase as a result of minimum support price fixed at a higher level than the free market equilibrium price. As a result of price support, they receive higher price than that which would prevail in the free market and also they produce and sell more than before. They sell a part of their larger production in the market and a part to the Government.

INCIDENCE OF INDIRECT TAXES

A significant application of demand-supply model is that it explains the problem of incidence of indirect taxes such as sales tax and excise duty on commodities. By incidence of taxes we mean who bear the money burden of taxes. For example, if sale tax is imposed on a commodity the question is whether the producers will bear the burden of the tax or the consumers who buy the commodity or the money burden of the sales tax would be distributed in some way between the producers and the consumers. We will confine ourselves to the explanation of incidence of indirect taxes, that is, taxes which are levied on either production or sale or purchase of commodities.

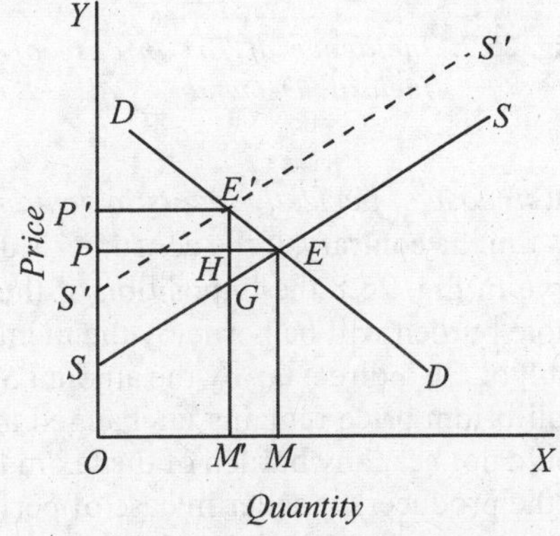

Fig. 25.6. *Incidence of Tax*

It is worthwhile to note that the price of a commodity is determined by demand and supply only when perfect competition prevails in the market. Supply curve of a commodity slopes upward as it is assumed that marginal cost of production increases with the increase in output by the firms. The upward sloping supply curve implies that as the price of a commodity rises the producer would offer more quantity for sale in the market. If no tax is levied on the commodity, the seller or producer will receive the whole amount of the price. Now, if the sales tax is imposed equal to ₹ 5 per unit then the *supply price* of each unit of the quantity offered for sale in the market will rise by ₹ 5. In this case, the producer would receive the market price minus the amount of the tax per unit. Thus, if the producer is to receive the same amount of price as prior to the imposition of the sales tax, then ; *the supply price of each unit of the commodity sold will rise by the full amount of the tax*. This implies that the supply curve of the commodity will now shift upward by the amount of the tax as a result of the imposition of the sales tax. Consider Fig. 25.6 where demand and supply curves of a commodity are shown. Before the imposition of any indirect tax, demand and supply curves intersect at point *E*, and accordingly, equilibrium price *OP* and equilibrium amount *OM* are determined. Now suppose that the sales tax equal to the amount *SS′* is imposed on the commodity. As explained above, the imposition of the sales tax will shift the supply curve vertically upward. The new supply curve *S′ S′* has been drawn which depicts the supply position after the imposition of the sales tax. It will be seen from the Fig. 25.6 that the new supply curve *S′ S′* intersects the given demand curve *DD* at point *E′*. Thus, as a result of the imposition of the sales tax the price of the commodity has riseen from *OP* to *OP′*. It means that the consumer will have to pay price of a commodity which is higher by the amount *PP′* than before. Obviously, the burden of the tax borne by the consumer is equal to *PP′*(=*E′ H*). This is the incidence of tax fallen on the consumer.

It will be seen from the diagram that the quantity sold in the market would now be *OM′* and the Government would receive *E′G* per unit of it as tax. Since *E′ H* will be paid by the consumer, the rest of the tax equal to the amount *GH* per unit will be borne by the producer or seller. Thus, a part of the tax has been passed on to the consumer through a higher price and a part has been borne by the producer himself. It is worthwhile to note that *the incidence of taxes borne by the producer and the consumer will depend upon the elasticity of demand as well as elasticity of supply. The lower the elasticity of demand, the greater will be the incidence of tax borne by the consumer.*

If the demand for a commodity is perfectly inelastic the whole of the burden of the commodity tax will fall on consumer. This is shown in Fig. 25.7. In this figure demand curve *DD* is a vertical straight line showing that demand for the commodity is completely inelastic. As a result of the intersection of the demand and supply curves, price *OP* is determined. If now the tax equal to *SS′* is imposed on the commodity, the supply curve will shift vertically upward to the dotted position *S′S′*. It will be seen that the new supply curve *S′ S′* intersects the demand curve *DD* at point *E′* and the new equilibrium price *OP′* is determined. It will be noticed from Fig. 25.7 that in this case the price of the commodity has risen by *PP′* or *EE′* which is equal to the full amount of the tax *SS′*. It means that producers pass on the full tax to the consumers and they themselves do not bear any incidence. It, therefore, follows that *in case of perfectly inealstic demand, the whole incidence of the tax falls on the consumer.*

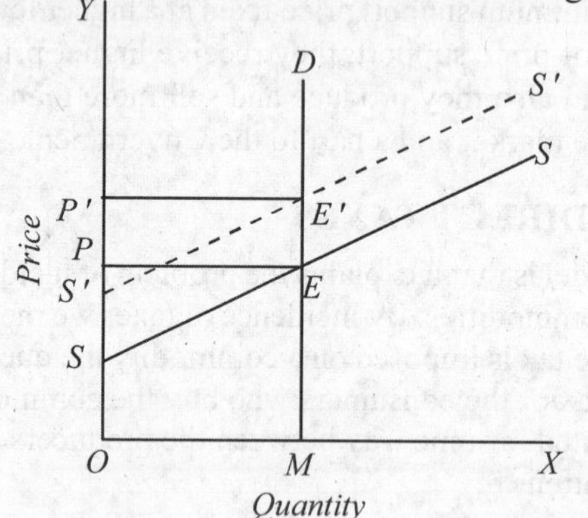

Fig. 25.7. *Incidence of Tax in case of Perfectly Inelastic Demand*

On the contrary, if the consumer's demand for a quantity is perfectly elastic, as is shown by *DD* curve in Fig. 25.8 the imposition of the tax on it will not cause any rise in price. In this case, the whole burden will be borne by the manufacturers or sellers. It will be seen from Fig. 25.8 that as a result of the indirect tax by the amount *SS′* and the resultant upward shift in supply curve to *S′ S′* the equilibrium price remains unchanged at the leve *OP*. Since the price has not risen, the consumer would not bear any burden of the tax in this case. Therefore, the whole incidence of the tax will fall on the producer, or seller in case of perfectly elastic demand.

It should be noted that the more inelastic the demand for a commodity, the greater the rise in the price paid by the consumer and the vice versa. In order to show this clearly we have drawn two demand curves—one inelastic and the other relatively more elastic in Fig. 25.9. Supply curve *SS* has been drawn which intersects the two demand curves *DD* and *D′D* at the same point *F*. Now before the imposition of any tax, the quantity sold and purchased is *OM* and the price of the commodity is OP_1. Now, if the sales tax tax is imposed, supply curve shifts upward to *S′ S′* by the amount of tax per unit imposed. It will be noticed from Fig. 25.9 that the new supply curve *S′ S′* intersects the inelastic curve DD at point S according to which equilibrium price OP_3 is determined. In this case of inelastic demand price has risen by P_1P_3 which is the burden borne by the consumers. Now, the new supply curve $S'S'$ intersects the relatively elastic demand curve $D'D'$ at point R according to which market price OP_2 is determined. Thus, in the case of elastic demand curve $D'D'$, the rise in price as a result of the same tax, is equal to P_1P_2 which smaller than P_1P_3 borne by the consumer in case of inelastic demand. Therefore, *the extent to which the incidence of the tax will fall on the consumer, depends on their elasticity of the demand for the commodity in question.*

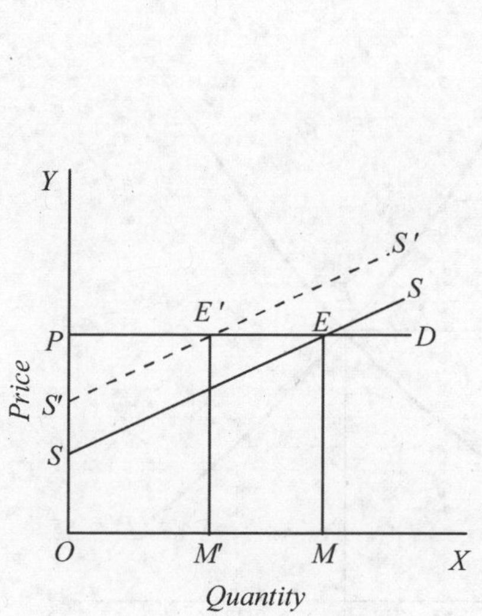

Fig. 25.8. *Incidence of tax in case of perfelectly elastic demand*

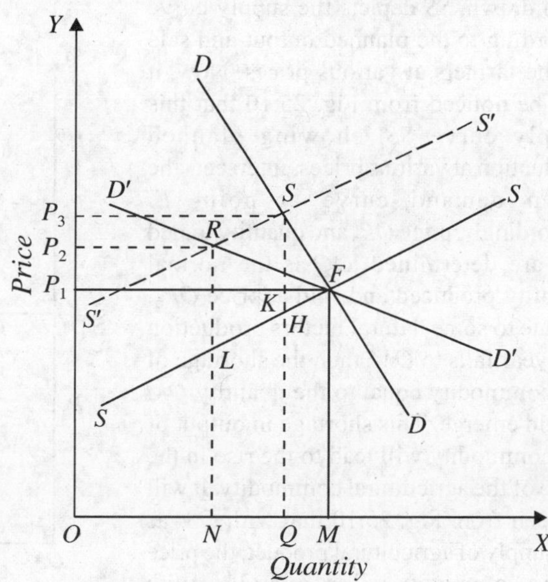

Fig. 25.9. *Incidence of an indirect tax in case of elastic and inelastic demand*

The predictions about the incidence of taxes borne by the consumers and the producers have been generally found true in the real world situation when the commodities on which taxes are imposed are sold under competitive conditions.

PRICE AND INCOME STABILISATION IN AGRICULTURE

The demand and supply analysis has important applications in understanding the problems of agriculture and framing suitable policies to stabilise agricultural prices and incomes. The nature of demand and supply of agricultural produce is also peculiar. *The demand for agricultural product is relatively inelastic*. This means that whether the price goes up or comes down the quantity demanded of the agricultural product changes very little. This is due to the fact that agricultural products belong to the 'necessaries group' which cannot be dispensed with, nor have they any good substitutes except that in periods of hardship inferior products may be substituted for superior ones or that in the periods of prosperity the farmer may himself give up the consumption of an inferior product, say, millets for superior products like wheat and rice. As will be seen below, the relatively inelastic demand curve for agricultural product has significant implications for agricultural prices and incomes.

The supply of agricultural products is also of a peculiar nature. This is because the production of the agricultural product is subject to quite large variations due to natural factors which are beyond the control of human beings. For instance, when monsoon fails in India or does not come in time the agricultural production falls down to a good extent. On the other hand, when there is good monsoon, bumper crop is reaped and as a result the supply of agricultural products in the market increases. Similarly, invasion of pests and insecticides, occurrence of floods are other natural factors which cause unplanned variations in agricultural output. *The unplanned fluctuations in agricultural output caused by natural factors greatly affect the prices and incomes of the farmers*. Now, how the demand and supply of agricultural products determine the prices of agricultural products and how the unplanned variations in agricultural output affect agricultural prices and incomes are explained below.

Effect of Unplanned Fluctuations in Agricultural Supply on Prices

Consider Fig. 25.10 where relatively inelastic demand curve *DD* of an agricultural product has been drawn. *SS* depicts the supply curve according to the planned output and sale by the farmers at various prices. Now, it will be noticed from Fig. 25.10 that this supply curve *SS*, showing planned production at various prices, intersects the given demand curve at point *E*. Accordingly, price OP_2 and quantity traded *OQ* are determined. *OQ* is the normal quantity produced and sold at price OP_2. If due to some natural factors production in a year falls to *OM*, then the shortage of the commodity equal to the quantity *QM* would emerge. This shortage in output of the commodity will lead to the rise in the price of the agricultural commodity. It will be seen from Fig. 25.10 that with *OM* as the supply of agricultural product, the price of the commodity will rise to OP_3. With this rise in price to OP_3 quantity demanded will become equal to the available output *OM*.

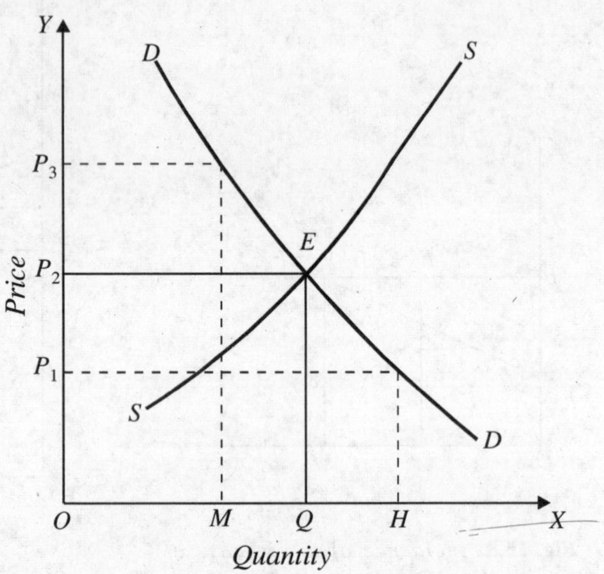

Fig. 25.10. *Fluctuations in agricultural prices due to unplanned fluctuations in production*

Now suppose that due to favourable weather unplanned production of the agricultural product increases to *OH*. This is greater than the normal output *OQ* by the amount *QH*. Accordingly, as a result of the unplanned increase in output, surplus equal to the amount *QH* would emerge. This surplus would lead to the fall in price of the agricultural product. It will be seen from Fig. 25.10 that with *OH* as the production of the agricultural product, price would fall to the level OP_1 so as to bring quantity demanded equal to the level of the available output. We thus see that *as a result of the unplanned variations in the agricultural output, prices of agricultural products vary a good deal*. It is also clear that unplanned fluctuations in agricultural output will cause changes in price in the opposite direction to the changes in output, that is, the larger the production, the lower the price and vice versa.

Effect of Unplanned Fluctuations in Production on Incomes of the Farmers

Now, these unplanned changes in agricultural production and consequent changes in prices significantly change the incomes earned by the farmers. In this connection, elasticity of demand for agricultural product plays a significant role, as it determines the changes in the incomes of the farmers. If the demand for an agricultural product is unitary elastic, then the changes in prices brought about by changes in unpalnned variations in agricultural output will offset the output effect and the result would be that income earned by a famrer would not change. If on the other hand, demand for a given agricultural product is elastic, the unplanned increase in the agricultural output will only cause slight reduction in price so that the incomes made by the farmers would increase. On the other hand, if in the face of elastic demand curve, the production of the agricultural product falls, the rise in price would be little as compared to the fall in agricultural output with the result that income earned by the farmer would decline.

If the demand for agricultural commodity is inealstic (*i.e.* less than unity), then farmer's income will decrease with the fall in price caused caused by an unplanned increase in the production, and on the other hand, the farmer's income will increase with the rise in price brought about by the unplanned fall in the production. Therefore, we arrive at the following predictions about the changes in income following the changes in production :

1. Unplanned fluctuations in agricultural production can cause a good deal of changes in farmers' income through changes in prices.
2. When the demand for an agricultural product is elastic, unplanned increase in production will raise the farmers' income and the unplanneded decrease in production will reduce farmers' income.
3. When the demand for an agricultural product, is inelastic, the increase in its output will reduce farmers' income.
4. The greater the divergence of elasticity of demand from unity (in either direction), the greater will be the fluctuations in farmers' income following the unplanned changes in agricultural production.

There is a good amount of empirical evidence from the real world to confirm the above conclusions based on demand and supply analysis as applied to agriculture. Unplanned changes in the production of agricultural products do occur frequently. As the prices of most of agricultural products are determined by the perfectly competitive markets, large changes in prices of agricultural products generally occur. Further, *as the demand for most of the agricultural products is quite inelastic, variation in production causes a very large change in prices. The inelastic nature of demand for many agricultural products gives rise to a peculiar phenomenon of* "poverty amidst plenty". This happens when there is a bumper crop which is expected to bring prosperity to the farmers but due to the inelastic demand this does not occur. The bumper crop leads to a very large fall in price and due to inelastic nature of demand the incomes earned by the farmers fall. Thus plenty of production lowers the incomes of the farmers rather than raising them. On the other hand, when due to the unkind weather crop production falls somewhat, it will lead to the rise in incomes of the farmers. This is because the decline in crop production will cause price to rise appreciably and due to inelastic demand this will raise the incomes of the farmers. From the point of view of the farmers lower production with higher prices is better than the larger production with lower prices. But it is larger production with lower prices which the consumers would like to have. Thus the interests of the farmers and consumers instead of being consistent conflict with each other.

Stabilisation of Agricultural Prices and Incomes

Owing to the peculiar nature of demand for and supply of agricultural products, their prices and farmers' income fluctuate very much if left to be determined by the free market. These agricultural fluctuations occur either due to the unplanned changes in agricultural production or due to changes in demand for farm products. In order to safeguard the interests of the farmers, the governments in many countries intervene to stabilise agricultural prices and incomes. The government may also aim at raising incomes of the farmers. But, as will be seen below, *the policy of stabilisation of agricultural prices and the policy of raising farmers' income often conflict with each other.* In what follows we shall analyse policies for stabilisation of agricutlural prices and income, and also for raising farmers' income in the context of the unplanned fluctuations in supply of agricultural products. A similar analysis can be made in the case of changes in demand.

Stabilisation of agricultural prices is illustrated in Fig. 25.11 where demand and supply curves of a certain agricultural product are drawn. The supply curve *SS* drawn in Fig. 25.11 shows the amount which the farmers will plan or desire to produce at each given price. Now, if the production can be planned with certainty, then *SS* will represent normal supply curve. And this supply curve *SS* intersects the given demand curve *DD* at point *E* and the free market price will be settled down at the equilibrium level OP_2 and OQ represents the planned production which would be bought and sold at the equilibrium price OP_2.

But actual production will generally differ from the planned equilibrium output OQ due to the vagaries of weather. Let us assume that actual production fluctuates around OQ between the amounts OL and OH. With OL as the actual supply and, given the demand curve DD, the free market price would settle at OP_3 and, with OH as actual supply, the free market price will settle at OP_1. Thus with the fluctuations of actual production between OL and OH, the free market price will fluctuate between OP_3 and OP_1. Fluctuations in actual production and resultant fluctuations in prices will cause the farmers' income to fluctuate. As the farmers' income is equal to the amount produced multiplied by the price, with planned production OQ, the farmers' income will be equal to the area of rectangle OP_2EQ. Similarly, with OL as the actual production and OP_3 as the price, farmers' income will be equal to the area of rectangle OP_3KL and with OH as the actual production and OP_1 as the free market price, farmers' income will be equal to the rectangle OP_1TH. Thus the fluctuations in actual production and consequent changes in free market price will cause fluctuations in farmers' income.

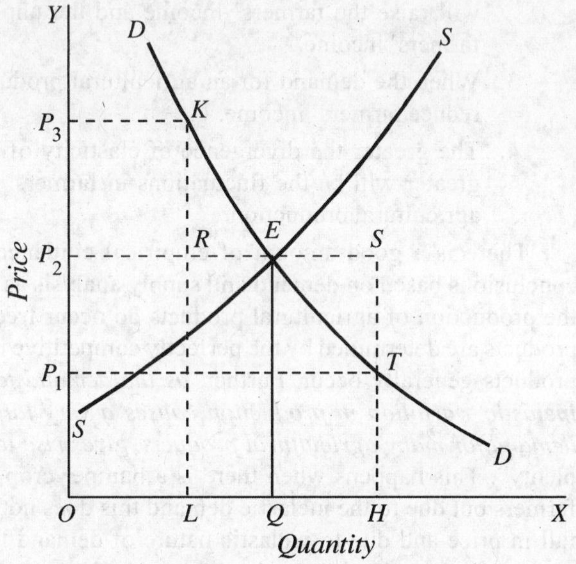

Fig. 25.11. *Stabilisation of Agricultural Prices*

The farmers can avoid the fluctuation in their prices and incomes, if they form a producers' association which regulates the supply *(i.e.*, amount offered for sale) in the market. The producers'association can successfully stabilise the price at the level OP_2 and income indicated by the area of the retangle OP_2EQ. Now, the pertinent question is what policy should the producers' association pursue for stabilising this price and income. In order to stabilise price and income, if the actual production exceeds OQ, then producers' association will have to withhold some quantity from the market and store it and if the actual production falls short of OQ, the producers' association will have to sell from its stock to make up the deficiency. For example, if the actual production is OH, then to ensure to the farmers price equal to OP_2, the producers' association will have to take away from the farmers the amount QH (=ES) and store it. On the other hand, if the actual production is OL, the association will sell the amount LQ in the market from its own stock in order to ensure price OP_2 and income equal to OP_2EQ to the farmers. We thus see *if the farmers form their association, they can stabilise their prices and incomes.*

It may be noted that the farmers' asociation can continue successfully this policy of keeping price fixed at OP_2 only if the average level of production over years is equal to OQ, the sales of which have to be maintained for ensuring stabilisation of price and farmers' income. On the other thand, if the price is fixed at a higher level, the sales would be smaller than the average level of production. In this case addition to the stocks of the commodity during the periods of bumper crop will exceed the sales out of it during lean periods with the result that stocks with the association over a number of years will accumulate. This will create problems and will prevent the sales to be kept constant at OQ. In the context of price being stabilised at the level of OP_2, the income of the farmers can be stabilised at OP_2EQ only if sales over a number of years are kept constant at OQ.

Buffer Stock Operations by the Government

Now, if instead of farmers forming their association, the Government intervenes to safeguard their interests and makes an attempt to stabilise their prices and incomes. The Government can do so through what has been called buffer stock operations which means buying the commodity by the Government from the open market in times of bumper crop and selling it in times of smaller crop production when there is shortage. In order to conduct the buffer stock operations successfully the first thing to be decided by the Government is whether it aims to stabilise prices, or to stabilise the farmers' income. This is because the *government cannot achieve stabilisation of both agricultural prices and incomes*.

Consider Fig. 25.11 again. Suppose the government wishes to keep price stable at the level OP_2. If the production in any year is *OH*, then to keep price at the level OP_2, the Government would buy *QH* quantity from the open market and store it. The farmers would get the price for the whole amount *OH* sold, *OQ* to the public and *QH* to the Government. Therefore, farmers' income will be $OP_2 SH$. On the other hand, if the production in any year falls to the level *OL*, then in order to keep the price at OP_2, the Government would sell *LQ* amount in the open market out of its stocks. But since the farmers would only sell the amount *OL* produced by them, income earned by them will now be equal to OP_2RL. This income OP_2RL earned is much smaller than that OP_2SH earned when their production was *OH*. With price kept fixed at the level OP_2, farmers' income will vary as their production changes, since farmers will sell whatever is produced, a part to the public and a part to the Government. It is therefore clear that the Government can successfully stabilise prices through buffer stock operations but this price stabilisation policy will not ensure stable income to the farmers. Indeed, with the price kept fixed by the Government *the farmers' income will change in direct proportion to the change in production*, the ten per cent rise in production will lead to the ten per cent increase in farmers' income and *vice versa*. Thus though the Government's intervention in the above way will ensure price stability but it will not eliminiate fluctuations in farmers' income. Thus the objective of price stability conflicts with the objective of stability of farmers' income.

Income Stabilisation. Now, the important question is what the Government should do if it wants to stabilise farmers' income. The stabilisation of farmer's income can be achieved with moderate changes in prices. In order to stabilise farmers' income, the Government should allow changes in prices exactly in proportion to unplanned changes in production. In order to ensure income stability, when there is 10 per cent increase in crop production, the Government should permit only 10 per cent fall in price, and when there occurs 10 per cent decline in crop production, the Government should permit only 10 per cent price rise in price.

Consider Fig.. 25.12 where *DD* is the demand curve for the commodity and *SS* represents the planned production by the farmers. The two intersect at point *E* which shows that the market price will be *OP* and the planned equilibrium production will be *OQ*. Let us assume that actual unplanned production varies between *OL* and *OH*. If this variation in production is left free to affect the price, then the market price will vary between *OR* and *OU*. In order to show graphically to what extent price should be allowed to vary in the face of variation in actual production to ensure stability in farmers' income, we have drawn a unitary elastic demand curve $D' D'$ (dotted) passing through point *E*. It will be recalled that on the unitary elastic demand curve, outlay made by the consumers on the commodity remains constant. Thus at each price-quantity combination represented on this unitary elastic demand $D' D'$, the expenditure (and hence the income earned by the farmers) will remain constant and will be equal to the area of the rectangle *OPEQ*.

Now suppose that the actual production in any year is *OH*. If the market system is allowed to work freely, then, with production *OH*, the price will fall to *OR* and the income earned by the farmers will be equal to the area of rectangle *ORGH*. It will now be clear from Fig. 25.12 that with production *OH*, the price should be allowed to fall from *OP* to the *OS* level only if income of *OSJH* (which is equal to the area *OPEQ*) is to be ensured to the farmers. It will be seen from Fig. 25.12 that with price *OS*, the public purchase *SZ* quantity and the remaining quantity *ZJ* will have to be bought by the Government to keep price at *OS*.

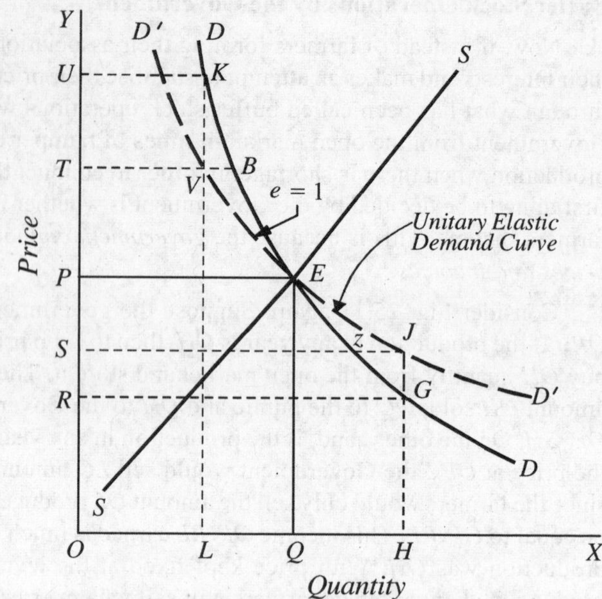

Fig. 25.12. *Stabilisation of Farmers' Income by the Government*

On the other hand, if the actual production falls to *OL*, the market price will rise to *OU*. But to keep farmers' income equal to the area *OPEQ* or *OSJH*, the price should be allowed to rise only to the *OT* level. With production *OL* and price equal to *OT*, the income of the farmers will be *OTVL* which is equal to *OPEQ* or *OSJH as points V, E and J lie on the unitary elastic demand curve D′ D′* (dotted). It may be noted that at price *OT*, the consuming public will buy *TB* amount of the commodity while the production of farmers is only *TV* (=*OL*). Therefore, *VB* amount will have to be sold to the public by the Government out of its buffer stock.

If the above policy of the income stabilisation is pursued with efficiency it will have the following beneficial effects. First, *there will be moderate changes in prices as compared to too large changes in price when left free to be determined by the free market mechanism*. Second, the income of the farmers will be stabilised at a certain level in the face of wide fluctuations in production. Third, the policy of income stabilisation will be self-financing and, indeed, it may give net profit to the Government. This is because, as will be seen from the Fig. 25.12, the Government buys at a lower price and sells at a higher price. If the gross profits made are equal to the costs of storing the commodity, then the scheme will be completely self.-financing. Thus whether from this scheme there will be net profits or net losses to the Government, depends on the magnitude of costs incurred on storage. This policy of income stabilisation with moderate changes in price is better than the policy of complete price stability explained earlier. For, in case of policy of complete price stabilisation, the Government buys and sells at the same price and consequently cost of storage will not be met.

QUESTIONS FOR REVIEW

1. What would happen if the government imposes a ceiling on the retail price of a commodity, say petrol. Why do advocates of price control also recommend rationing ? *(D.U.B.A.(Hons) 1994)*
2. Explain the effects on a market of setting (*a*) a maximum controlled price for a commodity and (*b*) a minimum support price for a commodity. *(D.U.B.A.(Hons) 1991)*
3. Explain the effect of price control on a commodity ? Why is it necessary to introduce rationing when price control is imposed on a commodity.
4. What is black market ? Why do black markets often arise when price control and rationing of a commodity is introduced.

5. What would happen if government imposes a ceiling on the retail price of a commodity, say patrol. Why do advocates of price control also recommend rationing ?
6. Explain the effect of setting a minimum support price for an agricultural product.
7. *Examine the effects of setting by government procurement price of wheat higher than the free market price. Explain with special reference to rising stocks of foodgrains in India.*
8. Explain how rent control leads to the shortage of housing units available on rent.
9. Do rent controls help those who wish to find cheap housing ?

Incidence of Taxes

10. Analyse the incidence of a per unit tax imposed on a commodity with perfectly inelastic demand. Assume that the conditions in the market are perfectly competitive. *(D.U. B.A.(Hons) 1980)*
11. A unit excise duty is imposed on a firm in a competitive market. Examine its short run and long-run effects on industry and firm. *(D.U.B.A.(Hons.) 1996)*
12. Analyse the incidence of a per unit tax imposed on a commodity with (*i*) perfectly elastic demand and (*ii*) perfectly inelastic demand. *(D.U.B.A.(Hons.) 1987)*
13. Explain the circumstances under which rise in price of a commodity as a result of the imposition of a per unit tax on it is more than the tax. *(D.U.B.A.(Hons.) 1997)*
14. How is a specific duty imposed on a commodity is shared by the consumers and producers? Show that the lower the price elasticity of demand, the higher is the incidence of a specific commodity tax on the consumers.

Applications of Demand Supply Model to Agriculture

15. How much wheat the Government should buy from the open market depends upon whether it wants to stabilise the price of wheat or income of the farmers? *(D.U.B.A. (Hons.) 1987)*
16. Explain the effects of setting a minimum support price for a commodity. *(D.U.B.A. (Hons.) 1991)*
17. Income stabilisation of farmers and price stabilisation of agricultural products can be inconsistent. Explain. *(D.U.B.A.(Hons.)1991)*
18. A policy goal of governments in developing countries is to stabilise agricultural incomes. Demonstrate that in stabilising agricultural incomes the Government must let agricultural prices to vary.
19. What are buffer-stock operations by the Government? Using demand-supply model, explain the role of buffer stock operations to stabilise agricultural prices.

Chapter 26

PRICE AND OUTPUT UNDER MONOPOLY

In the last two chapters, we discussed the equilibrium of the firm and the industry under perfect competition and as a consequence the determination of price and output under it. Monopoly is another important market from which merits detailed study in respect of output equilibrium of the firm as well as how price is determined under it. About seventy years ago it was thought that the existence of monopoly was an exceptional case. But now monopoly form of market structure extensively prevails in capitalist economies of the world, including that of India. Monopolistic market structure prevails in many large-scalc manufacturing industries and public utility services. Therefore, the analysis of price and output determination under monopoly has assumed vital importance. As has been pointed out earlier, monopoly is an extreme form of imperfect competition. Pure or perfect competition and pure or absolute monopoly are the two extreme opposite cases and between them the various intermediate market situations lie, which differ from each other in respect of the degree of monopoly. In this chapter we shall explain the principles which underlie the price and output equilibrium under monopoly and will compare them with those of perfectly competitive equilibrium.

MONOPOLY: ITS MEANING AND CONDITIONS

What is monopoly? *Monopoly is said to exist when one firm is the sole producer or seller of a product which has no close substitutes*. Three points are worth noting in this definition. First, there must be a single producer or seller of a product if their is to be monopoly. This single producer may be in the form of an individual owner or a single partnership or a joint stock company. If there are many producers producing a product, either perfect competition or monopolistic competition will prevail depending upon whether the product is homogeneous or differential. On the other hand, when there are few producers or sellers of a product, oligopoly is said to exist. If then there is to be monopoly, there must be one firm in the industry. Even literally monopoly means one seller. 'Mono' means one and 'poly' means seller. Thus monopoly means one seller or one producer.

But to say that monopoly means one seller or producer is not enough. A second condition which is essential for a firm to be called monopolistic is that *no close substitutes for the product of that monopolistic firm should be available in the market*. If there are some other firms which are producing close substitutes for the product in question there will be competition between them. In the presence of this competition a firm cannot be said to have monopoly. Monopoly implies absence of all competition. For instance, there is one firm in India which produces 'Binaca' toothpaste but this firm cannot be called monopolist since there are many other firms which produce close substitutes of Binaca toothpaste such as Colgate, Promise, Forhans, Meclean etc. These various brands of toothpaste compete with each other in the market and the producer of any one of them cannot be said to have a monopoly. Prof. Bober[1] rightly remarks, "The privilege of being the only seller of a product does not by itself make one a monopolist in the sense of possessing the power to set the price. As the one seller, he many be a king without a corwn."

1. M.M. Bober, *Intermediate Price and Income Theory*, revised edition (1962), p. 237

We can express the second condition of monopoly in terms of cross elasticity of demand also. Cross elasticity of demand shows a change in the demand for a good as a result of change in the price of another good. Therefore, if there is to be monopoly *the cross elasticity of demand between the product of the monopolist and the product of any other producer must be very small.*

The fact that there is one firm under monopoly means that other firms for one reason or other are prohibited to enter the monopolistic industry. In other words, strong barriers to the entry of firms exist wherever there is one firm having a sole control over the production of a commodity. The barriers which prevent the firms to enter the industry may be economic in nature or else of institutional and artificial nature. In case of monopoly, barriers are so strong that prevent entry of all firms except the one which is already in the field.

From above it follows that for the monopoly to exist, three conditions are necessary:

1. There is a single producer or seller of a product.
2. There are no close substitutes for the product.
3. Strong barriers to the entry into the industry exist.

THE NATURE OF DEMAND AND MARGINAL REVENUE CURVES UNDER MONOPOLY

It is important to understand the nature of the demand curve facing a monopolist. The demand curve facing *an industrial firm* under perfect competition, as explained in a previous chapter, is a horizontal straight line, but the demand curve facing the *whole industry* under perfect competition is sloping downward. This is so because the demand is by the consumers and the demand curve of consumers for a product usually slopes downward. The downward-sloping demand curve of the consumers faces the whole competitive industry. But an individual firm under perfect competition does not face a downward-sloping demand curve. This is because an individual firm under perfect competition is one among numerous firms constituting the industry so that it cannot affect the price by varying its individual level of output. A perfect competitive firm has to accept the ruling price as given and constant for it. It can sell as much as it likes at the ruling price of the product. Therefore, the demand curve facing an individual firm under perfect competition is a horizontal straight line at the level of prevailing price of the product. A perfectly competitive firm is a mere quantity adjuster; it has no influence over price.

But in the case of monopoly one firm constitutes the whole industry. Therefore, the entire demand of the consumers for a product faces the monopolist. Since the demand curve of the consumers for a product slopes downward, the monopolist faces a downward sloping demand curve. If he wants to increase the sale of his good, he must lower the price. He can raise the price if he is prepared to sacrifice some sales. To put it in another way, a monopolist can lower the price by increasing his level of sales and output, and he can raise the price by reducing his level of sales or output. A perfectly competitive firm merely adjusts the quantity of output it has to produce, price being a given and constant datum for him. But the monopolist encounters a more complicated problem. He cannot merely adjust quantity at a given price because each quantity change by him will bring about a change in the price at which the product can be sold. Consider Fig. 26.1. *DD* is the demand curve facing a monopolist. At price *OP* the quantity demanded is *OM*, therefore he would be able to sell *OM* quantity at price *OP*. If he wants to sell a greater quantity *ON*, then price the to *OL*. If would he restricts his quantity to *OG*, fall price will rise to *OH*. Thus every quantity change by him entails a change in price at which the product can be sold. Thus the problem faced by a monopolist is to choose a price-quantity combination which is optimum for him, that is, which yields him maximum possible profits.

Demand curve facing the monopolist will be his average revenue curve. Thus, the average revenue curve of the monopolist slopes downward throughout its length. Since average revenue curve slopes downward, marginal revenue curve will lie below it. This follows from usual average-

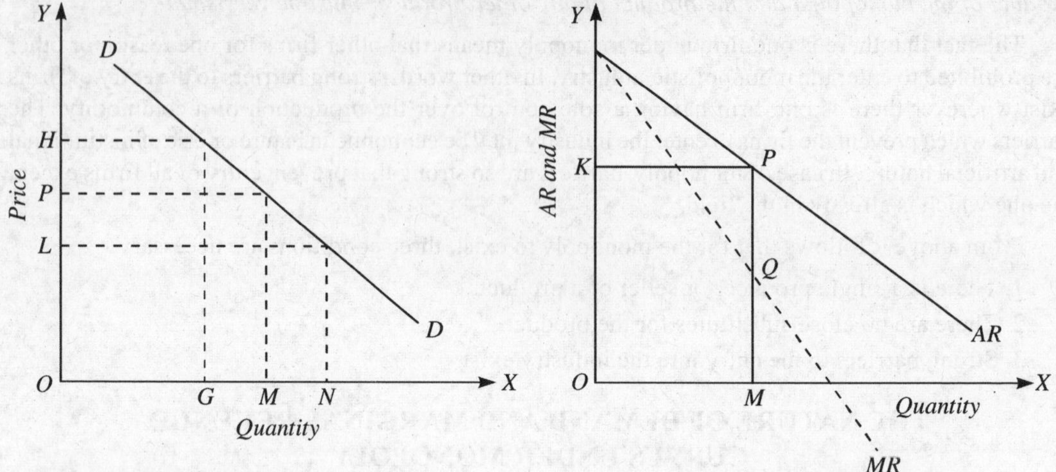

Fig. 26.1. *Demand curve of the monopolist slopes downward.*

Fig. 26.2. *Average and Marginal Revenue Curves under Monopoly*

marginal relationship. The implication of marginal revenue curve lying below average revenue curve is that the marginal revenue will be less than the price or average revenue. When monopolist sells more, the price of his product falls; marginal revenue therefore must be less than the price. In Fig. 26.2 *AR* is the average revenue curve of the monopolist and slopes downward. *MR* is the marginal revenue curve and lies below *AR* curve. At quantity *OM*, average revenue (or price) is *MP* and marginal revenue is *MQ* which is less than *MP*. In an earlier chapter we have explained that average and marginal revenue at a quantity are related to each other through price elasticity of demand and in this connection we derived the following formula :

$$MR = AR\frac{(e-1)}{e}, \text{ where } e \text{ stands for price elasticity.}$$

Since AR is the same thing as price

Therefore, $$MR = \text{price}\frac{(e-1)}{e}$$

or $$\text{price} = MR\frac{e}{e-1}$$

Since the expression $\frac{e-1}{e}$ will be less than unity, *MR* will be less than price, or price will be greater than *MR*. The extent to which *MR* curve lies below *AR* curve depends upon the value of the fraction $\frac{e-1}{e}$.

The monopolist has a clearly distinguished demand curve for his product, which is identical with the consumers' demand curve for the product in question. It is also worth mentioning that, unlike oligopolist or a firm under monopolistic competition, monopolist does not consider the repercussions of the price change by him upon those of other firms. Monopoly, as defined here, requires that the gap between the monopoly product and those of other firms is so sharp that changes in the price policies of the monopolist will not affect other firms and will therefore not evoke any readjustments of the policies by these firms.

PRICE–OUTPUT EQUILIBRIUM UNDER MONOPOLY

Monopolist, like a perfectly competitive firm, tries to maximize his profits. Profit maximization assumption on which is based the equilibrium analysis of the perfectly competitive firm is also taken to be the most valid assumption about the behaviour of the monopolist too. The motive of monopolist is the same as the motive of the perfectly competitive firm, that is, both aim at maximizing money profits. We thus do not attribute any more sinister motive to the monopolist. If the results of monopolist' behaviour on the basis of profit maximisation motive are different from that of the firm under perfect competition, it is not due to any more sinister motive of monopolist but due to the circumstances and situation in which he is placed. A firm under perfect competition faces a horizotal straight-line demand curve and marginal revenue is equal to average revenue (or price), but a monopolist faces a downward-sloping demand (or *AR*) curve and his marginal revenue curve lies below the average revenue curve. The difference in the demand conditions facing the monopolist and the perfectly competitive firm makes all the difference in the results of their equilibrium, even though both work on the basis of the same profit maximisation motive.

Monopoly equilibrium is depicted in Fig. 26.3. The monopolist will go on producing additional units of output so long as marginal revenue exceeds marginal cost. This is because it is profitable to produce an additional unit if it adds more to revenue than to cost. His profits will be maximum and he will attain equilibrium at the level of output at which marginal revenue equals marginal cost. If he stops short of the level of output at which *MR* equals *MC*, he will be unnecessarily forgoing some profits which otherwise he could make. In Fig. 26.3, marginal revenue is equal to marginal cost at *OM* level of output. The firm will be earning maximum profits and will therefore be in equilibrium when it is producing and selling *OM* quantity of the product. If he increases his output beyond *OM*, marginal revenue will be less than marginal cost, that is, additional units beyond *OM* will add more to cost than to revenue. Therefore, the monopolist will be incurring loss on the additional units beyond *OM* and will thus be reducing his total profits by producing more than *OM*. Thus he is in equilibrium at *OM* level of output at which marginal cost equals marginal revenue ($MC = MR$).

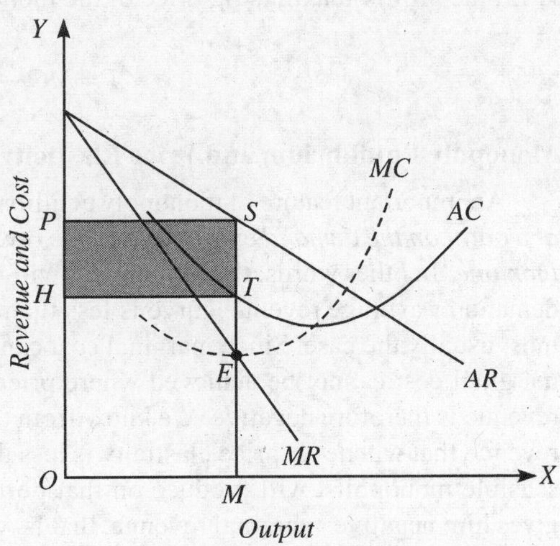

Fig. 26.3. *Firm's Equilibrium under Monopoly: Maximisation of Profits*

It will be seen from the *AR* curve in Fig. 26.3 that he will be getting the price *MS* or *OP* by selling *OM* quantity of output. The total profits earned by him are equal to the area *HTSP*. There is here a significant difference between monopoly and perfect competition. *The price under perfect competition is equal to marginal cost, but under monopoly price is greater than marginal cost.* The monopolist, unlike perfectly competitive firm, faces a downward-sloping average revenue curve and his marginal revenue lies below average revenue curve. Therefore, in monopoly equilibrium when marginal cost is equal to marginal revenue, it is less than price (or average revenue). From Fig. 26.3 it will be noticed that at equilibrium output *OM*, marginal cost and marginal revenue are equal and both are here equal to *ME*, while price fixed by monopolist is *MS* or *OP*. It thus follows that price under monopoly is greater than marginal cost.

$$MR = P\left(\frac{e-1}{e}\right)$$

where *MR* stands for marginal revenue, *P* for price and *e* for price elasticity of demand at the equilibrium output.

Since in equilibrium, *MR* = *MC*, therefore

$$P\left(\frac{e-1}{e}\right) = MC$$

or $$P = MC\frac{e}{e-1} \qquad \ldots (2)$$

Equation (2)* provides us with *rule of thumb for pricing* by the monopolist. If the knows marginal cost for his product and value of price elasticity of demand at or near the equilibrium output, he can easily calculate what price he should fix to maximise profits. In Equation (2) since $\frac{e}{e-1}$ is greater than one, *P* > *MC*. Further, it shows that *price is inversely* related to elasticity of demand. The greater the price elasticity of demand, the smaller the price fixed by the monopolist and vice-versa. Thus, if price elasticity of demand for the firm's product is equal to – 4 marginal cost of production is 12, the profits-maximising price of the monopolist will be

$$P = MC\frac{e}{e-1} = 12\frac{4}{4-1} = 16$$

Monopoly Equilibrium and Price Elasticity of Demand

An important feature of monopoly equilibrium is that the *monopolist will never be in equilibrium at a point on the demand curve or average revenue curve at which price elasticity of demand is less than one*. In other words, the monopolist will never fix his level of output at which elasticity of the demand or average revenue curve is less than one, provided the marginal cost is positive which is most usually the case. Since marginal cost can never be negative, equality of marginal revenue and marginal cost cannot be achieved where price elasticity of demand is less than one and marginal revenue is therefore negative. We know from the relationship between price elasticity and marginal revenue that whenever price elasticity is less than one, marginal revenue is negative. Therefore, no sensible monopolist will produce on that portion of the demand or average revenue curve which gives him negative marginal revenue, that is, which reduces his total revenue, while the production of additional marginal units of output adds to his total cost.

That the equilibrium of the monopolist will never be at the level of output at which the elasticity of demand curve or average revenue curve is less than one is illustrated in Fig. 26.4. It will be seen from Fig. 26.4 (upper panel) that upto *ON* level of output, *MR* is positive and total revenue is increasing because upto this output level, price elasticity of demand on the demand or average revenue curve is greater than one. *Monopoly equilibrium will always lie where price elasticity is greater than one if marginal cost is positive.* We know that at the middle point *R* of the straight-line demand or *AR* curve, elasticity is equal to one and corresponding to this unit elasticity point, marginal revenue is equal to zero. Below the middle point *R* on the average revenue curve, elasticity is less than one and marginal revenue is negative. The equilibrium of the monopolist, will never lie below the middle point of the average revenue curve *AR* as over this range, marginal revenue becomes negative and total revenue (*TR*) decreases as is evident from the falling the *TR* curve beyond *ON* output in the bottom part of Fig. 26.4. Thus, given that *MC* is positive, the equilibrium cannot lie below the middle point of the average revenue curve where elasticity is less than one. It will always lie above the middle point of the average revenue curve where elasticity is greater than one. The precise point on which equilibrium point lies depends, as already explained, upon the position of marginal cost curve and its intersection point with the marginal revenue curve.

Monopoly Equilibrium in Case of Zero Marginal Cost

There are, however, some cases where marginal cost is zero, that is, it costs nothing to produce additional units of output. For instance, in case of mineral spring, cost of production of mineral water is zero. Furthermore, in the very short period when a product is already on hand in excessive amount, it is not relevant to consider cost of production while determining the quantity of output to sell. *In these cases where cost of production is either zero or is irrelevant to consider the monopoly equilibrium will lie at a unit elasticity point on the demand curve.* This is because in such cases monopolist has only to decide at which output the total revenue will be maximum. And total revenue is maximum at the output level at which marginal revenue is zero. When marginal cost is zero, the condition of profit maximisation, that is, the equality between marginal cost and marginal revenue is achieved only at the output where the latter is zero. In Fig. 26.4 if marginal cost is zero, monopoly equilibrium will be achieved at *ON* level of output at which *MR* is zero. The price set by him in this situation will be *NR* or *OP*. *ON* quantity of output will yield maximum total revenue since beyond this marginal revenue becomes negative and total revenue will, therefore, start declining. Since cost of production is zero, the whole revenue will represent profits and because total revenue is maximum at *ON* output, the total profits will be maximum at this output. As at *ON* level of output, *MR* is zero and, as already seen, corresponding to zero marginal revenue, elasticity of the demand on the average revenue curve is equal to one or unity. We therefore conclude that *when cost of production is zero, monopoly equilibrium will be established at a level where price elasticity of demand is one*. If the marginal cost is positive, then, as explained above, monopolist will be in equilibrium at a point where elasticity on average revenue curve is greater than one.

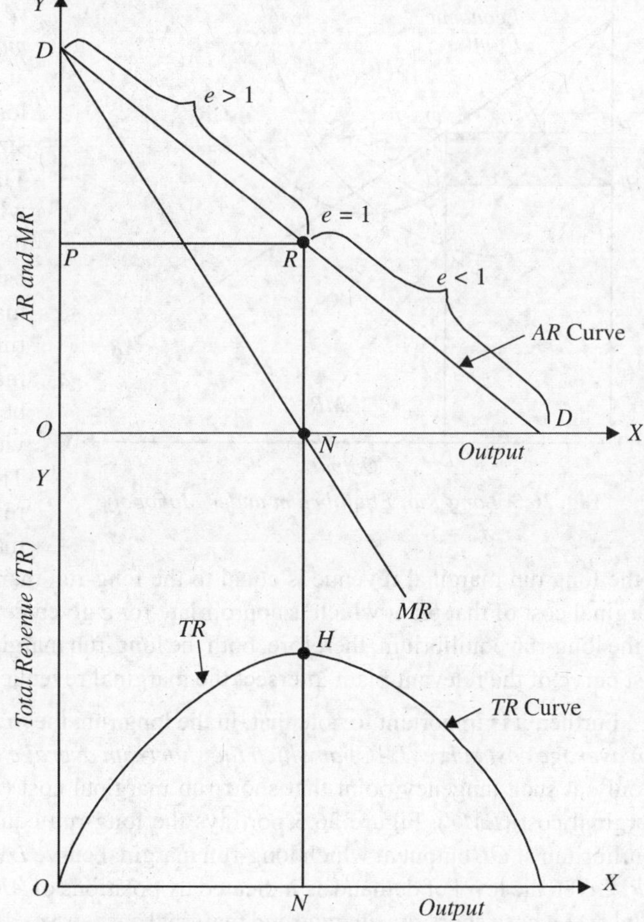

Fig. 26.4. *Monopolist cannot be in equilibrium at a level of output where elasticity of demand is less than one.*

LONG-RUN EQUILIBRIUM UNDER MONOPOLY

In the long run monopolist would make adjustment in the size of his plant. The long-run average cost curve and its corresponding long-run marginal cost curve portray the alternative plants, i.e., various plant sizes from which the firm has to choose for operation in the long-run. The monopolist would choose that plant size which is most appropriate for a particular level of demand. In the short run the monopolist adjusts the level of output while working with a given existing plant. His profit-maximising output in the short run will be where only the short-run marginal cost curve (*i.e.*, marginal cost curve with the existing plant) is equal to marginal revenue. But in the long run he can further increase his profits by adjusting the size of the plant. *So in the long run he will be in equilibrium at the level of output where given marginal revenue curve cuts the long run marginal cost curve.* Fixing output level at which marginal revenue is equal to long-run marginal cost shows that the size of the plant has also been adjusted. That is, a plant size has been chosen which is most optimal for a given demand for the product. It should be carefully noted that, in the long run, marginal revenue is also equal to short-run marginal cost. But this short-run marginal cost is of the plant which has been selected in the long run keeping in view the given demand for the product. Thus while, in the short run, marginal revenue is equal only to the short-run marginal cost of the given existing plant, in the long run marginal revenue is equal to the long-run marginal cost as well as to the short-run marginal cost of that plant which is appropriate for a given demand for the product in the long run. In the long-run equilibrium, therefore, both the long-run marginal cost curve and short-run marginal cost curve of the relevant plant intersect the marginal revenue curve at the same point.

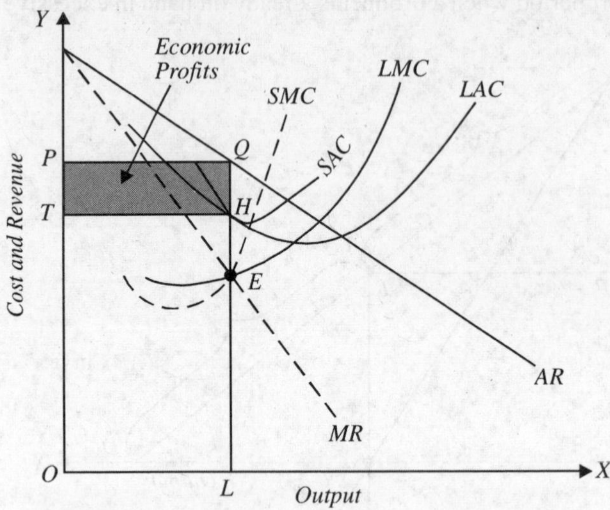

Fig. 26.5. *Long-Run Equilibrium under Monopoly*

Further, it is important to note that, in the long run, the *firm will operate at a point on the long-run average cost curve (LAC) at which the short-run average cost is tangent to it.* This is because it is only at such tangency point that short-run marginal cost (*SMC*) of a plant equals the long-run marginal cost (*LMC*). Figure 26.5 portrays the long-run equilibrium of the monopolist. He is in equilibrium at *OL* output at which long-run marginal curve *LMC* intersects marginal revenue curve *MR*. Given the level of demand as indicated by positions of *AR* and *MR* curves he would choose the plant size whose short-run average and marginal cost curves are *SAC* and *SMC*. He will be charging price equal to *LQ* or *OP* and will be making profits equal to the area of rectangle *THQP*.

It therefore follows that for the monopolist to maximise profits in the long run, the following conditions must be fulfilled

$$MR = LMC = SMC$$

$$SAC = LAC$$

$$P \geq LAC$$

The last condition implies that in *long-run monopoly equilibrium* price of the product should be either greater than long-run average cost or at least equal to it. The price cannot fall below long-run average cost because in the long run the monopolist will quit the industry if it is not even able to make normal profits.

Long-run Equilibrium Adjustment under Monopoly

In order to understand fully the difference between the short-run equilibrium and long-run adjustment under monopoly, it is necessary to show short-run equilibrium and long-run equilibrium in one figure. This has been done in Fig. 26.6 which shows that for a given level of demand, the monopolist will be in short-run equilibrium at point E or at output OQ_1 if he has plant size SAC_1 at that time. But in the long run he would not be in equilibrium at E since in the long run he can also change the plant and will employ that plant which is most appropriate for a given level of demand. In the long run he will be in equilibrium at point F where marginal revenue curve cuts his long-run marginal cost curve (LMC). But every point of the long-run marginal cost curve corresponds to a point of some short-run marginal cost curve. Long-run equilibrium point F at which marginal revenue curve cuts long-run marginal cost curve is also the point on short-run marginal cost curve SMC_2 which corresponds to the short-run average cost curve SAC_2. It means that, in long-run equilibrium position, monopolist has chosen the plant with short-run average and marginal cost curves SAC_2 and SMC_2. The plant having short-run cost curves SAC_2 and SMC_2 is optimal for him in the long run, given the level of demand as given by AR and MR curves. It is now clear that the monopolist who was in equilibrium at E in the short run with the given plant having cost curves SAC_1 and SMC_1 has shifted to the plant having cost curves SAC_2 and SMC_2 in his long run adjustment, level of demand being given. It will be noticed that, in the long run, the output has increased from OQ_1 to OQ_2 and price has fallen from OP to OJ. Profits have also increased in the long run; area $GHKJ$ is larger than the area $TRSP$.

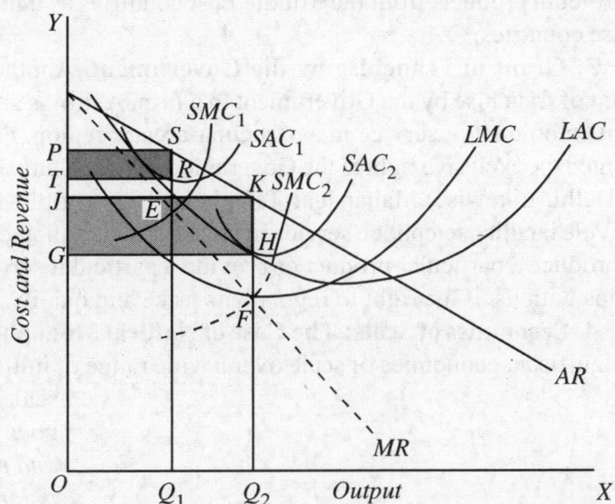

Fig. 26.6. *Shift from a Short-Run to Long-Run Equilibrium Position under Monopoly*

SOURCES OR REASONS OF MONOPOLY POWER

There are five major reasons or sources of monopoly. It is because of these reasons that monopolist enjoys a high degree of monopoly power. These sources relate to the factors which prevent the entry of new firms into an industry. Thus, these factors serve as barriers to the entry of new firms. We have explained earlier that strong barriers to the entry of new firms is an essential condition for the existence of monopoly. We explain below some of the important factors that serve as barriers to the entry of new firms and therefore constitute sources of monopoly.

1. Patent or Copyright. First important source of monopoly is that a firm may possess a patent or copyright which prevents others to produce the same product or use a particular production process. Generally, when the firms introduce *new products*, they get patent rights from the Governments so that others cannot produce them. These patent rights is granted for a certain period of time. For example, when copying machine was invented, its inventor '*Xerox*' Company had monopoly in its production based on a patent granted to it by the Government. Likewise, when a new medicine is invented by a medical company, it gets patent right from the Government so that it retains monopoly power over its production. Patents and copyrights constitute strong barrier to the entry of potential competitors.

2. Control over an Essential Raw-Material. Another source of monopoly is a control by a particular firm over an essential raw material or input used in the production of a commodity. For example, before World War II, Aluminium Company of America exercised exclusive control over almost every source of bauxite, which is an essential input for the manufacture of aluminium, had the monopoly power over the production of aluminium in the USA. Likewise OPEC (Organisation of Petroleum Exporting Countries) exercises monopoly power in the world over the supply of petroleum products from the Middle East countries as it has control over the supply of Crude Oil of these countries.

3. Grant of Franchise by the Government. Another important reason for monopoly is the grant of franchise by the Government to a firm. A firm is granted the exclusive legal right to produce a given product or service in a particular area or region. For example in Delhi the Government has granted the exclusive right to the Government owned Company DVB to provide or distribute electricity in Delhi. Likewise, Mahanagar Telephone Nigam Ltd. (MTNL) enjoys exclusive legal rights to provide landline telephone service in Delhi. It may be noted that when a company is given a franchise to produce a particular product or provide a particular service by the Government, the Government keeps with itself the right to regulate its price and quality.

4. Economies of Scale: The Case of Natural Monopoly. Another important source of monopoly is significant economies of scale over a wide range of initial output. When significant economies of scale are present, *average cost of production goes on falling over a wide range of output and reaches a minimum at an output rate that is large enough for a single firm to meet the entire market demand at a price that is profitable.* In such a situation if more than one firm operate to produce the product each firm must be producing the product at a higher than minimum cost per unit. In such a situation each firm is inclined to cut price to increase its output and reduce average cost of the product. This leads to price warfare and one who survives in this economic warfare emerges as monopolist.

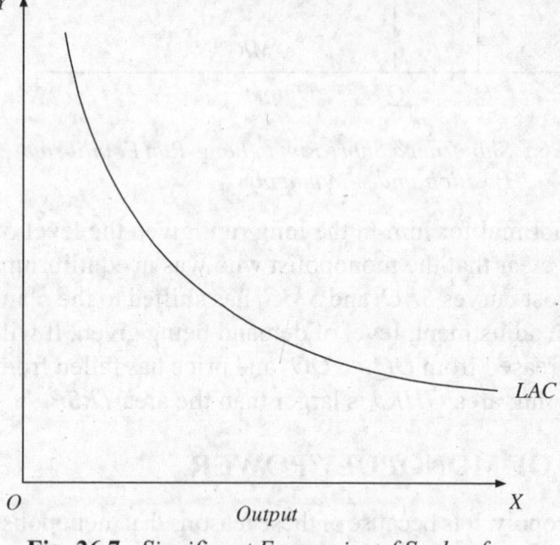

Fig. 26.7. *Significant Economies of Scale of a Natural Monopoly*

When there are significant economies of scale and as a result average cost decreases over a wide range of output, *natural monopoly is said to exist.* Natural monopolies are regulated by the Government so that they should not charge high prices and exploit the consumers.

5. Advertising and Brand Loyalties of the Established Firm. Another important reason that prevents the entry of new competitors in the industry is strong loyalties to the brands of the established firms and their heavy advertising campaigns to promote their brand. For example, strong loyalty of the consumers for 'Britannia Bread' makes it difficult for the potential competitors to enter. Further, for a long time in the USA the firm producing Coca Cola was a well-established firm to produce a famous cold drink and no one dared to enter this field till Pepsi Cola, another giant, came and broke its monopoly. Huge advertising campaigns and customer service programmes are often undertaken to enhance the market power of the producer and prevent the entry of potential competitors. Besides, if well-established firms are expecting new potential competitors, they cut prices of their products so that potential competitors find it unprofitable to enter the industry.

MONOPOLY EQUILIBRIUM AND PERFECTLY COMPETITIVE EQUILIBRIUM COMPARED

We have explained, in the preceding chapters, equilibrium of a perfectly competitive firm and industry and also how price was determined under. In the present chapter we have explained above the equilibrium under monopoly. It is now in the fitness of things to make a comparative study of the two. Only similarity between the two is that a firm under both perfect competition and monopoly is in equilibrium at the level of output at which marginal revenue equals marginal cost. But there are many important points of difference which we spell out below.

A significant difference between the two is that while *under perfect competition price equals marginal cost at the equilibrium output, under monopoly equilibrium price is greater than marginal cost*. Why? Under perfect competition average revenue curve is a horizontal straight line and therefore marginal revenue curve coincides with average revenue curve and as a result marginal revenue and average revenue are equal to each other at all levels of output. Therefore, at the equilibrium output marginal cost not only equals marginal revenue but also equals average revenue, that is, price. On the other hand, average revenue curve contronting a monopolistic firm slopes downward and marginal revenue curve therefore lies below it. Consequently, under monopoly, average revenue (or price) is greater than marginal revenue at all levels of output. Hence, at the equilibrium output of the monopolist where marginal cost equals marginal revenue, price stands higher than marginal cost. Thus, under perfectly competitive equilibrium, price = *MR* = *MC*. In monopoly equilibrium, price > *MC*.

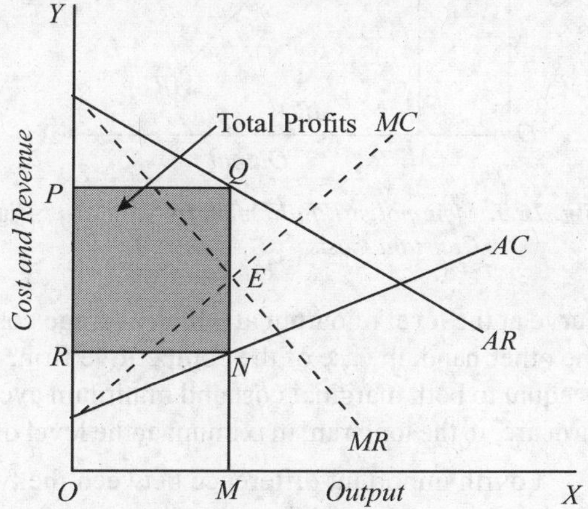

Fig. 26.8. *Monopoly Equilibrium in Case of Rising Marginal Cost*

A second important difference between the two is that while *under perfect competition, equilibrium is possible only when marginal cost is rising at the point of equilibrium, but monopoly equilibrium can be reached whether marginal cost is rising, remaining constant or falling at the equilibrium output*. This is so because the second order condition of equilibrium, namely, *MC* curve should cut *MR* curve from below at the equilibrium point, can be satisfied in monopoly in all the three cases, whether *MC* curve is rising, remaining constant or falling, whereas in perfect competition the second order condition is fulfilled only when *MC* curve is rising. Since in perfect competition the marginal revenue curve is a horizontal straight line, marginal cost curve can cut the marginal revenue curve from below only when it (*MC* curve) is rising. But, under monopoly, marginal revenue curve is falling downward and, therefore, marginal cost curve can cut the marginal revenue curve from below whether it (*MC* curve) is rising, remaining at constant level, or falling.

The equilibrium of the monopolist in these three cases is shown in Figs. 26.8, 26.9 and 26.10. Fig. 26.8 illustrates the equilibrium of the monopolist when marginal cost curve is rising at the equilibrium output. Fig. 26.9 shows monopoly equilibrium when marginal cost is constant at and near the equilibrium output. In Fig. 26.10 monopolist is in equilibrium when marginal cost is falling at and near the point of equilibrium. In all these three cases, *OP* represents the price determined, *OM* represents the equilibrium output, and *RNQP* represents total positive profits made, though their amount differs in different cases.

Another significant difference between the two is that *whereas a perfectly competitive firm is in long-run equilibrium at the minimum point of the long-run average cost curve, monopolistic firm is generally in equilibrium at the level of output where average cost is still declining* and has not yet reached its minimum point. In other words, whereas a perfectly competitive firm tends to be of optimum size in the long run, a monopolist firm stops short of the optimum size. This is so because it pays a competitive firm to expand production so long as average cost is falling since average revenue and marginal revenue remain constant but it does not pay a monopolist firm to expand production to the minimum point of the average cost curve because it is often not worthwhile for it to do so. More frequently, the marginal revenue curve of the monopolist intersects the marginal cost

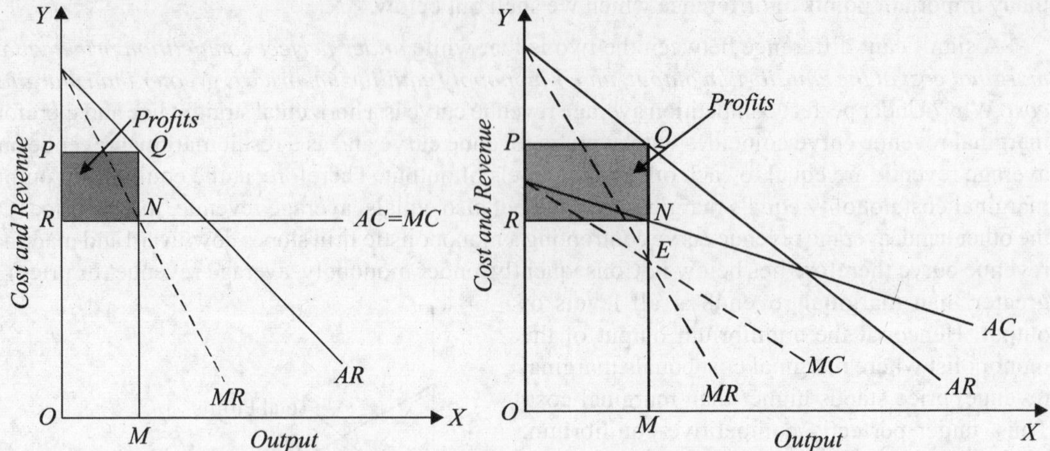

Fig. 26.9. *Monopolist Equilibrium in Case of Constant Costs*

Fig. 26.10. *Monopoly Equilibrium in Case of Falling Cost*

curve at the level of output at which average cost is still falling, as will be seen from Fig. 26.10. On the other hand, in case of the competitive firm, marginal revenue or price, in long-run equilibrium, is equal to both marginal cost and minimum average cost. In other words, profits of the competitive firm are, in the long run, maximum at the level of output at which long-run average cost is minimum.

Fourth important difference between the two is that *while the perfectly competitive firm is, in the long run, able to make only normal profits, a monopolist can make supernormal profits even in the long run*. Under perfect competition, if firms in the short run are making profits above normal, the new firms will enter the industry to compete away the profits. But under monopoly the firm continues earning supernormal profits even in the long run since there are strong barriers to the entry of new firms in the monopolistic industry. It should not however be understood from this that monopoly always guarantees super-normal profits. As already explained above, *the monopoly can even make losses, though only in the short run.* These short-run losses are not due to competition from any new firms but due to low demand and high costs. The point is, if a monopolist in the short run is making profits above normal, they cannot he eliminated by the entry of new firms in the long run with the result that these supernormal profits will persist in the long run provided that the demand and cost situations are nor changed unfavourably.

Another important difference between monopoly equilibrium and perfectly competitive equilibrium is that *under monopoly price is higher and output smaller than under perfect competition*, assuming cost conditions in the two cases to be the same. Suppose a number of firms are producing homogeneous products and pure or perfect competition in the sense so that no one can individually affect the price exists among them. Price and output will be determined at the level where demand and supply curves intersect each other. Suppose that all the firms constituting the competitive industry combine together so as to form a monopoly. We assume that no economies, internal or external, accrue when the firms combine together so that the cost or supply conditions remain unchanged. Note that the price and output under monopoly are determined by equality between marginal cost and marginal revenue and not by the intersection of demand and supply curves.

Price-output equilibrium under perfect competition and monopoly are graphically shown in a single diagram (Fig. 26.11). Curves *DD* and *SS* are respectively demand and supply curves of the perfectly competitive industry. The two curves intersect each other at point E and as a result price OP and output OM are determined under perfect competition. The marginal revenue curve MR is drawn corresponding to the demand curve DD. Now, what is the marginal cost curve of the monopoly? The curve SS which is the supply curve of perfectly competitive industry will be the marginal cost curve under monopoly. It will be seen from Fig. 26.11 that the marginal revenue curve MR cuts the marginal cost (MC) curve SS of the monopolist at point F and as a consequence monopoly price $O'P'$ and monopoly output OM' are determined. It is thus clear that if cost conditions remain unchanged, the combination of firms to form a monopoly has resulted in a higher price and a lower output of the product. As a result of monopoly coming into existence price has risen from OP to OP' and output has fallen from OM to OM' *Monopoly thus restricts output to raise price.*

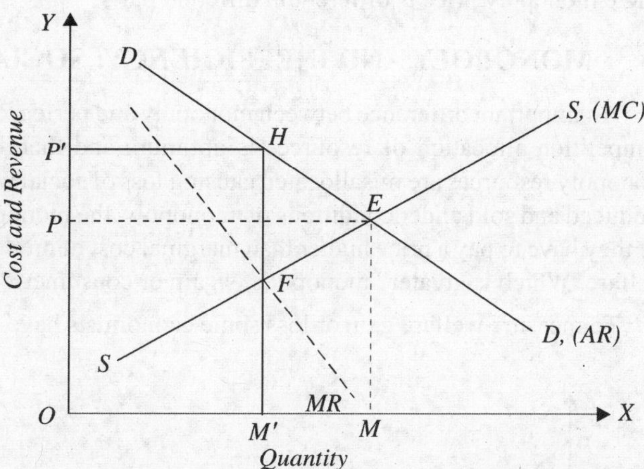

Fig. 26.11. *Under monopoly price is higher and output smaller than under perfect competition.*

Now, a pertinent question is whether cost conditions are likely to remain the same when a number of firms combine to become a monopoly? In other words, whether some extra economies will not accrue when the monopoly comes into existence and works on a large scale than that of a large number of small firms under perfect competition. There are two opinions on this issue. One view is that monopolist can introduce various economies such as greater specialisation in the work by bringing about suitable lateral and vertical integration, organization of sales on a large scale, buying raw materials and machinery on a big order, getting cheap credit, discovering and introducing new methods of production, and undertaking various measures which are generally associated with modernisation. These economies, it is held, will lower the cost of production of the monopolist with the result that the marginal cost curve (SS) in Fig. 26.11 will shift to a downward position when monopoly comes into existence. If these economies are large enough so that there is substantial fall in the cost curve, the monopoly price may be lower than the competitive price MP and monopoly output will be larger than the competitive output.

But in view of Mrs. Joan Robinson, "*perfect competition would bring about all the economies which monopoly could introduce.*"[2] She holds therefore that the cost curve will not shift downward significantly as a result of the establishment of monopoly in place of a large number of firms working under perfect competition. Thus, according to her, monopoly price is always higher and monopoly output always smaller than that under perfect competition.

Last significant difference between monopoly and perfect competition is that while a *monopolist can discriminate prices for his product, a perfect competition cannot.* The monopolist will be increasing his total profits by discriminating prices if he finds that elasticities of demand at the single monopoly price are different in different markets. But it is not possible for a firm under perfect competition to charge different prices from different buyers. This is because a seller under perfect competition confronts a perfectly elastic demand curve at the level of going market price. Therefore, if he tries to charge a bit higher price than the going market price from some buyers, they will turn to other sellers

2. Joan Robinson, *Economics of Imperfect Competition*, p. 168

and purchase the same product at the going market price. But the monopolist has the sole control over the supply of a product which has no close substitutes and therefore the demand curve of his product is very much less elastic. If the monopolist is able to break up his market into different parts on one basis or the other, it will be possible for him to discriminate prices in different parts of the market. But it will be profitable for him to charge prices in the different parts only if the elasticity of demand at the single monopoly price is different in different parts.

MONOPOLY AND INEFFICIENCY: SOCIAL COST OF MONOPOLY

An important difference between monopoly and perfect competition is that whereas under perfect competition allocation of resources is optimum and therefore social welfare is maximum, under monopoly resources are misallocated causing loss of social welfare. As seen above, when a product is produced and sold under conditions of monopoly, the monopolist gains at the expense of consumers, for they have to pay a price higher than marginal cost of production. This results in loss of consumers' welfare. Which is greater? monopolist's gain or consumers' loss.

To measure welfare gain or loss some economists have used the concept of consumer's surplus. Consumer's surplus, it will be recalled, is the surplus of price which consumers are prepared to pay for a commodity over and above what they actually pay for it. The dead-weight loss in consumer's welfare due to monopoly can be shown through Fig. 26.12 where *TD* is the demand curve for the monopolsist product, *MR* is the corresponding marginal revenue curve. It is assumed that the industry is a constant cost industry so that average cost (*AC*) remains the same as output is increased and marginal cost is equal to it.

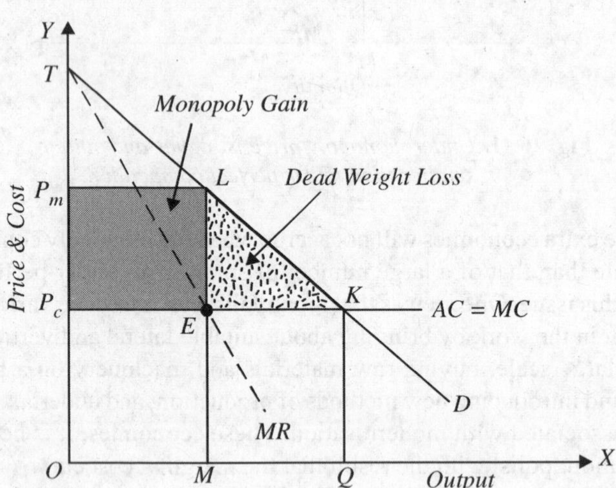

Fig. 26.12. *Dead-weight Loss or Social Cost of Monopoly*

Under perfect competition firms equate price with marginal cost and industry's output is determined by demand for and supply of the product. Since we are considering a constant cost industry, a horizontal line (*AC* = *MC*) is the supply curve of the industry. It will be seen from Figure 26.12 that under perfect competition price determined is equal to *QK* (or OP_C) and output *OQ* is being produced. Firms will be equating price OP_C with their marginal cost. The consumer surplus enjoyed by the consumers is equal to the area TKP_C. It may be noted that consumer surplus reflects social welfare as it is excess of what consumers are willing to pay (that is, the utility that they obtain) over and above what they actually pay.

Now, the monopolist would not produce OP_C output as he equates marginal revenue (*MR*) with marginal cost (*MC*) to maximise his profits. It will be seen from Fig. 26.12 that marginal revenue and marginal cost are equal at output level *OM*. Therefore the monopolist will produce *OM* output and charge *ML* or OP_m price. Thus, monopolist has restricted output to *OM* and raised price to OP_m. As a result monpolist makes profits equal to the area P_mLEP_c. On the other hand, as a result of rise in price to P_m, the consumers' surplus has been reduced to the area TLP_m and they suffer a loss of consumer surplus equal tot he area P_cKLP_m. Thus there is a redistribution of income from consumers to the monopolist, but it is important to note that loss of consumer surplus P_cKLP_m which is greater than the profits made by the monopolist by the area of triangle *LKE*.

The loss in consumer surplus can be divided into two components. First part is the profits equal to the area P_mLEP_c made by the monopolist at the expense of the consumers. This component of loss in consumer surplus is suffered by those who are still purchasing the product. The second component of the loss of consumer surplus is equal to the area of triangle *LKE* which is due to allocative inefficiency caused by the monopolist by reducing output of the product and raising its price. This second component of loss in consumer surplus as measured by the area of triangle *LKE* is called *dead-weight loss* of welfare caused by the monopolist. This represents **social cost of monopoly**. It will be seen from Fig. 26.12 that price which the last existing consumer is willing to pay for *M*th unit is M_L while the marginal cost which has to be incurred by the society is *ME* and therefore from *M*th unit, consumer enjoys consumer surplus equal to *EL*. In other words, consumer values the product more than the opportunity cost of production as measured by the marginal cost (*MC*). Likewise, the price which the consumers are willing to pay for additional *MQ* units exceeds the marginal cost (*MC*) to the society and therefore generates additional consumer surplus.

Thus, the social welfare or consumer surplus would be increased if output is extended to the point *Q*. It will be seen from Fig. 26.12 that consumers would gain additional consumer surplus equal to the area of triangle *LKE* if output is increased to *OQ*. But monopolist would not extend output to point *Q* because his profits are maximised at *OM* output. Since for extra units from *M* to *Q*, marginal cost exceeds marginal revenue of the monopolist, he will not produce them. *But from the social point of view, the extension of output to the point Q is desirable as it increases consumer surplus gained by the consumers.* Thus, monopoly causes a net loss of consumer welfare equal to area of triangle *LKE*. This is called *a dead weight loss of welfare* because though consumers suffer a loss of welfare, no one else, not even monopolist, gains from it. This is loss of welfare caused by allocative inefficiency of the monopoly. It may be noted that the gain of profits by the monopolist equal to the area P_mLEP_c has been made possible by the loss of consumer surplus of those who are still purchasing the *OM* quantity of the product. *Even if these profits gained by the monopolist are redistributed among the consumers, the loss of consumer surplus represented by the triangle LKE would still remain and would not be eliminated because, as explained above, they have been caused by allocative inefficiency of monopoly*. Hence it is appropriate to call it a *dead-weight loss* of welfare.

With lower production of the product by the monopolist, relatively less resources are allocated to its production. For optimum allocation of resources *OQ* amount of the product, at which marginal cost equals price should have been produced and resources allocated accordingly. To conclude, *monopoly causes misallocation of resources and dead-weight loss of welfare.* This is also called social cost of monopoly.

Dead - Weight Loss (Social Cost) under Monopoly in Case of Increasing Marginal Cost

In our above analysis of dead-weight welfare loss (or, in other words, social cost of monopoly) due to reduction in output and hike in the price by a monopolist as compared to the perfectly competitive equilibrium, it has been assumed that marginal cost curve is a horizontal straight line. When marginal cost curve is a horizontal straight line, the loss in welfare occurs only in consumer surplus. But when marginal cost curve is rising, the loss in welfare due to reduction in output by the monopolist will occur not only in reduction in consumer surplus but also in producer surplus. Producer surplus, it will be recalled, is the total revenue earned over and over all the opportunity costs (explicit and implicit) represented by the marginal cost curve. It may be noted that *maximum social welfare or economic efficiency is achieved when the sum of consumer surplus and producer surplus is the maximum.* In a perfectly competitive equilibrium where quantity demanded equals quantity supplied or price equals marginal cost, the sum of consumer surplus and producer surplus is maximum and therefore perfect competition ensures maximum social welfare or economic efficiency. But to maximize profits monopolist does not equate price with marginal cost. Instead, he equates marginal revenue with marginal cost and therefore reduces output and raises price and thereby causes loss of welfare. Loss in welfare as measured by the reduction in the sum of consumer surplus and producer surplus is illustrated in Fig. 26.12(a).

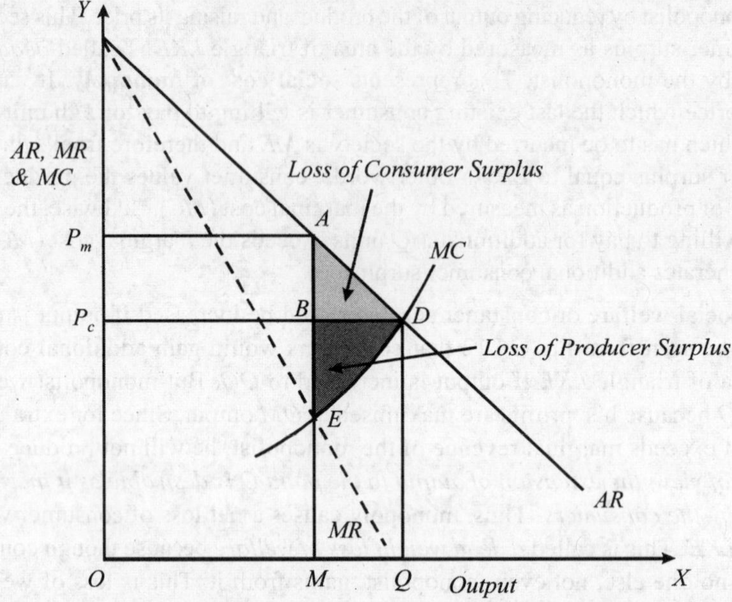

Fig. 26.12(a). *Dead-Weight Loss in Welfare in Case of Increasing Marginal Cost.*

It will be seen that, under perfect competition, equilibrium will be at point *D* where price is equal to marginal cost (*MC*) and *OQ* output is being produced and P_C price lis being charged. Now, if monopoly comes into existence, the monopolist-producer will maximize profits by producing lower output *OM* and will charge higher price P_m. It will be observed that the loss in consumer surplus suffered suffered by the buyers is equal to area P_cDAP_m. Due to the higher price charged by the monopolist, his gain in profits or producer surplus equals the rectangel P_cBAP_m. (It may be noted that this gain in producer surplus by the monopolist occurs at the expense of consumers who suffer a loss in consumer surplus and is equal to the price differential $P_m - P_c$ or *AB* multiplied by the monopoly output *OM*). Thus the gain in producer surplus represented by the area P_cBAP_m is just a transfer of income from the consumers to the monopolist. Net loss of consumer surplus or welfare is therefore the area of the triangle *ABD*.

However, in the situation depicted in Fig. 26.12(a) where marginal cost is rising, apart from the net loss of consumer surplus, there is also a loss of producer surplus due to reduction in output by *OM* amount under monopoly. It will be seen from Fig. 26.12(a) that under perfect competition with price equal to OP_c or *QD*, the extra profits or producer surplus earned over and above the marginal costs in the region of *MQ* output equals the area *BDE* which is lost due to the reduction in output equal to *QM* by the monopolist. This loss in producer surplus *BDE* is also a *dead weight loss* caused by the inefficiency or lower production due to monopoly because this has not benefited any other. Thus, the total dead weight loss of welfare caused by the monopoly is equal to the whole area *AED* which is the sum of net loss of consumer surplus (*ABD*) and the loss of producer surplus equal to *BDE* represents *social cost of monopoly*.

It follows from the partial equilibrium approach to the measurement of loss of welfare that monopoly is economically inefficient and causes misallocation of resources as it does not extend production of a product to the level at which the sum of consumer surplus and producer surplus is the maximum.

ABSENCE OF SUPPLY CURVE UNDER MONOPOLY

An important feature of the monopoly is that, unlike a competitive firm, *the monopolist does not have the supply curve.* It is worth noting that the supply curve shows how much output a firm will produce at various *given prices* of a product. The supply curve of a product by a firm traces out the *unique price-output relationship,* that is, against a given price there is a particular amount of output which the firm will produce and sell in the market. As explained in a previous chapter, the concept of supply curve is relevant only when the firm exercises no control over the price of the product and therefore takes it as given. Therefore, it is perfectly competitive firm which is a price taker and demand curve facing it is a horizontal straight line that a unique price-output relationship is established. For a perfectly competitive firm, marginal revenue *(MR)* equals price and therefore to maximize profits, the firm equates price (= *MR)* with marginal cost. As price changes due to the shift in demand, the competitive firm equates the new higher price *(i.e.* new *MR)* with its marginal cost at higher level of output. In this way under perfect competition, marginal cost curve becomes the supply curve of the firm. To quote professor Baumol, "The supply curve is strictly speaking, a concept which is usually relevant only for the case of pure or perfect competition . . . . The reason for this lies in its definition . . . the supply curve is designed to answer question of the form, "How much will firm A supply if it encounters a price which is at *P* dollars. But such a question is most relevant to the behaviour of firms that actually deal with price over whose determination they exercise no influence."[3]

But a monopolist does not take the price as given and exercises control over the price of the product as he is the sole producer of the product. Further, for a monopoly firm demand curve slopes downward and marginal revenue *(MR)* curve lies below it. Therefore, a monopolist in order to maximise

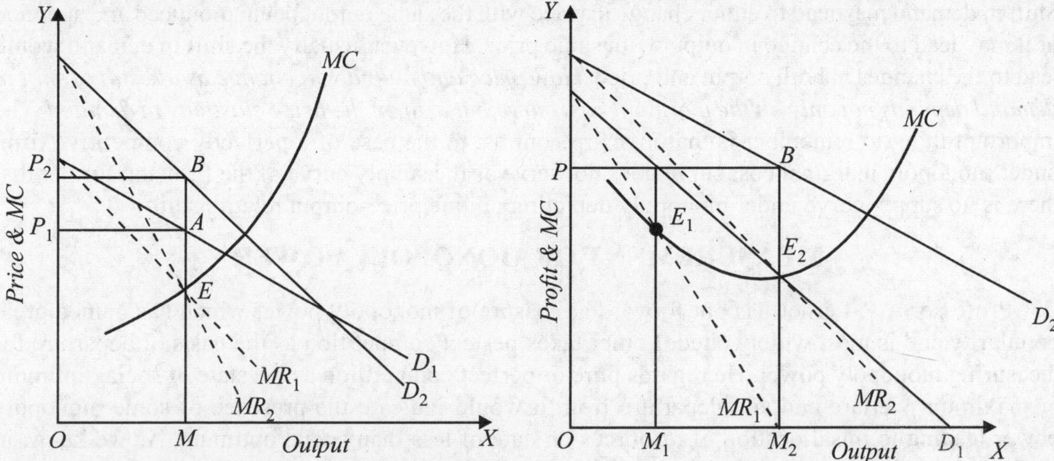

Fig. 26.13. *Under monopoly a shift in demand leading to the same output being supplied at two different prices.*

Fig. 26.14. *Under monopoly a shift in demand may lead to a change in output being supplied at the same price.*

profits does not equate price with marginal cost; instead he equates marginal revenue with marginal cost. As a result, shifts in demand causing changes in price do not trace out a unique price-output series as happens in case of a perfectly competitive firm. In fact, under monopoly shifts in demand

3. W.J. Baumol, *Economic Theory and Operations Analysis*, 2nd edition, p.342

can lead to a change in price with no change in output or a change in output with no change in price or they can lead to changes in both price and output. This renders the concept of supply curve inapplicable and irrelevant under conditions of monopoly. Thus, there is no unique price-quantity relationship, since quantity supplied by a firm under monopoly is not determined by price but instead by marginal revenue, given the marginal cost curve. Thus, Joan Robinson writes, "When competition is not perfect, the demand curve for the output of each individual producer is not perfectly elastic and each producer will sell that output at which his marginal cost is equal to his marginal receipts Marginal revenue will not be equal to price, it is marginal revenue and not price that determines output of the individual producer."[4]

That, under monopoly, we do not get a series of unique price-output relationship or supply curve of a product is illustrated in Fig. 26.13. Suppose the demand curve is initially D_1, corresponding to which MR_1 is the marginal revenue curve. Given the marginal cost curve *MC*, monopolist is in equilibrium at *OM* level of output and charging price OP_1. Now, suppose that demand curve shifts to the position D_2 corresponding to *which* MR_2 is the marginal revenue curve. It will be seen from Fig. 26.13 that the new marginal revenue curve MR_2 also intersects the given marginal cost curve *MC* at the same level of output *OM* as before the shift in the demand curve but price has risen to OP_2. Thus we see that under monopoly, a shift in demand may lead to the production and *supply of the same output at two different prices.* This clearly shows that there is no unique price-output relationship which is essential for the concept of supply curve to be applicable.

Fig. 28.14 illustrates another special case where shift in demand leads to the *different levels of output being supplied at the same* price. Initially, with D_1 and MR_1as the demand and marginal revenue curves respectively, the monopolist maximises his profits by producing output OM_1 and charging price *OP.* With the shift in demand curve to D_2 and the marginal revenue curve to MR_2, the marginal cost curve *MC* cuts the new MR_2 curve at E_2 and it will be observed from Fig. 28.14 that in the new equilibrium, the monopolist produces higher quantity OM_2 at the same price *OP.* This again shows that under monopoly there is no any *specific* quantity of the product supplied at a price.

To sum up, under monopoly, there is no supply curve associating a unique output with a price. Shift in demand may lead to either change in price with the same output being produced and supplied or it may lead to the change in output with same price. However, usually the shift in demand would lead to the changes in both output and price. *How price and output will change as a result of shift in demand depends not only on the marginal cost curve but also on the price elasticity of demand.* The important thing to remember is that in sharp contrast to the case of a perfectly competitive firm, under monopoly marginal cost curve does not serve as the supply curve of the firm and further that there is no supply curve under monopoly depicting unique price-output relationship.

MEASUREMENT OF MONOPOLY POWER

Professor A.P. Lerner has put forward a measure of monopoly power which has gained great popularity and is most widely cited. Lerner takes perfect competition as the basis of departure for measuring monopoly power. He regards pure or perfect competition as the state of social optimum or maximum welfare and any departure from it would indicate the presence of some monopoly power leading to misallocation of resources or state of less than social optimum. As we know, in perfect competiton price is equal to marginal cost of the product in the equilibrium position. And it is this equality of price with marginal cost under perfect competition that ensures maximum social welfare or optimum allocation of resources.

Now, when competition is less than pure or perfect the demand curve facing a firm will be sloping downward and marginal revenue curve will lie below it. Consequently, when competition is

4. Joan Robinson, *The Economics of Imperfect Competition*, p. 66

less than pure (perfect), that is, when it is imperfect, in a seller's equilibrium position, marginal cost will be equal to marginal revenue but price will stand higher than marginal cost or marginal revenue. *This divergence between price and marginal cost, according to Professor Lerner, is the indicator of the existence of monopoly power.* The greater this divergence between price and marginal cost, the greater the degree of monopoly power possessed by the seller. Based on this, Lerner has given the following precise index of the degree of monopoly power:

$$\text{Degree of monopoly power} = \frac{P - MC}{P}$$

where P denotes price and MC denotes marginal cost at the equilibrium level of output.

When competition is pure or perfect, price *(P)* is equal to marginal cost and therefore Lerner's index of monopoly power is equal to zero indicating no monopoly power at all, for when price is equal to marginal cost, $P - MC$ will be equal to zero and the above formula will yield the value of index as zero.

Thus, under perfect competition, Lerner's index of monopoly power $\left(\frac{P-MC}{P}\right) = \frac{0}{P} = 0$. On the other hand, when the monpolised product entails no cost of production, that is, when the product is a free good whose supply is controlled by one person, the marginal cost will be equal to zero and Lerner's index of monopoly power $\left(\frac{P-MC}{P}\right)$ would be equal to one or unity. Thus when MC is equal to zero $\frac{P-MC}{P} = \frac{P-0}{P} = \frac{P}{P} = 1$.

It is thus clear that Lerner's index of monopoly power can vary from zero to unity. Within this range, the greater the value of the index $\left(\frac{P-MC}{P}\right)$, *the greater the degree of monopoly power possessed by the seller.* For instance, if the price of a product is equal to ₹ 15 per unit and its marginal cost is ₹ 10, then the value of index of monopoly power will be $\frac{15-10}{15} = \frac{5}{15} = 1/3$ and when the price is equal to ₹ 20 and marginal cost is equal to 10, the index of monopoly power will be equal to $\frac{20-10}{20} = \frac{10}{20} = 1/2$.

Monopoly Power and Price Elasticity of Demand

Now, it has been shown that Lerner's index of monopoly power is equal to the inverse of the price elasticity of demand. We can prove this as follows:

Since at the equilibrium level, marginal cost is equal to marginal revenue, we can substitute in the above formula marginal revenue for marginal cost.

Thus

$$\text{Lerner's index of monopoly power} = \frac{P-MC}{P}$$

Since in equilibrium, $MC = MR$

$$\text{Lerner's index of monopoly power} = \frac{P-MR}{P} \quad \ldots \text{(i)}$$

We know that $MR = P\left(1 - \frac{1}{|e|}\right)$ where e is absolute value of the price elasticity of demand at the equilibrium output. Thus, putting $P\left(1-\frac{1}{|e|}\right)$ in place of *MR* in (i) above we get,

$$\text{Lerner's index of monopoly power} = \frac{P-MC}{P} = \frac{P-P\left(1-\frac{1}{|e|}\right)}{P} = 1-1+\frac{1}{|e|} = \frac{1}{|e|}.$$

It therefore follows that Lerner's index of monopoly power is equal to the inverse of price elasticity of demand. Thus degree of monopoly power can be judged by merely knowing the price elasticity of demand at the equilibrium output. *The degree of monopoly power varies inversely with the absolute value of price elasticity of demand for the good.* It is worth noting that the price elasticity of demand in Lerner's index refers only to the price elasticity at the equilibrium output.

Critique of Lerner's Measure of Monopoly Power

Lerner's measure has many shortcomings. First, the chief shortcoming of Lerner's measure is its inability to measure the strength of monopoly and competitive elements in the *non-price competition* and in cases of product differentiation as found in monopolistic competition and differentiated oligopoly. Lerner's index at the best measures the strengths of monopoly and competitive elements when the competition between the sellers is on the basis of price. When instead of competing on the basis of price the sellers in monopolistic competition and oligopoly compete on the basis of product variation, advertising and other of selling costs, Lerner's index fails to indicate truly the degree of monopoly and competition involved in such market structures. Suppose Lerner's monopoly index for a firm selling a differentiated product yields a high figure. But this does not necessarily mean that the seller will be possessing a high degree of monopoly power and facing less competition. It may be that the sellers of the various varieties of the product may not be competing on the basis of price and instead may be engaged in highly intense competition in product variation and advertising and other forms of selling costs to promote the sales of their products.

On the basis of the above non-price factors some firms may enjoy greater monopoly control over their products than others. Thus Professor Chamberlin rightly says that elasticity and Lerner's index measures "*pass over completely the important problems of competition and monopoly in the non price area; quality and other aspects of the product including location and advertising and other forms of selling costs*."[5]

Secondly Lerner's measure is based upon only one aspect of monopoly, namely, its control over price which depends upon the *availability and effectiveness of existing substitutes*. It ignores the restraints on monopoly power put by the *potential substitutes* which would come to exist with the entry of new firms into the industry as a powerful factor limiting be monopoly power of the existing sellers. To quote Chamberlin again, "Neither elasticity nor the Lerner's index measures anything the effectiveness of existing substitutes; it gives no indication as to potential substitutes (the important problem of entry)."[6]

Thirdly, Lerner's index is criticised on the ground that monopoly power does not express itself only in the divergence of price from marginal cost, it also expresses itself in the restriction of output.

THE RULE OF THUMB FOR PRICING

We have explained above Lerner's index of monopoly power according to which degree of monopoly is given by

5. E.H. Chamberlin, Measuring the Degree of Monopoly and Competition, printed in his "*Towards a General Theory of Value.*"
6. E.H. Chamberlin, *op. cit.*

$$\frac{P-MC}{P}=\frac{1}{e}$$

where $\frac{P-MC}{P}$ is the *mark-up* over marginal cost as a proportion of price. According to the above equation, this mark-up over price is equal to inverse of the absolute value of the price elasticity of demand for the product. It follows from above that there are following relations of mark-up over marginal cost with price elasticity of demand :

(*a*) The lower the price elasticity of demand, the greater the mark-up as a proportion of price and greater the monopoly power.

(*b*) The greater the price elasticity of demand, the smaller the mark-up as a proportion of price and less the monopoly power.

For example, if marginal cost of production at the equilibrium output is 9 and price elasticity of demand is 4. What will be the price ? Note that absolute value of price elasticity of demand is 4. Therefore,

$$\text{Mark up}=\frac{P-MC}{P}=\frac{P-9}{P}=\frac{1}{4}$$

$$P=4P-36$$

$$3P=36$$

$$\text{Thus price or } P=\frac{36}{3}=12$$

Now if at the equilibrium output with $MC = 9$, price elasticity of demand is 2, then

$$\frac{P-MC}{P}=\frac{P-9}{P}=\frac{1}{2}$$

$$P = 2P-18$$

$$2P-P = 18$$

$$P=18$$

It follows from above that given the marginal cost, the lower absolute value of price elasticity of demand, the greater the mark-up over marginal cost and, therefore, higher the level of price fixed by the producer.

Diagrammatic Illustration

The relation of price mark-up over marginal cost with monopoly power and price elasticity of demand is illustrated in Figure 26.14(A). In panel (*a*) of Fig. 26.14(A) price elasticity of demand at the equilibrium output *OQ* is relatively more, and therefore the power of the producer to raise price above marginal cost is less and as a result the mark-up (*P–MC*) is small. On the other hand, in panel (*b*) of Figure 26.14(A) price elasticity of demand at equilibrium output is less and therefore the greater the power of the producer to set price above marginal cost. As a result, price mark up $\left(\frac{P-MC}{P}\right)$ is much higher in panel (b) Fig. 26.14(A).

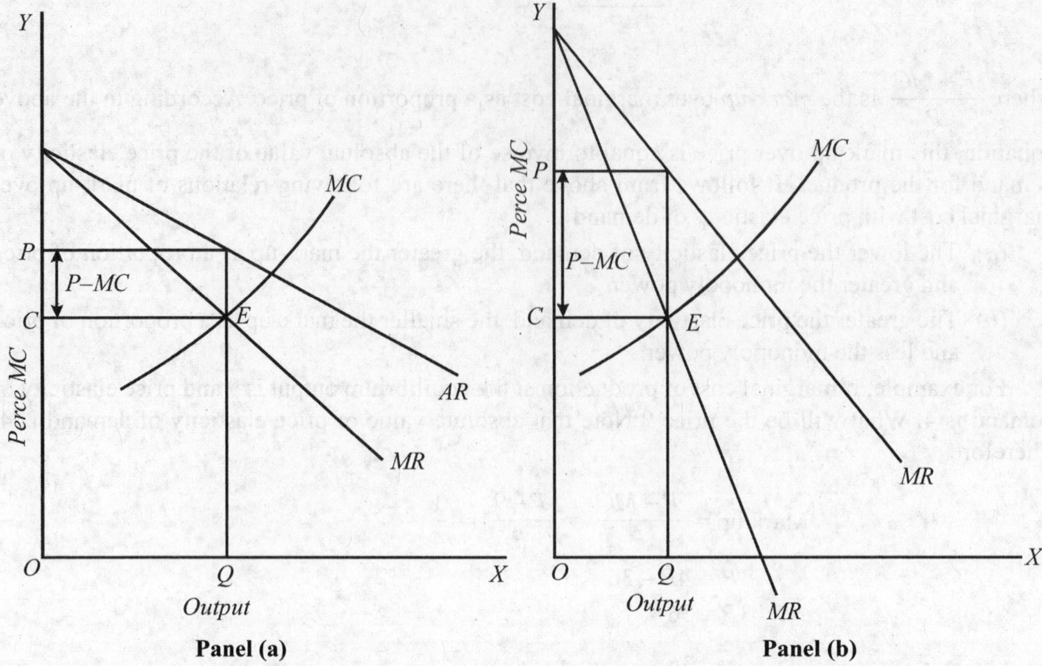

Fig. 26.14(*A*). *Price Elasticity of Demand and Price Mark-up*

Thus, the relationship $\frac{P - MC}{P} = \frac{1}{e}$ provides us a rule of thumb for price fixation by a firm enjoying monopoly power. It has to fix price as a mark-up over marginal cost which is determined by the absolute value of the price elasticity of demand

REGULATION OF MONOPOLY

It has been seen above that monopolists restrict output and raise price of their products, In this way they are not only generally able to make supernormal profits and increase inequalities in income distribution but also cause inefficiency in the allocation of resources of the society. It has therefore been felt to regulate monopoly with a view to achieving two objectives. First, it is regulated to improve income distribution and prevent exploitation of consumers by the monopolists. Secondly, monopoly is regulated so as to ensure economically efficient allocation of resources. Further, monopoly can be regulated either through suitable taxation or through fixation of the maximum price it can charge for its product. It is the *price regulation* of monopoly that we will discuss below.

Consider Fig. 26.15 which gives cost and demand conditions faced by a monopolist. Without regulation of price charged by him, he is in equilibrium by producing the level of output *OQ* at which his profits are maximum, and are equal to the area *PKLT.* He has been able to make these supernormal profits by producing a level of output less than what is socially desirable and also charging a price higher than what would be prevailing under perfect competition.

Marginal Cost Pricing

Suppose in order to improve allocation of resources or distribution of income the Government: decides to regulate the price charged by the monopoly. Obviously, the Government will fix the maximum price *(i.e.,* price ceiling) at the level below its profit maximising price *OP.* There are two types of pricing rules which have often been proposed for price regulation of monopoly. First, since social welfare is maximum when price of a commodity is fixed at the level where it equals marginal cost of production of the commodity, it is proposed that the maximum price for the monopoly should be fixed equal to the marginal cost. This is therefore known as *marginal cost pricing.* Consider Fig. 26.15 where *AR* represents the average revenue or demand curve of the commodity produced by the monopolist. It will be seen from the demand or average curve *AR* in Fig. 26.15 that when monopolist expands output on *ON*, price falls and becomes equal to the marginal cost at point *S*. Thus, if the objective of regulation of monopoly is to ensure economic efficiency, that is, maximum social welfare, the maximum price to be charged by the monopolists should be fixed at the level *OH* corresponding to the level of output *ON*.

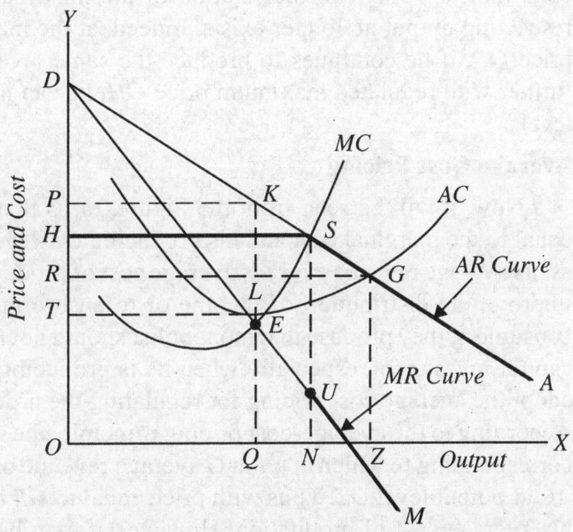

Fig. 26.15. *Regulation of Price Charged by a Monopoly*

It is important to note that with the maximum price *OH* fixed by the Government the monopolist will sell all units of output upto *N* at the same price *OH*. Therefore, with *OH* as the maximum regulated price for the monopolist, the part of the average revenue curve *DS* now becomes irrelevant. This is because he can sell any amount upto point *N* at the given regulated price *OH*. Therefore, his new average revenue curve upto the level of output *ON* becomes the horizontal straight line *HS*. Since average revenue curve remains constant over the range of output *ON*, marginal revenue curve would now coincide with the average revenue curve *HS* which has been thickened. Therefore, with regulation of price at *OH*, part *DU* of the marginal revenue curve also becomes irrelevant. With regulated price at *OH*, the portion of demand or average revenue curve *SA* lying below the price *OH* would remain intact and unaffected and marginal revenue curve corresponding to this part of the average revenue curve is the thickened portion *UM*. Thus, with maximum regulated price *OH*, the average revenue curve for the monopolist becomes *HSA* and the marginal revenue curve is *HSUM* with the broken or discontinues portion *SU*.

It should be noted that with the given marginal cost curve *MC* and the new marginal revenue curve *HSUM* formed after the fixation of regulated price *OH*, the monopolist will be maximising his profits by producing *ON*. This is because if with the regulated price *OH* the monopolist produces more than *ON*, its marginal revenue will fall below the marginal cost showing losses on the extra units produced. And if with the regulated price *OH*, he produces less than *ON* level of output, marginal revenue exceeds marginal cost indicating the possibility of increasing profits by expanding output to the level *ON*. Thus, with the regulated price *OH* fixed on the basis of marginal cost principle, the monopolist is in equilibrium and maximising profits by producing *ON* level of output.

It is interesting to observe that in case of monopoly, when maximum regulated price *OH* is fixed below its free market profit-maximizing price *OP*, the monopolist expands his output from *OQ* to *ON* and buyers demand this much output at price *OH* and therefore no shortage of output is created as a result of the fixation of regulated price below its free-market profit maximising price *OP* of the monopolist. This is in sharp contrast to the price regulation of an industry working under perfect competition in which case when maximum price is fixed below the free-market equilibrium price at which the quantity demanded equals the quantity supplied, at the lower regulated price the quantity demanded exceeds the quantity supplied which results in shortage of the commodity. However, in monopoly, fixation of a lower price can lead to a greater output up to the point of equality of price with marginal cost. This is because if left free a monopolist restricts output to raise the price of the commodity. Now, with the fixation of the maximum price which he can charge, the reason for restricting output no longer exists. Indeed, at the maximum regulated price below his free-market price *OP*, if he continues to produce the same level of output *OQ*, his profits will decline. And further with regulated maximum price *OH*, he can increase his profits by raising output to the *ON* level.

Average Cost Pricing

Now, it will be seen from the Figure 26.15 that even with the imposition of maximum price equal to the marginal cost and his producing *ON* level of output, he is making supernormal profits, as his average revenue exceeds average cost of production. Those who want to regulate monopoly to improve the distribution of income or to ensure that lowest possible price be charged from the consumers, they propose that monopolist should not be allowed to make more than *fair return* on his capital investment, especially when he is producing an essential commodity. Their proposal is to adopt the average cost pricing for regulating the product-price of monopolist. Consider Fig. 26.15. According to the average-cost pricing principle, the maximum price should be fixed at the level *OR* corresponding to which at point *G* average revenue or demand curve *DA* cuts the average cost curve *AC* at output level *OZ*. Thus with price equal to *OR* and output *OZ*, the monopolist is just covering his average cost of production. However, it may be noted that his *average cost includes normal profits or fair return on his capital investment*. This normal profits or fair return on capital is the opportunity cost of his capital, that is, the earnings which he can make elsewhere if he invests his capital in some other industry. Of course, what exactly is the fair return on capital has been a subject of severe controversy and Governments which regulate monopoly often appoint committees to decide about the fair return on capital investment which then become a part of the cost of production. To conclude, when average cost pricing principle is followed, it actually ensures fair return on capital.

Price Regulation of Natural Monopoly

Price regulation in case of a natural monopoly presents some problems. *A natural monopoly is said to exist when there occur economies of scale over a large expansion of output due to which average cost of production steadily declines and the extent of market demand for a commodity is such that it can support only one big optimum-size firm*. Under these circumstances a large-sized firm enjoying economies of scale and therefore having lower average cost of production can compete away the small-sized firms having higher average cost through setting a lower price. Thus, in case of occurrence of economies of scale upto quite a large output, one firm tends to dominate and succeed in establishing its monopoly.

Now, for a regulation of natural monopoly through marginal cost pricing a difficult problem is to faced. Fig. 26.16 illustrates the case of marginal cost pricing in case of natural monopoly. Owing to economies of scale average cost is steadily declining throughout and marginal cost curve lies below it. *AR* and *MR* are the average and marginal revenue curves representing the demand conditions. It will be seen that marginal cost curve intersects the average revenue curve at point *E*, that is, if marginal cost pricing is followed, then price will be fixed at the level OP_m at which OQ_m quantity of the commodity will be sold. A glance at the Fig. 26.16 will reveal that price OP_m is less than the average cost of production equal to *OL* at OQ_m level of output. This means that if marginal cost pricing is imposed on the monopolist he will have to incur losses equal to $LCEP_m$ and therefore will go out of business. In order that he continues to produce with price regulated at the marginal cost level, *he will have to be provided subsidies equal to the area* $LCEP_m$. This amount of subsidy will ensure him fair return on his capital investment, while producing OQ_m level of output.

However, if the Government does not want to give subsidies because of its financial constraint, average cost pricing policy may be adopted. It will be seen from Fig. 26.16 average cost curve cuts the demand or average revenue curve AR at point S and therefore according to the rule of average cost pricing, price should be fixed at the level of OP_a at which the monopolist will produce OQ_a. With price equal to the average cost, the monopolist will make fair return on capital (included in average cost). It is worth noting that unregulated monopoly price will be equal to OP and if left free and unregulated, the monopolist will produce output equal to OQ and well therefore make profits equal to $PTKJ$. Thus, *even regulation of monopoly through average cost pricing leads to the expansion in output by the monopolist to OQ_a level and thereby cause expansion of output towards economically more efficient output* (the consumer's surplus or welfare will increase by expanding output from OQ to OQ_a). In other words, loss in welfare or economic efficiency suffered in case of unregulated monopoly will be reduced under average cost pricing due to the expansion in output and lowering of price. It should be further noted that with average cost pricing, the producer would not be making any monopoly profits; he will be earning only normal profits and fair return on his capital investment. Of course, as mentioned above, full economic efficiency or maximum social welfare is reached if marginal cost pricing is adopted. But, as seen above, in a case of natural monopoly, marginal cost pricing requires provision of subsidy by the Government if he is to be induced to stay in business.

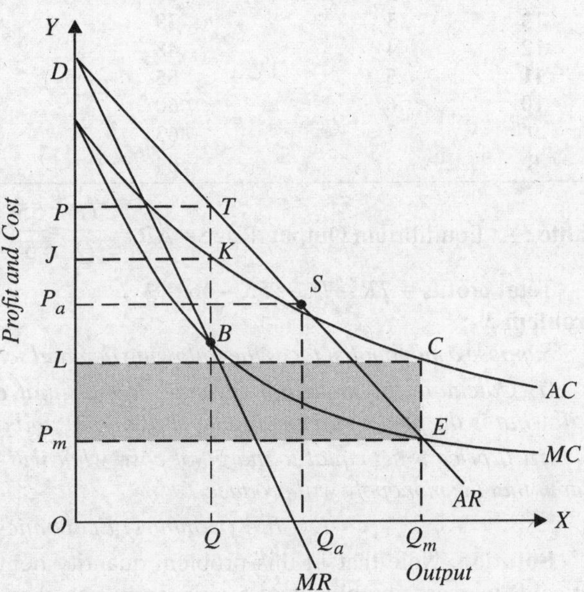

Fig. 26.16. *Regulating the Price of a Natural Monopoly*

NUMERICAL PROBLEMS

Problem 1. *The following demand function and total cost function of a monopolist are given. Calculate his marginal revenue and marginal cost. At what level of output, the monopolist will be in equilibrium. What price will be set at the equilibrium output and calculate total profits made by him.*

Solution. In order to find marginal revenue (MR) we have first to calculate total revenue (TR). $MR = TR_n - TR_{n-1}$ and marginal cost which is $TC_n - TC_{n-1}$ The monopolist will be in equilibrium at the level of output at which $MR = MC$. Calculating TR, MR and MC we have the following table.

Price (₹)	Quantity sold	Total cost (₹)
15	1	12
14	2	22
13	3	31
12	4	39
11	5	46
10	6	54
9	7	64
8	8	75

It will be seen from the table given below that $MC = MR$ when 5 units of output are sold. Therefore, to be in equilibrium the monopolist will produce 5 units to maximise his profits.

Finding Equilibrium Output

Price (P) ₹	Quantity sold (Q)	TR ($P \times Q$)	MR $TR_n - TR_{n-1}$	TC ₹	MC $TC_n - TC_{n-1}$
15	1	15	15	12	12
14	2	28	13	22	10
13	3	39	11	31	9
12	4	48	9	39	8
11	**5**	**55**	**7**	**46**	**7**
10	6	60	5	54	8
9	7	63	3	64	10
8	8	64	1	75	11

Table : At Equilibrium Output, Price = AR = $\dfrac{TR}{Q} = \dfrac{55}{5} = 11$

Total profits = $TR - TC = 55 - 46 = 9$

Problem 2

Suppose a monopolist faces the following demand schedule

(i) Calculate the marginal revenue. If marginal cost is ₹ 50 what is the profit-maximising level of output and price.

(ii) If price is set equal to marginal cost, what will be the output that the monopoly will produce ?

DU B.A (Honours) Economics 2008

Demand Schedule

Price, P	Quantity, Q
100	0
90	5
80	10
70	15
60	20
50	25
40	30
30	35

Solution. Note that in this problem quantity demanded (or sold) increases by blocks of 5 units and price changes by blocks of ₹ 10. This should be kept in mind while calculating marginal revenue.

Price (P) ₹	Quantity demanded or sold (Q)	Total Revenue (TR) ($P \times Q$)	$MR\left(=\dfrac{\Delta TR}{\Delta Q}\right)$
100	0	0	0
90	5	450	$\dfrac{450}{5}=90$
80	10	800	$\dfrac{350}{5}=70$
70	15	1050	$\dfrac{250}{5}=50$
60	20	1200	$\dfrac{150}{5}=30$
50	25	1250	$\dfrac{50}{5}=10$
40	30	1200	$\dfrac{-50}{5}=-10$
30	35	1050	$\dfrac{-150}{5}=-30$

To obtain marginal revenue we have first to calculate total revenue (*TR*) which is equal to $P \times Q$. Then marginal revenue $(MR) = \dfrac{\Delta TR}{\Delta Q}$. We have calculated *MR* above.

Given marginal cost equal to ₹ 50, the profit-maximising condition of $MC = MR$ is satisfied when 15 units of the product are sold. It will be seen from the above table that with 15 units of output sale price of the product is ₹ 70 per unit.

(*ii*) If price is set equal to marginal cost, that is, ₹ 50, then as will be seen from the above table, 25 units of output will be produced and sold by the monopolist.

Problem 3. *A monopolist faces a demand curve, $P = 100 - 2Q$. If marginal cost is constant and is equal to 20. What is the amount of profits made by the monopolist? What is dead weight welfare loss on account of monopoly?*

Solution. For monopoly equilibrium, $MR = MC$. The given demand curve is $P = 100 - 2Q$

$$TR = PQ = 100Q - 2Q^2$$

$$MR = \frac{d(TR)}{dQ} = \frac{d(PQ)}{dQ}$$

$$= 100 - 4Q$$

Equating *MR* with $MC(= 20)$ we have

$$100 - 4Q = 20$$

$$4Q = 100 - 20 = 80$$

$$Q = \frac{80}{4} = 20$$

To obtain equilibrium price we substitute $Q = 20$ in the given demand function. Thus

$$P = 100 - 2 \times 20$$

$$= 100 - 40 = ₹\ 60$$

Welfare is maximised. when at the output produced price equals marginal cost as under conditions of perfect competition. Thus, equating price with marginal cost we have

$$P = MC$$

$$100 - 2Q = 20$$

$$2Q = 100 - 20 = 80$$

$$Q = 40$$

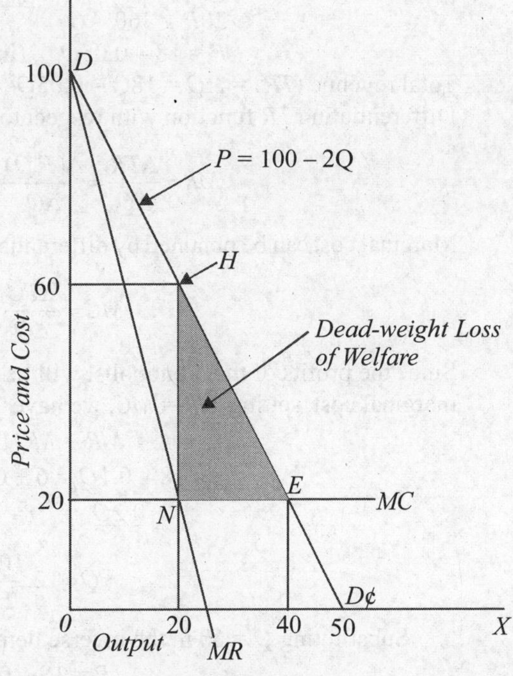

Fig. 26.17.

In Fig. 26.17 consumer's surplus with output equal to 40 and price equal to 20 is equal to the area *DPE*. Monopoly restricts output to 20 and raises price to 60 or *OP′*. Therefore, under monopoly, consumer surplus is reduced to the area *DP′H*. Thus, consumers suffer a loss of welfare (*i.e.,* consumer's surplus) equal to the area *PP′HE*. The monopolist's profits as a result of restriction of output from 40 units to 20 units and rasing of price of the product from 20 to ₹ 60, is equal to the area *PNHP′* (*i.e.,* 40 × 20 = ₹ 800). But consumers suffer a greater loss of consumer surplus equal to the area *PP′HE* , that is, *NHE* more than the gain in profits by the monopolist. The welfare loss of consumers equal to the area *NHE* represents the **dead weight loss** of welfare and is equal to

$$= \frac{1}{2}(40 \times 20) = \frac{1}{2}(800)$$
$$= 400$$

This dead-weight loss represents social cost of monopoly.

Problem 4. *Suppose the following demand and total cost functions of a monopolist are given (demand function)*

$Q = 360 - 20P$

$TC = 6Q + 0.05Q^2$ (*cost function*)

Determine equilibrium output of the monopolist. What price will be charged in this equilibrium solution ?

Solution. In order to find the profit-maximising solution we have to derive the marginal revenue and marginal cost from the demand and cost equations given above. In order to find out the marginal revenue we have to first obtain the total revenue function.

Rearranging to obtain the *inverse* demand function

$$Q = 360 - 20P$$
$$20P = 360 - Q$$
$$P = 18 - 0.05Q \quad \text{(inverse demand function)} \qquad \ldots \text{(i)}$$

Total revenue (TR) = $P.Q = 18Q - 0.05Q^2$

Differentiating TR function with respect to output Q, we can get MR. Thus,

$$MR = \frac{\Delta TR}{\Delta Q} = \frac{\Delta(P.Q)}{\Delta Q} = 18 - 0.1Q \qquad \ldots \text{(ii)}$$

Marginal cost can be obtained by differentiating the given total cost function ($TC = 6Q + 0.05Q^2$)

$$MC = \frac{\Delta TC}{\Delta Q} = 6 + 0.1Q \qquad \ldots (iii)$$

Since the profits of the monopolist will be maximised when he equates marginal revenue with marginal cost, setting $MR = MC$, we have

$$MR = MC$$
$$18 - 0.1Q = 6 + 0.1Q$$
$$0.2Q = 18 - 6 = 12$$
$$Q = 12.\frac{10}{2} = 60$$

Substituting $Q = 60$ in the inverse demand function (i) we have

$$P = 18 - 0.05 \times 60 = 15$$
$$TR = P.Q = 15 \times 60 = 900$$
$$TC = 6Q + 0.05Q^2 = 6 \times 60 + 0.05\,(60)^2$$
$$= 360 + \frac{5}{100} \times 3600 = 540$$

Profits = $TR - TC = 900 - 540 = 360$

Thus, output (Q) = 60; price (P) = 15 and profits = ₹ 360

Problem 5. *A monopolist has the following total cost function* $TC = 10 + 5Q$

(1) *If the price elasticity of demand for his products is – 2, find out pirce he will fix for his product.*

(2) *If the price elasticity of demand for his product changes to –4, how will he change his price ?*

Solution.

$$TC = 10 + 5Q$$

$$MC = \frac{\Delta TC}{\Delta Q} = 5 \qquad \ldots (1)$$

The relationship between *MR*, price and price elasticity of demand (*e*) is

$$MR = P\left(1 + \frac{1}{e}\right)$$

Given that price elasticity of demand is – 2,

$$MR = P\left(1 + \frac{1}{-2}\right) = P\left(1 - \frac{1}{2}\right)$$

$$MR = \frac{1}{2}P \qquad \ldots (2)$$

In equilibrium

$$MR = MC$$

$$\frac{1}{2}P = 5$$

$$P = 5 \times 2 = ₹\,10$$

Given that price elasticity of demand, $e = -4$

$$MR = P\left(1 + \frac{1}{-4}\right) = P\left(1 - \frac{1}{4}\right) = \frac{3}{4}P$$

In equilibrium

$$MR = MC$$

$$\frac{3}{4}P = 5; \quad P = 5 \times \frac{4}{3} = 6.67$$

Thus, with the increase in absolute value of price elasticity of demand, monopolist will reduce the price of his product.

Problem 6. *Suppose the price elasticity of demand for the product of a monopolist is –2.0. Show that price fixed by him will be twice the marginal cost of production.*

Solution. In price-output equilibrium of the monopolist

$$MR = MC$$

Since

$$MR = P\left(1 - \frac{1}{|e|}\right)$$

In equilibrium,

$$MC = P\left(1 - \frac{1}{|e|}\right)$$

If $e = 2.0$

$$MC = P\left(1 - \frac{1}{2}\right) = \frac{1}{2}P, \quad \text{or} \quad P = 2MC$$

Problem 7. *Given the following linear demand and cost functions, show that monopolist will produce half the output under perfect competition*

$Q = 300 - 2P$ (Linear demand function)

$TC = 150 + 10\,Q$

Solution.

$$TC = 150 + 10Q$$

$$MC = \frac{\Delta TC}{\Delta Q} = 10 \qquad \ldots (i)$$

Now, the given linear demand function is

$$Q = 300 - 2P$$
$$2P = 300 - Q$$
$$P = 150 - 0.5Q \quad \ldots (ii)$$
$$TR = PQ = 150Q - 0.5Q^2$$
$$MR = \frac{\Delta(PQ)}{\Delta Q} = 150 - Q \quad \ldots (iii)$$

Ouput under perfect competition is determined where $MC = P$

Thus, in equilibrium under perfect competition

$$10 = 150 - 0.5Q$$
$$0.5Q = 150 - 10 = 140$$

Hence, $$Q_{PC} = 280 \quad \ldots (iv)$$

In equilibrium under monopoly,

$$MR = MC$$

From (*iii*) we know $MR = 150 - Q$ and from (*i*) we know that $MC = 10$

Thus, in equilibrium under monopoly

$$150 - Q = 10$$
$$Q = 150 - 10 = 140$$

Thus, $$Q_m = 140 \quad \ldots (v)$$

Comparing (*iv*) nad (*v*) we find that output under monopoly is half of that under perfect competition.

QUESTIONS FOR REVIEW

1. What is monopoly? Explain the three conditions necessary for the existence of monopoly.
2. Explain the relationship between average revenue, marginal revenue and price elasticity of demand under the conditions of monopoly.
3. Explain the equilibrium of a monopoly firm? Show that price fixed by the monopolist is less than marginal cost of production.
4. How are price and output determined under monopoly? Show that under monopoly price is higher and output smaller than under perfect competition
5. *Show that a monopolist will always operate on the elastic part of the demand curve.*
   D.U. B.A. (Hons.) 2002
6. A monopolist will never be in equilibrium on the *inelastic* ($e_p < 1$) portion of the demand curve. Explain. *D.U.B.A.(H) 2001*
7. A monopolistic firm has the following total cost function and demand function.

Price (Rupees)	*Quantity demand (Units)*	*Total cost (₹)*
8	5	20
7	6	21
6	7	22
5	8	23
4	9	24
3	10	25

Explain what price will be charged and what output will be produced.

[**Hint**: From the given data of price and quantity demanded, first find out total revenue ($TR = P.Q$). From TR, marginal revenue can be obtained at different levels of quantity demanded or sold. From total cost data, MC can be obtained. Then the output level at which $MC = MR$, will be the equilibrium output and price at that will be equilibrium price. In the above table, MC is constant at Re. 1. When 6 units of sold, MR =2 and when 7, units are sold, $MR = 0$, therefore the firm should produce 6 units of output to maximise profits].

8. *How is monopoly power measured ? State Lerner's measure of degree of monopoly power. Show that degree of monopoly power is inverse of the price elasticity of demand.*

9. A simple monopolist is in equilibrium. At the point of equilibrium the coefficient of price elasticity is - 2.0 and the marginal cost is ₹ 4.0. Calculate his equilibrium price. How will this price be effected by an increase in the fixed cost of the monopolist. *D.U. B.A. (Hons). 1996.*

[**Hints** : In equilibrium, $MC = MR$

Since $MR = P(1 - \frac{1}{|e|})$

Given $MC = 4$ and e = – 2

i.e. $4 = P(1 - \frac{1}{2})$

$P = 4 \times \frac{2}{1} = 8$]

10. When a monopolist is maximising profits price is greater than marginal cost. Thus consumers will be paying more for additional units of output than it costs to produce. So why does not the monopolist produce more ? *D.U. B.A (Hons) 1988*

11. A monopolist plans to produce 100 units of a commodity and observes that the price elasticity of demand at this output is 0.5. What advice would you give to the monopolist ? *D.U. B.A. (Hons) 1993*

[**Hints.** Since MR is negative at an output level where price elasticity is less than one, the monopolist will not be making maximum profits. In fact, he may be making losses if he pursues his plan. Therefore, the mopolist should restrict output of the commodity to the level where price elasticity of demand is greater than one and where his $MC = MR$]

12. A monopolist will never sell its product at a price less than its average total cost. Is it true or false ? Give reasons. *D.U. B.A. (Hons) 1993*

13. Explain why marginal revenue of a monopolist is less than the price charged. *D.U. B.A. (Hons) 1993*

14. Explain why a perfectly competitive firm will never operate when marginal cost is declining but a monopoly firm can do so. *D.U. B.A. (Hons) 1993*

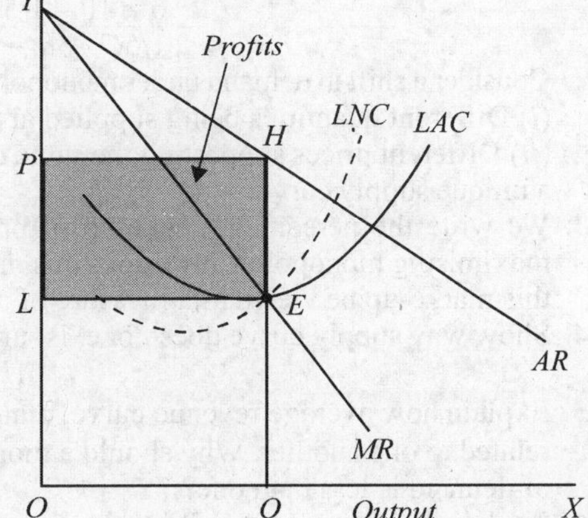

Fig. 26.18. *Monopolist working at the minimum point of LAC*

15. If a price making monopolistic firm wants to maximise its sales revenue, it should :

(*a*) set the highest price it can get.

(*b*) set the lowest price it can get.

(*c*) choose a selling price at which the elasticity of demand for its product is unity.

(*d*) choose a selling price where the extra revenue received from the last unit sold exceeds the extra cost of making that unit. *Tick the right answer.*

16. Suppose that at the profit maximising output, a monopolist price is twice as high as his marginal cost. What is the price elasticity of demand. *D.U. B.A. (Hons) 1991*

17. Does a monopolist always set his price above marginal cost ? Will he ever produce at the minimum point of long-run average curve.
[**Hints.** As explained in the text a monopolist always sets his price above marginal cost of production. Further, in the long run he usually produces less than the minimum point of LAC, that is, he produces with excess capacity in the long run. *However, this is not necessary* as it depends on the demand conditions facing him. If demand for his product is sufficiently large he may produce at or near the minimum point of long-run average curve, that is, produces with a optimal scale of plant and uses it at its full capacity (see Fig. 26.18). However, there is no certainty that monopolist will produce at this optimal level, as is the case under perfect competition. This is because unlike in competitive market conditions, no market forces operate under monopoly that compel the monopolist to work wtih optimum plant size and uses it at its full capacity.

18. A monopolist has attained equilibrium on a point on the demand curve where the coefficient of price elasticity is – 2.5 and the equilibrium price is ₹ 20.00. Calculate his marginal revenue.

19. What do you understand by 'dead-weight loss' of monopoly ? If the gains to producers be redistributed to consumers, would the "dead-weight loss" be eliminated ?
D.U. B.A (Hons) 2000

20. What is a social cost of monopoly ? If the gains to the monopolist could be redistributed to consumers, would the social cost of monopoly be eliminated. Explain briefly.
D.U. B.A. (Hons) 1995

21. Let the demand function for an industry's output be $Q = 50 - 5P$, where P is price and Q is the output. Assume constant costs at ₹ 6 per unit of output.
(*i*) How much would a profit-maximising monopolist produce ? What would be the equilibrium price ?
(*ii*) Now assume the commodity is produced by a public sector undertaking which follows the *marginal* cost *pricing principle*. What would be the optimal quantity produced ?
[**Hints** (*ii*) A public sector undertaking following marginal cost-pricing principle would set the price equal to marginal cost of production. Thus

$$Q = 50 - 5P \quad \ldots (1)$$
$$5P = 50 - Q$$
$$P = 10 - 1/5\,Q \quad \ldots (2)$$

Setting price (P) equal to marginal cost we have

$$6 = 10 - 1/5\,Q$$
$$1/5Q \text{ or } = 10 - 6 = 4, \quad Q = 20]$$

22. Consider a shift in demand under monopoly. In two separate diagrams, show that this can result in:
(*i*) Different quantities being supplied at the same price.
(*ii*) Different prices supporting the same quantity. Hence show that a monopolist does not have a unique supply curve. *D.U. B.A. (Hons) 1999*

23. We write the percentage make-up of price over marginal cost as $(P - MC)/P$. For a profit maximising monopolist, how does this mark-up depend on the elasticity of demand ? Why can this marke-up be viewd as a measure of monopoly power ? *D.U. B.A. (Hons) 1995*

24. Show why supply curve does not exist under monopoly.
[D.U.B. Com. (H), 1997, C.U., B.Com., (H), 1998, 95]

25. Explain how average revenue curve, marginal revenue curve and elasticity of demand are related to one another. Why should a monopoly firm not sell below the price where elasticity of demand is less than one? *D.U.B. Com (Hons) 2000*

26. Explain the long-run equilibrium under monopoly. Explain the adjustment process how a monopoly firm with shift from the short-run equilibrium to the long-run equilibrium position. Why is monopoly profits not likely to be eliminated ?

27. Show how monopoly causes misallocation of resources and thereby causes loss of social welfare
OR
Monopoly is said to be economically inefficient. Explain and show how it causes dead weight loss. *[D.U. B.Com. (H), 1996]*

CHAPTER 27
PRICE DISCRIMINATION

MEANING OF PRICE DISCRIMINATION

Price discrimination refers to the practice of a seller of selling the same product at different prices to different buyers. A seller makes price discrimination between different buyers when it is both possible and profitable for him to do so. If the manufacturer of a refrigerator of a given variety sells it at ₹ 5000 to one buyer and at ₹ 5,500 to another buyer (all conditions of sale and delivery being the same in two cases), he is practising price discrimination.

Price discrimination, as defined above, is not a very common phenomenon. It is very difficult to charge different prices for the identical product from the different buyers. More often, the product is slightly differentiated to practise successfully price discrimination. Thus, the concept of price discrimination can be broadened to include the sale of the various varieties of the same good at prices which are not proportional to their marginal costs. Thus, Prof. Stigler defines price discrimination as *"the sales of technically similar products at prices which are not proportional to marginal costs."*[1] On this definition, a seller is doing price discrimination when he is charging different prices from different buyers for the different varieties of the same good if the differences in prices are not the same as or proportional to the differences in the costs of producing them. For example, if a book costs the publisher ₹ 58 per unit and its deluxe edition ₹ 65 per unit, then he will be practising price discrimination if he sells the ordinary edition at ₹ 70 per unit and the deluxe edition at ₹ 130 per unit. In this case, he is said to be practising price discrimination because the price difference between the two editions (₹ 130 – 70 = 60) is greater than the cost difference between them (₹ 65 – 58 = 7).

Though this second case of price discrimination is very relevant, but is more complicated. Therefore, for the purpose of analysis given below, we shall restrict ourselves to the simple case of price discrimination—the sale of the same product at different prices to different buyers. But the conclusions arrived at in this simple case will be generally valid in case of the more complicated case mentioned above.

Three types of price discrimination may be noted. Price discrimination may be (*a*) *personal*, (*b*) *local*, or (*c*) *according to use or trade*. Price discrimination is personal when a seller charges different prices from different persons. Price discrimination is local when the seller charges different prices from people of different localities or places. For instance, producer may sell a commodity at one price at home and at another price abroad. Discrimination is according to use when different prices of a commodity are charged according to the uses to which the commodity is put. For example, the electricity is usually sold at a cheaper rate for domestic uses than for commercial purposes.

Degrees of Price Discrimination

Prof. A.C. Pigou has distinguished between the following three types of price discrimination on another ground : (*i*) price discrimination of the first degree; (*ii*) price discrimination of the second degree; and (*iii*) price discrimination of the third degree.

1. G.J. Sitgler, *The Theory of Price*, revised edition 1952, p. 215.

Price discrimination of the first degree involves maximum possible exploitation of each buyer in the interest of a seller's profits. *Price discrimination of the first degree is also called perfect price discrimination. Price discrimination of the first degree or perfect price discrimination is said to occur when the monopolist is able to sell each separate unit of the output at a different price.* Thus under discrimination of the 'first degree' every buyer is forced to pay the price which is equal to the maximum amount he would be willing to pay rather than do without the good altogether. In other words, under perfect price discrimination, the seller leaves no consumer's surplus to any buyer.

Price discrimination of the *second degree* would occur when a monopolist is able to charge *separate prices for different blocks or quantities of a commodity from buyers* and in this way he takes away a part, but not all of consumer surplus from them . Thus, under the second degree price discrimination a monopolist may charge a high price for first block of say 10 units, the medium price for the additional block of 10 units, and a lower price for additional units of a commodity. For example, a monopolist may charge from a buyer a price of ₹ 50 per unit for the first 10 units, ₹ 40 per unit for the next 10 units and ₹ 30 per unit for the additional units of the commodity.

Price discrimination of the *third degree* is said to occur *when the seller divides his buyers into two or more than two sub-markets or groups depending on the price elasticity of demand and charges a different price in each sub-market.* The price charged in each sub-market depends upon the output sold in that sub-market and the demand conditions of that sub-market. Price discrimination of the third degree is most common. A common example of such discrimination is found in the practice of a manufacturer who sells his product at a higher price at home and at a lower price abroad. Again, the third degree price discrimination is found when an electric company sells electric power at a lower price to the households and at a higher price to the manufacturers who use it for industrial purposes. In our analysis of price discrimination made below, we will assume only third degree price discrimination since this is usually more practicable as well as most commonly found in case of the real world.

WHEN IS PRICE DISCRIMINATION POSSIBLE?

Two fundamental conditions are necessary for the price discrimination to become possible. First, price discrimination can occur only *if it is not possible to transfer any unit of the product from one market to another.* In other words, a seller can practise price discrimination only when he is selling in different markets which are divided in such a way that product sold by him in the cheaper market cannot be resold in the dearer market. Price discrimination by the original seller will break down if his buyers in the cheaper market purchase the product from him and resell it to the buyers of the dearer market. Buyers in the dearer market of the original seller will instead of buying from him will buy the product from the buyers of his cheaper market. Thus, a seller can charge different prices in the two markets when there is no possibility of the product being transferred from the cheaper market to the dearer market.

Second essential condition for price discrimination to occur is that *it should not be possible for the buyers in the dearer market to transfer themselves into the cheaper market to buy the product* or *service at the lower price.* For instance, if a doctor is charging a smaller fee from the poor than from the rich, then his price discrimination will break down if a rich man can pretend to be poor and pay a poor man's charges to the doctor.

It is clear from above that for the price discrimination to become practicable, neither the unit of the good, nor the unit of demand (*i.e.*, buyer) can be transferred from one market to the other. In other words, there should not be any seepage or communication between the two markets. Thus price discrimination depends upon the ability of the seller to keep his two markets quite separate. If he is not able to keep the different markets separate, the price discrimination by him will break down. Price discrimination is possible in the following cases.

1. **The Nature of the Commodity.** The nature of the commodity or service may be such that there is no possibility of transference from one market to the other. The most usual case is the sale of direct personal services like that of a surgeon or lawyer. The surgeons usually charge different fees from the rich and the poor for the same kind of operation. This is possible for them since the service has to be delivered personally by the surgeon and therefore it cannot be transferred. Neither is it possible for the rich men to assume to be poor so easily in order to pay the smaller fee.

2. **Long Distances or Tariff Barriers.** Discrimination often occurs *when the markets are separated by long distance or tariff barriers* so that it is very expensive to transfer goods from a cheaper market to be resold in the dearer market. A monopolist manufacturer at Chennai may sell his product in one town, say Kolkata, at ₹ 20 and in another town, say Delhi, at ₹ 15. If the transport cost between Delhi and Kolkata is greater than ₹ 5 per unit it will not be worthwhile for the buyers in Delhi to transfer the goods to Kolkata on their own. Similarly, if a seller is selling his good in two different markets, say, in a home market which is protected by a traiff and in a foreign market without a tariff, he can take advantage of the tariff barrier and can raise the price of his product in the home market (which is protected by the tariff). As a result, he will be selling the product in the foreign market at a lower price than at home. *This practice of selling the product at cheaper rates abroad than at home is known as dumping.*

3. **Legal Sanction.** In some cases there may be *legal sanction for price discrimination.* For example, an electricity company sells electricity at a lower price if it is used for domestic purposes and at a higher price if it is used for commercial purposes. In this case customers are liable to be fined if they use electricity for commercial purposes if the sanction has been granted for domestic purposes only. The same is the case with railways which charge different fares for travelling in First Class, and Second Class compartments. Though the service of carrying rendered in two classes of compartments slightly differs in each case but the differences in fares are out of porportion to the differences in comforts provided. So this is a clear case of price discrimination by legal sanction. It is unlawful and a criminal offence to travel in the first class with a ticket for the second class.

4. **Preferences or Prejudices of the Buyers.** Price discrimination may become possible due to *preferences or prejudices of the buyers*. The same good is generally converted into different varieties by providing different packings, different names or labels in order to convince the buyer that certain varieties are superior to others. Different prices are charged for different varieties, although they differ only in name or label. In this way the producers are usually able to break up their market and sell the so-called superior varieties to the rich people at higher prices and the so-called inferior varieties to the poor people. Sometimes there is some actual difference in the various varieties of the good, for instance, generally there is a difference in the paper used and quality of the binding between the deluxe edition and ordinary edition of a book, but the difference in prices of the two kinds of editions is more than proportional to the extra costs incurred on the de luxe edition. So, this is a clear case of price discrimination based on the preferences or prejudices of the various buyers of the product. It is worth quoting Joan Robinson in this connection. "Various brands of a certain article which in fact are almost exactly alike may be sold as different qualities under names and labels which induce rich and snobbish buyers to divide themselves from poor buyers, and in this way the market is split up and the monopolist can sell what is substantially the same thing at several prices.[2]"

Another case of price discrimination falling in this category is that when some people prefer to buy goods in a particular locality at a higher price. For example, if a seller has two shops, one in Cannaught Place which is the most fashionable shopping centure in Delhi and another at Sadar Bazar which is very congested and ugly locality in Delhi, he may be selling the same product at a higher price in Connaught Place and at a lower price in Sadar Bazar. It is the fashionable and rich people who usually buy goods in Connaught Place and they will be prepared to pay a higher price rather than go for shopping in the congested and ugly locality of Sadar Bazar.

2. Joan Robinson, *Economics of Imperfect Competition*, pp. 180-81.

5. **Ignorance and Laziness of Buyers.** Price discrimination may become possible due to *ignorance and laziness of buyers*. If a seller is discriminating between two markets but the buyers of the dearer market are quite ignorant of that fact that the seller is selling the product at a lower price in another market, then price discrimination by the seller will persist. Price discrimination will also persist if the buyers of the dearer market are aware of the seller's act of selling the same product at a lower price in another market but due to laziness may not go for shopping in the cheaper market. In these cases if the ignorance is removed or laziness is given up, the price discrimination will break down.

6. Price discrimination may become possible *when several groups of buyers require the same service for clearly differentiated commodities*. For example, railways charge different rates of fare for the transport of cotton and coal. In this case price discrimination is possible since bales of cotton cannot be turned into loads of coal in order to take advantage of the cheaper rate of transport for coal.

Under Which Market Structure Price Discrimination is Possible? We have seen above those conditions under which price discrimination is possible. Now, the question arises under what market form a seller can practise price discrimination. It is obvious that *under perfect or pure competition no seller can charge different prices from different buyers for the same product*. Under perfect or pure competition, there are many sellers selling the homogeneous product. If any seller tries to charge from some buyers a higher price than the prevailing market price, they will refuse to buy from him and will buy the same product at the prevailing price from other sellers. It is worth noting that under conditions of perfect or pure competition price discrimination cannot prevail even if the market can be easily divided into separate parts. This is so because if conditions of perfect or pure competition prevails in each part of the whole market, then will confront a perfectly elastic demand curve in each part and will like to sell the whole of his output in that part of the market in which the highest price prevails. But the attempt by all sellers to do so would force down the price to the competitive level so that a single price will prevail throughout the whole market. But if all sellers under perfect competition, combine or arrive at some understanding, then they can discriminate prices. "*So long as market is perfect it is only if all sellers are combined or are acting in agreement that they can take advantage of the barriers between one part of a market and another to charge different prices for the same thing*."[3] However, it may be pointed out that if all sellers combine or enter into an agreement regarding price discrimination, perfect competition ceases to exist. We thus see that price discrimination is not possible under perfect competition.

Under imperfect or monopolistic competition, price discrimination can occur. The degree of price discrimination practised depends upon the degree of imperfection in the market. The imperfect or monopolistic competition prevails when the product is differentiated and every seller has some attached customers who will not move so readily from one seller to another. Therefore, if imperfect or monopolistic competition exists and also the market can be divided into different parts by a seller, then price discrimination becomes possible. It should be noted that in this case an individual seller may not produce a single variety of the product but may produce various varieties of his product and thus may break up his market into different parts and charge different prices for different varieties of his good. Price discrimination will occur only if extra prices charged for the so-called superior varieties are not proportional to the extra costs incurred on them.

But *price discrimination is more likely to occur when there is monopoly of the product by a single seller or when there is agreement among the various sellers selling the same product or service*. Monopoly exists when there are no other sellers selling the same good or its close substitutes. Therefore, monopolist is in a position to charge different prices from different buyers for the same good. Price discrimination also usually occurs when there are various sellers selling the same product or same service but there is agreement among them for charging different prices from different

3. Joan Robinson. *Economics of Imperfect Competition*, p. 179.

groups of buyers. For instance, doctors have generally some understanding with each other to charge higher fees from the rich and lower fees from the poor.

WHEN IS PRICE DISCRIMINATION PROFITABLE?

We have seen above under what conditions price discrimination is possible. Price discrimination may be possible yet it may not pay the monopolist to discriminate prices in the separate markets. In other words, the monopolist may be able to discriminate prices but it may not be profitable for him to do so. We have to see now under what conditions it is profitable for the monopolist to discriminate prices between the two markets. *Price discrimination is profitable only if elasticity of demand in one market is different from elasticity of demand in the other.* Therefore, the monopolist will discriminate prices between two markets only when he finds that the price elasticity of demand of his product is different in the different sub-markets. We shall analyse below this condition for the profitability of price discrimination.

(*a*) *When Demand Curves in the Separate Markets are Iso-elastic.* If the demand curves in the two markets are iso-elastic so that at every price the elasticity of demand in the two markets is the same, then it will not pay the monopolist to charge different prices in the two markets. Why? When elasticity of demand is the same in the two markets, it follows from the formula, $MR = AR\frac{e-1}{e}$ that marginal revenues in the two markets *at every price* (i.e. every *AR*) of the good will also be the same. Now, if marginal revenue at every price of the product is the same in the two markets, it will not be profitable for the monopolist to transfer any amount of the good from one market to the other and thus to charge different prices of the good in the two markets.

(*b*) *When Elasticity of Demand is Different in Various Markets at the Single Monopoly Price.* It will be to the advantage of the monopolist to set different prices if price elasticities of demand in the two markets at the single monopoly price are not the same. In fact, if he wants to maximise profits he must discriminate prices if the price elasticities of demand in the two markets *at the single monopoly price* are different. If the producer regards the two markets as one and charges a single monopoly price on the basis of aggregate marginal revenue and marginal cost of the output, he would not be maximising profits if elasticities of demand in the two markets at the single monopoly price are different. If price elasticity of demand is the same in the two markets at the single monopoly price, it will not pay the monopolist to discriminate between the two markets, even if the elasticities are different at other prices.

Suppose on the basis of aggregate marginal revenue and marginal cost, a monopolist fixes a single price (which is called the single monopoly price) and charges the same price in both the markets. If he now finds that price elasticity of demand at this single monopoly price is different he can increase his total profits by discriminating prices between the two markets. How is it profitable for the monopolist to charge different prices in the two markets when price elasticities of demand in them at the single monopoly price are different? This follows from the formula, $MR = AR\frac{e-1}{e}$. When average revenue in both the markets is the same, that is, when the monopolist charges a single monopoly price in both the markets, but price elasticities are different in the two markets, then marginal revenues in the two markets will be different. Suppose the single monopoly price is ₹ 15 and price elasticity of demand in markets *A* and *B* is 2 and 5 respectively. Then,

$$MR \text{ in market } A = AR_a\frac{e_a - 1}{e_a}$$

$$= 15\frac{2-1}{2} = 15 \times \frac{1}{2} = 7.5$$

$$MR \text{ in market } B = AR_b \, \frac{e_b - 1}{e_b}$$

$$= 15 \times \frac{5-1}{5}$$

$$= 15 \times \frac{4}{5} = 12$$

It is thus clear that marginal revenues in the two markets are different when elasticities of demand at the single monopoly price are different. Further, from the above numerical example, it is evident that the marginal revenue in the market in which price elasticity is higher is greater than the marginal revenue in the market where price elastiticy is lower. Now, it is profitable for the monopolist to transfer some amount of the product from the market *A* where elasticity is less and therefore marginal revenue is low to the market *B* where elasticity is higher and, therefore, marginal revenue is larger. In this way, the loss of revenue by reducing sales in market *A* by some marginal units will be smaller than the gain in revenue from increasing sales in market *B* by those units. Thus, in the above example, if one unit of the product is withdrawn from market *A*, the loss in revenue will be ₹ 7.5, while with the addition to sales by one more unit of the product in market B, the gain in revenue will be about ₹ 12. It is thus clear that the transference of some units of the product will be profitable, when there is difference in price elasticities of demand and hence in marginal revenues.

It is worth mentioning that when some units of the product are transferred from market A to market *B*, price in market *A* will rise and price in the market *B* will fall. This means that the monopolist will now be discriminating prices between the two markets.

But here a relevant question arises: how long will it be profitable for the monopolist to continue shifting his product from the market with lower elasticity of demand to the market with higher elasticity of demand? It is worthwhile for the monopolist to go on transferring units from market *A* (with lower elasticity of demand) to market *B* (with higher elasticity of demand) until the marginal revenues in the two markets become equal. This is because as long as marginal revenue in market *B* is greater than that in market *A*, he will be making addition to revenue in market *B* by selling an additional unit of the product more than the loss he will be incurring in market *A* from reducing sales by one unit. When the marginal revenues in the two markets become equal as a result of transference of some units of output, it will no longer be profitable to shift more units of output from market *A* to market *B*. When the position of equality of the marginal revenues in the two markets is reached, he will be charging different prices in the two markets—a higher price in market A with lower elasticity of demand and a lower price in market *B* with a higher elasticity of demand.

EQUILIBRIUM UNDER PRICE DISCRIMINATION

We have explained above the conditions under which price discrimination is possible and profitable. We now turn to analyse the equilibrium of a discriminating monopolist. Under simple monopoly, a single price is charged for the whole output; but under price discrimination the monopolist will charge different prices in different sub-markets. First of all, therefore, the monopolist has to divide his total market into various sub-markets on the basis of differences in price elasticity of demand in them. The monopolist can divide his total market into *several sub-markets* according to the differences in demand elasticity, but for the sake of making our analysis simple we shall explain the case when the total market is divided into *two sub-markets*.

In order to reach the equilibrium position, the discriminating monopolist has to take two decisions: (1) how much total output should be produced; and (2) how the total output should be distributed between the two sub-markets and what prices he should charge in the two sub-markets.

The same marginal principle will guide the decision of the discriminating monopolist to produce a total output as that which guides a perfect competitior or a simple monopolist. In other words, the discriminating monopolist will compare the marginal revenue with the marginal cost of the output. But he has to find out first the aggregate marginal revenue (*AMR*) of the two sub-markets taken together and then compare this aggregate marginal revenue with the marginal cost of the total output. Aggregate marginal revenue curve is obtained by summing up laterally the marginal revenue curves of the sub-markets. Consider Fig. 27.1. MR_a is the marginal revenue curve in sub-market *A* corresponding to the demand curve D_a. Similarly, MR_b is the marginal revenue curve in sub-market *B* corresponding to the demand curve D_a. Now, the aggregate marginal revenue curve *AMR*, which has been shown in diagram (*iii*) of Fig. 27.1, has been derived by adding up laterally MR_a and MR_b. This aggregate marginal revenue curve depicts the total amount of output that can be sold in the two sub-markets taken together corresponding to each value of the marginal revenue. Marginal cost curve of the monopolist is shown by the curve *MC* in Fig. 27.1 (*iii*).

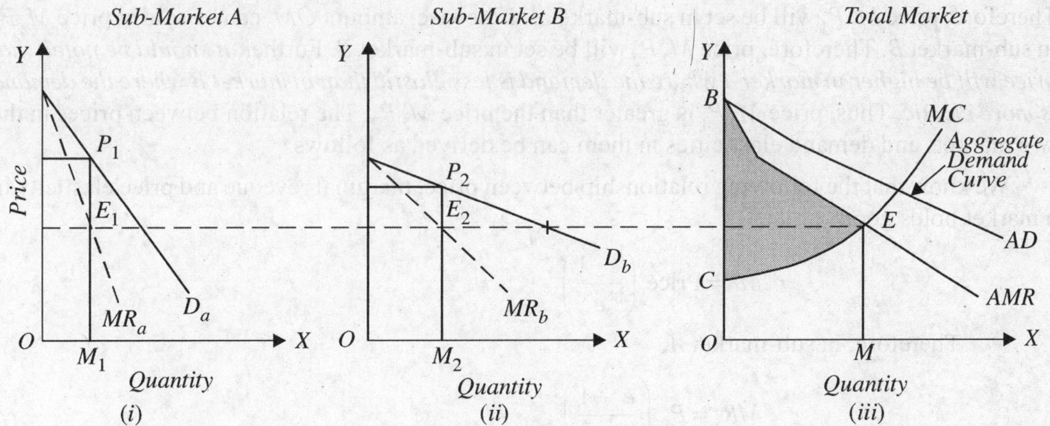

Fig. 27.1. *Determination of total output and different prices in the two markets*

The discriminating monopolist will maximise his profits by producing the level of output at which marginal cost curve *MC* intersects the aggregate marginal revenue curve *AMR*. It will be seen from Fig. 27.1 (*iii*) that profit-maximising output is *OM*, for only at *OM* aggregate marginal revenue (*AMR*) is equal to the marginal cost (*MC*) of the total output. Thus the discriminating monopolist will decide to produce *OM* level of output.

Once the total output to be produced has been determined the next task for the discriminating monopolist is to distribute the total output between the two sub-markets. He will distribute the total output *OM* in such a way that marginal revenues in the two sub-markets are equal. Marginal revenues in the two sub-markets must be equal if the profits are to be maximised. If he is so allocating the output in the two markets that the marginal revenues in the two are not equal, then it will pay him to transfer some amount of output from the sub-market in which the marginal revenue is less to the sub-market in which the marginal revenue is greater. Only when the marginal revenues in the two markets are equal, it will be unprofitable for him to shift any amount from one market to the other.

But for the discriminating monopolist to be in equilibrium it is essential not only that the marginal revenues in the two markets should be the same but that they should also be equal to the marginal cost of the whole output. Equality of marginal revenues in the two markets with marginal cost of the whole output ensures that the amount sold in the two markets will together be equal to the whole output *OM* which has been fixed by equalising aggregate marginal revenue with marginal cost. It will be seen from Fig. 27.1 [diagram (*iii*)] that at equilibrium output *OM*, marginal cost is *ME*. Now, the output *OM* has to be distributed in the two markets in such a way that marginal revenue in them should be equal to the marginal cost *ME* of the whole output. It is clear from the diagram (*i*) that OM_1 should be sold in sub-market *A*, because marginal revenue M_1E_1 at amount OM_1 is equal to marginal cost *ME*. Similarly, OM_2 should be sold in sub-market *B*, since marginal revenue M_2E_2 at amount OM_2 is equal to the marginal cost *ME* of the whole output. To conclude, demand and cost conditions being given, the discriminating monopolist will produce total output *OM* and will sell amount OM_1 in sub-market *A* and amount OM_2 in sub-market *B*. It should be carefully noted that the total output *OM* will be equal to $OM_1 + OM_2$.

Thus, for the discriminating monopolist to be in equilibrium, the following conditions must be fulfilled:

1. $AMR = MC$
2. $MR_1 = MR_2 = MC$

Another important thing to know is what prices will be charged in the two markets. It is clear from the demand curve that amount OM_1 of the good can be sold at price M_1P_1 in sub-market *A*. Therefore, price M_1P_1 will be set in sub-market *A*. Likewise, amount OM_2 can be sold at price M_2P_2 in sub-market *B*. Therefore, price M_2P_2 will be set in sub-market *B*. Further, *it should be noted that price will be higher in market A where the demand is less elastic than in market B where the demand is more elastic*. Thus, price M_1P_1 is greater than the price M_2P_2. The relation between prices in the two markets and demand elasticities in them can be derived as follows.

We know that the following relationship between price, marginal revenue and price elasticity in a market holds good.

$$MR = \text{Price}\left(\frac{e-1}{e}\right)$$

Therefore, in sub-market *A*,

$$MR_a = P_a\left(\frac{e_a-1}{e_a}\right) \qquad ...(i)$$

where P_a stands for price, MR_a for marginal revenue and e_a for price elasticity in market *A*.

Likewise, in sub-market *B*,

$$MR_b = P_b\left(\frac{e_b-1}{e_b}\right) \qquad ...(ii)$$

where P_b stands for price, MR_b for marginal revenue and e_b for price elasticity in market *B*.

Since in equilibrium under price discrimination, $MR_a = MR_b$, from (*i*) and (*ii*) we get

$$P_a\left(\frac{e_a-1}{e_a}\right) = P_b\left(\frac{e_b-1}{e_b}\right)$$

$$\frac{P_a}{P_b} = \frac{\frac{e_b-1}{e_b}}{\frac{e_a-1}{e_a}} = \frac{\left(1-\frac{1}{e_b}\right)}{\left(1-\frac{1}{e_a}\right)}$$

Suppose absolute value of price elasticity in market *A* is equal to 2 and in market *B* it is equal to 3, then

$$\frac{P_a}{P_b} = \frac{\frac{3-1}{3}}{\frac{2-1}{2}} = \frac{\frac{2}{3}}{\frac{1}{2}}$$

$$= \frac{2}{3} \times \frac{2}{1} = \frac{4}{3}$$

Thus, when elasticities in markets A and B are 2 and 3 respectively, the prices in the two markets will be in the ratio of 4 : 3.

From the forgoing analysis it follows that the following two conditions are required to be satisfied for the equilibrium of a discriminating monopolist :

(1) Aggregate Marginal Revenue (*AMR*) = Marginal Cost (*MC*) of the total output.

(2) $MR_a = MR_b = MC$.

Equilibrium under Price Discrimination in the Dumping Case

A special case of price discrimination occurs when a producer is selling in two markets, one in which he faces perfect competititon, while in the other he has a monopoly. The demand curve for the product will be perfectly elastic for him in the market in which he faces perfect competition, while the demand curve will be sloping downward in the market in which he enjoys monopoly position. Such situation might occur when a producer sells his product in his home country in which he has a monopoly and in the world market which is perfectly competitive. Equilibrium in this situation is depicted in Fig. 27.2. In the home market in which the producer has a monopoly, demand curve or the average revenue curve AR^H is sloping downward. So does the marginal revenue curve MR^H. In the international or world market in which he faces perfect competition, the demand for his product is perfectly elastic. The average revenue curve AR^W of the producer in the world market is therefore a horizontal straight line and marginal revenue curve MR^W coincides with it. *MC* is the marginal cost curve of output. Aggregate marginal revenue (*AMR*) curve in this case is the composite curve *BFED* which is the lateral summation of MR^H and MR^W. The marginal cost curve *MC* intersects the aggregate marginal revenue curve *BFED* at point *E* and equilibrium output *OM* is determined. The total output *OM* is to be distributed between the home market and the world market in such a way that marginal revenue in each market is equal to each other and to the marginal cost *ME*. It is clear from Fig. 27.2 that when amount *OR* is sold in the home market, the marginal revenue is *RF* which is equal to marginal cost *ME*. Thus, out of total output *OM*, amount *OR* will be sold in the home market. From the curve AR^H, it is clear that price OP^H will be charged in the home market.

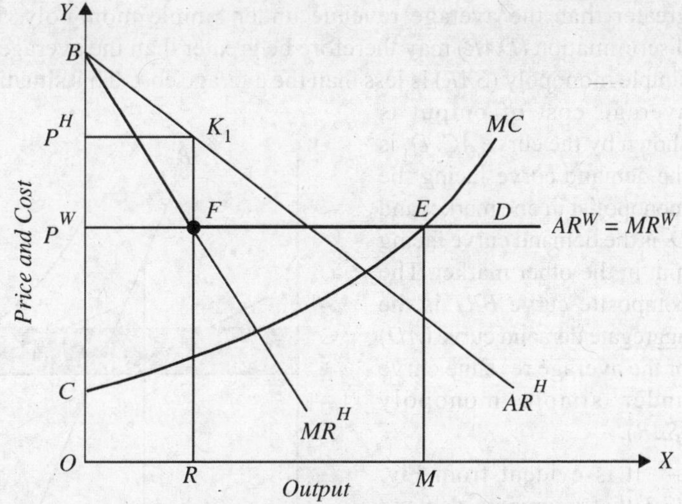

Fig. 27.2. *Equilibrium of the discriminating monopolist when he has a monopoly in the home market and faces perfect competition in the world market.*

Rest of the amount *RM* will be sold in the world market at price OP^W. Area *CEFB* represents the total profits earned by the producer from both the markets. Price in the world market OP^W is lower than the price OP^H in the home market. *When a producer charges a lower price in the world market than in the home market, he is said to be dumping in the world market.*

COMPARING BETWEEN SIMPLE MONOPOLY OUTPUT AND DISCRIMINATING MONOPOLY OUTPUT

It is now proper to discuss whether the total output of the product under price discrimination will be greater than, equal to or smaller than output under simple monopoly in which a single price for the product is charged. There is no single rule in regard to the effect of price discrimination on output. Whether the price discrimination will increase output or reduce output or leave output unchanged depends upon the various conditions.

First of all, there are cases in which no output will be produced at all under simple monopoly. In such cases, output of the product is possible only under price discrimination. The fact that the average revenue under price discrimination is greater than the average revenue under simple monopoly has an important bearing over this question. If the average cost curve of a product lies above the demand curve for it throughout its length, then it will not be profitable for the simple monopolist to produce any output at all. He may find that if he breaks up the market and charges different prices in the various separate markets, it may be profitable for him to produce some output of the product. This is so because the average revenue obtained by discriminating prices in the two separate markets is greater than the average revenue under simple monopoly. The average revenue under price discrimination (*DAR*) may therefore be greater than the average cost, when average revenue under simple monopoly (*SAR*) is less than the average cost. This situation is depicted in Fig. 27.3 in which average cost of output is shown by the curve *AC*. D_1 is the demand curve facing the monopolist in one market and D_2 is the demand curve facing him in the other market. The composite curve *BRT* is the aggregate demand curve (*AD*) or the average revenue curve under simple monopoly (*SAR*).

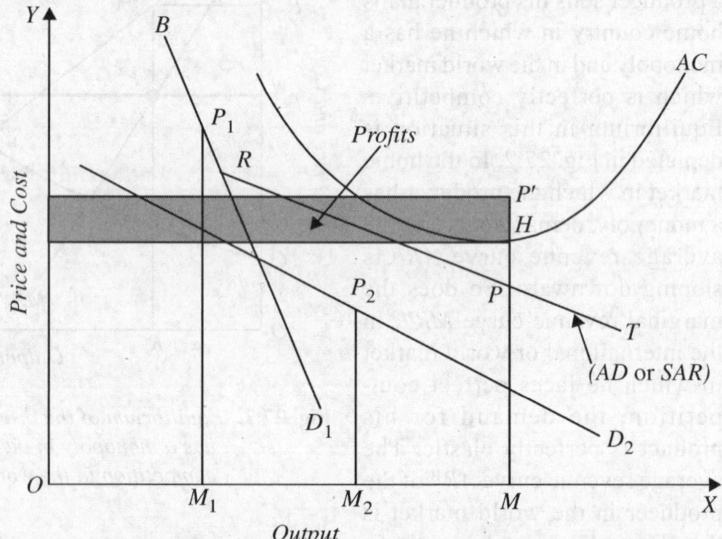

Fig. 27.3. *The case where production is possible only under price discrimination.*

It is evident from Fig. 27.3 that aggregate demand curve (*AD*) lies below the average cost curve *AC* throughout its length. In other words, average cost is greater than average revenue at all levels of output when a single price is charged for the product. For instance, suppose a simple monopolist decides to produce *OM* level of output, the single price which he will be able to charge from both markets is *MP* which is less than the average cost *MH*. So it will not pay the simple monopolist to produce *OM* output and charge the single price. But it will be profitable for him to produce *OM* output if he discriminates prices in the two markets. Under discrimination he may sell output OM_1 in one market at price M_1P_1 and output OM_2 in the other market at price M_2P_2,

($OM_1 + OM_2 = OM$). By discriminating prices in this way, suppose the average revenue obtained by him is equal to MP'. It will be seen from Fig. 27.3. That MP' is greater than the average cost MH of output OM. Thus, wereas it is not profitable for the simple monopolist to produce any output under single price system, it pays him to produce output by discriminating prices in the two markets.

We have discussed above the extreme case when there will be no output without price discrimination. In less extreme cases, output under price discrimination can be larger than the output without price discrimination.

PRICE DISCRIMINATION AND SOCIAL WELFARE

Whether price discrimination promotes social welfare or not is difficult to say, Judgement as to whether price discrimination promotes social welfare partly depends upon whether with the adoption of price discrimination, total output of the product increases or not. If we put aside the output effect of price discrimination and consider the *distribution of a given output* in case when price discrimination is being practised, then economist's verdict would be that the price discrimination will adversely affect the social welfare since it will inefficiently allocate the output between the various consumers.

A necessary condition for the achievement of consumption efficiency is that marginal rate of substitution (*MRS*) between any two goods must be the same for all consumers consuming the goods. This condition is violated under price discrimination. This is because under price discrimination, different consumers face different prices and consequently the consumers will not have the same marginal rates of substitution between the goods in their equilibrium positions. In such a case satisfaction or welfare of some consumers can be increased without causing any reduction in the satisfaction or welfare of any other consumer. Thus, *for any given level of output*, the third-degree price discrimination will have an adverse effect on the consumption efficiency, that is, efficiency in the distribution of goods between the consumers and will therefore reduce welfare of the consumers. Thus we conclude that the condition for maximum social welfare that the marginal rate of substitution (*MRS*) of different consumers between any two goods should be the same, is violated by price discrimination. G.J. Stigler rightly writes : "*given commodities and the buyers money incomes, all buyers can gain by the elimination of price discrimination, for price discrimination prevents them from reaching of the contract curve.*"[4]

But the above argument against price discrimination is based on the assumption that the quantity of commodities produced of the goods is given. The question of their optimum distribution is analysed, by taking the given outputs of the goods. On the other hand, the total output that the monopolist sells in the two markets may increase or decrease as a result of price discrimination. Even monopolist who charges a single price produces less than the efficient or socially optimum output. And when as a result of price discrimination output falls, the inefficiency of monopoly is made even worse. However, when price discrimination leads to the increase in output, the effect of this increase in output on social welfare is positive. *Thus when the price discrimination leads to the increase in output, this extra output made possible by price discrimination has socially beneficial effect flowing from it and which should be put against the possible loss of social welfare due to misallocation of goods between the two individuals*. As we explained above, price discrimination in some cases leads to the increase in output. That is, output of a commodity in some cases is greater under price discrimination than under simple monopoly. Thus, from the point of view of output, especially when a society prefers a greater output to a smaller output, *price discrimination in all those cases where it leads to the increase in output can be held to be promoting social welfare and is therefore justified.* We agree with Mrs Joan Robinson when she writes:

4. G.J. Stigler, *The Theory of Price,* revised edition, p.93, 1952.

"From the point of view of society as a whole, it is impossible to say whether price discrimination is desirable or not. From one point of view, therefore, *price discrimination must be held to be superior to simple monopoly in all those cases in which it leads to an increase of output*, and these cases are likely to be more common. But against this advantage must be set the fact that *price discrimination leads to maldistribution of resources as between different uses...* Before it is possible to say whether discrimination is desirable or not it is necessary to weigh up the benefit from the increase in output against this disadvantage. In those cases in which discrimination will decrease output, it is undersirable on both counts."[5]

There is another important reason for which in some cases price discrimination is socially justified. This is the case when under a single uniform price, no output of a commodity is produced and only under price discrimination production of a commodity becomes profitable. Rail transportation is a case in point. It has been observed by many that if railway authorities are not permitted to charge higher fare from the rich people who travel in first or AC class, then it may not be profitable for the authorities to run the railways on a single uniform fare from all, rich and poor.

We have already graphically explained above with the aid of Fig. 27.3 that when aggregate demand curve (*AD*) for curve a commodity lies below the average cost curve throughout and therefore at no level of output average revenue (when single price is charged) is greater than the average cost, the commodity would not be produced at all with a single uniform price, because it is not profitable to do so. When in such cases, price discrimination is practised the average revenue under price discrimination (*DAR*) may become higher than the average cost of production, it becomes worthwhile or profitable to produce a commodity (for full explanation of this point, see above the explanation given in connection with Figure 27.3). In those cases, where under single uniform price no production is done and under price discrimination, production becomes possible (to be more exact, becomes profitable), then price discrimination is socially desirable and justified, if the production of that commodity is considered essential and important for the society.

Price Discrimination and Equity. Finally, price discrimination may be socially justified on grounds of equity. We have discussed above the desirability of price discrimination from the point of view of *efficiency criterion*. As regards *optimum distribution of goods*, (that is, distributive efficiency), we have found that price discrimination leads to the maldistribution of the commodities between the individuals and thus violates the criterion of distributive efficiency. But this whole analysis of distributive efficiency is based on the given present distribution of income. If the present distribution of income is not considered equitable, then the mere distributive efficiency is not enough. *Under price discrimination when price is raised for the rich and is lowered for the poor, it has a redistributive effect; the poor are benefited at the expense of the rich*. Therefore, in order to reduce inequalities of personal real incomes, Government often itself practises price discrimination or when it controls prices in the private sector, it may permit or even encourage price discrimination. We thus see that equity criterion may outweigh the efficiency criterion and may make the price discrimination socially justified.

We, therefore, conclude that from the point of view of distributive efficiency and optimum allocation of resources, given the present distribution of income, price discrimination is not socially desirable. But from the point of view of expansion in output as well as for making the distribution of real incomes more equitable, price discrimination is socially justified. Thus, to pass judgement on the social desirability of a particular case of price discrimination, all these various considerations have to be weighed against each other.

NUMERICAL PROBLEM

Problem. *Suppose a discriminating monopolist is selling a product in two separate markets in whcih demand functions are*

5. John Robinson, *The Economics of Imperfect Competition, p.206.(italics added)*

$$P_1 = 12 - Q_1$$
$$P_2 = 20 - Q_2$$

The monopolist's total cost fucntion is

$$TC = 3 + 2Q$$

As the economic adviser you are asked to determine the prices to be charged in the two markets and amount of output to be sold in each market so that profits are maximised. You are also asked to calculate the total profits to be made from the strategy of price discrimination. What advise will you give?

Solution. As profits in case of price discrimination are maximised when $MR_1 = MR_2 = MC$. Therefore, we have to calculate the marginal revenue in the two markets from the given demand functions of the two markets.

Total revenue in market $1 = P_1Q_1 = 12Q_1 - Q_1^2$ . . . (1)

MR_1 in market $1 = \dfrac{\Delta(P_1Q_1)}{\Delta Q_1} = 12 - 2Q_1$

Total revenue in market $2 = P_2Q_2 = 20Q_2 - Q_2^2$ . . . (2)

MR_2 in market $2 = \dfrac{\Delta(P_2Q_2)}{\Delta Q_2} = 20 - 2Q_2$

We can derive the marginal cost from the total cost function

$$TC = 3 + 2Q$$

$$MC = \frac{\Delta TC}{\Delta Q} = 2$$

Profit-maximising amounts of output to be sold in the two markets are determined by applying the equilibrium condition $MR_1 = MR_2 = MC$ and solving the following equations:

$$MR_1 = MC$$
$$12 - 2Q_1 = 2$$
$$2Q_1 = 12 - 2 = 10$$
$$Q_1 = 5$$
$$MR_2 = MC$$
$$20 - 2Q_2 = 2$$
$$2Q_2 = 20 - 2$$
$$Q_2 = 18/2 = 9$$

Substituting these equilibrium outputs, Q_1 and Q_2 in the demand functions, we obtain the profit-maximising prices:

$$P_1 = 12 - Q_1 = 12 - 5 = 7$$
$$P_2 = 20 - Q_2 = 20 - 9 = 11$$

Total profits can be obtained in the usual way

Total profits $\pi = (TR_1 + TR_2) - TC$

$$= (P_1Q_1 + P_2Q_2) - (3 + 2Q)$$

$$= 7(5) + 11(9) - [3 + 2(5 + 9)]$$

$$= (35 + 99) - 31 = 103$$

QUESTIONS FOR REVIEW

1. A monopolist is able to spearate two markets. In one market, the demand can be expressed as $Q_1 = 14 - P_1$. In the second market the demand is $Q_2 = 20 - 2P_2$.
   The monopolist's *MC* (Marginal Cost) equals 4. Find the profit maximising output, its allocation between the two markets and the prices charged. What happens to profit ?
2. Define price discrimination. Under what conditions can a monopolist discriminate between different buyers in fixing the price of his product? Under what conditions price discrimination is profitable ? *(C.U., B.Com. (H), 1996)*
3. Explain the conditions under which monopolistic price discrimination is both possible and profitable. *(D.U.B.A.(Hons)1990, C.U., B.Com., (H), 1994)*
4. Discuss the equilibrium of a monopolist, if one of the markets in which he operates is perfectly competitive.
   [**Hint.** This is the situation just like the dumping case discussed in the text]
5. A discriminating monopolist finds price elasticity of demand of his product –2.0 in one market and –1.5 in the other. What would be the ratio of prices charged by him in the two markets ?
6. When will a monopolist discriminate prices between buyers of his product ? *Is price discrimination socially justified* ? *D.U. B.Com .(Hons)* 2000
7. Suppose a monopolist sells his product in the home market and also exports a part of it. The foreign elasticity of demand for its product is below that of the domestic market. If consumer arbitrage betwen the two markets is impossible, how do the domestic and foreign prices compare ?

CHAPTER 28

PRICE UNDER MONOPOLISTIC COMPETITION

IMPERFECT COMPETITION: MONOPOLISTIC COMPETITION AND OLIGOPOLY

Perfect competition and monopoly are rarely found in the real world and thus they do not represent, for the most part, the acutal market situations. Therefore, the conclusions which follow from the theories of pure competition were found to be inapplicable to the behaviour of business firms in the actual world. For instance, in the real world, firms were found to be enjoying 'internal economies of scale' which were incompatible with the theory of perfect competition. The urgent need was therefore felt to reformulate the theory of price so as to bring it nearer to the actual world. This was accomplished by E.H. Chamberlin and Joan Robinson who worked quite independently and brought out simultaneously "*The Theory of Monopolistic Competition*" and "*The Economics of Imperfect Competition*" respectively. Joan Robinson got the clues for her theory of imperfect competition from who in his article "The Laws of Returns under Competitive Conditions" in *Economic Journal* asserted, "*It is necessary, therefore, to abandon the path of free competition and turn in the opposite direction, namely, towards monopoly.*"

Monopolistic competition theory of Prof. Chamberlin and imperfect competition theory of Joan Robinson, though similar in various ways differ in some important respects. The nutshell of these theories, especially of the theory of monopilstic competition, is that the pure competition and pure monopoly are the two opposite limiting cases, lying between which is a series of intermediate cases, which differ from each other in relative strengths of monopoly and competitive elements, or in other words, in *degrees of imperfection*. It may be noted that the extreme limit of monopoly is reached when a seller does not face competition from any substitute product. Since every good has to complete with others for buyer's money income, every good is a substitute of others to some extent. It follows that the extreme limit of monopoly, which Sraffa denotes as *pure monopoly*, will be reached only when a single person or agency comes to have a "control over the supply of all economic goods". At the other extreme is pure or perfect competition in which case an individual seller has to compete with the products (of rival sellers) which are perfect substitutes of his own product, since products of all sellers are completely identical or homogeneous. Between these two extremes of pure monopoly and pure competition, there are all gradations in which both the monopolistic and competitive elements are present. In the terminology of Robinson and Chamberlin, pure monopoly in the sense of a single seller of a product which has got no close substitutes is an extreme form of imperfect competition. *The two important forms of imperfect competition are : (a) monopolistic competition, (b) oligopoly.*

The fundamental distinguishing feature of imperfect competition is that unlike that under perfect or pure competition, the demand curve confronting an individual firm under it slopes down-ward. As a result, the marginal revenue curve lies below it. Marginal revenue curve plays such a crucial role because of its definite relation with price and elasticity ($MR = \text{Price}\ \frac{e-1}{e}$). It is the nature of this relation which distinguishes a state of competititon that is pure from one that is impure or imperfect.

The difference between mrginal revenue and price at an output level depends upon the size of price elasticity of demand at it. Under perfect competition, price is equal to marginal revenue since price elasticity of demand is infinite, while under imperfect competition, price is greater than marginal revenue since price elasticity of demand is less than infinite. Therefore, the difference between the price and marginal revenue (or marginal cost) at equilibrium output is regarded as the *degree of imperfection or degree of monopoly*. Thus, the relative magnitudes of price and marginal revenue at equilibrium output help us to distinguish between different degrees of imperfection or monopoly power in various market structures. The greater the difference between the price and the marginal revenue, the greater the degree of imperfection or monopoly element and *vice versa*.

PRODUCT DIFFERENTIATION AND MONOPOLISTIC COMPETITION

The concept of monopolistic competition put forth by Chamberlin is a true revolutionary as well as more realistic than either perfect competition or pure monopoly. Before Chamberlin, monopoly and competition were regarded as two mutually exclusive alternatives; one would be absent when the other exists. On the other hand, according to Chamberlin, in most of the real world economic situations, both monopoly and competitive elements are present. Chamberlin's concept of monopolistic competition is thus a blending of competition and moncpoly.

The distinguishing feature of monopolistic competition which makes it as a blending of competition and monopoly is the *differentiation of the product*. This means that the products of various firms are not homogeneous but differentiated though they are closely related to each other. Product differentiation does not mean that the products of various firms are altogether different. They are only slightly different so that they are quite similar and serve as close substitutes of each other. When there is any degree of differentiation of products, monopoly element enters the situation. And, the greater the differentiation, the greater the element of monopoly involved in the market situation. When there is a large number of firms producing differentiated products, each one has a monopoly of its own product but is subject to the competition of close substitutes. Since each is a monopolist and yet has competitors, we have a market situation which can be aptly described as "monopolistic competition." It is thus clear that monopolistic competititon involves both the monopoly and competitive elements. Here, it is worth quoting Chamberlin himself "With differentiation appears monopoly and as it proceeds further, the element of monopoly becomes greater. Where there is any degree of differentiation whatever, each seller has an absolute monopoly of his own product, but is subject to the competition of more or less imperfect substitutes. Since each is a monopolist and yet has competitors we may speak of them as 'competing monopolists' and with peculiar appropriateness, of the forces at work as those of monopolistic competition."[1]

It is thus clear from above that in monopolistic competition, products are not identical as in perfect competition, but neither are they remote substitutes as in monopoly. The products of various sellers are fairly similar (but not the same) and serve as close substitutes of each other. Every seller has a monopoly of his own differentiated product but he has to face a stiff competition from his rival sellers which are selling close substitutes of his product.

Many examples of product differentiation can be given from the Indian scene. For instance, in India there are various manufacturers of toothpaste which produce different brands such as Colgate, Binaca, Forhans, Pepsodent, Signals, Neem etc. Thus, the manufacturer of 'Colgate' has a monopoly of producing it (nobody else can produce and sell the toothpaste with the name 'Colgate') but he faces competition from the manufacturers of Forhans, Binaca, Pepsodent etc. which are close substitutes of Colgate. The manufacturer of Colgate cannot therefore decide about its price-output policies without considering the possible reactions of rival firms producing close substitutes. Other examples of monopolistic competition are the producers of bathing soaps (Lux, Godrej, Breeze, Hamam, Palmolive,

1. E.H. Chamberlin, *The Theory of Monopolistic Competition*, 7th ed., 1956, p. 904,

Jai etc.); the manufacturers of tooth brushes (Colgate, Dr. West's, Wisdom, Binaca, etc.); retailers' shops in the towns; barbers' shops in the towns, etc. We thus see that monopolistic competititon corresponds more to the real world economic situation than perfect competition or monopoly.

A general class of product is differentiated if a basis exists for preferring goods of one seller to those of others. Such a basis for preference may be real or fancied, it will cause differentiation of the product. When such differentiation of the product exists, even if it is slight, buyers will be paired with sellers not in a random fashion (as in perfect competition) but according to their preferences, There are, broadly speaking, two bases of product differentiation. Firstly, differentiation may be based upon certain *characteristic of the product itself* such as exclusive patented features, trade marks and trade names, peculiarities of packages or wrappers if any, or difference in quality, design, colour or style. Real qualitative differences like those of materials used, design and workmanship are no doubt important means of differentiating products. But imaginary differences created through advertising, the use of attractive packets, the use of trade marks and brand names are more usual methods by which products are differentiated, even if physically they are identical or almost so. Secondly, Differentiation may be based upon the *conditions surrounding the sale of the product*. This means that product is differentiated if the services rendered in the process of selling the product by one seller or firm are not identical with those rendered by any other seller or firm. Thus, in retail trade to take only one instance, the conditions surrounding the sale of the product include "the convenience of the seller's location, the general tone or character of his establishment, his way of doing business, his reputation for fail dealing, courtesy, efficiency, and all others which attach his customers either to himself or to those employed by him".[2] If these and other intangible factors surrounding the sale of the product are different in case of different sellers, the product in each case will be different since the buyers take these intangible factors into consideration while making purchases. These factors like the patents, trade marks etc. serve as a basis for preference.

If the above two aspects of differentiation are borne in mind, then it will be quite manifest that virtually all products in the real world are more or less differentiated. It should be carefully noted that even when product differentiation is caused by the circumstances surrounding the sale of the products, both monopolistic and competitive elements are present. In retail trade, to take the above example, every product is rendered differentiated by the individuality of the establishment in which it is sold including the location factor, the general tone, reputation, goodwill of the establishment, way of dealing with customers etc. Each retail trader has complete and absolute control over the supply of his 'product' when this is taken to include location factor, general tone and reputation, way of dealing with the customers etc; this is the monopolistic aspect. Again, every retail trade is subject to the competition of other 'products' sold under different conditions and at different locations; this is the competitive aspect. Thus, as in the case of patents and trade marks, both monopoly and competition are present in the case of 'products' differentiated by the conditions surrounding their sale.

Under monopoly, the demand curve for the product of the monopolist is given. But the demand curve for the product of an individual producer or seller under monopolistic competition cannot be taken as given since there exist, in monopolistic competition, close competitive interrelationships among various producers selling close substitutes. The demand curve for the product of an individual depends upon the nature and prices of its closely competing substitues. Thus, according to Chamberlin, "Monopolistic competition concerns itself not only with the problem of an *individual equilibrium* (the ordinary theory of monopoly), but also with that of a *group equilibrium* (the adjustment of economic forces within a group of competing monopolists, ordinarily regarded as merely a group of competitors). In this, it differes both from the theory of competition and from the theory of monopoly".[3]

2. *Ibid.*, p. 69,
3. E.H. Chamberlin, *op. cit.*, p. 68

This leads us to Chamberlin's concepts of group and group equilibrium. Chamberlin has used the word group instead of industry in order to draw distinction between pure competition and monopolistic competition. The word industry is generally used in economics in the context of pure or perfect competition and *refers to a collection of firms that produce homogeneous products*. But, as explained above, under monopolistic competition we have product hetrogeneity. Therefore, the word 'industry' loses its significance under monopolistic competition. Therefore, Prof. Chamberlin has used the word 'group'. *By group he means "a number of producers whose goods are fairly close substitutes*". We thus see that the Chamberlin's 'group' refers to a collection of firms that produce closely related but not identical products. It became necessary for Prof. Chamberlin to evolve the concept of 'group' as distinct from industry because in the case of group where products are differentiated there are special problems which are absent in the case of industry where products are homogeneous. The various firms in a given group produce products which are close substitutes of each other and therefore compete with each other in the market. The demand for the product of one producer depends upon the price and the nature of the products of his close rivals. These competitive inter-relationships among firms are absent in the case of an industry under pure or perfect competition since all of them produce homogeneous products and each of them can sell his entire supply of the product at the going market price.

An individual producer in the 'group' under monopolistic competition cannot therefore be treated in complete isolation from his rivals producing close substitutes, since the demand for his product depends upon the nature and price of the close substitutes. To quote Prof. Chamberlin: "From our point of view, each producer within the group is a monopolist, yet his market is interwoven with those of his competitors, and he is no longer to be isolated from them"[4].

Lastly, it may be noted that the sense in which the word 'competition' is generally used in the business world is in conformity more with the monopolistic competition than with the pure competition. The phrases such as price cutting, under-selling, unfair competition, meeting competition, cut-throat competition, securing market etc. often used in the business world are irrelevant in the context of pure competition. This is because each seller under so-called 'pure competition' accepts the going market price and can dispose of his entire supply of the product without affecting price. For a pure competitor there is no problem of choosing a price policy, problem of advertising in order to attact customers. Thus, the competition in the ordinary sense of the competitive phrases, just noted above, has meaning only in the context of monopolistic competition (and also of oligopoly) where products of various sellers are differentiated but closely related. Sellers really compete in the ordinary sense of the word only under the monopolistic competition and oligopoly.

IMPORTANT FEATURES OF MONOPOLISTIC COMPETITION

It is important to understand the important characteristics of monopolistic competition. The knowledge of these features will enable the students to know how this form of market structure is different from perfect competition and oligopoly. We explain below its important features.

1. A large number of firms. The first important feature of monopolistic competition is that under it there are a relatively large number of firms each satisfying a small share of the market demand for the product. Because there is a large number of firms under monopolistic competition, there exists stiff competition between them. Unlike perfect competition these large number of firms do not produce identical products. Instead, they produce and sell differentiated products which are close substitutes of each other. This makes the competition among firms real and tough.

4. E.H. Chamberlin, *op. cit.* p.8.

Further, the fact that there is a large number of firms under monopolistic competition, size of each firm will be relatively small. This is unlike oligopoly where there are a few firms of big size.

2. Product differentiation. The second important feature of monopolistic competition is that the products produced by various firms are not identical but are slightly different from each other. Though different firms make their products slightly different from others, they remain close substitutes of each other. In other words, the products of various firms working under monopolistic competition are not the same but are similar. Therefore, their prices cannot be very much different from each other. It is because of the fact that their products are similar and close substitutes of each other that the various firms under monopolistic competition compete with each other.

3. Some influence over the price. Each firm under monopolistic competition produces a product variety which is close substitute of others. Therefore, if a firm lowers the price of its product variety, some customers of other product varieties will switch over to it. This means as it lowers the price of its product variety, quantity demanded of it will increase. On the other hand, if it raises the price of its product, some of its customers will leave it and buy the similar products from its competing firms. This implies that demand curve facing a firm working under monopolistic competition slopes downward and marginal revenue curve lies below it. This means that under monopolistic competition an individual firm is not a price taker but will have some influence over the price of its product. If it fixes a higher price, it will be able to sell a relatively smaller quantity of output. And if it fixes a lower price, it will be able to sell more. *Thus under monopolistic competition, a firm has to choose a price-output combination which will maximise its profits*.

4. Non-price competition: Expenditure on advertisement and other selling costs. An important feature of monopolistic competition is that firms incur a considerable expenditure on advertisements and other selling costs to promote the sales of their products. Promoting sales of their products through advertisement is an important example of non-price competition. The expenditure incurred on advertisement is prominent among the various types of selling costs. The advertisement and other selling outlays by a firm change the demand for its product as well as its costs. Like the adjustments of price and product, a seller under monopolistic competition will also adjust the amount of his advertisement expenditure so as to maximise his profits. This problem of adjusting one's selling outlay is unique to monopolistic competition, because the firm under perfect competition has not to incur any expenditure on advertisement. The advertisement expenditure by a purely competitive firm will be without purpose since it can sell as much amount as it pleases at the going market price without any advertisement expenditure. The rival firms under monopolistic competition keenly compete with each other through advertisement by which they change the consumers' wants for their products and attract more customers. Thus, a full explanation of the equilibrium under monopolistic competition must also involve equilibrium in regard to the amount of expenditure on advertisement and other sales promotion activities.

5. Product variation. Another form of non-price competition which a firm under monopolistic competition has to face is the variation in products by various firms. A firm, under perfect competition, does not confront this problem, for the product is homogeneous under perfect competition. The problem of product variation under monopolistic competition exists because there is differentiation of products of various firms. The firm will try to adjust its product so as to conform more to the wishes of the buyers. The variation of the product may refer to a change in the quality of the product itself, technical changes, a new design, better materials, and it may mean only a new package or container. It may also mean more prompt or courteous service, and a different way of doing business. The amount of the product which a firm will be able to sell in the market depends in part upon the manner in which its product differs from others. Where the possibility of product differentiation exists, sales depends upon the skill with which a product is distinguished from others and made to appeal to a particular group of buyers. The profit maximisation principle applies to the choice of the nature of the product as to its price. In other words, a firm will choose that nature of the product, given its price, which gives it maximum profits. Therefore, in a full explanation of the firm's equilibrium under monopolistic competition we have also to explain product equilibrium in addition to price equilibrium and selling-costs equilibrium.

6. Freedom of entry and exit. This is another important feature of monopolistic competition. In a monopolistically competitive industry it is easy for the new firms to enter and the existing firms to leave it. Free entry means that when in the industry existing firms are making super-normal profits, the new firms enter the industry which leads to the expansion of output. As a result, price of product tends to fall in the long run. However, it may be noted that under monopolistic competition entry may not be as easy or free as under perfect competition. Whereas under perfect competition the new firms which enter the industry can produce identical products, but under monopolistic competition, the new firms can produce only new brands or product varieties which may initially find it difficult to compete with the already well-established brands and product varieties.

The Nature of Demand and Marginal Revenue Curves under Monopolistic Competition

It is important to understand the nature of the demand curve facing an individual firm under monopolistic competition. As has been explained in previous chapters, demand curve facing a firm working under perfect competition is perfectly elastic at the ruling market price since it has absolutely no control over the price of the product. On the contrary, a firm working under monopolistic competition enjoys some control over the price of its product since its product is somewhat differentiated from others. If a firm under monopolistic competition raises the price of its product, it will find some of its customers going away to buy other products. As a result, the quantity demanded of its product will fall. On the contrary, if it lowers the price, it will find that buyers of other varieties of the product will start purchasing its product and as a result the quantity demanded of its product will increase. It therefore follows that the demand curve facing an individual firm under monopolistic competition slopes downward.

If a firm working under monopolistic competition wants to increase the sales of its product, it must lower the price. It can raise the price if it is prepared to sacrifice some sales. To put it in another way, a firm working under monopolistic competition can lower the price by increasing its level of sales and output. A purely competitive firm merely adjusts the quantity of output it has to produce, price being a given and constant datum for it. But a firm working under monopolistic competition faces a more complicated problem. It cannot merely adjust quantity at a given price because each quantity change by it will bring about a change in the price at which the product can be sold. Consider Fig. 28.1. *DD* is the demand curve facing an individual firm under monopolistic competition. At price *OP* the quantity demanded is *OM*. Therefore, the firm would be able to sell *OM* quantity at price *OP*. If it wants to sell greater quantity *ON*, then it will have to reduce price to *OL*. If it restricts its quantity to *OG*, price will rise to *OH*. Thus, every quantity change by it entails a change in price at which the product can be sold. Thus *the problem faced by a firm working under monopolistic competition is to choose the price-quantity combination which is optimum for it, that is, which yields it maximum possible profits*.

Demand curve facing a firm will be his average revenue curve. Thus the average revenue curve of the monopolistically competitive firm slopes downward throughout its length. Since average revenue curve slopes downward, marginal revenue curve lies below it. This follows from usual average-marginal relationship. The implication of marginal revenue curve lying below average revenue curve is that the marginal revenue will be less than the price or average revenue. When a firm working under monopolistic competition sells more, the price of its product falls; marginal revenue therefore must be less than price. In Fig. 28.2 *AR* is the average revenue curve of the firm under monopolistic competition and slopes downward. *MR* is the marginal revenue curve and lies below *AR* curve. At quantity *OM*, average revenue (or price) is *OP* and marginal revenue is *MQ* which is less than *OP*. In an earlier chapter we have explained that average and marginal revenues at a level of output are related to each other through price elasticity of demand and in this connection we derived the following formula:

$$MR = AR\left(\frac{e-1}{e}\right), \text{ where } e \text{ stands for price elasticity of demand.}$$

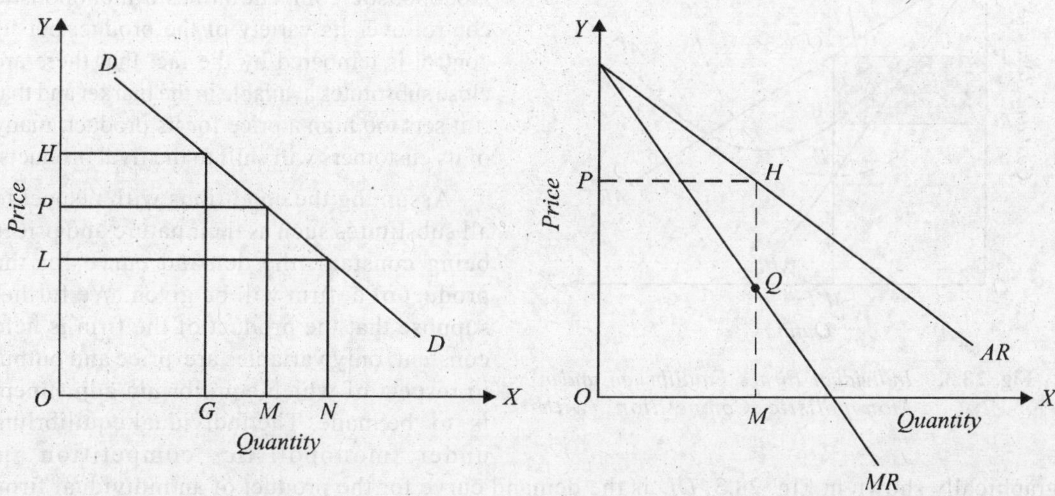

Fig. 28.1. *Demand Curve Facing a Monopolistically Competitive Firm*

Fig. 28.2. *Average and Marginal Revenue Curves under Monopolistic Competition*

PRICE-OUTPUT EQUILIBRIUM UNDER MONOPOLISTIC COMPETITION

A firm under monopolistic competition has to face various problems which are absent under perfect competition. Since the market of an individual firm under perfect competition is completely merged with the general one, it can sell any amount of the good at the ruling market price. But, under monopolistic competition, individual firm's market is isolated to a certain degree from those of its rivals with the result that its sales are limited and depend upon (1) its price, (2) the nature of its product, and (3) the advertising outlay it makes. Thus, the firm under monopolistic competition has to confront a more complicated problem than the perfectly competitive firm. Equilibrium of an individual firm under monopolistic competition involves equilibrium in three respects, that is, in regard to the price, the nature of the product, and the amount of advertising outlay it should make.

Equilibrium of the firm in respect of three variables simultaneously—price, nature of product, selling outlay—is difficult to discuss. Therefore, the method of explaining equilibrium in respect of each of them separately is adopted, keeping the other two variables given and constant. Moreover, as noted above, the equilibrium under monopolistic competition involves "individual equilibrium" of the firms as well as "group equilibrium". We shall discuss these two types of equilibrium first in respect of price and output and then in respects of product and advertising expenditure adjustments.

Individual Firm's Equilibrium under Monopolistic Competition

The demand curve for the product of an individual firm, as noted above, is downward sloping. Since the various firms under monopolistic competition produce products which are close substitutes of each other, the position and elasticity of the demand curve for the product of any of them depend upon the availability of the competing substitutes and their prices. Therefore, the equilibrium adjustment of an individual firm cannot be defined in isolation from the general field of which it is a part. However, for the sake of simplicity in analysis, conditions regarding the availability of substitute products produced by the rival firms and prices charged for them are held constant while the equilibrium adjustment of an individual firm is considered in isolation. Since close substitutes for its product are available in the market, the demand curve for the product of an individual firm working under conditions of monopolistic competition is fairly elastic. Thus, although a firm under monopolistic competition has a monopolistic control over its variety of the product but its control is tempered by the fact that there are close substitutes available in the market and that if it sets too high a price for its product, many of its customers will shift to the rival products.

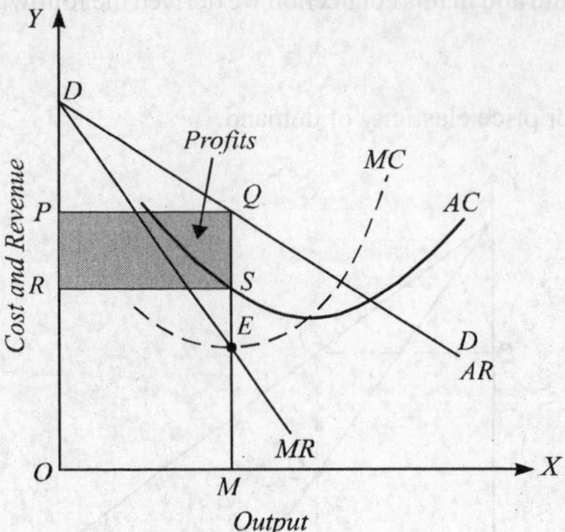

Fig. 28.3. *Individual Firm's Equilibrium under Monopolistic Competition (with*

Assuming the conditions with respect to all substitutes such as their nature and prices being constant, the demand curve for the product of a firm will be given. We further suppose that the product of the firm is held constant, only variables are price and output in respect of which equilibrium adjustment is to be made. The individual equilibrium under monopolistic competition is graphically shown in Fig. 28.3. *DD* is the demand curve for the product of an individual firm, the nature and prices of all substitutes being given. This demand curve *DD* is also the average revenue (*AR*) curve of the firm. *AC* represents the average cost curve of the firm, while *MC* is the marginal cost curve corresponding to it. It may be recalled that average cost curve first falls due to internal economies and then rises due to internal diseconomies.

Given these demand and cost conditions a firm will adjust his price and output at the level which gives it maximum total profits. Theory of value under monopolistic competition is also based upon the profit maximisation principle, as is the theory of value under perfect competition. Thus a firm in order to maximise profits will equate marginal cost with marginal revenue. In Fig. 28.3, the firm will fix its level of output at *OM*, for at *OM* output marginal cost is equal to marginal revenue. The demand curve *DD* facing the firm in question indicates that output *OM* can be sold at price *MQ* = *OP*. Therefore, the determined price will evidently be *MQ* or *OP*. In this equilibrium position, by fixing its price at *OP* and output at *OM*, the firm is making profits equal to the area *RSQP* which is maximum. It may be recalled that profits *RSQP* are in excess of normal profits because the normal profits which represent the minimum profits necessary to secure the entrepreneur's services are included in average cost curve *AC*. Thus, the area *RSQP* indicates the amount of supernormal or economic profits made by the firm.

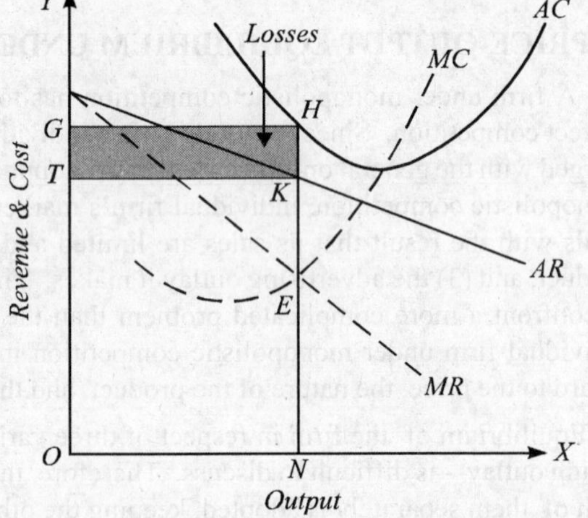

Fig. 28.4. *Individual Firm's Equilibrium under Monopolistic Competition (with*

In the short-run, the firm, in equilibrium, may make supernormal profits, as shown in Fig. 28.3 above, but it may make losses too if the demand conditions for its product are not so favourable relative to cost conditions. Fig. 28.4 depicts the case of a firm whose demand or average revenue curve *DD* for the product lies below the average cost curve throughout indicating thereby that no output of the product can be produced at positive profits. However, the firm is in equilibrium at output *ON* and setting price *NK* or *OT*, for by adjusting price at *OT* and output at *ON*, it is rendering the losses to the minimum. In such an unfavourable situation there is no alternative for the firm except to make the best of the bad bargain.

We thus see that a firm in equilibrium under monopolistic competition, as under pure or perfect competition, may be making supernormal profits or losses depending upon the position of the demand curve relative to the position of the average cost curve. Further, a firm may be making only normal profits even in the short run if the demand curve happens to be tangent to the average cost curve.

It should be carefully noted that in individual equilibrium of the firm both in Fig. 28.3 and Fig. 28.4, the firm having once adjusted price at *OP* and *OT* respectively will have no tendency to vary the price any more. If it varies its price upward, the loss due to fall in quantity demanded will be more than made up by the higher price. If it cuts down its price, the gain due to the increase in quantity demanded will be less than the loss due to the lower price. Hence, price will remain stable at *OP* and *OT* in the two cases respectively.

Long-Run Firm's Equilibrium and Group Equilibrium under Monopolistic Competition

We may now turn to see how the 'group' comes to be in equilibrium. In other words, We have now to explain how the equilibrium adjustment of prices and outputs of a number of firms whose products are close substitues comes about. As explained above, each firm within a group has monopoly of its own particular product, yet its market is interwoven with those of his competitors who produce closely related products. The price and output decisions of a firm will affect his rival firms who may in turn revise their price and output policies. This dependence of the various producers upon one another is aprominent feature of monopolistic competition. The question now is: what characterises the system of relationship into which the group tends to fall as a result of producers's influence on each other?

A difficulty faced in describing the group equilibrium is the vast diversity of conditions which exist in respect of many matters between the various firms constituting the group. The product of each firm has special characteristics and adapted to the tastes and preferenes of its customers. The qualitative differences among the products lead to the large variations in cost and demand curves of the various firms. The demand curves of the products of various firms differ in respect of elasticity as well as position. Similary, cost curves of the various firms differ in respect of shape and position. As a result of these heterogeneous conditions surrounding each firm, there will be differences in prices, in outputs (scales of production) and profits of the various firms in the group. Putting this matter in another way, E.H. Chamberlin says: "The differentiation of the product is not, so to speak, 'uniformly spaced'; it is not distributed homogeneously among all the products which are grouped together. Each has its own individuality, and the size of its market depends on the strength of the preference for it over other varieties".[5]

In order to simplify the analysis of equilibrium, Chamberlin ignores these diverse conditions surrounding each firm and takes an assumption what has been called "*uniformity assumption*". Thus Chamberlin says: "We, therefore, proceed under the heroic assumption that both demand and cost curves for all the 'products' are uniform throughout the group."[6] Chamberlin points out that by taking this assumption we do not reduce the differentiation of products. Under the

5. *Op. cit.* p.82
6. *Ibid.*, p. 82

uniformity assumption, "It is required only that *consumer's preferences be evenly distributed among the different varieties*, and that differences between them be not such as to give rise to differences in cost."[7]

Further, to facilitate exposition of his theory, Chamberlin introduces a further assumption which has been called '*symmetry assumption*' by Prof. Stigler. It is that the number of firms being large under monopolistic competition, an individual's actions regarding price and output adjustment will have a negligible effect upon its numerous competitors so that they will not think of retaliation for readjusting their prices and outputs. He thus says, specifically, we assume for the present that any adjustment of price of a "product" by a single producer spreads its influence over so many of his competitors that the impact felt by any one is negligible and does not lead him to any readjustment of his own situation. A price cut, for instance, which increases the sales of him who made it draws inappreciable amounts from the markets of each of his many competitors, achieving a considerable result for the one who cut, but without making incursions upon the market of any single competitor sufficient to cause him to do anything he would not have done any way."[8]

Given the above assumptions, we proceed to explain how under monopolistic competition an individual firm and a group of firms producing close substitutes come to be in equilibrium position. To begin with, suppose that the demand and cost curves of each of the firms in the group are *DD* and *AC* as depicted in Fig. 28.3. Each firm will set price *OP* at which marginal cost is equal to marginal revenue and hence profits are maximum. Although all firms are making supernormal profits, there is no reason for any one to cut down price below *OP* because the sales gained thereby will be insufficient to make up the loss due to the lower price. These supernormal profits will, however, attract new firms into the field in the long run.

Here it may be pointed out that full freedom of entry cannot prevail under monopolistic competition. Entry can be fully free only if the new firms who propose to enter the field can produce exactly identical products as those of the existing firms. But under monopolistic competition, this is not possible. Therefore, entry in the full and strict sense cannot exist under monopolistic competition. However new entrants are free to produce closely related products which are very similar to the products of the existing firms. Thus, *under monopolistic competition there can be freedom of entry only in the sense of a freedom to produce close substitutes.*

Turning to the above argument, when the new firms lured by the abnormal profits enjoyed by the existing firms enter the field, the market would be shared between more firms and a result the demand curve (or average revenue curve) for the product of each firm will shift downward *i.e.*, to the left. This process of entry of new firms and the resultant shift in the demand (average revenue) curve to the left will continue until the average revenue curve becomes tangent to the average cost curve and the abnormal profits are completely wiped out. This is shown in Fig. 28.5 where average revenue curve is tangent to average cost curve at point *T*. Marginal cost and marginal revenue curves intersect each other exactly vertically below *T*. Therefore, *the firm is in long-run equilibrium by setting price QT or OP and producing OQ quantity of its product*. Because average revenue is equal to average cost, the firm will be making only normal profits. Since all firms are alike in respect of demand and cost curves (by assumption), the average revenue of all will be tangent to their average cost curves and all firms will, therefore, be earning only normal profits. Because only normal profits are accruing to the firms there will be no more tendency for the new competitors to enter the field and the group as a whole will, therefore, be in equilibrium.

An important point is worth nothing here. It is that a *firm in long-run equilibrium under monopolistic competition makes only normal profits, as in perfect competition, but its price is*

7. *Ibid*, p. 83 (*italics added*)
8. *Ibid., p.* 83

higher and output smaller than under perfect competition. Under perfect competition, long-run equilibrium of the firm is established at the minimum point of the average cost curve. In other words, a firm under perfect competition tends to be of the optimum size. But a firm under monopolistic competition, as is evident from Fig. 28.5, stops short of the optimum point and operates at the point at which average cost is still falling. In Fig. 28.5 the firm produces output *OQ*, while a firm under perfect competition would have produced outut *OR* at which average cost is minimum. The firm under monopolistic competition can reduce its cost of production by expanding output to the point *R* but it will not do so because by expanding output beyond *OQ* it will be reducing price more than the average cost. It is, therefore, clear that by producing *OQ* instead of *OR*, the firm under monopolistic competition does not use its capacity fully. (The firm would be using its capacity fully, if it produces optimum or full capacity output *OR*). Thus capacity equal to *QR* is lying unused in the firm's long-run equilibrium under monopolistic competition. This *unused capacity is called excess capacity which is a prominent feature of long-run equilibrium under monopolistic compe-tition.*

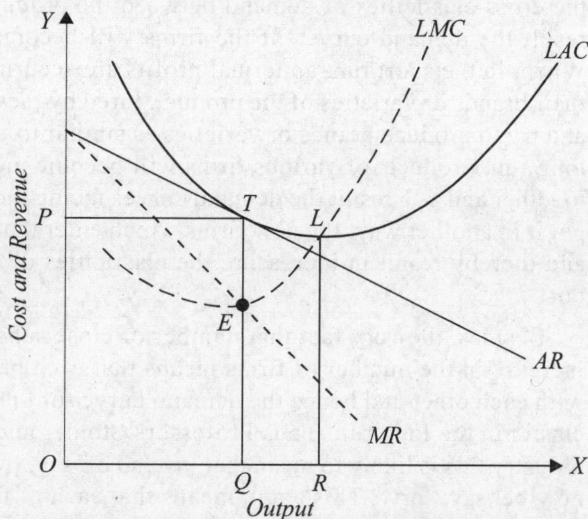

Fig. 28.5 *The Group is in equilibrium when average revenue curves of the firms are tangent to their long-run average cost curves.*

Further, it may be noted that in the long-run equilibrium, firms under monopolistic competition make only normal profits as under pure competition, but the price set under monopolistic competition is higher than competitive price. In Fig. 28.5, price set in long-run equilibrium under monopolistic competition is *QT* while competitive price would have been equal to *RL*. This higher price under monopolistic competition is due to monopoly element contained in it. The monopoly element involved in monopolistic competition makes the demand or average revenue curve facing an individual firm downward sloping and a downward-sloping average revenue curve can be tangent to the average cost curve only to the left of its minimum point.

Thus price under monopolistic competition will be higher than the competitive price due to the monopoly element in monopolistic competition. But, in spite of the higher price, a firm under monopolistic competition will not be making profits above normal in the long run. We may, therefore, say that a *firm under monopolistic competition, in the long-run equilibrium, charges higher price without enjoying monopoly profits*. A significant result follows from this. It is that the non-existence of abnormal profits is no indicator of the absence of monopoly element. In long-run equilibrium under monopolistic competition, as seen above, the firm has monopoly power (it has a sole control over its own differentiated product with the result that demand curve for it slopes downward) but does not make any supernormal profits. Further, the existence of abnormal or excessive profits does not necessarily mean the presence of monopoly power. Under perfect competition, a firm in the short run may enjoy excessive supernormal profits due to the increase in demand for the product. Thus "*he skates on thin ice who identifies profits with monopoly and monopoly with profits.*"[9]

9. M.M. Bober, *Intermediate Price and Income Theory.*

Another noteworthy point about long-run equilibrium under monopolistic competition is that with the increase in the number of firms in the long run *demand curves facing individual firms will become more elastic in the sense that they will flatten out*. Although Chamberlin denies it, it is now commonly believed that as the number of firms gets larger in the long run, the cross elasticities of demand between the products of various firms will increase and as a result the demand curves of the firms will become more elastic (or graphically less steep). When, in the short run, abnormal profits are accuring to the firms due to the large demand for their brands or varieties of the product, lured by these abnormal profits the new firms will enter and try to produce brands or varieties as similar to the existing brands as possible. Thus, in the long run, products of various firms will become more similar or in other worlds, come closer together and as a result the demand curves facing individual firms will become more elastic. To put it in another way, the new firms which enter the group will come "in between" the old ones and thereby result in increasing the elasticities of demand curves (that is, make them flatten out).

Besides, the very fact that number of close substitutes in the long run increases due to the increase in the number of firms means that each brand will become more closely competitive with each other and hence the demand curve for the product of any one firm will become more elastic in the long run. Thus Professors Stonier and Hague write: "If new producers enter the industry this is likely to mean that instead of, say, twenty similar cars being produced there will now be, say, forty. This again means that each of the cars is likely to be more similar to each other than it was previously. And the more closely competitive substitutes there are, the more elastic the demand for the product of any one firm in the group will be."[10]

SELLING COSTS AND ADVERTISING

For determining price and output under monopolistic competition and oligopoly, firms often compete through incurring selling costs or advertisement expenditure. For firms working under conditions of monopolistic competition and oilgopoly, besides adjustments of price, output and product, an important decision that has to be taken is that of how much selling costs or advertisement expenditure it should undertake so as to achieve its aim of profit maximisation. The first problem that is encountered in connection with selling costs is how do they differ from production costs. The other important question which arises is why firms under monopolistic competition and oligopoly incur selling costs and not the firms working udner conditions of perfect competition and monopoly. Further, how the firm will decide about the optimum level of selling costs or advertisement expenditure. And, lastly, what is the influence of selling costs on price and output of the product. We shall discuss all these questions concerning selling costs below.

Selling Costs Distinguished from Production Costs

The term "selling costs" is broader than advertisement expenditure. Whereas advertisement expenditure includes costs incurred only on getting product advertised in newspapers and magazines, on radio and television, selling costs include the salaries and wages of salesmen, allowances to retailers for the purpose of getting their product *displayed* by them and so many other types of promotional activities besides advertisement. Chamberlin who introduced the analysis of selling costs in price theory distinguished selling costs from production costs. According to Chamberlin, cost of production includes all those expenses which are incurred to manufacture and provide a product to the consumer to meet his *given demand or want*, while the selling costs are those which are incurred to *change, alter or create the demand for a product*. Costs of production therefore include manufacturing costs, transportation costs, and cost of handling, storing and delivering a product to the consumers, since all of these activities add utilities to a commodity. And the addition

---

10. Stonier and Hague, *A Textbook of Economic Theory*, 4th edition, 1972, p.209

or creation of utilities to satisfy the given wants is called production in economics. To quote Chamberlin, "Cost of production includes all expenses which must be met in order to provide the commodity or service, transport it to the buyer, and put it into his hands ready to satisfy his wants. *Cost of selling includes all outlays made in order to secure a demand or market* for the product. The former costs create utilities in order that given demands may be satisfied; the *latter create and shift the demands themselves*. A simple criterion is this: of all the costs incurred in the manufacture and sale of a given product, *those which alter the demand curve for it are selling costs*, and those which do not are costs of production."[11] The selling costs, according to Chamberlin, include "advertising in its many forms, salaries of salesmen and the expenses of sales departments and sales agencies (except where these agencies actually handle the goods), window displays, and displays and demonstration of all kinds."[12]

It should be noted that transportation should not be construed as *increasing the demand*, as it apparently appears. This is because the transportation does not really increase the demand; it merely enables the producer to meet the demand of the consumer which is already there whether the transport cost is incurred by the producer or by the consumer himself. Likewise, a high site rent for a shop in a well-located area will increase the sales of the firm but cannot be considered as a part of selling costs, since in this the firm is meeting the given or existing demand for the product more accurately or exactly and not *altering* the demand for the product. By paying a high rent for a shop or a factory building in the well-located area, the producer is merely adapting the product or himself more exactly to the given demand and not altering the demand or adapting his customers. Therefore, Chamberlin while drawing the distinction between production costs and selling costs writes that those costs which are "*made to adapt the product to the demand are costs of production; those made to adapt the demand to the product are costs of selling.*".[13]

It should, however, be noted that the distinction between production costs and selling costs cannot always be sharply made and there are cases where it cannot be said whether product is being adapted to meet the given demand, or the demand is being adapted to sell the product. For instance, it is difficult to say whether the extra cost on attractive packaging is production cost or selling cost. However, as far as advertisement expenditure is concerned, there is little doubt about its being a selling cost, since purpose of advertisement is to increase or create the demand for the product. Thus Chamberlin's distinction is quite applicable so far as advertisement expenditure is concerned. Because advertising expenditure is the most important and dominant form of selling costs, we in our analysis below shall use them interchangeably and discuss the various questions concerning selling costs by taking the case of advertisement expenditure.

Role of Selling Costs under Perfect Competition, Monopoly and Imperfect Competition

As has been explained in previous chapters, there is no need for a firm working under perfect competition to undertake advertisement expenditure or to incur other types of selling costs, since by assumption, the product produced by all firms in the perfectly competitive industry is homogeneous, and an individual firm can sell as much quantity of the product as it likes at the given price. If a perfectly competitive firm advertises for the product, the consumers who are influenced by it may purchase the product from other firms in the industry, since all are selling homogeneous products. Of course, the whole perfectly competitive industry, that is, all firms together or their association may advertise to promote the sales of their product at the expense of the *products of other industries*. Such advertising is known as *promotional advertising*, as compared to competitive advertising with which we are here concerned. In India, there has been advertising by *Terene* industry producing terylene fabrics for increasing the demand for its product at the cost of other kinds of fabrics. We,

11. Edward, H. Chamberlin, *The Theory of Monopolistic Competition*, 6th edition, p. 123 (italics added).
12. *Op. cit.*, p. 124
13. *Ibid.*, p. 125

therefore, conclude that under perfect competition, there can be promotional advertising by the whole industry but not competitive advertising by individual firms to snatch away the customers from each other.

Under monopoly also, there is no competitive advertising since, by definition, a monopolist produces a product which has no close substitutes. The monopolist only needs to inform or remind the buyers that his product exists and he need not emphasise the competitive nature of its product. Of course, the monopolist may advertise to promote his sales or demand but it will not be at the expense of its rivals, since no rivals producing close substitutes are there under monopoly. Hence the advertisement by the monopolist is informative and promotional and not competitive.

It is under conditions of imperfect competition, that is, monopolistic competition and oligopoly with product differentiation that advertisement and other selling costs become important as a competitive weapon at the disposal of one firm to increase its sales at the expense of others. This is because differentiated products produced by different firms under monopolistic competition and differentiated oligopoly are close substitutes of each other. Therefore, each firm under monopolistic competition tries to convince the buyers that its product is better than those of others in the industry. A firm under monopolistic competition and differentiated oligopoly may keep its price and product design constant and seek to increase the demand for its product by increasing the amount of advertisement expenditure and through it persuading the buyers that its brand of the product is of superior quality than others. Thus, this is competitive advertising which is aimed at attracting the customers to their product and weaning them away from the closely related products of the rivals. Thus, "*the fundamental aim of all 'competitive' advertising is to attract the customers' attention and to imprint the name of a particular product on his mind; the aim is to persuade the consumer to put his hand in his pocket and buy the product in question........... the main aim is to increase the sales of one firm at the expense of others and not to increase the sales of the 'group' as a whole.*"[14] For instance, we all know that all tooth pastes are based upon the same chemical formula recommended by the medical science. But the firm producing Colgate through its radio and television commerical programmes has been propagating that the tooth paste of 'Colgate' variety is very much better than others and has special and superior qualities which are absent in other brands of toothpaste. The fundamental aim of Colgate advertisment is not to increase the aggregate demand for toothpaste in the country but to increase the demand for 'Colgate toothpaste' by competing away the buyers from other brands of toothpaste. Similarly, the manufactures of other brands of toothpaste such as Binaca, Pepsodent, Oral–B, Signal, Forhans etc. are also incurring expenditure on advertisement through various means and trying to convince the buyers that their particular brand of tooth paste is better than others. Such competitive advertisement by a firm often proves to be successful in its objective of increasing the demand for a particular brand of the product. Thus, as a result of advertisement, demand curve facing an individual firm shifts to the right which indicates that at a given price, a greater quantity of the product can be sold.

It follows from above that in the presence of selling costs or advertisement, demand curve for a product cannot be taken as an objective fact given by the tastes or wants of the consumers. A firm can alter or shift the demand curve for its product through its own efforts by incurring advertisement expenditure and other forms of selling costs.

Effect of Selling Costs (Advertising Expenditure) on Demand

The purpose and effect of successful advertising is to increase the demand, that is, to shift the demand curve for the product to the right. However, these selling costs or advertising outlays are subject to the varying returns. That is, equal increments in advertising outlay first yield increasing returns and then eventually diminishing returns in terms of its effect on demand for the product. In the beginning, increases in advertising outlay will bring about increasing returns in raising demand

14. Stonier and Hague, *op. cit.*, p. 222. (italics added)

for the product for two reasons. Firstly, increase in advertising outlay (or selling costs) permits a firm to *repeat many times* the advertisement for the product. And this repetition of advertisement produces favourable effects on demand. "It is well established that repetition is essential if advertising is to make an impact on the consumer's mind. A single advertisement seen once will have at the most a negligible, and probably no effect on the consumer. The outlay for it is wasted. But continued advertising over a period of time and in different media is far more likely to impinge on the consumers' thoughts and consequent consumption choices."[15]

Second reason for occurrence of increasing returns as the advertisement outlay is increased in the beginning is the *economies of large scale selling operations or advertising*. The main economy or adventage is the specialisation which is made possible by large-scale selling or advertising activity. To quote Prof. Hibdon again, "Large-scale activities permit the *use of specialised personnel with greater expertise and effectiveness*. There may also be *economies in the use of advertising media*. Greater total spending permits a shift in the technique and media that are used in the selling effort as well as the use of combinations of media."[16] As a result of the increasing returns from advertising outlay in the beginning, the demand increases more than proportionately to the equal increases in advertising outlay.

But as the advertising outlay is stepped up, diminishing returns are likely to set in eventually. Firstly, this is because potential buyers differ in tastes, income and wealth. These differences among the potential buyers of the product mean that they will vary in their responses to selling operations or advertising by a firm. Initial advertisements will cause large increases in demand as the more susceptible buyers respond to advertisement. But further increase in advertisements is likely to bring about relatively less increases in demand as they fail to influence other buyers who do not prefer the good so much.

Second reason for the ultimate occurrence of diminishing returns from the increases in advertisement is that the existing buyers may not further increase the demand for the product as a result of more advertising by the firm. This is because as the consumer buys more of a product, its marginal utility to him falls. Further, in order to purchase one good more, he has to give up more of some other goods, his income being limited. As a result, marginal utility of 'other goods' increases. Thus, on the one hand, marginal utility of a good declines as it is purchased more under the influence of advertisement, and, on the other hand, the opportunity cost of buying this good increases on account of the rise in the marginal utility of 'other goods'. This makes the existing buyers of a product reluctant to buy more units of it when advertising effort by a firm is greatly increased. Thus this also causes diminishing returns to advertising outlay.

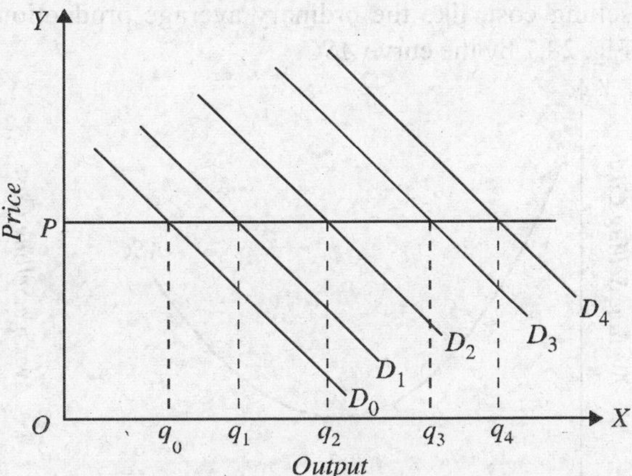

Fig. 28.6. *Effect of Selling Cost on Demand*

The effect of selling costs on the demand for a product and the varying returns in this connection are illustrated in Fig. 28.6. Demand curve before any advertisement expenditure is undertaken is D_0. Now, equal increments in advertisement expenditure successively bring about rightward shift in the demand curve to D_1, D_2, D_3 and D_4 respectively. We have assumed that the shift in the demand curve is parallel, while in the real world it may not be so. Because in the beginning there are increasing returns and then after a point

15. James E. Hibdon, *Price and Welfare Theory,* McGrawHill, 1969, p. 302
16. *Op., cit.*, p. 302.

diminishing returns occur, successive shifts in the demand curve differ in magnitude. At the given price *OP*, as a result of equal increments in advertising outlay, the quantity demanded increases from q_0 to q_1, q_1 to q_2, q_2 to q_3 and q_3 to q_4. It will be seen from the Fig. 28.6 that after D_3, diminishing returns to extra advertising outlay occur.

The Curve of Average Selling Costs

The concept of the curve of average selling cost should be fully understood. There are two concepts regarding average selling cost, and average selling cost curves drawn according to these two concepts are different. Since the shape of the average selling cost curve is different in these two concepts, for analysing selling cost and its effect on equilibrium price and output, it must, therefore, be known according to which concept the curve of average selling cost is drawn. A firm can choose among various amounts of advertising outlay (*i.e.*, selling costs). Therefore, the firm may at the outset treat the selling costs as variable amounts. Thus, with varing amounts of selling costs (advertising outlay) incurred in a period, the average selling cost per unit will depend upon the output sold as a result of the rightward shift in the demand curve brought about by a particular amount of selling costs. This average selling cost is obtained by dividing the amount of selling costs incurred by the quantity of output sold. And as a firm plans to increase the amount of selling costs incurred in a period, the average selling cost will change depending, on the one hand, upon the increase in selling costs and, on the other, upon the resultant increase in output demanded (or sold) at a *given price*. As explained above, it is generally believed that *selling costs* (*advertisement outlay*) *is subject to varying returns*. In the beginning, increasing returns to selling costs are obtained, that is, equal increases in advertisement outlay cause more than proportionate increase in the amount demanded of the product at the *given price*. In other words, selling costs per unit will fall in the beginning. After a point diminishing returns to selling costs set in and increases in advertisement outlay cause less than proportionate increases in the amount demanded of the product. In other words, after a point, average selling cost will rise. Hence, with varying amounts of selling cost average selling cost in the beginning falls due to increasing returns, reaches the minimum level and then rises due to diminishing returns. Thus with varying amounts of selling cost, a curve of average selling cost, like the ordinary average production-cost curve, is U-shaped, which is shown in Fig. 28.7 by the curve *ASC*.

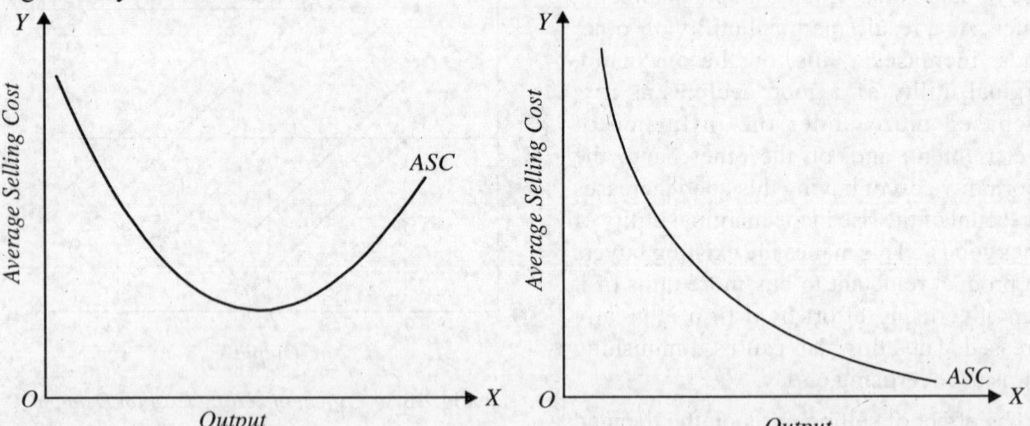

Fig. 28.7. *Average Selling Cost Required to Sell Various Amounts of Output*

Fig. 28.8. *Curve of Average Selling Cost with a Fixed Amount of Total Selling Costs*

However, the average selling cost curve *ASC* drawn in Fig. 24.7 should be carefully interpreted. It does not mean how the average selling cost per unit changes as output is increased. But *it means the average selling cost per unit which is required to be undertaken to sell various amounts of output.* Ultimately the average selling cost curve *ASC* will become vertical. This is because quite often saturation point regarding the effect of extra selling costs on raising the demand for the product is reached, beyond which no further increases in selling costs can cause any expansion in amount demanded of the product.

We have discussed above the nature of the average selling cost curve when the total selling costs are treated as variable magnitudes and, in fact, the firm treats them as such when it has to plan for an amount of selling or advertisement cost it should incur in a period so as to maximise profits. However, *once a firm commits itself to a particular amount of selling costs or advertisement outlay to be incurred in a period, it may regard that cost as a fixed cost during that period.* In other words, when a *given fixed amount* of total selling costs is decided or considered to be incurred, then greater the level of output sold, the selling cost per unit would continue to decline. In other words, when a firm commits itself to a given amount of selling costs, it will be then treat them as a fixed cost. Given a fixed amount of selling costs, average selling cost will change in the same manner as does the average fixed cost. Like the average fixed cost curve, this type of average selling cost curve will have a shape of rectangular hyperbola, as shown in Fig. 28.8.

Optimum Level of Advertising Outlay (Selling Costs): With Price and Product Design as Constants

An important question is how much selling costs (advertisement outlay), a firm will undertake so as to maximise profits. In other words, what is the *optimum amount* of advertisement expenditure for a firm. The determination of optimal advertising outlay (selling costs) for the firm can be explained by the average and marginal cost curves. For explaining the optimal amount of advertising expenditure with average and marginal costs, we have to use the concept of average selling cost when advertisement outlay is taken to be variable. Consider Fig. 28.9 where *ASC* and *APC* are average selling cost and average production cost curves respectively. Average selling cost curve *ASC* has been superimposed over the average production cost curve *APC* to obtain average total curve *AC* (*AC* = *APC* + *ASC*). It should, therefore, be noted that the vertical distance between the *AC* and *APC* curves measures the average selling cost. *MC* is the marginal cost curve to the average total cost curve *AC*. We assume that price *OP* has already been fixed by the firm which is kept constant. Further, the nature of the product is also held unchanged and it is only the advertising expenditure which is varied and consequently demand shifts to the right and output sold increases.

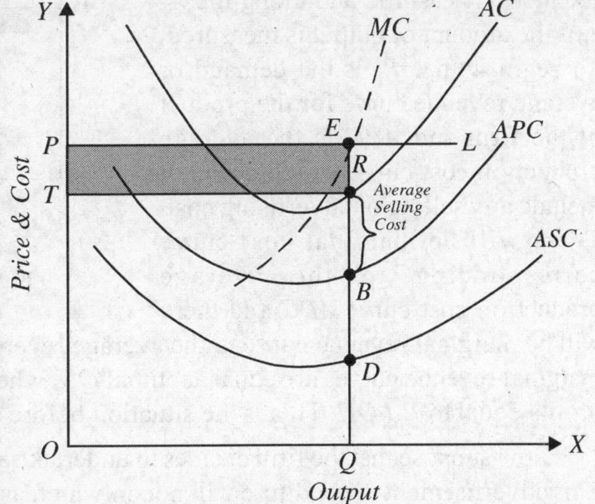

Fig. 28.9. *Optimal Level of Advertising Expenditure (with price and Product Design as Constants)*

Since price of the product remains fixed at *OP*, the horizontal line *PL* can be viewed as if it were a marginal revenue curve. This is because through increase in advertisement expenditure, a firm can sell more quantity of the product without lowering price. *If the firm aims to maximise profits, then it will be in equilibrium regarding advertising outlay where the marginal cost (which is inclusive of both production cost and selling cost) is equal to the marginal revenue, i.e., given price OP.* It will be seen from the Fig. 28.9 that marginal cost is equal to marginal revenue (or price) at *OQ* level of output at which profits will be maximised. With *OQ* as the output produced and sold, total profits made by the firm are equal *PERT* and, as is evident from the figure, average selling cost incurred by the firm in its equilibrium position is equal to *QD* or *BR*. Therefore, the optimal amount of advertisement outlay, incurred by the firm will be equal to (*QD* or *BR*) multiplied by the output *OQ*.

Optimal Level of Advertising Expenditure with both Price and Output Variable

We have discussed above how much advertising expenditure a firm will undertake so as to maximise its profits when the price as well as nature of the product remain constant. We shall now explain the case of optimal level of advertising expenditure when price also varies. In other words, we have to explain optimal combination of advertising expenditure, price and output, only the physical make-up of the product will not be varied. We shall illustrate our analysis with two dimensional diagram. Here we shall be taking *a given sum of advertising expenditure and then analyse its effect on demand, output, price and profits*. With a given sum of advertising expenditure, average selling cost per unit will go on declining as more output is produced. It is because of this that the distance between the two successive average cost curves will become smaller and smaller as output is expanded. However, it should be noted that we will be increasing this given sum of advertising expenditure and assess its effect on demand and profits to obtain the optimal levels of advertising expenditure and output.

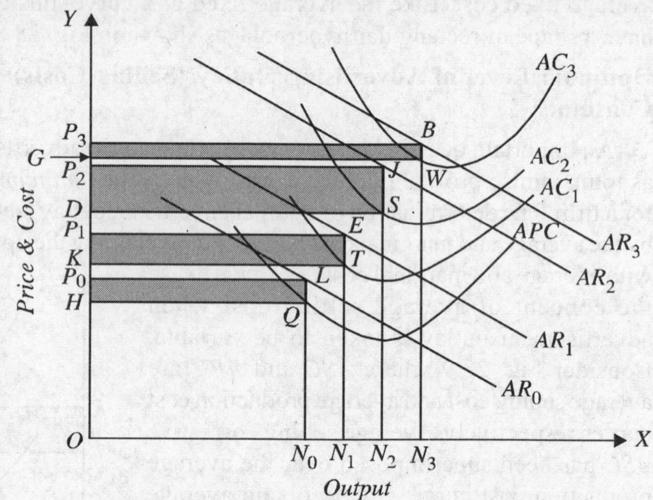

Fig. 28.10. *Optimal Advertising Expenditure*

The optimal amount of advertising outlay and the choice of price-output combination by a firm is illustrated in Fig. 28.10. It should be noted that we continue to assume that the firm aims to maximise profits. Along the *Y*-axis in Fig. 28.10 the price and cost of the product is measured and along the *X*-axis the amount of output is measured. To begin with, AR_0 is the demand or average revenue curve for the product of the firm and *APC* is the average production cost curve which does not include any selling or advertising cost. There will be marginal cost curve corresponding to the average production cost curve *APC* and there will be marginal revenue curve to the average revenue curve AR_0. Suppose these marginal cost and marginal revenue curves are equal at output ON_0 where price P_0 is determined and the firm is making profits equal to P_0LQH. This is the situation before any advertisement expenditure is undertaken.

Now suppose that the firm decides to undertake advertisement expenditure equal to 1,000 rupees. This advertisement expenditure will not only increase demand by shifting the average revenue curve to the right, but will also add to costs. Assume that with advertisement expenditure of ₹ 1,000, the demand curve shifts to the curve AR_1 and the new average cost (inclusive of advertising cost) is AC_1. Suppose with AR_1, as the new demand curve and AC_1 as the average cost curve, equilibrium is at output ON_1 and price P_1 (this price-output equilibrium is determined by the equality of new marginal revenue and marginal cost curves which are not drawn in the figure for sake of avoiding confusion). It will be seen from the figure that profit has now risen to the area P_1ETK. It should be noted that profits will increase only when the addition to *net* revenue earned from advertising outlay is greater than the advertising outlay incurred. It should be further noted that as a result of advertising outlay, output increases and therefore total production cost also increases. And the addition *to net* revenue attributable to advertising outlay is obtained by subtracing the addition to production cost from the addition to gross revenue.

Since as a result of advertisement expenditure, profits have increased, firm will be tempted to undertake further advertisement expenditure. Suppose the firm incurs additional advertising expenditure of ₹ 1,000 (i.e., now total ₹ 2,000) and with this the demand (or average revenue curve) shifts to the position AR_2 and average cost curve to AC_2. Now, the new equilibrium position is reached at output ON_2 and price P_2, and profits are further increased to P_2JSD. It may again be mentioned that profits increase only when additional revenue generated (net of production cost) is greater than the additional advertising cost. "*Indeed it will pay the firm to go on increasing selling costs in this way so long as each increment of advertising expenditure adds more to revenue than to costs. Only when the additional revenue generated (net of production costs) equals the extra (marginal) amount spent in order to generate that net revenue, will profit be at the highest possible level.*"[17]

In Fig. 28.10 now assume that lured by the increase in profits to P_2JSD, firm undertakes additional advertising expenditure of ₹ 1,000 (that is, now total advertising expenditure is raised to ₹ 3,000). With this, the average revenue curve shifts to AR_3 and average cost curve to AC_3. It will be seen from the figure that now profits earned by the firms are equal to the area P_3BWG. It will be observed that profits P_3BWG are less than the previous profits P_2JSD. The decline in profits must be due to the additional net revenue (*i.e.*, additional revenue net of increase in production costs) made possible by the additional advertising expenditure being smaller than the extra cost of one thousand rupees on advertisement.

It is thus clear that the firm will not undertake the third additional outlay of one thousand rupees and will be in equilibrium by incurring advertising expenditure of two thousands rupees at which its profits are largest (that is, equal to P_2JSD) and additional net revenue made possible by the advertising is equal to the additional advertising outlay. Thus in the situation depicted in Fig. 28.10 two thousand rupees per period is the optimal advertising expenditure for the firm.

It should be noted for each level of advertising expenditure, that is, for one thousand, two thousand and three thousand rupees there is profit-maximising price and output. Various levels of profits P_0LQH, P_1ETK, P_2JSD, and P_3BWG are all maximum with different levels of advertising expenditure. The task for the firm is to choose the combination of advertising expenditure, price and output which yields maximum out of the various maxima of profits. That is, the firm has to determine the combination of advertising expenditure, price and output which yields him *maximum-maximorum* level of profits. And we have seen above that in the situation depicted in Fig. 28.10 this *maximum-maximorum* equilibrium is reached when advertising expenditure is two thousand rupees, price is P_2 and output is ON_2.

Effect of Advertising (Selling Costs) on Elasticity of Demand

We have seen above that as a result of advertising expenditure demand for the product increases, that is, demand curve shifts to the right. For the sake of convenience, we have assumed that the new demand curves after advertising is undertaken are parallel to the old one, though in actual practice it need not be so. However, in this connection it is useful to consider whether when the demand increases and demand curve shifts to the right, elasticity of demand at each price remains the same, declines or rises. The purpose of competitive advertising or other forms of selling costs is to influence the consumers to buy a particular brand of the product rather than other substitute brands of it. The intention of the producer who advertises for his brand of the product is to differentiate his brand more from the viewpoint of the consumers and try to prove his brand to be superior to others. Thus if the purpose and intention of advertisement is achieved, then the consumers would begin to consider a particular brand of the product much superior to others. That is, they will now regard the other competitive brands as less closely substitutes than they were thinking before. This greater degree of differentiation and consequently fall in the elasticity of substitution will cause a decline in the price elasticity of demand for the product at each price as the demand curve shifts to the right under the influence of the advertisement. It is, therefore likely that price elasticity of demand should decline

17. Stonier and Hague, *op. cit.*, p. 224.

under the influence of advertising or other forms of selling costs. The extent to which price elasticity will decline is of course very uncertain. As we shall see below, changes in price elasticity of demand as a result of advertisement expenditure have significant implications for price-output equilibrium.

Effect of Advertising (Selling Costs) on Price and Output

The effect of advertising and other forms of selling costs on price and output are quite uncertain. This effect depends, on the one hand, upon the change in price elasticity of demand as a result of advertising expenditure and, on the other, upon the behaviour of average cost of production. If when the demand curve shifts to the right under the influence of advertising, price elasticity of demand at the current price remains the same and average cost of production is falling sharply with the expansion in output, then it may be profitable for the firm to set a lower price after advertisement. In this case profit-maximising price will be lower and output larger than those before advertisement. On the other hand, if price elasticity of demand declines very much as demand curve shifts to the right as a result of advertisement, and the average cost of production is rising sharply with the increase in production, then in order to maximise profits in the new situation after advertisement, the firm may raise the price and reduce its level of output. In this case, the advantage of the increase in demand due to advertisement expenditure will be enjoyed in the form of higher price of the product rather than of increased sales of the product. But the most likely case to occur is that price elasticity of demand declines after advertisement and the average cost of product is not rising very sharply, and as a result of these the firm may find it profitable to raise price of the product somewhat and also increase output after advertising expenditure is undertaken. Thus, according to Stonier and Hague, "The most probable result for a profit-maximising firm will be that the elasticity of demand will fall somewhat at each price, that the volume of demand will increase at each price and that price and output will both increase to some extent as a result of advertising campaign."[18] And this is the case which we have considered in our analysis of advertising expenditure as depicted in Fig. 28.10. It will be seen from Fig. 28.10 that with successive increments in advertising expenditure price has risen from P_0 to P_1, P_1 to P_2 and P_2 to P_3 and also output has increased from N_0 to N_1, N_1 to N_2 and N_2 to N_3. Thus in our analysis of Fig. 28.10 both price and output have increased as a result of advertisement outlay.

EXCESS CAPACITY UNDER MONOPOLISTIC OR IMPERFECT COMPETITION

Theories of Chamberlin's monopolistic competition and Joan Robinson's imperfect competition have revealed that a firm under monopolistic competition or imperfect competition in long-run equilibrium produces an output which is less than socially optimum or ideal output. This means that firms operate at the point on the falling portion of long-run average cost curve, that is, they do not produce the level of output at which long-run average cost is minimum. Long-run equilibrium of a firm under monopolistic competition is achieved when the demand curve (or average revenue curve) facing a firm becomes tangential to the long-run average cost curve so that it earns only normal profits. Under such circumstances a firm can reduce average cost (and hence price) by expanding output to the minimum level of long-run average cost, but it will not do so because its profits are maximised (equality of marginal revenue with marginal cost is attained) at the level of output smaller than that at which its long-run average cost is minimum.

Society's productive resources are fully utilised when they are used to produce the level of output which renders long-run average cost minimum. Thus a monopolistically competitive firm produces less than the socially optimum or ideal output, that is, the output corresponding to the lowest point of long-run average cost curve. This is in sharp contrast to the position of the firm in long-run equilibrium under perfect competition, which operates at the minimum point of the long-run average cost curve.

18. Stonier and Hague, *op. cit.*, p. 226.

The amount by which the actual long-run output of the firm under monopolistic competition falls short of the socially ideal output is a measure of *excess capacity* which means unutilised capacity.

The existence of excess-capacity under imperfect or monopolistic competition can be understood from Fig. 28.11 and 28.12. Figure 28.12 depicts the long-run position of a perfectly competitive firm which is in long-run equilibrium at the level of output *ON* corresponding to which long-run average

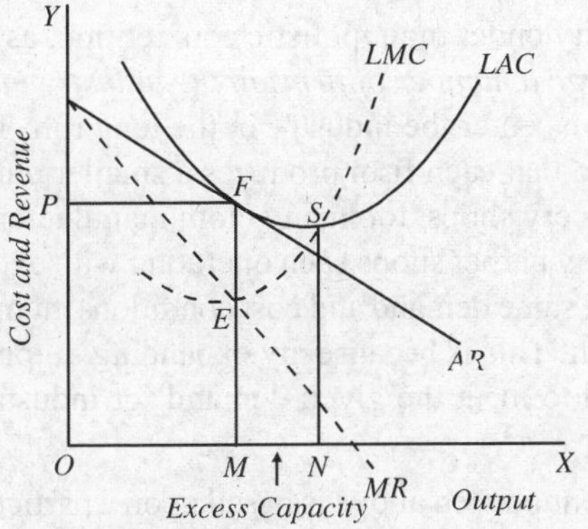

Fig. 28.11. *Excess Capacity under Monopolistic Competition*

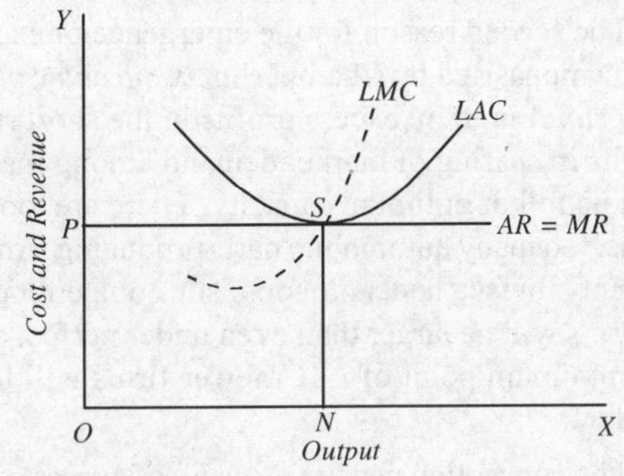

Fig. 28.12. *Ideal or Socially Optimum Output under Perfect Competition*

cost is minimum. It is at output *ON* that the double condition of long-run equilibrium, namely Price = *MC* = *AC* is fulfilled. It is thus clear that firms under perfect competition produce socially ideal output. On the other hand, a firm under monopolistic competition depicted in Fig. 28.11 is in long-run equilibrium at output *OM* at which its marginal revenue is equal to marginal cost and price is equal to average cost (Average revenue curve *AR* is tangential to average cost curve *LAC* at point *F* corresponding to output *OM*). It will be noticed that at output *OM* long-run average cost is still falling and goes on falling up to output *ON*. This means that the firm can expand its production up to *ON* and reduce his long-run average cost to the minimum. Ideal output is the output at an which long-run average cost is minimum. Therefore, the firm is *producing MN less than the ideal output. Thus MN output represents the excess capacity which emerges under monopolistic competition.* It is worth nothing that the concept of excess capacity refers only to the long run. This is because in the short run under any type of market structure (including perfect competition) there can be all sorts of departures from the ideal reflecting incomplete adjustment to the existing market conditions.

Causes of Excess Capacity

What factors are responsible for the existence of excess capacity under monopolistic competition? It is due to the existence of excess capacity that average cost of production and price of product are higher and output smaller monopolistic competition than under perfect competition. There are three main causes of the emergence of excess capacity under monopolistic competition. First, the most important cause of the eixstence of excess capacity under monopolistic competition is *downward-sloping demand curve* (*or average revenue curve*) *of the firm*. A downward-sloping curve can be tangent to a U-shaped average cost curve only at the latter's falling portion. It is only the horizontal demand curve or average revenue curve (as is actually found under perfect competition) which can be tangent to a U-shaped average cost curve at the latter's minimum point. From this, it also follows that the greater the elasticity of average revenue (or demand) curve confronting a monopolistically competitive firm, the less the excess capacity and *vice versa*. When the demand curve facing a firm is perfectly elastic, there is no ecess capacity, as is the case under perfect competition.

Now, demand curve facing individual firms under monopolistic competition slopes downward due to *product differentiation* found in it. Various firms produce different varieties and brands of product and each has a certain degree of monopoly power over the variety or brand it produces for fixing price and output. If products were homogeneous the demand curve would not have been downward sloping and equilibrium would have been established at the minimum point of *LAC* without there being any excess capacity.

The second reason for the emergence of excess capacity under monopolistic competition, as has been emphasised by Chamberlin, *is the entry of a very large a number of firms in the industry in the long run*. Lured by excess profits in the short run new firms enter the industry in the long run. This results in sharing of market demand among many firms so that each firm produces a smaller output than its full or optimum capacity. There are too many grocery shops, too many cloth manufacturing firms, too many automobile parts producing firms, too many barber shops each operating with excess capacity. In fact, under monopolistic competition, given the same demand and cost conditions, number of firms will be larger than even under perfect competition. This is because by expanding output to the minimum point of *LAC*, fewer firms will be required to meet the given demand for industry's product.

The conception and the measure of excess capacity as enunciated above is based upon a particular notion of ideal output. Marshall, Kahn[19], Harrod[20], Cassel[21] and Joan Robinson have regarded *ideal output or optimum size of the firm as that output at which its long-run average cost is minimum*. To quote Joan Robinson, "In a perfectly competitive industry each firm in full equilibrium will produce that output at which its average costs are minimum. Each firm will then be of the optimum size... If competition is imperfect, the demand curve for the output of the individual firm will be falling and the double condition of equilibrium can only be fulfilled for some output at which average cost if falling. The firms will, therefore, be of less than optimum size when profits are normal... It is only if conditions of perfect competition prevail that firms will be of the optimum size and there is no reason to expect that they will be of optimum size in the real world since in the real world competition is not perfect."[22]

Benefits of Excess Capacity. However, many modern economists are of the view that excess capacity under monopolistic competition is *desirable in some respects*. According to them, excess capacity under monopolistic competition provides some benefits which increase consumer welfare. As mentioned above, the excess capacity comes into existence mainly due to product differentiation under monopolistic competition. Now, this *product differentiation leads to product variety* which is highly beneficial to the consumers. The ability to choose among a wide variety of clothes, furniture, restaurant meals and other types of styles of product designs add greatly to the satisfaction or welfare of the consumers. Therefore, according to this view, social benefits of excess capacity should be weighed against the cost to the society of excess capacity.

EQUILIBRIUM UNDER MONOPOLISTIC COMPETITION: CHAMBERLIN'S ALTERNATIVE APPROACH

Concepts of Proportional and Perceived Demand Curves

The process of equilibrium adjustment under monopolistic competition has also been explained by an alternative approach put forward by Chamberlin. This alternative approach makes use of two

19. R.F. Khan, "Some Notes on Ideal Output", *Economic Journal*, XIV (1935), pp. 1-35.
20. R.F. Harrod, Doctrines of Imperfect Competition, *Quarterly Journal of Economics,* XLIX (1934-35), pp. 442-70.
21. J.M. Cassels, Excess Capacity and Monopolistic Competition, *Quarterly Journal of Economics,* LI (1936-37), pp. 426-43.
22. Joan Robinson, *Economics of Imperfect Competition*, pp. 96-97.

types of demand curves, namely, *perceived demand curve and proportional demand curve*. The demand curve facing an individual firm, as perceived by it, describes the demand for the product of one firm on the assumption that all other firms in the industry or group keep the *prices of their products constant*. The perceived demand curve shows the increase in quantity demanded of a product of a firm when a firm cuts down its price provided others keep their prices at the present level. On the contrary, it shows the fall in quantity demanded of the product of a firm which will occur if it raises its price on the assumption that others would not raise their prices. This type of demand curve in the present alternative approach, is known as *perceived demand curve* and is based upon the above important assumption. This type of demand curve is also known as *subjective or imagined or expected demand* curve because this is based on the individual firm deciding subjectively what it perceives or imagines its demand curve looks like. The number of firms being large in a product group under monopolistic competition, it is assumed that each firm is so small relative to the whole group that it thinks that the price change by it will have negligible impact upon its competitive firms with the result that they would not think of reacting to the change in price by it in retaliation.

The other type of the demand curve used in this approach is *proportional demand curve* facing an individual firm. This shows the demand or sales of the product of a firm when the prices of all firms in a product-group or industry *change, all of them simultaneously in the same direction and by the same amount* so that they charge same or uniform price. Obviously, the proportional demand curve of a firm will be less elastic than its perceived demand curve, since equal change in price by all firms of the industry (*i.e.*, group) will prevent the movement of customers from one seller to another. The proportional demand curve of each firm slopes downward because the market demand for the general class of the product increases as a result of the fall in price. In fact, the proportional demand curve facing an individual firm is a proportinate part of the total market demand curve for the general class of the product and will be of the same elasticity. Thus, since each firm gets some proportional share of the total market demand for the general class of the product, the proportional demand for each firm varies with the number of firms in the product group. The greater the number of firms in a product-group, the smaller the share of an individual firm at a given price. Therefore, the proportional demand curve facing an individual firm shifts to the left as more and more firms enter the product-group or industry. We designate the proportional demand curve as *DD′* curve and, the perceived demand curve as *dd′* curve. The two types of demand curves are graphically shown in Fig. 28.13. The two demand curves have been shown to be intersecting at point *A* corresponding to price *OP* and quantity demanded *OQ*. This means that at price *OP* of the product, the proportional share of each firm of the market demand for the industry product is equal to *OQ*. Since it is assumed that all firms charge the same price, all firms in the industry will be producing and selling *OQ* quantity and each charging *OP* price. Therefore, point *A* lies on the proportional demand curve. From the initial situtation at *A* on the proportional demand curve *DD′* (with price equal to *OP* and quantity demanded of each equal to *OQ*) we can draw the perceived demand curve of a firm. An individual firm believes that if it makes small alternation in price, it will have negligible effect on each of its many competitors so that they will not think of readjusting their prices. Prices of other firms remaining unchanged at *OP*, an individual firm thinks that if it reduces its price, the sales or quantity demanded of its product will greatly increase as it will attract customers of other firms. Thus, an individual firm perceives that its demand is more elastic

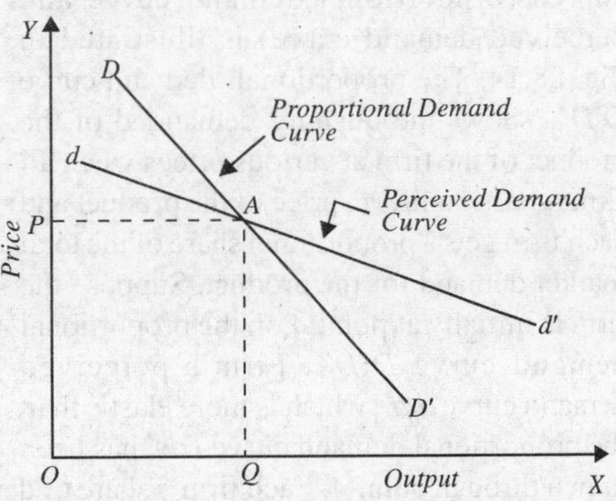

Fig. 28.13. *Chamberlin's Concepts of Perceived and Proportional Demand Curves*

than the proportinal demand curve and it can increase its profits by cutting down its price. Therefore, we construct the perceived demand curve dd' passing through point A as being more elastic than the proportinal demand curve DD'. However, it may be pointed out that since *each* firm in the product group will think independently that its price reduction will have a negligible effect upon each of his rivals and therefore assumes that others would keep their prices constant, each firm will cut its price and therefore the actual movement would not be along the perceived demand curve dd' but along the proportional demand curve DD' which shows actual sales by each firm when the prices of all change equally and are identical.

Short-Run Firm's Equilibrium under Monopolistic Competition

Price-output equilibrium of an individual firm in the short run in terms of two demand curves (proportional demand curve and perceived demand curve) is illustrated in Fig. 28.14. The proportional demand curve $D_0D'_0$ shows the quantity demanded of the product of the firm at various prices when all firms charge uniform price of the product and each firm gets a proportional share of the total market demand for the product. Suppose the firm is initially at point A on the proportional demand curve $D_0D'_0$. Firm's perceived demand curve $d_0d'_0$ which is more elastic than the proportional demand curve DD' has been drawn through point A. Each firm's share of demand for its product is equal to OQ_0 and all of them are charging uniform price OP_0.

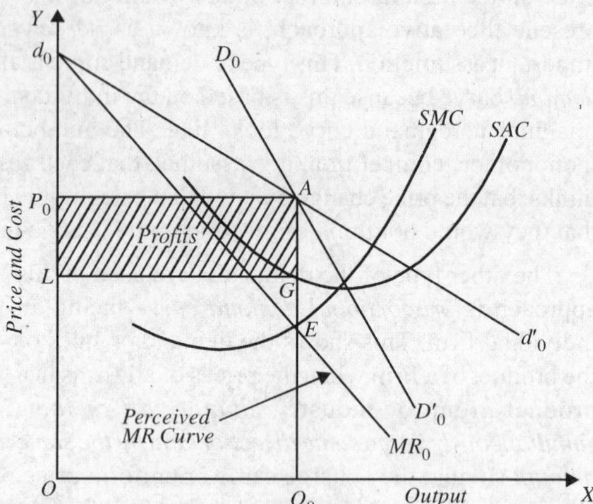

Fig. 28.14. *Short-run Equilibrium of the Firm under Monopolistic Competition: Proportional and Perceived Demand Curves Approach*

SAC and *SMC* are short-run average cost curve and short-run marginal cost curve respectively. Marginal revenue curve MR_0 corresponding the perceived demand curve dd'_0 has been drawn. *SMC* and MR_0 curves intersect at point E and accordingly output OQ_0 is equilibrium output of the firm and OP_0 is the equilibrium price charged with proportional and perceived demand curves cutting each other at point A in the present price-quantity equilibrium. It will be seen from Fig. 28.14. that price OP_0 exceeds average cost of production at OQ_0 output level and therefore the firm is making supernormal profits equal to the shaded area P_0AGL. Thus, according to Chamberlin's altenative approach short-run equilibrium under monopolistic competition is reached when the following two conditions are satisfied:

(1) The price-output combination is such at which *perceived marginal revenue curve intersects the marginal cost curve* so that $MR = MC$

(2) The price-output combination where $MR = MC$ is such that corresponds to the point at which *perceived demand curve dd' intersects the proportional demand curve D_0D_0'.*

Only when the above two conditions are fulfilled, the firm will be maximising its profits and will have no incentive to change its price. Thus firm's equilibrium in the short run occurs at the point on proportional demand curve and perceived demand curve which also corresponds to the output level where marginal revenue is equal to marginal cost.

Long-Run Adjustment

Chamberlin explains adjustment to long-run equilibrium at two stages. In the first stage he explains adjustment in the number of firms alone without price competition. In the second stage he discusses price competition among firms to reach long-run equilibrium position with zero economic profits. We have shown this adjustment to reach long-run equilibrium in Fig. 28.15. It has been assumed that firms face same demand and cost conditions. Lured by the supernormal profits made by the firms in the short run new firms enter the industry and as a result proportional market demand curve D_0D_0' starts shifting to the left as a given market demand is shared by more firms. Entry of the new firms will stop when the proportional market demand curve becomes tangent to the long-run average curve so that supernormal profits are wiped out. It will be seen from Fig. 28.15 that proportional demand curve facing firms in the absence of price competition has become tangent to the long-run average curve (*LAC*) at point *H*. However, this will not be long-run equilibrium situation if each of the existing firms working independently thinks that it can secure larger market share and thereby can increase its profits. From Figure 28.15 it is evident that *in its position at point H, a firm has more elastic perceived demand curve $d_0d'_0$ than its proportinate market demand curve $D_1D'_1$*.

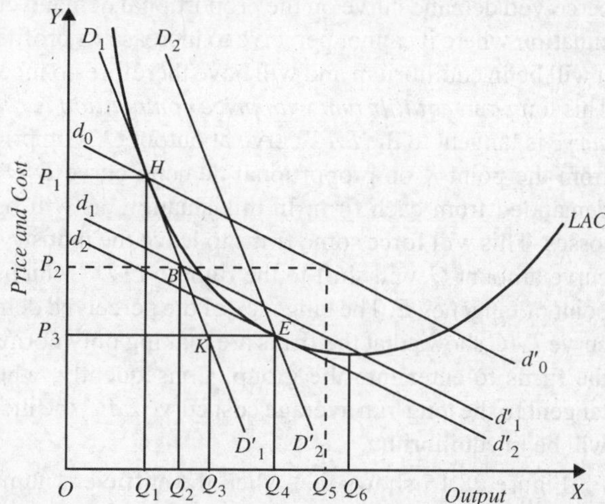

Fig. 28.15. *Long-run Equilibrium under Monopolistic Competition with Free Entry and Price Competition: Chamberlin's Alternative Approach*

Thus, with its position at point *H*, the firm perceives that it can increase its profits by cutting down price to P_2 ($AR > LAC$) by securing more than proporionate increase in demand for its product assuming that other firms will not react and would maintain their prices at the current level P_1. However each firm in monopolistically competitive industry will perceive, through independently of each other, that if it lowers its price below the current one, it can lure away customers from others assuming that others would keep their prices constant. Though this is a naive and myopic behaviour on the part of the firms, Chamberlin assumes it and thinks that in the real world firms do behave in this manner when they indulge in price competition.

But as a matter of fact since all firms working in this myopic manner cut down their prices, each firm *will not be in equilibrium* at point *H* producing Q_1 quantity of the product. Instead, the firms will find themselves at point *B* on the proportinoal demand curve $D_1D'_1$ getting a proportinale share of the increase in market demand at the lower price OP_2. Thus each firm will be working at point *B* on the proportinal demand curve $D_1D'_1$, and producing OQ_2 output. Thus, as a result of price cutting the perceived demand curve of each firm will slide down the proportinate demand curve to point *B*. In other words, from a new point *B* on the proportional demand curve $D_1D'_1$ at which the firms land themselves as a result of price reduction, the new perceived demand curve d_1d_1' has to be drawn. It will be seen from Figure 28.15 that with point *B* on the new perceived demand curve d_1d_1', a firm still thinks that it can increase its demand if it lowers its price below OP_2 provided other rival firms keep their prices constant. Thus, though its earlier attempt to increase profits by lowering price was frustrated by others also cutting down their prices by an equal amount, each firm again believes, though still suffering from illusion, that if it lowers the price, it can bring about increase in quantity demanded of its product and make profits.

However, since each firm perceives in the same way and cuts down its prices, though acting independently, it would not succeed in snatching away customers from its rival firms and its attempt to increase profits would again be frustrated. As a result, each firm instead of moving to a point on the perceived demand curve for increasing its profits it will land itself on the proportional demand curve $D_1D'_1$ getting the same proportionate share from the increase in quantity demanded of the product at the lower price. In this way the process of price-cutting competition and sliding down of perceived demand curve on the proportional demand curve would continue until a firm has reached a situation where it cannot perceive to increase its profits by reducing its price. In fact, In this situation it will be in equilibrium and will have therefore no incentive to lower further the price of the product. This long-run *equilibrium with price competition is reached at point E where the perceived demand curve* is tangent to the *LAC* curve at output Q_4 with price P_3 in Fig. 28.15. It will however be noticed from the point *K* on proportional demand curve $D_1D'_1$ that at price P_3 actually amount Q_3 will be demanded from each firm. In this siuation, as will be seen from Fig. 28.15 firms will be making losses. This will force some firms to leave the industry and as consequence the proportional demand curve amount Q_3 will shift to the right to $D_2D'_2$ which cuts the perceived demand curve $d_2d'_2$ at the point of tangency *E*. The tangency of the perceived demand curve $d_2d'_2$ with the long-run average cost curve *LAC* shows that the firms are making only normal profits. Hence there will be no tendency for the firms to enter into the group. Consequently, when the perceived demand curve *dd′* becomes tangent to the long-run average cost curve *LAC*, the individual firms and the product group as a whole will be in equilibrium.

Figure 28.15 shows that when the sufficient number of new firms have entered the group and price cutting has been made, the perceived demand curve has fallen to the position of tangency with the long-run average cost curve *LAC*. The proportional demand curve $D_2D'_2$ will intersect the d_2d_2' and *LAC* at the point of tangency. If too many firms have entered the group the perceived demand curve will fall to the position below that of the tangency and the firms will be making losses. The result will be that some firms will leave the group and the proportional demand curve along with perceived demand curve curve will shift to the right. For the achievement of long-run equilibrium of the firms and the group equilibrium, the two conditions are necessary.

(*i*) The perceived demand curve *dd′* must be tangent to the long-run average cost curve *LAC*.

(*ii*) The proportional demand curve *DD′* facing individual firm must intersect both the *dd′* and *LAC* curves at the point of tangency.

In the above discussion, the two competitive forces, namely, the entry of firms and the price cutting have been shown to be operating separately with the net result that the perceived demand curve *dd′* curve falls to the position of tangency with the *LAC* curve and the proportional demand curve *DD′* cuts both of them at the point of tangency. In actual world, the two competitive forces operate simultaneously but for analytical purposes, it is better to describe separately the working of these two forces.

It may be noted that, according to Chamberlin, with the entry of new firms, and *before the price competition taking place, that is, when firms operate* at point *H*, there exists excess capacity equal to Q_1Q_4 as according to him, OQ_4 is socially ideal output in the presence of product differentiation and price competition having been fully worked out. It may be further noted that Chamberlin does not regard Q_4Q_6 as excess capacity as according to him this much lower output is inevitable under monopolistic competition due to the existence of product differentiation and product variety.

Chamberlin's Concepts of Ideal Output and Excess Capacity

The view of ideal output described earlier rests upon the perfect competition and its associated product homogeneity which gives rise to the horizontal demand curve confronting an individual firm under it and consequently the establishment of the long-run equilibrium of the firm under perfect competition at the minimum point of the long-run average cost curve. The output at this minimum long-run average cost under perfect competition with product homogeneity is regarded as "ideal" from the viewpoint of social welfare. But Prof. Chamberlin argues that this "*Competitive Ideal*" cannot be considered as "ideal" under monopolistic competition.

Under monopolistic competition there is product differentiation which inevitably gives rise to the downward-sloping demand curve. Downward sloping demand curve along with free entry into the product group and active price competition necessarily involves long-run equilibrium to the left of minimum point of the long-run average cost curve. According to Chamberlin, product differentiation is desired *per se* and therefore the long-run equilibrium output of the monopolistically competitive firm under free entry and *active price competition* represents the "ideal output". The departure from the long-run equilibrium output under free entry and active price competition from the long-run minimum average cost does not prevent the output to be ideal because this departure is only due to product differentiation which is desired by the consumers for its own sake.

Chamberlin regards product differentiation a quality of the product which carries a cost just as any other quality. The cost of product differentiation is represented by production in the long run to the left of the minimum long-run average cost. This means that "the difference between the actual long-run equilibrium output and output at minimum cost is then a measure of the cost of producing differentiation rather than a measure of excess capacity"[23]. But according to Chamberlin, this is true only when effective price competition in the market is present, because it is only when effective price competition among the sellers is present that the buyers would have the degree of product differentiation they wish to purchase.

The ideal output as conceived by Chamberlin is illustrated in Fig. 28.15. According to Chamberlin, the long-run equilibrium under monopolistic competition is established at the point where the subjective or perceived demand curve $d_2d'_2$ is tangent to the long-run average cost curve and also proportional demand curve $D_2D'_2$ curve cuts this point of tangency. This is achieved at point E where $d_2d'_2$ is tangent to LAC and the proportional demand curve $D_2D'_2$ is cutting LAC at this point. This long-run equilibrium point has been achieved after adjustment in the number of firms in the 'group' has taken place and also price competition has fully worked itself out. Corresponding to this long-run equilibrium point E, the firm is producing output OQ_4. But point E lies on the falling portion of the long-run average cost curve. Chamberlin regards long-run equilibrium output OQ_4 corresponding to point E as the ideal output under monopolistic competition in which there is product differentiation. According to Chamberlin, output which corresponds to the minimum point of the long-run average cost curve LAC cannot represent ideal output under monopolistic competition involving product differentiation.

Thus with active price competition when long-run equilibrium is at pont E where perceived demand curve $d_2d'_2$ is tangent to the long-run average cost curve and the firm is producing OQ_4 output which represents ideal one under product differentiation, There is no excess capacity. Therefore, *according to Chamberlin, so long as there is active price competition in the market, monopolistic competition does not create excess capacity. In his view, excess capacity arises when there is free entry of firms but no price competition.* Thus it is the entry of firms in the long run in the monopolistically competitive group which, *in the absence of price competition*, gives rise to the excess capacity.

Chamberlin[24] has given various reasons for the lack of price competition between the firms in monopolistic competition. First, the business firms may follow the policy of "*to live and let live*" and therefore they may not indulge in price cutting. They may set prices with reference to cost (including normal profits) rather than demand and accordingly seek only normal profits rather than maximum profits; more or less taking it granted that they will enjoy usual share of the total market demand. Secondly, the business firms in monopolistic competition may have entered into formal or tacit agreements. They might have formed 'price association' which builds up the spirit of 'maintaining price' among the competitors and preventing them from price cutting. Thirdly, *business or professonal ethics*, also prevents them from disturbing the market by price cutting. It is generally considered unethical in the professions to compete on the basis of price. The fourth factor preventing price

23. C.F. Ferguson, *Microeconomic Theory* (1966), p. 262.
24. *Op. Cit.*, pp. 105-107.

cutting is the fear of business firms that the lower price may lead the consumers to regard the particular product as of inferior quality. It is generally seen in the real world that many consumers blindly link quality with price.

In the fifth place, the business firm may make disguised or hidden price cuts while maintaining the price openly. Since open price cuts generally bring about retaliation from the rivals, it is considered desirable to give certain non-price concessions, offer facilities free of charge, such as free coupons, premiums, etc., while displaying and charging the same price as competitors. All these extra non-price concessions or facilities are regarded as hidden price cuts. Finally, the prices may not be free to move at all, for they are often set by custom or tradition.

As pointed out above, if price competition is in fact absent individual firms will have no regard for the existence of perceived demand curve *dd′*. They will be concerned only with market demand curve *DD′* which represents the effects of price rise or price decline when all firms in the group simultaneously raise or reduce price. According to Chamberlin, the result of the absence of price competition among the firms in monopolistic competition "is excess productive capacity for which there is no automatic corrective. Such excess capacity may develop, of course, under pure competition owing to miscalculation on the part of producers or to sudden fluctuations in demand or cost conditions. *But it is the peculiarity of monopolistic competition* that it may develop over long periods *with impunity*, prices always covering costs, and may, in fact, become permanent and normal through the failure of price competition to function. The surplus capacity is never case off, the result is high prices and waste"[25]. He further says that his theory of monopolistic competition "affords an explanation of such wastes in the economic system—wastes which are usually referred to as "wastes of competition". In fact they could never occur under pure competition. They are wastes of monopoly—of the monopoly elements in monopolistic competition".[25]

PRICE-OUTPUT EQUILIBRIUM UNDER MONOPOLISTIC COMPETITION COMPARED WITH THAT UNDER PERFECT COMPETITION

It is useful to explain how price-output equilibrium under monopolistic competition differs from that under perfect competition.

1. Price is greater than MC under monopolistic competition. A significant difference between the two relates to the relation between price and marginal cost. Whereas in equilibrium under perfect competition, price is equal to marginal cost, in equilibrium under monopolistic competition price is greater than marginal cost. Since under perfect competition, an individual firm cannot influence the price of its product and takes price as given and constant, the demand or average revenue curve is a horizontal straight line and marginal revenue (*MR*) is equal to average revenue (*AR*) or price. Therefore, under perfect competition when a firm equates marginal cost with marginal revenue so as to maximise its profits, the former also becomes equal to price.

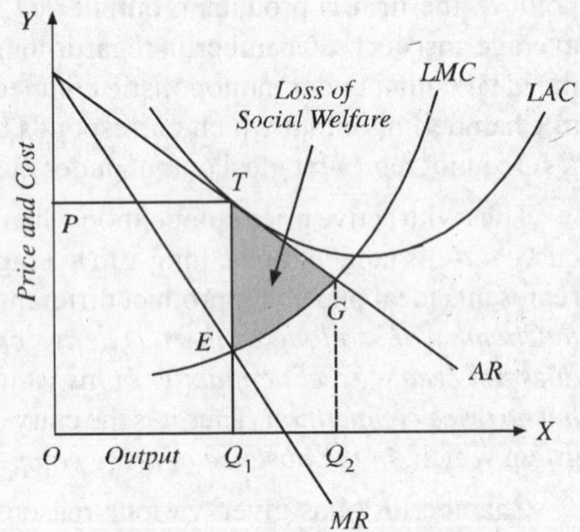

Fig. 28.16. *Loss of Social Welfare.*

On the other hand, under monopolistic competition, a firm exercises some control over the price of its product and the demand curve for it, representing prices at various quantities, slopes downward. As a result, marginal revenue (*MR*) curve lies below average revenue (*AR*) curve. Therefore, in order to maximise profits when a firm under monopolistic competition equates marginal cost with marginal

25. E.H. Chamberlin, *op. cit.*, p. 109.

revenue, price stands at a higher level than marginal cost. This is clear from Fig. 28.3 where, in equilibrium, price determined is equal to *MQ* or *OP*, which is greater than marginal cost (*ME*).

It should be noted that *producing level of output much less than at which marginal cost equals price implies a loss of social welfare.* It should be noted that social welfare is maximum when output is extended to the point where price is equal to long-run marginal cost. It will be seen from Fig. 28.16 that such point is *G* where price or *AR* = *LMC*. But as will be seen from Fig. 28.16 the firm working under conditions of monopolistic competition produces OQ_1 in the long-run equilibrium. Thus, the area *TEG* represents loss of social welfare which is also called *dead weight loss*. This indicates inefficiency of monopolistic competition.

2. Long-run equilibrium under monopolistic competition is established at less than optimum size. Another important difference between the equilibrium under monopolistic competition and perfect competition is that whereas a firm in long-run equilibrium under monopolistic competition produces less than its optimum size of output, under perfect competition long-run equilibrium of the firm is established at the minimum point of the long-run average cost curve. In other words, a firm under perfect competition tends to be of the optimum size. But a firm under monopolistic competition, as is evident from Fig. 28.5 stops short of the optimum point and operates in the long run at the point at which long-run average cost is still falling. In Fig. 28.5 the firm produces output *OQ*, while a firm under pure or perfect competition would have produced output *OR* at which long-run average cost is minimum. The firm under monopolistic competition can reduce its cost of production by expanding output to the point *R* but it will not do so because by expanding output beyond *OQ* it will be reducing price more than the average cost. It is, therefore, clear that by producing *OQ* instead of *OR* in the long run, the firm under monopolistic competition does not use its capacity fully. (The firm would be using its capacity fully if it produces optimum output *OR*). Thus, capacity equal to *QR* lies unused in the firm under monopolistic competition. This unused capacity is called excess capacity which is a prominent feature of long-run equilibrium under monopolistic competition.

3. Price under monopolistic competition is greater than competitive price. Further, it may be noted that in the long-run equilibrium, firms under monopolistic competition make only normal profits as under perfect competition, but the price set under monopolistic competition is higher than competitive price. In Fig. 28.5, price set in long-run equilibrium under monopolistic competition is *QT* while competitive price in the long run would have been equal to *RL*. This higher price under monopolistic competition is due to monopoly element contained in it. The monopoly element involved in monopolistic competition makes the demand or average revenue curve facing an individual firm downward sloping and a downward-sloping average revenue curve can be tangent to the average cost curve only to the left of its minimum point.

Thus price under monopolistic competition will be higher than the competitive price due to the monopoly element in monopolistic competition. But, in spite of the higher price, a firm under monopolistic competition will not be making profits above normal in the long run. We may, therefore, say that a firm under monopolistic competition, in the long-run equilibrium, charges higher price without enjoying monopoly profits. A significant result follows from this. It is that the *non-existence of abnormal profits is no indicator of the absence of monopoly element.* In long-run equilibrium under monopolistic competition, as seen above, the firm has monopoly power (it has a sole control over its own differentiated product with the result that demand curve for it slopes downward) but it does not make any super-normal profits in the long-run.

4. Excess capacity and other wastes of competition under monopolistic competition. Further, it is evident from the long-run equilibrium situation of a firm under monopolistic competition (see Fig. 28.5) that it occurs at the output level which falls short of the output *OR* at which the long-run average cost is minimum. This implies, that maximum economic efficiency is not achieved under monopolistic competition because the resources are not being used to produce the good at the minimum possible average cost. That is why economists such as Harrod, Cassel have called the output difference *QR* as *the excess capacity that tends to prevail under monopolistic competition which represents waste of resources*. However, it is worth noting that some economists do not consider *QR* as the excess capacity. According to them, this small loss of output and consequently higher average cost is price which people are paying for the product variety which they get under monopolistic competition due to product differentiation as compared to perfect competition. Further, *firms under monopolistic competition spend a lot of money on advertisement and other sales promotion activities which also represents* wastes of competition.

According to Chamberlin, *a lot of excess capacity is created under monopolistic as a result of entry of a large number of firms only if price competition does not prevail.* If price competition prevails then there is not much excess capacity under monopolistic competition. In the long-run too mainy firm enter the industry so that market is shared among many and in the absence of price competition each produces much less than its full capacity.

MONOPOLISTIC COMPETITION AND ECONOMIC EFFICIENCY

Perfect competition is said to be the ideal market form as it ensures maximum possible social welfare. In partial equilibrium analysis welfare is measuerd by the consumer surplus gained by the consumers and producer surplus earned by the producers. Under perfect competition, given the demand and cost conditions the maximum possible sum of consumer surplus and producer surplus is achieved so that social welfare is maximised. Besides, under perfect competition in the long run firms operate at the minimum point of the long-run average cost curve which ensures maximum efficiency in the use of resources. According to this, not only *optimum scale of plant* (*i.e*, plant with minimum *LAC*) is set up but also that plant is operated at its *optimum* or full *capacity*, that is, at the minimum point of short-run average cost curve of the plant.

On the other hand, firms operating in monoplistic competitive environment earn zero economic profits in their long-run equilibrium but they operate with *two types of inefficiencies*. First, as discussed above in detail, firms opeate in the long run with *excess capacity* which implies inefficiency or waste in the use of resources and impedes the achievement of maximum social welfare. In long-run equilibrium under monopolistic competition, firms operate at less than the scale of output at which long-run average cost is minimum. In this long-run equilibrium situation, neither the optimum scale of plant is set up, nor the plant actually set up is operated at its optimum capacity (*i.e.*, at minimum point of short-run average cost). This implies higher cost and price as well as lower output than if production had been done at minimum point of *LAC*. Thus, the existence of excess capacity under monopolistic competition causes inefficiency in the use of resources and loss of consumer welfare.

The second type of inefficiency that prevails under monopolistic competition is calle*d allocative inefficiency* which refers to the misallocation of resources among products which also causes loss of welfare. *Allocative efficiency is achieved if price of a product is fixed equal to the marginal cost of production*. However, under monopolistic competition firms are in long-run equilibrium at the level of output at which price exceeds marginal cost of production. Consider Fig. 28.16, firm is in long-run equilibrium at output OQ_1 at which *MR* equals *MC* but price fixed is Q_1T or *OP* which exceeds marginal cost Q_1E at the equilibrium output. The ideal output from the viewpoint of allocative efficiency is OQ_2 at which price equals *MC*. It will be observed from the *AR* curve in the figure that the price which consumers are willing to pay for additional units from Q_1 to Q_2 exceeds the marginal cost of production to be incurred by the society. If output were extended to Q_2 at which demand (*AR*) curve cuts the long-run marginal cost curve *LMC*, the total surplus (sum of consumer surplus and producer surplus) could be increased by an amount equal to the shaded area *ETG*. This shaded area *ETG* represents *dead weight loss* of welfare. This dead-weight loss accrues due to the monopoly element involved in monopolistic competition. It may be recalled that monopoly element is present in monopolistic competition because products of different firms are differentiated and each of them has some control over the price of its product. Expansion of output to Q_2 requires allocation of more resources to the production of the product. By restricting output to Q_1 and thereby raising price of the product, firms operating under monopolistic competition do not allocate resources efficiently which cause dead weight loss in welfare. This is called dead weight loss because there is no compensating gain by any one else.

Comments. Due to the existence of above two types of inefficiencies monopolistic competition is considered socially undesirable form of market structure by traditional economists. However, modern economists do not think so. This is for two reasons. First, the excess capacity that exists under monopolistic competition is very small because in the long run under monopolistic competition demand curve facing each firm becomes highly elastic as a result of the entry of new firms in the industry making available more close substitutes in the market.

Second, due to the demand (*AR*) curve becoming highly elastic in the long run monopolistic competition ensures that the difference between the price and marginal cost will also be very small which means whatever dead weight loss in welfare occurs is not very significant.

Above all there is gain or benefits to consumers because of *product variety* found in this market structure. Most consumers derive great satisfaction from the ability to choose among a wide variety of products and brands that exist under monopolistic competition. The gain in consumer welfare or benefits due to product diversity may be weighed against the cost of little inefficiency that prevails under monopolistic competition.

QUESTIONS FOR REVIEW

1. What is monopolistic competition? Explain the important features of monopolistic competition.
2. What is product differentiation? What role does it play in the determination of price and output under monopolistic competition?
3. Analyse the short-run and long-run equilibrium of a firm working under monopolistic competition. [*C.U., B.Com.*, (*H*), *1999*]
4. What is group equilibrium? Explain how group equilibrium is achieved under monopolistic competition.
5. (*a*) What is meant by excess capacity? Show that how a firm working under monopolistic competition works with excess capacity in the long run. [*D.U.B.Com.* (*H*) *2001*]
   (*b*) What causes excess capacity under monopolistic competition? Why is it undesirable?
6. *Are there any benefits of excess capacity associated with monopolistc competition*?
7. What are wastes of competition? How are they found under monopolistic competition?
8. Compare price-output equilibrium under monopolistic competition with that under (*a*) perfect competition, (*b*) monopoly.
9. Distinguish between selling costs and production cost? Explain the optimal level of advertising expenditure.
10. Explain Chamberlin's concepts of *proportional* and *perceived* demand curves of firms operating under monopolistic competition. Explain with their help short-run and long-run firm's equilibrium under monopolistic competition.
11. Explain Chamberlin's concept of excess capacity. What, according to him, is responsible for the emergence of excess capacity under monopolistic competition?
12. Explain the concept of economic efficiency. Evaluate econmic inefficiency of firms operating under monopolistic competition.
13. Distinguish between production cost and selling costs. How does monopolistically competitive firm determine the optimal amount of selling costs. [*D.U. B.A.*(*Hons*) *1997*]
14. "Monopolistically Competitive industries are characterised by too many firms each of which produces too little" [*D.U. B.A.*(*Hons*) *1987*]
15. Assuming that a firm operating under monopolistic competition does not engage in price competition and that it has already decided what kind and quality of product it will produce, how will the firm decide how much selling costs to incur in order to maximise its profits? [*D.U. B.A.*(*Hons*)]
16. Under monopolistic competition, which demand curve (perceived demand or proportionate demand curve) determines the firm's choice of output and price? Why is the firm's demand curve flatter than the total market demand curve? Why must the two demand curves intersect at the equilibrium point? [*D.U. (B. Com Hons) 2004*]

Chapter 29

PRICE AND OUTPUT UNDER OLIGOPOLY

We have studied price and output determination under three market forms, namely, perfect competition, monopoly, and monopolistic competition. However, in the real world economies we find that many of the markets or industries are oligopolistic. Oligopoly is an important form of imperfect competition. Oligopoly is said to prevail when there are few firms or sellers in the market producing or selling a product. In other words, when there are two or more than two, but not many, producers or sellers of a product, oligopoly is said to exist. Oligopoly is also often referred to as "*Competition among the Few*". *The simplest case of oligopoly is duopoly which prevails when there are only two producers or sellers of a product*. Analysis of duopoly raises all those basic problems which are confronted while explaining oligopoly with more than two firms.

Although there is no borderline between few and many but when the number of sellers of a product are two to ten, oligopoly situation is said to exist. When products of a few sellers are homogeneous, we talk of *Oligopoly without Product Differentiation or Pure Oligopoly.* On the other hand, when products of the few sellers or firms, instead of being homogeneous, are differentiated but close substitutes for each other, *Oligopoly with Product Differentiation or Differentiated Oligopoly* is said to prevail.

CHARACTERISTICS OF OLIGOPOLY

In oligopoly some special characteristics are found which are not present in other market structures. We discuss some of these characteristics below:

Interdependence. The most important feature of oligopoly is the interdependence in decision-making of the few firms which comprise the industry. This is because when the number of competitors is few, any change in price, output, product etc. by a firm will have a direct effect on the fortune of its rivals, which will then retaliate in changing their own prices, output or products as the case may be. It is, therefore, clear that the oligopolistic firm must consider not only the market demand for the industry's product but also the reactions of the other firms in the industry to any action or decision it may take. Since more than one reaction-pattern is possible from the other firms, we have to make some assumptions about the reaction of the others before we can provide a definite and determinate solution of price-output fixation under oligopoly.

Importance of advertising and selling costs. A direct effect of interdependence of oligopolists is that the various firms have to employ various aggressive and defensive marketing weapons to gain a greater share in the market or to prevent a fall in their market share. For this various firms have to incur a good deal of costs on advertising and on other measures of sales promotion. Therefore, there is a great importance of advertising and selling costs under conditions of market situation characterised by oligopoly. Prof. Baumol rightly says that "it is only under oligopoly that advertising comes fully into its own."[1] Under perfect competition, advertising by an individual firm is unnecessary in view of the fact that it can sell any amount of its product at the going price. A monopolist has also not to

1. William J. Baumol, *Economic Theory and Operations Analysis,* 3rd ed., p. 352.

make any competitive advertisement since he is the only seller of a product. A monopolist may perhaps advertise when he has to inform the public about his introduction of a new model of his product or he may advertise in order to attract potential consumers who have not yet tried his product. Under monopolistic competition advertising plays an important role because of the product differentiation that exists under it, but not as much important as under oligopoly. "Under oligopoly, advertising can become a life-and-death matter where a firm which fails to keep up with the advertising budget of its competitors may find its customers drifting off to rival products".[2]

In view of the fact that a firm in an oligopolistic industry competes by changing the advertisement costs, quality of the product, prices, output etc., *the presence of competitive conditions in it can hardly be denied.* To an oligopolist "Competition can consist not in the quiescent stalemate of perfect competition where there is no battle because there is never anyone strong enough to disturb the peace. Rather to him, true competition consists of the life of constant struggle, rival against rival, which one can only find under oligopoly (or, on a smaller scale, under conditions of monopolistic competition)."[3]

Group behaviour. Further, another important feature of oligopoly is that for the proper solution to the problem of determination of price and output under it, analysis of group behaviour is important. Theories of perfect competition, monopoly and monopolistic competition present no difficult problem for making suitable assumption about human behaviour. In cases of perfect competition and monopolistic competition (with a large number of firms), the economists assume that the business firms behave in such a way as to maximise their profits. Assumption of profit maximisation gives overall good results in these situations where mass of people are involved and there is no interdependence of firms. On the other end, the theory of monopoly deals with a sole individual and it is also appropriate to assume profit-maximising behaviour on his part.

But the theory of oligopoly is a theory of *group behaviour* not of mass or individual behaviour and to assume profit-maximising behaviour on the part of a producer of a group may not be very valid. There are few firms in a group which are very much interdependent. Given the present state of our economic and social science, there is no generally accepted theory of group behaviour. Do the members of a group agree to pull together in promotion of common interests or will they fight to promote their individual interests? Does the group possess any leader? If so, how does he get the others to follow him? These are some of the questions that need to be answered by the theory of group behaviour.

Indeterminateness of demand curve facing an oligopolist. Another important feature is the indeterminateness of the demand curve facing an oligopolist. The demand curve shows what amounts of its product a firm will be able to sell at various prices. Now, under perfect competition, an individual firm's demand curve is given and definite. Since a perfectly competitive firm is one among a large number of firms producing an identical product, it is incapable of influencing the price of its product by its own individual action. Therefore, a firm under perfect competition faces a perfectly elastic demand curve at the level of the going price in the market. On the other hand, a monopolist produces a product which has only remote substitutes. Therefore, a monopolist can safely ignore the effects of its own price changes on his distant rivals and therefore the monopolist faces a given and definite demand curve depending upon the consumer's demand for his product. Under monopolistic competition, where there is a large number of firms producing products which are close substitutes for each other, changes in price by an individual firm will have a negligible effect on each of its many rivals. Therefore, a firm under monopolistic competition can validly assume the prices of its rivals to remain unchanged when it makes changes in the price of its product. Thus, the demand curve for a firm under monopolistic competition can be taken as definite and is given by the buyers' preferences for its product.

---

2. *Ibid*, p. 352.
3. *Ibid.*

But the situation under oligopoly is quite different because of interdependence of the firms in it. Under oligopoly, a firm cannot assume that its rivals will keep their prices unchanged when it makes changes in its own price. As a result of this, the demand curve facing an oligopolistic firm loses its definiteness and determinateness because it goes on constantly shifting as the rivals change their prices in reaction to price changes by a firm.

ARE PRICE AND OUTPUT UNDER OLIGOPOLY INDETERMINATE?

We have explained above the various characteristics and problems of oligopoly. The readers will now like to know how the economists analyse the determination of price and output under oligopoly. Because of the interdependence of firms in oligopoly and the uncertainty about the reaction patterns of the rivals, the easy and determinate solution to the oligopoly problem is not possible. In other words, interdependence of firms in an oligopoly and consequently firm's reactions to each other's behaviour poses serious difficulties in establishing the theory of the determination of price and output in an oligopolistic market.

Interdependence of Firms in Oligopoly

A significant consequence of interdependence of firms in an oligopolistic market situation is that under it a *wide variety of behaviour patterns becomes possible*. "Rivals may decide to get together and co-operate in the pursuit of their objectives, at least so far as the law allows or, at the other extreme, they may try to fight each other to death. Even if they enter into an agreement it may last or it may break down. And the agreements may follow a wide variety of patterns."[4] Therefore, a large variety of models analysing price-output determination under oligopoly have been evolved by the economists depending upon the different assumptions about the behaviour of the oligopolistic group and the reaction patterns of rivals to a change in price or output by a firm.

Another difficulty that arises from interdependence of oligopolistic firms is the *indeterminateness of the demand curves facing individual firms*. As has been stated earlier, because of the interdependence an oligopolistic firm cannot assume that its rival firms will keep their price and quantities constant when it makes changes in its price. When an oligopolistic firm changes its price, its rival firms will retaliate or react and change their prices which in turn would affect the demand for the product of the former firm. Therefore, an oligopolistic firm cannot have sure and definite demand curve, since it keeps shifting as the rivals change their prices in reaction to the price changes made by it. Moreover, there is quite uncertainty about the rivals' reactions to a price change by one firm. That is, when an oligopoly firm cuts its price, whether its rivals will also cut their prices similarly, or whether they will keep their prices unchanged. If they cut their prices, whether they will cut their prices by the same amount, or by a smaller or greater amount. A definite and determinate demand curve for a firm can be drawn if its rivals' prices remain unchanged or if it is known beforehand that they will change their prices in a *certain particular way* in response to price changes by one firm. But under oligopoly there is no certainty about the reactions of the rivals to the price changes made by a firm. Hence the demand curve of an oligopolistic firm cannot be easily determined.

Now, when an oligopolist does not know the demand curve confronting him, what price and output he will fix cannot be ascertained by economic analysis. In other words, in view of the indeterminateness of demand curve for a firm under oligopoly, the solution for the determination of price and output under oligopoly cannot be provided by the economic theory. Under conditions of perfect competition, monopoly and monopolistic competition, an individual firm faces a determinate demand curve which has a corresponding definite marginal revenue curve. Then, on the basis of profit maximisation principle, the determinate solution for the price and output fixation under perfect competition, monopoly and monopolistic competition is found by the equality of marginal revenue with marginal cost. This solution cannot be applied to price and output determination under

4. William J. Baumol, *opcit.*, pp., 223–24.

oligopoly without qualification or making some additional assumptions because the individual firm's demand curve and therefore the marginal revenue curve is indeterminate or unknown. Thus even if profit maximisation assumption is considered as valid under conditions of oligopoly, no determinate solution for price-output fixation can be provided because of the indeterminateness of demand curve.

Even when the firms of an oligopoly do not enter into collusion, tacit or formal, or choose a leader from among themselves and instead try to compete with each other no single and simple solution is possible as to how a firm will fix its price and output. This is because of the uncertainty about the reaction patterns of the rivals to a move by one firm. This uncertainty about the reaction patterns of competitors poses a serious analytical difficulty in the way of providing a determinate solution for the oligopoly problem. Quoting Prof. Baumol again, "When a businessman wonders about his competitors' likely response to some move which he is considering, he must recognize that his competitors, too, are likely to take this interdependence phenomenon into account. The firms' attempts to outguess one another are then likely to lead to an *interplay of anticipated strategies and counterstrategies which is tangled beyond hope of direct analysis*."[5] Thus under oligopoly a firm is likely to imagine an infinite sequence of compounded hypotheses such as "If I make move A, he may consider making countermove B, but he may realise that I might then respond by making move C, in which case..., and so an *ad infinitum*."

Profit-Maximising Assumption Challenged in Case of Oligopoly

Again, a determinate solution to the price-output problem in other market forms (perfect competition, monopoly and monopolistic competition) is arrived at by assuming profit-maximising motive on the part of the firms. But some economists have challenged the validity of the profit-maximising hypothesis in oligopolistic situations. According to Prof. Rothschild, *oligopolists aim at maximising their security or achieving reasonable amount of stable profits* over a long period of time rather than maximising profits at a time.[6] On the other hand, Prof. Baumol thinks that in oligopolistic circumstances it is legitimate to assume *sales maximising objective* on the part of the firms.[7] Some other economists think that managers of oligopolistic firms maximise their own *utility function*. Still others like R.L. Morris think that firms try to maximise their *growth rate*. Finally, some economists assert that oligopolists do not maximise anything, they merely *satisfice*. In other words, they aim to obtain satisfactory profits rather than maximum profits. All this controversy about the real objective of the firms relates especially to the oligopolistic firms. This controversy about the most probable objective of the oligopolists further introduces indeterminacy in the analysis of price and output under oligopoly.

In view of above, there is no single determinate solution of the oligopoly problem but a wide variety of possible solutions, each depending upon different assumptions. It is worth noting as to what exactly economists mean by *indeterminacy*. When no single solution of a problem is possible, economists generally say that the problem has no determinate solution. Thus economists usually speak of indeterminacy where mathematicians would speak of a multiplicity of solutions. Prof. Fritz Machlup explains the meaning of indeterminancy as follows:

"In a general way, economists speak of indeterminacy if not enough information is available to give a safe and unambiguous answer to a question before them. If they wish to solve a problem—for example, how the price of a certain commodity will change under certain conditions—but find that the data which are assumed to be given *would permit of two or more* (*perhaps of an infinite number of*) answers, they will say that the problem has no determinate solution."[8]

---

5. *Op. cit.,* p. 224 (italics added).
6. See for Rothschild's view about security motive in chapter 22.
7. See for Baumol's view about sales maximization objective, chapter 30.
8. Fritz Machlup, The *Economics of Sellers'* p. 415 (italics added).

Various Approaches to Determination of Price and Output under Oligopoly

It may, however, be pointed out that in spite of what has been said above, economists have tried to provide a determinate solution to the oligopoly problem. But there is no any *single determinate solution, but a number of determinate solutions* depending upon different assumptions. The determinate solution to the oligopoly problem has been provided in the following ways.

Ignoring interdependence. First, for providing a determinate solution to the price-output determination, some economists have assumed that oligopolistic firms *ignore interdependence.* Now, when interdependence disappears from decision-making of the oligopolistic firms, the demand curve facing them becomes determinate and can be ascertained. With this, the standard analysis of the theory of the firm can be applied to provide a determinate solution for price and output problem of oligopoly. Classical models of duopoly put forward by Cournot and Bertrand fall in this category. In both those models, oligopolistic interdependence has been ignored. Cournot[9] in his famous, model assumed that each oligopolistic firm would set its output in the belief that its rival firms' output would remain constant. On the other hand, Bertrand[10] assumed that an oligopolistic firm would set its price in the belief that its rival firm would keep its price unchanged. But for providing a solution for price and output determination under oligopoly by ignoring the interdependence is a fundamentally mistaken approach. Rothschild rightly writes, "The determinate solution can be reached.... if it is assumed that the oligopolists do not take into account the effect of their action on the policy of their rivals as in the famous Cournot and Bertrand's solutions; ...But this type of approach is absolutely valueless, because it only solves the oligopoly problem by removing from the analysis its most essential differentiating aspect: the oligopolists' consciousness of their interdependence."[11]

Predicting reaction pattern and counter-moves of rivals. The second approach to provide a determinate solution to the price and output problem of oligopoly is to assume that oligopolistic firm is able to *predict the reaction patterns and counter moves of his rivals*. In this approach various oligopoly models based on different assumptions regarding the particular reaction patterns have been propounded. Chamberlin[12] while recognizing that oligopolistic firms are conscious of their interdependence took some assumptions regarding the reaction pattern and provided a determinate solution according to which under duopoly or oligopoly, monopoly output and price are determined so that *profits of duopolists (oligopolists) are jointly maximised.* P.M. Sweezy and Oxford economists Hall and Hitch assumed that an oligopolist firm considers that while the price increase by it will not be followed by its rival firms, the price reduction will be matched by them. With such an assumption about the reaction pattern, demand curve facing an individual oligopolistic firm is of a kinked type with a kink at the current price. We shall discuss the kinked demand curve approach to oligopoly problem in detail later in this chapter. Many other economists assuming other reaction patterns have propounded various other oligopoly models.

Forming a collusion to maximise joint profits. The third approach to oligopoly problem assumes that oligopolistic firms, realising their interdependence, will pursue their common interest and will form a collusion, formal or tacit, that is, will enter into agreement and work in the pursuit of their common interest. They will maximise joint profits and share profits, market or output as agreed to between them. A variant of this approach is that firms in an oligopoly would accept one firm as a leader, which may be a dominant or low-cost firm, and they will follow their leader in the fixation of price of the product. Both the collusive and price leadership oligopoly models will be critically examined later in this chapter. In case of collusion and price leadership, the problem of indeterminateness of demand curve is not encountered.

---

9. A. Cournot, *"Researches into the Mathematical Principles of the Theory of Wealth"*, translated by Nathaniel T. Bacon, Macmillar Co., New York, 1897.
10. J. Bertrand, "Theorie Mathematique de la Richesse Sociate", *Journal des Savants*, Paris, September 1893.
11. K.W. Rothschild. "Price Theory and Oligopoly", *Economic Journal,* Vol. 57, 1947, reprinted in *Reading in Price Theory* (AEA).
12. E.H. Chamberlin, *The Theory of Monopolistic Competition,* chapter 3 and Appendix A.

Game theory approach to oligopoly. Another significant approach to the oligopoly is that of the *theory of games*, put forward by Neuman and Morgenstern. In the theory of games, an oligopolistic firm does not guess at its rivals' reaction pattern, but *calculates the optimal moves by rival firms*, that is, their best possible strategies and in view of that adopts its own policies and counter moves. The explanation of the application of the theory of games to the oligopoly problem is an advanced-level study and will not be discussed in this book.

From our above analysis it follows that there is no single determinate solution to the price-output fixation under oligopoly. As said above, economists have developed a large number of models by taking different assumptions regarding the behaviour of the oligopolistic group (that is, whether they will cooperate together or fight with each other), regarding the objective they seek to achieve (that is, whether they are assumed to maximise individual or joint profits or they are assumed to maximise security or sales), and regarding the *different reaction patterns* of rival firms to price and output changes by one firm.

COLLUSIVE OLIGOPOLY: CARTELS

In order to avoid uncertainty arising out of interdependence and to avoid price wars and cut throat competition, firms working under oligopolistic conditions often enter into agreement regarding a uniform price-output policy to be pursued by them. The agreement may be either formal (open) or tacit (secret). But since formal or open agreements to form monopolies are illegal in most countries, agreements reached between oligopolists are generally tacit or secret. When the firms enter into such collusive agreements formally or secretly, collusive oligopoly prevails. But collusions are of two main types: (*a*) cartels and (*b*) price leadership. In a cartel type of collusive oligopoly, firms jointly fix a price and output policy through agreements. But under price leadership one firm sets the price and others follow it. The one which sets the price is a price leader and the others who follow it are its followers. The follower firms adopt the price of the leader, even though they have to depart from their profit-maximising position, as they think that it is to their advantage not to compete with their leader and between themselves. In this present section we confine ourselves to explain the determination of price and output under the cartel type of collusive oligopoly and in the next section we will explain price and output determination under price-leadership form of collusive oligopoly.

Originally, the term 'cartel' was used for the agreement in which there existed a common sales agency which alone undertook the selling operations of all the firms that were party to the agreement. But now-a-days all types of formal or informal and tacit agreements reached among the oligopolistic firms of an industry are known as cartels. Since these cartels restrain competition among the member firms, their formations have been made illegal in some countries by the Governments passing laws against them. For instance, the formation of a cartel is illegal in U.S.A. under the Anti-Trust Laws passed there. However, in spite of the illegality of cartels they are still formed in U.S.A. through secret devices and by adopting some means or the other shrewd businessmen are able to evade the anti-monopoly laws.

Formal collusion or agreement among the oligopolists may itself take various forms. An extreme form of collusion is found when the member firms agree to surrender completely their rights of price and output determination to a 'Central Administrative Agency' so as to secure maximum joint profits for them. Formation of such a formal collusion is generally designated as *perfect cartel*. Thus under perfect cartel type of collusive oligopoly, the price and output determination of the whole industry as well as of each member firm is determined by the common administrative authority so as to *achieve maximum joint profits for the member firms*.

The total profits are distributed among the member firms in a way already agreed between them. The share from total profits of each member firm is not necessarily in proportion to the output quota it has to supply and the cost it incurs on it. The output quota to be produced by each firm is decided by the central administrative authority in such a way that the total costs of the total output produced is minimum. In fact, under perfect cartel, the central authority determines the separate outputs to be produced by the various members and the price they have to charge in the same way as a monopolist operating multiple plants would do. Now, the question arises as to what outputs different firms in a cartel will be asked to produce so that the total cost is made minimum. Total cost will be minimised when the various firms in the cartel produce such separate outputs so that their marginal costs are equal. This is because if the marginal costs of the member firms are not equal, then the marginal units of output could be produced at a smaller cost by the firms with a lower marginal cost than by those with a higher marginal cost.

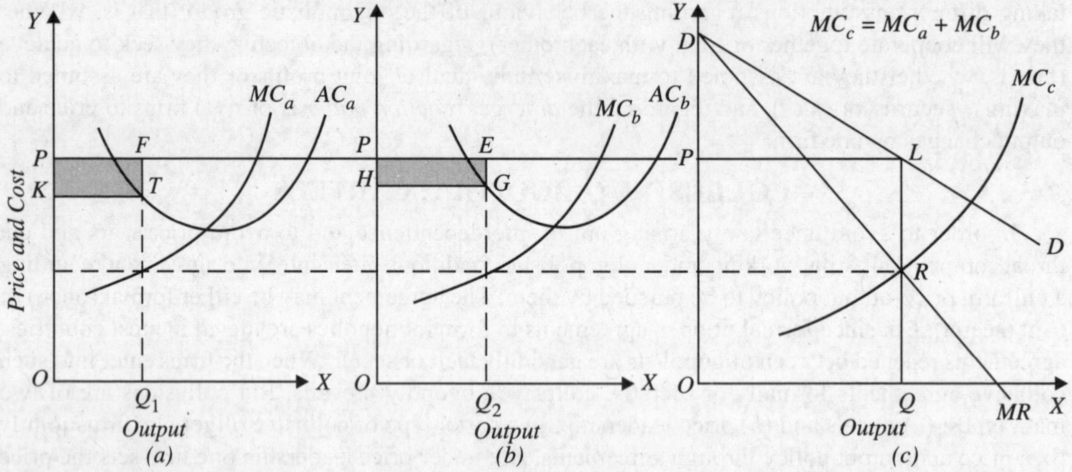

Fig. 29.1. *Price and Output Determination under Cartel: Joint Profit Maximization*

Let us now see how the cartel works and determines its price and output. Let us assume that two firms have formed a cartel by entering into an agreement. We assume that the *cartel will aim at maximising joint profits for the member firms*. First of all, the cartel will estimate the demand curve of the industry's product. As the demand curve facing a cartel will be the aggregate demand curve of the consumers of the product, it will be sloping downward as is shown by the curve *DD* in Fig. 29.1(*c*) Marginal revenue curve *MR* showing the addition to cartel's revenue for successive additions to its output and sales will lie below the demand curve *DD*. Cartel's marginal cost curve (MC_C) will be given by the horizontal addition of the marginal cost curves of the two firms. This has been done in Fig. 29.1(*c*) where MC_C curve has been obtained by adding horizontally marginal cost curves MC_a and MC_b of firms *A* and *B* respectively. It should be noted that cartel's marginal cost curve MC_c, obtained as it is by horizontal addition of marginal cost curves of the two firms, will indicate the minimum possible total cost of producing each industry output on it; each industry output being distributed among the two firms in such a way that their marginal costs are equal.

Now, the cartel will maximise its profits by fixing the industry's output at the level at which *MR* and *MC* curves of the cartel intersect each other. It will be seen in Fig. 29.1(*c*) that *MR* and *MC* curves cut each other at point *R* or output *OQ*. It will also be seen from the demand curve *DD* that the output *OQ* will determine price equal to *QL* or *OP*. Having decided the total output *OQ* to be produced, the cartel will allot output quota to be produced by each firm so that the marginal cost of each firm is the same. This can be known by drawing a horizontal straight line from point *R* towards the *Y*-axis. It will be seen from the figure that when firm *A* produces OQ_1 and firm *B* produces OQ_2, the marginal costs of the two firms are equal. The output quota of firm *A* will be OQ_1, and of firm *B* will be OQ_2. It is worth noting that the total output *OQ* will be equal to the sum of OQ_1 and OQ_2.

Thus, the determination of output *OQ* and price *OP* and the outputs OQ_1 and OQ_2 by the two firms *A* and *B* will ensure the maximum joint profits for the member firms constituting the cartel. It will be seen from Fig. 29.1(*a*) that with output *OQ* and cartel price *OP*, the profits made in firm *A* are equal to *PFTK* and with output OQ_2 and cartel price *OP* the profits made in firm *B* are equal to *PEGH*. The sum of the profits, that is, the joint profits made by the cartel will be maximum under the given demand and cost conditions as they have been arrived at as a result of equating combined marginal cost (MC_c) with the combined marginal revenue (*MR*). *The allocation of output quota to each of them is made on the grounds of minimissing cost and not as a basis for determining profit distribution*. Prof. J.S. Bain rightly says, "There is no particular reason for believing that the operating firms will retain just the profits resulting from the sale of their quotas, which are determined on cost grounds alone. Relative bargaining strengths will presumably determine the division of profits."[13]

Marketsharing Cartels

The formation of perfect cartels, as described above, has been quite rare in the real world even where their formation is not illegal. In a perfect cartel not only the price but also the output to be produced by each member of a cartel is decided by a central management authority and profits made in all of them are pooled together and distributed among the members according to the terms of a prior agreement. But when cartels are loose, instead of being perfect, the distribution of profits and fixation of outputs of individual firms are not determined in a manner perfect cartel does. In a loose type of cartel the market-sharing by the firms occurs. Further, there are two methods of market sharing: non-price competition and quotas.

Marketsharing by Non-price Competition. Under market sharing by non-price competition, *only a uniform price is set and the member firms are free to produce and sell the amount of outputs* which will maximise their individual profits. Though the firms agree not to sell at a price below the fixed price they are free to vary the style of their product and the advertising expenditure and to promote sales in other ways. That is, the price being a fixed datum, the firms compete on non-price basis. If the different member firms have identical costs, then the agreed uniform price will be the monopoly price which will ensure maximisation of joint profits. But when there are cost differences between the firms as is generally the case, the cartel price will be fixed by bargaining between the firms. The level of this price will be such as will ensure some profits to high-cost firms.

But with cost differences such loose cartels are quite unstable. This is because the low-cost firms will have an incentive to cut price to increase their profits and therefore they will tend to break away from the cartel. However, they may not openly charge lower price than the fixed one and instead cheat the other firms by giving secret price concessions to the buyers. However, as the rivals gradually lose their customers, the cheating by the low-cost firms will be ultimately discovered and consequently open price war may commence and cartel breaks down.

Marketsharing by Output Quota. The second type of market-sharing cartel is the agreement reached between the oligopolistic firms regarding quota of output to be produced and sold by each of them at the agreed price. If all firms are producing homogeneous product and have same costs, the monopoly solution (that is, the maximisation of joint profits) will emerge with the market being equally shared by them. However, when costs of member-firms are different, the different quotas for various firms will be fixed and therefore their market shares will differ. The quotas and market shares in case of cost differences are decided through bargaining between the firms. During the bargaining process, two criteria are usually adopted to fix the quotas of the firms. One is the past level of sales of the various firms and the second is the productive capacity of the firms. However, the 'past-period sales' and 'productive capacity' of various firms are not very firm criteria as they can be easily manipulated. Ultimately the quotas fixed for various firms depend upon their bargaining power and skill.

---

13. J.S. Bain, *"Pricing, Distribution and Employment,"* 1953, p. 286.

The second common basis for the quota system and market sharing is the division of market regionwise, that is, the geographical division of the market between the cartel firms. In this arrangement, price and also style of the product of cartel firms may vary.

It is worth noting that all types of cartels are unstable when there exist cost differences between firms. The low-cost firms always have a tendency to reduce price of the product to maximise their profits which ultimately results in the collapse of the collusive agreement. Further, if the entry of firms in the oligopolistic industry is free, the instability of the cartel is intensified. The new entrants may not join the cartel and may fix a lower price of the product to sell a large quantity. This may start a price war between the cartel firms and the new entrants. We thus see that the stability of the cartel arrangement is always in danger.

PRICE LEADERSHIP

As mentioned above, price leadership is an important form of collusive oligopoly. Under it, one firm sets the price, others follow it. Price leadership also comes into existence either through tacit or formal agreement. But as the formal or open agreement to establish price leadership are generally illegal, price leadership is generally established as a result of informal and tacit understanding between the oligopolists. The competing oligopolists in an informal meeting choose a leader and agree to follow him in setting price.

Types of Price Leadership

Price leadership is of various types. Firstly, there is a *price leadership by a low-cost firm.* In order to maximise profits the low-cost firm sets a lower price than the profit-maximizing price of the high-cost firms. Since the high-cost firms will not be able to sell their product at the higher price, they are forced to agree to the low price set by the low-cost firm. Of course, the low-cost price leader has to ensure that the price which he sets must yield some profits to the high-cost firms—their followers.

Secondly, there is a *price leadership of the dominant firm*. Under this one of the few firms in the industry may be producing a vary large proportion of the total production of the industry and may therefore dominate the market for the product. This dominant firm wields a great influence over the market for the product, while other firms are small and are incapable of making any impact on the market. As a result, the dominant firm estimates its own demand curve and fixes a price which maximises its own profits. The other firms, which are small having no individual effects on the price of the product, follow the dominant firm accepting the price set by it and adjust their output accordingly.

Thirdly, there is a *barometric price leadership* under which an old, experienced, largest or most respected firm assumes the role of a custodian who protects the interests of all. He assesses the change in the market conditions with regard to the demand for the product, cost of production, competition from the related products etc. and makes changes in price which are best from the viewpoint of all the firms in the industry. Naturally, other firms follow him willingly.

Fourthly, there is *exploitative or aggressive price leadership* under which a very large or dominant firm establishes its leadership by following aggressive price policies and thus compels the other firms in the industry to follow him in respect of price. Such a firm will often initiate a move threatening to compete the others out of market if they do not follow him in setting their prices.

Price-output Determination under Low-cost Price Leadership

Economists have developed various models concerning price-output determination under price leadership making different assumptions about the behaviour of price leader and his followers. We shall first explain price-output determination under price leadership by a low-cost firm. In order to simplify our analysis we make the following assumptions:

(1) There are two firms, *A* and *B*. The firm *A* has a lower cost of production than firm *B*.

(2) The product produced by the two firms is homogeneous so that the consumers have no preference between them.

(3) Each of the two firms has equal share in the market. In other words, demand curve facing each firm will be the same and will be half of the total market demand curve of product.

Given the above assumptions, price and output determination under price leadership is illustrated in Fig. 29.2. Each firm is facing demand curve *Dd* which is half of the total market demand curve *DD* for the product. *MR* is the marginal revenue curve of each firm. AC_a and MC_a are the average and marginal cost curves of firm *A* and AC_b and MC_b are the average and marginal cost curves of firm *B*. Cost curves of firm *A* lie below the cost curves of firm *B* because we are assuming that firm *A* has a lower cost of production than firm *B*.

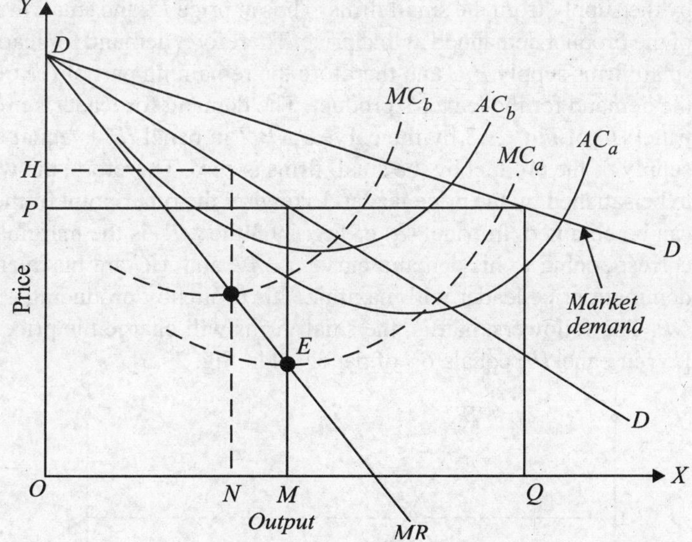

Fig. 29.2. *Price-Output Determination under Low-cost Price Leadership*

The firm *A* will be maximising its profits by selling output *OM* and setting price *OP*, since at output *OM*, its marginal cost is equal to the marginal revenue. Firm *B*'s profits will be maximum when it fixes price *OH* and sells output *ON*. It will be seen from the figure that profit-maximising price *OP* of firm *A* is lower than the profit-maximising price *OH* of firm *B*. Since the two firms are producing a homogeneous product, they cannot charge two different prices. Because the profit-maximising price *OP* of firm *A* is lower than the profit-maximising price *OH* of firm *B*, firm *A* will dictate the price to the firm *B* or, in other words, firm *A* will win if there is price war between the two and will emerge as a price leader and firm *B* will be compelled to follow. Given these facts, the agreement reached between them, even though tacit it may be, will require that the firm *A* will act as the price leader and firm *B* as the price follower.

It should be noted that firm *B* after having accepted firm *A* as the price leader will actually charge price *OP* and produce and sell *OM*. This is because at price *OP*, it can sell *OM* output like firm *A* because the demand curve facing each firm is the same. Thus both the firms will charge the same price *OP* and sell the same amount (*OM*). Note that the total output of the two firms will be $OM + OM = OQ$ which will be equal to the market demand for the good at price *OP*. But there is an important difference between the two. While firm *A*, the price leader, will be maximising its profits by selling output *OM* and charging price *OP*, the firm *B* will not be making maximum profits with this price-output combination because its profits are maximum at output *ON* and price *OH*. Profits earned by firm *B* by producing and selling output *OM* and charging price *OP* will be smaller than those of firm *A* because its costs are greater.

When the products of the price leader and his price-followers are differentiated, then the price charged by them will be different but the prices charged by the followers will be only slightly different either way from that of the price leader and they will conform to a definite pattern of differentials.

Price-output Determination under Price Leadership by the Dominant Firm

We now proceed to explain the determination of price and output when there exists price leadership by a dominant firm which is having a large share of the market with a number of small firms as followers each of which has a small share of the market. To explain this we assume that the dominant firm knows the total market demand curve for the product. Further, the dominant firm also knows the marginal cost curves of the smaller firms whose lateral summation yields the total supply of the product by the small firms at various prices. This implies that from his past experience the dominant firm can estimate fairly well the likely supply of the product by the small firms at various prices. With this information, the leader can obtain his demand curve. Consider panel (*a*) of Fig. 29.3 where *DD* is the market demand curve for the product S_m is the supply curve the product of all the small firms taken together. At each price the leader will be able to sell the part of the market demand not fulfilled by the supply from the small firms. Thus at price P_1, the small firms supply the whole of the quantity of the product demanded at that price. Therefore, demand for leader's product is zero. At price P_2, the small firms supply P_2C and therefore the remaining part of *CT* of the market demand will constitute the demand for the leader's product. The demand for leader's product has been separately shown in panel (*b*) of Fig. 29.3 by the curve d_L. P_2Z in panel *(b)* is equal to *C T* in panel *(a)*. At price P_3, the supply of the product by the small firms is zero. Therefore, the whole market demand P_3U will have to be satisfied by the price leader. Likewise, the other point of the demand curve for the price leader can be obtained. In panel (*b*) of Fig. 29.3 the MR_L is the marginal revenue curve of the price leader corresponding to his demand curve d_L. *AC* and *MC* are his average and marginal cost curves. The dominant price leader will maximize his profits by producing output *OQ* (or *PH*) and setting price *OP*. The followers, that is, the small firms will charge the price *OP* and will together produce *PB*. [*PH* in panel (*b*) equals *BS* of panel (*a*) in Fig. 29.3].

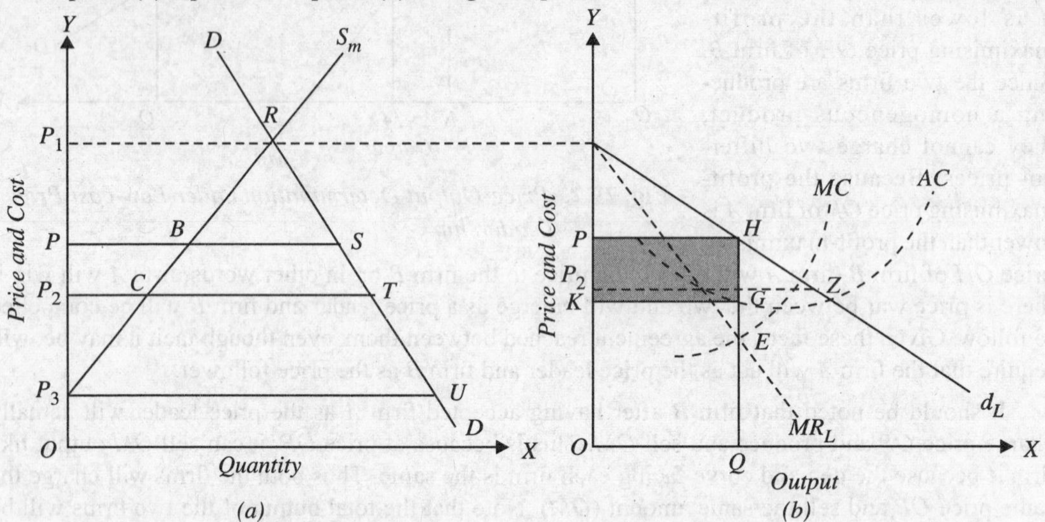

Fig. 29.3. *Price Leadership by the Dominant Firm*

It is worth noting that in order that profits of the leader are maximised it is not enough that followers should charge profit-maximizing price *OP* set by him, he will also have to ensure that they produce output *PB*. If the followers produce more or less than this, given the market demand *DD*, the leader will be pushed to a non-profit maximising position. This implies that if price-leadership is to remain, there must be some definite market-sharing agreement tacit though it may be.

Difficulties of Price Leadership

Price leadership involves many difficulties in the real world. First, the success of price leadership of a firm depends upon the correctness of his estimates about the reactions of his followers. If his estimates about the reactions of his rivals to price changes by it prove to be incorrect, then not only the success of his price policy but also his leadership in the market will be jeopardised. Secondly, when a price leader fixes a higher price than the followers would prefer, there is a strong tendency for the followers to make hidden price cuts in order to increase their shares of the market without openly challenging the price leader. A good number of devices which amount to secret price cutting are used by business firms. Some of these secret price-cutting devices are the offer of rebates, favourable credit terms, 'money back' guarantees, after-delivery free services, sale on the payment of price in easy installments with low rates of interest etc., and liberal entertainment of the buyers. Price leaders are generally fed up with the increasing number of concessions granted by their rivals and they make an open price cut to prevent further fall in their share of the market. In such circumstances price leadership becomes infructuous.

Another important difficulty of maintaining price leadership is the tendency on the part of the rivals to indulge in *non-price competition* to increase sales while go on charging the price set by the price leader. The devices used under 'non-price competition' include advertising and other methods of the sales promotion, like improvement of the quality of the product, in addition to the hidden price-product concessions mentioned above. While charging the same price, the rivals try to increase their share of the market by increasing the advertisement expenditure. As a result of this non-price competition, the price leader has also to adopt similar devices to prevent the fall in its sales or has to make outright cut in price in order to achieve his objective. In view of these facts, the price leader may not be able to maintain his leadership for a long time.

Further, there is a great limitation on the price leader to fix a high price of his product. This is because the high price will induce the rivals to make secret price cuts which will adversely affect the sales of the price leader. Moreover, a high price fixed by the price leader will attract new competitiors into the industry which may not accept his leadership. Lastly, differences in costs also pose a problem. If the price leader has higher costs, then the high price fixed by him will, as mentioned above, induce the rivals to undercut price or will attract the entry of new firms into the industry. If the price leader has lower costs than his rivals, he will set a low price which will antagonise his rivals who will disturb him quite frequently.

THE KINKED DEMAND CURVE THEORY OF OLIGOPOLY

It has been observed that many oligopolistic industries exhibit an appreciable degree of price rigidity or stability. In other words, in many oligopolistic industries prices remain sticky or inflexible, that is, there is no tendency on the part of the oligopolists to change the price even if the economic conditions undergo a change. Many explanations have been given of this price rigidity under oligopoly and most popular explanation is the so-called kinked demand curve hypothesis. The kinked demand curve hypothesis was put forward independently by Paul M. Sweezy[14], an American economist, and by Hall and Hitch[15], Oxford economists.

It is for explaining price and output *under oligopoly with product differentiation*, that economists often use the kinked demand curve hypothesis. This is because when under oligopoly products are differentiated, it is unlikely that when a firm raises its price, all customers would leave it because some customers are intimately attached to it due to product differentiation. As a result, demand curve facing a firm under differentiated oligopoly is not perfectly elastic. On the other hand, under oligopoly without product differentiation, when a firm raises its price, all its customers would leave it so that demand curve facing an oligopolist producing homogeneous product may be perfectly elastic. Further, under oligopoly without product differentiation, there is a greater tendency on the part of

---

14. Paul M. Sweezy, "Demand under Conditions of Oligopoly". *Journal of Political Economy*, Vol. XL VIII, August 1939, reprinted in American Economics Association, *Readings in Price Theory*.
15. R.L. Hall and C.J. Hitch, "Price Theory and Business Behaviour", Oxford Economic Papers, No. 2, May, 1939.

the firms to join together and form a collusion, formal or tacit, and, alternatively, to accept one of them as their leader in setting their price. No doubt, kinked demand curve has a special relevance for differentiated oligopoly, but it has also been applied for explaining price and output under oligopoly without product differentiation.

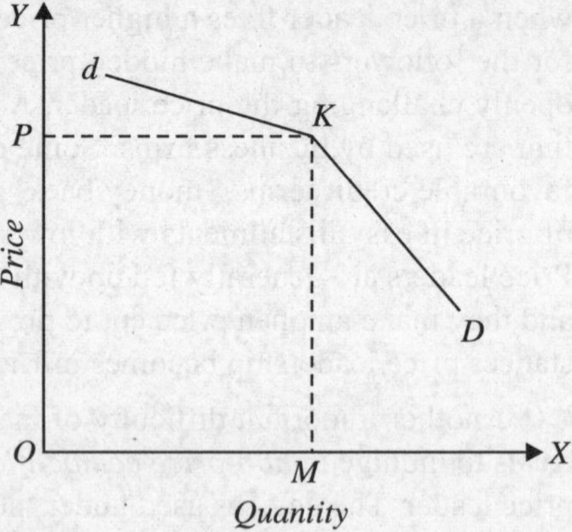

Fig. 29.4. *Kinked Demand Curve under Oligopoly*

The demand curve facing an oligopolist, according to the kinked demand curve hypothesis, has a 'kink' at the level of the prevailing price. The kink is formed at the prevailing price level because the segment of the demand curve above the prevailing price level is highly elastic and the segment of the demand curve below the prevailing price level is inelastic. A kinked demand curve *dD* with a kink at point *K* has been shown in Fig. 29.4. The prevailing price level is *OP* and the firm is producing and selling the output *OM*. Now, the upper segment *dK* of the demand curve *dD* is relatively elastic and the lower segment *KD* is relatively inelastic. This difference in elasticities is due to the particular competitive reaction pattern assumed by the kinked demand curve hypothesis.

The competitive reaction pattern assumed by the kinked demand curve oligopoly theory is as follows:

Each oligopolist believes that if he lowers the price below the prevailing level, his competitors will follow him and will accordingly lower their prices, whereas if he raises the price above the prevailing level, his competitors will not follow his increase in price.

In other words, each oligopolistic firm believes that though its rival firms will not match his increase in price above the prevailing level, they will indeed match its price cut. These two different types of reaction of the competitors to the increase in price on the one hand and to the reduction in price on the other make the portion of the demand curve above the prevailing price level relatively elastic and the lower portion of the demand curve relatively inelastic. This is explained below:

(*a*) **Price reduction.** If the oligopolist reduces its price below the prevailing price level *OP* in order to increase his sales, the competitors will fear that their customers would go away from them to buy the product from the former oligopolist which has made a price cut. Therefore, in order to retain their customers they will be forced quickly to match the price cut. Because of the competitors quickly following the reduction in price by an oligopolist, he will gain in sales only very little. (His sales will increase not at the expense of his competitors but because of the rise in total quantity demanded due to the reduction in price of the good. In fact each will gain in sales to the extent of a proportionate share in the increase in total demand). Very small increase in sales of an oligopolist following his reduction in price below the prevailing level means that the demand for him is inelastic below the prevailing price. Thus the segment *KD* of the demand curve in Fig. 29.4 which lies below the prevailing price *OP* is inelastic showing that very little increase in sales can be obtained by a reduction in price by an oligopolist.

(*b*) **Price increase.** If an oligopolist raises his price above the prevailing level, there will be a substantial reduction in his sales. This is because as a result of the rise in his price, his customers will withdraw from him and will go to his competitors who will welcome the new customers and will gain in sales. These happy competitors will have therefore no motivation to match the price rise. The oligopolist who raises his price will be able to retain only those customers who either have a strong preference for his product (if the products are differentiated) or who cannot obtain the desired quantity of the product from the competitors because of their limited productive capacity. Large reduction in sales following an increase in price above the prevailing level by an oligopolist means that demand with respect to increases in price above the existing one is highly elastic. Thus, in Fig. 29.4 the segment *dK* of the demand curve which lies above the current price level *OP* is elastic showing a large fall in sales if a producter raises his price.

It is now evident from above that each oligopolist finds himself placed in such a position that while, on the one hand, he expects his rivals to match his price cuts very quickly, he does not expect his rivals to match his price increases on the other. Given this expected competitive reaction pattern, each oligopolist will have a kinked demand curve *dKD* with the upper segment *dK* being relatively elastic and the lower segment *KD* being relatively inelastic.

Why Price Rigidity Under Oligopoly?

From what has been said above, it is easy to see why an oligopolist confronting a kinked demand curve will have no incentive to raise its price or to lower it. *Since the oligopolist will not gain a large share of the market by reducing his price below the prevailing level, and will have substantial reduction in sales by increasing his price above the prevailing level, he will be extremely reluctant to change the prevailing price*. In other words, each oligopolist will adhere to the prevailing price seeing no gain in changing it. Thus, rigid prices are explained in this way by the kinked demand curve theory. In Fig. 29.4, the prevailing price is *OP* at which kink is found in the demand curve *dKD*. The price *P* will tend to remain stable or rigid as every member of the oligopoly will not see any gain in lowering it or in increasing it. It should be noted that if the prevailing price *OP* is greater than average cost, more than normal profits will be made.

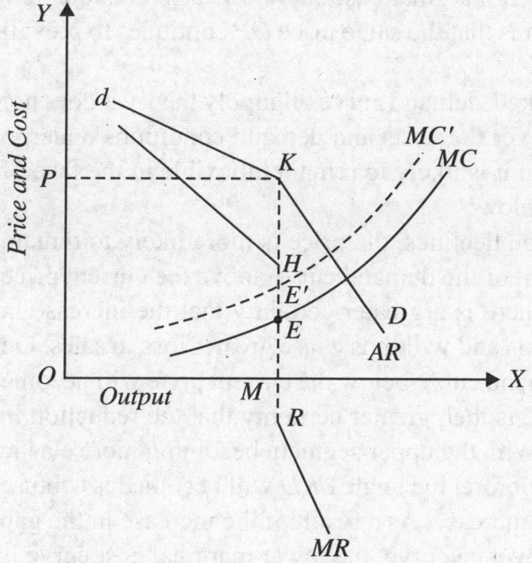

Fig. 29.5. *Changes in costs within limits do not affect the oligopoly price.*

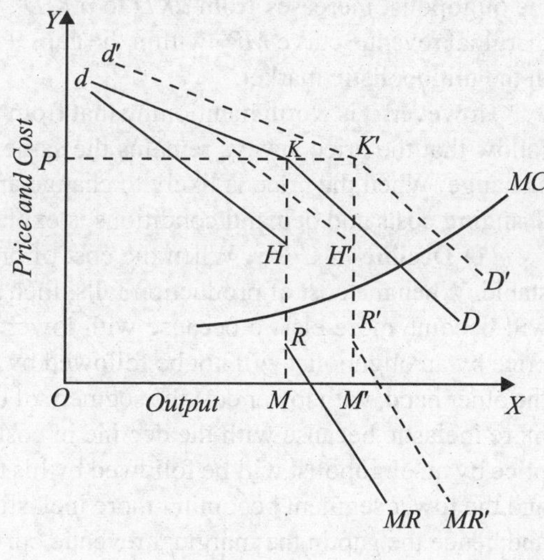

Fig. 29.6. *Changes in demand do not affect the oligopoly price.*

Further, it is worth mentioning that the oligopolist confronting a kinked demand curve will be maximising his profits at the current price level. For finding the profit-maximizing price-output combination, marginal revenue curve *MR* corresponding to the kinked demand curve *dKD* has been drawn. It is worth mentioning that the marginal revenue curve associated with a kinked demand curve is discontinuous, or in other words, it has a broken vertical portion. The length of the discontinuity depends upon the relative elasticities of two segments *dK* and *KD* of the demand curve at point *K*. The greater the difference in the two elasticities, the greater the length of the discontinuity. In Fig. 29.5 marginal revenue curve *MR* corresponding to the kinked demand curve *dKD* has been drawn which has a discontinuous portion or gap *HR*. Now, if the marginal cost curve of the oligopolist is such that it passes anywhere, say from point *E*, through the discontinuous portion *HR* of the marginal revenue curve *MR*, as shown in Fig. 29.5, the oligopolist will be maximizsing his profits at the prevailing price level *OP*, that is, he will be in equilibrium at point *E* or at the prevailing price *OP*. *Since the oligopolist is in equilibrium, or in other words, maximising his profits at the prevailing price level, he will have no incentive to change the price.*

Furthermore, even if there are changes in costs, the price will remain stable so long as the marginal cost curve passes through the gap *HR* in the marginal revenue curve. In Fig. 29.5 when the marginal cost curve shifts upward from *MC* to *MC′* (dotted) due to the rise in cost, the equilibrium price and output remain unchanged since the new marginal cost *MC′* also passes from point *E′* through the gap *HR*.

Likewise, the kinked demand curve theory explains that even when the demand conditions change, the price may remain stable. This is illustrated in Fig. 29.6 in which when the demand for the oligopolist increases from *dKD* to *d′K′D′*, the given marginal cost curve *MC* also cuts the new marginal revenue curve *MR′* within the gap. This means that the same price *OP* continues to prevail in the oligopolistic market.

However, it is worth mentioning that from the kinked demand curve oligopoly theory it does not follow that the price always remains the same whenever the costs and demand conditions undergo a change. When the price is likely to change and when it is likely to remain inflexible in the face of changing costs and demand conditions is explained below:

(1) **Decline in Costs.** When the cost of production declines, the price is more likely to remain stable. When the cost of production falls, then segment of the demand curve above the current price will become more elastic because with lower costs there is a greater certainty that the increase in price by an oligopolist will not be followed by his rivals and will thus cause greater loss in sales. On the other hand, with lower cost the segment of the demand curve below the current price will become more inelastic because with the decline in costs, there is then greater certainty that the reduction in price by an oligopolist will be followed by his rivals. With the upper segment becoming more elastic and the lower segment becoming more inelastic than before, the angle *dKD* will become less obtuse and hence the gap in the marginal revenue curve will increase. As a result of the increase in the gap (that is, the length of discontinuity) in the marginal revenue curve, the lower marginal cost curve is likely to pass through this gap showing that the price and output remain the same as before.

(2) **Rise in Cost.** *If there is a rise in the cost of the oligopolistic industry, the price is not likely to stay rigid.* When there is a rise in cost of industry an oligopolist can reasonably expect that his increase in price will be followed by the others in the industry. Consequently, the segment of the demand curve above the prevailing price will become less elastic and thereby make the angle *dKD* more obtuse and this will narrow down the gap in the marginal revenue curve. With the smaller gap in the marginal revenue curve, the *higher marginal cost curve is likely to cut it above the upper point H* indicating that the equilibrium price will rise and the equilibrium output will fall. Thus it follows from the kinked demand curve theory that price is not likely to remain stable in the event of rise in cost.

(3) **Decrease in Demand.** In case of decrease in demand, the price is very likely to remain inflexible and will not fall. When the demand decreases, it becomes more certain that if one oligopolist initiates the reduction in price, others will follow with the result that the lower segment of the demand curve will become more inelastic. On the other hand, in the face of a decline in demand it is very certain that the increase in price by one oligopolist will never be followed by others. As a result, the upper segment of the demand curve becomes more elastic, that is, it becomes more nearly horizontal. With the increase in the elasticity of the upper segment and the decrease in the elasticity of the lower segment, the gap in the marginal revenue curve becomes wider and there it is most likely that the given marginal cost curve will cross the marginal revenue curve inside the gap when

the demand curve *dKD* shifts downward. This indicates that the price will remain unchanged in the case of decrease in demand.

(4) **Increase in Demand.** *When the demand increases, the price is unlikely to remain stable, instead the price is likely to rise*. In the event of increase in demand, an oligopolist can expect that if he initiates the increase in price, his competitors will most probably follow him. Therefore, the upper segment *dK* of the demand curve will become less elastic and the angle *dKD* will become more obtuse. As a result, the gap *HR* in the marginal revenue curve will decrease and if this gap decreases much, it is very likely that the marginal cost curve crosses the marginal revenue curve above the upper point *H*, that is, above the gap, indicating that the price will rise above *OP*.

From above, it is clear that the kinked demand curve analysis of oligopoly explains stability in price in the face of falling costs or declining demand, whereas, prices are likely to rise when either the costs rise or demand increases. M.M. Bober thus rightly writes:

"*The kinky demand curve analysis points to the likelihood of price rigidity in oligopoly when a price reduction is in order and of price flexibility when conditions warrant a rise in price. There is hardly any disposition to lower the price when there is decline in demand or in costs, but the price may be raised in response to increased demand or to rising cost.*"[16]

Critical Appraisal of Kinked Demand Curve Theory

1. We saw above how the kinked demand curve theory of oligopoly provides an explanation of price rigidity under oligopoly. But there is a major drawback in the theory. It only explains why once an oligopoly price has been determined it would remain rigid or stable, *it does not explain how the price has been determined*. There is nothing in the kinked demand theory which explains how the price which is prevailing is determined. In other words, whereas this theory shows why price tends to stay where it is, it tells us nothing about why the price is where it is. In Fig. 29.4 the kink occurs at the price *OP* because *OP* happens to be the prevailing or established price. The theory does not explain how the price got to be equal to *OP*. Commenting upon kinked demand curve theory Prof. Silberston rightly writes, "The most interesting question is not 'why are prices sticky in the short run?' (if they are), but who decides what the price is to be and on what principles."[17]

However, it may be mentioned that the above criticism applies especially to P.M. Sweezy's version of the kinked demand curve analysis. Hall and Hitch's version of kinked demand curve analysis also explains the determination of oligopoly price. According to Hall and Hitch, equilibrium price is determined by average cost (including normal profits), that is, by the tangency between average cost curve and the demand curve, as shown in Fig. 29.7. However, Hall and Hitch version runs into difficulty when the average cost curves of the various firms in an oligopolistic industry are different.

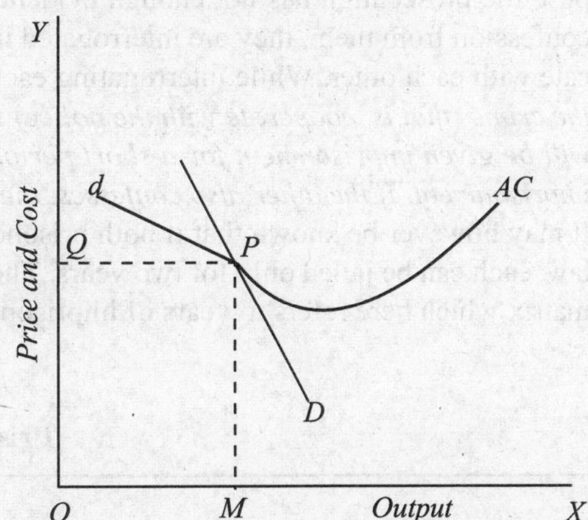

Fig. 29.7. *Full-Cost Pricing and Kinked Demand Curve.*

2. Another shortcoming of the kinked-demand oligopoly theory is that it does not apply to the oligopoly cases of price leadership and price cartels which account for quite a large part of the oligopolistic markets. When price leadership and price cartels exist in oligopolistic markets there is concerted behaviour in regard to the price changes and hence there is no kink in the demand curve in these cases.

16. M.M. Bober, *Intermediate Price and Income Theory*.
17. Aubery Silberston, "Price Behaviour of Firms", *Economic Journal*, September 1970.

3. Finally, even in the case of pure oligopoly (*i.e.*, oligopoly with homogeneous products), the kinked demand curve theory does not furnish a complete explanation for price rigidity observed in oligopolistic markets. As explained earlier, from the kinked demand curve analysis it follows that prices are likely to remain stable when demand or cost conditions decrease, whereas under pure oligopoly prices are likely to rise in the case of increase in cost or demand.

4. Finally, it has been rightly asserted that explanation of price stability by Sweezy's kinked demand curve theory applies only to depression periods. In periods of depression, demand for the products decreases. As has been explained above, in the context of decreased demand, price in kinked demand curve theory is likely to remain sticky. But in periods of boom and inflation when the demand for the product is high and increasing, the price is likely to rise rather than remaining stable.

We, therefore, conclude that from Sweezy as well as Hall and Hitch's versions of kinked demand curve, it follows that prices are likely to remain stable during depression periods but not during boom and inflationary periods. Our analysis shows that whether we use kinked demand curve of the type postulated by Sweezy, or Hall and Hitch prices are unlikely to be stable during the boom periods.

PRISONER'S DILEMMA AND OLIGOPOLY THEORY

The Prisoner's Dilemma

The firms working in oligopolistic markets make decisions in face of uncertainty about how their rivals will react to their moves. The game theory is a mathematical technique of analaysing the behaviour of rival firms with regard to changes in price, output and advertisement expenditure in the situations of conflicts of interest among individuals or firms. An important game model that has significant implications for the behaviour of the oligopolists is popularly known as prisoner's dilemma. Model of prisoner's dilemma explains how rivals behaving selfishly act contrary to their mutual or common interests. We first explain prisoner's dilemma with the examples which were originally given while propounding this model.

Suppose two persons, Billa and Ranga, have been caught for committing a bank robbery. Suppose the prosecution has not enough evidence for their committing the crime. In order to procure confession from them, they are interrogated in two separate rooms so that they could not communicate with each other. While interrogating each accused, the police offers to Billa *"If you confess to the crime (that is, cooperate with the police) while the other keeps silent (i.e., does not confess) you will be given imprisonment for a short period, say, 1 year only but punish the other with 10 years imprisonment. If the other also confesses, then both of you would be sentenced to jail for 5 years."* It may however be known that if both prisoners do not confess, then according to the provisions of law, each can be jailed only for two years. The choices open to each accused are presented in payoff matrix which here refers to years of imprisonment.

Table 29.1

Prisoner's Dilemma

		Ranga's Choice	
		Confesses	Doesn't confess
Billa's Choice :	Confesses	Ranga : 5 years Billa : 5 years	Ranga : 10 years Billa : 1 year
	Doesn't confess	Ranga : 1 year Billa : 10 years	Ranga : 2 years Billa : 2 years

It will be seen from the payoff matrix that the outcome (*i.e.*, length of sentences to each) is determined by the specific strategy (that is, choice) adopted by each prisoner. The two strategies (choices) refer to (*i*) confesses and (*ii*) does not confess. If both Ranga and Billa confess, each gets 5 years imprisonment. If one confesses, but the other does not, the one who confesses (*i.e.*, cooperates with the police) gets a very light punishment, namely, imprisonment for 1 year only, and the one who does not confess is sentenced for 10 years imprisonment. It will be further seen from the table that if both don't confess (that is, they remain loyal and faithful to each other and do not cooperate with the police), both are sentenced to 2 years imprisonment.

Now, each prisoner faces an uncertain situation regarding how the other person will behave, that is, whether or not he will confess. Though each person has to make an independent choice whether to confess or not but the outcome, *i.e.*, payoff depends on what the other does.

Now, under these circumstances what choice will be made by the prisoners when they cannot communicate with each other and have to choose between the two alternatives independently. *The model of prisoner's dilemma suggests that both behaving selfishly and working in self-interest will confess to the crime and cheat each other*. Since both confess, each will get imprisonment for 5 years. Why do they make this choice and confess can be shown as follows. Take Ranga first. Most probably, he would confess when he does not know how his co-accused will act. Ranga would reason like this. If I do not confess it is very likely that I will be imprisoned for 10 years as the other prisoner will most probably confess. If I confess, I will get 5 years imprisonment if the other one also confesses and only one year imprisonment even if he does not confess. So in the presence of uncertainty about the other prisoner's choice, and behaving in self-interest, Ranga is likely to confess. Billa too reasoning similarly would confess. As a result, both prisoners confess and therefore would be sentenced for 5 years, though they would have received a light sentence of only two years if they had not confessed and remained loyal to each other. However, it is self-interest which leads each prisoner to confess and prevents them from attaining the best solution for themselves (2 years imprisonment) if both do not confess to the crime and remain loyal to each other. *But the decision of each prisoner in favour of confession is quite rational because each person works in self-interest and tries to make the "best" of the "worse outcomes" in an uncertain situation.*

Prisoner's Dilemma and Oligopolistic Behaviour: Instability of a Cartel

The game of prisoner's dilemma is of important relevance to the oligopoly theory. The incentive to cheat by a member of a cartel (*i.e.*, in the model of collusive oligopoly) and eventual collapse of cartel agreement is better explained with the model of prisoner's dilemma. Instead of two prisoners we take the two firms *A* and *B* which have entered into a cartel agreement and have fixed the price of the product each has to charge and output each has to produce and sell (*i.e.*, share of the market). The choice problem facing each member firm of the cartel is whether to cooperate and abide by the

Table 29.2

Payoff Matrix for Cartel Members

		Firm *A*	
		Cheat	Cooperate
Firm *B*	Cheat (lower the price)	*A* : 5 lakhs *B* : 5 lakhs	*A* : 2 lakhs *B* : 25 lakhs
	Cooperate	*A* : 25 lakhs *B* : 2 lakhs	*A* : 15 lakhs *B* : 15 lakhs

agreement and thus sharing the joint monopoly profits or to cheat the other and try to make higher individual profits. But if both cheat and violate the agreement, the cartel would break down and profits would fall to the competitive level. We will show that though both would lose by cheating others but as seen in case of prisoner's dilemma, their selfish behaviour would lead them to cheat others. The payoff matrix for two member firms of a cartel from the various combination of choices to be made by them is presented in Table 29.2.

It will be seen from the above payoff matrix, that if both firms cooperate and abide by cartel agreement, they share monopoly profits; 15 lakhs to each of them (right hand bottom). If both firms cheat and thus violate the agreement, profits to each firm fall to the competitive level, ₹ 5 lakh to each firm (left hand top). If firm *A* cheats, while firm *B* cooperates, *B*'s profits drop to low level of ₹ 2 lakhs and *A*'s profits rise to ₹ 25 lakhs (left hand bottom). On the other hand, if firm *B* cheats and firm *A* adheres to the agreement, profits of *A* declines to ₹ 2 lakhs and *B*'s profits shoot up to ₹ 25 lakhs (left hand top).

It is evident from the payoff from the different choices that each firm has a strong incentive to cheat. Under the prevailing circumstances. *A*'s best strategy is to cheat, rather than cooperate. The same is true for firm *B* whose best strategy is also to cheat. Again, *it is the pursuit of self-interest rather than common interest that prompts the firms to cheat each other*. Thus, both firms will cheat and this will bring about the breakdown of the cartel.

Graphical Illustration. The strong incentive to cheat on the part of cartel members and consequently causing a breakdown of a cartel is graphically illustrated in Fig. 29.8, where *DD* is market demand curve facing the cartel consisting of two firms *A* and *B*. MC_a in panel (*i*) is the marginal cost curve of the firm *A*. Summing up horizontally marginal cost curves of the two firms forming a cartel we get the combined marginal cost curve MC_{a+b}. The cartel maximises its profits by equating *MR* with MC_{a+b} and accordingly output *OQ* and price *OP* are fixed. The output share of each firm will be fixed where marginal cost of each firm equals the combined marginal cost of the cartel members. Accordingly, the output share of firm *A* shown in panel (*i*) is Oq_a at which marginal cost of firm *A* equals the combined marginal cost *OT* or *QE* in the equilibrium situation. Oq_a will be the agreed share of output of firm *A*. To simplify our analysis we have not shown the cost situation of firm *B*, the other member of the cartel.

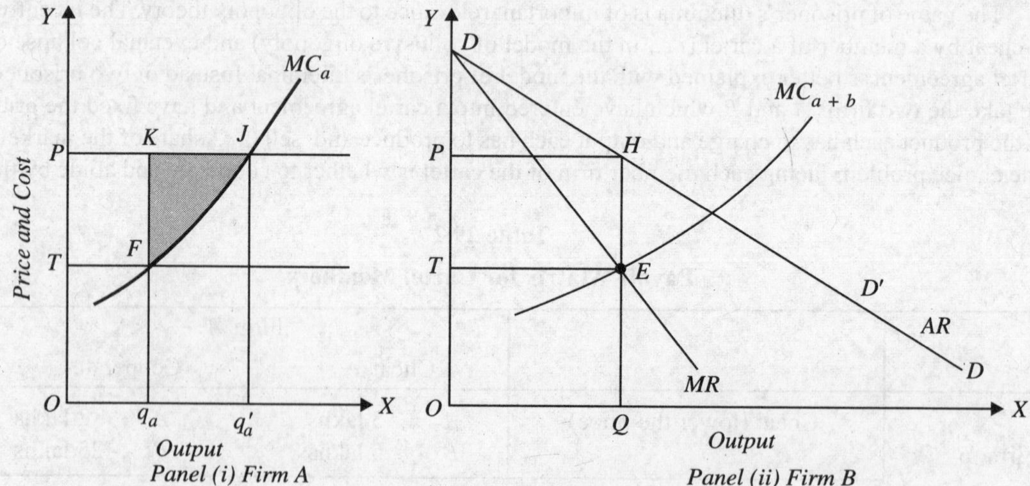

Fig. 29.8. *Instability of a Cartel Due to Cheating by a Member*

Thus, under the cartel agreement, firm *A* will be producing Oq_a, and charging price *OP*. Now, a glance at panel (*i*) in Fig. 29.8 reveals that given the price *OP*, fixed by the cartel, if the firm *A* increases its output from agreed share Oq_a to Oq'_a, it can increase its profits by the shaded area *FKJ*. This means that in order to increase its profits the firm *A* will have incentive to cheat by trying to produce and sell more at the agreed price *OP*. Similarly, the firm *B*, other member of the cartel (not shown in Fig. 29.8), will also find that it can increase its profits by violating the cartel agreement by producing and selling more than its agreed output share. To sum up, *it is due to the promotion of self-interest by cartel members that accounts for the instability of cartel arrangements and frequent price wars*.

THE THEORY OF CONTESTABLE MARKETS

Baumol, Panzor and Willing have developed a new theory of markets known as the theory of contestable markets.[18] According to this theory, even when there prevail monopoly and oligopoly in the product markets, the firms may find it most profitable to behave in setting price of its product as if they were working in a perfectly competitive market. This is because a firm thinks that if it sets a price higher than competitive market, other firms will enter the market and compete for the customers and thereby push down the price. Hence, the threat of new entrants in the markets forces the existing firm or firms to charge no more than the competitive price and therefore make only normal profits.

Important conditions for contestable markets to prevail are :

1. The new firms can enter the market with the same cost conditions as the existing firm or firms. If the existing firm enjoys some cost advantage as compared to the potential entrants in the market, then the new firms that enter the market cannot succeed in competing with the existing one which can afford to lower the price and inflict losses on the new entrants.
2. The second condition for the contestable markets is that firms should be able to leave or exit the market (*i.e.* industry) without incurring any capital loss. If the production in an industry involves much capital investment which is specific to that industry and therefore of no use elsewhere, leaving the industry will entail losses equal to the sunk costs. Recall that *sunk costs are the costs incurred on capital assets which are specific to an industry and cannot be recovered by selling them or which cannot be deployed or used elsewhere for commercial or production purposes*. Sunk costs act as an important barrier to the entry of new firms and are therefore deterrent to the existence of contestable markets.
3. The third condition for contestable markets to exist is that the potential entrants must be at no disadvantage as compared to the existing firms with regard to the *production technology or product quality as perceived by the consumers*. Any lack of access to the same production technology as used by the existing firms prevents the new entrants from competing the existing firms on the basis of cost or product quality. This would work to reduce the threat by potential entrants and enable the existing firms to charge higher than competitive price and earn supernormal profits.
4. The last condition for the contestable markets to exist is that the new entrants must be able to engage in **'hit and run'** tactics. That is, entry should be free and costless so that the new entrants enter the market or industry and make profits and exit the industry before the existing firms adjust their prices downward. Hit and run tactics can succeed when the existing firms can adjust their prices with some time-lags whereas the consumers immediately respond to buy from new entrants who offer same or similar products at slightly lower prices.

It follows from above that fewer the barriers to the entry in a market, the more contestable the market is. So to the extent monopolistic and oligopolistic markets are contestable depends on the barriers to the entry of new firms into a market or industry.

---

18. See William J. Baumol, John C. Panzar and Robert D. Willing, *Contestable Markets and the Theory of Industrial Structure* (New York: Harcourt, Brace, Jovanovich, 1982).

When there are absolutely no barriers to the entry of new firms, the perfectly contestable markets would exist even though the existing firms may be working in monopolistic or oligopolistic market structures. In perfectly contestable markets, the existing firms are forced to keep their prices equal to average cost and therefore make only normal profits.

Price setting in a contestable market is illustrated in Fig. 29.9. Suppose there are two firms in an industry, that is, there is duopoly in the product market. *DD* is the demand curve for the industry's product. The average and marginal cost curves of each firm are shown. The threat of entry of new firms leads each duopolist to charge a price *OP* (which is equal to their minimum average cost) and produces output equal to *Oq*. The total output *OQ* will be equal to two times *Oq*. (That is $OQ = 2Oq$). Each firm earns zero economic profits.

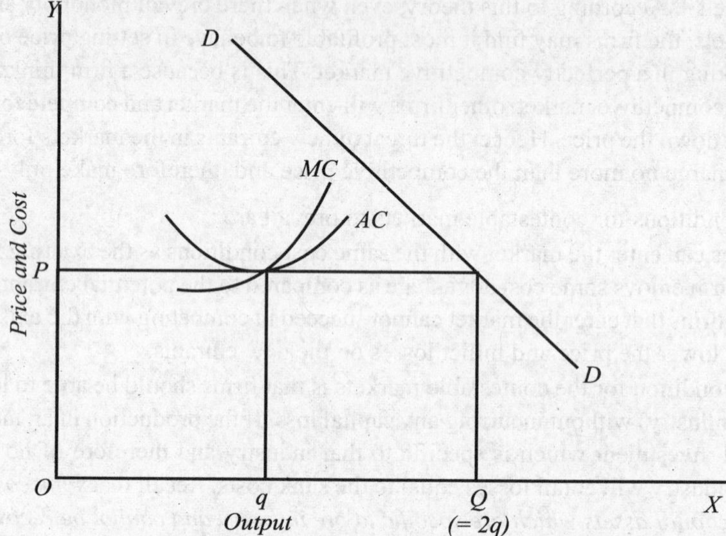

Fig. 29.9. *A Constestable Market : The threat of potential entrants leads the duopolists to behave like competitors.*

Though the two firms can enter into tacit collusion and push up the price and earn supernormal profits, they do not do so because they think new entrants will enter industry quickly and undercut price and in this way inflict losses on them. To prevent entry of new firms and avoid competiting with them each duopolist produces *OQ* output and charges *OP* price. By producing at the lowest minimum average cost they are able to enjoy all the economies of scale. Thus, in producing the level of output at the minimum average cost and charging a price equal to it, the two firms behave like a perfectly competitive firm.

The theory of contestable markets has been criticised, among others by Shepherd[19] who points out that it is based on extremely unrealistic assumptions regarding entry and exit of new firms. According to him, in the real world entry is not free and exit is not costless. He contends that most monopolistic and oligopolistic markets are not contestable because there exist a good amount of sunk costs in most of the industries. These sunk costs act as barrier to the entry of potential entrants as they cannot exit the market without much loss. Because some of the investment involves sunk cost, the new competitors would have to set a price high enough to cover all their costs, whereas the existing firms could set a slightly lower price and make more than normal profits.

To conclude, the threat of new entrants can lead to the existing monopolists and oligopolists to behave like a competitive firm in some situations but when there exist substantial sunk investment

19. William G. Shepherd, "Contestability vs Competition", *American Economic Review,* Sept. 1984, pp. 572–587.

costs, the existing firms have an edge over the new entrants and charge higher than competitive price and make large supernormal profits. We thus see that contestable market theory does not apply to all monopolistic and oligopolistic market situations.

QUESTIONS FOR REVIEW

1. What is oligopoly? Explain the important features of oligopoly. Do you think price and output under oligopoly is indeterminate?
2. There is no unique solution to the problem of determination of price and output under oligopoly. Discuss.
3. What is Collusive Oligopoly? How are price and output determined under it?
4. What is Cartel? Explain how a cartel determines price and output of a product to maximise joint profits.
5. Why are cartels generally unstable? Why is there a tendency on the part of members of cartels to cheat?
6. Explain Sweezy's kinked demand curve model of oligopoly. How does it explain price rigidity under oligopoly?
7. What is non-price competition ? What role does it play in oligopolistic market situation? [**Hints**: Selling costs (*e.g.*, advertisement expenditure etc.) and product variation by the firms are two important forms of non-price competition.]
8. What alternative objectives to profit maximisation have been proposed which the firms working under oligopoly seek to achieve? Briefly explain them.
   [**Hints** : Three alternative objectives to profit maximisation are: (*a*) to seek normal profits which is assumed in the theory of mark-up pricing, (*b*) sales maximisation which has been emphasised by Prof. Baumol and (*c*) security motive or to achieve maximum profits in the long run.]
9. What are the different types of price leadership that may be established in oligopolistic market situation? Assuming that there are two firms producing homogeneous products. Explain how price and output are determined where there is price leadership by
   (1) the lower cost firm, and
   (2) the dominant firm
10. Explain determination of price and output under price leadership when the two firms have :
    (*a*) equal market share
    (*b*) unequal market share.
11. W*hat is kinked demand curve? How does it help in explaining price rigidity under oligopoly*? [*D.U. B.A*(*Hons*.)]
12. What is kinked demand curve? How does the kinked demand curve hypothesis explain price stability under oligopoly? [*D.U. B. Com.* (*Hons*)]
13. Using the kinked demand curve model explain how the increase and reduction in marginal cost need not lead to any change in price or output. [*D.U. B.A. (Hons), 1993, 1997*]
14. What is meant by price rigidity? Why are prices rigid under oligopoly? Explain with the help of a model. [*D.U. B.Com.* (*Hons*) 1998, 1999]
15. Can a kinked demand curve arise in a duopoly market with homogeneous products? If so, what would be the equilibrium price and output? [*D.U. B.A* (*Hons*.), 1991]
16. What is prisoner's dilemma? How does it help to explain the likelihood of firms opting for sub-optimal solution in oligopoly? [*B.Com.* (*Hons*.) *D.U.* 2002]

17. What are the chief difficulties of maintaining price leadership in oligopolistic market situation? Examine the view that price leadership is likely to break down as there is generally a tendency on the part of the rival firms to indulge in *non-price competition* to promote their sales.

18. What is meant by interdependence of firms in oligopoly ? How does it affect price-output equilibrium of a firm?

19. Explain briefly the various approaches that have been offered to explain the determination of price and output under oligopoly.

20. Which of the following statements are true and which false?

(*a*) Interdependence of firms is an important feature of an oligopolistic market.
(*b*) Advertisement expenditure does not play any role in an oligopolistic market.
(*c*) Firms working in oligopolistic market have kinked demand curve.
(*d*) Demand curve facing a firm working in an oligopolistic market is indeterminate.
(*e*) There is no unique solution to the determination of price and output under oligopoly.
(*f*) Price rigidity is one of the features of oligopoly.
(*g*) It is the pursuit of self-interest by members that is often responsible for breakdown of a cartel.

[**Hint.** (*a*) Correct. (*b*) Incorrect. (*c*) Correct (according to the kinked demand curve theory) (*d*) Correct. (*e*) Correct. (*f*) Correct. (*g*) Correct.]

21. An oligopolistic firm suspects that if it cuts price, its rivals will follow suit. How will this action on the part of the rivals affect the elasticity of demand curve facing the firm?

(*a*) It will make the firm's demand curve more elastic.
(*b*) It will make the firm's demand curve less elastic.
(*c*) It will not affect the elasticity of the firm's demand curve.
(*Which is correct answer and Why*?) [*D.U. B.A.* (*Hons*) 1989]

22. Explain Prisoner's dilemma. How does it help to explain that cartels are quite unstable ?

23. (*a*) Use the prisoner's dilemma to show why firms attempt to form cartels and yet cheat.
(*b*) How does a cartel determine the joint profit maximising quantity and allocate it among the firms. [*D.U.B. Com* (*Hons*) *2004]*

Appendix to Chapter 29

CLASSICAL MODELS OF OLIGOPOLY

In this appendix, we will discuss two important models of oligopoly. A model of oligopoly was first of all put forward by Cournot, a French economist, in 1838. Cournot's model of oligopoly is one of the oldest theories of the behaviour of the individual firm and relates to non-collusive oligopoly. In Cournot's model, it is assumed that an oligopolist thinks that his rival will keep his output fixed regardless of what he might do. That is, each oligopolist does not take into account the possible reactions of his rivals in response to his actions.

Another important model of non-collusive oligopoly which we will discuss below was put forward by E H Chamberlin in his famous work *The Theory of Monopolistic Competition*. Chamberlin made an important improvement over the classical models of oligopoly, including that of Cournot. In sharp contrast to Cournot and other classical models, Chamberlin assumes in his model that oligopoly firms recognise their interdependence while fixing their output and price. Through his model, Chamberlin arrives at a monopoly solution of pricing and output under oligopoly wherein oligopolistic firms in an industry jointly maximise their profits.

COURNOT'S DUOPOLY MODEL

As said above, Augustin Cournot, a French economist, published his theory of duopoly in 1838. But it remained almost unnoticed until 1880's when Walras called the attention of the economists to Cournot's work. Cournot dealt with the case of duopoly. Let us first state the assumptions which are made by Cournot in his analysis of price and output under duopoly. Firstly, Cournot takes the case of two identical mineral springs operated by two owners who are selling the mineral water in the same market. Their waters are identical. Therefore, *his model relates to the duopoly with homogeneous products.* Secondly, it is assumed by Cournot, for the sake of simplicity, that the owners operate mineral springs and sell water without incurring any cost of production. Thus, in Cournot's model, cost of production is taken as zero; only the demand side of the market is analysed. It may be noted that the assumption of zero cost of production is made only to simplify the analysis. His model can be presented when cost of production is positive. Thirdly, it is assumed the duopolists fully knew the market demand for the mineral water; they could set every point on the demand curve. Moreover, the market demand for the product is assumed to be linear, that is, market demand curve facing the two producers is a straight line.

Lastly, Cournot assumes that *each duopolist believes that regardless of his actions and their effect upon market price of the product, the rival firm will keep its output constant,* that is, it will go on producing the same amount of output which it is presently producing. In other words, the duopolist will decide about the amount of output which is most profitable for him to produce in the light of his rival's *present* output and assumes that it will remain constant. In other words, for determining the output to be produced, he will not take into account reactions of his rival in response to his variation in output and thus decides its level of output *independently.*

Cournot's Approach to Equilibrium of the Duopolists

Suppose the demand curve confronting the two producers of the mineral water is the straight line *MD* as shown in Fig. 29A.1. Further suppose that $ON = ND$ is the maximum daily output of each mineral spring. Thus, the total output of both the springs is $OD = ON + ND$. It will be seen from the figure that when the total output OD of both the springs is offered for sale in the market, the price will be zero. It may be noted here that if there was a perfect competition, the long-run equilibrium price would have been zero and actual output produced equal to OD. This is because cost of production being assumed to be zero, price must also be zero so as to provide a zero profit long-run equilibrium under perfect competition[1].

Assume for a moment that one producer A of the mineral water starts the business first. Thus, to begin with he will be the monopolist. He will then produce daily ON output because his profits will be maximum at output ON' and will be equal to $ONKP$ (since the costs are zero, the whole of total revenue $ONKP$ will represent profits). The price which that producer will charge will be OP. Suppose now that the owner of the other spring enters into the business and starts operating his spring, This new producer B sees that the former producer A is producing ON amount of output. According to the assumption made by Cournot, the producer B believes that the former producer A will continue producing ON ($= \frac{1}{2}$ OD) amount of output, regardless of what output he himself decides to produce. Given this belief, the best that the new producer B can do is to regard segment KD as the demand curve confronting him. With his demand curve KD, and corresponding marginal revenue curve MR_B, the producer B will produce NH ($= \frac{1}{2}$ ND) amount of output. The total output will now be $ON + NH = OH$, and as a result the price will fall to OP' or HL per unit. The total profits made by the two producers will be $OHLP'$ which are less than $ONKP$. Out of total profits $OHLP'$, profits of producer A will be $ONGP'$ and profits of producer B will *be* $NHLG$. Thus, entry into the market by producer B and producing output NH by him, the producer A's profits has been reduced. A will therefore reconsider the situation. But he will assume that producer B will continue to produce output NH. With producer B producing output NH, the best that the producer A can do is to produce $\frac{1}{2}$ (OD $-NH$). He, will, therefore, reduce his output.

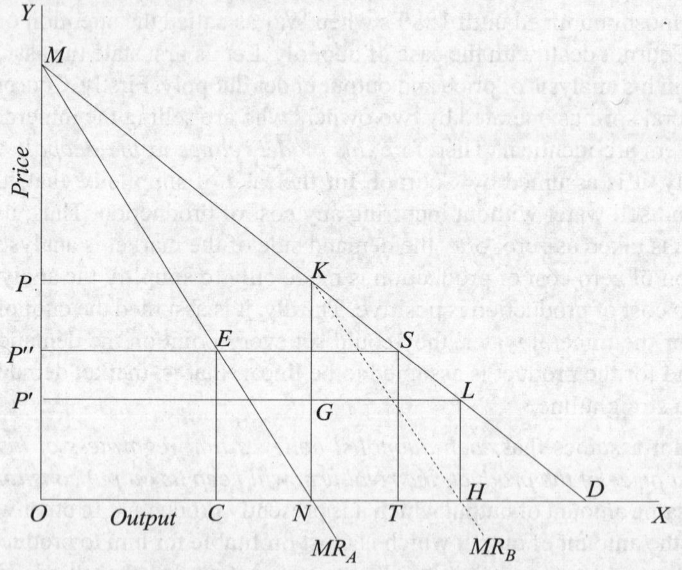

Fig. 29A.1 *Cournot's Duopoly Solution*

1. When cost of production is zero, the marginal cost will also be zero. With demand curve MD as a straight line, the marginal revenue curve will pass through the middle of OD, that is, through N so that $ON = ND$. Thus, at ON, output MR will be zero. Therefore, a monopolist with zero marginal cost will be in equilibrium at output ON where marginal revenue and marginal cost will be equal (both being zero at output ON). In such a situation, the monopolist will make maximum profits by producing output ON, given the linear demand curve MD.

Now that the producer *B* has been surprised by the reduction of output by producer *A* and will also find that his share of total profits is less than that of producer *A,* he will reconsider his situation. Learning nothing from his earlier experience and believing that producer *A* will continue producing its new current level of output, the producer *B* will find that he will now be making maximum profits by producing output equal to $\frac{1}{2}$ (*OD* – New output of *A*). Producer *B,* accordingly, will increase his output. With this move of producer *B,* producer *A* will find his profits reduced. The producer *A* will therefore again reconsider his position and will find that he can increase his profits by producing output equal to $\frac{1}{2}$ (*OD* – Current output of producer *B*). This process of adjustment and readjustment will continue and producer *A* will be forced gradually to reduce his output and producer *B* will be able to increase his output gradually until the total output *OT* is produced ($OT = \frac{2}{3}$ *OD*) and each is *producing the same amount of output* equal to $\frac{1}{3}$ *OD.* In this final position, producer *A* produces *OC* amount of output and producer *B* produces *CT* amount of output, and *OC* = *CT.* Throughout this process of adjustment and readjustment, each producer assumes that the other will keep his output constant at the present level and then always finds his maximum profits by producing output equal to $\frac{1}{2}$ (*OD* – the present output of the other).

As seen above, producer *A* starts by producing $ON = (\frac{1}{2}$ *OD*) and continuously reduces his output until he produces *OC.* The final output *OC* of producer *A* will be equal to $\frac{1}{3}$ *OD* (= $\frac{1}{2}$ *OT*). On the other hand, producer *B* begins by producing $\frac{1}{4}$th of *OD* and continuously increases his output until he produces *CT.* His final output *CT* will be equal to $\frac{1}{3}$ *OD* (= $\frac{1}{2}$ *OT*). Thus, the two producers together will produce total output equal to $\frac{1}{3}$ *OD* + $\frac{1}{3}$ *OD* = $\frac{2}{3}$ *OD* (= *OT*).

Cournot's Duopoly Equilibrium

It will be seen from Fig. 29A.1 that when each producer is producing $\frac{1}{3}$ *OD* (that is, when producer *A* is producing *OC* and producer *B* equal to *CT*), the best that his rival can do is to produce $\frac{1}{2}$ (*OD* – $\frac{1}{3}$ *OD*) which is equal to $\frac{1}{3}$ *OD* = *OC* = *CT.* Thus, *when each producer is producing* $\frac{1}{3}$ *OD so that the total output of the two together is* $\frac{2}{3}$ *OD, no one will expect to increase his profits by making any further adjustment in output. Thus, in Cournot's model of duopoly, stable equilibrium is reached when total output produced is* $\frac{2}{3}$ *rd of OD and each producer is producing* $\frac{1}{3}$ *rd of OD.*

It will be useful to compare Cournot' s duopoly equilibrium with the monopolistic and the purely competitive equilibriums. If the two producers had combined and formed a coalition, then the output produced by them together will be the monopoly output *ON* and, therefore, the price set will be the monopoly price *OP.* Monopoly output *ON* produced in case of coalition is much less than the output *OT* produced in Cournot's duopoly equilibrium. Further, the monopoly price *OP* charged in case of coalition is much greater than the price *OP''* determined in Cournot's duopoly equilibrium. In case of coalition, they will enjoy the monopoly profits *ONKP* which are *maximum possible joint profits*, given the demand curve *MD.* These monopoly or maximum joint profits can be shared equally by them. It will be seen from Fig. 29A.1 that these monopoly profits *ONKP* made in case of coalition are much greater than the total profits *OTSP''* made by them in Cournot's duopoly equilibrium. It is thus clear that in case of the duopolists competing with each other as conceived by Cournot's duopoly solution, the price and the profits are lower and output is greater than if they had combined together and formed a monopoly.

On the other hand, if the markets were perfectly competitive, the output would have been *OD* and price would have been zero. This is because with assumed marginal cost being equal to zero, perfectly competitive equilibrium will be reached at the output level where price is equal to zero. That is, perfectly competitive solution would have resulted in greater output and lower price than under Cournot's duopoly equilibrium.

To sum up, under Cournot's duopoly equilibrium, output is two-thirds of the maximum possible output (*i.e.,* perfectly competitive output) and price is two-thirds of the most profitable price (*i.e.,* monopoly price).

Following Cournot, the cost of production in the above discussion of Cournot's oligopoly solution has been taken to be zero. However, it should be noted that above conclusions will not change if the cost curves with positive cost of production are introduced into the discussion.

Reaction Functions and Cournot's Duopoly Solution

Cournot's solution of duopoly problem can also be obtained with *reaction functions* of the two firms. *An output reaction function depicts the profit-maximising output of a firm, on the assumption that the other firm's output remains constant.* We have seen above that the profit-maximising output of a Cournot's duopolist is one-half of the difference between the other firm's output and the market demand for output at which price equals marginal cost. This is called reaction function of a firm. This output at which price equals marginal cost (MC) is the maximum output which can be produced because any output beyond this will cause the price to go below marginal cost (which is equal to *AC* under constant cost conditions) and will therefore not be worthwhile to produce.

The following example will make clear the concept of reaction functions. Let's assume the market demand function is: $Q = 100 - P$ and marginal cost is ₹ 10. *In order to determine reaction functions of two duopolist firms, we set price equal to the given marginal cost to determine market demand at price (P) = MC.* Thus, from the given demand function

$$P = 100 - Q \qquad ....(i)$$

Setting it equal to *MC* we have

$$100 - Q = 10$$

Or $$Q = 100 - 10 = 90$$

Thus, the reaction function of firm *A* is :

$$Q_a = \frac{90 - Q_b}{2} \qquad ....(ii)$$

where Q_a and Q_b are the outputs of firm *A* and *B* respectively.

Similarly, reaction function of firm *B* is :

$$Q_b = \frac{90 - Q_a}{2} \qquad ....(iii)$$

The above two equations (*ii*) and (*iii*) can be solved simultaneously to determine Q_a and Q_b. To do so, we substitute the value of $Q_b = \frac{90 - Q_a}{2}$ in equation (*ii*) and have :

$$Q_a = \frac{90 - \left(\frac{90 - Q_a}{2}\right)}{2}$$

$$Q_a = \frac{90 - 45 + \frac{Q_a}{2}}{2} = 45 - 22.5 + \frac{1}{4}Q_a$$

$$Q_a - \frac{1}{4}Q_a = \frac{3}{4}Q_a = 22.5$$

$$Q_a = 22.5 \times \frac{4}{3} = \frac{90}{3} = 30$$

With $Q_a = 30$,

$$Q_b = \frac{90 - Q_a}{2} = \frac{90 - 30}{2} = 30$$

$$Q = Q_a + Q_b = 60$$

Solving equation (*i*) for *P* we get

$$P = 100 - Q = 100 - 60 = 40$$

Cournot Equilibrium as Nash Equilibrium

John F. Nash, a noted American Mathematician and a Nobel Prize winner in Economics, has put forward the concept of equilibrium known as *Nash Equilibrium*. Cournot duopoly equilibrium is an example of Nash equilibrium. *According to Nash equilibrium, competing firms reach their equilibrium state when each of them thinks that it is doing its best, that is, maximising its profits in response to the given strategy adopted by others which think they are also maximising their profits with the given strategies*. As a result, no one has a tendency to change its strategy. Therefore, we have a stable equilibrium. Since in Cournot duopoly equilibrium each firm chooses to produce an output level that maximises its profits, given the profit-maximising level of output of the other firm, Cournot duopoly is generally called *Cournot-Nash duopoly equilibrium.*

Cournot's Duopoly Equilibrium Explained with the Aid of Reaction Curves

Some economists have employed the reaction curves to explain Cournot's duopoly equilibrium. The reaction curves may be output reaction curves or price reaction curves depending upon whether it is the output or the price which is the adjustment viable. Since, in Cournot's model, it is the output which is subject to the adjusting variation, output reaction curves are relevant. It should be carefully noted that these *reaction curves refer not to the reactions which a seller expects will be forthcoming from his rivals but to the sellers' own reactions to the moves of his rival.*

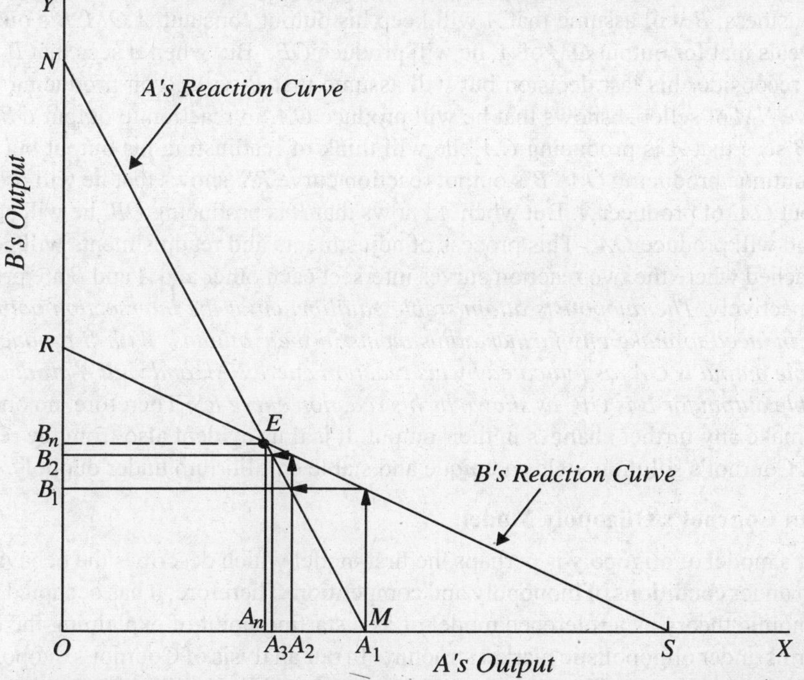

Fig. 29A.2. *Output Reaction Curves According to Cournot's Model*

In Fig. 29A.2, output reaction curves of two producers (sellers) *A* and *B* are shown; *MN* is the output reaction curve of *A* and *RS* is the output reaction curve of *B.* The output reaction curve *MN* of seller *A* shows how *A* will react to any change in output by *B,* that is, *A's* output reaction curve shows how much output *A* will decide to produce for each given output of producer *B.* In other words, *A's* output reaction curve *MN* indicates the most profitable output for *A* for each given output of *B.* Likewise, *B's* output reaction curve *RS* shows how much output *B* will decide to produce (that is, what will be *B*'s most profitable output) for each given output of *A.* For example, if *B* produces output OB_1, *A's* output reaction curve *MN* shows that *A* will produce output OA_2 in response to *B's* output OB_1. Similarly, for all other outputs. On the other hand, if *A* produces OA_2, *B's* output reaction curve shows that *B* will produce OB_2 and so forth for all other outputs.

It will be seen from Fig. 29 A.2 that output reaction curves have been drawn to be straight lines. This is because we are assuming that market demand curve for the product of duopolist is a straight line and that the marginal costs of production of both producers *A* and *B* are constant (at zero). It should be noted that output *OM* is the monopoly output since producer *A* will produce output *OM* if producer *B's* output is zero. In other words, producer *A* will produce and sell output *OM* if he were the monopolist. On the other hand, *A* will produce zero output if *B'S* output is *ON*. Given the marginal cost equal to zero, a producer will be forced to produce zero output when the price has fallen to zero and, therefore, production is no longer profitable. Output *ON* will be produced under conditions of perfect competition since at output *ON,* the price will be zero and therefore equal to marginal cost which is assumed to be zero in the present case. Thus, *while OM is the monopoly output, ON is the perfectly competitive output*. We assume the two producers *A* and *B* to be completely identical, *OR* will, therefore, be equal to *OM,* and *OS* will be equal to *ON.*

Output reaction curves, as interpreted above, can be used to explain Cournot's duopoly equilibrium. Each producer, as before, assumes that his rival will continue producing the same amount of output regardless of what he might himself decide to produce. To begin with, suppose producer *A* goes into business first and is therefore initially a monopolist. Therefore, in the beginning *A* will produce output *OM* which is a monopoly output as output by the firm *B* is zero. Suppose now *B* also enters into business, *B* will assume that *A* will keep his output constant at *OM. B*'s output reaction curve *RS* reveals that for output *OM* of *A*, he will produce OB_1. But when *A* sees that *B* is producing OB_1 he will reconsider his last decision but will assume that *B* will go on producing OB_1, Output reaction curve *NM* of seller *A* shows that he will produce OA_2 in reaction to output OB_1, of firm *B.* Now when *B* sees that *A* is producing OA_2; he will think of readjusting his output but will assume that *A* will continue producing OA_2. *B*'s output reaction curve *RS,* shows that he will produce output OB_2 for output OA_2 of producer *A.* But when *A* knows that *B* is producing OB_2 he will again readjust his output and will produce OA_3, This process of adjustments and readjustments will continue until point *E* is reached where the two reaction curves intersect each other and *A* and *B* are producing OA_n and OB_n respectively. *The duopolists attain stable equilibrium at the intersection point, since they will not feel induced to make any further adjustments in their outputs. With B producing* OB_n, *A's most profitable output is* OA_n *as indicated by his reaction curve NM, and with A producing* OA_n, *the most profitable output for B is* OB_n *as shown by his reaction curve RS,* Therefore, no one will have a tendency to make any further changes in their output. It is thus evident also from the reaction curve analysis that Cournot's solution yields a unique and stable equilibrium under duopoly.

A Critique of Cournot's Oligopoly Model

Cournot's model of oligopoly is perhaps the first model which describes the behaviour of an individual firm under conditions of monopoly and competition. Therefore, it has occupied an important place in economic theory as a reference model or as a starting point of explaining the behaviour of individual firms under oligopolistic market structure. In our analysis of Cournot's duopoly model, we have seen that he makes an important assumption, namely, while deciding about his output policy, each duopolist believes that his rival will hold output constant at the present level whatever output he himself might produce. Further, a producer remains unshaken in this erroneous belief even when he constantly finds himself to be proved incorrect since after his action, the rival does react and change his output. This is a chief logical error in Cournot's model.

Furthermore, by assuming that duopolist (oligopolist) will think that his rival will continue producing the current level of output, *Cournot model ignores the mutual interdependence* between the duopolist which is the chief characteristic of oligopoly, Thus, Cournot model provides solution for oligopoly problem by removing from it its most important feature.

NUMERICAL PROBLEMS

Problem 1. *Let the market for telecommunications equipment be represented by a duopoly where the two firms produce outputs q_1 and q_2, respectively. The inverse market demand function is represented by $P = 100 - 2Q$ (where $Q = q_1 + q_2$). The marginal cost that each firm faces is 4:*

(*i*) *What are the reaction functions of the Cournot duopolists?*

(*ii*) *Calculate each firm's output and the market price at Cournot equilibrium.*

[*D.U. B.A.* (*Hons.*)]

Solution:

We first find out market demand for output (Q) when $P = MC$. For this, setting the given inverse demand function equal to MC, we have

$$100 - 2Q = 4$$
$$Q = 48$$

Reaction function of firm B :

$$Q_a = \frac{48 - Q_b}{2} \qquad ...(i)$$

Reaction function of firm A :

$$Q_b = \frac{48 - Q_a}{2} \qquad ...(ii)$$

To determine Q_a we substitute the value of Q_b in (*i*) and have

$$Q_a = \frac{48 - \left(\frac{48 - Q_a}{2}\right)}{2}$$

$$= \frac{48 - 24 + \frac{1}{2}Q_a}{2}$$

$$Q_a = 24 - 12 + \frac{1}{4}Q_a$$

$$\frac{3}{4}Q_a = 12$$

$$Q_a = 12 \times \frac{4}{3} = 16$$

To obtain Q_b we substitute the value of Q_a in (*ii*).

$$Q_b = \frac{48 - Q_a}{2} = \frac{48 - 16}{2} = 16$$

Total output (Q) under Cournot duopoly equilibrium = 16 + 16 = 32

Substituting the value of Q in the inverse demand function we have

$$P = 100 - 2Q = 100 - 2 \times 32$$
$$= 36$$

Problem 2. *Suppose the market demand curve for a commodity is $P = 70 - Q$ and that there are two firms each with constant MC of ₹ 10. Assuming they behave as Cournot duopolists, what will be the price and total industry output? Compare with the outcome under pure monopoly and perfect competition.* [*D.U. B.A.* (*Hons.*) 1989]

Solution:

Market demand for output at which $P = MC$:

$$70 - Q = 10$$
$$Q = 60$$

Let the output of the two firms A and B are Q_a and Q_b respectively

$$Q_a = \frac{60 - Q_b}{2} \quad \text{...}(i)$$

$$Q_b = \frac{60 - Q_a}{2} \quad \text{...}(ii)$$

Substituting the value of Q_b in (i) we have

$$Q_a = \frac{60 - \left(\dfrac{60 - Q_a}{2}\right)}{2} = \frac{60 - 30 + \dfrac{1}{2} Q_a}{2}$$

$$= 30 - 15 + \frac{1}{4} Q_a$$

$$\frac{3}{4} Q_a = 15$$

$$Q_a = 15 \times \frac{4}{3} = 20$$

$$Q_b = \frac{60 - Q_a}{2}$$

$$= \frac{60 - 20}{2} = 20$$

Total output (Q) under Cournot duopoly equilibrium = $Q_a + Q_b = 40$

Price under Cournot duopoly equilibrium:

$$P = 70 - \text{Q}$$
$$= 70 - 40 = 30$$

MR under Pure Monopoly

$$MR = \frac{dTR}{dQ} = \frac{d(PQ)}{dQ} = \frac{d(70Q - Q^2)}{dQ}$$

$$= 70 - 2Q$$

Now, in equilibrium output under monopoly : $MR = MC$

$$70 - 2Q = 10$$
$$2Q = 60$$
$$Q = 30$$

Price under Pure Monopoly

$$P = 70 - Q = 70 - 30 = 40$$

Output and Price under Pefect Competition

For equilibrium output under perfect competition:

$$\text{Price} = MC$$
$$70 - Q = 10$$
$$Q = 70 - 10 = 60$$

Price under Perfect Competition:

$$P = 70 - Q$$
$$= 70 - 60 = 10$$

It is evident from above that output (40) under Cournot duopoly is $\frac{2}{3}$ rd of output (60) under perfect competition and monopoly output (30) is half of the output under perfect competition. Price under perfect competition (₹ 10) is lower than those under Cournot's duopoly and pure monopoly.

Problem 3. *In a constant cost industry having zero costs, the inverse demand function is given by $P = 12 - Q$. Estimate the equilibrium value of the price and output for monopoly, Cournot's duopoly and perfect competition.* [*D.U. B.A.* (*Hons.*) 1999]

Solution:

Monopoly:

For price and output under monopoly:

$$MR = MC$$

$$MR = \frac{d\,(TR)}{dQ} = \frac{d\,(PQ)}{dQ} = \frac{12\,(12Q - Q^2)}{dQ}$$

$$= 12 - 2Q$$

$$\text{Since } MC = 0 \text{ (given)}$$
$$12 - 2Q = 0$$
$$Q = 6$$

Price under monopoly

$$P = 12 - Q = 12 - 6 = 6$$

Cournot's Duopoly:

Setting price equal to marginal cost (which is zero) we have $12 - Q = 0$ or $Q = 12$

$$\text{Output of Firm 1 } (Q_1) = \frac{12 - Q_2}{2} \qquad ...(i)$$

$$\text{Output of Firm 2 } (Q_2) = \frac{12 - Q_1}{2} \qquad ...(ii)$$

Substituting the value of Q_2 in (*i*) we have

$$Q_1 = \frac{12 - \left(\dfrac{12 - Q_1}{2}\right)}{2}$$

$$Q_1 = \frac{12 - 6 + \frac{1}{2}Q_1}{2}$$

$$Q_1 = 6 - 3 + \frac{1}{4}Q_1$$

$$\frac{3}{4}Q_1 = 3 \text{ or } Q_1 = 3 \times \frac{4}{3} = 4$$

Similarly $$Q_2 = \frac{12 - Q_1}{2} = \frac{12 - 4}{2} = 4$$

Cournot duopoly output, $Q_1 + Q_2 = 4 + 4 = 8$

Price under Cournot duopoly :

$$P = 12 - Q = 12 - 8 = 4$$

Under Perfect Competition :

Perfectly competitive output : Price = MC

$$12 - Q = 0 \text{ or } Q = 12$$

Perfectly competitive price : $P = 12 - Q = 12 - 12 = 0$

BERTRAND'S DUOPOLY MODEL

Joseph Bertrand, a French mathematician, criticized Cournot's duopoly solution and put forward a substitute model of duoply. According to Bertrand, there was no limit to the fall in price since each producer can always lower the price by underbridding the other and increasing his supply of output until the price becomes equal to his unit cost of production. There are some important differences in assumptions of Bertrand and Cournot's models of duopoly. In Bertrand's model, producers do not produce any output and then sell whatever price it can bring in. Instead, the producers first set the price of the product and then produce the output which is demanded at that price. Thus, *in Bertrand's model adjusting variable is price and not output.*

In Cournot's model, each producer adjusts his output believing that rival will continue to produce the same output as he is doing at present, but in *Bertrand's model each producer believes that his rival will keep his price constant at the present level whatever price he might himself set.* Thus, in Bertrand's adjusting variable is price and not output.

Furthermore, in Bertrand's model, it is not very important that the producers should know the correct market demand of their product, or should have identical view about the market demand. It is enough for each producer to know that he can capture the whole market by undercutting his rival. The other assumptions of Bertrand's model are the same as those of Cournot's model though their implications may be somewhat different. Thus, *in Bertrand's model the products produced and sold by the two producers are completely identical and in no way differentiated.* Its implication is that if a producer underbids the other, it can conquer the whole market (that is, snatch away all the customers from his rival). Further, the two producers have identical costs and also work under condition of constant marginal cost. Moreover, the productive capacity of the producers is unlimited, that is, there is no limit to their increase in the supply of output up to the maximum requirement of demand.

Bertrand's duopoly model is illustrated through Fig. 29A.3. Let there be two producers A and B. Market demand curve for the product produced by them is given by linear curve DD'. Suppose that producer A goes into business first. Because A is the only producer at present he sets the price at the monopoly level, which is the most profitable for him. This monopoly price is P_m and producer A produces monopoly output ON which is half of perfectly competitive output Q assuming constant average and marginal cost equal to OG. Now, suppose that B also enters into the business and starts producing the same product as produced by A. But B assumes that A will go on charging the same price P_m which he is doing at present, irrespective of whatever price he himself might set. Further B finds that he can capture the whole market by slightly undercutting the price and thereby make substantial amount of profits. Accordingly, B sets a price slightly lower than A's price P_m and as a result gets the entire demand of the product. *A*'s sales, for the moment, falls to zero. Now, threatened with the loss of his entire business, producer *A* will reconsider his price policy. But while deciding about his new price policy he assumes that *B* will continue to charge the same price which he is doing at present. There are two alternatives open to him. First, he may match the price cut made by *B*, that is, he may charge the same price as *B* is now charging. In this case, he will secure half the market, the other half going to the producer *B*. Secondly, he may undercut *B* and set a slightly lower price than that of *B*. In this case, *A* thinks he will seize the entire market. Evidently, the latter course looks more profitable and thus, *A* undercuts *B* and sets a price lower than *B*'s price.

But with the above move of *A*, producer *B* finding himself deprived of all his sales will react and think of changing his price. Since *B* also assumes *A*'s price to remain fixed at the present level, whatever price he himself might set. Producer *B* has similarly two alternatives : he may match *A*'s price or undercut him. Finding the undercutting more profitable, *B* will set a bit lower price than *A* and thus seize the whole market.

But again, *A* will be forced to undercut *B*. This *price war* (*i.e.* the process of undercutting) will go on until the price falls to the competitive levels, that is, equal to average or marginal cost of production. Once the price has fallen to the level of average or marginal cost of production, neither of them will like to cut the price further because in that case total cost would exceed total revenue and will therefore bring losses to the duopolists. Also, neither of them would like to raise the price, since in doing so each of them would be afraid of losing his entire business given the belief that the other will go on charging the same lower price. Thus, when the price has fallen to the competitive level of average cost of production, neither of the duopolists would have any incentive to lower the price further, or to raise it and, therefore, the equilibrium has been achieved. *In Bertrand's model equilibrium is achieved when as a result of price war market price has fallen to the average cost of production and the combined equilibrium output of the two duopolists is equal to the competitive output.*

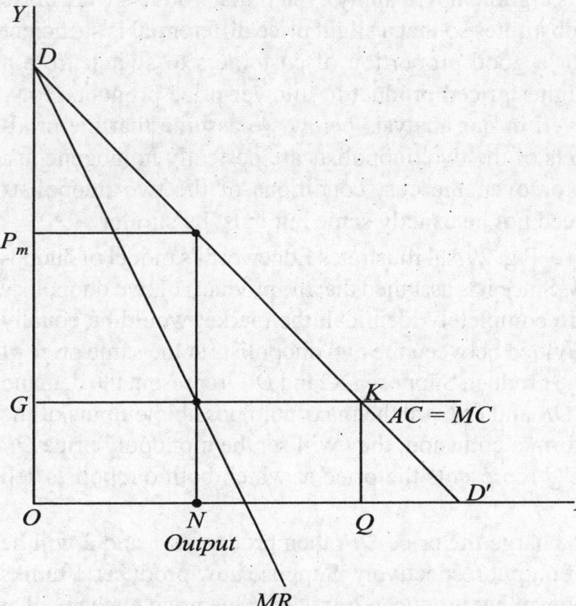

Fig. 29A.3. *Bertrand's Duopoly Model*

It is evident from the above analysis of the Cournot and Bertrand's models of duopoly that the fundamental assumption about the behaviour of the duopolists in the two models is similar. The duopolists in both models have erroneous and incorrigible belief that the rival will continue to do what he is presently doing regardless of what he himself might do. However, the basic assumption in the two models is not exactly the same. In Cournot's model, the basic assumption relates to the output policy, but in Bertrand's model, it relates to the price policy. Therefore, the two models yield different results. According to Cournot's model, equilibrium output is less than the perfectly competitive output and, therefore, under it price is higher than the perfectly competitive price. But, according to Bertrand's model, output and price under duopoly are equal to those under pure competition.

EDGEWORTH'S DUOPOLY MODEL

F.Y. Edgeworth, a famous French economist, also attacked Cournot's duopoly solution. He criticised Cournot's assumption that each duopolist believes that his rival will continue to produce the same output irrespective of what he himself might produce. According to Edgeworth (as in Bertrand's model), each duopolist believes that his rival will continue to charge the same price as he is just doing irrespective of what price he himself sets. With his assumption, and taking the example of Cournot's mineral wells' with zero cost of production, *Edgeworth showed that no determinate equilibrium would be reached in duopoly.*

The main difference between Edgeworth's model and Bertrand's model is that whereas in Bertrand, productive capacity of each duopolist is practically unlimited so that he could satisfy any amount of demand but in Edgeworth's model, the productive capacity of each duopolist is limited so that neither duoplist can meet entire demand at the lower price ranges. Each duopolist accepts as much demand of the product at a price as he can meet. It is not essential in Edgeworth's model that the products of duopolists should be perfectly homogeneous, his argument will apply even if the products were close substitutes so that a slight price differential is sufficient for a good proportion of customers to switch from a higher priced product to a lower-price product. However, in our analysis below we assume that the products of the two duopolists are perfectly homogeneous. Moreover, the cost conditions of the two duopolists need not be exactly same but must be similar.

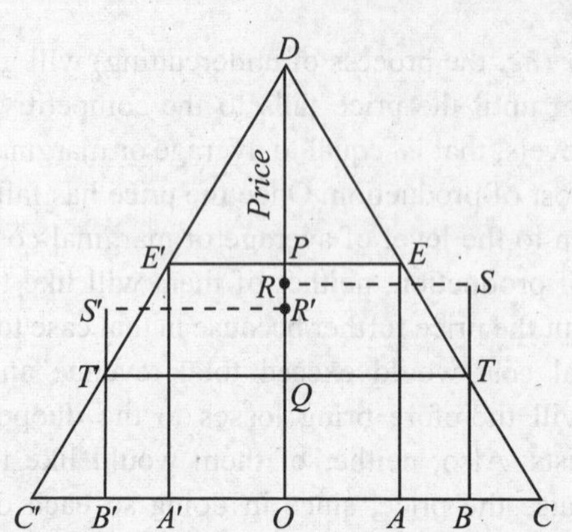

Fig. 29A.4. *Edgeworth's Duopoly Solution.*

Fig. 29A.4 illustrates Edgeworth's model of duopoly. Since it is assumed that the products of two duopolists are completely identical, the market would be equally divided between the two duopolists at the same price of the product. Suppose *DC* and *DC'* represent the demand curves facing each duopolist. Further suppose *OB* and *OB'* are the maximum possible outputs of the two duopolists respectively. If the duopolists form a collusion, they will set the monopoly price *OP* and will make maximum joint profits. Price *OQ* represents the price at which both duopolists sell their maximum possible outputs.

Assume that the two duopolists happen to charge the price *OP*, then producers 1 and 2 will be producing and selling *OA* and *OA'* amounts of output respectively. Suppose now producer 1 thinks of revising his price policy. Producer 1 will believe that producer 2 will keep his price unchanged at *OP* regardless of whatever price he himself might charge. With producer 2's price remaining fixed at *OP*, producer 1 realises that if he sets the price slightly lower than *OP*, he will be able to attract a sufficient number of producer 2's customers so that he can sell his whole maximum output which he can produce. This would yield greater profits to producer 1 than he is making at present. Thus in Fig. 29A.4 if producer 1 lowers his price from *OP* to *OR*, he will be able to sell his entire maximum and will be earning profits equal to the area *OBSR* which are greater than *OAEP*. Thus *A* would increase his profit by lowering his price.

But when producer 1 reduces his price, producer 2 will find most of his customers deserting him and his sales considerably reduced. Profits of producer 2 will therefore fall considerably. As a result, producer 2 will think of making a counter move, but he too will assume that producer 1 will hold his price constant at *OR*. Producer 2 sees that if he cuts his price slightly below producer 1's price *OR*, say he fixes *OR'* he can take away enough customers of *A* to sell his entire maximum possible output *OB'*. Thus when producer 2 cuts his price to *OR'*, he sells his entire output *OB'* and makes profits equal to *OR'S'B* which are greater than the profits he was making before. As a result of this, sales and profits of producer 1 will greatly decline. Producer 1 will then react and will think that if he reduces his price a bit below *OR*′, he will be able to sell his whole maximum possible output *OB* by attracting customers of producer 2, still believing that producer 2 will keep his price fixed at *OR*′. Thus when producer 1 reduces his price, his profits will rise for a moment. But producer 2 will then react and reduce his price further in order to increase his profits. In this way, according to Edgeworth, the price cutting by two producers will continue until the price falls to the level *OQ* at which both producers sell their entire maximum possible outputs.

It will be seen in Fig. 29A.4 that at price *OQ*, producers 1 and 2 are selling *OB* and *OB*′ respectively (*OB* = *OB*′) and are making profits equal to *OBTQ* and *OB*′*T*′*Q* respectively. When the price has been bid down to the level *OQ*, none of the producers will see any advantage to cut the price further. Since at price *OQ* each is selling the entire output he can produce, he will not be able to increase his profits because of his inability to increase his output further. But, according to Edgeworth, equilibrium is not attained at price *OQ*. Edgeworth argues that each producer will have no incentive to lower the price below *OQ*, but each will have incentive to raise it above *OQ*. Thus, Edgeworth says: "At this point it might seem that equilibrium would have been reached,. Certainly it is not in the interest of either monopolist to lower the price still further. But it is in the interest of each to raise it." At price *OQ*, one of the two producers, say producer 1, may realise that his rival producer 2 is selling his entire possible output *OB*′ and serving half of the customers and cannot increase his output further to serve more customers. Thus producer 1 realises that he can serve the other half of the customers at the price which is most profitable for him and he will accordingly raise the price to *OP* at which he sells *OA* and earns profits *OAEP* which are larger than profits *OBTQ* at price *OQ*. Thus knowing that his rival has done his worst by putting his entire possible output on the market and that producer 2 cannot attract any of his *OA* units of demand because of his inability to produce more, producer 1 raises the price to *OP* and thereby increases his profits.

But when producer 1 has raised the price to *OP*, producer 2 will realise that if he sets his price slightly below *OP*, he would still be able to sell *OB*′ by attracting enough customers of producer 1 who is charging the price *OP* and will therefore increase his profits. Accordingly, producer 2 raises his price to the level slightly below *OP*. But producer 1 then finding his customers deserting him and sales being reduced will believe that he can increases his profits by reducing his price slightly below producer 2's level. When he does so, then producer 2 will react, and so on. Thus, once again the process of competitive price cutting starts and the price again ultimately reaches the level *OQ*. But once the price has reached *OQ*, any of the producers will again raise it to *OP* and so on. In this way, price will oscillate between *OP* and *OQ*, gradually downward but upward in a jump. As said above, price *OP* is the monopoly price and price *OQ* is the competitive price. It follows from above that Edgeworth's duopoly solution is one of perpetual disequilibrium, price constantly oscillating between the monopoly price and competitive price. Thus no determinate and unique equilibrium of duopoly is suggested by Edgeworth's duopoly model.

Comments over Classical Models of Duopoly (Oligopoly)

In the analysis of three classical models of duopoly, it was seen that one common assumption among them is that the duopolists have zero conjectural variation, that is, while deciding about the output or price policy, each duopolist believes that his rival will hold output or price constant at the present level whatever he himself might do. Further, a producer remains unshaken in this erroneous belief even when he constantly finds himself to be proved incorrect since after his action, the rival does react and change his output or price. This is a chief logical error in classical models.

Furthermore, by assuming zero conjectural variation on the part of the duopolists (oligopolists), classical models ignore the mutual interdependence which is the chief characteristic of oligopoly. Thus, classical models provide solution for the oligopoly problem by removing from it its most important feature.

CHAMBERLIN'S OLIGOPOLY MODEL

In his now famous work *"The Theory of Monopolistic Competition"* Chamberlin made an important contribution to the explanation of pricing and output under oligopoly. His oligopoly model makes an advance over the classical models of Cournot, Edgeworth and Bertrand in that, in sharp contrast to above classical models, *his model is based on the assumption that the oligopolists recognise their interdependence and act accordingly*. Chamberlin criticises the behavioural assumption of Cournot, Bertrand and Edgeworth that the oligopolists behave independently in the sense that they ignore their mutual dependence and while deciding about their output or price assume that their rivals will keep their output or price constant at the present level. According to him, oligopolists behave quite intelligently as they recognise their interdependence and learn from the experience when they find that their action in fact causes the rivals to react and adjust their output level. This *realisation of mutual dependence on the part of the oligopolists leads to the monopoly output being produced jointly and thus charging of the monopoly price*. In this way, *according to Chamberlin, maximisation of joint profits and stable equilibrium are achieved by the oligopolists even though they act in a non-collusive manner*. Given identical costs, they will also equally share these monopoly profits.

Chamberlin's Approach to Stable Joint Profit-Maximising Equilibrium under Oligopoly

The process by which stable equilibrium under oligopoly is reached in Chamberlin's oligopoly model is illustrated in Fig. 29A.5. Chamberlin considers the case of a duopoly with zero cost of production of the two producers, *A* and *B*. Like Cournot he also assumes that the market demand curve for the product is linear. In Fig. 29A.5, *MD* represents this linear market demand curve for the homogeneous product of the duopolists. As in Cournot's model, suppose producer *A* is the first to start production. He will view the whole market demand curve *MD* facing him and corresponding to it MR_a is the marginal revenue curve. In order to maximise his profits he will equate marginal revenue with marginal cost (which is here taken to be equal to zero). It will be seen from Fig. 29A.5 that he will be in equilibrium by making $MR = MC$ when he produces *OQ* output (*i.e.* half of *OD*) which is in fact the monopoly output, and will fix price equal to *OP*.

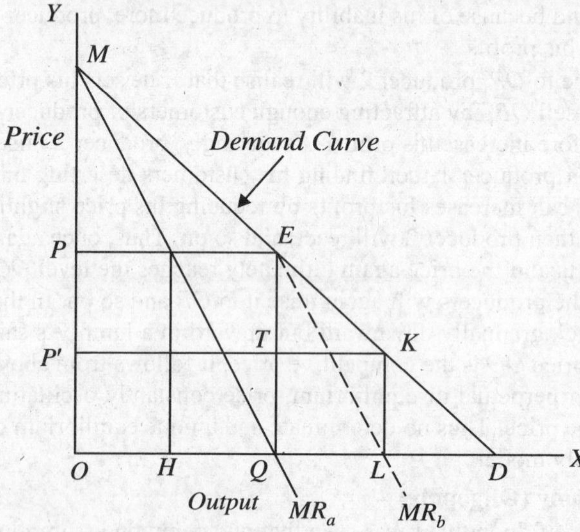

Fig. 29A.5. *Chamberlin's Stable Model of Duopoly with Mutual Dependence Recognised*

Now, suppose producer *B* enters the market. He thinks, as in Cournot' s model, that producer *A* would continue to produce *OQ* output and therefore views *ED* portion of the market demand curve as the relevant demand curve facing him and corresponding to it MR_b is the marginal revenue curve. With marginal cost being equal to zero, for maximising profits he will produce half of *QD,* that is, *QL* or at point *L* at which his marginal revenue curve *MR* intersects the *X*-axis along which output is measured. With aggregate output *OL* (*OL* = *OQ* of *A* + *QL* of *B*), price will fall to the level *LK* or *OP′* with the result that profits earned by producer *B* will be equal to the area of rectangle *QLKT,* and due to the fall in price the profit of producer *A* will decrease from *OPEQ* to *OP′ TQ*.

However, from this point onward Chamberlin's analysis deviates from Cournot's model. Whereas in Cournot's model, the firm *A* will readjust his output and will continue to assume that his rival will keep his output constant at *QL* level, but in Chamberlin's model producer *A* learns from his experience that they are interdependent. With the realisation of mutual dependence, producer *A* decides to produce output *OH* equal to output *QL* of producer *B* and half of monopoly output *OQ* so that the aggregate output of both of them is the monopoly output (*OQ* = *OH* of *A* + *QL* of *B*). With *OQ* as the aggregate output level, price will rise to *QE* or *OP.* Firm *B* also realises that in view of interdependence it is in the best interest for both of them to produce half of monopoly output and will therefore maintain output at the *QL* or *OH* level which is half of the monopoly output. Thus, *each producer producing half of monopoly output, will result in maximisation of joint profits though they do not enter into any formal collusion*. In this way Chamberlin explains that duopolists behaving intelligently and realising their interdependence reach a stable equilibrium and together produce monopoly output and charge monopoly price each sharing profits equally.

A Critical Evaluation. Chamberlin's model is an advance over the classical models in that the firms behave intelligently and recognise their interdependence. *Their behaviour leads them to the monopoly solution of output and pricing which ensures maximisation of joint profits though they do not formally collude*. This implies that firms have full information about the market demand curve and quickly learn from the experience and realise that the ultimate consequence of alternative chain of adjustments to rival's moves will be less profitable than sharing the monopoly profits equally with him.

Further, it is assumed in Chamberlin's model that the oligopolists know fully the costs of production of their rivals which enable them to arrive at a monopoly output and price which is in the best interest of all of them. Thus, unless all oligopolists have identical costs and demands, it seems impossible that the oligopolists will be able to reach monopoly solution, that is, maximisation of joint profits without collusion. It may be noted that *even in a formal collusion there is always incentive on the part of rival firms to cheat by under-cutting price to increase their individual profits*. In Chamberlin's model of oligopoly without collusion, incentive for the firms to undercut price to increase their share of profit will be relatively more. Besides, Chamberlin's model has another great flaw as *it ignores the entry of new firms* and is thus a closed model. Due to the attraction of monopoly profits jointly earned by the existing firms, the new firms are likely to enter the industry. With the entry of new firms the attainment of stable equilibrium of oligopoly is unlikely to occur.

QUESTIONS FOR REVIEW

1. Explain how equilibirum output and price are determined in Cournot's duopoly model. State the underlying assumptions.
2. How is Cournot-Nash equilibrium determined? Why is it stable? Why don't the duopolists set the output at joint profit maximizing level by tacit collusion? *[D.U. B.A.* (*Hons*) 2001]
   [**Hint:** Cournot equilibrium is also called Cournot–Nash equilibrium.]
3. (*a*) Explain Cournot's duopoly model. What should the firms do to maximise their profits? [*D.U. B.Com.* (*Hons.*) 1996]
   (*b*) Bring out precisely the difference between Cournot equilibrium and profit-maximising behaviour. [*D.U. B.A.* (*Hons.*) 1988]
4. Explain Chamberlin's model of duopoly. How does it differ from Cournot's solution of duopoly problem? [*D.U. B.Com.* (*Hons*) 2001]
5. Let the market for telecommunications equipment be represented by a duopoly where the two firms produce outputs q_1 and q_2, respectively. The inverse demand function is represented by $P = 100 - 2Q$ (where $Q = q_1 + q_2$). The marginal cost that each firm faces is 4:
   (*i*) What are the reaction functions of Cournot's duopolists?
   (*ii*) Calculate each firm's output and the market price at Cournot's equilibrium. [*D.U. B.A. (Hons) 2003*]
6. (*a*) Explain Bertrand's model of oligopoly taking the case of two firms. How does it differ from Cournot's model?
   (*b*) Show that in Bertrand's model a price war between the two rival firms will drive the price to the competitive level.
7. Explain Edgeworth's duopoly model of pricing. Show that in Edgeworth's model price is indeterminate, goes on moving perpetually between monopoly price and competitive price and no stable equilibrium is reached.
8. According to Chamberlin, maximisation of joint profits and stable equilibrium are achieved under oligopoly even though they act in a non-collusive manner. Discuss.

Chapter 30

SALES MAXIMISATION MODEL OF OLIGOPOLY

Sales maximisation model of obligopoly is another important alternative to profit maximization model. This has been propounded by W.J. Baumol, an American economist. In an earlier chapter we have explained sales maximisation objective and saw how Baumol challenged the profit maximisation assumption regarding business behaviour in these days of manager-dominated corporate form of business organisation and showed how sales maximisation was more valid and realistic assumption of business behaviour. Further, we pointed out there that sales maximisation was quite consistent with *rationality* assumption about business behaviour. It may also be noted that sales maximisation model represents one of the managerial theories of the firm because in it, the great importance has been given to the manager's role and to his pursuing self-interest in making price, output and advertising policies. Prof. Baumol thinks that *managers are more interested in maximizing sales than profits*.

Sales Maximisation vs Profit Maximisation

It should be noted that by sales maximisation Baumol does not mean the maximisation of the *physical volume* of sales but the maximisation of *total revenue from sales*, that is, the rupee value of the sales made. Therefore, his theory is also known as revenue maximisation model. Further, as has been mentioned in a previous chapter, Prof. Baumol does not ignore profit motive altogether. He argues that there is a *minimum acceptable level of profits* which must be earned by the management so as to finance future growth of the firm through retained profits and also to induce the potential shareholders for subscribing to the share capital of the company. Thus, according to him, management of oligopolistic firms *seeks to maximise sales or, in other words, total revenue subject to this minimum profit constraint*.

He thus writes: "My hypothesis then is that oligopolists typically seek to maximize their sales subject to a minimum profit constraint. The determination of the minimum just acceptable profit level is a major analytical problem and I shall only suggest here that it is determined by long-run considerations. Profits must be high enough to provide the retained earnings needed to finance current expansion plans and dividends sufficient to make future issues of stocks attractive to potential purchasers"[1]

Sales Maximisation: Price-output Determination

It is better to explain graphically price-output determination in Prof. Baumol's sales or total revenue maximization model. Consider Fig. 30.1 where on the *Y*-axis we measure total revenue, total cost and total profits in terms of rupees and on the *X*-axis we measure the total output. *TR* and *TC* are respectively total revenue and total cost curves. Since total cost curve *TC* starts from the origin, it means the diagram refers to the long-run cost-revenue situation. *TP* is the total profit curve

1. W.J. Baumol, On the Theory of Oligopoly, *Economica*, new series, Vol. 25, 1958.

which first rises and then after a point falls downward. Since total profits are the difference between total revenue and total costs at various levels of output, the total profit curve measures the vertical difference between the *TR* and *TC* curves at various levels of output.

If the firm aims at maximising profits, it will produce *OA* output. This is because corresponding to *OA* output, the highest point of *TP* curve lies. But, as we have seen above, according to Prof. Baumol, the firm does not seek maximisation of profits. On the other hand, if the firm wants to maximise sales (or total revenue) it will fix output at *OC* level which is greater than *OA*. At output *OC* total revenue is CR_2, which is maximum in the diagram. At this total revenue (sales) maximising output level *OC*, the firm is making total profits equal to *CG* which are less than the maximum attainable profits *AH*. It will be clear from the figure that sales (or total revenue) maximisation output *OC* is larger than profit-maximising output *OA*. Prof. Baumol contends that the business firms aim at total revenue (sales) maximisation subject to a minimum profit constraint. Now, if *OM* is the minimum total profits which a firm wants to obtain, then *ML* is the minimum profit line. Now, this minimum profit line *ML* cuts the total profit curve *TP* at point *E*. Therefore, if the firm wants total revenue (sales) maximi-sation subject to the minimum profits of *OM*, as has been contended by Prof. Baumol, then it will produce and sell output *OB*. At output *OB*, the firm will be having total revenue equal to BR_1 which is less than the maximum possible total revenue of CR_2. But the total revenue BR_1 is the maximum obtainable revenue to earn the minimum desirable profits *OM*.

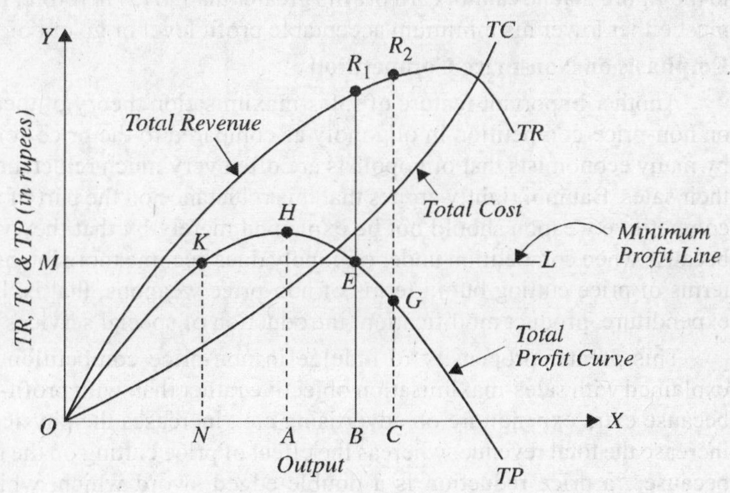

Fig. 30.1. *Prof. Baumol' Sales Maximisation Hypothesis.*

It should be noted that the firm can earn the minimum profit *OM* even by producing *ON* output. (Minimum profit line *ML* also cuts the total profits curve *TP* at point *K*). But the total revenue at output *ON* is much less than at output *OB*. Therefore, given the firm's objective of maximising the total revenue subject to the minimum profit constraint, the firm will not produce *ON* output or at point *K*. It will be noticed from the figure that the output *OB* lies in between *OA* and *OC*, that is, it is larger than profit-maximising output *OA* but smaller than the total revenue-maximising output *OC*. Thus, in Prof. Baumol's model, oligopolistic firm will be in equilibrium at output *OB* and will be earning profits *BE* (or *OM*).

It should be carefully noted that the objective of total revenue (or sales) maximisation subject to the minimum profit constraint leads to a greater output and lower price than does profit maximisation. Price will be lower under revenue maximisation because the output under it, as seen above, is greater and given that demand or average revenue curve is sloping downward, the price will be less when the output is larger. To quote Prof. Baumol "The profit-maximizsing output *OA* will usually be smaller than the one which yields either type of sales maximum, *OC* or *OB*. This can be proved with the aid of the standard rule that at the point of maximum profit marginal cost must equal marginal revenue. For marginal cost is normally a positive number (we can't usually produce more of a good for nothing). *Hence marginal revenue will also be positive when profits are at a maximum, i.e.* a further increase in output will increase total sales (revenue). Therefore, if at the point of maximum profit the firm earns more profit than the required minimum, it will pay the sales maximizer to lower his price and increase his physical output."[2]

The price charged at output OB will be equal to $\frac{\text{total revenue}}{\text{output}}$, that is, $\frac{BR_1}{OB}$

Now suppose the minimum acceptable profits are equal to AH (which are maximum possible profits under the given cost-revenue situation), then even under total revenue maximisation objective subject to the minimum profit constraint, the firm will produce the profit-maximising output OA. But it will produce output OA not to maximise profits but to maximise total revenue, given the minimum profit constraint AH. Now, suppose the minimum acceptable profits for an entrepreneur are larger than AH, then it is obvious from the figure that given the cost-revenue situations depicted in the figure that he cannot earn profits greater than AH. Therefore, in this situation the entrepreneur must either lower his minimum acceptable profit level or go out of the industry.

Emphasis on Non-price Competition

Another important feature of sales-maximisation theory of the firm of Baumol is its emphasis on non-price-competition in oligopoly as compared to the price competition. It has been observed by many economists that oligopolists are often very much reluctant to use price cutting to promote their sales. Baumol rightly argues that this reluctance on the part of the oligopolists to use price as a competitive weapon should not be explained merely by that they want to live in *quiet life*. This is because when competition under oligopoly does become more intense and vigorous, it may not be in terms of price cutting but in terms of non-price weapons, that is, in the form of more advertising expenditure, product modification, introduction of special services for the customers etc.

This greater propensity to indulge in non-price competition under oligopoly can be better explained with sales-maximisation objective rather than with profit-maximisation objective. This is because extra expenditure on advertising etc., increases the physical volume of sales, it must also increase the total revenue, whereas the effect of price cutting on the total revenue is doubtful. This is because, "a price reduction is a double-edged sword which, while it serves as an influence to increase total revenue in that it usually adds to the number of units which can be sold, simultaneously it also works in the opposite direction by reducing the revenue on each unit sold. In other words, as the economists know so well, depending on whether demand is or is not elastic, *price cutting is an uncertain means for increasing dollar sales*."[3] The effect of price cutting on profits is more uncertain because if it fails to raise total revenue, it will most probably reduce profits because the increase in output as a result of reduction in price, will increase total costs. On the other hand, while the profitability of advertising, product modification, improved service is doubtful, their favourable effect on the sales is quite certain. Thus, according to Baumol, "the effect of advertising, improved services, etc., on sales is fairly sure while, very often, their profitability may be quite doubtful. Thus, *sales maximization makes for greater presumption that the businessman will consider non-price competition to be more advantageous alternative*."[4]

We now turn to explain that, according to Baumol's sales maximisation model, how much advertising expenditure a firm will undertake.

Sales Maximisation Model: Optimal Advertising Expenditure

We know that firms in oligopolistic market conditions compete not only in terms of price but also in terms of advertising expenditure, product variation and special services offered to the buyers. We shall discuss here the question of optimal advertising expenditure to be incurred by an oligopolist and the conclusions reached in this connection will apply equally to the questions of optimal product adjustment and the optimal amount of special services to be provided by an oligopolist when he chooses to maximize sales (total revenue).

---

2. William J. Baumol, *Economic Theory and Operations Analysis*, 3rd. ed., pp. 326-297.
3. Baumol, *op. cit.*, p.266 (*italics added*)
4. *Ibid*, p. 266-67 (*italics added*)

The important question in regard to the advertising is how much advertising expenditure a firm will make so as to achieve its objective. How much advertising outlay will be incurred by a firm is greatly influenced by the objective of the firm as to whether it seeks to maximise sales or profits. This optimal advertising expenditure from the viewpoints of both sales maximisation and profit maximisation is illustrated in Fig. 30.2 in which advertising outlay is measured along the X-axis and total cost, total revenue and total profits on the Y-axis. Baumol takes an important assumption in connection with the effect of advertising outlay on total revenue or sales. He assumes, and he quotes empirical evidence for this, that the increase in advertising outlay by a firm will always raise the physical volume of sales, though after a point these sales will increase at a diminishing rate.

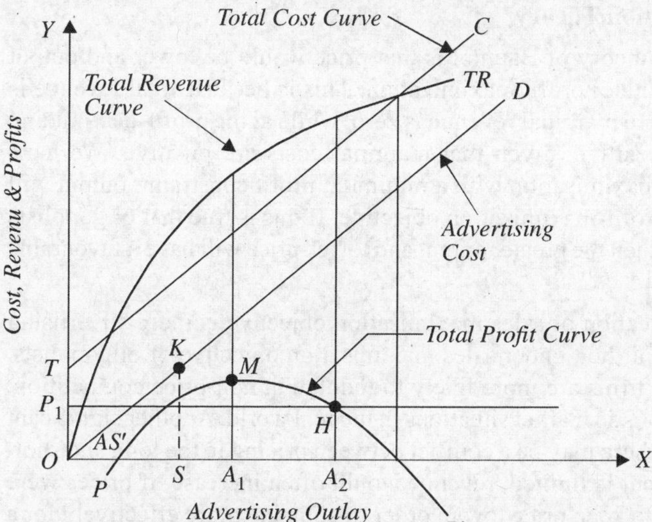

Fig. 30.2. *Optimal Advertising Outlay with Sales Maximisation and Profit Maximisation*

Now, given the price of product, the total revenue (*i.e.*, monetary value of the sales) will increase in proportion to the increase in the physical value of sales as a result of the increase in advertising outlay. Therefore, the increase in advertising outlay will always cause the total revenue to increase, though after a point diminishing returns are likely to set in. In Fig. 30.2, *TR* is the total revenue curve which represents the change in total revenue as the advertising outlay is raised, given the price of the product. Curve *OD* represents the advertising cost and has been so drawn as to make 45° angle with *X*-axis. This is because we have simply transferred the advertising outlay shown on the *X*-axis to the vertical axis as advertising cost, (for instance $OS = SK$). The *other costs* of the firm incurred on fixed and variable factors are taken to be independent of the amount of advertising outlay. Therefore, by adding a fixed amount of other costs (equal to *OT*) to the advertising curve *OD*, we obtain the total cost curve *TC*. Finally, by taking out the difference between total revenue curve (*TR*) and total cost curve (*TC*) we draw the total profit curve *PP′*.

Now, it will be seen from Fig. 30.2 that if the firm seeks to maximise its profits, it will incur advertising outlay equal to OA_1, at which the profit curve reaches its maximum point *M*. On the other hand, if OP_1 is the minimum profit constraint and the firm chooses to maximize its total revenue with OP_1 as the minimum profit constraint, it will spend OA_2 on advertisement which is greater than OA_1. We thus see that objective of constrained revenue maximization leads to a greater level of advertising outlay than the objective of profit maximization. In this connection, it should be noted that here there is no possibility of *unconstrained sales or revenue maximum*, as is there corresponding to output *OC* in the previous Fig. 30.1. This is because, unlike a reduction in price, *increase in advertising outlay always raises total revenue or sales* (by assumption). As a result, Baumol concludes that "it will always pay the sales maximizer to increase his advertising outlay until he is stopped by the profit constraint—until profits have been reduced to the minimum acceptable level. This means that sales maximizer will normally advertise no less than, and usually more than, do profit maximizers. For, unless the maximum profit level A_1M is not greater than the required minimum OP_1 it will be possible to increase advertising somewhat beyond the profit-maximizing level OA_1 without violating the profit constraint. Moreover, this increase will be desired since, by assumption, it will increase physical sales, and with them, dollar sales will rise proportionately."[5]

Critical Appraisal of Sales Maximisation Theory

Implication of sales maximisation theory of Baumol is that price would be lower and output greater under sales maximisation than under profit maximization. This is because total revenue is maximised at the price-output level where marginal revenue is zero, while at the profit-maximizing level of output marginal revenue is positive, given that marginal costs are positive. We have explained above that even under sales maximisation with a minimum profit constraint, output will be greater and price lower than under profit-maximization objective. If this is true that oligopolists seek to maximise sales or total revenue, then the greater output and lower-price will have a favourable effect on the welfare of the people.

As explained above, another implication of sales maximisation objective is more advertising expenditure will be incurred under it. Further, under sales maximisation objective of oligopolists, price is likely to remain sticky and the firms are more likely to indulge in non-price competition. This is what actually happens in oligopolistic market situations in the real world. Another significant implication of Baumol's model is that "there may be a conflict between pricing in the long and short run. In a short-run situation where output is limited, revenue would often increase, if prices were raised: but in the long-run it might pay to keep price low in order to compete more effectively for a large share of the market. This price policy to be followed in the short run would then depend on the expected repercussions of short-run decisions on long-run revenue."[6]

But sales maximisation model has not been without its critics. Shepherd[7] has asserted that an oligopolist confronts kinked demand curve and that if the kink is quite large, total revenue (*i.e.*, sales) and total profits would be maximised at the same level of output. But Hawkins[8] has shown that Shepherd's conclusions are invalid if the oligopolistic firms indulge in any form of non-price competition such as advertising, product variation, improvement in service, etc. and normally in the real world they do so.

An important and convincing criticism against sales maximisation model has been made by Hawkins.[9] As has been noted above, according to Baumol, sales-maximising firm will generally produce and advertise more than the profit-maximising firm. But Hawkins has shown that this conclusion is generally invalid. According to him, in case of single-product firms as compared with profit maximizing firm, whether a sales maximising firm will produce greater, smaller or same output and incur a greater, smaller or same advertising outlay depends upon the responsiveness of demand or total revenue to advertising expenditure as compared with the responsiveness of demand or total revenue to price cuts. As regards multi-product firms, which are generally found these days in the real world, in the *static model*, both sales maximisation and profit maximisation arrive at the same conclusion regarding choice of output and input combinations.[10] But besides a static model, Baumol has also developed a growth model of a sales maximising firm which, as has been shown by Williamson, that different results follow as compared with profit maximisating firm.

In spite of the above cirticisms, the present author is of the view that Prof. Baumol's sales maximisation model is a significant alternative to profit maximisation and brings us closer to reality,

5. *Op. cit.*, p.262
6. A. Silberston, Price Behaviour of Firms, *Economic Journal*, 1970
7. W.G. Shepherd, On Sales Maximising and Oligopoly Behaviour, *Economica*, 1962
8. C.J. Hawkins, 'On the Sales Revenue Maximization Hypothesis, *Journal of Industrial Economics*, April 1970.
9. C.J. Hawkins, "The Revenue Maximisation Oligopoly Model", Comment, *American Economic Revew*, March, 1971
10. Sec C.J. Hawkins, On the Sales Revenue Maximization Hypothesis', *Journal of Industrial Economics*, April 1970

for in many cases as we brought out above in the explanation of the model, it explains the business behaviour in the real world better than profit maximisation. Even if in certain cases, sales and profit maximisation yield same or similar results, even then by providing intersting insight into managerial motivation in these days of manager-dominated big business corporations Baumol has made an important contribution to the theory of firm. Further, by explicitly incorporating advertising and other forms of non-price competition in his model, Baumol has made a significant contribution to our price theory.

QUESTIONS FOR REVIEW

1. "In corporate form of business organisation managers are more interested in maximising sales rather than profits" Discuss.
2. What is meant by sales-maximisation ? Explain how, a firm determines optimal price and output when it pursues the goal of sales maximisation.
3. Explain Baumol's sales maximisation model using a suitable diagram. Campare it with profit, maximisation objective of the firm. *D.U. B.Com (Hons) 2001.*
4. What is meant by non-price competition ? Explain that sales maximisation objective brings out that businessmen will consider non-price competition to more advantageous than price competition. Show, through sales maximisation model, the optimal amount of advertising expenditure.
5. If a price making monopolist firm wants to maximise its sales revenue, it should —

   (*a*) set the highest price it can get.

   (*b*) set the lowest price it can get.

   (*c*) choose a selling price at which the elasticity of demand for its product is unity.

   (*d*) choose a selling price where the extra revenue received from the last unit sold exceeds the extra cost of making that unit.

   (Tick the right answer) *D.U. B.A.(Hons) Economics*

   [**Hint.** (C) is correct. This is because total revenue is maximum at the point on a demand curve at which price elasticity of demand is equal to unity.
6. Assuming that a firm pursues the objective of sales maximisation. How much advertising expenditure it will incure to maximise value of sales subject a minimum profit constraint. Compare it with the amount of advertising expenditure incurred for maximising profits.

Chapter 31
THEORY OF GAMES AND COMPETITIVE STRATEGY

Introduction

So far we have discussed those models of oligopoly which assume profit maximizing behaviour on the part of the oligopolists. There are other models of oligopoly which explain the price and output under oligopoly by assuming objectives other than profit maximization. One such model is to be found in the application of the theory of games to the oligopoly problem. Professors Neumann and Morgenstern in their book "*The Theory of Games and Economic Behaviour*" which was first published in 1944 provided a *new approach to many problems involving conflicting situations*. The game theory has been applied not only to the oligopoly but also to other economic questions like demand when uncertainty is present. Not only that, the game theory has been applied to the problems of subjects other than economics such as business administration, sociology, psychology, political science, military planning. The theory of games examines the outcome of a situation of interactions between the parties when they have *conflicting* interests. Basically, the game theory seeks to explain what is the rational course of action for an individual who is faced with an uncertain situation, the outcome of which depends not only upon his own actions but also upon the actions of others who too confront the same problem of choosing a rational strategic course of action. We shall describe below how the game theory explains his fundamental question. We shall confine ourselves only to the oligopoly problem. According to Professors Neumann and Morgenstern,[1] in an oligopolistic market situation, individual oligopolist is faced with a problem of choosing a rational course of action which is often called a strategy, keeping in view the possible reactions of his rivals whose reactions in turn would affect him. Thus, he confronts a problem similar to that of the player of any other game.

In a simple form of the game theory, the player has to choose among many possible courses of action which are called strategies. *A strategy is thus a course of action or a policy which a player or a participant in a game will adopt during the play of the game*. There are many possible strategies open to an individual among which he has to choose one at a time. In the case of oligopoly, the various alternative possible strategies which are relevant are : (*a*) changing the price, (*b*) changing the level of output, (*c*) increasing advertisement expenditure, and (*d*) varying the product. Changing the price may itself be divided into three strategies : (1) lowering the price, (2) raising the price, and (3) keeping the price unchanged. Similarly, the strategy of output may be (1) to increase the level of output, (2) to decrease output, and (3) to keep output constant. Likewise, increasing advertisement expenditure may further be divided into various strategies depending upon the various forms of advertisement, for instance, advertisement on radios, on television, in newspapers, in magazines, through handbills, through posters etc. Likewise, varying the product can be subdivided into various strategies depending upon the nature of the product to be chosen such as whether the colour of the package or the type of the package, or the quality of the product should be changed.

---

1. John Von Neumann and Oskar Morgenstern, *Theory of Games and Economic Behaviour,* 3rd ed., New York, Wiley, 1964.

A basic feature of oligopoly is that each firm must take into account its rival's reactions to its own actions. For example, Maruti Udyog cannot ignore the effect of an increase in price of its product on the prices and profits of its rival firms and how they will respond to its move of rise in price of its product. Thus, it is clear that oligopolistic behaviour has some of the characteristics of a game where a player must know how his move will affect his rival and how, assuming that he is rational, his rival will react to his move.

Game theory highlights that, in an oligopolistic market, a *firm behaves strategically,* that is, it adopts *strategic decision-making which means that while taking decisions regarding price, output, advertising etc. it takes into account how its rivals will react to its decisions and assuming them to be rational it thinks that they will do their best to promote their interests and take this into account while making decisions*. Game theory sheds new light on some of the important issues faced in explaining decision-making by firms operating in oligopolistic markets. It explains why an individual firm decides to cheat on a cartel agreement. Further, it explains why and how firms operating in oligopolistic markets prevent the entry of new firms in the industry.

Cooperative and Non-cooperative Games

Games which firms play can be either cooperative or non-cooperative. A game is cooperative if the firm (*i.e.* players in the game) can arrive at an enforceable or *binding contract* that permits them to adopt a strategy to maximise joint profits. Suppose making of a carpet costs ₹ 500 but the buyers value it at ₹ 1000. Fixation of price between ₹ 500 and 1000 per carpet will yield profits. In this case, two firms producing carpet can cooperate with each other and adopt a joint price strategy to maximise their joint profits rather than competing with each other. *If the two firms can sign a binding contract to share the profits between them from the production and sale of carpets, the game is called a cooperative game.*

On the other hand, a non-cooperative game is one where because of *conflict of interests* two firms cannot sign a binding contract. In most of the oligopolistic market situations binding contracts, that is, contracts that are enforceable cannot be negotiated. Therefore, in oligopoly in most cases we find examples of non-cooperative games. *In a situation of non-cooperative games while the competing firms take each other's actions into account but they take decisions independently and adopt strategies* regarding pricing, advertising, product variation to promote their interests.

It should be noted that a basic difference between a cooperative and non-cooperative game lies in the possibility of negotiating an enforceable contract. In cooperative games, negotiating binding or enforceable contracts are possible, in non-cooperative games they are not. In this chapter while explaining firms' decisions regarding pricing, advertising, we will be concerned mostly with non-cooperative games.

Note that there are games where the players move simultaneously. Each firm chooses a strategy before observing any action or strategy chosen by the rival firms. Not all games are of this type. In some games one player goes first and after which the other player reacts.

DOMINANT STRATEGY

How can firms decide about the optimal choice of a strategy? Some strategies may be successful (that is, more profitable) if competitors make a *particular* choice, that is, take a *certain* decision but will not be successful if competitors make *other* choices. On the other hand, *a dominant strategy is one which will be successful or optimal for a firm regardless of what others do, that is, no matter what strategy the rival firms adopt.*

Let us illustrate the dominant strategy in case of duopoly in the choice of whether to '*advertise*' *or not*. In this case, deciding in favour of advertising by a firm to promote its sales and hence profits or deciding not to advertise are the two strategies. Thus, '*Advertising*' or '*Not Advertising*' are the two strategies between which each firm has to make a choice. We assume there are two firms, *A* and

B which have to make a choice between the two strategies. The outcome (or profits made) from the various combinations of two strategies chosen by the two firms are presented in the following table in the form of payoff matrix. It should be noted that outcome or profits made by a firm by adopting a strategy is influended by the choice of a particular strategy by the rival firm.

Table 31.1. Payoff Matrix for Advertising Game

(in ₹ Crores)

		FIRM B	
		Advertising	*Not Advertising*
FIRM A	*Advertising*	A : 10 B : 5	A : 15 B : 0
	Not Advertising	A : 6 B : 8	A : 10 B : 2

It will be seen from the payoff matrix that if *both firms* adopt the strategy of '*Advertising*', the firm *A* will make profits of 10 crores and firm *B* will earn profits of 5 crores. If firm *A* decides to advertise and firm *B* decides not to advertise, profits of firm *A* are 15 crores and of firm *B* are zero. Similarly, if firm *A* decides not to advertise but firm *B* decides in favour of advertising, firm *A* makes profits of 6 crores and *B* of 8 crores. Further, if both firms go in for '*not advertising*' profits of *A* are 10 crores and of *B* are 2 crores.

Now, the question is what strategy each firm should choose. It is assumed that each firm is rational and will adopt a strategy which will ensure it more profits. Let us first consider choices and their outcome available for firm *A*. If the firm *B* adopts a strategy of '*Advertising*', profits of firm *A* are 10 crores if it also chooses the strategy of advertising but only 6 crores if it chooses not to advertise. On the other hand, if firm *B* adopts strategy of '*Not Advertising*', profits of firm *A* are 15 crores if it opts for 'Advertising' and profits of 10 crores if it also chooses strategy of '*Not Advertising*'.

It is thus clear from the payoff matrix that choice of strategy of '*Advertising*' by firm *A* is better or optimal since it ensures more profits whether firm *B* adopts strategy of '*Advertising*' or the strategy of '*Not Advertising*'. Thus, in the present payoff matrix, whatever strategy firm B adopts, for firm *A* strategy of '*Advertising*' is optimal. *When payoff matrix of a game is such that a choice of one strategy is better regardless of whatever strategy the other firm chooses, the strategy is known as dominant strategy.* In the present case choice of strategy of '*Advertising*' is a dominant strategy for firm *A*.

From the payoff matrix of the advertising game given in Table 31.1, the similar conclusion can be drawn for the optimal strategy to be adopted by firm *B*. Let us state the choices that are open to firm *B*. If firm *A* adopts strategy of 'Advertising', the firm *B* makes profits of 5 crores if it also chooses strategy of 'Advertising' and zero if it chooses strategy of 'Not Advertising'. Thus, choice of strategy of 'Advertising' by firm *B* is better, if firm *A* opts for the strategy of 'Advertising'. On the other hand, if firm *A* chooses strategy of 'Not Advertising', profits of firm *B* are 8 crores if it chooses strategy of 'Advertising' and 2 crores if it adopts strategy of 'Not Advertising'. Thus, in this case too, choice of strategy of 'Advertising' by firm *B* is optimal whatever strategy the firm *A* adopts. Thus, strategy of '*Advertising*' is a dominant strategy for firm *B*.

Since it is assumed that both firms behave rationally, each of them will choose the strategy of 'Advertising' and the outcome will be profits of ₹ 10 crores for firm A and Rs 5 crores for firm B.

It is important to note that all games do not have a dominant strategy for each player. To make it clear we make some changes in the payoff matrix and present them in Table 31.2. The payoff matrix in Table 31.2 differs from the previous payoff matrix in that profits shown in the bottom right hand corner are different, they are ₹ 20 crores for firm *A* and ₹ 2 crores for firm *B* in case both adopt the strategy of 'Not Advertising'.

Table 31.2. Payoff Matrix for Advertising Game

		FIRM B	
		Advertising	*Not Advertising*
FIRM A	*Advertising*	A : 10 B : 5	A : 15 B : 0
	Not Advertising	A : 6 B : 8	A : 20 B : 2

Note. *Numbers in the above table represent profits and are in ₹ crores.*

As will be seen from payoff matrix in Table 31.2, if firm *B* chooses strategy of 'Advertising' profits of firm *A* are ₹ 10 crores if it also opts for strategy of 'Advertising' and are ₹ 6 crores if it opts for strategy of 'Not Advertising'. Clearly, choice of strategy of 'Advertising' by firm *A* yields more profits and is therefore optimal if firm *B* adopts strategy of 'Advertising'. Now, if firm *B* chooses strategy of 'Not Advertising', profits of firm *A* are ₹ 15 crores, if it decides in favour of 'Advertising' strategy and its profits are ₹ 20 crores if it too adopts strategy of 'Not Advertising'. Thus, in this case given that firm *B* chooses stragegy of 'Not Advertising' choice of strategy of 'Not Advertising' by firm *A* is optimal.

It follows from above that in the payoff matrix presented in Table 31.2, *optimal strategy for firm A depends on which strategy the firm B adopts.* Choice of strategy of 'Advertising' is optimal for firm *A*, given that the firm *B* adopts the strategy of 'Advertising'. On the other hand, choice of strategy of 'Not Advertising' by firm *A* is better, given that the *B* adopts strategy of 'Not Advertising'. Thus, in this case there is no dominant strategy for firm *A*. The choice of an optimal strategy by firm *A* in the present case, that is, when dominant strategy does not exist, will be easier if firm *B* adopts a strategy before the firm *A* has to make its choice. But how a firm makes an optimal decision regarding choice of strategy if both firms must choose their strategies simultaneously, that is, at the same time. This is explained below.

Choice of an Optimal Strategy in the Absence of a Dominant Strategy

To decide about the optimal strategy by firm *A* when the choice of strategy by it depends on what strategy the other firm *B* adopts, the *firm A must put itself in firm B's place.* For this the firm *A* has to know what strategy is the best from firm *B's* point of view and further that it should assume that the firm *B* is rational and will therefore adopt the best strategy. From the payoff matrix given in Table 31.2, it will be seen that given that the firm *A* chooses strategy of 'Advertising' the firm *B* will make profits equal to ₹ 5 crores if it adopts strategy of advertising and its profits will be equal to zero if it opts for 'Not Advertising' strategy. Further, if firm *A* chooses strategy of 'Not Advertising' profits of firm *B* will be ₹ 8 crores if it decides to advertise and only ₹ 2 crores if it decides not to advertise. Thus, for firm *B*, strategy of 'Advertising' is better no matter firm *A* adopts strategy of 'Advertising' or 'Not Advertising' and therefore the firm *A* can safely conclude that firm *B* will adopt this strategy of 'Advertising'.

Now, given that firm *B will adopt strategy of 'Advertising'* the firm *A* will choose its strategy. If firm *A* adopts strategy of 'Advertising' its profits will be ₹10 crores and if it adopts strategy of 'Not Advertising', its profits will be ₹ 6 crores. Thus, given the firm *B's* strategy of 'Advertising', the optimal strategy of firm *A* is that of 'Advertising' too. In this way both firms will reach the equilibrium *state by choosing strategy of 'Advertising' and will have no incentive to deviate from it.* It is quite logical outcome of the game because firm *A* is choosing the best strategy it can, given firm *B's* strategy, and firm *B* is choosing the best strategy, given firm *A's* strategy.

The Nash Equilibrium

We may refer here to the concept of Nash equilibrium which has been already explained in connection with Cournot's duopoly equilibrium. Nash equilibrium is named after John F. Nash, an American mathematician and economist. We have explained above that in many games we do not have dominant strategies, but still the firms achieve equilibrium in the adoption of their strategies. The application of the concept of Nash equilibrium is quite relevant here. Nash equilibrium is a more general concept of equilibrium that is widely applicable and highly appealing. In the second advertising game whose payoff matrix is given in Table 31.2 and in which firm *A* has no dominant strategy we reached the conclusion that the equilibrium state is reached when firm *A* adopts strategy of 'Advertising' *given that the firm B will choose the strategy of 'Advertising'*. That is, firm *A* is making the best choice, given the choice by its rival firm *B* and *B* is choosing the best strategy, given the strategy of firm *A*. Therefore, they have no incentive to change their strategies. Hence, there exists an equilibrium, called Nash equilibrium.

Nash equilibrium describes a set of strategies where each player believes that it is doing the best it can, given the stragy of the other playe or players.

In our above example of game 2 of advertising where firm *A* has no dominant strategy, each firm promotes its own interests and makes a best choice of strategy, given the other firm's strategy. In the above game, both firms *A* and *B* adopt strategy of 'Advertising' which is optimal for them. Since each is doing the best, given other's strategy and no one has a tendency to change it unilaterally, there exists Nash equilibrium. As no one has a tendency to deviate from the Nash equilibrium state, strategies chosen by them *are stable*.

Dominant Strategy and Nash Equilibrium. It is important to compare Nash equilibrium and equilibrium reached where each firm has a dominant strategy. Whereas a *dominant strategy equilibrium describes an optimal or best choice regardless of what strategy the other player adopts,* in Nash equilibrium each player adopts a strategy that is the best or optimal, *given the strategy the other player adopts.* However, it may be noted that in some games we do not have Nash equilibrium and that in some we have more than one Nash equilibrium.

THE PRISONER'S DILEMMA AND OLIGOPOLY THEORY

The firms working in oligopolistic markets make decisions in the face of uncertainty about how their rivals will react to their moves. As explained above, game theory is a mathematical technique of analysing the behaviour of rival firms with regard to changes in prices, output and advertisement expenditure in the situations of conflicts of interest among individuals or firms. An important game model that has significant implications for the behaviour of the oligopolists is popularly known as *prisoner's dilemma*. Model of prisoner's dilemma explains how rivals behaving selfishly act contrary to their mutual or common interests. We have explained prisoner's dilemma in connection with the instability of a cartel.

Now, under these circumstances what choice will be made by two prisoners, say Ranga and Billa, when they cannot communicate with each other and have to choose between the two alternatives independently. *The model of Prisoner's Dilemma suggests that both behaving selfishly and working in self-interest confess to the crime and cheat each other.* Since both confess, each will get maximum imprisonment under the law. Why do they make this choice and confess can be shown as under. Take Ranga first, most probably, he would confess when he does not know how his co-accused will act. Ranga would reason like this : If I don't confess it is very likely that I will be imprisoned for 10 years as the other prisoner will most probably confess. If I confess, I will get 5 years imprisonment if the other one also confesses, and only one year imprisonment if he does not confess. So in the presence of uncertainty about the other person's choice, and behaving in self-interest, Ranga is likely to confess. Billa too reasoning similarly would confess. As a result, both prisoners would be sentenced for 5 years, though they would have received a lighter sentence of only one year if both of them had not confessed and remained loyal to each other. However, *it is self-interest which leads prisoners to confess and prevents them from attaining the best solution* for themselves (1 year imprisonment) if both do not confess to the crime and remain loyal to each other. But the decision of each prisoner in favour of confession is quite rational because each person works in self-interest and tries to make the "*best*" of the "*worst outcomes*" in an uncertain situation.

Similarly, in case of a cartel we have seen that to increase their own profits the member firms have incentive to cheat by trying to produce and sell more at the agreed price. It is due to the working of inner pressures and promotion of self-interest by cartel members that accounts for the instability of cartel arrangements.

REPEATED GAMES AND TIT-FOR-TAT STRATEGY

In our analysis of Prisoner's Dilemma it was assumed that game was played just once. While applying prisoner's dilemma type game to the case of a cartel we concluded that oligopolists like the prisoners lacking trust in each other and behaving selfishly cheated each other. This resulted in bad outcome (*i.e.* lower or no profits) for them. However, the firms facing prisoner's dilemma can increase their profits if they coopeate with each other. But *such cooperation is unlikely to occur in a prisoner's dilemma game played only once*. In this game of prisoner's dilemma the players have only a single opportunity to play a game (*i.e.* to confess or not). But in the real world oligopolists have to *play games repeatedly* as they have to set price and output over and over again.

In the case of working of a cartel at every point of time each firm has to decide whether to cheat or not. Behaving selfishly and having no trust in others, all member firms of a cartel cheat (i.e. undercut price) and as a result make only small profits. However, in case of repeated games the oligopolists may adopt a cooperative behaviour which enables them to earn large profits. Thus when oligopolists play a repeated game, the analysis of prisoner's dilemma type game played only once may not be correct.

In case of a game played repeatedly the players come to know how the others react to their moves and this in turn changes their strategic behaviour. Thus in case of a repeated game, one firm has the opportunity to penalise the other for his previous bad behaviour. In this context it has been suggested that **tit-for-tat strategy** is the optimal strategy that will ensure cooperative behaviour of the players participating in a game.[2] Let us suppose an oligopolist firm *A* adopts a cooperative behaviour and charges a high price. Tit-for-tat strategy means that firm *A* will continue to charge high price so long as its rival firm *B* will also continue to do so (*i.e.* adopts cooperative behaviour). But if firm *B* cheats and undercuts its price in a round, then in the next round firm *A* will retaliate and will also set a low price. Thus the firm *B* knowing that the firm *A* is adopting a tit-for-tat strategy will have to take into account the possibility of the rival firm *A* retaliating in the next round. In case of a repeated game, this tit-for-tat strategy results in cooperative behaviour among oligopolists.

However, whether tit-for-tat strategy will be viable depends on whether the repeated game is played indefinitely or a finite number of times. Let us first explain the outcome when repeated game is played indefinitely. We assume that there are two oligopolistic firms *A* and *B* and there are two possible strategies, namely, (1) charging a high price, and (2) charging a low price. The firms adopt a tit-for-tat strategy. It may be mentioned again that, according to tit-for-tat strategy, what one firm does in the current period, the other firm will do in the next period. In the case of game of prisoner's dilemma which is played only once, if one firm cheats the retaliation by the other firm in the next period does not arise as the game is over in the first round itself. However, in case of a repeated game, the other player (firm in our case) can penalise the other firm in the next period for any cheating by any player in the current period. It is assumed that a firm knows that its rival firm is adopting tit-for-tat strategy.

2. See R. Axelord, "*The Evolution of Cooperation*" (New York : Basic Books, 1984).

How tit-for-tat strategy is an optimal strategy and will result in cooperative behaviour on the part of the oligopolists is illustrated in payoff matrix given in Table 31.3.

Table 31.3. Pay-off Matrix

(Figures in millions)

		FIRM B's Strategy	
		Low Price	*High Price*
FIRM A's Strategy	*Low Price*	A's Profits : 10 B's Profits : 10	A's Profits : 100 B's Profits : –50
	High Price	A's Profits : –50 B's Profits : 100	A's Profits : 50 B's Profits : 50

If game is played only once as in the above case of Prisoner's Dilemma, both firms will cheat and charge low price and as will be seen from payoff matrix each firm will earn profits of only ₹ 10 million (see top left side box), while if they had cooperated and charged high price they could have earned ₹50 million each (see bottom right side box). Under tit-for-tat strategy in case of repeated game played for an indefinite period, suppose firm *A* starts with charging a high price and decides to continue charging the high price so long as the other firm also does likewise. But when the firm *B* cheats, that is, charges a low price, *B*'s profits rise to 100 million in that round while the firm *A*'s profits have become negative (–50 million). Now, under the tit-for-tat strategy, the firm *A* will retaliate in the next round and set a low price. When both charge low price profits of each are 10 million (see upper left box of Table 31.3). As the game is repeated indefinitely, round after round, the cumulative loss of profits suffered by firm *B* will outweights his gain of profits in the round when it undercuts price. Thus cheating (*i.e.* undercutting price in the present example) when the rivals are pursuing tit-for-tat strategy is not a profitable proposition. In this way the firms will learn that cooperative behaviour is the best course of action when each firm is pursuing the tit-for-tat strategy.

When both cooperate and charge a high price, each firm will earn profits of ₹50 million in each round (see bottom right hand box in payoff matrix of Table 31.3). Thus, Hal Varian writes, "The tit-for-tat strategy does very well because it offers an immediate punishment for defection. It is also a forgiving strategy. It punishes the other player only once for each defection. If he falls into line and starts to cooperate, then tit-for-tat will reward the other player with cooperation. It appears to be a remarkably good mechanism for the efficient outcome in a prisoner's dilemma that will be played an indefinite number of times."[3]

Let us now consider the case when the game is repeated a *finite number of times*, say in 10 rounds. Both players know that the game will be played 10 times and also that each is pursuing tit-for-tat strategy. Let us first consider the 10th round which by assumption is the last round when the game will be played between the two firms. Whether they will cooperate, each charging a high price or cheat each charging a low price. If firm *B* believes that its rival firm is rational will reason like this : Even knowing that the firm *A* is playing tit-for-tat strategy, the firm *B* will think that since 10th round is the last round of playing the game and after that since the game is over, the firm *A* will have no chance to retaliate. Therefore, firm *B* will charge the high price for the first nine rounds but will choose to cheat, that is, will charge the low price and make a large profit in the last 10th round (this is shown in the bottom left-side box of the payoff matrix of Table 31.3).

However, firm *A* will also reason likewise and will charge high price in the first 9 rounds but will plan to cheat (charge low price) in the last 10th round and will hope to make a large profit in the last 10th round thinking that firm *B* will have no chance to retaliate thereafter. Thus, both thinking rationally will decide to charge a low price in the last 10th round and will not cooperate with each

3. Hal R. Varian, *Intermediate Microeconomics : A Modern Approach* (New York : W.W. Norton & Co., Fourth Edition, 1997), p. 487.

other. Hal Varian rightly writes' "Players cooperate because they hope that cooperation will induce further cooperation in the future. But this requires that there will always be the possibility of future play. Since there is no possibility of future play in the last round, no one will cooperate then".[4]

But what about next to the 10th round, that is, 9th round. Firm *B* will reason that it should charge low price in this next to the last round because in any case there will be no cooperation between the two in the last round. But of course, the firm *A* being equally rational will also reason likewise and will plan to charge the low price in the 9th round (*i.e.* next to the last round). The same reasoning can be repeated by both the firms for undercutting price, that is, for charging a low price in the 8th earlier rounds as well, that is, for the rounds 8th, 7th, 6th etc. till the first round. *Thus, when the game is played a finite number of times, even while pursuing tit-for-tat strategy the two firms will opt for non-cooperative behaviour*. Thus, even with tit-for-tat strategy in case of repeated games to be played a finite number of times, we are stuck in prisoner's dilemma without the outcome of the cooperative behaviour.

But cooperative outcome can come about even when the game is to be played finite number of times if a firm has a doubt about its competitor's rationality in pursuing tit-for-tat strategy and its ability to reason out the logical implications of a finite time horizon as explained above. Thus, if competing firms have doubts about whether the other firm is playing tit-for-tat or playing tit-for-tat blindly, this will make cooperative behaviour a good strategy. Besides, in the case of finite number of times the game is to be played, cooperative behaviour can be regarded as a good strategy by the competing firms if the time is long enough and the firms are uncertain about how long they will be competing. Most managers don't know how long they and their firms will be competing with their rivals and this also serves to make cooperative behaviour a good strategy. Although the number of months that the firms compete is probably finite, managers are unlikely to know just what the number is. As a result, the unravelling argument that begins with a clear expectation of undercutting in the last month no longer applies. As with an infinitely repeated game it will be rational to play tit for tat.[5] Thus in view of the fact that in most oligopolist markets the game is in fact repeated over a long period and uncertain length of time and managers have doubts about how rationaly their competitors behave in case of the repeated game for a finite number of times, the prisoner's dilemma can have cooperative outcome.

STRATEGIC MOVES

In the earlier sections it has been emphasised that oligopolists must realise that their own profits depend not only on their own decision and behaviour but also on the decision and behaviour of their rivals. This shows the importance of strategic moves by the oligopolists to enhance their profits. By making certain strategic moves an oligopolist can gain competitive advantage in the market. Thomas Schelling of Harvard University who has make an important contribution to the theory of strategic decision making defines the concept of strategic move in the following words: "*A strategic move is one that influences the other person's choice in a manner favourable to one's self, by affecting the other person's expectations on how one's self will behave*"[6]. For example, if Maruti Udyog threatens to retaliate by cutting the price of their cars to a level that would cause losses to its rival firms which reduce their prices, this move of Maruti Udyog is a strategic move. This is because this threat is intended to ensure that the rival firms do not cut the prices of their cars.

Threat, Commitment and Credibility

For the strategic move of giving threat to be successful, there must be commitment that the firm making a threat will definitely carry it out. Only when there is commitment to carry out a threat

4. Hal R. Varian, *op. cit.*, p. 486
5. See Pindyck and Rubinfeld, *op. cit.*, p. 466. This result is based on the original research work done by David Kreps, Paul Milgrom, John Roberts, and Roberts Wilson, "Rational Coopeation in the Finitely Repeated Prisoner's, Dilemma", *Journal of Economic Theory*, 1982, pp. 245-252.
6. Thomas C Schelling , *The Strategy of Conflict,* (New York: Oxford University Press, 1960). Another work is M. Porter, *Competitive Strategy* (New York : Free Press, 1980).

that it becomes credible. If there is no commitment to carry out the threat, it will be an empty threat and will therefore not have the desired effect on the behaviour of the rivals. If a firm can convince its rival firms that it is committed to a particular move that it is making, then the rivals may cooperate without retaliating because they may think that they would lose more than they would gain from a long period of conflict with the firm making a move.

When a threat is credible is illustrated in the payoff matrix of firms *A* and *B* given in Table 31.4 where the profits of the two firms making different brands of cars are shown when they charge low price or high price for their cars. This payoff matrix shows that charging a high price is a dominant strategy for firm *A*, that is, whatever strategy (whether of charging a high price or low price) the rival firm *B* pursues, the strategy of charging a high price is optimal for firm *A*. Thus, if firm *B* charges a low price, the firm *A* will earn profits of ₹ 20 lakhs if it charges a low price and ₹ 30 lakhs if it charges a high price. On the other hand, if the firm *B* charges a high price, the firm *A* will earn ₹ 20 lakhs if it charges a low price and ₹ 50 lakhs if it charges a high price. Thus, whether the firm *B* plays a low price strategy or a high price strategy, for firm *A* high price strategy is the optimal strategy to adopt. It will be seen from the payoff matrix of Table 31.4 that when the firm *A* will charge a high price, the firm *B* will opt for charging a low price and in this way will earn ₹ 40 lakhs instead of ₹ 30 lakhs if it charges a high price.

Table 31.4. Payoff Matrix for Pricing Game

(in lacs)

		FIRM B	
		Low Price	*High Price*
FIRM A	*Low Price*	20, 20	20, 10
	High Price	30, 40	50, 30

(**Note :** *The first number in the pair shows profits of firm A, the second number shows profits of firm B*)

Under these circumstances if firm *A* threatens firm *B* that it will charge a low price, this threat will be incredible or empty because the firm *B* knows that by charging a low price, the firm *A* will cause its profits to fall to ₹ 20 lakhs. Being an incredible threat the firm *B* will not take it seriously. As explained above, one way to make the threat credible is to make it binding and irreversible. Thus, if a firm threatens to enter a particular market it can make its threat credible if the potential firm buys a plant rather than lease it or enters into a long-term contract for buying raw materials. This shows that the firm which gives a threat to enter has made an irreversible commitment and will therefore enter the market, come what way, and this makes the threat credible. Take another example. If a firm commits to a price reduction then if its rival firm lowers its price, then to make its commitment credible, it can make a verbal or written agreement with the customers that it will match any price cut by its rival. On the other hand, if a particular firm has the image that it will easily ignore its particular commitment that it makes, then the commitment is not credible and its competitors will not pay much attention to the commitment made by it.

Another way to make the threat credible is to build a reputation of irrationality for carrying out its threat even if it has to lose some profits or even incur losses. This irrational reputation is developed when a firm has actually carried out its threat several times in the past (even at the expense of its profits). Thus, the threat of a firm with reputation of irrationality is a credible threat and its rivals will take serious note of it. Consider payoff matrix of Table 31.4 again. If the firm *A* is charging a high price and the firm *B* is charging the low price, they are earning profits of ₹ 30 and 40 lakhs respectively (see bottom left hand corner of Table 31.4), then if the firm *A* has a reputation for irrationality and gives a threat to lower its price to carry out its threat to lower price, then *B* will be induced to charge high price and as a result both firms will be charging high price and firm *A*'s profit will rise to ₹50 lakhs, but profit of the firm *B* will fall to ₹ 40 lakhs (see bottom right hand corner). It is important to note though profits of the firm *B* has fallen because under threat from the firm *A* it has decided to cooperate, it is still greater than profits of ₹ 20 lakhs that it would have earned if firm *A* had actually carried out its threat and both charged the low price.

In addition to what has been said above about credible commitment, it may be noted that for a threat to be credible, the firm's commitment must be backed up with assets, skills and expertise, financial and technological powers to carry out its commitment. Besides, a firm's commitments are more credible if it has a reputation and a long history of adhering to its commitments.

However, for tit-for-tat strategy to be successful certain conditions must be fulfilled. First, a reasonably stable set of players (that is, firms) is required for the successful working of tit-for-tat strategy. If the players (firms) change quite frequently, cooperative behaviour between them is not likely to develop. Second, in tit-for-tat strategy for cooperative behaviour to be achieved, there must be a small number of players (firms). In case of a large number of competing firms, it is difficult to know what each firm is doing. As a result, cooperation cannot be enforced and generally breaks down when there are many firms confronting each other. Third, for the success of tit-for-tat strategy to induce cooperative behaviour it is assumed that each firm can quickly detect cheating by others and is able and willing to retaliate if rivals do cheating. Fourth, the demand and cost conditions must remain stable for the success of tit-for-tat strategy. The failure to cooperate is quite often the result of changing demand or cost conditions. Uncertainties about demand or costs make it difficult for the firms to arrive at an implicit understanding of what cooperative behaviour requires. Lastly, tit-for-tat strategy to induce cooperative behaviour, the game is to be played either indefinitely or for a long uncertain number of times.

ENTRY DETERRENCE

The existing firms, especially the monopolists, try to prevent the entry of new firms as the entry of new firms reduces the profits of the existing firms. An important strategy for the existing firm to deter the entry of new firms in the market is giving threat to lower price and thereby inflict loss on the potential entrant. However, such a threat by the existing firm will work only if it is credible. To illustrate whether threat is credible or not consider the payoff matrix shown in Table 31.5. From this payoff matrix in Table 31.6 it follows that the threat by the existing firm *A* that if the potential firm *B* enters the market it will lower the price and impose loss on *B*, is not credible. It will be seen from the payoff matrix of Table 31.5 that before entry of firm *B* the firm *A* is charging high price and is making profits of ₹10 lacks (see bottom right hand corner). Now, on entry by firm *B* in the market, *if the existing firm charges* a higher price, the profits of the existing firm *A* are 7 lakhs and of the new firm *B* are 2 lakhs and on the other hand, if the existing firm *A* lowers the price to carry out its threat, profits of firm *A* are 4 lakhs and the new firm *B* incurs a loss of 2 lakhs. This shows that despite the entry of new firm *B*, it is profitable for the existing firm to charge a high price and earn profits of ₹7 lakhs which are greater than of ₹ 4 lakhs[7] in case it lowers the price in accordance with the threat held out. This shows that threat is not creditable. And since the potential entrant knows this, the threat will not work and will not prevent firm *B* from entering the market. Unless the firm *A* is irrational, it will not lower the price on the entry of new firm *B*.

Table 31.5. Payoff Matrix

(Profits in lakhs)

	Firm B (Potential Entrant)	
	Enter	*Stay Out*
Low Price	4, –2	6, 0
High Price	7, 2	10, 0

To make the threat credible the existing firm has to commit itself to resist the entry of the new firm *B* even at the loss of its profits. One way to make a credible commitment to resist the entry of the potential firm is the expansion of its capacity by the existing firm before it is needed, that is,

7. Note that fall in profits of the existing firm *A* is due to (1) the market will now be shared between the two firms, and (2) lowering of price of the product to deter entry.

by building excess capacity.[8] Since for building of excess capacity, the existing firm will incur costs, there will be a change in the payoff matrix. The new payoff matrix is given in Table 31.6.

Table 31.6. Payoff Matrix with Credible Commitment

		Firm B (Potential entrant)	
		Enter	*Stay Out*
Firm A	Low Price	4, –2	6, 0
(the existing firm)	High Price	3, 2	5, 0

Table 31.6 is the same as the previous Table 31.5 except that after building excess capacity, firm *A*'s profits are ₹ 3 lakhs if it continues to charge high price and ₹ 4 lakhs if it lowers the price. The profits of the existing firm *A* are now smaller even with the high price charged because of the costs incurred on building new capacity and sharing of the market with the new entrant. On the other hand, in case of low price being charged on the entry of new firm the profits of firm *A* are the same, namely, 4 lakhs as in the previous payoff Table 31.5. This is because at the low price, the sales of the existing firm will increase resulting in greater revenue and as a result it will be able to utilise a part of the extra capacity built. Thus, with low price the increase in revenue may cancel out the increase in costs due to the addition of extra capacity and therefore profits of the existing firm *A* remains the same by charging a low price. However, charging a low price by firm *A* will cause losses of ₹ 2 lakhs to the new entrant (see upper left hand corner). Thus realising that by entering the market it will suffer a loss, the firm *B* will not enter the market and stay out. Thus building of excess capacity before it is needed the existing firm commits itself to lowering price if the firm *B* dares to enter the market and this makes its threat credible and deters the entry of the potential firm. Having built excess productive capacity, the existing firm *A* will charge a low price and make a profit of ₹ 4 lakhs instead of ₹ 3 lakhs if it charges a high price. Since the new firm *B* on entry will have to sell the product at the low price, it will suffer a loss ₹ 2 lakhs if it entered the market. Therefore, the firm *B* would not enter the market and stay out. Thus, the existing firm *A* has succeeded in deterring entry by holding out a credible threat.

An alternative to buidling excess capacity is creating a reputation of irrationality for preventing entry of potential firms in the market even if its causes decline in profits of the existing firms for quite a long time. Thus, when a firm has good reputation for behaving irrationally then, even given the payoff matrix of Table 31.5, it will succeed in deterring entry. The reputation of irrationality of the existing firm creates a credible threat of price warfare if the potential firms enter the industry. As a matter of fact, in the real world it is reputation for irrationality that seems to work in deterring entry.

QUESTIONS FOR REVIEW

1. Distinguish between cooperative game and non-cooperative game. Give examples.
2. Explain a dominant strategy. How is stable equilibrium reached when a firm pursues its dominant strategy?
3. What is Nash Equilibrium? Does equilibrium in a Prisoner's Dilemma game represent Nash Equilibrium?
4. What is prisoner's dilemma? How is it used to explain the instability of a cartel?
5. Explain Neumann-Morgenstern theory of game? On what grounds has it been criticised?
6. What is a tit-for-tat strategy ? Why is it a rational strategy when Prisoner's Dilemma type game is infinitely repeated ?
7. Explain the Prisoner's Dilemma game when it is repeated 15 times and both players are rational. Is tit-for-tat strategy optimal in this case?
8. What is a "strategic move"? What are the conditions required for strategic move to be successful?
9. What is a credible threat? How can a threat be made credible?
10. How does an incumbent firm succeed in deterring entry into the market by potential competitors?
11. Can a threat of a price war deter entry by new firms? What actions might a firm take to make this threat credible?

8 . For an elaborate analysis of the use of excess capacity to prevent entry, see J. Tirole, *The Theory of Industrial Organisation* (Cambridge, MA : MIT Press, 1988).

PART V

FACTOR PRICING AND INCOME DISTRIBUTION

- Factor Pricing and Income Distribution: A General View
- Wage Determination in Competitive and Imperfectly Competitive Markets
- The Theory of Rent
- Alternative Theories of Interest
- The Theory of Profits

CHAPTER 32
FACTOR PRICING AND INCOME DISTRIBUTION: A GENERAL VIEW

Functional vs. Personal Distribution

So far we have discussed how the prices of products are determined in different market forms. The pricing of factors of production is the subject matter of this part. The theory of factor of prices is popularly known as the *theory of distribution*. The distribution may be functional or personal. The distribution theory with which we are concerned in this book is the theory of functional distribution. The concept of functional distribution should be carefully distinguished from that of personal distribution. *Personal distribution* of national income or what is known as '*size distribution of incomes*' means the distribution of national income among various individuals or persons in a society. As is well known, national income is not equally distributed among various individuals in the country. Some are rich, while others are poor. In fact, there are large inequalities of income between various individuals. Thus the theory of personal distribution studies how personal incomes of individuals are determined and how the inequalities of income emerge. On the other hand, in the theory of *functional distribution* we study how the various factors of production are rewarded for their services or functions performed in the production process.

Factors of production have been classified by economists under four major heads, *viz.*, land, labour, capital and entrepreneur. Thus in theory of functional distribution we study how the relative prices of these factors of production are determined. The prices of land, labour, capital and entrepreneurship are called rent, wages, interest and profit respectively. Thus in the theory of functional distribution we discuss how the rent of land, wages of labour, interest on capital and profits of entrepreneur are determined. To be brief, *theory of functional distribution means the theory of factor pricing*. To conclude, in the words of Professor Jan Pen, "In functional distribution, we are no longer concerned with individuals and their individual incomes, but with factors of production: labour, capital, land and something else that may perhaps best be called 'entrepreneurial activity'. The theory of functional distribution examines how these factors of production are remunerated. It is primarily concerned with the price of a unit of labour, a unit of capital, a unit of land, and being an extension of price theory it is sometimes called the theory of factor prices."[1]

The question that now arises is : Is it not the functional distribution that determines the personal distribution of national income? Personal distribution of income *only partly* depends upon functional distribution. How much income an individual will be able to get depends not only on the *price* of a particular factor he has but also on the *amount* of that factor he owns as well as the *prices* and amounts of other productive factors which he may possess.

Thus the personal income of a landlord depends not only on the rent but also on the amount of land he owns. Given the rent per acre, the greater quantity of land he owns, the greater will be his income. Further, the landlord may have lent some money to others for which he may be earning interest. The total income from interest on money will also add to his personal income. Thus a person

1. Jan Pen, *Income Distribution*, Penguin Books, 1971, p. 16.

may be getting income from several sources *i.e.*, from the earnings of various factors of production. The earnings from all the sources will constitute his personal income. Thus if our landlord does not do any other work and owns no other factor of production, his personal income will depend on the rates of rent and interest and also on the amount of land he owns which he has given on rent and the amount of money he has lent out. Thus, "*personal distribution* (or the size distribution of incomes) relates to individual persons and their income. The *way in which* that income has been acquired often remains in the background. What matters is how much one earns, not so much whether income consists of wage, interest, profit, pension or whatever."[2] Thus, total income from rent and interest will make up his personal income. Theory of personal distribution of income has, therefore, to explain not only how prices or rewards for factors such as rent of land, interest on capital are determined but also *how various people happen to own different quantities of these productive factors*. The theory of functional distribution, or the theory of factor prices, as it is often called, is thus only a part of the theory of personal distribution which covers the former. In this book we are primarily concerned with functional distribution of income or the pricing of factors of production.

Before we proceed further we would like to remove one confusion. It is that when we say how the prices of factors are determined, we do not really mean the prices of the factors *themselves*. By the prices of the factors, we mean the prices of their *services or use* for a period of time. Thus in factor pricing we do not study how the price or value of land as such is determined, instead we study how the *price for the use of land*, which is called rent, is determined. Similarly, in the theory of factor prices, we do not explain how the price of labourer as such is determined (the labourer never sells himself, he sells only his labour or *service* for a period), but we study how the price for the use or service of labour for a period which is called the wage rate is determined. It is only for the sake of convenience and brevity that we speak of pricing of factors while we actually mean the pricing of their *services* or *uses*.

Theory of Distribution as a Special Case of the Theory of Price

It is worth mentioning here that in modern economic theory, theory of distribution is only a special case of the theory of price. As the prices of products are explained with the interaction of the demand for and supply of them, similarly distribution is conceived as the determination of prices of the factors which are also explained with interaction of demand for and supply of them. The income which a factor will obtain depends on the price determined by market, *i.e.,* demand and supply and the amount that will be used or employed of that factor. In other words, it is the forces of free market, that is, demand and supply, that go to determine the prices and incomes of various factors and not any institutional framework such as ownership of property. Further, the association of various factors with particular social classes, such as land with land owning class, capital with capitalists, and labour with the working class is also not emphasised. In fact, the factors are conceived merely as productive agents and distribution of income among them as merely functional rewards for their contribution to production. In other words, the contemporary theory of distribution merely explains functional distribution of income and not personal distribution of income.

Prof. A.K. Das Gupta describes the nature of contemporary distribution theory very clearly. He remarks, "Distribution appears an extension of the theory of value ... *being just a problem of pricing of factors of production*. The two aspects of the economic problem are then integrated into a unified and logically self-consistent system. Value of a commodity is derived in the ultimate analysis from utility, and value of factors derived from productivity imputed by the commodities which they help in producing. The old tripartite division of factors into land, labour and capital is retained but their old

2. *Op. cit.*, p. 15.

association with social classes is lost. Factors are conceived as just productive agents independently of the institutional framework within which they operate."[3]

In the opinion of the present author, the contemporary theory of distribution is on the wrong track. Distribution of income, that is, who gets what share of the national cake in a society cannot be explained merely by the mechanism of the impersonal forces of the market, by the equilibrium between the demand for and supply of factors. The production relations governed by the ownership of property or means of production, power structure in the society play a vital role in the distribution of national income. The contemporary theory by asserting that everybody or every factor under conditions of perfect competition is remunerated according to the value of its marginal product (that is, what it contributes to general output of industry) is tacit approval of the present distribution of income as just and right. But this is far from truth, since the present-day highly skewed distribution of income found in the capitalist countries (including India) has been greatly determined by the uneven ownership of property, production relations based on it and the power structure in the society. This is not to say that marginal productivity as a determinant of incomes is quite unimportant but the importance of the institutional factors just mentioned above cannot be glossed over.

Micro and Macro Theories of Distribution

Even the 'functional distribution' of income can be studied from two viewpoints, *viz.*, micro and macro. The theory of micro distribution explains how the *rates of reward* for various factors of production are determined. In other words, the theory of micro distribution deals with the determination of *relative prices* of productive factors. It thus studies how the wage rate of labour, rate of rent on land, rate of interest on capital are determined. On the other hand, the theory of macro distribution deals with the problem of the determination of *aggregate rewards* of various factors in national income. In other words, the macro distribution means the relative shares of various factors in national income. Therefore, the theory of macro distribution is also known as the *theory of distributive shares*. Thus the theory of macro distribution or distributive shares tells us as to how the share of labour in the national income (*i.e.,* the total amount of all wages of all labourers in the country) is determined. Similarly, the macro theory of distribution explains how the share of profits in the national income (*i.e.*, total amount of profits earned by all entrepreneurs as a ratio of national income of the country) are determined. To quote Professor Jan Pen again, "The theory of distributive shares attempts to explain the share of the total national income that each factor of production receives. It enquires into the percentage that labour receives of the whole, and also into the shares of interest, rent and profit. Now individual income recipients disappear beyond the horizon".[4]

MARGINAL PRODUCTIVITY THEORY OF DISTRIBUTION

We now turn to the question as to what determines the prices of factors of production A theory which tries to answer this question and which has been fairly widely held by professional economists is known as *marginal productivity theory of distribution.* It may, however, be pointed out that in recent years its popularity has somewhat declined due to bitter criticisms levelled against it. The essence of this theory is that price of a factor of production depends upon its marginal productivity. It also seems to be very fair and just that price of a factor of production should get its reward according to the contribution it makes to the total output, *i.e.*, its marginal productivity.

Marginal productivity theory was first put forward to explain the determination of wages, *i.e.*, reward for labour but later on prices of other factors of production such as land, capital etc. also were explained with marginal productivity. The origin of the concept of marginal productivity can be traced to Ricardo and West. But both Ricardo and West applied the marginal productivity doctrine only to land. The concept of marginal productivity is implicit in the Ricardian theory of rent. But

3. A.K. Das Gupta, *Tendencies in Economic Theory*, Presidential Address to the 43rd Annual Conference of the Indian Economic Association, held at Chandigarh, December, 1960.
4 Jan Pen, *Op. cit.,* p. 18.

the idea of marginal productivity did not gain much popularity till the last quarter of 19th century, when it was re-discovered by economists like J.B. Clark. Jevons, Wicksteed, Walras and later Marshall and J.R. Hicks popularised the doctrine of marginal productivity. Since marginal productivity theory has been mainly evolved for the determination of reward for labour, we shall discuss below its application to wage determination. But it should be understood to apply equally to the rewards of other factors of production.

Clark's Version of Marginal Productivity Theory

J.B. Clark, an American economist who developed marginal productivity theory of distribution in a number of articles and later on presented it in a complete form as an explanation for "*The Distribution of Wealth*". In order to bring out the fundamental factors at work in the mechanics of income distribution Clark assumed a *completely static society*, free from the disturbances caused by economic growth or change. In other words, he assumed a constant population, a constant amount of capital and unchanging techniques of production. Besides the assumption of a static economy, he also assumed perfect competition in the factor market and perfect mobility on the part of both labour and capital. Further, it was assumed by him that the total stock of capital remains constant. Clark also supposed that the *form* of capital can be varied at will. In other words, physical instruments of production can be adapted to varying quantities and abilities of available labour. Further, he treats labour as a homogeneous factor by taking identical labour units and discusses how the wage rate of labour is determined.

Every rational employer or entrepreneur will try to utilise his fixed amount of capital so as to maximise his profits. For this he will hire as many labourers (labour units) as can be profitably put to work with a given amount of capital. For an individual firm or industry, marginal productivity of labour will decline as more and more workers are added to the fixed quantity of capital. He will go on hiring more and more labour units as long as the addition made to the total product by an extra labour unit is greater than the wage rate he has to pay for it. The employer will reach equilibrium position when the wage rate is just equal to the marginal product of labour.

Just consider Fig. 32.1 where units of labour are represented on the *X*-axis and the marginal product of labour on the *Y*-axis. Then the *MP* curve shows the diminishing marginal product of labour. If the prevailing wage rate which an employer must pay is equal to *OW*, then it will be profitable for the employer to go on employing additional workers until the marginal product of labour becomes equal to the prevailing wage rate *OW*. It will be evident from Fig. 32.1 that if the prevailing wage rate is *OW*, then the employer will employ *OL* units of labour since the marginal product of labour is equal to *OW* at *OL* employment of labour. He would not employ more than *OL* amount of labour as the marginal product of labour falls below the wage rate *OW* and he would therefore be incurring losses on the employment of additional workers beyond *OL*. Thus, an employer would be maximising his profits by equalising the marginal product of labour with the wage rate *OW*. Since perfect competition is assumed to be prevailing in the labour market, an individual firm or industry will have got no control over the wage rate. An individual firm or industry has, therefore, to determine only the number of factor units

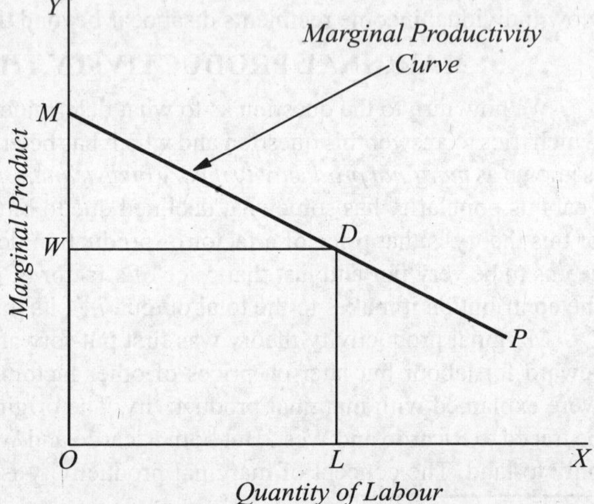

Fig. 32.1. *Wage rate is equalised with marginal productivity of labour.*

(labour in the present case) to which it has to give employment at the prevailing wage rate. Thus at micro level (*i.e.* for individual firm or industry) marginal productity theory is the theory of employment.

A marginal-product schedule or curve shows a particular wage-employment relationship. Since Clark has assumed a stationary state, he takes the total supply of labour available for employment in the whole economy as given and constant. In other words, *in Clarkian analysis, aggregate supply curve of labour has been assumed to be perfectly inelastic*. Given the total supply of labour in the economy, the wage rate will be determined by the marginal product of the available amount of labour assuming that all labour get employment. Given the aggregate amount of labour that is seeking employment, the wage rate that the labourers will get will be equal to the *addition made to the total product* by the employment of the marginal unit of labour. In other words, if the total quantity of labour seeking employment is '*n*' units, then each unit of labour will get wage which will be equal to the difference between the total production when *n* labour units were employed and that when *n* – 1 labour units were employed. In other words, in the competitive labour market the wage rate is *determined by the marginal product of a given quantity of labour force*.

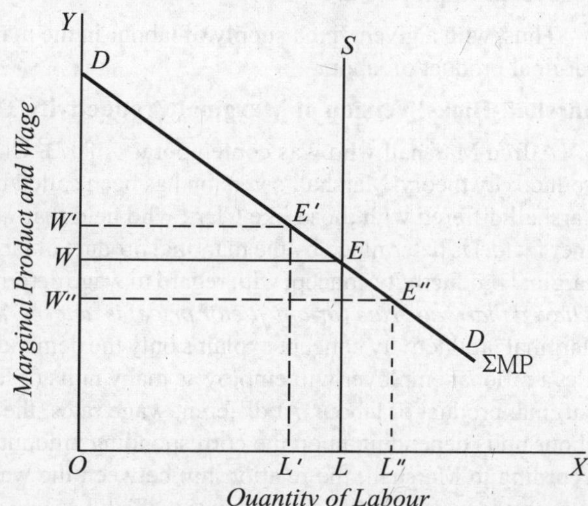

Fig. 32.2. *Wage rate is determined by marginal productivity.*

If the labourers compete with each other for obtaining jobs, they will bid down the wage rate if some of them find themselves unemployed. The employers will bid the wage rate up if the prevailing wage rate is smaller than the marginal product of available labour force. This is so because at the wage rate lower than marginal product the employers' demand for labour force will be more than the available number of labourers. Consider Fig. 32.2. In this figure *DD* curve represents the demand curve for labour by all employers and has been obtained by summing up horizontally marginal product curves (Σ *MP*) of all employers demanding labour. Marginal product of labour diminishes as more units of labour are employed in the economy, *assuming the quantities of other factors used as unchanged*. Now, if the available supply of labour force is *OL* in the whole economy, the marginal product of *OL* quantity of labour is *LE*. The wage rate will be determined by this marginal product *LE* and therefore equilibrium wage rate which will settle down in the market will be equal to *LE* or *OW*. At a higher wage rate *OW'* the employers will employ *OL'* amount of labour leaving *LL'* amount of labour unemployed. Unemployed workers in their attempt to get employment will bring the wage rate down to the level *OW* (=*LE*) at which all are employed.

On the other hand, at a lower wage rate than *OW*, say *OW''* , the employers will demand *OL''* amount of labour since at this their profits will be maximium but the available of labour is *OL*. Thus, at a lower wage rate than *OW* the demand for labour by the employers will be greater than the available quantity of labour. In their bid to get more labour, competition among employers will push the wage rate up to *OW* at which the employers' demand for labour is just the same amount of labour which is actually available.

Thus given the quantity of labour in the country, wage rate is *determined* by marginal productivity of labour. *One assumption which is implicit in the Clarkian marginal productivity theory as applied to the economy as a whole is that of full employment of labour and further that the supply curve of labour is perfectly inelastic at this full-employment level.* In other words, it is assumed that all the existing number of workers in the economy are employed. To sum up, in Clark's presentation, the marginal productivity of a given quantity of available labour determines its wage rate when we consider the market as a whole. In the disaggregated picture, however, where a single employer finds the wage level determined by the forces beyond his control the marginal product of labour determines the level of employment.

Thus, with a given fixed supply of labour in the market the wage rate will be determined by the marginal product of labour.

Marshall-Hicks' Version of Marginal Productivity Theory

Alfred Marshall who was contemporary of J.B. Clark gave a different version of the marginal productivity theory. Marshall's version has been called by many as the marginal productivity theory. Marshall differed with those like Clark who held that wage rate (or for that matter, the price of any other factor) is determined by the marginal product of labour. Marshall said it was wrong to regard the marginal productivity concept with regard to wage determination as a *wage theory. This is because he believed that wage rate (or any factor price) is determined by both demand for and supply of labour.* Marginal productivity concept explains only the demand side of the problem. That is, given the wage rate, a rational employer will employ as many units of labour as will equalise the wage rate with the marginal product of labour. At different wage rates, the employer will employ different amounts of labour units depending upon the corresponding amount of the value of the marginal product. Thus, according to Marshall, the relationship between the wage rate and the marginal productivity of labour provides us with the demand curve of labour. In a complete theory for an *explanation of wage determination, the upward-sloping supply curve of labour has also to be introduced into the analysis*. The wage rate at which the supply curve of labour cuts the demand curve of labour (governed by the marginal productivity) will be *determined. However, the wage rate thus determined by demand and supply will be equal to the value of marginal product of a factor*.

It is clear from above that Marshall considered marginal productivity principle as one of the two forces that *determine* wages, the other force being the supply of labour. Marshall and Hicks believed that wages would tend to be equal to the marginal product, but they emphasised several times that the *wages are not determined* by marginal product, since like all other, marginal quantities, marginal product, *together with the price* (wage) is determined by the interaction of demand and supply. Further, *they considered the supply curve of labour as upward sloping to the right.*

Furthermore, Marshall pointed out that the marginal productivity doctrine plus the competitive conditions in the labour market would in the long run tend to make the *wages of labour in different industries or uses equal to each other* and to the marginal product of labour (assuming of course that labour is homogeneous). Further, Marshall drew the distinction between the marginal productivity principle which determines the demand for a factor and the marginal productivity theory as a complete theory of determination of factor prices. Thus, according to Marshall, *price of a factor such as wage rate of labour, rent of land are determined by demand for and supply of the factor and is equal to the marginal productivity of the factor*. While the marginal productivity principle determines the demand for a factor, namely, how much quantity of the factor is demanded at its various prices, the marginal productivity theory explains how through the interaction of demand and supply, price of a factor, say wage rate of labour, is determined.

It may however be noted that in our view the difference between Clark's version and Marshall-Hicks version of marginal productivity theory is not that whereas Clark considered the demand side (*i.e.* marginal productivity) of a factor and ignored the supply of labour, Marshall and Hicks considered the roles of both of them as the determinants of wage rate. The real difference in our view is that *while Clark considered supply curve of labour as perfectly inelastic at full-employment level, Marshall and Hicks considered it as upward-sloping* to the right indicating as wage rate (*i.e.*, price of the factor) rises, its quantity supplied increases. This is how we have explained above Clark's version of marginal productivity theory.

A detailed analysis how wage rate (and also prices of other factors) are determined by demand for and supply of the factor will be explained in detail in the next chapters.

CRITICAL EVALUATION OF MARGINAL PRODUCTIVITY THEORY

Marginal productivity theory has been a pillar in the traditional theory of distribution and till today it continues to be an important factor in modern analysis of factor pricing. We discuss below the various objections raised against marginal productivity theory.

1. It has often been argued that *marginal productivity theory takes too many assumptions which are quite unrealistic*. Therefore, it is concluded that theory has no validity. The theory of marginal productivity (Clarkian version) assumes a stationary state, perfect competition, perfect mobility of factors, equal bargaining power of buyers and sellers, and perfect knowledge which are all far away from the actual conditions of the real world. World is not static. Instead, developments are continually taking place making the actual world a dynamic one. Competition is not perfect; instead there are large imperfections in the factor markets which make any analysis of factor pricing based on the assumption of perfect competition quite useless. Bargaining powers of buyers and sellers of factor services, for instance, of employers and labourers are not equal and thus make the exploitation of the weaker party possible.

2. Another significant criticism levelled against marginal productivity theory (both Clarkian and Marshall-Hick's versions) is that being based upon the assumption of perfect competition both in product and factor markets, *it is unable to explain the determination of factor process under conditions of imperfect competition in the factor and product markets*. As we shall see later in this chapter, following the developments of imperfect and monopolistic competition theories by Joan Robinson and Chamberlin, there emerged two concepts of marginal productivity, namely, marginal revenue product (*MRP*) and value of the marginal product (*VMP*). Thus, when there prevails imperfect competition in the product market (assuming perfect competition in the factor market) a factor of production would not get remuneration equal to the value of the marginal product as is asserted in marginal productivity theory. Under imperfect competition in the product market, a factor of production is remunerated according to a different principle, namely, marginal revenue product (*MRP*) which is less than the value of marginal product (*VMP*). According to Joan Robinson[5], a factor is exploited if it is paid less than value of its marginal product, whereas in marginal productivity theory, as it was presented in neo-classical economic thought, there is just and fair distribution of total product; every factor getting equal to its contribution to the total production.

If imperfect competition or *monopsony prevails in the factor market, a factor will not even get its reward equal to its marginal revenue product*. Under imperfect competition or monopsony or oligopsony in the factor market, the firm to be in equilibrium will equate marginal wage of labour with the marginal revenue product of labour and, as shall be seen later, this marginal wage is greater than the average wage rate which will be paid to labour. We are therefore of the view that in the context of imperfect competition in the product markets and factor markets marginal productivity theory needs to be modified.

3. Another serious shortcoming of marginal productivity theory is that it cannot explain the rewards of the factors which are to be used in *fixed proportions*. Marginal productivity theory takes it granted that a good degree of elasticity of substitution exists between the factors of production so that increase in one factor, keeping other factors constant, leads to the increase in the addition to the total product, that is, it has a positive marginal productivity and therefore gets positive reward for its contribution to production. But when the factors are used in fixed proportions, increase in one factor,

5. Joan Robinson, *Economics of Imperfect Competition*, p. 283.

keeping the others constant, will not lead to any increase in total production at all. That is to say, *in case of fixed proportions or fixed relations between the factors, marginal productivity of the factors is zero*. In view of their zero marginal productivity, then according to the marginal productivity theory, their rewards or prices should also the zero. But this is quite absurd, for the factors of production, even when they have fixed relations with each other, obtain positive rewards.

4. Still another serious drawback of marginal productivity theory is that, in its original and rigid version, *trade unions or collective bargaining cannot raise the wages of labour without creating unemployment*. Thus, according to this theory, trade unions are superfluous and collective bargaining by them is a futile activity. Given the downward sloping nature of the marginal productivity curve, at a higher wage secured by the union, the employer will demand or employ less number of workers than before so that some labour will be rendered unemployed. But, as we shall see in a later chapter, increase in wage rate by the union does not always cause unemployment. Indeed, we shall study there that under conditions of monopsony in the labour market the increase in the wage rate by the union may be accompained by the increase in employment rather than the creation of unemployment. As noted above, under conditions of imperfect competition in the product and factor markets, labour (or any other factor) is exploited, that is, paid less than the value of its marginal product or less than its marginal revenue product. In this context trade unions can play a useful role of removing the exploitation of labour by getting the wage rate raised to the level of value of marginal product or marginal revenue product.

5. Marginal productivity theory also *ignores the positive interrelation between rewards of the factors and their productivity, especially between wages and the efficiency or productivity of labour*. It has been pointed out that rise in wages has a favourable effect on the efficiency and productivity of labour. With higher wages workers can afford to have better standard of living and better health which will raise their productivity and efficiency. This positive relationship between wages and labour efficiency especially holds good in case of the developing countries like India where wage rates in many industries are even below the minimum subsistence level. With wages even below the subsistence level, workers remain underfed and undernourished and, as a result, unhealthy and inefficient. If, following the rise in wages, efficiency and productivity of labour improve, then it may be worthwhile from the viewpoint of employers to raise wages. It has therefore sometimes been asserted that " *higher wages are economical*" or there is a '*economy of high wages*'. But, as mentioned above, marginal productivity theory completely ignores this favourable effect of higher wages on productivity of labour.

Now, if the favourable effect of higher wages on labour productivity is recognised, then the unique level of wage-equilibrium arrived at in the marginal productivity theory is not valid. With every rise in the wage rate there will be a different curve of marginal productivity of labour and a different wage-employment equilibrium. Thus, there are various possible positions of wage-employment equilibrium depending upon the productivity and efficiency and there is a choice for a firm or the industry to select among them. That there is a *unique* wage equilibrium, as has been asserted by the strict and rigid version of marginal productivity theory, is therefore not acceptable.

6. Like the neo-classical theory of pricing of products, marginal productivity theory of distribution which has been developed by neo-classical economists is also a *marginalist approach* to the problem and therefore *assumes that the entrepreneurs or employers seek to maximise profits*. It is only if the entrepreneurs are maximisers of profits that they will equate wage with marginal product of labour. If they do not seek to maximise profits, then they may employ the amount of labour at which marginal product of labour stands higher than the wage. Likewise, they may employ the number of workers at which the marginal product lies below the wage rate. Just as Hall and Hitch have criticised the marginalist approach as it applies to product pricing on the ground that entrepreneurs do not maximise

profits. R.A. Lester[6], an American economist, has criticized the marginal productivity theory of distribution on the ground that entrepreneurs do not behave as maximisers of profits and therefore wages in the real world will differ from the marginal product of labour. Lester has provided empirical evidence in support of his view. But Machlup[7] and Pen[8] have defended the marginal productivity theory of distribution and the profit maximization assumption on which it is based. To quote Pen, "It is not necessary for every entrepreneur to be able to find the exact point of equilibrium, some will overshoot the equality of wage and marginal product, others will remain below it. However the *trend is towards equality*. In this sense the theory gives only a rough approximation of equality, but as such it is probably not bad."

7. Another basic objection that has been raised against marginal productivity theory is that various factors are jointly demanded for the production of a commodity. That is, production of a commodity is the end result of co-operation of various factors and their *individual productivities cannot be estimated separately*. Labour without the assistance of capital goods can produce almost nothing and capital without the assistance of labour will not produce anything at all. Now, when one cannot speak of the individual productivities of the factors at all or when we cannot calculate their individual productivities at all, then the question of rewarding the factors according to their marginal product cannot arise at all. This argument was forcefully advanced by English literary figures such as Bernard Shaw and Bertrand Russel. Bernard Shaw writes, "When a farmer and his labourers sow and reap a field, nobody on earth can say how much of the wheat each of them has grown."[9] Likewise, Bertrand Russel writes, "In an industrial system a man never makes the whole of everything, but makes the thousandth part of a million things. Under these circumstances it is totally absurd to say that a man has the right to the produce of his own labour. Consider a porter on a railway whose business it is to shunt goods trains. What proportions of the goods carried can be said to represent the produce of his own labour? The question is wholly insoluble."[10]

It may however be noted that many economists believe that there exists a good degree of elasticity of substitution between the factors and therefore one factor can be varied by a small amount keeping the other factors constant. In this way they argue that marginal products of various factors can be estimated separately.

8. A controversial problem concerning the marginal productivity theory is that if the various factors are remunerated in accordance with their marginal products, whether the total product would be just exhausted. Suppose there are two factors of production, labour and capital, which are required for production of a commodity (ignore other factors). Now the question is when wages of the labour are paid equal to its marginal product, whether the remaining total product would be equal to capital's marginal product or it will be less or more than it. This difficulty is called the *adding up problem* or *product exhaustion problem* of the marginal productivity theory of distribution. Whether payment to factors equal to their marginal products does or does not exactly exhaust the total product depends on the form of the production function. If the production function is homogeneously linear, or in economic terms, constant returns to scale prevail, then with the aid of Euler's theorem of mathematics it has been proved that payments to factors equal to their marginal products would just exhaust the total product. But, "The practical question is whether constant returns to scale do or do not occur in reality; once again, that differs entirely for the different branches of industry. At some places,

6. R.A. Lester, "Shortcomings of Marginal Analysis for Wage Employment Problems", *American Economic Reivew*. 1946.
7. F. Machlup, "Marginal Analysis and Empirical Research", *American Economic Review*, Vol. 36, 1946.
8. J.Pen, *Income Distribution*, Penguin Books, 1971.
9. Bernard Shaw, *Intelligent Woman's Guide to Socialism,* p. 21, quoted by Dennis Robertson, *Principles of Economics*, The Fontana Library Edition, p. 186.
10. Bernard Russel, *Prospects of Industrial Civilisation*, p. 146, quoted by Dennis Robertson, *op. cit.*, pp. 186-87.

something will be left. That too shows that the marginal productivity theory gives only a rough approximation of reality.[11]

9. Another important criticism of marginal productivity theory of distribution is that it does not explain the remuneration of entrepreneurs, that is, profits. Marginal productivity of a factor can be known if it can be varied by keeping other factors fixed. But the entrepreneur in a firm is only one and a fixed factor and variation in it is not possible. Therefore, marginal productivity of entrepreneur from the viewpoint of a firm is meaningless. If the single entrepreneur is withdrawn from the firm, keeping all other factors constant, the whole production process of the firm will collapse. And there is no meaning of adding one entrepreneur to a firm. The new entrepreneur will mean the establishment of altogether a new firm. It is because of this that in the neo-classical theory of distribution profits are shown as surplus or residual income and not as determined by marginal productivity.

10. Finally, the marginal productivity theory of distribution *does not give any importance to the power structure, social conventions, social status, and prestige of a group of workers in the determination of remuneration of various groups or classes of labour force*. Professor Pen rightly writes that marginal productivity theory based on perfect competition "does not explain discrimination between men and women, between races and between social classes; it does not make it clear why top executives earn as much as they do and why unions can push up wages"[12]. According to Pen, the high salaries drawn by the top executives of the firm cannot be explained by marginal productivity theory, since the concept of marginal productivity in their case is utterly vague and further that their remunerations can be explained only by the power structure. To quote him again, "Remuneration of executives and staff workers are fixed in another way: social conventions, the power structure, considerations of prestige and status play a much larger part than marginal productivity. And that also holds good for the remuneration of the people not working in industry: of teachers, for instance (what is the marginal productivity of a contribution to the knowledge of economics ?) and doctors (what is the marginal productivity of a human life?). These are the sectors in which other laws apply than the derivatives of production. Economists often forget this".[13]

Conclusion

We have discussed above the various criticisms levelled against marginal productivity theory of distribution. Marginal productivity theory of distribution does not explain fully the determination of all factor prices. *But marginal productivity of a factor is the most important economic factor governing the prices of factors*. Other factors such as power structure, social conventions, status and prestige do play a part in fixation of remunerations but the economic factor of marginal productivity does exercise an important influence on the determination of factor rewards.

CONCEPTS OF FACTOR PRODUCTIVITY

Before turning to the detailed study of the various determinants of factor prices, it will be helpful for the proper understanding of the subject if we first explain the various concepts of productivity. The knowledge of these various concepts will greatly help in understanding the modern theory of factor prices. At the very outset it is desirable to make it clear why economists are interested at all in the productivity of a factor. *We are concerned with productivity since the price which a factor will be able to get depends upon its productivity*. Why? This is because the factors are demanded not because they directly satisfy the wants. The factors are purchased to put them to work for producing consumer goods which satisfy human wants. Other things being equal, the greater the contribution made to the production of goods by a factor unit, the greater the price which it will be able to command in the market. Thus productivity is an important determinant of the price of a factor.

11. J.Pen, *Op. cit.*, p. 85.
12. *Ibid* p. 80.
13. *Ibid* p. 86.

We have already explained the concepts of average physical productivity and marginal physical productivity in a previous chapter. To repeat here, *average physical product of a factor is the total production divided by the number of units of a factor employed.*

$$\text{Average physical product } (APP) \text{ of a factor} = \frac{\text{Total output}}{\text{Total No. of units of a factor used}}$$

Marginal physical product (MPP) of a factor is the increase in total output caused by employing an additional unit of the factor, quantity of other factors remaining fixed. The fixed factors are, however, conceived to be adjusted or adapted in such a way that increased amount of the variable factor can be used with them.

Before the development of the theory of product pricing under imperfect competition (monopoly and monopolistic competition) by Joan Robinson and Chamberlin, the assumption of perfect competition in the product market was usually made while discussing the pricing of factors. Under perfect competition in the product market conversion of the marginal physical product into money terms merely involves multiplying the marginal physical product with the price of the product since the price of the product of an individual firm under perfect competition is a given and constant quantity. Money value of the marginal physical product under perfect competition thus means the marginal physical product multiplied by the price of the product. However, with the development of imperfect competition theory in explaining product prices the two distinct concepts of marginal productivity have been evolved. They are :

(*i*) Marginal Revenue Product (*MRP*)

(*ii*) Value of Marginal Product (*VMP*)

Marginal Revenue Product (MRP)

Marginal revenue product is the increment in the total value product caused by employing an additional unit of a factor, the expenditure on other factors remaining unchanged. In other words, marginal revenue product is the marginal physical product of the factor multiplied by the marginal revenue

$$MRP = MPP \times MR$$

It is the marginal revenue product which is often termed as marginal product or marginal productivity.

Table 32.1
Value of Marginal Product (VMP) and Marginal Revenue Product (MRP) when there is Perfect Competition in the Product Market

I	II	III	IV	V	VI	VII
Units of a Factor	*Total Output (Q)*	*Marginal Physical Product (MPP)*	*Price of Product (P)*	*Value of Marginal Product (VMP)*	*Total Revenue (PQ)*	*Marginal Revenue Product (MRP)*
1	25	25	2	50	50	50
2	70	45	2	90	140	90
3	110	40	2	80	220	80
4	145	35	2	70	290	70
5	172	27	2	54	344	54
6	191	19	2	38	382	38
7	199	8	2	16	398	16
8	199	0	2	0	398	0

Value of Marginal Product (VMP)

It means the marginal physical product of the factor multiplied by the price of the product (*i.e.* average revenue).

$$VMP = MPP \times \text{Price (or } AR)$$

Since under perfect competition the demand curve of the product facing an individual firm is perfectly elastic and therefore price and marginal revenue are equal, the value of marginal product (*VMP*) and marginal revenue product (*MRP*) will be equal to each other as is shown in Table 32.1 and Fig. 32.3. But since in monopoly or imperfect competition in the product market, average revenue (or demand curve) is falling downward and marginal revenue curve lies below the average revenue curve, price is not equal to marginal revenue. Therefore, in monopoly or in other forms of imperfect competition in the product market, marginal revenue product will not be equal to the value of marginal product. Since price is higher than marginal revenue under monopoly or monopolistic competition in the product market, the value of marginal product (*VMP*) will be larger than the marginal revenue product (*MRP*) and the marginal revenue product (*MRP*) curve will lie below the value of marginal product (*VMP*) curve as is shown in Fig. 32.4. Thus, in perfect competition *MRP* and *VMP* have identical meaning but in monopoly and imperfect competition in the product market they diverge.

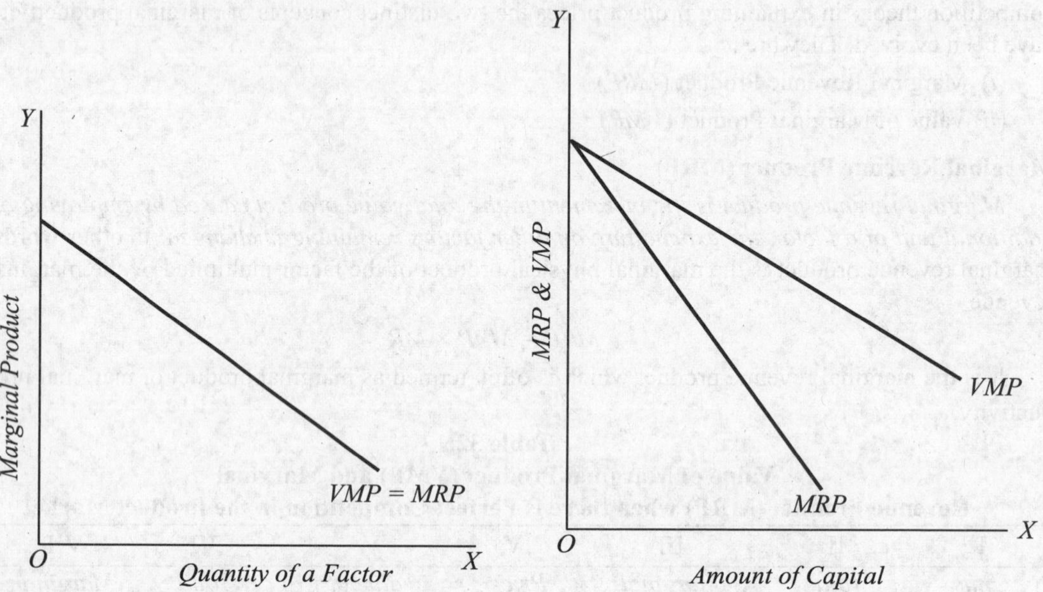

Fig. 32.3. *VMP and MRP under Perfect Competition in the Product Market*

Fig. 32.4. *VMP and MRP under Imperfect Competition in the Product Market.*

The distinction between Marginal Revenue Product (*MRP*) and the Value of Marginal Product (*VMP*) can be better understood from Table 32.1. In the above Table 32.1 perfect competition has been assumed to be prevailing in the product market. Therefore, price of the product (₹ 2) for an individual firm remains the same whatever the level of its output. As more units of labour are employed, total output is increasing but at a diminishing rate. That is to say, marginal physical product of the factor is declining (law of diminishing marginal returns has been assumed to be operating). Since *VMP* is equal to *MPR* × Price, *VMP* (Col. V) can be found by multiplying Col. III by Col. IV. Thus when one labour unit is employed the marginal physical product is 25. Since the price of the product is ₹ 2, the value of marginal product (*VMP*) will be equal to 25 × 2 = 50 and so for the subsequent units of labour. It shall be noticed from Col. V that value of marginal product is declining as more units of labour are employed after the second unit. This is so because marginal physical product is declining due to the operation of law of diminishing returns.

Since when there is perfect competition in the product market *MR* is equal to price (*P*), Marginal Revenue Product (*MRP*) also can be found out by multiplying Col. III by Col. IV. Thus under perfect competition value of marginal product (VMP) will be equal to marginal revenue product (*MRP*) (compare Col. VII with Col. V). Since marginal revenue product (*MRP*) of a factor can also be defined as the increment in the total revenue of a firm by employing an additional unit of a factor, it can also be directly found out from Col. VI which shows the total revenue at the various levels of output.

MRP can be obtained by taking out the difference between the two successive total revenues. The difference in the two successive total revenues occurs due to the employment of an extra unit of a factor. Thus when two units of labour are employed, total revenue is ₹ 140 which is obtained by selling 70 units of output produced by 2 units of labour at price of ₹ 2 per unit. When another labour unit is employed, the total output is 110 and total revenue obtained is ₹ 220. Thus this additional unit of labour has added ₹ 80 (₹ 220 – ₹ 140) to the firm's total revenue, *MRP* of labour at this stage is therefore ₹ 80. Likewise, *MRP* of the subsequent units of labour can be found out in two ways: firstly, by multiplying marginal physical product *MPP* by *MR* (in the present case *MR* is equal to price); and secondly by taking out the difference between the two successive total revenues caused by employing an additional unit of labour.

VMP and MRP When there is Imperfect Competition in the Product Market

Whereas values of marginal product and marginal revenue product are equal under perfect competition, they differ if there is monopoly or imperfect competition in the product market. Under monopoly or imperfect competition in the product market, average revenue curve or demand curve facing an individual firm slopes downward. In other words, as the firm increases its output (and sales) by employing more units of labour, the price of the product declines (see Col. IV in Table 32.2). Since

Table 32.2
VMP and MRP under Imperfect Competition

I	*II*	*III*	*IV*	*V*	*VI*	*VII*
Units of Labour	*Total Output (Q)*	*Marginal Physical Product (MPP)*	*Price of Product (P)*	*Value of Marginal Product (VMP)*	*Total Revenue (II × IV) (Q.P.)*	*Marginal Revenue Product (MRP)*
1	25	25	2.00	50.00	50	50
2	70	45	1.80	81.00	126	76
3	110	40	1.50	60.00	165	39
4	145	35	1.30	45.50	188.50	23.50
5	172	27	1.20	32.40	206.40	17.90
6	191	19	1.15	21.85	219.65	13.25
7	199	8	1.13	9.04	224.87	5.22
8	199	0	1.13	0.00	224.87	0.00

the average revenue curve (*i.e.* price curve) falls downward under monopoly and imperfect competition in the product market, *MR* curve will lie below it. In other words, *MR* will be less than *AR* (*i.e.*, price) of output. Since *VMP* is equal to *MPP* × Price and *MRP* is equal to *MPP* × *MR*, the two will not be equal to each other when monopoly or imperfect competition prevails in the product market. *VMP* at various levels of labour employment which is shown in Col. V of Table 32.2 is obtained by multiplying *MPP* (Col. III) with price of the product which goes on falling (Col. IV). Thus when two units of labour are engaged, marginal physical product (*MPP*) is 45 and price of the product is ₹ 1.80. By multiplying 45 with ₹ 1.80 we get *VMP* equal to ₹ 81 which we write in Col. V corresponding to two units of labour. Likewise, *VMP* for other levels of labour employment can be found out.

In order to obtain *MRP*, we have first to find out the total revenue at various levels of employment. The total revenue (Col. VI) is obtained by multiplying total output (Col. II) with price of product (Col. IV). By finding out the the difference between two successive total revenues, *MRP* can be obtained. Thus when two units of labour are employed, total revenue is ₹ 126 and when 3 units of labour are employed the total revenue is ₹ 165. Increment in total revenue caused by the third unit of labour is thus equal to ₹ 165–126 = ₹ 39. Thus *MRP* of labour when three units of it are employed is ₹ 39. Similarly, *MRP* of other levels of labour employment is found out and recorded in Col. VII. By comparing Col. V and Col. VII, it will be clear that *VMP* and *MRP* are not the same and further that *MRP* is less than *VMP*. This is so because, as explained above, price and *MR* are not equal (*MR* is less than the price) for a firm working under monopoly or imperfect competition in the product market. To sum up, *under monopoly or imperfect competition in the product market, MRP and VMP will diverge and MRP curve will lie below VMP curve.*

A COMPETITIVE FIRM'S EQUILIBRIUM REGARDING FACTOR EMPLOYMENT: GENERAL CONDITIONS

In the previous parts we have discussed the conditions of firm's equilibrium in the product market in order to show what level of output it will produce so as to maximise its profits. In other words, we have explained there firm's equilibrium in regard to the level of output it has to produce. At the present moment, we are concerned with the firm's equilibrium in the factor market. That is, we are here concerned with the equilibrium of the firm with respect to the amount of a factor it has to employ. Of course, we explained in a previous chapter firm's equilibrium in regard to factor combination and found that in order to produce a given level of output at the minimum cost possible with given factor prices, a firm will be in equilibrium when it is combining the various factors in such a way that marginal rate of technical substitution between any two factors is equal to their price ratio $\left(MRTS_{ab} = \frac{P_a}{P_b}\right)$. In terms of marginal products of factors we can say that a firm is in equilibrium in regard to the usage of factors so that

$$\frac{MP_a}{P_a} = \frac{MP_b}{P_b} = \frac{MP_c}{P_c}$$

where *a*, *b* and *c* stand for various factors of production, *MP* for marginal products of the factors and *P* for factor prices.

We shall explain below how a firm determines the absolute amount of a single variable factor it has to use or employ and what factors govern its decision about this. There can be various possible situations in both the factor and product markets in which a firm may find itself. In the present analysis we assume a firm is facing perfect competition in both the factor and product markets. Further, in our analysis we assume that the entrepreneur who has to hire other factors of production is rational, *i.e.,* he tries to maximise his profits. In deciding about the number of units of a factor which will be employed, the rational entrepreneur will be guided by the change in cost as a result of employment of a factor unit and the change in revenue made by the sale of the additional product produced by that factor unit. The entrepreneur will compare the change in revenue with the change in cost of utilising an extra unit of a factor. If the change in revenue resulting from the employment of an additional factor unit exceeds the change in cost of hiring it, it will be profitable for the firm to employ that unit.

Change in revenue resulting from the employment of an extra unit of the factor is called *Marg Revenue Product* (*MRP*) and the change in cost brought about by hiring an extra unit of the fa called *Marginal Factor Cost* (*MFC*). *The firm will go on employing more and more units of as long as marginal revenue product of the factor exceeds the marginal factor cost of it* will not employ an extra unit of a factor if its marginal factor cost is greater than its margi product, because it will not be profitable for the firm to do so. The profits of the entrep maximum when it is employing such a number of units of a factor that its marginal r equals its marginal factor cost. Therefore, the firm will achieve equilibrium in the factor when the following condition is satisfied:

Marginal Revenue Product = Marginal Factor Cost

$$MRP = MFC$$

The equality of *MRP* with *MFC* of the factor is necessary but not a su equilibrium with regard to the employment of the factor. For the firm a second order condition must also be fulfilled. This second order cor *enue curve must cut the marginal factor cost curve from above at t* words, this second order condition states that for employment marginal revenue product must be less than marginal factor co equilibrium level, the marginal revenue product must be great for the equilibrium level of employment the following two

(1) *MRP* = *MFC*
(2) *MRP* curve must cut the *MFC* curve *from ab*

O'

For employment greater than the equilibrium

for employment less than the equilibrium c

When there is perfect competition in th the price of the product by changing its ov the product in the market as given and Since for a perfectly competitive firm constant the average revenue curve f coincides with it. For it, Price = *MF* competition in the product market revenue product of the factor (equal to the value of margina factor (*VMP*). Thus under p in the product market, *M* will be the same.

Perfect competit market when :

(1) the facto are identical fr

(2) bot fect inforr the mar

14.

(3) the number of buyers and sellers is very large; and

(4) both buyers and sellers are free to enter or leave the market.

When a firm is confronting perfect competition in the factor market, it cannot affect the price of the factor by varying its level of factor employment. A single firm's demand for the factor is sc small compared with the total demand for the factor that any change in its demand for the factor will not affect its price. It has to take the price of the factor as given and constant for it. The firm can employ as many number of factors. Therefore, *the supply curve of the factor for a single firm under perfect competition in the factor market will be perfectly elastic (horizontal straight line) at the level of the prevailing price of the factor*. As a result, for a firm working under perfect competition in the factor market, the extra cost of hiring an extra unit of the factor will be equal to the price of the factor which remains unchanged. Thus marginal factor cost under perfect competition in the factor market is equal to the price of the factor ($P_F = MFC$). This is shown in Fig. 32.5. In this figure, ruling price of the factor in the market is *OP*, which the firm has to accept as given and constant. Supply curve of the factor (or *AFC* curve) is a horizontal straight line and *MFC* curve coincides with it.

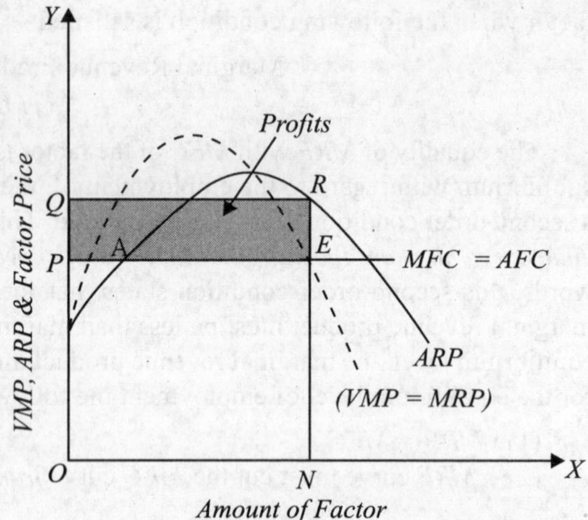

Fig. 32.6. *Employment of a Factor Under Conditions of Perfect Competition*

As explained above, the firm will continue employing more units of a factor as long as the marginal revenue product is greater than marginal factor cost and will achieve equilibrium position where marginal revenue product is equal to marginal factor cost. As in the present case when perfect competition prevails in both the factor and product markets, *MRP* is equal to *VMP* and *MFC* = *P*, we can state that a firm which is working under conditions of perfect competition will attain equilibrium at the level of factor employment where *VMP* is equal to the price of the factor. In Fig. 32.5 the firm is in equilibrium at *ON* level of employment (or at point *E*) at which *VMP* (or *MRP*) is equal to the price of the factor (or *MFC*). At point *E*, the second order condition is also satisfied since the *VMP* (or *MRP*) curve is cutting *MFC* curve from above. It will be noticed from Fig. 32.5 that marginal revenue product (*MRP*) is also equal to *MFC* at point *A*. But *A* cannot be the position of equilibrium since second order condition is not fulfilled. At point *A*, *VMP* or *MRP* curve is cutting *MFC* curve from below and it will be profitable for the firm to increase employment further. Thus, *E* is the point of equilibrium where *MRP* = *MFC* or *VMP* = Price of the factor as well as *MRP* curve is cutting *MFC* curve from above.

To sum up, in equilibrium under perfect competition in both the factor and product markets :

$$MRP = MFC = P_F$$

Since MRP and VMP are equal when there is perfect competition in the product market, therefore, $MRP = VMP = MFC = P_F$

Thus, we conclude that *under perfect competition in both the markets, a factor of production will get the price equal to the marginal revenue product of the factor which under perfect competition in both the factor and product markets is equal to value of its marginal product.*

In order to know whether the firm is earning profits or losses we have also to draw average revenue product curve (*ARP*) as is shown in Fig. 32.6. It will be seen from this figure that when the firm is in equilibrium at *ON* level of employment (*i.e.* at point *E*) average revenue product which is equal to *RN* is greater than average factor cost (*AFC*) which is equal to *EN*. Therefore, *ER* is the profit per unit of the factor. Rectangle *PERQ* is the total profit earned by the firm. Thus, with factor price *OP* and equilibrium at *E*, the firm is making supernormal profits by employing *OM* of the factor. Firm can earn such supernormal profits in the short run. But this cannot happen in the long run since by lure of these profits other firms in the long run will be attracted into the industry which is using that factor.

To sum up, while in the short run, the firms may make supernormal profits, but in the long run adjustments will take place in the number of firms so that only normal profits are being earned.

DEMAND CURVE FOR A FACTOR

We shall explain the demand curve for a factor under condition of perfect competition in the factor market in the following three cases. In all three cases we assume that perfect competition prevails in the product market. Under these conditions, *MRP* and *VMP* of the factors are equal.

1. A competitive firm's demand curve for a single variable factor.
2. A competitive firm's demand curve for a factor when other factors are also variable
3. A competitive industry's market demand curve for a factor.

Competitive Firm's Demand Curve for a Single Variable Factor

We have explained above how much amount of a factor a firm, under conditions of perfect competition in the factor market will employ and demand both in the short run and long run. We are now in a position to derive the *demand curve* for a factor of production under conditions of perfect competition, which with the intersection of supply curve of that factor determines its price. The derivation of demand curve for factor is illustrated in Fig. 32.7. It always pays an entrepreneur to go on hiring more and more units of a factor until its marginal revenue product equals its marginal factor cost. If there is a perfect competition in the factor market, then marginal factor cost (*MFC*) of the factor will be equal to the market price of the factor which remains unaltered. As already explained, a perfectly competitive buyer in the factor market will employ a factor of production to a point where its *MRP,* which is equal to *VMP,* equals its market price. In doing so, he will be maximising his profits and will thus be in equilibrium position. Thus, in Fig. 32.7, if the market price of a factor is *OP*, then the employer will hire or employ *ON* units of the factor since at *ON* units *MRP* of the factor is equal to its price *OP*. Thus, at market price *OP*, *ON* amount of the factor will be demanded by the producer. If now the market price of the factor falls to *OP′* the amount demanded of the factor will rise to *ON′* where marginal revenue product of the factor is equal to the new market price *OP′* . If the factor price further falls to *OP′ ′*, the *ON′ ′* quantity of the factor will be demanded by the firm.

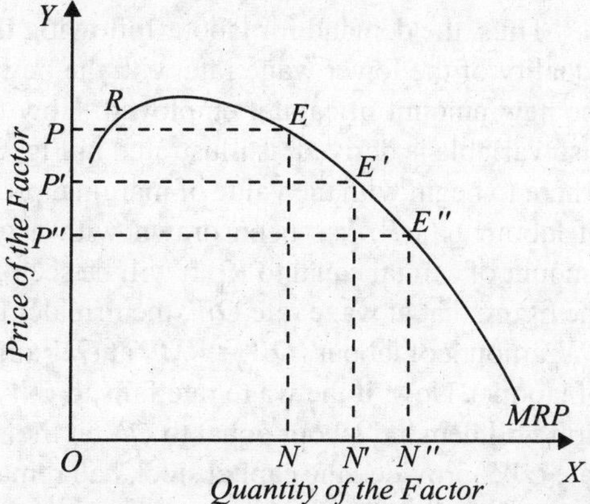

Fig. 32.7. *Marginal revenue product curve constitutes the demand curve for a single variable factor*

It is thus obvious that given the market price of the factor, we can read the quantity demanded of the factor by the firm from the marginal revenue product curve of the factor. The marginal revenue product curve of a variable factor usually first rises upward to a point and then slopes downward. But it should be noted that only the downward-sloping portion of the *MRP* curve forms the demand curve for that factor. This is so because the entrepreneur cannot be in equilibrium at the rising part of the *MRP* curve. For instance, with market price *OP*, the entrepreneur will not be in equilibrium at point *R*, since at *R*, marginal revenue product curve is cutting marginal factor cost curve (which is here same as factor price curve) from below. The firm will be in equilibrium at point

E where *MRP* curve is cutting the factor price curve from above and will employ or demand *ON* amount of the factor. We thus conclude *that, under perfect competition in the factor market, downward sloping part of the marginal revenue product curve of the factor is the demand curve for that factor.*

We have derived the demand curve for the factor of an individual competitive firm. We now turn to derive the demand curve for the factor of the whole competitive industry. To make our analysis simple, if we assume that the price of the product does not change as the industry employs more of the factor and thus expands its output. If this assumption holds, the demand curve for the factor of production of the whole competitive industry would be obtained by the lateral summation of the *MRP* or *VMP* curves of the firms (*i.e.* demand curves of the firms) in the industry.

Firm's Demand Curve for a Factor: With all Factors (inputs) Variable

In showing value of marginal product (*VMP*) curve as a firm's demand curve for labour we assumed that quantities of other factors such as capital, raw materials fixed. Thus when wage rate declines and as a result more labour is employed, a firm moves along the given *VMP* or demand curve. However, a fall in the price of a variable factor, such as wage rate of labour, generally leads to the increase in employment not only of that factor but also of other factors as well, at least in the long run. Note that the two factors are said to be complementary if increase in the amount of one raises the marginal productivity of the other. Thus, when due to the decline in wage rate the employment of labour is increased, it also leads to the increase in the quantity of capital.[15] Further, this increase in the quantity of capital will in turn raise the marginal productivity of labour. As a result of this adjustment in capital, value of marginal product (*VMP*) curve of labour will shift to the right. Thus when we consider that a firm can vary several factors, the firm's demand curve for a factor is no longer its value of marginal product (*VMP*) curve. This is because a change in price of a factor will result in change in the quantities of other factors. Thus, when several other factors such as capital, raw materials are variable, (and wage rate of labour) falls, it will not only lead to the increase in the demand for labour but will also require adjustment in the amounts of other factors such as capital, raw materials in the production process. Considering that the two factors, labour and capital, are complementary to each other, when there is increase in the employment of labour consequent to a fall in wage rate, it would bring about increase in the quantity of capital (and other factors), at least in the long run.

Thus, the demand for labour following the reduction in the wage rate will be determined by the equality of the lower wage rate with the new value of marginal product of labour corresponding to the new amount of capital employed. How the long-run demand curve of a factor when capital is also variable is derived is illustrated in Fig. 32.8, where to begin with the value of marginal product of labour (*VMP*) has been drawn with a given amount of capital equal to K_1. It will be seen from the figure that at wage rate *OW*, the firm demands ON_0 amount of labour ($OW = VMP$ at ON_0 amount of labour). Now, if the wage rate falls to *OW'* the firm will demand labour equal to ON_1 at the wage rate *OW'* provided the capital stock had remained constant at K_1. However, as explained above, the increase in the quantity demanded of labour at the lower wage rate *OW'* would raise the marginal product of capital and therefore in the long run the firm would adjust the amount of capital. Suppose that by adjusting the amount of capital the firm raises it to the new level K_2. The increase in capital

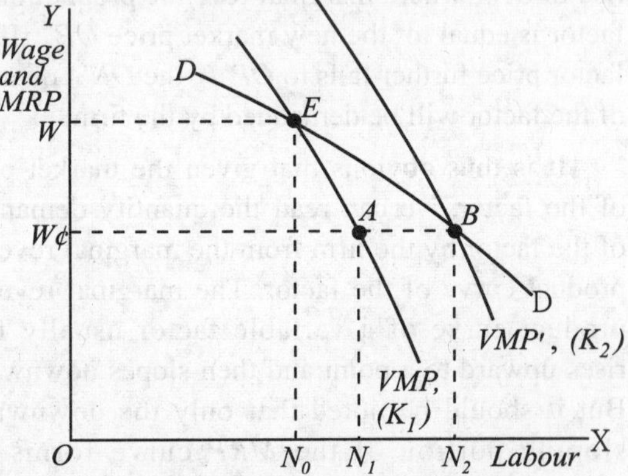

Fig. 32.8. *Demand for a Factor (Labour) with Capital Adjustment in Capital*

15. We are here assuming that labour and capital are complementary to each other.

to K_2 in the production process causes a shift in the value of marginal product (*VMP*) curve of labour to the new position *VMP*′. With this new marginal value product curve of labour, *VMP* ′ obtained after adjusting the amount of capital to the level K_2, at the lower rate *OW* ′ the firm will demand ON_2 quantity of labour. Thus by joining points like *E* and *B*, we obtain the firm's demand curve *EB* for labour when capital is also variable and adjusted appropriately. It will be seen from Fig. 32.8 that firm's demand curve for labour *EB* obtained when other factors such as capital are varied is more elastic than either of the two *VMP* curves.

COMPETITIVE INDUSTRY'S DEMAND CURVE FOR A FACTOR

We now turn to explain competitive industry's demand curve for labour. In a previous section we obtained the competitive industry's demand curve for labour by the horizontal summation of the *VMP* curve (*i.e.* demand curves) of the factor of all firms assuming that price of the product remains constant or unchanged. This is however not realistic. When following the reduction in the wage rate all the firms in the industry employ more labour in the production of a commodity, the output of commodity and therefore its supply in the market increases. This would result in lowering the price of the product.

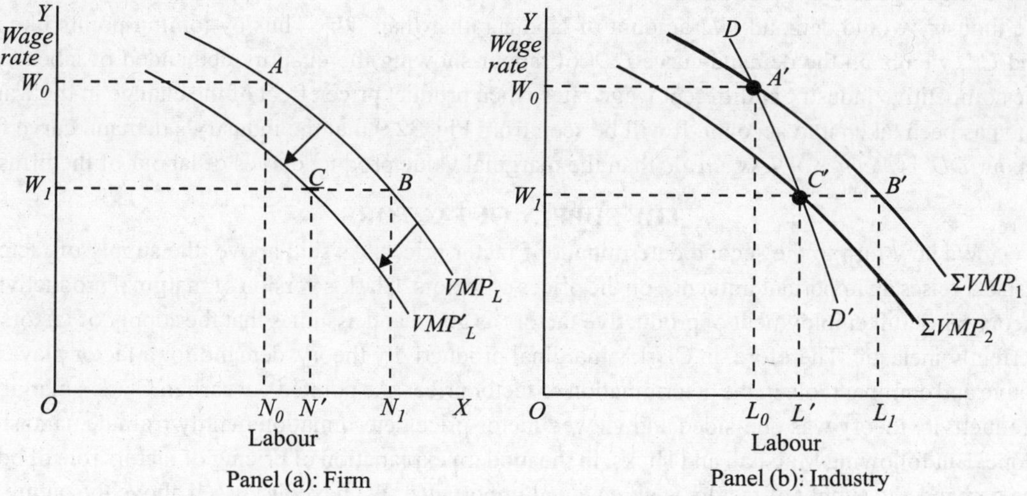

Fig. 32.9. *The Competitive Industry (Market) Demand Curve for Labour*

It may be recalled that the value of marginal product (VMP_L) of labour under perfect competition in the product market equals marginal physical product (*MPP*) of labour multiplied by the price of the product (under perfect competition $MRP_L = VMP_L = MPP_L \times$ Price of Product). Now, following the reduction in wage rate, as more labour is employed by all the firms in the competitive industry output expands bringing about a fall in the price of the product. This will cause a shift in the value of marginal product curve of labour (VMP_L) to the left as shown in panel (*a*) of Fig. 32.9 where when the price of the product falls from P_0 to P_1, the value of marginal product curve of labour shifts to the left to its new position VMP'_L and as will be seen from panel (*a*) of Fig. 32.9 that at the lower wage rate W_1 the firm demands and employs labour *ON*′ rather than ON_1 amount of labour which it would have demanded at the lower wage rate W_1 if the product price had remained constant at P_0. This represents the *product price effect* on the demand for labour. It will be seen that when this price effect is taken into account, with the reduction in the wage rate from W_0 to W_1 the firm's demand for labour increases from ON_0 to *ON* ′ and not to ON_1.

Having explained how a firm's demand for labour is affected by the change in price of the product following the reduction in the wage rate, we can now derive the market demand curve for labour of the competitive industry. This is illustrated in panel (*b*) of Fig. 32.9. At the wage rate OW_0, a competitive firm employs ON_0 amount of labour by equating wage rate OW_0 with the value of marginal product of labour (VMP_L), price of the product equal to P_0 being given. (See panel (*a*) in Fig. 32.9). At the wage rate OW_0 the demand for labour of the industry as a whole can be obtained by the horizontal summation of the quantity demanded of labour by the given number of firms. Suppose there are 100 firms in the industry. By adding up horizontally the quantity demanded of labour by all the firms in the industry, that is, 100 ON_0, given the price of the product equal to P_0, we get a point A' in panel (*b*) on ΣVMP_1 curve which shows that at wage rate OW_0 the industry demand for labour will equal OL_0 (*i.e.*, $OL_0 = ON_O \times 100$). Now, when the wage rate falls to W_1 and with expansion in output or supply, price of the product falls to P_1 and consequently *VMP* curve of labour shifts to the left to the new position VMP_L, a firm demands ON' quantity of labour. With this, the horizontal summation of the VMP'_L curves to obtain the industry's demand curve for labour would also shift to the left to the new position ΣVMP_2 with the price of the product equal to P_1. As will be seen from the figure that at the wage rate OW_0 the industry will demand OL_0 amount of labour and at the wage rate OW_1, the industry would demand OL' amount of labour rather than OL_1. Thus by joining points like A' and C', we obtain the demand curve DD' of labour showing the quantity demanded of labour by the competitive industry at different wage rates when product price effect of the change in the wage rate has been taken into account. It will be seen from Fig. 32.9 that the industry's demand curve for labour DD' is *steeper or less elastic* than the marginal value product curves of labour of the firms.

THE SUPPLY OF FACTORS

We now turn to the second determinant of factor prices. As said above, the supply of factors also exercises an important influence on the prices of factors. Clark's version of marginal productivity theory takes full employment of productive factors as given and assumes that the supply of factors is perfectly inelastic. Therefore, in Clark's marginal productivity theory, demand for a factor plays an active and dominant role in the determination of factor prices. As pointed out earlier, Clark's marginal productivity theory was one-sided and viewed factor-price determination mainly from demand side alone. But following Marshall and Hicks, in the modern explanation of pricing of factors role of both the demand and supply of factors is given equal importance. We have discussed above the nature of demand for factors and the factors determining it and its price elasticity of demand. In what follows we shall spell out the nature of supply of various factors. It is worth mentioning at the very outset that, unlike the demand for factors, the nature and behaviour of supply of various factors is not uniform. We shall explain in detail the nature of supply of various factors when we make analysis of the price determination of each factor separately. Here we shall indicate only the broad features of their supply.

It may be noted that the supply represents the different quantities of that factor which are offered for sale at various alternative prices. From the viewpoint of the nature of supply, factors may be divided into two classes : *original and produced factors*. Produced factors are the inermediate physical inputs such as capital equipment (*i.e.* machinery, tools, components etc.), steel, cement, fertilizers, fuels which are themselves produced in some industries and are used as inputs or productive factors in other industries for production of other products. So far as the supply curve of these produced factors is concerned, it is governed by the same laws of production which apply to the the production and supply of goods produced by various industries. Therefore, the supply curves of produced inputs (factors) depend upon the changes in marginal cost of production. So long as marginal cost rises as output is expanded in industries such as machine-making steel, fertilizers, marginal cost curve slopes upward and therefore the supply curve of these material inputs slopes upward. However, if the industries producing certain intermediate physical inputs experience decreasing costs as their outputs expand, their supply curve will be sloping downward.

Supply of Land

Land and labour are called original or primary factors of production as they are not produced in the industries. Land is a free gift from nature and therefore its quantity is fixed by nature. More land cannot be produced in response to greater demand for it. Whatever the rent, high or low, for the use of land, its supply to the economy as a whole remains unchanged. In other words, the supply of land to the entire economy does not depend on the price, *i.e.,* rent for its use. Hence, from the standpoint of the whole economy, the supply of land (which includes natural resources) is *perfectly inelastic.* Since supply of land is a free gift from nature and not a produced factor, cost of production has no relevance for its supply. For the society as a whole, land has got no cost of production, since society did not produce it; it got it free from nature.

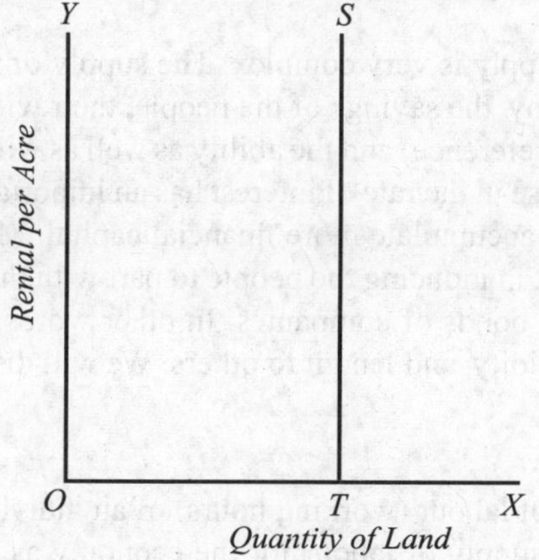

Fig. 32.10. *Perfectly Inelastic Supply Curve of Land for the Economy as a Whole*

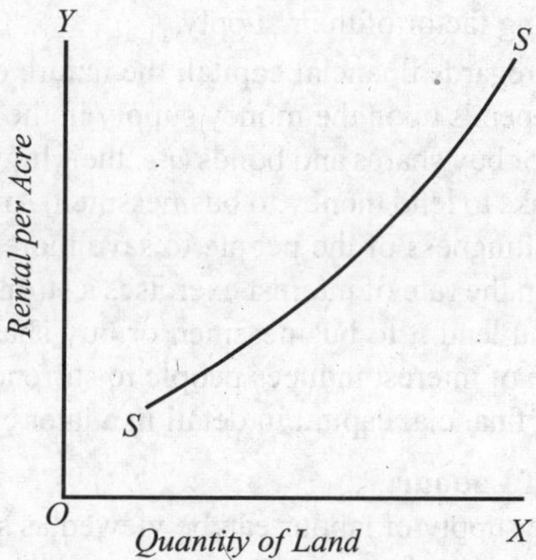

Fig. 32.11. *Somewhat Elastic Supply Curve of Land for a Particular Crop or Industry*

But the supply of land to a *single use or to a particular* industry is not perfectly inelastic. The supply of land to a particular use or industry can be increased by shifting of land from other uses or industries. By offering attractive rents, the supply of land for a particular use can be increased by taking it away from other competitive uses.

The supply of land for a particular crop or industry can be better explained with the help of the concept of *transfer or alternative earnings. The transfer earnings of a factor may be defined as the earnings in the next best alternative use.* Thus the transfer earnings of a piece of land under wheat is the amount of money earned by it if it is put to the cultivation of cotton, assuming that, after wheat, growing of cotton is the next best use of that piece of land. If that piece of land earns ₹ 500 under wheat and ₹ 400 under cotton, then ₹ 400 is the transfer earnings of that piece of land. Now, a piece of land will be supplied to a particular use if at least its transfer earnings are paid to it. In our present example, if the earnings of the prices of land fall below ₹ 400, then the price of land (*i.e.* rent) increases in one particular use, the land from other uses would be attracted towards it, so that the supply of land to the particular use in question would increase. Therefore, the *supply curve of land to a particular use, crop or industry is elastic and slopes upward from left to right.* This is shown in Fig. 32.11.

Supply of Capital

We now turn to the supply of capital which is a produced factor and occupies these days a paramount place in the productive process. Here a distinction must be made between *real or physical capital, i.e.,* capital goods on the one hand and *financial capital* or *money capital* on the other, since the nature of their supply is quite different. Capital goods are produced factors as compared to the primary factors like land and labour which are not produced. Capital goods are produced by firms on the same basis as consumer goods. The nature of supply of consumer goods has been discussed in Part IV. The supply of capital goods is determined by the same factors as those which determine the supply of consumer goods. Since capital goods are reproducible, the cost of production exercises a significant influence over their supply. If the industry producing a capital good is subject to increasing costs, the supply curve of that capital good will be upward sloping indicating that more of it will be supplied at a higher price. And if the industry producing a certain capital good is working under constant cost conditions, the more of that capital good will be supplied at the same price and therefore its supply curve will be a horizontal straight line. However, once the durable capital goods have been produced, their supply is independent of their cost. But, over a period of time, cost is, no doubt, a determining factor of their supply.

As regards financial capital, the nature of its supply is very complex. The supply of financial captial depends upon the money supply in the economy, the savings of the people, their willingness to lend it or buy shares and bonds (*i.e.* their liquidity preference) and the ability as well as willingness of the banks to lend money to businessmen. An increase in the rate of interest has an important effect on the willingness of the people to save more and to accumulate more financial capital. Moreover, increase in the rate of interest exercises a strong effect in inducing the people to part with the money capital and lend it to businessmen or buy shares and bonds of companies. In other words, the rise in the rate of interest induces people to surrender liquidity, and lend it to others. We will discuss the supply of financial capital in detail in a later chapter.

Supply of Labour

The supply of labour can be viewed as supply of labour (working hours) of an individual, the supply of labour for an industry or occupation, and supply of labour for the economy as a whole. We begin with the analysis of supply of labour (working hours) by an individual. As explained in Chapter 33, supply of labour by an individual depends on his choice between work and leisure. Alternative or opportunity cost of working for a period is the sacrifice of leisure by the individual for that time period. It should be noted that leisure is a desirable object which provides satisfaction to the individuals. On the other hand, work provides income to the individual with which he can buy goods and services to satisfy his wants. How much leisure an individual will be willing to sacrifice, that is, how many hours of work he will do depends on the wage rate. We discuss in the next chapter their choice between work (income) and leisure and therefore the supply of labour in detail with the help of indifference curves.

DETERMINATION OF FACTOR PRICES UNDER PERFECT COMPETITION

According to the neo-classical theory, under conditions of perfect competition in the factor and product markets, it is both demand for and supply of factors which determine their prices. It is therefore essential to understand first the nature of demand for factors of production. Demand for a factor differs in certain respects from the demand for consumer goods or products. Products or consumer goods are demanded because they satisfy the wants of the people directly. People demand food to satisfy the pangs of their hunger, they demand clothes to satisfy their want of providing a cover to their bodies and so forth. These products possess utility which directly satisfies the desires of the people and who are therefore willing to pay price for these products.

Derived Demand

But unlike the products, the factors of production do not satisfy the wants of the people directly. The factors of production are demanded not because they directly satisfy the wants of the people who wish to buy them. Instead, they are demanded because they can be used to produce consumer goods which then directly satisfy human wants. Therefore, demand for factors of production is called *derived demand*. It is derived from the demand for the product they help to make. Thus, the demand for a factor ultimately depends upon the demands for goods it helps to produce. The greater the demand for goods a particular type of factor helps to make, the greater the demand for that type of factor. Just as demand for a consumer good depends upon its utility, the demand for a factor depends upon the marginal revenue productivity of the factor. In fact, as has beem explained earlier, the marginal revenue productivity curve of the factor is the demand curve for that factor. The entrepreneur's demand for a factor of production is governed by the marginal productivity of the factor.

Determination of a Factor Price

According to Marshall-Hicks'version of marginal productivity theory of distribution, price of a factor is determined by demand and supply of a factor. Marshall and Hicks held that the price of a factor of production is determined by both the demand for and supply of the factor, but it is to be equal to the marginal revenue product of the factor. Thus, in their view, price of the factor is not *determined* by the marginal revenue product but is, in equilibrium, *equal* to the marginal revenue product of the factor. We will discuss below the various determinants of the demand for a factor of production. Further, we have seen above how the demand for a factor of production depends upon its marginal revenue product. We have also derived the demand curve for a factor of production of an industry. The supply curve of a productive factor is given by the curve showing the amounts of factor offered by the owners of the factor at various factor prices and it slopes upward to the right. The supply curve of a factor for an industry depends upon the transfer earnings of the various units of the factor. *The price of a factor is determined by the intersection of these demand and supply curves of the factor*. In other words, given the demand and supply curves of a factor, the price of the factor will adjust to the level at which the amount of the factor supplied is equal to the amount demanded. This is shown in Fig. 32.12, where *DD* is the demand curve and *SS* is the supply curve of the factor. Only at price *OP*, quantity demanded is equal to the quantity supplied. The price *OP* is thus determined. The price of a factor cannot be determined at a level higher than or lower than price *OP*, *i.e.*, other than the price where amount demanded is equal to the amount supplied. For example, the price cannot be established at the level *OP*′, since at price *OP*′ the quantity offered to supply (*P*′ *H*) of the factor is greater than the quantity demanded (*P*′*S*) of it. As a result, the competition between the owners of the factor will force down the price to the level *OP* where the quantity supplied is equal to the quantity demanded. Likewise, the price of the factor cannot be determined at the level *OP*″, since at price *OP*″ the quantity demanded of the factor is greater than the quantity offered to supply of it. Consequently, the competition among the producers or entrepreneurs demanding the factor of production will push up the price to the level *OP*.

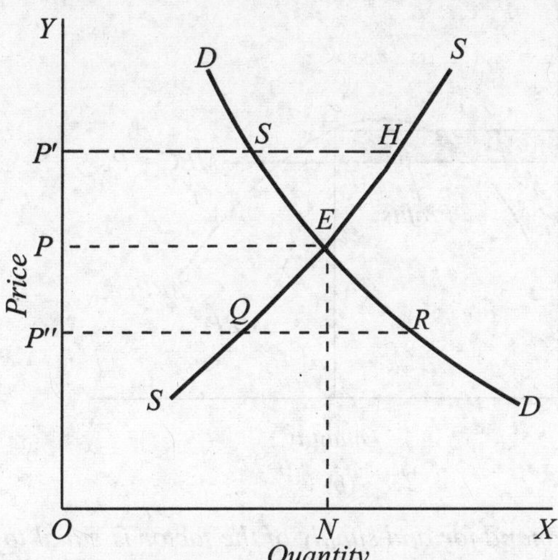

Fig. 32.12. *Equilibrium between demand for and supply of a factor determines its price.*

Though price of a factor is determined by demand for and supply of the factor, it is equal to the marginal revenue product of the factor. This is illustrated by Fig. 32.13. It will be seen from Fig. 32.13 (*a*) that equilibrium price *OP* of the factor is determined in the market and *ON* is equilibrium quantity demanded and supplied of the factor. An individual producer or firm who demands that factor will take the factor price *OP* as given. It will now be seen from Fig. 32.13 (*b*) which depicts the position of a single firm or entrepreneur that at price *OP* the firm will employ or use *OM* quantity of the factor. This is so because in order to maximise its profits, the firm will equalise the price of the factor with the *MRP* of the factor, and at *OM*, the price of the factor is equal to the marginal revenue product of the factor. If the firm employs fewer than *OM* units of the factor, then the *MRP* of the f actor will be greater than the price of the factor which will imply that there is still a scope for earning more profits by increasing the use of the factor. If, on the other hand, the firm employs more than *OM* units of the factor, *MRP* of the factor will be less than the price paid for it. As a result, the firm will incur losses on the marginal units and it will therefore be to the advantage of the firm to reduce the employment of the factor. Thus, the firm maximises its profits and is in equilibrium when it is employing *OM* amount of the factor at which *MRP* of the factor is equal to the price of the factor. *To sum up, price of a factor is determined by the demand for and supply of the factor and is equal to the marginal revenue product of the factor.*

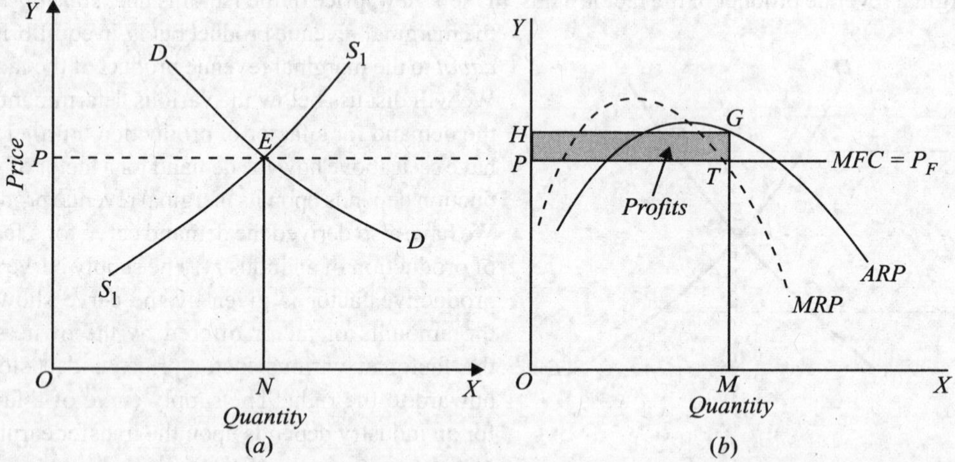

Fig. 32.13. *Factor price which is determined by demand for and supply of the factor is equal to marginal revenue product of the factor.*

As is evident from Fig. 32.13, at price *OP*, the firm is earning supernormal profits, since in equilibrium *ARP* of the factor is greater than the price of factor. This can happen in the short run, but not in the long run. If firms are earning supernormal profits, more entrepreneurs will enter the market in the long run to purchase that particular factor to produce the products made by that particular type of the factor. Entry of more entrepreneurs to the factor market will compete away the supernormal profits. As a result, the demand for factors will rise and the demand curve for the factor in Fig. 32.13 (*a*) will shift outward to the right. This shift in demand curve due to rise in demand for the factor is shown in Fig. 32.14. With this increase in demand, the price of the factor willl rise to *OP*′. It is evident from Fig. 32.14 that with factor price *OP* ′, the firm will be in equilibrium at *H* when it is employing *OM*′ amount of the factor. At *OM*′ amount of the factor the price of the factor is equal to *MRP* as well as *ARP* of the factor. Since at *OM*′ the price of the factor *OP*′ is equal to *ARP* of the factor, the firm is neither making supernormal profits, nor having losses. It is earning only normal profits.

If, in the short run, firms are having losses, some entrepreneurs will leave and stop purchasing the factor. As a result, the demand for the factor will decrease. The demand curve will shift downward and to the left so that the price of the factor will fall to a level at which price the firms earn only normal profits. Thus, in the long run, under perfect competition in the factor market, price of the factor is equal to both *MRP* and *ARP* of the factor.

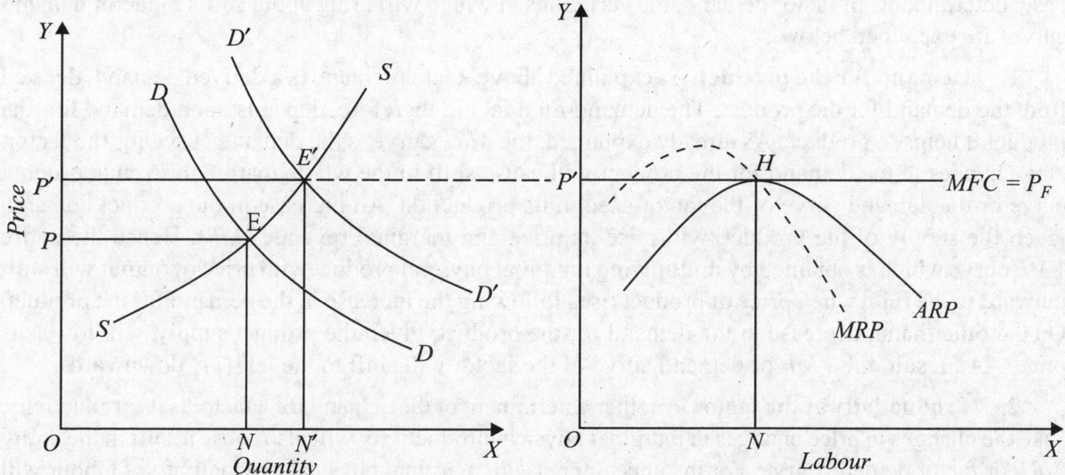

Fig. 32.14. *Factor Price Determination in the Long Run*

To sum up, in the long run, the equilibrium between demand for and supply of the factor is established at the level where the price of the factor is equal to both the MRP and ARP of the factor and thus the firms earn only normal profits.

We have seen above that when demand for a factor of production increases, given the supply curve of the factor, the factor price will rise. Now, what happens when the supply of a factor increases, given the demand curve of the factor. When the supply of a factor increases, the supply curve will shift to the right. This new supply curve will intersect the given demand curve at a lower price. Thus, with the increase in the supply of a factor, its price will tend to fall. On the other hand, when the supply of a factor decreases, the supply curve will shift to the left and, given the demand curve, the price of the factor will rise.

As regards the policy of factor owners, two results follow from our analysis. First, if the owners of a factor want to raise the price of the service of their factor, they should try to increase the demand for their factor service. The demand for a factor will rise if demand for and price of the product rise, or the price of the substitute factor rises, or there is increase in the productivity of the factor due to the improvement in technology. Second, if factor owners want to maintain the price of their factor service, *i.e.*, to prevent the price from falling, they should not allow their supply to increase.

The above modern theory of factor pricing under conditions of perfect competition is based upon Marshall-Hicks' version of marginal productivity theory. In this, marginal productivity of a factor is an important economic force which determines the price of the factor. Therefore, this is subject to many criticisms which have been levelled against the marginal productivity theory and which we have already discussed above in the present chapter.

DETERMINANTS OF THE DEMAND FOR FACTORS

We shall now explain those causes which bring about changes in the factor demand. In other words, we shall discuss those factors which cause shifts in the whole *MRP* curve or demand curve of the factor. These shifts or changes in factor demand are caused not by the change in the price of the factor itself but by other influences which work upon factor demand. These changes or shifts in *MRP* curve or demand curve for a factor are effected by the changes in the basic determinants of factor demand. Just as the demand curve of a product changes or shifts as a result of changes in income or changes in tastes or preference of the consumers, similarly the demand curve for the factor (*i.e., MRP* curve) will shift following the changes in the basic determinants of the factor demand. These basic determinants of factor demand, the variations in which will bring about shifts in factor demand curve, are explained below :

1. **Demand for the product**. As explained above, factor demand is a derived demand, derived from the demand for the product. The demand for a factor, therefore, depends upon demand for the product it helps to produce. As already explained, the *MRP* curve is the demand curve for the factor. Any changes in the demand for the product will cause shift in the whole marginal revenue product curve or the demand curve of the factor used in its production. An increase in the product demand, given the supply of the product, will raise its price and marginal revenue (*MR*). Hence the entire *VMP* curve which is obtained by multiplying marginal physical product with price of output will shift outward to the right when price of product rises following the increase in the demand for the product. On the other hand, decrease in the demand for the product, given the product supply, will lower its price. As a result, the *VMP* or demand curve of the factor will shift to the left (*i.e.* downward).

2. **Productivity of the factor**. Another determinant of the demand for a factor is its productivity. Like the changes in price changes in marginal physical productivity will also cause a shift in the entire *MRP* or factor demand curve. For instance, increase in marginal physical productivity of labour will shift the *MRP* or demand curve of labour to the right (*i.e.* outward). It may also be pointed out that historically we have experienced only *increase* in the physical productivity of the factors. *Decrease* in physical productivity, except under exceptional circumstances, has not been noticed. There are mainly three ways in which the marginal physical productivity of the factor can be increased. First, the *quality of the factor* may be improved. For instance, the labour productivity can be increased by making labour more educated and more skilful. Thus any improvement in the quality of the factor by enhancing its marginal productivity will cause outward shift in the factor demand curve.

Secondly, the marginal physical producty of any factor depends upon the *quantity of the fixed co-operating factors used* with it. For example, the position of *VMP* or *MPP* curve of labour is dependent on the *quantity of capital* used with it. The greater the quantity of fixed capital used with labour, the higher the level of marginal physical product curve of the factor. Thus increase in the marginal physical productivity of a factor as a result of increase in the quantity of capital (likewise, other fixed factors) will raise the *MRP* or factor demand curve outward to the right.

Thirdly, the marginal physical productivity is increased by the *advancement in technology*. The technological progress raises productivity by bringing about improvement in techniques of production. Hence as a result of advance in technology, *MRP* curve or the demand curve of the factor will shift to the right.

3. **Prices of other factors**. Just as the demand for a commodity depends upon the prices of other related commodities, demand for a factor also depends upon the prices of other related factors. But the effect of the changes in the price of related factors on the demand for a given factor would have different effects depending upon whether these related factors are substitutes or complements for the given factor. We first take the case of substitutes.

Suppose two factors, labour and machinery, are used in production of a commodity. Now, labour and machinery are largely substitutes. We want to consider the effect on the demand for labour as a result of change in the price of machinery. If the price of machinery falls so that machinery becomes relatively cheaper than labour, there would be large substitution of machinery for labour. Since machinery is now relatively cheaper than labour, it will pay the employer to use more machinery and less labour. Therefore, as a result of change in the price of machinery, demand for labour will fall and machinery would be used in place of labour. Thus the change in the price of machinery will have what is called *substitution effect* on the demand for labour. As a result of the substitution effect of the fall in the price of machinery, the demand curve for labour will shift to the left. The extent to which the demand for labour will fall depends upon the extent to which it is possible to substitute machinery for labour. Since machinery and labour are close substitutes, such substitution takes place on a large scale.

But the change in the price of factor has not only a substitution effect but also *expansion or output effect*. The prices of factors govern the cost and price of the product. When, in our above example, the price of machinery falls, the cost of production of the product will decline and hence there would be fall in the price of the product. With the fall in the price of the product, more of it will be demanded. In response to greater demand, more output of the product would be made. The fall in the price of the machinery will thus lead to greater output of the product. To produce more output, more of labour as well as more of other factors including machinery will be required. This expansion in output due to the fall in the price of machinery, that is, output effect, would tend to increase the demand for labour. Thus the fall in price of machinery has two opposite effects: first, substitution effect which tends to reduce the demand for labour: and second, output effect (expansion effect) which tends to increase the demand for labour.

The net effect on the demand for labour would depend upon the relative strengths of these two opposite effects. But in the present case, substitution effect would outweigh the output effect and, therefore, there would be a net decrease in the demand for labour as a consequence of fall in the price of machinery.

But the change in the price of a complementary factor would have a different effect on the demand for a given factor. Suppose two factors *A* and *B* are being used in the production of a commodity. Further suppose that factors *A* and *B* bear complementary relation to each other. We are interested to know what happens to the demand for factor *A*, when the price of factor *B* changes. If the price of factor *B* falls, its quantity demanded will rise. Since *B* is complementary to *A*, when *B* will be used more due to fall in its price, it will necessitate more employment of factor *A* also. Thus the demand for factor *A* will rise as a consequence of the fall in the price of its complementary factor *B*. But this is not all. *Output effect will also exercise its influence*. When the price of *B* falls, cost of production will decline. As a result, the price of the product will fall which will bring about an increase in the quantity demanded of the product. Consequently, output of the product would be expanded. This increase in output will require more of both factors *A* and *B*. Thus the demand for factor *A* will rise due to not only complementary effect but also output effect in the case of complementary factors which work in the same direction, thereby reinforcing each other.

EULER'S THEOREM AND PRODUCT EXHAUSTION PROBLEM

As soon as it was propounded that the factors of production are paid equal to their marginal products, a difficult problem cropped up over which there has been a serious debate among the famous economists. The difficult problem which has been posed is that if all factors were paid rewards equal to their marginal products, would the total product be just exactly exhausted ? In other words, if each factor is rewarded equal to its marginal product, the total product should be disposed of without any surplus or deficit. The problem of proving that the total production will be just exhausted if all factors are paid rewards equal to their marginal products has been called "*Adding-up Problem*" or *Product Exhaustion Problem.*

The two solutions to the problem of product exhaustion have been put forward. First, important solution was put forward by P.H. Wicksteed who assumed the operation of constant returns to scale in production (that is, the first degree homogeneous production function) and applied Euler theory to prove the product exhaustion problem. The second important solution has been provided by J.R. Hicks and P.A. Samuleson who used perfect competition model of determination of product and factor prices to prove the product exhaustion problem. We discuss below these solutions of product exhaustion problem.

Wicksteed's Solution of Product Exhaustion Problem with Euler's Theorem

Philip Wicksteed was one of the first economists who posed this problem and provided a solution for it. Wicksteed applied a mathematical proposition called Euler's Theorem[16] to prove that the total product will be just exhausted if all the factors are paid equal to their marginal products. Let Q stand for the total output of the product, a stands for the factor labour and b stands for the factor capital and c stands for land. Assuming that there are only three factors employed for production. Then, the adding up problem implies that,

$$Q = MP_a \times \mathrm{a} + MP_b \times b + MP_c \times c$$

That is, the marginal product of factor a multiplied by the amount of factor a plus the marginal product of factor b multiplied by the amount of factor b plus the marginal product of factor c multiplied by the amount of factor c equals the total product of the firm. Marginal products of various factors can be expressed as partial derivatives. Thus, the marginal product of labour (*i.e.* factor a) can be expressed as $\frac{\partial Q}{\partial a}$, and the marginal product of capital (factor b) as $\frac{\partial Q}{\partial b}$, and the marginal product of land (factor c) as $\frac{\partial Q}{\partial c}$, then for the adding-up problem (*i.e.* product exhaustion problem) to be fulfilled, the following equation must hold good:

$$Q = a\frac{\partial Q}{\partial a} + b\frac{\partial Q}{\partial b} + c\frac{\partial Q}{\partial c},$$

where $a\frac{\partial Q}{\partial a}$ represents share of the total product going to labour.

$a\frac{\partial Q}{\partial b}$ represents share of the total product going to capital

$a\frac{\partial Q}{\partial c}$ represents share of the total product going to land.

Now, Euler's Theorem states that if production function is a homogeneous function of the first degree, that is, if in $Q = f(a, b, c)$ for any increase in the variables a, b and c by the amount n, the output Q also increases by n, then Q will be equal to the total sum of the partial derivatives of production function with respect to various factors multiplied by the amounts of the factors respectively. The homogeneous function of the first degree or linear homogeneous function is written in the following form :

$$nQ = f(na, nb, nc)$$

Now, according to Euler's theorem, for this linear homogeneous function:

$$Q = a\frac{\partial Q}{\partial a} + b\frac{\partial Q}{\partial b} + c\frac{\partial Q}{\partial c}$$

Thus, if production function is homogeneous of the first degree, then according to Euler's theorem the total product is :

$$Q = a\frac{\partial Q}{\partial a} + b\frac{\partial Q}{\partial b} + c\frac{\partial Q}{\partial c}.$$

16. This is after the name of its author 'Leonhard Euler' (1707-1783), a Swis mathematician.

where Q represents the total product and $\frac{\partial Q}{\partial a}$, $\frac{\partial Q}{\partial b}$, $\frac{\partial Q}{\partial c}$ are partial derivatives of the production function and therefore represent the marginal products of labour, capital, and land respectively. It follows therefore that if production function is homogeneous of the first degree (that is, where there are constant returns to scale), then, according to Euler's Theorem, if the various factors a, b and c are paid rewards equal to their marginal products, the total product will be just exhausted, with no surplus or deficit.

We thus see that Euler's Theorem is able to explain product exhaustion when production function is homogeneous of the first degree. In this way, Wicksteed assuming constant returns to scale and applying Euler's Theorem, proved the adding-up problem, that is, demonstrated that if all factors are paid equal to their marginal products, the total product will be just exactly exhausted.

A Critique of Euler's Theorem and Wicksteed's Solution

Wicksteed's solution was critcized by Walras, Barone, Edgworth and Pareto. It was asserted by these writers that production function was not homogeneous of the first degree, that is; returns to scale are not constant in the actual world. Thus Edgeworth satirically commented on Wicksteed's solution, "There is a magnificence in this generalisation which recalls the youth of philosophy. Justice is a perfect cube, said the ancient sage ; and rational conduct is a homogeneous function, adds the modern savant". Critics pointed out that production function is such that it yields a U-shaped long-run average cost curve. The U-shape of the long-run average cost curve implies that up to a point increasing returns to scale occur and after it diminishing returns to scale are obtained. In case a firm is still working under increasing returns to scale, then if all factors are paid equal to their marginal products, the total factor rewards would exceed the total product. On the other hand, if a firm is working under diminishing returns to scale, and if all factors are paid equal to their marginal products the total factor rewards would not fully exhaust the total product and will therefore leave a surplus. It follows that Euler's Theorem does not apply and therefore the adding-up problem does not hold good when either there is increasing returns to scale or decreasing returns to scale.

Another drawback pointed out in Wicksteed's solution is that when there is constant returns to scale, the long-run average cost curve of the firm is a horizontal straight line which is incompatible with perfect competition. (Under horizontal long-run average cost curve, the firm cannot have a determinate equilibrium position). But perfect competition was essential to the marginal productivity theory and therefore to Wicksteed's solution. Thus Wicksteed solution leads us to two contradictory things.

Wicksell, Walras and Barone's Solution of Production Exhaustion Problem

After Wicksteed, Wicksell, Walras and Barone, each independently, advanced more satisfactory solution to the problem that marginally determined factor rewards would just exhaust the total product. These authors assumed that the typical production function was not homogeneous of the first degree, but was such that yielded U-shaped long-run average cost curve. They pointed out that in the long run under perfect competition, the firm was in equilibrium at the minimum point of the long-run average cost curve. At *the minimum point of the long-run average cost curve, the returns to scale are momentarily constant*, that is, returns to scale are constant within the range of small variations of output. Thus the condition required for the marginally determined rewards to exhaust the total product, that is, the *operation of constant returns to scale, was fulfilled at the minimum* point of the long-run average cost curve, where a perfectly competitive firm is in long-run equilibrium. Thus in the case of perfectly long-run equilibrium, Euler Theorem can be applied and if the factors are paid rewards equal to their marginal products, the total product would be just exactly exhausted.

Hicks-Samuelson's Solution to the Product Exhaustion Problem

After Wicksell, Walras and Barone, J.R. Hicks[17] and P.A. Samuelson[18] provided more satisfactory solution to the problem of product exhaustion. The basic point to note in their solution is that it is the market conditions of perfect competition with its important feature of zero economic profits in the long run and not the first degree-homogeneous production function that ensures that if factors are paid rewards equal to their marginal products, total value product would be just exhausted.[19]

In a perfectly competitive market structure, firms make neither economic profits, nor make losses. Thus the solution of product exhaustion problem in case of the firms working in competitive factor markets where factors are paid equal to their marginal products, the existence of perfect competition in the product markets will ensure zero economic profits in the long run. Consider Fig. 32.15 where a perfectly competitive firm is in long-run equilibrium at the minimum point of the long-run average cost curve (*LAC*) producing level of output *OQ* at price *OP*.

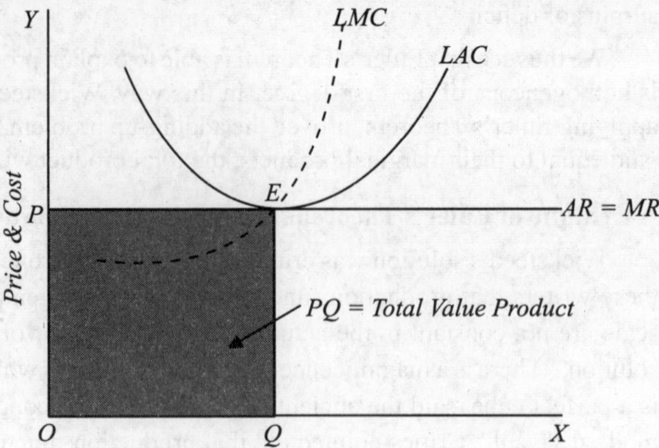

Fig. 32.15. *Hicks–Samuelson Solution to Product Exhausion Problem.*

The total value product produced by the firm in this long-run equilibrium is equal to the area *OPEQ*. Since price *OP* is equal to average cost (*AC*) at this long-run equilibrium output with zero pure profits, total value product (*P.Q.*) will be equal to the total cost (*TC*). Thus

In long-run competitive equilibrium

$$\text{Total Value Product } (P.Q.) = w.L + K.r. \quad ...(1)$$

Now, marginal productivity theory of distribution requires that

$$w = VMP_L = P.\ MPP_L \quad ...\ (2)$$

$$r = VMP_K = P.\ MPP_K \quad ...\ (3)$$

where w and r are prices of labour and capital respectively and MPP_L and MPP_K are marginal physical products of labour and capital respectively and P is the price of the product.

Substituting the values of w and r into equation (1) we have

$$P.Q = L.\ (P.\ MPP_L) + K.\ (P.\ MPP_K)$$

Dividing both sides by P we have

$$Q = L.\ MPP_L + K.\ MPP_K$$

That is, if labour and capital are paid equal to their marginal physical products, total output will be just exhausted.

It is important to note that in contrast to the solutions of Wicksteed and of Wicksell, Walras and Barone, the solution furnished by Hicks and Samuelson proves the product exhaustion theorem

17. John R. Hicks, *The Theory of Wages.*
18. Paul A. Samuelson, *The Foundations of Economics.*
19. According to Hicks, in long-run competitive equilibrium with zero profits, entrepreneur has no function of decision making to perform and hence his economic profits are zero.

without assuming constant returns to scale (*i.e.*, first-degree homogeneous production function) and without using Euler theorem. They prove it by just assuming conditions of perfect market structure.

The merit of the Hicks-Samuleson solution is that it highlights when conditions of perfect competitive market do not hold, that is, when there is either monopoly or imperfect competition in the product market or monopsony or imperfect competition in the factor market, the hired factors do not get rewards equal to the value of their marginal products and are therefore exploited by the entrepreneurs who may enjoy large economic profits.

PRICING OF FACTORS IN IMPERFECTLY COMPETITIVE MARKETS

We have explained above how the price of a factor of production is determined when there prevails perfect competition both in the product and factor markets. Before the theories of imperfect competition and monopolistic competition were introduced in economic theory no distinction was made between value of marginal product (*VMP*) and marginal revenue product (*MRP*). We have seen above that when there is imperfect competition (*i.e.* monopoly, oligopoly or monopolistic competition) in the product market, marginal revenue differs from the price of the product. As a result, under conditions of imperfect competition in the product market, *marginal revenue product* (*MRP*) of the factor differs from *value of the marginal product (VMP)*. This affects the demand for a factor and the price it will get under conditions of imperfect competition Determination of Factor Price when there exists Monopoly (or Imperfect Competition) in the Product Market but Perfect Competition in the Factor Market or monopoly in the product.

Determination of Factor Price when there is Imperfect Competition (or Monopoly) in the Product Market and Perfect Competition in Factor Market

In what follows we will explain the determination of prices and employment of factors under imperfect competition in the product and factor markets in general. In the next chapter, we will explain in detail the determination of wages and employment of labour under conditions of imperfect competition and monophony in the labour market.

We will explain below the employment of a factor by a firm and the price it will pay to a factor when the firm is working under conditions of imperfect competition or monoply in the product market. *However we assume in this section that as far as factor market is concerned perfect competition prevails in it*. Since perfect competition is assumed to be prevailing in the factor market, price of the factor will be determined by demand for and supply of the factor of production, as explained above. But now the demand for the factor of production is determined not by the value of the marginal product (*VMP*) but by the marginal revenue product (*MRP*) of the factor. As we will see below, in this case price of the factor, which is determined by demand for and supply of the factor, will be equal to the marginal revenue product, but will be less than the value of the marginal product (*VMP*) of the factor.

The conditions of firm's equilibrium in factor market developed above will also apply in the present case. The firm working under *perfect competition in factor market but monopoly or imperfect competition in the product market* would also be in equilibrium position where *MRP* = *MFC*, and *MRP* curve cuts *MFC* curve from above. But there are some differences between this case and the case explained above.

Since in this case, as in the previous, the firm is working under perfect competition in the factor market it will not be able to affect the price of the factor and factor-cost line will be a horizontal straight line. Therefore, the firm will be in equilibrium, that is, will be maximising profits when *MRP* = *MFC* = Price of the factor. But because the firm in the present case is working under conditions of monopoly or imperfect competition in the product market, it will be able to exercise some influence or control over the price of the product. *AR* curve for it will slope downward and *MR* curve will be below it. Consequently, *MRP* which is equal to *MPP* × *MR* will not be equal to *VMP* which is equal to *MPP* × price of the product. Since *MR* is less than the price of the product under monopoly or imperfect competition, *MRP* would be less than *VMP*. In symbolic terms:

$$MRP = MPP \times MR$$
$$VMP = MPP \times \text{Price of the product}$$

Since, under imperfect competition or monopoly in the product market, *MR* < Price of the product, therefore

$$MRP < VMP$$

In equilibrium in the factor market, the firm will make

$$P_F = MRP$$

Therefore, $$P_F = MRP < VMP$$

It is, therefore, concluded that *under conditions of monopoly or imperfect competition in the product market, assuming perfect competition in the factor market, the factor will get price less than the value of its marginal product.*

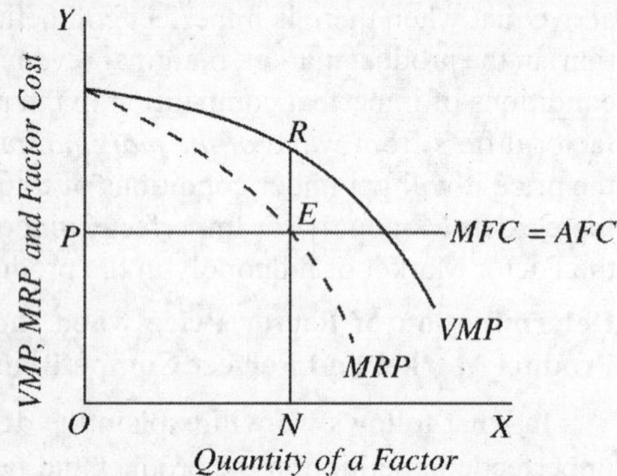

Fig. 32.16. *Employment of a Factor by a Firm under Conditions of Perfect Competition in the Factor Market and Monopoly or Imperfect Competition in the Product Market*

The equilibrium of the firm when it is working under conditions of perfect competition in the factor market and monopoly or imperfect competition in the product market is shown in Fig. 32.16. Since *VMP* is greater than *MRP* when there is imperfect competition in the product market, *VMP* curve will be above *MRP* curve (for the sake of convenience, we have drawn only the downward-sloping portions of *MRP* and *VMP* curves). The firm will be in equilibrium at *E*, where *MRP* = *P*. The equilibrium employment of the factor is *ON*. It will be noticed from the figure that the price of the factor *OP* is, in equilibrium, equal to marginal revenue product *EN* but is less than its value of marginal product which is equal to *RN*. Therefore, factor gets *RE* less than the value of its marginal product.

Meaning of Factor Exploitation

It follows from above that *price of a factor will be less than the value of the marginal product of the factor under conditions of monopoly and imperfect competition in the product market.* According to Joan Robinson[20] *a factor is exploited when it is paid less than the value of its marginal product (VMP).* Therefore, according to Joan Robinson, when imperfect competition prevails in the product market, labour and other factors, (*i.e.*, factors other than the entrepreneur) are exploited by the entrepreneur. But many economists, especially E.H. Chamberlin, do not agree with Robinson's definition of exploitation of labour. According to Chamberlin, a factor is exploited only when it is paid less than the marginal revenue product (*MRP*). As explained above, when there prevails imperfect or monopolistic competition (including monopoly and oligopoly) with perfect competition in the factor market, price of a factor is equal to the marginal revenue product, though it is less than the value of the marginal product. Therefore, according to Chamberlin, there is no any exploitation of labour or any other factor by the entrepreneur when imperfect competition exists in the product market if there is perfect competition in the factor market. We shall discuss this question of exploitation of labour in detail in the next chapter.

20. See Joan Robinson, *Economics of Imperfect Competition,* p. 283.

DETERMINATION OF FACTOR PRICES UNDER MONOPSONY

What is monopsony ? Monopsonist means a single buyer. There can be a single buyer in the product market as well as in the factor market. *Where there is a single buyer of a product, he is monopsonist in the product market*. However, we are concerned here with monopsonist in the factor market. *Monopsony in the factor market is said to exist when there is a single buyer of a specific factor of production*. For instance, when in a particular area, there is only one employer of specific type of labour, he is monopsonist of that labour. Monopsony in the labour market also comes into existence when various employers of labour in an area form a collusion so far as recruitment of labour is concerned. While monopsony is very rare in product markets, it is more often found in factor markets.

While a monopolist in the product market faces a downward sloping demand curve and can influence the price of the product by varying the level of his output, *a monopsonist in the factor market faces an upward sloping supply curve of the factor and can affect the price of the factor by varying the level of its employment*. By restricting employment it can lower the price of the factor. And if he wants to buy more amount of the factor, he will have to raise its price. Therefore, *the supply curve of the factor or average cost (AFC) curve of the monopsonist will be rising upward to the right*. As the *AFC* curve will be rising, *MFC* curve will be above it. *MRP* curve, as usual, will have an inverted U-shape. As we are assuming that perfect competition prevails in the product market, the value of marginal product (*VMP*) and marginal revenue product (*MRP*) would be equal. The *AFC* and *MFC* curves and *MRP* or *VMP* curves under these conditions are shown in Fig. 32.17.

Firm's Equilibrium when there is Monopsony in Factor Market and Perfect Competition in Product Market.

The general condition for the firm's equilibrium also applies to this case. The firm will be in equilibrium when *MRP* is equal to *MFC* and *MRP* curve cuts *MFC* curve from above. But here there is an important difference. Under monopsony in the factor market *MFC* and *AFC* (or P_F) will not be equal, the former being greater than the latter. *MFC* curve lies above the *AFC* curve. Therefore, in equilibrium when *MRP* is equal to *MFC*, it is greater than the *AFC*, that is, price of the factor.

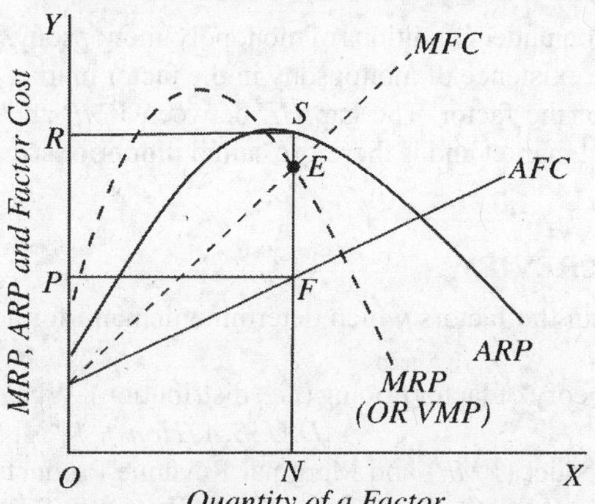

Fig. 32.17. *Determination of Price and Employment of a Factor under Monopsony when there is Perfect Competition in Product Market*

The equilibrium of the firm under monopsony in a factor market assuming perfect competition in the product market is presented in Fig. 32.17. The firm is in equilibrium at point *E* (or at *ON* level of employment) where *MRP* = *MFC*. But the *AFC* at *ON* level of employment is *NF*. In other words, the factor is being paid *NF* (or *OP*) price, while its marginal revenue product is *NE*. The difference *EF* between the marginal revenue product (*NE*) and the price paid to the factor of production (*NF*) is called **monopsonistic exploitation of factor** *because this has arisen due to the monopsony in the factor market*. *EF* is monopsonistic exploitation per unit of the factor. *The degree of monopsony power can be measured as the difference between MRP and price of the factor (AFC) at the equilibrium employment*. This is similar to measuring monopoly power as the difference between the price and marginal cost of a product at the equilibrium output in the product market. It should be noted that in Fig. 32.17, the firm is making profits equal to *PFSR* which is the difference between the total revenue product and the total factor cost.

When the firm is working under perfect competition in the factor market, $MFC = P_F$ so that in equilibrium $MRP = MFC = P_F$. But if the firm is a monopsonist, $MFC > P_F$ so that in equilibrium.

$$MRP = MFC > \text{Price of the factor } (P_F)$$

Thus a factor under conditions of monopsony will get less than its *MRP*. As we are assuming perfect competition in the product market, *VMP* will be equal to *MRP*. Therefore, we can write complete condition for firm's equilibrium in the present case.

$$VMP = MRP = MFC > \text{Price of the factor}$$

Firm's Equilbrium when there is Monopsony in Factor Market and Monopoly or Imperfect Competition in the Product Market.

The same general equilibrium conditions will also apply when there is monopsony in the factor market and monopoly in the product market. Under these conditions, as in others the firm will be in equilibrium where $MRP = MFC$. But since there is monopoly in the product market, *MRP* and *VMP* will not be equal; the latter will be greater than the former. *MFC* curve will be above *AFC* curve, because the firm has monopsony in the factor market. Equilibrium of a monopolist-monopsonist firm is shown in Fig. 32.18. The firm's profits will be maximized at *E* (or at *ON* level of employment) where $MRP = MFC$. But the price which is paid to the factor of production by the firm is *FN* which is less than both *MRP* and *VMP*. Thus the conditions of firm's equilibrium can be stated as:

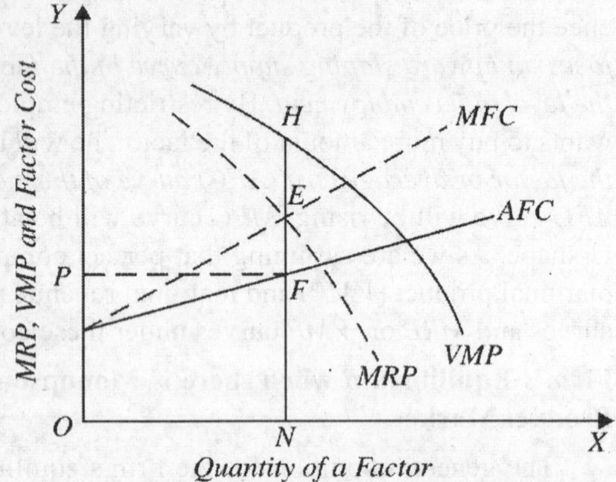

Fig. 32.18. *Determination of Price and Employment of a Factor under Monopsony.*

$$VMP > MRP = MFC > P_F$$

The factor is, therefore, subject to double exploitation under conditions of monopoly-monopsony. The gap *EF* between *MRP* and *AFC* or P_F is due to the existence of monopsony in the factor market and is therefore called **monopsonistic exploitation** of the factor. The gap *HE* between *VMP* and *MRP* is due to the existence of monopoly in the product market and is therefore called **monopolistic exploitation** of the factor.

QUESTIONS FOR REVIEW

1. What is meant by derived demand ? Explain the factors which determine demand for a factor or input.
2. Briefly explain the Marginal Productivity Theory of factor pricing (*i.e.*, distribution). What are main shortcomings? [*D.U. B.A (Hons), 1984*]
3. Distinguish between Value of Marginal Product (*VMP*) and Marginal Revenue Product (*MRP*) of an input. How are they related to each other when there prevails (*a*) perfect competition in the product market, (*b*) monopolistic competition or monopoly in the product market?
4. How is reward of a factor (say labour) determined when both the factor market and product market are perfectly competitive? Explain. [*D.U. B.Com (Hons) 2002*]
5. Explain the determination of optimal use of a variable input (*e.g.*, labour) by a firm when there prevails perfect competition in the input market.
6. Under conditions of perfect competition in the labour market value of marginal product (*VMP*) curve of labour is firm's demand curve for it. Explain.

7. What factors determine the demand for a factor input ? Explain.

8. The economy-wide demand curve for a factor will be more elastic, the greater the elasticity of the consumer demand for goods in the production of which the factor is employed. Explain

9. A theory of factor pricing is only a special case of general theory of pricing. Discuss.

10. A perfectly competitive firm can hire labour at ₹ 50 per day. The firm's production function is given below:

Number of days/labour	*Number of units of output*
0	0
1	16
2	30
3	42
4	52
5	60

If the output is sold at price ₹ 5 per unit, how many labour days should the firm hire? [**Hints:** From the production data, we have to find out marginal physical product of labour days (MPP_P). We then multiply marginal physical products by price of output (= ₹ 5) to obtain VMP_L at different levels of labour use. The firm will hire the number of labour days at which $VMP_L = 50$.]

11. Wages under perfect competition in labour market are equal to the marginal revenue product; but under imperfect competition (*e.g.* monopsony) in labour market wages are not equal to the marginal revenue product of labour. Explain with the help of diagrams.

12. Derive graphically the competitive firm's demand curve for a variable factor when several variable factors are used. [*D.U. B.A. (Hons) 1998*]

13. How is market demand for a variable factor of production derived from the demand of individual firms?

14. A monopolist's demand curve for labour will be downward sloping even if marginal physical product of labour is constant. Explain. [*D.U. B.A. (Hons) 1997*]

15. What is meant by 'Product Exhaustion Problem' (or 'Adding up Problem')? Is total product fully exhausted according to marginal productivity theory of distribution?

16. What is Euler theorem ? How is it used to solve the product exhaustion problem in the theory of distribution?

17. If each factor is rewarded equal to its marginal product, the product will be just exhausted. Critically examine.

18. What is meant by factor exploitation? Distinguish between monopsonistic exploitation and monopolistic exploitation. Under what conditions do they prevail?

CHAPTER 33
WAGE DETERMINATION IN COMPETITIVE AND IMPERFECTLY COMPETITIVE MARKETS

Introduction

In the previous chapter we have explained the determination of prices of factors in general. In the present chapter we shall explain in the detail how wages are determined under conditions of perfectly and imperfectly competitive markets. It is here worth noting some *special characteristics of labour* First, unlike land and capital, labour (that is, workers) are usually found to be combining together to form trade unions for demanding higher wages and better working conditions from the entrepreneurs. Second, labour cannot be separated from labourer himself, whereas land and capital are distinct from their owners (*i.e.*, landlords and capitalists). Thirdly, workers are, within certain limits, free to choose whether or not they will work on a particular day as they can take leave or abstain from work. Fourthly, unlike other factors, workers can decide whether or not to increase their supply by working a large number of hours per week.

Nominal and Real Wages

While nominal wage is wage in terms of money, the real wage is measured in terms of goods and services it will purchase. When prices in the economy rise, the same money wage will buy less goods and services. Therefore, the rise in prices, money wage remaining constant, will lead to the fall in real wages. Thus, *real wages are obtained after adjusting for inflation, that is, for rise in prices in the economy.* Thus real wage is equal to money wage divided by the price level, that is, $W_r = \frac{W}{P}$ stands for real wage, W for money wage rate and P for the general price index number.

The term real wage is also sometimes used to mean the *money wages plus the non-monetary benefits* which workers receive in some jobs. Thus college professors with equal qualifications and skill may prefer lower money wage rates as compared to the persons employed as business executives in private firms because the teaching in a university or college offers some *non-monetary benefits* such as flexible working hours, pleasant surroundings, opportunities for advancement etc. Thus real wages of a college professor will be his money salary plus the above mentioned non-monetary benefits which he gets in the teaching profession.

Like the prices of other factors of production, wage rate depends on demand for and supply of labour. For explanation of wage determination it is therefore essential to understand first the nature of demand for labour. Demand for labour like other factor prices differs in certain respects from the demand for consumer goods. Consumer goods are demanded because they satisfy the wants of the people directly. People demand food to satisfy the pangs of their hunger, they demand clothes to satisfy their want of providing a cover to their bodies and so forth. These products possess utility which directly satisfies the wants of the people who are therefore willing to pay price for these products.

Demand for labour is a derived demand. But, unlike the goods, the labour does not satisfy the wants of the people directly. The labour is demanded not because it directly satisfies the wants of the people who wish to buy them. Instead, labour is demanded because it can be used to produce consumer goods which then satisfy human wants. Therefore, demand for labour, like that of other factors of production, is called derived demand. It is derived from the demand for the products it helps to produce. Thus, the *demand for labour ultimately depends upon the demand for goods it helps to produce.* The greater the demand for goods which a particular type of labour helps to make, the greater the demand for that type of labour. Just as demand for goods depend upon their utility, the *demand for labour depends upon marginal revenue product of labour.* In what follows we first explain supply of labour.

SUPPLY CURVE OF LABOUR

It is important to know how many hours a worker will be willing to work at different wage rates. When the real wage rate increases, the individual will be pulled in two opposite directions. As mentioned above, the real wage rate is the relative price of leisure which has to be given up for doing work to earn income. As real wage rate rises, leisure becomes relatively more expensive (in terms of income foregone) and this induces the individual to substitute work (or income) for leisure. This is called *substitution effect* of the rise in real wage and induces the individual to work more hours (i.e. supply more labour) to earn more income.

But the increase in the real wage rate also makes the individual richer, that is, his income increases. This increase in income tends to make the individual to consume more of all commodities including leisure. This is called *income* effect of the rise in wage rate which tends to increase leisure and reduce number of work hours (*i.e.* reduce labour supply. The economists generally believe that substitution effect of a rise in real wage is larger than its income effect and therefore individuals work for more hours (that is, supply more labour) at a higher wage rate. However, beyond a certain higher real wage and number of hours worked, leisure becomes more desirable and income effect outweights substitution effect, and as a result supply of labour decreases beyond a certain higher wage rate. In what follows we shall explain how we derive a supply curve of labour of an individual and of the economy as a whole in all these circumstances. Thus, whether an individual will supply more work-effort or less as a result of the rise in the wage rate depends upon the relative strengths of the income and substitution effects.

The changes in the work-effort or labour supplied by an individual worker due to the changes in the wage rate is illustrated in Fig. 33.1(*a*) To begin with, the wage line is AW_1; the slope of the wage line indicates the wage rate per hour. With wage line AW_1, the individual is in equilibrium at point Q on indifference curve I_1 and is working AL_1 hours in a week. Suppose the wage rate rises so that the new wage line is AW_2 With wage line AW_2, the individual is in equilibrium at point R on the indifference curve I_2 and is now working AL_2 hours which are more than before.

If the wage rate further rises so that the new wage line is AW_3, the individual moves to the point S on indifference curve I_3 and works AL_3 hours which are more than AL_1 or AL_2. Suppose the wage rate further rises so that the wage line is AW_4. With wage line AW_4, the individual is in equilibrium at point T and works AL_4 hours. If points Q, R, S and T are connected, we get what is called *wage offer curve which shows the number of hours that an individual offers to work at various wage rates*. It should be noted that the wage offer curve, strictly speaking, is not the supply curve of labour though it provides the same information as the supply curve of labour. The supply curve of labour is obtained when the wage rate is directly represented on the Y-axis and labour (*i.e.* work effort) supplied at various wage rates on the X-axis reading from left to right. In Fig. 33.2 the supply curve of labour has been drawn from the information gained from Fig. 33.1. Let the wage line AW_1 represent the wage rate equal to P_1, wage line AW_2 represents wage rate P_2, wage line AW_3 represents wage rate P_3, and wage line AW_4 represent wage rate P_4. It will be seen that as the wage rate rises from P_1 to P_4

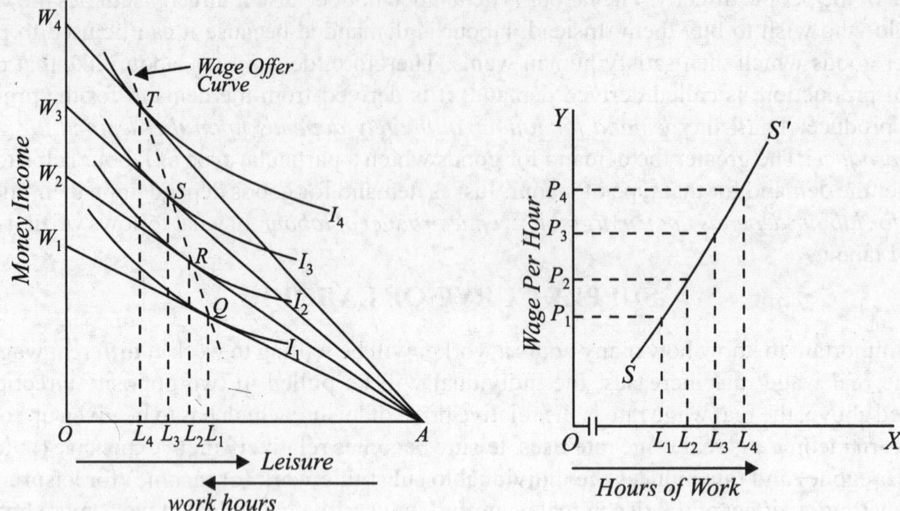

Fig. 33.1. Wage Offer Curve **Fig. 33.2.** *Upward Sloping Supply Curve of Labour*

and as a result the wage line shifts from AW_1 to AW_4 the number of hours worked, that is, the amount of labour supplied increases from AL_1 to AL_4. As a result, the supply curve of labour in Fig. 33.2 is upward sloping. *The indifference map depicted in Fig. 33.1 is such that the substitution effect of the rise in the wage rate is stronger than the income effect of the rise in the wage rate so that the work-effort supplied increases as the wage rate rises.*

Backward-sloping Supply Curve of Labour

But the supply curve of labour is not always upward sloping. When an individual prefers leisure to income, then the supply of labour (number of hours worked) by an individual will decrease as the wage rate rises. This is because in such a case income effect which tends to reduce the work effort outweighs the substitution effect which tends to increase the work effort. In Fig. 33.3 such an indifference map is shown which yields a backward sloping supply curve of labour which indicates that the number of hours worked per week decreases as the wage rate rises. AW_1, AW_2, AW_3 and AW_4 are the wage lines when the wages rates are P_1, P_2, P_3 and P_4 respectively. *Q, R, S* and *T* are the equilibrium points with the wage lines AW_1, AW_2, AW_3 and AW_4 respectively. It will be noticed from Fig. 33.3(*a*) that when the wage rate rises and as a consequence the wage line shifts from AW_1 to AW_4. the number of hours worked per week decreases from AL_1 to AL_4. In Fig. 33.3(*b*) supply curve of labour is drawn with *Y*-axis representing the hourly wage rate and *X*-axis representing number of hours worked per week at various wage rates. It will be seen from Fig. 33.3(*b*) as the wage rate rises from P_1 to P_4 the supply of labour (*i.e.*, number of hours worked per week) decreases from OL_1 to OL_4. In other words, the *supply curve of labour slopes backward,* that is, *slopes upward from right to left.* It should be noted that it is the nature or pattern of indifference curves between income and leisure that yields backward sloping supply curve. A glance at Fig. 33.3(*a*) and Fig. 33.3(*a*) will reveal that the nature of indifference curves in the two is different. As said above, the nature of indifference curves depend upon the relative preference between income and leisure. In Fig. 33.3(*a*) *indifference curves between income and leisure are such that the individual's preference for leisure*

is relatively greater than for income. In this case, when the wage rate rises the individual enjoys more leisure and accordingly reduces the number of hours worked per week.

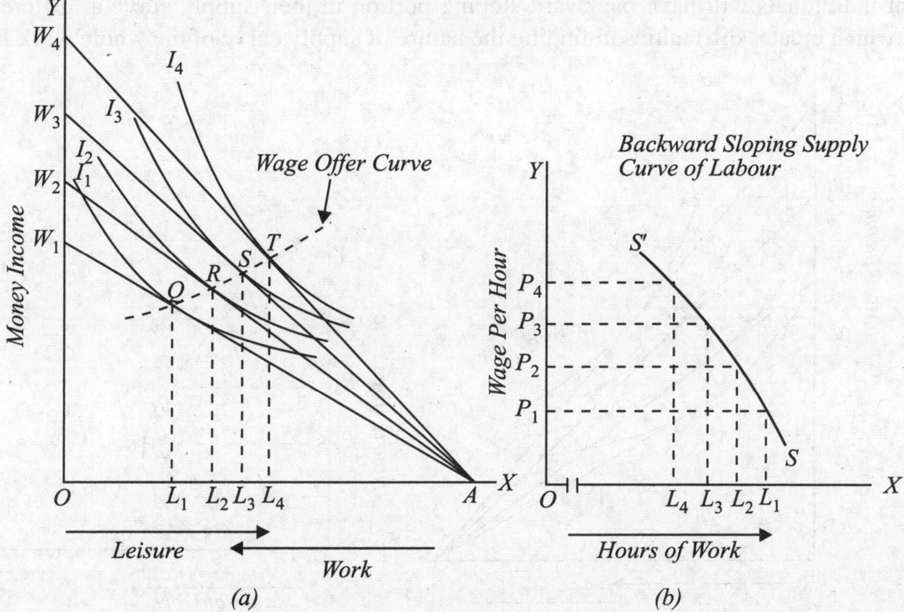

Fig. 33.3. *Backward Bending Supply Curve of Labour*

But it sometimes happens that as the hourly wage rate rises from a very low level to a reasonably good level, the number of hours worked per week rises and as the hourly wage rate rises further, the number of hours worked per week decreases. This may be the case of an individual who has some more or less fixed minimum wants for goods and services which he can satisfy with a certain money income. When the wage rate is so low that he is not earning sufficient money income, then to satisfy his more or less fixed minimum wants for goods and services, his preference for income will be relatively greater than that for leisure and, therefore, when the wage rate rises the individual will work more hours per week. When the wage rate has risen to a level which is sufficient to yield a sufficient money income for satisfying his fixed minimum wants, then for further increases in wage rate the number of hours worked per week will decrease because now the individual can afford to have more leisure and also earn an income sufficient to meet his minimum wants for goods and services.

It follows from above that up to a certain wage rate the supply curve will slope upward from left to right and then for further increases in the wage rate the supply curve of labour will slope backward.

In Fig. 33.4(*a*) an indifference map along with a set of wage lines AW_1, AW_2, AW_3, AW_4 (showing wage rates P_1, P_2, P_3, P_4 respectively) are shown. As the wage rate rises to P_2 and hence the wage line shifts to AW_2 the number of hours worked by the individual per week increases but when the wage rate further rises to P_3 and P_4 and hence the wage line shifts to AW_3 and AW_4, the number of hours worked by the individual decreases. From Fig. 33.4(*b*) it will be explicitly seen that the supply curve of labour slopes upward to the wage rate P_2 (that is, point *K*) and beyond that it slopes backward.

Supply Curve of Labour for the Economy as a Whole

The supply curve of labour of a group of individuals or of the whole working force in the economy can be derived by summing up horizontally the supply curves of individuals. It may be noted that the supply curve of labour for the economy as a whole will be upward sloping or backward sloping depending upon whether the relative number of individuals having upward sloping supply curves is greater or less than those having backward sloping supply curves of labour. Further, different individuals will have backward sloping portion in their supply curve at different wage ranges, which creates difficulties in finding the nature of supply curve of the whole work force.

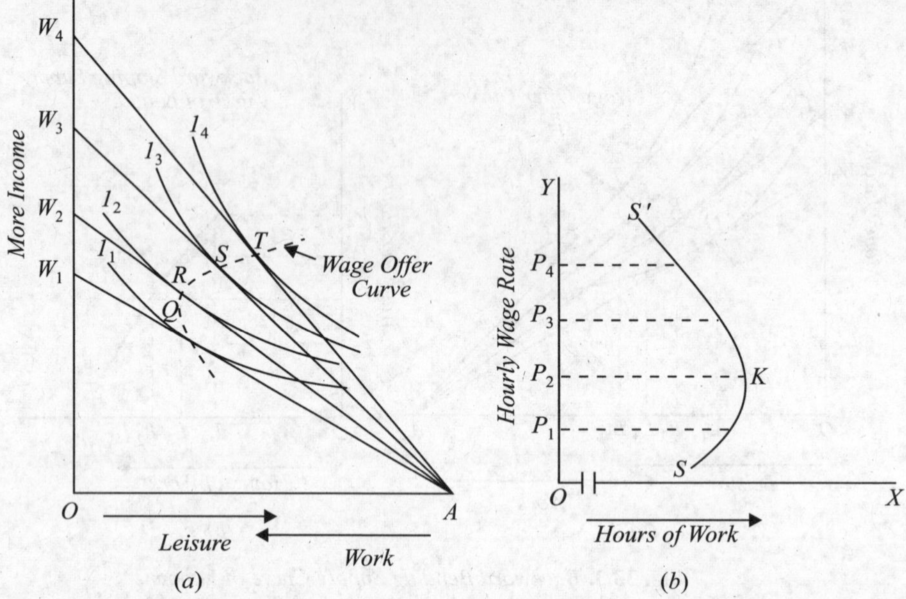

Fig. 33.4. *Work effort increases with the rise in the wage rate when the substitution effect is greater than the income effect.*

It is generally found that when the wage rate rises from the initially low level to a sufficiently good level, the total supply of labour to the economy as a whole increases (that is, supply curve for the economy as a whole slopes upward to a certain wage rate) and for further increases in the wage rate, the total supply of labour to the economy as a whole decreases (that is, beyond a certain wage rate the total supply curve of labour slopes backward). Thus, the total supply curve of labour for the economy as a whole is generally believed to be the shape depicted in Fig. 33.3(*b*).

WAGE DETERMINATION UNDER PERFECT COMPETITION IN THE LABOUR MARKET

We have explained in the Chapter 32 how the price of factor is determined under conditions of perfect competition in the factor market. The analysis of wage determination under conditions of perfect competition is exactly the same as given there. In the case of wage determination, it should be remembered that average factor cost (*AFC*) becomes average wage (*AW*) and marginal factor cost becomes marginal wage (*MW*). As explained there, when there prevails perfect competition in the labour market, wage rate is determined by the equilibrium between the demand for and supply of labour. Demand for labour is governed by marginal revenue product of labour (*MRP*). Wage rate determined by demand for and supply of labour is equal to the marginal revenue product of labour. Thus, under perfect competition in labour market, a firm will employ the amount of labour at which wage rate = *MRP* of labour.

As regards the supply of labour, it may be pointed out that supply of labour to the whole economy depends upon the size of population, the number of workers available for work out of a given population, the number of hours worked, the intensity of work, the skills of workers and their willingness to work. The size of population depends upon a great variety of social, cultural, religious and economic factors among which wage rate the size of population rises or falls with a rise or fall respectively in the wage rate, and from this they had deduced a law called "Iron Law of Wages". But the history has shown that rise in the wage rate may have just the opposite effect on the size of population from what the subsistence theory of wages conceives.

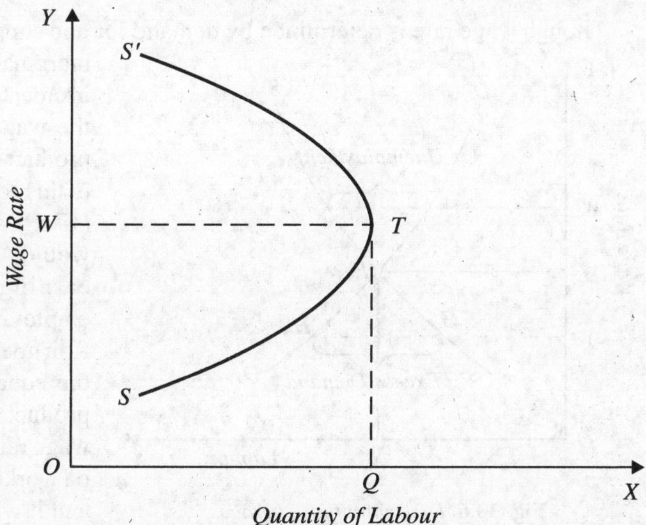

Fig. 33.5. *Beyond a certain wage rate, supply curve of labour is backward sloping.*

Moreover, the historical experiences have revealed that the size of population is dependent upon the great variety of social, cultural, religious and economic factors among which wage rate plays only a minor determining role. However, the willingness to work may be influenced greatly by the changes in the wage rate. On the one hand, as wages rise, some persons will do not work at lower wages may now be willing to supply their labour. But, on the other hand, as wages rise, some persons may be willing to work fewer hours and others like women may withdraw themselves from labour force, since the wages of their husbands have increased. Thus there are two conflicting responses to the rise in wages and therefore the exact nature of supply curve of labour is difficult to ascertain. *It is, however, generally held that the total supply curve of labour rises up to a certain wage level and after that it slopes backward.* This is shown in Fig. 33.5. As wage rate rises up to *OW*, the total quantity supplied of labour rises, but beyond *OW*, the quantity supplied of labour decreases as the wage rate is increased.

But so far as supply of labour to a particular industry is concerned it slopes upward. As the wages in an industry are increased labourers from other industries will shift to this industry. The elasticity of the supply curve of labour to an industry will also depend upon the transfer earnings of labourers. Similar is the case of supply of workers to a particular occupation. If wages in one occupation go up, some persons from other similar occupations would be attracted to it and thus the supply of labour to that occupation will increase. Thus because of occupational shifts, the supply curve of labour to a particular occupation is elastic and rises upwards. The long-run supply curve of labour is more elastic than the short-run supply curve since, in the long-run, besides the occupational shift in the labour force, new entrants in the labour market (who are now children) can also adopt the occupation by getting training for it in the very first instance.

How the wage rate is determined by demand for and supply of labour is shown in Fig. 33.6 where *DD* represents the demand curve for labour and *SS* represents its supply curve. The two curves intersect at point *E*. This means that at wage rate *OW*, quantity demanded of labour is equal to quantity supplied of it. Thus, given the demand for and supply of labour wage rate *OW* is determined and at this wage rate labour market is cleared. All those who are willing to work at the wage rate *OW* get employment. This implies that there is no involuntary unemployment and full employment of labour prevails. It is important to note that there will be no equilibrium at any wage rate higher or lower than *OW*. For example, at a higher wage *OW′*, supply of labour exceeds quantity demanded of it and as a result involuntary unemployment equal to *UT* emerges. Given the competition among labourers, this unemployment would push down the wage rate to *OW*. On the other hand, at a lower wage rate *OW″*, the demand for labour exceeds the amount of labour which people are willing to supply. In view of the excess demand for labour, the wage rate will go up to *OW* where the demand for labour equals the amount supplied of it. Thus wage rate *OW* will finally settle in the labour market.

Though wage rate is determined by demand for and supply of labour, it is equal to the value of marginal product of labour. This is so because in order to maximise its profits, a firm will equalise the wage rate with the value of the marginal product (*VMP*) of labour. If the firm stops short of this equality, the value of the marginal product (*VMP*) will be greater than the wage rate which would imply that there was still scope for earning more profits by increasing the employment of labour. On the other hand, if the firm goes beyond and employs more labour than the equality point, the value of the marginal product of labour will become smaller than the wage rate. As a result, the firm will incur losses on workers employed beyond the equality point and it will therefore be to the advantage of the firm to reduce the employment of labour. Thus in order to maximise profits and be in equilibrium the firm working under conditions of perfect competition in the factor and product markets will employ so much labour that the wage rate is equal to the value of marginal product (or marginal revenue product) of labour. It will be seen from Fig. 33.6 that the *firm* working in perfect competition in the labour market will take the wage rate *OW* as given and equates it with value of marginal product (*VMP*) and employs *OM* labour. To sum up, *the wage rate is determined by demand for and supply of labour, but is equal to the value of marginal product (or marginal revenue product) of labour*.

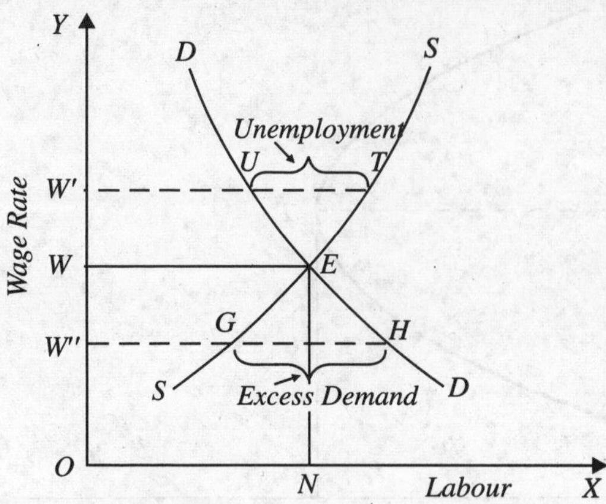

Fig. 33.6. *Determination of Wages*

It is worth mentioning that when the firms are in equilibrium by equating value of marginal product of labour to the wage rate, they may be making profits or losses in the short run. Consider Fig. 33.7 which depicts the equilibrium position of the firm in the short run. It will be seen from Fig. 33.7 that at the wage rate *OW*, the firm is in equilibrium when it is employing *OM* amount of labour. It will be further seen that the firm is making super-normal profits since in equilibrium employment *OM*, average revenue product of labour (*ARP*) which is equal to *RM* is greater than the

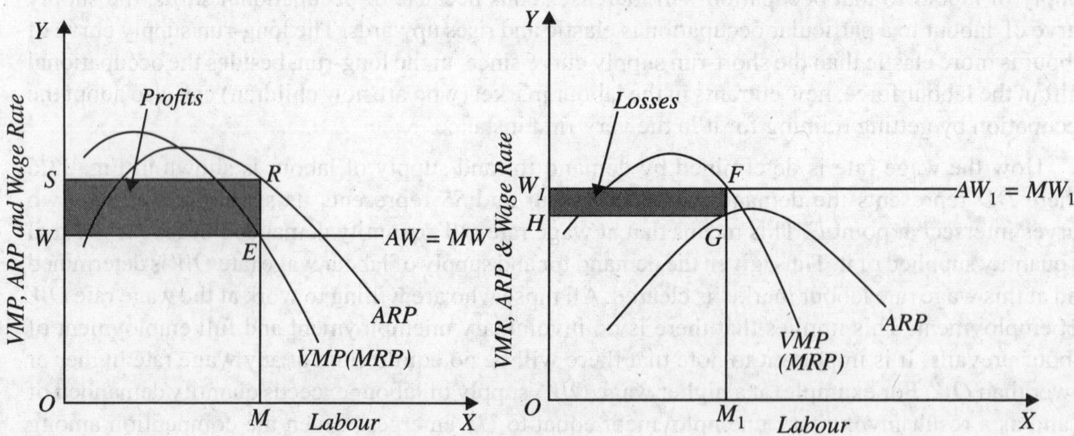

Fig. 33.7. *Equilibrium of the Firm with Super-Normal Profits*

Fig. 33.8. *A firm will not employ labour at wage rate* OW_1.

wage rate $OW(=ME)$. This can happen in the short run, but not in the long run. When firms are earning super-normal profits in the short run more entrepreneurs will enter the market in the long run to purchase labour to produce the products made by it. Entry of more entrepreneurs to the labour market will compete away the super-normal profits. As a result, the demand for labour will rise and the demand curve for labour will shift outward to the right, which will raise the wage rate and will eliminate the profits.

It should be carefully noted that a firm will not employ labour if wage rate exceeds average product of labour. Unlike machines labour is a variable factor and if its employment is not sufficient to recover its wages, it will be laid off even in the short run. Consider Fig. 33.8 at wage rate OW_1, a firm will be incurring losses if it employs ON_1 amount of labour at which wage rate $OW_1 = VMP = MRP$. Therefore, at wage rate OW_1, the firm will not employ labour.

To sum up, in the long run, the equilibrium between demand for and supply of labour is established at the level where the wage rate of labour is equal to both the VMP (MRP) and ARP of labour and thus the firms earn only normal profits. The long-run equilibrium position of the firm working under perfect competition is depicted in Fig. 33.9 where it will be seen that the firm is in equilibrium at *ON* level of employment (*i.e.*, at point *T*) at which wage rate is not only equal to value of marginal product but also average revenue product of labour. Given the *ARP* and *VMP* curves, if the wage rate is lower than OW' (= $N.T.$), the number of firms employing labour will change causing changes in demand for labour. As a result of this, the wage rate will ultimately settle at the level *OW* or *NT*.

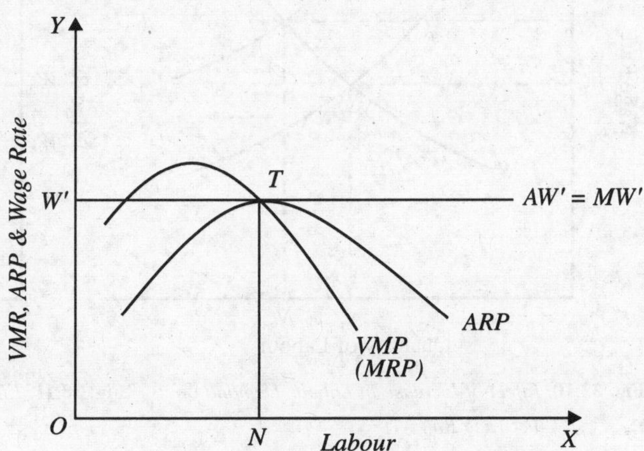

Fig. 33.9. *Long-Run Equilibrium of the Firm*

Changes in Equilibrium Wage Rate

We have explained above how through interaction of demand for and supply of labour determines the market wage rate. Now, if any of the factor causes a shift either in demand curve or in supply curve of labour, the equilibrium will be disturbed causing a change in the wage rate. Both demand for and supply of labour can shift.

Shift in Demand Curve for Labour. Demand for labour increases if its productivity increases, say through technological improvement. This will cause a rightward shift in the demand curve for labour and as shall be seen from Fig. 33.10 this will bring about a rise in the wage rate. Similarly, if the demand for a product, say of a textile cloth, increases, the demand for textile workers being a derived demand will also go up. This too will cause an upward shift in the demand for textile workers causing a rise in their wage rate. Further, if the price of a textile cloth rises, it will increase the value of marginal product, ($VMP = Price \times MPP$) of textile workers. With this higher value of marginal product, it will become profitable for the producer to hire more workers. As a result, demand for textile workers will increase causing a rise in their wage rate.

Conversely, if the demand for a product decreases or its price falls, it will induce a reduction in the demand for labour. Given the supply curve, decrease in demand for labour will bring about reduction in wage rate.

Shift in Labour Supply Curve. If the factors determining labour supply undergo a change, the supply curve of labour will shift causing a change in the equilibrium wage rate. The supply of labour to a given occupation or industry will decrease *if the wages in alternative occupations or industries go up.* In this case at every wage rate less labour will be offered to a given occupation or industry. This will cause a shift in the supply curve of labour to the left and, given the demand curve for labour, result in rise in the wage rate. This is shown in Fig. 33.11.

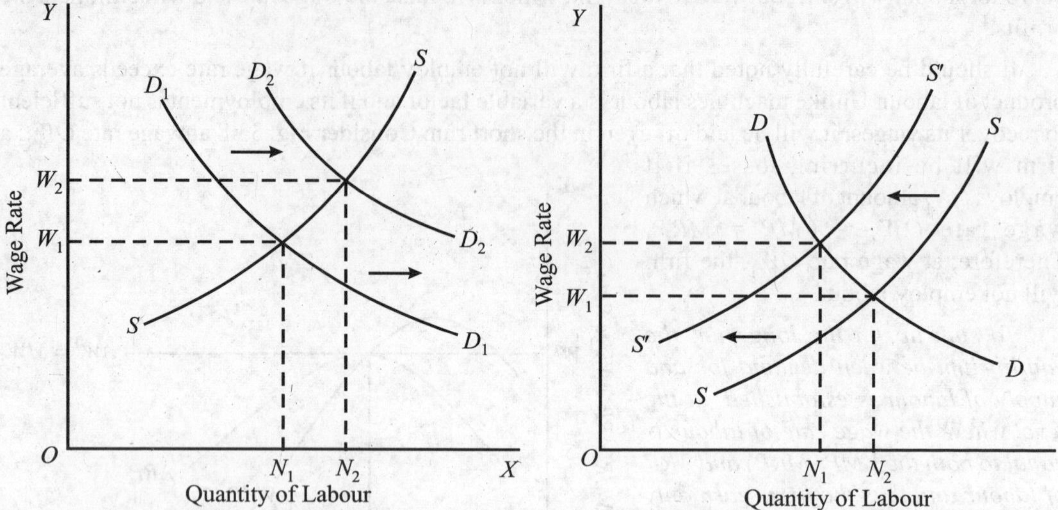

Fig. 33.10. *Effect of Increase in Labour Demand on the Wage Rate*

Fig. 33.11. *Effect of Decrease in Labour Supply on Wage Rate*

Similarly, if workers valuation of their leisure time changes, the supply curve of labour will shift. If most workers start attaching a higher value to their leisure time spent with their families, the less labour will be supplied to an occupation or industry. This will cause a shift in the labour supply curve to the left resulting in higher wage rate as is illustrated in Fig. 33.11.

Conversely, if for any reason, the wage rate in alternative occupation fall or workers' preferences for leisure declines, supply of labour to a given occupation or industry will increase at every wage rate. This will cause a shift in the supply curve of labour to the right and result in fall in the wage rate.

WAGE DETERMINATION UNDER MONOPSONY

Monopsony in the labour market is said to exist when there is a single buyer of labour. Under monopsony in the labour market , a single buyer faces a large number of workers who are unorganised and whose geographical mobility of labour is very much limited. Monopsony may prevail when a big employer hires a proportionately very large number of a given type of labour so that he is in a position to influence the wage rate or it may prevail when various employers have an understanding not to compete for labour and thus act as one in hiring labour. Thus, in the actual world, monopsony exists in the labour market when a large single employer or, various employers acting as one confronts a large number of workers who are unorganised (*i.e.*, non-unionized) and who lack geographical mobility.

It should be noted that non-organisation of labourers into unions is an essential condition for the existence of monopsony, for when the labourers organise themselves into trade union, then the supply of labour is channelled through the trade union and the trade union therefore becomes a sole seller of labour. When a single buyer-the employer, faces a single seller—the trade union, the market situation is one of bilateral monopoly and not of monopsony. Thus monopsony will prevail when the workers are not organised into trade unions. Likewise, immobility of labour is also an essential condition for the existence of monopsony in the labour market. If the workers are sufficiently mobile so that they will move to places or industries where wages are higher, then the single employer in a local market will not possess a determining influence in hiring labour.

It is clear from above that non-organisation of workers and lack of mobility on their part are the essential conditions on the side of the workers for the existence of monopsony in the labour market. We shall now explain how wage rate is determined in a monopsonistic market situation. In a proper analysis of the same, we must know what type of market situation is confronted by the monopsonist in selling the product produced by the labour employed by him. We shall first explain the wage determination in the case where monopsony in the labour market is found with perfect competition in the product market. After this we shall explain wage determination when monopsony in the labour market is found with monopoly in the product market, that is, where the monopsonist in the labour market is also the monopolist in the product market.

Wage Determination when there is Monopsony in the Labour Market but Perfect Competition in the Product Market: Monopsonistic Exploitation

The demand curve of labour of the monopsonist, as of perfect competitor, is given by the curve of marginal revenue productivity. It should be noted that when perfect competition prevails in the product market, marginal revenue product of labour will be equal to the value of its marginal product. Therefore, in this case, the curve of marginal revenue product of labour will coincide with its curve of value of the marginal product.

What will be the nature of supply curve of the monopsonist? The supply curve of labour to a monopsonist is usually drawn as a rising curve, as S_L curve in Fig. 33.12 which indicates that to get more labour the monopsonist must pay higher wages. It should be noted that the supply curve of labour S_L in Fig. 33.12 is the same as the average factor cost (*AFC*) curve described in Chapter 32. As wage rate rises, the greater quantity of labour is supplied. Therefore, the supply curve of labour S_L is sloping upward. Since when wage rate rises and more labour is hired, marginal factor cost of labour (*MFC*) of labour is greater than average factor cost. Therefore, marginal factor cost curve of labour (MFC_L) lies above *AFC* of labour (*i.e.* the labour supply curve S_L). Note that marginal factor cost curve MFC_L is also called marginal expenditure curve.

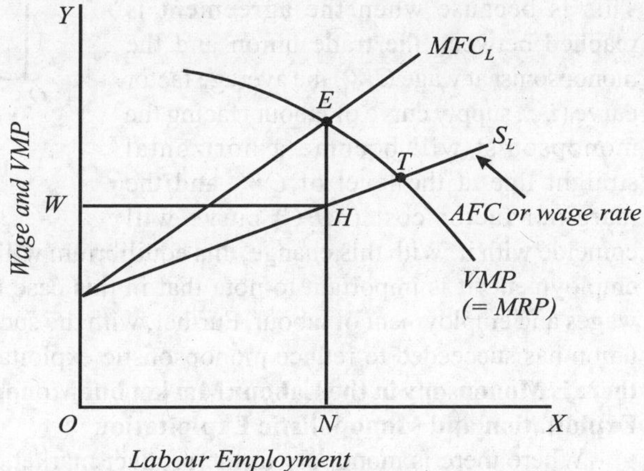

Fig. 33.12. *Wage Determination under Monoposony*

Now, the monopsonist will be maximising his profits and therefore will be in equilibrium where the marginal factor cost of labour (MFC_L) is equal to its marginal revenue product. It will be seen from Fig. 33.12 that the corresponding to the point *E* at which marginal revenue product curve of labour cuts the marginal factor cost curve (MFC_L) the level of labour employment is ON_0. *Further, it is evident from the figure that ON* amount of labour is forthcoming at *NH* or *OW* wage rate. Thus the monopsonist, in equilibrium, will employ *ON* amount of labour and will pay *OW* wage rate to the labourers. It is clear from the figure that the wage rate *NH* (= *OW*) determined under monopsony is smaller than the value of marginal product (*VMP*) which is equal to the *NE*. Thus each worker gets *EH* less than the value of its marginal product. According to Joan Robinson, to pay a worker less than his value of marginal product (*VMP*) is to exploit him.[1] Therefore, in Fig. 33.12 exploitation of each worker done by the monopsonist is equal to *EH*. Because this exploitation is due to the existence of monopsony in the labour market, Joan Robinson calls it as '*monopsonistic exploitation.*

1. Joan Robinson, *Economics of Imperfect Competition.*

It should be noted that monopsony results in a lower wage rate and lower level of labour employment than under perfect competition in the labour market. Given Fig. 33.12 if there was perfect competition in the labour market, the equilibrium would have been at the point *T* where the demand curve or *VMP* curve for labour intersects the supply curve, S_L. Therefore, under perfect competition in the labour market, the higher wage rate and larger employment would have been determined.

A very important conclusion can be derived from the above analysis of wage-employment equilibrium under monopsony in the labour market. It is that *under conditions of monopsony in the labour market, trade unions can raise the wage rate without creating unemployment*. In fact, it can be shown that a wage increase secured by trade unions under such circumstances may result in greater employment. Thus, in Fig. 33.13 if the wage rate is raised to OW_1 as a result of trade union's bargaining with the monopsonist, the employment offered by the monopsonist will increase from ON_0 to ON_1. This is because when the agreement is reached between the trade union and the monopsonist at wage OW_1, the average factor curve (*i.e.*, supply curve of labour) facing the monopsonist will become a horizontal straight line at the level of OW_1 and the marginal factor cost (*MFC*) curve will coincide with it. With this change, his equilibrium will be at point *C* corresponding to ON_1 level of employment. It is important to note that in this case trade union has succeded is increasing both wages and employment of labour. Further, with the success in getting a higher wage rate OW_1, trade union has succeeded to reduce monopsonstic exploitation of labour. **Wage Determination when there is Monopsony in the Labour Market but Monopoly in the Product Market: Monopsonistic Exploitation and Monopolistic Exploitation**

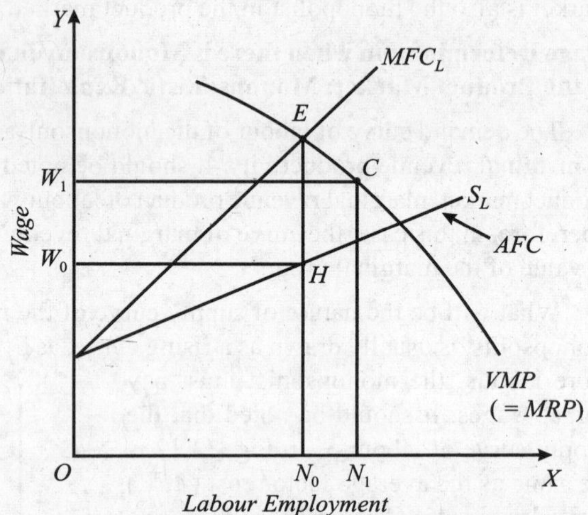

Fig. 33.13. *Effect of Increase in Wages by Trade Union*

Where there is monopoly in the product market, the curve of marginal revenue product will differ from the curve of the value of marginal product. The marginal revenue product curve (*MRP*) will lie below the curve of the value of marginal product (*VMP*), for the marginal revenue is less than the average revenue when there is monopoly in the product market. The supply curve of S_L labour (average factor cost curve *AFC*), as in the above case, slopes upward and the marginal factor cost curve lies above it. The equilibrium of the monopsonist will be where the marginal revenue product of labour equals marginal factor cost of labour (*MFC*). In Fig. 33.14 the equilibrium of the monopsonist is at point *E* according to which wage *NH* or *OW* is determined and labour *ON* is employed. It will be seen from Fig. 33.14 that wage rate *NH* is not only less than marginal revenue

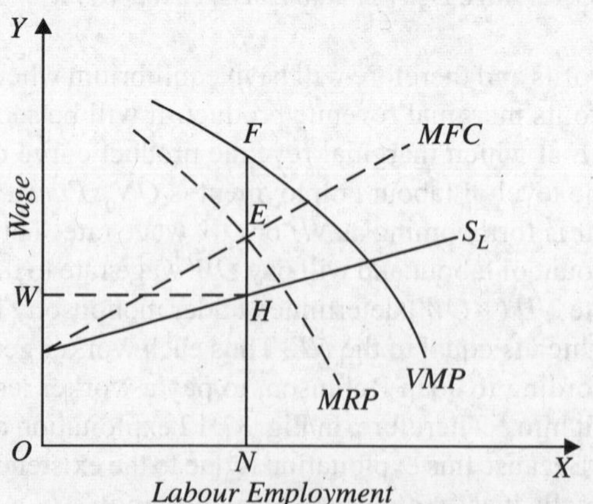

Fig. 33.14. *Wage Determination under Monopsony in Labour Market and Monopoly in the Produt Market*

product (*NE*) but is also less than the value of the marginal product (*NF*). The difference *EH* between marginal revenue product *NE* and the wage rate *NH* is due to the existence of monopsony in the labour market and has, therefore, been called by Joan Robinson as *monopsonistic exploitation*. But the difference *FE* between the value of the marginal product (*VMP*) which is equal to *NF* and the marginal revenue product (*MRP*) which is equal to *NE* is due to existence of monopoly in the product market. The worker get *FE* amount less than his value of the marginal product because of the fact of monopoly in the product market and has, therefore, been designated as '*monopolistic exploitation*' by Joan Robinson. Under monopsony-monopoly market situation, the worker is subjected to double exploitation; because of monopoly he gets less than value of its marginal product and because of monopsony he gets less than his marginal revenue product.

Exploitation of Labour

We have explained above that labour is exploited when there prevails imperfect competition in the product market as well as when there is imperfect competition or monopsony in the labour market. In the former case it is called monopolistic exploitation and in the latter it is called monopsonistic exploitation. In our above analysis we have followed Joan Robinson's approach to the exploitation of labour. However, Robinson's definition of exploitation of labour is not accepted as valid by some economists, especially Chamberlin. It is, therefore, useful to discuss the concept of exploitation of labour and various meanings and interpretations of exploitation of labour that have been provided.

It should be noted that Pigou-Robinson concept of exploitation of labour, namely, labour being paid less than the value of its marginal product (*VMP*), assumes perfect competition as the 'ideal' and the wage rate determined under it as just, fair and proper. Any deviation from this perfectly competitive wage is regarded as exploitation. Thus, according to Rothschild, "Professor Pigou, and following him Joan Robinson have made the *ideal of perfect competition* their starting point. Under the system the worker would get, as we saw, a wage equal to the value of the marginal physical product. Every deviation from this is regarded as exploitation."[2]

The above concept of exploitation can be looked at from two viewpoints. Firstly, it can be viewed as a *purely technical way* of describing the wage payments which are less than those which would have prevailed under conditions of perfect competition. In this purely technical viewpoint of exploitation, any evil design or sinister motive of the entrepreneur is not implied. Secondly, the Pigou-Robinson's concept of exploitation can be viewed as the *concept which regards perfect competition wages as the just, fair or right wages which ought to be paid by the entrepreneur to the labour*. In this second sense emotional colouring is given to the concept of exploitation and sinister motive on the part of the entrepreneur is implied. Further, it is also implied in this second sense of exploitation that market price of a product reflects its social value. Pigou and to somewhat lesser extent Robinson regarded exploitation in the second sense, that is, from the viewpoint of emotional colouring and ethical standard. In the opinion of present author to regard "deviations from perfect competition" as unethical is unjustified and unwarranted. We, therefore, agree with Rothschild who writes: "this usage of exploitation....may be regarded as an implicit assumption that perfect competition wages are the 'right' or 'proper' wages of labour. There can be no doubt that such a moral understone is part of Professor Pigou's definition, who stood under the influence of the newly developed marginal productivity theory; and to a lesser extent this is probably also true of Joan Robinson. It should be clear, however, that there is *no any scientific reason to warrant such an assumption*. By calling deviations from the perfect-competition equilibrium 'exploitation', it does not follow that *this equilibrium has any superior ethical or political qualities*."[3]

---

2. K.W. Rothschild, *The Theory of Wages*, Augustus M. Kelley, New York, 1966
3. *Op. cit*., p. 103

How can Labour Exploitation be Removed?

We have discussed above the two concepts of exploitation and also highlighted the conditions under which labour exploitation arises. Now, an important question is how this labour exploitation can be removed. That is, whether trade unions or Government can remove exploitation by raising the wages of workers or some other steps have to be taken to remove exploitation.

As far as *monopolistic exploitation* depicted in Fig. 33.14 by *EF* which has arisen due to the imperfect competition in the product market, is concerned, it cannot be removed by raising wages by the trade unions. This is because, in this situation, if the trade unions succeed in raising wages, the employer will employ smaller amount of labour so as to equate the new high wage rate with the marginal revenue product (*MRP*) of labour. But the important point to note is that with lower employment and higher wage rate, labour would still be exploited, for in this new wage position also, value of the marginal product (*VMP*) will be greater than the marginal revenue product (*MRP*) with which new higher wage will be equated by the employer. We thus see that *monopolistic exploitation of labour* as conceived by Joan Robinson cannot be removed through raising wages by trade unions or government. Monopolistic exploitation can only be removed by *creating the conditions of perfect competition in the product market*. The government can take measures for removing monopolistic conditions or imperfections from the product market.

But so far as monopsonistic exploitation of labour is concerned it can be, as explained above, removed by raising wages through trade unions or government. We have already explained this in this chapter.

ROLE OF TRADE UNIONS AND COLLECTIVE BARGAINING IN RAISING WAGES

For a long time economists believed that trade unions and collective bargaining could not play an important role in raising the wages of workers or effecting improvements in their economic conditions. In other words, they thought that trade unions as an instrument to raise wages of workers or to improve their overall economic conditions are *ineffective and superfluous* and thus the collective bargaining by them is a *futile undertaking*. Therefore, in the leading nineteenth century wage theories such as subsistence theory of wages and wage fund theory. the role played by trade unions and collective bargaining in the determination of wages was entirely neglected.

According to the Marginal Productivity Theory of Wages with its assumptions of perfect competition and the given supply of labour, trade unions cannot succeed in raising wages or cannot succeed in raising wages without creating unemployment. According to this theory, marginal revenue productivity (*MRP*) curve is the employers' demand curve. Consider Fig. 33.15 where *MRP* is the marginal revenue productivity curve of labour. If *ON* is the available supply of labour, *OW* is the equilibrium wage rate. Now, if the wage rate is increased to *OW′* by the collective bargaining of trade unions, *NN′* amount of workers would be rendered unemployed. If these unemployed workers are free to compete, they would press down the wage back to *OW.* If these *NN′* workers are, for one reason or the other, not free to compete, they will continue to remain

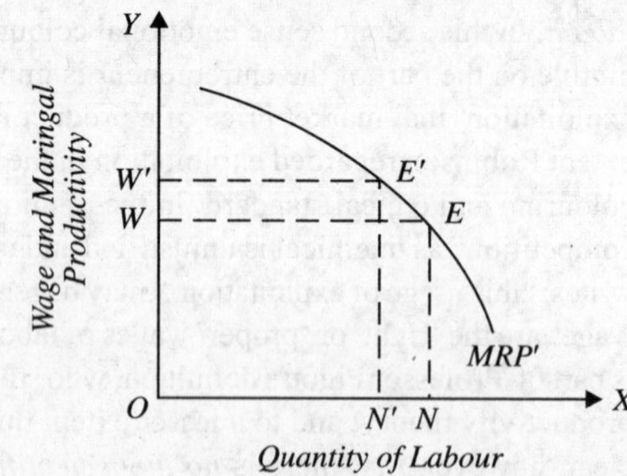

Fig. 33.15. *Marginal Productivity Theory: Trade unions cannot enhance wages without creating unemployment.*

unemployed. It is thus clear that, even according to marginal productivity theory, trade unions are unable to raise wages without creating unemployment.

It is evident from above that in the *static* marginal productivity theory there was no room for trade unions and collective bargaining in fixing and improving wages of the workers. The view about the futility of trade union also prevailed in the early twentieth century, even though many theorists expressed doubts about the correctness of this view. It was only in the 'thirties' of the 20th century that role of trade unions and collective bargaining was introduced into the economic theory and significant place was accorded to them in fixing wages of the workers. The realisation of the fact that perfect competition only prevailed in some exceptional cases in the real world led to the reconsideration of whole price and wage theory. Though the marginal productivity approach to the wage fixation was maintained; it was however extended to the conditions of imperfect competition wherein scope for trade unions and collective bargaining in raising wages was shown.

Moreover, there has been the emergence of two kinds of theories known as *institutional and psychological theories of wage determination under trade unions*. These theories do not try to reconcile the marginal productivity approach with the role of collective bargaining in the context of imperfect competition, but instead assign an eminent role to trade unions and collective bargaining in the wage determination, regardless of the marginal productivity doctrine. Thus in these institutional and psychological theories, trade unions and collective bargaining do not enter through the back door but occupy a central place right from the beginning in the determination of wages. In recent years there has been a due recognition of the role which the trade unions and collective bargaining play in the fixation of wages and various theories of bargaining have been advanced. Rather than being alternative explanations of wage determination under collective bargaining, these different bargaining theories of wages in fact bring out various factors which play a determining role in the bargaining process; different theories laying stress on different factors.

Bargaining Approach to Wages and Trade Unions

As explained above, there was not much room for the bargaining by trade unions in the marginal productivity theory in its earliest static versions. First of all, a bargaining approach to wages was developed which indicated a scope for collective bargaining within the framework of marginal productivity theory. This was done by giving up some of the assumptions of marginal productivity theory which had been essential parts of its earlier versions. Then, with the emergence of the theories of imperfect competition, monopsony and oligopoly, *bargaining approach within the framework of marginal productivity principle was considerably widened.*

Rise in Wage Rate and Labour Efficiency. To begin with, static assumptions of marginal productivity theory have been dropped. It is accepted that marginal productivity curve is employer's demand curve and that, given the marginal productivity curve, the increase in the wage rate by trade-union power will lead to the creation of unemployment. But it is pointed out that when the wage increase is achieved through a successful bargaining, the marginal productivity curve may not remain the same but may shift above due to the rise in efficiency of workers brought about by the higher wage. When the efficiency and therefore the marginal productivity curve rises due to the increase in wages of the workers the unemployment may not be created as a result of

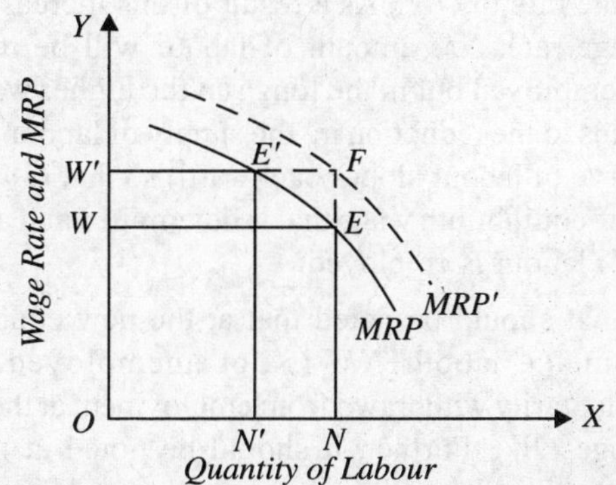

Fig. 33.16. *Increase in Marginal Revenue Productivity or Efficiency as a result of Rise in Wage Rate*

the increase in the wage rate secured by the trade union. This is shown in Fig. 33.16. Initially the marginal productivity curve is *MRP* and the equilibrium wage is *OW* and employment is *ON*. Now suppose the wage is raised to *OW′* by the successful collective bargaining of the trade union. If the marginal productivity curve *MRP* remains unchanged, then at *OW′* wage rate, *ON′* men will be employed, which means that *NN′* number of men will be rendered unemployed. But if the rise in wage brings about a sufficient increase in efficiency and productivity so that the marginal productivity curve shifts upward to the dotted position, then unemployment will not be created. It will be seen from the Fig. 33.16 that with the marginal revenue productivity curve *MRP′*, *ON* men are employed at the higher wage rate *OW′*.

We thus see that if we consider the effect of the wage bargain on the increase in efficiency or marginal productivity, then the trade unions can succeed in raising wages without creating unemployment. Again, the increase in wages may force the employers to improve the efficiency of production process in which case also the marginal productivity curve shifts upward and as a result at the increased wage rate the same men may be employed. Again, the marginal revenue productivity curve may also be shifted above, if the increase in the wage rate brought about by collective bargaining is passed on to the consumers in the form of higher price of the product. In this case also the danger of unemployment being created will not be very much there and the higher wage rate will become the equilibrium wage rate equal to the new higher marginal revenue productivity. Therefore, the supporters of bargaining approach to wage fixation maintain that it is not so much the movements along the marginal productivity curve as it is the shifting of this curve that has to be considered when the impact of a wage bargaining is to be known.

Impact of Wage-hike on Labour Supply. Similarly, when the wage rate is raised by collective bargaining, the supply of labour may fall so that the increased wage becomes the equilibrium of wage without involuntary unemployment. Supply of labour can fall because when men are earning higher wage their women can stop working (that is, women withdraw themselves from labour force and stay at home) and children can be kept at school for a longer time. Further, at higher wages individuals may work less hours in a week or smaller number of days in a month or year. Because of the reduction in the supply of labour at the higher wage rate, the higher wage rate may become the 'equilibrium' wage rate. Consider Fig. 33.17 where *SS′* is the supply curve of labour and *DD′* is the demand curve for labour. Suppose the initial equilibrium is at point *P* where *ON* labour is employed at the wage rate *OW* or *NP*. Suppose the trade union through collective bargaining succeeds in raising the wage rate to *OW′*. As a result of this increase in the wage rate, *NM* amount of labour will be rendered unemployed but in the long run the higher wage rate leads to the reduction in the supply of labour (supply curve of labour slopes backward) so that *OW′* is the new equilibrium wage that is determined and at which *OM* labour is employed.

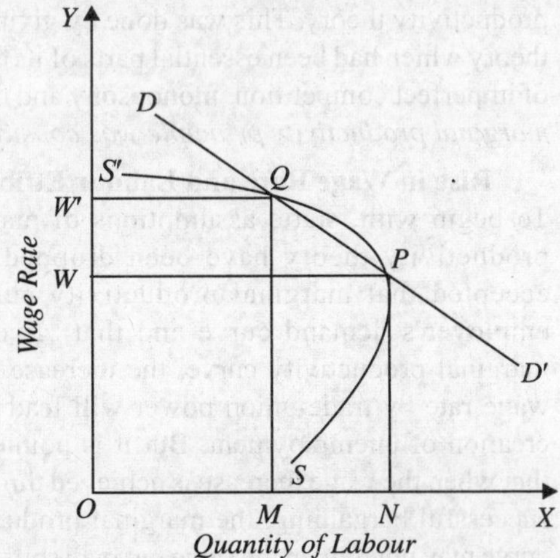

Fig. 33.17. *A higher wage secured by the union can become an equilibrium wage without involuntary unemployment if a part of the supply of labour withdraws itself from the labour force.*

It should be noted that at the new equilibrium point *Q*, labour *NM* is not unemployed, it has voluntarily withdrawn from employment at the higher wage *OW′*. Further, it should be noted at the new higher-level equilibrium where the supply of labour is less than before, the workers would get a large proportion of the new higher national income and enjoy higher standards of leisure and education. (The national income will increase if the increase in output due to higher productivity at the increased wage rate is greater than the loss in output due to the reduction in the supply of labour.)

It is clear from above that when the increase in wage rate is attained by trade unions through collective bargaining, there are such changes in the factors determining wage employment situation that the increased wage becomes the equilibrium wage without creating unemployment. Prof. Rothschild rightly remarks, "*The imposition of a higher wage may lead initially to some unemployment but may then produce such a change in the determinants of the wage-employment situation that the unemployment disappears and the higher wage rate becomes an equilibrium wage*".[4]

Role of Trade Unions in Raising Wages Under Monopsony. It is evident from above that under perfect competition and in the framework of the marginal productivity theory conceived in dynamic terms, there is a good room for collective bargaining to raise wages. Further, the realisation that it is not the perfect competition, but the market forms of imperfect competition, oligopoly, monopoly, monopsony, oligopsony, *etc.* which mostly prevail in the real world, opened up new vistas for combining the marginal productivity analysis with the bargaining approach to the wage determination. The most striking case in these various forms of imperfect markets is of monopsony when there is a single buyer of labour. The monopsonist, as explained before, working on the marginal productivity principle equates marginal revenue product of labour with the marginal wage to be in equilibrium posit labour. In such a situation wage rate (*i.e.*, average wage) determined is less than the marginal revenue productivity. It will be seen from Fig. 33.18 that, under monopsony, wage rate *OW* and employment *ON* is determined. Under such circumstances if workers organise themselves into trade unions, they can achieve increase in the wage rate without creating unemployment, indeed the employment will increase for some increases in the wage rate.

When the trade unions come into existence, the supply of labour is channelled through it and the bargaining with the employer is on the basis of '*all or nothing*' at a particular wage rate demanded, that is, no supply of labour will be offered below the demanded and/or mutually agreed wage rate and the whole supply of labour will be offered at the mutually agreed wage. This means, in other words, that the supply curve of labour under trade union becomes perfectly elastic at the demanded or mutually agreed wage rate. It will be seen from Fig. 33.18 that if the trade union succeed in getting higher wage rate *OW′*, the supply curve of labour becomes horizontal or prefectly elastic shape at the agreed wage level *OW′*, the new supply supply curve of labour will coincide with the marginal factor cost curve. It will be seen that, given the wage rate equal to *OW′* and the labour supply curve *W′ F*, the employer's equilibrium will be at point *F*, at which employment *ON′*, which is greater than *ON*, will be offered by the employer.

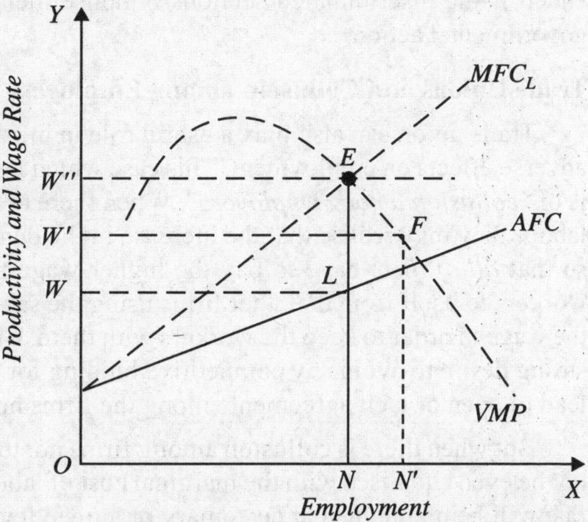

Fig. 33.18. *When there is monopsony in the labour market, increase in the wage rate can be secured by the labour union without creating unemployment.*

It should be carefully noted that a powerful trade union can raise the wage rate even up to

4. K.W. Rothschild, Approaches to the Theory of Bargaining, published in *The Theory of Wage Determination*, edited by J.T. Dunlop, p. 284

OW″, that is equal to the marginal value product *NE* at the original level of employment *ON*. When the wage rate *OW″* is fixed under collective bargaining and as a consequence the supply curve of labour or margunal factor cost curve *W″E* becomes perfectly elastic at the level *OW″*, the employer's equilibrium will be at *ON*, the original level of employment. Thus, under conditions of monopsony, a strong trade union can raise the wage rate to the level of value of marginal product *NE* without the fear of creating unemployment. In the absence of trade union, the monopsonist would exploit the workers to the extent of *LE* or *WW″*. It is, therefore, clear that the workers by organising themselves into trade unions and thereby collectively bargaining with the employer, can raise the wage rate to remove monopsonistic exploitation by the monopsonist.

Trade Unions and Monopsonistic Discrimination

The role of trade union to improve the lot of the workers is also important when there prevails the conditions of *monopsonistic discrimination*. Monopsonistic discrimination is said to prevail when the monopsonist pays different wages to the different workers. Monopsonistic discrimination is quite common in the actual world where workers are unorganised and where due to fear of unemployment they have to accept the essential minimum wage. Under such circumstances, if the workers organise themselves into strong trade unions, they can force the monopsonist to stop discrimination and pay the same wage to all the workers of a given type and thereby can reduce his excess profits earned because of discrimination.

In developed countries and well established industries, the organisation of labourers into powerful trade unions and thereby bargaining collectively with the employers, perfect monopsonistic discrimination (that is, discrimination between individual and individual) has vanished. But "group discrimination" is still very common in all countries, developed as well as underdeveloped. By group discrimination is meant the discrimination between different, groups or sections of the population. Thus, for the same work women are generally paid lower wages than men in many countries. In some countries, coloured people are likewise paid less wages than white men for the same work. Boys are also sometimes paid smaller wages than men. The group discrimination can be ended if the discriminated sections organise themselves into powerful trade unions and also by governmental action.

Trade Unions and Collusion among Employers

Trade union can also play a useful role in improving the wages of the workers without causing adverse effects on employment. This case which is intensely associated with the idea of monopsony is of "*collusion among employers*". When there are few large firms competing for the same kind of labour they may realise that the increase in the demand for the labour by one firm may raise wages so that *all of them* have to pay the higher wage rate. If one firm offers a higher wage to attract workers to itself from the other firms using the same type of labour the others too will have to raise the wage in order to keep the workers with them. Under such circumstances the firms will develop a strong desire to avoid any competitive bidding for labour and spoiling the labour market. This may lead to open or tacit agreement among the firms not to raise wages.

But when there is collusion among firms not to raise wages, then the marginal productivity will not be even equalised with the marginal cost of labour. Under collusion among employers, the wage rate will be maintained at customary or agreed level even though the marginal revenue product of labour will stand higher than the wage rate paid. Although the firm can increase its profits by expanding employment to the point where the marginal revenue product equals marginal cost of labour, but in doing so the firm will have to increase the wage rate which is prohibited under collusion. Under such conditions of collusion, the formation of trade union by workers can force the employers to pay the wage rate equal to the marginal revenue product of labour. Such a rise in the wage rate to the level of marginal productivity under pressure of trade union would not create any unemployment, because such a rise in the wage rate will only fill up the gap between the marginal revenue productivity and marginal cost of labour, and will not raise the latter above the former.

Trade Unions and Oligopoly in Product Market

Furthermore, even in the case of oligopoly in the product market which so extensively prevails in the capitalist countries, the increase in the wage rate by the trade union may be achieved without creating unemployment. It is generally believed that the oligopolist confronts a 'kinked' demand curve (having a kink at the prevailing price of the product), corresponding to which the marginal revenue curve has a discontinuous or broken portion vertically below the kink. In such a case, when the increase in the wage rate occurs due to the collective bargaining by the trade union, the marginal cost curve will shift above, but for a moderate increase in the wage rate it will still cut the marginal revenue curve through its discontinuous portion, indicating thereby that the output remains unchanged despite the increase in the wage rate and consequently the rise in cost. No change in output as a result of the increase in the wage rate means that the employment will remain the same, provided the employer does not substitute machinery for labour.

It follows, therefore, that in this case too the collective bargaining can succeed in raising the wage rate without adversely affecting the employment level. The oligopolistic market situation imposes a certain price and output policy on the employer due to which he is forced to swallow the whole increase in the wage bill following a moderate increase in the wage rate. Thus in the oligopolistic market situation the trade union can succeed in raising wages by making inroads into the profits of the capitalist employer.

Conclusion

It is clear from the above analysis that trade unions are not superfluous and collective bargaining by them not a futile activity. In fact, the trade union can play a useful role in raising the wage of workers and can save them from the exploitation of the employers. Indeed, in many cases, the trade unions can raise wages without creating unemployment.

WAGE DETERMINATION UNDER BILATERAL MONOPOLY WHEN MONOPSONIST BUYER FACES A MONOPOLIST SELLER

Collective bargaining by trade union with an employer or, if it is industry-wide bargaining, with the employers' association *represents a situation where a single seller (i.e., monopolist) faces a single buyer*. Trade union of the firm or industry acts as a single voice representing the workers so that trade union becomes a single seller of labour to the employer. In other words, trade union has the monopoly of selling labour. On the other hand, the employer, if he is monopsonist, or the employers' association is a single buyer of labour. Thus in such a bilateral monopoly a single buyer of labour faces a single seller of labour. Therefore, we are confornted here with a special case of bilateral monopoly and wage determination under trade unions. In this case collective bargaining becomes a special case of price determination under bilateral monopoly. *Analysis of wage determination under bilateral monopoly as of product under bilateral monopoly does not theoretically lead us to a certain particular wage at which agreement will be reached*. Analysis of bilateral monopoly only brings out two limits—the upper limit sought by the union and the lower limit set by the employer—within which range the wage will be fixed. Actual wage rate fixed in a bilateral monopoly will lie within that range and whether it will be more nearer to the upper limit or to the lower limit depends upon the relative bargaining strengths of the union and the employer. Apart from pointing to the range within which the wage will be determined, economic theory cannot lead us to conclude at which particular wage the agreement will be reached. Theoretically, the wage determination under bilateral monopoly or collective bargaining within the range between upper and lower limits is indeterminate; wage may be set at any level within the range between the two limits.

However, it is useful to analyse the two limits or the range within which wage rate will be fixed under collective bargaining. A difficulty which crops up in the beginning of the analysis is in regard to the union behaviour. Just as economists have built up many models of oligopoly depending upon the assumptions in regard to the behaviour of the oligopolists, similarly many models of collective bargaining have been constructed depending upon the different assumptions in regard to the goal and behaviour of the union. A basic question in this regard is whether the trade unions are economic or political or politico-economic institutions. Again, even if aims of trade unions are purely economic, what economic goal will they pursue or what quantity will they seek to maximise. Thus, will the trade union may seek a wage rate which could *maximise the incomes of its members* or will the trade union try to seek the wage which will *maximise the number of its members* ? Again, will the trade union strive to increase the wage rate to as high a level as possible without regard to the employment effect of its action, or will it seek some optimum combination of wage rate and employment. These are the different alternative objectives of trade unions which imply different patterns of their behaviour in their negotiation or bargaining with the employer.

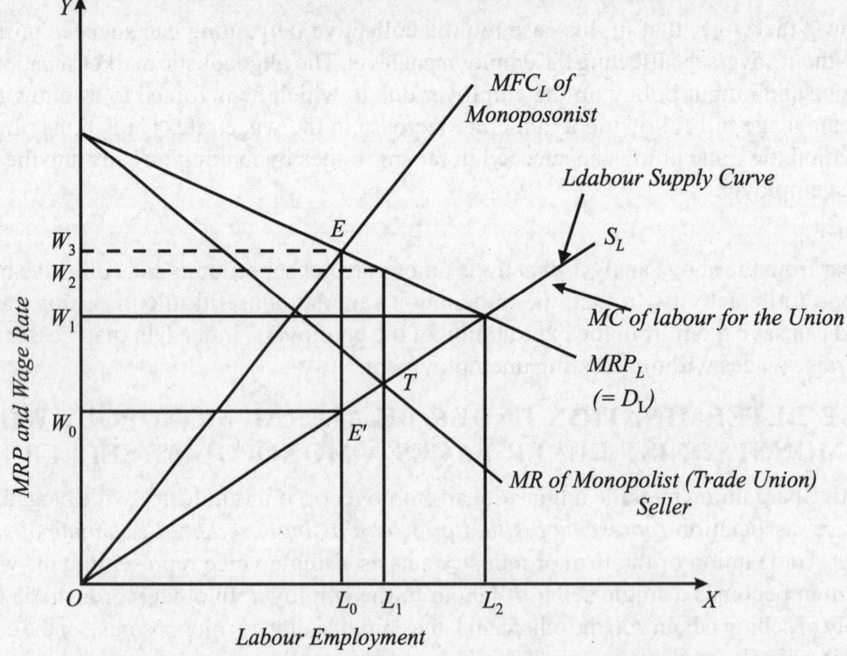

Fig. 33.19. *Indeterminacy of Wage Determination in Bilateral Monopoly*

In modern economic theory wage determination under collective bargaining is explained in terms of bilateral monopoly model. *Bilateral monopoly is a market form where a monopolist, the single seller of a product or service, sells to the monopsonist who is a single buyer of that product or service.* In the present case the trade union is a single seller of labour, whereas the firm, if it is a monopsonist is a single buyer of labour. The range of wage rates within which a particular wage rate will be settled can be easily explained with the Fig. 33.19 where the bargaining about wage rate between a trade union and a single employer is explained. It will be seen from this figure that the employer (monopsonist) would maximise his profits where marginal factor cost (*MFC*) on labour equals marginal revenue product (MRP_L) of labour. As will be seen from Fig. 33.19 that the employer will be maximising his profits by setting wage rate equal to OW_0 and employing OL_0 labour. Note that marginal factor cost (*MFC*) curve of labour intersects the marginal revenue product curve (*MRP*) of labour at point *E* and accordingly wage rate OW_0 will be paid by him. Thus, OW_0 is the minimum limit below which wage rate cannot fall.

The trade union (that is, the single seller of labour – a monopolist) faces *MRP* curve of labour which acts as a demand curve (D_L) for labour. This is because, given a wage rate, the monopsonist (*i.e.* employer) will equate the wage rate with marginal revenue product (*MRP*) of labour. The downward sloping *MRP* curve of labour means that as the number of workers employed increases, wage rate falls. In fact, the trade union has to choose a point on the MRP_L curve ($= D_L$) of labour. Since the wage rate paid to workers falls as more workers are hired, marginal revenue (*MR*) curve which describes the additional wages obtained by the union for its workers as the number of workers hired increases lies below MRP_L (*i.e.* D_L) curve of labour. On the other hand, the supply curve of labour S_L represents the minimum wage necessary to induce the union workers to offer their labour to the buyer. Therefore, the trade union views the supply curve S_L of labour as the marginal cost of labour. *If the labour union aims at maximizing net revenue or economic rent (i.e. revenue over and above the opportunity costs of labour) it will press for wage rate equal to* W_2 at which OL_1 quantity of labour will be employed. It should be noted that this economic rent or surplus is maximised at the level of employment L_1 (in Fig. 33.19) at which supply curve of labour (S_L) representing marginal opportunity costs of workers intersects the marginal revenue curve (*MR*) of the union. Note that if trade union wants to maximise employment of labour, it will agree to wage rate W_1 corresponding to which supply curve of labour S_L intersects the MRP_L (*i.e.*, D_L) curve of labour and at which employment is OL_2.

Thus W_2 is the upper limit or the wage rate sought by the union whereas W_0 is the lower limit. At which wage rate and employment, settlement will be reached between the two parties depends on their bargaining powers and strategies. If the union can make a strong threat to strike, it might succeed in achieving a wage closer to W_2. On the other hand, if the employer makes a credible threat to declare lockout or hire non-union labour, it might secure a wage rate closer to W_0. The result which depends on the relative bargaining powers of the two parties is indeterminate.

WAGE DIFFERENTIALS

Under conditions of perfect competition the identical workers doing the same type of jobs would get same wages. However, in the real world it is seen that different wages are paid to workers because of the following three factors:

1. *Workers differ* in quality, skill and training.
2. *Jobs differ*; some jobs are dangerous and others pleasant, some require more education and training than others.
3. Some institutional factors cause imperfections in labour markets such as discrimination against some workers, such as black race in America, women in many parts of the world, Scheduled Castes and Scheduled Tribes in India.

It should be noted that differences in wages of workers do not make demand–supply analysis of wage determination invalid. It is differences in demand and supply conditions in various labour markets that cause differences in wage rates. In other words, there is not one labour market but many—each with its different demand-supply conditions and therefore different equilibrium wage rates. Wages of the workers for whose services demand is relatively high and supply is relatively small are high. On the other hand, wages of workers whose supply is large but demand relatively weak are low. Several factors operate to cause the differences in demand and supply conditions of different types of workers. We explain below these different factors.

1. Differences in Abilities, Skill and Training

The first important factor that causes differences in workers and therefore wages earned by them is that *various workers differ in abilities, skill and training*. An example will make this clear. Let us consider the wages of computer engineers and unskilled workers. To become computer engineers one requires a lot of education and training to acquire the skill. On the other hand, unskilled workers do not have to spend time and money for obtaining education and training. The result is that not only demand for computer professionals is high but also their supply is relatively small. This is illustrated in

Fig. 33.20. In panel (*A*) of this figure determination of wages of computer engineers is shown. In this panel (*A*) demand for computer engineers D_1D_1 is high and supply S_1S_1 relatively small. As is seen from Fig. 33.20 the wage rate of computer engineers determined by these demand and supply curves is OW_1 which is much higher than the wage rate OW_2 of unskilled workers shown in panel (*B*). In panel (*B*) the demand curve D_2D_2 represents the demand for unskilled workers which is low and supply for them depicted by S_2S_2 is relatively large. Therefore, wage rate OW_2 of the unskilled workers is low. It should

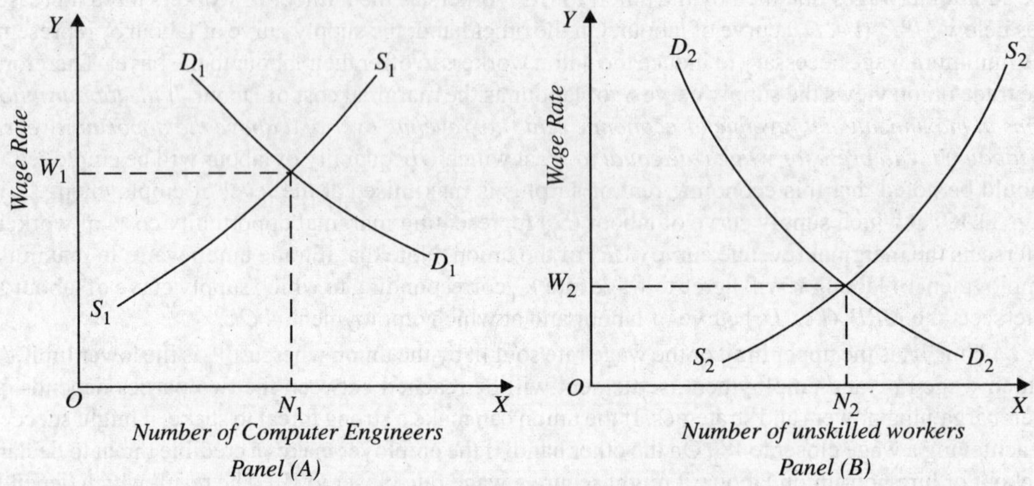

Fig. 33.20. *Difference in Wage Rates*

be noted again that demand for unskilled workers is small because due to lack of skill, education and training their marginal productivity is low and their supply is large because those who cannot spend time and money in acquiring education and training can get employment as unskilled workers. It is thus clear that the difference in wages can be explained through demand-supply analysis.

2. Differences in Jobs or Occupations: Compensating Wage differentials

The second important factor that causes differences in wages is the differences in the nature of jobs. Some jobs are more dangerous, risky and dirty than others. For example, jobs of miners in coal mines is quite dangerous; a blast or some other accident can cause even one's life. The workers working in coal mines are therefore paid higher wages than in a manufacturing industry, say in the textile industry, where there does not exist much risk of life. In the U.S.A. coal miners are generally paid 25 per cent more wages than textile workers. It should be noted that the reason for the differences in wages in this case lies on the supply side. At each wage rate, the quantity supplied of workers is smaller for work in coal mines than in a textile industry. This is illustrated in Fig. 33.21 where demand curve *DD* is assumed to be the same for miners and textile workers, considering that all workers are identical. However, the supply curves of workers are different in them. S_1S_1 is the supply curve of

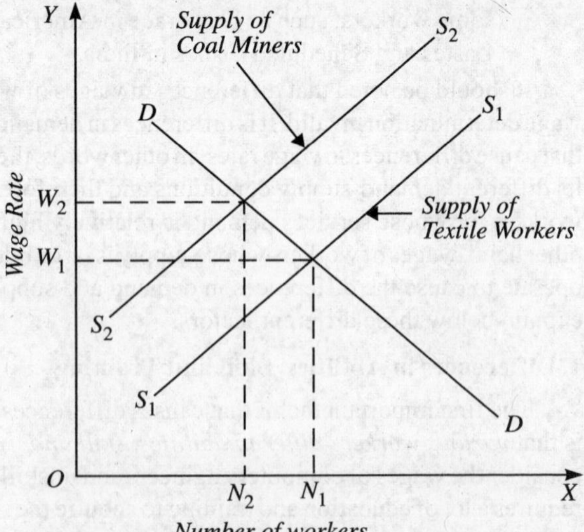

Fig. 33.21. *Difference in Wage Rates*

workers for the textile industry, whereas S_2S_2 is the supply curve of workers for coal mines. The supply curve S_2S_2 for coal mine workers shows that at each wage rate, the smaller quantity of workers is supplied to the coal mines due to the dangerous nature of the job in them as compared to the supply of textile workers depicted by the supply curve S_1S_1. It will be seen from this Fig. 33.21 that the intersection of demand curve DD and supply curve S_2S_2 of coal miners determines higher wage rate OW_2 whereas the intersection of demand curve DD and supply curve of S_1S_1 of textile workers determines a lower wage rate equal to OW_1. *Higher wage rate paid to coal miners is to compensate them for bearing higher risk for working in coal mines.*

Similarly, in the developed countries sanitation workers are paid higher wages than clerical workers because of dirty and unpleasant nature of the sanitation work. *The differences in wages which arise due to the differences in jobs arising due to more dangerous, more dirty, more harsh climate, higher cost of living are called compensating wage differentials.* Thus compensating wage differentials show higher wages that must be paid to the workers to compensate them for undesirable job characteristics.

3. Institutional Factors Causing Imperfect Labour Market Conditions

Institutional factors such as discrimination on the basis of sex, race, colour of skin etc. make labour market imperfect and give rise to differences in wages. In the U.S.A. black workers (Negros) are generally paid less wages than the white workers for the same type of work on account of discrimination practiced between them. In many countries, including India, women are paid less wages than men for the same work. Therefore, demand for equal pay for equal work both for men and women has been raised in recent years. Similarly, in several parts of India, rural workers belonging to Scheduled Caste and Scheduled Tribes are paid less than high caste workers.

We have explained above only some of the factors causing differences in wage rates. *The differences in natural abilities, differences in non-monetary benefits such as job satisfaction, pleasant atmosphere, freedom to choose one's work* schedule as in case of a college professor and some other factors account for differences in wage rates.

NUMERICAL PROBLEMS

Problem 1. *Suppose a firm is a monopsonist in the market for labour it hires. The firm can hire the quantity of labour at $W = 10 + 4L$. If MRP of labour $= 100 - L$, how many number of workers the firm will hire and what wage rate will it pay?*

Solution. To maximise profits the firm will hire labour until $MFC_L = MRP_L$. Therefore, we have to find out marginal factor cost of labour (MFC_L) from the given wage function. Thus

$$W = 10 + 4L \qquad \text{... }(i)$$

To find total cost of labour we multiply (*i*) by the quantity of labour (*L*). Thus

$$TFC_L = W \cdot L = 10L + 4L^2$$

$$MFC_L = \frac{d\,(TFC)}{d\,L} = 10 + 8L$$

For profit maximisation :

$$MFC_L = MRP_L$$

$$10 + 8L = 100 - L$$

$$9L = 90$$

$$L = 10$$

Thus the 10 workers will be employed.

To find wage rate paid ,we substitute *L* in the wage function (*i*). Thus,

$$W = 10 + 4L = 10 + 4 \times 10$$

$$= ₹\ 50$$

The wage rate of ₹ 50 will be paid by him.

Problem 2. *A firm has a monopsony in labour market. Its MRP of labour is 14 – L and the wage rate is W = 2 + L.*

(a) How much labour the firm will hire and what wage rate it will pay?

(b) How much labour it will hire and use if a minimum wage of ₹ 10 is imposed on it.

Solution. The firm will hire labour until $MFC_L = MRP_L$

The supply curve of labour is :

$$W = 2 + L \qquad \text{... (i)}$$

To obtain *TFC* of labour we multiply (*i*) by *L*

$$TFC = W \cdot L = 2L + L^2$$

$$MFC_L = \frac{d(TFC_L)}{dL} = 2 + 2L$$

and the given $MRP_L = 14 - L$

Equating MFC_L and MRP_L we have

$$2 + 2L = 14 - L$$

$$3L = 14 - 2 = 12$$

$$L = 4$$

Thus 4 units of labour will be employed. To obtain wage rate (*w*) we substitute $L = 4$ in equation (*i*). Thus

$W = 2 + L = 2 + 4 = 6$

Wage rate of ₹ 6 will be paid.

(*b*) When the minimum wage rate of ₹ 10 is imposed on it, it will equate *MRP* of labour with this minimum wage rate. Thus

$$10 = 14 - L$$

$$L = 14 - 10 = 4$$

Thus at the minimum wage rate of ₹ 10, 4 units of labour input will be hired. (Note that by imposing minimum wage rate of ₹ 10, the government has succeeded in removing labour exploitation without adversely affecting employment).

QUESTIONS FOR REVIEW

1. Explain the effect of increase in wage rate on supply of labour (*i.e.*, work effort) by an individual in terms of income effect and substitution effect.
2. Derive the supply curve of labour using indifference curves between income and leisure. Why does supply curve of labour generally slope upward? When is the supply curve of labour effort backward bending? Explain. *D.U., B.Com.* (*Hons*) 1999
3. "Will an increase in the wage rate always lead to an increase in the number of hours worked"? Explain. *D.U., B.A.* (*Hons*) 2002
4. How is wage rate determined in a perfectly competitive labour market? Show that the wage rate determined in such a market is equal to the value of marginal (*VMP*) of labour.
5. Show that wage rate in a perfectly competitive labour market will be equal to both the value of marginal product and average revenue product of labour in the long run.
6. How are wages determined in a perfectly competitive labour market? If in such a market trade unions succeed in getting wage rate higher than the competitive equilibrium wage rate, what will be its consequences?
7. A trade union enforces a floor on minimum wage which lies above the market clearing wage. What are economic implications?
8. What is monopsonistic exploitation of labour? How can collective bargaining by trade unions help us increasing wages and employment both? Explain with the help of diagrams. *D.U. B.Com.* (*Hons*)

9. Given the following production function of a firm, where L is the number of workers hired per day (the only variable input) and Q is the quantity of a commodity produced per day and the constant commodity price of $P =$ ₹ 5.00 is assumed.

L	0	1	2	3	4	5
Q	0	10	18	24	28	30

(*a*) Find the marginal revenue product of labour.

(*b*)How many workers per day will the firm hire, if the wage rate is ₹ 40.00?

[**Hints.** First find the *Total Revenue* by multiplying output (Q) with price (₹ 5). Then find the successive differences in *TR* to find *MRP* of labour. Firm will employ labour where MRP_L = ₹ 40. (**Ans.** 2 workers).

10. Distinguish between monopolistic and monopsonistic exploitation of labour and illustrate diagrammatically. *D.U. B.A. (Hons)*, 1996

11. How far are labour unions successful in raising wages and employment under monopsonistic conditions? Explain *D.U.B.A. (Hons)* 1989, 1997

12. What is monopsony? Explain how wages and employment are determined in a monopsonistic labour market.

13. What is meant by exploitation of labour? Show how labour is exploited under monopsony in a labour market. *D.U. B.Com. (Hons)* 2000

14. What would be effect of increase in wage rate secured by labour union in monopsonistic labour market? Would it lead to unemployment? *D.U. B.Com. (Hons)* 2000

15. Discuss the circumstances under which a trade union can raise wages in an industry without affecting level of employment adversely. *D.U. B.A. (Hons)* 1986., *B.Com. (Hons)* 1996

16. Explain the role of trade unions in raising wages of labour. Does it necessarily lead to unemployment? Explain.

17. Is it possible for a labour union to raise wages without reducing employment under conditions of:

(*i*) Perfect competition in the labour market.

(*ii*)Monopsony in the labour market.

18. Explain the alternative union strategies for raising wages of labour.

[**Hints:** There are mainly four strategies which trade unions can adopt to improve wages of workers. First, *using its monopoly power over supply of labour*, trade union can raise wages, without bothering about employment. Secondly, the union can raise wages by *increasing labour demand* through raising productivity or efficiency of labour and also by promoting the demand for the product which labour helps to produce. Thirdly, it can improve wages by restricting the supply of labour. Fourthly, when it faces a monopsonist it can remove exploitation by raising wages without any adverse effect on employment]

19. If a trade union of labour faces a single buyer of labour, how would the wage rate and level of employment is determined? Explain. *D.U.B.A. (Hons)* 1994

20. Wages and employment are indeterminate when a monopsonistic employer faces a monopolistic union. Explain. *D.U. B.A. (Hons)*, 1998

21. Explain the role of trade unions in raising wages in competitive labour market. Does raising wage necessarily lead to unemployment of labour?

22. Define and contrast 'equalising' and non-equalising differences in wages. Which concept is involved when high wages contained 'pure economic rent'? *D.U.B.A. (Hons)* 1988

23. If wages are determined by marginal productivity, trade unions are superfluous. Do you agree? Discuss.

24. If the government wants to intervene in a monopsonistic labour market, at what level would you suggest the minimum wages be fixed? Why. *DUBA (Hons.) 2000*

APPENDIX TO CHAPTER 33

AN APPLICATION OF THE THEORY OF WAGES

In the previous chapters we have explained the determination of prices of factors. Thus we have explained the determination of wages of labour,under conditions of perfect competition and monopsony in labour market. In the present chapter we will discuss an important application of the theory wages, namely, fixation of minimum wages.

FIXATION OF MINIMUM WAGES

The way in which labour market works greatly influences the distribution of income in a society. Some workers, especially unskilled ones are poor because they are paid low wages. The skilled workers are relatively rich because they receive high wages. Therefore, the objective of fixation of minimum wages by the Government is to remove the poverty of the unskilled workers by fixing a minimum wage rate at a level higher than the one determined by the working of free labour market. However, it has been argued by some economists that fixation of minimum wages of the unskilled workers by the Government may not improve the economic conditions of the unskilled workers as a whole. The consequences of fixation of minimum wages by the Government is illustrated in Fig. 33A.1 where determination of wages of unskilled workers is shown. *DD* and *SS* are demand and supply curves of labour respectively. It will be seen from the figure that in the free-market equilibrium wage rate is determined at the level OW_0 where quantity demanded of labour is equal to its quantity supplied. If this free-market wage rate OW_0 is considered to be too low, the Government may intervene and fix the minimum wage at a higher level OW_1. As a result of the imposition of minimum wage legislation, the employers cannot pay wage rate below this minimum wage rate OW_1.

It is important to note that minimum wage rate is fixed *above the equilibrium wage rate* so as to enable unskilled workers to earn a decent living. The fixation of minimum wage below the free-market equilibrium rate will not serve any purpose and will also have no effect on labour market. When the minimum wage rate is fixed at a level higher than the equilibrium wage rate OW_0, that is, at OW_1 in Fig. 33A.1, although it will increase the wage rate of low-wage workers, it will have other important effects on the labour market and on the economic conditions of the working class. We discuss below the various important effects of fixation of minimum wages.

Labour Employment. An important effect of the minimum wage rate is on the level of labour employment which can be easily seen from Fig. 33A.1. When the minimum wage rate is fixed at the level OW_1, that is, higher than the equilibrium wage rate OW_0, the producers will reduce the quantity demanded of labour to ON_1. This means the number of workers equal to N_0N_1 who were already employed will now be rendered unemployed. In addition to this, as will be seen from Fig. 33A.1, at the higher minimum wage rate OW_1, the quantity supplied of labour increases to ON_2 (or W_1L). This means at the higher minimum wage rate OW_1, the additional N_0N_2 workers offer themselves for work, that is, seek employment. With this total number of workers equal to N_1N_2 or KL ($N_1N_2 = N_0N_1 + N_0N_2$) will become unemployed at the higher minimum wage rate OW_1. It may be emphasised again that the labour surplus or unemployment of workers has come into existence because the minimum wage rate OW_1 has been fixed at a level higher than the equilibrium rate (employers employ or demand smaller amount of labour ON_1) and, secondly, because, at the higher wage rate OW_1, the additional number of workers are now willing to work and therefore add to the number of employment seekers who are unable to find jobs.

Since fixation of minimum wages has resulted in the emergence of surplus labour, there will be pressure for wage cutting because the workers who are rendered unemployed will be willing to offer their labour at lower wages than the legally fixed minimum wage rate OW_1. However, an important point to note is that, *unlike the case of maximum price control, this case of minimum price fixation will not lead to the emergence of a group of black-market operators who buy at the controlled price (i.e., minimum wage rate) and sell at the lower free-market equilibrium price (i.e., wage rate OW_0) because nothing will be gained by this.*

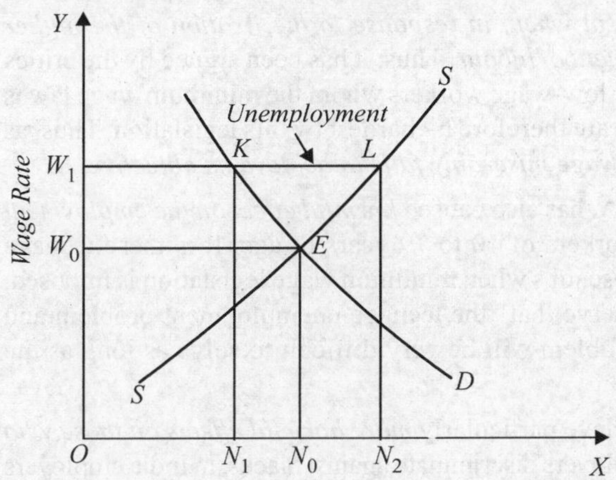

Fig. 33A.1. *Fixation of Minimum Wages*

From the above analysis of the effect of minimum wage rate on employment of labour in case perfect competition prevails in the labour market we reach the following conclusions:

1. The fixation of minimum wage rate will reduce employment of labour (by the amount $N_0 N_1$ in Fig. 33A.l)

2. The fixation of minimum wage will create labour surplus or unemployment. Some workers will be willing to obtain employment but cannot get it.

3. Some workers who are rendered unemployed will be tempted to evade the law and offer to supply their labour at the wage rate below the legally fixed one.

4. The fixation of minimum wages will not lead to the emergence of black marketers who buy labour at the controlled wage rate and sell it in a black market.

Effect on Income of Workers. If the minimum wage law is effectively enforced, it will raise the incomes of those workers who remain employed. But an important question is to understand to what extent the minimum wage legislation benefits the workers as a whole group. The workers who are able to retain their employment when the minimum wage rate is fixed will definitely become better off. They receive higher wages because minimum wage rate is fixed at a level higher than the market determined wage rate. But those unskilled workers who lose their jobs because at a higher minimum wage rate, employers reduce employment of labour and are unable to find new employment will become worse off as a result of minimum wage legislation. It has been argued by some economists that if the total income of the unskilled workers as a whole group increases, then the whole group of unskilled workers gain from the fixation of minimum wages. But even this cannot be said with certainty. This is because the *total income of the unskilled workers as a group will increase only when the demand for unskilled workers* is *inelastic.* In Fig. 33A.1 with the free market determined wage rate OW_0 and employment of unskilled workers ON_0 at this wage rate, the total income of the workers equals the area OW_0EN_0. With the higher minimum wage rate OW_1 and reduced employment equal to N_1, the total earnings of the workers will be OW_1KN_1 which will be greater if the elasticity of demand for labour is less than unity. On the contrary, if the demand for labour is *elastic,* the employment of labour falls greatly with the result that the new total income OW_1KN_1. of the workers as a group will decrease.

However, even if the demand for unskilled labour is inelastic and they as a whole group gain from fixation of minimum wage rate, the fact remains that a good number of workers who are rendered unemployed as a result will individually suffer a loss.

Another important thing to note is that all unskilled workers are not identical. Some unskilled workers earn wages only slightly lower than the minimum wage rate fixed while others earn much less than the minimum wage rate fixed. The employment of those unskilled workers will decline most who are initially paid very low wages because the fixation of a higher minimum wage rate sharply increases the cost of hiring these workers. Thus, *it is the most poor among the unskilled workers who are likely to lose their employment when, in response to the fixation of the higher minimum wage rate, employers reduce employment of labour*. Thus, it has been argued by the critics of minimum wage fixation that some of the poor low-wage workers whom the minimum wage law is intended to help find themselves out of jobs and are therefore hit hardest by this legislation. Thus, as *an anti-poverty measure, fixation of minimum wage miserably fails to achieve its objective.*

Fixation of minimum wage rate in the U.S.A. has also caused *harmful effect on the employment of teenage workers,* that is, employment of workers of 10 to 19 years of age. It is these teenage workers, who are less productive also, mostly lose jobs when minimum wage legislation is imposed. Therefore, Professors Banmol and Blinder observe that "the teenage unemployment problem and especially the black teenage unemployment problem will be very difficult to solve as long as the minimum wage remains effective."[1]

Besides, fixation of minimum wages may have particularly *more harmful effects on those who are the victims of discrimination.* In America employers discriminate against blacks, in India employers often discriminate against workers belonging to schedule castes and schedule tribes. When minimum wages are imposed, the employers of unskilled workers have more applicants seeking jobs than the available employment opportunities. As a result, the employers pick and choose among the applicants and often discriminate against blacks in the U.S.A. and against Schedule castes and tribes in India. It is worth nothing that it is the workers belonging to these weaker sections who because of the past discrimination have not been able to acquire skills required for being employed in higher-paid jobs. In case of the USA, most economists believe that the fixation of minimum wage has contributed greatly to the increase in unemployment of black teenage workers.

It is important to understand as to who bears the cost of fixing higher minimum wages. It is generally assumed that the cost of paying higher minimum wages are borne by the employers, that is, the capitalist class. However, this is not true. When higher minimum wages are imposed, the employers often pass on these higher wages to the consumers in the form of higher prices of goods produced by them. Moreover, the employers often try to shift the cost of higher wages to the suppliers of other inputs which are complementary factors of labour in the form of lower prices paid for these inputs. It follows, therefore, that the costs of higher minimum wages are more widely spread through the society as a whole and not exclusively borne by the employers or business class.

An important effect of fixation of minimum wages is worth mentioning. In our above analysis we have assumed that all workers are covered by minimum wage law. This is however not correct. Minimum wage law does not apply to or cover all jobs or industries. For example, in case of the USA, it has been estimated that 15 per cent of workers are not covered by the minimum wage legislation. In India, the percentage of uncovered workers or jobs is much greater. This has an important consequence. When the minimum wage law is imposed in some sectors or industries and as a result employment of workers is reduced in these covered sectors, supply of labour to the uncovered sectors or jobs increases which depresses wages in these uncovered sectors. This means the workers who remain employed in the covered sectors benefit at the expense of the workers employed in the uncovered jobs. Therefore, whether workers as a group will receive net gain from minimum wage legislation depends on how large is the uncovered sector.

Another potential effect of minimum wage is *the reduction* in *the fringe benefits enjoyed by the workers.* Fringe benefits are the non-monetary benefits such as medical insurance, sickness benefits, pensionary benefits which workers receive in addition to the money wage. As a matter of fact, real

1. Baumol and Blinder, *Economics, Principles and Policy*, Fifth edition, 1992, p.774

wage rate is money wage plus these fringe benefits. Now, when as a result of minimum wage law, employers have to pay highestmoney wage rate, they may reduce fringe benefits offered to their workers so that real wages may not rise much. To the extent the employers are able to reduce the fringe benefits and real wages do not rise as much as money wages, the fall in employment as a result of fixation of minimum wages will be less. Since fixation of minimum wages creates unemployment of labour, workers cannot resist reduction in their fringe benefits.

Lastly, the fall in employment of labour as a result of minimum wage may not fully reflect itself in the reduced number of workers employed but in the form *of reduction in the number of hours worked per day or per week by each worker*. In other words, as a result of reduced employment opportunities, instead of two out of ten workers becoming unemployed, all workers remain employed but each worker now works 80 per cent of the hours which he worked before the fixation of minimum wages. It has however been pointed out by some economists quoting some empirical evidence that since hiring each worker is associated with increase in some overhead costs, the employers prefer to cut back on employment of workers rather than to reduce hours per worker.

Conclusion. The analysis of adverse effect of minimum wage fixation leads us to the conclusion that it cannot be said with certainty that fixation of minimum wages benefits the workers as a whole. Fixation of minimum wages reduce employment opportunities and thereby lead to unemployment of labour, especially of the teenage and young workers. Further, it reduces the fringe benefits of workers and promote discrimination against the blacks, women, Schedule castes and tribes. Of course, some workers who retain employment benefit from minimum wages. But the main objective of minimum wage legislation is to provide decent living to the low-wage or poor workers. It is believed by many Western economists that *minimum wage fails to achieve its objective of removal of poverty of these workers*.

Case for Minimum Wage

In the opinion of the present author, the above analysis of the adverse effects of minimum wage legislation is based upon the assumptions of wage determination by demand for and supply of labour in a competitive labour market. Further, the above analysis assumes static conditions. If we remove these assumptions, the beneficial effects of the fixation of minimum wage legislation can be easily shown and the unfavourable effect of increase in unemployment of the fixation of minimum wage rate can be prevented. This is because *fixation of higher minimum wage rate may increase labour productivity.* The increase in labour productivity will cause a shift in the labour demand curve to the right. This will prevent the increase in unemployment which fixation of higher minimum wage rate may create. But the pertinent question is how the fixation of higher minimum wage rate for unskilled workers can result in increase in labour productivity. First, it is argued that *higher minimum wage for low-wage unskilled workers may produce shock effect upon employers*. It has been generally observed that firms using low-wage unskilled workers tend to be inefficient in the use of labour. The imposition of higher minimum wage will provide a shock to them, that is, prompt them to improve labour efficiency so that it is profitable for them to keep employing them. Second, it has been pointed out that the *fixation of higher minimum wages leads to the increase in real incomes of these unskilled workers who with their higher wages may improve their health and vigour by meeting their basic needs*. This is true especially in the case of poor and developing countries like ours where unskilled workers are paid very low wages which are quite inadequate even to meet basic needs. Besides, the higher real income *may increase the motivation of these workers* to do more work. The improved health, greater vigour and higher motivation to do more work resulting from higher minimum wages may result in rise in their efficiency or productivity. This will cause a shift in the demand curve for labour to the right which will tend to offset the tendency to reduce labour employment when the higher minimum wage is imposed. This is generally referred to as *economy of high wages*.

In the case of the existence of monopsony in the labour market, workers are exploited by the employees by restricting employment of labour and paying low wages. *Under the monopsonistic conditions in labour market, imposition of minimum wages can lead to the increase in wage rate without creating unemployment.* As a matter of fact, fixation of higher minimum wage can lead to the increase in employment of labour by doing away with monopsonistic employer's motive of restricting employment. This is illustrated in Fig. 33A.2 where the curve *VMP* represents value of marginal product of labour. *SL* is the supply curve of labour and *MW* is marginal wage or marginal factor cost curve of labour of the monopsonist. The monopsonist is in equilibrium when he employs *OL* labour and pays *OW* or *LK* wage rate which is less than the value of marginal product of labour *LE.* Thus, he is exploiting labour by restricting employment.

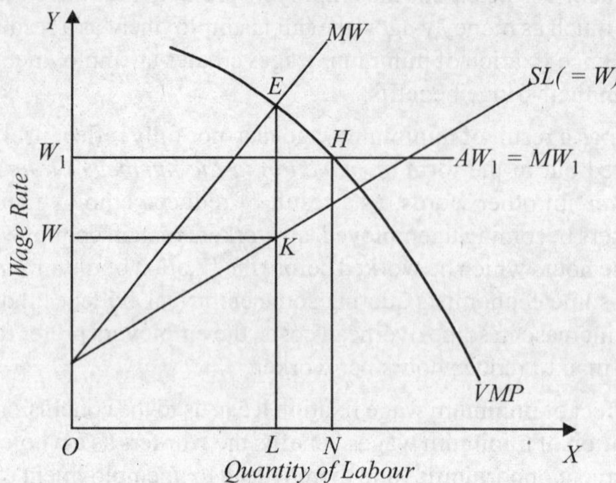

Fig. 33A.2. *Effect of Minimum Wage in Case of Monopsonistic Labour Market*

Now, if minimum wage rate OW_1 is fixed, the monopsonist must pay wage rate OW_1 since payment of wage rate below it will not be permitted by law. With the given minimum wage rate OW_1 the supply curve of labour will now become horizontal straight line and marginal wage curve will coincide with it. With the wage rate OW_1, the monopsonist will maximise profits by employing *ON* labour which is greater than *OL*. Thus, it is clear that in monopsonistic labour market, fixation of minimum wage rate has led to the increase in employment of labour. What is true of monopsonistic labour market generally applies to oligopsonistic labour market.

Conclusion

Two opposite views regarding effects of minimum wages have been explained above. Whether or not fixation of minimum wage by the Government has a net beneficial or adverse effect on unskilled workers cannot be decided on the basis of theoretical grounds alone. This 'is primarily an empirical issue which can be clinched by undertaking research studies. A large number of research studies concerning the effect of minimum wages have been made in the case of the developed countries such as USA. These studies generally conclude that minimum wage legislation causes some increase in unemployment of unskilled workers, especially of teenage workers and black ones.

QUESTIONS FOR REVIEW

1. Explain what is the effect of the fixation of minimum wages in a competitive labour market. Does it necessarily lead to unemployment of labour? *D.U. B.A. (Hons) 2000*
2. As an anti-poverty measure fixation of minimum wages fails to achieve its objective. Do you agree? Discuss.
3. What is meant by *economy of high wages*? How would you explain the beneficial effect of fixation of minimum wages on the basis of the hyphothesis of economy of high wages.
4. Show that the fixation of minimum wages in a monopsonistic labour market may lead to the increase in both wages and employment of labour. *D.U. B.Com Hons)*
5. Explain the effect on empoyment if a minimum wage is applied to a monopsonic labour market. *D.U. B.A. (Hons) 2001*
6. "Rent Control of houses reduces the availability and quality of houses in the long run" Discuss.

CHAPTER 34
THE THEORY OF RENT

CONCEPTS OF RENT

Having discussed the determination of reward for labor, we now turn to explain the determination of rent which is the price paid for the use of land. However, in modern economic theory, the term rent is used not only in the sense of reward for the use of land, but also in the sense of surplus earnings of the factors over their transfer earnings. In fact, in the later sense the concept of rent has been generalised so that it applies to *surplus return over and above the transfer earnings of all factors of production*, so that it is no longer peculiarly associated with land. We shall first discuss the determination of land rent and will then explain the concept of rent as surplus return over transfer earnings of the factor.

A distinguishing feature of land is that no human effort or sacrifice has been necessary to make it available to the society. Land to society is a free gift of nature. Society has not incurred any cost to obtain the land. Further, since land is not producible by man, *its supply is absolutely inelastic*, although its productivity can be increased by various improvements such as clearing, draining, introduction of irrigation facilities etc. These improvements are made by the efforts of man and therefore constitute capital goods. As the quantity of land available for use is scarce relative to demand, a price must be paid for its use. This price for the use of land, or what is commonly called *land rent*, is obtained by those people in the society in whom the ownership of land is vested. Since these private owners of land have not incurred any real cost to bring land into existence, the rent which they obtain is a surplus payment to them. The whole of the earnings of land *i.e.*, land rent (excluding, of course, the return on capital investment in the form of improvements made on land by the owners) are *surplus*, since land is there in any case and does not require any costs or human efforts to be made to bring it into existence. Thus, the term rent which was originally employed for the price paid for the use of land came to be used for *the surplus earnings of any factor of production in excess of the cost incurred to obtain its service*. Since the land in its entirety to the whole society, being free gift of nature, does not require any cost to be paid in order to make it available to the society for use in production, the whole earnings of land are regarded as surplus. Thus, the whole earnings of land from the view point of the society become economic rent. We shall discuss this concept of economic rent in greater detail when we take up the explanation of the concept of rent as a surplus over transfer earnings to all factors of production.

It should be noted that rent as a payment to the landlord for hiring or use of land by the tenant and the concept of economic rent as surplus over transfer earnings which applies to all factors of production are altogether different concepts and therefore should not be confused with each other. Modern economists generally call payment for hiring of land as *land rent* and surplus over transfer earnings of the factors as *economic rent*.

It is worthwhile to note here that Marshall extended the concept of rent to cover the earnings (net of depreciation and interest charges) of fixed capital equipment like machinery in the short run. The distinguishing characteristic of land is the fact that its supply is perfectly inelastic and therefore its earnings depend mainly upon the demand for it. But, in the short run period, fixed capital equipment like machinery is likewise perfectly inelastic in supply and cost of production is not relevant once it has been installed for production. Thus, in the short period, the earnings of fixed capital equipment depend mainly upon the demand conditions and are thus similar to land rent and have therefore been called rent by Marshall. Since these capital equipments are not permanently in fixed supply like land, and instead their supply is very much elastic in the long run. Marshall preferred to call their earnings in the short period as *Quasi-Rent rather than rent*.

Now we propose to discuss the above three concepts of rent as mentioned above. We start with the Ricardian theory of rent which discusses the problem of payment for the use of land and other natural resources.

RICARDIAN THEORY OF RENT

The Ricardian theory of rent follows from the views of classical writers about the operation of law of diminishing returns in agriculture. Classicical authors, West, Torrents, Malthus and Ricardo, each of them independently formulated the theory of differential rent. However, the classical theory of rent in the form presented and elaborated by David Ricardo has become more popular, though the ideas of all of them concerning the land rent is fundamentally same. Ricardo gave credit to West and Malthus as his forerunner in the development of the theory of rent.

Ricardo defined rent as follows : "*Rent is that portion of the produce of earth which is paid to the landlord for the use of the original and indestructible powers of soil.* "It should be noticed that land rent, according to Recardian definition, is a payment for the use of only land and is different from contractual rent which includes the return on capital investment made by the landlord in the form of hedges, drains, wells and the like. When return on the capital investment made by the land owner is deducted from the contractual rent, what is left is pure land rent which is the price for the use of only land or "*the original and indestructible powers of the soil*".

Assumptions of Ricardian Theory

It will greatly help in the understanding of the Ricardian model of rent determination, if we clearly state the various assumptions made by him. Firstly, Ricardo considers the supply of land from the viewpoint of the whole society and takes the quantity of land as completely fixed. No amount of higher price for the use of land can call forth an increased supply of it. Thus the total supply of land is perfectly inelastic and unresponsive to any changes in rent. Secondly, *he does not take into account the various alternative uses to which land can be put*. He assumes the land to be used for growing a single composite crop 'corn'. Thus land has been taken to be completely specific to one crop, *i.e.*, corn. In this way, in Ricardian model, either land is to be used for growing of corn or alternatively it has be left idle. There are only two alternative uses of land: its use for growing of corn or no use at all. Thus *he takes the transfer earnings of land as zero*. No land owner would like to leave the land idle and therefore every land owner will be would prepared to give it for any rent however little it may be provided that perfect competition prevails.

Thirdly, *he assumes that land differs in quality*. There are various grades of land, differing from each other in respect of fertility and location. Some pieces of land are more fertile than others and, as compared to others, some are more well located or near to the market centres.

Fourthly, *he assumes that there is perfect competition in the market for land*. In other words, there are many land owners who are to give their land on rent and there are many farmers who are to get land on rent for the purpose of growing corn. Further, each individual land owner and farmer has no influence over rent *i.e.*, the price for the use of land.

Given the above assumptions, according to the Ricardian theory, rent arises due to two reasons. Firstly, if land is homogeneous, *i.e.*, of uniform quality and same location, the scarcity of land relative to demand will give rise to rent. Ricardo calls it a *scarcity rent*. Second, when land differs in quality, *i.e.*, in fertility and location, the scarcity of superior grades of land will give rise to *differential rents*.

We shall discuss below the emergence of scarcity and differential rents, as conceived in the Ricardian theory.

Scarcity Rent

The emergence of land rent in the classical theory can be easily explained by imagining that a new island is discovered and some people come to settle there. We suppose that all land in this island is completely homogeneous or is of uniform quality. In other words, all pieces of land in this island are equally fertile and equally well-situated. The quantity of land available for cultivation on this island is fixed and is therefore completely inelastic to changes in the price for its use. Land is to be used for the cultivation of a single crop "*corn*". Land is assumed to be having no other alternative uses. When the people come to settle on this island, they will use the land for producing corn by applying labour and capital on it. When all the available land is not yet put in use, the price of the corn will be equal to the *average cost of output incurred on labour and capital*, with the farmers working at the minimum point of the average cost (exclusive of land rent). The price of the corn must at least be equal to the average cost (exclusive of land rent) in the long run if the use of labour and capital is to be worthwhile. Since we are assuming perfect competition in the market for corn, the farmer's equilibrium will be established at the lowest point of long-run average cost curve (exclusive of rent).

As long as some land is idle, the production of corn will be increased by bringing new land under cultivation. Thus until land is not scarce, *i.e.*, some land is yet idle, the price of corn cannot rise *permanently* above the average cost of labour and capital cost. Since the price of corn is, in long-run equilibrium, equal to the average cost of only labour and capital, as long as all land is not yet in use, *there will be no surplus left to be earned as rent on land*. In other words, it means that so long as there is some available land which is not yet brought into use, farmers will not have to pay any rent to the landlords for the use of their land.

Provided the competition among landlords is perfect (as is the case we are assuming here), the rent will not arise when there is still surplus land for use because the demand for land is relatively less than the supply of it. In other words, land is yet not scarce relative to demand. Price of any things arises only when it is scarce in relation to demand. If any landlord tries to charge any rent when there is still some land lying idle with other landlords, farmers will go to take up that land for cultivation. The landlord need not be paid rent for the use of land since its only alternative use is keeping it idle. To sum up, so long as land is not scarce, rent cannot arise, since price will equal minimum average (labour and capital) cost. Suppose that the population continues increasing so that the demand for corn becomes so large that all available land is brought under cultivation. If the population of the island further increases beyond this, it will raise the demand for the product which will bring about rise in the price level above the minimum average (labour and capital) cost per unit of output giving rise to rent on land. Since it has arise in due to scarcity of land, it has been called scarcity rent.

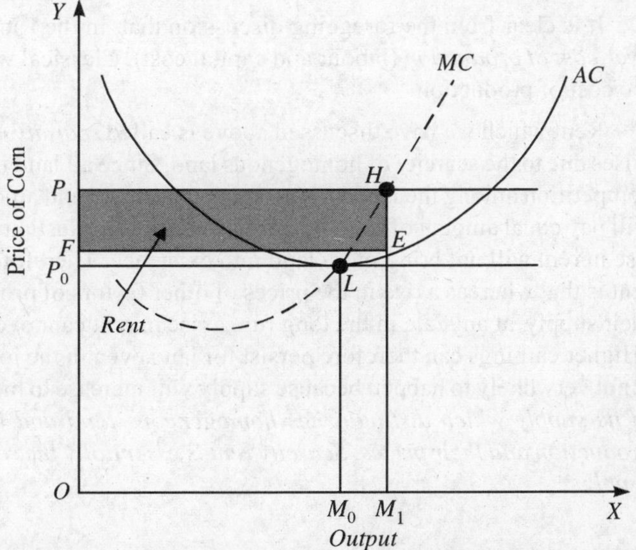

Fig. 34.1. *In Ricardian theory scarcity rent arises as surplus over cost of production.*

Recardian concept of scarcity rent is illustrated in Fig. 34.1. Where *AC* and *MC* curves show average and marginal cost per unit output of corn incurred on labour and capital. Price of corn must be equal to OP_0 if land is to be cultivated at all. Note that price OP_0 is equal to the minimum average cost per unit of corn output on labour and capital. At price OP_0 there is no surplus over cost of production and therefore no rent accrues to the land. In other words, supply of land is not scarce in relation to demand for it upto price of corn equal to OP_0.

Now, if due to the expansion in population, demand for corn increases and as a result price of corn rises to OP_1 land will be more intensively cultivated. With price of corn equal to OP_1, the equilibrium of the farmer is at point *H* or at output OM_1 as price of corn is equal to marginal cost at output level OM_1. It will be seen that with price OP_1, surplus over cost of production equal to P_1HEF (shaded area) has emerged. This surplus over cost will be given to the landlord. The price of corn rises abvoe the minimum average cost of production only when the demand for corn has greatly increased and as a result land has become scarce in relation to the derived demand for it. Since all pieces of land are homogeneous, the same amount of rent will accrue on all pieces of land.

It is evident from Fig. 34.1 that a difference between the price of the corn and the average cost on labour and capital has arisen. In other words, farmer earns more than the labour and capital cost incurred by him. While the average labour and capital cost incurred by him is M_1E, the price of the corn is $M_1H(= OP)$. Thus the differential *EH* between the price and the average labour and capital cost has arisen. This *EH* is the rent per unit of output which will be paid by the farmer to the landlord. Total rent to be paid by the farmer to the landlord will be $FEHP_1$. This rent (difference between price and cost) cannot be competed away by the entry of more farmers in production since all land is already being employed for production. This rent has arisen because of the scarcity of land. In other words, rent arises due to the niggardliness of nature; nature has not provided land large enough to meet the level of demand by producing on the minimum level of average cost. The demand for corn has increased so much that the required output cannot be produced with total available supply of land at the minimum average cost (exclusive of rent). In order to meet the increased demand, output has to be expanded to the point *H* of marginal cost curve so that the new price OP_1 equals marginal cost. The marginal cost M_1H which is equal to price OP_1 exceeds average cost M_1E by *EH*. Thus a differential *EH* between price and average cost emerges, which is rent of land required to be paid to the landlord.

It is clear from the foregoing discussion that, in the Ricardian theory, *rent emerges as surplus over cost of production* (labour and capital cost). Classical writers did not consider rent as a part of the cost of production.

Rent which we have discussed above is called *scarcity rent*. It is called scarcity rent because it arises due to the scarcity of homogenous land. Since all land is homogeneous and there exists perfect competition among the land owners on the one hand and among the tenants on the other all farmers will pay equal amount of rent. Because land has zero elasticity of supply *i.e.*, its quantity is fixed, the rise in rent will not bring more land into existence. Therefore, the essential feature of pure scarcity rent is that whereas a rise in the prices of other factors of production will bring about an increase in their supply, at any rate in the long run, a rise in rent cannot cause an increase in the supply of land. "Higher earnings can therefore persist for land even in the long run, whereas with other factors this is not very likely to happen because supply will increase to meet the increased demand. *It is the fixity of its supply which distinguishes homogeneous land and its scarcity rent from other factors of production and their prices. Scarcity rent is essentially the result of the fact that land is in inflexible supply.*"[1]

1. Stonier and Hague, *A Textbook of Economic Theory*, 4th edition, 1972, p.283

Differential Rent

In the discussion of scarcity rent above, we have assumed that all land is homogeneous, *i.e.*, equally well fertile and equally well-situated. This is, however, not a realistic assumption. In fact, Ricardo was most interested in showing the emergence of rent when the land differs in quality *i.e.*, in fertility and situation. Some pieces of land are more fertile than others. Again, some pieces of land are more favourably situated than others. That is, they are located near to the market centres where produce has to be sold, than others.

Fertility of tracts of land varies primarily because of the differences in the nature of the soil , temperature, rainfall and other climatic factors. With a given application of labour and capital, some pieces of land will yield more output per acre than others. Thus the differences in fertility will bring about differences in the costs of production (exclusive of rent) of various farmers operating on the different grades of land. The farmers working on the superior or more fertile grades of land will have their average cost curve at a lower level than those working on the inferior or less fertile grades of land. Likewise, differences in location cause differences in costs of various farmers because of the differences in transportation costs. In practice, land will be of numerous grades, shading off gradually from the best to the poorest. To simplify our analysis, we however, assume that in our island there are three grades of land. Land *A* being the superior most and *C* the poorest, *B* grade of land lies between *A* and *C*.

When people first come to island, they will take up the best grade land *A* for the production of corn. So long as some of grade *A* land is yet lying idle, there will be no rent. When with the increase in the population of the island or with the development of the island, the demand for corn increases, the whole of the grade *A* land will be put into use for the production of corn. At this stage each of the many farmers who will be using the grade *A* land will work at the lowest point of the average cost curve as shown in Fig. 34.2(*a*). When once the whole of grade *A* land is brought into use and the demand for corn still further increases due to either growth of population or the development of the island, *two courses of action will be adopted.* First, grade *B* land will also be taken up for cultivation. Secondly, grade *A* land will be *more intensively used i.e.*, more doses of labour and capital will be applied to the pieces of grade *A* land.

Now, the grade *B* land can be taken for use only when the price sufficiently rises so that it covers the average cost of production on grade *B* land. In other words, price must be high enough to cover the minimum average cost (exclusive of rent) on grade *B* land, otherwise it will not be worthwhile to cultivate it. In other words, if the price is lower than the lowest average cost on grade *B* land, its cultivation will not pay back even the labour and capital cost incurred and therefore it will not be

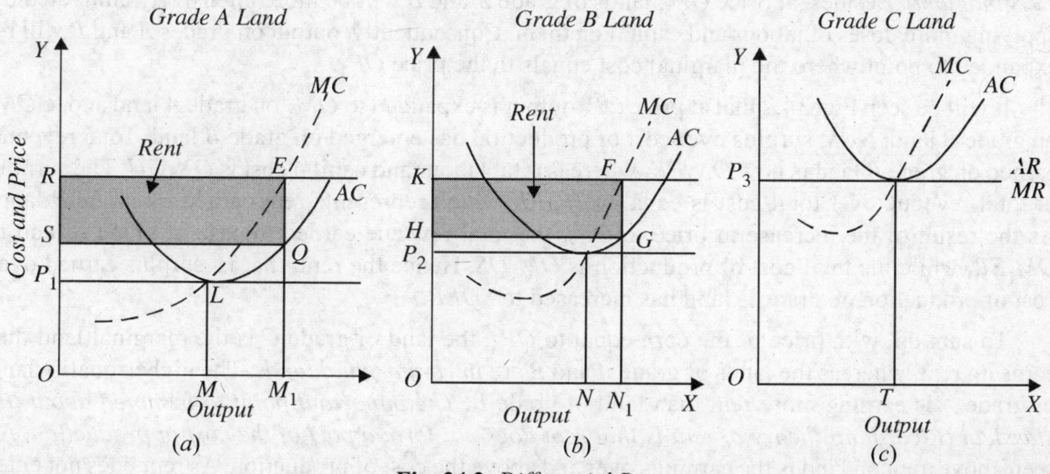

Fig. 34.2. *Differential Rent*

brought under cultivation. It is evident from Fig. 34.2 (*b*), that price must rise to OP_2 if the grade *B* land is to be taken up for production. Now suppose that demand for corn has risen so much that price of corn is OP_2 and therefore grade *B* land has been brought under cultivation. Thus, margin of cultivation has been extended to grade *B* land. In other words, grade *B* land is now on the *margin of extensive cultivation*. Every farmer cultivating the grade *B* land will operate on the lowest point of average cost curve *AC* in Fig. 34.2(b). Since the price of OP_2 is equal to average labour and capital cost on grade *B* land, there is no surplus over cost of production and hence grade *B* land does not earn any rent. But because price OP_2 stands higher than the lowest average cost on grade *A* land, surplus over cot of production would appear on grade *A* land. This surplus is rent which will be paid to the landlord of grade A land.

It should be noticed that besides extending the margin of cultivation to grade *B* land, there will also be side by side more intensive cultivation of grade *A* land by applying more does of labour and capital on it. In other words, *margin of intensive cultivation* will also be pushed forward. In terms of Fig. 34.2, it will mean that the farmers operating on grade *A* will not produce at the lowest average cost, they will also expand output to meet the increased demand. With the expansion in output, the marginal costs on farms of grade *A* land will rise. The price must rise to cover this increase in marginal cost, if the extra costs incurred on additions to capital and labour for expanding output are to be recovered. In Fig. 34.2(*a*) when the farmers of grade *A* land extend the margin of their intensive cultivation in response to increased demand, their new equilibrium position will be where the marginal cost is equal to new higher price OP_2.

It should be noticed that rent on grade *A* land would have arisen even if no more intensive cultivation was done and the output was restricted to *OM* level since the price OP_2 stands higher than the lowest average cost *ML* on grade *A* land. But, in practice, both the extensive and intensive margins are pushed further in order to meet the increase in demand and the surplus over cost of production *i.e.*, land rent on intra-marginal lands arises because of both the more extensive and intensive cultivation. At this stage grade *B* land is *marginal land* which earns no rent and grade *A* land is *intra-marginal land*, which will earn rent.

Now suppose that population of the island further increases which brings about further increases in demand for the produce of land so that the price of corn further rises to the level OP_3. As a result of this, the grade *C* land will also be brought under cultivation and lands of grade *A* and *B* will be more intensively cultivated. Price OP_3 is equal to the minimum average cost on grade *C* land. There will be no surplus earned over cost of production on grade *C* land and hence grade *C* land does not earn any rent. Grade *C* land is now on the margin of extensive cultivation. Thus, grade *C* land is the *marginal land.* Besides, at price OP_3, lands of grade *A* and *B* will be more intensively cultivated by applying more doses of labour and capital on them. Consequently, output on grades *A* and *B* will be expanded to point where the marginal cost equals to the price OP_3.

It will be seen Fig. 34.2 that at price OP_3 output is expanded to OM_1 on grade *A* land and to ON_1 on grade *B* land. Now, surplus over cost of production has emerged on grade *B* land. Total revenue earned on grade *B* land is now ON_1FK, whereas total labour and capital cost is ON_1GH. The surplus of total revenue over total cost is equal to *HGFK* which represents rent earned by grade *B* land. As the result of the increase in price to OP_3, the total revenue earned in case of grade A land is OM_1ER, while the total cost of production is OM_1QS. Hence the rent, that is, surplus earned over cost of production on grade *A* land has increased to *SQER*.

To sum up, with price of the corn equal to OP_3, the land of grade *C* is the marginal land that earns no rent, whereas the lands of grade *A* and *B* are *intra-marginal lands*. The higher-quality land of grade *A* is earning more rent than land of grade *B*. *The important point to be noted about the classical (Ricardian) theory of rent is that rent does not form a part of the cost of production*. As seen above, rent on land is the earnings over and above the cost of production. As rent does not enter into cost of production, it therefore does not determine price. Price of corn (or produce of the land) must be equal to the minimum average cost of production of the marginal land, but the marginal land earns no rent. It is thus clear that in Ricardian Theory, rent is not *price determining*. In fact, in this theory rent is price determined, that is, it is price of corn which determines rent, and not other way around. The quote Ricardo, "*Corn is not high because a rent is paid, but a rent is paid because corn is high*."

CRITICAL EVALUATION OF RICARDIAN THEORY OF RENT

So far as the determination of land rent is concerned and the forces which influence it, modern economists agree with Ricardian theory of rent. Like Ricardo, modern economists are also of the view rent of land arises because of its scarcity. Although Ricardo explained the determination of land rent through a '*differential return' approach* and not the basis of *direct* demand for and supply of land, and accordingly did not employ demand and supply curves to depict the determination of land rent, yet in Ricardian theory it is the forces of demand for and supply of land which determine the rent of land. Like modern economists, Ricardo too believed that demand for land is derived demand, it is derived from the demand for the produce of land, that is, what Ricardo called 'corn'.

In the Ricardian theory with its differential approach, increase in the population of a country raises the demand for the corn and brings about rise in the land rent. In the modern approach based on direct interaction of demand and supply, the increase in population will shift the demand curve for land to the right and thereby will push up the rent. Thus *the demand-supply approach (or in other words, marginal productivity approach) of the modern economists and differential return approach of Ricardo are alternative explanation of the same phenomenon and are in no sense contradictory*; in both approaches forces of demand and supply play a crucial role in determination of rent. But the use of Ricardian differential approach does not provide any better understanding of the nature of rent than that of modern approach based on the use of direct demand and supply concepts. However, the adoption of Ricardian differential approach often leads to misunderstanding, for its suggests that rent of land requires a special theory for its explanation, that is, it may lead one to conclude that whereas rent of land can be explained with differential principle, the other factor rewards of labour, interest on capital, etc. can be explained with demand and supply or marginal productivity theory. Obviously, such a conclusion would not be correct. We thus see that whereas there is not contradiction between differential theory of rent propounded by Rciardo and the modern theory based on direct use of demand and supply and the marginal productivity principle, there seems to be a little reason for the special theory of rent (*i.e.*, differential theory). Unity of economic thought and similarity of treatment requires that land rent should also be explained directly with demand and supply curves as other factor prices. We shall explain later in this chapter how rent of land is determined by demand for and supply of land.

Ricardian theory has also been citicised by some modern economists for its laying undue emphasis on the "*original and industructible powers of the soil*". According to them, fertility of the soil gets destroyed after a few years of continuous cultivation. Further, fertility of soil can be greatly improved by using modern techniques of production. Thus, according to Stonier and Hague, "In these days of nuclear physics and atomic energy it is very dangerous to assert that anything is indestructible..... It is not reasonable to regard the powers of the land as indestructible."[2] They further write "it seems more reasonable to attribute the payment of rent not to the original and indestructible powers of the soil but rather to the fact that land is a factor of production which is in almost completely inelastic supply to changes in its price."[3]

Briggs and Jordan criticise the Ricardian differential theory of rent as a mere truism. According to them, it just seeks to prove that a superior grade land would earn more than an inferior land. They

2. Stonier and Hague, A Textbook of Economic Theory, 4th edition, p.276
3. Ibid, p.267

thus write "Fundamentally, all that the Ricardian theory of rent amounts to is the truism that the better article will always command the higher price. A more fertile acre will be worth more than less fertile one simply because they are different things." However, in our view, criticism of Briggs and Jordan is misplaced. Through his defferential theory of rent Ricardo tried to show that productivity of land like the marginal productivity of other factors is an important factor in determining demand for land and therefore rent for its use.

Another disagreement of modern economists with Ricardo relates to the role of rent in determining price of product (*i.e.*, corn in Ricardo's case). As seen above, Recardo was of the view that rent is a surplus over cost of production and therefore does not enter into cost of producing corn and hence does not determine its price. In fact, according to him, *rent is price-determined and not price determining*. Thus, according to him, price of corn determines rent of land and not the other way around. As we shall explain late in detail, from the point of view of an individual firm or individual industry rent of land or at least transfer earnings of land is a necessary part of the cost of production and therefore contributes to the determination of the price of corn or product of land. Thus modern economists do not agree with Ricardo that land rent does not determine price of corn.

Land's Share in National Income and Stagnation of Economic Growth. Another aspect of Ricardian theory with which modern economists do not agree is the prediction by Ricardo, on the basis of his rent theory, that process of economic growth would come to an end and there will be economic stagnation. He argued that as the population increased, the demand for land would also increase. As a result, both the extensive and intensive margins would be pushed further and due to this rent of land will go up. According to him, in this process, share of land rent in the toal national product will go up and share of profits will decline. Decline in profits means not much money will be available for financing the industry. Further, decline in the rate of profit will discourage inducement to invest. Consequently, further investment and growth process would come to a halt. But in actual practice things have not worked out in accordance with the prediction of Ricardo. To quote Jan Pen again, "the grandiose theory developed by D. Ricardo at the beginning of the nineteenth century.... amounts to the fact that population growth and shortage of land will force up the share of land rent, as a result of which no money will be left for the financing of industry. Ricardo predicted stagnation of economic growth, but this prediction did not work out, nor has the increase in the share of rent of land occurred.... There is in fact a clear shortage of land at many places, and especially in the cities this may lead to high prices and rents. This situation is not without it problems, but it is nothing like Ricardo's predictions

EXPLAINING DETERMINATION OF LAND RENT THROUGH DEMAND AND SUPPLY

Ricardian differential approach, demand and supply of land determine rent, through they are not directly shown to do so. The Ricardian theory can be expressed and explained directly in terms of demand for and supply of land. In the Ricardian treatment of rent, the supply of land for the economy as a whole is fixed, and the demand for land is derived from the demand for the "corn" produced by land. Thus, according to Lipsey, "The modern student of Economics will recognize in the Ricardian arguments the idea of derived demand. Landlords, Ricardo was saying, cannot just charge any price they want for land; the prices they get will depend on demand and supply. The supply of land is pretty well fixed and the demand depends on the price of corn. The higher the price of corn, the more profitable will it be to grow corn, higher will be the demand for corn land, and the higher will be the price for its use."[4]

In the Ricardian model, it is assumed that land has only one use, that is, of growing corn on it. Therefore, no price is required to be paid to prevent land from transferring itself to other uses, that is, to uses other than growing corn. Further, the land has been considered for the economy as a

4. *Op. cit.*, p. 349

whole. Therefore, the supply of land is a given and fixed quantity. Given these assumptions, supply curve of land is *perfectly inelastic*. Scarcity rent will arise only when the available quantity of land is scarce in relation to demand for it, which is derived from the demand for the corn. The Ricardian model of the determination of scarcity rent through demand and supply curves has been illustrated in Fig. 34.3 where SS' is the supply curve of land with OS as the available quantity of land. It is assumed that all land is homogeneous and therefore no differences in fertility or location exist. It will be seen from the Fig. 34.3 that various demand curves such as D_0D_0, D_1D_1, D_2D_2, D_3D_3 and D_4D_4 for land depend upon the various levels of demand for the product *i.e.*, corn.

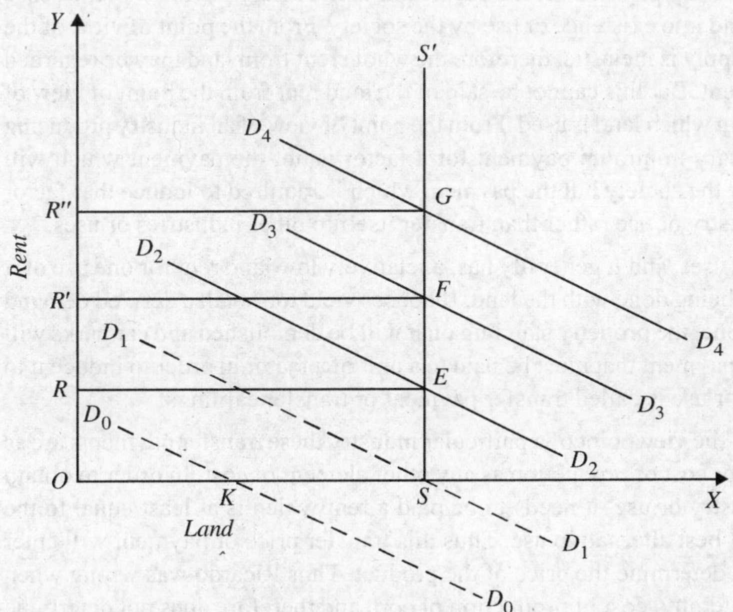

Fig. 34.3. *Ricardian Model: Determination of Land Rent through Demand and Supply Curves*

It should be noted that aggregate demand curves for land have been obtained by summing up the marginal revenue product curves of all farmers cultivating the land. Thus, as the demand for other factors, demand for land, is also determined by its marginal revenue productivity. If, to begin with, demand curve for land is given by the curve D_0D_0, *then no rent on land will be charged* on hiring it. This is because with D_0D_0 as the demand curve, the available land OS is abundant in relation to the demand for it. With D_0D_0 as the demand curve, various farmers will use OK amount of land and as a result KS amount of land will remain idle or uncultivated. If the demand for land increases to D_1D_1, even then the rent of land will be nil, but now the whole amount of available land OS will be cultivated. If the demand for land increases to D_2D_2, the demand for and supply of land will be in equilibrium at point E and rent OR is determined. With the increase in demand for land to D_3D_3, rent rises to OR' and with the further increase in demand for land to D_4D_4, rent rises to OR'', whereas the supply of land remains fixed at OS. Demand for land can increase as a result of population growth which raises the consumption of corn or agricultural produce. This increased demand for corn or agricultural produce causes the demand curve for land to shift upward and thereby brings about a rise in the rent of land.

We thus see that Ricardian model of determination of land rent can be better explained with demand and supply curves. Therefore, there is no need for the adoption of special differential return theory of rent, while other factor prices are explained on the demand and supply basis and marginal productivity principle.

Demand Supply Analysis and Differential Rents. With demand-supply approach we can also explain differential rents arising on different kinds of lands between which there are differences of fertility and location. In our above analysis of rent determination through demand and supply, we have assumed that all land is homogeneous. Therefore, a single uniform rent is determined. When there are differences in land, each kind of land would have its own separate demand and supply curves and consequently different rents depending on their demand and supply will be determined.

Thus, on demand and supply basis, even differential rents on different grades of land in respect of fertility and location can be explained.

LAND RENT, COST AND PRICE

Ricardo explained the determination of land rent from the standpoint of the economy as a whole. He did not consider the question of rent payment from the point of view of a single industry or use. He did not consider the *various different uses or industries* for which land can be used. Since he confined himself to the whole land and a single use of it, he concluded that whole land rent is a surplus, superfluous and unnecessary payment. It is a surplus or unnecessary payment in the sense that it is not required to bring land into existence or use by the society. From the point of view of the economy as a whole, the total supply is inelastic, therefore the whole rent from land may be regarded as surplus or unnecessary payment. But this cannot be said of the land rent from the point of view of the individual industries or uses in which land is used. From the point of view of an industry producing a particular product, the necessary minimum payment for a factor is not the payment which will bring it into existence or use for the society but the payment which is required to induce that factor to remain in that particular industry or use rather than transfer itself to other industries or uses.

If land has various possible uses, and it generally has, a relatively low land rent for one use of it will lead to something different being done with the land. If houses yield too small a derived demand for land, and car parks a higher one, the property standing on it will be demolished and car parks will appear.[5] The minimum price or payment that must be paid to a unit of a factor in order to induce it to work in the particular industry or use is called transfer payment or transfer earnings.

It should be noted that from the viewpoint of a particular industry these transfer payments are as much a part of the supply price or cost of production as any other element of cost. In order to obtain the land for any particular industry or use, it needs to be paid a rent which is at least equal to the income it could earn in the next best alternative use. Thus this transfer price or payment will enter into cost of production and will determine the price of the product. Thus Ricardo was wrong when he asserted that rent does not enter into cost of production of corn and therefore does not determine the price of the corn. Thus, according to modern economists, at least that part of land rent which is the transfer cost or transfer earnings enters into the cost of production of the commodity produced by land and therefore determines the price of the commodity as any other element of cost. Prof. Robertson expresses the modern viewpoint in regard to the land rent entering into the cost of production of a commodity in the following words: "the old phrase about rent not entering into cost . . . . conjures upto my mind a stately temple labelled costs of production with poor Rent standing disconsolately on the mat outside . . . . from the point of view not only of the individual firm but of the individual industry the price paid for the use of land is exactly on all fours with the price paid for the use of the other factors of production."[6]

That the land rent enters into the cost of production and thus determines price of a commodity for the production of which land is used is quite evident if we consider the supply of land from the viewpoint of the individual farmer or any other individual uses of land. For an individual farmer, the supply of land is perfectly elastic at the given rate of land rent. If an individual farmer does not pay this land rent, the land will get transferred to other farmers who will be ready to pay the current rate of land rent. It is therefore quite clear that for the individual farmer the whole of land rent is a necessary payment and he therefore must include it in his cost of production on the basis of which price of the commodity will be determined. Jan Pen rightly writes: "Ricardo was right when he wrote: corn is not high because a rent is paid, but a rent is paid because corn is high. But that 'right'

5. Jan Pen, *Income Distribution*, Penguin Books, 1971, p.128
6. D.H. Robertson. *Lectures on Economic Principles*, Fontana Library Edition, p. 204.

must be interpreted solely in the sense of inelastic supply curves that are confronted with a derived demand. If the latter is at a high level because end-product is expensive, a high land rent results from the market, that rent is not a cause but a consequence. However, *one may not deduce from this that rent for the individual farmer is not a part of the costs*. Not a single farmer would believe that, and rightly so. He simply has to pay for the land as for labour and for capital."[7]

We therefore see that Ricardo's view that rent of land does not enter into cost of production and therefore does not take part in the determination of the price of corn is quite wrong. He formed this view by considering land from the point of view of the whole society for which land is completely inelastic and has zero transfer cost. As a result, from the point of view of the society as a whole, the whole land rent is surplus and unnecessary payment. But, as Chamberlin writes, "Although rents may be surpluses from certain points of view or for certain purposes, or subject to certain interpretation, they are to the individual producer no different from any other money expense. They do not arise as a surplus from his own operations; they are a cost rigidly imposed upon him by the competition of his rivals for the use of rent-yielding property. They figure in the same way as do the wages of labour and the interest of capital in his computations ."[8]

To sum up, from the point of view of the individual industry and farmer or producer, land rent enters into cost of production and therefore determines price. Ricardo was wrong in holding an opposite view that rent does not enter into cost of production and therefore does not determine price.

MODERN THEORY OF RENT

Economic Rent as a Surplus over Transfer Earnings

In the modern theory of rent the concept of economic rent has been generalised and extended to the surplus payments made to other factors of production besides land. As we saw above, Ricardo regarded land as a free gift of nature and considered the whole earnings of land as the economic rent. Later on, this Ricardian concept of rent has been extended to designate a part of earnings of other factors of production—labour, capital and entrepreneurial ability—over and above the minimum necessary income required to induce the factors to do their work. Thus, in accord with Ricardo's concept, from the social point of view Joan Robinson say, "The essence of the conception of *rent* is the conception of a surplus earned by a particular part of a factor of production over and above the minimum earnings necessary to induce it to do work."[9]

According to Benham, economic rents are "the sum paid to the factors which need not be paid in order to retain the factors *in the industry*."[10] Likewise, according to Boulding, economic rent is the payment to a factor "in excess of the minimum amount necessary to keep that factor in *its present occupation*."[11] Joan Robinson also discusses the question of surplus or economic rent from the viewpoint of individual industries or uses. She thus say, "from the point of view of an industry producing a particular commodity the necessary minimum payment for a factor is not the payment which will cause that factor to exist, but the payment which will cause it to take service in that particular industry rather than in another."[12] And "the difference between the earnings actually received by a certain unit of a factor and its transfer price is its rent from the point of view of the industry."[13]

7. Jan Pen, *op, cit.* p. 128 (*italics added*)
8. E.H. Chamberlin, *The Theory of Monopolistic Competition*, 1950
9. *Economics of Imperfect Competition* (London 1948), p.102.
10. Fredrick Benham, *Economics*, 6th edition (London), p. 227.
11. K.E. Boulding, The Concept of Economic Surplus, *American Economic Review*, Vol. 35 December 1945.
12. *Economics of Imperfect Competition*,p.104.
13. *Op. cit.*, p.110.

Likewise, according to George Stigler, the rent of a factor is "the excess of its return in the best use over its possible return in other uses."[14]

It is clear from above that other factors of production, labour, capital, entrepreneur may also be found to be earning economic rent when they are getting payment greater than what is required to induce them to work in the present industry occupation or use. Thus, in modern economic theory, *economic rent is not merely confined to land, it also refers to the surplus payments made to some units of other factors over and above what is necessary to keep them in the present industry or use.* Thus, the units of other factors may also earn economic rent.

ECONOMIC RENT ARISES WHEN THE SUPPLY OF A FACTOR IS LESS THAN PERFECTLY ELASTIC

Boulding[15] and Joan Robinson[16] emphasized that whenever the supply of factor units to an industry or economy is not perfectly elastic, a part of the earnings of a factor will consist of surplus or economic rent, since the full price they get are not necessary to make all the factor units available. If the supply is not perfectly elastic, some factor units will be available at a price lower than the price they will actually receive, the difference between the actual price and the one necessary to make them available is surplus or economic rent. We shall now explain below the concept of economic rent with special reference to land and will bring out the conditions under which it arises.

As explained above, in the Ricardian theory, the rent was conceived as a payment made to a factor of production whose total supply was completely inelastic or fixed. When the total supply of a factor is completely fixed or inelastic, no price is needed to be paid in order to induce it to be available for production. Land is the main example of the factor of production whose total supply is fixed and completely inelastic to the economy as a whole. This is because land is a free gift of nature and is non-reproducible. Thus supply of land cannot be increased when the demand for it and hence the price for its use rises. On the contrary, even if the price of its use falls to zero, the total supply of land will remain unaffected. Land is there in any case and does not require to be paid any price in order to exist and become available for production. To put it in other words, the transfer earnings of land for the economy as a whole are zero. Thus, to Ricardo and other classical economists, the whole earnings of land are functionless surplus, that is, whatever price the land happens to earn is not required to be paid for it in order to keep it in existence or to make it available to the society. Therefore, they regarded the whole price or earnings for land as *economic rent defined as the surplus payment over and above what is required to be paid to make it available to society.* Thus the term rent which in ordinary usage means the price or payment for hiring land came to be used by economists as the *title for the surplus earnings which the land receives.*

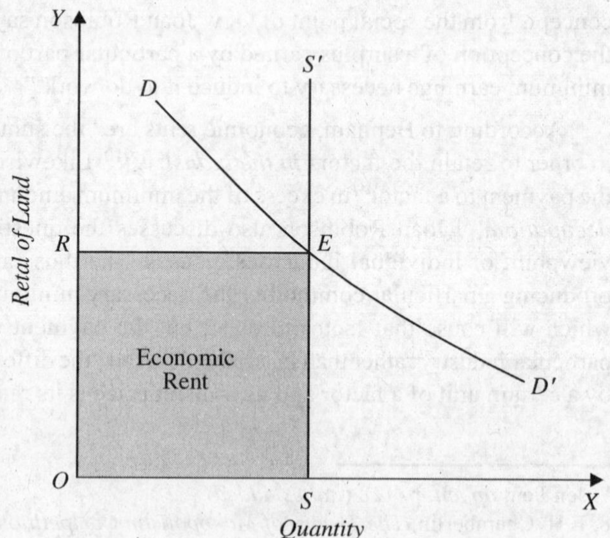

Fig. 34.4. *Whole Earnings are Economic Rent.*

14. George Stigler, *The Theory of Price*, 2nd edition p.99
15. K.E. Boulding. The Concept of Economic Surplus, *American Economic Review*, vol. 35, Dec. 1945.
16. *Economics of Imperfect Competition*, Ch. 8.

Since the whole price or payment made to land is surplus earnings in the sense defined above, therefore the rent as a price for the use of land and rent as surplus earnings of land have often been used interchangeably, though they are conceptually different. Figure 34.4 illustrates that whole of the earnings of land are economic rent from the viewpoint of the society. Suppose *OS* is the amount of land available to the society. The society then confronts a perfectly inelastic supply curve *SS′* of land. This perfectly inelastic curve *SS′* indicates that even if the price for the use of land falls to zero, its supply remains *OS*. This is because the transfer earnings of land for the society are zero. Curve *DD′* represents the demand for land of the society as a whole. As a result of the intersection of demand and supply curves of land, the rental *OR* per unit for the use of land is determined. Since transfer earnings are zero, the whole rental *OR* will be the economic rent earned per unit of land. The total earnings of land will be *ORES* and whole of these earnings of land will represent economic rent, since the transfer earnings are zero.

The whole earnings of a factor can be surplus only if its supply is perfectly inelastic. Land is such a factor because of its being a free gift of nature and society having incurred no cost to make it available for production. Therefore, *economic rent has also been defined as a payment for any factor whose supply is perfectly inelastic*. Thus, economic rent refers to the payment for the use of land and other natural resources since it is the supply of land and other natural resources which are permanently in fixed supply.

According to the modern economists, it is only from the standpoint of the economy as a whole that land has perfectly inelastic supply. The supply of land for a particular use or industry is not perfectly inelastic. There are various uses or industries competing for the use of land. A unit of land being employed to produce a particular crop will be transferred to the production of other crops, if its earnings in the present use fall below the possible earnings in the production of any other crop. It is therefore clear that from the viewpoint of a particular use or industry a payment has to be made for a unit of land so as to prevent it from being transferred to some other use or industry.

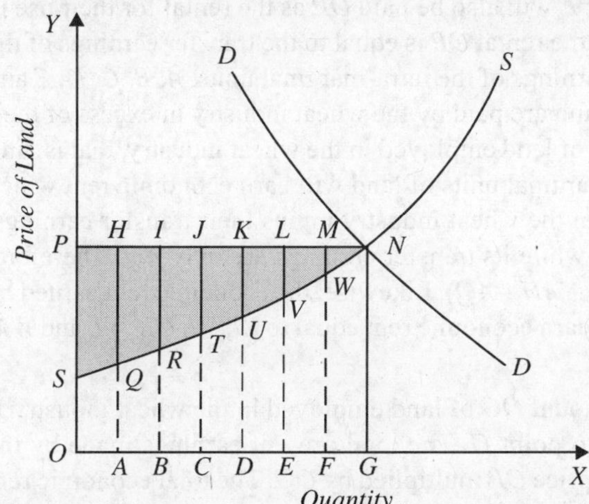

Fig. 34.5. *When the supply curve of a factor is less than perfectly elastic, a part of the earnings is economic rent and a part transfer earnings.*

The payment of price which is necessary to keep a unit of factor (land in the present case) in a certain use or industry is called transfer earnings or transfer price, because a payment of a price below this will cause it to be transferred elsewhere. Thus, *transfer earnings are the minimum earnings which must be paid to a unit of factor in order to induce it to remain in the present use or industry*. The transfer earnings of a unit of a factor may also be defined as the amount of earnings which it can obtain in the *next best alternative use, occupation or industry*. This is because reduction in the earnings of a unit of factor below its earnings in the next best alternative use will cause it to transfer itself from the present use to its next best use. For instance, if a wheat-growing industry has to keep land with it, then it must pay for the land at least its transfer earnings. Suppose, an acre of land currently employed in the production of wheat has its next best alternative of being used for the production of potato where it can earn ₹ 500. Thus the transfer earnings of this acre of land is ₹ 500. Then, in order to keep this particular acre of land in the wheat industry it must be paid at least ₹ 500, otherwise it will transfer itself to the potato-growing industry. Now, if actually this acre of land is being paid ₹ 600 in the wheat industry, it will be earning economic rent of ₹ 100 which is the difference between actual earnings of ₹ 600 in the present use and its transfer earnings of ₹ 500.

As the wheat industry uses more land, it will draw into it the land with successively higher supply price (*i.e.* transfer earnings). Thus, in Fig. 34.5, *A*th unit of land has a supply price equal to *AQ*. In other words, *AQ* must be paid to the *A*th unit of land in order to keep it in the wheat industry. Thus *AQ* represents the transfer earnings of *Ath* unit of land. Similarly, *B*th unit of land must be paid *BR* if it is to be attracted into or retained in the wheat industry. *BR* is therefore the transfer earnigs of the *Bth* unit of land. Likewise, transfer earnings of *Cth*, *Dth*, *Eth*, *Fth* and *G*th units of land are *CT*, *DU*, *EV*, *FW* and *GN* respectively. Demand curve *DD* of wheat industry for land and supply curve *SS* of land intersect at point *M* and thus determine *OP* as price for the use of land.

Suppose all units of land are equally productive in respect of wheat production. Then all acres of land will get the same price as determined by demand for land for wheat production and supply of land to it. But if various units of land are different in respect of other uses, they will have different supply prices to the wheat industry, that is, they need to be paid different prices to attract them to the wheat industry. The units of land having larger earnings in other uses will need to be paid higher prices to attract them to the wheat industry and those with smaller earnings in other uses will be required to be paid relatively smaller prices to draw them into the wheat industry. Under such conditions, therefore the supply curve of land to the wheat industry will be upward sloping, as is shown by the curve *SS* in Fig. 34.5 where price for the use of land determined is *OP* and equilibrium amount of land employed in the wheat industry is *OG*.

It will be seen from Fig. 34.5 the *G*th is the marginal unit of land employed in the wheat industry and further that the rental *OP* per unit of wheat land is equal to transfer earnings *GN* of the marginal unit *G*th of land. But, since we are assuming that all units of land are identical, *i.e.*, equally productive in respect of wheat, every unit of land employed in the wheat industry must get *OP* as the price for its use. Thus intra-marginal units, *A*, *B*, *C*, *D*, *E*, *F* will also be paid *OP* as the rental for their use in wheat industry. But it will be seen that whereas the rental *OP* is equal to the transfer earnings of the marginal unit *G*, it is greater than the transfer earnings of the intra-marginal units, *A*, *B*, *C*, *D*, *E* and *F*. That is to say, these intra-marginal units of land are paid by the wheat industry in excess of their transfer earnings. Thus, while the marginal unit of land employed in the wheat industry, that is, unit *G* does not earn any economic rent, the intra-marginal units of land will earn economic rent which will be equal to the price or earnings they get in the wheat industry minus their transfer earnings. Thus *A*th unit of land obtains price ($AH = OP$) while its transfer earnings are only *AQ*. Therefore, Ath unit of land earns *QH* as economic rent ($QH = AH - AQ$). Likewise, the economic rent earned by the *Bth*, *Cth*, *Dth*, *Eth* and *Fth* units of land will earn economic rent equal to *RI*, *TJ*, *UK*, *VL* and *WM* respectively.

The total transfer earnings of the whole amount *OG* of land employed in the wheat industry is *OSNG* — the area under the supply curve up to point *G*. The total present earnings made by the whole amount *OG* of land is *OPNG*, that is, the price *OP* multiplied by *OG*. The total economic rent earned by the whole amount *OG* of land employed in the wheat industry is, therefore, the area *SNP* which is equal to the whole earnings *OPNG* minus the transfer earnings *OSNG*. It is now clear that *whereas in the case of society as a whole, the whole earnings of land represents a surplus or economic rent, from the viewpoint of a single industry or use, a part of the earnings are transfer earnings (which must be paid in order to retain the land in the given use or industry) and only the remaining amount of earnings (i.e., surplus over transfer earnings) represents economic rent.*

A careful study of Fig. 34.5 will reveal that economic rent as a surplus over transfer earnings is different from the rent as price for the use of land. In Fig. 34.5 whereas the price (per unit) for the use of land is *OP* and the total price (that is, the total rent for the use of land) of the whole land employed in the given industry is the area *OPNG*, the total economic rent (*i.e.*, the total rent as surplus over transfer earnings) earned is the area *SNP*.

Furthermore, Fig. 34.5 makes it clear that all units of land do not earn equal amounts of economic rent; different units of land earn different amounts of economic rent depending upon their transfer earnings. Thus, economic rent earned by *A*th unit is *QH*, *B*th unit is *RI*, Cth unit is *TJ* and so on. It should be noted that these differences in economic rent earned by different units of land are not due to the differences in their productivity in respect of wheat but due to the differences in their ability to earn in their next best uses.

When the supply of a factor is perfectly elastic, no economic rent is earned

Now, suppose that all units of land used in the wheat industry are not only identical in respect of wheat production but are also identical in respect of all other uses. In such circumstances, each unit of land employed in the wheat industry will have the same transfer earnings and therefore the supply curve of land for the wheat industry will be perfectly elastic as shown by *SS* curve in Fig. 34.6. In this, therefore, the transfer earnings of every unit of land employed in the wheat industry is *OS*. In other words, every unit of land needs to be paid equal to *OS* in order to retain it in the wheat industry. It will be seen in Fig. 34.6 that price per unit for the use of land in the wheat industry as determined by demand for and supply of land is also *OS*. Thus, every unit of land in the wheat industry gets *OS* and its transfer earnings are also *OS*. Therefore, in this case no unit of land in the wheat industry earns more than its transfer earnings, i.e., no unit of land in the *wheat industry earns any economic rent. It is, therefore, clear that when the supply of any factor is perfectly elastic to an industry, no unit of this factor will earn any economic rent*. Further, it has been seen above, that whereas to the economy or society as a whole, the supply of land is perfectly inelastic and, therefore, the whole of its earnings are *economic rent*, the supply of land to an industry (the wheat industry in the above case) may be perfectly elastic and earns no economic rent at all.

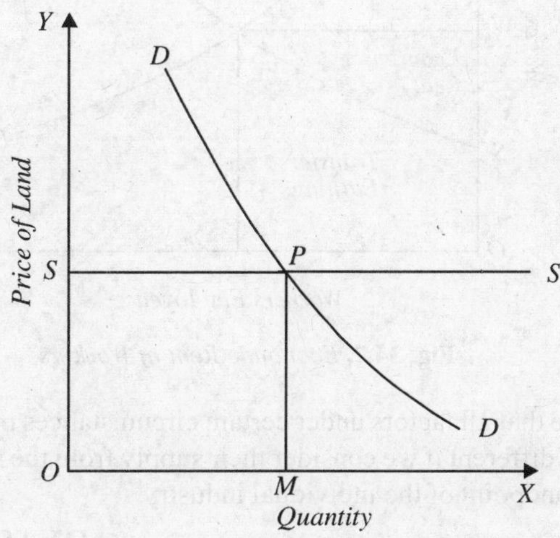

Fig. 34.6. *No economic rent is earned when the supply of a factor is perfetly elastic.*

Economic Rent can accrue to all factors

We have seen above that under certain circumstances some units of land from the viewpoint of industry may earn economic rent, *i.e.*, earnings in excess of their transfer earnings, and under some other conditions, land may not earn any economic rent at all. The same kind of analysis can be made for other factors as well. As pointed out above, the concept of *economic rent is not merely confined to land, it applies to other factors of production as well. Thus labour, capital and entrepreneurs may also be earning economic rent when their supplies are less than perfectly elastic*. Now, consider the supply of a particular kind of workers in a given industry, all of which are being paid a certain uniform wage rate by the industry. But all these workers may be heterogeneous from the point of view of the other industries. As a result of this, they would have different transfer earnings and the supply curve of this kind of workers will be upward sloping (*i.e.*, less than perfectly elastic). Since all are getting the same wage rate in the given industry which is determined by demand and supply. Now, workers whose transfer earnings are less than the wage rate will be getting economic rent or surplus. It will be seen from Fig. 34.7 that only the marginal worker (*i.e.*, *L*th worker) employed would not be obtaining any economic surplus or rent. In Fig. 34.7 the total economic rent earned by all the intra-marginal workers is equal to *WES*.

Likewise, entrepreneurs may be able to earn different amounts of profits in different industries. An entrepreneur will work in the industry in which he is able to get greater amount of profits and if his profits fall below a certain level, he will shift himself to the next best industry. But the additional amount of profits which he is able to earn in an industry over and above his transfer earnings (that is, the profits in the next best profitable industry) are surplus or economic rent from the point of the given industry.

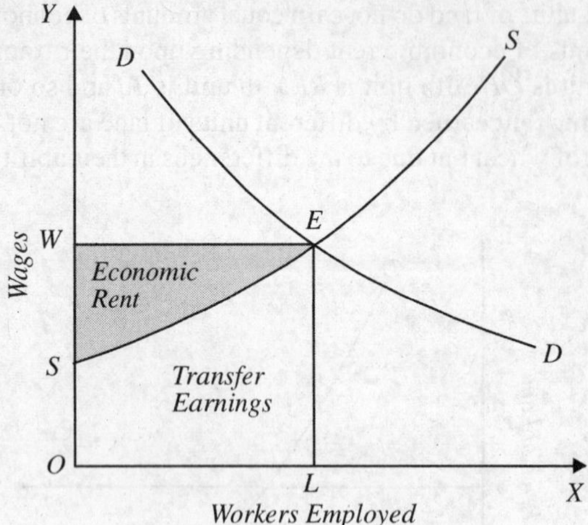

Fig. 34.7. *Economic Rent of Workers*

Similarly, a particular kind of capital equipment is not equally efficient from the viewpoint of all industries. It will earn higher income for which it is best suited and will earn over and above the necessary income that is required to keep that in the industry. This additional income, which owner of capital equipment will get, will be the economic rent from the standpoint of that industry. We thus see that all factors under certain circumstances may earn economic rent. But this economic rent will be different if we consider their supply from the standpoint of the economy as a whole than from the standpoint of the individual industry.

QUASI RENT

The concept of quasi rent was introduced in economic theory by Marshall Marshall's concept of quasi-rent is the extension of the Ricardian concept of rent to the short run earnings of the capital equipment (such as machinery, building etc.) which are in inelastic supply in the short run. The distinguishing characterisitc of land is the fact that its supply is perfectly inelastic to changes in its price and therefore its earnings depend mainly upon the demand for it. But, in the short run, the fixed capital equipment such as machinery is likewise perfectly inelastic in supply and cost of its production is not relevant once it has been produced. During the short period, the earnings of specialised capital equipment depend mainly upon the demand conditions and are thus similar to land rent and have therefore been called rent by Marhsall. Since the capital equipment is not permanently in fixed supply like land and instead their supply is very much elastic in the long run. Marshall preferred to call their earnings in the short period as Quasi rent rather than rent.

The *quasi-rent is only temporary surplus* which is enjoyed by the owner of the capital equipment in the short run due to the increase in demand for it and this will disappear in the long run due to the increase in the supply of capital equipment in response to the increased demand. In the short run, specialised machinery has no alternative use and therefore its supply will remain fixed in the short run even if its earnings fall to zero. Thus, the transfer earnings of the capital equipment or machinery in the short run are zero. Therefore, the *whole of the earnings of the machinery in the short run are surplus over transfer earnings and therefore represent rent*. It may, however, be pointed out that some maintenance costs are required to be incurred in the short run to keep the machinery in the running order. Therefore, more precisely, the quasi rent may be defined as *the short run earnings of a machine minus the short run cost of keeping it in running order*.

There is every reason to believe that quasi rents will be generally earned in the short run by the capital equipment like machinery, building etc. This is because, however keen the competition between entrepreneurs may be, the supply of capital equipment cannot be increased in the short run. Consequently, when very high earnings are being made from capital equipment they will not be competed away in the short run. But in the long run the position regarding the supply of capital equipment (*e.g.*, machines) is quite different. Capital equipment are man-made instruments of production and therefore their supply can be increased in the long run to meet the increased demand for them. Thus, as a result of the increase in the supply of machines in the long run their excessive earnings will be competed away. In the long run, therefore, the competitive equilibrium is reached when the earnings from the capital equipment are just sufficient to maintain them in running order and provide only normal profits to entrepreneur. Thus in the long run no surplus over cost of production is earned by the machines. Therefore, quasi rent will disappear in the long-run competitive equilibrium. Professors Stonier and Hague rightly remark, "The supply of machines is fixed in the short run whether they are paid much money or little so they earn a kind of rent. In the long run this rent disappears for it is not a true rent, but only an ephemeral reward—a 'quasi-rent".[17]

But the case of land is quite different. The supply of land being a free gift of nature and non-reproducible, its supply is perfectly inelastic in the short run as well as in the long run. Thus the surplus earnings or rent earned by land persist in the long run also. It is thus clear that the earnings of land and of capital equipment (machines etc.) are similar only in the short run. The analogy between the two does not hold in the long run because of the difference in the nature of their long-run supply. To quoto Professors Stonier and Hague again, "In the long period, machines will stand on a very different footing from land or natural ability. For machines are produced by human effort while land or human ability are gift of nature. In the long run, therefore, the supply of land will not respond to an increase in demand for it, the supply of machines will. In the long-run, therefore, land will earn rent but machines will, assuming competition, earn only just enough to make their existence worthwhile".[18]

Quasi Rent as Surplus over Variable Costs

Production of a good is possible when a fixed factor is combined with some variable factors. The amount of variable factors used depends upon the level of output produced, while the quantity of the fixed factor remains unchanged during the short period. The variable costs must be recovered in the short run, otherwise the production would be stopped. Whatever excess earnings over and above the total variable costs are made are ascribed to the machines (*i.e.*, fixed factor). Therefore, quasi-rent has also been defined as the excess of total revenue in the short run over and above the total variable costs. Thus,

Quasi Rent = Total Revenue Earned –Total Variable Costs

Since in long run, all costs are variable and, in long run competitive equilibrium, total receipts are equal to total costs (including normal profits to the entrepreneur), no excess earnings over and above the costs will accrue to the machines and therefore no quasi rent will be earned by the machines. The earnings of quasi

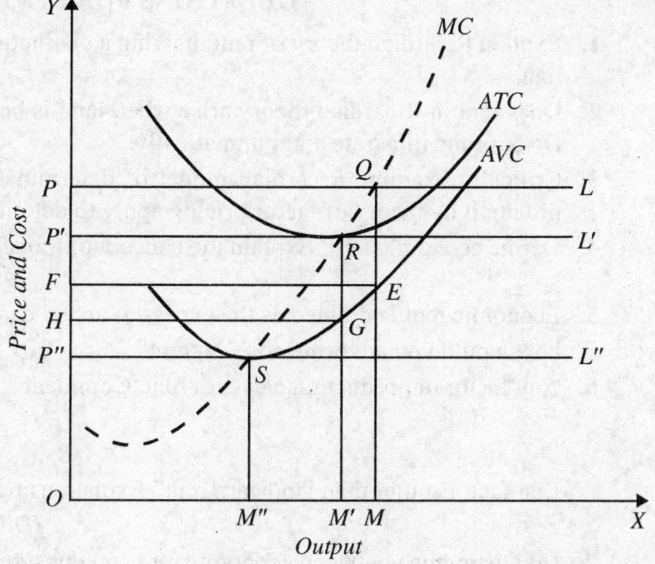

Fig. 34.8. *Quasi rent*

17. A.W. Stonier and D.C. Hague: *A Textbook of Economic Theory*, Fourth edition, 1972, p. 329.
18. *Op. cit.*, p.329

rent in the short run and its disappearance in the long run is illustrated in Fig. 34.8 wherein, as usual, output is measured on the *X*-axis and price and cost are measured on the *Y*-axis. *ATC* and *AVC* represents the average total cost and average variable cost curves respectively in the short run. It should be noted that the average variable cost, *AVC* includes the cost incurred per unit of output on the variable factors such as labour, raw materials etc. as well as the cost per unit of output for keeping the machinery in the working order during the short period. (As said above, the cost of maintaining the machinery in the working order during the short period is a part of variable costs).

Now suppose that the demand for the product is such that the price *OP* the is determined. With price of the product *OP* the price line faced by an individual entrepreneur is *PL* which represents the marginal revenue as well as the average revenue. With price line *PL*, the entrepreneur is in equilibrium at point *Q* and is producing *OM* level of output. It will be seen from the figure that the total revenue earned is *OMQP*, while total variable costs incurred is *OMEF*. The area *FEQP* represents the surplus of total revenue earned over the total variable costs (*FEQP* = *OMPQ* – *OMEF*). Thus *FEQP* is quasi rent, that is, the short-run earnings of the machinery.

If now the demand for the product declines so that the price for the product falls to *OP′*. With price *OP′*, the price line is *P′L′* and equilibrium of the entrepreneur is at point *R* with output *OM′*. Now, the total revenue earned is *OM′RP′* and the total variable cost is *OM′GH*. Thus the quasi rent earned by the machinery is now *HGRP′* (= *OM′R-P′* – *OM′GH*). If the demand for the product further declines and the price falls to *OP″*, the price line confronting the entrepreneur will be *P″L″* and he will be in equilibrium position at point *S* – the minimum point of the curve *AVC*. At point *S*, the total revenue earned is just equal to the total variable costs and quasi rent earned by the machinery has fallen to zero. The entrepreneur will close down production if the price falls below *OP″*, for at a price below *OP′*, it will not be realising fully even the variable costs. It is, therefore that clear the *quasi rent cannot be negative*. But in the long run all factors are variable and all costs including costs incurred on capital equipment such as machinery must be recovered if the firm is to stay in the industry. In the long run competitive equilibrium, price must equal long run average cost. Consequently no quasi rent is earned in the long run.

QUESTIONS FOR REVIEW

1. Explain Ricardian theory of rent drawing a distinction betwen scarcity rent and differntial rent.
2. Does rent in Ricardian theory arise when land is homogenous and equally well-situated? Discuss and illustrate diagrammatically.
3. Critically examine Ricardian model of determination of land rent. Does not marginal productivity theory of factor pricing apply to determination of land rent.
4. Define economic rent. Explain the concept of transfer earnings ? How is it related to rent? *[D.U. B.Com (Hons.) 2000]*
5. Economic rent is defined as the excess of actual earnings over transfer earnings. Explain how would you advocate a tax on rent? *[D.U. BA. (Hons.) 1991]*
6. Any factor of production can earn rent. Comment *[D.U. B.A. (Hons.) 2000]*

   OR

   Can factors other than land earn rent? Explain using diagrams. *[D.U. B.Com. (Hons.) 1997, 1998,* 2001]
7. (*a*) Distinguish between economic rent and quasi rent . *[D.U. B.Com (Hons) 1999, B.A. (Hons. 1996, 1997, 2001]*

   Can quasi rent be negative? *[D.U. B.A. (Hons) 1996]*

(*b*) In case the supply function of labour is 45° line passing through the origin, what would be the percentage of economic rent in the toal earnings of a factor.

[D.U. B.A. (Hons.) 2001]

[**Hints.** In figure given below *DD* and *SS* are the demand and supply curves of a factor and through their intersection determine *OP* or *NE* as the price of the factor and *ON* as the quantity employed of the factor. Total earnings of the factor are equal to the area *OPEN*. The shaded area *OPEN* represents economic rent as surplus over transfer earnings. Transfer earnings of the factor are equal to *OEN*. Since supply curve *SS* makes 45° angle with the *X*-axis, $\angle EON = \angle EOP$, both being equal to 45°. Further, $ON = PE$, both being equal to the quantity of the factor employed. Besides, $\angle ENO = \angle EPO$ both being right angles. Thus triangle *OPE* is congruent to the triangle *ONE* and therefore each would be half of the total earnings *OPEN*. Thus, economic rent *OPE* is half of total earnings *OPEN*]

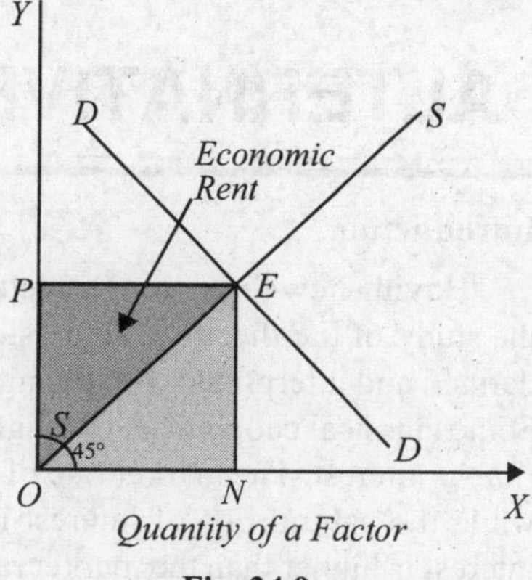

Fig. 34.9

8. Is rent price determined or price determining ? Explain *[D.U. B.Com (Hons) 2001]*

OR

Explain the relationship between land rent, cost of production and price of the production of the land. Show that from the point of view not only of the individual firm but of the individual industry the price paid for the use of land is exactly on *all fours* with the price paid for the use of the other factors of production.

9. Explain the concept of Quasi rent. Must it always be non-negative.

[D.U. B.A. (Hons) 1999]

10. What is economic rent ? Explain that economic rent arises when the supply of a factor is less than perfectly elastic.

11. The more elastic the supply curve the less the amount of payment to factors that is rent and the more that are transfer earnings. *[D.U. B.A. (Hons.) 1999]*

12. Division of a given factor payment between transfer earnings and economic rent depends on the type of transfer earnings considered. *[D.U. B.A (Hons.) 1989]*

13. Distinguish economic rent earned by a factor from the viewpoint of (1) the society as whole, (2) an individual industry and (3) the individual firm.

14. While the whole earnings of land are economic rent from the viewpoint of the society as a whole, an individual farmer earns no economic rent. Explain.

15. Explain the concept of economic rent. How is it related to transfer earnings ? Explain with the help of diagrams. *[D.U. B.Com (Hons.) 2nd Year 2006, 2007]*

16. Any excess that a factor earns over its reservation price is called its economic rent. Explain with the help of diagrams. *[D.U., B.Com (Hons.) 2007, 2008]*

[**Hint.** Note that reservation price of a factor is its transfer earnings].

CHAPTER 35

ALTERNATIVE THEORIES OF INTEREST

Introduction

Having now discussed the determination of rent of land and wages of labour, we now pass on to the study of the theory of interest. But what do we mean by interest ? Interest has been variously defined and interpreted. Firstly, interest is conceived by economists as the rate of return on capital. Some classical economists distinguished between the *natural or real rate* of interest and the *market rate of* interest. The market rate of interest is the rate at which funds can be borrowed in the market, while the natural rate of interest is rate of return on capital investment. When the natural rate of interest is higher than the market rate of interest, then there will be greater investment in capital with the result that the natural rate of interest (*i.e.*, the rate of return on capital) will fall. The equilibrium will be established when the natural rate of interest becomes equal to the market rate of interest.

Since physical capital has to be purchased with monetary funds, rate of interest becomes the rate of return over money invested in physical capital. But since money to be invested in physical capital has to be saved by some one, interest also becomes rate of return or saving. Thus, there are two concepts of interest which are related to each other. First, *the term interest is used to express a rate of return earned on capital as a factor of production. The second concept of interest refers to the price which is paid by the borrowers to lenders for the use of their saving funds*. When the borrowers are entrepreneurs or businessmen, who use the saving funds for investment in capital for them rate of return on physical capital is highly significant. However when we consider things from the viewpoint of lenders who save money to slnd it to others, the concept of interest as a price for borrowed funds is of crucial importance therefore, *interest is generally defined as price for the use of borrowed funds*.

It is also important to note the difference between the real rate of interest and nominal rate of interest. *The real rate of interest is the nominal rate of interest corrected for inflation* (*i.e.* rise in the general price level) in the economy. Thus

Real rate of interest = Nominal rate of interest – rate of inflation.

In our analysis in this chapter we will be concerned with real rate of interest.

For explaining the determination of rate of interest, the three theories have been put forward. First theory of interest is *Classical Theory of Interest* which explains interest as determined by saving and investment. Secondly, neo-classical economists such as Wicksell, Ohlin, Haberler, Robertson, Viner developed what is known as *Loanable Funds or Neo-Classical Theory of interest*. These writers consider the interplay of monetary and non-monetary forces in the determination of the rate of interest. At their hands, interest theory ceased to be purely real or non-monetary theory. In their view, monetary factors along with the real factors determine the rate of interest. The loanable funds theory is partly a monetary theory of interest.

But monetary theory gained more recognition with the publication of *Keynes' General Theory*. According to Keynes, interest is a purely a monetary phenomenon and as such it is determined by the demand for money (*i.e.*, liquidity preference) and the supply of money. According to him, interest is a price not for the sacrifice of waiting or time preference but for parting with liquidity.

Since he emphasised the role of *liquidity preference* in the determination of the interest rate, his theory is known as liquidity preference theory of interest. Keynesian theory is a purely *monetary theory*.

It is worth noting that all these theories of interest seek to explain the determination of the rate of interest through the equilibrium between the forces of demand and supply. In other words, all these theories are *demand and supply theories*. The difference between the various theories of interest lies in the answer to the question : *demand for what and supply of what*. According to the classical theory, rate of interest is determined by *demand for savings to make investment and the supply of savings*. Loanable-funds theory seeks to explain the determination of the rate of interest through the equilibrium between *demand for and supply of loanable funds*. Besides savings, loanable funds consist of funds derived from other sources as well. Keynesian theory of interest explains the determination of interest through the equilibrium between *demand for and supply of money*.

Another point worth mentioning is that a theory of interest has to explain two things. First, why does interest arise ? Secondly, how rate of interest is determined ? All the three theories mentioned above explain both these aspects of interest.

CLASSICAL THEORY OF INTEREST

As mentioned above, this theory seeks to explain the determination of the rate of interest through the interaction of the demand for savings to make investment and the supply of savings. Since this theory explains the determination of the rate of interest by real forces such as thriftiness, time preference and productivity of capital, it is also called the *real theory of interest*. Various classical writers differ a good deal from each other in respect of their views about interest. Some of them laid emphasis on the forces governing the supply of savings. Thus they considered interest as a price for abstinence or waiting or time preference. Some others like J.B. Clark and F.H. Knight thought the marginal productivity of capital, which is a force that operates on the demand side of savings, determines the rate of interest.

Fisher and Bohm-Bawerk explained the interest with both types of factors. There is a basic assumption that is common to all classical writers. It is that all of them assume full employment of resources. In other words, in their models if more resources are to be devoted to investment, that is, to the production of capital goods, some resources have to be withdrawn from the production of consumers' goods. According to this theory, money which is lent out to the entrepreneurs for investment in capital goods is to be made available by those who save out of their incomes. By abstaining from consumption they release resources for the production of capital goods. In order to induce people to save and refrain from consuming a part of their incomes, they must be offered some interest as a reward. To persuade them to save more, the higher rate of interest has to be offered. So far the various classical economists agreed but they differed in detail about the nature of interest. We shall discuss below the views of some of them.

Interest is a price for abstinence or waiting. It was Nasau Senior who first pointed out that saving involved a sacrifice of abstinence and interest is a price for this sacrifice. Any one who saves some money and is therefore able to lend it to others abstains from consuming a part of his income and in order to induce him to do so, he must be paid interest by the borrower. Thus, according to Senior, interest arises because of the abstinence involved in the act of saving. Without giving him the interest as compensation, the individuals will not like to undergo the sacrifice of abstaining from consumption. The idea of abstinence was criticised by some economists, in particular by Karl Marx, who pointed out that the rich people who are the main source of savings are able to save without making any real sacrifice of abstinence. They save because something is left over after they have indulged in consumption to their heart's desire. In order to avoid this criticism Marshall substituted the word *waiting* for "abstinence". According to him, when a person saves money and lends it to others, he does not abstain from consumption for all time ; *he merely postpones consumption*.

But the individual who lends his savings has to wait until he gets back the money. Thus, the person who saves money and lends it to others undergoes the sacrifice of waiting. To induce people to save and wait some price has to be paid to them as compensation for making this sacrifice. According to this view, interest is a price for waiting.

Interest is paid because of time preference (Fisher's Theory). The views of Irving Fisher, an eminent American economist, are very popular among economists. According to him, enjoyment interst needs to be paid to the lenders because people prefer present enjoyment of goods to future of them. Fisher laid greater emphasis on time preference as a cause of interest. But along with time preference he also considered the role of marginal productivity of capital for which he used the term '*rate of return over cost*' as a factor that also determines interest.

Rate of interest arises because people prefer present satisfaction to future satisfaction. They are therefore impatient to spend their incomes in the present. According to Fisher, interest is a compensation for the time preference of the individual. The greater the impatience to spend money in the present, that is, the greater the preference of individuals for the present enjoyment of goods to future enjoyment of them, the higher will have to be the rate of interest to induce them to lend money.

The degree of impatience to spend income in the present depends upon the size of the income, the distribution of income over time, the degree of certainty regarding enjoyment in the future and the temperament and character of the individual. The people whose incomes are large are likely to have their present wants more fully satisfied. Therefore, these rich people will discount the future at a relatively lower rate of interest (that is, their time preference will be less) and will be required to be paid a relatively lower rate of interest. As regards distribution of income over time, three kinds of situation are possible. The income may be uniform throughout one's life or may increase with age or decrease with age. If it is uniform the degree of impatience to spend in the present will be determined by the size of the income and the temperament of the individual. If income increases with age, it means the future is well provided for and the degree of impatience to spend money in the present (that is, time preference) will be greater. On the other hand, if income decreases with age, the degree of impatience to spend money at present will be less.

As regards certainty of enjoyment in the future, if the individual is sure of enjoyment of income in the future, other things remaining the same, the impatience to spend money in the present will be less, that is, the degree of time preference will be smaller. Finally, the character and the temperament of the individual will also determine his time preference. A man of foresight will be less impatient to spend income in the present, that is, his rate of time preference will be less as compared to that of a spendthrift. The rate of time preference is also influenced by expectation of life. If a man expects to live long, his preference for spending income in the present will be comparatively low.

It is clear from the above analysis that Fisher, like Bohm Bawerk, regarded the rate of interest as an agio on the present goods exchanged for future goods of the same kind. Fisher based his explanation of the rate of interest on his concept of income. According to him, interest is the link between expected future income values and the present capital values based on them. The present capital value of the stream of expected income in future years depends on the rate interest (*i.e.*, rate of time preference) at which they have to be discounted. He says, "The value of the orchard depends upon the value of its crops and in this dependence lurks implicitly the rate of interest itself."[1]

As said above, Fisher also regarded productivity of capital which he called *rate of return over cost*, as a determinant of interest. According to him, several different uses of capital which may yield different income streams are open to the owner of capital. The greater the expected income stream from use of capital, the greater will be the rate of interest. Another point worth mentioning is

1. Irving Fisher, *Rate of Interest*, (1907), p. 13.

that Fisher introduced risk and uncertainly in his explanation of interest. According to him, individual has a choice of any *one* of a number of uncertain income streams so that instead of a single rate of interest representing the rate of exchange between this year and next year we now find a great variety of rates according to the risks involved.

Determination of the Rate of Interest in the Classical Theory

According to the classical theory rate of interest is determined by the supply of savings and demand for savings to invest. We have explained above the forces working on the side of supply of saving. Some classical economists laid stress on the abstinence or waiting involved in the act of savings and supply of them and some others emphasised the role of time preference as a determinant of the supply of savings. According to this theory, the money which is to be used for purchasing capital goods is made available by those who save from their current income. By postponing consumption of a part of their income they release resources for the production of capital goods. It is assumed in this theory that *savings are interest elastic. The higher the rate of interest, the more the savings which people will be induced to make*. Besides, at higher rate of interest, savings would be forthcoming from those persons whose rates of time preference are more strongly weighed in favour of present satisfaction. The supply curve of savings will, therefore, slope upward to the right.

On the other hand, the demand for savings comes from the entrepreneurs or firms which desire to invest in capital goods. Capital goods are demanded because they can be used to produce further goods which can be sold to earn income. Thus capital goods have a revenue productivity like all other factors. For any given type of capital asset, *e.g.*, a machine , it is possible to draw a marginal revenue productivity curve showing the addition made to total revenue by an additional unit of a machine at various levels of the stock of that machine.

As said above, like other factors of production, capital has marginal revenue productivity. But the marginal revenue productivity of capital is a more complex concept than that of other factors because capital has a life of many years. A capital asset continues to yield returns for many years. But the future is quite uncertain. Therefore, the entrepreneurs have to judge the uncertainties of the future and estimate prospective yield or income from a capital asset after making allowance for

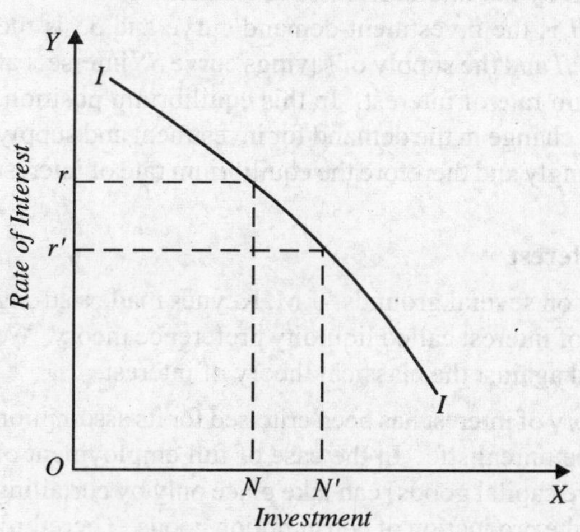

Fig. 35.1. *Investment Demand Curve*

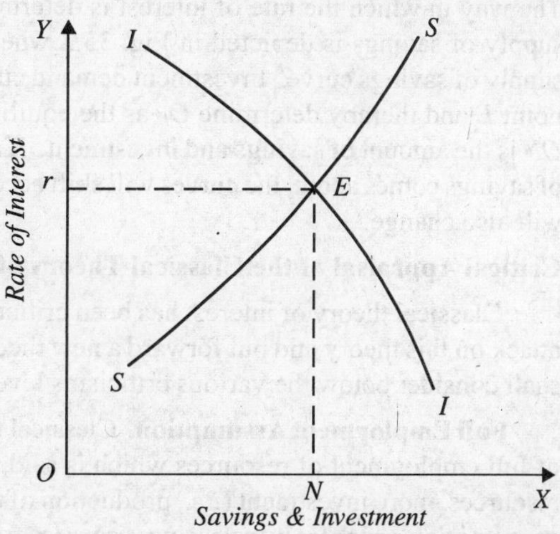

Fig. 35.2. *Classical Theory: Determination of Interest*

maintenance and operating costs. In other words, they have to find out the *net expected return* of a capital asset. This net expected return is expressed as percentage of the cost of capital asset. The more capital assets of a given kind there are, the less income will be expected to accrue from a marginal unit of it. Therefore, the marginal revenue productivity curve of capital slopes downward to the right.

We have seen in the previous chapters that a firm in a perfectly competitive factor market will hire a factor upto the point at which the price of that factor equals the marginal revenue product of the factor. Now, the marginal revenue product of a capital asset can be regarded as marginal revenue product of money invested in that capital asset. The *price* of money invested in capital assets is the rate of interest which a person has to pay on the borrowed funds. An entrepreneur will continue making investment in capital assets as long as the expected net rate of return, or in other words, marginal revenue produc of capital or investment is greater than the rate of interest. He will be in equilibrium position when marginal revenue product of capital (or expected rate of return) falls to the level of the rate of interest. Since the marginal revenue product curve of capital slopes downward, it will become profitable to purchase more capital goods as the rate of interest falls, *i.e.*, with the fall in the rate of interest more money will be demanded for investment. Thus, the investment demand curve connecting the rate of interest with the investment demand will be downward sloping. In other words, investment demand is assumed to be interest-elastic.

The way in which the investment demand increases as the interest falls is illustrated in Fig.35.1 where *II* is the investment demand curve showing the falling marginal revenue productivity of capital which, in other words, indicates the declining marginal net expected return as more investment is undertaken. When the rate of interest is *Or*, the entrepreneurs will make investment upto *ON* because the marginal net expected return is equal to *Or* rate of interest when *ON* investment is made. Now, it the rate of interest falls to *Or′*, then more capital projects will become profitable to be undertaken. Therefore, as a result of the fall in the rate of interest to *Or′*, investment increases to *ON′*

It is clear from the above analysis that investment demand curve slopes downward to the right. With the change in the rate of interest, investment will change.

Equilibrium between Demand and Supply. As seen above, according to the classical theory, the rate of interest is determined by the intersection of the investment demand curve and the supply of savings curve – the curves showing the relation of investment and savings to the rate of interest. The way in which the rate of interest is determined by the intersection of investment demand and supply of savings is depicted in Fig. 35.2 where *II* is the investment demand curve and *SS* is the supply of savings curve. Investment demand curve *II* and the supply of savings curve *SS* intersect at point *E* and thereby determine *Or* as the equilibrium rate of interest. In this equilibrium position, *ON* is the amount of savings and investment. If any change in the demand for investment and supply of savings comes about, the curves will shift accordingly and therefore the equilibrium rate of interest will also change.

Critical Appraisal of the Classical Theory of Interest

Classical theory of interest has been criticised on several grounds. J.M. Keynes made a strong attack on this theory and put forward a new theory of interest called liquidity preference theory. We shall consider below the various criticisms levelled against the classical theory of interest.

Full Employment Assumption. Classical theory of interest has been critcised for its assumption of full employment of resources which is said to be unrealistic. In the case of full employment of resources, more investment (*i.e.*, production of more capital goods) can take place only by curtailing consumption and there by releasing resources from the production of consumption goods. Therefore, when full employment of resources prevails, people have to be paid interest so as to induce them to abstain from consumption so that more resources should be devoted to the production of capital goods. But, when unemployed resources are found on a large scale there is no need for paying people to abstain from consumption or to postpone consumption and wait in order that more savings for undertaking investment should take place. More investment then can be undertaken by employing the unemployed or unutilised productive resources. Prof. Dillard rightly remarks : "Within the framework of a system of theory built on the assumption of full employment, the notion of interest as reward for waiting or abstinence is highly plausible. It is the premise that resources are typically fully employed that lacks plausibility in the contemporary world."[2]

Changes in Income Level Ignored. By assuming full employment the classical theory has ignored the changes in income level and their effect on savings and investment. Classical theory establishes a direct functional relationship between interest rate and the volume of savings. As the rate of interest goes up, more savings will take place. But at the higher rate of interest investment demand will be less with the result that interest will tend to fall to the level where savings and investment are in equilibrium. But this is not so realistic firstly, because the direct functional relationship between savings and the rate of interest is doubtful, and secondly, because when more savings take place as a result of the rise in the rate of interest, these more savings should lead to more investment, as according to classical theory investment is governed by savings. But, in the whole process of adjustment, change in income is not at all considered by the classical theory. As a matter of fact, when the rate of interest rises and investment shrinks as a consequence, income will decline. With the decline in income, the savings will decline. Therefore, the equality between savings and investment are brought about not so much through changes in the rate of interest but through changes in income.

Now, take the opposite case. If the rate of interest falls, according to classical theory, the investment demand will increase. But because of the lower rate of interest the greater supply of savings would not be forthcoming. Therefore, in classical theory more investment cannot take place even at lower rates of interest, because of the paucity of savings at lower rates of interest. But this is not what actually happens. At a lower rate of interest, more investment will be undertaken and increase in the investment will lead to the increase in income via multiplier. And out of increased income more would be saved. Again the tendency to equalise savings and investment is brought about by changes in income. Thus, the lower rate of interest through the increase in investment and income leads to the rise in savings. But this is quite contrary to the classical theory wherein at the lower rate of interest small savings are made.

From the above analysis it follows that by neglecting the changes in income the classical theory is led into the error of viewing the rate of interest as the factor which brings about he equality of savings and investment. The classical theory ignores the changes in income level because it assumes full employment of resources. When the resources are fully employed, income level will remain constant, production techniques being given. Now, it was Keynes who abandoned the assumption of full employment and, therefore, considered the change in the income level and its relation with savings and investment. Quoting Prof. Dillard again, "*The difference between the traditional theory of interest and Keynes' money theory of interest is a fundamental aspect of the difference between the economics of full employment and the economics of less than full employment.*"[3]

Disincentive Effect of Lesser Consumption on Investment Ignored. According to the classical theory, more investment can occur only by cutting down consumption. More the reduction in consumption, the greater the increase in investment in capital goods. But as we know the demand for capital goods is a derived demand; it is derived from the demand for consumer goods. Therefore, the reduction in consumption, which means decrease in demand for consumer goods, will adversely affect the demand for capital goods and will thus reduce the inducement to invest. The disincentive effect of the fall in consumption on investment is glossed over by the classical theory.

---

2. D. Dillard, *Economics of J.M. Keynes*, p. 162.
3. *Op. cit.*, p. 160.

As we shall see later, in Keynes' theory more investment does not occur at the expense of consumption. In Keynes' theory, in view of the unemployment of resources, more investment is possible by utilising the unemployed and under-employed resources. When investment increases, it leads to the increase in the income level. With the increase in incomes, people will consume more. Thus in Keynesian analysis more investment leads to more consumption, or in other words, investment and consumption go together. Keynesian analysis is more realistic in the context of unemployment of resources prevailing in the economy.

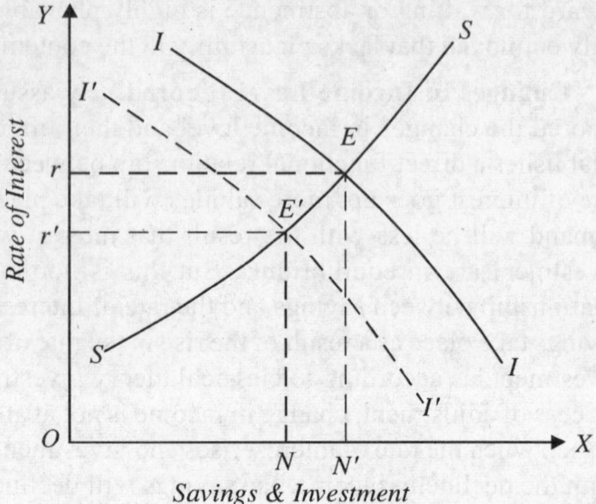

Fig. 35.3. *According to Classical Theory, investment schedule can change without affecting the savings schedule.*

Independence of Savings Schedule from Investment Schedule Assumed. Another implication of assuming full employment and constant level of income by classical theory of interest is that investment demand schedule can change without causing a change in the savings schedule. For instance, according to classical theory, if investment demand curve *II* shifts downward to the dotted position *I′I′* (Fig. 35.3) because the profit prospects have lessened, then according to classical theory, the new equilibrium rate of interest is *Or′* where the new investment demand curve *I′I′* intersects the supply curve *SS* which remains unaltered. But this is quite untenable. As a result of the fall in investment, income will decline. Since the supply curve of savings is drawn with a given level of income, when income falls, there will be less savings than before and as a result savings curve will shift to the left. But the classical theory does not take into account changes in the income level as a result of changes in investment and regards the savings schedule as independent of investment schedule which is not correct and realistic.

Indeterminateness. Finally, the classical theory, as pointed out by Keynes, is indeterminate. Position of the savings curve varies with the level of income. There will be different savings schedules for different levels of income. As income rises, the savings curve will shift to the right and as income falls the savings curve will shift to the left. Thus, we cannot know the position of the savings curve unless we already know the level of income, and if we do not know the position of the savings curve, we cannot know the rate of interest. It follows therefore that we cannot know what the rate of interest will be unless we already know what the income level is. But we cannot know the income level without already knowing the rate of interest because with the changes in the rate of interest investment will change which will in turn bring about changes in the income level. The classical theory, therefore, offers no determinate solution to the problem of interest rate determination and is indeterminate.

Savings out of current income is not the only source of supply of funds. As we have seen, the classical theory considered only savings out of current income as constituting the supply of funds in the market. But savings out of current income is not the only source of capital supply. People have usually past hoarded savings, which they may dishoard in a period adding to the supply of funds in the market. Further, now-a-days bank credit has become a very important source of investible funds which are also not taken into account by the classical theory.

We have critically explained the classical theory of interest. Some of the shortcomings of the classical theory were removed by the loanable funds theory which we now turn to discuss.

LOANABLE FUNDS THEORY OF INTEREST

Another school of thought developed what is called loanable funds theory of interest. Among the principal economists who contributed to the development of loanable funds theory mention may be made of Wicksell, Bertil Ohlin, Robertson, Myrdal, Lindahl, Viner, etc. According to this theory, real forces, such as thriftiness, waiting, time-preference and productivity of capital alone do not go to determine the rate of interest, monetary forces such as hoarding and dishoarding of money, money created by banks, monetary loans for consumption purposes also play a part in the determination of the rate of interest. Thus the exponents of the loanable funds theory saw the interplay of monetary and non-monetary forces in the determination of the rate of interest. We, therefore, see that loanable funds theory is a monetary theory of interest, although it is only partly monetary since it also recognises the importance of real forces such as thriftiness and productivity of capital in the determination of the rate of interest.

According to this theory, rate of interest is determined by demand for and supply of loanable funds. The supply of loanable funds consists of savings out of disposable income, dishoarding, money created by the banks and disinvestment (*i.e.*, disentangling of fixed and working capital). The demand for loanable funds is composed of the demand for investment, demand for consumption and demand for hoarding money. We shall discuss below in detail these several sources of supply and demand of loanable funds.

Supply of Loanable Funds

Savings. Savings by individuals and households constitute the most important source of the supply of loanable funds. In the loanable funds theory savings are considered in either of the two ways. Firstly, in the sense of *ex-ante* savings, that is, savings planned by individuals and households in the beginning of a period in the hope of expected incomes and anticipated expenditures on consumption, and secondly, in the sense of Robertsonian savings, that is, the difference between the income of the preceding period (which becomes disposable in the present period) and consumption of the present period. In both these senses of savings it is assumed that the amount of savings varies with rate of interest. More savings will be forthcoming at higher rates of interest and *vice versa.* It is granted that savings by individuals and households primarily depend upon the size of their income. But, given the level of income, savings vary with the rate of interest; the higher the rate of interest, the greater the volume of savings. Therefore, supply curve of savings slopes upward to the right.

Like individuals, business concerns also save. When the business is of the type of single proprietorship or partnership, a part of the income from the business is used for consumption purposes and a part is kept for further expansion of the business. When the business is of the type of joint stock company, a part of the earnings is distributed as dividends to the shareholders and a part of the earnings of the company is retained undistributed which constitutes the *corporate savings*. Business savings depend partly upon the current rate of interest. A higher rate of the interest is likely to encourage business savings as a substitute for borrowing from the loan market. These savings are mostly used for investment purposes by the business firms themselves and, therefore, most of them do not enter into the market for loanable funds. But these savings influence the rate of interest since they serve as substitute for borrowed funds and, therefore, reduce the market demand for loanable funds. In Fig. 35.4, the curve labelled as S indicates the supply curve of savings which slopes upward to the right.

Dishoarding. Dishoarding of the past accumulated savings constitutes another source of supply of loanable funds. Individuals may possess idle cash balances hoarded from the incomes of the previous periods which they may dishoard in a period. When people dishoard, the idle cash balances become active cash balances in the present period and thus add to the supply of loanable funds. People hoard money because of their preference for liquidity. When the rate of interest rises or when the prices of securities decline, they may like to take advantage of these market movements and thus dishoard money for lending it to others or for purchasing securities. At a higher rate of interest, the individuals possessing idle cash balances will be induced to dishoard more money. At very low rates of interest, their parting with liquidity will not be rewarded sufficiently and, therefore, they will hold on to money. It is evident that *dishoarding is interest-elastic and*, therefore, the curve of dishoarding slopes upwards to the right as is shown in Fig. 35.4 by curve *DH*.

Bank Money. The banking system is another important source of the supply of loanable funds. The commercial banks by creating credit money advance loans to the businessmen and industries for investment. Banks can also reduce the supply of loanable funds by contracting their lending. Banks also purchase and sell securities and thereby affect the supply of loanable funds. The supply curve of funds provided by banks is to some degree interest-elastic. Generally speaking, the banks would like to lend more money at higher rates of interest than at lower ones. Therefore, supply curve of bank money also slopes upward to the right as is shown by the curve *BM* in Fig. 35.4.

Disinvestment. Disinvestment is another source of the supply of loanable funds. Disinvestment means disentangling of the present fixed and working capital. Usually a good amount of depreciation reserves are kept so as to replace the fixed capital when it is completely worn out. When there is a declining tendency in certain industries due to some structural changes in the economy, the entrepreneurs may not like to remain tied to those industries and therefore they may allow the existing stock of machines and other equipment belonging to those industries wear out without replacement. As a result, they may bring the depreciation reserves in the market for loanable funds. Similarly, working capital invested in business may be withdrawn gradually and made available as loanable funds. When disinvestment is decided to be undertaken, then not only the depreciation reserves but also a part of the revenue earned from the sale of the output instead of going into capital replacement flows into the market for loanable funds. At higher rates of interest, the entrepreneurs will generally contemplate a greater amount of disinvestment. Prof. M.M. Bober rightly remarks, "Disinvestment is encouraged somewhat by a high rate of interest on loanable funds. When the rate is high, some of the current capital may not produce a marginal revenue product to match this rate of interest. The firm may decide to let this capital run down and to put the depreciation funds in the loan market."[4] It is, therefore, clear that disinvestment curve will also slope upward to the right, as is indicated by the curve *DI* in Fig. 35.4.

By lateral summation of the curves of savings (*S*), dishoarding (*DH*), bank money (*BM*) and disinvestment (*DI*) we get the total supply curve of loanable funds *SL* which slopes upward to the right showing that a greater amount of loanable fund will be available at higher rates of interest and vice versa.

Demand for Loanable Funds

Having now explained the sources of supply of loanable funds, we turn to explain the factors which determine the demand for loanable funds. Loanable funds theory also differs from the classical theory in its explanation of the demand for funds. Whereas the classical theory considers only the demand for funds for investment purposes, the loanable funds theory also considers the demand of loans for consumption and demand for hoarding money, apart from the demand of funds for investment. In considering the hoarding of money, loanable funds theory incorporates in itself the factor of liquidity preference on which Keynes later laid great stress as an important determinant of interest. We shall explain below these different sources of demand for loanable funds.

Investment Demand. Demand for investment constitutes an important factor working on the side of demand for loanable funds. Investment demand includes businessmen's borrowings for purchasing or making of new capital goods including the building up of inventories. The price of

4. M.M. Bober, *Intermediate Price and Income Theory* (1955), p. 371.

obtaining the loanable funds required to purchase or make in capital goods is obviously the rate of interest. It will pay businessmen to demand and undertake investment of loanable funds upto the point where the expected net rate of return on investment equals the rate of interest. In the loanable

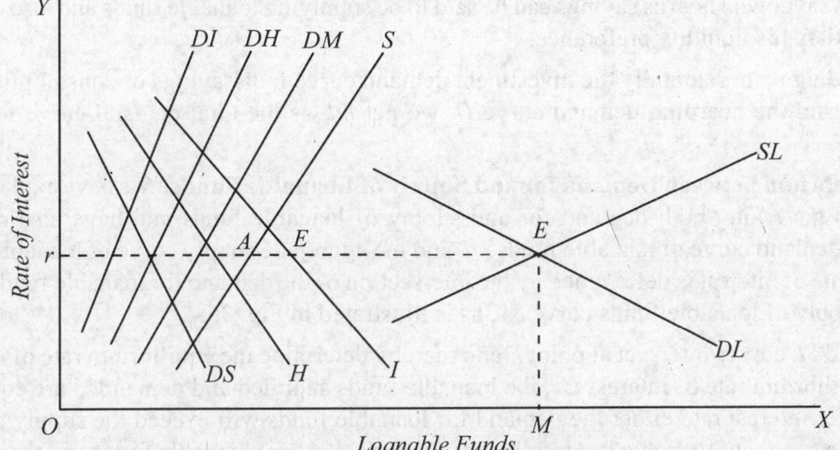

Fig. 35.4. *Determination of rate of interest*

funds theory, demand for investment depends upon the marginal revenue productivity of capital (or the marginal rate of return) in the same way as in the classical theory. When the rate of interest falls, businessmen will find it profitable to increase investment in capital goods with the result that their demand for loanable funds will rise. We thus see that demand of loanable funds for investment is interest-elastic ; at a low rate of interest, there will be greater investment demand and vice versa. Therefore, the curve of investment demand for loanable funds slopes downward to the right as is shown by the curve *I* in Fig. 35.4.

Consumption Demand. Another important source of demand for loanable funds are the loans desired to be taken by the people for consumption purposes. Loans for consumption purposes are demanded by the people when they wish to make purchases in excess of their current incomes and idle cash resources. The loans for consumption purposes are demanded generally for buying durable use goods such as houses, automobiles, refrigerators, television sets, air conditioners etc. Whereas a lower rate of interest will induce people to borrow more for consumption, the higher rates of interest will discourage borrowing for consumption. Therefore, consumption demand for loanable funds slopes downward to the right and is shown by the curve labelled as *DS* in Fig. 35.4.

Demand for Hoarding. Lastly, demand for money to hoard is another important factor determining demand for loanable funds. Demand for hoarding money arises because of people's preference for liquidity, *i.e.*, for cash balances. Hoarded money represents idle cash balances. People save money when they do not spend all their disposable income on consumption. Now the people can lend out their savings to others or purchase securities with their savings or invest their savings in real capital assets. *Another alternative use of income saved (i.e., income not spent on consumption) is to hoard them, that is, to hold them as idle cash balances*. People can also hoard money when they sell securities or assets owned by them and not spending the proceeds obtained therefrom. Now the question arises as to why people hoard money when they can earn some income by lending it to others or investing it in securities or capital assets. An important reason for the demand for hoarding money is that people like to take advantage of the changes in the rate of interest or changes in the prices of securities in the future. At higher current rates of interest, people will hoard less money because much of the money will be lent out to take advantage of the higher rates of interest, and people will hoard more money at lower rates of interest because loss suffered in hoarding money in this case will not be very much. It follows therefore that the curve of demand for hoarding money will slope downwards as is shown by the curve *H* in Fig. 35.4. An important point to be noted here is that the person who has a demand for funds to hoard supplies the funds to himself for that purpose. A saver who hoards savings can be said to be supplying loanable funds and also demanding them to satisfy his liquidity preference.

By adding up horizontally the investment demand curve *I*, dissavings or consumption demand curve *DS*, and the hoarding demand curve *H*, we get *DL* as the total demand curve for loanable funds.

Equilibrium between Demand for and Supply of Loanable Funds. We have explained above the factors governing both demand for and supply of loanable funds and have also derived the aggregate demand curve of loanable funds *DL* and the aggregate supply curve of loanable funds *SL*. Now, the rate of interest is determined by the intersection of the demand for loanable funds curve *DL* and the supply of loanable funds curve *SL*, as is illustrated in Fig. 35.4.

DL and *SL* curves intersect at point *E* and thereby determine the equilibrium rate of interest *Or*. At the equilibrium rate of interest *Or*, the loanable funds supplied and demanded are equal to *OM*. At any other interest rate either the demand for loanable funds will exceed the supply of loanable funds or the supply of loanable funds will exceed the demand for loanable funds and therefore there will be a change in the rate of interest until it reaches the level or where demand for and supply of loanable funds are equal.

It should be noted that *at the equilibrium rate of interest where aggregate demand for and supply of loanable funds are equal, planned savings and investment may not be equal*, as is the case in Fig. 35.4. It will be seen from this figure that at equilibrium rate of interest *Or*, while investment is equal to *rE*, the savings are equal to *rA*. As a result of investment being greater than savings, income will increase. With the increase in income, the savings curve *S* and therefore the aggregate supply curve *SL* will shift to the right. And this shift in the savings and supply of loanable funds curve *SL* will cause a change in the rate of interest. We thus see that the *rate of interest as determined by the demand for and supply of loanable funds will not be a stable one if there is inequality between savings and investment at that rate*. This inequality will bring about a change in the income and thereby a change in the savings and supply of loanable funds. As a result, the rate of interest will tend to change. A *stable equilibrium rate of interest will be achieved where intersection of the aggregate supply and demand curves of loanable funds determines the equilibrium rate of interest at which the savings and investment are also equal*. But in a single period, the rate of interest at which demand and supply of loanable funds are equal will prevail in spite of the inequality of savings and investment, although the equilibrium rate of interest in such a situation will tend to change over time through changes in income.

We have explained above the various components of demand for and supply of loanable funds and have shown how the equilibrium rate of interest is determined. We have taken the savings (*S*) on the supply side, dissavings (*DS*) on the demand side ; dishoarding (*DH*) on the supply side and hoarding (*H*) on the demand side ; investment (*I*) on the demand side and disinvestment on the supply side. We can further simplify our analysis of components of demand for and supply of loanable funds and bring out the conditions for the equilibrium rate of interest in a better way if we use *net of savings* (*i.e.*, savings *minus* dissavings), *net* of hoarding (*i.e.*, hoarding *minus* dishoarding) and *net* of investment (*i.e.*, investment *minus* disinvestment). This will become clear from the following :

We know that equilibrium rate of interest is determined where

Supply of loanable funds = Demand for loanable funds

or $S + DH + BM + DI = I + DS + H$

By taking DS to the left hand sidc and DH and DI to the right hand side we get

$(S - DS) + BM = (1 - DI) + (H - DH)$

Or $Net\ S + BM = Net\ I + Net\ H$

Or, Net savings + Bank money = Net Investment + Net hoarding

Thus we see that the rate of interest will be at the equilibrium level where the supply of net savings and bank money will be equal to the demand for investment and net hoarding. This is the essence of the loanable funds theory of interest.

Critical Evaluation of Loanable Funds Theory

Loanable funds theory is superior to classical theory of interest. It has greatly improved our understanding of the forces working on the supply of and demand for loanable funds. It makes quite comprehensive analysis of the determination of the rate of interest and takes into consideration all the relevant factors which have a bearing on the rate of interest, namely, saving or thriftiness, investment demand, hoarding and bank credit. However, loanable funds theory has been criticised by Keynes and Keynesians.

First, it was asserted by Keynes that the concept of hoarding as used in loanable funds theory is quite dubious. This is so because hoarding simply cannot increase or decrease as long as the amount of money remains the same. Money in circulation in an economy has to be in somebody's cash balances at any time. According to him, if the quantity of money remains the same, then the total amount of cash balances in the beginning and at the end of a period will be the same ; the greater hoarding of money by one person must have been offset by the dishocarding of any other person. But this criticism of loanable funds theory is misplaced. As a matter of fact, the effective supply of money in a society does not merely depend upon the quantity of money, it also depends upon the velocity of circulation of money. And it is this velocity of circulation which changes as a result of hoarding or dishoarding and, therefore, involves the changes in the effective supply of money, although the amount of money in existence may have remained the same.

Thus, we see that the hoarding can occur even if the quantity of money in circulation remains constant during a period and, therefore, Keynes's objection against the loanable funds theory on this ground is not valid. As a matter of fact, Keynes himself introduced distinction between '*active*' and '*idle*' balances. Now, increase in idle balances at the expense of active balances is hoarding and results in a reduction in the velocity of circulation of money. If the time duration of idleness of money (*i.e.*, period of rest between the two transfers) increases, it will mean hoarding which will reduce the supply of loanable funds in the market and thus affect the determination of the rate of interest.

Keynes also criticised the loanable funds theory on the ground that like classical theory it *did not provide a determinate solution to the interest-rate determination* and involved what was called circular reasoning. According to him, since savings is an important constituent of the supply of loanable funds, the supply of loanable funds curve will vary with the level of income which determines savings. We, therefore, cannot know the rate of interest unless we know what the level of income is. And we cannot know the level of income unless we know the rate of interest since rate of interest affects investment which in turn determines the level of income. Following Keynes, Hansen also disapproves loanable funds theory and maintains that "the schedule of loanable funds is composed of savings plus net additions to loanable funds from new money and dishoarding of idle balances. But since the savings portion of the schedule varies with the level of disposable income, it follows that the total supply schedule of loanable funds also varies with income making the rate of interest indeterminate."[5]

5. A.H. Hansen, *Guide to Keynes*, p. 141.

Keynes was correct in criticising the classical theory for its ignoring the effect of changes in the level of income upon the supply of savings but his criticism against loanable funds theory is not valid. This is because *loanable funds theory seeks to explain the interest rate determination through period analysis with a lag of one period,* which makes the theory quite determinate. In loanable funds theory, the supply of savings is regarded as being determined by the income of the preceding period and savings so determined along with other components of supply and the demand for loanable funds determine the rate of interest in the current period. The current rate of interest so determined affects the level of income in the succeeding period through investment. Prof. Halm rightly maintains that "It is not circular reasoning to say that income is influenced by investment, investment by rates of interest, rates of interest by the supply of loanable funds, the supply of loanable funds by savings, and savings in turn, by the income received in the last period."[6] We, therefore, conclude that charge against loanable funds theory that it is indeterminate is untenable. In fact, it is Keynes's own liquidity preference theory of interest, as we shall see later, which is indeterminate.

Another charge against the loanable funds theory is that it is based upon the assumption of full employment level of income which does not hold in the real world. And the superiority of Keynes's theory is sought to be proved on the basis of its being based upon realistic assumption of less than full employment. But this is not correct interpretation of loanable funds theory. As we have seen above in the explanation of the loanable funds theory that it takes into account the increases in the level of income as a result of investment and their influence on savings. According to loanable funds theory, stable equilibrium regarding rate of interst will be reached at the level where not only demand for and supply of loanable funds are equal also saving and investment are equal. If the full-employment were income the assumption, how the income could increase ?

As a matter of fact, loanable funds theory is a synthesis between the classical theory and Keynes's liquidity preference theory since it takes into account the savings and investment demand of the classical theory as well as liquidity preference of Keynes's theory. By incorporating hoarding and dishoarding it considers the liquidity preference on which Keynes laid a great stress as an important factor determining the rate of interest. Besides, loanable funds theory has been described as dynamic. Thus Prof. H.G. Johnson[7] has suggested that the Keynesian theory is 'static' seeking only to explain the state of affairs in a short period equilibrium and how changes in circumstances will alter the equilibrium values, while the loanable funds theory is dynamic and seeks to explain precisely how interest and income move from one equilibrium level to another when circumstances have changed.

KEYNES'S LIQUIDITY PREFERENCE THEORY OF INTEREST

In his epoch-making book, "*The General Theory of Employment, Interest and Money*" J.M. Keynes gave a new view of interest. According to him, "*interest is the reward for parting with liquidity for a specified period.*"[8] A man with a given income has to decide first how much he is to consume and how much to save. The former will depend on, what Keynes calls, the propensity to consume. Given this propensity to consume, the individual will save a certain proportion of his given income. He then has to make another decision. Should he hold his savings ? How much of his resources will he hold in the form of ready money (cash or non-interest-paying bank deposits) and how much will he part with or lend depends upon what Keynes calls his "*liquidity preference*". Liquidity preference means the *demand for money to hold or the desire of the public to hold cash.*

Demand for Money or Motives for Liquidity Preference

Liquidity preference of a particular individual depends upon several considerations. The question is : Why should the people hold their resources liquid or in the form of ready money, when they can

6. G.N. Halm, *Monetary Theory* p.147 (italics added).
7. "Some Cambridge Controversies in Monetary Theory", *Review of Economic Studies*, Vol. XIX. No. 49 (1951-52).
8. Op. cit., p. 167.

get interest by lending such resources ? The desire for liquidity arises because of three motives; (*i*) the transactions motive, (*ii*) the precautionary motive, and (*iii*) the speculative motive.

The **Transactions Motive.** The transactions motive relates to the demand for money or need for cash for the current transactions of individuals and businessmen. Individuals want to hold cash in order "to bridge the interval between the receipt of income and its expenditure". This is called the '*Income Motive*'. Most of the people receive their incomes by the week or the month, while the expenditure goes on day by day. A certain amount of ready money, therefore, is kept in hand to make current payments for goods and services to be purchased. This amount will depend upon the size of the individual's income, the interval at which the income is received and the methods of payment prevailing in the society.

The businessmen and the entrepreneurs also have to keep a proportion of their resources in ready cash in order to meet current needs of various kinds. They need money all the time in order to pay for raw materials and transport, to pay wages and salaries and to meet all other current expenses business. This Keynes calls the '*Business Motive*' for keeping money. It is clear that the amount of money held under this business motive will depend to a very large extent on the turnover (*i.e.*, the volume of trade of the firm in question). The larger, the turnover, the larger in general, will be the amount of money needed to cover current expenses.

Precautionary Motive. Precautionary motive for holding money refers to the desire of the people to hold cash balances for unforeseen contingencies. People hold a certain amount of money to provide for the danger of unemployment, sickness, accidents, and the other uncertain emergencies. The amount of money held under this motive will depend on the nature of the individual and on the conditions in which he lives.

Speculative Motive. The speculative motive relates to the desire to hold one's resources in liquid form in order to take advantage of market movements regarding the future changes in the rate of interest (or bond prices). The notion of holding money for speculative motive is a new typically Keynesian idea. *Money held under the speculative motive serves as a store of value* as money held under the precautionary motive does. But it is a store of money meant for a different purpose. The cash held under this motive is used to make speculative gains by dealing in bonds[9] whose prices fluctuate. If bond prices are expected to rise, which, in other words, means that the rate of interest is expected to fall, businessmen will buy bonds to sell when their prices actually rise. If, however, bond prices are expected to fall, *i.e.*, the rate of interest is expected to rise, businessmen will sell bonds to avoid capital losses. Nothing being certain in this dynamic world, where guesses about the future course of events are made on precarious basis, businessmen keep cash balances to speculate on the probable future changes in bond prices (*i.e.* the rate of interest) with a view to making profits.

Given the expectations about the changes in the rate of interest in future, less money will be held under the speculative motive at a higher current or prevailing rate of interest and more money will be held under this motive at a lower current rate of interest. The reason for this inverse relation between money held for speculative motive and the prevailing rate of interest is that at a lower rate of interest less is lost by not lending money or not investing it, that is, by holding on to money, while at a higher rate of interest holders of cash balances would lose more by not lending or investing.

Thus, the demand for money under speculative motive is a function of the current rate of interest, increasing as the interest rate falls and decreasing as the interest rate rises. Thus, demand for money under this speculative motive is a decreasing function of the rate of interest. This is shown in Fig.35.5.

---

9. All securities and other such papers that yield a fixed and known rate of interest over a period of time are known as bonds.

Along the X-axis is represented the speculative demand for money and along the Y axis the rate of interest. The liquidity preference curve LP is a downward sloping towards the right signifying that the higher the rate of interest, the lower the demand for speculative motive, and *vice versa*. Thus, at the high current rate of interest Or, a very small amount OM is held for speculative motive. This is because at a high current rate of interest much money would have been lent out or used for buying bonds and therefore less money will be kept as inactive balances. If the rate of interest falls to Or', then a greater amount OM' is held under speculative motive. With the further fall in the rate of interest to Or'', money held under speculative motive increases to OM''. It will be seen in Fig. 35.5 that the liquidity preference curve LP becomes quite flat *i.e.*, perfectly elastic at a very low rate of interest. It is horizontal line beyond point E'' towards the right. This perfectly elastic portion of liquidity preference curve indicates the position of absolute liquidity preference of the people. That is, at a very low rate of interest people will hold with them as inactive balances any amount of money they come to have. This *portion of liquidity preference curve with absolute liquidity preference is called liquidity trap by some economists*.

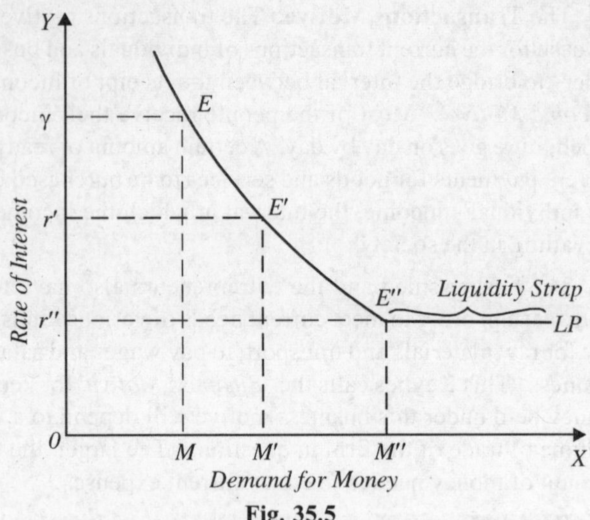

Fig. 35.5

But the demand for money to satisfy the speculative motive does not depend so much upon what the current rate of interest is, as on expectations of changes in the rate of interest. If there is a change in the expectations regarding the future rate of interest, the whole curve of liquidity preference for speculative motive will change accordingly. Thus, if the public on balance expect the rate of interest to be higher (*i.e.*, bond prices to be lower) in the future than had been previously supposed, the speculative demand for money will increase and the whole liquidity preference curve for speculative motive will shift upward.

If the total supply of money is represented by M, we may refer to that part of M held for transactions and precautionary motives as M_1 and to that part held for the speculative motive as M_2. Thus $M = M_1 + M_2$. According to Keynes, the money held under the transactions and precautionary motives, *i.e.* M_1, is completely interest-inelastic unless the interest rate is very high. The amount of money held as M_1, that is, for transactions and precautionary motive, is mainly a function of the size of income and business transactions together with the contingencies growing out of the conduct of personal and business affairs. We can write this in a functional form as follows :

$$M_1 = L_1(Y) \qquad \ldots (i)$$

where Y stands for income, L_1 for liquidity preference function, and M_1 for money held under the transactions and precautionary motive. The above function implies that money held under the transaction and precautionary motive is a function of income.

On the other hand, money demanded for speculative motives, *i.e.*, M_2, as explained above, is primarily a function of the rate of interest. This can be written as :

$$M_2 = L_2(r) \qquad \ldots (ii)$$

where r stands for the rate of interest, L_2 for liquidity preference function for speculative motive.

Since total supply of money $M = M_1 + M_2$, we get from (*i*) and (*ii*) above

$$M = L_1(Y) + L_2(r)$$

Determination of the Rate of Interest: Interaction of Liquidity Preference and Supply of Money

According to Keynes, the demand for money, *i.e.*, the liquidity preference and supply of money determine the rate of interest. It is in fact the liquidity preference for speculative motive which along with the quantity of money determines the rate of interest. We have explained above the speculative demand for money in detail. As for the supply of money, it is determined by the policies of the Government and the Central Bank of the country. The total supply of money consists of coins plus notes plus bank deposits. How rate of interest is determined by the equilibrium between the liquidity preference for speculative motive and the supply of money is shown in Fig. 35.6.

In Fig. 35.6, *LP* is the curve of liquidity preference for speculative motive. In other words, *LP* curve shows the demand for money for speculative motive. To begin with, *ON* is the quantity of money available for satisfying liquidity preference for speculative motive. Rate of interest will be determined where the speculative demand for money is in balance or equal to the fixed supply of money *ON*. It is clear from the figure that speculative demand for money is equal to *ON* quantity of money at *Or* rate of interest. Hence *Or* is the equilibrium rate of interest. Assuming no change in expectations, an increase in the quantity of money (say through open market operations by central bank of a country) for the speculative motive will lower the rate of interest. In Fig. 35.6, when the quantity of money increases from *ON* to *ON′*, the rate of interest falls from *Or* to *Or′* because the new quantity of money *ON′* is in balance with the speculative demand for money at *Or′* rate of interest. In this case we move down the curve. Thus, given the curve of liquidity preference for speculative motive, an increase in the quantity of money brings down the rate of interest.

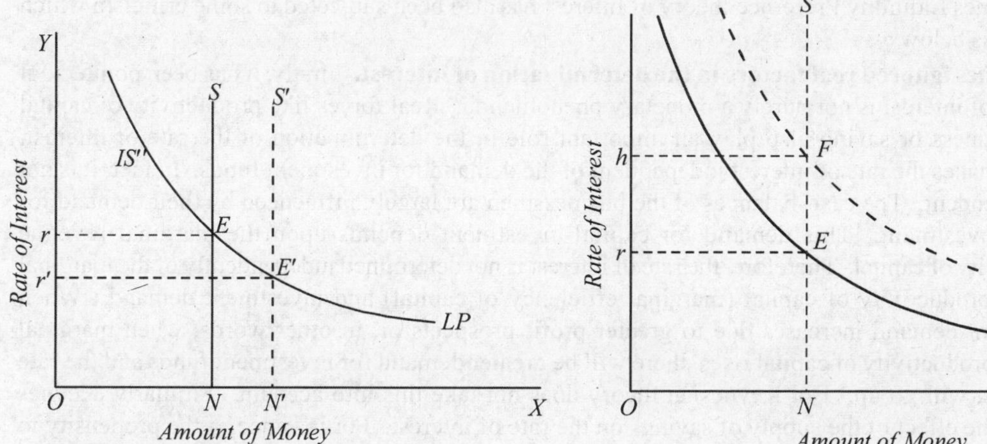

Fig. 35.6. *Determination of Rate of Interest and Effect of Expansion in Money Supply*

Fig. 35.7. *Effect of Increase in the Lignidity prefereance on Rate of Interest*

But the act of increase in the quantity of money may cause a change in the expectations of the public and thereby cause an upward shift in liquidity preference curve for speculative motive bringing the rate of interest up. But this is not certain. "New developments may only cause wide differences of opinion leading to increased activity in the bond market without necessarily causing any shift in the aggregate speculative demand for money[10].

10. Hansen, *A Guide to Keynes*, p. 133.

It is worth mentioning that shift in liquidity preference schedule or curve can be caused by many other factors which affect expectations and might take place independently of changes in the quantity of money by the central bank. Shifts in the liquidity function may be either downward or upward depending on the way in which the public interprets people a change in events. If some change in events leads the people on balance to expect a higher rate of interest in the future than they had previously supposed,theliquidity preference for speculative motive will increase which will bring about an upward shift in the curve of liquidity preference for speculative motive and will raise the rate of interest. In Fig. 35.7, assuming that the quantity of money remains unchanged at *ON*, the rise in the liquidity preference curve from *LP* to *L′*P′, the rate of interest rises from *Or* to *Oh* because at *Oh*, the new speculative demand for money is in equilibrium with the supply of money *ON*. It is worth noting that when the liquidity preference for speculative motive increases from *LP* to *L′P′*, the amount of money hoarded does not increase; it remains *ON* as before. Only the rate of interest rises from *Or* to *Oh* to equilibrate the new liquidity preference for speculative motive with the available quantity of money *ON*.

Thus we see that Keynes explained interest in terms of purely monetary forces and not in terms of real forces like productivity of capital and thriftiness times which formed the foundation-stones of both classical and loanable fund theories. According to him, demand for money for speculative motive together with the supply of money determines the rate of interest. He agreed that the marginal revenue product of capital tends to become equal to the rate of interest but the rate of interest is not determined by marginal revenue product of capital. Moreover, according to him, interest is not a reward for saving or thriftiness or waiting but for parting with liquidity. Keynes asserted that it is not the rate of interest which equalises saving and investment. But this equality is brought about through changes in the level of income.

Critical Appraisal of Keynes's Liquidity Preference Theory of Interest

Keynes Liquidity Prefernce theory of Interest has also been subjected to some criticism which we discuss below :

Keynes ignored real factors in the determination of interest. Firstly, it has been pointed out that rate of interest is not purely a monetary phenomenon. Real forces like productivity of capital and thriftiness or saving also play an important role in the determination of the rate of interest. Keynes makes the rate of interest independent of the demand for investment funds. In fact, it is not so independent. The cash-balances of the businessmen are largely influenced by their demand for capital investment. This demand for capital-investment depends upon the marginal revenue productivity of capital. Therefore, the rate of interest is not determined independently of the marginal revenue productivity of capital (marginal efficiency of capital) and investment demand. When investment demand increases due to greater profit prospects or, in other words, when marginal revenue productivity of capital rises, there will be greater demand for investment funds and the rate of interest will go up. But Keynesian theory does not take this into account. Similarly, Keynes ignored the effect of the supply of savings on the rate of interest. For instance, if the propensity to consume of the people increases, savings would decline. As a result, supply of funds in the market will decline which will raise the rate of interest.

2. **Keynesian theory is also indeterminate.** Now exactly the same criticism applies to Keynesian theory itself on the basis of which Keynes rejected the classical and loanable funds theories. Keynes's theory of interest is also indeterminate. According to Keynes, rate of interest is determined by the speculative demand for money and the supply of money available for satisfying speculative demand. Given the total money supply, we cannot know how much money will be available to satisfy the speculative demand for money unless we know how much the transactions demand for money is. And we cannot know the transactions demand for money unless we first know the level of income Thus the Keynesian theory, like the classical, is indeterminate. "In the Keynesian case the supply and demand for money schedules cannot give the rate of interest unless we already know the

income level; in the classical case the demand and supply schedules for saving offer no solution until the income is known. Precisely the same is true of loanable-fund theory. Keynes' criticism of the classical and loanable-fund theories applies equally to his own theory."[11]

3. **No liquidity without Savings.** According to Keynes, interest is a reward for parting with liquidity and in no way a compensation and inducement for saving or waiting. But without saving how can the funds be available to be kept as liquid and how can there be question of surrendering liquidity if one has not already saved money. Jacob Viner rightly maintains, "*Without saving there can be no liquidity to surrender*". Therefore, the rate of interest is vitally connected with saving which is neglected by Keynes in the determination of interest.

It follows from above that Keynesian theory of interest is also not without flaws. But importance Keynes gave to liquidity preference as a determinant of interest is correct. In fact, the exponents of loanable funds theory incorporated the liquidity preference in their theory by laying greater stress on hoarding and dishoarding. We are inclined to agree with Prof. D. Hamberg when he says, "Keynes did not forge nearly as *new* a theory as he and others at first thought. Rather, his great emphasis on the influence of hoarding on the rate of interest constituted an invaluable addition to the theory of interest as it had been developed by the loanable funds theorists who incorporated much of Keynes's ideas into their theory to make it more complete."[12]

QUESTIONS FOR REVIEW

1. **Define interest.** Distinguish between nominal rate of interest and real rate of interest
2. Critically examine the following views about why rate of interest arises (or is paid).
   (*a*) Interest is a reward for abstinence
   (*b*) Interest arises because of time preference
   (*c*) Interest is a reward for parting with liquidity
3. Explain how saving and investment are related to rate of interest. Critically examine the classical theory that rate of interest is determined by saving and investment.
4. Explain how Keynes showed that Classical theory of interest was indeterminate. The same criticism of indeterminacy has been made against Keynes' own theory of interest. Discuss
5. Explain the components of demand for and supply of loanable funds. Show that rate of interest is determined by demand for and supply of loanable funds. Show that rate of interest is determined by demand for and supply of loanable funds. Is the rate of interest determined through equality of demand for and supply of loanable funds stable?
6. How does loanable funds theory of interest differ from (i) Classical theory, and (2) Keynes' theory of interest
7. Explain the Keynesian theory of interest. How is the Keynesian theory of interest different from the Classical theory. *[D.U. B.Com (Hons) 1999]*
8. What is meant by liquidity preference? How does it explain why interest is paid?
9. Explain Keynes' liquidity preference theory of interest. Why is it considered to be indeterminate? *[D.U. B.Com (Hons) 1998]*
10. In what ways is the Keynesian theory of interest a departure from the classical theory? Discuss *[D.U. B.A. (Hons), 1995]*
11. Explain why demand for money is considered to be declining function of rate of interest. *[D.U. B.A. (hons) 1999]*
12. The deference between the classical theory of interest and Keynes' money theory of interest is a fundamental aspect of the difference between the economics of full employment and the economics of less than full employment (D. Dillard). Discuss.

---

11. Hansen, *Guide to Keynes*, p. 141.
12. D. Hamberg, *Business' Cycles*, p. 183.

Chapter 36
THE THEORY OF PROFITS

Introduction

After having discussed the determination of rent of land, wages of labour and interest on capital, we now pass on to the study of profits which are said to be the reward for enterprise, the fourth factor of production. No doubt profits are associated with entrepreneur and his functions but the economists from time to time have expressed diverse and conflicting views about the nature, origin and role of profits. Till today, there is no complete agreement among economists about the true nature and origin of profits. As a matter of fact, there has been perhaps no topic in the whole economic theory which has been in such a confused and tangled state as the theory of profit.

A part of the confusion in the theory of profit is due to the lack of agreement among economists about the true or proper function of the entrepreneur. Some have held the view that the function of the entrepreneur is to organise and co-ordinate the other factors of production. According to them, entrepreneur earns profits for his performing this function. On this view, enterprise is a special type of labour and profits a special form of wages. Some others have described the entrepreneur as performing the functions of bearing risk and uncertainty as he controls the business and takes price and output decision. The entrepreneur earns profits because he bears risk and uncertainty because as his price and output policies may prove to be incorrect in view of the future business movements. Schumpeter has assigned to the entrepreneur the role of an innovator and profits as a reward for his introducing innovations. Lastly, F.H. Knight has emphasised uncertainty in the economy as a factor which gives rise to profits and bearing uncertainty is the task of the entrepreneur.

Besides, some economists have described profits as *non-functional income*. Thus, J.M. Keynes expressed the view that profits resulted from the favourable movements of the general price level. Joan Robinson, E.H. Chamberlin and M. Kalecki have associated profits with imperfect competition and monopoly. According to them, the greater the degree of imperfection or, in other words, the greater the degree of monopoly power, the greater the profits made by the entrepreneur. Thus, profits has been connected by F.H. Knight with uncertainty, by Schumperter with innovations, by Hawley with risk-bearing, and by Joan Robinson. E.H. Chamberin and M. Kalecki with the degree of monopoly power. As a matter of fact, profits arise from all these sources. Therefore, no single explanation or theory of profits is adequate; each omits some crucial factors and fails to bring out some important economic factors having a relation with profits. B.S Keristead, therefore, expresses the view that profits originate from monopoly, successful innovations and a correct estimate of uncertain future. He thus says, "Profits may come to exist as a result of monopoly or monopsony, as a reward for innovation, as a reward for the correct estimate of uncertain factors, either particular to the industry or general to the whole economy."[1]

Profits as a Residual Income

It is worth mentioning that profits are *residual income* left after the payment of the contractual rewards to other factors of production. The entrepreneur while engaging other factors of production enters into contract with them. He thus pays wages to the workers, rent on the land employed,

1. B.S. Keirstead, *Capital, Interest and Profits*, (1959), p. 6.

interest on the loans taken at the rates already fixed by contracts. In fact, the entrepreneur makes payment to these factors much in advance of the realisation of the values of output produced after sale of the product. What is left after paying the contractual rewards to other factors employed is the profits of entrepreneur. Thus profits are *non-contractual income* and therefore they may be positive or negative, whereas the contractual income of other factors such as wages, rent and interest are always positive and never negative. It should be further noted that *pure profits* of the entrepreneur are found by subtracting from the gross residual income the *imputed values* of rent and interest on self-owned land and capital employed by the entrepreneur and also the imputed *wages* for his work of routine management.

PROFITS AS A DYNAMICS SURPLUS: CLARK'S DYNAMIC THEORY OF PROFITS

A popular conception of profits is that they arise in a dynamic economy, that is, in an economy where changes are taking place. In a static economy where nothing changes there can be no profits. It was J.B. Clark who first propounded that profits are a dynamic surplus. He argued that in a stationary state where no changes in conditions of demand and supply are occurring, the prices paid to the factors on the basis of their marginal productivity would exhaust the total value product and therefore no profits would accrue to the entrepreneur. Profits result when selling prices of the goods exceed their cost of production. Now, in a competitive long-run equilibrium, price equals average cost of production (including normal profits which are in fact wages for routine supervision and management) and therefore no pure profits are made. Now, if no changes either in the conditions of demand or in the conditions of supply occur, competitive equilibrium will persist and therefore no pure profits will be earned by the entrepreneur.

On the contrary, if due to the changes in either demand or supply, price exceeds cost of production, profits will emerge. If due to these changes, price falls below the cost of production, negative profits, that is, losses will accrue to the entrepreneur. It is evident that changes distrub the long-run competitive equilibrium achieved and thereby give rise to profits. In other words, profits arise due to disequilibrium caused by the changes in demand and supply conditions. Prof. Stigler rightly says, "Firms in a competitive industry may receive profits....because of a state of disequilibrium....these profits can arise even if all entrepreneurs are identical, for *disequilibrium* can characterize a whole industry. If prices are higher, or costs lower than were anticipated, entrepreneurs will receive a return in excess of the alternative product of their resources. If prices were lower or costs higher than were anticipated, entrepreneurs will receive less than the alternative product of their resources *i.e.*, negative profits. Positive profits may persist for a long time if firms outside the industry are slow to enter the industry and negative profits can persist as long as specialised equipment yields more than when used in the industry that when used elsewhere, say as scrap."[2]

It should be noted that these disequilibrium profits arise from unanticipated changes in demand or cost conditions. If the changes could have been foreseen in advance, then suitable adjustments could have been made according to the anticipated changes so that forces of competition would have driven profits to zero.

The Various Types of Changes

Now, the question is what changes occur in the economy and give rise to profits. Clark mentioned five changes that occur in a dynamic economy and which give rise to profits. These five changes are: (1) The changes in the quantity and quality of human wants, (2) changes in methods or techniques of production, (3) changes in the amount of capital (4) changes in the forms of business organisation, and (5) the growth of population. These changes are constantly taking place and bring about the

2. G.J. Stigler. *The Theory of Price* (revised edition, 1952), p. 181.

divergence between price and cost and thereby give rise to profits, positive or negative. If the demand for a commodity increases due to the increase in population or increase in the incomes of the people or due to the increase in consumers' preferences for the commodity, the price of the commodity will rise, and if cost remains the same, profits would accrue to the entrepreneurs producing the commodity. On the other hand, cost of production may go down as a result of the adoption of a new technique of production, or as a result of cheapening of the raw material, and if price remains constant or does not fall to the same extent, the profits would emerge.

Apart from the five changes mentioned by Clark, there are other changes also which occur in the economy. All the changes which take place and as a result of which profits arise in a dynamic economy may be classified into two types: (1) *innovations* and (2) *external changes*. We explain below these changes in some detail. Innovations represent changes which are introduced by individual entrepreneurs themselves.

Innovations. The entrepreneur earns large profits from introducing innovations such as a new product, a new and cheaper method of production, a new method of marketing the product, a new way of advertisement. The innovational changes may either reduce cost or increase the demand for the product and thereby bring profits into existence. Those entrepreneurs who introduce successful innovations earn large profits. But as the innovation gets known to other entrepreneurs and they also adopt similar other innovations, profits which arise because of a particular innovation tend to disappear. But new innovations are being continuously introduced by the entrepreneurs and profits continue to arise out of them.

External Changes. External changes refer to those changes which are external to the firms or industries in an economy. These changes affect all firms in an industry or sometimes all industries in the economy. Examples of external changes are breaking out of wars, occurrence of sometimes periods of inflation and rising prices and sometimes business depression and falling prices, changes in the monetary and fiscal policies of government affecting favourably or unfavourably, changes in the technology of production, changes in tastes and preferences of the consumers, changes in income and spending habits of the people, changes in the availability of substitute products, alteration in the legislative and legal environment affecting the industries, and changes in preference between income and leisure. All these changes affect either the cost or demand for the products and give rise to profits, positive or negative as the case may be. For instance, during wars when prices of goods mount up, and costs lag behind, the entrepreneurs make a lot of profits. Similarly, when inflation occurs due to the increased demand for goods caused by the rising incomes, increasing population and expansion in the money supply, huge profits accrue to the firms. On the contrary, when period of depression comes due to the fall in effective aggregate demand, firms suffer huge losses and some may go into liquidation. During periods of depression all prices, rents, wages, and interest tend to fall but because of the non-contractual nature, profits fall sharply and may even become negative.

Knight's Views on Dynamic Theory

Here the views of Prof. F.H. Knight about dynamic changes giving rise to profits are worth mentioning. According to him, "Dynamic changes give rise to a peculiar form of income only in so far as the changes and their consequences are unpredictable in character....It cannot, then, be the change which is the cause of profits, since if the law of change is known, as in fact is largely the case, no profits can arise. The connection between change and profits is uncertain and always indirect. Change may cause a situation out of which profit will be made if it brings about ignorance of the future....It is not dynamic change, nor any change as such which causes profits, but divergence of actual conditions from those which have been expected and on the basis of which business arrangements have been made. For a satisfactory explanation of profit we seem to be thrown back from the dyanmic theory to the *Uncertainty of the Future*."[3]

3. F.H. Knight, *Risk, Uncertainty and Profits*, pp. 37-38.

As far as unpredictable, unforeseen changes and uncertainty about the future giving rise to profits are concerned, there can be no disagreement with Prof. Knight. But with his assertion that dynamic changes as such are not the cause of profits, one can differ. Against Knight's view it may be pointed out if there is no change there will be no uncertainty about the future and therefore no profits. Thus the factor of change is fundamental in bringing profits into existence. Professors Stonier and Hague rightly maintain, "*in an economy where nothing changes, there can be no profits. There is no uncertainty about the future, so there are no risk and no profits.*"[4]

SCHUMPETER'S INNOVATIONS THEORY OF PROFITS

Successful innovations as important dynamic changes and as source of profit have been, in brief, explained above. But since innovations have been singled out as a very important factor responsible for the occurrence of profits to the entrepreneurs it requires to be dealt with separately. It has been held by Joseph Schumpeter that the main function of the entrepreneur is to introduce innovations in the economy and profits are reward for his performing this function. Now, what is innovation ? Innovation, as used by Schumpeter, has a very wide connotation. Any new measure or policy adopted by an entrepreneur to reduce his cost of production or to increase the demand for his product is an innovation. Thus innovations can be divided into two categories.

First type of innovations are those which reduce cost of production, or in other words, which change the production functions. In this first type of innovations are included the introduction of a new machinery, new and cheaper technique or process of production, utilisation of a new source of raw material, a new and better method of organising the firm, etc. Second type of innovations are those which increase the demand for the product, or in other words, which change the demand or utility function. In this category are included the introduction of a new product, a new variety or design of the product, a new and superior method of advertisement, discovery of new markets etc. If an innovation proves successful, that is, if it achieves its aim of either reducing the cost of production or enhancing the demand for a product, it will give rise to profit. Profits emerge because due to successful innovations either cost falls below the prevailing price of the product or the entrepreneur is able to sell more and at a better price than before. It should be noted that profits accrue not to him who conceives innovation, nor to him who finances it but to him who introduces it. Further, whenever any new innovation is to be introduced, it always calls for a new combination of factors or reallocation of resources.

It is here worth mentioning that *profits caused by a particular innovation are only temporary* and tend to be competed away as others imitate and also adopt that. An innvation ceases to be new or novel, when others also come to know of it and adopt it. When an entrepreneur introduces a new innovation, he is first in a monopoly position, for the new innovation is confined to him only. He therefore makes large profits. When after some time others also adopt it in order to get a share, profits will disappear. If the law allows and the entrepreneur is able to get his new innovation *e.g.*, new product patented, then he will continue to earn profits.

But in a competitive economy and without patent laws, the existing competitors or the new firms will soon adopt any successful innovation and profits would be eliminated. But in a competitive and progressive economy the entrepreneurs always continue to introduce new innovations and thus profits continue emerging out of them. Thus Prof. Stigler writes, "Unless one can construct a permanent monopoly, such profits as are realised by successful innovations are essentially transitional and will be eliminated by the attempts of other firms to share them. But these profits may exist for a considerable time because of the ignorance of other firms of their existence or because of the time required for the

4. Stonier and Hague, *A Textbook of Economic Theory*, 2nd edition, p. 327.

entry of new firms. More important, the successful innovator can continuously seek new disequilibrium profits since the horizon of conceivable innovations is unlimited."[5]

We have seen above that innovations are important source of profits. Obtaining profits is a necessary incentive for the entrepreneurs to conceive and introduce innovations which help the economic development of the country. Since innovations, if successful, yield profits and profits is also the motive to introduce innovations, *profits are both the cause and effect of innovations*.

RISK, UNCERTAINTY AND PROFITS: KNIGHT'S THEORY OF PROFITS

An important theory associates profit with risk and uncertainty. According to F.H. Knight, profit is a reward for uncertainty bearing. Even before Knight, F.B. Hawley and A.C. Pigou had pointed out that entrepreneurs earn profits because they have to bear the risks of production. But Knight has greatly developed the theory of profits based on uncertainty. He has distinguished between risk and uncertainty on the one hand and predictable and unpredictable changes on the other. According to him, dynamic changes give rise to profits only if changes and their consequences are of unpredictable character. Only those changes whose occurrence cannot be known before hand give rise to profits.

Profits, Unpredictable Changes and Uncertainty

As we have noted above, if there were no changes or if the changes were foreseen and predictable, there would have been no uncertainty about the future and therefore no profits. Profits arise because of the uncertainty of future. If the future conditions could be completely foreknown in the present, then competition would certainly adjust things to the ideal state where all prices would equal costs and profits would not emerge. Thus it is our ignorance about the future and uncertainty of it that give rise to profits. In other words, it is the divergence of actual conditions from those which have been expected and on the basis of which business contracts have been made that give rise to uncertainty and profits. Prof. A.K Dass Gupta rightly maintains, "uncertainty is thus a permanent feature of economic system. It is one of the limitations of human ingenuity that it cannot unearth the contents of the future. Trained insticts of businessmen coupled with statistical information may go a long way, but in so far as the course of nature (both physical and human) is any thing but rhythmical, the future would always remain more or less of mystery."[6] He further writes, "so long as entrepreneurs start operations with imperfect knowledge about the state of the market and so long as the anticipated marginal product of the hired factors deviates from their actual product, so long a surplus would persist."[7]

We thus see that entrepreneurs have to undertake the work of production under conditions of uncertainty. In advance they have to make estimates of the future conditions regarding demand for the product and other factors which affect price and costs. In view of their estimates and anticipations, they make contract with the suppliers of factors of production in advance at fixed rates of remuneration. They realise the value of the output produced by the hired factors after it has been produced and sold in the market. But a good deal of time is spent in the process of producing and selling the product. It follows, therefore, that a good time gap elapses between the contracts made by the entrepreneur with the factors of production at fixed rates and the realisation of sale proceeds from the output made by them. As mentioned before, these contracts are based upon anticipations about the future conditions. But between the times of contracts and sale of the output many changes may take place which may upset anticipations for good or for worse and thereby give rise to the profits, positive and negative. Now, if the conditions prevailing at the time of sale of output could be known or predicted when the

5. G.J. Stigler, *op. cit.*, p. 182.
6. A.K. Dass Gupta, *The Conception of Surplus in Theoretical Economics*, p. 188.
7. *Ibid.*, p. 188.

entrepreneurs enter into contractual relationships with the factors of production about their rates of remuneration, there would have been no uncertainty and, therefore, no profits. *Thus uncertainty, that is, ignorance about the future conditions of demand and supply, is the cause of profits*. It should be noted that positive profits accrue to those entrepreneurs who make correct estimate of the future or whose anticipations prove to be correct. Those whose anticipations prove to be incorrect will have to suffer losses.

We thus see that profit is a residual and non-contractual income which accrues to the entrepreneurs because of the fact of uncertainty. The entrepreneur is unhired factor; he hires others for work of production. It is, therefore, *entrepreneur who bears uncertainty and earns profits as a reward for that*. J.F. Weston who has been a prominent exponent and supporter of uncertainty theory of profit explains the emergence of profits in the following way: "Under uncertainity total product may not be equal to total costs (explicit and implicit) *because plans are not fulfilled*. How this occurs is briefly indicated. Two classes of owners of productive services are distinguished. First, those with rates of compensation fixed in advance of the determination of the results of operations, are called *hired factors* and receive contractural returns. Second, those with rates of compensation dependent upon the results of operation are referred to as unhired factors who receive non-contractual or residual returns. Whatever the basis upon which contractual relationships have been entered, actual results will not have been accurately foreseen because of uncertainty. Hence *whatever the basis upon which contractual commitments have been made events actually do not turn out that way*. This is the significance of economic profit. *It is not possible to plan in advance exactly what total product or total costs will be*."[8]

What Causes Uncertainty?

Now, the question is what changes cause uncertainty. As has been explained earlier, there are *two types of changes* which take place and are responsible for conditions of uncertainty. First type of changes refer to innovations (for example, introduction of a new product or a new cheaper method of production etc.) which are introduced by the entrepreneurs themselves. These innovations not only create uncertainty for the rivals or competitors who are affected by them but they also involve uncertainty for the entrepreneur who introduces them because one cannot be certain whether a particular innovation will be definitely successful. The second type of changes which cause uncertainty are those which are external to the firms and industries. These changes are: (1) changes in tastes and fashions of the people, (2) changes in Government policies and laws especially taxation, (3) wage and labour policies and laws, (4) movements of prices as a result of inflation and depression, (5) changes in income of the people, (6) changes in production technology etc. All these changes cause uncertainty and bring profits, positive or negative, into existence.

Insurable and Non-Insurable Risks. We have seen above that entrepreneurs work under conditions of uncertainty and that they bear uncertainty and earn profits as a reward for that. Here a distinction drawn by F.H. Knight between *insurable and non-insurable risk* is worth mentioning. Because of the changes that are continuously occurring in the economy entrepreneur has to face many risks. But all risks do not cause uncertainty and give rise to profits. It is only *non-insurable risks* that involve uncertainty and the entrepreneur earns profits for bearing these non-insurable risks. Now, the question arises as to what kind of risks are insurable and what non-insurable. The entrepreneur faces risks like fire, theft, accident etc. which may cause him huge losses. But these risks of fire, theft, accident etc. can be insured against on the payment of a fixed premium. Insurance premium is included in the cost of production. Thus no uncertainty arises due to insurable risks as far as individual entrepreneurs are concerned and therefore they cannot give rise to profits.

---

8. J. Fred Weston, 'A Generalized Uncertainty Theory of Profit', *American Economic Review*, XL (March 1950), (italics added).

Now, only those risks can be insured the probability of whose occurrence can be calculated. Thus an insurance company knows by its calculation on the basis of past statistics that how much percentage of the factories will catch fire in a year. On the basis of this information, it will fix the rate of premium and is able to insure the factories against the risk. But there are risks which cannot be insured and therefore they have to be borne by the entrepreneurs. These non-insurable risks relate to the outcomes of the price-output decisions taken by the entrepreneurs. Whether it will pay him to increase output, reduce output and what will be the outcome in terms of profits or losses as a result of his particular output decision. Again, whether it will pay him to lower price or to raise it and when he takes a particular price decision whether he would make profits or losses. Similarly, he has to face risks as a result of his decisions regarding mode of advertisement and outlay to be made on it, product variation etc. For taking all these decisions he has to guess about demand and cost conditions and always there is risk of suffering losses as a result of decisions. No insurance company can insure the entrepreneurs against commercial losses which may emerge out of decisions regarding price, output, product variation and also against the losses which may fall upon the entrepreneurs due to the structural, cyclical and other exogenous changes which take place in the economy. It is, therefore, clear that it is non-insurable risks that involve *uncertainty and give rise to profits*. To quote Knight, "*It is 'uncertainty' distinguished from insurable risk that effectively gives rise to the entrepreneurial form of organisation and to the much condemned 'profit' as an income form*."[9]

Conclusion. All the theories of profits explained above have some element of truth. No single theory can adequately explain the existence of profits in all cases. Thus, economic profits can arise as a result of disequilibrium caused by dynamic changes in the economy and frictions in the instantaneous adjustment to the new conditions. They can arise due to the existence of monopoly in the product and factor markets, due to the introduction of innovations by the entrepreneurs, due to higher risk and correctly estimating the uncertain future and due to higher managerial efficiency and skills. B.S. Keirstead rightly writes, "Profits may come to exist as a result of monopoly or monopsony as a reward for innovation, as a reward for the correct estimate of uncertain factors either particular to the industry or general to the whole economy"[10].

Role and Functions of Profits

Profits play an important role in a free market economy. Profits perform three important primary roles in such an economy. First, profits serve as a signal to change the rate of output or for the firms to enter or leave the industry. Second, profits play a critical role in providing incentive to introduce innovations and increase productive efficiency and to take risks. Third, profit serves as source of saving which can be invested in building up more productive capacity. Even corporate firms do not distribute all profits earned by them among shareholders as dividends but retain a good part of them as undistributed profits which are ploughed back into investment for expansion of their productive capacity by buying new machines, equipment and buildings. By serving as a source of saving and investment profits make a significant contribution to economic growth of a country.

As regards the first role, high economic profits being earned in an industry serve as a *signal* that consumers want more of the commodity being produced by that industry. These profits indicate to the firm to expened output of the commodity and for the new firms to enter the industry to gain a share of economic profits that exist in the industry. As a result, more resources will be allocated to the output of that industry. On the other hand, below normal profits in an industry serve as a signal that either less output of the industry is demanded by the consumers or inefficient production methods are being used by the firms. In response to the lower demand for the product the firms will reduce their output and also some firms will leave the industry. As a result, some productive resources will be released

---

9. Social Economic Policy, *Canadian Journal of Economic and Political Science*, (26 Feb. 1960), p.31
10. B.S. Keirstead, Capital, Interest and Profits (1959), p.6.

from that industry and made available for the production of other goods. If the lower profits are due to the inefficient production and organisation, this will induce firm to improve efficiency by changing the production methods or making organisational changes to reduce costs.

As explained above, profit motive drives a free-market economy. Although it has been observed that sometimes managers and entrepreneurs in a free market system are swayed by greed and avarice, and break laws to make money or profits by exploiting the consumers or workers but in general profits perform useful function of sending signals for changing levels of output of various products and for reallocation of resources among them.

Secondly, above normal rate of profits in a free enterprise system is an essential reward for introducing innovations and taking risks. No entrepreneur will introduce new products or more efficient production methods or undertake investment in risky projects unless there is prospect of making profits. Some firms continue to earn above-normal rate of profit year after year as they are continually introducing new products, new production methods and providing good customer services.

In the economy changes in demand for the product often occur due to cyclical and structural changes. Besides, new strategies of rival firms also affect the demand for the product of a firm. All these uncertain and unanticipated changes involve a good deal of risk. An important function of economic profits is to reward entrepreneurs for taking these risks involved in making investment and organising factors for the production of products.

However, in some cases firms are also able to make supernormal profits by virtue of their having monopoly power. Their monopoly power may be due to some legal patent and license obtained from the government, the economies of large scale production, exclusive control over essential raw materials which prevent the other firms from producing the same product or service. These enable the monopoly firms to charge higher prices and thereby make large economic profits. Therefore, even in free-market economies steps are taken to prevent the emergence of monopolies through anti-trust laws or Competition Acts as recently enacted in India. Of course, monopolies are legally permitted if they are needed in public interest. For example, in several cities Government grants license to some private firms to provide public utility services such electricity, gas, telephone etc. In these cases of legal monopolies government regulates them and fix reasonable prices to be charged by them from the public but at the same time ensures fair return or normal profits to the private firms on their investment.

Social Responsibility of Business and Profit Motive

It has been argued by some that business firms should be socailly responsible and works in a way that provides maximum benefits to the society. This implies that in their efforts to maximise private profits business firms should not harm social welfare. This implies that they should take steps to reduce pollution of environment caused by their activities and adopt measures for the safety and health of the workers and fulfill other social obligations. The experience of free market economices of the USA, Great Britain, Japan, Canada over the past several decades reveals that contribution of profit-driven business firms to their economic growth and increase in social welfare has been quite significant.

Despite this it is difficult to agree with the viewpoint of a noted American economist, Milton Friedman who states, "Business has only one social responsibility to make profits (so long as it stays within the legal and moral rules of the game established by the society). Few trends could so thoroughly undermine the very foundations of our society as the acceptance by corporate officials of a social responsibility other than to make as much money for their shareholders as possible".

This is an extreme view. Friedman has been an advocate of unregulated and unfettered capitalist system. Though, as stated above, profits in general play a useful role for allocating resources efficiently and promoting rapid economic growth . But it is difficult for us to agree with Freedman's view that business firms should not be guided by social responsibility. It has been seen that some business firms in their single-minded pursuit of maximising their private profits regardless of their harmful effect on social well-being behave in a socially irresponsible way. Thus, to make money at the expense of consumers, workers and society at large is not a socially responsible behaviour. That is why at present in all free-market economies some activities of business firms are regulated by government so that their decision making is consistent with the overall social goals.

Dr. Manmohan Singh, the present Prime Minister of India who has been instrumental in initiating economic reforms aimed at promoting leberalisation and privatisation in his inaugural address at the Annual Session of CII (Confederation of Indian Industries) on May 24, 2007 was surprised to note that in India top corporate companies worked in oligoplastic markets and form Cartles which exploit the consumers by charging higher prices. To quote him, "The operation of cartels by groups of big corporate companies in India to keep prices high must end. It is unacceptable to obstruct the forces of competition from having free play. It is even more from distressing in a country where poor are severely affected by rising commodity prices. Cartels are a crime and go against, the grain of an open economy "*More importantly, profits should be within the bounds of decency and greed*".

QUESTIONS FOR REVIEW

1. Profits arise due to innovations and entrepreneurs are rewarded for introducing innvations. Discuss
2. Profit is the reward for making innovations. Do you consider this as an adequate explanation for profits? *[D.U. B.A. (Hons) 1989]*
3. Profit is defined as a functional reward to the entrepreneur. *[D.U. B.A (Hons) 1991]*
4. Explain why profits are zero in a competitive economy. *[D.U. B.A. (Hons) 1995]*
5. "In a purely competitive static economy economic profits are eliminated" Discuss. *[D.U. B.A. (Hons), 1993]*

   OR

   "There can be no profits in the stationary state" Discuss *[D.U. B.A. (Hons) 1986]*
6. Profits are a dynamic surplus. Discuss. *[D.U. B.A. (Hons) 1987]*
7. Profits are a reward for uncertainity bearing. *[D.U. B.A. (Hons) 1996]*
8. "Profits may come to exist as a result of monoposoly or monopsony, as a reward for innovation, as a reward for making a correct estimate of uncertain factors either particular to the industry or general to the economy as a whole" Discuss.
9. What is the role of profit in a free market economy? Is there a conflict between profit earning and social welfare? *[D.U. B.A. (Hons) 1990]*
10. Distinguish between risk and uncertainty. What causes uncertainty in an economy ? Critically examine the view that entrepreneurs is rewarded in the form of profits for bearing uncertainity.

PART VI

GENERAL EQUILIBRIUM ANALYSIS AND WELFARE ECONOMICS

- General Equilibrium Analysis
- Welfare Economics: An Introduction
- Economic Efficiency and Pareto Optimality
- Market Failures, Externalities and Public Goods
- New Welfare Economics: Compensation Principle
- Social Welfare Function

Chapter 37
GENERAL EQUILIBRIUM ANALYSIS

Partial Equilibrium Analysis and General Equilibrium Analysis

In the previous chapters we have focussed on explaining the determination of price and quantity of a commodity or a factor and the working of its market viewed in isolation of what happens to other commodities and factors. We have analysed how equilibrium price and quantity of a commodity or a factor is determined through demand and supply, assuming prices of other commodities and factors would remain the same when changes occur in the price of the commodity under consideration. That means the effect, if any, of the changes in price of a commodity on the demand for other commodities is ignored. *This type of analysis where we do not take into account the interrelationship or inter-dependence between prices of commodities or between prices of commodities and factors of production is called partial equilibrium analysis.* In this partial equilibrium analysis each product or factor market is considered as independent and self-contained for the proper explanation of the determination of price and quantity of a commodity or a factor.

However, *partial equilibrium analysis is not useful and relevant to apply when there is inter-relationship between commodities or between factors*. Thus when markets for various commodities and factors are interdependent, that is, when changes in the price of a commodity or a factor have important repercussions on the demand for other commodities or factors, partial equilibrium analysis would not yield correct results. In such cases when there is significant inter-relationship between various markets or that the changes in one market would significantly affect others, we should employ general equilibrium analysis which considers *simultaneous equilibrium* of all markets taking into account all effects of changes in price in one market over others. It may be mentioned that both types of equilibrium analysis are useful, each being valuable in its own way. Partial equilibrium analysis is useful when the changes in conditions in one market have little repercussions on other markets. However, when the changes in conditions in one market have significant effects on other markets, general equilibrium analysis should be used. Thus, in partial equilibrium analysis when we consider the determination of market price of a commodity we assume that prices of other goods do not change.

For example, the rise in price of petrol following imposition of a tax on it would cause little effect on the prices of goods such as wrist watches, drapers, bowling balls, and in turn there would be negligible feedback effect of changes in prices of these goods on the demand and price of petrol. If prices of petrol and of only these commodities are to be considered and since there are little repercussions of changes in prices of petrol on these other commodities, the use of partial equilibrium analysis of price determination of petrol would be quite reasonable. However, when market for automobiles is considered, the rise in price of petrol would have an important effect on their demand and price. Therefore, the assumption of partial equilibrium analysis that prices of automobiles would remain constant, when the price of petrol changes would be seriously wrong. This is because petrol and automobiles being *complements* to each other, their markets are inter-related and mutually inter-dependent and changes in their prices would significantly affect each other. In such cases when there exist inter-relationship and inter-dependence of the markets for goods (whether they are complements or substitutes), the general equilibrium analysis should be used. In *general equilibrium analysis, all prices are considered variable and the analysis of simultaneous determination of equilibrium in all markets is made.*

In fact when we look at the economic system as a whole, there is a great deal of inter-relationship and inter-dependence among various markets for commodities and factors and there are a large number of decision making agents—consumers, producers, workers, (who supply labour) and other resource owners. All these agents are self-interested and would behave to maximise their goals; consumers would maximise their utility, and producers would maximise their profits. A comprehensive analysis of the economic system when prices and quantities of all commodities and factors are considered as variable and which would take into account all inter-relationships and inter-dependence could be made only through general equilibrium analysis. The general equilibrium would occur when markets for all commodities and factors and all decision-making agents, consumers, producers, resource owners are simultaneously in equilibrium.

To sum up, *partial equilibrium analysis focuses on explaining the determination of price and quantity in a given product or factor market when one market is viewed as independent of other markets. On the other hand, general equilibrium analysis deals with explaining simultaneous equilibrium in all markets when prices and quantities of all products and factors are considered as variables.*

Thus, in general equilibrium analysis inter-relationship among markets of all products and factors are explicitly taken into account.

In what follows we shall explain the conditions of general equilibrium of exchange and production in an economy. We shall concentrate on the conditions of general equilibrium with regard to the following three aspects :–

1. the distribution of goods and services for consumption among individuals in the society;
2. the allocation of productive factors to the production of various goods and services; and
3. the composition of production (or output mix) together with the distribution of consumption.

In this chapter, we shall discuss *whether a general equilibrium exists at all.* Although the question of existence of general equilibrium is an abstract one, it is very important because many propositions of economics rest on the existence of general equilibrium. We shall also explain that in case general equilibrium exists, whether it is *unique*.

GENERAL EQUILIBRIUM OF EXCHANGE AND CONSUMPTION: DISTRIBUTION OF GOODS BETWEEN INDIVIDUALS

First, we shall *explain general equilibrium in a pure exchange economy*. In this pure exchange system, we assume *that there is no production.* That is, we consider the case when two goods are provided to the individuals in the economy from outside the system. To keep our analysis simple we assume that there are (1) *two goods*, a specific bundles of which have been made available to the individuals for consumption; and (2) there are *two individuals* between which exchange of goods has to take place and equilibrium reached with regard to the distribution of the specific amounts of these two goods.

Edgeworth Box and General Equilibrium of Exchange. In this two goods, two individuals (2x2) model of pure exchange, the famous Edgeworth Box diagram has been employed to explain the general equilibrium of distribution of two goods between two individuals. In what follows we first explain the concept of Edgeworth Box and then analyse the general equilibrium in this pure exchange system. Consider Fig. 37.1 where a box with a certain fixed dimensions has been drawn. Along the X-axis we measure the commodity X and along the Y-axis, the commodity Y. The total available amount of commodity X is OX_0 and of commodity Y is OY_0. The available amounts of the two commodities, OX_0 and OY_0 determine the dimension of the box. The quantity of X available with the individual A is measured from left to right along the X-axis with bottom left-hand corner O_A

as the origin. And, quantity of commodity Y available with the individual A is measured along the Y-axis from bottom upwards with the origin O_A. For individual B, the top right hand corner O_B has been taken as the origin and with the given quantities of X and Y, the quantity of X available for consumption for individual B is measured, right to left, with the origin O_B and the quantity of Y available for B is measured, from top to bottom, from the origin O_B.

It follows from above that Edgeworth Box has fixed dimensions representing the maximum available quantities of X and Y to be distributed between the two individuals. We further assume that the two individuals between them will entirely consume all the available quantities of the two goods. It may be noted that a point in the Edgeworth Box represents a particular distribution pattern of two goods between the two consumers. This implies that if the two individuals trade goods with each other and accordingly move from one point in the Edgeworth Box to another, the quantities purchased and sold of each good would be equal. Thus, with trade or exchange of goods, it is the distribution or consumption of two goods of the two individuals that will change, the total quantities of the two goods remaining constant.

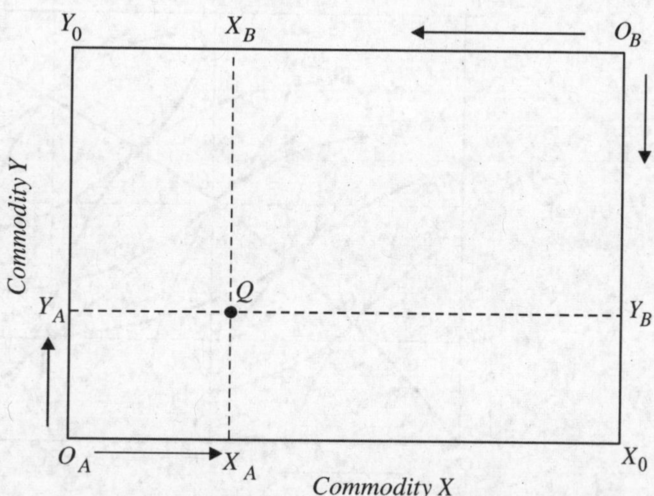

Fig. 37.1. *Edgeworth Box*

In the Edgeworth Consumption Box we also draw the indifference curves of the two individuals A and B depicting their scale of preferences between the two goods. As we move upward from bottom-left to top right, the satisfaction of individual A increases and that of B decreases, that is, A-moves to successively higher indifference curves and individual B to successively lower indifference curves.

We can now show that the *general exchange equilibrium would lie somewhere on the contract curve*, that is, the curve QT in Fig. 37.2 which passes through the tangency points of indifference curves of two individuals. At these tangency points of indifference curves, MRS_{xy} of individual A equals that of individual B. Thus, the general equilibrium of exchange will occur when the following condition holds good :-

$$MRS^A_{XY} = MRS^B_{XY}$$

Since a point on the contract curve lies within the Edgeworth box with the fixed quantities of the two goods, the equilibrium reached after exchange or trading between the two individuals implies that the distribution for consumption of the two goods between the two individuals would just exhaust the available quantities of the two goods.

From the above it cannot be known at which specific point or location of the contract curve, the general equilibrium of exchange will be reached. This is because the equality of MRS_{xy} of the two individuals exists at all points of the contract curve. However, if we know the initial distribution of two goods between the two individuals we can pinpoint the boundaries within which the general equilibrium of exchange would lie. Consider Figure 37.2. If the *initial distribution* of two goods between the two individuals is represented by point C where individual A has X_{AI} amount of good X and Y_{A1} amount of good Y. The remaining quantity of good X, that is, $X_0 - XA_1 = X_{BI}$ *would be allocated to* individual B and the remaining Y_{BI} amount of good Y would go to individual B. At this initial distribution of goods X and Y between the two individuals A and B the indifference curves of two individuals are intersecting. Now, this initial distribution at point C cannot be the position of

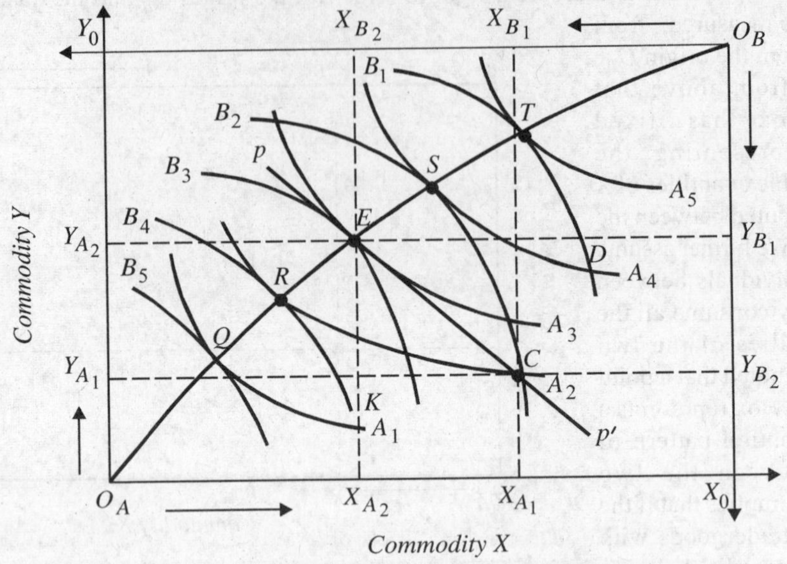

Fig. 37.2. *General Equilibrium of Exchange*

equilibrium for the two individuals, since the two individuals can gain in welfare or, in other words, can become better off if they exchange some amounts of the goods possessed by them and move to the contract curve. If the individuals think that they can benefit from trading or exchange, they will trade with each others. As long as they think there are possibilities of becoming better off, they will exchange goods and end up at the contract curve.

With the initial distribution of two goods as implied by point C, if the two individuals through exchange of goods between them move to the point R on the contract curve, individual B reaches on his higher indifference curve B_4 and therefore becomes better off and A is no worse off as he remains on the same indifference curve A_2 as on the initial distribution point C. On the other hand, if through exchange they move to point S on the contract curve, individual A becomes better off and individual B no worse off as compared to the initial position C. And if through exchange of goods they move to any point between R and S on the contract curve both the individuals will gain from exchange of goods as they will be reaching their respective higher indifference curves.

With initial distribution at point C and through exchange of goods nearer they move to point R on the contract curve, individual B will benefit more and nearer they move to point S on the contract curve, the individual A will gain more as compared to the initial distribution position C. *Where exactly on the contract curve, their equilibrium position of exchange will lie depends upon the bargaining power of each individual*. With their almost equal bargaining power, their equilibrium position of exchange on the contract curve may lie at point E where the two individuals gain almost equally as a result of exchange. Thus, if the initial distribution of two individuals is not on the contract curve, there will be tendency on the part of individuals to trade or exchange goods between themselves and to move to a point on the contract curve because in doing so they will be increasing their satisfaction.

It is evident from the foregoing analysis that the position of exchange equilibrium can be somewhere between R and S on the contract curve. *On all points between R and S, the exchange equilibrium can exist. Although equilibrium will exist at a point on the contract curve, there is no unique position of exchange equilibrium ; all points between R and S on the contract curve are possible equilibrium positions*. If point E on the contract curve is the position of exchange equilibrium actually reached, then individual A has exchanged the amount of commodity X equal to CK for the amount of commodity Y equal to KE Since point E lies on the contract curve which is the locus of the tangency points of indifference curves of the two individuals, marginal rate of substitution between the two goods (MRS_{XY}) of individual A equals marginal rate of substitution between the two commodities (MRS_{XY}) of individual B. Thus exchange of CK of commodity X for KE of commodity Y has been settled between them at the equilibrium position E. The general equilibrium of exchange attained at point E on the contract curve has the following important features :

1. Individuals maximise their satisfaction by equating their MRS_{XY} *subject to their initial endowments of goods*.
2. Since the equilibrium point E lies within the Edgeworth Box, drawn with the given amounts of two goods, the exchange of goods between the two individuals when they move to the equilibrium point E on the contract curve would imply that quantity sold of each good equals the quantity purchased of the good. That is, markets for the two goods would clear. This implies that on moving to the equilibrium position E, individual A relative to his initial endowment of goods is selling good X and buying good Y. The opposite is true of individual B who buys good X and sells good Y. The quantity sold and purchased of each good must equal each other. If this does not happen the two markets will not clear and shortages or surplus would emerge.
3. The general exchange equilibrium determines not only the final distribution of two goods between the individuals *but also a certain exchange rate (i.e. relative prices of the two goods)*. Thus at the equilibrium position E, the exchange rate CK of X for KE of Y has been settled between them. It is at this price ratio that exchange of goods takes place between the individuals..
4. The general equilibrium of exchange does not lead to the determination of *absolute prices* of goods but only *relative prices* of goods. This will be explained in detail later in this chapter.
5. The general equilibrium of exchange must lie on the contract curve, and given the initial distribution implied by point C, it must lie between the point R and S on the contract curve. *The general exchange equilibrium cannot be at a point in the Edgeworth Box which is not on the contract curve*. This is because at a point which is not on the contract curve, indifference curves of two individuals will intersect each other and therefore their MRS_{XY} will not be equal to each other.
6. The equilibrium can lie *anywhere* between R and S on the contract curve, that is, general equilibrium of exchange in this bargaining is *not unique*.

GENERAL EQUILIBRIUM OF PRODUCTION

We now extend our analysis of general equilibrium to the sphere of production. Production of goods requires the use of inputs or factors of production. The level of production of goods depends upon the allocation of resources to them. As emphasised in the beginning, the general equilibrium analysis takes into account the mutual inter-dependence of markets. In it, we are not only concerned with the mutual inter-dependence of markets for goods between themselves but also between product markets and factor markets. To keep our analysis simple we shall assume that two factors or inputs, labour and capital, are required for the production of two goods X and Y. This analysis of general equilibrium by taking two goods and two factors will enable us to capture the essential characteristics of the general equilibrium of production, Besides taking 2×2 model, we make the following assumptions:

1. All units of labour are homogeneous so that they receive equal remunerations for their contribution to the production of goods. So is the case with all units of capital.

2. The available quantities of the two factors, labour and capital, are fixed in the economy and both of them are fully employed and utilised.

3. There is smooth production function for each good so that production factors, labour and capital, can be freely transferred from one good to the other.

4. Technology is given which together with the factor endowments limits the production possibilities.

It may be mentioned here that there are four markets in the model considered here : two factor markets of labour and capital, and two product markets of goods *X* and *Y*.

With the above assumptions we shall analyse the general equilibrium of production under conditions of perfect competion in all the markets. It may be emphasised again that the various markets are inter-related. For example, if more labour is employed in the production of good *X*, then, given its fixed supply, some labour will have to be withdrawn from the production of good *Y*. Changes in labour allocation between the production of the two goods, would also have repercussions on the use of capital in the production of two goods. It is indeed the task of general equilibrium analysis to determine the pattern of relative prices and quantities of goods and factors at which all markets clear together, that is, prices and quantities at which demand and supply in each of the four markets are brought into equilibrium simultaneously.

Edgeworth Production Box and General Equilibrium of Production

As in case of general equilibrium of exchange or consumption, the general equilibrium of production can also be analysed with the help of Edgeworth Box diagram. In case of Edgeworth Production Box, dimensions of the box represent the available fixed quantities of the two factors, labour and capital. Thus, in the Edgeworth Production Box as shown in Fig. 37.3, along the horizontal

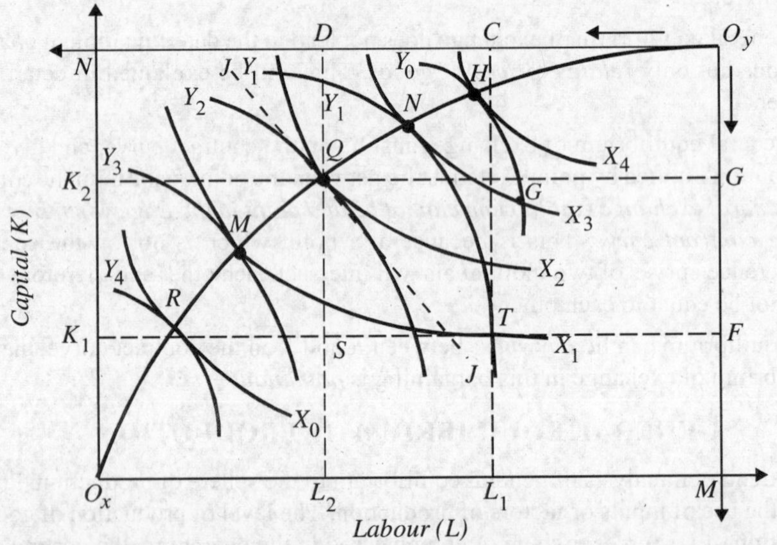

Fig. 37.3. *General Equilibrium of Production*

axis we measure the quantity of labour and along the vertical axis we measure the quantity of capital. In this box various isoquants are also drawn for the products *X* and *Y*. For good *X*, various isoquants representing successively higher levels of output such as X_0, X_1, X_2 etc. are drawn with the bottom left hand corner O_x as their origin. For good *Y*, various isoquants such as Y_0, Y_1, Y_2 representing-successively higher levels of output of *Y* are drawn with top right hand corner O_y as the origin. The isoquant map of each good exhibits its production function. As constructed, isoquants for good *X* imply that the good *X* is relatively capital-intensive and isoquants for good *Y* imply that good *Y* is relatively labour-intensive. This means that for any given factor price ratio, that is, ratio of price of labour to the price of capital (*w/r*), minimisation of cost for *a given level of output* of good *X* requires higher capital-labour ratio (*K/L*) than that of good *Y*.

It is important to note that any point in the Edgeworth Box represents a *particular allocation of labour and capital* between the two industries, one producing good *X* and the other producing good *Y*. Various points in the box represent different alternative allocations of factors between the two commodities. For example, point *T* in the box shows that O_xL_1 (or K_1T) amount of labour and O_xK_1 (or L_1T) amount of capital are allocated to the production of *X* and the remaining amount of labour *TF* and the remaining amount of capital *TC* are allocated to the production of good *Y*. It will be seen from Fig. 37.3 that the points *M*, *Q*, *N* etc. represent different allocations of factors from the point *T*.

Now, a smooth curve *RH* joining the tangency points of isoquants of *X* and *Y* has been drawn. This is called the *production contract curve*. It can be shown that the general equilibrium of production under competitive conditions in the factor markets would lie somewhere on this production contract curve. The allocation of factors implied by the point away from the contract curve such as point *T* cannot be the possible competitive equilibrium position. This is because from the point *T* where *X*-isoquant and *Y*-isoquant are intersecting, the economy can move by re-allocating resources between the two goods to a point *M* or *N* on the contract curve where the output of one good increases without the reduction in output of the other. And, if through reallocation of factors, the economy moves to any point between *M* and *N* on the contract curve, say to point *Q* the outputs of both the goods *X* and *Y* would be higher than at *T*. Thus, resource allocation implied by a point on the contract curve leads to greater output than those not on the contract curve. Since the production contract curve is the locus of the tangency points of the isoquants of *X* and *Y*, slopes of isoquants of the two goods are equal to each other at various points on it. Since slope of an isoquant measures marginal rate of technical substitution between the two factors (MRS_{LK}), on the various points on the contract curve, $MRTS^X{}_{LX} = MRTS^S{}_{LK}$. It is at the production efficient points of the contract curve that general equilibrium of production would lie.

Now, what ensures that the general equilibrium of production would lie at a point on the contract curve. If the two firms or industries find themselves away from the contract curve, they will trade or exchange the factors and move to a point on the contract curve because in doing so they will be increasing their output and will have therefore incentive to move to the contract curve.

It should be noted that if the firms produce the goods at a point on the contract curve, one firm can increase output only at the cost of output of the other firm. Thus, if in Fig. 37.3 the firms move from point *M* to *N* on the contract curve, the production of good *X* increases, the production of good *Y* decreases. It is worth mentioning that *point of general equilibrium of production is not unique because it may occur at any point on the contract curve, depending on the starting point*, that is, the initial allocation of factors (labour and capital to the production of goods. For example, if the initial allocation of factors is denoted by point *T* in Fig. 37.3, then with reallocation through trading of factors, the two firms can move to and attain general equilibrium at any point on the segment *MN* of the production contract curve. Further, with a different initial allocation of a given amount of factors, say point *G* in the Edgeworth Box, there will be a different point of general equilibrium of production anywhere on the segment *HN* of the contract curve.

Note that an initial point such as *T* or *G* shows not only the allocation of factors between the goods but also the amounts of the two goods produced. Further, it may also be noted that in the equilibrium situation *how much amount of goods are produced depends on the demands for them which in turn are determined by the initial ownership of the factors by the two individuals* and the factor prices together with their preferences of the two goods. Ownership of the factors and factor prices determine the income of the people which determine the demand for the two goods.

It will be observed from Fig. 37.3 that given the initial factor allocation point T, if the general equilibrium of production occurs at point Q on the contact curve RH, it will determine not only allocation of factors between the two goods but also the equilibrium amount of the two goods produced. Thus, if the general equilibrium of production occurs at point Q, it means that output X_2 of good X and output Y_2 of good Y are determined. Besides, the *general equilibrium of production determines ratio of factor prices (i.e. exchange rate of labour for capital*). It will also be seen from Fig. 37.3 that in moving from the starting point T, if the general equilibrium of production occurs at point Q, TS amount of labour has been exchanged for QS amount of capital.

The exactly at which point on the production contract curve the general equilibrium of production would lie depends upon the demand for goods X and Y. If the demand for good X is relatively higher than that of good Y, the equilibrium of production would lie on a some upper point such as N on the contract curve RH where a larger quantity of good X and smaller quantity of good Y are produced. On the other hand, the lower the demand for X and greater the demand for Y, the production equilibrium would lie at a lower point on the contract curve which indicates smaller output of X and larger output of Y. Thus, the exact point of equilibrium on the contract curve depends upon the specific consumer' demand for goods X and Y. And *individuals demand for good is determined by the initial ownership of factors and the factor price received for them*.

From the foregoing analysis it follows that the general equilibrium of production requires the *simultaneous equilibrium of the two factor markets determining* allocation of resources between the production of the two goods. For an economy with many goods and many factors, the general equilibrium of production requires that the marginal rate of technical substitution *between any pair of factors* is the same for all goods and all producers using the same pair of factors.

Transformation Curve and General Equilibrium of Production

In our above analysis we have shown the output of two goods X and Y through isoquants in the *factor space* of the Edgeworth Production Box. To bring outputs of two goods directly into the picture and to clearly show the equilibrium in markets for goods, we require to display outputs of the two goods in the *output space*. This is done through a familiar concept of *transformation curve* which is also known as production possibility curve. A transformation curve shows the alternative combinations of two goods that can be produced with the given fixed amounts of the factors. The transformation curve is derived from the contract curve of production in the space by mapping or plotting the various output combinations of two goods directly in *output space* corresponding to the various points of the contract curve. In Fig. 37.4 a transformation curve has been drawn from the production contract curve in Fig. 37.3. For example, corresponding to point Q on the contract curve RH in Fig. 37.3 indicating X_2 level of output of good X and Y_2 level of output of good Y, we plot the point Q' in the output space of Fig. 37.4. Likewise, point N' corresponding to point N on the contract curve, point M' corresponding to the point M of the contract curve are plotted and on joining such points as N', Q', M' we get the transformation curve TT'. A transformation curve shows how one good is transformed into another by transferring resources from the production of one good into the production of the other.

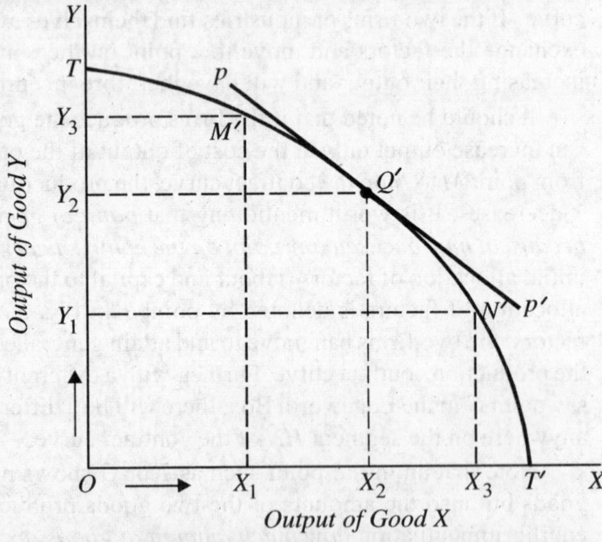

Fig. 37.4. *Transformation Curve*

Three aspects of the transformation curve are worth noting. First, *it is concave to the origin* which implies that the amount of commodity Y which has to be given up or sacrificed in order to produce an additional unit of good X goes on increasing as we produce more of good X. This implies the operation of diminishing returns to scale in the production of X. In case returns to scale increase, the transformation curve obtained would be convex to the origin. We will get a straight line transformation curve when returns to scale are constant. Thus the shape of the transformation curve depends upon the nature of the production function of a commodity.

Secondly, the slope of the transformation curve at a point measures the marginal rate of transformation between the two goods. Marginal rate of transformation of good X for good Y ($MRTxy$) indicates the amount by which the production of good Y has to be reduced to produce one more unit of good X with the help of resources released from the reduction in output of good Y, that is, MRT_{xy} is the rate at which one good is transformed into another and is equal to the ratio of marginal costs of production of the two goods. Thus

$$MRT_{XY} = \frac{MC_X}{MC_Y}$$

Thirdly, every point on the production possibility or transformation curve is a point of general equilibrium of production. This is because we have derived the transformation curve by mapping the production contract curve from input space to output space.

GENERAL EQUILIBRIUM OF PRODUCTION AND EXCHANGE (CONSUMPTION)

There are several points on the transformation curve; each point will indicate a different price ratio of the two goods and a different production equilibrium indicating a different output-mix of the two goods. The general equilibrium of production can occur at any of the points on the given transformation curve depending upon the prevailing price ratio of the two goods.

Now, an important question is which of these price ratios of commodities or MRT_{xy} will be the equilibrium ratio. The general equilibrium level of price ratio $\left(\frac{P_X}{P_Y}\right)$ or MRT_{XY} will be one which on the one hand maximises profits of the firms and on the other maximises satisfaction of consumers. To analyse the equilibrium price ratio determining output-mix, we have to introduce in our analysis the consumers preference pattern or demand for goods. It can be easily shown that if MRS_{xy} of the consumers is not equal to the MRT_{xy} in the production, consumer satisfaction will not be maximised with the result that further changes in the price ratio of the two goods and output-mix will tend to occur. Since the marginal rate of transformation (MRT_{xy}) shows the rate at which one good is 'transformed' into another in the production process and marginal rate of substitution measures the rate at which consumers are willing to exchange one good for the other, equilibrium cannot be reached unless the two rates are equal. Thus, only when MRT_{xy} equals MRS_{xy}, planned pattern of production will be consistent with the preferences of consumers and ensure general equilibrium of production and exchange. This can be made clear with a simple numerical example. For instance, suppose at a given output-mix of two goods X and Y, MRT_{xy} of the producers of the economy is $3Y/lX$ and MRS_{xy} of consumers equals $2Y/1X$. Thus, in this case MRS_{xy} of consumers is less than MRT_{xy} of producers. That is, the economy can producer 3 units of Y by foregoing one unit of good X, while the consumers are willing to exchange or buy 2 units of Y for one unit of X. MRS_{xy} of consumers being less than MRT_{xy} of producers implies that relatively greater quantity of commodity X and the smaller quantity of commodity Y is being produced than desired by the consumers. Obviously, if the economy redues the production of X by one unit and product 3 units more of Y and give them to the consumers, their satisfaction or welfare will increase as they are willing to get 2 units of Y for the sacrifice of one unit of X. Thus consumers' satisfaction can be raised by expanding the production of Y and reducing the production of X until the two rates, namely, MRS_{xy} of consumers and MRT_{xy} of producers become

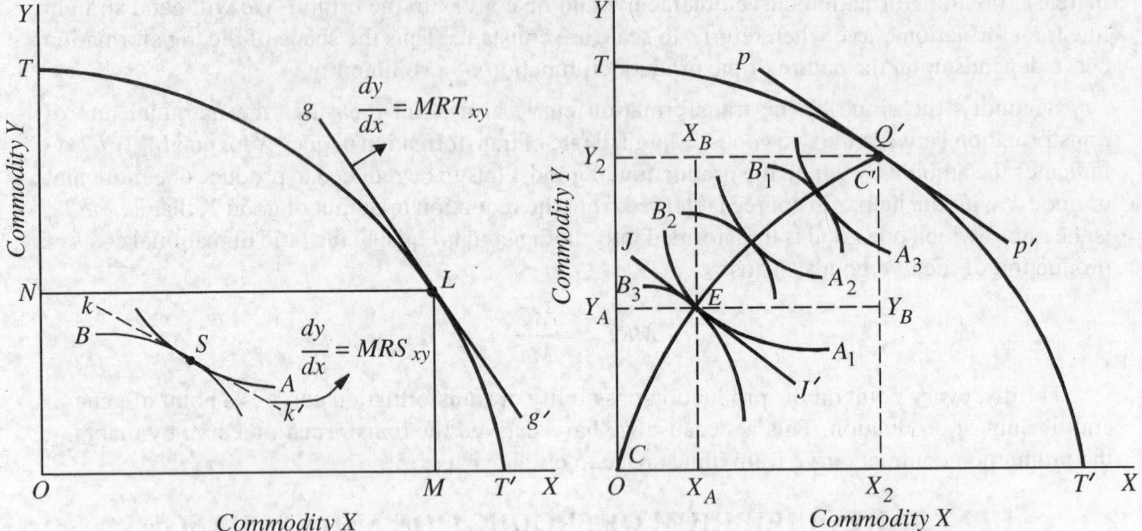

Fig. 37.5. *General Disequilibrium* **Fig. 37.6.** *General Equilibrium*

equal. General equilibrium will thus be reached if with adjustment in output-mix of the two commodities, MRS_{xy} equals MRT_{xy}. In a free market economy, forces of competition would ensure such adjustment which will bring about product-mix that will equate MRS_{xy} of consumers with MRT_{xy} of producers and thus ensuring maximum consumers' satisfaction.

The general equilibrium of production together with the general equilibrium of exchange (or consumption) requires that the marginal rate of transformation *(MRT_{xy})* be equal not only to marginal rate of substitution *(MRS_{xy})* of the consumers but also MRS_{xy} of the two consumers be equal to each other.

Thus, for the achievement of general equilibrium of production and exchange simultaneously we arrive at the following condition:

$$MRT_{XY} = MRS^{A}_{XY} = MRS^{B}_{XY}$$

The overall general equilibrium of production and exchange is illustrated in Fig. 37.5 and 37.6. Consider Fig. 37.5 where a transformation curve TT' has been drawn. Let us consider the point L on the production transformation curve TT'. At point L on it OM of good X and ON of good Y are being produced. With the given preference pattern and resource endowments of the two consumers, their indifference curves are tangent to each other at point S and their MRS_{xy} is indicated by the slope of the tangent line kk'. It will be seen from Fig. 37.5 that MRS_{xy} of consumers is less than MRT_{xy} of producers indicating that production pattern is inconsistent with consumers preferences. Thus, the system is in disequilibrium indicating that relatively greater quantity of X and smaller quantity Y is being produced than desired by the consumers. In response to this disequilibrium situation and to maximise their profits producers would tend to produce more of Y and less of X and this process of adjustment in output mix will continue until the MRT_{xy} in production is brought into equality with MRS_{xy} of consumers. With this equality, general equilibrium of production and exchange (consumption) would be attained. This general equilibrium is shown in Fig. 37.6 where at position Q' on the transformation curve TT', OX_2 level of output of X and OY_2 level of output of Y are being produced and MRT_{xy} at point equals MRS_{xy} at E (slopes of JJ' and PP' are equal). Note that there is general equilibrium of production at point Q' because all points of transformation curve corresponds to the points of the production contract curve. There is general equilibrium of exchange or consumption as MRS of, the two individuals are the same (their indifference curves being tangent to each other at point E). There is joint equilibrium of production and exchange as the MRT_{xy} in production is equal to MRS_{xy} of the two individuals at the consumption equilibrium point E. Thus, with point Q on the transformation curve TT' and point E on the consumption contract curve drawn in the Edgeworth Box made from point Q', the following condition of general equilibrium of production and consumption holds good:–

$$MRT_{XY} = MRS^{A}_{XY} = MRS^{B}_{XY}$$

General equilibrium of production determines total output X_2 of. commodity X and total output Y_2 of commodity. It is with OX_2 and OY_2 as the dimensions, Edgeworth Box has been drawn and indifference curves of two individuals depicting their scale of preferences are drawn with CC' as the contract curve. Consumption equilibrium point E reveals that out of total output X_2 of good X the amount X_A is being consumed by individual A and the remaining amount of X goes to individual B for consumption. Out of total output Y_2 of commodity Y, the amount Y_A is consumed by A and the remaining amount by the individual B.

Conclusion. It follows from above that in our two goods, two persons (2 × 2 × 2) model of general equilibrium, we arrive at the conclusion that the position of general equilibrium can exist. However, this general equilibrium position is *not unique.* There can be *several points* of general equilibrium on the production possibility curve IT' in Figure 37.6 and consumption contract curve in it depending upon the T', *distribution* of income which (among other things) is determined by ownership of resources. A general equilibrium position at point Q', on the production possibility curve IT' in Fig. 37.6 indicates that factor markers of labour and capital and product markets of goods X and Y are simultaneously in equilibrium and determine the following things :—

1. relative prices of factors (w & r) as indicated by the exchange rate of ST of labour for SQ of Y in Fig. 37.3
2. Allocation of factors between products X and Y as indicated by point Q in Fig.37.3.
3. relative prices of products X and Y as measured by the slope of tangent PP' in Fig. 37.6.
4. the product-mix, that is, the levels of output of goods X and Y as given by point Q' in Fig. 37.6.
5. the distribution of goods X and Y between the two individuals as indicated by point E in Fig. 37.6.

GENERAL EQUILIBRIUM AND PERFECT COMPETITION

In our foregoing analysis we have explained the general equilibrium of exchange and production by taking two individuals, two factors and two goods case (*i.e.* 2 × 2 × 2 model) through the process of bargaining by the individuals. It was *emphasised that exactly at what point on the contract curve the equilibrium occurs depends on the bargaining strengths of the two individuals.* It was further pointed out that equilibrium arrived at through bargaining between the two parties is not unique, it can occur anywhere on the consumption contract curve or the production contract curve as the case may be. We shall now explain *how general equilibrium is established when perfect competition prevails in the product and factor markets.* In other words, we shall now show that competitive equilibrium is consistent with general equilibrium.

It may be noted that in perfectly competitive market there are many buyers and sellers of products and factors. If a buyer does not like the terms of exchange of one seller, he can look for another seller who is willing to exchange on better terms, Therefore, in competitive markets no individual buyer or seller of a product or factor is in a position to influence the price. They take the price prevailing in the market as given and constant and are therefore mere *price takers.* At the given price, they decide how much to buy or sell of a good or factor. We shall explain how general equilibrium of exchange and production, both separately and simultaneously, is achieved under perfect competition. It will be known from our analysis made below that *while in case of the bargaining between the two parties, the general equilibrium is indeterminate or not unique, under competitive conditions general equilibrium is unique.*

Perfect Competition and General Equilibrium of Exchange

We will explain first general equilibrium of exchange under competitive conditions. With perfect competition in the buying and selling of goods no individual can influence the prices of goods; he takes the prices as given and constant. Thus, with given prices of goods X and Y, consumer A maximises his satisfaction by equating his marginal rate of substitution *(MRS_{xy})* with the given price ratio. Thus

$$MRS^{A}_{XY} = \frac{P_x}{P_y} \qquad \text{... }(i)$$

Similarly, the individual B would be maximising his satisfaction when

$$MRS^{B}_{XY} = \frac{P_x}{P_y} \qquad \text{...}(ii)$$

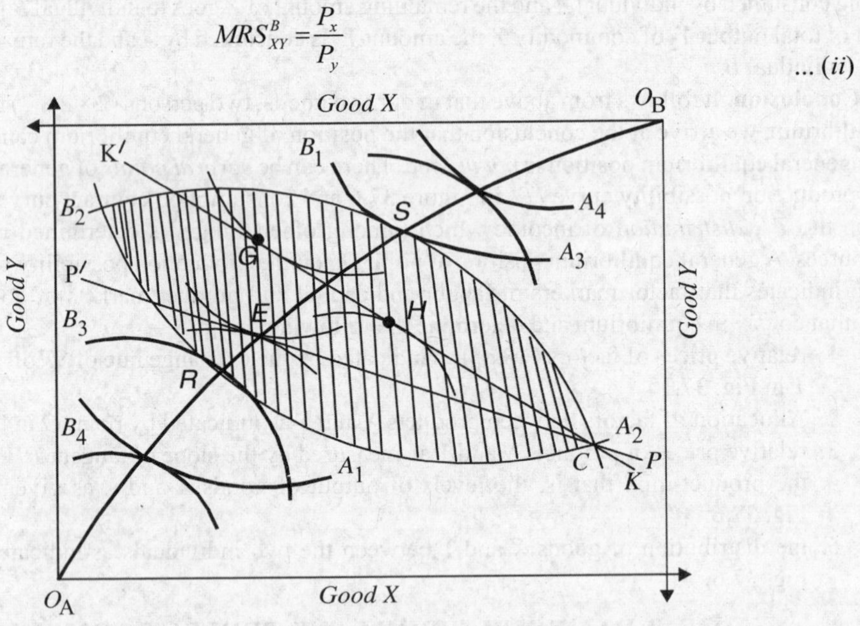

Fig. 37.7. *General Equilibrium of Exchange Under Perfectly Competitive Conditions*

Since under perfect competition both the individuals will be facing the same set of prices, from (*i*) and (*ii*) we get

$$MRS^{A}_{XY} = MRS^{B}_{XY} = \frac{P_X}{P_Y}$$

Now consider Fig. 37.7. Suppose the initial endowments of individuals A and B is given by point C. Further suppose that the prevailing perfectly competitive price ratio $\frac{P_x}{P_y}$ is denoted by the slope of the price line PP' which is drawn through the initial endowment point C. In fact, the price line PP' passing through initial endowment point C is the budget line for the two consumers. With the initial endowment of the two individuals represented by point C and the price-ratio line PP' the two individuals in their attempt to maximise satisfaction will trade goods with each other and move to a *unique point E* on the contract curve where their indifference curves are tangent to the given price line PP'. In this way, by adding the assumption of perfect competition in the sale and purchase of goods and starting from initial endowment point *C* we have reached the unique point *E* where the general equilibrium of exchange occurs on the segment *RS* of the contract curve.

With position *C* as the starting point and point *E* where they reach the equilibrium position, individual *A* has sold good *X* and purchased good *Y*, whereas individual *B* has purchased good *X* and sold good *Y*. Price ratio of the two goods is given by the slope of the budget line *PP'*. At equilibrium at *E* with the budget line *PP' (i.e.* the given price ratio) the quantity of good *X* demanded by *B* equals the quantity of good *X* supplied by A and, similarly, quantity demanded of good *Y* equals the quantity supplied of good *Y*. As a result, markets for the two goods are in equilibrium and therefore general equilibrium of exchange exists. Note that a *general equilibrium refers to a set of prices at which quantity demanded equals the quantity supplied.* We thus that *competitive equilibrium is consistent with perfect competition.*

However, it may be noted that not all prices are consistent with general equilibrium under perfect competition. For instance, price ratio of the two goods represented by the price line *KK′* which also passes through the initial endowment point *C*. With this price ratio or price line *KK′* individual *A* is in equilibrium at point *G* and individual *B* is in equilibrium at point *H*, the markets of goods *X* and *Y* are not in equilibrium as their MRS_{XY} of the two individuals are not equal to $\frac{P_x}{P_y}$ at the same point. While individual *B* demands less of good X, the individual *A* is willing to supply more of it, that is, surplus of good *X* emerges in the market. As regards good *Y*, with price ratio line *KK′*, individual A demands more of *Y*, than individual *B* is willing to supply it Thus, with price line *KK′*, the markets of two goods are not in equilibrium.

It should be emphasised again that the equilibrium of exchange on the segment *RS* of the contract curve in Fig. 37.7 is reached by assuming that *C* is the initial endowment of goods (i.e. initial income) of the two individuals. With a different initial endowment point in the Edgeworht box, we will obtain a different point of exchange equilibrium on the contract curve.

Perfect Competiton and General Equilibrium of Production

General equilibrium of production requires the determination of allocation of resources among the firms which use the factors for production of goods. We have seen above that general equilibrium of production occurs at a point on the production contract curve where the $MRTS_{LK}$ in the production of commodity *X* equals the $MRTS_{LK}$ in the production of commodity *Y*. With perfect competition prevailing in the factor markets factor-price ratio $\left(\frac{w}{r}\right)$ will be given for each firm. In order to minimise cost for producing a given output, a firm will be in equilibrium when $MRTS_{LK} = \frac{w}{r}$, where w stands for wage rate (*i.e.* price of labour and γ stands for rate of interest (*i.e.* price of captial).

Now, what ensures that the general equilibrium of production would occur at a point on the contract curve. As we are assuming factor markets are competitive, it is the competitive forces that would bring the economy in equilibrium at a point on the contract curve. If the two firms find themselves away from the contract curve, they will trade or exchange the factors and move to a point on the contract curve because in doing so they will be increasing their output and will have therefore incentive to move to the contract curve. To produce a given level of output of a commodity, each firm will minimise its cost of production by equating marginal rate of technical substitution between factors ($MRTS_{LK}$) with the *same factor price ratio* of labour and capital $\frac{w}{r}$ where *w* stands for the wage rate of labour and *r* for the rental price of capital. To conclude, under competitive conditions general equilibrium of production would occur at a point on the contract curve where the following condition holds :

$$MRTS_{LK}^{X} = MRTS_{LK}^{Y} = \frac{w}{r}$$

Consider Fig. 37.3 again, competitive forces will bring the economy to be in general equilibrium at point Q on the production contract curve in Figure 37.3. It should be noted that JJ' is the price line which passes through the initial endowment point T representing the given factor-price ratio $\left(\frac{w}{r}\right)$.

Perfect Competition and General Equilibrium of Production and Exchange

In the two factors, two goods and two persons (2 × 2 × model) with the exchange conducted through bargaining between the two persons. We have explained above that the simultaneous equilibrium of production and exchange occurs at a point where the following condition is fulfilled :

$$MRTS_{XY} = MRS^A_{XY} = MRS^B_{XY}$$

As explained earlier, marginal rate of transformation between two goods, MRT_{XY} is equal to the ratio of marginal costs of production of two commodities. Thus

$$MRT_{XY} = \frac{MC_X}{MC_Y}$$

Now, if perfect competition is prevailing in the goods markets, then the individual firms producing the goods would have no influence on their prices and will, therefore, equate the given price of each good with its marginal cost of production. Thus

$$MC_X = P_X \text{ and } MC_Y = P_Y$$

The slope of the production transformation curve measures the marginal rate of transformation (MRT_{xy}). Thus, given the market prices of goods for the firms as is the case under perfect competition, the general equilibrium of production will occur at a point on the production transformation curve at the point where marginal rate of transformation equals the ratio of prices of goods. Thus, in equilibrium under perfect, competition.

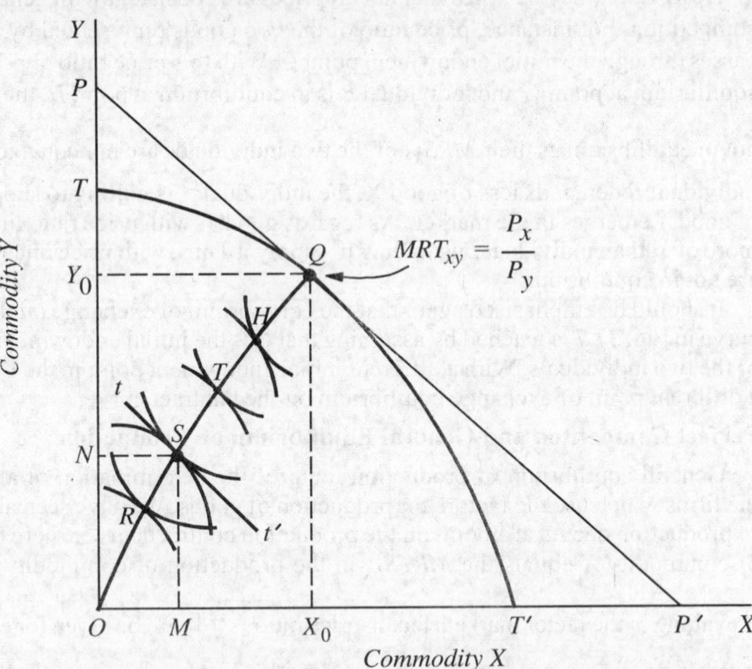

Fig. 37.8. *General Equilibrium of Production and Exchange*

$$MRT_{XY} = \frac{MC_X}{MC_Y} = \frac{P_X}{P_Y} \qquad \text{... (i)}$$

Such general equilibrium of production is represented by point Q on the transformation curve TT in Fig. 37.8 where the price line PP' is tangent to it. Note that the slope of the price line PP' measures the given ratio of competitive prices of the two goods $\left(\frac{P_x}{P_y}\right)$. This general equilibrium of production at point Q on the production transformation curve TT' determines the product mix; X_0 of good X and Y_0 of good Y.

It follows from above that, given the factor endowments of the economy and the nature of production function (*i.e.*, technology), under perfect competition the general equilibrium of production reaches a unique position on the transformation curve of the economy and a unique position on the corresponding contract curve in the Edgeworth Production Box.

Now, we have to see how in general equilibrium under perfect competition, this equilibrium product mix is to be distributed for consumption between the two individuals. We know that in order to maximise satisfaction, consumers equate their marginal rate of substitution between the two goods (MRS_{xy}) with the ratio of the given competitive prices of the two goods. Thus, in equilibrium under perfect competition.

$$MRS^{A}_{XY} = \frac{P_x}{P_y} \text{ and } MRS^{B}_{XY} = \frac{P_x}{P_y} \qquad \text{... (ii)}$$

Since the consumers and producers (firms) *face the same price ratio of the two goods*, from conditions *(i)* and *(ii)* above we have

$$MRT_{xy} = MRS^{A}_{xy} = MRS^{B}_{xy} = \frac{P_x}{P_y}$$

This general equilibrium of production and exchange is shown in Figure 37.8 where from point Q at the transformation curve we have made Edgeworth Box with dimension X_0 of good X and Y_0 of good Y. It will be observed from Fig. 37.8 that at its point S on the contract curve, the slope of the indifference curve of the two individuals measuring the marginal rate of substitution (MRS_{xy}) between two goods is equal to the slope of the price line PP'. Thus, *with perfect competition we are able to determine a unique position of general equilibrium of production and exchange at a point on the contract curve.*

It may however be emphasised again that even this position of general equilibrium depends on the *initial allocation of fixed amounts* of the two factors to the two goods which in turn is determined by the *initial* endowments or distribution of goods (*i.e.* distribution of real income) between the two individuals. With a different initial allocation of factors to the two goods or different initial endowments of the goods of the two individuals even with perfect competition, we would have reached a different position. of general equilibrium of production and exchange.

GENERAL EQUILIBRIUM DETERMINES ONLY RELATIVE PRICES

It is important to note that micro-economic theory is concerned with analysing the factors that determine *relative prices* and not *absolute prices.* Relative price means prices of a good as compared to the price of another good or a factor. In other words, by relative prices we mean the ratio of prices of goods. That micro-economic theory is concernsed with relative prices is made quite clear in the analysis of general equilibrium analysis. In our analysis of general equilibrium when we talked about the determination of factor prices, it was the ratio of factor prices (*i.e.,* the ratio of price of labour to the price of capital, $\frac{w}{r}$) as measured by slope of the tangent line JJ' to the relevant isoquants that was determined. In fact, we did not explain how absolute prices, namely, wage rate of labour, w and price of capital, r were determined (see Fig. 37.3). Likewise, in general equilibrium of exchange we did not concern ourselves with the determination of absolute price of good X (*i.e.,* P_x) and absolute price of good Y (*i.e.* P_Y) separately; we only showed how general equilibrium determine the *ratio of prices* of two goods *i.e.* $\left(\frac{P_x}{P_y}\right)$. as measured by the slope of prcie line PP' which is tangent to the indifference curces of the two individuals at point E (see Fig. 37.8). Thus we found that in Figure 37.6 general equilibrium is established at point Q on the transformation curve TT' where $MRT_{XY} = MRS^{A}_{XY} = MRS^{B}_{XY}$ and saw that this yields relative prices of goods X and Y, that is, the ratio $\frac{P_X}{P_Y}$ which is measured by the slope of the transformation curve at that equilibrium point Q' and not the absolute prices.

An important thing to note is that because it is the relative and not absolute prices that result from the general equilibrium analysis that doubling or halving of *all prices* so that relative prices remain the same would not bring any change in the equilibrium position. For example, suppose the relative prices of two goods *X* and *Y i.e.* $\frac{P_X}{P_Y}$ as measured by the marginal rate of transformation at point Q' in Fig. 37.6 is 2/1 and the relative price of labour and capital (w/r) as measured by the slope of the isoquants at point Q in Fig.37.3 is 3/1. Further suppose price of good *X* is ₹ 20 and the price of good *Y* is 10, so that relative product prices are 20/10 = 2/1 and the price of labour (w) is ₹ 30 and the price of capital (r) is ₹ 10. So the relative factor prices $\frac{w}{r}$ equals 3/1. Now, if all prices are doubled, P_x = ₹ 40, P_y = ₹ 20, w = ₹ 60 and r = ₹ 20, then the relative prices of products $\frac{P_x}{P_y} = \frac{40}{20} = \frac{2}{1}$ and relative factor prices $\frac{w}{r} = \frac{60}{20} = \frac{3}{1}$ since all relative prices have remained the same despite the change in their absolute levels, the position of general equilibrium would remain unaltered. In real terms nothing would change. Consumers would have twice the payments or income but would buy exactly the same quantities of goods *X* and *Y*. Each firm would receive double the total revenue than before but would buy and employ the same amounts of factors at double the prices. Thus, *it is the relative prices that are significant* in determining the real outcome or production, employment of factors, income distribution and the distribution of goods between the individuals. That is why micro-economic theory lays stress on the determination of relative prices. To conclude, general equilibrium analysis yields relative and not absolute prices. If absolute prices of all products and all factors change by a same percentage, the real situation would remain unchanged.

QUESTIONS FOR REVIEW

1. Distinguish between partial and general equilibrium. What does the contract curve indicate in an Edgeworth Box diagram?
   [D.U. B.Com (Hons) 1996, 1999, 2000 D.U. B.A. (Hons) 19991, 1993]
2. In what way is general equilibrium analysis more useful? *[B.Com. (Hons) D.U. 1999]*
3. In an economy characterised by two consumers, two commodities show how general equilibrium of exchange is attained. State your assumption clearly.
   [B.A. (Hons) Delhi 1992]
4. What is the marginal rate of transformation ? Explain why the marginal rate of transformation of one good for another is equal to the ratio of the marginal costs of producing the two goods. *[D.U. B.A. (Hons) 2001]*
5. Point out the inadequacies of partial equilibrium analysis. Under what conditions can the existence of a meaningful general equilibrium solution be guaranteed?
   [B.A. (Hons.) D.U. 2001]
6. Define a contract curve. What determines the final location of the two parties on the contract curve. In what sense is the contract curve a locus of optimal locations?
7. Define and illustrate general equilibrium in an economy with perfectly competitive factor and product markets. State and explain clearly the conditions under which an equilibrium will exist for such an economy. *[D.U. B.A. (Hons.) 1997]*
8. "General equilibrium of exchange occurs at a point on the exchange contract curve." Explain. Can there be two such points? *[D.U. B.A. (Hons.) 1994]*
9. "Absolute prices are indeterminate in a model of general equilibrium analysis" True or false? Explain. *[D.U. B.A. (Hons.) 1973]*
10. "General equilibrium analysis is solely concerned with relative prices" Explain.
    [D.U. B.A. (Hons.) 1984, 1956]

CHAPTER 38
WELFARE ECONOMICS: AN INTRODUCTION

WHAT WELFARE ECONOMICS IS ABOUT

So far we have been concerned with the analysis of what has been called positive economics which explains merely "how it is". That is, we have been concerned with explaining how prices of products and factors are determined and further on the basis of these prices how the allocation of resources is made in a free private enterprise economy. Now, in this part of the book we shall discuss whether any allocation of resources is efficient or not. *By efficiency in economics we mean whether any state or situation regarding resource allocation maximises social welfare.* In welfare economics attempt is made to establish criteria or norms with which to judge or evaluate alternative economic states and policies from the viewpoint of efficiency or social welfare. These criteria or norms serve as a basis for recommending economic policies which will increase social welfare. Thus the norms established by welfare economics are supposed to guarantee the optimal allocation of economic resources of the society. Putting it more specifically, Prof. Baumol writes, "Welfare Economics has concerned itself mostly with policy issues which arise out of the allocation of resources, with the distribution of inputs among the various commodities and the distribution of commodities among various consumers."[1] And it may be emphasised again that allocation of resources is efficient or optimum when social welfare is maximum.

The inter-relationship among various parts of the economy means that certain particular change in one part of the economy affects resource allocation in all other parts of it. Thus, a central problem in welfare economics relates to whether a particular change in resource allocation will increase or decrease social welfare. However, an almost insurmountable difficulty which is faced in welfare economics is that it is not possible to measure social *welfare objectively*, for it involves making interpersonal comparison of utilities or welfares of different individuals comprising the society. In order to avoid making interpersonal comparison of utility, whose scientific nature has been challenged, among others, by Lord Robbins, economists have mostly used what is known as *Pareto-optimality criterion* for evaluating whether social welfare increases or decreases as a result of a specific change in economic state, situation or policy. According to Pareto criterion of optimality or efficiency, *any change that makes at least one individual better off without making any other worse off is an improvement in social welfare.* Of course, when a certain change makes every one in the society better off, social welfare will undoubtedly increase. On the other hand, social welfare will decrease if a certain change makes no individual better off while it makes at least one individual worse off. With the aid of this criterion we can define the state of maximum social welfare or what is known as *Pareto optimality or economic efficiency. Economic state or situation is said to be Pareto-optimal or efficient in which allocation of resources is such that by any rearrangement of them it is impossible to make any individual better off without making any other worse off.* This concept of Pareto-optimality or economic efficiency is the basis of welfare economics and has a large number of applications in applied economics. We shall explain the conditions which ensure Pareto optimality and also critically evaluate Pareto's welfare criterion in the next chapter at length.

---

1. W.J. Baumol, *Economic Theory and Operations Analysis*, 4th edition, 1978, p. 406.

Pareto criterion and the concept of Pareto optimality does not embrace those changes in economic state which make some better off and others worse off. This involves interpersonal comparisons of utility which were ruled out by Pareto and his followers. Kaldor and Hicks propounded a welfare criterion, which is known as *compensation principle*, to judge those changes in situation which make some better off and others worse off. They have claimed that their welfare criterion does not involve interpersonal comparison of utility and value judgements. It is asserted that Kaldor and Hicks rehabilitated welfare economics from the damaging criticism of Lord Robbins and founded a "*New Welfare Economics*" free from value judgments or interpersonal comparison of utility. In the development of new welfare economics, Scitovsky and Little have also made significant contributions. The critical evaluation of "New Welfare Economics" forms the subject matter of a separate chapter.

After discussing new welfare economics we shall study the concept of the *social welfare function* propounded by Bergson and Samuelson. According to this social welfare concept, any attempt to establish propositions in welfare economics without the introduction of explicit value judgements is sterile. Bergson-Samuelson social welfare function as well as Arrow's analysis of how to obtain social welfare function from preferences of individuals shall be discussed at the end of this part. Welfare economics has been a controversial subject in recent years because it involves value judgements about which there is a sharp difference of opinion among economists.

Now-a-days the establishment of a welfare state is the fundamental objective of modern democratic Governments. In order to achieve this objective the State attempts to satisfy the wants of each and every individual of the society. Non-satisfaction of wants gives pain to a man and its satisfaction, a pleasure. The word 'pleasure' is associated with welfare. Some wants always remain unsatisfied and give pain to an individual whereas some are always satisfied or are in the process of being satisfied and thus give rise to welfare. The fact remains that the welfare is the result of the satisfaction of wants. It can be concluded from this fact that to increase the welfare of a man his wants must be satisfied. According to Prof. J.K. Mehta, if we compare the two periods of time, then "a given man has greater welfare in which a larger number of his wants have been satisfied. Conversely, we can say that the smaller the number of wants (of given intensities) that remain unsatisfied the greater is the welfare."[2]

Three Concepts of Social Welfare

Prof. Graff has distinguished three concepts of social welfare.[3] The first concept of social or group welfare is the *paternalist* one which describes the views of a paternalist authority or state and not of the individuals in the society. According to this concept, the preferences of the individual members of the society may be ignored and the state or a paternalist authority or a dictator uses its own ideas about social welfare; social welfare increases when that paternal authority or the dictator thinks it so.

The second concept of social welfare is one which has been used by V. Pareto and his followers. According to this *Paretian concept*, welfare of the society is simply the sum total of the welfare of different individuals comprising it. If some persons are made better off and none worse off, social welfare increases and if some are made worse off and none better off, it decreases. But if some are made better off and some worse off, then, according to Paretian concept, we cannot know what has happened to the welfare of the society. The Paretian concept of social welfare rules out interpersonal comparison of utility or welfare and rests only on a generally accepted ethical view that "it is good to make somebody better off while making nobody worse off". But since in most cases of changes in economic organisation and policy, at least some people are made worse off, the Paretian concept of social welfare is of limited value for most of the real world economic problems.

---

2. J.K. Mehta and Mahesh Chand, *A Guide to Modern Economics*, Somaiya Publications, Bombay, 1970, p. 180.
3. J. De V. Graff. *Theoretical Welfare Economics*, Cambridge University Press, 1957, pp. 7-11.

The third concept of social welfare involves interpersonal comparison of utility which is to be made by introducing explicit value judgements. This concept of social welfare has been propounded by Bergson and Samuelson in their now well-known theory of *social welfare function*. Thus they have described the utility functions of the various persons in the society with the help of a social welfare function. These economists are of the opinion that changes in social welfare cannot be assessed without making interpersonal comparisons of utility and therefore without making value judgements. It is because of this that this concept of social welfare is able to judge the welfare implications of even those changes in economic organisation and policies that make some people better off and others worse off.

Various concepts of social welfare have been referred to above. Economists generally do not accept the paternalist or dictatorial concept of social welfare. Some economists such as Robbins and his followers tried to separate economics from ethics but now-a-days there is consensus among economists that welfare economics cannot be separated from ethics. Bergson, Samuelson, Little, Arrow and others are of the opinion that value judgements are most important in welfare economics. But the fact remains that social welfare and changes in it cannot be measured accurately due to heterogeneity of the interests of the various individuals in a society.

ROLE OF VALUE JUDGEMENTS IN WELFARE ECONOMICS

It is important to explain the role of value judgements in welfare economics. Since welfare economics is concerned with the desirability or otherwise of economic policies, the value judgements play a crucial role. As mentioned above, by value judgements or values we mean the conceptions or ethical beliefs of the people about what is good or bad. These conceptions regarding values of the people are based on ethical, political, philosophical and religious beliefs of the people and are not based on any scientific logic or scientific law. There is a great controversy regarding whether value judgements should have any role to play in welfare economies. Robbins and his followers have been asserting that the inclusion of value judgements would make our subject unscientific and therefore, according to them, economists should refrain from making value judgements. On the other hand, majority of modern economists are of the view that economists should not fight shy of making value judgements if there is a wide consensus about them among the community. Using his knowledge of economics together with these value judgements he should comment upon the desirability or otherwise of economic policies and issues. Professor Paul Streeten rightly says, "Economists cannot and should not refrain from making value judgements if their studies are to be more than a purely formal technique of reasoning, an algebra of choice. The technique, the algebra, is important and ought to be as scientific as possible, but it is significant only as a means to study of wealth and welfare and of the way to improve them".[4]

It should be noted that as far as the welfare of individual is concerned, though difficult to measure in cardinal terms, economists can measure it in ordinal terms and by observing the act of choice of the individual. For instance, if an individual chooses A rather than B, it shows that his welfare is greater in A than in B. Thus, choice by an individual is an objective test for knowing and comparing his welfare in different economic states. Therefore, what promotes individual welfare or not can be tested and verified. However, when welfare economics has to judge the social welfare or group welfare, it encounters difficulties because the measurement of social welfare is not an easy task and involves value judgements and interpersonal comparisons of utility. This is because the society or group whose welfare we have to judge cannot be regarded as an organic whole, having its own mind. Therefore, social welfare, unlike individual welfare, is not something which resides in the mind of the society. We cannot derive propositions of social welfare from choice of individuals comprising the society, because individuals choose differently and, therefore, there is no unanimous

4. Paul Streeten, Economics and Value Judgements, *Quarterly Journal of Economics*, 1950, p. 595.

social choice. Individual choices differ because various individuals have different tastes, preferences and ethical beliefs and therefore different value judgements. The vital issues in welfare economics are concerned with social welfare and devising certain criteria to judge the social welfare. Therefore, welfare economics cannot be purely objective or free from value judgements.

It is worth noting that *Pareto evolved the concept of social welfare which is said to be free from any value judgements*, because it is not based upon any interpersonal comparison of utility. According to Pareto, the social welfare depends upon the welfare of the individuals comprising the society and, according to him, if at least one individual is made better off by certain economic reorganisation and no one being made worse off, the social welfare increases, that is, if any economic reorganisation increases the welfare of one without reducing the welfare of any other, then the social welfare increases. When such an economic state is reached *that through any reorganisation it is not possible to make at least one individual better off with no other being worse off, this is called the state of maximum social welfare or Pareto optimum*. However, Paretian concept of social welfare is confined to only limited issue of welfare economics. Generally, when any economic reorganisation increases the welfare of some, it would reduce the welfare of some others and therefore, in this case, *Pareto criteria* will not apply. Following Robbins some economists object to making inter-personal comparison of utility to derive welfare propositions, since, according to them, inter-personal comparison of utility is based upon value judgements.

However, Kaldor and Hicks by propounding a compensation principle laid the foundations of *New Welfare Economics* which is supposed to be free from value judgements. According to this compensation principle, if a change in economic organisation increases the welfare of some and reduces the welfare of others, but those who gain in welfare are able to compensate the losers and still be better off than before, then the change in economic organisation will increase the social welfare. However, Kaldor-Hicks welfare criterion has been subjected to swvere criticisms. The claim of Kaldor and Hicks that their criterion is free from value judgements or ethical assumptions has been contested. To quote Professor Baumol, "*Both the Kaldor and the Scitovsky tests operate on the basis of an implicit and unacceptable value judgement. By using a criterion involving potential money compensation, they set up a concealed inter-personal comparison on a money basis*."[5] He further writes, "It is no answer to this criticism to say that these criteria are just designed to measure whether production, and hence, potential welfare, are increased by a policy change that these criteria disentangle the evaluation of a production change from that of the distribution change by which it is accompanied. Consider a change in *production* which increases gin output but reduces the output of whiskey. If *X* likes highballs but *Y* prefers martins, the question whether this is an increase in production is inextricably tied in with the question of the distribution of these beverages between *X* and *Y*."[6]

In the end we may note that Professor Bergson has pursued a different line of approach to welfare economics. He has propounded the concept of social welfare function in which a set of value judgements is *explicitly* introduced and with this social welfare function, the economists can judge the desirability of certain economic reorganisations or policy changes. *These value judgements, according to Bergson, "must be determined by its compatibility with the values prevailing in the community the welfare of which is being studied*."[7] Followers of Bergson like Samuelson and I.M.D. Little are of the view that welfare economics cannot be separated from value judgements, because any statement about increase or decrease of social welfare necessarily involves value judgements. On the rightness of Bergson's social welfare function, and his introduction of explicit value judgements in it, Prof. Baumol writes, "*Essentially the Bergson criterion must be judged right, if not very*

5. Nicholas Kaldor, Welfare Propositions in Welfare Economics, *Economic Journal*, 1939 and J.R. Hicks, Foundations of Welfare Economics, *Economic Journal*, 1939.
6. Op. cit., p. 403-4.
7. A. Bergson, On the Concept of Social Welfare, *Quarterly Journal of Economics*, 1954.

helpful. To decide whether B is better than A, we must certainly employ some value judgements, and unless these judgements are explicit they must be treated with suspicion."[8] Likewise, Professor K. E. Boulding writes: "One must admit that the task of making value judgements explicit is very important. It is obviously preposterous to suppose that one can set up criteria for judgement which are somehow independent of ethical norms".[9]

Thus, according to several modern economists such as Samuelson, Little, Boulding welfare economics cannot be purged of value judgements. In fact, the study of welfare economics has been developed to make policy recommendations to promote social welfare. And for doing so economists cannot escape from introducing ethical norms or value judgements since we all take interest in the question concerning happiness and welfare of the society. "*Welfare economics and ethics cannot, then, be separated.* They are inseparable because the welfare terminology is a value terminology.....*Getting rid of value judgements would be throwing the baby away with the bath water.* The subject is one about which nothing interesting can be said without value judgements for the reason that we take a moral interest in welfare and happiness".[10]

It should not be gathered from above that the explicit introduction of value judgements makes the study of welfare economics unscientific. In spite of the explicit introduction of value judgements in welfare studies, the economist's approach can still be scientific in the sense that he scientifically deduces the welfare propositions from the given value judgements.

QUESTIONS FOR REVIEW

1. In welfare economics attempt is made to establish criteria or norms to judge the social desirability of economic policies. Explain.
2. What is meant by social welfare? Is it possible to measure social welfare without interpersonal comparison of utility?
3. What is meant by economic efficiency? Is it right to ignore equity in distribution when it clashes with the objective of economic efficiency?
4. What is new welfare economies? Explain its approach to social welfare.
5. Explain three concepts of social welfare mentioning in each case the approach to the inter-personal comparision of utility.
6. What is meant by value judgements? Explain their role in welfare economics
7. "Welfare economics and ethics cannot be separated.... Getting rid of value judgements would be throwing the baby away with the bath water" (I.M.D. Little) Examine critically.
8. "Wheras the normal way of testing a theory in positive economics is to test its conclusions, the normal way of testing a welfare proposition is to test its assumptions" (J. de. V. Graaff) Discuss
9. "The interest attaching to a theory of welfare depends almost entirely upon the realism and relevance of assumptions".

---

8. *Op.cit.*, p.404
9. K.E. Boulding, 'Welfare Economics' in B.E. Haley (ed) A.E.A., *A Survey of Contemporary Economics*, Richard D. Irwin 1952
10. I.M.D. Little, *A Critique of Welfare Economics*, Oxford University Press, 2nd Edition, 1957. (*Italic added*)

CHAPTER 39
ECONOMIC EFFICIENCY AND PARETO OPTIMALITY

Notion of Pareto Optimality and Economic Efficiency

We saw in the previous chapter that neo-classical and earlier economists defined social welfare as a sum total of cardinally measurable utilities of different members of the society. An optimum allocation of resources was one which maximised the social welfare in this sense. V. Pareto was the first to part with this traditional approach to social welfare in two important respects. First, he rejected notion of cardinal utility and its additive nature and, second, he detached welfare economics from the inter-personal comparisons of utilities. Pareto's concept of maximum social welfare which is based upon ordinal utility and is also free from value judgements occupies a significant place in modern welfare economics. Pareto optimum may not be sufficient condition for attaining maximum social welfare but it is a necessary condition for it. We have already explained the concept of Pareto optimality in Chapter 38. To repeat, *Pareto optimum (often called Economic Efficiency) is a position from which it is impossible to make anyone better off without making someone worse off by any reallocation of resources or distribution of outputs.* Thus, in the Pareto "optimum position the welfare of any individual of the society can not be increased without decreasing the welfare of another member. Before explaining the conditions of achieving Pareto optimality, we shall explain Pareto criterion of evaluating changes in social welfare because the concept of Pareto optimality or maximum social welfare is based upon Pareto criterion of welfare.

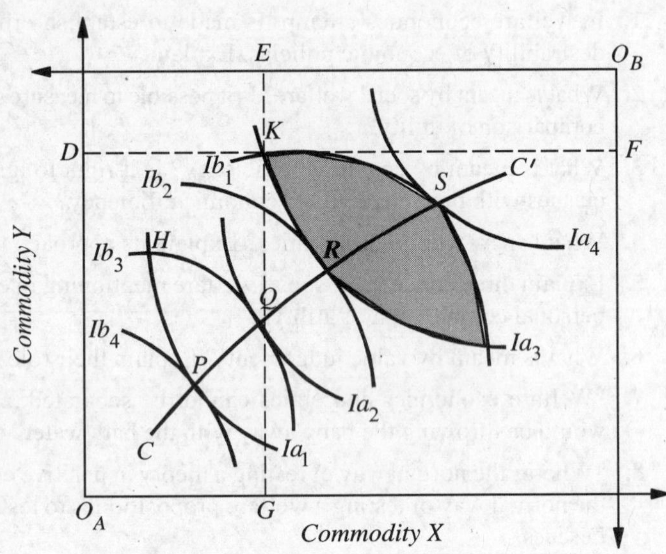

Fig. 39.1 *Pareto Criterion and Pareto Optimality.*

Pareto Criterion of Social Welfare

The concept of Pareto optimum or economic efficiency stated above is based on a welfare criterion put forward by V. Pareto. Pareto criterion states that if any reorganisation of economic resources does not harm anybody and makes someone better off, it indicates an increase in social welfare. If any reorganistaion or change makes everybody in a society better off, it will, according to Pareto, undoubtedly mean increase in social welfare. Thus, in the words of Prof. Baumol "any change which harms no one and which makes some people better off (in their own estimation) must be considered to be an improvement."[1] Parto criterion can be explained with the help of Edgeworth Box diagram which is based on the assumptions of ordinal utility and non-interpersonal comparison of utilities. Suppose two persons *A* and *B* form the society and consume two goods *X* and *Y*. The various levels of their satisfaction by consuming various combinations of the two goods have been represented by their respective indifference curves.

In Fig. 39.1 O_a and O_b are the origins for the utilities of two persons *A* and *B* respectively. I_{a1}, I_{a2}, I_{a3}, I_{a4} and I_{b1}, I_{b2}, I_{b3}, I_{b4} are their successively higher indifference curve. Suppose the initial distribution of goods *X* and *Y* between the members of the society, *A* and *B*, is represented by point-*K* in the Edgeworth Box. Accordingly, individual *A* consumes O_AG of $X + GK$ of *Y* and is at the level of satisfaction represented by indifference curve I_{a3}. Similarly, individual *B* consumes *KF* of $X +$ *KE* of *Y* and gets the satisfaction represented by indifference curve Ib_1. Thus the total given volume of goods *X* and *Y* is distributed between *A* and *B*. In this distribution, individual *A* consumes relatively larger quantity of good *Y* and individual *B* of good *X*. Now, it can be shown with the aid of Pareto's welfare criterion that a movement from the point *K* to a point such as *S* or *R* or any other point in the shaded region will increase social welfare.

Any movement from *K* to *S* through redistribution of two goods between two individuals increases the level of satisfaction of *A* without any change in the satisfaction of *B* because as a result of this *A* moves to his higher indifference curve I_{a4}, and *B* remains on his same indifference curve I_{b1} (*K* and *S* lie on *B*'s same indifference curve I_{b1}). In other words, as a result of the movement from *K* to *S*, individual *A* has become better off whereas individual *B* is no worse off. Thus, according to Pareto criterion, social welfare has increased following the movement from *K* to *S* and therefore *K* is not the position of economic optimum. Similarly, the movement from *K* to *R* is also desirable from the point of view of social welfare because in this individual *B* becomes better off without any change-in-the satisfaction of individual *A*. Therefore, both the positions *S* and *R* are better than *K*. The tangency points of the various indifference curves of the two individuals of the society are the Pareto optimum points and the locus of these points is called '*contract curve*'.

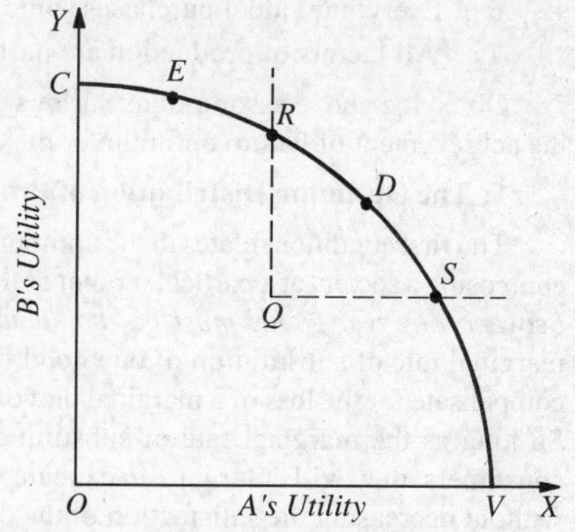

Fig. 39.2. *Pareto Criterion Explained with Utility Possibility Curve*

Pareto criterion can also be explained with the help of Samuelson's utility possibility curve. *Utility possibility curve is the locus of the various combinations of utilities obtained by two persons from the consumption of a particular bundle of goods*. In Fig. 39.2, *CV* is a utility possibility curve which shows the various levels of utilities obtained by two individuals *A* and *B* of the society resulting from the redistribution of a fixed bundle of goods and its consumption by them. According to Pareto criterion, a movement from *Q* to *R*, or *Q* to *D*, or *Q* to *S* represents the increase in social welfare because in such movements the utility of either *A* or *B* or both increases. *A* movement from *Q* to *R* implies that the utility or welfare of *B* increases, while that of *A* remains the same. On the other hand, a movement from *Q* to *S* implies that while *A* has become better off, *B* is no worse off. And a movement from *Q* to *D* or any other point on the segment between *R* and *S* will mean increase in welfare or utility of both the individuals. Thus points *R*, *D* and *S* are preferable to *Q* from the point

1. W.J. Baumol, *Economic Theory and Operations Analysis*, Prentice Hall, 4th edition,1978, p. 527.

of view of social welfare. But unfortunately Pareto criterion does not help us in evaluating the changes in welfare if the movement as a result of redistribution is from the point Q to a point outside the segment RS; such as point E on the utility possibility curve CV. As a result of the movement from point Q to E, the utility of A decreases while that of B increases. In such circumstances, Pareto criterion can not tell us as to whether social welfare increases or decreases.

MARGINAL CONDITIONS OF PARETO OPTIMALITY

Pareto concluded from his criterion that competition leads the society to an optimum position but he had not given any mathematical proof of it, nor he derived the marginal conditions to be fulfilled for achievement of the optimum position. Later on, Lerner and Hicks derived the marginal conditions which must be fulfilled for the attainment of Pareto optimum. These marginal conditions are based on the following important assumptions:

1. Each individual has his own ordinal utility function and possesses definite amount of each product and factor.
2. Production function of every firm and the state of technology is given and remains constant.
3. Goods are perfectly divisible.
4. A producer tries to produce a given output with the least-cost combination of factors.
5. Every individual wants to maximise his satisfaction.
6. Every individual purchases some quantity of all goods.
7. All factors of production are perfectly mobile.

Given the above assumptions various marginal conditions (first-order conditions) required for the achievement of Pareto optimum or maximum social welfare are explained below:

1. The Optimum Distribution of Products among the Consumers: Efficiency in Exchange

The first condition relates to the optimum distribution of the goods among the different consumers composing a society at a particular point of time. The condition says: "*The marginal rate of substitution between any two goods must be the same for every individual who consumes them both.*"[2] The marginal rate of substitution of one good for another so as is the amount of one good necessary to compensate for the loss of a marginal unit of another so as to maintain a constant level of satisfaction. So long as the marginal rate of substitution (*MRS*) between two goods is not equal for any two consumers, they will enter into an exchange which would increase the satisfaction of both or of one without decreasing the satisfaction of the other.

This condition can be better explained with the help of the Edgeworth Box diagram. In Fig. 39.3, goods X and Y, which are consumed by two individuals A and B composing a society are represented on the X and Y axes respectively. O_A and O_B are origins for A and B respectively. I_{a1}, Ia_2, Ia_3 and Ib_1, Ib_2, Ib_3 are the indifference curves showing successively higher and higher satisfaction of consumers A and B respectively. CC' is the contract curve passing through various tangency points Q, R, S of the indifference curves of A and B. The marginal rates of substitution (*MRS*) between the two goods for individuals A and B are equal on the various points of the contract curve CC'. Any point outside the contract curve does not represent the equality of *MRS* between the two goods for two individuals A and B of the society.

Let us consider point K where indifference curves I_{a1} and I_{b1} of individuals A and B respectively intersect each other instead of being tangential. Therefore, at point K marginal rate of substitution between two goods X and Y (MRS_{xy}) of individual A is not equal to that of B. With the initial distribution of goods as represented by point K, it is possible to increase the satisfaction of one

2. M.W. Reder, *Studies in the Theory of Welfare Economics*, Columbia University Press, New York, 1947, p. 24.

individual without any decrease in that of the other or to increase the satisfaction of both by redistribution of the two goods *X* and *Y* between them. A movement from *K* to *S* increases the satisfaction of A without any decrease in *B*'s satisfaction. Similarly, a movement from *K* to *Q*

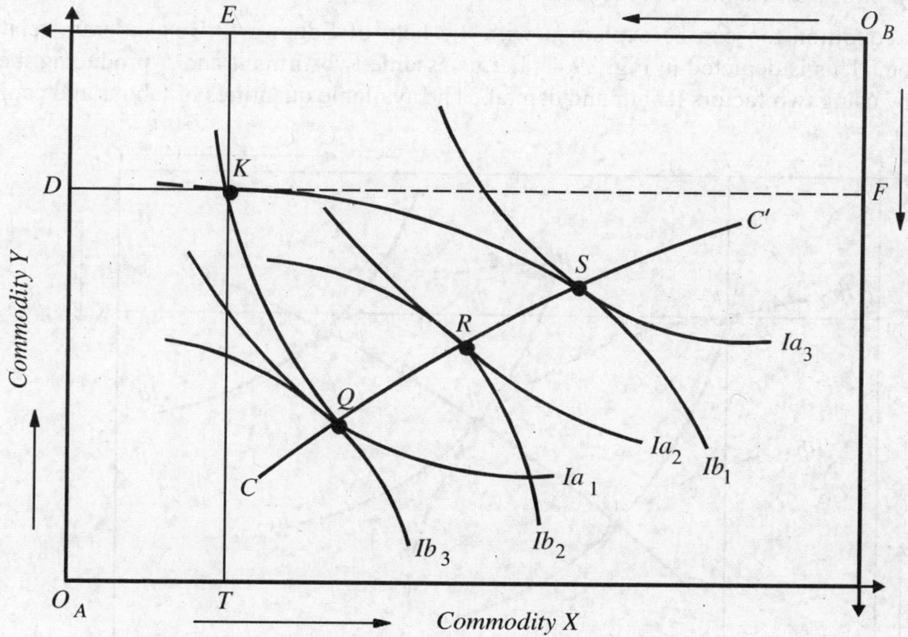

Fig. 39.3. *The Optimum Distribution of Goods.*

increases *B*'s satisfaction without any decrease in *A*'s satisfaction. The movement from *K* to *R* increases the satisfaction of both because both move to their higher indifference curves. Thus, a movement from *K* to *Q* or to *S* or to any other point on the segment *SQ* of the contract curve will, according to Pareto criterion, increase the level of social welfare.

From above it follows that movement from any point away from the contract curve to a point on the relevant segment of the contract curve will mean increase in social welfare. At any point away from the contract curve in the Edgeworth box, the indifference curves of the two individuals will intersect which will mean that MRS_{xy} of two individuals is not the same. And, as explained above, this indicates that through exchange of some units of goods between them, they can move to some point on the contract curve where the social welfare (that is, welfare of two individuals taken together) will be higher.

Since the slope of an indifference curve represents the marginal rate of substitution (MRS_{xy}) at any point of the contract curve, which represents tangency points of the indifference curves, MRS_{xy} of the two individuals are equal. Therefore, points on the contact curve represent the maximum social welfare. However, a movement along the contract curve in either direction will make one individual better off and the other worse off since it will put one individual on his successively higher indifference curves and the other on his successively lower indifference curves. Thus, every point on the contract curve denotes maximum social welfare in the Paretian sense but we can not say anything about the best of them with the help of Pareto criterion.

2. Optimum Allocation of Factors: Pareto Efficiency in Production

The second condition for Pareto optimum requires that the available factors of production should be utilised in the production of products in such a manner that it is impossible to increase the output of oen firm without a decrease in the output of another or to increase the output of both the goods by any re-allocation of factors of production. *This situation would be achieved if the marginal technical rate of substitution between any pair of factors must be the same for any two firms producing two different products and using both the factors to produce the products.*[3]

This condition too can be explained with the help of Edgeworth Box diagram relating to production. This is depicted in Fig. 39.4. Let us assume two firms *A* and *B* producing the same product by using two factors labour and capital. The available quantities of labour and capital are

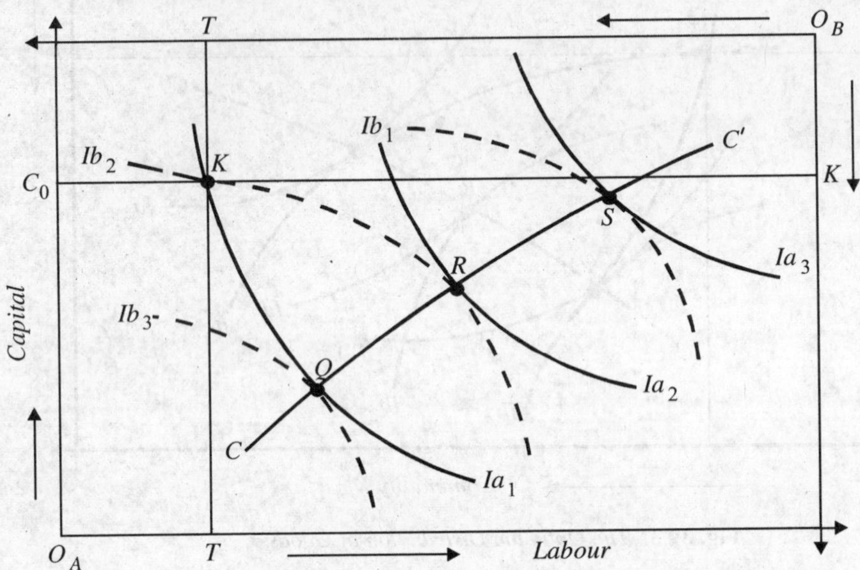

Fig. 39.4. *The Optimum Allocation of Factors.*

represented on *X* and *Y* axes respectively. O_A and O_B are the origins for firms *A* and *B* respectively. Isoquants I_{a1}, I_{a2}, I_{a3} and I_{b1}, I_{b2}, I_{b3} of firms *A* and *B* respectively represent successively higher and higher quantities of output which they can produce by different combinations of labour and capital. The slope of the iosquants, which are convex to the origin, represents the marginal rate of technical substitution (*MRTS*) between two factors. *MRTS* of one factor for another is the amount of one factor necessary to compensate for the loss of the marginal unit of another so that the level of output remains the same. So long as the *MRTS* between two factors for two firms is not equal, total output of a product can be increased by transfer of factors from one firm to another. In terms of the above diagram any movement from *K* to *S* or to *Q* raises the output of one firm without any decrease in the output of the other. The total output of the two firms increases when through redistribution of factors between the two firms, a movement is made from the point *K* to the point *Q* or *S* on the contract curve. A glance at Fig. 39.4 will reveal that movement from point *K* outside the contract curve to the point *R* on the contract curve will raise the output of both the firms individually as well as collectively. Therefore, it follows that corresponding to a point outside the contract curve there will be some points on the contract curve production at which will ensure greater total output of the two firms. As the contract curve is the locus of the tangency points of the isoquants of two firms, the marginal rate of substitution of the two firms is the same at every point of the contract curve *CC'*. It, therefore, follows that on the contract curve at every point of which *MRTS* between the two factors of two firms is the same, the allocation of factors between the two firms is optimum. When the

3. M.W. Reder, *op. ci*, p. 29.

allocation of factors between the two firms is such that they are producing at a point on the contract curve, then no re-allocation of factors will increase the total output of the two firms taken together.

But it is worth mentioning that there are several points on the contract curve and each of them represents the optimum allocation of labour and capital as between the two firms. But which one of them is best cannot be said on the basis of Pareto criterion because movement along the contract curve in either direction represents such factor reallocation which increases the output of one and reduces the output of another firm.

3. Optimum Direction of Production: Efficiency in Product Mix

This condition relates to the pattern of production. The fulfilment of this condition determines the optimum quantities of different commodities to be produced with the given factor endowments. This condition states that "*the marginal rate of substitution between any pair of products for any person consuming both must be the same as the marginal rate of transformation (for the community) between them.*"[4] According to this condition, for the attainment of maximum social welfare goods should be produced in accordance with consumer's preferences. Let us explain this with the help of Fig. 39.5[5]

In Fig. 39.5 commodities *X* and *Y* have been represented on the *X* and Y axes respectively. *AB* is a community's transformation curve between any pair of goods *X* and *Y*. This curve represents the maximum amount of *X* that can be produced for any quantity of *Y*, given the amounts of other goods that are produced and fixed supplies of available resources. IC_1 and IC_2 are the indifference curves of a consumer the slope of which at a point represents the marginal rate of substitution between the two goods of the consumer. The MRT_{xy} of the community and MRS_{xy} of the consumer are equal to each other at point *R* at which the community's transformation curve is tangent to the indifference curve IC_2 of a representative consumers, Point *R* represents optimum composition of production in which commodities *X* and *Y* are being produced and consumed in *OM* and *ON* quantities. This is because of all the points on the community's transformation curve, point *R* lies at the highest possible indifference curve IC_2 of the consumer. For instance, if a combination of goods *X* and *Y* represented by *S* is being produced and consumed, the consumer would be at a lower level of welfare because *S* lies on his lower indifference curve IC_1 which intersects the community's transformation curve instead of being tangential to it. As a result, at point *S*, MRS_{xy} of the consumer is not equal to the MRT_{xy} of the community. With the situation at *S* there is a possibility of moving the consumers to a higher indifference curve by changing the direction (*i.e.* composition) of production *i.e.* by increasing the production of *X* and reducing the production of *Y*. Thus, the

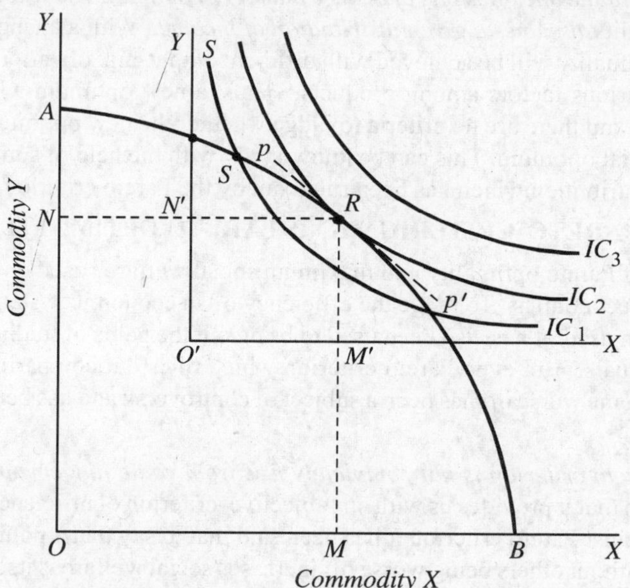

Fig. 39.5. *Optimum Direction of Production :Optimum Product Mix.*

4. Reder, *op. cit.*, p. 30.
5. Reder, *op. cit.*, p. 37.

optimum direction of production is established at point R where community's transformation curve is tangent to the indifference curve of a consumer in the society.

The Second Order and Total Conditions of Pareto Optimality

The marginal or the first order conditions explained above are '*necessary*' but not sufficient for the attainment of maximum social welfare because the marginal conditions by themselves do not guarantee maximum welfare. The marginal conditions can be fulfilled even at the level of minimum welfare. To attain the maximum social welfare position second-order conditions together with the marginal conditions must be satisfied. *The second order conditions require that all indifference curves must be convex to the origin and all transformation curves concave to it in the neighbourhood of any portion where marginal conditions are satisfied.*

But even the satisfaction of both (first and second order conditions) does not ensure the largest maximum welfare because even when marginal conditions (first and second order) are fulfilled, it may still be possible to move to a position where social welfare is greater. To attain the maximum social welfare, another set of conditions which are called by J.R. Hicks as the '*total conditions*' must also be satisfied. The total conditions state, "that if welfare is to be a maximum, it must be impossible to increase welfare by producing a product not otherwise produced or by using a factor not otherwise used."[5] If it is possible to increase welfare by such activities the optimum position is not determined by marginal conditions alone.

Therefore, welfare will be really maximum if the marginal as well as total conditions are satisfied. *But such a social optimum too is not a unique one. It is one of a large number of optima.* The whole analysis of conditions of Pareto optimality *assumes a given distribution of income*. With a change in the distribution of income Pareto optimality will be achieved with different output-mix of various products and different allocation of various factors among products. Thus, a new optimum will emerge due to redistribution of income and there are no criteria to judge whether the new optimum is better or worse than the previous social optimum. This can be known only with the help of some value judgements regarding income distribution which has been ruled out by the Pareto criterion.

A CRITICAL EVALUATION OF PARETO CRITERION AND PARETO OPTIMALITY

Pareto criterion and the concept of Pareto optimality and maximum social welfare based on it occupies a significant place in welfare economics. To judge the efficiency of an economic system, the notion of Pareto optimality has been used. It has also been used to bring out the gains of trading or exchange of goods between individuals. But even Pareto criterion which rules out comparing those changes in policies which make some worse off has been a subject of controversy and has been criticised on several grounds.

First, it has been alleged that *Pareto criterion is not completely free from value judgements.* The supporters of Pareto criterion claim that it provides us with an 'objective' criterion of efficiency. However, this has been contested. Against Pareto criterion it has been said that to say that a policy change which makes some better off without others being worse off increases social welfare is itself a value judgement. This is because *we recommend such changes which pass Pareto criterion.* The implication of this assertion will become obvious when the persons who gain as a result of policy change are the rich and those who remain where they were before are poor. Therefore, to say on the basis of Pareto criterion that whenever any policy change which, without harming anyone, benefits some people regardless of whoever they may be, increases social welfare is a value judgement which may not be accepted by all.

Second, an important limitation of Pareto criterion is that *it cannot be applied to judge the social desirability of those policy proposals which benefit some and harm others.* Such policy changes are quite rare which do not harm at least some individuals in the society. Thus, Pareto criterion is of limited applicability as it cannot be used to pronounce judgements on a majority of policy proposals which involve a conflict of preferences of two individuals. Thus, according to Prasanta K. Patnaik, "Pareto criterion fails seriously when it comes to comparing alternatives. *Whenever there is conflict of preferences of two individuals with respect to two alternatives, the criterion fails to rank those two alternatives no matter what the preferences of the rest of individuals in the society might be*".[6] To evaluate social desirability of those policy changes which benefit some and harm others, we need to make interpersonal comparison of utility which Pareto criterion refuses to do. Thus, "Pareto criterion works by sidestepping the crucial issue of inter-personal comparison and income distribution, that is, by dealing only with cases where no one is harmed so that the problem does not arise".[7]

Another shortcoming of Pareto criterion and notion of maximum social welfare based on it is that *it leaves a considerable amount of indeterminacy in the welfare analysis* since every point on the contract curve is Pareto-optimal. For instance, in Fig. 39.1, every point such as *P*, *Q*, *R*, *S* on the contract curve is Pareto-superior to any point such as *K* and *H* which lies outside the contract curve. Movement from one point on the contract curve to another as a result of change in economic policy, that is, through re-allocation of resources that makes one individual better off and the other worse off, that is, one gains at the expense of the other. This means that on the basis of Pareto criterion, social alternatives lying on the contract curve cannot be compared since with any movement on the contract curve one individual gains and the other loses, that is, it involves redistribution of income or welfare. Therefore, to compare various alternatives lying on the contract curve and to choose between them, inter-personal comparison and value judgements regarding proper distribution of income need to be made. However, Pareto refused to make value judgements and sought to put forward a value-free or objective criterion of welfare.

It, therefore, follows that on the basis of Pareto criterion where the change from an alternative lying outside the contract curve to an alternative on the contract curve is judged to increase social welfare but this cannot be said of the change from one position on the contract curve to another on it. But as there are infinite number of points on the contract curve all of which are Pareto optimal, no choice can be made out of them on the basis of Pareto criterion. To remove this indeterminacy and to choose among the alternatives lying on the contract curve one needs to make some additional value judgements beyond what is implied in the Pareto criterion. Henderson and Quandt hold a similar view when they assert, "*The analysis of welfare in terms of Pareto optimality leaves a considerable amount of indeterminacy in the solution : there are infinite number of points which are Pareto optimal.*" They further remark that, `The indeterminacy is the consequence of considering an increase in welfare to be unambiguously defined only if an improvement in one individual's position is not accompanied by a deterioration of the position of another. The indeterminacy can only be removed by further value judgements."[8]

Above all, *a chief drawback of Pareto-optimality analysis is that it accepts the prevailing income distribution and no attempt is made to find an optimal distribution of income*, since it is thought that there does not exist any objective, value-free and scientific way of finding optimal distribution of income. Thus, Pareto optimality analysis remains either silent or biased in favour of status quo on the issue of income distribution. Further, Pareto optimality analysis may lead to recommend the prevailing income distribution where a majority of the population lives on the subsistence level or below the poverty line while a few live a life of affluence. Thus, "Ultimately, the Paretian approach can be considered the welfare economists' instrument par excellence for the circumvention of the issue of income distribution."[9]

---

6. Prasanta K. Patnaik, Some Aspects of Welfare Economics, *The Indian Economic Journal*, Conference Number, 1974, p. 68C.
7. W. J. Baumol, *Economic Theory and Operations Analysis*, 4th edition, 1977 Prentice Hall, p.527.
8. J. M. Henderson, and R. E. Quandt, *Microeconomic Theory*, 2nd edition, 1971, p. 265.
9. Baumol, *op.cit*. p. 503.

It may also be mentioned that for any initial distribution of income (that is, for any given distribution of goods) between the individuals, there will be several Pareto optimal positions. Consider Fig. 39.1. Corresponding to point, *K*, the points on the segment *RS* on the contract curve *CC′* will all represent Pareto optimal positions. Likewise, corresponding to a given distribution of income (*i.e.* distribution of goods) as represented by point *H*, the points on the segment *PQ* of the contract curve *CC′* will be Pareto-optimal. Thus corresponding to a different distribution of income, there will be different Pareto optima. In the Paretian analysis there is no way of evaluating whether one pattern of income distribution is better than the other.

Prof. Amartya Sen's Critique of Pareto Optimality

Further, criticising Pareto criterion Prof. Amartya Sen has pointed out that the success that the criterion of Pareto optimality has achieved in judging the desirability of a social state or a policy change is very limited. To quote him, "A Social state is described as Pareto optimal if and only if no-one's utility can be raised without reducing the utility of someone else. This is a very limited kind of success and in itself may or may not guarantee much. *A state can be Pareto optimal with some people in extreme misery and others rolling in luxury, so long as the miserable cannot be made better off without cutting into the luxury of the rich*."[10] So, according to him, this is not a good and adequate criterion for juding social welfare.

Further, Prof. Sen has criticised Pareto optimality on the basis that *it identifies well-being with utility* and captures the efficiency aspects only of utility-based accounting. It may be noted that utility is interpreted in two ways, Firstly, it is said to mean '*happiness*'. Secondly, it is interpreted in the sense of '*desire-fulfilment*'. He is of the view that utility does not always reflect well-being. To quote him, "*To judge the well-being of a person exclusively in metric of happiness or desire-fulfilment has some obvious limitations*. These limitations are particularly damaging in the context of interpersonal comparison of well-being. Since the extent of happiness reflects what one would expect and how the social 'deal' seems in comparison with that."[11] He is of the view that *people living a life of great misfortune with little hope and opportunities may get more utility or happiness even from small gains*. But that should not be interpreted that there is a significant improvement in their well-being. The measure of utility in the sense of happiness may not reveal the true picture about the state of his deprivation. He thus writes The hopeless beggar, precarious landless labourers, the dominated housewife, the hardened unemployed or the over-exhausted coolie may all take pleasure in small mercies, and manage to suppress intense suffering for the necessity of continued survival, but it would be ethically deeply mistaken to attach correspondingly small value to the loss of their well-being because of their survival strategy."[12]

According to Prof. Sen, even in case of desire-fulfilment, the same problem arises, because "*the hopelessly deprived lack the courage to desire much, and their deprivations are muted and deadened in scale of desire-fulfilment*." The sum and subsistence of Sen's criticism is that the concept of utility used in for indjudging Pareto optimality whether it is interpreted in terms of happiness or desire-fulfilment is seriously inadequate and insufficient for judging a person's well-being. To quote him, well-being is ultimately a matter of valuation, and while happiness and fulfilment of desire may well be valuable for the person's well being, they cannot on their own or even together adequately reflect the value of well-being."[13]

---

10. Amartya Sen, *On Ethics and Economics*, Oxford University Press, 1990, pp. 31-32
11. *Ibid*
12. Amartya Sen, *op, cit., pp*. 45-46
13. *Ibid.*, p. 46

It is thus clear that welfare or well-being of individuals depends on a wide range of variables that those associated with utility which is derived from the consumption of goods and services and amount of leisure enjoyed. *Welfare or well-being also depends on such variables as political and environmental factors, personal and political freedom individuals enjoy, disposition of their neighbours*. For an adequate measure of well-being these variables cannot be ignored. "In comparing different economic systems or comparing different ways of organising a given economy, the possibility that some of these variables might be affected cannot be ignored. Thus, *a reorganisation that gives everyone more income and leisure might not improve the welfare of the community if at the same time it limits individual freedoms or requires the abandonement of cherished cultural traditions*."[14]

In the end, it may be pointed out that Pareto criterion is not altogether unless. It is useful in the sense that, "by throwing out the Pareto inoptimal alternatives, it reduces the range within which socially best alternatives are to be looked for, and therefore does serve as a useful first step. The trouble arises if one gets so fascinated with this first step that one does not try to go any further, but that can hardly be called a defect of Pareto criterion."[15] Moreover, as has been pointed out above, Pareto analysis has been used to bring out the gains from trading or exchange of goods between the two individuals.

PERFECTLY COMPETITIVE EQUILIBRIUM AND PARETO OPTIMALITY

In our above analysis we have explained the various marginal conditions of attaining Pareto optimality or, in other words, optimum allocation of resources. It has been claimed by several economists that perfect competition is an ideal market form which ensures the attainment of Pareto optimality or maximum social welfare as it fulfils all the marginal conditions required for the purpose. In what follows we shall show how perfectly competitive equilibrium satisfies all the marginal conditions required for the achievement of Pareto optimum. We shall further explain what are the major obstacles in the way of maximising social welfare or achieving Pareto optimality.

Perfect Competition and Optimal Distribution of Goods or Efficiency in Exchange

The condition for Pareto optimality with regard to the distribution of goods among consumers requires that the marginal rate of substitution (*MRS*) between any two goods, say *X* and *Y*, must be the same for any pair of consumers. Let *A* and *B* be the two consumers between whom two goods *X* and *Y* are to be distributed. Under perfect competition prices of all goods are given and same for every consumer. It is also assumed that consumers try to maximise their satisfaction subject to their budget constraint. Now, given the prices of two goods, consumer A will maximise his satisfaction when he is buying the two goods *X* and *Y* in such amounts that:

$$MRS^A_{XY} = \frac{P_X}{P_Y} \quad \ldots (i)$$

Likewise, the consumer *B* will also be in equilibrium (maximise his satisfaction) when he is purchasing and consuming the two goods *X* and *Y* in such amounts that:

$$MRS^B_{XY} = \frac{P_X}{P_Y} \quad \ldots (ii)$$

Since this is essential condition of perfect competition that prices of goods are the same or uniform for all consumers, the price ratio of the two goods $\left(\frac{P_x}{P_y}\right)$ in equations (*i*) and (*ii*) above will

14. Else, P. Curwen, *Principles of Microeconomics*, Unwin, Hyman, London, 1990, p. 324.
15. Prasanta K. Patnaik, *Op.cit.*

be the same for consumers *A* and *B*. It, therefore, follows from equations (*i*) and (*ii*) above that under conditions of perfect competition marginal rate of substitution between two goods *X* and *Y* will be equal for the two consumers. That is,

$$MRS_{XY}^{A} = MRS_{XY}^{B}$$

This result will hold good between any pair of goods for any pair of consumers.

Perfect Competition and Optimal Allocation of Factors: Production Efficiency

The second marginal condition for Pareto optimality relates to the optimal allocation of factors in the production of various goods. This condition requires that for the optimal allocation of factors marginal rate of technical substitution (*MRTS*) between any two factors, say labour and capital, must be the same in the production of any pair of products. This condition is also satisfied by perfect competition. For a producer working under perfect competition prices of factors he employs are given and constant and he is in equilibrium (that is, minimises his cost for a given level of output) at the combination of factors where the given isoquant is tangent to an iso-cost line. As is well known, the slope of the isoquant represents marginal rate of technical substitution between the two factors and the slope of the iso-cost line measures the ratio of the prices of two factors. Thus, under perfect competition, a cost-minimising producer producing goods will equate *MRTS* between labour and capital with the price ratio of these two factors. Thus under perfect competition :

$$MRTS_{LK}^{X} = \frac{P_L}{P_K} \qquad \ldots (i)$$

where P_L and P_K are the prices of labour and capital respectively and and $MRTS_{LK}^{x}$ is the marginal rate of technical substitution between labour and capital in the production of good *X*. Similarly, producer *B* producing good *Y* and working under perfect competition will also equate his marginal rate of technical substitution between the two factors with their price ratios. Thus

$$MRTS^{Y}{}_{LK} = \frac{P_L}{P_K} \qquad \ldots (ii)$$

Since, under perfect competition, prices of factors are the same for all the producers, each producer will adjust the use of factors in such a way that his marginal rate of technical substitution (*MRTS*) between labour and capital in the production of goods is equal to the same factor price ratio. In other words, $\left(\frac{P_L}{P_K}\right)$ will be the same for all of them and to this $MRTS_{LK}$ of the producers will be made equal. It, therefore, follows from (*i*) and (*ii*) above that under perfect competition :

$$MRTS_{LK}^{X} = MRTS_{LK}^{Y}$$

We thus see that perfect competition ensures optimal allocation of resources as between different firms using these resources for production of commodities.

Perfect Competition and Optimum Direction of Production (Product-mix): Allocative Economic Efficiency

The most important condition for the attainment of Pareto optimum is one which refers to the optimum direction or composition of production. In other words, this condition requires how much amounts of different goods should be produced and resources allocated accordingly. This refers to the general condition for optimum allocation of resources which has also been called the condition for *General Economic Efficiency* and *General Pareto Optimum*. This condition states that marginal rate of substitution between any two commodities for any consumer should be the same as the marginal rate of transformation for the community between these two commodities.

Under conditions of perfect competition, each firm to be in equilibrium produces so much output of a commodity that its marginal cost is equal to the price of the commodity. Thus, for firms in perfect competition, $MC_x = P_x$, $MC_y = P_y$, where MC_z and MC_y are marginal costs of production of commodities X and Y respectively and P_x and P_y are prices of commodities X and Y. Therefore, it follows that firms working in perfect competition will be in equilibrium when they are producing commodities in such quantities that

$$\frac{MC_x}{MC_y} = \frac{P_x}{P_y} \qquad \ldots (i)$$

The ratio of marginal costs of two commodities represents the marginal rate of transformation between them. Therefore, for firms producing under perfect competition :

$$MRT_{xy} = \frac{MC_x}{MC_y} = \frac{P_x}{P_y} \qquad \ldots (ii)$$

When there prevails perfect competition on the buying side, each consumer maximises his satisfaction and is in equilibrium at the point where the given budget line is tangent to his indifference curve. In other words, each consumer is in equilibrium when :

$$MRS_{xy} = \frac{P_x}{P_y}$$

Since, under perfect competition, *the ratio of prices of two commodities* $\left(\frac{P_x}{P_y}\right)$ *is the same for a consumers and producers* it follows from (i) and (ii) above that

$$MRS_{xy} = MRT_{xy}$$

Likewise, this will hold good for any other pair of commodities. Thus, perfect competition satisfies the marginal condition required for the Pareto optimal composition or direction of production.

We thus see that all first order marginal conditions required for the attainment of Pareto-optimality or maximum social welfare are fulfilled under perfect competition. It is in this sense that perfect competition represents economic optimum from the viewpoint of social welfare.

Fundamental Theorem of Welfare Economics and Its Critique

It has been shown above that *perfectly competitive equilibrium is Pareto optimal*. This is called fundamental theorem of welfare economics. This is also called the **invisible hand theorem**. The belief that competitive market economy provides an efficient means of allocating scarce resources goes back to Adam Smith who argued in his famous book "*Wealth of Nations*" that *individuals who pursue their self-interest, they operating through market promote the welfare of others and welfare of the society as a whole*. Thus individual consumers seek to maximise their own satisfaction and producers pursue to maximise their own profits. Even though promoting the interests of the society as a whole is not a part of their intention but they are led by the invisible forces of market system to promote the interest of the society as a whole.

We have proved above that perfect competition in the market satisfies Pareto's optimum condition of exchange, that is (1) MRS_{xy} of any pair of individuals under it is the same, (2) Pareto's optimum condition of production, that is, $MRTS_{LK}$ of any pair of firms using the two factors for producing products under it is the same, and (3) Pareto's condition for optimal direction of production (*i.e.* optimum product mix), namely, MRT_{xy} in production equals MRS_{XY} of consumers (For details regarding proofs of these and other conditions see the above section).

However, the conditions under which a perfect competitive market system achieves Pareto-optimality or what is also called *economic efficiency* are quite restrictive. One important condition for the achievement of Pareto optimality is that the *general competitive equilibrium exists*. This requires that all markets concerned are in equilibrium simultaneously. If one market is not in equilibrium for some reason, the condition for Pareto optimality would be violated which would leave unutilised the opportunities for Pareto improvement.

The second important requirement for the validity of the fundamental theorem of welfare economics is that *second order conditions for equilibrium must be fulfilled*. This implies that consumer preferences (or indifference curves) are convex and also producers' production sets (i.e. isoquants) are convex. This implies that consumers' marginal rate of substitution and producers' marginal rate of technical substitution ($MRTS_{LK}$) must be diminishing at or near the equilibrium point. Further, the second order condition also requires *that production transformation curves must be concave in the relevant region*. The existence of perfect competition does not guarantee that these second order conditions will be fulfilled. In this context it may be noted that many areas of production there prevail *increasing returns* to scale. In case of increasing returns to scale equilibrium of competitive firms is not possible. This would ensure that general competitive equilibrium will not exist which will lead to the violation of the condition.

The third condition required for the fulfilment of fundamental theorem of welfare economics is that externalities in production and externalities in consumption do not exist. The assumption of the absence of production externalities implies that consumption production choices by any firm do not affect the production possibilities of other firms. Similarly, the assumption of the absence of consumption externalities implies that consumption decisions of a consumer do not affect the consumption possibilities of the other consumers. In case these externalities in production and consumption exist, the competitive equilibrium will not achieve Pareto-optimality from the social point of view. How externalities prevent the achievement of Pareto optimality and lead to loss of social welfare will be explained in detail in a later chapter.

Lastly, it is important to note that the competitive equilibrium under the conditions mentioned above ensures Pareto optimality or efficiency in use and allocation of resources. It has nothing to do with desirable distribution of welfare. In other words, *it ensures Pareto efficiency not justice*. Pareto optimality analysis assumes the initial factor endowment as given. Initial inequalities in the ownership of assets or factor-endowments causes inequalities which leads to non-optimal distribution of goods and services and therefore loss of social welfare.

Further, it may be noted that perfectly competitive equilibrium achieves Pareto optimality when the second order conditions of equilibrium are satisfied. These second order conditions require that at or near the equilibrium point indifference curves of consumers and isoquants of producers are convex and production transformation curves are concave to the origin. However, *perfect competition does not guarantee that second order conditions required for the achievement of Paret optimality will also be fulfilled*. Besides, when externalities, that is, external economies and diseconomies in production and consumption are present, perfect competition will not lead to Pareto optimality. When external economies and diseconomies either in production or consumption are present, social marginal cost (or benefit) will diverge from private marginal cost (benefit). Now, *perfect competition only ensures the equality of price of a product with the private marginal cost and not with the equality of price with the social marginal cost*. Thus, the existence of externalities will prevent the achievement of Pareto optimality or efficient allocation of resources even when perfect competition prevails in the economy. How the occurrence of externalities obstructs the achievement of maximum social welfare will be explained at length in a later chapter.

Further, even if the above two factors, namely, non-fulfilment of the second order conditions and the existence of externalities are not actually found, the perfect competition will not lead to economic efficiency or Pareto optimality (that is, optimum allocation of resources) if the given distribution of income is not optimal from the viewpoint of social welfare. As mentioned before, analysis of Pareto optimality accepts the prevailing income distribution which may be far from the optimum distribution. There is nothing in perfect competition which ensures optimum distribution of income. That the distribution of income is an important factor determining social welfare is now widely recognised by the economists.

Finally, there is another factor which prevents the achievement of Pareto-optimality or maximum social welfare even when perfect competition prevails in the economy. This factor relates to the employment or utilisation of available resources. *Pareto-optimality will not be attained if the available resources are not fully employed or utilised.* This is because if some of the available resources are unemployed or unutilised, then the society could produce more of a commodity by employing the unemployed resources and therefore without cutting down the production of any other commodity. Now, when it is possible to produce more of a commodity without reduction in the output of another, then the society could make either all individuals better off or at least some better off without making *others worse off.*

If the economy is operating at a point inside its transformation curve (*i.e.* production possibility curve), it would not then he employing or utilising its resources fully and it would then be possible to increase the output of both commodities (represented on the two axes) or to increase the output of one commodity without reducing the output of the other. Thus, any position of working inside the transformation or production possibility curve cannot be a position of Pareto optimality. Therefore, for the economy to achieve Pareto-optimality it must work at some point on the given production possibility curve implying full employment of resources. Full employment of available resources is therefore a necessary condition for the attainment of Pareto optimality. But it is important to note that perfect competition does not guarantee full employment of resources and therefore does not necessarily lead to the achievement of Pareto-optimality.

It follows from above that *perfect competition though a necessary condition is not a sufficient condition for Pareto-optimality*. Therefore, that a free enterprise economy characterised by perfect competition ensures efficient allocation of resources or maximum social welfare cannot be accepted without some qualifications. And these qualifications are : (1) the second order conditions are satisfied, (2) the externalities in production and consumption are absent, (3) prevailing distribution of income is optimal from the social point of view, and (4) available resources are fully employed. It may also be noted that in present-day free enterprise capitalist economies perfect competition is an exception rather than the rule. In the present-day capitalist economies, it is monopolies, oligopolies and monopolistic competition which largely prevail and, as shall be demonstrated in the next chapter, these market forms serve as a great obstacle for the achievement of Pareto-optimality or optimum allocation of resources.

QUESTIONS FOR REVIEW

1. Define economic efficiency? Explain Pareto's three conditions for achievement of economic efficiency
2. State Pareto's Criterion of Social welfare. Explain Pareto's Criterion of social welfare using Edgeworth Box diagram.
3. Explain the concept of utility possibility curve. How would you explain Pareto's. criterion of social welfare with utility possibility curve?
4. What is Pareto Optimality? State the conditions of Pareto optimality regarding (1) distribution of goods between individuals, (2) allocation of resources between firms and (3) direction of production.
5. Pareto Critierion does not give us a sufficient basis for ordering states from the viewpoint of social welfare. Examine critically *[B.A. (Hons) D.U. 1984]*
6. State and explain Pareto optimality criterion. How far is it useful in suggesting policies that will make the community better off? *[B.A. (Hons) D.U. 1986]*

7. Perfect Competition will tend to yield an optimal allocation of resources. Discuss. *[B.A. (Hons.) D.U. 1986]*

8. State and explain diagrammatically the marginal conditions for a Pareto-optimal distribution of goods and resources in a two factors, two commodities, two consumers model. *[B.A. (Hons.) D.U. 1987]*

9. State the conditions of Pareto optimality. How are these conditions affected in the presence of (*i*) social costs and (*ii*) public goods. Explain *[B.A. (Hons.) D.U. 1988, 1999, 1991]*

   [**Hints** (*i*) *Social Costs* are brought about when productive activities of private firms generate external diseconomics and external economies. In case of the emergence of external diseconomies, marginal social cost will be higher than marginal private cost which will lead to more output of a product to be produced under conditions of perfect competition than is socially desirable from the viewpoint of Pareto Optimality. (For details see next chapter)

   (*ii*)*Public goods* will also not be produced in a competitive market conditions in optimal quantities as required by Pareto optimality. Public good are not rival in consumption. A person using a public good (say enjoying a walk in a public park) does not affect the utility derived by other persons. Similarly, in case of public goods it is not possible to exclude from their consumption those persons who do not pay for it. If the production of the public goods is left to be produced by the market driven private firms, they will not be produced in Paretian optimal quantities (for details see the next chapter).

10. "Pareto criterion works by sidestepping the crucial issue of interpersonal comparison and income distribution" (Baumol) Discuss.

11. Pareto's analysis leaves a considerable amount of indeterminacy in maximisation of social welfare. Discuss

12. Pareto optimality is necessary but not a sufficient condition for social welfare maximisation. Discuss. *[B.A. (Hons.) (D.U.), 1991]*

13. The collapse of the communist regime in Eastern Europe and Soviet Union can be attributed to economic inefficiencies. Discuss. *[B.A. (Hons.) D.U. 1992]*

14. Show that a general competitive equilibrium is also Pareto optimal. *[B.A. (Hons.) (D.U.) 1994]*

15. There is only one Pareto-optimal allocation in an economy. True or false ? Explain *[B.A. (Hons.) D.U. 1998]*

16. When perfectly competitive markets are in equilibrium, all the three conditions for economic efficiency are also satisfied *[B.A. (Hons.) D.U. 1999]*

17. Perfect competition is an ideal market form as it ensures the attainment of Pareto optimality. Examine critically.

CHAPTER 40
MARKET FAILURES, EXTERNALITIES AND PUBLIC GOODS

We have explained in the previous chapter conditions for the achievement of pareto optimality or maximum social welfare. We have also briefly discussed that with some qualifications perfect competition ensures the achievement of Pareto optimality or maximum social welfare. Now, the pertinent question is what are the factors which hinder the attainment of pareto-optimality and maximum social welfare. The main obstacles are :

1. the existence of monopoly or imperfect competition;
2. the presence of externalities, *i.e.* external economies and diseconomies in production and consumption; and
3. the consumption of Public Goods, and

We shall explain below these obstacles at some length.

MONOPOLY (MARKET POWER) AS AN OBSTACLE TO THE ATTAINMENT OF PARETO OPTIMALITY

An important complaint against monopoly (and as a matter of fact against all forms of imperfect competition) is that it causes misallocation of productive resources and thus hinders the achievement of maximum social welfare. The crucial condition required for Pareto optimality is that marginal rate of transformation (*MRT*) of the community between any two commodities should be equal to the marginal rate of substitution (*MRS*) between these commodities of every consumer. We saw above that perfect competition satisfies this condition of Pareto optimality. But under monopoly (or any other form of imperfect competition such as oligopoly or monopolistic competition) the marginal rate of transformation of the community between two commodities is not equal to the marginal rate of substitution between them of the consumers. Consequently, monopoly does not ensure optimum allocation of resources and serves as an obstacle to the attainment of maximum social welfare. The reason for this is that a monopolist does not equate price of his product with marginal cost of production; he restricts output and charges higher price than marginal cost.

Assume that there are two commodities, X and Y, and further that the commodity X is being produced under conditions of monopoly whereas the commodity Y is being produced under conditions of perfect competition. Since the commodity X is being produced under conditions of monopoly, the price (P_x) of commodity X will be grater than marginal cost (MCx) of its production. Thus $P_x > MC_x$. But since the commodity Y is being produced under conditions of perfect competition, price (Py) of commodity Y will be equal to the marginal cost (MC_y) of production. That is, $P_y = MC_y$. It therefore follows that :

$$\frac{MC_x}{MC_y} < \frac{P_x}{P_y}$$

Since the ratio of marginal costs of two commodities represents the marginal rate of transformation (MRT_{xy}) between them, therefore,

$$MRT_{xy} < \frac{P_x}{P_y}$$

But the consumers in order to be in equilibrium will equate their marginal rate of substitution between two commodities (MRS_{xy}) with the price ratio of the two commodities $\left(\frac{P_x}{P_y}\right)$. This is because each individual consumer will take the prices of commodities as given and constant for him. Thus, for consumers,

$$MRS_{xy} = \frac{P_x}{P_y} \quad \ldots (ii)$$

From (*i*) and (*ii*) above it follows that under conditions of monopoly in the production of *X*

$$MRT_{xy} < MRS_{xy}$$

or

$$MRS_{xy} > MRT_{xy}$$

That is, when monopoly exists in the production of a commodity marginal rate of substitution between commodities will be greater than the marginal rate of transformation. In other words, consumers would like the commodity under monopoly production to be produced more but monopolists would not be producing the desired quantity of the commodity and will, therefore, be causing loss of satisfaction and misallocation of resources.

That monopoly causes loss of welfare and misallocation of resources will become very clear by considering Fig. 40.1. It will be seen from this figure that the transformation curve of the community *AB* is tangent to the community indifference curve IC_3 at point *E*. Therefore, at point *E* marginal rate of transformation (MRT_{xy}) of the community between two commodities is equal to the marginal rate substitution (MRS_{xy}) of the community. Thus *E* represents maximum possible level of social welfare and the combination of two commodities being produced (*i.e.* *OM* of *X* and *ON* of *Y*) represents optimum allocation of resources.

But when the commodity *X* is being produced under conditions of monopoly, the equilibrium

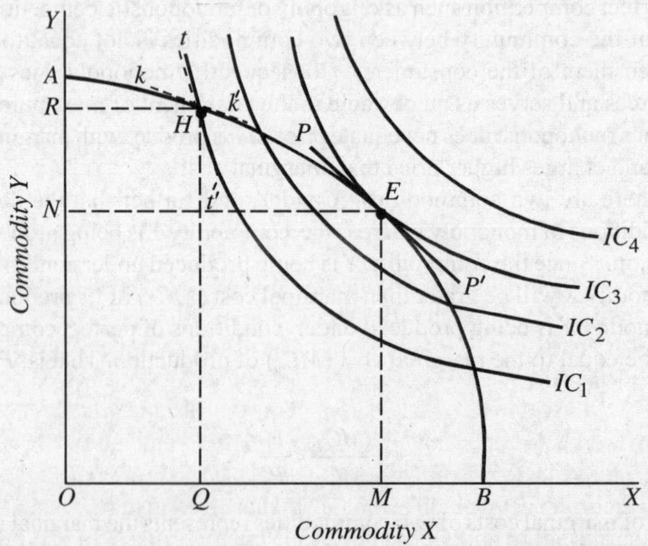

Fig. 40.1. *Monopoly as an Obstacle to Maximisation of Social Welfare.*

will not be at point *E*, but instead it will be at point *H*. This is because, under monopoly, producers would be equating marginal rate of transformation (*MRT*) or ratio of marginal costs $\left(\frac{MC_x}{MC_y}\right)$ with the ratio of marginal revenues $\left(\frac{MR_x}{MR_y}\right)$ and not with the ratio of prices of two goods $\left(\frac{P_x}{P_y}\right)$. Since consumers would be equating marginal rate of substitution (MRS_{xy}) with the price ratio of two goods, the marginal rate of transformation in the equilibrium position at point *H* will not be equal to the marginal rate of substitution. This is quite obvious from Fig. 40.1 where at point *H* transformation curve *AB* and consumers' indifference curve IC_1, are intersecting each other. This implies that slopes of transformation curve at point *H*, which indicates marginal rate of transformation, and the slope of consumers' indifference curve IC_1, which indicates marginal rate of substitution, are not the same. It will be observed from Fig. 40.1 that at point *H* marginal rate of substitution between two goods (*MRSxy*) is greater than the marginal rate of transformation (*MRTxy*) between them as tangent *tt′*, *drawn at point H* on the indiffrence curve IC_1 is greater than the slope of the langent *kk′* drawn at point *H* on the transformation curve *AB*. This means that consumers' preferences require that good *X* should be produced more but because of the existence of monopoly in the production of commodity *X*, it is not being produced equal to the desired quantity. As a result, the level of satisfaction or welfare of the consuming community is at a lower level than possible under the given production conditions. Consuming community's satisfaction will be greater at point *E* which lies on indifference IC_3 but under conditions of monopoly in the production of *X*, equilibrium is at point *H* which lies on the lower indifference curve IC_1.

Thus monopoly causes loss of satisfaction or welfare. This loss of satisfaction or welfare is due to the fact that monopoly is not optimally allocating its resources between the production of two commodities according to the consumers' preferences. Given the transformation curve and consuming community's indifference map, the optimal production pattern is represented by the point *E* where *OM* amount of commodity *X* and *ON* amount of commodity *Y* are being produced. But under conditions of monopoly in the production of *X*, the equilibrium is established at point *H* where smaller quantity *OQ* of commodity *X* and larger quantity *OR* of commodity *Y* are being produced. Thus *monopoly has caused misallocation of resources.*

EXTERNALITIES AND MARKET FAILURE

The existence of externalities is an important factor which prevents the achievement of Pareto-optimality (or economic efficiency) and thus cause market failure even when perfect competition prevails. *Externalities refer to the beneficial and detrimental effects of an economic unit (a firm, a consumer or an industry) on others*. The beneficial externalities created by a firm or consumer for others are also known as *external economies* and detrimental externalities imposed on others by a productive firm or a consumer are known as *external diseconomies*. To be more precise, *when an economic unit creates benefits for others for which he does not receive any payment, there exist beneficial externalities (or external economies). On the other hand, detiremental externalities or external diseconomies occur when an economic unit inflicts costs on others for which he does not make any payment*.

It is noteworthy that the term "externalities" covers both the external economies and external diseconomies. When for a productive firm there exist external economies, that is, beneficial external effects, then the private marginal cost of the firm will be higher than the social marginal cost, since the firm will not take into account benefits external to it (*i.e.* benefits created for others). And the market price fixed on the basis of private marginal cost will not reflect the social marginal cost which will be lower when external economies occur. Similarly, when with the expansion of a firm external diseconomies (detrimental externalities) occur, then the private marginal cost will be lower than the social marginal cost, since the firm will not take into account the harms it causes to others by its activity. Thus when a firm expands its production, it increases the smoke it emits and thereby harms other's health who live in the surrounding areas. The others may suffer from respiratory diseases and have to incur extra medical expenditure. Besides, living of the people become quite unpleasant in the surrounding areas due to cloud of noxious fumes emitted and huge noise created. Therefore, *when a firm's activities create detrimental externalities its private marginal cost would be lower than* the social marginal cost as external costs it imposes on others are not counted in calculating its private marginal cost. Thus, when external diseconomies occur, price fixed on the basis of private marginal cost will be lower than that determined on the basis of social cost.

It follows from above that *in the absence of externalities*, all costs incurred and all benefits received by producers and consumers will be reflected in market prices and that there will not be any divergence between private and social costs (or benefits). But, *when externalities (external economies and diseconomies) occur, market prices determined on the basis of private costs and benefits will not truly reflect social costs and therefore divergence is caused between private and social costs (or benefits)*.

It follows from above that externalities arise primarily due to the fact that the effects regarding costs, output, employment, labour skills, technological capabilities of the activities of a producer or consumer on others or society as a whole are not reflected in market prices and therefore *market prices do not truly reflect social costs*. We shall give below a few examples of externalities, that is, external economies and diseconomies in production and consumption. The basic idea behind the belief that a competitive price system is optimal is based on the fact that a producer benefits himself only by benefitting the society because he makes available certain goods and services to the society. In other words, by promoting his own interests he promotes the interests of society as well. But there are so many cases in production and consumption "when members of the economy do things which benefit others in such a way that they can receive no payment in return or where their actions are detrimental to others and involve no commensurate cost to themselves."[1] Thus, due to externalities there arises the divergence between social and private costs and benefits.

Beneficial Externalities (External Economies) in Production

As the firm expands its scale of production, it becomes possible for the firm to produce a unit of product at a relatively lower cost due to internal economies of large-scale production. On the other hand, external economies (beneficial externalities) occur when the expansion of a firm's output creates benefits, part of which goes to others. A firm may create external benefits for others in two ways : (*a*) By expanding its production, the firm may render a direct service to others such as training the labourers by its manpower training programme and thus benefit the other firms by making available skilled and trained labourers when they have to pay no cost or only nominal cost. (*b*) By expanding its production a firm may make the supply of some inputs cheaper for all the firms in the industry.

For example, an expansion in the production of an engineering firm may increase the demand for steel. And if the steel production is subject to internal economies of large-scale production, the expansion of steel industry following the increase in its demand will lower its cost and price. Another example of external economies (beneficial externalities) is provided by the construction of a bridge or a highway which reduces transport cost and increases the land values in the neighbouring areas. Still another example of external economies of production is provided by the pumping of water from a mine. If a firm pumps out water from mine '*A*', it will lower the cost of pumping water from mine '*B*' owned by another firm. Similarly, bees of producers of honey create benefits for the owners of nearby orange groves, for the bees help pollinate oranges in the groves. On the other hand, orange

1. W.J. Baumol, *op. cit.* p.517

groves create external economies for the honey producers since the orange groves provide nectar for the bees producing honey.

In all these cases, a firm incurs cost in its expansion but the benefits arising out of it are also reaped by others who pay no price for them.

External Diseconomies or Detrimental Externalities in Production. Let us explain some external diseconomies (harmful externalities) of production. There are a good number of external diseconomies which may be created by the productive activity of a firm. *The pollution of air* by the factories through emitting smoke and the wastes of factories poured into streams or ocean create health hazard for men, especially those who live in the surrounding areas. For these external harms caused to the other members of the society, the firms are not required to pay any price. A factory owner pays nothing to the residents of the neighbouring colony who happen to be the victims of pollution by the factory. Another example of external diseconomies (detrimental externalities) is provided by a firm or industry which has to keep more trucks on the road to do its business. This will overcrowd the road which will increase the transportation costs of other firms or industries which had to carry their own goods by trucks. The expanding firm or industry does not pay any price for the higher transport cost incurred by others.

Externalities (External Economies and Diseconomies) in Consumption

External economies in consumption arise when the consumption of a person creates beneficial effects on others. Many examples of external economies can be given. For example, the satisfaction of a telephone owner increases with the increase in the number of telephone owners because he can now contact a larger number of persons on telephone. Likewise, if a person maintains a beautiful garden or lawn, he not only increases his own satisfaction but also that of others, especially his neighbours who also enjoy the look of his garden or lawn. Similarly, when a person maintains his car in such a way that it is quite safe to drive it and also does not emit any smoke, it will also improve others' safety and health and therefore welfare. In this category, we may also include the expenditure incurred by the parents on educating their sons. This will not only benefit them and their sons but also other members of the society. This is because the education makes a person civilised and a better citizen and therefore whoever comes in contact with him, derives satisfaction from him. Thus in the presence of external economies in consumption the social utility exceeds the private utility and therefore divergence between social and private benefits is caused when external economies in consumption prevail.

On the contrary, external diseconomies of consumption occur when a person's consumption creates unfavourable impact on other consumers. A good example of it is provided by the *conspicuous consumption* of a person who through demonstration effect causes a lot of dissatisfaction to friends and neighbourers who now feel themselves inferior to him. Likewise, when a person purchases candy bars for his children, it will make his neighbour's children unhappy because their parents cannot afford. Likewise, loud music played by your neighbour may disturb you and cause a lot of dissatisfaction. Also falls in this category the purchasing of a new Maruti Esteem by your friend because now your Maruti car 800 in your own eyes becomes outdated. More examples of such external diseconomies of consumption can be given.

HOW EXTERNALITIES CAUSE MARKET FAILURE

The existence of external economies and diseconomies (i.e., beneficial and detrimental externalities) explained above plays a significant role in determining the activities of production and consumption in the economy. A pertinent question is how these externalities can cause market failure and thereby act as on obstacle to the attainment of Pereto optimality or maximum social welfare. When externalities in production and consumption prevail and as a result divergence is caused between private and social costs and benefits, the economy guided by the market prices alone, even in the presence of perfect competition, will fail to achieve optimum allocation of resources (or, in other words, maximum social welfare). When external economies (beneficial externalities) of production occur private marginal cost will be greater than the social marginal cost and when external diseconomies (detrimental externalities) in production are present, private marginal cost will be lower than social marginal cost.

Under these circumstances, therefore, a firm which creates external benefits for others will not produce *its product* to the extent social interest requires. This is because equating price with the private marginal cost, which is higher than the social marginal cost, will result in *under-production of the product*. Thus in this case of the existence of *external economies in production*, output of the product determined on the basis of private marginal cost will be less than the socially optimal level of output. This is illustrated in Fig. 40.2 where *SS* represents the supply curve for the product of the industry which has been obtained by summing up the private marginal cost curves of firms. Due to the existence of external economies (beneficial externalities), social marginal cost will be smaller than the private marginal costs. Therefore, the supply curve *S′ S′* (dotted) of the product reflecting social costs will be lower than the supply curve *SS* based on private marginal costs. The supply curve reflecting social cost is lower because it takes into account external economies generated by the production in the industry, while private cost does not take into account these external economies. It will be seen from the Fig. 40.2 that the given demand curve and the supply curve *SS*, based upon the private cost of production, intersect at point *E* and thus determine *OQ* as the actual amount of the product produced. But the socially optimum output is *OM* at which the supply curve *S′S′* reflecting social cost intersects the given demand curve. It is thus evident that the product is being produced in smaller quantity than the socially optimum output *OM*. Thus, *the existence of external economies (beneficial externalities) results in under-production and loss of social welfare equal to the area EKT*. Thus this represents the case of market failure.

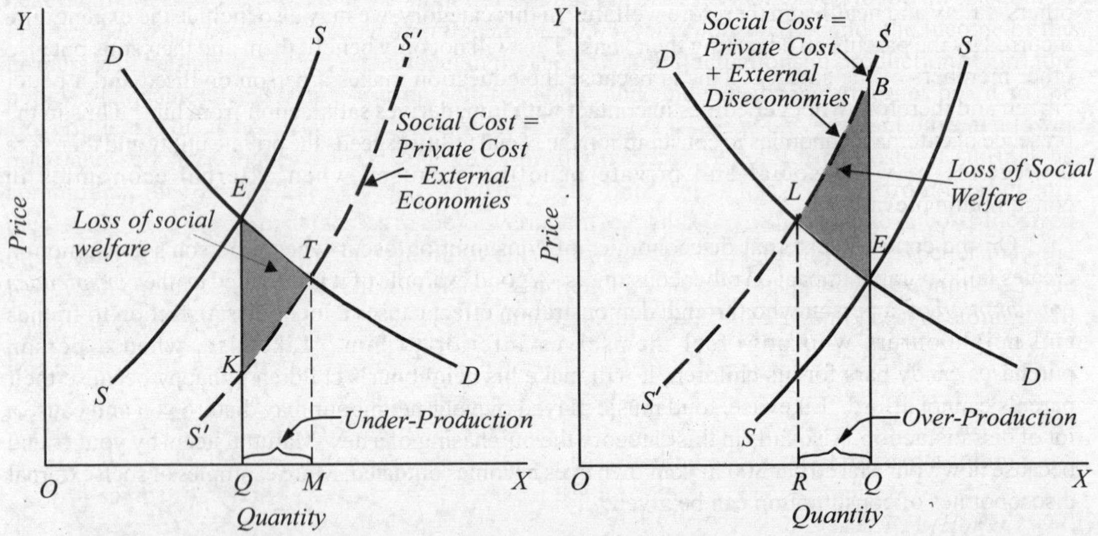

Fig. 40.2. *Under-Production in Case of External Economies (i.e., Beneficial Externalities)*

Fig. 40.3. *Over-Production in Case of External Diseconomies (i.e., Detrimental Externalities)*

On the other hand, when there exist external diseconomies in production, private marginal cost will be lower than the social marginal cost, since the former will not take into account costs or harms imposed on others. Therefore, *when external diseconomies are present, equating price with marginal cost will result in over-production of the product*, that is, more than socially optimum output will be produced. This case of market failure is illustrated in Fig. 40.3. It will be seen that the supply curve *SS* based on private marginal costs intersects the demand curve at point *E* and thus determines *OQ* amount of output. Supply *S′ S′*) (dotted) which takes into account external diseconomies and therefore reflects social cost lies at a higher level and intersects the demand curve at point *L* and therefore socially optimum output will be *OR*. Thus, it follows *when external diseconomies (detrimental externalities) are present, equating price with private marginal cost will result in over-production of the product, that is, more than socially optimum output will be produced and will cause a loss of social welfare equal to the area ELB*.

When there are external economies in consumption, then the demand curve for the product determined on the basis of private marginal utility will be lower than that based on social marginal utility, because the former will fail to reflect the external economies in consumption being generated. Therefore, in this case too, output determined on the basis of private marginal utility and demand will result in lower output than the socially optimum level. On the other hand, when there exist external diseconomies in consumption the private marginal utility will be higher than the social marginal utility, since the former will not take into account the external diseconomies. As a result, when external diseconomies in consumption are present, the output determined on the basis of private marginal utility (benefit) will be more than the socially optimum level.

Government Intervention and Externalities

Where there are a good amount of external benefits or external costs, the Government intervenes to take appropriate measures to promote social well-being. The imposition of taxes and provision of subsidies are the two important measures that Government can take to tackle the problems created by the presence of externalities. When there is excess production by firms due to the *creation of detrimental externalities* by their productive activity the Government can impose per unit tax so that private marginal cost (*MPC*) curve inclusive of per unit tax shifts above to the level of social marginal cost (*SMC*) so that equilibrium of the firms is at optimum level of output. Consider Fig. 40.3 with detrimental external effects the industry produces *OQ* output at which supply curve *SS* based on private marginal costs of the firms cuts demand curve *DD* at point *E* and is producing *RQ* more than the optimum level. If the Government imposes per unit tax equal to *EB*, private marginal cost (*PMC*) curves of the firm shown by the supply curve *SS* with shift up and coincide with social marginal cost curve (*SMC*) shown by the supply curve *S′S′* (dotted). With this tax per unit, the industry will be in equilibrium at point *L* where *S′S′* curve cuts the demand curve *DD* of the product and produces optimum output *OR*. In this way taxation solves the problem of over-production in case of detrimental externalities.

Provision of Subsidies

On the other hand, to promote social well-being the Government provides subsidies to those activities which are believed to generate external benefits. Thus education is subsidised by the Government not only because it creates equal opportunities for all people of a country but also because it generates beneficial externalities. For example, it has been found that educated people commit less crimes, an external benefit so that the Government has to spend less on prevention of crime. Besides, the academic research which is a by-product of the education system, benefits all people when it discovers useful things for them and also makes significant contribution to economic growth. If education is provided by profit-making enterprises, its private marginal cost will be greater than social marginal cost and it will be produced less than the optimal level as shown in Fig. 40.2 where without any subsidy *OQ* level of output is produced which is less than the optimal level *OM*. If Government provides subsidy equal to *EK* per unit the supply curve based on private marginal cost curves will shift below to the level of *S′S′* (dotted) so that the industry now produces optimal output *OM* at which supply curve based on *SMCs* cuts demand curve *DD*. Thus, with provision of subsidies the Government can correct market failure caused by beneficial externalities.

PUBLIC GOODS AND MARKET FAILURE

The existence of public goods provides us another important source of market failure. It should be noted that public goods though highly valuable from social viewpoint will, for reasons explained below, cannot be produced by the private sector firms driven by profit motive. It is due to the possession of certain properties that some goods are called public goods and therefore cannot be produced profitability by the private sector. It is thus very nature of some goods that makes it difficult, if not impossible, for the markets to achieve Pareto optimality or economic efficiency. Two essential characteristics of public goods are that they are *non-rival* and *non-exclusive* in consumption. Let us explain these characteristics of public goods in some details.

Non-Rivalry in Consumption

In order to explain what are non-rival goods, it is better to know first what are *rival goods* which are also called *private goods*. Private goods are rival goods in the sense that consumption of them by an individual reduces their availability for others to consume them. For example, if you use a gallon of petrol, *that, much less* of it will be available for others to use. Thus the consumption of a private good depletes the supply of it for others consumption. A rival or private good is one of which when one unit is consumed by an individual, that very unit cannot be consumed by another. For example, if Rekha consumes an apple, any other person, say Karishma cannot eat the same very apple. Of course, Karishma can get another apple for her consumption from the market by paying a price for it. Similarly, if Amit drinks Pepsi Cola, Bela cannot drink the same very pack of Pepsi Cola, that is, two individuals cannot consume the same very Pepsi Cola; its consumption by one individual excludes others to consume it. Thus goods like, apples, Pepsi Cola, shirts, machines and several such other goods, the consumption of which reduce their availability for other persons are called rival goods. Rival goods cannot be public goods, they are private goods.

On the other hand, *public goods are non-rival in consumption.* National defence, parks, television signals, flood control project, pollution control project, light house in the sea are some examples of public goods. Thus persons of a nation can enjoy (consume) equally the security provided by the national defence system. All persons of a city can benefit from the television signals and enjoy the programme telecast. The enjoyment provided by a park, if there is free access to it, can be obtained by all who visit it. National defence, parks, television signals and such other goods are non-rival goods as their consumption by one individual does not exclude its consumption by others. That is, the consumption of a non-rival good by an individuals does not reduce its amount available for others to consume. To conclude, public goods are non-rival.

Non-Excludiability

The other essential characteristic of a public good is non-excludiability in distribution of its consumption benefits. This non-exclusive nature of a public goods implies that it is difficult if not impossible, to exclude those from consuming them who are not willing to pay for them. In case of private rival goods such as shirts, cars, Pepsi Cola, apples those who do not pay for them can be easily prevented from consuming them or receiving benefits from them because the producer or seller simply does not provide them these goods, if they do not pay price for them. On the contrary, in case of public goods, either it is not possible or it is very costly to prevent those people from consuming them who do not pay for these goods. We will explain later that it is due to the feature of non-excludiability of public goods that accounts for the failure of market in case of these goods to ensure Pareto efficiency.

For example, national defence is a public good and is provided to all members of a society and its benefits are available to all equally irrespective of whether some people pay for it or not. It is difficult if not impossible to exclude those people from receiving benefits of security provided by national defence system who do not pay for it. Likewise, if a lighthouse is constructed in a sea, it provides light for all the ships whether any one of them pays for it or not and it is not possible to prevent those who do not pay from receiving light from the light house. This inability to exclude those who do not pay from receiving benefits also applies in case of other public goods such as television signals, pollution control project to provide clear air, flood control projects, parks etc.

Free - Rider's Problem and Public Goods

It is easy to show how non-excludiability of a public good can lead to the market failure, that is, failure of market to achieve pareto efficiency. As explained above, non-excludiability of public goods arises because producers are not able to prevent those from consuming these or enjoying benefits from these who do not pay their share of cost. There is a problem called a *free-rider's problem* which states that because people cannot be excluded from consuming public goods or enjoying benefits from them, there is incentive for persons in these situations to free ride and try to enjoy benefits from reduced pollution, parks, television signals, light house without paying for them. These persons want to get something for nothing and rely on others to make payment for public goods whose benefits they will also automatically get.

Due to this free-rider problem or inability of the producers of public goods to prevent those who do not pay for receiving benefits from them, that a profit-maximising firm will either not produce a public good or produce too little of it. This creates economic inefficiency or Pareto non-optimality. Let us take an example of this free rider's problem in case of public goods leading to economic inefficiency. Suppose the construction of a dam to check floods which cause a lot of damage in a city is required. This dam when built will protect equally all people of the city from the damages due to floods. However, some people of the city would not like to pay for the dam with the hope that others would pay for it and they because of non-excludiability would also enjoy its benefits. But in view of this incentive to free ride, adequate revenue to cover costs of building the dam cannot be provided, and therefore, no private entrepreneur would consider it worthwhile to construct the dam to control floods. Similarly, the production of other public goods such as lighthouse, television signals, pollution abatement projects would not be extended to the socially desirable level in view of the non-excludiability and incentive to free ride.

Public Goods and Pareto - Efficiency

Before explaining further how the free-rider problem results in less than socially optimal production of public goods, it is important to understand how Pareto optimal level of production of public good is determined. Because public goods are non-rival in consumption, some modifications are required in formulation of Pareto optimality conditions. To illustrate the conditions of Pareto - optimality in case of public goods we take the case of a society composed of two persons A and B and the public good is the pollution control project aimed to clean air which if produced would benefit both of them. But the two persons may not perceive to receive the same amount of marginal benefits from this pollution control measure. In other words, they may have different evaluation of the marginal benefits of pollution control measure. Each person will place some value on the pollution control. The marginal benefits they obtain or values they place on the different quantities of pollution abatement are depicted in Fig. 40.4. Due to differences in tastes or perceptions of two individuals, the curve showing marginal benefits from the pollution-free air are different, the curves MB_A and MB_B depict the marginal benefits obtained by individual A and B respectively from the varying quantities of pollution-free air. *The marginal benefit curve can also be interpreted as the price which the individuals are willing to pay for the different quantities of pollution free air*. Thus, it will be seen from the marginal benefit curves that individual A will be willing to pay price Q_1A_1 for OQ_1 quantity of pollution-free air, whereas individual B is willing to pay Q_1B_1 for the same OQ_1 quantity of the pollution-free air. Similarly, for OQ_2 quantity of the pollution-free air, the individual A is wiling to pay price equal to Q_2A_2 and individual B is willing to pay price equal to Q_2B_2. Therefore, the marginal benefit curves can be interpreted as the demand curves of the individuals for pollution-free air. It should also be noted that the marginal benefits or the price which the individuals are willing to pay depend on the values they place on the different quantities of the pollution-free air.

In order to determine the Pareto-optimal quantity of pollution-free air we need the total market demand curve or the aggregate marginal benefit curve of the individuals comprising the society. Market demand curve for a public good cannot be obtained in the way market demand curve is obtained in case of private goods. Since a private good is rival in consumption, the market demand curve of it is obtained by adding up sideways (*i.e.* horizontal addition) of the demand curves (*i.e.* marginal benefit curves) of the two individuals. But, as explained above, public goods are non-rival in consumption, that is, in case of public goods same units of output can be consumed by various people at the same time. Therefore, different individuals can pay for the same units of a public good at the same time. Thus, a pollution control project renders the air of a town free of pollution to some degree from which everybody in the town is benefited and should pay for it. Consider Fig. 40.4, where it will be seen that individual A is prepared to pay price Q_1A_1 for OQ_1 quantity of pollution-free air and the individual B is prepared to pay price Q_1B_1 for the same OQ_1 quantity of pollution-free air which he enjoys or consumes at the same time as individual A. Thus, for OQ_1 quantity of clean air, the total price which the two individuals are willing to pay equals $Q_1A_1 + Q_1B_1 = Q_1M$. Similarly, for the same OQ_2 quantity of pollution-free air, individual A is prepared to pay price equal to Q_2A_2, and individual B is prepared to pay price equal to Q_2B_2. Thus, the total price which the individuals together are willing to pay for the same OQ_2 quantity of the good is equal to the sum of these two prices, *i.e.* $Q_2A_2 + Q_2B_2 = Q_2N$. It therefore follows that in case of a public good market demand curve is derived by *summing up vertically* the demand curves of the individuals because each individual consumes the same units of the good at the same time.

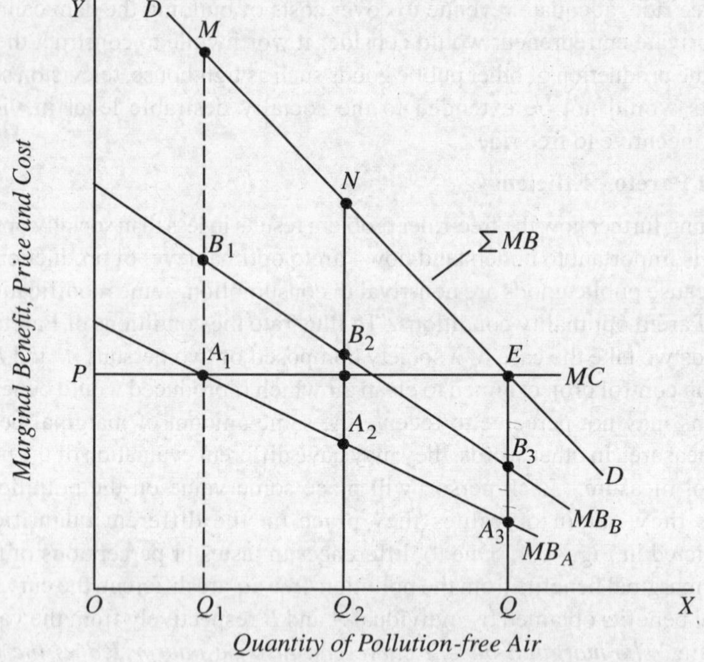

Fig. 40.4 *Public Goods and Pareto Optimality*

Having now obtained the market demand curve of a public good we can now show what will be the Pareto-efficient output of a public good. In this connection it should be noted that a society has to bear the costs it incurs on labour and materials to produce pollution-free air. Pareto efficient level of output is determined at which the price, which the individuals together are willing to pay for the good (that is, aggregate benefit), equals the marginal cost of production. Suppose the marginal cost of production is equal to *OP* per unit and remains constant. In Fig. 40.4, with *OP* as the constant marginal cost, *MC* is the marginal cost curve. It will be observed from the figure that price which the two individuals together are willing to pay equals marginal cost (*MC*) at *OQ* quantity of pollution free air. As said above, price which the persons are willing to pay indicates the aggregate marginal benefit. Thus, aggregate marginal benefit and marginal cost of production are equal at *OQ* level of output of pollution-free air. If resources are allocated to the pollution control project to the extent that *OQ* quantity of pollution-free air is produced at which price (marginal benefit) equals marginal cost incurred the social welfare (*i.e.*, the aggregate benefits of the two individuals) will be maximum. Thus *OQ* is Pareto-efficient level of output of the public good.

But a private firm will produce Pareto optimum output *OQ only if each individual pays a price equal to the marginal benefit*. At *OQ* output marginal benefit of pollution free air is QA_3 for individual *A* and QB_3 for individual *B*. If both are willing to pay prices equal to these marginal benefits, the aggregate price per unit which they together will pay for Pareto efficient quantity *OQ* amounts to *OP* or $QE = QA_3 + QB_3$. In this way total revenue collected by the private firm will cover the cost of pollution control project which cleans the air and therefore will be worthwhile for the private firm to undertake the pollution control project.

But, as explained above, due to inability of the producer of a public good to exclude those who do not pay and want to be free-riders, the costs of optimal level of output cannot be covered by a private producer. Therefore, in this situation too little or even none of the public good will be produced though the marginal benefits of additional units (*i.e.*, the value the individuals place on these additional units) exceeds the social marginal costs of producing these units. Thus, private production and functioning of market in case of public goods do not lead to Pareto efficiency in the provision of public goods.

It may be further noted that in case of two individuals composing a society, there may not be-much problem for an individual trying to be a free rider because of his being constantly watched and pressured by the other, but in the real world a society consists of *many* persons. There is incentive to the persons for misrepresenting the values they place (*i.e.* the benefits they receive) on public goods such as national defence, pollution control project, flood control programme, television signals, and apparently claiming that they have little interest in the provision of these public goods. Since a large number of individuals are involved, each one is likely to think that his not paying for the public good-will not make much difference to the overall revenue and the amount of the public good that will be produced and he will be able to enjoy its benefits without making any contribution. This, of course, would be true if one individual tries to be a free rider. But, as is likely the case, if many individuals and may even all of them thinking in a similar way try to be free riders, then as explained above in Figure 40.4, enough revenue cannot be collected to cover the cost of production of a public good. In this situation there will be no production of a public good at all, at least its Pareto optimal quantity would not be produced. Thus the production of highly important and useful public goods such as national defence, pollution control project, flood control project, television signals may not be undertaken at all if we rely on private sector and market. This is glaring example of the market failure.

An interesting way of explaining market failure to achieve Pareto efficiency in case of public goods is to emphasise that marginal cost of allowing a person to consume the public good is zero *once it is produced*, even if it is possible to prevent him from consuming the commodity. Thus, parks, television signals, flood control projects etc. having been produced, the cost of letting additional consumers to consume these goods or their services is zero. For example, within good limits a visitor to a public park who has not paid is not to affect the enjoyment of the park by those who have paid for it, and it costs the society or a private producer nothing for this additional person visiting the park and enjoying (consuming) it. In fact he would be made better off and no one would be worse off because no more resources of the society are used when the additional person is allowed to enjoy the park (i.e. marginal cost is zero). Now, if the marginal cost of permitting additional persons to consume the good is zero, then Pareto-efficiency requires that price of the public good should be zero. But the total cost of production of public goods is not zero; to produce public good is indeed very expensive. To meet these total costs of production private producer sets a positive price to cover cost. Consequently, price set will be higher than marginal cost and less than Pareto-optimal quantity will be consumed. *Thus production of a public good by the private sector does not lead to Pareto-optimality in allocation of the good.* In other words, market fails to achieve Pareto efficiency in the production of public goods.

QUESTIONS FOR REVIEW

1. What is meant by optimum allocation of resources? Show that monopoly leads to misallocation of resources.
2. What are the major sources of competitive market failure? Explain briefly in each case why the competitive market does not operate efficiently. *[B.Com (Hons) D.U. 1998]*
3. (*a*) In the case of production externalities the optimal pattern of production is independent of the assignment of property rights.

   (*b*) What are public goods? What limits the possibility of the private supply of public goods? *[B.A. (Hons) D.U. 1998]*
4. (*a*) What are the characteristics of a pure public good?

   (*b*) What do you understand by the *free rider problem*
5. What is public good? What are the economic implications of public good? *[B.Com (Hons) D.U. 2002]*
6. Write a note on the role of externalities and public goods in causing market failure?
7. Competitive markets fail due to market power, incomplete information, externalities and presence of public goods. Explain these concepts in support of your answer? *[B.Com (Hons) D.U. 1998]*
8. When external costs are associated with the production of some goods, should the outputs of the goods on efficiency grounds are limited to a level at which the external costs are zero? Why or why not *[B.A. (Hons) D.U. 1997]*
9. Show that when there is externality in consumption, the equality of marginal rate of substitution across individuals may not provide a Pareto optimum outcome. *[B.A. (Hons) D.U. 1996]*
10. Distinguish between beneficial and detrimental externalities. How do they cause market failures to achieve pareto optimality.
11. How do taxes and subsidies by the Government can correct market failures and achieve maximum social welfare?

CHAPTER 41
NEW WELFARE ECONOMICS: COMPENSATION PRINCIPLE

Pareto laid the foundation of the modern welfare economics by formulating the concept of social optimum which is based on the concept of ordinal utility and is free from interpersonal comparisons of utilities and value judgements. He aimed at formulating a value-free objective criterion designed to test whether a proposed policy change increases social welfare or not. Pareto critrion states simply that an economic change which harms no one and makes someone better off indicates an increase in social welfare. Thus, this criterion does not apply to those economic changes which harm some and benefit others. In terms of Edgeworth box diagram Pareto criterion fails to say as to whether or not social welfare increases as movement is made in either direction along the contract curve because it rejects the notion of interpersonal comparison of utility. As we have seen in the previous chapter, every tangency point of the two indifference curves on the contract carve represents a Pareto optimum. There is thus no any unique optimum position. This criterion does not tell us about changes in the level of social welfare if one moves on the contract curve from one tangency point to another because such movement harms one and benefits the other. Thus, the analysis of welfare in terms of Pareto optimality leaves a considerable amount of indeterminacy, for there are numerous Pareto optimum points on the contract curve.

Kaldor-Hicks Welfare Criterion: Compensation Principle

Economists like Kaldor, Hicks and Scitovsky have made efforts to evaluate the changes in social welfare resulting from any economic reorganisation which harms somebody and benefits the others. These economists have sought to remove indeterminacy in the analysis of Pareto optimality. They have put forward a criterion known as the '*compensation principle*' on the basis of which they claim to evaluate those changes in economic policy or organisation which makes some individual better off and others worse off. The 'compensation principle' is based on the following assumptions.

Assumptions

1. The satisfaction of an individual is independent of the others and he is the best judge of his welfare.
2. There exist no externalities of consumption and production.
3. The tastes of the individuals remain constant.
4. The problems of production and exchange can be separated from the problems of distribution. Compensation principle accepts the level of social welfare to be a function of the level of production. Thus it ignores the effects of a change in distribution on social welfare.
5. Utility can be measured ordinally and interpersonal comparisons of utilities are not possible.

Given the above assumptions, a criterion of compensation principle can be discussed. Kaldor, Hicks and Scitovsky have claimed to formulate a value-free objective criterion of measuring the changes in social welfare with the help of the concept of 'compensating payments'.

Nicholas Kaldor was the first economist to give a welfare criterion based on compensating payments. Kaldor's criterion helps us to measure the welfare implications of a movement in either direction on the contract curve in terms of Edgeworth box diagram. According to Kaldor's welfare criterion, if a certain change in economic organisation or policy makes some people better off and others worse off, *then a change will increase social welfare if those who gain from the change could compensate the losers and still be better off than before*. In the words of Prof. Baumol, "Kaldor's criterion states that a change is an improvement if those who gain evaluate their gains at a higher figure than the value which the losers set upon their losses."[1] Thus, if any policy change benefits any one section of the society (gainers) to such an extent that it is better off even after the payment of compensation to the other sections of the society (losers) out of the benefits received, then that change leads to increase in social welfare. In Kaldor's own words, "*In all cases.... where a certain policy leads to an increase in physical productivity and thus of aggregate real income.... it is possible to make everybody better off without making anybody worse off. It is quite sufficient.... to show that even if all those who suffer as a result are fully compensated for their loss, the rest of the community will still be better off than before.*"[2]

Prof. J.R. Hicks supported Kaldor for employing compensation principle to evaluate the change in social welfare resulting from any economic reorganisation that benefits some people and harms the others. This criterion states that, "if *A* is made so much better by the change that he could compensate *B* for his loss and still have something left over, then the reorganisation is unequivocal improvement."[3] In other words, *a change is an improvement if the losers in the changed situation cannot profitably bribe the gainers not to change from the original situation*. Hicks has given his criterion from the losers' point of view, while Kaldor had formulated his criterion from gainers' point of view. Thus the two criteria are really the same though they are clothed in different words. That is why they are generally called by a single name 'Kaldor-Hicks criterion'.

Kaldor-Hicks criterion can be explained with the help of the utility possibility curve. In Fig. 41.1 ordinal utility of two individuals *A* and *B* is shown on *X* and *Y* axes respectively. *DE* is the utility possibility curve which represents the various combinations of utilities obtained by individuals *A* and *B*. As we move downward on the curve *DE*, utility of *A* increases while that of *B* falls. On the other hand, if we move up on the utility curve *ED*, utility of *B* increases while that of *A* falls.

Suppose the utilities obtained by *A* and *B* from the distribution of income or output between them are represented by point *Q* inside the utility possibility curve *DE*. Let us assume that as a result of some change in economic policy, the two individuals move from point *Q* to point *T* on the utility possibility curve *DE*. As a result of this movement, utility of individual *B* has increased while the utility of *A* has declined, that is, *B* has been become better off and *A* has become worse off than before. Therefore, this movement from point *Q* to point *T* cannot be evaluated by means of Pareto criterion. Of course, points such as *R*, *G*, *S* or any other point on the segment *RS* of utility-possibility curve *DE* are socially preferable to point *Q* on the basis of Pareto criterion.

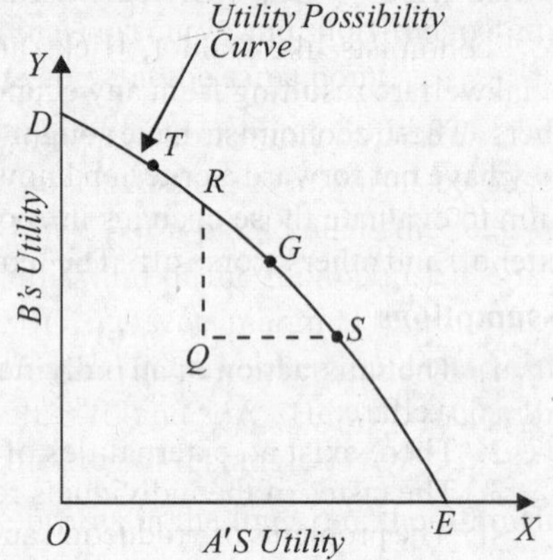

Fig. 41.1. *Kaldor-Hicks Criterion Explained with Utility Possibility Curve*

However, the compensation principle propounded by Kaldor-Hicks enables us to say whether or not social welfare has increased as a result of movement from *Q* to *T*. According to Kaldor-Hicks criterion, we have to see whether the individual *A* who gains with the movement from position *Q* to position *T* could compensate the individual *A* who is loser and still be better off than before. Now, it will be seen from Fig. 41.1 that utility possibility curve *DE* passes through points *R*, *G* and *S*. This means that by mere redistribution of income between the two individuals, that is, if individual *B* gives some compensation to individual *A* for the loss suffered, they can move from position *T* to the position *R*. It is evident from the figure that at position *R* individual *A* is as well off as at the position *Q* but individual *B* is still better off as compared to the position *Q*. It means due to a policy change and consequent movement from position *Q* to position *T*, the gainer (individual *B*) could compensate the loser

1. W. J. Baumol, *Economic Theory and Operations Analysis*, 3rd Ed. 1972, p. 402.
2. N. Kaldor, "Welfare Propositions of Economics and Interpersonal Comparison of Utility", *Economic Journal*, Sept. 1939.
3. J.R. Hicks, The Foundations of Welfare Economics, *Economic Journal*, Vol. XIX, December 1939.

(individual *A*) and is still better off than at *Q*. Therefore, according to Kaldor-Hicks criterion, social welfare increases with the movement from position *Q* to position *T*, because from *T* they could move to the position *R* through mere redistribution of income (*i.e.* compensation).

It is noteworthy that, according to Kaldor-Hicks criterion, compensation may not be actually paid to judge whether or not social welfare has increased. It is enough to know whether the gainer *could* compensate the loser for the loss of welfare and still be better off. Whether redistribution of income (that is, payment of compensation) should be actually made following the change in policy is left for the Government to decide. If it is *possible* for the gainer to compensate the loser and still be better off, the economists can say that social welfare has increased. It may be noted that gainer can compensate the losers and still be better off only when the change in economic policy leads to the increase in output or real income. That is why Kaldor and Hicks claim that they have been able to distinguish between *change* in output from *change in distribution*. When their criterion is satisfied by a change in the situation, it means that the economy has moved to a potentially more efficient position and as a result social welfare can be said to have increased. Now, whether redistribution of income is actually made through payment of compensation by the gainers to the losers, according to them, is a different matter.

Now, the implications of Kaldor-Hicks criterion become more clear if through redistribution the position of the two individual changes from *T* to *G* (see Fig. 41.1). It is quite manifest that at position *G* both the individuals *A* and *B* are better off than at the position *Q*. Thus, the position *T* to which the two individuals moved as result of a certain change in economic policy is superior to the initial position *Q* from the viewpoint of social welfare, since from position *T* movement can be made merely through redistribution of income to position *G* where both are better off as compared to the position *Q*.

It may be noted that in the situation depicted in Fig. 41.1, the change in economic policy brings about a movement from a position *inside* the utility possibility curve to a *point on it*. Now let us see what happens to social welfare if as a result of the adoption of a certain economic policy the utility possibility curve moves outward and the two individuals move from a point on a lower utility possibility curve to a point on a higher utility possibility curve. It can be shown that, according to Kaldor-Hicks criterion, such a movement causes an improvement in social welfare. Consider Fig. 41.2. *UV* is the original utility possibility curve and *Q* represents the position at which the two individuals are initially placed. Now, suppose utility possibility curve shifts outward to the new position, *U′V′*, and the two individuals are placed at point *R* on it. In movement from *Q* on the utility possibility curve *UV* to point *R* on the utility possibility curve *U′V′* the utility of *A* has increased and that of *B* has declined. But position *R* denotes greater social welfare on the basis-of Kaldor's criterion when compared to the position *Q* on the original utility possibility curve *UV* because with *U′V′* as the utility possibility curve it is possible to move through mere redistribution of income from position *R* to position *S* where the individual *B* has been fully compensated for his loss of utility, the individual *A* is still better off as compared to position *Q*. To conclude, any change in the economy that moves the individuals from a position on a lower utility possibility curve to a position on a higher utility possibility curve increases social welfare.

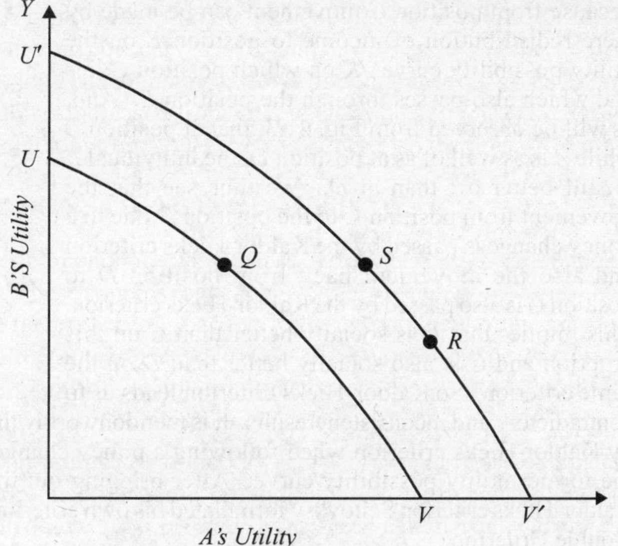

Fig. 41.2. *Kaldor-Hicks Welfare Criterion*

Scitovsky Paradox

Scitovksy pointed out an important limitation of Kaldor-Hicks criterion that it might lead to contradictory results. He showed that, if in some situation, position *B* is shown to be an improvement over position *A* on Kaldor-Hicks criterion, it may be possible that position *A* is also shown to be an improvement over *B* on the basis of the same criterion. For getting consistent results when position *B* has been revealed to be preferred to position *A* on the basis of a welfare criterion, then position *A* must not be preferred to position *B* on the same criterion. According to Scitovsky, Kaldor-Hicks criterion involves such contradictory and inconsistent results. Since Scitovsky was the first to point out this paradoxical result in Kaldor-Hicks criterion, it is known as '*Scitovsky Paradox*'.

How Kaldor-Hicks criterion may lead to contradictory results in some situation is depicted in Fig. 41.3. In this figure *JK* and *GH* are the two utility possibility curves which intersect each other. Now suppose that the initial position is at point *C* on *JK*. Further suppose that due to a certain policy change, utility possibility curve changes and takes the position *GH* and the two individuals find themselves at position *D*. Position *D* is superior to position *C* on the basis of Kaldor-Hicks criterion because from position *D* movement can be made through mere redistribution to position *F* at which individual *B* has been fully compensated but individual *A* is still better off as compared to the original position *C*. Thus movement from position *C* to position *D* satisfies Kaldor-Hicks criterion.

But, as has been pointed out by Scitovsky, reverse movement from position *D* on the new utility possibility curve *GH* to the position *C* on the old utility possibility curve *JK* also represents an improvement on Kaldor-Hicks criterion, that is, *C* is socially better than *D* on the basis of Kaldor-Hicks criterion. This is because from position *C* movement can be made by mere redistribution of income to position *E* on the utility possibility curve *JK* on which position *C* lies and which also passes through the position *E*. And, as will be observed from Fig. 41.3, that at position *E* while *A* is as well of as at position *D*, the individual *B* is still better off than at *D*. We thus see that the movement from position *C* to the position *D* due to a policy change is passed by the Kaldor-Hicks criterion and also the movement back from position *D* to position *C* is also passed by the Kaldor-Hicks criterion. This implies that *D* is socially better than *C* on this criterion and *C* is also socially better than *D* on the same criterion. So Kaldor-Hicks criterion leads us to contradictory and inconsistent results. It is mentionworthy that these contradictory results are obtained by Kaldor-Hicks criterion when following a policy change new utility possibility curve intersects the former utility possibility curve. After bringing out the possibility of contradictory results in Kaldor-Hicks criterion Scitovksy formulated his own criterion which is generally known as Scitovsky's Double Criterion.

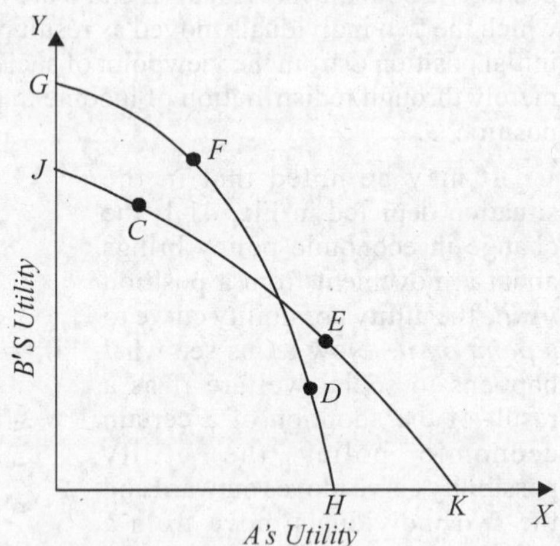

Fig. 41.3. *Scitvosky Paradox*

Scitovsky's Double Criterion of Welfare

To rule out the possibility of contradictory results in Kaldor-Hicks criterion Scitovsky formulated a double criterion which requires the fulfilment of Kaldor-Hicks criterion and also the fulfilment of the reversal test. It means that a change is an improvement if the gainers in the changed situation are able to persuade the losers to accept the change and simultaneously losers are not able to persuade the gainers to remain in the original situation. Scitovsky's double criterion can also be explained with the help of utility possibility curve. In Figure 41.4, *CD* and *EF* are the two utility possibility curves which do not intersect each other at any point. Suppose there is a change from position *Q* on the utility possibility curve *CD* to the position *G* on the utility possibility curve *EF* as a result of the adoption of a new economic policy. Such a movement is an improvement on Kaldor-Hicks criterion because G lies on the utility possibility curve EF passing through point R. From the position G, movement can be made to the position R simply by redistributing income between the two individuals. And R is better than Q because the utility of both the individuals is greater at R as compared to the position Q. Thus the Kaldor-Hicks criterion is satisfied and therefore change from Q to G will increase social welfare.

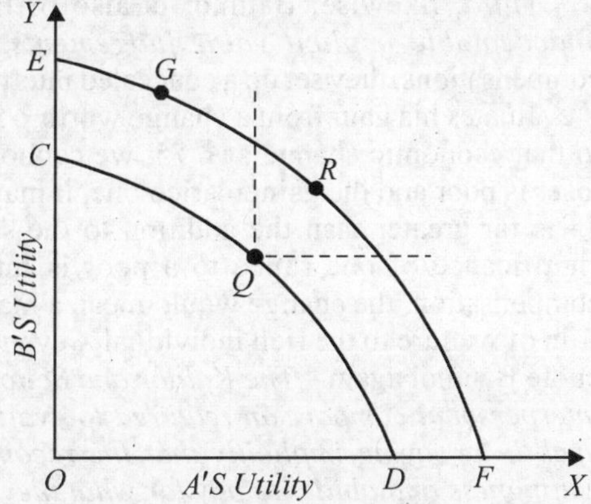

Fig. 41.4. *Scitovsky's Double Criterion*

Now, let us see, what happens to the reversal test. It must also be satisfied, if the Scitovsky double test is to be fulfilled. That is, a movement from the position G back to the original position Q *must not be* passed by Kaldor-Hicks criterion if Scitovsky's reversal test is to be satisfied. It is evident from Fig. 41.4 that from position R we cannot move to any position on the utility possibility curve CD merely through redistribution of income which is socially better than G (that is, which raises utility of either A or B, the utility of the other remaining constant or which raises the utility of both). We thus see that while movement from position Q to G is passed by Kaldor-Hicks criterion, reverse movement from position G to position Q is not passed by Kaldor-Hicks criterion. Hence, in Fig. 41.4 the movement from the position Q to G satisfies Scitovsky's criterion. Thus when the two utility possibility curves are non-intersecting and change involves movement from a position on a lower utility possibility curve to a position on a higher utility possibility curve, the change raises social welfare on the basis of Kaldor-Hicks-Scitovsky criterion. This happens only when a change brings about increase in aggregate output or real income.

A CRITIQUE OF THE COMPENSATION PRINCIPLE

The compensation principle as developed by Kaldor, Hicks and Scitovsky, has been a topic of much discussion in welfare economics since 1939. Prof. Kaldor was the first to give a criterion to judge the changes in social welfare when an economic change benefits some people and harms the others. Later Hicks also supported this criterion in 1940, though he put it in different words. Scitovsky tried to improve the Kaldor-Hicks criterion by formulating his own double criterion. These welfare economists have claimed that they have succeeded in developing a welfare criterion based on ordinal concept of utility and also which is free from any value judgements. But compensation principle has been bitterly criticised by the various welfare economists.

First, Little has pointed out that Kaldor did not formulate a new welfare criterion at all because he assumed welfare to be a function of increase in production or efficiency irrespective of the changes in distribution. Thus, according to Little, Kaldor has given only a definition of 'increase in wealth' or 'increase in efficiency'. Kaldor himself has interpreted the compensation principle in this sense as he says that, "when the production of wealth goes up, some income distribution could be found which makes some people better off, and no one worse off than before". However, as *desired income distribution via compensation is only hypothetical, therefore, according to Little, it is not a welfare test but a definition of 'economic efficiency' in terms of over-compensation*.

Second, compensation principle is *not free value judgements* as is claimed by its propounders. It involves, implicit value judgements. Prof. Baumol and Little are of the opinion that the contention of Prof. Kaldor that the changes which enable the gainers to compensate the losers and still be better off are good changes, is itself a value judgement. According to Little, to say that a policy which meets the Kaldor-Hicks criterion increases the output or "efficiency" of society is, in effect, to recommend it. According to him, Kaldor and Hicks have coined a *definition of* "efficiency" whose implicit ethical implications or value judgements will hardly find favour with many people. Compensation is after all only hypothetical; it is consistent with making the poor yet poorer. Thus, according to Little, if the value judgements implicit in Kaldor-Hicks criterion are made explicit, then the claim of Kaldor and Hicks that they have discovered a criterion of detecting increases in wealth, production or efficiency free from value judgements is hardly acceptable.

Third, likewise, Baumol is also of the view that *Kaldor-Hicks criterion is based upon unacceptable implicit value judgements*. "By using a criterion involving potential money compensations, they set up a concealed interpersonal comparison on a money basis"[4]. If an individual *A* evaluates his gain from a change worth ₹ 500 whereas another individual *B* evaluates his loss due to that economic change at ₹ 75, we cannot conclude that social welfare has increased; for if the loser is poor and the gainer a rich one, it may be possible that loss of satisfaction of the poor from ₹ 75 is far greater than the addition to the satisfaction of the rich by ₹ 500 because the marginal significance of one rupee to a poor is far greater than that of the rich. Thus without actual compensation, the change would mean a major loss of welfare to the poor individual *B* and a trivial gain of welfare to the rich individual *A* even if it passes the Kaldor criterion with flying colours. To quote Baumol again, "*The Kaldor and Scitovsky criteria have thus ducked the basic problem of the interpersonal comparison required to evaluate a policy change which harms X but aids Y. They duck it by saying implicitly that the recommendation should be based on X's and Y's relative willingness and ability to pay for what they want*".[5]

Fourth, *Kaldor-Hicks* have claimed that through compensation principle they have been able to separate a production change from the distribution change by which it is accompanied. For instance, as a result of a policy change output of Coca-Cola increases and that of whisky decreases. Now, if individual *X* prefers Cola Cola but *Y* prefers whisky, the question whether there has occurred any increase in production is inseparably connected with the distribution of these beverages between *X* and *Y*. In many cases it is, therefore, difficult to say whether or not production has increased without considering how the output or real income is being distributed.

Moreover, Kaldor and Hicks think that the *level of production is the main determinant of social welfare and the distribution a secondary one*. But this is quite untenable. A lower total output equitably distributed ensures greater social welfare than larger output, inequitably distributed. They essentially accept the existing distribution of income and wealth and ignore its impact on individual utilities and well-being.

Fifth. Prof. Baumol, Little and Arrow point out another major flaw in compensation principle that it does not envisage social welfare. This principle proves the sociol desirability of change in the social state on the basis of the criterion that gainers *could compensate* the losers and still be better off than before. These critics are of the opinion that policy changes which would increase social welfare when accompanied by actual compensation need not lead to improvement in social welfare if compensation is not actually made. Dr. Rothenberg has given a very good example to illustrate this. He supposes an initial social state in which a firm adopts a new invention and as a result the cost of production of the firm is reduced but it throws the competitors out of industry and the workers become unemployed. Let us suppose that the gainer firm from the invention can compensate the losers out of its increased income and still be better off. If the compensation is not actually made in the changed situation, social welfare will decrease as the welfare loss suffered by the workers rendered unemployed will be very large indeed. As a matter of fact, there is no guarantee that compensation will be actually made in such cases. Thus, so long as compensation is hypothetical a change might make the rich richer and the poor yet poorer and therefore reduce social welfare.

It follows from above that a basic flaw in Kaldor-Hicks compensation principle is that it refers to *potential welfare* rather than *actual welfare* since it does not envisage that compensation should be actually made. In the absence of actual compensation one cannot say whether or not actual social welfare has increased as a result of a certain policy change unless one is prepared to make some value judgements. Therefore, making value judgements, especially that concerning distribution of income or welfare, is quite indispensable in welfare economics. And economists should not fight shy of making those value judgements which are widely accepted by the people.

It may also be noted that if compensation is actually made then Kaldor-Hicks criterion is quite unnecessary, for in that case only Pareto criterion will be sufficient to judge the effect of a policy change on social welfare.

---

4. Baumol, *op. cit*, p. 530.
5. *Op.cit.* p.530

Sixth, compensation principle does not take into account the external effects on consumption and production. The exponents of compensation principle are of the opinion that an individual's welfare depends solely upon his own level of production and consumption and is not affected by the production and consumption activities of the others. But this is not a realistic assumption because a person's level of satisfaction (or dissatisfaction) depends to a large extent upon the consumption of goods and services by other persons. A person is more satisfied as his *relative economic* position in the society is improved. Thus, if an economic change leaves a person where he was before but makes some other individuals better off, he will not feel as well off as in the original situation, that is, his level of welfare will fall. The gains by some individuals from a policy change have usually unfavourable external effects on the welfare position of those whose position is said to have remain unchanged.

Bergson Social Welfare Criterion

Bergson pointed out that for comparing utility levels of different individuals resulting from changes in economic policy, making interpersonal comparison of utility cannot be avoided. But he stressed that these comparision of interpersonal utilities should be made explicit. Thus Bergson put forward a social welfare criterion in which he incorporates explicitly such comparison as utility of individual *A* from Re 1= utility of individual *B* from Re 0.20. On the basis of such explicit value judgements or interpersonal comparison of utilities he gives ranking to alternative social states from the viewpoint of social welfare. Bergson along with Samuelson developed the concept of social welfare function which incorporates explicit value judgements for evaluating the welfare implications of policy changes and also finding out a unique social optimum. We shall discuss Bergson – Samulson's social welfare function in the next Chapter.

QUESTIONS FOR REVIEW

1. What is compensation principle of judging change in social welfare brought about by the adoption of an economy policy? In compensation considered in this principle real or hypothetical.
2. State Kaldor-Hicks criterion of judging social welfare. Explain with the help of utility possibility curve how it judges the social welfare impact of an economic policy which makes some better off and others worse off.
3. Show how Kaldor-Hicks criterion helps in solving the problem of indetermincy involved in Pareto's analysis of maximisation of social welfare.
4. "Pareto criterion does not give us sufficient basis of ordering states. The Kaldor-Hicks – Scitovsky criteria do not carry us much further" Examine critically. *[B.A. (Hons) D.U. 1984]*
5. Does Kaldor-Hicks-Scitovsky criterion give us sufficient basis for ordering states from the viewpoint of social welfare? Give reasons for your answer. *[B.A. Hons (D.U) 1993]*
6. In proposing compensation criterion Kaldor and Hicks claimed that they succeded in giving a welfare criterion which was free from value judgements. Do you agree? Discuss.
7. In what way is Scitovsky criterion an improvement over Kaldor-Hicks criterion ? Do you agree with the view that Scitovsky criterion does not remove all the weaknesses of Kaldor - Hicks criterion? Discuss. *[B.A. (Hons) D.U. 1996]*
8. How far can Kaldor - Hicks Scitovsky criteria be considered an improvement over Pareto's criterion of social welfare? Discuss.
9. Show that alternative *A* being better than *B* on Kaldor-Hicks criterion does not rule out the possibility that alternative *B* is better than *A* on Kaldor - Hicks criterion.
10. The Kaldor and Scitovsky criteria have ducked the basic problem of the interpersonal comparison of utility required to evaluate a policy change which *harms X* but aids *Y*" (Baumol). Do you agree? Discuss.
11. It is generally felt that a rupee means more to a poor man than to a rich man. Hence an economic system that merely reacts to the number of rupees spent in different directions is grossly unsatisfactory in the optimal allocation of resources and goods. Discuss in the context of Kaldor-Hicks-Scitovsky compensation principle.

Chapter 42
SOCIAL WELFARE FUNCTION

The concept of 'Social Welfare Function' was propounded by A. Bergson in his article '*A Reformulation of Certain Aspects of Welfare Economics*' in 1938. Prior to it various concepts of social welfare had been given by different welfare theorists but they failed to provide a satisfactory solution to the problem of maximisation of social welfare and measurement. Bentham talked of welfare in terms of '*the greatest happiness of the greatest number.*' Neo-Classical welfare theorists discussed the problem of social welfare on the basis of cardinal measurability of utility and interpersonal comparison of utility. Analysis of Pareto optimality maximises social welfare by satisfying various marginal conditions of production, distribution and allocation of resources among products. But unfortunately they are not fulfilled due to the existence of various externalities and imperfections in the market. Moreover, Pareto optimality analysis fails to measure the changes in welfare resulting from any change which benefits one section of society and harms the other. Compensation principle as given by Kaldor-Hicks-Scitovsky attempts to measure the changes in social welfare resulting from such economic changes which harm some and benefit others through hypothetical compensating payments. Compensation theorists claimed to give a value-free objective criterion based on ordinal concept of utility but, as seen in a previous chapter, this is based upon implicit value judgements and does not evaluate changes in social welfare satisfactorily.

By providing the concept of social welfare function Bergson and Samuelson have attempted to provide a new approach to welfare economics and have succeeded in rehabilitating welfare economics. They have put forward the concept of social welfare function that considers only the ordinal preferences of individuals. They agree to Robbins' view that interpersonal comparison of utility involves value judgements but they assert that without making some value judgements, economists cannot evaluate the impact of changes in economic policy on social welfare. Thus, according to them, welfare economics cannot be separated from value judgements. According to them, welfare economics is essentially a normative study. But the approach to study it must be scientific despite the fact that the use of value judgements in it is unavoidable.

BERGSON-SAMUELSON SOCIAL WELFARE FUNCTION

Social welfare function is an ordinal index of society's welfare and is a function of the utility levles of all individuals constituting the society. Bergson-Samuelson social welfare function can be written in the following manner :

$$W = W(U_1, U_2, U_3 \ldots\ldots, U_n)$$

where W represents the social welfare $U_1, U_2, U_3, \ldots\ldots, U_n$ represent the ordinal utility indices of different individuals of the society. The ordinal utility index of an individual depends upon the goods and services he consumes and the magnitude and kind of the work he does. The important thing to note about social welfare function is that in its construction *explicit value judgements* are introduced. Value judgements determine a form of the social welfare function; with a different set of value judgements, the form of social welfare function would be different. Value judgemtns are essentially ethical notions which are introduced from outside economics. The value judgements required to construct a social welfare function may be obtained through democratic process with voting by individuals or it may have to be imposed on the society in a dictational manner. Whatever the case may be, the form of social welfare function depnds upon the value judgements of those who decide about them since it expresses their views regarding the effect which the utility level of each individual has on the social welfare. In the worlds of Prof. Scitovsky. "The social welfare function can be thought of as a funtion of each individual's welfare which in turn depends both in his personal well being and on his appraisal of the distribution of wlefare among all members of the community".[1]

Since the value judgements requaired for the formation of social welfare function are not of the economist himself and instead they are introduced from outside economics they are not obtained through any scientific method. It has been claimed that social welfare function has solved the basic problem of welfare economics, since it thinks unnecessary for the economists themselves to make value judgements concerning what is a desirable distribution of welfare as between indindividuals constituting the society. In other words, economist need not himself decide about what is the most desirable distribution of wlefare. He can take value judgements regarding distribution as given from outside economics.

Bergson's social welfare function is supposed to be dependent on changes in economic events that have a direct effect on individual welfares. The ordinal utility level of an individual is a function of his own consumption of goods and services and not of others. Morever, the utility level of an individual depends on his own value judgments regarding the composition of different goods and services consumed which depends upon his tastes. An individual may derive more utility from the consumption of liquor whereas another individual may derive very nominal utility or no utility at all from it.

Social Welfare Function and Value Judgements

So far we have been mainly concerned with the value judgements of individuals regarding their utility levels. From the view point of social welfare function, the value judgements regarding the welfare of the society as a whole are relevant. The formulation of a welfare function for the society as a whole is a very difficult task because utility being a mental phenomenon cannot be measured or estimated accurately by any person or institution entrusted to furnish value judgements regarding the changes in social welfare, Moreover, addition and substraction of utilities of different individuals by an authorised person or institution too is a very difficult task.

The social welfare function and its form depends upon the value judgements of the person or institution whom the society has authorised to decide. The authorised person or institution may be any body but for true value judgements regarding the social welfare *he must be unbiased* because changes in social welfare will depend upon his value judgements. "These judgements as to what constitutes justice and virtue in distribution may be those of the economist himself or those set up by the legislature, by some other governmental authority or by some other unspecified person or group."[2] A social welfare function can be attained by common consensus or it may be forced upon the society by a dictator.

Since the forms of social welfare functions are known by value judgements about social welfare, therefore there arises the problem of finding an authority who could give purely unbiased value judgemtents. Bergson and Samuelson have assumed a "Superman" who provides a value judgements about changes in social welfare. Superman alone can take decisions about the solution of various problems of the economy. What goods and services should be produced and supplied in the society ? How much of various goods should be produced ? What should be the quality and kind of goods ? What should be capital intensity of producing a particular type of good ? What should be the pattern of distribution of national income among different sections of the society ? Which wants should be satisfied at present and which at a future date and so on. All these questions can be answered by the

1. T. Scitovsky, *The State of Welfare Economics,* printed in his *Papers on Welfare and Growth*, George Allen and Unwin Ltd., 1964, p.184
2. W. J. Baumol, *Economic Theory and Operations Analysis,* p. 531.

superman alone in accordance with his views about the determinants of social welfare. The society would have to accept the solutions of all these questions provided by him assumpting that he will give any value judgements which aim to achieve maximum social welfare rather than maximum self-interest. Thus we are free from the addition, substraction, measurement and interpersonal comparisons of utilities by assumpting the existence of a superman.

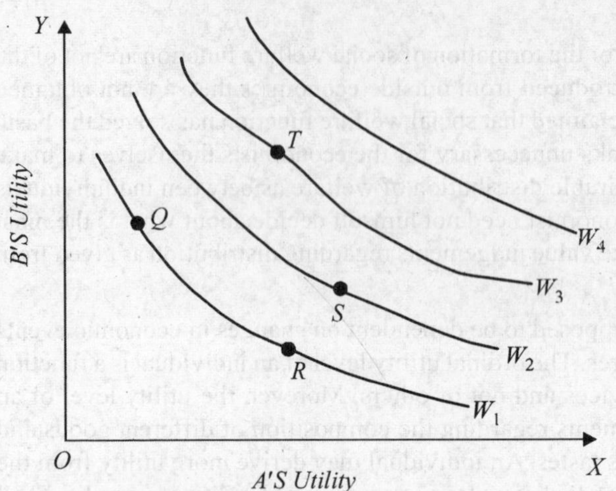

Fig. 42.1. *Social Indifference Curves depicting Social Welfare Function*

In modern age of democratic governments people elect their representatives who constitue the Government. The political party in majority forms the Government and rules the country. The representatives' Government formed by the majority rule formulates various policies on the basis of value judgements and it is expected that all the policy decisions by the Government will aim at maximising social welfare rather than maximising the welfare of an individual or a particular section of the soceity.

Bergson and Samuelson expressed the view that *all value judgements used to construct the social welfare function must be consistent which implies* that if in a given situation *A* is perferred to *B* and *B* is preferred to *C* then *A* must be preferred to *C*. This is nothing new to the students of economics as this is the well know assumption of transitivity in social choice among various alternatives. We can explain the social welfare function wilth the help of *social indifference curves or welfare frontiers*. Let us assume a society of two persons. In such a case social welfare function can be represented with the help of social indifference curves.

In Fig. 42.1 the utilities of individuals *A* and *B* have been represented on the horizontal and vertical axes respectively. W_1, W_2 and W_3 are the social indifference curves representing successively higher levels of social welfare. A social indifference curve is a locus of various combinations of utilities of *A* and *B* which result in an equal level of social welfare. The properties of social indifference curves are just like that of individual consumer's indifference curves. Given a family of social indifference curves, the effect of a proposed change in policy on social welfare can be evaluated. In terms of Fig. 42.1 any policy change that moves the economy from *Q* to *T* is an improvement. Similarly, a movement from *Q* to *S* or from *R* to *S* also represents an improvement in social welfare, and a movement from *T* to *Q* or *T* to *S* represents a decrease in social welfare. A movement along the same social indifference curve represents no change in the level of social welfare.

Analysis of Pareto optimality failed to provide a 'unique optimum solution' which represents maximum social welfare. There are a large number of solutions which are optimum on the basis of Pareto criterion. In terms of Edgeworth-box diagram every point on the contract curve represents the optimum position. In terms of Grand Utility Possibility Frontier, all points on it are Pareto optimal or economically efficient. But Pareto criterion does not tell us the best of them. Thus, Paretian analysis leaves us with a lot of indeterminacy in the choice of maximum social welfare point. Now, the significance of social welfare function is that it enables us to obtain a unique optimum position regarding social welfare. This unique optimum position is best of all the Pareto optima and therefore ensures the maximum social welfare. By including the concept of grand utility possibility frontier along with Bergson-Samuleson social welfare function we are able to obtain a unique optimum position or maximum social welfare position which is explained below.

GRAND UTILITY POSSIBILITY FRONTIER AND POSITION OF CONSTRAINED BLISS

As shall be explained below, a grand utility possibility forntier is a locus of the vairous physically attainable utility combinations of two persons when the factor endowment, state of technology and preference orders of the individuals are given.[3] In other words, *every point on the grand utility posibility curve represents the optimum position with regard to the allocation of the products among the consumers, allocation of factors among different products and the direction of production.* Thus every point on the grand utility possibility curve represents a Pareto optimum and as we move from one point to another on it the utility of one individual increases while that of the other falls.

Now, let us superimpose grant utility possibility curve on the social indifference curves representing social welfare function to find a unique optimum position of social welfare. In Fig. 42.2 social indifference curves W_1, W_2, W_3 and W_4 representing the social welfare function have been drawn along with the grand utility possibility curve VV'. Social indifference curve W_3 is tangent to the grand utility possibility curve VV' at point Q. Thus, point Q represents the maximum possible social welfare given the factor endowments, state of technology and preference scales of the individuals. Point Q is known as the point of **constrained bliss** since, given the constraints regarding factor endowments and the state of technology. Q is the highest possible state of social welfare which the society can attain. Social welfare represented by the social indifference curve W_4 is higher than social indifference curve W_3 passing through Q but it is not possible to attain it, given the technology and factor endowment. Thus, from among a large number of Pareto optimum points on the grand utility possibility curve, we have a unique optimum point Q at which the social welfare is the maximum. The point of constrained bliss represents the *unique* pattern of production of goods, *unique* distribution of goods between the individuals and unique combination of factors employed to produce the goods.

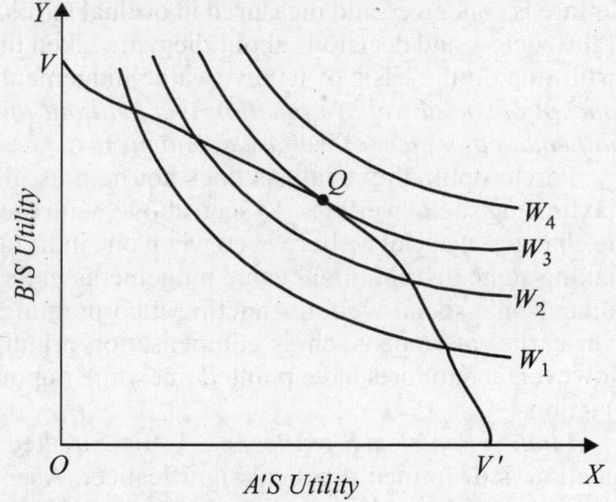

Fig. 42.2. *Social Welfare Function and Position of Constrained Bliss.*

The following features of the Bergson-Samuelson Social Welfare function are worth noting :

1. The Bergson-Samuelson social welfare function is based on explicit value judgements and involves interpersonal comparisons of utility in ordinal terms.
2. Bergson-Samuelson social welfare function, the maximum social welfare position is completely determined as a result of the introduction of value judgements regarding distribution of welfare among individuals.
3. The social welfare function is not based on any unique value judgements. Instead, any set of value judgements can be used by a welfare economist to construct a social welfare function. Thus, it is not any unique function but changes with the change in value judgements.
4. Once the social welfare function has been decided upon by value judgements, the maximisation technique is used to obtain the maximum social welfare position at which

3. How Grand Utility Possibility Curve is derived see Author's book *Advanced Economic Theory*, 14th edition, S. Chand & Co.

allocation of resources is Pareto optimum and also the distribution of goods and services is equitable. Thus, both efficiency and equity are achieved so that social welfare may be maximised.

5. Used along with the Pareto optimality analysis the concept of social welfare function enables us to find a unique optimum solution which combines economic efficiency with distributive justice.

A CRITICAL EVALUATION OF BERGSON-SAMUELSON SOCIAL WELFARE FUNCTION

The main aim of welfare economics has been to find an acceptable social welfare function which could measure the changes in social welfare resulting from a change in economic and non-economic variables. Bergson and Samuelson solved this problem by formulating a social welfare function which is based on explicit value jdugements. This function can incorporate the various economic and non-economic determinants of the welfare of individuals. In this function utility or welfare is conceived and measured in ordinal terms. Preferences or utilities of different individuals of the society and decisions about them are taken through a democratic method or by an authorised institution on the basis of its own value judgements. Even according to its bitter critic Little, the *concept of social welfare function is a brilliant theoretical construct which completes the formal mathematical system of welfare economics*.

Pareto optimality analysis does not help us in providing a unique solution to the problem of maximising social welfare. As seen above, with the help of social welfare function we can measure the changes in social welfare even when one individual becomes better off and another worse off by making some distributional value judgements in the form of social welfare function. The Bergson-Samuelson's social welfare function incorporating explicit value judgements is an improvement over earlier attempts such as compensation principle advanced by Kaldor, Hicks and Scitovsky. However, economists have pointed out some important drawbacks in the concept of social welfare function.

Limited Practical Significance. Little, Streeten and Baumol have pointed out that social welfare function is of limited practical significance. According to Little, the social welfare function can neither be used in a democratic state, nor even in a totalitarian one because in them there would be as many vague social welfare functions as there are individuals. Social welfare function, to quote Little is only "a formal device necessary to a perfectly general abstract system of 'welfare', which is devoid of any practical significance."[4]

Likewise, Paul Streeton also thinks that *social welfare function is a highly formal concept which has hardly any relation with the important facts of social life and choice*. To quote him, "No political programme or individual value standard would fit the model of a social welfare function of the required type"[5] Prof. Baumol is also of the opinion that the concept of social welfare is of limited practical value as it does not tell us how to get the value judgements which it requires for its construction. Though Bergson criterion of social welfare function, writes Baumol, "provides us with a highly useful frame of reference, unfortunately it does not come equipped with a kit and a set of instructions for collecting the welfare judgements which it requires. Thus it still leaves us with the difficult part of the job unsolved"[6]

Welfare depends on a wider range of variables than those associated with utility. Social welfare function approach is based on the utility which an individual derives from economic variables such as consumption of goods and services. Apart from these economic variables, welfare or well-being of individuals depends on a whole range of political and environmental variables such as enjoyment of human rights, political freedom, pollution-free environment. Thus, "in comparing different economic systems or in comparing different ways of organising a given economy, the possibility that some of these variables might be affected cannot be ignored. Thus a reorganisation that gives everyone

---

4. I.M.D. Little, *A Critique of Welfare Economics*, p. 122
5. Paul Streeton, Appendix to Gunnar Myrdal's *The Political Elements in the Development of Economic Theory*, p. 216.
6. Baumol, *op. cit.*, p. 531.

more income and leisure *might not improve the welfare of the community if at the same time it limits individual freedom or requires the abandonment of cherished cultural traditions."*[7]

Impossibility of Constructing a Social Welfare Function from Individual Preferences. A highly damaging drawback of social welfare function has been pointed out by K.J. Arrow who has shown that social welfare function cannot be constructed on the basis of value judgements arrived at through democratic process of majority rule in group decision-making. Arrow has proved that the *majority rule leads to contradictory results or intransitivity of social choices when individuals are asked to make a choice from among more than two alternatives available to them*. Therefore, Prof. Arrow concludes that a social welfare function which is based on mere ordinal preferences cannot in principle be constructed from the preferences of all the individuals comprising a society. Of course, social welfare function can be set up on the basis of value judgements of an individual who can impose his will on the society but that will reflect the aims and aspirations of an absolute dictator.

Prof. Amartya Sen's Critique : Judging welfare or well-being in terms of utility is of limited significance.

Prof. Amartya Sen has criticised modern welfare economics covenng both Pareto efficiency and social welfare function on the ground that utility is not a true indicator of well-being. To quote him, "A difficulty with welfarism arises from the particular interpretation of well-being that utility provides. To judge the well-being of a person exclusively in the metric of happiness or desire-fulfilment has some obvious limitations. These limitations are particularly damaging in the context of interpersonal comparisons .of well-being."[8] He further adds, "A person who has had a life of misfortune, with very little opportunities, and rather little hope, may be more easily reconciled to deprivations than others reared in more fortunate and affluent circumstances. *The metric of happiness may, therefore, distort the extent of deprivation, in a specific and biased way*. The hopeless beggar, the precarious landless labourer, the dominated housewife, the hardened unemployed or the over-exhausted coolie may all take pleasures in small mercies and manage to suppress intense suffering for the necessity of continuing survival, but *it would be ethically deeply mistaken to attach a correspondingly small value to the loss of their well-being because of the survival strategy*".[9]

It follows from above that Amartya Sen has criticised the concept of social welfare based on utility which means psychological reactions of individuals to goods and services which they consume. Further, Prof. Sen shifts the focus on promoting *positive freedoms* of individuals for assessing the change in their welfare following a change in organisation or public policy. He defines freedom as '*capabilites to function*' as to what persons can do or cannot do. It is capabilities to function that reflect freedom in the positive sense and determine well-being or welfare of the people.[10]

QUESTIONS FOR REVIEW

1. Show diagramtically how socially optimum allocation of consumption and resources can be obtained from the *Bliss Point*. What is the underlying assumption? *[D.U., B.A. (Hons) 2001]*
2. How is Grand Utility Possibility Frontier derived? Examine its significance in the attainment of maximum social welfare. *[D.U., B.A. (Hons)1990]*
3. What is social welfare function? Is it analogous to an individual's utility function ?
4. What is Bergson-Samuelson's social function? How does it incorporate value judgements to evaluate changes in social welfare?
5. Explain how Bergson-Samuelson's concept of Social Welfare Function helps us to solve the problem of indeterminacy of optimality found in Pareto's analysis of social welfare?
6. What is meant by the point of constrained bliss? How do the concepts of grand utility possibility curve and social welfare function help us to achieve it?

---

7. P. Else and P. Curwin, *Principles of Microeconomics*, Unwin Hyman, London, 1990. p. 324 (Italics added).
8. Amartya Sen, *On Ethics and Economics*, Oxford University Press, Delhi, 1990. pp. 45-46. (italics added).
9. *Ibid*.
10. See his well-known article, "*The Concept of Well-Being*" a Silver Jubilee Lecture at Institute of Economic Growth, Delhi University.

[illegible]

Impossibility [illegible] social [illegible] individual preferences [illegible] Arrow [illegible]

[illegible]

[illegible] Arrow [illegible]

[illegible]

[illegible]

QUESTIONS [illegible]

1. [illegible]
2. [illegible]
3. [illegible] social welfare function [illegible]
4. [illegible]
5. [illegible]

[illegible]

PART VII

THEORY OF INTERNATIONAL TRADE

- International Trade: Comparative Cost Theory
- Heckscher-Ohlin's Theory of International Trade
- Terms of Trade

Chapter 43
INTERNATIONAL TRADE: COMPARATIVE COST THEORY

In one of our previous chapters we discussed personal specialisation or division of labour. We also discussed the great part which personal specialisation plays in the functioning of the modern economic organisation. Equally important are the roles of the regional and international specialisation. *Regional specialisation means that various regions or areas in a country specialise themselves in the production of different products. International specialisation means that different countries of the world specialise in producing different goods*. Factors which determine regional specialisation are more or less the same as those which determine international specialisation. A country which produces surplus of a good, *i.e.*, produces more than its requirements, will export it to other countries in exchange for the surplus products of those countries.

In the modern world, there is mutual interdependence of the various national economies. Today it is hard to find the example of a closed economy. All economies of the world have become open. But the degree of openness varies from one country to another. Thus, *in the modern world no country is completely self-sufficient*. Self-sufficiency, in the sense used here, means the proportion of the goods and services consumed to their total output produced within a country. But the degree of self-sufficiency varies from one country to another.

The Theory of Absolute Advantage

The theory of absolute advantage was put forward by Adam Smith who argued that different countries enjoyed absolute advantage in the production of some goods which formed the basis of trade between the countries. Consider Table 43.1 where man-hours required to produce a unit of wheat or cloth in the U.S.A. and India are given.

Table 43.1
Man-hours required to produce one unit of wheat and cloth

	U.S.A.	*India*
Wheat	3	10
Cloth	6	4

It will be seen from the above table that to produce one unit of wheat in the U.S.A. 3 man-hours and in India 10 man-hours are required. On the other hand, to produce one unit of cloth, in the U.S.A. 6 man-hours and in India 4 man-hours are required. Thus the U.S.A. can produce wheat more efficiently (that is, at a lower cost), while India can produce cloth more efficiently. To put it in other words, while the U.S.A. has an *absolute advantage in the production of wheat, India has an absolute advantage in the production of cloth*. Adam Smith showed that the two countries would benefit and world output will increase if the two countries specialise in the production of goods in which they have absolute advantage and trade with each other. How such specialisation and trade would lead to gain in output and would be mutually beneficial for the two countries is shown in Table 43.2.

Table 43.2
Gain in output when labour resources are transferred

	U.S.A.	*India*	*World Output*
Gain in Wheat	+ 2 Wheat	–1Wheat	+ 1 Wheat
Gain Cloth	– 1 cloth	+ 2.5 Cloth	+ 1.5 Cloth

Suppose to specialise in the production of Wheat, the U.S.A. withdraws 6 man-hours from the production of cloth and devote them to the production of wheat, it will lose 1 unit of cloth and gain 2 units of wheat. Similarly, to specialise in the production of cloth if India withdraws 10 hours of labour from wheat and use them for the production of cloth, it will lose one unit of wheat but gain 2.5 units of cloth. In this way, transfer of labour resources to the goods in which they have absolute advantage, will result in the net gain of one unit of wheat and 2.5 units of cloth. The gain in output can be distributed between the two countries through voluntary exchange.

It is also clear from above that without any increase in productive resources international division of labour and trade leads to the expansion in world output and wealth. *According to Adam Smith, given perfect competition in the industries and free trade between the countries, it is the market forces that would ensure specialisation and trade on the lines of absolute advantage.*

THEORY OF COMPARATIVE COSTS OR COMPARATIVE ADVANTAGE

The fundamental cause of international specialisation and hence international trade is the difference in costs of production. It is the relative differences in costs which determine the products to be produced by different countries. Considering climatic conditions, availability of mineral and other resources and differences in costs arising from them, every country seems to be better suited for the production of certain articles rather than for others. A country tends to specialise in the *production of those goods for which it has got relative or comparative advantage*. A country can produce numerous goods. But it will not produce all of them since it will simply not be paying to do so. It will be much better if after comparing the costs of the various articles that it can produce, it selects those in which the comparative costs are lower or in which it enjoys relative advantage. It will be to the advantage of each country as well as of the world as a whole that each country specialises in the production of those commodities for which it has comparative advantage. In this way the productive resources of the country and of the whole world will be optimally utilised.

The theory of comparative costs is simply an application of the principle of division of labour to different countries. An individual can do a number of jobs but he cannot do them all alike. Take the case of a doctor. Undoubtedly, he can do the dispensing work better than his dispenser. But still he employs a dispenser, and he himself specialises in examining the patients. He knows that the time which he will spend in dispensing can be more remuneratively employed by examining patients. The work of dispensing can be done by a low-paid person, while he can earn much more as a doctor.

Similarly, a professor may be able to teach his own son who is reading in a lower class much better than any school teacher. But if the hour that he devotes to the teaching of his son is devoted to coaching of a student for the degree examination, he will get much more payment than he has to pay a tutor whom he may employ for coaching his son. This specialisation is very gainful. This is the application of the theory of comparative costs to the case of individuals. Each individual compares the cost and the income of the various jobs that he could take up and of these he selects that one which is most profitable.

Now, let us see how this principle applies to international specialisation. The fundamental cause as to why international specialisation occurs is the differences in costs, which result from the differences in the availability of the amount and the quality of resources, the prices of these resources or factors and the method of their use. Considering the differences in costs of producing different goods, every country seems to be better suited for the production of certain goods rather than the others. *A country tends to specialise in the production of those goods in which it has got comparative advantage or lower comparative cost*. A country can produce many goods. But it will not produce all of them since it will simply not be paying to do so. It will be much better if after comparing the costs of the various goods which it can produce it selects those in which the comparative costs are lower or in which it enjoys comparative advantage. It will be to the advantage of each country as well as of the world as a whole that each country specialises in the production of those commodities for which it has comparative advantage. In this way the productive resources of the country will be optimally utilised. It is worth mentioning that *specialisation necessitates trade or exchange of goods with other countries*. Under specialisation, a country will import a good it does not produce and exports the good in which it specialises. The advantage of having a country specialise in the production of the good which it can produce more efficiently and exporting some of it in exchange for the imports of the good it can produce relatively less efficiently is that such specialisation increases the total supply of goods. In this way both countries are able to increase their level of consumption beyond what is possible in the absence of specialisation.

Let us take two countries U.S.A. and India and two goods, wheat and cloth. We shall explain what would be the basis of trade between these two countries and how the two would gain from specialising and trading with each other on the basis of comparative advantage or comparative cost. In order to simplify our analysis, we make the following assumptions :—

1. There are no transport costs between the two countries.
2. Conditions of perfect competition prevail in both countries.
3. Labour is the only resource of production and prices of products are equal to their relative labour costs.
4. There is free trade between the countries.

Specialisation and Trade with Comparative Advantage

When each country has an absolute advantage over the other in the production of a commodity, the gain from specialisation and trading between countries is quite obvious. However, a pertinent question is if the U.S.A. can produce both the commodities wheat and cloth more efficiently than India, would she gain from specialisation and trading with India. In fact, it was this question which was raised by David Ricardo, a classical economist, who put forward the theory of comparative costs (advantage) as an explanation of the potential gain from international trade.

Let us illustrate the theory of comparative cost (or comparative advantage) with a numerical example. Taking two countries, two commodities model as used by Ricardo we give in Table 43.3 *labour requirements per unit of cloth and wheat in the U.S.A. and India.*

Table 43.3.
Man-hours Required to Produce One Unit of Wheat and Cloth

Country	*Wheat*	*Cloth*	*Domestic Exchange Ratio between Wheat and Cloth*
U.S.A.	3	6	1 Wheat = 0.5 Cloth
India	12	9	1 Wheat = 1.33 Cloth

It will be seen from Table 43.3 that the U.S.A. is more efficient (or has absolute advantage) in the production of both wheat and cloth. To produce one unit of wheat the U.S.A. requires 3 man-hours, while India requires 12 man-hours. In case of cloth, to produce one unit of it 6 hours of labour are needed in the USA, while 9 hours are needed in India. It is thus evident that the U.S.A. is more efficient in the production of both the commodities as it produces them at a lower labour cost than India. That is, U.S.A. has an absolute advantage in the production of both the commodities. However, Ricardo argued that the two countries can still gain from specialising and trading between them if they produce according to their comparative advantage. It will be seen from above table that the U.S.A. is four times more efficient in the production of wheat as compared to India, while its efficiency in the production of cloth is 1.5 times greater than India. Thus, while U.S.A. has absolute advantage in the production of both wheat and cloth, it has a *comparative advantage* in the production of wheat. On the other hand, India is less efficient in the production of both wheat and cloth, its inefficiency is comparatively less in cloth. *India is said to have comparative advantage in the production of cloth.* Both the countries will be better off if the U.S.A. specialises in the production of wheat and exports it to India for import of cloth and India specialises in the production of cloth and exports it to the U.S.A. and import wheat from it.

It might appear that the U.S.A. which can produce both wheat and cloth more efficiently than India has nothing to gain by trading with India which is comparatively inefficient in the production of both wheat and cloth. Even though the U.S.A. is more efficient in the production of both wheat and cloth, she will still gain by having specialisation and trading with India. Let us see how the two countries will gain if they specialise and trade on the lines of comparative advantage. In the absence of trade between the U.S.A. and India, depending on their labour costs, one unit of wheat will be exchanged for 0.5 units of cloth in the U.S.A. and 1.33 units of cloth in India. When the U.S.A. and India specialise in wheat and cloth respectively and trade takes place between them, the U.S.A. will gain if it has to give less than can get more than 0.5 units of cloth from India for one unit of wheat and India will gain if it 1.33 units of cloth to the U.S.A. for import of one unit of wheat. Hence any exchange ratio between 0.5 and 1.33 units of cloth against one unit of wheat represents a gain for both the countries. The actual exchange rate settled between them will be determined by the *reciprocal demand* of the two countries for wheat and cloth.

Now, the question is what will be the source of gain from specialisation in the present case. The answer to this is that whereas the U.S.A. has an absolute advantage in the production of both wheat and cloth, its *margin of advantage* is greater in case of wheat as compared to cloth. Likewise, although India has absolute disadvantage in the production of both wheat and cloth, the extent of its disadvantage is less in case of cloth. Thus, with the U.S.A. specialising in the production of wheat and India in cloth, through reallocation of labour between wheat and cloth, the total production will increase. How does the total joint output of the two countries increases if U.S.A. specialises in wheat for which it has comparative advantage and India in cloth for which it enjoys comparative advantage ? We show below in Table 43.4 the gain in wheat output that occurs when U.S.A. shifts its labour resources by reducing the production of cloth by one unit and India shifts its labour to cloth by reducing the production of wheat by one unit (the above data of man-hours cost of wheat and cloth are used).

Table 43.4.
Gain in Output Resulting From Shift of Labour

	U.S.A.	*India*	*World output*
Wheat	+ 2 Wheat	– 1 Wheat	+ 1 Wheat
Cloth	– 1 Cloth	+ 1.33 cloth	+ 0.33 Cloth

It will be seen from Table 43.3 and 43.4 that if U.S.A. reduces the production of cloth by one unit 6 man-hours of labour will be released and if these are used for the production of wheat it will gain 2 units of wheat production. On the other hand, if India reduces production of one unit of wheat, 12 hours of labour will be released which on using for cloth production will result in gain of 1.33 units of cloth. It will be seen from Table 43.4 that total world output (*i.e.* joint output of the two countries) will rise by 1 unit of wheat and 0.33 units of cloth as a result of the above shift of man-hours to the products of their comparative advantage.

Thus, specialisation, according to the comparative advantage, would lead to the increase in production of both wheat and cloth and the two countries would gain from trading with each other by exporting the goods in which they speicalise.

COMPARATIVE COST THEORY: OPPORTUNITY COST APPROACH

Comparative cost theory explained above is based upon labour theory of value. But this labour theory of value has been abandoned by the modern economists. However, comparative cost theory is still believed to be valid and important basis of international trade. The labour theory of value does not hold good because the wages of labour are not brought to equality throughout the country. Further, labour is not homogeneous and the wages of different non-competing groups do not tend to be equal at least in the short run. Another serious objection against labour theory of value has been that goods are not produced by labour alone but by various combinations of different factors of production, land, labour and capital. Therefore, to compare the values of the two commodities by taking into account only their labour content would give a wrong view of the relative values because one commodity may require more capital than labour, another may require more land than labour. Therefore, variable proportions of factors used in the production of different commodities make the labour theory of value inapplicable in determining comparative cost of commodities. Therefore, a noted economist Haberler has explained comparative cost doctrine in terms of opportunity costs. As explained in an earlier chapter, the opportunity cost of a commodity, say wheat, is amount of another commodity, say cloth, which a country has to give up to produce an additional unit of wheat. Thus opportunity cost measures the ratio of marginal costs of the two commodities. In the explanation of comparative cost theory, the concept of opportunity cost is generally illustrated through production possibility curve.

Constant Opportunity Costs

We shall first explain the case when opportunity costs of the commodities in each country have been assumed to be constant. Suppose, with given resources, India can produce 20 kgs. of wheat or 40 meters of cloth, whereas with the same resources, the U.S.A. can produce 60 kgs of wheat and 80 metres of cloth. This is given in the Table 43.5.

Table 43.5
Production Possibilities in U.S.A. and India

Country	*Wheat*		*Cloth*
U.S.A	60 kgs	or	80 Metres
India	20 Kgs	or	40 Metres

From these production possibilities, it follows that U.S.A. is more efficient in the production of both wheat and cloth. However, in case of wheat its efficiency is three times greater and in case of cloth two times greater as compared to India. From these prodution possibilities we can calculate the opportunity cost of wheat and cloth in the two countries. In U.S.A. opportunity cost of 60 kg. of wheat is 80 metres of cloth, that is, 1 kg. of wheat has opportunity cost of 80/60 or 1.33 metres of cloth. Likewise, 80 metres of cloth in U.S.A. has the opportunity cost of 60 kg. of wheat, that is, 1 metre of cloth has the opportunity cost of 0.75 kg. of wheat in U.S.A. In India one kg. of wheat has the opportunity cost of 2 metres of cloth and one metre of cloth has 0.50 kg. of wheat. In Table 43.6 opportunity costs of the two countries are given :

Table 43.6. Opportunity Costs of Wheat and Cloth in U.S.A. & India

Country	*Wheat*	*Cloth*
U.S.A.	1.33 metres of cloth	0.75 kg of wheat
India	2.00 metres of cloth	0.50 kg of wheat

It is evident from the above table that in U.S.A. opportunity cost of wheat is less than that of India, while in India opportunity cost of cloth is less than that of U.S.A. *These opportunity costs reflect comparative advantage*. Thus, it would be to the advantage of U.S.A. to specialise in the production of wheat and of India in the production of cloth.

Production possibilities between wheat and cloth of U.S.A. has been show in Fig. 43.1 where on the *X*-axis wheat and on the *Y*-axis cloth have been measured, Now, line *AB* represents the production possibility curve in India between wheat and cloth. As constant costs are assumed, a straight line production possibility curve has been drawn. Similarly, in Fig. 43.2 production possibility curve between wheat and cloth of India is *CD*. In order to explain the comparative cost theory and the gain from trade we have to superimpose the production possibility curve of one country over that of the other. It will be seen from the slopes of the production possibility curves of the two countries that while India can produce cloth at the lower comparative cost, the U.S.A.can produce wheat at a comparatively cheaper cost. Therefore, it would be advantageous for India and U.S.A. to specialise

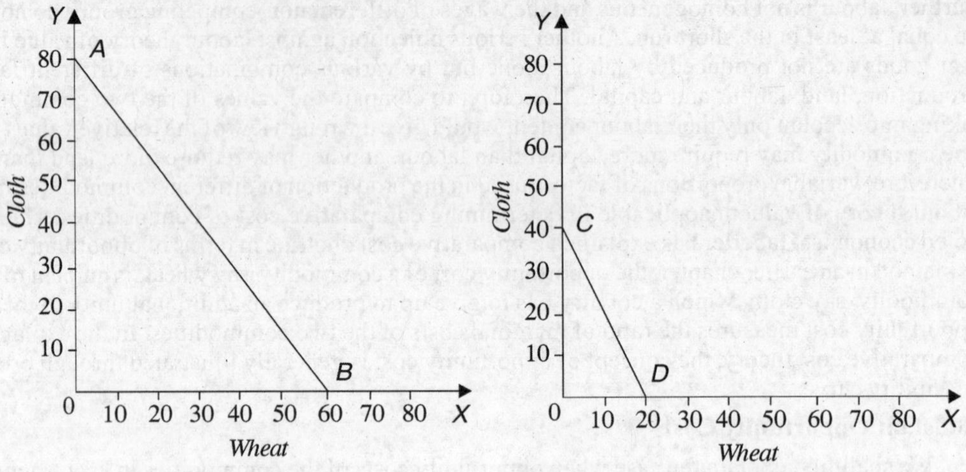

Fig. 43.1. *Opportunity Cost of U.S.A.* **Fig. 43.2.** *Opportunity Cost of India*

in cloth and wheat respectively. This can be shown by superimposing the opportunity cost curve *CD* of India over the opportunity cost of U.S.A., in such a way that point *D* is joined with point *B* of U.S.A. This has been shown in Fig. 43.3 the product possibility. Now, suppose that before trade U.S.A. is producing and consuming at point *E* on her production possibility curve *AB*. The slope of the production possibility curve *AB* shows the comparative cost ratio of the two commodities in U.S.A. If U.S.A. can trade with another country, say India, at a different price-ratio than this, it will then gain from the trade. It will be seen in Fig. 43.3 that the product possibility curve *AB* of U.S.A. lies below India's production possibility curve *C′D′*. The terms of trade which will be settled between the two will lie between the production possibility curve *C′D′* of India and the production possibility curve *AB* of *U.S.A*. Suppose, *D′T′* is the terms of trade line showing the price ratio settled between the two countries. Since U.S.A. has a comparative advantage in the production of wheat it will specialise in wheat and would produce *OB* or 60 units of wheat, whereas India has comparative advantage in the production of cloth and will specialise in cloth production. After trade the equilibrium of U.S.A. would lie at some point on the terms of the trade line *D′T*, depending upon its demand for goods. Suppose *R* is such point on the terms of trade line *D′T*. Thus U.S.A. would now get the combination of the goods indicated by point *R* for its consumption while it would be producing only wheat. It will be seen From Fig. 43.3 that U.S.A.

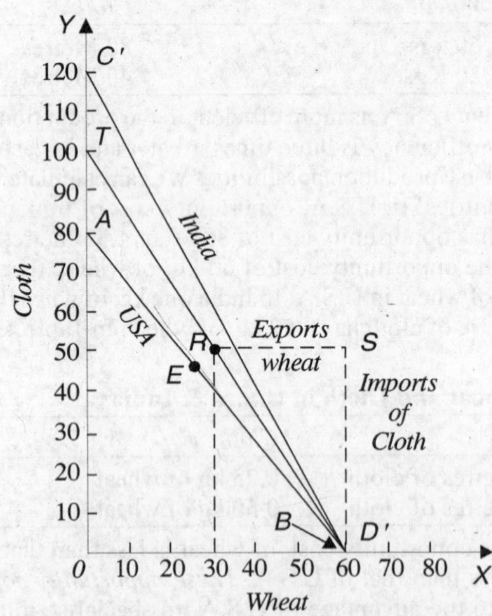

Fig. 43.3. *Comparative Cost Theory: Constant opportunity costs*

would gain from specialisation and trade as point *R* lies at a higher level than point *E*. At point *R*, U.S.A. would be consuming more of both wheat and cloth than at point *E* which is the position before trade. At point *R*, U.S.A. would be exporting *RS* amount of wheat and would be getting in return for it *SD′* amount of cloth from India. Similarly, it can be shown that India will also gain from specialising in cloth and exchanging it for wheat with U.S.A.

It may be noted that in case of constant opportunity costs, there is complete specialisation, that is, of the two goods a country produces only one commodity, that is, either wheat or cloth after specialisation and trade.

Comparative Cost Theory: Increasing Opportunity costs

In the above analysis of opportunity cost version of the comparative cost theory we have assumed that opportunity costs are constant. The constancy of opportunity costs implies that the various resources are equally suitable for the production of each of the two goods. However, this does not represent the real situation, where all resources do not produce equally well both of the two commodities. Some resources may be equally suitable for the production of both wheat and cloth but all resources are not of this kind. Resources such as land, villagers trained for the art of agriculture, are more suited for the production of wheat. On the other hand, resources such as spindles, looms, technicians are suitable for producing cloth. When resources are not equally efficient in the production of the two commodites we may have the situation of increasing costs. In case of increaisng opportunity costs, the production possibility curve instead of being a straight line is concave to the origin, as shown in Fig. 43.4. This indicates as we shift resources from the production of wheat to the production of cloth, marginal opportunity cost of cloth (that is, the amount of wheat forgone for a unit of cloth) goes on increasing and vice versa.

Under conditions of increasing opportunity costs, the production possibility curve is not identical with a price curve as in case of constant opportunity costs. The domestic price-ratio (that is, the rate of which two goods would be exchanged in the absence of trade) are not determined by the production possibility curve alone. Instead, this domestic exchange rate is determined by the production possibilities (i.e. supply or cost conditions) and the demand for the goods. In the equilibrium situation, the two goods would be produced in such quantities where :

$$MRT_{CW} = \frac{P_C}{P_W}$$

where

MRT_{CW} stands for marginal rate of transformation between the two goods

P_C denotes price of cloth

P_W denotes price of wheat.

In Fig. 43.4 suppose the price-ratio line between wheat and cloth, as determined by demand and supply conditions, is *pp′*. The production equilibrium in the absence of trade will be reached at point *R* where the price line is tangent to the production possibility curve so that the marginal rate of transformation between the two goods is equal to the price ratio between them.

Now, if foreign trade possibilities are such that price of cloth is relatively more in foreign market than at home, it would be advantageous for the country to enter into trade and increase the production of cloth and reduce that of wheat. Suppose the price-ratio line in the foreign market (or what is also called *terms of trade* line) is given by the price line *tt* in Fig. 43.4. It will be seen from this figure that slope of the price-ratio line *tt* indicates higher price for cloth and lower one for wheat as compared to domestic price-ratio line *pp′* since according to *tt* more wheat can be obtained for a given amount of cloth.

Given these trade possibilities, it would be advantageous for India to increase the production of cloth. (or so to say specialise in cloth) by shifting to it some resources from wheat. With price-ratio (or terms of trade) line *tt*, India would be in equilibrium at point *R′* where its production possibilities curve *AB* is tangent to the terms of trade line *tt*. However, the consumption of two goods in a country depends on the tastes or demand pattern for the goods. In the modern international trade theory, tastes or demand for goods in a country are shown through the medium of community indifference curves, the use of which we are avoiding at this stage. It will suffice to mention here that consumption of the two goods in India would take place at a certain point on the terms of trade line *tt* to the left of point R, say at point C. Of course, this consumption point C would be determined by the demand conditions in India. With production at R' and consumption at C in India after it enters into trade with U.S.A. it follows that India would produce $OH(=JR')$ of cloth, OJ (or HR') of wheat, and would consume OG (or WC) of cloth and OW (= GC) of wheat. Further India will export QR' of cloth and import QC of wheat. It is evident from Fig.43.4 that *trade has enabled India to consume more of both cloth and wheat than it could produce at home with its own production possibilities.*

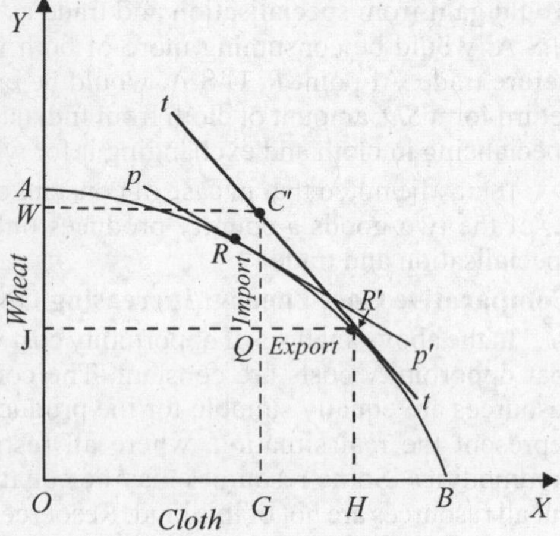

Fig. 43.4. *Comparative Cost Theory: Increasing Opportunity Costs.*

It is noteworthy that unlike in case of constant opportunity costs, in case of increasing opportunity costs, specialisation is not likely to be complete. This is because of the occurrence of diminishing returns or increasing costs as the production of one commodity is stepped up at the expense of the other commodity. When production of cloth is expanded productive resources less suited to the production of cloth are drawn into that industry. This raises per unit cost of cloth. On the other hand, as more factors of production are drawn from wheat for allocating them to the production of cloth, per unit cost of wheat falls. The opposite will happen in the trading partner who will increase the production of wheat and reduce that of cloth by shifting resources out of the latter into the former. Cost ratios in the two countries may become equal *before* either country completely specialises in the production of a single commodity.

The law of comparative costs when freed from the labour theory of value and expressed in terms of opportunity costs is still believed to be true by the modern economists. In a two-commodity world when one country can produce both of them at a lower cost than another, it will pay to it to specialise in the production of a commodity which it can produce at comparatively lower cost and import the other commodity for which it has a comparatively higher costs. It gains from trade as it is able to get a higher price for the commodity in which it specialises because of its relatively lower cost of production and also because it pays a relatively lower price for the commodity for which its efficiency is relatively not so high.

CRITICAL APPRAISAL OF COMPARATIVE COST THEORY

Theory of comparative cost which is the important doctrine of classical economics is still valid and widely acclaimed as the correct explanation of international trade. Most of the criticisms that have been leveled against this doctrine relate to the Ricardian version of comparative cost theory based on labour-theory of value. Haberler and others broke away from this labour-cost version and reformulated the comparative cost theory in terms of opportunity costs which takes into consideration all factors. The basic contention of the theory that a country will specialise in the production of a commodity and export it for which it has a lower comparative cost and import a commodity which can be produced at a lower comparative cost by others, is based on a sound logic. The theory correctly explains the gain from trade accruing to the participating countries if they specialise according to their comparative costs. These merits of the theory have led Professor Samuelson to remark, "*If theories, like girls, could win beauty contests, comparative advantage would certainly rate high in that it is an elegantly logical structure.*"[1] He further writes, "the theory of comparative advantage has in it a most important glimpse of truth.... A nation that neglects comparative advantage may have to pay a heavy price in terms of living standards and potential rates of growth"[2] Despite the sound logical structure and vivid explanation of gains from trade, the comparative cost theory, especially the Ricardian version based on labour theory of value has been criticized. The following criticisms have been levelled against this theory :–

1. In the first place, Ricardian version of comparative cost theory has been attacked on the

1. *P.A.* Samuelson, *Economics*, McGrawHill, 8th Edition, 1973.

ground that being *based on labour theory of value*, it considers only labour cost to measure the comparative costs of various goods. It has been pointed out that labour is not the only factor needed for the production of commodities, other factors such as capital, raw materials, land also contribute to production. Therefore, it is the total money costs incurred on labour as well as other factors that should be considered for assessing comparative costs of various commodities.

Taussig tried to defend Ricardo by pointing out that even if labour theory of value was defective and even if other factors made important contributions to the production of goods, comparative costs could still be based on labour cost alone, if it is assumed that the trading countries are at the same stage of technological development. This is because, he argued that given the same technological development, the proportions in which other factors could be combined with labour would be the same. In view of this he asserted that other factors could be validly ignored and for purpose of comparative costs relative efficiency of labour alone of different countries could be considered. However, Taussig's defense of Ricardian version of comparative cost theory is poor and invalid. The various trading partners are not at the same stage of technological development and therefore the factor proportions used for the production of commodities in different countries are vastly different. Hence, it is quite unrealistic and improper to consider relative efficiency of labour alone.

However, as stated earlier, Haberler rescued the comparative cost theory from labour theory of value and reformulated it in terms of opportunity cost which covers all factors. The opportunity cost version of comparative cost theory has already been explained above.

2. The comparative cost theory explained that different countries would specialise in the production of goods on the basis of comparative costs and that they would gain from trade if they export those goods in which they have comparative advantage and import those goods from abroad in respect of which other countries enjoyed comparative advantage. But it could not provide a satisfactory explanation *of why comparative costs of producing commodities in various countries differ*. Ricardo thought comparative costs of producing commodities in various countries differed due to the differences in efficiency of labour. But this begs the question why labour efficiency is different in various countries.

The credit of providing an adequate and valid answer to this question goes to Heckscher and Ohlin who explained that comparative costs of different commodities in the two countries vary because of the following factors :

1. The various countries differ in respect of factor endowments suited for the production of different commodities.

2. The different commodities require different factor proportions for their production.

Thus Heckscher and Ohlin supplemented the comparative costs theory by providing valid reasons for differences in comparative costs in various countries. We shall explain in detail the Heckscher-Ohlin theory of international trade in the next chapter.

3. Against the Ricardian doctrine of comparative cost it has also been said that it is based on the constant cost of production in the two trading countries. This assumption of constant costs leads them to conclude that different countries would *completely specialise* in the production of a single product on the basis of their comparative costs. Thus, of the two commodities cloth and wheat, if India has a comparative advantage in the production of cloth, it will produce all cloth and no wheat. On the other hand, if U.S.A. has a comparative advantage in the production of wheat, it will produce all wheat and no cloth. But the pattern of international trade shows that this is far from reality. As a matter of fact, a stage comes when it is no longer advantageous for India to import wheat from U.S.A. (because of increasing costs in producing wheat). Further, in the real world it is found that countries do not have complete specialisation. Indeed, a country produces a certain commodity and also imports a part of it.

However, it may be noted that even if the phenomenon of increasing costs is taken into account, foreign trade can still be explained in terms of differences in comparative costs. Only in the situation of increasing costs, countries would not have complete specialisation. As explained earlier, opportunity cost version of comparative costs theory does consider the case of increasing costs.

4. The Ricardian theory of comparative costs, has also been critcized for its not going into the question *what determines the terms of trade* between the countries. Voicing this criticism Elseworth

2. *Ibid.*

remarks, 'the comparative costs theorem, the way in which Ricardo set up his illustration, tended to obscure the problem of the terms of trade."[3] Ricardian theory of comparative costs explains what commodity a country will export and what commodity it will import but it does not investigate at what rate it will exchange its exports for imports (*i.e.* terms of trade). However, the fixation of terms of trade is a vital issue, for on it a country's share of gains from trade depends.

It is worthwhile to note that J.S. Mill, another noted classical economist, removed this shortcoming of the comparative cost theory by supplementing it with *Reciprocal Demand Theory* which explains the determination of terms of trade. We shall explain Mill's reciprocal demand theory in a later chapter.

5. Ohlin attacked the comparative cost theory for its assumption that *factors of production were perfectly mobile within a country but immobile between countries*. He pointed out that immobility of factors between countries could not serve as a basis for international trade, since immobility of factors is not peculiar to the relations between countries but is also present between different regions of the same country. He further expressed the view that comparative cost doctrine applied not only to international trade but also to inter-regional trade. Indeed, according to him, *international trade is only a special case of inter-regional trade*. He further criticized the classical theory of comparative cost for its emphasis on supply conditions as an explanation of international trade and its neglect of the importance of demand conditions in determining the pattern of international trade. He writes, "The comparative cost reasoning alone explains very little about international trade. It is indeed nothing more than an abbreviated account of the conditions of supply"[4] According to him, prices of different goods and their quantities produced and consumed depend on both supply and demand conditions. He therefore, propounded a new theory of international trade based on general equilibrium theory of value.

It may be mentioned here that Ohlin's criticisms do not invalidate comparative cost theory. Indeed, he only refined and modified it. Even in his theory, popularly known as factor-proportions theory of international trade, comparative costs serve as a basis of international trade. His contribution lies in his inquiring into the question why comparative costs of commodities in different countries differ and offering a satisfactory explanation of it in terms of different factor-proportions required for the production of various goods. He further improved the comparative cost theory by incorporating in his analysis the demand aspect as he based his international trade theory on the general equilibrium theory of value.

6. It is alleged that comparative costs theory is *static in character as it is based on fixed supplies of factors of production, the given technology, and the fixed and identical production functions in the trading countries*. Its conclusions cannot therefore be applied in the context of a dynamic economy, especially in the present-day developing countries where resources are being developed, technology is being improved, production functions are undergoing a change. Indeed, structural changes are being brought about in these economies. In view of the changes in factor supplies and technology in developing countries, comparative costs of producing different commodities are also changing. In this dynamic context, a developing economy may have a comparative disadvantage in producing a certain commodity but may attain a comparative advnatage after a certain stage of its development.

Note that this criticism about the static character of the comparatice cost theory does not invalidate it. It only pinpoints the need for reformulating and refining it so as to make it applicable to the dynamic conditions of the developing countries.

Conclusion

To sum up, bereft of the labour theory of value and expressed in terms of opportunity costs comparative cost theory is still a valid explanation of international trade. It highlights the need for removal of artificial restrictions in the form of tariffs and other means on foreign trade so that various countries specialise on the basis of their comparative costs and derive mutual benefits from trade. This theory has been a victim of undue criticisms such as that it assumes the absence of transport costs, the existence of perfect competition and full employment, and further that it considers two commodities, two countries model. These are only simplifying assumptions and do not invalidate

3. P.T. Elseworth *The International Economy*.
4. B. Ohlin, *Interregional and International Trade*, p. 586

its conclusions in a substantial way. Indeed, every theory makes some such simplifying assumptions in order to bring out the economic forces that have an important bearing on the subject under investigation.

QUESTIONS FOR REVIEW

1. Explain how international trade extends a country's consumption possibility beyond its production possibility frontier.
2. Taking an example of two countries, two goods and one factor of production, illustrate the Ricardian theory of comparative advantage. *[B.A. (Hons) D.U. 2000]*
3. Explain the theory of comparative cost with the opportunity cost approach. *[B.Com (Hons) D.U. 1996]*
4. In the Ricardian model of trade do you think introduction of trade would lead to complete specialisation. Explain briefly. *[B.A. (Hons) D.U. 1999]*
5. Distinguish between absolute difference in cost and comparative difference in cost. How does free trade between countries benefit consumers in both countries ? *[B.Com (Hons) D.U. 2001]*
6. Suppose that labour is the only factor of production in two countries *A* and *B*, who produce only two goods *X* and *Y*. Country *A* is endowed with 6000 units of labour and country *B* is endowed with 3000 labour units. The labour units required to produce one unit of a good in each country are as follows:

Labour Units required in

	Country A	Country B
1 unit of X	4	1
1 unit of Y	3	2

Answer the following questions :

(*a*) Which country has absolute advantage over the other in which good/goods?
(*b*) Which country has comparative advantage over the other in which good/goods?
(*c*) Is trade possible between *A* and *B* ? What will be the basis of trade?
(*d*) Before trade, what will be the price of good *X* in terms of good *Y* in country *A* and *B*?
(*e*) In which price range will international trade be beneficial to both the countries, assuming no transport costs?
(*f*) Draw the production possibility frontier of country *A* and *B* before and after trade. *[B.Com (Hons) D.U. 2002]*

[**Hints.** The country *B* has absolute advantage in the production of both goods *X* and *Y* over country *A*. However, country *B* is comparatively more efficient in the production of good *X*, whereas country *A* is relatively less inefficient in the production of *Y*. Thus country *B* has comparative advantage in the production of *X* and the country *A* has comparative advantage in the production of good *Y*.
The comparative advantage is the basis of trade between the two countries).

7. Show that gain from trade arises partly from the opportunity to trade at the international price ratio and partly from the specialisation in the product in which the country has comparative advantage.
8. Identify the conditions that may give rise to trade between two nations. How can two countries simultaneously gain from trade? *[B.A. (Hons) D.U. 2001]*
9. (*a*) There are two countries *A* and *B*, each producing two goods *X* and *Y*, with full employment of its resources, country *A* can produce a maximum of 18 units of *X* or 12 units of *Y*; country *B* can produce a maximum of 6 units of *X* or 12 units of Y. In both countries the marginal rates of transformation are constant.

(*i*) If trade is allowed, which countries specialise in which goods?
(*ii*) In the pre-trade situation, country A produces 9 units of *X* and 6 units of *Y*, and country *B* produces 4 units of *X* and 4 units of *Y*. With trade the international rate of exchange becomes $P_x = P_y$. Calculate the gains from trade for each country. *[B.A. (Hons.) D.U. 1997]*

CHAPTER 44
HECKSCHER–OHLIN THEORY OF INTERNATIONAL TRADE

The classical comparative cost theory did not satisfactorily explain why comparative costs of producing various commodities differ as between different countries. The new theory propounded by Heckscher and Ohlin went deeper into the underlying forces which cause differences in comparative costs. They explained that it is differences in factor endowments of different countries and different factor-proportions needed for producing different commodities that account for difference in comparative costs. This new theory is therefore-called Heckscher-Ohlin theory of international trade. Since there is wide agreement among modern economists about the explanation of intenational trade offered by Heckscher and Ohlin this theory is also called *modern theory of international trade*. Further, since this theory is based on general equilibrium analysis of price determination, this is also known as *General Equilibrium Theory of International Trade*.

It is worthwhile to note that, contrary to the viewpoint of classical economists, Ohlin asserts that there does not exist any basic difference between the domestic (inter-regional) trade and international trade. Indeed, according to him, *international trade is only a special case of inter-regional trade*. Thus, Ohlin asserts that it is not the cost of transport which distinguishes international trade from domestic trade, for transport cost is present in the domestic inter-regional trade. Trade because currencies of different countries are related to each other through foreign exchange rates which determine the value or purchasing power of different currencies. Ohlin, therefore, regards different nations as mere regions separated from each other by national frontiers, different languages and customs, etc. But these differences are not such that prevent the occurrence of trade between nations. He, therefore, asserts that general theory of value which can be applied to explain inter-regional trade can also be applied equally well to explain international trade.

According to general equilibrium theory of value, relative prices of commodious are determined by demand for and supply of them. In the long-run equilibrium under conditions of perfect competition, relative prices of commodities, as determined by demand and supply, are equal to average cost of production. The cost of production of a commodity, as is well-known, depends upon the prices paid for the factors of production employed in the production of that commodity. Factor prices in turn determine the incomes of the factor owners and hence the demand for goods. Thus there is mutual inter-dependence between prices of commodities and prices of factors and the exchange of goods and factors between different individuals in a region or country. This is how general equilibrium theory of value explains prices of commodities and factors between different individuals in a region or a country. However, according to Ohlin, the classical analysis presumes it to apply to a single market in a country and ignores the space factor whose intrdocution is crucial for explanation of trade between regions. The factors which explain the trade between different regions also explain the trade between different nations or countries as well.

Heckscher-Ohlin Theory

According to Ricardo and other classical economists, international trade is based on differences in comparative costs. It is important to note that Heckscher and Ohlin agreed with this fundamental proposition and only elaborated this by explaining the factors which cause differences in comparative costs of commodities between different regions or countries. Ricardo and others who followed him explained differences in comparative costs as arising from differences in skill and efficiency of labour alone. This is not a satisfactory explanation of differences in comparative cots. Ohlin pointed out more significant factors, namely, differences in factor endowments of the nations and difference in factor proportions of producing different commodities, which account for differences in comparative costs and hence from the ultimate basis of inter-regional or international trade. Thus, Heckscher-Ohlin theory does not contradict and supplant the comparative cost theory but supplements it by offering sufficiently satisfactory explanation of what causes differences in comparative costs. According to Ohlin, the underlying forces behind differences in comparative costs are two-fold :

1. T*he different regions or countries have different factor endowments.*
2. *The different goods require different factor-proportions for their production.*

It is a well-known fact that various countries (regions) are differently endowed with productive factors required for production of goods. Some countries posses relatively more capital, some relatively more labour, and some relatively more land. The factor which is relatively abundant in a country will tend to have a lower price and the factor which is relatively scarce will tend to have a higher price. Thus, according to Ohlin, factor endowments and factor prices are intimately associated with each other, Suppose K stands for the availability or supply of capital in a country, L for that of labour and PK for price of capital and PL for the price of labour. Further, take two countries A and B; in country A capital is relatively abundant and labour is relatively scarce. The reverse is the case in country B. Given these factor-endowments, in country A capital will be relatively cheaper. In symbolic terms :

$$\text{Since } \left(\frac{K}{L}\right)_A > \left(\frac{K}{L}\right)_B$$

$$\left(\frac{PK}{PL}\right)_A < \left(\frac{PK}{PL}\right)_B$$

Thus the differences in factor endowments cause differences in factor prices and therefore account for differences in comparative costs of producing different commodities.

Together with the difference in factor-endowments, differences in factor proportions required for the production of different commodities also constitute an important force underlying differences in comparative costs as between different countries. Some commodities are such that their production requires relatively more capital than other factors ; they are therefore called capital-intensive commodities. Still other commodities require relatively more land than capital and labour and are therefore called land-intensive commodities. These differences in factor-productions (or what is also called differences in *factor-intensities*) needed for the production of different commodities account for differences in comparative costs of producing different commodities. The differences in comparative costs of producing different commodities lead to the differences in market prices of different commodities in different countries.

It follows from above that some countries have a comparative advantage in the production of a commodity for which the required factors are found in abundance and comparative disadvantage in the production of a commodity for which the required factors are not available in sufficient quantities. Thus a country A which has a relative abundance of capital and relative scarcity of labour will have a comparative advantage in specialising in the production of capital-intensive commodities and in return will import labour-intensive goods. This is because $\left(\frac{PK}{PL}\right)_A < \left(\frac{PK}{PL}\right)_B$.

On the other hand, a labour-abundant country *B* with a scarcity of capital will have a comparative advantage in specialising in the production of labour-intensive commodities and export some quantities of them and in exchange for import capital-intensive commodities. This is because in this country $\left(\frac{PL}{PK}\right)_B < \left(\frac{PL}{PK}\right)_A$.

If factor endowments in the two countries are the same and factor-productions used in the production of different commodities do not differe there will be no differences in relative factor prices $\left[i.e. \left(\frac{PK}{PL}\right)_A = \left(\frac{PK}{PL}\right)_B\right]$ which will mean differences in comparative costs of producing commodities in the two countries will be non-existent. In this situation the countries will not gain from entering into trade with each other.

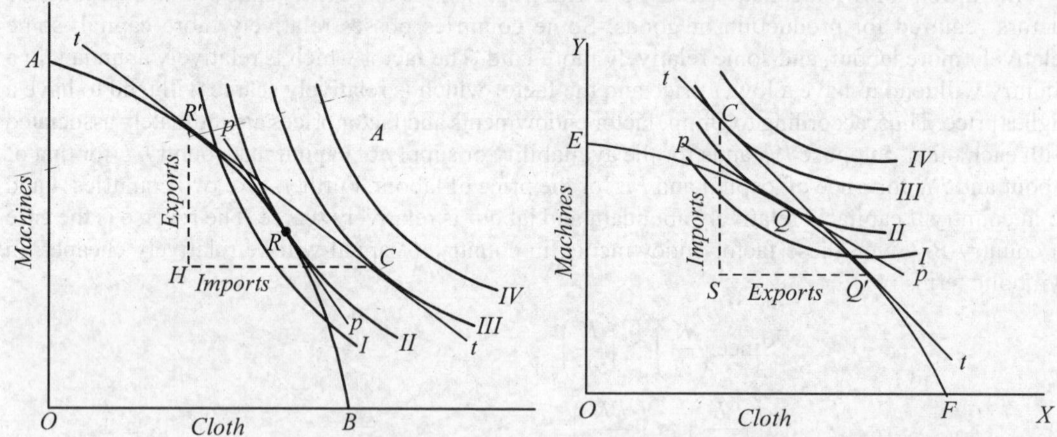

44.1. *Exports and Imports of U.S.A.* **Fig. 44.2.** *Exports and Imports of India*

Let us graphically explain the Heckscher-Ohlin theory of international trade. Take two countries U.S.A. and India. Assume that there is a relative abundance of capital and scarcity of labour in U.S.A. and, on the contrary, there is a relative abundance of labour and scarcity of capital in India. (This is the real situation as well). Given these factor endowments we have drawn the production possibility curves (also known as transformation curves) between two commodities, cloth and machines of the two countries, U.S,.A. and India in Fig. 44.1 and 44.2 respectively. Since the two countries have different factor endowments their production possibility curves will differ. As will be seen from Fig. 44.1, the production possibility curve *AB* of U.S.A. shows that given its factor endowments, U.S.A. can produce relatively more of capital-intensive commodity machines and relatively less of labour-intensive commodity cloth. On the contrary, as will be seen from Fig. 44.2 with given factor endowments, India can produce relatively more of labour-intensive commodity cloth and relatively less of capital-intensive commodity machines. In the absence of foreign trade, equilibrium in each country would be determined by the following rule :

$$MRT_{MC} = MRS_{MC} = \frac{PM}{PC}$$

Where MRS_{MC} stands for a marginal rate of transformation of machines into cloth, MRS_{MC} for marginal rate of substitution of machines for cloth and $\frac{PM}{PC}$ for the price ratio between the machines and cloth.

In the geometric terms, the above rule implies that in the absence of foreign trade the production and consumption in the two countries would take place at the tangency point of the given production possibility curve with the highest possible community indifference curve.

It will be observed from Fig. 44.1 that in the absence of trade, U.S.A. will be in equilibrium position with production and consumption at point *R* where its production possibility curve *AB* is tangent to its community indifference curve *II*. The tangent *pp* to the production possibility curve *AB* and the community indifference curve *II* at point *R* indicates the ratio of price of the two commodities (*i.e.* the domestic rate of exchange) before trade in U.S.A.

As regards India, as shown in Fig. 44.2 before trade, she will be in equilibrium with production and consumption at point *Q* at which its production possibility curve is tangent to its community indifference curve *II*. Tangent *pp* at point *Q* to its production possibility curve *EF* and the community indifference curve *II* shows the domestic rate of exchange of two commodities before foreign trade.

It will be seen from Fig. 44.1 and 44.2 that the price ratio (rate of exchange) of the two commodities in the two countries differs (slopes of tangents *pp* in them vary). It will therefore pay the two countries to enter into trade with each other. Suppose, the terms of trade, that is, the ratio of exchange of goods between the two countries is given by the line *tt*. It will be observed that with terms of trade line *tt*, U.S.A. will be in equilibrium from the view point of production at point *R′* at which the terms of trade line *tt* is tangent to its production possibility curve *AB*. However, its consumption point after trade is *C* which is determined by the tangency of the terms of trade line *tt* with the community indifference curve *III*. It will be seen from Fig. 44.1 that in U.S.A. the consumption point *C* after trade lies at a higher indifference curve than before trade indicating the gain from trade it obtains. The consumption point *C* of U.S.A. as compared to its production point *R′* after trade reveals that U.S.A. produces *HR′* more of machines and *HC* less of cloth than it consumes domestically. Thus U.S.A. will export *HR′* of machines and import *HC* of cloth.

As regards India, it is evident from Fig. 44.2 that as a result of trade its production point will shift to point *Q′* where its product possibility curve *EF* is tangent to the terms of trade line *tt*. After trade, consumption in India will take place at point *C* at which the terms of trade line *tt* is tangent to its community indifference curve *III*. As a result of trade India has also gained as she has reached a higher community indifference curve. Thus after trade with the production point *Q′* and consumption point *C*, India will produce *SQ′* more of cloth and *SC* less of machines than it consumes at home. Thus India will export *SQ′* of cloth and import *SC* of machines.

It follows from above that due to differences in factor endowments in U.S.A. and India and also due to different factor proportions required for the production of different commodities the basis for trade between the two countries exists and both would gain from trade by specialising in the production of commodities which require factors in respect of which they are well endowed and will import those commodities which need factors which are relatively in scarce supply.

EQUALISATION OF FACTOR PRICES

As explained above, trade takes place when relative prices of commodities differ between countries due to difference in comparative costs. The volume of trade goes on expanding until the differences in relative prices of commodities (ignoring transport costs) are eliminated. Thus, in the absence of transport costs and tariffs, the effect of trade would be to equalise the relative commodity prices in the trading countries. For instance, in the example given above, as a result of trading between U.S.A. and India under the given conditions, the exports of machines to India by the U.S.A. will lower the prices of machines in India and raise them in U.S.A. This is because before opening up of trade, the price of machines was relatively high in India as compared to that in U.S.A. Now, with the increase in supply of machines in India due to their imports from U.S.A., their prices will fall. On the other hand, prices of machines in U.S.A. will rise due to the reduced supply in the domestic market than before as a result of their exports to India. Thus the prices of machines would tend to rise in U.S.A. and those in India would tend to fall. The flow of trade would continue expanding untill the prices of machines in the two countries (in the absence of transport costs and tariffs) would become equal. Likewise, the price of cloth which before trade is lower in India than in U.S.A. would tend to become equal in the two countries following the opening up of trade between them.

Equalisation of factor prices

It is important to note that *trade tends to equalise not only commodity prices but also factor prices*. The exports of a product using the abundant factor by a country will cause the demand for that factor to increase and thereby make it relatively less abundant and raise its price. On the other hand, the imports of a product embodying large amounts of relatively scarce factor would make it less scarce and tend to lower its price. Thus changes in factor prices following the trade of commodities between the two countries result in equalisation of factor prices in them.

Let us take an example. As noted above, in U.S.A. capital is relatively abundant and cheap whereas labour is relatively scarce and expensive. On the countrary, in India labour is relatively abundant and cheap whereas capital is scarce and expensive. With these factor endowments it will pay India to export labour-intensive commodity cloth which can produce it at a cheaper price and in exchange to import capital-intensive commodity machines from U.S.A. which can produce them at a lower price. As a result of this trade, the demand for labour in India would increase and its price would tend to increase. Now, with the imports of labour-intensive commodity cloth by U.S.A. and concentrating its more resources on production of capital-intensive machines, the demand for labour in U.S.A. would decrease and its price would tend to fall. Thus, other things remaining the same, the price of labour in India and U.S.A. would tend to become equal after opening up of trade between the two countries. The same applies to the price of capital. To sum up, *according to Heckscher-Ohlin theory, free trading of commodities between the two countries results in equalisation of factor prices*. If factors were mobile between countries, then the free movement of factors from one country to another would have equalised their prices.

But in actual practice factors lack interregional and international mobility. Therefore, in absence of trade of commodities, factor prices would not tend to the become equal in the different countries. Thus Ohlin argues that what would have been accomplished through free movement of factors between countries is indirectly accomplished through movement of commodities embodying different factor-proportions. Indeed, according to Ohlin, *international trade in commodities serves as a substitute for international mobility of factors*.

It is worthwhile to note that trade would achieve complete factor price equalisation only when some conditions and assumptions are fulfilled. It is also realised that these conditions and assumptions are quite restrictive so that in actual practice differences in factor-prices are not completely eliminated. The conditions and assumptions underlying the factor-price equalisation theorem are :

1. Tastes, that is, demand pattern for commodities, are the same
2. It is the supply conditions of the factors which are different in different countries and no qualitative differences prevail in them. This implies that the level of technological progress is the same in the different countries.
3. Production function of each commodity is the same in the different countries and is of a simple character, that is, it is either capital-intensive, or labour-intensive. In other words, production functions of commodities provides a limited degree of factor substitution.
4. There are no restrictions on trade in the form of tariffs and quotas in the trading countries.

5. There are no transport costs.
6. There exists perfect competition in the commodity market as well as in factor markets in each region or country.

Since in the real world, above conditions are not fulfilled, complete factor price equalisation does not take place. However, this does not invalidate the factor price equalisation theorem. Indeed, every theory is based upon some assumptions. What this theory asserts is that, given the above conditions, factor prices would become equal. To the extent these conditions do not exist, factor prices will remain unequal even after trade takes place between the countries.

Critical Evaluation of Heckscher-Ohlin Theory of International Trade

Heckscher and Ohlin theory has made invaluable contributions to the explanation of international trade. Though this theory accepts comparative costs as the basis of international trade, it makes several improvements in the classical comparative cost theory. First, it rescued the theory of international trade from the grip of labour theory of value and based it on the general equilibrium theory of value according to which both demand and supply conditions determine the prices of goods and factors. Second, Heckscher-Ohlin theory removes the difference between intenational trade and inter-regional trade, for the factors determining the two are the same. Third, a significant improvement is the explanation offered for difference in comparative costs of commodities between trading countries. Ricardo thought that the differences in labour efficiency alone accounted for the differences in comparative costs. According to Heckscher and Ohlin, as seen above, the differences in factor-endowments of the countries and also the differences in factor proportions required for producing various commodities explain differences in comparative costs and hence from the ultimate basis of intenational trade. These reasons advanced by Heckscher and Ohlin for differences in comparative costs of commodities in different countries are considered to be broadly true.

Fourth, as has been pointed out by Prof. Lancaster, Heckscher-Ohlin model provides a satisfactory picture of the future of foreign trade. According to the Ricardian theory, international trade exists because of differences in skill and efficiency of labour alone. This implies that as there is transmission of knowledge between the countries so that they master the techniques and skills of each other, then differences in comparative costs would cease to exist and as a result international trade would come to an end. But this is not likely to occur despite the fact that transmission of knowledge and techniques has greatly increased these days. Heckscher and Ohlin explain that international trade is due to the differences in factor-endowments (*i.e.* differences in supplies of all factors and not only of labour efficiency) and different factor-proportions required for different commodities. Since the factors such as land and other natural resources lack mobility, international trade would not cease to exist even if there is perfect transmission of knowledge between the countries.

Despite the above merits of Heckscher-Ohlin theory, it has some shortcomings which are briefly discussed below :

1. In the Heckscher-Ohlin theory it has been assumed that relative factor prices reflect the relative supplies of factors. That is, a factor which is found in abundance in a country will have a lower price and vice versa. This means that in the determination of factor-prices supply outweighs demand. But if demand for factors prevails over supply, then factor prices so determined would not conform to the supplies of factors. Thus, if in a country there is abundance of capital and scarcity of labour in physical terms but there is relatively much greater demand for capital, then the price of capital would be relatively higher to that of labour. Then, under these circumstances, contrary to its factor-endowments, the country many export labour-intensive goods and import capital-intensive

goods. Perhaps it is this which lies behind the empirical findings by Leontief that though America is a capital abundant and labour-scarce country, in the structure of its imports capital-intensive goods are relatively greater whereas in the structure of its exports labour-intensive goods are relatively greater.[1] As this is contrary to the popularly held view, this is known as **Leontief Paradox**.

2. **Differences in preferences or demands for goods.** Against Hecksher-Ohlin theorem, it has also been pointed out that differences in tastes and preferences for goods or, to put it in other words,

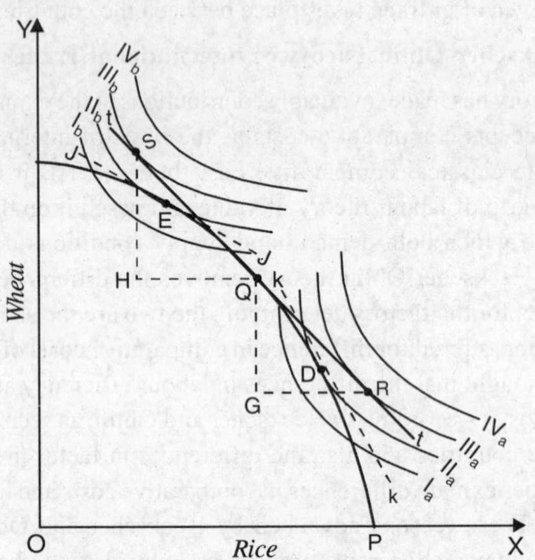

Fig. 44.3. *Differences in Consumption Pattern and International Trade*

differences in pattern of demand also give rise to trade between the countries. This is because under differences in demand or preferences for goods, the commodity price-ratios would not conform to the cost-ratios based on factor endowments. Let us take an extreme example. Suppose there are two countries *A* and *B* with same factor-endowments. According to Heckscher-Ohlin theorem, with same factor endowments cost-ratio of producing the two commodities and hence the commodity price ratio would be the same. Hence there is no possibility of trade between the two countries on the basis of Heckscher-Ohlin theorem. However, trade between the two countries is possible if the demand pattern or preferences of the people of the two countries for wheat and rice greatly differ. This is illustrated in Fig. 44.3. Suppose people of country A prefer rice to wheat and people of country *B* prefer wheat consumption. With these preferences for wheat and rice we have drawn the community indifference curves of the two countries *A* and *B*. Since factor-endowments in the two countries are the same, *PP* is the production possibility curve in both of them. In the absence of trade country *A* is in equilibrium at point *D* where its community indifference curve II_a is tangent to the production possibility curve *PP*. *kk* is the commodity price-ratio line determined in country *A*. Country *B* is in equilibrium at point *E* where its community indifference curve II_b is tangent to the production possibility curve *PP*. *JJ* is the commodity price-ratio determined in country *B*. Thus, due to difference in demand (preferences) for rice and wheat in the two countries different commodity price ratios

1. See W.W. Leontief. 'Factor Proportions and Structure of American Trade, *Review of Economic Statistics,* 1956.

have come to prevail in the two countries though their factor endowments (and therefore the cost-ratios of the two commodities) are the same. Because of the differences in commodity price ratios the trade is possible and will be mutually beneficial. The opening of trade would equalise commodity prices in the two countries and would therefore change their pattern of production giving rise to exports and imports by the two countries. Let *tt* be the commodity price ratio (*i.e.* terms of trade) that is settled between the two countries. It will be noticed in Fig. 44.3 that with terms of trade *tt* each country would produce at point *Q*. But the consumption of the two commodities would be different in the countries. After trade whereas country *A*'s consumption would be at *R* where terms of trade line *tt* is tangent to its indifference curve III_a; country *B*'s consumption would be at point *S* on its indifference curve III_b. It is evident that country *A* would be importing *GR* rice and exporting *GQ* wheat, whereas country *B* would be exporting *QH* rice and importing *HS* wheat. It will be further noticed that through trade both have reached at their higher indifference curves indicating higher level of welfare than before trade.

GAINS FROM TRADE

Foreign trade confers a good deal of benefits on the trading countries. If different countries specialise on the basis of their comparative costs, it would enable them to make optimum use of their resources and thereby add to their output, income and welfare of their people. Gains from trade are broadly divided into two types:

(*i*) *Static Gains*

(*ii*) *Dynamic Gains*.

Static gains from trade refer to the increase in utility or welfare of the people of the trading countries as a result of the optimum utilisation of their given factor-endowments, for they specialise on the basis of their comparative costs. On the other hand, dynamic gains refer to the contributions which foreign trade makes to the overall economic growth of the trading countries. We shall explain below in detail these two types of gains.

Static Gains from Trade

As stated above, static gains from trade are measured by the increase in the utility or level of welfare when there is opening of trade between the countries. Note that in modern economics increase

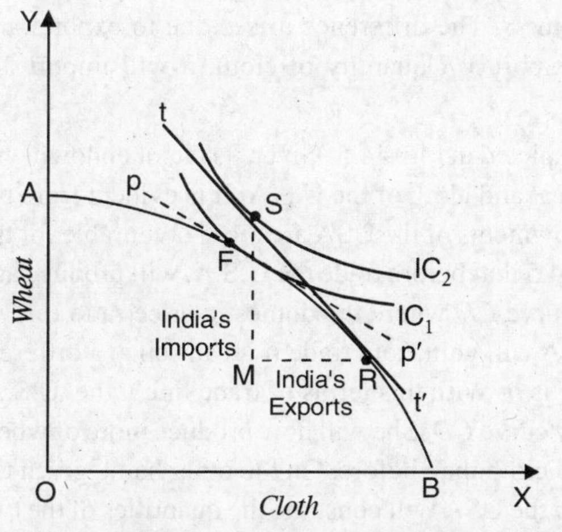

Fig. 44.4. *Gain from Trade: India.*

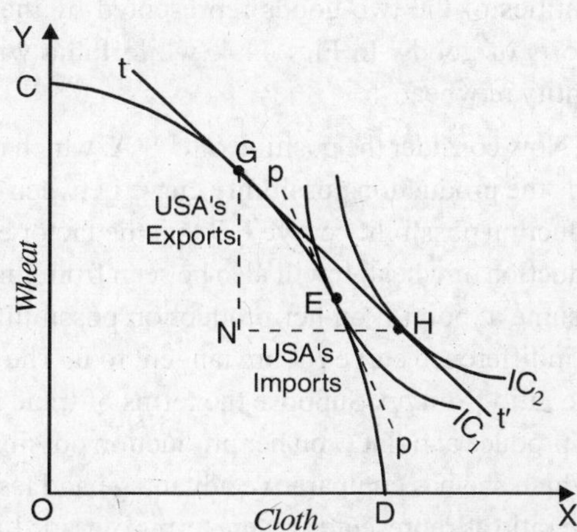

Fig. 44.5. *Gain from Trade: U.S.A.*

in utility or welfare is measured through indifference curves. When as a result of foreign trade, a country moves from a lower indifference curve to a higher one, it implies that the welfare of the people has increased. To show the static gains from trade, let us take an example. Suppose two commodities cloth and wheat are produced in two countries, India and U.S.A., before they enter into trade. Their production possibility and indifference curves are shown in Fig. 44.4 and 44.5. It will be seen from Fig. 44.4 that before trade India would be in equilibrium at point *F* (*i.e.* producing and consuming at point *F*) where the price line *pp′* is tangent to both production possibility curve *AB* and indifference curve IC_1. The slope of the price line *pp′* shows the price-ratio (or cost ratio) of the two commodities in India. India can gain if international price-ratio (terms of trade) is different from the domestic price-ratio represented by *pp′*. Suppose the terms of trade settled are such that we get *tt* as the terms of trade line showing the price ratio at which goods can be exchanged between India and the U.S.A. Now, with *tt′* as the given terms of trade line (*i.e.* new price-ratio line). India would produce at point *R* at which the terms of trade line *tt* is tangent to her production possibility curve. It will be seen from Fig. 44.4 that at point *R*, India will produce more of cloth in which it has comparative advantage and less of wheat than at *F*. Though India will produce at point *R* on his production possibility curve, where the terms of trade line *tt* is tangent to her production possibility curve *AB* it will not consume the quantities of wheat and cloth represented by the point *R*. Given the new price-ratio represented by the terms of trade line *tt* the consumption of the goods will depend upon the pattern of demand of the country. To incorporate this factor we have drawn social indifference curves IC_1, IC_2 of the country. These social indifference curves represent the demands for the two goods, or, in other words, the scale of preferences between the two goods of the society.

It will be seen from Fig. 44.4 that the terms of trade line *tt* is tangent to the social indifference curve IC_2 of India at point *S*. Therefore, after trade India will consume the quantities of cloth and wheat as represented by point *S*. It is therefore clear that as a result of specialisation and trade India has been able to shift from point *F* on indifference curve IC_1 to the point *S* on the higher indifference curve IC_2. This is the gain obtained from specialisation and trade and implies that trade enables a country to increase her consumption beyond her production possibility curve. (It will be seen that point *S* lies beyond the production possibility curve *AB* of India).

It is also worth noting that when specialisation and trade occur, the quantities of the two goods consumed by a country will differ from the quantities of the two goods produced by her. In Fig. 44.4 whereas India produces the quantities of two goods represented by point *R*, it will consume the quantities of the two goods represented by the point *S*. The difference arises due to exports and imports of goods. In Fig. 44.4, while India will export *MR* quantity of cloth, it will import *MS* quantity of wheat.

Now consider the position of U.S.A. which is depicted in Fig. 44.5. Given its factor endowments *CD* is the production possibility curve between wheat and cloth of the U.S.A. It is evident from the production possibility curve *CD* that the factor endowments of the USA are more favourable for the production of wheat. It will also be seen from Fig. 44.5 that before trade the U.S.A. will produce and consume at point *E* on her production possibility curve *CD* where the domestic price ratio line *pp* and indifference curve IC_1 are tangent to it. The USA will gain from trade if it can sell at a different price ratio from *pp*. Suppose the terms of trade line is *tt*. With this terms of trade line *tt* the U.S.A. will produce at point *G* on her production possibility curve *CD*. She will now produce more of wheat in which she has comparative advantage and less of cloth than before. On the other hand, given the price ratio as represented by the terms of trade line *tt* the USA will consume the quantities of the two goods given by the point *H* where the terms of trade line is tangent to her indifference curve IC_2. It is therefore clear that the specialisation and consequently trade with India has enabled the U.S.A. to shift from her lower indifference curve IC_1 to her higher indifference curve IC_2. This is the gain which she obtains from trade.

By comparing the production and consumption points of the U.S.A. it will be observed that the U.S.A. will export *NG* amount of wheat and import *NH* amount of cloth.

It is worth remembering that while in case of constant opportunity cost, each country attains complete specialisation, that is, it produces one of the two goods after trade, in case of increasing opportunity cost specialisation is not complete. In case of increasing opportunity cost, a country produces only a relatively large amount of the good in which it has comparative advantage.

Dynamic Gains from Trade: International Trade and Economic Growth

We have seen above in the discussion of comparative cost theory that specialisation followed by international trade makes it possible for the countries to have more of both commodities than before. This additional production of commodities is the gain which flows from specialisation by different countries in the production of different goods and then trading with each other. Specialisation by different countries in the production of different goods according to their efficiency and resource endowment brings about an increase in the total world production by increasing the level of their productivity. It is this trade that makes possible the division and specialisation of labour on which higher productivity of different countries is so largely based. If the various countries could not exchange the products of their specialised labour, each of them would have to be self-sufficient (*i.e.*, each of them would have to produce all goods it requires, even those which it could not produce efficiently) with the result that their productivity and standard of living will go down. Thus, according to Professor Haberler, "*International division of labour and international trade, which enable every country to specialise and to export those things which it can produce cheaper in exchange for what others can provide at a lower cost, have been and still are one of the basic factors promoting economic well being and increasing national income of every participating country*."[2]

We thus see that the main gain from specialisation and trade is the increase in national production, income and consumption of the participating countries. But the above explanation of gains from trade in terms of comparative cost theory deals only with *static* gains from trade, that is, the gains which accrue to a country from reallocation of a given amount of resources. We shall now discuss, *dynamic gains* from trade, that is gains from trade which accrue to a country in terms of promotion of its economic growth.

Dennis Robertson described foreign trade as "*an engine of growth.*" With greater income and production made possible by specialisation and trade, greater savings and investment become possible and as a result higher rate of economic growth can be achieved. Through promotion of exports, a developing country can earn valuable foreign exchange which it can use for the imports of capital equipment and raw materials which are so essential for economic development. Therefore, Professor Haberler argues that since international trade raises the level of income, it also promotes economic development. He thus remarks : "What is good for the national income and the standard of living is, at least potentially, also good for economic development; for the greater the volume of outputs the greater can be the rate of growth—provided the people individually or collectively have the urge to save and to invest and economically to develop. The higher the level of output, the easier it is to escape the "vicious circle of poverty" and to "take off into self-sustained growth" to use the jargon of modern development theory. Hence, *if trade raises the level of income, it also promotes economic development*[3].

---

2. G. Haberler, *International Trade and Economic Development*
3. G. Haberler, *Op, cut*. (emphasis added)

As pointed out above, the importance of and gain from international trade follows from the theory of comparative cost. Specialisation by different countries according to their production efficiency and factor endowments *ensures optimum use and allocation of resources* of the countries. Differences in production possbilities and costs of production of various products between different countries of the world are so great that tremendous gain in terms of additional output and income accrues to the world community from in ternational specialisation and trade. For instance, the relative differences in cost of production of industrial products and food and raw materials between developed and developing countries are almost infinite in the sense that either type of these countries cannot produce what they buy from the other.

But the theory of comparative cost is static. it indicates only those gains which accure to the trading countries as a result of the differences in *given cost* of production and *given production possibilities* of various products at *a given point of time.* As pointed out above, besides the static gains indicated by comparative cost theory, international trade bestows very important indirect gains and benefits, which are generally described as *dynamic gains,* upon the participating countries. These dynamic gains also promote economic growth in the participating countries. It is worth noting that both developed and developing countries have obtained benefits from trade. The international trade has contributed a good deal to the economic development of countries. To quote Professor Haberler again "If we were to estimate the contribution of international trade to economic developent especially of the under-developed countries solely by the static gains from trade in any given year on the usual assumption of *given production capabilities,* we would indeed grossly under-rate the importance of trade. For over and above the direct stastic gains dwelt upon by the traditional theory of comparative cost, trade bestows very important indirect benefits upon the participating countries".[4] Dynamic gains which' accrue to the developing countries from international trade are as follows:

Firstly, through foreign trade, developing countries get material means of production such as capital equipment, machinery and raw materials which are so essential for economic growth of these countries. There has been rapid technological progress in the developed countries. This advanced and superior technology is incorporated or embodied in various types of capital goods. It is thus clear that developing countries derive tremendous gains from technological progress in the developed countries through the imports of capital goods such as machinery, transport equipment, vehicles, power generation equipment, road building machinery, medicines, chemicals. It is worth mentioning here that the pattern of import trade of the developing countries has changed in the last several years and now consists of greater quantity of various forms of capital goods and less of textiles.

Secondly, even more important than the importation of capital goods is the transmission of technical know-how, skills, managerial talents., entrepreneurship through foreign trade. When the developing countries come to have trade relationship with the developed countries, they also often import technical know-how, with all their skills, managers, etc., from them. With this they are also able to develop their own technical know-how, managerial and entrepreneurial ability. The growth of technical know-how, skill and managerial ability is an important requisite for economic development of developing countries. Professor Haberler rightly says: "The late-comers and successors in the process of development and industrialization have always had the great advantage that they could learn from the experinces, from the successes as well as from the failures and mistakes of the pioneers and forerunners. . . Today the developing countries 'have a tremendous, constantly growing store of technical know-how to draw from. True, simple adoption of methods, developed for the conditions of

---

4. *Op. cit.*

the developed countries is often not possible. But adaptation is surely much easier than the first creation. . . . Trade is the most important vehicle for the transmission of technological know-how ... Today there are a dozen industrial centres in Europe, the U.S., Canada, and Japan, and Russia which are ready to sell machinery as well as engineering advice and know-how."[5]

QUESTIONS FOR REVIEW

1. What is the Heckscher-Ohlin theorem of trade? Clearly state its underlying assumptions. *[D.U. B.A. (Hons) 1991, 1998, 2000]*
2. Using the Heckscher-Ohlin model of trade show that free trade is better than no trade. Explain graphically the gains from trade to the countries.
3. Show that under Heckscher-Ohlin framework international trade would normally lead to equalisation of relative factor prices. *[D.U. B.A. (Hons) 1995]*
4. Under what conditions are factor prices equilised by trade between two countries. *[D.U. B.A. (Hons) 1993]*
5. Two nations A and B have identical production possibility curves. Can trade between them still be beneficial to both? *[D.U. B.A. (Hons) 1998]*

   OR

   Show that in a two country, two commodity, two factor world, trade can take place even if factor endowments are different? *[D.U. B.A. (Hons) 1993]*
6. "Commodity trade acts as a substitute for international mobility of factors of production in its effect on factor prices" Discuss, *[D.U. B.A. (Hons) 2001]*
7. What is meant by factor intensity reversal ? Is factor intensity non-reversal a necessary condition for Heckscher-Ohlin model of trade to be valid. *[D.U. B.A. (Hons) 2000]*
8. What do you understand by factor-intensity reversal? What are its implications for factor price equalisation? *[D.U. B.A. (Hons)1998]*
9. What are the necessary and sufficient conditions for the factor price equalisation theorem to hold ? Explain *[D.U. B.A. (Hons) 1992]*
10. Do you agree with the view that difference in factor endownments are neither a necessary nor a sufficient basis for international trade? Give arguments in support of your answer?
11. Free trade is better than no trade Discuss.

---

5. G, Haberler, *International Trade and Economic Development*

CHAPTER 45

TERMS OF TRADE

Gains from Trade and Terms of Trade

In the previous chapter we have seen that the countries would gain from specialising in the production of commodities and exporting them in which they have lower comparative cost. *How the gain from international trade would be shared by the participating countries depends upon the terms of trade*. The terms of trade refer to the rate at which one country exchanges its goods for the goods of other countries. Thus, terms of trade determine the international values of commodities. Obviously, the terms of trade depend upon the prices of exports a country and the prices of its imports. When the prices of exports of a country are higher as compared to those of its imports, it would be able to obtain greater quantity of imports for a given amount of its exports. In this case terms of trade are said to be favourable for the country as its share of gain from trade would be relatively larger. On the contrary, if the prices of its exports are relatively lower than those of its imports, it would get smaller quantity of imported goods for a given quantity of its exports. Therefore, in this case, terms of trade are said to be unfavourable to the country as its share of gain from trade would be relatively smaller. In what follows we first explain the various concepts of the terms of trade and then explain how they are determined.

CONCEPTS OF TERMS OF TRADE

Net Barter Terms of Trade. The most widely used concept of the terms of trade is what has been caned the net barter terms of trade which refers to the relation between prices of exports and prices of imports. In symbolic terms:

$$T_n = \frac{Px}{Pm}$$

where

T_n stands for net barter terms of trade,
P_x stands for price of exports *(x)*,
P_m stands for price of imports *(m)*.

When we want to know the changes in net barter tends of trade over a period of time, we prepare the price index numbers of exports and imports by choosing a certain appropriate base year and obtain the following ratio:

$$\frac{Px_1}{Pm_1} : \frac{Px_0}{Pm_0}$$

where Px_o and Pm_0 stand for price index numbers of exports and imports in the *base year* respectively, and Px_1) and Pm_1) denote price index numbers of exports and imports respectively in the *current year*.

Since the prices of both exports and imports in the base year are taken as 100, the terms of trade in the base year would be equal to one

$$\frac{Px_0}{Pm_0} = \frac{100}{100} = 1$$

Suppose in the current period the price index number of exports has gone upto 165, and the price index number of imports has risen to 110, then terms of trade in the current period would be:

$$\frac{165}{110}:\frac{100}{100}=1.5:1$$

Thus, in the current period, terms of trade have improved by 50 per cent as compared to the base period. Further, it implies that if the prices of exports of a country rise relatively greater than those of its imports, terms of trade for it would improve or become favourable. On the other hand, if the prices of imports rise relatively greater than those of its exports, terms of trade for it would deteriorate or become unfavourable. Thus, net barter terms of trade is an important concept which can be applied to measure changes in the capacity of exports of a country to buy the imported products. Obviously, if the net barter terms of trade of a country improve over a period of time, it can buy more quantity of imported products for a given volume of its exports.

But the concept of net barter terms of trade suffers from some important limitations in that *it shows nothing about the changes in the volume of trade.* If the prices of exports rise relatively to *those* of its imports but due to this rise in prices, the volume of exports falls substantially, then the gain from rise in export prices may be offset or even more than offset by the decline in exports. This has been well described by saying, "We make a big profit on every sale but we don't sell much". In order to overcome this drawback, the net barter terms of trade is weighted by the volume of exports. This has led to the development of another concept of terms of trade known as the income terms of trade which shall be explained later. Even so, the net barter terms of trade is most widely used concept to measure the power of the exports of a country to buy imports.

Gross Barter Terms of Trade. This concept of the gross terms of trade was introduced by F.W. Taussig and in his view this is an improvement over the concept of net barter terms of trade as it directly takes into account the volume of trade. Accordingly, the gross barter trms of trade refer to the relation of the volume of imports to the volume of exports. Thus,

$$T_g=\frac{Om}{Qx}$$

where

Tg = gross barter terms of trade, Qm = quantity of imports
Qx = quantity of exports

To compare the change in the trade situation over a period of time, the following ratio is employed:

$$\frac{Om_1}{Qx_1}:\frac{Qm_0}{Qx_0}$$

where the subscript 0 denotes the base year and the subscript I denotes the current year.

It is obvious that the gross barter tenns of trade for a country will rise *(i.e.,* will improve) if more imports can be obtained for a given volume of exports. It is important to note that *when the balance of trade is in equilibrium (that is, when value of exports is equal to the value of imports), the gross barter terms of trade amount to the same thing as net barter terms of trade.* This, can be shown as under :

Value of imports = price of imports × quantity of imports = $Pm.Qm$
Value of exports = Price of exports × quantity of exports = $Px.Qx$
Therefore, when balance of trade is in equilibrium.

$$Px.\ Qx = Pm.\ Qm$$

$$\frac{Px}{Pm}=\frac{Qm}{Qx}$$

However, when balance of trade is not it equilibrium, the gross barter terms of trade would differ from net barter terms of trade.

Income Terms of Trade

In order to improve upon the net barter terms of trade G.S. Dorrance developed the concept of income terms of trade which is obtained by weighting net barter terms of trade by the volume of exports. Income terms of trade therefore refer to the index of the value of exports divided by the price of imports. Symbolically, income terms of trade can be written as

$$T_y = \frac{Px.Qx}{Pm}$$

where

T_y = Income terms of trade
P_x = Price of exports
Q_x = Volume of exports
P_m = Price of imports

Income terms of trade yields a better index of the capacity to import of a country and is, indeed, sometimes called '*capacity to import*. This is because in the long run balance of payments must be in equilibrium, the value of exports would be equal to the value of imports. Thus, in the long run:

$$Pm, Qm = Px, Qx$$

$$Qm = \frac{Px.Qx}{Pm}$$

It follows from above that the volume of imports *(Qm)* which a country can buy (that is, capacity to import) depends upon the income terms of trade *i.e.*, $\frac{Px.Qx}{Pm}$ Since income terms of trade is a better indicator of the capacity to import and since the developing countries are unable to change *Px* and *Pm*. Kindleberger' thinks it to be superior to the net barter terms of trade for these countries, However, it may be mentioned once again that it is the concept of net barter terms of trade that is usually employed.

DETERMINATION OF TERMS OF TRADE: THEORY OF RECIPROCAL DEMAND

As seen above, the share of a country from the gain in international trade depends on the terms of trade. The terms of trade at which the foreign trade would take place is determined by reciprocal demand of each country for the product of the other countries. The theory of reciprocal demand was put forward by JS. Mill and is thought to be still valid and true even today. *By reciprocal demand we mean the relative strength and elasticity of the demand of the two trading countries for each other's product.* Let us take two countries *A* and *B* which on the basis of their comparative costs specialise in the production of cloth and wheat respectively. Obviously, country *A* would export cloth to country *B,* and in exchange import wheat from it. Reciprocal demand means the strength and elasticity of demand of country *A* for wheat of country *B,* and the intensity and elasticity of country *B's* demand for cloth from country *A.* If the country *A* has inelastic demand for wheat of country *B,* she will be prepared to give more of cloth for a given amount of wheat. In this case terms of trade will be unfavourable to it and consequently its share of gain from trade will be relatively smaller. On the contrary, if country *A's* demand for import of wheat is elastic, it will be willing to offer a smaller quantity of its cloth for a given quantity of the imports of wheat. In this case terms of trade would be favourable to country *A* and its share of gain from trade will be relatively larger. *The equilibrium terms of trade would settle at a level at which its reciprocal demand, that is, quantity of its exports which it will be willing to give for a given quantity of its imports is equal to the reciprocal demand of the other country.*

Note that the equilibrium terms of trade are determined by the intensity of reciprocal demand of the two trading countries but *they will lie in between the comparative costs (i.e., domestic exchange ratios) of the two countries*. This is because no country would be willing to trade at a price which is lower than at which it can produce at home. Let us return to the example of the two countries *A* and *B* which specialise in the production of two commodities cloth and wheat respectively, and exchange them with each other. Production conditions in the two countries are given below :–

Table 45.1. Production of one man per week

	Country A	*Country B*
Wheat	4 Bushels	12 Bushels
Cloth	12 Yards	20 Yards

It will be seen from above table that before trade production conditions in country *B* are such that 12 bushels of wheat would be exchanged for 20 yards of cloth. in it, that is, the domestic exchange ratio is 12 : 20 (or 3 : 5). On the other hand, in country *A* production conditions are such that 4 bushels of wheat would be exchanged for 12, yards of cloth, that is, the domestic exchange ratio is 4 : 12 or 1 : 3. Obviously, after trade, terms of trade will be settled within these domestic exchange ratios of the two countries. The domestic exchange ratios of the two countries set the limits beyond which terms of trade would not settle after trade. It is evident that country *B* will be unwilling to offer more than 12 bushels of wheat for 20 yards of cloth since by sacrificing 12 bushels of wheat it can produce 20 yards of cloth at home. Likewise, country *A* would not accept less than 6.66 bushels of wheat for 20 yards of cloth, for this is the domestic exchange rate cloth of wheat for (1 :3) determined by production or cost conditions at home in country *A*. It is within these limits that terms of trade will be settled between the two countries as determined by the strength of reciprocal demand of the trading countries. It also follows that it is not mere demand but also the comparative production costs *(i.e.,* the supply conditions) that go to determine the terms of trade. Indeed, the law of reciprocal demand, if properly understood, considers both the forces of demand and supply as determinants of the terms of trade.

Critical Evaluation of the Reciprocal Demand Theory

The reciprocal demand theory of the terms of trade is based upon two countries, two commodities model. It assumes that full-employment conditions prevail in the economy and also that there is perfect competition in both the product and factor markets in the economies of the various countries. It also assumes the governments of the various countries follow free trade policy and impose no restrictions on foreign trade by imposing tariffs or adopting other means to restrict imports. Further, this theory grants that there is free mobility of factors domestically within the economies of the two countries. To the extent these assumptions do not hold in the real world, the terms of trade would not conform to those determined by the reciprocal demand. However, as stated above, every theory makes some simplifying assumptions. The soundness of a theory depends upon whether the deductive logic it employs is flawless and the conclusions it draws about the impact of economic forces on the subject investigated are correct or not. On this test the reciprocal demand theory fares very well as reciprocal demand is undoubtedly an important factor which influences terms of trade.

F.D. Graham criticized this theory by pointing out that it is applicable only to trade in antiques and old masters which are found in fixed supplies and therefore in their case demand plays a crucial role in the determination of terms of trade. He stressed that the theory of reciprocal demand was not relevant in case of goods produced currently since their international values *(i.e.,* terms of trade) were determined by comparative production costs *(i.e.,* the supply conditions). In his view, reciprocal demand theory grossly exaggerates the role of reciprocal demand and neglects the importance of comparative cost conditions. However, Graham's criticism is not valid. The reciprocal demand or offer curve embodies both demand and production costs. In reply to Graham's criticism, Viner writes that "The terms of trade can be directly influenced by the reciprocal demands and by nothing else.

The reciprocal demands in turn are ultimately determined by the cost conditions together with the basic utility function."[1]

We, therefore, conclude that terms of trade are determined by reciprocal demands of the trading countries. Reciprocal demands in turn are governed by both the demand and supply (cost) conditions. Thus, the intensity of demand by others for exports of a country and the intensity of its demand for imports from the other country are the important factors that determine the terms of trade. Besides, comparative cost conditions of the products exported and those imported have also an important role in the determination of terms of trade.

DETERMINATION OF TERMS OF TRADE AND OFFER CURVES

The theory of reciprocal demand has been explained graphically with the help of the concept of offer curves developed by Edgeworth and Marshall. *The offer curve of a country shows the amounts of a commodity it offers at various prices for a given quantity of the commodity produced by the other country*. To understand how offer curves are derived and how with their help determination of the terms of trade is explained, we shall first explain how a country reaches its equilibrium position about the amounts of goods to be produced and consumed. For this purpose, modern economists usually employ the tools of production possibility curve and the community indifference curves. The production possibility curve represents the combinations of two commodities which a country, given its resources and technology, can produce. A community indifference curve shows the combinations of two goods which provide same satisfaction to the community as a whole. *A* map of community indifference curves portrays the tastes and demand pattern of a community for the two goods. A production possibility curve *TT'* and a set of community indifference curves IC_1 IC_2 and IC_3 of country *A* have been drawn in Fig. 45.1. The country *A* reaches its equilibrium position with regard to production and consumption of cloth and wheat at the point *Q* where the production possibility curve *TT'* is tangent to the highest possible indifference curve IC_2 at which marginal rate of transformation of cloth for wheat (MRT_{CW}) equals marginal rate of substitution of cloth for wheat (MRS_{CW}) as well as

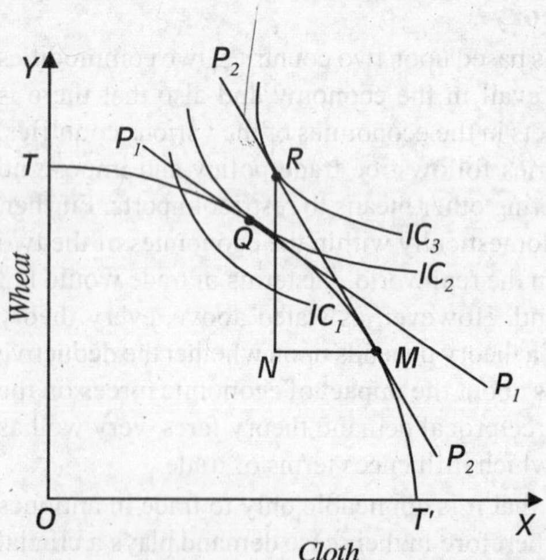

Fig. 45.1. *Production and Consumption of Country A in the absence of Trade and with Trade.*

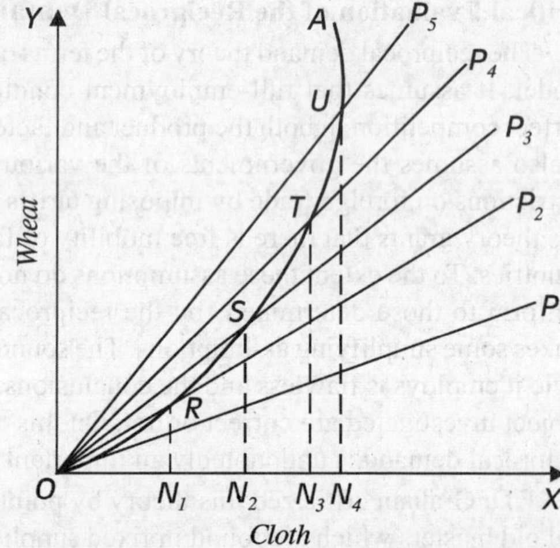

Fig. 45.2. *Offer Curve of Country A.*

the price ratio of the two commodities *Pc/Pw* as shown by the slope of the price line $P_1 P_1$. Thus, tangency point *Q* in Fig. 45.1 depicts the equilibrium position of country *A* in the absence of trade. Suppose country *A* enters into trade relation with country *B* and price of cloth rises relative to wheat

1. Jacob Viner, *Studies in the Theroy of International Trade*, p. 549

so that new price-ratio line becomes P_2P_2. It will be observed from Fig. 45.1 that with price-ratio line P_2P_2 production equilibrium of country *A* is at point *M,* its consumption equilibrium is at point *R.* This shows that with price-ratio line PP_2 country *A* will offer or export *MN* of cloth for *RN* imports of wheat. Similarly, if price of cloth further rises relative to wheat, price-ratio line will become more steep, then for the same quantity offered of export of clothm, the or import of wheat will increase. With such information gathered from Fig. 45.1, we can derive offer curve of country *A* In Fig. 45.2.

The tangent line P_1P_1 in Fig. 45.1 shows the domestic price ratio of the two commodities and has a negative slope. *In the analysis of the offer curve, the price line is drawn with a positive slope from the origin.* This is is because in the drawing of an offer curve we are interested only in knowing the quantity of one commodity which can be exchanged for a certain quantity of another commodity. In other words, in the analysis of terms of trade what we are really interested is the absolute slope of the curve, *i.e,* the price ratio. In Fig. 45.2 the positively sloping price line OP_1 from the origin, which in absolute terms, has the same slope as P_1P_1 of Fig. 45.1 has been drawn. In Fig. 45.2 at price ratio line O_1P_1 no trade occurs. When price of cloth rises and price ratio line shifts to OP_2, as will be from Fig. 45.2, country *A* offers ON_1 of cloth (exports) for RN_1 of wheat (imports). (Note that at a given price ratio how much quantity of a commodity, a country will offer for imports from the other country is determined by production possibility curve and commnity's indifference curves as illustrated in Fig. 45.1).

Suppose the price of cloth further rises relatively to that of wheat causing the price line to shift to the position OP_3. It will be seen that with the price line OP_3, country *A* is willing to offer for export ON_2 quantity of cloth for SN_2 of wheat. Likewise, Fig. 45.2 portravys the exports and imports of the country *A* as price of cloth in terms of wheat increases further and consequently price line shifts further above to OP_4 and OP_5 and the new offers of export of cloth for import of wheat are determined by equilibrium points *T* and *U.* If points such as *R, S, T* and *U* representing the country *A's* offers of cloth for wheat are joined we get its offer curve.

It is important to note that the offer curve may be regarded as the supply curve in the international trade as it shows amounts of cloth which the country *A* is willing to offer for certain amounts of imports of wheat at various price ratios. Another important point to be noted is that the offer curve cannot go below the price line OP_1 which represents the domestic exchange ratio determined by the tangency point *Q* of production possibility curve and community indifference curve of country A as shown in Fig. 45.1. This is because, as stated above, no country will be willing to export its product for the quantity of the imported product which is smaller than that it can produce at home.

Likewise, we can derive the offer curve of country *B.* Figure 45.3 portrays the derivation of the offer curve of country *B,* representing quantities of wheat which it is willing to exchange for certain quantities of cloth from country *A* at various prices. Note that so long as country *B* is importing a smaller quantity of cloth, it will be willing to offer relatively more wheat for cloth. But as the quantity of imported cloth is increased, it would be prepared to offer relatively less wheat for the given quantity of imports of cloth. In Fig. 45.3 whose *Y*-axis represents wheat, the origin for indifference curves of country *B* will be the North-West Corner Price lines. OP_7, OP_6, OP_5, OP_4 etc., express successively higher price ratios of wheat for cloth. *Price line* OP_7 represents the

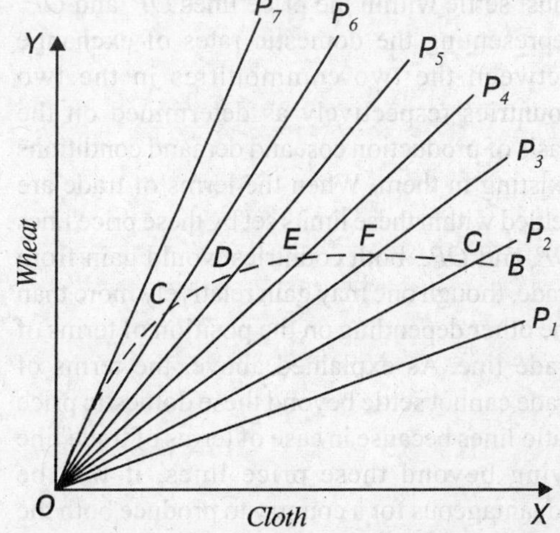

Fig. 45.3. *Offer Curve of Country B*

domestic price ratio in country *B* in the absence of trade. The points C, *D, E, F,* G which has been obtained from the equilibrium or tangency points between the community indifference curves of country *B* and the various price-ratio lines show the equilibrium offers of wheat by country *B* for cloth of country *A* at various prices. By joining together points, C, *D, E, F* and G we obtain the offer curve of country *B* indicating its demand for cloth of country *A* in terms of its own product wheat.

It would be observed from Fig. 45.2 and 45.3 that offer curves *OA* and *OB* of the two countries have been drawn with the same origin *O* (*i.e* South-West Corner) as the basis. *These offer curves, as explained earlier, represent reciprocal demand of the two countries for each other's product in terms of their own product*. The offer curves *OA* and *OB* of the two countries have been brought together in Fig. 45.4.

The intersection of the offer curves of the two countries determines the equilibrium terms of trade. It will be seen from Fig. 45.4 that the offer curves of two countries cross at point *T.* By joining point *T* with the origin we get the price-ratio line *OT* whose slope represents the equilibrium terms of trade which will be finally settled between the two countries. At any other price-ratio line the offer of a product by country *A* in exchange for the product of the other would not be equal to the reciprocal offer and demand of the other country *B*. For instance, at price-ratio line OP_5, country *B* would offer *OM* wheat for *MH* or *ON* of cloth from country A (*H* lies on *B*'s offer curve corresponding to price-ratio line OP_5). But at this price-ratio line OP_5 country A would demand much greater quantity of wheat *UW* for *OU* of cloth as determined by point *W* at which the offer curve of country *A* intersects the price ratio line OP_5. This will result in rise in price of wheat and the price-ratio line will shift to the right until it reaches the equilibrium position *OT* or OP_4. On the other hand, if price ratio line lies to the right of *OT* (for instance, if it is OP_3), then, as will be observed from Fig. 45.4, it cuts the offer curve of country *A* at point *L* implying thereby that the country A would offer *OR* of cloth in exchange for *RL* of wheat. However, with terms of trade implied by the price ratio line OP_3, the country *B* would demand *OZ* of cloth for *ZS* quantity of wheat as determined by point *S*. It therefore follows that only at the terms of trade implied by the price ratio line *OT* (*i.e.,* OP_4) that the offer of a product by one country will be equal to its demand by the other. We therefore conclude that the intersection of the offer curves of the two countries determines the equilibrium terms of trade.

As explained above, the offer curves of the two countries are determined by their reciprocal demand. Any change in the strength and elasticity of reciprocal demand would cause a change in the offer curves and hence in the equilibrium terms of trade.

It is worthwhile to note that terms of trade must settle within the price lines OP_1 and OP_7 representing the domestic rates of exchange between the two commodities in the two countries respectively as determined on the basis of production cost and demand conditions existing in them. When the terms of trade are settled within these limits set by these price lines OP_1 and OP_7, both countries would gain from trade, though one may gain relatively more than the other depending on the position of terms of trade line. As explained above, the terms of trade cannot settle beyond these domestic price ratio lines because in case of terms of trade line lying beyond these price lines, it will be advantageous for a country to produce both the goods (wheat and cloth) domestically rather than entering into foreign trade.

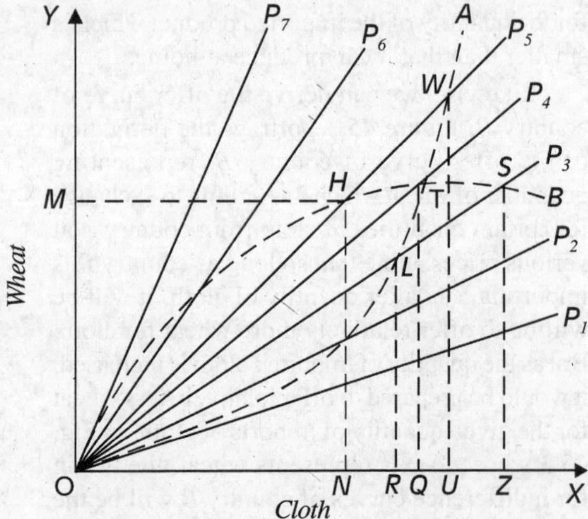

Fig. 45.4. *Determination of Terms of Trade*

EFFECT OF TARIFF ON TERMS OF TRADE

The various countries of the world have imposed tariffs (*i.e.,* import duties) to protect their domestic industries. It has been said in favour of tariffs that through them a country can provide not only protection to its industries but under appropriate circumstances it can also improve its terms of trade, that is, tariffs under favourable circumstances enable a country to get its imports cheaper. These favourable circumstances are: (1) the demand for the exports of the tariff-imposing countries is both large and inelastic (2) the demand for the imports by the country is quite elastic. Under these circumstances, as a result of the imposition of tariff by that country, the imports of the country will decline since the price of the imported commodity will rise. But this is not the end of the story. The decline in imports of the tariff-imposing country would reduce the export earnings of its trading partner as it will lead to the decrease in demand for its exported commodity. The decrease in demand for the exported commodity in the trading partner would result in lowering its domestic price. As a result of the fall in the domestic price of the exported commodity and in order to maintain its export earnings the exporting country is likely to reduce the price of its exports. This means that the tariff-imposing country would now be able to get its imports at a relatively lower price than before. Given the demand and price of its exports, the fall in its prices of imports of the tariff-imposing country

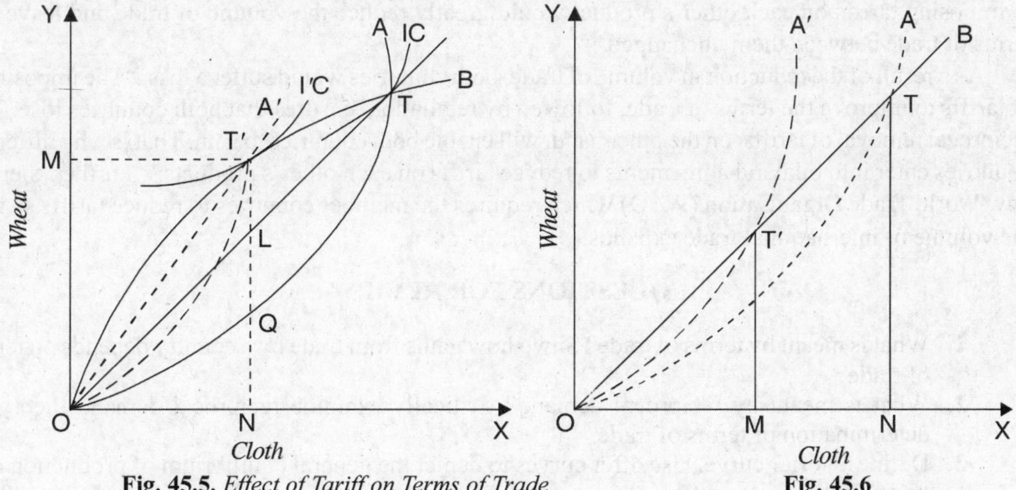

Fig. 45.5. *Effect of Tariff on Terms of Trade* **Fig. 45.6**

would imply the improvement in its terms of trade. It is worth mentioning that the improvement in the terms of trade through tariff depends upon the changes in price and resultant changes in quantity demanded of imports and exports of the trading countries which in turn depends upon the elasticities of their reciprocal demand.

The effect of tariff on the terms of trade can be explained through the geometrical device of offer curves. In Fig. 45.5 the offer curves *OA* and *OB* respectively of the two countries *A* and *B* are shown. These offer curves intersect at *T* implying thereby that terms of trade equal to the slope of *OT* are determined between them. Now, suppose that country *A* imposes import duty on wheat from country *B.* As a result of this imposition of tariff, the offer curve of country *A* will shift to a new position *OA'* (dotted). This implies, for instance, that, before tariff, country *A* was prepared to offer *ON* of cloth for *NQ* of wheat, but after imposition of tariff it requires *NT' of wheat for ON* of cloth and collects *QT'* as import duty.

It will be noticed from Fig. 45.5 that the new offer curve *OA'* (dotted) of country *A* intersects the offer curve *OB* of country *B* at point *T'* and thereby the terms of trade changes from *OT* to *OT'*. Note that the slope of the terms of trade line *OT′* is greater than that of *OT′*. Thus terms of trade for country A have improved consequent upon the imposition of tariff by country *A.* For instance, whereas according to terms of trade line *OT* country *A* was exchanging *ON* of cloth for *N L* imports of wheat, it is now exchanging *ON* of cloth for *NT'* of wheat. The following three things are worth nothing about the impact of tariffs on terms of trade:

1. The gain in terms of trade from imposing a tariff depends on the elasticity of the offer curve of the opposite trading country. If the offer curve of the opposite trading country is perfectly elastic, that is, when it has constant costs so that offer curve is the straight line *OB* from the origin with slope equal to that of *OT* as shown in Fig. 45.6, the imposition of tariff would reduce the volume of trade between them, the terms of trade remaining the same. For example, if in the situation depicted in Fig. 45.6 country *A* imposes a tariff on imports of a wheat from the country *B* and as a result the offer curve of *A* shifts upward to the new position *OA'* (dotted), the terms of trade remain constant as measured by the slope of the terms of trade line *OT*. It will be seen from Fig. 45.6 that in this case only volume of trade has declined from *ON* to *OM*

2. The gain in terms of trade from imposing a tariff will finally accrue to a country *A* only in the absence of rataliation from the trading country *B*. But when one country can play a game to improve its position, the other can retaliate and play the same game. That is, on country *A* imposing a tariff on its imports from country *B* in a bid to improve its terms of trade, the latter can also impose a tariff on the imports from the former and thereby cancels out the original gain by country *A*. Such competition in imposing tariffs on each other's product would greatly reduce the volume of trade and leave the terms of trade between them unchanged.

As a result of the reduction in volume of trade, both countries would suffer a loss. "The imposition of tariffs to improve the terms of trade, followed by retaliation, ensures that both countries lose. The reciprocal removal of tariffs, on the other hand, will enable both countries to gain. That is why different countries enter into bilateral agreements to reduce tariffs on each other's products." Further, there is now World Trade Organisation (WTO) which requires the member countries to reduce tariffs so that the volume of international trade expands.

QUESTIONS FOR REVIEW

1. What is meant by terms of trade? Shwo how gains from trade by a country depends on terms of trade?
2. What is meant by reciprocal demand? Critically examine reciprocal demand theory of determination of terms of trade.
3. Define an offer curve. Use offer curves to depict the general equilibrium of production and trade, in the two countries. *[B.A. (Hons) D.U. 1997]*
4. What are offer, curves? Explain the determination of terms of trade through these offer curves? *[B.A. (Hons.) D.U. 2002]*

3. See C.P. Kindleberger, *International Economics*, 1973, p. 40.
4. Jacob Viner, *Studies* in *the Theory of International Trade*,p.549.

VOLUME 2
MACROECONOMICS AND ECONOMICS OF GROWTH AND DEVELOPMENT

PART VIII

Macroeconomics: Theory of Income and Employment

- Nature and Scope of Macroeconomics
- Circular Flow of Income and National Income Accounting
- Classical Theory of Income and Employment
- Keynes's Theory of Employment and Unemployment
- Determination of National Income: Keynes's Basic Two-sector Model
- Determination of National Income in an Open Economy with Government and Foreign Trade: Three Sector and Four Sector Models
- Consumption Function
- Post-Keynesian Theories of Consumption
- Investment Demand
- Theory of Multiplier
- Aggregate Demand and Aggregate Supply Model (with Price Flexibility)
- Unemployment, Full Employment and Price-wage Flexibility
- Unemployment in Developing Countries with Special Reference to India

CHAPTER 1
NATURE AND SCOPE OF MACROECONOMICS

In the first volume of this book we have studied microeconomics which is concerned with the behaviour of individual consumers, factory owners, firms and individual industries and markets. It is through their interaction that microeconomic theory explains how prices of products and factors are determined and how resources are allocated between various products. In microeconomic theory, it is assumed that full employment of resources such as labour and capital prevails and analyses how they are allocated among different products. Further, it explains whether resource allocation achieved is economically efficient. In this second volume, we will study macroeconomic theory and its applications to the formulation of economic policies. In this chapter, we shall briefly explain what macroeconomics is about and why the British economist J.M. Keynes laid stress on macroeconomic analysis as a separate study from microeconomic analysis. Besides, we shall explain the various issues that are studied in macroeconomics. In the end, we shall briefly explain the various post-Keynesian developments in macroeconomics. We shall also discuss the importance of the study of macroeconomics.

WHAT IS MACROECONOMICS

Microeconomics deals with the analysis of small individual units of an economy such as individual consumers, individual firms, individual industries and markets and explains how prices of products and factors are determined. On the basis of these prices microeconomics explains how resources are allocated among various products and how income distribution among factors is determined. *Macroeconomics is concerned with the analysis of the behaviour of the economic system in totality. Thus, macroeconomics studies how the large aggregates such as total employment, national product or national income of an economy and the general price level are determined. Macroeconomics is therefore a study of aggregates. Besides, macroeconomics explains how the productive capacity and national income of the country increase over time in the long run.*

Professor Gardner Ackley makes the distinction between macroeconomics and microeconomics more clearly when he says, "Macroeconomics concerns itself with such variables as the aggregate volume of the output of an economy, with the extent to which its resources are employed, with the size of the national income, with the 'general price level'. Microeconomics, on the other hand, deals with the division of total output among industries, products and firms and the allocation of resources among competing uses. It considers problem of income distribution. Its interest is in relative prices of particular goods and services."[1]

It is evident from above that the subject matter of macroeconomics is to explain what *determines the level of total economic activity* (that is, the size of the national income and employment) and fluctuations (*i.e.*, ups and downs) in it in the short run. It also explains what causes the general price level to rise and determines the rate of inflation in the economy. Besides, modern macroeconomics analyses those factors which determine the increase in productive capacity and national income in the long run. The problem of increasing productive capacity and national income over time in the long run is called the problem of *economic growth*. Thus, what determines rate of growth of an economy is also the concern of macroeconomics.

1. Gardner Ackley, *Macroeconomic Theory*, 1961, p. 4.

Thus, why is national income higher today than it was in 1950? Why does rate of unemployment in a free market economy go up in a period and falls in another period? Why do some countries have high rates of inflation, while others maintain price stability? What causes alternating periods of depression and boom (generally described as business cycles)? Why should government intervene in the economy and what policy it should adopt to check inflation, control business cycles, raise level of national income, reduce unemployment and restore equilibrium in the balance of payments are some of the important questions which macroeconomics attempts to answer.

SAY'S LAW AND KEYNESIAN REVOLUTION

Classical and neoclassical economists[2] assumed that full-employment of labour and other resources always prevailed in the economy and concentrated mainly upon explaining how the resources were allocated to the production of various goods and services and how the relative prices of products and factors were determined. It is mainly because of their full-employment assumption that they concentrated on the problem of determination of prices, outputs and resource employment in the individual industries. They believed that if there were departures from full employment, a free market economy would automatically work in a way that would restore full employment of resources. They argued that involuntary unemployment and underutilisation of the productive capacity could not occur in capitalist economies if market mechanism were allowed to work freely without any interference by the trade unions. Thus, according to them, even when deficiency of aggregate demand arises as it happens during the times of recession or depression, prices and wages would change in such a way that employment, output and national income would not decline. The belief of classical economists that full employment of labour and capital stock will always exist was based on *Say's Law of Markets*. According to Say's Law, *supply creates its own demand* and, therefore, the problem of lack of demand for supply of goods and services does not arise. Factors which produce goods and services for supply in the market get rewards (wages, interest and rent) for their contributions to the production of goods and services produced. Thus, incomes earned become expenditure made on goods and services. Therefore, the problem of deficiency of demand does not arise. They thus could not provide adequate explanation of the occurrence of huge unemployment that prevailed during depression of 1930s in the capitalist economies. What is worse, the classical economists, particularly A.C. Pigou, tried to apply the economic laws that hold good in the case of an individual industry to the case of the behaviour of the whole economic system and macroeconomic variables.

For instance, Pigou asserted that involuntary unemployment existing at the time of depression in the 1930s was due to the obstacles put up by trade unions and government and further that this involuntary unemployment could be eliminated and employment expanded by cutting down the wages. A noted British economist J.M. Keynes challenged this classical viewpoint during the early nineteen thirties when severe depression took place in the capitalist countries such as Britain and the U.S.A. While the cut in wages may expand employment in an individual industry, the reduction in wages throughout the economy will result in fall in incomes of the working classes and is ikely to cause decrease in the level of aggregate demand. The fall in aggregate demand will tend to lower the level of employment rather than expand it.

Though there were pre-Keynesian theories of business cycles and the general price level that were "macro" in nature, but it was J.M. Keynes, an eminent British economist, who laid stress on macroeconomic analysis and put forward a general theory of income and employment in his revolutionary book, *A General Theory of Employment, Interest and Money*, published in 1936. Keynes's theory made a genuine break from the classical and neoclassical economics and produced such a fundamental and drastic change in economic thinking that his macroeconomic analysis has earned the names '*Keynesian Revolution*' and '*New Economics*'. Keynes in his analysis made a frontal

2. Keynes called all those economists who preceded him as classical economists. Thus, neoclassical economists are included in the term 'classical economists' as used by Keynes.

attack on the classical '*Say's Law of Markets*' which was the basis of full employment assumption of classical view that involuntary unemployment could not prevail in a free private enterprise economy. He showed how the equilibrium level of national income and employment was determined by aggregate demand and aggregate supply and further that due to lack of aggregate effective demand equilibrium level of income and employment might well be established at far less than full-employment level in a free-market capitalist economy. This causes involuntary unemployment of labour on the one hand and excess productive capacity (*i.e.*, underutilisation of the existing capital stock) on the other. His macroeconomic model reveals how consumption function, investment function, liquidity preference function, conceived in aggregate terms, interact to determine the level of national income and employment.

Further, the fall in employment in the Great Depression of 1930s and emergence of huge unemployment of labour which in some countries went up to 25 per cent of labour force gives unmistakable evidence that aggregate demand is not always large enough to ensure full employment of labour and full use of productive capacity.

After Keynes, there have been significant developments in macroeconomics. Apart from what determines the level of employment of labour and use of productive capacity, modern macroeconomics is concerned with many more issues such as the problems of inflation, economic growth, business cycles, stagflation, the balance of payments and the exchange rate. The analysis of these six major issues describes the scope of macroeconomics. We shall be explaining all these six issues of macroeconomics in detail in later chapters. Here we shall explain only briefly what these problems are.

MAJOR ISSUES IN MACROECONOMICS

As explained above, Keynes in his book, *Theory of Employment, Interest and Money,* explained how levels of income and employment were determined in a free market economy. During the Second World War period, he extended his macro-theory to explain inflation. However, after Keynes, economists have further developed and extended macroeconomics. We briefly explain below the main issues in macroeconomics.

Determination of National Income and Employment. The first major issue in macroeconomics is to explain what determines the level of employment and national income in an economy and therefore what causes involuntary unemployment. Why is level of national income and employment very low in times of depression as in 1930s in various capitalist countries of the world. This will explain the cause of huge unemployment that emerged in these countries. As mentioned above, classical economists denied that there could be involuntary unemployment of labour and other resources for a long time. They thought that with changes in wages and prices, unemployment would be automatically removed and full employment established. But this did not appear to be so at the time of depression in the thirties and after. Keynes explained that level of employment and national income is determined by aggregate demand and aggregate supply. With aggregate supply curve remaining unchanged in the short run, it is the deficiency of aggregate demand that causes underemployment equilibrium with the appearance of involuntary unemployment. According to him, it is the changes in private investment that cause fluctuations in aggregate demand and is, therefore, responsible for the problems of cyclical unemployment. We will explain Keynes's theory of employment and income and its various aspects in detail in the subsequent chapters.

General Price Level and Inflation. Another macroeconomic issue is to explain the problem of inflation. Inflation has been a major problem faced by both the developed and developing countries in the last fifty years. Classical economists thought that it was the quantity of money in the economy that determined the general price level in the economy and, according to them, rate of inflation depended on the growth of money supply in the economy. Keynes criticised the '*Quantity Theory of Money*' and showed that expansion in money supply did not always lead to inflation or rise in price level. Keynes who before the Second World War explained that involuntary unemployment and depression were due to the deficiency of aggregate demand, during the war period when prices rose very high he explained in a booklet, '*How to Pay for War*', that just as unemployment and depression were caused by the deficiency of aggregate demand, inflation was due to the excessive aggregate demand. Thus, Keynes put forward what is now called *demand-pull theory of inflation*. After Keynes, theory of inflation has been further developed and many theories of inflation depending upon various causes have been put forward. *Cost-push* and *structuralist* theories of inflation have been put forward. To analyse the problem of inflation is an important issue of macroeconomics.

Business Cycles. Throughout history market economies have experienced what are called business cycles. Business cycles refer to fluctuations in output and employment with alternating periods of boom and recession. In boom periods both output and employment are at high levels, whereas in recession periods both output and employment fall and as a consequence large unemployment comes to exist in the economy. When these recessions are extremely severe, they are called depressions. What are the causes of these business cycles or ups and downs in market economies is an important macroeconomic issue which has been highly controversial. The objective of macroeconomic policy is to achieve economic stability with equilibrium at full-employment level of output and income. We shall discuss various theories of business cycles and also monetary and fiscal policies to control business cycles and achieve economic stability.

Stagflation. How to control business cycles and achieve economic stability has been a very difficult problem for the economies to solve. But during the decade of 1970s and in some later times in the subsequent decades market economies have experienced a still more intricate problem which has been described as stagflation. While in business cycles recession or depression is accompanied by high unemployment and falling prices, *in the seventies recession or stagnation was accompanied by not only high unemployment but also rapid inflation*. Since in that period high unemployment and recession (or stagnation) co-existed with high inflation, this problem was given the name stagflation. This stagflation could not be explained with the Keynesian theory which focuses on the demand side. Therefore, a new economic thought which is called *Supply-side Economics* emerged which explained stagflation by laying stress on the supply-side of economic activity. Stagflation is an important issue of modern macroeconomics. Stagflation along with supply-side economics will be explained at length in a separate chapter.

Economic Growth. Another important issue in macroeconomics is to explain what determines economic growth in a country. *Theory of economic growth* or what is simply called *growth economics* which has been recently developed a good deal is an important branch of macroeconomics. The problem of growth is a long-run problem and Keynes did not deal with it. In fact, Keynes is said to have once remarked that "in the long run we are all dead". From this remark of Keynes it should not be understood that he thought long run to be quite unimportant. By this remark he simply emphasized the importance of short-run problem of fluctuations in the level of economic activity (involuntary cyclical unemployment, depression). It was Harrod[3] and Domar[4] who extended the Keynesian analysis to the long-run problem of growth with stability. They laid stress on the dual role of investment—one of *income-generating*, which Keynes considered, and the second of *capacity-creating* which Keynes ignored because of his preoccupation with the short run. In view of the fact that an investment adds to the productive capacity (*i.e.*, capital stock), then if growth with stability (*i.e.*, without stagnation or inflation) is to be achieved, income or demand must be increasing at a rate large enough to ensure the full utilisation of the increasing capacity. Thus, macroeconomic models of Harrod and Domar have explained the rate of growth of income that must take place if the steady growth of the economy is to be achieved. These days growth economics has been further developed and extended a good deal and new theories of growth have been put forward by Solow, Meade, Kaldor and Joan Robinson.

3. R.F. Harrod, *Towards a Dynamic Economics* (1948).
4. E.D. Domar, "Expansion and Employment", *American Economic Review* (March 1947).

Since the above growth theories apply particularly to the present-day developed economies, special theories which explain the causes of underdevelopment and poverty in less developed countries (LDCs) and which also suggest strategies for initiating and accelerating growth in them have also been propounded. These special growth theories relating to less developed countries (LDCs) are generally known as *Economics of Development*. We shall discuss some theories of development such as *balanced growth theory* of Nurkse, *unbalanced growth theory* of Hirschman, *'theory of economic development with unlimited supplies of labour'* of Arthur Lewis in this book.

Balance of Payments and Exchange Rate. Balance of payments is the record of economic transactions of the residents of a country with the rest of the world during a period. The aim to prepare such a record is to present an account of all receipts on account of goods exported, services rendered and capital received by the residents of a country and the payments made for goods imported, services received and capital transferred to other countries by residents of a country. There may be deficit or surplus in balance of payments. Both create problems for an economy. An important effect is that the transactions in balance of payment are influenced by the exchange rate. The exchange rate is the rate at which a country's currency is exchanged for foreign currencies. The instability in exchange rate has been a major problem in recent years which has given rise to serious balance of payments problems. During the 12 years period (1901-2002), Indian rupee depreciated to a large extent in terms of U.S. dollar giving rise to serious problems. In the year 2003, the Indian rupee started appreciating against the U.S. dollar. During the two years 1997 and 1998, currencies of many South-East Asian countries and Japan rapidly depreciated in terms of U.S. dollar. This creates the situation of economic crisis which needs to be overcome. Both interrelated problems of balance of payments and instability of foreign exchange rate will be analysed in a separate part of this book.

POST-KEYNESIAN SCHOOLS OF MACROECONOMIC THOUGHT

There have been significant developments of macroeconomic ideas in the post-Keynesian era. These developments came about as a critique of Keynesian macroeconomics. We briefly explain below the following four main developments in macroeconomics after Keynes:

(1) Monetarism
(2) Supply-side Economics
(3) Rational Expectations Theory
(4) New Keynesian Economics

Monetarism

American economist Milton Friedman, a Nobel Laureate in economics, criticised Keynes's macroeconomics and put forward a new view or idea. Milton Friedman along with Anna Schwartz published an important work, "*A Monetary History of the United States*", in which they argued that monetary policy is the prime engine in causing fluctuations in economic activity by bringing about changes in aggregate demand. Milton Friedman made a sharp criticism of Keynes's view that monetary policy was quite ineffective instrument in bringing about economic stability. In fact, he asserted that monetary policy caused or contributed to almost all recessions he studied. He stressed that even the Great Depression of 1930s was primarily caused by the tight monetary policy adopted at that time. He further argued that the Great Depression of the 1930s did not show the failure of free-market system, but the failure of Government's interventionist policy, especially monetary policy of the Central Bank which landed the U.S. economy into depression. According to Friedman, it is the excessive contraction of money supply by the Federal Reserve Bank System (which is the Central Bank of the U.S.A.) that caused depression in the U.S. economy. There are differences between the monetarists and Keynesians in respect of two important issues. First issue relates to the relationship between money supply and inflation. The second relates to the role of Government in the economy.

Monetarists led by Friedman believe that *inflation is always and everywhere a monetary phenomenon*. According to them, inflation is caused by the rapid expansion of money supply in the economy. Friedman and his followers restated the classical quantity theory of money and made improvements in it but retained its essence that it is the greater increase in money supply relative to growth of output that causes inflation. In order to control inflation, they suggest a constant growth rate of money supply in the economy.

An important feature of monetarism is that to achieve economic stabilisation it is opposed to the adoption of *activist role* by the Government. As against this, Keynesian economists emphasise that active role should be played by the Government to control business cycles and achieve economic stability. Like classical economists, monetarists also believe that a free-market economy is inherently stable and if the economy departs from the state of full employment, full-employment equilibrium is restored through automatic adjustments in it. Therefore, they even argue that Government or its central bank should not adopt *active* discretionary monetary policy, rather it *should pursue a policy of stable rate of growth of money supply*.

In sharp contrast to the monetarists' view, Keynes and his followers advocate the adoption of activist role by the Government. In this respect Keynes lays stress on the adoption of discretionary fiscal and monetary policies.[5] Besides, Keynesian economists believe that expansion in money supply does not always cause inflation in the economy. According to them, whether or not increase in money supply will lead to inflation depends on the possibility of expansion in output. When the economy is in the grip of depression, the increase in money supply is likely to lead to the large expansion in output of goods which will prevent the rise in prices. Similarly, monetarists are opposed to the fiscal policy of budget deficits and public debt. They argue for reducing taxes and lowering public expenditure so that the role of Government in the economy is restricted.

Supply-side Economics

In the 1970s and early eighties the problem of stagflation appeared in which high inflation was accompanied by high rate of unemployment. This problem of stagflation demonstrated that Keynes's theory which focused on fluctuations in aggregate demand as responsible for either high unemployment or high inflation could not explain the problem of stagflation in which both high unemployment and high inflation occurred together. The failure of Keynesian economics to deal with stagflation led some economists to believe that problem was on the supply side of economic activity. The problem looked quite strange and Keynesian policies were incapable of solving it. Following Keynesian policy, if expansionary fiscal and monetary measures were taken to raise aggregate demand to remove stagnation or high unemployment, it accelerated inflation further. On the other hand, if steps were taken to lower aggregate demand to lower the inflation rate, it would have further increased the already high unemployment rate. Supply-side economists pointed out that it was *supply-shocks*, delivered among others by reduction in oil supplies and *increase in oil prices,* that caused the problem of stagflation. As a result of contraction in supply due to the adverse supply shocks, given the aggregate demand curve, price level and inflation rate could rise on the one hand and aggregate output could fall giving rise to more unemployment on the other.

Advocates of supply-side economics argued that for the expansion in aggregate supply and thereby increase in employment opportunities, incentives to work, save and invest more were required to be promoted. According to them, more work or labour, and higher investment will lead to the increase in aggregate supply. The increase in aggregate supply, given the aggregate demand curve, will lead to the increase in employment on the one hand and reduction of inflation on the other. According to them, high rates of income tax serve as disincentives to work, save and invest

5. Keynes himself advocated for the adoption of only discretionary fiscal policy to cure depression and control inflation but his followers generally called *'New Keynesians'* emphasise both fiscal and monetary policies to attain economic stability.

more. Therefore, to encourage more saving, work and investment, they advocated for the reduction in the then prevailing high rates of income tax. As a result of more work and investment, aggregate supply will increase which will not only cause employment to rise and unemployment to decrease but also lower the rate of inflation.

Besides, in the opinion of the supply-side economists the reduction in tax rates will increase income and output to such an extent that even with lower rates of taxes, the Government revenue will increase which would tend to reduce Government's budget deficit. In this respect, the concept of *Laffer curve* has been put forward, according to which when rate of a tax is increased from zero upward, Government's revenue from it initially rises, but after a point further hike in the rate of a tax brings about decrease in revenue for the Government. Therefore, they are of the view that higher rates of personal taxes are responsible for low tax revenue. Therefore, they assert that with lowering of tax rates, not only national income and employment will increase through greater labour supply and investment but this will also reduce Government's budget deficit by increasing tax revenue. We will critically examine in detail the supply-side economics in a separate chapter.

Rational Expectations Theory

Recently a new macroeconomic theory has been put forward which is also opposed to Keynesian macroeconomic theory and policy which focused on aggregate demand for goods and services. According to this new classical macroeconomic theory, consumers, workers and producers behave rationally to promote their interests and welfare. On the basis of their *rational expectations*, they make quick adjustments in their behaviour. Therefore, according to the supporters of rational expectations theory, involuntary unemployment cannot prevail. They argue that producers and consumers collect all the necessary information about economic situation and policies and determine their behaviour according to the rational expectations formed on the basis of all information collected. According to them, people do not make any mistakes in establishing correct relationship between economic events and Government's policies on the one hand, and the results that follow from them on the other. In other words, they always make correct predictions from the Government's policies and changes in the economic environment. For example, when Government makes a deficit budget they will expect that rates of interest will rise. Therefore, they will attempt to take loans now when the rates of interest are lower so as to save themselves from paying higher interest rates in the future. Unfortunately, because of this behaviour the interest rates rise immediately rather than in future.

A significant difference between the Keynesian theory and rational expectations theory may be noted here. In the Keynesian theory deficit in Government budget leads to increase in aggregate demand and will therefore promote private investment. On the other hand, according to rational expectations theory, budget deficit will cause rate of interest to rise which will discourage private investment. Thus, according to rational expectations theory, increase in aggregate demand as a result of budget deficit is offset by decrease in private investment so that national output, income and employment remain unaffected.

Similarly, according to rational expectations theory, if central bank of a country increases the money supply, consumers, producers and workers will expect rationally that price level will rise. On the basis of these rational expectations, workers would get their wages raised, landlords will raise their rent, lenders and bankers will increase the rate of interest, and producers will raise their profit margins. As a result of these adjustments by various persons, the effect of expansion in money supply on these persons will get cancelled. According to the rational expectations theorists, since the consumers, workers and producers themselves make adjustments to save them from the adverse effects of economic events and policies there is no need for the Government to intervene in the economy through adoption of proper macroeconomic policy. Thus, like Friedman and other monetarists, supporters of rational expectations theory are opposed to the activist role by the Government. According to them, it is very difficult to implement an activist policy successfully. They are of the view that market is usually in full-employment equilibrium and people make self-adjustments in their behaviour to protect and promote their interests. The Government cannot achieve any success in improving economic situation through its activist policy. As compared to the Government individuals themselves are in a better position to take corrective measures to safeguard their interests.

New Keynesian Economics

New classical economics based on rational expectations put forward by Lucas and Robert Barro criticised the traditional Keynesian model by expressing doubts on the validity of its assumption of *price stickiness* (*i.e.,* failure of prices to adjust quickly as a result of changes in demand or supply conditions) by pointing out that it was not based on the firm foundations of microeconomics based on rational and profit-maximising principles. In response to the critique of Lucas that the original Keynesian explanation of determination of aggregate output and employment level and fluctuations in them was not based on rationality assumption of microeconomics, a number of Keynesian economists explained short-run price stickiness with the help of microeconomics and still keeping the conclusions of traditional Keynesian economics intact. That is why they are called *new* Keynesians and the models they propounded are called *New Keynesian Economics.* Now, the question arises how do these new Keynesian economists explain price stickiness with microeconomic theory. There are mainly two reasons given by new Keynesian economists for short-run stickiness of prices: (1) First, according to them, imperfect competition (monopolistic competition and oligopoly) prevails in the real world product markets and prices are set by the firms working in these markets. while Keynes's original model was based on perfect competition. Second, the other reason for not changing prices by the firms is due to the *menu costs* that they have to incur.

As regards price stickiness under imperfect competition, N. Gregory Mankiw and other new Keynesian economists explain that it is quite rational for the imperfectly competitive firm which has a control over its price of the product not to adjust or lower its price when there is decrease in aggregate demand, because though it will lose some of its customers but not all of them. If all firms hold on to their prices when demand declines, it will lose some sales as a result but still retain much of its sales to be profitable for it. If all firms stick to their initial price, no individual firm will lose much of its sales.

According to the second reason of not changing or adjusting prices by the firms when demand for their products changes is that they have to incur *menu costs* if they change their prices when demand for the products changes. *Menu costs refer to the costs the firms have to incur for making adjustments in prices. To change prices, a firm has to print a new catalogue and send them to its customers*. Though these costs are small, they have a large effect on the economy.*

We will explain New Keynesian model put forward by N. Gregory Mankiw and how he explains recession and involuntary unemployment with price stickiness in the short run. With its model of price stickiness he also explains unemployment and inflation trade-off involved in the short-run Phillips Curve.

Further, New Keynesian economics with its focus on price stickiness can explain the important result of original Keynesian economics that, contrary to classical macroeconomics, *money is non-neutral in its effect on real variables such as real national income (i.e., aggregate output), level of employment and real rate of interest.* If nominal prices remain sticky when there is increase in money supply, that is, do not adjust quickly to offset the changes in money supply, then, according to New Keynesian economists, it will lead to the adjustment in real variables. Thus, when money supply is increased by the Central Bank, it will cause rate of interest to fall. Fall in rate of interest will lead to more investment and more investment will raise the levels of real national income and employment by raising the level of aggregate demand, prices remaining sticky. This shows that money is non-neutral.

* For details see Chapter 16.

WHY A SEPARATE STUDY OF MACROECONOMICS

Now an important question which arises is why a separate study of the economic system as a whole or its large aggregates is necessary. Can't we generalise about the behaviour of the economic system as a whole or about the behaviour of large aggregates such as aggregate consumption, aggregate saving, aggregate investment from the economic laws governing the behaviour patterns of the individual units found by microeconomics? In other words, can't we obtain the laws governing the macroeconomic variables such as total national product, total employment and total income, general price level, etc by simply adding up, multiplying or averaging the results obtained from the behaviour of the individual firms and industries? The answer to this question is the behaviour of the economic system as a whole or the macroeconomic aggregates is not merely a matter of addition or multiplication or averaging of what happens in the various individual parts of the whole. As a matter of fact, in the economic system what is true of parts is not necessary true of the whole. Therefore, the application of micro-approach to generalise about the behaviour of the economic system as a whole or macroeconomic aggregates is incorrect and may lead to misleading conclusions. Therefore, a separate macro-analysis is needed to study the behaviour of the economic system as a whole in respect of various macroeconomic aggregates. When laws or generalisations are true of constituent individual parts but untrue and invalid in case of the whole economy, paradoxes seem to exist. Boulding has called these paradoxes as *macro-economic paradoxes*. It is because of the existence of these macro-economic paradoxes that there is a justification for making macro-analysis of the behaviour of the whole economic system or its large economic aggregates. Thus Professor Boulding rightly remarks, "*It is these paradoxes more than any other factor, which justify the separate study of the system as whole, not merely as an inventory or list of particular items, but as a complex of aggregates.*"[6]

Professor Boulding further elaborates his point by liking the economic system with a forest and the individual firms or industries with the trees in the forest. Forest, he says, is the aggregation of trees but it does not reveal the same properties and behaviour pattern as those of the individual trees. It will be misleading to apply the rules governing the individual trees to generalise about the behaviour of the forest.

Various examples of macro-paradoxes (that is, what is true of parts is not true of the whole) can be given from the economic field. We shall give two such examples of savings and wages, on the basis of which Keynes laid stress on evolving and applying macroeconomic analysis as separate and distinct approach from microeconomic analysis.

Paradox of Thrift. Savings are generally good for individuals. People save for some motives in view. They save for meeting their needs in old age, for education of their children, for purchasing durable goods such as houses and cars and thereby raise their standards of living. Further, they save for making investment or depositing in banks which raise their income in future years. But an interesting paradox arises which shows importance of macroeconomic analysis as distinct from microeconomic analysis. The paradox of thrift arises because it so happens that when *all people* in a society try to save more but in fact they are not only unable to do so but actually attempt to save more by *all people* cause their income or standard of living to decline. Keynes's macroeconomic analysis helps to explain this paradox, Keynes pointed out that efforts to save more, especially at times of depression, will lower the consumption demand of the people and will therefore adversely affect aggregate demand in the economy. *The decline in aggregate demand will cause national output and income to fall and unemployment to increase*. At the lower level of national income, savings will fall to the original level but the consumption of the people will be less than before which implies that people would become worse off as a result of their decision of saving more. Thus decision to save more will deepen the economic depression. It goes to the credit of Keynes's theory of effective demand as determinant of

6. K.E. Boulding, *A Reconstruction of Economics* (1952), p. 173.

national income and his multiplier theory (which shows that income and employment fall more than the original rise in saving) that paradox of thrift has been resolved. It clearly shows that at times of depression more savings deepen the economic crisis.

Wage-Employment Paradox. Another common example to prove that what is true for the individual may not be true for the society as a whole is the wage-employment relationship. As pointed out above, classical and neoclassical economists, especially A.C. Pigou, contended that the cut in money wages at times of depression and unemployment would lead to the increase in employment and thereby eliminate unemployment and depression. Now, it is true that a cut in money wages in an individual industry will lead to more employment in that industry. It is quite commonplace conclusion of microeconomic theory that, given the demand curve for labour, at a lower wage more men will be employed. But for the society or economy as a whole this is highly misleading. If the wages are cut all round in the economy, as was suggested by Pigou and others on the basis of wage-employment relationship in an individual industry, the aggregate demand for goods and services in the society will decline, since wages are incomes of the workers who constitute majority in the society. The decline in aggregate demand will mean the decrease in demand for goods of many industries. Because the demand for labour is a derived demand, *i.e.*, derived from the demand for goods, the fall in aggregate demand for goods will result in the decline in demand for labour which will create more unemployment rather than reduce it.

Fallacy of Composition

We thus see that the laws or generalisations which hold good for the behaviour of an individual consumer, firm or industry may be quite invalid and misleading when applied to the behaviour of the economic system as a whole. There is thus *a fallacy of composition*. This is so because what is true of individual components is not true of the collective whole. As mentioned above, these are called macroeconomic paradoxes and it is because of these paradoxes that a separate study of the economic system as a whole is essential.

Macroeconomic analysis takes account of many relationships which are not applicable to individual parts at all. For instance, an individual may save more than that he invests or he may invest more than he saves, but for economy as a whole it is one of the important principles of Keynesian macroeconomics that actual savings are always equal to actual investment. Likewise, for an individual or a group of individuals, expenditure may be more or less than the income but the national expenditure of the economy must be equal to the national income. In fact, the national expenditure and national income are two identical things. Similarly, in the case of full employment, an individual industry may increase its output and employment by bidding away the workers from other industries, but the economy cannot increase its output and employment in this way. Thus what applies to an individual industry, does not do so in case of the economic system as a whole.

We therefore conclude that a separate and distinct macroeconomic analysis is essential if we want to understand the true working of the economic system as a whole. As a matter of fact, microeconomics and macroeconomics are complementary to each other rather than being competitive. The two types of theories deal with different subjects, one deals mainly with the explanation of relative prices of goods and factors and the other mainly with the short-run determination of income and employment of the society and its long-run economic growth. Professor Samuelson rightly says, "There is really no opposition between micro-and macro-economics. Both are absolutely vital. And you are only half-educated if you understand the one while being ignorant of the other."[7]

7. Paul A. Samuelson, *Economics*, 10th edition, 1976.

QUESTIONS FOR REVIEW

1. What is macroeconomics? How will you distinguish it from microeconomics?
2. What are the important problems that constitute the subject matter of macroeconomics? Briefly explain them.
3. What are macroeconomic paradoxes? Explain any two of them. Explain how these paradoxes limit the applicability of microeconomic theories to explain the behaviour of the economy as a whole.
4. In what sense did Keynes bring about revolution in economics?
5. Macroeconomics has been described as a study of aggregates. Name some important 'aggregates' or variables whose behaviour is analysed in macroeconomics. Explain them briefly.
6. Why is there a need for a separate theory of macroeconomics? Explain.
7. What is involuntary unemployment? Why did classical economists believe that involuntary unemployment cannot exist in a free-market economy?
8. Explain briefly the following models of macroeconomics. In what fundamental ways do they differ from the Keynesian model?
   1. Monetarism
   2. Supply-side economics
   3. Rational expectations theory
9. Keynesian macroeconomics is said to be irrelevant for developing countries like India. Do you agree? Explain briefly.
10. Name which of the following problems fall within the purview of macroeconomics?
    1. How is general price level determined?
    2. Determination of output of a cotton textile industry.
    3. How is wage rate of computer engineer determined?
    4. The problem of unemployment in the economy.
    5. On what factors does economic growth of a country depend?
    6. How is level of national income determined?
    7. Deficit in the balance of payments of a country.
11. Explain the importance of the study of macroeconomics.
12. What is New Keynesian Economics? Explain. Why was there a need for putting forward New Keynesian Economics? How is it different from Keynesian Economics?

CHAPTER 2
CIRCULAR FLOW OF INCOME AND NATIONAL INCOME ACCOUNTING

MEANING OF NATIONAL INCOME

The labour and capital of a country acting on its natural resources produce annually a certain amount of goods and services. This is called national income of the country. *National income of a country can be defined as the total market value of all final goods and services produced in the economy in a year*. Two things must be noted in regard to the meaning of national income. First, it measures the market value of annual output. In other words, national income is a *monetary measure*. This is because there is no other way of adding up the different sorts of goods and services except with their money prices. But in order to know accurately the changes in physical output, the figure for national income is adjusted for price changes. Secondly, for calculating national income accurately all goods and services produced in any given year must be counted only once, and not more than once. Most of the goods go through a series of production stages before reaching a market. As a result, parts or components of many goods are bought and sold many times. Hence, in order to avoid counting several times the parts of goods that are sold and resold, national income only includes the market value of all *final goods* and ignores the transactions involving *intermediate goods*.

The above way of explaining national income is only one way of interpreting it. In fact, the concept of national income has three interpretations. It represents (1) a total value of production (as explained above); (2) it represents a receipts total and (3) it represents an expenditure total. It is an obvious fact that every expenditure is at the same time a receipt. In other words, amount spent by the buyers is equal to the amount received by the sellers. But if goods and services are valued at their market prices, we have three-fold identity, namely, that the value received equals the value paid equals the value of goods and services produced and sold.

To explain the above idea, let us take an economy where there are only two agents: households and firms. Firms are required to produce goods. To produce them, they require services of factors of production. Factors of production are paid the rewards for their contribution to the production of goods. Thus incomes of these factors arise in the course of production. The sales value of net production must equal the sum total of payments made by the firms to the factors of production in the course of production. The sales value of net production must equal the sum total of payments made by the firms to the factors of production in the form of wages, rents, interest and profits. These incomes in turn become the sources of expenditure. Thus income flows from the firms to the households in exchange for productive services. This income again returns to the firms when expenditure is made by the households on the goods produced by the firms.

From above it follows that: *National Income = National Product = National Expenditure*. In other words, there are three measures of national income of a country. They are:

(*a*) the sum of values of all final goods and services produced;

(*b*) the sum of all incomes, in cash and kind, accruing to factors of production in a year; and

(*c*) the sum of consumers' expenditure, net investment expenditure and government expenditure on goods and services.

Sum of all income, sum of values of all final production, and sum of all expenditures will be the same, but the significance of each arises from the fact that they reflect the three basic activities of the nation's economy, *viz*., production, distribution and expenditure.

CIRCULAR FLOW OF INCOME

The modern economy is a monetary economy. In the modern economy, money is used in the process of exchange. Money has facilitated the process of exchange and has removed the difficulties of the barter system. Thus money acts as a medium of exchange. The households supply the economic resources or factors to the productive firms and receive in return the payments in terms of money. In is thus clear that, in the monetary economy, there will be flows of money corresponding to the flows of economic resources and the flows of goods and services. But each money flow is in opposite direction to the real flow.

Circular Income Flow in a Two-Sector Economy

To begin with, to explain the circular flow of income and expenditure we assume that all incomes which households receive are spent on consumer goods and services and thus there is no savings by them. Similarly, we assume that there is no investment by business firms. Money flows of income and expenditure corresponding to the real flows in terms of goods, services and productive factors are shown in Fig. 2.1. In the upper loop of this figure, the resources such as land, capital and entrepreneurial ability flow from households to business firms as indicated by the arrow mark. In opposite direction to this, money flows from business firms to the households as factor payments such as wages, rent, interest and profits. In the lower part of the figure, money flows from households to firms as consumption expenditure made by the households on the goods and services produced by the firms, while the flow of goods and services is in opposite direction from business firms to households. Thus we see that money flows from business firms to households as factor

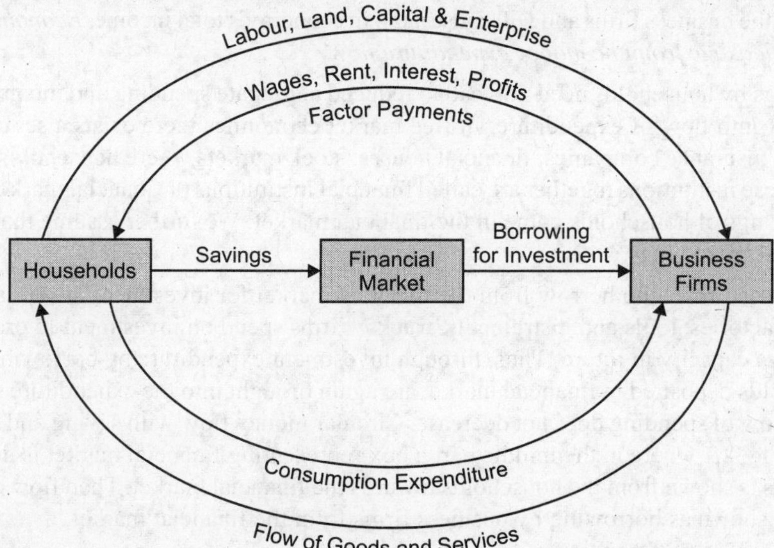

Fig. 2.1. *Circular Flow of Income in a Simple Two-Sector Economy*

payments and then it flows from households to firms. Thus there is, in fact, *a circular flow of money or income*. This circular flow of money will continue indefinitely week by week and year by year. This is how the economy functions.

It may, however, be pointed out that this flow of money income will not always remain the same in volume. In other words, the flow of money income will not always continue at a constant level. In years of depression, the circular flow of money income will contract, *i.e.*, will become lesser in volume, and in years of prosperity it will expand, *i.e.*, will become greater in volume. This is so because the flow of money is a measure of national income and will, therefore, change with changes in the national income. In years of depression, when national income is low, the volume of the flow of money will be small and in years of prosperity when the level of national income is quite high, the flow of money will be large.

In order to make our analysis simple and to explain the central issues involved, we take many assumptions. In the first place, we assume that neither the households save from their incomes, nor the firms save from their profits. We further assume that the government does not play any part in the national economy. In other words, the government does not receive any money from the people by way of taxes, nor does the government spend any money on the goods and services produced by the firms or on the resources and services supplied by the households. Thirdly, we assume that the economy neither imports goods and services, nor exports anything. In other words, in our above analysis we have not taken into account the role of foreign trade. In fact we have explained above the flow of money that occurs in the functioning of a closed economy with no savings and no role of government.

Circular Money Flow with Saving and Investment

In our above analysis of the circular flow of money we have assumed that all income which the households receive, they spend it on consumer goods and services. As a result, circular flow of money spending and income remains undiminished. We will now explain if households save a part of their income, how their savings will affect money flows in the economy. When households save, their expenditure on goods and services will decline to that extent and as a result money flow to the business firms will contract. With reduced money receipts, firms will hire fewer workers (or lay off some workers) or reduce the factor payments they make to the suppliers of factors such as workers. This will lead to the fall in total incomes of the households. Thus, savings reduce the flow of money expenditure to the business firms and will cause a fall in economy's total income. *Economists therefore call savings a leakage from the money expenditure flow.*

But savings by households need not lead to reduced aggregate spending and income if they find their way back into flow of expenditure. In free market economies there exists a set of institutions such as banks, insurance companies, financial houses, stock markets where households deposit their savings. All these institutions together are called financial institutions or financial market. We assume that all the savings of households come in the financial market. We further assume that there are no inter-household borrowings.

It is business firms who borrow from the financial market for investment in capital goods such as machines, factories, tools and instruments, trucks. Firms spend on investment in order to expand their productive capacity in future. Thus, through investment expenditure by borrowing the savings of the households deposited in financial market are again brought into the expenditure stream and as a result total flow of spending does not decrease. Circular money flow with saving and investment is illustrated in Fig. 2.1 where in the middle part a box representing financial market is drawn. Money flow of savings is shown from the households towards the financial market. Then flow of investment expenditure is shown as borrowing by business firms from the financial market.[1]

Condition for the Constancy of Circular Money Flow

Now the question arises: what is the condition for the flow of money income to continue at a steady level so that it makes possible the production and subsequent flow of a given volume of goods

1. It may be noted that business firms also save when they do not distribute among shareholders all the profits made by them and keep the undistributed profits as reserves. We are ignoring them in our money flow analysis as firms themselves use their savings for buying machinery and making other investment expenditure.

and services at constant prices? To explain this we have to introduce saving and investment in the analysis of circular flow of income. As mentioned above, saving a part of income means it is not spent on consumer goods and services. In other words, saving *is withdrawal of some money from the income flow*. On the other hand, investment means some money is spent on buying new capital goods to expand production capacity. In other words, *investment is injection of some money in circular flow of income*. For the circular flow of income to continue unabated, the withdrawal of money from the income stream by way of saving must equal injection of money by way of investment expenditure. Therefore, planned savings must be equal to planned investment if the constant money income flow in an economy is to be obtained. Now, what will happen if planned investment expenditure falls short of the planned savings? As a result of fall in planned investment expenditure, income, output and employment will fall and therefore the flow of money will contract. If the equality between planned savings and planned investment is disturbed by increase in savings, then the immediate effect will be that the stocks of goods lying in the shelves of the shops will increase (as some of the goods will not be sold due to the fall in consumption, *i.e.,* increase in savings). Owing to the deficiency of demand for goods and the accumulation of stocks, retailers will place small orders with the wholesalers. Consequently, smaller amount of goods will be produced and therefore fewer capital goods like machinery will be needed with the result that fixed investment will tend to fall. Thus the ultimate effect of either the fall in planned investment or the increase in planned savings is the same, namely, the fall in income, output, employment and prices with the result that the flow of money will contract.

On the other hand, if the equality between planned savings and planned investment is disturbed by the increase in investment demand, the result will be increase in income, output and employment. Consequently, the flow of money income will expand.

It is thus clear from the above analysis that the *flow of money income will continue at a constant level only when the condition of equality between planned saving and investment is satisfied*. It was believed by classical economists that financial market provides a mechanism which coordinates the savings of households and the investment expenditure by the firms. Rate of interest, which is the price for the use of savings, is determined by saving and investment. If savings exceed investment expenditure, rate of interest falls so that, at a lower rate of interest, investment increases and both become equal. On the contrary, if investment expenditure is greater than savings, rate of interest will rise so that at a higher rate of interest savings increase and become equal to planned investment expenditure.

However, an eminent British economist, J.M. Keynes, refuted the above argument that changes in rate of interest will cause saving and investment to become equal. According to him, since in a free market capitalist economy investment is made by business enterprises and savings are mostly done by households and for different reasons, there is no guarantee that planned investment will be equal to planned savings and thus fluctuations in income, output and employment are inevitable. As a result, circular flow of income does not continue at a steady level in a free-enterprise capitalist economy unless certain corrective and preventive steps are taken by the government to maintain stability in the economy.

Saving–Investment Identity in National Income Accounts

Despite the fact that people who save are different from the business firms which primarily invest, in national income accounts savings are identical or always equal to investment in a simple two-sector economy having no roles of Government and foreign trade. This is a basic identity in national income accounts which needs to be carefully understood. Of course, in our above analysis of circular flow of income, we explained that planned investment by business firms can differ from savings by household. But in that analysis we referred to *planned* or *intended* investment and savings which often differ and affect the flow of national income. However, in national income accounts we are concerned with *actual saving* and *actual investment*. It is these actual or realised saving and investment that are identical in national income accounts. We can prove their identity in the following way.

In a simple economy which has neither government, nor foreign trade, the value of *output produced* which we denote by Y is equal to the value of *output sold*. Since the value of output sold in a simple two-sector economy is equal to the sum of consumption expenditure and investment expenditure, we have

$$Y \equiv C + I \qquad \text{...}(i)$$

where Y = Value of aggregate output, C = Consumption expenditure and I = Investment expenditure.

A pertinent question which arises here is what happens to the unsold output. The unsold output leads to the *increase in the inventories of goods* and in national income accounting increase in inventories of goods is treated as a part of actual investment. This may be considered as the firms selling the goods to themselves to add to their inventories. Thus, gross national product (GNP) produced is used either for consumption or for investment.

Now, look at the gross national product or income in the simple economy from the viewpoint of its allocation between consumption and saving. Since national income (which is equal to GNP) can be either consumed or saved,[2]. we have

$$Y \equiv C + S \qquad \text{...}(ii)$$

From the identities (*i*) and (*ii*) we get

$$C + I \equiv Y \equiv C + S \qquad \text{...}(iii)$$

The left-hand side of the identity (*iii*), namely $C + I = Y$ shows the components of aggregate demand (that is, aggregate expenditure on goods and services produced) and the right-hand side of the identity (*iii*), namely, $Y = C + S$ shows the allocation of national income to either consumption or saving. Thus, the identity (*iii*) shows that the value of output produced or sold is equal to the total income received. It is income received that is spent on goods and services produced.

Now subtracting the consumption (C) from both sides of the identity (*iii*) we have

$$I = S$$

Thus, in *our two-sector simple economy with neither government, nor foreign trade, investment is identically equal to saving.*

Circular Income Flow in a Three-Sector Economy with Government Sector

In our above analysis of money flow, we have ignored the existence of government for the sake of making our circular flow model simple. This is quite unrealistic because government absorbs a good part of the incomes earned by households. Government affects the economy in a number of ways. Here we will concentrate on its taxing, spending and borrowing roles. Government purchases goods and services just as households and firms do. Government expenditure takes many forms including spending on capital goods and infrastructure (highways, power, communication), on defence goods, and on education and public health and so on. These add to the money flows which are shown in Fig. 2.2 where a box representing Government has been drawn. It will be seen that government purchases of goods and services from firms and households are shown as flow of money spending on goods and services.

Government expenditure may be financed through taxes, out of assets or by borrowing. The money flow from households and business firms to the government is labelled as tax payments in Fig. 2.2. This money flow includes all the tax payments made by households less transfer payments received from the Government. Transfer payments are treated as negative tax payments.

2. Note that decisions about saving are made both by households and business firms. In national income accounts they are generally put together and referred to as private sector savings.

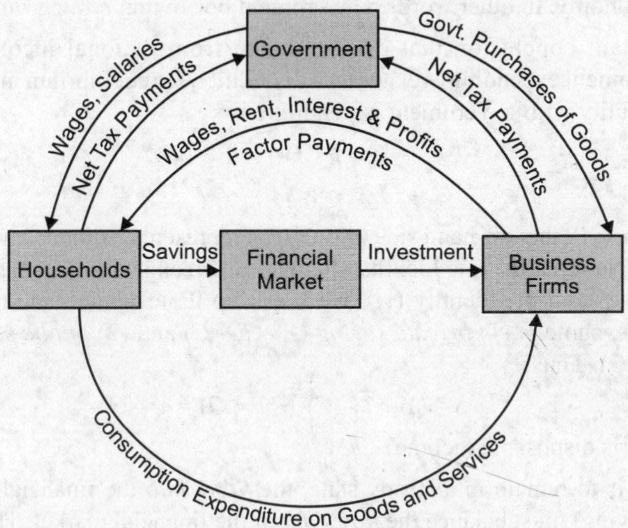

Fig. 2.2. *Circular Money Flow Model with Government*

Another method of financing Government expenditure is borrowing from the financial market. This can be represented by the money flow from the financial market to the Government and is labelled as Government borrowing (To avoid confusion we have not drawn this money flow from financial market to the Government). Government borrowing increases the demand for credit which causes rate of interest to rise. The government borrowing through its effect on the rate of interest affects the behaviour of firms and households. Business firms consider the interest rate as cost of borrowing and the rise in the interest rate as a result of borrowing by the Government lowers private investment. However, households who view the rate of interest as return on savings feel encouraged to save more.

National Income Account Identity with Government

It follows from above that the inclusion of the Government sector significantly affects the overall economic situation. Total expenditure flow in the economy is now the sum of consumption expenditure (denoted by C) , investment expenditure (I) and Government expenditure (denoted by G). Thus,

$$\text{Total expenditure } (E) = C + I + G \qquad \text{...(i)}$$

Total income (Y) received is allocated to consumption (C), savings (S) and taxes (T). Thus

$$Y = C + S + T \qquad \text{...}(ii)$$

Since expenditure (E) made must be equal to the income received (Y), from equations (i) and (ii) above we have

$$C + I + G = C + S + T \qquad \text{...}(iii)$$

Since C occurs on both sides of the equation (iii) and will therefore be cancelled out, we have

$$I + G = S + T \qquad \text{...}(iv)$$

By rearranging we obtain

$$G - T = S - I \qquad \text{...}(v)$$

Equation (v) is very significant as it depicts what would be the consequences if government budget is not balanced. If Government expenditure (G) is greater than the tax revenue (T), that is, $G > T$, the government will have a budget deficit. To finance the budget deficit, the Government will borrow from the financial market. For this purpose, the private investment (I) by business firms must be less than the savings (S) of the households. Thus Government borrowing reduces private investment in the economy. In other words, Government borrowing *crowds out* private investment.

Another important conclusion that can be drawn from national income account identity incorporating Government expenditure relates to the condition for equilibrium in the financial market. National income identity with government expenditure is

$$Y = C + I + G$$

$$\text{or} \qquad Y - C - G = I \qquad \ldots(vi)$$

In the expression (*vi*), the left-hand side ($Y - C - G$) represents national saving or simply saving (S). Note that in this National Income Identity all government expenditure is treated as consumption expenditure. To understand the identity (*vi*), we break up its left-hand side representing national saving into two parts, namely, (1) *private saving* ($Y - T - C$) and (2) *public saving* (*i.e.*, saving of the government ($T - G$). Thus

$$S = (Y - T - C) + (T - G) = Y - C - G$$

(Note that $Y - T$ is disposable income)

If the economy is to remain in a steady state, the *flows into* the financial market (*i.e.*, private saving and public saving) must balance the *flows out* of the financial market. Thus, for the economy to remain in a steady state

$$Y - T - C + (T - G) = I$$

That is, for the economy to remain in a steady state, the sum of private saving and public saving must equal investment.

MONEY FLOWS IN THE FOUR-SECTOR OPEN ECONOMY: ADDING FOREIGN SECTOR

We now turn to explain the money flows that are generated in an open economy, that is, economy which has trade relations with foreign countries. Thus, the inclusion of the foreign sector will reveal to us the interaction of the domestic economy with foreign countries. Foreigners interact with the domestic firms and households through exports and imports of goods and services as well as through borrowing and lending operations through financial market. Goods and services produced within the domestic territory which are sold to the foreigners are called exports.

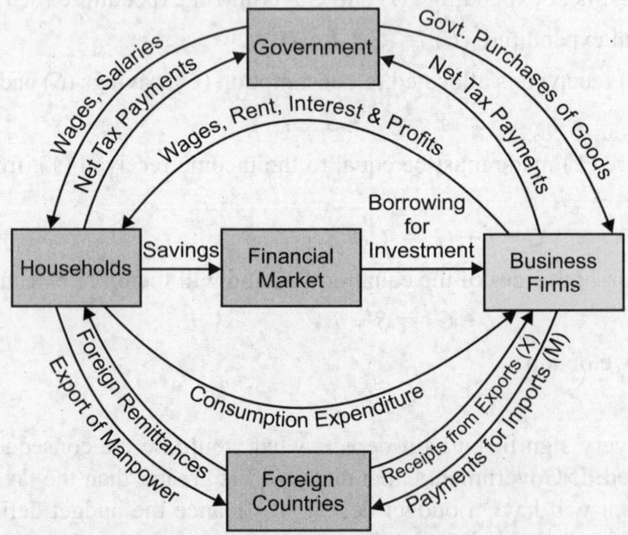

Fig. 2.3. *Circular Flow of Income in an Open Economy with Government and Foreign Sector*

On the other hand, purchases of foreign-made goods and services by domestic households are called imports. Fig. 2.3 illustrates additional money flows that occur in the open economy when exports and imports also exist in the economy. In our analysis, we assume it is only the business firms of the domestic economy that interact with foreign countries and therefore export and import goods and services. A flow of money spending on imports has been shown to be occurring from the domestic business firms to the foreign countries (*i.e.*, rest of the world) . On the contrary, flow of money expenditure on exports of a domestic economy has been shown to be taking place from foreign countries to the business firms of the domestic economy. If exports are equal to the imports, then there exists a balance of trade. Generally, exports and imports are not equal to each other. If value of exports exceeds the value of imports, *trade surplus* occurs. On the other hand, if value of imports exceeds value of exports of a country, *trade deficit* occurs.

In the open economy there is interaction between countries not only through exports and imports of goods and services but also through borrowing and lending funds or what is also called financial market. These days financial markets around the world have become well integrated. When there is a trade surplus in the economy, that is, when exports (X) exceed imports (M), *net capital outflow* will take place. By net capital outflow we mean foreigners will borrow from domestic savers to finance their purchases of our exports. In this way as a result of net capital outflow domestic savers will lend to foreigners, that is, acquire foreign financial assets.

On the contrary, in case of *import surplus*, that is, when imports are greater than exports, trade deficit will occur. Therefore, in case of trade deficit, domestic consumer households and business firms will borrow from abroad to finance their excess of imports over exports. This means there will be capital inflow in our economy. As a result, foreigners will acquire domestic financial assets.

Saving–investment Identity in the Open Economy

From the circular flows that occur in the open economy the national income must be measured by aggregate expenditure that includes net of exports, that is, $X–M$ where X represents exports and M represents imports. Imports must be subtracted from the total expenditure made by foreigners on our domestically produced goods and services exported to them to get the value of net exports. Thus, in the open economy

$$\text{National Income} = C + I + G + NX$$

where NX represents net exports, $X–M$.

Since national income can be either consumed, saved or paid as taxes to the Government we have

$$C + I + G + NX = C + S + T$$

Since C is common on both sides of the above equation, we have

$$I + G + NX = S + T \qquad \ldots(vii)$$

The above equation (*vii*) shows that in the open economy sum of private investment (I), Government expenditure (G) and net exports (NX) is equal to the sum of savings and tax revenue.

Relation of Aggregate Domestic Output and Expenditure with Trade Balance

The national income account identity of the open economy can be used further to show how aggregate domestic output (Y) and aggregate expenditure ($C + I + G$) and net exports (*i.e.*, trade balance) are related. Rearranging the national income of the open economy ($Y = C + I + G + NX$), we have

$$NX = Y – (C + I + G)$$

where $C + I + G$ is aggregate domestic expenditure. Thus

Net Exports = National Domestic Product (Y) *– Aggregate Domestic Expenditure*

From above it follows that in an open economy aggregate domestic expenditure need not be equal to the aggregate domestic output of goods and services (*i.e.*, Y). If aggregate domestic output (Y)

exceeds aggregate domestic expenditure ($C+I+G$), there is export-surplus and we export the excess of domestic output, that is, net exports (NX) are positive ($NX > 0$). On the other hand, if domestic output is less than domestic expenditure, we import this shortfall (*i.e.*, $NX < 0$).

Foreign Capital Flows and Trade Balance

As in the case of a closed economy, goods market is intimately related to the financial market in the open economy. To show this, let us rearrange the national income accounts identity ($Y = C + I + G + NX$) of the open economy.

$$Y - C - G = I + NX \quad \ldots(viii)$$

The left-hand side of the above equation (*i.e.*, $Y - C - G$) represents national saving (S). Note that all Government expenditure is here treated as consumption expenditure. Thus, it follows from above that

$$S = I + NX$$

or

$$S - I = NX \quad \ldots(ix)$$

The national income identity of expression (*ix*) shows that the *economy's net exports (NX) must always be equal to the difference between saving and investment. The net exports are also called trade balance because it shows the difference between exports and imports of an economy*. The trade balance (NX) may be positive or negative. If exports exceed imports, trade balance (NX) is positive and if exports fall short of imports, the trade balance (NX) is negative.

But the important result that follows from national income identity of the open economy relates to the link between *international capital flows* and the *goods market*. If $S - I$ in (*ix*) is positive, that is, if the economy's national saving exceeds domestic investment, it will be lending the excess funds to foreigners, that is, there will be *net capital outflow* from the economy. On the other hand, if $S - I$ is negative, that is, if domestic investment exceeds domestic saving, the economy will be borrowing from abroad to finance the excess investment. That is, there will be *net capital inflow* into the economy to finance higher capital formation in the economy.

It follows from above that net capital flows ($S - I$) always equal the trade balance or to current account balance which is the broader term that includes *invisibles* (that is, exports and imports of services) also.

NATIONAL INCOME AND NATIONAL PRODUCT

The sum of all incomes of the people of a country is called national income. This national income is greatly related to the national product. In fact in a two-sector economy without the Government and its imposition of indirect taxes and grant of subsidies and also assuming no depreciation national income and national product are one and the same thing. The incomes which different people of the society get are obtained by them for their contribution of labour, land, capital and entrepreneurial services to the national production. Hence the income which the labourers get are wages for the productive services which labourers lend to the various firms which undertake the work of production. Similarly, the owners of land get income as rent because of their contribution of land to the productive firms; the capitalists get interest for lending their money capital to the entrepreneurs for undertaking any work of production or business. The entrepreneurs get profits for starting and organising the work of production and bearing risk and uncertainty involved in it. It is thus clear that the different individuals of a country obtain their income either as wages of their labour, or as interest on their money capital, or as rent for their land, or as profits for their enterprise. The sum of incomes obtained as wages, rent, interest and profits is the national income of the country.

Various households obtain their income from the productive firms or businesses which utilise their labour, land, capital and other services for the production of goods and services. The incomes earned by the various households and individuals from the work of production are in fact costs of production of the goods produced. The total value of all final goods and services produced by various productive firms or businesses in a year is known as national product. Therefore, the national product of a country can be estimated by multiplying the total output of final goods and services with their market prices.

Hence the total national product in terms of rupees or value which is produced by various productive firms in a year will be distributed among the various productive factors which have contributed to its production. Therefore, out of this national production (in value terms) wages will be paid to those households which have sold their labour to the productive firms. Out of this total value of the national product, landowners would get rent for the contribution of the services of land and capitalists would get interest for lending money capital to the productive firms. After the payment of wages, rent, interest, what is left is the profit of the entrepreneurs who set up productive firms, organise their work and bear risk and uncertainty and because of these services they get profits. It is thus clear that the national product (value of total output of final goods and services in a year of a country) is distributed among wages, rents, interests and profits. As we have stated above, the sum of total wages, total interest, total rents and total profits is national income. Hence national product will be equal to the national income. The above conclusion that in a two-sector economy without Government and foreign trade national income equals national product can be expressed in the form of the following equation:

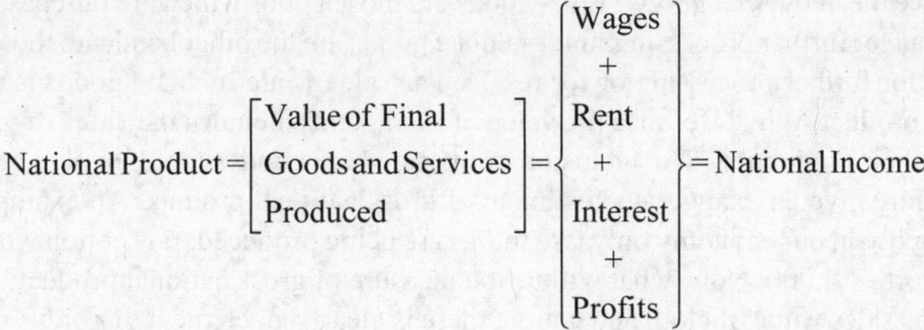

$$\text{National Product} = \begin{bmatrix} \text{Value of Final} \\ \text{Goods and Services} \\ \text{Produced} \end{bmatrix} = \left\{ \begin{matrix} \text{Wages} \\ + \\ \text{Rent} \\ + \\ \text{Interest} \\ + \\ \text{Profits} \end{matrix} \right\} = \text{National Income}$$

From the above analysis it is evident that national income and national product are one and the same thing. Professor J.R. Hicks rightly writes, "*The value of the net social product of the community and the sum of the incomes of its members are exactly equal. The net social product and the social income are one and the same thing.*"[3]

It needs to be emphasised that the above conclusion regarding equality of national product and national income is valid only in our simple two-sector economy where we have assumed no depreciation funds are kept and also no role of Government (and therefore no indirect taxes collected by it and no subsidies granted by it). In case of a real economy, firms keep a part of some value created as depreciation allowance which therefore do not become a part of the income of the factors. When role of Government is brought into our analysis, a part of the value of output is taken away by the Government as indirect taxes and therefore the incomes of factor owners are reduced to that extent. Thus, only in the absence of depreciation funds and net indirect taxes, national product will be equal to national income. In our simple economy we assume away these depreciation funds and net indirect taxes levied by the Government on business firms to emphasise the point that national income does not come out of thin air, it is in fact generated by the production of goods and services. Therefore, growth in national income can be achieved by the increase in production of goods and services.

3. J.R. Hicks, *The Social Framework*.

CONCEPTS OF NATIONAL INCOME

There are various concepts of national income. Which shall study then one by one.

Gross Domestic Product (GDP)

This is the basic national accounting measure of the total output or aggregate supply of goods and services. *Gross Domestic Product is defined as the total market value of all final goods and services produced in a year in the domestic territory of a country*. Two things must be noted in regard to gross national product. First, it measures the market value of annual output. In other words, GDP is a *monetary measure*. There is no other way of adding up the different sorts of goods and services produced in a year except in terms of their money prices. But in order to know accurately the changes in physical output, the figure for gross national product is adjusted for price changes.

Secondly, for calculating gross domestic product accurately, all goods and services produced in any given year must be counted once, and not more than once. Most of the goods go through a series of production stages before reaching a market. As a result, parts or components of many goods are bought and sold many times. Hence to avoid counting several times the parts of goods that are sold and resold, gross national product includes the market value of only final goods and services and ignores transactions involving intermediate goods.

Final Goods and Intermediate Goods

What do we mean by *final goods*? Final goods are those goods which are purchased for final use and not for resale or further processing. Intermediate goods, on the other hand, are those goods which are purchased for further processing or for resale. The sale of intermediate goods is excluded from gross national product. Why? Because the value of final goods includes the value of all intermediate goods used in their production. The inclusion of intermediate goods would involve double counting and will therefore give an exaggerated estimate of gross national product. An example will clarify this point. Suppose in our economy only two things are being produced, raw cotton worth ₹ 1000 and cotton cloth worth ₹ 2,000. Now, what will be the measure of gross national product? For finding it, if we add up the sales value of cloth and cotton, there is clearly an element of *double counting* in the sense that we have added the value of cotton *twice*—one as the sales value of cotton and secondly when we added to it the value of cloth. Actually, the value of cloth includes also the value of cotton which having been accounted for already should not be added second time.

Further, it is worth noting that Gross Domestic Product (GDP) includes only *currently produced* goods and services in a year. It is a *flow measure* of output of goods and services per time period, say in a year. In it only goods and services produced in a particular year are included. Market transactions involving goods produced in previous periods such as old houses, cars, factories built earlier are not included in GDP of the current year. Similarly, purchase and sale of assets such as stock and bonds do not involve current production of goods and are therefore excluded from GDP of the year.

Lastly, Gross Domestic Product (GDP) refers to the market value of goods and services currently produced by *normal residents of a country working within its domestic territory*. These residents may be national or non-national companies. Thus, many foreign companies have set up plants in India. They are owned by non-nationals but work to produce goods and services within the domestic territory of India and generate incomes for the Indian people employed in them. These foreign companies can only send back to their home countries profits earned by them.

Gross domestic product has the following components:

1. Value of *final consumer goods and services* produced in a year and consumed by the households which is denoted by *consumption* (C) by households.
2. Value of *new capital goods produced and addition to the inventories of goods* such as raw materials, unfinished goods and consumer goods produced but not sold during a year. This

is called *Gross Private Investment* (I).

3. Value of output of General Government which is taken to be equal to the value of *purchases of goods and services by the Government* which we denote by G.
4. Net Exports (NX) which is equal to value of goods *exported* minus the value of goods *imported* (M).

We can obtain Gross Domestic Product (GDP) at market prices by adding up the above four items. Thus,

$$GDP_{MP} = C + I + G + NX$$
$$\text{where } NX = (X - M)$$

We have graphically depicted GDP_{MP} by the second bar in Fig. 2.4

Gross National Product (GNP)

Another important concept of national income is gross national product (GNP). Gross national product is the market value of all final goods and services produced by normal *residents working in the domestic territory of a country and also includes net factor income earned from abroad.* Thus difference between gross domestic product (GDP) and gross national product (GNP) at market prices arises due to the existence of '*net factor income from abroad* '. Gross domestic product does not include net factor income from abroad, whereas gross national product includes it. Therefore,

Gross National Product (at market prices) or

$$GNP_{MP} = GDP_{MP} + \text{net factor income from abroad}$$

Net Factor Income from Abroad. Now, what does net factor income from abroad stand for? The sum of factor incomes such as wages and salaries (*i.e.*, compensation to employees), rent, interest and profits generated within the domestic territory of a country in a year is called domestic factor income. It includes factor incomes generated both by residents and non-residents working in the domestic territory of a country. For example, non-residents (*i.e.*, foreigners) work in the domestic territory of India and earn wages and salary. Thus, foreign individuals and companies from the USA, Great Britain and other countries have acquired property such as factories, offices, buildings, places and have also acquired financial assets such as bonds and shares of Indian companies. This generates incomes in the form of rent and interest to them. In addition to this, foreign residents—individuals and companies—have set up industrial plants and factories producing goods and services from which they earn profits.

On the other hand, Indians go abroad and work in the territories of other countries and earn wages and salaries. Likewise, some Indian individuals and corporate companies have acquired assets such as buildings, factories, commercial space, and have also invested in bonds, bank deposits of foreign countries and thus receive rent and interest. Some Indian companies have set up factories abroad and earn profits.

Now, the *net factor income from abroad is the difference between factor income received from abroad by normal residents of India for rendering factor services in other countries on the one hand and the factor incomes paid to the foreign residents for factor services rendered by them in the domestic territory of India on the other.*

Net factor incomes earned from abroad have therefore the following three components:

1. *Net* compensation to employees.
2. *Net* income from property, *i.e.*, rent, interest and income from entrepreneurship (that is, profits and dividends).
3. *Net* retained earnings of the resident companies working in foreign countries.

(**Note** that '*Net*' in the above means the difference between the relevant income of normal residents earned from abroad and the same type of incomes paid to non-residents working in the domestic territory of a country)

It should however be noted that net factor income from abroad should not be confused with income from *net exports* which are a part of both gross national product (GNP) and gross domestic product (GDP). Net exports equal total exports minus imports. Gross exports are the total value of domestically produced goods and services sold to foreign buyers in a year and are therefore a part of Gross Domestic Product (GDP) and not a factor income from abroad. On the other hand, imports are the value of imported goods and services purchased by the domestic buyers. Therefore, imports do not form a part of GDP or GNP. In national income accounting, we subtract the value of imports from the value of exports to arrive at net exports which are a part of GDP and therefore also of GNP. Thus, earnings from net exports are quite distinct from net factor income from abroad.

We have graphically depicted GNP_{MP} by the first bar in Fig. 2.4.

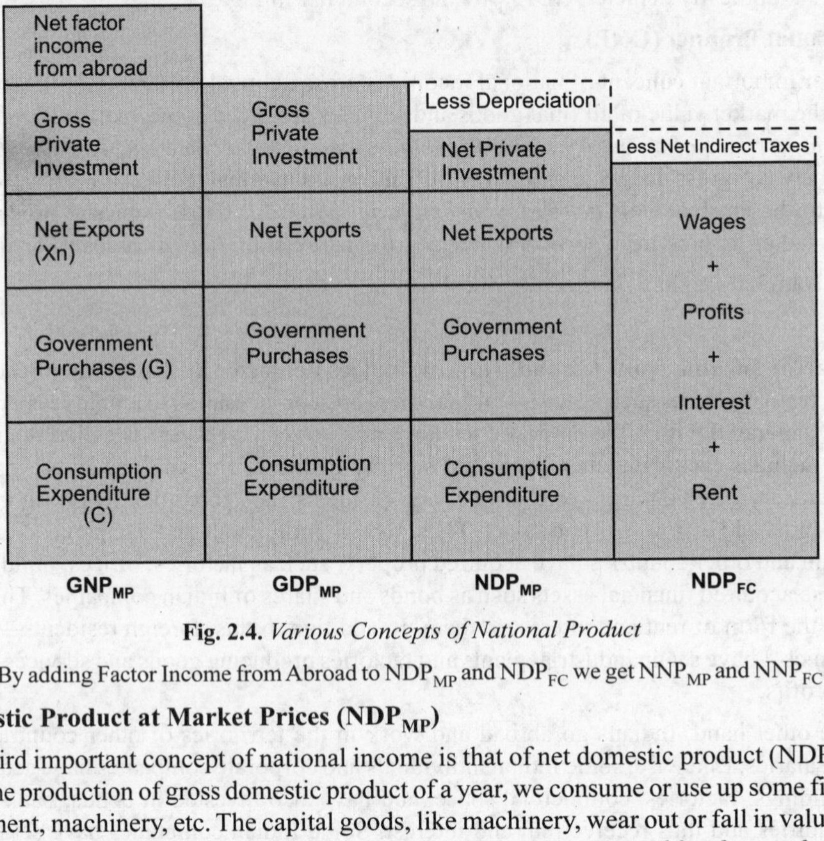

GNP_{MP}	GDP_{MP}	NDP_{MP}	NDP_{FC}
Net factor income from abroad			
Gross Private Investment	Gross Private Investment	Less Depreciation	
		Net Private Investment	Less Net Indirect Taxes
Net Exports (Xn)	Net Exports	Net Exports	Wages + Profits + Interest + Rent
Government Purchases (G)	Government Purchases	Government Purchases	
Consumption Expenditure (C)	Consumption Expenditure	Consumption Expenditure	

Fig. 2.4. *Various Concepts of National Product*

(**Note:** By adding Factor Income from Abroad to NDP_{MP} and NDP_{FC} we get NNP_{MP} and NNP_{FC} respectively.

Net Domestic Product at Market Prices (NDP_{MP})

The third important concept of national income is that of net domestic product (NDP) at market prices. In the production of gross domestic product of a year, we consume or use up some fixed capital, *i.e.*, equipment, machinery, etc. The capital goods, like machinery, wear out or fall in value as a result of its consumption or use in the production process. *This consumption of fixed capital or fall in the value of fixed capital due to wear and tear is called depreciation.* When charges for depreciation are deducted from the gross national product we get net national product. Clearly, it means the *market value of all final goods and services* produced in a year after providing for depreciation. Therefore, it is also called '*domestic product or income at market prices*'. Therefore,

$$\left.\begin{array}{c}\text{Net Domestic Product}\\ \text{at Market Prices}\end{array}\right\} = \text{Gross Domestic Product at market prices} - \text{Depreciation}$$

or $$NDP_{MP} = GDP_{MP} - \text{Depreciation}$$

Net National Product at Factor Cost (NNP_{FC}) or National Income

Net National Product at factor cost which is also simply called *national income* means the sum of all incomes earned by resource suppliers for their contribution of land, labour, capital and entrepreneurial ability which go into the year's net production. In other words, national income (or national income at factor cost) shows how much it costs society in terms of economic resources to produce net output. It is really the national income at factor cost for which we use the term National Income. The difference between national income (or national income at factor cost) and net national product (national income at market prices) arises from the fact that *indirect taxes* and *subsidies* cause market prices of output to be different from the factor incomes resulting from it. Suppose for instance, a metre of mill cloth sold for ₹ 200 includes ₹ 25 on account of the excise and the sales tax. In this case while the market price of the cloth is ₹ 200 a metre, the factors engaged in its production and distribution would receive ₹ 175 a metre. The value of cloth at factor cost would thus be equal to its value at market price less the indirect taxes on it. On the other hand, a subsidy causes the market price to be less than the factor cost. Suppose handloom cloth is subsidised at the rate of ₹ 10 per metre and it sells at ₹ 90 per metre. Then while the consumer pays ₹ 90 per metre, the factors engaged in the production and distribution of such cloth will receive ₹ 100 per metre (₹ 90 + 10 = ₹ 100). The value of handloom cloth at factor cost would thus be equal to its market price plus the subsidies paid on it. It follows, therefore, that the national income (or national income at factor cost) is equal to net national product minus indirect taxes plus subsidies.

$$\left.\begin{array}{l}\text{National Income} \\ \qquad\text{or} \\ \text{Net National Product at} \\ \text{Factor Cost}\end{array}\right\} = \begin{array}{c}\text{Net National Product (National Income at Market Prices)} - \\ \text{Indirect Taxes} + \text{Subsidies}\end{array}$$

Net of indirect taxes and subsidies is called Net Indirect Taxes. Therefore,

NNP_{FC} or National Income = NNP_{MP} – Net Indirect Taxes

Personal Income (PI)

Personal Income is the sum of all incomes actually received by all individuals or households during a given year. National income, that is, total incomes earned and personal income, that is, total incomes received must be different because some incomes which are earned such as social security

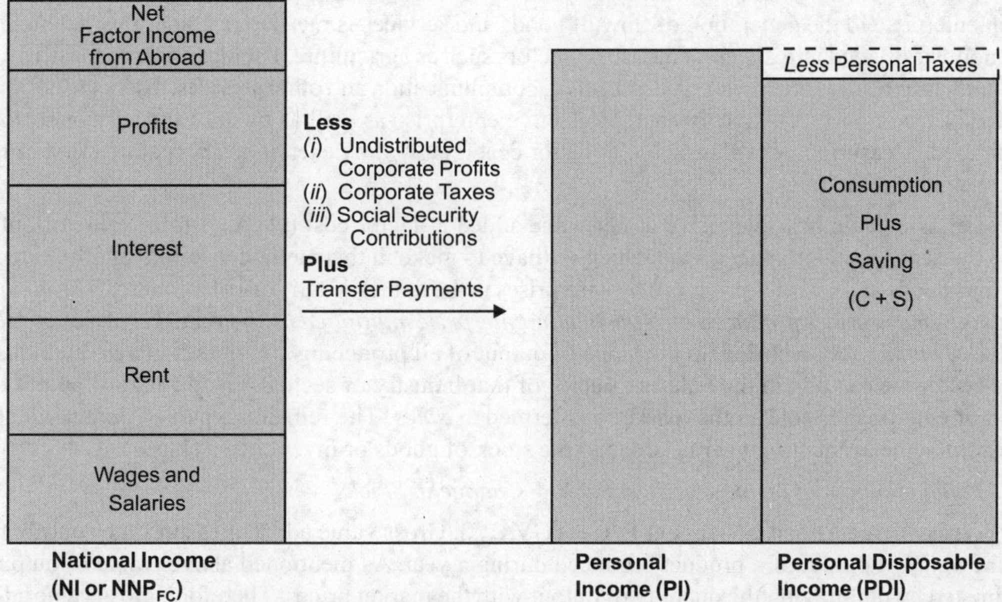

Fig. 2.5. *National Income to Personal Disposable Income*

contributions, corporate income taxes and undistributed corporate profits are not actually received by households, and conversely, some incomes which are received like transfer payments are not currently earned (examples of transfer payments are old-age pensions, unemployment compensation, relief payments, interest payments on the public debt, etc.). Obviously, in moving from national income as an indicator of income earned to personal income as an indicator of income actually received, we must subtract from national income those three types of income which are earned but not received and add those incomes which are received but currently not earned. Therefore,

Personal Income = National Income – Social Security Contributions – Corporate Income Taxes – Undistributed Corporate Profits + Transfer Payments.

Personal Disposable Income (PDI)

Even whole of the incomes which are actually received by the people are not available to them for consumption. This is because governments levy some personal taxes such as income tax, personal property taxes. Therefore, after a part of personal income is paid to government in the form of *personal taxes* like income tax, personal property taxes, etc., what remains of personal income is called *personal disposable income*. Therefore,

Personal Disposable Income (PDI) = Personal Income – Personal Taxes.

Personal Disposable Income can either be consumed or saved. Hence,

Personal Disposable Income = Consumption + Saving.

How do we get personal income and personal disposable income from national income is illustrated in Fig. 2.5.

MEASUREMENT OF NATIONAL INCOME

Since factor *incomes* arise from the production of goods and services, and since incomes are spent on goods and services produced, three alternative methods of measuring national income are possible.

1. Value Added Method

This is also called *output method* or *production method.* In this method the contribution of each enterprise to the generation of flow of goods and services is measured. Under this method, the economy is divided into different industrial sectors such as agriculture, fishing, mining, construction, manufacturing, trade and commerce, transport, communication and other services. Then, the net value added at factor cost (NVA_{FC}) by each productive enterprise as well as by each industry or sector is estimated. Measuring net value added at factor cost (NVA_{FC}) by each industry requires first to find out the value of output.

Let us explain how we arrive at net value added at factor cost (NVA_{FC}) from value of output step by step. An important estimate which we have to make in this method is to find out the value of various goods and services produced by enterprises in the domestic territory of a country. *The quantity of goods and services produced by a particular enterprise multiplied by their market prices is called value of output.* By summing up the value of output of all producing enterprises in a given industry or a sector we can obtain the value of output of that industry or sector. A major part of output of a firm or enterprise is sold in the market and termed as *sales*. The remaining part of output which is *not sold* in the accounting year is added to the stock of goods or inventories. Thus.

Value of output of an enterprise = Sales + Change in Stocks

Gross Value Added at Market Prices (GVA_{MP}). Gross value added measures the contribution to the value of output of a product produced during a year. As mentioned above, value of output is estimated by multiplying the quantity of output with the market prices. Therefore, gross value added at market prices by a production unit is obtained by deducting the value of intermediate consumption (that is, the value of intermediate goods such as raw materials used from the value of output (at market prices) produced. Thus,

Gross value added at market prices (GVA_{MP}*) = Value of Output – Intermediate Consumption*

A firm disposes of its value added at market prices (GVA_{MP}) among the three items : (1) making depreciation provision for consumption of fixed capital during the year, (2) making payments of indirect taxes such as excise duties, sales tax, import duty to the government, and (3) making factor payments such as wages, interest and profits to the factors of production whose services have been used for the production of a good.

Net Value Added at Market Prices (NVA_{MP}). When from gross value added at market prices we deduct depreciation on account of consumption of fixed capital during the production process of a good during the year, we get Net Value Added at market prices (NVA_{MP}). Thus,

$$\text{NVA}_{MP} = \text{GVA}_{MP} - \text{Consumption of Fixed Capital } \textit{i.e.} \text{ Depreciation}$$

Thus net value added is net of depreciation or consumption of fixed capital.

Net Value Added at Factor Cost (NVA_{FC}). When adjustment is made in net value added at market prices (NVA_{MP}) for the payment of net indirect taxes, (that is, indirect taxes such as sales tax, excise duty, customs duties to the government minus subsidies received from the government), we get Net Value Added at Factor Cost (NVA_{FC}). This is because after the subtraction of depreciation amount and net indirect taxes what remains is used for making payments to factors of production such as wages to labour, interest on capital borrowed, rent for land and building hired from others and profits and dividends to entrepreneur. In other words, NVA_{FC} measures the value of factor cost a firm has to incur.

Summing up the net values added at factor cost (NVA_{FC}) by all productive enterprises of an industry or sector gives us the net value added at factor cost of each industry or sector. We then add up net values added at factor cost by all industries or sectors to get *net domestic product at factor cost* (NDP_{FC}). Lastly, to the net domestic product we add the *net factor income from abroad* to get *net national product at factor cost* (NNP_{FC}) which is also called national income. Thus,

$$\text{NI or NNP}_{FC} = \text{NDP}_{FC} + \text{Net factor income from abroad}$$

This method of calculating national income can be used where there exists a census of production for the year. In many countries, the data of production of only important industries are known. Hence this method is employed along with other methods to arrive at the national income. The one great advantage of this method is that it reveals the relative importance of the different sectors of the economy by showing their respective contributions to the national income.

Precautions. The following precautions should be taken while measuring national income of a country through value added method:

1. *Imputed rent values* of self-occupied houses should be included in the value of output. Though these payments are not made to others, their values can be easily estimated from prevailing values in the market.
2. Sale and purchase of *second-hand goods* should not be included in measuring value of output of a year because their values were counted in the year of output of the year of their production. Of course, commission or brokerage earned in their sale and purchase has to be included because this is a new service rendered in the current year.
3. *Value of production for self-consumption* are to be counted while measuring national income. In this method, the production for self-consumption should be valued at the prevailing market prices.
4. *Value of services of housewives* are not included because it is not easy to find out correctly the value of their services.

5. Value of intermediate goods *must not* be counted while measuring value added because this will amount to double counting.

2. Income Method

This method approaches national income from distribution side. In other words, this method measures national income at the phase of distribution and appears as income paid and/or received by individuals of the country. Thus, under this method, *national income is obtained by summing up the incomes of all individuals of a country*. Individuals earn incomes by contributing their own services and the services of their property such as land and capital to the national production. Therefore, national income is calculated by adding up the rent of land, wages and salaries of employees, interest on capital, profits of entrepreneurs (including undistributed corporate profits) and incomes of self-employed people. This method of estimating national income has the great advantage of indicating the distribution of national income among different income groups such as landlords, owners of capital, workers, entrepreneurs. Measurement of national income through income method involves the following main steps:

1. Like the value added method, the first step in income method is also to *identify* the productive enterprises and then *classify* them into various industrial sectors such as agriculture, fishing, forestry, manufacturing, transport, trade and commerce, banking, etc.

2. The second step is to classify *the factor payments*. The factor payments are classified into the following groups:

1. *Compensation to employees* which includes wages and salaries, employers' contribution to social security schemes.
2. *Rent and also royalty*, if any.
3. *Interest*.
4. *Profits*: Profits are divided into three sub-groups:
   (*i*) Dividends
   (*ii*) Undistributed profits
   (*iii*) Corporate income tax
5. *Mixed income of the self-employed*: In India, as in other developing countries, there is fifth category of factor income which is termed as *mixed income of self-employed*. In India a good number of people are engaged in household industries, in family farms and other unorganised enterprises. Because of self-employment nature of the business it is difficult to separate wages for the work done by the self-employed from the surplus or profits made by them. Therefore, the incomes earned by them are mix of wages, rent, interest and profit and are, therefore, called *mixed income of the self-employed*.

3. The third step is to *measure factor payments*. Income paid out by each enterprise can be estimated by gathering information about the number of units of each factor employed and the income paid out to each unit of every factor. Price paid out to each factor multiplied by the number of units of each factor employed would give us the factor's income.

4. The *adding up of factor payments* by all enterprises belonging to an industrial sector would give us the incomes paid out to various factors by a *particular* industrial sector.

5. By summing up the incomes paid out by *all industrial sectors* we will obtain domestic factor income which is also called net domestic product at factor cost (NDP_{FC}).

6. Finally, by adding net factor income earned from abroad to domestic factor income or NDP_{FC} we get net national product at factor cost (NNP_{FC}) which is also called *national income*.

Income approach to measurement of national income is shown through bar diagrams in Fig. 2.6.

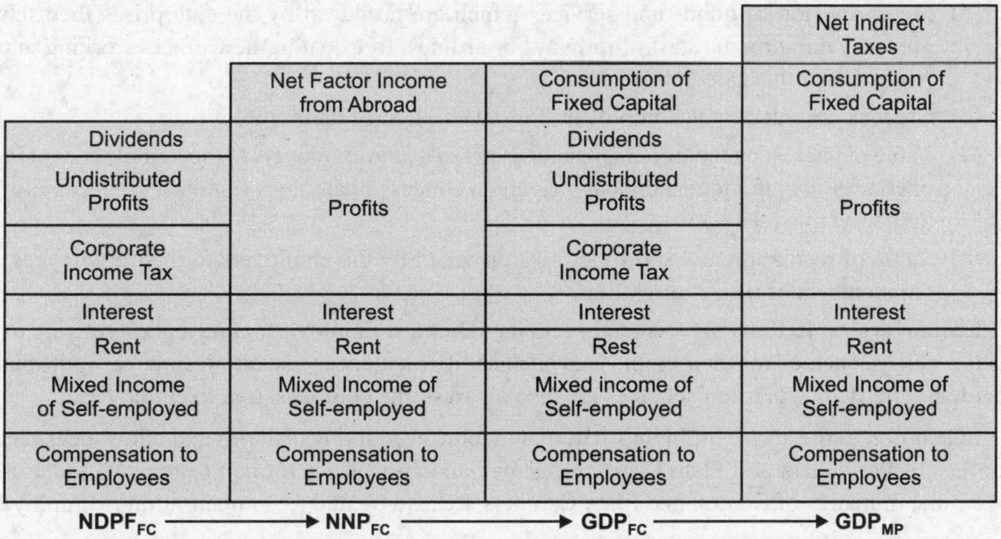

$NDPF_{FC}$ →	NNP_{FC} →	GDP_{FC} →	GDP_{MP}
			Net Indirect Taxes
	Net Factor Income from Abroad	Consumption of Fixed Capital	Consumption of Fixed Capital
Dividends	Profits	Dividends	Profits
Undistributed Profits		Undistributed Profits	
Corporate Income Tax		Corporate Income Tax	
Interest	Interest	Interest	Interest
Rent	Rent	Rest	Rent
Mixed Income of Self-employed	Mixed Income of Self-employed	Mixed income of Self-employed	Mixed Income of Self-employed
Compensation to Employees	Compensation to Employees	Compensation to Employees	Compensation to Employees

Fig. 2.6. *Income Approach to National Income*

Compensation to Employees

From the above items of income categories compensation to employees requires further explanation. Compensation to employees by the producers is the sum of wages and salaries, paid both in cash and kind, and contribution to social security schemes for the employees made by the employers. Thus, it has the following components:

Wages and Salaries: These include all payments made by the employers to their employees, both in cash and kind, for the services they render to their employers.

Wages and Salaries Payable in Cash. They include the following:

(1) Wages and salaries received in cash by the employees from their employers.

(2) Payments received by the employees for overtime work done by them.

(3) Travelling allowance received by the employees for going to and coming home from their workplaces.

(4) Bonuses, if any, receivable by the workers.

(5) Dearness allowance paid to the employees to neutralise the rise in cost of living.

(6) Vacation allowance and sick leave allowance.

(7) Leave travelling allowance (LTC).

(8) Commission provided, if any, to the sales staff on the sales.

Wages and Salaries in Kind. These are the remunerations in kind received in the form of goods and services by the employees for their use by themselves or by the members of their households. The following are some important types of remuneration received in kind:

(1) Housing accommodation provided free of cost.

(2) Free meals and drinks (such as tea, coffee, cold drinks) provided free to the employees when they are on duty.

(3) Uniforms and special clothing, if any, received free of cost from the employer.

(4) The free services of vehicles (cars, scooters, vans etc.) provided by the employers to their employees.

(5) Free provision of goods and services which are produced by the enterprises themselves. Free travelling by the staff of railways or airlines, free coal to the workers working in coal mines fall in this category.

(6) Creches provided by the employers for the children of their employees.

(7) Value of interest on the free-interest loans given by the employers to their employees or value of concessions in interest on loans given on concessional rates of interest by the employers to their employees.

(8) Value of recreation and sport facilities provided by the employers to their employees and the members of their households.

Employers' contributions to social security schemes. Employers' contribution relating to the social security schemes for their employees such as life insurance, casualty insurance, contributory provident fund (CPF), pension schemes are also a part of the compensation to employees.

In addition to the above, in India's national income accounting, salaries and allowances paid to members of Parliament and State Legislatures, pay and allowances to the President of India, State Governors, ministers of Central and State Cabinets are also treated as compensation to employees.

Precautions. While estimating national income through income method the following precautions should be taken:

1. *Transfer payments* are not included in estimating national income through this method.

2. *Imputed rent* of self-occupied houses are included in national income as these houses provide services to those who occupy them and its value can be easily estimated from the market value data.

3. *Illegal money* such as *hawala money, money earned through smuggling* etc. are not included as they cannot be easily estimated.

4. *Windfall gains* such as prizes won, lotteries are also not included.

5. Corporate profit tax (that is, tax on income of the companies) should not be separately included as it has already been included as a part of profits.

6. *Death duties, gift tax, wealth tax, tax on lotteries*, etc., are paid from past savings or wealth and not from current income. Therefore, they should not be treated as a part of national income of a year.

7. The receipts from the *sale of second-hand goods* should not be treated as a part of national income. This is because the sale of second-hand goods does not create new flows of goods and services in the current year.

8. Income equal to the *value of production used for self-consumption* should be estimated and included in the measure of national income.

3. Expenditure Method

Expenditure method arrives at national income by adding up all expenditures made on goods and services during a year. Income can be spent either on consumer goods or capital goods. Again, expenditure can be made by private individuals and households or by government and business enterprises. Further, people of foreign countries spend on the goods and services which a country exports to them. Similarly, people of a country spend on imports of goods and services from other countries. We add up the following types of expenditure by households, government and by productive enterprises to obtain national income:

1. Expenditure on consumer goods and services by individuals and households. This is called *final private consumption expenditure*, and is denoted by C.
2. Government's expenditure on goods and services to satisfy collective wants. This is called *government's final consumption expenditure*, and is denoted by G.

3. The expenditure by productive enterprises on capital goods and inventories or stocks. This is called *gross domestic capital formation*, or gross domestic investment and is denoted by I or GDCF. Gross domestic capital formation is divided into two parts:
   (*i*) Gross fixed capital formation
   (*ii*) Addition to the stocks or inventories of goods
4. The expenditure made by foreigners on goods and services of a country exported to other countries which are called exports and are denoted by X. We deduct from exports (X) the expenditure by people, enterprises and government of a country on imports (M) of goods and services from other countries. That is, we have to estimate net exports (that is, exports–imports) or (X—M).

Thus, we add up the above four types of expenditure to get final expenditure on gross domestic product at market prices (GDP_{MP}). Thus,

GDP_{MP} = Private final consumption expenditure + Government's final consumption expenditure + Gross domestic capital formation + Exports – Imports or

$$GDP_{MP} = C + G + I + (X - M)$$
$$= C + G + I + NX$$

On deducting consumption of fixed capital (*i.e.*, depreciation) from gross domestic product at market prices (GDP_{MP}) we get net domestic product at market prices (NDP_{MP}).

In this method, we then subtract net indirect taxes (that is, indirect taxes – subsidies) to arrive at net domestic product at factor cost (NDP_{FC}),

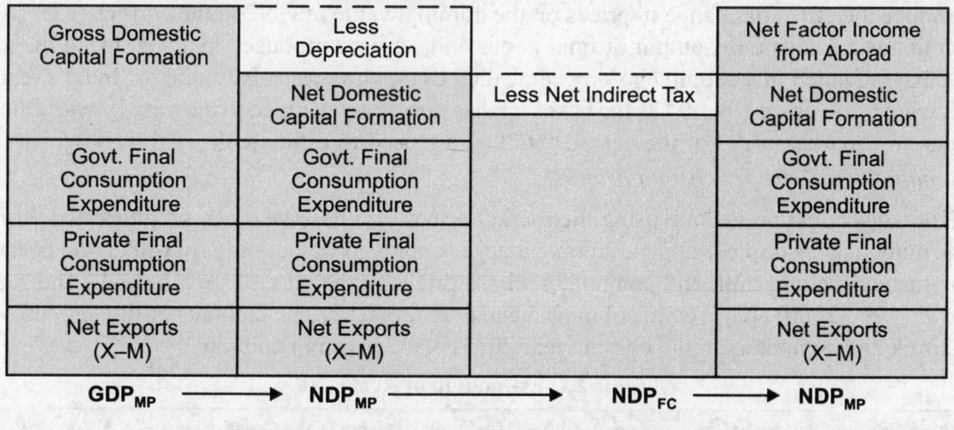

GDP_{MP} →	NDP_{MP} →	NDP_{FC} →	NDP_{MP}
Gross Domestic Capital Formation	Less Depreciation		Net Factor Income from Abroad
	Net Domestic Capital Formation	Less Net Indirect Tax	Net Domestic Capital Formation
Govt. Final Consumption Expenditure	Govt. Final Consumption Expenditure		Govt. Final Consumption Expenditure
Private Final Consumption Expenditure	Private Final Consumption Expenditure		Private Final Consumption Expenditure
Net Exports (X–M)	Net Exports (X–M)		Net Exports (X–M)

Fig. 2.7. *Expenditure Approach to National Income Concepts*

Lastly, we add 'net factor income from abroad' to obtain net national product at factor cost (NNP_{FC}), which is called national income. Thus,

NNP_{FC} = GDP_{MP} – Consumption of Fixed capital - Net Indirect taxes + Net Factor Income from Abroad.

Expenditure approach to national income is shown through bar diagram in Fig. 2.7.

(**Note:** To any concept of the domestic product, if we add '*Net Income from Abroad*' we will get corresponding National Product)

Precautions. While estimating Gross Domestic Product through expenditure method or measuring final expenditure on Gross National Product, the following precautions should be taken:

1. *Second-hand goods*: The expenditure made on second-hand goods should not be included because this does not contribute to the current year production of goods and services.

2. *Purchase of shares and bonds*. Expenditure on purchase of old shares and bonds from other people and from business enterprises should not be included while estimating Gross Domestic Product through expenditure method. This is because bonds and shares are mere financial claims and do not represent expenditure on currently produced goods and services.

3. *Expenditure on transfer payments* by government such as unemployment benefits, old-age pension should also not be included because no goods or productive services are produced in exchange by the recipients of these payments.

4. *Expenditure on intermediate goods* such as fertilisers and seeds by the farmers and wool, cotton and yarn by manufacturers of garments should also be excluded. This is because we have to avoid double counting. Therefore, for estimating Gross Domestic Product we have to include only *expenditure on final goods and services*.

NOMINAL GNP AND REAL GNP

Nominal GNP, as explained above, is the money value of all final goods and services produced in a year. This money value is obtained using current year market prices of final goods and services produced. However, Nominal GNP does not truly indicate the real performance or economic growth of a country over time if prices are changing. While nominal GNP may be increasing due to rise in price level, the quantity of goods and services produced may remain constant. Therefore, in order to estimate to what extent the amount of final goods and services produced has increased we have to eliminate the effect of change in prices on the nominal value of GNP in the current year. To do so the economists evaluate the output of final goods and services produced in a year using the market prices that prevailed in a certain chosen year, called as the *base year*. Estimates of India's real GNP which were made using 1993-94 as the base year has now been changed to the year 1999-2000. Thus, in India now market prices of the year 1999-2000 are used to estimate the real GNP which is also called *national income at constant prices*.

How this valuation is done using the market prices of a base year can be illustrated through a simple numerical example. Suppose an imaginary economy produces only four final goods, namely, wheat, standard cloth, milk and computers whose prices in the current year 2013-14 and the base year (*i.e.*, 1999-2000) are given in columns 3 and 5 of Table 2.1. The amounts of final goods, wheat, cloth, milk and computers in the current year 2013-14 are given in column 2.

Table 2.1. Estimation of Real GNP

Final Goods	*Quantities produced in 2013-14*	*Market Prices in 2013-14* (₹)	*Value of Final Goods at Current Market Prices of 2013-14* (₹)	*Market Price in base year (1999-2000)* (₹)	*Value of Final Goods at base year 1999-2000 prices* (₹)
1	2	3	4	5	6
Wheat (quintals)	650	700 per quintal	4,55,000	550	3,57,500
Standard cloth (metres)	345	120 per metre	41,400	100	34,500
Milk (litres)	520	15 per litre	7,800	12	6,240
Computers (Nos.)	200	2,500 per unit	50,00,000	18,000	3,66,0000
			55,04,200		39,98,240

In order to obtain the *nominal GNP* for the current year 2013-14 we multiply the quantities of four final commodities by the market prices of the current year (2013-14) and the results are given in column 4. By adding up the market values of all the final goods at current prices we get the total value equal to ₹ 55,04,200 for the year 2013-14. ₹ 55,04,200 is the nominal GNP for the current year (2013-14).

For obtaining the value of *real GNP* for the current year 2013-14 we multiply the outputs of final goods produced in the current year (2013-14) by the prices of the base year (1999-2000), and record the estimates in column 6. Summing up the values of four final goods evaluated at base-year prices we get the total estimate of ₹ 39,98,240 which is the real GNP (or gross national income at constant prices) for the current year (2006-07).

It may be noted that real GNP and nominal GNP for the base year will be the same.

It will be noticed from Table 2.1 that for the current year 2013-14, whereas the value of nominal GNP is equal to ₹ 55,04,200, the real GNP is only ₹ 39,98,240. This is because market prices have risen between the base year (1999-2000) and the current year (2006-07).

The calculation of real GNP helps us to know whether the availability of real goods and services in the economy has increased over time. *Besides, by calculating the percentage change of real GNP in a year enables us to measure rate of economic growth in a year*. Further, it is real GNP that is often used for making comparisons of international standards of living and rates of economic growth of various countries.

Finally, with the help of nominal GNP and real GNP we can measure rate of inflation that has taken place over time. This measure of rate of inflation is known as **GNP deflator**. This is calculated as follows :

$$\text{GNP Deflator} = \frac{\text{Nominal GNP of a year}}{\text{Real GNP of the year}} \times 100$$

LIMITATIONS OF GDP CONCEPT AS A MEASURE OF SOCIAL WELFARE

National income, or GNP, as explained in the last chapter, measures the value of aggregate output of goods and services produced in a year. Since goods and services satisfy the wants of the people, national income or GNP has often been used as a measure of satisfaction or economic welfare of the people. The greater the magnitude of national income, the greater the level of economic welfare. Besides, economic progress or what is now generally called economic growth has been measured in terms of increase in national income (that is, increase in GNP or NNP) in terms of total or per capita income. In recent years doubts have been expressed about the validity of national income or gross national product (GNP) as a measure and index of economic welfare.

It has been asserted by several modern economists that national income as it is usually defined is not a satisfactory measure of economic welfare. According to them, in order to obtain a true measure of economic welfare, some adjustments both in the form of additions and subtractions have to be made in the aggregate of national income. National income, as it is usually conceived and measured, includes some things that do not increase welfare of the people. Therefore, such things ought to be excluded in order to get a true measure of economic welfare. This true measure of economic welfare is now often called "*Net Economic Welfare*" or simply *NEW*. On the other hand, the usual concept of national income excludes some goods and services which increase satisfaction of the people and therefore ought to be included in any good index of Net Economic Welfare.

As regards the things that ought to be added to obtain the index of Net Economic Welfare (NEW), the first important thing is the *value of satisfaction that people derive from leisure*. The usual concept of national income (or GNP) does not attach any significance to the amount of leisure people enjoy. However, the individuals derive satisfaction not only from the consumption of goods and services but also from leisure they have. Therefore, for constructing any index of net economic welfare, the value of leisure which the people enjoy must also be included. For example, if the average working hours are reduced, this is likely to reduce national production or national income but may raise welfare of the people by enabling them to enjoy more leisure.

The other important items that ought to be added to obtain a true measure of welfare are the *non-marketed personal services* (that is, the personal services which are not sold and purchased in the market) which also greatly raise the satisfaction and welfare of the people. For instance, services rendered by the housewives to the family members greatly add to their welfare but they are not recorded in national income accounting. Likewise, personal services rendered by the individuals to themselves such as gardening, painting one's own house significantly raise their welfare but do not get registered in the national income or GNP. Hence in order to get any true index of economic welfare, the value of non-market activities such as personal services which increase welfare ought to be incorporated.

As regards the items that have to be deducted from the national income, mention may be made of those harmful effects which result from increase in output. As is well known, the production of modern industries pollutes the environment such as polluting air, water and calmness which significantly reduce welfare of the people. Though modern industrialization has greatly increased national income of the countries but by causing *air pollution, water pollution and noise* has tended to reduce the welfare of the people. Therefore, for preparing a measure of net welfare, negative values ought to be assigned to the *environment pollution* that results from the production of goods and services. The various forms of pollution of environment have often been referred to as *costs of economic growth*, which, like other costs, have to be deducted to obtain the index of net economic welfare.

Apart from environment pollution, certain other deductions on account of wasteful and non-productive expenditure such as Government expenditure on police and law courts so as to maintain law and order and on defence to protect the country from external aggression. These have been called "*regrettable costs*", because economists consider them regrettable necessities expenditure on which does not lead to increase in welfare of the people. Hence expenditure on them should be excluded from GNP to arrive at the measure of net economic welfare.

To sum up, the relation between national income (GNP) and Net Economic Welfare (NEW) can be represented as follows:

Real GNP
– Depreciation
+ Value of leisure
+ Value of non-market activities (*i.e.*, services of housewives and personal services)
– Environment pollution
– Regrettable costs
= Net Economic Welfare

But it is worth mentioning that even the measure of Net Economic Welfare (NEW) as defined above does not truly indicate the welfare enjoyed by the people. There are other things which play a significant role in determining welfare of a nation and which do not get registered in national income or in Net Economic Welfare. Thus, *composition of national output* as between wage goods and luxuries and also the *distribution of goods between individuals* determine welfare to a great extent. If with the increase in total national income and per capita income the rich are getting richer and the poor getting poorer, then this growth in national income and per capita income cannot be said to promote welfare. Professor Hicks rightly remarks, "*The national income only measures the total volume of goods and services at the disposal of the community during the year; it can tell us nothing, for example, about the way in which the national income is divided up between rich and poor.*"[4]

4. J.R. Hicks, *The Social Framework*.

Further national income and per capita income are also not true measures of welfare because they do not consider the composition of output. A country's national income and per capita income may be very high but the well-being of the people may be very low because the national output consists of a larger quantity of war material. This is a very important factor because it is a well known fact that various countries are spending quite a large part of national budgets on the manufacture of war materials and the greater the expenditure on the defence forces, the smaller will be the actual well-being of the people, given the size of national and per capita income.

Similarly, the well-being of the people also depends upon the relative proportions of wage-goods (necessities) and luxuries in the composition of national product. It may be noted that in a country if luxuries are being produced in relatively greater quantities than the wage-goods, then in that case while the few rich will be rolling in luxuries, the poor will be deprived of even sufficient quantities of necessities of life. Likewise, national and per capita income do not accurately reflect improvements in the quality of products. This is a shortcoming because improvements in the quality of products affect the economic welfare of the people as much as the increase in quantities of goods.

Lastly, the increase in national income does not truly indicate the increase in welfare of the people because it does not take into account *how the national income is being produced and how the increase in it has been brought about*. If increase in national income has been brought about by making workers work longer hours which impair their health and efficiency, then this increase in national income will not promote welfare but will adversely affect it. Similarly, if the increase in national income has been obtained by introducing labour-saving machinery throwing out a large number of workers out of employment, this growth in national income increases unemployment and therefore cannot lead to the increase in social welfare.

THE CONCEPT OF GREEN GDP

We have explained above the limitations of GDP or GNP as a measure of welfare. In recent years the economists have realised that in national income accounting two aspects or costs of producing GDP need to be incorporated in estimating GDP or national income so that it should truly reflect welfare of the people. These real costs are of two types. First, *when producing GDP firms pollute the air and water which adversely affects the welfare of the people.* Thus, it has been found that while producing goods the firms pollute the river water by dumping their wastes in local rivers. Similarly, the factories in the urban areas using oil and coal emit smoke and gas that pollute the air which harms those who live in the surrounding area. Therefore, ***in estimating real GDP or national income the costs of pollution of air and water by the firms in the production process of goods must be subtracted to arrive at what has been called green GDP* or green *GNP*.**

Another important aspect of producing GDP which requires adjustment for calculating green GDP is **depletion of natural resources** such as oil, forests, coal and natural gas. Conventional national income accounting does not take into account the depletion of natural resources. For example, when oil drilling machines pump out oil from an underground field, it leads to the depletion of the non-renewable resources but no deduction is made to account for the *depletion of oil reserves* in national income accounting. In fact, the depletion of natural resources should be treated as a type of *negative inventory investment* which if accounted for in national income accounting would tend to lower the GDP estimates, other things remaining the same. Likewise, measuring the true cost of timber from cutting trees of the forests which causes not only destruction of forests—a natural resource—but also soil erosion. The damage caused by cutting trees and thereby destroying forests and causing soil erosion be assessed and accordingly proper deduction be made on this count in estimation of real GDP. In this connection the loss due to destruction of forests can be better understood from the damage and devastation caused by floods and landslides in May 2013 in Uttarakhand in India which was the result of cutting of trees in the forests nearby the rivers passing through them. It follows from above that "*ideal accounting system, the economic costs of environment degradation would be subtracted in the calculation of a firm's contribution to output and all activities that improve the environment because they provide real economic benefits to the people be added to output.*"[5]

In this regard some western economists have pointed out that some developing countries in their attempt to speed up economic growth and raising their GDP as rapidly as possible have overexploited the natural resources and caused a good deal of environment pollution and natural resources degradation, they therefore suggest that costs of environment pollution and natural resources degradation are incorporated in their estimates of growth of GDP and their growth policies need to be modified. However, in our view this is not entirely correct to blame the developing countries who want higher GDP growth rate to reduce poverty in their countries. As a matter of fact, it is the developed countries which in the past for achieving rapid industrial growth have contributed a lot to the emission of harmful gases such as carbon dioxide that have significantly contributed to the global warming affecting the welfare of the people of developing countries. As a matter of fact, they should not only provide financial assistance to the developing countries for the harm done to the global environment in their growth process in the past but also transfer technology that ensures protection of environment from the growth processes in poor developing countries.

However, apart from what the developed countries say, it needs to be emphasized that India needs **green growth**, that is, growth in green GDP so as to ensure environment sustainability and thereby to promote the welfare not only of the current generation but also of future generations. India's quest for growth with the objective to pull out millions of its people from poverty is a necessary and legitimate pursuit. But so is the pursuit of clean and safe environment and conservation of natural resources. No growth process can afford to neglect the environmental consequences of economic activity, or allow unsustainable depletion and deterioration of natural resources. For the last over one decade (2003-14) India is the second fastest growing economy in the world, next only to China. The growth cannot take place without additional energy. The Indian economy heavily depends on coal and hydropower to meet its energy needs and the growth of each of these energy sources involves the issue of protection of environment and depletion of its natural resources. Therefore, the pursuit of our growth objective has to be reconciled with the objective of protection of environment. Unfortunately, the experience of growth in many countries and our own experience suggests that environmental pollution and unsustainable depletion and deterioration of natural resources occur due to laxity in environment monitoring. This has to be avoided in the future to achieve clean and sustainable growth. Therefore, we have to explore the sources and practices that use less of the polluting agents and more of clean sources of energy such as solar energy and nuclear energy. A good start has already been made in India in both these alternative sources of energy.

It growth of green GDP is adopting the pursued, then the growth will promote welfare of the people and protect them from health hazards. For such growth to occur, environmental and ecological consequences of growth activities must be taken into account. The current estimates of growth based on GDP and national income as conventionally measured do not reflect a true and genuine growth in the sense of green and sustainable growth. Kenneth Arrow and Parthadas Gupta find India's growth rate to be 2.5 to 3 per cent lower than the reported average of 7.6 per cent achieved in the last 11 years (2003-14). Therefore, we conclude that a green economy will make growth more inclusive and sustainable. Neglecting the ecological consequences of growth adversely affects the welfare of the people in the long run.

---

5. Andrew B. Abel, Ben S. Bernanke and Dean Croushore, *Macroeconomics*, 7th edition, Pearson, 2011, p. 29.

SOME EXAMPLES OF NATIONAL INCOME CALCULATION

We solve below some numerical problems of calculation of national income by three methods, namely, expenditure method, income method and value added method.

Problem 1. *Calculate (1) GDP at market prices and (2) national income from the following information.*

	₹
Personal consumption expenditure	***6,500***
Indirect taxes less subsidies	***150***
State government consumption and investment expenditure	***500***
Central government consumption and investment expenditure	***2,000***
Change in business inventories	***100***
Gross private domestic fixed investment	***1,200***
Exports	***900***
Net factor payments to rest of the world	***– 100***
Imports	***1,200***
Depreciation	***200***

DU BA(Hons.) 2006

Solution.

(*i*) GDP_{MP} = (Private Consumption Expenditure + Gross private investment (both fixed and inventories) + Govt. Expenditure (both State and Central) + Net exports (X–M))

$$= 6500 + (1200 + 100) + (500 + 2000) + (900 - 1200)$$
$$= 6500 + 1300 + 2500 - 300$$
$$= 10{,}000$$

(*ii*) *National Income*. National income is net national product at factor cost (NNP_{FC}). Therefore, to find out *net* domestic product at factor cost we have to deduct depreciation and net indirect taxes from GDP_{MP} and to get *national* income we have to deduct net factor payments to the rest of the world. Thus

NI or NNP_{FC} = GDP_{MP} – Depreciation – net Indirect taxes (*i.e.*, indirect taxes less subsidies) – net factor payments to the rest of the world.

$$NI = 10{,}000 - 200 - 150 - 100$$
$$= 9550$$

Problem 2. *Calculate the national income and personal disposable income from the following information*:

	(₹)
GDP_{MP}	***6,000***
Receipts of factor income from the rest of the world	***150***
Payments of factor income to the rest of the world	***225***
Depreciation	***800***
Indirect taxes minus subsidies	***700***
Corporate Profits	***1,200***

Dividend	***600***
Transfer payments to persons	***1,300***
Personal Taxes	***1,500***

D.U. BA (Hons.) 2007

Solution.

(*i*) National Income is Net National Product at factor cost

$$\text{NI or } NNP_{FC} = GDP_{MP} - \text{Depreciation} - \text{Net Indirect taxes} + \text{Net Factor income from abroad}$$
$$= 6000 - 800 - 700 + (150 - 225)$$
$$= 6000 - 1500 - 75$$
$$= 6000 - 1575$$
$$= 4425$$

(*ii*) Personal Disposable Income = National Income – Retained Corporate Profits + Transfer Payments to Persons – Personal taxes

$$PDI = 4425 - (1200 - 600) + 1300 - 1500$$
$$= 4425 - 600 + 1300 - 1500$$
$$= 3625$$

(**Note:** Retained corporate profits equal corporate profits minus dividends paid to shareholders)

Problem 3. *Calculate (1) Net Domestic Product at Market Prices and (2) National Income from the following data of all the enterprises in an economy.*

		₹ in Crore
(i)	***Subsidies***	***10***
(ii)	***Sales***	***1000***
(iii)	***Closing stock***	***100***
(iv)	***Indirect taxes***	***50***
(v)	***Intermediate consumption***	***300***
(vi)	***Opening stock***	***200***
(vii)	***Consumption of fixed capital***	***150***
(viii)	***Net factor income from abroad***	***10***

Solution.

NDP_{MP} and National Income be obtained through *value added method.* Net Domestic Product at Market Prices is the sum of net value added by all enterprises in the economy.

$$\text{Value of output} = \text{Sales} + \text{Change in stock}$$
$$= 1000 + (\text{Closing Stock} - \text{Opening Stock})$$
$$= 1000 + (100 - 200)$$
$$= 900 \text{ crore}$$

Net value added (NVA) at market prices (NDP_{MP}) = Value of output–Intermediate Consumption – Consumption of fixed capital

$$= 900 - 300 - 150$$
$$= 450 \text{ crore}$$

National Income (NNP_{FC}) = NDP_{FC} Net Indirect taxes + Net factor income from abroad
= 450 – (50 – 10) + 10
= 450 – 40 + 10
= 420 crore

Problem 4. *On the basis of the following data about the economy which consists of only two firms, find out:*

(*a*) *Value added by firms A and B.*

(*b*) *Gross Domestic Product at market prices and factor cost*

		₹ *in Crore*
(i)	***Sales by firm A***	***200***
(ii)	***Purchases from B by firm A***	***80***
(iii)	***Purchases from A by firm B***	***120***
(iv)	***Sales by firm B***	***400***
(v)	***Closing stock of firm A***	***40***
(vi)	***Closing stock of firm B***	***70***
(vii)	***Opening stock of firm A***	***50***
(viii)	***Opening stock of firm B***	***90***
(ix)	***Indirect taxes paid by both firms***	***60***

Solution.

Firms	Output	Intermediate cost	Value added
Firm A:	Sale = 200 Change in stock = 40 – 50 = –10 Total Value of Output = 200 – 10 = 190	Purchases From B = 80	190 – 80 = 110
Firm B:	Sales = 400 Change in stock = 70 – 90 = –20 Total Value of Output = 400 + (–20) = 380	Purchases From A = 120	380 – 120 = 260

GDP_{MP} is sum total of GVA_{MP} of all firms

Gross domestic product at market prices (GDP_{MP}) = ΣGVA_{MP} = 110 + 260 = 370 crore

Gross domestic product at factor cost (GDP_{FC}) = GDP_{MP} – Net Indirect taxes
= 370 – 60
= 310 crore

Problem 5. *Calculate Gross Domestic Product at Market Prices and Net National Product at Factor Cost and Gross Domestic Product at Factor Cost from the following data:*

		₹ *in Crore*
(i)	***Net indirect taxes***	***38***
(ii)	***Consumption of fixed capital***	***34***
(iii)	***Net factor income from abroad***	***(–) 3***
(iv)	***Rent***	***100***
(v)	***Profits***	***125***

(vi)	***Interest***	***20***
(vii)	***Mixed Income of self-employed***	***120***
(viii)	***Wages and salaries***	***170***
(ix)	***Employers' contribution to social security schemes***	***30***

Solution.

Note that (*viii*) wages and salaries and (*ix*) Employer's contribution to social security schemes together constitute compensation to employees

Net national product at factor cost (NNP_{FC}) = Wages and salaries + Employers' contribution to social security schemes + Rent + Profit + Interest + Mixed Income of Self-employed + Net factor income from abroad

$$= 170 + 30 + 100 + 125 + 20 + 120 - 3$$
$$= 562 \text{ crore}$$
$$GDP_{MP} = NNP_{FC} + \text{Net Indirect taxes} + \text{Consumption of fixed capital} - \text{Net factor income from abroad}$$
$$= 562 + 38 + 34 - (-3)$$
$$= 562 + 38 + 34 + 3$$
$$= 637 \text{ crore}$$
$$GDP_{FC} = GDP_{MP} - \text{Net Indirect Taxes}$$
$$= 637 - 38 = 599 \text{ crore}$$

Problem 6. *Calculate gross domestic product at market prices (GDP_{MP}), gross domestic product at factor cost (GDP_{FC}) and net national product at factor cost (NNP_{FC}) from the following data:*

		₹ in Crore
(i)	***Private final consumption expenditure***	***290***
(ii)	***Government's final consumption expenditure***	***50***
(iii)	***Subsidies***	***20***
(iv)	***Gross domestic fixed capital formation***	***105***
(v)	***Indirect taxes***	***70***
(vi)	***Depreciation (i.e., Consumption of Fixed Capital)***	***45***
(vii)	***Net factor income from abroad***	***(–) 5***
(viii)	***Net addition to stock***	***15***
(ix)	***Net exports***	***(–) 5***

Solution :

Gross domestic product at market prices (GDP_{MP}) = Private final consumption expenditure + Government's final consumption expenditure + Gross domestic fixed capital formation + Change in stock + Net exports

$$= 290 + 50 + (105 + 15) + (-5)$$
$$= 290 + 50 + 105 + 15 - 5$$
$$= 455 \text{ crore}$$
$$GDP_{FC} = GDP_{MP} - \text{Net Indirect Taxes}$$
$$= 455 - (\text{Indirect Taxes} - \text{Subsidies})$$
$$= 455 - (70 - 20) = 405 \text{ crore}$$
$$NNP_{MP} = GDP_{MP} - \text{Depreciation}$$
$$= 455 - 45$$
$$= 410 \text{ crore}$$
$$NDP_{FC} = NDP_{MP} - \text{Indirect taxes} + \text{Subsidies}$$

$$= 410 - 70 + 20$$
$$= 360 \text{ crore}$$
$$NNP_{FC} = NDP_{FC} + \text{Net factor income from abroad}$$
$$= 360 + (-5)$$
$$= 360 - 5 = 355 \text{ crore}$$

QUESTIONS FOR REVIEW

1. Explain the circular flow of income in an economy. What does it measure?
2. Describe the circular flow of income in a two-sector economy with households and business firms. What determines the magnitude of circular flow of income and expenditure?
3. What is meant by saving and investment? How do they affect circular flow of income in a free market economy?
4. Despite the fact that motives to save are different from the motives of businessmen to invest, in national income accounts savings are always equal to investment. Explain.
5. How does the addition of Government in a two-sector economy affect circular flow of income? If Government's budget is deficit, how will it affect circular flow of income?
6. Show with a circular income flow model of a two-sector economy that national product equals national income.
7. In a three-sector economy with firms, households and Government show with a circular income flow model that the sum of private investment and Government expenditure equals the sum of saving and taxes.
8. What is meant by withdrawals and injections? How do they affect the size of circular flow of income and expenditure in an economy?
9. If investment by business firms falls short of savings, how will it affect the circular flow of income?
10. How will circular income flow in an economy be affected if a country has foreign trade transactions?
11. In an open economy, explain with a circular income flow model that the sum of investment, Government expenditure and net exports must equal the sum of savings and taxes.
12. How are aggregate domestic output and expenditure related to trade balance?
13. Explain saving-investment identity in an open economy.
14. Show that the difference between saving and investment of an economy always equals trade balance.
15. Explain how foreign capital flows are related to domestic saving and investment.
16. Define gross domestic product. Are value of exports and imports included in it?
17. Distinguish between final goods and intermediate goods. Why are intermediate goods not included in gross domestic product of a country?
18. What are the components of gross domestic product? Explain them briefly.
19. Distinguish between gross national product (GNP) and gross domestic product (GDP). In this connection explain 'net factor income from abroad'.
20. Define *'national income'*. How is it different from gross domestic product (GDP)? Explain.
21. Distinguish between Net Domestic Product at Market Prices (NDP_{MP}) and Net Domestic Product at Factor Cost (NDP_{FC}).
22. What is meant by *'value added'*? Explain how national income is measured by value added method.

23. Explain the following terms:
(*i*) Net value added
(*ii*) Net factor income from abroad
(*iii*) Transfer Payments by the Government

24. Distinguish between *nominal* GDP and *real* GDP. Which concept is used for measuring economic growth of a country?

25. Explain how national income is measured through expenditure method. What types of expenditure are included in measuring national income of a country?

26. Explain 'income method' of measuring national income. Which of the following terms will be included while calculating national income of a country through income method?
(*i*) Transfer Payments by the government to individuals in a year
(*ii*) Imputed rent of self-occupied houses
(*iii*) Hawala money
(*iv*) Windfall gains
(*v*) The receipts from the sale of second-hand goods.

27. Is gross domestic product (GDP) a true indicator of welfare of the society? Explain.

28. What is meant by *net economic welfare?* How is it derived or obtained from the estimate of GNP?

29. On the basis of the following data about an economy which consists of only two firms, find out:
(*a*) Value added by Firms A and B.
(*b*) Gross domestic product at market prices.

		₹ *in Crore*
(i)	***Exports by Firm A***	***40***
(ii)	***Imports by Firm A***	***100***
(iii)	***Sales to households by Firm A***	***180***
(iv)	***Sales to Firm B by Firm A***	***80***
(v)	***Sales to Firm A by Firm B***	***60***
(vi)	***Sales to households by Firm B***	***120***

(**Hint.** Note that value of exports by firm A will be a part of value of output, whereas imports by it will be a part of intermediate cost of production.)

(**Ans.** GDP_{MP} = 240 crore)

30. From the following data calculate Gross National Product and Net National Product at market prices by (*i*) income method, and (*ii*) expenditure method.

		₹ *in Crore*
(i)	***Mixed income of self-employed***	***400***
(ii)	***Compensation to employees***	***500***
(iii)	***Private final consumption expenditure***	***900***
(iv)	***Net factor income from abroad***	***(–)20***
(v)	***Net direct taxes***	***100***
(vi)	***Consumption of fixed capital***	***120***
(vii)	***Net domestic capital formation***	***280***
(viii)	***Net Exports***	***(–)30***
(ix)	***Profits***	***350***
(x)	***Rent***	***100***
(xi)	***Government final consumption expenditure***	***300***

Chapter 3
CLASSICAL THEORY OF INCOME AND EMPLOYMENT

ECONOMY IN THE LONG RUN: THE CLASSICAL FULL-EMPLOYMENT MODEL

Before explaining the Keynesian macro-theory of income and employment, it will be in the fitness of things to explain classical theory regarding income and employment determination. The study of classical theory of income and employment is essential because some of the aspects of classical theory are more relevant to the conditions prevailing in the developing countries and this theory highlights those factors which govern income and employment in these countries. While the Keynesian theory emphasises the role of effective demand in the determination of income and employment, classical economists believed that in a free-market economy there was always a tendency toward the establishment of full employment of labour and there was sufficient demand for the output produced. Note that Keynes called all economists who preceded him (including Marshall and Pigou) as classical economists.

CLASSICAL THEORY OF EMPLOYMENT: AN INTRODUCTORY ANALYSIS

Classical theory of employment and output is based on the following two basic notions:

1. Say's Law
2. Wage-price flexibility

We explain below these two notions of classical theory

Say's Law and Classical Theory

According to the classical theory propounded by Ricardo and Adam Smith, levels of income and employment are governed by fixed capital stock on the one hand and wage-goods fund on the other. It may be noted in the beginning that the classical theory believes in full employment or near full employment prevailing in the economy. This belief of classical theory regarding the existence of full employment in the economy is based on Say's Law put forward by a French economist, J.B. Say. According to J.B. Say's law, *"Supply creates its own demand"*. This implies that every increase in production made possible by the increase in the productive capacity or the stock of fixed capital will be sold in the market and there will be no problem of lack of demand. Thus, classical economists rule out the possibility of overproduction; there being no problem in selling the output produced. According to Say's Law, greater production automatically leads to a greater money income which creates the market for the greater flow of goods produced. Thus, deficiency in demand being no problem, the process of capital accumulation and expansion of productive capacity will continue till all people are employed and there is no reason why the productive capacity created remains unutilized or underutilized. According to this theory, the income which is not spent on consumer goods and thus saved will become investment expenditure. Therefore, investment equals saving. Thus, the leakage caused by the saving in the income flow is made up by the investment expenditure. In this way, a given productive capacity continues to be fully utilized and no problem of deficiency of demand arises.

Classical economists thought that if price mechanism in a capitalist economy is allowed to work freely without any interference by the Government, there is always a tendency towards full employment in it. Of course, they admitted that in advanced capitalist economies often certain circumstances arise due to which they are is not in full-employment equilibrium. But they firmly believed that there was always a tendency towards full-employment in the economy and certain economic forces automatically operate so as to move the economy towards full employment. Therefore, according to the classical economists, whenever there are lapses from full-employment level, these are removed automatically by the working of free price mechanism. The modern economists do not regard this aspect of classical theory of employment as valid and correct description of the real world. J.M. Keynes bitterly criticised the classical theory of automatic full-employment.

The classical theory of employment was based upon two basic assumptions. The first assumption is that there is always enough expenditure or aggregate demand to purchase the total production at full-employment level of resources. In other words, in this theory the classical economists disregarded the problem of deficiency of demand for purchasing goods produced at full-employment level of resources. The second assumption is that even when deficiency of aggregate expenditure or demand arises, the prices and wages would change in such a way that real production, employment and income will not decline. The classical thought that there was no problem of deficiency of expenditure and demand was based upon Say's law of markets. J.B. Say has been the famous French economist of the 19th century. Say's law is based upon the fact that every production of goods also creates incomes equal to the value of goods produced and these incomes are spent on purchasing these goods. In other words, production of goods itself creates its own purchasing power. Therefore, Say's law is expressed as "*supply creates its own demand*", that is, the supply of goods produced creates demand for it equal to its own value with the result that the problem of general overproduction does not arise. In this way in Say's law, the possibility of lack of aggregate demand has not been visualised.

Say's law expresses an important fact about the working of a free enterprise economy. The fact is that the source of demand for goods is the incomes earned by various factors of production employed for their production. All unemployed and idle labourers and other resources when employed for production, create their own demand because the total incomes which they earn create equal market demand for the goods produced by their employment. When a new entrepreneur employs some factors of production and pays them their monetary rewards, he not only increases the supply of goods but also at the same time creates the demand for them. Therefore, it is the production which creates market or demand for goods. Production is the only source of demand. Dillard rightly writes that *"Say's Law of Markets"* is the denial of the possibility of deficiency of aggregate demand. Therefore, employment of more resources will always be profitable and will take place to the point of full-employment, subject to the limitation that the contributors of the resources are willing to accept rewards no greater than their physical productivity justifies. There could be no general unemployment, according to this view, if workers will account what they are worth."[1]

We thus see that according to Say's law aggregate expenditure or demand will always be such that all resources are fully employed. The factors which participate in productive activity and earn incomes from it, they spend a good part of their incomes on consumer goods and some part they save. But, according to classical economists, the savings by the individuals are actually spent on investment or capital goods. Since saving when invested also becomes expenditure or demand, in classical theory the whole income is spent, partly on consumption and partly on investment. There is thus no reason for any leakage in the income stream and hence supply creates its own demand. Now, a question arises as to how in classical theory saving becomes equal to investment expenditure. According to the classical theory, it is the rate of interest which makes investment equal to saving. When savings of the people increase, the rate of interest declines. As a result of fall in the rate of interest, demand for investment rises and in this way investment becomes equal to the increased savings. Hence, accord-

1. D. Dillard, *Economics of J.M. Keynes*.

ing to the classical economists, it is the interest rate mechanism that brings about equality between saving and investment and therefore Say's law applies in spite of saving by the people. This guarantees full employment in the economy. In other words, it is changes in the rate of interest due to which the withdrawal of some money from the income stream as a result of savings automatically comes back to it in the form of investment expenditure and therefore income flow continues unchanged and supply goes on creating its own demand.

Wage-Price Flexibility and Full-Employment

The classical economists also proved the validity of the assumption of full-employment with another fundamental logic. According to them, the amount of production which the business firms can supply does not depend only on aggregate demand or expenditure but also on the prices of products. If the rate of interest temporarily fails to bring about equality between saving and investment and as a result deficiency of aggregate expenditure arises, even then the problem of general over-production and unemployment will not arise. This is because they thought that the deficiency in aggregate expenditure would be made up by changes in the price level. When due to the increase in the savings of the people, the expenditure of the people declines, it will then affect the prices of products. As a result of fall in aggregate expenditure or demand, the prices of products would decline and at reduced prices their quantity demanded will increase and as a result all the quantity produced of goods will be sold out at lower prices. In this way, they expressed the view that in spite of the decline in aggregate expenditure caused by the increase in savings, the real output, income and employment will not fall provided the fall in prices of products is proportionate to the decline in aggregate expenditure.

Classical economists thought that a free-market capitalist economy actually worked in that way. Owing to the intense competition between the sellers of products as a consequence of the fall in expenditure, the prices will decline. This is because when aggregate expenditure on goods or demand for them declines, the various sellers and producers reduce the prices of their products so as to avoid the excessive accumulation of stocks of goods with them. Hence, according to the classical logic, increased saving will bring down the prices of products and not the amount of production and employment. But now a question arises to what extent the sellers or producers will tolerate the decline in prices. However, to make their business profitable they will have to reduce the prices of the factors of production such as labour. With a fall in wages of labour, all workers will get employment. If some workers do not want to work at reduced wages, they will not get any job or employment and therefore will remain unemployed. But, according to classical economists, those workers who do not want to work at lower wages and thus remain unemployed are only *voluntarily unemployed*. This voluntary unemployment is not real unemployment. According to the classical thought, it is *involuntary unemployment* which is not possible in a free-market capitalist economy. All those workers who want to work at the wage rate determined by market forces will get employment.

During the period 1929-33 when there was a great depression in capitalist economies, a renowned neoclassical economist Pigou suggested a cut in wage rates in order to remove huge and widespread unemployment prevailing at that time. According to him, the cause of depression or unemployment was that the Government and trade unions of workers were preventing the free working of the capitalist economies and were artificially keeping the wage rates at high levels. He expressed the view that if the wage rates were cut down, the demand for labour would increase so that all would get employment. It was at this time that J.M. Keynes challenged the classical theory and put forward a new theory of income and employment. He brought about a fundamental change in economic thought regarding the determination of income and employment in a developed capitalist economy. Therefore, it is often said that *Keynes brought about a revolution in our economic theory*.

CLASSICAL THEORY OF EMPLOYMENT AND OUTPUT (INCOME)– A FORMAL MODEL

Classical economists such as Adam Smith and Ricardo maintained that the growth of income and employment depends on the growth of the stock of fixed capital and inventories of wage goods. But, in the short run, the stock of fixed capital and wage goods inventories are given and constant. According to them, even in the short run full-employment of labour force would tend to prevail as the economy would not experience any problem of deficiency of demand. On the basis of their theory they denied the possibility of the existence of *involuntary unemployment* in the economy. The short-run classical theory of income and employment can be explained through the following three stages:

1. *Determination of income and employment when there is no saving and* investment;
2. *Determination of income and employment in an economy with saving and investment; and*
3. *Determination of income and employment : Role of money and prices.*

Determination of Income and Employment in the Short Run Without Saving and Investment

According to the classical theory, the magnitude of national income and employment depends on the aggregate production function and the supply and demand for labour. To show this let us assume that the economy produces one homogeneous and divisible good, say corn. Let symbol Y stand for the output of this good. To produce this good we require two factors of production : (1) labour which we denote by N and (2) capital which we denote by K. Thus we have the following aggregate prodcution function

$$Y = F(K, N) \qquad ...(i)$$

In the short run the stock of capital (i.e. plant and equipment) is assumed to be fixed. The state of technology and the population are also assumed to be constant in the short run. Thus, rewritting the aggregate production function we have

$$Y = F(\bar{K}, N) \qquad ...(ii)$$

The bar over the symbol K for capital indicates that stock of capital is fixed. It is worth noting that change in technology will cause a shift the production junction.

Therefore, with a fixed capital stock and a given and constant technology, the output Y (or what is also the real income) would increase only when employment of labour N increases. That is, employment of labour and output (income) rise or fall together. Now, according to classical theory, with a fixed capital stock as employment of labour increases, marginal product of labour would diminish. This is the famous law of diminishing returns of the classical economics.

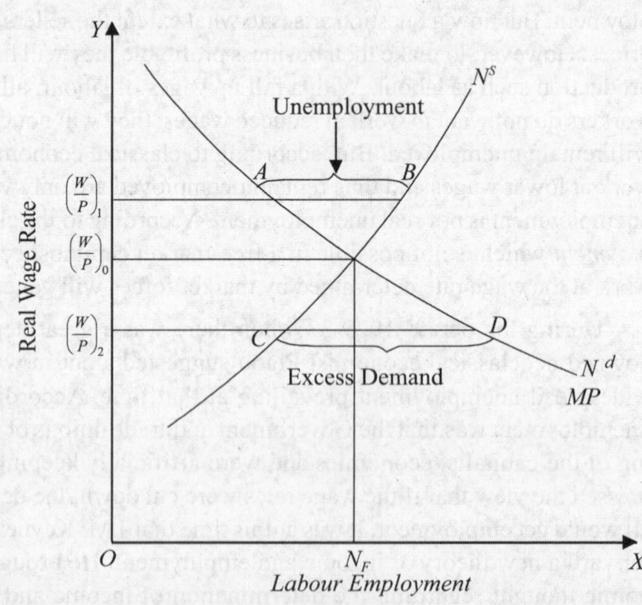

Fig. 3.1. *Labour-Market Equilibrium – Determination of Employment and Wages*

The demand for labour is derived from this short-run production function, that is, diminishing marginal product of labour. The classical theory assumes perfect competition in both the factor and product markets . Further, assuming that the firms which undertake the task of production attempt to maximise profits, they will employ labour until the marginal product of labour is equal to the given real wage rate. It may be noted that real wage rate is given by nominal wage rate divided by the general price, level, that is, real wage rate $= \frac{W}{P}$ where W is the nominal or money wage rate and P is the average price level. Thus, a firm will employ so much labour at which

$$\frac{W}{P} = MP_N$$

where MP_N stands for marginal product of labour

At a lower real wage rate, more labour will be demanded or employed by the firms and *vice versa*. Thus, the demand curve for labour is derived from the marginal product curve of labour. In fact, the former coincides with the latter. Thus demand function for labour can be written as

$$N^d = f\left(\frac{W}{P}\right) \qquad \ldots (iii)$$

Consider Fig. 3.1 where *MP* curve depicts the diminishing marginal product of labour with a given stock of fixed capital and a given state of technology. As explained just above, marginal prouct (*MP*) curve of labour also represents the demand curve of labour (N^d).

On the other hand, the *supply of labour by the households in the economy depends on their pattern of preference between income and leisure*. The classical theory assumes that in the short run when population does not vary, supply curve of labour slopes upward. Now, what is the rational behind the upward-sloping supply curve of labour. This is based on the assumption that households or individual workers maximise their utility or satisfaction in their choice of work (which yields them income) and leisure. When real wage rate rises, two effects work in opposite direction. It may be noted that real wage is the opportunity cost or relative price of lesiure. When real wage rate rises leisure becomes relatively more expensive, that is, opportunity cost or price of leisure in terms of income forgone by not working goes up. This induces the individual to work more (i.e. supply more labour hours) and thereby substitutes income for leisure. This is the substitution effect. On the other hand, with a rise in real wage rate individuals become relatively richer than before, and this induces them to consume more of all commodities (including leisure which is regarded as a normal commodity). This is income effect of the rise in real wage rate which tends to increase leisure and reduce labour-hours supplied. The classical economists believed that substitution effect is larger than income effect of the rise in real wage rate and as a result supply of labour increases with the rise in wage rate. Thus the supply function of labour can be written as

$$N^s = g\left(\frac{W}{P}\right) \qquad \ldots (iv)$$

This implies that at a higher wage rate, more labour would be supplied and *vice versa*. It will be seen from Fig. 3.1 that supply and demand for labour are in equilibrium at the real wage rate $\left(\frac{W}{P}\right)_0$. Hence, given the supply and demand curves, the wage rate $\left(\frac{W}{P}\right)_0$ is determined. It will be seen that ON_F labour is employed in this equilibrium situation. Thus, *in classical theory level of employment is determined by labour market equilibrium*. This *equilibrium between supply and demand for labour at the real wage rate* $\left(\frac{W}{P}\right)_0$ *implies that all those who offer their labour services at this wage rate are in fact employed.* There is neither excess supply of labour, nor excess demand for labour. In other words, there is no involuntary unemployment of labour in this equilibrium situation. If somehow real wage rate in the labour market is higher than this equilibrium wage rate $\left(\frac{W}{P}\right)_0$, say it is equal to $\left(\frac{W}{P}\right)_1$, then it will be observed from Fig. 3.1 that excess supply of labour equal to *AB* would emerge. In other words, at real wage rate $\left(\frac{W}{P}\right)_1$, *AB* workers will be unemployed.

But given the competition among workers, the excess supply of labour at wage rate $\left(\frac{W}{P}\right)_1$ would cause the wage rate to fall to the equilibrium level $\left(\frac{W}{P}\right)_0$ at which the labour market is cleared. On the contrary, if somehow real wage rate in the labour market is $\left(\frac{W}{P}\right)_2$, the firms would demand more labour than is offered at this real wage rate. As a result of the competition among the firms to hire labour desired by them, the wage rate would go up to the equilibrium level $\left(\frac{W}{P}\right)_0$. At $\left(\frac{W}{P}\right)_0$ to repeat, all those who offer their labour services are in fact demanded and employed. *It therefore follows that at the real wage* $\left(\frac{W}{P}\right)_0$, *there is no involuntary unemployment, or, in other words, full-employment of labour prevails. Further, it is the wage flexibility (i.e., changes in the wage rate) which ultimately brings about this full-employment situation.*

To clarify further the restoration of full employment of labour due to quick adjustment of real wage rate let us consider the decrease in demand for labour following the fall in aggregate demand for output as it happens when depression or recession occurs in the economy. Consider Fig. 3.1(A) where following the *decrease in aggregate demand for output* labour demand curve shifts to the left to N_1^d so that at the initial wage rate $\frac{W_0}{P_0}$ fewer workers will be demanded than the number of workers who are willing to supply their labour at this wage rate. As a result (as is seen from Fig. 3.1A) the ex-

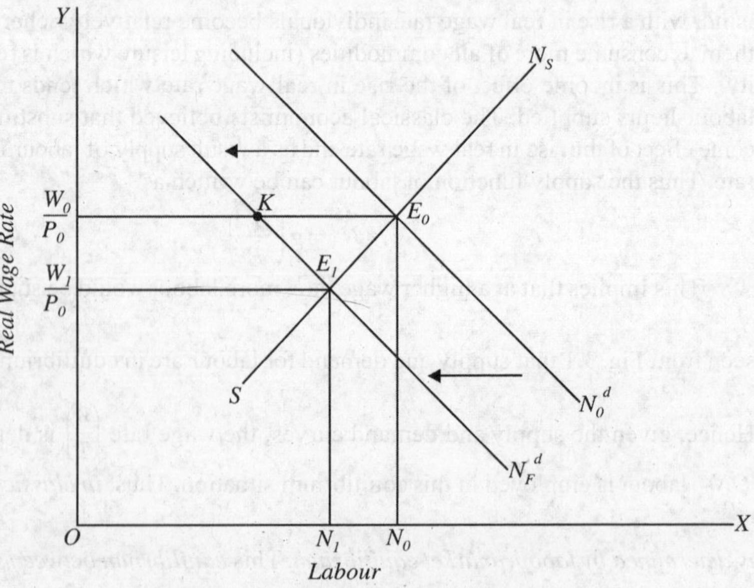

Fig. 3.1A. *Adjustment of Real Wage Rate*

cess supply of labour equal to KE_0 will emerge at this initial real wage rate $\frac{W_0}{P_0}$. However, in the classical full employment model this excess supply of labour (i.e. unemployment of workers) will cause real wage rate to fall to $\frac{W_1}{P_0}$ (where $W_1 < W_0$) at which new equilibrium between demand for and supply of labour is again established at point E_1. Note that even in this new labour market equilibrium at lower real wage rate $\frac{W_1}{P_0}$ full employment of labour prevails as all those who are willing to work at this real wage rate find employment. Of course, N_0N_1 workers have voluntarily withdrawn themselves from labour force and therefore no one remains involuntarily unemployed.

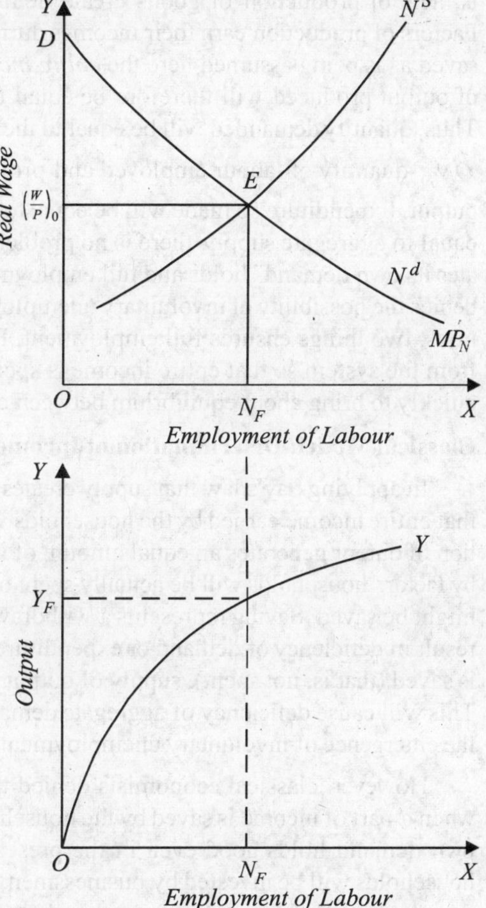

Fig. 3.2. *Classical Theory : Determination of Employment and Output*

How much output will be produced in this full employment situation can be readily known from the aggregate production function. We depict this in Fig. 3.2 where in addition to the supply of and demand for labour, the aggregate production function (*OY*) representing the relation between employment of labour (*N*) and total output (*Y*) is shown. It will be seen from the lower panel of Fig. 3.2 that, given the stock of fixed capital and the state of technology, employment of ON_F labour produces OY_F output. This output OY_F of corn will constitute the income of the society and will be distributed between wages and profits. Thus sum of wages as reward for labour and total profits as reward for capital would constitute the total income of the society and would be equal to the national output OY_F produced.

It follows from above that the quick changes in the real wage rate upward or downward ensures that neither excess supply of labour, nor excess demand for labour will persist and thus equilibrium will be reached with full employment of labour in the economy. Further, given the stock of capital and the state of technology with this full employment of labour, total output or income of the economy equal to OY_F is determined.

Now, an important question to enquire is what guarantees that output produced by the full employment of labour and the level of capital (assumed as fixed in the short run) will be actually demanded. If this does not happen, then the problem of insufficient demand for the output (i.e., corn) will emerge which will ultimately lead to reduction in output and employment and hence to the emergence of involuntary unemployment.

Say's Law and No Deficiency of Demand. In the absence of saving and investment which we are assuming here, classical economists ruled out the possibility of deficiency of aggregate demand on the basis of Say's law. Say's law, as mentioned above, states that *supply creates its own demand*, that is, acts of production of goods create demand equal to the value of output of goods produced. Factors of production earn their incomes during the process of production. Since no part of income is saved as is being assumed here the *entire income will be spent on consumer goods produced*. Value of output produced will therefore be equal to the income generated in the process of production. Thus, quantity demanded will be equal to the supply of output produced. In Fig. 3.2, wages earned by ON_F quantity of labour employed and profits earned by the entrepreneurs will be spent on OY_F output. Expenditure so made will be equal to the value of output produced. Aggregate demand being equal to aggregate supply, there is no problem of deficiency of demand. Say's law that "supply creates its own demand" holds and full employment of labour is guaranteed. In this way classical theory denies the possibility of involuntary unemployment. It needs to be emphasised that under such conditions, two things ensures full employment. First, it is because saving and investment are excluded from the system so that entire income is spent on consumer goods. Second, real wage rate changes quickly to bring about equilibrium between demand for and supply of labour.

Classical Model: Determination of Income and Employment with Saving and Investment

In applying Say's law that supply creates its own demand an invalid assumption was made above that entire income earned by the households will be actually spent. Although it is correct that production of output generates an equal amount of income but what is the guarantee that all income earned by factors/households will be actually spent on goods and services produced. In fact, a part of income might be saved. Saving represents a withdrawal of some income from the expenditure flow. This will result in deficiency of demand or expenditure on output of goods produced. Thus, if a part of income is saved (that is, not spent), supply of output produced would not create sufficient demand for itself. This will cause deficiency of aggregate demand which will cause fall in output and employment and the emergence of involuntary unemployment.

However, classical economists denied the possibility of deficiency of aggregate demand even when a part of income is saved by the households. They showed that Say's law that supply creates its own demand holds good even in the presence of saving. They argued that every rupee saved by households will be invested by businessmen, that is, investment expenditure will be equal to savings done by households. In fact, output produced consists of consumer goods and capital goods. Income earned from production will be partly spent on consumer goods and partly on investment in capital goods. What is not spent on consumer goods is saved and investment expenditure on capital goods made by businessmen equals this savings. Therefore, there is no deficiency of demand or expenditure and circular flow of income goes on undisturbed. Thus, supply goes on creating its own demand and Say's law applies.

Now the pertinent question is what is the guarantee that investment expenditure will be equal to savings of the households. *According to classical economists, it is the changes in the rate of interest that brings about equality between saving and investment.* Further, according to them, rate of interest is determined by supply of savings and demand for investment. The investment demand is stipulated to be decreasing function of the rate of interest. At the lower rate of interest, more would be borrowed for investment. On the other hand, the savings of the people are taken to be the increasing function of the rate of interest, that is, higher the rate of interest, the larger the savings and *vice versa*. The loan market will be in equilibrium at the rate of interest at which the demand for investment is equal to the supply of savings. The changes in rate of interest would cause investment and supply of saving to become equal. This is illustrated in Fig. 3.3 (*a*). It will be seen that intersection of investment demand cuve *II* and the supply of savings curve *SS* determines the rate of interest *i*. At a higher rate of interest i_2, the investment demand is less than the intended supply of savings. Due to the excess supply of savings, the rate of interest would fall to *i*. On the contrary, at a lower rate of interest, say i_1, the demand for investment exceeds the supply of savings. Now, due to the excess demand for investment in the loan market rate of interest would go up. Thus, it is at rate of interest *i* that loan market is in equilibrium, *i.e.*, investment is equal to savings ($I = S$).

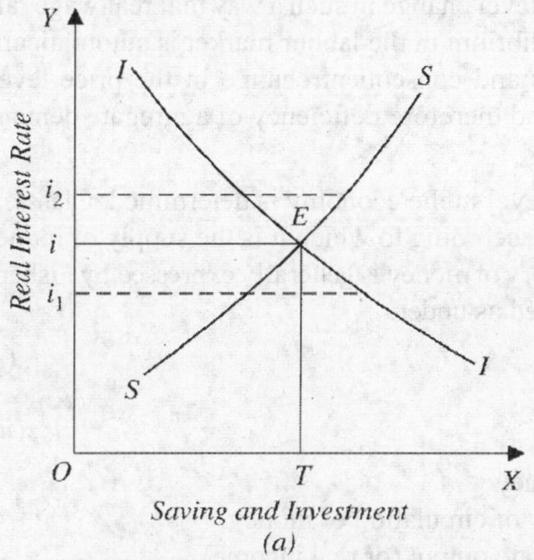

Fig. 3.3. *Changes in Rate of Interest bring about equality between saving and investment.*

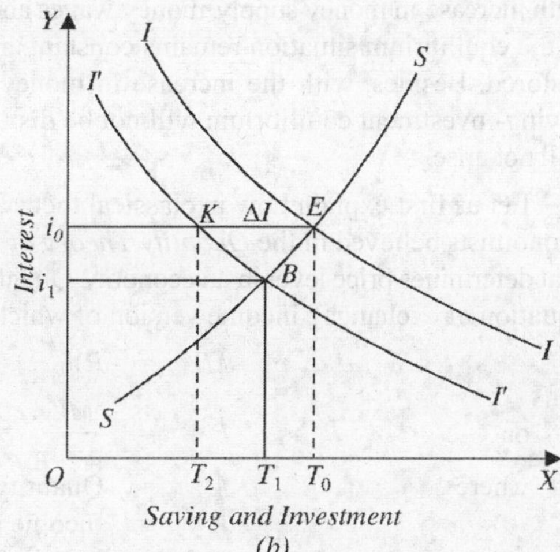

Fig. 3.4. *Decrease in investment demand does not disturb full-employment equilibrium.*

Now an important thing to know about classical theory is when due to decline in profit expectations of business firms if investment falls as it happens at times of recession or depression how it then explains that demand deficiency problem would not arise and equilibrium will continue to remain at full employment. This is illustrated in Figure 3.4, where initially saving and investment are in equilibrium at rate of interest i_0. Now suppose that due to fall in profit expectations investment by business firms decreases by ΔI or EK causing a shift in the investment curve to the left to the new position $I'I'$. With this at the initial rate of interest i_0, the supply of savings exceeds investment by KE. This excess supply of savings will put downward pressure on the rate of interest and as result interest will fall to i_1, at which saving and investment are again equal. According to classical theory, the lower interest induces more investment and therefore as a result of fall in interest to i_1, investment increases from OT_2. Besides, with the fall in interest rate from i_0 to i_1, savings decline by T_0T_1, which implies consumption demand will increase by T_0T_1. Thus, shift in investment demand curve to the left results in lowering of rate of interest which leads to more investment and consumption demand so that aggregate demand is not affected. It is thus clear that due to adjustment in interest rate even decline in investment does not give rise to demand deficiency problem and full-employment continues to prevail.

It follows from above that the equality between investment and saving, brought about by changes in the rate of interest, would guarantee that aggregate demand for output would be equal to aggregate supply of output. Thus, the problem of deficiency of aggregate demand would not be faced and full employment of labour will prevail.

Classical Theory of Income and Employment – Aggregate Demand, Money and Prices

Now, we shall examine how full employement of labour is assured in the classical theory even when money is introduced in the system. The introduction of money does not affect the result of the classical theory that problem of deficiency of aggregate demand would not be experienced by the free-market system and therefore full employment of labour is guaranteed. *The quantity of money, according to the classical theory, determines only the price level of output and in no way affects the real magnitudes of saving and investment. Further, since quantity of money determines the price level of output, it also affects real wage rate*, that is, the ratio of money wages and the price level, or $\frac{W}{P}$. But with increase in money supply, money wages and price level change in such a way that real wage rate in the equilibrium situation remains constant and equilibrium in the labour market is automatically restored. Besides, with the increase in money supply and consequent change in the price level, saving-investment equilibrium will not be disturbed and therefore deficiency of aggregate demand will not arise.

Let us first explain how in classical theory price level in the economy is determined. Classical economists believed in the *Quantity Theory of Money* according to which it is the supply of money that determines price level in an economy. Quantity theory of money is generally expressed by Fisher's equation of exchange[2], income version of which is stated as under:

$$MV = PY \qquad \text{...}(i)$$

or

$$P = \frac{MV}{Y} \qquad \text{...}(ii)$$

where
M = Quantity of money
V = Income velocity of circulation of money
Y = Level of aggregate output (or real income)
P = Price level of goods and services

Velocity of money is defined as the number of times a unit of money is used for purchase of final goods and services in a period, say during a year. In classical theory velocity is assumed to be constant. Besides, since in classical theory level of aggregate output is determined by the supply of productive resources, (*i.e.*, capital stock, availability of labour, land etc.) and the state of technology which do not change in the short run. Further, due to operation of Say's law and wage-price flexibility full employment of resources occur in the economy. Thus, with a given amount of productive resources and constant technology and with further assumption that they are fully utilised and employed, the aggregate output (Y) is held constant at full-employment level of output in the short run. Therefore, in Fig. 3.5 we have shown aggregate supply curve as a vertical straight line which shows that whatever the price level, aggregate output remains constant. How aggregate supply curve is derived in classical theory and why it is perfectly inelastic at full employment level of output will be explained later in this chapter. From equation (*ii*) above it follows that, with V and Q remaining constant, increase in money supply will cause proportionate increase in the price level. The effect of increase in the quantity of money is graphically shown in Fig. 3.5. It may be noted that MV in the quantity theory of money represents aggregate expenditure on goods and services made in a year. For instance, if money supply in an economy equals ₹ 500 crores and velocity of circulation is 4, then 500 × 4 = 2000 crores will be aggregate expenditure. In other words, in Figure 3.5 MV represents aggregate expenditure or aggrigate demand (AD) curve which slopes downward to the right. This is because at a lower price level, given the ag-

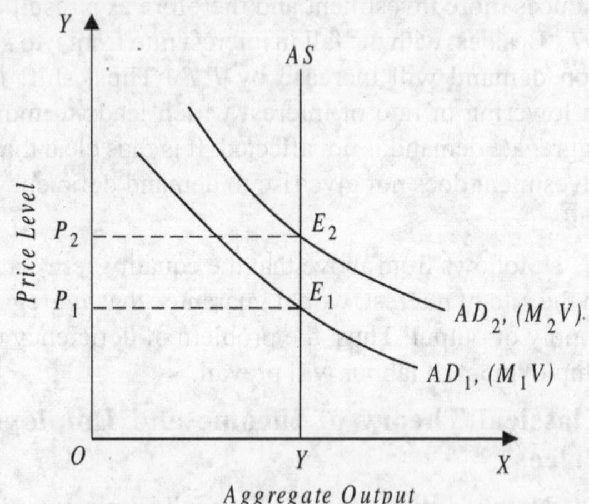

Fig. 3.5. *Increase in the quantity of money causes proportionate rise in the price level.*

2. Quantity theory of money and Fisher's Equation of Exchange will be explained and critically examined in detail in a separate chapter.

gregate expenditure as determined by MV, more quantity of goods and services will be purchased.

Suppose quantity of money in the economy is equal to M_1. With velocity of money V, aggregate expenditure on final goods and services will be equal to M_1V and corresponding to this aggregate demand curve AD_1 has been drawn in Fig. 3.5. With aggregate supply curve AS and aggregate demand curve AD_1, price level OP_1 is determined. Now, with increase in money supply to M_2, velocity of money V remaining constant, aggregate expenditure will rise to M_2V and aggregate demand curve will shift to the right to AD_2 position. Intersection of aggregate demand curve AD_2 and aggregate supply curve AS determines a higher price level OP_2. It therefore follows that increase in the quantity of money causes price level to rise. Indeed, the rise in price level will be proportional to the increase in quantity of money.

Classical Aggregate Supply Curve

Aggregate supply curve describe the relationship between aggregate supply of output with price level. Classical theory regards aggregate supply curve to be perfectly inelastic. Now, an important question is why in classical model, aggregate supply curve is perfectly inelastic. As explained above, aggregate output Y_F is determined by the equilibrium level of employment N_F, given the aggregate production function. Equilibrium level of employment along with real wage rate is determined by labour market equilibrium, that is, equilibrium between demand for the supply of labour. Thus, in classical theory aggregate supply curve is determined by supply-side factors, namely, preferences of households or individuals regarding work and leisure, the stock of capital (and other factor endowments), the state of technology. Supply of labour, as seen above, is determined by indvidual preferences between work and leisure and demand curve for labour is determined by marginal product of labour.

Thus in classical model aggregate supply curve reflects supply-determined nature of output and does not depend on the aggregate demand and price level. The classical aggregate supply curve is shown in Fig. 3.6. The pertinent questions is how with changes in price level, which in the classical theory depends on the quantity of money, leave level of employment and output unaffected. The reason for this is that changes in price level causes equal proportionate changes in money wage rate with the result that the equilibrium real wage rate which is given by $\frac{W}{P}$ remains constant and therefore equilibrium level of employment does not get affected. The adjustment process works in the following way.

If due to the increase in supply of money price level rises, with a given money wage rate (W), real wage rate, which is equal to $\frac{W}{P}$, will fall. At a real wage rate lower than the equilibrium real wage rate, the quantity demanded of labour will exceed the supply of labour. This disequilibrium between labour demand and supply will cause money wage rate to rise to the level so that original real wage rate determined by labour market equilibrium is restored. Suppose that in labour-market equilibrium money wage rate is W_1 and given the price level equal to P_1, and the equilibrium real wage rate will be $\frac{W_1}{P_1}$. Now, if price level is

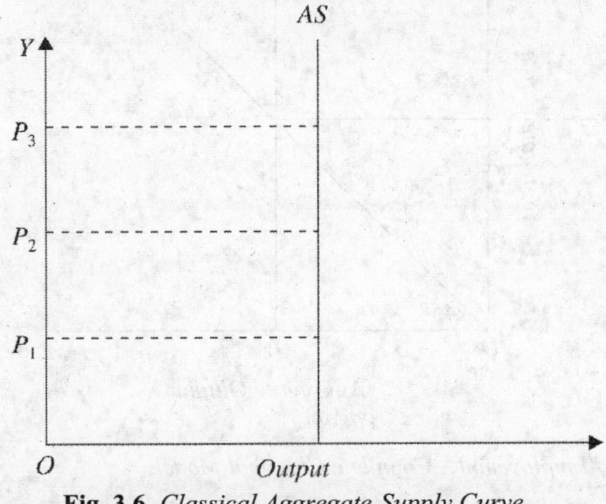

Fig. 3.6. *Classical Aggregate Supply Curve*

doubled to $2P_1$ and money wage rate rises to $2W_1$, then the equilibrium real wage rate will become equal to $\frac{2W_1}{2P_1} = \frac{W_1}{P_1}$. Thus, with equal proportionate increase in money wage rate as a result of rise in price level, equilibrium real wage rate and level of employment will remain unaffected. Thus, with rise in price level, level of employment remains unchanged and, given the aggregate production function, level of output will remain constant. This implies that aggregate supply curve of output is perfectly inelastic. Thus whatever the price level, money wage rate changes in such a way that equilibrium real wage rate, level of employment and therefore output remain constant. Thus in classical theory aggregate supply of output is determined by supply-side real variables and does not depend on money and prices.

CLASSICAL THEORY OF OUTPUT AND EMPLOYMENT – COMPLETE CLASSICAL MODEL*

We now illustrate the complete classical model of income and employment determination in an economy in Fig. 3.7. In panel (*a*) of this figure labour market equilibrium is shown wherein it will be seen that the intersection of demand for and supply of labour determines the real wage rate $\left(\frac{W_0}{P_0}\right)$. At this equilibrium real wage rate the amount of labour employed is N_F and, as explained above, this is full employment level. As depicted in panel (*b*) of the figure this full employment level of labour N_F produces Y_F level of output (or income). In panel (*c*) of Figure 3.7 we have drawn 45° line that is used to transfer the

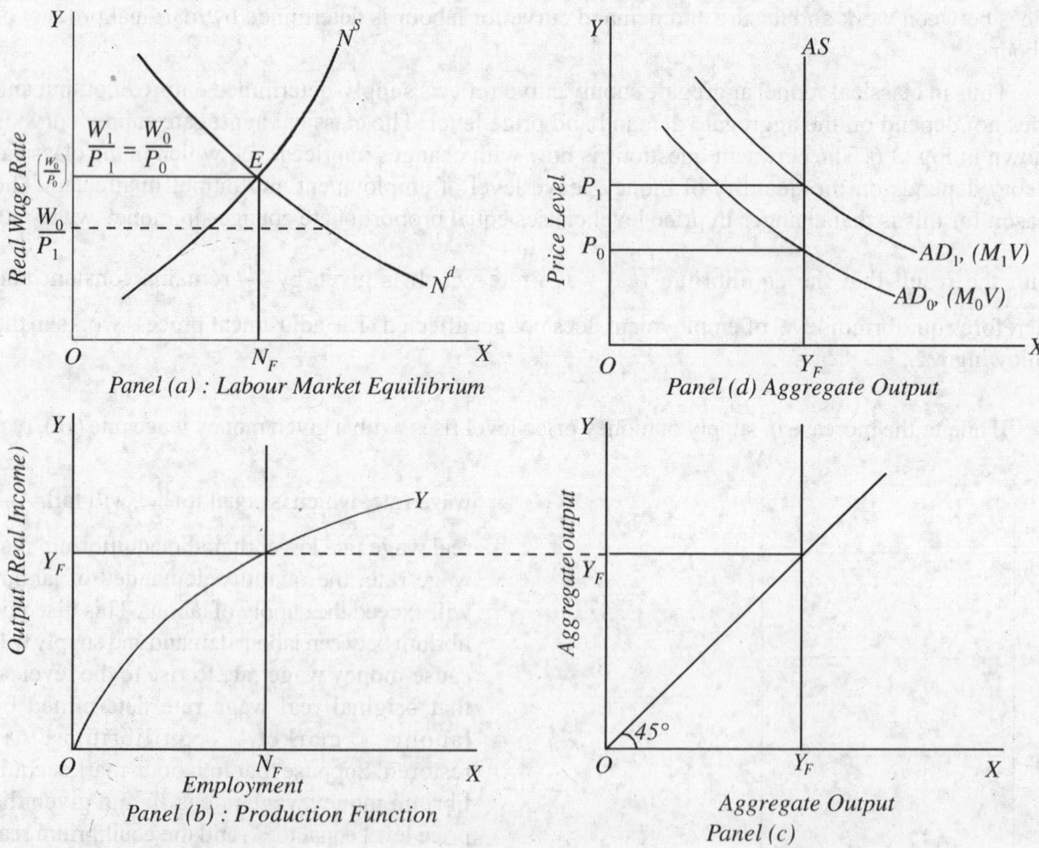

Fig. 3.7. *Determination of Income and Employemnt – Complete Classical Model*

* This section is meant for higher level courses in macroeconomics.

level of output on the vertical axis in panel (*b*) to the horizontal axis of panel (*c*). In panel (*d*) we have shown the determination of price level through intersection of the curves of aggregate demand for and aggregate supply of output, as explained by the quantity theory of money. In the classical theory, aggregate supply curve AS is a vertical straight line at full-employment level of output Y_F. Thus, given constant velocity of money V, the quantity of money M_0 will determine the expenditure or aggregate demand equal to M_0V according to which aggregate demand curve (with flexible prices) is AD_0. It will be seen from panel (*d*) of Fig. 3.7 that intersection of vertical aggregate supply curve AS at fully-employment level output Y_F and aggregate demand curve AD_0 determines the price level P_0. With price level at P_0, the money wage rate is W_0 so that $\left(\frac{W_0}{P_0}\right)$ is the real wage rate as determined by the intersection of demand for and supply of labour [see panel (*a*) of Fig. 3.7].

Now, a relevant question is how this equilibrium level of real wage rate, prices, employment and output (income) will change following the increase in the quantity of money. Suppose the quantity of money increases from M_0 to M_1. With the given capital stock (as we are considering the short-run case) and the labour force being already fully employed, the output cannot increase. Therefore, as depicted in panel (*d*) following the increase in money supply to M_1 aggregate demand or expenditure will increase to M_1V and thereby causing aggregate demand curve to shift to AD_1. As a result, price level rises from P_0 to P_1.

However, as explained above, with the given money wage rate W_0, the rise in price level from P_0 to P_1 will cause a fall in real wage rate. As will be seen from panel (*a*), with the rise in price level to P_1, real wage rate falls to $\frac{W_0}{P_1}$. This will cause temporary disequilibrium in the labour market. At the lower real wage rate $\frac{W_0}{P_1}$, more labour is demanded than is supplied. Given the competition among the firms, this excess demand for labour will cause the money wage rate to rise to W_1 level so that the real wage is bid up to the original level $\frac{W_1}{P_1} = \frac{W_0}{P_0}$. With the real wage rate being quickly restored to the original level, employment of labour N_F and total output or income Y_F will remain unaffected. To sum up, *the result of increase in money supply is to raise money wages and prices in equal proportion, leaving real wages, employment and output unaffected.* The results of decrease in money supply can be similarly worked out.

NEUTRALITY OF MONEY AND CLASSICAL DICHOTOMY

An important conclusion which follows from the classical theory of output and employment is that *changes in the quantity of money affect only nominal variables (i.e. money wages, nominal GNP, money balances), and have no influence whatsoever on the real variables of the economy* such as real GNP (*i.e.* output of goods and services produced), level of employment (*i.e.* number of labour - hours or number of workers employed), real wage rate (*i.e.* wage rate in terms of its purchasing power). Actually, according to classical theory, the nominal variables move in proportion to changes in the quantity of money, while real variables such as GNP, employment, real wage rate, real rate of interest remain unaffected. Classical economists explained that real variables such as GNP, employment, real wage rate are determined by real factors such as stock of capital, the state of technology, marginal physical product of labour, households' preferences regarding work and leisure. In the classical model based on flexibility of prices and wages, changes in money supply only affect the price level and nominal magnitudes (*i.e.* money wages, nominal interest rate, while the real variabiles such as levels of labour employment and output, saving and investment, real wages, real rate of interest remain unaffected. *This independence of real variables from changes in money supply and nominal variables is called classical dichotomy.*

The neutrality of money can be graphically illustrated with the help Fig. 3.7 and 3.8. Suppose to begin with, the stock of money in the economy is equal to M_0. With this, as will be seen from Panel (*d*) of Figure 3.7, aggregate demand curve for output is AD_0 which with interaction with aggregate supply curve AS determines price level P_0. Given the price level P_0, labour-market equilibrium determines money wage rate W_0 and real wage rate equal to $\frac{W_0}{P_0}$ and level of employment N_F in Panel (*a*) of Fig. 3.7. The level of employment N_F, given the production function, determines aggregate output Y_F in Panel (*b*) of Fig. 3.7.

Now suppose there is expansion in money supply from M_0 to M_1 which causes an upward shift in the aggregate demand curve from AD_0 to AD_1 [see Panel (*d*) of Fig. 3.7]. As a result of this upward shift in the aggregate demand curve from AD_0 to AD_1, price level rises from P_0 to P_1. Now, as will be seen from Panel (*a*) of Fig. 3.7, with money wage rate W_0 and price level equal to P_1, real wage rate falls to $\frac{W_0}{P_1}$ at which demand for labour exceeds supply of labour. This will cause, according to classical theory, money wage rate to rise to W_1 in equal proportion to the rise in price level so that real wage is restored to the original level $\left(\frac{W_1}{P_1}=\frac{W_0}{P_0}\right)$ and labour-market equilibrium determines the original level of employment N_F. With the same level of labour employment aggregate output (*i.e.* GNP) will not be affected. Thus, we see that with the expansion in money supply, nominal wage rate and price level have risen, but real wage rate, level of employment and output remain constant. Hence it shows that money is neutral in its effect on real variables.

Changes in Money Supply, Saving-Investment Equilibrium and Neutrality of Money. According to the classical theory, money performs the function of merely a medium of exchange of goods and services and is therefore demanded only for transaction purposes. This means alternative to holding money is the purchase of goods and services. Therefore, demand for and supply of money in the classical system does not determine the rate of interest. When the quantity of money increases, it will leave the *real rate of interest* unchanged and hence the amount of output saved and allocated to investment (*i.e., real saving and investment*) will remain the same as shown in Fig. 3.8. This means the increase in money supply does not disturb the capital market equilibrium or saving-investment equality and consequently the continuation of full-employment equilibrium. However, it may be noted that the higher level of prices of commodities would mean that investment expenditure in money terms will increase in the same proportion as the rise in prices even though the output of commodities allocated for investment purposes remains the same. But this increase in *monetary expenditure* for investment is matched by the equal increase in *monetary saving* brought about by the rise in prices. The higher prices of commodities also mean a proportionate increase in the amount of money received from the sale of commodities so that savers are willing to provide proportionately larger amount of saving at a given rate of interest. Thus, with the increase in quantity of money, the supply curve of nominal saving and investment demand curve will

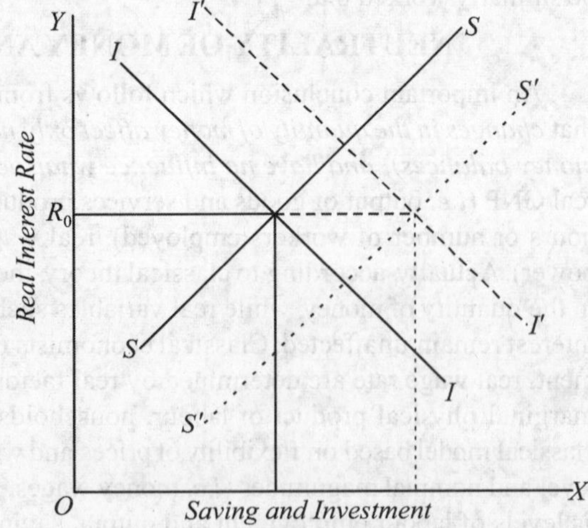

Fig. 3.8. *Capital Market Equilibrium*

shift to the right as shown by dotted $S'S'$ and $I'I'$ curves by the same proportion so that the same real rate of interest is maintained and the same amounts of real saving and investment in terms of commodities are made at the higher price level.

A serious limitation of the classical concept of neutrality of money may be noted. As seen above, *the neutrality of money is a basic result reached in the classical full-employment model based on flexibility of prices and wages*. If increase in money supply and consequent rise in prices has no real effects, then inflation would not be a matter of concern. However, we know that inflation is a matter of serious concern as it lowers standards of living of the people and also adversely affects economic growth. Therefore, efforts are made to control inflation and achieve price stability in the economy.

KEYNES'S CRITIQUE OF SAY'S LAW AND CLASSICAL THEORY

Keynes in his renowned book "*General Theory*" severely criticised the classical theory of employment.We explain below various criticisms of classical theory made by Keynes.

Keynes challenged Say's Law. Keynes criticised Say's Law and proved that it was quite invalid. As we have said above, according to Say's Law, every supply or production creates its own demand and therefore problems of over-production and unemployment do not arise. It is, of course, true that supply creates demand for goods because the various factors which are employed in a productive activity earn incomes from it, which in turn are spent on goods. For example, when factors of production are employed in producing cloth, then the incomes in the form of wages, rent, interest and profits accrue to them which they spend on various goods. But from this it does not follow that the supply of production will create its own entire demand. The incomes earned by the various factors of production are equal to the value of output produced, but this does not mean that the whole income received by the factors of production will be spent on goods and services. A part of the income is saved and the saved part does not necessarily create demand for goods and services. If entrepreneurs do not invest equal to the desired savings, then aggregate demand which, without government intervation, consists of demand for consumer goods and capital goods, will not be enough to purchase the available supply of output. Hence, if aggregate demand is not sufficient to purchase the available supply, the producers would be unable to sell their whole output due to which their profits would decline and as a result they would reduce their level of production giving rise to involuntary unemployment in the economy.

In a given period, consumers spend a part of their income on consumption and the rest they save. Likewise, in a period, the entrepreneurs plan to spend on factories and machines, that is, they plan to invest. Aggregate demand is the sum of consumption demand and investment demand. But in a free market capitalist economy, the persons who save are often different from those who invest and further that the factors that determine savings are different from the factors which determine investment by the entrepreneurs. People save to provide for their old age, to accumulate money for education and marriage of their children and also save and hold money balances for speculative motive, that is, to buy stocks and bonds in future to earn profits. But investment by entrepreneurs depends upon marginal efficiency of capital (that is, expected rate of profit), rate of interest, population growth and technological progress. Keynes also explained that the equality between saving and investment cannot be brought about by changes in interest rate as saving mainly depends on income and it was changes in income that bring about equality between saving and investment rather than changes in rate of interest. But classical economists ignored the changes in level of income because of their assumption of full-employment.

To conclude, savers and investors are different people with different motives. Much of the economy's saving is done by households while investment is mostly done by business firms on the basis of profit expectations and the amount of investment they want to make fluctuates widely from year to year and is unlikely to be equal to savings which households want to do. This affects aggregate demand and causes fluctuations in income, output and employment in capitalist economies. We thus see that there is no any mechanism in a free market economy which guarantees that investments made by the entrepreneurs are equal to the saving by the people. If the desired investment by entrepreneurs falls short of the amount of saving at full-employment level of income, the equilibrium of the economy will be at less than full employment level and as a result involuntary unemployment will emerge in the economy. In this way, according to Keynes, there is no reason that sum of consumption expenditure and investment expenditure is necessarily equal to the value of output produced. In other words, there is no guarantee that aggregate demand will be equal to aggregate supply forthcoming at full-employment level of resources. Hence, it is not necessary that the economy will be in equilibrium at the level of full employment. This invalidates Say's law, since according to it over-production and unemployment cannot occur.

Keynes proved Pigou's view that price-wage flexibility would automatically restore full-employment as fallacious. Keynes also criticised Pigou's view that a general fall in wages and prices in times of depression will remove unemployment and automatically restore full empooyment in the economy if market mechanism is allowed to work freely without any obstruction by trade unions and Government. According to Keynes, a general fall in wages will not bring about increase in employment because the reduction in wages will reduce aggregate demand for goods. Keynes put forward the view that *wages are not only the cost of production, they are also incomes of the workers* which constitute the majority of the population of a country. As a result of a *general fall* in wages, the incomes of the workers will fall due to which aggregate demand will decline. As a result of decline in aggregate demand, level of production will have to be reduced and less labour will be employed than before. This will create more unemployment rather than reducing it. No doubt, as a result of a general cut in wages, cost of production of industries will fall but with the fall in costs, the demand for the products will not increase because due to the *all-round cut in wages,* purchasing power of the working class will decrease. Hence all-round cut in wages will reduce the level of employment by reducing aggregate demand and will thus deepen the depression.

There is a fundamental difference between Keynes and Pigou in respect of the relationship between wages and employment. Pigou thought that level of employment in an economy depends upon the level of money wages and therefore reduction in money wages will promote employment. On the other hand, Keynes thought that the level of employment depends upon the aggregate demand and aggregate demand declines as a result of an all-round cut in money wages. According to Keynes, even if wage rates are perfectly flexible, unemployment will prevail in the economy if the aggregate demand is deficient.

Classical economists thought that a general cut in wages would reduce the cost of production of various industries but they ignored the fact that a general cut in wages will also reduce the incomes of the workers. In view of the fall in incomes and aggregate demand, how will manufacturers be able to sell their whole output? It is the sales of output that makes the wheel of trade, output and employment going. However, note that the classical theory is valid in case of an individual industry. With the decline in wages, the cost of the industry decreases and as a result the price of its product falls. The industry will be able to sell a larger amount of output at a lower price because it is not necessary that the goods produced by the industry are to be purchased by the workers employed in that industry whose wages have been reduced. But in the case of the economy as a whole, this is not valid because a general cut in wages will reduce the incomes of the working class and as a result enough demand will not be there for the output produced by the whole economy. This deficiency in demand will reduce demand for workers as a result of which unemployment will spread among them. Although it is true that a reduction in real wages (i.e. money wages relative to the general price level, $\frac{W}{P}$) in a single firm or industry is not be likely to affect the *overall demand* for that product, it is quite wrong to assume that a *general economy-wide reduction in wages* of *all workers* have no effect on aggregate demand. Pigou and other classical economists *committed a logical falacy* in their thinking by applying the analysis which is true for a particular firm or industry to the economy as a whole. Thus, the fundamental flaw in Pigou and other classical economists is that they applied partial equilibrium analysis, which is valid in the case of an individual industry, to the determination of income and employment in the whole economy. The determination of the level of aggregate income and employment in the economy should be explained with the aid of general equilibrium analysis rather than with partial or particular equilibrium analysis of microeconomics.

Price Flexibility and Unemployment. A basic idea of classical economists is that in a free market economy full employment is the normal state of affairs and any deviation from it will be automatically corrected through quick adjustment in prices and wages. As explained above, when during the period of great repression, 25 per cent of labour force in the USA was unemployed A.C. Pigou wrote, "With perfectly free competition, there will always be a strong tendency toward full employment. Such *unemployment as exists at any time is due wholly to the frictional resistances that prevent the appropriate wage and price adjustments being made instantaneously.*"[3]

On the contrary, Keignes explained that unemployed that prevailed during depression was due to fall in aggregate demand and he argued that prices and wages were inflexible downward and fall in aggregate demand causes decline in real output and employment. As a result, involuntary unemployment emerges.

The Classical and Keynessian viewpoints are illustrated in Fig. 3.9 through *AS-AD* model. According to classical economists, aggregate supply curve is vertical at full employment output Y_F and is represented by *AS*. Keynes's short run aggregate supply curve is given by the horizontal line *SAS*. Suppose, to begin with, aggregate demand curve is AD_2 which intersects aggregate supply curve *AS* at point *E* with price level equal to P_2. Now suppose that aggregate demand declines due to the fall in investment demand or due to the contraction in money supply and as a result aggregate demand curve shifts leftward to the new position AD_1 (dotted). According to the classical economists, prices and wages would adjust quickly so that equilibrium will be achieved at point *T* at the lower price level P_1, level of national output remaining unchanged at full-employment output level Y_F. Thus, in the classical framework, if market system is allowed to work freely, even with the fall in aggregate demand, full employment tends to prevail and no involuntary unemployment can exist.

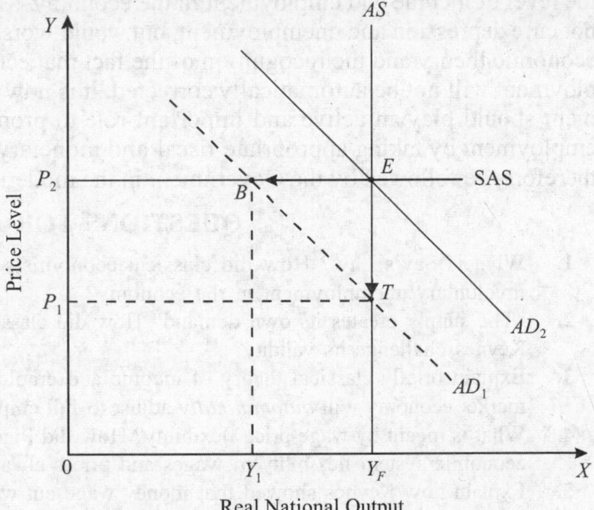

Fig. 3.9. *Wage-Price Flexibility: Keynes vs. Classics*

On the other hand, according to Keynes, *prices and wages are sticky* and therefore Keynes's short-run aggregate supply curve is flat as is represented by *SAS* in Fig. 3.9. Therefore, when there is leftward shift in aggregate demand due to decline in desired investment, the real national output will fall by *EB* or Y_FY_1, price level and money wages remaining incharged.[4] We shall discuss in detail in a later chapter the Keynesian and Classical theories regarding wage-price flexibility and employment.

3. A.C. Pigou, *The Theory of Unemployment*, 1933 (Italics added)
4. Note that modern Keynesians consider the short-run aggregate supply curve to be gently sloping upward. Therefore, when aggregate demand falls, there is a small fall in price but, a large fall in real national output.

Because of the above-mentioned shortcomings of the classical theory, there was a need for development of new theory which could provide a correct explanation of the determination of income and employment in the economy. A capitalist economy cannot automatically attain a state of full employment. Keynes in his famous work "*General Theory of Employment, Interest and Money*" not only criticised the classical theory but also propounded a new one which is still regarded as substantially valid and correct.

Conclusion

We have discussed above Say's law of classical economics. This is a basic law of the classical economics. In brief this law states that supply creates its own demand. From this, it has been concluded that in a free-enterprise capitalist economy, there is always a tendency towards full employment. According to them, if sometimes unemployment appears in the economy, then wages would decline, the rate of interest and prices would also fall. As a result, employment of labour would increase and unemployment will be automatically removed, provided the economy is allowed to work freely without any interference by Government and trade unions. Hence a state of full employment will be established. In this way due to the *flexibility of wages, prices and interest rate*, there can neither be general overproduction, nor unemployment in the economy for a long time. Therefore, classical and neoclassical economists thought that there was always a tendency toward full-employment provided no restrictions were placed in the working of free and perfect competition. Thus, according to them, the Government need not interfere in the working of the economy and should follow a *laissez fair policy*. But Keynes proved this as invalid not only theoretically but also practically. Keynes put forward a new theory of income and employment which is the correct explanation of the phenomenon in a developed capitalist economy. For this purpose, Keynes invented new concepts such as propensity to consume, marginal efficiency of capital, liquidity preference which affect the level of income and employment in the economy. Keynes also proved that a cut in wages would not cure depression and unemployment, but would worsen them. Following Keynesian revolution in economic theory and the recognition of the fact that economic fluctuations or lapses from full-employment will not be automatically corrected, it is now believed by many economists that Government should play an active and important role to promote economic stability at the level of full-employment by taking appropriate fiscal and monetary measures. *Laissez faire* policy should not, therefore, be followed by the Government in the modern world.

QUESTIONS FOR REVIEW

1. What is Say's Law? How did classical economists use this law to show that there could not be involuntary unemployment in the economy?
2. "The supply creates its own demand" How did classical economists justify this argument? How did Keynes challenge its validity?
3. Explain briefly classical theory of income and employment. How does this theory show that a free market economy will *automatically* adjust to full employment level?
4. What is meant by wage-price flexibility? How did Pigou show that given a free enterprise competitive economic system flexibility of wages and prices always ensures full employment.
5. Explain how Keynes showed that money wage cut would fail to increase employment.
6. What is aggregate demand? How classical economists showed that there would always be sufficient aggregate demand even when there is saving by the households?
7. What is meant by *neutrality of money*? How did classical economists explain that money had no effect on real income, output and employment.
8. How is the level of output determined in a classical model? What will happen to output if investment demand falls due to the decline in profit expectations of business firms?

   [**Hint:** Output means the same thing as real income. The classical theory of income and employment is the classical theory of output].
9. What are the main factors according to Keynes that may prevent the attainment of full-employment when wages and prices are flexible.
10. Explain briefly Keynes's critique of classical theory of income and employment
11. What mechanism in classical theory of output and employment ensures equality between saving and investment at full employment? How did Keynes prove it as fallacious?
12. What is meant by neutrality of money. Show that why money is neutral under the classical model of output and employment .

CHAPTER 4
KEYNES'S THEORY OF EMPLOYMENT AND UNEMPLOYMENT

KEYNESIAN REVOLUTION

Classical economists were of the view that there is always full employment in the economy or there is always a tendency towards full employment in the economy. This view of them was based upon their belief in *Say's Law of Markets*. They thought that when there is unemployment in the economy, then, given the free and perfect competition in the economy, certain economic forces automatically operate in such a way that the condition of full employment is restored. During the period 1929-33, there occurred great depression in the capitalist countries which caused huge unemployment of labour and other resources in those countries and as a result level of national income fell down. Due to this depression many factories were closed in these countries and factories which were working were also not being used to their full productive capacity. In other words, there appeared a good deal of excess productive capacity in these economies. As a result of unemployment, low income and production created by depression, people had to undergo a good deal of sufferings.

This condition of depression and unemployment did not seem to disappear automatically. Therefore, people's belief in the classical economic thought regarding the tendency to full employment was shaken. Thus, the classical theory of full employment was proved empirically wrong. In this background, Keynes wrote his book, *"General Theory of Employment, Interest and Money"*, in which he challenged the validity of the classical theory of employment. Not only did he criticise the classical theory of full employment and proved it wrong but also presented a new theory of income and employment which is generally believed to be correct and valid by modern economists. Keynes brought about such a fundamental and important change in our economic thought that the Keynesian theory was generally known as *new economics at that time*. Being impressed by the fundamental and revolutionary nature of change in our economic theory by Keynes, many economists called his *"General Theory of Employment, Interest and Money"* as the *Keynesian Revolution*. We explain below the outline of the Keynes's theory of employment.

We have critically examined the classical model of full employment and Say's Law of Markets on which the classical theory is based. Keynes in his eminent work, *"General Theory of Employment, Interest and Money"*, not only criticised the classical Say's law but also propounded a new theory of income and employment. Keynes in his work put forward a more systematic and realistic analysis of the determinants of employment in an advanced capitalist economy and the factors which lead to unemployment. Keynes tried to prove that full employment is not the normal feature of an advanced capitalist economy and that underemployment equilibrium is its normal feature. Keynes also invented new tools and concepts of economic analysis in terms of which he propounded his theory of income and employment. These new tools and concepts are propensity to consume, multiplier, marginal efficiency of capital and liquidity preference. We will explain these concepts in detail in later chapters. Here we will explain only the outline of the Keynesian theory of income and employment.

It is important to note that Keynesian theory of income and employment is a short-run theory because Keynes assumes that the amount of capital, the size of population and labour force, technology, efficiency of labourers, etc., do not change. That is why in Keynesian theory, the amount of employment depends upon the level of national income and production. This is because, given the amount of capital, technology and labour efficiency, increase in income and output can be obtained by employment of more labour. Therefore, in the Keynesian short run, the higher the level of national income, the greater the amount of employment, and lower the level of national income, the lower the amount of employment. Hence, Keynes's theory of employment determination is also Keynes's theory of income determination. But, for simplicity, in this chapter we shall explain the Keynesian theory in terms of employment and all the diagrams we will draw will show explicitly the determination of employment. In the next chapter, we will explain Keynesian theory in terms of national income through *income expenditure analysis* and the diagrams which we will draw there will explicitly show the income determination. But it is important to note that the factors that determine the level of income are the same as those that determine the level of employment; only the diagrams used to show their determination are different. It is important to note that Keynes thought that *prices and wages do not adjust quickly to balance demand and supply*. Therefore, in his theory of income and employment he *assumes that prices remain constant*.

KEYNES'S THEORY OF EMPLOYMENT – PRINCIPLE OF EFFECTIVE DEMAND

At the outset, it may be noted that in Keynesian theory of income and employment determination, principle of effective demand occupies a significant place. In an advanced capitalist economy, the level of employment depends upon the level of aggregate effective demand, the greater the level of effective demand, the greater the amount of employment in the economy. How many men will be employed by an individual firm depends upon the number of persons employed which will make maximum profits. Likewise, in the whole economy, how many men will be employed by firms or entrepreneurs in the economy depends upon the fact that they make their individual profits. In the whole economy, the amount of employment is determined by aggregate supply and aggregate demand. We will now discuss below these concepts of aggregate supply and aggregate demand functions and will show how they determine the equilibrium level of employment.

Aggregate Supply Function

When entrepreneurs employ some people they incur some cost of production. If the proceeds obtained from the sale of output produced by a certain number of people employed is greater than the cost of production incurred, it will be worthwhile to employ them. The cost of production incurred on the employment of a certain number of labourers must be received by the entrepreneur, otherwise they will not produce and provide employment to labour. At any given level of employment of labour, *aggregate supply price is the total amount of money which all the entrepreneurs in the economy taken together must expect to receive from the sale of the output produced* by the given number of labourers employed. In other words, *aggregate supply price is the total cost of production* incurred by employing a certain given number of labourers. It is obvious that if the cost of production incurred by the entrepreneur in employing a certain number of labourers is not covered, they will reduce the amount of employment offered. As the amount of employment of labour increases, the total cost of production will also increase. Therefore, aggregate supply price will rise as more labour is employed to produce goods and services. *Keynes's aggregate supply function (curve) shows the relationship between the number of workers employed and the receipts which all firms in the economy must get if it is just worth employing them, prices remaining constant.* In other words, Keynes's aggregate supply function (curve) shows the relationship between the number of workers employed and the aggregate supply price. Thus we can construct *a schedule or curve of aggregate supply showing aggregate supply price at different levels of employment*.

Let us now explain in detail the factors on which aggregate supply curve depends. *The aggregate supply curve depends ultimately on the physical and technical conditions of production* (that is, capital stock, the state of technology and the nature of production function). However, physical and technical conditions remain constant in the short run. Hence, given these technical conditions, the level of output can be increased only by increasing employment of labour. But, when output and employment are increased, more cost of production is incurred. Whether production is subject to the law of increasing, diminishing or constant returns, when more workers are employed to increase production, more cost has to be incurred since additional workers are to be paid wages. Hence, more workers will be employed only if the entrepreneurs *must expect to receive* greater revenue so as to cover the rise in cost incurred. Therefore, the aggregate supply curve slopes upward to the right.

The *slope* of the aggregate supply curve (function) depends on the physical or technical conditions of production. If the technical conditions are such that with the increase in output, marginal cost of production does not rise, the aggregate supply curve will be a straight line. On the other hand, if technical conditions are such that diminishing returns occur with the increase in employment of labour, marginal cost of production will rise with the increase in the level of output. This will make the slope of aggregate supply curve to increase with the increase in employment of labour. Besides, if with the increase in employment of labour, wage rate of labour rises, the slope of aggregate supply curve will increase with the employment of more labour. It is however worth noting that Keynes thought in conditions of depression/recession when huge unemployment of labour prevailed in the economy, with the increase in labour employment to produce more, wage rate will remain constant. Assuming that marginal cost of production rises with the increase in employment of labour, upward rising aggregate supply curve *AS* with increasing slope as more labour is employed is shown in Fig. 4.1. This aggregate supply curve (function) depicts rising aggregate supply price at various levels of labour employment.

We have said that the aggregate supply is determined by the physical and technical conditions prevailing in the economy, that is, the quantity and quality of labour, stock of capital and raw materials available in the economy and the state of technology. As these changes, or as production technology is improved, *AS* curve will also change. But in the analysis of the determination of employment in the advanced capitalist economies in the short-run aggregate supply curve can be assumed to be given and constant. This is for the simple reason that in times of depression the main problem of advanced capitalist economies is how to employ idle manpower and capital resources to increase production by raising demand and not that how the productive capacity be raised by augmenting the stock of capital or by improving the techniques of production. That is why Keynes assumed the *AS* curve to be constant and paid greater attention to the factors determining aggregate demand. The need of the moment in times of recession is to increase aggregate demand so that the equilibrium be achieved at full-employment level. When aggregate demand is increased, the aggregate demand curve will shift upward and it will intersect the *AS* curve more on the right, *i.e.*, the number of men employed will increase.

When the full-employment level has been reached, then, given the stock of fixed capital and the prevailing technology, output and employment cannot be further increased by increasing aggregate demand. Under such conditions increase in aggregate demand will only result in inflation. In such a situation, it becomes necessary to cause a rightward shift in the aggregate supply curve by making addition to the stock of capital and effecting improvements in production technology. This will contain the inflationary pressures. In other words, efforts should be made to increase aggregate supply when full-employment level has already been achieved and the economy is in the grip of inflation.

The curve of aggregate supply price *AS* starts from the point of origin and slopes upward to the right. In the beginning aggregate supply price curve *AS* rises slowly, afterwards it rises rapidly. This curve *AS* shows that as the number of men employed is increased, the aggregate supply price rises slowly in the beginning and rapidly afterwards. This is because cost of production rises as more people are employed and further due to the operations of law of diminishing returns total cost of production increases at an increasing rate. Once all the men willing to get employment are employed, then we have a state of full employment. When the state of full employment is reached further increases in aggregate demand or expenditure will be unable to increase employment further since output of goods and services cannot be increased further as no more labour is available for production after full-employment level is reached. Therefore, aggregate supply curve assumes a vertical shape after full employment is reached. In Fig. 4.1, ON_F is the level of full employment at which aggregate supply curve assumes a vertical shape.

Aggregate Demand Function

It is aggregate demand function which plays a more important role in the determination of employment. Aggregate demand function (curve) shows for each possible level of employment the total sum of money (proceeds) which all the firms or entrepreneurs in the economy *actually expect* to receive from the sale of output produced by those workers employed in the economy. *The amount of expenditure actually expected when a given number of workers are employed to produce goods and services is called aggregate demand price.* Like the aggregate supply price, aggregate demand price also varies at different levels of employment. This is because at different levels of employment different income levels would be generated and at different income levels, expenditure, especially consumption expenditure, would be different. Aggregate demand has the following 4 components: (1) *Consumption demand,* (2) *Investment demand,* (3) *Government expenditure, and* (4) *Net Exports (that is, exports-imports).* In this introduction to the theory of employment, we confine ourselves to the consumption demand and investment demand and will introduce Government expenditure and net exports in a later chapter of this book.

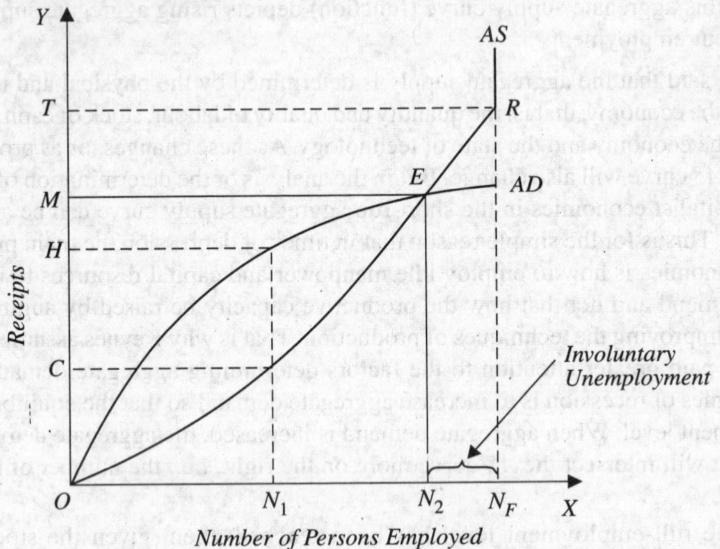

Fig. 4.1. *Determination of Employment*

Thus the factors which determine aggregate demand are the factors which determine consumption demand and investment demand. Consumption demand depends on disposable income on the one hand and propensity to consume on the other. Propensity to consume depends on some subjective factors such as willingness to save, desire to imitate others' higher levels of living and objective factors such as price level, taxation policy of the Government, rate of interest. Given these factors, changes in which cause a shift in the entire consumption function, the higher the level of disposable income, the greater the amount of consumption demand. As regards investment demand, according to Keynes, it is determined by rate of interest on the one hand and marginal efficiency of capital on the other. While rate of interest is more or less sticky, it is changes in *marginal efficiency of capital* (*i.e., expected rate of return*) that cause frequent changes in inducement to invest. Marginal efficiency of capital means expected rate of profit by the entrepreneurs from the investment they propose to undertake. When prospects for profit making in future are bright, there will be more investment. If investors become pessimistic about profit earning in future, they will undertake less new investment.

But what are the factors on which business expectations about opportunities of making profits depend? The expected rate of profits (which Keynes calls marginal efficiency of capital) of business men depends on their estimates of consumer demand for goods, taxation policy of the Government, expectations regarding changes in technology. It is worth noting that it is due to the frequent changes in business expectations that investment demand is volatile. When business men become pessimistic about profit earning, investment declines which lowers aggregate demand. On the other hand, when entrepreneurs become bullish or optimistic, they undertake new investment on a large scale which raises the level of aggregate demand of the economy. On the other hand, *consumption function*, according to Keynes, remains stable in the short run. However, *amount of consumption demand* increases as the income rises in the short run. Therefore, we can construct a schedule or curve of aggregate demand showing different aggregate demand prices at different levels of employment. The curve of aggregate demand is shown by the curve *AD* in Fig. 4.1. The aggregate demand price curve also rises from left to right. It will be seen in Fig. 4.1 that when ON_1 number of men are employed, the aggregate demand price is *OH* and when ON_2 men are employed, aggregate demand price is *OM*.

Determination of the Equilibrium Level of Employment by Effective Demand

In Fig. 4.1 we have shown together aggregate supply curve and aggregate demand curve. The amount of employment is measured along the *X*-axis and the receipts or proceeds obtained at various levels of employment are measured along the *Y*-axis. As said above, aggregate supply curve shows the revenue or receipts *which must be received* by the entrepreneurs so as to provide employment to different numbers of workers, whereas aggregate demand curve shows proceeds or receipts which entrepreneurs *actually do expect* to receive at different levels of employment and production. These aggregate demand and the aggregate supply curves determine the level of employment in the economy. Given that the perfect competition prevails in the economy, then so long as opportunities to earn profits or make money exist, the entrepreneurs will increase the level of employment. The opportunities to make profits exist if aggregate demand price is greater than the aggregate supply price for a given number of employment. For example, in Fig. 4.1 at ON_1 number of persons employed aggregate demand price *OH* exceeds aggregate supply price *OC*. Therefore, it is profitable to offer employment to ON_1 workers. Therefore, so long as aggregate demand price exceeds aggregate supply price, the entrepreneurs will go on employing extra workers. When at a level of employment aggregate demand price becomes equal to aggregate supply price, then after this it will be no more profitable to employ workers. Since beyond this point aggregate supply price will exceed aggregate demand price, the cost of production incurred on employing a certain number of people will not be covered. Therefore, when aggregate demand price falls short of aggregate supply price, employment of labour will fall.

Equilibrium level of employment is determined by the intersection of aggregate demand curve and the aggregate supply curve, where the amount of money which the entrepreneurs actually expect to receive from employing a certain number of workers is equal to the amount of money which they must receive. In other words, *the employment of labour will be in equilibrium at the level at which aggregate demand price equals aggregate supply price*. It will be seen from Fig. 4.1 that aggregate supply curve and aggregate demand curve intersect at point *E* and therefore ON_2 level of employment is determined. It will be noticed that *at less than* ON_2 level of employment, aggregate demand curve *AD* lies above the aggregate supply curve *AS* showing that it is profitable to expand the amount of employment. However, beyond ON_2 amount of employment, the aggregate demand curve *AD* lies below aggregate supply curve *AS*, which shows that it is no more profitable to employ extra workers beyond ON_2. We, therefore, conclude that ON_2 is the equilibrium level of employment which will be determined by aggregate demand curve *AD* and aggregate supply curve *AS*.

Effective Demand and Determination of Employment

We are now in a position to explain more clearly what effective demand means and how it is important for determination of employment and output in the economy. We have seen that the magnitude of employment in the economy is determined by the equilibrium between aggregate demand and aggregate supply. There is an aggregate demand function for an economy which shows aggregate demand price at varying levels of employment.

But of these varying levels of employment the *aggregate demand which at a level of employment is also equal to the aggregate supply is called effective demand.* In other words, effective demand is that aggregate demand price which becomes '*effective*' because it is equal to aggregate supply price and thus represents a position of short-run equilibrium. There are several other points on the aggregate demand curve but what distinguishes effective demand from all these points is that at this point the aggregate demand price is equal to aggregate supply price. On all other points, aggregate demand price is either more or less than aggregate supply price.

It is thus clear that employment in the economy in the short run is determined by effective demand. The higher the level of effective demand, the greater is the volume of employment, and vice versa. Unemployment is due to the deficiency of effective demand and the basic remedy to remove this unemployment is to raise the level of the effective demand on the basis of Say's law. The classical economists believed that effective demand was always large enough to ensure full employment. But Keynes proved that it was not so and that is why the phenomenon of unemployment is common in free-market capitalist economies.

Underemployment Equilibrium – the Problem of Demand Deficiency

It is not necessary that the equilibrium level of employment is always at full employment. Equality between aggregate demand and aggregate supply does not necessarily indicate the full employment level. The economy can be in equilibrium at less than full employment or, in other words, an underemployment equilibrium can exist. The classical economists denied that there could be an equilibrium at less than full employment, because they believed in *Say's Law of Markets* that supply would always create its own demand and therefore no problem of deficiency of aggregate effective demand would be experienced. Keynes demolished the classical thesis of full employment both on theoretical grounds and on the basis of illustrations from real life.

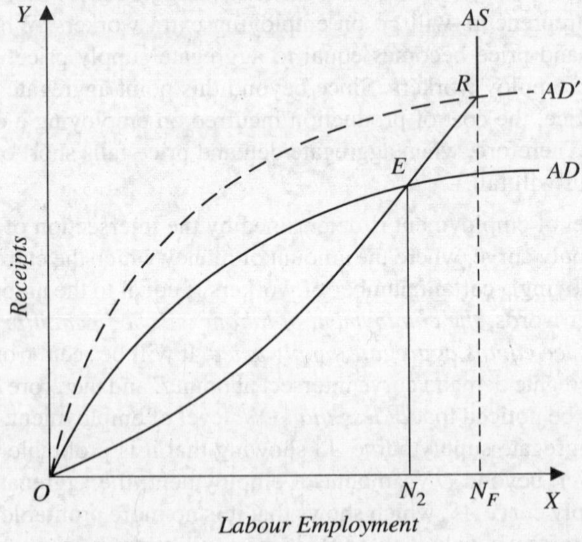

Fig. 4.2. *Raising Aggregate Demand Curve to Achieve Full Employment*

It will be seen from Fig. 4.2 that in the situation of equilibrium at the employment level ON_2, the N_2N_F persons remain unemployed. Thus equilibrium at *E* represents an underemployment equilibrium (or, in other words, less than full-employment equilibrium). It is important to note that N_2N_F

persons are *involuntarily unemployed*: they are willing to work at the existing wage rates but are unable to find jobs. It is important to remember that, according to Keynes, this ***unemployment is due to deficiency of aggregate demand.***

Now, an important question arises that when unemployment of labour emerges due to the fall in aggregate demand, why wage rate does not fall so that at lower wage rate there is more demand for labour resulting in expansion in employment. Why does wage rate not go on falling until all unemployed workers get employment and full employment again occurs. Keynes argued that for the reasons explained later in this chapter that *money wages are rigid downward.* Furthermore, according to Keynes, even if wages are reduced throughout in the economy, it will not help to restore full employment as wages constitute incomes of the working class and reduction in their money wages will cause a decline in aggregate demand for goods and services which will further worsen the employment situation rather than improving it.

This unemployment will be removed and full-employment equilibrium will be reached if through increase in investment demand or increase in consumption, or increase in both, aggregate demand curve shifts upward so that it intersects the aggregate supply curve at point R as depicted in Fig. 4.2. It will be seen that with the intersection of aggregate demand and aggregate supply curves at point R, equilibrium is established at full-employment level ON_F.

According to Keynes, *aggregate demand price and aggregate supply price will be equal at full employment only if investment demand is sufficient to cover the gap between the aggregate supply price corresponding to full-employment level and the consumption expenditure out of income at the full-employment level*. The view of Keynes is that *when investment demand falls short of this gap between full-employment income and consumption, recession occurs resulting in the emergence of involuntary unemployment*. According to Keynes, while propensity to consume is stable in the short run, it is fluctuations in investment demand that cause changes in aggregate demand which in turn cause changes in level of employment and income of the two determinants of investment, namely, rate of interest and marginal efficiency of capital (*i.e.*, expected rate of profits) rate of interest is sticky in the short run, it is changes in marginal efficiency of capital that brings about changes in rate of investment. When inducement to invest in capitalist countries declines due to the fall in marginal efficiency of capital (*i.e.*, expected rate of profit), investment demand falls so that equilibrium is established at less than full-employment level. This causes unemployment in the economy bringing about fall in level of output and income of the community. Keynesian unemployment which occurs due to the deficiency of aggregate demand brought about by the decline in investment demand is also called *cyclical unemployment*. This is because when boom conditions appear investment increases and pull up aggregate demand, both levels of employment and income increase and as a result unemployment disappears.

SUMMARY OF KEYNES'S THEORY OF EMPLOYMENT

After explaining Keynes's theory of employment at some length, we are now in a position to describe it in a summary form bringing out relationship between various elements or factors that go to determine the equilibrium level of employment.

1. Level of output or income of a country depends on the level of employment. Given the capital stock and technology, greater the employment of labour, higher the level of aggregate output or national income.
2. The level of employment depends on the magnitude of effective demand, which is the sum of consumption demand and investment demand, at the point where aggregate supply curve intersects the aggregate demand curve.

3. Aggregate supply of an economy depends on physical and technical conditions of production. Since these factors do not change much in the short run, aggregate supply curve remains constant in the short run. Aggregate supply curve slopes upward to the right as level of employment increases. This is because with the increase in labour employment, the greater cost has to be incurred.
4. Aggregate demand in a simple Keynesian model consists of consumption demand and investment demand. Since the consumption demand increases with the increase in labour employment, aggregate demand curve also slopes upward to the right. In Keynes's model, *investment demand is regarded as autonomous of changes in income or employment*.
5. Consumption demand depends on propensity to consume on the one hand and disposable income on the other. Propensity to consume of a community does not change much in the short run. Therefore, consumption function which relates consumption demand with the level of income remains stable in the short run.
6. Investment demand depends on the rate of interest and marginal efficiency of capital. According to Keynes, rate of interest is determined by supply of money and the state of liquidity preference. Marginal efficiency of capital (*i.e.*, expected rate of profit) depends on the expected future yields or *profit expectations* of entrepreneurs on the one hand and *replacement cost of capital* on the other.
7. According to Keynes, while rate of interest is more or less sticky it is frequent changes in profit expectations of the entrepreneurs, that is, changes in marginal efficiency of capital that cause a great deal of fluctuations in investment by entrepreneurs. Investment demand is thus highly volatile and causes recession or depression when it falls, and boom and prosperity when it increases significantly.

We summarise below the various determinants of employment and income (output) in a tabular form.

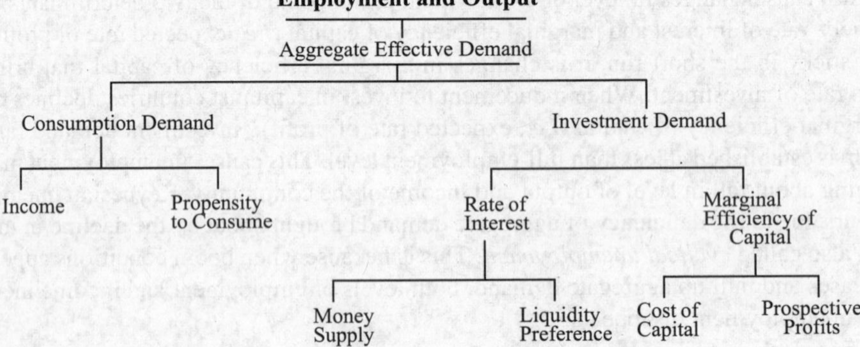

INVOLUNTARY UNEMPLOYMENT – KEYNES'S MONEY-WAGE RIGIDITY MODEL

According to Keynes, money-wage rigidity, that is, downward inflexibility of money wages results in involuntary unemployment of labour. The workers are rendered unemployed because at a given wage rate supply of labour exceeds demand for labour. Keynes believed that money wage would not change sufficiently in the short run to keep the economy at full employment. Classical economists believed that money wage rate is perfectly flexible and adjusts to bring demand for and supply of labour in equilibrium and keeps the economy at full employment level. To understand money-wage rigidity which results in unemployment we have to examine why labour market does not clear through reduction in money wages, Keynes gave three reasons for the stickiness of money wage rate. It may be noted that stickiness or rigidity of money wage implies that money wage rate will not quickly change, especially in the downward direction to keep equilibrium at full employment level.

Causes of Money Wage Rigidity

1. **Money Illusion.** The first reason why firms fail to cut wages despite an excess supply of labour is that workers will resist any move for cut in money wages though they might accept fall in real wages brought about by rise in prices of commodities. Keynes attributed this to money illusion on the part of the workers. *By money illusion it is meant that workers fail to realise that value of money, that is, its purchasing power in terms of commodities, changes when prices rise.* They regard money such as a rupee as something which has a stable value or purchasing power, that a rupee is a rupee and a dollar is a dollar with fixed real purchasing power. Therefore, while they would strongly oppose and resist any cut in money wages, they would not resist much if their real wages are reduced through rise in prices of commodities with money wages remaining constant. Thus Keynes wrote, "*Whilst workers will usually resist a reduction of money wages, it is not their practice to withdraw their labour whenever there is a rise in the price of wage goods.*"

There are two reasons for existence of money illusion.

(*i*) First reason for the existence of money illusion is that workers of a firm or industry think that though rise in prices reduces their real wages, but that this rise in prices equally affects workers in other industries so that their relative wages as compared to those employed in other industries remain the same. Therefore, workers who are more concerned with their relative position with other workers will strongly resist the cut in their money wages, while they will not oppose so strongly their cut in real wages through rise in the general price level.

(*ii*) The second reason for strong resistance to cut in money wages is that the workers blame their own employers for this, whereas they think that a cut in real wages through rise in prices in general is the outcome of the working of general economic forces over which strikes in an industry would have little effect. However, it does not necessarily mean that trade unions remain silent spectators if they feel that changes in Government policy adversely affect their economic interests.

From the above two reasons given for money illusion it follows that if additional employment can be created by lowering real wages, it is more practical to do so *through bringing about rise in general price level rather than by cutting money wages*.

2. **Wage Fixation through Contracts.** In most of the free market economies such as those of USA and Great Britain, wages are fixed by the firms through contracts made with the workers for a year or two. There is little possibility of changing money wages fixed through contracts when the situation of either labour surplus or shortage emerges. For workers organised into trade unions wages are even rigid. Through collective bargaining by trade unions with the employers wage scales are fixed for 3 to 4 years by contract. Money wages cannot be changed when either surplus or shortage of labour emerges during the period of the contract. Trade unions of workers never accept wage cuts even if some of union workers remain unemployed. Thus the sticky or rigid money wages lead to the existence of involuntary unemployment. This means that labour market does not clear in the short run.

3. **Minimum Wage Laws.** Another reason for money wage rigidity or, what is also called money-wage stickiness, is the intervention by the Government in fixing minimum wages below which employers are not permitted to pay wages to the workers.

4. **Efficiency Wages.** Another factor which accounts for money-wage rigidity is that employers themselves are not interested in lowering wages as high wages make workers more efficient and productive. The adverse effect of lower wages on workers' efficiency may explain the unwillingness on the part of employers to cut money wages despite the excess supply of or unemployment of workers at higher money wages.

We have explained above the practical difficulties pointed out by Keynes and his followers which are faced by firms in reducing wages and which therefore explain money-wage rigidity or stickiness. The sticky or rigid money wages above the equilibrium level cause unemployment of labour.

Price Flexibility and Rigid Money Wage: Keynes's View of Involuntary Unemployment

In Keynes's contractual view of labour market, it is assumed that whereas prices are free to vary, the money wage is fixed. It is important to note that Keynesians do not believe that money wage rate is completely fixed or sticky. What do they actually mean by sticky wages is that money wages do not fall quickly to bring demand for and supply of labour in equilibrium at full employment. In their view money wages are very slow to adjust sufficiently to ensure full employment of labour when there is a decline in aggregate demand resulting in lowering of prices of products. As a consequence, involuntarily unemployment comes into existence. It may be further noted that Keynes was particularly concerned with downward rigidity of money wages at which the demand for labour exceeds the supply of labour and consequently unemployment or excess supply of labour emerges. It is important to note that Keynes accepted the classical theory of labour demand according to which firms demand labour up to the point at which real wage rate (that is, money wage rate divided by the price level or, W/P) is equal to the marginal product of labour. At a higher real wage rate, less amount of labour will be demanded and, at a lower real wage rate, more labour will be demanded or employed. In other words, demand curve of labour is downward sloping. Keynes's theory of involuntary unemployment based on price flexibility and money wage rigidity is depicted in Fig. 4.3

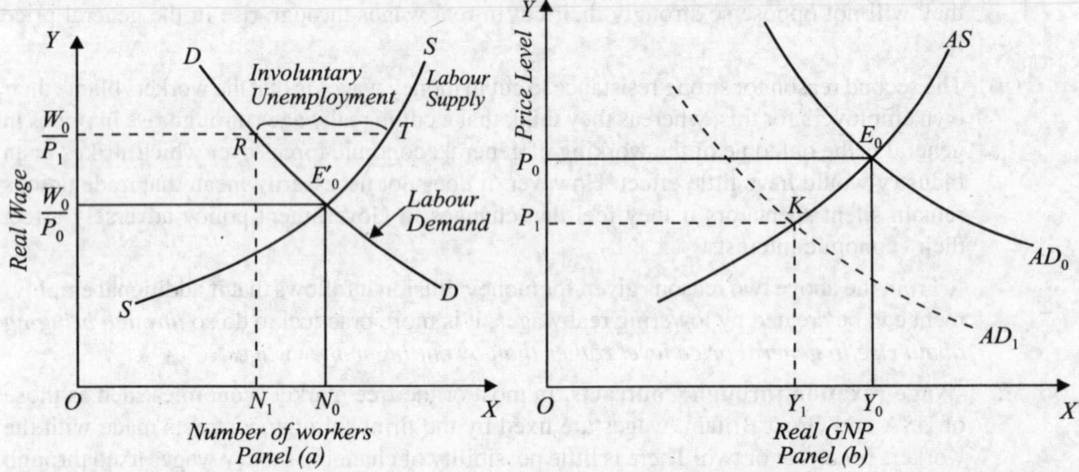

Fig. 4.3. *Keynes's Rigid Money Wage and Flexible Price Model: The Emergence of Involuntary Unemployment*

In panel (*b*) of Fig. 4.3 short-run aggregate supply curve *AS* and aggregate demand curve AD_0 have been drawn and through their interaction determine price level P_0 and the level of real *GNP* equal to Y_0.[1] It is important to note that short-run aggregate supply curve *AS* has been drawn with a given fixed money wage rate, say W_0. In panel (*a*) of Fig. 4.3 the level of labour employment N_0 shows the number of jobs when the economy is producing Y_0 level of national output in panel (*b*) corresponding to the equilibrium between aggregate supply *AS* and aggregate demand AD_0 at price level P_0, with a fixed money wage and the level of *GNP* equal to Y_0. The labour market must be in equilibrium at

1. Students should note that concepts of aggregate demand curve (*AD*) and aggregate supply curve (*AS*) used in this Fig. 4.3 are different from those used earlier in this chapter as they are based on price-flexibility assumption. Aggregate demand curve drawn in this Fig. 4.3, means the *total quantity of goods and services households, firms and Government are willing to buy at various price levels*. Similarly, aggregate supply curve (*AS*) here means the total output of goods and services that all firms in the economy are willing to produce and supply at different price levels. These concepts of aggregate demand and aggregate supply curves will be explained in a later chapter.

point E_0 or real image rate W_0/P_0 at which N_0 workers are demanded and employed. All those who are willing to get jobs at the real wage rate W_0/P_0 are in fact demanded and employed. Thus, equilibrium at E_0' or at level of employment N_0 represents full-employment equilibrium.

Now consider again panel (*b*) of Fig. 4.3. Suppose due to fall in marginal efficiency of capital there is reduction in investment demand which along with its multiplier effect causes a leftward shift in the aggregate demand curve *AD*. Since Keynes believed that with a fixed money wage rate aggregate supply curve *AS* is given and remains unchanged, it will be seen from panel (*b*) of Fig. 4.3 that new aggregate demand curve AD_1 and the fixed aggregate supply curve *AS* intersect at point *K* determining new equilibrium lower price P_1 and smaller real *GNP* equal to Y_1. Keynes asserted that the economy would remain stuck at point *K* with less than full-employment level of output Y_1 and lower price level P_1. Now, a glance at panel (*a*) of Fig. 4.3 shows that with fixed money image W_0 and lower price level P_1 ($P_1 < P_0$), the real wage rate rises to W_0/P_1. It will be seen from panel (*a*) of Fig. 4.3 that at this higher real wage rate W_0/P_1, the smaller amount of labour N_1 will be demanded and employed by all firms in the economy. However, at this higher wage rate W_0/P_1 (with money wage rate fixed at W_0), *RT* number of workers are rendered unemployed. It is in this way that Keynes explained that with money wage rate remaining fixed at the level W_0 and with flexible prices, the fall in aggregate demand results in persistent involuntary unemployment. Thus, by explaining the emergence of persistent involuntary unemployment Keynes made a fundamental departure from the classical view of a free market economy which denied the existence of involuntary unemployment except for a short time.

QUESTIONS FOR REVIEW

1. Explain Keynes's theory of employment. How does it differ from classical theory?
2. In what respects do the Classical and Keynesian theories of employment differ? Spell out the reasons for underemployment equilibrium in the Keynesian model.
3. Explain how, according to Keynes, *involuntary unemployment* is caused by lack of effective demand. How is the money wage-rigidity responsible for the emergence of involuntary unemployment?
4. The logical starting point of Keynes's theory of employment is the principle of effective demand. Explain.
5. What do you understand by the term 'money illusion' in the Keynesian theory of employment? Is wage-price rigidity a consequence of money illusion?
6. What are the policy implications of Keynesian theory of income and employment? Have they been successfully used to overcome recession?
7. Define effective demand. How does it determine level of employment in a country?

CHAPTER 5
DETERMINATION OF NATIONAL INCOME: KEYNES'S BASIC TWO-SECTOR MODEL

KEYNES'S INCOME-EXPENDITURE APPROACH

We have explained in the previous chapter how the level of employment is determined in an advanced capitalist economy. We explained there Keynesian theory with special reference to the determination of employment. It is worth noting here that the Keynesian theory is relevant in the context of the short run only since the stock of capital, techniques of production, efficiency of labour, the size of population, forms of business organisation have been assumed to remain constant in this theory. Further, in his model of income determination *Keynes assumed that price level in the economy remains unchanged.* Therefore, in the Keynesian theory which deals with the short run, the level of income of the country will change as a result of changes in the level of labour employment. Thus, in free market economy in the short run, when capital stock and technology remain unchanged, income is a function of labour employment. In fact, both income and employment go together. The higher the level of employment, the higher the level of income. As level of employment is determined by aggregate demand and aggregate supply, the level of income is also determined by aggregate demand and aggregate supply. In this chapter, we shall explain how the equilibrium level of national income is determined through *Keynes's income-expenditure analysis*. This analysis explains determination of national income by relating income (output) to aggregate expenditure on goods and services. The aggregate expenditure shows aggregate demand for goods and services. Keynesian theory of income determination can be explained by assuming two sectors in the economy, namely, households and business firms. Keynes focussed on this simple two-sector model of determination of national income and derived conclusions regarding policy formulation from this basic model. Analysis of determination of national income can be extended to incorporate Government and foreign trade. We start with the analysis of determination of national income by *taking a simple two-sector economy with a fixed price level.*

AGGREGATE EXPENDITURE (WITH A FIXED PRICE LEVEL)

Aggregate expenditure is the total expenditure which at given fixed prices all households and business firms want to make on goods and services in a period at various levels of national income. Though J.M. Keynes used the term *aggregate demand*, in modern macroeconomics, the term *aggregate expenditure* is generally used. The terms aggregate demand and aggregate supply are now generally used in the model with variable price level. In this chapter however we use the term aggregate demand and aggregate expenditure interchangeably but assume that price level remains constant. In a two-sector closed economy, aggregate expenditure or aggregate demand consists of two components: First, there is consumption demand, and secondly, there is a demand for capital goods, which is called investment demand. Thus, by aggregate expenditure, we mean how much expenditure the households and the entrepreneurs are willing to undertake on consumption and investment. Therefore,

Aggregate Demand = Consumption Demand + Investment Demand

$$AD = C + I$$

where AD stands for aggregate demand, C for consumption demand and I for investment demand.

Using the term aggregate expenditure instead of aggregate demand we have

Aggregate Expenditure = Consumption Expenditure + Investment Expenditure

or $$AE = C + I$$

Consumption Demand. As for consumption demand, it depends upon the propensity to consume of the community and the level of national income. Given the propensity to consume, as income increases, consumption demand will also increase. In other words, given the propensity to consume, consumption demand is a function of income. Consumption function can take several forms. The most common form of short-run consumption function is

$$C = a + bY$$

where a is the intercept term of the function and represents *autonomous consumption* whereas b represents the slope of the consumption function. According to Keynes's theory current consumption expenditure depends primarily on current income. Further, according to Keynes, the chief factor that determines consumption expenditure is disposable income, that is, income available after taxes. Increase in personal taxes reduces personal disposable income and therefore consumption expenditure.

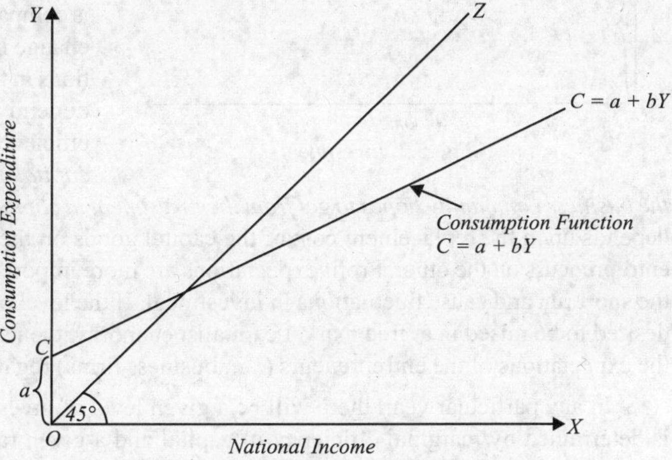

Fig. 5.1. *Consumption Function*

Consider Fig. 5.1 in which national income is measured along the X-axis and consumption demand (C) is shown on the Y-axis. In this figure, a straight line OZ which makes 45° angle with the X-axis has been drawn. This straight line OZ with 45° angle with the X-axis represents the *reference income line* to measure the difference between consumption and level of income. This is also often called *income line*. This 45° line represents national income in money terms. In fact, national product and income are the same things. In this figure a curve C has also been drawn which represents consumption function, $C = a + bY$ of the community. This curve of consumption function slopes upward from left to right, which shows that as income increases the amount of consumption demand also increases. As income line OZ makes 45° angle with the X-axis, the *gap between the consumption function curve C and the income line OZ represents the saving of the community*. The reason for this is that a part of the income is consumed and a part is saved, *i.e.*, National Income = Consumption + Saving. This is also written as $Y = C + S$, where Y represents income, C consumption and S saving. It will be seen from Fig. 5.1 that the gap between the consumption function curve C and the income line OZ goes on increasing as income increases. In other words, the amount of saving or saving gap increases as income increases.

It is worth mentioning that in the short run consumption function does not change. This is because the propensity to consume, that is, the whole consumption function curve C depends upon the tastes, preferences, the income distribution in the society, the population level, wealth of the people, etc., which do not undergo much change in the short run. The implication of the stability of consumption function is that the consumption demand is primarily determined by the level of current national income.

However, through changes in monetary policy and fiscal policy by the Government the consumption function can be shifted. For example, when rate of interest, an instrument of monetary policy, is reduced, the people borrow more for durable consumer goods such as cars, airconditioners, houses and with this at a given level of income consumption demand increases causing upward shift in the consumption function curve. Similarly, when income tax, an instrument of fiscal policy, is reduced disposable income of the households increases and as a result at a given level of national income (GDP), consumption demand rises leading to the upward shift in consumption function.

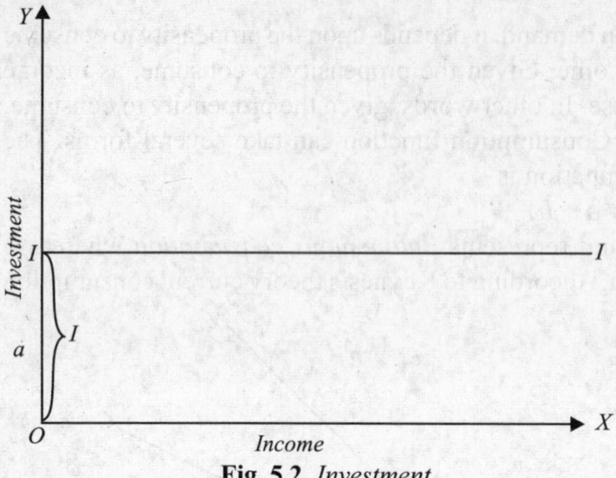

Fig. 5.2. *Investment*

Investment Demand. The other component of the aggregate demand is investment which is a crucial factor in the determination of equilibrium level of national income. Investment demand depends upon two factors: (1) marginal efficiency of capital and (2) rate of interest. Of these two factors, rate of interest is comparatively stable and does not frequently change in the short run. Therefore, the fluctuations in the level of investment demand chiefly depend upon the changes in the marginal efficiency of capital. *The marginal efficiency of capital means the expected rate of profit which the business community hopes to get from the investment in capital assets*. Marginal efficiency of capital depends upon the replacement cost of the capital goods on the one hand, and profit expectations of entrepreneurs on the other. Profit expectations are more important because they often change even in the short run and cause fluctuations in investment. If the level of national income and employment is desired to be raised in a free market capitalist economy, then steps should be taken which will raise the expectations of the entrepreneurs (*i.e.*, business firms) regarding profit-earning from investment.

In any particular year, there will be a given level of investment demand which, as seen above, is determined by marginal efficiency of capital and a given rate of interest. However, in Keynes's theory investment being determined by marginal efficiency of capital and rate of interest does not depend on the level of income. Thus, in Keynesian theory of income determination, investment does not vary with change in income. In other words, in *Keynes's income-expenditure analysis investment is treated as autonomous of income, that is, investment does not change with a change in the level of income*. This type of investment is shown in Fig. 5.2. This is the assumption which Keynes made and in this chapter we will also do likewise. In actual practice when the level of income rises, the demand for goods will also rise and this will favourably affect the expectations of the entrepreneurs regarding making of profits. Rise in the profit expectations will raise the marginal efficiency of capital which in turn will increase the level of investment. But it is quite clear that investment demand does not directly depend upon income; it is only affected indirectly by changes in income. Therefore, in our Fig. 5.3 we have taken a given amount of investment demand independent of the level of income and added it to upward sloping consumption function curve to get aggregate expenditure curve $C + I$. The distance between the C curve and the $C + I$ curve is parallel to the C curve throughout which indicates that the level of investment is constant and does not change with the change in income. It may however be noted that

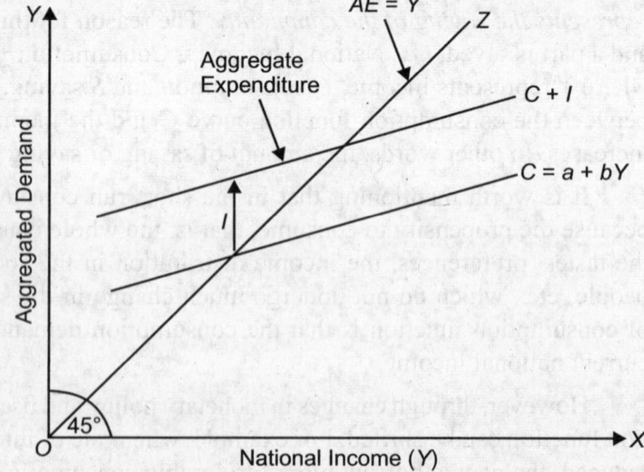

Fig. 5.3. *Aggregate Expenditure or Demand Curve*

with either a change in the rate of interest or marginal efficiency of capital investment will change. Therefore, in income-expenditure diagram as shown in Fig. 5.3, *a different new amount of investment* will have to be taken.

Aggregate Output

As mentioned above, in the short run the level of national income and employment in a free-market economy depends upon the equilibrium between aggregate expenditure and aggregate output. We have also explained above the various components of aggregate expenditure on goods and services. We now turn to explain the aggregate supply and factors on which it depends. It is important to note again that Keynes in his analysis assumed that prices and wages remain constant in the short run.

In an economy without any role of Government, national income *means the total money value of goods and services produced in an economy in a year*. There are two important constituents of aggregate output:

1. the supply or output of final consumer goods and services in a year; and
2. the output of capital goods which are also called producer goods because they help in producing further goods.

National income is the same thing as national product as both represent the value of output of final goods and services produced. In fact, aggregate supply or money value of national product of goods and services is distributed among the various factors of production as wages, rent, interest and profits as rewards for their contribution to the national product.

The aggregate output which is also sometimes referred to as aggregate supply of goods of an economy depends upon the stock of capital, the amount of labour used and the state of technology. In the short run, stock of capital remains constant and therefore output can be increased by increasing the amount of labour employed. Like the classical economists Keynes also thought that aggregate output or national income depends on the short-run production function with a given capital stock and constant technology. Thus, we have the following production function in the short run:

$$Y = (N, \overline{K}, \overline{T})$$

where Y is national output, $\overline{K}$ is the constant amount of capital stock, $\overline{T}$ is the constant state of technology and N is the labour employed which is a variable factor.

It may be noted that the amount of production which a given amount of labour can produce depends upon the given stock of capital and state of technology. The higher the level of technology, more can be produced with the help of given resources. Since J.M. Keynes in the nineteen thirties and we at present are concerned with the determination of income and employment *in the short run*, the stock of capital and the state of technology are assumed to be constant. For similar reasons, size of population is also assumed to be constant. However, the amount of labour employed out of this given size of population can vary depending upon the demand for labour.

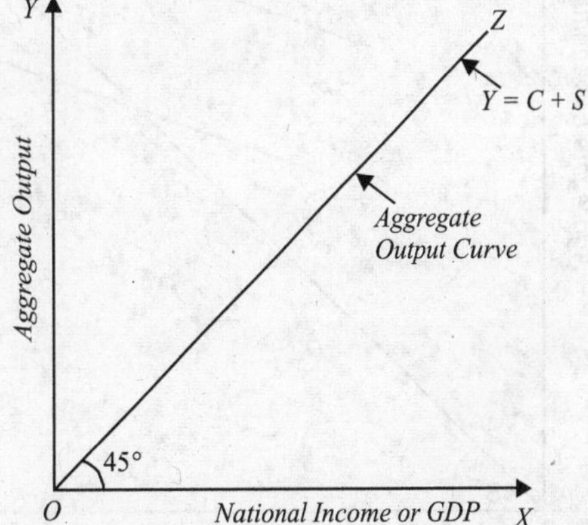

Fig. 5.4. *45° line Showing Aggregate Output*

45° Line as Aggregate Output Curve (With Fixed Prices)

In the income-expenditure analysis with which we are presently concerned we need to compare Gross Domestic Product (or National Income) with aggregate expenditure (*AE*), also called aggregate demand (*AD*), which is represented on the vertical axis. For this purpose we draw a 45° line from the origin which helps us to *transfer* Gross Domestic Product or real National Income (*i.e.* gross supply of output at constant prices) from the horizontal axis to the vertical axis for comparing it with total expenditure on goods and services. At every point of this 45° line from the origin, the vertical distance equals horizontal distance which measures real national income or GDP which is generally denoted by Y. This 45° line is also called income line along which $Y = C + S$ because income can either be consumed or saved.

The essential feature which emerges from all this is that more output will be produced and supplied at the given price level in response to increase in aggregate demand or expenditure. In other words, whatever the aggregate demand, more output of goods will be produced and supplied at the same price before full employment of resources is reached. This type of aggregate supply curve has been shown by 45° OZ line drawn against the X and Y axes in Fig. 5.4 where along the X-axis real national income (GDP) and along the Y-axis aggregate supply of output are measured. However, as noted above, aggregate output and real national income (GDP) are identical.

It follows from above that 45° line shows two things. First, it shows *varying* levels of aggregate production or the supply of goods (both consumer and capital goods) that will be offered for sale at the given price level at various levels of aggregate expenditure. This shows that up to the level of full employment of resources *any amount* of aggregate supply of output will be forthcoming at the given price level depending on the aggregate demand or expenditure. The greater the aggregate demand or expenditure, the greater the aggregate supply of output at the given price level. Secondly, it represents national income. In fact, as you would have read in national income accounting, national product and national income are the same things.

EQUILIBRIUM LEVEL OF NATIONAL INCOME

Now, we shall explain how through the intersection of aggregate demand and aggregate supply curves the equilibrium level of national income is determined in Keynes's two-sector model. $C + I$ curve represents the aggregate expenditure and 45° OZ line represents aggregate supply of output.

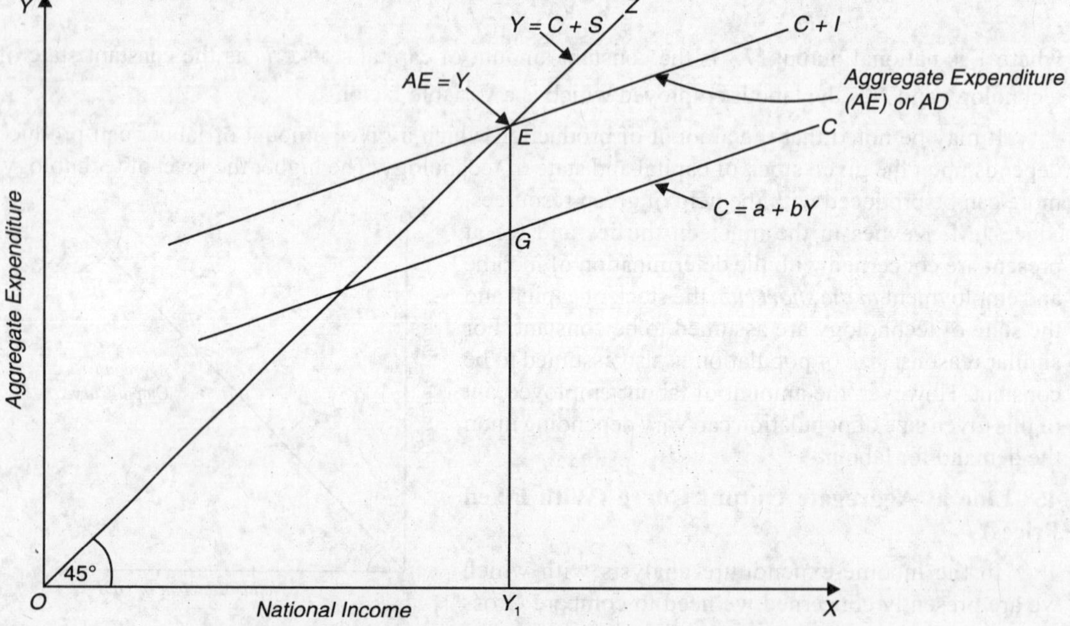

Fig. 5.5. *Determination of National Income : Basic Keynesian Model*

Normally the goods and services are produced by firms when they think they can sell them in the market. There will be equilibrium in the goods market when total output of goods and services produced will be equal to the total demand for output. Aggregate demand for them is represented by aggregate expenditure. In equilbrium aggregate expenditure (which is denoted by *AE*) must equal aggregate output (GDP). Since aggregate output or GDP equals national income (*Y*) we have the following condition for equilibrium:

$$AE = GDP = Y$$

It will be seen in Fig. 5.5 that aggregate expenditure curve (*AE*) or $C + I$ curve intersects 45° line at point *E* which satisfies the equilibrium condition. That is, at point *E* which corresponds to the income level OY_1, aggregate expenditure is equal to aggregate output. Therefore, *E* is the equilibrium point and OY_1 represents the equilibrium level of national income. Now, income cannot be in equilibrium at levels smaller than OY_1, since aggregate expenditure exceeds aggregate supply of output as $C + I$ curve which depicts aggregate expenditure of output lies above 45° line. This excess demand will be met by the firms selling goods from their stocks or inventories of goods kept by them. This leads to the decline in inventories of goods below the desired levels. This unintended fall in inventories will induce the firms to expand their output of goods and services to meet the extra demand for them and keep their inventories of goods at the desired levels. Thus when at a given level of national income, aggregate expenditure (*i.e.*, aggregate demand) exceeds aggregate output, national income will increase. With this increase in national income or output, employment of labour will also rise to produce the increment in output. This process of expansion in output under the pressure of excess demand will continue till national income OY_1 is reached.

On the contrary, the equilibrium level of national income cannot be greater than OY_1 because at any level greater than OY_1 aggregate expenditure or demand ($C + I$) falls short of aggregate output. This will cause the increase in inventories of goods with the firms beyond the desired levels. To this situation of the unintended increase in inventories of goods, the firms will respond by cutting down production to keep their inventories at the desired levels. Thus, deficiency in aggregate demand relative to the aggregate supply of output will lead to the fall in national income and output until the level OY_1 is reached where aggregate expenditure ($C + I$) is equal to the value of aggregate output. Thus, OY_1 is the equilibrium level of national income.

PRINCIPLE OF EFFECTIVE DEMAND

We have seen above that the equilibrium level of national income is determined by aggregate demand and aggregate supply of output. Consider Fig. 5.5 where it will be seen that, the prices remaining constant, aggregate expenditure (*AE*) curve $C + I$ *shows varying levels of aggregate demand at various levels of national income. The particular aggregate demand which is equal to aggregate output and therefore determines the equilibrium national income is called effective demand.* In other words, effective demand is that level of aggregate demand (aggregate expenditure) which becomes effective in determining equilibrium level of income because it is equal to aggregate supply of output. This is called *Keynesian principle of effective demand.* In Fig. 5.5, the effective demand is equal to Y_1E. Note that the level of national income OY_1, which has been determined, equals the effective demand Y_1E ($OY_1 = Y_1E$).

There are several other points on the aggregate demand (expenditure) curve but what distinguishes effective demand from all these points is that at this point aggregate demand is equal to aggregate output. On all other points aggregate demand is either more or less than aggregate output. Thus, the level of national income is determined by and equal to effective demand. In a two-sector Keynesian model, we can express the principle of effective demand in symbolic terms as under:

$$Y = AE^* \text{ or } AD^*$$

$$AD^* = C + I$$

$$Y = AD^* = C + I$$

where Y = National Income
AD^* = Effective demand
C = Consumption demand
I = Investment

Note that *star mark* on *AE* or *AD* shows the aggregate demand which becomes effective.

Importance. It is thus clear that national income and employment in the short run are determined by effective demand. The higher the level of effective demand, the greater are the levels of national income and employment and *vice versa*. *Unemployment is due to deficiency of effective demand and the basic remedy to remove this unemployment is to raise level of effective demand.* Due to their belief in Say's Law and in quick adjustment of wages and prices to balance between demand and supply the classical economists believed that effective demand was always large enough to ensure full employment. But Keynes proved that it was not so and that is why the problem of involuntary unemployment was common in the free-market capitalist economies.

EQUILIBRIUM IS NOT NECESSARILY ESTABLISHED AT FULL-EMPLOYMENT LEVEL

The most important contention of J.M. Keynes in his now famous work "General Theory of Employment, Interest and Money" was that in a capitalist economy it is not necessary that the equilibrium level of income is established at the level of full employment. This view of Keynes was quite contrary to the views of the classical economists who believed that the economy always tended towards full employment. Keynes refuted this important idea of the classical economists. To understand the Keynesian viewpoint consider Fig. 5.6. Suppose the full-employment level of labour corresponds to the level of national income equal to OY_F. Thus OY_F represents the level of income at which there is full employment in the economy. But, as seen above, the equilibrium level of income is not established at OY_F but at OY_1 at which some number of labour will be unemployed. Thus equilibrium at OY_1 is known as *under-employment equilibrium*. It is worth noting that in the situation depicted in Fig. 5.6 the equilibrium level of national income will be established at full-employment level only when investment demand is sufficiently large to fill the saving gap between income and

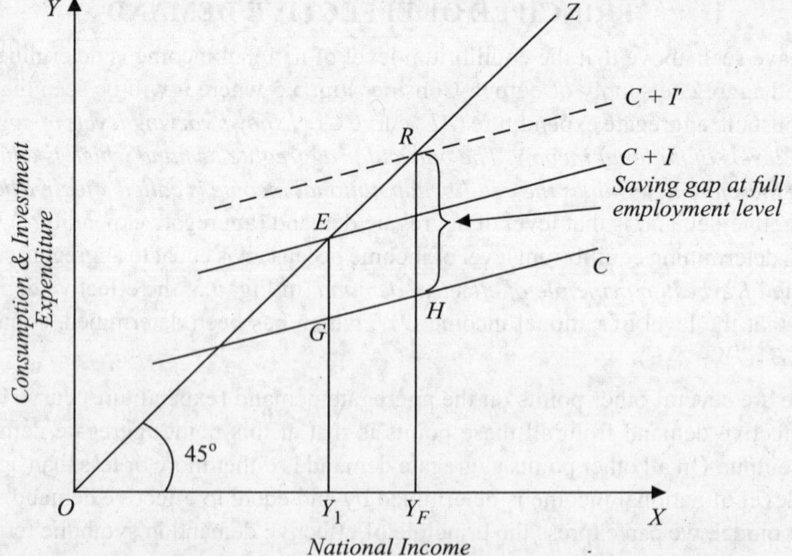

Fig. 5.6. *Under-employment Equilibrium due to Insufficiency of Investment Expenditure*

consumption at the full-employment level of income OY_F. It will be seen from Fig. 5.6 that corresponding to the full-employment level of income OY_F, RH is the saving of the community. Therefore, if the equilibrium is to be established at OY_F level, then investment demand must be equal to level of saving RH at full-employment level of income Y_F. But the prevailing investment demand is only equal to EG which is less than RH. With RH as the level of investment and the C curve as the level of propensity to consume, $C + I$ curve intersects the 45° aggregate output curve OZ at point R and therefore determines equilibrium level of income equal to OY_F at which labour will be fully employed.

But there is no guarantee that the investment demand will be equal to the saving gap corresponding to full-employment level of income. There are various reasons for this. First, the people who save are not necessarily those who make investment; while it is the general public which save but only few people constituting the entrepreneurial class undertake investment. Secondly, the factors which determine saving are quite different from those which determine investment. While the people save for providing for the education, marriages of their children and also for contingency purposes such as diseases and unemployment periods. They also save in order to acquire durable goods such as houses, gold and jewellery. They also save to accumulate sufficient funds for their old age. But the level of investment depends upon marginal efficiency of capital and rate of interest in the short run and on the level of population, technological progress in the long run. Therefore, it is not essential that investment should be equal to the amount of savings by the people at full-employment level. If due to the adverse changes in the profit expectations of the entrepreneurs level of investment falls, the equilibrium level of national income will also decline.

It is also worth noting that if the aggregate expenditure income is beyond the $C + I'$ in Fig. 5.6 it will mean that the economy is spending and demanding more than it can produce. The result of this will be rise in prices of the commodities. In other words, if the investment demand in any year is higher than HR, then the inflationary process will start. In this situation, national income will rise beyond OY_F only in money terms but not in real terms.

DETERMINATION OF EQUILIBRIUM LEVEL OF NATIONAL INCOME: ALGEBRAIC ANALYSIS

A study of how the level of national income is determined will become more clear by using simple mathematics. As has been explained above, the level of national income is in equilibrium at which aggregate demand equals aggregate supply of output. In a simple model of income determination in which we do not consider the impact of Government expenditure and taxation and also exports and imports, the national income is the sum of consumption demand (C) and investment demand (I), that is,

$$Y = C + I$$

where Y stands for the level of national income.

Suppose the consumption function is of the following form :

$$C = a + bY$$

a is the intercept term in the consumption function and therefore represents the autonomous consumption expenditure which does not very with income. b is a constant which represents the marginal propensity to consume (mpc $\Delta C/\Delta Y$). Thus total consumption demand is equal to the sum of autonomous consumption expenditure (a) and the induced consumption expenditure (bY).

Now suppose that investment demand equals I_a. This investment I_a is autonomous because this does not depend on income. Thus, we get the following three equations for the determination of the equilibrium level of national income:

$$Y = C + I \qquad ...(i)$$

$$C = a + bY \qquad ...(ii)$$

$$I = I_a \qquad ...(iii)$$

Substituting the values of C and I in equation (*i*) we have

$$Y = a + bY + I_a \quad ...(iv)$$

$$Y - bY = a + I_a$$

$$Y(1-b) = a + I_a$$

$$Y = \frac{1}{1-b}(a+I_a)$$

The equation (*v*) shows the equilibrium level of national income when aggregate demand equals aggregate supply. *The equation (v) reveals that autonomous consumption and autonomous investment* $(a + I_a)$ *generate so much expenditure or aggregate demand which is equal to the income generated by the production of goods and services*. From the equation (*v*) it also follows that the equilibrium level of national income can be known from multiplying the elements of autonomous expenditure (that is, $a + I_a$) by the term $\frac{1}{1-b}$ which is equal to the value of multiplier.

The multiplier tells us how much increase in income occurs when *autonomous investment* (*or consumption*) increases by ₹ 1, that is, multiplier is $\frac{\Delta Y}{\Delta I}$ and its value is equal to $\frac{1}{1-b}$ where b stands for marginal propensity to consume (*MPC*). Thus, multiplier $= \frac{\Delta Y}{\Delta I} = \frac{1}{1-b}$. Further, it also follows that equilibrium level of income is higher, the greater the marginal propensity to consume (*i.e. b*) and autonomous investment (I).

Now, higher the marginal propensity to consume (b), the greater the value of multiplier. For example, if marginal propensity to consume (b) is 0.8, investment multiplier is

$$\frac{\Delta Y}{\Delta I} = \frac{1}{1-08} = \frac{1}{0.2} = 1 \times \frac{10}{2} = 5$$

If *MPC* or $b = 0.75$, multiplier is

$$= \frac{\Delta Y}{\Delta I} = \frac{1}{1-0.75} = \frac{1}{0.25} = \frac{100}{25} = 4$$

We can find out the increase in income (ΔY) resulting from a certain increase in investment (ΔI) by using this multiplier relationship. Thus

$$\frac{\Delta Y}{\Delta I} = \frac{I}{1-b}$$

$$\Delta Y = \Delta I . \frac{1}{1-b}$$

If marginal propensity to consume is equal to 0.8, with the increase in investment by ₹ 100 crore the increase in income will be :

$$\frac{\Delta Y}{\Delta I} = \frac{1}{1-b}$$

$$\Delta Y = \Delta I \times \frac{1}{1-b} = 100 \times \frac{1}{1-0.8}$$

$$100 \times \frac{1}{0.2} = 100 \times 5 = 500 \text{ crore}$$

Illustrations of Equilibrium Level of Income with Numerical Examples

A few numerical examples will make it clear how the equilibrium level of national income is determined.

Problem 1. *Suppose in an economy, autonomous investment (I) is ₹ 600 crore and the following consumption function is given :*

$$C = 200 + 0.8Y$$

Given the above, find out the equilibrium level of income.

Solution. The equilibrium level of income is

$$Y = C + I$$

$$C = 200 + 0.8Y \quad \ldots (i)$$

$$I = 600$$

Putting the values of C and I in the equilibrium (i) we have

$$Y = 200 + 0.8Y + 600$$

$$(Y - 0.8Y) = 200 + 600$$

$$Y(1 - 0.8) = 800$$

or $$Y = \frac{800}{0.2} = \frac{800}{\frac{2}{10}} = ₹\ \mathbf{4000}$$

Problem 2. *Suppose the consumption function of an economy is C = 0.8 Y. Planned investment by entrepreneurs for a year is ₹500 crore. Find out what will be the equilibrium level of income.*

Solution. $$Y = C + I$$

$$C = 0.8Y$$

$$I = \text{Rs. } 500 \text{ crore} \quad \ldots (i)$$

Substituting the values of C and I in (i) we have

$$Y = 0.8Y + 500$$

$$Y - 0.8Y = 500$$

$$Y(1 - 0.8) = 500$$

$$0.2 = 500$$

$$y = 500 \times \frac{10}{2} = ₹\ \mathbf{2500\ crore}$$

Problem 3. *Suppose the consumption of an economy is given by*

$$C = 20 + 0.6Y$$

The following investment function is given :

$$I = 10 + 02\ Y$$

What will be the equilibrium level of national income ?

Solution. Note that in this problem, investment varies with income. However, this will not change our method of determining equilibrium level of income.

$$Y = C + I$$

$$C = 20 + 0.6Y \quad \ldots (i)$$

$$I = 10 + 0.2Y$$

Substituting the values of C and I in (i) we have

$$Y = 20 + 0.6Y + 10 + 0.2Y$$

$$Y = 30 + 0.8Y$$

$$Y - 0.8Y = 30$$

$$Y(1 - 0.8Y) = 30$$

$$0.2Y = 30$$

$$Y = 30 \times \frac{10}{2} = 150$$

Thus, we find the equilibrium level of income to be equal to 150.

HOW TO OVERCOME RECESSION: SHIFTING AGGREGATE EXPENDITURE CURVE UPWARD

Now, an important question is what measures can be adopted to overcome recession or involuntary unemployment which comes into existence as a result of deficiency of aggregate demand brought about by fall in investment.

From our above analysis it is clear that increase in equilibrium level of national income can be brought about by increases in any components of aggregate demand, namely, consumption demand (C), private investment demand (I), Government expenditure (G), and net exports (NX). In order to increase national income through upward shift in the consumption function, *the Government can reduce personal income taxes*. In 1964, the reduction in income tax by John Kennedy Government in USA was quite successful in boosting consumption demand and thereby raising aggregate output. As a result, more income and employment were generated. The rate of unemployment sharply declined and the American economy was lifted out of depression. Recently, in 2002 and again in early 2008, President George W. Bush made a significant cut of 3.5 billion dollars in income tax to revive the American economy. He refunded good amount of income tax collected from the people in these years. The cut in direct taxes increases the disposable income of the people which tends to raise their consumption demand. However, there is a limitation of cut in direct taxes for raising aggregate demand because a part of the increase in disposbale income can be saved rather than being spent on goods and services. This happened in the United States in the year 2008. To boost aggregate demand, indirect tax can also be reduced. In India in December 2008, to correct the sagging demand for industrial products under the fiscal stimulus package 4 per cent across the board cut in central excise duty was made. This too has a limitation because benefit of this cut in excise duty may not be fully passed on to the consumers by reducing prices.

Secondly, the equilibrium level of national income (GNP) and employment can be increased by *raising the rate of private investment* (I). Through monetary policy measures businessmen can be induced to invest more by lowering the rate of interest and increasing the availability of credit. We know that lower the rate of interest, the higher will be the level of private investment. Alternatively, the Government may encourage private investment by reducing tax on profits so that post-tax rate of profit will be higher than before. The higher level of investment will shift the aggregate demand curve ($C + I + G$) upward and determine a higher level of national income and employment.

Thirdly, the recession can be overcome and national income (GNP) and employment can more efficiently be increased by *increasing Government expenditure* (G). It was the increase in Government expenditure on Public Works Programme which was the main recommendation of Keynes to raise the level of national output and income to restore equilibrium at full-employment level. Recently in 1993-94, President Clinton stepped up public expenditure on public works in the USA to overcome recession in the American economy and reduce unemployment. Now in 2009 President Obama has planned to give a fiscal stimulus of over $ 800 billion the major part of which is the increase in Government expenditure to raise aggregate demand to overcome recession in the American economy. In India also fiscal stimulus packages were announced in 2008-09 under which government expenditure, especially on infrastructure, was increased to revive slowing Indian economy.

Lastly, the *expansion in positive net exports* (NX) will also cause an increase in equilibrium level of national income and employment. Exports can be promoted through tax concession on profits earned through exports as well as making credit available at lower interest rate for purpose of exporting goods. Another common measure adopted to promote exports is the depreciation of national currency. Depreciation makes the exports cheaper and imports costlier which raise net exports (NX_n). However, when there is global recession depreciation of its currency by each country will offset the effect of others.

It follows from above that a mix of fiscal and monetary policy measures are adopted to increase aggregate demand and thereby to lift the economy out of recession.

DETERMINATION OF NATIONAL INCOME: SAVING-INVESTMENT APPROACH

We have seen how equilibrium level of national income is determined by the interaction of aggregate demand and aggregate supply. The equilibrium level of national income is established at the point where aggregate demand equals aggregate supply. But there is an alternative method for the explanation of the determination of national income. This alternative method explains the determination of national income directly by intended saving and investment.

Look at Fig. 5.5 again. In this figure, at the equilibrium level of national income OY_1, saving and investment are equal to GE. Given the aggregate demand curve $C + I$, the amount of saving at income greater than OY_1 exceeds investment and for income less than OY_1 investment exceeds saving. It is obvious that intended saving and investment are equal only at the equilibrium level of national income and when intended saving and investment are not equal, national income will not be in equilibrium. Let us see why it is so and how national income is determined by intended saving and investment.

When at a certain level of national income, intended investment by the entrepreneurs is more than intended saving by the people, this would mean that aggregate expenditure is greater than aggregate supply of output. This will lead to decline in inventories below the desired level. This would induce the firms to increase production, raising the level of income and employment. The result will be that national output will be increased on account of which national income will go up. Further, when at any level of income investment is less than saving, it means that aggregate demand is less than aggregate supply. As a result, the entrepreneurs will not be able to sell their entire output at given prices. The result will be that output will be reduced which will result in the reduction of national income.

Thus, when, at any level of national income, investment demand of the entrepreneurs is less than the intended savings of the people, the national income will decrease. It will come down to the level at which investment spending is just equal to the planned savings by the community. But, when at any level of national income intended investment demand on the part of the enterpreneurs is equal to intended savings of the people, it means that aggregate demand is equal to the total output or aggregate supply and therefore the national income will be in equilibrium. Hence, equilibrium level of national income will be determined at the level at which the amount of intended investment by the entrepreneurs is equal to the amount of intended savings by the people.

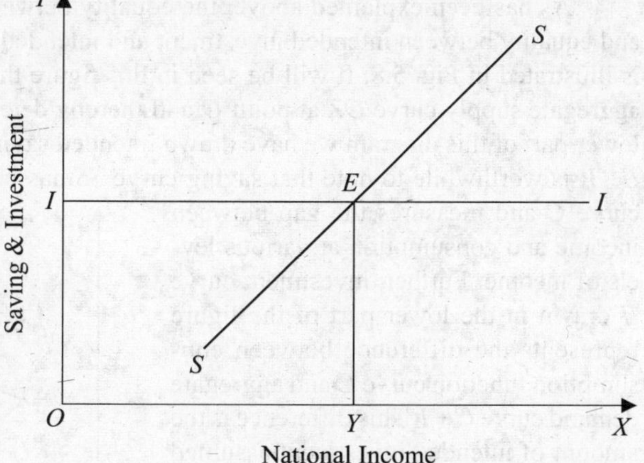

Fig. 5.7. *Determination of National Income: Saving-Investment Approach*

We can explain the determination of national income through saving and investment in another way. *Saving represents withdrawal of some money from the income stream. On the other hand, investment represents injection of money into the income stream.* Now, if intended investment is greater than intended saving, it means that more money has been put into the income stream than has been taken out of it. As a result, income stream, *i.e.*, flow of national income would expand. On the contrary, if investment is less than intended saving, it means that less amount of money has been put into the income stream than has been taken out of it. The result would be that national income would decrease. But when investment is just equal to saving, it would mean that as much money has been put into the income stream as has been taken out of it. The result will be that the national income will neither increase nor decrease, *i.e.*, it will be in equilibrium. It is thus clear that the equilibrium level of national income will be determined at the level at which the intended investment is equal to intended saving.

The determination of national income by investment and saving is illustrated in Fig. 5.7. In this figure, national income is shown along the *X*-axis. *SS* is the saving curve which shows intended saving at different levels of income, *II* curve shows investment demand *i.e.*, intended investment. The *II* investment curve has been drawn parallel to the *X*-axis. This is done on the assumption that in any year the entrepreneurs intend to invest a certain amount of money. This implies that we assume that investment is independent of change in income, that is, investment does not change with income.

The saving curve *SS* and the investment curve *II* intersect each other at *E*. That is, the intended investment and intended saving are equal at the *OY* level of income. Hence, *OY* is the equilibrium level of income. It will be seen from Fig. 5.7 that at the level of income less than *OY*, the amount of intended investment is more than intended savings. As a result, income will increase. On the contrary, at the level of income greater than *OY*, the amount of intended investment is less than the intended savings with the result that income will decrease. The decrease in income will continue until it becomes equal to *OY*. At the level of *OY* income, intended investment and intended saving are equal, so that there is neither the tendency for income to increase, nor to decrease. Hence, national income *OY* is determined. It is thus clear that national income is determined by the investment and saving.

Synthesis between the Two Approaches

The determination of national income has been explained above by two methods. The equilibrium level of national income is determined where two conditions are fulfilled :

(*i*) *Aggregate Expenditure or Demand = Aggregate Supply of Output*, and

(*ii*) *Intended Investment = Intended Saving*.

As has been explained above, the equality between aggregate demand and aggregate supply and equality between intended investment and intended saving in reality mean the same thing. This is illustrated in Fig. 5.8. It will be seen in this figure that aggregate demand curve *C* + *I* intersects aggregate supply curve *OZ* at point *Q* and thereby determines national income equal to *OY*. In the lower part of this diagram we have drawn intended savings curve *SS* and intended investment curve *II*. It is worthwhile to note that saving curve *SS* has been derived from the consumption function curve *C* and measures the gap between income and consumption at various levels of income. Further, investment curve *II* drawn in the lower part of the figure represents the difference between consumption function curve *C* and aggregate demand curve *C* + *I*. This difference is the amount of intended investment assumed in this figure. We thus see that both saving and investment curves drawn in the lower part are embedded in the upper part showing aggregate demand and aggregate supply. It is because of this that intended saving and intended investment curves also determine the same level of national income *OY* which is determined through the equality of aggregate demand (*C* + *I*) and aggregate supply.

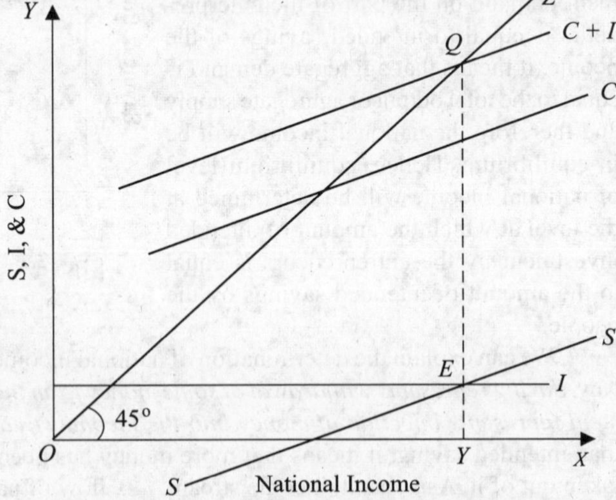

Fig. 5.8. Reconciliation between Income Expenditure Approach and Saving-Investment Approach

Saving-Investment Approach : Algebraic Analysis

That the level of national income is determined by the equality of planned saving and planned investment can be derived from the equilibrium condition explained above that level of income is equal to effective demand.

That is,

$$Y = AD^* \quad \text{or} \quad AE^* \qquad \ldots (i)$$

where AD^* represents effective aggregate demand

Now in our simple economy, effective aggregate demand (AD^*) is equal to the sum of consumption expenditure and investment expenditure. Thus

$$AE^* \text{ or } AD^* = C + I$$

or $$Y = C + I \qquad \ldots (ii)$$

Now $$C = Y - S$$

Substituting $Y - S$ for C in equation (*ii*) we have

$$Y = Y - S + I$$

$$Y - Y + S = I$$

or $$S = I \qquad \ldots (iii)$$

Thus, we reach at the alternative condition of equilibrium of national income, namely, planned (intended) saving be equal to planned (intended) investment.

We can further extend saving–investment approach to show the determination of national income as a multiple of autonomous factors. Since investment is treated as autonomous of national income, its amount is taken to be a fixed amount determined *exogenously*. Therefore we have

$$I = \overline{I}$$

Saving function as derived from the consumption function ($C = a + bY$) can be written as**

$$S = -a + (1 - b)\,Y$$

In equilibrium

$$\overline{I} = -a + (1 - b)Y$$

$$(1 - b)Y = \overline{I} + a$$

$$Y = \frac{1}{1-b}(\overline{I} + a) \qquad \ldots (iv)$$

where $\frac{1}{1-b}$ is the value of multiplier and b = marginal propensity to consume and therefore $1 - b$ = marginal propensity to save. The equation (*iv*) describes the same condition which we derived earlier. Note that a in equation (*iv*) is autonomous consumption which is intercept term in the consumption function.

Numerical Problem on Saving-Investment Approach

Problem 4. *Suppose the level of autonomous investment in an economy is ₹200 crore. The following saving function is given* :

$S = -80 + 0.25\,Y$

Find the equilibrium level of income.

Solution. According to the saving-investment approach, the level of income is in equilibrium at which

$$S = I \qquad \ldots (i)$$

Given, $$S = -80 + 0.25Y, \text{ and}$$

** This can be derived as follows

Since $C = Y - S$, consumption function can be written as

$Y - S = a + bY$

$S => -a - bY + Y$ or $S = -a + (1 - b)Y$

$$I = ₹\ 200 \text{ crore}$$

Substituting the values of S and I in the equilibrium equation we have

$$-80 + 0.25Y = 200$$

$$0.25Y = 200 + 80 = 280$$

$$Y = 280 \times \frac{100}{25} = 1120 \text{ crore}$$

Note that we can also use the equation (*iv*) derived above to obtain the equilibrium level of national income. Thus

$$Y = \frac{1}{1-b} = (I + a)$$

Now, $b = 1 - 0.25 = 0.75$, $a = 80$ and $I = 200$

Thus

$$Y = \frac{1}{1-0.75} = (200+80)$$

$$= \frac{1}{0.25} \times 280 = 1120 \text{ crore}$$

WHAT CAUSES DEPRESSION OR RECESSION: KEYNES'S VIEW

During 1929-33, capitalist economies of the world found themselves in the grip of depression. Such was the severity of this depression that in the United States, the rate of unemployment rose from 3.2 per cent in 1929 to 25% of the labour force in 1933. Besides, there was a drastic decline in Gross National Product (GNP) which fell from 315 billion dollars in 1929 to 222 billion dollars in 1933, that is, national income declined by 30 per cent during these four years period. This reflects a dismal picture of the American economy during the period of Great Depression. In England as well as in other European countries also such a grave situation of severe recession and huge unemployment prevailed during this period. Before 1929-33 and even after it, recessions have occurred in these economies but they have not been as severe as that took place during 1929-33.

Classical economists had no valid explanation of such a severe depression and large-scale cyclical unemployment of labour. A.C. Pigou and other economists of his view attributed this situation to the high wage rates kept by trade unions and Government. Pigou, therefore, suggested all-round cut in wages to increase employment and to remove depression and unemployment. However, this solution of the problem was neither logically sound nor practical to be implemented. Keynes challenged this view of the classical economists and put forward a different explanation of depression and cyclical unemployment which was accepted by many as logical and correct. He explained his viewpoint in his now noted work, '*General Theory of Employment, Interest, and Money*'. He not only gave a sound and valid explanation of depression and its associated problem of cyclical unemployment but also suggested effective policy measures to cure them.

According to Keynes, the cause of depression and cyclical unemployment in the industrialised capitalist countries was sharp decline in private investment due to the adverse business expectations about profit making. There was a wave of pessimism prevailing among investors. The decline in private investment due to fall in marginal efficiency of capital (that is, expected rate of return) caused a fall in aggregate demand and resulted in less than full-employment equilibrium. Consequently, level of output and employment fell drastically and involuntary unemployment came to prevail on a large scale. The involuntary unemployment that prevails in times of recession/depression is called cyclical unemployment and, as we have seen above, according to Keynes, it is due to deficiency of aggregate demand.

Large fluctuations in investment, according to Keynes, is due to the uncertain basis of profit expectations on which investment decisions are made. To quote him, "*We have to admit that our basis of knowledge for estimating the yield ten years hence of a Railway, a copper mine, a textile factory, the goodwill of a patent medicine ... amounts to little and sometimes to nothing.*"[1] In view of this uncertainty of future. Keynes stressed that investment decisions were greatly influenced by how optimistic or pessimistic investors feel. He used the term *animal spirits* to describe these pessimistic or optimistic expectations of the investors about profit earning from investment projects. The term animal spirits implies that there may be no good or intelligent basis for expectations on which investors base their decisions. Thus the cause of depression or cyclical involuntary unemployment is the deficiency of aggregate demand due to fall in private investment that results from adverse business expectations and loss of confidence of the investor in a market economy.

It will be useful to explain through a diagram how a fall in investment causes a decline in level of output and employment and results in depression. This is illustrated in Fig. 5.9. It will be seen from Fig. 5.9 (*a*) that due to adverse profit expectations or pessimism of investors, investment demand curve shifts to the left from *II* to *I′I′*. With this it will be seen that investment falls from I_1 to I_0 at a given rate of interest. This fall in investment demand by I_1I_0 causes a downward shift in the aggregate expenditure curve from *AE* to AE_0 (See Fig. 5.9 (*b*). As a result, the equilibrium between aggregate expenditure and aggregate output which was initially at full employment level Y_F (which corresponds to point E_1 where the aggregate expenditure curve AE_1 intersects 45° line) falls to the new equilibrium level of income OY_0. Therefore, the *cause of depression or recession is deficiency of aggregate demand.* It should also be noted that the decline in the level of GNP or national income following the fall in investment and aggregate demand also results in decrease in level of employment in the economy and causes the emergence of large-scale cyclical unemployment.

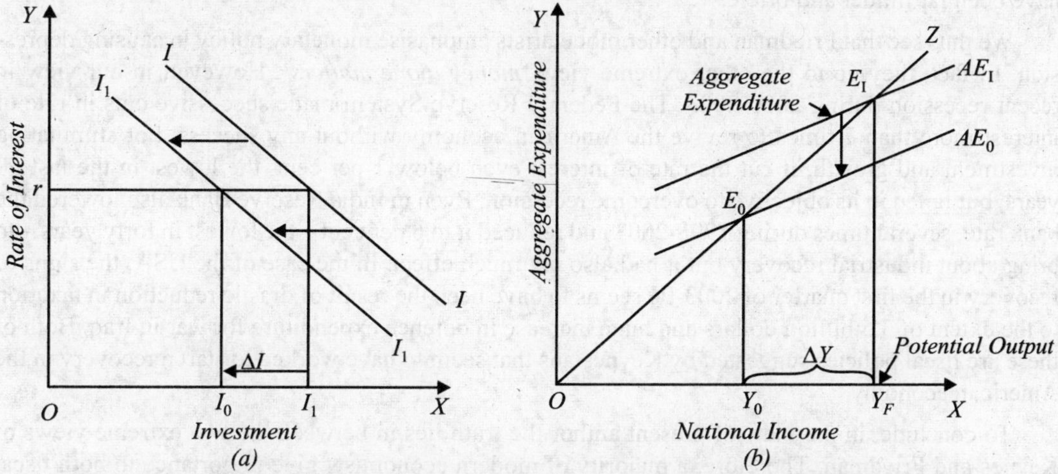

Fig. 5.9. *What Causes Depression and Cyclical Unemployment : Keynes's Explanation.*

It is also important to note that with help of *his theory of multiplier* Keynes showed that the fall in the level of national income was much greater than that of decline in investment. This can be known by comparing decline in investment demand (Δ*I*) and fall in national income. (Δ*Y*) in Fig. 5.9 (*b*).

Explaining Great Depression (1929-33) : Monetarist View

We have explained above Keynes's view that great depression of 1929-33 was caused by fall in investment caused by the drastic fall in the profit expectations of business class. The decline

1. J.M. Keynes, *Op. cit.* pp. 149-150.

in investment, according to him, caused aggregate demand to fall which along with the reverse working of multiplier caused conditions of depression (*i.e.* high rate of unemployment, substantial decrease in GNP and fall in prices) Keynes and his followers emphasise the role of fiscal policy (increase in Government expenditure and cut in taxation) to stimulate investment. They downplayed the role of monetary policy and in fact pointed out that monetary policy in the thirties during the depression period proved to be unimportant and ineffective. As a matter of fact, they asserted that monetary policy was tried during the depression period to revive the economy but miserably failed. At times of depression, according to them, businessmen are in the grip of pessimism about future profitability of investment and more money made available through easy and liberal monetary policy would not be effective in reviving the economy. It was in this context that some Keynesian economists remarked that '*money does not matter*' and also referred to the policy of making more credit available for investment by businesses – 'you can bring horse to water but cannot make it drink'. Thus Keynesian view prevailed during the forties and fifties almost unchallenged.

Keynes's view was attacked by Milton Friedman, a Nobel Laureate in Economics, in the middle of fifties. Friedman and Schwartz in their important work '*A Monetary History of the United States*' written in 1963 quoted statistical evidence from the United States, and argued that "a re-examination of the Great Depression of the 1930s demonstrated that, contrary to general belief, it had been a tragic testament to the great power of monetary policy, not to its impotence." They pointed out that drastic decline in the quantity of money was made by Federal Reserve System (the Central Banks of the USA) from 1929 to 1933 which was responsible for severe depression in the American economy. They argued that the failure of the Federal Reserve System to prevent bank failures and the drastic decrease in money supply during 1930-1933 increased the intensity of depression. To quote them, "The US Federal Reserve System could have prevented the *decline of one-third* in the quantity of money that occurred from 1929 to 1933. Had it done so, the evidence indicated, the depression would have been far milder and briefer"[2].

We thus see that Friedman and other monetarists emphasise monetary policy in causing depression. In fact, they go to the other extreme view '*money alone matters*'. However, in our view in recent recession during 2000-2003. The Federaly Reserve System made successive cuts in rate of interest more than 7 times to revive the American economy without any success. For stimulating investment and growth, it cut the rate of interest even below 1 per cent, the lowest in the last 50 years, but failed in its objective to overcome recession. Even in India Reserve Bank also lowered the bank rate, several times during 1998-2003 and reduced it to 6 per cent – the lowest in forty years – to bring about industrial recovery but it had also not much effect. In the case of the USA, the signs of recovery in the first quarter of 2003-04 seems to have been the result of drastic reduction in taxation to the extent of 3.5 billion dollars and huge increase in defence expenditure for war in Iraq. Both of these are fiscal policies suggested by Keynesians that seem to have worked to start a recovery in the American economy.

To conclude, in the view of present author the truth lies in between the two extreme views of Keynes and Friedman. Therefore, a majority of modern economists give importance to both fiscal and monetary policies for influencing investment and growth in GNP. Both can be complements to each other. Furthermore, profit expectations of entrepreneurs and cost and availability of credit play a significant role in determining national income, employment and price level in the economy.

POLICY IMPLICATIONS OF KEYNES'S THEORY OF INCOME AND EMPLOYMENT

Keynes's theory has important policy implications for raising the level of employment and income in the economy. When the equilibrium of the economy is at less than full-employment level and as

2. *Ibid.*

a consequence there is depression or recession in the economy and with it huge unemployment of labour prevails. According to Keynes, this occurs due to lack of aggregate effective demand. There are two types of policies which may be adopted to overcome this situation of recession and unemployment. They are (1) monetary policy and (2) fiscal policy. Under monetary policy, through expansion in money supply rate of interest can be lowered which will encourage private investment. With the increase in private investment, aggregate demand will increase (that is, aggregate demand curve will shift upward) which will raise the equilibrium level of employment. As a result, the economy will be lifted out of recessionary conditions and unemployment will be removed.

However, Keynes had serious doubts about the effectiveness of monetary policy. He was of the view that in times of depression rate of interest is already very low and at this liquidity preference curve of the community (that is, the curve of demand for money to hold) is *absolutely elastic, that is, it is of horizontal shape*. Therefore, in this situation when money supply is increased, rate of interest will not fall. Thus, with no fall in the rate of interest, private investment will not pick up further.

Importance of Fiscal Policy

In view of ineffectiveness of monetary policy, Keynes laid stress on the role of fiscal policy in curing recession/depression and removing involuntary unemployment. Under the fiscal policy, a major measure is the *increase in expenditure by the Government on several types of public works* in times of depression. The increase in Government expenditure will cause an upward shift in the aggregate demand curve. This increase in aggregate demand will bring about increase in employment and output. If the increase in Government expenditure and as a result rise in aggregate demand is sufficient, it will help in achieving equilibrium at full-employment level. As a result, depression and involuntary unemployment will be eliminated. It is worth noting that in this regard Keynes put forward a theory of multiplier which strengthened the case for raising Government expenditure on public works in times of depression or when large-scale unemployment prevailed in the economy. We shall discuss the theory of multiplier at length in later chapter. Briefly, theory of multiplier implies that increase in Government expenditure will raise aggregate demand and therefore output and employment not by the amount of increase in Government expenditure but by a multiple of it.

Another important measure of fiscal policy to raise employment and output is the *reduction in taxes*. When rates of personal taxes such as income tax are reduced, disposable income of the people increases which brings about rise in their demand for consumption. The rise in consumption demand pushes up aggregate demand curve and helps in removing recession and unemployment. It may be noted that on the recommendations of Keynesian economists American President John Kennedy made a drastic cut in income taxes in 1964. This had a great success as output and employment in the USA increased significantly and, as a result, recession and unemployment were removed. Cut in personal taxes to boost aggregate demand has also been applied in later years in the USA as well as in Great Britain. Recently in 2003, the President of the USA George W. Bush has made a 3.5 billion cut in taxes to revive the American economy. .

In the end, it is important to note that modern economists though agree with Keynes's view about the effectiveness of fiscal policy in curing recession, they do not share Keynes's view about the ineffectiveness of monetary policy. It may however be noted that, according to the modern Keynesians and others, liquidity preference curve of the community is fairly elastic even in times of recession which implies that expansion in money supply will cause a decline in interest rate and will therefore stimulate private investment. Further, according to them, investment demand curve is also fairly elastic which implies that fall in interest rate will have significant impact on private investment. Thus, according to a majority of modern economists, both monetary and fiscal policies are important instruments whereby aggregate demand of the economy can be changed. We shall discuss in detail the roles of monetary and fiscal policies for attaining economic stabilisation in later chapters.

That the free market system does not guarantee economic stability and full employment has been proved by the recent ***global financial crisis*** (2007-09) which started in the US as a result of collapse of financial system following bursting of housing bubble and spread to other countries (including India) through global linkages of free trade and capital flows and caused global slowdown. The market system which was believed by the classical economists-monetarists, led by Milton Friedman and new classical economists led by Robert Lucas to correct itself automatically and recover from the crisis failed to deliver. As a result, there were loud protests about the failure of the free market system to ensure economic stability and full employment.

Again it were Keynesian policies which came to the rescue of the US, India and other countries which were gripped by severe recession. These countries adopted Keynesian expansionary fiscal and monetary policies which provided stimulus to the economies by raising aggregate demand so as to get out of recessionary conditions. These counter cyclical Keynesian policies worked and in the recession-ridden economies of the US, European countries, Australia, Japan, China and India, economic recovery started and by the early 2010 they seem to have returned to the path of growth.

CONCEPTS OF INFLATIONARY AND DEFLATIONARY GAPS

Inflationary Gap

It is useful and important to understand the concept of inflationary gap because with it we are able to know the main cause of the rise in general level of prices. The equilibrium of an economy is established at the level of full employment when aggregate demand or total expenditure is equal to the level of income corresponding to full employment. This happens when the amount of investment is equal to the saving gap corresponding to full employment level of income. Consider Fig. 5.10 where OY_F is national income corresponding to the level of full employment. Equilibrium at national income OY_F would be established only when aggregate demand or total expenditure $(C+I+G)$ is equal to YF_E (YF_E is equal to OY_F). Real national income cannot increase beyond OY_F because when all means of production including labour are fully employed, there is no possibility of further rise in production or real national income. Thus when aggregate demand is greater than the aggregate demand YF_E which is required to establish the equilibrium at OY_F level of national income, the equilibrium would not be established at OY_F.

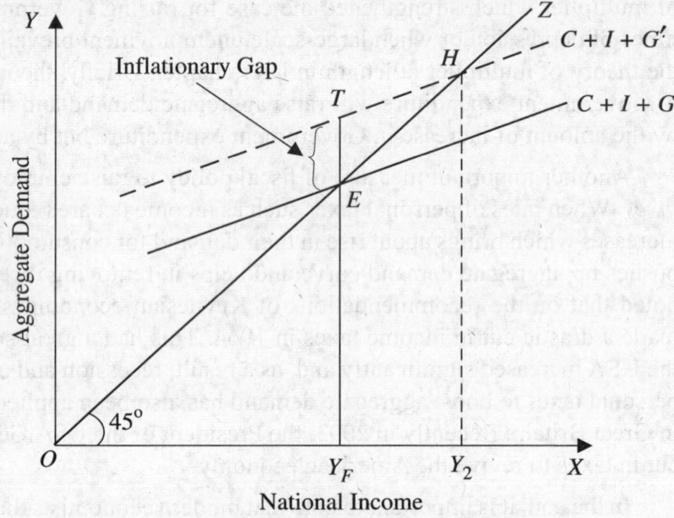

Fig. 5.10. *Inflationary Gap*

It would be seen from Fig. 5.10 that aggregate demand Y_FT is greater than aggregate demand Y_FE which is required to maintain the equilibrium at OY_F. Thus with the level of aggregate demand $(C + I + G')$, which is obtained by adding expenditure ET to the aggregate demand curve $C + I + G$, equilibrium would not be established at OY_F which corresponds to full employment level of income. The actual aggregate demand being greater than YF_E by the amount ET the level of national income would be greater than OY_F. Since OY_F is full-employment level of national income, actual production cannot increase beyond that but there would be rise in prices which would raise the money value of OY_F production. The amount by which the actual aggregate demand exceeds the level of national income corresponding to full employment is known as inflationary gap because this excess of aggregate demand causes inflation or rise in prices in the country. In Fig. 5.10 this excess of aggregate demand or inflationary gap is equal to *ET*. It would be seen from Fig. 5.10 that the aggregate demand curve $C + I + G'$ intersects 45° line (*OZ* line) at *H* so that equilibrium level of national income would be OY_2. It should be carefully understood that there is no differencc between OY_F and OY_2 in terms of real income or actual production; only as a result of rise in price level, national income has increased from OY_F to OY_2 in money terms. Inflationary gap represents excess demand in relation to aggregate production or supply of output which brings about demand-pull inflation. J.M. Keynes in his revolutionary book "*General Theory of Employment, Interest and Money*" did not discuss the concept of inflationary gap because he was then preoccupied with the analysis of the state of depression and deflation. During the Second World War when the problem of inflation cropped up, then Keynes applied his macroeconomic analysis to explain inflation as well and in this connection he put forward the concept of inflationary gap.

Deflationary Gap

In the theory of income and employment, the concept of deflationary gap occupies an important place, since in a capitalist economy unemployment and depression occur due to this gap. According to Keynesian theory of income and employment, equilibrium at the level of full employment is established when aggregate demand consisting of consumption demand plus investment demand plus government demand ($C + I + G$) is equal to the national income at the level of full employment. This happens when investment and government demand is equal to the saving gap at full employment level of national income. If aggregate demand is less than the full-employment level of national income, i.e., when investment and government demand is less than the saving gap at full employment level of income, the deficiency of aggregate demand occurs due to which national income and employment will fall below the full employment level causing unemployment and depression in the economy. The concept of deflationary gap is illustrated in Fig. 5.11 in which along the *X*-axis national income is measured and along the *Y*-axis level of aggregate demand is measured. Suppose national income at the level of full employment is equal to OY_F. Now the equilibrium level of income and employment would be established at OY_F when aggregate demand (consumption demand plus investment demand) is equal to YF_E (which is equal to national income OY_F). But in the real world if aggregate demand is less than the full-employment level of income OY_F or it is less than YF_E, then the problem of deficiency of aggregate demand will arise. Therefore, *EK* in Fig. 5.11 represents deflationary gap. Hence, deflationary gap represents the difference between the actual aggregated demand and the aggregate demand which is required to establish the equilibrium at full-employment level of income. It should be carefully understood that due to the deflationary gap *EK*, the level of national income and employment will decline. The decline in national income and employment will not only be equal to the deflationary

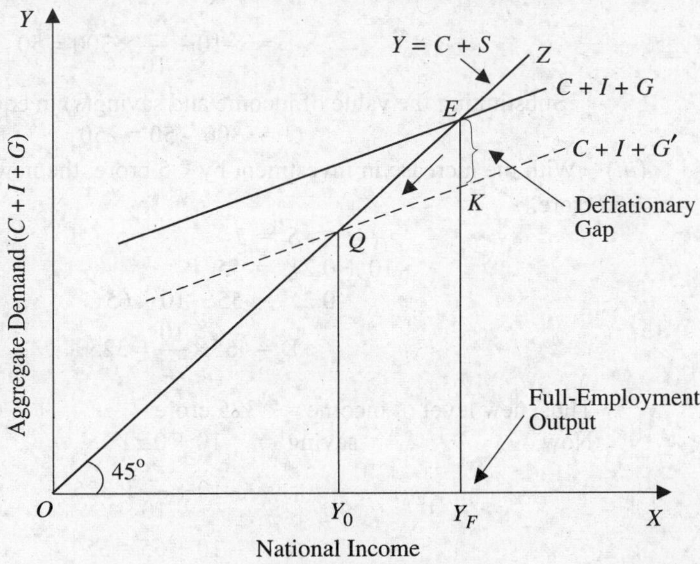

Fig. 5.11. *Deflationary Gap*

gap EK but it will be much greater than this. The decline in national income is determined by the value of the multiplier. In Fig. 5.11 when aggregate demand curve is $C + I + G'$ deflationary gap is equal to EK. The aggregate demand curve $C + I + G'$ (dotted) which cuts the 45° line at point Q as a result of which equilibrium is established at OY_1 level of national income. It will be seen from Fig. 5.11 that OY_1 is less than full-employment level of income OY_F. Deflationary gap represents the situation of deficient demand in the economy. This deficiency in aggregate demand causes fall in national output and level of employment. As a result, involuntary unemployment in the economy emerges. The depression of 1929-33 in capitalist countries was caused by the emergence of deflationary gap or by demand deficiency in these economies.

NUMERICAL PROBLEMS

Problem 1. *Given the saving function $S = -10 + 0.2y$ and autonomous investment, I = ₹ 5 crore.*

(*i*) *Find the equilibrium level of income.*

(*ii*) *Find the level of consumption.*

(*iii*) *If investment increases permanently by ₹ 5 crore, what will be the new levels of income and consumption?*

Solution: (*i*) According to saving-investment approach, equilibrium level of national income is determined by equality between planned saving and investment. Thus,

$$S = I \qquad ...(i)$$

$$-10 + 0.2Y = 50$$

$$0.2\,Y = 50 + 10 = 60$$

$$Y = 60 \times \frac{10}{2} = 300$$

(*ii*)

$$C = Y - S \qquad ...(ii)$$

$$S = -10 + 0.2Y$$

$$= -10 + \frac{2}{10} \times 300 = 50$$

Substituting the value of income and saving(*s*) in equation (*ii*), we have

$$C = 300 - 50 = 250$$

(*iii*) With the increase in investment by ₹ 5 crore, the new investment will become equal to ₹ 55 crore.

$$S = I$$

$$-10 + 0.2Y = 55$$

$$0.2Y = 55 + 10 = 65$$

$$Y = 65 \times \frac{10}{2} = 325$$

Thus, new level of income = ₹ 325 crore

Now,

$$\text{saving} = -10 + 0.2Y$$

$$= -10 + \frac{2}{10} \times 325$$

$$= -10 + 65 = 55$$

$$\text{consumption } (C) = Y - S$$

$$= 325 - 55$$

$$= 270$$

Problem 2. *The following consumption function of an economy is given:*

$40 + 0.8Y$ where Y is national income

If the planned level of investment in a year equals ₹ 75 crore, what will be the equilibrium level of national income and consumption?

Solution : Equilibrium level of national income is determined where

$$Y = C + I$$

Now, $C = 40 + 0.8\,Y$

$$I = 75$$

$$Y = 40 + 0.8Y + 75$$

$$Y - 0.8Y = 40 + 785 = 115$$

$$0.2Y = 115$$

$$Y = 115 + \frac{10}{2} = 575$$

$$C = Y - I = 575 - 75 = 500$$

Problem 3. *For an economy following consumption function is given* :

$$C = 60 + 0.75\,Y$$

(*a*) *If investment in a year is ₹ 35 crore what will be the equilibrium level of income or output?*

(*b*) *If full-employment level of income (i.e. level of potential output) is ₹ 460 crore what investment is required to be undertaken to ensure equilibrium at full employment.*

Solution : (*a*)

$$Y = C + I$$

$$Y = 60 + 0.75Y + 35$$

$$Y - 0.75Y = 60 + 35 = 95$$

$$Y(1 - 0.75) = 95$$

$$0.25Y = 95$$

$$Y = 95 \times \frac{100}{25} = 380$$

(*b*) To ensure full-employment equilibrium, investment should be equal to the saving gap at full-employment income. With the given full-employment income equals to ₹ 460 crore.

$$S_F = Y_F - C_F = Y_F - (60 + 0.75Y_F)$$

$$= 460 - 60 - (0.75 \times 460)$$

$$= 400 - 345 = 55$$

Thus, investment required for full-employment equilibrium is 55 crore.

QUESTIONS FOR REVIEW

1. Explain how equilibrium level of national income is determined by level of effective demand in a simple two-sector Keynesian model. Is this equilibrium necessarily established at full employment level ?

2. Explain the equilibrium level of national income by consumption and investment. When will the equilibrium be established at full employment ?

3. (*i*) Saving and Investment are *always equal.*

(*ii*) Saving and Investment are *equal only in equilibrium.*

How would you reconcile the above two statements ?

4. Explain Keynes's income-expenditure approach to the determination of national income. Why can this equilibrium be reached below full-employment level of output ?

5. Define planned (desired) saving and planned (desired) investment. Explain how equilibrium level of national income is determined by planned saving and planned investment. Why is this equilibrium not always attained at full-employment level ?

6. What do you understand by the equilibrium level of national income? Explain that both saving-investment approach and aggregate demand-aggregate supply approach yield the same level of equilibrium.

7. Explain what, according to Keynes, causes depression or recession. What is Friedman's view about the cause of depression or recession?

8. Explain the Keynesian concept of aggregate demand. How is the Keynesian aggregate demand function different from the classical demand function?

9. Explain the concept of deflationary gap. What are the causes and consequences of this gap?

10. (a) What is meant by underemployment equilibrium? Explain the factors which cause it.

(b) Explain the various measures that can be taken to increase the equilibrium level of income and employment. Is it possible to increase the level of income without increasing the level of investment?

11. Explain the concept of inflationary gap. What are its causes and consequences?

12. Explain the role of monetary and fiscal policies to overcome recession in the economy.

13. What, according to Keynes, is the cause of recession or depression in the economy? What measures can be adopted to overcome recession in the economy?

14. The following consumption of an economy is given:

$$C = 75 + 0.8\,Y$$

(*i*) What will be the equilibrium level of national income if investment equals 40 crore?

(*ii*) Find the equilibrium level of consumption.

(*iii*) How much will national income increase if investment increases to 50 crore?

[**Hint:** Answer to (*iii*) can be directly found by applying the multiplier principle, that is, $\Delta Y = \Delta I\left(\frac{1}{1-MPC}\right)$. In the given problem $\Delta I = 50 - 40 = 10$ crore and $MPC = 0.8$.]

15. Given the following saving function of the economy;

$$S = -15 + 0.25\,Y$$

(*i*) If planned investment in a year of the economy equals ₹ 20 crore, what will be the equilibrium level of national income?

(*ii*) What is the equilibrium level of consumption?

Appendix to Chapter 5

KEYNES'S COMPLETE MACROECONOMIC MODEL: INTEGRATING GOODS MARKET WITH MONEY MARKET

Linking Goods Market with Money Market

As explained in the previous chapter, classical economists thought that supply of money only determines the price level and nominal variables and has no effect on the real variables of the economy such as level of real national income (output) and employment. However, in Keynes's theory money plays an important role in determination of real income and employment. In fact *Keynes brought about integration between money market and goods market.* Keynes was able to do so because he explained that rate of interest is determined by demand for money (which he called liquidity preference) and supply of money, that is, through money market equilibrium, while classical economists thought that rate of interest was determined by saving and investment, that is, through capital market equilibrium. It is the rate of interest which is crucial channel through which money market is linked with goods market. Changes in money supply by monetary policy can bring about changes in rate of interest which determines investment. And, according to Keynes, it is investment which plays a crucial role in determining level of real income and employment. For example, increase in the supply of money will lower the rate of interest. Since investment depends on the rate of interest and marginal efficiency of capital, at the lower rate of interest more investment will be undertaken. The higher investment will lead to the greater level of real national income and employment. In fact, Keynesian analysis brings out that monetary policy can play significant role in influencing the real variables and controlling business cycles.

However, Keynes pointed out that *effect of increase in money supply depends on interest elasticity of demand for money (i.e. liquidity preference curve) and interest-sensitivity of investment demand.* The greater the elasticity of demand for money and lower the elasticity of investment demand curve, the smaller will be the effect of changes in money supply on national income and employment. Though during the period of great depression (1929-33), Keynes pointed out that demand for money was highly elastic and therefore he argued that effect of monetary policy would not be enough to lift the economy out of depression and therefore he emphasised that along with monetary policy expansionary fiscal policy should be adopted to raise aggregate demand and thereby to increase the levels of real income and employment. However, according to him, in *normal times*, monetary policy is quite effective in determining real income and employment.

Empirical evidence after Keynes has revealed that even during recessions demand for money is not absolutely elastic and also investment demand is somewhat elastic. Therefore, the modern Keynesian economists emphasise the roles of both monetary and fiscal policies to lift the economy out of recession and achieve economic stability.

KEYNES'S COMPLETE MACROECONOMIC MODEL

In Fig. 5A.1 we have graphically depicted Keynes's complete macroeconomic model that shows how Keynes integrates money market and real goods market. In Panel (*c*) in Fig. 5A.1 money market equilibrium is shown. Given the money supply equal to M_0 money supply curve MS_0 intersects money demand curve *LP* at rate of interest r_0. It will be seen from Panel (*d*) that at rate of interest r_0, investment is I_0. Then in Panel (*b*) with the consumption function curve C and investment equal to I_0, aggregate demand curve is $C + I_0$ which intersects the 45° income line OZ at point E and determines level of real national income equal to Y_0. Panel (*a*) of the figure shows that Y_0 level of national income generates N_0 level of labour employment.

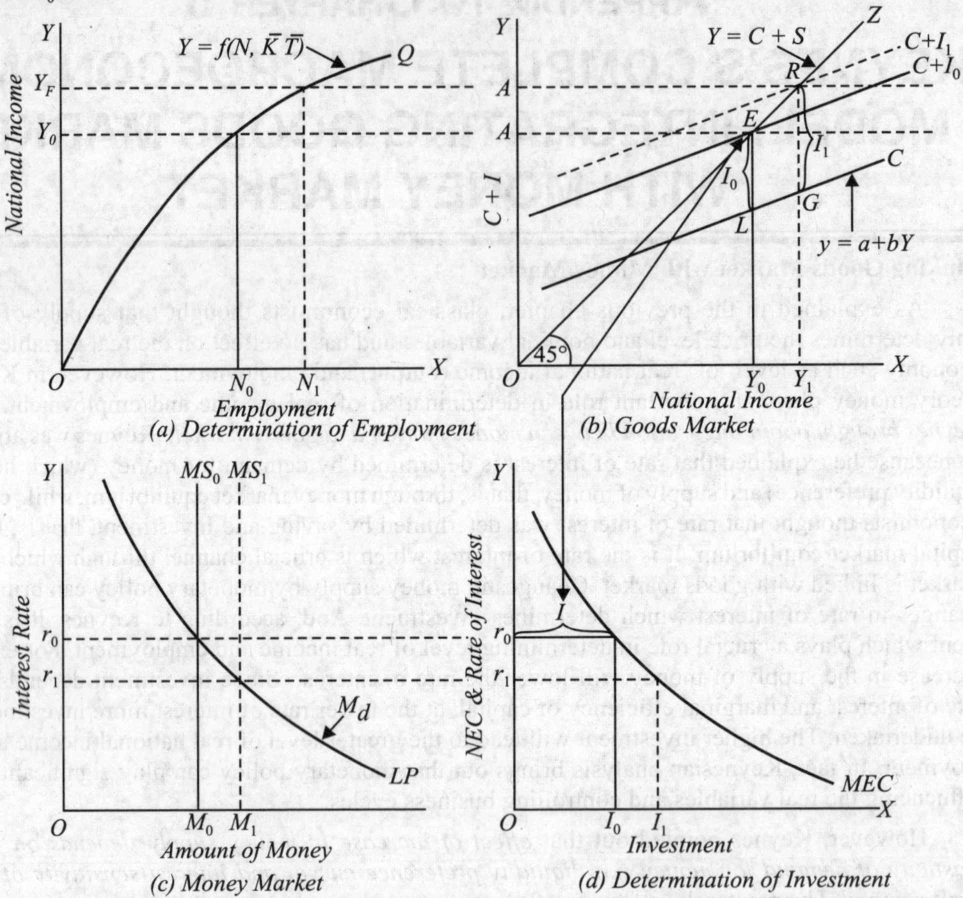

Fig. 5A.1. *Keynes's Complete Macroeconomic Model: Integrating Money Market with Goods Market*

Now suppose that there is increase in money supply from M_0 to M_1 which causes shifts in money supply curve from MS_0 to MS_1. As a result, rate of interest falls from r_0 to r_1 of Panal (*d*). Fig. 5A.1 shows that at a lower interest rate r_1 investment increases from I_0 to I_1. With this increase in investment (ΔI), aggregate demand curve shifts above to the dotted position $C + I_1$. With this increase in aggregate demand, level of real national income increases from Y_0 to Y_1. Corresponding to this increase in real national income the level of labour employment in Panel (*a*) rises from ON_0 to ON_1.

It is evident from above that Keynes succeeded in integrating money market with goods market and with this he showed that money supply plays an important role in determining real variables such as real investment (*i.e.,* addition to the stock of capital), real national output and employment.

It follows from above that liquidity preference, money supply, rate of interest, consumption function and investment are crucial elements in Keynes's model which determine level of income and employment. Fig. 5A.1 also shows *how changes in rate of interest through their effect on investment shifts aggregate expenditure (aggregate demand) curve and thereby affects level of income and employment.* For example, if money supply is increased from M_0 to M_1, it will be seen from Panel (*c*) that rate of interest falls from r_0 to r_1. In Panel (*d*) at lower rate of interest r_1, investment increases to I_1. Now, it will be seen from Panel (*b*) that with higher investment I_1 aggregate expenditure curve shifts upward to $C + I_1$ (dotted). With this new expenditure curve $C + I_1$, equilibrium level of national income increases to Y_1 which causes labour employment to rise from N_0 to N_1. The increase in money supply has led to the increase in real income and employment. In Chapter 5 in Fig. 5.3 where Keynes's simple model of income determination is shown we have taken investment as *exogenously* fixed amount autonomous of income. However, as seen in Fig. 5A.1, rate of interest is *endogenously* determined in Keynes's complete macroeconomic model and it is through the channel of change in interest rate that Keynes was able to successfully integrate money market with goods market Keynes's complete macroeconomic model can be represented in tabular form as under:

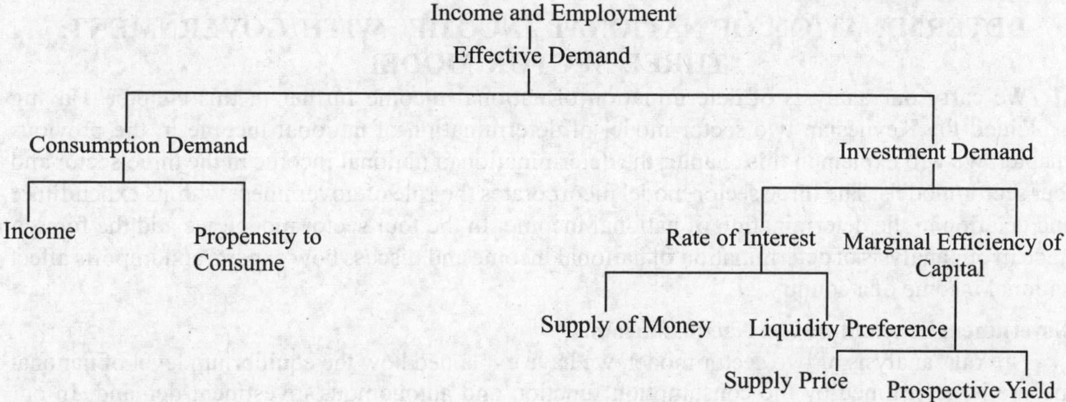

Policy Implications

The policy implications of Keynes's model of the determination of income and employment may be noted. When the equilibrium is at less than full employment and the economy is in a state of depression, two types of economic policies, monetary and fiscal, may be adopted. Monetary policy will seek to increase the supply of money to reduce the rate of interest. This lower interest rate will tend to raise private investment and consequently the levels of income and employment as has been seen above. More than the cheap money policy Keynes strongly advocated for the adoption of *fiscal policy* to overcome recession. Under fiscal policy Government may give tax concessions to the private sector so that their profit may rise resulting in the upward shift in the marginal efficiency of capital (MEC) curve. This will raise the amount of private investment and will help in raising national income and employment. Besides, under fiscal policy Government may itself increase its expenditure on public works which will increase income and employment not only directly but also indirectly through the working of multiplier.

QUESTIONS FOR REVIEW

1. Explain how Keynes links money market with goods market. What are its policy implications? [*DU. BA(Hons.)*]
2. Show how changes in interest rate affect aggregate demand and level of income and employment.
3. In Keynes's macroeconomic model money is not neutral and affects the magnitudes of real variables. Explain
4. What type of interest rate policy is adopted to promote economic growth?
5. What type of interest rate policy is adopted to control inflation. Explain the transmission process in this regard.

CHAPTER 5A
DETERMINATION OF NATIONAL INCOME IN AN OPEN ECONOMY WITH GOVERNMENT AND FOREIGN TRADE: THREE SECTOR AND FOUR SECTOR MODELS*

DETERMINATION OF NATIONAL INCOME WITH GOVERNMENT: THREE SECTOR MODEL

We carry our analysis of determination of national income further in this chapter. Having explained the Keynesian two sector model of determination of national income in the previous chapter, we will explain in this chapter the determination of national income in the three sector and four sector models. The three sector model incorporates the role of government with its expenditure and taxation in the determination of national income. In the four sector model, we add the foreign trade in our analysis of determination of national income and discuss how exports and imports affect national income of a country.

Government Expenditure and National Income

In our analysis of two sector model, we have explained how the equilibrium level of national income is determined by the consumption function and autonomous investment demand. In our analysis, we did not take into account the role of Government expenditure in the determination of national income. However, in all economies today including the free-market capitalist economies such as those of U.S.A., Britain, Japan, the Government expenditure on goods and services plays an important role in the determination of national income and therefore it should also be included in the analysis of income determination. It is important to note here that the magnitude of Government expenditure on such things as highways, public parks, education, health services is governed by the consideration of promoting social welfare, employment and growth in the economy and does not depend on the level of income of the economy. Therefore, in the simple Keynesian model of income determination, the Government expenditure is treated as *autonomous expenditure*. We further assume for the moment that Government spends on goods and services but imposes no taxes. We denote Government expenditure by *G*. Thus in the three sector economy when we take into account the income generating effects of Government expenditure, we get the following equation for the equilibrium level of national income.

$$Y = C + I + G \qquad \ldots (i)$$

where Y is national income or output and $C + I + G$ represents the level of aggregate demand including Government expenditure, G.

Since consumption function $C = a + bY$ *where b stands for marginal propensity to consume* and a for autonomous consumption expenditure, we can rewrite the equilibrium level of national income as under :

$$Y = a + bY + I + G$$

$$Y - bY = a + I + G$$

$$Y(1 - b) = a + I + G$$

$$Y = \frac{1}{1-b}\,(a + I + G) \qquad \ldots (ii)$$

Thus it is clear that the *equilibrium level of income (when impact of Government expenditure is*

*This chapter is meant for higher level courses in macroeconomics

also considered) is equal to the sum of three types of fixed autonomous expenditure, namely, autonomous consumption, autonomous investment and Government expenditure ($a + I + G$) multiplied by the value of the multiplier [$1/(1 - b)$].

Graphic Illustration

The impact of Government expenditure on goods and services on the equilibrium level of national income is illustrated in Fig. 5A.1. The $C + I$ is the aggregate expenditure curve which has been obtained by adding autonomous investment equal to I to the consumption functional curve. On top of the consumption curve C, we have added the Government expenditure on goods and services (G) and investment demand (I) to get the aggregate expenditure (or aggregate demand) curve $C + I + G$. With this, the equilibrium national income OY is determined at the level at which the aggregate expenditure curve ($C + I + G$) intersects the 45° line, that is, aggregate demand ($C + I + G$) equals aggregate supply of output. The reason for the determination of equilibrium level of income at OY when Government expenditure is included in the aggregate demand curve is the same as stated earlier. It will be observed from Fig. 5A.1 that at a level of national income *less* than OY aggregate expenditure $C + I + G$ exceeds aggregate output. This implies unintended decline in the inventories of goods with the firms which will induce them to expand the level of aggregate output to the level OY. On the contrary, at the level of national income *greater* than OY, the aggregate expenditure is less than aggregate output. This deficiency of aggregate demand for goods will cause unintended accumulation of inventories. The firms would respond to this by cutting back production of goods which will lead to the reduction of national income to OY.

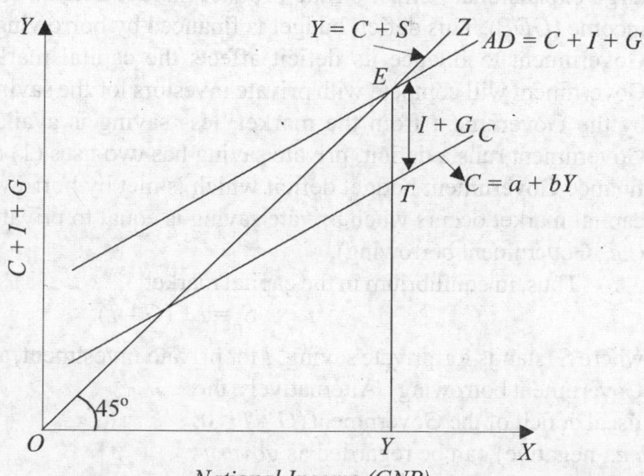

Fig. 5A.1. *Determination of National Income: The Three-Sector Model*

It is important to note that *in the presence of Government expenditure, national income is determined at the level at which the saving gap is equal to the sum of private investment and Government expenditure* ($I + G$).

This can be shown as under:

In equilibrium, $Y = C + I + G$

or $\quad Y - C = I + G$

Since $\quad Y - C = \text{Saving } (S)$

Therefore, in equilibrium $S = I + G$ . . . (*iii*)

Government Expenditure Multiplier

It may be further noted from equation (*ii*) that just as investment expenditure has a multiplier effect on increase in income *Government expenditure too has a multiplier effect. It will be seen from equation (ii) above, Government expenditure (G) multiplier, like that of autonomous consumption (a) and investment (I) multiplier, is equal to* $\dfrac{1}{1+b}$

GOVERNMENT EXPENDITURE, BUDGET DEFICIT AND CAPITAL MARKET

We have explained above how the government expenditure affects the product market by bringing about increase in aggregate demand. Now, an important question is how the Government expenditure is financed. Imposition of taxes is an important way of financing Government expenditure.

But taxes reduce aggregate demand in the product market by lowering disposable income of the community and thereby causing a decline in its consumption expenditure. Therefore, if Government expenditure equals tax revenue collected ($G = T$), then Government has a balanced budget and much of the expansionary effect of government expenditure is offset by taxes levied. If Government increases its expenditure without raising taxes, (*i.e.*, $G > T$), then it has a *budget deficit* which has a large expansionary effect on the product market and, as seen above, leads to increase in national income (*GDP*). This deficit budget is financed by borrowing from the public. But borrowing by the Government to finance its deficit affects the capital market. This is because borrowing by the Government will compete with private investors for the saving of the public. As a result of borrowing by the Government from the market, less saving is available for private investment. When the Government runs a deficit, private saving has two uses (1) to finance private investment and (2) to finance Government budget deficit which is met by borrowing from the public. Equilibrium in the capital market occurs when private saving is equal to private investment plus budget deficit, $G - T$, (*i.e.*, Government borrowing).

Thus, in equilibrium in the capital market

$$S_p = I + (G - T)$$

where S_p stands for private saving, I for private investment, and $(G - T)$ stands for budget deficit (*i.e.*, Government borrowing). Alternatively, the fiscal deficit of the Government ($G - T < 0$, *i.e.*, negative) can be regarded as *government dissaving*. Therefore, in case of fiscal deficit or negative *Government savings*, national saving will decrease and national saving curve will shift to the left. How budget deficit or Government borrowing reduces or crowds out private investment is illustrated through Fig. 5A.2 where to begin with vertical saving curve S_0 intersects the downward sloping investment curve II at point E_0 and determines real interest rate r_0 at which private investment equal to I_0 takes place. Now as a result of Government borrowing to finance its budget deficit $(G - T)$ national saving available for investment falls and the national saving curve shifts to the left[1] to the new position S_1. Thus when Government borrows to finance budget deficit, Government is said to dissave and therefore in this case Government or public saving is negative ($G - T = -S_g$). Note further that national saving = private saving and Government saving ($S = S_p + S_g$). As a result of fiscal deficit and consequent leftward shift in the national saving curve to S_1, the capital market is now in equilibrium at point E_1 and rate of interest rises to r_1. At this higher interest rate r_1, private investment falls to I_1. Thus, borrowing by the Government to finance its budget deficit has led to crowding out of private investment. As mentional above, a part of private saving is used to finance private investment and a part to finance budget deficit of the Government.

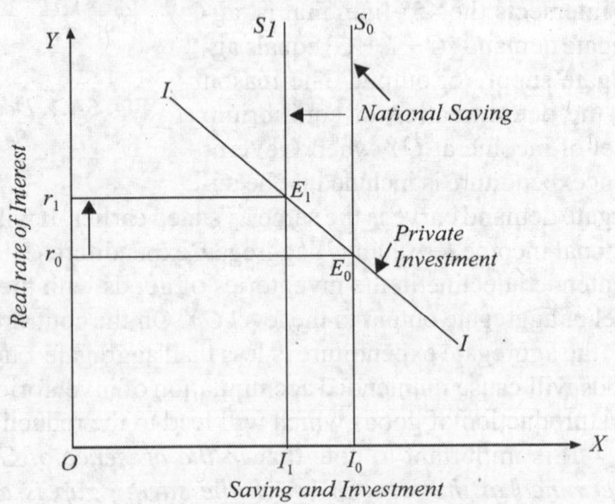

Fig. 5A.2. *Effect of Budget Deficit and Government Borrowing on Capital Market*

It is worth noting that when Government expenditure is fully matched by increase in taxes (*i.e.*, $G = T$) we have a balanced budget but it will still have an expansionary effect on national income, though, as shall be explained below, the multiplier effect of a balanced budget is equal to one.

1. Note that all expenditure by the Government is here treated as consumption expenditure.

However, since in case of balanced budget increase in Government expenditure is matched by increase in taxes which reduces disposable income $Y_d = Y - T$. And reduction in disposable income (Y_d) leads to decrease in both consumption and saving ($Y_d = C + S$). This decrease in private saving will cause rate of interest to rise and thereby crowds out private investment.

Further, when the Government has a *budget surplus* (i.e. $T > G$), it has a *positive saving* which will lead to the increase in national saving and make more funds available for private sector to borrow. This will cause real rate of interest to fall. As a result, higher private investment will occur in case of surplus budget.

It is worth noting that in our above analysis Government expenditure has been treated as consumption expenditure on goods and services purchased to satisfy collective wants. However, this is strictly true in case of a purely free-market economy such as that of the US where the Government is not a major investor. But in case of India's mixed economy the Government is an important investor and a good part of its borrowing is used for investment purposes, especially for investment in infrastructure projects such as transport, highways, roads, ports etc. Such Government borrowing for investment purposes cannot be considered as consumption expenditure or *dissaving*. Therefore, it has a large expansionary effect on national income with a usual multiplier effect. Viewed as such Government borrowing for investment, especially for infrastructure projects in developing countries, may *crowd in* private investment rather than crowding it out. This is so because lack of adequate infrastructure hinders private investment.

Determination of National Income with Government Expenditure and a Lump Sum Tax

Now, in a three sector model we analyse the impact of Government expenditure when to finance this expenditure the government levies a lump sum tax. Note that a lump-sum tax does not vary with income. In this case if government expenditure equals revenue from lump-sum tax, its budget will be in *balance*. On the other hand, if government expenditure exceeds tax revenue, there will be a *budget deficit*. Further, if tax revenue is more than its expenditure, government will have a *surplus budget*. We are thus concerned here with the impact of government expenditure and taxation together on the determination of national income. Let us assume *the government follows a balanced budget policy* in which case government expenditure equals tax revenue from lump-sum tax. Let us assume that there are no *transfer payments* made by the government.

The simplest kind of a tax is the *lump-sum tax* in which a given fixed amount of revenue is collected irrespective of the level of income. Suppose government has decided to raise extra ₹ 500 crore by way of imposing lump-sum tax on the public.

Every one will have to pay the fixed amount out of his income as his share of lump sum tax. In the aggregate, the disposable income of the economy will be reduced by this fixed amount of lump-sum tax *i.e.*, (500 crore) whatever the level of national income (GNP). Consequently, people will consume less after the levy of lump-sum tax. A lump-sum tax of ₹ 500 crores imposed by the Government will reduce the disposable income by the same amount, that is, by ₹ 500 crores. As a result, the people will consume less (and also save less) at each level of national income (GNP). However, it is worth mentioning that *consumption will not fall by the full amount of the tax* because a part of the reduction in disposable income due to the tax was being consumed and a part of it was being saved prior to taxation. Therefore, reduction in disposable income due to the lump-sum tax will partly cause a decline in consumption and partly a decline in saving. Therefore, the amount by which consumption will fall due to the introduction of lump-sum tax will be equal to the amount of tax multiplied by the marginal propensity to consume. If ΔY_d stand for change in disposable income, T for tax imposed and MPC for marginal propensity to consume, then the decline in consumption ($-\Delta C$) due to the imposition of lump sum tax is given by

$$-\Delta C = T.MPC$$

Since a lump-sum tax causes a change in disposable income (Y_d)

Therefore tax, $T = \Delta Y_d$

$-\Delta C = \Delta Y_d . MPC$

Let us take an example. If the aggregate amount collected by way of lump sum tax is ₹ 500 crore and *MPC* is 0.75, then the decline in consumption ($-\Delta C$) due to lump sum tax will be ₹ 500 × 0.75 = ₹ 375 crore.

On the contrary, *if lump-sum tax is reduced*, the consumption will increase by the amount of reduction in the lump sum tax multiplied by the marginal propensity to consume ($+\Delta C = +\Delta Y_d . MPC$).

The impact of imposition of lump-sum tax on consumption function is graphically shown in Fig. 5A.3. It will be seen from this figure that introduction of lump sum tax shifts the consumption function curve *C* downward to the new position *C'* by the amount given by the tax amount multiplied by the *MPC*. Similarly, the reduction in lump sum income tax shifts the consumption function curve upward to *C''*.

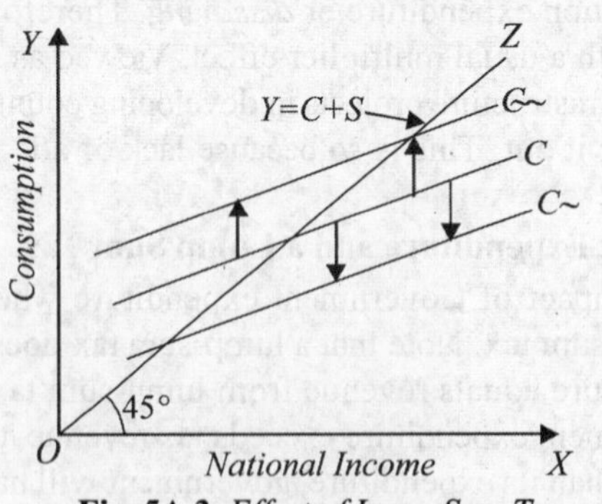

Fig. 5A.3. *Effect of Lump Sum Tax on Consumption Function*

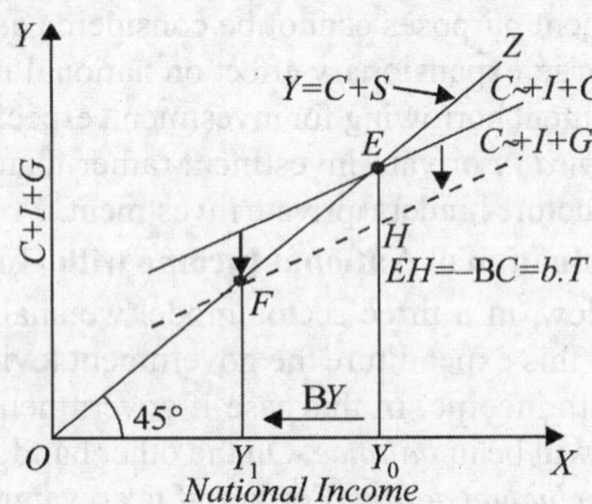

Fig. 5A.4. *Effect of Lump Sum Tax on National Income*

Fig. 5A.4 shows the effect of imposition of lump sum tax on the equilibrium level of national income (GNP). It will be observed from Fig. 5A.4 that due to lump sum tax equal to *T*, and consequently downward shift in aggregate expenditure curve to the new dotted position $C' + I + G$, the equilibrium level of national income falls from OY_0 to OY_1. An important point worth mentioning is that the fall in the level of income ($-\Delta Y$) is not equal to the amount of tax (*T*) collected but by a multiple of it, tax multiplier being equal to $\frac{-b}{1-b}$ where *b* stands for marginal propensity to consume.

Thus, $\frac{-b}{1-b}$ or $\frac{-MPC}{1-MPC}$ is **tax multiplier of a lump sum tax**. We will explain this tax multiplier in detail in a later section. It will be seen that tax multiplier is negative.

It will be seen from Fig. 5A.3 that *if Government reduces lump sum income tax*, the consumption function curve shifts upward to the new position C''. From this it follows that *reduction in tax by the government will increase the disposaible income of the society and raise consumption demand. The rise in consumption demand will raise the equilibrium level of income by a multiple amount depending on the marginal propensity to consume.*

Algebraic Analysis

Three-sector model with the above specifications can be developed as follows:

$$Y = AD = C + \overline{I} + \overline{G}$$

or $$Y = C + \overline{I} + \overline{G} \qquad ...(i)$$

Now, consumption function can be stated as :

$$C = a + bY_d$$

Where a is constant autonomous consumption expenditure, Y_d is disposable income. In this three sector model with lump-sum tax and no transfer payments, consumption function can be written as

$$C = a + b\,(Y - T)$$

where T = lump-sum tax

or $$C = a + bY - bT$$

Substituting the value of C in income equation (i) we have

$$Y = a + bY - bT + \bar{I} + \bar{G}$$

$$Y - bY = a - bT - \bar{I} + \bar{G}$$

$$Y(1-b) = a - bT + \bar{I} + \bar{G}$$

$$Y = \frac{1}{1-b}\left(a - bT + \bar{I} + \bar{G}\right) \qquad ...(ii)$$

The equation (ii) furnishes us the equilibrium level of national income in the three sector economy when there is both government expenditure and lump sum tax. Therefore, for an economy if the data regarding consumption function, investment, Government expenditure and lump-sum tax are given we can compute the equilibrium level of national income. A numerical example will clarify the determination of equilibrium level of income in this case.

Numerical Examples

Problem 1. *The following data characterises the macroeconomic conditions of a hypothetical economy*

$$C = 50 + 0.8\ Y_d$$
$$I = 100$$
$$G = T = 75$$

where C, I, and Y_d are consumption, investment and disposable income respectively. Calculate equilibrium income of the economy. What is the value of multiplier?

Solution: Note that since $G = T$, budget of the Government is balanced.

$$Y = C + I + G$$
$$C = a + bY_d = a + b(Y - T)$$

Substituting the values of C, I and G we have :

$$Y = 50 + 0.8\,(Y - 75) + 100 + 75$$
$$= 50 + 0.8Y - 60 + 100 + 75$$
$$Y = 165 + 0.8Y$$
$$Y - 0.8Y = 165$$
$$Y(1 - 0.8) = 165$$
$$0.2\ Y = 165$$

$$Y = 165 \times \frac{10}{2} = 825$$

Thus, equilibrium level of income is 825.

Value of multiplier of autonomous consumption (50) and autonomous investment (100) is

$$\frac{1}{1-b} = \frac{1}{1-mpc}$$

$$= \frac{1}{1-mpc} = \frac{1}{1-0.8} = \frac{1}{0.2}$$

$$= \frac{10}{2} = 5$$

Since in the present case $G = T$, there is a balanced budget, therefore

value of balanced budget multiplier $= \frac{1}{1-b} + \frac{-b}{1-b}$

$$= \frac{1}{1-0.8} + \frac{-0.8}{1-0.8} = \frac{1}{0.2} + \frac{-0.8}{0.2} = 5 - 4 = 1$$

Problem 2. *You are given the following data for an economy*

Consumption $C = 20 + 0.8\,Y_d$

Investment $I = 50$

Government expenditure $G = 20$

Taxes $T = 10$

Find (a) equilibrium level of income (b) If lump-sum taxes increase by 10 (i) what is the equilibrium level of income, and (ii) lump-sum tax multiplier

(c) If government expenditure decreases by 10, what is (i) equilibrium level of income, (ii) Government expenditure multiplier

(d) Can you explain the difference between in the answer for (b) and (c)

DU. BA(Hons.) 2006

Ans. (*a*) $Y = C + I + G$

$$C = 20 + 0.8\,Y_d = 20 + 0.8\,(Y - T)$$
$$= 20 + 0.8\,(Y - 10)$$
$$I = 50 \text{ and } G = 20, \text{ and } T = 10$$
$$\therefore \quad Y = 20 + 0.8\,(Y - 10) + 50 + 20$$
$$= 20 + 0.8\,Y - 8 + 50 + 20$$
$$Y - 0.8Y = 82$$
$$Y(1-0.8) = 82$$
$$Y = \frac{82}{1-0.8} = \frac{82}{0.2} = 82 \times \frac{10}{2} = 410$$

(*b*) (*i*) With increase in lump sum tax by 10, the change in equilibrium level of income will be

$$\Delta Y = \Delta T \times \text{Lump-Sum Tax Multiplier}$$
$$= 10 \times \frac{-b}{1-b} \quad \text{where } b = MPC = 0.8$$
$$= 10 \times \frac{-0.8}{1-0.8} = 10 \times \frac{-0.8}{0.2} = -40$$

Therefore, new equilibrium level of income is $410 - 40 = 370$

(*b*) (*i*) Lump-Sum Tax Multiplier $= \frac{-b}{1-b} = \frac{-0.8}{1-0.8} = \frac{-0.8}{0.2} = -4$

(*c*) (*i*) With decrease in government expenditure by 10 the change in equilibrium level of income

$$\Delta Y = -\Delta G \times \frac{1}{1-b} = -10 \times \frac{1}{1-0.8} = -10 \times \frac{1}{0.2}$$
$$= -10 \times \frac{10}{2} = -50$$

Therefore, new equilbrium level of income = 410 – 50 = 360

(*ii*) Government expenditure multiplier $= \dfrac{1}{1-b} = \dfrac{1}{1-0.8} = \dfrac{1}{0.2} = 5$

(*d*) *Explaining the difference in the Impact of decrease in Government Expenditure and increase in lump sum tax.*

A decrease in Government expenditure causes a decrease in aggregate expenditure by the same amount and will therefore have a greater impact on income. On the other hand, increase in lump sum tax which will cause reduction in disposable income by the same amount but reduction in expenditure will be smaller because households will cut their consumption expenditure by the amount of tax multiplied by marginal propensity to consume. Thus, in this case decrease in consumption expenditure ($-\Delta C$) will be equal to $-\Delta T \times MPC$, that is, $-\Delta C = -\Delta T \times MPC$. Thus, if lump sum tax is increased by ₹ 10, then, given $MPC = 0.8$, the decrease in consumption expenditure will be $10 \times 0.8 =$ ₹ 8, that is, less than ₹ 10.

Proportional Income Tax and Equilibrium Level of National Income

Now, instead of lump sum tax, if Government levies proportional income tax how it will affect the consumption function curve and the equilibrium level of income. With proportional income tax, a *fixed percentage* of income is collected as tax, irrespective of the level of income. If an income tax at the rate of 25 per cent of income is levied regardless of the level of income, it will be an example of a proportional income tax. As in case of lump sum tax, the introduction of proportional income tax will reduce the disposable income. This will bring about decline in the amount of consumption and saving at each level of real national income. However, in case of proportional income tax consumption will not fall by the same amount at each level of income but instead it will fall by the *same percentage of the level of income*. Therefore, as a result of proportional income tax, the consumption function curve rotates downward (with point *a* as a pivot) to the right, so that the fall in amount of consumption varies but fall as a percentage of income level remains the same. This is shown in Fig. 5A.5 where the vertical distance between the consumption function curve *C* before tax and the consumption function curve *C′* after tax is equal to the fall in consumption due to the proportional income tax. It is however worth noting that in case of proportional income tax at *each level of real national income, consumption falls by the tax revenue multiplied by the marginal propensity to consume.*

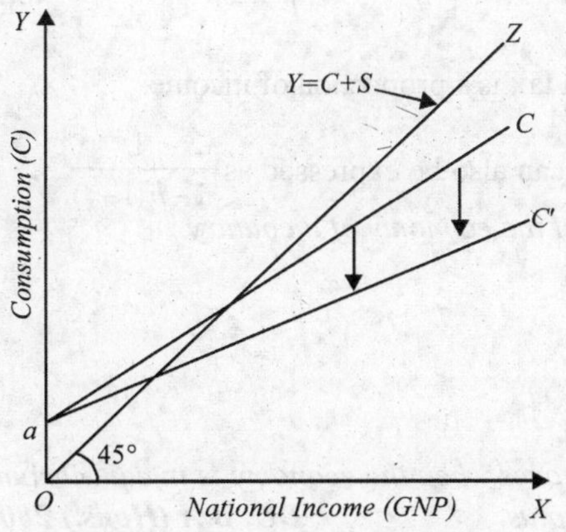

Fig. 5A.5. *Effect of Proportional Income Tax on Consumption Function*

Fig. 5A.6. *Effect of Proportional Income Tax on Aggregate Demand and National Income*

In Fig. 5A.6 we have shown the effect of proportional income tax on the equilibrium level of national income tax. As a result of imposition of proportional income tax, the aggregate demand curve $C+I+G$ rotates downward due to the downward rotation of consumption function curve. It will be seen from Figure 5A.5 that due to imposition of prooprotional income tax, the equilibrium level of national income falls from OY_0 to OY_1. It may be noted that other things remaining the same, the higher rate of proportional income tax, the greater the fall in consumption and, therefore, the greater the decline in equilibrium level of national income.

Three Sector Model Further Extended: (Incorporating Proportional Income Tax)

So far in our above analysis of determination of equilibrium national income in a three sector economy we have assumed autonomous constant investment, apart from consumption function of households and Government expenditure on goods and services and transfer payments. In the further extension of the model we also include *proportional income tax*, that is, a tax levied as a fixed percentage or proportion of income irrespective the level of income . Mathematically, proportional income tax is therefore expressed as tY where t is the rate or proportion of income which is payable as a tax. In a real economy proportional income tax may be imposed along with any lump sum tax. Besides, Governments also makes **transfer payments** such as unemployment allowance, pension and are denoted by TR. While imposition of taxes reduces disposable income, transfer payments (TR) increase disposable income. Thus, the total tax (T) can be expressed as

$$T = tY$$

where t = rate or proportion of income tax

Y = Income

Equilibrium income, $Y = C + I + G$ ...(*i*)

where $C = a + bY_d$

$$= a + b(Y - tY + TR)$$

Substituting the consumption function in equation (*i*) for equilibrium income we have

$$Y = a + b(Y - tY + TR) + I + G$$

$$= a + bY - btY + bTR + I + G$$

$$Y - bY + btY = a + bTR + I + G$$

$$Y(1 - b + bt) = a + bTR + I + G$$

$$Y = \frac{1}{1-b+bt}(a + bTR + I + G) \quad ...(ii)$$

Note that $\frac{1}{1-b+bt}$ represents tax multiplier when tax is a proportion of income.

Tax multiplier with tax as a proportion of income can also be expressed as $\frac{1}{1-b(1-t)}$.

Problem 1. *Given the following information about the economy of a country.*

Consumption function, $C = 85 + 0.5\,Y_d$

Investment function, $I = 85$

Government spending $G = 60$

Net Taxes, $T = 40 + 0.25\,Y$

(i) *Solve for equilibrium income and consumption*

(ii) *How much does the government collect in net taxes when the economy is in equilibrium*

(iii) *What is the government's budget deficit, or surplus.* ***DU, B.A (Hons.) 2007***

Solution.(*i*) Equilibrium income (Y) = $C + I + G$

$$C = 85 + 0.5\,Y_d$$

Since $Y_d = Y - T$

$$C = 85 + 0.5[Y - (40 + 0.25Y)]$$

$$= 85 + 0.5Y - 20 - 0.125\,Y$$

$$= 65 + 0.375\,Y$$

$$Y = C + I + G$$

$$= 65 + 0.375\,Y + 85 + 60$$

$$Y - 0.375Y = 210$$

$$Y(1 - 0.375) = 210$$

$$Y = \frac{210}{1-0.375} = \frac{210}{0.625} = 210 \times \frac{1000}{625}$$

$$= 336$$

(*i*) Equilibrium income is 336

and Equilibrium consumption $= 65 + 0.375\,Y$

$$= 65 + 0.375 \times 336$$

$$= 65 + 126$$

$$= 191$$

(*ii*) Tax money collected by the Government

$$T = 40 + 0.25Y$$

$$= 40 + 0.25 \times 336$$

$$= 40 + 84$$

$$= 124$$

(*iii*) *Government budget deficit or surplus*, Since government expenditure (G) is less than its tax revenue, there will be surplus budget.

Budget surplus $= 124 - 60 = 64$

Numerical Example 2. *For an economy, the following data is given:*

Consumption (C) = 100 + 0.8 Y.

Investment (I) = 150

Government Expenditure (G) = 75

Rate of income tax (T) = 0.25Y

(a) Find the equilibrium level of income.

(b) What is the value of multiplier with tax and without tax.

Ans. Reduced form for the equilibrium level of income with proportionate income tax is

$$Y = \frac{1}{1-b+bt}(a+I+G)$$

Given $b = 0.8$, $t = 0.25$, $a = 100$, $I = 150$ and $G = 75$. Substituting these values in equation for equilibrium income we have

$$Y = \frac{1}{1-0.8+08\times0.25}(100+150+75)$$

$$= \frac{1}{1-0.8+0.2}(325)$$

$$= \frac{1}{1-0.6}(325) = \frac{1}{0.4}(325)$$

$$Y = 2.5\,(325)$$

$$= 812.5$$

The value of proportional tax multiplier $= \dfrac{1}{1-b+bt}$

$$= \frac{1}{1-0.8+0.8\times 0.25}$$

$$= \frac{1}{1-0.6} = \frac{1}{0.4}$$

$$= 2.5$$

The value of multiplier without tax $= \dfrac{1}{1-b} = \dfrac{1}{1-0.8} = \dfrac{1}{0.2} = 5$

THE LUMP-TAX MULTIPLIER

We have explained the multiplier effect of change in Government expenditure on goods and services and of change in transfer payments (*TR*) on the equilibrium level of income. We will now explain the multiplier effect of *change in tax*. Reduction in taxes has often been used by the policy makers to increase aggregate demand of the economy and thereby to lift the economy out of recession. For example, in 1964 John F Kennedy, the president of the United States made a substantial cut in personal and corporate taxes to stimulate expenditure on consumption and investment to bring about expansion in income and employment in the economy. This policy of cut in taxes to revive the American economy had a grand success. There was a large growth in Gross Domestic Product (GDP) of the United States in 1964 and 1965. There was a substantial fall in unemployment as a result of the policy of cut in taxes.

Recently when George Bush was elected president in 2001 the American economy found itself in the grip of recession. He also made a drastic cut in income tax to stimulate aggregate demand so as to overcome recession. Again this policy of tax cut was greatly successful and the American economy recovered in 2003-04.

The change in lump sum income tax and proportional income tax affect equilibrium income differently, but both have a multiplier effect on equilibrium income. We have already explained the effect of lump-sum income tax and proportional income tax on the equilibrium level of income. We are now concerned with analysing the effect of change in lump sum tax and proportional income tax. For the sake of simplicity and convenience we explain the multiplier effect of lump sum income tax. Note that while a cut in tax will have a positive effect on equilibrium income, increase in tax will reduce equilibrium income.

Suppose the government *increases lump-sum income tax* by ΔT we are interested to know its multiplier effect on income. As found above, in the three sector economy with Government expenditure and lump-sum tax :

$$Y = \frac{1}{1-b}(a - bT + I + G) \qquad \text{... (1)}$$

where T is lump sum income tax. Now, a decrease in lump-sum income tax by ΔT will cause a change in income (that is, ΔY). Increase in tax will reduce the disposable income and therefore the consumption demand. As a result of increase in tax, consumption demand will decrease by the change in tax times the marginal propensity consume, that is, $\Delta Y = b\Delta T$. Let us incorporate change in the tax (ΔT) in the three sector equation (1) for equilibrium income and see how much change in income (ΔY) it brings about. Thus

$$Y + \Delta Y = \frac{1}{1-b}[a - b(T + \Delta T) + I + G]$$

$$Y+\Delta Y=\frac{1}{1-b}\left(a-bT-b\Delta T+I+G\right) \qquad \ldots(2)$$

Now subtracting equation (1) from equation (2) we have

$$\Delta Y=\frac{1}{1-b}\left(-b\Delta T\right)$$

Dividing both sides by ΔT we have

$$\frac{\Delta Y}{\Delta T}=\frac{-b}{1-b} \qquad \ldots(3)$$

$\frac{\Delta Y}{\Delta T}$ is the expression for tax multiplier. It will be seen that tax multiplier is *negative*. It will be seen that increase in tax by ΔT will *reduce* equilibrium income by a multiple. For example, if marginal propensity to consume is 0.75, then the tax multiplier is

$$\frac{\Delta Y}{\Delta T}=\frac{-b}{1-b}=\frac{-0.75}{1-0.75}=\frac{-0.75}{0.25}=-3$$

This means that increase in tax by Re. 1 will cause a decline in equilibrium income by 3.

Note that we can similarly prove that a *cut in tax* will *raise* the equilibrium income through the tax multiplier. Therefore, in case of *reduction in lump sum income tax*, the tax multiplier is

$$\frac{\Delta Y}{\Delta T}=\frac{b}{1-b}$$

With 0.75 as marginal propensity to consume, tax multiplier of cut in tax is,

$$\frac{\Delta Y}{\Delta T}=\frac{0.75}{1-0.75}=\frac{0.75}{0.25}=3.$$

Problem. *The following specifications are given for an economy*
Consumption, $C = 250 + 0.75\ Y_d$ *where* Y_d *is disposable income*
Government expenditure $G = 150$
Investment $I = 80$
Taxes $T = 200$

(i) Find the equilibrium level of income (Y), Consumption (C) and Private Sector Saving (S_p)

(ii) Using the value of tax multiplier how much will income increase if taxes are reduced by 30 Crore. *DU BA (Hons.) 2009*

Ans. Equilibrium

$$Y = C + I + G$$
$$C = 250 + 0.75\ Y_d$$
$$= 250 + 0.75\ (Y - T)$$
$$= 250 + 0.75\ (Y - 200)$$
$$Y = 250 + 0.75\ (Y - 200) + 80 + 150$$
$$= 250 + 0.75Y - 150 + 80 + 150$$
$$(Y - 0.75Y) = 250 - 150 + 80 + 150$$
$$Y(1 - 0.75) = 330$$
$$Y = \frac{330}{1-0.75}=\frac{330}{0.25}=330\times\frac{100}{25}=1320$$
$$\text{Consumption, } C = 250 + 0.75\ (Y - T)$$
$$= 250 + 0.75\ (1320 - 200)$$

$$= 250 + \frac{75}{100} \times 1120$$

$$= 250 + 840$$

Therefore, $C = 1090$

Private Sector Saving (S_p) = Disposable income – Consumption

Disposable Income $= Y - T$

Private Saving $S_p = Y - T - C$

$$= 1320 - 200 - 1090$$

$$= 30$$

(*ii*) The *decrease* in taxes will lead to *increase* in income by raising consumption expenditure.

Tax multiplier when taxes are reduced $= \frac{b}{1-b}$

Decrease in Taxes $= 30$

Increase in Income $= \Delta T \times \frac{b}{1-b}$

$$= 30 \times \frac{0.75}{1-0.75}$$

$$= 30 \times \frac{0.75}{0.25} = 90$$

BALANCED BUDGET MULTIPLIER

We now analyse the impact of balanced budget of the government. In case of a balanced budget, the government's expenditure is equal to its tax revenue, that is, $G = T$ and therefore when the government increases its expenditure by ΔG, then it also raises taxes to get more tax revenue (ΔT) so that $\Delta T = \Delta G$. The effect of a balanced budget on the equilibrium income is described by the *balanced budget multiplier* which is also called *balanced budget theorem*. According to this, the *balanced budget multiplier is always equal to one*. We prove it below by assuming a lump sum tax.

In order to analyse the effect of a balanced budget multiplier, we reproduce below the equation for equilibrium income for a three-sector economy.

$$Y = \frac{1}{1-b}\left(a - b\overline{T} + I + G\right) \qquad ...(1)$$

where $\overline{T}$ is constant lump sum tax levied by the government.

Let us incorporate increase in government expenditure (ΔG) fully financed by additional lump sum tax (ΔT), that is, $\Delta G = \Delta T$. By doing so we get

$$Y + \Delta Y = \frac{1}{1-b}\left[a - b(\overline{T} + \Delta T) + I + G + \Delta G\right] \qquad ...(2)$$

Subtracting equation (1) from equation (2) we have

$$\Delta Y = \frac{1}{1-b}(-b\Delta T + \Delta G)$$

Since in a balanced budget, $\Delta T = \Delta G$

$$\Delta Y = \frac{1}{1-b}(-b\Delta G + \Delta G)$$

$$\Delta Y(1-b) = \Delta G(1-b)$$

or $$\Delta Y = \Delta G \quad ...(3)$$

Dividing both sides of the above equation by ΔG we have

$$\frac{\Delta Y}{\Delta G} = \frac{\Delta G}{\Delta G} = 1$$

Thus, balanced budget multiplier is equal to one. This implies that when $\Delta G = \Delta T$ and budget is therefore balanced, national income increases by the same amount as the increase in government expenditure.

Alternative Proof

That the balanced budget multiplier is equal to one can be proved in an alternative way. Balance budget multiplier is the sum of government expenditure multiplier and lump-sum tax multiplier.

Government expenditure multiplier $(G_m) = \frac{1}{1-b}$

where b represents marginal propensity to conseve

Lump-sum tax multiplier $(T_m) = \frac{-b}{1-b}$

Balanced budget multiplier = $G_m + T_m$

$$= \frac{1}{1-b} + \frac{-b}{1-b}$$

$$= \frac{1-b}{1-b} = 1$$

Thus, the balanced budget multiplier is equal to one.

NUMERICAL EXAMPLES

Problem. *Suppose we have an economy characterised by the following functions :*

$$C = 100 + 0.8\, Y_d$$
$$\overline{I} = 100$$
$$\overline{G} = 100$$
$$\overline{T} = 100$$

(*Figures are in crores*)

(*a*) *Find the equilibrium level of income.*

(*b*) *How much increase in income will take place if government expenditure on goods and services increases by ₹60 crores.*

(*c*) *Find the tax multiplier and balanced budget multiplier*

(*d*) *Find the equilibrium level of national income if* $T = \overline{T} + tY = 100 + 0.25Y$.

Solution :

(*a*) Equilibrium $Y = C + \overline{I} + \overline{G}$

$$Y = a + bY_d + \overline{I} + \overline{G}$$

Therefore, $Y = a + b(Y - \overline{T}) + \overline{I} + \overline{G}$

Substituting the values of *a*, *b*, $\overline{T}$, $\overline{I}$ and $\overline{G}$ we have

$$Y = 100 + 0.8\,(Y - 100) + 100 + 100$$

$$= 100 + 0.8Y - 80 + 100 + 100$$

$$Y - 0.8Y = 220$$

$$Y(1 - 0.8) = 220$$

$$0.2Y = 220$$

$$Y = 220 \times \frac{10}{2} = 1100$$

(*b*) Government expenditure multiplier $(G_m) = \frac{1}{1-b} = \frac{1}{1-0.8} = 5$

Increase in national income if government expenditure increases by 60 crores.

$$\Delta Y = G_m . \Delta G$$
$$= 5 \times 60 = 300$$

(*c*) Tax multiplier, $T_m = \frac{-b}{1-b} = \frac{-0.8}{1-0.8} = \frac{-0.8}{0.2}$

$$= -4$$

Balanced Budget Multiplier $= G_m + T_m$

$$= 5 + (-4)$$
$$= 1$$

(*d*) The equilibrium level of national income when $T = \bar{T}) \; tY \;\; 100 + 0.25Y$

Note that $\bar{T}$ (= 100) is a fixed lump sum tax that does not vary with income while tY will vary with income (Y) depending on the value of t.

$$Y = \frac{1}{1-b+bt}\left(a - b\bar{T} + \bar{I} + \bar{G}\right)$$

$$= \frac{1}{1-0.8+0.2}(100 - 0.80 \times 100 + 100 + 100)$$

$$= \frac{1}{1-0.6}(100 - 80 + 100 + 100)$$

$$= \frac{1}{0.4}(220)$$

$$= 2.5\,(220) = 550$$

DETERMINATION OF NATIONAL INCOME IN AN OPEN ECONOMY: THE FOUR SECTOR MODEL

In modern times economies of different countries are integrated with each other though some are more integrated than others. An open economy is one which has not only trade relations with other countries but has financial capital flows between it and other economies of the world. It is because of such openness and integration of the economies of the world that recent financial crisis that occurred in the United States in 2007 has significantly affected other economies (including that of India) causing global meltdown through its affect on trade and capital flows. Under capital flows, funds flow from one country to another. Foreign trade of goods and services affect product market and influences the determination of national income of a country as net exports (*i.e.* exports-imports) is a part of aggregate demand for output produced by an economy. On the other hand, capital flows in an open economy provide funds for financing domestic investment or for financing government budget deficits. If the demand for funds in an economy exceeds domestic national saving, the private sector and Government of an open economy can get them from abroad either through borrowing or direct investment by foreign investors. We start with the analysis of impact of inclusion of foreign trade on the determination of national income of a country.

Foreign Trade and National Income in an Open Economy

In the four sector-model of determination of national income, we add the foreign trade sector to the three sectors, namely, households, firms, and Government. Foreign trade, that is, volume of exports and imports of a country also affects the level of national income of a country. We now extend the Keynesian model of determination of national income by including the effect of exports and imports of a country on the generation of income. For example, the exports of India represent the foreign demand and generate income for the Indian people. On the other hand, imports represent the demand for foreign goods by the Indians and generate incomes for the people of other countries. Therefore, imports tend to reduce the domestic aggregate expenditure. It therefore follows that national income will depend on the net exports, that is, exports minus imports $(X-M)$ where X stands for exports and M for imports.

The exports and imports of a country depend to a great extent on the level of economic activity (that is, the level of output and income of the country). Thus, when the growth of industrial output in India is rapid, it will generate greater demand for imported materials. On the other hand, the higher industrial growth would also cause our exports to rise provided there is demand for our goods abroad. However, in the Keynesian model of income determination, exports and imports are considered as autonomous, that is, independent of income, being determined outside the model. Further, as mentioned above, the increase or decrease in aggregate expenditure or demand due to exports and imports depends on the *net exports*, that is, $X–M$ which we may write as NX. If the net exports are positive, there will be addition to the aggregate demand or expenditure of a country. On the other hand, if the net exports are negative, there will be decrease in aggregate expenditure.

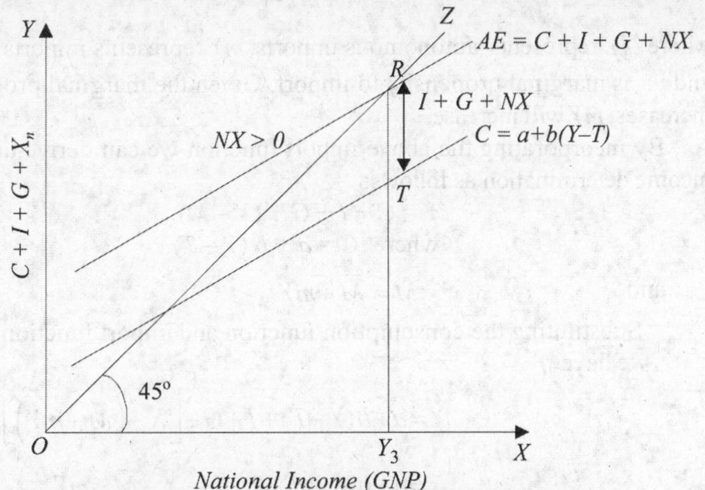

Fig. 5A.7. *Determination of National Income with Foreign Trade: Four-Sector Model*

When we include net exports (NX) in our analysis we get the following equation for the equilibrium level of income.

$$Y = C+I+G+(X-M)$$

where
$$C = a+b(Y-T) = a+bY-bT$$

$$Y = a+bY-bT+I+G+NX$$

or
$$Y(1-b) = a-bT+I+G+NX$$

$$Y = \frac{1}{1-b}(a-bT+I+G+NX)$$

Thus, the equilibrium level of income is the sum of all fixed autonomous expenditures (*i.e.*,

$a - bT + I + G + NX$) times the value of multiplier [$1/1 - b$]. It is worth noting that in the four-sector model, national income is determined at the level at which saving gap between consumption and income is equal to the sum of investment, Government expenditure and net exports. (that is, $I + G + NX$)

Graphic Illustration. In Fig. 5A.7, we have depicted the determination of national income when there are positive net exports (that is, when exports exceed imports, $X - M > 0$. To obtain the aggregate expenditure curve incorporating the positive net exports (NX) we add the $I + G + NX$ to the consumption function curve C to get the higher aggregate expenditure curve $C + I + G + NX$ which intersects the 45° line at point R and determines a level of income OY. If the net exports NX were negative, that is, imports exceed exports, $X - M < 0$, the aggregate demand curve incorporating net exports would lie at a lower level than $C + I + G$ curve and determine a lower level of income.

Four-Sector Model Further Extended: Foreign Trade Multiplier

So far in our analysis of four sector model of income determination we have assumed that both exports and imports are constant autonomous items, that is, independent of income. In our later chapters concerning the external sector we shall explain in detail what are the factors that determine exports and imports of a country. In the model of income determination while exports are regarded as constant autonomous factor *determined exogenously* (that is, $X = \overline{X}$), imports are treated as function of two variables, namely, (1) autonomous imports (*i.e.*, independent of income), and (2) level of income. Thus, import function used in the model can be written as

$$M = \overline{M} + mY$$

where $\overline{M}$ represents autonomous imports, mY represents imports dependent on level of income (Y) and m is marginal propensity to import. Given the marginal propensity to import (m), as income increases, mY will increase.

By incorporating the above import function we can derive the complete four-sector model of income determination as follows:

$$Y = C + I + G + (X - M) \quad \ldots(i)$$

$$\text{where } C = a + b\,(Y - T)$$

$$\text{and} \quad M = \overline{M} + mY$$

Substituting the consumption function and import function in the income equation (*i*) we have

$$Y = a + b(Y - T) + I + G + \left[X - \left(\overline{M} + mY\right)\right]$$

$$= a + bY - bT + I + G + X - \overline{M} - mY$$

$$Y - bY + mY = a - bT + I + G + X - \overline{M}$$

$$Y(1 - b + m) = a - bT + I + G + X - \overline{M}$$

$$Y = \frac{1}{1 - b + m}(a - bT + I + G + X - \overline{M})$$

where the term $\frac{1}{1 - b + m}$ is known as *foreign trade multiplier* whose value is determined by marginal propensity to consume (b) and marginal propensity to import (m). *Foreign trade multiplier is also known as open economy multiplier.* Note that change in any autonomous factor of the model such as X, G, I, a, $\overline{M}$ will cause a change in national income by the value of the foreign trade multiplier $\left(\dfrac{1}{1-b+m}\right)$ times the change in the amount of the factor. Thus, if exports increase by ΔX, the national income will increase by

$$\Delta Y = \frac{1}{1-b+m} . \Delta X$$

The value of foreign trade multiplier (*i.e.*, open economy multiplier) is less than the closed economy investment or Government expenditure. This is because imports represents the demand for foreign goods and services and generate income for people of *other countries*. Like saving, imports represent leakage from the domestic income stream.

Incorporating Proportional Income Tax in the Four Sector Model

If in the above four sector model of income determination proportional income tax is incorporated, then only the term of foreign trade multiplier will change, the other terms of the model remaining the same. Thus, if income tax is of form $T = \overline{T} + tY$ where T is constant lump-sum, t is the proportion of income that is taken as tax. With incorporation of proporational income tax, the value of foreign trade multiplier becomes :

$$= \frac{1}{1-b(1-t)+m}$$

or
$$= \frac{1}{1-b+bt+m}$$

where t is the proportional income tax rate.

With this proportional income tax, the above equilibrium income equation can be written as

$$Y = \frac{1}{1-b(1-t)+m}\left(a - bT + I + G + X - \overline{M}\right)$$

TRADE BALANCE (NET EXPORTS) AND FOREIGN CAPITAL FLOWS

In an open economy as in the closed economy product markets and capital markets are closely related. To know this relationship let us consider the equilibrium level of national income in an open economy which is :

$$Y = C + I + G + (X - M) \quad \ldots (1)$$

where $X - M$ are net exports (N^X)

Therefore, rearranging the above equation and writing NX for $X - M$ we have

$$Y - C - G = I + NX \quad \ldots (2)$$

Now Y–C–G represents national saving[2] which we represent by S.

Writting S for national saving (Y–C–G) in equation (2) we have

$$S = I + NX \quad \ldots (3)$$

or
$$S - I = NX \quad \ldots (4)$$

From equation (4) it is evident that difference between domestic saving and domestic investment ($S - I$) equals net exports (NX). *Since difference between domestic saving and investment ($S - I$) represents capital flows*, equation (4) reveals that capital flows between an economy and the foreign ones are related to net exports (NX) which are also called *trade balance*. If for a country domestic saving exceeds domestic investment ($S > I$) then the country would have *positive* net *capital outflows* which means residents of the country would be lending or investing abroad more than the foreigners would be lending or investing in the country concerned. The equation (4) above

2. Note that national saving $Y - C - G$ is the sum of private saving ($Y - T - C$) and Government saving ($T - G$). Thus $Y - T - C + T - G = Y - C - G$

shows that these *positive net capital outflows* imply that they will be equal to *positive net exports* (*NX*), that is, the country will be exporting more than it will be importing. Foreigners will need financial loans to buy that country's more goods and services than they are exporting to it. This has been the relationship with Japan and the US. Japan has a trade surplus (i.e. positive net exports) with the US and to enable the US to buy more Japanese goods than Japan is importing from it, it has been lending to the US which has been running a trade deficit, that is, there has been net capital outflows from Japan to the US.

Now, when domestic saving of a country is less than the domestic investment ($S < I$ or $S - I$ is negative), the equation (4) above reveals that the net exports (*NX*) will be negative (i.e., the country has a trade deficit) and, therefore, there will *net capital-inflow*. The net capital inflow implies that the country's will be borrowing from abroad or receiving foreign investment. The country having negative net exports or deficit trade balance requires foreign capital funds either through borrowing from foreigners or foreign investment so as to enable it to import more goods and services than it is exporting them. For example, in India we have a *deficit in the current account of balance of payments* (which has the same meaning as the concept of trade balance involving both goods and services) for several years now so that foriegn saving or capital inflow have contributed to the increase in investment in India. Though the share of capital inflows (foriegn savings) as a per cent of its GDP in India are relatively very small.

It follows from above that *international flow of capital to finance investment and international flow of goods and services are two sides of the same coin.*[3]

NUMERICAL PROBLEMS

Problem 1. *An economy is characterised by the following equations:*

C (*Consumption*)	=	$60 + 0.9Y_d$
I (*Investment*)	=	10
G (*Government expenditure*)	=	10
T (*Tax*)	=	0
X (*Exports*)	=	20
M (*Imports*)	=	$10 + 0.05\ Y$

(a) *What is the equilibrium income*?

(b) *Calculate trade balance.*

(c) *What is the value of foreign trade multiplier*?

Solution : (*a*) *National Income*

$$Y = C + I + G + Nx$$
$$= (60 + 0.9\ Y_d) + 10 + 10 + 20 - (10 + 0.05Y)$$
$$= 60 + 0.9\ (Y - T) + 10 + 10 + 20 - 10 - 0.05Y$$
$$= 60 + 0.9Y - 0 + 30 - 0.05Y$$
$$Y = 90 + .85Y$$
$$Y - 0.85Y = 90$$
$$0.15Y = 90$$
$$Y = 90 \times \frac{100}{15} = 600$$

(*b*) *Trade Balance* = $X - M$

$= 20 - (10 + 0.05Y)$

Substituting the value of *Y* we have

3. See N. Gregory Mankiw, *Macroeconomics*, Worth Publishers, New York 5th edition, 2003, p. 118

$$\text{Trade balance} = 20 - \left(10 + \frac{5}{100} \times 600\right)$$

$$= 20 - 10 - 30 = -20$$

Trade balance is in deficit.

(*c*) *Value of foreign trade multiplier* $= \dfrac{1}{1-b+m}$

Where b is marginal propensity to consume, and m is marginal propensity to import.

$$\text{Foreign trade multiplier} = \frac{1}{1-0.9+0.05}$$

$$= \frac{1}{1-0.85} = \frac{1}{0.15}$$

$$= \frac{100}{15} = 6.66$$

EXERCISES

1. How does Government expenditure on goods and services affect level of national income? Does it have a multiplier effect? What is the value of government expenditure multiplier.
2. Show that with Government expenditure in a three sector economy (households, firms and Government) equilibrium of national income is established at the level where sum of investment and Government expenditure ($I + G$) is equal to the amount of saving.
3. Show how the increase in Government expenditure without raising taxes (*i.e.*, incurring budget deficit) affects national income, real rate of interest and private investment. Illustrate diagrammatically.
4. It is often asserted that budget deficit crowds out private investment. Do you agree ? Explain.
5. Explain what will be the effect of cut in government expenditure without reduction in taxes (creating budget surplus in this way) on equilibrium national income, real rate of interest and private investment in the economy.
   [**Hints.** A cut in government expenditure without reducing taxes leads to increase in public saving (*i.e.*, $T - G$ increases). This causes a shift in national saving curve to the right and leads to the fall in rate of interest and increase in private investment.
6. Show how does a cut in taxes without being balanced by cut in government expenditure affect real interest rate and investment.
7. What is effect of a balanced budget (*i.e.*, increasing government expenditure matched by increase in equal amount of taxes) on national income, real interest rate and investment
8. Explain how balanced budget has an expansionary effect. What is value of balanced budget multiplier? Explain.
9. How does cut in direct taxes lead to increase in national income? What is the size of lump-sum tax multiplier?
10. Assuming full-employment in an economy balanced budget *increase* in taxes and Government expenditure will reduce consumption and investment and raise the real interest rate? Explain.
11. Assuming that full-employment prevails in an economy, increased government expenditure matched by taxes crowds out private investment.
12. When the government runs a budget deficit, how is the capital market affected?

13. What are the components of aggregate expenditure in an open economy. With Government spending on goods and services show that equilibrium level of national income in an open economy is established at the level at which $I + G + NX$ equals saving.

14. Explain foreign trade multiplier (the open economy multiplier). What is its value? Why is it smaller than closed economy expenditure multiplier? ***DU B.A (Hons.) 2007***

15. You are given the following information about an economy.

Consumption function	$C = 80 + 0.8Y$
Investment function	$I = 80$
Government spending	$G = 60$
Net Taxes	$T = 60$
Exports	$X = 20$
Imports	$M = 0.2Y_d$

(*i*) What is the current account balance for this economy

(*ii*) What is the value of the open economy multiplier? ***DU. B.A.(Hons.) 2007***

(**Note**: Since in this question no data of '*invisibles*' are given, current account balance equals *trade balance*).

16. You are given the following information about an economy–

Consumption function, $C = 1000 + 0.5\,(Y - T)$

Investment, I = ₹2,000 crores.

Government expenditure = ₹1,000 crores

Taxes = ₹1,000 crores

(*i*) Find the equilibrium level of GDP without taxes.

(*ii*) Find the equilibrium level of GDP with taxes.

(*iii*) Use your answers in (*i*) and (*ii*) above to calculate the tax multiplier.

(*iv*) Explain the working of the tax multiplier intuitively. ***DU B.A., (Hons.) 2008***

[**Hints.** Find the equilibrium level of GDP (*i.e.* national income Y), first by laking $T = 0$ and then find the equilibrium GDP *i.e.*, (Y) by taking $T = 1000$. Then tax multiplier is $\dfrac{\Delta Y}{\Delta T}$. Note that $\Delta T = 1000$.)

17. Explain the relationship between trade deficit and inflow of foreign capital.

18. International flow of capital to finance investment and international flow of goods and services are two sides of the same coin. Explain.

CHAPTER 6
CONSUMPTION FUNCTION

INTRODUCTION

In the previous chapter we have explained how the level of national income is determined. In the present chapter and the succeeding one we shall study the factors that determine consumption demand and investment demand – the two important components of aggregate demand. As we have seen, given the aggregate supply, the level of income or employment is determined by the level of aggregate demand; the greater the aggregate demand, the greater the level of income and employment and vice versa. Keynes was not interested in the factors determining the aggregate supply since he was concerned with the short run and the existing productive capacity. We will also not explain in detail the factors which determine the aggregate supply and will confine ourselves to explaining the determinants of aggregate demand. In the present chapter we will explain the consumption demand and the factors on which it depends and how it changes over a period of time. The investment demand will be explained in the next chapter. Consumption demand depends upon the level of income and the propensity to consume.

CONCEPT OF CONSUMPTION FUNCTION

As the demand for a good depends upon its price, similarly consumption of a community depends upon the level of income. In other words, consumption is a function of income. *The consumption function relates the amount of consumption to the level of income.* When the income of a community rises, consumption also rises. How much consumption rises in response to a given increase in income depends upon the marginal propensity to consume. It should be borne in mind that the *consumption function is the whole schedule which describes the amounts of consumption at various levels of income.* We give in Table 6.1 on the next page such a schedule of consumption function.

Consumption function should be carefully distinguished from the *amount of consumption.* By consumption function is meant the whole schedule which shows consumption at various levels of income, whereas *amount of consumption means the amount consumed at a specific level of income.* The schedule given in Table 6.1 reflects the consumption function of a community *i.e.,* it indicates how the amount of consumption changes in response to the changes in income. From the schedule given in Table 6.1 it will be seen that at the level of income equal to ₹ 1200 crore, the amount of consumption is ₹ 1090 crores. As the national income increases to ₹ 1500 crore, the amount of consumption rises to ₹ 1300 crore. Thus, with a given consumption function, amount of consumption is different at different levels of income.

The schedule of consumption function in Table 6.1 reveals an important fact that when income rises, consumption also rises but not as much as the income. This fact about consumption function was emphasised by Keynes who first of all evolved the concept of consumption function. The reason why consumption rises less than income is that a part of the increment in income is saved. Therefore, we see that when income increases from ₹ 1000 crore to ₹ 1100 crore, the amount of consumption rises from ₹ 950 crore to 1020 crore. Thus, with the increase in income by ₹ 100 crore, consumption rises by ₹ 70 crore; the remaining ₹ 30 crore are saved. Similarly, when income rises from ₹ 1100 crore to ₹1200 crore, the amount of consumption increases from ₹ 1020 crore to ₹ 1090 crore.

Table 6.1. Linear Consumption Function

Income (₹ in crores) Y	Consumption (₹ in crores) C	Average Propensity to Consume $\left(\frac{C}{Y}\right)$	Marginal Propensity to Consume $\left(\frac{\Delta C}{\Delta Y}\right)$
1000	950	$\frac{950}{1000} = 0.950$	—
1100	1020	$\frac{1020}{1100} = 0.927$	$\frac{70}{100} = .70$
1200	1090	$\frac{1090}{1200} = 0.908$	$\frac{70}{100} = .70$
1300	1160	$\frac{1160}{1300} = 0.892$	$\frac{70}{100} = .70$
1400	1230	$\frac{1230}{1400} = 0.878$	$\frac{70}{100} = .70$
1500	1300	$\frac{1300}{1500} = 0.867$	$\frac{70}{100} = .70$
1600	1370	$\frac{1370}{1600} = 0.856$	$\frac{70}{100} = .70$

Here also, as a result of increase in income by ₹ 100 crore, the amount of consumption has risen by ₹ 70 crore and the remaining ₹ 30 crore have been saved. The same applies to further increases in income and consumption. We shall see later that Keynes based his theory of multiplier on the proposition that consumption increases less than income and this theory of multiplier occupies an important place in macroeconomics.

In the Keynesian consumption function given in Table 6.1 while marginal propensity to consume $\left(\frac{\Delta C}{\Delta Y}\right)$ remains constant at ₹ 70 crore, average propensity to consume is $\left(\frac{C}{Y}\right)$ is falling with the increase in income. It should be noted that Keynesian consumption function given in Table 6.1 is linear as marginal propensity to consume which measures the slope $\left(\frac{\Delta C}{\Delta Y}\right)$ of the consumption function curve is constant. It is noteworthy that while marginal propensity to consume $\left(\frac{\Delta C}{\Delta Y}\right)$ in the consumption function given in Table 6.1 is constant, average propensity to consume $\left(\frac{C}{Y}\right)$ falls with increase in national income (Y). *The fall in average propensity to consume with the increase in income has an important implication that increase in consumption is not proportional to increase in income*. Besides, as will be seen from Table 6.1 in Keynesian consumption function, $MPC < APC$ at various levels of income. We shall explain the importance of these two features of Keynesian consumption function later.

Consumption demand depends on income and propensity to consume. Propensity to consume depends on various factors such as price level, interest rate, stock of wealth and several subjective factors. *Since Keynes was concerned with short-run consumption function he assumed price level, interest rate, stock of wealth etc. constant in his theory of consumption.* Thus with these factors being assumed constant in the short run, Keynesian consumption function considers consumption as a function of *current* income. Thus

$$C=f(Y)$$

In a specific form, Keynesian consumption function can be written as:

$$C = a + bY$$

where *a* and *b* are constants. While *a* is intercept term of the consumption function, *b* stands for slope of the consumption function and therefore represents marginal propensity to consume and *Y* represents the level of *current income*.

Keynesian consumption function, namely, $C = a + bY$ has been depicted by *CC′* curve in Fig.6.1 in which along the *X* -axis national income and along the *Y*-axis the amount of consumption are measured. In this figure, a line *OZ* making 45° angle with the *X*-axis, has been drawn. Because line *OZ* makes 45° angle with the *X*-axis every point on it is equidistant from both the *X*-axis and *Y*-axis. Therefore, if consumption function curve coincides with 45° line *OZ* it would imply that the amount of consumption is equal to the income at every level of income. In this case, with the increase in income, consumption would also increase by the same amount. As has been said above, in actual practice consumption increases less than the increase in income. Therefore, in actual practice the curve depicting the consumption function will deviate from the 45° line. If we represent the above consumption schedule by a curve, we would get the propensity to consume curve such as *CC′* in Fig. 6.1.

It is evident from Fig. 6.1 that the consumption function curve *CC′* deviates from the 45° line *OZ.* At lower levels of income, the consumption function curve *CC′* lies above the *OZ* line, signifying that at these lower levels of income consumption is greater than the income. It is so because at lower levels of income, a nation may draw upon its accumulated savings to maintain its consumption standard or it may borrow from abroad. As income increases, consumption also increases and at the national income level OY_0, consumption is equal to income. Beyond this, with the increase in income,

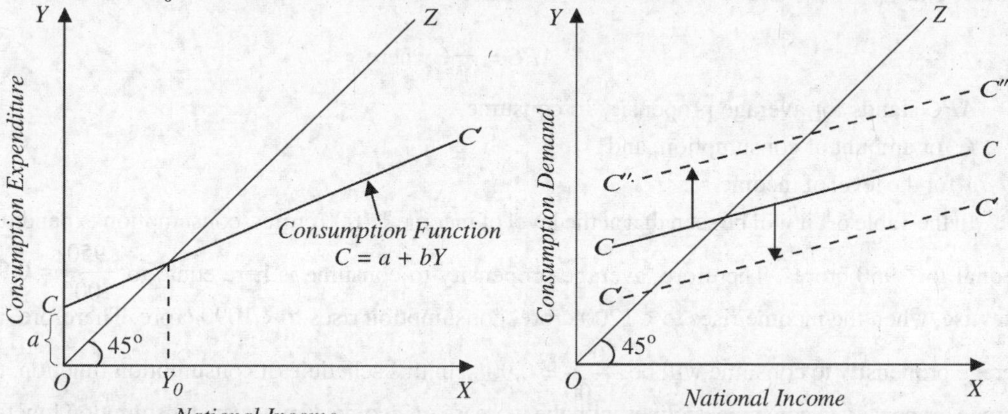

Fig. 6.1. *Keynesian Linear Consumption Function* **Fig. 6.2.** *Shift in Consumption Function*

consumption increases but less than the increase in income and therefore, consumption function curve *CC′* lies below the 45° line *OZ* beyond Y_0. An important point to be noted here is that beyond the level of income OY_0, the gap between consumption and income is widening. The difference between consumption and income represents savings. Therefore, *with the increase in income, saving gap widens* and, as we shall see later, this has a significant implication in macroeconomics.

It is useful to point out here that *when the consumption function of a community changes, the whole consumption function curve changes or shifts*. When propensity to consume increases, it means that at any given level of income more is consumed than before. Therefore, as a result of increase in propensity to consume of the community, the whole consumption function curve shifts upward as has been shown by the upper curve *C′′C′′* in Fig. 6.2. On the contrary, when the propensity to consume of the community decreases, the whole consumption function curve shifts downward signifying that at any given level of income, less is consumed than before.

AVERAGE AND MARGINAL PROPENSITY TO CONSUME

There are two important concepts of propensity to consume, the one being average propensity to consume and the other marginal propensity to consume. They should be carefully distinguished, for they are equal in some cases but different in others. Consider Table 6.1, where we have calculated the average and marginal propensity to consume in columns 3 and 4. As seen above, consumption changes as income changes. Now, how much consumption changes in response to a given change in income depends upon the average and marginal propensity to consume. Thus, propensity to consume of a community can be known by the average and marginal propensity to consume. *Average propensity to consume is the ratio of the amount of consumption to total income*. Therefore, average propensity to consume is calculated by dividing the amount of consumption by the total income. Thus,

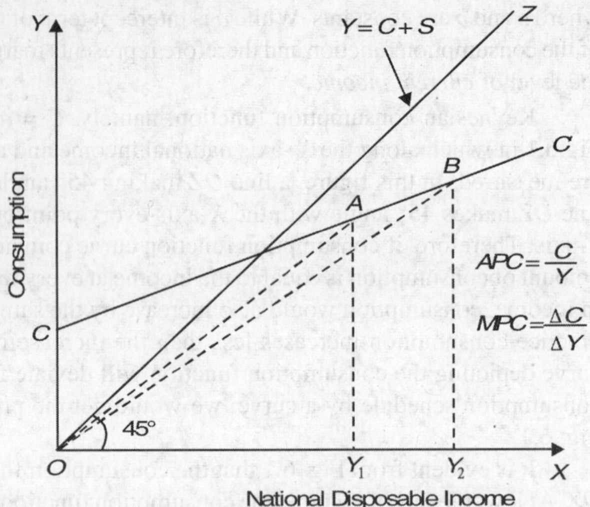

Fig, 6.3. *Consumption Function: Declining Average Propensity to Consume and Constant Marginal Propensity to Consume*

$$APC = \frac{C}{Y}, \text{ where}$$

APC stands for average propensity to consume,

C for amount of consumption, and

Y for the level of income.

In the Table 6.1 it will be seen that at the level of income ₹ 1000 crores, consumption expenditure is equal to ₹ 950 crores. Therefore, average propensity to consume is here equal to $\frac{950}{1000} = 0.95$. Likewise, when the income rises to ₹ 1200 crore, consumption rises to ₹ 1090 crore. Therefore, the average propensity to consume will be $\frac{1090}{1200} = 0.908$. In this schedule of consumption function, the average propensity to consume declines with the increase in income. Keynesian consumption function CC' is shown in Figures 6.1 and 6.3. Average propensity to consume at a point on the consumption function curve can be obtained by measuring the slope of the ray from the origin to that point. For example, at income level OY_1, corresponding point on the consumption function curve is *A*. Therefore, at OY_1 income level, average propensity to consume (*APC*) is the slope of the ray *OA*. Similarly, at income level OY_2, average propensity to consume is the slope of the ray *OB*. It will be observed from Fig.6.3 that slope of *OB* is less than that of *OA*. Therefore, average propensity to consume at income level OY_2 is less than that at income level OY_1. In other words, average propensity to consume has declined with the increase in national disposable income. However, marginal to consume (*MPC*) which measures the slope $\left(\frac{\Delta C}{\Delta Y}\right)$ of the consumption function curve CC' in Fig. 6.3 remains constant with the increase in income throughout because the slope of a straight line is constant.

Non-Linear Consumption Function : Average and Marginal Propensity to Consume

In the consumption function depicted in Fig. 6.3, though average propensity to consume $\left(\frac{C}{Y}\right)$ declines, marginal propensity to consume which equals $\frac{\Delta C}{\Delta Y}$ remains constant since consumption function curve CC' is a straight line and therefore its slope $\left(\frac{\Delta C}{\Delta Y}\right)$ is constant. But, according to some economists, it is not necessary that marginal propensity to consume should be the same at all levels of income. We have constructed below another schedule of consumption function in which marginal propensity to consume declines with the increase in income.

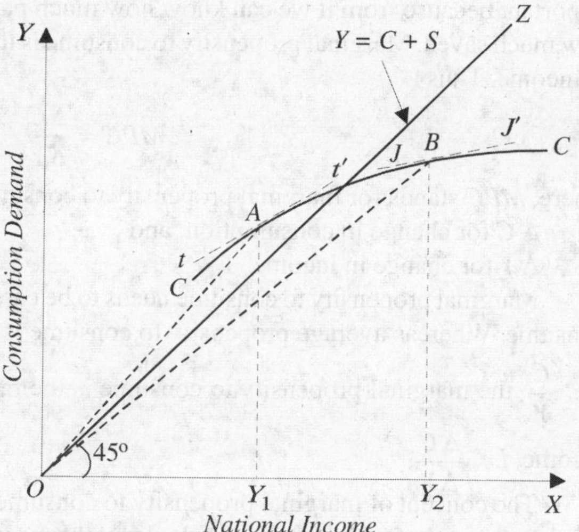

Fig. 6.4. *Non-Linear Consumption Function: Declining both marginal and Average Propensity to Consume*

It will be seen from Table 6.2 that at the level of income of ₹ 1000 crores, marginal propensity to consume is 0.9, and when income rises to ₹ 1500 crores, the marginal propensity to consume has declined to 0.5. When with the increase in income both marginal propensity to consume and average propensity to consume decline, then the curve of consumption function is not a straight line but has a concave shape as shown in Fig. 6.4. From any point on the propensity to consume curve CC we can find out average propensity to consume by joining that point with the point of origin by a straight line whose slope will measure the average propensity to consume. In Fig. 6.4, if we have to find out average propensity to consume at point A on the consumption function

Table 6.2. Non-Linear Consumption Function: Average and Marginal Propensity to Consume

Income (*₹ in crores*) Y	*Consumption* (*₹ in crores*) C	*Average Propensity to Consume* $\left(\frac{C}{Y}\right)$	*Marginal Propensity to Consume* $\left(\frac{\Delta C}{\Delta Y}\right)$
1000	950	$\frac{950}{1000} = 0.950$	–
1100	1040	$\frac{1040}{1100} = 0.945$	$\frac{90}{100} = 0.9$
1200	1120	$\frac{1120}{1200} = 0.933$	$\frac{80}{100} = 0.8$
1300	1190	$\frac{1190}{1300} = 0.915$	$\frac{70}{100} = 0.7$
1400	1250	$\frac{1250}{1400} = 0.893$	$\frac{60}{100} = 0.6$
1500	1300	$\frac{1300}{1500} = 0.866$	$\frac{50}{100} = 0.5$

curve CC, we connect point A with the origin by a straight line. Now, the slope of the line OA *i.e.*, AY_1/OY_1 will indicate the average propensity to consume. Similarly, at point B of the given consumption function CC, the average propensity to consume will be given by the slope of the line OB which is equal to BY_2/OY_2.

The glance at the figure will show that the slope of the line OB is smaller than the slope of the line OA. Therefore, average propensity to consume at point B or at income level OY_2 is less than at point A or income level OY_1.

Marginal Propensity to Consume. The concept of marginal propensity to consume is very important because from it we can know how much part of the increment in income is consumed and how much saved. Marginal propensity to consume is the ratio of change in consumption to the change in income. Thus :

$$MPC = \frac{\Delta C}{\Delta Y}$$

where, MPC stands for marginal propensity to consume,

ΔC for change in consumption, and

ΔY for change in income.

Marginal propensity to consume needs to be carefully distinguished from average propensity to consume. Whereas average propensity to consume is the ratio of total consumption to total income, *i.e.*, $\frac{C}{Y}$, the marginal propensity to consume is the ratio of change in consumption to the change in income, *i.e.* $\frac{\Delta C}{\Delta Y}$.

The concept of marginal propensity to consume can be easily understood with the aid of Table 6.2, in column 4 of which we have calculated the marginal propensity to consume at various levels of income. In this schedule when income rises from ₹ 1000 crores to ₹ 1100 crores, the consumption increases from ₹ 950 crores to ₹ 1040 crores. Here the increment in income is ₹ 100 crores and the increment in consumption is ₹ 90 crores. Therefore, marginal propensity to consume which is $\frac{\Delta C}{\Delta Y}$ is here equal to $\frac{90}{100}$ or 0.9. Similarly, when national income rises to ₹ 1200 crores and as a result consumption increases from ₹ 1040 crores to ₹ 1120 crores, the marginal propensity to consume is now equal to $\frac{80}{100}$ or 0.8. In Table 6.2, it will be seen that marginal propensity to consume declines as the income rises. It is worth noting that when with the increase in income average propensity to consume declines, marginal propensity to consume is less than average propensity to consume. This is in accordance with the usual relationship between the average and marginal quantities, $MPC < APC$. This is evident from Table 6.2.

Marginal propensity to consume can be estimated by drawing the tangent at a point on the consumption function curve. Consider Fig. 6.4 where curve CC depicting the consumption function has been drawn. Marginal propensity to consume at point A on this will be equal to the slope of the tangent tt' drawn at this point. Similarly, marginal propensity to consume at point B on it is given by the slope of the tangent JJ' drawn at this point. It will be seen that slope of the tangent JJ' is less than the slope of the tangent tt'. Therefore, marginal propensity to consume at point B on the consumption function CC in Fig. 6.4 is smaller than the marginal propensity to consume at point A on this consumption function. Thus, marginal propensity to consume declines with the increases in income in the non-linear consumption function curve CC in Fig. 6.4. Thus *when marginal propensity to consume declines with the increase in income, consumption function is non-linear whose slope declines as income rises.* Non-linear consumption function is shown in Fig. 6.4 where the slope of the propensity to consume curve CC declines as income increases. In Fig. 6.1 and Fig. 6.3 propensity to consume curve is a straight line *i.e.,* the slope of the consumption function curve remains constant. Therefore, marginal propensity to consume which is given by the slope of the consumption function curve remains constant in Figures 6.1 and 6.3

It is worth noting that marginal propensity to consume is neither zero nor equal to one. It has been found by empirical studies *that marginal propensity to consume varies between zero and unity.* If the marginal propensity to consume was zero, then the whole of the increment in income would have been saved and the consumption function curve would have a horizontal shape. As we have seen before, this is not so realistic. On the other hand, if the marginal propensity to consume was equal to unity, then the whole of the increment in income would be consumed and in that case consumption function curve would have coincided with 45° line.

SAVING FUNCTION

As mentioned above, consumption increases as income increases but less than the rise in income. We will now explain what happens to saving when income increases. Saving is defined as the part of income which is not consumed because disposable income is either consumed or saved.

Thus,

$$Y = C + S$$
$$S = Y - C$$

where Y = Disposable income, C = Consumption, S = Saving

Like consumption, saving is also a function of income. Thus, saving function can be written as

$$S = f(Y)$$

Saving function is a counterpart of a consumption function, Therefore, given a particular consumption, function, we can derive the corresponding saving function. Let us take the Keynesian consumption, namely, $C = a + bY$. We can derive saving function corresponding to it.

$$\text{Since } Y = C + S$$
$$S = Y - C \quad \text{... }(i)$$

Now, substituting the Keynesian consumption function for C in (i) we have

$$S = Y - (a + bY)$$
$$= Y - a - bY$$
$$= -a + Y - bY$$
$$= -a + (1 - b)\,Y \quad \text{... }(ii)$$

Note that $(1- b)$ in the above saving function in *(ii)* is the value of marginal propensity to save where b is the value of marginal propensity to consume. Let us give a numerical example. Suppose the following consumption function is given.

$$C = 150 + 0.80\,Y$$
$$S = Y - C$$

Substituting the given consumption function for C we have

$$S = Y - 150 - 0.80\,Y$$
$$= -150 + Y - 0.80\,Y$$
$$= 150 + (I - 0.80)\,Y$$
$$= -150 + 0.20\,Y$$

Note that 0.20 represents marginal propensity to save. It also follows from above that the sum of marginal propensity to consume and marginal propensity to save is equal to one $(MPC + MPS = 1)$. It is important to distinguish between average propensity to save and marginal propensity to save.

Average Propensity to Save (*APS*)

An important relationship between income and saving is described by the concept of average propensity to save. *(APS)*. Average propensity to save is the proportion of disposable income that is saved *(i.e.* not consumed). Mathematically

$$APS = \frac{\text{Savings}}{\text{Disposable Income}} = \frac{S}{Y}$$

Like the average propensity to consume *(APC)* average propensity to save also generally varies as income increases. As seen above, average propensity to consume *(APC)* falls as income increases. This implies that average propensity to save will increase as income rises.

Let us derive an important relationship between average propensity to consume and average propensity to save. Restating below the relation that income is either consumed or saved:

$$C + S = Y$$

Dividing both sides by disposable income Y we have

$$\frac{C}{Y} + \frac{S}{Y} = \frac{Y}{Y} = 1$$

Since $\frac{C}{Y}$ is average propensity to consume and $\frac{S}{Y}$ is average propensity to save, we have

$$APC + APS = 1$$

or $$APS = 1 - APC$$

This means, for example, that if a society consumes 75 per cent of its disposable income, that is, $APC = 0.75$, then it will save 25 per cent of its disposable income or its average propensity to save (*APS*) will be 0.25 $(1 - 0.75 = 0.25)$.

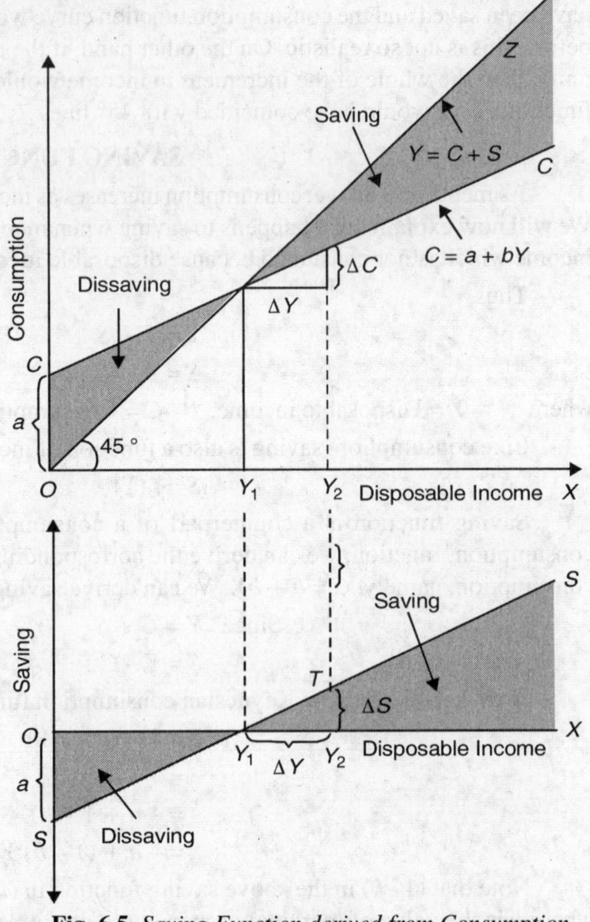

Fig. 6.5. *Saving Function derived from Consumption Function*

In Fig. 6.6 we have drawn the saving curve *SS* in the panel at the bottom. The saving curve *SS* shows the gap between consumption curve *CC* and the income curve *OZ* in the upper panel of Fig. 6.5. It will be seen that upto income level OY_1, consumption exceeds income, that is, there is dissaving. Beyond income level OY_1, there is positive saving. It is worth mentioning that as average propensity to consume *(APC)* falls with the increase in income in the upper panel, average propensity to save rises as income increases. Thus in Fig. 6.6 with the increase in income not only the absolute amount of saving increases, the average propensity to save also increases.

Marginal Propensity to Save (*MPS*)

Whereas average propensity to save indicates the proportion of income that is saved, marginal propensity to save represents how much of the *additional* disposable income is devoted to saving. The marginal propensity to save is therefore change in saving induced by a change in the disposable income.

Thus,

$$MPS = \frac{\Delta S}{\Delta Y}$$

For example, if disposable income increases from rupees 10,000 to 12,000 and this causes planned savings to increase by ₹ 500 crores, marginal propensity to save is:

$$MPS = \frac{500}{2000} = \frac{1}{4} = 0.25$$

Since the additional income is either consumed or saved, *the sum of marginal propensity to consume and marginal propensity to save is equal to one.*

$$MPC + MPS = 1$$

This can be mathematically proved as under

From $C + S = Y$, it follows that any change in income (ΔY) must induce either change in consumption (ΔC) or change in saving (ΔS). Thus.

$$\Delta C + \Delta S = \Delta Y$$

Dividing both sides by ΔY we have

$$\frac{\Delta C}{\Delta Y} + \frac{\Delta S}{\Delta Y} = \frac{\Delta Y}{\Delta Y} = 1$$

$$MPC + MPS = 1$$

The concept of marginal propensity to save is graphically shown at the bottom of Fig. 6.6. It will be seen from this figure that when disposable income increases from OY_1 (say ₹ 10,000) to OY_2 (say ₹ 12,000), that is, ΔY = ₹ 2000, the saving increases by Y_2T, (₹ 500), that is, ΔS is ₹ 500. Thus marginal propensity to save (MPS) is

$$\frac{\Delta S}{\Delta Y} = \frac{Y_2T}{Y_1Y_2} = \frac{500}{2000} = \frac{1}{4} = 0.25$$

KEYNES'S THEORY OF CONSUMPTION

Keynes in his *"General theory"*, published in 1936, laid the foundations of modern macroeconomics. The concept of consumption function plays an important role in Keynes' theory of income and employment. Keynes mentioned several subjective and objective factors which determine consumption of a society. However, according to Keynes, of all the factors it is the *current level of income* that determines the consumption of an individual and also of society. Since Keynes laid stress on the *absolute size of current income* as a determinant of consumption, his theory of consumption is also known as **absolute income theory of consumption.** While Keynes recognized that many subjective and objective factors including interest rate and wealth influenced the level consumption expenditure, he emphasised that it is the **current level of income** on which the consumption spending of an individual and the society depends.

Keynes's Psychological Law of Consumption. Further, Keynes put forward a *psychological law of consumption,* according to which, as income increases consumption increases but not by as much as the increase in income. In other words, marginal propensity to consume is less than one but greater than zero. That is,

$$1 > \frac{\Delta C}{\Delta Y} > 0$$

Thus implies that out of an increment in income a part is consumed and a part saved. To quote Keynes, "*The amount of aggregate consumption depends mainly on the amount of aggregate income. The fundamental psychological law, upon which we are entitled to depend with great confidence both a priori from our knowledge of human nature and from (the detailed facts of experience is that men (and women, too) are disposed, as a rule and on an average to increase their consumption as their income increases, but not by as much as the increase in their income.*"[1] This psychological law of consumption plays a significant rule in macroeconomic theory as the size of investment multiplier depends on it. For example, if *MPC* is 0.75, the value of multiplier will be $\frac{1}{1-MPC}=\frac{1}{1-0.75}=\frac{1}{0.25}=4$. The higher the marginal propensity to consume, the greater the value of investment multiplier.

[1]Keynes's consumption theory has three features. First, he suggests that consumption expenditure depends mainly on absolute income of the current period, that is, consumption is a positive function of the *absolute level of current income*. The more income in a period one has, the more is likely to be his consumption expenditure in that period. In other words, in any period the rich people tend to consume more than the poor people do. Secondly, he propounds a psychological law of consumption, according to which marginal propensity to consume is less than one but greater than zero. Thirdly, Keynes points out that *consumption expenditure does not have a proportional relationship with income*. According to him, as income increases, a smaller proportion of income is consumed. The proportion of consumption to income is called average propensity to consume (*APC*). Thus, the third property of Keynes's consumption function is that *average propensity to consume (APC) falls as income increases.*

The Keynes' consumption function can be expressed in the following form

$$C = a + bY_d$$

where C is consumption expenditure and Y_d *is* the real disposable income which equals gross national income minus taxes, a and b are constants, where a is the intercept term, that is, the amount of consumption expenditure at zero level of income. Thus, a is *autonomous consumption.* The parameter b is the marginal propensity to consume (*MPC*) which measures the increase in consumption spending in response to per unit increase in disposable income. Thus, $b = MPC = \frac{\Delta C}{\Delta Y}$

The behaviour of consumption expenditure as perceived by Keynes implies that the marginal propensity to consume is less than average propensity to consume ($MPC < APC$). This is because as income increases consumption does not rise proportionately and as income falls consumption does not fall proportunately. This is because people protect their consumption standards by not reducing their consumption proportionately to the reduction in their income. Implication of the proposition that $MPC < APC$ as income increases is that the ratio of consumption demand to income falls. This can be seen from Figure 6.3 where the slope of consumption function curve is *marginal propensity to consume (MPC) which remains constant* and the slope of the lines such as *OA* and *OB* from the origin to points *A* and *B* on the consumption function line *CC′* measure *APC*, that is, ratio *C*/*Y* at these points is falling showing that consumption demand does not increase proportionately to increase in national income. This Keynesion consumption function is shown in figure 6.1 and 6.3.

In Fig. 6.3 we have shown a linear consumption function with an intercept term. In this form of linear consumption function, though marginal propensity to consume ($\Delta C/\Delta Y$) is constant, average propensity to consume is declining with the increase in income as indicated by the slopes of the lines *OA* and *OB* at levels of income Y_1 and Y_2 respectively. The straight line *OB* .drawn from the origin indicating average propensity to consume at higher income level Y_2 has a relatively less slope than the straight line *OA* drawn from the origin to point *A* at lower income level Y_1. The decline in average propensity to consume as the income increases implies that the *proportion of income that is saved increases with the increase in national income of the country.* This result also follows from the studies of family budgets of various families at different income levels. The fraction of income spent on consumption by the rich families is lower than that of the poor families. In other words, the rich

1. J.M. Keynes, *The General Theory of Employment, Interest and Money,* (New York: Harcourt), First Edition, 1936, p. 96.

families save a higher proportion of their income as compared to the poor families.

The assumption of diminishing average propensity to consume is a significant part of Keynesian theory of income and employment. This implies that as income increases, a progressively larger proportion of national income would be saved. Therefore, to achieve and maintain equilibrium at full-employment level of income, increasing proportion of national income is needed to be invested. If sufficient profitable opportunities for investment are not available, the economy would then run into trouble and in that case it would not be possible to maintain full-employment because aggregate demand will fall short of full-employment output. On the basis of this increasing proportion of saving with the increase in income and consequently the emergence of the problem of demand deficiency, some Keynesian economists such as Hanson based the theory of *secular stagnation* on the declining propensity to consume.

Important Features of Keynes's Consumption Function

In macroeconomics, Keynes's consumption function plays a highly important role. Therefore, it is essential to state its important features. The following are the important features of Keynes's consumption function:

1. First, *absolute level of current income is the important factor that determines consumption* of the community. Increase in national income causes an increase in consumption. On the other hand, classical economists thought that it was rate of interest that primarily determined saving and consumption of the community. A rise in rate of interest induces the people to save more and thus to reduce their level of consumption. According to Keynes, though rate of interest is one of the factors that determine consumption of the community, he did not consider it a very important determinant of it. By considering level of *current income* as the most important factor determining consumption and saving, Keynes made a significant contribution to the macroeconomic theory.
2. The second important feature of Keynes' consumption function is that *marginal propensity to consume is less than one* but greater than zero ($0 < MPC < 1$). As has been explained above, the feature of Keynes's consumption function that marginal propensity to consume is less than one is known as Keynes's psychological law of consumption. According to this law, as income increases, consumption increases but not as much as the increase in income. We will explain in a later chapter that Keynes's theory of multiplier is based on the marginal propensity to consume being less than one but greater than zero.
3. In *Keynes consumption function, namely,* $C = a + by$, *as income increases, average propensity to consume* (*APC*) *falls.* Keynes was of the view that rich people relatively save a higher proportion of their income so that at higher levels of income average propensity to consume (*APC*), that is, proportion of total consumption to national income falls as national income rises.
4. Another important feature of consumption function as put forward by Keynes is that *it remains stable in the short run.* Consumption function, according to Keynes, depends on various institutional factors such as distribution of income and wealth and psychological factors such as willingness to save. Since there cannot be much changes in these institutional and psychological factors, consumption function remains stable in the short run, that is, it does not shift upward or downward. Therefore, Keynes in his theory explains the determination of income and employment in the short run by considering that the consumption function is stable.

DETERMINANTS OF PROPENSITY TO CONSUME

The important question is on what factors does the consumption demand of a community depend. In other words, what are the factors that determine the level and position of the propensity to consume or the consumption function. Keynes divided the factors determining the propensity to consume into two groups: one group of factors was called by him as objective factors and the second group was named by him as subjective factors. We shall explain below in detail these objective and subjective factors which affect the consumption function of a community.

Objective Factors

The following six types of objective factors influence the consumption function:

(1) Changes in the General Price Level : Real Balance Effect. The general price level is an important factor which influences the consumption of a community. When the general price level increases or, in other words, when inflation occurs, the consumption function shifts downward. This is because the rise in the general price level, real value (that is, purchasing power) of people's money balances and financial assets with fixed monetary values declines. This causes a downward shift in the consumption function. This is called *real balance effect.* Similarly, when the general price level falls, real value of money balances and financial assets increases. This will induce people to consume relatively more out of their current income. This will cause an upward shift in the consumption function.

(2) Fiscal Policy. Fiscal policy of the Government, especially taxation policy affects the propensity to consume of the country. By levying excise duties, sales tax, the Government can cut down the consumption and thereby increase savings of the community. Likewise, when the Government reduces taxes, consumption of the people increases and this raises the propensity to consume. Rationing and price control by the Government also affects the propensity to consume, as was witnessed during the Second World War. In the modern times, pursuing of the welfare state policy by the Government under which progressive taxes have been levied on the rich people and the revenue obtained from them have been spent to provide many social security benefits and amenities to the poor people, has tended to raise the consumption function.

(3) Rate of Interest. Rate of interest also affects the propensity to consume and save. It is generally believed that higher rate of interest induces the people to save more and this results in reducing their propensity to consume. But this is not true in the case of all the people. Some individuals are of such a type who want a certain fixed income in the future. And when the rate of interest rises these individuals consume more and save less because with higher rate of interest they can obtain the given fixed income with lesser savings. Therefore, when the rate of interest rises such individuals save less than before. Thus, it cannot be said with certainty whether with the changes in the rate of interest the propensity to consume of the whole community will change or not.

(4) Stock of Wealth. The stock of wealth owned by the households in the economy is also an important factor that determines propensity to consume. In wealth we include not only *real assets* such as land, houses, automobiles but also *financial assets* such as cash balances, saving and fixed deposits with banks, stocks and bonds possessed by households. The greater the amount of wealth accumulated by households in the economy, the greater is generally the propensity to consume *(i.e.* the greater the amount of consumption out of any level of current income). The important motive of the people to save is to accumulate wealth. Generally speaking, the greater the wealth which people have accumulated, the weaker is the incentive to save further. In other words, the other things remaining the same the increase in wealth generally causes an upward shift in the consumption function and decrease is wealth causes a downward shift in the consumption function.

An important example which is often cited to emphasise the importance of wealth as a determinant of consumption is the stock market crash of 1929 in England (*i.e.* drastic fall in share prices) which substantially reduced the financial wealth of the households overnight resulting in shifting the consumption function downward.

(5) Credit Conditions and Consumer Indebtedness. The availability of easy credit causes an increase in consumption and shifts the consumption function upward. It is now a common experience in India that in recent years lowering of lending interest rates by Indian banks on loans for houses, cars, computers and other durable consumer goods has greatly increased the consumption of the people and shifted consumption function upward. On the other hand, tightening of credit produces an opposite effect, that is, causes a downward shift in the consumption function, Furthermore, the recent increase in facilities of Credit Cards by banks and their acceptance buy sellers of consumer goods have also worked to shift the consumption function upward in India.

Similarly, the level of consumer indebtedness also greatly affects the propensity to consume of the people. If the households are heavily indebted, say 25 to 30 per cent of their current, they are committed to save *(i.e.* consume less) to that extent so that they are able to pay their instalments of previous credit taken. Thus, the greater the degree of indebtedness of households 'in the economy, the higher will be the consumption function curve and *vice versa.*

(6) Income Distribution. Lastly, distribution of income in a society also determines the level of consumption function. If national income is more unequally distributed, the lower will be the propensity to consume. This is because propensity to consume of the rich is relatively less as compared to that of the poor. Therefore, if inequalities in income distribution increase, this reduces the consumption out of any given level of national income and thus causes a downward shift in the consumption function.

(7) Windfall Gains and Losses. Windfall gains and losses also affect the propensity to consume. When the prices of the shares go up, the shareholders begin to think themselves better off and this raises their consumption. On the other hand, when the prices of the shares go down, the shareholders have to suffer sudden losses and they begin to think themselves relatively poorer than before. This induces them to reduce their consumption. We thus see that the windfall gains and losses affect the propensity to consume.

(8) Change in Expectations. Changes in the expectations of the people also influence the propensity to consume. When people expect that war will break out in the near future and they expect prices to go up, then they will try to spend more on goods so as to meet the needs of the immediate future. This raises the consumption function in the current period. On the other hand, when people expect the prices to fall they reduce their current consumption so that they should spend more when the prices actually fall.

Subjective Factors

Among the subjective factors are included those factors which induce and prompt people to save some part of their income. First, people save because they want to provide for unforeseen contingencies, such as illness, unemployment, accidents, etc. Secondly, people are induced to save because they want to provide for the expected future needs such as education of the children, marriages of their children, etc. Thirdly, several people wish to save from their current incomes so that they may be able to use accumulated savings for investment which will increase their future income. Investments will bring them more income in the form of more profits and interest. Fourthly, people are motivated to save so that they can accumulate large wealth which will increase their social status. With increased wealth they would think themselves to be economically more independent and they could buy many things with more wealth. Further, many individuals also save so that they can use them for speculative purposes and other business projects. Besides, several people are prompted to save for the sake of leaving a good fortune for their heirs and children. Lastly, many people save because of their miserly instinct and habits. The accumulation of more wealth gives them a great psychic satisfaction.

The above subjective factors increase the propensity to save and therefore reduce the propensity to consume. These subjective factors play a crucial rule in determining the level and shape of the consumption function. However, Keynes pointed out and rightly so that some subjective factors raise the propensity to consume. The desire for ostentation generally leads to greater consumption expenditure. People have a natural instint to imitate others' consumption habits. As pointed out by Duesenberry, people in lower and middle income ranges imitate the consumption standards of the higher income groups and this increases their propensity to con-sume. This has been called *demonstration effect* which is a great subjective or psychological force that works in raising the propensity to consume.

Subjective factors also lead the business firms to save much or little from their incomes. Many of the subjective factors which influence the savings of the firms are the following ; (1) *Enterprise*—Many business industrial firms desire to save a part of their current income so that they can make investment in new enterprises and carry out expansion in the future. Business firms generally save a good part of their income for their further expansion. (2) *Liquidity*—Business firms also are induced to save so that they can face contingencies in the future. If they have good amount of liquid wealth in their hands, they would be able to meet the emergent situations more successfully. More cautious and farsighted firms will save more than others on this count. (3) *Successful Management*—Many managers of the business firms are motivated to save more because they want to prove themselves successful managers. With the investment of the saved money, the income of a firm increases and their managers are regarded as successful. (4) *Financial Prudence*—Business firms desire to save for making up the depreciation in plant and machinery. Since after some years business firms have to replace their plant and machinery, if a good part of their current income is not saved, it would not be possible for them to replace plant and machinery. If the firms put aside a greater part of their income for depreciation or replacement purposes, they would pay lower dividends to the shareholders and this will generally lower the propensity to consume of a community. On the other hand, if the firms keep a relatively small amount for depreciation, they will pay larger amounts as dividends to the shareholders and this will generally increase the propensity to consume of the community. Lastly, firms also want to save because they have to repay their debts.

We have explained above various subjective and objective factors which taken together determine the consumption function of a community. It is worth noting that propensity to consume does not generally change in the short run, because it depends more on psychological and institutional factors which change only in the long run. *The institutional factors which determine the distribution of income in the society are important forces determining the consumption function.* And these institutional factors do not change in the short run. Therefore, Keynes was of the view that consumption function remains stable in the short run.

In the next chapter, we shall critically examine Keynes's consumption function and compare it with other theories of consumption, function.

LIFE CYCLE AND PERMANENT CONSUMPTION THEORIES OF CONSUMPTION

Since Keynes propounded his theory of consumption there have been significant developments in this field and some alternative theories of consumption have been put forward. However, all these post-keynesian theories of consumption *relate consumption to some form of income*. First, J.S. Duesenberry expressed the view that consumption expenditure of individuals depends not on their own absolute incomes but on their *incomes relative to the incomes of others.*[2] Therefore, Duesenberry's theory of consumption is called, *Relative Income Theory of consumption.* Second, Modigliani and Ando put forward a theory known as '*Life Cycle Theory of Consumption* according to which individuals plan their consumption profile for their while life time which depends not so much on their current income but on *their expectations of whole life time income*. Further, a famous American economist Milton Friedman advanced an hypothesis regarding consumers' behaviour called permanent income hypothesis according to which *consumption of individuals depends on their permanent incomes rather their current levels of income*. We will critically examine these post-keynsion theories of consumption in the next chapter. Here we will explain here briefly life cycle theory of consumption and permanent Income Hypothesis.

In Keynes's theory of consumption function, *current* consumption expenditure depends mainly on *current income*. After Keynes some economists such as Franco Modigliani and Miltion Friedman pointed out that individuals while making their consumption decisions take into account not only the current income which was emphasised by Keynes but also the income they expect to receive in the future. Therefore, the consumption theories of Franco Modigliani and Milton Friedman, have been

2. See J. Stiglitz, *Economics,* 2002, p. 634

called *Future-Oriented Consumption theories* by Joseph Stiglitz.[2] Consumption theory put forward by Modigliani and Ando is called *Life-Cycle Theory of Consumption* as they lay stress that in deciding about consumption out of income in the working years of their life individuals attempt to save for the post-retirement period. Their motive for doing so is to smooth their consumption in their entire life span. That is, they save from the current income of working period of their life so that will not have to cut down consumption level after they retire.

Like Modigliani, and Ando, Miltion Friedman also thinks that *expectations about future income affect, consumption* in a current period. He points out that people save in good years so that they have not to curtail consumption in bad years. In other words, in Friedman's view consumption of individuals or households depends on *permanent income* in contrast to *current income* level as emphasised by Keynes. *By permanent income he means average long-run income*. Therefore, Friedman's theory is known as *permanent consumption hypothesis*. According to this view, temporary changes in income such as bonus received in a year, or large capital gains received due to rise in prices of shares or real estate would not have the same effect as the permanent changes in income, that is, changes in long-run average income. According to Friedman, transitory changes in income will have little effect on consumption while permanent changes in income will have large effect on consumption. Thus, according to Friedman's permanent consumption theory, if households could predict their future income level, they will adjust their consumption levels according to their average long-run income.

Milton Friedman emphasises that consumption of individuals is affected by changes in individuals' wealth which is a stock variable in contrast to the current income which is a flow variable. Friedman's permanent consumption theory suggests that individuals' consumption depends on how well-off they are and that is better measured by their wealth. Individuals' wealth change when the value of their assets such as prices of stock shares and real estate change. When their prices rise people enjoy capital gains which make them wealthier and as a result their consumption out of a given income rises. With this consumption function curve shifts above. Commenting on the future-oriented theories of consumption and saving of Modigliani and Friedman Joseph Stiglitz write, "Both Modigliani and Friedman stressed that individuals generally try to stabilise their consumption using savings to smooth consumption so that it does not fluctuate dramatically from year to year. They also stressed that individuals will be forward looking, basing current consumption decisions on their expectations about future income"[3].

Implications. There are three important implications of the above two future-oriented theories of consumption and saving. First important implication is that *changes in expectations about future income* effect the current consumption of individuals even though their current income has not changed. Consider the case when an economy is currently experiening recession as has been the case in 2008-09 afflicted by global financial crisis and global meltdown, people became quite pessimistic about their future income prospects which led them to curtail consumption and save more even though their current income did not change. This causes the consumption function curve to shift downward.

Another implication is that one-time or temporary changes in income will cause only a small change in consumption. Suppose a person has a windfall gain of ₹ 10 lakhs in a year due to his winning a lottery. Given his marginal propensity to consume equal to 0.8, the Keynesian consumption function predicts that his consumption will increase by ₹ 8 lakhs and the remaining ₹ 2 lakhs will be saved by him in that year. On the other hand, the future-oriented theories of Modigliani and Freidman predict that the consumption of extra income gained by the lottery winner during a year will be spread over his life time or over a long-period of time. Similarly, a more important case from the viewpoint of economic policy is the *temporary reduction in the direct tax* at the time of recession by the government to boost consumption demand as had been done even recently by the US government in 2007-08 and 2008-09 to fight recession will have only limited effect on consumption. The above two

3. Stiglitiz, *op. cit.*, p.634

future-oriented theories predict that contrary to the prediction of the Keynesian model the temporary reduction in direct taxes will be much less effective in boosting consumption demand as due to uncertain economic prospects much of the income gained through transitory tax reduction will be saved. The evidence from the US shows this has in fact been the case.

The third related implication of the future-oriented theories is that since changes in income that individuals expect to be only temporary will have only small effect on consumption spending, marginal propensily to consume will be lower and the multiplier effect of such changes in taxes will be smaller. To conclude, leaving aside the cases of temporary changes in income and the concern of the forward looking people to ensure same consumption standard in the future, the Keynesian theory of consumption is right in visualising that individuals' consumption spending is more dependent on current income. To quote Stiglitz again, "*The permanent income and life-cycle saving hypothesis contain large elements of truth. Families do save for their retirement and they do smooth their consumption between good years and bad.* Even so, *households spending appear to be more depended on current income than either future-oriented theory would suggest*"[4].

CONSUMPTION FUNCTION PUZZLE: KEYNES'S CONSUMPTION FUNCTION AND KUZNETS' FINDINGS

Empirical studies of long-term times series data of the US economy for the period 1869-1938 made by the noted American economist Kuznets estimated a consumption function which contradicts Keynes' consumption function which was found to be correct on the basis of cross-section studies of household budget data and short-term time series data. *This contradiction between Kuznets empirical findings and Keynes' consumption function has been called con-sumption function puzzle.* Efforts have been made by several economists to resolve this puzzle and new theories of consumption function have been put forward to resolve the conflict between Keynes's consumption function and Kuznets's findings. These alternative theories of consumption function will be discussed in the next chapter.

To compare the Keynes' and Kuznets's consumption functions, it will be useful to write them in algebraic form and graphically represent them. Keynes's consumption function can be algebraically written as below:

$$C = a + bY \qquad ...(1)$$

where a is a positive intercept term which is also called *autonomous consumption* as it does not vary with income. The constant a shows that even when income is zero, a certain consumption is present. This is possible when in any year a community can live either on its past savings or borrow from other communities. Keynes' consumption function is shown in Fig. 6.6.

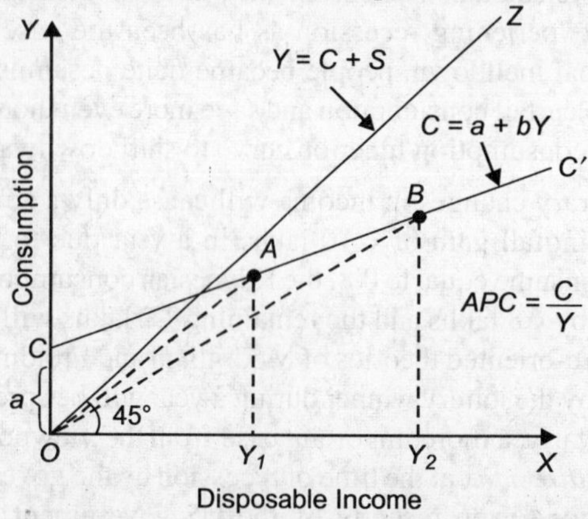

Fig. 6.6. *Keynes Consumption Function: Falling Average Propensity to Consume*

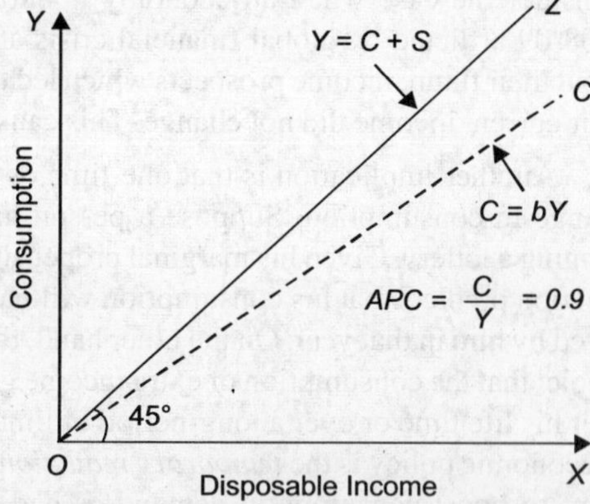

Fig. 6.7. *Kuznets's Consumption Function: Constant APC*

4. Stiglitz *op.* cit. p. 635 (italics added)

Secondly, b in the consumption function represents marginal propensity to consume $\left(\frac{\Delta C}{\Delta Y}\right)$. The above mentioned Keynes's consumption function ($C = a + bY$) shows that average propensity to consume $\left(\frac{C}{Y}\right)$ falls as income increases. This can be known by comparing slopes of the rays *OA* and *OB* at income levels Y_1 and Y_2 respectively in Fig. 6.6.

Kuznets's Consumption Function

On the other hand, Kuznets found that consumption function is of the following form:

$$C = bY \quad ...(2)$$

In Kuznets' consumption function there is no intercept term (that is, autonomous consumption). This is shown in Fig. 6.7 where it will be seen that Kuznets consumption function curve starts from the origin and is very near to 45° line depicting that the propensity to consume *(b)* is very high. From his empirical study Kuznets estimated that average propensity to consume was nearly 0.9. Besides, by dividing the entire period (1869-1933) into three overlaping 30 years sub-periods Kuznets found that the proportion of consumption to income (that is, average propensity to consume) was nearly the same and equal to about 0.87 in all the three sub-periods. Thus Kuznets concluded that there was no tendency for the average propensity to consume to decline as disposable income rises. Thus, rounding off Kuznets estimated propensity to consume is equal to 0.9. His consumption function presented in equation (2) can be rewritten as

$$C = 0.97 \quad ...(3)$$

From the above discussion it follows that implication of Keynes's consumption function ($C = a + bY$) and Kuznets consumption function ($C = bY$) are different. *Whereas in Keynes' consumption function APC falls as income rises, in Kuznets s function it remains constant over a long period.* Further, the value of marginal propensity to consume which is less than one is much higher in Kuznets's function as compared to that of Keynes. The reconciliation between two types of consumption functions has been made by some economists by pointing out that whereas Keynes's function is short-run consumption function *Kuznets s function is concerned with long run and is referred to as long-run consumption function.* In the long run, short run consumption function curve shifts above and therefore in the long run consumption function, propensity to consume is higher as compared to that in the short run. *Further, Friedman's permanent Income Hypothesis and Modigliani's Life Cycle Hypothesis* have also tried to reconcile the two functions by referring to the short-run and long-run consumption behaviour of the people. We will discuss these consumption hypothesis in the next chapter.

IMPORTANCE OF CONSUMPTION FUNCTION

The concept of consumption function is greatly important both in theory and actual practice. To remove unemployment and to control economic fluctuations in the economy, it is very essential to adopt a proper macroeconomic policy. In the formation of such a policy, understanding of the concept of propensity of consume is very essential. Therefore. Prof. A.H. Hansen has remarked that "*consumption function is epoch-making contribution of Keynes to economic theory*". We shall explain below some of the theoretical and practical importance of this consumption function.

(1) **The concept of consumption function helps us to invalidate Say's law of classical economics.** In fact, Keynes relied on his consumption function for proving the invalidity of Say's law. According to Say's law, every supply creates its own demand and therefore there is no problem of deficiency of aggregate demand. Therefore, general overproduction and unemployment in the economy is not possible because adequate amount of aggregate demand is ever present. Now, according to Keynes's psychological law of consumption, when income increases consumption increases but less than the increase in income (that is, *MPC* is less than one) and therefore saving gap emerges between

income and consumption. This saving gap implies that all output produced may not be sold in a period and the problem of deficiency of demand will arise unless this saving gap is matched by an equal amount of investment demand. There is no guarantee that the saving done will be automatically invested or, in other words, it is not necessary that investment demand will be equal to the saving gap. Thus, the contention of the Say's law that every supply creates its equal demand is not valid. No doubt, every supply or production creates income equal to the output produced. But since all income is not consumed and there is no guarantee that investment will be equal to the saving so emerged, Say's law is proved invalid. When investment is less than the saving gap corresponding to full-employment level of income, the aggregate demand is not sufficient to provide full-employment to the people and other resources. Thus, the problem of deficiency of effective demand and hence general unemployment and overproduction arises in a free enterprise capitalist economy.

(2) The concept of propensity to consume is also important because **it brings out crucial significance of investment demand** for determination of the level of income and employment in a capitalist economy. From the concept of propensity to consume we know that consumption increases less than the increase in income and as a result gap emerges between income and consumption. To maintain a certain given level of income and employment, gap between income and consumption at that level must be bridged by investment expenditure otherwise it will not' be possible to maintain that level of income and employment because aggregate demand would not be large enough. This indicates the crucial importance of the investment demand in the determination of income and employment. To prevent the establishment of underemployment equilibrium in the economy or, in other words, for achieving full-employment equilibrium, investment demand must be equal to the saving gap at the level of full-employment. Keynes also showed that consumption function remains stable in the short run and therefore economic fluctuations in a capitalist economy are largely due to the fluctuations in investment demand. Thus from the concept of propensity to consume it follows that investment demand is vitally important in determining the level of income and employment. If it were possible to raise the propensity to consume in the short run, then without raising investment, we could have raised the level of income and employment. Since propensity to consume at a given level of income generally remains stable, we have to increase investment for achieving full employment in the economy.

(3) Another crucial importance of the Keynesian concept of consumption function is that **we derive the theory of multiplier from it which has great practical importance in the formulation of macro-economic policy,** especially of public works in times of depression. As we shall study in the next chapter, the magnitude of multiplier is equal to the reciprocal of one minus marginal propensity to consume $\left(K = \dfrac{1}{1 - MPC}\right)$, where K stands for multiplier and MPC for marginal propensity to consume. According to Keynes's psychological law of consumption marginal propensity to consume is less than one. According to this concept of multiplier, when investment increases, income, output and employment increase by a multiple amount, depending upon the size of the multiplier. Income increases manifold than the original investment because of the nature of consumption function. When some investment in some projects is undertaken, it leads to the increase in income of those employed in the projects but the process does not stop here. The increases in income are further spent on consumption and this leads to further increase in income and so the chain of increases in income and consumption continues and the ultimate increase in income and employment is multiple of the original increment in investment. If the marginal propensity to consume were equal to zero, then all increments in income brought about by additional investment would have been saved and therefore multiplier process would not have worked. Since the marginal propensity to consume is greater than zero, the increase in net investment have a multiplier effect on income, output and employment. Thus, the effect of investment on income depends on the size of the multiplier which depends on the value of

the marginal propensity to consume. The greater the marginal propensity to consume, the greater the size of the multiplier.

(4) From the concept of consumption function, we can also explain ***why there is a tendency for the marginal efficiency of capital to decline.*** The declining tendency of the marginal efficiency of capital is due to the nature of the consumption function. Two features of consumption function are important. First, the marginal propensity to consume is less than one which implies that as income increases, consumption increases less than this. Secondly, consumption function is stable in the short run *i.e.,* it does not shift much in the short run. As we know that the level of investment is a crucial factor in the determination of income and employment, fluctuations in the levels of income and employment depend primarily on the fluctuations in investment. The investment demand in the short run is determined by the rate of interest on the one hand and marginal efficiency of capital on the other. Since the rate of interest is relatively sticky, it is the marginal efficiency of capital which greatly affects the level of investment in the short run. Marginal efficiency of capital is nothing but the expected rate of profit on investment in the future. Thus, the marginal efficiency of capital is determined by the expectations of the entrepreneurs regarding the earning of profits from capital assets in the future.

Now, the most important fact that affects the entrepreneurs' expectations regarding profit prospects and thereby the marginal efficiency of capital is the level of future consumption demand for goods and services. Their estimate of future consumption demand depends on, among others, on the population growth. If population growth of a country is expected to fall as was estimated in the early thirties when Keynes wrote his book (*General Theory of Employment, Interest and Money*), this would adversely affect future consumption demand which in turn would adversely affect investment in the long run, Besides, according to Keynes, average propensity to consume (*APC*) falls as income of a community increases overtime. This also adversely affects inducement to invest. If there does not occur capital-using technological change, this will result in decline in investment opportunities in the long run, causing secular stagnation. Thus we see that in the Keynesian scheme of things level of investment depends upon the level of consumption demand in the long run.

Since marginal propensity to consume is less than one and also the consumption function is stable when income increases, consumption does not increase proportionately. As a result, the aggregate demand becomes deficient and the marginal efficiency of capital declines. The decline in the marginal efficiency of .capital adversely affects investment which stops rising. As a result, the growth process stops and economic recession occurs. In this way, Keynes himself and later important Keynesian economist, Prof. A.H. Hansen developed the theory of ***secular stagnation*** (i.e., prolonged period of depression of indefinite future) for the mature capitalist economies. This secular stagnation theory is based upon the assertion that investment opportunities in a capitalist economy will be exhausted soon due to the absence of the possibilities of increasing consumption demand. The meagre opportunities of increasing investment in the mature capitalist economies, according to them, were partly due to the constancy of consumption function and declining average propensity to consume which caused the marginal efficiency of capital to decline.

(5) As has been explained above, Keynesian concept of propensity to consume also **helps us in explaining the *turning points of a business cycle.*** The economy swings down from the peak it reaches because with marginal propensity to consume being less than one and average propensity to consume falling with increase in income consumption demand does not increase as much as the increase in income and output. Over time this causes deficiency in aggregate demand causing fall in marginal efficiency of capital which adversely affects investment demand by private business men. Likewise, economy's downward movement stops and it starts recovery because marginal propensity to consume being less than one and average propensity to consume (*APC*) rises as income decreases people do not reduce their consumption as much as the reduction in their income. Even during depression people try to maintain their previous level of consumption. This ultimately induce

investment for replacement of capital goods which wear out over a period of business cycle. With the working of Keynesian investment multiplier recovery from recession gathers momentum.

QUESTIONS FOR REVIEW

1. What is consumption function? Explain the objective and subjective factors which determine consumption expenditure in the economy. What are the three forms of consumption function? How do *APC* and *MPC* behave in these forms of consumption function? Which form of the consumption function seems to be realistic?
2. Explain Keynes's psychological law of consumption. Explain its importance in the determination of income and employment in the economy.
3. Explain and graphically represent Keynesian consumption function $C = a + bY$. How does marginal propensity to consume (*MPC*) and average propensity to consume (*APC*) change with increase in income in this consumption function?
4. What is consumption function? Explain the factors that cause a shift in the consumption function.
5. Distinguish between Keynes' consumption function and Kuznet's consumption function. How have these two consumption functions been reconciled?
6. What is consumption function puzzle? How has it been resolved?
7. What is consumption function? Explain its importance in the theory of determination of income and employment
8. (*a*) What is a saving function? Derive a saving function from the consumption function, $C = a + bY$

   (*b*) Show that the sum of *APC* and *APS* is equal to one .
9. "Consumption function is an epoch making contribution to the tools of economic analysis analogous to but even more important than Marshall's discovery of the demand function" (Hansen) Discuss.
10. Why does the *APC* differ from *MPC*? What factors other than income are likely to be most important in determining consumption?
11. Given the data of the disposable income (Y_d) and amount of consumption at the initial level of income (₹ 100). Assuming that marginal propensity consume is 50 per cent. Complete the Table and draw the graphs of consumption and saving functions.

Y_d	*C*	*S*	*APC*	*MPC*	*APS*	*MPS*
₹ 100	150					
200						
300						
400						
500						
600						

12. Explain the important features of Keynesian Consumption function.
13. Explain briefly 'Life Cycle theory of consumption' and 'permanent income theory of consumption'. How do they differ from Keynesian theory of consumption?

CHAPTER 7
POST-KEYNESIAN THEORIES OF CONSUMPTION*

INTRODUCTION

In the previous chapter we explained the concept of consumption function which plays an important role in the theory of income and employment. Keynes mentioned several subjective and objective factors which determine consumption of a society. However, according to Keynes, of all the factors it is the *current level of income* that determines the consumption of an individual and also of society. Since Keynes lays stress on the *absolute size of income as a determinant of consumption,* his theory of consumption is also known as *absolute income theory*. Further, Keynes put forward a *psychlological law of consumption, according to which, as income increases consumption increases but not by as much as the increase in income. In other words, marginal propensity to consume is less than one.*

$$1 > \frac{\Delta C}{\Delta Y} > 0$$

Since Keynes propounded his theory of consumption there have been significant developments in this field and several alternative theories of consumer behaviour have been put forward. First, Duesenberry has propounded that consumption expenditure depends on income of an individual relative to incomes of others rather than the absolute size of his own income. His theory is therefore called *Relative Income Theory of Consumption*. Secondly, Modigliani put forward a theory known as *life cycle hypothesis*, according to which an individual plans his even consumption profile in his lifetime which depends not so much on his current income but on his expectations of income in the whole lifetime. Further, a famous American economist Friedman has advanced a hypothesis regarding consumption behaviour, called *permanent income hypothesis*, according to which consumption of an individual depends on permanent income rather than current level of income.

It is important to mention here an important puzzle about consumption function pointed out by Kuznets, a nobel prize-winner in economics. Contrary to Keynes's proposition that proportion of income spent on consumption declines as income increases (that is, average propensity to consume falls with the increase in income), Kuznets found from a statistical empirical study of consumption of the economy of the USA that average propensity to consume had remained constant over a long period despite the substantial increase in income. How the average propensity to consume has remained stable despite the substantial increase in income has been a great puzzle in consumption theory. We shall study below how modern theories of consumption such as Duesenberry's relative income theory of consumption life cycle hypothesis and Friedman's permanent income theory succeed in resolving this puzzle.

RELATIVE INCOME THEORY OF CONSUMPTION

An American economist J.S. Duesenberry put forward the theory of consumer behaviour which lays stress on relative income of an individual rather than his absolute income as a determinant of his consumption. Another important departure made by Duesenberry from Keynes's consumption theory

*This chapter is meant for higher level courses in macroeconomics.

is that, according to him, the consumption of a person does not depend on his current income but on a certain previously reached income level.

According to Duesenberry's relative income hypothesis, consumption of an individual is not the function of his absolute income but of his relative position in the income distribution in a society, that is, his consumption depends on his income relative to the incomes of other individuals in the society. For example, if the incomes of *all individuals* in a society increase by the *same percentage*, then his relative income would remain the same, though his absolute income would have increased. According to Duesenberry, because his relative income has remained the same the individual will spend the same proportion of his income on consumption as he was doing before the absolute increase in his income. That is, his average propensity to consume (APC) will remain the same despite the increase in his absolute income.

As mentioned above, empirical studies based on time-series data made by Kuznets reveal that over a long period the average propensity to consume remains almost constant. Now, Duesenberry's relative income hypothesis suggests that in the long run the community would continue to consume the same proportion of income as its income increases. According to Duesenberry, saving as a proportion of income of the individuals with relatively low incomes would not rise much with the increase in their incomes. That is, their savings would not rise to the same proportion of income as was being done by the individuals who had the same higher income prior to the present increase in income. This is because with the increase in incomes of all individuals by the same proportion, the relative incomes of the individuals would not change and therefore they would consume the same proportion of their income. This applies to all individuals and households. It therefore follows that assuming that relative distribution of income remains the same with the growth of income of a society, its average propensity to consume (APC) would remain constant. Thus, this conclusion of the relative income hypothesis differs from the Keynesian theory of consumption according to which, as seen above, as absolute income of a community increases, it will devote a smaller proportion of its income to consumption expenditure, that is, its APC will decline.

It is important to note that relative income theory implies that with the increase in income of a community, the relative distribution of income remaining the same, does not move along the same aggregate consumption function, but its consumption function shifts upward. Since as income increases, movement along the same consumption function curve implies a fall in average propensity to consume, *Duesenberry's relative income hypothesis suggests that as income increases consumption function curve shifts above so that average propensity to consume remains constant.*

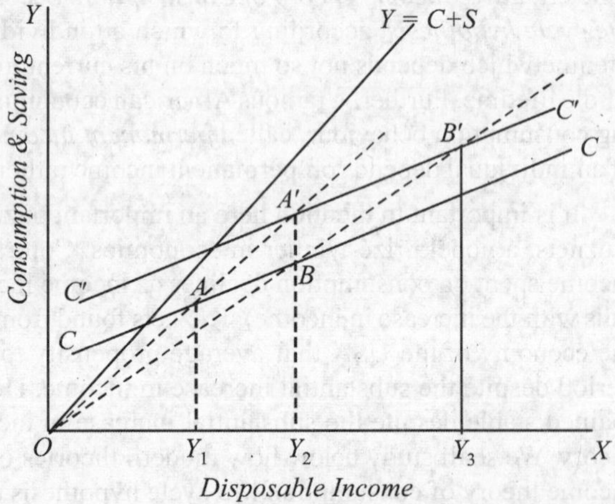

Fig. 7.1. *Duesenberry's Relative Income Theory of Consumption*

This is illustrated in Figure 7.1. Suppose a family A' has Y_1 level of income and is spending Y_1A' on consumption. Suppose its income level rises to Y_2. Now, its consumption would not rise only to Y_2B (*i.e.* equal to the consumption of the family B at Y_2 income level) but to Y_2A' where A' lies on the same ray from the origin as the previous point A of consumption. *This implies that the consumption expenditure of family A has risen in the same proportion as its income with the result that its average propensity to consume remains constant.*

Likewise, if income of family B which is having consumption expendiutre Y_2B at income level

Y_2, rises to Y_3, its consumption expenditure will increase to Y_3B' where B' lies on the same ray from the origin as B. This again means that the proportion of income devoted to consumption by family B (*i.e.* its *APC*) remains constant as there is increase in its absolute income. Thus, *if the proportion of income* devoted to consumption of the average family at each income level remains the same as its income increases, the aggregate consumption of the community as proportion of its income will also remain constant though its absolute consumption and absolute savings will increase with the absolute increse in income.

As income increases and a society moves along the same consumption function curve, its average propensity to consume falls. But Duesenberry's relative income hypothesis suggests that *as income increases consumption function curver shifts above* so that average propensity to consume remains constant. In Figure 7.1 it will be seen that if points A′ and B′ are joined together, we get, a new consuption function curve C′C′.

Demonstration Effect. By emphasising relative income as a determinant of consumption, the relative income hypothesis suggests that individuals or households try to imitate or copy the consumption levels of their neighbours or other families in a particular community. This is called demonstration effect or *Duesenberry effect*. Two things follows from this. First, the average propensity to consume does not fall. This is because *if incomes of all families increase in the same proportion, distribution of relative incomes would remain unchanged and therefore the proportion of consumption expenditure to income which depends on relative income will remain constant*.

Secondly, a family with a given income would devote more of his income to consumption if it is living in a community in which that income is regarded as relatively low because of the working of demonstration effect. On the other hand, a family will spend a lower proportion of its income if it is living in a community in which that income is considered as relatively high because demonstration effect will not be present in this case. For example, the recent studies of household expenditure made in India reveal that the families with a given income, say ₹ 5000 per month spend a larger proportion of their income on consumption if they live in urban areas as compared to their counterparts in rural areas. The higher propensity to consume of families living in urban areas is due to the working of demonstration effect where families with relatively higher income reside whose higher consumption standards tempt others in lower income brackets to consume more.

Ratchet Effect. The other significant part of Duesenberry's relative income hypothesis is that it suggests that when income of individuals or households falls, their consumption expenditure does not fall much. This is often called a *ratchet effect*. This is because, according to Duesenberry, the people try to maintain their consumption at the highest level attained earlier. This is partly due to the demonstration effect explained above. People do not want to show to their neighbours that they no longer afford to maintain their high standard of living. Further, this is also partly due to the fact that they become accustomed to their previous higher level of consumption and it is quite hard and difficult to reduce their consumption expenditure when their income has fallen. They maintain their earlier consumption level by reducing their savings. Therefore, the fall in their income, as during the period of recession or depression, does not result in decrease in consumption expenditure very much as one would conclude from family budget studies.

This is illustrated in Figure 7.2 where on the X-axis we measure disposable income and on the *Y*-axis the consumption and savings. Starting with disposable income of zero, we assume that there is steady growth of disposable inocme till it reaches Y_1. The linear consumption function C_{LR} is the long-run consumption function. It will be seen from the figure that at Y_1 level of disposable income, the consumption expenditure equals Y_1C_1. Now suppose with initial income level Y_1, there is recession in the economy with the result that disposable income falls to the level Y_0. According to Duesenberry, consumption would not fall greatly to the level Y_0C_0 as the long-run consumption function curve C_{LR} would suggest. In their bid to maintain their consumption level previously reached people would now save less and reduce their consumption level only slightly to $Y_0C'_0$ whereas point C'_0 is on the short-run consumption function curve C_{SR}. Since $Y_0C'_0 > Y_0C_0$, the average propensity to consume at income level Y_0 is greater at C'_0 than at C_1 at income level Y_1 (A ray drawn from the origin to the point C'_0 will have greater slope than that of OC_1). When the economy recovers from recession and disposable income increases, the economy would move along the short-run consumption function curve C_{SR} till the consumption level C_1 is reached at income level Y_1. Beyond this, with the growth of income the consumption will increase along the long-run consumption function curve CL_R.

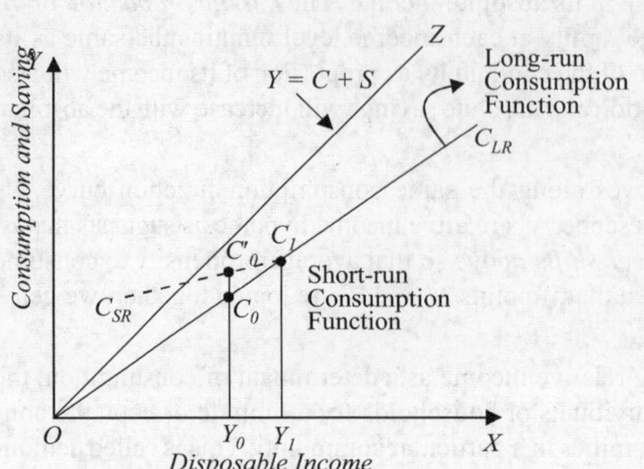

Fig. 7.2. *Duesenberry's Ratchet Effect*

Aggregate consumption function of the community. From the analysis of demonstration and ratchet effects it follows that Duesenberry's relative income hypothesis provides an explanation for why *aggregate consumption function of the community may be flatter than the family budget studies would suggest.* Duesenberry emphasizes that it is relative income rather than absolute income which determines consumption expenditure of households. When income of the community increases, relative income remaining constant, the proportion of consumption expenditure to income will not increase much because relative incomes of the households remain the same (Note that this implies that saving ratio will not rise much). Due to demonstration effect every household will increase its expenditure in the same proportion as the increase in income. On the other hand, if the income of the community decreases, the consumption expenditure would not decline much due to the ratchet effect according to which people try to maintain their previously attained higher level of consumption. This makes the consumption function of the community flatter than suggested by the cross-sectional family budget studies.

Further, it also follows from the Duesenberry relative income hypothesis that *short-run aggregate consumption function of the community is linear rather than curved.* As stated above, if, in the short run, the level of income increases, the proportion of consumption expenditure to income is not likely to increase much due to the operation of demonstration effect and with the fall in income the proportion of consumption to income is not likely to decline much due to the ratchet effect. This makes the short-run aggregate consumption function of the community linear. It is worth noting that Duesenberry's theory assumes that relative distribution of income does not change much. This is in accord with the facts of the real world situation where changes in income distribution do not take place in the short run. Thus Duesenberry's theory provides a convincing explanation in terms of demonstration and ratchet effects why aggregate consumption function is linear rather than non-linear.

LIFE CYCLE THEORY OF CONSUMPTION

An important post-Keynesian theory of consumption has been put forward by Modigliani and Ando[1] which is known as life cycle theory. According to life cycle theory, the *consumption in any*

1. See Ando, A., Modigliani, Franco "The Life Cycle Hypothesis of Saving: Aggregate Implications and Tests," *American Economic Review*, March 1963, pp, 55-84 and Modigliani, "The Life Cycle Hypothesis of Saving and the Demand for Wealth and Supply of Capital," *Social Research*, June 1966, pp. 160-217.

period is not the function of current income of that period but of the whole lifetime expected income. Thus, in life cycle hypothesis the individual is assumed to plan a pattern of consumption expenditure based on expected income in their entire lifetime. It is further assumed that individual maintains a more or less constant or slightly increasing level of consumption. However, this level of consumption is limited by his expectations of lifetime income. A typical individual in this theory in his early years of life spends on consumption either by borrowing from others or spending the assets bequeathed from his parents. It is in his main working years of his lifetime that he consumes less than the income he earns and therefore makes net positive savings. He invests these savings in assets, that is, accumulates wealth which he consumes in the future years. In his lifetime after retirement he again dissaves, that is, consumes more than his income in these later years of his life but is able to maintain or even slightly increase his consumption in the lifetime after retirement.

Life cycle hypothesis has been depicted in Fig. 7.3. It is assumed that a typical individual knows exactly at what age he will die. In Fig. 7.3 it is taken that the individual would die at the age of 75 years. That is, years 75 is his expected lifetime. It is further assumed in the life cycle theory that net savings in the entire lifetime is zero, that is, the savings done by the individual in his working years of his life is equal to the dissavings made by him in his early years of life before he is able to earn income as well as the dissavings which he makes after retirement. It is also assumed for the sake of simplicity that interest paid on his assets is zero. The curve *YY′* shows income pattern of the whole life-time of the individual whereas *CC′* is the curve of consumption which is assumed to be slightly increasing as the individual grows old. It is assumed that our individual enters into labour force (*i.e.*, working life) at the age of 15 years. It will be noticed from Fig. 7.3 that upto the age of 25 years his income, though increasing, is less than his consumption, that is, he will be dissaving during the first 13 years of his working life. To finance his excess consumption over his income, he may be borrowing from others.

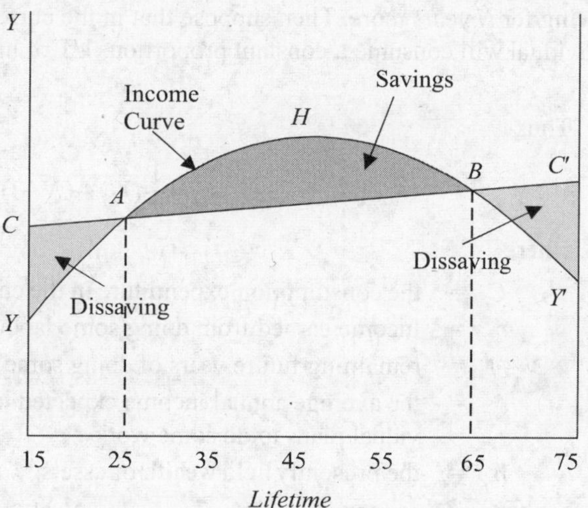

Fig. 7.3. *Life Cycle Theory of Consumption*

Beyond the age of 25 or point A on the income and consumption curves and upto the age of 65 years his income exceeds his consumption, that is, he will be saving during this period of his working life. With these savings he will build up assets or wealth. He may use these savings or wealth to pay off his debt incurred by him in the early stage of his working life. Another important motive of his savings and building up assets or wealth is to provide for his consumption after retirement when his income drops below his level of consumption. It will be observed from the Fig. 7.3 that beyond point B (that is, after retirement at 65 years) his current income falls short of his consumption and therefore he once again dissaves. He would be using his accumulated assets or wealth from his earlier working years to meet the dissavings after retirement at the age of 65. It is important to note that we assume that he does not intend to leave any assets for his children. Given this assumption, his net savings over his lifetime will be zero. Therefore, in Fig. 7.3 his savings during the period when he earns more than his consumption expenditure, that is, the *shaded* area *AHB* will be equal to the two areas of dissavings, *CYA* + *BC′Y′*. Thus he dies leaving behind no assets or wealth. He has planned his consumption expenditure over the years that his net savings at the time of death are zero. However, this assumption can be relaxed if he wishes to leave some assets or wealth for his children.

Some important conclusions follow from the life cycle theory of consumption. The fundamental idea of the life-cycle hypothesis is that people make their consumption plans for their entire lifetime and further that they make their lifetime consumption plans on the basis of their expectations of lifetime income. Thus in the life cycle model *consumption is not a mere function of current income but on the expected lifetime income. Besides, in life cycle theory the wealth presently held by individuals also affects their consumption.*

How the consumption of an individual in a period depends on these factors highlighted by life cycle theory can be expressed in the form of an equation. To do so let us consider an individual of a given age with an *additional life* expectancy of T years and intends to retire from working after serving for N years more. Then suppose that in the current period and thereafter in his life span the individual will consume a constant proportion, 1/T of his life-time income in equal instalments per year.

Thus

$$C_t = \frac{1}{T}(Y_{Lt} + (N-1)Y^e{}_L + W_t)$$

where

C_t = the consumption expenditure in the current period t

Y_{Lt} = Income earned from doing some labour in the current period t

$N - 1$ = remaining future years of doing some labour or work

Y^e_L = the average annual income expected to be earned over N-1 years for which individual plans to do some work

W_t = the presently held wealth or assets

It will be observed from the above equation that life cycle hypothesis suggests that consumption in any period does not depend only on current income but also on expected income over his entire working years. Besides, consumption in any period also depends on his presently owned wealth or assets which are built up during the prime working years of one's life when income exceeds savings.

The general consumption behaviour as suggested by Ando-Modigliani life cycle hypothesis can be expressed in the following functional form:

$$C_t = b_1 Y_{Lt} + b_2 Y^e_L + b_3 W_t$$

where C_t = Consumption expenditure in a period t.

Y_{Lt} = Income earned from doing some labour in the current period t.

Y^e_L = the average annual income expected to be earned from labour during the further years of working life.

W_t = wealth currently owned

b_1 represents marginal propensity to consume out of current income

b_2 is marginal propensity to consume out of expected lifetime income, and

b_3 is the marginal propensity to consume out of wealth.

It is significant to note that consumption would not be much responsive to changes in current income (*i.e.*, Y_{Lt}) unless it also changes expected future lifetime income (Y^e_L). A one time or temporary change in income, say, by ₹ 1000, will affect consumption in the same way as the increase in wealth. The consumption of these ₹ 1000 will be spread over the entire lifetime in a planned consumption flow per period. With 50 years of future life, increase of ₹ 1000 of transient or temporary income will raise the consumption by 1000/50 = ₹ 20 per period. This implies that consumption function curve will shift above. A permanent increase in income that is expected to persist throughout the working years, which implies that in future expected lifetime income also rises, will produce a large effect on consumption in each of the remaining period of one's lifetime. Further, the increase in wealth will shift the consumption function upward, that is, will increase the intercept term of the consumption function.

To estimate behaviour of the consumer on the basis of life cycle hypothesis, one is required to make some assumptions how people form their expectations regarding labour income over their life time. In the study of consumption function for the United States, Ando and Modigliani made the assumption that the expected future labour income is simply a multiple of current labour income. Thus, according to this assumption,

$$Y_L^e = \beta Y_{LT}$$

where β is a multiple of current labour income. This assumption implies that people revise their expected labour income of future by a certain multiple of the change in current labour income. With this assumption, aggregate consumption function for the community can be expressed as under

$$C_1 = (b_1 + b_2\beta)Y_{LT} + b_3 W_t$$

This function has been estimated taking time-series data for the U.S.A. and the following estimates have been obtained :–

$$C_t = 0.72\ Y_{LT} + 0.06\ W$$

According to these estimates, if current labour income increases by Rs. 100 along with assumed effect on expected future income, consumption will increase by Rs. 72 per period. Besides, the increase in wealth by Rs. 100 will raise the consumption expenditure by Rs. 6. It therefore follows that according to life cycle hypothesis the relationship between income and consumption is non-proportional, increase in labour income by Rs. 100 crore leads to increase in consumption by Rs. 72. Further, the increase in wealth will shift the consumption function upward, that is, will increase the intercept term of the consumption function.

The consumption function based on life cycle hypothesis is illustrated in Fig. 7.4 where along the X-axis we measure disposable income and along the *Y*-axis the consumption expenditure. The short-run consumption function is shown by the curve C_{SR} which has a slope of 0.6 which is the marginal propensity to consume out of labour income in the short run. This short-run consumption function is linear and has an intercept term indicating that average propensity to consume declines as labour income increases and $MPC < APC$. The intercept of the short-run consumption function measures the effect of wealth on consumption. Since wealth increases over a period of time due to savings in the prime working years, the short-run consumption function will be shifting upward, that is, the intercept of the short-run consumption function will be increasing as wealth grows in the long run. Overtime the shift in the short-run consumption function may trace a series of points on a long-run consumption function C_{LR} passing through the origin.

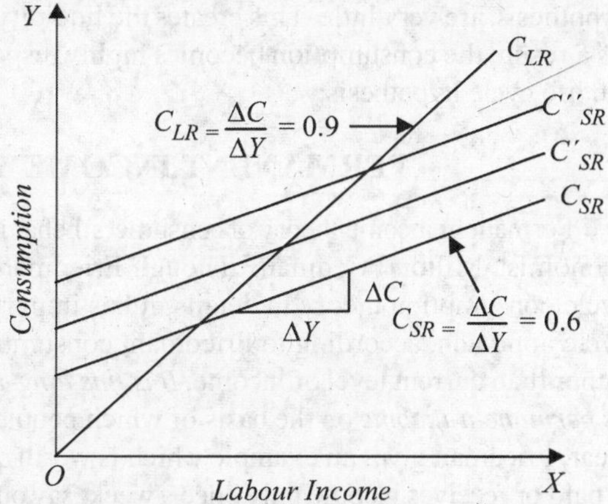

Fig. 7.4. *Life Cycle Consumption and Income : Short Run and Long Run*

Since the ratios of wealth and labour income are constant over time, the life cycle consumption function is in accord with the conclusion arrived at by Kuznets from the long-run time series data that the long-run consumption function is proportional, with average propensity to consume (*APC* or *MPC*) remaining constant and being equal to nearly 0.9. These facts are quite consistent with the long-run consumption function of life cycle hypothesis and thus help in resolving the Kuznets puzzle.

Life cycle hypothesis also explains the non-proportional relationship between consumption and income found in the cross-sectional family budget-studies. It has been found in these studies that high income families consume a smaller proportion of their income, that is, their average propensity to consume (*APC*) is relatively lower than those of the low-income families. This can be easily explained by life-cycle hypothesis. Suppose we choose a random sample of families from the population and rank them according to their incomes. The families with higher incomes are expected to be middle-aged income earners who are in the prime working years of their lifetime and therefore earn more than they consume (*i.e.*, their *APC* will be relatively lower). On the other hand, the families with lower incomes are likely to have relatively high proportion of new entrants into the labour force and the old people who have retired and, as seen above, they consume more than their current income and their *APC* being quite high pushes up the *APC* of the low income families.

Shortcomings. Although life cycle theory has provided an explanation of various puzzles about consumption function, it is not without critics, Gardner Ackley[2] has criticized the assumption of life cycle hypothesis that in making consumption plans, households have "a definite and conscious vision."

According to Ackley, the possession of this vision on the part of households sounds unrealistic. Further, according to him, to assume that a household has complete knowledge of "family's future size, including the life expectancy of each member, entire lifetime profile of income of each member, the extent of credit available in the future, future emergencies, opportunities and social pressure which have a bearing on consumption spending" is quite unrealistic.

Life cycle theory has also been criticized that it fails to recognize the importance of *liquidity constraints* in determining the response of consumption to income. According to critics, even if a household has a concrete vision of future income, the opportunities to borrow from the capital markets for quite a long period on the basis of expected future income, as has been visualised by life cycle hypothesis, are very little. This creates the liquidity constraints for deciding about consumption plans. As a result, the consumption becomes highly responsive to current income which is quite contrary to the life cycle hypothesis.

PERMANENT INCOME THEORY OF CONSUMPTION

Permanent income theory of consumers' behaviour has been put forward by a well-known American economist, Milton Friedman.[3] Though Friedman's permanent income hypothesis differs from life cycle consumption theory in details, it has important common features with the latter. Like the life cycle approach, according to Friedman, consumption is determined by long-term expected income rather than current level of income. *It is this long-term expected income which is called by Friedman as permanent income* on the basis of which people make their consumption plans. To make his point clear, Friedman gives an example which is worth quoting. According to Friedman, an individual who is paid or receives income only once a week, say on Friday, he would not concentrate his consumption on one day with zero consumption on all other days of the week. He argues that an individual would prefer a smooth consumption flow per day rather than plenty of consumption today and little consumption tomorrow. Thus consumption in one day is not determined by income received on that particular day. Instead, it is determined by average daily income received for a period. This is on the line of life cycle hypothesis. Thus, according to him, people plan their consumption on the basis of

2. Gardner Ackley, "Discussion of a Paper by James Tobin and W. Dolde in Consumer Spending and Monetary Policy, The Linkages, Federal Reserve Bank of Boston, 1971.
3. Milton Friedman, *A Theory of Consumption Function*, Princeton University Press, 1957.

expected average income over a long period which Friedman calls permanent income.

It may be noted that permanent income or expected long-term average income is earned from both "human and non-human wealth". The income earned from human wealth which is also called human capital refers to the return on income derived from selling household's labour services, that is, efforts and abilities of its labour. This is generally referred to as labour income. Non-human wealth consists of tangible assets such as saved money, debentures, equity shares, real estate and consumer durables. It is worth noting that Friedman regards consumer durables such as cars, refrigerators, airconditioners, television sets as part of households' non-human wealth. The imputed value of the flow of services from these consumer durables is considered as consumption by Friedman.

Relationship between Consumption and Permanent Income

Now, what is the precise relationship between consumption and permanent income (that is, the expected long period average income). According to permanent income hypothesis, Friedman thinks that consumption is proportional to permanent income

$$C^p = kY^p$$

where Y^p is the permanent income

C^p is the permanent consumption

k is the proportion of permanent income that is consumed.

The proportion or fraction k of permanent income that is consumed depends upon the following factors:

1. Rate of interest (*i*). At a higher rate of interest the people would tend to save more and their consumption expenditure will decrease. The lowering of rate of interest will have opposite effect on the consumption.

2. The proportion of non-human wealth to human wealth. The relative amounts of income from physical assets (*i.e., non-human wealth*) and income from labour (i.e., *human wealth*) also affects consumption expenditure. This is denoted by the term w in the parmanent consumption function and is measured by the ratio of non-human wealth to income. In his permanent income hypothesis Friedman suggests that consumption expenditure depends a good deal on the wealth or assets possessed by the people. The greater the amount of wealth or assets held by an individual, the greater would be its propensity to consume and vice-versa.

3. Desire to add to one's wealth. Lastly, households' *preference for immediate consumption as against the desire to add to the stock of wealth or assets* also determines the proportion of permanent income to be devoted to consumption. The desire to add to one's wealth rather than to fulfil one's wants of immediate consumption is denoted by u.

Thus rewriting the consumption function based on Friedman's permanent income hypothesis we have

$$C^p = k\,(i, w, u)\; Y^p$$

The above function implies that permanent consumption is function of permanent income. The proportion of permanent income devoted to consumption depends on the rate of interest (i), the ratio of non-human wealth to labour income (w) and desire to add to the stock of assets (u).

Permanent and transitory income. In addition to permanent income, the individual's income may contain a transitory component that Friedman calls as a transitory income. *A transitory income is a temporary income that is not going to persist in future periods.* For example, a clerk in an office may get a substantial income from overtime work in a month which he thinks cannot be maintained. Thus, this large overtime income for a month will be transitory component of income. According to Friedman, transitory income is not likely to have much effect on consumption. Thus, income of an individual consists of two parts, permanent and transitory, which we may write as under:

$$Y^m = Y^p + Y^t$$

where Y^m is measured income in a period, Y^p is the permanent income and Y^t is transitory income.

Measuring permanent income. To make the permanent income hypothesis operational we need to measure permanent income. Permanent income, as is generally defined is *"the steady rate of consumption a person could maintain for the rest of his or her life, given the present level of wealth and income now and in the future."*[4] However, it is very difficult for a person to know what part of any change in income is likely to persist and is therefore permanent and what part would not persist and is therefore transitory. Friedman has suggested a simple way of measuring permanent income by relating it to the current and past incomes. According to him, permanent income is equal to the last year's income plus a *proportion of change in income* occurred between the last year and the current year. Thus, permanent income can be measured as under:

$$Y^p = Y_{t-1} + a(Y_t - Y_{t-1}) \quad o < a < 1 \qquad \ldots(1)$$

$$Y^p = aY_t + (1 - a)\, Y_{t-1} \qquad \ldots(2)$$

Let us illustrate this with an example. Suppose, the proportion of change in income in the last year and the current year equals 0.6 and the last year's income (Y_{t-1}) is Rs. 80,000 and the current year's income (Y_t) is Rs. 85,000, then from above equation permanent income can be estimated as under.

$$\begin{aligned} Y^p &= 0.6\,(85{,}000) + (1 - 0.6)\,80{,}000 \\ &= 51{,}000 + 32{,}000 \\ &= 83{,}000 \end{aligned}$$

It is worthwhile to note the two features of the above equations estimating permanent income. First, if $Y_t = Y_{t-1}$, it implies that current year's income is equal to last year. This further means that last year income is being maintained and therefore the individual would expect to earn the same income in the future also. In this case then permanent income is equal to the current or last year's income. Secondly, when income of an individual increases in the current year as compared to the last year, the permanent income will be less than the current year's income. This is because individual is not sure whether the increase in income will persist in the future and therefore does not immediately revise his estimate of permanent income by the full amount of the increase in his income in the current year.

Permanent Income, Long-Run and Short-Run Consumption Functions

Now, having known the meaning of permanent income and permanent consumption we can describe the precise relationship between consumption and income both in the short run and the long run as under.

$$C = kY^p = kaY_t + k\,(1 - a)\, Y_{t-1}$$

In the above consumption function *ka* is the marginal propensity to consume in the short-run which is obviously less than the long-run marginal propensity to consume which is equal to *k*. Thus, according to Friedman's permanent income hypothesis, short-run marginal propensity to consume differs from long-run marginal propensity to consume, the latter being greater than the former. Further, $k(1-a)Y_{t-1}$ is the intercept of the short-run consumption function.

4. Dornbusch, R. and Fischer, S., *Macroeconomics*, McGraw-Hill, Fifth Edition, 1990, p. 279.

Friedman's permanent income hypothesis is illustrated in Figure 7.5. It shall be seen from this figure that permanent consumption function is represented by the long-run consumption function curve C_{LR}, ($C_{LR} = kY^p$). This long-run consumption function shows the proportional relationship between consumption and income and is a straight line passing through the origin which implies that *APC* is constant and is equal to *MPC*.

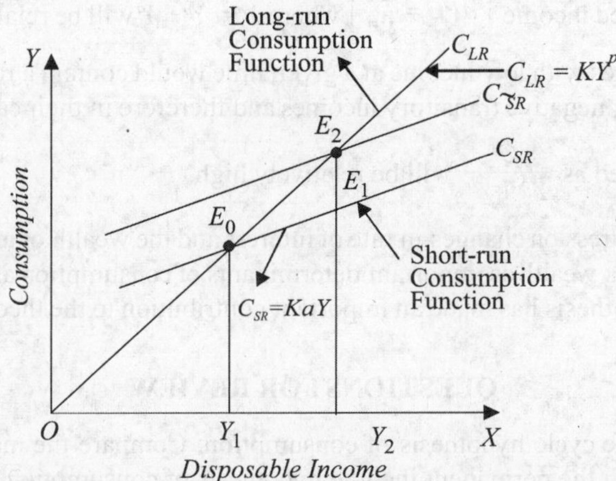

Fig. 7.5. *Permanent Income Hypothesis: Long-Run and Short-Run Consumption Functions*

In accordance with permanent income hypothesis, short-run consumption function curves are flatter as compared to the long-run consumption function curve indicating that the short-run marginal propensity to consume is lower than long-run marginal propensity to consume. The reason for this is that the individual is not sure whether the increase in income will persist over the longer period which determines the consumption plans of individuals. Therefore, the individuals do not fully adjust their consumption expenditure according to their higher current income than would be the case if the current increase in income is expected to be permanent. If the rise in income happens to be permanent, that is, if the next year's income is equal to the higher income of the current year, the individual will fully adjust his consumption expenditure to the higher income level.

It is important to note that in our above analysis we have assumed that full adjustment of consumption expenditure to change in income takes place in two years time. In this case permanent income is the average of the two years incomes. However, *in real world permanent income depends on expectations of income for a much longer period depending upon the vision of the individual.* In case of longer vision adjustment of consumption expenditure will take place slowly over a long period. However, if the individual is sure that the increase in income is permanent he will adjust his consumption quickly to higher current income. It, therefore, follows that whereas in the short run average propensity to consume falls as income increases because people are not sure whether the increase in income will persist or not. But when they actually find that the increase in income is permanent, they fully adjust their consumption to their higher permanent income as reflected in the long-run consumption function.

Conclusion

Permanent income hypothesis is similar to life cycle hypothesis and differs only in details. Like the life cycle hypothesis, permanent income hypothesis can explain the puzzle about the relationship between consumption and income, namely, whereas in the long-run time series data, consumption-income ratio (*i.e.*, *APC*) is constant, in the short run it declines with the increase in income as we have seen above. The permanent income hypothesis is quite consistent with the constancy of *APC* in the long run and its variation in the short run.

Permanent income hypothesis is also consistent with the evidence from the cross-sectional budget studies that high income families have low average propensity to consume than that of low-income families. A sample of high income families at a given time is likely to contain a relatively larger number of families who are having positive transitory increase in incomes. Since the consumption depends on permanent income, the average propensity to consume computed as the ratio of consumption to measured income $\left[APC = \frac{C}{Y^m}\right]$ where $Y^m = Y^p + Y^t$ will be relatively low. On the other hand, a sample of families with low income at a given time would contain a relatively larger number of families experiencing negative transitory incomes and therefore in their case the average propensity to consume estimated as $\frac{C}{Y^p + Y^t}$ will be relatively high.

Further, by laying stress on changes in rate of interest and the wealth or assets held by the people and desire to add to one's wealth as important determinants of consumption and savings, Friedman's permanent income hypothesis has made an important contribution to the theory of consumption and saving.

QUESTIONS FOR REVIEW

1. Explain the life cycle hypothesis of consumption, Compare the major differences of this hypothesis with the permanent income hypothesis of consumption.
2. Explain the cyclical relationship between aggregate consumption and national income in terms of the relative income hypothesis.
3. Explain relative income hypothesis of consumption behaviour. How does it reconcile long-run (sector) consumption function with the cyclical consumption function?
4. What is relative income hypothesis? How does it differ from Keynes's absolute income hypothesis. Explain demonstration effect and rachet effect" in the relative income hypothesis of consumption. Explain the shortcomings of relative income hypothesis.
5. Explain permanent income hypothesis. How does it differ from Keynes' absolute income hypothesis? How is permanent income measured?
6. How does permanent income hypothesis explain that the short-run marginal propensity to consume is lower than the long-run marginal propensity to consume ? Do you think that the short-run *MPC* of someone whose income is highly variable will be relatively low? Why?
7. Explain the permanent income hypothesis of consumption. Explain its weaknesses.
8. What is money illusion? How does it affect the consumption of those who suffer from it?

CHAPTER 8
INVESTMENT DEMAND

INTRODUCTION

Levels of national income and employment in the short run depend upon the level of aggregate demand. The aggregate demand consists of two constituents—consumption demand and investment demand. In the previous chapter we have studied the consumption function and the various factors which determine it. In the present chapter we shall discuss the factors which determine the investment demand or inducement to invest. Since consumption function is more or less stable in the short run, investment demand is of crucial importance in the determination of income and employment. The greater the level of investment, the greater the level of income and employment. According to Keynes, the equilibrium of a capitalist economy is generally not established at the level of full employment, primarily because investment demand is insufficient to fill up the saving gap corresponding to the level of full employment. Therefore, underemployment equilibrium, on which Keynes laid a great stress, is due to the lack of investment demand as compared to the amount of saving at full-employment level of income, there emerges an *inflationary gap* in the economy which brings about the rise in general price level. We thus see that investment demand plays a vital role in the determination of national income, employment and prices in the country.

MEANING OF INVESTMENT

It is useful to make the meaning of investment clear. When a person buys shares, bonds or debentures of a public limited company from the market, it is generally said that he has made investment. But this is not the real investment which determines income and employment in the country and with which we are here concerned. Buying of existing shares and bonds by an individual is merely a financial investment. When one individual purchases the shares or bonds, some other one would sell them. Thus, the purchase and sale of the shares merely represents the change in the ownership of assets which already exists rather than the creation of new capital assets. It is the new addition to the stock of physical capital such as plant, machines, trucks, new factories and so on that creates income and employment. Therefore, *by real investment we mean the addition to the stock of physical capital*. Thus, in economics, investment means the new expenditure incurred on addition of capital goods such as machines, buildings, equipment, tools, etc. The addition to the stock of physical capital, i.e., net investment-raises the level of aggregate demand which brings about addition to the level of income and employment in the economy. Keynes and many other economists also include the increase in the *inventories* of consumer goods in the capital of the country and therefore in the investment.

AUTONOMOUS INVESTMENT AND INDUCED INVESTMENT

After Keynes two types of investments have been distinguished. One is *autonomous investment* and the second is *induced investment*. By autonomous investment we mean the investment which does not change with the changes in the income level and is therefore independent of income. Keynes thought that the level of investment depends upon marginal efficiency of capital and the rate of interest. He thought changes in income level will not affect investment. This view of Keynes is based upon his preoccupation with short-run problem. He was of the opinion that changes in income level with affect investment only in the long run. Therefore, considering as he was the short-run problem he treated investment as independent of the changes in the income level. In fact the distinction between autonomous investment and induced investment has been made by post-Keynesian economists. *Autonomous investment refers to the investment which does not depend upon changes in the income level.* This autonomous investment generally takes place in houses, roads, public undertakings and in other types of economic infrastructure such as power, transport and communication. This *autonomous investment depends more on population growth and technical progress than on the level of income.* Most of the investment undertaken by Government is of the autonomous nature. The investment undertaken by Government in various development projects to accelerate economic growth of the country is of autonomous type. The autonomous investment is depicted in Fig. 8.1 where it will be seen that whatever the level of national income, investment remains the same at I_a. Therefore the autonomous investment curve is a horizontal straight line.

On the other hand, induced investment is that investment which is affected by the changes

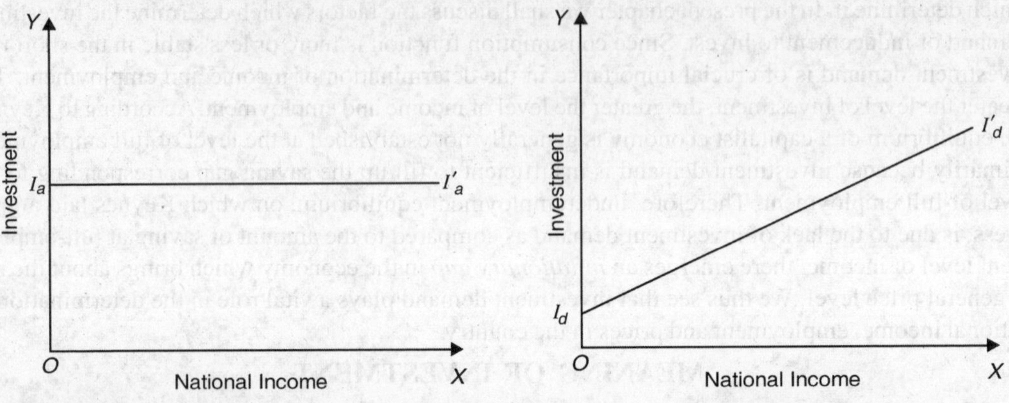

Fig. 8.1. *Autonomous Investment* **Fig. 8.2.** *Induced*

in the level of income. The greater the level of income, the larger will be the consumption of the community. In order to produce more consumer goods, more investment has to be made in capital goods so that greater output of consumer goods becomes possible. Keynes regarded rate of interest as a factor determining induced investment but the empirical evidence gathered so far suggests that induced investment depends more on income than on the rate of interest. Induced investment is shown in Fig. 8.2 where it will be seen that with the increase in national income, induced investment is increasing. Increase in national income implies that demand for output of goods and services increases. To produce greater output, more capital goods are required to produce them. To have more capital goods more investment has to be undertaken. This induced investment is undertaken both in fixed capital assets and in inventories. The essence of induced investment is that greater income and therefore greater aggregate demand affects the level of investment in the economy. The induced investment underlines the concept of the prin-ciple of accelerator, which is highly useful in explaining the occurrence of trade cycles.

DETERMINANTS OF INVESTMENT

Inducement to invest or investment demand depends upon two factors: (1) expected rate of profits to which Keynes gives the name *Marginal Efficiency of Capital,* and. (2) the rate of interest. It can be easily shown that investment is determined by expected rate of profit and the rate of interest. A person having an amount of savings has two alternatives before him. Either he should invest his savings in machines, factories, etc., or he can lend his savings to others on a certain rate of interest. If investment undertaken in machines or factories is expected to fetch 25% rate of profit, while the current rate of interest is only 15%, then it is obvious that the person would invest his savings in machinery or factory.

It follows from above, if investment is to be profitable then the expected rate of profit must not be less than the current market rate of interest. If the expected rate of profit is greater than the market rate of interest, new investment will take place. If an entrepreneur does not invest his own savings but has to borrow from others, then it becomes much clear that the expected rate of profit from investment in any capital asset must not be less than the rate of interest he has to pay. For instance, if an entrepreneur borrows funds from others at 15% rate of interest, then the investment proposed to be undertaken will actually be undertaken only if the expected rate of profit from it is more than 15 per cent. We thus see that investment depends upon marginal efficiency of capital on the one hand and the rate of interest on the other. Investment will be undertaken in any given form of capital asset so long as expected rate of profit or marginal efficiency of capital falls to the level of the current market rate of interest. The equilibrium of the entrepreneur is established at the level of investment where expected rate of profit or marginal efficiency of capital is equal to the current rate of interest. Therefore, the theory of investment is also based upon the assumption that the entrepreneur tries to maximize his profits.

Of the two determinants of inducement to invest, marginal efficiency of capital or expected rate of profit is of comparatively greater importance than the rate of interest. This is because rate of interest does not change much in the short run; it is more or less sticky. But changes in the expectations of profits greatly affect the marginal efficiency of capital and make it very much unstable and volatile. As a result of changes in marginal efficiency of capital, investment demand is greatly affected which causes aggregate demand to fluctuate very much. The changes in aggregate demand bring about economic fluctuations which are generally known as trade cycles. When profit expectations of businessmen are good, large investment is undertaken which causes aggregate demand to rise and bring about conditions of boom and prosperity in the economy. On the other hand, when expectations regarding profits are adverse, the rate of investment falls which causes decline in aggregate demand and brings about depression, unemployment and excess productive capacity in the economy. Thus, the changes in the marginal efficiency of capital play a crucial role in causing changes in the investment level and economic activity.

We have read in the theory of interest that, according to Keynes, the rate of interest is determined by liquidity preference and the supply of money. The greater the liquidity preference, the greater the rate of interest. Given the liquidity preference curve, the greater the supply of money, the lower will be the rate of interest. We have already studied how the rate of interest is determined. We explain below in detail the concept of marginal efficiency of capital and describe the factors on which it depends.

Marginal Efficiency of Capital

As has been pointed out above, the concept of marginal efficiency of capital refers to the rate of profit expected to be made from investment in certain capital assets. The rate of profit expected from an extra unit of a capital asset is known as marginal efficiency of capital. Now, the question is how the marginal efficiency of capital is measured. Suppose an entrepreneur invests money in a certain machinery, how will he estimate the expected rate of profit from it. To estimate the marginal efficiency of capital, the entrepreneur will first take into consideration how much price he has to pay for the particular capital asset. The price which he has to pay for the particular capital asset is called its *supply price* or *cost of capital.* The second thing which he will consider is that how much yield he expects to obtain from investment from that particular capital asset. A capital asset continues to produce goods and yield income over a long period of time. Therefore, an entrepreneur has to estimate the prospective yield from a capital asset over his whole life period. Thus, the supply price and the prospective yields of a capital asset determine the marginal efficiency of capital. By deducting the supply price from the prospective yield during whole life of a capital asset the entrepreneur can estimate the expected rate of profit or marginal efficiency of capital. Keynes has defined the marginal efficiency of capital in the following words: "I define the marginal efficiency of capital as being equal to that rate of discount which would make the present value of the series of annuities given by the returns expected from the capital asset during its life just equal to its supply price." Thus, according to Keynes, marginal efficiency of capital is the rate of discount which renders the prospective yields from a capital asset over its whole life period equal to the supply price of that asset. Therefore, we can obtain the marginal efficiency of capital in the following way:

$$\text{Supply Price} = \text{Discounted Prospective Yields}$$

$$C = \frac{R_1}{1+r} + \frac{R_2}{(1+r)^2} + \frac{R_3}{(1+r)^3} \ldots\ldots + \frac{R_n}{(1+r)^n}$$

In the above formula, C stands for Supply Price or Replacement Cost and $R_1, R_2, R_3, .. R_n$ etc., represent the annual prospective yields from the capital asset, r is that rate of discount which renders the annual prospective yields equal to the supply price of the capital asset. Thus, r represents the expected rate of profit or marginal efficiency of capital.

The measurement of marginal efficiency of capital can be illustrated by an arithmatical example. Suppose it costs 3000 rupees to invest in a certain machinery and the life of the machinery is two years, that is, after two years it becomes quite useless, having no value. Suppose further that in the first year the machinery is expected to yield income of ₹1100 and in the second year ₹2420/-. By substituting these values in the above formula, we can calculate the value of r, that is, the marginal efficiency of capital.

$$\text{Supply Price} = \text{Discounted Prospective Yields}$$

$$C = \frac{R_1}{(1+r)} + \frac{R_2}{(1+r)^2}$$

$$3000 = \frac{1{,}100}{1+r} + \frac{2{,}420}{(1+r)^2}$$

On calculating the value of r in the above equation it is found to be equal to 10. In other words, marginal efficiency of capital is here equal to 10 per cent. If we put the value of r, that is, 10 in the above equation, we obtain the following:

$$3{,}000 = \frac{1{,}100}{1.10} + \frac{2{,}420}{(1+10)^2}$$

$$= 1{,}000 + 2{,}000 = 3{,}000$$

Marginal Efficiency of Capital in General. We have explained above the marginal efficiency of a *particular type* of capital asset. But we also require to know the marginal efficiency of capital *in general.* It is the marginal efficiency of capital in general that will indicate the scope for investment opportunities at a particular time in any economy, *At a particular time in an economy the marginal efficiency of that particular capital asset which yields the greatest rate of profit, is called the marginal efficiency of capital in general.* In other words, marginal efficiency of capital in general is the greatest of all the marginal efficiencies of various types of capital assets which could be produced but have not yet been produced. Thus, marginal efficiency of capital in general represents the highest expected rate of return to the community from an extra unit of a capital asset which yields the maximum profit which could be produced.

Keeping in view the existing stock of a capital asset, we can always calculate the marginal efficiency of any particular capital asset. The marginal efficiency of capital will vary when more is invested in a given particular capital asset. In any given period of time marginal efficiency of capital from every type of capital asset will decline as more investment is undertaken in it. In other words, marginal efficiency of a particular type of capital asset will be sloping downward as the stock of capital increases. The main reason for the decline in marginal efficiency of capital with the increase in investment in it is that the prospective yields from capital asset fall as more units are installed and used for production of a good. Prospective yields decline because when more quantity of a good is produced with the greater amount of a capital asset prices of goods decline. The second reason for the decline in the marginal efficiency of capital is that the supply price of the capital asset may rise because the increase in demand for it will bring about increase in its cost of production.

Rate of Interest and Investment Demand Curve

We have explained above how the marginal efficiency of capital is estimated. We can represent diminishing marginal efficiency of capital in general by a curve which will slope downward. This has been done in Fig. 8.3 in which along the *X-axis* investment in capital assets is measured and along the *X*-axis marginal efficiency of capital in general is shown. It will be seen from the figure that when investment in capital asset is equal to 07,, marginal efficiency of capital is i,. When the investment is increased to OI_2, marginal efficiency of capital falls to i_2. Likewise, when investment rises to OI_3, marginal efficiency of capital further diminishes to i_3.

We have seen above that inducement to invest depends upon the marginal efficiency of capital

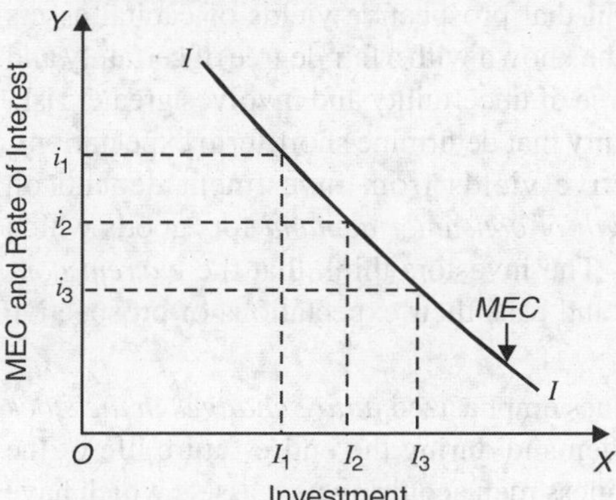

Fig. 8.3. *Investment Demand Curve*

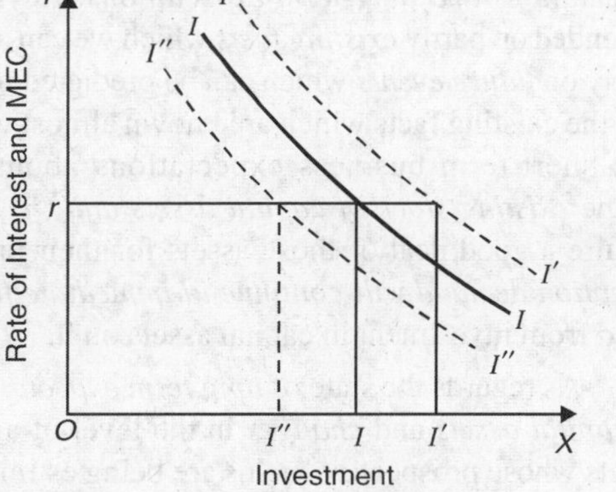

Fig. 8.4. *Shifts in Investment Demand Curve due to Changes Marginal Efficiency of Capital*

and the rate of interest. With the given particular rate of interest and given the curve of marginal efficiency of capital in general we can show what will be the equilibrium level of investment in the economy. If in Fig. 8.3 along the *Y*-axis, rate of interest is also shown, then level of investment can be easily determined. The equilibrium level of investment will be established at the point where marginal efficiency of capital becomes equal to the given current rate of interest. Thus, if the rate of interest is i_1 then I_1 investment will be undertaken. Since at OI_1 level of investment marginal efficiency of capital is equal to the rate of interest i_1. If the rate of interest falls to i_2, investment in the capital assets will rise to OI_2 since at OI_2 level of investment the new rate of interest i_2 is equal to the marginal efficiency of capital. Thus, we see that the curve of marginal efficiency of capital in general shows the demand for investment or inducement to invest at various rates of interest. Hence marginal efficiency of capital curve represents the investment demand curve. This investment demand curve shows how much investment will be undertaken by the entrepreneurs at various rates of interest. If the investment demand curve is less elastic, then investment demand will not increase much with the fall in the rate of interest. But if the investment demand curve or marginal efficiency of capital curve is very much elastic, then the changes in the rate of interest will bring about large changes in investment demand.

Business Expectations and Investment Demand Curve

We have studied above that the marginal efficiency of capital depends upon the supply price of capital on the one hand and prospective yields on the other. It is important to note here that the prospective yields are greatly affected by the expectations of the entrepreneurial class regarding profit making. These expectations generally change very frequently. Indeed, it is the profit expectations of the entrepreneur which determine the level of investment. When the expectations of the entrepreneur regarding profit making become dim, the marginal efficiency of capital declines and as a result demand for investment falls. The occurrence of depression is mainly due to the pessimistic expectations of the entrepreneurial class regarding profit making. On the contrary, when the expectations of the entrepreneurs regarding profit opportunities increase, their inducement to invest rises. As a result of the increase in investment, aggregate demand in the economy increases and levels of income and employment increase. We thus see that profit expectations of entrepreneurs greatly affect investment demand and consequently the level of income and employment.

We have explained above that, according to Keynes, marginal efficiency of capital along with rate of interest determines the level of investment and therefore of the levels of income and employment. A significant contribution of Keynes is that he emphasized the important role that business expectations play in determing the prospective yield from assets on which marginal efficiency of capital and therefore the level of investment depends. He drew distinctions between *short-term expectations and long-term expectations*. Keynes thought that prospective yields of capital assets depended on partly *existing facts* which we can assume to be known with a fair degree of certainty and partly on *future events* which can be predicted with a degree of uncertainty and involves greater risk. It is the existing facts which are known almost with certainty that determine short-term expectations. The short-term business expectations about prospective yields from investment depend on (*i*) the *existing stock of capital assets and* (*ii*) the *strength of consumer demand* for goods which require a good deal of those assets for their production. The investors think that the *current consumption demand will continue in immediate future* too and base their expectations of prospective yield from investment in capital assets on it.

As regards the state of *long term expectations*, Keynes emphasized *future changes in the stock of capital assets* and *changes* in the level of aggregate demand during the entire future life of the assets whose prospective yields are being estimated. Business men acquiring new assets which have long life for production of goods are more concerned with long-run forces on which future earnings from capital assets depend. It is these long-run forces which make long-run expectations about future yields that make them quite uncertain. Changes in these long term expectations make investment quite volatile. Therefore, Keynes gave great importance to the state of confidence for determining investment.

It should be carefully noted that when the expectations regarding prospective yields change, the whole curve of the marginal efficiency of capital will shift. If profit expectations fall, the marginal efficiency of capital curve, that is, investment demand curve will shift downward to left, as shown by the shift of the curve from *II* to *I″I″* in Fig. 8.4. On the other hand, when the profit expectations of the entrepreneurial class become better than before, the marginal efficiency of capital curve will shift upward to the right, as shown by the marginal efficiency of capital curve *I′I′*. Downward shift in the marginal efficiency of capital curve indicates that at the given rate of interest, less investment will be undertaken than before. And upward shift in the marginal efficiency of capital curve indicates that more will be invested at the given rate of interest than before. In Fig. 8.4 when in the beginning, investment demand curve, that is, marginal efficiency of capital curve is represented by *II*, at the rate of interest *r*, demand for investment is *OI*, As a result of downward shift in the marginal efficiency of capital curve to *I″I″*, investment demand at the rate of interest *r* falls to *OI″*. When the marginal efficiency of capital shifts upward to *I′I′*, investment demand at the given rate of interest *r* rises to *OI′*.

It follows, therefore, that the rate of investment depends on marginal efficiency of capital and the rate of interest. If investment demand is interest-elastic, a reduction in rate of interest tends to stimulate investment. But it may fail to do so, if marginal efficiency of investment is already lower than the rate of interest (as may well happen during a depression). Of the two determinants of the rate of investment, marginal efficiency of investment is more volatile than the rate of interest.

The rate of interest is usually 'sticky' in the short run, while marginal efficiency of investment can fluctuate from one extreme to another. If there is a divergence between the two, usually the marginal efficiency of investment will adjust to the rate of interest. If, for example, the marginal efficiency of investment is 20 per cent, while the current rate of interest is 12 per cent, forces will be set in motion so as to bring the former to the level of the latter. In such a situation, there will be more investment because marginal efficiency of investment is greater than the rate of interest, and with the increase in investment, marginal efficiency of investment will fall. At the point where it is just reduced to the level of the current rate of interest, further investment will cease.

FACTORS CAUSING SHIFT IN INVESTMENT DEMAND CURVE

Investment is highly volatile and fluctuates greatly. It is fluctuations in investment that is mainly responsible for fluctuations in aggregate demand. We have seen above that investment demand depends on rate of interest and marginal efficiency of capital, that is, expected rate of profit in investment. Now, an important question is what are the factors that determine marginal efficiency of capital which determines investment demand. There are several factors which cause shift in the marginal efficiency of investment curve and brings about change in investment demand. We discuss these factors below :

1. Changes in Marginal Efficiency of Capital due to Changes in Expectations. The Keynesian theory gives importance to profit expectations of business for determining marginal efficiency of capital and therefore investment in the economy. It is worth mentioning here the concept of **animal spirits** of Keynes. Keynes pointed out that it was not possible to make rational calculations about the future profitability of investment. He believed that the swings of optimism and pessimism of the business class were more important driving forces in the stock market that determine investment. It is these swings of optimism and pessimism about future profitability of investment which he referred to as *animal spirits.* According to Keynes, it is these animal spirits rather than economic fundamentals that determine investment by business firms.

We now term to explain **fundamental economic factors** that determine the profitability of investment on which investment demand depends.

2. Expected Demand for Products. Expected net return on investment by a business firm depends to a large extent on the demand for its product it anticipates. For the economy as a whole the marginal efficiency of investment (*i.e.* expected net return on investment) depends significantly on the consumption expenditure of households on the products produced in the economy. As studied in the previous chapter, apart from income, consumption expenditure on goods and services depends on several factors such as propensity, to save of the people, price level, rate of interest, stock of wealth, taxation policy of the government. Changes in these factors would cause a change in consumption of households and demand for the products.

3. Technology and Innovations. In the recent years advances in technology and introduction of new products and processes have been important determinants of investment. The introduction new improved technology and new products makes it necessary to build new plants or install new capital equipment. This stimulates investment in new capital goods. An important recent example of this is the introduction of micro-computers for use in various fields of industries and services. This has spurred new investment in computers and other related equipments. Another important recent example is the introduction of mobile cellular telephones which has boosted investment in telecommunications by business class.

In addition to the above, advances in technology and new inventions *make investment more productive.* Increase in productivity of capital causes the firms to invest more at a given rate of interest. As a result, curve of marginal efficiency of capital shifts to the right.

4. User Cost of Capital. The user cost of capital also greatly influences the rate of return on capital. The user cost of capital depends on the real rate of interest, rate of depreciation, corporate income taxes, investment tax credit, if any. The lower real rate of interest, lower corporate income taxes, higher investment tax credit, the lower will be the user cost of capital which will increase the rate of return on capital and stimulate investment. On the other hand, the higher real rate of interest, the higher corporate income tax will discourage investment and shift the curve of marginal efficiency of capital to the left. It may be noted that real rate of interest is the nominal rate of interest minus expected rate of inflation.

5. Availability of Credit. Besides cost of credit, availability of credit also determines investment in the economy. Investment is financed in three ways : (1) Borrowing from the banks and other financial institutions, (2) raising resources through issue of equity capital that is, selling shares in the stock market and (3) internal savings of the companies (*i.e.* retained earnings of the companies). If the Central Bank of a country follows tight monetary policy [*i.e.,* higher bank rate, the higher cash reserve ratio (CRR)], credit availability for investment will be less which will decourage private investment. If the Central bank adopts expansionary or easy monetary policy credit availability from banks increases which is likely to encourage investment.

6. Fiscal Policy. Fiscal policy of the Government, especially expenditure and taxation policy of it can work in two ways. If Government increases investment and finance it by borrowing from the open market, this will raise the rate of interest which will crowd out private investment. However, if expansionary monetary policy is adopted simultaneously rate of interest will not rise and therefore crowding out effect will not occur. But there are beneficial effects of expansionary fiscal policy of the Government as well. Increase in Government expenditure, say on infrastructure such as power, communication, highways, roads will raise the incomes of the people which will through the multiplier effect cause a significant increase in aggregate demand for goods produced by the private sector and raise the profit expectations of business firms. This will encourage private investment. Many Indian economists believe that increase in Government expenditure, especially if it is made on infrastructure *crowds in* private investment rather than *crowds it out.*

ACCELERATOR THEORY OF INVESTMENT

We have explained in a previous chapter the Keynesian concept of multiplier which states that as the investment increase, income increases by a multiple amount. On the other hand, there is a concept of accelerator which was not taken into account by Keynes has become popular after Keynes, especially in the discussions of theories of trade cycles and economic growth. The acceleration principle describes the effect quite opposite to that of multiplier. According to this, *when income or consumption increases, investment will increase by a multiple amount.* When income and therefore consumption of the people increases, the greater amount of the commodities will have to be produced. This will require more capital to produce them if the already given stock of capital is fully used. Since in this case, investment is induced by changes in income or consumption, this is known as *induced investment.* The accelerator is the numerical value of the relation between the increase in investment resulting from an increase in income. The net induced investment will be positive if national income increases and induced investment may fall to zero if the national income or output remains constant.

To produce a given amount of output, it requires a certain amount of capital. If Y_t output is required to be produced and v is capital-output ratio, the required amount of capital to produce Y_t output will be given by the following equation:

$$K_t = vY_t \qquad \text{... } (i)$$

where K, stands for the stock of capital

Y_t for the level of output or income, and

v for capital-output ratio.

This capital-output ratio v is equal to $\frac{K}{Y}$ and in the theory of accelerator this capital-output ratio is assumed to be constant. Therefore, under the assumption of constant capital-output ratio, changes in output are made possible by changes in the stock of capital. Thus, when income is Y_t then required stock of capital $K_t = vY_t$, when output or income is equal to Y_{t-1}, then required stock of capital will be $K_{t-1} = vY_{t-1}$.

It is clear from above that when income increases from Y_{t-1} in period *t*–1 to Y_t in period, *t,* then the stock of capital will increase from K_{t-1} to K_t. As seen above, K_{t-1} is equal to vY_{t-1} and K_t is equal to vY_t. Hence, the increase in the stock of capital in period *t* is given by the following equation:

$$K_t - K_{t-1} = vY_t - vY_{t-1}$$

Since increase in the stock of capital in a year $(K_t - K_{t-1})$ represents investment in that year, the above equation (*ii*) can be written as below:

$$I_{t1} = V(y_t - y_{t-1})$$

Equation (*iii*) reveals that as a result of increase in income in any year *t* from a previous year $t-1$, increase in investment will be *v* times more than the increase in income. Hence, it is *v i.e.,* capital-output ratio which represents the magnitude of the accelerator. If the capital-output ratio is equal to 3, then as a result of a certain increase in income, investment will increase three times more *i.e.,* accelerator here will be equal to 3.

It thus follows that *investment is a function of change in income.* If income or output increases over time, that is, when Y_t is greater than Y_{t-1} then investment will be positive. If income declines, that is, Y_t is less Y_{t-1} then disinvestment will take place. And if the income remains constant, that is, $Y_t = Y_{t-1}$ the investment will be equal to zero.

An arithmetical example will make clear the working of the accelerator. This has been represented in the accompanying table. We have made the following assumptions in making this table:

(*i*) Capital-output ratio remains constant and is equal to 3.

(*ii*) The depreciation that takes place in the stock of capital is equal to one-fifth of the stock existing in the *previous year.* Therefore, one-fifth of the stock of capital is to be replaced every year.

Table 8.1. Explanation of the Accelerator

Period	*Output (income)*	*Required Stock of Capital*	*Investment*		
			Capital Replacement	*Net Investment*	*Gross Investment*
(1)	(2)	(3)	(4)	(5)	(6)
$t-1$	500	1,500	300	0	300
t	510	1,530	300	30	330
$t+1$	525	1,575	306	45	351
$t+2$	550	1,650	315	75	390
$t+3$	575	1,725	330	75	405
$t+4$	575	1,725	345	0	345
$t+5$	560	1,680	345	–45	300
$t+6$	550	1,650	336	–30	306
$t+7$	500	1,500	330	–150	180
$t+8$	400	1,200	300	–300	0
$t+9$	400	1,200	240	0	240

In the table, it is supposed that in period $t-1$ and several periods before it, output or income is equal to ₹500. Given that the capital-output ratio is equal to 3, then to produce ₹500 worth of output, ₹1500 worth of capital will be required. [$K = vY$; $1500 = 3(500)$] which is written in column (3). Since depreciation of capital occurred in period $t-1$, will be one-fifth of the stock of capital existing in the previous period (which is also ₹1500). Therefore, replacement investment in period $t-1$ will be equal to ₹300. Since as compared to the previous period, there is no change in output in period $t-1$, the net investment in period $t-1$ will be equal to zero. As a result, the gross investment in period $t-1$ will be equal to ₹300.

Now suppose that production in the period *t* rises to ₹ 510 crores as a result of increase in

Government expenditure or autonomous investment. To produce output worth ₹ 510 crores, total capital worth ₹ 1530 is required [$K_t = vY_t$ 1530 = 3(510)] which is written in column (3). Thus, as a result of increase in output (income) by ₹ 10, net investment has increased by ₹ 30, that is, 1530 – 1500 = 30 which means that accelerator is here equal to 3. In period *t* the depreciation equation equal to l/5th of the capital stock of period *t* – 1 will occur, that is, capital depreciation of ₹ 300, (1/5 × 1500 = 300) will occur in period *t*. Therefore, capital replacement investment in period *t* will be equal to ₹ 300. Thus, gross investment in period *t* will be equal to 30 + 300 = 330. In this way, if output (or income) increases by ₹ 15 in period *t* + 1, ₹ 25 in period *t* + 2, and also ₹ 25 in period *t* + 3, the net investment will increase by three times the increment in output (or income), that is, net investment will increase by ₹ 45 in period *t* + 1, ₹ 75 in period *t* + 2 and also ₹ 75 in period *t* + 3. It will be further observed from Table 6.1 that when output falls in period *t* + 5 by ₹ 15, the net, investment will decline by 3 times of it, that is, equal to ₹ 45. Likewise, from changes in output in different periods we can find out net investment that will take place in any period and with the capital replacement investment we can obtain the gross investment that will occur in any period.

A glance at columns 2, 5 and 6 will show that with a change in output, investment will increase by a multiple of it. This shows that acceleration principle is a powerful destabilising force working in the economy. If the accelerator is the only force at work, then we shall have too much of instability in the economy—more than is actually found. In real life, we find that there are limits to instability, both in the upward as well as the downward direction, so that fluctuations in economic activity or what are called business cycles must have a peak as well as a bottom.

Criticism of the Accelerator Theory

The principle of acceleration has come in for a good deal of criticism in recent years. For example, it has been pointed out by Kaldor that we cannot assume a constant value of the accelerator throughout the trade cycle, that is, it is not true that an increase in output or income by an amount must always give rise to a multiple increase in investment. This is because, if already, some machines are lying idle, we shall try to use them before rushing in for new equipment. Also, if expectation of entrepreneurs is that the rise in demand brought about by increase in income or output is only a temporary one, they well try to meet it by overworking the existing machinery rather than installing a new plant. Thus, *in the theory of accelerator it has been assumed that there is no excess capacity existing in consumer goods industries.* In other words, it has been assumed that no machines are lying idle and no extra shift working is possible. If there had been excess capacity and extra shift working was possible, the supply of goods could be increased with the existing equipment and the accelerator would not come into play.

Further, in the principle of acceleration principle it *has also been assumed that in the capital goods industries, there exists surplus productive capacity.* If there is no excess capacity in the machine-making industries, increased demand for machines caused by the requirement for additional output would not lead to increase in the supply of machines. In the absence of supply of machines, investment cannot increase in the short run. It is thus assumed in the accelerator theory that the machine-making industry is capable of increasing its output for the time being at least. The supply can be increased by reducing stocks of finished machines, by working extra shifts, and so on. But stocks cannot be reduced below zero and working double shifts or adoption of other experiments is found to be expensive. Only when the demand has increased permanently, will the entrepreneurs find it worthwhile to increase investment in machine-making industries.

The size of the accelerator does not remain constant over time. Its value will be affected by the businessmen's calculation regarding the profitability of installing new plants to make more machines on the basis of their probable working life. It is also assumed that the demand for machines will remain stable in future, although the increase in demand has suddenly cropped up. However, in spite of the "above limitations of acceleration principle, it points out an important force which causes economic fluctuations in the economy. Economists like Samuelson,

Hicks and Dusenberry have shown how accelerator combined with mulltiplier provides an adequate and satisfactory theory of trade cycles that occur in the capitalist economies.

RELATIONSHIP BETWEEN SAVING AND INVESTMENT

An important controversy in macroeconomics relates to the relationship between saving and investment. Many economists before J.M. Keynes were generally of the view that saving and investment are generally not equal; they are equal only under condition of equilibrium. Besides, they thought that equality between saving and investment is brought about by changes in the rate of interest. Keynes in his famous work "*General Theory of Employment, Interest and Money*" put forward the view that *saving and investment are always equal*. This gave rise to a severe controversy in economics as to whether saving and investment are always equal or they are generally unequal. This controversy has now been resolved, and there is general agreeement among the economists about the correct relationship between saving and investment. Moddem economists use the concepts of saving and investment in two different senses. In one sense, saving and investment are always equal, equilibrium or no equilibrium. In the second sense, saving and investment are equal only in equilibrium; they are unequal under conditions of disequilibrium. We shall explain below in detail the relationship between saving and investment in these two different senses.

It will be useful to recall here the meaning of saving and investment. In the beginning of this chapter we mentioned that investment is addition made to the stock of capital during a period. In the stock of capital are included new machines, raw materials, buildings, tools and implements and addition to the inventories of raw materials, goods in process, finished consumer goods. When in a certain year there is net addition to the stock of capital, investment is said to have taken place. It is worth mentioning here that *by investment we do not mean the stock of capital but the net addition to the stock of capital i.e., investment is a flow concept*. Of course, addition to the stock of capital is made through the flow of investment. In every year stock of capital expands through net investment.

On the other hand, by saving we mean the part of the income which has not been spent on consumer goods and services. In other words, saving is the difference between income and consumption expenditure. It is worth noting that in consumption expenditure all types of expenditure are not included. If an individual spends a part of his income on providing irrigation facilities, on buying tools and machinery, then that expenditure is not the consumption expenditure, it is in fact an investment expenditure. In order to obtain the saving, we have only to deduct the consumption expenditure from income and not the investment expenditure. When an individual makes investment expenditure he is deemed to spend his saved income on investment. For instance, if a farmer's annual income is ₹ 10,000 and he spends ₹ 6,000 on consumer goods and services and spends ₹ 1,000 on the construction of a well for his fields, and another ₹ 1,000 on building a drainage system for his fields and providing fencing, then his saving would be 10 – 6 = ₹ 4 thousands. The expenditure of ₹ 2,000 on well, drainage and fencing will be included in the saving and will not constitute the consumption expenditure. If Y represents the national income of a country and C the total consumption, then the saving of the country will be equal to $Y - C$. Thus,

$$S = Y - C$$

Ex-post Savings and Ex-post Investment are always Equal

Pre-Keynesian economists were of the view that savings and investment are generally not equal. This is firstly because saving and investment are made by two different classes of people. While investment is undertaken by entrepreneurial class of the society, saving is done by the general public. Secondly, saving and investment depend upon different factors and are made for different purposes and motives. Therefore, it is not inevitable that savings and investment of a society MUST always be equal. Besides, some pre-Keynesian economists pointed out that investmist expenditure is also undertaken by borrowing money from the banks which create new credit for this purpose. It was thus pointed out that more amount of investment than savings is possible because excess of investment over savings is financed by new bank cJ;edit. But Keynes expressed a totally opposite view that saving and investment are always equal. *The sense in which savings and investment are always equal refers to the actual savings and actual investment made in the economy during a year.* They are also called ex-post saving and ex-post investment. If we have to calculate that during the year 2002-03, how much actual savings and investment have been made in India, we will have to deduct the total consumption expenditure made by the citizens of India during that year from the national income. Likewise, the real investment during the year 2002-03 of the Indian economy will be obtained by summing up the investments actually made by the Indian people during that year. In fact, national income estimates of savings and investment are made in this actual or ex-post sense.

The second sense in which saving and investment words are used is that in a certain year how much saving or how much investment people of the country desire or intend to do. Thereefore, saving and investment in this sense are known as desired, intended or planned savings and investment. They are also called *ex-ante* saving and *ex-ante* investment.

Keynes in his book, "*General Theory of Employment, Interest and Money*" showed that in spite of the fact that saving and investment are done by two different classes of people and also for different purposes and motives, actual saving and actual investment are always equal. Thus, he used the word saving and investment in the ex-post or actual sense and proved the equality between saving and investment in the following way:

Income of a country is earned in two ways: (1) by producing and selling conSUll\er goods and services, and (2) by producing and selling capital goods. That is, national income of a country is composed of the value of consumer goods and services and the value of capital goods. This can be expressed in the form of the following equation:

National Income = Consumption + Investment

or $$Y = C + I \qquad \ldots (i)$$

where Y stands for national income, C for consumption and I for investment.
The above equation represents the production or earning side of the national income. The second aspect of national income is the expenditure side. The total national income can be fully consumed but generally it does not happen so. In actual practice, a part of the total income is spent on consumption and the remaining part is saved. From this we get the following equation:

National Income = Consumption + Saving

or $$Y = C + S \qquad \ldots (ii)$$

where Y stands for national income, C for consumption and S for saving.

In the above two equations (*i*) and (*ii*) it is clear that national income is equal to the sum of consumption and investment and also equal to the sum of consumption and saving. From this it follow& that:

Consumption + Saving = Consumption + Investment

$$C + S = C + I \qquad \ldots (iii)$$

In equation (iii) above, since C occurs on both sides of the equation, we get:

Saving = Investment

or $$S = I$$

From the foregoing analysis, it follows that saving and investment are defined in such a way that they are necessarily equal to each other. In equation (i), investment is that part of national income which is obtained from the production of goods other than those consumed and in equation (ii) saving is that part of national income which is not spent on consumption. Hence in the actual or expost sense, saving and investment by definition are equal.

It is worth mentioning that in macroeconomics, saving and investment do not refer to the saving and investment by an individual; they refer to the saving and investment of the whole community or economy. Saving and investment by an individual can differ but in the ex-post sense, the saving of the whole country must always be equal to the investment.

Now the question arises, why *ex-post saving* and *ex-post investment* are always equal. For instance, when more investment is undertaken by the entrepreneurs how actual saving becomes equal to this larger investment and if the saving falls how investment will become equal to smaller savings. In this connection it is worth mentioning that *modern economists, as did Keynes, include the addition to the inventories of consumer goods in investment.* Now, when saving increases, it implies that consumption will be less. The decline in consumption would result in the addition to the inventories of consumer goods with the shopkeepers and manufacturers, which were not planned or intended by them. This addition to inventories, though unintended, will raise the level of actual investment. Thus unintended increase in inventories will raise the level of investment and in this way investment will increase to become equal to the greater saving. On the other hand, if in any year saving declines, it will result in the unplanned decline in the inventories of consumer goods with the. traders and manufacturers. This unintended decline in inventories will mean the fall in actual investment. In this way, investment will decline to become equal to the lower savings.

Ex-ante Saving and Ex-ante Investment are Equal only in Equilibrium

As said above, in the desired, planned or ex-ante sense, saving and investment can differ. In fact planned or ex-ante saving and investment are generally not equal to each other. This is due to the fact that the persons or classes who save are different from those who invest. Savings are done by general public for various objectives and purposes. On the other hand, investment is made by the entrepreneurial class in the community and is generally governed by marginal efficiency of capital on the one hand and rate of interest on the other hand. Therefore, savings and investment in planned or ex-ante sense generally differ from each other. But through the mechanism of change in the income level, there is tendency for ex-ante saving and ex-ante investment to become equal. When in a year planned investment is larger than planned saving, the level of income rises. At a higher level of income, more is saved and therefore intended saving becomes equal to intended investment On the other hand, when planned saving is greater than planned investment in a period, the level of income will fall. At a lower level of income, less will be saved and therefore planned saving will become equal to planned investment. We thus see that *planned or ex-ante saving and planned or ex-ante investment are brought to equality through changes in the level of income*. When ex-ante saving and ex-ante investment are equal, level of income is in equilibrium *i.e.*, it has no tendency to rise or fall. It is thus clear that whereas realised or ex-post saving is equal to realised or expost investment, intended, planned or ex-ante saving and investment may differ; intended or ex-ante saving and investment have only a tenndency to be equal and are equal only at the equiilibrium level of income.

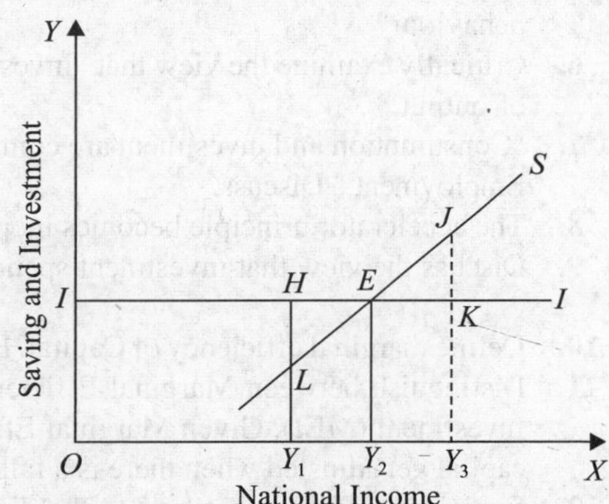

Fig. 8.5. *Equality between Saving and Investment in the ex-ante sense*

That the planned or intended saving is equal to intended investment only at the equilibrium level of income can be easily understood from Fig. 8.3. In this figure, national income is measured along the X-axis while saving and investment are measured along the Y-axis. SS is the saving' curve which slopes upward indicating thereby that with the rise in income, saving also increases. II is the investment curve. Investment curve II is drawn as horizontal straight line because, following Keynes, it has been assumed that investment is independent of the level of income i.e., it depends upon factors other than the current level of income. It will be seen from the Fig. 8.5 that saving and investment curves intersect at point E. Therefore, OY is the equilibrium level of income. If the level of income is OY_1, the intended investment is Y_1H whereas the intended saving is $Y_1 L$. It is thus clear that at OY_1 level of income, intended investment is greater than intended saving. As a result of this, level of income will rise and at higher levels of income more will be saved. It will be seen that with the rise in income to OYz, saving rises and becomes equal to investment. On the other hand, if in any period, level of income is OY_3 intended investment is $Y_3 K$ and intended saving is Y_3J. As a result of this, level of national income will fall to OY_2 at which ex-ante saving and ex-ante investment are once again equal and thus level of national income is in equilibrium

To sum up, whereas ex-post savings and ex-post investment are always equal, ex-ante saving and ex-ante investment are equal only in equilibrium.

QUESTIONS FOR REVIEW

1. What do you understand by investment function? Explain the factors that influence the level of investment in the economy.
   Distinguish between autonomous investment and induced investment. Explain the factors that determine the level of investment in the economy [*D. V., B.Com., 2001*]
2. What is meant by marginal efficiency of capital? How is it calculated? What are the factors that determine marginal efficiency of capital in the economy?
3. What is marginal efficiency of investment ? What role do expectations of entrepreneurs play in determining marginal efficiency of investment or capital?
4. How do changes in the rate of interest influence the decision to invest? Is investment interest elastic?
5. Explain accelerator theory of investment? Is it a good representation of investment behaviour?
6. Critically examine the view that "investment spending depends on the changes in the level of output."
7. "Consumption and investment are complementary during deflation, but competitive at full Employment." Discuss.
8. The accelerator principle becomes inoperative during periods of excess capacity. Discus
9. Discuss the view that investment spending depends on change in the level of income. [*D.V. B.A. (Hons) 2000*]
10. Define marginal efficiency of Capital. How does it determine the volume of investment?
11. Distinguish between Marginal Efficiency of Captial (MEC) and Marginal Efficiency of Investment (MEI). Given Marginal Efficiency of Capital (MEC) how would the stock of capital get adjusted when there is a fall in the rate of interest?
12. Explain the factors that lead to the discrepancy between the actual and desered stock of capital. [*D.V. B.A. (Hons) 1997*]
13. Given the saving function S = - 10 + 0.2Y and autonomous investment I = ₹ 50 cores
    (*i*) Find the level of consumption
    (*ii*) If investment decreases permanently by ₹ 5 crore, what is the new level of consumption?

CHAPTER 9

THEORY OF MULTIPLIER

THE CONCEPT OF MULTIPLIER

The theory of multiplier occupies an important place in the modern theory of income and employment. The concept of multiplier was first of all developed by F.A. Kahn in the early 1930s. But Keynes later further refined it. F.A. Kahn developed the concept of multiplier with reference to the increase in employment, direct as well as indirect, as a result of initial increase in investment and employment. Keynes, however, propounded the concept of multiplier with reference to the increase in total income, direct as well as indirect, as a result of original increase in investment and income. Therefore, whereas Kahn's multiplier is known as *'employment multiplier'*, Keynes's multiplier is known as *investment* or *income multiplier.* The essence of multiplier is that total increase in income, output or employment is manifold the original increase in investment. For example, if investment equal to ₹ 100 crores is made, then the income will not rise by ₹ 100 crores only but a multiple of it. If as a result of the investment of ₹ 100 crores, the national income increases by ₹ 300 crores, multiplier is equal to 3. If as a result of investment of ₹ 100 crores, total national income increases by ₹ 400 crores, multiplier is 4. The multiplier is, therefore, the ratio of increment in income to the increment in investment. If ΔI stands for increment in investment and ΔY stands for the resultant increase in income, then multiplier is equal to the ratio of increment in income (ΔY) to the increment in investment (ΔI). Therefore

$k = \dfrac{\Delta Y}{\Delta I}$ where k stands for multiplier.

Now, the question is why the increase in income is many times more than the initial increase in investment. It is easy to explain this. Suppose Government undertakes investment expenditure equal to ₹ 100 crores on some public works, say the construction of rural roads. For this Government will pay wages to the labourers engaged, prices for the materials to the suppliers and remunerations to other factors who make contribution to the work of road-building. The total cost will amount to ₹ 100 crores. This will increase incomes of the people equal to ₹ 100 crores. But this is not all. The people who receive ₹ 100 crores will spend a good part of them on consumer goods. Suppose marginal propensity to consume of the people is 4/5 or 80%. Then out of ₹ 100 crores they will spend ₹ 80 crores on consumer goods, which would increase incomes of those people who supply consumer goods equal to ₹ 80 crores. But those who receive these ₹ 80 crores will also in turn spend these incomes, depending upon their marginal propensity to consume. If their marginal propensity to consume is also 4/5, then they will spend ₹ 64 crores on consumer goods. Thus, this will further increase incomes of some other people equal to ₹ 64 crores. In this way, the chain of consumption expenditure would continue and the income of the people will go on increasing. But every additional increase in income will be progressively less since a part of the income received will be saved. Thus, we see that the income will not increase by only ₹ 100 crores, which was initially invested in the construction of roads, but by many time more.

Derivation of Investment Multiplier

How much increase in national income will take place as a result of an initial increase in investment can be expressed in the following mathematical form:

Increase in income Or

$$\Delta Y = 100 + 100 \times 4/5 + 100(4/5)^2 + 100(4/5)^3 + 100(4/5)^4 ...$$
$$= 100[1 + (4/5) + (4/5)^2 + (4/5)^3 + (4/5)^4 ...]$$

But the above series is one of geometric progression. Therefore, increase in income (ΔY)

$$= 100 \frac{1}{1-4/5}$$
$$= 100 \frac{1}{1/5}$$
$$= 100 \times 5$$
$$= 500$$

It is thus clear that if the marginal propensity to consume is 4/5, the investment of ₹ 100 crores leads to the increase in the national income by ₹ 500 crores. Therefore, multiplier here is equal to 5. We can express this in a general formula. If ΔY stands for increase in income, ΔI stands for increase in investment and *MPC* for marginal propensity to consume, we can write the equation (i) above as follows:

$$\Delta Y = \Delta I \frac{1}{1 - MPC}$$

$$\frac{\Delta Y}{\Delta I} = \frac{1}{1 - MPC}$$

$\frac{\Delta Y}{\Delta I}$ measures the size of the multiplier. Therefore,

$$\text{Size of multiplier or } k = \frac{1}{1 - MPC} \qquad ...(ii)$$

It is clear from above that the size of multiplier depends upon the marginal propensity to consume of the community. The multiplier is the reciprocal of one minus marginal propensity to consume. However, we can express multiplier in a simpler form. As we know that saving is equal to income minus consumption, one minus marginal propensity to consume will be equal to marginal propensity to save, that is, $1 - MPC = MPS$. Therefore, multiplier is equal to $\frac{1}{1 - MPC} = \frac{1}{MPS}$.

Algebraic Derivation of Multiplier

The multiplier can be derived algebraically as follows:

Writing the equation for the equilibrium level of income we have

$$Y = C + I \qquad ...(1)$$

As in the multiplier analysis we are concerned with changes in income induced by changes in investment, rewriting the equation (1) in terms of changes in the variables we have

$$\Delta Y = \Delta C + \Delta I \qquad ...(2)$$

In the simple Keynesian model of income determination, change in investment is considered to be autonomous or independent of changes in income while changes in consumption are function of changes in income.

In the consumption function,

$$C = a + bY$$

where a is a constant term, b is marginal propensity to consume which is also assumed to remain constant. Therefore, change in consumption can occur only if there is change in income. Thus

$$\Delta C = b\Delta Y \qquad ...(3)$$

Substituting (3) into (2) we have

$$\Delta Y = b\Delta Y + \Delta I$$

$$\Delta Y - b\Delta Y = \Delta I$$

$$\Delta Y\,(1 - b) = \Delta I$$

$$\Delta Y = \frac{1}{1-b}\,\Delta I$$

$$\frac{\Delta Y}{\Delta I} = \frac{1}{1-b}$$

As b stands for marginal propensity to consume

$$\frac{\Delta Y}{\Delta I} = \frac{1}{1-MPC} = \frac{1}{MPS}$$

This is the same formula of multiplier as obtained earlier. Note that the value of multiplier $\frac{\Delta Y}{\Delta I}$ will remain constant as long as marginal propensity to consume remains the same.

Calculating the Size or Value of Multiplier

It follows from above that the *size or value of multiplier is the reciprocal of marginal propensity to save.* Therefore, we can obtain the value of multiplier if we know the marginal propensity to consume or the marginal propensity to save of the community. Given the size of multiplier form the net increase in investment, we can find out the total increment in income that will occur as a result of investment.

If the marginal propensity to consume of a community is equal to $\frac{2}{3}$, we can find out the size of multiplier as under:

$$\text{Multiplier, } k = \frac{1}{1-MPC}$$

$$= \frac{1}{1-\frac{2}{3}} = \frac{1}{\frac{1}{3}} = 3$$

Likewise, if the marginal propensity to consume is equal to $\frac{1}{2}$ or 0.5, then the multiplier:

Two Limiting Cases of the Value of Multiplier. There are two limiting cases of the multiplier. One limiting case occurs when the marginal propensity to consume is equal to one, that is, when the whole of the increment in income is consumed and nothing is saved. In this case, the size of multiplier will be equal to infinity, that is, a small increase in investment will bring about a very large increase in income and employment so that full employment is reached and even the process goes beyond that. "In such circumstances, the Government would need to employ only one road builder to raise income indefinitely, causing first full employment and then a limitless spiral of inflation."[1] However, this is unlikely to occur since marginal propensity to consume in the real world is less than one. The other limiting case occurs when marginal propensity to consume is equal to zero, that is, when nothing out of the increment in income is consumed, and the whole increment in income is saved. In this case, the value of the multiplier will be equal to one. That is, in this case, the increment in income will be equal

1. Stonier and Hague, *A Textbook of Economic Theory,* 4th edition, 1972.

to the original increase in investment and not a multiple of it. But in actual practice the marginal propensity to consume is less than one but more than zero $\left(1 > \frac{\Delta C}{\Delta Y} > 0\right)$. Therefore, *the value of the multiplier is greater than one but less than infinity.*

Assumptions of Multiplier Theory

In our above explanation of multiplier, we have made many simplifying assumptions. First, we have assumed that the *marginal propensity to consume remains constant throughout* as the income increases in various rounds of consumption expenditure. However, the marginal propensity to consume may differ in various rounds of consumption expenditure. But this constancy of marginal propensity to consume is a realistic assumption, since all available empirical evidence shows that marginal propensity to consume is very stable in the short run. Secondly, we have assumed that there is a net increase in investment in a period and no further indirect effects on investment in that period occur or if they occur they have been taken into account so that there is a given net increase in investment. Further, we have assumed that *there is no any time-lag* between the increase in investment and the resultant increment in income. That is, increment in income takes place instantaneously as a result of increment in investment. J.M. Keynes ignored the time-lag in the process of income generation and therefore his multiplier is also called *instantaneous multiplier.* In recent years, the importance of time lag has been recognised and concept of dynamic multiplier has been developed on that basis. But in an ele-mentary study as the present one the time lags will be ignored as was done by Keynes.

Another important assumption in the theory of multiplier is that *excess capacity exists in the consumer goods industries* so that when the demand for them increases, more amounts of consumer goods can be produced to meet this demand. If there is no excess capacity in consumer goods industries, the increase in demand as a result of some original increase in investment will bring about rise in prices rather than increases in real income, output and employment. As we shall see later, Keynes's multiplier was evolved in the context of advanced capitalist economies which were in grip of depression and in times of depression and there did exist excess capacity in the consumer goods industries due to lack of aggregate demand. The Keynesian multiplier effect is very small in developing countries like India since there is not much excess capacity in consumer goods industries.

In our above analysis of the multiplier process we have taken a closed economy, that is, we have not taken into account imports and exports. If ours were an open economy, then a part of the increment in consumption expenditure would have been made on imports of goods from abroad. This would have caused increment in income in foreign countries rather than within the country. This will reduce the value of the multiplier. *Imports are important leakage from the multiplier process* and we have ignored them in our above analysis for the purpose of simplicity.

It is worth noting that multiplier not only works in money terms but also in real terms. In other words, multiple increment in income as a result of a given net increase in investment does not only take place in money terms but also in terms of real output, that is, in terms of goods and services. When incomes increase as a result of investment and these increments in income are spent on consumer goods, the output of consumer goods is increased to meet the extra demand brought about by increased incomes. Therefore, real income or output, increases by the same amount as the increment in money incomes, since *the prices of goods have been assumed to be constant.* Of course, we have assumed, as has been mentioned above, that there exists excess productive capacity in the consumer goods industries so that when the demand for consumer goods increases, their production can be easily increased to meet this demand. However, if due to some bottlenecks output of goods cannot be increased in response to increasing demand, prices will rise and as result the real multiplier effect will be small.

Diagrammatic Representation of Multiplier

We have already explained that the level of national income is determined by the equilibrium between aggregate demand and aggregate supply. In other words, the level of national income is fixed at the level where $C + I$ curve intersects the 45° income curve. With such a diagram we can explain the multiplier. The multiplier is illustrated in Fig. 9.1. In this figure C represents marginal propensity to consume. Marginal propensity to consume has been here assumed to be equal to $\frac{1}{2}$ *i.e.,* 0.5. Therefore, the slope of the curve C of marginal propensity to consume curve C has been taken to be equal to 0.5. $C + I$ represents aggregate demand curve. It will be seen from Fig. 7.1 that the aggregate demand curve $C + I$ which intersects the 45° line at point E so that the level of income equal to OY_1 is determined. If investment increases by the amount EH we can then find out how much increment in income and occur as a result of this. As a consequence of increase in investment by EH, the aggregate demand curve shifted upward to the new position $C + I'$. This new aggregate demand curve $C + I'$ intersect income line at point F so that the equilibrium level of income increases to OY result of net increase in investment equal to EH, the income has increased by Y_1Y_2. seen from the figure that Y_1Y_2 is greater than EH. On measuring will be found that Y_1Y_2 is *twice* the length of EH. This is as it is expected because the market propensity to consume is here equal to $\frac{1}{2}$ and therefore the size of multiplier will be equal to 2.

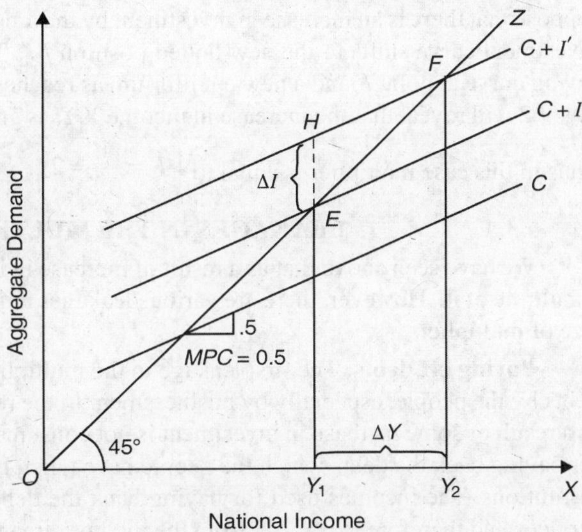

Fig. 9.1. *Keynes's Income Multiplier*

The multiplier can be illustrated through saving investment diagram also. In a previous chapter we explained the determination of national income also through saving the investment. Therefore, the multiplier can also be explained with the help of saving investment diagram, as has been shown in Fig. 9.2. In this figure SS is the saving curve indicating that as the level of income increases, the community plans to save more. II is the investment curve showing the level of investment planned to be undertaken by the investors in the community. The investment has been taken to be a constant amount

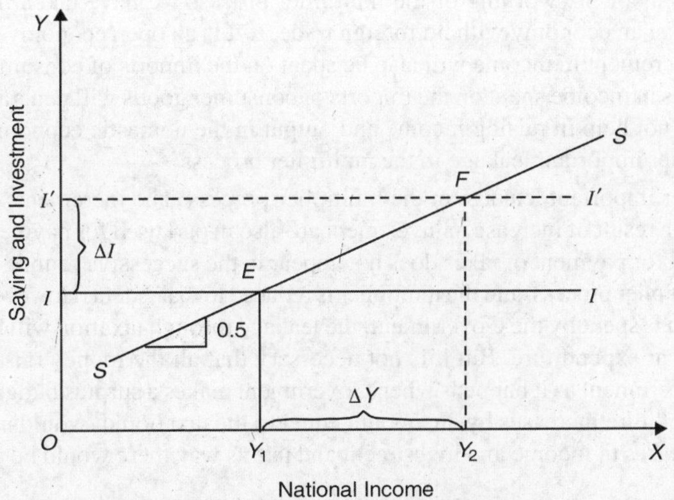

Fig. 9.2. *Multiplier Explained with the Aid of Savings-Investment Diagram*

and autonomous of changes in income. This investment level OI has been determined the marginal efficiency of capital and the rate of interest. Investment being autonomous of income means that it does not change with the level of income. Keynes treated investment as autonomous of income and we will here follow him. It will be seen from Fig. 9.2 that saving and investment curves intersect at point E, that is, planned saving and planned investment are in equilibrium at the level of income OY_1. Thus, with the given saving and investment curves level of income equal to OY_1 is determined. Now suppose that there is an increase in investment by the amount II'. With this increase in investment, the investment curve shifts to the new dotted position $I'I'$. This new investment curve $I'I$ intersects the saving curve at point F and a new equilibrium as reached at the level of income OY_2. A glance at the Fig. 9.2 will reveal that the increase in income Y_1Y_2 is greater than the increase in investment by II''. Thus in this case multiplier is equal to $\left(K = \frac{1}{0.5} = 2\right)$.

LEAKAGES IN THE MULTIPLIER PROCESS

We have seen above that as a result of increase in investment, the level of income increases by a multiple of it. However, there are various leakages in income generation process which reduce the size of multiplier.

Paying off debts. The first leakage in the multiplier process occurs in the form of payment of debts by the people, especially by businessmen. In the real world, all income received by the people as a result of some increase in investment is not consumed. A part of the increment in income is used for paying back the debts which the people have taken from moneylenders, banks or other financial institutions. The incomes used for paying back the debts do not get spent on consumer goods and services and therefore leak away from the income stream. This reduces the size of the multiplier. Of course, when incomes received by the moneylenders, banks or institutions are again lent back to the people, they come back to the income stream and enhance the size of multiplier. But this may or may not happen.

Holding of idle cash balances. If the people hold a part of their increment in income as idle cash balances and do not use it for consumption, they also constitute leakage in the multiplier process. As we have seen, people keep part of their income for satisfying their precautionary and speculative motives, money kept for such purposes is not consumed and therefore does not appear in the successive rounds of consumption expenditure and therefore reduces the increments in total income and output.

Imports. In our above analysis of the working of the multiplier process we have taken the example of a closed economy, that is, an economy with no foreign trade. If it is an open economy as is usually the case, then a part of increment in income will also be spent on the imports of consumer goods. The proportion of increments in income spent on the imports of consumer goods will generate income in other countries and will not help in raising income and output in the domestic economy. Therefore, imports constitute another important leakage in the multiplier process

Taxation. Taxation is another important leakage in the multiplier process. The increments in income which the people receive as a result of increase in investment are also in part used for payment of taxes. Therefore, the money used for payment of taxes does not appear in the successive rounds of consumption expenditure in the multiplier process, and the multiplier is reduced to that extent. However, if the money raised through taxation is spent by the Government, the leakage through taxation will be offset by the increase in Government expenditure. But it is not necessary that all the money raised through taxation is spent by the Government as it happens when Government makes a surplus budget. No doubt, if the Government expenditure increases by an amount equal to the taxation, it would not have any adverse effect on the increases in income and investment and in this way there would be no leakage in the multiplier process.

Increase in Prices. Price inflation constitutes another important leakage in the working of the multiplier process in real terms. As we have noted above, the multiplier works in real terms only when as a result of increase in money income and aggregate demand, output of consumer goods is also increased. When output of consumer goods cannot be easily increased, a part of the increases in the money income and aggregate demand raises prices of the goods rather than their output. Therefore, the multiplier is reduced to the extent of price inflation. In developing countries like India the extra incomes and demand are mostly spent on foodgrains whose output cannot be increased so easily. Therefore, the increments in demand raise the prices of goods to a greater extent than the increase in their output. Besides, in developing countries like India, there is not much excess capacity in many consumer goods industries, especially in agriculture and other wage-goods industries. Therefore, when income and demand increase as a result of increase in investment, it generally raises the prices of these goods rather than their output and therefore weakens the working of the multiplier in real terms. Thus, it is often asserted that Keynesian theory of multiplier is not very much relevant to the conditions of developing countries like India.[2]

The above various leakages reduce the multiplier effect of the investment undertaken. If these leakages are plugged, the effect of change in investment on income and employment would be greater.

IMPORTANCE OF THE CONCEPT OF MULTIPLIER

Multiplier is one of the most important concepts developed by J.M. Keynes to explain the determination of income and employment in an economy. The theory of multiplier has been used to explain the cumulative upward and downward swings of the trade cycles that occur in a free-enterprise capitalist economy. When investment in an economy rises, it has a multiple and cumulative effect on national income, output and employment. As a result, economy experiences rapid upward movement. On the other hand, when due to some reasons, especially due to the adverse change in the expectations of the business class, investment falls, then backward working of the multiplier causes a multiple and cumulative fall in income, output and employment and as a result the economy rapidly moves on downswing of the trade cycle. Thus, Keynesian theory of multiplier helps a good deal in explaining the movements of trade cycles or fluctuations in the economy.

The theory of multiplier has also a great practical importance in the field of fiscal policy to be pursued by the Government to get out of the depression and achieve the state of full employment. To get rid of depression and remove unemployment, Government investment in public works was recommended even before Keynes. But it was thought that the increase in income will be limited to the amount of investment undertaken in these public works. But the importance of public works is enhanced when it is realised that the total effect on income, output and employment as a result of some initial investment has a multiplier effect. Thus, Keynes recommended Government investment in public works to solve the problem of depression and unemployment. The public investment in public works such as road building, construction of hospitals, schools, irrigation facilities will raise aggregate demand by a multiple amount. The multiple increase in income and demand will also encourage the increase in private investment. Thus, the deficiency in private investment which leads to the state of depression and underemployment equilibrium will now be made up and a state of full employment will be restored. If the multiplier had not worked, the income and demand would have risen as a result of some public investment but not as much as they rise with the multiplier effect. Inspired by the Keynesian theory of multiplier. Expansionary fiscal policy of increase in Government expenditure and reduction in income tax was have been adopted by President John Kenedy and press president George W. Bush in the United State of America to remove involuntary unemployment and depreciation. This had a great success in removing unemployment and, depression and therefore, Keynesian theory of multiplier was vindicated and as a result people's belief in it increased.

A Numerical Problem on Multiplier

Problem. *Suppose the level of autonomous investment in an economy is ₹ 200 crores*

2. See, V.K.R.V. Rao, Investment, Income and the Multiplier in an Underdeveloped Economy printed in A.N. Agarwala and S.P. Singh (edited). *The Economics of Underdevelopment,* Oxford, 1958.

and consumption function of the economy is :

$$C = 80 + 0.75Y$$

(*a*) *What will be the equilibrium level of income?*

(*b*) *What will be the increase in national income if investment increases by ₹ 25 crores.*

Solution. (*a*) For equilibrium level of income,

$$Y = C + I \quad ...(i)$$

where $C = 80 + 0.75$

$$I = 200 \text{ crores}$$

Substituting the values of C and I in (1) we have

$$Y = 80 + 0.75K + 200$$

$$(Y - 0.75Y) = 80 + 200 = 280$$

$$0.25\,Y = 280$$

$$Y = 280 \times \frac{100}{25} = 1120$$

Equilibrium level of income is therefore equal to 1120 crores.

(*b*) How much increase in income will occur as a result of increase in investment by ₹ 25 crores depends on the size of multiplier. The size of multiple is determined by the value of marginal propensely to consume. In the given consumption function ($C = 80 + 0.75Y$) marginal propensity to consume is equal to 0.75 or $\frac{3}{4}$. Thus,

$$\text{multiplier} = \frac{1}{1 - MPC} = \frac{1}{1 - \frac{3}{4}} = 4$$

Thus, with increase in investment by ₹ 25 crores, national income will rise by $25 \times 4 = 100$ crores.

THE PARADOX OF THRIFT

An interesting paradox arises when all people in a society try to save more but in fact they are unable to do so. The multiplier theory of Keynes helps a good deal in explaining this paradox. According to this paradox of thrift, the attempt by the people as a whole to save more for hard times such as impending period of recession or unemployment may not materialise and in their bid to save more the society infact may not only end up with the same savings (Or, even lower savings) but also

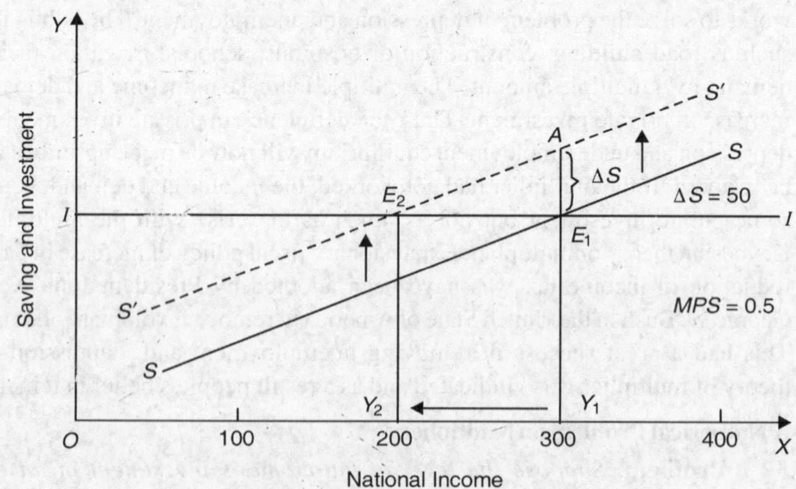

Fig. 9.3. *The Paradox of Thrift*

in the process cause their consumption or standard of living to decline. Thrift (*i.e.*, the desire to save more) is considered to be a virtue in most of the societies and it is regarded as an act of prudence on the part of individuals to save for a rainy day. According to a proverb, "a penny saved is a penny earned". Further, according to classical economists, savings determine investment which plays a crucial role in accelerating the rate of economic growth. However, the paradox of thrift shows that the efforts to save more, especially in times of depression, may actually deepen the economic crisis and cause output to fall and unemployment to increase. It goes to the credit of Keynes that with his multiplier theory he was able to resolve the paradox of thrift. Keynesian explanation of paradox of thrift has been shown in Fig. 9.3. According to the Keynesian theory, the saying "penny saved is penny earned" is quite inappropriate for the economy as a whole when it is working at underemployment equilibrium, that is, when there prevails recession or depression. Keynes has showed that if *all* people in a society decide to save more, they may actually fail to do so but nevertheless reduce their consumption. This is because, according to Keynes, the effort to save more by all in a society will lower the aggregate demand for goods and services resulting in a drop in the level of national income. At the lower level of national income, the saving fall to the original level but consumption will be less than before which implies that the people would become worse off.

Consider Fig. 9.3, where *SS* is the saving curve with a slope equal to 0.5, and *II* is the planned investment curve. It will be seen that saving and investment curves intersect at point E_1 and determines level of income equal to Y_1 or ₹ 300 crores. Now suppose that expecting hard times ahead all people try to save more by the amount of ₹ 50 crores which would cause an autonomous downward shift in the consumption function. This downward shift in the consumption function brings about an upward shift by ₹ 50 crores or E_1A in the saving function curve to *S′S′*. This new saving function curve *S′S′* cuts the planned investment curve *II* at point E_2 according to which new equilibrium level of income falls to Y_2 or ₹ 200 crores. It is important to note that level of income does not drop only by the amount (E_1A or ₹ 50 crores), that is, by the extent of reduction in consumption due to more saving but by a multiple of it. With marginal propensity to save (*MPS*) being equal to 0.5 or $\frac{1}{2}$, the value of multiplier would be $1/MPS = 1\frac{1}{2} = 2$. Further, the decline in consumption due to more saving would cause the multiplier to work in reverse, that is, the multiplier would operate to reduce the level of consumption and income by a magnified amount. The decline in consumption expenditure of the people by ₹ 50 crores in the first instance due to more saving by them implies that the producers and sellers of goods and services will find their income to fall by ₹ 50 crores. But the reverse process will not stop here. Given the marginal propensity to consume being equal to 0.5 or $\frac{1}{2}$, the producers/sellers of goods and services in turn would spend ₹ 25 crores less when they find their income has fallen by ₹ 50 crores. It will be observed from Fig. 9.3 that this process of reduction of the level of income will continue till the new saving is equal to investment at the lower level of income Y_2 (₹ 200 crores), that is, the level of income has declined by ₹ 100 crores (50 × 2) from its initial equilibrium level of income Y_1 of ₹ 300 crores. Thus the attempt by all people to save more has led to the decline in the equilibrium level of income to Y_2 or ₹ 200 crores at which, with marginal propensity to consume remaining unchanged at 0.5 or $\frac{1}{2}$, saving of the society will fall to the initial level of Y_1E or ₹ 100 crores ($200 \times 0.5 = 100$). This is clearly depicted in Fig. 7.3. With the decrease in planned saving by ₹ 50 crores at every level of income the saving function (*SS*) shifts upward. This sets in motion the operation of the multiplier in the reverse and as will be seen from the Fig. 9.3, the new equilibrium is reached at the new lower level of income Y_2 (₹ 200 crores). It is important to observe that the saving which had risen to Y_1A (₹ 150 crores) has once again fallen to the original level of ₹ 100 crores ($Y_2E_2 = Y_1E_1$) due to reduction in consumption expenditure inducing the working of multiplier in the reverse which causes a decline in the equilibrium level of income from K_t (₹ 300 crores) to Y_2 (₹ 200 crores). In other words, the increases in saving by ₹ 50 crores has led to the fall in income by ₹ 100 crores because the multiplier is equal to 2. This explains the paradoxical feature of an economy gripped by recession. *This is paradoxical because in their attempt to save more the people have caused a decline in their income and consumption with no increase in the saving of the society at all.*

In our analysis we have assumed that the planned investment is fixed, that is, determined outside the model. In other words, the investment has been assumed to be *autonomous of income,* that is, it does not vary with income.

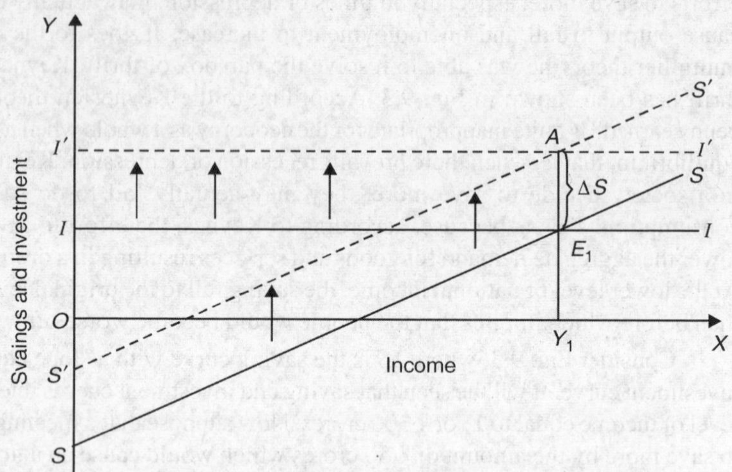

Fig. 9.4. *The Situation when paradox of thrift does not hold*

Can We Avert the Paradox of Thrift?

Paradox of thrift holds good when a free market economy is in the grip of recession or depression and investment demand is inadequate due to lack of profit opportunities. However, it has been pointed out by some economists that paradox of thrift can be averted if the extra savings that the people do for a rainy day are somehow channeled into additional investment through financial markets. Indeed, the classical economists argued that the increase in the supply of savings would lead to the fall in the rate of interest which would induce increase in planned investment. If this happens, then in our saving-investment diagram the investment curve *II* would shift up to *I′I′* and as will be seen from Fig. 9.4 the new equilibrium level of income may not fall and therefore the paradox of thrift is averted. In Fig. 9.4, initially the saving curve (S_1S_1) and investment curve (*II*) intersect at point E_1 and determine Y_1 level of income. Now, if the people of the society expecting difficult times ahead, desire to save E_1A more. If these extra savings, for reasons mentioned above, result in more investment, the investment curve will shift to *I′I′′*, the new equilibrium will be at point *A* corresponding to the original level of income Y_1. In this way the paradox of thrift has been averted.

However, according to the modern economists, especially the followers of Keynes, the empirical evidence does not support the above argument of averting the paradox of thrift. This is because at times of recession or depression, the prospective yields from investment are so small that no possible reduction in the rate of interest will induce sufficient increase in investment. Thus, according to them, in a free-market and private enterprise economy without Government intervention paradox of thrift cannot be averted. Of course, if the Government intervenes as it does even in the present-day predominantly private enterprise economies of the USA and Great Britain, it can mobilise the extra savings of the people and invest them in some worthwhile projects and thus prevent aggregate demand and income from falling. This can happen because the Government undertakes investment because it is not motivated by profit motive but by the considerations of promoting social interest and economic growth. It is because of this that the role of the Government has greatly increased for overcoming recession in the capitalist countries.

THE KEYNESIAN EXPLANATION OF GREAT DEPRESSION: THE IMPACT OF MULTIPLIER

During the 1930s the capitalist economies experienced severe depression which caused widespread involuntary unemployment, substantial loss of output and income and crushing hunger and poverty among the working classes. As explained earlier, the classical economists attributed this unemployment and depression to the higher was rates maintained by the trade unions and the Government. However, this explanation did not prove to be valid. It was English economist J.M. Keynes who radically departed from the classical thought and put forward the view that it was the large decline in investment that caused the depression and substantial increase in involuntary unemployment. According to Keynes, the investment was highly volatile and it was a drastic decline in it due to the pressimistic expectations of the entrepreneurs about the prospective profits from investment that brought about a decline in aggregate demand (expenditure) which through working of the multiplier in the reverse caused a magnified fall income (output) and employment. For example, during the first four years (1929-33) of depression in the USA the unemployment which was only 3.2 per cent in 1929 soared to 25 per cent in 1933, that is, one out of four in the labour force in the United States became unemployed. The level of national income dropped from \$ 315 billion in 1929 to \$ 222 billion in 1933 at 1972 prices, a decline of \$ 93 billion in just four years. According to Keynes, this was caused by a drastic fall in investment from \$ 56 billion in 1929 to \$ 8.5 billion in 1993.

The huge decline in national income and the emergence of unemployment in the USA, UK and other industrialised capitalist countries during the period of depression is graphically shown in Fig. 9.5. It is assumed that to begin with, say in 1929, the aggregate demand curve C + I_2 intersects 45° line at point H and determines equilibrium level of income at full-employment or potential output level OY_1. The sharp decline in investment by the amount *HT* due to the fall in profitability of investment following a crash in stock markets in 1929 and other unfavourable events caused a downward shift in the aggregate demand curve to C + I_1 (where $I_1 < I_2$). This new aggregate demand curve $C + I_1$ intersects the 45° line at point *E* and accordingly determines equilibrium level of income OY_1 which is much lower than full-employment level OY_F and thus represents a state of depression with a large unemployment of workers. *The important point made by Keynes was that income would not fall merely equal to the decline in investment but by a multiple of it.* In fact, during the depression period of 1930s, it actually happened so and is evident from the Table 9.1. It will be seen from the Figure 9.5 that the decline in national income Y_FY_1 is not equal to the fall in investment by *HT* by out by a multiple of it. Y_FY_1 is twice that of *HT*. As explained above, this is due to the working of multiplier in the reverse. Further note that after taking into all leakages in the multiplier process it has been assumed that marginal propensity to consume is equal to 0.5 which yields the value of multiplier 1/1-*MPC* = 1/1-1/2 = 2, This is why fall in income by Y_FY_1 is twice the decline in investment by *HT.*

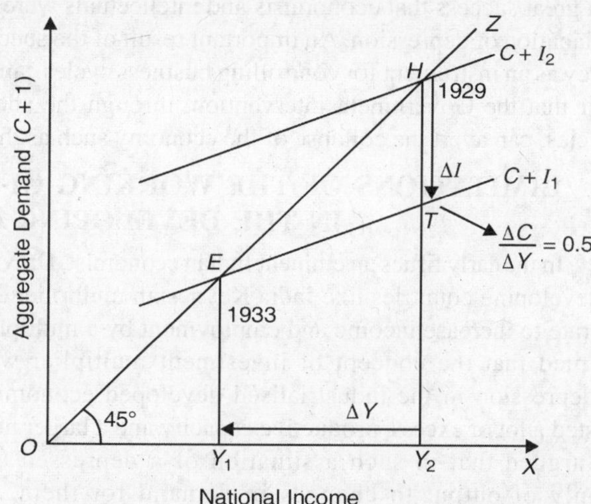

Fig. 9.5. *Depression of 1930s Caused by a Decline in Investment and Working of Multiplier in Reverse*

Now, the historical record of this period about the various components of aggregate demand of the US economy shows that changes in net exports and Government expenditure were quite small and they mostly offset each other during the period 1929-33. The drastic drop in private investment appears to be the basic reason for the huge fall in aggregate demand or spending. The private investment which was \$ 56 billion in 1920 fell to only \$8.5 billion in 1933 in the U.S.A., the decline

Table 9.1

	1929	1933	(in billion US $) Change
Investment	56	8.5	47.5
National Income	315	222	93
Unemployment (% of labour force)	3.2	25	21.8

of $ 47.5 billion in four years. But, as has been explained by Keynes, the decrease in aggregate expenditure was not merely equal to $ 47.5 billion, but by a multiple amount due to the operation of the multiplier in the reverse. It has been estimated that taking into account all leakages in the multiplier process, the value of the multiplier was around 2 during the period. Therefore, as a result of sharp decline in investment by $ 47.5 billion and consequently operation of the multiplier in the reverse there was a fall in the induced consumption expenditure. Thus, as a result of the sharp drop in private investment and resultant fall in induced consumption due to working of multiplier caused much bigger decrease ($ 93 billion) in the level of aggregate effective demand, income and output. This is in accordance with the value of multiplier being equal to around 2. Multiplier is here equal to

$$\frac{\Delta Y}{\Delta I} = \frac{93}{47.3} = 1.96$$

Thus, the Keynesian theory of income determination provides a fairly accurate explanation of the first four years of the great depression. This looks rather simple but during the early 1930s it was not understood at all. Only after the Keynesian prescription toward off depression and involuntary unemployment, namely, launching by the Government public works programme financed by the deficit budgets to raise aggregate demand, such as adopted under New Deal Policy in the U.S.A. proved to be a great success that economists and intellectuals were convinced about the validity of Keynes's explanation of depression. An important result of the success of the Keynesian model was that fiscal policy as an instrument for controlling business cycles came into prominence. Further, it now became clear that the Government intervention, through the adoption of appropriate fiscal and monetary policies, can avert the collapse of the economy such as that happened during 1929-33.

LIMITATIONS OF THE WORKING OF KEYNESIAN MULTIPLIER IN THE DEVELOPING COUNTRIES

In the early fifties an eminent Indian economist, Dr. V.K. R.V. Rao, and some others explained that in developing countries like India Keynesian multiplier did not work in real terms, that is, does not operate to increase income and employment by a multiple of the initial increase in investment.[3] He claimed that the concept of investment multiplier was valid in the context of the situation of depression in the industrialised developed economies of the UK and the USA where there existed a lot of excess productive capacity and a larger number of open involuntary unemployment. He argued that in such a situation of a depressed economy there was a high elasticity of supply of output to changes in demand for them. Therefore, in the developed capitalist economies ridden with depression increase in investment leading to successive rounds of consumption expenditure raises aggregate demand. Due to the existence of large excess capacity and involuntary unemployment under conditions of depression aggregate supply of outputs highly elastic, increase in aggregate demand brings about increase in real income, output and employment which is a multiple of original increase in investment.

3. Dr. V.K.R.V. Rao, "Investment, Income and Multiplier in an Underdeveloped Economy", *Indian Economic Review,* February 1952.

On the other hand, they claimed that in underdeveloped countries there was little excess capacity in consumer goods industries and therefore supply of output was inelastic. Therefore, when there is injection of investment, and as a result through successive rounds of the operation of multiplier, aggregate demand for consumer goods increases, it results mainly in rise in money income brought about through rise in prices and not an increase in real national income.

The second condition, according to Dr. Rao and his followers for the working of multiplier in raising national income and employment was that the supply of raw materials, financial capital must be sufficiently elastic so that when aggregate demand increases as a result of multiplier effect of increase in investment the supply of output could be increased adequately to meet this higher demand for goods and services. They argued that in underdeveloped countries like India due to under-developed nature of their economies, there was acute scarcity of raw materials, other intermediate goods such as steel, cement and financial capital which put great obstacles for the working of multiplier in real terms.

The third condition required for the working of multiplier in real terms was that there should be involuntary open unemployment so that when aggregate demand for goods increases as a consequence of new investment, the adequate supply of workers must be forthcoming to be employed in the production processes of various industries. They argued this condition too was not fulfilled in the under-developed countries where there existed disguised unemployment, especially in the agricultural sector. The disguisedly unemployed workers who are supported by joint family system could not be easily shifted to be employed in the industries for expansion of output to achieve the multiplier effect.

Lastly, it was pointed out that the under-developed countries like India had predominantly agricultural economies and income elasticity of demand for foodgrains was very high in these economies. In view of this when increase in investment leads to the rise in money incomes of the people, a large part is spent on foodgrains. But the supply of agricultural products is inelastic because their production is subject to uncertain natural factors like Monsoon and climate and further there was lack of irrigation facilities, improved seeds, fertilizers etc. Hence it was difficult to increase agricultural production in response to the increase in demand through the multiplier effect of increase in investment.

It follows from above that the Keynesian assumptions for the working of multiplier in real terms, namely (1) the supply of output of goods is elastic due to the existence of large excess capacity, (b) the supply of raw materials and others intermediate goods can be adequately increased, (c) there exists involuntarily unemployed workers searching for work and, (d) sufficiently elastic agricultural output. In view of the earlier economists these assumptions for realising the multiplier effect in terms of rise in real income and employment were not valid in case of under-developed countries. Therefore, according to them, Keynesian multiplier did not operate in real terms in under-developed countries and actually leads to the rise in price or inflationary conditions in them.

The Relevance of Multiplier in Developing Countries : The Modern View

We have explained above the views of some eminent Indian economists, such Dr. V.K.R.V. Rao, Dr. A.K. Dass Gupta expressed during the early fifties regarding non-operation of the Keynesian multiplier in the under-developed countries. But the situation in the present day developing countries has substantially changed in the last 50 years. There has been a lot of economic growth and structural transformation in the Indian economy during the last half a centuary so that supply conditions today have become significantly elastic. So in the present state of the Indian economy and also of some other developing economies, it cannot be said that Keynesian multiplier is not applicable in real terms in them. However, it may be noted that even in the fifties and early sixties the view that Keynesian multiplier did not work in the under-developed countries did not go entirely unchallanged. Thus commenting on Dr. Rao's article Dr. K.N. Raj remarked that "*Discarding the Keynesian thesis as altogether inoperative in under-developed countries is really throwing the baby away with the bath water*".[4] Similarly Dr. D.R. Khatkhate wrote, "In conclusion we may state that the multiplier can operate in an under-developed economy when it is associated with a carefully designed pattern of investment. The theory that the multiplier works in a backward economy only with reference to the money income is based on static assumptions and is, therefore, not correct"[5]. Likewise, Dr. Ashok Mathur write. "Our main objection against the view that Keynesian multiplier does not operate in the under-developed countries is that it views the operation of multiplier process in a completly static setting and as a purely short-period concept, whereas the very rationale of economic development is long run dynamic change. Once we relax these two restrictive assumptions, the essential content of the Keynesian multiplier, that is increase of investment results in an increase in output which is much in excess of the original outlay on investment holds true in case of the developing as much as in the developed economies"[6]

At present, in the beginning of the new millenium as a result of economic growth both in the industrial and agricultural sectors the Indian economy has a widely diversified structure and supply of output has become quite elastic, at least in the industrial sector. Besides, at times there is a lot of excess or unutilised capacity in several industries in India due to the deficieny of aggregate demand. The potential for increasing raw materials and intermediate products such as cement, steel and fertilizers has significantly increased to meet the rising demand for them.

If there is injection of investment it will result in manifold increase in output or real income and employment through the working of the multiplier. If the injection of new investment package is quite diversified and balanced, as is generally planned in our Five Years Plans, the investment and growth in several industries simultaneously will create not only additional demand for each other as was visualised by Nurkse[7] but will also create productive, capacities in them which will ultimately over a period of time will result in multiple increase in output an employment.

It is worth noting that in India today there is not only a lot of preexisting excess production capacity in the Indian industries but new investment every year also creates additional production capacity which with some time lag will result in increase in real income or output, if adequate aggregate demand is forthcoming for its utilisation. Harrod-Domar in their famous dynamic growth models emphasised that investment not only creates demand but also new productive capacity[8]. Thus, if we look at increment in investment from the viewpoint of dynamics of development and take a longer time horizon, multiplier effect of new investment in the developing countries can become a reality. It is true that increase in money incomes and demand may tend to occur ahead of the increase in real income but subject to some time lag between investment and consequent increase in production capacity, the latter would tend to catch up with the former. The significant point to note is that investment not only creates demand but it also creates production capacity. Ultimately there is no reason as to why multiplier effect of new investment on real income or output may not materialise, though the actual period required for realisation of the multiplier effect depends on various time lags in the process of income generation and capacity creation. The wider the range ot industries over which initial investment is undertaken the greater will be the multiplier effect. This is because monetary demand or expenditure generated by investment in any one industry would be easily met by the increase in production capacity in a variety of industries. In this way increase in demand resulting from investment would not lead to rise in prices but will cause real output to rise.

Agriculture and Multiplier. As mentioned above, the argument for non-operation of multiplier in underdeveloped countries was also partly based on the inelastic nature of supply of agricultural

4. Dr. K.N. Raj, "Dr. Rao on Investment, Income and the Multiplier in an Under-developed Economy A Comment" *Indian Economic Review,* p. 1952.
5. D.R. Khatkhate "The Multiplier Process in Developing Economies" *Indian Economic Journal,* Oct. 1954.
6. Ashok Mathur, On Throwing the Baby away with the Bath Water, *Indian Economic Journal,* April-June; 1965
7. Ragnar Nurkse, *Problems of Capital Formation in Underdeveloped Countries,*
8. See E. Domar, Capital Expansion, Rate of Growth and Employment, *Economica* Vol. 14, 1948.

output especially foodgrains as it was pointed out that a large part of monetary demand or money incomes generated by investment would be spent on foodgrains. Inability to meet the rise in demand for foodgrains would cause rise in price level or inflation in the economy rather than increase in real output. It may be pointed out that thanks to the spread of green revolution technology expansion in irrigation facilities in various states of India, foodgrains production can be adequately increased in response to rising demand for foodgrains.

Furthermore, it was asserted by Dr. Rao, the existence of disguised unemployment in underdeveloped countries instead of Keynesian type involuntary open unemployment also prevented the working of multiplier in real terms. In the Indian economy today there are a large number of involuntarily unemployed workers crying out for employment. So this argument for failure of multiplier to work in real terms no longer hold good in the present economic situation.

To conclude, in the present economic situation of the Indian economy with a lot of excess production capacity in several consumer goods industries and a large potential for expanding agricultural production, increase in investment would produce a real multiplier effect on increasing real income and output without causing inflationary pressures in the economy. Multiplier effect of new investment can be further increased, if investment package is quite diversified covering a large number of industries (including agriculture) so that monetary demand and income generated by any one industry can be adequately met by increase in output capacity in other industries. Further, even when their is no preexisting excess capacity in the industries increase in investment leads to the increase is demand for consumption goods which in turn cause further rise on investment to meet that consumption demand. The effect of increase in consumption demand on expansion in investment in generally referred to as *accelerator.* Indeed, the combined working of multiplier and acceleration, which is called *super-multiplier,* leading to manifold increase in output can take place in the growth process in the developing countries like India[9].

QUESTIONS FOR REVIEW

1. What is investment multiplier? How is it related to marginal propensity to consume? [**Hints:** Keynes's investment multiplier is also called Income multiplier.]
2. Wht is income multiplier ? Find out the value of multiplier given the following: *(a)* $MPC = 0.80$, *(b)* $MPC = 0.75$, *(c)* $MPS = 0.20$, *(d)* $MPS = 0.50$.
3. Explain the concept of investment multiplier what is the value of multiplier when marginal propensity to save is 0.1
4. Explain the Keynesian theory of investment multiplier. What are possible leakages in the multiplier mechanism?
5. Keynesian investment multiplier does not operate in the developing countries like India." Do you agree? Discuss.
6. "Multiplier does not work in real terms in developing countries" *(Dr. V.K.R.V Rao)* What prevents the multiplier to work in real terms in India?
7. Even though marginal propensity to consume is quite high in LDCs, an increase in government expenditure may have little expansionary effect. Discuss *(D.U., B.A. (Hons) 1995.)*
8. What is paradox of thrift? How would you explain it.
9. Suppose in a country investment increases by ₹ 100 and consumption is given by $C = 10 + 0.6Y$ (where C = consumption and Y = income). How much increase will take place in income? *[C.U., B.Com. (H), 1997]*
10. Explain the relevance of multiplier in developing countries like India.

9. See V.N. Pandit, Macroeconomic Characteristics of the Indian Economy, in Prabhat Patnaik, ed. *Macro-economics,* Oxford University Press, p. 205.

CHAPTER 10

AGGREGATE DEMAND AND AGGREGATE SUPPLY MODEL (WITH PRICE FLEXIBILITY)

AD-AS MODEL WITH FLEXIBLE PRICES

Keynes in his income-expenditure analysis of employment assumed that price level remains constant. As explained in the previous chapters, Keynes in his macroeconomic analysis related aggregate demand and supply to the levels of national income. Concerned as he was with the problem of an economy under the grip of depression characterised by demand deficiency and excess capacity in the economy he assumed the price level to remain constant.

On the other hand, classical economists thought national output or income was determined by real factors such as capital stock, state of technology, labour supply and in no way affected by the general price level which was determined by the quantity of money. This classical doctrine is generally referred to as **classical dichotomy**. Thus AD-AS model with flexible price level highlights the breakdown of **classical dichotomy**. The aggregate demand and aggregates supply model, which is generally referred to as AD-AS model, is used to explain fluctuations in output, price level and rate of inflation in the economy. In what follows, we explain the concepts of aggregate demand and aggregate supply with flexible price level and analyse how the interaction between the two determines jointly the aggregate output (i.e. real GDP) and the general price level.

AGGREGATE DEMAND (AD)

Let us first explain aggregate demand. *Aggregate demand is the total desired quantity of goods and services that are bought by consumer households, private investors, government and foreigners at each possible price level, other things held constant.* Thus aggregate demand is not any quantity demanded at a particular price level but is a schedule of total output demanded at various price levels and is represented by a curve. Thus aggregate demand has four components: consumption demand, private investment demand, Government purchases of goods and services and net exports. *Thus, a aggregate demand curve depicts the total output of goods and services which households, firms, and Government are willing to buy at each possible price level.*

Thus, aggregate demand curve shows the relationship between the total quantity demanded of goods and services and general price level. It is worth noting that aggregate demand curve (*AD*) defiers from the ordinary demand curve of an individual commodity with which we are concerned in microeconomics though both slope downward to the light. In case of a demand curve of an individual commodity when price of a commodity rises, it will tend to be substituted by other commodities which are its close substitutes resulting in fall in the quantity demanded of the commodity at a higher price. Thus, the slope of demand curve of an individual commodity depends mainly on the possibility of substitution between the commodities. On the other hand, as we shall see below, the slope of aggregate demand curve depend on factors totally different from those that cause demand curve of an individual commodity to slope downward. We have drawn an *AD* curve in Figure 10.1 where on the horizontal axis we measure level of aggregate output and on the vertical axis we measure the general price level. As will be seen from Figure 10.1 aggregate demand curve *AD* slopes downward to the right.

It is important to note that the AD-AS model is unlike that of market demand-supply model of microeconomic theory. When we consider demand and supply in a particular market, say for cotton cloth, rise in its price will cause increase in its quantity supplied but in doing so resources will be withdrawn from other goods. Such reallocation of resources between products and consequently rise in output of one product and fall in that of the other is not considered in macroeconomics where we are concerned with determination of aggregate output, total employment of resources in the economy as a whole. Similarly, in macro-ecnomic model of aggregate demand and aggregate supply we study the determination of general price level and does not explain the relative prices of various products.

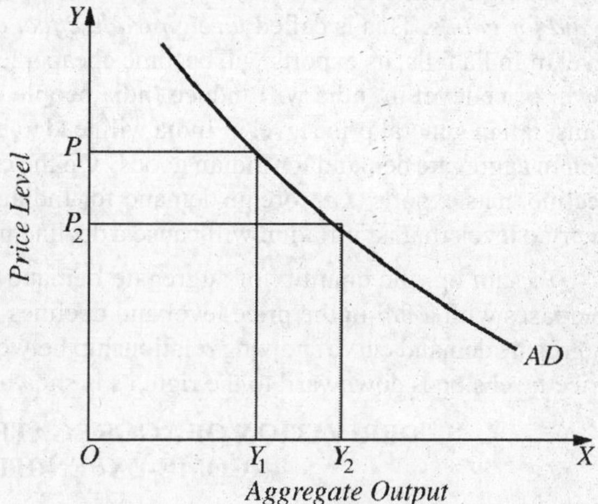

Fig. 10.1. *Aggregate Demand Curve with Varying Price Level*

We explain below in detail the concepts of aggregate demand (*AD*) and aggregate supply (*AS*) curves and their likely shape and factors determining them. We will also discuss some important controversial issues of macroeconomics with this AD-AS model.

Why Aggregate Demand Curve Slopes downward ?

The aggregate demand curve shows graphically the relationship between total spending and price levels and it slopes downward to the right. This means that at higher price levels, the total spending or quantity of aggregate output purchased or demanded is less and at lower price level the total spending or total purchases of aggregate output of goods is higher.

This is depicted in Figure 10.1. Aggregate output demanded per period of time is measured along the *X*-axis, and the general price level along the *Y*-axis. It should be carefully understood why aggregate demand for output or total spending falls at higher aggregate price level and increases at lower price levels, or, in other words, why aggregate demand (*AD*) curve slopes downward. The following factors are responsible for this.

Real Balance Effect. First, the changes in the general price level affect the *real value* or *purchasing power of money balances and monetary assets* with fixed nominal values (such as bank deposits, bonds, etc.) held by the people. With the rise in the general price level, the real value of these monetary assets will fall making people feel poorer than before. This induces them to consume less and therefore lead to the decline in quantity of output purchased by them. Conversely, if the price level falls, the real value of their monetary assets increases inducing them to buy more. This is called *real balance effect* of the change in the price level.

Rate of Interest Effect. Secondly, another important reason for the downward sloping nature of aggregate demand curve is *the effect of change in general price level on the rate of interest and through it on investment demand.* At a higher price level, the people will require more money for conducting a given amount of transactions. This will lead to the increase in the demand for money for making transactions. Given the money supply, the increase in demand for money for transaction purposes will cause the rate of interest to go up. At a higher rate of interest, demand for investment in new capital goods (*i.e.* plant, machinery and equipment) will decrease. On the contrary, if aggregate price level falls, demand for money for transaction purposes will decline and, the money supply being given, this will lead to the fall in the rate of interest. At the lower rate of interest, investment demand will increase. To conclude, *the investment demand and the general price level are also inversely related.*

Foreign Trade Effect. Thirdly, changes in the price level also causes a change in *foreign demand for goods*. This is called *foreign trade effect of the change in price level*. If a general price level in India falls, its exports will become cheaper leading to their increase. On the other hand, the lower price level in India will induce India people to buy Indian goods instead of imported ones. Thus, fall in general price level in India will lead to more exports and lesser imports causing expansion in aggregate demand for Indian goods. On the contrary, a rise in price level in India will lead to decline in its exports (*i.e.* foreign demand for Indian goods) and increase in its imports. Thus, rise in price level, that is, inflation will cause a decline in net exports.

To sum up, the quantity of aggregate demand for consumption, investment and net exports increases with a fall in the price level and declines with a rise in the price level. This means that aggregate demand curve showing relationship between aggregate output demanded and the general price level slopes downward to the right as is shown in Fig. 10.1.

DERIVATION OF AGGREGATE DEMAND CURVE FROM INCOME-EXPENDITURE ANALYSIS

We can now derive the aggregate demand curve using Keynesian income-expenditure framework and incorporating price level into the model. It should be noted that Keynesian aggregate expenditure curve ($C+I+G+X_n$) shows planned aggregate expenditure at various levels of national income (*i.e.*, real *GNP*), the aggregate demand curve (*AD*), which we are considering here shows equilibrium aggregate expenditure (*i.e.* equilibrium quantity of aggregate output demanded) at various price levels. In order to derive this aggregate demand curve with flexible prices we ask the question what is the effect of change in the price level on the economy's aggregate expenditure ($C+I+G+X_n$) function.

As has been explained above, a change in the price level causes a change in the quantity demanded through producing three effects, namely, *real balance effect*, *interest rate effect*, and *foreign trade effect*. Let us suppose the price level falls. As explained above, with a lower price level real purchasing power of the money balances or financial assets with fixed nominal values held by the people will increase. As a consequence, the people will start feeling themselves richer. Thus lower price level will induce people to consume more at each level of national income. That is, consumption function curve in the income expenditure model will shift above which in turn will cause upward shift in the aggregate expenditure ($C+I+G+X_n$) curve. This is illustrated in Figure 10.2. In panel (*a*) at the top of this figure, we have shown the determination of equilibrium level of real national income (*i.e.* equilibrium level of aggregate output demanded). Initially, at a price level P_0, the aggregate expenditure function curve ($C+I+G+X_n$) intersects the 45° line at point E_0 according to which Y_0 is the equilibrium quantity of real *GNP* or aggregate output demanded. At the level Y_0 of GNP, planned aggregate expenditure is equal to the value of aggregate national output. Thus, at the initial price level P_0 the equilibrium quantity of aggregate output demanded is Y_0. Therefore, in the panel (*b*) of Fig. 10.2 at the bottom we represent aggregate output Y_0 directly against price level P_0.

Suppose the price level falls from P_0 to P_2. With this real purchasing power of the money balances and financial assets held by the people will increase and they will be induced to consume more than before. As a result, consumption function curve will shift above causing upward shift in the aggregate expenditure curve to the new higher position ($C_2+I+G+X_n$). It will be seen from panel (a) of Figure 10.2 that this new aggregate expenditure function curve $C_2+I+G+X_n$ intersects the 45° line at point E_2 yielding greater quantity of aggregate output demanded Y_2. Thus, in panel (*b*) at the bottom we show aggregate output Y_2 against the lower price level P_2. This shows that at a lower price level, more aggregate output is demanded.

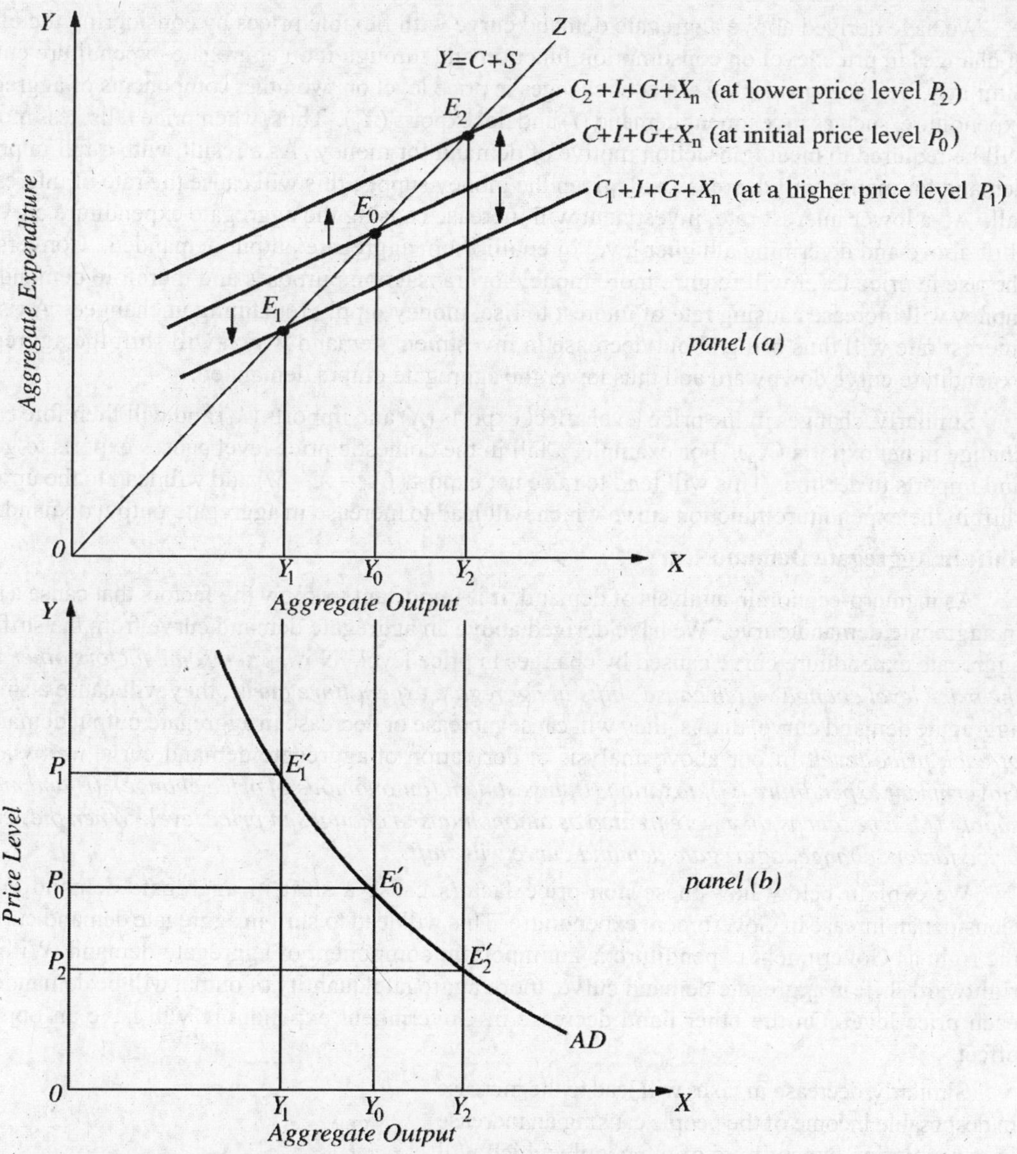

Fig. 10.2. *Derivation of Aggregate Demand Curve*

Now, suppose instead of fall in price level the price level rises from P_0 to P_1. At a higher price level, the real value of money balances and financial assets with fixed nominal value will decrease. As a result, people will feel poorer than before causing them to spend less on consumption than before at each level of national income. This will cause a downward shift in the consumption function causing the whole aggregate expenditure function curve to shift downward to the new lower level $C_1 + I + G + X_n$. It will be seen from the upper panel (a) of Fig. 10.2 that the new aggregate expenditure curve $C_1 + I + G + X_n$ intersects line at point E_1 and determines a lower level equilibrium aggregate output demanded and produced. Accordingly, against a higher price OP_1, we plot a smaller aggregate output demanded Y_1. It will be seen from the panel (*b*) at the bottom of Fig. 10.2 that aggregate demand curve obtained by plotting various equilibrium quantities of aggregate output demanded at different price levels slopes downward to the right.

We have derived above aggregate demand curve with flexible prices by considering the effect of changes in price level on consumption function and through it on aggregate expenditure curve. Similarly, we can consider the effect of changes in price level on two other components of aggregate expenditure, such as investment demand (*I*) and net exports (X_n). Thus, when price falls, less money will be required to meet transaction motive of demand for money. As a result, with a fall in price, demand for money will decrease and, given the money supply, this will cause the rate of interest to fall. At a lower interest rate, investment will increase causing the aggregate expenditure curve to shift above and determine a higher level of equilibrium aggregate output demanded. Conversely, the rise in price level will require more money for transaction purposes and therefore demand for money will increase causing rate of interest to rise, money supply remaining unchanged. A rise in interest rate will thus bring about decrease in investment demand which will shift the aggregate expenditure curve downward and thus lower the aggregate output demanded.

Similarly, changes in the price level affect exports (*X*) and imports (*M*) and will therefore cause change in net exports (X_n). For example, a fall in the domestic price level causes exports to go up and imports to decline. This will tend to raise net exports ($X_n = X - M$) and will lead to the upward shift in the expenditure function curve which will lead to increase in aggregate output demanded.

Shift in Aggregate Demand Curve

As in micro-economic analysis of demand, it is important to know the factors that cause a shift in aggregate demand curve. We have derived above an aggregate demand curve from the shifts in aggregate expenditure curve caused by changes in price level. Now, *when some factors other than the price level change which cause shifts in aggregate expenditure curve*, they will cause a shift in aggregate demand curve, that is, they will cause increase or decrease in aggregate output demanded *at each price level*. In our above analysis of derivation of aggregate demand curve *we assumed Government expenditure (G), taxation (I), investment (autonomous of price change) (I) and money supply (M) constant as they were treated as autonomous of changes in price level. When these non-price factors change, aggregate demand curve will shift.*

We explain below how these non-price factors cause a shift in aggregate demand curve. Consider an inrease in Government expendiutre. This will lead to shift in aggregate demand curve to the right as Government expenditure is an important component of aggregate demand. With this rightward shift in aggregate demand curve, more aggregate quantity of output will be demanded at each price level. On the other hand decrease in Government expenditure will have an opposite effect.

Similarly, decrease in taxes will lead to the increase in desposable income of the people causing an increase in consumption expenditure of the people which will bring about rightward shift in aggregate demand. Further, expansion in money supply as a part of expansionary monetary policy will also cause a rightward shift in aggregate demand curve as shown Fig. 10.3. This happens because expansion in money supply leads to fall in rate of interest which induces *increase in investment demand,* an important component of aggregate demand. Likewise, to fight inflation the Central Bank of a country adopts a contractionary monetary policy and raises interest rate which causes investment demand to decrease resulting in leftward shift in aggregate demand curve so that at any price level less aggregate output is demanded than before.

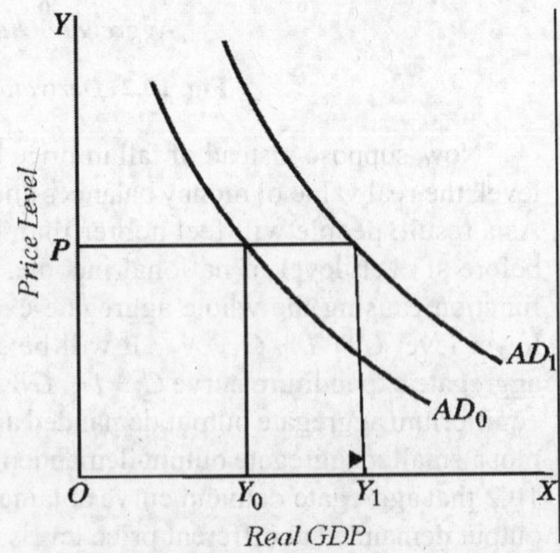

Fig. 10.3. Shift in Aggregate Demand Curve

Similarly, if expectations or sentiments of the entrepreneurs about profit prospects favourably change they will cause increase in investment demand leading the increase or rightward shift in aggregate demand curve.

Another important factor causing a shift in demand curve is changes in *net exports* (i.e. exports –imports). If there is increase in incomes abroad, the demand for our exports will increase. Given our imports, the increase in net exports will cause a shift in aggregate demand curve to the right. Similarly, if the rate of exchange rate of our currency (rupee) falls, that is, if there is depreciation of our currency our exports will become cheaper and imports dearer than before. This will cause an increase in our net exports leading to the shift in aggregate demand curve to the right.

Similarly, *increase in money supply* (*M*) will cause a rightward shift in aggregate demand curve. In the derivation of a given aggregate demand curve, money supply in the economy is held constant. If at a given price level, money supply is increased, the interest rate will fall. The fall in interest rate will cause investment demand to increase. Aggregate output demanded will thus be greater at the given price level. Thus, expansion in money supply brings about shift in aggregate demand curve to the right.

The same logic applies to the effect of contractionary fiscal and monetary policies. A decrease in Government expenditure or increase in taxes will cause *AD* curve to shift to the left. Similarly, decrease in investment autonomous of any change in the price level will also shift the *AD* curve to the left. Similarly, contraction in money supply (*M*) at a given price level will lead to the leftward shift in the aggregate demand curve.

It will be recalled that increase in Government expenditure or investment has a multiplier effect on aggregate output demanded depending on the size of multiplier. We will explain this multiplier effect in AS-AD model in a later section

$$\Delta Y = \Delta G \frac{1}{1-mpc}$$

Where *mpc* is the marginal propensity to consume and $\frac{1}{1-mpc}$ is the value of multiplier. As in case of increase in Government expenditure, *reduction in taxes* will also increase aggregate output demanded at each price level and will therefore cause a shift in aggregate demand curve.

AGGREGATE SUPPLY CURVE

Aggregate supply is the total output of goods and services that firms want to produce at each possible price level. Thus, like aggregate demand, *aggregate supply is the whole schedule of total quantities of aggregate output that firms in the economy are willing to produce and can be represented by an aggregate supply curve*. It is worth noting that aggregate supply is the outcome of the decisions of all producers in the economy to hire workers and buy other inputs for production of goods and services for selling them to consumers, other producers, government as well as for exporting them to other countries. It may be noted while drawing aggregate supply curve that depicts relationship been quantities produced for sale in the market of aggregate output that are by the firms in the economy at various price levels, all other factors that affect aggregate supply beeing held constant.

There is a lot of disagreement among economists about the shape of aggregate supply curve. The classical economists assumed that there normally prevailed full employment of resources in the economy. According to them, if at any time there is deviation from this full employment level, the wages, interest and prices quickly and automatically adjust and change to restore equilibrium at the full employment level. Thus, in the classical theory, the aggregate supply curve of output is perfectly inelastic (*i.e.* a vertical straight line) at the output level corresponding to full-employment level of resources. This aggregate supply curve relating aggregate supply with price level of the classical theory of income and employment is shown in Figure 10.4 by a vertical *AS* curve.

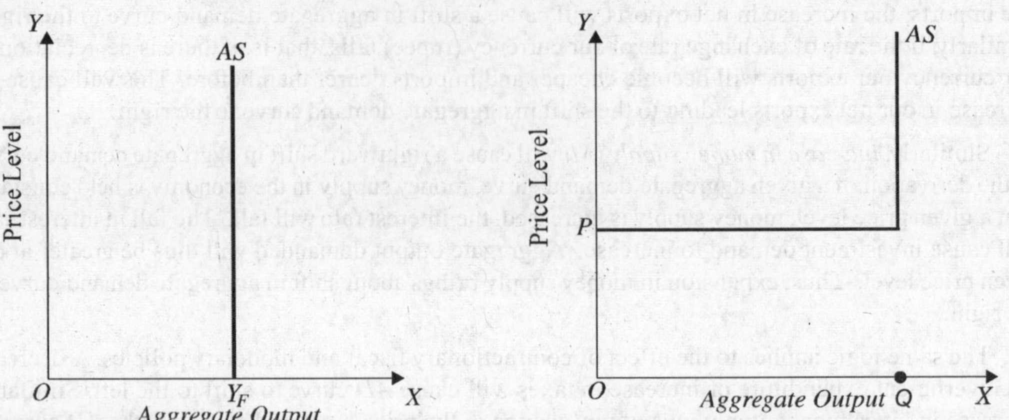

Fig. 10.4. *Aggregate Supply Curve : Classical View* **Fig. 10.5.** *Keynes's Aggregate Supply Curve*

As mentioned above, Keynes considered the situation of economic depression when the economy was operating before the level of full employment of resources. He further believed that in such a situation money wage rates were sticky *i.e.* remain constant. He further assumed that average and margined products of labour remain constant when more of it is employed following the increase in aggregate demand. With these assumptions, more aggregate output is produced and supplied at the given price level in response to increase in aggregate demand. But when full employment of labour and capital stock is attained and aggregate demand further increases, aggregate supply curve being unable to increase any more, it is the price level that will rise in response to the increase in aggregate demand. The Keynes's aggregate supply curve depicting the relationship between price level and the aggregate production (supply) during the period of depression and involuntary unemployment when there is a lot of excess capacity in the economy is shown in Figure 10.5 where it will be seen that aggregate supply is a horizontal straight line (*i.e.* perfectly elastic) upto full-employment output *Q* showing thereby that more is produced and supplied at the same price level *OP* but corresponding to full-employment level point *Q* it becomes vertical.

It may however be noted that Keynes recoganised that as the aggregate supply approaches close to the full-employment level the cost of output per unit tends to rise due to the rise in wage rate and also due to diminishing returns to extra factors employed. But, according to Keynes, the rise in price full-employment level level before full employment or less than capacity output will not be much.

It is evident from above that shape of aggregate supply curve is a highly controversial issue. As mentioned above, aggregate supply curve indicates the total output which the producers are willing and able to sell at different price levels. It is important to note that as the average price level rises above the current marginal cost of production, producers find it profitable to expand output. When the economy is working substantially below capacity, that is, at times of depression or severe recession, the more can be produced without much rise in marginal cost of production and therefore the aggregate supply curve is nearly flat. With the given stock of capital (*i.e.* plant and equipment) when output is expanded the diminishing returns and rising marginal costs occur which ultimately cause the aggregate supply curve to slope upward gently. But as the firms in the economy approach near their capacity output their marginal costs sharply rise which causes sharply rising aggregate supply curve. At the level of capacity-output, that is, when the given resources of the economy are fully employed, aggregate supply curve (*AS*) becomes a vertical straight line as classical economists assumed.

Modern View. There is now consensus among economists that short-run aggregate supply curve has segments or ranges-: (1) the horizontal range, (2) the intermediate (or upward sloping) range, and (3) the vertical range. This aggregate supply curve having three distinct segments is shown in Figure 10.6. Let us now give reasons for these three ranges of aggregate supply curve.

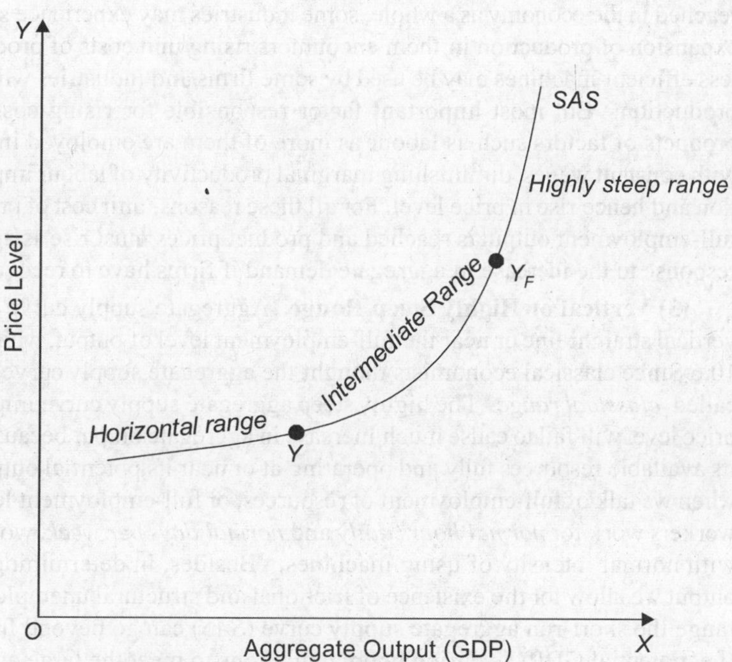

Fig. 10.6 *Short-run Aggregate Supply Curve with three Ranges*

(1) **Horizontal Range.** Because Keynes and his early followers thought aggregate supply curve was horizontal, this range is also called *Keynesian range*. It may be noted that the shape of aggregate supply curve is governed by what happens to unit cost of production as national output expands or contracts. It will be noticed from Figure 10.6 that we have taken Y_F as the full-employment level of output, that is, Y_F is the capacity production. This represents the level of output which is produced when available resources of the economy are fully employed or used. Therefore, this full-employment or capacity output is also called *potential output*. Now, it will be seen that horizontal or Keynesian range comprises levels of output that are substantially less than the full capacity production Y_F. This implies that during this horizontal range of output upto *Y*, the economy can expand its production without facing much rise in unit cost of production. This is in fact the stage when the economy is in the grip of severe recession or depression with a lot of idle capacity in the form of unemployed labour, unutilised machinery and other capital equipment. These idle or unused resources can be put to use without causing any rise in unit cost of production and therefore without any *upward pressure* on the price level. The unemployed labour would be willing to work at the current wage rate (that is, without any need for offering higher wage rate to them). Since other resources are lying unused, no shortages or bottlenecks will be experienced causing higher cost per unit or rise in the price level, if aggregate output is expanded in this range.

(2) **Intermediate Range.** It will be seen in Figure 10.6 that in this intermediate range between the output levels *Y* and Y_F, increase in aggregate output brings about rise in price level, that is, aggregate supply curve slopes upward in this range. There are more than one reasons for this upward-sloping range of *AS* curve. First, the national economy consists of several industries and resource markets and full-employment is not attained simultaneously in all industries and in respect of all types of workers. For example, as the national production expands in this range some industries, say electronics and computer hardware may experience shortage of skilled engineers engaged

in these industries, while some industries such as textile industries may be still facing quite large unemployment. Due to the shortage of engineers for some crucial industries, bottlenecks in production may arise which push up the cost in these industries while the economy as a whole is still operating below full-employment level. Secondly, in order to overcome shortage of skilled workers such as engineers, as mentioned above, less efficient or less skilful workers may be employed resulting in rise in unit cost of production. Similarly, even before full employment of resources, is reached in the economy as a whole, some industries may experience shortage of raw materials due to expansion of production in them encounters rising unit costs of production. Further, the older and less efficient machines may be used by some firms and industries when they reach close to capacity production. But most important factor responsible for rising cost is the diminishing marginal products of factors such as labour as more of them are employed in expanding production. Even with constant wages, diminishing marginal productivity of labour implies rise in unit cost of production and hence rise in price level. For all these reasons, unit cost of production increases even before full-employment output is reached and product prices must rise as aggregate output is expanded in response to the increase in aggregate demand if firms have to recover the rising costs.

(3) Vertical or Highly Steep Range. Aggregate supply curve in this range is highly steep or vertical straight line or near the full-employment level of output, which is designated by Y_F in Figure 10.6 Since classical economists thought the aggregate supply curve was vertical, this range is also called *classical range*. The highly steep aggregate supply curve implies that any further rise in the price level will fail to cause much increase in aggregate output because the economy is already using its available resources fully and operating at or near its potential output. It is important to note that when we talk of full-employment of resources, or full-employment level of output, we consider that workers work for *normal hours daily* and *normal days per week*, working of usual number of shifts with normal intensity of using machines. Besides, in determining full-employment or capacity output we allow for the existence of frictional and structural unemployment. Therefore, in this third range the short-run aggregate supply curve (*SAS*) can go beyond full-employment level of output (i.e. potential GDP) as during boom conditions to meet the large aggregate demand the firms can increase production by extending the work-week (say having six-days work-week instead of 5 days work-week, by having workers work overtime or arranging night shift for workers for expanding output. Besides during boom conditions when wages are high even women and children can enter labour force and offer them for work. Therefore, in the third range aggregate supply curve, though steep, can go beyond the potential GDP level (i.e. full employment level of output).

Shifts in Aggregate Supply Curve

In explaining the upward-sloping nature of short-run aggregate supply we stated that aggregate supply curve depicts the relationship between price level and aggregate output (i.e. real GDP), other factors that determine aggregate supply being held constant. Now, it is the *changes* in these other determining factors that cause a *shift* in the aggregate supply curve. We explain below the factors that cause a shift in aggregate supply curve.

Wages. Change in wage rate is an important factor that causes a shift in aggregate supply curve. In the short run labour is available at fixed wage rate. There are many reasons for this. Workers and their employing firms generally enter into long-term contracts under which money wages are generally fixed up to 3 years. Even when there is no explicit agreement money wages of workers are changed once in a year. During the interim period money wages remain fixed or sticky. *It is important to note that a short-run aggregate supply is drawn assuming money wages remain fixed.*

Now when wage rates of workers increase, they cause a leftward shift in short-run aggregate supply curve as shown in Fig. 10.7. This is because increase in wages raises cost per unit of output. With a given price of output, higher wage rate means profit per unit of output will decline with production becomes less profitable, it is likely that the firms will cut back on production and supply less output.

Prices of Inputs. What is true of the effect of rise in wage rate on aggregate supply is true of the effect of prices of other inputs. Changes in prices of other inputs such as energy (for example crude oil) and raw materials also bring about a shift in short-run aggregate supply curve. If is well known that increase in price of crude oil by OPEC in 1973 and again in 1979 affected aggregate supply by raising cost per unit of production. This caused a leftward shift in short-run aggregate supply curve as shown in Fig. 10.7

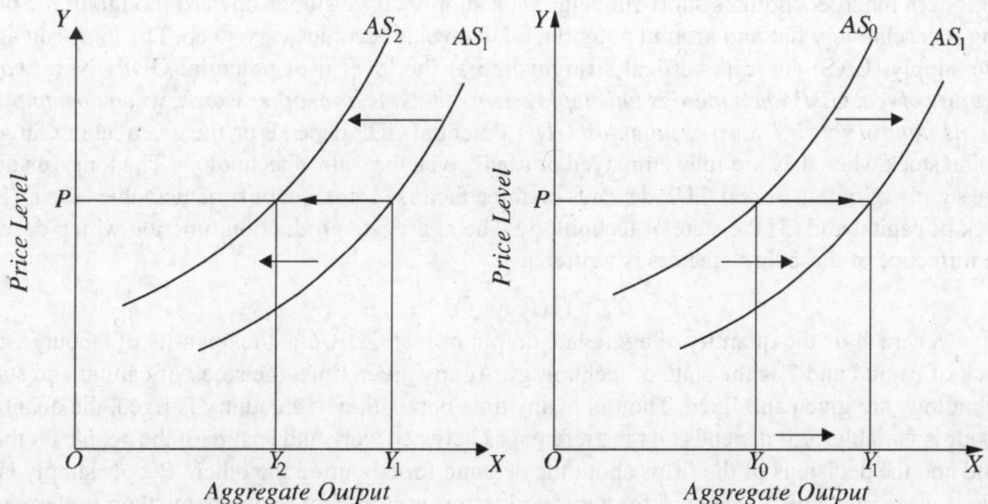

Fig. 10.7. *Leftward Shift in Aggregate Supply curve* **Fig. 10.8.** *Rightward Shift in Aggregate Supply Curve*

This leftward shift in the aggregate supply curve implies that at any given price level less output is supplied than before.

On the other hand when price of crude oil falls as has happened at several occasions in the past, aggregate supply curve shifts to the right as is shown Fig. 10.8 indicating that at any given price level more output will be produced and supplied than before.

Change in Technology. The change in technology is another important factor that causes a shift in aggregate supply curve. When there is improvement in technology productivity of factors rises causing a fall in the unit cost of production. This brings about a rightward shift in aggregate in supply curve showing that at any given price level more will be produced and supplied than before.

It is worth noting that in India at present it is being emphasized by many economists that increase in productivity in agriculture by effecting improvement in technology is crucial for increase in supply of agricultural output to solve the problem of food inflation.

Business Taxes and Subsidies. Increase in rates of business taxes such as excise duty, sales tax, customs duties increase per unit cost of production just as rise in wage rate. (Note that tax is considered as cost of production as it raises the supply price of output). Thus by levying business taxes or increasing their sales tax causes a leftward shift in the aggregate supply curve. On the other hand, lowering of taxes as happened during the recent global slowdown (2007-09) caused by bursting of sub-prime housing loans bubble in United States when to prevent the deepening of recession many countries (including India lowered the rate of these business taxes.

By contrast, provision of subsidies on products of various industries causes a shift in the aggregate supply curve to the right.

Available Supply of Resources. Lastly, a very important factor determining the position of aggregate supply curve is the available quantity of resources. When the available supplies of resources such as labour and capital increase, the short-run aggregate supply curve will shift to the right. As labour force grows and supply of capital is increased through investment, the short-run short-run aggregate supply curve will shift to the right implying that more output will be produced for sale at any given price level.

Long-run Aggregate Supply Curve

In our above analysis we did not distinguish between the short-run and long-run supply curves. In modern macroeconomics short-run aggregate supply curve slopes upward though in the beginning it is relatively flat and around potential GDP level it becomes very steep. The long-run aggregate supply (LAS) curve is vertical straight line at the level it of potential GDP. Note *that the quantity of real GDP when there is full employment of labour or in other words, when unemployment is at its natural rate is known as potential GDP*. Potential GDP depends on the given labour force and capital stock when they are fully employed or used, given the state of technology.The long-run aggregate supply of output or real GDP depends on three factors (1) the quantity of available labour (2) the stock of capital and (3) the state of technology. The aggregate production function which describes the influence of these three factors is written as :

$$Y_F = F(L, K, T)$$

Where Y_F is the quantity of aggregate output or real GDP, L is the quantity of labour, K is the stock of capital and T is the state of technology. At any given time, the stock of capital and state of technology are given and fixed. Though at any time population of a country is fixed, the quantity of labour is variable as it depends on the preferences between work and leisure of the people on the one hand and the decisions of the firms about the demand for labour on the other. People supply labour only if wage rate which is reward for their work effort is sufficient to overcome their preference for leisure. The higher the wage rate, the greater is the quantity of labour supplied. On the other hand, the firms demand labour if it is profitable to use it for production. The lower the wage rate, which is the cost of labour, the greater is the quantity of labour used. The equilibrium wage rate and the quantity of labour employed are determined by labour-market equilibrium. The labour market is in equilibrium at the wage rate at which the quantity of labour demanded equals the quantity of labour supplied. *At this equilibrium wage rate, all those who are willing to supply their labour are is fact demanded and employed. Therefore, at this equilibrium wage rate, full employment of labour is said to prevail.*

However, even at full employment, there are always some workers searching for jobs and some firms are looking for workers to offer them employment. This is because of two factors, frictional and structural. Every week some workers leave their old jobs and search for new better jobs more suited to their skills and ability. But due to lack of information, it takes time to find the new jobs even through these are available. This represents what is called *frictional unemployment*. Second, every week or month some industries are declining due to changes in technology or preferences of the people for goods while others are expanding. In this case while some people are laid off from the declining industries but sometime is required before they acquire new skills and training needed for employment in the expanding industries. Therefore, for some time they remain unemployed though job vacancies for them exist. This second type of unemployment is called *structural unemployment*.

Thus, at any time some frictional and structural unemployment inevitably exist in a free market economy. Therefore, in modern macroeconomics, the amount of frictional and structural unemployment is called the ***natural rate of unemployment***. Around 4 to 5 per cent of labour force in the developed free-market economies represents the natural rate of unemployment. And full employment is said exist in spite of the existence of frictional and structural unemployment. As mentioned above, *the quantity of real GDP produced and supplied when there is full employment (that is, when there exists only natural rate of unemployment) is called potential GDP*. It may be noted again that potential GDP depends on full employment of labour, the full use of the existing stock of capital and the available technology.

The long-run aggregate supply describes the relationship between the quantity of real GDP and the price level in the long run when real GDP equals potential GDP. The long-run aggregate supply curve is a vertical line (at potential GDP level) as shown by *LAS* in Figure 10.9. The long-run aggregate supply curve is vertical because potential GDP does not vary with price level, that is, it is independent of the price level. Suppose given a long-run supply *LAS* in Fig. 10.9 and a given level of aggregate demand curve AD_1 the price level determined is OP_1. Now suppose the aggregate demand falls while potential *GDP* remains unchanged at Y_F. The reason for the independence of potential GDP from the price level is that the movement along the long-run aggregate supply curve involves not only the change in price level of goods but also change in prices of factor inputs such as wages of labour etc. For example, when there is 5 per cent decrease in the prices of goods and services, this is matched by the same (*i.e.* 5 per cent) decline in wage rate and other factor prices so that relative prices and real wage rate remain unchanged. This explains why it is profitable to produce the same quantity of real GDP at lower price level of goods and services. When price level of goods and services falls, the cost also falls as wage rate and other factor prices fall by the same percentage. Therefore, aggregate supply of output (*i.e.* real GDP) in the long run also remains constant at potential GDP level.

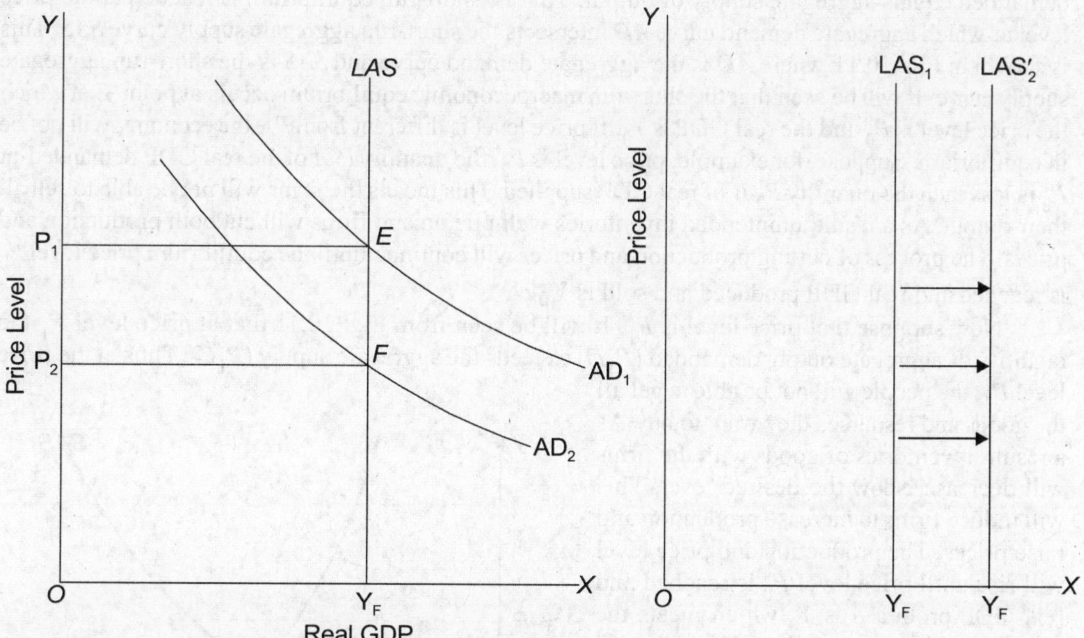

Fig. 10.9. *Long-Run Aggregate Supply Curve* **Fig. 10.10.** *Shift in Long-run Aggregate Supply Curve*

Changes in Long-run Aggregate Supply Curve

We have seen above that long-run aggregate supply curve is a vertical straight line at the level of potential GDP. Changes in price level bring about a movement along the long-run aggregate supply, but the quantity of aggregate supply remains fixed at the level of potential GDP. It is changes in potential GDP that causes a shift in the long-run aggregate supply curve. The following factors cause a change in potential GDP resulting in a rightward shift in the long-run supply curve *LAS* as shown in Fig. 10.10.

1. The increase in labour force.
2. Increase in the stock of capital.
3. The progress in technology.

As the increase in GDP is known as economic growth it is clear that when economic growth takes place, long-run aggregate supply by curve shifts to the right.

DETERMINATION OF GDP AND PRICE LEVEL : AS-AD MODEL

Macroeconomic Equilibrium. Having explained the concepts of aggregate demand and aggregate supply with variable price level. Now we shall explain how macro equilibrium is reached between the aggregate supply and aggregate demand to determine the amount of real GDP and the price level. As there is difference between the long-run and the short run aggregate supply curves, the long-run equilibrium differs from the short-run macroeconomic equilibrium. While long-run equilibrium is the state towards which the economy is moving, short-run equilibrium is the actual state of the economy in the short run as it fluctuates around potential GDP. The purpose of AS-AD model is to explain how the changes in various factors, fiscal and monetary policies bring about changes in real GDP and the price level. We explain below both the short-run and long-run macroeconomic equilibrium.

Short-run Macroeconomic Equilibrium

Short-run macroeconomic equilibrium occurs at the price level at which aggregate output demanded equals aggregate supply of output. That is, short-run equilibrium is reached at the price level at which aggregate demand curve *AD* intersects the short-run aggregate supply curve *SAS*. This is shown in Fig. 10.11 where *AD* is the aggregate demand curve and *SAS* is the short-run aggregate supply curve. It will be seen that the short-run macroeconomic equilibrium occurs at point *E* at which the price level is P_0 and the real GDP is Y_0. If price level is different from P_0, the economy will not be in equilibrium. Suppose, for example, price level is P_2, the quantity P_2A of the real GDP demanded at P_2 is less than the quantity P_2B of real GDP supplied. This means the firms will not be able to sell all their output. As a result ,unintended inventories well pile up and firms will cut both production and prices. The process of cutting production and prices will continue until the equilibrium price level P_0 is reached and real GDP produced and sold is Y_0.

Now suppose that price level is P_1. It will be seen from Fig. 10.11 that at price level P_1 the quantity of aggregate output demanded (P_1D) exceeds the aggregate supply (P_1C). Thus at the price level P_1, the people will not be able to get all the goods and resources they want to buy. As a result, inventories of goods with the firms will decrease below the desired level. This will induce firms to increase production and raise prices. The production and price level will rise until price level P_0 is reached and real GDP produced is Y_0 which meets the demands of the people fully at the price level P_0. Thus, the price level P_0 and real GDP equal to Y_0 represent the short-run macroeconomic equilibrium.

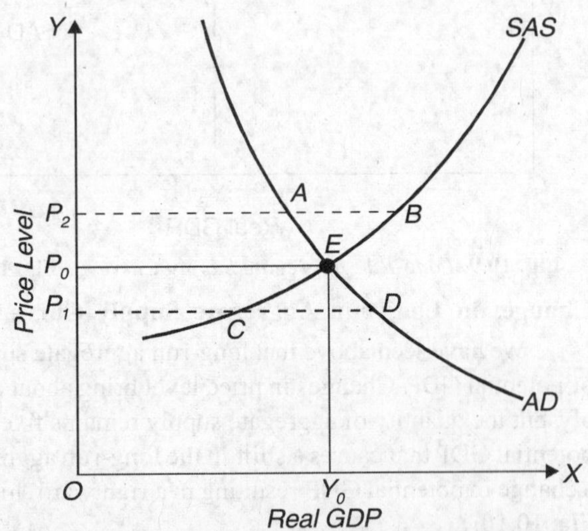

Fig. 10.11. *Short-run Macroeconomic Equilibrium : Joint Determination of GDP and Price Level*

It is worthwhile to note that in the short run the money wage rate is fixed. *It does not adjust to bring macro-equilibrium at full-employment level of real GDP.* Thus, in the short run macro-equilibrium can be attained with real GDP less than or greater than potential GDP (*i.e.* the level of GDP at which there is full employment of labour) depending on the level of aggregate demand. It is only in the long run when money wage rate adjusts that equilibrium is restored at potential GDP.

It is worth noting that as Keynes emphasized that equilibrium between aggregate demand and aggregate supply may not necessarily be at full-employment level. This short-run equilibrium between *AD* and *AS* may occur at less full employment as shown in Fig. 10.12 where aggregate demand curve intersects short-run aggregate supply curve SAS_0 at point *E* and equilibrium is at GDP level Y_0 which is less than potential or full-employment output level $\bar{Y}$. This *less than full employment level is due to deficiency of aggregate demand relative to potential aggregate supply of output.* This results in recession and involuntary unemployment of labour.

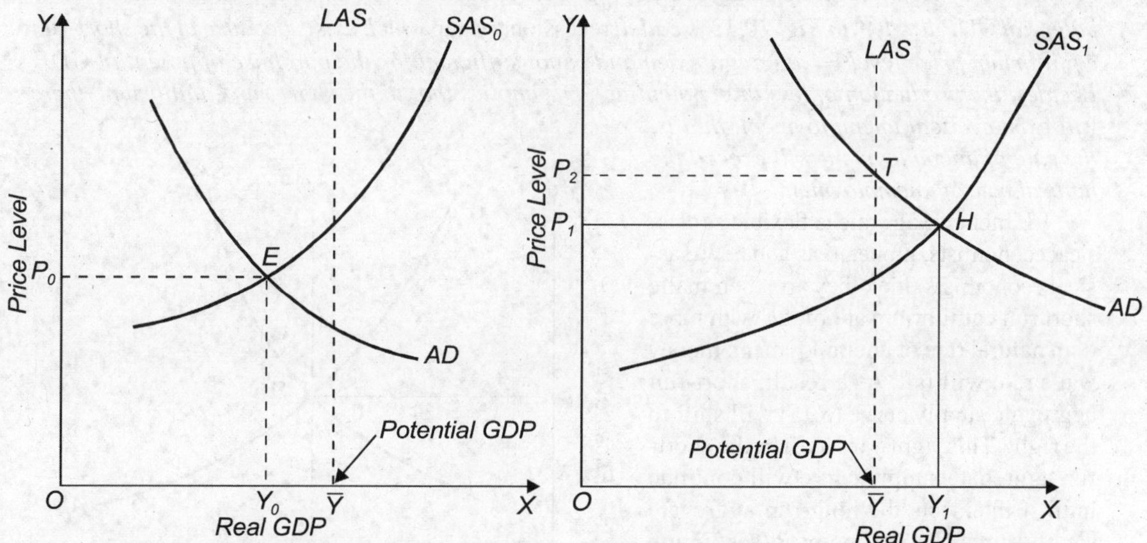

Fig. 10.12. *Short-run Equilibrium at less than full-employment (or Potential GDP).*

Fig. 10.13. *Short run Equilibrium above full employment with Inflation.*

On the other hand, the short-run equilibrium between aggregate demand and aggregate supply may reach at more than potential GDP (or full-employment) level, as is shown in Fig. 10.13 where aggregate demand curve *AD* cuts short-run aggregate supply curve SAS_1 and determine Y_1 level of GDP which exceds potential GDP level $\bar{Y}$. As mentioned above, this implies unemployment has gone down below natural rate of unemployment. The workers may also be doing overtime or larger number days in a week to produce more than the potential GDP. Besides, the economy is experiencing boom and inflationary conditions. The distance $\bar{Y}Y_1$ is called *inflationary, gap* as it exerts inflationary pressures.

Long-Run Macroeconomic Equilibrium

Further, it may be noted that fluctuations in the economy occur due to changes in the factors that cause changes in either aggregate demand or aggregate supply. For example, changes in money supply, the government expenditure, taxes, investment demand by business firms or consumption demand of households will bring about changes in aggregate demand and cause shift in the short-run macroeconomic equilibrium. On the other hand, changes in money wage rate and other resource prices such as oil price stock will cause a shift in aggregate supply curve and bring about a change in the short-run macroeconomic equilibrium.

The long-run macroeconomic equilibrium occurs at the price level where the aggregate demand curve intersects the long run aggregate supply curve which is vertical at the potential GDP level.

Thus, long-run equilibrium occurs *when real GDP equals potential GDP. But this long-run equilibrium of price level and real GDP is reached when money wage rate adjusts so that the short-run aggregate supply curve shifts to intersect the long-run aggregate supply curve (LAS) at the point at which aggregate demand curve intersects the latter. The long-run macro equilibrium and how it is reached is illustrated in Figure 10.14. From Fig. 10.12 it will be seen that the short-run aggregate supply curve SAS_0 intersects the given aggregate demand curve AD at point E and determines the price level equal to P_0 and real GDP equal to Y_0 in the short run. Thus in this short-run macroeconomic equilbrium, real GDP which is equal to Y_0 is less than the potential GDP (remember that the potential GDP corresponds to full employment of labour, the full use of the given stock of capital and the given state of technology). The difference $Y_0\bar{Y}$ between the short-run equilibrium GDP and the potential GDP level $\bar{Y}$ in Fig. 10.12 is called* recessionary gap *which exists because at the short-run equilibrium price level* P_0 *aggregate demand is not sufficient for the purchase of potential GDP. Besides, the production of less than potential GDP implies that in the short-run equilibrium, there will exist cyclical unemployment, that is, the rate of unemployment will exceed the natural rate of unemployment.*

If money wage rate is flexible as classical economists, monetarists and new classical economists think they are, then in the short run equilibrium at point *E* with more than natural rate of unemployment, money wage rate will fall. As a result, short-run aggregate supply curve (*SAS*) will shift to the right. This rightward shift in the short-run aggregate supply curve will continue until it intersects the long-run aggregate supply curve (*LAS*) at point *B* (see Figure 10.14) where the given aggregate demand curve cuts the latter. At the new equilibrium point *B* price level has fallen to *P* and aggregate quantity of output demanded has increased to the potential GDP level $\bar{Y}$. Point *B* in the Figure 10.14 represents the long-run equilibrium at full-employment level, that is, the level at which only natural rate of unemployment prevails.

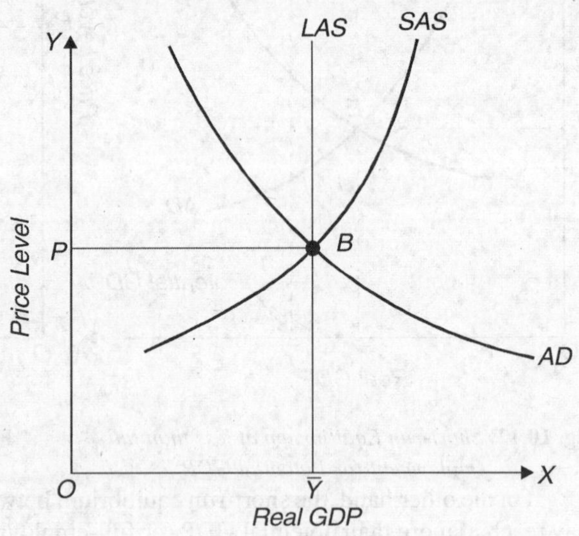

Fig. 10.14. *Long-run Equilibrium at Potential GDP*

Let us consider the opposite case when the short-run equilibrium is initially at more than potential GDP, that is, above full employment level. Suppose initially with aggregate demand curve AD_1 and the short-run aggregate supply curve SAS_1, the short-run equilibrium is at point *H* at which the price level is P_1 and real GDP is Y_1 which is greater than potential GDP equal to $\bar{Y}$. (see Fig. 10.13). The amount $\bar{Y}\ Y_1$ by which real GDP = Y_1) exceeds the potential GDP level $\bar{Y}$ is called *inflationary gap* as this gap creates infaltionary pressures in the econmy. In a free market economy this represents the situation when unemployment has gone below the natural rate of unemployment as due to boom jobs are in plenty and labour is in great demand. Many firms may find it difficult to find enough labour to produce expanded output. They will try to lure away labour from others by offering higher wages. This will set in motion the process of rise in wages. The rise in wages will cause shift in short-run aggregate supply curve *SAS* upward until it reaches point *T* at which long-run equilibrium is reached and price level has gone up to P_2 (see Fig. 10.13).

MULTIPLIER WITH CHANGES IN PRICE LEVEL : AS-MD MODEL

In our analysis of multiplier in Keynes's income - expenditure approach, it is assumed that price level remains constant and the firms are willing to supply more output at a given price. How much national income or GNP increases as a result of any autonomous expenditure such as government expenditure, investment expenditure, net exports is determined by a shift in aggregate demand curve which depends on the size of marginal propensity to consume (*MPC*) when price level is fixed. The price level remaining fixed implies a horizontal short-run aggregate supply curve. However, as studied above, short-run aggregate supply curve slopes upward as the firms are willing to supply additional output in the short run only at a higher price level. With short-run aggregate supply curve sloping upward, a rightward shift in aggregate demand curve raises new equilibrium GNP level not equal to the horizontal shift in the aggregate demand curve but less than it. Consequently, the size of multiplier is smaller than that of simple Keynesian multiplier with a given fixed price level. This is because a part of expansionary effect of GNP of the increase in autonomous investment or government expenditure is offset by rise in the price level.

The multiplier effect in case of upward sloping curve is shown in Fig. 10.15. To begin with, in the top panel of Fig. 10.15 aggregate expenditure curve AE_0 intersects 45° line at point B and determines Y_0 equilibrium level of GNP. In the panel at the bottom of Fig. 10.15 the corresponding aggregate demand curve AD_0 and the short-run aggregate supply curve SAS to intersect at B', at the above determined GNP level Y_0. Now suppose autonomous investment expenditure (which is independent of changes in price level) increases by I. As a result, aggregate expenditure curve AE shifts upward to AE_1 and determines new equilibrium GNP level equal to Y_2. In the lower panel (*b*), due to the upward shift in aggregate expenditure curve, aggregate demand curve shifts rightward from AD_0 to AD_1. The horizontal shift in the aggregate demand curve *at a given price level* is determined by the increase in investment expenditure multiplied by the simple Keynesian multiplier at the given fixed price level ($B'H$ or $\Delta Y = \Delta I \dfrac{1}{1-MPC}$)

But given the upward sloping short-un aggregate supply curve SAS with new aggregate demand curve AD_1, price level does not remain fixed. As will be seen from the lower panel (*b*) of Fig. 10.15, the aggregate demand curve AD_1 intersects the short-run aggregate supply curve SAS at point R' and as a result price level rises to P_1.

Now, with this rise in price level to P_1, aggregate expenditure curve in the upper panel (*a*) will not remain unaffected but will shift downward. This fall in aggregate expenditure curve is due to the adverse effects on wealth or real balances, interest rate and net exports. Much of wealth is held in the form of bank deposits, bonds and shares of companies and other assets. With the rise in price level, *real value or purchasing power* of wealth possessed by the people declines. This induces them to spend less. As a result, consumption expenditure declines due to this wealth effect. Secondly, the rise in price level reduces the *supply of real money balances* (M^s/P) that causes a shift in money supply curve to the left. Given the demand function for money (M^d), the decline in the real money supply will cause rate of interest to rise. Now, the rise in interest will induce private investment expenditure to decline. Lastly, rise in price level in the domestic economy will adversely affect exports of a country causing net exports to fall.

Thus, as a *result of negative effects of rise in price level on real wealth, private investment and net exports, in the upper panel (a) of Fig. 10.15 aggregate expenditure curve shifts downward to* AE_1' (dotted) so that it determines GNP level Y_1 corresponding to the intersection of aggregate demand curve AD_1 and short-run aggregate supply curve SAS at point R' in the lower panel (*b*) of this Fig. 10.15.

Thus with the upward sloping short-run-aggregate supply curve SAS, the effect of increase in autonomous investment expenditure (or for that matter increase in any other autonomous expenditure such as Government expenditure, net exports, autonomous consumption) on the GNP level can be visualised to occur in two stages. First, increase in investment expenditure shifts aggregate expenditure curve *AE* upward in the upper panel (*a*) of Fig. 10.15 and correspondingly aggregate demand curve in the lower panel (*b*) shifts to the right to AD_1 and bring about increase in GNP level from Y_0 to Y_2 with the given fixed price level P_1. In the second stage due to the upward-sloping short-

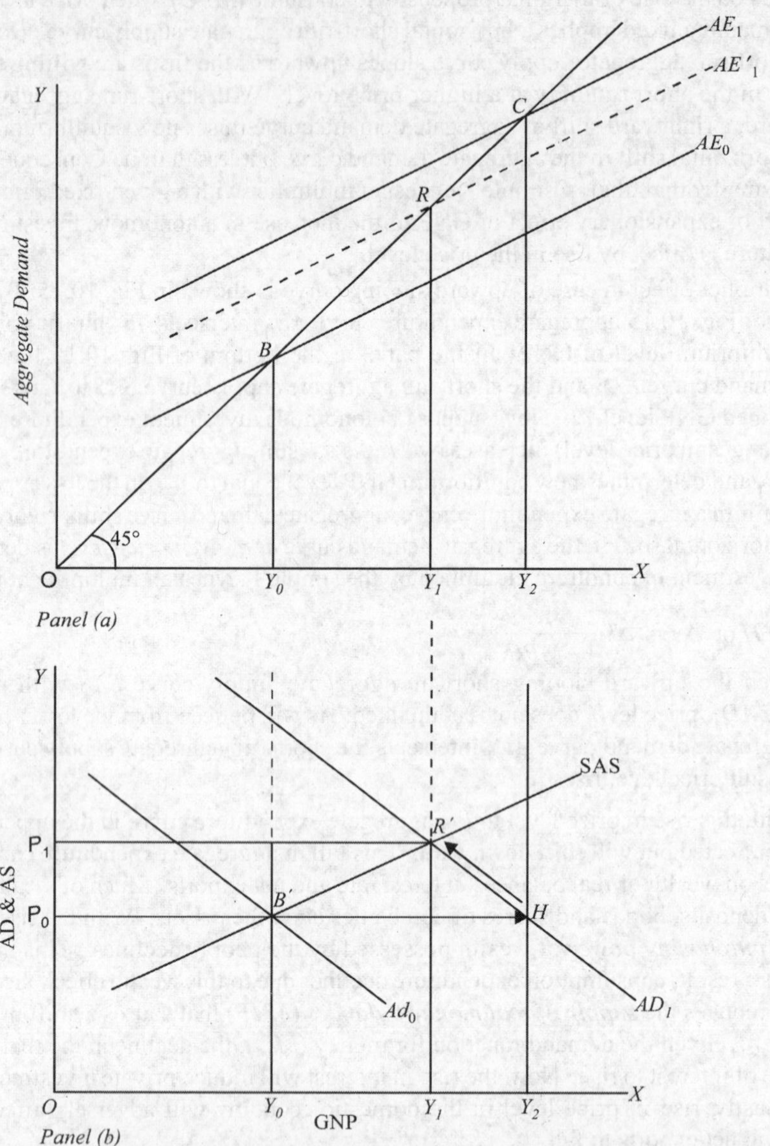

Fig. 10.15. *Multipler Effect with Changes in Price Level*

run aggregate supply curve *SAS*, the rightward shift in the aggregate demand curve causes price level to rise from P_0 to P_1 and causes decrease in GNP from Y_2 to Y_1.

However, as shall be seen from Fig. 10.15, when price level effect is taken into account, the increase in investment expenditure has still a multiplier effect on real GDP but this effect is smaller than it would be if price level remained fixed. It may be further noted that steeper the slope of the short-run aggregate supply curve, the greater is the increase in the price level and smaller is the effect on real GNP.

With this aggregate demand-aggregate supply model, popularly known as *AD-AS* model, we can explain the effects of fiscal and monetary policies on aggregate output (*i.e., GNP*) and price level in the economy. For example, if under expansionary fiscal policy the Government steps up its expenditure without increasing taxes, this will cause aggregate demand curve *AD* to shift to the-right and thereby will lead to the increase in Gross-Domestic Product (*i.e.*, National Income) and the price level as is illustrated in Figure 10.15. Likewise, reduction in taxes without reducing Government expenditure will cause increase in disposable income of the people leading to the increase in consumption expenditure of the people also result in shift in the aggregate demand curve to the right causing price level and national output to increase.

However, to what extent GDP will increase as a result of expansionary fiscal policy depends on the *slope of short-run aggregate supply curve*. When short-run aggregate supply (*SAS*) is relatively flat as many Keynesian believes and shown in Fig. 10.16, the effect of increase in aggregate demand (*AD*) will be more to increase GDP as compared to the rise in price level. On the other hand, if short-run aggregate supply curve (*SAS*) is steep as shown in Fig. 10.17(given on the next page), the effect of increase in demand and as a result of expansionary fiscal policy will be to cause more rise in price level (i.e. inflation) than increase in GDP. It is important to note that when aggregate demand increases due to either expansionary fiscal policy or expansionary monetary policy or due to increase in private investment expenditure and as a result price level increases this is called *demand pull inflation*. Demand-pull inflation arises due to the excess of aggregate demand over aggregate supply at the current price level.

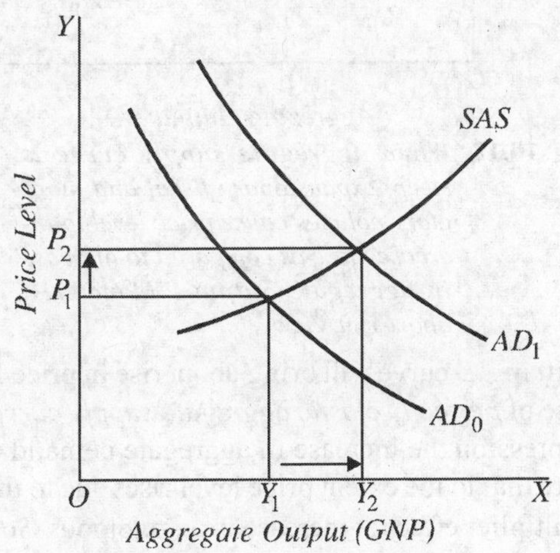

Fig. 10.16. *Expansionary Fiscal and Monetary Policies lead to more Increase in GNP and less rise in Price Level: Keynesian View*

Similarly, *increase in money supply* (*M*) will cause a rightward shift in aggregate demand curve. In the derivation of a given aggregate demand curve, money supply in the economy is held constant. If at a given price level, money supply is increased, the interest rate will fall. The fall in interest rate will cause investment demand to increase. Aggregate output demanded will thus be greater at the given price level. Thus, expansion in money supply brings about shift in aggregate demand curve to the right and will bring about rise in price level and GDP.

The same logic applies to the effect of contractionary fiscal and monetary policies. A decrease in Government expenditure or increase in taxes will cause *AD* curve to shift to the left. Similarly, decrease in investment autonomous of any change in the price level will also shift the *AD* curve to the left. Similarly, contraction in money supply (*M*) at a given price level will lead to the leftward shift in the aggregate demand curve.

An important result follows from this *AD*–AS model with flexible prices when under the influence of expansionaries policies aggregate demand curve shifts to the right. *To what extent it will affect the national output and the price level depends on the elasticity of aggregate supply curve*. Keynesian economists *i.e.* followers of Keynes are of the view that the aggregate supply curve during a period of recession is relatively elastic (*i.e.* flat) and therefore most of the response to expansion in aggregate demand will be the increase in output rather than rise in the price level as is shown in Figure 10.16. On the other hand, moneterists think that the supply curve is *steep as is shown* in Fig. 10.17 and in this case the expansionary fiscal and monetary policies causing outward

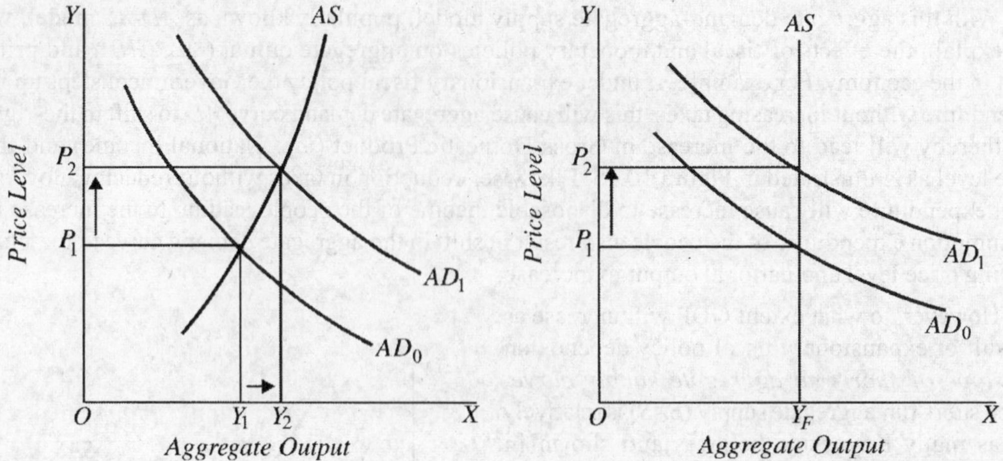

Fig. 10.17. *When aggregate supply curve is steep, expansionary fiscal and monetary policies cause price level to increase more as compared to increase in aggregate output : Moneterist Short-run View*

Fig. 10.18. *With full-employment of resources aggregate supply curve is a vertical straight line, an increase in aggregate demand will cause price level to rise without change in aggregate output : Classical View*

shift in *AD* curve will bring about rise in price level more than the increase in output. In the extreme case of *perfectly elastic aggregate supply curve* (Fig. 10.5) which Keynes assumed in his analysis of depression the increase in aggregate demand will not affect the price level. It is important to note here that to the extent price level rises due to the upward-sloping aggregate supply curve, the size of multiplier effect of increase in autonomous Government expenditure or investment will be reduced.

Lastly, when the economy is operating at the level of capacity output (*i.e.* full complement level) as the classical economists assumed, increase in aggregate demand in this case will lead only to higher price level or inflation in the economy without affecting aggregate output (national income). The view of classical economists is illustrated in Figure 10.18 where aggregate supply curve *AS* is a vertical straight line at the full-employment level of output OY_F. It will be seen from this figure that when as a result of expansionary fiscal and monetary policies aggregate demand curve shifts outward to the right from AD_0 to AD_1, price level rises from P_1 to P_2 whereas aggregate output remains constant at OY_F level.

SHIFT IN AGGREGATE SUPPLY AND STAGFLATION

Let us now turn to examine the effect of changes in aggregate supply, aggregate demand remaining constant. With the advent of *supply-side economics and new classical macroeconomics* embodying rational expectations theory in recent years economists have also concerned with shifts in aggregate supply curve. Important factors that cause shift in supply are changes in factor-prices such as prices of crude oil and raw materials and availability of resources, change in productivity and expectations about future inflation. Institutional factors such as Government regulations that affect resource use efficiency also cause shift in aggregate supply.

An important cause of shift in aggregate supply curve is the *rise in prices of resources such as large increase in oil price by OPEC* in 1973-74 and again in 1979-1980. The rise in oil prices causes the aggregate supply curve to shift to the left as shown in Figure 10.19 where due to the higher per unit resource cost aggregate supply curve has shifted leftward from AS_0 to AS_1. The aggregate demand curve *AD* remaining constant, with the leftward shift in aggregate supply curve from AS_0 to AS_1, leads to the new macro-economic equilibrium being established at point *T*, at which price level is higher and aggregate output smaller than before.

The emergence of rise in price level or inflation due to the leftward shift in aggregate supply curve is called *cost-push inflation*. There is important differences between deman pull inflation cost-push inflation. *Whereas in case of demand-pull inflation, price rises along with rise in GDP, in case of cost-push inflation price level rises but GDP declines.*

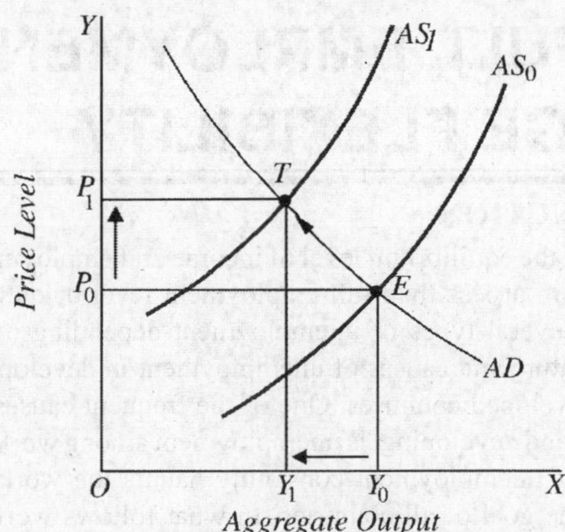

Fig. 10.19. *Emergence of stagflation due to leftward shift in aggregate supply curve*

Thus, leftward shift in the AS curve leads to rise in the price level from P_0 to P_1, while national output falls from Y_0 to Y_1. *When both inflation and recession occur simultaneously, economists call this situation as stagflation.* We thus see that increase in costs or resource prices are both inflationary and recessionary.

It may be noted that stagflation is also said to occur when instead of absolute fall in aggregate output there is *slowdown in economic growth rate* when inflation rate remains steadly high and also unemployment levels are quite high.

It is evident from above that stagflation occurs as a result of supply shock, that is, rise in price of essential inputs such as crude oil which causes a shift in the short-run aggregate supply curve to the left. Given the aggregate demand curve, the new equilibrium is reached at a higher price level (i.e. inflation) and at the same time reduced aggregate output (i.e. GDP) which generates unemployment in the economy.

We have explained above stagflation as a result of sharp rise in price of crude oil effected by restriction of oil output by callusive agreement of OPEC. In India stagflation can arise when there is decline in agricultural output due to failure of monsoon. The decline in agricultural output causes *food inflation* which in turn leads to rise in wages of workers which cause a shift in the short-run aggregate supply curve to the left. Besides, drastic decline in agricultural output also leads to the rise in prices of raw materials provided by agricultural sector to industries. The rise in raw material prices for industries shifts aggregate supply curve to the left causing both inflation and reduction in output.

Solution. To solve the problem of stagflation is a difficult task. In the seventies when the problem of stagflation was first faced efforts were made to solve it through demand management policies. However, when to reduce inflation, steps were taken to reduce aggregate demand by cutting expenditure, the result was decline in aggregate output which further worsened the unemployment problem. On the other hand, when to reduce unemployment attempt was made to increase aggregate expenditure through expansionary fiscal and monetary policies the result was acceleration in inflation.

The proper solution of stagflation is to shift the short-run aggregate supply curve to the right by reducing cost through raising productivity levels. The productivity in both industry and agriculture can be raised through technological improvements and more capital accumulation. Besides rightward shift in short-run aggregate supply can be made by reducing indirect taxes such as excise duties, sales tax, customs duties not only on final products but also on essential inputs used in the production of commodities.

CHAPTER 11
UNEMPLOYMENT, FULL EMPLOYMENT AND PRICE-WAGE FLEXIBILITY

INTRODUCTION

In the last few chapters, we have explained how the equilibrium level of income and employment is determined. When the economy is in equilibrium at less than full-employment level of output, unemployment comes into existence. There are several types of unemployment depending upon the different factors which cause it. Further, the nature and causes of unemployment in developing countries like India are different from those of developed countries. One of the frequent causes of labour unrest in moderns societies, both developed and developing, is unemployment among workers and the insecurity of employment. The specter of unemployment constantly haunts the working classes. Full employment has therefore become the goal of all countries. In what follows we will explain the various types of unemployment and their causes. We shall also explain the concept of full employment and what policies should be pursued to eliminate unemployment and achieve full employment.

MEANING OF UNEMPLOYMENT

Technically speaking, *unemployment is defined as a state of affairs when in a country there are a large number of able-bodied persons of working age who are willing to work but cannot find work at the current wage levels.* People who are either unfit for work for physical or mental reasons, or don't want to work, *e.g., sadhus*, are excluded from the category of the unemployed.

Mere engagement in some productive occupation does not necessarily mean absence of unemployment. People, who are only partially employed or are engaged in inferior jobs, though they can do better jobs, are not adequately employed. It is called a state of *underemployment* which is equally bad for the prosperity of a country. Under capitalism, it seems some amount of unemployment is inevitable. The best that can be done is to keep the number of the unemployed as low as possible.

TYPES OF UNEMPLOYMENT

There are three main types of unemployment: (1) frictional unemployment, (2) structural unemployment; and (3) cyclical unemployment. In order to understand the concept of full employment the difference between these types of unemployment needs to be grasped.

Frictional Unemployment

There is always some minimum amount of unemployment that prevails in the economy among workers who have voluntarily quit their previous jobs and are searching for new better jobs or looking for employment for the first time. They are said to be *between jobs*. They are not able to get jobs immediately because of frictions such as lack of market information about availability of jobs and lack of perfect mobility on the part of workers. The distinguishing feature of frictionally unemployed persons is that the number of job vacancies equal to them are available in the economy. They remain unemployed for a relatively short period of time before they are able to get new jobs. This is because some time is required for job searchers to have information about the availability of jobs. When presently fractionally unemployed persons get jobs, the new frictionally unemployed persons come into existence and thus there is always some frictional unemployment prevailing in the economy due to imperfections or lack of market information about the availability of jobs. Since frictionally unemployed are those who have either quit their old jobs voluntarily and are looking for better jobs or those who have entered into the labour force *for the first time* and searching for jobs according to their acquired skills or those who had re-entered the labour force for some time, for example, after having children, the frictional unemployment is considered as voluntary unemployment.

Structural Unemployment

Structural unemployment is another type of unemployment which always exists to some extent in a growing economy. Structural unemployment refers to the mismatch between the unemployment persons and the demand for specific types of workers for employment occurs because whereas demand for one kind of labour is expanding, the demand for another kind of labour is declining either due to the changes in the structure (*i.e.*, composition) of demand for the industrial products or due to the changes in technology that take place in an economy. *The distinguishing feature of structural unemployment is that the unemployed workers lack skills required by the expanding industries.* Structural unemployment also occurs because of the mismatch between the location of job-vacancies of the expanding industries and the present location of the unemployed workers.

Let us give an example of structural unemployment. In a growing economy techniques of production are constantly changing with the result that people are likely to lose their jobs when these are replaced by newer and more efficient techniques. The unemployment created in this way is structural unemployment and is inevitable consequence of technological progress. These structurally unemployed persons will need to acquire new skills and training before they will be absorbed in the new technologically superior jobs. Take another example, when computerisation of banks, offices occurred in India recently some workers were rendered unemployed. But computerisation created new job opportunities. On being given training in computer operations some of them again got jobs. But for some time they remain unemployed. Further, due to changes in demand for some goods, the output of industries producing them declines rendering some people unemployed. On the other hand, those industries the demand for whose products are increasing, new jobs are created and unemployed workers are absorbed in them but before getting jobs in the expanding industries they remain unemployed.

Structural unemployment tends to last much longer than frictional unemployment because more time is required for people to get new training or acquire new skills or to move to new locations of expanding industries. Structural unemployment is more serious than frictional unemployment. This is because frictionally unemployed are likely to get jobs in a relatively short period of time as job vacancies exist for them. On the other hand, structurally unemployed have no immediate job prospects as they have to get training or acquire new skills required for the jobs available. The structurally unemployed persons can be expected to get only low paying unskilled types of jobs in the immediate future. Some accept such jobs while others prefer to remain unemployed and go on searching for better jobs which match their skills.

It may be noted that frictional and structural types of unemployment together constitute what is called **natural rate of unemployment** which may be of the order of 4 to 5 per cent of labour force in free-market economies. Thus, natural rate of unemployment arises due to labour market frictions and structural changes in a free market economy. Note that when unemployment is equal to the natural rate of unemployment, full employment is said to exist. In the late nineties, natural rate of unemployment was estimated to be 5.2 per cent of labour force in the United States. In 2003 actual unemployment rate rose to 6.2 percent in the United States. This higher than natural rate of unemployment is due to recessionary conditions prevailing in the American economy. *Thus full employment is said to prevail despite the existence of natural rate of unemployment which is unavoidable in a changing economy.*

Cyclical Unemployment

It is unemployment due to deficiency of effective demand. It is also called cyclical '*Keynesian unemployment'*. Advanced capitalist countries have been suffering from time to time from this type of unemployment. This type of unemployment greatly increases during periods of recession or depression. Since recession or depression is one phase of the business cycles that generally occur in the industrialised developed economies this type of unemployment is called cyclical unemployment. This type of unemployment is due to the fact that the total effective demand of the community is not sufficient to absorb the entire production of goods that can be produced with the available stock of capital. In a private enterprise economy, production takes place in response to the profit motive. When businessmen cannot sell their entire output, their profit expectations are not fulfilled so that their reaction in the next period is to reduce their output. The factors of production earn their incomes because of their participation in the process of production and when entrepreneurs decide to reduce their production, some factors of production become unemployed. Since employment is the major source of income for a great majority of the people, a fall in employment signifies a fall in their incomes also.

The deficiency of aggregate demand as an explanation of involuntary unemployment in the industrially advanced countries is associated with the name of J.M. Keynes. Classical and neo-classicial economists often denied the very existence of the problem. In the view of classical economists, symbolised in *Say's Law of Markets*, a freely competitive, capitalist economy could never suffer from a deficiency of aggregate effective demand. The argument advanced by Say (a French economist) and accepted implicity by many others in support of this contention was that there could never be a problem of general over-production and that if there was any unemployment it was because of hindrances placed in the working of the freely competitive price system by artificial monopolistic action on the part of trade unions, employers or the government.

Now, it is true that supply does create a demand for goods and services because various factors of production earn their incomes in the process of production by helping to create additional supply of output. When factors of production are employed to produce a good they get their reward in the form of wages, rents, interest and profit. But from this it does no follow that the entire supply of national output will always be demanded by them. The incomes of the factors of production are necessarily equal to the value added in the productive process but it does not mean that the entire income will be automatically spent on goods and services created in a given time period. A part of income will be saved so that this part of income is not available to create demand for goods and services. Unless investors are willing to invest to an equivalent extent of intended savings, the total effective demand which consists of demand for consumers' goods and producers' good $(C + I)$ will not be sufficient to absorb the entire available supply of output. And if it happens like this, producers will not be able to sell their entire output, their profits will fall and they will reduce their production and this will create unemployment.

In a given period, consumers are planning to spend a given part of their income and save the rest. Similarly, entrepreneurs are planning to invest in factories, machines, etc., to a given extent. The total effective demand is the sum of the consumption and the investment demand. Savers are saving for different reasons than the investors whose investment is determined by different factors and in a private enterprise economy there is no mechanism to ensure that what savers are planning to save is just equal to what investors are planning to invest. If there is any discrepancy between planned saving and investment, output, income and employment will change to correct this discrepancy. If planned investment is greater than planned saving, the current output will not be sufficient to meet the emerging demand and hence income, output and employment will increase and vice versa. Thus, we see that the basic weakness of Say's law arises because of lack of any agency to ensure that intended savings are just equal to intended investment and that since savings and investment are undertaken by different persons and for different reasons a discrepancy between the two is bound to arise and when it arises the necessary mechanism to correct it is through changes in the volume of employment and income.

Cyclical Unemployment and Labour Market Equilibrium

Let us explain through labour market equilibrium how Keynes explains the emergence of cyclical unemployment of *workers who are seeking work or willing to work at the prevailing wage rate but are unable to get jobs*. According to Keynes, this involuntary unemployment happens to prevail in an economy on a large scale in times of depression due to fall in aggregate demand for goods and services. Classical economists thought that at times of recession wages and prices would fall to eliminate involuntary unemployment and ensure full employment. Keynes challenged this Classical viewpoint on two grounds. First, he pointed out that *money wages are rigid and fail to adjust even when there exists involuntary unemployment due* to lack of aggregate demand for output. Secondly, even if money wages are flexible downward or there is all-round cut in money wages, involuntary unemployment will not be removed. Consider Figure 11.1 where we have shown labour market equilibrium. *DD* is the demand curve of labour which is sloping downward to the light. This shows that at lower wage rates more labour is demanded and employed and vice versa. *SS* is the supply curve of labour which is sloping upward to the right. This shows that in higher wages more labour is supplied and vice versa. These demand and supply curves of labour intersect at point *E* at which real wage rate $\frac{W}{P}$ and labour employment of ON_0 are determined. At the wage rate $\frac{W}{P}$, all those who are willing to work and supply their labour services are employed. Thus, in this labour market equilibrium, there is full employment of labour and involuntary unemployment does not prevail.

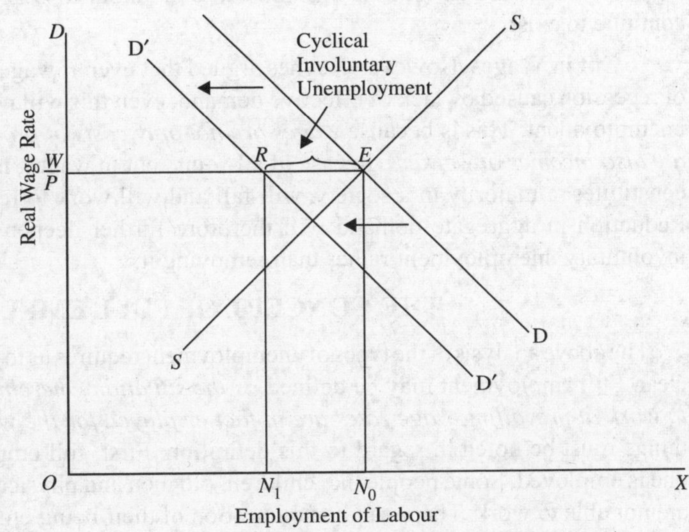

Fig. 11.1: *Labour Market Equilibrium and Involuntary Unemployment*

Now suppose aggregate demand for output of goods and services declines, as it happens in times of recession. *Demand for labour is a derived deman.*[1] Decline in aggregate demand for output causes demand curve of labour to shift to the left to *D′D′*. Now, at the given wage rate, $\frac{W}{P}$ the quantity ON_1 labour will be demanded, whereas supply of labour at this wage rate remains ON_0. Thus, *RE* number of workers will be rendered unemployed who are willing to work at the prevailing wage rate but their services are not demanded by the economy.

According to classical economists, this involuntary unemployment cannot persist for long. The excess supply of labour and competition among unemployed workers will bring down the wage rate and at lower wage rate more labour will be demanded and employed. In this way, according to classical economists, due to downward adjustment in wages as well as prices of goods involuntary unemployment of labour will be eliminated and full employment will be restored.

Rigid Money Wages. However, Keynes contended that *money wages are rigid* and do not fall and as a result involuntary unemployment continue to prevail. There are a number of reasons for rigidity of money wages. First, workers strongly resist a reduction in money wages. Second, there is usually a wage contract between workers (or their trade unions) with the employers for a period of two to three years during which wage rates cannot be changed. Thirdly, there are minimum wages fixed by the government below which employers are not permitted to pay wages to the workers. Lastly, employers themselves may not be interested in reducing wages, as lower wages are likely to reduce efficiency or productivity of labour. Thus, if in Figure 11.1, wages remained rigid at $\frac{W}{P}$ despite the leftward shift in the demand curve of labour to $D'D'$, involuntary unemployment would continue to exist.

Cut in Wages. However, Keynes argued that even if wages are cut all round during the period of recession caused by lack of effective demand, even this will not ensure elimination of involuntary unemployment. This is because *wages are not only costs of production for firms and industries but are also incomes of workers.* In case of all-round cut in wages, incomes of the working class, which constitutes a majority in a society, will fall and will work to reduce aggregate demand for output. Reduction in aggregate demand will therefore further deepen the depression and result in more involuntary unemployment rather than removing it.

THE CONCEPT OF FULL EMPLOYMENT

The above analysis of the types of unemployment requires us to make the concept of full employment clear. Full employment may be defined *as the situation wherein all those who are willing and able to work at prevailing wage rates are in fact employed for the work in which they are trained.* Two things must be noted in regard to this definition. First, full employment does not mean that every one is employed. Some people like children, old men and physically or mentally handicapped people are not able to work. There can be no question of their being employed. In fact these people are not included in the labour force of the country. Full employment will exist in spite of their not working. Secondly, some people called 'idle rich' though able to work are not willing to work because they get enough unearned incomes to live. These people are also not included in the labour force of the country.

From above, it follows that *the unemployed people are those who are involuntarily idle. They are able and willing to work but the economy does not provide them jobs*.

Another important qualification to the above definition is worth noting. It is that, at any given time, there is bound to be some frictional and structural unemployment owing to labour-saving technological improvements or a change in consumer tastes for the product of some individual industries, some workers are in the process of voluntarily changing jobs. Some time may elapse before they get a new job after leaving the old one and, therefore, they remain temporally unemployed. Similarly, some amount of seasonal unemployment is also in available. Experience indicates that the total amount of frictional and structural unemployment may amount to 4 to 6 per cent of the labour force.

Thus, *full employment is said to exist in the economy even if there is prevailing some amount of frictional and structural unemployment in the economy.*[1] If the number of unemployed is greater than frictional and structural unemployment, only then we shall say that full employment does not prevail.

Keynesian full employment is, by definition, the maximum level of employment that private enterprise countries can attain without experiencing strong inflationary pressure. But for purposes of practical policy it is necessary to reduce the concept to quantitative terms. It should be possible to say precisely when employment is less than full so that remedial action is called for. The concept, as defined above, raises two quantitative problems, namely, how to determine (1) the amount of employment sought by those who at the ruling wage-rates wish to be employed, and (2) the inevitable minimum amount of frictional and structural unemployment? The first, in turn, depends upon (a) the number of people able and willing to work for wages, and (b) the average number of hours of work which each of them wants to be employed. The number of wage-employment seekers depends upon the following: the size of the population of working age, the number who, having sufficient

1. Pigou, A.C., *Lapses from Full Employment,* 1945, p. 45.

unearned incomes, choose to remain idle; the number who, being in command of the requisite means of production, are self-employed; and the prevailing wage-rates and other incentives provided. Normally, apart from the size of the population of working age, these factors are likely to be more or less stable over short periods. For instance, the span of life regarded as falling within the working age, while subject to variation as a result of changes in the period of schooling or in the normal age for retirement consequent on changes in health standards and longevity, is likely to remain unchanged in the short period.

Again, since most workers have no large unearned incomes or accumulated savings, they are compelled to try to remain continuously in employment. They cannot, therefore, throw their labour power in the market or withhold it therefrom in some unpredictable manner. Accordingly, the number who choose to remain idle is unlikely to vary much in any short period. And so on. Moreover, even when some of these factors change somewhat, the net quantitative effect may not be important. *Higher wages, for instance, may induce some of the older workers to postpone their retirement while these may impel some of the married women workers, now that their husbands are better off, to relinquish their jobs. The net effect of a rise in a wages will not, therefore, be quantitatively important.*[1] *The same applies to other of factors.* The size of the population of working age too, though not invariant, is measurable as it changes according to definite trends. It follows that the number of wage-seeking population can be determined with a great measure of accuracy. And since the average number of hours of work which each wage-employment seeker wants to put in is likely to be more or less stable in any short period, the total amount of wage-employment sought is quite precisely measurable.

Similar observations may be made regarding the permissible allowance for structural and frictional unemployment. It varies from year to year, in accordance with the magnitude of the structural shifts in demand and production that together with immobility of labour and lack of adequate information about the availability and locations of new jobs cause structural and frictional unemployment. It may, therefore, be better defined as a range rather than as a precise figure. Since normally structural shifts in demand and production are unlikely to be violent or spasmodic and since in industrialised countries the incidence of frictional, unemployment is bound to be quite low, the allowance for frictional and structural unemployment needs to be very small. A US study has suggested that this allowance need not be beyond a range of 4 to 6 per cent of the available labour force.

The possibility of measuring the size of the available labour force and the inevitable minimum of frictional and structural unemployment makes full employment a determinate quantity. As suggested by a UN study, as a necessary step in the effective implementation of full-employment policies, each country should fix a full-employment target expressed in terms of the permissible range of frictional and structural unemployment. Unemployment in excess of the fixed target would indicate a lapse from full employment calling for a remedial action. The fixation of such a target would help to reduce the chances of government inaction and vacillation in the face of growing unemployment. It may also help to maintain confidence among businessmen whose pessimism ordinary plays a notable part in magnifying the downswing. There is, of course, the danger of the government acting on a false scent. For at times unemployment may exceed the level of frictional and structural unemployment due to causes other than insufficient demand. This danger may be minimized by allowing the government the discretion to disregard the signal if it has clear evidence that rise in unemployment is not due to the demand deficiency.

WAGE CUT AND EMPLOYMENT

The relationship between wages and employment has been a highly controversial issue between Keynes and classical economists. As seen earlier, classical economists believed that a cut in money wages would increase employment and help in removing unemployment. Keynes pointed out that workers would strongly resist any attempt to cut *money wages*, though they might be prepared to agree to cuts in *real wages* caused by rise in prices. Therefore, according to him, money wages remain rigid and any cut in money wages for the purpose of promoting employment was not a practical policy proposal. However, though wage cuts may not be practically possible, it is important to know whether it is a theoretically valid proposition that wage cuts or will promote employment and help in removing unemployment. In fact, Keynes made two-pronged attack on the classical view of cutting wages to remove unemployment. First, he challenged the classical view that wage cuts would promote employment at times of depression on grounds of practical feasibility. Second, he showed that even theoretically downward wage-price flexibility would not promote employment and attain full employment of labour. We discuss below both classical and Keynesian viewpoints about the relationship between wage-price flexibility and employment.

Wage Cut and Unemployment: Classical View

For analysing the effect of wage cuts on employment the classical economists applied partial or micro-economic analysis to the macro level. They argued that if wages fell, the prices of products made by labour would also fall. The price of products will fall because reduction in wages causes reduction in marginal costs of production. Since they used partial equilibrium analysis they assumed that the demand for the output of the industries would remain unaffected when cut in wages is made. They argued that with a fall in prices of products consequent to the cut in wages, the amount of products produced would increase and new price - output equilibrium will be established at lower prices and larger quantities of products.

The expansion in output of the products will lead to the increase in employment of labour and other inputs. The extent of expansion in output (and therefore increase in employment) following the cut in wages and consequent fall in product prices depends on the elasticity of demand for output made by labour. This is illustrated in Figure 11.2 where *DD* is the demand curve for the product of labour and *SS* is its supply curve. Initially, the demand and supply curves of the product intersect at point E_0 and determine output OQ_0 of the product. Suppose cut in wage rate is made which lowers marginal cost. As a result of decline in marginal costs supply curve of output shifts to right and at the new equilibrium between demand and supply at point E_1 price of product falls from P_0 to P_1 and consequently output of the product increases from Q_0 to Q_1. This increase in output will depend on the elasticity of demand curve of output ; the higher the demand elasticity, the greater the expansion in output. This expansion in output will lead to the increase in demand for or employment of labour. Similarly, on the basis of partial equilibrium analysis of the impact of wage cut on price, output and employment in an industry the classical economists applied this result to the impact of all-round cut in wages on the increase in output and employment of labour in the economy.

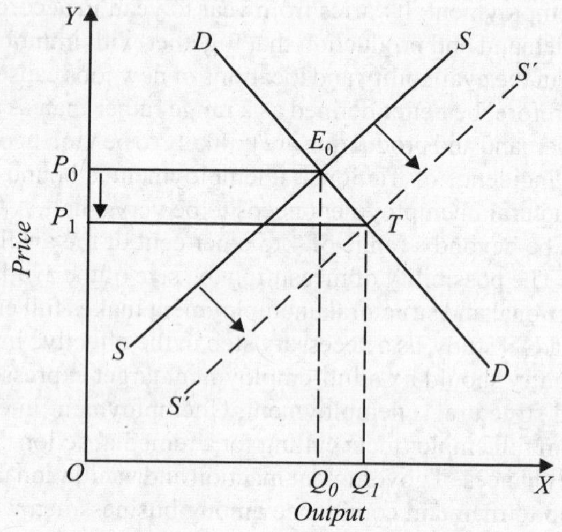

Fig. 11.2. *Fall in Price and Increase in output due to cut in Money Wages.*

Keynes's Critique of the Classical View

Keynes challenged the classical viewpoint regarding the impact of all-round cut in money wage on output and employment of labour. According to him, while in case of the analysis of price and output determination of an individual industry, it is justified to assume that a cut in wages by the industry would not significantly affect the demand for the product of that industry because most of the demand for the product of that industry comes from the workers and persons employed in other industries. However, to assume that demand curve for output of all industries will remain unchanged when *cut in wages in all industries together* is made is not valid. In other words, to apply the result of partial equilibrium analysis of the determination of price, output and employment to the economy as a whole is quit misleading and invalid. This is because wages are not only costs from the viewpoint of individual industries they constitute incomes of the workers and these incomes determine the demand for the products of various industries. When all-round cut in wages is made in all industries, it will reduce the aggregate demand for the products because workers would have now less incomes and therefore would spend less on goods and services. With reduced demand for the products of industries smaller output will be produced. Therefore, smaller amount of labour will be demanded and employed. To quote Stonner and Hague, *"When there is general unemployment, a general cut in wages in all industries cannot be assumed to leave demand unchanged, for part of that demand results from spending out of wages. It is thus quite clear that a general cut in wages will merely cause a reduction and will not in itself remove unemployment"*.[2] Thus, we see that classical economists neglected the adverse effect of all-around cut in wages on the level of aggregate demand.

The effect of *all-around reduction in wages* on aggregate output is illustrated in Fig. 11.3 where short-run aggregate curve supply curve SAS_0 and aggregate demand curve AD interest at point E and determine aggregate output Y_o which is less than full employment level of output Y_F. LAS is the long-run aggregate supply curve at full-employment level of output or national income. According to classical viewpoint, the sufficient cut in wages will cause a downward shift in aggregate supply curve, say to the new position SAS_1 with the result that new equilibrium is established at full employment

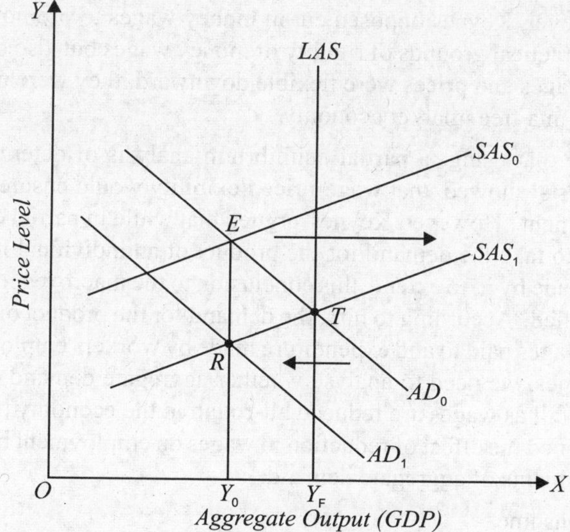

Fig. 11.3. *Impact of Wage-Cut on Aggregate Output and Employment*

output Y_F. But, according to Keynes, all-round cut in wages will reduce incomes of the workers and therefore expenditure made by workers which will cause a leftward shift in the aggregate demand curve AD, say to the new position AD_1. If the decrease in aggregate demand is proportional to the increase in aggregate supply, then equilibrium might be reached at the original aggregate output Y_o. Consequently, level of employment will remain below the full-employment level. It may however be noted that the impact of a wage cut on aggregate demand is more complicated than described above. As a matter of fact, as a result of all-round cut in wages the extent to which aggregate demand will decline depends on several factors such as its effect on propensity to consume, rate of interest, marginal efficiency of capital etc. But the classical analysis that there is a direct link between wage cut and rising employment through reduction in cost of production is also not valid because it neglects the role of aggregate effective demand when all-round cut in money wages is made. Professors Stonier and Hague rightly write, "*A general equilibrium analysis of wage cuts is futile unless it pays attention to the problem of whether aggregate demand will rise or fall as wages are reduced.*"[3]

2. Stonier and Hague, *A Textbook of Economic Theory*, Fourth Edition, 1977, p. 397.
3. Stonier and Hague, *A Textbook of Economic Theory*, Fifth edition 1980., p. 578.

Whether employment will rise or fall when wages are cut depends on whether aggregate demand in terms of money will remain constant, will it fall less than in proposition to the cut in money wages or will it fall more than in proportion to cut in money wages. In Keynesain macroeconomic analysis, the impact of cut in money wages is explained through its effect on the three main determinants of aggregate demand, namely, (1) propensity to consume, (2) rate of interest and (3) marginal efficiency of capital.

A General Equilibrium Analysis Wage Cut and Employment

We have explained in an earlier chapter Keynes' rigid money-wage rate but flexible price model of involuntary unemployment. Now, *we shall explain if both money wages and prices are flexible, will they ensure full employment of labour*. The viewpoint of classical economists, especially A.C. Pigou, the renowned British economist, was that wage-price flexibility would ensure full employment and therefore they recommended cut in money wages to increase employment and restore full-employment equilibrium. We have seen above that classical economists' suggestion of cutting money wages can be rarely put into practice, although workers might accept cut in real wages through rise prices. We have further discussed above various reasons for money wage-rigidity or sticky wages which explains the persistence of involuntary unemployment despite the fact that labour market is in disequilibrium. However, Keynes opposed cut in money wages to promote employment during depression not only on practical grounds of rigidity of money wages but also on theoretical grounds. He argued that even if wages and prices were flexible downward, they were not likely to lead to the increase in employment in a free market economy.

We have seen above that using a partial equilibrium analysis of determination of wages and prices classical economists showed that wage-price flexibility would ensure automatic adjustment to establish full-employment. However, Keynes argued that while in partial equilibrium analysis, it was theoretically right to take the demand for the product of an individual industry as given when cut in money wages is made by it, to extend this conclusion to the macro level or general equilibrium framework was not justified. According to him, the demand for the product of an individual industry depends mostly on the wages paid to and expenditure made by workers employed in other industries. Thus, according to Keynes, we need to analyse whether aggregate demand or aggregate monetary expenditure will rise or fall as wages are reduced all-round in the economy (that is, in all industries together). Keynes examined the effect of reduction in wages on employment by considering its effect on the following determinants of aggregate demand:

1. Propensity to consume
2. The rate of internest
3. The marginal efficiency of capital

After having examined the effect of cut in money wages on the above determinants of aggregate demand he reached the conclusion that when all things are considered, cut in money wages is likely to affect aggregate effective demand adversely and will therefore be unable to remove unemployment in the economy. In any case, according to him, favourable effect of wage-price flexibility on aggregate demand and employment occurs through reduction in rate of interest which gives a boost to investment demand. But reduction in rate of interest can be easily achieved through expansion in money supply in the economy and therefore for that purpose it is not a sound policy to adopt the policy of cutting money wages of workers. We examine below the effect of wage cuts on three determinants of aggregate demand.

Propensity to Consume. The effect of reduction in money wages is likely to reduce consumption demand by adversly affecting proensity to consume. All-round cut in money wages will reduce prices of products since wages are an important element of cost. The greater the ratio of wage costs to total production costs, the greater the fall in prices of products. The reduction in money wages and fall in prices of products are likely to lead to the redistribution of real income from wage earners to those sections of the society whose money incomes have not been reduced. Since propensity to consume of wage earners is higher than the non-wage earners who happen to be relatively richer sections of a society, this redistribution of income is likely to reduce propensity to consume causing a decline in aggregate consumption demand.

Another type of redistributive effect of falling prices is that redistribution of income from debtors to creditors occurs which also reduces consumption expenditure. The debit repayments are usually fixed in money terms so that fall in prices due to all-round reduction in wages which was not expected when debts were incurred causes an increase in the real value of mortgage and instalment payments by debtors to creditors which usually happen to be individual savers and banks. Even banks receive funds from individuals and companies who save and deposit their savings. This redistribution of income also reduces aggregate consumption demand, because creditors spend a relatively small part of the addition to their income due to fall in prices whereas debtors have to reduce their consumption greatly to meet their higher real interest payments caused by fall in prices.

It follows from above that cut in money wages will have an adverse effect on propensity to consume and will therefore reduce aggregate demand.

Marginal Efficiency of Capital. The effect of cut in money wages on marginal efficiency of capital is quite uncertain. This effect depends on what has been called *expectation effect*. If entrepreneurs expect that although money wages have been reduced in the present, they will rise again in the future, this will have a favourable effect on marginal efficiency of capital in the present. This is because higher money wages in the future will ensure better prospects of future demand for goods which will raise marginal efficiency of capital. Further, the fact that entrepreneurs expect money wages and therefore prices to rise in future, the people will prefer to buy goods in the present rather than in the future. This will improve prospects of making profits temporarily in the present and will therefore raise marginal efficiency of capital. But this favourable effect is unlikely to be significant.

But the expectation effect of fall in wages and prices is likely to have more adversely affect marginal efficiency of capital. If fall in wages and prices lead people to expect that they will fall even further in the future, it will reduce consumption demand. This will make entrepreneurs to become pessimistic about future economic prospects. As a result, marginal efficiency of capital will fall which will cause a reduction in investment demand.

It follows from above that the effect of a wage cut and fall in price on the marginal efficiency of capital is uncertain because much depends in expectations of entrepreneurs. However, cut in money wages is likely to reduce consumption demand which ultimately determines the future prospective yields from investment. It is believed by Keynesians that cut in wages and falling prices make future prospects of selling consumer goods less rosy and therefore adversely affect marginal efficiency of capital.

Rate of Interest : Keynes' Effect. The cut in money wages and consequent fall in prices is likely to reduce the rate of interest which will favourably affect investment demand. This is generally referred to as Keynes's effect as compared to Pigou effect which traces out the favourable effect of money wage cuts on consumption demand. According to Keynes, when cut in money wages is made and consequently prices fall, transactions demand for money will decline which will cause increase in the money supply for speculative motive. This increase in money supply for speculative motive will cause a reduction in the rate of interest. At a lower rate of interest, more investment will be forthcoming which will raise the level of aggregate demand and employment. It is important to notice here that in Keynesian analysis the hope of stimulus to investment following the cut in money wages and resultant fall in prices lies in the lowering of rate of interest. However, Keynes asserted that rate of interest can be easily lowered through expansion in money supply. Therefore, according to Keynes, theoretically "*We can produce the same effect on the rate of interest by increasing the quantity of money while keeping the level of wages unchanged*".[4] Further, according to him, expansionary

4. J.M. Keynes, The *General Theory of Employment, Interest and Money*

monetary policy to reduce rate of interest is preferable to the policy of reducing money wages since the latter will not only be strongly resisted by workers but also an all-round wage cut will reduce consumption demand for goods and will therefore adversely affect employment. Keynes scarstically remarks that while the flexible wage policy and monetary policy produce the same effect on investment demand through lowering rate of interest but there is a "World of difference between them in practice only a foolish person would prefer a flexible wage policy to a flexible monetary policy"[5]

However, Keynes emphasised, that expansionary monetary policy (and for that matter flexible wage policy so far its effect on rate of interest is concerned) will not necessarily ensure full employment. To what extent increase in money supply will succeed in lowering rate of interest and boost investment demand depends first on interest-elasticity of the liquidity preference curve and secondly on interest-elasticity of investment demand curve. If the liquidity preference curve is highly elastic, the increase in quantity of money will not bring about much fall in the rate of interest. Again, if the investment demand curve is relatively inelastic, then even the fall in rate of interest as a consequence of expansion in money supply may not be able to achieve much increase in investment.

Thus, Keynes was not hopeful of securing adequate increase in employment through either flexible wage policy or flexible monetary policy at times of depression.

Real Balance Effect or Pigou Effect

All-round cut in wages and consequent fall in prices produces a favourable effect on aggregate demand and employment in another may which was first pointed out by Pigou and is, therefore, kn(wn after his name as Pigou effect. According to Pigou effect, when wage cut causes fall in prices, real value of money balances increases as a result of which people become richer which induces them to increase their consumption expenditure which raises aggregate demand and employment. Though Keynes and his followers admitted that real balance effect operated but they expressed doubts about its quantitative significance. Of course, Pigou effect or real balance effect is theoretically possible but whether it is strong enough to raise aggregate demand sufficiently to ensure full employment is highly doubtful.

Patinkin and Real Balance Effect. It may be noted that 'Pigou effect' has been subsequently developed by Don Patinkin under the title of real balance effect. Don Patinkin[6] tried to revive the classical economics and tried to show that Keynes was wrong for advancing the view that the free market economy did not have a mechanism to restore full-employment equilibrium after an initial disturbance. Like the Pigou effect, real balance effect means that when the economy goes into recession or depression, wages and prices fall and as a result real value or purchasing power of money balances increases. This enables people to buy more than they could before. At lower prices they will actually purchase more than they did before. This increase in demand will lead to rise in output and employment. Thus, the operation of real balance effect, according to Patinkin, will bring the economy back automatically to full employment.

Of course, there may be some increase in output and employment as a result of the working of the real balance effect. But the crucial question is whether it is quantitatively strong enough to get the economy out of depression and restore full employment. As mentioned above, the real balance effect is not very significant to ensure the return to full employment. Thus Stewart writes, "*The process of falling prices was limited in its extent, it was not possible for the expansionary effect of its purchasing power to be more than a minor effect to the contractionary effect of falling incomes, consumption and investment, particularly when account was taken of the multiplier.*"[7]

---

5. Keynes, *op. cit.*, pp. 267-68.
6. Don Patinkin, *Money, Interest and Prices*, Evanston, 1956.
7. Michael Stewart, *Keynes and After,* 2nd edition, 1972, Penguin Books, pp. 139-40.

Summing up, the debate between Keynes and the classical economists on pursuing a flexible wage policy to ensure full employment. In the recent years the quantitative importance of Keynes's effect and real balance effect has been hotly debated. Many modern economists doubt the importance of real balance effect though its theoretical possibility is not denised. Again, as regards the choice between flexible wage policy and flexible monetary policy for increasing investment demand through lowering the rate of interest is concerned one is likely to achieve much less success by reducing the level of money wages than by increasing the supply of money. Besides, there are strong practical objectives to following a flexible wage policy as an alternative to monetary policy. First, as has been mentioned above, workers and their trade unions will strongly oppose all-round cut in money wages for every section of workers. It is much easier to lower interest rate through expansionary monetary policy than by cuts in money wages. Secondly, it is inequitable and unreasonable to expect that workers alone should accept cut in their money wages while other sections of the society continue enjoying higher money incomes. Thus, to increase the supply of money to reduce interest and thereby to reduce unemployment during depression is preferable to cutting wages. Thus, cuts in money wages are unlikely to bring about expansion in employment and output and therefore cannot become a practical policy of achieving full employment. Further, what can be achieved by cuts in money wages can be easily achieved through increase in money supply and reduction in interest rate to raise aggregate demand and employment.

INSIDER-OUTSIDER MODEL

A recent approach about wage-employment relation known as insider-outsider model is worth mentioning[8]. According to this approach, those who become unemployed as a result of fall in aggregate demand resulting in recession in the economy lose their influence on the wage-setting process and do not participate in wage negotiations with the employers. This is because the unemployed often lose their status as union members and therefore neither the employers nor the labour union care about them. As a result, while the unemployed workers would want firms to cut wages so as to create more jobs which would enable them to get back employment but the firms ignore them and negotiate with the workers who have jobs and are members of unions. It may be noted that it is quite expensive for the firms to often turn over the workers as costs have to be incurred for hiring new workers, training them and when not required retrenching them. Therefore, insiders have advantage over the outsiders who are seeking employment. Therefore, it is difficult for the firms to cut wages of the insiders (*i.e.*, their employed labour force), especially if their labour-union is strong. Even threats by the firms to the workers that they will be fired and rendered jobless if they do not accept wage cuts may not succeed.

Even if the workers yield to these threats and accept cut in wages to retain their employment it may not be in the interest of the firms themselves to lower wages as it is likely to adversely affect the morale, effort and productivity of workers. The firms will therefore think it better for them to deal with the insiders and pay them good wages even if there are unemployed workers outside who are eagerly seeking employment even at lower wages. Thus, "*The insider-outsider model predicts that wages will not respond substantially to unemployment and thus another reason why we do not quickly return to full employment once the economy experiences recession.*"[9]

Some economists believe that the insider-outsider model provides an important explanation of the persistent high unemployment in the European countries for the last about three decades starting in early 1980s. This is because labour unions in the European countries are quite strong and force the firms to bargain with the insiders and ignore the outsiders, the unemployed. This keeps the wages at high levels despite the existence of large unemployment in the European countries. However, if the

8. See Assar Lindech and Dennis Snower, *The Insider-Outsider Theory of Employment and Unemployment* (Cambridge, M.A: MIT Press 1988).
9. Dornbusch, Fischer and Startz, *Macroeconomics*, Tata McGraw-Hill, 8th edition, 2002, p.116.

labour unions were not strong, the firms might like to hire the outsiders (*i.e.*, unemployed) at lower wages and thereby reduce unemployment and the new firms might be set up to take advantage of cheap labour and provide employment. This would reduce European unemployment but the strong unions would resist lowering of wages. As a result, there is persistent high unemployment in European countries.

ANTI-RECESSIONARY POLICY TO REMOVE CYCLICAL UNEMPLOYMENT

To pull the economy out of recession or depression, the adoption of measures to raise aggregate demand is called for. We spell out below the various measures for raising aggregate demand so as to correct demand deficiency.

1. Deficit Budgeting: Expanding Aggregate Demand

An important fiscal measure to correct demand deficiency is to plan for deficit budgets. Deficit budgets can be attained in two ways. First, government should increase its expenditure on goods and services without being matched by imposition of fresh taxes or without raising the rates of old taxes. In 1930s when all capitalist economies of the world suffered from severe depression Keynes recommended the increase in government expenditure on *public works programmes* such as constructions of roads, railways, canals, school buildings, irrigation works, electrification of villages etc., to tackle the problem of demand deficiency. It should be noted that increase in public expenditure will have a multiplier effect on raising income and employment, that is, income and employment will not rise to the extent of increase in government expenditure but by much more than that depending upon the magnitude of multiplier.

2. Reduction in Taxes: Increasing Disposable Income

Another fiscal measure to raise aggregate demand is cutting down taxes. With reduction in taxes, people's disposable income will rise and consequent to this, their spending on goods and services will rise. If corporate taxes are reduced, this will increase the profitability of investment due to which business class will be induced to invest more. Thus, with reduction in taxation the two components of aggregate demand, namely, consumption demand and investment demand, will increase. This would help in taking the economy out of depression. In 1964 on the advice of the Keynesian economists government in US made significant cut in income tax. And this did a miracle in reviving the American economy and soon output and employment increased substantially. The recession was thus overcome with the help of fiscal policy. In June 2003, present President of the United States announced a tax cut of 3.5 billion dollars to revive the American economy which was quite successful.

3. Monetary Policy: Expanding Credit

Monetary policy can also be used to overcome depression by removing demand deficiency. The following three types of monetary measures may be taken under monetary policy to raise aggregate demand. Whereas the first measure works to raise the *cost of credit*, the other two increase the *availability of credit*. When there is a depression in the economy, the central bank tries to overcome it by lowering the bank rate. As a result of the fall in the bank rate, the market rates of interest also fall. With the fall in the market rates of interest, investment is stimulated which brings about the increase in aggregate demand and helps to remove depression.

Another method to stimulate expansion in credit supply is lowering of *Cash Reserve Ratio* (CRR). With lower cash reserve ratio banks will have more available funds which they will be willing to lend to business firms for investment purposes. This will raise aggregate demand in the economy leading to the expansion in income and employment.

4. Export Promotion to Raise Aggregate Demand

Deficiency in aggregate demand can also be corrected through promotion of exports of goods. When domestic demand or market for goods is limited or has shrinked due to the operation of some adverse factors, the goods can be sold abroad. But growth of exports of a country can depend upon several factors. The first important factor is the prices of the goods to be exported. We can export those goods which can be produced comparatively at lower costs and therefore can be sold abroad at competitive prices. That is, expansion in exports of those products can be achieved in which the country has a comparative advantage. To promote exports, foreign exchange note of the domestic currency can be reduced.

QUESTIONS FOR REVIEW

1. What is meant by involuntary unemployment? Explain Keynes's theory of involuntary unemployment.

2. Explain the following types of employment. What factors are responsible for their existence?
1. Frictional Unemployment
2. Structural Unemployment
3. Cyclical Unemployment

3. Explain the concept of full employment. Is it consistent with the existence of natural rate of unemployment? Explain the measures that can be adopted to achieve it.

4. What is natural rate of unemployment? Why full employment is said to exist even when there is natural rate of unemployment?

5. What is full employment? Discuss the different ways of attaining equilibrium level of full employment

6. Distinguish between fiscal policy and monetary policy. How is monetary policy helpful in generation of employment opportunities in the economy?

7. Explain Keynes effect and Pigou effect. Which do you think is more effective in lifting the economy out of depression?

8. What is real balance effect? Is it strong enough to get the economy out of depression and restore full employment?

9. Whether the policy of making cuts in money-wages or adopting an expansionary monetary policy to tackle the problem of unemployment in an economy suffering from recession.

10. Explain the following measures to increase employment in a depressed economy.
1. Fiscal policy of deficit budgeting
2. Expansionary monetary policy
3. Reduction in Taxes
4. Devaluation to promote exports

Chapter 12
UNEMPLOYMENT IN DEVELOPING COUNTRIES WITH SPECIAL REFERENCE TO INDIA

UNEMPLOYMENT IN DEVELOPING COUNTRIES: NATURE AND CAUSES

As stated above, Keynesian theory is mainly concerned with cyclical unemployment which emerges in the developed capitalist countries, especially in times of depression due to the fall in aggregate demand. During the period 1929-33, the developed capitalist economies suffered from sever depression which caused huge magnitude of unemployment. Keynes analysed this cyclical type of unemployment and asserted that it was caused by deficiency of aggregate demand. *The nature of unemployment in developing countries such as India is however quite different; rather than being cyclical it is of chronic and long-term nature*. Caused by the operation of structural factors, namely, the rate of capital formation has not been large enough to absorb increase in labour force in productive employment. It is now almost universally recognised that the chronic unemployment and underemployment in India are not due to the lack of aggregate effective demand which, according to J.M. Keynes, was responsible for unemployment in developed countries in times of depression. *Rather it is stated to be due to the lack of land, capital and other complementary resources in relation to the total population and labour force*.

Besides, extreme unequal distribution of land resulting in lack of access of land to a vast majority of rural population and use of capital-intensive technology in industries has accentuated the problem of unemployment and underemployment in developing countries. In the phenomenon examined by Keynes, not only labour force but also capital equipment was unemployed due to the deficiency of aggregate effective demand. In other words, in the Keynesian scheme, both the labour force and capital equipment was crying out for full employment which could be achieved by raising the level of aggregate monetary expenditure. Thus, according to Joan Robinson: "*Keynes's theory has little to say, directly, to the underdeveloped countries, for it was framed entirely in the context of an advanced industrial economy, with highly developed financial institutions and a sophisticated business class. The unemployment that concerned Keynes was accompanied by underutilisation of capacity already in existence. It had resulted from a fall in effective demand. The unemployment of underdeveloped economies arises because capacity and effective demand never have been great enough*"[1].

In the modern world, man by himself can hardly produce anything. Even the primitive man needed some elementary tools like bow and the arrow to engage in hunting for the earning of his livelihood. With the growth of technology and specialisation, he needs much more capital with which to engage him in the productive activity. If he is an agriculturist, he needs not only a piece of land but also a plough, a pair of oxen, seeds and some foodgrains and other necessities of life to sustain himself during the period of sowing to the reaping of the harvest. In the industrial sector, he needs factories to work in and machines to work with. All these aids to production belong to the community's stock of capital. Now, if the working force grows faster than the stock of capital of a country, the entire addition of labour force cannot be absorbed in productive employment—because not enough instruments of production would be there to employ them. The resulting unemployment is known as the *long-term or chronic unemployment as against Keynesian unemployment which is short-term cyclical unemployment*.

1. Joan Robinson, *Economic Philosphy*, p. 119.

A nation's stock of capital can be enlarged by increased investment which, in the absence of any unutilised resources, requires additional savings on the part of the community. The concern of the classical economists was to ensure that the rate of capital formation was kept sufficiently high so that employment opportunities were successively enlarged to absorb the additions to the working force of a country as a result of population growth. This is also the problem that developing countries like India are facing today. In recent years (between 1999-2000 and 2011-2012), the labour force in India has been growing at an annual rate of 2.2 per cent but our *rate of investment expressed as a percentage of our stock of capital* has not been growing fast enough rate to keep pace with the growth of population and therefore the country's ability to offer productive employment to the new entrants in the labour market has been severely limited. This manifests itself in two things—first, the prevalence of large-scale open unemployment in both the rural and urban areas as evidenced by the statistics of National Sample Surveys; second, it manifests itself in the form of underemployment as well as disguised unemployment in agriculture.

It is a common knowledge that with minor changes in organisation and with the existing techniques, our agriculture can be looked after by a much smaller number of workers and they can be withdrawn from agriculture without reduction of agricultural output if alternative employment opportunities are available. Since employment opportunities in the formal non-agricultural sector have not been growing rapidly, the new entrants to the workforce are compelled to remain in agriculture on self-employment basis which perpetuates the phenomenon of *disguised unemployment* which means that people are engaged in occupations where their marginal productivity is very low (if not zero or negative) and that a shift to alternative occupations will increase their marginal productivity and add to national income of the country.

Thus, *the standard explanation for the existence of labour surplus or unemployment and under-employment in less developing countries like India is that as compared with the magnitude of population and labour force there is limited availability of capital or complementary resources which include land, factories, machines, tools and implements–the means with which labour produces*. Now, if the population grows faster than the stock of capital of a country, the entire addition to the labour force cannot be absorbed in productive employment because not enough instruments of production would be there to employ them. Since in less developed countries, the stock of capital has not been growing at a rate fast enough to keep pace with the growth of population, the ability to offer productive employment is not adequate. This has resulted in surplus labour which is manifested in the existence of huge magnitude of open unemployment, underemployment and disguised unemployment in both the rural and urban areas.

Capital as the major bottleneck to growth of employment was made popular by Harrod-Domar model of economic growth in which capital accumulation plays a pivotal role and according to which rate of growth of output depends upon the proportion of national income saved, divided by the capital-output ratio ($g = s/v$, where g stands for growth rate, s for the proportion of income saved and v for the capital-output ratio). This model assumes capital coefficients (*i.e.,* capital-output ratio and capital-labour ratio) to remain constant. With constant capital-output and capital-labour ratios, the more the capital, the more will be both output and employment. Therefore, when adapted to the less developed countries, this model suggests that the *rate of growth of output and therefore of employment is determined by the growth of capital stock.*

WAGE GOODS CONSTRAINT AND UNEMPLOYMENT IN DEVELOPING COUNTRIES

It is worth mentioning a dissenting view regarding the cause of unemployment and under-employment in developing countries. This dissenting view has heen put forward by Prof. P.R. Brahmananda and C.N. Vakil of Bombay University. According to them, the basic cause of unemployment in developing countries is the deficiency of the availability of essential consumer goods, often called *wage goods.* They point out that when the unemployed people or disguisedly unemployed people who are withdrawn from agriculture are engaged in some public works, they will have to be supplied with wage-goods so that employed labourers can subsist. If the wage-goods are not sufficiently available, their employment in capital-creation works cannot be sustained. Given the real wage rate, a particular number of people can be employed in the economy, depending upon the supply of wage-goods in the economy. Now, the total quantity of wage-goods required to employ all the disguisedly unemployed workers in agriculture, according to them, would exceed the actual available supply of wage-goods even when the release of wage-goods by the withdrawal of disguised unemployed is taken into account. Thus, according to Prof. Brahmananda and Vakil, there exists a *wage-goods* gap which is the fundamental cause of unemployment in labour-surplus developing countries.[2]

A prominent Indian economist, Prof. Amartya Sen, has also emphasised the supply of wage-goods in determining employment in developing countries. According to Prof. Sen, the quantum of wage employment in the economy depends on the total supply of wage-goods on the one hand and the real wage rate on the other. If E represents the quantum of employment, M represents the supply of wage-goods and W, the real wage-rate, then the employment which can be provided will be given by the following equation :[3]

$$E = \frac{M}{W}$$

It is thus evident from Sen's above equation that if the supply of wage-goods (M) is less than their required supply to provide them to all labour force, then all workers cannot be fully employed, which will result in the emergence of unemployment. Thus, to generate enough employment and solve the problem of unemployment and underemployment, the wage-goods industries, especially agriculture, must be accorded a high priority in the strategy of economic development.

Most of the unemployment in developing countries is therefore of a different nature from that in advanced developed countries. A major part of unemployment in present-day developed countries is of cyclical nature which is due to *deficiency of aggregate effective demand.* But most of the unemployment in developing countries is not cyclical. Thus, in developing countries, there is not much Keynesian type short-term unemployment. Instead, it is a chronic problem.

Causes of Unemployment in the Indian Economy

We have explained above the two **basic** explanations of unemployment and underemployment prevailing in the developing countries. We now explain below in some detail the various causes which account for unemployment and underemployment that prevails in the developing countries.

1. Lack of the Stock of Physical Capital. The major cause of unemployment and underemployment in developing countries like India is *the deficiency of the stock of capital* in relation to the needs of the growing labour force. In the modern world, man by himself can hardly produce anything. Even the primitive man needed some elementary tools like the bow and arrow to engage in hunting for the earning of his livelihood. With the growth of technology and specialisation, he needs much more capital with which to engage in the productive activity. If he is an agriculturist, he needs a piece of land and also a plough, a pair of oxen, fertilizers, means of irrigation to produce agricultural products. In the industrial sector, he needs factories to work in and machines to work with. All these aids to production belong to the community's stock to capital. Now, if the working force grows faster than the stock of capital of a country, the entire addition of labour force cannot be absorbed in productive employment because not enough instruments of production would be there to employ them. The resulting unemployment is known as the *long-term* or *chronic unemployment.*

A nation's stock of capital can be enlarged by increased investment which, in the absence of any unutilised resources, requires additional saving on the part of the community. The concern of the

2. Amartya Sen, *Technology and Employment,* Oxford University Press.
3. *Ibid.*

classical economists was to ensure that the rate of capital formation be kept sufficiently high so that employment opportunities were successively enlarged to absorb the additions to the working force of a country as a result of population growth. This is also the problem that the developing countries like India are facing today. In recent times, the labour force in India has been growing at more than 2 per cent per year, yet our rate of investment *expressed as a percentage of our stock of capital* has not been growing at a fast enough rate so as to keep pace with the growth of population. As a result, the country's ability to offer productive employment to the new entrants in the labour force has been severely limited. This manifests itself in two things: first, the prevalence of large-scale open unemployment in the urban areas, and second, it manifests itself in the form of open unemployment and underemployment in the rural areas as well as disguised unemployment in agriculture.

2. Use of Capital-Intensive Techniques. An important factor responsible for slow growth of employment has been the use of capital-intensive techniques of production, even in consumer goods industries where alternative labour-intensive techniques are available. Even before 1991, under the Industrial Policy Resolution 1956, the development of consumer goods industries was left open for the private sector. However, private sector prefers to invest in highly capital-intensive plants and equipment on the basis of technology developed in labour-scarce western countries. It is argued by them that the alternative labour-intensive techniques have low productivity and low surplus-generating capacity. However, the important reason for the use of capital-intensive techniques has been the availability of cheap capital. Even firms in modern small industry sector which were expected to generate large employment opportunities have also tended to use capital-intensive techniques of production. Thus, Prof. J.C. Sandesara states, "*The availability of cheap capital has tended to encourage the modern small-scale industries sector to over-capitalise and use more capital-intensive methods of production and thus reduce employment potential*".[4]

In agriculture, reckless mechanisation of various agricultural operations despite the existence of surplus labour has reduced the employment-augmenting effect of new high-yielding technology involving the use of HYV seeds, fertilisers and pesticides. This has prevented the generation of enough employment opportunities in rural areas.

Now, a pertinent question is why capital-intensive techniques are used in industries despite the condition of labour abundance in the economy. First reason, as mentioned above, is the *relatively low price of capital*. Relatively low price of capital has been caused by (*a*) lower rate of interest, (*b*) liberal depreciation allowance on capital investment permitted in the taxation system of the country, (*c*) relatively cheap capital equipment imported from abroad. Second, higher wages of labour in the organised sector relative to their productivity under pressure from trade unions. Thirdly, *rigid labour* laws also discourage the employment of labour. It is difficult to retrench labour even when it is not required in case an industrial unit becomes sick and proposes to close down or exit. Fourthly, Research and Development (R & D) activity has not been adequately directed to discover and identify *labour-intensive appropriate techniques* to be used in industries which, though labour-intensive, have also reasonably good productivity.

3. Inequitable Distribution of Land. Another cause of unemployment prevailing in the developing countries like India is inequitable distribution of land so that many agricultural households have *no adequate access to land* which is an important asset for agricultural production and employment. Sub-division of land holdings under the pressure of rapid population growth since 1951 has further reduced access to land for several agricultural households. As a result, many persons who were self-employed in agriculture have become landless agricultural labourers who suffer from acute unemployment and underemployment.

---

4. J.C. Sandesara, *Small-scale Enterprises in Industrial Development. The Indian Experience* (1998), Sage Publications, Delhi.

4. Rigid Protective Labour Legislation. Another reason for the slow growth of employment in the organised sector has been the existence of unduly rigid protective labour legislation which makes it very difficult to retrench a worker who has been employed for 240 days. Labour legislation is so much rigid that it is even difficult to close down the unit and quit the industry. Thus, this excessively protective labour legislation induces private entrepreneurs to prefer the maximum use of capital in place of labour.

5. Neglect of the Role of Agriculture in Employment Generation. An important factor responsible for slow growth of employment opportunities is the neglect of agriculture for generating employment opportunities. The general perception, as existed in the first three five year plans in India (1951-65) as well as in the theoretical models of growth for dualistic economies such as Lewis' "*Economic Development with Unlimited Supplies of Labour*" was that agriculture already contained surplus labour and it was required to withdraw this surplus labour from agriculture and employ them in the modern industrial sector. By the mid-sixties it was realised that not to speak of employing new entrants to the labour force year after year, the modern industrial sector could not absorb productivity even a fraction of the then existing unemployed persons in the foreseeable feature.

Agriculture though containing surplus labour can generate employment opportunities if proper strategy for its development is adopted. For instance, the empirical evidence shows that on an irrigated hectare of land the number of manhours employed is almost twice that on the unirrigated hectare. Irrigation requires more labour input for watering the fields, and also since output per hectare on irrigated land is much higher, more labour is used for harvesting and threshing the crop. Besides, irrigation makes the *adoption of double cropping* possible which greatly raises the employment potential of agriculture.

It is worth noting that new agricultural technology, commonly called green revolution technology, involving the use of HYV (High Yielding Varieties) seeds, greater use of fertilisers and pesticides along with water is highly labour absorptive. What is equally important, this new green revolution technology is **size-neutral**, that is, it can be equally well adopted by small farmers. Further, HYV seeds are of short-duration type, that is, they mature in a short time making multiple cropping more feasible. The use of double or multiple cropping greatly enhances the opportunities of employment generation in agriculture. The experience of Punjab, Haryana and Western U.P. is a shining example of large employment generation in agriculture through green revolution technology. What is needed for the generation of large employment opportunities in agriculture is that the new green revolution technology should be widely diffused and adopted in the backward and lagging agricultural regions in India.

6. Lack of Infrastructure. We have explained above lack of physical capital with which labour is equipped for productive employment as the cause of unemployment prevailing in the developing countries like India. By capital we generally mean machines, plant and equipment, factory buildings etc. But a similar factor responsible for huge unemployment prevailing in these countries is lack of infrastructure such as roads, power, telecommunications, highways, irrigation facilities in agriculture. Inadequate availability of infrastructure is a great obstacle for the generation of opportunities for productive employment. Expansion in infrastructure facilities generate employment *directly as well as indirectly* in the production of industrial products and services as they have a large *forward linkages*.

It follows from above that unemployment and underemployment prevailing in India and other developing countries is *not cyclical Keynesian type of unemployment caused by decline in aggregate demand. Unemployment and underemployment in India are caused by more basic structural factors such as lack of capital, use of capital-intensive technologies, lack of access to land for agricultural households, lack of infrastructure, rapid growth of population resulting in large annual increments in labour force year after year.*

MEASUREMENT OF UNEMPLOYMENT IN THE INDIAN ECONOMY

The surplus labour in a developing economy such as ours manifests itself in (*a*) open unemployment, (*b*) underemployment and (*c*) disguised unemployment. Therefore, a realistic assessment of the magnitude of surplus labour that exists in the Indian economy should contain information about all these three forms of unemployment. However, the measurement of unemployment in an economy such as ours where self-employment is predominant and the bulk of productive activity is carried on within household enterprises is a different task. A slight difference in the definitions and criteria adopted to measure unemployment often makes enormous change in its magnitude. For example, late Professor Raj Krishna suggested the following four criteria to measure unemployment:

(*a*) *Time Criterion*: This refers to identifying those persons as unemployed who have no gainful work for less than some normal standard of hours in a week or less than a certain normal standard days in a month or a year.

(*b*) *Willingness Criterion*: This calls a person unemployed when he is either seeking or is available for work at the prevailing wage rate.

(*c*) *Income Criterion*: This considers those persons unemployed or underemployed who have income (or expenditure) less than a certain '*poverty norm*' generally fixed on the basis of minimum calories required for subsistence.

(*d*) *Productivity Criterion*: This considers those persons as unemployed or underemployed who may be working but they have less than some normal standard of productivity.

Thus, a change from time and willingness criteria to productivity criterion changes the estimate of rural unemployment for the year 1972-73 from 8.21 per cent of rural labour force (on daily status basis) to 33.7 per cent.[5]

Note that both willingness and time criteria have been used in India in the various rounds of National Sample Survey to measure unemployment in the economy. Productivity criterion has been used by Ashok Mathur, Shakuntala Mehra, Ashok Rudra and Amartya Sen to measure the extent of disguised unemployment in the Indian agriculture. Further, using household expenditure data from earlier rounds of NSS, Dandekar-Rath in their study of "*Poverty in India*" adopted income criterion to measure the magnitude of unemployment in the Indian economy.

Measurement of Unemployment in India: Dantwala Committee Approach

Let us now discuss the actual approaches that have been adopted in India to measure open unemployment and underemployment in India. On the basis of time and willingness criteria open unemployment and underemployment have been estimated using the following three approaches which were recommended by an expert committee headed by Prof. M.L. Dantwala:

(*i*) **Usual Status Approach.** This approach records only those persons as unemployed who had no gainful work for a *major time* during the 365 days preceding the date of survey and are seeking or are available for work. Thus, the estimates of unemployment obtained on the basis of usual status approach are expected to capture chronic or long-term open unemployment.

(*ii*) **Weekly Status Approach.** In this approach current activity status relates to the week preceding the date of survey is recorded and those persons are classified as unemployed who did not *have gainful work even for an hour on any day in the preceding week and were seeking or were available for work*. The persons who may be employed on usual status approach may however become intermittently unemployed during some seasons or parts of the year. Thus, unlike the usual status approach, *weekly status approach would capture not only open chronic unemployment but also seasonal unemployment*. This approach provides us a weekly average rate of unemployment.

5. 33.7 per cent of labour force was Sen's estimate of rural unemployment for the year 1972-73 on the basis of his productivity criterion.

(*iii*) **Daily Status Approach.** As noted above, the weekly status approach records a person employed *even if he works only for an hour on any day of the whole week.* It is thus clear that the weekly status approach would tend to underestimate unemployment in the economy because it does not appear to be proper to treat all those who have been unemployed for the whole week except an hour as employed. Indeed, the demand for labour in farming and non-farming households often fluctuates over a small period within a week. Hence the need for the use of daily status approach to measure the magnitude of unemployment and underemployment in India. In the daily status approach current activity status of a person with regard to whether employed or unemployed or outside labour force is recorded *for each day in reference week.* Further, for estimating employment and unemployment, *half-day has been adopted as a unit of measurement*. A person who works for 4 hours or more up to 8 hours in a day is recorded as employed for the full day and one who works for an hour or more but less than 4 hours on a day is recorded as employed for half-day. Accordingly, persons having no gainful work even for one hour on a day are described as unemployed for full day provided that they are either seeking or are available for work.

Thus, the daily status approach would capture not only the unemployed days of those persons who are usually unemployed but also the unemployed days of those who are recorded as employed on weekly status basis. Hence daily status concept of unemployment is more inclusive than those of usual status and weekly status approaches and would yield an average number of unemployed person-days per day in the year indicating the magnitude of both open unemployment and underemployment. They are also referred to as *person-years unemployed* so as to distinguish them from persons unemployed.[6]

The estimates of unemployment on the basis of above three approaches are presented in Table 12.1. It may be observed from this table that according to NSS 1999-2000 round, 2.2 per cent of labour force were unemployed on usual status basis and 4.4 per cent of labour force were unemployed on weekly status basis and 7.3 per cent of labour force on current daily status (CDS) basis. In 2004-05 (60th Round) unemployment rate further increased. On current weekly status basis, unemployment rate was estimated at 4.4 per cent in 2004-05 for India as a whole, but on current daily status basis, in 2004-05 rate of unemployment rose to 8.2 per cent in case of All India. In absolute numbers, Planning Commission estimated unemployment to be of the order of 35 million persons on current daily status basis in April 2002 in the beginning of the 10th Plan.

On the basis of 60th round, number of persons unemployed on the current daily status basis rose to 34.7 millions in 2004-05. However, according to 66th round of National Sample Survey (NSS) for the year 2009-2010, current daily status (CDS) unemployment rate as per cent of labour declined to 6.8 in rural areas and to 5.8 in urban areas giving us unemployment rate as 6.6% of labour force for whole India. In 2011-12, while the usual status unemployment as a per cent of labour force rose to 2.2 per cent, on current daily status unemployment further declined to 5.6 per cent of labour force. However, in our view, this is mainly the result of special employment guarantee scheme, MGNREGA, started in 2007-08 under which people are employed in construction activities for a short period of time and does not represent the *stable growth* of employment opportunities.

Table 12.1. Magnitude of Unemployment in the Indian Economy (% of Labour Force)

Approach	1972-1973	1977-1978	1983	1987-1988	1993-1994	1999-2000	2004-2005	2009-2010	2011-2012
Usual Principal Status	1.61	4.23	2.77	3.77	2.58	2.2	2.3	2.0	2.2
Current Weekly Status	4.32	4.48	4.51	4.80	3.6	4.4	4.4	3.6	–
Current Daily Status	8.35	8.18	8.28	6.09	6.0	7.3	8.2	6.6	5.6

Source: Planning Commission, *Eighth Five Year Plan* 1992-97, Vol. I, p. 134 and *Tenth Five Year Plan.*

6. See CSO, *National Sample Survey,* 32nd round (1977-78), Report No. 298, p. 20.

Note that, contrary to what could be expected, weekly status unemployment is not much greater than that of the usual status approach. This is for two reasons. First, a person to be identified as unemployed on weekly status approach has to pass through a severe test, namely, he did not work even for only an hour in the whole reference week. Now, in an economy where there is predominance of self-employed households engaged in family-based enterprises, it is quite easy for a person to have worked for an hour in a week even during the lean periods of the year and therefore get registered himself as employed rather than unemployed. Secondly, during the slack seasons of the year some persons, mainly women, withdraw themselves from the workforce. Therefore, instead of being counted as unemployed, they are treated as 'outside labour force', for they generally report as 'not available for work'. As a result, whereas weekly status employment estimates are relatively greater than that of usual status, their unemployment estimates are not much higher.

Further, it is noteworthy that both the usual and weekly status approaches reveal a rather small magnitude of surplus labour. This is because, as mentioned above, *both these approaches measure open unemployment in the country*. However, in India with predominance of self-employed households and absence of social security system to support the people, they can hardly afford to remain absolutely unemployed. The new hands that are added to the labour force every year as a result of increasing population do not remain openly unemployed; they share employment and work with others so that there is more of underemployment and disguised unemployment rather than open unemployment.

Thus, quite expectedly, the daily status approach which seeks to capture both open unemployment and underemployment reveals that quite a larger magnitude of surplus labour exists in India. It will be seen from Table 12.1 that 8.2 per cent of labour force was found unemployed in 2004-05 on daily status basis. It was estimated that the absolute magnitude of daily status unemployment increased to 36.7 million person-years in 2006-07.

Critical Evaluation of Dantwala Committee's Measures of Unemployment

Note that, contrary to what could be expected, usual status unemployment is very low. This is for two reasons. First, a person to be identified as unemployed on usual status approach has to pass through a severe test, namely, he did not work for a *major time* during the 365 days preceding the date of survey. Now, in an economy where there is predominance of self-employed households engaged in family-based enterprises, it is quite easy for a person to have worked for some days even during the lean periods of the year and therefore get registered as employed rather than unemployed. Secondly, during the slack seasons of the year some persons, mainly women, withdraw themselves from the workforce. Therefore, instead of being counted as unemployed, they are treated as 'outside labour force', for they generally report as 'not available for work'. As a result, usual status unemployment estimates are relatively very low.

However, even the daily status approach does not fully measure the magnitude of surplus labour in the Indian economy. Evidently, this also would have a tendency to underrecord unemployment as it considered a person working for only 4 hours (1/2 of the standard day) as fully employed and a person who worked for an hour (1/8th of the standard day) as employed for half-day. With this we are not able to measure disguised unemployment from daily status approach. In fact, as an alternative, hourly status approach measuring employment and unemployment in terms of man-hours would be more appropriate.

Further, the daily status approach-like the usual and weekly status approach, cannot fully capture disguised unemployment of those persons whose intensity of work and consequently productivity is very low even though they are employed for a long time. In India, as in other labour-surplus developing countries, people are generally found to be working longer hours and larger number of days than one is really required to do the given work by *stretching or spreading the work*. Therefore, for the proper assessment of the magnitude of surplus labour that currently exists in the Indian economy, unemployment on the basis of daily/hourly status ought to be supplemented by a measure of disguised unemployment based on a norm of productivity or work-intensity.

Despite the above limitations it is clear that daily status unemployment is a relatively better measure of the magnitude of surplus labour as it covers open unemployment as well as some amount of underemployment. It is evident from above that three measures of unemployment are based *mainly* on time criterion as stated by late Dr. Raj Krishna. The time criterion has been used in the above three measures of unemployment as they are based on the number of days in a year persons remain unemployed. Though the willingness criterion has been sought to be incorporated by asking the question from unemployed persons whether they are seeking to work or available for work. As far as seeking work is concerned it would have been relevant only if enough employment exchanges would have been there. In the rural India, employment exchanges do not exist. Even in urban areas, the unemployed people do not go to register their names for employment as it is a common knowledge that employment exchanges can arrange employment only for skilled and education and that too in a very small number of cases. Therefore, 'seeking for work' is not very relevant in labour surplus developing countries. Especially, women in rural India generally do not seek work outside their villages.

The other question relating to '*willingness to work*' is also sought to be obtained by asking question from unemployed whether they are "*available for work*". However, *without reference* to at what wages the work will be available is not very meaningful. Moreover, many female workers would neither be seeking nor available for work outside their village where they reside. Thus the three measures of unemployment suggested by Dantwala Committee which are being used by Planning Commission and Government are mainly based on time criterion regardless of earnings they make in their employment which can be described as gainful. They should be described as acutely underemployed or disguisedly unemployed if their employment in some work does not yield enough earnings to meet their basic needs.

DISGUISED UNEMPLOYMENT

A good number of unemployment in a labour-surplus country such as India manifests itself in the form of disguised unemployment, especially in the agricultural sector. It is a common knowledge that with minor changes in organisation and with the existing techniques, our agriculture can be looked after by a much smaller number of workers and they can be withdrawn from agriculture without reduction of agricultural output if alternative employment opportunities are available. Since employment opportunities in the non-agricultural sector have not been increasing large enough to absorb the new entrants of the labour force a good number of additional workers are compelled to remain in agriculture and related activities and this perpetuates the phenomenon of disguised unemployment. To be precise *disguised unemployment is said to prevail when so many workers are employed in agriculture that the marginal productivity of workers or labourers is zero*. Therefore, if they are withdrawn from agriculture, techniques of cultivation remaining the same, agricultural output will not fall. Thus, Prof. Ragnar Nurkse defines disguised unemployment in underdeveloped countries in the following way. "*These countries suffer from large-scale unemployment in the sense that even with the unchanged techniques of agriculture a large part of population engaged in agriculture could be removed without reducing agricultural output. The same farm output could be obtained with a smaller labour force. In technical terms, the marginal productivity of labour is zero*"[7]

It is worth noting that the concept of disguised unemployment is applicable to self-employed labourers of extended family system in agriculture and does not apply to wage employment which usually takes place on the large-sized capitalist farms. Thus the significance of the "term" 'disguisedly' is that it is applied only to persons who are not normally engaged in wage employment but are self employed.

The measurement of disguised unemployment is very difficult. However, it is generally believed that around 25 per cent of labour force in labour-surplus developing countries such as India is disguisedly unemployed.

7. R. Nurkse, *Problems of Capital Formation in Underdeveloped Countries*, p.32

Prof. Amartya Sen's Analysis of Disguised Unemployment

Though the concept of disguised unemployment visualises production aspect of unemployment, it mainly manifests itself in idleness, that is, time not worked. In other words, disguised unemployment appears in the form of reduction in the hours worked in a day or a week and days in a week, or a month or year. As long as more work opportunities are available on the family farms, the additional labour is employed for that and as a result total output increases, though at a diminishing rate. When all the work opportunities on the farm are exhausted, no more addition to labour input will bring about increase in output. When the size of the families continues to increase due to the rapid growth of population and when alternative employment opportunities outside their families are not available, the additional members have to be absorbed on the family farms. Since on the family farms a point reaches beyond which no increase in labour input leads to the rise in total output, additional workers are absorbed by sharing of the work among more workers rather than by increasing the quantum of labour input employed on the farms. In other words, *the same quantity of labour is now applied by more workers. As a result, the number of working hours performed by all labourers in a day, or week and days in a month or year are reduced.*[8]

Distinction between marginal productivity of labourers and marginal productivity of labour. In explaining the concept of disguised unemployment, Amartya Sen has drawn the distinction between ***marginal product of labourers*** and ***marginal product of labour***. He explain that it is the marginal productivity of labourers over a wide range which is zero and not the marginal productivity of labour (*i.e.*, labour input) over a wide range of labour used. In fact, marginal productivity of labour is just equal to zero at the margin. Therefore, the distinction drawn between labour and labourers by Prof. Amartya Sen is significant since it helps in understanding the nature of disguised unemployment and the form in which it appears as well as in removing the apparent contradiction between the disguised unemployment and rational behaviour. Thus, Prof. Amartya Sen writes :

"*This confusion arises because of not distinguishing between labour and labourer*. It is not that too much labour is being spent in the production process, but that too many labourers are spending it. Disguised unemployment thus normally takes the form of a smaller number of working hours per head per year; for example, each of three brothers shepherding the sheep every third day. It is thus, the marginal productivity of the labourer, so to say that is nil over a wide range and the productivity of labour may be just equal to zero at the margin. *It may also take the form of lower intensity of work with people 'taking it easy'*, e.g. the peasant having time to watch the birds while working. If a number of labourers went away, the others would be able to produce about the same output working longer and harder. There is not contradiction between disguised unemployment will naturally put on this disguise. A piece of land that can be cultivated fully by two, may actually be looked after by four, if a family of four working men having no other employment opportunity happens to own it."[9]

It is evident from Professor Sen's quotation given above, the disguised unemployment appears not only in the form of the idle man hours (that is, all workers work for lesser hours of work than those considered standard or normal and remain idle for the rest) but also in the reduced intensity of work-hours performed. In other words, the disguised unemployment also creeps into work hours performed by labourers. This happens through what has been called work-spreading or work-stretching devices.

It is worth noting that not all sizes of farms suffer from disguised unemployment. As mentioned above, large-sized farms generally use hired labour (or are cultivated through tenancy system) and granting that landlords of those farms operate on the principle of profit maximisation, no disguised unemployment will exist on those farms. As the size of the farms goes on declining, man-land ratio on them will be increasing. As the man-land ratio rises with the decrease in the size of farms, the average and marginal products of the labourers will diminish due to the operation of diminishing returns. Since a certain minimum ratio of land to labour is required even in the most labour-intensive

8. Amartya Sen, *Choice of Techniques*, Basil Blakwell, 1957.
9. *Op. cit.*, p. 3

agricultural processes, a point will reach when the further reduction in the size of the farm and consequently further increase in the man-land ratio will result in zero marginal product of labour.

Graphic Illustration of Sen's Concept of Disguised Unemployment

Let us graphically explain the concept of disguised unemployments as has been clarified by Prof. Amartya Sen. Consider Fig. 12.1 where along the *Y*-axis total output is measured and along the *X*-axis, labour (say labour hours) is measured. *TP* represents total product curve which in the beginning rises and after a point it becomes horizontal indicating thereby that total output remains constant after OL_1 amount of labour is spent. Now, according to Amartya Sen, it will be irrational on the part of the family to put in more labour on the farm beyond OL_1 since extra labour spent beyond OL_1 does not add to total output. Thus if the family goes beyond OL_1 and devotes OL_2 labour, then L_1L_2 labour input has been unnecessarily spent, for it has not brought any increment in output. Marginal product of labour is zero when OL_1 amount of labour is put in. Therefore, a rational family will put in OL_1 amount of labour.

But, according to Prof. Amartya Sen, the disguised unemployment occurs due to the fact that a given amount of labour is spent by too many labourers. Consider Fig. 12.1. again. Along the south, number of labourers are measured. Suppose in the beginning OL_1 amount of labour is being put in by ON_1 labourers who are working normal hours in a week. If now the size of the family expands and the number of labourers increases from ON_1 to ON_2, the amount of land, other factors and technique of production remaining the same. As stated above, it will be futile and irrational to put in more labour beyond OL_1 even though the number of labourers in the family has increased. If the additional labourers do not get employment outside the family farm, the reaction of the family will be that the same amount of labour input OL_1 be shared by ON_2 labourers. As a result, the number of work hours devoted by each labourer will be reduced.

The total output will not increase since the same amount of labour input OL_1 is being put in. It follows therefore that the increase in the number of labourers from ON_1 to ON_2 has not resulted in any increment in total output: only the same amount of labour is being shared by ON_2 labourers instead of ON_1. Thus number of labourers N_1N_2 represents disguised unemployment since marginal productivity of labourers over this range is zero. Now if N_1N labourers are withdrawn from the agricultural family, total output will not decline. With the withdrawal of N_1N_2 workers, the remaining workers will put in OL_1 labour so that each of them will work for normal hours.

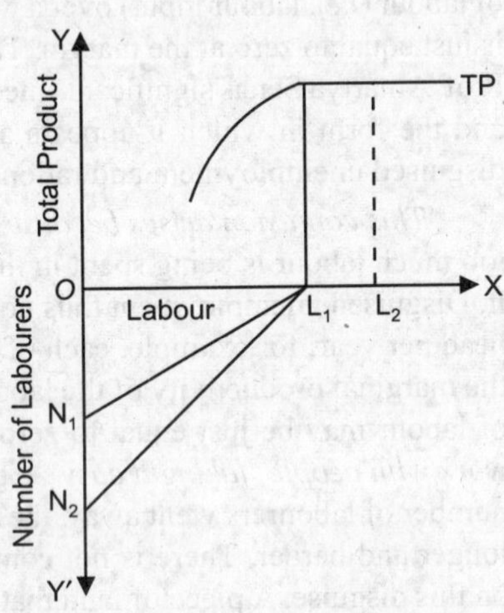

Fig. 12.1. *The Emergence of Disguised Unemployment: Sen's View*

The magnitude of disguised unemployment has been increasing in underdeveloped countries since the rate of capital accumulation has not kept pace with the rapid growth of population. It has been estimated that 15 to 30 per cent of agricultural labour force is disguisedly unemployed in various labour-surplus underdeveloped countries. This type of unemployment can be relieved by the development of non-agricultural sectors and by increasing the rate of capital formation that will open new avenues of employment. But the economies of such over-populated agricultural countries have remained generally stagnant and economic development has been slow to generate sufficient employment opportunities to absorb surplus farm labour. This has led to the emergence of the phenomenon of continuously mounting pressure of disguised unemployment. The increasing pressure of population leads to subdivision and fragmentation of holdings with all their attendant evils. The farm income goes down with every generation. Low consumption of the unemployed presses down level of economic activity. Surplus available for investment is meagre. Thus mass disguised unemployment is a drag on economic progress. The problem can be solved only by stepping up the rate of investment above the rate of population growth. The central problem of economic growth in such countries is thus to create conditions under which rate of investment can be increased above that of growth of population.

Solution. The basic solution to the problem of disguised unemployment is faster rate of capital formation and economic growth that creates a large number of employment opportunities. For this purpose rate of saving must be raised for increasing the rate of investment. In the developing countries investment incentives for the private sector are very low and the government can assist in the process of capital formation directly as well as indirectly. Through a fiscal policy which encourages savings and investment and an easy monetary policy it can do much to encourage investors. The government itself can participate in the process of capital formation by building infrastructure such as roads, power, ports, airports as undertaking of such development activities do not attract the private investors. Therefore, the Government has got to assume a special role in speeding up the rate of economic development and employment generation.

The other line of attack has got to be on the rate of population growth. If population grows at a rapid rate, then to maintain the people even at their existing levels, large amounts of capital are needed which could otherwise have been used to raise the amount of capital available per man and hence to raise the living standards at a faster rate. Apart from accumulation of capital (*i.e.*, machines, factory buildings, tools and equipment), we should increase investment in agriculture and expand irrigation facilities to generate opportunities for productive employment. Besides, we should give incentives to *small and medium enterprises* which being labour-intensive will generate a lot of employment. Further, by adopting suitable fiscal and monetary policies we should discourage the use of capital-intensive techniques. Lastly, land reforms initiated in the earlier periods of planning should be effectively implemented to ensure access to land for agricultural households. We explain below in general what measures should be taken as a part of employment strategy to create employment opportunities.

EMPLOYMENT STRATEGY TO REDUCE UNEMPLOYMENT

The following measures should be adopted if unemployment problem is to be solved

Use of Labour-Intensive Technology. Both the organised and unorganised sectors must adopt *labour-intensive technology* if sufficient employment opportunities are to be generated in both the rural and urban sectors of the economy. The decline in employment elasticity of output growth[10] in recent years in India is primarily due to the *increasing trend in capital intensity* in the organised industrial sector as well as in agriculture. Increasing mechanisation of agriculture in various states has lowered the employment elasticity of growth of agricultural output. Therefore, for raising labour intensity, suitable monetary and fiscal measures need to be adopted to discourage the use of capital-intensive techniques. Of course, the use of labour-intensive techniques with lower productivity of workers in the industry and agriculture may lower the growth of output. Thus, there might be some *trade off between employment and growth of output*. In our view due to the seriousness of unemployment problem some output growth should be sacrificed for the sake of more employment.

Accelerating Investment in Agriculture. Second, an important reason for slow growth of employment in agriculture and rural sector has also been a *shortfall in investment or capital formation in agriculture*. Both the public and private sector investment in agriculture has declined since the early nineties. Of special importance from the viewpoint of employment generation is investment in irrigation, rural roads, flood control projects, power generation and other infrastructure. It is worth noting that investment not only generates employment directly but also has a multipier effect which operates through backward and forward linkages. Therefore, stepping up of investment in agriculture and rural infrastructures will greatly help in employment generation. The announcement by the government *to furnish more credit to farmers at lower than market rates of interest* from commercial banks will also ensure that the small and medium farmers will be able to buy fertilizers,

10. Note that employment elasticity with respect to output growth means percentage growth in employment resulting from one per cent growth in output.

other high-yielding inputs, and arrange for their irrigation. This will raise their productivity and tend to reduce underemployment and disguised unemployment. Despite more than six decades of planned development in India, agriculture continues to be principal source of employment in the Indian economy. Though the share of GDP from agriculture has come down to around 14 per cent, still about 50 per cent of labour force continues to be employed in agriculture. In fact, agriculture continues to be the parking lot of the unemployed in the country. But a good number of persons engaged in agriculture and related activities are not productively employed. In fact there is widespread underemployment and disguised unemployed. Economic reforms initiated in 1991 has by and large neglected agriculture which even now does not get enough credit from commercial banks.

No wonder that there has been a fall in investment or capital formation in agriculture both by the private and public sectors. This is an important reason why employment opportunities in agriculture have not risen much causing increase in rural unemployment. What is needed is the substantial investment in agriculture so as to generate enough employment opportunities in this sector.

Diversification of Agriculture. Besides there is urgent need for diversification of agriculture. For example, there is an urgent need for a relative shift from growing of crops to horticulture, vegetable production, floriculture, animal husbandry, fisheries etc. which are more labour absorbing and higher income-yielding. In addition to this, promotion of agro-processing industries for export purposes has a large employment potential.

For the rapid growth of employment opportunities in agriculture and related sectors and agro-processing industries, investment in infrastructure including irrigation, rural reads, power supply, and development of agricultural markets are required.

Labour-Intensive Industrial Growth. For solving unemployment problem in the urban areas, the organised industrial sector must also absorb a sufficient number of workers. The failure of the organised industrial sector to generate enough employment opportunities is due to the use of capital-intensive technologies imported from abroad. The tendency to use the higher capital-intensive technology by the Indian industries in the post reform period has been intensified because they are trying *to improve competitiveness to face competition from imported commodities*. The other factor responsible for the use of higher capital intensive technologies is the **factor-price distortions** such as cheap capital and relatively higher wages of workers who have strong labour unions. Capital has become relatively cheap due to various fiscal concessions such as liberal depreciation and investment allowance and low interest rates on borrowed funds for investment. Consequent to the use of high capital-intensive technologies in our industries the labour productivity has been increasing while employment has been growing at a snail pace. Therefore, to encourage the use of relatively more labour-intensive technologies and thereby to generate more employment opportunities, fiscal and monetary concessions mentioned above on use of capital must be withdrawn. In this case there is bound to be some loss of workers' productivity. Therefore, this will involve some trade-off between employment and output. In our view, because unemployment problem is very serious and greatly hurts people's welfare, some growth of output is worth sacrificing for more employment opportunities.

Services and Employment Growth. The growth of services in India has a large employment potential. In 1993-94 to 1999-2000, next to the construction sector the employment elasticity of output growth in services was higher and was estimated at about 0.50. Of special mention are software services and BPO (Business Processing Outsourcing) which have a lot of employment potential, especially for the educated youth of the country. The other services such as finance, insurance, trade, hotels, and restaurants have also a lot of employment potential. But an important thing to note is that growth of these services is dependent on industrial and agricultural growth in the economy.

Education, Health and Employment Generation. Lastly, the expansion of education and health care not only promotes accumulation of human capital and thereby contributes to growth of output, it will also generate a good deal of employment opportunities. More schools, hospitals, health care clinics in the rural and urban areas will not only create employment during their construction but also, and more important, when they start working to provide education and health services. Their working provides employment to both the educated and unskilled persons. Therefore, a higher allocation of resources is required to be made for them in government budgets and in our future plans.

PART IX

Monetary Demand and Supply

- Money: Nature, Functions and Role
- Credit and Commercial Banking
- Central Banking
- Objectives and Instruments of Monetary Policy
- Role of Monetary Policy in Economic Growth
- Supply of Money and its Determinants
- Demand for Money and Keynes's Liquidity Preference Theory of Interest
- Post-Keynesian Theories of Demand for Money

CHAPTER 13
Money: Nature, Functions and Role

DIFFICULTIES OF BARTER SYSTEM AND INVENTION OF MONEY

Money was not used in the early history of man. Exchanges were few since each family was self-sufficient. Whatever exchanges there were, they took the form of barter, that is, exchange of goods for the other goods. Various difficulties were faced by the people in the barter economy. There was no acceptable means of payment for the direct purchase of goods and services in the barter economy. In other words, in a purely barter system, there was no generally acceptable medium of exchange in the form of a particular good or asset which could be used to buy goods and services and do other types of transactions. The following are the main difficulties which were found in the barter system.

1. **Double Coincidence of Wants.** Owing to lack of generally acceptable medium of exchange, a difficult problem of double coincidence of wants was faced by the persons who wanted to sell and buy goods. For exchange of goods persons desiring to exchange goods must specifically want those goods what others offer in exchange. Thus, an individual who wants to have a good he must locate another person who offers to give up the good wanted by him and who is willing to accept in exchange the good offered by him. Thus, under barter system only when wants for buying and selling goods of different persons coincided the exchange of goods was possible. A good deal of time was spent by a person in searching for a man with whom wants coincided. Halm rightly says, "It is next to impossible that all wishes of bartering individuals should coincide as to the kind, quality and quantity and value of things which are mutually desired, especially in a modern economy in which on a single day millions of persons may exchange millions of goods and services."[1]

2. **Lack of a Standard Unit of Account.** A barter economy lacked not only a common medium of exchange but also a standard unit of account in which prices could be measured and quoted. In the absence of a common unit of account, the number of exchange ratios (that is, prices of goods expressed in terms of each other) between goods would be very large. For example, two cows for one horse, one cow for two quintals of wheat, one pen for three pencils and so on. Thus, lack of a standard unit of account with which to measure values of different goods and services made exchange or trade difficult.

3. **Impossibility of Subdivision of Goods.** Another problem faced under the barter system for exchange of goods was impossibility of subdivision of goods without loss of their value. For instance, if a person has a cow and wants to have 5 kg of wheat, obviously, it is too costly to give one cow for 5 kg of wheat he requires. Then, to do this transaction cow has to be divided. But cow cannot be divided or cut into pieces because cow will lose much of its value if it is divided. Thus, impossibility of division of goods for the purpose of exchange posed a great difficulty and obstructed the growth of trade.

4. **Lack of Information.** Another problem found in the barter system was that in it traders required a good deal of information for exchange of goods. For example, if Amit wants to have a saw in exchange of a wooden table which he has made. Not only should Amit be able to assess the value of saw but the maker of a saw should also be able to determine the value of the wooden table which Amit wishes to exchange. All this required a lot of information about goods for which people

1. G.N. Halm, *Monetary Theory*, 1946, p. 1.

must spend a good deal of time and resources to obtain such information. If there exists a medium of exchange, it will solve half the problem. However, Amit will still have to determine the value of the table in terms of the medium of exchange. Thus, if there exists a medium of exchange, with well-known characteristics, it will reduce the information costs of trading. Without the medium of exchange information cost will indeed be very large.

5. **Production of Large and Very Costly Goods not Feasible.** Another problem of barter economy relates to the production of large, costly goods. Suppose an individual who has technical skill and equipment to manufacture a car will not have much incentive to manufacture it in the barter economy. This is because he can exchange a car with a person who has enough goods having a value equal to a car so that their exchange with a car can take place. The car maker must obtain food, clothing and several other commodities of day-to-day consumption in exchange for a car. It will be very difficult, almost impossible, to find a prospective buyer who has enough of these goods and services to give in return for a car.

It is evident from above that barter system could work in a primitive economy where life was simple and man was self-sufficient. As man made some economic progress, division of labour or specialisation and large-scale production came to exist, barter system could not fulfil the increasing needs for exchange of goods. Due to the difficulties of exchange barter economy would have no large-scale production, no advantage of the use of capital-intensive specialised machinery and no easy and cheap means in which wealth could be stored. *The range of goods produced must be much smaller than those produced in the modern developed economies. To meet the needs for a common unit of account and also as a generally accepted medium of exchange and thereby to overcome the difficulties faced under the barter system, money was invented.*

EVOLUTION OF MONEY

Commodity Money. Today when in India we think of money, it is generally in terms of rupee notes, in the USA it is in terms of dollar notes and in Great Britain it is in terms of Pound Sterling, and all of these are mostly made of paper. However, in the beginning it was commodities which was selected as a medium of exchange and thus came to be used as money. Bows, sea shells, beads, arrows, furs and skin etc. were adopted as money at different times in the early stages of development, especially in the hunting stage of human development. With further development and in the pastoral stage, animals such as sheep and cattle (goats and cows) were started being used as money, that is, as a medium of exchange of goods. The use of cattle and other animals as money suffered from certain disadvantages. Since all cows and goats were not identical, they could not serve as a standard unit of measurement. Secondly, the supply of animals such as cows, sheep and goats was subject to large and abrupt fluctuations. Thirdly, ordinary commodities and animals cannot serve as a satisfactory store of value.

Metallic Money. In view of the above limitations of ordinary commodities and animals for being used as money and with further progress of human civilisation, ordinary commodities and animals were replaced by precious metals such as gold and silver for being used as money. Crowther thus sums up the superiority of precious metals like gold and silver for monetary use: "They are easily handled and stored, they do not deteriorate, they have just the right degree of scarcity and they can be relied upon neither to increase nor to diminish in quality except gradually."[2] But with invention of coinage, it was coins made of gold and silver which began to be widely used as money rather than simple and plain bits of gold and silver whose values were difficult to be ascertained. It is important to note that precious metals were used as money not because they were valuable but because they were scarce. *For a thing to serve as money scarcity is more important than value*. These days, *it is the scarcity of paper money which is responsible for its efficiency as money*; its lack of value is no hindrance for it to serve as money.

2. G. Crowther, *An Outline of Money*.

Paper Money. As it came to be firmly realised that for a sound money, scarcity is more important than value, precious metals were replaced by *paper money*. In the beginning paper money, that is, paper notes were simple claims to and substitutes for metallic money. But in the course of time paper money came to be regarded as money itself. Paper money took the form of bank notes which were not mere substitutes but were considered as an addition to the supply of money. At first, notes could be issued by all commercial banks but with the passage of time when paper money became inconvertible into metallic money issuing of notes became the monopoly of the Central Bank of a country. Central Bank of India is named as Reserve Bank of India. We will explain later in this chapter how paper money system has been changed from being convertible into gold to become inconvertible.

Emergence of Bank Deposits as Money. Finally, there was further development in the form of money. In the developed countries the main type of money is not paper notes issued by the central bank but the *bank deposits* (especially *demand deposits*) which people hold with the commercial banks and against which cheques can be drawn. In India too *bank money* (*i.e.*, bank deposits) or which is also called credit money has become a significant part of the total money supply. It is worth mentioning that chequeable bank deposits held by the public in the banks serve as money because through drawing cheques on them we can use them for making payments for purchase of goods and services and assets.

It is worthwhile to note that it is not only the currency deposited by the people that constitutes a part of money supply in the economy, but, as will be explained later, *banking system as a whole itself creates deposits depending on the reserve requirements and they also form a part of money supply in the economy*.

It is also worth noting that *deposits in the banks are money* and *not the cheques drawn on them*. Through a cheque a person instructs the bank to transfer some specified amount of money from his deposit account to another person or institution On receipt of the cheque bank debits the specified amount from the account of the drawer and credits that amount in the deposit account of the drawee. It is therefore the person's deposits that are used as a medium of exchange and therefore serve as money.

Similarly, credit card is not money. Presenting your credit card to pay for something, the credit card company gives you loan at the moment you buy a thing. Later the credit card company sends you the bill and you have to pay the credit card company either paper currency or drawing a cheque on your bank and also generally pays interest on the loan taken for a certain period. Unless you have currency or bank deposits in the bank you cannot make the necessary payments to the credit card company. Thus, it is the currency or bank deposits that serve as money and not the credit card. In fact, credit card is just like an identity card for providing you loan by the credit card company.

An essential prerequisite of money which needs to be emphasised is that it should be generally acceptable in a society as a medium of exchange. You accept paper notes in payment for goods or services you sell because you are confident that others will accept them from whom you wish to buy goods or services. If people lose confidence in any money, they will not accept it in payment for goods and services. Thus, when people lose confidence in any money as, for instance, when its value is fast depreciating, the money ceases to be generally acceptable. Indeed, in that case it ceases to be money. In the early twenties this happened to 'Mark' in Germany.

DEFINITION OF MONEY

To give a precise definition of money is a difficult task. Various authors have given different definitions of money. Professor D.H. Robertson defines money as "*any thing which is widely accepted in payment for goods or in discharge of other kinds of business obligations*". According to Crowther, "*Money can be defined as anything that is generally acceptable as a means of exchange and that at the same time acts as a measure and a store of value.*"

Two important things about money emerge from the above two definitions of money. First, money has been defined in terms of the functions it performs. That is why some economists have said that money is what money does. That is, money is anything which performs the functions of money. Secondly, *an essential requirement of any kind of money is that it must be generally acceptable to every member of the society*. Money has a value for 'A' only when he thinks that 'B' will accept it in exchange for the goods. And money is useful for 'B' only when he is confident that 'C' will accept it in settlement of debts. But the general acceptability is not the physical quality possessed by a good. General acceptability is a social phenomenon and is conferred upon a good when the society by law or convention adopts it as a medium of exchange.

FUNCTIONS OF MONEY

There are four main functions of money which are summed up in the following couplet:

"Money is a matter of functions four,
A medium, a measure, a standard, a store"

We explain below these four functions of money in some detail.

1. **Medium of Exchange.** The most important function of money is that it serves as a medium of exchange. In the barter economy a great difficulty was experienced in the exchange of goods as the exchange in the barter system required double coincidence of wants. Money has removed this difficulty. Now a person A can sell his goods to B for money and then he can use that money to buy the goods he wants from others who have these goods. As long as money is generally acceptable, there will be no difficulty in the process of exchange. By serving as a very convenient medium of exchange money has made possible the complex division of labour or specialisation in the modern economic organisation.

2. **Measure of Value.** Another important function of money is that it serves as a common measure of value or a unit of account. Under barter economy there was no common measure of value in which the values of different goods could be measured and compared with each other. Money has also solved this difficulty. Money serves as a yardstick for measuring the value of goods and services. As the value of all goods and services is measured in a standard unit of money, their relative values can be easily compared.

3. **Standard for Deferred Payments.** Another function of money it that it serves as a standard for deferred payments. Deferred payments mean those payments which are to be made in the future. If a loan is taken today, it would be paid back after a period of time. The amount of loan is measured in terms of money and it is paid back in money. A large number of credit transactions involving huge future payments are made daily. Money performs this function as standard for deferred payments because its value remains more or less stable.

If the prices are falling, *i.e.*, the value of money is rising, the creditors will gain in real terms and the debtors will lose. Conversely, if the prices are rising (or value of money is falling) creditors will be the losers. Thus if the money is to serve as a fair and correct standard for deferred payments, its value must remain stable. In case the value of money is changing very much, the creditors or debtors will be put to much loss and sufferings. Thus when there is severe inflation or deflation, money ceases to serve as a standard for deferred payments.

4. **Store of Value.** Lastly, money acts as store of value. Money being the most liquid of all assets is a convenient form in which to store wealth, that is, *money can be held as an asset*. Thus store of value function is also called *asset function of money*. It is, therefore, essential that the good chosen as money should be such as can be easily stored without deterioration or wastage. That is why gold was popular in the past as money material. Gold could be kept safely without deterioration. But in the modern times even paper money can be kept as deposits in banks to serve as asset. Of course, there are other assets like houses, factories, bonds, shares etc., in which wealth can be stored. But money performs the store of value function with a difference. Money being the most liquid of all

assets has the advantage that an individual or a firm can buy with it anything at any time. But this is not the case with other assets. Other assets like houses and shares have to be sold first and converted into money and only then they can be used to buy other things. Money would perform the store of value function properly if it remains stable in value.

From the above it is clear that money has removed the difficulties of barter system. It has facilitated trade and has made possible the complex division of labour and specialisation of the modern economic system. Without the use of money, modern economy will not run smoothly and the production will get a serious setback. For these reasons, monetary economy is preferred to barter economy.

FORMS OF MONEY

Money of Account. Money of account is the monetary unit in terms of which the accounts of a country are kept and transactions settled, *i.e.*, in which general purchasing power, debts and prices are expressed. The rupee is, for instance, our money of account, sterling is the money of account of Great Britain and mark that of Germany. Money of account need not however be actually circulating in the country. During 1922-24 the mark in Germany depreciated to such an extent that it ceased to be money of account. The paisa in India has not been in circulation and yet it is money of account. In the words of Keynes, "Money of account is the description of little title money is the thing which answers to the description."

Limited and Unlimited Legal Tender: Coins may be limited legal tender or unlimited legal tender. *A legal tender currency is one in terms of which debts can be legally paid.* It is an offence to refuse to accept payment in legal tender money. *A currency is unlimited legal tender when debts up to any amount can be paid through it.* It is limited legal tender when payments only up to a given limit can be made by means of it. For instance, rupee coins and rupee notes are unlimited legal tender in India. And so is the half-rupee coin. But coins of lower denominations are only limited legal tender. They are legal tender only up to ten rupees.

When a coin is worn out and becomes light beyond a certain limit, then it ceases to be a legal tender. When one rupee and half-rupee coins are more than 20% below the standard weight they are no longer legal tender. A Government may take away the legal tender quality of a currency. For instance, the old rupee coins of one hundred and eighty grains 11/12 fine are no longer legal tender in India. In 1978 the Government declared currency notes of rupees one thousand denomination and above as no longer legal tender. It is usually done to bring such currency out of the hoards. However, recently notes of ₹ 1000 have again been issued by Reserve Bank of India and have unlimited legal tender status.

Standard Money. Standard money is that in which the value of goods as well as all other forms of money are measured. Thus, in India all prices of goods are measured in terms of rupees. Moreover, the other forms of money such as two-rupee notes, ten-rupee notes, hundred-rupee notes and one -half rupee coins are expressed in terms of rupees. Thus rupee is the standard money of India. Standard money is always made the unlimited legal tender money. In old days the standard money was a *full-bodied money, i.e.*, its face value was equal to the real or intrinsic worth of the metal it contained. But now-a-days in almost all countries of the world, even the standard money is only a *token money, i.e.*, the material contained in it is very much less than the face value written on it. Thus, the rupee in India is available in the form of paper notes which have no intrinsic worth. Even the rupee coin has a metallic value much less than its face value. Rupee coin has been called a '*note printed on nickel*' (Rupee coin these days is made of nickel).

Token Money. As explained above, the token money is that the metallic value of which is much less than the real or intrinsic worth of the metal it contains. Rupee and all other coins in India are all token money.

Bank Money. Demand deposits of banks are usually called bank money. Bank deposits are created when somebody deposits money with them. Banks also create deposits when they advance loans to the businessmen and traders. Today these demand deposits are the important constituent of the money supply in the country.

It is important to note that bank deposits are generally divided in two categories : *demand deposits* and *time deposits*. Demand deposits are those deposits which are payable on demand through cheques and without serving any prior notice to the banks. On the other hand, time deposits are those deposits which have a fixed term of maturity and are not withdrawable on demand and also cheques cannot be drawn on them. Clearly, it is only the demand deposits which serve as a medium of exchange, for they can be transferred from one person to another through drawing a cheque on them as and when desired by them. However, since even time or fixed deposits can be withdrawn by foregoing some interest and can be used for making payments, they are included in the concept of broad money, generally called *M*3. It may be noted that latest addition to the forms of money are *credit cards* issued by the banks which are these days extensively used for making purchases.

MODERN PAPER CURRENCY SYSTEM OR MANAGED CURRENCY STANDARD

Paper money has now come to occupy a very important place in the modern monetary system of almost all the countries of the world. The term paper money only applies to Government notes and the notes issued by the Central Bank of the country. In early times when notes were introduced they were backed by an exactly equal amount in gold or silver kept in reserve by the issuing authority. Such notes could be exchanged for gold or silver coins when demanded and did nothing more than representing metallic coins. Such paper money or notes are therefore called *representative paper money*. Paper money now-a-days is not backed by gold or silver. Representative paper money is now not found anywhere in the world.

For a long time, paper money remained a convertible paper money. Under this, money is convertible into standard coins made of gold or silver. Under it the paper currency issued by the Central Bank was fully backed by the reserves of gold and silver of equal value kept by it. Therefore, this paper currency system was called "***Full Reserve System***". But with the passage of time it was thought that a cent per cent reserve against paper currency issued was not needed and instead only proportion of 30 to 50 per cent was enough to convert the notes presented for conversion into gold. Therefore, ***proportional reserve system*** was adopted. According to this, the issuing authority was called upon to keep a 30 to 50% of the total amount of notes issued as gold reserves. A percentage of 30 to 50 was considered enough to honour the notes when they were presented for exchange into gold. It was based upon the fact that people found notes very convenient and they seldom thought of presenting them to the issuing authority. Therefore, full backing of gold was not required. In India, this proportional reserve system was adopted in 1927 and continued till 1957.

Inconvertible Paper Money. Thus, now-a-days paper money is of *inconvertible type*. Under the incovertible paper money system, money is not convertible into gold or other precious metals. Thus, when paper money is inconvertible, the issuing authority is not responsible to convert the paper notes into gold or gold coins. The currency notes that are issued by Reserve Bank of India are **'fiat' paper money**, that is, they are issued by the fiat (*i.e.*, order) of the Government. As they are legal tender, they are generally acceptable in exchange for goods and services and for payment of debt. It may be noted that '***promises to pay***' written on the currency notes are not 'promises to pay' something else. For these notes, only other paper notes can be given whose value would be equal to the face value of the note you present for payment

Minimum Reserve System. But as time passed even proportional reserve system was thought to be inadequate for the monetary needs of a growing economy. Therefore, it was abandoned in India in 1957 and was replaced by the minimum reserve system. According to this, Reserve Bank was required to keep only a minimum amount of gold and other approved securities (such as dollar and Pound Sterling) of the value of ₹ 200 crores. Out of these reserves it was required that gold must not be less than the value of ₹ 115 crores. On the basis of these minimum reserves, Reserve Bank of India could issue *any number of notes or currency subject to the economic conditions of the country.* Thus at present, the government or Central Bank of the country can create new money, that is, print new notes without being backed by any new addition to its stock of gold.

However, one important disadvantage of incovertible paper currency is that government is tempted to over-issue the currency to meet its ever increasing expenditure. This causes serious inflationary pressures in the economy. Desired benefits from incovertible paper currency can be obtained if the government properly manages it and new money is printed if the economic conditions require it. Therefore, present inconvertible paper money has to be *managed properly* if its benefits are to be obtained. Therefore, the present inconvertible paper currency system is called *managed currency system*. If the adoption of paper currency system causes instability in the economy, the Government is responsible for it and not the paper currency system. The paper currency system is very elastic and its main advantage is that the quantity of money can be increased or decreased according to the requirements of the economy. But if improper use of its elastic nature is made by the Government by creating excessive quantity of money, the Government should be blamed and not the standard. Furthermore, as explained above, in the modern economy paper money has been replaced by *bank deposits* which can be transferred merely by issuing cheques without paper money being used in the process. However, the creation of bank deposits is also managed by the Central Bank of the country. Besides, the use of *credit cards* and *debit cards* issued by the bank has further restricted the use of paper money.

Managed Currency Standard: Its Advantages

Paper currency standard or managed currency system which prevails in the modern economy has several advantages and disadvantages. Its main advantage is that under it the quantity of money can be easily increased according to economic needs of the country. When any country is under gold standard, it cannot increase the quantity of money unless the quantity of gold with it increases, despite the fact that the country needs expansion in money supply badly for purpose of promoting economic growth. Therefore, paper or fiat currency standard, which delinks the creation of money from gold, facilitates expansion in the quantity of money for promoting economic growth. The process of economic growth brings about an increase in output of goods and services in agriculture, industry, transport and communication. As a result, there is need for financing these activities so that transactions involving a larger amount of goods and services take place easily. But this requires the availability of the sufficient amount of money. If the required amount of money is not forthcoming, as is the case under gold and silver standards, then the process of economic growth will be retarded.

In times of depression there is need to increase the aggregate demand in the economy so that the state of full employment is reached. The best way to increase aggregate demand is the creation of new money by the Government and spending it on various public works. Following the increase in aggregate demand, production will expand and employment for workers will increase causing their incomes to rise. But, as we have seen above, only under paper currency standard, the quantity of money can be increased easily, whereas under gold standard, without the increase in the quantity of gold, money supply cannot be increased. In this way we see that paper currency standard is helpful to overcome depression and to achieve the objective of full employment.

On the other hand, when inflation occurs in the economy the Government can make a surplus budget and in this way it can withdraw the extra money from the public. This will lower the aggregate demand which will help in controlling inflation. With the appropriate monetary and fiscal policies price stability can be achieved. But these appropriate monetary and fiscal policies are possible only under paper currency standard. J.M. Keynes also supported the managed currency standard on the ground that, as compared to gold standard or any other monetary standard, this is very elastic and under it the quantity of money can be varied according to the needs of the economy.

Unlike the gold and silver standard, paper currency is not a fair weather friend. This standard is very helpful to the Government in situations of crisis such as war, drought, etc. when it needs extra quantity of money. By creating new paper money the Government can conduct the war successfully. With the newly created money the Government can buy real resources and goods when sufficient quantity of money cannot be procured through taxes. Under gold and silver standards, money cannot be increased unless the quantity of gold or silver increases. This is true that under paper currency standard Governments are tempted to issue excessive notes which results in inflation. But if proper controls are applied and production of goods expands, the issue of more money within reasonable limits to match the extra output does not result in inflation.

Besides, paper currency standard is very economical. As compared to the prices of metals, the price of paper is very low. Even those countries can adopt this standard which have no or few resources of gold and silver. Besides this, with the adoption of paper currency standard the gold and silver are saved which can be used for industrial and productive purposes. Lastly, under the paper currency standard, rate of foreign exchange automatically changes according to the conditions of demand and supply and therefore disequilibrium in the balance of trade is automatically removed. According to a foreign exchange specialist, Agnas, "*the beauty of paper system is that as soon as any disequilibrium between demand and supply occurs, sharp fluctuation in the price of foreign exchange immediately takes place which by its immediate action on both exports and imports rapidly restores equilibrium.*"

Disadvantages of Paper or Managed Currency Standard

But paper currency standard has several disadvantages too. First, it creates foreign exchange instability. Due to excessive changes in the foreign exchange rate, the prices of exports and imports of a country go on changing rapidly which affects foreign exchange earnings of the country. As a result of this, the country has to continually face foreign exchange instability and this instability is very dangerous for those countries which predominantly depend on foreign trade. These days when most of the countries of the world have floated their currencies (*i.e.*, foreign exchange rate is left to be determined by demand and supply), foreign exchange instability has assumed serious proportions. Therefore, many suggestions have been put forward to reform the international monetary system so that the problem of foreign exchange instability is avoided. Secondly, it is also asserted that even internal economic stability cannot be achieved with paper currency standard. The over-issue of paper currency can lead to inflation in the economy. This has been actually so in India where too much new paper money has been created to finance ever increasing Government expenditure. This has caused serious inflationary pressures in the Indian economy. Thus, the paper currency standard can give rise to both internal and foreign exchange instability. However, for this paper currency system should not be blamed. Desired benefits from paper currency system can be obtained if the Government properly manages it. If the adoption of paper currency system causes instability in the economy, the Government is responsible for it and not the paper currency system. The paper currency system is very elastic and its main advantage is that the quantity of money can be increased or decreased according to the requirements of the economy. But if improper use of its elastic nature is made by the Government by creating excessive quantity of money, the Government should be blamed and not the standard.

IMPORTANCE OF MONEY

Classical economists thought that money is a just veil and does not affect the real volume of goods and services in the economy, that is, the level of national product. According to them, money only affects the price level in the economy with no effect on the levels of real income and employment. They contended that what we really want was goods and services which satisfy the wants of the people and promote their welfare. Therefore, according to them, we must go behind the veil of money to see changes in the total output of goods and services or in real income. It is true that what we really want is not money but the goods and services which money can buy. It is the goods and services which satisfy the wants of the people and not money as such. But it must not be concluded from this that money is of no real importance. Modern economic life based on the complex division of labour or specialisation which has added so greatly to the productivity of the economy will not be possible without money. The following are some of the advantages of money which show that money plays a vital and significant role in the modern economy.

1. **Money has Facilitated Exchange and Promoted Trade.** In the first place by serving as a medium of exchange and a common measure of value, money has removed the difficulties of the barter system and promoted trade in the economy. The difficulties of the barter system, namely, lack of double coincidence of wants, lack of division and lack of common measure of value are well known. By removing these difficulties, money has greatly facilitated the process of exchange. In the absence of money, trade and exchanges must have been few and far between and entailed a great waste of time and energy.

2. **Money Promotes Division of Labour and Productivity.** Money is of great importance as it promotes division of labour and productivity in the modern economies. Since under the barter system, exchange was difficult, a man had to be self-sufficient, that is, produced most of the goods for himself. In the absence of money there were great difficulties in exchanging goods and services. This worked as an obstacle to the division of labour and specialisation among various individuals and nations. Long ago Adam Smith clearly brought out in his now well-known book, "*Wealth of Nations*", how the division of labour and specialisation enhance productivity and efficiency of labour force. It is this division of labour and specialisation that has made the use of more efficient machines and advanced technology possible for production of goods. Indeed, it is because of the output-augmenting effect of division of labour that *Adam Smith regarded increase in the division of labour as progress in technology*. By opening up the opportunities of effecting division of labour, and through facilitating exchange and trade of goods and services money contributes to the expansion of production. Therefore, money enables the increase in the amount of production and variety of goods and services produced. "*If a modern economy were somehow deprived of the monetary mechanism and were driven back to a system of barter, the level of output would be much lower and the variety of goods and services much smaller than is enjoyed with a money system.*"[3]

3. **Money Promotes Saving.** A great significance of money is that it contributes a great deal to the increase in saving of the economy. Money has made saving easier than in the barter system. Increase in saving leads to the increase in investment which determines economic growth of a country. Further, money has made easier the acts of borrowing and lending and has given birth to various financial institutions which promote saving. Investment which is made possible by saving raises the rate of capital formation in the economy. The higher level of output in the modern economy is mainly due to the extensive use of capital in the production process.

4. **Money helps in Maximising Satisfaction or Profits by Consumers and Producers.** Fifthly, money is of immense advantage both to the consumers and producers. To the consumers money represents a general purchasing power. He can buy anything with money he has and at any time convenient to him. Since the values of goods are expressed in terms of common measure *i.e.*, money, the consumer can easily compare the relative money prices of the goods and expected utilities from them. A consumer can easily spend his given money income on various goods in such a way that marginal utilities goods are proportional to their prices. With this he will be maximising his satisfaction from his given income. Thus the existence of money helps the consumer to maximise his satisfaction by acting on the principle of equip-marginal utility.

5. **Money helps the Producer too.** The producer can easily compare the money cost and money

3. A.A. Walters, *Money and Banking*, Penguin, Readings, pp. 8-9.

income of the different levels of output. He can thus easily decide about the level of output which maximises his profits by equating marginal cost with marginal revenue (*MC* = *MR*). He can also easily decide about the units of a factor to be employed by comparing its marginal revenue product with the money payment he has to make to it. To maximise profits he can easily equate marginal revenue product (*MRP*) of a factor with the marginal factor cost (*MFC*) on that factor. Moreover, the workers engaged by a producer want goods and services to satisfy their wants. They want food, clothing, shelter and so many other things. The producer cannot supply them all these things easily. But what be can do is to pay them in money with which they buy the commodities they like at their own convenience and when they need them.

6. **Money can Help in Reviving the Economy from Recession or Depression.** According to modern economists, money may play an important role in bringing about real changes. They point out that at times of recession or depression when there exists a lot of idle productive capacity and unemployment of labour, expansion of money supply, say through Government's deficit budget financed by the creation of new money, will cause aggregate expenditure or demand to shift upward. This increase in aggregate expenditure or demand will cause output and employment in the economy to rise and thus will help the economy to recover from recession or depression.

Beneficial effect on output and employment of expansion in money supply can be obtained even if Central Bank of a country takes steps to expand money supply in the economy say through reduction in Cash Reserve Ratio (*CRR*) and through buying securities in the open market. These steps by the central bank aim at expansion in credit availability for the businessmen. The greater credit availability will lead to more private investment which through multiplier process will cause income, output and employment to rise.

It follows from above that *money is not neutral in its effect on real income and employment*. In fact, as seen just above, *it plays an important role in the determination of the level of real economic activity*.

ROLE OF MONEY IN ECONOMIC DEVELOPMENT OF THE DEVELOPING COUNTRIES

Economic development is generally believed to be dependent on the growth of real factors such as capital accumulation, technological progress, and increase in quality and skills of labour force. This view does not adequately stress the role of money in the process of economic development. It is said that money is a mere veil and instrinsically unimportant. What matters is the real goods and productive factors which money buys. However, this extreme view about the unimportance of money as such is no longer believed. Not only is money an important factor without which modern complex economic organisation is impossible, but it is also an important factor for promoting economic development. We discuss below the importance of money in the process of economic development.

In the economy today money performs several functions. Money serves as a standard of value in which other values are measured. Money is a store of value, that is, the means in which wealth can be held. It acts as a standard for deferred payments. However, the most important function of money which distinguishes it from other goods is that it serves as a medium of exchange. That is, money is a means of payment for goods and services. It is this use of money that distinguishes a monetary economy from a barter economy. A monetary economy is one in which goods are sold for money and money is used to buy goods.

Money Promotes Productivity and Economic Growth. Barter system was full of difficulties of exchanging goods and services between individuals. In the absence of easy exchange of goods and services the barter system worked as an obstacle to the division of labour and specialisation among individuals which is an important factor for increasing productivity and economic growth. Further, the process of economic growth leads to the expansion of production of goods and services and consequential rise in incomes of the people. As a result, volume of transactions in the developing economy increases. This raises the demand for money to finance the increased transactions brought about by the expanded level of economic activity. Thus, the process of economic growth would be held in check if adequate supply of money is not forthcoming to meet the requirements of increase in the level of economic activity.

Money Promotes Investment. From the viewpoint of development another important role of money lies in making the magnitude of investment independent of the current level of savings. In a barter system, the goods not consumed constitute the savings as well as investment. That is, investment is not different from current savings. The greater the current savings, the greater the investment. However, in a modern economy, this is not so. Whereas it is households which save in the form of money, it is the firms which invest money in capital goods. Therefore, investment can differ from saving because investment activity is separated from the act of saving. More importantly, investment in a monetary economy can exceed the current level of savings. This excess of investment over savings is possible because new money can be created by the Government in the form of currency or by banks in the form of bank deposits. And this is what is important for the purpose of economic development.

In the developed countries in times of depression when idle productive capacity exists, the increase in investment made possible by creation of new money by the Government or banks would lead to the increase in aggregate demand for goods and services. In such times the supply of goods and services is elastic due to the existence of excess capacity. Therefore, increase in aggregate demand generated by the investment financed by created money brings about expansion in output of goods and services and thereby causes an increase in the level of employment.

In developing countries, the created money can play a useful role in promoting economic development. Rapid economic development can be achieved by stepping up the rate of investment or capital formation. But additional resources are required to increase the rate of investment. But in a country where a majority of the people are living at the bare subsistence level, voluntary savings, taxation, Government borrowing cannot by themselves provide sufficient investible resources for development. The government therefore attempts to increase the volume of investible resources beyond what is possible on the basis of current level of savings through creating new money. The newly created money can be spent on investment projects both in the industrial and agricultural fields which would lead to the increase in output, income and employment.

Money and Investment in Quick-Yielding Projects. It is widely believed that any increase in the supply of money in developing countries would lead to the rise in prices or to the emergence of inflationary pressures. However, this is not always true. A reasonable amount of newly created money helps the development of the economy by raising the level of investment. In the developing economies a lot of natural and human resources lie unutilised and underutilised which can be employed for productive purposes. If the newly created money is used for investment in those projects such as small irrigation works, land reclamation schemes, flood control and anti-soil erosion measures, cottage industries which yield quick returns, then the danger of inflation will not be there. These quick-yielding projects will increase the production of essential consumer goods in the short run and will therefore prevent the rise in prices. Further, if development strategy is such that a higher priority is assigned to agriculture and other wage goods industries and further that organisational and institutional reforms are undertaken to provide all farmers with irrigation facilities, fertilizers and high-yielding varieties, agricultural output can be raised in the short period. In this framework, new money can be created to increase the level of investment without much adverse effect on prices.

Monetisation and Economic Growth. Further, as is well known, most underdeveloped countries have a large non-monetised (*i.e.* barter) sector where production is for the purpose of subsistence only. To break the subsistence nature of economic activity and thus generate new forces for economic growth, its monetisation is required. The introduction of money helps in bringing it in contact with the modern sector. This contact of the subsistence sector with the modern sector will lead to the expansion of its output. In order to obtain the products of the modern industrial sector, the people engaged in the subsistence sector will make efforts to raise their output. Thus, a surplus of output over their self-consumption will be generated in this way which will ultimately break their subsistence nature. It is supported by the past history of the developing countries. During the colonial period, the monetisation of the peasant sector led to the expansion in exports in exchange for the imported industrial products. This stepped up their agricultural development to a good extent.[4]

Similar to the growth of production for exports the introduction of money in the subsistence agricultural sector and its contact with the modern sector would lead to the increase in marketable surplus of foodgrains and other agricultural products which is an important factor in economic development. If some rise in agricultural prices occurs as a result of increase in investment financed by the created money, as is likely the case, it would serve as an incentive to produce more foodgrains and supply it to the market. The rise in agricultural incomes will increase demand for industrial products and would therefore accelerate their growth.

Further, the monetisation of the subsistence sector will also help in raising the volume of savings. Monetisation will bring this sector in contact with the financial institutions such as commercial and cooperative banks and insurance companies. The opportunities of earning more income through interest on saving will raise the propensity to save of the people in the present-day subsistence sector. If proper monetary policies are pursued, then instead of consuming or hoarding all their incomes, these people can deposit a part of them in the financial intermediaries.

QUESTIONS FOR REVIEW

1. What is money? Show how it has removed the difficulties of barter system.
2. Define money. Describe its functions in a modern economy.
3. What is the necessary condition which makes something money? How is the store of value function of money derived from its function as a medium of exchange?
   [*D.U. B.Com (Hons). 2002*]
4. "Money is what money does". Discuss
5. Define money. Explain its important functions.
6. Explain the following forms of money:
   (*a*) Demand Deposits
   (*b*) Paper money
   (*c*) Legal Tender money
   (*d*) Convertible and inconvertible money.
   (*e*) Fiat money.
7. What is money? Would you include time deposits in money supply?
8. Distinguish between money and near money. Why are time deposits considered as near money whereas demand deposits are regarded as money?
9. Discuss the various functions of money. What would happen to an economy if money supply suddenly disappeared?
10. Explain the paper monetary system as it prevails today in the modern economies. For the issue of new money should the Government or Reserve Bank of India has to keep proportionate amount of gold as reserves? Explain the advantages and disadvantages of paper monetary system.
11. Explain the significance of money in the economy for promoting (a) division of labour, (b) saving, (c) income and employment.
12. Explain the importance of money for economic development of a country.
13. "Money which is a source of so many blessings to mankind becomes, unless we control it, a source of peril and confusion". Discuss.

---

4. See H. Myint, *The Economics of the Developing Countries*, Hutchinson University Library, London, 3rd edition, 1967, Ch. 3.

CHAPTER 14
CREDIT AND COMMERCIAL BANKING

CREDIT CREATION BY BANKS: MEANING

In the previous chapter we have explained that chequable bank deposits are treated as money as by drawing a cheque on his bank a person transfers some amount of funds to another person or institution. In developed countries most money is in the form of bank deposits. However, its may be noted that deposits are created not only by the public depositing their currency funds with the banks. The banks themselves create deposits when they give loans to the businessmen. *The creation of bank deposits themselves by the banks while giving loans to businessmen is called credit creation* by banks. The banks create credit or bank deposits much more than the currency funds deposited by the public with them. Banks also create deposits or money when they buy securities of the government or the private corporate sector.

The creation of bank deposits in excess of the currency reserves with them has become possible because of the **fractional reserve system.** The fraction of banks' total deposits that are held in currency reserves is called *reserve ratio.* Suppose 10 per cent is currency reserve ratio against its deposits that a bank maintains. Then, if a person deposits currency notes of ₹ 1000 with the bank, then the bank will keep 10% of ₹1000, that is, ₹ 100 as reserves and give loans of the remaining amount of ₹ 900 to business men. When the bank gives loans to the businessman of ₹ 900, it generally credit this amount in chequable deposits of his customer businessman. Before it is withdrawn the chequable deposits of the bank rises to ₹ 1900. The extra ₹ 900 of the deposits which the bank itself creates will also be treated as money. Thus bank has created extra money of ₹ 900 as if out of thin air. But the process of creation of credit or bank deposits does not stop here. As will be explained later in detail when ₹ 900 will be transferred by the bank's customer to another person by issuing a cheque to him, this may be deposited in another bank whose reserves will increase by ₹ 900. With 10 % as reserve ratio, the other bank will keep 10% of ₹ 900 (that is, ₹ 90) as reserves and the remaining ₹ 810 will be given as loan to his customer by crediting ₹ 800 to his chequable deposits. This process will go on and creation of bank deposits will go on increasing until they rise by $\frac{1}{r}$ where r is reserve ratio. $\frac{1}{r}$ represents deposit multiplier. If reserve ratio is 10 per cent, then the deposit multiplier will be equal to $\frac{1}{0.10} = 10$. Thus, with the original currency deposit of ₹ 1000 will cause total deposits to rise by 10 × 1000 = ₹10,000. Thus, the fractional reserve system enables the banking system as a whole to create much more demand deposits than the amount of the original deposit of currency.

Functions of Bank Credit

The main function of credit is to relieve the constraint imposed by balanced budgets on economic agents, that is, to meet the financial requirements of investors who have to spend more on trade and investment than their own savings. Accordingly, it is through the means of credit that the surplus funds with some individuals and institutions are made available to the deficit spenders, that is, those who are required to spend more than their resources, for instance, traders, companies, investors. It is the performance of this function, that is, the transferring surplus funds of some to meet the spending by the businessmen and investors, that the banking and other segments of the financial system are able to promote savings and investment, to ensure better allocation of financial resources and thereby to encourage economic growth in the economy.

But, for the credit system to perform these functions, it needs to be efficiently managed and controlled. If credit is not efficiently managed, it can cause inflation or deflation, recession and unemployment in the economy. Besides, the mismanagement of credit can lead to misallocation of investible resources and thereby hinder economic growth. It can also cause concentration of economic power in a few hands and exploitation of weaker sections and thus work against the achievement of social justice.

PURPOSE OR END-USES OF CREDIT

Credit is required for different purposes and by all sectors of the economy. Therefore, there is need for the proper allocation of credit between different uses and sectors if the society is to achieve its objectives. When credit is demanded and used for productive purposes, it may be used to finance the needs of working capital or for fixed investment (*i.e.* capital equipment, machinery etc.). The broad categories of economic activity for which credit for productive purposes is demanded are (*a*) agriculture, (*b*) industry, (*c*) construction, and (*d*) trade, both domestic and foreign. Further, in each of these categories, the allocation of credit between different users is of crucial importance from the viewpoint of economic growth and social justice. Thus, in agriculture allocation of credit between large landowners and small farmers has been a matter of serious debate. Likewise, the allocation of credit between the large-scale industries and small-scale industries has been a major concern in India's credit policy. Credit policy in India in recent years has emphasised certain priority sectors such as agriculture, small-scale industries, export and weaker sections of society such as small and marginal farmers, young entrepreneurs for whom greater amount of credit is to be made available.

ORIGIN AND EVOLUTION OF COMMERCIAL BANKING

A Commercial Bank is a business organisation which deals in money; it borrows and lends money. In this borrowing and lending of money, it makes profit. The distinction between money lender and a commercial bank may be noted. Whereas money lender only lends money to others and that too from his own sources; a commercial bank does both the lending and borrowing businesses. A commercial bank raises its resources through borrowing from the public in the form of deposits and lends them to the businessmen. Its lending rate of interest is greater than that it pays to its depositors. It is because of this difference in lending and borrowing rates of interest that it is able to make profits.

Now, a curious student will ask the question how the institution of banking originated and evolved. The banking owes its origin to the activities of goldsmiths in England. In the Seventeenth Century, goldsmiths in England possessed strongest safe vaults where valuable goods such as gold, silver, diamonds could be kept safely. Therefore, the rich people in England at that time used to keep their valuable things such as gold and silver with the goldsmiths and obtained receipts from them for their deposits of gold and silver. At a later stage it was found that these deposits with goldsmiths could serve as money, that is, a medium of exchange. The person who had a deposits of gold and silver with a goldsmith could pay off his debt simply by endorsing the deposit receipt in favour of the person to whom he has to pay back the loan. In this way gold and silver got automatically transferred to another person even though they remained in the safe custody of the goldsmiths. It was soon discovered that there was no need for endorsing the whole deposit receipt in favour of the other person. With a deposits of gold and silver a person could issue a letter to his goldsmith instructing him to transfer a certain amount of gold or silver to another person. Obviously from this practice emerged the present-day system of payment through cheques.

Another important development took place when goldsmiths realised that a good amount of gold deposited with them remained idle for long as only the ownership of them changed through transfer. From this they concluded that they could issue more deposit receipts than the actual deposits of gold and silver with them to meet the financial needs of some persons and started charging some interest on them. In this practice one can see the beginning of the creation of deposits or credit by the modern banks which will be described in detail later in this chapter.

BALANCE SHEET OF A BANK: LIABILITIES AND ASSETS STRUCTURE

Balance sheet of a bank is of great importance for understanding the sources of funds it possesses and the uses to which these funds are put. As is well known, a balance sheet of an institution indicates its liabilities and assets. The liabilities of a bank shows the sources of its funds and assets show its uses by it. The balance sheet of Bank of Baroda as on 31st March 1997 is given below :

(₹ *in crores*)

Liabilities		*Assets*	
Capital	254.3	Cash and Balances with RBI	3,048.2
Reserves and Surplus	1,733.4	Balances with other banks and money at call and short notice	4,683.4
Deposits	32,156.7	Investment in Govt. and other securities	10,927.2
Borrowings	339.9	Advances	16,531.6
Other Liabilities and Provisions	3,155.4	Fixed Assets	570.1
		Other Assets	1,872.2
Total	37,639.7	Total	37,639.7

Liabilities

It will be observed from the balance sheet of a bank given above that deposits constitute a very large proportion of the total funds available with a bank. It should be borne in mind that these deposits are bank's liabilities as they are depositors' claim against the bank. Deposits are mainly of two types: (1) demand deposits, (2) time deposits. Demand deposits are payable on demand and therefore can be withdrawn by the public through cheques. On the other hand, the time deposits are repayable by the bank only after a certain fixed period of time. In addition to these, in India there is another type of deposits called Savings Bank Deposits. Though money in such deposits can be withdrawn through cheques, there are limits to the amount withdrawable in a week or month.

Banks also borrow from the Central Bank of the country and these borrowings also constitute its liabilities and source of funds. In the balance sheet, these borrowings from the Reserve Bank of India (*i.e.,* Central Bank of India) are included in other items of the liabilities. When supply of money is very tight, the borrowings from Central Bank is of immense help to the banks.

Assets Structure : Liquidity Vs. Profitability

The asset side of the balance sheet of a bank shows for what purposes it has used the funds obtained from the depositors. As pointed out above, a viable bank has to operate so as to make reasonable profits. On the other hand, in order to meet the demands for withdrawals by the public and thus to retain faith and creditibity, it has to keep some ready cash with it, that is, it has to ensure some liquidity. Profitability and liquidity are the two major considerations that weigh with the commercial banks in deciding about the composition of its assets. If all of its deposits are kept by the bank in the form of cash, it will have perfect liquidity in this case but will make no profits at all. But if the bank advances all its deposits as long-term loans to the industries it will lose liquidity and will be unable to honour the demands of withdrawals by the depositors. Therefore, a bank has to keep such an asset structure (that is, a combination of different types of assets) that strikes a balance between liquidity and profitability.

A glance at the balance sheet of a commercial bank given above shows that *cash in hand* and with other banks, which is a liquid asset, constitutes about 8 per cent of the total assets of a bank. There is another quite a good liquid asset, namely, *money at call and short notice,* which is about 12 per cent of the total assets. Another important liquid asset is the *investment in government and other securities.* Investment in government securities and other securities by the bank are also liquid as they can be sold in a short time and cash realised therefrom. It may be noted that in India the banks are by law required to invest a given percentage of their deposits in government securities. But legal requirements apart, the investment in government securities by the banks ensures their liquidity position such as these can be readily converted into cash. It will be seen from the balance sheet of the bank that investment in Government and other securities by the bank constitutes about 29 per cent of its assets.

Loans and advances by the banks to industries and traders are the most profitable item of asset. It is against these profitable assets that liquid assets, mentioned above, have to be balanced. It will be noticed that loans and advances account for about 44 per cent in the asset structure of the banks.

It is worthwhile to note that asset structure of different banks will vary depending upon the composition of their deposits. A bank that has relatively more demand deposits, will have to keep a greater proportion of its assets in liquid form. On the other hand, if a bank possesses more time deposits, it will need to hold a relatively smaller proportion of its assets in liquid form.

FUNCTIONS OF COMMERCIAL BANKS

The main functions of the commercial banks can be summed up in one sentence : *the banks borrow to lend.* To receive deposits and to advance loans are thus the two main functions of all commercial banks. They borrow in the form of deposits and lend in various types of advances. Besides, there are other miscellaneous functions of commercial banks which have developed according to the needs of the society.

(a) Accepting Deposits

The banks borrow in the form of deposits. This function is important because banks mainly depend on the funds deposited with them by the public. The deposits received by the banks may be of the following types.

(*i*) **Demand Deposits or Current Account Deposits.** If a depositor deposits money in the bank in the current account (*i.e.,* demand deposits), he can withdraw it in part or in full at any time he likes without notice. These accounts are generally kept by businessmen whose requirements of making business payments are quite uncertain. Usually no interest is paid on them, because the bank cannot utilise these short-term deposits for lending purposes and must keep almost cent per cent reserve against them. But in return for these current account deposits, the banks offer some facilities or concessions to the account holders. The most important is the cheque facility made available to them, that is, the account holders make payment to the parties through cheques on these accounts. Further, on behalf of the holders of current account deposits, banks collect cheques, drafts, dividend warrants, postal orders, etc.

(*ii*) **Fixed Deposits or Time Deposits.** These deposits are made for a fixed period of time, which varies from fifteen days to a few years. These deposits cannot, therefore, be withdrawn before the expiry of that period. However, a loan can be taken from the bank against the security of these deposits within that period. A higher rate of interest is paid on the fixed deposits. As the fixed deposits carry a good rate of interest, they are a good source of investment by the people who are in a position to save. In India, the increase in the rates of interest on fixed deposits of various durations has attracted a good deal of savings in the banks. Since June, 1998, the interest on these deposits for 6 months and above but less than 1 year is 8 per cent, on deposits for one year and above is not regulated by RBI but varies from 9 per cent to 12 per cent depending on the period.

(*iii*) **Savings Bank Deposits.** In this case the depositor can generally withdraw money usually once a week. Sometimes there are also restrictions as to the total amount that can be withdrawn at one time and the total amount that can be placed in one deposit. These deposits are generally made by the people of small means, usually, people with fixed salaries, for holding their short-term savings. Like the current account deposits, the saving bank deposits are payable on demand and also they can be drawn upon through cheques. But in order to discourage people to use the savings bank deposits very frequently, there are some restrictions on the number of times withdrawals (through cheque or otherwise) that can be made from these accounts. The saving deposits carry lower rate of interest than the fixed deposits. Since March 1, 1996, the interest on savings bank deposits in India is 4.5 per cent per annum.

(*b*) Advancing Loans

Another function of the bank is to give loans to others. If the bank does not lend the deposited money to others, how can it pay the interest on the deposits to depositors? Banks give loans to businessmen and firms usually for short periods only. This is so because the bank must keep itself ready to meet the demands of the people who have deposited money for short period only. In advancing loans, the bank has to shoulder a heavy responsibility. The bank makes profit by advancing loans. But the bank deals in other people's money and it has to keep some ready cash to meet the depositors' demands. Hence a great care has to be exercised in the matter of lending and keeping resources. The bank must strike a fine balance between liquidity and profitability. If it keeps its assets in too liquid a form, it loses profit and if it tries to make too much profit, it may not be able to meet the depositor's demand. It must aim at both liquidity and profitability. Banks advance loans in the following ways.

1. By allowing an overdraft. Those people who keep current account with the banks are sometimes given the right to overdraw their accounts. In other words, people make arrangements with the banks that if a cheque has been drawn by them which is not covered by the deposit, then the bank should grant the overdraft and honour the cheque. Thus under overdraft arrangements, people can get more than they have deposited but they have to pay interest on the extra amount which has to be paid back within a short period. Overdraft facilities are generally granted to businessmen who can pay off the money after the sale of the goods.

2. Cash-Credit Loan. Under the cash-credit system, borrower is sanctioned a credit limit up to which he can borrow from the bank. But before granting a credit limit the bank satisfies itself about the credit-worthiness of the borrower. However, the actual utilisation of the credit limit by the borrower depends on his withdrawing power. Withdrawing power of a borrower is determined by his current assets which consists of stocks of goods, that is, stocks of raw materials, semi-finished goods and payments receivable from others. The borrower has to submit periodically a list of his assets showing his borrowing power. The interest payable by the borrower is calculated on the amount of the credit limit actually drawn.

3. Demand Loans. Demand loans granted by a bank are those loans which can *be recalled on demand by the bank any time.* The entire sum of demand loan granted to a borrower is paid in lump sum by crediting it to his account. Therefore, the interest is payable on the entire sum of demand loan granted. Usually demand loans are granted to stock-brokers whose need for credit fluctuates from day to day.

4. Short-Term Loan. Short-term loans are given as personal loans against security. Car loans, computer loans, housing loans are some examples of short-term loans. Short-term loans are also given to finance needs for working capital of businessmen. In case of short-term loans also the whole amount of loan granted is paid in lump-sum by crediting it to the loan account of the borrower. The interest is payable on the entire sum of loan granted.

(c) Discounting Bills of Exchange or Hundies

A very important function of a modern bank is to discount bills or hundies of businessmen. It is like this, a businessman buys goods and is granted credit, say, for a month. The seller of the goods draws a bill of exchange which the purchaser is asked to sign. The bill orders the purchaser to pay a certain sum after the expiry of one month. If the seller goes on selling goods on this basis, he will soon find that all his stock is gone and he has got only these hundies in his cash box. Unless these hundies are changed into cash, his business will come to a standstill. He, therefore, does not keep these hundies with him till they mature for payment. But he takes them to the bank and gets the present worth of the hundies, leaving the bank to realise them when the date of payment comes. This is called discounting a bill. It is obvious that the bank has advanced money to the businessman for the period of the currency of the bill.

Bill discounting is considered a very suitable investment for a bank. The bills are certain to be paid on maturity. They can be rediscounted with the central bank, if necessary. They set up a regular flow of incomings and outgoings of cash. Being negotiable instruments, the bills do not create any difficulty at the time of payment and do not involve the bank in any litigation. It represents a short-term investment, which suits the bank very well because most of its deposits are also of short term. For all these reasons this form of advance represents the most suitable investment from the bank's point of view. That is why it is sometimes remarked that a good banker knows the difference between a bill and a mortgage.

(*d*) Transfer of Money

Banks transfer money from one place to another for their customers. Banks remit the funds of the people by means of a bank draft or a cheque. This is a cheap as well as safe method of transferring money from one place to another.

(*e*) Miscellaneous Functions

A bank now-a-days serves its customers in various other ways. It has lockers or 'safe deposit vaults'. They are meant to keep the valuables of customers in safe custody. Further, a bank collects interest on behalf of its customers as well as pays dividends on behalf of joint-stock companies. It purchases and sells stocks and shares of companies for its clients. It pays insurance premium on behalf of their customers from their deposits. It executes the wills of deceased customers and acts for them as a trustee.

From the above discussion it follows that these days the functions of banks are many and varied. Banks are not merely the *traders of money* but they are also the *manufacturers of money.* In other words, they not only deal in money, *i.e.,* borrow and lend money but also manufacture or make money. They create money through credit creation. Bank deposits these days are as much money as any other form of money.

ROLE OF COMMERCIAL BANKS IN ECONOMIC DEVELOPMENT

As is well known, the rate of economic development depends to a large measure on the rate of capital formation. The rate of capital formation in turn depends on the rate of savings and investment and the proper allocation of investible funds among different sectors and users. The banking system helps economic growth in all these ways, that is, by (*a*) promoting savings, *(b)* mobilizing savings, and (*c*) allocating savings among alternative uses and users. Let us explain how commercial banking, which is an important part of the financial system of the country, performs these functions.

(*a*) Promotion of Savings

People save for various reasons. Thus people save to provide for future needs such as periods of unemployment, old age, sickness, to provide for education and marriage of their children, to own property such as real estate, houses etc. in future, and to purchase durable consumer goods. But they require assets in the form of which they should keep their savings in safe custody and earn a rate of return as well. Commercial banks promote savings by providing a wide range of deposits with varying combinations of liquidity and rate of interest to suit the needs and preferences of different savers. It has been found that with the growth of commercial banking in unbanked and under-banked regions, the household savings go up. As a store of value, bank deposits enjoy certain advantages over tangible assets (physical capital, inventories of commodities) and other financial assets. The bank deposits are convenient to hold as store of value and are more safe and more liquid, *i.e.,* they can be converted into cash easily. They are also greatly divisible and less risky. With all these advantages, they earn varying rates of interest depending upon the type of deposits into which the savers put their savings. These advantages of bank deposits induce households to save more and encourage the habit of thrift.

It is worth noting that the commercial banks would perform well this function of promoting savings in the framework of price stability. If the prices of goods are rising, that is, inflation grips the economy, then the savers would prefer to use their savings to buy gold and silver, other commodities, and real estate whose prices are also rising. This is because the real rate of interest on bank deposits goes down to the extent of rise in prices. This underlines the importance of keeping prices stable, if savings by households are to be promoted.

(*b*) Mobilisation of Savings

Not only do the banks encourage savings but they also mobilise savings done by serveral households and make them available for the purposes of production and investment to the entrepreneurs in various sectors of the economy. This function of mobilizing savings is of crucial importance because in the modern monetary economy, the act of saving has been separated from the act of real investment. Savings are done by millions of households and firms, whose individual savings may be very small, savings of some may be of short-term and of others of long-term nature. Banks and other financial intermediaries collect or mobilise these savings before these can be made available to the producers or investors. Without the banks (and other financial intermediaries), these savings would have remained scattered and also idle, that is, would not have been utilised for productive and investment purposes. As pointed out above, banks mobilise savings of households and firms through offering variety of deposits to suit the needs and preference of different households possessing surplus funds.

It follows from above that the *commercial banks, like other financial institutions, provide a link between those who have savings (i.e. surplus funds) and those who are in need of such funds to use them for production and investment purposes.* If commercial banks and other financial intermediaries were not there, those with surplus funds would have to search for appropriate borrowers and strike individual bargains with them and bear risks of lending them. The existence of commercial banks makes the tasks of lenders easy and with the control over commercial banks by the Government or central bank of the country, the risks of depositors have almost been eliminated. This enables the banks to mobilise more resources for production and investment purposes.

It is evident from above that banks act as a financial intermediary between lender and borrower. The financial assets can be classified into two categories (1) primary securities, (2) secondary securities. Equity shares, debentures, and company deposits of corporate firms represent primary securities. When households buy these securities they directly invest or lend money to the investor and bear risk of such investment. On the other hand, bank deposits represent secondary securities and when households go in for them, they provide their savings to banks who allocate them among competing borrowers—traders, producers and investors. In this way it is the banks who bear risks of lending, whereas the depositors' money and interest are safe and certain. Those savers who are risk averters find the secondary securities (bank deposits) more acceptable than the primary securities.

(*c*) Allocation of Funds

Allocation of funds or economic surplus among different sectors, users and producers so as to make maximum social return and thus to ensure optimum utilisation of savings is an other important function performed by the banks. Whereas the corporate firms can raise resources through sale of equity shares and debentures, the noncorporate firms and borrowers depend greatly on banks for financing the needs of both working capital and fixed capital. Through the lending rates of interest determined by market mechanism or fixed by the central bank of the country credit advanced by the banks get rationed among various potential borrowers and sectors. Further, before lending banks take into account the credit-worthiness or capacity to pay back the loans. Thus the banks are in a better position to judge the returns or productivity from the uses for which the funds are lent out. This helps in maximisation of returns from scarce financial resources.

However, it may be mentioned, that commercial banks do not always work and allocate resources in the way that maximises production or social welfare. For instance, before nationalisation in 1969 the commercial banks in India in their allocation of funds neglected socially highly desirable sectors such as agriculture, small-scale industries and the weaker sections of the society such as small and marginal farmers, the young entrepreneurs seeking self-employment. On the contrary, they preferred to invest funds collected from the public in the business concerns of the big business houses which controlled these banks. Therefore, it was thought necessary to nationalise them so that they should allocate resources in socially desirable directions.

(*d*) Promotions of Trade, Production and Investment

By encouraging inducement to save and also mobilising savings from the public, banks help to increase the aggregate rate of investment in the economy. It may also be noted that banks not only mobilise the saved funds from the public, but also themselves create deposits or credit which serve as money. The new deposits are created by the banks when they lend money to the investors or other users. These deposits are by the banks in excess of the cash reserves they obtain through deposits by the public. These days, the bank deposits, especially demand deposits, are as much good money as the currency issued by the Government or Reserve Bank of India. This creation of credit, if it is used for productive purposes, greatly larges production and investment and thus promotes economic growth.

CREDIT CREATION BY BANKS: BALANCE SHEET APPROACH

Creation of credit is one of the most outstanding functions of a modern bank. A bank has sometimes been called a factory for the manufacture of credit. How credit is created? It is an open secret that the banks do not keep cent per cent reserves against deposits in order to meet the demands of depositors. The bank is not a cloak room where you can keep your currency notes or coins and claim those very notes or coins back when you desire. It is generally understood that money received by the bank is meant to be advanced to others. A depositor has to be content simply with the bank's promise or undertaking to pay him whenever he makes a demand. Thus the banks are able to do with a very small reserve, because all the depositors do not come to withdraw money simultaneously; some withdraw, while others deposit at the same time. The bank is thus enabled to erect a vast superstructure of credit on the basis of a small cash reserve. The bank is able to lend money and charge interest without parting with cash, as the bank loan creates simply a deposit or it creates a credit for the borrower. This is what is meant by creation of credit.

Similarly, the bank buys securities and pays the seller with its own cheque which again is no cash; it is just a promise to pay cash. The cheque is deposited in some bank and a deposit is created or credit is created for the seller of the securities. This is credit creation. The term 'credit creation' implies a situation, to use Benham's words, when "a bank may receive interest simply by permitting a customer to overdraw their accounts or by purchasing securities and paying for them with its own cheques, thus increasing the total bank deposits."

Let us see the actual process. Let us assume that there is only one bank in the country. Suppose a customer deposits ₹ 1,000 in the bank. The bank has to pay him interest. Therefore, the bank must seek a safe and profitable investment for this amount. It must lend it to somebody. But this amount is not actually paid out to the borower; it is retained by the bank to meet its obligations, *i.e.*, to pay to those of its depositors who need cash and draw cheques for the purpose. The bank's experience tells him that for this purpose only a certain percentage of cash reserves to total liabilities need be kept. In countries like England, they keep nearly 10 per cent. The ratio of cash reserves to liabilities is much higher in countries like India, where banking habit has yet to develop.

Suppose the bank, in which a depositor has deposited ₹ 1,000, keeps 20 per cent cash reserve to meet the demand of depositors. This means that, as soon as the bank has received ₹ 1,000 it will make up its mind to advance loans up to the amount of ₹ 5,000 (only one-fifth reserve is kept). When, therefore, a businessman comes to the bank with a request for a loan of ₹ 5,000, he may be sure of being granted accommodation to this extent, provided of course, his credit is good. The bank lends ₹ 5,000, although it has only ₹ 1,000 in cash. It is here that credit comes in. This transaction is rendered possible because the borrower is not given the loan in cash; only an account is opened in his name and the amount is credited to his account. He is simply given the cheque book, *i.e.,* the right to draw cheques as and when he needs money. Even when he withdraws cash, it will be deposited in the bank by the recipients, because businessmen do not raise funds to keep them locked up in a cash box but to run their business and to make payments to their creditors. When this particular businessman draws cheques on this bank to pay his creditors, these cheques are passed on by them to their own banks, where the amount is deposited in their account. Cash is seldom withdrawn. The banks adjust their mutual obligations through a system of bank clearing. Thus the bank has succeeded in creating a credit of ₹ 5,000 against a cash reserve of ₹ 1,000.

The bank also creates credit when it purchases securities. The bank can purchase securities without paying any cash. It issues its own cheque to pay the purchase price. The cheque is deposited in this bank or some other bank and the small cash reserve which the bank keeps is sufficient to meet an obligation arising from this transaction too. It is thus that, on a small cash foundation, a vast superstructure of credit is built up.

Let us now understand the process of credit creation when there are *several* banks in the country, as they are in the real world. In the case of several commercial banks in the country, one individual bank cannot create all the credit as described above. But what no single bank can do individually, the banking system as a whole can do, *i.e.,* create credit. We shall explain the process of credit creation or the expansion of money supply in the country by the banking system collectively with the help of *balance sheets* of the banks. We shall illustrate how deposit of ₹ 1,00,000 of currency in a commercial bank enables the banking system as a whole to expand deposits by another ₹ 4,00,000, that is, deposits of ₹ 1,00,000 in currency leads to a total deposits of ₹ 5,00,000 in the banking system.

Banks, as other business firms, show their financial condition on a balance sheet. A simple balance sheet has two columns, its left column represents all the *assets* of a bank and its right column represents all the *liabilities* of a bank. Assets are all the things or claims a bank owns, liabilities, on the other hand, are claims against those assets; some of the claims are of creditors and some of them are of owners of the banks themselves. Because assets show everything that a bank owns and because liabilities represent claims against those assets, the two sides of the balance sheet, that is, assets and liabilities must equal each other.

Let us suppose that an individual or a firm deposits ₹ 1,00,000 in cash with a bank A. Ignoring everything else in the balance sheet, let us know how the balance sheet of bank A will look like with this fresh deposit of ₹ 1,00,000 in currency with it. The cash of ₹ 1,00,000 which the bank A will receive will become its assets, and at the same time individual's deposits of ₹ 1,00,000 will be its liabilities, the assets and liabilities of bank A will therefore be equal to each other.

BANK A
Balance Sheet

Assets		*Liabilities*	
Cash	₹ 1,00,000	Deposits	₹ 1,00,000

Let us assume that cash reserve ratio is 20%. Now the bank does not require all the ₹ 1,00,000 in cash against the deposits of ₹ 1,00,000. The bank A requires only 20% of it, that is, ₹ 20,000 cash against its deposits of ₹ 1,00,000. The bank can lend or invest in securities the remaining amount of ₹ 80,000. Actually if the bank does not lend or invest it will suffer a loss, since it will pay the interest to the depositor with no profit from the cash it possesses. Therefore, the bank A will lend ₹ 80,000 to the business firms or individuals whom it finds creditworthy. Now, when a bank lends to a person or firm it does not give him cash immediately. The bank makes deposits in the name of the person whom he lends the money and gives him the right to draw cheques against it when required. It is a new deposit, one that did not exist before. The person or firm getting loans from the bank will, however, after some time completely withdraw the money through cheques from his deposits. When the loan of ₹ 80,000 has been sanctioned to a person, but before that person starts withdrawing his money, the balance sheet of bank A will look like as follows:

BANK A
Balance Sheet
(When the bank sanctions loan of ₹ 80,000 but before loan is cashed)

Assets		*Liabilities*	
Cash	₹ 1,00,000	Deposits	₹ 1,00,000
Loan	₹ 80,000	New Deposits (Created)	₹ 80,000
	₹ 1,80,000		₹ 1,80,000

In this balance sheet, loan of ₹ 80,000 becomes asset of the bank, while the new deposits created constitute the liability of the bank, since the person getting the loan has the right to draw upon these deposits.

Now, when the person wholly withdraws his deposits through cheques and the recipients of these cheques deposit them in some other bank, say bank B, then the bank A will have to surrender to bank B cash money equal to ₹ 80,000. After the whole newly created deposits of ₹ 80,000 have been thus withdrawn, the balance sheet of bank A will now look like as follows:

BANK A
Balance Sheet

Assets		*Liabilities*	
Cash	₹ 20,000	Original Deposits	₹ 1,00,000
Loan	₹ 80,000	New Deposits	0
Total	₹ 1,00,000	Total	₹ 1,00,000

As said above, cheques worth ₹ 80,000 against Bank A are deposited in Bank B, for the Bank B these will constitute new cash money and will therefore become the assets of bank B. But ₹ 80,000 will also be the liability of the bank B in the form of deposits in the name of those persons who have deposited the cheques with it. Ignoring other assets and liabilities of bank B and taking into account only this above transaction, the balance sheet of bank B will be as follows:

BANK B

Balance Sheet

Assets		*Liabilities*	
Cash	₹ 80,000	Deposits	₹ 80,000

Now against the deposits of ₹ 80,000 bank B requires to keep its 20%, that is, ₹ 16,000 and it can lend or invest the remaining amount of ₹ 64,000. When bank B lends ₹ 64,000 to a firm, it will create deposits for that firm. Before the firm draws upon that deposits, the balance sheet of bank B will took like:

BANK B

Balance Sheet

Assets		*Liabilities*	
Cash	₹ 80,000	Deposits	₹ 80,000
Loan	₹ 64,000	New Deposits Created	₹ 64,000
Total	₹ 1,44,000	Total	₹ 1,44,000

Now, when the firm which has got loan from bank B completely withdraws ₹ 64,000 through cheques, the balance sheet of bank B will be as follows. As a result of the firm spending the loan money of ₹ 64,000, the bank B will transfer cash of ₹ 64,000 to another bank, say C, in which the cheques drawn by the firm are deposited. As a result of this, the cash with bank B will fall to ₹ 16,000 (₹ 80,000-₹ 64,000 = ₹ 16,000).

BANK B

Balance Sheet

Assets		*Liabilities*	
Cash	₹ 16,000	Deposits	₹ 80,000
Loan	₹ 64,000	New Deposits	0
Total	₹ 80,000	Total	₹ 80,000

Now when the bank C will get ₹ 64,000, it will also require to keep 20% of it *(i.e.,* ₹ 12,800) and the remaining amount of ₹ 51,200 will be lent out or invested by it.

From the foregoing analysis it is clear that the currency deposits of ₹ 1,00,000 led to the creation of deposits of ₹ 80,000 by Bank A, ₹ 64,000 by Bank B, ₹ 51,200 by Bank C. But the process of expansion of deposits will not stop here, it will go on as the money lent out by one bank is spent through cheques and these cheques are deposited in other banks, till the total deposits of ₹ 5,00,000 in all the banks (including original deposits of ₹ 1,00,000) are created. But it should be remembered that at each stage the new deposits created by a bank goes on declining. This is because at each stage a bank is required to keep 20% of the money it receives as cash reserves and therefore lends and creates deposits equal only to the remaining amount. Thus bank A created deposits of ₹ 80,000, bank B created deposits of ₹ 64,000, bank C created deposits of ₹ 51,200 and so on.

We are now in a position to state how much deposits have been created by the banking system out of the currency deposits of ₹ 1,00,000.

Total Deposits = ₹ 1,00,000 + 80,000 + 64,000 + 51,200 + ... = ₹ 5,00,000

Out of the total deposits of ₹ 5,00,000, the deposits of ₹ 1,00,000 in cash was made in the banking system, the remaining deposits have been created by the banking system itself, as if out of thin air.

$$\left.\begin{array}{r}\text{Deposits created by the}\\ \text{banking system}\end{array}\right\} \begin{array}{l}= ₹\ 5{,}00{,}000 - 1{,}00{,}000\\ = ₹\ 4{,}00{,}000\end{array}$$

It should be further noted that the *total expansion of deposits by the banking system depends upon the cash reserve ratio* (CRR). The smaller the cash reserve ratio the large the expansion of deposits or credit. Thus, in the above case, we noted that, given the cash reserve ratio of 20%, the total deposits expansion from the cash deposits of ₹ 1,00,000, was equal to ₹ 5,00,000. Thus the total deposits expanded to was five times the original cash deposits. Therefore the deposits of cash in the banking system leads to multiple expansion in the total deposits. This is known as *deposits* or *credit multiplier.* In our above case, the deposits or credit multiplier is 5. It should be remembered that the *magnitude of deposits multiplier depends on the cash reserve ratio.*

$$\text{Deposit multiplier } d_m = \frac{\Delta D}{\Delta R} = \frac{1}{r}$$

where r stands for cash reserve ratio.

Thus deposit multiplier is the reciprocal of cash reserve ratio (CRR) which we have denoted by r in the measure of deposit multiplier.

Thus, when the cash reserve ratio is 20%, that is, 0.20 or 1/5, the deposit multiplier.

$$= \frac{1}{\frac{1}{5}} = 5$$

Now, if the cash reserve ratio is raised by Reserve Bank to 25 per cent, that is, 0.25, the deposit multiplier,

$$d_m = \frac{1}{0.25} = \frac{1}{1/4} = 4$$

Thus, *the greater the cash reserve ratio, the lower will be the value of deposit multiplier.* In other words, increase in the cash reserve ratio *(CRR* or *r)* will lead to the contraction of credit created by the banks, and vice versa

Similarly, if the cash reserve ratio is 10%, *i.e.,* 1/10, then the deposits multiplier = 1/1/10 = 10.

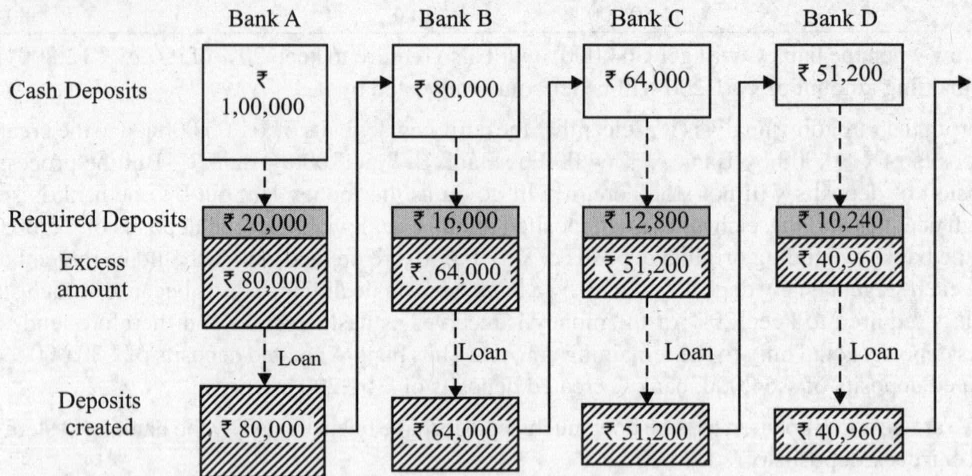

Fig. 18.1. *Credit Creation by Banks*

In this case initial deposits of some cash amount in the banking system will lead to ten times expansion in the total deposits. Figure 18.1 makes clear the process of deposits expansion by the banking system.

Deposit Multiplier and Credit Multiplier

Some economists distinguish between deposit multiplier and credit multiplier. In our above example, original or primary deposits of ₹ 100,000 made by the public in the banking system, given the cash reserve ratio (r) equal to 20 per cent (*i.e.* 0.20), resulted in the increase in total deposits in the banking system equal to ₹ 500,000. If we denote total increase in deposits by ΔD and original increase in cash deposits as ΔR, then the deposit multiplier can be written as

$$d_m = \frac{\Delta D}{\Delta R} \qquad ...(i)$$

or, in our first example, $$d_m = \frac{\Delta D}{\Delta R} = \frac{1}{r} = \frac{1}{0.20} = 5 \qquad ...(i)$$

LIMITATIONS ON THE CREDIT CREATING POWER OF THE BANKS

From the foregoing account of credit or deposits creation by the banks, it would seem that the banks reap where they have not sown. They advance loans or buy securities without actually paying cash. But they earn interest on the loans they give or earn dividends on the securities they purchase all the same. This is very tempting. They make profits without investing cash. They would, of course, like to make as much profit like this as they can. But they cannot go on expanding credit indefinitely. In their own interest, they have to apply the brake, and they do actually apply it, for it is well-known that the profits made by the banks are not very high. The overriding limitation arises from the obligation of the banks to meet the demands of their depositors.

Benham has mentioned three limitations on the powers of the banks to create credit:

(*i*) The amount of cash in the country;

(*ii*) The amount of cash which the public wishes to hold; and

(*iii*) The minimum percentage of cash to deposits, called cash revenue ratio which the banks have to maintain.

(*iv*) The amount of money which the public wants to hold as deposits in the banks.

As for (*i*), it may be said that credit can be created on the basis of cash. *The larger the cash (i.e., legal tender money) the larger the amount of credit that can be created.* But the amount of cash that a bank may have is subject to the control of the central bank. The central bank has the monopoly of issue of cash. It may increase it or decrease it, and credit will expand or contract accordingly. The power of the central bank to control currency helps it to control the extent of credit that the banks have the power to create.

The second limitation arises from *the habit of the people regarding the use of cash i.e. currency.* If people are in the habit of using cash and not cheques, as in India, then as soon as credit is granted by the bank to a borrower, he will draw the cheque and get cash. When the bank's cash reserves is thus reduced, its power to create credit is correspondingly reduced. On the other hand. If people use cash only for very small and odd transactions, then the cash reserve of the banks is not much drawn upon and their power of creating credit remains unimpaired. This is the case in advanced countries like the U.S.A., England and other European countries. There the banks hardly keep 10 per cent cash reserve.

The third limitation is the most important. It arises from the *cash reserve ratio cash, which the banks must maintain* to ensure the safety of the bank and to retain the degree of liquidity that is considered desirable. It is clear that when a bank creates a credit or grants a loan, it undertakes a liability. There is an increase in its liabilities and there is correspondingly a fall in cash reserve ratio. The bank will not let the cash reserve ratio fall below a certain minimum. When that minimum is reached, the power of the bank to create credit comes to an end. To grant any further credit will be risky unless the bank's experience is reassuring enough to permit the adoption of a lower percentage. Then that would be the limit.

The other important limitation on the credit creating power of banks is *the amount of money which the public choose to hold as deposits in banks.* The more money which the public deposits with the banks, the more reserves banks would have and therefore more credit they will be able to create and vice versa. It may noted that public can use their saved money in more than one way. The public can buy shares or debenture of the companies, it can invest in mutual funds of both public and privates sectors. But the credit creation by the banks depends on the money the public deposits in them. It is important to note than rate of interest paid by the banks on deposits determines to a good extent the amount of money deposits with them by the public. Other things being equal, the higher the rate of interest, the greater the amount of money the public will deposit money with the banks.

To these may be added the fourth limitation. *The bank cannot create credit without acquiring some asset.* An asset is a form of wealth. Thus the bank only turns immobile wealth into mobile wealth. Hence, as Crowther observes, "the bank does not create money out of thin air, it transmutes other forms of wealth into money." However, banking system today has become quite advanced. These days banks give credit on the basis of personal goodwill rather than on the basis of any form of wealth.

QUESTIONS FOR REVIEW

1. Explain credit creation. Describe the factors which determine the volume of credit which can be created.
2. Explain the process of credit creation through the 'Balance Sheet Approach'. What factors limit credit creation?
3. What are the functions of a Commercial Bank
4. Explain the role of commercial bank in the economic development of India.
5. Explain the process of creation of credit by commercial banks. What are the limitation on the credit creating power of banks?
6. Explain how commercial banks create credit. Are there any limit to credit creation by them? If cash deposits ratio changes from 0.1 to 0.2, will there be any change in the credit creating power of banks?
7. Explain the functions of Commercial Banks. How is Commercial bank different from the Central Bank?
   (*a*) Explain why commercial banks should be concerned about liquidity.
   (*b*) Is there a conflict between the profit motive of a bank and the need for liquidity?
   (*c*) Which assets are a bank's principal sources of profitability

CHAPTER 15
CENTRAL BANKING

INTRODUCTION

In the monetary system of all countries, the central bank occupies an important place. The central bank is an apex institution of the monetary system which seeks to regulate the functioning of the commercial banks of a country. The central bank of India is called the Reserve Bank of India which was set up in 1935. As explained in the previous chapter, the commercial banks keep only a fraction of their deposits in cash and the rest they lend out to the traders and investors. Therefore, the commercial banking is often known as *fractional reserve system.* In view of the fact that commercial banks keep only a fraction of their deposits in cash, they will run into difficulties if at a time there is rush of depositors to withdraw their money. This indicates the need for an institution which should come to the rescue of the commercial banks and provide them the money required to meet the excessive demand of the depositors. The central bank fulfils this need. However, in the modern times, the central bank not only provides monetary aid to the commercial banks in time of crisis but also performs many other functions. Indeed, the control over cost and availability of credit in the economy and regulation of the growth of money supply are special responsibilities of the central bank.

PRINCIPLES OF CENTRAL BANKING

The central bank of a country enjoys a special status in the banking structure of the country. The principles on which a central bank is run differ from the ordinary banking principles. An ordinary bank is run for profits. A central bank, on the other hand, is primarily meant to promote the financial and economic stability of the country. "The guiding principle of a central bank", says DeKock, "is that it should act only in the public interest and for welfare of the country and without regard to profit as primary consideration". Earning of profit for a central bank is thus a secondary consideration.

The central bank is thus not a profit hunting institution. It does not act as rival of other banks. In fact, it is a monetary authority of the country and has to function in a manner so as to promote economic stability and development. The functions of the central bank, especially the Reserve Bank of India, have increased enormously in recent years. Not only does the Reserve Bank of India regulate credit and money supply in the country, but it also promotes economic development and price stability. Guiding principles of the Reserve Bank are to operate its most instruments in a way that serves the objectives of economic policy laid down by the Government and Planning Commission.

FUNCTIONS OF CENTRAL BANK

The following are the main functions of a central bank:

1. It acts as a note issuing agency.
2. It acts as the banker to the state.
3. It acts as the bankers' bank.
4. It controls credit.
5. It acts as the lender of the last resort.
6. It manages exchange rate.

Note Issuing Agency

The central bank of the country has the monopoly of issuing notes or paper currency to the public. Therefore, the central bank of the country exercises control over the supply of currency in the country. In India with the exception of one rupee notes which are issued by the Ministry of Finance of the Government of India, the entire note issue is done by the Reserve Bank of India. In the past the central bank of various countries used to keep as reserves some gold and foreign exchange securities against the notes issued. The percentage of the reserves to be kept against the total amount of notes issued was fixed by law and is subject to change by the Government. Theoretically, there is no need of the backing of gold reserves against the notes issued. It may be pointed out that paper notes these days cannot be converted into gold or some other precious metals; they are inconvertible. This is called ***proportional reserve system***. Before 1956 in India also, there was proportional reserve system of issuing currency or notes. According to this, the Reserve Bank was required to keep as reserves 40 per cent of the total notes issued in the form of gold and foreign exchange securities.

Since 1956 has been given up and instead ***minimum reserve system*** has been adopted according to which the Reserve Bank is required to keep only a *minimum amount of reserves* in the form of gold and foreign exchange securities and given this minimum reserve it can issue notes as much as it thinks desirable in view of the needs and conditions of the economy. There is even no need for backing of gold for the currency issued by the Government or the central bank. From the economic point of view what matters is the production of real goods and services and not the amount of gold supporting a currency. The real value of a currency depends on how much it can buy goods and services and not how much gold or silver is kept as reserve against it.

Thus, ultimately the credibility of the currency of a country depends not upon whether it is convertible into gold or silver but upon to what extent it is possible to maintain the stability of its value by suitable monetary control.

Banker to the Government

Another important function of the central bank is to act as the banker to the Government. All the balances of the Government are kept with the central bank. On these balances the central bank pays no interest. The central bank receives and makes all payments on behalf of the Government. Further, the central bank has to manage the public debt and also to arrange for the issue of new loans on behalf of the Government. The central bank also provides short-term loans to the Government. This is usually done through the central bank discounting the Government treasury bills either directly or when presented by other banks. Thus the central bank performs a number of services to the Government. In fact, the central bank is the fiscal agent of the Government and advises the latter in matters relating to currency and exchange as well as finance.

Bakers' Bank

Broadly speaking, the central bank acts as a bankers' bank in three capacities: (*i*) as the custodian of the cash reserves of the commercial banks; (*ii*) as the lender of the last resort; and (*iii*) as bank of central clearance, settlement and transfers. All other banks in the country are bound by law to keep a fixed portion of their total deposits as reserves with the central bank. These reserves help the central bank to control the issue of credit by commercial banks. They in return can depend upon the central bank for support at the time of emergency. This help may be in the form of a loan on the strength of approved securities or through rediscounting of bills of exchange. Thus the central bank is the lender of the last resort for other banks in difficult times because on such occasions there is no hope of getting help from any competing institution.

In India, scheduled banks have to keep deposits with the Reserve Bank not less than 5% of their current demand deposits and 2% of their fixed deposits as reserves. In return they enjoy the privilege of rediscounting their bills with the Reserve Bank as well as securing loans against approved securities when in need.

Clearing function is also performed by the central bank for the banks. Since banks keep cash reserves with the central bank, settlement between them may be easily effected by means of debits and credits in the books of the central bank. If clearings go heavily against some bank, its cash reserves with the central bank will fall below the prescribed limit and therefore the bank concerned will have to make up the deficiency.

Control of Credit

The chief objective of the central bank is to maintain price and economic stability. Price instability—both inflation and deflation—has harmful effects. Moreover, fluctuations in overall economic activity, that is, trade cycles, entail a lot of human sufferings. Main reason for the fluctuations in prices as well as in overall economic activity is the changes in aggregate demand. Aggregate demand, especially the investment demand, depends upon the supply of money. And credit these days is the important constituent of the money supply. Thus the supply of credit greatly affects the prices, national income and employment through changes in investment demand.

Now it is the responsibility of the central bank of a country to guide the money market, *i.e.,* the commercial banks, regarding supply of credit so as to maintain stability in prices as well as in overall economic activity. To overcome inflation it has to restrict the supply of credit and to prevent or get rid of depression and deflation it has to expand the credit. There are various methods by which the central bank can control the supply of credit in the economy. These methods are: *(a)* varying the bank rate; (*b*) engaging in open market operations; (*c*) changing the reserve ratio, and *(d)* exercising selective credit controls. We shall explain these methods of credit in detail in later sections of this chapter.

It is through controlling the supply of credit and cost of credit *(i.e.,* rate of interest on it) that the central bank of a country tries to bring about stability in prices as well as in overall level of economic activity. The central bank is the monetary authority of the country and monetary policy is one of the important measures which are taken to avoid and cure both depression and inflation. To remedy inflation the central bank tries to restrict the supply of credit by raising the bank rate and using other weapons of credit control. To overcome depression it tries to expand credit by lowering the bank rate and cash reserve ratio and also by buying securities from the open market.

In India Reserve Bank which is the central bank of the country has been making important contribution to the achievement of the objective of price stability. To achieve price stability Reserve Bank has been setting forgets of expansion in money supply which are consistent with the growth of output. Control of inflation by checking excessive expansion in credit supply has been the major concern of monetary policy imposed by Reserve Bank of India.

Lender of the Last Resort

As mentioned above, the commercial banks operate on the basis of fractional reserve system. Therefore, even a well-managed commercial bank can run into difficulty if there is a great rush of demand for cash by the depositors, because with a fraction of its deposits in cash, it will not be able to meet a sudden and large demand for cash. The central bank must therefore come to their rescue at such times. Thus, central bank is the last source of supply of credit. It is the duty of the central bank to meet demand.for cash by a bank at the time of emergency when panic prevails among the public and people's confidence has been shaken and when other banks have refused to supply credit. The central bank boldly steps forward to supply cash and allay panic. The central bank must meet situation of high liquidity preference.

To Promote Economic Growth

A very important function of central bank these days in developing countries like India is to promote economic growth. It can help in both agricultural and industrial growth in the country. The central bank can promote agricultural and industrial growth by providing finance or credit to agriculture and industry. Central bank adopts such a monetary policy as is conducive to economic growth. In order to accelerate the rate of investment or capital formation, the central bank takes steps to make more credit available for investment at lower lending rates of interest. In the developing countries, the role of central bank as promoter of economic development is very important.

Thus, in India apart from regulatory function, Reserve Bank of India has been playing a promotional role. RBI has been making important contribution for building up appropriate financial institutions such as Industrial Finance Corporation of India, State Finance Corporations to promote saving and investment. By ensuring adequate supply of agricultural credit, term finance to industries, credit for exports, RBI has performed a useful promotional role in encouraging economic growth.

Managing Exchange Rate of the National Currency

Another important function of a central bank is to maintain the exchange rate of the national currency. For example, the Reserve Bank of India has the responsibility of maintaining the exchange value of the rupee. When a country has adopted flexible exchange rate system under which value of a currency is determined by demand for and supply of a currency, the value of a currency, that is, its exchange rate with other currencies, is subject to large fluctuations which are harmful for the economy. Under these circumstances, it is the duty of the central bank to prevent large depreciation or appreciation of the national currency. Since 1991 when the rupee has been floated, the value of Indian rupee, that is, its exchange rate with US dollar and other foreign currencies, has been left to be determined by market forces, RBI has been taking several steps from time to time to stabilise the exchange rate of rupee, especially in terms of US dollar.

There are several ways by which RBI can manage or maintain the exchange rate of the rupee. First, if due to speculative activities of foreign exchange operators, the rupee starts depreciating rapidly, RBI can intervene in the market. It can use its reserves of dollars and supply dollars in the market from its own reserves. With the increase in supply of dollars, the rupee will be prevented from depreciation. It may however be noted that the success of this step depends on the amounts of dollar reserves with Reserve Bank of India. This is illustrated in Fig. 15.1 where we have depicted demand and supply curves of US dollars which intersect at point E and determine exchange value of rupee equal to ₹ 65 to a US dollar. Now suppose that demand for dollars by Indian traders and companies importing goods from abroad and market operators increases so that demand curve for dollar shifts to the right to $D' D'$ position. It will be seen that the intersection of this new demand curve for the US dollars $D'D'$ intersects the supply curve SS of dollars at point H and determines the exchange rate of rupee for the US dollars equal to ₹ 66 per dollar. Thus with increase in demand for dollars rupee has depreciated (and the US dollar appreciated). Now, if RBI intervenes and from its foreign exchange reserves supplies extra dollars equal to EB, the supply curve of

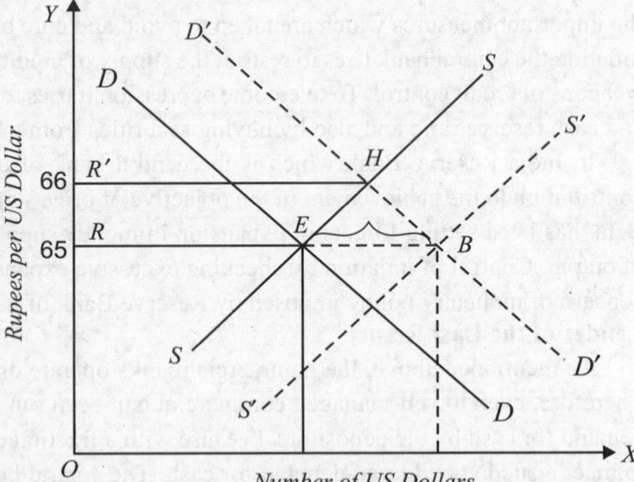

Fig. 15.1. *Intervention by Central Bank to Maintain Exchange Rate*

dollars in the foreign exchange market will shift to the right to the position *S′S′* which intersects the higher demand curve *D′D′* of dollars at point *B* so that again ₹ 65 becomes the equilibrium exchange rate of rupees for the US dollar. In this way by its intervention and supplying extra dollars from its foreign exchange reserves, RBI can succeed in maintaining the exchange rate of rupee at ₹ 65 to a US dollar. In actual practice in recent years, when rupee was depreciating RBI intervened and succeeded in preventing the rapid depreciating rupee against US dollar.

Another method by which RBI can manage the exchange rate of rupee is adopting measures which will reduce the demand for dollars. Some importers, foreign investors, foreign exchange operators try to avail of cheap credit facilities of banks and borrow rupee funds from the banks and try to convert them into dollars. This raises the demand for dollars and leads to the depreciation of the Indian rupee. Such a situation occurred in July-September 1998. RBI intervened and raised the Cash Reserve Ratio (CRR) and increased its repo rate. This succeeded in mopping up the excess liquidity with the banks and reduced their lending capacity. This led to the reduction in the demand for dollars and helped in preventing the rupee from depreciating.

On the contrary, if rupee is appreciating against the US dollar and it is thought desirable to check undue appreciation of the Indian rupee, RBI can intervene to check its further appreciation. For this purpose, it can buy dollars from the market. This will raise the demand for dollars in the foreign exchange market and appreciation of Indian rupee will be checked.

METHODS OF CREDIT CONTROL

The central bank of a country has the responsibility of controlling the volume and direction of credit in the country. Bank credit has become these days an important constituent of the money supply in the country. The volume and direction of bank credit has, therefore, an important bearing on the level of economic activity. Excessive credit will tend to generate inflationary pressures in the economy, while deficiency of credit supply may tend to cause depression or deflation. Lack of the availability of cheap credit may also hinder economic development of a country. At times of depression, there is a need to expand credit and at times of boom there is need to contract credit. For promoting economic development, expansion of cheap credit (credit at low rates of interest) is desirable. In order to prevent booms and depressions (*i.e.,* to maintain economic stability) and to promote economic growth, central bank seeks to control credit in accordance with the needs of the situation.

Broadly speaking, there are two types of methods for controlling credit.

1. Quantitative or General Methods. These methods seek to change the total quantity of credit in general. These are three in number: (*i*) changing the bank rate; (*ii*) open market operations; (*iii*) changing the cash reserve ratio.

2. Qualitative or Selective Control Methods. These methods aim at changing the volume of a specific type of credit. In other words, the selective control method affects the use of credit for particular purposes.

We will discuss these methods in detail.

Bank Rate Policy

Bank rate is the minimum rate at which the central bank of a country provides loans to the commercial banks of the country. Bank rate is also called discount rate because in the earlier days central bank used to provide finance to the commercial banks by rediscounting bills of exchange. Through changes in the bank rate, the central bank can influence the creation of credit by the commercial banks. Bank credit these days is an important component of the supply of money in the economy. Changes in money supply affect aggregate demand and thereby output and prices. For instance, when the central bank raises the bank rate, the cost of borrowings by the commercial banks from the central bank would rise. This would discourage the commercial banks to borrow from the central bank. Further, when the bank rate is raised, the commercial banks also raise their lending rates. When the lending rates of interest charged by the commercial banks are higher, businessmen and industrialists would feel discouraged to borrow from commercial banks. This would tend to contract bank credit and hence would result in the reduction of money supply in the economy. The reduction in the supply of money would reduce aggregate demand or money expenditure. This would reduce prices and check inflation in the economy. Thus when the economy is gripped by inflation rising prices, the bank rate is generally raised to contract credit creation by the bank.

On the other hand, when there is recession or depression in the economy, the bank rate is lowered to overcome it. A fall in the bank rate will cause the reduction in lending interest rates of the commercial banks. With credit or loans from banks becoming cheaper, businessmen would borrow more from the commercial banks for investment and other purposes. This would lead to the increase in aggregate demand for goods and services and help in overcoming recession and bringing about recovery of the economy.

If a country permits the free flow of funds in and out of the country, then the changes in the bank rate will also have an effect on the external flows. For instance, as said above, when the bank rate is raised all interest rates in the market would also generally rise. With the rise in the deposit rates of interest of the banks, funds from outside would be attracted to the banks of the country. Further, with the rise in deposit rates, outflow of funds from the banks to the other countries would be prevented. Therefore, these effects on the inflow and outflow of funds of the rise in bank rate will have a favourable effect on the balance of payments of the country.

It is important to note that the *change in bank rate affect the credit creation by banks through altering the cost of credit*. Changes in the cost of credit affect borrowings by the commercial banks from the central bank and also affect the demand for credit by businessmen from the commercial banks.

Limitations of Bank Rate Policy

Bank rate policy does not have always the desired effect on investment, output and prices. There are certain conditions which must be met for the successful working of the bank rate policy. These conditions are:

(1) All other rates should follow the bank rate in its movement so that bank credit should expand or contract as desired. This will not happen if the commercial banks have considerable reserves of their own at their disposal and, therefore, their dependence on borrowed funds from the central bank may be very less. Further, for the changes in the bank rate to cause changes in all other interest rates in the market, a well-organised money market is needed. If well-organised money market does not exist as is the case in India where indigenous bankers contribute a good part of the money market, changes in the bank rate would not be followed by the appropriate changes in all rate.

(2) The second important condition for the successful working of the bank rate policy is the responsiveness of businessmen to the changes in the lending rates of interest. If businessmen and investors reduce their borrowings when the bank rate and hence the lending rates of commercial bank are raised, and increase their borrowings for investment when the bank rate and lending rates of the banks are reduced, the changes in the bank rate will have the effect, predicted in the bank rate theory. However, empirical studies conducted recently have shown that rate of interest does not exercise a strong influence on borrowings for investment and other purposes. When there is inflationary situation in the economy, the rate of interest will have to be raised very high in order to have desirable effects. Similarly, the response of investment to the fall in the rate of interest is never vigorous. Demand by businessmen for loans for investment purposes from the banks depends on the economic situation

prevailing in the economy. When the economy is gripped by severe depression and consequently prospects for making profits are bleak, businessmen would be reluctant to borrow for investment even though the lending rates of interest have been lowered considerably to induce them to borrow. It has been aptly remarked that 'you can bring the horse to water but you cannot make it drink'.

The present belief is that bank rate policy will play only a subsidiary role in the monetary control. The bank rate at the best could check a boom and inflation but could not bring about a recovery if the country was suffering from recession or depression. Thus it has much greater potential effectiveness if it is desired to arrest credit expansion than when it is sought to stimulate credit. But used in conjunction with other methods it can still play an useful role. The bank rate has the value as a signal. If the traditions are that a signal must be obeyed, the bank rate policy can fulfil its purpose since it will be in keeping with respectable conduct.

Open Market Operations

Open market operations are another important instrument of credit control, especially in the developed countries. The term open market operations means the purchase and sale of securities by the central bank of the country. The theory of open market operations is like this: The sale of securities by the central bank leads to the contraction of credit and the purchase thereof to credit expansion. When the central bank sells securities in the open market, it receives payment in the form of a cheque on one of commercial banks. If the purchaser is a bank, the cheque is drawn against the purchasing bank. In both cases the result is the same. The cash balance of the bank in question, which it keeps with the central bank, is to that extent reduced. With the reduction of its cash, the commercial bank has to reduce its lending. Thus, credit contracts. When the central bank purchases securities, it pays through cheques drawn on itself. This increases the cash balance of the commercial banks and enables them to expand credit. "Take care of the legal tender money and credit will take care of itself" is the maxim.

The open market operations method is sometimes adopted to make the bank rate policy effective. If the member-banks do not raise lending rates following the rise in the bank rate due to surplus funds available with them, the central bank can withdraw such surplus funds by the sale of securities and thus compel the member-banks to raise their rates. Scarcity of funds in the market compels the banks directly or indirectly to borrow from the central bank through rediscounting of bills. If the bank rate is high, the market rate of interest cannot remain low.

Limitations of the Open Market Operations. It is obvious that this method will succeed only if certain conditions are satisfied. The limitations are discussed below:

(1) According to the theory of open market operations, when the central bank purchases securities, the cash reserves of the member-banks will be increased and, conversely, the cash reserves will be decreased when the central bank sells securities. This, however, may not happen. The sale of securities may be offset by return of notes from circulation and hoards. The purchase of securities, on the other hand, may be accompanied by a withdrawal of notes for increased currency requirements or for hoarding by the public. In both the cases, therefore, the cash reserves of the member-banks may remain unaffected.

(2) But even if the cash reserves of the member-banks are increased or decreased, the banks may not expand or contract credit accordingly. The percentage of cash to credit is not rigidly fixed and can vary within quite wide limits. The banks will expand and contract credit according to the prevailing economic and political circumstances and not merely with reference to their cash reserves.

(3) The third condition is that when the commercial bank's cash reserves increase, the demand for loans and advances should increase too and *vice versa*. This may not happen. Owing to economic or political uncertainty, even cheap money rate may not attract borrowers. Conversely, when there is boom in the economy and prospects of profits bright, entrepreneurs would borrow even at high rates of interest.

(4) Finally, the circulation of bank credit should have a constant velocity. But the velocity of bank deposits is rarely constant. It increases in periods of rising business activity and decreases in periods of depression. Thus, a policy of contracting credit may be neutralised by increased velocity of circulation, and *vice versa*.

Open market operations as an instrument of credit control have proved to be quite successful in regulating the availability of credit in developed countries. This is for two reasons. First, when the central bank purchases or buys securities, the reserves of the commercial banks automatically expand or contract. The changes in the reserves of the banks directly affect their capacity to lend. Unlike in the case of instrument of bank rate, the success of the market operations does not depend on the attitude or responsiveness of the commercial banks; their capacity to provide credit is affected automatically. Secondly, the open market operations can be managed in a way that the cash reserves of the banks increase or decrease to the desired extent.

Since for the success of market operations it is necessary that there should be broad and active market in short and long-term government securities, and such markets exist only in the U.S.A. and Great Britain and other developed countries, this method of credit control has been most effectively used in these countries.

Open market operations which did not play a significant role as an instrument of credit control in India in the past have now become important tool and are being used extensively by RBI. This is because market for Government securities in India has now become quite wide and there is often excess liquidity in the banking system. General public do not buy more than a fraction of government securities. It is banks and other financial institutions that buy or sell government securities. When there is excess liquidity in the banking system, banks willingly invest in Government securities more than the statutory limits.

Changing the Cash Reserve Ratio (CRR)

Another monetary tool to vary the quantity of credit is to change the required cash reserve ratio. By law, banks have to keep a certain amount of cash reserves with RBI as reserves against the demand and time deposits. If the legal minimum cash reserve ratio, for example, is 20%, the bank will have to keep ₹ 4,000 as reserves with RBI against its deposits of ₹ 20,000. Now the central bank of a country has the authority to vary the cash reserve ratio. If now the central bank increases the cash reserve ratio from 20% to 25%, then the reserves of ₹ 4,000 could support only deposits of ₹ 10,000 and therefore the banks having reserves of ₹ 4,000 will have to reduce their deposits from ₹ 20,000 to ₹ 16,000. This will cause the contraction of credit.

On the other hand, if the central bank reduces the cash reserve ratio from 20% to 10% then the reserves of ₹ 4,000 could support the deposit of ₹ 40,000. Therefore the bank having reserves of ₹ 2,000 can increase their deposits, *i.e.,* expand credit to ₹ 40,000. To sum up, increase in the legal cash reserve ratio leads to the contraction of credit and decrease in the legal cash reserve ratio leads to the expansion of credit. *An increase in the legal cash reserve ratio will succeed in contracting credit only when banks have no excess reserves.* If banks are holding excess reserves, the increase in the legal reserve ratio will not bring about the contraction of credit. On the other hand, the reduction in the cash reserve ratio will have the desired effect of expanding credit only if the borrowers respond favourably. As a result of reduction in the cash reserve ratio, the availability to lend by the banks increases and the banks tend to make more credit available to borrowers, or to make it available at lower rates. Now, if the borrowers for one reason or another do not respond favourably, *i.e.,* are not prepared to borrow, credit will not be expanded.

SELECTIVE CREDIT CONTROLS

The methods of credit control described above are known as quantitative or general methods as they are meant to control the availability of credit in general. Thus, bank rate policy, open market operations and variations in cash reserves ratio expand or contract the availability of credit for all purposes. On the other hand, selective credit controls are meant to regulate the flow of credit for particular or specific purposes. Whereas, the general credit controls seek to regulate the total available quantity of credit (through changes in the high powered money) and the cost of credit, the selective credit control seeks to change the distribution or allocation of credit between its various uses. The selective credit controls have both the positive and negative aspects. In its positive aspect, measures are taken to stimulate the greater flow of credit to some particular sectors considered as important. Thus in India, agriculture, small and marginal farmers, small artisans, small-scale industries are the priority sectors to which greater flow of bank credit has been sought to be encouraged by the Reserve Bank of India. In its negative aspect, several measures are taken to restrict the credit flowing into some specific activities or sectors which are regarded as undesirable or harmful from the social point of view. The selective credit controls generally used are:

(1) Changes in the minimum margin for lending by banks against the stocks of specific goods kept or against other types of securities.
(2) The fixation of maximum limit or ceiling on advances to individual borrowers against stocks of particular sensitive commodities.
(3) The fixation of minimum discriminatory rates of interest chargeable on credit for particular purposes.
(4) Prohibition of the discounting of bills of exchange involving sale of sensitive commodities.

The selective credit controls originated in U.S.A. to regulate the flow of bank credit to the stock market *(i.e.,* the market for shares). They were also used to restrict the volume of credit available for purchasing durable consumer goods during the period of the Second War. However, in India, the selective credit controls are being used by the Reserve Bank to prevent speculative hoarding of commodities so as to check the rise in prices of these commodities. The selective credit controls in India are being used in the case of foodgrains, oilseeds, vegetable oils, cotton, sugar, gur and khandsari.

Though all the above techniques of selective credit control are used, in India it is the first technique, namely, the changes in the minimum margin against stocks of commodities or other securities, that has been mostly used. It may be noted that the central bank of a country has the power to vary the minimum margin requirements against the security of stocks of commodities. While lending advances to businessmen, the commercial banks leave a margin of the value of stock kept as security to be financed by the businessmen from their own sources and lend funds to them equal to the remaining amount of the value of the stock. This minimum requirement of the value of the stock left to be financed by the borrowers themselves is known as margin. Suppose the margin fixed for a stock of particular commodity is 60 per cent. In this case the businessmen can borrow up to the value of 40 per cent of the stock of that commodity while the 60 per cent of the value of stock will be financed by the businessmen themselves. Now, if the bank raises the margin to 70 per cent, then businessmen can borrow from the bank to the extent of 30 per cent of the value of the stock of that commodity. This will lead to the contraction of credit for holding the stock of the commodity by the businessmen. If the businessmen are not able to finance the holding of 10 per cent extra stock of the commodity, they will be forced to sell that in the market and thus raising market supply of the commodity. This will lower the prices of the commodity, other things remaining the same.

In developed countries, the selective credit controls are generally used to prevent excessive speculation in share market. In the share market, the buyers purchase a good amount of shares by making a small payment and the remaining value of shares are paid by brokers through borrowing from the banks against the shares thus purchased. When the central bank raises the margin, the buyers of shares have to pay a larger sum of money for the shares purchased and as a result bank credit contracts and the speculative activity in the stock market is discouraged.

Conditions Necessary for Success of Selective Credit Controls

Some conditions are necessary for the successful operations of selective credit controls of commodities. However, the clever businessmen can obtain credit from the banks by offering other securities and use the funds so obtained to finance the speculative holdings of the stocks of sensitive commodities. Therefore, if the selective credit controls are to succeed in preventing the rise in prices of sensitive commodities, they have to be accompanied by general credit controls aimed at reducing the capacity of banks to lend money. It also follows from above that end-use or purpose of all credit ought to be taken into account by the banks and credit advanced accordingly if selective credit controls are to be effective.

In India the selective credit controls have been in operation since 1956 to check the rise in prices of sensitive commodities. However, the success of selective credit controls in achieving its objective of restraining the rise in commodity prices depends upon certain conditions explained below:

1. The Use of Quantitative Credit Controls. Selective credit controls are effective only when they are accompanied by the general quantitative credit controls such as variation in bank rate and cash reserve ratio. This is because the selective credit controls operate through regulating credit against particular securities or stocks.

2. The Availability of Non-Bank Finance. The success of the selective credit control also depends upon the extent to which the funds from non-bank sources *(i.e.,* from their own funds and also from the unregulated credit market) is available to the businessmen. When the bank credit for a particular purpose is reduced, the businessmen can use their own funds or borrow from non-regulated markets to indulge in speculative holding of inventories. In India today the businessmen have large quantities of black money with them which they generally use for speculative holding of inventories of sensitive commodities and in this way succeed in defeating the purpose of selective credit controls.

QUESTIONS FOR REVIEW

1. What are the main functions of a Central Bank? How does a Central Bank control the *direction of credit*?
2. How are the main functions of a Central Bank different from the functions of a commercial bank?
3. Explain the quantitative methods of credit control. Do you think these methods are adequate for effective control of credit supply?
4. How fare are the quantitative methods of credit control effective in developing countries like India.
5. Discuss the functions of a Central Bank. Between bank rate and open market operations which is more effective as an instrument of credit control.
6. Explain the various methods of selective credit control. Explain their significance in an economy like India.
7. Explain the quantitative methods of Credit Control adopted by a Central Bank. Which of the methods – quantitative or qualitative – is better for India.
8. Explain how Reserve Bank India tries to mountain exchange rate of rupee. Illustrate diagrammatically.

CHAPTER 16
OBJECTIVES AND INSTRUMENTS OF MONETARY POLICY

INTRODUCTION

Monetary policy is concerned with the measures taken to regulate the supply of money, the cost and availability of credit in the economy. Further, it also deals with the distribution of credit between uses and users and also with both the lending and borrowing rates of interest of the banks. In developed countries the monetary policy has been usefully used for overcoming depression and inflation as an anti-cyclical policy. However, in developing countries it has to play a significant role in promoting economic growth. As Prof. R. Prebisch writes, "The time has come to formulate a monetary policy which meets the requirements of economic development, which fits into its framework perfectly." Further, along with encouraging economic growth, the monetary policy has also to ensure price stability, because the excessive inflation not only adversely affects distribution of income but also hinders economic growth.

It is important to understand the distinction between *objectives* or *goals*, *targets* and *instruments* of monetary policy. Whereas **goals** of monetary policy refer to its objectives which, as mentioned above, may be price stability, full employment, exchange rate stability or economic growth, ***targets*** *refer to the variables such as supply of money or bank credit, interest rates which are sought to be changed through the tools or instruments of monetary policy so as to attain these objectives*. The various tools of monetary policy are changes in the supply of currency, variations in bank rate and other interest rates, open market operations, selective credit controls, and variations in reserve requirements.

OBJECTIVES OF MONETARY POLICY

Before explaining in detail the monetary tools undertaken by the Central Bank to regulate credit and growth of money supply, it is important to explain the objectives of monetary policy pursued by RBI which is India's central bank in formulation of its policy. Since monetary policy is one instrument of economic policy, its objectives cannot be different from those of overall economic policy. The three important objectives of monetary policy are:

1. ensuring *price stability,* that is, containing inflation.
2. to encourage *economic* growth.
3. to ensure *stability of exchange rate of the rupee,* that is, exchange rate of rupee with the US dollar, pound sterling and other foreign currencies.

Let us explain below these objectives in some detail.

Price Stability or Control of Inflation. It may be noted that each instrument of economic policy is better suited to achieve a particular objective. Monetary policy is better suited to the achievement of price stability, that is, containing inflation. To quote C. Rangarajan, a former Governor of Reserve Bank of India, "Faced with multiple objectives that are equally relevant and desirable, there is always the problem of assigning to each instrument the most appropriate target or objective. Of the various objectives, *price stability is perhaps the one that can be pursued most effectively by monetary policy.* In a developing country like ours, acceleration of investment activity in the context of supply shocks in the agricultural sector tends to be accompanied by pressures on prices and, therefore, monetary policy has much to contribute in the short-run management."[1]

Thus, achieving price stability has remained the dominant objective of monetary policy of Reserve Bank of India. It may however be noted that price stability does not mean absolutely no

1. C. Rangarajan, "Issues in Monetary Management" printed in *Indian Economy: Essays on Money and Finance,* 1998, p. 6.

change in price at all. In a developing economy like ours where structural changes take place during the process of economic growth, some changes in relative prices do occur that generally put upward pressure on prices. Therefore, some changes in price level or, in other words, a certain rate of inflation is inevitable in a developing economy. Thus, *price stability means reasonable rate of inflation.* A high degree of inflation has adverse effects on the economy. Firstly, inflation raises the cost of living of the people and hurts the poor most. Therefore, inflation has been described as enemy No. 1 of the poor. Inflation sends many people below the poverty line. Secondly, inflation makes exports costlier and, therefore, discourages them. On the other hand, due to higher prices at home people are induced to import goods to a large extent. Thus, inflation has an adverse effect on the balance of payments. Thirdly, when due to a higher rate of inflation value of money is rapidly falling, people do not have much incentives to save. This lowers the rate of saving on which investment and economic growth depend. Fourthly, a high rate of inflation encourages businessmen to invest in the unproductive assets such as gold, jewellery, real estate, etc.

An expert committee on monetary reforms headed by Late Prof. S. Chakravarty suggested 4 per cent rate of inflation as a reasonable rate of inflation and recommended that monetary policy by RBI should be so formulated that ensures that rate of inflation does not exceed 4 per cent per annum.

Economic Growth. Promoting economic growth is another important objective of the monetary policy. In the past Reserve Bank has been criticised that it pursued the objective of achieving price stability and neglected the objective of promoting economic growth. Monetary policy can promote economic growth through ensuring adequate availability of credit and lower cost of credit. There are two types of credit requirements of businesses. Firstly, they have to finance their requirements of working capital and for importing needed raw materials and machines from abroad. Secondly, they need credit for financing investment in projects for building fixed capital. Easy availability of credit at low interest rate stimulates investment and thereby quickens economic growth.

However, during the seventies, eighties and the first half of nineties, Reserve Bank followed a tight monetary policy under which Cash Reserve Ratio (CRR) and Statutory Liquidity Ratio (SLR) were continually raised to restrict the availability of credit for private sector. Besides, lending rates of interest were kept at high levels which discouraged private investment. This tight monetary policy worked against promoting growth. However, in the opinion of Prof. Rangarajan, there is no conflict between the objectives of price stability and growth. Price stability, according to him, is a means to ensure economic growth. To quote him, "*It is price stability which provides the appropriate environment under which growth can occur and social justice can be ensured.*"[2] In our opinion, this may be true in the long run but *in the short* run *there* exists *trade-off between growth and inflation.* To ensure higher economic growth the adequate expansion of money supply and greater availability of credit at a lower rate of interest is needed. But large expansion in money supply and bank credit leads to the increase in aggregate demand which tends to cause a higher rate of inflation. This raises the issue of what is *acceptable trade-off* between growth and inflation, that is, what rate of inflation is acceptable to promote growth through appropriate monetary policy. Expert Committee on monetary policy headed by Late Prof. Chakravarty suggested a target of 4 per cent as "the acceptable rise in prices". According to it, the growth of money supply and availability of credit should be so regulated that rate of inflation does not exceed 4 per cent per annum. However, C. Rangarajan, former Governor of Reserve Bank, fixed a higher target, namely, 5 to 6 per cent rate of inflation in the context of objective of achieving 6 to 7 per cent rate of economic growth.[3]

It may be noted that in the context of the openness of the economy and floating exchange rate system, as is the case of the Indian economy today, the objective of achieving higher rate of economic growth through monetary measures may also conflict with objective of exchange rate stability, that

2. C. Rangarajan, "*The Changing Context of Monetary Policy*", Address to 31st Convocation of the Indian Statistical Institute, March 31, 1997.
3. *Op. cit.*

is, value of rupee in terms of the U.S dollar and other foreign currencies. Whereas prevention of the depreciation of rupee requires tightening of monetary policy, that is, raising of interest rate and reducing liquidity of the banking system so that banks restrict their credit supply, the promotion of growth objective requires lower lending rates of interest and greater availability of credit for encouraging private investment. It is this dilemma of conflicting objectives of achieving higher economic growth rate or price stability which is being presently faced in India (2013-14).

Exchange Rate Stability

Until 1991, India followed fixed exchange rate system and only occasionally devalued the rupee with the permission of IMF. The policies of floating exchange rate and increasing openness and globalisation of the Indian economy, adopted since 1991, have made the exchange rate of rupee quite volatile. The changes in capital inflows and capital outflows and changes in demand for and supply of foreign exchange, particularly US dollar, arising from the imports and exports cause great fluctuations in the foreign exchange rate of rupee. In order to prevent large depreciation and appreciation of foreign exchange rate Reserve Bank has to take suitable monetary measures to ensure foreign exchange rate stability. Owing to the fixed exchange rate system prior to 1991 the concern about foreign exchange rate had not played a significant role in the formulation of monetary policy.

Today, the exchange rate of rupee is determined by demand for and supply of foreign exchange (say, U.S dollar). When there is mismatch between demand for and supply of foreign exchange, external value of rupee changes. For instance, from August 2011 to December 2011 depreciation of rupee as against U.S dollar had been caused by the increase in demand for dollars from (1) the corporate sector for financing their imports, (2) Foreign Institutional Investors (FII) who wanted to take out their dollars from India (*i.e.,* capital outflow) to the U.S. had risen and (3) increase in demand for U.S dollar by the Indian banks on the instructions of the public sector undertakings for financing necessary imports from abroad. Since export earnings and capital inflows which determine the supply of dollars had not risen adequately, mismatch between dollars and supply of dollars had arisen causing the depreciation of rupee as against the U.S dollar.

Thus when there is large current account deficit (CAD) that exceeds comfortable level (which in Indian situation is estimated at 2.5 per cent of GDP as it can be financed by normal capital inflows), it tends to depreciate the Indian rupee. In 2011-12 there was current account deficit (CAD) of 4.2 per cent and for 2012-13 it is expected that current account deficit (CAD) will be around 5 per cent of GDP. This is too high current account deficit which is putting downward pressure on the exchange rate of rupee. The depreciation of rupee raises the costs of imports which adds to the inflationary pressures in the Indian economy. Therefore, reducing current account deficit (CAD) to the comfortable level is not only essential for exchange rate stability but also for controlling inflation and achieving price stability. According to D. Subbarao, the present Governor of RBI, the widening of current account deficit in India during the last two years (2011-12 and 2012-13) has emerged as a major constraint on easing of monetary policy as causing lower lending interest rates and large availability of funds with banks will lead to more imports which will add to the current account deficit.

To arrest the fall in value of rupee Reserve Bank raises the bank rate and repo rate and thus sending signals to the banks to raise their lending rates. It also raises cash reserve ratio (CRR) to reduce the liquidity in the banking system and thus reducing lendable resources of the banks. Thus, through rise in the cost of credit and reduction in the availability of credit, borrowings from the banks are discouraged which is expected to reduce the demand for dollars. The higher interest rates in India would also discourage foreign institutional investors and Indian corporates to invest abroad. This will also work to reduce the demand for dollars which will prevent the fall in the value of the rupee.

Alternatively, to prevent the depreciation of the rupee, Reserve Bank can release more dollars from its foreign exchange reserves. The release of more dollars by Reserve Bank will increase the supply of U.S. dollars in the foreign exchange market and will therefore tend to correct the mismatch between demand for and supply of the U.S. dollars. This will help in stabilising the exchange rate of the rupee.

It is clear from above that in the context of flexible exchange rate system, Reserve Bank has to intervene frequently to achieve stability of exchange rate at a reasonable level.

INSTRUMENTS OF MONETARY POLICY

The central bank of a country has the responsibility of controlling the volume and direction of credit in the country. Bank credit has become these days an important constituent of the money supply in the country. The volume and direction of bank credit has, therefore, an important bearing on the level of economic activity. Excessive credit will tend to generate inflationary pressures in the economy, while deficiency of credit supply may tend to cause depression or deflation. Lack of the availability of cheap credit may also hinder economic development of a country. At times of depression, there is a need to expand credit and at a time of inflation there is need to contract credit. For promoting economic development, expansion of cheap credit (credit at low rates of interest) is desirable. In order to prevent booms and depressions (*i.e.,* to maintain economic stability) and to promote economic growth, central bank seeks to control credit in accordance with the needs of the situation.

Broadly speaking, there are two types of tools or instruments of monetary policy for controlling credit:

(1) Quantitative or General Tools. These tools seek to change the total quantity of credit in general. These are three in number: (*i*) changing the bank rate and repo rate; (*ii*) open market operations; and (*iii*) changing the cash reserve ratio.

(2) Qualitative or Selective Control Tools. These tools aim at changing the volume of a specific type of credit. In other words, the selective credit control tools affect the use of credit for particular purposes.

We discuss below these tools in detail.

Bank Rate Policy

Bank rate is the minimum rate at which the central bank of a country provides loans to the commercial banks of the country. Bank rate is also called *discount rate* because in the earlier days central bank used to provide finance to the commercial banks by rediscounting bills of exchange. Through changes in the bank rate, the central bank can influence the creation of credit by the commercial banks. Bank credit these days is an important component of the supply of money in the economy. Changes in money supply affect aggregate demand and thereby output and prices. For instance, when the central bank raises the bank rate, the cost of borrowings by the commercial banks from the central bank would rise. This would discourage the commercial banks to borrow from the central bank. Further, when the bank rate is raised, the commercial banks also raise their lending rates. When the lending rates of interest charged by the commercial banks are higher, businessmen and industrialists would feel discouraged to borrow from commercial banks. This would tend to contract bank credit and hence would result in the reduction of money supply in the economy. The reduction in the supply of money would reduce aggregate demand or money expenditure. This would reduce prices and check inflation in the economy. Thus when the economy is gripped by inflation or rising prices, the bank rate is generally raised to contract credit creation by the banks.

On the other hand, when there is recession or depression in the economy, the bank rate is lowered to overcome it. A fall in the bank rate will cause the reduction in lending interest rates of the commercial banks. With credit or loans from banks becoming cheaper, businessmen would borrow more from the commercial banks for investment and other purposes. This would lead to the increase in aggregate demand for goods and services and help in overcoming recession and bringing about recovery of the economy.

If a country permits the free flow of funds in and out of the country, then the changes in the bank rate will also have an effect on the external capital flows. For instance, as said above, when the bank rate is raised all interest rates in the market would also generally rise. With the rise in the deposit rates of interest of the banks, capital flows from outside would be attracted to the banks of the country for investment in debt and equity markets. Further, with the rise in deposit rates, outflow of funds from the banks to the other countries would be prevented. Therefore, these effects on the capital flows of the rise in bank rate will have a favourable effect on the balance of payments of the country as it will help to meet the current account deficit of the country. At present in India we are facing a huge current account deficit in our balance of payments as our import bill of crude oil, petroleum products and gold has gone up while our exports have slowed down.

Capital outflows create a lot of problems. When foreign capital is withdrawn from our debt and equity markets, not only our stock market crashes but also they raise the demand for U.S. dollars to send abroad. As a result of increase in demand for dollars, the exchange rate of rupee per U.S. dollar falls, that is, Indian rupee depreciates. Though depreciation of a currency is generally considered as desirable as it tends to raise exports and reduce imports. However, because our demand for imports is inelastic, their prices rise as a result of depreciation, but the volume of imports remains unaffected. The rise in prices of imports therefore contributes to inflation in the economy and worsens it further. To prevent this, rate of interest must be kept high.

It is important to note that the *changes in bank rate affect the credit creation by banks through altering the cost of credit.* Changes in the cost of credit affect borrowings by the commercial banks from the central bank and also affect the demand for credit by businessmen from the commercial banks.

Repo Rate in India

However, changes in bank rate as a tool of monetary policy in India has been quite limited. Before 1991, bank rate remained unchanged at 10 per cent in the whole decade of 1981-1991. This was because prior to 1991, interest rates charged by banks were administered by the Reserve Bank and therefore could not be used as a reference rate by the banks for the purpose of fixing their lending rates. In the post-reform period from July 1991, to control inflationary pressures in the Indian economy bank rate was raised from 10 per cent to 11 per cent. However, since then bank rate has been rarely used. This is for two reasons. First, the bank rate is meant **for a longer intermediate period for regulating interest rates charged by the banks** from their customers and thereby influences the cost of credit. However, in the Indian context what is needed is to regulate liquidity of the banks for a very short period, even on daily basis. Secondly, even now when lending rates of interest have been freed from control by RBI, **there is not much refinance** being made available at the bank rate and therefore the banks can ignore this as a reference in setting their own lending rates.

However, the need for influencing the lending rates of banks and liquidity of the Indian banks is very much needed. Therefore, for this purpose in 1997 RBI introduced under its **liquidity adjustment facility (LAF),** a rate of interest called **repo rate.** Under *repo rate system* banks could get overnight short-term loans from the RBI to meet their needs of liquidity if they arise. Through repo system RBI buys securities from the banks and thereby provides loans to them. *The rate of interest at which the RBI lends overnight funds to the banks against securities is called repo rate.* Repo rate is short-run interest rate and bank rate is the long-term interest rate at which banks could borrow from the RBI. The introduction of repo rate helps the Reserve Bank to increase the liquidity of the banks by purchasing from them Government securities and meeting their needs for liquidity when they arise. There is also ***reverse repo*** mechanism in which the banks park their surplus funds with the RBI on which they earn interest. The interest rate which RBI pays to the banks for the funds they keep with it for the short term under this scheme is called *reverse repo rate.* Through reverse repo system, the RBI mops up liquidity from the banking system.

Repo refers to agreement for a transaction between RBI and banks through which RBI supplies funds immediately against Government securities and simultaneously banks agree to repurchase the same or similar securities after a specified time which may be one day to 14 days. Through *reverse repo,* RBI mops up liquidity from the banking system by selling them securities when it finds that there is excess liquidity in the system which may create inflationary pressures. Under the repo agreement banks can get short-term loans from RBI against the securities to meet the needs of liquidity when they arise. Thus, *Repo and Reverse Repo operations are useful tools of short-run liquidity management by RBI.* Repo and Reverse Repo operations should not be confused with open market operations as the latter involve sale and purchase of Government securities in the market for a longer intermediate period and not for one day or few days. Besides, Repo and Reverse Repo rates of interest are fixed by RBI while in open market operations rate of interest varies and depends on demand and supply conditions of funds.

As a tool of monetary policy when Reserve Bank of India wants to check inflation, **it raises the repo rate.** When the commercial banks find that borrowing from the RBI has become costly and if they have not much excess reserves with them they will raise their lending rates. At higher lending rates businessmen will be discouraged for taking loans from the banks to finance their investment projects. Likewise, persons who want loans for building or purchasing houses and cars will also be discouraged to get loans from the banks. This will result in contraction of credit which will tend to lower aggregate demand and thereby help in controlling inflation.

On the other hand, when there are recessionary conditions in the economy, the RBI lowers the repo rate. With fall in the repo rate the banks can borrow from the banks at lower short-term rate of interest. In turn they lower their lending rates to businessmen for investment and to the consumers who seek home loans and car loans. This raises investment and consumption demand which through the operation of multiplier helps to raise the level of income and employment. As a result, the economy is lifted out of recession. Such a situation prevailed in India in 2007-09 when following the global financial crisis there was decline of our exports landing the Indian economy into industrial slowdown. Then RBI lowered its repo rate from 9 per cent to 4.5 per cent during the period from Oct. 2008 to March 2009. This helped to raise investment and consumption demand which led to recovery of the Indian economy in 2009-10 and 2010-11.

Limitations of Bank Rate and Repo Rate Policy

Both bank rate and repo rate are interest rate policy of RBI and therefore both have common limitations. Bank rate or repo rate policy does not have always the desired effect on investment, output and prices. There are certain conditions which must be met for the successful working of the bank rate or repo rate policy. These conditions are:

(1) All other rates should follow the bank/repo rate in its movement so that bank credit should expand or contract as desired. This will not happen if the commercial banks have considerable reserves of their own at their disposal and, therefore, their dependence on borrowed funds from the central bank may be very less. Further, for the changes in the bank rate to cause changes in all other interest rates in the market, a well-organised money market is needed. If well-organised money market does not exist as is the case in India where indigenous bankers contribute a good part of the money market, changes in the bank rate or repo rate would not be followed by the appropriate changes in all rates.

(2) The second important condition for the successful working of the bank/repo rate policy is the responsiveness of businessmen to the changes in the lending rates of interest. If businessmen and investors reduce their borrowings when the bank rate/repo rate and hence the lending rates of commercial bank are raised, and increase their borrowings for investment when the bank rate and lending rates of the banks are reduced, the changes in the bank/repo rate will have the desired effect, predicted in the bank rate theory. However, empirical studies conducted recently have shown that rate of interest does not exercise a strong influence on borrowings for investment and other purposes. When there is inflationary situation in the economy, the rate of interest will have to be raised very high in order to have desired effects. Similarly, the response of investment to the fall in the rate of interest is never vigorous. Demand by businessmen for loans for investment purposes from the banks depends on the economic situation prevailing in the economy. When the economy is gripped by severe depression and consequently prospects for making profits are bleak, businessmen would be reluctant to borrow for investment even though the lending rates of interest have been lowered considerably to induce them to borrow. It has been aptly remarked that '*you can bring the horse to water but you cannot make it drink*'.

The present belief is that interest rate policy will play only a subsidiary role in the monetary control. The bank rate or repo rate at the best could check a boom and inflation but could not bring about a recovery if the country was suffering from recession or depression. Thus it has much greater potential effectiveness if it is desired to arrest credit expansion than when it is sought to stimulate credit. But used in conjunction with other methods, it can still play a useful role. The bank rate and repo rate have the value as a signal. If the traditions are that a signal must be obeyed, the bank rate or repo rate policy can fulfil its purpose since it will be in keeping with respectable conduct.

Open Market Operations

Open market operations are another important tool of monetary control, especially in the developed countries. The term open market operations means the purchase and sale of securities by the central bank of the country. The theory of open market operations is like this: The sale of securities by the central bank leads to the contraction of credit and the purchase thereof to credit expansion. When the central bank sells securities in the open market, it receives payment in the form of a cheque on one of commercial banks. If the purchaser is a bank, the cheque is drawn against the purchasing bank itself. In both cases the result is the same. The cash reserves of the bank in question, which it keeps with the central bank, is to that extent reduced. With the reduction of its cash reserves, the commercial bank has to reduce its lending. Thus, credit contracts. On the other hand, when the central bank purchases securities, it pays through cheques drawn on itself. This increases the cash balance of the commercial banks and enables them to expand credit. "Take care of the legal tender money and credit will take care of itself " is the maxim.

The open market operations method is sometimes adopted to make the bank/repo rate policy effective. If the member-banks do not raise lending rates following the rise in the bank or repo rate due to surplus funds available with them, the central bank can withdraw such surplus funds by the sale of securities and thus compel the member-banks to raise their rates. Scarcity of funds in the market compels the banks directly or indirectly to borrow from the central bank through rediscounting of bills or through repo rate system of liquidity. If the bank rate or repo rate is high, the market rate of interest cannot remain low.

It may be noted that to check inflation the RBI sells government securities to the banks through open market operations and receives in return cash from them. This reduces the cash reserves with the banks and thus reducing their liquidity. This causes contraction of credit which helps in fighting inflation.

On the other hand, when RBI faces recessionary problem in the economy, it will seek to expand credit or money supply in the economy. This can be done through open market operations by buying securities from the open market. With this purchase of securities from the market the RBI pays cash funds to the banks and thus raising their cash reserves or liquidity. Against this increase in cash reserves, banks are able to create credit or bank deposits through providing loans by a multiple of the increase in their cash reserves depending on the value of deposit multiplier.

Limitations of the Open Market Operations. It is obvious that this method will succeed only if certain conditions are satisfied. The limitations are discussed below:

(1) According to the theory of open market operations, when the central bank purchases securities, the cash reserves of the member-banks will be increased and, conversely, the cash reserves will be decreased when the central bank sells securities. This, however, may not happen. The sale of securities may be offset by return of notes from circulation and hoards. The purchase of securities, on the other hand, may be accompanied by a withdrawal of notes for increased currency requirements or for hoarding by the public. In both the cases, therefore, the cash reserves of the member-banks may remain unaffected.

(2) But even if the cash reserves of the member-banks are increased or decreased, the banks may not expand or contract credit accordingly. The percentage of cash to credit is not rigidly fixed and can vary within quite wide limits. The banks will expand and contract credit according to the prevailing economic and political circumstances and not merely with reference to their cash reserves.

(3) The third condition is that when the commercial bank's cash reserves increase, the demand for loans and advances should increase too and *vice versa*. This may not happen. Owing to economic or political uncertainty, even cheap money rate may not attract borrowers. Conversely, when there is boom in the economy and prospects of profits bright, entrepreneurs would borrow even at high rates of interest.

(4) Finally, the circulation of bank credit should have a constant velocity. But the velocity of bank deposits is rarely constant. It increases in periods of rising business activity and decreases in periods of depression. Thus, a policy of contracting credit may be neutralised by increased velocity of circulation, and *vice versa*.

Open market operations as an instrument of credit control have proved to be quite successful in regulating the availability of credit in developed countries. This is for two reasons. First, when the central bank purchases or buys securities, the reserves of the commercial banks automatically expand or contract. The changes in the reserves of the banks directly affect their capacity to lend. Unlike in the case of instrument of bank rate, the success of the market operations does not depend on the attitude or responsiveness of the commercial banks; their capacity to provide credit is affected automatically. Secondly, the open market operations can be managed in a way that the cash reserves of the banks increase or decrease to the desired extent.

Since for the success of market operations it is necessary that there should be broad and active market in short and long-term government securities, and such markets exist only in the U.S.A. and Great Britain and other developed countries, this method of credit control has been most effectively used in these countries.

Open market operations which did not play a significant role as an instrument of credit control in India in the past have now become important tool and are being used extensively by RBI. This is because market for Government securities in India has now become quite wide and there is often excess liquidity in the banking system. General public do not buy more than a fraction of government securities. It is banks and other financial institutions that buy or sell government securities. When there is excess liquidity in the banking system, banks willingly invest in Government securities more than the statutory limits.

Changing the Cash Reserve Ratio (CRR)

Another monetary tool to vary the quantity of credit is to change the required cash reserve ratio. By law, banks have to keep a certain amount of cash reserves with RBI as reserves against the demand and time deposits. If the legal minimum cash reserve ratio, for example, is 10%, the bank will have to keep ₹ 2,000 as reserves with RBI against its deposits of ₹ 20,000. Now the central bank of a country has the authority to vary the cash reserve ratio. If now the central bank increases the cash reserve ratio from 10% to 20%, then the reserves of ₹ 2,000 could support only deposits of ₹ 10,000 and therefore the banks having reserves of ₹ 2,000 will have to reduce their deposits form ₹ 20,000 to ₹ 10,000. This will cause the contraction of credit.

On the other hand, if the central bank reduces the cash reserve ratio from 20% to 10% then the reserves of ₹ 2,000 could support the deposit of ₹ 20,000. Therefore the bank having reserves of ₹ 2,000 can increase their deposits, *i.e.,* expand credit to ₹ 20,000. To sum up, increase in the legal cash reserve ratio leads to the contraction of credit and decrease in the legal cash reserve ratio leads to the expansion of credit. *An increase in the legal cash reserve ratio will succeed in contracting credit only when banks have no excess reserves.* If banks are holding excess reserves, the increase in the legal reserve ratio will not bring about the contraction of credit. On the other hand, the reduction in the cash reserve ratio will have the desired effect of expanding credit only if the borrowers respond favourably. As a result of reduction in the cash reserve ratio, the availability to lend by the banks increases and the banks tend to make more credit available to borrowers, or to make it available at lower rates. Now, if the borrowers for one reason or another do not respond favourably, *i.e.,* are not prepared to borrow, credit will not be expanded.

Reserve Bank of India has often used this monetary tool to influence the availability of bank credit depending on the conditions prevailing in the Indian economy. For example, when there were recessionary conditions in the Indian economy following the global financial crisis in 2007-09, RBI over several stages sharply reduced the cash reserve ratio (CRR) to overcome economic slowdown and bring about recovery. With the reduction in cash reserve ratio, a lot of cash reserves were released for the banks which was used by the banks to provide more loans to businessmen for investment purposes and to individuals for buying houses, cars and scooters. This led to the increase in aggregate demand in the economy which helped in recovery of the Indian economy.

On the other hand, when during 2011-12 and 2012-13, a serious problem of inflation (*i.e.,* rise in general price level) prevailed in the Indian economy, Reserve Bank of India raised the cash reserve ratio (CRR). As a result, banks had to keep more cash reserves with the RBI resulting in decline in the credit availability of the banks. This helped to curtail consumption and investment demand in the economy due to lack of credit. This increase in cash reserve ratio for the banks helped a lot in reducing aggregate demand in the economy and bringing about fall in the WPI inflation rate to around 5 per cent in April 2013.

SELECTIVE CREDIT CONTROLS

The methods of credit control described above are known as quantitative or general methods as they are meant to control the availability of credit in general. Thus, bank rate policy, open market operations and variations in cash reserves ratio expand or contract the availability of credit for all purposes. On the other hand, selective credit controls are meant to regulate the flow of credit for particular or specific purposes. Whereas the general credit controls seek to regulate the total available quantity of credit (through changes in the high powered money) and the cost of credit, the selective credit control seeks to change the distribution or allocation of credit between its various uses. The selective credit controls have both the positive and negative aspects. In its positive aspect, measures are taken to stimulate the greater flow of credit to some particular sectors considered as important. Thus in India, agriculture, small and marginal farmers, small artisans, small-scale industries are the priority sectors to which greater flow of bank credit has been sought to be encouraged by the Reserve Bank of India. In its negative aspect, several measures are taken to restrict the credit flowing into some specific activities or sectors which are regarded as undesirable or harmful from the social point of view. The selective credit controls generally used are:

(1) Changes in the minimum margin for lending by banks against the stocks of specific goods kept or against other types of securities.

(2) The fixation of maximum limit or ceiling on advances to individual borrowers against stocks of particular sensitive commodities.

(3) The fixation of minimum discriminatory rates of interest chargeable on credit for particular purposes.

(4) Prohibition of the discounting of bills of exchange involving sale of sensitive commodities.

The selective credit controls originated in U.S.A. to regulate the flow of bank credit to the stock market *(i.e.,* the market for shares). They were also used to restrict the volume of credit available for purchasing durable consumer goods during the period of the Second War. However, in India, the selective credit controls are being used by the Reserve Bank to prevent speculative hoarding of commodities so as to check the rise in prices of these commodities. The selective credit controls in India are being used in the case of foodgrains, oilseeds, vegetable oils, cotton, sugar, gur and khandsari.

Though all the above techniques of selective credit control are used, in India it is the first technique, namely, the changes in the minimum margin against stocks of commodities or other securities, that has been mostly used. It may be noted that the central bank of a country has the power to vary the minimum margin requirements against the security of stocks of commodities. While lending advances to businessmen, the commercial banks leave a margin of the value of stock kept as security to be financed by the businessmen from their own sources and lend funds to them equal to the remaining amount of the value of the stock. This minimum requirement of the value of the stock left to be financed by the borrowers themselves is known as margin. Suppose the margin fixed for a stock of particular commodity is 60 per cent. In this case the businessmen can borrow up to the value of 40 per cent of the stock of that commodity while the 60 per cent of the value of stock will be financed by the businessmen themselves. Now, if the bank raises the margin to 70 per cent, then businessmen can borrow from the bank to the extent of 30 per cent of the value of the stock of that commodity. This will lead to the contraction of credit for holding the stock of the commodity by the businessmen. If the businessmen are not able to finance the holding of 10 per cent extra stock of the commodity, they will be forced to sell that in the market and thus raising market supply of the commodity. This will lower the prices of the commodity, other things remaining the same.

In developed countries, the selective credit controls are generally used to prevent excessive speculation in share market. In the share market, the buyers purchase a good amount of shares by making a small payment and the remaining value of shares are paid by brokers through borrowing from the banks against the shares thus purchased. When the central bank raises the margin, the buyers of shares have to pay a larger sum of money for the shares purchased and as a result bank credit contracts and the speculative activity in the stock market is discouraged.

Conditions Necessary for Success of Selective Credit Controls

Some conditions are necessary for the successful operations of selective credit controls of commodities. However, the clever businessmen can obtain credit from the banks by offering other securities and use the funds so obtained to finance the speculative holdings of the stocks of sensitive commodities. Therefore, if the selective credit controls are to succeed in preventing the rise in prices of sensitive commodities, they have to be accompanied by general credit controls aimed at reducing the capacity of banks to lend money. It also follows from above that end-use or purpose of all credit ought to be taken into account by the banks and credit advanced accordingly if selective credit controls are to be effective.

In India the selective credit controls have been in operation since 1956 to check the rise in prices of sensitive commodities. However, the success of selective credit controls in achieving its objective of restraining the rise in commodity prices depends upon certain conditions explained below:

1. The Use of Quantitative Credit Controls. Selective credit controls are effective only when they are accompanied by the general quantitative credit controls such as variation in bank rate and cash reserve ratio. This is because the selective credit controls operate through regulating credit against particular securities or stocks.

2. The Availability of Non-Bank Finance. The success of the selective credit control also depends upon the exent to which the funds from non-bank sources *(i.e.,* from their own funds and also from the unregulated credit market) is available to the businessmen. When the bank credit for a particular purpose is reduced, the businessmen can use their own funds or borrow from non-regulated markets to indulge in speculative holding of inventories. In India today the businessmen have large quantities of black money with them which they generally use for speculative holding of inventories of sensitive commodities and in this way succeed in defeating the purpose of selective credit controls.

QUESTIONS FOR REVIEW

1. What are the objectives of monetary policy, especially of Reserve Bank of India? Is there a trade-off between the objectives of controlling inflation and promoting economic growth? Discuss.
2. What are the instruments of monetary policy? Which tools of monetary policy are mostly used by the Reserve Bank of India?
3. What are repo and reverse repo rates? How are they used by RBI to regulate liquidity in the Indian banking system?
4. What is Bank Rate? What is the effect of increase in bank rate? What is the effect of reducing the bank rate? What are the limitations of using the tool of changes in bank rate to influence the level of economic activity?
5. What are open market operations? If Reserve Bank of India wants to promote economic growth, how will it use the tool of open market operations?
6. If Reserve Bank of India seeks to control inflation, how will it conduct its open market operations? Trace the events that follow from the sale of securities by the Reserve Bank of India in this context.
7. What is cash reserve ratio? How has a reduction in cash reserve ratio by Reserve Bank of India a multiplier effect on the aggregate demand?
8. How does increase in cash reserve ratio help to check inflation in the economy?
9. How does the open market purchase of government securities by the Reserve Bank of India help to overcome recession in the economy? What are its limitations?
10. Prior to 1991, what monetary policy tools were used by Reserve Bank of India to neutralise the inflationary impact of expansionary fiscal policy? Explain.

CHAPTER 16A
ROLE OF MONETARY POLICY IN ECONOMIC GROWTH

INTRODUCTION

Economic growth implies the expansion in productive capacity or capital stock in the economy so that increases in real national output or income are attained. As is well known, economic growth can be speeded up by accelerating the rate of savings and investment in the economy. This requires the following steps:

(*a*) Increase in the aggregate rate of savings in the economy;
(*b*) Mobilisation of these savings so that they are made available for the purpose of investment and production;
(*c*) Increase in the rate of investment; and
(*d*) Allocation of investment funds for productive purposes and priority sectors of the economy.

Proper monetary policy can help in producing favourable effects on the above requirements of economic growth. We explain below the appropriate monetary policy to promote growth in developing countries.

MONETARY POLICY AND SAVINGS

Several monetary measures can be adopted to raise the aggregate rate of saving. First, a high interest rate policy can promote savings. In the fifties and sixties of the 20th Century, it was widely believed that interest rate reflected the price of capital and since capital was scarce in developing countries, the real interest rates should be kept at higher levels to promote saving so that capital accumulation is speeded up. However, this argument in favour of higher real interest rates in developing countries was based upon the belief or assumption that *savings were positively elastic or sensitive to the interest rate*.

It is worthwhile to mention an opposite viewpoint regarding interest rate policy which has gained large support in recent years. According to this view (emphasised by Keynes in his monetary theory), rate of interest represents the cost of investment, and lower the rate of interest, the greater will be inducement to invest. However, Keynes was of the view that investment was not much sensitive or elastic to lower rate of interest Dr. K.N. Raj argued that *since investment is an important factor determining economic growth in developing countries, it should be promoted by lowering the interest rates*. However, this is not fully correct. This is because whereas inducement to invest may be promoted by lowering the interest rates, the adequate amount of savings or resources needed to finance larger investment may not be forthcoming at lower interest rates. Further, a low interest rate policy in developing countries like India is likely to promote more investment in inventories and luxury consumer goods such as cars, airconditioners, luxury houses rather than that in fixed capital goods.

In our view, therefore, in a developing economy *a reasonably positive real interest rate policy* should be pursued to provide incentives for more savings so that large sources are made available for investment in fixed capital stock. If monetary policy in a developing country is to promote economic growth it must aim at raising the rate of saving. It may be recalled that *real rate of interest is nominal rate of interest minus rate of inflation*. In our view real rate of interest should be positive and preferably should not be allowed to fall below 5 per cent per annum, if it is to provide reasonable rate of return on savings. Therefore, with the rise in rate of inflation nominal rate of interest should also be raised to keep incentives for saving intact. It may be noted that in recent years (2011-12 and 2012-13) in India low deposit rates offered to the public by banks have prompted the people to invest their savings in unproductive assets such as gold, jewellery and real estate. *This led to the fall in financial saving* required for productive investment. This led the Finance Minister, Mr. Chidambaram, in his budget for 2013-14 to announce issuing of **inflation-indexed bonds** on which nominal interest rate will automatically rise with the increase in inflation rate.

Besides, the monetary policy can play a strategic role in increasing national savings by promoting the expansion of banking facilities and other financial intermediaries in the developing countries, especially in their rural areas. With more bank branches in under-banked and underdeveloped regions, the people who consume away their surplus incomes will be induced to save them in the form of bank deposits which are quite safe as a store of value. The commercial banking encourages thrift or propensity to save by offering a return on savings in the form of interest rate on bank deposits. It also induces more savings by providing outlets for mutual funds which can be equity linked and other saving schemes for fruitful investment of the savings by the people who would have otherwise put them to unproductive or wasteful uses such as buying land, real estate, gold and jewellery. But to tap and raise savings sufficiently and to prevent its unproductive use, banks and other financial institutions need to be numerous and dispersed throughout the economy in both the urban and rural areas. Speaking from the development experience of various countries, Prof. Lewis, a development economist, has emphasised that the volume of savings depends partly upon how widespread banking and other financial institutions are. To quote him, "*If the saving institutions are placed under the individual's nose, people save more than if the nearest saving institution is some distance away.*"[1]

Thus growth in saving in the form of bank deposits will be greater, if a reasonable real rate of interest policy is pursued. Similarly, with the expansion of other financial institutions, the people will feel induced to save more for buying financial assets. It is generally agreed that the rapid expansion of banking facilities after the nationalisation of major commercial banks in 1969 promoted by the Reserve Bank of India has contributed considerably to the growth of aggregate savings. While the aggregate saving rate has gone up from 16.5 per cent in 1969-70 to 22.7 per cent in 1982-83, bank deposits, which accounted for 8.7 per cent of total domestic savings in 1969-70, accounted for 22.5 per cent of these savings in 1982-83. The number of commercial bank offices increased from 8,262 in July 1969 to 44,521 in 1984 and to 62,350 in 1995. Although it is difficult to establish a precise quantitative relationship between expansion of bank branches and savings in the form of bank deposits, its significant contribution to the promotion of savings cannot be denied. Therefore, the future branch expansion policy must fully take into account the untapped saving potential of under-banked and underdeveloped regions.

It may be noted that in order to facilitate mobilisation of an increasing proportion of savings by the banking system, it is essential to maintain a reasonable price stability. Dr. Manmohan Singh, when he was Governor of Reserve Bank of India, pointed out, "*It is only under conditions of reasonable price stability that people will have less attraction for savings in the form of physical assets such as gold, real estate and excessive accumulation of inventories*." Further, if banks are to mobilise adequate amount of savings in the form of bank deposits, interest rates on bank deposits must remain positive in real terms. That is, the rate of interest be maintained at a higher level than the rate of inflation. If through excessive rise in prices real rate of interest becomes negative, people will be discouraged to save.

MONETARY POLICY AND INVESTMENT

Monetary policy has an important role to play in boosting up the level of investment by making available savings or resources mobilised by banks for purposes of investment and production. The banks fulfil this task by offering bank credit for investment to business and industry.

Cost of Credit. It may be noted that Keynesian theory of monetary policy emphasises that the effect of a change in the supply of money on the level of production and investment operates through the changes in the rate of interest. The increase in the supply of money by the monetary authority will cause the market rate of interest to fall. At lower rate of interest, the entrepreneurs will be induced to invest more. Accordingly, Keynesian monetary theory has been described as "*cost of credit*" theory. However, Keynes himself and several economists after him believed that this effect of changes in money supply on investment which operates through rate of interest is not very vigorous. It has been asserted that investment in fixed capital is interest inelastic. That is why Keynes did not have much faith in the effectiveness of monetary policy and instead emphasised the role of fiscal policy in influencing the level of economic activity. However, it is now generally believed by the economists that since interest is an important cost of investment, *investment demand is fairly interest elastic*. Therefore, to promote investment, the easy monetary policy should be adopted.

It may be noted that in the post-economic reforms period from 1996 to 2003 easy or low interest rate policy was adopted to promote private investment so as to revive the industrial sector whose growth slowed down from 1996-97 to 2002-03. Following the soft interest rate policy of Reserve Bank of India prime lending rates of interest of commercial banks, which varied between 15 and 16 per cent prior to 1996, were reduced over a period of time to 9 to 10 per cent in 2003. Again when in September 2008 global financial crisis occurred as a result of bust of housing bubble in the U.S, there was worldwide (including India) economic slowdown, Reserve Bank of India in October 2008

1. Arthur Lewis, *Theory of Economic Growth*, George Allen and Unwin, 1960.

reduced its repo interest rate from 9 per cent in July 2008 to 7 per cent in October 2008 and further to 4.75 per cent on March 4, 2009, to promote private investment so as to prevent sharp economic slowdown in India.

However, the soft interest rate policy does not always, especially during recession, achieve much success in stimulating private investment. This is because when there are sluggish demand conditions, the existing fixed capital is underutilised, it is not profitable to make more investment for expanding fixed capital stock, private enterprises are therefore not willing to make further investment despite lower interest rates. It is for such economic conditions that it has been said that '*you can bring the horse to water but you cannot make it drink*'. However, in normal circumstances cheap interest rate policy promotes private investment and therefore helps to achieve higher economic growth.

Availability of Credit. It is noteworthy that in the recent times monetary theory emphasises the **credit-availability effect on investment** of the changes in the supply of money. According to this, an increase in the supply of money causing expansion in the reserve money with the banks directly enlarges the availability of bank credit for investment purposes and thereby raises the level of investment in the economy. On the contrary, contraction in money supply will directly reduce the availability of credit and hence tend to decrease investment in the economy.

Further, in this context it is noteworthy that in developing countries, Government or public investment plays a crucial role in the development of their economies. Therefore, monetary policy requires to promote not only private investment but also public investment by making available adequate amount of credit for it.

Monetary Policy and Private Investment. If the private sector plays an important role in the development process, as it does in India after economic reforms undertaken since 1991, monetary policy has also to ensure that the needs for bank credit for investment and production in the private sector are fully met. The banks must provide adequate bank credit to meet at least essential working capital requirements of industry and agriculture. Both large-scale and medium industries require funds for investment in fixed capital, working capital and for maintaining inventories. Subject to appropriate norms fixed for inventory holdings, the credit needs of the private sector should be met if existing capacities in the private sector are to be fully used and also more productive capacity is to be built up. In India banks have been asked by the Reserve Bank of India to give specific percentage of loans at concessional rates of interest to some priority sectors such as agriculture and small and medium enterprises.

Monetary Policy and Public Investment. Let us first explain the promotion of the public investment through monetary policy. Monetary policy has to ensure that the banking system contributes to the financing of the planned public investment. For this a good part of bank deposits mobilised by the banks should be invested in Government and other approved securities so that the government should be able to finance its planned investment, especially in infrastructure. At present in India, there is lack of infrastructure such as power, roads, highways, ports, irrigation facilities which is obstructing economic growth. Besides, when industry faces demand recession, public investment is an ideal tool to develop infrastructure and to increase the demand for industrial products through the operation of multiplier. This will stimulate private investment. Therefore, in our view, policy of raising public investment will *crowd in* private investment rather than *crowding it out*. In India a new technique of monetary policy has been designed to secure larger resources from the banking system for financing public investment. This technique has been described as *Statutory Liquidity Ratio* (SLR). According to this, *in addition to cash reserve requirements* banks in India are required to keep a minimum proportion of their total demand and time deposits in the form of specified liquid assets. And the most important specified liquid asset for this purpose is the investment in Government and other securities. To raise the lendable resources of the banks, cash reserve ratio of the banks must be kept at low level.

ALLOCATION OF INVESTMENT FUNDS

Mobilisation of savings alone would not do. Proper canalisation of these into suitable directions of investment is or perhaps more important than mobilisation itself. The monetary policy should restrict the growth of wasteful lines of investment which are inimical to economic growth. It should be able to direct investment in productive channels. In this regard, the monetary policy has to play a selective or qualitative role in so far as it is possible through its operations to *discriminate between productive and unproductive outlays. It should give a fillip to the former and stint* the growth of the latter. Further, it has to be so designed as to influence the specific sectors and industries which are most significant to affect the growth of the economy. In fact, the particularised application of credit can greatly activise the process of economic growth. Therefore, it is necessary to operate the *selective credit rationing with a view to influence the pattern of investment*. However, in the gamut of monetary policy, there are selective credit control measures of both general and specific nature. In the general category are included measures such as voluntary credit restriction and moral suasion. In the category of specific measures there exist measures to control credit institutions.

For the purpose of regulating the individual credit institutions, the method of varying the reserve requirements is quite effective. On the other hand, in order to divert the financial resources into desired channels, the method of **credit rationing** should be used by the Central Bank. This may be done by fixation of a ceiling on the aggregate portfolio of the commercial banks, thereby making it incumbent that loans and advances do not exceed the fixed ceiling. Alternatively, it may be done by directly allocating funds that can be granted and used. Besides, policies such as the selective rediscount policy, prior deposit requirement policy and the fixation of deposit requirement policy can also be adopted to achieve similar goals. However, it may be pointed out that the potency of credit planning depends upon the extent of the area over which it operates. The developing economies must strive for an extension of credit planning over wider areas.

Credit control measures such as noted above promote growth by directing the stream of domestic savings into the desired lines of investment. Measures such as lengthening the periods of repayment, lowering of margin requirements, providing rediscounting facilities at rates below the bank rate, provision of special loans by commercial banks to be used for specific purposes or the setting up of special Investment institutions such as Industrial Development Bank can provide the required inducement to channelise savings in the desired directions. In the indirect sense, these measures may prove conducive to growth in two ways. Firstly, the qualitative credit control measures prevent the savings from being wasted in unproductive channels. Through their application it becomes possible to deny or discourage certain lines of investment that are inimical to the growth of the economy. However, the extent to which such measures can help providing resources for investment in the desired directions depends on the extent to which the flow of credit towards the undesirable channels can be prevented.

Secondly, such measures may go a long way in galvanising the process of growth by restraining inflation and its adverse effects. When inflationary tendencies set in, generally the bank advances to businessmen tend to rise. In this way certain undesirable and unproductive enterprises may grow and flourish. For instance, activities such as speculative demand for building up inventories, accumulation of precious metals for purchase of foreign exchange and real estate get a fillip. Growth of these and such other unproductive activities can be held in check by raising the margin requirements for the blackballed collaterals. Further, the monetary authority can fix a ceiling by holding the loans and advances of the value of the collateral. Thus, it *serves the dual purpose of curbing inflation and protecting certain essential productive forms of investment from being restricted.*

In the context of the planned development of the underdeveloped countries, the use of methods of selective credit controls and credit rationing is not only necessary but also essential. They greatly widen the horizons of development along predetermined and desired directions. They are not only helpful in preventing inflation but also *act as a positive means of directing the process of economic development on the desired lines*. Furthermore, the policy of selective credit controls is especially suited to the needs of underdeveloped countries where the orthodox monetary techniques have limited applicability. As it is, the structure of these economies is not very conducive to the general methods of credit control.

The huge investment expenditure made by the government in the public sector in its endeavour to accelerate the process of growth is not amenable to control by the monetary authority. The monetary authority is, in fact, subversive to the wishes of the government in the matter of providing resources for development. Further, in order to help government borrowings, the bank rate policy of the Reserve Bank tends to become more or less inflexible. Besides, for the sake of supporting government loans, the institutions such as SEBI (Stock Exchange Board of India) is also required to stabilise the securities market.

QUESTIONS FOR REVIEW

1. Economic growth depends on saving and investment. Explain the role of monetary policy in promoting saving and investment in a developing country such as India.
2. Explain the dilemma faced by Reserve Bank of India to achieve its to objectives of price stability and economic growth.
3. Role of monetary policy in India is to achieve economic growth with stability. Discuss.
4. Explain the role of monetary policy to ensure proper allocation of funds among uses and sectors to ensure *inclusive economic growth*.

CHAPTER 17
SUPPLY OF MONEY AND ITS DETERMINANTS

IMPORTANCE OF MONEY SUPPLY

Money supply plays a crucial role in the determination of price level and interest rate. In the present chapter we shall explain what determines the money supply in an economy. In economic analysis it is generally presumed that money supply is determined by the policy of Central Bank of a country and the Government. However, this is not fully correct as in the determination of money supply, besides Central Bank and Government, the public and commercial banks also play an important role. There are various measures of money supply depending upon which types of deposits of banks and other financial institutions are included in it. We will explain various measures of money supply later in this chapter.

Growth of money supply is an important factor not only for acceleration of the process of economic development but also for the achievement of price stability in the economy. What constitutes the money supply and what factors causes variation and growth in money supply will be explained in the present chapter. There must be controlled expansion of money supply if the objective of development with stability is to be achieved. A healthy growth of an economy requires that there should be neither inflation nor deflation. Inflation is the greatest headache of a developing economy. A mild inflation arising out of the creation of money by deficit financing may stimulate investment by raising profit expectations and extracting forced savings. But a runaway inflation is highly detrimental to economic growth. The developing economies have to face the problem of inadequacy of resources in initial stages of development and it can make up this deficiency by deficit financing. But it has to be kept strictly within safe limits. Thus, increase in money supply affects vitally the rate of economic growth. In fact, it is now regarded as a legitimate instrument of economic growth. Kept within proper limits it can accelerate economic growth but exceeding of the limits will retard it. Thus, management of money supply is essential in the interest of steady economic growth.

THE CONCEPT OF MONEY SUPPLY AND ITS MEASUREMENT

By money supply we mean the total stock of monetary media of exchange available to a society for use in connection with the economic activity of the country. According to the standard concept of money supply, it is composed of the following two elements:

1. Currency with the public,
2. Demand deposits with the public.

Before explaining these two components of money supply two things must be noted with regard to the money supply in the economy. First, the money supply refers to the total sum of money available to the public in the economy *at a point of time*. That is, money supply is a *stock concept* in sharp contrast to the national income which is a flow representing the value of goods and services produced *per unit of time*, usually taken as a year. Secondly, money supply always refers to the amount of money *held by the public*. In the term public are included households, firms and institutions other than banks and the government. The rationale behind considering money supply as held by the public is to separate the producers of money from those who use money to fulfil their various types of demand for money. Since the Government and the banks produce or create money for the use by the public, the money (cash reserves) held by them are not used for transaction and speculative purposes and are excluded from the standard measures of money supply. This *separation of producers of money from the users of money is important from the viewpoint of both monetary theory and policy.*[1]

Let us explain the two components of money supply at some length.

Currency with the Public

In order to arrive at the total currency with the public in India we add the following items:

1. Currency notes in circulation issued by the Reserve Bank of India.
2. The number of rupee notes and coins in circulation.
3. Small coins in circulation.

It is worth noting that *cash reserves* with the banks has to be deducted from the value of the above three items of currency in order to arrive at the total currency with the public. This is because cash reserves with the banks must remain with them and cannot therefore be used for making payments for goods or by any commercial banks' transactions. It may further be noted that these days paper currency issued by Reserve Bank of India (RBI) are not fully backed by the reserves of gold and silver, nor it is considered necessary to do so. Full backing of paper currency by reserves of gold prevailed in the past when gold standard or silver standard type of monetary system existed. According to the modern economic thinking the magnitude of currency issued should be determined by the monetary needs of the economy and not by the available reserves of gold and silver. As mentioned in the last chapter, as in other developed countries, since 1957 Reserve Bank of India follows Minimum Reserve System of issuing currency. Under this system, minimum reserves of Rs. 200 crores of gold and other approved securities (such as dollars, pound sterling, etc.) have to be kept and against this any amount of currency can be issued depending on the monetary requirements of the economy.

As stated earlier, RBI is not bound to convert notes into equal value of gold or silver. In the present times currency is inconvertible. The word written on the note, say 100 rupee notes and signed by the governor of RBI that 'I promise to pay the bearer a sum of 100 rupees' is only a legacy of the past and does not imply its convertibility into gold or silver. Another important thing to note is that paper currency or coins are *fiat money*, which means that currency notes and metallic coins serve as money on the bases of the *fiat* (*i.e.* order) of the Government. In other words, on the authority of the Government no one can refuse to accept them in payment for the transaction made. That is why they are called *legal tender*.

Demand Deposits with the Public

The other important component of money supply are demand deposits of the public with the banks. These demand deposits held by the public are also called *bank money* or *deposit money*. Deposits with the banks are broadly divided into two types: demand deposits and time deposits. Demand deposits in the banks are those deposits which can be withdrawn by drawing cheques on them. Through cheques these deposits can be transferred to others for making payments from whom goods and services have been purchased. Thus, cheques make these demand deposits as a medium of exchange and therefore make them to serve as money. It may be noted that demand deposits are *fiduciary money proper*. Fiduciary money is one which functions as money on the basis of *trust of* the persons who make payment rather than on the basis of the authority of Government. Thus, despite the fact that demand deposits and cheques through which they are operated are not legal tender, they functions as money on the basis of the trust commanded by those who draw cheques on them. They are money as they are generally acceptable as medium of payment.

1. S. B. Gupta, *Monetary Economics : Institutions, Theory and Policy*, S. Chand & Co., New Delhi, 3rd Edition, 1997, p. 272.

Bank deposits are created when people deposits currency with them. But far more important is that banks themselves create deposits when they give advances to businessmen and others. On the basis of small cash reserves of currency, they are able to create a much larger amount of demand deposits through a system called *fractional reserve system* which will be explained later in detail.

In the developed countries such as USA and Great Britain deposit money accounted for over 80 per cent of the total money supply, currency being a relatively small part of it. This is because banking system has greatly developed there and also people have developed banking habits. On the other hand, in the developing countries banking has not developed sufficiently and also people have not acquired banking habits and they prefer to make transactions in currency. However in India after 50 years of independence and economic development the proportion of bank deposits in the money supply has risen to about 50 per cent.

FOUR MEASURES OF MONEY SUPPLY

Several definitions of money supply have been given and therefore various measures of money supply based on them have been estimated. First, different components of money supply have been distinguished on the basis of the different functions that money perform. For example, demand deposits, credit card and currency are used by the people primarily as a medium of exchange for buying goods and services and making other transactions. Obviously, they are money because they are used as a medium of exchange and are generally referred to as M_1. Another measure of money supply is M_3 which includes both M_1 and time deposits held by the public in the banks. Time deposits are money that people hold as store of value.

The main reason why money supply is classified into various measures on the basis of its functions is that effective predictions can be made about the likely affects on the economy of changes in the different components of money supply. For example, if M_1 is increasing fastly it can be reasonably expected that pepole are planning to make a large number of transactions. On the other hand, if time-deposits component of money-supply measure M_3 which serves as a store of value is increasing rapidly, it can be validly concluded that people are planning to save more and accordingly consume less. Therefore, it is believed that for monetary analysis and policy formulation, a single measure of money supply is not only inadequate but may be misleading too. Hence various measures of money supply are prepared to meet the needs of monetary analysis and policy formulation.

Recently in India as well as in some developed countries, four concepts of money supply have been distinguished. The definition of money supply given above represents a narrow measure of money supply and is generally described as M_1. From April 1977, the Reserve Bank of India has adopted four concepts of money supply in its analysis of the quantum of and variations in money supply. These four concepts of measures of money supply are explained below.

1. Money Supply M1 or Narrow Money. This is the narrow measure of money supply and is composed of the following items:

$$M1 = C + DD + OD$$

where C = Currency with the public

DD = Demand deposits with the public in the Commercial and Cooperative Banks.

OD = Other deposits held by the public with Reserve Bank of India.

The money supply is the most liquid measure of money supply as the money included in it can be easily used as a medium of exchange, that is, as a means of making payments for transactions.

Currency with the public (C) in the above measure of money supply consists of the followings:

(*i*) Notes in circulation.

(*ii*) Circulation of rupee coins as well as small coins

(*iii*) Cash reserves on hand with all banks.

Note that in measuring demand deposits with the public in the banks (*i.e.*, *DD*), inter-bank deposits, that is, deposits held by a bank in other banks are excluded from this measure.

In the other deposits with Reserve Bank of India (*i.e.*, *OD*) deposits held by the Central and State Governments and a few others such as RBI Employees Pension and Provident Funds are excluded. However, these other deposits of Reserve Bank of India include the following items.

(*i*) Deposits of Insititutions such UTI, IDBI, IFCI, NABARD etc.

(*ii*) Demand deposits of foreign Central Banks and Foreign Governments.

(*iii*) Demand deposits of IMF and World Bank.

It may be noted that other deposits of Reserve Bank of India constitute a very small proportion (less than one per cent).

Money Supply *M2*

M_2 is a broader concept of money supply in India than M_1. In addition to the three items of M_1, the concept of money supply M_2 includes savings deposits with the post office savings banks. Thus,

$M2 = M1$ + Savings deposits with the post office savings banks.

The reason why money supply M_2 has been distinguished from M_1 is that saving deposits with post office savings banks are not as liquid as demand deposits with Commercial and Co-operative Banks as they are not chequable accounts. However, saving deposits with post offices are more liquid than time deposits with the banks.

Money Supply *M3* or Broad Money

M3 is a broad concept of money supply. In addition to the items of money supply included in measure M_1, in money supply M_3 time deposits with the banks are also included. Thus

$M3 = M1$ + Time Deposits with the banks.

It is generally thought that time deposits serve as store of value and represent savings of the people and are not liquid as they cannot be withdrawn through drawing cheque on them. However, since loans from the banks can be easily obtained against these time deposits, they can be used if found necessary for transaction purposes in this way. Further, they can be withdrawn at any time by forgoing some interest earned on them.

It may be noted that recently *M3* has become a popular measure of money supply. The working group on monetary reforms under the chairmanship of Late Prof. Sukhamoy Chakravarty recommended its use for monetary planning of the economy and setting target of the growth of money supply in terms of *M3*. Therefore, recently RBI in its analysis of growth of money supply and its effects on the economy has shifted to the use of M_3 measure of money supply. In the terminology of money supply employed by the Reserve Bank of India till April 1977, this *M3* was called *Aggregate Monetary Resources* (AMR).

Money Supply M4

The measure *M4* of money supply includes not only all the items of M_3 described above but also the total deposits with the post office savings organisation. However, this excludes contributions made by the public to the national saving certificates. Thus, $M4 = M3$ + *Total Deposits* with *Post Office Savings Organisation*.

Let us summarise the four concepts of money supply as used by Reserve Bank of India in the following tabular form:

Money Supply

$M1$	$M2$	$M3$	M_4
=	=	=	=
Currency (C)	$M1$	$M1$	$M3$
+	+	+	+
Demand Deposits (D)	Savings Deposits with Post-Office Savings Banks	Time Deposits with the Banks	Total Deposits of Post Office Savings Organisation (excluding NSC)
+			
Other Deposits of RBI (OD)			

DETERMINANTS OF MONEY SUPPLY

In order to explain the determinants of money supply in an economy we shall use M_1 concept of money supply which is the most fundamental concept of money supply. We shall denote it simply by M rather than M_1. As seen above this concept of money supply is composed of currency held by the public (C_p) and demand deposits with the banks (D). Thus

$$M = C_p + D \qquad ...(1)$$

where
M = Total money supply with the public
C_p = Currency with the public
D = Demand deposits held by the public

The two important determinants of money supply as described in (1) are (*a*) the amounts of high-powered money which is also called *Reserve Money* by the Reserve Bank of India and (*b*) the size of money multiplier. We explain below the role of these two factors in the determination of money supply in the economy.

1. **High-Powered Money (H).** The high-powered money which we denote by H consists of the currency (notes and coins) issued by the Government and the Reserve Bank of India. A part of the currency issued is held by the public, which we designate as C_p and a part is held by the banks as reserves which we designate as R. A part of these currency reserves of the banks is held by them in their own cash vaults and a part is deposited in the Reserve Bank of India in the Reserve Accounts which banks hold with RBI. Accordingly, the high-powered money can be obtained as sum of currency held by the public and the part held by the banks as reserves. Thus

$$H = C_p + R \qquad ...(2)$$

where
H = the amount of high-powered money
C_p = Currency held by the public
R = Cash Reserves of currency with the banks.

It is worth noting that Reserve Bank of India and Government are *producers of the high-powered money* and the commercial banks do not have any role in producing this high-powered money (H). However, commercial banks are *producers of demand deposits* which are also used as money like currency. But for producing demand deposits or credit, banks have to keep with themselves cash *reserves of currency* which have been denoted by R in equation (2) above. Since these cash reserves with the banks serve as a basis for the multiple creation of demand deposits which constitute an important part of total money supply in the economy, it provides high poweredness to the currency issued by Reserve Bank and Government. A glance at equations (1) and (2) above will reveal that the difference in the two equations, one describing the total money supply and the other high-powered money is that whereas in the former, demand deposits (D) are added to the currency held by the public, in the later it is cash reserves (R) of the banks that are added to the currency held by the public. In fact, it is against these cash reserves (R) that banks are able to create a multiple expansion of credit or demand leposits due to which there is large expansion in money supply in the economy

The theory of determination of money supply is based on the supply of and demand for high-powered money. Some economists therefore call it '*The H Theory of Money Supply*[2]. However, it is more popularly called '*Money-multiplier Theory of Money Supply*' because it explains the determination of money supply as a certain multiple of the high-powered money. How the high-powered money (H) is related to the total money supply is graphically depicted in Fig. 17.1. T*he base of this figure shows the supply of high-powered money (H), while the top of the figure shows the total stock of money* supply. It will be seen that the total stock of money supply (that is, the top) is determined by a multiple of the high-powered money (H). It will be further seen that whereas currency held by the public (C_p) uses the same amount of high-powered money, that is, there is one-to-one relationship between currency held by the public and the money supply. In sharp contrast to this, bank deposits (D) are a multiple of the cash reserves (R) of the banks which are part of the supply of high-powered money. That is, one rupee of high-powered money kept as bank reserves gives rise to much more amount of demand deposits. Thus, the relationship between money supply and the high-powered money is determined by the money multiplier. The money multiplier which we denote by m is the ratio of total money supply (M) to the stock of high-powered money, that is, $m = \dfrac{M}{H}$. The size of money multiplier depends on the preference of the public to hold currency relative to deposits, (that is, ratio of currency to deposits which we denote by K) and banks' desired cash reserves ratio to deposits which we call r. We explain below the precise multiplier relationship between high-powered money and the total stock of money supply.

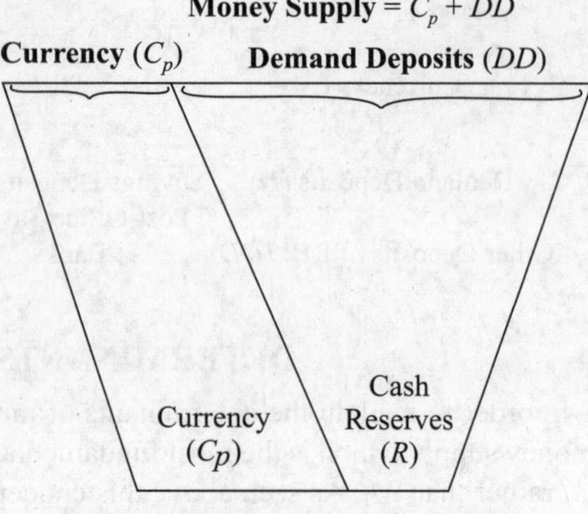

Fig. 17.1. ***The High Powered Money and the Stock of Total Money-Supply***

It follows from above that if there is increase in currency held by the public which is a part of the high powered money with demand deposits remaining unchanged, there will be a direct increase in the money supply in the economy because this constitutes a part of the money supply. If instead currency reserves held by the banks increase, this will not change the money supply immediately but will set in motion a process of multiple creation of demand deposits of the public in the banks. Although banks use these currency reserves held by the public which constitutes a part of the high-powered money to give more loans to the businessmen and thus create demand deposits, they do not affect either the amount of currency or the composition of high-powered money. The *amount of high-powered money is fixed by RBI by its past actions. Thus, changes in high-powered money are the result of decisions of Reserve Bank of India or the Government which own and controls it.*

2. **Money Multiplier.** As stated above, money multiplier is the degree to which money supply is expanded as a result of the increase in high-powered money. Thus

$$M = H, m$$

Rearranging we have, $$M = H.m \quad ...(3)$$

2. See S.B. Gupta, *Monetary Economics: Institutions Theory and Policy*, 3rd Edition, 1992, S. Chand & Co., New Delhi.

Thus money supply is determined by the size of money multiplier (m) and the amount of high-powered money (H). If we know the value of money multiplier we can predict how much money will change when there is a change in the amount of high-powered money. As mentioned above, change in the high-powered money is decided and controlled by Reserve Bank of India, the money multiplier determines the extent to which decision by RBI regarding the change in high-powered money will bring about change in the total money supply in the economy.

Size of Money Multiplier

Now, an important question is what determines the size of money multiplier. It is the cash or currency reserve ratio r of the banks (which determines deposit multiplier) and currency-deposit ratio of the public (which we denote by k) which together determines size of money multiplier. We derive below the expression for the size of multiplier.

From equation (1) above, we know that total money supply (M) consists of currency with the public (C_p) and demand deposits with the banks. Thus

$$M = C_p + D \qquad \ldots (1)$$

The public hold the amount of currency in a certain ratio of demand deposits with the banks. Let this currency-deposit ratio be devoted by k,

$$C_p = kD$$

Substituting kD for C_p in equation (1) we have

$$M = kD + D = (k + 1)D \qquad \ldots (2)$$

Now take equation which defines high powered money (H) as

$$H = C_p + R \qquad \ldots (3)$$

where R represents cash or currency reserves which banks keep as a certain ratio of their deposits and is called cash-reserve ratio and is denoted by r. Thus

$$R = rD$$

Now substituting rD for R and kD for C_p in equation (3) we have

$$H = kD + rD$$

$$H = (k + r)D \qquad \ldots (4)$$

Now money multiplier is ratio of total money supply to the high powered money, therefore we divide equation (2) by equation (4), to get the value of multiplier, which we denote by m. Thus

$$m = \frac{M}{H} = \frac{(k+1)D}{(k+r)D} = \frac{k+1}{k+r}$$

or, $$\text{Money multiplier } (m) = \frac{M}{H} = \frac{1+k}{r+k}$$

or $$M = H\frac{1+k}{r+k} \qquad \ldots (5)$$

where r = Cash or Currency-reserve ratio of the banks

k = Currency-deposit ratio of the public.

From above it follows that money supply is determined by the following :

1. H, that is, the amount of high powered money.
2. r, that is, *cash reserve ratio* of banks (*i.e.*, ratio of currency reserves to deposits of the banks). This cash reserve ratio of banks determines the magnitude of deposit multiplier.

3. k, that is, *currency deposit ratio of the public.*

From the equation (5) expressing the determinants of money supply, it follows that *money supply will increase* :

1. When the supply of high-powered money (*i.e.*, reserve money) H increases;
2. When the currency-deposit ratio (k)[3] of the public decreases; and
3. When the cash or currency reserves-deposit ratio of the banks (r) falls.

Cash Reserve Ratio and the Deposit Multiplier

In a later chapter we will see that because of *fractional reserve system*, with a small increase in cash reserves with the banks, they are able to create a multiple increase in total demand deposits which are an important part of money supply. The ratio of change in total deposits to a change in reserves is called the *deposit multiplier* which depends on cash reserve ratio. The value of deposit multiplier is the reciprocal of cash reserve ratio, $\left(d_m = \frac{1}{r}\right)$ where dm stands for deposit multiplier. If cash reserve ratio is 10 per cent of deposits, then $d_m = \frac{1}{0.10} = 10$. Thus deposit multiplier of 10 shows that for every ₹ 100 increase in cash reserves with the banks, there will be expansion in demand deposits of the banks by ₹ 1000 assuming that no leakage of cash to the public occurs during the process of deposit expansion by the banks.

Currency Deposit Ratio and Multiplier

However, in the real world, with the increase in reserves of the banks, demand deposits and money supply do not increase to the full extent of deposit multiplier. This is for two reasons. First, the public does not hold all its money balances in the form of demand deposits with the banks. When as a result of increase in cash reserves, banks start increasing demand deposits, the people may also like to have some more currency with them as money balances. This means during the process of creation of demand deposits by banks, some currency is leaked out from the banks to the people. This drainage of currency to the people in the real world reduces the magnitude of expansion of demand deposit and therefore the size of money multiplier. Suppose the cash reserve ratio is 10 per cent and cash or currency of ₹ 100 is deposited in a bank *A*. The bank *A* will lend out ₹ 90 and therefore create demand deposits of ₹ 90 and so the process will continue as the borrowers use these deposits for payment through cheques to others who deposit them in another bank *B*. However, if borrower of bank *A* withdraw ₹ 10 in cash from the bank and issue cheques of the remaining borrowed amount of ₹ 80, then bank *B* will have only ₹ 80 as new deposits instead of ₹ 90 which it would have if cash of ₹ 10 was not withdrawn by the borrower. With these new deposits of ₹ 80, bank *B* will create demand deposits of ₹ 72, that is, it will lend out ₹ 72 and keep ₹ 8 as reserves with it $\left(\text{Note that } MC = \frac{W}{MPP_L}\right)$. The drainage of currency may occur during all the subsequent stages of deposit expansion in the banking system. The greater the leakage of currency, the lower will be the money

It will be noticed that in the equation $M = H\frac{1+k}{r+k}$, k appears both in the numerator and denominator. While in the numerator k is added to unity, in the denominator it is added to fraction. Mathematically, it means that a fall in k will affect the denominator relatively more than the numerator. Therefore, with fall in k, the value of multiplier term $\frac{1+H}{r+k}$, will increase. Thus, with decrease in k, the money supply will increase and *vice versa*.

multiplier. We thus see the currency-deposit ratio, which we denote by k is an important determinant of the actual value of money multiplier.

It is important to note that deposit multiplier works both ways, positively when cash reserves with banks increase, and negatively when the cash reserves with the banks decline. That is, when there is a decrease in currency reserves with the banks, there will be multiple contraction in demand deposits with the banks.

Excess Reserves

In the explanation of the expansion of demand deposits or deposit multiplier we assumed that banks do not keep currency reserves in excess of the required cash reserve ratio. The ratio r in the deposit multiplier is the *required cash reserve ratio* fixed by Reserve Bank of India. However, banks like to keep with themselves some excess reserves, the amount of which depend on the extent of liquidity (*i.e.* availability of cash with them) and profitability of making investment and rate of interest on loans advanced to business firms. Therefore, the *desired reserve ratio* is greater than the statutory minimum required reserve ratio. Obviously, the holding of excess reserves by the banks also reduce the value of deposit multiplier.

Conclusion

Theory of determination of money supply explains how a given supply of high-powered money (which is also called *monetary base or reserve money*), leads to multiple expansion in money supply through the working of money multiplier. We have seen above how a small increase in reserves of currency with the banks leads to a multiple expansion in demand deposits by the banks through the process of deposit multiplier and thus causes growth of money supply in the economy. Deposit multiplier measures how much increase in demand deposits (or money supply) occur as a result of a given increase in cash or currency, reserves with the banks depending on the required cash reserve ratio (r) if there are no cash drainage from the banking system. But in the real world drainage of currency does take place which reduce the extent of expansion of money supply following the increase in cash reserves with the banks. Therefore, *the deposit multiplier exaggerates the actual increase in money supply from a given increase in cash reserves with the banks*. In contrast, money multiplier takes into account these leakages of currency from the banking system and therefore measures actual increase in money supply when the cash reserves with the banks increase. *The money multiplier can be defined as increase in money supply for every rupee increase in cash reserves (or high powered money), drainage of currency having been taken into account. Therefore, money multiplier is less than the deposit multiplier*.

It is worth noting that rapid growth in money supply in India has been due to the increase in high-powered money H, or what is also called Reserve Money (Lastly Reserve Bank of India, the money multiplier remaining almost constant.

The money supply in a country can be changed by Reserve Bank of India by undertaking open market operations, changing minimum required currency reserve-deposit ratio, and by varying the bank rate. The main source of growth in money supply in India is creation of credit by RBI for Government for financing its budget deficit and thus creating high-powered money. Further, though the required currency reserve-deposit ratio of banks can be easily varied by RBI, the actual currency reserve-deposit ratio cannot be so easily varied as reserves maintained by banks not only depend on minimum required cash reserve ratio but also on their willingness to hold excess reserves.

Lastly[3], an important noteworthy point is that though money multiplier does not show much variation in the long run, it can change significantly in the short run causing large variations in money supply. This unpredictable variation in money multiplier in the short run affecting money supply in the economy prevents the Central Bank of a country from controlling exactly and precisely the money supply in the economy.

3. Note that Reserve Bank of India calls the high-powered money as Reserve Money (RM).

FACTORS DETERMINING MONEY SUPPLY: RBI'S APPROACH

In its analysis of factors determining money supply in India and sources of variation in it, Reserve Bank of India does not follow any explicit theory of money supply such as money multiplier theory explained above. It provides only purely *accounting or ex-post analysis* of variations in money supply and the factors or sources causing these variations. Although Reserve Bank provides figures of the high powered money in its analysis, it virtually clubs high-powered money with the ordinary money to calculate the total money supply in the country and therefore does not give due importance to the high-powered money as an important factor causing variation in money supply in the economy. Further, *Reserve Bank also does not lay emphasis on the two behavioural ratios, namely, desired currency-deposit ratio (k) of the public and desired cash reserve ratio* (*r*) of the banks as determinants of money supply, though it provides *ex-post* or realised figures of these ratios. We explain below Reserve Bank's analysis of sources of variation in money supply.

Reserve Bank of India classifies factors determining money supply into the following categories:

(a) Government borrowing from the banking system;
(b) Borrowing of the private or commercial sector from the banking system;
(c) Changes in net foreign assets held by the Reserve Bank of India caused by changes in balance of payments position; and
(d) Government's currency liabilities to the public.

Bank Credit to the Government. When the Government expenditure exceeds government revenue and there is deficit in government's budget, then *it may* resort to borrowing from Reserve Bank of India which creates new currency notes for the purpose. This creation of new currency for financing the deficit of the Central Government Budget is known as *monetisation of deficit*. It was previously called *deficit financing*. Monetisation of deficit is an important source of change in money supply in the economy. It may be noted here that since 1995 when Fiscal Responsibility and Budget Management (FRBM) Act was passed Central Government fiscal deficit is financed by borrowing through selling Government bonds to the banks. This has caused a stoppage of RBI's credit to the Government through monetisation of fiscal deficit. However, for the years 2008-09 and 2009-10 the Government is thinking of *directly borrowing from RBI* by selling its bonds to it and thus monetising a part of its fiscal deficit which has greatly risen due to large increase in Government expenditure incurred to prevent slowdown in economic growth. Thus in recent years budget deficit of the Central Government has been largely financed by borrowing from the commercial banks and other financial institutions who buy government securities or bonds.

The commercial banks also lend money to the Government for purchase of foodgrains by the Food Corporation of India. The creation of credit by the banks when they lend to the Government leads to the increase in money supply in the economy. When the Government uses this bank credit for meeting its expenditure, money supply with the public increases.

Bank Credit to the Commercial or Private Sector. The private sector also borrows from the banking system when its own resources are less than its total expenditure. This also adds to the money supply with the public because when the banks lend they create credit or deposits. Bank deposits, as we have seen above, is a part of money supply in the economy. This also affects the money supply in the same manner as the Government borrowing from the banking system. There is, however, an important difference. However, the Central Bank can influence the credit supply by the banks to the private sector by changing liquidity with the banks through making changes in cash reserve ratio.

Changes in Net Foreign Exchange Assets. Changes in the foreign exchange assets held by the Reserve Bank can also bring about a change in the money supply. The change in the net foreign assets may be caused by balance of payment situation. Suppose there is deficit in balance of payments and therefore available foreign exchange is less than the country needs to pay for its imports, both visible and invisible. In order to meet this deficit in balance of payments the country will have to dispose of some of its foreign exchange assets. If there is a net deficit in balance of payments, rupees would flow into the Reserve Bank which pays out foreign exchange reserves in return for the rupee currency. This would have the effect of reducing the *Reserve Money* (*i.e.*, the high-powered money) in India and the contraction of money supply with the public. Opposite result would follow when there is a net surplus in balance of payments of a country.

It follows from above that a deficit in the balance of payments decreases the supply of rupee currency (that is, high-powered or reserve money) in the economy and thereby causes contraction in money supply with the public. On the contrary, a surplus in the balance of payments will increase the foreign exchange assets and thereby will lead to the expansion in reserve money and money supply in the economy.

It may also be noted that apart from surplus in balance of payments foreign exchange reserves or assets may also come through *capital inflows* by either foreign aid or deposits in Indian banks by NRI or foreign direct investment made by foreign companies in India or by portfolio investment by foreign institutional investors (FII). For example, in years 2005 to 2008 there had been a large-scale inflow of foreign exchange through investment made by foreign companies in India. As a result, our foreign exchange reserves substantially went up which resulted in the issue and expansion of rupee currency in circulation. But RBI neutralised its monetary impact to a large extent by mopping up liquidity of the banks through open market operations by selling them Government bonds.

The plenty of foreign exchange received posed a problem of its optimal use during 2005-2008. One proposal was to use foreign exchange reserves to finance the infrastructure projects which are crucial for accelerating economic growth. But argument against this was that the use of foreign exchange reserves will lead to expansion in money supply which would cause higher rate of inflation. The other proposal was that these foreign exchange reserves should be used to import goods in short supply which help in lowering inflation rate. But since September 2008, the opposite problem of *capital outflows* arose due to foreign institutional investors selling shares in Indian stock market and repatriating dollars abroad following the global financial crisis. These capital outflows increased the demand for dollars and caused depreciation of Indian rupee. In order to prevent the rapid depreciation of rupee, RBI sold dollars from its reserves and in return got rupees resulting in decline in the number of rupees in circulation, that is, decrease in money supply.

Government's Currency Liabilities to the Public. Changes in money supply in the economy are also brought about by Government's currency liabilities to the public. Coins and one-rupee notes represent Government's currency liabilities to the public. On January 28, 2005, there were ₹ 7374 crores of coins and one-rupee notes as compared to ₹ 7296 crores on March 31, 2004. If Government's currency liabilities increase, the money supply also increases.

FISCAL DEFICITS AND MONEY SUPPLY

A budget deficit is also an important source of expansion of money supply in the economy. There are two possible links between budget deficit and growth in money supply. First, when following an expansionary fiscal policy the government raises its expenditure without financed by extra taxation and thereby causing a budget deficit, it will tend to raise interest rate. This happens when budget deficit is financed through borrowing from the market. As a result, demand for money or loanable funds increases which, given the supply of money, causes interest rate to rise. Rise in interest rate tends to reduce or crowd out private investment. If the Central Bank is following the policy of a fixed interest rate target, when the government resorts to borrowing to finance the budget deficit, then to prevent the rise in interest rate the Central Bank will take steps to increase the money supply in the economy.

The second link between budget deficit and expansion in money supply is direct. This occurs when the Central Bank itself purchases government securities when the government resorts to borrowing.[4] The Central Bank is said to *monetise* budget deficit when it purchases government securities as it prints new notes for the purpose and gives it to the government for meeting public expenditure. In some countries such as the U.S, Federal Reserve (which is the Central Bank of the USA) enjoys a good deal of independence from the Treasury (*i.e.,* the Government) and voluntarily decides when and how much to purchase government securities to finance its budget deficit.

Central Bank's Dilemma

The Central Bank of a country faces a dilemma in deciding whether or not to monetise budget deficit. If the Central Bank does not monetise budget deficit to meet its increased expenditure, the government will borrow from the market and in the absence of any accommodating monetary policy this will tend to raise interest rate and thereby *reduce or crowd out private investment.* Referring to the policy of Federal Reserve of the United States, Dornbusch, Fischer and Startz write, " There is accordingly a temptation for the Federal Reserve to prevent crowding out by buying government securities thereby increasing the money supply and hence allows an expansion in income without a rise in interest rates."[5]

But the policy of monetisation of budget deficit by the Central Bank involves a risk. If the economy is working near full-employment level, that is, at near full-production capacity, monetisation of budget deficit will cause inflation in the economy. However, if the economy is in the grip of a severe depression, the risk of causing inflation through monetisation of budget deficit and consequent growth in money supply is not much there.

It follows from above that in any particular case the Central Bank, if it enjoys freedom from the Government, has to judge whether it should adopt accommodatory monetary policy to achieve its goal of interest-targeting or allow fiscal expansion through monetisation of budget deficit accompanied by the tight monetary policy to check inflation. It is the latter course of action that was adopted by Reserve Bank of India before 1995 when government's fiscal deficit was high and a good part of it was monetised by it.

MONEY SUPPLY AND THE OPEN ECONOMY

The transactions of an open economy also affect the growth of money supply in it. In the open economy there is free flow of goods and services through trade with foreign countries. Besides, in the open economy there are flows of capital between countries. The impact of transactions of an open economy on the money supply can be better understood from national income identity of an open economy. National income of the open economy is written as

$$Y = C + I + G + NX \qquad \ldots (1)$$

or

$$NX = Y - (C + I + G) \qquad \ldots (2)$$

where NX stands for net exports or trade balance. In the trade balance if we also include exports and imports of services (*i.e.,* invisibles), then NX can be taken as current account balance.

The current account balance (NX) can be either positive or negative. If in equation (2) above aggregate expenditure ($C + I + G$) exceeds national output (Y), current account balance or NX will be negative, that is, imports will be greater than exports. In other words, there will be deficit in current account of the balance of payments. On the other hand, if aggregate expenditure is less than national income [$(C + I + G) < NX$], there will be surplus in current account balance of payments. This implies that our exports will be greater than imports.

Now, if in a year there is deficit in current account, that is, NX is negative, it means our demand for foreign exchange, say the US dollars, for imports of goods and services will exceed the supply of foreign exchange. This situation is depicted in Fig. 19.2 where the curve DD represents demand curve for foreign exchange

4. It is important to note that when government borrows from the market, it does so through sale of its securities (*i.e.* bonds)
5. Dornbusch, Fischer, and Startz, *op.cit*; 9th edition, 2004, p.275.

(US \$) and *SS* is the supply curve of foreign exchange (US \$) at exchange rate (₹ per US dollar), *i.e.*, *OR* and *LK* represents deficit in current account.

If the economy is under flexible exchange rate regime and the Central Bank of the country does not intervene at all, the exchange rate will change to *OR′* and as a result deficit in current account balance will be eliminated and equilibrium restored at the new exchange rate. If there is such a situation, there is no impact on the money supply. However, if the Central Bank wants to maintain the exchange rate at *OR*, then current account deficit equal to *LK* has to be met. If there are no capital inflows, then to maintain the exchange rate at *OR*, the Central Bank of the country has to supply foreign exchange equal to *LK* out of the reserves held by it. But when the Central Bank (RBI in case of India) pays out foreign exchange from its reserves, *it will receive money (i.e., rupees in India)* from *importers of goods and services in return for foreign exchange paid to them to meet the deficit*. Thus some money (say Indian rupees) will flow into the Central Bank and thus withdrawn from circulation. As a result of Central Bank intervention to meet the current account deficit and to maintain the exchange rate money supply in the economy decreases. It is important to note that the Central Bank of the country cannot go on supplying foreign exchange reserves, year after year, for a long time because foreign exchange assets with the Central Bank are available in limited amount.

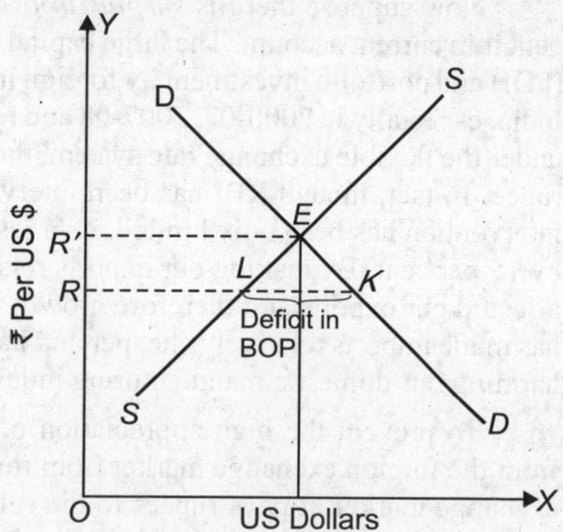

Fig. 19.2. *Deficit in Balance of Payments and Foreign Exchange Market*

The above analysis of expansion in money supply as a result of use of foreign exchange reserves to meet the current account deficit is based on two assumptions. First, it is assumed that there are no capital flows to meet the deficit in current account balance. Second, it is assumed the exchange rate is not allowed to change as a result of imbalance between demand and supply of foreign exchange due to current account deficit.

Capital inflows. However, if there are sufficient net capital inflows accruing from the capital account of balance of payments, then deficit in current account (*i.e.*, negative *NX*) can be met by these capital inflows. In this case there will be no impact of deficit in current account balance of payments on money supply in the economy.

Now take the opposite case of surplus in current account balance (*i.e.*, when *NX* is positive). This implies that the supply of foreign exchange exceeds demand for it. In the absence of capital outflows this excess supply of foreign exchange will have to be purchased by the Central Bank if exchange rate is to be maintained. The Central Bank (RBI) will print new notes to pay for the purchase of foreign exchange. This will lead to the increase in money supply in the economy. However, if exchange rate is allowed to change, as is the case under flexible exchange rate system, the exchange rate will adjust to bring supply and demand for foreign exchange in equilibrium.

Overall Balance of Payments and Capital Inflows

When in an open economy with flexible exchange rate regime there is *deficit in overall balance of payments* (*i.e.*, on both current and capital accounts), it means that capital inflows are insufficient to bridge the gap in the balance of payments, then, in case of India, this has to be met with use of foreign exchange reserves by the Reserve Bank of India. When Reserve Bank of India pays foreign exchange (*e.g.* US \$) to finance the deficit in overall balance of payments, it gets rupees in return. Thus rupee currency flows into the RBI. As a result, money supply (rupee currency) in the economy will decline. However, as explained above, under flexible rate system, if RBI does not intervene, the deficit in overall balance of payments will cause rupee to depreciate.

Now suppose there is *surplus in overall balance of payments* as capital inflows exceed the deficit in current account. The large capital inflows can occur due to heavy foreign direct investment (FDI) and portfolio investment by foreign institutional investors (FII) as it happened in some years in India, especially in 2006-07, 2007-08 and recently in 2010-11. In the absence of intervention by RBI under the flexible exchange rate system, these large capital inflows will cause appreciation of Indian rupee. In fact, though RBI has been intervening in foreign exchange market from time to time, its intervention has been only limited. As a result, between Oct. 2006 and Oct. 2007, rupee appreciated by 15 per cent. By making our exports relatively expensive the appreciation of rupee has adversely affected our exports and therefore growth in GNP and employment. Besides, appreciation of rupee has made imports relatively cheaper and has led to large imports of goods and materials and thereby harming our domestic manufacturing industries.

To prevent the high appreciation of the Indian rupee RBI has been purchasing US dollars from the foreign exchange market from time to time. When RBI purchases dollars from the foreign exchange market, it pays rupees to the sellers of foreign exchange. To do so more rupee currency is printed by RBI to pay for US dollars purchased by it. In this way more rupee currency (*i.e.,* high-powered money) comes into existence in the economy. Thus intervention by RBI to prevent appreciation of rupee results in increase in money supply in the economy.

On the other hand, in 2011 and again in 2013 the RBI faced the opposite problem of depreciation. When, after August 2011, there was net *large capital outflow* from India due to uncertainty caused by European debt crisis and economic slowdown in the U.S. the FIIs started selling Indian equity and bonds and converting rupees into US dollars. This led to the increase in demand for dollars resulting in appreciation of US dollar and *depreciation of Indian rupee*. The value of rupee which was around ₹ 45 to a US dollar in the first week of September 2011 depreciated to around ₹ 53 in the second week of November 2011 and further to ₹ 54 in mid-December 2011. This depreciation of rupee will make our imports constlier which will tend to raise inflation if not matched by fall in international commodity prices. To prevent sharp depreciation of rupee the RBI intervened in the foreign exchange market by selling dollars in the market. As a result in the first week of January 2012 it moved in the range of ₹ 51 to ₹ 52 to a US dollar. Again its intervention was only limited. In fact, the RBI has no fixed target for maintaining exchange rate of rupee at any level and instead its policy is to *allow exchange rate of rupee to fluctuate within a band. In fact, RBI faces a dilemma which we discuss below.*

RBI Dilemma: External Balance and Internal Balance

RBI faces a dilemma because if it does not intervene in the face of large capital inflows, rupee will appreciate much which will adversely affect our exports and therefore growth of GNP and employment in our economy. On the other hand, if it intervenes and purchases enough US dollars from the market to prevent any appreciation of rupee, it will cause large increase in money supply that would cause higher rate of inflation. A major objective of RBI is to control inflation. Therefore, RBI has to strike a balance between the two alternatives. It has been intervening in the foreign exchange market to prevent large appreciation of rupee. But it cannot buy inflows of foreign exchange indiscriminately as it leads to higher inflation. RBI has also resorted to *sterilization* of increase in money supply by selling government securities to the banks and thereby getting back the money issued by it. But there is a limit to this sterilization operation as it has not unlimited amount of government securities to sell them to the banks. Hence the dilemma faced by it. We explain the sterilization operations by RBI later.

It follows from above that the two objectives of external balance and internal balance clash with each other. External balance occurs when balance of payments is in equilibrium or close to it. When external balance does not exist the Central Bank will either go on losing foreign exchange reserves which it cannot do so for long or it will be gaining foreign exchange reserves which also poses a problem as it leads to increase in money supply and causes inflationary pressures in the economy.

On the other hand, internal balance exists when the economy is in equilibrium at full employment or full productive capacity level without any inflationary pressures. Thus, to ensure internal balance requires that money supply should not be allowed to increase much. Since the two require different types of policy measure by the Central Bank, they clash with each other. Hence, the dilemma faced by the Central Bank.

Sterilization by the Central Bank

Sterilization provides a way out of the problem of clash between the goals of external balance and internal balance. *Sterilization refers to the action by the Central Bank of a country to offset or cancel the impact of its foreign exchange market intervention on the money supply through open market operations.* The sterilization measures can be used both to offset the reduction in money supply when in case of current account deficit the Central Bank of the country sells foreign exchange in the market and also when the Central Bank offsets the effect of increase in money supply when it buys foreign exchange from the market in case of surplus in balance of payments or when large capital inflows are coming into the economy.

Let us first explain sterilization operation by the Central Bank in case of deficit in current account of the balance of payments. As seen above, the deficit in current account balance requires the Central Bank to sell foreign exchange from its reserves to prevent the depreciation of domestic currency (that is, to maintain the exchange rate constant). The sale of foreign exchange in foreign exchange market by the Central Bank causes money supply in the economy to decrease that has deflationary effect on the economy. To avoid this adverse effect, the Central Bank buys government securities (*i.e.,* bonds) through open market operations. When it does so the Central Bank prints domestic currency to pay for the bonds it purchases. In this way money supply in the economy increases which offsets the decrease in money supply brought about by the Central Bank when it sells foreign exchange to prevent the depreciation of the domestic currency. Thus, provided it has enough foreign exchange assets, with sterilization operations by the Central Bank *persistent* deficit in balance of payments is possible because it *insulates* the money supply changes in the domestic economy from the Central Bank intervention in the foreign exchange market.

Sterilization Operations in Case of Surplus in Balance of Payments or Large Capital Inflows

Now, we take up the opposite case when there is surplus in balance of payments or when large capital inflows are taking place. This situation requires that Central Bank intervenes in the foreign exchange market and buys foreign exchange inflows from the market to maintain the foreign exchange rate or to prevent the appreciation of domestic currency. As mentioned above, in the two years (2006-08) due to large net capital inflows in the Indian economy there has been a quite large appreciation of the Indian rupee against US dollar that has produced undesirable effects. Therefore Reserve Bank has intervened in the foreign exchange market by buying US dollars to prevent too much appreciation of the Indian rupee. The purchase of foreign exchange (US dollars) from the foreign exchange market by the Reserve Bank has led to the increase in money supply in the Indian economy that has caused inflationary pressures. To sterilize the effect of this increase in money supply RBI has undertaken open market operations by *selling* government securities to the banks which have paid rupees to it. In this way through the sale of government securities some rupee currency has been withdrawn from the economy. In this way inflationary pressures created by the original increase in money supply through intervention in foreign exchange market have been offset.

QUESTIONS FOR REVIEW

1. Explain the concept of money supply. What are the factors responsible for the rapid increase in money supply?
2. Explain how supply of money depends upon the high-powered money, the currency deposit ratio and cash reserve ratio.
3. Explain the impact of budget deficit on the supply of money in the economy. Explain the Central Bank's dilemma in this connection.
4. Explain the impact of (1) deficit in balance of payments, and (2) large capital inflows on the money supply in the open economy. Explain sterilization operations by the Central Bank of the country in each case.
5. Explain the Reserve Bank's dilemma in the face of large capital inflows in the Indian economy.

CHAPTER 18

DEMAND FOR MONEY AND KEYNES'S LIQUIDITY PREFERENCE THEORY OF INTEREST

INTRODUCTION

Why people have demand for money to hold is an important issue in macroeconomics. The level of demand for money not only determines the rate of interest but also prices and national income of the economy. *Classical economists considered money as simply a means of payment or medium of exchange*. In the classical model, people therefore demand money in order to make payments for their purchase of goods and services. In other words, they want to keep money for transactions purposes. *On the other hand, J.M. Keynes laid stress on the store of value function of money*. According to him, money is an asset and people want to hold it so as to take advantage of changes in the price of this asset, that is, the rate of interest. Therefore, Keynes emphasised another motive for holding money which he called speculative motive. As will be explained in detail below, under speculative motive people demand to hold money balances to take advantages from the future changes in the rate of interest, or what means the same thing, from the future changes in bond prices.

An essential point to be noted about people's demand for money is that what people want is not *nominal* money holdings but *real money balances* (This is also referred to as simply *real balances*). This means that people are interested in the purchasing power of their money holdings, that is, the value of money balances in terms of goods and services which they could buy. Thus, people would not be interested in merely nominal money holdings irrespective of the price level, that is, the number of rupee notes and bank deposits. If with the doubling of price level, nominal money holdings are also doubled, their *real money balances* would remain the same. If people are merely concerned with nominal money holdings irrespective of the price level, they are said to suffer from *money illusion*.

The demand for money has been a subject of lively debate in economics. Interest in the study of demand for money has been due to the important role that monetary demand plays in the determination of the price level, interest and income. Till recently there were three approaches to demand for money, namely, transactions approach of Fisher, Cash-balance approach of Cambridge economists, Marshall and Pigou and Keynes theory of demand for money. However, in recent years Baumol, Tobin and Friedman have put forward new theories of demand for money. We critically examine below all these theories of demand for money.

FISHER'S TRANSACTIONS APPROACH TO DEMAND FOR MONEY

In his theory of demand for money Fisher[1] and other classical economists laid stress on the medium of exchange function of money, that is, money as a means of buying goods and services. All transactions involving purchase of goods, services, raw materials, assets require payment of money as value of the transaction made. If accounting identity, namely value paid must equal value received is to occur, value of goods, services and assets sold must be equal to the value of money paid for them. Thus, in any given period, the value of all goods, services or assets sold must equal to the number of transactions T made multiplied by the average price of these transactions. Thus, the total

1. Irving Fisher, *Purchasing Power of Money,* Macmillan, 1911.

value of transactions made is equal to *PT.*

On the other hand, because value paid is identically equal to the value of money flow used for buying goods, services and assets, the value of money flow is equal to the nominal quantity of money supply *M* multiplied by the average number of times the quantity of money in circulation is used or exchanged for transactions purposes. The average number of times a unit of money is used for transactions of goods, services or assets is called *transactions velocity of circulation* and denote it by *V.*

Symbolically, Fisher's equation of exchange is written as under :

$$MV = PT \qquad ...(1)$$

where M = the quantity of money in circulation

V = transactions velocity of circulation

P = Average price

T = the total number of transactions.

The above equation (1) is an identity, that is true by definition. However by taking some assumptions about the variables *V* and *T*, Fisher transformed the above identity into a theory of demand for money.

According to Fisher, the nominal quantity of money *M* is fixed by the Central Bank of a Country (note that Reserve Bank of India is the Central Bank of India) and is therefore treated as an exogeneous variable which is assumed to be a given quantity in a particular period of time. Further, the number of transactions in a period is a function of national income; the greater the national income, the larger the number of transactions required to be made. Further, since Fisher assumed that full employment of resources prevailed in the economy, the level of national income is determined by the amount of the fully employed resources. Thus, with the assumption of full employment of resources, the volume of transactions *T* is fixed in the short run.

But most important assumption which makes Fisher's equation of exchange as a theory of demand for money is that velocity of circulation (*V*) remains constant and is independent of *M*, *P* and *T*. This is because he thought that *velocity of circulation of money (V) is determined by institutional and technological factors involved in the transactions process.* Since these institutional and technological factors do not vary much in the short run, the transactions velocity of circulation of money (*V*) was assumed to be constant.

As we know that for money market to be in equilibrium, nominal quantity of money supply must be equal to the nominal quantity of money demand. In other words, for money market to be in equilibrium

$$M_s = M_d = M$$

where *M* is fixed by the Central Bank of a country.

With the above assumptions, Fisher's equation of exchange can be rewritten as

$$M_d = \frac{PT}{V}$$

or

$$M_d = \frac{1}{V}.PT \qquad ...(2)$$

Thus, according to Fisher's transactions approach, demand for money depends on the following three factors:

(1) the number of transactions (*T*)

(2) the average price of transactions (*P*)

(3) the transaction velocity of circulation of money

It has been pointed out that Fisher's transactions approach represents some kind of a mechanical relation between demand for money (M_d) and the total value of transactions (PT). Thus Prof. Suraj Bhan Gupta says that in Fisher's approach the relation between demand for money M_d and the value of transactions (PT), "betrays some kind of a mechanical relation between it (*i.e.* PT) and M_d as PT represents the total amount of *work to be done* by money as a medium of exchange. This makes demand for money (M_d) a technical requirement and not a behavioural function"[2].

In Fisher's transactions approach to demand for money some serious problems are faced when it is used for empirical research. First, in Fisher's transactions approach, not only transactions involving current production of goods and services are included but also those which arise in sales and purchase of capital assets such as securities, shares, land etc. are also included. Due to frequent changes in the values of these capital assets, it is not appropriate to assume that T will remain constant even if Y is taken to be constant due to full-employment assumption.

The second problem which is faced in Fisher's approach is that it is difficult to define and determine a general price level that covers not only goods and services currently produced but also capital assets just mentioned above.

Cambridge Cash-Balance Theory of Demand for Money

Cambridge Cash-Balance theory of demand for money was put forward by Cambridge economists, Marshall and Pigou[3]. This Cash-Balance theory of demand for money differs from Fisher's transaction approach in that it *places emphasis on the function of money as a store of value or wealth* instead of Fisher's emphasis on the use of money as a medium of exchange. It is worth noting that the exchange function of money eliminates the need to barter and solves the problem of double coincidence of wants faced in the barter system. On the other hand, the function of money as a store of value lays stress on holding money as a general purchasing power by individuals over a period of time between the sale of a good or service and subsequent purchase of a good or service at a later date. Marshall and Pigou focussed their analysis on the factors that determine individual demand for holding cash balances. Although, they recognised that current interest rate, wealth owned by the individuals, expectations of future prices and future rate of interest determine the demand for money, they however believed that changes in these factors remain constant or they are proportional to changes in individuals' income. Thus, they put forward a view that individual's demand for Cash-balances (*i.e.* nominal money balances) is proportional to the nominal income (*i.e.* money income). Thus, according to their approach, aggregate demand for money can be expressed as

$$M_d = kPY$$

where Y = real national income

P = average price level of currently produced goods and services

PY = nominal income

k = proportion of nominal income (PY) that people want to hold as cash balances

Cambridge Cash-balance approach to demand for money is illustrated in Fig. 18.1 where on the X-axis we measure nominal national income (PY) and on the Y-axis the demand for money (M_d). It will be seen from Fig. 18.1 that *demand for money (M_d) in this Cambridge Cash-balance approach is a linear function of nominal income*. The slope of the function is equal to k, that is, $k = \frac{M_d}{P_y}$. Thus important feature of Cash-balance approach is that *it makes the demand for money as function of*

2. S.B. Gupta, *Monetary Economics, Theory Policy and Institutions*, 4th edition, 1997, p. 183.
3. A.C. Pigou, Value of Money, *Quarterly Journal of Economics*, Vol. 32, 1917, pp. 38-65.

money income alone. A merit of this formulation is that it makes the relation between demand for money and income as behavioural in sharp contrast to Fisher's approach in which demand for money was related to total transactions in a mechanical manner.

Although, as mentioned above, Cambridge economists recognised the role of other factors such as rate of interest, wealth as the factors which play a part in the determination of demand for money but these factors were not systematically and formally incorporated into their analysis of demand for money. In their approach, these other factors determine the proportionality factor k, that is, the proportion of money income that people want to hold in the form of money, *i.e.* cash balances. It was J.M. Keynes who later emphasised the role of these other factors such as rate of interest, expectations regarding future interest rate and prices and formally incorporated them explicitly in his analysis of demand for money. Thus, Glahe rightly writes, "Cambridge approach is conceptually richer than the transactions approach, the former is incomplete because it does not formally incorporate the influence of economic variables just mentioned on the demand for cash balances... John Maynard Keynes first attempted to eliminate this shortcoming."[4]

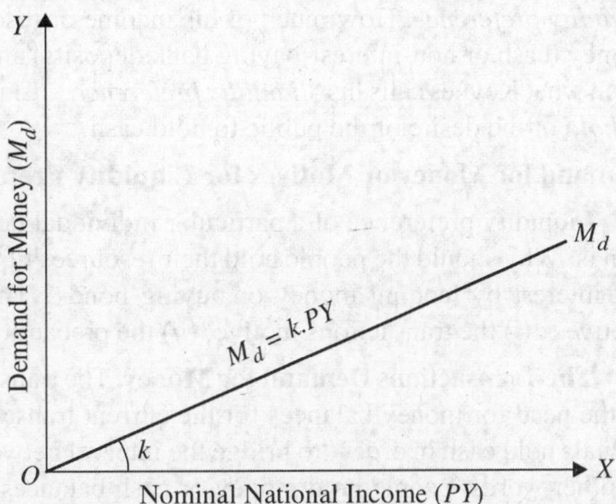

Fig. 18.1 : *Demand for Money : Cambridge Cash Balance Approach*

Another *important feature of Cambridge demand for money function is that the demand for money is proportional function of nominal income* ($M_d = kPY$). Thus, it is proportional function of both price level (P) and real income (Y). This implies two things. First, income elasticity of demand for money is unity and, secondly, price elasticity of demand for money is also equal to unity so that any change in the price level causes equal proportionate change in the demand for money.

Criticism. It has been pointed out by critics that other influences such as rate of interest, wealth, expectations regarding future prices and rate of interest have not been formally introduced into the Cambridge theory of the demand for cash-balances. These other influences remain in the background of the theory. "It was left to Keynes, another Cambridge economist, to highlight the influence of the rate of interest on the demand for money and change the course of monetary theory."[5]

Another criticism levelled against this theory is that income elasticity of demand for money may well be different from unity. Cambridge economists did not provide any theoretical reason for its being equal to unity. Nor is there any empirical evidence supporting unitary income elasticity of demand for money. Besides, price elasticity of demand is also not necessarily equal to unity. In fact, changes in the price level may cause non-proportional changes in the demand for money. However, these criticisms are against the mathematical formulation of cash-balance approach, namely, $M_d = kPY$. They do not deny the important relation between demand for money and the level of income. Empirical studies conducted so far point to a strong evidence that there is a significant and firm relation between demand for money and level of income.

4. Fred R. Glahe, *Macroeconomics: Theory and Policy*, 2nd Edition, 1977, Harcourt, p. 164.
5. S.B. Gupta, *op. cit.*, p. 184.

KEYNES'S THEORY OF DEMAND FOR MONEY

In his well-known book[6], Keynes propounded a theory of demand for money which occupies an important place in his monetary theory.

It is also worth noting that for demand for money to hold Keynes used the term what he called *liquidity preference*. How much of his income or resources will a person hold in the form of ready money (cash or non-interest-paying bank deposits) and how much will he part with or lend depends upon what Keynes calls his "*liquidity preference*." Liquidity preference means the *demand for money to hold* or the desire of the public to hold cash.

Demand for Money or Motives for Liquidity Preference : Keynes's Theory

Liquidity preference of a particular individual depends upon several considerations. The question is: Why should the people hold their resources liquid or in the form of ready money when he can get interest by lending money or buying bonds? The desire for liquidity arises because of three motives : (*i*) the transactions motive, (*ii*) the precautionary motive, and (*iii*) the speculative motive.

The Transactions Demand for Money. The transactions motive relates to the demand for money or the need for money balances for the current transactions of individuals and business firms. Individuals hold cash in order "to bridge the interval between the receipt of income and its expenditure". In other words, people hold money or cash balances for transactions purposes, because receipt of money and payments do not coincide. Most of the people receive their incomes weekly or monthly while the expenditure goes on day by day. A certain amount of ready money, therefore, is kept in hand to make current payments. This amount will depend upon the size of the individual's income, the interval at which the income is received and the methods of payments prevailing in the society.

The businessmen and the entrepreneurs also have to keep a proportion of their resources in money form in order to meet daily needs of various kinds. They need money all the time in order to pay for raw materials and transport, to pay wages and salaries and to meet all other current expenses incurred by any business firm. It is clear that the amount of money held under this business motive will depend to a very large extent on the turnover (*i.e.*, the volume of trade of the firm in question). The larger the turnover, the larger, in general, will be the amount of money needed to cover current expenses. It is worth noting that money demand for transactions motive arises primarily because of the use of money as a medium of exchange (*i.e.* means of payment).

Since the transactions demand for money arises because individuals have to incur expenditure on goods and services during the receipt of income and its use of payment for goods and services, money held for this motive depends upon the level of income of an individual. A poor man will hold less money for transactions motive as he spends less because of his small income. On the other hand, a rich man will tend to hold more money for transactions motive as his expenditure will be relatively greater.

The demand for money is a demand for *real* cash balances because people hold money for the purpose of buying goods and services. The higher the price level, the more money balances a person has to hold in order to purchase a given quantity of goods. If the price level doubles, then the individual has to keep twice the amount of money balances in order to be able to buy the same quantity of goods. Thus the demand for money balances is demand for *real* rather than *nominal* balances.

According to Keynes, the transactions demand for money depends only on the real income and is not influenced by the rate of interest. However, in recent years, it has been observed empirically and also according to the theories of Tobin and Baumol transactions demand for money also depends on the rate of interest. This can be explained in terms of *opportunity cost* of money holdings. Holding one's asset in the form of money balances has an opportunity cost. The cost of holding money

6. J.M. Keynes, *General Theory of Employment, Interest and Money*, Macmillan, 1936.

balances is the interest that is foregone by holding money balances rather than other assets. The higher the interest rate, the greater the opportunity cost of holding money rather than non-money assets. Individuals and business firms economise on their holding of money balances by carefully managing their money balances through transfer of money into bonds or short-term income yielding non-money assets. Thus, at higher interest rates, individuals and business firms will keep less money holdings at each level of income.

Precautionary Demand for Money. Precautionary motive for holding money refers to the desire of the people to hold cash balances for unforeseen contingencies. People hold a certain amount of money to provide for the danger of unemployment, sickness, accidents, and the other uncertain perils. The amount of money demanded for this motive will depend on the psychology of the individual and the conditions in which he lives.

Speculative Demand for Money. The speculative motive of the people relates to the desire to hold one's resources in liquid form in order to take advantage of market movements regarding the future changes in the rate of interest (or bond prices). The notion of holding money for speculative motive was a new and revolutionary Keynesian idea. Money held under the speculative motive serves as a store of value as money held under the precautionary motive does. But it is a store of money meant for a different purpose. The cash held under this motive is used to make speculative gains by dealing in bonds[7] whose prices fluctuate. If bond prices are expected to rise which, in other words, means that the rate of interest is expected to fall, businessmen will buy bonds to sell when their prices actually rise. If, however, bond prices are expected to fall, *i.e.*, the rate of interest is expected to rise, businessmen will sell bonds to aboid capital losses. Nothing is certain in the dynamic world, where guesses about the future course of events are made on precarious basis, businessmen keep cash to speculate on the probable future changes in bond prices (or the rate of interest) with a view to making profits.

Given the expectations about the changes in the rate of interest in future, less money will be held under the speculative motive at a *higher current* rate of interest and more money will be held under this motive at a *lower current* rate of interest. The reason for this inverse correlation between money held for speculative motive and the prevailing rate of interest is that at a lower rate of interest less is lost by not lending money or investing it, that is, by holding on to money, while at a higher current rate of interest holders of cash balance would lose more by not lending or investing.

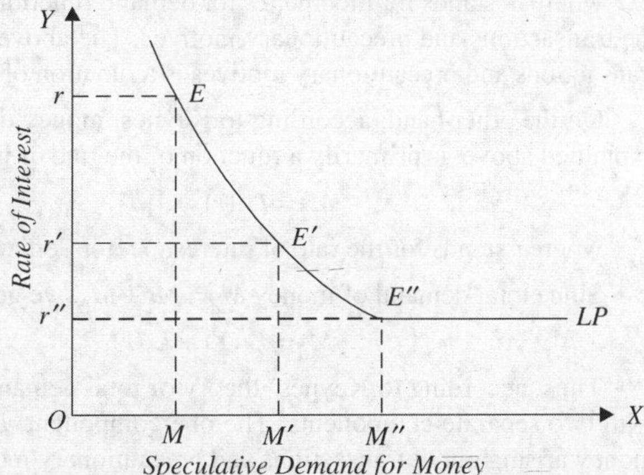

Fig. 18.2. *Liquidity Preference Curve and Liquidity Trap*

Thus the demand for money under speculative motive is a function of the current rate of interest, increasing as the interest rate falls and decreasing as the interest rate rises. Thus, demand for money under this motive is a decreasing function of the rate of interest. This is shown in Fig. 18.2. Along *X*-axis we represent the speculative demand for money and along the *Y*-axis the current rate of interest The liquidity preference curve *LP* is a downward sloping towards the right signifying that the higher the rate of interest, the lower the demand for money for speculative motive, and vice versa. Thus at the high current rate of interest Or, a very small amount *OM* is held for speculative motive. This is because at a high current rate of interest more money would have been lent out or used for buying bonds and therefore less money would be kept as inactive balances. If the rate of interest falls to *Or′*, then a greater amount of money *OM′* is held under speculative motive. With the further fall in the rate of interest to Or', money held under speculative motive increases to OM.

Liquidity Trap. It will be seen from Fig. 18.2 that the liquidity preference curve LP becomes quite flat *i.e.*, perfectly elastic at a very low rate of interest; it is horizontal line beyond point E'' towards the right. This perfectly elastic portion of liquidity preference curve indicates the position of *absolute liquidity* preference of the people. That is, at a very low rate of interest people will hold with them as inactive balances any amount of money they come to have. This portion of liquidity preference curve with absolute liquidity preference is called *liquidity trap* by the economists because expansion in money supply gets trapped in the sphere of liquidity trap and therefore cannot affect rate of interest and therefore the level of investment. According to Keynes, it is because of the existence of liquidity trap that monetary policy becomes ineffective to tide over economic depression.

But the demand for money to satisfy the speculative motive does not depend so much upon what the current rate of interest is, as on expectations about changes in the rate of interest. If there is a change in the expectations regarding the future rate of interest, the whole curve of demand for money or liquidity preference for speculative motive will change accordingly. Thus, if the public on balance expect the rate of interest to be higher (*i.e.*, bond prices to be lower) in the future than had been previously supposed, the speculative demand for money will increase and the whole liquidity preference curve for speculative motive will shift upward.[7]

Aggregate Demand for Money : Keynes's View

If the total demand of money is represented by M_d we may refer to that part of M held for transactions and precautionary motive as M_1 and to that part held for the speculative motive as M_2. Thus $M_d = M_1 + M_2$. According to Keynes, the money held under the transactions and precautionary motives, *i.e.*, M_1, is completely interest-inelastic unless the interest rate is very high. The amount of money held as M_1, that is, for transactions and precautionary motives, is mainly a function of the size of income and business transactions together with the contingencies growing out of the conduct of personal and business affairs. We can write this in a functional form as follows:

$$M_1 = L_1(Y) \qquad ...(i)$$

where Y stands for income, L_1 for demand function, and M_1 for money demanded or held under the transactions and precautionary motives. The above function implies that money held under the transactions and precautionary motives is a function of income.

On the other hand, according to Keynes, money demanded for speculative motive, *i.e.*, M_2, as explained above, is primarily a function of the rate of interest. This can be written as:

$$M_2 = L_2(r) \qquad ...(ii)$$

where r stands for the rate of interest, L_2 for demand function for speculative motive.

Since total demand of money $M_d = M_1 + M_2$, we get from (*i*) and (*ii*) above

$$M_d = L_1(Y) + L_2(r)$$

Thus, according to Keynes' theory of total demand for money is an *additive demand function* with two separate components. The one component, $L_1(Y)$ represents the transactions demand for money arising out of transactions and precautionary motives is an increasing function of the level of money income. The second component of the demand for money, that is, $L_2(r)$ represents the speculative demand for money, which depends upon rate of interest, is a decreasing function of the rate of interest. Keynes' additive form of demand for money function has now been rejected by the modern economists. It has been pointed out that money represents a single asset, and not the several ones.

7. All securities and other such papers yield a fixed and known rate of interest over a period of time and are known as bonds.

There may be more than one motive to hold it and the same unit of money can serve several motives. Therefore, the demand for money cannot be divided into separate compartments independent of each.

Further, as has been argued by Tobin and Baumol, the transactions demand for money also depends upon the rate of interest. Others have explained that speculative demand for money is an increasing function of the total assets or wealth. If income is taken as a proxy for total wealth then even speculative demand for money will depend upon the size of income, apart from the rate of interest. In view of all these arguments, the Keynesian total demand for money function is written in the following modified form

$$M_d = L(Y, r)$$

where it is conceived that demand for money function (M_d) is increasing function of the level of income, it is a decreasing function of the rate of interest. The presentation of the demand for money function in the above revised and modified form, $M_d = L(Y, r)$ has been a highly significant development in monetary theory.

Critique of Keynes's Theory. By introducing speculative demand for money, Keynes made a significant departure from the classical theory of money demand which emphasized only the transactions demand for money. However, as seen above, Keynes' theory of speculative demand for money has been challenged. The main drawback of Keynes' speculative demand for money is that *it visualises that people hold their assets in either all money or all bonds*. This seems quite unrealistic as individuals hold their financial wealth in some combination of both money and bonds. This gave rise to portfolio approach to demand for money put forward by Tobin, Baumol and Freidman. The portfolio of wealth consists of money, interest-bearing bonds, shares, physical assets etc. Further, while according to Keynes' theory, demand for money for transaction purposes is insensitive to interest rate, the modern theories of money demand put forward by BaUmol and Tobin show that money held for transaction purposes is interest elastic. We will discuss in the next chapter the Post-Keynesian theories of demand for money put forward by Tobin, Baumol and Friedman.

Further, *Keynes additive form* of demand for money function, namely, $M_d = L_1(Y) + L_2(r)$ has now been rejected by the modern economists. It has been pointed out that *money represents a single asset*, and not the several ones. There may be more than one motive to hold money but the same units of money can serve several motives. Therefore, the demand for money cannot be divided into two or more different departments independent of each other.

In view of all these arguments, the Keynesian total demand for money functions is written in the following modified form

$$M_d = L(Y, r)$$

where it is conceived that demand for money function (M_d) is increasing function of the level of income, it is a decreasing function of the rate of interest. The presentation of the demand for money function in the above revised and modified form, $M_d = L(Y, r)$, has been a highly significant development in monetary theory.

KEYNES'S LIQUIDITY PREFERENCE THEORY OF RATE OF INTEREST

In his epoch-making book "*The General Theory of Employment, Interest and Money*", J.M. Keynes gave a new view of interest. According to him, the rate of interest is a purely monetary phenomenon and is determined by demand for money and supply of money. According to him "*interest is a reward for parting with liquidity for a specified period.*" Since people prefer liquidity or want to hold money to meet their various motives, they need to be paid some reward for surrendering liquidity or money. And this reward is the rate of interest that must be paid to them in order to induce them to part with liquidity or money. Further, according to Keynes, rate of interest is determined by liquidity preference or demand for money to hold and the supply of money known as *Liqudity Preference Theory.*

The Demand for Money in a Two Asset Economy. In order to explain the demand for money and interest-rate determination. Kenyes assumed a simplified economy where there are two assets which people can keep in their portfolio balance. These two assets are : (1) money in the form of currency and demand deposits in the banks which earn no interest, (2) long term bonds. It is important to note that rate of interest and bond prices are inversely related. When bond prices go up, rate of interest rises and vice versa. The demand for money by the people depends upon how they decide to balance their portfolios between money and bonds. This decision about the portfolio balance can be influenced by two factors.

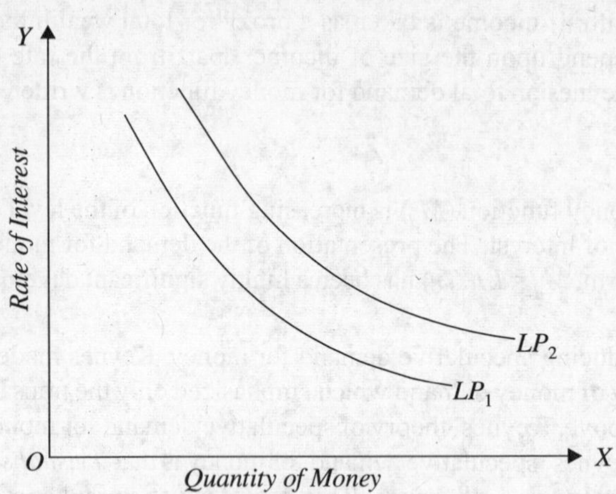

Fig. 18.3. *Demand for Money (Liquidity Preference) Rate of Interest and Level of income*

First, the higher the level of nominal income in a two asset economy people would want to hold more money in their portfolio balance. This is because of transactions motive according to which at the higher level of nominal income, the purchases by the people of goods and services in their daily life will be relatively larger which require more money to be kept for transactions purposes.

Second, the higher the nominal rate of interest, the lower the demand for money for speculative motive. This is firstly because a higher nominal rate of interest implies a higher opportunity cost for holding money. At higher rate of interest holders of money can earn more incomes by holding bonds instead of money. Secondly, if the current rate of interest is higher than what is expected in the future, the people would like to hold more bonds and less money in their portfolio. On the other hand, if the current rate of interest is low (in other words, if the bond prices are currently high), the people will be reluctant to hold larger quantity of bonds (and instead they could hold more money in their portfolio) for the fear that bond prices would fall in the future causing capital losses to them.

Money Demand Curve. It follows from above that quantity of money demanded increases with the fall in the rate of interest or with the increase in level of nominal income. At a given level of nominal income, we can draw a money demand curve showing the quantity of money demanded at various rates of interest. As demand for money is inversely related to the rate of interest, the money demand curve at a given level of income say, will be downward-sloping as is shown by the curve LP_1 in Figure 18.3. When the level of money income increases, suppose from Y_1 to Y_2, the curve of demand for money shifts upward to the new position LP_2.

Determination of Rate of Interest: Equilibrium in the Money Market

The rate of interest, according to J.M. Keynes, is determined by demand for money (Liquidity Preference) and supply of money. The factors which determine demand for money has been explained above. The supply of money, at a given time, is fixed by the monetary authority of the country. In Figure 18.4 *LP* is the demand curve for money at a given level of nominal income. *MS* is the money supply curve which is a vertical straight line showing that 200 crores of rupees is the money supply fixed by the monetary authority. It will be seen that quantity demanded of money equals the given money supply at 10 per cent rate of interest. So the money market is in equilibrium at 10 per cent rate of interest. There will be disequilibrium if rate of interest is eighter higher or lower than 10 per cent.

Suppose the rate of interest is 12 per cent. It will be seen from Figure 21.3 that at 12 per cent rate of interest, supply of money exceeds the demand for money. The excess supply of money reflects the fact that people do not want to hold as much money in their portfolio as the monetary authority has made it available to them. The people holding assets in the present two-asset economy would react to this excess money supply with them by buying bonds and thus buying bonds and thus replace some of money in their replace some of money in their portfolios with bonds. Since the *total money supply* at a given moment remains fixed, it cannot be reduced by buying bonds by individuals. What the bonds-buying spree would lead to the rise in prices of bonds. The rise in bond prices mean the fall in the rate of interest. As will be seen from the Figure with the fall in the interest rate from 12 per cent to 10 per cent, quantity demand of money has increased to be once again equal to the given

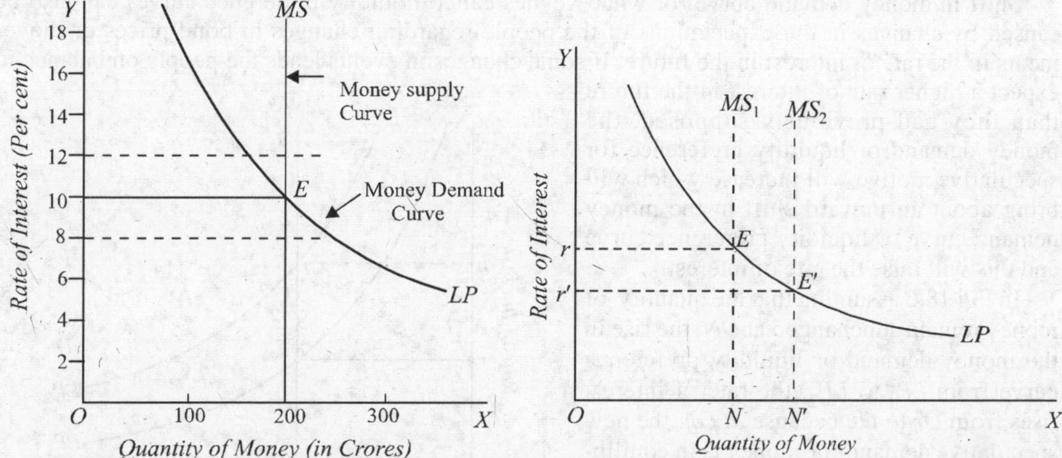

Fig. 18.4. *Money Market Equilibrium*

Fig. 18.5. *Effect of Increase in Money Supply on the Rate of Interest.*

supply of money and the excess supply of money is entirely eliminated and money market is in equilibrium.

On the other hand, if the rate of interest is lower than the equilibrium rate of 10 per cent, say it is 8 per cent, then as will be seen from Figure there will emerge *excess demand for money*. As a reaction to this excess demand for money, people would like to sell bonds in order to obtain a greater quantity of money for holding at lower rate of interest. The stock of money remaining fixed, the attempt by the people to hold more money balances at a rate of interest lower than the equilibrium level through sale of bonds will only cause the bond prices to fall. The fall in bond prices implies the rise in the rate of interest. Thus, the process started as a reaction to the excess demand for money at an interest rate below the equilibrium will end up with the rise in the interest rate of the equilibrium level.

Effect of an Increase in Money Supply

Let us now examine the effect of increase in money supply on the rate of interest. In Fig. 18.5, *MD* is the demand for money for satisfying various motives. To begin with, *ON* is the quantity of money available. Rate of interest will be determined where the demand for money is in balance or equal to the fixed supply of money *ON*. It is clear from Fig. 18.5 that demand for money is equal to *ON* quantity of money at *Or* rate of interest. Hence *Or* is the equilibrium rate of interest. Assuming no change in expectations and nominal income, an increase in the quantity of money (through buying securities by the Central Bank of the country from the open market), will lower the rate of interest. In Fig. 18.5, when the quantity of money increases from *ON* to *ON*, the rate of interest falls from *Or* to *Or′* because the new quantity of money *ON′* is in balance with the demand for money at *Or′* rate of interest. In this case we move down on the curve. Thus, given the money demand curve or curve of liquidity preference, an increase in the quantity of money brings down the rate of interest.

Let us see how increase in money supply leads to the fall in the rate of interest. With initial equilibrium at *Or*, when the money supply is expanded from *ON* to *ON′*, there emerges excess supply of money at the initial *Or* rate of interest. The people would react to this excess quantity of money supplied by buying bonds. As a result, the bond prices will go up which implies that the rate of interest will decline. This is how the increase in money supply leads to the fall in rate of interest.

Shifts in Money Demand or Liquidity Preference Curve

The position of money demand curve depends upon two factors: (1) the level of nominal income, (2) the expectations about the changes in bond prices in the future which implies changes in rate of interest in future. As has been explained above, a money demand curve is drawn by assuming a certain level of nominal income. With the increase in nominal income, money demand for transactions and precautionary motives increase causing an upward shift in the money demand curve.

Shift in money demand curve (or what Keynes called liquidity preference curve) can also be caused by changes in the expectations of the people regarding changes in bond prices or movements in the rate of interest in the future. If some changes in events leads the people on balance to expect a higher rate of interest in the future than they had previously supposed, the money demand or liquidity preference for speculative motive will increase which will bring about an upward shift in the money demand curve or liquidity preference curve and this will raise the rate of interest.

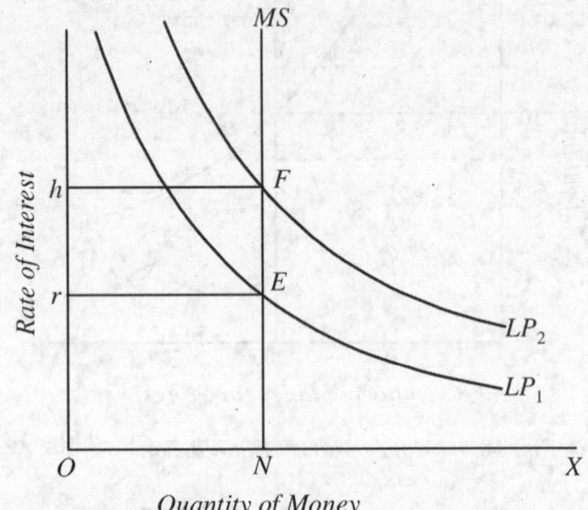

Fig. 18.6. *Effect of Increase in Liquidity Preference on the Rate of Interest.*

In Fig 18.6 assuming that the quantity of money remains unchanged at *ON*, the rise in the money demand or liquidity preference curve from LP_1 to LP_2, the rate of interest rises from *Or* to *Oh* because *at Oh*, the new speculative demand for money is in equilibrium with the supply of money *ON*. It is worth noting that when the liquidity preference curve rises from LP_1 to LP_2, the amount of moneyheld does not increase; it remains *ON* as before. Only the rate of interest rises from *Or* to *Oh* to equilibrate the new liquidity preference or money demand with the available quantity of money *ON*.

Thus we see that Keynes explained interest in terms of purely monetary forces and not in terms of real forces like productivity of capital and thrift which formed the foundation-stones of both classical and loanable fund theories. According to him, demand for money for speculative motive together with the supply of money determines the rate of interest. He agreed that the marginal revenue product of capital tends to become equal to the rate of interest but the rate of interest is not determined by marginal revenue productivity of capital. Moreover, according to him, interest is not a reward for saving or thriftiness or waiting but for parting with liquidity. Keynes asserted that it is not the rate of interest which equalises saving an investment. But this equality is brought about through changes in the level of income.

Critical Appraisal of Keynes's Liquidity Preference Theory of Interest

1. *Keynes ignored the role of real factors in the determination of interest.* Firstly, it has been pointed out that rate of interest is not purely a monetary phenomenon. Real forces like productivity of capital and thriftiness or saving also pay an important role in the determination of the rate of interest. Keynes makes the rate of interest independent of the demand for investment funds. In fact, it is not so independent. The cash-balances of the businessmen are largely influenced by their demand for capital investment. This demand for Capital investment depends upon the marginal revenue productivity of capital. Therefore, the rate of interest is not determined independently of the marginal revenue productivity of capital (marginal efficiency of capital) and investment demand. When investment demand increases due to greater profit prospects or, in other words, when marginal revenue productivity of capital rises, there will be greater demand for investment funds and the rate of interest will go up. But Keynesian theory does not account for this. Similarly, Keynes ignored the effect of the availability of savings on the rate of interest. For instance, if the propensity to consume of the people increases, savings would decline. As a result, supply of funds in the market will decline which will raise the rate of interest.

2. *Keynesian theory is also indeterminate.* Now, exactly the same criticism applies to Keynesian theory itself on the basis of which Keynes rejected the classical and loanable funds theories. Keynes's theory of interest, like the classical and loanable funds theories, is indeterminate.

According to Keynes, rate of interest is determined by liquidity preference (*i.e.* demand for money) and supply of money. However, as we have seen, liquidity preference, especially demand for money for transactions motive depends on level of income. Now, when income increases, liquidity preference curve (that is, money demand curve will shift to the right and, given the supply of money, new equilibrium rate of interest will be obtained. Thus at different levels of income there will be different liquidity preference curve or money demand curve. As a result, at different levels of income, therė will be different equilibrium rates of interest. Thus, we cannot know the rate of interest unless we know the liquidity preference curve, and also we cannot know the liquidity preference curve unless know the level of income. However, we cannot know the level of income unless we first know the rate of interest. This is because rate of interest influences investment which in turn determines the level of income. Thus, Keynes's theory is indeterminate, that is, we are not able to arrive at a single determinate rate of interest; rate of interest various as income varies.

Thus, Keynes's analysis at the most help as to obtain *LM* curve which shows what will be the rate of interest at different levels of income and not any unique or particular rate of interest. Thus, the Keynesian theory, like the classical theory, is indeterminate. "In the Keynesian case the supply and demand for money curves cannot give the rate of interest unless we already know the income level, in the classical case the demand and supply schedules for saving offer no solution until the income is known. Precisely the same is true of loanable funds theory. Keynes' criticism of the classical and lonable funds theories applies equally to his now theory[8].

3. *No liquidity without savings.* According to Keynes, interest is a reward for parting with liquidity and in no way a compensation and inducement for saving or waiting. But without saving how can the funds be available to be kept as liquid and how can there be the question of surrendering liquidity if one has not already saved money. Jacob Viner rightly maintains, "*Without saving there can be no liquidity to surrender.*" Therefore, the rate of interest is vitally connected with saving which is neglected by Keynes is the determination of interest.

It follows from above that Keynesian theory of interest is also not without flaws. But importance Keynes gave to liquidity preference as a determinant of interest is correct. A valid and an adequate explanation of interest must incorporate this important factor of demand for money to hold.

QUESTIONS FOR REVIEW

1. Explain Fisher's transactions approach to demand for money. Classical economists such as Fisher considered money as simply a medium of exchange. Examine critically.
2. Critically examine the Cambridge Cash-Balance Theory of Demand for money. How does it differ from Fisher's transactions approach to demand for money?
3. What, according to Keynes, are the motives for holding money. What are the factors that determine the money held under these motives?
4. Why do people hold money when they can earn interest by lending it to others or by buying interest yielding bonds of the companies?
5. In what ways is the Keynesian theory of demand for money a departure from the classical theory? Discuss. *[D.U. B.Com. (Hons) 1998]*
6. Explain the speculative motive for holding money? Why is it interest elastic?
7. What is meant by liquidity preference? Explain various motives of liquidity preference. What role does it play in determining rate of interest?
8. Explain the Keynesian theory of interest. How is the Keynesian theory of interest different from the classical theory? *[D.U. B.Com (Hons) 1999]*
9. Explain Keynes's liquidity preference theory of interest. Why is it considered to be indeterminate? Show that at best we can get *LM* curve from it. *[D.U. B.Com (Hons), 1998]*
10. What is meant by liquidity trap? How does its existence make the expansionary monetary policy ineffective at times of depression?

8. A. H. Hansen, *Guide to Keynes*. p. 141

CHAPTER 19
POST-KEYNESIAN THEORIES OF DEMAND FOR MONEY*

INTRODUCTION

By introducing speculative demand for money, Keynes made a significant departure from the classical theory of money demand which emphasized only the transactions demand for money. However, as seen above, Keynes' theory of speculative demand for money has been challenged. The main drawback of Keynes' speculative demand for money is that *it visualises that people hold their assets in either all money or all bonds*. This seems quite unrealistic as individuals hold their financial wealth in some combination of both money and bonds. This gave rise to portfolio approach to demand for money put forward by Tobin, Baumol and Freidman. The portfolio of wealth consists of money, interest-bearing bonds, shares, physical assets etc. Further, while according to Keynes' theory, demand for money for transaction purposes is insensitive to interest rate, the modern theories of money demand put forward by Baumol and Tobin show that money held for transaction purposes is interest elastic. We discuss below the Post-Keynesian theories of demand for money put forward by Tobin, Baumol and Friedman.

TOBIN'S PORTFOLIO APPROACH TO DEMAND FOR MONEY

An American economist James Tobin[1], in his important contribution explained that rational behaviour on the part of the individuals is that they should keep a portfolio of assets which consists of both bonds and money. In his analysis he makes a valid assumption that people prefer more wealth to less. According to him, an investor is faced with a problem of what proportion of his portfolio of financial assets he should keep in the form of money (which earns no interest) and interest-bearing bonds. The portfolio of individuals may also consist of more risky assets such as shares. *According to Tobin, faced with various safe and risky assets, individuals diversify their portfolio by holding a balanced combination of safe and risky assets*. According to Tobin, individual's behaviour shows *risk aversion*. That is, they prefer less risk to more risk at a given rate of return. In the Keynes' analysis an individual holds his wealth in either all money or all bonds depending upon his estimate of the future rate of interest. But, according to Tobin, individuals are uncertain about future rate of interest. If a wealth holder chooses to hold a greater proportion of risky assets such as bonds in his portfolio, he will be earning a high average return but will bear a higher degree of risk. Tobin argues that a risk averter will not opt for such a portfolio with all risky bonds or a greater proportion of them.

On the other hand, a person who, in his portfolio of wealth, holds only safe and riskless assets such as money (in the form of currency and demand deposits in banks) he will be taking almost zero risk but will also be having no return and as a result there will be no growth of his wealth. Therefore, people generally prefer a mixed diversified portfolio of money, bonds and shares, with each person opting for a little different balance between riskiness and return.

It is important to note that a person will be unwilling to hold all risky assets such as bonds unless he obtains a higher average return on them. In view of the desire of individuals to have both safety and reasonable return, they strike a balance between them and hold a mixed and balanced

* This chapter is meant for Higher Level Courses in Macroeconomics.

1. James Tobin, Liquidity Preference as Behaviour Towards Risk, Review of Economic Studies Vol. 25 (1958)

portfolio consisting of money (which is a safe and riskless asset) and risky assets such as bonds and shares though this balance or mix varies between various individuals depending on their attitude towards risk and hence their trade-off between risk and return.

Tobin's Liquidity Preference Function

Tobin derived his liquidity preference function depicting relationship between rate of interest and demand for money (that is, preference for holding wealth in money form which is a safe and "riskless" asset[2]. He argues that with the increase in the rate of interest (*i.e.* rate of return on bonds), wealth holders will be generally attracted to hold a greater fraction of their wealth in bonds and thus reduce their holding of money. That is, at a higher rate of interest, their demand for holding money (*i.e.*, liquidity) will be less and therefore they will hold more bonds in their portfolio. On the other hand, at a lower rate of interest they will hold more money and less bonds in their portfolio. This means, like the Keynes's speculative demand for money, in Tobin's portfolio approach demand function for money as an asset (*i.e.* his liquidity preference function curve) slopes downwards as is shown in Fig. 19.1, where on the horizontal axis asset demand for money is shown. This downward-sloping liquidity preference function curve shows that the asset demand for money in the portfolio increases as the rate of interest on bonds falls. In this way Tobin derives the aggregate liquidity preference curve by determining the effects of changes in interest rate on the asset demand for money in the portfolio of individuals. Tobin's liquidity preference theory has been found to be true by the empirical studies conducted to measure interest elasticity of the demand for money. As shown by Tobin through his portfolio approach, these empirical studies reveal that aggregate liquidity preference curve is negatively sloped. This means that most of the people in the economy have liquidity preference function similar to the one shown by curve M_d in Fig. 19.1.

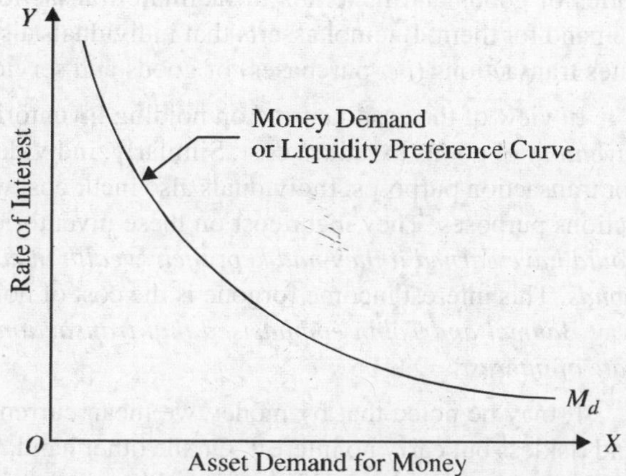

Fig. 19.1. *Tobin's Liquidity Preference Curve*

Evaluation

Tobin's approach has done away with the limitation of Keynes' theory of liquidity preference for speculative motive, namely, individuals hold their wealth in *either all money or all bonds*. Thus, Tobin's approach, according to which individuals simultaneously hold both money and bonds but in different proportion at different rates of interest yields a *continuous* liquidity preference curve. Further, Tobin's analysis of simultaneous holding of money and bonds is not based on the erroneous Keynes's assumption that interest rate will move only in one direction but on a simple fact that individuals do not know with certainty which way the interest rate will change. It is worth mentioning that Tobin's portfolio approach, according to which liquidity preference (*i.e.* demand for money) is determined by the individuals attitude towards risk, can be extended to the problem of asset choice when there are several alternative assets, not just two, of money and bonds.

2. Note that riskless is written in quotes because strictly speaking even money defined as currency and demand deposits is also not completely riskless in view of the frequent changes in prices of commodities. If inflation occurs, it reduces the value of money and thus inflicts a capital loss on its holders.

BAUMOL'S INVENTORY APPROACH TO TRANSACTIONS DEMAND FOR MONEY

Instead of Keynes's speculative demand for money, Baumol concentrated on transactions demand for money and put forward a new approach to explain it. Baumol explains the transaction demand for money from the viewpoint of the inventory control or inventory management similar to the inventory management of goods and materials by business firms. As businessmen keep inventories of goods and materials to facilitate transactions or exchange in the context of changes in demand for them, Baumol asserts that individuals also hold *inventory of money* because this facilitates transactions (*i.e.* purchases) of goods and services.[3]

In view of the cost incurred on holding inventories of goods there is need for keeping *optimal inventory* of goods to reduce cost. Similarly, individuals have to keep *optimum inventory of money* for transaction purposes. Individuals also incur cost when they hold inventories of money for transactions purposes. They incur cost on these inventories as *they have to forgone interest which they could have earned if they had kept their wealth in saving deposits or fixed deposits or invested in bonds*. This interest income forgone is the cost of holding money for transactions purposes. In this way *Baumol and Tobin emphasised that transaction demand for money is not independent of the rate of interest*.

It may be noted that by money we mean currency and demand deposits which are quite safe and riskless but carry no interest. On the other hand, bonds yield interest or return but are risky and may involve capital loss if wealth holders invest in them. However, saving deposits in banks, according to Baumol, are quite free from risk and also yield some interest. Therefore, Baumol asks the question why an individual holds money (*i.e.* currency and demand deposits) instead of keeping his wealth in saving deposits which are quite safe and earn some interest as well. According to him, it is for convenience and capability of it being easily used for transactions of goods that people hold money with them in preference to the saving deposits. Unlike Keynes both *Baumol and Tobin argue that transactions demand for money depends on the rate of interest*. People hold money for transaction purposes "*to bridge the gap between the receipt of income and its spending*."[4] As interest rate on saving deposits goes up people will tend to shift a part of their money holdings to the interest-bearing saving deposits.

Individuals compare the costs and benefits of funds in the form of money with the interest-bearing saving deposits. According to Baumol, the cost which people incur when they hold funds in money is the *opportunity cost of these funds*, that is, interest income forgone by not putting them in saving deposits.

Baumol's Analysis of Transactions Demand

Baumol analyses the transactions demand for money of an individual who receives income at a specified interval, say every month, and spends it gradually at a steady rate. This is illustrated in Fig. 19.2. It is assumed that individual is paid ₹ 12000 salary cheque on the first day of each month. Suppose he gets it cashed (*i.e.* converted into money) on the very first day and gradually spends it daily throughout the month. (₹ 400 per day) so that at the end of the month he is left with no money. It can be easily seen that his average money holding in the month will be ₹ $= \frac{12000}{2} =$ ₹ 6,000 (before 15th of a month he will be having more than ₹ 6,000 and after 15th day he will have less than ₹ 6,000). Average holding of money equal to ₹ 6,000 has been shown by the dotted line. Now, the

3. See James Tobin, "The Interest Elasticity of Transactions Demand for Cash,'" *Review of Economics and Statistics*, Vol. 38 (1956).
4. W.J. Baumol, The Transactions Demand for Cash: an Inventory Theoretic Approach, *Quarterly Journal of Economics*, Vol. 66 (Nov. 1952).

question arises whether it is the optimal strategy of managing money or what is called optimal cash management. The simple answer is no. This is because the individual is losing interest which he could have earned if he had deposited some funds in interest-bearing saving deposits instead of withdrawing all his salary in cash on the first day. He can manage his money balances so as to earn some interest income as well. Suppose, instead of withdrawing his entire salary on the first day of a month, he withdraws only half of it *i.e.* (₹ 6,000 in cash and deposits the remaining amount of ₹ 6,000 in saving account which gives him interest of 5 per cent, his expenditure per day remaining constant at ₹ 400. This is illustrated in Fig. 19.3. It will be seen that his money holdings of ₹ 6,000 will be reduced to zero at the end of the 15th day of each month. Now, he can withdraw ₹ 6,000 on the morning of 16th of each month and then spends it gradually, at a steady rate of 400 per day for the

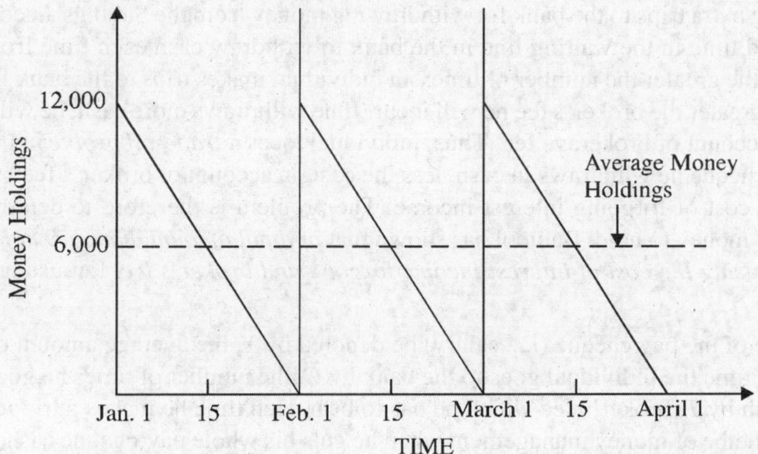

Fig. 19.2. *Stream of Cash Payments and Transactions Demand for Money*

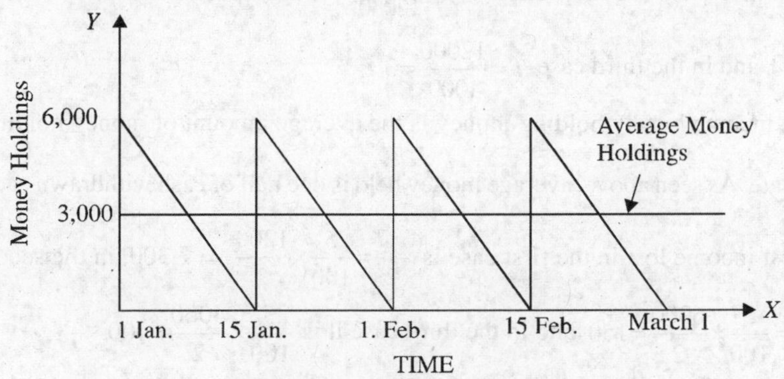

Fig. 19.3. *Transactions Demand for Money and Stream of Cash Payments*

next 15 days of a month. This is a better method of managing funds as he will be earning interest on ₹ 6,000 for 15 days in each month. Average money holdings in this money management scheme is

₹ $= \frac{6000}{2} = 3000.$

Likewise, the individual may decide to withdraw ₹ 4,000 (*i.e.*, 1/3rd of his salary) on the first day of each month and deposits ₹ 8,000 in the saving deposits. His ₹ 4,000 will be reduced to zero, as he spends his money on transactions, (that is, buying of goods and services) at the end of the 10th day and on the morning of 11th of each month he again withdraws ₹ 4,000 to spend on goods and services till the end of the 20th day and on 21st day of the month he again withdraws ₹ 4,000 to spend steadily till the end of the month. In this scheme on an average he will be holding ₹ $\frac{4000}{2}$ = 2000 and will be investing remaining funds in saving deposits and earn interest on them. Thus, in this scheme he will be earning more interest income.

Now, which scheme will he decide to adopt? It may be noted that investing in saving deposits and then withdrawing cash from it to meet the transactions demand involves cost also. Cost on brokerage fee is incurred when one invests in interest-bearing bonds and sells them. Even in case of saving deposits, the asset which we are taking for illustration, one has to spend on transportation costs for making extra trips to the bank for withdrawing money from the Savings Account. Besides, one has to spend time in the waiting line in the bank to withdraw cash each time from the saving deposits. Thus, the greater the number of times an individual makes trips to the bank for withdrawing money, the greater the broker's fee he will incur. If he withdraws more cash, he will be avoiding some costs on account of brokerage fee. Thus, individual faces a *trade-off problem*; the greater the amount of pay cheque he withdraws in cash, less the cost on account of broker's fee but the greater the opportunity cost of forgoing interest income. The problem is therefore to determine an optimum amount of money to hold. Baumol has shown that *optimal amount of money holding is determined by minimising the cost of interest income forgone and broker's fee*. Let us elaborate it further.

Let the size of the pay cheque (*i.e.* salary) be denoted by *Y*, the average amount of the cash he withdraws each time the individual goes to the bank by *C*, the number of times he goes to the bank to withdraw cash by *T*, broker's fee which he has to bear each time he makes a trip to the bank by *b*. In the first scheme of money management when he gets his whole pay-cheque cashed on the first day of every month he incurs broker's fee only once since he makes only a single trip to the bank. Thus *T* in our first case is equal to one $T = \frac{Y}{C} = \frac{12000}{12000} = 1$ because in this case C = Y. In the second case, $T = \frac{12000}{6000} = 2$ and in the third case $T = \frac{12000}{4000} = 3$.

Interest income lost by holding money is the average amount of money holding multiplied by the interest rate. As seen above, average money held is one half of cash withdrawn each time $\left(i.e, \frac{C}{2}\right)$.

Thus, interest income lost in the first case is $\frac{rC}{2} = \frac{5}{100} \times \frac{1200}{2}$ = ₹ 300, in the second case interest lost $= r.\frac{C}{2} = \frac{5}{100} \times \frac{6000}{2} = 150$ and in the third case it is $\frac{5}{100} \times \frac{4000}{2} = 100$.

Thus the total cost incurred on broker's fee and interest income forgone is given by

$$\text{Total Cost } = bT + \frac{r.C}{2}$$

where b stands for broker's fee.

$$\text{As seen above, } T = \frac{Y}{C}$$

Therefore,

$$\text{Total Cost } = \frac{Y}{C}b + \frac{r.C}{2}$$

Baumol has shown that average amount of cash withdrawal which minimises cost is given by

$$C = \sqrt{\frac{2bY}{r}}$$

This means that average *amount of cash withdrawal which minimise cost is the square root of the two times broker's fee multiplied by the size of individual's income (Y) and divided by the interest rate*. This is generally referred to as *Square Root Rule*. For this rule, it follows that a higher broker's fee will raise the money holdings as it will discourage the individuals to make more trips to the bank. On the other hand, a higher interest rate will induce them to reduce their money holdings for transaction purposes as they will be induced to keep more funds in saving deposits to earn higher interest income. That is, at a higher rate of interest transactions demand for money holdings will decline.

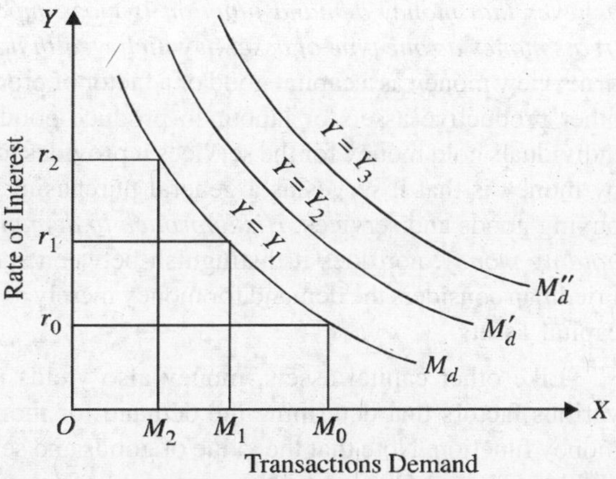

Fig. 19.4. *Transactions Demand for Money : Baumol-Tobin Approach*

As we have studied earlier, Keynes thought that transactions demand for money was independent of rate of interest. According to him, transactions demand for money depends on the level of income. However, Baumol and Tobin have shown that transactions demand for money is sensitive to rate of interest. As explained above, *interest represents the opportunity cost of holding money instead of bonds, saving and fixed deposits*. The higher the rate of interest, the greater the opportunity cost of holding money (*i.e.* the greater the interest income forgone for holding money for transactions). Therefore, *at a higher rate of interest people will try to economise the use of money and will demand less money for transactions*. At a lower interest rate on bonds, saving and fixed deposits, the opportunity cost of holding money will be less which will prompt people to hold more money for transactions. Therefore, according to Baumol and Tobin, transactions demand curve for money slopes downward as shown in Fig. 19.4. At higher interest rates, bonds, savings and fixed deposits are more attractive relative to money holding for transactions. Therefore, at higher interest rates people tend to hold less money for transaction purposes. On the other hand, when the rates of interest are low, opportunity cost of holding money will be less and, as a consequence, people will hold more money for transactions. Therefore, the curve of transaction demand for money slopes downward.

It will be observed from the square root rule given above that transactions demand for money varies directly with the income (*Y*) of the individuals. Therefore, the higher the level of income, the greater the transactions demand for money at a given rate of interest. In Fig. 19.4. the three transactions demand curves for money M_d, M_d' and M_d'', for three different income levels, Y_1, Y_2, Y_3 are shown. It will be known from the square root rule that optimum money holding for transactions will increase less than proportionately to the increase in income. Thus, transactions demand for money, according to Baumol and Tobin, is function of both rate of interest and the level of income.

$$M_{td} = f(r, Y)$$

where M_{td} stands for transactions demand for money, *r* for rate of interest and *Y* for the level of income.

FRIEDMAN'S THEORY OF DEMAND FOR MONEY

A noted monetarist economist Friedman put forward demand for money function which plays an important role in his restatement of the quantity theory of money and prices. In this section we shall be concerned with Friedman's money demand function and will explain in a later chapter money which shows how the changes in money supply affect the price level in the economy. *Friedman believes that money demand function is most important stable function of macroeconomics. He treats money as one type of asset in which wealth holders can keep a part of their wealth.* Business firms view money as a capital good or a factor of production which they combine with the services of other productive assets or labour to produce goods and services. Thus, according to Friedman, individuals hold money for the services it provides to them. It may be noted that the service rendered by money is that it serves as a general purchasing power so that it can be conveniently used for buying goods and services. *His approach to demand for money does not consider any motives for holding money*, nor does it distinguish between speculative and transactions demand for money. Friedman considers the demand for money merely as an application of a general theory of demand for capital assets.

Like other capital assets, money also yields return and provides services. He analyses the various factors that determine the demand for money and from this analysis derives demand for money function. Note that the value of goods and services which money can buy represents the real yield on money. Obviously, this real yield of money in terms of goods and services which it can purchase will depend on the price level of goods and services. Besides money, *bonds* are another type of asset in which people can hold their wealth. Bonds are securities which yield a stream of interest income, fixed in nominal terms. Yield on bond is the *coupen rate of interest* and also anticipated capital gain or loss due to expected changes in the market rate of interest.

Equities or Shares are another form of asset in which wealth can be held. The yield from equity is determined by the dividend rate, expected capital gain or loss and expected changes in the price level. The fourth form in which people can hold their wealth is the *stock of producer and durable consumer commodities*. These commodities also yield a stream of income but in kind rather than in money. Thus, the basic yield from commodities is implicit one. However, Friedman also considers an explicit yield from commodities in the form of expected rate of change in their price per unit of time.

Friedman's nominal demand function (M_d) for money can be written as

$$M_d = f\left(W, h, r_m, r_b, r_e, P, \frac{\Delta P}{P}, U\right)$$

As demand for real money balances is nominal demand for money divided by the price level, demand for real money balances can be written as

$$\frac{M_d}{P} = f\left(W, h, r_m, r_b, r_e, \frac{\Delta P}{P}, U\right)$$

where M_d stands for *nominal* demand for money and $\frac{M_d}{P}$ for demand for real money balances, W stands for wealth of the individuals, h for the proportion of human wealth to the total wealth held by the individuals, r_m for rate of return or interest on money, r_b for rate of interest on bonds, r_e for rate of return on equities, P for the price level, $\frac{\Delta P}{P}$ for the change in price level (*i.e.* rate of inflation), and U for the institutional factors.

1. **Wealth** (W): The major factor determining the demand for money is the wealth of the individual

(W). In wealth Friedman includes not only *non-human wealth* such as bonds, shares, money which yield various rates of return but also *human wealth* or *human capital*. By human wealth Friedman means the value of an individual's present and future earnings. Whereas non-human wealth can be easily converted into money, that is, can be made liquid. Such substitution of human wealth is not easily possible. Thus human wealth represents illiquid component of wealth and, therefore, the proportion of human wealth to the non-human wealth has been included in the demand for money function as an independent variable.

Individual's demand for money directly depends on his total wealth. Indeed, the total wealth of an individual represents an upper limit of holding money by an individual and is similar to the budget constraint of the consumer in the theory of demand. The greater the wealth of an individual, the more money he will demand for transactions and other purposes. As a country, becomes richer, its demand for money for transaction and other purposes will increase. Since as compared to non-human wealth, human wealth is much less liquid, Friedman has argued that as the proportion of human wealth in the total wealth increases, there will be a greater demand for money to make up for the illiquidity of human wealth.

2. Rates of Interest or Return (r_m, r_b, r_e). Friedman considers three rates of interest, namely, rm, r_b and r_e which determine the demand for money. r_m is the own rate of interest on money. Note that money kept in the form of currency and demand deposits does not earn any interest. But money held as saving deposits and fixed deposits earns certain rates of interest and it is this rate of interest which is designated by r_m in the money demand function. Given the other rates of interest or return, the higher the own rate of interest, the greater the demand for money.

In deciding how large a part of his wealth to hold in the form of money the individual will compare the rate of interest on money with rates of interest (or return) on bonds and other assets. As mentioned earlier, *the opportunity cost of holding money is the interest or return given up by not holding these other forms of assets*. As rates of return on bond (r_b) and equities (r_e) rise, the opportunity cost of holding money will increase which will reduce the demand for money holdings. Thus, the *demand for money is negatively related to the rate of interest (or return) on bonds, equities and other such non-money assets.*

3. **Price Level (*P*).** Price level also determines the demand for money balances. A higher price level means people will require a larger *nominal money balances* in order to do the same amount of transactions, that is, to purchase the same amount of goods and services. *If income (Y) is used as proxy for wealth (W) which, as stated above, is the most important determinant of demand for money, then nominal income is given by Y.P* which becomes a crucial determinant of demand for money. Here Y stands for real income (*i.e.* in terms of goods and services) and P for price level. As the price level goes up, the demand for money will rise and, on the other hand, if price level falls, the demand for money will decline. As a matter of fact, people adjust the nominal money balances (M) to achieve their desired level of real money balances $\left(\frac{M}{P}\right)$.

4. **The Expected Rate of Inflation** $\left(\frac{\Delta P}{P}\right)$. If people expect a higher rate of inflation, they will reduce their demand for money holdings. This is because inflation reduces the value of their money balances in terms of its power to purchase goods and services. If the rate of inflation exceeds the nominal rate of interest, there will be negative rate of return on money. Therefore, when people expect a higher rate of inflation they will tend to convert their money holdings into goods or other assets which are not affected by inflation.

On the other hand, if people expect a fall in the price level, their demand for money holdings will increase.

5. **Institutional Factors (U).** Institutional factors such as mode of wage payments and bill payments also affect the demand for money. Several other factors which influence the overall economic environment affect the demand for money. For example, if recession or war is anticipated, the demand for money balances will increase. Besides, instability in capital markets, which erodes the confidence of the people in making profits from investment in bonds and equity shares will also raise the demand for money. Even political instability in the country influences the demand for money. To account for these institutional factors Friedman includes the variable U in his demand for money function.

Simplifying Friedman's Demand for Money Function

A major problem faced in using Friedman's demand for money function has been that due to the non-existence of reliable data about the value of wealth (W), it is difficult to estimate the demand for money. To overcome this difficulty Friedman suggested that since the *present value of* walth or $W = \frac{Y_P}{r}$ (where Y_p is the permanent income and r is the rate of interest on money.), permanent income Yp can be used as a proxy variable for wealth. Incorporating this in Friedman's demand for money function we have

$$M_d = \left(Y_p, h, r_m, r_b, r_e \frac{\Delta P}{P}, U \right)$$

If, we assume that no price change is anticipated and institutional factors such as h and U remain fixed in the short run and also all the three rates of interest return are clubbed into one, Friedman's demand for money function is simplified to

$$M_d = f(Y_p r)$$

QUESTIONS FOR REVIEW

1. Explain Tobin's Portfolio Approach to demand for money. Why according to Tobin, asset demand for money increases as the rate of interest on bonds falls?
2. How does Tobin's approach to liquidity preference differ from Keynes's liquidity prefernce function?
3. How does Baumol-Tobin model explain that the transaction demand for money is a function of interest rate and level of income. *[D.U. B.A. (Hons), 1997]*
4. Explain why the transactions demand for money depends not only on the level of income but also on the rate of interest.
5. Explain Friedman's theory of demand for money. How does it differ from Keynes' theory of demand for money? *[D.U. B.Com (Hons) 1997, 1999]*
6. Why, according to Freidman, are the main factors that determine demand for money. How do the stock of wealth and rate of interest influence the demand for money?
7. State and explain Friedman's Quantity Theory of Demand for Money. Friedman thinks that his theory is closely related to Cambridge version of quantity theory of money while Don Patinkin holds that Friedman provides us a most elegant and sophisticated statement of Keynesian theory demand for money. Discuss.

PART X

Money, Prices and Inflation

- Money and Prices: Quantity Theory of Money
- Keynesian Monetary Theory: Money, Income and Prices
- Monetarism and Friedman's Modern Quantity Theory of Money
- Inflation and Hyperinflation: Causes, Social Costs, Effects and Cure

Chapter 20
MONEY AND PRICES: QUANTITY THEORY OF MONEY

VALUE OF MONEY AND THE PRICE LEVEL

Value in economics means value-in-exchange. The value of a commodity means what it can be exchanged for or what it can buy. It means the purchasing power of a commodity. Similarly, the value of money means its purchasing power in terms of goods and services in general. The purchasing power of money obviously depends on the prevalent price level. If the prices are high, money will buy less and its value will be low. Conversely, if the price level is low, the value of money is high. The value of money is thus inversely proportional to the price level, or the value of money is the reciprocal of price level.

QUANTITY THEORY OF MONEY: FISHER'S TRANSACTIONS APPROACH

We are now in a position to explain how the general level of prices is determined, that is, why at sometimes the general level of prices rises and sometimes it declines. Sometime back it was believed by the economists that the quantity of money in the economy is the prime cause of fluctuations in the price level. The theory that increase in the quantity of money leads to the rise in the general price was effectively put forward by Irving Fisher.[1] They believed that the greater the quantity of money, the higher the level of prices and vice versa. Therefore, the theory which linked prices with the quantity of money came to be known as quantity theory of money. In the following analysis we shall first critically examine the quantity theory of money and then explain the modern view about the relationship between money and prices and also the determination of general level of prices.

The quantity theory of money seeks to explain the value of money in terms of changes in its quantity. Stated in its simplest form, the quantity theory of money says that the level of prices varies directly with quantity of money. "Double the quantity of money, and other things being equal, prices will be twice as high as before, and the value of money one-half. Halve the quantity of money and, other things being equal, prices will be one-half of what they were before and the value of money double." The theory can also be stated in these words: *The price level rises proportionately with a given increase in the quantity of money. Conversely, the price level falls proportionately with a given decrease in the quantity of money, other things remaining the same.*

There are several forces that determine the value of money and the general price level. The general price level in a community is influenced by the following factors:

(*a*) the volume of trade or transactions;

(*b*) the quantity of money;

(*c*) velocity of circulation of money.

The first factor, the volume of trade or transactions, depends upon the supply or amount of goods and services to be exchanged. The greater the amount or supply of goods in an economy,

1. Irving Fisher, *Purchasing Power of Money*, Macmillan, 1911.

the larger the number of transactions and trade, and *vice versa*. But the classical and neoclassical economists who believed in the quantity theory of money assumed that *full employment* of all resources (including labour) prevailed in the economy. Resources being fully employed, the total output or supply of goods (and therefore the total trade or transactions) cannot increase. Therefore, those who believed in the quantity theory of money assumed that the total volume of trade or transactions remained the same.

The second factor in the determination of general level of prices is the quantity of money. It should be noted that the quantity of money in the economy consists of not only the notes and currency issued by the Government but also the amount of credit or deposits created by the banks.

The third factor influencing the price level is the velocity of circulation. A unit of money is used for exchange and transactions purposes not once but several times in a year. During several exchanges of goods and services, a unit of money passes from one hand to another. Thus, if a single rupee is used five times in a year for exchange of goods and services, the velocity of circulation is 5. Hence, the velocity of money is the *number of times* a unit of money changes hands during exchanges in a year. The work done by one rupee which is circulated five times in a year is equal to that done by the five rupees which change hands only once each.

Let us illustrate the quantity theory of money. Suppose in a country there is only one good, wheat, which is to be exchanged. The total output of wheat is 2,000 quintals in a year. Further suppose that the government has issued money equal to ₹ 25,000 and no credit is issued by the banks. We further assume that one rupee is used four times in a year for exchange of wheat. That is, velocity of circulation of money is four. Under these circumstances, 2,000 quintals of wheat are to be exchanged for ₹1,00,000, (25,000 × 4 = 1,00,000). The price of wheat will be $\frac{1,00,000}{2,000}$ = ₹50 per quintal. Suppose the quantity of money is doubled to ₹50,000, while the output of wheat remains at 2,000 quintals. As a result of this increase in the quantity of money, the price of wheat will rise to $\frac{2,00,000}{2,000}$ = ₹100 per quintal. Thus with doubling of the quantity of money, the price has doubled. If the quantity of money is further increased to ₹75,000, the amount of wheat remaining constant, the price level will rise to $\frac{3,00,000}{2,000}$ = ₹150 per quintal. It is thus clear that if the volume of transactions, *i.e.*, output to be exchanged remains constant, the price level rises with the increase in the quantity of money.

Fisher's Equation of Exchange

An American economist, Irving Fisher, expressed the relationship between the quantity of money and the price level in the form of an equation, which is called 'the equation of exchange'. This is:

$$PT = MV \qquad \ldots (1)$$

or

$$P = \frac{MV}{T} \qquad \ldots (2)$$

where P stands for the average price level:

T stands for total amount of transactions (or total trade or amount of goods and services, raw materials, old goods etc.)

M stands for the quantity of money; and

V stands for the transactions velocity of circulation of money.

The equation (1) or (2) is an accounting identity and true by definition. This is, because MV which represents money spent on transactions must be equal to PT which represents money received from transactions.

However, the equation of exchange as given in equations (1) and (2) has been converted into a theory of determination of general level of prices by the classical economists by making some assumptions. First, it has been assumed that *the physical volume of transactions is constant* because it is determined by a given amount of real resources, the given level of technology and the efficiency with which the given available resources are used. These real factors determine a level of aggregate output which necessitates various types of transactions. Another crucial assumption is that *transactions velocity of circulation* (V) *is also constant*. The quantity theorists accordingly believed that velocity of circulation (V) depends on the methods and practices of factor payments such as frequency of wage payments to the workers, and habits of the people regarding spending their money incomes after they receive them. Further, velocity of circulation of money also depends on the development of banking and credit system, that is, the ways and speed with which cheques are cleared, loans are granted and repaid. According to them, these practices do not change in the short run. This assumption is very crucial for the quantity theory of money because when the quantity of money is increased this may cause a decline in velocity of circulation of money, then MV may not change if the decline in V offsets the increase in M. As a result, increase in M will not affect PY.

The quantity theorists believed that the volume of transactions (T) and the changes in it were largely independent of the quantity of money. Further, according to them, changes in velocity of circulation (V) and price level (P) do not cause any change in volume of transactions except temporarily. Thus classical economists who put forward the quantity theory of money believed that the number of transactions (which ultimately depends on aggregate real output) does not depend on other variables (M, V and P) in the equation of exchange. Thus we see that the assumption of constant V and T converts the equation of exchange ($MV = PT$), which is an accounting identity, into a theory of the determination of general price level.

The quantity of money is fixed by the Government and the Central Bank of a country. Further, it is assumed that quantity of money in the economy depends upon the monetary system and policy of the central bank and the Government and is assumed to be autonomous of the real forces which determine the volume of transactions or national output.

Now, with the assumptions that M and V remain constant, the price level P depends upon the quantity of money M; the greater the quantity of M, the higher the level of prices. Let us give a numerical example. Suppose the quantity of money is ₹ 5,00,000 in an economy, the velocity of circulation of money (V) is 5; and the total output to be transacted (T) is 2,50,000 units, the average price level (P) will be:

$$P = \frac{MV}{T}$$

$$= \frac{5,00,000 \times 5}{2,50,000} = \frac{2,500,000}{2,50,000}$$

$$= ₹\ 10 \text{ per unit.}$$

If now, other things remaining the same, the quantity of money is doubled, *i.e.*, increased to ₹ 10,00,000 then:

$$P = \frac{10,00,000 \times 5}{2,50,000} = ₹\ 20 \text{ per unit}$$

We thus see that according to the quantity theory of money, price level varies in direct proportion to the quantity of money. A doubling of the quantity of money (M) will lead to the doubling of the price level. Further, since changes in the quantity of money are assumed to be independent or autonomous of the price level, the changes in the quantity of money become the cause of the changes in the price level.

Quantity Theory of Money: Income Version

Fisher's transactions approach to quantity theory of money described in equation (1) and (2) above considers such variables as total volume of transaction (T) and average price level of these transactions are conceptually vague and difficult to measure. Therefore, in later years quantity theory was formulated in income form which considers real income or national output (*i.e.*, transactions of final goods only) rather than all transactions. As the data regarding national income or output is readily available, the income version of the quantity theory is being increasingly used. Moreover, the average price level of output is a more meaningful and useful concept. Indeed, in actual practice, the general price level in a country is measured taking into account only the prices of final goods and services which constitute national product. It may be noted that even in this income version of the quantity theory of money, the function of money is considered to be a means of exchange as in the transactions approach of Fisher. In this approach, the concept of *income velocity of money* has been used instead of transactions velocity of circulation. *By income velocity we mean the average number of times per period a unit of money is used in making payments involving final goods and services*, that is, national product or national income. In fact, income velocity of money is measured by $\frac{Y}{M}$ where Y stands for real national income and M for the quantity of money.

In view of the above, the income version of quantity theory of money is written as under:

$$MV = PY \qquad \ldots(3)$$

$$P = \frac{MV}{Y} \qquad \ldots(4)$$

where

M = Quantity of money

V = Income velocity of money

P = Average price level of final goods and services

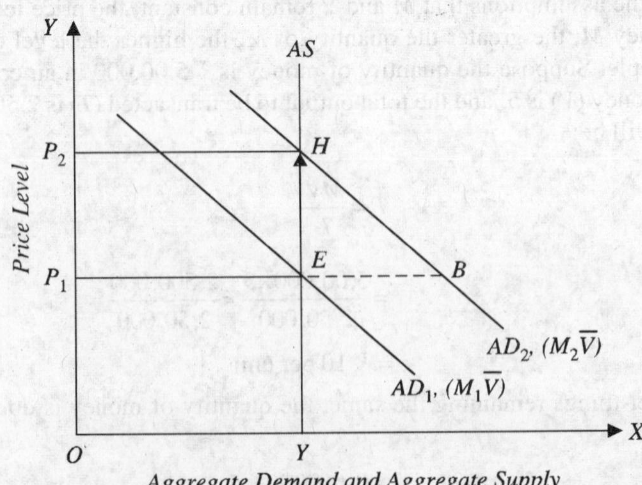

Aggregate Demand and Aggregate Supply

Fig. 20.1. *Quantity Theory of Money*

Y = Real national income (or aggregatc output)

Like that in the transactions approach, in this new income version of the quantity theory also the different variables are assumed to be independent of each other. Further, income velocity of money (V) and real income or aggregate output (Y) are assumed to be given and constant during a short period. More specifically, they do not vary in response to the changes in M. In fact, real income or output (Y) *is assumed to be determined by the real sector forces such as capital stock, the amount and skills of labour, technology etc*. But as these factors are taken to be given and constant in the short run, and further full employment of the given resources is assumed to be prevailing due to the operation of Say's law and wage-price flexibility supply of output is taken to be inelastic and constant for purposes of determination of price level.

It follows from equations (3) and (4) above that with income velocity (V) and national output (Y) remaining constant, price level (P) is determined by the quantity of money (M).

Classical quantity theory of money is illustrated in Fig. 20.1 through aggregate demand and aggregate supply model. It is worth noting that the quantity of money (M) multiplied by the income velocity of circulation (V), that is, MV gives us aggregate expenditure in the quantity theory of money. Now with a given quantity of money, say M_1 and constant velocity of money $\bar{V}$, we have a given amount of monetary expenditure ($M_1 \bar{V}$). Given this aggregate expenditure, at a lower price level more quantities of goods can be purchased and at a higher price level, less quantities of goods can be purchased. Therefore, in accordance with classical quantity theory of money aggregate demand representing $M_1 \bar{V}$ slopes downward as shown by the aggregate demand curve AD_1 in Fig. 20.1. If now the quantity of money is increased, say to M_2, aggregate demand curve representing new aggregate monetary expenditure $M_2 \bar{V}$ will shift upward.

As regards, aggregate supply curve, due to the assumption of wge-price flexibility, it is perfectly inelastic at full-employment level of output as is shown by the vertical aggregate supply curve AS in Fig. 20.1. Now, with a given quantity of money equal to M_1, aggregate demand curve AD_1 cuts the aggregate supply curve AS at point E and determines price level OP_1. Now, if the quantity of money is increased to M_2, the aggregate demand curve shifts upward to AD_2. It will be seen from Fig. 20.1 that with the increase in aggregate demand to AD_2 consequent to the expansion in money supply to M_2, excess demand equal to EB emerges at the current price level OP_1. This excess demand for goods and services will lead to the rise in price level to OP_2 at which again aggregate quantity demanded equals the aggregate supply which remains unchanged at OY due to the existence of full employment in the economy.

QUANTITY THEORY OF MONEY: THE CAMBRIDGE CASH–BALANCE APPROACH

The equation of exchange has been stated by Cambridge economists, Marshall and Pigou, in a form different from Irving Fisher. Cambridge economists explained the determination of value of money in line with the determination of value in general. Value of a commodity is determined by demand for and supply of it and likewise, according to them, the value of money (*i.e.*, its purchasing power) is determined by the demand for and supply of money. As studied in cash-balance approach to demand for money Cambridge economists laid stress on the store of value function of money in sharp contrast to the medium of exchange function of money emphasised by in Fisher's transactions approach to demand for money. According to cash balance approach, the public likes to hold a proportion of nominal income in the form of money (*i.e.*, cash balances). Let us call this proportion of nominal income that people want to hold in money as k. Then cash balance approach can be written as:

$$M^d = kPY \quad ...(1)$$

where
Y = *real* national income (*i.e.*, aggregate output)
P = the price level
PY = *nominal* national income
k = the proportion of nominal income that people want to hold in money
M^d = the amount of money which public want to hold

Now, for the achievement of money-market equilibrium, demand for money must equal worth the supply of money which we denote by M. It is important to note that the supply of money M is exogenously given and is determined by the monetary policies of the central bank of a country. Thus, for equilibrium in the money market.

$$M = M^d$$

As $\quad M^d = kPY$

Therefore, in equilibrium $\quad M = kPY \quad ...(2)$

Monetary equilibrium Cambridge cash balance approach is shown in Fig. 20.2 where demand for money is shown by a rising straight line kPY which indicates that with k and Y being held constant demand for money increases proportionately to the rise in price level. As price level rises people demand more money for transaction purposes.

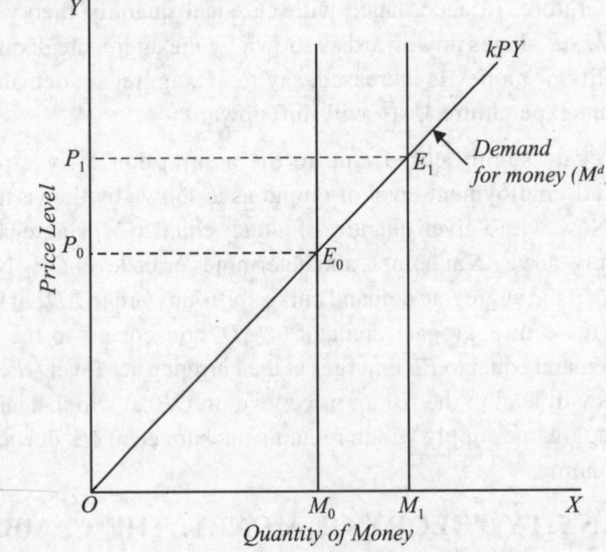

Fig. 20.2. *Determination of Price Level : Cambridge Cash Balance Approach*

Now, if supply of money fixed by the Government (or the Central Bank) is equal to M_0, the demand for money kPY equals the supply of money, M_0 at price level P_0. Thus, with supply of money equal to M_0 equilibrium price level P_0 is determined. If money supply is increased, how the monetary equilibrium will change? Suppose money suply is increased to M_1, at the initial price level P_0 the people will be holding more money than they demand at it. Therefore, they would want to reduce their money holding. In order to reduce their money holding they would increase their spending on goods and services. In response to the increase in money spending by the households the firms will increase prices of their goods and services. As prices rise, the households will need and demand more money to hold for transaction purposes (*i.e.* for buying goods and services). It will be seen from Fig. 20.2 that with the increase in money suply to M_1 new equilibrium between demand for money and suply of money is attained at point E_1 on the demand for money curve kPY and price level has risen to P_1.

It is worth mentioning that k in the equations (1) and (2) is related to velocity of circulation of money V in Fisher's transactions approach. Thus, when a greater proportion of nominal income is held in the form of money (*i.e.*, when k is higher), V falls. On the other hand, when less proportion of nominal income is held in money, V rises. In the words of Crowther, "The higher the proportion of their real incomes that people decide to keep in money, the lower will be the velocity of circulation, and vice versa."[2]

It follows from above that $k = \frac{1}{V}$. Now, rearranging equation (2) we have cash balance approach in which P appears as dependent variable. Thus, on rearranging equation (2) we have

$$P = \frac{1}{k} \cdot \frac{M}{Y} \qquad \text{...(3)}$$

Like Fisher's equation, cash balance equation is also an accounting identity because k is defined as:

$$\frac{\text{Quantity of Money Supply}}{\text{National Income}}, \text{ that is, } \frac{M}{PY}$$

Now, Cambridge economists also assumed that k remains constant. Further, due to their belief that wage-price flexibility ensures full employment of resources, the level of real national income was also fixed corresponding to the level of aggregate output produced by full employment of resources.

Thus, from equation (3) it follows that with k and Y remaining constant price level (P) is determined by the quantity of money (M); changes in the quantity of money will cause proportionate changes in the price level.

Some economists have pointed out similarity between Cambridge cash-balance approach and Fisher's transactions approach. According to them, k is reciprocal of V $\left(k = \frac{1}{V} \text{ or } V = \frac{1}{k}\right)$. Thus in equation (2) if we replace k by , we have

$$M = \frac{1}{V} PY$$

or

$$MV = PY$$

Which is income version of Fisher's quantity theory of money. However, in spite of the formal similarity between the cash balance and transactions approaches, there are important conceptual differences between the two which makes cash balance approach superior to the transactions approach. First, as mentioned above, Fisher's transactions approach lays stress on the medium of exchange function of money, that is, according to it, people want money to use it as a means of payment for buying goods and services. On the other hand, cash balance approach emphasizes the store-of-value function of money. They hold money so that some value is stored for spending on goods and services after some lapse of time.

---

2. Crowther, *Outline of Money*, p. 150.
3. S.B. Gupta, *op. cit.*, p. 206.

Further, in explaining the factors which determine velocity of circulation, transactions approach points to the mechanical aspects of payment methods and practices such as frequency of wages and other factor payments, the speed with which funds can be sent from one place to another, the extent to which bank deposits and cheques are used in dealing with others and so on. On the other hand, k in the cash balance approach is *behavioural in nature*. Thus, according to Prof. S.B. Gupta, "*Cash-balance approach is behavioural in nature: it is build around the demand for money, however simple. Unlike Fisher's V, k is a behavioural ratio. As such it can easily lead to stress being placed on the relative usefulness of money as an asset*."[3]

Thirdly, cash balance approach explains determination of value of money in a framework of general demand-supply analysis of value. Thus, according to this approach value of money (that is, its purchasing power is determined by the demand for and supply of money).

To sum up cash balance approach has made some improvements over Fisher's transactions approach in explaining the relation between money and prices. However, it is essentially the same as the Fisher's transactions approach. Like Fisher's approach if considers substitution between money and commodities. That is, if they decide to hold less money, they spend more on commodities rather than on other assets such as bonds, shares, real property, durable consumer goods. Further, like Fisher's transactions approach it visualises changes in the quantity of money causes proportional changes in the price level. Like Fisher's approach, cash balance approach also assumes that full-employment of resources will prevail due to the wage-price flexibility. Hence, it also believes the aggregate supply curve as perfectly inelastic at full-employment level of output.

An important limitation of cash balance approach is that it also assumes that the proportion of income that people want to hold in money, that is, k, remains constant. Note that . In practice it has been found that proportionality factor k or velocity of circulation has not remained constant but has been fluctuating, especially in the short run.

Besides, cash-balance approach falls short of considering *demand for money as an asset*. If demand for money as an asset were considered, it would have a determining influence on the rate of interest on which amount of investment in the economy depends. Investment plays an important role in the determination of level of real income in the economy. It was left to J.M. Keynes who later emphasised the role of demand for money as an asset which was one of the alternative assets in which individuals can keep their income or wealth. Finally, it may be mentioned that other criticisms of Fisher's transactions approach to quantity theory of money discussed above equally apply to the Cambridge cash balance approach.

Keynes's Critique of the Quantity Theory of Money

The quantity theory of money has been widely criticised. The following criticisms have been levelled against the quantity theory of money lay by Keynes and his followers.[4]

1. **Useless truism.** With the qualification that velocity of money (V) and the total output (T) remain the same, the equation of exchange ($MV = PT$) is a *useless truism*. The real trouble is that these

4. For detailed analysis of Keynes's theory of money and prices, See Chapter 21.

things seldom remain the same. They change not only in the long run but also in a short period. Fisher's equation of exchange simply tells us that expenditure made on goods (*MV*) is equal to the value of output of goods and services sold (*PT*).

2. **Velocity of money is not stable.** *Keynesian economists has challenged the assumption that velocity of money remains stable*. According to them, velocity of money changes inversely with the change in money supply. They argue that increase in money supply, demand for money remaining constant, leads to the fall in the rate of interest. At a lower rate of interest, people will be induced to hold more money as idle cash balances (under speculative motive). This means velocity of circulation of money will be reduced. Thus, *if a decline in interest rate reduces velocity, then increase in the money supply will be offset by reduction in velocity*, with the result that price level, need not rise when money supply is increased.

3. **Increase in quantity of money may not always lead to the increase in aggregate spending or demand.** Further, according to Keynes' the quantity theory of money is based upon two more *wrong assumptions*. Basically, for the quantity theory to be true, the following two assumptions must hold:

(*i*) An increase is money supply must lead to an increase in spending, that is, aggregate demand *i.e.*, no part of additional money created should be kept in idle hoards.

(*ii*) The resulting increase in spending or aggregate demand must face a totally inelastic output.

Both the assumptions according to Keynes, lack generality and, therefore, if either of them does not hold, the quantity theory cannot be accepted as a valid explanation of the changes in price level.

Let us take the first assumption. Under this assumption, the entire increase in the quantity of money must express itself in the form of increased spending. If spending does not increase, there is no question of a change in prices or output. But, is it valid to make such an assumption? Obviously, *there is no such direct link* between the increase in the quantity of money and the increase in the volume of total spending or aggregate demand. No one is going to increase his expenditure simply because the government is printing more notes or the banks are more liberal in their lending policies. Thus, if the demand for money is highly interest-elastic, the increase in money supply will not lead to any appreciable fall in the rate of interest. With no significant fall in rate of interest, the investment expenditure and expenditure on durable consumer goods will not increase much. As a result, increase in money supply may not lead to increase in expenditure or aggregate demand and therefore price level may remain unaffected.

This is not to say, however, that changes in the quantity of money have no influence whatsoever on the volume of aggregate spending. As we shall show below, changes in the quantity of money are often capable of inducing changes in the volume of aggregate spending. What Keynes and his followers deny is the assertion that there exists a direct, simple, and more or less a proportional relation between variation in money supply and variation in the level of total spending.

4. **Assumption of constant volume of transactions or constant level of aggregate output is not valid.** Keys asserted that the assumption of constant aggregate output valid only under conditions of full employment. It is only then that we can assume a totally inelastic supply of output, for all the

available resources are being already fully utilised. In conditions of less than full employment, the supply curve of output will be elastic. Now, if we assume that aggregate spending or demand increases with an increase in the quantity of money, it does not follow that prices must necessarily rise. If the supply curve of output is fairly elastic, it is more likely that effect of an increase in spending will be more to raise production rather than prices. Of course, at full-employment level every further increase in spending or aggregate demand must lead to the rise in the price level as output is inelastic in supply at full-employment level. Since full-employment cannot be assumed to be a normal affair, we cannot accept the quantity theory of money as a valid explanation of changes in the price level in the short run.

QUESTIONS FOR REVIEW

1. Explain the Quantity Theory of Money. Will doubling of money supply always lead to a doubling of general price level?
2. Does increase in money supply always lead to a proportional increase in prices? Give reasons in support of your answer.
3. Critically examine the Quantity Theory of Money. Does an increase in money supply always lead to a proportionate increase in prices?
4. Explain Keynes's criticism of the Quantity Theory of Money.
5. Discuss Fisher's and Keynes's approach to the relationship between the quantity of money and price level.
6. Explain Fisher's equation of exchange. It is asserted by some that Fisher's equation of exchange is a '*useless truism*' Do you agree? Discuss.
7. Explain Cambridge Cash Balance Approach of the Quantity Theory of Money. How does it differ from Fischer's transactions approach?
8. Cash Balance Approach of the Quantity Theory of Money is behavioural in nature and is built around the demand for money. Discuss.

CHAPTER 21

KEYNESIAN MONETARY THEORY: MONEY, INCOME AND PRICES

INTRODUCTION

The main thrust of Keynes's criticism of classical quantity theory of money was directed at its conclusion that (*i*) velocity of circulation is constant, and (*ii*) full employment of resources is the natural state of a free market economy. Keynes believed that velocity of circulation was volatile and there often existed underemployment of resources due to recessionary conditions in the economy. As seen in the previous chapters, classical economists believed that people demanded money only for transactions purpose and money balances held for transactions purposes were proportional to nominal income. Keynes challenged this viewpoint and held that people could hold income-earning assets such as bonds instead of holding money balances. To the transactions motive for holding money: Keynes added precautionary motive and speculative motive (*i.e.*, demand for money as an asset) for holding money. Income or interest earned on assets such as bonds is the opportunity cost of holding money. The higher the rate of interest on these assets, the less money will be held by the public. It is worth noting that people have adjust their money holdings until what they demand equals what they actually have. If people have more money than what they demand, they will spend either on consumer goods and services, or invest more. On the other hand, if their demand for money to hold is greater than what they presently have, they will try to acquire more money either by reducing expenditure on goods and services or selling some of their assets such as bonds and shares. Keynes laid stress on financial investment in bonds as a major way to reduce one's money holdings.

The task of a monetary theory is to explain the influence of changes in money supply on the level of economic activity (*i.e.*, levels of real income, output and employment) and the price level. *Keynes's monetary theory explains the effect of variation in money supply on the level of economic activity through its effect on the rate of interest which determines investment in the economy*. In what follows we first explain the impact of expansion in money supply on that levels of real income and employment. In the second stage of our analysis of Keynes's monetary theory we show how changes in money supply affect the price level in the economy.

KEYNES'S MONETARY THEORY: INTEGRATING MONEY MARKET WITH GOODS MARKET

According to Keynes, rate of interest is determined by equilibrium between demand for money and supply of money (*i.e.*, through money market equilibrium). The effect of money supply on rate of interest and the effect of rate of interest on aggregate demand provides a mechanism through which changes in money supply affect the goods market which determines level of economic activity in the economy, that is, level of output and employment. We know from the study of money market that monetary policy has a profound effect on the rate of interest. Thus, if rate of interest is reduced as a result of an increase in money supply, the rate of investment will rise and the increase in investment will lead to increase in income and employment via the multiplier. Thus, when in times of recession, money supply in the economy is increased it will cause investment to increase and as result, there will be an increase in aggregate expenditure (*i.e.*, aggregate demand) which will lead to the increase in real national income (aggregate output) and employment will increase. In this way Keynes succeed in integrating money market with goods market.

Transmission Mechanism. How, according to Keynes, the change in money supply leads to the increase real income output and employment is shown in the following scheme:

$$M^S \xrightarrow{(1)} r \xrightarrow{(2)} I \xrightarrow{(3)} AD \text{ (or } C + I + G + X_n) \xrightarrow{(4)} Y \text{ (i.e., GNP)}$$

The first link in the transmission mechanism is the effect of expansion in money supply on the rate of interest which depends on how far demand for money holdings is sensitive (*i.e.*, elastic) to the changes in rate of interest. The expansion in money supply (M^S) causes the rate of interest to fall. The second step in the transmission mechanism is the influence of change in rate of interest on the rate of investment, which is determined by the elasticity of investment with respect to rate of interest. The fall in rate of interest leads to the increase in investment in the economy. The next step in the process is the effect of increase in investment on aggregate demand and therefore on national income (aggregate output) and employment in the economy. The effect of investment on income, output and employment is determined by the size of multiplier.

We explain below at length the above factors and show how the increase in money supply affects the level of economic activity. It may be noted that expansion in money supply which leads to the increase in aggregate demand will affect both the real national income (*i.e.*, *GNP*) and the price level jointly. However for better understanding of the subject by the students we shall explain the Keynesian monetary theory with regard to the relation between money supply and price level separately as well.

Money Supply and Rate of Interest. Rate of interest, according to Keynes, is a purely monetary phenomenon. Demand for money to hold depends on the level of income and rate of interest. At a higher current rate of interest, less money is demanded by the people to hold and vice versa. Therefore, money demand curve (M^d) or what Keynes calls liquidity preference curve slopes downward as shown by M^d curve in Fig. 21.1. Rate of interest is determined by demand for money and supply of money. This is shown in Fig. 21.1 where, the quantity of money fixed

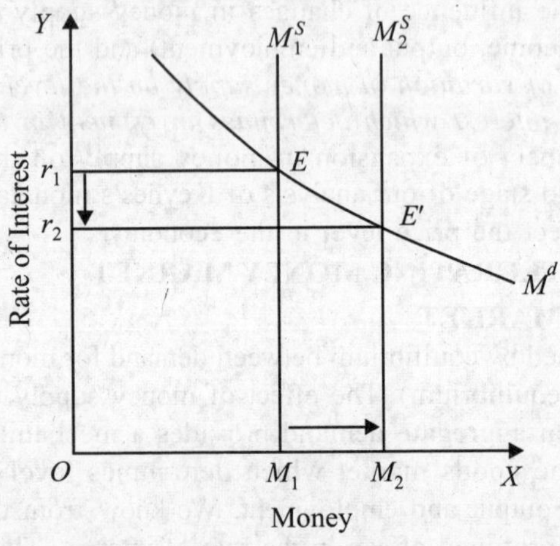

Fig. 21.1. *Money Supply and Rate of Interest*

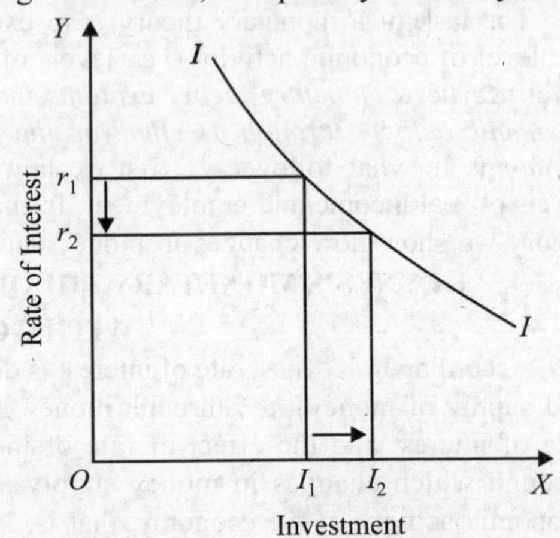

Fig. 21.2. *Rate of Interest and Investment*

by the Government is OM_1 so that money supply curve is M_1^S. The intersection of demand for money curve M^d and the supply of money curve M_1^S determines r_1 rate of interest. Thus at rate of interest r_1, demand for money to hold is equal to the available supply of money M_1. Now, observing that there is unemployment of labour and other resources and recessionary conditions prevail in the economy, the central bank takes steps to raise money supply. The central bank can raise money supply by purchasing Government securities from the market (that is undertaking open market operations) or lowering cash-reserve ratio (CRR) of the banks. Suppose ultimately these steps lead to the expansion in money supply to M_2. It will be seen from Fig. 21.1 that with the increase in money supply from M_1 to M_2 rate of interest falls to r_2 at which demand for money holdings equals the increased supply of money M_2.

It may however be noted that the extent to which rate of interest falls as a result of expansion in money supply depends on the elasticity of demand for money holdings with respect to the rate of interest. The higher the elasticity of demand for money with respect to the rate of interest, the smaller the fall in rate of interest as a result of increase in money supply by the central bank of a country.

Rate of Investment. The next link in the chain of causation is the effect of change in rate of interest on rate of investment in the economy. In the Keynesian system, investment in the economy depends on the rate of interest on the one hand and marginal efficiency of investment (*MEI*) on the other. Marginal efficiency of investment (*i.e.*, expected rate of profit), it may be emphasised, depends on the expectations of entrepreneurs.

The determination of investment is shown in Fig. 21.2 where *II* is the 'investment demand curve whose position depends on the profit expectations of the entrepreneurs which determine marginal efficiency of investment. At the rate of interest r_1 investment equal to I_1 will be made. Now, if the expansion in money supply results in fall in rate of interest to r_2 investment increases to I_2. It is worth noting that the *increase in investment as a result of change in the rate of interest depends on the responsiveness* (*i.e., elasticity*) *of investment demand to the changes in rate of interest.* The higher the elasticity of investment expenditure to the changes in the rate of interest, the greater will be the increase in investment for a given fall in the rate of interest.

Investment and Aggregate Demand. Next step in the transmission mechanism of the effect of money supply on the national income and price level is concerned with the impact of increase in investment on aggregate demand. Aggregate demand which we may write as *AD* is determined in the Keynesian theory by the sum of private consumption expenditure, private investment expenditure (*I*), Government's expenditure on goods and services (*G*) and net exports (X_n) *i.e.*, excess of exports over imports. Thus

$$AD = C + I + G + X_n$$

When the rate of interest is reduced as a result of expansion in money supply and causes investment to increase, it will shift the aggregate demand curve upward. This is depicted in Fig. 21.3 where initially with investment equal to I_1 along with other variables, aggregate demand curve is AD_1 or $C + I_1 + G + X_n$. When due to the expansion in money supply and resultant fall in rate of interest investment increases from I_1 to I_2, aggregate demand curve shifts upward to the new position $C + I_2 + G + X_n$. The upward shift in aggregate demand curve is equal to the increase in investment (ΔI) from I_1 to I_2.

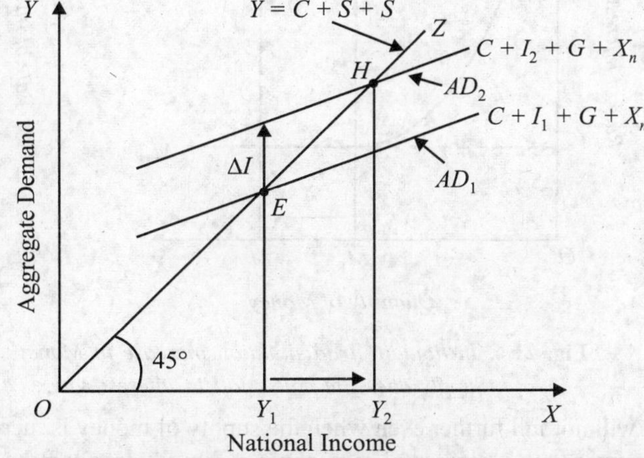

Fig. 21.3. *The Effect of Increase in Investment on Aggregate Demand and National Income*

Multiplier and National Income. Finally, the effect of increase in investment and aggregate demand on real national income (GNP) depends on the size of multiplier. The size of multiplier depends on the marginal propensity to consume (*MPC*) of a community. The higher the marginal propensity to consume, the greater the size of the multiplier $\left(\text{Multiplier} = \dfrac{1}{1-MPC}\right)$. Thus increase in national income (GNP) following the increase in investment by ΔI will be equal to $\Delta I \dfrac{1}{1-MPC}$. This is illustrated in Fig. 21.3 where we have assumed marginal propensity to consume being equal to 0.5 and the size of multiplier equal to 2. Thus when investment increases from I_2 to I_2 (*i.e.*, *M*), national increases from Y_1 to Y_2 (*i.e.*, ΔY) which is twice the increase in investment.

It follows from above that the *effect of expansion in money supply on national income and employment depends on the elasticity of demand curve for money holding, on the elasticity of investment to changes in rate of interest, and on the size of multiplier.*

Ineffectiveness of Monetary Policy: Keynes'View

It may however be noted that Keynes and early Keynesians were not very optimistic about the success of expansionary monetary policy in lifting the economy out of depression. They point out various weak links in the chain of causation in the process of increase in money supply leading to the rise in aggregate demand. The first weak link occurs at the stage of increase in money supply causing lowering of the rate of interest. Keynes was considering an economy which was in the grip of depression and rate of interest. prevailing then was already at a low level and therefore they did not expect any further fall. Whatever expansion in money supply is made at that time, all of it would be held by them rather than investing it in bonds. In other words, according to Keynes and his earlier followers, the demand for money (*i.e.*, liquidity preference curve) was perfectly elastic at a very low rate of interest as shown in Fig. 21.4 where at rate of interest r_3 demand for money becomes perfectly elastic. It is said that at a very low rate of interest economy gets caught up in the **liquidity trap** in times of depression.

Thus, if the rate of interest is already very low, as is usually the cause in times of depression, the expansion in the quantity of money will not be able to reduce it still further. As shown in Fig. 21.4 when the rate of interest is already at a low level so that people demand any amount of money (*i.e.*, the economy is caught in the liquidity trap) at it, the rate of interest

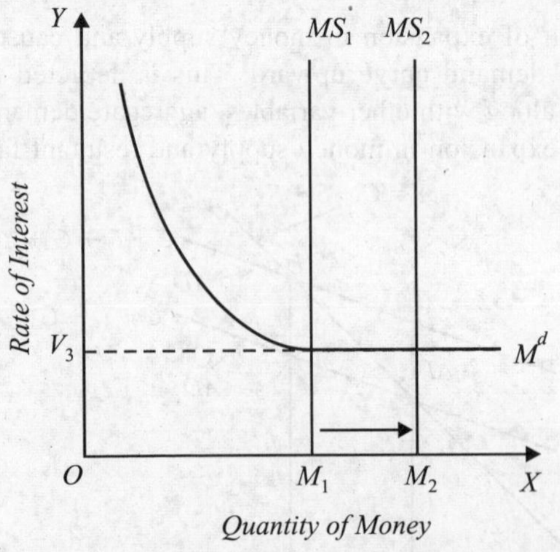

Fig. 21.4. *In liquidity trap situation increase in Money Supply does not cause fall in interest rate.*

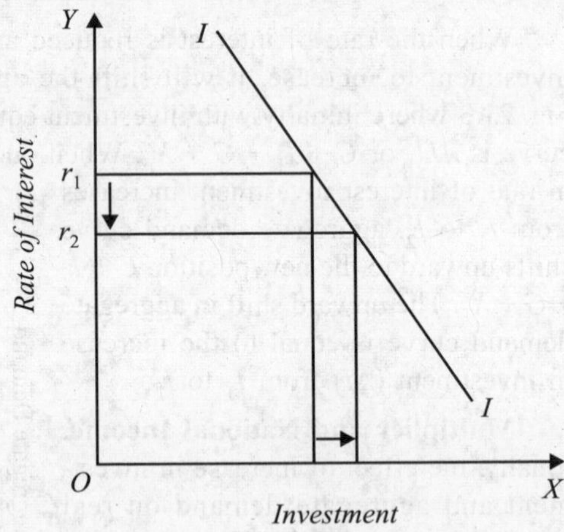

Fig. 21.5. *Investment is interest Inelastic.*

will not fall further even when the supply of money is increased. The curve M^d is the money, demand curve with a horizontal segment (*i.e.*, liquidy trap) on the right side. Suppose the quantity of money is initially MI' With the given money demand curve M^d and the money supply M_1, the rate of interest

Or_3 is determined which is very low. Now, if the money supply is increased to M_2, the rate of interest does not fall; all the additional money is held by the people with themselves and not lent out or used for buying bonds or debentures. Thus, when we are operating along the perfectly elastic part of the liquidity preference or money demand curve the rate of interest cannot be reduced by the increase in money supply. With no fall in the rate of interest, investment demand curve remaining the same: the rate of investment will not increase and if investment does not increase, aggregate demand and expenditure will not increase.

Another weak link in the transmission mechanism of curs in the effect of change in the rate of interest on the investment. Keynes and early Keynesians believed that investment demand curve is not much sensitive to the rate of interest, that is, interest-elasticity of investment demand is very low. If this is so them even if there is a large fall in the rate of interest brought about by increase in money supply by the central bank, investment will not increase much. This is illustrated in Fig. 21.5 where we have drawn an inelastic investment demand curve II. It will be seen from Fig. 21.5 that when rate of interest falls from r_1 to r_2, there is only small increase in investment because investment demand curve is not sensitive to the rate of interest. Therefore, when investment does not increase much even when there is a large fall in rate of interest as a result of expansion in money supply, aggregate demand or expenditure will not increase much.

Thus, there are circumstances, especially when recessionary conditions prevail in the economy with large scale unemployment and excess capacity in the economy, expansionary monetary policy may fail to increase the level of aggregate demand or expenditure. With no increase in aggregate demand, real national income, that is, GNP will not increase. Besides, in this Keynesian system, when due to ether highly interest-elastic nature of the money demand curve (*i.e.*, due to liquidity trap) or due to the interest-inelastic nature of the investment demand curve, increase in money supply fails to raise, aggregate demand, it will not cause rise in the price level.

It is thus clear from above that the Keynesian theory traces the effect of the increase in money supply on the level of economic activity (i.e., income, output and employment) via its effect on the rate of interest. Thus, in the Keynesian monetary theory, the relationship between money-supply and national income is not direct but it is thought to be much more indirect and uncertain than is assumed in the Friedman's modem quantity theory of money.

KEYNES THEORY OF MONEY AND PRICES

We know turn to explain the Keynesian monetary theory with regard to the relationship between the supply of money and the price level. Keynes believed that changes in money supply could being about changes in the price level but contrary to the classical economists' view he explained that there was no any direct and proportionate relationship between the quantity of money and the price level. He showed that changes in money supply indirectly affect the price level through its effect on the rate of interest. When money supply is increased, given the demand for money holdings curve, it leads to the fall in the rate of interest depending upon how far money demand curve is sensitive to the rate of interest. A change in the rate of interest affects investment which through multiplier process affects aggregate expenditure or demand. It is then the magnitude of aggregate demand relative to aggregate supply of output that causes price level to change. Thus, relation between money and the price level far from direct and proportionate is only indirect.

The Keynesian theory emphasises that the price level is in fact a consequence of aggregate demand or expenditure relative to aggregate supply rather than of quantity of money. The real cause of fluctuations in price level is to be found in fluctuations in the level of aggregate expenditure. Therefore, changes in the quantity of money can bring about changes in the level of prices only if they change aggregate demand in relation to the supply of output. Unless aggregate expenditure increases, there can be no increase in demand for goods. And if demand for goods does not increase, the question of rise in price level does not arise. However, even if aggregate demand or expenditure does increase, prices may still not rise if the supply curve of output is fairly elastic. Therefore, the effects of a change in quantity of money on the price-level depend on the following factors:

(*i*) effect of changes in money supply on the level of aggregate demand or spending;

(*ii*) relation between aggregate spending and the volume of production.

As regards the volume of aggregate expenditure or aggregate demand, in the keynesian theory it depends on the following:

(*a*) *rate of interest* which is determined by the demand for money and the supply of money;

(*b*) *the investment demand curve* which determines the increase in investment demand following a fall in the rate of interest and

(*c*) *the propensity to consume* which determines the magnitude of the multiplier effect of increase in investment.

(*d*) *supply of money*.

Transmission Mechanism. The mechanism through which increase in money supply can lead to the rise in aggregate demand and price level can be represented by the following scheme:

$$M^S \uparrow \longrightarrow r \downarrow I \uparrow \longrightarrow AD \uparrow \longrightarrow Y \uparrow \text{ and } P \uparrow$$

where M^S = supply or money

r = rate of interest

I = amount of investment

AD = aggregate demand

Y = level of national income, that is, aggregate output

P = price level

The above scheme represents the chief proposition of Keynesian monetary theory. According to this, expansion in money supply (M^S) causes the rate of interest to fall. Then, given the investment demand function, at a lower rate of interest, there is more demand for investment. This higher rate of investment boosts up the level of aggregate demand or expenditure through multiplier process. However, as we will see below, *whether or not this increase in aggregate demand or expenditure will cause the rise in the price level depends upon the nature of aggregate supply curve.*

It is evident from above that in the Keynesian theory the general price level is determined by the same forces which determine the level of national income and employment, that is, the level of aggregate demand and aggregate supply. It needs to be emphasised that price level and national income (*i.e.*, aggregate output) are determined jointly by aggregate demand and aggregate supply. However, in the analysis of determination of real national income in the framework of 45° line diagram, we have used the concept of aggregate demand with a given fixed price level. But to explain the Keynesian theory of money and prices, we need to use the concept of *aggregate demand with varying price level*. Aggregate demand (with varying price level) is the sum of total expenditure, which consumers, businessmen, Government and foreigners are willing to make on aggregate output of goods and services at different price levels during a given period. That is, aggregate demand (with flexible prices) shows how much output, the consumers, businesses, Government and foreigners are willing to buy at various price levels. Thus aggregate demand curve in Keynesian theory is $C + I + G + X_n$ at various price levels.[1] At higher price levels, aggregate output demanded or purchased is less at a higher price level and it increases at a lower price level. In other words, aggregate demand ($C + I + G + X_n$) curve with variable price level slopes downward as shown in Fig. 21.6. Three factors account for the downward sloping nature of the aggregate demand curve. These

1. Note that in sharp contrast to Keynes' theory, in the quantity theory of money aggregate demand at various price levels is determined by MV.

three factors are *real balance effect, interest effect and foreign trade effect* of the change in the price level which have been explained in an earlier chapter concerning *AD-AS* model (with flexible prices).

As explained in an earlier chapter concerning *AD–AS* model (with flexible prices), the sum of aggregate demand for consumption, investment and net exports increases with a fall in the price level and declines with a rise in the price level. This means that aggregate demand curve showing relationship between aggregate output demanded and the general price level slopes downward to the right as is shown in Fig. 21.6.

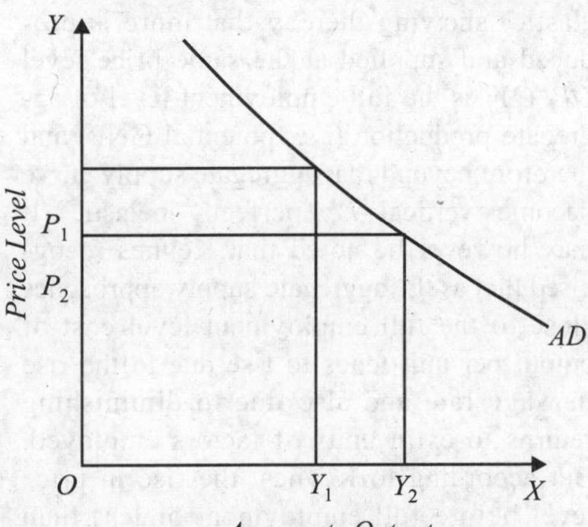

Fig. 21.6. *Aggregate Demand Curve with Varying Price Level*

Aggregate demand curve is derived from changes in aggregate expenditure caused by changes in the price level. If there is a change in the *non-price factors such as money supply, investment demand, Government expenditure, taxation*, the aggregate demand curve will shift. Thus, when money supply is increased by the central bank of a country, it will lower the rate of interest. As seen above, at a lower rate of interest more investment is undertaken depending on the elasticity of investment demand curve. Further, with more investment expenditure the entire aggregate demand curve (with variable price level) depicting $C + I + G + X_n$ will shift to the right implying thereby that at each price, the aggregate quantity demanded will increase.

Money Supply, Aggregate Demand and Price Level

But what happens to the equilibrium price level and real national product as a result of change in money supply, we must consider aggregate supply as well. Thus, even if aggregate demand or expenditure increases it does not follow that prices must necessarily rise. Whether or not the increase in aggregate demand will cause a rise in price level depends upon the nature of aggregate supply curve. If the supply curve of output is fairly elastic, it is more likely that the effect of increase in aggregate demand or spending will be more to raise production or real income than prices. In the model of determination of general price level, the concept of aggregate supply curve with variable price level is used. Precisely speaking, in this sense, *aggregate supply curve shows the various amounts of aggregate supply which the producers in the economy are willing to produce and sell in the market at various price levels*. The classical economists assumed that there normally prevailed full employment of resources in the economy. According to them, if at any time there is deviation from this full employment level, the wages, interest and prices quickly and automatically adjust or change to restore equilibrium at the full employment level. Thus, in the classical theory, the aggregate supply curve of output is perfectly inelastic (*i.e.*, a vertical straight line) at the output level corresponding to the full employment level of resources.

As mentioned above, Keynes considered the situation of economic depression when the economy was operating at less than full-employment level of resources. He further believed that in such a situation money wage rates were sucky, *i.e.*, remain stable. He further assumed that average and margined products of labour remain constant when more of it is employed following the increase in aggregate demand. With these assumptions, more aggregate output is produced and supplied at the given price level in response to increase in aggregate demand. But when full employment of labour and capital stock is attained and aggregate demand further increases, aggregate supply curve being unable to increase any more, it is the price level that will rise in response to the increase in aggregate demand. The Keynes's aggregate supply curve depicting the relationship between price level and the aggregate production (supply) is shown in Fig. 21.7 where it will be seen that up to the level of aggregate output OY_F aggregate supply curve is a horizontal straight line (*i.e.*, perfectly elastic) showing thereby that more is produced and supplied at the same price level OP. OY_F is the full employment level of aggregate production (*i.e.*, potential GNP) and therefore beyond that aggregate supply curve becomes vertical (*i.e.*, perfectly inelastic). It may however be noted that Keynes recognised that as the aggregate supply approaches close to the full employment level cost of output per unit tends to rise due to the rise in wage rate and also due to diminishing returns to extra units of factors employed. But according to Keynes, the rise in price level before full employment or less than capacity output will not be much.

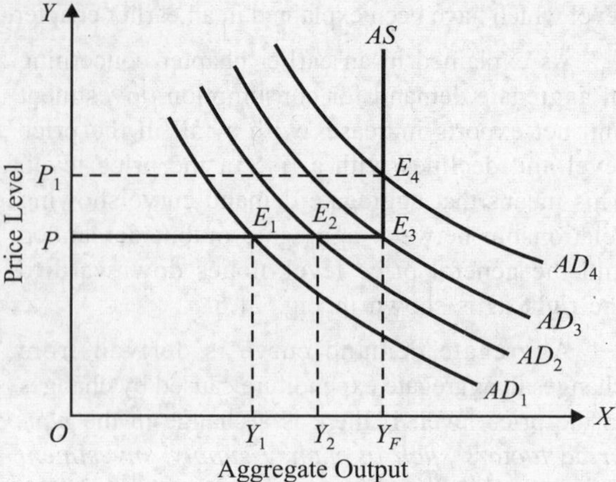

Fig. 21.7. *Effect of Increase in Money Supply on Price Level: Keynesian Model*

Suppose the economy is in state of depression so that a lot of resources including labour are lying idle. Initially, the aggregate demand curve is AD_1 which cuts the aggregate supply curve AS at point E_1 and as a result price level OP is determined (see Fig. 21.7). Now suppose that expansion in money supply succeeds in the raising aggregate demand curve to AD_2. It will be seen from Fig. 21.7 that is new aggregate demand curve stills cuts the aggregate supply curve in its horizontal range at point E_2. Thus increase in aggregate demand to AD_2 caused by expansionary monetary policy has led to the increase in aggregate output (*i.e.*, real national income) from Y_1 to Y_2 without any rise in general price level. If money supply is further increased and as a result aggregate demand curve shifts upward to AD_3 and cuts aggregate supply curve AS at point E_3, even then only aggregate output rises to its full employment level Y_F, price level remaining unchanged. Thus with aggregate demand curve AD_3 economy is operating at full employment level of output Y_F. Now, if there is further expansion in money supply causing aggregate demand curve to rise above AD_3, say to AD_4 in Fig. 21.7, resources being already fully utilised supply of output will not respond to the increased demand and will cause the price level to rise to P_1. This happens when firms try to hire more workers and other resources for expanding output to meet the increased demand. In view of full employment of resources having been already achieved, they will only bid up wages and their cost of production. Hence the price level will rise. Thus when full employment is prevailing increase in aggregate demand for goods brought about by increase in money supply leads only to higher price level and not to higher output.

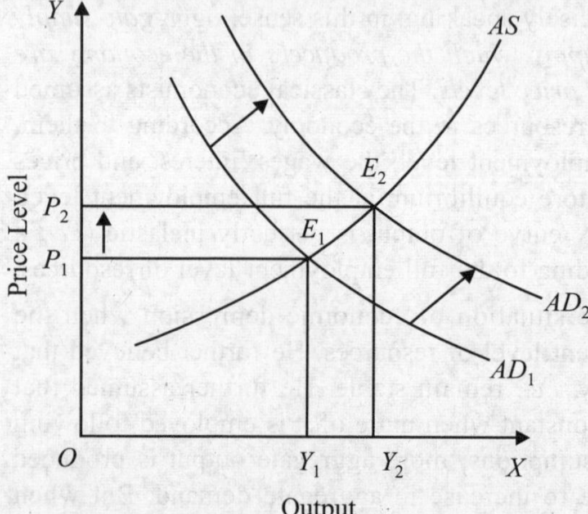

Fig. 21.8. *Impact of Increase in Aggregate Demand on Price and Output*

It is clear from above that *it is not necessary that even if expansion in money supply succeeds in raising aggregate demand, price level must rise*. The effect of increase in aggregate demand depends on whether the economy is operating at less than full employment level when there are recessionary conditions in the economy or the economy is working at full employment level at which aggregate supply curve is perfectly inelastic. It may however be noted that the modem Keynesians believe that in normal times the short-run aggregate supply curve slopes upward and is elastic as shown in Fig. 21.8. In this case increase in aggregate demand say from AD_1 to AD_2 in Fig. 21.8 caused by expansion in money supply will cause both output and price level to rise, the extent of their rise depends on the elasticity of the aggregate supply curve.

QUESTIONS FOR REVIEW

1. Explain briefly Keynes' monetary theory. Show how, according to Keynes, expansion in money supply can lead to the increase in national income and employment.
2. What, according to Keynes, are the factors which determine the effect of changes in money supply on the level of income and employment in the economy? Explain with the help of Keynes' monetary theory.
3. Explain how Keynes showed that monetary policy is ineffective in tackling recession/ depression and thereby revive the economy. What policy did he suggest for this purpose?
4. Explain the relationship between money supply and price level as established by Keynes. Explain how he proved that there is no direct and proportionate increase in the general price level as a result of increase in money supply in an economy.
5. According to J.M. Keynes price level is in fact the consequence of aggregate expenditure rather than of quantity of money. Discuss.
6. What is real balance effect? How does it affect aggregate demand? Is it strong enough to, cancel out the adverse effect on employment of all-round cut in wages in the economy?

CHAPTER 22

MONETARISM AND FRIEDMAN'S MODERN QUANTITY THEORY OF MONEY

MONETARISM: AN INTRODUCTION

The quantity theory of money as put forward by classical economists emphasised that increase in the quantity of money would bring about an equal proportionate rise in the price level. The quantity theory of money had come into disrepute, together with the rest of classical economists as a result of the Great Depression of the 1930s. J.M. Keynes criticised quantity theory of money and brought out that expansion in money supply did not always cause the price level to rise. Keynes and his *early* followers, often called *early* Keynesians[1], believed that money was not important in influencing the level of economic activity and that depression was not caused by contraction in money supply by the central banks of the countries. Keynes and his early followers argued that demand for money at a low rate of interest was almost infinetely elastic (that is, liquidity trap existed in the demand for money) so that increase in money supply would not succeed in lowering the rate of interest. As a result, investment would not increase following the expansion in money supply. They further argued that investment demand was very much less interest-elastic which made it doubly sure that expansionary monetary policy was quite ineffective in giving boost to investment and thereby the level of economic activity. Briefly speeking Early Keynesians thought that money was unimportant or *"Money does not matter"*.

However, in the fifties and sixties, the new thinking emerged under the influence of Milton Friedman, an eminent American economist and a Nobel Laureate in economics, who laid stress on the importance of money not only in determining the general price level but more importantly in influencing the level of economic activity. Friedman asserted that events of 1930s had been wrongly assessed and did not in fact offer evidence against the quantity theory of money. He however realised that there was a need to restate or *reformulate the quantity theory of money* which should re-establish the importance of money determining the level of economic activity and the price level. However, in his restatement of the quantity theory of money, he took account of Keynes's contribution to monetary theory, especially his emphasis on the demand for money as an asset. We may describe Friedman's monetarism into the following three propositions:

1. The level of economic activity in current rupee terms, that is, the level of *nominal income* is determined *primarily* by the stock of money.
2. In the long run, the effect of expansion in money supply is primarily on the price level and other *nominal* variables. In the long run, the level of economic activity in real terms, that is level of real output and employment are determined by the real factors such as stock of capital goods, the state of technology, the size and quality of labour force.
3. In the short run price level as well as the level of real national income (*i.e.*, real output) and employment are determined by the supply of money. In the short-run changes in the quantity of money are the dominant factor causing cyclical fluctuations in output and employment.

1. Note that later Keynesians, often called modern or *New Keynesians*, hold slightly different views from the *early Keynesians*.

We shall explain below the above three propositions of Friedman's monetary theory. The above conclusions derived by Friedman depends on the restatement of the quantity theory of money. It is important to note that Friedman's modern quantity theory of money is in fact based on his theory of demand. Therefore, in our analysis below we start from Friedman's theory of demand for money. We then explain how his theory of money demand explains the determination of the level of economic activity and the price level, both in the short run and long run.

DEMAND FOR MONEY AND FRIEDMAN'S RESTATEMENT QUANTITY THEORY OF MONEY

Friedman's modern quantity theory of money is very close to the Cambridge's cash balance approach. Friedman and other modern monetarists have emphasised that *k* in Cambridge approach should be interpreted as proportion of nominal income that people *desire* or *demand to hold* in the form of money balances. Interpreting *k* in this *desired or ex-ante* sense helps to convert the Cambridge equation of exchange into a *theory of nominal income*. Thus, rewriting Cambridge equation as demand for money (M^d) we have:

$$M^d = kPY \qquad ...(1)$$

where *k* is assumed to be constant. PY is the nominal income obtained by multiplying the real income (*Y*) with the price level (*P*). Like Cambridge economists, Friedman regards the quantity of money being fixed exogenously by the central bank of the country. If *M* represents the quantity of money set exogenously by the central bank we have the equation which describes the Cambridge theory of determination of nominal income.

$$M = M^d = \bar{k}PY \qquad ...(2)$$

or

$$M.\frac{1}{\bar{k}} = PY \qquad ...(3)$$

According to Cambridge equation (3) nominal income is determined by the supply of money (*M*) multiplied by the reciprocal of constant *k*. Now, Friedman *introduced changes in the above Cambridge theory of money demand* incorporating important aspects of Keynes's theory of demand for money. Keynes emphasised the *role of money as an asset* apart from its role in meeting transactions demand.

In studying the factors that determine demand for money as an asset relative to other assets, he simplified his analysis by lumping together all non-monetary assets under a single category '*bonds*'. He then examined what determined people's allocation of their wealth between money and bonds. According to him, the level of income and rate of interest determined the allocation of wealth between money and bonds. According to Keynes, rate of interest was the most important determinant of *k* in the Cambridge cash balance approach. It may be recalled that *k* in Cambridge theory indicates how much proportion of income people hold as money. However, Keynes regarded *k* as the proportion of income that people want to hold in money as an asset. *Friedman accepted Keynes's emphasis on the role of money as an asset* and presented his own theory of demand for money. In his analysis of determinants of money demand, Friedman included not only level of income and rate of interest on bonds but also rates at return on other assets such as equity shares, durable goods including real property. Thus, Friedman's theory demand for money can be written as follows:

$$M^d = F(P, Y, r_B, r_E, r_D)$$

where
P = price level
Y = level of real income
r_B = rate of interest on bonds
r_E = rate of return on equity shares

$$r_D = \text{rate of return on durable goods}$$

It will be seen from Friedman's money demand function that the product of the first two variables, namely, P and Y give us the level of nominal income. It therefore follows that, in Friedman's function, demand for money depends on nominal income. The higher the level of nominal income, the greater the demand for money. It is worth mentioning that for a given level of nominal income, in Friedman's money demand function as in that of Keynes, demand for money depends on the rates of return on non-monetary alternative assets. But, unlike Keynes, *Friedman regards money demand function is stable*. Stability of money demand function implies that it will not shift erratically and that variables in the function will determine the quantity of money that will be demanded. Friedman's theory of demand for money can be used to restate the Cambridge money demand equation so as to bring out the important role of money demand function in the determination of the level of economic activity. Thus Friedman money demand function can be restated as follows:

$$M^d = k\,(r_B, r_E, r_D)\,PY \qquad ...(4)$$

It is worth noting that whereas in Cambridge cash balance equation, k, that is, the proportion of income that is held in money, is mainly dependent on the transactions demand for money in Friedman's theory, it has been taken to be a function of rates of return on alternative non-monetary assets such as bonds, equity shares, durable goods. Any rise in the rates of return on these alternative assets, it will cause k to fall showing the increased desirability of alternative non-monetary assets. "In these terms Friedman can be seen to have restated the quantity theory, providing a systematic explanation of k, an explanation that takes account of the Keynesian analysis of money's role as an asset."[2]

Money Market Equilibrium: Friedman's Analysis. In terms of Friedman money demand function, the condition for equilibrium in the money market can be stated as under:

$$M = M^d = k\,(r_B, r_E, r_D)PY$$

where M stands for the supply of money which is determined by the central bank policies. k is the proportion of nominal income (PY) that is held in the form of money and is determined by the rates of return (r_B, r_E, r_D) on alternative non-monetary assets such as bond, equity shares, durable goods respectively. Friedman assumes that money demand function is stable. *Given the stable money demand function, any increase in money supply by the central bank will cause either increase in nominal income (PY) or decline in rates of return so that k rises or alternatively some increase in nominal income (PY) and some rise in k.* Friedman believes that much of the effect of exogenous increase in money supply will be to bring about increase in nominal income (PY) rather than k. Thus Friedman concludes that quantity of money is an important determinant of the level of economic activity, that is, output, employment and prices. Here the short-run and long-run effects of exogenous increase in money supply must be distinguished. In the short run, the increase in money supply will lead to a change partly in real income (*i.e.*, real aggregate output or Y) and partly in the price level (P). That is, in the short run the effect of increase in money supply will be distributed between change in real income (Y) and change in price level (P) depending upon the elasticity of the aggregate supply curve. Friedman recognised that at times of depression, aggregate supply curve was fairly elastic so that the effect of the increase in money will be more in the form of expansion in real output and less in the form of rise in the price level. But, in the long run, the aggregate supply curve is perfectly inelastic at full-employment level and, therefore, the exogenous increase in money supply will be reflected in the rise in price level.

DETERMINATION OF NOMINAL INCOME : FRIEDMAN'S APPROACH

Let us show how in Friedman's modern quantity theory nominal income is determined. To show this Friedman makes some strong assumptions about the behaviour of k. Friedman converts his money demand function into a theory of nominal income by assuming that variables in his money demand

2. Richard Froyen, *Macroeconomics: Theories and Policies*, 3rd edition, Macmillan, New York, p. 278.

function other that nominal income (PY), that is, r_B, r_E, r_D have little effect on k. With this assumption, money held as a proportion of income will be nearly constant. It may be noted by taking this assumption Friedman's theory comes very close to Cambridge theory of determination of nominal income. With this assumption Friedman's theory of nominal income can be derived as follows:

$$M^d = k\,(r_B, r_E, r_D)\,PY$$

With variables r_B, r_E, r_D having little or no effect on k, k can be assumed to be constant. Therefore,

$$M^d = kPY$$

Rearranging we have

$$PY = \frac{1}{k}M^d$$

since $\frac{1}{k} = V$, we can write

$$PY = VM^d$$

In equilibrium demand for money (M^d) equals money supply (M) which is exogenously given, we have

$$PY = VM$$

or

$$PY = \frac{1}{k}M$$

The last condition shows that *nominal income is determined by the supply of money, k being remaining constant*. The determination of nominal income is graphically shown in Fig. 22.1 where we measure nominal income (PY) on the Y-axis, and the supply and demand for money on the X-axis. Suppose central bank sets money supply equal to OM. With OM as the supply of money, supply curve of money is a vertical straight line M^S. It will be seen that at the level of nominal income P_0Y_0, money market is in equili-brium as at it demand for money (M^d) equals the supply of money (OM). Let us explain how this equilibrium level of nominal income is determined. Suppose the level of nominal income is P_1Y_1. It will be seen from Fig. 22.1 that at nominal income P_1Y_1, money supply OM exceeds money demand (M^d). Given the stable money demand function and constant k, Friedman argues that this excess money supply will be spent by the people on goods and services so as to bring money supply in equilibrium with the demand for money. As a result, nominal income will increase to the level P_0Y_0

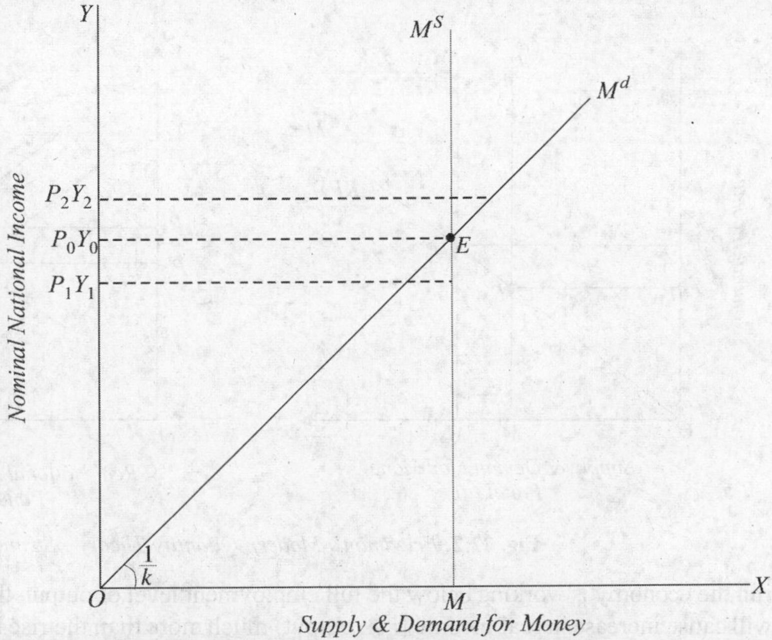

Fig. 22.1. *Determination of Nominal Income : Monetarist Theory*

at which money demand becomes equal to money supply (*OM*). If the period we are considering is the short run when there prevails depression and unemployment in the economy, the increase in nominal income will be more in terms of real income (*i.e.*, physical output) and less in terms of price level (*P*). On the other hand, if money supply (M^S) is less than the demand for money ($M^S < M^d$) as it is at nominal income P_2Y_2 in Fig. 22.1. The people will react to such a situation by spending less on goods and services so that demand for money by them decreases to become equal to the given supply of money *OM*. The decline in expenditure by the people on goods and services will cause nominal income to decrease to the level P_0Y_0. Again this decline in nominal income from P_2Y_2 to P_0Y_0 may occur mostly in real income (*Y*) and less in terms of fall in price (*P*) if we are considering the short-run situation when there is excess capacity and unemployment in the economy due to recession. It follows from above that nominal income is in equilibrium at the level where demand for money equals available quantity of money supply.

Increase in Money Supply and Changes in National Income Price Level

Let us explain in more detail how in Friedman's monetarist approach increase in money supply will affect the price level and national income. We will also spell out the transmission mechanism through which increase in money supply operates to bring about rise in the price level and income. Consider again the equation for money equilibrium,

$$kPY = M^S$$

$$PY = \frac{1}{k}M^S \quad \text{...(5)}$$

Now, if *k* and *Y* are constant, then it follows from (5) above that price is directly proportional to money supply. This is the viewpoint of the old quantity theory of money which thought velocity of money (*V*) or in other words, *k* [which equals $\frac{1}{v}$] and real national income or output remained constant. Friedman and his followers, in their strong version of quantity theory of money or demand for money assume *k* to be stable, the effect of expansion in money supply will be distributed between the increase in real income (*i.e.*, *Y*) and increase in price level (*P*). They further explain that if in the short

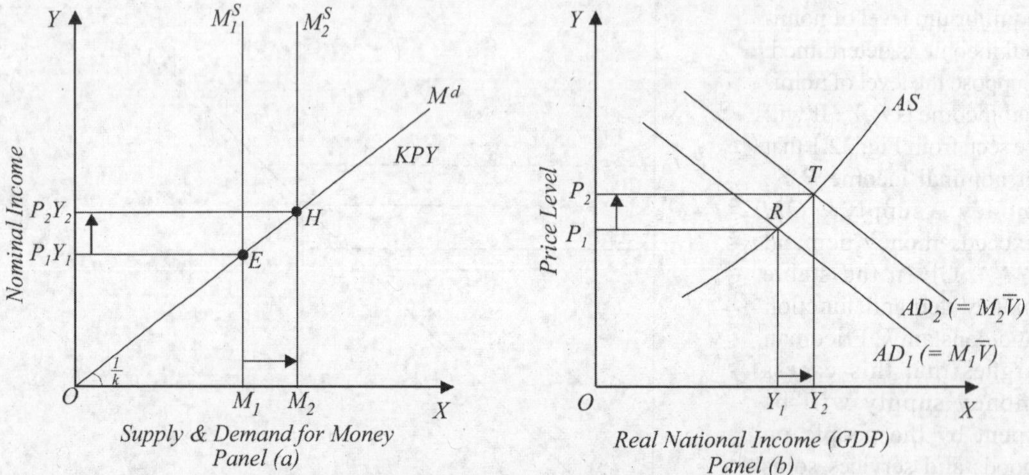

Fig. 22.2. *Friedman's Modern Quantity Theory : Strong Version*

run the economy is working below the full employment level of output, the increase in money supply will cause increase in real income (*i.e.*, output) much more than the rise in price level. Thus, according to them, *if in the short run the economy is in recession, then the effect of increase in money supply, with k or V remaining constant, is likely to lead to the increase in output or real income* (*i.e.*, *Y*) *more than rise in the price level* (*P*). It is only in the long run that at the full employment level of output when *Y* and *k* remain constant the increase in money supply will raise the price level or cause inflation in the economy.

To fully considered Friedman's new quantity theory, consider panel (a) in Fig. 22.2. Let us suppose the supply of money is raised from M_1^S to M_2^S. As will be seen from the figure, the expansion in money supply causes the nominal income to rise. In panel (a) of Fig.22.2 with M_1 as the supply of money and M^d as the curve of demand for money ($M^d = kPY$), money market is initially in equilibrium at point *E* with P_1Y_1 as the equilibrium level of nominal income. When the supply of money is increased from M_1 to M_2 and as a result money supply curve shifts above to M_2^S, then at nominal income P_1Y_1, money supply becomes excessive relative to the demand for money to hold at it. As a result, people will be induced to spend the excess money supply and thereby increase the aggregate demand for goods and services. This increase in aggregate demand for or expenditure on goods and services leads to the increase in nominal income. With this increase in nominal income demand for money increases and the new money market equilibrium is reached at point *H* at the new nominal income level P_2Y_2. How much increase in nominal income will occur in terms of increase in real income (*i.e.* physical output) and how much in terms of rise in price level can be explained in framework of aggregate demand (*AD*) and aggregate supply (*AS*) curves with flexible prices. This is shown in panel (*b*) of Fig. 22.2 where on the *X*-axis we measure aggregate demand and aggregate supply, on the *Y*-axis price level is shown. The concept of aggregate demand curve in this framework needs to be spelled out. In the monetarist view aggregate demand curve is determined by the quantity of money supply (*M*) and the velocity of circulation (*V*). That is, aggregate demand or aggregate expenditure in this approach is obtained from multiplying a given quantity of money supply by the velocity of circulation (*V*) which is assumed to remain constant. Note that velocity of circulation (*V*) is equal to $\frac{1}{k}$. Thus, if ₹4000 crores is the quantity of money supply and velocity of circulation is 5, the aggregate expenditure or aggregate demand (*MV*) will be equal to ₹20,000 crores. Now, if money supply is raised to ₹5,000 crores aggregate demand will increase to ₹25,000 crores, velocity of circulation remaining constant at 5. Now, with a given aggregate expenditure ($M\overline{V}$), aggregate demand curve slopes downward because at a lower price level, more quantity of goods and services can be purchased.

Now, a glance at panel (*b*) of Fig. 22.2 reveals that with money supply equal to M_1, the aggregate demand curve is AD_1 with the constant velocity $\overline{V}$. Aggregate demand curve AD_1 and aggregate supply curve *AS* determines the price level OP_1 and *real* national income equal to Y_1. Now, when in panel (*a*), money supply is increased from M_1 to M_2, and as a result nominal income rises to P_2Y_2, in panel (*b*) aggregate demand curve based on $M_2\overline{V}$ shifts upward to AD_2 and its intersection with the given aggregate supply curve *AS*, determines higher level of real national income equal to OY_2 and new higher price level P_2. It is thus evident that increase in nominal income to $P_2\,Y_2$ as a result of expansion in money supply from M_1 to M_2 is distributed or absorbed as increase in real national income (*i.e.*, physical output) by the amount Y_1Y_2 and as rise in price level from P_1 to P_2. According to Friedman, and other monetarists, this occurs in the short run, especially when the economy is operating in recessionary conditions and therefore less than full-employment level of national income. Thus, to what extent real income as a result of expansion in money supply increases depends on the elasticity of aggregate supply curve *AS*. Thus, Friedman writes, "*I regard the description of our position as 'money is all that matters for changes in nominal income and for short-run changes in real income as an exaggregation but one that gives the right flavour to our conclusion.*"[3]

3. Milton Friedman, "A Theoretical Framework for Monetary Analysis" in Robert Gordon, ed., *Milton Friedman's Monetary Framework*, University of Chicago Press, 1974, p. 27

As seen above, according to the monetarist view, the impact of increase in the money supply on prices and real income is direct. The expansion in money supply is seen as putting more money in the hands of both consumers and investors. Since monetarists believe that demand for money to hold (M_d) is a stable function of nominal income, expansion in money supply leads to the excess money supply or, in other words, excess cash balances with the people relative to their demand for money holdings. The excess money balances are soon spent leading to the increase in aggregate demand for goods and services. Unlike the transmission mechanism of Keynesian analysis wherein increase in money supply first reduces the rate of interest and thereby raises investment and aggregate demand and then affects prices, in the analysis of Friedman and other modern monetarists, the expansion in *money supply has a direct impact on prices and income through increase in aggregate demand.* Liquidity trap and inelastic nature of investment demand are not involved in this adjustment process which could prevent or curtail the effect of monetary expansion on prices and national income.

However, many modern monetarists (including Friedman) aggree to the Keynesian idea that expansionary macroeconomic policy may lead to the increase in national output and employment and this enables the economy to come out of recession quickly. It may be recalled that Keynes laid stress on the use of expansionary fiscal policy (expansion in Government expenditure and lower taxation) as the proper instrument to tide over recession. Keynes believed that in view of the liquidity trap and interest-unresponsive nature of investment demand, expansionary monetary policy will not help to get the economy out of recession. However, monetarists emphasize that since expansion in money supply directly causes increase in aggregate demand through people spending the excess money supply, expansionary monetary policy can help in curing recession and achieving full employment.

Long-Run Impact of Expansion in Money Supply: Views of Friedman and Other Modern Monetarists

Monetarists believe that perfectly elastic aggregate supply curve envisaged by Keynes actually represents the recessionary state of the economy in the short run. According to the monetarists, though short run aggregate supply curve of output at times of recession is elastic, the long-run aggregate supply curve of output is perfectly inelastic, that is, it is a vertical straight line. The monetarist view is portrayed in Fig. 22.3. Assume that aggregate demand curve is initially at AD_1 corresponding to the available money supply equal to ₹ 4000 crores, and a given velocity of circulation (V), cuts the short-run aggregate supply curve SAS at point T and thus determines the price level equal to P_1 and the income level Y_1 which is less than full employment level of output Y_F. In such an equuilibrium situation idle or unutilized productive capacity would prevail causing unemployment of labour. Thus, in equilibrium at point T, the economy will find itself in the grip of recession. However, according to Friedman and his followers modern monetarists, this only represents a short-run situation. They argue that if in such a situation the money supply and aggregate demand curve were held constant, then in view of the prevailing unutilized capacity and unemployment, wages would eventu-

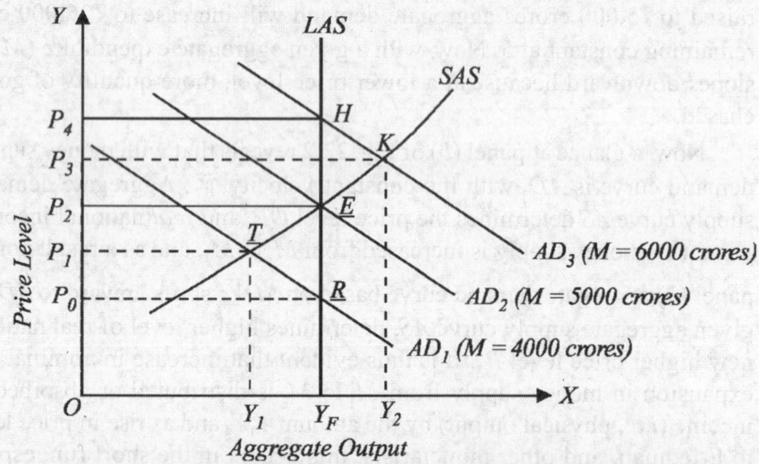

Fig. 22.3. *Long run Impact of Expansion in Money Supply on the Price Level*

ally fall causing a shift in the short-run aggregate supply curve (*SAS*) to the right so that long-run equilibrium is established at point *R* on the long-run aggregate supply curve *LAS*. It will be seen from Fig. 22.3 at point *R* where as price has fallen, aggregate output is restored at full-employment level Y_F though at lower wages.

However, Friedman and other modern monetarists accept that if instead of waiting for the downward adjustment in prices and wages, money supply is increased, say from ₹ 4000 crores to ₹ 5000 crores raising the aggregate demand curve from AD_1 to AD_2, the new equilibrium is attained at point *E* where the effect of increase in money supply has been divided into rise in price level from P_1 to P_2 and increase in output from Y_1 to Y_F. The extent to which price level will rise and output will increase depends on the elasticity of supply in the short run. Further, if money supply is increased further to ₹ 6000 crores so that aggregate demand curve rises to AD_3 the price level will rise to P_3 and output will increase to Y_2 in the short run. However, since in equilibrium at point *K* aggregate output exceeds full employment level of output, this implies unemployment with fall below natural rate of unemployment and as a result causing shortage of labour. The labour shortage will cause a shift in short-run aggregate supply curve *SAS* to the left to be in equilibrium at point *H* on the long-run aggregate supply curve *LAS*. With this while price level has risen to P_4, the equilibrium is restored at full-employment level of output Y_F.

Monetarists are Opposed to the Active Discretionary Monetary Policy

It is however worthnoting that most of the modern monetarists are opposed to the use of active monetary policy to tide over recession since they believe that prices and wages are quite flexible and downward adjustment in them will take place fairly rapidly to restore equilibrium at full-employment level of output. Second, monetarists are opposed to pursuing active monetary policy because they feel very much concerned about inflation as a result of *discretionary and active monetary policy*. They think than in its bid to overcome recession monetary authority is likely to *overshoot* and expand money supply much greater than that required to achieve equilibrium at full employment level of national income and thus causing inflation in the economy. As will be seen from Fig. 22.3, when aggregate demand increases to AD_3 level, as a result of the expansion of money supply to ₹ 6000 crores when only ₹ 5000 crores were needed to achieve equilibrium at full employment level, price level rises to P_4 and thus inflation emerges in the economy as a result of discretionary active monetary policy.

KEY PROPOSITIONS OF MONETARISM

We state below the key propositions of monetarism and in a later section we shall compare them with Keynesian proposition of macroeconomics.

1. **Money Demand Function.** Miltion Friedman, the pioneer of modern monetarism, rests his monetary theory on the demand function for money which is written as under

$$M^d = k\,(r_B, r_E, r_D)\,PY$$

That is, demand for money is a function of nominal income (*PY*) and *k* which is proportion of nominal income held in the form of money and this proportion is determined by the rate of returns on alternative non-monetary assets such as bonds, equity consumer durables. Together with stock of money supply this money demand function determines price level, output and rates of return on assets.

2. **The quantity of money supply**, *given the money demand function, determines the level of nominal income* (*PY*). If the quantity of money increases rapidly, so will nominal income. The velocity of money circulation or *k* in the money demand function is fairly stable and predictable. However, it takes time for changes in money supply to affect nominal income. This time lag between growth of money supply and its effect on nominal income, according to Friedman, is about 6 to nine months.

3. **Increase in money supply affects output in the short run only.** This short run, according to Friedman, may mean five to ten years. Over the long run, according to monetarists, the old quantity theory of money holds. That is, in the long run increase in money supply affects only prices. Thus, according to a fundamental proposition of monetarism, in the long run *inflation is always and everywhere a monetary phenomenon*. This means that in the long run inflation cannot occur without a *more rapid increase in the quantity of money than in output*. In fact, monetarists claim that sustained increase in the growth rate of money will, in the long run when all adjustments have taken place, lead to an equal increase in the rate of inflation.

4. **Money alone matters.** In sharp contrast to the views of early Keynesians, as seen above, Friedman and other modern monetarists argue that money has significant effect on price level or inflation in the economy in the long run and have real effects on output and employment in the short run. Thus, contrary to the views of early Keynesians who thought that "*money does not matter*" in view of the above effects of the growth of money supply, monetarists think '*Money alone matters*'.

5. **The output or real national income in the long run is determined by the real factors** such as stock of capital, propensity to save, supplies of other resources such as labour, natural resources, state of technology. An increase in the money supply has no long run effects on the level of output. Thus, according monetarists, there is no long-run trade off between inflation and output or for that matter, between inflation and unemployment. According to monetarists, in the long run Philips Curve is vertical. However, monetarists adviets that there is trade off in the short run and accordingly, short run Phillips Curve slopes downward to the right. (This proposition of monetarists will be explained at length in a later chapter concerning this issue).

6. *Milton Friedman and other modern monetarists oppose the use of active counter cyclical monetary policy to control economic fluctuations in the economy*. Instead they recommend a *rule of constant money growth rate* to ensure economic stability in the economy. Monetary policy should be conducted in a manner that should ensure money supply to grow at a constant rate. However, monetarists have suggested different constant money growth rate which varies from 2 to 5.

This monetary policy of constant money growth rate implies that monetary policy should not respond to disturbances in the economy unless these are major disturbances such as the Great Depression that occurred in industrialised countries during 1929–33. To quote Friedman, "*By setting itself a steady course and keeping to it, the monetary authority could make a major contribution to promoting economic stability. By making that course one of steady but moderate growth in the quantity of money, it would make a major contribution to avoidance of either inflation or deflation of prices.*"[4]

It therefore follows from above that Friedman and other monetarists oppose activist policy of "*Fine tuning*". By *fine tuning* we mean policy variables are continuously adjusted in response to small disturbances in the economy so as to ensure economic stability at full-employment level of output. Dornbusch and Fischer explain the monetarists position on fine tuning in the following words.

"The major" argument of monetarist against fine tuning is that in actual practice policy makers do not in fact behave as suggested—making only small adjustments in response to small disturbances. If allowed to do anything they may do too much. Instead of merely trying to offset disturbances, policy makers shift towards fine tuning to keep the economy always exactly at full employment, with the risk of overdoing a good thing. The 'full employment bias in policies risks creating inflation".[5]

7. **Money Supply and Interest Rate Targets.** Another controversial issue between monetarists and Keynesians is what should be the target variable of monetary policy, that is, whether monetary authority should have fix targets of money supply growth or target of interest rate as an appropriate

4. Milton Friedman, "The Role of Monetary Policy", *American Economic Review*, March 1968.
5. Dornbusch, Fisher & Startz, *Macroeconomics,* Tata McGraw Hll, 8th edition 2002, p. 176

monetary variable for formulating monetary policy. In the fifties and sixties in the USA, emphasis of monetary policy was entirely on controlling interest rates — a policy suggested by Keynesians Monetarists are opposed to it and support the money stock target for a monetary policy. They argue that increase in the money supply leads ultimately to inflation and therefore only by keeping in money supply growth moderate, inflation in the economy can be avoided. They point out that while the Central Bank of a country keeps its eye on interest rate target, the money supply growth rate and inflation rate often tend to rise.

As a result of the above monetarist critique of interest rate as a target variable of monetary policy, in recent years, not only in the USA but in other countries as well emphasis has been shifted to proper *money supply growth rate* as target variable of monetary policy. In India however, since 1996, both money supply growth and interest rate targets are being used as policy variables of monetary policy.

8. **Inherent Stability of the Private Sector.** Monetarists believe that private sector is inherently stable but the fluctuations in the economy are caused by wrong policies and mistakes of the Government or the Central bank of the country. They try to show that even occurrence of Great depression of 1929–33, especially its severity was due to failure of Federal Reserve System (which is the name of Central Bank of the United States) to prevent bank failures and reduction of the money supply from 1929 to 1933. Likewise, major inflation occurs due to the more rapid expansion in money supply than growth in output. In addition to this, they argue that left to itself and without interference by the Government and trade unions the price wage flexibility will ensure equilibrium at natural rate of unemployment *i.e.* full employment level) in the long run. They are therefore opposed to the active role of government in the economy and adoption of counter-cyclical discretionary monetary and fiscal policies.

A CRITIQUE OF MONETARISM

The observation of data belonging to the United States of the sixties and seventies and upto late eighties indicates that inflation rate and growth rate of money supply *broadly moved together*. Until the late seventies upward trend in both the rate of inflation and rate of money growth was observed and during the seventies downward trend in both had been seen. From this it may appear that Friedman's monitary theory has proved to be correct. But, as has been rightly pointed out by Dornbusch and Fischer, the relationship between growth of money and rate of inflation has been *very rough with large gaps between their growth lines* that persist for several years and this means change in output growth and/or velocity changes have affected inflation.[6]

Thus, it shows that monetarism has not been fully correct in explaining rate of inflation in terms of growth of money supply alone. Examining the relationship between rate of inflation and rate of money growth *after adjusting for growth in output* they observe that during the sixties and seventies and to a lesser extent during the eighties hold somewhat but *during the nineties the relationship between growth in M2 and rate of inflation seems to have broken down*" in the united states. Further, they find that relationship between money growth and inflation is much looser if we *measure growth of money supply in terms of* $M1$. "To obtain a stable relationship between money growth and inflation we need a stable real money demand function or equivalently stable velocity."[7] It is this velocity of money circulation which has been found to be largely fluctuating as will be observed from Figure 22.4 where on the Y-axis we have measured k which stands for the proportion of income which people are willing to hold in cash balances to (Recall that $k ; \frac{1}{v}$). It will be seen from the graph of k in Figure 22.4 that velocity of money has been greatly fluctuating that has

6. Dornbusch, Fischer and Startz, *Macroeconomics*, Tata McGraw Hill, 2002, p. 422
7. Op. cit. p. 423.

disturbed the close relationship between money growth and rate of inflation. It will be seen from the graph that proportion of national income that people want to hold as money balances (*k*) has fallen to a very low level which provides explanation why relationship between money growth and inflation is not close.

It will be seen from Fig. 22.4 that the propensity to hold money has fluctuated considerably since 1890. In the early part of the century it was relatively stable. During the 1930s and 1940s it increased dramatically, reaching a peak of nearly 25 weeks of income in 1946. Since the end of World War II it has been falling. This vindicates the viewpoint of Keynesians that *k* or *v* is not stable and may move in opposite direction to the growth in money supply and this has disturbed the close relation between the two.

Recent Experience. During the period 2000-03, the United States economy experienced recession and in order to revive the American economy and stimulate investment money supply was increased by Reserve System (Central Bank of the United States) and also in the last two years, it reduced bank rate more than seven times and finally in July 2003 it was reduced to as low as 0.75 per

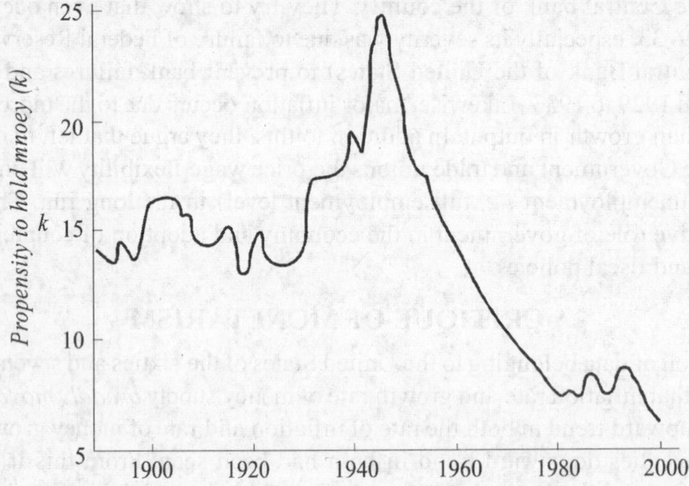

Fig. 22.4. *The Propensity to Hold Money (k) in the United States*

cent per annum, but without much success. Thus, monetary policy has been tried again in the United States to overcome recession but has failed as was predicted by Keynesians. This again shows the failure of monetarism. That is why in the United States fiscal policy measures such as reduction in taxes to the tune of 3.5 billion US dollars in 2001-02 and increase in Government expenditure were adopted to get rid of recession.

The conclusion of ineffectiveness of monetary policy also appears to hold in India too. Since 1996 Reserve Bank of India has been following expansionary and liberal monetary and credit policy by reducing significantly the cost of credit and also making more bank credit available but it has achieved little success in stimulating private investment. Bank rate which was 9 per cent in 1995 has been reduced over successive periods to 6 per cent in April 2003 (the lowest in the last four decades). As a result, prime lending rates of bank which were around 16.5 per cent have now been reduced to 10 to 10.5 per cent (August 2003). Thus more credit at very soft interest rates have been made available but credit off take has not significantly picked up. Besides, low interest rate and liberal credit policy has failed to spur private investment. This again shows the ineffectiveness of monetary policy to stimulate private investment and accelerate industrial growth.

Besides, during the seventies in America and other Western industrialised countries a ticklish problem of stagflation arose. *Stagflation means while prices and inflation rose, output declined.* This was brought about by cost-push inflation caused by oil price shock during the seventies. Cost-push inflation and stagflation cannot be explained by monetarism according to which "*Inflation is always and everywhere a monetary phenomenon and is caused by rapid growth in the quantity of money than output*" Lastly, an important challenge to monetarism is that monetarism *direction of causation* is not always valid. According to monetarists, rapid growth of money supply causes inflation and not the other way around. Monetarists consider rate of money growth as an exogenous variable determined by Government or the Central Bank. Critics have pointed out that when wages of workers are raised by the efforts of trade unions or when prices of raw materials, especially energy inputs (such as coal, crude oil, and petroleum products) are pushed up, they are passed on to the buyers as price hikes resulting in cost-push inflation. Critics rightly point out that the increase in the money supply that is often observed to accompany inflation can also be explained as *accommodating response* of Central bank that increases money growth in order to avoid recession or prevent decline in GDP. In this case money growth is the *effect* of inflation, not its *cause*.

MONETARISM AND KEYNESIANISM COMPARED

The old monetarists like Irving Fisher put forward quantity theory of money which explained that changes in money supply had a direct and proportionate relationship with the price level. However, modern monetarists led by Prof. Milton Friedman of Chicago University have put forward an alternative macroeconomics to the Keynesian macroeconomic theory. Accordingly, they not only explain the changes in the general price level but also changes in output and employment. Monetarists hold that the money supply is the prime determinant of nominal GNP in the short run and changes in price level in the long run. In what follows we compare the two theories in detail and in the end will bring out how in recent years the difference between the two approaches has quite narrowed down.

It is important to note that the main difference between the monetarists and Keynesians lies in their approaches to the determination of aggregate demand. While Keynesians hold that many different factors such as consumption investment Government expenditure, taxes, exports and money determine aggregate demand, monetarists argue it is the changes in money supply that are primary factor in determining aggregate demand which affect both output and prices. Besides, there is important ideological differences between the monetarists and Keynesians. Monetarists believe that a private market economy is inherently stable and if left free will automatically adjust itself to full-employment level of output. Therefore, they argue that there is no need for Government intervention in the economy. In fact, they point out that it is the discretionary monetary and fiscal policies pursued by the monetary authorities and Government that are responsible for much instability in the private economies. On the other hand, Keynesians believe that private economy is inherently unstable and for its stabilisation and growth, the Government should play an active role by adopting proper discretionary fiscal and monetary policies.

1. Two Different Approaches to Aggregate Demand

Keynesians focus on aggregate demand as a determinant of income and employment and view it as a sum of consumption, investment, Government expenditure and net exports (that is, exports–imports). In symbolic terms Keynesian aggregate demand is written as

$$AD = C + I + G + (X - M)$$

Keynesians equate it with value of goods and services produced to determine equilibrium level of national income or national product. Thus,

$$C + I + G + X - M = Y$$

where Y is national income or Gross National Product (GNP).

On the other hand, monetarists focus on only money supply as a primary determinant of aggregate demand. The fundamental equation of monetarists is the equation of exchange which is as under :

$$MV = PQ$$

Where M is the quantity of money
V is income velocity of money
P is general price level
Q is level of physical output of goods and services

MV in the above equation represents aggregate expenditure or aggregate demand in the monetarist approach. Thus, in monetarist approach aggregate demand is simply the sum of money multiplied by its velocity. *PQ* in the above equation of exchange represents nominal income or nominal GNP. Thus, in monetarist model,

$$MV = \text{Nominal GNP}$$

Since monetarist believe that velocity (*V*) is stable and predictable, there is direct relationship between money supply and nominal income or GNP. With *V* as stable, when money supply increases, consumers and businesses find themselves with excess money balances which they spend on goods and services. As a result of this increase in aggregate demand caused by the expansion in money supply, *PQ* or nominal GNP rises.

It is clear from above that though aggregate demand may be equal in the two approaches, the fundamental difference lies in the conception of aggregate demand. Keynesians view it as $C + I + G + (X - M)$, monetarists view it as simply *MV*. The question which has been debated which approach provides a better explanation of determination of income, employment and prices and which therefore can be made a basis for formulation of economic policy.

2. Growth of Money Supply is the Prime Determinant of Growth in Nominal GNP

A fundamental principal of monetarism is that "*Only money matters*". As seen above, like the Keynesian approach, monetarism is basically a theory of aggregate demand. They view aggregate demand as being equal to *MV*. With their belief that *V* is stable, aggregated demand is influenced primarily by changes in money supply. From the equation of exchange, $MV = PQ$, if *V* is stable the only force that can affect nominal income *PQ* is supply of money (*M*). According to monetarists, changes in money play a more and direct role in determining nominal income or GNP. They emphasise that while fiscal measures such as changes in amount and pattern of Government expenditure and taxation is important for determining how much is allocated to defence, private consumption or investment but the important macroeconomic variables such national output, employment and prices are determined mainly by money supply.

On the other hand, as seen above, Keynesians view aggregate demand as $C + I + G + X - M$ which determines the level of national income, output and employment. In the Keynesian model, the effect of increase in money supply on national income and employment is very indirect and operates through its effect on investment component of aggregate demand. According to it, expansion in money supply causes a fall in rate of interest. At a lower interest rate, more private investment is undertaken. When the economy is in the grip of depression and there is a lot of idle productive capacity, more investment through multiplier process leads to the increase in aggregate demand and therefore in national income and employment. It may however be noted that early Keynesians regarded this effect of changes in money supply through its effect on interest and investment very weak because they thought money demand curve (*i.e.* liquidity preference curve) at low rates of interest is *quite flat* and investment demand curve *quite steep*. Therefore, the early Keynesians thought that monetary policy did not play an important role in reviving the depressed economy. However, modern Keynesians have veered round to the view that changes in money supply can play a significant role in raising national output and employment. But unlike monetarists they trace the effect of changes in money supply on the real macroeconomic variables through its effect on interest and investment.[8]

3. Differences Regarding Shape of Aggregate Supply Curve

Another important difference between monetarist and Keynesian theories revolves around the shape of economy's aggregate supply curve. It is important to note that in monetarist approach it is the elasticity or steepness of the aggregate supply curve that determines how changes in nominal

8. This is explained in detail in a later section of this chapter.

GNP (*i.e., PQ*) will be divided between the change in output (*i.e.,* real income) and change in the price level. Monetarists believe that short-run aggregate supply curve is relatively steep so that when aggregate demand increases consequent to the expansion in money supply, it results in rise in price level (P) much more than the expansion in output (Q).

In other words, in case of steep aggregate supply curve, increase in nominal income leads more to rise in price than to increase in output. This is illustrated in panel (*a*) of Fig. 22.5. It will be seen from panel (*a*) where a relatively steep short-run aggregate supply is drawn that when due to the expansion in money supply (ΔM) aggregate demand increases from AD_1 to AD_2, price level rises sharply from P_1 and P_2 whereas *real* GNP (that is, aggregate output) increases causes relatively much less from Q_1 and Q_2. Thus, it is clear how division of *nominal GNP* into change in price level and change in output depends on the steepness of aggregate supply curve.

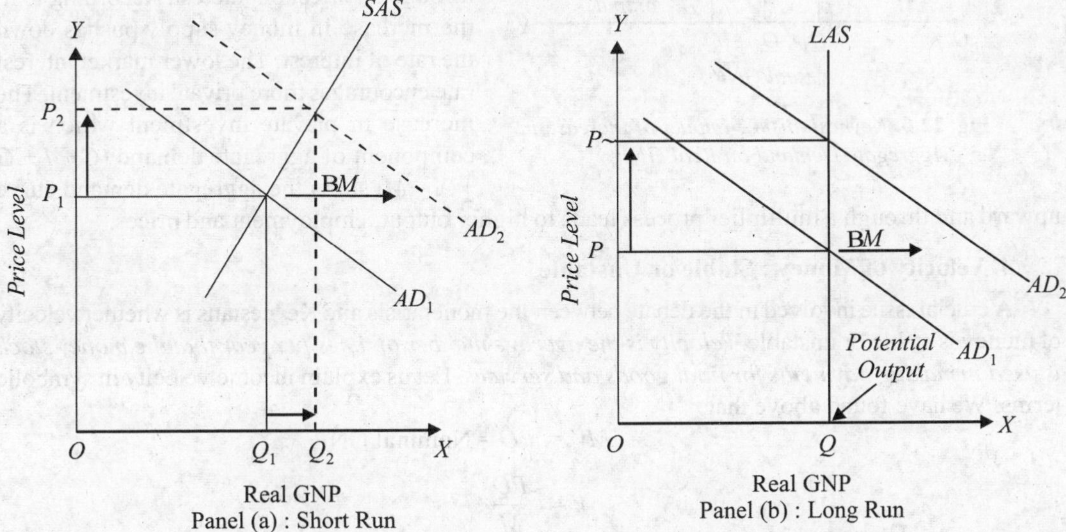

Fig. 22.5. *Monetarism : Effect of Expansion in Money Supply in the Short Run and Long Run.*

It is important to note that, according to monetarists, long-run aggregate supply is vertical because they believe that *due to wage-price flexibility,* long-run equilibrium is established at the level of potential output (that is, full-employment level of output). In panel (*b*) of Fig. 22.5 vertical straight *LAS* represents long-run aggregate supply curve. Now, in this case of long-run vertical aggregate supply curve when aggregate demand shifts upward from AD_1 to AD_2 due to the expansion in money supply (ΔM), price level rises sharply from P to P' whereas real GNP remains constant. Thus, according to monetarists, in the long run expansion in money supply only causes price level to rise, level of output remaining unaffected.

On the other hand, Keynesians believe that the short-run aggregate supply curve is quite flat (in the extreme cause it is a horizontal straight line), the effect of increase in aggregate demand is more on raising real GNP with little rise in the price level. This is shown in Fig. 22.6 where it will be observed that the part *SR* is a short-run aggregate supply curve which is quite flat. According to Keynesians, this flat short-run aggregate supply curve represents the situation of an economy which is having depression or recession and therefore has a lot of idle capacity and a large unemployment of labour in the economy. In such a situation when aggregate demand increases, say through increase in investment, from AD_1 to AD_2, it leads to a large expansion in real GNP with only a small rise in price level from P_1 to P_2. Similarly, when aggregate demand increases from AD_2 to AD_3, real GNP increases by a large amount with only a relatively small rise in price from P_2 to P_3.

However, Keynesians' point out that when the economy is operating at the level of full-employment or potential output, aggregate supply curve takes a vertical shape. Thus, with vertical

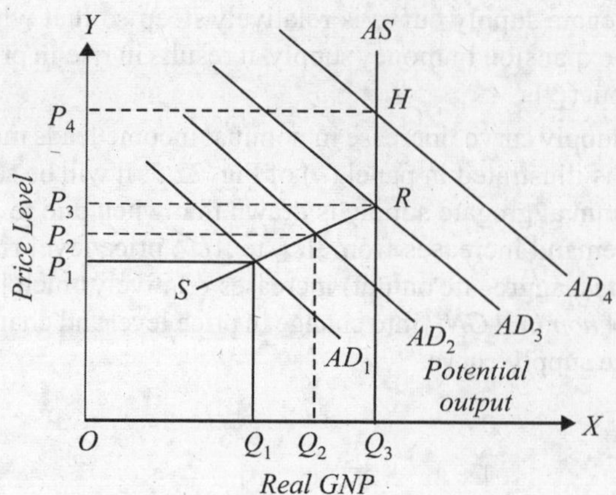

Fig. 22.6. *Keynesianism : Impact of Increase in Aggregate Demand on Real GNP.*

aggregate supply curve at full-employment level of output, any increase in aggregate demand in this case from AD_3 to AD_4 will cause a sharp rise in price level from P_3 to P_4 with no effect at all on real GNP which remains constant at Q_3.[9]

It is important to note that, in Keynesian theory, the effect of changes in money supply on aggregate demand and consequently on output and prices is quite indirect and operates through its effect on interest. According to it, the increase in money supply pushes down the rate of interest. The lower market interest rate encourages more private investment. The increase in private investment which is a component of aggregate demand ($C + I + G + X - M$) shifts the aggregate demand curve upward and through a multiplier process leads to higher output, employment and prices.

4. Velocity of Money : Stable or Unstable

A crucial issue involved in the debate between the monetarists and Keynesians is whether velocity of money is stable or unstable. *Velocity is the average number of times per year that the money stock is used in making payments for final goods and services.* Let us explain income velocity in symbolic terms. We have found above that

$$MV = PQ = \text{Nominal GNP}$$

$$V = \frac{PQ}{M}$$

Since PQ = nomimal GNP

$$V = \frac{\text{Nominal GNP}}{M}$$

Thus, income velocity of money (V) is the average number of money stock (M) used to buy nominal GNP which is the value of final goods and services produced in a year. It follows from the above definition of velocity that, according to monetarists, given the money stock M, if we can *predict* the income velocity of money V, we can predict the level of nominal GNP (that is, nominal income). Further, if velocity was *constant,* changing the money supply would result in proportionate changes in nominal income.

Note that modern monetarists do not assert that velocity is constant but that it is stable and predictable. Velocity depends on the methods of payment, duration of pay periods, use of more rapid means of making payments such as cheques and credit cards. Changes in these factors alter the velocity of money. Monetarists argue that the *factors which alter velocity change gradually and predictably.* Hence, changes in velocity from year to year can be easily anticipated. Moreover, monetarists contend that velocity does not change in response to changes in supply of money.

The assumption of stable and predictable V is crucial to the monetarist theory. If V is stable, it follows from the equation of exchange that there is a direct predictable relationship between money

9. Some Keynesians believe that short-run aggregate supply curve during periods of depression is a horizontal straight line. In this case when aggregate demand is raised, it leads only to expansion in real GNP with no effect on price level at all.

supply and nominal GNP. According to monetarists.

$$MV = \text{Nominal GNP}$$

If V is stable and predictable we can predict the growth in nominal GNP from the growth in money supply. For instance, if in a year $M = 100$, $V = 1$, the nominal GNP will be $100 \times 1 = 100$. Suppose velocity annually increases at a stable rate of 2 per cent. Now, using the equation of exchange, if there is 5 per cent in increase inmoney supply, then it can be correctly predicted that nominal GNP will grow by 7 Per cent.

On the other hand, if V is unstable and unpredictable, the relation between money supply and nominal GNP (that is, between M and PQ) will be losse and uncertain. Thus, in the context of variable V, a steady growth in money supply (M) will not necessarily translate into a steady growth in nominal GNP.

Now, if velocity is stable or not is an empirical question. In the United States, velocity increased predictably (steadily at the rate of 3.4% per annum) in the 1960s and 1970s and during this period many economists became monetarist in their approach not only in the USA but in many other countries as well. But then during the 1980s and 1990s in the USA and other countries there was a sharp fall in velocity and it behaved quite erratically and unpredictably many economists abandoned monetarism. However still the debate between monetarists and Keynesians go on as there is a difference of opinion about the proper interpretation of the data from which change in velocity is estimated.

Keynesian View : V is Unstable. Quite opposite to the monetarists' view Keynesians believe that *velocity of money is variable and unpredictable.* Moreover, according to Keynesians, velocity of money is affected by changes in money supply and rate of interest. According to Keynesians, money is demanded not only for transaction purposes but also for *holding it as an asset,* which was called speculative demand for money by Keynes. Money demanded for transactions purposes is termed as *active* money because it changes hands and circulates through income-expenditure stream during a period. In other words, money meant for transaction purposes has a certain *positive velocity,* say, five in a year. But money demanded and held as an asset *remains idle* and does not flow through the transactions involving final goods and services and, therefore, its velocity is zero. It follows therefore that *overall velocity of money depends on how public divides the total money supply between transaction and asset balances.* The greater the extent to which money is demanded for transaction motive, the larger will be the overall veloxity of money. Conversely. the greater the proportion of money held as an asset, the lower will the velocity of money.

Keynesians also challenge the transmission mechanism of changes in moiney affecting changes in GNP. They argue that if a large proportion of expansion in money supply goes into asset demand for money, velocity of money will fall. In the extreme case when all additional money is held as an asset balances, then the increase in money supply will be accompanied by the offsetting decline in velocity so that in that case the size of nominal GNP will remain unaffected.

Keynesians viewpoint can be better understood with their notion of demand for money. According to Keynesians, total demand for money is a function of rate of interest and level of national income. In symbolic terms,

$$M_d = L(i, Y)$$

where i denotes interest, Y denotes income and L stands for function

Now rewriting the velocity of money by replacing money supply (M_s) with M_d we have[10]

$$V = \frac{\text{Nominal GNP}}{M_s} = \frac{\text{Nominal GNP}}{M^d}$$

10. In fact, in equilibrium money supply (M_s) is equal to demand for money (M_d).

Let us write Y for nominal GNP and $L(i, Y)$ for M_d we have

$$V = \frac{Y}{L(i, Y)}$$

Now, *if increase in money supply lowers the interest rate,* it will lead to the public holding *more money* as an asset. This is because lower interest rate means cost of holding money is less. This will induce the public to hold large amount of money as asset balances which have a zero velocity. As a result, *expansion* in money supply will cause *decline* in velocity of money and vice-versa. Thus, *in the Keynesian view velocity of money varies (1) inversely with the supply of money (2) and directly with the rate of interest.*

In view of the variable velocity of money, Keynesians hold that stable relationship between money supply (M) and nominal GNP as visualised by monetarists' transmission mechanism *does not hold good as* V varies as money supply and interest change.

5. Price-Wage Flexibility

Another important difference between Keynesians and monetarists revolves around wage-price flexibility. According to monetarists, money is the prime determinant of nominal GNP and prices and wages are fairly flexible so that wages and prices are so adjusted in the long run that the long-run equilibrium is established at potential GNP or full-employment level of output. According to them, though changes in money supply affect *both output and prices in the short run,* in the long run equilibrium is achieved through appropriate changes in wages and prices at full-employment level of output. Thus, according to monetarists, long-run aggregate supply curve is vertical at full-employment level of output. In the long run, according to monetarists, as a result of changes in money supply *only price level changes,* equilibrium level of GNP remains constant at full-employment or potential level.

On the other hand, Keynesian economists think that *wages and prices are sticky,* especially in the downward direction. Therefore, according to them, wages and prices do not adjust fully to restore the equilibrium at full-employment level of output. The difference in the two approaches is illustrated in Fig. 22.7 where SAS_0 and LAS are short-run and long-run aggregate supply curves respectively. To begin with, economy is in equilibrium at point R where LAS and SRS_0 intersect the aggregate demand curve AD_0 at point R. As a result, equilibrium exists at the level of potential or full-employment output Q_F. Suppose due to contraction in money supply aggregate demand curve shifts downward to the new position AD_1. From Fig. 22.7, it will be seen that new short-run equilibrium is established at GNP level Q_1 or point T where SAS_0 intersects the new aggregate demand curve AD_1. Thus, in the short run, GNP has declined to Q_1 and price level to P_1. This fall in GNP in the short run will cause the emergence of involuntary unemployment. Keynesian economists stop at this point because they contend that wages cannot fall and therefore price level cannot decline further.

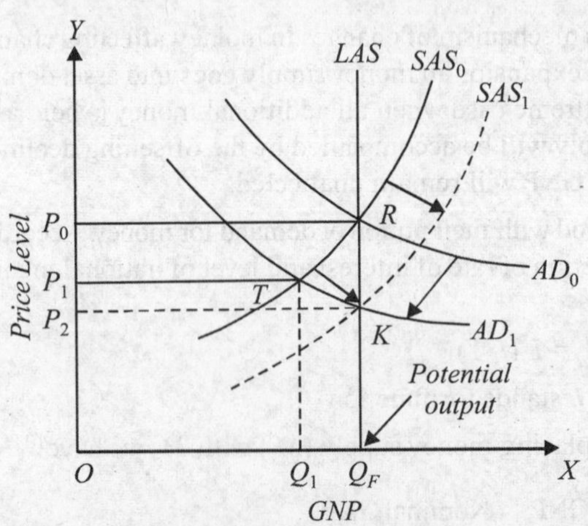

Fig. 22.7. *Wage-Price Flexibility : Monetarists vs. Keynesians.*

However, monetarists believe that pressure of unemployment will force the workers to bid down wages. When wages fall, cost of production will decrease causing short-run aggregate supply curve to shift below till it reaches the position SAS_1 where it intersects the aggregate demand curve AD_1 and LAS at point K. At point K, new long run equilibrium is established at potential level Q_F of GNP and price level has further fallen to P_2 level. Thus, according to monetarists, in the long run wages and prices change so that equilibrium is achieved at full-employment level. To conclude, according to monetarists in the long run only wages and prices change, while GNP remains unchanged at potential level.

6. Role of Monetary Policy

The difference in the theoretical viewpoints about the effect of money supply on output and prices leads to their divergent views about the effectiveness of monetary policy. On the basis of their monetary theory, monetarists believe that monetary policy can play a much more important role in stabilisation of the economy than do the Keynesians. While Keynesians agree that *money also matters* but they contend that expansionary monetary policy alone will be quite ineffective in lifting the economy out of depression. Keynesians argue that during recession demand for money curve (that is, liquidity preference curve) is quite flat so that expansion in money supply does not reduce very much market rate of interest. Besides, they believe that private investment is relatively insensitive to interest rates (that is, investment demand curve is steep) as it depends primarily on business expectations about possibilities of profit makeing. This suggests that a slight fall in interest rate consequent to the expansion in money supply will not lead to much increase in investment. As a result, aggreagate demand will rise only by a small amount. Hence, expansionary monetary policy will be ineffective in curing recession. This is illustrated in Fig. 22.8 where Pnael (*a*) depicts the determination of rate of interest where demand for money is shown by the curve M_d which is quite flat at lower rates of interest. When money supply is increased from M_1 to M_2, there is only a small fall in interest from r_0 to r_1, investment increases by only a small amount. As a result, aggregate demand will not rise much to product any significant effect on output and employment. Thus, according to Keynesians, monetary policy as an instryment of stabilisation is quite ineffective.

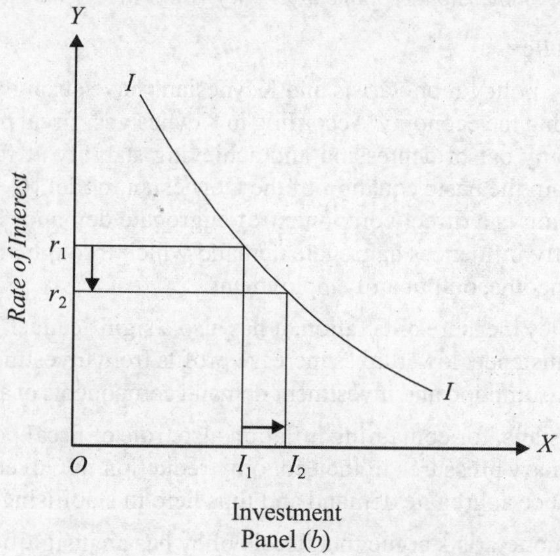

Fig. 22.8. *Ineffectiveness of Monetary Policy : Keynesian View.*

On the other hand, monetarists believe tha demand for money curve is quaite steep (that is, demand for money is relatively insensitive to interest rates) so that small expansion in money supply causes a large reduction in interest rate. Besides, according to monetarists, investment is very sensitive to interest rates (that is, investment demand curve is relatively flat so that a fall in interest rate causes a large increase in investment demand with a significant effect on otuput and employment. Thus, monetarists argue that monetary policy is very effective and can became an important tool for stabilisation. This monetarists' view is graphically shown in Fig. 22.9.

It will be seen from Fig. 22.9 panel (*a*) that small expansion in money supply from M_1 to M_2 causes a relatively large drop in interest rate from r_0 to r_1. With this full in interest rate, there is a substantial increase in investment from I_0 to I_1 as investment demand is quite interest-elastic. This large increase in investment demand will have appreciable effect on aggregate demand, output and employment. Thus, monetarists argue that expansionary monetary policy is quite effective in curing recessionm or depression.

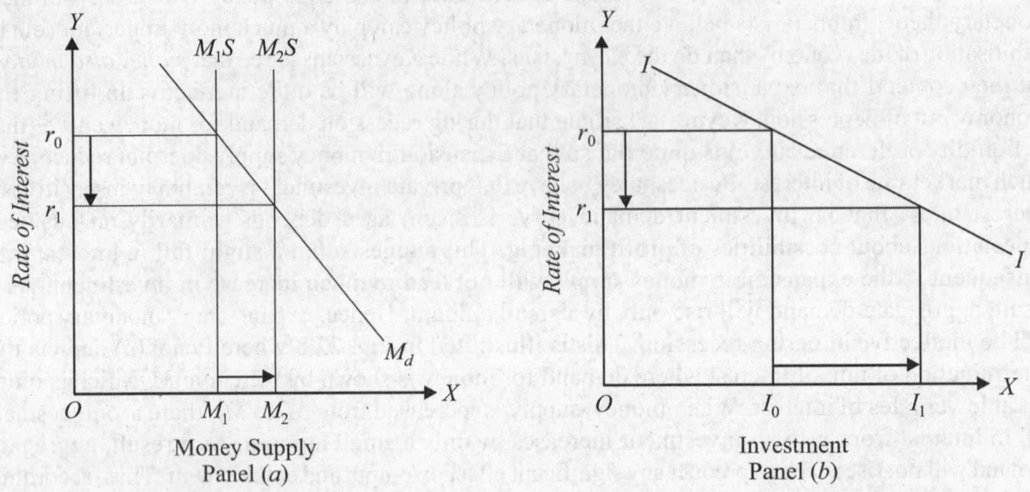

Fig. 22.9. *Expansionary Monetary Policy is Effective : Monetarist's View.*

7. Role of Fiscal Policy

Like that of monetary policy monetarists and Keynesians have strongly debated about the role of fiscal policy in stabilising the economy. According to Keynesians, fiscal policy is a more powerful tool for lifting the economy out of depression and achieving stability at full-employment level of output. this is evident from the basic equation of the Keynesian model $[C + I + G + (X - M) = Y]$ where investment spending is a direct component of aggregate demand. Increase in government spending therefore directly influences aggregate demand which through a multiplier effect pushes up the level of national income, output and employment.

As regards fiscal policy measure of taxation, it has also a significant effect on consumption and investment demand. for instance, lower taxes increase profits from investment by businesses. Thus, lower taxes boost both consumption and investment demand components of aggregate demand curve.

According to Keynesians, for containing inflation also, role of fiscal policy plays an important role. At times of inflationary pressures in the economy reduction in Government expenditure and increase in taxes will reduce aggregate demand and thus help in stabilising prices.

On the other hand, monetarists argue that fiscal policy has an insignificant effect on aggregate demand and therefore does not serve as an important tool of stabilisation. They believe that fiscal policy is quite weak and ineffective because of the *crowding out effect* of Government expenditure. Monetarists point out that when Government makes a deficit budget and finances it through borrowing from the market, it will be completing with private businesses for funds. As a result, demand for money will increase which, given the money supply, will push up the interest rate. higher interest rate will cause a large reduction in private investment due to investment demand being highly sensitive to the interest rates. Thus, increase in Government expenditure through its large borrowing from the market *crowds out* substantial amount of private investment which other wise would have been [possible. Monetarists therefore contend that due to crowding out effect increase in government expenditure simply replaces private investment with little effect on aggregate demand and consequently levels of output and employment will remain unaffected. To quote Friedman, the leader of monetarist school "...in my opinion the state of budget by itself has not significant effect on the course of nominal income, on inflation, or an cyclical fluctuations."[11]

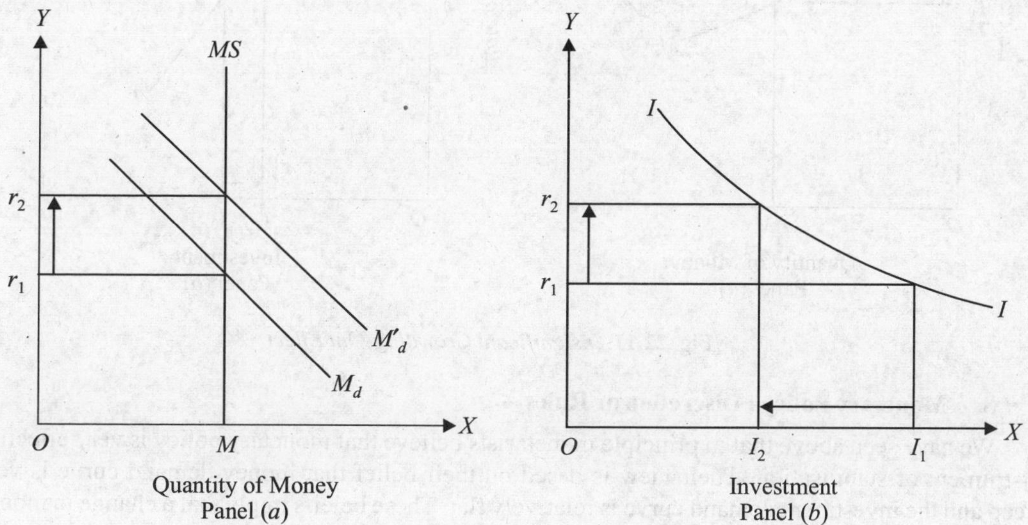

Fig. 22.10. *Crowding Out Effect of Rise in Government Expenditure*

Crowding out effect of Government expenditure is illustrated in Fig. 22.10. From increase in Government expenditure pushes up the demand curve for money from money from M_d to M'_d and, given the omeny supply *OM*, raises the interest rate form r_1 to r_2. From Panel (*b*) of Fig. 22.10 it will be seen that rise in interest rate from r_1 to r_2 causes a reduction in private investment from I_1 to I_2. Thus, a good amount of private investment has been crowded out as a result of expansion in government expenditure financed by borrowing from the market.

Of course, if budget deficit or increase in Government expenditure is financed by issuing new money, it will have no effect on demand for money and interest rate. As a result there will be no crowding out effect. But monetarists argue that this would not be due to the fiscal policy of increase in Government expenditure but due to the *creation of additional money*.

It may however be noted that most of the Keynesians admit that some crowding out effect of fiscal policy may be there but they strongly contend that the *net effect* of expansionary fiscal policy on real national income and employment will be substantial. Keynesians argue that demand for money curve is relatively flat because demand for money is highly sensitive to changes in interest rate [see Panel (*a*) of Fig. 22.11] so that increase in aggregate demand curve, given the money supply, does not lead much to the rise in rate of interest. Besides, according to them, investment demand is relatively steep because investment is insesitive to changes in interest rate [see Panel (*b*) of Fig. 22.11]. Under these circumstance, increase in Government expenditure will cause only a little rise in interest rate which due to the steep investment demand curve will result in a very small decrease in private investment. Thus, Keynesians believe that crowding-out effect of expansionary fiscal policy is quite insignificant.

---

11. Milton Friedman and Walter Heller, *Monetary vs. Fiscal Policy* (New York : W.W. Norton and Company, (1969) p. 51

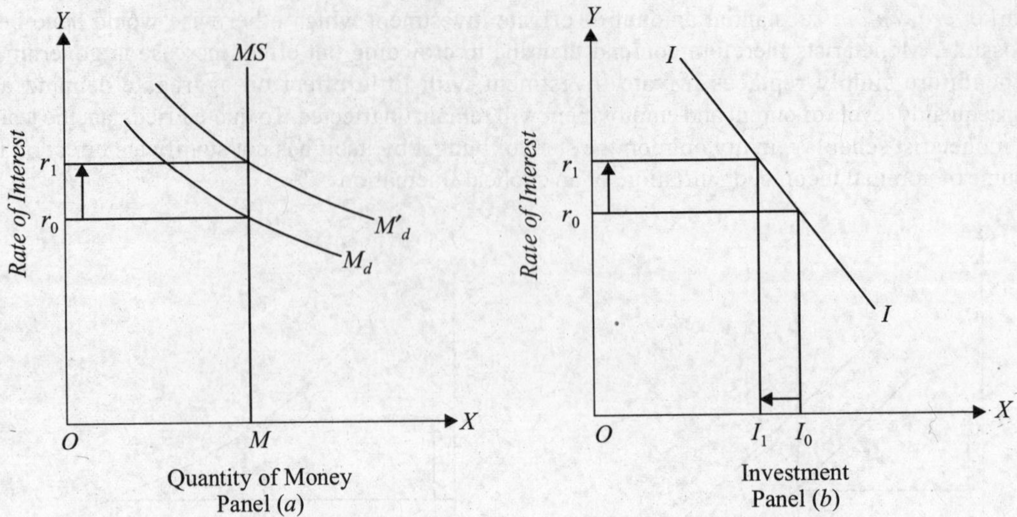

Fig. 22.11. *Insignificant Crowding-Out Effect.*

8. Monetary Policy : Discretion or Rules

We have seen above that in principle monetarists believe that monetary policy is very effective instrument of stabilisation. Their view is based on their belief that money demand curve is very steep and the investment demand curve is relatively flat. These beleifs imply that a change in money supply has a powerful effect on equilibrium level of *nominal* income. However, most monetarists do not recommend *active* use of monetary policy to stabilise the economy. This is because they believe that private economy is inherently stable and central bank of a country through its erratic and improper discretionary monetary policy tends to destabilise rather than stabilise the economy. Friedman, on the basis of historical evidence from the United States contended that discretionary changes in money supply made by the monetary atuhorities had in fact been a destabilising influence on the American economy.

Moreover, monetarists argue that *changes in money supply affect output only after long, irregular and variable lags* which make it very difficult to formulate a proper monetary policy *at right time* to achieve stability. Therefore, monetarists advocate the adoption of *Constant monetary rule* in preference to the discretionary monetary policy, According to the constant monetary rule which they recommend *money supply should be increased each year at the same annual rate as the potential growth of real* GNP. For instance, if the annual potential growth of real GNP is 3 per cent, money supply should be expanded also by 3 per cent; if potential growth in real GNP is 5 per cent money supply should also be increased every year by 5 per cent.

The rationale behind constant monetary rule can be explained with the equation of exchange ($MV = PQ$) behind constant monetary rule can be explained with the equation of exchange ($MF = PQ$). If the money supply (M) grows at the same rate as our ability to produce (Q) and if velocity (V) is constant, then the price level will remain stable. This can be shown mathematically. In growth terms equation of exchange can be written as

$$\frac{\Delta M}{M} + \frac{\Delta V}{V} = \frac{\Delta P}{P} + \frac{\Delta Q}{Q}$$

where $\frac{\Delta M}{M}$ stand for growth of money supply, $\frac{\Delta P}{P}$ for growth in price level and $\frac{\Delta Q}{Q}$ for growth in real output.

Now, if $\frac{\Delta M}{M} = \frac{\Delta Q}{Q}$ and $\frac{\Delta V}{V} = 0$ it follows from above equation that $\frac{\Delta P}{P} = 0$.

Monetarists think that the constant monetary rule will do away with the major cause of instability which, according to them, is the pursuit of erratic and capricious changes in discretionary monetary policy. they believe that as long as the money supply grows at a constant rate each year, be it 3, 4 or 5 per cent emergence of recession in the economy will be only temporary. the expansion in money supply provided by a constatnly growing money supply will cause aggregate demand to increase and prevent recession to last long. Similarly, if the supply of money is allowed to rise at a more than average growth rate of GNP, any inflationary pressures caused by excess aggregate demand will get itself eliminated due to lack of money supply.

On the other hand, Keynesians think it would not be prudent to follow a fixed or constant monetary rule. they argue that in view of the fact that velocity of money is variable both cyclically and secularly, a constant annual rate of growth of money supply would in fact cause substantial fluctuations in aggregate demand and would therefore promote instability.

QUESTIONS FOR REVIEW

1. Explain the *restatement of Quantity Theory of Money* as made by Friedman. How does it differ from Fisher's version of the Quantity Theory of Money?
2. Discuss briefly the relation between the supply of money and price level in terms of Friedman's Quantity Theory of Money. What are the main shortcomings of this theory? [*D.U. B. Com (Hons)*]
3. According to Friedman's modern quantity theory of money, Changes in the quantity of money would affect output as well as prices. Discuss?
4. Referring to his restatement of the theory Friedman says, "the quantity theory of money is not the tautology of the equation, it is rather an analysis of the factors determining the quantity of money that the community wishes to hold". Discuss?
5. "The substantial changes in prices or nominal income are *almost invariably the result* of changes in the nominal supply of money". (Friedman) Discuss?
6. 'Inflation is always and everywhere a purely monetary phenomenon" (Friedman). Discuss?
7. What is monetarism? How does it explain depression and inflation in the economy? Discuss.
8. Critically examine Fredman Monetarist theory of macroeconomy
9. What are the essential features of monetarism Compare monetarism with Keynesianism.

Chapter 23

INFLATION AND HYPERINFLATION: CAUSES, SOCIAL COSTS, EFFECTS AND CURE

INFLATION: MEANING AND CAUSES

Having discussed the relationship between quantity of money and price level in the previous chapter, we are now in a position to explain the nature and causes of inflation. By inflation we mean *a general rise in prices.* To be more correct, *inflation is a persistent rise in the general price level rather than a once-for-all rise in it.* On the other hand, *deflation represents persistently falling prices.* Inflation or persistently rising prices is a major problem in India today. When price level rises due to inflation, the value of money falls. When there is a persistent rise in price level, the people need more and more money to buy goods and services. To enable the people to meet their daily needs of consumption of goods and services when their prices are rising, their incomes must rise if they have to maintain their standard of living. For government employees, their dearners allowance is increased. Wages and salaries employed in the organised private sector are also raised, though after some time lag. But people with fixed incomes and those who are self-employed are unable to raise their prices suffer a lot due to inflation. The poor suffer the most from persistent rise in prices, especially of foodgrains and other essential items. Rate of inflation during the seventies and eighties was very high as compared to the rates of inflation experienced earlier during previous periods. In India, in recent years, 2010-11, 2011-12 and 2012-13, rate of inflation as measured by consumer price index (CPI) has been in double digit Fig.s. Prior to Jan. 2013, even WPI inflation was quite high which compelled Reserve Bank of India to adopt tight monetary policy.

Causes of Inflation

Let us understand how the inflation originates or what causes it. Depending upon the specific causes, three types of inflation have been distinguished: (1) *Demand-pull inflation,* (2) *Cost-push inflation*, and (3) *Structuralist inflation*. An important cause of demand-pull inflation is the excessive growth of money supply in the economy. We will explain this cause of inflation in the *Monetarist Theory of Inflation*. We will explain and discuss below these three types of inflation.

DEMAND-PULL INFLATION

This represents a situation where the basic factor at work is the increase in aggregate demand for output either from the households or the entrepreneurs or government organised. The result is that the pressure of demand is such that it cannot be met by the currently available supply of output. If, for example, in a situation of full employment, the government expenditure or private investment goes up, this is bound to generate inflationary pressures in the economy. Keynes explained that inflation arises when there occurs an *inflationary gap* in the economy which comes to exist when aggregate demand for goods and services exceeds aggregate supply at full employment level of output. Basically, inflation is caused by a situation whereby the pressure of aggregate demand for goods and services exceeds the available supply of output (both being counted at the prices ruling at the beginning of a period). In such a situation, the rise in price level is the natural consequence.

Now, this imbalance between aggregate demand and supply may be the result of more than one force at work. As we know aggregate demand is the sum of consumers' spending on consumer goods and services, government spending on goods and services and net investment being contemplated by the entrepreneurs. *When aggregate demand for all purposes—consumption, investment and government expenditure—exceeds the supply of goods at current prices, there is a rise in price level. Since inflation is a continuous increase in the price level, not a one time rise in it, sustained inflation requires continuous increase in aggregate demand.* In the modern macroeconomics, inflation is explained with AD-AS model. Inflation can be explained by increase in aggregate demand (called "***demand shock***") or decrease in aggregate supply or rise in cost of production generally called "***supply shock***". Demand-pull inflation occurs when there is increase in aggregate demand when supply shocks are absent.

As stated above, demand-pull inflation occurs when there is increase in any component of aggregate demand, namely, consumption demand by households, investment by business firms, increase in government expenditure unmatched by increase in taxes (that is, deficit spending by the government financed by either creation of new money by the central bank or borrowing by the government from the market. To illustrate the cause of demand-pull inflation, let us assume the government adopts expansionary fiscal policy under which it increases its expenditure without levying extra taxes to finance its increased expenditure by borrowing from the Reserve Bank of India.

To illustrate the above point, let us assume that the government adopts expansionary fiscal policy under which it increases its expenditure on education, health, defence and finances this extra expenditure by borrowing from Reserve Bank of India which prints new notes for this purpose. This will lead to increase in aggregate demand ($C + I + G$). If aggregate supply of output does not increase or increases by a relatively less amount in the short run, this will cause demand-supply imbalances which will lead to demand-pull inflation in the economy, that is, general rise in price level.

Similarly, an inflationary process will be initiated if business firms anticipating the opportunities of making profits decide to invest more and to finance the new investment projects by borrowing from the banks being unable to get sufficient funds throught savings out of profits and savings invested by the public in them. This new investment by the firms leads to the increase in aggregate demand for goods and services. However, inflation will occur by this new investment if aggregate supply of output does not increase adequately in the short run to match the increase in aggregate demand. Therefore, *demand-pull inflation generally occurs when the economy is already working at full-employment level of resources or what is now generally called when there is natural rate of unemployment*. This is because if aggregate demand increases beyond the full-employment level of output, output of goods cannot be increased adequately without much increase in cost. Note that in developing countries such as India, there are difficulties of measuring employment, unemployment and full employment. Therefore in the Indian context, instead of full-employment level of output, we use ***full capacity output of the economy beyond which supply of output cannot be generally increased.***

It is important to note that Keynes in his booklet, *How to Pay for the War*, published during the Second World War explained inflation in terms of excess demand for goods relative to the aggregate supply of their output. His notion of the *inflationary gap* which he put forward in his booklet represented excess of aggregate demand over full-employment output. This inflationary gap, according to him, leads to the rise in prices. Thus, Keynes explained inflation in terms of demand-pull forces. Therefore, the theory of demand-pull inflation is associated with the name of Keynes. Since beyond full-employment level of aggregate supply of output cannot increase in response to increase in demand this results in rise in prices under the pressure of excess demand. Aggregate supply curve, according to him, is vertical at full-employment level.

Demand pull inflation is shown in Fig. 23.1. In the modern macoreconomics distinction is drawn between *long-run aggregate supply cure (LAS) and short-run aggregate supply curve* (*SAS*). The long-run aggregate supply curve (*LAS*) is a *vertical line drawn at the full-employment level* (*i.e.* at natural rate of unemployment) or in the Indian context *full-capacity output level.* This full-employment level or full-capacity output is also called ***potential output.*** The short-run aggregate supply curve SAS_0 slopes upward to the right which shows more supply of output is forthcoming at a higher price and *is drawn with a constant wage rate*. This short-run aggregate supply curve (*SAS*) slopes upward because with the increase in employment of labour, diminishing returns to labour occur which raises marginal cost of production. This short-run aggregate supply curve slopes upward to the right even beyond full-employment or potential level of output. This is because, as explained earlier, even at full-employment level, some unemployment occurs due to frictional and structural factors and therefore beyond full-employments level, employment of labour can increase with reduction in natural unemployment under the pressure of aggregate demand. It will be seen from Fig. 23.1 that short-run aggregate supply curve SAS_0 cuts long-run aggregate supply curve *LAS* at point E_0. To begin with, aggregate demand curve AD_0 intersects both short-run aggregate supply curve SAS_0 and long-run aggregate supply curve (*LAS*) at point E_0 and this shows that at point E_0 there is long-run equilibrium at potential or full-employment GDP level $\overline{Y}$ and at price level P_0.

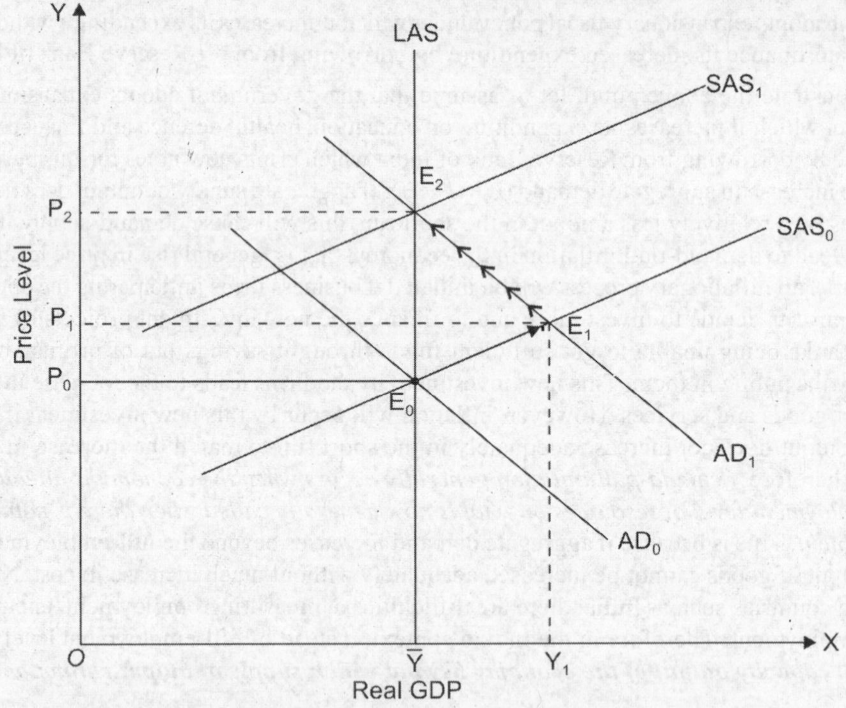

Fig. 23.1. *Demand-Pull Inflation*

Now suppose without increasing taxes government increases its expenditure on goods and services and finances it by borrowing from the Reserve Bank which in turn prints money for this purpose. As a reuslt of increase in government expenditure, aggregate demand curve shifts to the right to AD_1 which intersects the short-run aggregate supply curve SAS_0 at point E_1 and as a result price level rises to P_1 and real GDP to Y_1. It needs to be emphasized that price level has risen as a result of rightward shift in aggregate demand curve to AD_1 because aggregate supply has not increased to the extent of increase in aggregate demand and thereby creating demand-supply imbalances. If short-run aggregate curve had been horizontal line through point E_0, the aggregate supply would have increased by an equal amount to the increase in aggregate demand when aggregate demand curve shifts to right to AD_1. Thus, *the price level has risen because aggregate demand has increased relatively more than the aggregate supply, that is, due to demand-supply imbalances*.

However, the response to the initial increase in aggregate demand to AD_1 does not stop at point E_1. It may be recalled that in drawing short-run aggregate supply curve wage rate of labour is kept constant. Now, the rightward shift in aggregate demand curve to AD_1 has caused the price level to rise from P_0 to P_1 and real gross domestic output (GDP) has risen to Y_1. This rise in price level, the money wage rate remaining constant, would cause decline in workers' real wage rate, $\frac{W}{P_1} = \frac{W}{P_0}$ where $P_1 > P_0$. When workers realise that their real wage has fallen as a result of increase in aggregate demand causing rise in price level, they will demand higher wages in their negotiations with their employers which are likely to be conceded to. Further, as pointed out above, equilibrium to the right of potential GDP level $\bar{Y}$ implies that unemployment has fallen below the natural rate of unemployment and will cause shortage of labour which implies wage rate will rise. When wages are raised by the firms short-run aggregate supply curve will shift upward and this process of shifting short-run aggregate supply curve upward will continue until it reaches SAS_1 which cuts the new aggregate demand curve AD_1 at point E_2 that lies at the long run supply curve *LAS* and as a result pri ce level further rises to P_2. Equilibrium between aggregate demand and aggregate supply with price as P_2 and at point E_2 on the long-run average supply curve *LAC* the wage rate has risen by E_0E_2 equal to the rise in price level by P_0P_2. Thus, real wage has been restored at the level prior to increase in aggregate demand from AD_0 to AD_1. Note that equilibrium and wage rate at point E_1 on the short-run average supply curve will move to point E_2 at the long-run average supply curve not with a single jump but through various steps of wage adjustment and upward shifting of short-run aggregate supply curve. That is why we depicted this movement from point E_1 to point E_2 on the *LAS* through *various* arrows. Note that not only there is shifting upward of short-run average supply curve to resotre previous level of real wage rate but also real GDP has returned to potential or full employment GDP level $\bar{Y}$.

Demand-Pull inflation Process to Cause Continuous Rise in Price Level

We have explained above the process of demand-pull inflation when for a given increase in aggregate demand, *price level has eventually gone up to price level* P_2 and wages have risen enough to restore the real wage to its previous full-employment level. But, as noted above, *inflation is a continuous increase in the price level and not a one-time jump in the price level in response to a given increase in aggregate demand.* For sustained or persistent inflation to take place the continuous increase in aggregate demand must occur. In our example aggregate demand can persistently increase if a government has a large persistent budget deficit that it finances by borrowing from the Central Bank or alternatively borrowing from the market year after year. Besides, continuous increase in aggregate demand can occur if quantity of money is persistently increased by the Central Bank of the country. In these two cases aggregate demand increases year after year causing price level to rise persistently. This continuous rise in price level is shown in Fig. 23.2 in which we have assumed that government runs the budget deficit year after year and finances it by borrowing from the Central Bank (that is, government sells its bonds to the Central Bank which prints new money to pay for these bonds).

As a result, aggregate demand increases year after year bringing about continuous rise in price level. Initially, the equilibrium is at point E_0 at which an aggregate demand curve AD_0 intersects long run aggregate supply curve (*LAS*) as well as short-run aggregate supply curve SAS_0. Now suppose that due to increase in government expenditure financed by borrowing from the Central Bank shifts aggregate demand curve to AD_1 which intersects the short-run aggregate supply curve SAS_0 at point E_1 raising the price level to P_1. As explained above, with rise in price level to P_1 real wages of labour will fall and they will demand higher wages. As a result of rise in wage rate over a number of stages or steps short-run aggregate supply curve (*SAS*) shifts to the left until it intersects the new aggregate demand curve AD_1 at point E_2 that lies at the long-run aggregate supply curve (*LAS*). With this equal rise in price level and wage rate, real wage rate of workers is restored.

Now suppose next year the government again incurs a budget deficit and finances its further increase in expenditure by borrowing more from the Central Bank. With this aggregate demand curve shifts to the right to AD_2 and cuts the short-run aggregate supply curve SAS_1 at E_3 and causes price level to rise further to P_3 as shown in Fig. 23.2. The rise in price level will cause the fall in real wage rate of workers who will demand increase in their money wage rate further. When higher money wage rate is conceded to, short-run aggregate supply curve will start shifting to the left and the process will eventually end when short-run aggregate supply curve has shifted to the position SAS_2 which intersects aggregate demand curve AD_2 at point E_4 at which price level has risen to P_4 and with that real wage rate of workers is restored at full-employment level with potential GDP equal to $\bar{Y}$. If government further increases its expenditure next year and finances it by borrowing from the Central Bank, price level will further rise. In this way there is continuous inflation triggered by increase in government expenditure and the operation of wage price spiral under the pressure of increase in aggregate demand. In Fig. 23.2 we *have traced the rise in price level through arrows as aggregate demand increases and short-run aggregate supply curve shifts to the right as a result of increase in money wage rate.*

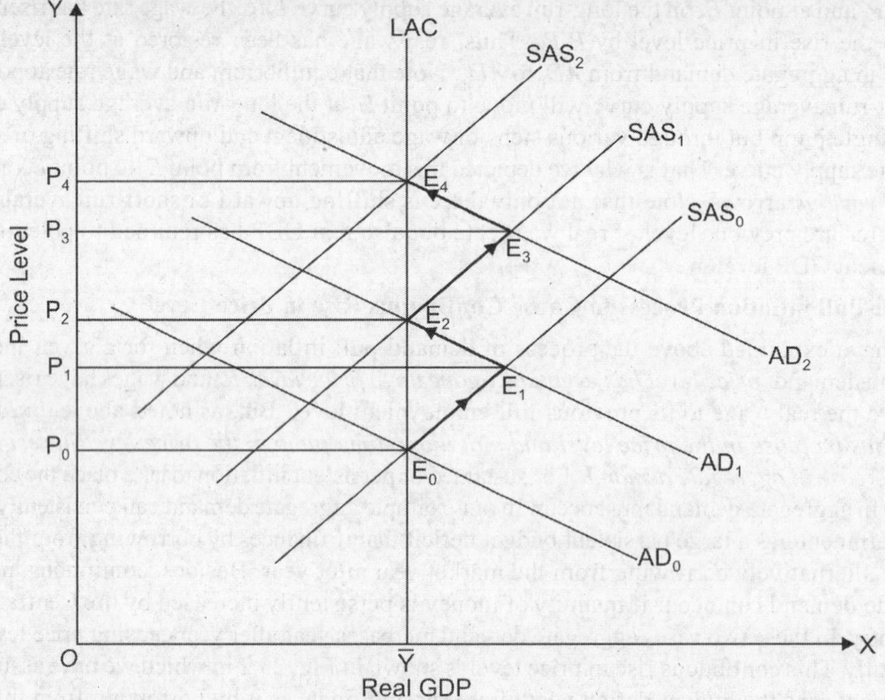

Fig. 23.2. *Demand-Pull Inflation Process: Persistent Increase in Price Level*

Inflationary Expectations

Inflationary expectations are an important cause of inflation. The expectations of future prices play a significant role in decision-making by firms regarding price and output. If a firm expects that its rival firms will raise their prices, it may also raise its own price in anticipation. Suppose inflation has been occurring at a rate of 8 per cent per annum in the past, a firm will expect this inflation rate to continue in future too and therefore it will raise its price by 8 per cent. If every firm expects that other firms will raise their prices by 8 per cent, every one will raise its prices by 8 per cent. As a result, inflation rate of about 8 per cent will occur. This is how inflationary expectations cause inflation. Elaborating on this, Case and Fair write, "*Expectations can lead to an inertia that makes it difficult to stop an inflationary spiral. If prices have been rising and if people's expectations are adaptive, that is, if they form their expectations on the basis of past pricing behaviour – that firms may continue raising prices even if demand is slowing.*"[1] Therefore, to check inflation, steps should be taken to break the inflationary expectations.

1. Karl E. Case and Race C. Fair, *Principles of Economics*, Pearson Education, 6th Edition, 2002,

COST-PUSH INFLATION

We can visualise situations where even though there is no increase in aggregate demand, prices may still rise. This may happen if there is initial increase in costs *independent of any increase in aggregate demand.* The four main autonomous increases in costs which generate cost-push inflation have been suggested :

1. *Oil Price Shock*
2. *Farm Price Shock*
3. *Import Price Shock*
4. *Wage-Push inflation*

Cost-push inflation is also called *supply-side inflation.*

1. Oil Price Shock. In the seventies the supply shocks causing increase in marginal cost of production became more prominent in bringing about cost-push inflation. During the seventies, *rise in prices of energy inputs* (hike in crude oil price made by OPEC resulting in rise in prices of petroleum products). The sharp rise in world oil prices during 1973-75 and again in 1979-80 produced significant supply-shocks resulting in cost-push inflation.

The sharp rise in the price of oil leads to inflation in all oil-importing countries. The rise in oil price also occurred in 1990, 1999-2000 and again in 2003-08 which resulted in rise in rate of inflation in oil-importing countries such as India. In recent years, there have been a good deal of fluctuations in oil prices; in some periods they go up and in some others they go down. It may be noted that rise in oil prices not only gives rise to the increase in inflation, but also adversely affects the balance of payments raising current account deficit of the oil-importing countries such as India.

2. Farm Price Shock. Cost-push inflation can also come about from increase in prices of other raw materials, especially farm products in economies such as that of India where they are of greater importance. In India when monsoon is not adequate or comes very late or when weather conditions are quite unfavourable, they reduce the supply of agricultural products and raise their prices. These farm products are raw materials for various industries such as sugar industry, other agro-processing industries, cotton textile industry, jute industry and as a result when prices of farm products rise they lead to rise in prices of goods which use the farm products as raw materials. This is farm price shock causing cost-push inflation. Even rise in food prices or what is called *food inflation* is caused by supply-side factors such as inadequate rainfall or untimely monsoon and other adverse weather conditions and inadequate availability of fertilizers which lead to reduction in output of foodsgrains is the example of supply-side inflation.

3. Import Price Shock. These days currencies of most countries of the world are *flexible*, that is, determined by demand for and supply of a currency and they can appreciate or depreciate every month in terms of the US dollar. For example, when the Indian rupee depreciates, more rupees are required to buy one US dollar and therefore in terms of rupees, imports become costlier. The Indians who import raw materials for industries such as petroleum products, coal, machines and other equipment, oilseeds, fertilizers, Indian consumers who import gold, cars and other final products have to pay higher prices in terms of rupees when Indian rupee depreciates against US dollar. This raises the cost of production of the producers who in turn raise the prices of final products produced by them. *This inflation is the result of import price shock.* Thus depreciation of rupee causes cost-push inflation. For example, in the months of June and 2013, there was sharp depreciation of the Indian rupee. The value of rupee fell by about 9.5 per cent in these months from about ₹ 56 to a US dollar in the first week of June 2013 to around ₹ 61 to a dollar in the second week of August 2013.

4. Wage Push Inflation. It has been suggested that the growth of powerful trade unions is responsible for the spread of inflation, especially in the industrialized countries. When trade unions push for higher wages which are not justifiable either on grounds of a prior rise in productivity or of cost of living they produce a cost-push effect. The employers in a situation of high demand and employment are more agreeable to concede to these wage claims because they hope to pass on these rises in costs to the consumers in the form of hike in prices. If this happens we have cost-push inflation. It may be noted that as a result of cost-push effect of higher wages, short-run aggregate supply curve of output shifts to the left and, given the aggregate demand curve, results in higher price of output.

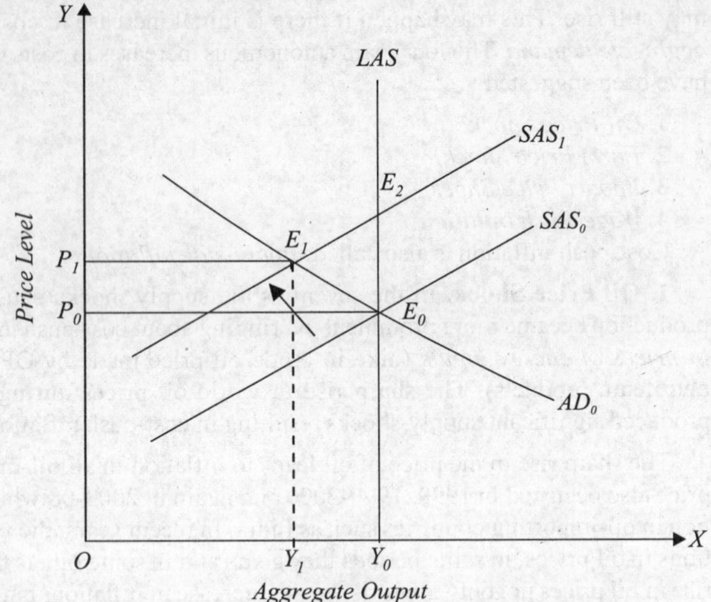

Fig. 23.3. *Cost-Push Inflation and Stagflation*

Let us consider Fig. 23.3 where to begin with aggregate demand curve AD_0 and short-run aggregate supply curve SAS_0 intersect at point E_0 and determine price level P_0 and output level Y_0. Further suppose that Y_0 is the full capacity (*i.e.*, full-employment) level of output and therefore long-run aggregate supply curve *LAS* is vertical at Y_0 level of output. Suppose there is increase in oil prices which causes shifts in short-run aggregate supply curve to the left from SAS_0 to SAS_1. As a result, price level rises to P_1 but output falls from Y_0 to Y_1. With decline in output unemployment will also increase. *Thus cost-push inflation not only causes rise in price level (or inflation) but also brings about fall in GDP level. The rise in price level or inflation and simultaneously fall in GDP level is called* ***stagflation****. Thus cost-push inflation* ***results in stagflation***.

Cost-Push Inflation Spiral

When the real GDP decreases and as a result we have cost-push inflation in the first stage, the unemployment will also rise. When unemployment emerges there is a huge hue and cry by the workers who are rendered unemployed. In such a situation either the Central Bank will respond by increasing the money supply to raise aggregate demand or the government will *increase its expenditure to provide fiscal stimulus to aggregate demand.* As a result of either of these responses, aggregate demand curve shifts to the right from AD_0 to AD_1 in Fig. 23.4. With this the economy moves from equilibrium position E_1 to the equilibrium position E_2. As will be seen from Fig. 23.4, as a result of this increase in aggregate demand while real GDP has returned to the potential GDP level Y_0, price level has further risen to P_2 (that is, more inflation will occur). But the inflationary process of cost-push inflation will not stop at equilibrium point E_2. If there is further rise in any of the cost-push factors such as rise in oil price, increase in price of farm output, import-price shock or wage rate increase takes place short-run aggregate supply curve *SAS* will shift further to the left of SAS_1 and intersects the aggregate demand curve AD_1 to the left of the long-run aggregate supply curve *LAS* causing further rise in price level or inflation above P_2 and fall in GDP from the potential level Y_0 *resulting again in unemployment of workers*. To restore full employment and raise GDP to the potential level Y_0, the Central Bank will further increase the money supply or the government will further increase its expenditure. As a result, aggregate demand curve will again shift to the right of AD_1. With this, *GDP* level will be back at the potential level Y_0 and full employment of workers will be restored but price level or inflation will further rise. In this way cost-push inflation spiral will work to cause sustained or persistent inflation.

Many economists think *inflation in the economy is generally caused by the interaction of the demand-pull and cost-push factors*. These interaction between demand-pull inflation and cost-push inflation are shown in Fig. 23.2 and 23.4. The inflation may be started in the first instance either by cost-push factors or by demand pull factors both work and interact to cause sustained inflation over time. Thus, according to Machlup, "there cannot be a thing as cost-push inflation because without an increase in purchasing power and demand, cost increases will lead to unemployment and depression, not to inflation"[2] Likewise, Cairncross writes, "*there is no need to pretend that demand and cost inflation do not interact or that excess demand does not aggrevate wage inflation, of course it does.*"[3]

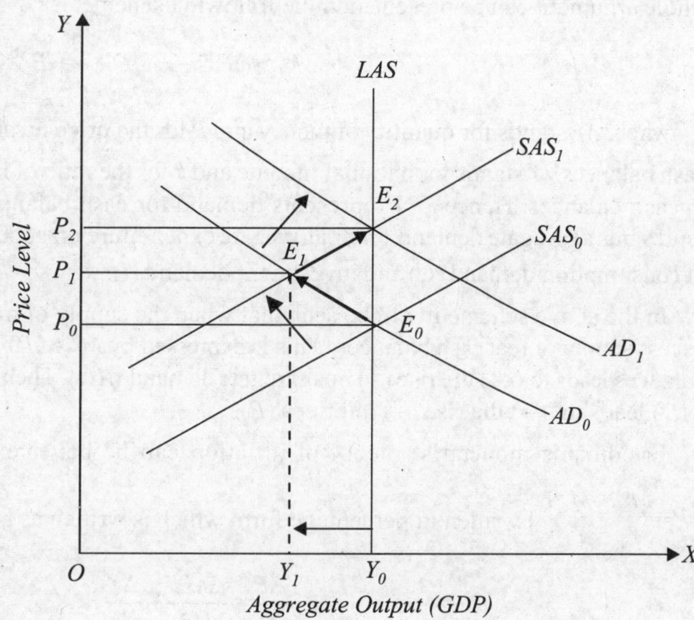

Fig. 23.4. *Cost-Push Inflation Spiral*

MONETARIST THEORY OF INFLATION

We have explained above the Keynesian theory of demand-pull inflation. It is important to note that both the original quantity theorists and *the modern monetarists, prominent among whom is Milton Friedman, also explain inflation in terms of excess demand for goods and services*. But there is an important difference between the monetarist view of demand-pull inflation and the Keynesian view of it. Keynes explained inflation as arising out of real sector forces. In his model of inflation excess demand comes into being as a result of *autonomous* increase in expenditure on investment or consumption, or increase in government expenditure on goods and services, that is, the increase in aggregate expenditure or demand occurs independent of any increase in the supply of money.

On the other hand, monetarists explain the emergence of excess demand and the resultant rise in prices on account of the increase in money supply in the economy. To quote Friedman, *"Inflation is always and everywhere a monetary phenomenon........ and can be produced only by a more rapid increase in the quantity of money than in output"*.[4]

Friedman holds that when money supply is increased in the economy, then there emerges an excess supply of real money balances with the public over the demand for money. This disturbs the equilibrium. In order to restore the equilibrium, the public will reduce the money balances by increasing expenditure on goods and services. Thus, according to Friedman and other modern

2. Fritz Machlup, "Another View of Cost-Push and Demand-Pull Inflation" in *Review of Economics and Statistics,* Vol. 42. 1960.
3. Alec Cairncross, *Inflation Growth and International Trade*, George Allen and Unwin, 1975.
4. See N.G. Mankiw, *Principles of Economics*, Second Edition, p., 641

quantity theorists, the excess supply of real monetary balances results in the increase in aggregate demand for goods and services. If there is no proportionate increase in output, then extra money supply leads to excess demand for goods and services. This causes inflation or rise in prices. The whole argument can be presented in the following scheme:

$$M^S > kPY \rightarrow AD\uparrow \rightarrow P\uparrow \qquad \ldots (1)$$

where M^S stands for quantity of money and P for the price level. Therefore, $\frac{M^S}{P}$ represents real cash balances. Y stands for national income and k for the ratio of income which people want to keep in cash balances. Hence kPY represents demand for cash balances (*i.e.*, demand for money), AD represents aggregate demand for or aggregate expenditure on goods and services which is composed of consumption demand (C) and investment demand (I).

In the above scheme it will be seen that when the supply of money (M^S) is increased, it creates excess supply of real cash balances. This is expressed by $M^S > kPY$. This excess supply of real money balances leads to (→) the rise (↑) in aggregate demand (AD). Then increase (↑) in aggregate demand (AD) leads to (→) the rise (↑) in prices (P).

Friedman's monetarist theory of inflation can be better explained with quantity equation $\left(P = \frac{MV}{Y} = \frac{M}{Y} \cdot \frac{1}{k}\right)$ written in percentage form which is written as below taking V or k as constant

$$\frac{\Delta P}{P} = \frac{\Delta M^S}{M^S} - \frac{\Delta Y}{Y} \qquad \ldots (2)$$

where $\frac{\Delta P}{P}$ is the rate of inflation, $\frac{\Delta M^S}{M^S}$ is the rate of growth of money supply and $\frac{\Delta Y}{Y}$ is the rate of growth of output. Thus, according to equation (2), rate of inflation $\left(\frac{\Delta P}{P}\right)$ is determined by growth of money supply $\left(\frac{\Delta M^S}{M}\right)$ and rate of growth of output $\left(\frac{\Delta Y}{Y}\right)$, with velocity of circulation (V) or k remaining constant. Friedman and other monetarists claim that *inflation* is predominantly a monetary phenomenon which implies that changes in velocity and output are small.

It thus follows that when money supply increases, it causes disturbance in the equilibrium, that is, $M^S > kPY$. According to Friedman and other monetarists, the reaction of the people would be to spend the excess money supply on goods and services so as to bring money supply in equilibrium with the demand for money. This leads to the increase in aggregate demand or expenditure on goods and services which, k remaining constant, will lead to the increase in *nominal* national income (PY). They further argue that the real national income or aggregate output (*i.e.*, Y in the demand for money function stated above) remains stable at full employment level in the long run due to the flexibility of wages. Therefore, according to Friedman and his followers (modern monetarists), in the long run, the increase in nominal national income (PY) brought about by the expansion in money supply and resultant increase in aggregate demand will cause a proportional increase in the price level.

However, in the short run, like Keynesians, they believe that the economy may be working at less than full employment, that is, in the short run there may prevail excess capacity and unemployment of labour so that expansion in money supply and consequent increase in nominal income partly induces expansion in real income (Y) and partly results in rise in the price level as shown in Fig. 23.5. To what extent price level increases depends upon the elasticity of supply or aggregate output in the short run. It will be seen from Fig. 23.5 that effect of increase in money supply from M_0 to M_1 and resultant increase in aggregate demand curve for goods and services from AD_0 to AD_1 is split up into the rise in price level (from P_0 to P_1) and the increase in real income or aggregate output (from Y_0 to Y_1).

It should be noted that Friedman and other modern quantity theorists believe that in the short run full employment of labour and other resources, may not prevail due to recessionary conditions and, therefore, they admit the possibilities of increase in output. But they emphasise that in the short run when the growth in money supply is greater than the growth in output, the result is excess demand for goods and services which causes rise in prices or demand-pull inflation.

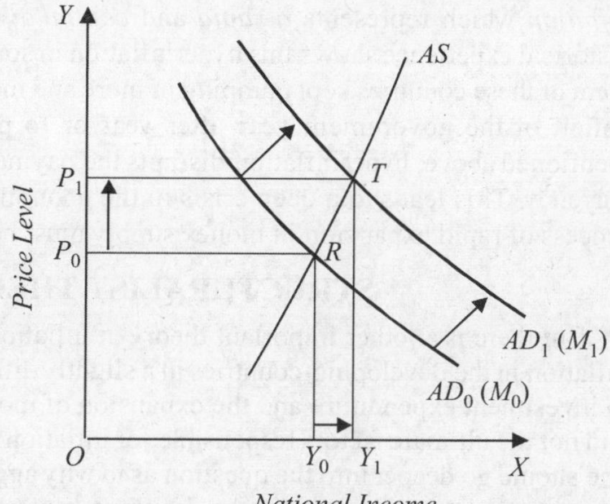

Fig. 23.5. *The Effect of expansion in money supply in the short run is split up into price rise and increase in Real National Income: Friedman's Monetarist Approach.*

It follows from above that both Friedman and Keynesians explain inflation in terms of excess demand for goods and services. Whereas Keynesians explain the emergence of excess demand due to the increase in *autonomous expenditure,* independent of any increase in money supply. Friedman explains that inflation is caused by *proportionately greater increase in money supply than the increase in aggregate output.* In both views inflation is of demand-pull variety.

Money and Sustained Inflation

Many economists believe in the monetarist view of inflation. Increase in money shifts the aggregate demand curve to the right and if the economy is operating at full capacity (*i.e.* along the vertical part of the aggregate supply curve), the upward shift in aggregate demand curve will cause price level to rise. A big drawback of this approach is that it assumes that supply of output does not increase sufficiently to counter this effect of expansion in money supply on aggregate demand. In this context there is a need to distinguish between a *one-time increase* in the price level and *sustained inflation* which occurs when the general price level continues to rise over a long period of time. It is generally believed by most of the economists that whatever be the initial cause of inflation (demand-pull, cost-push or inflationary expectations), for the price level to continue rising, period after period, it must be *accommodated* by expansion in money supply. *Sustained inflation is therefore considered as a purely monetary phenomenon.* It is not possible for the price level to continue rising if the money supply remains constant. The increase in money supply continues shifting the aggregate demand curve to the right; if aggregate supply does not increase sufficiently to match the increase in aggregate demand, price level will continue rising.

Sustained inflation can be better understood when Government increases its expenditure without raising taxes. This leads to the increase in aggregate demand which, *aggregate supply remaining constant*, will cause a rise in price level. It is important to know what happens when the price level rises. The higher price level raises the demand for money to rise for transactions purposes. With supply of money remaining constant, the greater demand for money causes interest rate to rise. The rise in interest rate *crowds out* private investment. If the Central Bank of a country wants to prevent the fall in the private investment, it will expand the money supply to keep the interest constant. But this expansion in money supply through its effect on aggregate demand will cause the price level to rise further if increase in more supply of output is not possible. This further rise in price level will again cause greater demand for money leading to higher interest rate. And the Central Bank, it is committed to keep the interest rate constant so that private investment does not decline, if will further expand the money supply which will cause further inflation. This process could lead to *hyper inflation* which represents *a rapid* and *continuous rise in price level*, period after period. The historical experience shows this hyperinflation in some countries when the Central Bank or Government of these countries kept pumping in more and more money either to finance its persistent budget deficit of the government year after year or to prevent the interest rate to rise. However, as mentioned above, hyperinflation disrupts the payment system and people's loss of credibility of the currency. This leads to a deep crisis in the economy. If hyperinflation is to be avoided, then the process of rapid expansion in money supply must be halted.

STRUCTURALIST THEORY OF INFLATION

But there is another important theory of inflation known as *structural inflation* which explains inflation in the developing-countries in a slightly different way. The structuralists argue that increase in investment expenditure and the expansion of money supply to finance it are the only proximate and not the ultimate factors responsible for inflation in the developing countries. According to them, one should go deeper into the question as to why aggregate output, especially of foodgrains, has not been increasing sufficiently in the developing countries to match the increase in demand brought about by the increase in investment expenditure, and money supply. Further, they argue why investment expenditure has not been fully financed by voluntary savings and as a result excessive deficit financing has been done.[5]

Structural theory of inflation has been put forward as an explanation of inflation in the developing countries especially of Latin America. The well-known economists, Myrdal and Streeten who have proposed this theory have analysed inflation in these developing countries in terms of structural features of their economies. Recently Kirkpatrick and Nixon have generalised this structural theory of inflation as an explanation of inflation prevailing in all developing countries.[6]

Myrdal and Streeten have argued that it is not correct to apply the highly aggregative demand-supply model for explaining inflation in the developing countries. According to them, there is a lack of balanced integrated structure in them where substitution possibilities between consumption and production and inter-sectoral flows of resources between different sectors of the economy are not quite smooth and quick so that the inflation in them can not be reasonably explained in terms of aggregate demand and aggregate supply. In this connection it is noteworthy that Prof. V.N. Pandit of Delhi School of Economics has also felt the need for distinguishing price behaviour in the Indian agricultural sector from that in the manufacturing sector.[7]

Thus, it has been argued by the exponents of structural theory of inflation that economies of the developing countries of Latin America and India are structurally underdeveloped as well as highly fragmented due to the existence of market imperfections and structural rigidies of various types. The result of these structural imbalances and rigidies is that whereas in some sectors of these developing countries, we find shortages of supply relative to demand, in others under-utilisation of resources and excess capacity exist due to lack of demand. According to structuralists, these structural features of the developing countries make the aggregate demand-supply model of inflation inapplicable to them. They therefore argue for analysing disaggregative and sectoral demand-supply imbalances to explain inflation in the developing countries. They mention various sectoral constraints or bottlenecks which generate the sectoral imbalances and lead to rise in prices. Therefore, to explain the origin and propagation of inflation in the developing countries, the forces which generate these bottlenecks or imbalances of various types in the process of economic development need to be

5. G.Myrdal, *Asian Drama, An Enquiry into the Poverty of Nations*, 1968, Penguin Books.
6. The expansionary monetary and fiscal policies adopted in response to recessionary conditions are called *accomadating polcies.*
7. P. Streeten, *The Frontiers of Development Studies*, 1972, London, MacMillan.

analysed. A study of these bottlenecks is therefore essential for explaining inflation in the developing countries. These bottlenecks are of three types : (1) Agricultural bottlenecks which make supply of agricultural products inelastic, (2) resources constraint or Government budget constraint, and (3) foreign exchange bottleneck. Let us explain briefly how these structural bottlenecks cause inflation in the developing countries.

Agricultural Bottlenecks. The first and foremost bottlenecks faced by the developing countries relate to agriculture and they prevent supply of food grains to increase adequately. Of special mention of the structural factors are disparities in land ownership, defective land tanure system which act as disincentives for raising agricultural production in response to increasing demand for them arising from increase in people's incomes, growth in population and urbanisation. Besides, use of backward agricultural technology also hampers agricultural growth. Thus, in order to control inflation, these bottlenecks have to be removed so that agricultural output grows rapidly to meet the increasing demand for it in the process of economic development.

Resources Gap or Government's Budget Constraint. Another important bottleneck mentioned by structuralists relate to the lack of resources for financing economic development. In the developing countries planned efforts are being made by the Government to industrialise their economies. This requires large resources to finance public sector investment in various industries. For example, in India, huge amount of resources were used for investment in basic heavy industries started in the public sector. But socio-economic and political structure of these coun tries is such that it is not possible for the Government to raise enough resources through taxation, borrowing from the public, surplus generation in the public sector enterprises for investment in the projects of economic development. Revenue raising from taxation has been relatively very small due to low tax base, large scale tax evasion, inefficient and corrupt tax administration. Consequently, the government has been forced to resort to excessive deficit financing (that is, creation of new currency) which has caused excessive growth in money supply relative to increase in output year after year and has therefore resulted in inflation in the developing countries. Though rapid growth of money supply is the proximate cause of inflation, it is not the proper and adequate explanation of inflation in these economies. For proper explanation of inflation one should go deeper and enquire into the operation of structural forces which has caused excessive growth in money supply in these developing economies. Besides, resources gap in the private sector due to inadequate voluntary savings and under-development of the capital market have led to their larger borrowings from the banking system which has created excessive bank credit for it. This has greatly contributed to the growth of money supply in the developing countries and has caused rise in prices. Thus, Kirkpatrick and Nixon write, "the increase in the supply of money was a permissive factor which allowed the inflationary spiral to manifest itself and become cumulative—it was a system of the structural rigidities which give rise to the inflationary pressures rather than the cause of inflation itself."[8]

Foreign Exchange Bottleneck. The other important bottleneck which the developing countries have to encounter is the shortage of foreign exchange for financing needed imports for development. In the developing countries ambitious programme of industrialisation is being undertaken. Industrialisation requires heavy imports of capital goods, essential raw materials and in some cases, as in India, even foodgrains have been imported. Besides, imports of oil on a large scale are being made. On account of all these imports, import expenditure of the developing countries has been rapidly increasing. On the other hand, due to lack of export surplus, restrictions imposed by the developing countries, relatively low competitiveness of exports, the growth of exports of the developed countries has been sluggish. As a result of sluggish exports and mounting imports, the developing countries have been facing balance of payment difficulties and shortage of foreign exchange which at times has assumed crisis proportions. This has affected the price level in two ways. First,

---

8. C. Kirkpatrick and F., Nixon, 1987 op. cit.

due to foreign exchange shortage domestic availability of goods in short-supply could not be increased which led to the rise in their prices. Secondly, in Latin American countries as well as in India and Pakistan, to solve the problem of foreign exchange shortage through encouraging exports and reducing imports devaluation in the national currencies had to be made. But this devaluation caused rise in prices of imported goods and materials which further raised the prices of other goods as well due to cascading effect. This brought about cost-push inflation in their economies.

Physical Infrastructural Bottlenecks. Further, the structuralists point out various *bottlenecks* such as lack of infrastructural facilities *i.e.*, lack of power, transport and fuel which stands in the way of adequate growth in output. At present in India, there is acute shortage of these infrastructural inputs which are hampering growth of output. Sluggish growth of output on the one hand, and excessive growth of money supply on the other have caused what is now called *stagflation*, that is inflation which exists along with stagnation or slow economic growth.

According to the structuralist school of thought, the *above bottlenecks and constraints are rooted in the social, political and economic structure of these countries*. Therefore, in its view a broad-based strategy of development which aims to bring about social, institutional and structural changes in these economies is needed to bring about economic growth without inflation. Further, many structuralists argue for giving higher priority to agriculture in the strategy of development if price stability is to be ensured. Thus, we see that structuralist view is greatly relevant for explaining inflation in the developing countries and for the adoption of measures to control it. Let us further elaborate the causes of inflation in the developing countries.

INFLATION IN DEVELOPING COUNTRIES: DEMAND-PULL OR COST-PUSH INFLATION

Now, the question arises as to which theory of inflation can explain inflation in developing countries. Of course, the rise in prices has come about as a result of excess of aggregate demand over aggregate supply. In other words, inflation in the developing countries is mainly of demand-pull variety. However, how this excess demand for goods and services has been caused is issue at dispute. In our view both the Keynesian and Friedman's views are relevant to explain the emergence of excess demand for goods. To promote rapid economic growth a hunge investment expenditure has been made in India in successive development plans. If this increase in investment expenditure had been financed by raising resources through taxation, Government borrowings from the public, profits of public undertakings, then the increase in investment expenditure would have been matched by increase in savings with the result that excess demand for goods had not arisen. As a matter of fact, a good deal of increase in investment expenditure has been financed by deficit financing *i.e.*, through creating of new money by the central bank. Besides, the increase in private investment expenditure has also been financed to a good extent by expansion of credit or demand deposits by the commercial banks. This increase in investment expenditure for promoting economic growth has been made possible by the expansion in money supply, which as emphasized by Friedman and other monetarists, causes prices to rise through creating excess demand for goods and services. Thus, it is hard to distinguish whether

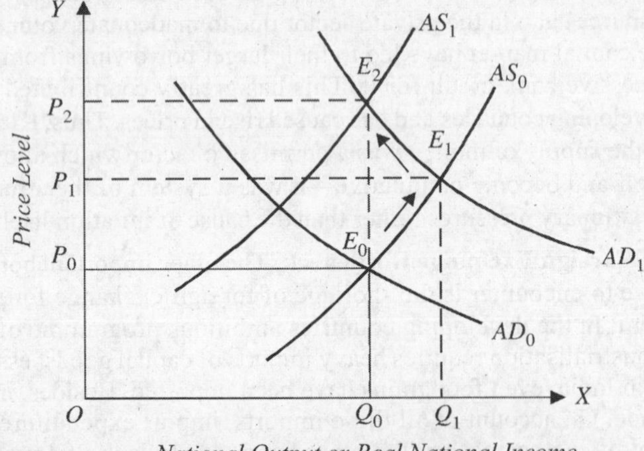

Fig. 23.6. *Rise in General Price Level as a Result of Joint Impact of Demand-Pull and Cost-Push Factors*

it is increase in investment expenditure as such or expansion in money supply to finance it which has caused demand-pull inflation in the face of slow growth of output of goods, especially foodgrains and other essential consumer goods. Indeed, both have played a part in causing inflationary situation in India.

What has caused the inflationary pressures in India's developing economy? To think that only demand-pull factors or excess demand generated by huge budget deficits and huge growth in money supply are responsible for the problem of inflation facing the Indian economy would not be fully correct. As a matter of fact, both types of factors, namely, demand-pull and cost push factors have operated to cause inflation in India, which crossed the two-digit Fig. in some years. As a result of rapid growth in Government expenditure without corresponding increase in mobilisation of resources, the Government has resorted to heavy deficit financing (*i.e.*, the creation of new money). This has further created the conditions of excess demand in the economy which have led to the rise in general price level. Besides, the increase in petroleum prices, many times during the last few years, the hike in administered prices of steel, cement, coal, fertilizers, increase in indirect taxes on several commodities, increase in railway fares and freights have all led to the *cost-push inflation* or what is also called *supply-side inflation*.

Now, when due to the demand pull inflation, the general price-level rises, the demand for rise in wages and dearness allowance are made by the working classes and they have to be conceded to in view of the rising cost of living. The rise in wages and dearness allowance of the workers raises the cost of production. The increases in cost of production due to higher wages and dearness allowance also cause the aggregate supply curve to shift to the left and brings about cost-push or supply-side inflation. Thus there has been persistent rise in prices under the combined impact of demand-pull factors caused by the large budget deficits and cost-push factors due to the hike in administrated prices of petroleum, steel, cement, coal etc., and increase in indirect taxes on commodities such as sugar, cloth, cooking gas, soaps, cold-drink, etc.

Let us illustrate graphically how the operation of the various demand-pull and cost-push factors resulting in persistent rise in general price level. Consider Fig. 23.6, where AD_0 represents aggregate demand curve and AS_0 the aggregate supply curve and the two intersect at point E_0 and the price level P_0 is determined. With huge fiscal deficits and large expansion in money supply generating conditions of excess demand cause upward shift in the aggregate demand curve to AD_1. The general price level rises to P_1 corresponding to the new equilibrium point E_1. Now, the increase in cost of production brought about by either independently of the rise in price caused by increase in aggregate demand or by rise in wages and dearness allowances induced by the demand-pull inflation in the first instance, aggregate supply curve shifts to the left to AS_1. With this the price level further rises to P_2 under the influence of cost-push factors. The price equilibrium moves from E_0 to E_1 and then to E_2. This process of joint operation of demand-pull and cost-push factors has been going on year after year and is responsible for the persistent rise in prices in the Indian economy. Therefore, to check and moderate this inflationary process not only the fiscal deficits have to be drastically cut but also steps should be taken to prevent the increase in cost of production by avoiding the frequent hike in administered prices and indirect taxes.

It is worth noticing that it is the food prices which first start rising rapidly in a developing economy. Rise in food prices is then followed by the rise in the prices of other consumers' goods. This is so because a greater part of the increase in demand generated by the investment expenditure is spent on the foodgrains such as wheat and rice and further the supply of foodgrains cannot be sufficiently increased in the short run due to the failure of monsoons, lack of irrigation facilities, HYV seeds, fertilizers and inefficient techniques of cultivation. Besides, in India's five year plans adequate investment in agriculture has not been made and as a result, growth rate in agriculture has been low. On the other hand, *income elasticity of demand for food is very high because the vast majority of the people are undernourished. Thus, as a result of huge investment expenditure by the government and private sector there is a sharp rise in demand for foodgrains, leading to increases in food prices.*

In developing countries like India, which are predominantly agricultural, the prices of agricultural commodities, especially of foodgrains, hold the key position in the price structure of the country. Any rise in agricultural prices leads to a distortion in the whole price structure. A steep rise in food prices increases the cost of living of the people. Consumers are hit hard as their incomes do not increase so easily to offset the increase in prices. Workers whose cost of living rises press for higher wages. When wage increases are conceded to, the cost of production of manufactured articles rises and this, in turn, increases their prices, and so on. Moreover, some agricultural products are raw materials for industries and increase in their prices will directly increase the cost of production of industrial goods. Hoarding of and speculation in both agricultural and industrial products add fuel to the inflationary fire. Thus *once the prices of agricultural goods rise they are likely to cause inflationary spiral in the economy*.

A factor which deserves special mention in this connection is the *mode of financing development plans*. The developing countries, especially India, are not in a position to finance their plans fully through voluntary savings of the people and taxes by the government. They have often made resort to fiscal deficit (*i.e.*, borrowing by the Government from the Central Bank and Commercial banks) as a method of financing their development plans. In the past in India up till 1996, the Government in India borrowed heavily from the Central Bank (i.e. RBI) which often led to the large increase in high powered money. Increase in the high-powered money as a result of borrowing from RBI, which was then called *deficit financing was highly inflationary*.

Deficit financing which is now called *money financing* or *monetisation of fiscal deficit* to some extent is good and can be absorbed by the economy without experiencing inflation. This is so because as the economy grows, its monetary sector expands and also there is an increase in production as a result of economic growth for which extra money is needed. But owing to acute shortage of finance the developing countries had often indulged in deficit financing to an excessive degree. Sharp increase in the money supply with the public as a result of excessive deficit financing added greatly to the level of aggregate demand for consumers' goods. On the other hand, the supply of consumers' goods, especially of food products, could not be increased rapidly and sufficiently. The pressure of demand, therefore, led to the inflationary rise in prices. In recent years in India (after 1996) the Government does not generally borrow directly from RBI but to finance its fiscal deficit, it borrows from commercial banks through sale of its bonds to them. A high degree of fiscal deficit increases greatly government expenditure and leads to excess demand conditions which exert inflationary pressures in the economy.

It may, however, be pointed out that investment expenditure made by the government under the development plans not only generates the additional demand for goods, it also increases the productive capacity. *Investment has a dual effect. On the one hand, it generates demand or income, on the other it increases the productive capacity*. As a result of increase in productive capacity, more output of goods can be obtained which will counteract inflationary tendencies. But, in the earlier stages of development, investment expenditure is largely made on huge dams, steel plants and other heavy and basic industries which have long-gestation periods. In other words, the long-term projects can help in increasing the supply of consumers' goods only in the long run. In the short-run, prices generally shoot up under the pressure of excessive demand for goods. And once inflationary spiral starts operation, it is difficult to control it. However, if increase in Government expenditure takes place on financing its consumption expenditure which does not lead to any durable productive assets, it has a large inflationary potential

It is often said that inflation in India is due to the emergence of imbalance between demand and supply. As we have said above, increase in aggregate demand is due to the increase in Government expenditure, which has been financed largely by either by money financing or borrowing from the market. Besides, commercial banks have also created a good amount of bank money or credit to finance private investment. As a result, the aggregate expenditure increased very much and has brought about a huge expansion in aggregate demand. On the other hand, the production of foodgrains and other consumer goods industries did not increase adequately and this brought about a rapid rise in prices.

It may however be noted that *strategy of economic development*, adopted in our Five Year Plans, prior to 1991 gave high priority to basic heavy industries and relatively neglected agriculture and consumer goods industries. This is the structural reason why output of foodgrains and other essential consumers goods did not increase adequately to meet the rising demand for them resulting in inflation. After 1991 economic reforms undertaken also gave higher priority to industrialisation and did not accord due importance to agriculture. As a result, there has been slower growth of agriculture even after 1991.

SOCIAL COSTS OF INFLATION

Having discussed the theories of inflation we proceed to explain in detail the social cost and effects of inflation. Apart from reducing the purchasing power of people's incomes, inflation inflicts some other costs on the society. To explain such costs of inflation it is necessary to distinguish between anticipated inflation and unanticipated (i.e. unexpected) inflation. As noted above, in case of anticipated inflation, the expected rise in price level is taken into account while making economic transactions, for example, in negotiating wage rate of labour etc. Above all in case of India inflation hurts the poor people most.

Costs of Anticipated Inflation

Suppose in an economy there has been annual inflation rate of 5 per cent for a long time in the past and every body expects that this 5 per cent rate of inflation will continue in the future too. In such a case all contracts made by the people such as loan agreements with borrowers, wage contracts with labour, property lease contracts will provide for 5 per cent annual rise in rates of interest, wages, rent to compensate for inflation of that order. That is, in any contract in which passage of time is involved 5 per cent rate of inflation will be taken into account and rates will be agreed to rise per period equal to the anticipated rate of inflation. If rates of interest, wages, rent etc are agreed to rise at the anticipated rate of inflation, then there will be no cost of inflation except the following two types of costs – shoe-leather costs and menu costs which are not very high. We explain below both these types of costs.

1. **Shoe-leather Costs.** This type of cost occurs because on account of inflation cost of holding money in the form of currency *i.e.*, notes and coins, rises with the increase in inflation rate. Such cost arises because no interest is paid on holding currency, while money kept in deposits with the bank or used for keeping bonds earns interest. When inflation rate rises, the nominal interest rate on bank deposits rises, the interest lost by holding currency by the people therefore increases. In order to reduce the cost of holding currency people will tend to reduce their holdings of currency for transaction purposes. Accordingly, at a time people will hold less currency with them and keep as long as possible greater amount of money in bank deposits that yield interest. Therefore, rather than withdrawing a large amount of currency from banks at a time, they will withdraw less money which is sufficient for meeting daily expenses for a few days, say for a week. But for doing so the people will make more trips to withdraw cash. More trips to a bank in a month involves greater cost to the people. These costs have to be incurred on spending on petrol if car is used for making trips, more wear and tear of car, the time spent for making a trip. These costs of making more trips to the bank for withdrawing currency is metaphorically called *shoe-leather costs* of inflation, as walking to banks more often, one's shoes wear out more rapidly and one has to spend money on new shoes more often.

2. **Menu Costs.** The second type of anticipated inflation is menu costs, a term derived from a restaurant's cost of printing a new menu. Menu costs arise because high inflation requires them to change their listed prices more often. Changing prices is somewhat more expensive because the firms have to print new catalogues listing new prices and distribute them among their customers. They have even to incur expenditure on advertisements to inform the public about their new prices.

3. **Macroeconomic Inefficiency in Resource Allocation.** A third cost of inflation arises because firms having menu costs change their prices quite infrequently. Given the reluctance to change prices frequently, the higher the rate of inflation, the greater the variability in relative prices of a firm. Suppose a firm issues a new catalogue listing prices of its products once in a year, say in the month of January of every year. If during the year inflation occurs, there will be change in the relative prices of a firm to the general price level. If inflation rate of one per cent per month lakes place in a year the firm's relative prices to the general price level will *fall* by 12 per cent by the end of the year. As a result, his sales will tend to be lower in the early part of the year (when its prices are relatively high) and higher in the later part of the year (when its prices are relatively low). Thus when due to inflation relative prices of a firm vary during a year as compared to the overall price level, it causes destortion in production and therefore leads microeconomic inefficiencies in resource allocation.[9]

4. **Inconvenience of Living.** Lastly, another social cost of inflation is the *inconvenience of living* in a world with a changing price level. Money is the yardstick with which we measure the value of transactions. When inflation is taking place the value of money changes and as a result it becomes difficult to correctly estimate the value of transactions in real terms every time a transaction is made during a year. The rising price level makes it difficult to make optimal decisions about saving and investment and thus do the rational financial planning covering a long period of time. To quote Mankiw, "A dollar saved today and invested at a fixed nominal interest rate will yield a fixed dollar amount in the future. Yet the real value of that dollar amount – which will determine the retiree's living standard – depends on the future price level. *Deciding how much to save would be much simpler* if people could count on the price level in 30 years being similar to its level today."[10]

Conclusion. Taking account of all costs of anticipated inflation one finds that the costs of anticipated inflation are quite small or trivial and if these alone are considered, it is then surprising why inflation is a matter of serious concern for the policy makers and politicians. In our opinion the above view of costs of inflation does not consider the true cost of inflation which, as mentioned above, refers to the reduction in purchasing power or real-incomes of the people which lowers their standard of living. Besides, there is ample cross-country evidence that *high rates of inflation lead to low rate of sustanced economic growth* [11].

It may be further noted that in the above analysis of cost of inflation it is assumed that there is only small to moderate inflation rate, say a single digit rate of inflation occurs so that it does not disrupt the payment system. With such a low to moderate inflation, costs of inflation are small. The hyperinflation has more harmful effect as it disrupts the payment system which leads to the collapse of the economy.

Social Cost of Unanticipated Inflation

Unanticipated inflation has a more substantial and harmful effect as compared to the cost of anticipated infllation rate. The significant effect of unanticipated inflation is that it *arbitrarily redistributes wealth* among individuals. Consider the value of assets fixed in nominal terms. Between 1995 and 2006, price level in India rose by about 100 per cent. This implies that those who held *claims on assets fixed in nominal terms in 1996*, their real value in terms of purchasing power would have declined significantly. Thus, a person who bought government bound of 10 years maturity with a face value of ₹ 1000 bearing 8 per cent nominal interest rate in 1996 will find that

---

9. Mankiw, *op. cit*., p. 98.
10. Stanley Fischer, Macroeconomic Factors in Economic Growth, *Journal of Monetary Economics*, Dec. 1993.
11. See D. Felix, "Profit Inflation and Industrial Growth, "*Quarterly Journal of Economics, August* 1956, pp. 441-63 and E.J. Hamilton, Prices as a Factor in Business Growth, *Journal of Economic History*, December 1952, pp. 325-49.

₹ 1000 he gets bank in 2006 has far less value than when he purchased the bond in 1996.

Similarly, unanticipated inflation harms the individuals, who retire on pensions fixed in rupee terms. After some years of inflation, the real value or purchasing power of the fixed nominal pension will greatly decline and will therefore reduce his standard of living in his old age. Thus inflation hurts individuals with fixed pensions. Workers and the private firms often agree on fixed nominal pension payable to the workers after retirement. Workers are greatly harmed when inflation is higher than anticipated. Likewise, higher than anticipated inflation rate hurts the creditors who give loans to the others and get back the principal amount after the stipulated period. Thus *inflation redistributes wealth in favour of debtors*.

EFFECTS OF INFLATION

Inflation is a very unpopular happening in an economy. Opinion surveys conduced in India, the U.S.A. and other countries reveal that inflation is the most important concern of the people as it badly affects their standard of living. The political fortunes of many political leaders (Prime Ministers and Presidents) and Governments in India and abroad have been determined by how far they have succeeded in tackling the problem of inflation. So much so that some American presidental candidates called '*inflation as enemy number one*'. Same is the case in India where inflation is the most hotly debated issue during the general elections for the Parliament and Assemblies. A high rate of inflation makes the life of the poor very miserable. It is, therefore, described as anti-poor. It redistributes income and wealth in favour of some and greatly harms others. By making the rich richer and the poor poorer, it militates against social justice. Besides, inflation lowers national output and employment and impedes long-run economic growth, especially in developing countries like India. We shall discuss below all these effects of inflation.

Anticipated and Unanticipated Inflation

The difference between anticipated inflation and unanticipated inflation is of crucial importance as the effects of inflation, especially its redistributive effect, depend on whether it is anticipated or not. If rate of inflation is anticipated, then people take steps to make suitable adjustments in their contracts to avoid the adverse effects which inflation could bring to them. For example, if a worker correctly anticipates the rate of inflation in a particular year to be equal to 10 per cent and if his present wage rate is ₹ 5000 per month, he can enter into contract with the employer that to compensate for the 10 per cent rise in prices his money wage per month next year be raised by 10 per cent so that next year he gets ₹ 5500 per month. In this way he has been able to prevent the erosion of his real income with the automatic revision of his money wage depending on the anticipated rate of inflation.

Take another example. You lend ₹ 10,000 to a person at a rate of 10 per cent per annum. After a year you will receive ₹ 11,000. But if it is anticipated that during the year there will be 8 per cent rate of inflation, then 8 per cent of your income will be offset by the rise in prices that would occur so that you will get only 2 *per cent real rate of interest*. Therefore, in order to receive 10 per cent real rate of interest, in view of 8 per cent anticipated inflation rate you must demand 18 per cent nominal rate of interest.

On the other hand, effects of unanticipated inflation are unavoidable because in this case you do not know what would be the rise in the price level. That is, unanticipated inflation catches you by surprise. In what follows we shall examine the *effects of unanticipated inflation*. The effects of inflation can be divided into three categories:

1. Effect on real income;
2. Effect on distribution of income and wealth;
3. Effect on output; and
4. Effect on long-run economic growth.

Inflation Erodes Real Incomes of the People

To examine the effect of inflation it is important to note the difference between money income and real income. It is the change in the general price level that creates the crucial difference between the two. Money income or what is also called nominal income means the income such as wages, interest, rent *received in terms of rupees.* On the other hand, *real income* implies the amount of goods and services which you can buy. In other words, real income means the purchasing power of your income. If your money or nominal income increases at a lower rate than the rate of rise in the general price level (*i.e.*, the rate of inflation), you will be able to buy less goods and services, that is, your real income will decline. Real income will rise only if nominal income rises faster than the rate of inflation. For illustration, take the case of workers who enter into contract with their employer at an agreed wage rate of ₹ 5000 per month for the period, say, 5 years. Now, suppose the rate of inflation is 10 per cent per annum. This means after a year, with money wage rate of ₹ 5,000 workers will be able to buy less goods and services. That is, their real income will decrease and therefore their standard of living will fall.

Take another example. Suppose you deposit your saving of ₹ 100 in a savings account which carries 5 per cent rate of interest. After a year you will receive ₹ 105. However, if during that year rate of inflation has been 12 per cent, you will be a loser in real terms. In fact your real interest-income will be negative as, with 12 per cent rate of inflation, ₹ 105 after a year will buy less goods and services than what you can purchase with ₹ 100 today.

The above two examples clearly show that inflation reduces the purchasing power of money and thereby adversely affects real income of the people.

Effect on Distribution of Income and Wealth

An important effect of inflation is that is redistributes income and wealth in favour of some at the cost of others. Inflation adversely affects those who receive relatively fixed incomes and benefits businessmen, producers, traders and others who enjoy flexible incomes.[13] Inflation brings windfall profits for the producers and traders. Thus, all do not lose as a result of inflation, rather some gain from it. We examine below how inflation redistributes income and wealth and thereby harms some people and benefits others.

Creditors and Debtors. Unanticipated inflation harms creditors and benefits debtors and in this way redistributes income in favour of the latter. As explained above, value of money declines due to inflation. For creditors (including financial institutions such as banks and insurance companies) who enter into agreement with the borrowers to provide loans at fixed nominal rate of interest, the real value of money in terms of goods and services which they will receive at the end of the period would be much less if during the period prices rise sharply. Thus, the debtors or borrowers gain because they would return the loan-money when its real value has declined greatly due to the unexpected rapid rate of inflation.

Fixed Income Groups. Those who get fixed incomes stand to lose from inflation. Workers and salaried people who earn fixed wages and salaries are hit hard by unanticipated inflation. These people often enter into contract with the employers regarding wages or salaries fixed in nominal terms. When inflation occurs, the purchasing power of their nominal incomes falls greatly causing a decline in their levels of living. Thus, when inflation persists for some years there are demands for revision of wages and salaries. It may be mentioned that now-a-days workers and other salaried people get dearness allowances to compensate them for the rise in cost of living due to inflation. However, these dearness allowances do not fully neutralise the rise in price level and therefore they also demand revision of wages and pay scales.

Pensioners. They also come in the category of the people who get income in fixed nominal

13. Robert J. Gordon, *op. cit.* p.337

terms. For the people who retired in 1984 with the monthly pension of ₹ 2000, the real value of their pension in Oct. 1998 would been reduced to one third as compared to 1984 as there has been more than 300 per cent rise in the price level during this period. It may also be noted that in order to reduce the hardship of the pensioners, some dearness allowance is also provided on pension. But effects of inflation on the real value of their pension are only partly offset in this way.

Businessmen : Producers and Traders. Businessmen, that is, entrepreneurs and traders, stand to gain by inflation. During periods of inflation, the prices of goods produced by entrepreneurs rise relatively faster than the cost of production because wages lag behind the rise in prices of goods. Consequently, *inflation increases the profits of businessmen.* The value of the inventories or stocks of goods and materials kept by the entrepreneurs and traders increases due to rise in prices of goods which brings about an increase in their profits.

Wealth Holders of Cash, Bonds and Debentures. Inflation also adversely affects wealth holders who hold their wealth in the form of cash money, demand deposits, saving and fixed deposits and interest-bearing bonds and debentures. These wealth holders are severely hurt by inflation as inflation reduces the real value of their wealth. Savings and demand deposits, bonds and debentures represent assets whose value is fixed in terms of money. The rise in prices reduces the purchasing power of these fixed-value money assets such as savings and time deposits, bonds and debentures which bear a fixed nominal rate of interest.

Inflation, therefore reduces the real rate of interest earned by them. Consequently, it has been observed that during periods of rapid inflation people try to convert their holdings of money and near money into goods and physical property so as to avoid the loss due to inflation. It may also be noted that if inflation is anticipated and all expect equal rates of inflation the nominal rates of interest are adjusted upward so as to obtain targeted real rate of interest. Thus, if creditors want real rate of interest equal to 10 per cent and anticipate rate of inflation is equal to 8 per cent, they will try to have nominal rate of interest fixed at 18 per cent. This is known as **Fisher effect** *which states that market or nominal rate of interest is equal to the real rate of interest (based on productivity of capital and rate of time preference) plus the anticipated rate of inflation.* Thus nominal rate of interest includes what is called *inflation premium* to prevent the erosion of purchasing power due to inflation.[14]

Effect of Inflation on Output

There is a good deal of uncertainty and also disagreement as to whether inflation will adversely or favourably affect national output. The effect of inflation on output also depends on whether it has been caused by demand-pull or cost-push factors. Further, the effect of inflation on output depends on whether it is moderate or very rapid or whether it is anticipated or unanticipated. Let us examine the likely effects of inflation in each of these cases.

Demand-pull Inflation and Output. Till recently a large number of economists favoured a mild or modest inflation and they argued that rising prices are caused by increase in aggregate demand which are accompanied by expansion in output. Indeed, according to them, rising prices create a tonic effect on the level of investment and employment of labour and other resources by raising the marginal efficiency of capital (*i.e.*, expected rate of profits). We know the level of output and employment depend on aggregate demand, given the aggregate supply curve. In a previous chapter we have explained that short-run aggregate supply curve has three ranges: in range 1 aggregate supply curve is horizontal, in range 2 when the economy is near full employment level or potential level of output, aggregate supply curve mildly slopes upward and in range 3 aggregate supply curve becomes vertical. To start with, aggregate demand is low. For instance, if in Fig. 23.7 it is AD_0, the economy will produce level of output Y_0 and will be far below the full-employment

14. Robert Gordon, *Op. cit.*, P. 337

level of output Y_3. Now, if aggregate demand increases to AD_1, the price level will remain stable and output will rise to Y_1. But if aggregate demand further increases to AD_2, it will intersect the AS curve in range 2, where it is sloping upward mildly. In this range 2, price level will increase but output and employment will also increase. Even if aggregate demand further increases upto AD_3, aggregate output increases to Y_3 and is accompanied by rise in the price level to P_2. Thus, in range 2 of aggregate supply curve until the full-employment output Y_3 is reached, increase in aggregate demand causes moderate inflation and also raises aggregate output and employment. It is only when aggregate demand increases beyond full-employment output that it leads to a higher rate of inflation without affecting output. Thus, so far as economy finds itself in range 2, moderate inflation has to be tolerated if higher levels of output and employment are to be achieved. Therefore, some economists have argued that there exists trade-off between output and inflation or between employment and inflation. Some inflation has to be accepted if you want higher output (and therefore less unemployment). However, this viewpoint has been criticised in recent years and it has been asserted that any trad-off between rate of inflation and output (or unemployment) is only a short-run phenomenon and there is no such trade-off in the long run.

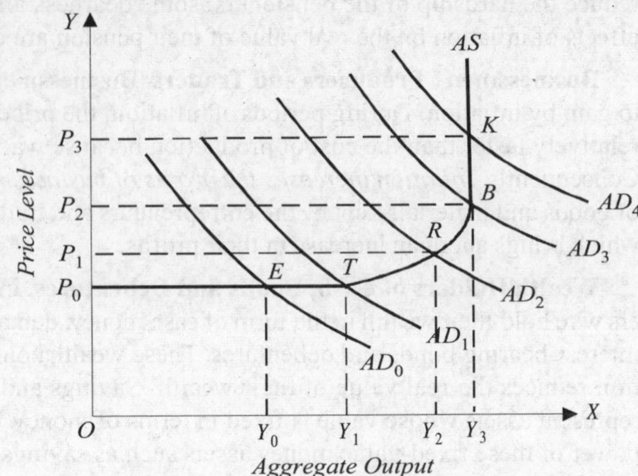

Fig. 23.7. *Effect of Inflation on Output*

Cost-Push Inflation and Output. But, as seen above, the economic events in recent years have thrown up another type of inflation which is generally described as cost-push inflation which is caused by leftward shift in the aggregate supply curve due to rise in prices of crucial inputs such as petroleum oil, wages of labour etc. In the cost-push inflation, rise in price level is associated with a fall in aggregate output as will be seen from Fig. 23.8. In the early years of 1970s and again in 1979–80 when oil price shock given by *OPEC* created cost-push inflation, prices rose but at the same time output fell and unemployment went up. Such a situation has been described as *stagflation* which implies occurrence of inflation with static or lower output and higher unemployment.

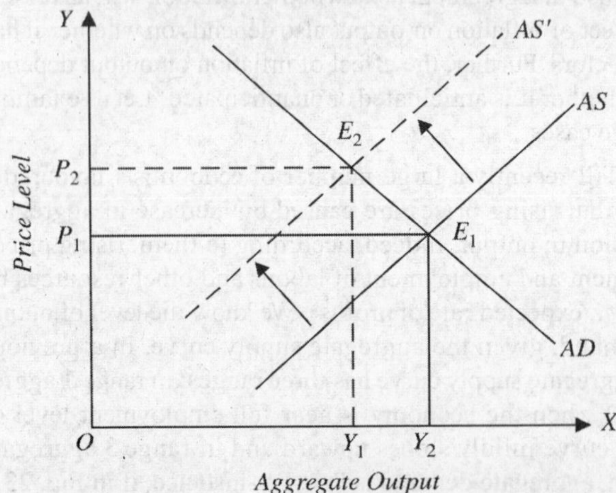

Fig. 23.8. *Effect of Cost-Push Inflation an Output*

From the above it is clear that there is no necessary relationship between inflation and output. Inflation may come about with either a higher or a lower level of output and employment. To quote Samuelson, "*Today, macroeconomists believe that there is no necessary relationship between prices and output. An increase in aggregate demand will increase both prices and output, but a supply shock shifting up aggregate supply curve will raise prices and lower output.*"[12]

EFFECTS OF INFLATION ON LONG-RUN ECONOMIC GROWTH

Some Economists have argued that inflation of a creeping or mild variety has a tonic effect on the long-run economic growth. In their support they give the example of today's industralised countries in the Eighteenth and Nineteenth Centuries when the rate of growth of output had been more rapid during long periods of inflation witnessed in these countries. The driving force in the process of economic growth, according to them, has been *high profit margins* created by inflation. They argue that wages lag behind the rise in general price level and thus creating higher profit margins for businessmen and industrialists. This tends to increase the profit share in national income. The businessmen and industrialists who receive profits as income belong to the upper income brackets whose propensity to save is higher as compared to the workers. As a result, savings go up which ensures higher rate of investment. With greater rate of investment more accumulation of capital is made possible. More rapid capital accumulation generates a higher rate of long-run economic growth. Looking at the problem from an alternative angle, with wages lagging behind rise in prices, inflation causes a large shift of resources away from the production of consumer goods for the wage earners to the production of capital goods. The higher rate of expansion in capital stock raises the growth of productive capacity of the economy and productivity of labour. This generates rapid economic growth.

Adverse Effects of Inflation on Economic Growth

However, it is now widely recognised that, far from encouraging savings and generating higher rate of economic growth, inflation slows down the rate of capital accumulation. There are several reasons responsible for this. First, as seen above, when due to rapid inflation value of money is declining, people will not like to keep money with themselves and will, therefore, be eager to spend it before its value goes down heavily. This raises their consumption demand and therefore lowers their saving. Besides, people find that the rapid inflation will erode the real value of their savings. This discourages them to save. Thus, *inflation or rapid rise in prices serves as a disincentive to save.*[15] Further, as a consequence of the rise in prices, a relatively greater part of the income of the people is spent on consumption to maintain their level ofin living of living and therefore little is left to be saved. Thus, not only does inflation reduce the willingness to save, it also slashes their ability to save.

Secondly, *inflation or rising prices lead to unproductive form of investment* in gold, jewellery, real estate, construction of houses etc. These unproductive forms of wealth do not add to the productive capacity of the economy and are quite useless from the viewpoint of economic growth. Thus, inflation may lead to more investment but much of this is of unproductive type. In this way economic surplus is frittered away in unproductive investment.

Thirdly, a highly undesirable consequence of inflation, especially in developing countries, is that *it accentuates the problem of poverty in these countries*. It is often said inflation is enemy number one of the poor people. Due to rising prices poor people are not able to meet their basic needs and maintain minimum subsistence level of consumption. Thus inflation sends many people to live below the poverty line with the result that the number of people living below the poverty line increases. Besides, due to inflation, consumption of a large number of poor people is reduced much below what may be regarded as *productive consumption*, that is, essential consumption required to maintain health and productive efficiency. In India, rapid inflation in recent years is as much responsible for the mounting number of people below the poverty line as the lack of employment opportunities.

Fourthly, *inflation adversely affects balance of payments* and thereby hampers economic growth, especially in the developing countries. When prices of domestic goods rise due to inflation, they cannot compete abroad and as a consequence exports of a country are discouraged. On the other

15. Samuelson, P.A. & Nordhaus, W.D., *Economics,* Fourteenth edition, McGraw-Hill, 1992

hand, when domestic prices rise relative to prices of foreign goods, imports of foreign goods increase. Thus, falling exports and rising imports create disequilibrium in the balance of payments which may, in the long run, result in a foreign exchange crisis. The shortage of foreign exchange prevents the country to import even essential materials and capital goods needed for industrial growth of the economy. The Indian experience during 1988–92 when foreign exchange reserves declined to abysmally low level and created an economic crisis in the country, shows the validity of this argument.

There is no agreement among economists whether or not moderate or mild inflation encourages saving and therefore ensures higher rate of capital accumulation and economic growth. However, there is complete unanimity that a very rapid inflation discourages saving and hinders economic growth. However, barring the special case of hyperinflation, whether or not saving is encouraged by inflation depends on whether there exists *wage lag.*[16] While there is sufficient evidence in the industrialised countries such as the U.S.A. Great Britain, France etc., about the existence of wage lag in the period before the World War II, in the period after this there is no solid evidence of it. In the present wages quickly catch up with the rising prices. Indeed, there is evidence in some developed countries that the share of profits in national income has declined and that of wages has going up during the post World War II period. Therefore, "To the extent that rate of long-run economic growth depends on the rate of capital accumulation, a major basis for the conclusion that inflation promotes rapid economic growth is undermined given that *wages no longer lag during inflation as they apparently did in time of past*."[17]

However, it may be noted that in the developing countries like India where labour is mostly unorganised and trade unions of labour are not strong and further there is a lack of information which causes wages lagging behind prices during periods of inflation. This itself will cause greater proportion of national income going to profits and other business incomes which should ensure higher saving rate. However, in India, businessmen are prone to make unproductive investment in speculative activities, gold, jewellery, real estate and palatial houses whose prices rise rapidly during periods of inflation. Such kind of investment is not only counter-productive and anti-growth but is repugnant to social justice as it further accentuates inequalities in the distribution of income and wealth.

It follows from above that rising prices as a goal of monetary policy are full of disastrous consequences for the economy and the people and therefore cannot be recommended as a desirable goal for the economic policy. Rising prices often get out of hand and hyperinflation might set in which will shake the confidence of the people in the monetary and fiscal system of the country.

HYPERINFLATION AND ECONOMIC CRISIS

When inflation is extremely rapid, it is called hyperinflation. *A hypherinfaltion is generally defined as inflation at the rate of 50 per cent or more per month.* Some economists, prominent among them is R. Dornbusch have defined *hyperinflation as rise in price level at the rate of 1000 per cent in a year.*[18] There are several examples of hyperinflation that occurred in European countries after World Wars I and II. The most important episode of hyperinflation took place in Germany between Aug. 1922 and Nov. 1923. German inflation occurred at the rate of 322 per cent per month during this period and in the final months rate of inflation in Germany reached the highly extreme rate of 32000 per cent per month.[19] It may be noted that German hyperinflation was caused by

16. See D. Felix, "Profit Inflation and Industrial Growth, "*Quarterly Journal of Economics, August* 1956, pp. 441-63 and E.J. Hamilton, Prices as a Factor in Business Growth, *Journal of Economic History*, December 1952, pp. 325-49.
17. E. Shapiro, *Macroeconomic Analysis,* Fourth Edition, 1978, p. 443
18. See R. Dornbusch, et. al, "Extreme Inflation, Dynamics and Stabilisation", *Brookings Paper on Economic Activity*, 1977.
19. Robert J. Gordon, *Macroeconomics,* 12th edition, 2012, p. 14

deficit financing by the government (*i.e.*, creation of new money to finance the budget deficit during World War I when it had to pay massive reparations to Britain and France. Argentina, Brazil, Peru and Poland all suffered from inflation rate of 1000 per cent a year for one or more years in late 1980 or 1990. The effect of hyperinflation on national output and employment turns out to be devastating. *Thus hyperinflation is generally caused when Government issues too much currency which greatly adds to the money supply in the economy.*

However, some economists are of the view that even mild or moderate inflation may ultimately lead to hyperinflation. They argue that when prices go on creeping upward for some time, people start expecting that prices will rise further and value of money will depreciate. In order to protect themselves from the fall in the purchasing power of money in the future, they try to spend money now. That is, they try to beat the anticipated price increases. This raises the aggregate demand for goods in the present. Businessmen too increase their purchases of capital goods and build up larger than normal inventories if they anticipate rise in prices. Thus, inflationary expectations raise the pressure on prices and in this way inflation feeds on itself. Further, the rise in prices and the cost of living under the influence of rising aggregate demand, prompts the workers and their unions to demand higher wages to compensate them for the rise in the prices. During the periods of boom these demands of the workers for hike in wages are generally conceded. But rise in labour costs due to higher wages are recovered by business firms from the consumers by raising the prices of their products. This increase in prices gives rise to demand for further increases in wages resulting in still higher costs. Thus, *cumulative wage-price inflationary spiral starts operating which may culminate in hyperinflation.*

The major reason for hyperinflation is rapid growth of money supply as a result of financing budget deficits through creation of new money. During wars government expenditure increases more than the revenue. But hyperinflation does not occur during the war period as controls are imposed on prices during the war period and therefore inflation remains suppressed. When controls are lifted after the war, hyperinflation takes places as a consequence of excessive deficit financing and monetary growth duirng the war period.

Some economists attribute hyperinflation to the *accommodation of adverse supply shocks* that may take place as a result of rise in wage rates or rise in oil prices. As these supply shocks cause inflation on the one hand and unemployment or fall in GDP on the other. To increase employment opportunities to get rid of unemployment, the government or its Central Bank increases the supply of money to raise aggregates demand. This higher aggregate demand succeeds in eliminating unemployment but generates still higher inflation. This gives rise to *wage-price spiral* under which every time when the price level rises, money wages are raised to restore real wages. This process goes on and ultimately ends in hyperinflation.

The Costs of Hyperinflation

The costs of hyperinflation is much greater than the costs of moderate inflation. For example, when prices are rising at an extremely high rate, the incentives to minimise holding of currency become quite strong that result in enormous *shoe-leather costs*. In the situation of hyperinflation workers are paid much more frequently, even on daily basis and people rush out to spend their wages on goods (or to convert their currency in some other form such as a foreign currency or into gold and silver) before the prices rise even more. The greater time and energy are spent in getting rid of currency as soon as possible. Thus hyperinflation involves lot of wastes of resources and leads to disruption of production.

Hyperinflation not only has disruptive redistributive effect, it also brings about economic crisis and may even cause collapse of the economic system. Hyperinflation encourages speculative activity on the part of people and businessmen who shy away from productive activities, as they find it highly profitable to hoard both finished goods and materials on the basis of expectations of further rise in prices. But such hoardings of goods and materials restrict the supply and availability of goods and tend to intensify the inflationary pressures in the economy. Instead of making productive investment, people and businesses tend to invest in unproductive assets such as gold and jewellery, real estate, houses etc., as a means of protecting themselves from inflation.

In the extreme when as a result of issuing too much money supply to finance government deficit or working of wage-price spiral, inflation becomes extremely rapid or what economists called hyper-inflation, normal working of the economy collapses. In this situation, "prices are so rapidly rising and consequently purchasing power of money is so much dwindling that businessmen do not know what to charge for their products and consumers do not know what to pay. Resource suppliers will want to be paid with actual output rather than with rapidly depreciating money. Creditors will avoid debtors to escape the rapayment of debts with cheap money. Money becomes virtually worthless and ceases to do its job as a measure of value and a medium of exchange. The economy may be literally thrown into a state of barter. Production and exchange grind towards a halt and the net result is economic, social and political chaos." [20]

Such grim and gloomy situation created by hyperinflation did occur in the Germany during 1920s and in Hungary and Japan in the forties. At that time, money depreciated so much that for some time barter system came to prevail and after some time, the new currency had to be issued. It is therefore desirable that appropriate anti-inflationary measures be taken so that inflation should not go out of control and get transformed into hyperinflation.

Measures to End Hyperinflation

It is important to note that all hyperinflation episodes lasted for one to 2 years. The governments succeeded in controlling them through the adoption of proper *stabilisation policy*. The stabilisation measures adopted were the following:

1. The most important measure was to reduce sharply budget deficit *by cutting down government expenditure and subsidies and raising more resources through taxes*. This resulted in slowdown in growth of money and helped in controlling inflation.

2. The Central Bank made creditable announcement that it would not automatically monetise the budget deficits of the government. The government on its part also assured that it would finance its expenditure through taxes and would not resort to deficit financing. Once the people were convinced about the commitments by the Central Bank and the government they did not rush to spend the money income they earned and kept it with them to spend when needed. This helped to curb inflationary expectations.

As suggested by some economists, to control hyperinflation some type of control on wages often called **incomes policy** was needed. This visualised reducing the frequency of wages-indexation or agreement among firms and labour unions not to raise wages until hyperinflation was brought under control. Besides, this also involved commitments by the firms not to raise their profit margins. However, the experience of several countries which succeeded in ending hyperinflation suggest that, "stopping hyperinflation is a complex and difficult task. Much depends on the credibility of the government, that is, public belief that budget deficit and monetary growth are really going to stop. It may take several dramatic actions all at once to achieve credibility."[21]

20. Robert J. Gordon, *Op. cit.*, p. 337.
21. Robert Gordon, *Op. cit.* p. 337

MEASURES TO CONTROL DEMAND-PULL INFLATION

As has been explained above, inflation occurs due to the emergence of excess demand for goods and services relative to their supply of output at the prevailing prices. Inflation of this type is called *demand-pull inflation*. Various fiscal and monetary measures can be adopted to check this inflation. We discuss below the efficacy of the various policy measures to check demand-pull inflation which is caused by excess aggregate demand.

1. Fiscal Policy: Reducing Fiscal Deficit

The budget deals with how a Government raises its revenue and spends it. If the total revenue raised by the Government through taxation, fees, surpluses from public undertakings is less than the expenditure it incurs on buying goods and services to meet its requirements of defence, civil administration and various welfare and developmental activities, there emerges a fiscal deficit. It may be noted here that the budget of the government has two parts: (1) *Revenue Budget*, (2) *Capital Budget*. In the revenue budget on the receipts side revenue raised through taxes, interests, fees, surpluses from public undertakings are given and on the expenditure side consumption expenditure by the government on goods and services required to meet the needs of defence, civil administration, education and health services, subsidies on food, fertilizers and exports, and interest payments on the loans taken by it in the previous years are important items. In the capital budget, the main items of receipts are market borrowings by the government from the banks and other financial institutions, foreign aid, small savings (*i.e.*, Provident Fund, National Savings Schemes etc.). The important items of expenditure in the capital budget are defence, funding of public enterprises for developmental purposes, especially those relating to infrastructure projects and loans to States and Union territories.

Fiscal deficit can be financed in two ways. First, borrowing by the government from the central bank (RBI in case of India) against its own securities. This leads to the creation of more money supply and therefore gives rise to inflationary pressures in the economy. Some years ago, this was called *deficit financing* which was held to be the main cause of demand-pull inflation in India. Now, we call it *monetisation of fiscal deficit*. However, in recent years *fiscal deficit in India is financed mainly through borrowing by the Government through sale of its bonds which are generally purchased by banks, insurance companies, mutual funds and corporate firms*. The increase in government expenditure made possible by borrowing without being matched by extra taxation causes aggregate demand to increase. The net increase in Government expenditure leads to multiple increase in aggregate demand through *Government expenditure* multiplier.

In the opinion of many economists, the expansion in aggregate expenditure caused by fiscal deficit leads to the excess aggregate demand and inflationary pressures in the economy, especially when aggregate supply of output is inelastic. To some extent increase in aggregate demand may not generate demand-pull inflation because if the aggregate output increases, especially of essential consumer goods such as foodgrains, cloth, the extra demand arising out of increase in expenditure would be matched by extra supply of output.

But when increase in expenditure occurs and the economy is already utilising its production capacity fully or working near to it, this leads to demand-pull inflation. Therefore, to tackle the problem of demand-pull inflation fiscal deficit should be reduced. In its recommendation for India, IMF has suggested that fiscal deficit must be reduced to 3 per cent of GDP if inflation has to be kept under check.

To reduce fiscal deficit the Government can mobilise more resources through taxes, both direct and indirect. In India there is a lot of scope for raising resources through taxation. In both personal and corporate income taxes, there are a large number of unnecessary exemptions which lower the effective rate of income tax. Thus it has been found that though there is 33 per cent tax on corporate income but because of several exemptions which have outlived their utility, effective rate of corporate income tax in India is only about 24 per cent. Thus with withdrawal of these exemptions can lead to significant increase in revenue of the Government. Besides, improving efficiency of tax-collecting administration tax evasion which occurs on a large scale in India and generate a lot of black money can *yield* large revenue for the Government. On the other hand, it can reduce fiscal deficit by curtailing its wasteful and inessential expenditure, especially subsidies to the non-poor people. In India, it is often argued that there is a large scope for pruning down non-plan expenditure on defence, police and general administration and on subsidies being provided on food, fertilizers and fuel oil. Though it is easy to suggest cutting down of Government expenditure, it is difficult to implement it in practice. However, in our view, there is a large-scale inefficiency in resource use and also a lot of corruption involved in spending by the Government which can be curtailed to a good extent. Thus, both by greater resource mobilisation on the one hand and pruning down of wasteful and non-essential Government expenditure on the other, the fiscal deficit can be reduced. This will help in controlling inflation.

2. Monetary Policy to Control Inflation : Monetary Tightening

Monetary policy refers to the adoption of suitable policy regarding *interest rate* and the *availability of credit*. Monetary policy is an important measure for reducing aggregate demand to control inflation. As an instrument of demand management, monetary policy can work in two ways. First, it can affect the *cost of credit* and second, it can influence the *credit availability* for private business firms. Let us first consider the cost of credit. The higher the rate of interest, the greater the cost of borrowing from the banks by the business firms. As anti-inflationary measure, the rate of interest has to be raised to discourage businessmen to borrow more so that less bank credit is created. The cheap credit policy (*i.e.*, lower interest rates) is recommended on the ground that lower rate of interest will promote more private investment which is an important factor determining economic growth. Keeping in view this consideration, cheap money policy was adopted in India up to 1972 and accordingly bank rate was kept low.

The dear money policy (that is, *higher interest rate policy*) has often been used in India to curb the inflationary pressures in the Indian economy. In India, bank rate has not been generally used to check inflation. Instruments like *repo rate* and *reverse repo* rate have often been used to manage aggregate demand. Repo rate is the interest rate at which Reserve Bank of India lends funds to the commercial banks *for a short period*. To curb inflation *repo rate* is raised. Hike in repo rate raises the cost of funds for the banks which will, if they do not have excess reserves, raise their *lending rates*. The higher lending rates will lower the demand for bank credit for investment and for purchases of cars (auto loans) and housing (housing loans). Under these circumstances to mobilise funds banks raise their deposit rates. The higher rate of interest on savings and fixed deposits will induce more savings by the households and help in cutting down aggregate consumption expenditure. Besides, higher rates of interest will discourage more investment in inventories and consumer durables and will help in reducing aggregate demand.

However, rise in repo rate will lead to the decline in credit growth if ***monetary transmission mechanism*** works. This will happen only if banks are short of liquid funds. If the banks have excess liquid reserves with them they would not raise their lending rates when the RBI raises repo rate. Therefore, for monetary transmission mechanism to work liquidity with the banks must be curtailed. This can be done by the RBI by raising cash reserve ratio (CRR) and by *open market operation* through sale of Government securities.

It is noteworthy that a recent monetary theory emphasizes that it is the changes in the ***credit availability*** rather than ***cost of credit*** (*i.e.*, rate of interest) that is a more effective instrument of regulating aggregate demand. There are several methods by which credit availability can be reduced. Firstly, it is through *open market operations* that the central bank of a country can reduce the availability of credit in the economy. Under open market operations, the Reserve Bank sells Government securities. Those, especially banks, who buy these securities will make payment for them in terms of cash reserves. With their reduced cash reserves, their capacity to lend money to the business firms will be curtailed. This will tend to reduce the supply of credit or loanable funds by the banks which in turn would tend to reduce aggregate demand.

However, in the past in India open market operations did not play a significant role as an instrument of credit control to fight against inflationary situation. This was because market for Government securities was narrow as well as captive. However, with financial reforms this is no more the situation. At present RBI often uses open market operations to influence liquidity of the banking system. General public do not buy more than a fraction of Government securities. It is the institutions such as commercial banks, LIC, GIC and Provident Funds which are required by law to invest a certain proportion of their funds in buying Government securities. The RBI can reduce the liquidity with the banks by selling Government securities to them through open market operations.

In India, it is the ***Cash Reserve Ratio* (*CRR*)** which can be effectively used to curb inflation. By law banks have to keep a certain proportion of cash money as reserve against their demand and time deposits. This is called cash reserve ratio. To reduce liquidity with the banks and thereby to contract credit availability Reserve Bank can raise this ratio. In recent years to squeeze credit for checking inflation, cash reserve ratio in India has been raised from time to time.

***Statutory Liquidity Ratio* (*SLR*).** Another instrument with Reserve Bank of India for affecting credit availability is the statutory liquidity ratio, in addition to *CRR*, banks have to keep a certain minimum proportion of their deposits in the form of specified liquid assets. And the most important specified liquid asset for this purpose is the Government securities. To mop up extra liquid assets with banks which may lead to undue expansion in credit availability for the business class, the Reserve Bank has often raised statutory liquidity ratio.

Limitations of Tight Monetary Policy

However, tight monetary policy for controlling inflation is not without its limitations. First, monetary transmission mechanism may be weak and raising of short-term interest rates by the RBI may not actually lead to the restriction of bank credit. First, the banks may have surplus liquidity (*i.e.*, cash reserves) with them and therefore they may not follow tight monetary policy and raise their lending rates. As a result, supply of credit by banks will not be restricted. Secondly, if the economic environment is such that boom conditions prevail in the economy and aggregate demand for products is quite high, demand for credit may not be much affected by higher lending rates. As emphasized by J.M. Keynes, investment is determined more by marginal efficiency of capital (that is, expected rate of return) rather than rate of interest. Thirdly, at present in India corporate firms are more easily able to borrow from foreign capital markets (*i.e.*, external commercial borrowing, ECB) especially when rates of interest in the US, European zone and Japan are extremely low. Therefore, unless this debt-capital inflow (*i.e.* ECB) is checked RBI's monetary policy may not be effective to check the supply of credit to control inflation in the economy.

Thirdly, when the stock market prices are rising and aggregate demand is quite high, the corporate sector is able to raise funds itself more easily from the capital market. This will also offset the impact of tight monetary policy of the RBI to control inflation. Fourthly, if adequate internal funds are available with the corporate firms as a result of *retained profit earnings* by the companies they can use them to finance their expansion plans and thereby add to the aggregate expenditure or demand. This will also nullify the tight monetary policy of the RBI to curb inflation.

It may be noted that in recent years to control inflation the Reserve Bank of India raised its *repo rate* 13 times from March 2010 to Nov. 2011 (from 4.75% in March 2010 to 8.50% in Nov. 2011) but inflation as measured by the WPI remained at an elevated level. It was estimated at 10 per cent (YoY) in September 2011, 9.73% in Oct. 2011 and 9.11 per cent in Nov. 2011. After Nov. 2011,

WPI inflation declined because of dip in food inflation due partly to the base effect and partly to easing of supply position of some food articles. Thus tight monetary policy failed to check inflation despite 3.75 per centage points increase in repo rate. The RBI explained failure of its tight monetary policy to curb inflation by blaming the *large fiscal deficit* of the Government in 2011-12. Due to large fiscal deficit, aggregate demand was increasing which was feeding inflation in the Indian economy. According to the RBI, unless fiscal deficit is brought down by the Government, tight monetary policy alone will not succeed in checking inflation. Besides, RBI blamed *supply-side factors* responsible for food inflation which has contributed to overall rise in WPI inflation.

Lastly, if inflation in the economy has originated from the **supply-side factors**, for example, if production of foodgrains and other essential food articles such as milk, vegetables, fruits etc. in a year declines or does not increase adequately to meet the growing demand for them due to certain supply-side bottlenecks, this will cause the *demand-supply imbalances* leading to the rise in inflation. Such supply-side inflation cannot be checked by raising interest rates under the tight monetary policy. This happened in 2009-10 when due to the shortage of monsoon rainfall, drop in agricultural production was expected, inflationary pressures emerged in the Indian economy raising food-inflation to around 20 per cent in December 2009. This food inflation continued to prevail at double digit level till November 2010. Tight monetary policy which is aimed at management of aggregate demand rather than augmentation of supply is ineffective in tackling this supply-side inflation as in case of food inflation in 2010. Likewise, in 2010-11 and 2011-12 *shortage of protein-based food* products such as pulses, milk, fruits and vegetables, eggs, meat and fish, contributed a good deal to the persistence of food inflation. It is only after Oct. 2011 when there was a drop in food-inflation, that headline WPI inflation also declined. Similarly, supply-side inflation arises when prices of petroleum products rise which is passed on to the domestic consumers and cannot be tackled with use of tight monetary policy. In addition to fuel, if output growth in *core industries* such as steel, cement, coal declined as they did during 2011-12 in India, they create supply-side bottlenecks in various industries leading to supply-side inflation.

Now to control food inflation what is required is to take long-term measures to augment supply of foodgrains and other food articles by raising agricultural productivity and undertaking appropriate technological changes and land-reform measures. In the short run to control food inflation what is needed is to release food stocks through public distribution system (PDS) which should be properly monitorad to check black market. Besides, to control rise in food prices in the short run release of food stocks for sale in the open market should be made. However, this presupposes enough food stocks with the Government. Further, to check food-inflation, food grains, pulses, oilseeds and other feed articles in short-supply can be imported.

It may however be noted that when inflationary expectations arise as a result of emergence of shortage of supply of some essential commodities there is tendency on the part of traders to hoard stocks of goods in short supply for speculative purposes. To discourage such speculative hoarding of goods in short supply monetary policy of hgih interest rates can be helpful. Besides, as explained above, selective credit controls of monetary policy can also be used to check excessive hoarding

Selective Credit Controls

By far the most important anti-inflationary measure in India is the use of selective credit control. The methods of credit control described above are known as quantitative or general methods as they are meant to control the availability of credit in general. Thus, bank rate policy, open market operations and variation in cash reserves ratio expand or contract the availability of credit for all purposes. On the other hand, selective credit controls are meant to regulate the flow of credit for particular or specific purposes. Whereas the general credit controls seek to regulate the total available quantity of credit (through changes in the high powered money) and the cost of credit, the selective credit control seeks to change the distribution or allocation of credit between its various uses. These selective credit controls are also known as *Qualitative Credit Controls*. The selective credit controls have both the positive and negative aspect. In its positive aspect, measures are taken to stimulate the greater flow of credit to some particular sectors considered as important.

(1) Changes in the minimum margin for lending by banks against the stocks of specific goods kept or against other types of securities.
(2) The fixation of maximum limit or ceiling on advances to individual borrowers against stock of particular sensitive commodities.
(3) The fixation of minimum discriminatory rates of interest chargeable on credit for particular purposes.

3. Supply Management through Imports

To correct excess demand relative to aggregate supply, the latter can also be raised by importing goods in short supply. In India, to check the rise in prices of foodgrains, edible oils, sugar etc., the Government has often taken steps to increase imports of goods in short supply to enlarge their available supplies. When inflation is of the type *of supply-side inflation*, imports are increased to augment the domestic supplies of goods. To increase imports of goods in short supply the Government reduce customs duties on them so that their imports become cheaper and help in containing inflation. For example in 2008-09 the Indian Government removed customs duties on imports of wheat and rice and reduced them on oilseeds, steel etc. to increase their supplies in India.

At times of inflationary expectations, there is a tendency on the part of businessmen to hoard goods for speculative purposes. The attempt by the Government to import goods in short supply would compel the hoarders to release their hoarded stocks. This will have a favourable impact on prices of these goods. However, the country can sufficiently increase the imports of goods if there are either enough *foreign exchange reserves* which can be used to spend on imports or if sufficient *foreign aid* is available to import the goods in short supply.

4. Incomes Policy: Freezing Wages

Another anti-inflationary measure which has often been suggested is the avoidance of wage increases which are unrelated to improvements in productivity. This requires exercising control over wage-income. As seen in the previous chapter, it is through wage-price spiral that inflation gets momentum. When cost of living rises due to the initial rise in prices, workers demand higher wages to compensate for the rise in cost of living. When their wage demands are conceded to, it gives rise to cost-push inflation. And this generates inflationary expectations which add fuel to the fire. To check this vicious circle of wages-chasing prices, an important measure will be to exercise control over wages. However, if wages are raised equal to the increase in the productivity of labour, then it will have no inflationary effect. Therefore, the proposal has been to *freeze wages* in the short run and wages should be linked with the changes in the level of productivity over a long period of time. According to this, wage increases should be allowed to the extent of rise in labour productivity only. This will check the net growth in aggregate demand relative to aggregate supply of output.

However, freezing wages and linking it with productivity only irrespective of what happens to the cost of living has been strongly opposed by trade unions. It has been validly pointed out why freeze wages only; to ensure social justice, the other kinds of income such as rent, interest and profits should also be freezed similarly. Indeed, an effective way to control inflation will be to adopt a broad-based incomes policy which should cover not only wages but also profits, interest and rental incomes.

QUESTIONS FOR REVIEW

1. Define demand-pull inflation. How is it caused by the Central Bank insistence of keeping the market rate of interest low so as to stimulate investment.
2. What are the main factors that cause cost-push inflation? Suggest measures to control cost-push inflation.
3. Distinguish between demand-pull inflation and cost-push inflation. How are they often intertwined?
4. Inflation may originate because of cost-push but it cannot be sustained for long unless it is supported by demand-pull inflation.
5. Critically examme monetarist theory of inflation.
6. What is stagflation ? Explain how cost-push inflation leads to stagflation.
7. An inflationary process may begin on the supply side but it will not continue for long unless there is increase in demand. Discuss.
8. Distinguish between demand-pull and cost-push inflation. How can demand-pull inflation be controlled?
9. Inflation is always and everywhere a purely monetary phenomenon. Critically examine.
10. Define inflation. Explain its effects on (*a*) creditors and debtors and (*b*) persons of fixed income group.
11. Explain the effects of inflation on economic growth and distribution of income.
12. How does inflation distort the pattern of production and distribution in an economy?
13. Define inflation. What measures would you suggest to control it?
14. What are the main causes of inflation in India? Which policies of government have helped to check rise in prices?
15. What is *inflation fallacy* with regard to cost of inflation? How far is it valid? Discuss.
16. Explain the social cost of (*a*) anticipated inflation and (*b*) unanticipated inflation.
17. How does inflation adversely affect long-term economic growth of a country?
18. What is hyperinflation? How does it lead to economic crisis?

PART XI

IS-LM Curves Model, Phillips Curve Concept and Business Cycles

- The IS-LM Curve Model
- Inflation and Unemployment: Phillips Curve and Rational Expectations Theory
- Stagflation and Supply-side Economics
- Business Cycles
- Friedman's Monetarist and New Classical (Rational Expectations) Theories of Business Cycles
- Economic Stabilisation: Fiscal Policy
- Economic Stabilisation: Monetary Policy

CHAPTER 24

THE IS-LM CURVE MODEL

THE GOODS MARKET AND MONEY MARKET: LINKS BETWEEN THEM

The Keynes in his analysis of national income explains that national income is determined at the level where aggregate demand (i.e., aggregate expenditure) for consumption and investment goods (C + I) equals aggregate output. In other words, in Keynes' simple model the level of national income is shown to be determined by the *goods market equilibrium*. In this simple analysis of equlibirum in the goods market Keynes considers investment to be determined by the rate of interest along with the marginal efficiency of capital and is shown to be independent of the level of national income. The rate of interest, according to Keynes, is determined by money market equilibrium by the demand for and supply of money. In this Keynes' model, changes in rate of interest either due to change in money supply or change in demand for money will affect the determination of national income and output in the goods market through causing changes in the level of investment. In this way changes in money market equilibrium influence the determination of national income and output in the goods market. However, there is apparently one flaw in the Keynesian analysis which has been pointed out by some economists and has been a subject of a good deal of controversy. It has been asserted that in the Keynesian model whereas the changes in rate of interest in the money market affect investment and therefore the level of income and output in the goods market, *there is seemingly no inverse influence of changes in goods market i.e., (investment and income)* on the money market equilibrium.

It has been shown by J.R. Hicks and others that with greater insights into the Keynesian theory one finds that the changes in income caused by changes in investment or propensity to consume in the goods market also influence the determination of interest in the money market. According to him, the level of income which depends on the investment and consumption demand determines the transactions demand for money which affects the rate of interest. Hicks, Hansen, Lerner and Johnson have put forward a complete and integrated model based on the Keynesian framework wherein the variables such as investment, national income, rate of interest, demand for and supply of money are interrelated and mutually interdependent and can be represented by the two curves called the *IS* and *LM* curves. This extended Keynesian model is therefore known as *IS-LM* curve model. In this model they have shown how the level of national income and rate of interest are jointly determined by the simultaneous equilibrium in the two interdependent goods and money markets. Now, this *IS-LM* curve model has become a standard tool of macroeconomics and the effects of monetary and fiscal policies are discussed using this *IS* and *LM* curves model.

GOODS MARKET EQUILIBRIUM: THE DERIVATION OF THE IS CURVE

The *IS-LM* curve model emphasises the interaction between the goods and money markets. The goods market is in equilibrium when aggregate demand is equal to income. The aggregate demand is determined by consumption demand and investment demand. In the Keynesian model of goods market equilibrium we also now introduce the rate of interest as an important determinant of investment. With this introduction of interest as a determinant of investment, the latter now becomes an endogenous variable in the model. When the rate of interest falls the level of investment increases and vice versa. Thus, changes in the rate of interest affect aggregate demand or aggregate expenditure by causing changes in the investment demand. When the rate of interest falls, it lowers the cost of investment projects and thereby raises the profitability of investment. The businessmen will therefore undertake greater investment at a lower rate of interest. The increase in investment demand will bring about increase in aggregate demand which in turn will raise the equilibrium level of income. *In the derivation of the IS Curve we seek to find out the equilibrium level of national income as determined by the equilibrium in goods market by a level of investment determined by a given rate of interest. Thus IS curve relates different equilibrium levels of national income with various rates of interest*. As explained above, with a fall in the rate of interest, the planned investment will increase which will cause an upward shift in aggregate demand function ($C + I$) resulting in goods market equilibrium at a higher level of national income.

The lower the rate of interest, the higher will be the equilibrium level of national income. Thus,

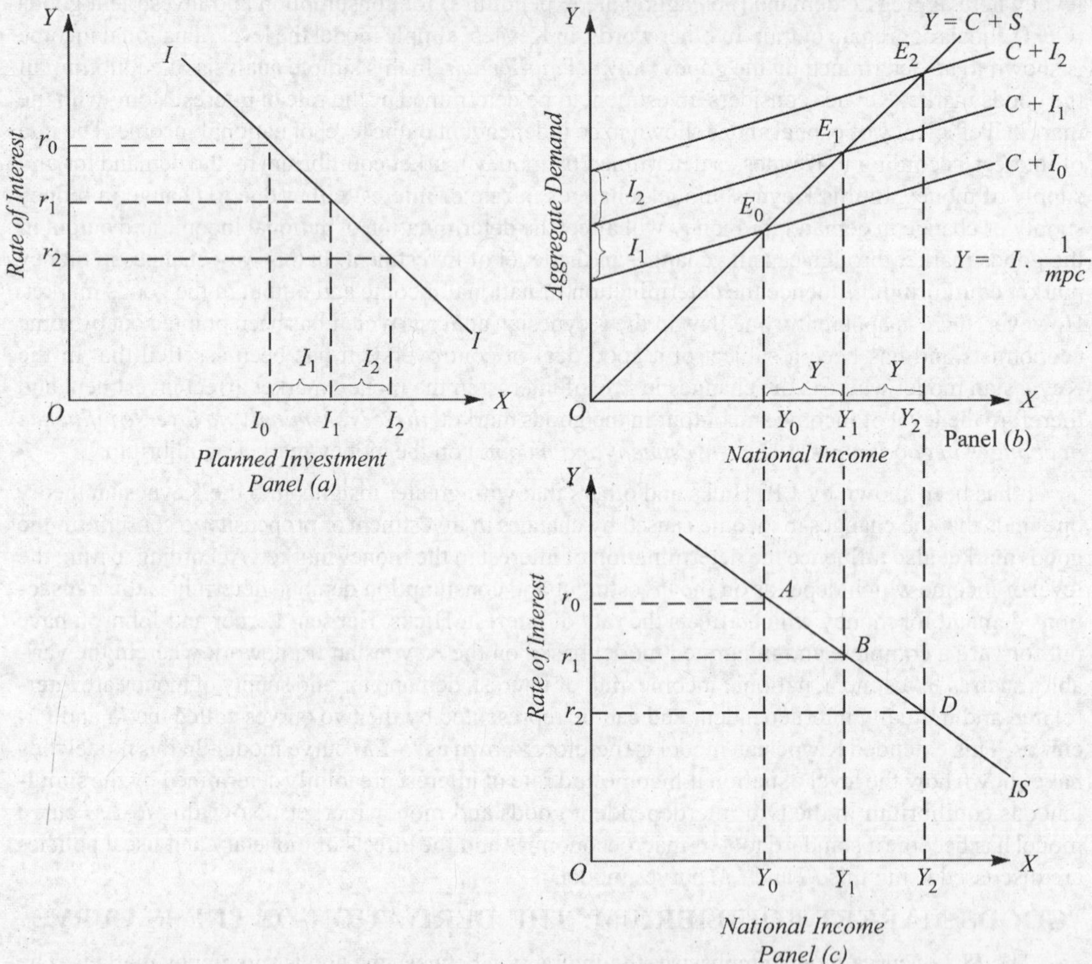

Fig. 24.1. *Derivation of IS Curve: Linking Rate of Interest with National Income through Investment and Aggregate Demand*

the *IS* curve is the locus of those combinations of rate of interest and the level of national income at *which goods market is in equilibrium*. How the *IS* curve is derived is illustrated in Fig. 24.1. In panel (a) of Fig. 24.1 the relationship between rate of interest and planned investment is depicted by the investment demand curve *II*. It will be seen from panel (*a*) that at rate of interest Or_0 the planned investment is equal to OI_0. With OI_0 as the amount of planned investment, the aggregate demand curve is $C + I_0$ which, as will be seen in panel (*b*) of Fig. 24.1 equals aggregate output at OY_0 level of national income. Therefore, in the panel (*c*) at the bottom of the Fig. 24.1, against rate of interest Or_0, level of income equal to OY_0 has been plotted. Now, if the rate of interest falls to Or_1, the planned investment by businessmen increases from OI_0 to OI_1 [see panel (*a*)]. With this increase in planned investment, the aggregate demand curve shifts upward to the new position $C + I1$ in panel (*b*), and the goods market is in equilibrium at OY_1 level of national income. Thus, in panel (c) at the bottom of Fig. 24.1 the level of national income OY_1 is plotted against the rate of interest, Or_1. With further lowering of the rate of interest to Or_2, the planned investment increases to OI_2 (see panel *a*). With this further rise in planned investment the aggregate demand curve in panel (*b*) shifts upward to the new position $C + I_2$ corresponding to which goods market is in equilibrium at OY_2 level of income. Therefore, in panel (*c*) the equilibrium income OY_2 is shown against the interest rate Or_2. By joining points A, B, D representing various interest-income combinations at which goods market is in equilibrium we obtain the *IS* Curve. It will be observed from Fig. 24.1 that the *IS* Curve is downward sloping (*i.e.*, has a negative slope) which implies that when rate of interest declines, the equilibrium level of national income increases.

Why does IS Curve Slope Downward?

What accounts for the downward-sloping nature of the *IS* curve. As seen above, the decline in the rate of interest brings about an increase in the planned investment expenditure. The increase in investment spending causes the aggregate demand curve to shift upward and therefore leads to the increase in the equilibrium level of national income. Thus, a lower rate of interest is associated with a higher level of national income and vice-versa. This makes the *IS* curve, which relates the level of income with the rate of interest, to slope downward.

Steepness of the IS curve depends on (1) the elasticity of the investment demand curve, and (2) the size of the multiplier. The elasticity of investment demand signifies the degree of responsiveness of investment spending to the changes in the rate of interest. Suppose the investment demand is highly elastic or responsive to the changes in the rate of interest, then a given fall in the rate of interest will cause a large increase in investment demand which in turn will produce a large upward shift in the aggregate demand curve. A large upward shift in the aggregate demand curve will bring about a large expansion in the level of national income. Thus when investment demand is more elastic to the changes in the rate of interest, the investment demand curve will be relatively flat (or less steep). Similarly, when investment demand is not very sensitive or elastic to the changes in the rate of interest, the *IS* curve will be relatively more steep.

The steepness of the *IS* curve also depends on the magnitude of the multiplier. As has been explained in a previous chapter the value of multiplier depends on the marginal propensity to consume (*mpc*). It may be noted that the higher the marginal propensity to consume, the aggregate demand curve $(C + I)$ will be more steep and the magnitude of multiplier will be large. In case of a higher marginal propensity to consume (*mpc*) and therefore a higher value of multiplier, a given increment in investment demand caused by a given fall in the rate of interest will help to bring about a greater increase in equilibrium level of income. Thus, the higher the value of multiplier, the greater will be the rise in equilibrium income produced by a given fall in the rate of interest and this makes the IS curve flatter. On the other hand, the smaller the value of multiplier due to lower marginal propensity to consume, the smaller will be the increase in equilibrium level of income following a given increment in investment caused by a given fall in the rate of interest. Thus, in case of smaller size of multiplier the *IS* curve will be more steep.

Shift in *IS* Curve

It is important to understand what determines the position of the IS curve and what causes shifts in it. It is the level of autonomous expenditure which determines the position of the IS curve and changes in the autonomous expenditure cause a shift in it. *By autonomous expenditure we mean the expenditure, be it investment expenditure, the Government spending or consumption expenditure which does not depend on the level of income and the rate of interest.* The government expenditure is an important type of autonomous expenditure. Note that the Government expenditure which is determined by several factors as well as by the policies of the Government, does not depend on the level of income and the rate of interest. Similarly, some consumption expenditure has to be made if individuals have to survive even by borrowing from others or by spending their savings made in the past year. Such consumption expenditure is a sort of autonomous expenditure and changes in it do not depend on the changes in income and rate of interest. Further, autonomous changes in investment can also occur.

In the goods market equilibrium of the simple Keynesian model the investment expenditure is treated as autonomous or independent of the level of income and therefore does not vary as the level of income increases. However, in the complete Keynesian model, the investment spending is thought to be determined by the rate of interest along with marginal efficiency of investment. Following this complete Keynesian model, in the derivation of the *IS* curve we consider the level of investment and changes in it as determined by the rate of interest along with marginal efficiency of capital. However, there can be changes in investment spending autonomous or independent of the changes in rate of interest and the level of income. For instance, growing population requires more investment in house construction, school buildings, roads, etc., which does not depend on changes in level of income or rate of interest. Further, autonomous changes in investment spending can also take place when new innovations come about, that is, when there is progress in technology and new machines, equipment, tools etc., have to be built embodying the new technology. Besides, Government expenditure is also of autonomous type as it does not depend on income and rate of interest in the economy. As is well-known government increases its expenditure for the purpose of promoting social welfare and accelerating economic growth. Increase in Government expenditure will cause a rightward shift in the *IS* curve.

MONEY MARKET EQUILIBRIUM: DERIVATION OF LM CURVE

Derivation of the *LM* Curve

The *LM* curve can be derived from the Keynesian theory from its analysis of money market equilibrium. According to Keynes, demand for money to hold depends upon transactions motive and speculative motive. It is the money held for transactions motive which is a function of income. The greater the level of income, the greater the amount of money held for transactions motive and therefore higher the level of money demand curve.

The demand for money depends on the level of income because they have to finance their expenditure, that is, their transactions of buying goods and services. The demand for money also depends on the rate of interest which is the cost of holding money. This is because by holding money rather than lending it and buying other financial assets, one has to forgo interest.

Thus demand for money (M^d) can be expressed as:

$$Md = L(Y, r)$$

where M^d stands for demand for money, *Y* for real income and *r* for rate of interest. Thus, we can draw a family of money demand curves at various levels of income. Now, the intersection of these various money demand curves corresponding to different income levels with the supply curve of money fixed by the monetary authority would gives us the *LM* curve. The *LM* curve relates the level of income with the rate of interest which is determined by money-market equilibrium corresponding to different levels of demand for money. The *LM* curve tells what the various rates of interest will be (given the quantity of money and the family of demand curves for money) at different levels of income. But the money demand curve or what Keynes calls the liquidity preference curve alone cannot tell us what exactly the rate of interest will be. In Fig. 24.2 (*a*) and (*b*) we have derived the *LM* curve from a family of demand curves for money. As income inicreases, money demand curve shifts

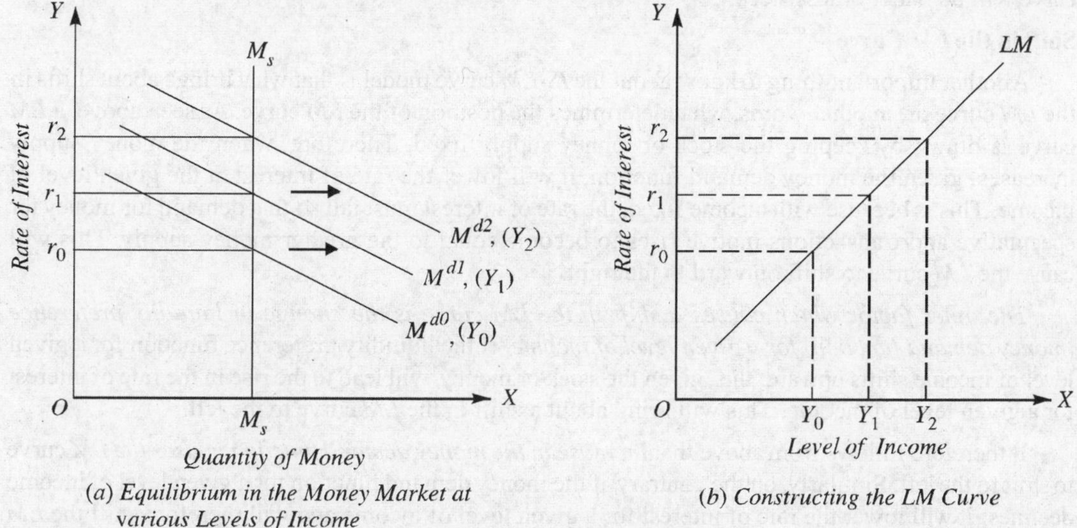

(*a*) *Equilibrium in the Money Market at various Levels of Income*

(*b*) *Constructing the LM Curve*

Fig. 24.2. *Derivation of LM Curve*

outward and therefore the rate of interest which equates supply of money with demand for money rises. In Fig. 24.2 (*b*) we measure income on the *X*-axis and plot the income level corresponding to the various interest rates determined at those income levels through money market equilibrium by the equality of demand for and the supply of money in Fig. 24.2 (*a*).

Slope of *LM* Curve

It will be noticed from Fig. 24.2 (*b*) that the *LM* curve slopes upward to the right. This is because with higher levels of income, demand curve for money (M^d) is higher and consequently the money-market equilibrium, that is, the equality of the given money supply with money demand curve occurs at a higher rate of interest. This implies that rate of interest varies directly with income. It is important to know the factors on which the slope of the *LM* curve depends. There are two factors on which the slope of the *LM* curve depends. First, *the responsiveness of demand for money (i.e., liquidity preference) to the changes in income*. As the income increases, say from Y_0 to Y_1, the demand curve for money shifts from Md_0 to Md_1, that is, with an increase in income, demand for money would increase for being held for transactions motive, M^d or $L_1 = f(Y)$. This extra demand for money would disturb the money market equilibrium and for the equilibrium to be restored the rate of interest will rise to the level where the given money supply curve intersects the new demand curve corresponding to the higher income level. It is worth noting that in the new equilibrium position, with the given stock of money supply, money held under the transactions motive will increase whereas the money held for speculative motive will decline. The greater the extent to which demand for money for transactions motive increases with the increase in income, the greater the decline in the supply of money available for speculative motive and, given the demand for money for speculative motive, the higher the rise in the rate of interest and consequently the steeper the *LM* curve. $r = f(M_2, L_2)$ where r is the rate of interest, M_2 is the stock of money available for speculative motive and L_2 is the money demand or liquidity preference for speculative motive.

The second factor which determines the slope of the *LM* curve is the *elasticity or responsiveness of demand for money (i.e., liquidity preference for speculative motive) to the changes in rate of interest*. The lower the elasticity of liquidity preference for speculative motive with respect to the changes in the rate of interest, the steeper will be the *LM* curve. On the other hand, if the elasticity of liquidity preference (money demand-function) to the changes in the rate of interest is high, the LM curve will be flatter or less steep.

Shift in the *LM* Curve

Another important thing to know about the *IS-LM* curve model is that what brings about shifts in the *LM* curve or, in other words, what determines the position of the *LM* curve. As seen above, a *LM* curve is drawn by keeping the stock or money supply fixed. Therefore, when the money supply increases, given the money demand function, it will lower the rate of interest at the given level of income. This is because with income fixed, the rate of interest must fall so that demand for money for speculative and transactions motive rises to become equal to the greater money supply. This will cause the *LM* curve to shift outward to the right.

The other factor which causes a shift in the LM curve is the change in liquidity preference (money demand function) for a given level of income. If the liquidity preference function for a given level of income shifts upward, this, given the stock of money, will lead to the rise in the rate of interest for a given level of income. This will bring about a shift in the *LM* curve to the left.

It therefore follows from above that *increase in the money demand function causes the LM* curve to shfit to the left. Similarly, on the contrary, if the money demand function for a given level of income declines, it will lower the rate of interest for a given level of income and will therefore shift the *LM* curve to the right.

The *LM* Curve: The Essential Features

From our analysis of the *LM* curve, we arrive at its following essential features:"

1. The *LM* curve is a schedule that describes the combinations of rate of interest and level of income *at which money market is in equilibrium.*
2. The *LM* curve slopes upward to the right.
3. The *LM* curve is flatter if the interest elasticity of demand for money is high. On the contrary, the *LM* curve is steep if the interest elasticity demand for money is low.
4. The *LM* curve shifts to the right when the stock of money supply is increased and it shifts to the left if the stock of money supply is reduced.
5. The *LM* curve shifts to the left if there is an increase in the money demand function which raises the quantity of money demanded at the given interest rate and income level. On the other hand, the *LM* curve shifts to the right if there is a decrease in the money demand function which lowers the amount of money demanded at given levels of interest rate and income.

SIMULTANEOUS EQUILIBRIUM OF THE GOODS MARKET AND MONEY MARKET

The *IS* and the *LM* curves relate the two variables: (*a*) income and (*b*) the rate of interest. Income and the rate of interest are therefore determined together at the point of intersection of these two curves, *i.e.*, *E* in Fig. 24.3. The equilibrium rate of interest thus determined is Or_2 and the level of income determined is OY_2. At this point income and the rate of interest stand in relation to each other such that (1) the goods market is in equilibrium, that is, the aggregate demand equals the level of aggregate output, and (2) the demand for money is in equilibrium with the supply of money (*i.e.*, the desired amount of money is equal to the actual supply of money). It should be noted that LM curve has been drawn by keeping the supply of money fixed.

Thus, the *IS-LM* curve model is based on: (1) the investment-demand function, (2) the consumption function, (3) the money demand function, and (4) the quantity of money. *We see, therefore, that according to the IS-LM curve model both the real factors, namely, saving and investment, productivity of capital and propensity to consume and save, and the monetary factors, that is, the demand for money (liquidity preference) and supply of money play a part in the joint determination of the rate of interest and the level of income. Any change in these factors will cause a shift in IS or LM curve and will therefore change the equilibrium levels of the rate of interest and income.*

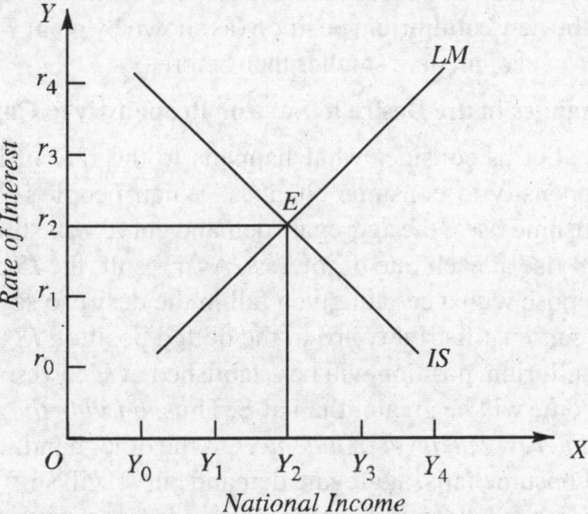

Fig. 24.3. *The IS and LM Curves Combined: The Joint Determination of the Interest Rate and the Income Level*

The *IS-LM* curve model explained above has succeeded in integrating the theory of money with the theory of income determination. And by doing so, as we shall see below, it has succeeded in synthesising the monerary and fiscal policies. Further, with the *IS-LM* curve analysis, we are better able to explain the effect of changes in certain important economic variables such as desire to save, the supply of money, investment, demand for money on the rate of interest and level of income.

Effect of Changes in Supply of Money on the Rate of Interest and Income Level

Let us first consider what will happen if the supply of money is increased by the action of the Central Bank. Given the liquidity preference schedule, with the increase in the supply of money, more money will be available for speculative motive at a given level of income which will cause the interest rate to fall. As a result, the *LM* curve will shift to the right. With this rightward shift in the *LM* curve, in the new equilibrium position, rate of interest will be lower and the level of income greater than before. This is shown in Fig. 24.4 where with a given supply of money, *LM* and *IS* curves intersect at point *E*. With the increase in the supply of money, *LM* curve shifts to the right to the position *LM*, and with *IS* schedule remaining unchanged, new equilibrium is at point *G* corresponding to which rate of interest is lower and level of income greater than at *E*. Now, suppose that instead of increasing the

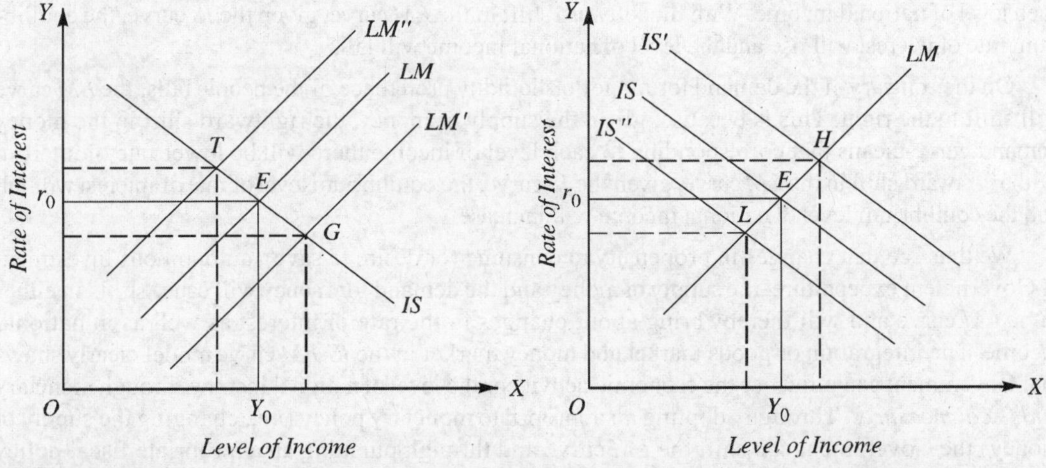

Fig. 24.4. *Impact of Change in Supply of Money* **Fig. 24.5.** *Impact of Change in Propensity to Save*

supply of money, Central Bank of the country takes steps to reduce the supply of money. With the reduction in the supply of money, less money will be available for speculative motive at each level of income and, as a result, the *LM* curve will shift to the left of *E*, and the *IS* curve remaining unchaned, in the new equilibrium position (as shown by point *T* in Fig. 24.4) the rate of interest will be higher and the level of income smaller than before.

Changes in the Desire to Save or Propensity to Consume

Let us consider what happens to the rate of interest when desire to save or in other words, propensity to consume changes. When people's desire to save falls, that is, when propensity to consume rises, the aggregate demand curve will shift upward and, therefore, level of national income will rise at each rate of interest. As a result, the *IS* curve will shift outward to the right. In Fig. 24.5 suppose with a certain given fall in the desire to save (or increase in the propensity to consume), the *IS* curve shifts rightward to the dotted position *IS* . With *LM* curve remaining unchanged, the new equilibrium position will be established at *H* corresponding to which rate of interest as well as level of income will be greater than at *E*. Thus, *a fall in the desire to save has led to the increase in both rate of interest and level of income*. On the other hand, if the desire to save rises, that is, if the propensity to consume falls, aggregate demand curve will shift downward which will cause the level of national income to fall for each rate of interest and as a result the *IS* curve will shift to the left. With this, and *LM* curve remaining unchanged, the new equilibrium position will be reached to the left of *E*, say at point *L* (as shown in Fig. 24.5) corresponding to which both rate of interest and level of national income will be smaller than at *E*.

Changes in Autonomous Investment and Government Expenditure

Changes in autonomous investment and Government expenditure will also shift the IS curve. If either there is increase in autonomous private investment or Government steps up its expenditure, aggregate demand for goods will increase and this will bring about increase in national income through the multiplier process. This will shift *IS* schedule to the right, and given the *LM* curve, the rate of interest as well as the level of income will rise. On the contrary, if somehow private investment expenditure falls or the Government reduces its expenditure, the *IS* curve will shift to the left and, given the *LM* curve, both the rate of interest and the level of income will fall.

Changes in Demand for Money or Liquidity Preference

Changes in liquidity preference will bring about changes in the *LM* curve. If the liquidity preference or demand for money of the people rises, the *LM* curve will shift to the left. This is because, greater demand for money, given the supply of money, will raise the rate of interest corresponding to each level of national income. With the leftward shift in the *LM* curve, given the *IS* curve, the equilibrium rate of interest will rise and the level of national income will fall.

On the contrary, if the demand for money or liquidity preference of the people falls, the *LM* curve will shift to the right. This is because, given the supply of money, the rightward shift in the money demand curve means that corresponding to each level of income there will be lower rate of interest. With rightward shift in the *LM* curve, given the *IS* curve, the equilibrium level of rate of interest will fall and the equilibrium level of national income will increase.

We thus see that changes in propensity to consume (or desire to save), autonomous investment or Government expenditure, the supply of money and the demand for money will cause shifts in either *IS* or *LM* curve and will thereby bring about changes in the rate of interest as well as in national income. The integration of goods market and money market in the *IS-LM* curve model clearly shows that Government can influence the economic activity or the level of national income through monetary *and fiscal measures*. Through adopting an appropriate monetary policy (*i.e.*, changing the supply of money) the Government can shift the *LM* curve and through pursuing an appropriate fiscal policy (expenditure and taxation policy) the Government can shift the *IS* curve. Thus both monetary and fiscal policies can play a useful role in regulating the level of economic activity in the country.

Critique of the IS-LM Curve Model

The *IS-LM* curve model makes a significant advance in explaining the simultaneous determination of the rate of interest and the level of national income. It represents a more general, inclusive and realistic approach to the determination of interest rate and level of income. Further, the IS-LM model succeeds in integrating and synthesising fiscal with monetary policies, and theory of income determination with the theory of money. But the *IS-LM* curve model is not without limitations. Firstly, it is based on the assumption that the rate of interest is quite flexible, that is, free to vary and not rigidly fixed by the Central Bank of a country. If the rate of interest is quite inflexible, then the appropriate adjustment explained above will not take place. Secondly, the model is also based upon the assumption that investment is interest-elastic, that is, investment varies with the rate of interest. If investment is interest-inelastic, then the *IS-LM* curve model breaks down since the required adjustments do not occur. Thirdly, Don Patinkin1 and Milton Friedman have cirticised the *IS-LM* curve model as being too artificial and over-simplified. In their view, division of the economy into two sectors – monetary and real – is artificial and unrealistic. According to them, monetary and real sectors are quite interwoven and act and react on each other. Further, Patinkin has pointed out that the *IS-LM* curve model has ignored the possibility of changes in the price level of commodities. According to him, the various economic variables such as supply of money, propensity to consume or save, investment and the demand for money not only influence the rate of interest and the level of national income but also the prices of commodities and services. Patinkin has suggested a more integrated and general equilibrium approach which involves the simultaneous determination of not only the rate of interest and the level of income but also of the prices of commodities and services.

IS-LM CURVE MODEL: EXPLAINING ROLES OF GOVERNMENT'S FISCAL AND MONETARY POLICIES

With the help of IS-LM curve model we can explain how the intervention by the Government with proper fiscal and monetary policies can influence the level of economic activity, that is, income and employment level. We explain below the impact of changes in fiscal and monetary policy on the economy in the IS-LM model.

Effect of Fiscal Policy

Let us first explain how *IS-LM* model shows the effect of increase in Government expenditure on level of income. This is illustrated in Fig. 24.6. As explained above, increase in Government expenditure which is of autonomous nature raises aggregate demand for goods and services and thereby causes an outward shift in *IS* curve, as is shown in Fig. 24.6 where increase in Government expenditure leads to the shift in *IS* curve from IS_1 to IS_2. Note that the horizontal distance between the two *IS* curves is equal to $\Delta G \times \dfrac{1}{1 - MPC}$ which shows the increase in income that occurs in Keynes's multiplier model. It will be seen from Fig. 24.6 that with the LM curve remaining unchanged, the new IS_2 curve intersects *LM* curve at point *B*. Thus, in *IS-LM* model with the increase in Government expenditure (ΔG), the equilibrium moves from point E to *B* and with this the rate of interest rises from r_1 to r_2 and income level from Y_1 to Y_2. Thus, *IS-LM* model shows that expanoinary fiscal policy of increase in Government expenditure raises both the level of income and rate of interest.

It is worth noting that in the *IS-LM* model increase in national income by Y_1Y_2 in Fig. 24.6 is less than *EK* which would occur in Keynes's model. This is because Keynes in his simple multiplier model (popularly called Keynesian cross model) assumes that investment is fixed and autonomous, whereas *IS-LM* model takes into account the fall in private investment due to the rise in interest rate that takes place with the increase in Government expenditure. That is, increase in Government expenditure crowds out some private investment.

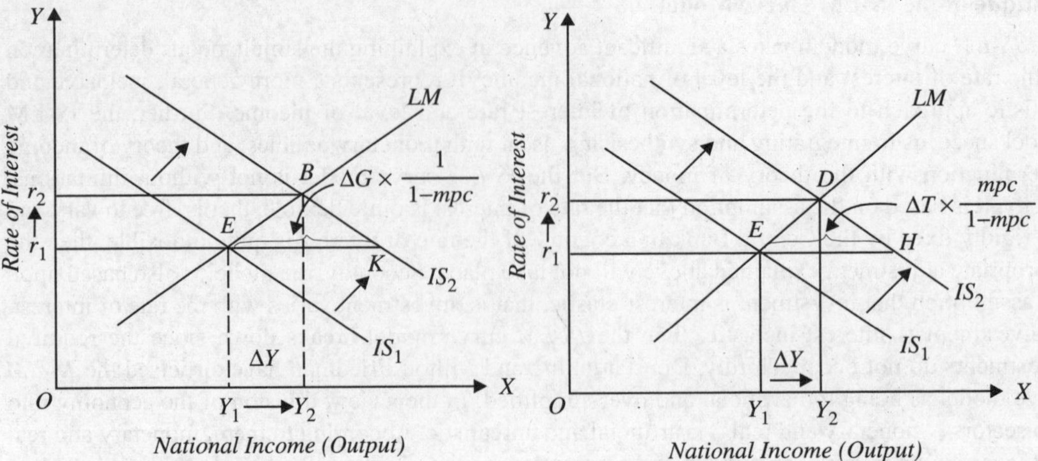

Fig. 24.6. *Impact of Increase in Government Expenditure on Interest rate and Income*

Fig. 24.7. *Effect of Cut in Taxes on Income and Interest Rate*

Likewise, it can be illustrated that the reduction in Government expenditure will cause a rightward shift in the IS curve, and given the *LM* curve unchanged, will lead to the fall in both rate of interest and level of income. It should be noted that Government often cuts expenditure to control inflation in the economy.

Reduction in Taxes. An alternative measure of expansionary fiscal policy which may be adopted is the reduction in taxes which through increase in disposable income of the people raises consumption demand of the people. As a result, cut in taxes causes a shift in the *IS* curve to the right as is shown in Fig. 24.7, from IS_1 to IS_2. It may however noted that in the Keynesian multiplier model, the horizontal shift in the *IS* curve is determined by the value of tax multiplier which is equal to $\Delta T \times \frac{MPC}{1-MPC}$ and causes level of income to increase by *EH*. However, in the *IS-LM* model, with the shift of the *IS* curve from IS_1 to IS_2 following the reduction in taxes, the economy moves from equilibrium point *E* to *D* and as is evident from Fig. 24.7, rate of interest rises from r_1 to r_2 and level of income increases from Y_1 to Y_2.

On the other hand, if the Government intervences in the economy to reduce inflationary pressures, it will raise the rates of personal taxes to reduce disposable income of the people. Rise in personal taxes will lead to the decrease in aggregate demand. Decrease in aggregate demand will help in controlling inflation. This case can also be shown by *IS-LM* curve model. In case of reduction in taxes, *IS* curve will shift to the left.

Effectiveness of Fiscal Policy

We have seen above the changes in fiscal policy shift the *IS* curve. Recall that the *IS* curve describes equilibrium in the goods market. We have explained that *IS* curve slopes downward because as the rate of interest falls investment spending increases causing a rise in aggregate demand that leads to the increase in real national income (*i.e.,*GDP). Expansionary fiscal policy may be either in the form of increase in government expenditure or cut in taxes. In both these forms of fiscal stimulus, the *IS* curve shifts to the right. In our previous Fig. 24.6. of *IS-LM* curve model we have explained that given the normal upward-sloping *LM* curve, increase in government expenditure leads to increase in output or real national income less than that under Keynesian government expenditure (*i.e.,* less than $\Delta G \times \frac{1}{1-MPC}$) because of the rise in interest rate. This increase in rate of interest causes private investment to fall, that is, increase in government expenditure *crowds out some private investment.*

When the *LM* curve is *more steep, that is, when interest-responsiveness of demand for money is less* a given increase in government expenditure will have large crowding-out effect as shown in Fig. 24.8. Initially the *IS*, and *LM* curves intersect at point E_1 and determines Y_1 level of national income and r_1 rate of interest (The given *LM* curve is relatively steep). Now suppose under the expansionary fiscal policy the government increases its expenditure equal to the horizontal distance $E_1 H$ shifting the IS_1 curve to the right to IS_2. This new IS_2 curve intersects the given steep *LM* Curve at point E_2 and, as will be seen from Fig. 24.8, rate of interest rises to r_2 and the real national income increases from Y_1 to Y_2. A larger income equal to Y_2Y_3 or *KH* has been wiped out due to crowding-out effect of rise in interest rate on investment. To conclude, in case of lower interest-responsiveness of demand for money expansionary fiscal policy is not very effective in bringing about a sufficient increase in real national income.

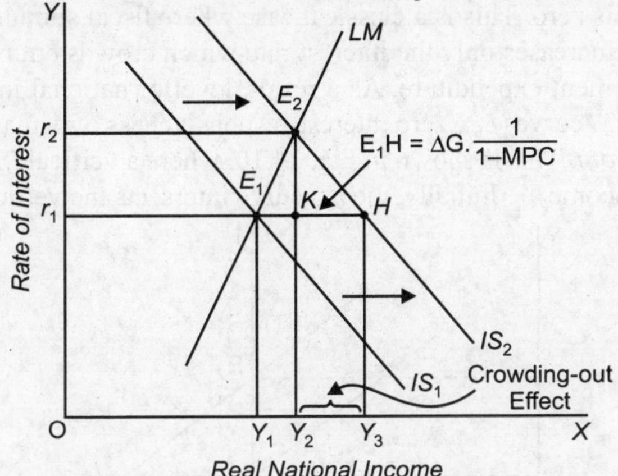

Fig. 24.8. *Large Crowding-out Effect in Case of Steep LM Curve and as a Result, Increase in Government Expenditure Causes only a Small Increase in Real National Income.*

Note that contrary to Fig. 24.8 where due to upward-sloping *LM* curve, increase in government expenditure on national income has less than full Keynesian multiplier effect on the equilibrium level of national income due to crowding-out effect of rise in rate of interest. However, when there is *infinite interest-responsiveness* of demand for money the *LM* curve is horizontal which occurs *when economy is in the liquidity trap.*

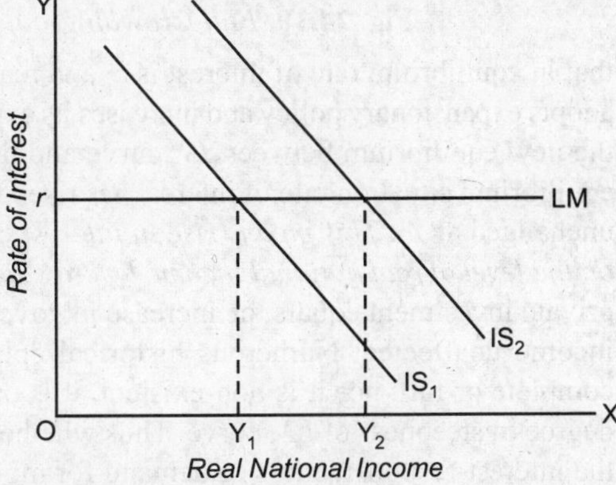

Fig. 24.9. *With a Horizontal LM Curve, When There is no Crowding-out Effect, Expansionary Fiscal Policy has Maximum Effect on National Income.*

In this case of horizontal *LM* curve shown in Fig. 24.9 increase in the government expenditure causes a shift in the *IS* curve to the right to IS_2 position. As will be see from Fig. 24.9, rate of interest remains fixed and as a result there is no crowding-out effect and the level of national income increases by the full multiplier effect, that is, by $\Delta G \dfrac{1}{1-MPC}$. It may however be noted that the horizontal *LM* curve depicting liquidity trap in the demand for money in which case there is no crowding-out effect of fiscal stimulus is an extreme case that may occur when there is severe depression in the economy.

The Classical Case of Zero Interest Responsiveness of Demand for Money and Crowding Out Effect. Expansionary fiscal policy, that is, increase in government expenditure or cut in taxes has no effect on the level of real income when the *LM* curve is vertical, that is, interest-responsiveness of demand for money is zero. This is a classical case where fiscal stimulus provided by increase in government spending increases only the interest rate which crowds out private investment equal to the increase in government expenditure. As a result, level of national income remains unaffected. Thus, with a vertical *LM* curve (*i.e.,* zero interest-responsiveness of demand for money), there is *full or complete crowding out.* This is shown in Fig. 24.10. where a vertical *LM* curve has been drawn at the level of national income Y_1. Initially, the IS_1 curve intersects the vertical *LM* curve at point E_1 so

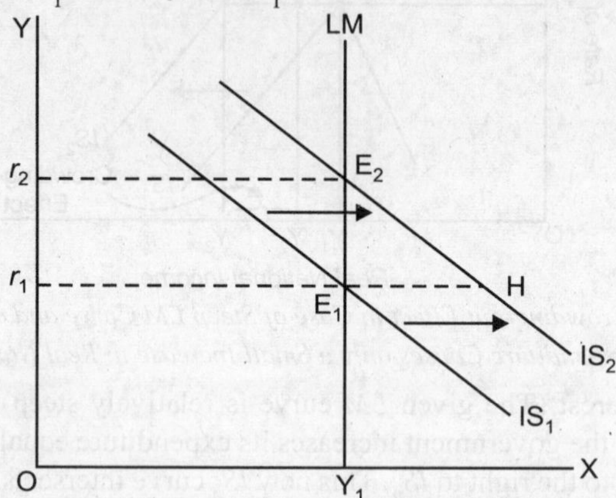

Fig. 24.10. *Full Crowding-out of Fiscal Stimulus: the Classical Case*

that in equilibrium rate of interest is r_1 and real national income is Y_1. Now suppose the government adopts expansionary policy and increases its expenditure by E_1H, shifting the *IS* curve to IS_2. However, the new equilibrium between IS_2 curve and the given vertical *LM* curve is at point E_2. In this new equilibrium situation rate of interest has risen from r_1 to r_2, the level of real national income remains unchanged at Y_1. *This means rise in interest rate has completely wiped out the expansionary effect on the level of real national income by crowding out private investment.* In this case crowding out of private investment equals the increase in government expenditure and therefore leaves real national income unaffected. Numerous historical episodes shows that the crowding-out effect is neither complete or full, nor it is non-existent, it is only partial as shown in Fig. 24.8 depending upon the degree of steepness of *LM* curve. Thus whether crowding out is zero, complete or partial depends on the interest-responsiveness of demand for money, that is, slope of the *LM* curve.

The Importance of Crowding Out Effect of Expansionary of Fiscal Policy

We have seen above that the increase in real national income (*i.e.,* multiplier effect) as a result of expansionary fiscal policy (*e.g.,* increase in government expenditure) depends on interest elasticity of demand for money (that is, slope of *LM* curve). Three points are worth considering about the effect of fiscal stimulus on real national income. First, in our analysis of *IS-LM* curves model, we have assumed that prices remain constant and the existing level of aggregate output (*i.e.,* real national income) is below the full-employment level. In this situation there is a scope for increase in output or real national income when the government expands its expenditure, causing increase in aggregate demand which induces firms to increase their output and employment. In this case the magnitude of fiscal multiplier is quite large.

However, in a fully- employed economy crowding out of fiscal stimulus occurs through a different route. When there is full-employment in the economy, increase in aggregate demand leads to the rise in price level as the economy moves up along a upward-sloping short-run aggregate supply curve. Now, the rise in price level, nominal money supply remaining constant, reduces the real money supply, that is, $\frac{M}{P}$ decreases. The reduction in money supply shifts the *LM* curve to the left raising the interest rate until the initial increase in aggregate demand as a result of expansion in government expenditure is fully wiped out This case is depicted in Fig. 24.11 where initially IS_1 and LM_1 curves

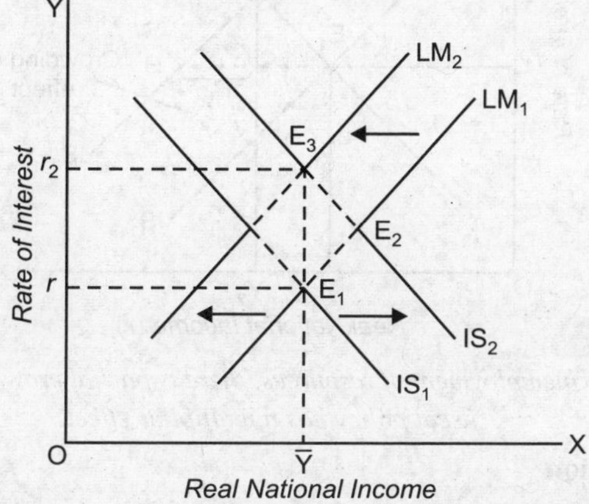

Fig. 24.11. *Full Crowding-out of Fiscal Stimulus in Case of Existing Full Employment of Resources*

interest at point E_1 and determine level of national income $\overline{Y}$ which is a full-employment level. Now, the increase in government expenditure causes *IS* curve to shift to the right to IS_2 and the economy moves to point E_2. Since with a shift in *IS* curve to IS_2 aggregate demand increases along a upward sloping short-run aggregate supply curve, this will lead to the rise in price level *resulting in decline in real money supply*. This decline in real money supply will bring about a leftward shift in the *LM* curve to the left to LM_2 position and *raise the interest rate to* r_2 so that the initial increase in national income is fully crowded out. As a result, expansionary fiscal policy fails to raise level of real national income and has therefore zero multiplier effect.

The second case occurs when there is unemployment of resources in the economy and the *LM* curve slopes upward to the right. In this case fiscal stimulus of increase in government expenditure will raise interest rate but level of real national income will also increase, Due to unemployment of resources, there will not be much increase in price level when aggregate demand increases. As a result, *crowding-out effect of fiscal stimulus is only partial*, and there is net increase in *national income*. Therefore, in this case there is a multiplier effect of expansionary fiscal policy though it is less than the Keynesian multiplier effect (*i.e.,* $\Delta G \cdot \frac{1}{1-MPC}$). This case is depicted in Fig. 24.12 where initially IS_1 and *LM* curves intersect at point E_1. Expansionary fiscal policy shifts *IS* curve to IS_2 equal to the horizontal distance, E_1H. With the given *LM* curve and the new IS_2 curve the new equilibrium is reached at point E_2 and, as will be seen from Fig. 24.12, the national income increases from Y_1 to Y_2, the income equal to *KH* has been wiped out due to crowding-out effect of rise in interest rate from r_1 to r_2. However, there is further effect of expansionary fiscal policy. With the net increase in national income from Y_1 to Y_2 resulting from shift in *IS* curve from IS_1 to IS_2 the *level of saving* will increase. This increase in saving enables the economy to finance a part of budget deficit with smaller amount of government borrowing which would ensure interest rate will not rise much and as a result crowding-out effect of expansionary fiscal policy on private investment will be smaller.

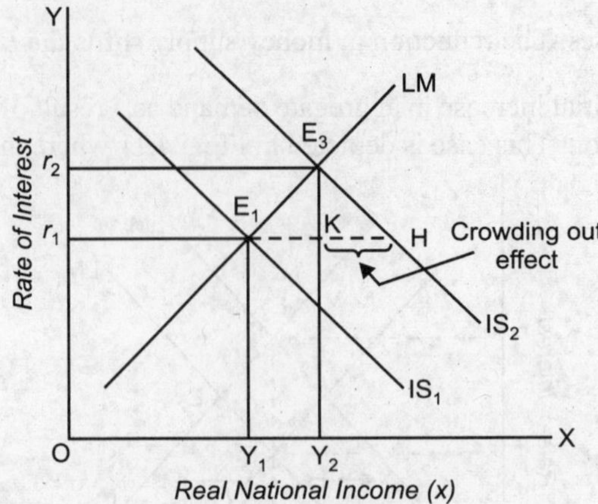

Fig. 24.12. *In case of Unemployment of resources, there is partial crowding-out and expansionary fiscal policy has a multiplier effect.*

Monetary Accommodation

The third case occurs when there is unemployment in the economy so that there is possibility of increase in output as a result of increase in aggregate demand. In this case interest rate need not rise when there is increase in government spending shifting *IS* curve to the right but at the same time the Central Bank of the country raises the money supply to prevent the rise in interest as a result of increase in government spending. This is called monetary accommodation by the Central Bank. Thus "*Monetary policy is accommodating when in the course of fiscal expansion, the money supply is increased to prevent interest rates from rising*"[1]. In case of sufficient monetary accommodation rate of interest does not rise and therefore there is no crowding-out effect on private investment, the expansionary fiscal policy brings about increase in national income equal to increase in government expenditure times the Keynesian multiplier (*i.e.,* $\Delta G \times \dfrac{1}{1-MPC}$). This case of monetary accommodation of fiscal expansion is depicted in Fig. 24.13 in which it will be seen that initially equilibrium is at point E_1, where IS_1 and LM_1 curves intersect to determine Y_1 level of income and r_1 rate of interest. Now, fiscal stimulus by the government shifts the *IS* curve to IS_2 and the given the LM_1 curve, equilibrium will be at point E_3 where rate of interest rises to r_2 which would crow out private investment. To prevent this crowding out, the Central Bank adopts the monetary accommodation policy and for this it increases money supply sufficiently so that *LM* curve shifts to the right to LM_2 which intersects IS_2 curve at point E_2 so that interest remains at the initial level r_1 and income increases to Y_2. Thus the increase in income equal to $E_1 E_2$ or $Y_1 Y_2$ that occurs equals the increase in government expenditure times the Keynesian multiplier (*i.e.,* $\Delta G.\dfrac{1}{1-MPC}$), crowding out

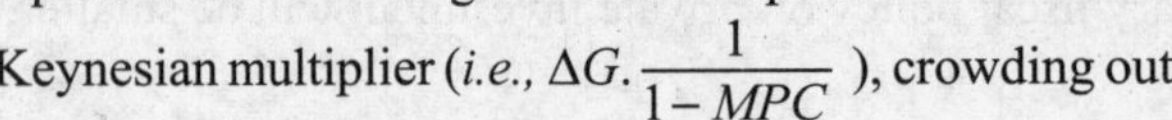

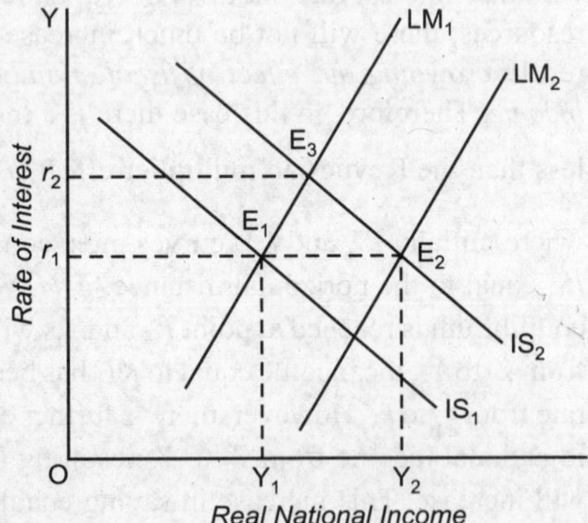

Fig. 24.13. *Monetary Accommodation of Expansionary Fiscal Policy*

1. Dornbusch, Fischer, and Startz, *Macroeconomics,* Tata MacGraw Hill, 8th Edition, p. 253

having been eliminated by expansion in money supply by the Central Bank.

It may be noted that in 2008-09 and 2009-10 when due to global financial crisis India faced the problem of large slowdown of the economy, the Indian government adopted fiscal stimulus measures such as raising its expenditure through borrowings on a large scale from the market and cut rates of many indirect taxes to prevent sharp slowdown of the Indian economy, the Reserve Bank of India adopted accommodative monetary policy so that rate of interest does not rise. To this end, the RBI greatly reduced its repo rate (the rate at which it lends to the commercial banks) and also lowered the cash reserve ratio (CRR) of the banks so that more funds are available with them to lend to the business firms for investment and consumption loans, such as housing loans, car loans at lower rate of interest. This policy succeeded and India achieved 6.7 per cent growth in the crisis year 2008-09 and 8.6 per cent in 2009-10.

IMPACT OF MONETARY POLICY

Through making appropriate changes in monetary policy the Government can influence the level of economic activity. Monetary policy may also be expansionary or contractionary depending on the prevailing economic situation. *IS-LM* model can be used to show the effect of expansionary and tight monetary policies. As has been explained above, a change in money supply causes a shift in the *LM* curve; expansion in money supply shifts it to the right and decrease in money supply shifts it to the left.

Suppose the economy is in grip of recession, the Government (through its Central Bank) adopts the expansionary monetary policy to lift the economy out of recession. Thus, it takes measures to increase the money supply in the economy. The increase in money supply, state of liquidity preference or demand for money remaining unchanged, will lead to the fall in rate of interest. At a lower interest there will be more investment by businessmen. More investment will cause aggregate demand and income to rise. This implies that with expansion in money supply *LM* curve will shift to the right as is shown in Fig. 24.14. As a result, the economy will move from equilibrium point *E* to *D* and with this the rate of interest will fall from r_1 to r_2 and national income will increase from Y_1 to Y_2. Thus, IS-LM model shows the expansion in money supply lowers interest rate and raises income. We have also indicated what is called monetary transmission mechanism, that is, how *IS-LM* curve model shows the expansion in money supply leads to the increase in aggregate demand for goods and services. We have thus seen that increase in money supply lowers the rate of interest which then stimulates more investment demand. Investment demand through multiplier process leads to a greater increase in aggregate demand and national inocme.

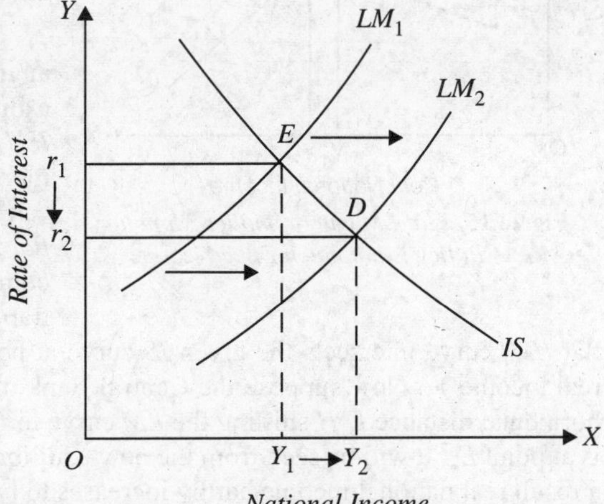

Fig. 24.14. *Effect of Expansion in Money Supply on Interest Rate and Income*

If the economy suffers from inflation, the Government will like to check it. Then its Central Bank should adopt tight or contractionary monetary policy. That is, it should reduce the money supply. *IS-LM model can be used to show, as we have seen above in case of expansionary monetary policy, that reduction in money supply will cause a leftward shift in LM curve and will lead to the rise in interest rate and fall in the level of income.*

Effectiveness of Monetary Policy

It is important to explain to what extent monetary policy is effective in influencing level of national output. Transmission of changes in money supply, say through open market operations, runs as follows. In the first step increase in money supply following the expansionary monetary policy leads to the fall in rate of interest. In the second step of transmission mechanism, fall in rate of interest causes increase in total spending or aggregate demand (especially, investment expenditure) Finally, the aggregate output adjusts to the changes in aggregate demand. However, some linkages in transmission process of the effect of changes in money supply may not work. First, the change in money supply may not lead to a change in rate of interest. J.M. Keynes pointed out the *liquidity trap* that may occur at a very low interest rate prevents the fall in rate of interest following the expansion in money supply. The liquidity trap is a situation in which the public prefers at a given rate of interest to hold whatever money is supplied, In this case demand for money is perfectly elastic and *LM* curve is a horizontal straight line. *With a horizontal LM curve, the 'increase in money supply does not cause a shift in it and therefore does not affect the rate of interest.* With rate of interest remaining unaffected, the expansion in money supply, say through open market operations, will not affect the aggregate spending (both consumption and investment demand). With no change in aggregate demand or spending, the level of national output will remain unchanged. Thus in the *situation of liquidity trap "monetary policy carried out through open market operations is powerless to affect the interest rate"*[2].

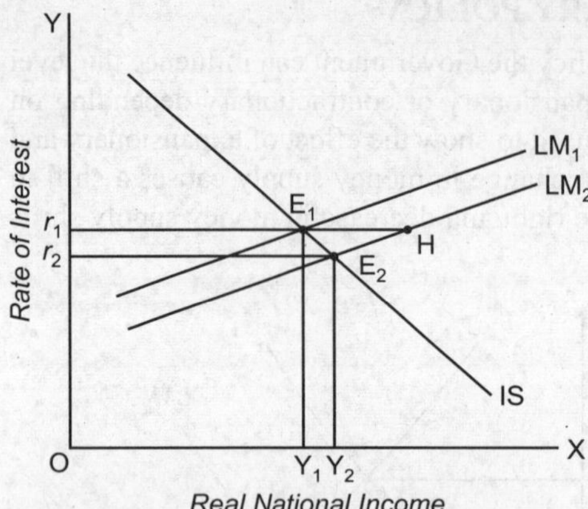

Fig 24.15. *Flat LM Curve implies Monetary Policy is not much Effective.*

The effect of increase in money supply on aggregate output in case of horizontal *LM* curve is a bit complicated to show diagrammatical through *IS-LM* curves model. However, the ineffectiveness of monetary policy in case of the liquidity-trap situation can be easily understood if we take the case of *relatively flat LM curve* (which can be considered as proxy for completely horizontal *LM* curve) caused by liquidity trap. This is depicted in Fig. 24.15 where a relatively flat LM_1 curve intersects the given *IS* curve at point E_1 determines rate of interest r_1 and level of real income Y_1. Now suppose the Central Bank of the country expands money supply equal to the horizontal distance E_1H shifting the *LM* curve to LM_2. With the given *IS* curve the new equilibrium is at point E_2. It will be seen from the new equilibrium that the interest rate falls only slightly and as a result real national income hardly increases to have any impact on recessionary conditions.

The second factor causing ineffectiveness of monetary policy occurs in the third step of transmission mechanism of changes in aggregate spending or demand to changes in interest rate. This happens when changes in rate of interest have insignificant effect on autonomous planned spending, especially investment expenditure. This situation occurs when business firms are so pessimistic about the future prospects of earning profits that business firms are reluctant to undertake any further investment, in response to lower interest rate. As a result, increase in money supply causing lower interest rates does not lead to the increase in real national income. Under these circumstances of unresponsiveness of investment to changes in interest rate the *IS* curve is a vertical straight line as shown in Fig. 24.16 where with the increase in money supply (ΔM) equal to E_1H does not raise real national income which is 'stuck' at Y_1 level.

Banks' Reluctance to Lend. The third case when monetary policy has only limited effect on investment spending and therefore on real national income occurs when banks are reluctant to increase lending for investment in response to lower interest rate. This occurred in the US in 1991 and then in 2008-09 and 2009-10 when global financial crisis occured. This situation also seems to have occurred in India in 2008-09 following the global financial crisis. In this case when the Central

2. Dornbusch, Fischer and Startz, op. 245

Bank of the US expanded money supply leading to lower interest rate, banks were reluctant to increase lending for the fear that lending might create bad loans with little possibility of being paid back. In the US. banks had made bad loans relating to real estate *i.e.*, housing. Instead of lending for private spending and investment, banks purchased government securities such as treasury bills which are quite safe investment by banks. If banks do not lend for private investment, the link in transmission mechanism that involves more private investment in response to lower interest rate breaks down to give boost to real national income. In India too, when the Reserve Bank of India lowered its repo rate and cash reserve ratio (*CRR*) in 2008-09 and 2009-10 for stimulating the economy the banks were not much enthusiastic for lending to private firms for fear of default by them in repaying the loans. Therefore, to earn some return on their excess cash reserves due to easy monetary policy, some banks opted for investing in government securities beyond what was required under statutory liquidity ratio.

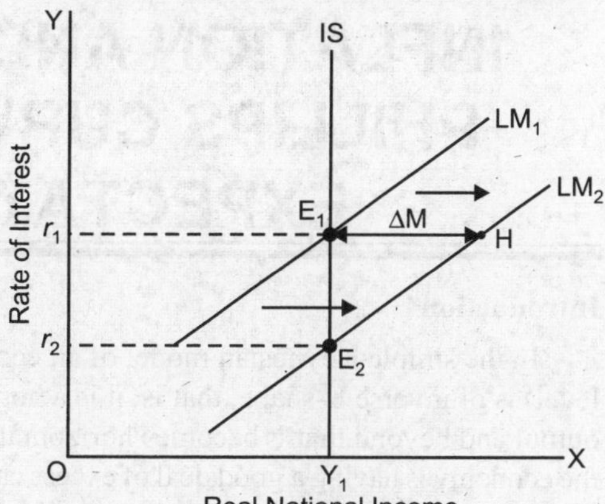

Fig. 24.16. *Zero interest-responsiveness of investment spending (i.e., vertical IS curve) implies expansionary monetary policy is ineffective in raising national income.*

However, both in the US in 1991 and 2008-09 and in India in 2008-09 ***further cuts*** *in interest rate and expansion in money supply,* brought about boost in lending for private investment and consumption for buying durable consumer goods leading to the recovery in the economies. In this context, it is worth mentioning the policy of *quantitative easing* (*QE*) under which the Federal Reserve of the US is attempting to revive the American economy under the readership of its governor Ben Bernanke. Under the policy of quantitative easing the Federal Reserve is buying government securities and pumping into the American economy more money, that is, the US dollars on a large scale keeping zero rate of interest. This policy of quantitative easing ultimately seems to have worked in raising the levels of output and employment in the US and thus achieving the recovery of the US economy in 2013 with rate of unemployment falling to 7.6 per cent from the peak of 10 per cent in the year 2009 and also growth of the US economy has risen to 2.5%.

QUESTIONS FOR REVIEW

1. What is IS curve? How is it derived from goods market equilibrium?
2. Define IS curve. What are the factors that determine the slope or steepness of IS Curve?
3. What is LM curve? How is it derived from money market equilibrium? On factors does its slope and position depend?
4. How do the product and money markets interact to determine the short-run equilibrium level of national income and the rate of interest? *[D.U., B.A. (H), 1986]*
5. Using IS-LM model analyse the impact of an increase in money supply on the equilibrium level of income and rate of interest. *[D.U. B.A. (H), 1984]*
6. Using IS-LM curve model analyse the impact of an increase in the money supply on the equilibrium level of national income and the rate of interest. *[D.U., B.A. (H)., 1990]*
7. Explain IS and LM curves. What is the role of these two curves in simultaneous determination of equilibrium rate of interest and level of income?
8. (a) Derive IS curve for an economy. What are the factors which determine slope of the IS curve?
   (b) Show how IS curve will change as a result of increase in Government expenditure.
9. Why does IS curve slope downward ? How is IS curve related to investment multiplier in the Keynesian Income theory? What are the factors which cause a shift in IS curve?
10. Explain with the aid of IS-LM curves role of monetary policy for stabilisation of the economy.
11. Explain with the help of IS-LM curve model, role of fiscal policy in overcoming recession in the economy.

CHAPTER 25

INFLATION AND UNEMPLOYMENT: PHILLIPS CURVE AND RATIONAL EXPECTATIONS THEORY

Introduction

In the simple Keynesian model of an economy, the aggregate supply curve (with variable price level) is of inverse L-shape, that is, it is a horizontal straight line up to the full-employment level of output and beyond that it becomes horizontal. This means that during recession or depression when the economy is having a good deal of excess capacity and large-scale unemployment of labour and idle capital stock, the aggregate supply curve is perfectly elastic. When full employment level of output is reached, aggregate supply curve becomes perfectly inelastic. With this shape of aggregate supply curve assumed in the simple Keynesian model, increase in aggregate demand before the level of full employment, causes increase in the level of real national output and employment with price level remaining unchanged. That is, no cost has to be incurred in the form of rise in the price level (*i.e.*, inflation rate) for raising the level of output and reducing unemployment. In the Keynesian model, once the full-employment level of output is reached and aggregate supply curve becomes vertical, further increase in aggregate demand caused by the expansionary fiscal and monetary policies will only raise the price level in the economy. That is, in this simple Keynesian model, inflation occurs in the economy only after full-employment level of output has been attained. Thus, *in the simple Keynesian model with inverse L-shaped aggregate supply curve there is no trade off or clash between inflation and unemployment*.

INFLATION-UNEMPLOYMENT TRADE-OFF: PHILLIPS CURVE

However, the actual empirical evidence did not fit well in the above simple Keynesian macro model. A noted British economists, A.W. Phillips[1] published an article in 1958 based on his good deal of research using historical data from the U.K. for about 100 years in which he arrived at the conclusion that *there in fact existed an inverse relationship between rate of unemployment and rate of inflation*. This inverse relation implies a trade-off, that is, for reducing unemployment, price in the form of a higher rate of inflation has to be paid, and for reducing the rate of inflation, price in terms of a higher rate of unemployment has to be borne. On graphically fitting a curve to the historical data Phillips obtained a downward sloping curve exhibiting the inverse relation between rate of inflation and the rate of unemployment and this curve is now named after his name as *Phillips Curve*. This Phillips curve is shown in Fig. 25.1 where along the horizontal axis the rate of unemployment and along the vertical axis the rate of inflation is measured. It will be seen that when rate of inflation is 10 per cent, the unemployment rate is 3 per cent, and when rate of inflation is reduced to 5 per cent per annum, say by pursuing contractionary fiscal policy and thereby reducing aggregate demand, the rate of unemployment increases to 8 per cent of labour force. The actual Phillips curve drawn from the data of sixties (1961-69) for the United States also shows the inverse relation between unemployment rate and rate of inflation (see Fig. 25.2). Such empirical data

1. A.W. Phillips, "Relation between Unemployment and Rate of Change in Money Wages in the United Kingdom, 1861-1957." *Economica* 25 (Nov. 1958), pp. 283-299.

pertaining to the fifties and sixties for other developed countries seemed to confirm the Phillips curve concept. On the basis of this, many economists came to believe that there existed a *stable Phillips curve* which depicted a predictable inverse relation between inflation and unemployment. Further, on the basis of a *stable Phillips curve* for a country, they emphasised the trade off that confronts the economic policy makers. This trade off presents a dilemma for the policy makers; should they choose a higher rate of inflation with lower unemployment or a higher rate of unemployment with a low inflation rate. In what follows we first explain the rationale underlying the Phillips curve, that is, how the inverse relationship between inflation and unemployment can be theoretically explained. We will further explain why this concept of stable Phillips curve depicting inverse relation between inflation and unemployment broke down during seventies and early eighties. During seventies a strange phenomenon was witnessed in the USA and Britain when there existed *a high rate of inflation side by side with high unemployment rate*. This was contrary to both Phillips curve concept and the simple Keynesian model. This simultaneous existence of both high rate of inflation and high unemployment rate (or low level of real national product) during the seventies and early eighties has been described as *stagflation* which has been explained in the previous chapter.

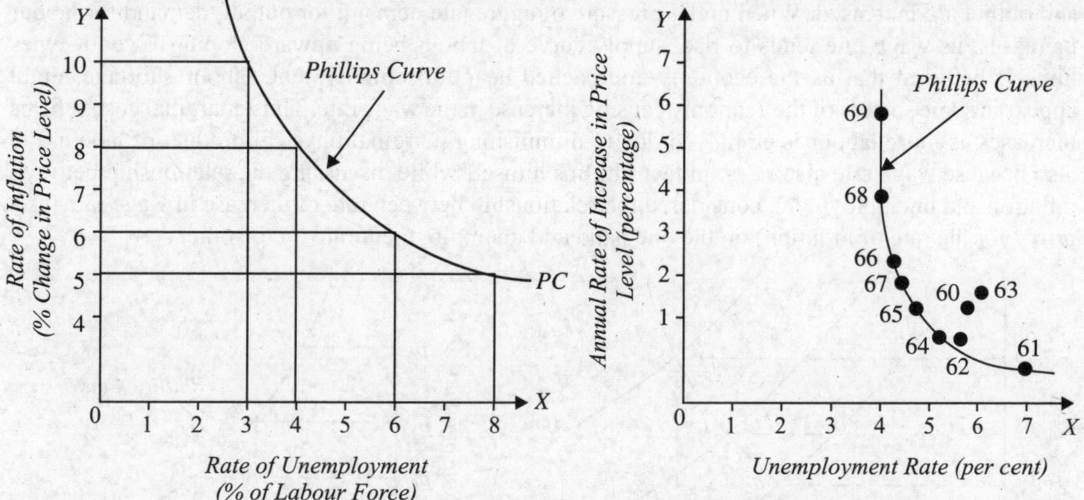

Fig. 25.1. *Phillips Curve showing Negative Relationship between Rates of Inflation and Unemployment*

Fig. 25.2. *Phillips Curve of the Sixties of the United States*

Explanation of Phillips Curve

Let us first provide an explanation for the Phillips curve. Both Keynesians and Monetarists agreed to the existence of the Phillips curve. The explanation of Phillips curve by the Keynesian economists is quite simple and is graphically illustrated in Fig. 25.3. It may be noted that Keynesian economists assume the upward-sloping aggregate supply curve. In fact, Keynes himself recognised that the curve *AS* is upward sloping in intermediate range, that is, as the economy approaches near full employment level, the aggregate supply curve slopes upward. According to Keynesian economists, aggregate supply curve is upward sloping for two reasons. First, as output is increased by the firms in the economy, diminishing returns to variable factors, especially to labour, accure resulting in fall in marginal physical product (MPP_L) of labour. With money wage rate (W) as given and fixed, the fall in the marginal physical product of labour causes the rise in the marginal cost (MC) of production $\left(\text{Note that } MC = \dfrac{W}{MPP_L}\right)$. With the fall in the MPP of labour, wage rate remaining constant, the term $\dfrac{W}{MPP_L}$ measuring marginal cost (MC) will rise.

The second reason for the marginal cost to go up is the rise in the wage rate as employment and output are increased. When under pressure of aggregate demand for output, demand for labour increases, its wage rate tends to rise, supply curve of labour being upward sloping. Even Keynes himself believed that as the economy approached near full employment, labour shortage might appear in some sectors of the economy causing increase in the wage rate. Thus, marginal cost of firms increases as more labour is employed due to diminishing marginal physical product of labour and also because wage rate also rises. In fact Phillips himself while discussing the relationship between inflation and unemployment, considered the relationship between rate of increase in wage rate (as a proxy for the rate of inflation) on the one hand and unemployment rate on the other.

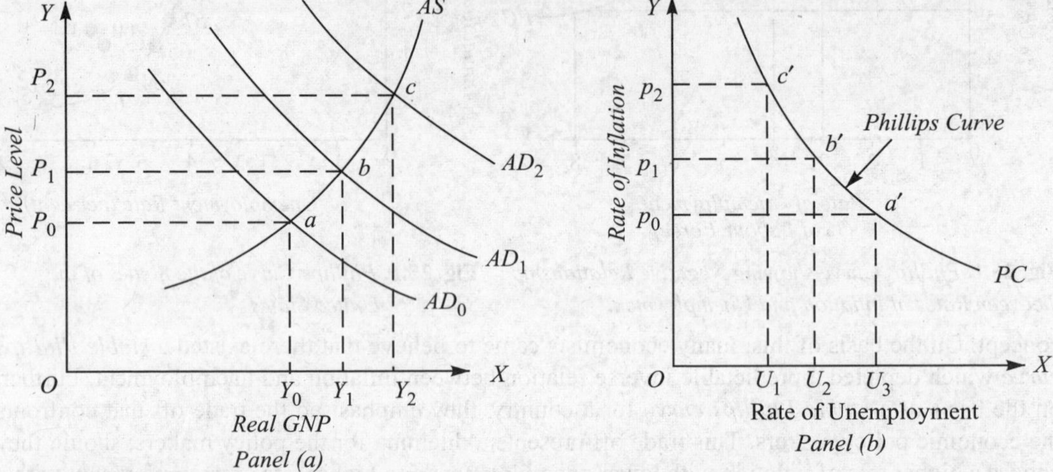

Fig. 25.3. *Keynesian Explanation of Phillips Curve*

Now, it will be seen from panel (*a*) of Fig. 25.3 that with the initial aggregate demand curve AD_0 and the given aggregate supply curve *AS*, the price level P_0 and output level Y_0 are determined. Now, suppose the aggregate demand curve increases from AD_0 to AD_1, it will be seen that price level rises to P_1 and aggregate national output increases from Y_0 to Y_1. Note that increase in aggregate national product means increase in employment of labour and therefore reduction in unemployment rate. Thus the rise in the price level from P_0 to P_1 (*i.e.*, occurrence of inflation) results in lowering of unemployment rate showing inverse relation between the two. Further, if aggregate demand increases to AD_2, the price level further rises to P_2 and national output increases to Y_2 which will further lower the rate of unemployment. The greater the rate at which aggregate demand increases, the higher will be the rate of inflation which will cause greater increase in aggregate output and employment resulting in much lower rate of unemployment. Thus, *a higher rate of increase in aggregate demand and consequently a higher rate of rise in price level is associated with the lower rate of unemployment and vice-versa.* This is what is represented by Phillips curve. Consider panel (*b*) of Fig. 25.3 where point *a′* on the downward sloping Phillips curve *PC* corresponds to point *a* of panel (*a*) of Fig. 25.3. In panel (*b*) of the Fig. 25.3 we have shown the rate of unemployment equal to U_3 corresponding to the price level P_0 of panel (*a*). When the aggregate demand shifts to AD_1, there is a certain rate of inflation and price level rises to P_1 and aggregate output expands to Y_1. As seen above, this increase in aggregate output leads to the increase in employment of labour bringing about decline in unemployment rate. Suppose the rate of rise in the price level (*i.e.*, the rate of inflation) when it increases from P_0 to P_1 in panel (*a*) following the increase in aggregate demand is greater than the rate of rise in the price level of the previous period, we obtain a lower rate of unemployment U_2 than before corresponding to a higher inflation rate p_1 in the Phillips curve *PC* in panel (*b*). With a still higher rate of inflation, say p_2, when price level rises from P_1 to P_2 in panel (*a*) following the increase in aggregate demand to

AD_2, we have a further lower rate of unemployment equal to U_1 in panel (*b*) corresponding to point c' on the Phillips curve *PC*. This gives us a downward-sloping Phillips curve *PC*.

It is clear from above the through increase in aggregate demand and upward-sloping aggregate supply curve, Keynesians were able to explain the downward-sloping Phillips curve showing the negative relation between rates inflation and unemployment.

Collapse of Phillips Curve (1971-91)

During the sixties Phillips curve became an important concept of macroeconomic analysis. The stable relationship described by it suggested that policy makers could have a lower rate of unemployment if they could bear with a higher rate of inflation. On the contrary, they could achieve a low rate of inflation only if they were prepared to reconcile with a higher rate of unemployment. But a stable Phillips curve could not hold good during the seventies and eighties, especially in the United States. Therefore, experience in the two decades (1971-91), has prompted some economists to say that the stable Phillips curve has disappeared. Fig. 25.4 shows that data regarding the behaviour of inflation and unemployment during the seventies and eighties in the United States which do not conform to a stable Phillips curve. *In these two decades we have periods when rates of both inflation and unemployment increased (that is, a high rate of inflation was associated with a high unemployment rate, which shows the absence of trade off.* We have shown the data of inflation rate and unemployment in case of the United States in Fig. 25.4. From the data it appears that instead of remaining stable, the Phillip curve shifted to the right in the seventies and early eighties and to the left during the late eighties. (see Fig. 25.4).

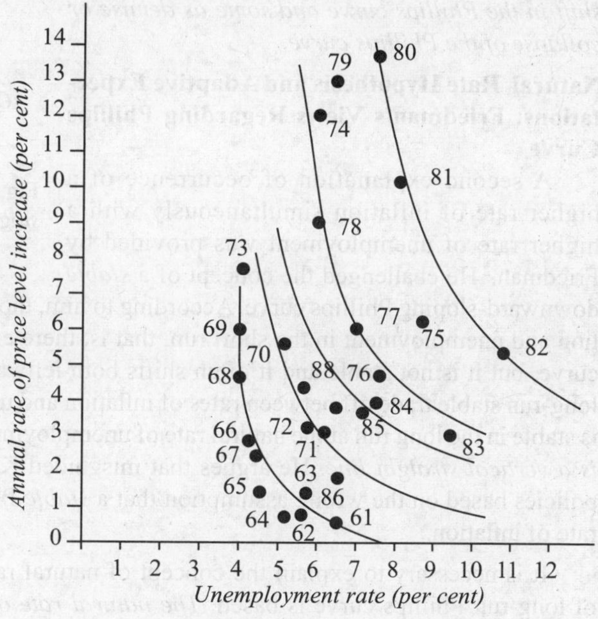

Fig. 25.4. *Shift in Phillips Curve: United States*

Causes of Shift in Phillips Curve

Now, what could be the cause of shift in the Phillips curve. There are two explanations for this. First, according to Keynesians, the occurrence of higher inflation rate along with the increase in unemployment rate witnessed during the seventies and early eighties was due to the *adverse supply shocks* in the form of four fold increase in the prices of oil and petroleum products delivered to the American economy first in 1973-74 and then again in 1979-80. Consider Fig. 25.5 where AD_0 and AS_0 are in equilibrium at point *E* and determine price level OP_0 and aggregate national output OY_0. The hike in price of oil by *OPEC*, the Cartel of Oil Producing Middle East Countries brought about a rise in the cost of production of several commodities for the production of which oil was used as an energy input. Further, the oil price hike also raised the transportation costs of all commodities. The increase in cost of production and transportation of commodities caused a shift in the aggregate supply curve upward to the left. This is generally described as adverse supply shock which raised the unit cost at each level of output. It will be seen from Fig. 25.5 that due to this adverse supply shock aggregate supply curve has shifted to the left to the new position AS_1 which intersects the given aggregate demand curve AD_0 at point

H. At the new equilibrium point *H*, price level has risen to P_1 and output has fallen to OY_1 which will cause unemployment rate to rise. Thus, we have a higher price level with a higher unemployment rate. This explains the rise in the price level with the rise in the unemployment rate, the phenomenon which was witnessed during the seventies and early eighties in the developed capitalist countries such as the U.S.A. Note that *this has been interpreted by some economists as a shift in the Phillips curve and some as demise or collapse of the Phillips curve.*

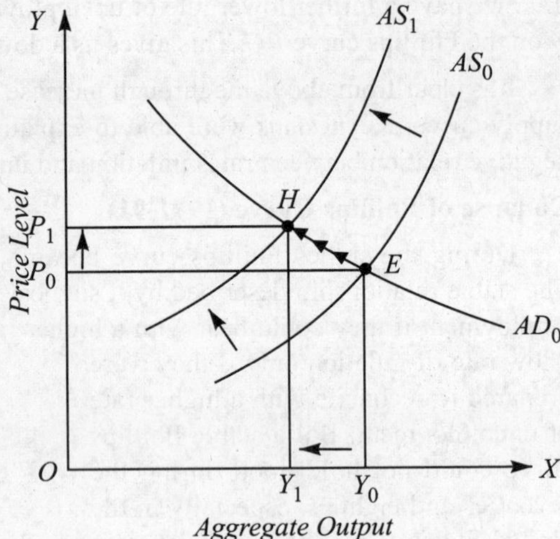

Fig. 25.5. *Adverse supply shock giving rise to stagflation and break down of Phillips Curve*

Natural Rate Hypothesis and Adaptive Expectations: Friedman's Views Regarding Phillips Curve

A second explanation of occurrence of a higher rate of inflation simultaneously with a higher rate of unemployment was provided by Friedman.[2] He challenged the concept of a *stable* downward-sloping Phillips curve. According to him, though there is a trade off between rate of inflation and unemployment in the short run, that is, there exists a short-run downward sloping Phillips curve, but it is not stable and it often shifts both leftward or rightward. He argued that there is no long-run stable trade off between rates of inflation and unemployment. His view is that the economy is stable in the long run at the natural rate of unemployment and therefore *the long-run Phillips curve is a vertical straight line*. He argues that misguided Keynesian expansionary fiscal and monetary policies based on the wrong assumption that a *stable* Phillips curve exists only result in increasing rate of inflation.

It is necessary to explain the concept of natural rate of unemployment on which the concept of long-run Phillips curve is based. *The natural rate of unemployment is the rate at which in the labour market the current number of unemployed is equal to the number of jobs available.* These unemployed workers are not employed for the frictional and structural reasons, though the equivalent number of jobs are available for them. For instance, the fresh entrants may spend a good deal of time in searching for the jobs before they are able to find work. Further, some industries may be registering a decline in their production rendering some workers unemployed, while others may be growing creating new jobs for workers. But the unemployed workers may have to be provided new training and skills before they are deployed in the newly created jobs in the growing industries. It is these frictional and structural unemployments that constitute the natural rate of unemployment. **Since the equivalent number of jobs are available for them, full employment is said to prevail even in the presence of this natural rate of unemployment**. It is presently believed that 4 to 5 per cent rate of unemployment represents a natural rate of unemployment in the developed countries.

Another important thing to understand from Friedman's explanation of shift in the short-run Phillips curve is that expectations about the future rate of inflation play an important role in it. Friedman put forward *a theory of adaptative expectations* according to which people form their expectations on the basis of previous and present rate of inflation, and change or adapt their expectations only when the actual inflation turns out to be different from their expected rate. According to this Friedman's theory of adaptive expectations, there may be a trade off between rates of inflation and unemployment in the short run, but there is no such trade off in the long run.

---

2. Milton Friedman, Inflation and Unemployment, *Journal of Political Economy*, 1977.

Long-run Phillips Curve: Natural Rate Hypothesis. The view of Friedman and his follower monetarists illustrated in Fig. 25.6. To begin with SPC_1 is the short run Phillips curve and the economy is at point A_0, on it corresponding to the natural rate of unemployment equal to 5 per cent of labour force. The location of this point A_0 on the short-run Phillips curve depends on the level of aggregate demand. Further, we assume that the economy is currently experiencing a rate of inflation equal to 5%. The other assumption we make is that nominal wages have been set on the expectations that 5 per cent rate of inflation will occur.

Now, suppose for some reasons the government adopts expansionary fiscal and monetary policies to raise aggregate demand. The consequent increase in aggregate demand will cause the rate of inflation to rise, say to seven per cent. Given the level of money wage rate which was fixed on the basis that the 5 per cent rate of inflation would continue to occur, the higher price level than expected would raise the profits of the firms which will induce the firms to increase their output and employ more labour. As a result of the increase in aggregate demand resulting in a higher rate of inflation and more output and employment, the economy will move to point A_1 on the short-run Phillips curve SPC_1 in Fig. 25.6, where unemployment has decreased to 3.5 per cent while inflation rate has risen to 7%. It may be noted from Fig. 25.6 that in *moving from point A_0 to A_1, on SPC_1 the economy accepts a higher rate of inflation at the cost of achieving a lower rate of unemployment. Thus, this is in conformity with the concept of Phillips curve explained earlier.* However, the advocates of natural unemployment rate theory interpret it in a slightly different way. They think that lower rate of unemployment achieved is only a temporary phenomenon. They think *when the actual rate of inflation exceeds the one that is expected unemployment rate will fall below the natural rate only in the short run. In the long run, the natural rate of unemployment will be restored.*

Long-run Phillips Curve and Adaptive Expectations. This brings us to the concept of long-run Phillips curve, which Friedman and other natural rate theorists have put forward. According to them, the economy will not remain in a stable equilibrium position at A_1. This is because the workers will realise that due to the higher rate of inflation than the expected one, their real wages and incomes have fallen. The workers will therefore demand higher nominal wages to restore their real income. But as nominal wages rise to compensate for the higher rate of inflation than expected, profits of business firms will fall to their earlier levels. This reduction in their profit implies that the original motivation that prompted them to expand output and increase employment resulting in lower unemployment rate will no longer be there. Consequently, they will reduce employment till the unemployment rate rises to the natural level of 5%. That is, with the increase is nominal wages in Fig. 25.6 the economy will move from A_1 to B_0, at a higher inflation rate of 7%. It may be noted that the higher level of aggregate demand which generated inflation rate of 7% and caused the economy to shift from A_0 to A_1 still persists.

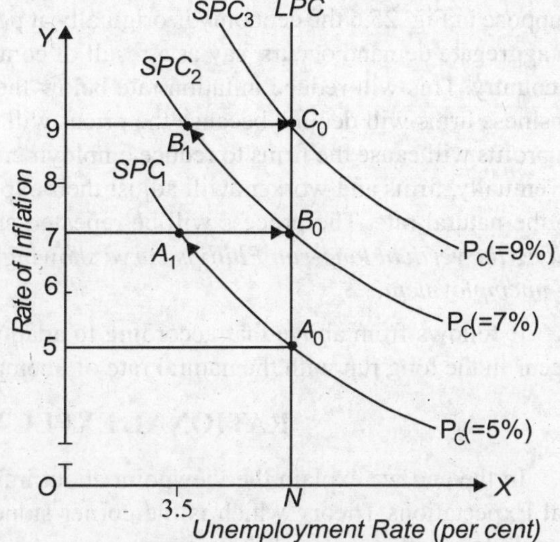

Fig. 25.6. *Shift in Short-run Phillips Supply Curve and Long-run Phillips Curve: Natural Rate Hypothesis*

Further, at point B_0, and with the actual present rate of inflation equal to 7 per cent, the workers will now expect this 7 per cent inflation rate to continue in future. As a result, the short-run Phillips curve *SPC* shifts upward from SPC_1 to SPC_2. It therefore follows, according to Friedman and other natural rate theoriests, the movement along a Phillips curve *SPC* is only a temporary or short-run phenomenon. In the long when nominal wages are fully adjusted to the changes in the inflation rate and consequently unemployment rate comes back to its natural level, a new short-run Phillips curve is formed at the higher expected rate of inflation.

However, the above process of reduction in unemployment rate and then its returning to the natural level may continue further. The Government may misjudge the situation and think that 7 per cent rate of inflation is too low and adopt expansionary fiscal and monetary policies to increase aggregate demand and thereby to expand the level of employment. With the new increase in aggregate demand, the price level will rise further with nominal wages lagging behind in the short-run. As a result, profits of business firms will increase and they will expand output and employment causing the reduction in rate of unemployment and rise in the inflation rate. With this, the economy will move from B_0 to B_1 along their short run Phillips curve SPC_2. After sometime, the workers will recognise the fall in their real wages and press for higher normal wages to compensate for the higher rate of inflation than expected. When this higher nominal wages are granted, the business profits decline which will cause the level of employment to fall and unemployment rate to return to the natural rate of 5%. That is, in Fig. 25.6 the economy moves from point B_1 to C_0. The new short run Phillips curve will now shift to SPC_2 passing through point C_0. The process may be repeated again with the result that while in the short run, the unemployment rate falls below the natural rate and in the long run it returns to its natural rate. But throughout this process the inflation rate continuously goes on rising. On joining points such as A_0, B_0, C_0 corresponding to the given natural rate of unemployment we get a vertical long run Phillips curve *LPC* in Fig. 25.6. Thus, in the adaptive expectations theory of the natural rate hypothesis while the short run Phillips curve is downward sloping indicating that trade-off between inflation and unemployment rate the short run, the long run Phillips curve is a vertical straight line showing that no trade off exists between inflation and unemployment in the long run.

It is important to remember that adaptive expectations theory has also been applied to explain the reverse process of disinflation, that is, fall in the rate of inflation as well as inflation itself. Suppose in Fig. 25.6 the economy is originally at point C_0 with 9% rate of inflation. Now, if a decline in aggregate demand occurs, say as a result of contraction of money supply by the Central Bank of a country. This will reduce inflation rate below the 9 per cent expected rate. As a result, profits of business firms will decline because the prices will be falling more rapidly than wages. The decline in profits will cause the firms to reduce employment and consequently unemployment rate will rise. Eventually, firms and workers will adjust their expectations and the unemployment rate will return to the natural rate. The process will be repeated and the economy in the long run will *slide down along the vertical long-run Phillips curve showing falling rate of inflation at the given natural rate of unemployment*.

It follows from above that according to adaptive expectations theory any rate of inflation can occur in the long run with the natural rate of unemployment.

RATIONAL EXPECTATIONS THEORY

In the end we explain the viewpoint about inflation and unemployment put forward by Rational Expectations Theory which is the corner stone of recently developed macroeconomic theory, popularly called *new classical macroeconomics*. As explained above, Friedman's adaptive expectations theory assumes that nominal wages lag behind changes in the price level. This lag in the adjustment of nominal wages to the price level brings about rise in business profits which induces the firms to expand output and employment in the short run and leads to the reduction in unemployment rate below the natural rate. But, *according to rational expectations theory, which is another version of natural unemployment rate theory, there is no lag in the adjustment of nominal wages consequent to the rise in price level.* The advocates of this theory further argue that nominal wages are *quickly* adjusted to any expected changes in the price level so that there does not exist Phillips curve showing trade-off between rates of inflation and unemployment. According to them, as a result of increase in aggregate demand, there is no reduction in unemployment rate. The rate of inflation resulting from increase in aggregate demand is *fully and correctly anticipated* by workers and business firms and get completely and quickly incorporated into the wage agreements resulting in higher prices of products. Thus, it is the price level that rises, the level of real output and employment remaining unchanged at the natural level. Hence, aggregate supply curve according to the rational expectations theory is a vertical straight line at the full-employment level.

Rational expectations theory rests on two basic elements. First, according to it, workers and producers being quite rational have a correct understanding of the economy and therefore correctly anticipate the effects of the Government's economic policies using all the available relevant information. On the basis of these anticipations of the effects of economic events and Government's policies they take correct decisions to promote their own interests.

The second premise of rational expectations theory is that, like the classical economists, *it assumes that all product and factor markets are highly competitive.* As a result, wages and product prices are highly flexible and therefore can quickly change upward and downward. Indeed, the rational expectations theory considers that new information is quickly assimilated (*i.e.*, taken into account) in the demand and supply curves of markets so that new equilibrium prices immediately adjust to the new economic events and policies, be it a new technological change or a supply shock such as a drought or act of OPEC Oil Cartel or change in Government's monetary and fiscal policies.

Fig. 25.7 illustrates the standpoint of rational expectations theory about the relation between inflation and unemployment. In this OY_F is the level of potential national output corresponding to the full-employment of labour (with a given natural rate of unemployment). *AS* is aggregate supply curve at *OQ* level of real national output. To begin with, AD_1 is the aggregate demand curve which intersects the aggregate supply curve *AS* at point *A* and determines price level equal to P_1. Suppose Government adopts an expansionary monetary policy to increase output and employment. As a consequence, aggregate demand curve shifts upward to the new position AD_2. According to rational expectations theory, people (*i.e.*, workers, businessmen, consumers, lenders) will correctly anticipate that this expansionary policy will cause inflation in the economy and they would take prompt measures to protect themselves against this inflation. Accordingly, workers would press for higher wages and get it granted, businessmen would raise the prices of their products, lenders would hike their rates of interest. All these increases would take place immediately. It is thus clear that the increase in aggregate demand (*i.e.*, aggregate expenditure) brought about by expansionary monetary policy will cause the price level to rise to P_2. Thus, the increase in aggregate demand or expenditure will be fully reflected in higher wages, higher interest rates and higher product prices, all of which will rise in proportion to the anticipated rate of inflation. Consequently, the levels of real national product and employment, wage rate, interest rate, levels of investment and consumption would remain unchanged. This can be easily understood with the help of

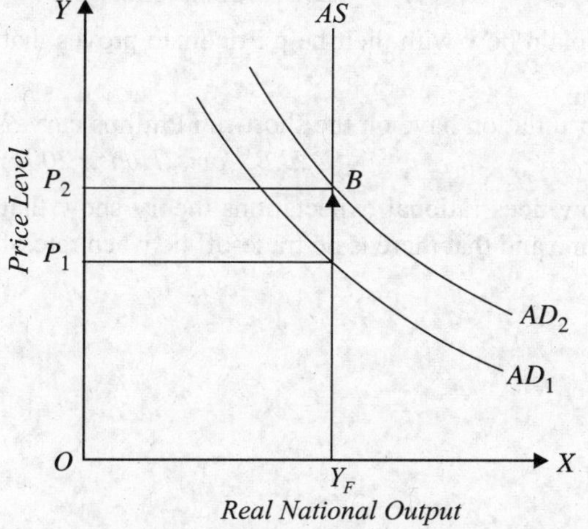

Fig. 25.7. *Inflation and Output: Rational Expectations Theory*

monetarist equation of exchange $P=\frac{MV}{O}$. Expansionary monetary policy leads to the increase in money supply *M*. As a result, aggregate expenditure, which in quantity theory is equal to *MV* increases. (Note that *V* is the velocity of circulation of money which remains stable). But people's anticipations or expectations of inflation causes an increase in *P* in equal proportion to the expansion in *MV*. This means that despite the increase in *MV*, real output *Q* and the level of employment will remain unchanged.

It is clear from above that people's anticipations or expectations of inflation and acting upon them in their decision making when expansionary monetary policy is adopted frustrate or nullify the intended effect (that is, increase in real output and employment) of Government's monetary policy. In other words, according to the rational expectations theory, the intended effect of expansionary monetary policy on investment, real output and employment does not materialise. As seen in Fig. 25.7 it is due to the anticipation of inflation by the people and quick upward adjustments made in wages, interest etc., by them that the price level instantly rises from P_1 to P_2, the level of output *Q* remaining constant. That is why, according to the rational expectations theory, aggregate supply curve is a vertical straight line. The vertical aggregate supply curve means that there is no trade off between inflation and unemployment, that is, downward-sloping Phillips curve does not exist. Thus, according to rational expectations theory, the increase in aggregate demand or expenditure as a consequence of easy monetary policy of the Government will fail to reduce unemployment and instead will only cause inflation in the economy.

QUESTIONS FOR REVIEW

1. What is the Phillips curve? What kind of trade-off between unemployment rate and inflation rate does it imply?
2. What is Phillips curve? What are its policy implications? *(D.U.B. Com. (Hons.)2000)*
3. Explain the factors causing downward slope of the Phillips Curve. What is the policy significance of the downward slope?
4. What is meant by trade-off between the rate of inflation and unemployment? Explain the Keynesian explanation of this trade-off. *(D.U.B.Com. (Hons.) 1999)*
5. What is meant by adaptive expectations? Explain how with their help Friedman proves that Phillips Curve is vertical in the long run.
6. What effect do expectations regarding rate of inflation have on the short-run Phillips curve? Can this help to explain stagflation? *(D.U.B.Com. (Hons.) 2001)*
7. What is meant by rational expectations? How does rational expectations theory show that aggregate supply curve is a vertical straight line and that there is no trade-off between rate of inflation and rate of unemployment.

CHAPTER 26

STAGFLATION AND SUPPLY-SIDE ECONOMICS

Introduction

Keynes put forward his theory of income and employment during the Great Depression of 1930s, when a large percentage of labour force was rendered unemployed (almost 25%) in today's developed capitalist economies such as U.K., U.S.A. In the fifties and sixties, Keynesian theory got great prominence when rate of inflation in these economies was a modest, about 2 to 3 per cent per annum, and the substantial unemployment rate prevailing at that time became a major concern of economic policy. The Keynesian policy to reduce this unemployment rate was to raise aggregate demand or expenditure. On the other hand, during the periods of high inflation and low unemployment, Keynesian economists recommended reduction in aggregate expenditure to fight inflation. Thus Keynesian economics emphasised management of aggregate demand through the adoption of proper fiscal and monetary policies. These policies could prove successful when there was either high inflation or high unemployment, that is, when high inflation and high unemployment did not exist simultaneously.

As mentioned in the last chapter, during the sixties Phillips curve concept describing inverse relation between inflation and unemployment became popular among economists. According to the Phillips curve, a high rate of inflation is accompanied by a low rate of unemployment or a low rate of inflation is accompanied by higher unemployment which shows that the goal of reduced inflation conflicts with the objective of reduced unemployment. This posed a great dilemma for policy makers. However, Keynesian economists advocated that Government should seek some socially acceptable short-run compromise. That is, it should try to achieve some combination of inflation and unemployment lying on the economy's Phillips curve. Monetarists led by Friedman recommend slow growth of money stock to fight inflation, while they thought unemployment would be automatically eliminated through the fee working of labour market.

The above was the general belief among economists until 1970s when stagflation appeared as a big problem for many economies of the world, especially those of USA and Britain. *Stagflation implies a high inflation rate prevailing simultaneously with a high unemployment rate*. Keynesian policy prescription of managing aggregate demand could not solve both high inflation and high unemployment existing simultaneously. If steps are taken to reduce aggregate demand to fight inflation, this would worsen the unemployment problem and, on the other hand, if measures aimed at increasing aggregate demand to reduce unemployment were taken, they would add fuel to the inflationary fire. Thus, the emergence of stagflation cast Keynesian theory into doubt. Some sounded the breakdown of Keynesian economics to tackle the problem of stagflation. Even monetarists could not provide any solution to reduce high inflation and high unemployment existing simultaneously. The search began for new ways of analysing and solving the twin problems of high inflation and high unemployment. This gave birth to a new economic thought which is now popularly called *supply-side economics* in contrast to the *demand-side* Keynesian economics. In what follows we will first explain in detail the meaning and causes of stagflation and then examine how supply-side macroeconomics offers a solution for this complicated problem. It may be noted that supply-side economics emphasises management of supply to fight against stagflation (*i.e.*, inflation and stagnation) rather than management of demand as recommended by Keynesian economics.

STAGFLATION

As mentioned above, stagflation refers to a situation when a high rate of inflation occurs simultaneously with a high rate of unemployment The existence of a high rate of unemployment means the reduced level of GNP. The term stagflation was coined in the seventies when several developed countries of the world, received a supply stock in terms of rapid hike in oil prices. In 1973, the Cartel of Oil Producing Countries OPEC raised the price of oil. There was a four times increase in the oil prices. In the United States during 1973-75 the higher costs of fuel-oil and other petroleum products brought about a sharp increase in the prices of manufactured goods. The rate of inflation went upto over 12 per cent during 1974 in USA. A severe recession, the worst since 1930s, also hit the American economy during the period 1973-75. The real GNP declined between the late 1973 and early 1975. As a consequence, the rate of unemployment shot upto nearly 9 per cent. Thus, both inflation and unemployment were unusually very high during this period (1973-75). This simultaneous occurrence of high inflation and high unemployment was also seen in case of other free market developed countries such as Britain, France and Germany. The recovery from recession began in 1975 and over the next few years GNP rose and unemployment declined. Inflation rate also declined from over 12 per cent to the range of 5 to 7 per cent.

But, again in 1979 when a revolution in Iran created a crises in world oil market, OPEC doubled the price of oil. This brought back stagflation again in 1979 in the developed countries. Real GNP fell at a rapid rate during 1979-81. Inflation rate again went upto over 10 per cent in these countries during this period.

India also could not escape from the oil price shocks in 1973 and 1979. But, in case of India, oil price triggered cost-push inflation but did not give rise to stagflation as the term is usually interpreted in 1973 and 1979. The public investment in India picked up from 1974 which generated economic growth.

Causes of Stagflation

Different explanations of stagflation have been given by eminent economists. It is worth noting that causes of stagflation in India during 1991-94 are different from those given by the economists for stagflation of 1973-75 and 1979-81 in the developed capitalist economies such as those of USA, Great Britain. We will first explain stagflation in USA, Great Britain and other developed capitalist countries during 1973-1975 and again in 1979-81 and then dwell on stagflation in India.

Adverse Supply Shocks. The main reason why typical stagflation arose in the developed capitalist economies during seventies and early eighties was the adverse supply shocks that occurred during these two periods. As mentioned above, there was four fold increase in oil prices by OPEC following Arab-Israel war in 1973 and then again doubling of oil prices by it in 1979 following the Iranian Revolution which pushed up the energy costs of the economies and resulted in higher product prices. In terms of aggregate supply curve, this cost-push factor delivered by oil price shock is interpreted as a decrease or leftward shift in the aggregate supply curve. How this adverse supply shock caused stagflation in the developed capitalist world is illustrated in Fig. 26.1 where initially aggregate demand curve AD_0 and aggregate supply curve AS_0 intersects

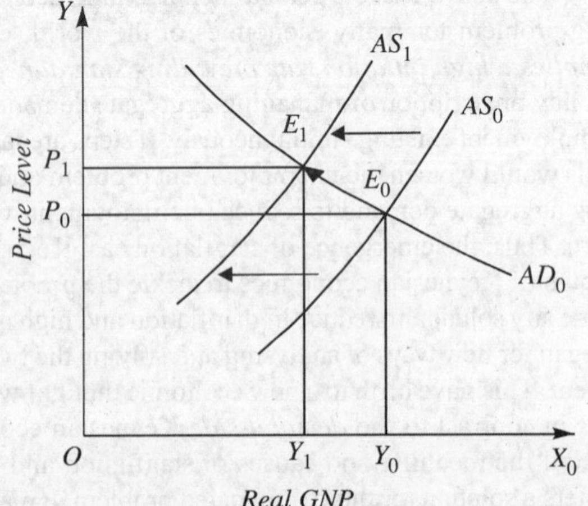

Fig. 26.1. *Stagflation Arising From an Adverse Supply Shock*

at E_0 and determines the price level equal to P_0. Since adverse supply shock delivered by hike in oil prices raises the cost per unit of production, aggregate supply curve shifts upward to the left to the new position AS_1. With the aggregate demand curve AD_0 remaining unchanged, the new aggregate supply curve AS_1 intersects it at E_1. It will be seen that in the new equilibrium position price level rises to P_1 and *GNP* falls to Y_1. Thus, adverse supply shock causes cost-puch inflation along with a reduction in the level of *GNP*. A reduction in *GNP* implies an increase in unemployment rate and occurrence of recession. Thus, an adverse supply shock causes both high inflation and high unemployment rate. It may be noted that in order to get out of recession and to reduce unemployment, if Government seeks to raise aggregate demand to the higher level AD_1 by adopting expansionary fiscal and monetary policies, the new equilibrium is reached at point E_2 (see Fig. 26.2) and as a result price level rises to P_2, while the real GNP comes back to the original higher level Y_0 where full employment of labour prevails.

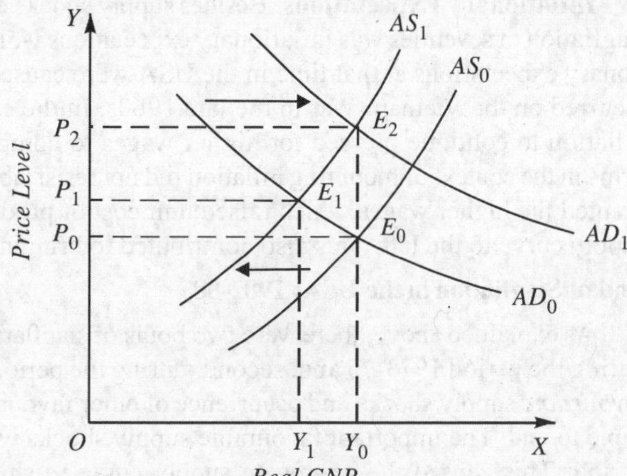

Fig. 26.2. *Further Surge in Inflation Rate if Aggregate Demand is Raised to Get Out of Stagflation.*

Thus in this context of stagflation in the economy attempts by the Government to raise aggregate demand to get out of recession and reduce unemployment result in further rise in the inflation rate. This shows that mere management of demand is quite inappropriate to solve the problem of stagflation. Though rise in oil prices has been a chief supply shock received by all the economies of the world which imported oil from Middle East Countries causing stagflation of 1970s and early 1980s, there are also other types of adverse supply shocks which occur. In different countries different types of supply shocks may occur bringing about rise in unit cost of production and causing a leftward shift in the aggregate supply curve. This has caused stagflation episodes from time to time. In case of USA, besides oil price shocks, the other supply shocks explained below also contributed to the stagflation of 1973-75.

An important supply shock operating in the USA was *the shortage in supplies of agricultural products* during this period. This happened because a good amount of American agricultural products had to be exported to Asia and the Soviet Union where severe shortfall in production occurred in 1972 and 1973. Larger exports reduced the domestic supplies of agricultural products used as raw material in the production of industries producing food and fibre products. This raised the unit cost of production of these commodities and their higher costs were passed on to the consumers as higher prices. This resulted in shifting of the aggregate supply curve to the left. It is important to note that the higher prices of agricultural commodities such as sugar cane, cotton, foodgrains which may occur due to either shortfall in production or due to the increase in their procurement prices has often been working also in the Indian economy which has resulted in higher costs to the industries processing these agricultural products.

Another adverse supply shock that occurred in the USA during the period of 1971-73 causing stagflation episode of 1973-75 was *depreciation of dollar*. Depreciation of dollar means that price of dollar in terms of foreign currencies was reduced. This raised the prices of American imports. To the extent the imports were used as inputs in American industries, unit production costs went up causing a shift in the aggregate supply curve to the left. In the period of 1973-75, *removal of wage and price controls* which had been imposed earlier also produced a supply stock to the American economy.

As these wages and price controls were lifted, workers got their wages increased and business firms pushed up the prices of their products. This also contributed to stagflation of 1973-75 in the USA.

Inflationary Expectations. Besides supply shocks explained above, another important cause of stagflation of seventies was inflationary expectations which were prevailing at that time. These inflationary expectations at that time in the USA were caused by greatly increased military expenditure incurred on the Vietnam War in the late 1960s. In the early seventies workers with expectations of inflation to continue pressed for higher wages to compensate for accelerating inflation. Business firms in the context of mounting inflation did not resist labour demand for higher nominal wages. They granted the higher wages which raised unit cost of production and resulted in shifting of aggregate supply curve to the left. This also contributed to bringing stagflation.

End of Stagflation in the USA: 1982-88

As explained above, there were two bouts of stagflation in the several countries of the world, first during the period 1973-75 and, second, during the period 1979-81. However, during 1982-88 due to *favourable* supply shocks and occurrence of other favourable factors, stagflation of the earlier period came to end. The important favourable supply shocks was the decline in oil prices by OPEC in this period. This caused the aggregate supply curve to shift to the right bringing about fall in both inflation and unemployment.

Another important factor contributing to the demise of stagflation in 1982-88 in USA was the deep recession that overtook the American economy in 1981-82 which was mainly caused by tight monetary policy pursued by Federal Bank. Such was the severity of recession that unemployment in USA rose to 9.7 per cent in 1982. Due to this high unemployment rate workers accepted smaller increases in their nominal wages or in some cases accepted even reduction in their wages. Further, due to still foreign competition and their eagerness to maintain relative shares in domestic and foreign markets, business firms were restrained to raise prices of their products. This also worked to bring stagflation to end.

It is important to note that while during the periods of stagflation in 1970s and early 1980s, both inflation and unemployment simultaneously increased, during the expansion period of 1982-88 when stagflation nearly subsided both inflation and unemployment rates fell simultaneously.

SUPPLY-SIDE ECONOMICS

Keynesian economics was born during the great depression of the 1930s, when a large percentage of labour force (about 25%) was rendered unemployed and also a good deal of productive capacity (*i.e.*, capital stock) lay idle resulting in a huge decline in Gross National Product (GNP) of the economies. The prices were actually falling during this depression period. When after the Second World War, problem of inflation rather than unemployment became the major concern of the economists. Keynesian economists explained it in terms of excess aggregate demand and therefore called it demand-pull inflation. As seen in the previous chapters that Keynes and his followers laid emphasis on the *management of aggregate demand* to bring about short-run stability in the economy. They recommended expansionary fiscal and monetary policies to raise aggregate demand to pull the economy out of depression or recession and thereby to reduce unemployment. On the other hand, to fight inflation, they advocated contractionary fiscal and monetary policies to reduce aggregate demand.

However, the problem of stagflation encountered in the USA and Great Britain during the seventies and early eighties when both high inflation and high unemployment prevailed simultaneously did not admit of easy solution through the Keynesian demand management policies. In fact, attempts to remedy the stagflation through Keynesian demand management worsened the situation.

Against this backdrop an alternative school of thought about macroeconomics was put forward. This alternative thought laid stress on supply-side of macroeconomic equilibrium, that is, *it focused on shifting the aggregate supply curve to the right* rather than causing a shift in aggregate demand curve. Thus, supply-side economics prefers to solve the problem of stagflation, that is, simultaneous existence of high inflation and high unemployment through *management of aggregate supply* rather than management of aggregate demand. Further, supply-side economics stresses the determinants of long-run growth instead of causes of short-run cyclical changes in the economy. Supply-side economists lay emphasis on the factors that determine the incentives to work, save and invest which ultimately determine the aggregate supply of output of the economy.

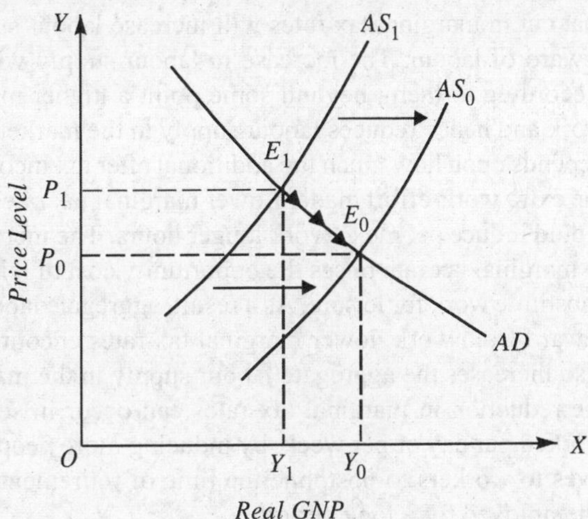

Fig. 26.3. *Supply-siders seek to shift aggregate supply curve to the right by stimulating incentives to work, save and invest.*

Difference in the approaches of Keynesian demand-side theory and alternative supply-side theory can be understood with reference to Fig. 26.3 which illustrates the emergence of stagflation as a consequence of a shift in the aggregate supply curve due to the cost-push factors and decline in productivity. Suppose aggregate supply curve shifts upward to the left from AS_0 to AS_1 due to some cost-push factors (*e.g.*, rise in oil price). As a result, it will be seen from Figure 26.3 that price level would rise to P_1 and output (*i.e.*, real GNP) would fall to Y_1 (which will cause increase in unemployment). This high inflation and high unemployment configuration represents the state of stagflation. Now, the supply-side economists argue that to get out of stagflation, aggregate supply curve should be shifted to the right. As is evident from the Fig. 26.3 with the rightward shift of the aggregate supply curve from AS_1 to AS_0, the economy moves from the equilibrium point E_1 to point E_0 showing that *while price level falls, aggregate national output increases* (*which will reduce unemployment*). Thus, in this way, through management of aggregate supply, the economy can be lifted out of stagflation.

It is worth mentioning that if to tackle this problem of stagflation Keynesian policy of increasing aggregate demand, that is, shifting the aggregate demand curve from AD_0 to AD_1 (see Fig. 26.2) through expansionary fiscal and monetary measures is adopted to reduce unemployment, it will cause price level to rise further to P_2 and will thus worsen the inflationary situation. On the other hand, if to tackle inflation, aggregate demand is reduced to AD_0, though i will cause the price level to fall, it will result in reduction in real aggregate output (GNP) causing unemployment to increase further and thus deepen the recession. Therefore, supply-side economists contend that Keynesian demand-management policy fails to provide a solution for the problem of stagflation. The supply-side economists, the prominent among whom is Arthur Laffer, are of the view that economic policy, especially of taxation, can be used to stimulate incentive to work, save and invest and task risks which cause increases in aggregate supply and yield higher growth in productivity. This leads to higher growth of real GNP and lower both rates of inflation and unemployment. We explain below the basic elements of supply-side economics and then critically evaluate it.

Basic Propositions of Supply-side Economics

As mentioned above, supply-side economists emphasise the importance of effects of tax incentives on labour supply, saving and investment for promoting growth of output. They further lay stress on the favourable effects of tax cuts on Government revenue and thereby to achieve reduction in budget deficit. The following are the basic propositions of supply side economics:

1. **Taxation and Labour-Supply.** The first important basic proposition of supply-side economics is that cut in marginal tax rates will increase labour supply or work effort as it will raise the after-tax reward of labour. The increase in labour supply will cause growth in aggregate supply of output. According to them, beyond some point a higher marginal tax rate reduces people's willingness to work and hence reduces labour supply in the market. They argue that how long individual will work depends upon how much the additional after tax income (i.e., after-tax wage rate) will be earned from the extra work-effort made. Lower marginal tax rates by increasing after tax earnings of extra labour would induce people to work longer hours. The increase in after-tax earnings as a result of reduction in marginal tax rate raises the opportunity cost of leisure and provide incentives to the individuals to substitute work for leisure. As a result, aggregate labour supply increases. Further, by ensuring higher reward from work, lower marginal tax rates encourage more people to enter the labour force. This also increases the aggregate labour supply in the market. Thus, increase in labour supply following the reduction in marginal tax rates can occur in several ways by increasing the number of hours worked per day or per week, by inducing more people to enter the labour force, by providing incentives to workers to postpone the time of retirement, and by discouraging workers from remaining unemployed for a long period.

Reduction in marginal tax rates on business income raises the after-tax return on labour employed. This will encourage businesses to demand and employ more labour. Thus reduction in marginal tax rates on incomes will increase both the supply of labour and demand for it.

2. **Incentives to Save and Invest.** The second basic proposition of supply-side economics is that reduction in marginal tax rates will increase the incentives to save and invest more. According to it, a high marginal tax rate on incomes reduces the after-tax return on saving and investment and therefore discourages saving and investment. Suppose an individual saves ₹ 1000 at 10 per cent rate of interest, he will earn ₹ 100 as interest income per annum. If marginal tax rate is 60 per cent, his after-tax interest income will be ₹ 40. This means that after-tax interest on his savings has fallen to 4 per cent $\left(\frac{40}{1000}\times 100 = 4\right)$.

Thus, whereas an individual might be willing to save at 10 per cent rate of return on his saving, he might prefer to consume more rather than save when the return he gets is only 4 per cent. To promote savings, it may be noted, is essential for raising investment and capital accumulation which in the long run determines growth of output. The supply-side economists emphasise lower marginal tax rates on income to encourage savings. They also argue for lower tax rates especially on income from investment such as business profits to induce businessmen and firms to invest more. It will be recalled that investment in an economy depends to a great extent on expected rate of profits (or what is called marginal efficiency of investment). A higher tax on business profits and corporate income discourages investment by reducing the after-tax net profit on investment. Thus, lower marginal tax rates on business profits will encourage saving and investment and step up capital accumulation. With more capital per worker, labour productivity will rise which will tend to reduce unit labour cost and lower the rate of inflation. Moreover, the higher rate of capital accumulation, will ensure greater growth of productive capacity. The lower unit labour cost and the higher rate of capital accumulation made possible by greater saving and investment will cause aggregate supply curve to shift to the right. This will lower the price level, increase the growth of output and reduce unemployment.

3. **Cost-Push Effect of the Tax Wedge.** Another important proposition of supply-side economics is that the substantial growth of public sector in the modern economies has necessitated a large increase in the tax revenue to finance its activities. The tax revenue has increased both absolutely and as a percentage of national income. The Keynesian economists view the tax revenue as withdrawal of money income from the people which operates to reduce aggregate demand. Thus, in the Keynesian

view the mobilisation of resources for the public sector through taxation have an anti-inflationary effect. On the contrary, supply-side economists think that sooner or later most of the taxes, especially excise duties and sales taxes, are incorporated in the business costs and shifted to the consumers in the form of higher prices of products. Thus, in their view, imposition of higher taxes, like higher wages, have a cost-push effect. Referring to the period of seventies and early eighties in the USA, which was plagued by the great stagflation, they point out that huge increases made in sales and excise taxes by the state and local Governments and substantial hike in payroll taxes by Fed Government in the USA during this period had greatly pushed up the business costs resulting in higher product prices. In fact, supply siders contend that many taxes constitute a wedge between the costs incurred on resources and the price of a product. With the substantial growth of the public sector, the funds required to finance it have greatly increased resulting in a greater tax wedge. This has worked to shift the aggregate supply curve to the left.

4. **Underground Economy.** Another important contention of supply-siders is that higher marginal tax rates encourage people to work in the underground economy (which in India is popularly called black or parallel economy) where their income cannot be traced by the income tax department. In India, this underground economy is very large. Not only individual businessmen evade income taxes, corporate firms also have devised several illegal ways to evade taxes on their profits. It is not only taxes on personal income and company profits but also excise duties and sales taxes which are not paid fully by individuals and companies. In line with the supply-side view, the former Finance Minister Dr. Manmohan Singh has often argued in favour of reduction in taxes. According to him, *lower tax rates would increase tax compliance* which increases the amount of income which people will report to the taxation authorities. Thus, the supply-side economists think that reduction in taxes will in fact raise the tax revenue by discouraging people from evading taxes and from operating in the underground economy.

5. **Tax Revenue and Laffer Curve.** By far the most important proposition of supply-side economics is that the lower marginal taxes will increase tax revenue. In seventies and eighties American economy was not only facing the problem of stagflation but also of large budget deficits of the Government. The supply-siders contend that through raising tax revenue, the reduction in tax rates will not only reduce inflation and unemployment by bringing about increase in aggregate supply but will also reduce budget deficits of the Government. An eminent supply-side economist, Arthur Laffer has argued that lower tax rates are quite consistent with the increase in tax revenue. He has shown the relationship between tax rates and the total tax revenue collected with the help of a curve named after him as Laffer Curve. *Laffer curve* shows that after a certain point increase in tax rates can reduce tax revenue as incentives to work, save and investment are adversely affected. The higher tax rates after a certain point prove to be counter-productive as they reduce supply of labour and capital accumulation through providing disincentives to work, save and invest. Therefore, these higher tax rates reduce national output and income. Remember that total tax revenue (TR) collected is equal to the tax rate, which we denote by t multiplied by the total income which we denote by Y. Thus, total tax revenue $TR = tY$. According to Laffer, when tax rate t is raised beyond a certain point the national output and

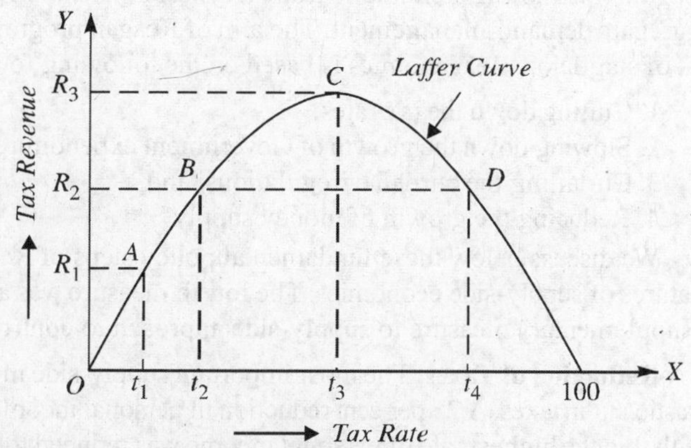

Fig. 26.4. *Laffer Curve: Relationship between Tax Rate and Tax Revenue Collected*

income Y which constitutes the base of taxation declines so much that total tax revenue t_Y falls. Laffer curve has been drawn in Fig. 26.4. The Laffer curve starts from the origin which means when the tax rate is zero, total tax revenue will also obviously be zero. Upto point C, Laffer curve is rising which shows that as tax rate rises to t_3, tax revenue collected increases. But if the tax rate is raised beyond t_3, Laffer curve slopes downward showing that tax revenue decreases as tax rate is increased above t_3 for the reasons explained above. At tax rate t_3, the tax revenue collected R_3 is the maximum. For instance, if tax rate is raised from t_3 to t_4, the tax revenue falls from R_3 to R_2. As explained above, when tax rate is raised beyond some point, tax revenue decreases. This is because *higher tax rates serve as disincentives to work, save and invest, innovate and take business risks and therefore the tax base*, (*that is, the level of national output, income and employment*) *decreases*. This can be easily understood if the tax rate is raised to 100 per cent. At 100 per cent tax rate no body would have any incentive to work, save and invest or engage in any productive activity and therefore the tax revenue will be reduced to zero. Production of goods and earning of income (i.e., tax base) would come to a halt at this consfiscatory (100 per cent) tax rate.

Most of the economists agree with Laffer that beyond a certain tax rate, the tax revenue will fall. However, *the moot point is at which point on Laffer curve the position of the economy is presently located*. For example, if the economy is presently at point D with tax rate t_4, that is, on the downward portion of the Laffer curve, the reduction in tax rate from t_4 to t_3 will raise the tax revenue from R_2 to R_3. If there is drastic cut in the tax rate from t_4 to t_2, the tax revenue remains unaffected.

It may be noted that with the reduction in tax rate, the tax revenue increases because of two additional reasons. First, as explained earlier, *the reduction in tax increases tax compliance and tends to reduce tax evasion and tendency to indulge in underground or black market activities*. Lower tax rates also reduce the inclination of people to avoid taxes through the use of various tax shelters (such as buying National Saving Certificates) investment in which is exempted from taxes. Second, the supply-side economists also argue that by promoting the growth of production and employment, lower tax rates will reduce Government transfer payments, such as unemployment allowances.

REAGANOMICS AND SUPPLY-SIDE ECONOMICS

Supply-side economics became popular when President Reagan of the USA actually put it into practice after winning the election in 1981. Reagan came into power at a time when the American economy was facing the problem of stagflation with higher rates of both inflation and unemployment. Though Reagan economic programme cannot be fully equated with supply-side economics, it contained the fundamentals of the latter. Reagan economic programme is generally described as *Reaganomics* to differentiate it from Keynesian and moneterist economics which were based on aggregate demand-management. The aim of Reagan programme was to pull the American economy out of stagflaion. Reaganomics is based on the following four pillars:

1. Cutting down the tax rates;
2. Slowing down the growth of Government expenditure;
3. Curtailing the burden of regulations; and
4. Reducing the growth of money supply.

We discuss below these fundamental policy steps of Reaganomics. The first three measures are features of supply-side economics. The fourth measure was adopted by Reagan to control inflation as a supplementary measure to supply-side approach to control stagflation.

Reduction of Taxes. The most important supply-side measure adopted by Reagan was to make a drastic cut in taxes. A 25 per cent reduction in personal income-tax rates over the three years was made. With this the highest rate of personal income was reduced to 33 per cent. Most of the Americans were brought in the bracket of 15 per cent tax rate. As seen above, such low tax rates are aimed at promoting incentives to work, save and invest and thereby increase aggregate supply of output. Further, in line with the view of supply-side economics, Reagan programme allowed a *higher depreciation allowance* to business firms to cover the cost of machinery and equipment installed. This virtually lowered the tax-burden on the companies which increased their incentives to invest in capital accumulation. Besides, another measure of tax reduction was *lowering of capital gains tax rates* to boost investment incentives. In 1986 tax reform, Reagan *reduced the corporation income tax*. This, according to supply-siders, promoted investment incentives by raising the profitability of investment and also making available more investible funds from within the internal resources of the companies.

As we will examine below, the reduction in taxes by Reagan became a highly controversial issue because of prevailing high inflation in USA at that time. Keynesian economists argued that these cuts in taxes will increase the disposable incomes with the individuals and companies and would bring about increase in aggregate demand resulting in higher inflation rate. Besides, they pointed out that tax cuts would lower the Government revenue and increase budget deficits.

Reducing the Burden of Government Regulation. Supply-siders argued that a high degree of Government regulation was also responsible for poor performance of the American economy in the past some decades. According to them, Government regulation of some industries such as transport and communication created monopolies and thus protected them from competition by rivals.

In the absence of competition, these monopolistic firms created through Government regulations tended to become inefficient which raised the cost of production in the regulated industries. Second, but more importantly, the supply-siders argue, there has been a substantial increase in the number of Government regulations and controls regarding protection of environment from pollution, safety of products (such as of insecticides, pesticides etc.) safety and health of workers, equal access to job opportunities. Supply-siders argue that these Government regulations and controls have raised the costs of production and of doing business. This have resulted in higher prices of goods and sluggish growth in output and has thus tended to give rise the problem of stagflation.

A Critical Appraisal of Supply-side Economics

As seen above, a central idea of supply-side economics is that the reduction in rates of certain type of taxes will increase aggregate supply of output by increasing both the supply of labour and capital. Taxes can certainly be cut in some ways that raise the rewards for working and saving and thereby provide incentives to work more and save more. If people actually respond positively to these incentives, then tax cuts would lead to the increase in supplies of labour and capital and cause a rightward shift in the aggregate supply curve. However, critics point out that there is no guarantee that tax cuts would induce people to actually work more and save more. According to them, those people whose aim is to earn a fixed targeted income to buy goods and services they want, at higher rewards for working resulting from tax cuts, they will be able to earn targeted incomes with fewer hours of work. Thus, the higher monetary reward per unit of work hour will enable them to enjoy more leisure and earn the same amount of income as before. As regards the people who behave positively to increased reward for working, it is pointed out that the increase in work-effort (*i.e.*, labour supply) obtained in this way may not be very large.

Similarly, cut in taxes raises the reward for saving, but we cannot certainly say whether this will cause people to save more. Those people who want to have a given amount of savings, they would find that their saving goals can be achieved by saving less when return on saving has been raised by reducing taxes. Thus, lower taxes can encourage some people to do greater saving and investment but at the same time they can also discourage others to save more. It has been pointed out by critics *that following tax cuts in USA by Reagan in 1981 the saving rate in the United States fell*. Commenting on the emperical evidence in this regard Baumol and Blinder write, "Most of the statistical evidence suggests that we should expect tax reduction to lead to only small increase in either labour supply or household savings"[1]

The Threat of Inflation: Demand-side Effects of Tax Cuts: The second important criticism made

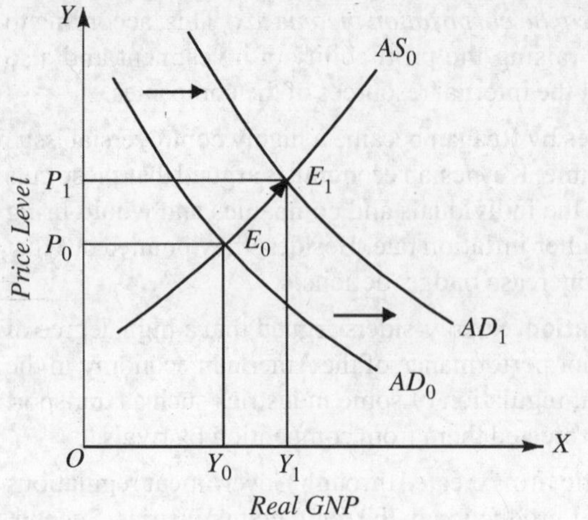

Fig. 26.5. *Effect of Tax Cut on Aggregate Demand and Inflation : Keynesian View*

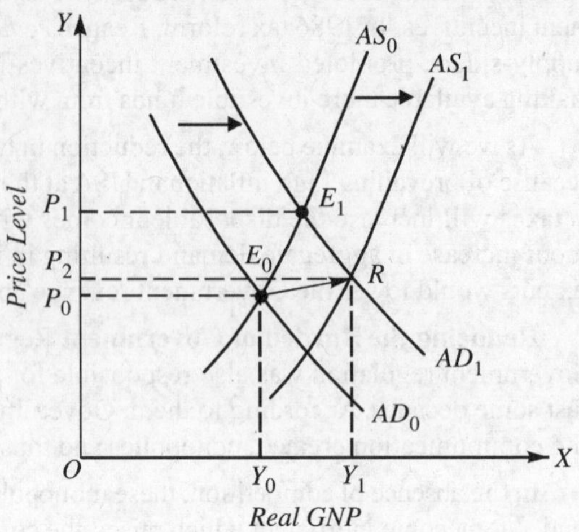

Fig. 26.6. *Effect of Tax Cut on Aggregate Supply and Demand, and Inflation Rate : Supply-Side View*

against supply-side economics is that it underestimates the effect of tax cuts on increasing aggregate demand and thus adding to the inflationary pressures in the economy. As we know that reduction in personal income tax increases the disposable income of the people and will therefore cause an increase in the demand for consumer goods. Similarly, the cuts in business taxes will raise the profitability of investment and thereby induce the firms to invest in expanding productive capacity. This will lead to the increase in demand for investment goods. Thus, reduction in taxes will cause aggregate demand for goods to increase which tend to raise the inflation rate. This is illustrated in Fig. 26.5 where initially aggregate demand curve AD_0 and aggregate supply curve AS_0 intersect at E_0 and determine price level P_0. If aggregate supply curve AS_0 remains constant, the increase in aggregate demand from AD_0 to AD_1 due to tax cuts establishes equilibrium at point E_1 and thereby determines new price level P_1 which is much higher than the initial price level P_0 (that is, rate of inflation is high).

On the other hand, supply-side economists argued that the increase in aggregate supply of goods stimulated by reduction in taxes will be large enough to counteract any inflationary pressures due to higher aggregate demand arising from tax cuts. Fig. 26.6 illustrates the supply-side view of the effect of tax cuts made to solve stagflation. Originally, aggregate demand and aggregate supply are in equilibrium at point E_0 where price level is P_0 and real GNP is equal to Y_0. This equilibrium situation is one of stagflation when there is high rate of inflation and lower GNP (and therefore high level of unemployment). According to supply-siders, the tax cuts through stimulating work effort, saving and investment substantially shift the aggregate supply curve to the right. Of course, it also shifts the aggregate demand curve upward. But the supply-side effect dominates resulting in *only small increase in the price level from P_0 to P_2 (that is, rate of inflation is quite low) and large expansion in real GNP from Y_0 to Y_1 which will cause unemployment to fall*. But, as we have pointed out above, this is a very optimistic view of the effect of tax cuts.

1. Baumol, William J. and Blinder, Allan S., *Economics : Principles and Policies*, Harcourt Publishers, Fifth Edition, 1991

Empirical evidence reveals that tax cuts have only a small effect on increasing aggregate supply, whereas they have substantial effect on raising aggregate demand and inflation. Figure 26.7 illustrates how a large effect on aggregate demand of tax cuts is stronger than their favourable impact on increasing the aggregate supply. With reduction in taxes, there is a large shift in aggregate demand curve from AD_0 to AD_2 but only a small increase in aggregate supply from AS_0 to AS_2. As a result, price level jumps to P_1, showing a higher rate of inflation (see Fig. 26.7).

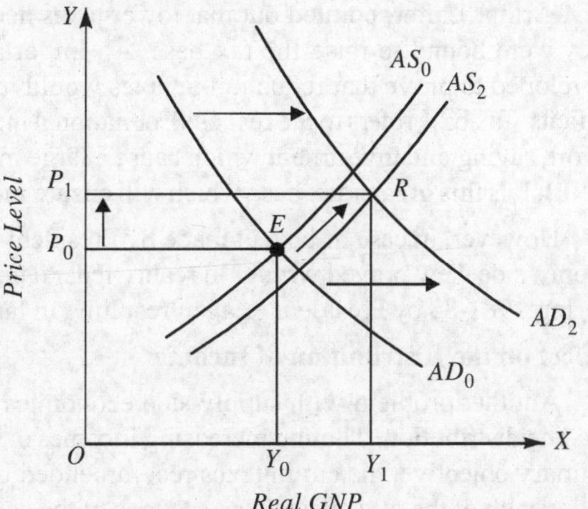

Fig. 26.7. *Whem the effect of tax cut an aggregate supply is much smaller than its effect on raising aggregate demand curve, tax cut causes a higher rate of inflation.*

It is worth mentioning here that perhaps recognising the demand-side effect of tax cuts and contrary to supply-side view President Reagan in his fiscal package, also made reduction in Government spending to cancel out the demand-side effect of generating inflationary pressures in the economy. As we know whereas tax cuts increases aggregate demand, reduction in Government expenditure reduces it and thus cancels out the effect of the former. By combining the tax cuts with reduction in Government expenditure of the right magnitude aggregate demand curve could be held constant which would make it possible to retain the favourable impact of tax cuts on aggregate supply. However, the problem with this fiscal strategy is that large cuts in Government expenditure are also to be made simultaneously, if aggregate demand is to be kept constant. However, in case of President Reagan's fiscal programme, many economists in the early 1980s felt that reduction in Government expenditure in Reagan's fiscal package was smaller than the tax cuts.

It follows from above that tax cuts of supply-side economics which was originally propounded to cure both inflation and stagnation, could be expected to make not more than a small dent in the inflation rate because of the demand-side effects they create. However, in case of USA, implementation of Reagan's economic programme did not actually lead to higher rate of inflation, despite the increase in budget deficits. But this was not because of successful effect of supply-side tax cuts but due to tight monetary policy adopted by Federal Reserve System of America to check inflation at that time.[2] It is this tight money policy that helped to contain aggregate demand by restraining the growth of money supply. Thus, credit for checking inflation must go to Federal Reserve's tight money policy rather than to tax cuts of supply-side economists.

Increase in Budget Deficits

Another important shortcoming of supply-side cuts in taxes is that they are likely to increase budget deficits. When a country is facing the problem of budget deficits, cut in taxes will cause reduction in Government revenue and will therefore raise budget deficits. It may be noted that in 1981 when on the advice of supply-side economists Reagan made large tax cuts, his critics argued that these would further increase the budget deficits, as he had made only small cuts in Government spending. Extreme supply-siders however denied that large tax cuts would raise budget deficits. As has been explained above, they argued that higher marginal tax rates were encouraging tax evasion

2. It should be noted that in the USA Federal Reserve System works independently of Government or President of the USA. Interestingly, the former Governor of Reserve Bank of India, Dr. C.R. Rangarajan also demanded independence of RBI from Government

and avoidance and also causing more and more activities to be done in the underground economy, and thus escaping from the tax net. Reduction in taxes, they argued, would increase tax compliance which will increase the Government revenue. Besides, they, especiailly an eminent supply-side economist Arthur Laffer, pointed out that lower taxes need not lead to reduction in tax revenues because they were bound to raise the tax base. An important graphic concept called as Laffer curve was developed to prove that reduction in taxes would increase tax revenue and help in reducing budget deficits. Tax base refers to the real GNP or national income. Tax cuts, according to them, stimulate work effort, saving and investment which cause a large increase in aggregate supply of goods (that is, real GNP). It is this greater tax base which will ensure increase in tax revenue.

However, in case of tax cut made by President Reagan during 1981-83 in the United States, the supply-side view proved wrong. In reality federal tax revenue fell sharply after the reduction in taxes during 1981-83 by President Reagan resulting in larger budget deficits.

Effect on the Distribution of Income

Another problem with supply-side economics is that it leads to the increase in inequilities of income distribution. Though increasing incomes of the richer sections of the society is not its explicit primary objective, the cut in taxes recommended by supply-side economics increases the incomes and wealth of the already affluent sections of the society. This is because it is the rich who earn most of capital gains, interest and dividend and tax cuts on them therefore greatly benefit them. Besides, it is the rich people who own the business corporations and tax cuts on corporate profits will also benefit them. Thus, the supply-side economics tilt income distribution towards the rich.

Conclusion

To sum up, it follows from above that supply-side theorists have too optimistic view of the favourable effect of tax cuts on inflation. Tax cuts have both supply-side and demand-side effects. Effects of tax cuts on raising aggregate supply through stimulating saving and business investment in plant and machinery *will accrue more slowly than their effect on increase in aggregate demand. Therefore, supply-side policies should not be considered as substitutes for short-run stabilisation policy which focuses on management of aggregate demand but rather they should be used to promote rapid growth of output in the long run*. Furthermore, the effect of tax cuts on raising aggregate demand is likely to be much greater than their effect on raising aggregate supply, at least in the short run. In view of the smaller supply-side effect as compared to the demand-side effect, tax cuts are unlikely to make any contribution to lowering the inflation rate. Further, the supply-side measures are likely to increase income inequalities as they benefit the rich more than the poor. Lastly, tax cuts of supply-side economics, if not accompanied by appropriate cuts in government spending would lead to larger and not smaller budget deficits.

Chapter 27
BUSINESS CYCLES

Introduction

Many free enterprise capitalist countries such as USA and Great Britain have registered rapid economic growth during the last two centuries. But economic growth in these countries has not followed steady and smooth upward trend. There has been a long-run upward trend in Gross National Product (GNP), but periodically there have been large short-run fluctuations in economic activity, that is, changes in output, income, employment and prices around this long-term trend. The period of high income, output and employment has been called the period of *expansion, upswing* or *prosperity*, and the period of low income; output and employment has been described as *contraction, recession, downswing* or *depression*. The economic history of the free market capitalist countries has shown that the period of economic prosperity or expansion alternates with the period of contraction or recession. *These alternating periods of expansion and contraction in economic activity have been* called business cycles. They are also known as *trade cycles*. J.M. Keynes writes, "A trade cycle is composed of periods of good trade characterised by rising prices and low unemployment percentages with periods of bad trade characterised by falling prices and high unemployment percentages."[1]

A noteworthy feature about these fluctuations in economic activity is that they are recurrent and have been occurring periodically in a more or less regular fashion. Therefore, these fluctuations have been called business cycles. It may be noted that calling these fluctuations as 'cycles' means they are periodic and occur regularly, though perfect regularity has not been observed. The duration of a business cycle has not been of the same length; it has varied from a minimum of two years to a maximum of ten to twelve years, though in the past it was often assumed that fluctuations of output and other economic indicators around the trend showed repetitive and regular pattern of alternating periods of expansion and contraction. However, actually there has been no clear evidence of very regular cycles of the same definite duration. Some business cycles have been very short lasting for only two to three years, while others have lasted for several years. Further, in some cycles there have been large swings away from trend and in others these swings have been of moderate nature.

A significant point worth noting about business cycles is that they have been very costly in the economic sense of the word. During a period of recession or depression many workers lose their jobs and as a result large-scale unemployment, which causes loss of output that could have been produced with full employment of resources, comes to prevail in the economy. Besides, during depression many businessmen go bankrupt and suffer huge losses. Depression causes a lot of human sufferings and lowers the levels of living of the people. Fluctuations in economic activity creates a lot of uncertainty in the economy which causes anxiety to the individuals about their future income and employment opportunities and involves a great risk for long-run investment in projects. Who does not remember the great havoc caused by the great depression of the early thirties of the present century? Even boom when it is accompanied by inflation has its social costs. Inflation erodes the real incomes of the people and makes life miserable for the poor people. Inflation distorts allocation of resources by drawing away scarce resources from productive uses to unproductive ones. Inflation redistributes income in favour of the richer sections

1. J.M. Keynes, *The General Theory of Employment; Interest and Money*, 1936.

and also when inflation rate is high, it impedes economic growth. About the harmful effects of the business cycles Crowther writes, "On the one hand, there is the misery and shame of unemployment with all the individual poverty and social disturbances that it may create. On the other hand, there is the loss of wealth represented by so much wasted and idle labour and capital."[2]

Phases of Business Cycles

Business cycles have shown distinct phases the study of which is useful to underlying causes. These phases have been called by different names by different economists. Generally, the following phases of business cycles have been distinguished:

1. Expansion (Boom, Upswing or Prosperity)
2. Peak (Upper Turning Point)
3. Contraction (Downswing, Recession or Depression)
4. Trough (Lower Turning Point)

The four phases of business cycles have been shown in Fig. 27.1 where we start from trough or depression when the level of economic activity, *i.e.*, level of production and employment is at

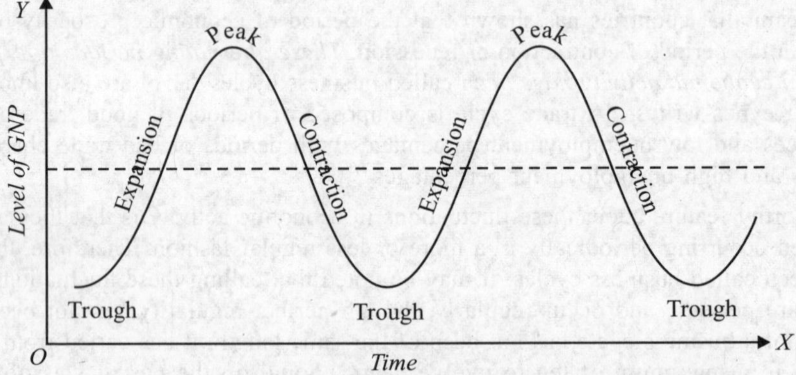

Fig. 27.1. *Four Phases of Business Cycles without Growth Trend*

the lowest level. With the revival of economic activity the economy moves into the expansion phase, but due to the causes explained below, the expansion cannot continue indefinitely, and after reaching peak, contraction or downswing starts. When the contraction gathers momentum, we have a depression. The downswing continues till the lowest turning point which

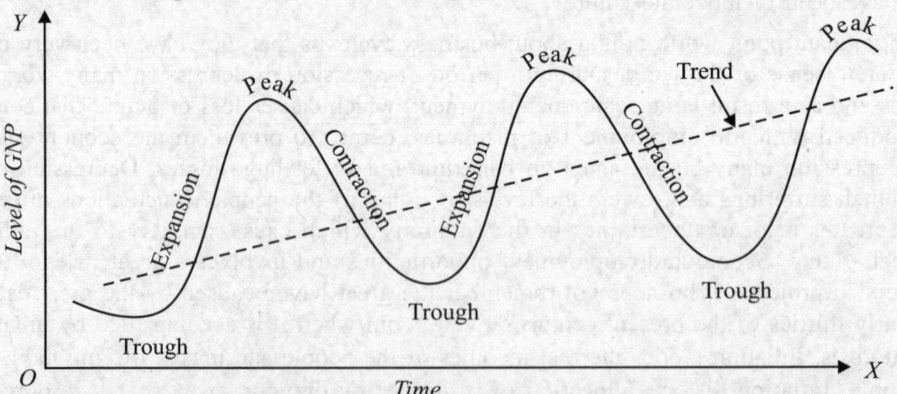

Fig. 27.2. *Cycles with Trend* (*i.e.,* *Growth*)

2. G. Crowther, *An Outline of Money* (London, 1962), p. 192.

is also called trough is reached. In this way cycle is complete. However, after remaining at the trough for some time the economy revives and again the new cycle starts. Haberler in his important work[3] on business cycles has named the four phases of business cycles as (1) Upswing, (2) Upper Turning point, (3) Downswing, and (4) Lower Turning Point.

There are two types of patterns of cyclic changes. One pattern is shown in Fig. 27.1 where fluctuations occur around a stable equilibrium position as shown by the horizontal line. It is a case of dynamic stability which depicts change but without growth or trend. The second pattern of cyclical fluctuations is shown in Fig. 27.2 where cyclical changes in economic activity take place around a growth path (*i.e.*, rising trend). J.R. Hicks in his model of business cycles explains such a pattern of fluctuations with long-run rising trend in economic activity by imposing factors such as autonomous investment due to population growth and technological progress causing economic growth on the otherwise stationary state. We briefly explain below various phases of business cycles.

Expansion and Prosperity. In its expansion phase, both output and employment increase till we have full employment of resources and production is at the highest possible level with the given productive resources. There is no involuntary unemployment and whatever unemployment prevails is only of frictional and structural types. Thus, when expansion gathers momentum and we have prosperity, the gap between potential GNP and actual GNP is zero, that is, the level of production is at the maximum production level. A good amount of net investment is occurring and demand for durable consumer goods is also high. Prices also generally rise during the expansion phase but due to high level of economic activity people enjoy a high standard of living.

Then something may occur, whether banks start reducing credit or profit expectations change adversely and businessmen become pessimistic about future state of the economy that bring an end to the expansion or prosperity phase. As shall be explained below, economists differ regarding the possible causes of the end of prosperity and start of downswing in economic activity. Monetarists have argued that contraction in bank credit may cause downswing. Keynes has argued that sudden collapse of expected rate of profit (which he calls marginal efficiency of capital, MEC) caused by adverse changes in expectations of entrepreneurs lowers investment in the economy. This fall in investment, according to him, causes downswing in economic activity.

Contraction and Depression. As stated above, expansion or prosperity is followed by contraction or depression. During contraction, not only there is a fall in GNP but also level of employment is reduced. As a result, involuntary unemployment appears on a large scale. Investment also decreases causing further fall in consumption of goods and services. At times of contraction or depression prices also generally fall due to fall in aggregate demand. A significant feature of depression phase is the fall in rate of interest. With lower rate of interest people's demand for money holdings increases. There is a lot of excess capacity as industries producing capital goods and consumer goods work much below their capacity due to lack of demand. Capital goods and durable consumer goods industries are especially hit hard during depression. Depression, it may be noted, occurs when there is a severe contraction or recession of economic activities. The depression of 1929-33 is still remembered because of its great intensity which caused a lot of human suffering.

Trough and Revival. There is a limit to which level of economic activity can fall. The lowest level of economic activity, generally called trough, lasts for some time. Capital stock is allowed to depreciate without replacement. The progress in technology makes the existing capital stock obsolete. If the banking system starts expanding credit or there is a spurt in investment activity due to the emergence of scarcity of capital as a result of non-replacement of depreciated capital and also because of new technology coming into existence requiring new types of machines and other capital goods.

3. See G. HaberJer, *Prosperity and Depression*: *A Theoretical Analysis of Cyclical Movements*, Harward University Press, 1960.

The stimulation of investment brings about the revival or recovery of the economy. The recovery is the turning point from depression into expansion. As investment rises, this causes induced increase in consumption. As a result industries start producing more and excess capacity is now put into full use due to the revival of aggregate demand. Employment of labour increases and rate of unemployment falls. With this the cycle is complete.

Features of Business Cycles

Though different business cycles differ in duration and intensity they have some common features which we explain below:

1. Business cycles occur *periodically*. Though they do not show same regularity, they have some distinct phases such as expansion, peak, contraction or depression and trough. Further the duration of cycles varies a good deal from minimum of two years to a maximum of ten to twelve years.
2. Secondly, business cycles are *Synchronic*. That is, they do not cause changes in any single industry or sector but are of all-embracing character. For example, depression or contraction occur simultaneously in all industries or sectors of the economy. Recession passes from one industry to another and chain reaction continues till the whole economy is in the grip of recession. Similar process is at work in the expansion phase, prosperity spreads through various linkages of input-output relations or demand relations between various industries and sectors.
3. Thirdly, it has been observed that fluctuations occur not only in level of production but also simultaneously in other variables such as employment, investment, consumption, rate of interest and price level.
4. Another important feature of business cycles is that investment and consumption of durable consumer goods such as cars, houses, refrigerators are affected most by the cyclical fluctuations. As stressed by J.M. Keynes, investment is greatly volatile and unstable as it depends on profit expectations of private entrepreneurs. These expectations of entrepreneurs change quite often making investment quite unstable. Since *consumption of durable consumer goods* can be deferred, it also fluctuates greatly during the course of business cycles.
5. An important feature of business cycles is that consumption of *non-durable goods and services* does not vary much during different phases of business cycles. Past data of business cycles reveal that households maintain a great stability in consumption of non-durable goods.
6. The immediate impact of depression and expansion is that on the inventories of goods. When depression sets in, the inventories start accumulating beyond the desired level. This leads to cut in production of goods. On the contrary, when recovery starts, the inventories go below the desired level. This encourages businessmen to place more orders for goods whose production picks up and stimulates investment in capital goods.
7. Another important feature of business cycles is that *profits fluctuate more than any other type of income*. The occurrence of business cycles causes a lot of uncertainty for businessmen and makes it difficult to forecast the economic conditions. During the depression period profits may even become negative and many businesses go bankrupt. In a free market economy profits are justified on the ground that they are necessary payments if the entrepreneurs are to be induced to bear uncertainty.
8. Lastly, business cycles are *international in character*. That is, once started in one country they spread to other countries through trade relations between them. For example, if there is a recession in the USA which is a large importer of goods from other countries, it will cause a fall in demand for imports from other countries whose exports would be adversely affected causing recession in them too. Depression of 1930s in USA and Great Britain engulfed the entire capital world.

THEORIES OF BUSINESS CYCLES

We have explained above the various phases and common features of business cycles. Now, an important question is what causes business cycles. Several theories of business cycles have been propounded from time to time. Each of these theories spells out the factors which cause business cycles. Before explaining the modem theories of business cycles we first explain below the *earlier* theories of business cycles as they too contain important elements whose study is essential for proper understanding of the causes of business cycles.

Sun-spot Theory

This is perhaps the oldest theory of business cycles. Sun-spot theory was developed in 1875 by Stanley Jevons. Sun-spots are storms on the surface of the sun caused by violent nuclear explosions there. Jevons argued that sun-spots affected weather on the earth. Since economies in the olden world were heavily dependent on agriculture, changes in climatic conditions due to sun-spots produced fluctuations in agricultural output. Changes in agricultural output through its demand and input-output relations affect industry. Thus, swings in agricultural output spread throughout the economy.

Other earlier economists also focused on changes in climatic or weather conditions in addition to those caused by sun-spots. According to them, weather cycles cause fluctuations in agricultural output which in turn cause instability in the whole economy. Even today weather is considered important in a country like India where agriculture is still important. In the years when due to lack of monsoon there are drought in the Indian agriculture, it affects the income of farmers and therefore reduces demand for the products of industries. This causes industrial recession. Even in USA in the year 1988 a severe drought in the farm belt drove up the food price around the world. It may be further noted that higher food prices reduce income available to be spent on industrial goods.

Critical Appraisal. Though the theories of business cycles which emphasise climatic conditions for business cycles contain an element of truth about fluctuations in economic activity, especially in the developing counties like India where agriculture still remains important, they do not offer an adequate explanation of business cycles. Therefore, much reliance is not placed on these theories by modem economists. Nobody can say with certainty about the nature of these sun-spots and the degree to which they affect rain. There is no doubt that climate affects agricultural production. But the climate theory does not adequately explain periodicity of the trade cycle. If there was truth in the climatic theories, the trade cycles may be pronounced in agricultural countries and almost disappear when the country becomes completely industrialised. But this is not the case. Highly industrialised countries are much more subject to business cycles than agricultural countries which are affected more by famines rather than business cycles. Hence variations in climate do not offer complete explanation of business cycles.

Hawtrey's Monetary Theory of Business Cycles

An old monetary theory of business cycles was put forward by Hawtrey.[4] His monetary theory of business cycles relates to the economy which is under gold standard. It will be remembered that economy is said to be under gold standard when either money in circulation consists of gold coins or when paper notes are fully backed by gold reserves in the banking system. According to Hawtrey, increases in the quantity of money raises the availability of bank credit for investment. Thus, by increasing the supply of credit expansion in money supply causes rate of interest to fall. The lower rate of interest induces businessmen to borrow more for investment in capital goods and also for investment in keeping more inventories of goods. Thus, Hawtrey argues that lower rate of interest will lead to the expansion of goods and services as a result of more investment in capital goods and inventories. Higher output, income and employment caused by more investment induces more spending

4. R.G. Hawtrey, *Good and Bad Trade*, Constable and Co., London, 1913.

on consumer goods. Thus, as a result of more investment made possible by increased supply of bank credit economy moves into the expansion phase. The process of expansion continues for some time. Increases in aggregate demand brought about by more investment also causes prices to rise. Rising prices lead to the increase in output in two ways. First, when prices begin to rise businessmen think they would rise further which induces them to invest more and produce more because prospects of making profits increase with the rise in prices. Secondly, the rising prices reduce the real value of idle money balances with the people which induces them to spend more on goods and services. In this way rising prices sustain expansion for some time.

However, according to Hawtrey, the expansion process must end. He argued that rise in incomes during the expansion phase induces more expenditure on domestically produced goods as well as more on imports of foreign goods. He further assumes that domestic output and income expand faster than foreign output. As a result, imports of a country increase more than its exports causing *trade deficit* with other countries. If exchange rate remains fixed, trade deficit means there will be *outflow of gold* to settle its balance of payments deficit. Since the country is on gold standard, outflow of gold will cause reduction in money supply in the economy. The decrease in money supply will reduce the availability of bank credit. Reduction in the supply of bank credit will cause the rate of interest to rise. Rising interest rate will reduce investment in physical capital goods. Reduction in investment will cause the process of contraction to set in. As a result of reduced order for inventories, producers will cut production which will lower income and consumption of goods and services. In this state of reduced demand for goods and services, prices of goods will fall. Once the prices begin to fall businessmen begin to expect that they will fall further. In response to it traders will cut order of goods still causing further fall in output. The fall in prices also causes real value of money balances to rise which induces people to hold larger money holdings with them. In this way contraction process gathers momentum as demand for goods start declining faster and with this economy plunges into depression.

But after a lapse of some time depression will also come to an end and the economy will start to recover. This happens because in the contraction process imports fall drastically due to decrease in income and consumption of households, whereas exports do not fall much. As a result, *trade surplus* emerges which causes *inflow of gold*. The inflow of gold would lead to the expansion of money supply and consequently availability of bank credit for investment will increase. With this, the economy will recover from depression and move into the expansion phase. Thus, the cycle is complete. The process, according to Hawtrey, will go on being repeated regularly.

Critical Appraisal. Hawtrey maintains that the economy under gold standard and fixed exchange rate system makes his model of business cycles self-generating as there is built-in tendency for the money supply to change with the emergence of trade deficit and trade surplus which cause movements of gold between countries and affect money supply in them. Changes in money supply influence economic activity in a cyclical fashion. However, Hawtrey's monetary theory does not apply to the present-day economies which have abandoned gold standard in 1930s. However, Hawtrey's theory still retains its importance because it shows how changes in money supply affects economic activity through changes in price level and rate of interest. In modem monetary theories of trade cycles this relation between money supply and rate of interest plays an important role in determining the level of economic activity.

Under-consumption Theory

Under-consumption theory of business cycles is a very old one which dates back to the 1930s. Malthus and Sismodi criticised Say's Law which states '*supply creates its own demand*' and argued that consumption of goods and services could be too small to generate sufficient demand for goods and services produced. They attribute over-production of goods due to lack of consumption demand for them. This over-production causes piling up of inventories of goods which results in recession.

Under-consumption theory as propounded by Sismodi and Hobson was not a theory of recurring business cycles. They made an attempt to explain how a free enterprise economy could enter a long-run economic slowdown. A crucial aspect of Sismodi and Hobson's under-consumption theory is the distinction they made between the rich and the poor. According to them, the rich sections in the society receive a large part of their income from returns on financial assets and real property owned by them. Further, they assume that the rich have a large propensity to save, that is, they save a relatively large proportion of their income and, therefore, consume a relatively smaller proportion of their income. On the other hand, less well-off people in a society obtain most of their income from work, that is, wages from labour and have a lower propensity to save. Therefore, these less well-off people spend a relatively less proportion of their income on consumer goods and services. In their theory, they further assume that during the expansion process, the incomes of the rich people increase relatively more than the wage-income. Thus, during the expansion phase, income distribution changes in favour of the rich with the result that average propensity to save falls, that is, in the expansion process saving increases and therefore consumption demand declines.[5]

According to Sismodi and Hobson, increase in saving during the expansion phase leads to more investment expenditure on capital goods and after some time lag, the greater stock of capital goods enables the economy to produce more consumer goods and services. But since society's propensity to consume continues to fall, consumption demand is not enough to absorb the increased production of consumer goods. In this way, lack of demand for consumer goods or what is called under-consumption emerges in the economy which halts the expansion of the economy. Further, since supply or production of goods increases relatively more as compared to the consumption demand for them, the prices fall. Prices continue falling and go even below the average cost of production bringing losses to the business firms. Thus, when under-consumption appears, production of goods becomes unprofitable. Firms cut their production resulting in recession or contraction in economic activity.

Karl Marx and Under-consumption. It is worth mentioning that Karl Marx, the philosopher of scientific socialism, had also predicted the collapse of the capitalist system due to the emergence of under-consumption. He predicted that capitalism would move periodically through expansion and contraction with each peak higher than its previous peak and each crash (*i.e.*, depression) deeper than the last. Ultimately, according to Marx, in a state of acute depression when the cup of misery of working class is full, they will overthrow the capitalist. class which exploits them and in this way the new era of socialism or communism would come into existence.

Like other under-consumption theorists, Marx argues that driving force behind business cycles is ever increasing income inequalities and concentration of wealth and economic power in the hands of the few capitalists who own the means of production. As a result, the poor workers lack income to purchase goods produced by the capitalist class resulting in under-consumption or over-production. With the capitalist producers lacking market for their goods, capitalist economy plunges into depression. Then the search for ways of opening of new markets is started. Even wars between capitalist countries take place to capture other countries to find new markets for their products. With the discovery of new methods of production of finding new markets, the economy recovers from depression and the new upswing starts.

Critical Appraisal. The view that income inequalities increase with growth or expansion of the economy and further that this causes recession or stagnation is widely accepted. Therefore, even many modem economics suggest that if growth is to be sustained (that is, if recession or stagnation is to be avoided), then consumption demand must be increasing sufficiently to absorb the increasing production of goods. For this deliberate efforts should be made to reduce inequalities in income distribution. Further, under-consumption theory rightly states that income redistribution schemes will reduce the amplitude of business cycles.

5. J.A. Hobson, *The Economics of Unemployment*, London: Allen & Unwin Ltd., 1922.

Besides, the suggested behaviour of average propensity to save and consume of the property owners and wage earners in this theory have been found to be consistent with the observed phenomena. Even in the theory of economic development the difference in average propensity to save (APS) of the property owners and workers has been widely used.

It is clear from above that under-consumption theory contains some important elements, especially the emergence of the lack of consumption demand as the cause of recession but it is regarded as too simple. There are many features other than growing income inequalities which are responsible for causing recession or trade cycles. Although under-consumption theory concentrates on a significant variable, it leaves too much unexplained.

Over-investment Theory

It has been observed that over time investment varies more than that of total output of final goods and services and consumption. This has led economists to investigate the causes of variation in investment and how it is responsible for business cycles. Two versions of over-investment theory have been put forward. One theory offered by Hayek emphasises monetary forces in causing fluctuations in investment. The second version of over-investment theory has been developed by Knut Wickshell which emphasises spurts of investment brought about by innovation. We explain below both these versions of over-investment theory. It is worth noting that in both the versions of this theory distinction between natural rate of interest and money rate of interest plays an important role. Natural rate of interest is defined as the rate at which saving equals investment and this equilibrium interest rate reflects marginal revenue product of capital or rate of return on capital. On the other hand, money rate of interest is the rate at which banks give loans to the businessmen.

Hayek's Monetary Version of Over-investment Theory. Hayek[6] suggests that it is monetary forces which cause fluctuations in investment which are prime cause of business cycles. In this respect Hayek's theory is similar to Hawtrey's monetary theory except that it does not involve inflow and outflow of gold causing changes in money supply in the economy.

To begin with, let us assume that the economy is in recession and businessmen's demand for bank credit is therefore very low. Thus, lower demand for bank credit in times of recession pushes down the money rate of interest below the natural rate. This means that businessmen will be able to borrow funds, that is, bank credit at a rate of interest which is below the expected rate of return in investment projects. This induces them to invest more by undertaking new investment projects. In this way, investment expenditure on new capital goods increases. This causes investment to exceed saving by the amount of newly created bank credit. With the spurt in investment expenditure, the expansion of the economy begins. Increase in investment causes income and employment to rise which induces more consumption expenditure. As a result, production of consumer goods increases. According to Hawtrey, the competition between capital goods and consumer goods industries for scarce resources causes their prices to rise which in turn push up the prices of goods and services.

But this process of expansion cannot go on indefinitely because the excess reserves with the banks come to an end which force the banks not to give further loans for investment, while demand for bank credit goes on increasing. Thus, the inelastic supply of credit from the banks and mounting demand for it cause the money rate of interest to go above the natural rate of interest. This makes further investment unprofitable. But at this point of time there has been over-investment in the sense that savings fall short of what is required to finance the desired investment, When no more bank credit is available for investment, there is decline in investment which causes both income and consumption to fall and in this way expansion comes to an end and the economy experiences downswing in economic activity.

---

6. F.A. Hayek, *Monetary Theory and Trade Cycle*, New York, Harcourt Brace, 1933 and also *Profits, Interest and Investment*, London, Routledge, 1935.

However, after a lapse of some time the fall in demand for bank credit lowers the money rate of interest which goes below the natural rate of interest. This again gives boost to investment activity and as a result recession ends. In this way alternating periods of expansion and contraction occur periodically.

Wicksell's Over-investment Theory. Over-investment theory developed by Wicksell[7] is of non-monetary type. Instead of focusing on monetary factors it attributes cyclical fluctuations to spurts of investment caused by new innovations introduced by entrepreneurs themselves. The introduction of new innovations or opening of new markets makes some investment projects profitable by either reducing cost or raising demand for the products. The expansion in investment is made possible because of the availability of bank credit at a lower money rate of interest. The expansion in economic activity ceases when investment exceeds saving. Again it may be noted that there is over-investment because the level of saving is insufficient to finance the desired level of investment. The end of investment expenditure causes the economy to go into recession. However, another set of innovations occurs or more new markets are found which stimulates investment. Thus, when investment picks up as a result of new innovations, the economy revives and moves into the expansion phase once again.

Appraisal. Though the over-investment theory does not offer an adequate explanation of business cycles, it contains an important element that fluctuations in investment are the prime cause of business cycles. However, it does not offer a valid explanation as to why changes in investment take place quite often. Many exponents of this theory point to the behaviour of banking system that causes diverges between money rate of interest and natural rate of interest. However, as Keynes later on emphasised, investment fluctuates quite often because of changes in profit expectations of entrepreneurs which depends on several economic and political factors operating in the economy. Thus, the theory fails to offer adequate explanation of business cycles.

KEYNES' S THEORY OF BUSINESS CYCLES

J.M. Keynes in his seminal work '*General Theory of Employment, Interest and Money*' made an important contribution to the analysis of the causes of business cycles. According to Keynes's theory, in the short run, the level of income, output or employment is determined by the level of aggregate effective demand. In a free private enterprise, the entrepreneurs will produce that much of goods as can be sold profitably. Now, if. the aggregate demand is large, that is, if the expenditure on goods and services is large, the entrepreneurs will be able to sell profitably a large quantity of goods and therefore they will produce more. In order to produce more they will employ a larger amount of resources, both men and materials. In short, a higher level of aggregate demand will result in greater output, income and employment. On the other hand, if the level of aggregate demand is low, smaller amount of goods and services can be sold profitably. This means that the total quantity of national output produced will be small. And a small output can be produced with a small amount of resources. As a result, there will be unemployment of resources, both labour and capital. Hence, the *changes in the level of aggregate effective demand will bring about fluctuations in the level of income, output and employment*. Thus, according to Keynes, the fluctuations in economic activity are due to the fluctuations in aggregate effective demand. Fall in aggregate effective demand will create the conditions of recession or depression. If the aggregate demand is increasing, economic expansion will take place.

Now the question arises : What causes fluctuations in aggregate demand? The aggregate demand is composed of demand for consumption goods and demand for investment goods. Thus aggregate demand depends on the total expenditure of the consumers on consumption goods and entrepreneurs on investment goods. Propensity to consume being more or less stable in the short run, fluctuations

7. Knut Wicksell, *Interest and Prices*, London: Macmillan, 1936.

in aggregate demand depend primarily upon the fluctuations in investment demand. Keynes shows that the fundamental cause of fluctuations in aggregate demand and hence in fluctuations in economic activity is the fluctuations in investment demand. *Investment demand is very unstable and volatile and brings about business cycles in the economy.*

Let us start from the phase of economic expansion to explain Keynes's theory of business cycles. We first explain how in Keynesian theory expansion comes to end and recession or depression sets in. During an economic expansion two factors eventually work to cause investment to fall. First, during the expansion phase increase in demand for capital goods due to large-scale investment activity leads to the rise in prices of capital goods due to rising marginal cost of their production. Higher prices of capital goods raise the cost of investment projects and thereby reduce marginal efficiency of capital (that is, expected rate of return). Secondly, as income rises during expansion phase, the demand for money increases which raises interest rate. Higher interest rate makes some potential projects unprofitable. Thus, fall in marginal efficiency of capital on the one hand and rise in interest rate on the other cause decline in investment demand. Declining trend of investment, according to Keynes, raises doubts about the *prospective yield on capital goods* which is more important factor determining marginal efficiency of capital than cost of investment projects and rate of interest. When among businessmen pessimism sets in about future profitability of investment projects stock prices tumble. The crash in stock prices worsens the situation and causes investment to fall even more. Besides, fall in prices of shares reduces wealth of households. Wealth, according to Keynes, is an important factor determining consumption. Thus, the decline in stock prices reduces autonomous consumption demand of households. With the fall in both investment and consumption demand aggregate demand declines which results in accumulation of unintended inventories with the firms. This induces the firms to cut production of goods.

It follows from above that besides the rise in cost of capital goods and rise in rate of interest towards the end of the expansion phase, it is the fall in expected prospective yield that reduces the marginal efficiency of capital and causes investment demand to fall. This induces a wave of pessimistic expectations among businessmen and speculators. These pessimistic expectations cause stock prices to tumble which work like adding fuel to the fire. They cause a further fall in the marginal efficiency of capital. *The turning point from expansion to contraction is thus caused by a sudden collapse in marginal efficiency of capital*. In terms of graph, a sudden fall in the marginal efficiency of capital causes a leftward shift in the investment demand curve, for example from I_0I_0 to I_1I_1 in Fig. 27.3 resulting in decline in investment from I_0* to I_1* at the given rate of interest. Note that decrease in investment does not automatically decrease the rate of interest to offset the fall in the marginal efficiency of capital.

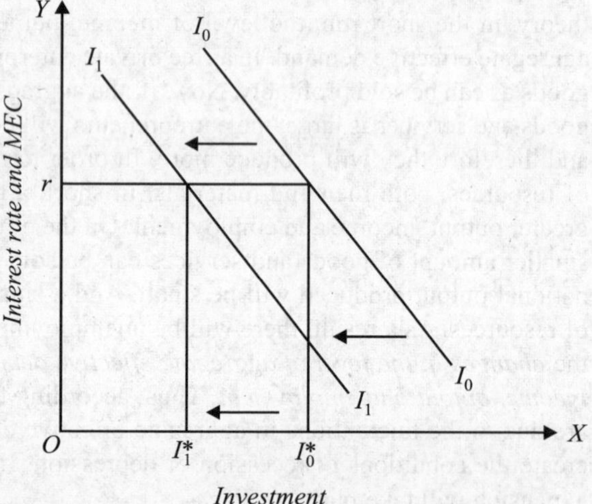

Fig. 27.3. *A Decline in Investment Expenditure due to Fall in Marginal Efficiency of Capital*

However, an additional factor that makes Keynes's business cycle theory potent is the working of multiplier which was an important discovery of J.M. Keynes. According to Keynes, a decrease in investment expenditure causes a decline in income which in turn reduces consumption expenditure. The reduction in consumption expenditure further reduces income and this process of reduction in income continues further. The total fall in income (ΔY) due to an initial decline in investment (ΔI) will be equal to $\Delta I \times \frac{1}{1 - MPC}$ where $\frac{1}{1 - MPC}$ is the value of multiplier. If marginal propensity to consume is 0.75, the multiplier will be equal to 4. Thus, a decline in investment by 100 crores will lead to the decline in income by 400 crores. Note that multiplier here works in reverse. Thus, the multiplier process magnifies the effect of decline in investment expenditure on aggregate demand and income and further deepens the depression. As income and output are falling rapidly under the multiplier effect, the employment also goes tumbling down. Thus, the Keynes theory of income multiplier plays a significant role in causing magnified changes in income, output and employment following a reduction in investment.

It is important to note that, in Keynes's view, wages and prices are not flexible enough to offset the decline in investment expenditure and thereby restore full employment. This is in sharp contrast to the classical theory where changes in wages and prices ensure continuous full employment. In Keynes model wages and prices are "sticky" downward which implies that though wages and prices do not remain constant but when demand falls wages and prices will fall but not sufficient to restore full employment in the economy.

Since wage and price flexibility does not ensure the recovery of the economy out of the state of depression, Keynes thinks that marginal efficiency of capital must rise to stimulate investment. During depression investment falls to a very low level, capital stock begins to wear out and requires replacement. Further, some existing capital equipment becomes technologically obsolete and has to be abandoned. This generates demand for replacement investment. A long period of time is necessary for existing capital to depreciate because most capital goods are durable as well as irreversible. By durability of capital goods we mean that they last for a long time and by irreversibility we mean that they cannot be used for purposes other than those for which they are meant.

Thus, just as the collapse of marginal efficiency of capital is the main cause of the upper turning point, similarly the lower turning point, *i.e.*, change from recession to recovery is due to the revival of the marginal efficiency of capital, that is, expected rate of profit. Restoration of business confidence is the most important, yet the most difficult factor to achieve. Even if the rate of interest is reduced, the investment will not increase. This is because of the fact that in the absence of confidence the profitability of investment may remain so low that no practicable reduction in the rate of interest will stimulate investment. The interval which will elapse between the upper turning point and the start of recovery is conditioned by two factors: (*i*) the time necessary for wearing out of durable capital assets, and (*ii*) the time required to absorb the excess stocks of goods left over from the boom. Just as the expected rate of profit was pushed down by the growing abundance of capital during the period of boom, similarly as the stocks of capital goods are depleted and there grows a scarcity of capital goods, then the expected rate of profit rises thereby inducing the businessmen to invest more. When the level of investment increases, income increases by a magnified amount due to the multiplier effect. So the cumulative process starts upward.

Thus, over time as depreciation of capital stock occurs without replacement and also some existing capital equipment becomes technologically obsolete, the size of capital stock declines. New investment must be undertaken even to produce reduced depression level of output. Thus with the emergence of scarcity of capital, marginal efficiency of capital rises which boosts investment. Once investment increases, it induces further rise in income and consumption demand through the multiplier process. Now, the multiplier works to magnify the effect of increase in investment on raising aggregate demand. The mood of bussinessmen changes from pessimism to optimism which drives up stock prices. All these factors work to lift the economy out of depression and put it on the road to prosperity.

However, it is noteworthy that the recovery process from depression takes a very long time. Keynes argued that Government should not wait for long for the natural recovery to occur. This is because persistence of depression creates a lot of human sufferings. He, therefore, advocated for the active intervention by the Government to raise aggregate demand through fiscal policy, that is, stepping up its expenditure or reducing taxes. Thus, he argued for the adoption of policy of deficit budget to boost aggregate demand so that economy is lifted out of depression.

It may be noted that Keynes's business cycle theory is self-generating. In it the economy passes through a long phase of expansion. But eventually some forces automatically work, for example, the growing abundance of capital stock, which reduces marginal efficiency of capital. Pessimism overtakes businessmen. This causes reduction in investment which is responsible for bringing about downswing in the economy. The idea that it is the fluctuations in investment that bring about the fluctuations in the level of economic activity is an important contribution made by Keynes. Of course, even before Keynes, it was believed that the fluctuations in the investment demand have something to do with the business cycles, but a systematic exposition was lacking. Keynes propounded a definite relationship between a change in investment and the resulting change in income and employment. This relationship is embodied in his famous theory of multiplier.

The Critical Appraisal of Keynes's Theory

J.M. Keynes has made three important contributions to the business cycle theory. First, it is fluctuations in investment that cause changes in aggregate demand' which bring about changes in economic activity (*i.e.*, income, output, and employment). Secondly, fluctuations in investment demand are caused by changes in expectations of businessmen regarding making of profits (that is, marginal efficiency of capital). Thirdly, Keynes put forward an important theory of multiplier which tells us how changes in investment bring about magnified changes in the level of income and employment.

But the Keynesian theory of multiplier alone does not offer a full and satisfactory explanation of the trade cycles. A basic feature of the trade cycle is its cumulative character both on the upswing as well as on the downswing, *i.e.*, once economic activity starts rising or falling, it gathers momentum and for a time feeds on itself. Thus, what we have to explain is the cumulative character of economic fluctuations. The theory of multiplier alone does not prove adequate for this task. For example, suppose that investment rises by 100 rupees and that the magnitude of multiplier is 4. From the theory of multiplier we know that national income will rise by 400 rupees and if multiplier is the only force at work, that will be the end of the matter, with the economy reaching a new stable equilibrium at a higher level of national income. But in real life this is not likely to be so, for a rise in income produced by a given rise in investment will have further repercussions in the economy. This reaction is described in the *principle of the accelerator*. According to the principle of acceleration, a change in national income will tend to induce changes in the rate of investment. *While multiplier refers to the change in income as a result of change in investment, the acceleration principle describes the relationship between a change in investment as a result of change in income.*

In the above example, when income has risen by 400 rupees, people's spending power has risen by an equivalent amount. This will induce them to spend more on goods and services. When the demand for goods rises, initially this will be met by overworking the existing plant and machinery. All this leads to an increase in profits with the result that businessmen will be induced to expand their productive capacity and will instal new plants, *i.e.*, they will invest more than before. Thus, *a rise in income leads to a further induced increase in investment*. The accelerator describes this relation between an increase in income and the resulting increase in investment. Thus, Samuelson combined the accelerator principle with the multiplier and showed that the interaction between the two can bring about cyclical fluctuations in economic activity.

SAMUELSON'S MODEL OF BUSINESS CYCLES: INTERACTION BETWEEN MULTIPLIER AND ACCELERATOR

We have briefly explained above that Keynes made an important contribution to the understanding of the cyclical fluctuations by pointing out that it is the ups and downs in investment demand, depending as it is on the profit expectations of the entrepreneurs, that causes changes in aggregate demand which affect the levels of income, output and employment. Further, by putting forward the theory of multiplier, Keynes has shown how the effect of increase and decrease in investment on output and employment gets magnified when multiplier is working during either the upswing or downswing of a business cycle. However, Keynes did not explain the cyclical and cumulative nature of the fluctuations in economic activity. This is because Keynes did not give any importance to the accelerator in his explanation of business cycles. Samuelson in his seminal paper[8] convincingly showed that it is the interaction between the multiplier and accelerator that gives rise to cyclical fluctuations in economic activity. The multiplier alone cannot adequately explain the cyclical and cumulative nature of the economic fluctuations. An autonomous increase in the level of investment raises income by a magnified amount depending upon the value of the multiplier. This increase in income further induces the increases in investment through the acceleration effect. The increase in income brings about increase in aggregate demand for goods and services. To produce more goods we require more capital goods for which extra investment is undertaken. Thus the relationship between investment and income is one of mutual interaction; investment affects income which in turn affects investment demand and in this process income and employment fluctuate in a cyclical manner.

We have shown below in Fig. 27.4 how income and output will increase by even larger amount when accelerator is combined with the Keynesian multiplier.

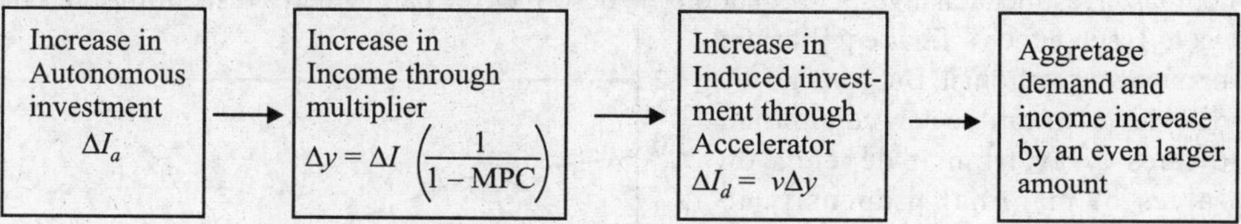

Fig. 27.4. Combining Accelerator with Keynesian Multiplier

where ΔI_a = Increase in Autonomous Investment

ΔY = Increase in Income.

$\frac{1}{1-MPC}$ = Size of Multiplier where MPC = Marginal Propensity to Consume.

ΔI_d = Increase in Induced Investment

v = Size of accelerator.

Fluctuations in investment are the main cause of instability in a free private-enterprise economy. This instability further increases due to the interaction of the multiplier and accelerator. The changes in any component of aggregate demand produce a multiplier effect whose magnitude depends upon the marginal propensity to consume. When consumption, income and output increase under the influence of multiplier effect, they induce further changes in investment and the extent of this induced investment in capital goods industries depends on the capital-output ratio, that is, the interaction between the multiplier and accelerator without any external shocks can give rise to the business cycles whose pattern differs depending upon the magnitudes of the marginal propensity to consume and capital-output ratio. The model of interaction between multiplier and accelerator can be mathematically represented as under:

---

8. Samuelson, P.A., "Interaction Between the Multiplier Analysis and the Principle of Acceleration" in *Review of Economic Statistics*, May 1939, p. 751.

$$Y = C + I \qquad ...(i)$$

$$C = C_a + c\,(Y_t - I) \qquad ...(ii)$$

$$I = I_a + v\,(Y_{t-1} - Y_{t-2}) \qquad ...(iii)$$

where Y_t, C_t, I_t stand for income, consumption and investment respectively for a period t, C_a stands for autonomous consumption, I_a for autonomous investment, c for marginal propensity to consume and v for the capital-output ratio or accelerator.

From the above equations it is evident that consumption in a period t is a function of income of the previous period Y_{t-1}. That is, one period lag has been assumed for income to determine the consumption of a period. As regards induced investment in period t, it is taken to be the function of the change in income in the previous period. This means that there is two periods gap for changes in income to determine induced investment. In the equation (*iii*) above, induced investment equals $v(Y_{t-1} - Y_{t-2})$ or $v(\Delta Y_{t-1})$. Substituting equations (*ii*) and (*iii*) in equation (*i*) we have the following income equation which states how changes in income are dependent on the values of marginal propensity to consume (c) and capital-output ratio v (*i.e.*, accelerator).

$$Y_t = C_a + c(Y_t - 1) + I_a + v(Y_{t-1} - Y_{t-2}) \qquad ...(iv)$$

In static equilibrium, the level of income determined will be:

$$Y = C_a + cY + I$$

This is due to the fact that in static equilibrium, given the data of the determining factors, the equilibrium level of income remains uncharged, that is, in this case, $Y_t = Y_{t-1} = Y_{t-2} = Y_{t-n}$ so that period lags have no influence at all and accelerator is reduced to zero. Thus, in a dynamic state when autonomous investment changes, the equation (*iv*) describes the path which a disequilibrium system follows to reach either a final equilibrium state or moves away from it. But whether the economy moves towards a new equilibrium or deviates away from it depends on the values of marginal propensity to consume (c) and capital-output ratio v (*i.e.*, accelerator).

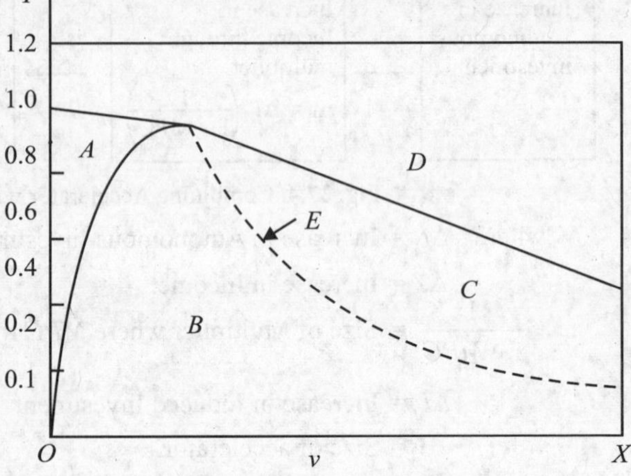

Fig. 27.5 *Combinations of c & v*

By taking different combinations of the values of marginal propensity to consume (c) and capital-output ratio (v), Samuelson has described different paths which the economy will follow. The various combinations of the values of marginal propensity to consume and capital-output ratio (which respectively determine the magnitudes of multiplier and accelerator) are shown in Fig. 27.5. The four paths or patterns of movements which the economic activity (as measured by gross national product or income) can have depending upon various combinations of the values of marginal propensity to consume (c) and capital-output ratio (v) are depicted in Fig. 27.6. When the combinations of the value of marginal propensity to consume (c) and capital-output ratio (v) lie within the region marked *A*, with a change in autonomous investment, the gross national product or income moves upward or downward at a decreasing rate and finally reaches a new equilibrium as is shown in panel (*a*) of Fig. 27.6. If the values of c and v are such that they lie within the region *B*, the change in autonomous investment or autonomous consumption will generate fluctuations in income which follow the pattern of a series of *damped cycles* whose amplitudes go on declining until the cycles disappear as is shown in panel (*b*) of Fig. 27.6.

The region C in Fig. 27.6 represents the combinations of c and v which are relatively high as compared to the region B and determine such values of multiplier and accelerator that bring about explosive cycles, that is, the fluctuations of income with successively greater and greater amplitude. The situation is depicted in panel (*c*) of Fig. 27.6 which shows that the system tends to explode and diverges greatly from the equilibrium level. The region D provides the combinations of c and v which cause income to move upward or downward at an increasing rate which has somehow to be restrained if the cyclical movements are to occur. This is depicted in panel (*d*) of Fig. 27.6. Like the values of multiplier and accelerator of region C, their values in region D cause the system to *explode* and diverge from the equilibrium state by an increasing amount.

In a special case when values of c and v (and therefore the magnitudes of multiplier and accelerator) lie in region E, they produce fluctuations in income of *constant amplitude* as is shown in panel (*e*) of Fig. 27.6.

It follows from above that region A and B are alike, they after a disturbance caused by a change in autonomous investment or consumption finally bring about stable equilibrium in the system. On the other hand, the values of c and v and therefore the magnitudes of multiplier and accelerator of region C and D resemble each other but are such that they cause great instability in the system as both of these values cause successively greater divergence from the equilibrium level and the system tends to explode. The case of region E lies in between the two as the combinations of values of c and v in it are such that cause cyclical movements of income which neither move toward nor away from the equilibrium.

It is worth noting that all the above five cases do not give rise to cyclical fluctuations or business cycles. *It is only combinations of c and v lying in regions B, C and E that produce business cycles*. The values of accelerator and multiplier in the region A are such that with a disturbance caused by a change in autonomous investment or autonomous consumption, the economic activity (as measured by the level of income or Gross National Product) moves smoothly from an initial equilibrium to a new equilibrium with no cyclical fluctuations or oscillations. On the other hand, the values of c and v (and therefore of multi-

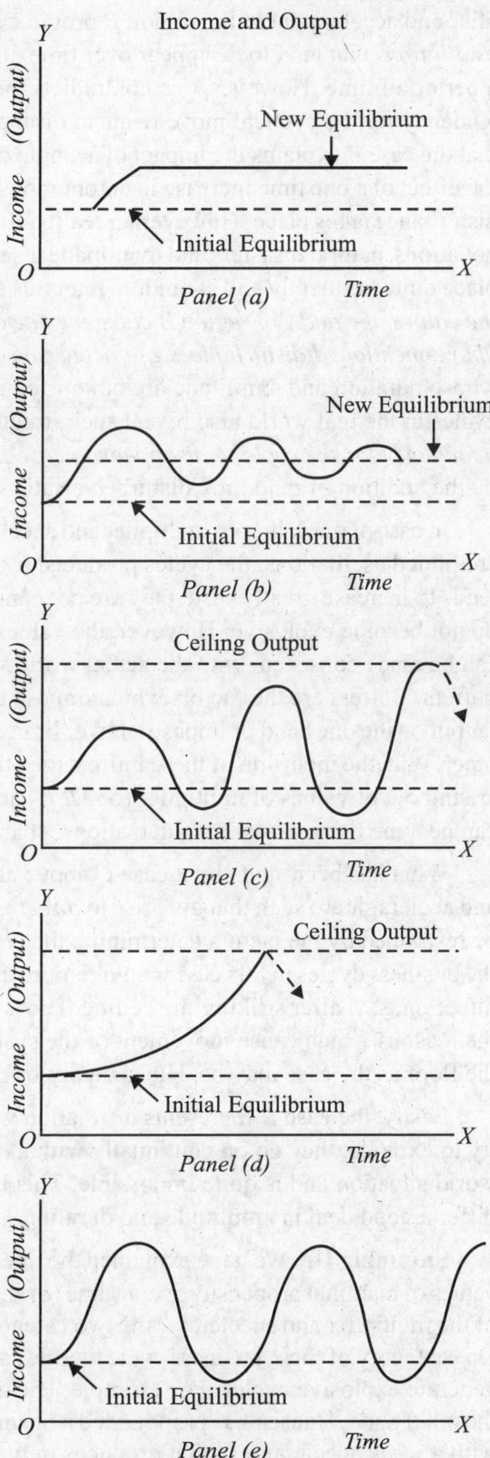

Fig. 27.6. *Interaction between Multiplier and Accelerator: Different Patterns of Income (output) movements for various values of c and v*

plier and accelerator) of the region *B* produce cyclical fluctuations which are of the type of *damped oscillations* that tend to disappear over time, that is, the amplitude of the cycles shrinks to zero over a period of time. However, this contradicts the historical experience which reveals that there is no tendency for the cyclical movements to disappear or die out over time. However, it is worth noting that the case *B* explains the impact of a single disturbance on income and employment. For example, the effect of a one time increase in autonomous investment goes on diminishing over time if no other disturbance takes place. However, in reality, further disturbances such as technological advances, innovations, natural disasters and man-made disasters such as security scam in India in 1991-92 do take place quite frequently and at random intervals and in a way they provide shocks to the system.[9] Thus, *the values of c and v of region B can generate cyclical fluctuations over time without dying out if the above-mentioned disturbances are occurring frequently at random.* This results in business cycles whose duration and amplitude are quite irregular and not uniform. As a matter of fact, the business cycles in the real world also reveal such irregular pattern. To sum up, "*what otherwise shows up as a tendency for the cycle to disappear in case B may be converted into unending sequence* of cycles by the addition of randomly disturbed erratic shock system."[10]

In case of the values of multiplier and accelerator falling within the region *C*, though they generate continued oscillations, the cycles produced by them tend to become 'explosive' (*i.e.* their amplitude tends to increase greatly). But they are not consistent with the real world situation where oscillations do not become explosive. However, the values of multiplier and accelerator falling within region *C* can be made consistent with the actual world situation by incorporating in the analysis the so called buffers. Buffers are the factors which impose upper limit or ceiling on the expansion of income and output on the one hand or impose a lower limit or floor on the contraction of output and income on the other. With the inclusion of these buffers the otherwise explosive upward and downward fluctuations arising out of values of multiplier (or *MPC*) and accelerator (or capital-output ratio) of the region *C* can become limited cyclical fluctuations, characteristic of the real world situation.

What has been said about case *C* above also applies to region *D* where the values of multiplier and acclerator are such that give rise to *directly explosive upward or downward movement* which can be restrained by the factors determining the ceiling and floor. However, the adequate explanation of the business cycles in this case would require the reasons why the system starts moving in the reverse direction, say, after striking the ceiling. Hicks in his famous theory of the business cycles provides the reasons which cause movement of the system in the reverse direction after it hits the ceiling or the floor as the case may be. Hicks theory of business cycles will be explained below at length.

Lastly, the case *E* represents a situation where the business cycles neither try to disappear, nor try to explode, they go on continually with a *constant amplitude*. This however contradicts the real world situation and is quite impossible. This is because in the real world situation, business cycles differ a good deal in amplitude and duration.

Summing Up. We have explained the interaction of multiplier and accelerator in case of various values of marginal propensity to consume (*c*) and capital-output ratio (*v*). On the basis of the interaction of the multiplier and accelerator the two categories of business cycle theories have been put forward. One category of these business cycle theories assumes the values of multiplier and accelerator which generate explosive cycles. For example, Hicks' theory of business cycles falls in this category. On the other hand, Hansen has propounded a business cycle theory based on the interaction of multiplier with a *weak* accelerator which produces only damped oscillations. Further, as indicated above, the

9. The concept of erratic shocks has been put forward by a Nobel Prize winner in economics, R. Frisch, "Propagation Problems and Impulse Problems in Dynamic Economics" in *Economic Essays in Honour of Gustav Cassel*, Allen and Unwin, 1993, pp. 171-205.
10. Shapiro Edward, *Macroeconomic Analysis*, Fourth Edition, 1978, p. 378.

interaction theories have been modified either by incorporating in the analysis erratic shocks or random disturbances or by including so called buffers which check the upward movement of income and output by imposing ceiling of expansion and checking a downward movement by imposing a floor on the contraction of output. One of the famous theories of business cycles based on the interaction of multiplier and accelerator which also incorporate buffers in his analysis of fluctuations is that put forward by the noted English economist J.R. Hicks. We discuss below his theory of business cycles in detail.

Table 27.1. A Multiplier-acceleration Interaction Model

Period	*Autonomous Investment*	*Induced Consumption (c = 2/3)*	*Induced Investment (v = 2)*	*Change in Income from the base period t*
1	2	3	4	5
	₹ in Crores	₹ in Crores	₹ in Crores	₹ in Crores
t	0	0	0	0
t + 1	10	0	0	10
t + 2	10	6.7	13.4	30.1
t + 3	10	20.0	26.6	56.6
t + 4	10	37.8	35.6	83.4
t + 5	10	55.6	35.6	101.2
t + 6	10	67.6	23.8	101.2
t + 7	10	67.6	0.2	67.8
t + 8	10	51.8	–10.0	51.8
t + 9	10	34.6	–10.0	33.8
t + 10	10	23.0	–10.0	23.0
t + 11	10	15.4	–10.0	15.4
t + 12	10	10.2	–10.0	10.2
t + 13	10	6.8	– 6.8	10.0
t + 14	10	6.6	+ 0.2	16.8

A Numerical Example of the Interaction of the Multiplier and Accelerator

How the interaction between the multiplier and accelerator gives rise to the cyclical movements in economic activity (as measured by income or output) will become clear from Table 27.1. In formulating this table we have assumed that marginal propensity to consume (c) being equal to 213 or 0.66 and capital-output ratio (v) or accelerator being equal to 2. Further, one period time lag has been assumed which implies that an increase in income in a period induces the increase in consumption in the next period. It is assumed that initially in period $t + 1$, autonomous investment is of ₹ 10 crores. In period $t + 3$, with autonomous investment being maintained constant at ₹ 10 crores, the deviation of total income in the period $t + 3$ as compared to the base period will be equal to 10 + 20 + 26.6 = ₹ 56.6 crores. Similarly, the changes in induced consumption and induced investment and hence in income brought about by the initial increase in autonomous investment of ₹ 10 crores which is maintained throughout, can be found out. It will be seen from column 5 of Table 27.1 that there are large fluctuations in income. Under the influence of the interaction between the multiplier and accelerator, the income increases up to the period $t + 6$. In other words, period up to $t + 6$ represents the expansion phase or upswing of the business cycle. Therefore, the period $t + 6$ is the upper turning point of the business cycle beyond which the contraction phase or downswing of the business cycle begins. It will be further observed that beyond the period $t + 13$, income again starts rising, that is, *recovery from the depression begins*. Thus, $t + 13$ represents the lower turning point of the business cycle. In this way we see that the interaction between the multiplier and accelerator can give rise to the cyclical movements of the economic activity and its various phases. It is worth mentioning that we have taken particular values of marginal propensity to consume (which determine the size of the multiplier) and capital-output ratio (which determines the size of the accelerator). The other values of multiplier and accelerator that have been explained above would give rise to the different patterns of fluctuations.

HICKS' THEORY OF BUSINESS CYCLES

Hicks[11] put forward a complete theory of business cycles based on the interaction between the multiplier and accelerator by choosing certain values of marginal propensity to consume (c) and capital-output ratio (v) which he thinks are representative of the real world situation. According to Hicks, the values of marginal propensity to consume and capital-output ratio fall in either region C or D of Fig. 27.5. As seen above, in case values of these parameters lie in the region C, they produce cyclical movements (*i.e.*, oscillations) whose amplitude increases over time and if they fall in region D they produce an *explosive upward movement* of income or output without oscillations. To explain business cycles of the real world which do not tend to explode, Hicks has incorporated in his analysis the role of buffers. On the one hand, *he introduces output ceiling* when all the given resources are fully employed and prevent income and output to go beyond it, and, on the other hand, he visualises *a floor or the lower* limit below which income and output cannot go because some autonomous investment is always taking place.

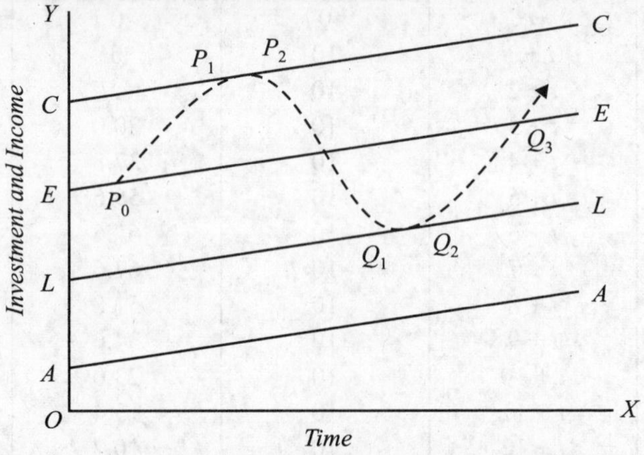

Fig. 27.7. *Hicks' Model of Business Cycles*

Another important features of Hicks' theory is that business cycles in the economy occur in the background of economic growth (*i.e.*, the rising trend of real income of output over time). In other words, cyclical fluctuations in real output of goods and services take place *above and below this rising line of trend* or growth of income and output. Thus in his theory he explains business cycles along with an equilibrium rate of growth. In Hicks' theory of long-run equilibrium growth that is determined by rate of increase of autonomous investment over time and, therefore, long-run equilibrium growth of income is determined by the autonomous investment and the magnitudes of multiplier and accelerator. Hicks assumes that autonomous investment, depending as it is on technological progress, innovations and population growth, grows at a constant rate. With further assumptions of stable multiplier and accelerator, equilibrium income will grow at the same rate as autonomous investment. It follows therefore that the failure of actual output to increase along the equilibrium growth path, sometimes to move above it and sometimes to move below it, determines the business cycles.

Hicks' theory of business cycles has been explained with the help of Fig. 27.7. In this figure, AA line represents autonomous investment. Autonomous investment is that investment which is not induced by changes in income and is made by entrepreneur as a result of technological progress or innovations or population growth. Hicks assumes that autonomous investment grows annually at a constant rate given by the slope of the line AA. Given the marginal propensity to consume, the simple multiplier is determined. Then the magnitude of multiplier and autonomous investment together

11. Hicks, J.R., *A Contribution to the Theory of Trade Cycles*, Oxford University Press, p. 50.

determine the equilibrium path of income shown by the line *LL*. Hicks calls this the floor line as this sets the lower limits below which income (output) cannot fall because of a given rate of growth of autonomous investment and the given size of the multiplier. But induced investment has not yet been taken into account. If national income grows from one year to the next, as it would move along the line *LL*, there is some amount of induced investment via accelerator. The line *EE* shows the equilibrium growth path of national income determined by autonomous investment and the combined effect of the multiplier and accelerator. *FF* is the full employment ceiling. It is a line that shows the maximum national output at any period of time when all the available resources of the economy are fully employed. Given the constant growth of autonomous investment, the magnitude of multiplier and the induced investment determined by the accelerator, the economy will be moving along the *equilibrium growth path line EE*. Thus starting from point *E*, the economy will be in equilibrium moving along the path *EE* determined by the combined effect of multiplier and accelerator and the growing level of the autonomous investment.

Suppose when the economy reaches point P_0 along the path *EE*, there is an *external shock*—say an outburst of investment due to certain innovation or jump in governmental investment. When the economy experiences such an outburst of autonomous investment it pushes the economy above the equilibrium growth path *EE* after point P_0. The rise in autonomous investment due to external shock *causes national income to increase at a greater rate than that shown* by the slope of *EE*. This greater increase in national income will cause further increase in induced investment through acceleration effect. This increase in induced investment causes national income to increase by a magnified amount through multiplier. So under the combined effect of multiplier and accelerator, national income or output will rapidly expand along the path from P_0 to P_1. Movement from P_0 to P_1 represents the upswing or expansion phase of the business cycle. But this expansion must stop at P_1 because this is the full employment output ceiling. The limited human and material resources of the economy do not permit a greater expansion of national income than shown by the ceiling line *CC*. Therefore, when point P_1 is reached the rapid growth of national income must come to an end. Prof. Hicks assumes that the full employment ceiling grows at the same rate as autonomous investment. Therefore, *CC* slopes gently unlike the very steep slope of the line from P_0 to P_1. When point P_1 is reached the economy must grow at the same rate as the usual growth in autonomous investment.

For a short time the economy may crawl along the full employment ceiling *CC*. But because national income has ceased to increase at the rapid rate, the induced investment via accelerator falls off to the level consistent with the modest rate of growth determined by the constant rate of growth of autonomous investment. But the economy cannot crawl along its full employment ceiling for a long time. The sharp decline in growth of income and consumption when the economy strikes the ceiling causes a sharp decline in induced investment. Thus with the sharp decline in induced investment when national income and hence consumption ceases to increase rapidly, the contraction in the level of the income and business actually must begin. Once the downswing starts, the accelerator works in the reverse direction. That is, since the change in income is now negative the inducement to invest must begin to decrease. Thus there is slackening off at point P_2 and national income starts moving toward equilibrium growth path *EE*. This movement from P_2 downward therefore represents the downswing or contraction phase of the business cycle. In this downswing investment falls off rapidly and therefore multiplier works in the reverse direction. The fall in national income and output resulting from the sharp fall in induced investment will not stop on touching the level *EE* but will go further down. The economy must consequently move all the way down from point P_2 to point Q_1. But at point Q_1 the floor has been reached.

Whereas the upswing was limited by the output ceiling set by the full employment of available resources, in the downswing the national income cannot fall below the level of output represented by the floor. This is because the floor level is determined by simple multiplier and autonomous investment growing at constant rate, while during the downswing after a point accelerator ceases to operate. It may be noted that during downswing the limit to negative investment (disinvestment) and therefore the limit to the contraction of output is set by the depreciation of capital stock. There is no way for the businessmen to make disinvestment at a desired rate higher than the depreciation. When during downswing such conditions arise, accelerator becomes inoperative. After hitting the floor the economy may for some time crawl along the floor through the path Q_1 to Q_2. In doing so, there is some growth in the level of national income. This rate of growth as before induces investment and both the multiplier and accelerator come into operation and the economy will move towards Q_3 and the full employment ceiling *CC*. This is how the upswing of cyclical movement again starts.

Critical Appraisal

But Hicks' theory of trade cycles is not without critics. A major weakness of Hicks' theory, according to Kaldor, is that it is based on the principle of acceleration in its rigid form.[12] If the rigid form of acceleration principle is not valid, then the interaction of the multiplier and accelerator which is the crucial concept of the Hicksian theory of trade cycles is not valid. Thus Duesenberry writes, "the basic concept of multiplier-accelerator interaction is important one but we cannot really accept to explain observed cycles by a mechanical application of that concept."[13] and, according to him, Hicks in his business cycle theory actually tries to do so.

It may be noted that Kaldor puts forward a theory of business cycles which does not make use of the rigid or strict form of the acceleration principle. In his trade cycle theory Kaldor provides for investment being *directly related* to the level of income and *inversely related* to the stock of capital. Thus Kaldor's approach which is also supported by Goodwin abandons the rigid and inflexible relation of investment to changes in income (output) as implied by the rigid acceleration principle [*i.e.*, $I_t = I_a + v(Y_{t-1} - Y_{t-2})$] and instead has used the following form of the investment function

$$I_t = I_a + gY_{t-1} - jK_t$$

where I_t stands for investment in period *t*, I_a for autonomous investment, Y_{t-1} for income in the previous period, K_t for the stock of capital, and *g* and *j* are constants.

A look at the above investment function used by Kaldor will reveal that investment is directly related to the income and inversely related to the stock of capital. Thus in Kaldor-Goodwin investment function, the increase in income, the capital stock remaining constant, will cause an increase in investment which will enlarge the stock of capital. On the other hand, according to this new investment function, if capital stock increases, output or income remaining constant, investment will fall due to its being negatively related to capital stock. Thus, Kaldor-Goodwin approach to investment while gives up the rigid acceleration principle but still retains the basic idea of investment related to income because in this approach investment will cause the capital stock to expand towards the stock of capital as desired for the production of output of the preceding year.

However, despite the shortcomings of Hicks' theory of business cycles, this is a valuable contribution to the theory of business cycles. Even its critics such as Kaldor though indicating some of its weaknesses acknowledge its merit. Thus Kaldor writes that "Hicks' theory of trade cycles provides us many brilliant and original pieces of analysis". Duesenberry considers it as an "ingenious piece of work".

---

12. See Kaldor N., "Hicks on the Trade Cycles" in *Economic Journal*, Dec. 1951, pp. 833-478.
13. Duesenberry, J.S. "Hicks on the Trend and Cycles" in *Quarterly Journal of Economics*, August 1950, pp. 464-76.

QUESTIONS FOR REVIEW

1. Explain briefly characteristics of a trade cycle. [*D.U., B.Com. (Hons), 1994, 96*]
2. Explain briefly the different phases of business cycle.
3. Explain Hawtrey's monetary theory of business cycle. Show that his model of business cycles in self-generating.
4. What is meant by under-consumption? What factors, according to Karl Marx, are responsible for it? How does under-consumption cause trade cycles in the economy?
5. Critically examine Hayek's over-investment theory of business cycles. Compare it with Wicksell's over-investment theory of business cycle.
6. Explain Keynes's contribution to business cycle theory. Is his theory a sufficient explanation of business cycles?
7. What is accelerator? What role does it play in explaining business cycles in the economy? [*D.U., B.A. (Honours), 1988*]
8. Explain how interaction of multiplier and accelerator explains cyclical fluctuations in the economy.
9. Critically examine Hicks' model of trade cycles. What determines floor and ceiling in his model?

CHAPTER 27A

FRIEDMAN'S MONETARIST AND NEW CLASSICAL (RATIONAL EXPECTATIONS) THEORIES OF BUSINESS CYCLES

We have explained in the previous chapters several theories of business cycles including Keynes's theory, Samuelson's theory of interaction between accelerator and multiplier, Hicks's theory, Kaldor's theory and Goodwins theory. The later theories are in the tradition of original Keynesian economics and make some improvement and modification in Keynes's explanation of trade cycles. However, in the post Keynesian era, Friedman led a revolution in macroeconomics and propounded a new theory called "*Monetarist Theory of Business Cycles*" and from the beginning of nineteen eighties Robert Lucas, a nobel laureate in economics, brought about another revolution in macroeconomics which dominated in the eighties and nineties. Lucas propounded a new theory of business cycles called '*New Classical Theory*' or '*Rational Expectations Theory*' of business cycles. To counter these some Keynesian economists put forward a new theory based on the original Keynesian economics but made some improvements and modifications in it and their theory is therefore called '*New Keynesian Theory of Business Cycles*'. In the present chapter we discuss these new theories of business cycles.

FRIEDMAN'S MONETARIST THEORY OF BUSINESS CYCLES

A different explanation of occurrence of business cycles has been propounded by Friedman and Schwartz [1] of Chicago University. They argue that *instability in growth of money supply is the source of most cyclical fluctuations in economic activity*. Therefore, their theory is called *monetarist theory of business cycles*. Friedman and his followers consider the free market economy as being inherently stable. According to them, it is the exogenous money shocks (*i.e.,* money supply changes) that affect aggregate demand which in turn causes cyclical changes in output and employment in the economy.

Friedman and Schwartz start from the historical data for the United States which show a high correlation between cyclical movements in economic activity and cyclical changes in money stock. They admit that *cyclical correlation does not prove direction of causality,* namely, changes in exogenous money supply leading to the cyclical movements in economic activity. Quoting the historical evidence from the US economy Friedman and Schwartz conclude that there is a causal linkage running from changes in money supply to the changes in levels of economic activity rather than other way around. Thus "Friedman and Schwartz adduce evidence from particular historical periods which suggests, first, that changes in activity have always been accompanied by changes in money stock while there have not been (major) disturbances in the money stock which have not been accompanied by changes in activity and, secondly, changes in money can be attributed to specific historical events rather than systematically to changes in activity"[2]

This raises two questions first, how is Friedman's monetarist theory opposed to the conventional cyclical theories, especially, the Keynesian one. Second, what is the *transmission mechanism* that links changes in money stock with occurrence of fluctuations in economic activity. Friedman, as opposed to Keynes, lays stress on the *exogenous changes* in *money stock* that causes fluctuations

1. M. Friedman and Anna Schwartz, Money and Business Cycles, *Review of Economics and Statistics* (Supplement), 1963.
2. M.T.C. Surrey, *Macroeconomic Themes,* Edited Readings in Macroeconomics, Oxford University Press, 1976, p. 255.

in aggregate demand and hence in economic activity while Keynes emphasized *autonomous changes in investment* that bring about changes in aggregate demand and hence in economic activity. Besides, Friedman and other monetarists put forward the *concept of money multiplier* which relates the statistical relationship between changes in money stock with changes in real income (*i.e.*, $\frac{\Delta Y}{\Delta M}$ is money multiplier where ΔY stands for change in national income and ΔM stands for change in money stock in a given period) as against *investment multiplier* of Keynes, which states the relationship between change in autonomous investment (ΔI) and resulting change in income (ΔY), that is, $\frac{\Delta Y}{\Delta I}$ measures the size of multiplier. According to Friedman and other monetarists, money multiplier shows greater stability than the Keynesian investment multiplier which varies depending on the various leakages such as the extent of imports and the degree of taxation of income. They assert that apparently greater stability of money multiplier shows *prima facie* evidence in favour of the *monetarists* explanation of cyclical fluctuations"[3]

Finally there is issue of *transmission mechanism* through which changes in money supply affect the level of national output and employment. Friedman explains it through *portfolio adjustment* consequent to changes in money stock where portfolios are considered to be comprising a *spectrum of assets ranging from money (liquid funds) through financial assets (bonds and equities to physical assets, durable producer and consumer goods*. When there is increase in money stock by the Central Bank of the country, individuals and firms would have temporarily more money balances (*i.e.*, cash or liquid funds) in their portfolio of assets than desired by them, they readjust their portfolio by spending some of the extra money balances on consumer goods and services; some part of extra money balances goes to purchasing new bonds and equity shares, some part to durable consumer goods such as houses. The new expenditure on bonds and shares raise their prices and lead to the fall in the rate of interest. The fall in interest and increase in wealth of the people will induce more investment and consumption demand. More investment will cause rise in demand for capital goods. Thus increase in money stock causes increase in aggregate demand for goods and services either directly through portfolio adjustment or indirectly through increase in wealth and fall in rate of interest caused in this adjustment process. The transmission process is quite complex as it involves not only readjustment of portfolio of the individuals but also of firms, banks and other financial institutions.

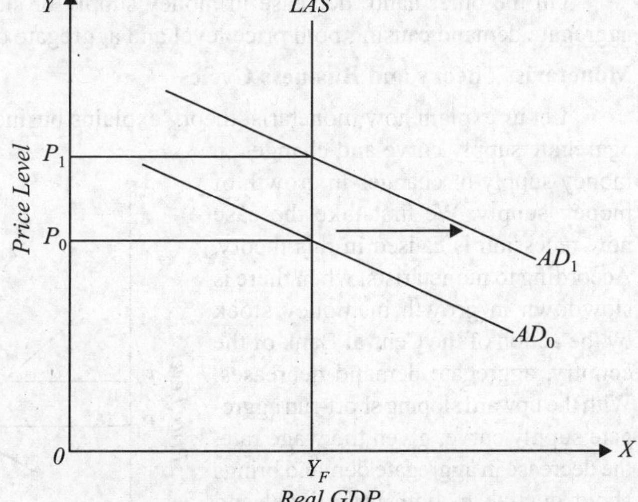

Fig. 27A.1. *Monetarist Theory: Changes in money supply do not affect real GDP in the long run; they affect only the price level.*

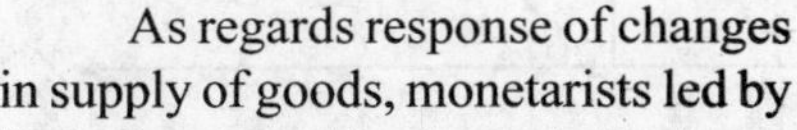

As regards response of changes in supply of goods, monetarists led by Friedman distinguish between short run and long run. The long-run aggregate supply of goods, according to them, is determined by the real factors such as the

3. M.J.C. Surrey, *op. cit.* p. 254.

availability of labour force, capital stock, infrastructure facilities energy resources and state of technology and is a vertical straight line at the level of potential GDP as shown by *LAS* as a vertical straight line in Fig. 27A.1. It is important to note that in the equation of exchange, $M = kPY$ generally used by the monetarists to explain the changes in price level, they assume that real national income (*i.e.,* Y) remains constant at potential GDP or full-employment level. Besides, they assume that k which is a proportion of income level held in the form of money to remain constant (note that k is inverse of income velocity of money, *i.e.,*) $k = \frac{1}{V}$. With k or V remaining stable (as is assumed by Friedman and other monetarists) and the real national income remaining constant at the potential *GDP* level (or full-employment level of output), increase in money supply brings about rise in price level only, real national income remaining constant.

Thus by making some assumptions about the behaviour of certain variables, namely, k and Y they turn equation of exchange into the theory of economic behaviour. Consider Fig. 27A.1 where long-run aggregate supply curve (*LAS*) is a vertical straight line at potential *GDP* level Y_F. The increase in money supply that causes aggregate demand curve to shift from AD_0 to AD_1 brings about rise in price level from P_0 to P_1, level of *GDP* remaining fixed at Y_F. But the monetarists explain business cycles on the one hand by the changes in money supply and, on the other hand, by the short-run supply curve which is assumed to be sloping upward. This upward-sloping short-run supply curve implies that both price level and aggregate output (real national income) change as aggregate demand curve shifts upward due to the increase in money supply. The short-run aggregate supply curve (*SAS*) slopes upward as money wages (W) are temporarily sticky in the short run and with the rise in price level (P), *real wage rate* $\left(\frac{W}{P}\right)$ *falls*. At lower real wage rate more labour is employed which produces more output or real income.

On the other hand, decrease in money supply or slowdown in its growth causes decline in aggregate demand causing both price level and aggregate output to fall.

Monetarist Theory and Business Cycles

Let us explain how monetarist theory explains business cycles with upward sloping short-run aggregate supply curve and changes in money supply or changes in growth of money supply. We first take the case how recession is caused in this theory. According to monetarists, when there is slowdown in growth in money stock by the action of the Central Bank of the country, aggregate demand decreases. With the upward sloping short-run aggregate supply curve, given the wage rate, the decrease in aggregate demand brings about decline in both price level and national output and employment causing unemployment in the economy. That is, the economy experiences recession. This is shown in Fig. 27A.2 where to begin with AD_0 is the aggregate demand curve which cuts both the vertical long run aggregate supply curve *LAS* and the upward-sloping

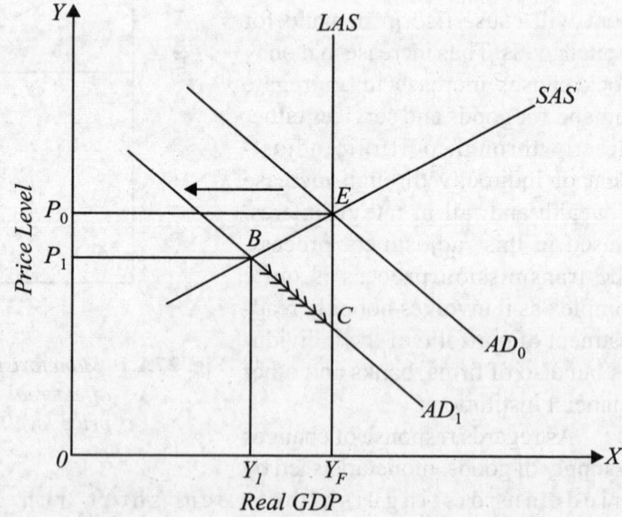

Fig. 27A.2. *Explaining Recession: Monetarist Theory.*

short-run aggregate supply curve *SAS* at point *E*. At point *E* the system is in long-run equilibrium. Now if there is a slowdown in growth of money supply causing a leftward shift in aggregate demand curve from AD_0 to AD_1. As a result, the economy moves to the new equilibrium point *B* at which aggregate demand AD_1 cuts the short-run aggregate supply curve *SAS*. It will be seen from Fig. 27A.2 that at point *B* aggregate output is smaller than the potential GDP level, unemployment in the economy will emerge. However, as mentioned above, monetarists believe that *wage rate is only temporarily sticky*. When aggregate demand decreases due to slowdown in growth of money supply and causes increase in unemployment, money wage rate will *eventually* begin to fall. As shown in Fig. 27A.2, with the fall in money wage rate short-run aggregate supply curve (*SAS*) shifts downward resulting in fall in price level. Short-run aggregate supply curve (*SAS*) goes on shifting downward until the equilibrium is reached at point *C* at the level of potential GDP level Y_F where full employment prevails. Thus, according to monetarist theory, through adjustment in money wage rate and price, the economy again reaches full-employment equilibrium and unemployment is eliminated. In this way monetarists explain how with the fall in growth of money supply, the economy goes into recession and then through adjustment in wage rate and price level new full-employment equilibrium is automatically achieved at *lower* wage rate and price level.

Explaining Expansion

Expansion phase of business cycle is explained by monetarists through a positive monetary shock, that is, *exogenous increase in money supply or higher rate of growth in money supply*. This is illustrated in Fig. 27A.3 where initially the full-employment equilibrium is at point *E* where aggregate demand curve AD_0 intersects *LAS* and SAS_0 at point *E*. Now, a speed up in growth in money stock causes an upward shift in aggregate demand curve to AD_2 resulting in increase in aggregate output (GDP) along the short-run aggregate supply curve SAS_0 which is drawn with a given wage rate.

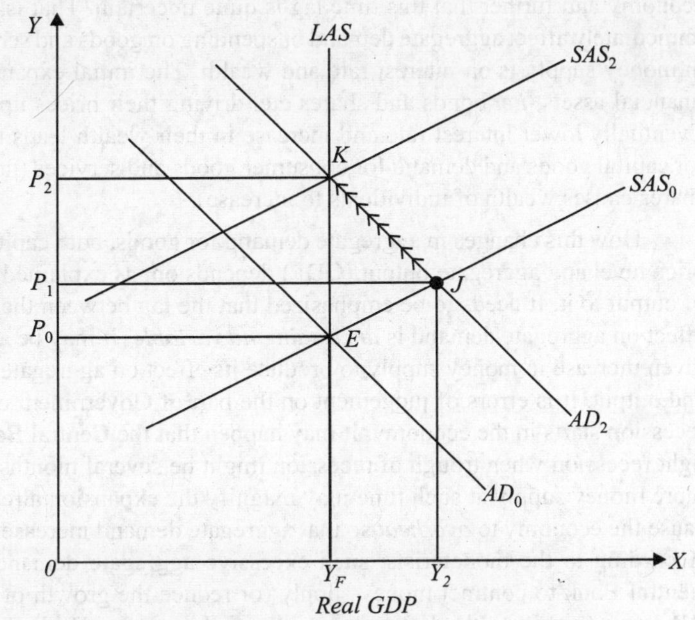

Fig. 27A.3. *Monetarist Explanation of Expansion*

It will be seen from Fig. 27A.3 that with the sticky wages, the higher aggregate demand curve AD_2 cuts SAS_0 curve at point *J* at which the new equilibrium is established at GDP equal to Y_2 which is greater than potential GDP equal to Y_F. Equilibrium at a level higher than potential GDP implies that *unemployment has fallen below the natural level of unemployment*. (It should be recalled that in full-employment situation at Y_F, the frictional and structural unemployments exist which together are called natural level of unemployment). However, the equilibrium at point *J* is only temporary. Eventually, due to the emergence of shortage of labour at GDP level Y_2 wage rate will start rising and cause shift in *SAS* to the left until it reaches SAS_2 which intersects AD_2 curve at point *K* where new full-employment equilibrium is reached at price level P_2 and the potential GDP level Y_F. Thus, in the long run, according to monetarists the economy automatically returns to the full-employment equilibrium (*i.e.*, potential GDP level) without any intervention by Government and Central Bank of the country. Prof. Parkin of the University of Ontario, US rightly says, "The monetarist business cycle is like a rocking horse. *It needs an outside force to get it going*, but once going it rocks back and forth (but just once). It does not matter in which direction the force initially hits. If it is a money growth slowdown, the economy cycles with a recession followed by expansion. If it is money growth speed up, the economy cycles with an expansion followed by recession."[4] It follows from above that monetarists believe that *changes in money supply* has a destabilising influence on the economy. They think that money shocks to the economy are the prime movers of cyclical fluctuations around the full-employment level of output in the economy.

Importance of Lags

If the monetarist theory would have stopped at explaining that changes in money supply are the major source of cyclical fluctuations in the economy, it would not have been much impressive as many Keynesian also believe that changes in money supply are important source of macroeconomic instability. There must be more in monetarist theory as it emerged as an important alternative to Keynesian explanation of cyclical fluctuations in the economy. However, *monetarists view of cyclical fluctuations has been founded as much on emperical evidence as on theoretical reasoning in terms of money shocks to the economy causing cyclical fluctuations*. Friedman and his followers quoted a lot of evidence of historical episodes in favour of their viewpoint but also showed that there is a time lag between changes in money supply and their actual effect on aggregate demand and the real economy and further that this time lag is quite uncertain. That is, changes in money supply do not immediately affect aggregate demand or spending on goods and services. The initial effect of changes in money supply is on interest rate and wealth. The initial expansion in money supply is spent on financial assets, *i.e.* bonds and shares etc. driving their prices up and thereby lowers interest rate. Eventually lower interest rate and increase in their wealth leads to increase in investment demand for capital goods and demand for consumer goods and services (note that rise in prices of bonds and shares causes wealth of individuals to increase).

How this changes in aggregate demand for goods, both capital and consumer goods, affect the price level and aggregate output (GDP) depends on, as explained above, on the response of supply of output to it. It needs to be emphasized that the lag between the increase in money supply and its effect on aggregate demand is *uncertain and variable*. It may be a few months, a year or more for a given increase in money supply to produce its effect on aggregate demand and hence on price level and output. It is errors of judgement on the part of Government or monetary authority about when recession starts in the economy. It may happen that the Central Bank injects more money supply to fight recession when trough of recession might be several months in the past. If so, the injection of more money supply at such time may magnify the expansion already in progress and possibly may cause the economy to *overheat* so that aggregate demand increases beyond the potential GDP level. According to the monetarists, such excessive aggregate demand pressure generally prompts the Central Bank to contract money supply (or reduce the growth of money supply) which will cause *AD* curve to shift to the left and ease demand pressure. However, according to them, the Central Bank may decide to contract the money supply when aggregate demand is already slowing down on its own. If such is the case, the contraction of money supply at this time could push the economy into recession. Thus, according to monetarists, intervention by the Central Bank through changes in money supply only aggravate the natural tendency of cyclical fluctuations to be of small amplitude. These small natural fluctuations, according to them, are caused by imperfect information, weather shocks and changes in international factors.

Critical Evaluation

Most economists, even Keynesians, believe that large money supply changes have a destabilising effect on the economy. But the monetarists view that money supply changes *alone* are responsible for

4. Michael Parkin, *Economics*, 4th Edition, Addison Wesley, 1998, p. 748.

cyclical fluctuations does not appear to be correct. As Keynes asserted even without any significant money supply changes, changes in investment demand due to changes in marginal efficiency of capital (*i.e.,* expected return on capital) are important factors that determine changes in aggregate demand and cause cyclical fluctuations in economic activity. Besides, it has been observed that when to pull the economy out of recession, the Central Bank takes steps to increase the money supply it may have no effect on aggregate spending due to lack of investors' confidence and bleak prospects for profit making. For example, in 2007-08 when worst ever recession took place since the Great Depression of 1930s, the Federal Reserve had pumped into the economy a large amount of money, reduced its lending rate to almost zero, but it did not lead to greater investment and consumption spending. In fact, those who got the extra money decided to hold on to it rather than spending it due to uncertain future economic environment. Because of the limitation of monetary expansion to revive the American economy, along with the expansionary monetary policy the adoption of fiscal stimulus policy measures helped the recovery of the US economy.

Further, monetarists think that economy is inherently stable and left to itself it would *automatically* correct itself through adjustment in wages and prices to restore equilibrium at full-employment (*i.e.* potential GDP) level. Therefore, they are of the view that Central Bank or Government of the Country should pursue non-interventionist policy. However, the experience of past (most recently recession of 2007-09) has abundantly showed that the automatic correction has failed to occur. Therefore, an important lesson from the past experience is that the roles of Government and Central Banks to fight recession and curb inflation are of paramount importance.

LUCAS' NEW CLASSICAL (RATIONAL EXPECTATIONS) THEORY OF BUSINESS CYCLES

Introduction

Rational expectations theory, also known as *New Classical Theory* was put forward by Nobel Laureate Robert E. Lucas of the University of Chicago. From the early eighties to 1997 Lucas New Classical Theory dominated macroeconomics. Lucas has been said to bring about a revolution in macroeconomics. He has plenty of differences with Keynes and leaned towards monetarist theory. However, he made an improvement over monetarist explanation of business cycles by introducing rational expectations in his analysis.

The cornerstone of Lucas new classical theory is the concept of rational expectations. *By rational expectations Lucas means that people use all available relevant information to make economic forecasts about price level.* This information includes not only explicit changes in money supply, Government's fiscal policy, international developments (which determine exports and prices of fuel, raw materials, and other commodities) but also economic theory about how the economy works.

According to rational expectations theory money wages are determined by rational expectations of the price level. Robert Lucas is of the view that **it is only *unanticipated* changes in aggregate demand that are the cause of cyclical fluctuations in the economy**. A greater than anticipated increase in aggregate demand causes expansion in the level of output and employment and less than anticipated increase in aggregate demand brings about recession and therefore decline in output and employment. Any factor that affects aggregate demand, for example, *larger than expected changes* in money stock, Government's fiscal deficit or changes in taxes or interest rate and unanticipated changes in international developments (that affect exports, prices of fuel and other commodities). According to Lucas, if changes in aggregate demand are anticipated, then money wages and prices would adjust so that equilibrium remains undisturbed. Thus, in his view, on the basis of all available information people estimate the future increase in money supply in forming their expectations and, if wages and prices are flexible, they are set on the basis of these expectations. Therefore, anticipated increase in aggregate demand based on these expectations in money supply changes would have no effect on the level of output and employment. This is because wages and prices would be raised in anticipation of increase in money supply and the short-run aggregate supply curve would shift to the left by the same amount as the rightward shift in aggregate demand. This is illustrated in Fig. 27A.4 where to begin with aggregate demand curve AD_0 intersects the long-run aggregate supply curve *LAS* and short-run aggregate supply curve *SAS* at point *E* and determine equilibrium price level P_0 and potential GDP level Y_F. Now, if on the basis of certain anticipated increase in money supply, aggregate demand curve shifts upward to AD_1, the anticipated price will be P_1. Now, the wages will be fixed immediately at a higher level in accordance with the new expected price level P_1. Now the wage rate will be immediately fixed at the higher level, *SAS* curve would also shift upward to SAS_1, by the same extent as the increase in aggregate demand curve to AD_1. With this, as will be seen from Fig. 27A.4, price level and wage rate have risen, the aggregate output remains at the potential GDP level. Therefore, according to Lucas, New Classical Theory based on rational expectations concept, *only unanticipated change in money supply* would affect output and employment as in the absence of adjustment in wage rate, in the short run, the response to the increase in aggregate demand the economy will move along the given short-run aggregate supply curve SAS_0. What is true of unanticipated changes in money supply and its impact on aggregate demand and output also applies to the effect of unanticipated changes in other factors.

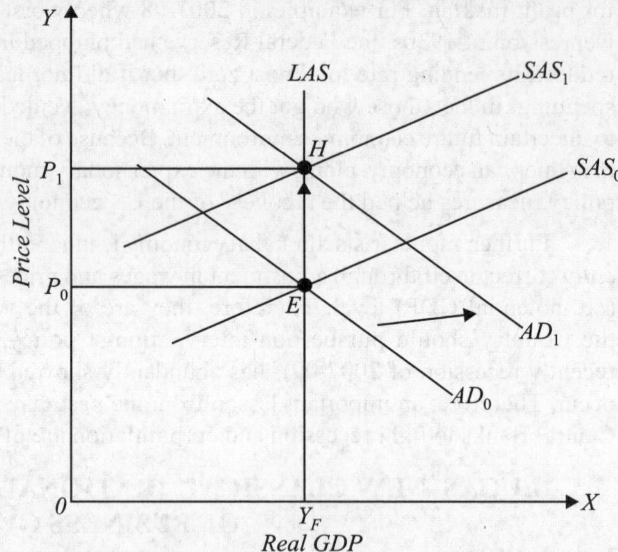

Fig. 27A.4 *Rational Expectation Theory: The Impact of Anticipated Increase in Aggregate Demand*

In what follows we explain how new classical theory explains cyclical fluctuations in economic activity (*i.e.,* levels of output and employment).

Rational Expectations: Explaining Recession

We first explain how new classical theory based on rational expectations explains the emergence of recession in the economy. Consider Fig. 27A.5 where *EAD* is expected aggregate demand curve which intersects the long-run aggregate supply curve *LAS* and short-run aggregate supply curve *SAS* and equilibrium is at potential *GDP* level Y_F with price level equal to P_0. Suppose there is unanticipated decrease in the aggregate demand due to unexpected decrease in money supply growth by the Central Bank of a country or due to unexpected imposition of a higher tax or unanticipated decline in demand for country's exports. Since this decline in aggregate demand is unanticipated, wage rate will not rise in the short run. As a result of unanticipated downward shift in

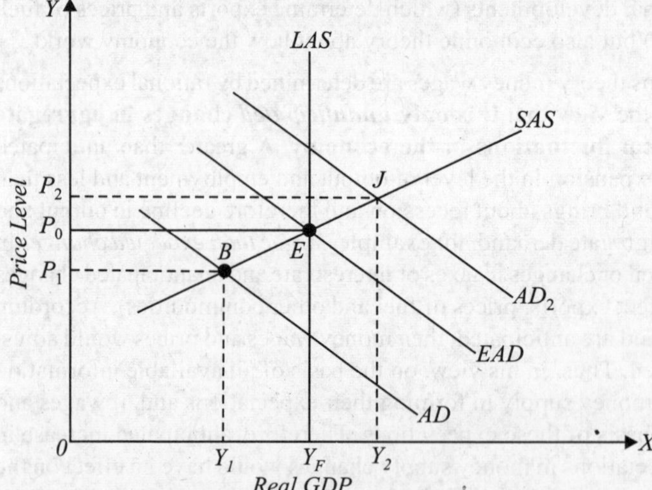

Fig. 27A.5. *Rational Expectations Theory of Business Cycles*

aggregate demand curved to AD_1, the economy will move along the given short-run aggregate supply curve SAS_0. With this the price level falls to P_1 and aggregate output (GDP) decreases to Y_1 causing unemployment in the economy. This indicates the situation of recession in the economy. Now, if the decline in aggregate demand is anticipated, price level will be expected to fall and therefore firms and workers will immediately agree to lower money wage rate. By doing so they will prevent the rise in real wage rate and therefore avoid the increase in unemployment. Thus it is only unanticipated decrease in aggregate demand that causes fall in price level and decline in real GDP below the full-employment level causing unemployment to rise. This recession persists until aggregate demand increases to the anticipated level *EAD*.

Rational Expectations: Explaining Expansion

Now, we proceed to explain the opposite case of expansion in economic activity with Lucas' rational expectations approach. As in case of recession, according to rational expectations theory, expansion in economic activity will occur when there is *unanticipated increase in aggregate demand.* Such an increase in aggregate demand may occur due to larger than expected increase in money supply or due to unexpected increase in exports or lowering of taxes causing increase in disposable income. This is also shown in Fig. 27A.5. *EAD* is the expected demand curve. Suppose there is unanticipated increase in aggregate demand to AD_2 due any of the factors mentioned above. Since increase in aggregate demand is unanticipated money wages will not rise, the economy will move along the given short-run aggregate supply curve *SAS*. The new aggregate demand curve AD_2 intersects short-run aggregate supply curve *SAS* at point *J* resulting in rise in price level to P_2 and real GDP level to Y_2. Since GDP has risen more than potential GDP level Y_F, unemployment will fall below the natural level of unemployment. This points to the emergence of expansion in economic activity. The new equilibrium will remain at point *J* until aggregate demand decreases to the level of expected aggregate demand curve *EAD*.

It follows from above that unanticipated fluctuations in aggregate demand such as changes from AD_1 to AD_2 around *EAD* cause changes in price level and GDP around the potential GDP level Y_F at which unemployment is at natural level (that is, at which full-employment prevails).

Critical Evaluation

Lucas' rational expectations theory has come under attack by New Keynesian economists assume that as soon as anticipated aggregate demand increases money wage rate will quickly rise. The new Keynesians point out that money wage rate does not rise quickly because labour-employer are locked in long-term contracts about money wage rate. It is only when new employer-labour contracts are renegotiated after the expiry of old ones that money wage rates can be raised. Thus, *according to new Keynesians, short-run aggregate supply curve SAS shifts only after sometime. In the meantime even anticipated increase in aggregate demand will lead to the rise in both the price level and GDP in the short run, SAS remaining unchanged.* On the contrary, according to supporters of new classical theory based on rational expectations, the employer-labour contracts are renegotiated immediately when conditions change. Therefore, they think employer-labour contracts do not impose any impediments to money wage rate flexibility. However, in our, view this can happen only if both parties agree to recognise the changed conditions.

Further, according to rational expectations theory, *anticipated policy change* brings about change in price level only with no change in real GDP and level of employment. This is because, as explained above, when policy changes are anticipated wage rate changes quickly causing *SAS* curve to shift immediately offsetting the effect of policy change on real GDP and employment. As seen above, this is not correct as there is always some time lag before change in money wage rate takes place.

Like monetarist theory, new classical theory also explain cyclical fluctuation on the basis of exogenous (*i.e.* outside) forces such as changes in money supply, fiscal (*e.g.* taxation), international; developments (e.g. changes in export demand for goods and services of a country) which affect aggregate demand. *In this theory there is no any endogenous mechanism to generate cyclical movements in economic activity.*

CHAPTER 28
ECONOMIC STABILISATION: FISCAL POLICY

Macroeconomic Policy and Stabilisation

The economy does not always work smoothly. There often occur fluctuations in the level of economic activity. At times the economy finds itself in the grip of recession when levels of national income, output and employment are far below their full potential levels. During recession, there is a lot of idle or unutilised productive capacity, that is, available machines and factories are not working to their full capacity. As a result, unemployment of labour increases along with the existence of excess capital stock. On the other hand, at times the economy is 'overheated' which means inflation (*i.e.,* rising prices) occurs in the economy. Thus, in a free market economy there is a lot of economic instability. The classical economists believed that an automatic mechanism works to restore stability, in the economy; recession would cure itself and, inflation will be automatically controlled. However, the empirical evidence during the 1930s when severe depression took place in the Western capitalist economies and also the evidence of post-World War II period amply shows that no such automatic mechanism works to bring about stability in the economy. That is why Keynes argued for intervention by the Government to cure depression and inflation by adopting appropriate tools of macroeconomic policy. *The two important tools of macroeconomic policy are fiscal policy and monetary policy*. According to Keynes, monetary po1icy was ineffective to lift the economy out of depression. He emphasized the role of fiscal policy as an effective tool for stabilising the economy. However, in view of the modern economists both fiscal and monetary policies play a useful role in stabilising the economy. In the present chapter we shall dwell upon the role of fiscal policy for stabilising the economy and will discuss the role of monetary policy in bringing about stabilisation in the next chapter.

Goals of Macroeconomic Policy

Stabilising the economy at a higher level of employment and national output is not the only goal of macroeconomic policy. Ensuring price stability is its another goal. Both inflation (that is, rising prices) and deflation (that is, falling prices) have bad economic consequences. It is therefore desirable to achieve price stability. Similarly, every nation wants to raise the level of living of its people which can be attained through bringing about economic growth which in turn depends on raising the rates of saving and investment and accumulating capital. Macroeconomic policies can play a useful role in raising the rate of saving and investment and therefore ensure rapid economic growth. Thus, three important goals or objectives of macroeconomic policy (both fiscal and monetary) are as follows:

1. Economic stability at a high level of output and employment.
2. Price stability.
3. Economic growth.

In the present chapter we shall confine ourselves to the discussion of the role of fiscal policy in achieving economic stability at full employment level and in controlling inflation and deflation and thus attaining price stability. The role of monetary policy for encouraging economic growth will be discussed at length in a later chapter.

DISCRETIONARY FISCAL POLICY FOR STABILISATION

Fiscal policy is an important instrument to stabilise the economy, that is, to overcome recession and control inflation in the economy. Fiscal policy is of two kinds: *Discretionary fiscal policy* and *Non-discretionaryl fiscal policy of automatic stabilisers*. By discretionary policy we mean deliberate change in the Government expenditure and taxes to influence the level of national output and prices. Fiscal policy generally aims at managing aggregate demand for goods and services. On the other hand, non-discretionary fiscal policy of automatic stabilisers is a built-in tax or expenditure mechanism that automatically increases aggregate demand when recession occurs and reduces aggregate demand when there is inflation in the economy without any special deliberate actions on the part of the Government. In this section we shall confine ourselves to the discussion of discretionary fiscal policy.

At the time of recession the Government increases its expenditure or cuts down taxes or adopts a combination of both. On the other hand, to control inflation the Government cuts down its expenditure or raises taxes. In other words, *to cure recession expansionary fiscal policy and to control inflation contractionary fiscal policy is adopted*. It is worth mentioning that fiscal policy aims at changing aggregate demand by suitable changes in Government spending and taxes. Thus, *fiscal policy is mainly a policy of demand management*. It should be further noted that when the Government adopts expansionary fiscal policy to cure recession, it raises its expenditure without raising taxes or cuts down taxes without changing expenditure or increases expenditure and cuts down taxes as well. With the adoption of any of these types of expansionary fiscal policy Government's budget will have a deficit. Thus expansionary fiscal policy to cure recession and unemployment is a deficit budget policy. If, on the other hand, to control inflation, Government reduces its expenditure or increases taxes or adopts a combination of the two, it will be planning for a budget surplus. *Thus policy of budget surplus, or at least reducing budget deficit is adopted to remedy inflation*. In what follows we will discuss fiscal policy first to cure recession and then to control inflation.

Fiscal Policy to Cure Recession

As we know, the recession in an economy occurs when aggregate demand decreases due to a fall in private investment. Private investment may fall when businessmen become highly pessimistic about making profits in future, resulting in decline in marginal efficiency of investment. As a result of fall in private investment expenditure, aggregate demand curve shifts down creating a *deflationary or recessionary gap*. It is the task of fiscal policy to close this gap by increasing Government expenditure, or reducing taxes. Thus there are two fiscal methods to get the economy out of recession:

(*a*) Increase in Government expenditure

(*b*) Reduction of taxes.

We discuss below both these methods.

(*a*) **Increase in Government Expenditure to Cure Recession.** For a discretionary fiscal policy to cure depression, the increase in Government expenditure is an important tool. Government may increase expenditure by starting public works, such as building roads, dams, ports, telecommunication links, irrigation works, electrification of new areas etc. For undertaking all these public works, Government buys various types of goods and materials and employs workers. The effect of this increase in expenditure is both direct and indirect. The direct effect is the increase in incomes of those who sell materials and supply labour for these projects. The output of these public works also goes up together with the increase in incomes. Not only that, Keynes showed that increase in Government expenditure also has an *indirect effect* in the form of the working of a multiplier. Those who get more incomes spend them further on consumer goods depending on their marginal propensity to consume. As during the period of recession there exists *excess capacity* in the consumer goods industries, the increase in demand for them brings about expansion in their output which further generates employment and incomes for the unemployed workers and so the new incomes are spent and re-spent further and the process of multiplier goes on working till it exhausts itself.

How large should be the increase in expenditure so that equilibrium is established at full employment or potential level of output. This depends on the magnitude of GNP gap caused by deflationary gap on the one hand and the size of multiplier on the other. It may be recalled that the size of the multiplier depends on the marginal propensity to consume. The impact of increase in Government expenditure in a recessionary condition is illustrated in Fig. 28.1. Suppose to begin with the economy is operating at full-employment or potential level of output Y_F with aggregate demand curve $C + I_2 + G_2$ intersecting 45° line at point E_2. Now, due to some adverse happening (say due to the crash in the stock market), investor's expectations of making profits from investment projects become dim causing a decline in investment. With the decline in investment, say equal to E_2B, aggregate demand curve will shift down to the new position $C + I_1 + G_1$ which will bring the economy to the new equilibrium position at point E_1 and thereby determine Y_1 level of output or income. The fall in output will create involuntary unemployment of labour and also excess capacity (*i.e.* idle capital stock) will come to exist in the economy. Thus emergence of deflationary gap equal to E_2B and the reverse working of the multiplier has brought about conditions of recession in the economy. It will be observed from Fig. 28.1 that, to overcome recession if the Government increases its expenditure by E_1H_1 the aggregate demand curve will shift upward to original position $C + I_2 + G_2$ and as a result the equilibrium level of income will increase to the full-employment or potential level of output Y_F and in this way the economy would be lifted out of depression. Note that the increase (ΔY) in national income or output by $Y_1 Y_F$ is not only equal to the increase in Government expenditure by ΔG or E_1H but a multiple of it depending on the marginal propensity to consume. Thus, increase in national income is equal to $\Delta G \times 1 \dfrac{1}{1 - MPC}$, where $\dfrac{1}{1 - MPC}$ is the value of multiplier.

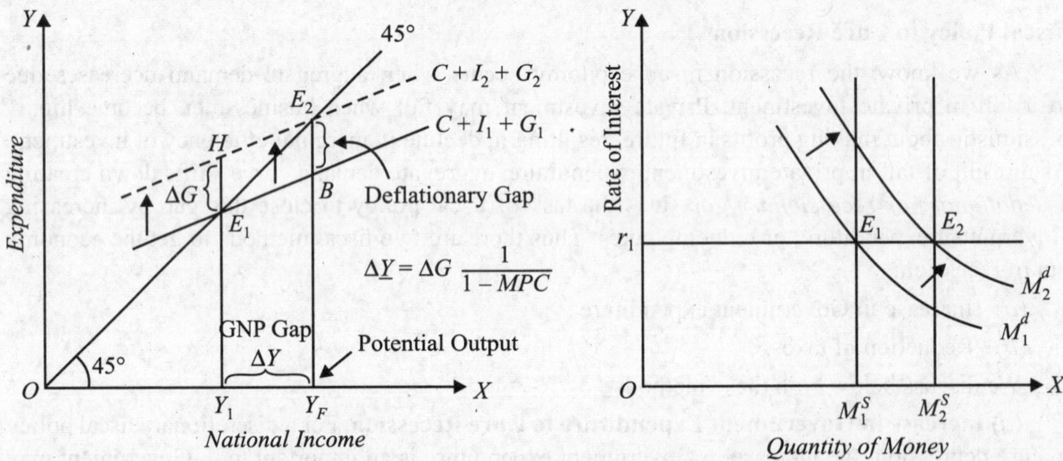

Fig. 28.1. *Increase in Goverment Expenditure to Cure Recession*

Fig. 28.2. *Expansion in Money Supply to Keep the Rate of Interest Unchanged*

It may also be further noted that increase in Government expenditure without raising taxes (and therefore the policy of deficit budgeting) will fully succeed in curing recession if rate of interest remains unchanged. With the increase in Government expenditure and resultant increase in output and employment demand for money for transactions purposes is likely to increase as is shown in Fig. 28.2 where demand for money curve shifts to right from $M_1{}^d$ to $M_2{}^d$ as a result of increase in transactions demand for money. Money supply remaining constant, with increase in demand for money rate of interest is likely to rise which will adversely affect the private investment. The decline in private investment will tend to offset the expansionary effect of rise in Government expenditure. Therefore, if fiscal policy of increase in Government expenditure (or of deficit budgeting) is to succeed in overcoming recession, the Central Bank of the country should also pursue expansionary monetary policy and take steps to increase the money supply so that increase in Government expenditure does not lead to the rise in rate of interest. It will be noticed from Fig. 28.2 that if money supply is increased from M_1^S to M_2^S, the rate of interest does not rise despite the increase in demand for money. With rate of interest remaining unchanged, private investment will not be adversely affected and increase in Government expenditure will have full effect on raising national income and employment.

Financing Increase in Government Expenditures and Budget Deficit

An important question is how to finance the increase in Government expenditure which is undertaken to cure recession. This increase in Government expenditure must not be financed by raising taxes because rise in taxes would reduce disposable incomes and consumers' demand for goods. As a matter of fact, rise in taxes would offset the expansionary effect of rise in Government spending. Therefore, proper discretionary fiscal policy at times of recession is to have the budget deficit if expansionary effect is to be realised.

Borrowing. A way to finance budget deficit is to borrow from the public by selling interest-bearing bonds to them. However, there is a problem in adopting borrowing as a method of financing budget deficit. When the Government borrows from the public in the money market, it will be competing with businessmen who also borrow for private investment. The Government borrowing will raise the demand for loanable funds which in a free market economy, if rate of interest is not administered by the Central Bank, will drive up the rate of interest. We know the rise in rate of interest will reduce or crowd out some private investment expenditure and interest-sensitive consumer spending for durable goods.

Creation of New Money. The more effective way of financing budget deficit is the creation of new money. By creating new money to finance the deficit, the crowding out of private investment can be avoided and full expansionary effect of rise in Government expenditure can be realised. Thus, creation of new money for financing budget deficit or what is called *monetisation of budget deficit* has a greater expansionary effect than that of borrowing by the Government.

(*b*) **Reduction in Taxes to Overcome Recession.** Alternative fiscal policy measure to overcome recession and to achieve expansion in output and employment is reduction of taxes. The reduction in taxes increases the disposable income of the society and causes the increase in consumption spending by the people. If tax reduction of ₹ 200 crores is made by the Finance Minister, it will lead to ₹ 150 crores in consumption, assuming marginal propensity to consume is 0.75 or 3/4. Thus reduction in taxes will cause an upward shift in the consumption function. If along with the reduction in taxes, the Government expenditure is kept unchanged, aggregate demand curve $C+I+G$ will shift upward due to rise in consumption function curve. This will have an expansionary effect and the economy will be lifted out of recession, and national income and employment will increase and as a result unemployment will be reduced. Note that reduction in taxes, with Government expenditure remaining constant, will also result in budget deficit which will have to be financed either by borrowing or creation of new money.

It is worth noting that *reduction in taxes has only an indirect effect on expansion and output* through causing a rise in consumption function. But, like the increase in government expenditure, the increase in consumption achieved through reduction in taxes will have a multiplier effect on increasing income, output and employment. As we have shown in a previous chapter the value of tax multiplier, as it is called, is given by

$$\Delta T \times \frac{MPC}{1-MPC} \quad \text{or} \quad \Delta C \times \frac{MPC}{1-MPC}$$

The effect of reduction in taxes in curing recession and in causing expansion in income and output can be graphically shown by a figure such as Fig. 28.1. In case of reduction in taxes, instead of increase in Government expenditure *G*, it is increase in consumption *C* which will cause upward shift in the aggregate demand curve ($C + I + G$) and will result in, through the working of multiplier, a higher level of equilibrium national income.

There are some instances in the history of the capitalist world, especially U.S.A when taxes were reduced to stimulate the economy. In 1964, President Kennedy reduced personal and business taxes by about \$12 billion to give a boost to the American economy when there was high unemployment and lower capacity utilisation in the American economy. This tax cut was quite successful in reducing unemployment substantially and expanding national income through full utilisation of excess capacity. Again, over the period 1981-84, President Reagan made a very large tax reduction to get out of recession and to achieve expansion in national income to reduce unemployment. There is some debate whether President Reagan's tax cut alone had positive effect on national income as some economists attribute the recovery in that period to the monetary expansion that took place. However, tax reduction by President Reagan did play a significant role in bringing about the recovery.

Fiscal Policy Options: Increase in Government Expenditure or Reduction in Taxes

Is it better to use Government expenditure or changes in taxes to stabilise the economy at full employment and potential level of output? The answer depends to a great extent upon one's view regarding the role of public sector. Those who think that public sector should play a significant role in the economy to meet various failures of a free market system will recommend the increase in Government expenditure during recession on public works to achieve expansion in output and employment. On the other hand, those economists who think that public sector is inefficient and involves waste of scarce resources would advocate for reduction in taxes to stimulate the economy.

The choice between tax reduction and increase in Government expenditure depends on the basis of another factor, namely, the magnitude of the effect of expenditure multiplier and tax multiplier. The value of tax multiplier is less than the Government expenditure multiplier. Ignoring the signs of the multipliers, it should be noted that whereas expenditure multiplier is equal to $\frac{1}{1-MPC}$, the tax multiplier equals $\frac{MPC}{1-MPC}$ or $MPC \times \frac{1}{1-MPC}$ which is less than $\frac{1}{1-MPC}$. Suppose marginal propensity to consume is 0.75 or 3/4 so that value of expenditure multiplier is 4. Increase in Government expenditure by ₹ 100 crores will raise national output by ₹ 400 crores. On the other hand, reduction in taxes by ₹ 100 crores will increase income and output by $100 \times \frac{MPC}{1-MPC} = 100 \times \frac{3/4}{1-3/4} =$ ₹ 300 crores.. Thus, the effect of reduction in taxes by an equal amount as the increase in Government expenditure has a smaller impact on national income than that of increase in Government expenditure. This difference in the effects of the two methods ot expanding output has implications for the size of the Government deficit. If we want to achieve expansion in income by the same amount, we need to cut taxes by more than we would need to increase Government expenditure because the size of the tax multiplier is less than that of expenditure multiplier. In other words, in case we adopt the policy of tax reduction, to achieve expansion by a given amount the budget deficit planned will to have to be much greater.

However, the size of expenditure multiplier relative to the size of the tax multiplier is not the sole deciding factor for the choice of a policy option. For example, reduction in taxes are greatly welcomed by the people as it directly increases their disposable incomes. Further, it is individual or households who themselves decide how to spend their extra disposable income made possible by a tax cut, while in case of increase in expenditure the Government decides how to spend it.

Fiscal Policy to Control Inflation

When due to large increases in consumption demand by the households or investment expenditure by the entrepreneurs, or bigger budget deficit caused by too large an increase in Government expenditure aggregate demand increases beyond what the economy can potentially produce by fully employing its given resources, it gives rise to the situation of *excess demand* which results in inflationary pressures in the economy. This inflationary situation can also arise if too large an increase in money supply in the economy occurs. In these circumstances *inflationary gap* occurs which tends to bring about rise in prices. If successful steps to check the emergence of excess demand or close the inflationary gap are not taken, the economy will experience a period of inflation or rising prices. For the last some decades, problem of demand pull inflation has been faced by both the developed and developing countries of the world[1]. An alternative way of looking at inflation is to view it from the angle of business cycles. After recovery from recession, when during upswing an economy finds itself in conditions of boom and becomes overheated prices start rising rapidly. Under such circumstances *anticyclical fiscal policy* calls for reduction in aggregate demand. Thus, *fiscal policy measures to control inflation are* (1) *reducing Government expenditure and* (2) *increasing taxes*. If in the beginning the *Government is having balanced budget*, then increasing taxes while keeping Government expenditure constant will yield budget surplus. The creation of budget surplus will cause downward shift in the aggregate demand curve and will therefore help in easing pressure on prices. If there is a balanced budget to begin with and the Government reduces its expenditure, say on defence, subsidies, transfer payments, while keeping taxes constant, this will also create budget surplus and result in removing excess demand in the economy.

It is important to mention that in the developing countries like India, the main factor responsible for inflationary pressures is *heavy budget deficit* of the Government for the last several years resulting in excess demand conditions. Rate of inflation can be reduced not necessarily by planning for budget surplus which is in fact impracticable but by *trying to take steps to reduce budget deficits*. It has been estimated that the aim should be to reduce fiscal deficit[2] to 3 per cent of GNP to achieve price stability in the Indian economy.

How the reduction in Government expenditure will help in checking inflation is shown in Fig. 28.3. It will be seen from this figure that an aggregate demand curve $C + I + G_1$ intersects 45° line at point E and determines equilibrium national income at full-employment level of income Y_F. However, if due to excessive Government expenditure and a large budget deficit the aggregate demand curve curve shifts upward to $C + I + G_2$, this will determine Y_2 level of income which is greater than full employment or potential output level Y_F. Since output cannot increase beyond Y_F, income will rise only in money terms through rise in prices, real income or output remaining unchanged. To put in other words, while the economy does not have labour, capital and other resources sufficient to produce Y_2 level of income or output, the households, businessmen and Government are demanding Y_2 level of output. This excess demand pushes up the price level so that level of only *nominal income* increases, real income or output remaining constant. It is thus clear that with the increase in aggregate demand beyond the full-employment level of output to $C + I + G_2$ causes excess demand equal to EA to emerge in the economy. It is this excess demand EA relative to full-employment output Y_F which causes the price level to rise and thus creates inflationary situation in the economy. This excess demand EA at full-employment level has therefore been called inflationary gap. *The task of fiscal policy is to close this inflationary gap by reducing Government expenditure or raising taxes.* With equilibrium at point H and nominal income equal to Y_2, if Government expenditure equal to

1. It is worth noting that in the mid-nineties in the developed countries inflation has become very moderate which is currently at the rate of about 4 per cent per annum.
2. There is some difference in the meaning of *budget deficit* and *fiscal deficit*. However, we shall not go into this issue and will be treating them to be of the same meaning.

HB (which is equal to inflationary gap AE) is reduced, aggregate demand curve will shift downward to $C + I + G_1$ which will restore the equilibrium at the full-employment level Y_F. The reduction in Government expenditure equal to HB through the operation of multiplier will result in a multiple decline in the level of national income or output. It will be seen from Fig. 28.3 that the decrease in Government expenditure by HB has led to a much bigger decline in output by Y_2Y_F. Ideally Government expenditure should cut down its expenditure on non-development or unproductive heads such as defence, unnecessary subsidies. It may however be noted that in India to control inflation the Government has been reducing capital expenditure which is mainly of development nature and has therefore been validly criticised

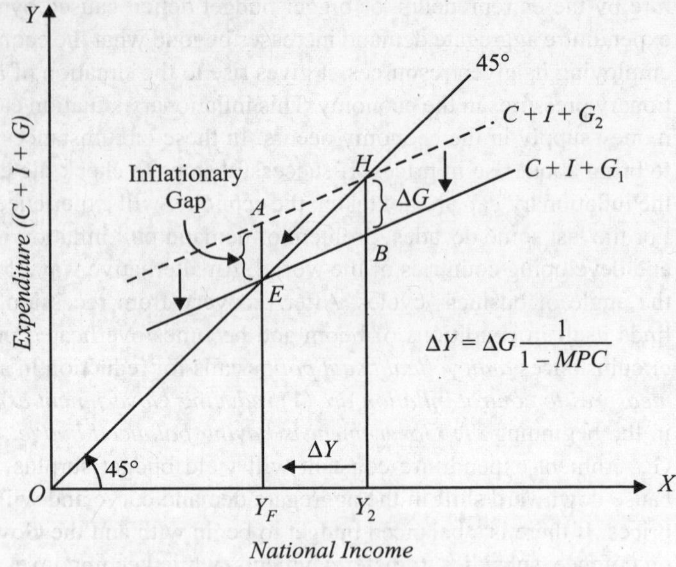

Fig. 28.3. *Reducing Expenditure to Check Inflation*

Raising Taxes to Control Inflation. *As an alternative to reduction in Government expenditure, the taxes can be increased to reduce aggregate demand.* For this purpose especially personal direct taxes such as income tax, wealth tax, corporate tax can be raised. The hike in taxes reduces the disposable incomes of the people and thereby force them to reduce their consumption demand. Note that in Fig. 28.3 as a result of hike in personal taxes it is the decrease in consumption-demand (C) component which will cause the aggregate demand curve $C + I + G_2$ to shift downward. Since, as shown above, the magnitude of tax multiplier is smaller than the expenditure multiplier, the tax revenue will be raised by a greater amount to achieve contraction in national income by Y_2Y_F.

Disposing of Budget Surplus. We have seen above to control demand-pull inflation, the Government either reduces its expenditure or raises taxes to lower aggregate demand for goods and services. Reduction in expenditure or hike in taxes results in decrease in budget deficits (if occurring before such steps) or in the emergence of budget surplus if the Government was having balanced budget prior to the adoption of anti-inflationary fiscal policy measures. Let us assume that anti-inflationary fiscal policy results in budget surplus. Anti-inflationary impact of budget surplus depends to a good extent on how the Government disposes of this budget surplus. There are two ways in which budget surplus can be disposed of : (1) reducing or retiring public debt and (2) impounding public debt.

We examine below the anti-inflationary effects of these two ways of disposing of the budget surplus

1. **Retiring Public Debt.** The budget surplus created by anti-inflationary policy can be used by the Government to pay back the outstanding debt. However, using budget surplus for retiring public debt will weaken its anti-inflationary effect. In paying off the debt held by the public, the Government will be returning the money to the public which it has collected through taxes. Further, this will also add to the money supply with the public. The general public will spend a part of the money so received which will raise consumption demand. Besides, retiring of public debt will result in the expansion of money supply in the

money market which will tend to lower the rate of interest. The lower rate of interest will stimulate consumption and investment demand while anti-inflationary policy requires that they should be reduced.

2. **Impounding of Public Debt.** To realise a large anti-inflationary effect of budget surplus it is desirable to impound the surplus fund. The impounding of surplus funds means that they should be kept idle. Thus by impounding the budget surplus, the Government shall be withdrawing some income or purchasing power from the income-expenditure stream and thus will not create any inflationary pressures to offset the deflationary impact of the budget surplus. To conclude, the impounding of budget surplus is a better method of disposing of budget surplus than of paying off public debt.

NON-DISCRETIONARY FISCAL POLICY: AUTOMATIC STABILISERS

There is an alternative to the use of discretionary fiscal policy which generally involves problems of lags in recognising the problem of recession or inflation and lag of taking appropriate action to tackle the problem. In this non-discretionary fiscal policy, *the tax structure and expenditure pattern are so designed that taxes and Government spending vary automatically in appropriate direction with the changes in national income.* That is, these taxes and expenditure pattern *without any special deliberate action* by the Government and Parliament automatically raise aggregate demand in times of recession and reduce aggregate demand in times of boom and inflation and thereby help in ensuring economic stability. These fiscal measures are therefore called *automatic stabilisers* or *built-in stabilisers.* Since these automatic stabilisers do not require any fresh deliberate policy action or legislation by the government, they represent non-discretionary fiscal policy. Built-instability of tax revenue and Government expenditure of transfer payments and subsidies is created because they vary with national income. These taxes and expenditure automatically bring about appropriate changes in aggregate demand and reduce the impact of recession and inflation that might occur in an economy at some time. This means that because of the existence of these automatic or built-in stabilisers recession and inflation will be shorter and less intense than otherwise be the case. Important automatic fiscal stabilisers are personal income taxes, corporate income taxes, transfer payments such as unemployment compensation, welfare benefits, corporate dividends.

We discuss below these taxes, revenue from which varies directly with the change in national income.

Personal Income Taxes. The tax rate structure is so designed that revenue from these taxes directly varies with income. Moreover, personal income taxes have progressive rates; the higher rates are charged from the upper income brackets. As a result, when national income increases during expansion and inflation, increasing percentage. of the people's income are paid to the Government. Thus, through causing a decline in their disposable income these taxes automatically reduce people's consumption and therefore aggregate demand. This decline in aggregate demand because of imposition of progressive personal income taxes tends to check inflation from becoming more severe. On the other hand, when national income declines at times of recession, the tax revenue declines as well which prevents aggregate demand from falling by the same proportion as the decline in income.

Corporate Income Taxes. Companies, or corporations as they are called now, also pay a percentage of their profits as tax to the Government. Like personal income taxes, corporate income tax rate is also generally higher at higher levels of corporate profits. As recession and inflation affect corporate taxes greatly, they have a powerful stabilising effect on aggregate demand; the revenue from them rises greatly during inflation and boom which tends to reduce aggregate demand, and revenue from them falls greatly during recession which tends to offset the decline in aggregate demand.

Transfer Payments: Unemployment Compensation and Welfare Benefits. When there is recession and as a result unemployment increases, the Government has to spend more on compensation for unemployment and other welfare programmes such as food stamps, rent-subsidies, subsidies to farmers. This hike in Government expenditure tends to make recession short-lived and less intense. On the other hand, when at times of boom and inflation national income increases and therefore unemployment falls, the Government curtails its programme of social benefits which result in lowering Government expenditure. The smaller spending by the Government helps to control inflation.

Corporate Dividend Policy. With economic fluctuations, corporate profits also rise and fall. However, corporations do not so quickly increase or reduce dividends in tune with fluctuations in profits and follow *a fairly stable dividend policy.* This permits the individuals to spend more during recession and spend less than would have the case if dividends were lowered' in time of recession and raised in conditions of boom and inflation. Thus, fairly stable dividends tend to cushion a recession and curb inflation by stabilising consumption expenditure.

Conclusion

It follows from above that automatic stabilisers reduce the intensity of business fluctuations, that is, both recession and inflation. However, the automatic or built-in stabilizers cannot alone correct the recession and inflation significantly. According to an estimate made for the U.S.A., automatic stabilizers have been able to reduce fluctuations in national income only by one-third. Therefore, the role of discretionary fiscal policy, namely, deliberate and explicit changes in tax rates and amount of Government expenditure are required to cure recession and curb inflation.

CROWDING OUT AND EFFECTIVENESS OF FISCAL POLICY

The critics of Keynesian theory have pointed out that expansionary effect of fiscal policy is not as much large as Keynesian economists suggest. In the Keynesian theory it is asserted that when Government increases its expenditure without raising taxes or when it reduces taxes without changing expenditure, it will have a large expansionary effect on national income. In other words, deficit budget would lead to the large increase in aggregate demand and thereby help to expand national output and income. However, it has been pointed that the above analysis of effect of expansionary fiscal policy of budget deficit ignores the effect of increase in Government expenditure or budget deficit on private investment. It has been argued that increase in Government expenditure or creation of budget deficit adversely affects private investment which offsets to a good extent the expansionary effect of budget deficit. This adverse effect comes about as increase in Government expenditure or reduction in taxes causes, rate of interest to go up. There are two ways in which rise in rate of interest is explained. First, within the framework of Keynesian theory increase in Government expenditure leads to the rise in national output which raises the transactions demand for money. Given the supply of money in the economy, the increase in transactions demand for money will cause the rate of interest to go up. Secondly, in order to finance its budget deficit the Government will borrow funds from the market. This will raise the demand for loanable funds which will bring about rise in the rate of interest.

Whatever the mechanism, the budget deficit or increase in Government expenditure to achieve expansion in national income and output will cause the rate of interest to go up. The rise in the rate of interest will discourage private investment. As we know from the theory of investment, at a higher rate of interest, private investment declines. Thus, *increase in Government expenditure or fiscal policy of budget deficit crowds out private investment.* This fall in private investment as a result of rise in rate of interest will offset or cancel out a part of the expansionary effect of increase in Government expenditure. The magnitude of this crowding-out effect depends on the elasticity of the investment demand. If investment demand is more elastic, the decrease in private investment

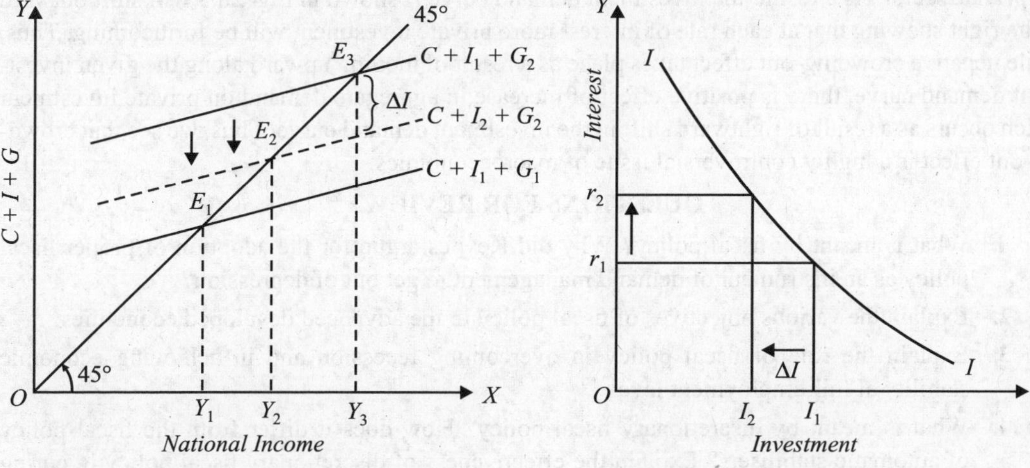

Fig. 28.4. *Effect of Crowding Out on National Output or Income*

Fig. 28.5. *Crowding Out of Private Investment*

consequent to the rise in rate of interest will be quite substantial and will greatly offset the expansionary effect of the increase in Government expenditure. On the contrary, if investment demand is relatively inelastic, the rise in rate of interest will lead to only a small decline in private investment and therefore crowding-out effect will be relatively small.

It, therefore, follows that the magnitude of crowding out weakens the effectiveness of fiscal policy. The crowding-out effect of expansionary fiscal policy and its effect on national output and employment is graphically shown in Fig. 28.4 and 28.5. To begin with the economy is in equilibrium at Y_1 level of income where aggregate demand curve $C + I_1 + G_1$ intersects the 45° line and determines Y_1 level of income. Let us assume that this is much below the potential or full-employment level of output. Suppose in order to raise the level of national income and output, the Government raises its expenditure from G_1 to G_2 so that the aggregate demand curve shifts upward to the new position $C + I_1 + G_2$ which intersects the 45° line at point E_3. With the increase in Government expenditure national income will rise by $\Delta G \times$ multiplier, that is, by $\Delta G \times \left(\frac{1}{1 - MPC}\right)$. In the absence of crowding-out effect, national income will rise to Y_3. This change in national income, ΔY or by $Y_1 Y_3$ is equal to the increase in Government expenditure (G) times the value of multiplier $\frac{1}{1 - MPC}$. However, as explained above, the increase in Government expenditure or the creation of budget deficit causes the rate of interest to rise, say from r_1 to r_2 (See Fig. 28.5). It will be seen from Fig. 28.5 that with the rise in interest from r_1 to r_2, private investment decreases from I_1 to I_2, Now, with the decline in private investment expenditure, aggregate demand curve in Fig. 28.4 shifts below to the new lower position $C + I_2 + G_2$ (dotted) and as a result new equilibrium is reached at Y_2 level of income. Thus, net result of increase in Government expenditure (DG) and crowding out of private investment equal to $I_1 I_2$ or ΔI is the expansion in national income equal to only $Y_1 Y_2$ which is relatively very small as compared to the rise in income by $Y_1 Y_3$ in the absence of crowding-out effect. Thus, the crowding-out effect has weakened the expansionary effect of fiscal policy.

However, in view of this author crowding-out effect has been blown out of all proportions by the critics of Keynes. As a matter of fact, when Government raises its expenditure, it leads to a large increase in aggregate demand for several goods and services through the working of multiplier. This rising aggregate demand improves investment climate by raising expectations of making profits by the private sector. As a result, the investment demand curve *II* shown in Fig. 28.5 will shift outward to the right showing that at each rate of interest more private investment will be forthcoming. Thus, while negative crowding-out effect takes place as a result of moving upward along the given investment demand curve, there is positive effect of increase in aggregate demand on private investment which occurs as a result of rightward shift in the investment demand curve. Thus, we see that crowding-out effect is a highly controversial issue of macroeconomics.

QUESTIONS FOR REVIEW

1. What is meant by fiscal policy? Why did Keynes argue for the adoption of proper fiscal policy as an instrument of demand management to get out of depression?
2. Explain the various objectives of fiscal policy in the advanced developed economies.
3. Explain the role of fiscal policy in overcoming recession and in achieving economic stability at full-employment level.
4. What is meant by discretionary fiscal policy? How does it differ from the fiscal policy of automatic stabilisers? Explain the effectiveness of discretionary fiscal policy in curing recession and attaining equilibrium at full employment level.
5. Expansionary fiscal policy and expansionary monetary policy are complementary to each other. Explain.
6. What is effective demand? How will it change if the Government covers its entire expenditure by (*a*) raising tax or (*b*) printing more currency?
7. Explain fiscal policy measures to control inflation. How far do you think they can be effective?
8. In an inflationary situation, the task of fiscal policy is to close the inflationary gap by reducing government expenditure or raising taxes. Discuss.
9. What is balanced budget multiplier? Explain that even balanced budget multiplier has an expansionary effect.
10. Derive mathematically to show that the balanced budget multiplier is equal to one.
11. What is meant by crowding-out effect? Show how the expansionary effect of fiscal policy is reduced by the crowding-out effect.
12. Increase in government expenditure, especially financed by borrowing from the open market, crowds out private investment. Discuss.
13. How does mechanism of crowding out of private investment work? Does it always occur?"
14. How is private investment affected if increase in Government spending is financed by sale of Government bonds? Do you think borrowing from the Central Bank to finance budget deficit is preferable if desired expansionary effect is to be realised?

CHAPTER 29
ECONOMIC STABILISATION: MONETARY POLICY

Introduction

Monetary policy is another important instrument with which objectives of macroeconomic policy can be achieved. It is worth noting that it is the Central Bank of a country which formulates and implements the monetary policy in a country. In some countries such as India the Central Bank (the Reserve Bank is the Central Bank of India) works on behalf of the Government and acts according to its directions and broad guidelines. However, in some countries such as the USA, the Central Bank (*i.e.*, Federal Reserve Bank System) enjoys an independent status and pursues its independent policy. Like the fiscal policy, the broad objectives of monetary policy are to establish equilibrium at full-employment level of output, to ensure price stability and to promote economic growth of the economy. Monetary policy is concerned with changing the supply of money stock and rate of interest for the purpose of stabilising the economy at full-employment or potential output level by influencing the level of aggregate demand. More specifically, at times of recession monetary policy involves the adoption of some monetary tools which tend to increase the money supply and lower interest rates so as to stimulate aggregate demand in the economy. On the other hand, at times of inflation, monetary policy seeks to contract the aggregate spending by tightening the money supply or raising the rate of interest.

It may however be noted that in a developing country such as India, in addition to achieving equilibrium at full-employment or potential output level, monetary policy has also to promote and encourage economic growth both in the industrial and agricultural sectors of the economy. Thus, in the context of developing countries the following three are the important goals or objectives of monetary policy:

(1) to ensure economic stability at full-employment or potential level of output;

(2) to achieve price stability by controlling inflation and deflation; and

(3) to promote and encourage economic growth in the economy.

In line with the above goals of monetary policy it has often been asserted by Governors of Reserve Bank of India that *growth with price stability is the goal of monetary policy of the Reserve Bank of India*. The role of monetary policy in achieving economic stability at a higher level of output and employment will be discussed below and its role in promoting economic growth in developing countries with special reference to India will be explained and critically examined at length in a later chapter.

Tools of Monetary Policy

There are four major tools or instruments of monetary policy which can be used to achieve economic and price stability by influencing aggregate demand or spending in the economy. They are:

1. Open market operations;
2. Changing the bank rate;
3. Changing the cash reserve ratio; and
4. Undertaking selective credit controls.

How these tools of monetary policy work to influence aggregate spending and economic activity has been explained in a previous chapter. We shall explain how these various tools can be used for formulating a proper monetary policy to influence levels of aggregate output, employment and prices in the economy. In times of recession or depression, *expansionary monetary policy* or what is also called *easy money policy* is adopted which raises aggregate demand and thus stimulates the economy. On the other hand, in times of inflation and excessive expansion, *contractionary monetary policy* or what is also called *tight money policy* is adopted to control inflation and achieve price stability through reducing aggregate demand in the economy. We discuss below both these policies.

EXPANSIONARY MONETARY POLICY TO CURE RECESSION OR DEPRESSION

When the economy is faced with recession or involuntary cyclical unemployment, which comes about due to fall in aggregate demand, the central bank intervenes to cure such a situation. Central Bank takes steps to expand the money supply in the economy and/or lower the rate of interest with a view to increase the aggregate demand which will help in stimulating the economy. The following three monetary policy measures are adopted as a part of an expansionary monetary policy to cure recession and to establish the equilibrium of national income at full-employment level of output.

1. The central bank undertakes open market operations and *buys securities in the open market.* Buying of securities by the central bank from the public, chiefly from commercial banks, will lead to the increase in reserves of the banks or amount of currency with the general public. With greater reserves, commercial banks can issue more credit to the investors and businessmen for undertaking more investment. More private investment will cause aggregate demand curve to shift upward. Thus buying of securities will have an expansionary effect.

2. The Central Bank *may lower the bank rate or what is also called discount rate*, which is the rate of interest charged by the central bank of a country on its loans to commercial banks. At a lower bank rate, the commercial banks will be induced to borrow more from the central bank and will be able to issue more credit at the lower rate of interest to businessmen and investors. This will not only make credit cheaper but also increase the availability of credit or money supply in the economy. The expansion in credit or money supply will increase the investment demand which will tend to raise aggregate output and income.

3. Thirdly, the *central bank may reduce the Cash Reserve Ratio (CRR) to be kept by the commercial banks*. In countries like India, this is a more effective and direct way of expanding credit and increasing money supply in the economy by the central bank. With lower reserve requirements, a large amount of funds is released for providing loans to businessmen and investors. As a result, credit expands and investment increases in the economy which has an expansionary effect on output and employment. In April 1996, when *Reserve Bank* lowered the CRR from 14 per cent to 13 per cent, it was estimated that this would release funds equal to ₹ 5,000 crores for the banks and thereby would significantly increase their lending capacity.

Similar to the Cash Reserve Ratio (CRR), in India there is another monetary instrument, namely *Statutory Liquidity Ratio* (SLR) used by the Reserve Bank to change the lending capacity and therefore credit availability in the economy. As explained in a previous chapter, according to Statutory Liquidity Ratio, in addition to the Cash Reserve Ratio (CRR) banks have to keep a certain minimum proportion of their deposits in the form of some specified liquid assets such as Government securities. To increase the lendable resources of the banks, Reserve Bank can lower this Statutory Liquidity Ratio (SLR). Thus, when Reserve Bank of India lowers statutory liquidity ratio (SLR), the credit availability for the private sector will increase.

It may be noted that the use of all the above tools of monetary policy leads to an increase in reserves or liquid resources with the banks. Such reserves are the basis on which banks expand their credit by lending, the increase in reserves raises the money supply in the economy, Thus, *appropriate monetary policy at times of recession or depression can increase the availability of credit and also lower the cost of credit.* This leads to more private investment spending which has an expansionary effect on the economy.

How Expansionary Monetary Policy Works: Keynesian Transmission Mechanism

Now, it is important to understand how expansionary monetary policy works to cause increase in output and employment and thus help the economy to recover from recession. In Keynes's theory, rate of interest is determined by the demand for and supply of money. According to Keynesian theory, expansion in money supply causes the rate of interest to fall . An this fall in the rate of interest will encourage businessmen to borrow more for investment spending. As is well known, *rate of interest is the opportunity cost of funds invested for purchasing capital goods.* As rate of interest falls, it becomes profitable to invest more in producing or buying capital goods. Thus, fall in the rate of interest rate raises the investment expenditure which is an important component of aggregate demand. The increase in aggregate demand causes expansion in aggregate output, national income and employment. How, according to Keynesian view, expansion in money supply can help to cure recession is illustrated in Fig. 29.1. In panel (*a*) of Fig. 29.1 it will be seen that when as a result of some measures taken by the central bank, the money supply increases from M_1 to M_2, the rate of interest falls from r_1 to r_2. It will be seen from panel (*b*) that with this fall in rate of interest, investment increases from I_1 to I_2. Now, in panel (*c*), it is shown how the increase in investment expenditure from I_1 to I_2 shifts the aggregate demand curve $(C+I_1+G)$ upward so that the new aggregate demand curve $C+I_2+G$ intersects the 40° line at point E_2 and thus establishes equilibrium at full-employment output level Y_F. It may be noted that with the given increase in investment how much aggregate output or national income will increase depends on the size of income multiplier which is determined by marginal propensity to consume. The greater the size of multiplier, the greater the impact of increment in investment on expansion of output and income.

From above, it is clear that monetary policy can play an important role in stimulating the economy and ensuring stability at full-employment level. But it is worth mentioning that there are several weak links in the full chain of increase in money supply achieving a significant expansion in economic activity. In fact, *Keynes himself was of the view that in times of depression, monetary policy will be ineffective in reviving the economy and therefore he laid stress on the adoption of fiscal policy to overcome depression.*

The first weak link in the above argument of expansionary monetary policy relates to the elasticity of money demand (*i.e.*, liquidity preference) curve M^d in panel (*a*) of Fig. 29.1. As shall be explained at length below, if demand for money curve M^d is nearly flat (*i.e.*, highly elastic), the increase in money supply by the central bank will not greatly affect the rate of interest and consequently further steps of significant expansion in investment and aggregate demand will not be realized. Besides, even if money demand curve is elastic and, therefore, expansion of money supply lowers the rate of interest significantly, the investment may not rise much. This is because if the investment demand curve is steep or inelastic, that is, investment is not sensitive to the changes in rate of interest, the fall in the rate of interest will fail to cause any significant increase in investment. As a result, aggregate demand curve will not change much and expansionary effect on output and employment will not be realised.

Further, the effect of increase in investment on output and employment depends on the size of multiplier. If there are several leakages in the multiplier process, even increase in investment may not bring about much change in output and employment. Thus, because of several weak links in the process or chain of expansion in money supply bringing about expansion, Keynes remarked that *there are many a slip between the cup and the lip*. Therefore, for all these reasons (especially because of the liquidity trap in the demand for money curve at lower rates of interest), Keynes was of the view that monetary policy is not an effective instrument in bringing about revival of the economy from the depressed state.

It may however be noted that modern Keynesians do not share the pessimistic view of the effectiveness of monetary policy. They think that liquidity preference curve is not flat and further that investment demand is fairly sensitive to the changes in the rates of interest. Therefore,

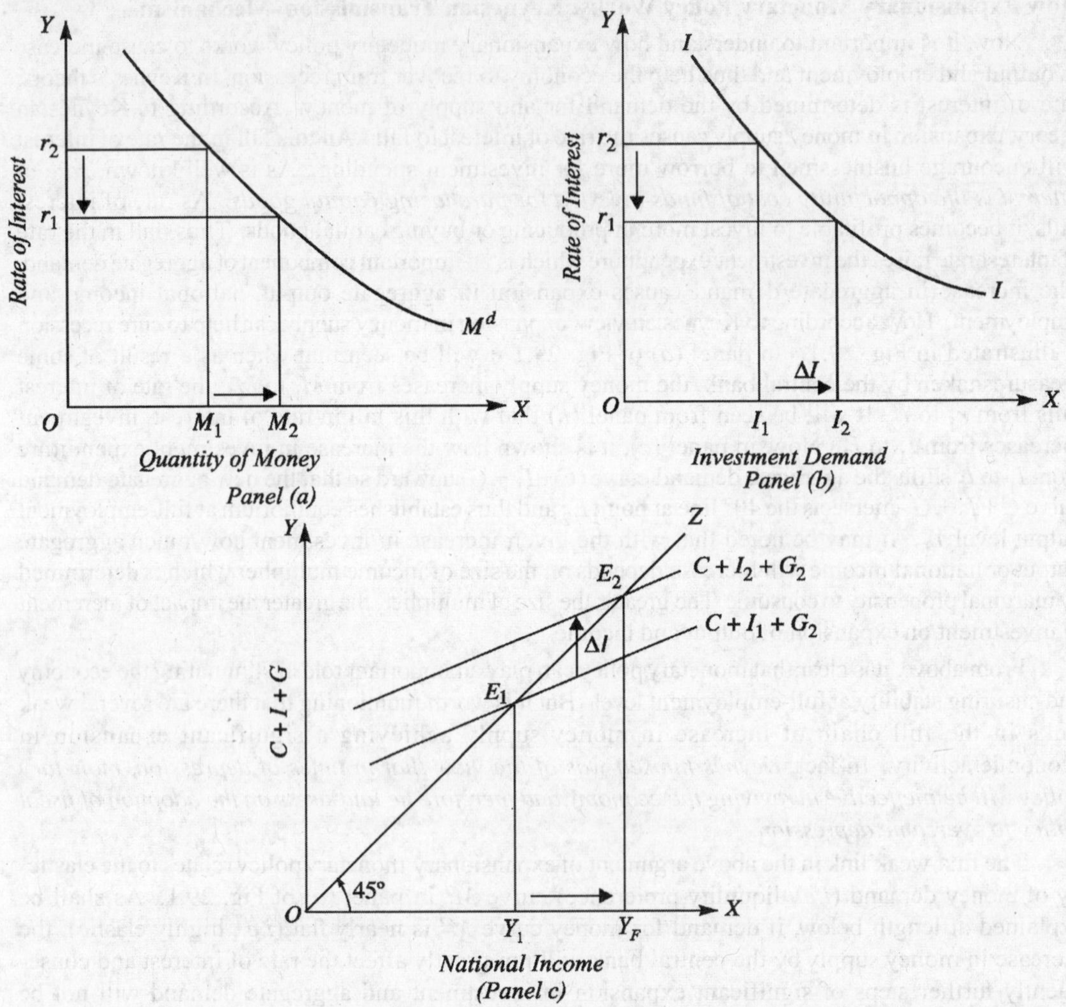

Fig. 29.1. *How Expansionary Monetary Policy Works: Keynesian View*

modern Keynesians equally advocate for the adoption of discretionary monetary policy as for the discretionary fiscal policy to get rid of recession.

TIGHT MONETARY POLICY TO CONTROL INFLATION

When aggregate demand rises sharply due to large consumption and investment expenditure or more imprantly, due to the large increase in Government expenditure relative to its revenue resulting in huge budget deficits, a demand-pull inflation occurs in the economy. Besides, when there is too much creation of money for one reason or the other, it generates inflationary pressures in the economy. To check the demand-pull inflation which has been major problem in India and several other countries in recent years the adoption of *contractionary* monetary policy which is popularly called tight monetary policy is called for. Note that tight or restrictive money policy is one which reduces the availability of credit and also raises its cost. The following monetary measures which constitute tight money policy are gen-erally adopted to control inflation:

1. *The Central Bank sells the Government securities to the banks, other depository institutions and the general public through open market operations.* This action will reduce the reserves with the banks and liquid funds with the general public. With less reserves with the banks, their lending capacity will be reduced. Therefore, they will have to reduce their demand deposits by refraning from giving new loans as old loans as old loans are paid back. As a result, money supply in the economy will shrink.

2. *The bank rate may also be raised which will discourage the banks to take loans from the central bank.* This will tend to reduce their liquidity and also induce them to raise their own lending rates. Thus this will reduce the availability of credit and also raise its cost. This will lead to the reduction in investment spending and help in reducing inflationary pressures.

It may be noted that in India changes in bank rate have not been generally used to bring stability in the economy. It is the instruments of **repo rate and reverse repo rate changes** that have often been used to fight inflation and recessionary situation. *Repo rate is the interest rate at which Reserve Bank of India lends funds to the commercial banks for a short period of time.* To control inflation repo rate is raised. Hike in repo rate raises the cost of funds for the banks which, if they do not have excess reserves or enough liquidity, would raise their lending rates. The rise in lending rates of banks which will raise the borrowing costs for businessmen will lead to the reduction in demand for bank credit and thus lower investment and consumpiton demand. The opposite happens when to fight recession the RBI lowers repo rate and thus reducing lending rate for the banks.

3. The most important anti-inflationary measure is *raising of statutory Cash Reserve Ratio (CRR).* Rise in cash reserve ratio (CRR) reduces cash reserves of banks. To meet the new higher reserve requirements, banks will reduce their lending, that is, credit availability. This will have a direct effect on the contraction of money supply in the economy and help in controlling demand-pull inflation. Besides Cash Reserve Ratio (CRR), the Statutory Liquidity Ratio (SLR) can also be increased through which excess reserves of the banks are mopped up resulting in contraction in credit.

4. Fourthly, an important anti-inflationary measure is the use of qualitative credit control, namely, *raising of minimum margins for obtaining loans from banks against the stocks of sensitive commodities* such as foodgrains, oilseeds, cotton, sugar, vegetable oil. As a result of this measure, businessmen themselves will have to finance to a greater extent the holding of inventories of goods and will be able to get less credit from banks. This selective credit control has been extensively used in India to control inflationary pressures.

How the Tight Monetary Policy Works: Keynesian Transmission Mechanism

It is important to understand how tight monetary policy works to check inflation. As explained above, tight monetary policy seeks to reduce the money supply through contraction of credit in the economy and also raising the cost of credit, that is lending rates of interest. The reduction in money supply itself raises the rate of interest. The higher interest rate reduces investment spending which results in lowering of aggregate demand curve ($C+I+G$). The decrease in aggregate demand tends to restrain demand-pull inflation. How tight money policy helps in checking inflation is graphically shown in Fig. 29.2. Let us assume that full-employment level of national income is Y_F as depicted in panel (*c*) of Fig. 29.2. Now, if due to a large budget deficit and excessive creation of money supply, aggregate demand curve shifts to $C+I_2+G_2$; inflationary gap of E_1H comes to exist at full-employment level. That is, the sum of consumption expenditure, private investment spending and Government expenditure exceeds the full-employment level of output by E_1H. This creates a demand-pull inflation causing rise in prices. Though with aggregate demand curve $C+I_2+G_2$ equilibrium reaches at point E_2 and as a result national income increases but only in money terms ; real income or output level remaining constant at OY_F

Now, it will be seen from panel (*a*) that if tight money policy succeeds in reducing money supply from M_2 to M_1, the rate of interest will rise from r_1 to r_2. Panel (*b*) of Fig. 29.2 shows that at a higher interest rate r_2, private investment falls from I_2 to I_1. This reduction in investment expenditure shifts aggregate demand curve $C + I_2 + G_2$ downward to $C + I_1 + G_2$ and in this way inflationary gap is closed and equilibrium at full-employment output level Y_F is once again established. In our figure it

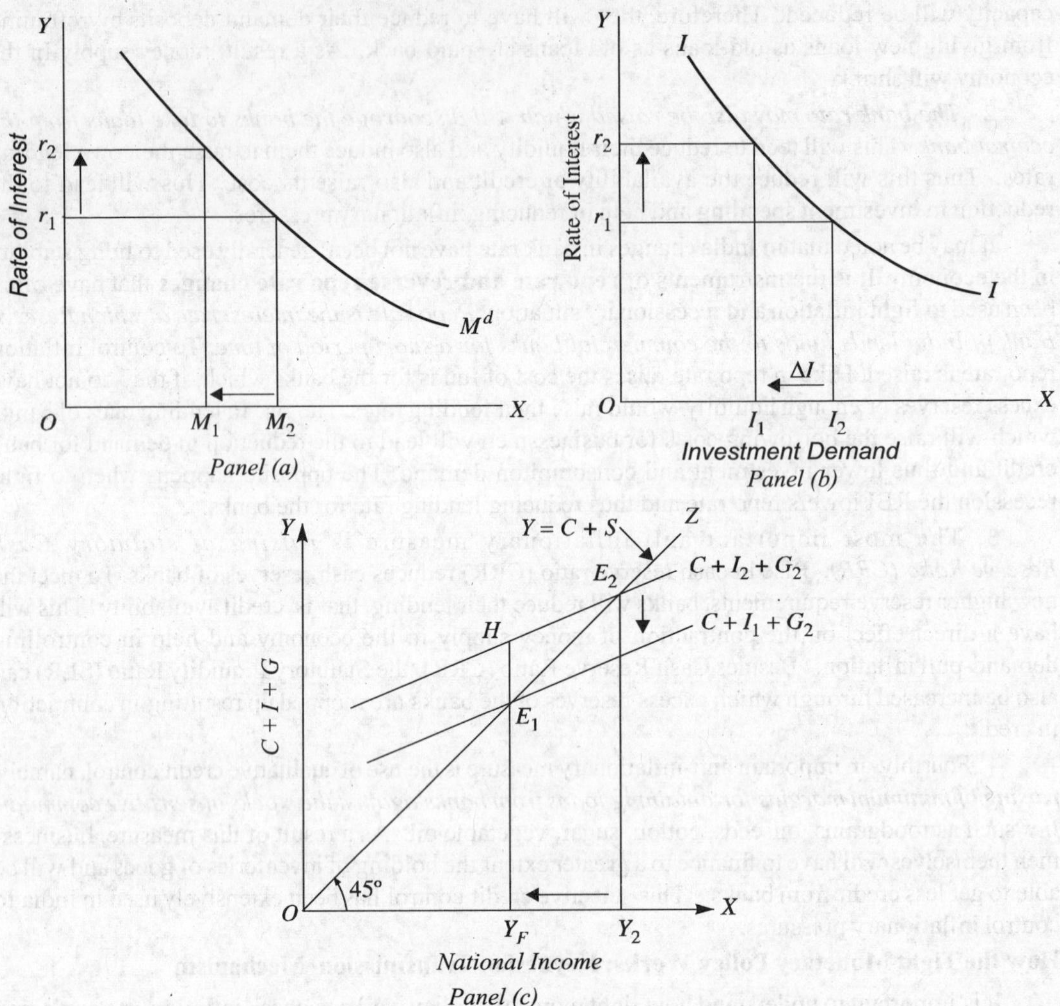

Fig. 29.2. *How Anti-Inflationary Monetary Policy Works: Keynesian View*

has been assumed that there is contraction of money supply from M_2 to M_1 and as a result rise in rate of interest from r_1 to r_2 is sufficient to reduce investment expenditure equal to $I_2 - I_1$ which is equal to inflationary gap and in this way macroeconomic equilibrium without any inflationary pressure is established at output level Y_F.

It should be further remembered that in our analysis of the successful working of the tight monetary policy it is assumed that demand for money curve (*i.e.*, liquidity preference curve) is fairly steep so as to push up the rate of interest from r_1 to r_2 and further that investment demand curve *II* in panel (*b*) of Fig. 29.2 is fairly elastic so that rise in rate of interest from r_1 to r_2 is sufficient to reduce investment by $I_2 - I_1$ or ΔI. If these conditions regarding the slopes of the money demand curve and investment demand curve represent the real world situation, then tight monetary policy will succeed in controlling inflation and ensuring price stability.

To sum up, Keynesian view of how expansionary and contractionary (tight) monetary policies work to achieve the twin goals of price stability and equilibrium at full-employment level of output is shown in the accompanying box.

Monetary Policy: Keynesian Transmission Mechanism

Expansionary Monetary Policy	Tight Monetary Policy
Problem: Recession and Unemployment	***Problem***: Inflation
Measures: (1) Central Bank buys securities through open market operations	***Measures***: (1) Central Bank sells securities through open market operations
(2) It reduces cash reserve ratio (CRR)	(2) It raises Cash Reserve Ratio (CRR) and Statutory Liquidity Ratio (SLR)
(3) It lowers bank rate	(3) It raises Bank Rate
	(4) It raises maximum margin against holding of stocks of goods.
↓ Money supply increases	↓ Money supply decreases
↓ Interest rate falls	↓ Interest rate rises
↓ Investment increases	↓ Investment expenditure declines
↓ Aggregate demand increases	↓ Aggregate demand declines
↓ Aggregate output increases by a multiple of the increase in investment	↓ Price level or inflation rate falls

Liquidity Trap and Ineffectiveness of Monetary Policy

Keynes and his early followers doubted the effectiveness of monetary policy in pulling the economy out of depression. They therefore emphasized the role of fiscal policy for fighting severe recession. According to Keynes and his followers, during severe recession people have a high elastic demand for money at low rates of interest. At low rates of interest banks hold on to whatever money

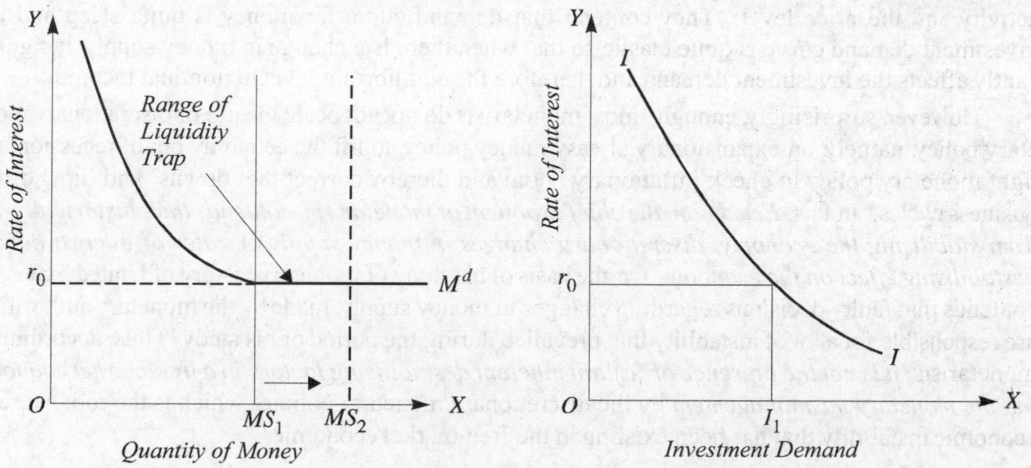

Fig. 29.3. *Liquidity Trap and Increase in Money Supply* **Fig. 29.4.** *Investment Demand*

reserves they happen to get and the people in general also hold on to whatever money they spare. According to Keynes, demand for money or what he calls liquidity preference is determined mainly by transactions and speculative motives. Whereas transactions demand for money is determined by the level of national income, the speculative demand for money depends on the expectations regarding future rates of interest. During depression, current rate of interest may fall so low that most of the people expect the interest rate to rise in future and therefore they hold on to their money for the present. This makes the demand for money absolutely elastic at a low rate of interest as is shown in Fig. 29.3. It will be seen from Fig. 29.3 that at a low rate of interest r_0 demand curve for money M^d is absolutely elastic showing people demand or hold on to all the increases in money supply beyond MS_1 for speculative purposes and not invest in bonds. Under these circumstances the economy is said to have fallen in a liquidity trap. A liquidity trap occurs when under conditions of depression the economy finds itself in a situation where people hold all the increments in the stock of money and money demand curve M^d takes a horizontal shape.

Suppose during a recession, stock of money is equal to MS_1 and money demand curve is given by M^d. The interaction between these two determines r_0 rate of interest. Suppose now, to pull the economy out of recession, the stock of money supply is expanded to MS_2. A glance at Fig. 29.3 reveals that expansion in money supply from MS_1 to MS_2 does not lower the rate of interest as the economy is operating in the range of liquidity trap. Now, Fig. 29.4 shows that with the rate of interest remaining unchanged at r_0, the level of investment does not rise. With level of investment remaining the same, there is no increase in aggregate demand and the economy remains in a state of depression. Thus, under these circumstances Keynes and his early followers contended that expansionary monetary policy did not help the economy in staging a recovery from recession.

It may, however, he noted that the concept of liquidity trap is not supported by empirical studies. The empirical studies show that demand for money (liquidity preference) never becomes flat and instead it falls throughout. Therefore, modern Keynesians and other economists now believe that monetary policy can play a useful role in stabilising the economy at full-employment level. However, as shall be discussed below, it is the *monetarists* led by Friedman who do not favour discretionary monetary policy to check cyclical instability.

MONETARY POLICY : MONETARIST VIEW

Though most of the modern economists regard both fiscal and monetary policies as important tools for stabilising the economy, there is a group of economists known as *monetarists* led by Friedman who think that changes in money supply are the key determinants in the level of economic activity and the price level. They contend that demand curve for money is quite steep and the investment demand curve is quite elastic so that when there is a change in money supply, it significantly affects the investment demand and therefore the equilibrium level of nominal income.

However, surprisingly, enough, most monetarists do not advocate the use of discretionary monetary policy, namely, an expansionary or easy money policy, to lift the economy out of recession and tight monetary policy to check inflationary boom and thereby correct the `downs' and `ups' of the business cycles. In fact, *Friedman, the chief exponent of monetarism, contends that, historically, far from stabilising the economy, discretionary changes in money supply or rates of interest have a destabilising effect on the economy.* On the basis of his study of monetary history of United States, he contends that faulty decisions regarding changes in money supply, made by the monetary authorities, are responsible for a lot of instability that prevailed during the period of his study. Thus, according to monetarists, *it is not the presence of certain inherent destabilising factors in a free-market economy but the monetary mismanagement* by the discretionary monetary policies which is the root cause of economic instability that has been existing in the free-market economies.

Source of Monetary Mismanagement

According to monetarists, there are two important sources of monetary mismanagement : (1) *variable time lags* concerning the effect of money supply on the nominal income and (2) *treating interest rate as the target of monetary policy* for influencing investment demand for stabilising the economy. We examine below both these sources of monetary mismanagement.

1. Variable Time Lags. First, there is a problem of variable long time lags that occur for changes in money supply to bring about desirable effect on nominal income. From his empirical studies Friedman concludes that it takes six months to two years for the changes in money supply to produce a significant effect on nominal income. Monetarists argue that since it is extremely difficult to know the time lag involved in a specific monetary policy measure adopted to tackle the problem, it is impossible to determine *when* a particular policy measure should be taken and *which policy measure*, expansionary or tight, is suitable under the given situation. In fact, according to the monetarists, in view of the uncertainty about the exact duration of time lags involved, the use of discretionary monetary policy to stabilise the economy may backfire and further intesify the cyclical instability. For example, if expansionary monetary policy is adopted because the various economic indicators show the situation of mild recession, then, due to the time lags involved, say six to eight months, for the policy to yield results, the economic situation might change and becomes reverse during that period and becomes one of mild inflationary situation. Expansionary monetary policy which produces the effect after 6 to 8 months may, therefore, acutally intensify the inflationary situation.

2. Interest Rate as a Wrong Target Variable. The second source of money mismanagement is the wrong target variable chosen by the monetary authorities. Monetarists have asserted that monetary authorities have tried to control the interest rates to stabilise the economy. It has been argued that Central Bank cannot simultaneously stabilise both the interest rate and money supply. By controlling the interest rate it has actually destabilised the economy. For example, if the economy is recovering from recession and is presently approaching full employment with aggregate demand, output, employment and prices all registering a rise, the transactions demand for money will increase. This increase in transactions demand for money will cause the rate of interest to rise. But if the monetary authorities have chosen to stabilise the interest rate, they would adopt tight monetary policy to prevent the interest rate from going up. This will have a recessionary effect and make the economy unstable. Similarly, when the economy is going into recession, it will result in lowering aggregate output and prices. This fall in aggregate output and prices will cause a decline in the transactions demand for money. And the decrease in transactions demand will lead to the fall in interest rate. To prevent this fall in interest rate, if money supply is increased, it will generate instability in the economy rather than removing it.

Monetary Rule: Monetary Policy Prescription

From the above analysis it follows that monetarists are not in favour of stabilising the interest rate, they advocate for the *adoption of a rule rather than pursuing discretionary monetary policy* to stabilise the economy. They prescribe a *rule for the growth of the money supply* to achieve economic grow with stability. According to the monetary rule suggested by Friedman, *money supply should be allowed to grow at the rate equal to the rate of growth of output*. If the economy is expected to grow annually at the rate of 3, 4 or 5 per cent, money supply should also grow at this rate. The grow of output of an economy will absorb the extra money supply created as per this rule, without generating inflationary or recessionary conditions, and will thus ensure stability in the economy. To quote Ritter and Silber, "such a rule would eliminate the major cause of instability in the economy—the capricious and unpredictable impact of counter cyclical monetary policy. As long as the money supply grows at a constant rate each year, be it 3, 4 or 5 per cent, any decline into recession will be temporary. The liquidity provided by a constantly growing money supply will cause the aggregate demand to expand. Similarly, if the supply of money does not rise at a more than average rate, any inflationary increase in spending will burn itself out for lack of fuel.

Monetary rule has been criticised by the Keynesian economists. They have argued that monetary rule will have a destabilising effect. Given that the velocity of money (V) is unstable or variable, increase in money supply (M), according to this rule, may not ensure growth of aggregate demand (which, according to monetarist theory, is equal to MV) equal to the rate of growth of output in a year which is difficult to predict. Thus, money supply increase may sometimes exceed the growth of output and sometimes fall short of it and as a result may cause sometimes demand-pull inflation and sometimes necessary conditions. Thus, according to Keynesian economists, policy of monetary rule does not guarantee economic stability and it may itself create economic instability.

THE POLICY OF QUANTITATIVE EASING (QE)

We have explained above the conventional monetary policy to bring stability in the economy. Now, in 2008 there was a severe economic crisis caused by bursting of housing bubble in the U.S.A. Prior to 2008, when the prices of houses were rising the Americans had made excessive investment in houses financed by sub-prime housing loans given on a large scale by the American banks at a low rate of interest. The American banks sold these sub-prime mortgaged housing loans to other financial institutions which converted them into securities that were traded in the market. In 2007-08 when the prices of houses declined sharply, the people who had got the sub-prime housing loans from the banks started defaulting on paying interest and the principal amount. This caused serious financial crisis as the prices of securities based on sub-prime mortgage housing loans fell almost to zero. As a result, the banks and financial institutions such as Lehman Brothers, Citi Bank which had made large investment in them came to the brink of bankruptcy. In July 2008, Lehman Brothers declared itself bankrupt. This caused great panic which ultimately resulted in severe recession in the U.S. economy which had world-wide repercussions. The recession in the U.S. economy was so severe that its growth rate came down to almost zero and unemployment shot up to 10 per cent of labour force.

It is to tackle such severe recessionary conditions the *non-conventional monetary policy* of quantitative easing (QE) was adopted by the Federal Reserve (which is the Central Bank of the U.S.). The interest rates were then so low that the conventional monetary policy of lowering interest rates had lost its ability to stimulate the economy. Therefore, non-conventional monetary policy of quantitative easing (QE) was adopted by the Federal Reserve keeping near zero rate of interest. Under the quantitative easing (QE) the Federal Reserve purchased bonds (both treasury bonds and private corporate bonds) and other financial assets on a large scale and *to purchase them created new money*. In this way, the Federal Reserve pumped a large liquidity (*i.e.*, cash) into the banking system and has continued this policy for several years now since 2008, though the scale of purchase of bonds has slowed down in recent years.

Now, selling bonds and other financial assets to Federal Reserve and getting a lot of cash allowed American commercial banks to provide loans to business firms and investors at low rate of interest to make more investment and thus increasing spending in the economy. It was hoped that more investment and spending would help the economy to get out of recession. Thus, quantitative easing (QE) was expected to stimulate the US economy and recover from recessionary conditions.

It may be noted that with the Federal Reserve purchasing bonds from the market on a large scale, their supply in the market was reduced which drived up the bond prices. Higher bond prices mean the yield from them declines and therefore investors are discouraged to buy them. The investors or business firms which got loans from the banks not only invested some money in building new factories or expanding their existing capacities but also used them to buy back their own stock or shares with the intention to shore up their prices. Since return on more investment in the US was very low, a good amount of US dollars created through quantitative easing also flowed into the emerging economies, especially India and China where returns from investment in shares and bonds were quite high. The inflows of capital (*i.e.*, US dollars) in India and China caused share prices in these countries to shoot up. In this way, American foreign institutional investors (FIIs) made a good amount of profits in investing in India and China.

With the policy of pumping a lot of liquidity into the banks which gave liberal loans at very low interest rates has succeeded in reviving the American economy whose annual GDP growth has on average reached 2.5 per cent and unemployment rate has also declined to 5 per cent labour force (June 2015). However, the recovery of the American economy is fragile. The Federal Reserve has quite often decided to withdraw completely the quantitative easing policy and raise rate of interest but each time it has postponed the implementation of its decision. It may be noted that when quantitative easing (QE) policy is completely withdrawn and rate of interest raised by the Federal Reserve, there will be *outflows of capital* (*i.e.*, dollars) by FIIs from India. When this happens the share prices in India will fall sharply, the Indian rupee would depreciate and thus creating a lot of economic instability in the Indian economy. This economic instability will also be seen in other emerging economies.

QUESTIONS FOR REVIEW

1. What is meant by monetary policy? Briefly explain the instruments of monetary policy.
2. Explain the objectives of monetary policy.
3. Explain monetary policy measures that should be adopted for curing recession (or depression) and reviving the economy. How does expansionary monetary policy work? How does Keynesian view in this regard differ from the monetarist view?
4. What are the factors that determine the effectiveness of monetary policy in reviving the economy from recession? Why Keynes thought that monetary policy was ineffective in reviving the economy from depression?
5. What is liquidity trap? How does it make the expansionary monetary policy ineffective in reviving the economy from recession?
6. Explain the main instruments of monetary policy that can be adopted to control inflation in the economy.
7. Explain the mechanism through which tight monetary policy works to check inflation. Illustrate diagrammatically.
8. Discretionary monetary policy can play an important role in stabilising the economy at full employment. Discuss.
9. Explain the views of Friedman and his follower modern monetarists regarding monetary policy as a tool of economic stabilisation. In this connection also comment on the monetarist view that interest rate is a wrong target variable of proper monetary policy.
10. What is meant by 'time lags'? How do they affect the efficiency of monetary policy to stabilise the economy?
11. What are the instruments of monetary policy? Explain the role of changes in bank rate and open market operations in controlling trade cycles.
12. "Monetary policy for its success depends on fiscal policy." Explain and critically examine this statement.

PART XII

Government and the Macroeconomy: Public Finance and Fiscal Policy

- Public Finance: Public Expenditure
- Public Revenue: Taxation and Budget
- Incidence of Taxation
- Fiscal Policy and Economic Growth in a Developing Country

Chapter 30
PUBLIC FINANCE: PUBLIC EXPENDITURE

What is Public Finance?

In public finance we study the finances of the Government. Thus, public finance deals with the question how the Government raises its resources to meet its ever-rising expenditure. As Dalton puts it, *public finance is "concerned with the income and expenditure of public authorities and with the adjustment of one to the other."*[1] Accordingly, effects of taxation, Government expenditure, public borrowing and deficit financing on the economy constitute the subject-matter of public finance. Thus, Prof. Otto Eckstein writes *"Public Finance is the study of the effects of budgets on the economy, particularly the effect on the achievement of the major economic objects—growth, stability, equity and efficiency."*[2] Further, it also deals with fiscal policies which ought to be adopted to achieve certain objectives such as price stability, economic growth, more equal distribution of income.

The Importance of Public Finance

Economic thinking about the role that public finance is expected to play has changed from time to time according to the changes in economic situation. Before the Great Depression that gripped the Western industrialised countries during the thirties, the role of public finance was considered to be raising sufficient resources for carrying out the Government functions of civil administration and defence from foreign countries. During this period, the classical economists considered it prudent to keep expenditure to the minimum so that taxing of the people is avoided as far as possible. Further, it was thought that Government budget must be balanced. Public borrowing was recommended mainly for production purposes. During a war, of course, public borrowing was considered legitimate but it was thought that the Government should repay or reduce the debt as soon as possible.

But under the impact of the Great Depression of thirties and the Keynesian explanation of it, the thinking about and role of public finance underwent a sea change. The classical view of public finance could not meet the requirements of the then prevailing situation. In order to increase aggregate effective demand and thereby raise the level of income and employment in the country, public finance was called upon to play an active role. Thus, according to Keynesian economists, to get the economy out of depression the government should increase its expenditure and cut taxes to raise aggregate demand. During the Second World War and after, the Western economies suffered from serious inflationary pressures which was attributed to the excessive aggregate demand. So, in such inflationary conditions, the public finance was expected to check prices through reducing aggregate demand. Thus the budget which was previously meant to raise resources for limited activities of the Government assumed a *functional role* to serve as an instrument of *economic stabilisation.*

It came to be realised that government's taxing and spending policies could go a long way in mitigating economic fluctuations. Balanced budgets are no longer considered sacrosanct and the governments can spend beyond their resources without offending canons of sound finance to restore the health of the economy. Public borrowing and consequent increase in public debt at the time of depression raises aggregate demand and thereby helps in raising the level of income and employment. Therefore, deficit budget and increase in public debt at such times is

1. Dalton, *Public Finance.*
2. O. Eckstein, *Public Finance,* Prentice Hall of India Private Ltd., 3rd Edition, p.53.

a thing to be welcomed. It was further demonstrated by Keynes that deficit spending by the Government could activise a depressed economy by creating income and employment much more than the original amount of deficit financing through the process of multiplier. Thus, after Keynesian revolution public finance assumed a functional role of *maintaining economic stability at full employment level.* Therefore, the present view of public finance is not one of mere resource-raising for the Government but one of serving as an instrument for maintaining stability through management of demand. Therefore, this present view of public finance has been described by A.P. Lerner as one of "*Functional Finance*".

In developing countries, public finance has to perform another important role. Whereas in the developed industrialised countries, the basic problem in the short run is to ensure stability at full employment level and in the long run to ensure steady rate of economic growth, that is, growth without fluctuations, the developing countries confront a more difficult problem of how to generate and sustain a higher rate of economic growth so as to tackle the problems of proverty and unemployment. Therefore, public finance has to play a special role of promoting economic growth in the developing countries besides maintaining price stability. Further, for developing countries mere economic growth is not enough; the composition of growing output and distribution of additional incomes ought to be such as will ensure removal of poverty and unemployment in the developing countries. Therefore, public finance has not only to augment resources for development and to achieve optimum allocation of resources, but also to promote fair distribution of income and expansion in employment opportunities. This is the functional view of public finance in the context of the developing countries.

PUBLIC EXPENDITURE

Of the two main branches of public finance, namely, public revenue and public expenditure, we shall first study the public expenditure. The classical economists did not analyse in depth the effects of public expenditure as public expenditure throughout the nineteenth century was very small owing to the very restricted Government activities. The Governments followed *laissez faire* economic policies and their functions were only confined to defend the country from foreign aggression and to maintain law and order within their territories. But now, the expenditure of Government all the world over have greatly increased. Therefore, the modern economists have started analysing the effects of public expenditure on production, distribution and the levels of income and employment in the economy.

Importance of Public Expenditure

Thanks to the macroeconomic theory advanced by J.M. Keynes, the role of public expenditure in the determination of level of income and its distribution is now well recognised. Keynesian macroeconomics provides a theoretical basis for recent developments in public expenditure programmes in the developed countries. The public expenditure can be used as a lever to raise aggregate demand and thereby to get the economy out of recession. On the other hand, through variation in public expenditure, aggregate demand can be managed to check inflation in the economy. Public expenditure can also be used to improve income distribution, to direct the allocation of resources in the desired lines and to influence the composition of national product. In the developing countries also, the role of public expenditure is highly significant.

In the developing countries, the variation in public expenditure is not only to ensure economic stability but also to generate and accelerate economic growth and to promote employment opportunities. The public expenditure policy in developing countries also plays a useful role in alleviating mass poverty existing in them and to reduce inequalities in income distribution. In what follows, we shall study the types of public expenditure, the causes of growth of public expenditure and its effects on production, distribution and economic growth in both the developed and the developing countries.

Classification of Public Expenditure

Revenue Expenditure and Capital Expenditure. Public expenditure has been classified into various categories. Firstly, Government expenditure has been classified into *revenue expenditure* and *capital expenditure.* Revenue expenditure is a current or consumption expenditure incurred on civil administration (*i.e.,* police, jails and judiciary), defence forces, public health and education. This revenue expenditure is of recurrent type which is incurred year after year. On the other hand, capital expenditure is incurred on building durable assets. It is a non-recurring type of expenditure. Expenditure incurred on building multipurpose river projects, power plants, ports, highways, steel plants etc., and buying machinery and equipment is regarded as capital expenditure.

Transfer Payments and Expenditure on Goods and Services. Another useful classification of public expenditure divides it into transfer payments and non-transfer payments. *Transfer payments refer to those kinds of expenditure against which there is no corresponding transfer of real resources (i.e., goods and services) to the Government.* Expenditure incurred on old-age pensions, unemployment allowance, sickness benefits, interest on public debt during a year etc., are examples of transfer payments because the Government does not get any service or goods against them in the particular year. On the other hand, *expenditure incurred on buying or using goods and services is a non-transfer payment as against such an expenditure, the Government receives goods or services.* It is therefore called expenditure on goods and services. It may be noted that expenditure on defence, education, health etc., are non-transfer expenditure as in return for these, Government obtains the services of army personnel, teachers, doctors etc., as well as some goods or equipments used in these activities. Investment expenditure is undoubtedly a non-transfer expenditure as through it Government obtains capital goods. It is worthwhile to mention that whereas in case of transfer payments, it is the beneficiaries who decide about the use of resources, in the case of non-transferable type of expenditure, the Government itself decides about the use of real resources, especially whether they are to be used for consumption or investment purposes.

Developmental and Non-Development Expenditure. Another useful classification of public expenditure rests on whether a particular expenditure by the Government promotes development. All those expenditures of Government which promote economic growth are called developmental expenditure. Expenditure on irrigation projects, flood control measures, transport and communication, capital formation in agricultural and industrial sectors are described as developmental. On the other hand, expenditure on defence, civil administration *(i.e.,* police, jails and judiciary), interest on public debt etc., are put into the category of non-development expenditure. It may be noted that, till recently, expenditure on education and health were regarded as non-developmental type. It has now been realised that the expenditure on education and public health promotes the growth of what is called *human capital* which promotes economic growth as much as physical capital, if not more. Therefore, these days, expenditure on education, research and health are generally regarded as developmental expenditure.

In Indian budgets government expenditure is often classified into *plan expenditure and non-plan expenditure.* Plan expenditure is government expenditure that is made on the schemes and projects included in Five Year Plans and is considered as development expenditure. On the other hand, non-plan expenditure is incurred on civil administration (police, law and order, judiciary), defence and is generally considered as non-developmental expenditure.

It is worth noting that division of Government expenditure into developmental or non-developmental is the modern counterpart of the distinction drawn by classical economists between *productive* and *unproductive public expenditure,* which has been a subject of great controversy. For instance, it has been pointed out that even Government expenditure on defence and civil administration helps to maintain conditions in which productive activity can be carried out. It is, therefore, claimed by some that, indirectly, expenditure on defence and civil administration is also productive. Thus, we see that what Government expenditure is developmental or productive and what non-developmental or unproductive is not based on any objective or fool-proof criteria and is therefore somewhat arbitrary.

GROWTH OF PUBLIC EXPENDITURE

Public expenditure has phenomenally increased all the world over. A pertinent question is what are the causes of this phenomenal growth in public expenditure. It will be useful to discuss this with reference to India. This is because the factors responsible for a large increase in public expenditure over time in India are generally applicable to other countries too. It will be interesting to mention here two laws about the growth of public expenditure.

Wagner's Law of Increasing State Activity

First, there is *Wagner's Law of Increasing State Activity.* According to Wagner, a German economist, there are inherent tendencies for the activities of the Government to increase both extensively and intensively. In other words, according to this law, as an economy develops over time, the activities or functions of the Government increase. With the development of the economy, new functions and activities are undertaken by the Government and old functions are performed more thoroughly. The expansion in the Government functions and activities leads to the increase in public expenditure. Though Wanger based his law on the historical evidence drawn from economic growth of Germany, this applies equally to other countries, both developed and developing ones.

Wiseman-Peacock Hypothesis

The second hypothesis about the growth of public expenditure has been put forward by Wiseman and Peacock in their study of public expenditure of U.K. According to this *Wiseman-Peacock hypothesis,* Government expenditure does not increase at a steady rate continuously but in jerks and step-like manner. However, in the view of the present author, both these factors, one making for a continuous increase in Government activity and consequently public expenditure as emphasised by Wagner and others like war and depression causing the public expenditure to rise by jerks as emphasised by Wiseman and Peacock have been responsible for the enormous increase in public expenditure. In what follows, we shall explain the factors responsible for the increase in public expenditure with special reference to the Indian economy.

1. Defence. An important factor responsible for public expenditure is the mounting defence expenditure incurred by countries all the world over. It is not only during actual wars that defence expenditure has been rising but even during peace time, the countries have to remain in the state of military preparedness demanding large defence expenditure. There is arms race going on between countries. A poor country like India has to safeguard its hard earned freedom and this involves a lot of expenditure on building up efficient and adequate armed forces. India is wedged in between two enemies, namely, expansionist China and aggressive Pakistan, which have been strengthening their armed forces. India had to fight three wars since independence. India has thus to remain in a state of military preparedness. Internally also in view of clash of linguistic, territorial and political interests, lot of expenditure has to be incurred on maintaining internal security.

2. Population Growth and Urbanisation. Another factor responsible for the increase in public expenditure is the growth in population and urbanisation of the economies. Population has been increasing in almost all countries of the world, though at varying rates. In India the population has been increasing at an alarming rate since independence. The population of India which was 36 crores in 1951 has now gone upto about 121crores in 2011. The scale of government activities such as providing education, public health, roads and transport facilities has to increase in harmony with the growth of population. Further, when population increases, more has to be spent on administrative services (police, jails, judiciary etc.) to maintain law and order in the country.

With the progress of the economy and the growth of population, the extent of urbanisation increases. In India, the proportion of urban population to the total population has risen from 11.3 per cent in 1921 to 25.5 per cent in 1991 and to 27.8 per cent in 2001. As a result of the increasing urbanisation, the existing towns expand and the new ones come up. Urbanisation calls for greater per capita expenditure on social and administrative services. Therefore, the increase in urbanisation in India has tended to increase the government expenditure.

3. Activities of a Welfare State. The Government activities and functions have been increasing due to the change in the nature of State. The modern States are no longer Police States concerned mainly with the maintenance of law and order. They have now become Welfare States. *A Welfare State is one which provides for social insurance of its citizens against old age, sickness, unemployment etc.* The modern Governments have therefore to incur a lot of expenditure on social security measures such as old age pensions, unemployment allowances, sickness benefits.

4. Maintaining Economic Stability. As pointed out by Wagner, state functions increase with the advancement and progress of the economy. In the nineteenth and early twentieth century, the Government followed *laissez-fair* policy. Now, need for active intervention of the Government has been increasingly felt. Thanks to J.M. Keynes whose macroeconomic theory has clearly brought out that the working of free-market mechanism does not ensure economic stability at full employment level. According to his theory, lapses from full employment or depressions are caused by deficiency of aggregate demand due to the slackened private investment activity. In order to compensate for this shortfall in private investment, the Government has to step up its expenditure on public works. The increase in Government expenditure raises aggregate demand manifold through the working of what Keynes has been called income multiplier. This helps to push the economy out of depression and to raise levels of income and employment. Now, this compensatory fiscal policy is being followed by all the world over, since achievement of full employment and maintenance of economic stability has become an important objective of the Government.

It is in line with the objective of employment that, in India, the Government has taken over several *private sick mills* and incurs a lot of expenditure on them so that workers employed in them are not rendered unemployed. Further, the Indian Governments, both of Centre and States, incur a lot of expenditure on relief public works in rural areas when drought and other natural calamities occurs. Besides, a lot of public expenditure is being incurred on *special employment and anti-poverty* schemes to promote the growth of employment in the economy.

5. Economic Growth and Development. The most important factor in developing countries such as ours that has led to a phenomenal increase in public expenditure is the expansion in developmental activities of the Government. In developing countries like India which have socialistic tendencies the public sector plays an important role in promoting economic growth and development. Not only public utility services such as water supply, electricity, post and petroleum and transport services are being provided by the public sector, but also the Government has invested a huge sum of resources in industrial and agricultural development of the economy. Several steel plants, multipurpose irrigation projects, fertilizer factories, coal mining, exploration and production of oil and petroleum, different kinds of machine-making industries and chemical plants have been started and are being operated in the public sector. On these a huge amount of expenditure is being incurred by the Government in India. Owing to these developmental activities of the Government in India, the proportion of developmental expenditure to the total Government expenditure has greately increased and is greater than that of non-developmental expenditure. In 2003-04, Central Governments plan expenditure, which is mainly development expenditure, was 122.3 thousand crores which rose to 137.4 thousand crores in 2004-05.

6. Mounting Debt Service Charges. The Governments in all developing countries (including India) has been borrowing heavily in recent years to finance their increasing activities. Not only the debt money has to be paid back when it matures, interest payments have also to be made annually to the creditors. These debt service charges have resulted in enormous increase in public expenditure. It should be noted that the Government in India has not only been borrowing from within the country but also from abroad through foreign aid or commerical loans from private capital markets to finance its development plans. It has been estimated that for the year 1998-99, ₹ 75,000 crore were spent on the interest payments which went up to ₹ 125,900 crore in 2004-05 and to 3,80,000 crore in 2013-14 by the Central Government.

7. Mounting Expenditure on Subsidies. Governments, both in the developed and developing countries, incur a lot of expenditure on subsidies to the various sections of population. In India, the Government has been providing subsidies on food, fertilizers, and oil expenditure on them has been increasing at a rapid rate which is the main cause of large fiscal deficit in India. For example while in 2004-2005, the Central Government expenditure on subsidies on these three items was of the order of ₹ 46.5 thousand crore and for the year 2013-14 it was estimated to go up to ₹ 244 thousand crores (RE). The expenditure of Central Government's expenditure on subsidies on food, fertilizers, oil now account for about 12 per cent of total budget expenditure of the Central Government. While the aim of giving food subsidy is to help the people below the poverty line, the aim of fertilizer subsidy is to promote the growth of agriculture and help small farmers.

8. Anti-Poverty Schemes. Another important cause of increasing public expenditure in India is huge expenditure which the Government is incurring on employment generating anti-poverty schemes. It has now been realised that economic growth alone will not eradicate poverty, at least in the short run. Therefore, various employment schemes have been started by the Government for the people living below the poverty line. Prominent among these anti-poverty schemes in India are Mahatma Gandhi National Rural Employment Gurantee Scheme (MGNREGS) and subsidies on food (food secunity) etc. Expenditure on these schemes has greatly risen in recent years.

Conclusion

We have seen above that several factors have been working to cause increase in public expenditure in all the economies of the world. Though most of the factors are common operating all the world over, there are some additional factors in developing countries like India in which public sector plays an important role in the process of socio-economic development. This calls for more rapid growth of public expenditure.

EFFECTS OF PUBLIC EXPENDITURE ON PRODUCTION AND DISTRIBUTION

Having studied the causes of large increase in public expenditure, it will be useful to explain the effects of public expenditure on the production and distribution in the economy. Public expenditure, if properly allocated and efficiently used, can have a wholesome effect on the economy. Public expenditure can augment productive capacity of the economy and improve productivity of its working class. It can also reduce inequalities in income distribution, if properly designed. In the following we shall spell out in detail the impact of public expenditure on production and income distribution in the economy.

Effect of Public Expenditure on Production

It is generally pointed out that all kinds of expenditure by Government is not productive. For instance, Government expenditure on defence and civil administration (police, jail and judiciary) is said to be non-development expenditure for it does not apparently add to the volume of production of the economy. It is true that public expenditure on defence and civil administration are unproductive *directly,* but even they can *under certain circumstances* in an indirect way promote production and employment. This will be made clear a bit later. Further, the effects of public expenditure on production may be different in the case of a developed economy from that of a developing economy, for the circumstances in them differ a good deal. Let us first take the case of a developed economy,

Effect of Public Expenditure on National Output at Times of Depression

Developed economies often find themseleves in the grip of a depression or recession caused by lack of aggregate effective demand. At certain times in the developed countries effective demand falls due to the decline in private investment. At times of depression in an industrialised developed economy, there is idle productive capacity on the one hand and unemployed manpower on the other.

Under these circumstances, the increase in Government expenditure on public works or any other type of investment or even expenditure on defence and civil administration will lead to the manifold increase in income and employment through the process of multiplier, as has been explained by J.M. Keynes. The increase in aggregate demand will cause fuller utilization of the existing productive capacity and unemployed manpower resulting in expansion of volume of production, employment and national income. This will become clear from Fig. 31.1 where along the *X*-axis we measure national product or income and on the *Y*-axis aggregate demand which comprises consumption demand (*C*) and investment demand (*I*). It will be seen from Fig. 31.1 that prior to the Government expenditure the aggregate demand curve cuts the 45° line representing aggregate supply curve at point *E*. Thus OY_1 is the equilibrium level of national product or income determined by aggregate demand and supply.

Now, as has been explained in the chapters relating to macroeconomic theory that this equilibrium level may not be established at full-employment level. Suppose full employment of labour and other resources corresponds to OY_2 level of national product. This implies that in equilibrium at OY_1, level of income aggregate demand is not sufficient to ensure full employment. If under these circumstances the Government undertakes extra expenditure, say equal to *EH* or G, then the aggregate demand curve shifts upward to *C* + *I* + *G* position (dotted). As will be seen from Fig. 31.1, the increase in Government expenditure by *EH* will cause the equilibrium level to shift to OY_2 level of national product resulting in higher level of employment. It will be further noticed that the increase in national output (*i.e.,* income) equal to Y_1Y_2 (*i.e.,* ΔY) will be greater than the Government expenditure (*G*) depending upon the magnitude of multiplier. It may be noted that the magnitude of Government expenditure multiplier depends upon the marginal propensity to consume of the people.

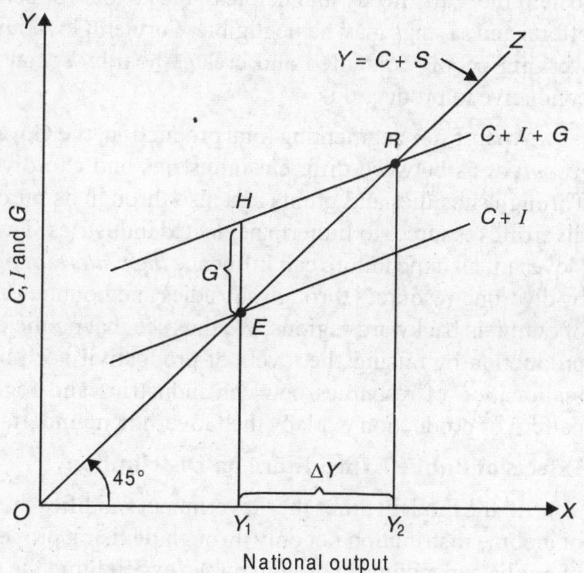

Fig. 31.1. *Effect of Government Expenditure on National Output and Income*

From the foregoing analysis we conclude that the increase in Government expenditure preferably financed by borrowings from the banks or printing new notes at times of depression will raise aggregate demand and thereby lead to the multiple increase in national output and employment. But once the level of full employment is attained the increase in Government expenditure cannot raise production through raising aggregate demand. Instead, prices will rise.

However, even in the developing countries such as India there there are ways in which government expenditure, if judiciously planned, can promote production. First, if the Government expenditure is incurred on *investment projects for capital formation,* for instance on building of canals, power, fertilizer, roads and highways, railways, and other infrastructural facilities, it will expand productive capacity and generate long-term economic growth. Secondly, if public expenditure is directed to *scientific research and development (R & D)*, it will ensure progress in technology and raise productivity or power to produce of workers. Thirdly, it has now been found that Government expenditure on education and public health helps in building "*human capital*" which also greatly enhances productivity or power to produce of the people. Against these, it may be pointed out that certain types of Government expenditure may adversely affect production. Government expenditure on social insurance like health insurance, unemployment insurance, and old age pensions is said to be of such a type. By insuring against their future and uncertain contingencies like sickness, unemployment and old age, they blunt the edge of the desire to work and save more. The social security expenditure by the Government makes the people indifferent towards the future and makes them neglect savings. This is bound to affect adversely productive efforts in the present. Because of the future security, people will work less and save less. However, in our view, if such social security expenditure is kept within proper limits and if it is used to help the really needy and helpless, the adverse effects of social security expenditure on productive efforts and savings may be negligible. Further, Government expenditure on social security makes the working people contented and creates healthy social environment and industrial peace which is conducive to production.

Apart from augmenting total production, the Government expenditure also affects *allocation of resources* as between different industries and can divert resources to socially desirable channels. Through subsidies and grants and also through its purchase policy, the Government may succeed in diverting resources to hitherto neglected industries. Thus, besides raising the level of production, the Government expenditure can influence *the pattern of production or composition of output*. Likewise, by diverting resources through subsidies and bounties to the backward regions it can promote growth of output in backward regions. We thus see that public expenditure, if wisely conducted, can promote production by raising the levels of productivity or powers to produce and save. It can also cause reallocation of resources between industries and regions and exert wholesome influence on the pattern of production work by the Government undertaken through public enterprises.

Effects of Public Expenditure on Distribution

In the modern times the Government modifies the free working of market mechanism in respect of income distribution not only through devising proper tax structure but also through various forms of public expenditure. Through public expenditure the Government redistributes income in favour of the poor. Too large inequalities in income distribution as produced by the free working of market system are not only socially unjust, but also not conducive to the maximisation of social output. Not all types of public expenditure reduce inequalities in income distribution. The following forms of public expenditure redistribute income in favour of the poor and thus reduce inequalities.

1. **Social Security Measures.** Expenditure on unemployment insurance, sickness benefits, old age pensions are some of the social security measures which help the people at times of contingencies. In India only in recent years some State Governments such as those of Haryana, Punjab, Delhi have introduced old age pension scheme. In capitalist countries such as U.S.A., Great Britain the social security system to help the people emerged with the idea of a Welfare State and in these countries Governments spend a large sum of money on the social security system.

2. **Expenditure on Subsidies.** Expenditure on various types of subsidies has also a redistributive effect. In India subsidies on foodgrain, sugar, kerosene oil, handloom cloth, fertilizers are provided to the people and Government spends a good part of its budget on these subsidies. Foodgrain, sugar, kerosene oil are sold through ration shops (*i.e.*, public distribution system) at prices below the market prices and the difference is borne by the Government as a subsidy. It may be noted that at present benefits of these subsidies are enjoyed not only by the poor but all those who are relatively well off. *If the public expenditure on subsidies is to have a real redistributive effect, these subsidies should be targeted to the poor.*

3. Expenditure on Social Infrastructure. Public expenditure by the Government on social infrastructure such as education, healthcare of the people, housing for the poor also tends to reduce income inequalities. With free or subsidised education, free or highly subsidised healthcare facilities the poor people's real income goes up. The modern governments spend a lot of money on schools, colleges, etc., to promote education. In most states in India education up to the middle class is free and for higher levels wards of the poor people are either given free education or charged only low fees.

Similarly, the Governments spend a lot on public hospitals and dispensaries to provide health care to the poor people. Likewise, in many countries poor people are given financial aid by the Government to build houses: In India under Indira Awas Yojna, poor people are being given aid to build their low-cost houses.

Expenditure on Anti-Poverty Programmes. An important step for increasing incomes of the poor people is starting of several employment generating anti-poverty schemes. Prominent among these anti-poverty schemes in India are Jawahar Rozgar Yojna, Employment Assistance Scheme (EAS), Integrated Rural Development Programme (IRDP). Recently in India, employment guarantee scheme under National Rural Employment Gurantee Act (NREGA) has been started which greatly helps the rural poor. By generating employment these schemes raise incomes of the poor.

Encouragement to Labour-Intensive Industries. The Indian Government gives various types of subsidies to the cottage and small-scale industries which adopt labour-intensive technique. Being labour-intensive, the growth of these industries generates a large number of employment opportunities which improve income distribution.

Negative Income Tax to Achieve More Equal Distribution of Income. Last but not the least, to achieve more equal distribution of income and make a big dent into the poverty problem, a proposal which has been recently put forward by some economists in the USA and Great Britian is introduction of negative Income Tax. Though the name shows it is a tax, but in fact it is a form of expenditure, called *transfer expenditure*. In this negative income tax scheme, payments are made by the Government to the poor to raise their incomes. Under this system, the Government has first to define poverty income standard for families of different sizes. Then it makes payment in the form of direct money transfer to bring the incomes of all families up to that standard or at least to close a large part of the gap between their income and poverty income standard.

But the use of public expenditure to reduce inequalities in income distribution has certain bad effects also. The first and foremost evil effect is that it adversely affects incentives to work and save of the people. Thus, when people know that in times of difficulties such as unemployment, old age, sickness, the Government will come to their help they will be less willing to work hard. The poor would not work more when they know whatever their income, the gap between the minimum income and theirs will be filled by the Government. Likewise, many people save to live comfortably in old age, in periods of sickness. But when all these difficult periods are taken care of by the Government their willingness to work more and save more will be badly hurt. Thus, *expenditure to reduce income inequalities will tend to affect adversely incentives to work, save and invest more. All these discourage production and growth.*

It is important to note that *redistributive effects of public expenditure must be considered in the light of how it is financed.* For example, if redistributive public expenditure is financed through taxation and if the tax system of the country is regressive, it will work against the desirable distributive effects of public expenditure. For example, if public expenditure is financed through deficit financing, as has been the case in India for several years, it causes inflation in the economy. This inflation hurts the poor most and will therefore cancel out the desired distributive effects of public expenditure. Thus, if reduction in income inequalities has to be achieved not only the pattern of public expenditure, but also method of financing it and the system of taxation have to be suitably designed and adjusted.

QUESTIONS FOR REVIEW

1. Explain Wagner's Law and Wiseman–Peacock Hypothesis of growth in public expenditure. How do they explain rising expenditure in the developing countries?
2. What are the canons of public expenditure? Describe the factors responsible for rising public expenditure in India.
3. Explain the main causes of rapid growth of public expenditure in India. Is this growth justifiable?
4. Explain the effects of public expenditure on production and distribution in an economy.
5. How does public expenditure reduce inequality in income distribution?
6. What are the principles of public expenditure? What are the effects of public expenditure on production and distribution of income?
7. What is meant, by functional finance? Explain the importance of public expenditure as a tool of functional finance for developing countries.
8. Distinguish between the following types of public expenditure.

   (*i*) Transfer payments and Government expenditure on goods and services.

   (*ii*) Development expenditure and non-development expenditure.
9. What is negative income tax? How can it be used to reduce income inequalities in the economy?

CHAPTER 31

PUBLIC REVENUE: TAXATION AND BUDGET

Government Budget Constraint

Having studied the increasing importance of public expenditure in both the developed and developing countries and its effects on production and distribution, we now turn to study of financing Government expenditure. The Government has to finance its expenditure including interest payments on accumulated public debt by using taxation, borrowing from the market (i.e. sale of new bonds) and use of printed money. Increase in rates of income and other taxes not only adversely affects incentives to work more, save and invest more but also causes tax evasion. Further, as Laffer curve concept shows that increase in rate of income tax beyond a point causes revenue from taxes to decline. Thus there are limits to increasing revenue from taxes to finance its increased expenditure fully through normal taxes. This means the government faces a resource constraint resulting in budget deficit which is called these days as *fiscal deficit*. The fiscal deficit is financed either through borrowing from the market or getting new money printed by the Reserve Bank of India. Both borrowing from the market and printing new money to finance increased government expenditure, as shall be explained later, have limitations. It is therefore said that government faces a *budget constraint*. **The government budget constraint refers to the limit placed on the government expenditure by the extent it can raise resources through taxation, borrowing from the market (i.e., through sale of its bonds) and getting new money printed**. The government has to make a choice between the magnitudes of taxation, borrowing from the market and using new printed money to finance its expenditure. Government's budget constraint equation can be written as

$$G = T + \Delta B + \Delta M \qquad \ldots (i)$$

Where G stands for Government expenditure, T for tax revenue, ΔB for new borrowing (i.e. issue of new bonds) and ΔM for new money created.

Rearranging government budget constraint (*i*) we have

$$G - T = \Delta B + \Delta M \qquad \ldots (ii)$$

When G exceeds T, we have a budget deficit. Budget constraint equation (*ii*) says, that budget deficit must be financed either by new borrowing by the Government (ΔB) or by using printed money (ΔM). In this chapter we will explain taxation as a mean of financing Government expenditure and use of borrowing (*i.e.* debt financing) and printed money (*i.e.* money financing) will be explained in the next few chapters.

What is Tax?

A tax is a *compulsory payment* levied on the persons or companies to meet the expenditure incurred on conferring *common benefits* upon the people of a country. Two aspects of taxes follow from this definition : (1) A tax is a compulsory payment and no one can refuse to-pay it. (2) Proceeds from taxes are used for *common benefits* or *general purposes* of the State. In other words, there is no direct *quid pro quo* involved in the payment of a tax. This implies that an individual cannot expect or demand that the Government should render him a specific service in return for the tax paid by him. However, this does not imply that Government does nothing for the people from whom it receives taxes. In fact. Government spends the tax money for the general or common benefits of all the people rather than conferring any special benefit on a particular tax payer. To quote Taussig. "The essence of a tax, as distinguished from the other charges by Government is the absence of any direct *quid pro quo* between the tax payer and the public authority."[1]

Tax should be carefully distinguished from a *fee.* Fee is also compulsory payment made by a person who receives in return a particular benefit or service from the Government. For paying fee on a television or radio, a person gets the benefits of programmes relayed by the Government on television or radio. Likewise, students who pay the education fee in schools and colleges, obtain the benefits of teaching arranged by the Government. The amount of fee is always less than the cost of service rendered by the Government in return and therefore covers only a part of the cost of service rendered. Thus, even in case of fee, there is a general public interest or common benefit of the service rendered by the Government. In this case, the Government undertakes a service for the common benefits of the citizens and obtains a fee from those who avail of that service to cover a part of the cost of service rendered.

CLASSIFICATION OF TAXES

The taxes have been variously classified. Taxes can be direct or indirect, they can be progressive, proportional or regressive, and indirect taxes can be *specific* or *ad-valorem.* We spell out below the meanings of these different types of taxes.

Direct and Indirect Taxes. The distinction between direct and indirect taxes is based on whether or not the burden of a tax can be shifted wholly or partly to others. If a tax is such that its burden cannot be shifted to others and the person who pays it to the Government has also to bear it, it is called a direct tax. Income tax, annual wealth tax, capital gains tax are examples of direct taxes. In case of a direct tax there is a direct contact between the tax payer and tax levying public authority.

On the other hand, indirect taxes are those whose burden can be shifted to others so that those who pay these taxes to the Government do not bear the whole burden but pass it on wholly or partly to others. For instance, excise duty on the production of sugar is an indirect tax because the manufactures of sugar include the excise duty in the price and pass it on to buyers. Ultimately, it is the consumers on whom the incidence of excise duty on sugar falls as they will pay higher price for sugar than before the imposition of the tax. Thus, though excise duties are on the production of commodities but they can be shifted to the consumers. Likewise, sales tax on commodities can also be passed on to buyers or consumers in the form of higher prices charged for the commodities. Therefore, excise duties and sales taxes on commodities are examples of indirect taxes. They are also known as *commodity taxes.* In the case of indirect taxes, there is an indirect relation, between the Government and those who ultimately bear the burden of the taxes.

Specific and Ad Valorem Taxes. Indirect taxes can be either *specific* or *ad-valorem.* A specific tax on a commodity is a tax *per unit* of the commodity, whatever its price. Thus the amount of total specific tax will vary in accordance with the changes in total output or sales of the commodity and not with the total value of output or sales. On the other hand, an ad-valorem type of an indirect tax is levied according to the value of the commodity. For instance, sales tax in India is an ad-valorem tax as the rate of sales tax in case of several commodities is 10 per cent of the value of sales of the commodities. Ad-valorem taxes are progressive in their burden on consumers whereas specific taxes are regressive.

Progressive, Proportional and Regressive Taxes. According to another classification, taxes can be progressive, proportional or regressive. In case of proportional tax, the *same rate* of the tax is charged, whatever be the magnitude of the base on which it is levied. For instance, if rate of income tax is 25 per cent whatever the size of income of a person, it will then be a proportional income tax. Likewise, if rate of wealth tax is 5 per cent, it will be porportional wealth tax. Thus, in case of proportional tax it is the *rate* which is fixed and not the *absolute* amount of the tax. Thus with the rate of 25 per cent proportional income tax, a person with income of ₹ 25,000 will pay ₹ 6,250 as the tax,

1. Taussig, *Principles of Economics.*

and a person with income of 50,000 will pay ₹ 12,500 as the tax. Thus, even under proportional income tax, a richer person has to pay greater amount of tax though rate of the tax is the same.

On the other hand, in case of a progressive tax, rate of the tax increases as the amount of the tax base (income, wealth or any other object) increases. The principle underlying a pro-gressive tax is that greater the tax base, the higher the tax rate. In India income tax, an important direct tax levied by the Central Government, is progressive. Its rate at present (1998-99) varies from 10 per cent in the slab of ₹ 40,000 to 60,000 to 30 per cent in the slab of income above ₹ 1,50,000. Under progressive income tax, the richer person pays not only absolutely more tax but also a higher rate of the tax. Thus, the burden of progressive tax falls more heavily on the richer persons as compared to proportional income tax.

A regressive tax is the opposite of a progressive tax. In case of a regressive income tax, the rate is lowered as the income rises. Thus, under regressive tax system, the burden of the tax is relatively more on the poor than on the rich. A regressive tax is therefore inequitable and no civilised Government in the world today will levy such a tax.

PRINCIPLES OR CANONS OF A GOOD TAX SYSTEM

A good tax system must fulfil certain principles if it is to raise adequate revenue and fulfil certain social objectives. Adam Smith had explained four canons of taxation which he thought a good tax must fulfil. These four canons are: of (1) *Equality,* (2) *Certainty,* (3) *Convenience,* and (4) *Economy.* These are still regarded as characteristics of a good tax system. However, there has been significant developments in economic theory and policy since Adam Smith wrote his book *"The Wealth of Nations".* Activities and functions of Government have enormously increased. Now, the Governments are expected to maintain economic stability at full employment level, they are to reduce inequalities in the income distribution, and they are also to perform the functions of a Welfare State. Above all, they are to promote economic growth and devel-opment, especially in the developing countries, not only through encouraging private enterprise but to undertake the task of production in some strategic industries. Thus, in order to devise a good tax system, these objectives and functions of Government's economic policy must be kept in view.

It may be noted that Adam Smith was basically concerned with how the wealth of nations or, in other words, production capacity of the economy can be increased and he thought that private enterprise working on the basis of free market mechanism would ensure efficient use of resources and, if left unfettered would bring about rapid economic growth. His ideas about public finance were influenced by his economic philosophy of virtues of free private enterprise. In proposing the above mentioned canons of taxation, he was guided only by the sole objective that Government should be able to raise sufficient revenue to discharge its limited functions of providing for defence, maintaining law and order, and, public utility services. However, as men-tioned above, both the objectives and functions of modern Governments have increased neces-sitating large resources. Therefore, the modern economists have added other principles or characteristics which taxation system of a country must satisfy if the objectives of modern Governments are to be achieved. In what follows we shall spell out in detail the principles and characteristics of a good tax system starting with the explanation of Smith ian canons of taxation.

1. Principle or Canon of Equality

The first canon or principle of a good tax system emphasised by Adam Smith is of equality. According to the canon of equality, every person should pay to the Government according to his ability to pay, that is in proportion of the income or revenue he *et jove onder* the protection of the State. Thus under the tax system based on equality principle the richer persons in the society will pay more than the poor. On the basis of this canon of equality or ability to pay Adam Smith argued that taxes should be proportional to income, that is, everybody should pay the same rate or percentage of his income as tax. However, modem economists interpret equality or ability to pay differently from Adam Smith. Based on the assumption of diminishing marginal utility of money income, they argue that ability to pay principle calls for progressive income tax, that is, the rate of tax increases as income rises. Now, in most of the countries, progressive system of income and other direct taxes have been adopted to ensure equality in the tax system.

We shall discuss at length later in this chapter the principle of ability to pay. It may, however, be mentioned here that there are two aspects of ability to pay principle. First is the concept of *horizontal equity*. According to the concept of horizontal equity, those who are equal, that is, similarly situated persons ought to be treated equally. This implies that those who have same income should pay the same amount of tax and there should be no discrimination between them. Second is the concept of *vertical equity*. The concept of vertical equity is concerned with how people with different abilities to pay should be treated for the purposes of division of tax burden. In other words, what various tax rates should be levied on people with different levels of income, A good tax system must be such as will ensure the horizontal as well as vertical equity.

2. Canon of Certainty

Another important principle of a good tax system on which Adam Smith laid a good deal of stress is the canon of certainty. To quote Adam Smith, 'The tax which each individual is bound to pay ought to be certain and not arbitrary. The time of payment, the manner of payment, the quantity to be paid ought all. to be clear and plain to the contributor and to every other person."[2] A successful function of an economy requires that the people, especially business class, must be certain about the sum of tax that they have to pay on their income from work or investment. The tax system should be such that sum of tax should not be arbitararily fixed by the income tax authorities. While taking a decision about the amount of work effort that a person should put in or how much investment should he undertake under risky circumstances, he must know with certainty the definite amount of the tax payable by him on his income. If the sum of tax payable by him is subject to much discretion and arbitrariness of the tax as-sessment authority, this will weaken his incentive to work and invest more. Moreover, lack of certainty in the tax system, as pointed out by Smith, encourages corruption'in the tax admini-stration. Therefore in a good tax system, "individuals should be secure against unpredictable taxes levied on their wages or other incomes. The law should be clear and specific; tax collectors should have little discretion about how much to assess tax payers, for this is a very great power and subject to abuse."[3] In the opinion of the present author the Indian tax system violates this canon of certainty as under the Indian income tax law a lot of discretionary powers have been given to the income tax officers, which have been absued with impunity. As a result, there is a lot of harassment of the tax payers and corruption is rampant in the income tax department.

3. Canon of Convenience

According to the third canon of Adam Smith, the sum, time and/manner of payment of a tax should not only be certain but the time and manner of its payment should also be convenient to the contributor. If a land revenue is collected at the time of harvest, it will be convenient since at this time farmers reap their crop and obtain income. In recent years efforts have made to make the Indian income tax convenient to the tax payers by providing for its payments in instalments as advance payments at various times during the year. Further, income tax in India is levied on the basis of *income received* rather than *income accrued* during a year. This also makes the income tax system convenient, However, there is a lot of harassment of the tax payers as they are asked to come to the income tax office several times during a year for clarifications of their income tax returns.

4. Canon of Economy

The Government has to spend money on collecting taxes levied by it- Since collection costs of taxes add nothing to the national product, they should be minimized as far as possible. If the collection costs of a tax is more than the total revenue yielded by it, it is not worthwhile to levy it. More

2. Adam Smith, *The Wealth of Nations,* Vol. *11,* p. 327.

3. Otto Eckstein, *Public Finance.* Prentice Hall of India Private Ltd., 3rd Edition, p. 53.

complicated a tax system, more elaborate administrative machinary will be employed to collect it and consequently collection costs will be relatively larger. Therefore, even for achieving economy in the tax collection, the taxes should be as simple as possible and tax laws should not be subject to different interpretations.

Other Principles or Characteristics of a Good Tax System

We have explained above the Smithian canons of a good tax system. Adam Smith viewed the problem of devising a good tax system chiefly from the viewpoint of devising a good tax payers. Taxation system should also be such that it meets the requirements of increasing state activity and achieves the objectives the society has placed before it. We explain below the characteristics and principles of a good tax system, especially in the context of the developing countries.

1. Productivity or Fiscal Adequacy: An important principle of a good tax system for a developing country is that it should yield adequate amount of resources for the Government so that it should be able to perform its increasing welfare and developmental activities. If the tax system fails to yield enough resources, the Government will resort to deficit financing. An excessive dose of deficit financing is bound to raise prices which are harmful for the society. To make the tax system sufficiently productive it should be broad-based and both direct and indirect taxes find place in it. Moreover, taxes should be progressive so that the revenue from them rises with the increase in income of tr)e people.

2. Elasticity of Taxation: Another principle of taxation suitable for the developing coun* tries is the principle of elasticity of taxation. According to the concept of elasticity of the taxation system, as national income increases as a result of economic growth, the Government revenue from taxes should also increase. In developing countries, the share of tax revenue as a proportion of national income is low as compared to the developed countries. This share of tax revenue will rise as national income increases, if the tax system is sufficiently elastic. Pro-gressive taxation of income and wealth provides this elasticity to the tax system. Imopositions of higher indirect taxes on luxury goods having a high income elasticity of demand also makes the tax system elastic.

3. Diversity: A good tax system should follow the principle of diversity. This implies that there should not be a single or a few taxes from which Government seeks to raise large revenue. This is because if a Government tries to get large revenue from a single tax or few taxes, it will have to raise the rates of taxation too high which will not only adversely affect the incentives to work, save and invest but also encourage evasion of taxes. Therefore, the tax system should be a multiple tax system with a large variety of taxes so that all those who can contribute to the public revenue should be made to do so. This calls for a mix of various direct and indirect taxes. With the diverse tax system, the principles of fiscal adequacy and equity will also be better satisfied. Commending a diversity in the tax system Arthur Young writes, "If I were to define a good system of taxation, it should be that of bearing lightly on an infinite nurpber of points, heavily on none". Similarly, another expert of public finance writes, "Excessive reliance on any one base may produce adverse economic effects because the rates may become too high. Therefore, a tax system may do less economic damage if it raises moderate amounts from several bases rather than large amounts from one or two."[4]

4. Taxation as an Instrument of Economic Growth: In a developing economy such as ours, taxation should serve as an instrument of economic growth. Economic growth is primarily a function of rate of capital formation. If in the development strategy public sector has been assigned an eminent place, then capital formation in the public sector must occur at a relatively higher rate. This calls for mobilization of resources by the Government so as to finance capital formation in public sector. Therefore, a good tax system for a developing country will be such as will enable the Government to mobilise adequate resources for capital formation or economic growth. This it can do in the following two ways.

4. T. Mathew, *Tax Policy: Some Aspects of Theory and Indian Experience*, p. 31.

(*a*) **Mobilisation of Economic Surplus.** An important principle for a developing country is that it should mobilise economic surplus found in the economy. Economic surplus is the "surplus of national income over *essential consumption*". It is the task of taxation system that it should restrain non-essential or unproductive consumption through appropriate system of pro-gressive direct and indirect taxes and thereby mobilize economic surplus. "In an underdeveloped economy, there are particular economic sectors and classes of people where the economic surplus is generally found and which therefore should receive the special attention of tax authorities in such countries.[4]

(*b*) **Increase in the Intermental Saving Ratio.** A good tax system not only tries to mobilise the existing economic surplus but also seeks to raise it with a view to mop up relatively greatei amount of increase in national income for the purpose of capital formation. Thus taxation in *s* developing economy has not only to restrain current unproductive consumption but also to check the large increases in consumption when with the increase in national income, economic surplus goes up. This will ensure rise in the incremental or marginal saving ratio which is a prime determinant of continuous economic growth. In other words, through the means of taxation consumption should not be allowed to increase in proportion to increase in incomes. Expansion in economic surplus accruing to the individuals should be mobilised and invested in the public sector for further growth. Progressive income tax and indirect taxes on goods with higher income elasticity will ensure this,

5. Taxation as an Instrument for Improving Income Distribution: A good tax system for a developing economy should also serve as an instrument for reducing economic inequalities. The purpose of a good tax system for a developing economy is not merely to raise revenue for the Government but also to ensure that burden of taxes falls more on the rich. This requires that the rates of progressive direct taxes on income, wealth, expenditure, capital gains etc., must be sufficiently high. This objective of reducing income inequalities will be better served if a good part of the ,tax revenue is used for proverty alleviation programmes.

6. Taxation for Ensuring Economic Stability: A tax system must also ensure economic stability. Economic fluctuations have been a big problem in the developed countries and for reducing these fluctuations taxation can play a useful role. For this purpose, tax system must have built-in-flexibility. To have built-in-flexibility, the taxation system must be progressive in relation in the changes in national income. This will ensure that when national income rises, an increasing part of the rise in income should automatically accrue to the Government. On the other hand, when national income falls, as in a recession or depression, the revenue obtained progressive from taxes will fall more rapidly than the decline in national income. Built-in-flexi-bility attained through progressive taxation ensures that when incomes are increasing* during the period of boom or inflation, the relatively greater amount of tax revenue accruing to Government will moderate the increase in purchasing power with the people and aggregate demand and thus help in keeping prices under check. Likewise, under progressive taxation at times of depression or recession, tax revenue will fall faster than the income so that purchasing power of the people does not fall as fast as their pre-tax income. This will serve to check decline in economic activity.

However, in developing countries, the problem is more of restraining inflation so as to achieve price stability. By discouraging or restraining consumption, especially of non-essential or unproductive type, taxation can pay a useful role in controlling inflation in the developing countries.

Conclusion

From the foregoing analysis, we conclude that in the world today taxation is called upon to achieve several socio-economic objectives. It is not just a means of raising revenue for the limited functions of the State. *Neutrality principle* of taxation, that is, leave them as you find them, no longer finds favour with the modern economists. Tax system today has to play a more positive role. It is intended to bring about rapid economic growth, reduce inequalities of incomes, promote stability and to achieve other socio-economic objectives.

PRINCIPLE OF EQUITY IN TAXATION

An important question widely discussed in public finance is what kind of tax system is fair, just or equitable. As seen above, equity in taxation was the first cajion of taxation on which Adam Smith laid a good deal of stress. A fair tax system is not merely an issue in pure economic analysis but also in social philosophy. There are two prominent theories put forward to devise a fair or equitable tax system. They are (1) Benefits Received Theory and (2) Ability to pay Theory. We discuss below these two theories of equity in taxation.

Benefits Received Theory

According to this theory of taxation, citizens should be asked to pay taxes in proportion to the benefits they receive from the services rendered by the Government. This theory is based upon the assumption that there is an exchange relationship or *quid pro quo* between the tax payer and Government. The Government confers some benefits on the tax payers by performing various services or providing them what are called social goods. In exchange for these benefits individuals pay taxes to the Government. Further, according to this theory, equity or fairness in taxation demands that an individual should be asked to pay a tax in proportion to the benefits he receives from the services rendered by the Government.

However, there are some difficulties in application of this theory. The most Crucial problem faced by benefits received approach is that it is *difficult to measure the benefits* received by an individual from the services rendered by the Government. For instance, how much benefit an individual tax payer derives from providing for national defence and education, and maintaining law and order by the Government cannot be measured with any objective criterion. Secondly, most of the Government expenditure is incurred on common indivisible benefits so that the division of benefits of Government expenditure is not possible.

Further, benefits received theory miltilates against the very notion of a tax. A tax is defined as a payment for general purposes of the State and not in return for a specific service. The benefit theory can have meaning if the benefits of the Government services to the community as a whole is considered. But this will only indicate how much total tax revenue the community should pay to the Government. This will not help us in dividing the tax liability among various individuals comprising the community. It may be noted that most important common benefits are peaceful enjoyment of life, liberty and property. So far as life and liberty are concerned, Government's protection is the same for all. This will indicate levying of a toll tax. But toll tax has long been discarded as it was found to be highly regressive and also a small yielder of revenue.

The benefit principle is applicable only in cases where the beneficiaries can be clearly identified. Thus benefit principle is applied to the collection of road tax from vehicle owners. This is also applied when local bodies collect special levies for the services such as construction of sewers and roads they render to the people of their locality. The benefit principle is also applied to social security programmes for workers. Social security contributions, or what are called payroll taxes, which are collected from workers are kept in reserves out of which benefit payments are made to them. To conclude, therefore, "at best the benefit principle can provide a partial solution to the problem of fairness in taxation."[5]

Ability to Pay Theory

The ability to pay is another criterion of equity or fairness in taxation. This theory requires that individuals should be asked to pay taxes according to their ability to pay. The rich have greater ability to pay, therefore they should pay more tax to the Government than the poor. Essentially, the ability to pay approach to fairness in taxation requires that *burden of tax* falling on the various persons should be the same. In the discussion of various characteristics of a good tax system, we mentioned about the two concepts of equity, namely horizontal equity and vertical equity based on the principle of ability

5. Otto Eckstein, *op. cit.*, p. 54.

to pay. According to the concept of horizontal equity, equals should be treated equally, that is, persons with the same ability to pay should be made to bear the same amount of tax burden. According to the vertical equity, unequals should be treated unequally, that is, how the tax burden among poeple with different abilities to pay be divided. In both these cencepts of equity, what excatly do we mean by ability to pay and what are the objective measures of ability to pay are crucial. Some have explained the ability to pay treating it as a subjective concept. Others have treated the ability to pay in terms of some objective bases such as income, wealth, consumption expenditure etc. We shall explain below both these approaches to the measurement of ability to pay.

Ability to Pay: Subjective Approach

In the subjective approach to tax paying ability, the concept of *sacrifice* undergone by a person in paying a tax occupies a crucial place. In paying a tax, a person feels a pinch or suffers from some disutility. This pinch or disutility felt by a tax payer is the sacrifice made by him. In this subjective approach to ability to pay, tax burden is measured in terms of sacrifice of utility made by the tax payers. The following three principles of sacrifice have been put forward by various authors:

1. The Principle of Equal Absolute Sacrifice;
2. The Principle of Equal Proportional Sacrifice; and
3. The Principle of Equal Marginal Sacrifice (or Minimum Aggregate Sacrifice).

The principle of equal absolute sacrifice implies that the tax burden in terms of utility sacrificed should be the same for all tax payers. If U stands for total utility, Y stands for income and T for the amount of tax paid, then the principle of equal absolute sacrifice requires that $U(Y) - U(Y - T)$ should be the same for all individuals. The term $U(Y)$ implies that total utility of a given income Y and $U(Y - T)$ implies the total utility of the post-tax income $(Y - T)$. If the equal absolute sacrifice principle is applied, none will be exempted from taxation and every body will pay same amount of the tax. Now, the pertient question is what type of tax, propor-tional or progressive, follows from this principle. If marginal utility of money income falls, as is generally believed, and if this fall in marginal utility of money income equals the rate of increase in income, then this principle will suggests proportional income tax. However, if the fall in marginal utility of income is greater than the rate of increase in income, then equal absolute sacrifice principle will suggest progressive income tax.

Equal Proportional Sacrifice. This principle requires that every person should be made to pay so much tax that the sacrifice of utility as a proportion of his income is the same for all tax payers. In terms of the notation used above, this implies that $\frac{U(Y) - U(Y - T)}{U(Y)}$ of all tax payers should be equal. If a person enjoying higher income is to bear same proportion of sacrifice, then given the falling marginal utility of income, he will have to pay income tax at a higher rate. This means progressive income tax.

Equal Marginal Sacrifice. According to this principle, tax burden should be so apportioned among various individuals that marginal sacrifice of utility of each person paying the tax should be the same. This approach seeks to minimize the aggregate sacrifice of the society as a whole. When all persons pay so much tax that their marginal sacrifice of utility is the same, the loss of total utility by the society will be minimum. Thus, the principle of equal marginal sacrifice looks at the problem of dividing the tax burden from the point of view of welfare of the whole society. The social philosophy underlying this principle is that the total sacrifice imposed by taxation on the community ought to be minimum. Assuming that marginal utility of income falls, the principle of equality of marginal sacrifice implies very high marginal rates of taxation. Indeed, in the extreme this principle can be used to recommend 100 per cent rate of tax on the people in highest income bracket in the society. Thus this principle recommends a highly progressive tax structure. This principle of taxation has been recommended among others by Edgeworth, Pigou and Musgrave who consider this as the ultimate principle of taxation. It is worthwhile to quote Edgeworth, a chief exponent of this principle. "The minimum sacrifice is the sovereign principle of taxation. If one is utilitarian and believes not only in the measurability of utility but also in the view that the law of diminishing utility is applicable to money also, then this principle would involve a high level of minimum exemption and a very steep pro-gression as income increases .... The less the aggregate sacrifice, the better the distribution of tax ourden in the community. The State exists to maximize human welfare. This it will be able to do by minimizing the sacrifices involved."

Comments. The whole subjective approach to ability to pay based on the sacrifice of utility has been termed as invalid because utility being a subjective entity cannot be measured in cardinal sense. Further, it is alleged there is no definite evidence that marginal utility of money income falls as income increases. Interpersonal comparison of utility which the sacrifice approach requires is held to be unscientific. However, in the view of the present author, these objections against ability to pay or sacrifice principle are not valid. We may not be able to measure utility of money income in exact absolute terms but a sufficiently good measure of utility of income can be obtained and this is enough for the application of this principle of ability to pay in terms of sacrifice. Observations in the world clearly indicate that people in the lower income brackets spend most of their incomes on buying necessaries, while people in the higher income groups spend a relatively greater proportion of their incomes on luxuries and non-essential goods. In view of this it is quite valid to assume diminishing marginal utility of money for purpose of taxation.

Ability to Pay: Objective Approach

The objective approach to the ability to pay principle considers what should be objective base of taxation which measures ability to pay correctly. There is even no agreement on this question also. However, income is generally considered to be the best measure of ability to pay. This is because a person's income determines a person's command over resources during a period to consume or to add to his wealth. However, it may be noted that ability to pay does not increase in direct proportion to money income. Ability to pay increases more than propor-tionately to the amount of income. Justification of progressive income tax is based on this. Further, in order to ensure equity in taxing income distinction ought to be made between earned income and unearned income, and considerations should also be given for a number of depend-ents on the person paying the tax.

Wealth of a person is another objective measure of ability to pay that has been suggested as a tax base. The ownership of the property or wealth of an individual determines how much resources he has accumulated. Saving from every year's income adds to his wealth. The wealth or property is therefore said to be a better index of taxable capacity.

It may however be noted that wealth alone is not considered to be adequate measure of taxable capacity and instead a combination of income and wealth taxes is regarded as a better measure from the viewpoint of ability to pay. Thus, according to Prof. Kaldor, *"Only a combination of income and property taxes can give approximation to taxation in accordance with ability to pay."*[6] Arguing the case for levying annual wealth tax in India, he writes "income taken by itself is an inadequate yardstick of taxable capacity as between income from work and income from property, and also as between different property owners". He further writes, "the ownership of property in the form of disposable assets endows the property-owner with a taxable capacity as such quite apart from the money income which that property yields."

It follows form above that a combination of income tax and wealth tax will be a better measure of ability to pay.

Prof. Kaldor has also advocated another base for taxation. He has been a strong advocate of *levying of expenditure tax* in both the developed and developing countries.[7] It should be noted that

6. See Nicholas Kaldor, *An Expenditure Tax,* Alien and Twin, 1955.
7. *Ibid.*

his expenditure tax is in fact a tax on consumption, that is, income minus saving. He claims that, IY *is consumption that is a fair or equitable base of taxation.* According to him, it is consumption that measures the resources that a person actually withdraws from the economy for his personal use. The part of his income not consumed, that is, savings lead to the increase in the capital stock and thus add to society's productive capacity. If a person consumes more than his income, he should the made to pay a higher tax because he reduces the capital stock of the country.[8] *The imposition of expenditure tax, according to Kaldor, is particularly relevant for developing countries where high consumption expenditure of richer classes reduces the rate of capital accumulation.* Imposition of expenditure tax will discourage consumption by taxing it heavily and promote savings *by* exempting it. In 1958 on the recommendation of Kaldor expenditure tax was levied in India. But after some years it was withdrawn on the ground that it was difficult to administer it and also that the revenue from it was very small.

Conclusion. In conclusion it may be said that it is *better to levy taxes on various bases instead of a single base.* As explained in the section of characteristics of a good tax system, *diversity of tax bases is preferable not only from the viewpoint of a measure of ability to pay but also because it will have less adverse economic effects.* Eckstein rightly points out, "Ex-cessive reliance on any one base may produce adverse economic effects because the rates may become too high. Therefore, a tax system may do less economic damage if it raises moderate amounts from several bases rather than larger amounts from one or two."[9] Perhaps it is because of these reasons that in actual practice, a good variety of taxes are levied. In India, income tax, corporation tax, wealth tax and union excise duties are the most important taxes levied by the central Government, while sales tax, land revenue, and certain excise duties are the important taxes at the state level.

DIRECT TAXES VS. INDIRECT TAXES

We have explained above the characteristics of a good tax system. A controversial issue in public finance is concerned with whether in tax structure of an economy direct or indirect taxes should be preferred. Indeed, both direct taxes and indirect taxes have their merits and demerits and therefore a good tax system should contain a proper mix of these two types of taxes. Direct taxes, it may be recalled, are those which are levied directly on the indi-viduals and firms and their burden is borne by those on whom they are levied. Persona! income tax, corporate income tax, wealth tax, expenditure tax, gift tax are some examples of direct taxes. On the other hand, indirect taxes are levied on production and sale of commodities and services. A small or a large part of the burden of an indirect tax is passed on to the consumers. Excise duties on the production of commodities, sales tax, service tax, tax on railway or bus fare are some examples of the indirect tax. The Indian tax system contains a combination of both direct and indirect taxes. We discuss below the merits and demerits of both direct and indirect taxes. It should be noted that tax has to be judged especially from the following viewpoints:

1. Equity;
2. Efficiency in allocation of resources;
3. Economic growth, that is, whether it promotes or discourages supply of work effort (labour), saving and investment on which economic growth depends;
4. Buoyancy, that is, to what extent revenue from tax increases with the increase in income; and
5. Economic stability.

Merits of Direct Taxes. 1. An important merit of direct taxes is their rate structure be *so*

---

8. Nicholas Kaldor, *Indian Tax Reforms.*
9. Otto, Eckstein, *op. cit.,* p. 56.

designed which is closely related to the ability jo pay principle. The direct taxes can be made progressive and the degree of progression among different income brackets can be deter-mined so as to satisfy the principle of ability to pay. The larger burden of these progressive taxes fall on the rich people who have capacity to bear them and the poor people with less ability to pay have to bear less burden.

2. Direct taxes are *important instrument of reducing inequalities of income and wealth.* A progressive income tax, imoposition of progressive wealth tax and estate duties have been used in different countries to reduce inequalities in income distribution. Large inequalities are harmful for maximization of social welfare. They are even bad from the viewpoint of achieving economic stability.

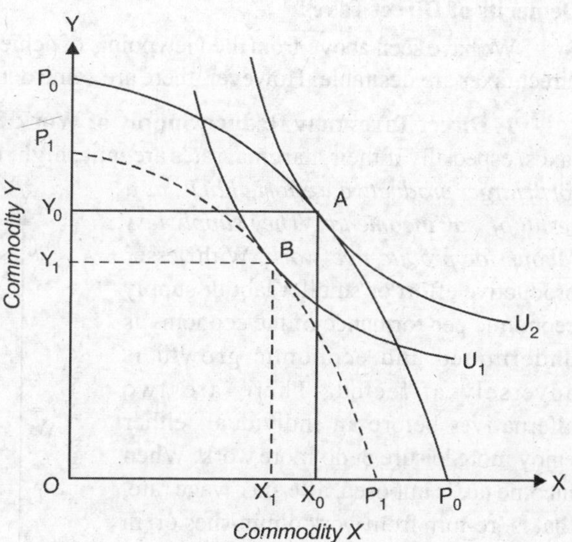

Fig. 31.1. *Direct Taxes do not Cause Inefficiency in Resource Allocation*

3. Another important merit of direct taxes is that unlike indirect taxes *they do not cause distortion in the allocation of resources.* As a result, they leave the consumers better off as compared to indirect taxes. Direct taxes such as general income and wealth taxes do not affect relative prices of goods and services and therefore leave the relative profitability of different industries unchanged. As a result, resource allocation among them remain unaffected. Therefore, from the viewpoint of efficiency in resource allocation direct taxes are preferred to indirect taxes.

That the direct taxes have no effect on allocation of resources except to withdraw resources from private use is illustrated in Fig. 31.1. Suppose there are two commodities X and Y. A production possibility curve P_0P_0 of a community is drawn. U_1, U_2 are the indifference curves of the community showing successively higher levels of welfare. In equilibrium and without a direct tax, the economy is producing X_0 of commodity X and Y_0 of commodity Y. Allocation of resources is efficient as by producing X_0 of good X and Y_0 of good Y, the community is reaching the highest possible indifference curve U_2. With this social welfare is maximum because marginal rate of transformation in production is equal to marginal rate of substitution of the community between the two commodities

Now suppose a certain amount of revenue is collected by leying a direct tax, say personal income tax. Since through this tax private purchasing power has been cut, some resources will be freed for Government use. The new production possibility curve for private economy shifts downward to P_1P_1. Now, the private economy will be in equilibrium at point B, where P_1P_1 is tangent to the community indifference curve U_1 and X_1 of good X and Y_1 of good Y are being produced. At point B, marginal rate of substitution between X and Y of the community (MRS_{xy}) equals marginal rate of transformation (MRT_{xy}) in production between the two goods. Thus, resource allocation between the two goods remain efficient. Of course, resources for the use of private sector have been reduced.

4. Fourthly, *revenue elasticity of direct taxes, especially if they are of progressive type is quite high.* As the national income increases, the revenue from these taxes also rises a great deal. It is due to this fact that progressive direct taxes serve as *automatic stabilizers.* When during upswing the economy gains momentum and boom conditions emerge, a relatively larger fraction of increasing income are siphoned of through the progressive direct taxes. This acts as a brake on the economy

getting into serious inflationary situation. Similarly, when during downswing the income and employment of the community is falling due to the progressive nature of the direct taxes, a relatively smaller part of income is taken away by way of these taxes. Therefore, the consumption demand does not fall to the same extent as the decrease in income. This prevents further slowdown of the economy which turns around and thereby recovery of the economy starts.

Demerits of Direct Taxes

We have seen above from the viewpoints of achieving equity, efficiency, economic stability that direct taxes are desirable. However, there are some demerits of direct taxes which we discuss below:

1. Direct Taxes may Reduce Supply of Work Effort. An important harmful effect of* direct taxes, especially if their marginal rates are quite high, is that they *reduce people's will-ingness to put forth larger productive y effort, that is, as a result of, say income tax, They supply less labour for produc-tive work.* With lesser productive effort or smaller labour supply economic per-formance of the economy is undermined and economic growth is adversely af-fected. There are two alternatives before an individual: either enjoy more leisure or do more work. When income tax is imposed, after-tax wage rate, that is, re-turn from work diminishes or, in other words, opportunity cost of leisure falls. This induces the individual to substitute leisure for work effort. As a result, sup-ply of work effort is reduced. To quote Eckstein, "If a worker has to pay taxes, say one half of his income from extra work, perhaps a lawyer working week-ends, a doctor taking on additional patients or a sales-person making evening calls—his incentives will certainly be reduced." He further writes about direct taxes, "if their high rates are diminishing effort substantially through-out the economy, we are giving up a lot of economic performance for the sake of equity."[10]

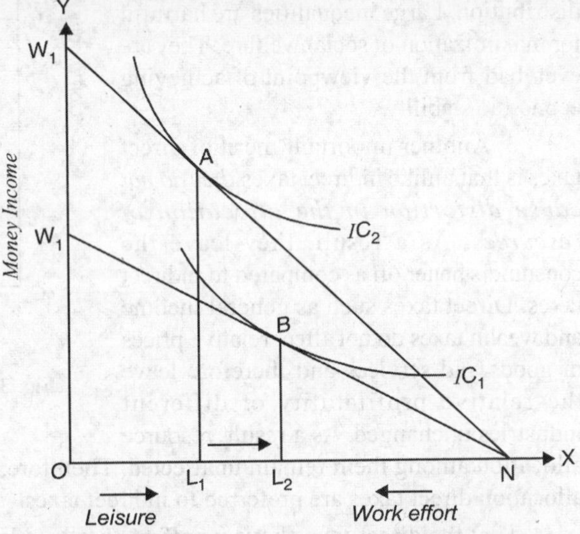

Fig. 31.2. *Effect of Income Tax on Supply of Work Effort*

Let us illustrate the harmful effect of a direct tax, say income tax, on supply of work effort (*i.e.*, labour). In Fig. 31.2 we have drawn the indifference curves of an individual between income and leisure. Suppose before taxation, the wage rate is depicted by the slope of the wage line NW_1. The individual is in equilibrium at point A on indifference curve IC_2 where he is enjoying OL_1 leisure and working NL_1 hours.

Suppose a proportional income tax at the rate of 50% is levied. As a result, after-tax wage rate is reduced and after-tax wage line shifts below to NW_2. The individual moves to equilibrium point B on indifference curve IC_1. In this equilibrium situation B the individual has increased his leisure to OL_2 and reduced hi- work effort to NL_2. Thus, as a result of levying of income tax and consequently lowering of the wage rate the individual has substituted leisure for work, resulting in the decrease of labour supply (*i.e.*, work effort). This is a normal behaviour. Of course, there are some exceptions to this normal behaviour.

10. Otto, Eckstein, *Public Finance,* Prentice Hall.

2. Direct Taxes Reduce Incentives to Save and Invest. A significant adverse effect of direct taxes is that *they reduce incentives to save and invest.* People save and invest their savings by sacrificing present consumption. Investment of savings yields a return. Here the *choice is between present consumption and saving for investment.* A direct tax, say income tax, reduces the rate of return on investment since a part of the return is taken away through tax. Thus, with the imposition of income tax, rate of return on investment is reduced which induces the individuals to substitute present consumption for saving and investment. As a consquence, *levying of income tax induces more consumption and discourages saving and investment.* The lower saving and investment retards economic growth. Empirical studies made in India and USA have shown that high rates of direct taxes have adversely affected savings. It is for this reason, budgets in India and the industrialised countries have provided for special tax treatment of some particular forms of saving and investment.

3. Corporate Income Tax Adversely Affects Investment. It is corporate income tax and its effect on investment and growth that has been widely discussed. Corporations, that is, joint stock companies finance their investment requirements by using their internal savings. It is these internal savings that is the chief source of capital accumulation by the companies. Imposition of corporate income tax reduces their investible funds and thereby ratards investment and growth. To overcome this demerit of corporate income tax, provisions for exempting depreciation funds from income tax and also for providing investment credit have been generally made in the budgets.

We have thus seen above there are both merits and demerits of direct taxes. In order to have minimal harmful effects of direct taxes some exemptions, special tax treatment of depre-ciation, saving and investment have been provided in income tax and wealth tax laws.

Indirect Taxes

Indirect taxes such as excise duties and sales tax are important source of revenue" for the Government. The revenue from indirect taxes has shown a great deal of buoyancy. In India revenue from indirect taxes has been continuously increasing. In Table 31.1 given below we give the tax revenue raised from direct and indirect taxes as percentage of gross tax revenue of the Central Government. It will be seen from the table that in 1995-96 indirect taxes, (customs and union excise duties) accounted for about 70 per cent of tax revenue of the Central Government while personal income tax and corporation tax accounted for about 28 per cent of total tax revenue. However, as a result of reduction in customs duties and excise duties by the Central Government in recent years, the proportion of indirect taxes to the total Central revenue has fallen to about 44.1 per cent and correspondingly the proportion of direct taxes has risen to 54.9 per cent in the fiscal year in 2011-12. The rise in proportion of direct taxes implies that tax-structure has become more progressive. Let us now explain the merits and demerits of indirect taxes.

Table 31.1. Tax Revenue as Percentage of Gross Tax Revenue of Central Government in India

	1994-95	1995-96		2007-08		2011-12	
Personal Income Tax	13.0	13.0	28.0	17.5	49.9	18.6	54.9
Corporation	15.0	15.0		32.3		36.3	
Customs	29.0	32.0	69.2	17.6	47.0	16.8	44.1
Excise Duties	40.5	37.2		20.8		16.3	
Service tax	–	–		8.6		11.0	

Source: GOI, *Economic Survey,* 2007-08, 2011-12 and 2012-13

Merits of Indirect Taxes

1. Convenient to Pay and Less Cumbersome. A significant merit of indirect taxes is *that they are convenient to pay and less cumbersome as compared to direct taxes.* In case of indirect taxes such as an excise duty or a costum duty people have to pay these taxes *bit by bit* when they produce, sell and buy these commodities. This piecemeal payment of an indirect tax makes it easier for the tax payers who do not feel its pinch in contrast to direct taxes. Indirect taxes are also not difficult to collect as they are collected from the firms and traders who pass them on to the buyers in the form "of higher prices of goods.

2. Large Revenue Potential. Another important merit of indirect taxes is that *they have a large revenue potential* for the Government. Through indirect taxes even the poor and low-income people can be brought in the tax net. Whoever, rich or poor, buys commodities ultimately bears the burden of indirect taxes. But direct taxes such as income tax cannot be levied on the poor as cost of collection from them will be much higher and also there is need for some exemption limit. For example in India with a population of over 900 million only about 8 million people pay income tax. On the other hand, a much large number of people pay indirect taxes such as excise duties, sales tax, tax on railway and bus fare. Thus through indirect taxes, *a large amount of resources can be raised* for meeting increasing requirements of the Government.

A large revenue potential of indirect taxes has a special importance for the developing countries like India where resources are needed by the Government not only to run the general administration and provide for defence but also for *investment and economic growth.* For ac-celerating capital formation and economic growth, the developing countries cannot rely on reve-nue from direct taxes alone. Revenue from indirect taxes can be increased substantially by bringing more commodities under their purview and increasing their rates. It has been found that revenue from indirect taxes show a great deal of *buoyancy,* that is, revenue from them increases very much with the growth of national income.

3. Important Instrument for Influencing Pattern of Production and Investment. An-other merit of indirect taxes is that it serves as an important tool for influencing pattern of production and investment in an economy. The working of market mechanism, left to itself, does not always ensure desirable or optimum pattern of productions and investment allocation. For example, left to itself market mechanism may lead to greater investment in producing,harm-ful goods such as alcohol, cigarettes, luxury goods such as airconditioners, washing machines, luxury cars. By imposing heavy excise and sales taxes, the production and investment in these industries can be discouraged. Similarly, indirect tax structure can be so designed as to encourage employment generation and the use of appropriate technology. For example, in India heavy excise duties have been levied on mill-made cloth while handloom cloth has been exempted from them. This has been done to promote the production of handloom cloth, a labour-intensive product, to promote employment generation. The point which is being emphasised is that *indirect faxes can be selective commoditywise and ratewise* so as to affect the prices of different com-modities to promote higher priority industries and discourage low priority ones.

Demerits of Indirect Taxes

Indirect taxes have been criticized for some of their shortcomings which we explain below

1. Indirect Taxes are Regressive. An important criticism of indirect taxes is that they are *regressive,* that is, people with lower incomes bear a higher burden as compared to the rich people. Rich or poor has to pay the same rate of excise duty or sales tax levied on commodities irrespective of their ability to pay. Thus, the fact that same rate of an excise duty on sugar, mill-made cloth etc., has to be paid both by the poor and rich, violates the principle of equity as poor people have lower ability to pay than the rich. However, the regressive nature of indirect taxes has been sought to be removed by levying higher taxes on luxuries which only rich people buy.

2. Indirect Taxes are Inflationary. Another important ground on which indirect taxes have been opposed is that they are *inflationary,* that is, they cause a rise in prices of commodities. In India before 1991 in order to raise more revenue rates of indirect taxes were often raised by Finance Ministers.

Even essential commodities such as sugar, cloth, petroleum products were subjected to heavy excise duties and sales taxes. These taxes were wholly or partly added to the prices of goods which caused *cost-push inflation.*

In this connection it is to important to mention the *cascading effect* of indirect taxes which add to their inflationary potential. Cascading effect refers to the effect of indirect taxes on the prices of inputs of the industrial commodities. Higher prices of inputs as a result i5f levying indirect taxes on them lead to the rise in prices of industrial products in whose production they are used as inputs. In this way through cascading effect much higher rise in price of industrial goods occur when *their several inputs* are subjected to indirect taxes.

3. Excess Burden of Indirect Taxes or Inefficiency in Resource Allocation. The most important harmful effect of indirect taxes is that they make the allocation of resources inefficient and cause *excess burden* on the consumers. As explained above, resource allocation is efficient when resources are so allocated to the production of goods as maximises social welfare. This is achieved when the pattern of production of goods is such that marginal rate of transformation (MRT_{xy}) between goods in production equals marginal rate of substitution between them of' consumers (MRS_{xy}). Indirect taxes distort the pattern of production or resource allocation from this ideal one and thereby cause inefficiency and loss of consumer welfare. The prices of com-modities on which indirect taxes are levied rise causing reduction in their quantities demanded and as a result less resources are devoted toDtheir production. Resources so released are used for the production of those goods which are not covered by indirect taxes. Thus indirect taxes cause distortion in the prices of goods and thereby reallocate resources away from the efficient or optimum state. Thus, consumer welfare decreases more as a result of indirect taxes than in case of direct taxes.

Graphic Illustration of Inefficient Resource Allocation of Indirect Taxes

Consider Fig. 31.3 where *AB* is the production possibility curve between two commodities *X* and *Y* before taxation. IC_1, IC_2, IC_3 are social indifference curves. Equilibrium is reached at point *H* where production possibility curve is tangent to social indifference curve IC_3 and the quantities of the two goods produced are indicated by the point *H.* At point *H,* the community is at the highest possible indifference curve and therefore social welfare is maximum. It should be noted that at the tangency point *H,* MRT_{xy} in production equals MRS_{xy} of consumers.

Now, suppose excise duty is levied on commodity *X,* while commodity *Y* remains untaxed. As a result, price of *X* rises, while price of *Y* remains unchanged. With this, the price line shifts to PL_2. Further, as a result of withdrawal of resources by the Government through indirect tax on *X,* the production possibility curve for private use shifts to the left to *CD.* Now, the equilibrium of the private economy will be at point *K* of the production possibility curve *CD* where the

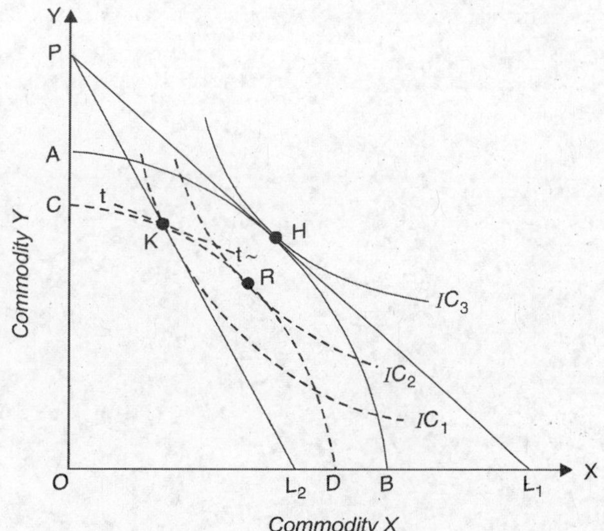

Fig. 31.3. *An Indirect Tax Causes Inefficiency in Resource Allocation.*

new price line PL_2 is tangent to the indifference curve IC_1. Thus, consumers suffer a loss of welfare as they have been shifted from a higher indifference IC_3 to the lower indifference curve IC_1 as a result of imposition of indirect tax on commodity X. Resource allocation at K is inefficient because slope of the price PL_2 (which is tangent to indifference curve IC_1) showing MRS_{xy} is greater than MRT_{xy} as measured by the slope of the tangent tt'. If a direct tax (say income tax) of *equal yield* is levied by the Government shifting the private production possibility curve to CD, the society will reach at point R on indifference curve IC_2. The social indifference curve IC_2 is higher than IC_1 which was reached by levying indirect tax on X. Thus indirect tax causes inefficient allocation of resources and reduces social welfare more than a direct tax. That is, an indirect tax causes an *excess burden* on consumers by distorting prices.

QUESTIONS FOR REVIEW

1. Describe the characteristics of a good tax system. How do you rate the Indian tax system in this context ?
2. Discuss the merits and demerits of direct and indirect taxes. Do you think indirect taxes and regressive ?
3. Distinguish between direct tax and indirect tax. Explain the role of indirect tax in the Indian economy.
4. Distinguish between progressive and proportional taxation. What are the effects of a progressive tax system ?
5. Explain the various canons of taxation. Discuss the principle of equity in taxation in a country like India.
6. Discuss the various objectives of taxation policy in India.
7. Explain the objectives and role of fiscal policy in a developing economy.
8. Discuss the role of fiscal policy in an economy. Do you think its role is expanding in the real world ?
9. Explain the main features of the Indian tax structure and examine the role of indirect taxes in a developing economy like India.

CHAPTER 32
INCIDENCE OF TAXATION

Meaning of Incidence

It is important to study who ultimately bears the burden of a tax. The incidence of taxation refers to this question of who and in what proportion bears the *final* burden *of* a tax. It is not necessary that a person or a firm who pays a tax to the Government or, in other words, bears the initial burden of a tax will also be one on whom the final burden of the tax rests. This is because a tax can be shifted or transferred to others. Therefore, in economics, we distinguish between the *impact* and *incidence* of a tax. Whereas the *impact* of a tax is said to be resting on the person or firm who pays the amount of the tax and thus receives the *initial burden,* the *incidence* of the tax rests on the person or firm who *ultimately bears the money burden* of the tax. If a person or a firm who pays the tax to the Government is also one who ultimately bears it, then the impact and incidence of the tax rests on the same person or firm. In such a case, there is no shift.

Shifting the Burden of a Tax

The burden of the tax can be transferred to others through a process of shifting. It may be noted that the whole burden of the tax may not be shifted to others. It may be that a part of the tax may be shifted to others and a part be borne by the one who initially pays the tax. As a matter of fact, a part of the tax burden rests on all the persons to a larger or smaller degree in the chain of transferring the burden so that at the ultimate end only a small burden rests. The process of shifting the burden of a tax goes on so long as different persons who come in the chain are able to pass on the burden to others till it ultimately rests on a person or a group of persons who cannot shift this unwelcome baby further. Theory of incidence of tax studies in what proportion the burden or incidence of a tax is shared among different persons. It may be noted that a tax can be shifted through a *process of exchange* or, in other words, an individual or a firm can shift the burden of the tax if there occurs exchange relations which are conducted on the basis of prices of goods and factors. The person who initially pays the tax can pass it on to the other either in the form of higher prices of goods he sells or in the form of lower prices of factors he buys.

Whether shifting can take place or if it does so how much tax burden can be shifted depends on a number of factors. They are briefly explained below:

1. The Nature of a Tax. The nature of a tax as to whether it is a tax on the production or sale of some commodities or it is a personal income or property tax. Tax shifting can easily take place in the case of taxes on the production and sale of commodities. The taxes on production or sale of commodities are called indirect taxes. The important example of indirect taxes are excise duties and sales tax. On the other hand, the burden of direct taxes such as income and wealth taxes cannot be shifted.

2. Market Conditions. Whether commodity is being produced under conditions of perfect competition, monopolistic competition or monopoly goes to determine the extent to which the burden of the tax can be shifted. A monopolist who has a full control over the supply of a commodity is in a better position to shift the burden of a tax on the commodity produced. Likewise, a producer working under monopolistic competition who produced a product somewhat different from others exercises a good deal of influence over the price of its product and therefore can pass on a part of the burden of the tax to the buyers. Even the firms working under perfect competition can shift the tax burden as the tax levied on a commodity raises its supply price for all of them. The difference in the three market forms lies in the extent to which the burden of the tax can be shifted.

3. Physical Conditions of Production. The shifting of the tax burden on a commodity also depends upon whether the commodity is being produced under increasing, constant or diminishing returns. This will be explained in detail a little later.

Factors Determining Incidence of Indirect (Commodity) Taxes

The questions of tax shifting especially arises in the case of indirect taxes, that is, taxes on the production and sale of goods such as excise duties and sales tax. In this regard, *whether and to what extent a tax on commodity can be shifted depends on the price elasticity of demand for and supply of a commodity.*

It is these elasticities of demand and supply that determine the bargaining strengths of the sellers and buyers of the taxed commodity. Sellers can shift the tax burden to the buyers if they are able to reduce the supply of the commodity and thereby raise its price. Thus, the power to shift the tax depends on the elasticity of supply of the taxed commodity. The elasticity of reducing supply of a commodity will be relatively smaller if there is excess capacity in the industry producing it. Further, the elasticity of supply of a commodity will'be larger in the long run than in the short run.

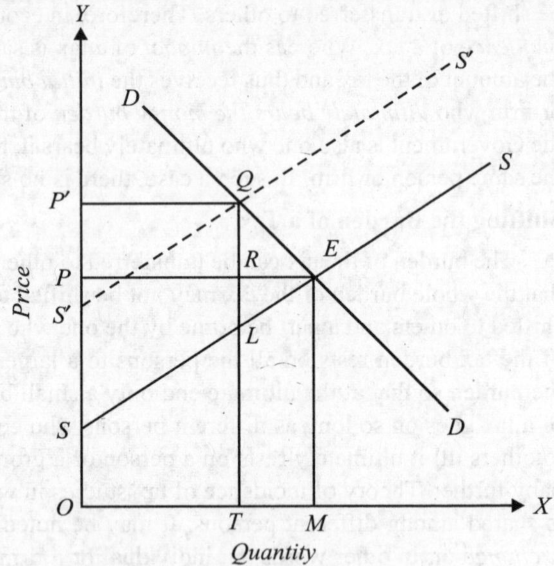

Fig. 32.1. *Incidence of an Indirect Tax*

Apart from the elasticity of supply, power to transfer the tax burden depends on the elasticity of demand for a commodity. The greater the elasticity of demand of the buyers, the smaller the extent to which the tax will be shifted to them. As shall be shown below, under conditions of perfect competition the incidence of a commodity tax is shared between the sellers and buyers in the ratio of the elasticity of demand and supply. Fig. 32.1. illustrates the incidence of an indirect tax under conditions of perfect competition.

DD is the demand curve for a commodity and *SS* is its supply curve before the imposition of tax on it. Interaction of these demand and supply curves determines price *OP* of the commodity and *OM* is the quantity sold and purchased. Suppose a umt sales tax, that is, sales tax per unit of the commodity equal to *SS'* or *LQ* is levied by the Government on the commodity in question. This will raise the supply price of the commodity by the sellers as the unit tax *SS'* will now be included by the sellers in their supply price. As a result the supply curve of the commodity will shift to the left by the magnitude of the tax *SS'*. The new supply curve *S'S'* intersects the demand curve *DD* at the point *Q* and determines the new price *OP'* and quantity exchanged *OT.* It will be seen from Fig. 32.1 that price for the buyers has risen by *PP'* or *RQ* whereas the tax per unit is *SS'* or *LQ* which the buyers will bear. It may be noted that the buyers will bear the burden of a tax to the extent that they have to pay the higher price than before. Thus the incidence of the tax borne by the buyers will be equal to *RQ.* The remaining part of the tax *RL* will be borne by the sellers. Thus, of the tax *SS'* or *LQ* per unit, *RQ* is incidence of the tax on the buyers and *RL* is the incidence on the sellers. Now, we can show that the incidence of the tax *RL* and *RQ* on the sellers and buyers respectively is equal to the ratio of the elasticity of demand and the elasticity of supply.

$$\frac{RL}{RQ} = \frac{\text{Incidence of tax on the sellers}}{\text{Incidence of tax on the buyers}}$$

$$\text{Price elasticity of demand} = \frac{\Delta q}{\Delta q} \cdot \frac{p}{q}$$

By referring to Fig. 32.1, elasticity of demand when the price rises from OP to OP' for the buyers will be

$$e_d = \frac{\Delta q_d}{\Delta p} \cdot \frac{p}{q} = \frac{MT}{RQ} \cdot \frac{OP}{OM} \quad ...(i)$$

Likewise, elasticity of supply when the price received by the sellers falls from OP or ME to TL and the quantity sold declines from OM to OT can be found out. Thus:

$$e_s = \frac{\Delta q_s}{\Delta p} \cdot \frac{p}{q} = \frac{MT}{RL} \cdot \frac{OP}{OM} \quad ...(ii)$$

$$\frac{e_d}{e_s} = \frac{\frac{MT}{RQ} \cdot \frac{OP}{OM}}{\frac{MT}{RL} \cdot \frac{OP}{OM}} = \frac{RL}{RQ}$$

Since RL is the incidence of the tax on the sellers and RQ is the incidence of tax on the buyers, we find that

$$\frac{RL}{RQ} = \frac{\text{Incidence on sellers}}{\text{Incidence on buyers}}$$

$$= \frac{\text{Price elasticity of demand}}{\text{Price elasticity of supply}} \quad ...(iii)$$

To conclude, to what extent the burden of the tax will be shifted and the proportion in which the buyers will share the incidence of a commodity tax depends on the elasticities of demand and supply. What exactly incidence of tax on buyers RQ and incidence of tax on sellers RL will be equal to is given below:

$$\frac{ed}{es} = \frac{RL}{RQ} = \frac{\text{Incidence of tax on sellers}}{\text{Incidence of tax on buyers}} \quad ...(iv)$$

If elasticity of demand for a commodity is zero, then from (*iv*) it follows that the incidence of tax on the sellers RL will be zero and from (*iv*) above it follows that incidence of tax RQ on buyers will be equal to whole tax t per unit. On the other hand, if the elasticity of supply is zero, then it follows from (*iv*) above that the incidence of tax on sellers RL will be equal to the whole amount of the tax t and as a result the incidence of tax on the buyers in this case will be zero in this case.

If elasticity of demand is infinity (ed. $= \infty$) then it also follows from above that $RL = t$, that is, the whole burden of the tax will fall on the sellers. On the other hand, if elasticity of supply is infinity, then it follows from above that $RQ = t$, that is, the whole burden of the tax will fall on the buyers. Let us graphically explain these cases of incidence of tax.

Incidence of Indirect Taxation and Elasticity of Demand

If demand for a commodity on which a unit tax has been imposed is elastic, then as a result of the rise in price caused by the imposition of a tax, the people will buy substitute goods and the quantity demanded of the good will fall greatly. In such a case, the incidence of the tax on the buyers will be relatively less. This is illustrated in Fig. 32.2. Take first, the demand curve DD which is relatively more elastic as compared with the demand curve $D'D'$ (dotted). SS is the supply curve which with the intersection of demand curve DD determines the price OP and quantity exchanged OM. With the imposition of a unit tax, the supply curve shifts to the left and intersects the demand curve DD at point Q determining new price OP' and the equilibrium amount ON. The distance between the two supply curves measures the tax per unit whith is equal to LQ. It will be seen from the figure that with the demand curve DD price has risen by PP' or RQ. The remaining tax LR will be borne by the sellers. It will be seen that in this case of elastic demand the incidence of the tax RQ fallen on the buyers is smaller than that on the sellers RL.

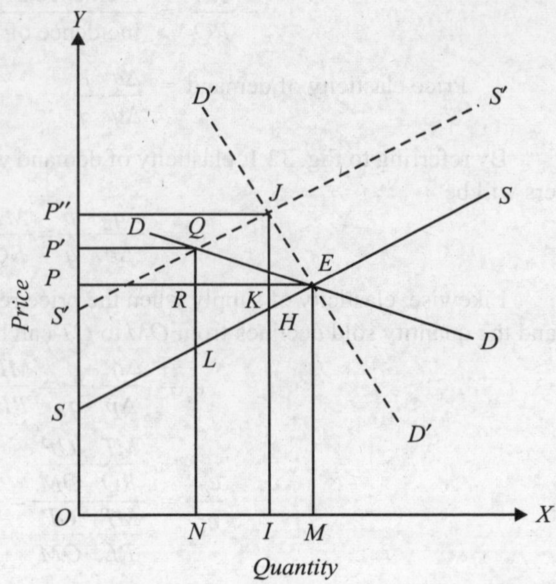

Fig. 32.2. *Elasticity of Demand and the Incidence of the Tax*

Now take the case of less elastic demand curve $D'D'$ (dotted). It will be seen from Fig. 32.2 that the supply curve $S'S'$ obtained after the imposition of the tax per unit intersects the less demand curve $D'D'$ at point J. In this case of less elastic demand curve, price has risen by a greater magnitude PP'' or KJ which will be the incidence on the buyers, whereas the sellers will receive HK less than before the imposition of tax. It will be thus noticed that in this case of less elastic demand curve, the incidence of the tax on the buyers (KJ) is much greater than the incidence on the sellers (HK).

It follows from above that *the less the elasticity of demand, the greater the incidence of tax on the buyers, and in the extreme case when the elasticity of demand is zero, the whole tax will be passed on to buyers*. On the other extreme, when demand for a commodity is perfectly elastic, the whole tax will be borne by the sellers of the commodity. These are illustrated in Figs. 32.3 and 32.4. Take Fig. 32.3 which illustrates the case where demand curve is completely inelastic (*i.e.,* $e_d = 0$) as actually happens in the case of necessities which cannot be substituted

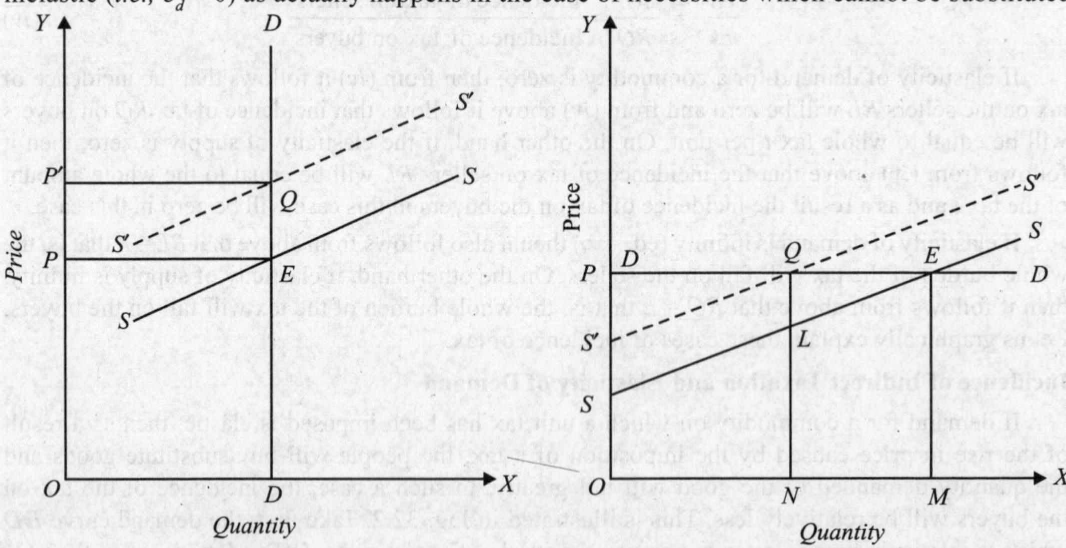

Fig. 32.3. *Incidence of a Tax in Case of Perfectly Inelastic Demand*

Fig. 32.4. *Incidence of a Tax in Case of Perfectly Inelastic Demand*

by some other goods and therefore their demand for them does not decline following the rise in the price after the imposition of a tax on them. It will be observed from Fig. 32.3 that in this case of completely inelastic demand price rises by the full amount of the tax when tax on the commodity is levied resulting in shift in the supply curve from *SS* to *S′S′*. Thus, in case of perfectly inelastic demand the whole incidence of tax rests on the buyers.

Fig. 32.4 illustrates the other extreme case of perfectly elastic demand curve for a commodity. In this case price for the buyers remains the same following the imposition of tax and shifting of the supply curve from *SS* to *S′S′*. Thus in this case of perfectly elastic demand, tax burden on the buyers will be nil. As will be seen from Fig. 32.4, the quantity demanded falls from *OM* to *ON* and the price received by the sellers falls by the full amount of the tax *QL*. Thus, in case of perfectly elastic demand, the whole burden of the tax rests on the sellers.

Elasticity of Supply and Incidence of a Tax

The division of the tax burden between the buyers md sellers also depends on the elasticity of supply. *Given the demand conditions, the greater the elasticity of supply, the greater the incidence of tax resting on the buyers (consumers) of a commodity.* Fig. 32.5 represents the case of a commodity with a relatively elastic supply. When the tax is levied on this commodity and supply curve shifts from *SS* to *S′S′*, the price rises from *OP* to *OP′*. The price increase is equal to *PP′* or *RQ* which will be the incidence of the tax on the buyers. Price received by the sellers is lower by *RL* and this will be the burden of the taxation on the sellers. It will be

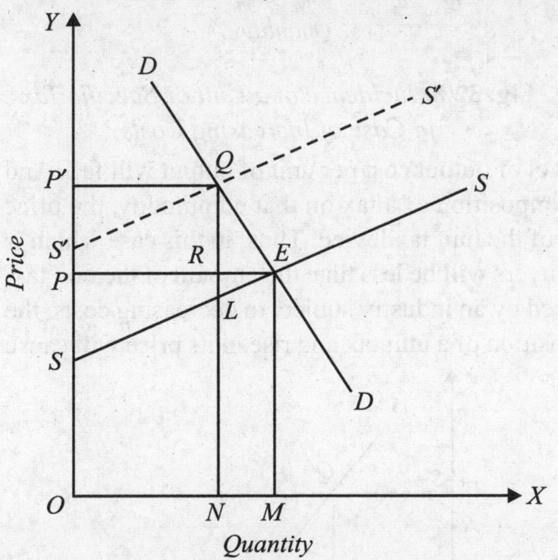

Fig. 32.5. *Incidence of Tax in Case of Elastic Supply*

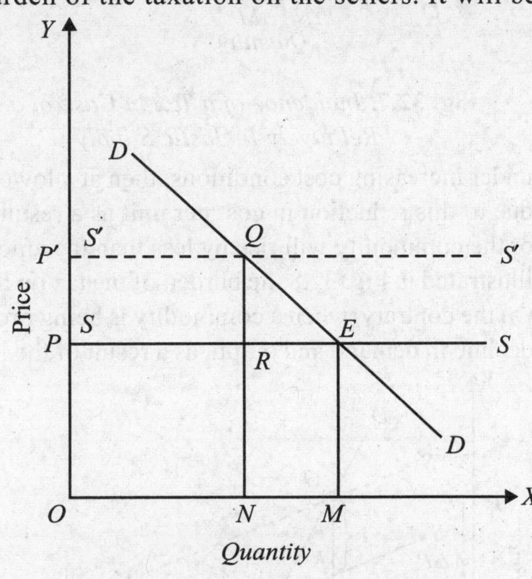

Fig. 32.6. *Incidence of Tax in Case of Perfectly Elastic Supply*

noticed that in this case of elastic supply curve, the incidence of tax on the buyers is greater than that on the sellers.

Fig. 32.6 shows the incidence of tax in case of commodity with a perfectly elastic supply. It will be seen from this that as a result of the imposition of tax on this commodity and shifting of the supply curve, price rises by the full amount of the tax. Thus, in case of *perfectly elastic supply the buyers have to bear the whole burden of the tax.*

Fig. 32.7 represents the incidence of tax in case of commodity with a relatively ineslastic supply. It will be observed that the incidence of a tax on this commodity will be less on the buyers (*RQ*) than on the sellers (*RL*).

Incidence of an Indirect Tax and Cost Conditions of the Industry

As mentioned above, cost conditions of the industry also determine the incidence of tax on buyers and sellers. From the viewpoint of cost, industries can be divided into three categories: (1) Increasing Cost Industries, (2) Constant Cost Industry, (3) Decreasing Cost Industry.[1]

Following the imposition of a tax on a commodity, its price rises and, given the demand curve, at a higher equilibrium price less is demanded and produced. If a commodity is being produced

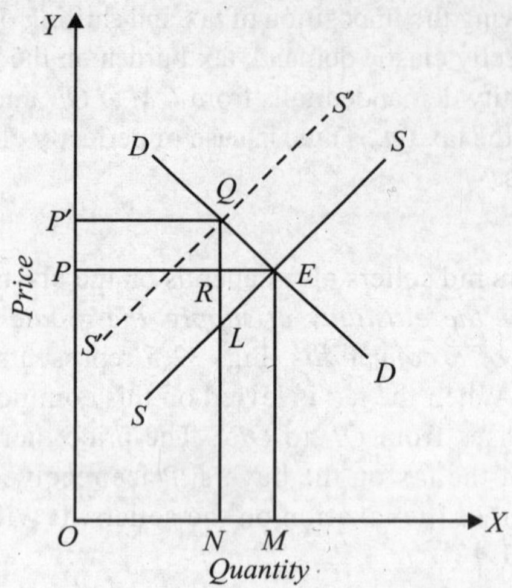

Fig. 32.7. *Incidence of a Tax in Case of a Relatively Inelastic Supply*

Fig. 32.8. *Incidence of a Unit or Specific Tax in Case of Increasing Costs*

under increasing cost conditions, then at a lower level of output cost per unit of output will fall. And due to this reduction in cost per unit as a result of imposition of a tax on that commodity, the price of the commodity will rise by less than the amount of the unit tax levied. Thus, in this case which is illustrated in Fig. 32.8 the burden of the tax on the buyers will be less than the amount of the unit tax. On the contrary, when a commodity is being produced by an industry subject to decreasing costs, the decline in demand and output as a result of the imposition of a unit tax and rise in its price will cause

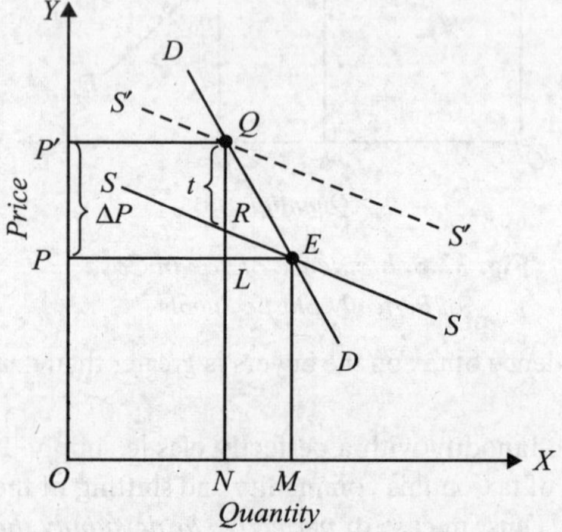

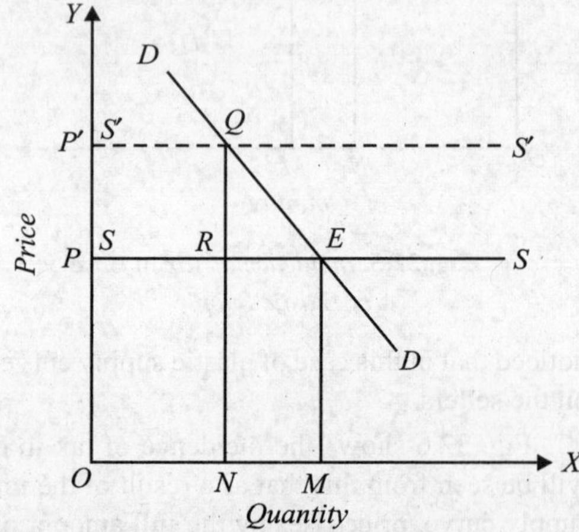

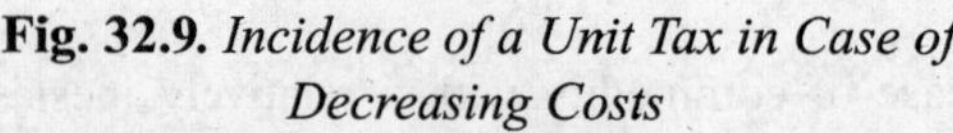

Fig. 32.9. *Incidence of a Unit Tax in Case of Decreasing Costs*

Fig. 32.10. *Incidence of a Unit Tax in Case of Constant Costs*

1. Since there is inverse relation between cost and physical returns, these industries can also be classed as (1) Diminishing Returns Industry, (2) Constant Returns Industry (3) Increasing Returns Industry.

cost per unit to rise. In this case, as shall be seen from Fig. 32.9, the *price will rise more than unit tax* on account of (1) imposition of tax on the commodity, and (2) rise in unit cost at a lower level of output. It will be observed from Fig. 32.9 that while tax imposed on the commodity equals *RQ*, the price has risen by *LQ* or *PP′* which is greater than *RQ*.

When the production of a commodity in an industry is subject to constant cost then decline in output following the imposition of a unit tax and rise in price does not cause change in unit cost of production. In this case, as shall be seen from Fig. 32.10, the price of a commodity rises by the same amount as the unit tax levied on it. Therefore, the whole burden of the tax will rest on the buyers.

Conclusion: From our above analysis of the impact of cost conditions or physical returns to scale on the incidence of a commodity tax on the buyers and sellers we arrive at the following conclusions :

1. The incidence of a unit tax on the buyers will be less than the amount of the tax when the commodity is being produced under increasing cost conditions.
2. The incidence of tax on the buyers will be greater than the tax when the commodity is being produced under decreasing cost conditions.
3. The incidence of tax on the buyers will be equal to the amount of the tax when the production of the commodity is subject to constant costs.

QUESTIONS FOR REVIEW

1. Distinguish between incidence and impact of a tax. How is the burden of a unit tax on a commodity shared between the producer and consumer under perfect competition?
2. Suppose a commodity is being produced under conditions of monopoly. Explain and illustrate how the burden of a unit tax on that commodity will be borne by the consumers and producer. Compare it with that under perfect competition.
3. Define the incidence of taxation. Analyse the incidence of a (*i*) unit tax and (*ii*) lump sum tax under conditions of perfect competition.
4. What do you mean by incidence of a tax? Explain the circumstances under which the incidence of a unit tax imposed under competitive conditions is passed wholly on the consumers.
5. Explain that the incidence of a unit tax on a commodity on the buyers is more, the greater the elasticity of supply.
6. Show that the less the elasticity of demand, the greater the incidence of a unit tax on the buyers.
7. Explain the impact of cost conditions of the industries on the incidence of an indirect tax (such as excise duty) on the buyers.

CHAPTER 33

FISCAL POLICY AND ECONOMIC GROWTH IN A DEVELOPING COUNTRY

Introduction

The most important instrument of government intervention in the economy today is that of fiscal or budgetary policy. Fiscal policy refers to the taxation, expenditure and borrowing by the Government. The economists now hold the government intervention through fiscal policy is essential in the matter of overcoming recession and inflation as well as of promoting and accelerating economic growth. Monetary policy alone will not do. There is no doubt that the Government budgetary or fiscal policy must be sound, keeping in view the needs and requirements of a developing economy.

Objectives of Fiscal Policy in Developing Countries

In developing countries, taxation, Government expenditure and borrowing have to play a very important role in accelerating economic development. In fact, fiscal policy is a powerful instrument in the hands of the Government by means of which it can achieve the objectives of development. There are several peculiar characteristics of a developing country which necessitate the adoption of a special fiscal policy which ensures a rapid economic growth. There are vast and diverse resources, human and material, which are lying underutilised. Such countries have weak infrastructure, *i.e*, they lack adequate means of transport and communications, roads, ports, highways, irrigation and power. They also lack technical know-how. Their population is increasing at an explosive rate which necessitates rapid economic development to meet the requirements of the rapidly growing population. Above all, these countries suffer from deficiency of capital. They are caught up in a vicious circle of poverty. In order to overcome these handicaps, a suitable fiscal and taxation policy is called for.

The principal objectives of fiscal policy in a developing economy are:

1. to mobilise resources for economic growth, especially for the public sector;
2. to promote economic growth in the private sector by providing incentives to save and invest;
3. to restrain inflationary forces in the economy in order to ensure price stability; and
4. to ensure equitable distribution of income and wealth so that fruits of economic growth are fairly distributed.

ROLE OF FISCAL POLICY FOR MOBILISATION OF RESOURCES IN DEVELOPING COUNTRIES

In developing economies, the Government has to play a very active role in promoting economic development and fiscal policy is the instrument that the state must use. Hence the great importance of public finance in underdeveloped countries desirous of rapid economic development. In a democratic society, there is an inherent dislike for direct (physical) controls and regulation by the state. The entrepreneurs would not like to be orderd about to produce this or that, how much to produce or where to produce. Fiscal incentives in the form of tax concessions, rebates or subsidies are, therefore, preferable. Similarly, the consumers would not like to be told directly to curtail their consumption or to consume this and not to consume that. Taxation of articles whose consumption is to be discouraged is therefore preferable. Hence, a democratic state must rely on indirect methods of control and regulation and this is done through fiscal and monetary policies. Thus, in democratic countries, fiscal policy is a powerful and least undesirable weapon on which the states can rely for promoting economic development.

Capital formation is of strategic importance in the matter of rapid economic development and the developing economies suffer from capital deficiency. It is, therefore, necessary to achieve a higher ratio of savings to national income. In early days of capitalism, payment of low wages and the existence of inequalities of income helped capital formation in the present-day developed countries. But no democratic country can adopt this method in modern times; the effort rather is to raise wages and reduce inequalities of income and wealth. Under a regime of socialist dictatorship, capital formation is brought about by ruthlessly curtailing consumption and keeping down the standards of living. But in modern democracies with every adult person having a right to cast vote very low levels of living for a long time is not feasible. *Hence the state must rely on instruments of fical policy to mobilise resources for economic development.* Taxation can be used to to raise collective savings for public investment and also at the same time to promote private investment. A well-conceived scheme of taxation is an important way of raising ratio of savings to national income which is one of the crucial determinants of the rate of economic growth. As Nurkse says, "*public finance assumes a new significance in the face of the problem of capital formation in underdeveloped countries.*"[1]

On the expenditure side, there is positive need for public investment, especially in those branches of economic activity where the private investments are not easily attracted, for example, the development of infrastructure such as power resources, means of transport and communications, basic heavy industries, education and research. Such investments are very often the very foundations of rapid economic advance. Thus, fiscal policy is of crucial importance in accelerating the pace of development in developing countries.

We explain below in detail how fiscal policy measures can be used to achieve the objectives of economic growth, more equal distribution of income and price stability in the developing countries.

Promoting Private Saving

As mentioned above, capital formation is an imoportant determinant of economic growth. For accelerating the rate of capital formation, savings and investment rate in the economy has to be stepped up. For this purpose savings have to be mobilised and channelled into productive investment. The alternative means other than fiscal policy available to promote savings and investment in the developing countries are not very effective in mobilising enough resources for investment and capital formation. The propensity to consume is very high in these countries. There exist large inequalities of income in these countries and this should ensure a large voluntary savings by the richer sections of the society. But the richer sections in them indulge in conspicious consumption such as building of luxury houses, indulging in five star culture, buying air-conditioners and such other things and, therefore, the volume of their voluntary savings is meagre. This propensity to indulge in conspicuous consumption is reinforced by demonstration effect which is operating strongly these days due to the development of electronic media and superior means of advertisement. Further, rich people tend to invest their rising incomes on unproductive investment such as gold and jewellery, real estate etc., which yield high profits due to their appreciation. In view of these, sufficient voluntary savings cannot be made available for raising significantly the rate of capital formation.

The alternative to taxation and Government borrowing to finance capital formation is to obtain *forced savings through creation of excessive creation of new money, often called money financing (i.e. monetisation of budget deficit) and resultant price inflation.* But this is not a desirable,

1. Ragnar Nurkse, *Problems of Capital Formation in Under-developed Countries.*

equitable and efficient way of obtaining the needed funds for development. First, the inflation tends to direct private investment into unproductive types of investment such as inventory holdings, purchase of real estates, gold and jewellery etc. Secondly, inflation raises the cost of public sector investment projects which lowers the rate of real investment. Thirdly, inflation reduces the voluntary savings of the people. This is because inflation reduces the real value of money which adversely affects willingness to save money. Fourthly, since inflation raises the incomes of the industrialists and traders on the one hand and reduces the real incomes of the general public on the other, it increases disparities of income in a society which runs counter to the objective of social justice. Therefore, most economists do not favour financing development through forced savings generated by deliberate inflation caused by excessive deficit financing.

Due to the severe limitations of alternative ways of mobilising resources for capital formation, the role of fiscal policy in performing this task assumes greater importance. Fiscal policy, if properly designed, is an efficient and equitable way of mobilising resources for augmenting public investment. Through it not only collective public savings can be raised for financing public investment but also at the same time private savings and investment can be encouraged. "In fact" *taxation is one of the effective means of increasing the total volume of savings and investments in an economy where the propensity to consume is normally high*." Further, the fiscal policy can be so devised that not only the objective of rapid capital accumulation or growth, but also other objectives of economic policy such as equitable distribution of income and wealth, price stability and promotion of employment opportunities can be achieved.

In what follows we shall explain how the various instruments of fiscal policy such as taxation and Government borrowing can be used to mobilise resources for economic development.

TAXATION AND MOBILISATION OF RESOURCES FOR ECONOMIC GROWTH

Taxation is an important instrument for fiscal policy which can be used for mobilising resources for capital formation in the public sector. To raise ratio of savings to national income and thereby raise resources for development, it is necessary that marginal saving rate be kept higher than the average saving rate. By imposition of direct progressive taxes on income and profits and higher rates of indirect taxes such as excise duties and sales tax on luxury goods for which income elasticity of demand is higher, the marginal saving rate can be made higher than the average saving rate. This will cause a continuous increase in the saving rate of the economy. An important merit of taxation is that it is not only a good instrument of resource moblisation for development but it also cuts down consumption of goods and thereby helps in checking inflation. Whereas direct taxes on income, profits and wealth reduce the disposable incomes of the people and thereby tend to reduce aggregate demand in the economy, indirect taxes directly discourages the consumption of the goods on which they are levied by raising their prices.

Taxation policy should be used to prevent this potential economic surplus from being wasted in conspicuous consumption and unproductive investment. It may be noted that this potential economic surplus is not a given amount but it increases in the very process of economic development. Firstly, economic development raises the incomes of the people, especially the business class and agriculturists and this augments the economic surplus. Secondly, some inflation is inherent in the process of development and this greatly benefits the traders, farmers and businessmen whose incomes rise much faster than others. The prices rise more than the costs and they get a good deal of profits. Further, due to the inflationary pressures market values of their investment in real estate, gold, shares, etc., go up and enormous capital gains accrue to them. All these enlarge economic surplus and therefore the taxation potential of the economy.

1. Baran, *Political Economy of Development.*

Direct Taxes and Mobilisation of Resources

Now the question arises what should be the taxation structure of a developing economy which will mobilise the potential economic surplus to the maximum, that is, what kinds of taxes be levied, how much progressive should be their rates and what should be the exemptions and concessions in various taxes. This is, however, a highly controversial issue. It has been suggested that an appropriate tax which would mobilise resources or mop up economic surplus is the progressive income tax. In India and the other developing countries income has been regarded as a good base for direct taxation. And the imposition of highly progressive income tax not only mops up relatively greater amount of resources but also tend to reduce inequalities of income. However, a progressive income tax with a high marginal rates of taxes adversely affects private saving and investment and also raises the propensity to evade the tax. In view of this, two proposels have been put forward to make the income tax both as an effective instrument of resource mobilisation for the public sector and of providing incentives to save and invest. First, Prof. Kaldor of Cambridge University, who in 1956 was invited by Government of India to suggest reforms in the Indian tax system for mobilising resources for development, suggested that whereas the marginal rate of income tax be reduced to, say, 45 to 50 per cent, expenditure tax be levied to discourage the people belonging to upper income brackets from dissipating their income in conspicuous consumption. According to him, this will also reduce the tendency to evade income tax on the one hand and promote private savings on the other. The second proposal to reform the income tax put forward by others is that whereas marginal rates of income tax be kept high but some exemptions for approved forms of saving and investment be allowed to the individuals. This will channel individual savings along desired lines and at the same time mobilise resources for development.

Apart from the income tax on individuals and companies, the imposition of other direct taxes such as wealth tax, gift tax, and estate duty are also needed to mobilise sufficient resources for capital formation. Unlike income tax, these capital taxes do not have any adverse effects on incentives to save and invest.[2] They are also important instruments of reducing inequalities of income and wealth. Because of these advantages, Professor Kaldor in his report of taxation reforms in India recommended the imposition of these capital taxes and this recommendation was accepted and the annual wealth tax and gift tax were levied in 1957 with estate duty having been already introduced in 1954.

Agricultural Taxation and Resource Mobilisation. A major part of national income in India and other developing countries originates in the agricultural sector which has substantial economic surplus which can be tapped for capital formation. This economic surplus mainly goes to rich farmers, landlords, merchants and other intermediaries and, in the absence of suitable taxation on agriculture, this is used for conspicuous consumption and for investing in unproductive activities such as buying gold, jewellery, real estate. Thus, according to Professor Kaldor, "*the taxation of agriculture by one means or another has a critical role to play in the acceleration of economic development.*"[3] Further, owing to economic growth in general and agricultural development in particular income of the agricultural class and therefore economic surplus enormously increases and therefore need to be mopped up for further development. Besides, the agricultural sector has to be taxed not only because it has a potential surplus but also to achieve maximium utilisation of land through devising a system of land taxation which would penalise poor use of good land. In this regard, a progressive land tax "the effective rates of taxation of which vary with the total value of landholdings of the family unit" may be suitable one. Besides this, progressive agricultural income tax with appropriate exemption may be levied to tap resources from affluent sections of the agricultural sector.

2. See, S. Gulati, *Capital Taxation in India*, Orient Longman, 1957.
3. N. Kaldor, The Role of Taxation in Economic Development, in E.A.G. Robinson (ed.), *Problems in Economic Development*, 1965, p. 177.

It may be noted that as compared to the non-agricultural sector, agricultural sector in India and other developing countries is quite under-taxed. Land revenue which was at one time the greatest source of revenue for government has now become an insignificant yielder of revenue. In the context of India, Agricultural Taxation Committee under the Chairmanship of Dr. K.N. Raj recommended Agricultural Holdings Tax (AHT) which was to be imposed at a progressive rate on the rateable value of an agricultural holdings of ₹5000 and above. It was expected to yield annually ₹200 crores. However, Agricultural Holdings Tax was found to be difficult to assess and administer. In our view, a *graded surcharge on the existing land revenue* will be far easier to assess and administer and make land revenue a more elastic source of revenue. However, it may be noted that in India agricultural taxation is a state subject and there is lack of political will on the part of the State Governments to raise the level of land taxation in a country. But if sufficient resources are to be mobilised for development, the level of agricultural taxation has to be raised.

Merits and Demerits of Direct Taxes for Resource Mobilisation, As seen above, as an instrument of resource mobilisation for development direct taxes enjoy several advantages. (1) They raise resources in a non-inflationary way. Indeed, they tend to check inflation by curtailing consumption demand. (2) They help to reduce inequalities of income and wealth, and (3) They discourage conspicuous and non-necessary consumption and thereby enlarge economic surplus. But the direct taxation of agricultural and non-agricultural sectors has its own limits. The coverage of direct taxes is quite narrow and difficult to expand. For instance, in India not more than one per cent of the population comes within the purview of income tax as incomes of the majority of the people fall below the exemption limit which has now been raised to the annual income of ₹40,000. Moreover, there is a considerable evasion of income tax. Yield from other direct taxes such as wealth tax, gift tax is quite meagre due to very small coverage, low rates and considerable evasion of them.

Role of Indirect Taxes in Resource Mobilisation

As a result of limitations of direct taxes, developing countries have resorted to extensive use of indirect taxes. *In India, almost all commodities have been brought within the net of indirect taxes such as excise duties and sales tax. Besides, there are custom duties* (*i.e., taxes on imports and exports*). Indirect taxation is an important source of development funds in a developing country. In the last five decades of planned development, revenue from several indirect taxes has been rising as a percentage of total revenue as well as of national income. It has been found that indirect taxes are better suited to the conditions obtaining in developing countries for reducing current consumption and mobilising resources for development. This is because in such countries quite a large proportion of national income tends to be diverted to current consumption instead of being productively invested. The average propensity to consume in such countries is much higher than is the case in advanced countries. Indirect taxes which reduce consumption must thus play a more important part. They will raise the rate of savings which are so essential for economic growth.

"High rates of taxes on commodities with a high income elasticity of demand are quite effective in siphoning a substantial proportion of increase in output into the resources of the public sector needed for development financing and a stiff rate of commodity taxes on luxury articles tends to introduce an element of progressiveness in an otherwise predominantly regressive tax structure in developing countries."[4] But in order to make sure that the resources raised through commodity taxes are adequate, it will be necessary to extend their coverage to include some articles of mass consumption. In the poor countries, it is not possible to exempt entirely goods of general and necessary consumption, because they are the only goods that provide a base broad enough to assure an adequate amount of resources. It may be noted that it is revenue from indirect takes that has substantially increased during the last five decades of planned development. From a small sum of ₹227

4. R.N. Tripathy, *Public Finance in Developing Countries*, 1966, p. 200.

crores in 1950-51, the revenue raised from indirect taxes rose to about 81,000 crores in 1998-98. Moreover the share of indirect takes to total tax revenue (direct plus indirect taxes) has go up from 64 per cent in 1950-51 to 70 per cent in 1998-99. It is thus clear that for mobilisation of resources for growth and capital formation indirect taxes have made an important contribution. It is worth mentioning that contrary to the trends in other countries of the world *custom duties* (*i.e.*, taxes on imports). Whereas in 1980-81, revenue of ₹3,400 crores was raised through custom duties, this rose to 48,000 crores in 1998-99. Revenue raised through union excise duties has also been quite substantial; revenue from them increased from ₹6,500 crores in 1980-81 to about ₹57,000 crores in 1998-99.

Limitations. But there are some limitations of raising resources through indirect taxes. First, they lead to *cost-push inflation*. The burden of indirect taxes are passed on to the consumers in the form of higher prices charged from them. In India excise duties on sugar, cloth, kerosene oil, petrol etc., has raised the prices and have contributed a good deal to cost-push inflation witnessed in last several years. Secondly, levying of customs duties on imports of capital goods, raw materials used for industrial production have accelerated cost-push inflation in the Indian economy. Thirdly, imposition of customs duties on imports of goods also protect inefficiency in domestic industries. This also contributes to high-cost production and promotes inefficiency. Lastly, indirect taxes are *regressive* in nature; both the rich and poor have to pay the same rate on the taxed commodities. The regressive character of indirect taxes have been attempted to be reduced by imposing higher rates of excise and custom duties on luxury items and smaller rates of duties on goods of mass consumption. However, the need for mobilising greater resources have forced the finance ministers to levy higher indirect taxes even on articles of mass consumption.

Role of Taxation in Promoting Private Savings and Investment

It is worth mentioning that in a mixed economy such as ours, there is need to raise not only the public savings and investment but also to promote private saving and investment so that the overall rate of saving and investment in the economy is stepped up. This implies that taxation measures should not impair incentives to save and invest of the people. Therefore, a developing economy encounters a crucial dilemma in augmenting larger resources for public investment on the one hand and to promote private savings and investment on the other. Thus, according to Prof. Heller, "Tax Policy faces a basic dilemma in its role as a an instrument of capital formation for economic development. On the one hand, high levels of taxes are necessary to finance that part of the development which falls in the government share and to mobilise for investment the private resources that might otherwise be dissipated. On the other hand, the lower the taxes the greater will be the inducement to private investment... The dilemma is worsened by the fact that those taxes which are most effective in capturing a large share of the gains from economic development for further capital formation are the ones most likely to affect the returns from private investment. *Only way of the dilemma may be to combine high rates of taxation in general, with preferential treatment for categories of desired development activity*."

Thus, while the policy of high rates of taxation on income and profit may be adopted but even with this private saving and investment can be promoted through preferential treatment of them. Rather than lowering the rate of income tax, private savings can be promoted in several other ways. First, interest on several types of private savings such as bank deposits, investment in units of Unit Trust of India and National Savings Certificates, and other approved forms of saving be totally or substantially exempted from taxation. This will encourage private savings through raising the rate of return on savings as the savers would earn interest income free of income tax. Secondly, voluntary savings in certain selective lines such as voluntary contributions to provident fund, life insurance premium, certain specific units (of UTI) and National Savings Certificates be substantially exempted from income tax. This will also stimulate private savings by raising the rate of return on these savings. For, people will not only earn rate of interest or dividend but will save a good amount of income tax on these approved forms of savings. Thirdly, some economists have suggested that if progressivity in tax rates is avoided (that is, if rates of income taxes do not increase much with the increase in income) private saving and investment will be boosted up. However, this is based upon a questionable assumption that richer people in countries like India have high propensity to save. In view of the present author, the lower marginal rates of taxes will place more disposable income in the hands of higher income brackets which will tend to raise their conspicuous consumption, especially when goods catering to their frivolous wants are available in abundance these days through both domestic production and imports. As mentioned above, this is reinforced by the international demonstration effect which exercises a strong influence on the consumption behaviour of the richer sections of the Indian society.

Besides private savings, private investment can be directly encouraged through taxation. Firstly, in order to stimulate private investment, the *retained profits* which are reinvested by the business firms instead of distributing them among the shareholders can be exempted from taxation or taxed at preferential rates. Secondly, to boost private investment *liberal depreciation and investment allowances* have also to be allowed to the business firms which will be used for purposes of investment in new plant, equipment and machinery. Further, "subsidies on investment can also be provided by the Government which generally prove to be a very effective way of promoting private investment".

Another important fiscal method of stimulating private investment in developing countries is *granting of tax holidays or relief from tax on the profits of new enterprises* for some specified period of time. Further, indirect taxes can also be manipulated in a variety of ways to promote private investment in certain selected fields of industrial activity. Thus exmption or lowering of sales tax or excise duties on some domestically produced important raw materials or lowering of import duties on raw materials and capital goods from abroad can provide a boost to private investment. Likewise, reduction of export duties may also bring about a healthy effect on investment outlook.

It needs to be stressed that various incentives to promote private savings and investment will prove effectively only if they are kept as simple, certain and stable as possible. A complicated schemes of incentives with frequent and arbitrary changes are likely to defeat the very purpose of these incentive schemes. Prof. Prest rightly points out, "*If the main purpose of investment incentives is to raise the long-term level of investment rather than offset cyclical downturns and this clearly is the case in underdeveloped countries ... the emphasis should be on the stability of these tax arrangements rather than their variability*."[5]

It is important to note that private investment depends upon several factors such as size of the market, cost of inputs, availability of infrastructural facilities such as power, coal, oil, transport, availability of technical know-how. Tax incentives will prove effective only if conditions regarding these other factors are not unfavourable. Further, unless there is an efficient and competent system of tax administration it is very likely that incentives become tax loopholes. The businessmen in the developing countries tend to take advantages of these tax exemptions on bogus basis without fulfilling and carrying out the underlying intent of such concessions. An honest and efficient administrative machinery, therefore, remains the basic precondition for the purposeful use of incentives.[6]

---

5. See Prest, A.R. Fiscal Measures and Capital Accumulation in Agarwala and Singh (ed.), *Accelerating Investment in Developing Countries*, Oxford University Press.
6. A.R. Prest. *op. cit., p.* 445.

GOVERNMENT BORROWING AND RESOURCE MOBILISATION FOR ECONOMIC DEVELOPMENT

Government borrowing is another fiscal method by which savings of the community may be mobilised for economic development. In developing economies, the Governments resort to borrowing in order to finance schemes of economic development. Government or what is also called public borrowing becomes necessary because taxation alone cannot provide sufficient funds for economic development. Besides, too heavy taxation has an adverse effect on private saving and investment. It may be noted that when there is fiscal deficit (i.e. budget deficit). It can be financed in two ways. First, by borrowing by the Government from the market and this borrowing leads to the increase in public debt. Borrowing as a means to finance the budget deficit is therefore also called **debt-financing of budget deficit**.

Under borrowing the government issues bonds and sells it to the public. Generally, sale of interest-bearing bonds to the public is indirect through financial intermediarises such as banks. Banks buy the bonds floated by the government with the currency deposits of the public. Therefore, *debt-financing of budget deficit* is also known as *bond-financing of budget deficit*. With the borrowed money in this way the government is able to expand its expenditure but at the same time it adds to public debt which has both short-run and long run consequences. It may also be noted that budget deficit also comes about when taxes are reduced, keeping government expenditure constant. This type of budget deficit can also be financed through incurring debt by selling bonds to the banks or public. The government has not only to pay annually interest on borrowed funds but has also to pay back the principal sum borrowed for which it may levy higher taxes in future.

The Keynesians have emphasised the expansionary effect of debt-financing of government expenditure or budget deficit. In the Keynesian model with a fixed price level, the increase in government expenditure through use of borrowed money causes an upward shift in aggregate expenditure $(C + I + G)$ curve. If the economy is working at less than full-employment level of national income so that output gap exists in the economy, the increase in debt-financed government expenditure will bring about expansion in output or income. With the increase in income at the given tax rate, tax revenue collected will rise which will over time reduce the budget deficit or even ultimately eliminate it so that budget becomes balanced.

However critics have pointed out that debt-financed government expenditure is largely offset by the *crowding-out effect* of debt financing on private investment. The crowding-out effect on private investment takes place in a variety of ways. First, it has been pointed out that government's borrowing funds to finance budget deficit will lead to the increase in demand for lendable funds which will cause the rate of interest to rise. The rise in interest rate will cause private investment to decline. Thus debt-financed increase in government expenditure crowds out private investment. According to this view, due to crowding-out effect on private investment net expansionary effect of increase in government expenditure is negligible. On the other hand, the society will have to bear the burden of increase in public debt as a result of debt-financed expansion in government expenditure. If budget deficit arises due to reduction in taxes, keeping government expenditure constant, this will also lead to rise in interest rate and will therefore cause crowding-out effect on private investment. This happens because reduction in taxes stimulates consumption expenditure of the people which reduces savings. The decline in savings causes interest rate to rise resulting in fall in private investment.

Government borrowing takes two forms : (*a*) *market loans* and (*b*) *small savings*. In the case of market borrowing, the Government sells to the public negotiable Government Securities of varying terms and duration and treasury bills. For financing capital projects long-term Government bonds are floated in the capital market. This form of public borrowing is more important for mobilising resources for development. The treasury bills which represent short-term borrowing are intended to meet only the current Government expenditure. New bonds may be issued for meeting old maturing bonds. The small savings represent public borrowing which are not negotiable and are not bought and sold in the capital market. For mobilising small savings, various types of saving certificates are issued, e.g., National Savings Certificates, National Development Certificates, Rural Development Bonds, Postal Certificates and Postal Accounts, Compulsory deposits, etc. A widespread campaign is necessary to attract small savings.

Borrowing is the quickest mode of raising funds. Tax finance is not so expeditious because passing of tax laws, assessment of taxes based on those laws and their collection involves considerable delay. Besides, *public debt does not involve any burden if it is devoted to productive works.* The subscribers to Government loans are able to find remunerative investment whereas the Government can pay the principal and interest out of the income yielded by investment financed from loans. Thus public borrowing is not only necessary but also desirable.

There is another advantage in Government borrowing. *Government borrowing is anti-inflationary.* The underdeveloped countries are victims of inflation since they have to resort to deficit financing for finding funds for economic development. Since deficit financing is inflationary, public borrowing is preferable to deficit financing. Public borrowing mops up the surplus purchasing power with the people. It thus checks consumption and so a rise in prices. At the same time idle balances are absorbed in productive activity.

Public borrowing generates additional productive capacity. The funds raised by public borrowing can be utilised for building up economic infrastructure for the economy through schemes for the development of irrigation, transport, power and communication. They can help also in building up of the agricultural and industrial base of the economy. All such investments facilitate further economic growth. There is no doubt, therefore, that the public borrowing plays a very useful and important role in accelerating economic development of underdeveloped economies. Public debt promotes saving and investment, the two most crucial determinants of economic growth.

But there exist a number of obstacles which hinder the success of borrowing policy in developing countries. In many such countries, there are no organised money and capital markets and in those where such markets exist, they constitute a very small proportion of total money market of the country. There may not, besides, exist any organic relationship between the organised and unorganised part of the money market.

Moreover, the resources of the organised money market may be too inadequate to fulfil the needs both of the private and public sectors. Further, in the financial market the competition for funds between the government and the private sector will raise rate of interest and this will have a highly disincentive effect on the expansion of investment in the private sector. More borrowing by the Government leads to the rise in market rates of interest. The higher interest rates tends to reduce private investment. Therefore, it is often asserted that the *higher borrowing target by the Government `crowds out' private investment*. In India the rates of interest on loans (on bonds or securities) of Government have been raised quite substantially. Since banks and others prefer to invest in Government securities because they are safe (*i.e.*, riskless). This has reduced funds for private investment.

For the success of government's borrowing policy, it is necessary that financial institutions be developed and extended into the rural areas of the economy in order to inculcate the habit of thrift in the population and to mobilise for productive purposes the amount of savings originating in this sector. Besides, for the mobilisation of savings it will be necessary to check and regulate the diversion of savings into unproductive investment such as real estate, gold and jewellery and inventory accumulation.

Suitable techniques of borrowing must also be devised. For example, bonds issued by the Government should be adjusted to the preferences of the general public; bonds of large denomination and long maturity may be offered to the institutional investors, whereas those of small denomination and short maturity may be reserved for the non-institutional public. If properly devised and conducted small savings campaign can mobilise a sizable amount of resources. Further, the mobilisation of the hoardings of gold and jewellery through government programmes constitutes a highly desirable source of resource mobilisation. Of course, suitable techniques of public borrowing for the mobilisation of these resources have to be evolved.

ROLE OF DEBT-FINANCING (*i.e.*, BORROWING) IN RESOURCES MOBILISATION: THE CASE OF INDIA

In India there has been a problem of persistent budget deficits which has been largely financed by borrowing from the banks and public and thus adding to the public debt. In 1991 as a measure of economic reforms IMF directed that fiscal deficit of the central government should be reduced to achieve economic stability. The economists associated with IMF and World Bank hold the view that large fiscal deficit raises inflation rate of the economy on the one hand and since it is financed largely by borrowing from the market, this raises rate of interest and therefore crowds out private investment and adversely affects economic growth. Besides, it has been pointed out that the persistent budget deficit year after year adds to public debt and as a result ratio of public debt to national income (Debt/GDP ratio) will greatly increase which will land the country in debt trap[13].

In view of the above it is appropriate to discuss the above issues relating to debt-financing of budget deficit in the light of the recent Indian experience. It is worth mentioning that from the years 1997 to 2002-03 and again in 2008 and 2009 the Indian economy experienced slowdown in economic growth due to lack of aggregate demand resulting mainly from the fall in public investment and poor performance in agriculture. In this situation *output gap* in the Indian economy emerged as the existing production capacity was not fully utilised. Besides, there has been lack of infrastructure facilities which stands in the way of raising private investment. Again in 2008-09 there was slowdown in the Indian economy due to the global financial crisis (2007-09) leading to a sharp fall in India's exports.

With their focus on reducing fiscal deficit the government reduces its capital expenditure (*i.e.*, investment expenditure) because it fails to curb its consumption expenditure and subsidies. Thus, Mihir Rakshit, a leading Indian macroeconomist, rightly states, "With their focus on fiscal deficit, policy makers seem to think that what matters primarily is curbing government expenditure ..... irrespective of changes in their composition. The period 1995-2000 was marked by 9.6 per cent and 12.6 per cent growth in government consumption and subsidies respectively, but government capital accumulation grew at snail's pace of 1.9 per cent and its two crucial constituents, namely, investment in agriculture and infrastructure actually declined by 5.6 per cent and 0.2 per cent respectively."[14]

The consequences of debt-financed budget deficit in the case of India depends on whether or not output gap exists due to lack of demand and infrastructural facilities. If the output gap exists, then increase in government expenditure on consumption and subsidies will raise aggregate demand and help in utilising more fully the existing production capacity and over time will cause private investment to rise. But this will not raise public saving and public investment and therefore will not result in much higher growth of the economy. This may also raise ratio of public debt to GDP since due to slow growth in GDP, tax collection will not appreciably rise. Mihir Rakshit aptly points out that, "The *Indian economy since 1997 has also been characterised by output gap, high fiscal deficit and growing debt/GDP ratio. The mainstream view in India is that despite the output gap fiscal deficits crowds out private investment and are unsustainable.*"[15]

13. For this view, see Shankar Acharya, former Chief Economic Advisor who puts forward this view in a forceful manner in 'How Fiscal Deficit Hurts Economic Growth'. We have explained in detail this mainstream view in the next chapter.
14. Mihir Rakshit, "Some Macroeconomics of India's Reforms Experience" in *India's Emerging Economy* (ed.) Kaushik Basu, Oxford University Press, 2004, p.104.
15. Mihir Rakshit describes the situation that prevailed during 1997-2002-03.

The mainstream view of debt-financed fiscal deficit is not correct and has been rightly challenged by some prominent Indian economists such as Mihir Rakshit, V.N. Pandit and Prabhat Patnaik, among others. Under the circumstances when there exists output-gap and under-utilisation of production capacity, the increase in government expenditure, especially of investment nature will *crowd in* private investment and speed up rate of economic growth. To quote Mihir Rakshit, "contrary to wide spread perception among commentators *budget deficits associated with closing the output gap in fact crowds in private investment*.......... crowding in will be much more pronounced in countries like India where government's infrastructural investment favourably affects private investment even while projects are being executed : the fact that highways are being built in some area or that it will soon have abundant supply of power, communication facilities is likely to induce private investors to start setting up new units and expending existing ones even in the shrot run"[16].

When there is outgap in a developing economy such as India, even increase in Government's consumption expenditure will crowd in private investment through the former's demand effect. But, as has been pointed out by Dr. Mihir Rakshit, crowding-in effect of government's investment expenditure will be greater because it expands the infrastructural facilities (*i.e.*, supply-side factors) badly needed for increase in private investment.[17]

In our view, the expansionary effect of government's investment expenditure and also its crowding-in effect on private investment in a developing country like India is not confined to the situation when output gap and under-utilisation of capital stock exists due to demand deficiency but also when such an economy functions at its normal capacity. Besides, even in normal times, debt-financed government investment expenditure which brings about rapid economic growth, budget deficits and debt/GDP ratio over time can be reduced. An important rule of public finance, which is sometimes described as **a golden rule** is that if borrowed money by the government is used for investment or developmental purposes, it leads to growth in GDP or national income. Given the tax rate, this increase in income will ensure collection of more tax revenue possible. As a result, budget deficit or what is called fiscal deficit will be reduced and debt/GDP ratio will decline. This has actually happened in India.

As a matter of fact long ago (in 1944) E.D. Domar had stated the condition for sustainability of persistent budget deficits and consequently of mounting public debit. According to Domar, debt-financed deficits *are sustainable if growth rate exceeds the rate of interest*. This is because economic growth means increase in income or GDP from which annual interest payments can be made and if growth in income exceeds the rate of interest, the part of increase in income may be used for retiring public debt. In this way with the growth of the economy budget deficit and debt/GDP ratio can be reduced Elaborating Domar's condition. Mihir Rakshit writes, "According to Domar, if the government finances part of its expenditure (amounting to a given fraction of full-employment output) through borrowing, in a growing economy public debt and the government interest outgo as proportion of GDP will be stable in the long run provided the growth rate exceeds the interest rate. The implication is that when this condition is satisfied, maintenance of full employment through debt financing of budget deficit does not erode fiscal viability or produce a debt trap"[18]

The viability and sustainability of debt-financed budget deficits in a growing economy must be based on the magnitude of Government investment expenditure (I_g) in the total expenditure ($I_g + G$). Suppose the economy is working at the full capacity output, and government increases its investment expenditure, say, in infrastructure and finances the budget deficit through incurring debt. This

16. Budget Deficit: Sustainability, Solvancy and Optimality, printed in Amaresh Bagchi, "*Readings in Public Finance*" Oxford University Press, 2005, pp. 342-343.
17. Mihir Rabshit, op. cit. p.345
18. Mihir Rakshit, op. cit. (2005), p. 340.

increase in government investment expenditure will raise income through both demand effect and capacity effect and will therefore stimulate private investment which depends on change in income (i.e. $Y_1 - Y_{t-1}$). This increase in both government and private investment will over time cause rapid growth in income which in turn will result in more tax revenue. Thus, in this way budget deficit will be reduced with favourable effect on public debt/GDP ratio, if rate of growth in income (GDP) exceeds rate of interest.

It is thus clear that sustainability of budget deficit and public debt depends on *shifting the composition of government expenditure towards investment*. To quote Mihir Rakshit, "Budgetary viability requires an improvement in revenue balances in the medium term and the long run....... public investment in agriculture aided by crowding-in effect, raises agricultural productivity and acts as a stimulant to a rural economy, brings down poverty and *enables the government to reduce over time expenditure on food or fertilizer subsidies and poverty alleviation measures*. An increase in government investment in rural and urban infrastructural facilities will through increased revenue receipts and decreased subsidies will strengthen budgetary position not only in the medium term and the long run but in the short run as well"[19]. It is interesting to note that when with effect from year 2003 the Government investment expenditure in India has been raised, tax revenue collection has significantly risen and also fiscal deficit, revenue deficit and debt/GDP ratio have declined. Besides, private investment has also gone up (i.e. crowed in) rather than crowding out by increase in debt-financed increase in government expenditure.

Thus, in the nineties when the focus in our strategy was to reduce fiscal deficit, no matter reducing capital expenditure of the government for this purpose, the Indian economy experienced slowdown in growth and our fiscal situation worsened. On the other hand, when from March 2003 we shifted the focus toward increase in government investment expenditure on power, ports, highways, rural roads in a big way, private investment also crowed in, generated higher economic growth which resulted in lower fiscal deficits and lower debt/GDP ratio.

MONEY FINANCING AND RESOURCE MOBILISATION FOR ECONOMIC GROWTH

Meaning of Money Financing. Money financing was earlier called **deficit financing** Deficit financing was variously defined but the one idea underlying all these definitions is clear viz., it refers to '*created money*', *i.e.*, creation of additional purchasing power in the form of currency notes. According to the Indian Planning Commission, deficit financing is equal to the net increase in the purchasing power of the economy arising out of the operations of the Government. Money financing or Deficit financing is said to have been practised whenever Government expenditure exceeds the Government receipts from taxes, fees, surpluses from public undertakings, borrowings from the public, etc. Such an excess of expenditure is financed by borrowing from the Central Bank.

When the Government borrows from the Central Bank which is a note-issuing authority, the Central Bank simply issues more notes and gives them to the Government against Government securities. Thus in the last analysis deficit financing means the creation of new currency. It may be noted that in India *Net Bank Credit from RBI* by the central Government is called deficit financing. In fact when Central Government borrows from RBI and the latter issues new currency it is called *monetisation of Government* deficit. It is this monetisation of debt that leads to the expansion in money supply due to Government's fiscal deficit that was earlier called deficit financing. However, in the modern terminology it is now called **monetisation of fiscal deficit**.

Need for Money Financing. The developing countries keen to promote rapid economic growth, the resources required for development far exceeds the amount which can be raised by normal means of resource mobilisation, viz., taxation, borrowing, surpluses from public enterprises, etc.

19. Mihir Rakshit, op. cit. 2005 p.105

The uncovered gap is made up by deficit financing. Rapid economic development can be achieved only by stepping up the rate of investment. But wherefrom are the developing countries to raise the additional resources? In the absence of sufficient foreign aid forthcoming from friendly countries and international organisations, the additional funds must come from domestic resources. For this purpose voluntary savings must be stepped up. These savings are then mopped up through national small savings schemes to add to the resources available to the government.

But in the country where a majority of the people are living at the subsistence level, the margin between income and consumption is very low so that voluntary savings, however welcome, cannot by themselves provide sufficient resources for development. The government may also attempt to increase the volume of resources by additional taxes. Yet because of extreme poverty of the great masses of the people, additional taxation beyond a point raises difficult problems, both economical and political. Therefore, in their anxiety to implement development schemes, the governments are compelled to resort to money financing.

Money Financing and Inflation. Since deficit financing involves expansion of money supply with the public, it has inflationary potential. But if the supply of output, especially that of consumer goods, is increasing when deficit financing is being done, there is no danger of inflation. Therefore, it should not be presumed that deficit financing necessarily leads to inflation. Deficit or money financing generates inflationary pressures only when the supply of output cannot be increased to match the increase in money supply which becomes available through deficit financing.

Money financing practised during the depression in the developed countries does not generate inflationary pressures because of the elastic nature of the supply curve of output. Additional government expenditure leads to an increase in effective demand, depending upon the magnitude of the multiplier. Similarly, when supply of output can be increased to match additional demand, there are no inflationary pressures. In the developed countries during the depression, there is a lot of excess capacity in the system so that an increase in output presents no difficult problems.

The situation in the developing countries is, however, different. Here an increase in investment does create additional demand but a corresponding increase in the supply of output cannot be taken for granted. First, a developing country suffers from the dearth of capital and there is no large excess capacity to be utilized for creating additional supply of consumer goods. Besides, the basic wage goods industry, *i.e.*, agriculture, suffers from an inelastic supply curve in the short run. The farmer's ability to increase his output is very limited. His holding is a tiny one and often lies scattered all over the village; he lacks resources to buy more and better seeds and fertilizers; irrigation facilities are lacking so that even with the best of will in the world, unaided by any outside agency, the farmer's efforts to increase his output may not bear fruit. In the absence of elastic supply of consumer goods, deficit financing carries with it great inflationary bias.

In the early stages of development, the inflationary danger is very real indeed. This is because of the urgent need to invest large sums in long-term projects such as building of big dams, steel plants, machines making industry and other types of heavy and basic industries as well as in the development of transport and communication system. Investment in such long-term projects does create additional demand, but these projects do not add to the supply of consumer goods in the short run. Thus in the short run prices generally shoot up under the pressure of excessive demand for consumer goods. And once inflationary spiral starts operating, it is difficult to control it.

Reasonable Money Financing Helps Development

From above it should not be understood that every and any amount of deficit financing is harmful. Money financing of a reasonable amount is very useful. A reasonable dose of deficit financing helps the development of the economy. In a developing country like India, a huge amount of natural and human resources are lying unutilised and underutilised. With the help of deficit financing these unutilised and underutilised resources can be employed for productive purposes. If deficit financing is used to invest in those projects of schemes such as small irrigation works, land reclamation schemes which yield quick results, then the danger of inflation is not there. These quick-yielding projects will increase the supply of consumer goods in the immediate period and will therefore prevent the creation of inflationary pressures. Further, if under the development plans, serious efforts are made by the government to help the farmers to increase agricultural production by providing the formers with HYV seeds, fertilizers and irrigation facilities in time and in adequate quantities, then the supply of agricultural goods can be made elastic. To the extent we are able to increase agricultural production, we can safely do deficit financing. Since the income elasticity of demand for foodgrains is very high in the developing countries like India the incomes generated by investment outlay under the development plans are mostly directed to the purchase of foodgrains. But if we succeed in raising agricultural production, especially foodgrains, then a reasonable dose of deficit financing will not result in inflation.

Further, as the economy develops, the monetary sector expands. In other words, *as the economy grows, it becomes more monetized*. As a result of the expansion of monetary sector, needs for money increase. If new money is created to match this extra need for money, that will not prove inflationary. Moreover, with the general growth of the economy under development plans, the output of goods and services rises. There is no harm in creating new money which will match this additional output made possible by growth of the economy. Thus, if economic growth occurs at the rate of 8 per cent per annum, then money supply through deficit financing can be increased by 8 per cent without the risk of inflation.

It is worth repeating here that the danger of deficit financing leading to inflation will be less only if resources raised through deficit financing are used for purposes of productive investment. In the long run, productive investments made under the plans will increase the productive capacity and will therefore bear fruit in the form of greater supply of consumer goods. Thus inflation created by productive investment in the short run is likely to be self-destructive in the long run. But there is real danger of short-run inflation becoming self-cumulative rather than self-destructive if deficit financing is used for providing subsidies or for financing consumption activities. If deficit financing is done to make investments in quick-yielding projects in both agriculture and industry, then the danger of inflation can be avoided. For instance, if new money is created to invest in quick-yielding schemes like building of wells. tube-wells, tanks and conservation of soil to get quick returns, it will not prove inflationary.

Thus within limits, deficit financing is not only desirable but essential as well. *It proves inflationary only when safe limit is crossed.* Deficit financing can serve as a useful instrument of economic development if a right dose of it is taken considering the growth of national output, especially that of foodgrains and if it is used for financing those schemes or projects which yield quick results.

FISCAL POLICY, EQUITY AND PRICE STABILITY

Fiscal Policy and Equitable Income Distribution

Let us now explain how fiscal policy can be used to achieve equitable distribution of income. Attainment of a wider measure of equality in incomes, wealth and opportunities must form an integral part of economic development and social advance. Existence of gross inequalities is a social evil and no measure of economic development will increase economic welfare unless an equitable distribution of the rising national product is assured. Instrument of taxation, therefore, must be used as a means of bringing about a redistribution of income in favour of the poorer sections of the society. A caution is, however, necessary. It is essential to strike a balance between the two objectives of lessening economic inequalities and that of sustaining and strengthening incentives to invest and increase production. Fiscal policy must maintain unimpaired the flow of savings and investment which makes for continued progress of productive enterprises. Larger production and greater equality are both objectives of high importance in relation to general economic and social policy. Higher incomes can be taxed without adverse repercussions on private productive effort and enterprise. Optimum rate of growth and maximum social welfare are not irreconcible. It will be possible to reconcile these two objectives by the formulation of a well-balanced fiscal programme.

But lessening inequalities through taxing higher incomes is only one form of fiscal operations. A better and complementary fiscal policy consists of *increasing public expenditure for promoting welfare of the less privileged classes*. Increasing public expenditure on anti-poverty programmes such as construction of rural public works, employment guarantee schemes will ensure equity in income distribution. Stepping up of public expenditure on primary education and public health will greatly improve the economic conditions of the poor people. In fact, international experience shows that active public expenditure policies aimed at raising the consumption of the poor are far more effective in promoting equity as compared to tax policies aimed at containing the incomes of the rich.

Fiscal Policy for Achieving Price Stability

India as well as other developing countries have been experiencing the problem of rising prices or inflation. Inflation in them has been of both demand-pull and cost-push types. The main cause of demand-pull inflation has been the *fiscal deficit* in Government's budgets which has arisen because they have not been able to finance the mounting public expenditure through revenue from taxes and public sector surpluses In India in the year 1990-91, fiscal deficit rose to 8.3 per cent of GDP (Gross Domestic Product) and after 14 years of efforts, the Indian Government has been able to achieve about 50 per cent reduction in fiscal deficit. In 2004-05, fiscal deficit is estimated to be of the order of 4.6 per cent of GDP. However, to achieve price stabilisation IMF recommended its reduction to 3 per cent of GDP. A huge fiscal deficit in Government is financed in two ways: (1) by Government borrowing from the market, (2) by monetising the deficit which is commonly called deficit financing in India. A high degree of fiscal deficit leads to excess market borrowing by the Government which causes expansion in bank credit to the Government and therefore increase in money supply in the economy causing prices to rise. Besides, excessive Government borrowing from the market leads to the rise in interest rate which discourages private investment. Further, a part of fiscal deficit is monetised through borrowing from Reserve Bank of India which issues new currency (which is reserve or high powered money) for the Government. This causes greater expansion in money supply through the process of money multiplier and generates inflationary situation in the economy. Thus, to check the rate of inflation, fiscal deficit has to be reduced through both raising revenue of the Government and reducing Government non-developmental expenditure. In what follows we first explain the concepts of fiscal deficit revenue deficit, primary deficit. We will then explain various measures in the Indian context that can be adopted to reduce fiscal deficit and thereby to reduce inflationary pressures in the economy.

FISCAL POLICY AND FISCAL DEFICIT IN INDIA

Government expenditure on goods and services and resources mobilised by it through taxes, etc., are important factors that determine aggregate demand in the economy. When there is a deficit in the budget of the government, it spends more than it collects resources through taxes and non-tax revenue. In recent years there have been huge budget deficits in India which have created large excess demand in the economy. This has resulted in inflation or sharp rise in the general level in prices. The deficit may occur either in the revenue budget or capital budget or in both taken together. When there is an overall budget deficit of the Government, it has to be financed by either borrowing from the market or from the Reserve Bank of India which is the nationalised central bank of the country. RBI has the power to create new money, that is, to issue new notes. Thus, to finance its budget deficit, the government may borrow from Reserve Bank of India against its own securities. This is only a technical way of creating new money because the government has to pay neither the rate of interest nor the original amount when it borrows from the Reserve Bank of India against its own securities. It is thus clear that budget deficit implies that government incurs more expenditure on goods and services than its normal receipts from taxes and non-tax revenue. This excess expenditure by the government is financed by either borrowing from the market or by newly created money which leads to the rise in incomes of the people. This causes the aggregate demand of the community to rise to a greater extent than the actual amount of deficit financing undertaken through the operation of what Keynes called *income multiplier.*

In the opinion of many economists, the expansion in money supply caused by monetisation of deficit, that is, deficit financing leads to the excess aggregate demand in the economy, especially when aggregate supply of output is inelastic. The excess aggregate demand causes rise in the price level or brings about inflation in the economy.

Revenue Deficit

As mentioned above, the budget has two parts : *Revenue Budget* and *Capital Budget.* Receipts in the revenue account come from such sources as various types of direct and indirect taxes and also from non-tax sources such as interest received on loans given, dividend and profits of public enterprises.

The items of expenditure in the revenue account are of the type that represents *collective consumption* of the society and therefore, creates no earning assets. The various types of expenditure on the revenue account are incurred on : (1) Interest payments on public debit, (2) Civil administration, (3) Defence, (4) Subsidies on food, fertilisers, exports and (5) Social services such as education, health, etc.

Revenue deficit refers to the excess of revenue expenditure over revenue receipts. Prudent management of public finance requires that receipts on revenue account should exceed expenditure on revenue account. In other words, there should be surplus on revenue account so that this surplus should be used for investment in development projects or building assets which yield return. In fact, surplus on revenue account of the budget represents *public or government savings* which can be used for financing developmental activities. However, in India, for the past several years, there has been deficit on the revenue account. That is, there has been *government dissaving* for the last some years in India. Therefore, borrowed funds from the capital account have been used to meet a part of consumption expenditure of the government. This has bad consequences for the economy. Revenue deficit as percentage of gross domestic product (GDP) had been quite high upto 2002-03. Revenue deficit of the central government of India for various years is given in Table 33.1.

Since revenue deficit represents negative savings of the Government, it lowers investment by the

Table 33.1: Trends in Budget Deficits of Central Government

Year	*Fiscal deficit*	*Revenue deficit*	*Primary deficit*	*Revenue deficit as per cent of fiscal deficit*
	(As per cent of GDP)			
1996-97	4.1	2.4	–0.2	58.2
1997-98	4.8	3.0	0.5	63.5
1998-99	5.1	3.8	0.7	74.8
1999-2000	5.4	3.5	0.7	64.6
2000-01	5.7	4.1	0.9	71.7
2001-02	6.2	4.4	1.5	71.1
2002-03	5.9	4.4	1.1	74.4
Enactment of FRBMA				
2003-04	4.5	3.6	0.0	79.7
2004-05	4.0	2.5	–0.1	62.3
2005-06	4.1	2.6	0.4	63.0
2006-07	3.4	1.9	–0.2	56.3
2007-08	2.7	1.4	–0.9	41.4
2008-09	6.0	4.5	2.6	75.2
2009-10	6.5	5.1	3.2	81.0
2010-11	4.8	3.5	1.8	67.5
2011-12	5.7	4.4	2.7	75.5
2012-13 (RE)	5.2	3.9	1.8	68.2
2013-14 (RE)	4.5	3.3	1.5	–

Source: Economic Survey, 2010-11, 2011-12 and 2012-13

government and, therefore, slows down economic growth of the country, every effort should be made to reduce revenue deficit. in 2003 Fiscal Responsibility Budget Management Act (FRBMA) was passed according to which revenue deficit was to be eliminated by 2008-09. It will be seen that after 2002-03 revenue deficit as per cent of GDP has come down to 1.4 in 2007-08 (RE).

Fiscal Deficit

Fiscal deficit in the budget is an important measure of deficit. IMF and the World Bank generally prescribe targets for budget deficit in terms of fiscal deficit. *Fiscal deficit is the excess of total expenditure* (*both on revenue and capital accounts*) *over revenue receipts and only non-debt type of capital receipts such as recoveries of loans.*

Thus,

Fiscal Deficit = (*Total Expenditure both on Revenue Account and Capital Account*) – (*Revenue Receipts + Non-debt Capital Receipts*)

Fiscal deficit is a more comprehensive measure of budgetary imbalances. Thus, we see that when the government's total expenditure, both revenue and capital expenditure, exceeds its revenue from taxes and other normal receipts, *fiscal deficit* is created. Now, this fiscal deficit can be financed in two ways. Firstly, through borrowing by the government from the market, both inside and outside the country. On this borrowing, the government has to pay rate of interest annually. Apart from that it has to pay back the internal and external debt taken. Secondly, the

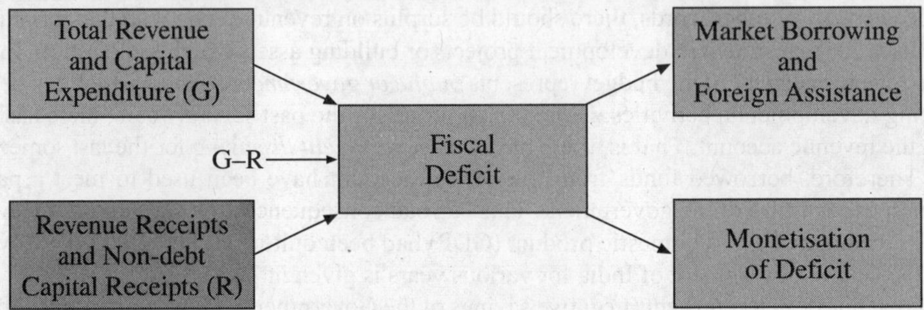

Fig. 1. *Financing of Fiscal Deficit*

government can finance the fiscal deficit by borrowing from the Reserve Bank of India which issues new notes against government securities. Thus, borrowing from the Reserve Bank results in expansion of high powered money in the economy and is popularly called deficit financing. It is in fact *monetisation* of fiscal deficit which, if undertaken to an excessive degree, leads to inflation in the economy. On the other hand, Government borrowing from the market to finance fiscal deficit adds to the public debt and increases burden on future generations on whom heavy taxes have to be imposed for repaying the loans.

In the Table 33A.1, fiscal deficit of the Central Government of India has been calculated for the last seventeen years. Fiscal deficit which was 4.7% of GDP in 1997-98 rose to 6.2% of GDP in 2001-2002 but it fell to 2.5% of GDP in 2007-08. It will be seen from this that in the Indian economy, fiscal deficit was quite large prior to 2003-04 which led to inflation or rise in prices in the economy on the one hand and increase in public debt on the other. Therefore, it was decided to reduce it to 3 per cent of GDP by the year 2008-09 and Fiscal Responsibility Budget Management (FRBM) Act was passed to discipline Government's fiscal policy and reduce fiscal deficit.

Accordingly, the government did not borrow directly from RBI through privately placing bonds with it since 2003 (That is, did not monetise the fiscal deficit since 2002). However, for the years 2008-09 and 2009-10, the Government planned to monetise the deficit to some extent as fiscal deficit greatly went up due to fiscal stimulus packages adopted to overcome recessionary conditions in the Indian economy created by global financial crisis of 2008-09 and it was not considered prudent to finance its fully through borrowing from the market which would keep interest rates higher and crowd out private investment apart from raising the burden of public debt. In the budget for 2007-08, fiscal deficit of the Central government was brought down to 2.5 per cent of GDP. In 1990-91, fiscal deficit of the Central Government was 7.85 per cent of GDP but came down to 4 per cent of GDP in 1996-97 but again rose to 5.7 per cent of GDP in 2000-2001 and to 6.2 per cent in 2001-2002. For 2008-09, actual fiscal deficit rose to 6 per cent of GDP and in 2009-10 to 6.3 per cent of GDP due to fiscal stimulus. In budget for 2010-11, the Finance Minister succeeded in to reducing it to 4.8 per of GDP (RE) (see Table 33.1). However, in 2011-12 due to adoption of several welfare schemes fiscal deficit again shot up to 5.7 per cent of GDP which caused much increase in aggregate demand and has been responsible for rise in inflation rate. For 2012-13 against the budget target of 4.6% of GDP it was feared that fiscal deficit would go up to even 6 per cent of GDP because of less than estimated collection of taxes due to slowdown of the economy and decline in estimated funds from 2G Spectrum and disinvestment of public sector. However, the Finance Minister succeeded in reducing government expenditure by cutting *planned expenditure* of over 91,000 crore though it was quite undesirable to do so. The States in India also have fiscal deficit of around 2.5 to 3 per cent of GDP.

Implications of Large Fiscal Deficit. Thus, there are three implications of fiscal deficit. A good part of it is financed through borrowing from within and outside the country. This *leads to the increase in public debt and its burden*. Secondly, a part of fiscal deficit is financed through deficit financing, that is, monetisation of fiscal deficit which leads to creation of new money and *rise in prices or inflation and increase in current account deficit (CAD)*. To check inflation and achieve price stability, IMF and World Bank have recommended that fiscal deficit in India should be reduced to 3 per cent of GDP in a phased manner.Thirdly, large deficits *adversely affects economic growth.* Due to large revenue deficit, a very large part of borrowed funds by the Government is used to finance current consumption expenditure of the Government. As result, a smaller amount of resources are left for productive investment in infrastructure and social capital (*i.e.* education and health) by the Government. This lowers the rate of economic growth. Lastly, more borrowing by the Government leaves less resources for private sector investment.

To reduce the central fiscal deficit to 3 per cent requires drastic cut in non-productive revenue expenditure which is a difficult task. *In India, attempts have been made to reduce fiscal deficit by curtailing capital expenditure which is incurred on projects of capital formation and other developmental activities.* This is not a desirable way of reducing fiscal deficit because this adversely affects economic growth. For sound management of Government finance there is need to cut revenue expenditure and raise revenue receipts by mobilising resources through taxation. Attempts to reduce fiscal deficit by curtailing capital expenditure, as has been the practice in India so far, adversely affects economic growth and should therefore be avoided.

To reduce fiscal deficit in his budget speech for 2013-14, the finance minister presented a roadmap for reducing fiscal deficit to 3% of GDP in 2016-17 as shown in Fig. 33.2. However, as shall be discussed below fiscal deficit, to a reasonable degree, it is not bad. In times of recessionary situation or slowdown in economic growth such as the one that prevails in 2008-09 and 2009-10, it is needed to fight recession and prevent slowdown in the economy. All over the world (including the US), the governments resort to a large dozes of fiscal deficit to stimulate the economy and prevent further deepening of recession and more losses of more jobs. It was for such a situation when private investment declines that J. M. Keynes recommended that Government should undertake deficit spending on a large scale to overcome recession and increase employment. In our view, though undertaking

1. See his article 'How Fiscal Deficits Hurt Economic Growth, *The Economic Times*, Nov-1, 2001 and 'Why Large Fiscal Deficits are bad for us, CNBC, January 2003
2. *Ibid.*

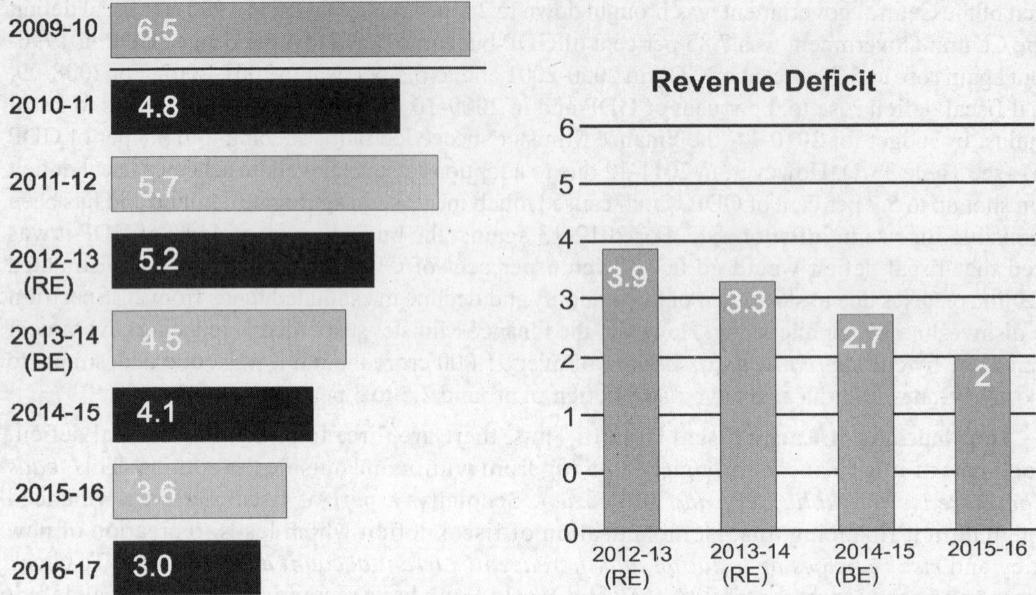

Fig. 33.2. *Roadmap of Reduction in Central Government's Fiscal and Revenue Deficits As Per Cent of GDP*

reasonability large fiscal deficit in recessionary situation or showdown in economic growth is desirable but in a developing country such as ours borrowed money should be used for ***productive purposes*** and not frittered away in spending on useless and wasteful projects.

Primary Deficit

Primary deficit is another important concept of budgetary deficit. Primary deficit is a measure of budget deficit which is obtained by deducting interest payments from fiscal deficit. Thus,

Primary Deficit = Fiscal Deficit – Interest Payments

Interest payments on public debt are transfer payments made by the government. The difference between the fiscal deficit and primary deficit shows the importance of interest payments on public debt incurred in the past. It will be seen from Table 3, that interest payments are very largely responsible for causing large fiscal deficit. A major part of Government borrowing to finance fiscal deficit is used for making interest payment on the past debt.

In India, in recent years, interest payments on public debt has increased very much. The higher interest payments on the past borrowings by the Government has greatly increased the fiscal deficit. Therefore, primary deficit is much lower than the fiscal deficit. In India, in the year 2001-02 primary deficit was 1.5 per cent of GDP, whereas fiscal deficit was 6.2 per of GDP. The primary deficit came down to zero in 2003-04 which implied that all the proceeds from government borrowing were used to spend on making interest payment on the past public debt. In the recent years 2006-07, 2007-08, primary deficit has been negative (– 0.2). This shows expenditure on interest payments exceeded borrowing by the Government. As mentioned above, interest payments is an item of liabilities in the revenue account of the budget. Therefore, if fiscal deficit has to be reduced, interest payments in the revenue account should be reduced which can be done if past public debt is quickly retired by mobilising more resources and curtailing non-developmental public expenditure.

Monetisation of Fiscal Deficit

It may be noted that budget deficit can financed by borrowing from the Reserve Bank of India by the central government against its own securities or treasury bills. The budget deficit is financed by borrowing from the Reserve Bank which issues new money or currency against government securities, which will lead to the expansion in money supply. In the old terminology, it was known as *deficit financing in India.* Financing the budget by issuing new money through borrowing from the Reserve Bank against securities in the new terminology is called *monetisation of fiscal deficit.* It may be noted that in the earlier years separate data regarding the budgetary deficit and therefore, the magnitude of deficit financing undertaken in a year were provided by the government. However, in recent years the data regarding budgetary deficit is not separately provided by the government and it has become a part of fiscal deficit which is shown to be financed by *government borrowing and other new liabilities.* Now the data regarding these deficits, namely, (1) revenue deficit, (2) fiscal deficit and (3) primary deficit are provided in the budget documents.

MEASURES FOR FISCAL CONSOLIDATION

By fiscal consolidation we mean raising more resource and reducing non-essential expenditure by the government so as to reduce fiscal deficit to a reasonable level. Reducing fiscal deficit is a formiddle challenge for control of inflation and achieving 8% growth in the 12the plan period (2012-17) large fiscal deficit has two bad consequences. First, it leads to excessive Government borrowing from the market which causes rise in market interest rate. Higher market interest rate tends to reduce private investment. Further, it reduces the resources available for private sector investment. Second, the extent to which a large fiscal deficit is financed by borrowing from the Reserve Bank of India which issues new currency (which is called reserve money or high-powered money) for the government. This causes greater expansion in money supply through the process of money multiplier and generates inflationary situation in the economy. Thus, to check the rate of inflation, fiscal deficit has to be reduced through both raising revenue of the government and reducing government expenditure. In the Indian context, the following measures can be adopted to reduce fiscal deficit and thereby to reduce inflationary pressures in the economy. We first spell out the measures which may be adopted to reduce government expenditure and then describe measures for raising Government revenue.

Measures to Reduce Public Expenditure. In the context of the Indian economy, the following measures can be adopted to reduce public expenditure for reducing fiscal deficit and thereby check inflation.

1. A drastic reduction in expenditure on major subsidies such as food, fertilisers, and petroleum products is required to curtail public expenditure. A huge sum of money is spent on major subsidies on food, fertilisers, petroleum products by the central government. Without a drastic cut in subsidies over time it is difficult to reduce public expenditure to an appreciable degree. In June 2012, the government rightly decontrolled petrol so that its price is aligned with global price level. Besides, to reduce its expenditure on subsidies diesel prices were increased in September 2012 by ₹ 5 per litre and now it has been decided to increase diesel price by around 50 paise every month. As regards checking leakages the policy of direct cash transfers to the beneficiaries has been adopted. To reduce subsidy on LPG, the supply of subsidised LPG cylinder to each consumer was capped at 6 per annum which together with hike in diesel prices was expected to lower government expenditure by ₹ 2300 crore. However, in the budget for 2013-14 expenditure on subsidies as per cent GDP is planned to be reduced marginally to 2 per cent of GDP compared to 2.2% of GDP for the years 2009-10 and 2010-11.
2. The huge sum of money is spent by the government on LTC (Leave Travelling Concessions), bonus, leave encashment etc. A reduction in expenditure on these is necessary if the government is determined to cut public expenditure.

3. Another useful measure to cut public expenditure is to reduce interest payments on past debt. In India, interest payments account for about 40 per cent of expenditure on revenue account of the central government. In our view, *funds raised through disinvestment in the public sector should be used to retire a part of old public debt rather than financing current expenditure.* Retirement of public debt quickly will reduce burden of interest payments in future.
4. Budgetary support to public sector enterprises other than infrastructure projects should be substantially reduced. Further, public sector enterprises should be asked to raise funds from the market and banks.
5. Austerity measures should be adopted to curtail non-plan expenditure in all government departments.

Increasing Revenue from Taxation

To reduce fiscal deficit and thereby check rise in inflation rate, apart from reducing government expenditure, government revenue has to be raised. However, in recent years tax-GDP ration has declined. In 2007-08, tax GDP ratio was 12 per cent which came down to 10.4 per cent for the year 2012-13. Continuing on the path of fiscal consolidation with a view to narrowing the gap in Government spending and resources, the tax-GDP ratio has been targeted at 10.9 per cent in the 2013-14 (BE) with a nominal growth rate of 19.1 per cent. However, this is not enough for fiscal consolidation to reduce fiscal deficit to 3% of GDP. The Finance Minister has not made genuine efforts to raise more resources through taxation. He has in his budget for 2013-14, has only levied a surcharge of 10% on income of those earning ₹ 1 crore or more. This is too low to yield enough tax revenue. With current income tax rate of 30% in this income bracket, it will means a super rich person will pay only ₹ 3 lakh more tax on his or her income of ₹ one crore per annum which is too small.

1. As regards mobilising resources to increase public revenue, it may be noted that the policy of moderate taxes with simplified taxation structure should be followed. This will help to increase public revenue rather than reduce it. High marginal rates of taxes should be avoided as they serve as disincentives to work more, save more and invest more. Further, high marginal rates of direct taxes cause evasion of taxes. However, *in our view, the present marginal rate of tax in India is 30 per cent for above 10 lakh income per annum.* This is too small. Besides, due to various exemptions effective rate has been estimated to be around 20 per cent. Therefore, there is need for introducing a higher income tax slab with 35% rate of income tax.
2. In India, the *tax base is narrow for both direct and indirect taxes*; only about 2.7 per cent of population pays income tax. To increase revenue from taxation, tax base should be broadened by taxing agricultural incomes and incomes derived from unorganised industrial and services sectors. *The various exemptions and deductions provided in the income and wealth taxes should be withdrawn to broaden the tax base and collect more revenue.* It may be noted that the Indian experience of the last 50 years reveals that these exemptions and deductions do not promote the intended objectives.
3. As is well known, there is a huge amount of black money in the Indian economy which has come into existence as a result of tax evasion. A good part of black money is parked in Swiss banks. In the last VDIS (Voluntary Disclosure Income Scheme) in 1997-98, more than, 10,000 crores of rupees were collected. However, there has been huge increase in the amount of black money since then. Not only the current black money has to be mopped up but also tax evasion that occurs every year has to be prevented by strict enforcement of the tax laws.

4. The past experience has shown that various tax concessions which have been given in income taxes and indirect taxes for promoting employment, industrial development of backward regions and for other such social objectives do not actually serve the intended purposes and are largely used for evading taxes. These *concessions should therefore be withdrawn to collect more revenue from taxes and the social objectives should be served by adopting more effective policy instruments.* Long-term capital gains on equity shares are exempted from income tax. In the context of boom conditions in the stock market, their exemption should be withdrawn to increase public revenue. They represent *unearned incomes* and canon of equity in taxation demands that this exemption should be withdrawn.

To sum up, with the adoption of above measures of reducing public expenditure and enhancing public revenue, it will be possible to reduce fiscal deficit to a safe limit. The reduction in fiscal deficit will prevent the emergence of excess demand in the economy and thereby help in controlling inflation and achieving price stability.

FISCAL DEFICIT AND ECONOMIC GROWTH

How does fiscal deficit affect economic growth is a hotly debated issue. A number of Keynesian economists argue that fiscal deficit promotes growth. In India where in the last some years since 1996, a good deal of industrial capacity has been lying idle due to lack of aggregate demand, and there has been enough stocks of foodgrains, it was asserted that fiscal deficit would stimulate demand and thereby ensure rapid economic growth. In the prevailing economic situation especially from 1996 to 2004 the Indian economy was a *demand-constrained economy* and therefore the increase in aggregate demand through larger fiscal deficit did not generate inflationary pressures in the economy. If the increase in public expenditure made possible by large fiscal deficit is used for productive investment, especially for investment in infrastructure and rural development, it will boost production and help increase employment opportunities in the economy. In fact, increase in public investment in infrastructure such as irrigation, roads, highways, is doubly beneficial from the viewpoint of accelerating economic growth. It helps in increasing aggregate demand on the one hand and helps to reduce supply constraints on economic growth on the other. So the focus should not be so much on reducing fiscal deficit but on reducing revenue deficit. Recall that revenue deficit is the excess of government revenue expenditure (*i.e.* current consumption expenditure) over revenue receipts. From the beginning of the nineties since economic reforms have been initiated, there has been large revenue deficit so that a large part of borrowings by the government has been used to bridge the revenue deficit. As a result, capital expenditure on investment in infrastructure and rural development as a per cent of GDP declined which has adversely affected economic growth.

Alternative View: Dr Shankar Acharya's View

However, from the above it should not be understood that any amount of fiscal deficit or government borrowing is good for economic growth and employment generation. Higher fiscal deficit, that is, borrowing by the government involves payment of interest and raises the burden of public debt. The large increase in public debt involving large interest payments year after year will not only make the process unsustainable but adversely affect economic growth through reducing investible resources for spending on infrastructure and social sectors. Reviewing the Indian growth scenario World Bank Study made in 2004[2] concludes: "Interest payments consumed less than 20% of total revenues in the pre-crisis period, compared with over 30% during the Ninth Plan period (1997-2002). Revenue deficits doubled from less than 3% in the second half of the 1980s to 6% during the Ninth

2. World Bank, *India Sustaining Reforms, Reducing Poverty*, Oxford University Press, 2004, p.14.

Plan period and beyond *representing a deterioration in the fiscal stance with spending on social and physical infrastructure crowded out by rising interest and other current payments.*"

As compared to the later half of the eighties (1985-86 to 1989-90), revenue and gross fiscal deficits (of the Centre and States combined) went up in the 9th Plan period (1997-98 to 2002), especially revenue deficit. As a result, interest payments as per cent of GDP greatly increased. The higher revenue deficits and large interest payments during the nineties caused a drastic decline in capital expenditure and public investment in physical infrastructure (irrigation + power + transport) as compared to the later half of the eighties (1985-86 to 1989-90).[3] The current expenditure on social services (such as education, health, housing, anti-poverty schemes) remained almost stagnant. This shows fiscal deficit (especially revenue deficit) has worked to lower economic growth.

Adverse impact of fiscal deficit on economic growth has been shown through its effect on saving and investment in the Indian economy. For example, Shankar Acharya, a former chief economic adviser to the Government of India, has contended that a *large* fiscal and revenue deficits during 1996-2003 as compared to 1995-96 slowed down economic growth in the second half of nineties and from 2000 to 2003 in India[4]. He argued that as a result of fiscal consolidation achieved between 1990-91 and 1995-96, combined fiscal deficit of Centre and States as percentage of GDP declined from 9.4 per cent in 1990-91 to 6.5 per cent in 1995-96 and revenue deficit fell from 4.2 per cent in 1990-91 to 3.2 per cent in 1995-96 (See Table 33.3). This freed resources for private investment and in fact brought about investment boom of 1993-96. Gross domestic investment rose from 22.5 per cent of GDP in 1991-92 (to which it had fallen during the crisis) to a peak of 26.8 per cent GDP in 1995-96 (See Table 33.3). Gross domestic saving also rose to a record level of 25.1 per cent of GDP in 1995-56. This increase in gross domestic saving and investment, according to Dr. Shankar Acharya, "helped propel India's growth to 7 per cent plus for three successive years in the mid-nineties" (1995-96, 1996-97 and 1997-98)."

After the mid-nineties, fiscal deficit and revenue deficit increased which caused the decline in gross domestic saving and investment and thereby contributed to slowdown in economic growth. To quote Dr. Shankar Acharya, "*It would be hard to find more telling evidence of the adverse impact of deficits on savings and investment*"[5]. Of course, other factors were also dampening investment, including the slowdown in reforms, the higher uncertainties associated with coalition government and a worsening international economic environment. But increased borrowing by government (rising fiscal deficits) and the associated high real interest rates clearly played a role in slowing private investment. The drop in aggregate savings and investment took its toll of economic growth, which slowed from 7 percent plus of the mid-nineties to around 5 percent in recent years"[6].

According to him, *high fiscal deficits during the eighties created balance of payments deficits as it was mainly financed by external commercial borrowing and therefore did not adversely affect economic growth.* But in the later half of nineties rising fiscal deficits were financed by domestic borrowings which were largely used to meet current consumption expenditure and resulted in *public dissavings* (see Table 33.3). It may be noted that the implementation (during 1997-2000) of Eighth Pay Commission report enhancing the salaries of Government staff both at the Centre and in States sharply increased the revenue expenditure of the governments and increased public dissavings. This adversely affected domestic rate of saving and investment and lowered rate of economic growth.

The high and rising fiscal deficits during the period from 1997 to 2003 which resulted in larger government borrowings from the market preempted the needed resources for investment by the private sector. This had an adverse effect on private investment and was mainly responsible for slowdown in economic growth.

3. *Op. cit.*
4. Shankar Acharya, 'High Fiscal Deficit Hurts Growth, *The Economic Times*, Nov. 1, 2003.
5. *Ibid*
6. *Ibid*

Role of Fiscal Deficit and Economic Growth: Our View

However, we do not fully agree with Dr. Shankar Acharya's view which is similar to that of World Bank. They consider only supply-side factors determining economic growth.The fall in growth rate during the period from 1997 to 2002 was largely due to demand recession in the economy caused by (1) reduction in the capital expenditure by Government as a matter of new economic policy initiated since 1991, (2) sluggish performance in agriculture during the period, and (3) world-wide recession during the period. Thus demand recession caused a slowdown in the Ninth Plan period (1999-2002) As a result, gross saving and investment rate and GDP growth slowed down. As will be seen from Table 33.5 of the previous chapter, capital expenditure of government increased substantially in 2003-04 and 2005-06 and with multiplier effects led to the increase in aggeregate demand. This increase in demand also caused increase in private investment and the two together ensured a higher rate of economic growth. In the two years (2005-07) even when Government's capital expenditure as per cent of GDP fell, under pressure of rise in aggregate demand it was made up by increase in private investment in the two years 2005-06 and 2006-07 which kept the momentum of higher economic growth.

Thus in our view both supply-side and demand-side factors play a role in determining economic growth. The growth experience of a country cannot be explained by supply-side factors alone. In fact, in 2008 and 2009 as a result of global financial crisis there was economic slowdown in India which caused huge job losses. To prevent the situation from worsening further the Government came out with three fiscal stimulus packages to keep the growth momentum. In these stimulus packages Government raised its expenditure, especially on infrastructure on the one hand and cut taxes on the other to stimulate aggregate demand. To increase its expenditure Government borrowed from the market about ₹ 300,000 crore in 2008-09 and ₹ 400,000 crore in 2009-10. As a result, its fiscal deficit went up from 2.7 per cent of GDP in 2007-08 to 6.0 per cent in 2008-09 and 6.4% of GDP in 2009-10. *But this did not produce bad results for the private sector investment, nor did it result in higher rate of interest*. Of course RBI reduced cash reserve ratio to increase liquidity in the banking system and lowered its repo rate to enable banks to lower the lending rates of interest. As a matter of fact, increased borrowing and expenditure by the Government in 2008-09 and 2009-10 led to the increase in demand for the products of private sector. Therefore, from May 2009 onward there were signs of revival of the Indian manufacturing sector. Manufacturing sector output grew at 9.6% in Sept. 2009 year-on-year basis and this growth further picked up to 17.6% year-on-year basis in Dec. 2009. Manufacturing sector registered growth of 13.5 per cent in March 2010, 16.5% in April 2010 and 11.1% in May 2010 year-on-year basis. This shows despite 6.0 per cent fiscal deficit in 2008-09 and 6.4 % in 2009-10 and consequently heavy borrowing by the government in these two years manufacturing output recorded high growth.

Besides, it is because of the fiscal stimulus packages and increase in government expenditure made possible by heavy borrowing and fiscal deficit that India could achieve 6.8 per cent rate of economic growth in 2008-09, 8.6% in 2009-10 and 9.3 % in 2010-11 which is quite high, especially when there were recessionary conditions in the US and Europe. So in our view fiscal deficit of more than 3% in a year is not always bad. All depends on the economic situation in the country and the purposes for which borrowed funds are spent. If there is recession or slowdown in the economy, the fiscal deficit of 6 to 7 per cent of GDP is necessary to lift the economy out of recession or to prevent sharp slowdown in the growth of the economy. If borrowed funds are spent by the Government for investment or building durable assets, it will lead to the expansion in productive capacity and therefore ensure sustained economic growth.

It is evident from above that larger fiscal deficit at the time of recession or economic slowdown far from reducing private saving and investment is helpful for fighting recession or keeping growth momentum of the economy. In fact, the present economic thinking is that even 3 per cent of fiscal deficit should be treated as *cyclically adjusted number*; that is, the fiscal deficit should go up at times of recession or economic slowdown and should come down in normal or boom period.

We have explained above the two alternative views regarding the impact of fiscal deficit on our economic growth. One view is of the Keynesian economists who think that in developing economies such as India when they suffer from demand constraint, fiscal deficit provides stimulus to economic growth by bringing about increase in aggregate demand. In the situation prevailing in India from 1997 to 2002, and then in 2008-2009 the Indian economy in fact suffered from lack of aggregate demand for manufactured products. This demand deficiency problem arose due to *bad agricultural performance in some years* (there was a negative agricultural growth in 1995-96, 1997-98, 1999-2000, 2000-01 and 2002-03 and also due to sharp *decline in public sector investment* which started in early nineties following the initiation of economic reforms since 1991.

Thus, during 1997-2002 there was a decline in public investment not only because a good deal of borrowing by the government was used to meet revenue deficit but also because of change in economic policy which aimed at reducing the role of public sector. Besides, public sector investment could not be raised also because of the fiscal policy stance of reducing fiscal deficit as recommended by IMF, World Bank and economists associated with these institutions. *In fact the focus of fiscal consolidation was reducing fiscal deficit rather than revenue deficit.* In fact, rates of income tax, corporation tax, excise and customs duties were reduced which resulted in fall in tax-GDP ratio till 2001-02 and caused revenue deficit to rise and therefore much of the borrowing by the government was used to bridge the revenue deficit. Thus, the fall in public investment was embedded in the new economic policy of reforms. *Fall in public sector investment resulted in decrease in aggregate demand on the one hand and shortage of infrastructure on the other which caused slowdown in economic growth during the period 1997-2002.* Again in 2008-09 and 2009-10 it was increase in Government expenditure made possible by heavy borrowing and large fiscal deficit that helped India to achieve GDP growth of 6.7% in 2008-09 and 8.6% in 2009-10.

Here it may be recalled the **Golden Rule of Public Finance**, namely, *borrowing by the government should be used only for investment purposes and not to meet excess current consumption expenditure except during times of recession.* In times of recession even government consumption expenditure will promote GDP growth. Thus, the focus on merely reducing fiscal deficit at all times and treat it as fetish is not right. In fact, in a demand-constrained economy, a moderate doze of fiscal stimulus is needed. Fiscal responsibility should not be conceived in ritualistic terms of reducing fiscal deficit alone regardless of its effect on the public investment and the economy. In fact, the increase in public investment in infrastructure will also stimulate private investment. The question of containing Government consumption expenditure, except during periods of recession or economic slowdown, is important but reduction in fiscal deficit should not be elevated to a dogma. A moderate amount of fiscal deficit and associated borrowing is good as long as it is used for increasing public investment in physical infrastructure, education and health of the people. Even foreign investment depends on our success in improving infrastructure.

PART XIII

OPEN ECONOMY MACROECONOMICS

- Balance of Payments
- Foreign Exchange Rate
- Free Trade Versus Protection
- Foreign Investments and Role of Multinational Corporations
- Globalisation, Commercial Policy and WTO

CHAPTER 34
BALANCE OF PAYMENTS

Balance of Payments: Meaning

In the modern world, there is hardly any country which is self-sufficient in the sense that it produces all the goods and services it needs. Every country imports from other countries the goods that cannot be produced at all in the country or can be produced only at an unduly high cost as compared to the foreign supplies. Similarly, a country exports to other countries the commodities which those countries prefer to buy from abroad rather than produce at home.

"*The balance of payments is a systematic record of economic transactions of the residents of a country with the rest of the world during a given period of time*." The record is so prepared as to provide meaning and measure to the various components of a country's external economic transactions. Thus, the aim is to present an account of all receipts and payments on account of goods exported, services rendered and capital received by residents of a country, and goods imported, services received and capital transferred by residents of the country. The main purpose of keeping these records is to know the international economic position of the country and to help the Government in reaching decisions on monetary and fiscal policies on the one hand, and trade and payments questions on the other.

Balance of Trade and Balance of Payments

Balance of trade and balance of payments are two related terms but they should be carefully distinguished from each other because they do not have exactly the same meaning. *Balance of trade refers to the difference in value of imports and exports of commodities only, i.e., visible items only*. Movement of goods between countries is known as visible trade because the movement is open and can be verified by the custom officials.

During a given period of time, the exports and imports may be exactly equal, in which case, the balance of payments of trade is said to be balanced. But this is not necessary, for those who export and import are not necessarily the same persons. If the value of exports exceeds the value of imports, the country is said to experience an *export surplus* or a favourable balance of trade. If the value of its imports exceeds the value of its exports, the country is said to have a *deficit or an adverse balance of trade*.

The terms "*favourable*" and "*unfavourable*" are derived from the mercantilist writers of the 18th century. In those days, settlements of the foreign transactions were made in gold. If India had exported ₹ 100 crore worth of goods but had imported ₹ 80 crore worth of goods, India would receive ₹ 20 crores worth of gold from the foreign countries. As gold was regarded as wealth and as the receipts of gold made a country wealthy, the mercantilist writers regarded exports surplus as being favourable to the country.

On the other hand, if India had exported ₹ 100 crores worth of goods, but imported ₹ 150 crores worth of goods, it had to pay ₹ 50 crores in gold to the foreigners. India would be losing gold and would be poorer to that extent. Therefore, an import surplus was regarded by the mercantilist writers as adverse balance. But in these days, the international transactions are not settled in terms of gold. Even then, the terms "*favourable*" and "*unfavourable*" balance of trades have continued to be used till today.

Exports and imports of a country are rarely equal. Balance of trade, in other words, will not balance. During any period a country may experience a favourable or an adverse balance of trade.

Distinction between Current Account and Capital Account

The distinction between the current account and capital account may be noted. The current account deals with payment for *currently produced goods and services*; it includes also interest earned or paid on claims and also gifts and donations. *The capital account, on the other hand, deals with payments of debts and claims*. The current account of the balance of payments affects the level of national income directly. For instance, when India sells its currently produced goods and services to foreign countries, the producers of those goods get income. In other words, current account receipts have the effect of increasing the flow of income in the country. On the other hand, when India imports goods and services from foreign countries and pays for them, money which would have been used to demand goods and services within the country flows out to foreign countries. The current account payments to foreigners involve reduction of the flow of income within the country and constitute a leakage. Thus, the current account or trade account of the balance of payments has a direct effect on the level of income in a country. The capital account, however, does not have such a direct effect an the level of income; it influences the volume of assets which a country holds.

It may be further noted that when there is a deficit in the current account, it has to be financed either by *using foreign exchange reserves* with Reserve Bank of Bank, if any, or by *capital inflows* (in the form of foreign assertance, funds flowing through FDI and portfolio investment by FIIs, commercial borrowing from abroad, non-resident deposits).

Determinants of Balance of Payments

There are several variables which determine the balance of payments position of a country, viz., national income at home and abroad, exchange rate of national currency, the prices of goods and factors, international oil and commodity prices, the supply of money, the rate of interest, etc. all of which determine exports, imports, and demand and supply of foreign currency. At the back of these variables lie the supply factors, production function, the state of technology, tastes, distribution of income, economic conditions, the state of expectations, etc. If there is a change in any of these variables and there are no appropriate changes in other variables, disequilibrium will be the result.

The main cause of disequilibrium in the balance of payments arises from imbalance between exports and imports of goods and services, that is, deficit or surplus in balance of trade. When for one reason or another exports of goods and services, of a country are smaller than their imports, disequilibrium in the balance of payments is the likely result. Exports may be small due to the lack of exportable surplus which in turn results from low production or the exports may be small because of the high costs and prices of exportable goods and severe competition in the world markets. Important causes of small exports are the inflation or rising prices in the country or over-valued exchange rate. When the prices of goods are high in the country, its exports are discouraged and imports encouraged. If it is not matched by other items in the balance of payments, disequilibrium emerges.

Balance of Payments on Current Account

Balance of payments is more comprehensive in scope than balance of trade. It includes not only imports and exports of goods which are visible items but also such invisible items as shipping, banking, insurance, tourism, interest on investments, gifts, etc.

A country, say India, has to make payments to the other countries not only for its imports of merchandise but also for banking, insurance and shipping services rendered by other countries; it has to pay further the royalties for foreign firms, expenditure of Indians in foreign countries, interest on foreign investments in India, or on loans obtained by India from other countries and such international organizations as the I.M.F., I.B.R.D., etc. These are debit items for India, since these transactions involve payments abroad. In the same way, foreign countries import goods from India, make use of Indian films and so on, for all of which they make payments to India. These are the credit items for India as the latter receives payments. Balance of payments thus gives a comprehensive picture of all such transactions including imports and exports of goods and services.

The Table 34.1 (given below) we give the position of India's balance of payments on current account for the years 2005-06 to 2012-13. In this table of balance of payments are given the visible as well as invisible items of trade. The *visible items* are export-import trade and the *invisible items of balance of payments on current account are travel, transportation and insurance, interest on loans given and other investment income and private transfers such as remittances from Indians living abroad.* Both visible and invisible items together make up the *current account.* Interest on loans, tourist expenditure, banking and insurance charges, etc., are similar to visible trade since receipts from selling such services to the foreigners are very similar in their effects to receipts from sales of goods; they provide income to the people who produce the goods or services.

Table 34.1. India's Balance of Payments on Current Account ***(in Billion US $)***

S.No. Items	2006-07	2007-08	2008-09	2009-10	2010-11	11-12	12-13	13-14
1. Exports	128.9	166.2	189.0	182.2	256.2	309.8	306.6	318.6
2. Imports	190.7	257.6	307.6	300.6	383.5	499.5	502.3	466.2
3. **Trade Balance**	–61.8	–91.5	–118.6	–118.4	–127.3	–189.9	195.7	–147.6
	(–6.5)	(–7.4)	(–9.8)	(–8.6)	(–7.8)	(–10.1)	(–10.6)	(N.A)
4. **Invisibles (Net)**	52.2	75.7	89.9	79.9	79.3	111.6	105.5	115.2
(*i*). Non-factor Services	29.5	38.9	49.6	35.7	44.1	.64.1	64.9	72.9
(*ii*). Investment Income	–7.2	–5.1	–4.0	–8.0	–17.9	–16.0	–21.5	–23.0
(*iii*). Private Transfers	29.8	41.7	44.6	52.3	53.1	63.5	64.0	65.3
5. **Goods and Services Balance**	–32.3	–52.6	–69.0	–83.0	–83.2	–125.7	–130.7	N.A
6. **Current Account Balance (Net)**	**– 9.6**	**–15.7**	**–28.7**	**–38.4**	**–48.1**	**–78.2**	**–88.2**	**–32.4**
	(–1.0)	**(–1.3)**	**(–2.4)**	**(–2.8)**	**(–2.7)**	**(–4.2)**	**(–4.7)**	**(–1.7)**

Source : RBI

Note : *Figures in brackets show per cent of GDP at market prices.*

It will be noted from Table 34.1 above that the most important item in the balance of payments on current account is balance of trade which refers to imports and exports of goods. In the Table 34.1 balance of trade does not balance and shows a deficit in all the seven years. In the years 2011-12 and 2012-13 trade deficit has substantially increased. Trade deficit was over 10 per cent of GDP in both these years. In fact, it is huge trade deficit in these two years that has caused huge current account deficit of over 4% of GDP in these two years Economic slowdown in advanced countries and its spill over effects in Emerging Market Economies coupled with high crude oil and gold prices were responsible for sharp increase in trade deficit. Due to *surplus in invisibles account*, there was a surplus on current account during 2001-2002, 2002-03 and 2003-04. In India's balance of payments on current account from 2004-05 onwards there has been a deficit. Contrary to popular perception, deficit on current account is not always bad provided it is within reasonable limits (which in case of India is estimated to be 2.5% of GDP) and can be easily met by non-debt capital receipts. In fact, *deficit on current account represents the extent of absorption of capital inflows in India during a year.*

It may be noted that when there is deficit on the current account, it is financed either by using *foreign exchange reserves* held by Reserve Bank of India or by *capital flows* that come into the country in the form of foreign direct investment (FDI) and portfolio investment by FIIs, external commercial borrowing (ECB) from abraod and by NRI deposits in foreign exchange account in our banks. However, due to global financial crisis in 2008-09, there was first slowdown and then decrease in exports. As a result, there was a large deficit of 2.4 of per cent of GDP on current account which could not be met by capital inflows as they were quite meagre ($ 8.6 billion) as a result of global financial crisis. Therefore, to finance the deficit on current account in 2008-09 we had to withdraw

U.S. $ 20 billion from our foreign exchange reserves. Again in the last two years 2011-12 and 2012-13 the current account deficit (CAD) has been quite high.

It may be noted that high current deficit tends to weaken the rupee by raising the demand for US dollars. In 2011-12, current account deficit tended to weaken the rupee by raising the demand for US dollars. In 2011-12, the current account deficit was 4.2 per cent of GDP. Since capital inflows in this year were not adequate to finance the current account deficit, RBI had to withdraw 12.8 million US dollars from its foreign exchange reserves to meet the demand for US dollars (see Table 34.2) . In the year 2012-13 the current account deficit has been estimated to the even higher at 4.8 per cent of GDP, capital inflows through portfolio investment by FIIs picked up in the latter half of 2012-13 but capital inflows through FDI had fallen. In 2013-14 current account deficit situation has worsened as FIIs have pulled out both from debt and equity markets since June 2013. This has led to the sharp depreciation of rupee by increasing demand for dollars. Thus current account deficit poses serious challenge to macroeconomic management of the economy. The dependence on volatile capital inflows through FIIs to meet the current account deficit is unsustainable as these capital flows go back when economic situation in their home countries improve and thereby causes sharp depreciation in exchange rate of rupee and crash in stock market prices.

Since in the recent years, 2011-12 and 2012-13 current account deficit has widened, which has increased the balance of payments vulnerability to sudden reversal of capital flows, especially when sizable flows are comprised of debt and volatile portfolio investment by FIIs. The priority has therefore been to reduce current account deficit (CAD) through improving trade balance. Efforts have been made to promote exports by diversifying the export commodity basket and export destinations. One way to limit imports is to bring prices up to the international level so that users bear the full cost. Accordingly, petrol has been decontrolled and diesel prices have been revised upward in Jan. 2013 to curtail subsidy on it. To discourage the imports of gold which has played a significant role in causing trade deficit, customs duty on its import was first raised from 6% to 8% and then to 10 per cent.

Further, to improve the current account deficit emphasis has been on facilitating remittances and encouraging software exports that have been responsible for surplus on the invisible account. In recent years this surplus has lowered the impact of widening trade deficit on current account deficit (CAD) significantly. The two components together met nearly two thirds of the trade deficit that was more than 10 per cent of GDP in 2011-12. Remittances particularly are known to exhibit resistance when the country is hit by external shock as was evident during the global crisis of 2008. Besides, several incentives have been given to NRI to deposit their foreign exchange in Indian banks.

Balance of Payments on Capital Account

In the balance of payments on capital account given in Table 34.2 important items are borrowings from foreign countries and lending funds to other countries. This takes two forms: (*i*) *External assistance* which means borrowing from foreign countries under concessional rate of interest; (*ii*) *Commercial borrowing* under which the Indian Government and the Private sector borrow funds from world money market at higher market rate of interest. Besides *non-resident* deposits are another important item in capital account. These are the deposits made by non-resident Indians (NRI) who keep their surplus funds with Indian Banks. Another important item in balance of payments on capital account is *foreign investment* by foreign companies in India. There are two types of foreign investment. First is *portfolio investment* under which foreign institutional investors (FII) *purchase shares (equity) and bonds* of Indian companies and Government. The second is *foreign direct investment* (FDI) under which foreign companies set up plants and factories on their own or in collaboration with the Indian companies. Still another item in capital account is other capital flows in which the important source of funds is remittances from abroad sent by the Indian citizens working in foreign countries. The Table 34.2 gives the position of India's capital account for the years 2004 -05 to 2009-10.

Table 34.2. India's Balance of Payments on Capital Account ***(in Billion US $)***

	2005-06	2006-07	2007-08	2008-09	2009-10	2010-11	2011-12
External Assistance (Net)	1.7	1.8	2.1	2.6	2.9	4.94	2.3
Commercial Borrowing (Net)	2.5	16.1	22.6	7.0	2.8	12.16	10.3
Non-Resident Deposits (Net)	2.8	4.3	0.2	– 4.3	2.9	3.14	11.9
Foreign Investment (Net) of which	15.5	14.8	43.3	3.5	50.4	42.13	39.2
(*i*) FDI net	3.0	7.7	15.9	17.5	18.0	11.83	22.1
(*ii*) Portfolio Investment (Net)	12.5	7.1	27.4	–15.0	32.4	30.3	17.2
(*iii*) Other capital flows (Net)	2.4	9.2	39.7	–9.7	–13.1	–10.48	–7.0
Capital Account Balance (Net)	**24.9**	**46.1**	**107.9**	**8.6**	**51.6**	**63.74**	67.8
Use of Exchange Reserves*	–15.0	–36.6	–92.2	+20.1*	–13.4*	(–13.1)*	(+12.8)*

* *Also includes errors and omissions*

Source: RBI Annual Report 2012-13 and 3013-14. Use of foreign exchange reserves (–indicates increase of reserves, and + indicates decrease of reserves on BOP basis.

Capital inflows in the capital account can be classified into ***debt creating*** and ***non-debt creating.*** Foreign investment (both direct and portfolio) represents non-debt creating capital inflows, whereas external assistance (i.e. concessional loans taken from abroad), external commercial borrowing (ECB) and nonresident deposits are debt-creating capital inflows.

It will be seen from Table 34.2, during 2007-08, there was net capital inflow of 43.3 billion US dollars on account of *foreign investment (both direct and portfolio).* The Table 34.2 gives the position of India's balance of payments in capital account for seven years, 2007-08, 2008-09 and 2009-10, 2010-11 and 2011-12, 2012-13 and 2013-14. When all items of balance of payments on capital account are taken into account we had *a surplus* of 107.9 billion US dollars in 2007-08. Taking into current account deficit of $ 15.7 billion on current account in year 2007-08 there was accretion to our foreign exchange reserves by $ 92.2 billion in 2007-08. Global financial crisis affected our capital account balance as there was ***reversal of capital flows after Sept. 2008*** with the result that ***we used*** $ 20.1 billion of our foreign exchange reserves in 2008-09 resulting in *decrease* of our foreign exchange reserves. That is, because we used our foreign exchange reserves equal to $ 20.1 billion, there was decline in our foreign exchange reserves by $ 20 billion in 2008-09. The situation improved in 2009-10 as foreign direct investment (FDI) and portfolio investment by FIIs picked up. As a result there was net capital account surplus of $ 51.6 billion in 2009-10 and after meeting the current account deficit of $ 38 billion there was addition to our foreign exchange reserves by $ 13.4 billion in 2009-10. In 2010-11 also there was surplus on capital account of $ 63.74 billion and after meeting current deficit we added $ 13.1 billion in our foreign exchange reserves in this year.

However, in 2011-2012 and 2012-13 the situation regarding capital flows changed significantly and capital flows were not sufficient to meet the large current account deficit (CAD). Consequently, in 2011-12 withdrawal from foreign exchange reserves of 12.8 billion US dollars was made. In 2012-13 and 2013-14, the situation improved and after meeting deficit on current account deficit, we added to our foreign exchange reserves.

Capital flows are driven by pull factors such as economic fundamentals of recipient countries and push factors such as policy stance of source countries. The capital flows have implications for exchange rate management, overall macroeconomic and financial stability including liquidity conditions. Capital account management therefore needs to emphasize promoting foreign direct investment (FDI) and reducing dependance on volatile portfolio capital. This would ensure that to the extent current account defect is bridged through capital surplus it would be better if it is done through stable and growth enhancing foreign direct investment flows. In the present international financial situation, reserves are the first line of defence against the volatile capital flows. However, the decline in reserves as a percentage of GDP is a source of concern.

Large Invisible Surpluses. At a micro-level the balance of payments profile reveals important trends. In the recent years deficits in the trade account have been made up by the large invisible surpluses resulting in a small deficit in current account balance. *The robust growth in invisibles is due to large inflows of private transfers and non-factors services*. Buoyant inflows of private transfers have been one of the main factors contributing to remarkable growth in invisibles. The current high level of private transfers which comprise largely inflows of remittances from Indians working abroad have made India the highest global recipient of remittances. The bulk of these remittances come in from expatriates in the USA and the Middle East.

In addition to private transfers, the contribution of *non-factor services* to the invisible account has been increasing steadily since 2001-02. This contribution increased further in 2006-07, 2007-08, 2008-09, 2009-10 and 2010-11. Among the non-factor services enhancing the invisible surplus is the *buoyant earnings from software services*. Net inflows from software services increased from US $ 6.9 billion in 2001-02 to US $ 10 billion in 2003-04. The software services (net) recorded a rise of 31.7 per cent during 2005-06, 30.4 per cent in 2006-07 and 28.3 per cent in 2007-08.

The high skill intensity of the Indian work force has given a decisive comparative advantage to Indian software services exports and has helped the country in carving out a niche in the global market for software services and IT-enabled services. Besides, in recent years, India has emerged as one of most preferred destination of outsourcing IT services from advanced economies. While software services exports more than trebled between 1998-99 and 2003-04, exports of IT -enabled and business process outsourcing (BPO) services have increased significantly in recent years.

On the other hand, the capital account surplus has not been growing strongly due to unsteady growth in *non-debt creating foreign investment flows*. Though *debt-creating capital inflows* through external commercial borrowing and external assistance have been significantly brought down and in fact in many years there were *net capital outflows* on these counts. However it is matter of serious concern that in recent years our dependence on volatile capital inflows through portfolio investment by FIIs has largely increased.

It is evident from above that trends indicate that the *fast growing invisibles and non-debt creating foreign investment flows* are the main factors behind large accumulation of foreign exchange reserves. But the situation in 2013-14 has changed for the worse. Due to large current account deficit and the announcement by Federal Reserve of the US about unwinding of quantitative easing (QE), there has been capital outflows from India since June 2013 by FIIs.

Does Balance of Payments must always Balance ?

It is often said that balance of payments must always balance. What does it mean ? The individuals and business firms of an economy have to pay for the imports from abroad. If exports are not sufficient to pay for the imports, then how the balance of payments will be in balance. For example, the balance of payments on current account of India had been in deficit for most of the years till 2000-01. Deficit on current account implies that the residents of a country are spending more on imports of goods and services than the incomes they are earning from exports of goods and services. For the overall balance of payments to be in balance, this deficit in the current account of the balance of payments must by financed by selling capital assets such as shares and bonds of companies or other assets such as gold or foreign exchange reserves of a country or by borrowing from abroad. Both by selling assets or by borrowing from abroad, foreign capital flows into the country as has been happening in the last several years in India. These foreign capital inflows are shown in the capital account of the balance of payments which must be in surplus to finance the deficit in the current account. Thus

current account deficit + capital account surplus = 0.

However, it is worthwhile to note that this "balance" is in the accounting sense only. Surplus on current account can lead to the grant of loans to the other countries by the Government or it can lead to the increase in the reserves of foreign exchange which show up themselves in the capital account. On the other hand, a deficit on current account can be met by capital inflows from abroad, that is, by obtaining foreign investment or getting foreign aid or by drawing from IMF or by running down country's foreign exchange reserves. Thus the surplus or deficit in the current account of a country's balance of payments gives rise to further financial transactions which show up themselves in capital account. In other words, if the balance of payments is used in the wider sense so as to include external assistance, drawing from IMF, drawing upon the country's reserves also as distinguished from its narrower sense, the balance of payments of a country must always balance—all the receipts taken together must be equal to all the payments taken together.

The above fact has an important lesson that must be borne in mind. If a country has no foreign currency reserves or it has no assets to sell to pay for the imports and if no body is willing to lend to it, it will have to cut down its imports which will reduce productive activity in the economy and adversely affect economic growth of the country. Such a crisis situation arose in India in 1991 when our foreign exchange reserves had fallen to a very low level and no one was willing to lend to us or give us aid. In fact, due to loss of confidence of foreign investors, *capital outflows* were taking place.

Therefore, in 1991 India had to mortgage its gold to Bank of England and Central Bank of Japan to get the necessary foreign exchange to pay for the needed imports. We had to accept the pre-conditions of IMF for providing us assistance to tide over the crisis. It is interesting to note this was done under the guidance of Dr. Manmohan Singh, the present Prime Minister of India, who was then the Finance Minister.

EQUILIBRIUM AND DISEQUILIBRIUM IN THE BALANCE OF PAYMENTS

Before we analyse the causes of disequilibrium in the balance of payments, we would like to explain what is meant by equilibrium in the balance of payments. We have noted above that when we add up all the demand for foreign currency and all the sources from which it comes, these two amounts are *necessarily equal* and thus the overall account of the balance of payments necessarily balance or must always be in *equilibrium*.

What then do we mean by when we say that the balance of payments of a country is 'in equilibrium or disequilibrium'. As a matter of fact, when we speak of equilibrium or disequilibrium in the balance of payments we refer to the balance on those parts of the account which do not include the accommodating items such as borrowing from the IMF, use of SDRs, drawing from the reserves of foreign currencies held by the Central Bank, etc. *When excluding these accommodating items there is neither deficit nor surplus in the overall balance of payments, it is said to be in equilibrium. When in this sense, there is either deficit or surplus, the balance of payments is said to be in disequilibrium.* The deficit in balance of payments can be financed by drawings from the IMF, use of SDRs, drawings from the reserves of foreign currencies and loan and aid received from abroad. For example in 2001-02, we added to our foreign exchange reserves to the tune of 11757 million US dollars. However for previous several years India's balance of payment on current account was in deficit. To finance the dificits India borrowed from IMF or from other countries or even resorted to commercial borrowing from abroad. But India's balance of payments for the year 2001-02 was favourable.

Basic Balance of Payments, Autonomous Items and Accommodating Items

However, a more important and popular concept of balance of payments equilibrium has been of *basic balance*. The concept of basic balance is based upon the idea of *autonomous items* in the balance of payments. The autonomous items in the balance of payments are those items which cannot be influenced or changed so easily or quickly by the Government and they are determined by some long-term factors. In this concept of basic balance, besides the items in the current account, *the long-term capital movements* both on private or Government account contained in the capital-account balance of payments are regarded as autonomous. On the other hand, in the capital account *short-term capital movements* such as borrowing from IMF or Central Banks of other countries, drawings from SDR, change in foreign exchange reserves are transitory and of *accommodating nature* and are therefore excluded from the concept of basic balance or of dequilibrium. The recourse has to be made to these accommodating items (also called *compensatory items*) so as to ensure equality between payments and receipts of foreign exchange. Changes in the compensatory items are made so as to offset the surplus or deficit in the autonomous items. Thus," *when autonomous movements cancel out over some appropriate time period and there is no need for compensatory movements, the balance of payments is in equilibrium.*[1] *Note that the equilibrium is a state of balance which can be sustained without intervention by the Government.*

The concept of balance of payments in the sense of basic balance can be represented by the following equation:

$$(X-M)+LTC=0$$

where X stands for exports including invisible items.

M stands for imports including invisible items.

LTC stands for long-term capital movements.

If $(X-M)$ is positive (*i.e.*, $X>M$), then for balance of payments to be in equilibrium, LTC will be negative and equal to $(X-M)$. This implies that there will be *net capital outflow*. On the other hand, if $M>X$, then for the balance of payments to be in equilibrium LTC must be positive (that is, there will be *net capital inflow* to offset the deficit in the current account).

When the balance of payments of a country is in equilibrium, the demand for the domestic currency is equal to its supply. The demand and supply situation is thus neither favourable nor unfavourable. If the balance of payments moves against a country, adjustments must be made by encouraging exports of goods, services or other forms of exports, or by discouraging imports of all kinds. No country can have a permanently unfavourable balance of payments. Total liabilities and total assets of nations, as of individuals, must balance in the long run.

This does not mean that the balance of payments of a country should be in equilibrium individually with every other country with which she has trade relations. This is not necessary, nor is it the case in the real world. Trade relations are multilateral. India, for instance, may have balance of payments deficit with the United States and surplus with the United Kingdom and/or other countries, but each country, in the long run, cannot receive more value than she has exported to other countries taken together.

Equilibrium in the balance of payments, therefore, is a sign of the soundness of a country's economy. But disequilibrium may arise either for short or for long periods. A continued disequilibrium indicates that the country is heading towards economic and financial bankruptcy. Every country, therefore, must try to maintain balance of payments in equilibrium. To know how this can be done involves the study of the causes of disequilibrium.

---

1. C.P. Kindleberger, *op. cit.*, p. 311.

CAUSES OF DISEQUILIBRIUM IN THE BALANCE OF PAYMENTS

There are several variables which join together to constitute equilibrium in the balance of payments position of a country, *viz.*, national income at home and abroad, the prices of goods and factors, the supply of money, the rate of interest, etc., all of which determine the exports, imports and demand and supply of foreign currency. At the back of these variables lie the supply factors, production functions, the state of technology, tastes, the distribution of income, the state of anticipations, etc. If there is a change in any of these variables and there are no appropriate changes in other variables, disequilibrium will be the result.

The main cause of the disequilibrium in the balance of payments arises from imbalance between exports and imports of goods and services. When for one reason or another exports of goods and services of a country are smaller than their imports, disequilibrium in the balance of payments is the likely result. Exports may be small due to the lack of exportable surplus which in turn result from low production or the exports may be small because of the high costs and prices of exportable goods and severe competition in the world markets. An important cause of small exports is the inflation or rising prices in the country. When the prices of goods are high in the country, its exports are discouraged and imports encouraged. If it is not matched by other items in the balance of payments, disequilibrium emerges.

Cyclical Disequilibrium

Cyclical disequilibrium is caused by the fluctuations in the economic activity or what are known as trade cycles. During the periods of prosperity, prices of goods fall and incomes of the people go down. These changes in incomes of the people and prices of goods affect exports and imports of goods and thereby influence the balance of payments."If prices rise in prosperity and decline in depression, a country with a price elasticity for imports greater than unity will experience a tendency for a decline in the value of imports in prosperity, while those for which imports price elasticity is less than one will experience a tendency for increase. These tendencies may be overshadowed by the effects of income changes, of course. Conversely, as prices decline in depression, the elastic demand will bring about an increase in imports, the inelastic demand a decrease."[2]

Secular or Long-run Disequilibrium

Secular (long-run) disequilibrium in balance of payments occur because of long-run and deep-seated changes in an economy as it develops from one stage of growth to another. The current account in the balance of payments follows a varying pattern from one stage to another. In the initial stages of development, domestic investment exceeds domestic savings and imports exceed exports. Disequilibrium arises due to lack of sufficient funds available to finance the import surplus, or the import surplus is not covered by available capital from abroad. Then comes a stage of growth when domestic savings tend to exceed domestic investment and export outrun imports. Disequilibrium may result because the long-term capital outflow falls short of the surplus savings or because surplus savings exceed the amount of investment opportunities abroad. At a still later stage of growth domestic savings tend to equal domestic investment and long-term capital movements are , on balance, zero.

Thus we see that a secular disequilibrium will occur when either the long-term capital movements get out of adjustment with deep-seated factors affecting savings and investment, or planned savings and investment change without an offsetting change in the movement of long-term capital. If investment adjusted itself readily to the amount of domestic savings plus foreign capital there could be no tendency for secular disequilibrium.

The balance-of-payments position will be in equilibrium, if the international capital flow falls into line with the requirements of domestic investment minus domestic savings. There is a tendency

2. C.P. Kindleberger, *op. cit.*, p. 526.

to secular disequilibrium, because of domestic savings and domestic investment are independent of the foreign capital flow and are of different magnitudes. There is a strong tendency for underdeveloped countries to over-invest and/or undersave. The underdeveloped countries are investing larger than their domestic savings and exports allow them because they are eager to acclerate the rate of economic growth. This tendency to over-invest causes a secular disequilibrium in the balance of payments.

Technological Disequilibrium

Technological disequilibrium in the balance of payments is caused by various technological changes. Technological changes involve inventions or innovations of new goods or new techniques of production. These technological changes affect the demand for goods and productive factors which in turn influence the various items in the balance of payments. Each technological change implies a new comparative advantage to which a country adjusts to. The innovation leads to increased exports if it is a new good and export-biased innovation. The innovation may lead to decline in imports if it is import-biased. This will create a disequilibrium. A new equilibrium will require either increased imports or reduced exports.

Structural Disequilibrium

Let us see how the structural type of disequilibrium is caused." *Structural disequilibrium at the goods level* occurs, when a change in demand or supply of exports alters a previously existing equilibrium, or when a change occurs in the basic circumstances under which income is earned or spent abroad, in both cases without the requisite parallel changes elsewhere in the economy."[3]

Suppose demand in foreign countries for Indian handicrafts falls. The resources engaged in the production of these handicrafts must shift to some other line or the country must restrict imports, otherwise the country will experience a structural disequilibrium. A change in supply may also cause a structural disequilibrium. Suppose Indian jute crop falls because of the change in the shift in the crop-pattern, Indian jute exports will fall and a disequilibrium will be created. Apart from goods, a loss of service income may also upset the balance-of-payments position on current account. Besides, the loss of income may arise because foreign investment has proved a failure or it has been confiscated or nationalised, *e.g.* nationalisation of Anglo-Iranian Company in Iran. A war also produces structural changes which may affect not only goods but also factors of production.

A deficit arising from a structural change can be filled by increased production or decreased expenditure, which in turn affect international transactions in increased exports or decreased imports. Actually, it is not so easy because the resources are relatively immobile and expenditure not readily compressible. Under such circumstances, more drastic steps are called for to correct a serious disequilibrium.

"Structural disequilibrium *at the factor level* results from factor prices which fail to reflect accurately factor endowments ... *i.e.*, when factor prices, out or line with factor endowments, distort the structure of production from the allocation of resources which appropriate factor prices would have indicated"[4] If, for instance, the price of labour is too high, it will be used more sparingly and the country will import highly capital-intensive equipment and machinerry. This will lead to disequilibrium in the balance of payments on the one hand and unemployment of labour on the other.

Conclusion

We have explained above four types of disequilibria-cyclical, secular and two kinds of structural disequilibria and how they are caused. In each case, the causes manifest themselves through changes in export of goods and services, making one exceed the other.

---

3. C.P. Kindleberger, *op. cit.*, p. 526.
4. See. C.P. Kindleberger, *International Economics*, 1978.

We have already detailed the various items that enter into the balance of payments. Any cause that leads to a persistently one-sided movement in those items may cause a disequilibrium. For instance, certain causes may lead to falling off in the export of merchandise, imports remaining unaffected or moving in the opposite direction. Falling off in exports may be due to all sorts of causes. Take, for example, the case of exports and imports of goods. Our exports may fall because of decreased production due to seasonal factors or other causes. The demand for our goods in the international market may fall off because of a fall in the purchasing power of consumers of such goods or because of a comparatively high cost of production in India which reduces our competitive strength in the international market. Our exports may become dear to foreigners because of an appreciation of our exchange rate, that is, a rise in the value of the rupee, say from ₹ 42.50 per US dollar to ₹ 40 per US dollar. If we persist in artificially keeping the value of the rupee at a higher level than justified by economic forces, unfavourable balance of trade and payments will tend to persist.

In the same way, disequilibrium may arise due to excessive imports or services neither balanced by exports nor import of capital, etc. Compulsory exports in the form of reparations or indemnities also cause international disequilibrium and obstruct the harmonious trade relations between the countries concerned.

HOW DISEQUILIBRIUM CAN BE CORRECTED

When serious disequilibrium arises in a country's balance of payments, steps must be taken to correct it, if the country's economy is to be kept in a sound condition. Obviously, the causes which are responsible for such a state of affairs must be removed. The 'classical' view of the adjustment mechanism is: "An active or passive balance, accompanied by an inflow or outflow of gold, was formally supposed to result in an expansion or contraction of the domestic money supply; and this expansion or contraction was expected to bring about a rise or fall in the level of domestic costs and prices tending, in the former case, to stimulate imports and discourage exports or, in the latter to discourage imports and stimulate exports. Gold flows, changes in the quantity of money and changes in relative levels thus appeared as the principal factors in the mechanism of adjustments"[5]

But gold standard has been abandoned long ago. These days different countries of the world are under managed currency standard. Therefore gold flows no longer occur to bring the balance of payments into equilibrium. Recent currency experience has, however, led to certain modifications in the classical theory. It is now thought that changes in the flow of income induced by balance of payments serve as an equilibrating factor."The main point is that any active or passive balance of current transaction....tends directly to expand or contract the total flow of money income within a given country... The changes induced by the balance of payments in the flows of income and expenditure affect, in turn, the demand for imported as well as home-produced goods and so react on the balance of payments to bring it into equilibrium".[6]

However, these automatic adjustments or corrections in the balance of payments through either changes in prices of goods and factors or through changes in the incomes do not always work due to the presence of various rigidities and immobilities in the system. When the causes of disequilibrium in the balance of payments are deeper and serious, then the automatic correction via price and income changes does not work at all. Under such circumstance, some concrete measures are required to be taken to restore the equilibrium in the balance of payments. Disequilibrium in the balance of payments may be corrected by the following methods by the Government.

There are four well-known methods of correcting in adverse balance of payments:

1. Trade Policy Measures: Expanding, Exports and Restraining Imports

Trade policy measures to improve the balance of payments refer to the measures adopted to

5. See C.P. Kindleberger, *International Economics,* 1978.
6. *Ibid.*

promote exports and reduce imports. Exports may be encouraged by reducing or abolishing export duties and lowering the interest rate on credit used for financing exports. Exports are also encouraged by grainting subsidies to manufacturers and exporters. Besides, on export earnings lower income tax can be levied to provide incentives to the exporters to produce and export more goods and services. By imposing lower excise duties, prices of exports can be reduced to make then competitive in the world markets.

On the other hand, imports may be reduced by imposing or raising tariffs (*i.e.*, import duties) on imports of goods. Imports may also be restricted through imposing import quotas, introducing licenses for imports. Imports of some inessential items may be totally prohibited.

Before the economic reforms carried out since 1991 India had been following all the above policy measures to promote exports and restrict imports so as to improve its balance of payments position. But they had not achieved nuch success in their aim to correct balance of payments disequilibrium. Therefore, India had to face great difficulties with regard to balance of payments. At several occasions it approached IMF to bail it out of the foreign exchange crisis that emerged as a result of huge deficits in the balance of payments. At long last, economic crisis caused by persistent deficits in balance of payments forced India to introduce structural reforms to achieve a long-lasting solution of balance of payments problem.

2. Expenditure-reducing Policies

The important way to reduce imports and thereby reduce deficit in balance of payments is *to adopt monetary and fiscal policies that aim at reducing aggregate expenditure* in the economy. The fall in aggregate expenditure or aggregate demand in the economy works to reduce imports and help in solving the balance of payments problem. The two important tools of reducing aggregate expenditure are the use of (1) tight monetary policy and (2) contrationary fiscal policy. We explain them below.

Tight Monetary Policy. Tight monetary is often used to check aggregate expenditure or demand by raising the cost of bank credit and restricting the availability of credit. For this bank rate is raised by the Central Bank of the coutnry which leads to higher lending rates charged by the commercial banks. This discoruages businessmen to borrow for investment and consumers to borrow for buying durable cosnumers goods. This therefore leads to the reduction in investment and consumption expenditure. Besides, availability of credit to lend for investment and consumption purposes is reduced by raising the cash reserve ratio (CRR) of the banks and also undertaking of open market operations (selling Government securities in the open market) by the Central Bank of the country. This also tends to lower aggregate expenditure or demand which will helps in reducing imports. But there are limitations of the successful use of monetary policy to check imports, especially in a developing country like India.This is because tight monetary policy adversely affects investment increase in which is necessary for accelerating economic grwoth. If a developing country is experiencing inflation, tight monetary policy is quite effective in curbing inflation by reducing aggregate demnad. This will help in reducing aggregate expenditure and, depending on the income propensity to import, will curtail imports. Besides, tight monetary policy helps to reduce prices or lower the rate of inflation. Lower price level or lower inflation rate will curb the tendency to import, both on the part of businessmen and consumers.

But when a developing country like India is experiencing recession or slowdown in economic growth along with deficits in balance of payments, use of tight monetary policy that reduces aggregate expenditure or demand will not help much as it will adversely affect economic growth and deepen economic recession. Therefore, in a developing country, monetary policy has to be used along with other policies such as a appropriate fiscal policy and trade policy to tackle the problem of disequilibrium in the balance of payments.

Contractionary Fiscal Policy. Appropriate fiscal policy is also an important means of reducing aggregate expenditure. An increase in direct taxes such as income tax will reduce aggregate expenditure. A part of reduction in expenditure may lead to decrease in imports. Increase in indirect taxes such as excise duties and sales tax will also cause reduction in expenditure. The other fiscal policy measure is to reduce Government expenditure, especially unproductive or non-developmental expenditure. The cut in Government expenditure will not only reduce expenditure directly but also indirectly through the operation of multiplier.

It may be noted that if tight monetary and contractionary fiscal policies succeed in lowring aggregate expenditure which causes reduction in prices or lowering the rate of inflation, they will work in two ways to improve the balance of payments. First, fall in doemstic prices or lower rate of inflation will induce people to buy doemstic products rather than imported goods. Second, lower doemstic prices or lower rate of inflation will stimulate exports. Fall in imports and rise in exports will help in reducing deficit in balance of payments.

However, it may be emphasised again that the method of reducing expenditure through contractionary monetary and fiscal policies is not without limitations. If reduction in aggregate demand lowers investment, this will adversely affect economic growth. Thus, correction in balance of payments may be achieved at the expense of economic growth. Further, it is not easy to reduce substantially government expenditure and impose heavy taxes as they are likely to affect incentives to work and invest and invite public protest and opposition. We thus see that correcting the balance of payments through contractionary fiscal policy is not an easy matter.

3. Expenditure-switching Policies: Devaluation

A significant method which is quite often used to correct fundamental disequilibrium in balance of payments is the use of expenditure-switching policies. Expenditure switching policies work through changes in relative prices. Prices of imports are increased by making domestically produced goods relatively cheaper. Expenditure switching policies may lower the prices of exports which will encourage exports of a country. In this way by changing relative prices, expenditure-switching policies help in correcting disequilibrium in balance of payments.

The important form of expenditure switching policy is the reduction in foreign exchange rate of the national currency, namely, devalaution. By devaluation we mean reducing the value or exchange rate of a national currency with respect to other foreign currencies. It should be remembered that devaluation is made when a country is under fixed exchange rate system and occasionally decides to lower the exchange rate of its currency to improve its balance of payments.

Under the Bretton Woods System adopted in 1946, fixed exchange rate system was adopted, but to correct fundamental disequilibrium in the balance of payments, the countries were allowed to make devaluation of their currencies with the permission of IMF. Now, Bretton Woods System has been abandoned and most of the countries of the world have floated their currencies and have thus adopted flexible exchange rate as determined by market forces of demand for and supply of them. However, even in the present flexible exchange rate system, the value of a currency or its exchange rate as determined by demand for and supply of it can fall. Fall in the value of a currency with respect to foreign currencies is described as depreciation. If a country permits its currency to *depreciate* without taking effective steps to check it, it will have the same effects as devaluation. Thus, in our analysis we will discuss effects of fall in value of a currency whether it is brought about through devaluation or depreciation. In July 1991, when India was under Bretton-Woods fixed exchange rate system devalued its rupee to the extent of about 20%. (From ₹ 20 per dollar to ₹ 25 per dollar) to correct disequilibrium in the balance of payments.

Now the question is how devaluation of a currency works to improve balance of payments. As a result of reduction in the exchange rate of a currency with respect to foreign currencies, the prices of goods to be exported fall, whereas prices of imports go up. This encourages exports and discourages imports. With exports so stimulated and imports discouraged, the deficit in the balance of payments will tend to be reduced. Thus policy of devaluation is also refered to as *expenditure switching policy* since as a result of reduction of imports, people of a country switches their expenditure on imports to the domestically produced goods. It may be noted that as a result of the lowering of prices of exports, export earnings will increase if the demand for a country's exports is price elastic (*i.e.*, ep > 1). And also with the rise in prices of imports the value of imports will fall if a country's demand for imports is elastic. If demand of a country for imports is inelastic, its expenditure on imports will rise instead of falling due to higher prices of imports.

Devaluation: Marshall Lerner Condition. It is clear from above that whether devaluation or depreciation will lead to the rise in export earnings and reduction in import expenditure depends on the price elasticity of foreign demand for exports and domestic demand for imports. Marshall and Lerner have developed a condition which states that devaluation will succeed in improving the balance of payments if sum of price elasticity of exports and price elasticity of imports is greater than one. Thus, according to Marshall-Lerner Condition, devaluation improves balance of payments if

$$e_x + e_m > 1$$

where e_x stands for price elasticity of exports

e_m stands for price elasticity of imports

If in case of a country $e_x + e_m < 1$, the devaluation will adversely affect balance of payments position instead of improving it. If $e_x + e_m = 1$, devaluation will leave the disequilibrium in the balance of payments unchanged.

Income Absorption Approach to Devealuation. Further, for devaluation to be successful in correcting disequilibrium in the balance of payments a country should have sufficient exportable surplus. If a country does not have adequate amount of goods and services to be exported, fall in their prices due to devaluation or depreciation will be of no avail. This can be explained through income-absorption approach put forward by Sidney S Alexander[7]. According to this approach, trade balance is the difference between the total output of goods and services produced in a country and its absorption by it. By absorption of output of goods and services we mean how much of them is used up for consumption and investment in that country. That is, absorption means the sum of consumption and investment expenditure on domestically produced goods and services. Expressing algebraically we have

$$B = Y - A$$

where B = trade balance or exportable surplus

Y = national income or value of output of goods and services produced

A = Absorption or sum of consumption and investment expenditure

It follows from above that if expenditure or absorption is less than national product, it will have positive trade balance or exportable surplus. To create this exportable surplus, expenditure on domestically produced consumer and investment goods should be reduced or national product must be raised sufficiently.

To sum up, it follows from above that *for devaluation or depreciation to be successful in correcting disequilibrium in the balance of payments, the sum of price elasticities of demand for a country's exports and imports should be high (that is, greater than one) and secondly it should have sufficient exportable surplus*. The devaluation will also not be successful in the achievement of its aim if other countries relaliate and make similar devaluation in their currencies and thus competitive devaluation of the exchange rate may start.

After Independence India devalued its currency three times, first in 1949, the second in June 1966 and third in July 1991 to correct the disequilibrium in the balance of payments. The devaluation of June 1966 was not successful for some time to reduce deficit in the balance of payments.

---

7. Sidney S. Alexander, "Effects of Devaluation on Trade Balance", *IMF Staff Papers*, Vol. 11 (April, 1952).

This is because the demand for bulk of our traditional exports was not very elastic and also we could not reduce our imports despite their higher prices. However devaluation of July 1991 proved quite successful as after it our exports grew at a rapid rate for some years and growth of imports remained within safe limits.

4. Exchange Control

Finally, there is the method of exchange control. We know that deflation is dangerous; devaluation has a temporary effect and may provoke others also to devalue. Devaluation also hits the prestige of a country. These methods are, therefore, avoided and instead foreign exchange is controlled by the government. Under it, all the exporters are ordered to surrender their foregin exchange to the central bank of a country and it is then rationed out among the licensed importers. None else is allowed to import goods without a licence. The balance of payments is thus rectified by keeping the imports within limits.

After the Second War World a new international institution 'International Monetary Fund (IMF)' was set up for maintaining equilibrium in the balance of payments of member countries for a short term. Member countries borrow from it for a short period to maintain equilibrium in the balance of payments. IMF also advises member countries how to correct fundamental disequilibrium in the balance of Payments when it does arise. It may, however, be mentioned here that no country now needs to be forced into deflation (and so depression) to root out the causes underlying disequilibrium as had to be done under the gold standard. On the contrary, the IMF provides a mechanism by which changes in the rates of foreign exchange can be made in an orderly fashion.

Conclusion. In short, correction of disequilibrium calls for a judicious combination of the following methods:

(*i*) Monetary and fiscal changes affecting income and prices in the country;
(*ii*) Exchange rate adjustment, *i.e.*, devaluation or appreciation of the home currency;
(*iii*) Trade restrictions, *i.e.*, tariffs, quotas, etc.;
(*iv*) Capital movement, *i.e.*, borrowing or lending aborad; and
(*v*) Exchange control.

No reliance can be placed on any single tool. There is room for more than one approach and for more than one device. But the application of the tools depends on the nature of the disequilibrium. There are, we have said, three types of disequilibrium: (1) cyclical disequilibrium, (2) secular disequilibrium, (3) structural disequilibrium (at the goods and the factor level). It is more appropriate that fiscal measures should be used to correct cyclical disequilibrium in the balance of payments. To correct structural disequilibrium adjustment in exchange rate should be avoided. Capital movements are needed to offset deep-seated forces in secular disequilibrium.

The main methods of desirable adjustment are, therefore, monetary and fiscal policies which directly affect income, and exchange depreciation (that is, devaluation) which affects prices in the first instance. Devaluation or depreciation of exchange rate can also have income effect through price effects. Monetary and fiscal policies affect relative prices also.

GLOBALISATION, CAPITAL FLOWS AND BALANCE OF PAYMENTS

Let us first explain what is meant by globalisation. Globalisation refers to the increased openness of an economy to the international trade, capital flows (both portfolio and foreign direct investment, FDI), transfer of technology and free movement of labour or people. Thus **globalisation means the integration of economies of the world resulting from free flows of trade, capital, labour (or people) and technology between countries.** The term globalisation signifies new international economic order which envisages the following :

(*a*) free flow of trade of goods and services between different countires of the world.
(*b*) free flow of capital among countires.
(*c*) free flow of technology among different countries of the world.

(*d*) free movement of labor or people internationally.

The growing interaction between different countries of the world has made the idea of globalisation increasingly popular in recent years.

The globalisation of the economy has an important consequence with regard to capital flows into the economy. Suppose India faces given prices of its imports and a given demand for its exports of goods and services. Under these circumstances, if domestic rate of interest (or rate of return on investment) is higher as compared to what exists abroad, then, given the mobility of capital, foreign capital will flow into the Indian economy to a very large extent. This principle can be expressed as follows.

$$BP = NX(Y_d, Y_f, R) + CF(i_f - i_d)$$

where BP = balance of payments, NX is net exports (*i.e.* exports–imports which is also called trade balance), Y_d stands for level of domestic income, Y_f for level of income of foreign countries, and R for foreign exchange rate CF stands for surplus in the capital account of the balance of payments, that is, capital flows. i_f represents rate of interest in foreign market and i_d represents rate of interest in the domestic economy.

The above equation reveals that trade balance (NX) is a function of level of domestic income (Y_d) and foreign income (Y_f) and real rate of exchange (R). An increase in the domestic income due to higher industrial growth or fall in real exchange rate of rupee will adversely affect the trade balance (NX) by increasing imports. $i_f - i_d$ in equation (*ii*) measures interest rate differential between the foreign country and domestic economy on which *i.e.* net capital flows depend. Further, the above equation (*ii*) shows that higher interest rate or rate of return in India as compared to that in the foreign country such as the United States will cause large capital inflows into India. Such capital inflows have actually taken place in India since 2003. Due to large capital inflows into the Indian economy our foreign exchange reserves crossed 100 billions US dollars in December 2003, reached 170 billion US $ in December 2006 and further increased to 247 billion US $ dollars in September 2007, and to 300 billion US $ in Sept. 2008. Due to reversal of capital flows, there was decline in our reserves in 2008-09 and in March 2009, they stood at US $252 billion. In the second half of 2009-10 there had been reversal of capital flows and capital inflows into Indian economy occurred on a large scale.

Globalisation has both advantages and disadvantages. It is through globalisation and openness of an economy that foreign trade of a country increases which promotes economic growth. Trade between countries help them to specialise in the production of goods and services according to their factor endowments and comparative efficiency. Besides, globalisation helps in the transfer of technology from the developed countries to the developing countries. Globalisation also promotes capital flows between countries which in general are welcome. Capital inflow in the form of foreign direct investment (FDI) adds to the productive capacity of a country by increasing capital accumulation. Capital inflows in the form of portfolio investment by FIIs (Foreign Institutional Investors), which are invested in buying equity shares and corporate and Government bonds lead to the development of financial markets. Besides, capital flows provide foreign exchange which can be used to meet the current account deficit which generally emerges in a developing country such as India due to large imports of fuel, machines and industrial raw materials required for industrial growth of the economy.

However, the global financial crisis of 2007-09 which originated in the sub-prime housing loan market in the US and enveloped the entire world has clearly brought out the adverse impact of globalisation. It adversely affected growth of exports of developing countries such as India which brought about large current account deficit. It also caused capital outflows from them which led to the stoek market crash. When capital flows are large and volatile they hurt macroeconomic stability. Large capital inflows as they occurred in India in 2010 lead to appreciation of exchange rate of the national currency (Rupee in case of India) which adversely affects growth of our exports and encourage imports. As a result they widen current account deficit.

Owing to these adverse effects of globalisation each country decides about the extent of openneses of its economy with which it is quite comfortable by imposing some restrictions on their foreign trade, especially in financial services. Besides, because of the adverse effects of large capital outflows India is trying to woo foreign investors and NRIs to make investment in India so that there are adequate capital inflows into India to meet our widening current account deficit (CAD).

Since current account deficit of India in 2008-09, was financed by withdrawing from foreign exchange reserves in 2008-09 but in the years 2009-10 and 2010-11 it was financed by capital inflows, and thus there was gainful absorption of foreign capital inflows. Therefore, if India wants to continue to grow at a high rate of 8% to 9% there will be deficit in current account of BOP since growth in our exports cannot keep face with our higher growth in imports due to rapid increase in our industrial activity. Current account deficit is a matter of concern only if growth is not high or if stable capital inflows represented by foreign direct investment (FDI) are not adequate to finance the current account deficit. It will be seen from Table 34.1 and 34.2 that from 2004-05 to 2010-11capital surplus [i.e. capital account (net)] exceeded our current account deficit, except for the years 2008-09 and 2011-12 with the result that *there was addition to our foreign exchange reserves* (see last row of Table 34.2). Note that minus (–) sign in the use of reserves for various years implies that instead of using foreign exchange reserves in these years we added to our foreign exchange reserves. In the year 2008-09 when there was global financial crisis which caused not only negative growth in our exports but also large capital outflows from the Indian economy with the result that capital surplus in 2008-09 was quite small and fell short of our current account deficit. Similarly, in 2011-12 our current account deficit could not be financed by capital inflows. As a result we withdrew $20 billion and $ 12.8 billion from our foreign exchange reserve to meet the current account deficit 2008-09 in 2008-09 and 2011-12 respectively. Hence the plus sign before US $ 20 in billion 2008-09 and $ 12.8 billion in 2011-12. However, in 2009-10 and 2010-11 there was revival of large capital inflows with capital account (net) equal to US $ 53.6 billion in 2009-10. With $ 38.4 billion deficit in current account we instead of withdrawing from our foreign exchange reserves we added to them in 2009-10. This means India successfully tackled global financial crisis and its consequences. However, it may be noted that in 2010-11 capital inflows into India through foreign direct investment declined and our current account deficit which was estimated to be 2.8 of our GDP had been financed mostly by *volatile capital flows by FIIs which is not desirable*. This is because these volatile capital flows affect our foreign exchange rate stability, stock market prices, price stability, and thus create macroeconomic instability.

QUESTIONS FOR REVIEW

1. Distinguish between balance of trade and balance of payments.
2. What is meant by balance of payments on current account? Explain the visible and invisible items on the current account.
3. Distinguish between current account and capital account of the balance of payments. Mention the various items on both these accounts of the balance of payments.
4. When is country's balance of payments said to be in equilibrium? Explain the factors that cause disequilibrium in the balance of payments.
5. Balance of payments must always balance. Is it true? Explain.
6. What is devaluation? Is it an effective way of correcting disequilibrium in the balance of payments.
7. Explain the various measures that can be adopted to correct disequilibrium.
8. What is expenditure reducing fiscal policy? How does it help to maintain the current account of the balance of payments in equilibrium?
9. What is meant by devaluation ? Under what conditions is it expected to improve the balance of payments position of an economy?
10. Discuss the absorption approach to devaluation as a measure of correcting balance of payments deficit.
11. Distinguish between Expenditure Reducing and Expenditure Switching policies to correct the balance of payments disequilibrium.

CHAPTER 35
FOREIGN EXCHANGE RATE

In the last chapter, we have discussed the concept of balance of payments which describes the transactions of goods and services and movements of financial capital between countries. The value of these transactions depends upon the prices of the currencies of various countries. For example, in India, the value of transactions involving balance of payments is measured either in rupee terms or in terms of US dollars. Given the volume of these transactions, their value in rupees or dollars would depend on the rate of exchange of rupee with US dollar.

FOREIGN EXCHANGE RATE AND FOREIGN EXCHANGE MARKET

It may be noted that the *foreign exchange is the name given to any foreign currency. Thus, US dollars* or *British pounds* are *foreign exchange for India.* Further, the exchange rate is the *price of a country's currency in terms* of another country's currency. Thus, at present (Aug. 30, 2013) one US dollar is exchanged for about 66.70 rupees of India (the spot exchange rate). Thus, 66.70 Indian rupees for one US dollar is the exchange rate of a dollar in terms of rupees.

Table 35.1 Foreign Exchange Rates of Indian Rupee

(Market Rate in ₹ Per Unit of Foreign Currency)

Foreign Currency	*Exchange rate (on Nov. 27, 2008)*	*Dec. 29, 2011*	*March 5 2013*	*August 30 2013*
US Dollar	49.44	52.71	54.9	66.70
British Pound	75.96	82.41	83.26	101.82
Canadian Dollar	40.35	51.75	53.56	62.36
Euro*	64.24	68.91	71.5	86.87
UAE (*Dirham*)	13.57	14.35	14.39	17.88
Swedish Kroner		7.57	8.21	9.93
Swiss Francs	–	56.10	57.8	70.69
Japanese 100 Yen	52.89	67.63	58.5	66.92
Australian Dollar	32.09	53.57	56.2	58.48
Singapore Dollar	32.81	40.78	44.0	51.53
New Zealand Dollar	–	40.62	44.6	50.79

*Euro is the Common Currency of European Union comprising a number of European countries.

In the above table, the exchange rates of Indian rupee in terms of currencies of some important countries of the world is given. It can be seen from the above table that there has been a lot of volatility in the exchange rate of rupee. There has been 15 per cent appreciation of Indian rupee vis-a-vis US Dollar between Oct. 2006 and Dec. 2007. Beyond Jan. 2008, rupee started depreciating under pressure of capital outflows by FIIs and it fell to ₹ 49.44 to a US dollar on Nov. 27, 2008. From later half of 2009-10 to Aug. 2011, as a result of return of capital flows to India, rupee started appreciating and its value rose to around ₹ 45 to a US dollar at the end of Aug. 2011. However, again from Sept. 2011 due to Eurozone sovereign debt crisis, there had been large capital outflows by FIIs from India (and other emerging economies) resulting in drastic depreciation of Indian rupee which fell to even ₹ 54 to a US dollar on Dec. 15, 2011. But due to intervention of RBI by way of selling dollars in the foreign exchange market from its foreign exchange reserves, the value of rupee fluctuated within range of ₹ 52 to ₹ 53 to a US dollar between December 16, 2011 and January 13, 2012. It will be seen that Indian rupee depreciated against currencies of all countries in March 5, 2013 except Japan. On March 5, 2013, value of rupee was 54.9 rupees per US dollar.

However, a very large depreciation of rupee has occurred since May 22, 2013. In the month of June 2013, the rupee fell from around ₹ 56 to a US dollar in the first week of June 2013 to ₹ 60 a dollar by the end of June 2013 and hit the level of ₹ 61.21 to a dollar on July 8, 2013. The RBI intervened by selling US dollars from its foreign exchange reserves and took steps to discourage the imports of gold so as to reduce the demand for dollars. The Finance Minister also raised the duty on gold, first from 6% to 8% and then to 10% to restrict the imports of gold to check further depreciation of rupee. However, despite these measures, depreciation of rupee continued and the value of rupee hit a record of ₹ 68.80 to a US dollar on Aug. 28, 2013. Then with aggressive intervention by the RBI through sale of dollars from its foreign exchange reserves and the value of rupee rose to ₹ 66.70 to a dollar on Aug. 30, 2013. That is, about 18% fall in the value of rupee from early June 2013.

Both domestic and external factors are responsible for this sharp depreciation of rupee since May 2013. The domestic factor responsible for this steep slide of rupee has been the emergence of large Current Account Deficit (CAD) which was estimated at $ 88 billion in 2012-13 (4.8% of GDP of the year) due to sluggish exports and rise in imports causing a net increase in demand for US dollars over their supply. The external factor which triggered the sharp depreciation of rupee was the announcement by the governor of the US Reserve System in early May 2013 that since the US economy had recovered, he would start unwinding of the policy of *Quantitative Easing* (QE) under which it would purchase bonds worth $ 85 billion per month from the market to revive the US economy. This triggered capital outflows by FIIs from the emerging economies including India. These capital outflows on a large scale from India contributed to the sharp slide in the exchange rate of rupee.

It may be noted that the exchange rate of rupee is now market determined and as shown in Table 35.1, it has shown great volatility in the last nine years (2004-2013).

FLOATING (FLEXIBLE) AND FIXED EXCHANGE RATE SYSTEM

Since exchange rate is a price, its determination can be explained through demand for and supply of currencies. Suppose we consider the transactions between two countries, India and USA, in this case, the demand for and supply of dollar is the demand for and supply of foreign exchange from the Indian perspective and the price of a US dollar in terms of Indian rupees or a number of dollars per Indian rupee is the exchange rate. *The system of exchange rate in which the value of a currency is allowed to adjust freely or to float as determined by demand for and supply of foreign exchange is called a flexible exchange system which is also known as floating exchange system.*

On the other hand, if the exchange rate instead of *being determined by demand for and supply of foreign exchange is fixed by the Government, it is called the fixed exchange rate system* which prevailed in the world under an agreement reached at Bretton Woods in New Hampshire in July 1944. It may be noted that under the fixed exchange rate system, exchange rate is not determined by demand for and supply of foreign exchange but is pegged at a certain rate. At the fixed exchange rate, if there is disequilibrium in the balance of payments giving rise to either excess demand or excess supply of foreign exchange, the Central Bank of the country has to buy and sell the required quantities of foreign exchange to eliminate the excess demand or supply.

In 1977, USA decided to float its dollar and switched over to the flexible exchange system resulting in the collapse of Bretton Woods System of fixed exchange rate. Both the floating (flexible) and fixed exchange rate system have their merits and demerits.

APPRECIATION AND DEPRECIATION OF CURRENCIES

It is important to explain the meanings of the term, appreciation and depreciation, of currencies which are often mentioned in the discussion of foreign exchange rate. Let us consider the exchange rate of rupee for dollar, *appreciation of a currency* is the increase in its value in terms of another foreign currency. Thus, if the value of a rupee in terms of US dollar increases from ₹ 45.50 to ₹ 44 to a dollar, Indian rupee is said to appreciate. This indicates strengthening of the Indian rupee. Note that when Indian rupee in dollar terms appreciates, the dollar would depreciate. On the other hand, if the value of Indian rupee in terms of US dollars falls, say from ₹ 45.5 to ₹ 46 to a dollar, the Indian rupee is said to depreciate which shows the weakening of Indian rupee. Thus, under a flexible exchange system, the exchange value of a currency frequently appreciates or depreciates depending upon the demand for and supply of a currency.

In a fixed exchange rate system the government has to buy or sell foreign exchange in order to maintain the rate at the controlled level. However, even under the fixed exchange rate system, the value of one's currency can be changed occasionally. For instance, in June 1966, the value of rupee in terms of US dollar and U.K's, pound sterling was lowered. Again in July 1991, India reduced its value of rupee in terms of dollar by about 20 per cent. Such a one-time lowering of value of its currency in terms of foreign exchange occasionally by a country is called *devaluation* as distinguished from depreciation which can often take place under the influence of changes in demand for and supply of a currency. On the other hand, with a fixed rate exchange system if a country raises the value of its currency in terms of foreign currency, it is called *revaluation.* It should be noted that since March 1993, India has also made its rupee convertible into a foreign currency and allowed its value to adjust freely depending upon the demand and supply forces.

DETERMINATION OF EXCHANGE RATE

We are now in a position to explain how in a flexible exchange system the exchange of a currency is determined by demand for and supply of foreign exchange. Let's assume that there are two countries, India and USA, the exchange rate of their currencies (namely, rupee and dollar) is to be determined. We explain below how the value of a dollar in terms of rupees (which will conversely indicate the value of a rupee in terms of dollars) is determined.

At present in both USA and India, there is floating or flexible exchange regime. Therefore, the value of currency of each country in terms of the other depends upon the demand for and supply of their currencies.

It is in the foreign exchange market that the exchange rate among different currencies is determined. The foreign exchange market is the market in which the currencies of various countries are converted into each other or exchanged for each other. In the case of the determination of exchange rate between US dollar and Indian rupee, the Indians sell rupees to buy US dollars (which is a foreign currency) and the Americans or others holding US dollars will sell dollars in exchange for rupees. It is the demand for and supply of a foreign currency or exchange that will determine the exchange rate between the two.

Demand for Foreign Exchange (US Dollars)

The demand for US dollars comes from the Indian people and firms who need US dollars to pay for the goods and services they want to import from the USA. The greater the imports of goods and services from the USA, the greater the demand for the US dollars by the Indians. Further, the demand for dollars also arises from Indian individuals and firms wanting to purchase assets in the USA, that is, desire to invest in US bonds and equity shares of the American companies or build factories, sales facilities or houses in the USA. The demand for dollars also arises from those who want to give loans or send gifts to some people in the USA. Thus, for whatever reasons the Indian residents need dollars they have to buy them in the foreign exchange market and pay for them with the Indian currency, the rupee. All of these constitute demand for dollars, the foreign exchange.

To sum up, the demand for dollars by the Indians arises due to the following factors:

1. The Indian individuals, firms or Government who import goods from USA into India.
2. The Indians travelling and studying in USA would require dollars to meet their travelling and educational expenses.
3. The Indians who want to invest in equity shares and bonds of the US companies and other financial instruments.
4. The Indian firms who want to invest directly in building factories, sales facilities, shops in the USA.

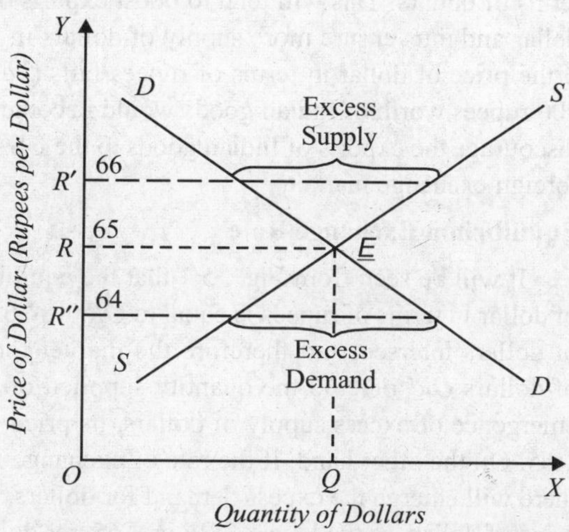

Fig. 35.1. *Determination of Exchange Rate of a Dollar in terms of Rupees*

An important thing to understand is how the demand curve for a foreign exchange would look like. *When there is a fall in the price of dollar in terms of rupees, that is, when the dollar depreciates, fewer rupees than before would be required to get a dollar*. With this, therefore, a dollar's worth of US goods could be purchased with fewer rupees, that is, the US goods would become cheaper in terms of rupees for Indians. This will induce the Indian individuals and firms to import more from the USA resulting in the increase in quantity demanded of dollars by the Indians. On the other hand, *if the price of US dollar rises (that is, U.S. dollar appreciates), a dollar's worth of US goods would now cost more in terms of rupees making American goods relatively expensive than before*. This will discourage the imports of US goods to India causing a decrease in quantity demanded of dollars for imports.

It therefore follows from above that at a lower price of dollars, the greater quantity of dollars is demanded for imports from the USA and at a higher price of dollar, the smaller quantity of dollars is demanded for imports from the USA by the Indians. This makes the demand curve for dollars downward sloping as shown by the *DD* curve in Fig. 35.1.

Supply of US Dollars (Foreign Exchange)

What determines the supply of dollars in the foreign exchange market? The individual firms and the Government *which export Indian goods to the USA will earn dollars from the American residents who would buy the Indian goods imported into the USA and pay their price in dollars*. Further, the Americans who travel in India and use the services of Indian transport, hotels etc., will also supply dollars to be converted into rupees for meeting these expenses.

Besides, the *American firms and individuals who want to buy assets in India, such as bonds and equity shares of Indian companies or wish to make loans to the Indian individuals and firms will also supply dollars*. There are Indians who are working in the USA and send their earnings in dollars to their relatives and friends. The supply of these dollars by the Indians working in the USA popularly called *remittances from the USA* also adds to the supply of dollars. Those holding dollars who have earned them from exports to the USA and the foreign firms and the individuals who want to invest in India or those who want to make loans to Indians or the American tourists travelling in India, and remittances from USA by the Indians working there will supply dollars in the foreign exchange market.

The supply curve of dollars plotted against the price of dollar in terms of rupees is positively sloping as shown in Fig. 35.1. What accounts for the upward sloping nature of the supply curve of the dollars? At a higher price of dollar in terms of rupees (or, in other words, lower value of the Indian rupee in terms of dollar), 100 rupees worth of Indian goods would be relatively cheaper in terms of dollars. This will tend to boost exports of the Indian goods to the USA at a higher price of dollar and thus ensure more supply of dollars in the foreign exchange markets. On the other hand, if the price of dollar in terms of rupees falls (*i.e.*, its exchange rate for Indian rupee declines) the 100 rupees worth of Indian goods would become relatively expensive in terms of dollars. This will discourage the exports of Indian goods to the USA and reduce the quantity supplied of dollars in the foreign exchange market.

Equilibrium Exchange Rate

It will be seen from Fig. 35.1 that the equilibrium exchange rate, that is, the equilibrium price of dollar in terms of rupees is equal to *OR* or ₹ 65 per dollar at which demand for and supply curve of dollars intersect and therefore the market for dollars is cleared at this rate. At a higher price of dollars *OR′* or ₹ 66 the quantity supplied of dollars exceeds the quantity demanded. With the emergence of excess supply of dollars, its price, that is, the exchange rate will again fall to *OR* or ₹ 65. On the other hand, if the rate of exchange is lower than *OR*, say it is *OR″* or ₹ 64 per dollar, there will emerge the excess demand for dollars. This excess demand for dollars would push up the price of dollars to the level of *OR* or ₹ 65 per dollar.

CHANGES IN EXCHANGE RATE

In our analysis of determination of exchange rate through demand and supply of foreign exchange, we have assumed that the underlying forces which determine demand for and supply of foreign exchange remain constant. In our foregoing analysis, it is only changes in the exchange rate that cause the quantity supplied and demanded of foreign exchange to change and bring about the equilibrium between demand for and supply of foreign exchange.

The equilibrium in the foreign exchange market will be disturbed if some changes occur in the underlying factors that determine the demand for and supply of foreign exchange. For example, if there is increase in incomes of the American people due to the boom conditions in the US economy, it will affect the equilibrium rate of exchange. The increase in incomes of the people of USA, will lead to the increase in demand for imported goods including those of India. Now, with this increase in demand for imported Indian goods, they will spend more dollars on the Indian goods. This will increase the supply of dollars to buy Indian goods in the foreign exchange market causing a rightward shift in the supply curve from *SS* to *S′S′*, as shown in Fig. 35.2. This increase in the supply of dollars in foreign exchange market will lower the price of dollar in terms of rupees from *OR* (₹ 65 per dollar) to *OR′* (₹ 64 per dollar). This implies that increase in imports by USA from India leading to more exports from India will cause the dollar to depreciate and the Indian rupee to appreciate. It will be seen from Fig. 35.2 that with the increase in supply of dollars and supply curve shifting to the right to *S′S′* position, at the original price *OR* (or ₹ 65 per dollar) there is excess supply of dollars by *EN* amount which will result in lowering the price of dollar to the level of *OR′* or ₹ 64 per dollar. This means increase in supply of dollars has led to the appreciation of rupee from ₹ 65 per dollar to ₹ 64 per dollar.

On the other hand, if due to the increase in incomes of the Indian people causing a rise in demand for American consumer goods or there is picking up of industrial activity in India requiring more imports of materials, machines, equipments and other capital goods from the USA, the Indian imports from USA will increase. The increase in imports from USA by India will have to be paid in dollars causing the demand for dollars to increase and as a result demand curve for dollar shifts to the right from *DD* curve to *D′D′* curve (see Fig. 35.3). This will upset the initial equilibrium at price *OR* (₹65 per dollar) as with the increase in demand for dollars following the increase in imports from USA there will emerge excess demand for dollar which will push up the price of dollar to *OR′ ′* (₹ 66 per dollar) so as to bring balance between demand for and supply of dollars (*i.e.*, foreign exchange). Note that rise in price of dollar in terms of rupees implies depreciation of the value of rupee from ₹ 65 to ₹ 66 per dollar.

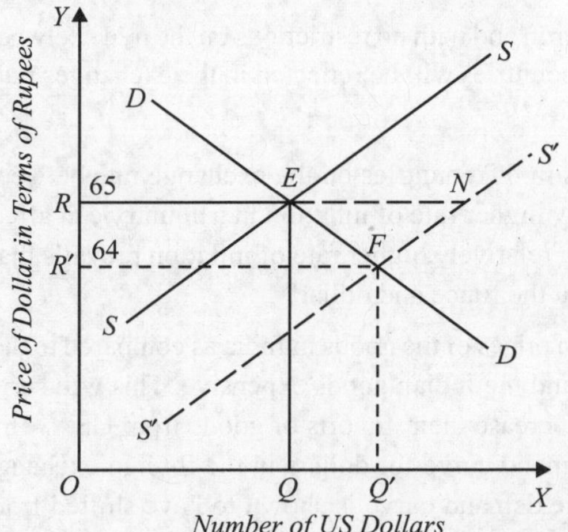

Fig. 35.2. *Increase in Incomes of US People Leading to more Exports from India will Lower the Price of Dollar and Increase the Indian Exports.*

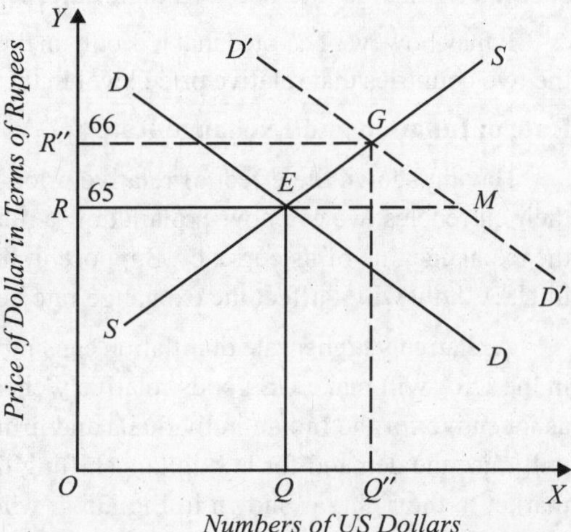

Fig. 35.3. *Increase in the Imports from USA by the Indians Causes Increase in Demand for Dollars and Lowers the Price of Rupee.*

It will be observed in Fig. 35.3 that as a result of adjustments following the increase in demand for dollars, the price of dollar has risen to *OR″* (₹ 66 per dollar) and the amount of dollars spent by Indians on imports from the USA increases to *Q.″*

FACTORS AFFECTING EXCHANGE RATE

Having now explained how the exchange rate between different currencies of various countries is determined, we can now explain the factors that are likely to affect the exchange rate between currencies. Changes in all those factors mentioned above which determine the demand for and supply of currencies can influence the exchange rate which will adjust according to the changes that have occurred.

Purchasing Power Parity: The Relative Price Levels

If there are no restrictions imposed on trade by the countries, the exchange rate between two national currencies is allowed to adjust freely and with the further assumption that costs of transport of goods between the countries is nil, then the exchange rate between the two currencies will reflect the differences in the price levels in the two countries. This is because with the above-mentioned assumptions, if the price of the same good BPL TV set is lower in Britain than in India, then it will pay traders to buy the BPL TV sets in Britain and sell them in India. This will reduce the supply of TV set in Britain pushing up its price there and increase the supply of TV sets in India and thus causing a decline in its price in India. This process would continue till the price differential of the same quality of TV sets is eliminated and the same price of TV set prevails in the two countries. Thus, if law of same price holds and each country consumes the same market basket of goods, the exchange rate between the two currencies would be determined by the relative price levels in the two countries. For example, if prices in Britain and India are such that a pair of same quality of shoes costs 5 pound in England and 350 rupees in India, then the exchange rate between rupees and pound would be 1 pound for 70 rupees. If the exchange rate between the two is different from this, it will be possible for the businessmen to make profits in sending pairs of shoes from a country to the other depending on the price levels in the two countries. We therefore conclude that price levels of commodities in the different countries influence the exchange rate between their currencies.

It may however be noted that it is only in the long run and with no restrictions on the trade between the two countries that relative price level in the two countries will be reflected in the exchanges rate.

Rate of Inflation and Exchange Rate

Having shown the effect of relative price levels in the countries on the exchange rate between their currencies, we can now explain how a relatively higher rate of inflation in a country can affect the exchange rate of its currency. Suppose in India, a relatively higher rate of inflation prevails than in USA, how will it affect the exchange rate between the rupee and dollar?

A relatively higher rate of inflation causing rise in prices of the goods in India as compared to that in the USA will make US goods relatively cheaper and the Indian goods expensive. This will serve as incentive for the Indian individuals and firms to increase their imports of goods from USA. This will raise the demand for US dollars shifting the demand curve for dollars in the foreign exchange market to the right, as shown in Fig. 35.4, where the demand curve is shown to have shifted from *DD* to *D′D′* under the influence of higher rate of inflation in India.

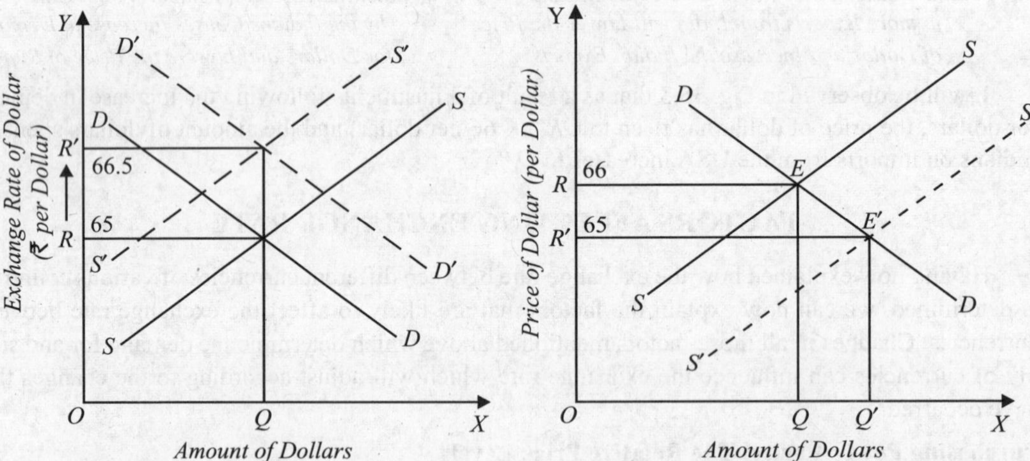

Fig. 35.4. *Effect of a Relatively Higher Rate of Inflation on the Exchange Rate of a Country*

Fig. 35.5. *Effect of a Relatively Higher Rate of Interest in India on the Exchange rate*

However, at the same time due to higher price level American people will find the Indian goods more expensive and as a result will reduce their imports of Indian goods. This will cause a decline in exports of goods from India to the USA, shifting the supply curve of dollars to the left to *S′S′*. Both these effects of a higher price level due to higher rate of inflation in a country, namely, rise in imports of US goods into India and the reduction in the Indian exports to the USA will cause the foreign exchange rate of dollar in terms of rupees to rise and the price of the Indian rupee in terms of dollar will fall. *Thus, as a result of higher rate of inflation in India, the US dollar will appreciate and the Indian rupee will depreciate* (See Fig. 35.4).

Interest Rates and Exchange Rate

Another important factor influencing the exchange rate is the interest rate in a country relative to interest rate of other countries with which it trades its goods. Suppose there are no restrictions imposed by the Governments on the flow of funds between the countries and assume that interest on securities or bonds in USA is 5 per cent whereas it is 8 per cent in India. The businessmen, firms, banks etc., with funds to invest would obviously have incentive to buy securities and bonds of US companies. That is, there will be *flight of capital* from the USA or capital inflows into India as the Americans business corporations, firms, banks, etc. will use their funds to purchase high-yielding Indian securities. In order to buy the Indian securities, they will have to convert dollars into rupees, and thus increasing the demand for dollars. Consequently, the supply of dollar curve will shift to the right. This will pull down the foreign exchange rate of dollar in terms of rupees (Conversely, the exchange rate of rupee in terms of dollar will appreciate). Thus, *a relatively higher interest rate in India as compared to India would lead to the depreciation of dollar and appreciation of rupee.* This is what actually happened in India in the years 2003 and 2004.

Fig. 35.5 illustrates the effect of higher interest rates in the USA on the dollar-rupee exchange rate. Initially, the demand for and the supply of US dollar are given by the curves *DD* and *SS* respectively. The balance between the two determines *OR* (or ₹ 65 per dollar) the exchange rate between the two. With the flow of United States funds into India to buy the Indian securities and shares and stocks, the supply curve for dollars shifts to the right to the new position *S′S′* (dotted) and lower the equilibrium exchange rate to *OR′* (₹ 66 per dollar) is determined.

It may be noted that during the early eighties, the Federal Reserve Bank of USA adopted a tight money policy to fight inflation which caused very high interest rates in USA. These high interest rates attracted a lot of foreign capital, particularly from Japanese firms which due to high saving rate in Japan had a large amount of funds to make investment in American bonds. To purchase American bonds, Japanese had to convert Yens into dollars. This caused the increase in demand for dollars and caused a sharp rise in the price of dollar. It is thus clear that *like the relative price levels, relative interest rates can also have a significant effect on the exchange rate.*

PURCHASING POWER PARITY THEORY

No country today is rich enough to have a free gold standard, not even USA. All countries now have paper currencies and these paper currencies of various countries are not convertible into gold or other valuable things. Therefore, these days various countries have paper currency standards. The exchange situation is difficult in such cases. In such circumstances, the ratio of exchange between the two currencies is determined by their respective purchasing powers.

The purchasing power parity theory was propounded by Professor Gustav Cassel of Sweden. According to this theory, rate of exchange between two countries depends upon the relative purchasing power of their respective currencies. Such will be the rate which equates the two purchasing powers. For example, if a certain assortment of goods can be had for £1 in Britain and a similar assortment with ₹ 80 in India, then it is clear that the purchasing power of £1 is equal to the purchasing power of ₹ 80. Thus, the rate of exchange, according to purchasing power parity theory, will be £1 = ₹ 80.

Let us take another example. Suppose in USA, one $ purchases a given collection of commodities. In India, same collection of goods costs ₹ 65. Then rate of exchange will tend to be

$1 = ₹ 65.

Now, suppose the price levels in the two countries remain the same but somehow exchange rate moves to $1 = ₹ 66.

This means that one US$ can purchase commodities worth more than ₹ 65. It will pay people to convert dollars into rupees at the rate ($1 = ₹ 66), purchase the given collection of commodities in India for ₹ 65 and sell them in USA for one dollar again, making a profit of

₹1 per dollar worth of transactions. This will create a large demand for rupees in the USA while supply thereof will be less because very few people would export commodities from USA to India. The value of the rupee in terms of the dollar will move up until it will reach $1 = ₹ 65. At that point, imports from India will not give abnormal profits. $1 = ₹ 65 is called the purchasing power parity between the two countries.

Thus while the value of the unit of one currency in terms of another currency is determined at any particular time by the market conditions of demand and supply, in the long run the exchange rate is determined by the relative values of the two currencies as indicated by their respective purchasing powers over goods and services. In other words, the rate of exchange tends to rest at the point which expresses equality between the respective purchasing powers of the two currencies. This point is called the purchasing power parity. Thus, under a system of autonomous paper standards, the external value of a currency is said to depend ultimately on the domestic purchasing power of that currency relative to that of another currency. In other words, exchange rates, under such a system, tend to be determined by the relative purchasing power parities of different currencies in different countries.

In the above example, if prices in India get doubled, the value of the rupee will be exactly halved. The new parity will be $1 = ₹ 130. This is because now ₹ 130 will buy the same collection of commodities in India which ₹ 65 did before. We suppose that prices in USA remain as before. But if prices in both countries get doubled, there will be no change in the parity.

In actual practice, however, the parity will be modified by the cost of transporting goods (including duties etc.) from one country to another.

Criticisms of Purchasing Power Parity Theory

The purchasing power parity theory has been subject to the following criticisms:

The actual rates of exchange between the two countries very seldom reflect the relative purchasing powers of the two currencies. This may be due to the fact that governments have either controlled prices or controlled exchange rates or imposed restrictions on import and export of goods. Moreover, the theory is true if we consider the purchasing power of the respective currencies in terms of goods which enter into international trade and not the purchasing power of goods in general. But we know that all articles produced in a country do not figure in international trade. Therefore, the rate of exchange cannot reflect the purchasing power of goods in general. For example, in India we may be able to get a dozen shirts washed with ₹ 65, but only 2 shirts with one dollar in the USA. Obviously, the purchasing power of one dollar in the USA is much less than the purchasing power of ₹ 65 in India. This is due to the fact that *dhobis* do not form an article of international trade. If *dhobis* entered into international trade and freely moved into the U.S.A., then in terms of clothes washed, the purchasing power of ₹ 65 may be equalised with the purchasing power of a dollar.

Further, it is very difficult to measure purchasing power of a currency. It is usually done with the help of index numbers. But we know that the index numbers are not infallible. Among the difficulties connected with index numbers are the following important ones: (*i*) Different types of goods that enter into the calculation of index numbers; (*ii*) Many goods which may enter into domestic trade may not figure in international trade; (*iii*) Internationally traded goods also may not have the same prices in all the markets because of differences in transport costs.

Besides, the theory of purchasing power applies to a stationary world. Actually the world is not static but dynamic. Conditions relating to money and prices, tariffs, etc., constantly go on changing and prevent us from arriving at any stable conclusion about the rates of exchange. The internal prices and the cost of production are constantly changing. Therefore, a new equilibrium between the two currencies is almost daily called for. As Cassel observes, "differences in two countries' economic situation, particularly in regard to transport and customs, may cause the normal exchange rate to deviate to a certain extent from the quotient of the currencies intrinsic purchasing powers." If a country raises its tariffs, the exchange value of its currency will rise but its price level will remain the same.

Besides, many items of balance of payments like insurance and banking transactions and capital movements are very little affected by changes in general price levels. But these items do influence exchange rates by acting upon the supply of and the demand for foreign currencies. The Purchasing Power Parity Theory ignores these influences altogether. Further, the theory, as propounded by Cassel, says that changes in price level bring about changes in exchange rates but changes in exchange rates do not cause any change in prices. This latter part is not true, for exchange movements do exercise some influence on internal prices.

The purchasing power parity theory compares the general price levels in two countries without making any provision for distinction being drawn between the price level of domestic goods and that of the internationally traded goods. The prices of internationally-traded goods will tend to be the same in all countries (transport costs are, of course, omitted). Domestic prices on the other hand, will be different in the two countries, even between two areas of the same country.

The purchasing power parity theory assumes that there is a direct link between the purchasing power of currencies and the rate of exchange. But in fact there is no direct relation between the two. Exchange rate can be influenced by many other considerations such as tariffs, speculation and capital movements.

According to Keynes, there are two basic defects in the purchasing power parity theory, *viz.* (*i*) it does not take into consideration the elasticities of reciprocal demand, and (*ii*) it ignores the influences of capital movements. In Keynes's view, foreign exchange rates are determined not only by the price movements but also by capital movements, the elasticities of reciprocal demand and many other forces affecting the demand for and supply of foreign exchange. "By elasticity of reciprocal demand is meant the responsiveness of one country's demand for another country's exports with respect to price or income." As for price elasticity, generally speaking, greater the proportion of luxuries and semi-luxuries in the exports demanded, the more elastic will be the country's demand for another country's exports. It will also be more elastic, when there is a greater number of alternative markets in which to buy and greater the capacity to produce the effective substitutes for goods imported. As for the income elasticity of demand for imports, changes in demand for goods and services and in the derived demand for foreign exchange is functionally related to the changes in national income. How far is a country's demand for another's exports responsive to a change in national income, will influence the rate of exchange. In other words, it is the character of the propensity to import out of a given income, that is, supposed to affect the exchange rate independently of international price movements. Technological improvement adding to the productivity of the country and making its goods cheaper and better, tariff changes and exports subsidies affect exchange rates via their influence upon reciprocal demand quite independently of international price movements.

Capital movements, both short-term and long-term, are the other important influences. There is hot money flying from a country trying to make profits or avoid loss on exchange fluctuations and there is a 'refugee capital' seeking safely and security abroad. An actual or expected change in the domestic price of a foreign currency may lead to inflow or outflow of 'hot money' causing a further change in the exchange rate without there being price changes in either country. The inflow tends to raise the exchange value of the currency of the capital receiving country and outflow will lower it. Long-term movement of capital also has a similar effect.

In view of the defects pointed out above, the purchasing power parity theory does not offer an adequate or satisfactory explanation of fluctuations in the rates of exchange. The determination of the exchange rate depends not only on international price relations but on many other factors as mentioned above. This leads to a more adequate explanation of the determination of foreign exchange rates through demand for and supply of foreign exchange viz. balance of payments theory. This theory has been explained above.

EFFECTS OF CHANGES IN EXCHANGE RATE (DEPRECIATION OR DEVALUATION) ON THE ECONOMY

Under the recent economic reforms in India, not only have we liberalised the Industrial sector but have also opened up the economy, made our currency convertible and allowed exchange rate to adjust freely. It is important to understand the full implications of opening up the economy and allowing our currency to 'float'.

It is worthwhile to note that under a fixed exchange rate system when citizens of a country spend some of their income on imports, it reduces the value of multiplier because imports, like savings and taxes, serve as a leakage from the circular flow of income. On the other hand, exports, like investment and Government expenditure, raise the aggregate demand for domestically produced goods and services and thereby cause an expansion in output through a multiplier process.

However, under a variable or floating exchange rate system, the effect of imports and exports on real output is highly complicated. First, the volume of imports and exports depends not only on income, price level, interest rate but also on the exchange rates themselves. Thus, when due to some factors, foreign exchange rate changes, it will have an effect on the level of GNP and the price level. Further, exchange rates themselves will adjust to the changes in the economy. We discuss below the effects of changes in the exchange rate, especially of depreciations and devaluation of the exchange rate, on exports, imports, national income, balance of payments and the price level in the economy.

Effects of Depreciation (or Devaluation) on Imports, Exports and Real National Income

From our foregoing discussion of determination of exchange rate through demand and supply curves of foreign exchange it follows that when a currency of a country, say Indian rupee, depreciates as a result of demand and supply conditions or is devalued by the Government, the prices of Indian exports in terms of foreign currency (say dollar) will fall. This will cause the increase in quantity demanded of Indian exports. As a result, Indian exports will increase. On the other hand, depreciation or devaluation of Indian rupee will make the imports from foreign countries more expensive in terms of rupees (for example, a dollar's worth of US goods will cost more in terms of the Indian rupee) when the Indian rupee depreciates or is devalued. Thus, higher prices of imports will induce individuals and firms in India to import less and they will make an attempt to substitute domestically produced goods for imports from abroad. Thus, *as a result of depreciation or devaluation and consequently increase in exports and decline in imports, net exports will rise and therefore the net aggregate demand for domestically produced goods will increase*. And if the level of output, especially industrial output is below full capacity output, the increase in net aggregate demand domestically will cause expansion in domestic output and will lead to increase in GDP or real national income. This is illustrated in Fig. 35.6 A *devaluation or depreciation can therefore serve as a stimulus to the economy*.

It will be recalled that in the Keynesian model of determination of real national income (GNP), the effect of net exports, that is, $X–M$, where X denotes value of exports and M denotes value of imports, is similar to investment in its effect on national income. Both raise national income through a multiplier process. It may be noted here that because of the favourable effect of depreciation or devaluation of national currency on exports, imports and real GNP, the countries are tempted at times to intervene in the foreign exchange market, devalue their currency to provide stimulus to their economies. However, if one country devalues its currency to stimulate its economy, the other countries can also do so. Now, when all countries attempt to devalue their currencies, there will be no gain in real income for any of them. Such a situation actually occurred during the early years of Great Depression (1929-33). In their attempt to maintain their exports and protect their level of income and employment, many countries devalued their currencies without bothering about its adverse effects on the economies of the other countries. Such policies of competitive devaluation are generally called "**beg-thy-neighbour**" policies. Since all followed this beg-thy-neighbour policy, no one could maintain their export sales and protect domestic employment and level of economic activity which were badly hurt by the severe depression that gripped their economies.

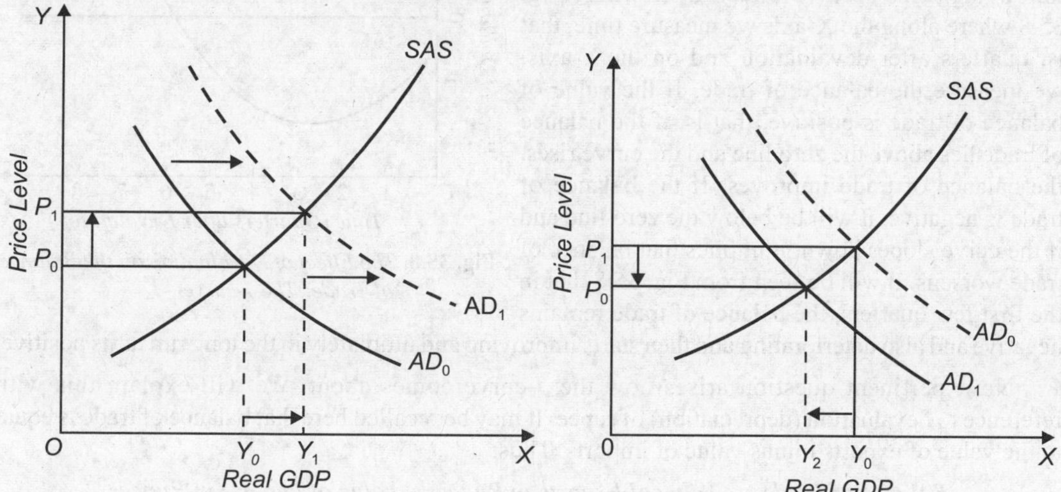

Fig. 35.6. *Depreciation Causes Net Exports to Rise which Leads to the Increase in both GDP and Price Level.*

Fig. 35.7. *Appreciation of National Currency Causes Fall in net Exports and Leads to Decline in both GDP and Price Level.*

On the contrary, the appreciation of a national currency will have opposite effect. When the currency of a country appreciates, its exports will become costlier causing a decline in them, whereas its imports will become cheaper resulting in increase in them. As a result, net exports of the country will decline leading to the decrease in net exports and will therefore cause leftward shift in the aggregate demand curve *AD* as is shown in Fig. 35.7. This will cause fall in both real GDP and price level. This can be seen from Fig. 35.7.

Devaluation or Depreciation and the Balance of Trade: The J-curve

As explained above, lowering of the value of a currency of a country tends to raise its exports by making its goods cheaper for foreigners. On the other hand, devaluation or depreciation makes the imports from abroad expensive in terms of domestic currency (rupees in case of India) and therefore the imports tend to fall. *With exports increasing and imports declining, it is expected that devaluation (depreciation) will reduce a country's trade deficit*. As a matter of fact, in recent years when a country experienced a severe disequilibrium in the balance of trade or balance of payments, it devaluated its currency to raise exports and reduce imports and thus restore equilibrium in the balance of payments. However, it may be noted that the effect of devaluation or depreciation on balance of trade is ambiguous and quite uncertain because a good deal depends on the price elasticity of exports and imports of a country. For example, if the *price-elasticity of exports* in terms of a foreign currency of a country is less than unity, the value of exports in terms of a foreign currency will fall as increase in physical volume of exports will be more than offset by the depreciation of the currency. On the other hand, if the demand for imports is inelastic, they will not decrease despite devaluation.

Many economists are of the view that devaluation is likely to worsen the balance of trade for the few quarters (probably three to six) after the initial devaluation. However, they think after a time lag, the balance of trade may improve. In fact, a concept called *J-curve effect* has been put forward. According to this, after the initial depreciation, the balance of trade moves according to the shape of the letter *J*. This means that in the first few quarters following devaluation the balance of trade becomes worse and after that it becomes positive and starts improving. This *J*-curve effect is shown in Fig. 35.8 where along the X-axis we measure time, that is, quarters after devaluation and on the Y-axis, we measure the balance of trade. If the value of balance of trade is positive, that is, if the balance of trade lies above the zero line and the curve rises, the balance of trade improves. If the balance of trade is negative, it will be below the zero line and if the curve slopes down, it implies that balance of trade worsens. It will be seen from Fig. 35.8 that in the first few quarters, the balance of trade remains negative and also deteriorating and then starts improving and ultimately in the long run turns positive.

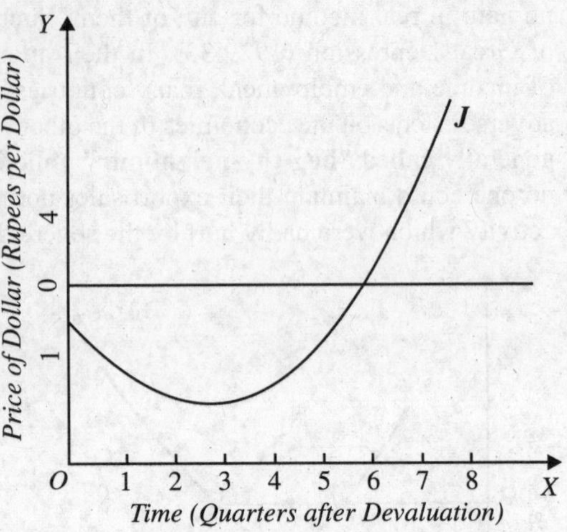

Fig. 35.8. *The Effect of Devaluation on the Balance of Trade: The J-curve*

Now, pertinent question arises how the *J*-curve comes about. We will explain this with reference to devaluation (depreciation) of rupee. It may be recalled here that balance of trade is equal to the value of exports minus value of imports. Thus;

Balance of Trade = Value of Exports in Rupees – Value of Imports in Rupees

It may be noted that value of both exports and imports is equal to the volume of exports or imports multiplied by the rupee price of exports and imports respectively. The depreciation (devaluation) of the currency affects both the volume and rupee price of exports and imports. First, depreciation (devaluation) of currency increases the volume of exports and reduces the volume of imports, both of which have a favourable effect on the balance of trade, that is, they will lower the trade deficit or increase the trade surplus. This has been explained above. Secondly, as a result of devaluation, rupee-price of exports is not likely to change much in the short run. The rupee-price of exports depends on the domestic price level and, in the short run, the devaluation (depreciation) of rupee will have only a very small effect on the domestic price level. On the other hand, the rupee-price of imports increases immediately after devaluation. Imports into India from abroad would be more costly because as a result of devaluation, a hundred rupee note will buy fewer US dollars and pound sterlings than before. Thus, *a rise in the rupee-price of imports has a negative effect on the balance of trade*, that is, it will tend to increase the trade deficit or reduce the trade surplus.

Price Effect and Quantity Effect of Devaluation

An example will make clear the negative effect of depreciation or devaluation on the balance of trade as a result of devaluation or depreciation. Suppose the rupee cost of a particular US machine goes up from ₹ 50,000 to 60,000 following the devaluation of rupee from ₹ 46 per dollar to ₹ 44 per dollar, with the rise in price of a US machine, Indians *will spend more* on a US machine than before. This is a *price effect*. But increase in the price of the US machine will lead to *the decrease in the quantity demanded* of US machines by the Indians. This is the *quantity effect*. Now, the net effect of devaluation on the *value of imports* depends on whether quantity effect is larger than price effect or vice versa. And this depends on the price elasticity of imports. It therefore follows that net effect of devaluation (depreciation) on the balance of trade could go either way. The historical experience shows that initially the negative effect predominates. This is because the effect of devaluation/depreciation on the price of imports is quite fast, whereas it takes some time for quantity of imports to decline in response to the rise in rupee-price of imports and the value of exports to increase in response to the fall in price of exports in terms of foreign currency. According to the *J-curve effect*, the initial effect of devaluation/depreciation on the balance of trade is negative and when in the long run imports and exports adjust to the changes in prices, the net effect on the balance of trade becomes positive. The more price elastic is the demand for exports and imports, the greater the improvement in the balance of trade in the long run.

Depreciation or Devaluation and Inflation

The devaluation or depreciation of currency tends to raise the price level in the country and thus increase the rate of inflation. This happens because of two reasons. Firstly, as a result of depreciation/devaluation, prices of imported goods rise. In case of imports of consumer goods, rise in their prices directly leads to the increase in the rate of inflation. In case of imports of capital goods and raw materials, the rise in their import prices will not only directly raise the price level but as they are used as inputs in the production of other goods, rise in their imports prices will also push up the cost of production of these other goods and thus will bring about cost-push inflation.

Secondly, as explained above, depreciation makes the exports cheaper and therefore more competitive in the world markets. This causes the exports of goods to increase and reduces the supply and availability of goods in the domestic market which tends to raise the domestic price level. Besides, due to higher prices of imported goods, people of a country tend to substitute domestically produced goods for the now more expensive imports. As a result, the aggregate demand or expenditure on domestically produced goods and services will increase causing either expansion in output of goods or rise in their prices or both. However, if the economy is working close to the capacity output, the effect will be more on raising prices of goods.

FIXED EXCHANGE RATE AND THE BRETTON WOODS SYSTEM

Until a few years ago, there was fixed exchange rate system introduced during the last year of the Second World War. This fixed exchange rates system under Bretton Woods System of foreign exchange is known as the Bretton Woods System as a group of economists from the United States and Europe met at the Bretton Woods, a town in New Hampshire to devise this system. According to the agreement made there, an international organisation called *International Monetary Fund* (IMF) was set up to administer the new fixed exchange rates system. According to the rules framed under the agreement, the USA was to fix a parity or par value for its dollar *in terms of gold*, whereas the other countries were required to fix parities for their currencies *in terms of dollars*. Since US dollar was tied to gold, the currencies of other countries with fixed exchange rate with US dollar automatically got pegged (that is, fixed) with a certain gold value. The Government of the USA made a commitment to maintain the convertibility between dollar and gold at a fixed rate, whereas other countries agreed to maintain the convertibility of their currencies with US dollar. The USA fixed $35 per ounce of gold as the convertible rate. The change in these fixed exchange rates (devaluation or revaluation) was to be made with the consent of IMF in case of *fundamental disequilibrium* in the balance of payments. The Bretton Woods agreement did not define the fundamental disequilibrium but in practice it came to mean *chronic and large* deficit in the balance of payments. Therefore, when a country was experiencing a *large and persistent* deficit in balance of payments, it was allowed by IMF to devalue its currency to the extent required to improve the balance of payments position. Thus, faced with the serious problem of balance of payments, India devalued its rupee by 36.5 per cent in 1966. Again in July 1991, India devalued its rupee by about 20 per cent to tide over the serious foreign exchange crisis which arose due to deficit in balance of payments.

IMF maintains funds or reserves with itself which are contributed by member countries. IMF was given power to give loans to the member countries when they had short-term or temporary deficit in the balance of payments from its reserves. In case of fundamental disequilibrium its function was to advise the countries to devalue their currency to tide over the problem of deficit in balance of payments. The purpose was to achieve a relatively fixed or stable exchange rate system which was required for promotion of world trade.

It may be noted that, under Bretton Woods System, *to maintain the exchange rate at the specified level, the Governments (or their Central Banks) were to keep with themselves reserves of internationally accepted foreign currencies*. Thus gold, which in the previous years served as an international currency was substituted by reserves of foreign currencies that were widely used in international dealings. The US dollar became the important currency in which the reserves were to be kept by the other countries to maintain the exchange rate of their currency. Pound Sterlings, German Marks, Japanese Yen were also kept for this purpose.

To maintain the fixed exchange rate, Bretton Woods System required the intervention by the Governments or their Central Banks. How they maintained the exchange value of their currency by using the reserves of foreign currencies is illustrated in Fig. 35.9 where the *demand and supply curves DD and SS* of US dollars are drawn. Along the *X*-axis, the quantity of US dollars and on the *Y*-axis, the value of US dollar in terms of Indian rupees is measured.

Suppose the Government of India is committed to maintain the exchange rate of its currency with US \$ at *OR* level which is also assumed to be the equilibrium exchange rate as determined by the demand for and supply of dollars. Now, suppose the demand of the USA for Indian exports greatly declines. This will reduce the supply of US dollars causing a shift to the left in the supply curve of US dollars to *S′S′*. Now, given the demand curve *DD* of dollars remaining unchanged, at the exchange rate *OR,* the quantity supplied of US dollars falls to *RL* whereas the quantity demanded of dollars continues to be *RE*. As a result, the excess demand of US dollars emerges at the fixed exchange rate *OR* and if left free to the market forces, the equilibrium exchange rate *will rise to OR′ level*. That is, price of *US* dollar in terms of rupees will appreciate or in other words the Indian rupee will depreciate. *In order to maintain the exchange rate of rupee in terms of dollar, and to prevent it from depreciating, the Government or RBI must sell US dollars from its foreign exchange reserves to increase the supply of US dollars in the market.* This is how the exchange rates of different countries remained fixed despite the frequent changes in demand for and supply of their currencies.

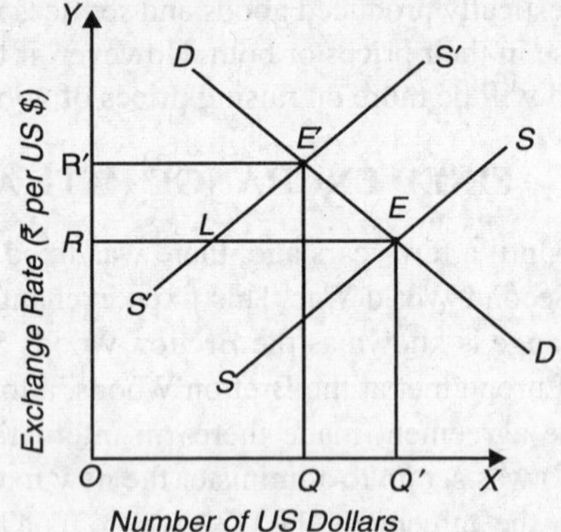

Fig. 35.9. *Government Intervention to Maintain the Exchange Rate at the Fixed Level*

Collapse of the Bretton Woods System

Fixed exchange rate system under the Bretton Woods arrangement came under heavy pressure during the sixties and it ultimately collapsed in the early seventies. As seen above, under the Bretton Woods System, the US dollar became an internationally accepted currency in which reserves of the countries were kept for international transactions. Before the Bretton Woods System gold was accepted as international medium of exchange. Under the Bretton Woods System the US dollar also assumed the role of international currency for three reasons. Firstly, after World War II, the United States emerged as the strongest economy. Secondly, the United States had accumulated a large stock of gold. Thirdly, during 1934 and 1971, the United States had been following the policy of selling and buying gold to other countries at a fixed exchange rate of \$35 per ounce. As under the Bretton

Woods System, the dollar was converted into gold at a fixed rate (that is, $35 per ounce), it was considered '*as good as gold*'.

Since the World's supply of gold did not increase adequately to meet the requirements of rapidly expanding international trade and finance, the dollar increasingly became the world currency in which international reserves were kept by various countries. But a serious problem arose which caused the demise of Bretton Woods system. The United States had a large and persistent deficit in balance of payments during the fifties and sixties. To meet these deficits, the US used its gold reserves for making payments to other countries, especially Germany and Japan. Obviously, this could not go on forever because its gold reserves would have eventually run out. In fact, this was a big flaw in the fixed exchange rate system adopted under the Bretton Woods arrangement. Further, as a result of surplus balance of payments with the United States, other countries, especially Germany and Japan, acquired not only gold but also a large quantity of the dollars from the US which were kept as reserves.

With the passage of time, reserves of dollars and gold held by other countries increased very much while the reserves of gold with USA declined a great deal, the doubts arose about the ability of the USA to fulfil its commitments of converting dollars into gold at the agreed exchange rate of $35 per ounce. Widespread fears emerged that there would be run on the dollar, that is, countries holding dollars would *en masse* try to exchange dollars for gold. If it had actually happened, it would have greatly hurt the international trade and finance. Therefore, in view of the large and persistent deficit in balance of payments, it was decided by the US Government to withdraw convertibility of dollar into gold at the rate of $35 per ounce on August 15, 1971. This brought about the collapse of the Bretton Woods System which was supported by the US commitment to convert dollar into gold at a fixed rate. With this, the link between gold and international value of the dollar ended and consequently the *dollar was floated*, that is, permitted its exchanged rate with other currencies to be determined by market forces.

FIXED AND FLEXIBLE EXCHANGE RATES

A study of economic history shows that three different exchange rate systems have been prevailing in the world economy. The first exchange rate system, popularly called *Gold Standard* prevailed over 1879-1934 period with the exception of World War I years. Under the gold standard, currencies of different countries were tied to gold (that is, the value of currency of each country was fixed in terms of a certain quantity of gold). With this, the exchange rate between different countries got automatically fixed. Thus, the gold standard represented fixed exchange rate system. As explained above, from the end of World War II to 1971, another fixed exchange rate system, generally known as Bretton Woods System prevailed. Under this, the US dollar was tied to a certain quantity of gold and the currencies of other countries were tied to dollar or in some countries directly to gold. However, as seen above, in 1971, due to large and persist deficit in balance of payments of the United States. Bretton Woods system also broke down. Since then, the *flexible* or what is also called *floating exchange rate system* has been existing. But even in this flexible or floating exchange rates, the Government (or the Central Bank) of the countries intervenes to keep the international value of their currencies (*i.e.*, exchanges rate) within certain limits. Therefore, even in the post 1971 period, the exchange rate system is not *purely* flexible and has therefore been called **Managed Float System**. We discuss below the merits and demerits of the fixed and flexible exchange rate system.

Merits of Fixed Exchange Rate System

Exchange Rate Stability. In defence of fixed exchange rate system, it has been pointed out that it ensures stability in exchange rate. Exchange rate stability, it is said, is necessary for orderly development of the international economy and rapid growth of world trade. If the exchange rate is unstable or variable, the exporters will not be certain about the price they would receive for the goods to be exported by them; the importers will not be certain about the price and the payments they have to make for their imports. These uncertainties involve risks for exporters and importers. This greatly harms the growth of world trade. Thus, the chief merit of fixed exchange rate system is that it eliminates the possibilities of such uncertainties and associated risks and thereby promotes foreign trade.

It is important to note that for the developing countries a stable or fixed exchange rate system has a special advantage. This is because the developing countries have a large and persistent balance of payment deficits. As a consequence, if the exchange rate is flexible and not fixed, due to these large persistent deficits, the exchange rate or international value of their currencies will be continuously falling. Thus, fixed exchange rate system prevents this continuous tendency for the depreciation of their currencies.

Promotes Capital Movements. Another advantage of fixed exchange rate is that it facilitates capital movement by private firms. A stable currency does not involve any uncertainties about capital loss on account of changes in exchange rate. Therefore, fixed exchange rate system would attract foreign capital investments. Foreign private firms would not be interested in making investment in those countries whose currency is not stable. Since foreign investment is an important source of economic growth, the fixed exchange rate system would promote rapid economic growth of the developing countries.

Prevents Capital Outflow. Further, flexible and unstable exchange rates may at times of difficult economic situation encourage *flight of capital*, as happened in the case of Mexico a few years ago causing a serious balance of payments problem. On the other hand, a stable exchange rate ensures that such capital outflow would not occur.

Prevents Speculation in Foreign Exchange Market. Another important merit of fixed exchange rate system is that it does away with speculation in foreign exchange markets. The advocates of fixed exchange rate system point out that the flexible and unstable exchange rate encourages speculation in foreign exchange market. The operators in foreign exchange market try to take advantage of the fluctuations in exchange rate as freely determined by demand and supply forces. They try to manipulate the exchange rate through their actions of buying and selling foreign exchange and try to obtain speculative gain in this way. Ragnar Nurkse in his study of *International Currency Experience* which he conducted for the period of twenties found that because of speculative activities, flexible exchange rate had a destabilising influence. It caused greater fluctuations than would have been the case otherwise. By eliminating speculation, fixed exchange rate system ensures stability in the exchange rate.

Serves as an Anchor against Inflation. Another important advantage of fixed exchange rate system is that it prevents the Government of the countries from adopting inflationary policies. Generally, Governments have often been tempted to pursue undue expansionary fiscal and monetary policies to lower unemployment and create boom conditions. However, inflation has several bad effects for the economies. It increases inequalities of income and wealth and hurts the poor most. Further, inflation hampers economic growth in the long run. Fixed exchange rate system forces the Governments to achieve price stability by taking effective anti-inflationary measures. This is because in the case of fixed exchange rate, inflation will cause balance of payments deficits and result in loss of international reserves. Therefore, this forces the Government to adopt measures to check inflationary pressures in the economy. To quote B. Soderston, "Fixed exchange rates can serve as an anchor. Inflation will cause balance of payments deficits and losses in reserves. Hence the authorities will have to take counter measures to stop inflation. Fixed exchange rates should therefore impose a discipline on governments and stop them from pursuing inflationary policies which are out of tune with the rest of the world."[1]

Promotes Economic Integration of the World. It has also been argued in favour of fixed exchange rate system that it is necessary for achieving economic integration of the world community. This is similar to a single common currency in a country which promotes economic integration of a nation in the sense that it facilitates communication, trade and free movement of finance between different regions of a country. It ensures free competition among all producers and consumers of various countries. Similarly, a common currency would promote economic integration of the world. Fixed exchange rate between the currencies of the various countries approximately serves the purpose of a single currency between different countries. Thus, fixed exchange rates among different currencies is a necessary condition for the purpose of *forming economic union* between various countries, that is, forming a regional grouping. To enjoy the advantages of economic unity in recent years, there is a proposal for issuing a single currency for European countries to strengthen the economic integration of European countries.

Promotes Growth of Internal Money and Capital Markets. Another big advantage of the fixed exchange rate system is that it promotes growth of internal money and capital markets. Since flexible exchange rates cause uncertainties about the future exchange rates, individuals, companies and institutions are reluctant to lend to and borrow from the internal money and capital markets.

In view of the above advantages, a fixed exchange rate system prevailed for a long time from 1944 to 1971. As we have seen above that the Bretton Woods aggreement adopted a fixed exchange rate system because of the above mentioned merits, IMF permitted a change in the exchange rate only in case of fundamental disequlibrium in the balance of payments.

Demerits of Fixed Exchange Rates

Fixed exchange rate had however a great flaw in that the countries with a large and persistent balance of payments deficits were losing gold and other foreign assets. This could not go on forever as evidently stock of gold and foreign currencies would have run out. That is why the USA abandoned Bretton Woods System which represented fixed exchange rate system. The other countries facing problem of balance of payment deficit found their international reserves dwindling which forced them to devalue their currency. The devaluation has an inflationary potential.

On the other hand, under the fixed exchange rates, the countries with a balance of payments (BoP) surplus will be providing its national currency and in return it will receive foreign currencies and assets such as US dollar, pounds, gold. The increase in foreign assets would lead to the expansion in money supply which is likely to create inflationary pressures in the economy. But remember an important difference between the countries with balance of payments deficits and the countries having balance of payment surplus. Whereas due to depletion of reserves of gold and foreign currencies, the countries with deficit balance of payments are forced to *devalue* their currency to overcome the deficit, the countries with a balance of payments surplus generally do not *revalue* their currency for the fear that it would discourage exports. Instead, the countries with surplus BoP face the problem of inflation and try to tackle it with other policies.

Another problem with the fixed exchange rate system is that *at what level exchange rate should be fixed*. If the exchange rate of a foreign currency with a national currency is fixed or pegged at the equilibrium level, that is, at the rate at which its quantity demanded equals quantity supplied, no problem arises. However, for the Government it is difficult to find such an equilibrium exchange rate of a foreign currency in terms of the national currency. Further, as underlying conditions change as they do so frequently, the equilibrium exchange rate will also change giving rise to the problems. If the exchange rate of a foreign currency in terms of the national

---

1. B.Soderston, *International Economics*, 2nd edition, Macmillan, 1980, p. 402.

currency is fixed at a lower level, there will be deficit in the balance of payments BoP. On the other hand, if the exchange rate of a foreign currency in terms of the national currency is fixed at a higher than the equilibrium level, there will be surplus in BoP. Consider Fig. 35.10, where demand and supply curves of dollars respectively for India at various prices of a dollar in terms of rupees are drawn. ₹ 45.50 to a dollar is the equilibrium exchange rate. If the Government or Reserve Bank of India fixes the exchange rate at ₹ 43 to a dollar as the exchange rate, the quantity demanded of dollars by Indians would exceed the quantity supplied so that there would be deficit in India's balance of payments. This means that at ₹ 43 to a dollar *Indian rupee is overvalued.* As a result, India's stock of foreign exchange (*i.e.*, dollars) and other foreign assets will decline. Sooner or later, India will be forced to take policy measures to correct this deficit in BoP. India will have to adopt contractionary fiscal and monetary policies to deflate the economy so that imports decrease or it will impose various restrictions on imports to reduce them. To reduce persistent deficit in BoP, it may even resort to devaluation.

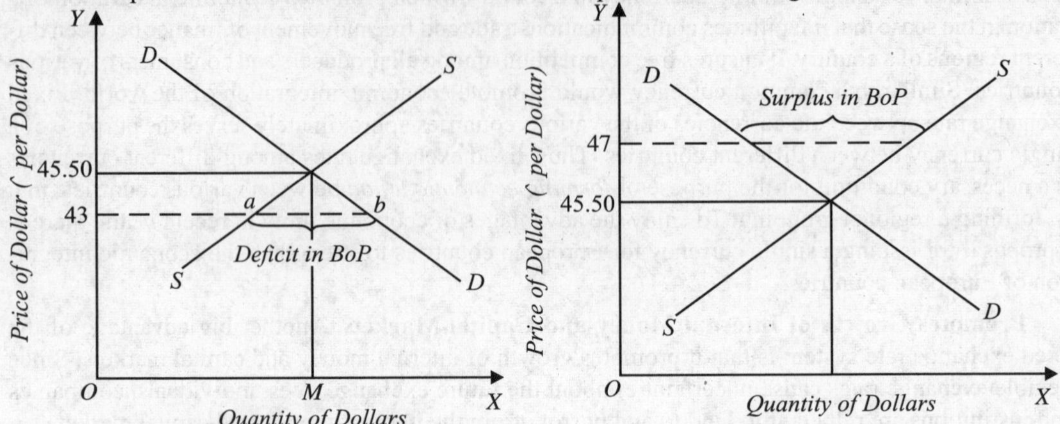

Fig. 35.10. *Overvaluation of a Currency Causing Deficit in BoP.*

Fig. 35.11. *Undervaluation of a Currency Leading to the Problem of Surplus in BoP.*

Now, consider Fig. 35.11, if the Government fixes the exchange rate of ₹ 47 to a dollar, the quantity supplied of dollars by other countries would exceed the quantity demanded by the Indians. As will be seen from Fig. 35.11, this causes *surplus in Balance of Payments* (BoP). This implies that at ₹ 47 to a dollar, the *Indian rupee is under-valued* (or, which is the same thing, *dollar is overvalued* in terms of rupees). Historical experience shows that countries with surplus in BoP have been reluctant to *revalue* their currency (that is, to raise the international value of their currency). Thus, a *surplus in balance of payments also gives rise to a problem*. As a result of this surplus, India will accumulate dollars (*i.e.*, foreign exchange) for which it would pay rupees, the national currency. The new rupee notes would be printed to pay for dollars. Thus, surplus in BoP leads to the expansion in money supply generating inflationary pressures in the Indian economy. To check inflation, the Indian Government might therefore take steps to eliminate the surplus in BoP by encouraging imports or restricting exports. Further, it may try to take steps to neutralise the impact of surplus in BoP on expansion in money supply. For this purpose, it might encourage *capital outflow* (*i.e.*, investment abroad) by private firms or provide financial assistance to other countries on a scale which is considered undesirable by its people.

In view of the above drawbacks and problems, the fixed exchange rate has been given up despite its various advantages explained above.

Merits of Flexible Exchange Rates

Under the flexible exchange rate system, exchange rate between different currencies, like the prices of commodities are freely determined by market forces, that is, by demand and supply forces. With the change in economic conditions underlying demand and supply, the exchange rate will automatically change without any intervention by the Government. That is why, it is called flexible or variable exchange rate system. It has the following merits.

Problems of Undervaluation and Overvaluation are Avoided. The advocates of flexible exchange rates contend that under it, the problems of undervaluation and overvaluation of currencies which are found in the fixed exchange rate system are avoided. Whenever there is deficit in balance of payments implying overvaluation of the national currency under the flexible exchange rates, it will depreciate (that is, its value will fall) which on one hand will make exports cheaper and thereby encourage them and on the other, will make the imports costlier than before which will tend to discourage them. Thus, increase in exports and decline in imports as a result of depreciation will lead to the automatic correction in the balance of payments.

On the other hand, whenever there is surplus in the balance of payments, the exchange rate will appreciate which will tend to reduce exports and raise imports. This again will tend to automatically restore the balance of payments equilibrium. This is how the flexible exchange system works to ensure the equilibrium in the balance of payments. To quote Prof. Soderston, "The same factors which under the fixed rates give rise to deficits and surpluses in the balance of payments would under floating rates make the exchange rate depreciate or appreciate. Thereby equilibrium would be preserved and the Government could be freed from consideration regarding external balance."

Promotes Growth of Multilateral Trade. The advocates of flexible exchange rates system are strongly of the view that as unlike fixed exchange rates system, this does not create serious and difficult problems, it will ensure rapid growth of multilateral world trade.[2] Further, they point out that promotion of world trade under the flexible exchange rates would not interfere in any way the adoption of policies to achieve domestic economic stability.[3]

Flexible Exchange Rates does not Necessarily Show Large Fluctuations. It has been pointed out in defence of the flexible exchange rates that the problems of undervalued or overvalued currency found under the fixed exchange rate regime are not found in the flexible exchange rate system. Further, it is contended that exchange rates being flexible does not necessarily mean there will be large fluctuations in them. Even under flexible exchange system there need not be large fluctuations in exchange rates. According to them, changes in exchange rates will occur only when economic conditions underlying demand for and supply of foreign currencies change. Further, according to them, random fluctuations around the normal exchange rates would be smoothened out through operations by private speculators. Thus, according to Soderston "if the currency appreciated above its equilibrium value, if its price fell in terms of foreign currency, speculators would buy the currency; and if it depreciated, speculators would sell the currency. Thereby they would smooth out fluctuations and help to keep the exchange rate stable. If the underlying conditions changed, however the price of foreign exchange would also change."

Ensures Individual Freedom. The system of floating exchange rate is advocated on the basis of philosophy that Government should intervene in the economy as little as possible and the individuals should be left free to pursue their economic interests. It has been noted above, the alternative to floating exchange rates system is the fixed exchange rates which require controls and restrictions which are implemented by experts who generally do not work under effective democratic controls. Generally, they fix the exchange rate out of consideration for other than economic objectives, for example for the purpose of retaining a certain market structure or influencing income distribution.

---

2. See Miltion Friedman, 'The Case for flexible Exchange Rate', AEA, *Readings in International Economics*, Vol. XI.

3 *Ibid.*

Frees the Government from Problems of Balance of Payments. A great merit of flexible exchange rates is that it frees the Government from problems of balance of payments. As has been seen above, the fixed exchange rates system leads either to deficit or surplus in balance of payments. Under this system, the Government remains preoccupied with the questions of devaluation or re-valuation of their currencies. Since floating exchange rates work automatically to restore balance of payments equilibrium, the Government need not pay any attention to the balance of payments. Instead, the Governments can concentrate on solving their domestic problems of economic stability, inflation and unemployment. "Flexible exchange rates should therefore give an increased freedom both to individuals and countries to pursue whatever aims they have for their economic policies."[4]

Demerits of Flexible Exchange Rates System

Flexible exchange rates system is also not without shortcomings. It has been opposed on some important grounds. Some important arguments given against flexible exchange rates are given below.

1. Flexible Exchange Rates Create a Situation of Instability and Uncertainty. An important argument against flexible exchange rates is that too frequent fluctuations in exchange rate under it create uncertainty about the exact amount of receipts and payments in foreign exchange transactions. *This instability hampers foreign trade and capital movements between the countries.* However, Friedman does not accept this charge against flexible exchange rates. He argues that flexible exchange rate is not necessarily an unstable exchange rate. According to him, it is the underlying economic conditions determining foreign trade that often change causing changes in the foreign exchange rate and for this, the flexible exchange rates should not be blamed.

2. Dampening Effect on Foreign Trade. Under the flexible exchange rates, the price of foreign exchange or international value of the national currency is quite uncertain. As a result, they are unable to take proper decisions regarding exports and imports of goods. Obviously, this has a dampening effect on the volume and growth of foreign trade. However, Friedman does not agree with this and points out that traders can always protect themselves against risks arising out of the changes in the exchange rate by hedging in the future market.[5] However, Soderston has challenged this view of Friedman and points out that no markets exist today that can protect traders against risks arising out of flexible exchange rates.[6]

It may however be noted that from the empirical evidence since 1971, when exchange rates have been flexible, it does not appear that flexible exchange rates have a dampening effect on the growth of foreign trade. In fact, the growth of world trade since 1971 has been fairly good.

3. Widespread Speculation with a Destabilising Effect. The system of flexible exchange rates has been opposed on the ground that under it there is widespread speculation regarding exchange rates of currencies which has a large destabilising effect on these rates. Friedman, on the other hand, contend that *speculation has a stabilising influence on exchange rates*. However, whether or not speculation has a destablising or stabilising effect is a highly controversial issue in economics which has so far remained unresolved.

4. Provides an Inflationary Bias to an Economy. Another shortcoming of the flexible, exchange rates is that they have an inflationary impact on the economy. It has been pointed out that whenever due to deficit in balance of payments, the currency depreciates, the prices of imports go up. The higher prices of imported materials raise the prices of industrial products and thus generate cost-push inflation. However, in defence of flexible exchange rates, it has been pointed out that when demand for imports decline due to appreciation of national currency, the *foreign suppliers are compelled to reduce their prices* to protect markets. Though this reduction in prices in the face of declining demand is theoretically possible, it may not happen in actual practice.

4. B. Soderston, *op. cit*.
5. Milton Friedman, *op. cit*
6. B. Soderston, *op. cit.*, p. 303.

To sum up, there are merits and demerits of flexible exchange rate system. Whether a flexible exchange rate systems suits an economy depends on circumstances. It depends on characteristics of the economy which adopts flexible exchange rate system. Further, it may be noted that since 1971 when Bretton Woods System representing the fixed exchange rate system was abandoned, it is not perfectly free flexible exchange rate system that has been adopted. Instead, it is what is called *Managed Float System* that has been prevailing. Under the managed float system, the exchange rates are flexible, that is, they change as a result of changes in demand and supply of currencies, but the Government or Central Bank intervenes through demand and supply to keep the variations in exchange rate within certain limits so as to ensure stability and certainty of the foreign exchange rates.

CURRENCY CONVERTIBILITY

For the rapid growth of world trade and capital flows between countries, convertibility of a currency is desirable. Without free and unrestricted convertibility of currencies into foreign exchange trade capital flows between countries cannot take place smoothly. Therefore, to achieve higher rate of economic growth and thereby to improve living standards through greater trade and capital flows, the need, for convertibility of currencies of different nations has been greatly felt. Under Bretton Woods system, fixed exchange rate system was adopted by various countries. In order to maintain the exchange rate of their currencies in terms of dollar or gold various countries imposed several controls over the use of foreign exchange. This required some restrictions on the use of foreign exchange and its allocation among different uses, the currency of a nation was converted into foreign exchange on the basis of officially fixed exchange rate.

When Bretton Woods system collapsed in 1971, the various countries switched over to the floating foreign exchange rate system. Under the floating or flexible exchange rate system, exchange rates between different national currencies are allowed to be determined through market demand for and supply of them. However, various countries still imposed restrictions on the free convertibility of their currencies in view of their difficult balance of payment situation.

Meaning of Currency Convertibility

Let us first understand what is exactly meant by currency convertibility. By convertibility of a currency we mean currency of a country can be freely converted into foreign exchange at *market determined rate of exchange*, that is, exchange rate as determined by demand for and supply of a currency. For example, convertibility of rupee means that those who have foreign exchange (*e.g.* US dollars, Pound Sterlings etc.) can get them converted into rupees and vice-versa at the market determined rate of exchange. Under convertibility of a currency there are authorised dealers of foreign exchange which constitute foreign exchange market. The exporters and others who recieve US dollars, Pound Sterlings, etc. can go to these dealers which are generally banks and get their dollars exchanged for rupees at the market determined rates of exchange. Similarly, under currency covertibility, importers and others who require foreign exchange can go to these banks dealing in foreign exchange and get rupees converted into foreign exchange.

Current Account and Capital Account Convertibility of Currency

A currency may be convertible on current account (that is, exports and imports of merchandise and invisibles) only. A currency may be convertible on both current and capital accounts. We have explained above the convertibility of a currency on current account only.

By capital account convertibility, we mean that in respect of capital flows, that is, flows of portfolio capital, direct investment flows, flows of borrowed funds and dividends and interest payable on them, a currency is freely convertible into foreign exchange and vice-versa at market determined exchange rate. Thus, convertibility of rupee on capital account means those who bring in foreign exchange for purchasing stocks, bonds in Indian stock markets or for direct investment in power projects, highways steel plants, etc. can get them freely converted into rupees without taking any permission from the government. Likewise, the dividends, capital gains, interest received on purchased stock, equity etc. profits earned on direct investment get the rupees converted into US dollars, Pound Sterlings at market determined exchange rate between these currencies and repatriate them. Since capital convertibility is risky and makes foreign exchange rate more volatile, it is introduced only some time after the introduction of convertibility on current account when exchange rate of currency of a country is relatively stable, deficit in balance of payments is well under control and enough foreign exchange reserves are available with the Central Bank.

Convertibility of Indian Rupee

In the seventies and eighties, many countries switched over to the free convertibility of their currencies into foreign exchange. By 1990, 70 countries of the world had introduced currency convertibility on current account, another 10 countries joined them in 1991.

As a part of the new economic reforms initiated in 1991, rupee was made partly convertible from March 1992 under the *Liberalised Exchange Rate Management* scheme in which 60 per cent of all receipts on current account (*i.e.* merchandise exports and invisible receipts) could be converted freely into rupees at market determined exchange rate quoted by authorised dealers, while 40 per cent of them was to be surrendered to the Reserve Bank of India at the officially fixed exchange rate. The 40 per cent exchange receipts on current account was meant for meeting Government needs for foreign exchange and for financing imports of essential commodities. Thus, partial convertibility of rupee on current account meant a *dual exchange rate system*. This partial convertibility of rupee on current account was adopted so that essential imports could be made available at lower exchange rate; to ensure that their prices do not rise much. Further, full convertibility of rupees at that stage was considered to be risky in view of large deficit in balance of payments on current account.

As even after partial convertibility of rupee, foreign exchange value of rupee remained stable, *full convertibility on current account* was announced in the budget for 1993-94. From March 1993, rupee was made convertible for *all trade in merchandise*. In March 1994, even indivisibles and remittances from abroad were allowed to be freely convertible into rupees at market-determined exchange rate. However, on capital account, rupee remained nonconvertible.

Advantages of Currency Convertibility

Convertibility of a currency has several advantages which are briefly discussed below:

1. **Encouragement to Exports.** An important advantage of currency convertibility is that it encourages exports by increasing their profitability. With convertibility, profitablility of exports increases because market foreign exchange rate is higher than the previous officially fixed exchange rate. This implies that from given exports, exporters can get more rupees against foreign exchange (*e.g.* US dollars) earned from exports. Currency convertibility especially encourages those exports which have low import-intensity.

2. **Encouragement to Import Substitution**. Since free or market determined exchange rate is higher than the previous officially fixed exchange rate, imports become more expensive after convertibility of a currency. This discourages imports and gives boost to import substitution.

3. **Incentive to send Remittances from Abroad.** Thirdly, rupee convertibility provided greater incentives to send remittances of foreign exchange by Indian workers living abroad and by NRI. further, it makes illegal remittance such as *hawala money* and smuggling of gold less attractive.

4. **A Self-balancing Mechanism.** Another important merit of currency convertibility lies in its self-balancing mechanism. When balance of payments is in deficit due to over-valued exchange rate, under currency convertibility, the currency of the country depreciates which gives boost to exports by lowering their prices on the one hand and discourages imports by raising their prices on the other. In this way, deficit in balance of payments gets automatically corrected without intervention by the Government or its Central bank. The opposite happens when balance of payments is in surplus due to the under-valued exchange rate.

5. **Specialisation in Accordance with Comparative Advantage.** Another merit of currency convertibility ensures production pattern of different trading countries in accordance with their

comparative advantage and resource endowment. It is only when there is currency convertibility that market exchange rate truly reflects the purchasing powers of their currencies which is based on the prices and costs of goods found in different countries. Since prices in competitive environment reflect that prices of those goods are lower in which the country has a comaparative advantage, this will encourage exports. On the other hand, a country will tend to import those goods in the production of which it has a comparative disadvantage. Thus, currency convertibility ensures specialisation and international trade on the basis of comparative advantage from which all the countries derive benefit.

6. **Integration of World Economy.** Finally, currency convertibility gives boost to the integration of the world economy. As under currency convertibility there is *easy access to foreign exchange*, it greatly helps the growth of trade and capital flows between the countries. The expansion in trade and capital flows between countries will ensure rapid economic growth in the economies of the world. In fact, currency convertibility is said to be a prerequisite for the success of globalisation.

Capital Account Convertibility of Rupee

As explained above, under capital account convertibility, any Indian or Indian company is entitled to move freely from the Rupee to another currency to convert Indian financial assets into foreign financial assets and back, at an exchange rate fixed by the foreign exchange market and not by RBI. In a way, capital account convertibility removes all the restraints on international flows on India's capital account. There is a basic difference between current account convertibility and capital account convertibility. In the case of current account convertibility, it is important to have a transaction – importing and exporting of goods, buying and selling of services, inward or outward remittances, etc. involving payment or receipt of one currency against another currency. In the case of capital account convertibility, a currency can be converted into any other currency *without any transaction*.

The Reserve Bank of India appointed in 1997 the Committee on Capital Account Convertibility with S S Tarapore, former Deputy Governor of RBI as its chairman. Tarapore Committee defined capital account convertibility as the freedom to convert local financial assets with foreign financial assets and vice-versa at market determined rates of exchange. In simple language, capital account convertibility allows anyone to freely move from local currency into foreign currency and back. The purpose of capital convertibility is to give foreign investors an easy market to move in and move out and to send a strong message that Indian economy was strong enough and that India had sufficient forex reserves to meet any flight of capital from the country to any extent.

Benefits of Capital Account Convertibility

The Tarapore Committee mentioned the following benefits of capital account convertibility to India.

1. Availibility of large funds to supplement domestic resources and thereby promote economic growth
2. Improved access to international financial markets and reduction in cost of capital
3. Incentive for Indians to acquire and hold international securities and assets
4. Improvement of the financial system in the context of global competition

Accordingly, the Tarapore Committee recommended the adoption of capital account convertibility. Under the system of capital account convertibility proposed by this committee, the following features are worth mentioning :

(*a*) Indian companies would be allowed to issue foreign currency denominated bonds to local investors, to invest in such bonds and deposits, to issue Global Deposit Receipts (GDRs) without RBI or Government approval to go in for external commercial borrowings within certain limits, etc.

(*b*) Indian residents would be permitted to have foreign currency denominated deposits with banks in India, to make financial capital transfers to other countries within certain limits, to take loans from non-relatives and others up to a ceiling of $ 1 million, etc.

(*c*) Indian banks would be allowed to borrow from overseas markets for short-term and long-term up to certain limits, to invest in overseas money markets, to accept deposits and extend loans denominated in foreign currency. Such facilities would be available to financial institutions and financial intermediaries also.

(*d*) All-India financial institutions which fulfil certain regulatory and prudential requirements would be allowed to participate in foreign exchange market along with Authorised Dealers (ADs) who are, at present, banks. In a later stage, certain select NBFCs would also be permitted to act as ADs in foreign exchange market.

(*e*) Banks and financial institutions would be allowed to operate in domestic and international markets and they would also be allowed to buy and sell gold freely and offer gold denominated deposits and loans.

Preconditions for Capital Account Convertibility

The Tarapore Committee recommended that, before adopting capital account convertibility (CAC), India should fulfil three crucial pre-conditions:

(i) Fiscal deficit should be reduced to 3.5 per cent. The Government should also set up a Consolidated Sinking Fund (CSF) to reduce Government debt.

(ii) The Governments should fix the annual inflation target between 3 to 5 per cent. This was called *mandated inflation target* — and give full freedom to RBI to use monetary weapons to achieve the inflation target.

(iii) The Indian financial sector should be strengthened. For this, interest rates should be fully deregulated, gross non-paying assets (NPAs) should be reduced to 5 per cent, the average effective CRR should be reduced to 3 per cent and weak banks should either be liquidated or be merged with other strong banks.

Apart from these the essential pre-conditions, the Tarapore Committee also recommended that :

(a) RBI should have a monitoring exchange rate band of 5 per cent around Real Effective Exchange Rate (REER) and should intervene only when the REER is outside the band :

(b) The size of the current account deficit should be within manageable limits and the debt service ratio should be gradually reduced from the present 25 per cent to 20 per cent of the export earnings :

(c) To meet import and debt service payments, forex reserves should be adequate and range between $ 22 billion and $ 32 billion; and

(d) The Government should remove all restrictions on the movement of gold.

It was generally agreed that full convertibility of the rupee, both on current account and capital account was a welcome measure and ***is necessary for closer integration of the Indian economy with the global economy.***

The major difficulty with the Tarapore Committee recommendations was that the current account convertibility was to be achieved in a 3-year period – 1998 to 2000. The period was too short and the pre-conditions and the macroeconomic indicators could not be achieved in such a short period. Basically, the Committee failed to appreciate the political instability in the country at that time, and the complete absence of political will and vision to carry forward the process of economic reforms and economic liberalisation. The outbreak of Asian financial crisis at this time was also responsible for shelving the recommendations of Tarapore Committee.

Conclusion

In a speech at RBI on March 18, 2006, former Prime Minister Dr Manmohan Singh stated: "Given the changes that have taken place during the last two decades, there is merit in moving towards fuller capital account convertibility with a transparent framework... I will, therefore, request the Finance Minister and RBI to revisit the subject and come out with a roadmap based on current realities." Promptly, within two days, RBI constituted the "**Comittee on Fuller Capital Account Convertibility**" with S S Tarapore again as chairman. This Tarapore Committee submitted its report in September 2006 (more commonly called the Second Tarapore Report or II).

The second Tarapore committee had drawn up a roadmap for 2011 as the target date for fuller capital convertibility of rupee and mentioned that the conditions were quite favourable. These conditions were:

1. Strong fundamentals of the Indian economy
2. A good amount of foreign exchange reserves of $ 165 million that existed in 2006
3. More liberalised use of foreign exchange already in place
4. A financial system better geared to deal with external capital flows

The fuller capital convertibility of rupee seemed to be desirable at the end of 2006 when the committee submitted its report. However, economic events, especially global financial crisis of 2007-09, brought about a sea change in the economic situation. The Indian economy would have been greatly affected by the global financial crisis if we had implemented the recommendations of the Tarapore Committee. We could not have coped with the extent of capital outflows that took place during 2008-09.

Problems

It may be noted that convertibility of currency can give rise to some problems. Firstly, since market determined exchange rate is generally higher than the previous officially fixed exchange rate, prices of essential imports rise which may generate cost-push inflation in the economy. Secondly, if current account convertibility is not properly managed and monitored, market exchange rate may lead to the *depreciation of domestic currency*. If a currency depreciates heavily, the confidence in it is shaken and no one will accept it in its transactions. As a result, trade and capital flows in the country are adversely affected.

Thirdly, convertibility of a currency sometimes makes it highly volatile. When due to speculative activity, a currency depreciates and confidence in it is shaken, there is capital flight from the country as it happened in 1997-98 in the case of South-East Asian economies such as Thailand, Malaysia, Indonesia, Singapore and South Korea. This adversely affects economic growth of the economy. In the context of heavy depreciation of the currency, not only there is capital, flight but inflow of capital in the economy is discouraged as owing to the resultant depreciation of the currency, profitability of investment in an economy is adversely affected.

QUESTIONS FOR REVIEW

1. What is meant by foreign exchange rate? How is it determined?
2. Distinguish between floating exchange rate and fixed exchange rate. What type of exchange rate system has been currently adopted by India?
3. Explain the factors that cause changes on the exchange rate of a currency. How will the exchange rate of rupee in terms of US dollar be affected if India has a deficit on current account of the balance of payments?
4. Explain the factors which determine demand for foreign exchange, say US dollars.
5. Explain the factors which determine supply of foreign exchange, say US dollar, for India.

6. Explain the fixed exchange rate system adopted under the Bretton Woods system. Why did it collapse?
7. Explain how a foreign exchange rate is determined. How will the following factors affect foreign exchange rate of rupee?
   (*a*) Higher rate of inflation in India
   (*b*) Higher rates of interest in India
8. Critically examine the purchasing power parity theory of foreign exchange. What shortcomings of this theory were pointed out by Keynes?
9. What is meant by appreciation of Indian rupee? How will it affect exports, imports and level of national income of India?
10. Distinguish between depreciation and devaluation. How will devaluation or depreciation affect balance of trade and the national economy? Explain in this connection the concept of *J-Curve*.
11. What is meant by appreciation of a currency? What will be the effect of appreciation of rupee against dollar on our balance of trade, foreign exchange reserves and price level in the Indian economy.
12. How is exchange rate determined in open market? What factors make exchange rate fluctuate? [*D.U. B.Com(Hons.) 1999*]
13. Explain the merits of a flexible exchange rate system. Discuss why Bretton Woods system had favoured relatively fixed exchange rates. [*D.U. B.Com.* (*Hons.*)]
14. A real depreçiation will worsen the trade balance in the short run, but will gradually improve in the long run. [*D.U. B.A.(Hons.) 1999*]
15. How do exchange rate changes help in the trade adjustment process? Discuss the role of monetary or domestic credit factors. [*D.U. B.A.(Hons.)1996*]
16. How far under flexible exchange rates can countries pursue their own national monetary and fiscal policies without having to worry about the balance of payments ? Is there a need for policy synchronization? [*D.U. B.A.(Hons.)1998*]
17. (*a*) What is meant by Currency Convertibility?
    (*b*) What are the advantages of currency convertibility and what problems are faced by making convertibility of a currency?
18. Distinguish between convertbility of currency on current account and capital account. Why has India not so far gone in for full convertibility on capital account.
19. What is capital account convertbility of a currency? What are its benefits and problems? Are the present conditions in India suitable for going in for full capital account convertibility?

CHAPTER 36

FREE TRADE VERSUS PROTECTION

Introduction

In this chapter we shall discuss commercial policy which is a highly important issue in economics. The commercial policy is concerned with whether a country should adopt the policy of free trade or of protection. If the policy of protection of domestic industries is adopted, the question which is faced whether protection should be granted through imposing tariffs on imports or through the fixation of quota or through licensing of imports. The commercial policy has been the subject of heated discussion since the time of Adam Smith who advocated for free trade and recommended that tariffs should be removed to avail of the advantages of free trade[1]. Even today, economists are divided over this question of commercial policy. Various arguments have been given for and against free trade. If the policy of protection of domestic industries is adopted, the question is whether for this purpose tariffs should be imposed on imports or quantitative restrictions through quota and licensing be applied. The readers should be knowing that a Bharatiya Janata Party in India has been demanding a policy of '*Swadeshi*' which in essence means that domestic industries should be protected against low-priced imports of goods from abroad, that is, free trade should not be allowed.

Besides Adam Smith, the other famous classical economist David Ricardo in his famous work "*On the Principles of Political Economy and Taxation*" also defended free trade to promote efficiency and productivity in the economy. We have seen in a previous chapter that Adam Smith and the other earlier economists thought that it pays a country to specialise in the production of those goods it can produce more cheaply than any other country and import those goods it can obtain at less cost or price than it would cost to produce them at home. This means they should specialise according to absolute cost advantage. However, Ricardo put forward the 'Theory of Comparative Cost' where he demonstrated that to obtain benefits from trade it is not necessary that countries should produce these goods for which their absolute cost of production is the lowest. He proved that it could pay a country to import a good even though it could produce that good at a lower cost, if its cost is relatively lower in the production of some other good. Ricardo's theory of trade rests on the idea of relative efficiency or comparative cost.

Despite the classical arguments for free trade to promote efficiency and well-being of the people, various countries have been following the protectionist policies which militate against free trade. By imposing heavy tariff duties on imports of goods or fixing quotas of imports they have prevented free trade to take place between countries. Several arguments have been given in favour of protection. In what follows we spell out this free trade vs. protection controversy.

CASE FOR FREE TRADE

The following arguments have been given in defence of free trade.

1. Gains in Output and Well-being from Specialisation

The case for free trade is fundamentally based on the gain in output and well-being a country obtains from specialising in the production of those goods in which it is relatively more efficient and

---

1. See Adam Smith, *The Wealth of Nations*, Modern Library Edition.

therefore export a part of them and in exchange gets those goods from other countries in production of which they are comparatively more efficient. Specialisation and trading in this way would achieve a more efficient allocation of resources and a higher level of output and well-being. To quote Prof. Haberler, "International division of labour and international trade which enable every country to specialise and to export those things which it can produce cheaper in exchange for what others can provide at a lower cost, have been and still are one of the basic factors promoting well-being and increasing national income of every participating country"[2]

2. Gains from Economies of Scale

An important gain from trade is that it enables the trading countries to benefit from the economies of scale. If a country does not trade with others, its firms will produce goods to meet the domestic demand for a product. If domestic demand for a product is small, each of them will produce at a higher cost since they would not be able to enjoy the benefits of the economies of large scale production. Accordingly, the production of goods will be inefficient. Trade allows a country to export goods with the result that level of output of goods in a country will exceed domestic demand within a country. Thus trade expands the market for goods and enables the producers to take advantage of the economies of scale. Adam Smith was the first economist who pointed out that specialisation was limited by the size of the market. Trade makes it possible for the producers to move beyond domestic market into international market and therefore makes it worthwhile to specialise and produce on a large scale and thereby to lower cost per unit.

For example, in a small country such as Ceylon domestic demand would not be sufficient to produce efficiently large luxury cars on a large scale at a lower cost. Their production on a large scale at lower cost requires wider international market for sale of luxury cars.

3. Long-run Dynamic Gains

Free trade also leads to dynamic gains being obtained from trade. Dynamic gains from trade refer to its stimulation of economic growth. Dennis Robertson described foreign trade as '*an engine of growth*'. The stimulation of growth through foreign trade are apparent from the rapid growth of such economies such as Japan, Taiwan, South Koria, Singapur, Hongkong and China. Free trade promotes economic growth through: (1) raising the rate of saving and investment; (2) import of capital goods, and (3) transfer of technology.

(*i*) *Raising rate of saving and investment*. As mentioned above, increase in national product or real national income of a country obtained through trade above the level that prevails in autarky leads to a higher level of saving. The higher level of saving ensures a higher rate of investment and capital formation which stimulates growth. Hence if trade raises the rate of saving, it also promotes economic growth. The higher rate of saving makes it easier for the developing countries to break 'the vicious circle of proverty' and to "take off into self-sustained growth"[3]

(*ii*) *Import of capital goods*. Besides trade permits a country to import capital goods in exchange for exports of consumer goods or surplus raw materials. and thereby accelerates industrial growth. Imports of capital goods adds to the capital stock in a country and raises its productive capacity more than it would have been possible without trade. Free trade also often enables a country to borrow from other countries to finance import of capital goods.

(*iii*) *Transfer of technology*. If different countries worked in isolation the new technology developed in one country would remain confined locally. Through trade technological progress tends to feed on each other. A technology discovered by one is improved by another and so technology goes on being improved successively. Imagine if every country had to invent a wheel, a steam engine,

2. G. Haberler, *International Trade and Economic Growth*.
3. *Ibid*.

electricity operating in an isolated manner, how slow would have been the progress in technology. The trade increases international diffusion of technology and in this way transfer of technology from the developed countries to the developing countries have been possible. In the modern times technology developed in one country by a firm is licensed to firms in other countries. Through this process, technology is transferred from country to country. In the absence of trade between countries such transfer of technology would not take place and as a result economic growth would be slower.

4. Promotes Competition and Prevents Monopoly

The case for free trade also rests on the fact that it promotes competition and prevents the emergence of monopolies in the domestic economy. In the absence of trade and therefore without facing any competition from foreign firms, domestic firms tend to become inefficient which causes rise in cost per unit of output and therefore higher prices of goods.

When trade is free, increased competition by foreign firms forces domestic firms to adopt measures to increase their efficiency and make efforts to reduce cost by employing lowest-cost production techniques. Free trade also compels them to be innovative and to improve the quality of their products. Further, free trade provides consumers a wide range of products from which to choose. The increase in efficiency and the adoption of improved technology not only lowers prices of products but also contributes to economic growth.

5. Political Gains from Free Trade

Free trade, as explained above, increases well-being or standared of living of the trading countries and this mutual welfare gains from trade make different nations economically dependent on each other. The economic interdependence raises the likelihood of reduced hostility between countries. Economic interdependence provides powerful incentives for peaceful solution of disputes. Trade between economically interdependent countries increases the potential losses from war and thus reduces the likelihood of armed conflict.

Despite the above gains from free trade, countries have put up various barriers to free trade flows. The important barriers to free trade are (1) the imposition of tariffs (i.e., duties on imports of goods), (2) the fixation of import quotas, (3) the licensing of imports. *The reasons for these trade barriers are that different nations want to protect their domestic industries, to increase employment opportunities, to improve their balance of payments and to achieve* other goals. We therefore discuss below the case for protection and then in a later section will examine the impact of trade barriers, especially tariffs on welfare and growth.

CASE FOR PROTECTION

Despite gains from free trade, many arguments have been given against free trade and in favour of protection. By protection we mean in order to safeguard the domestic industries from low-priced imports some barriers against import of foreign goods are imposed. Some arguments given in defence of protection are irrational and invalid, whereas some are valid. We critically examine below various arguments given in favour of protection (*i.e.*, against free foreign trade).

Nationalism

First argument for protection has been that nationalistic feeling or patriotism requires that people of a country should buy products of their domestic industries rather than foreign products. In the USA there has been a campaign '*Be American, buy American*' appealing people to buy American goods instead of imported foreign products. Similarly, in India recent campaign of '*Swadeshi*' appeals to the patriotic feeling of the Indian people that we should protect our indigenous industries and impose barriers on imports of foreign goods or provide subsidies to our industries. However, this argument is misplaced and invalid. Those policy makers who yield to such arguments deny the people of a country the gains from trade such as rise in productive efficiency and greater well-being, stimulus to growth through higher capital formation and spread of superior technology. Thus restrictions imposed on trade in the name of nationalism or *swadeshi* are actually contrary to our national interests because they promote inefficiency and prevents rapid economic growth.

Employment Argument

An important argument for protection is that it will lead to increase in domestic employment or at least preserves present domestic employment. It is often believed that imports of goods from abroad reduce domestic employment. Therefore, if instead of imports we produce those goods at home, employment in the country will increase. Besides, as prices of imported goods are lower, the domestic producers would not be able to compete with them and may be competed out of the market. This will destroy even present jobs in the domestic industries. It is therefore concluded that protection of domestic industries will lead to their expansion and therefore employment in them will increase.

In our view employment argument for protection is not logical and valid. This argument ignores the adverse effects of protection on our industries. An important economic principle is that exports must pay for imports. If imports are restricted by in, osing barriers, the exports cannot remain unaffected. For example, many raw materials and capital goods are imported to be used in industries which export goods. If imports are restricted, exports will therefore fall. This will lead to the decline in employment in export industries which will offset the increase in employment in the import-substituting industries.

Further, when you restrict imports to protect domestic industries so that they should expand, other countries are likely to retaliate and will impose restrictions on our exports which are imported by them. This too will reduce exports and cause reduction in employment in export industries. Thus net effect on employment of restricting imports for providing protection to domestic industries may not be positive.

Infant Industries Argument

A powerful argument given in support of protection, especially in the context of developing countries is infant industries should be provided protection from the competition of low-priced imports of the mature and well-established industries of the developed industrialised countries. Shortly after American Revolution, Alexander Hamilton argued that British industrial supremacy was due to its early start over American infant industries. He pointed out that these infant American industries required temporary protection for some time so that they should grow and achieve production efficiency and economies of scale before they could successfully compete with low-cost British goods. He thus argued that temporary protection of infant American industries was necessary for industrial development of America.

Similarly, the infant industry argument has been advanced for protecting infant industries of the developing countries from competition of the low-cost firms of the industrialised developed countries. Given some time, these infant industries will grow and will be able to benefit from the economies of scale and learn the techniques necessary to lower their cost of production. As a result, over a period of time their cost per unit will go down and will therefore be in a position to compete with the foreign imports. Therefore, for some time they should be protected otherwise they would be destroyed by foreign competition.

However, there are some lacuna in infant industry argument. First, it is assumed that protected infant industries will make efforts to lower cost when provided protection. However, actual experience shows that it is more likely that protected industries lose incentives to become efficient and lower cost. It is said "once an infant, always an infant." Secondly, even if an industry makes efforts to improve productivity and lower cost per unit when it is provided protection, it has been assumed in the argument that the Government is the best judge as to which industries will prove to be capable of competing low-priced foreign goods. It has been asserted in defence of free trade that selection of industries which will acquire competitive strength can be done better by private market mechanism. It is pointed out that when opening up the economy to foreign competition the domestic industries would try to increase their efficiency. As a result, only those industries will survive which are efficient and produce at a lower cost. Therefore, it is argued that it is better if the domestic industries are left to foreign competition and in this way they will have incentives to improve productivity to escape from losses. Only those domestic industries will survive and operate which are efficient and produce at a low cost per unit.

Indian Automobile industry is a shining example of an industry not making any efforts to become efficient even after given protection for more than three decades. Before the setting-up of Maruti Udyog with Japanese collaboration, Indian car industry was fully protected by heavy duties on imports of cars. The two domestic firms producing Ambassador and Fiat cars did not make any efforts to improve their efficiency, nor did they bring out any better models of their cars. It is only after 1991 that following the policy of liberalisation that new foreign firms such as Daewoo of South Korea, General Motors have come in India and producing new models and at relatively low prices. Even Maruti is now trrying to improve its efficiency further and brought out new models of Maruti such as Zen, Esteem.

However, it may be noted that in developing countries the Government is in a better position to protect certain industries such as steel, cement *which lead to an expansion of the infrastructure of the developing economies*. This is because these industries create external economies and the private firms will not be compensated for creating these external benefits.

Anti-dumping Argument

The other important argument for protection is that foreign producers compete unfairly by dumping the goods in another country. Dumping is a form of price discrimination when producers of a country sells goods in another country at lower prices than those charged at home. Of course, consumers in a country in which foreign goods are dumped are beneficiaries, the industries of that country suffer as they are unable to compete with the '*dumped goods*'. Besides, there is more harmful '*predatory dumping*' which implies that foreign firms try to sell goods in other countries even below cost to establish a worldwide monopoly by driving competitors out of the market. Once the local industries are competed out, they raise prices to obtain monopoly profits.

There is a lot of evidence that firms of USA and Japan often indulge in dumping of their goods in other countries to eliminate competition. But, in our view, instead of providing protection to domestic industries through tariffs or non-tariffs barriers, it will be a better policy to enact laws against dumping. Dumping should be prohibited by law declaring it illegal. In India such a law has been enacted but is not being properly implemented.

Correcting Balance of Payment Deficit

Correcting deficit in balance of payments is also mentioned as justification for imposing tariffs to restrict imports or fixing of quotas of imports. This appears to be a valid argument for providing protection. However, in our view the solution for fundamental disequilibrium in the balance of payments lies in the adoption of suitable adjustment in exchange rate, appropriate fiscal and monetary policies to lower domestic prices so as to encourage exports. The deficit in balance of payments can be reduced by ensuring rapid growth in exports of a country.

Redistribution of Income

Case for protection has also been built up on the ground that it can be used for making desirable redistribution of income from one section of society to another. Protection makes some people better off, while others worse off. By providing protection to domestic producers their profits can be raised at the expense of consumers who suffer a loss in consumer surplus as protection denies them consumption of low-priced imported goods. That is, protection redistributes income in favour of domestic producers.

Sometimes protection causes transfer of income from some factors to the others. For example, Heckscher-Ohlin Model of international trade shows that trade benefits the abundant factor and harms the scarce factor. It is therefore scarce factor that demands protection by the Government against imports so that its income may not decrease. This implies that the workers, the owners of labour, and capitalists tend to take opposite views with regard to protection. This is however not confirmed by empirical evidence.

In some countries one of the objectives of economic policy is to redistribute income from the rich to the poor. This can be done by imposing high tariffs on imports of goods considered to be luxury items and levying tariffs on exports of those goods which are considered as necessities. Higher import tariffs on luxuries will reduce the incomes of the rich as they would pay taxes to the Government. Similarly, higher taxes on exports of necessities ensure greater supplies of them in the domestic market which would lower their domestic prices and benefit the poor.

It may however be noted that direct taxes such as income tax are considered better methods of redistributing income among various sections of a society than the commercial policy. This is because as we shall see below import tariffs levied for protecting industries cause downweight loss of welfare which are avoided under the direct tax system.

Conclusion

We have critically examined the various arguments in favour of protection. Some of them are valid, other appears to be misplaced. Some people consider trade as a '*zero sum game*', that is, in trading if one gains, the other loses. This has given rise to the doctrine of exploitation. For example, it is believed by some that the developing countries like India are exploited by the developed countries such as the USA, Japan, Britain. That is, the developing countries are net losers in trading with the developed countries. However, in our view, this is wrong thinking. No trade can occur without expectations of gain. India would not have entered into trade relations with USA if it did not expect to gain from it. Trade occurs between two countries if it benefits both the trading partners, the developed and the developing countries. Therefore, in our view world trade should be promoted by lifting barriers put up by various countries based on wrong notions about effects of free trade. Some countries such as USA and Japan have resorted to protectionist measures as a retaliation against foreign countries who restrict imports into their countries. The retaliatory actions of imposing trade barriers have done great harm to the expansion of world trade. New international organisation WTO (World Trade Organisation) which has replaced earlier GATT has been set up. WTO has framed rules which every country should observe so that barriers to trade be removed and world trade be promoted without doing any injustice to the member countries. It may be noted that retaliatory activities of restricting imports from foreign countries generally lead to the depression in the economies of the world as it happened during the worldwide depression of 1930s. The retaliatory activities may cause another global depression.

TRADE BARRIERS

Despite many benefits of free trade, the various countries have put up barriers to trade. A number of instruments are used as barriers to free trade but most important are tariffs and quotas. Both tariffs and quotas can be imposed either an imports or exports but they are mostly imposed on imports. Barriers to exports are quite uncommon. We briefly explain below these tariff barriers.

Tariffs

Tariffs are excise duties imposed on imported goods. The objective of imposing tariffs may be either raising revenue for the Government or providing protection to the domestic industries. Therefore, two types of tariffs are distinguished: (1) revenue tariffs, and (2) protective tariffs. *Revenue tariffs* are usually imposed on the imports of those products which are not produced domestically. Rates of revenue tariffs are generally small but yields a good revenue for the Government. For example in USA, tariffs are imposed on tin, coffee and bananas which are not produced in that country. Their obvious purpose is to provide revenue to the Government.

Protective tariff, on the other hand, are imposed to provide protection to the domestic producers from foreign competition. The rates of these tariffs are not so high as to completly prohibit their imports into a country. Rise in prices of their products as a result of imposition of tariffs, foreign producers lose their superior competitive power.

Import Quotas

Import quotas are another instrument used to check free trade. Import quotas refer to the maximum quantities of goods which may be permitted to be imported during any period of time. They are also referred to as *quantitative restrictions* on imports. Quotas are more effective method of reducing trade than tariffs. A given commodity may be imported in a relatively large quantity despite high tariffs but low quotas totally stop the imports of a commodity beyond the fixed quota of the commodity. Since international negotiations to reduce trade barriers have tended to focus on tariffs, the various countries have resorted to non-tariff barriers to free trade. We discuss below the effects of tariffs and quotas.

EFFECTS AND EFFICIENCY COST OF A TARIFF

Let us now examine the economic effects of tariffs used as a trade barrier to protect domestic industries. We use partial equilibrium approach represented by supply and demand analysis to examine the effect of tariffs. Let us take a product, say computer, in which India has a comparative disadvantage. In Fig.36.1 we have drawn domestic demand and supply curve *DD* and *SS* respectively of computers in India. In the absence of foreign trade, domestic price OP_d is determined at which *OQ* quantity of computers is demanded and sold. Assume that the Indian economy is now opened up to trade with USA which has a comparative advantage in the production of computers. Suppose OP_w represents the world price at which USA sells computers. We assume that when the Indian economy is opened up to trade, it can import computers from the USA at this world price OP_w at which it has no influence. In other words, free trade price is OP_w. It will be seen from Fig. 36.1 that at free trade price OP_w, the domestic demand (or consumption) for computers is $OH = (P_wK)$ and the domestic producers are supplying *ON* or P_wL quantity. Thus, with free trade out of $OH (= P_wK)$ quantity of domestic consumption of computers, domestic production is *ON* and the quantity *NH* or *LK* of computers is being imported.

Thus with free trade price equal to OP_w and the quantity purchased equal to P_wK, consumers are getting consumer surplus equal to the area P_wKD and with domestic production equal to $ON (= P_wL)$ the domestic producers are obtaining producer surplus equal to the are P_wLS.

Consumption Effect. Now suppose that in order to protect its domestic computer industry India imposes a tariff of *T* per computer. As a result, price of computer in India will rise from OP_w to OP_t equal to tariff *T*. The imposition of tariff and consequently rise in the price of computers in India will have a variety of effects. First, as shall be seen from Fig. 36.1 that at a higher price OP_t, the consumption of computers in India will decline to *OQ* computers as the higher price causes buyers of computers to move up the demand curve *DD* from point to *K* to *F*. This is a *consumption effect* of the tariff. It follows that the Indian consumers of computers have been badly hurt by the imposition of tariff on computers. As a result of tariff, they pay P_wP_t more per computer which they now buy at the higher price. Besides, tariff induces them to buy fewer computers. Thus, with imposition of tariff and rise in price from P_w to P_t, they will now get consumer surplus equal to the area P_tFD. Thus, as a result of imposition of tariff the consumers surplus enjoyed by the consumers has declined by the area equal to the sum $a + b + c + d$. Thus, consumers suffer a loss of consumer surplus as a result of imposition of tariff, but who gains?

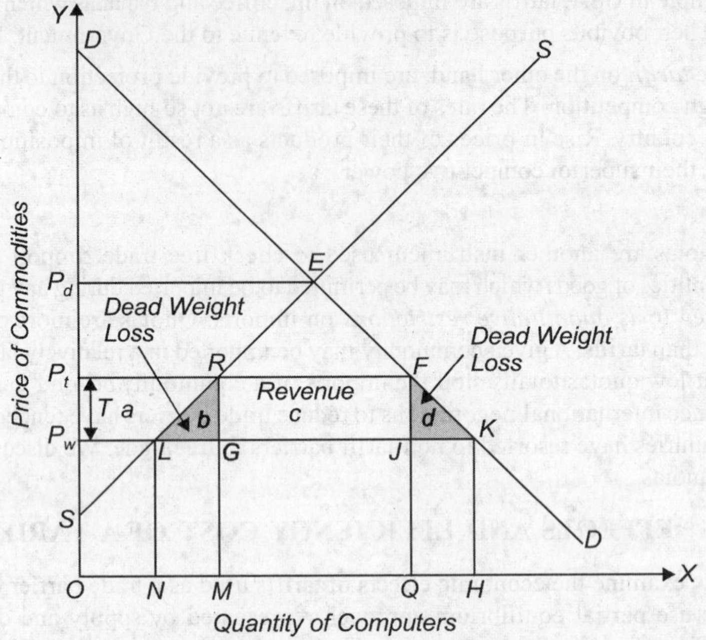

Fig. 36.1. *Effects and Efficiency Cost of Tariff*

Production Effect. Tariff benefits Indian producers of computers as they will now be able to sell their computers at a higher price OP_t instead of free trade price OP_w. Further, at a higher price OP_t, they will produce and supply more computers by moving up the domestic supply curve *SS*. It will be seen from Fig. 36.1 that at price OP_t, domestic producers of computers raises domestic production and quantity supplied from *ON* to *OM* and as a result their producer surplus increases to by area P_wLRP_t. It should be further noted that the increase in domestic production of computers by *NM* implies that some scarce resources will be bid away from other presumably more efficient industries.

Trade Effect. Third, as a result of imposition of tariff by India, American producers will be hurt. It may be noted that American producers would not get the higher price OP_t as the higher price is due to tariff which will be obtained by the Indian Government. For American producers price of computers will remain at OP_w. Since due to rise in price to OP_t, domestic production increases to *OM* and domestic consumption falls to *OQ*, the imports of computers fall from *NH* to *MQ*. This is trade effect of tariff.

Revenue Effect. Now, the important effect which is to be examined is whether economic well-being of the nation will increase as a result of imposition of tariff. The answer is in the negative. Of course, the Indian Government will gain from tariff equal to the revenue it collects from tariff. With rise in price by $P_w P_t$ per computer and the import of computers reduced to *MQ*, (or *GJ*), the total revenue of the Government from tariff will be equal to the rectangular area *RGJF*. This is the revenue effect of tariff. *This revenue from tariff obtained by the Government is "essentially a transfer of income from the consumers to government and does not represent any net change in the nation's well being. The result is that government gains a portion of what consumers lose."*[4]

4. Campbell R. McConnell and Stanley L. Brue, *Economics: Principles, Problems and Policies*, 11th edition, McGraw-Hill, New York, 1990, p. 784.

Effect on Welfare. Now the important question is what is the effect of imposition of tariff on welfare of the nation. As seen above, tariff has led to the rise in price of the computer from P_w to P_t resulting in the loss of consumer surplus of the consumers equal to the area P_wKFP_t (= the sum of the areas $a + b + c + d$). Out of the total loss of consumer surplus the producers gain equal to the area marked a as increase in producer surplus, government gets revenue equal to the area marked c. Thus, the consumers net loss of consumer surplus is equal to the shaded areas b and d. These shaded areas represent ***dead weight loss of welfare*** to the consumers of the society as a result of imposition of tariff.

But the effects of tariffs go beyond what has been explained above on the basis of partial equilibrium analysis of demand and supply. The imposition of tariff on computers will reduce export earnings of American computer industry—the industry in which it has a comparative advantage. Because of lower exports of computers, the production of computers will be reduced in the USA. This will cause the resources to be shifted from relatively more efficient computer industry to relatively inefficient industries of the USA in which it has a comparative disadvantage. Thus *tariffs cause misallocation of resources from the viewpoint of the world economy*. To conclude in the words of Professors McConnel and Brue, "*specialisation and unfettered world trade based on comparative advantage would lead to the efficient use of world resources and an expansion of the world's real output. The purpose and effect of protective tariffs are to reduce world trade. Therefore, aside from their specific effects upon consumers, foreign and domestic producers, tariffs diminish the world's real output.*[5]

EFFECTS AND DEAD-WEIGHT LOSS OF QUOTAS

Quotas are quantitative restrictions on the quantity or value of a commodity to be imported in a country during a period. Since quota limits the imports of a commodity, it reduces supply of a commodity in a country as compared to the case with a free trade. Like tariffs, quotas raise the prices of imported goods and encourage domestic production of those goods. But in case of quotas, the government does not collect any revenue. Quotas may be imposed against imports from all countries or used against the imports of only a few countries.

Economic effects of quota are graphically shown in Fig. 36.2 where D_M and S_M are domestic demand and supply curves of a commodity respectively. In the absence of trade price of the commodity in the country is P_A. Suppose the world price of the product is P_w. Under free trade, at price P_w of the commodity the domestic producers of country will produce OQ_1 quantity but as domestic demand of the product at price P_w is OQ_3, the quantity $Q_1 Q_3$ represents the imports at the world price P_w. Now assume that the Government imposes a quota that fixes the quantity of the product equal to Q_1Q_2 to be

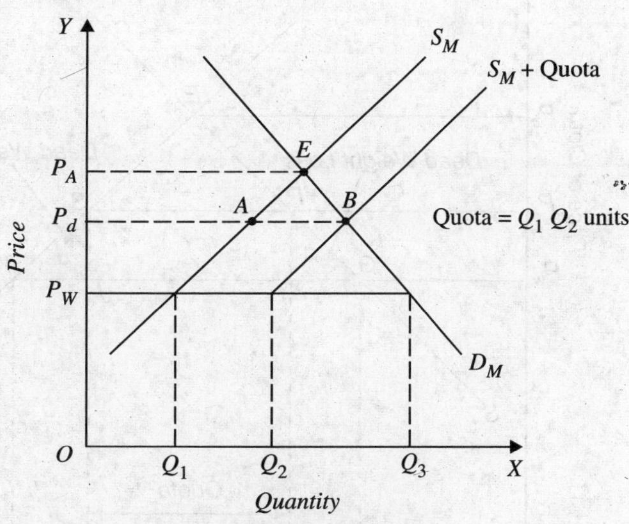

Fig. 36.2. *Effects of Quota*

5. *Ibid.*, p. 784.

ported. With this the total supply of the product in the domestic market will be away from the domestic supply S_M equal to the distance $Q_1 Q_2$. Incorporating the quota equal to $Q_1 Q_2$, we draw a new supply curve S_M+ Quota, which is to the right of the free-trade supply curve S_M. It will be seen from Fig. 36.2 that interaction of the supply curve (S_M+ Quota) with the domestic demand curve D_M determine higher price P_d than the world price P_w. It will be seen from Fig. 36.2 that difference *AB* between demand and domestic supply at price P_d is exactly equal to the fixed quota of Q_1Q_2 quantity of imports. It is thus dear that like tariffs fixation of quota has served to limit trade and raise price. It will therefore have same effects as we have explained in case of tariff. It may however be noted that unlike tariff in case of quota Government would not collect any revenue.

Efficiency Cost of Quota

Quotas are quantitative restrictions on the quantity or value of a commodity to be imported in a country during a period. Since quota limits the imports of a commodity, it reduces supply of a commodity in a country as compared to the case with a free trade. Like tariffs, quotas raise the prices of imported goods and encourage domestic production of these goods. But in case of quotas, the government does not collect any revenue. Quotas may be imposed against imports from all countries or used against the imports of only a few countries.

Economic effects of quota are graphically shown in Fig. 36.3 where *DD* and *SS* are domestic demand and supply curves of cloth. In the absence of trade with other countries, price determined by domestic demand and supply curves of cloth in the country is P_c. Suppose the world price of cloth is P_w over which India exercises no control

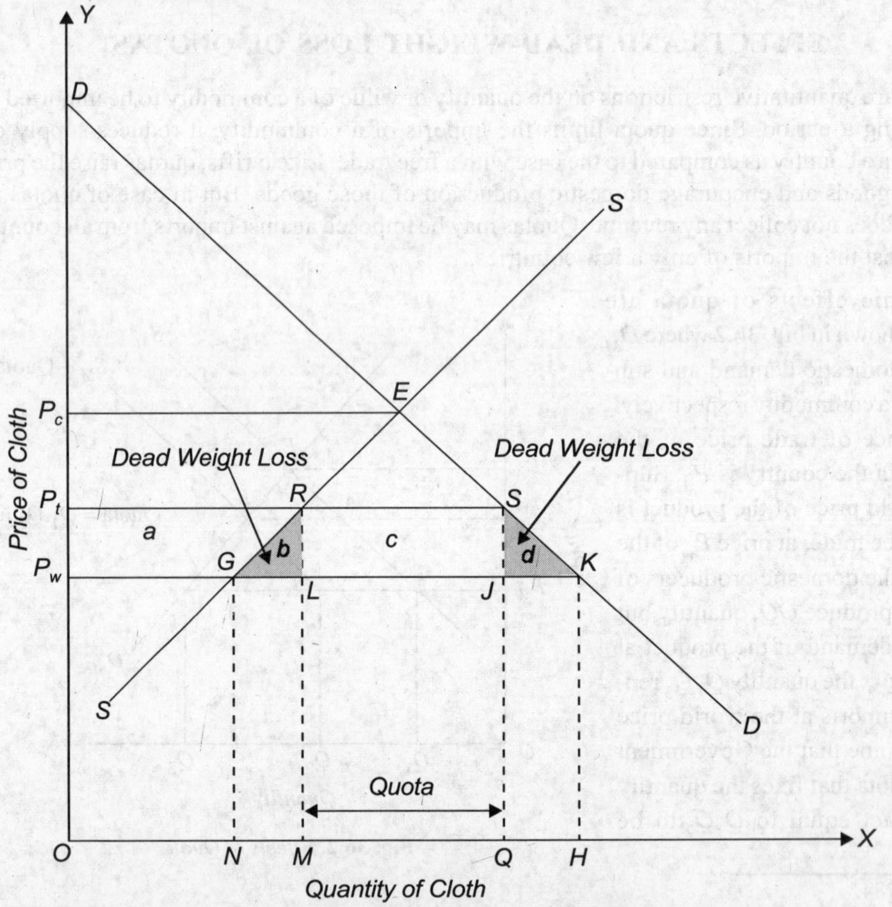

Fig. 36.3 *Efficiency Cost of Quota*

P_w at which *OH* quantity of the good is purchased by the consumers. At world price P_w of cloth, domestic producers produce *ON* (or P_wG) of cloth and the remaining *NH* or *GK* is imported to meet the demand of consumers.

Now suppose that the Government imposes an import quota equal to *MQ* quantity of cloth to protect the interests of domestic producers. Due to this reduction in supply of imported cloth, price of cloth will rise until a new equilibrium is reached at price P_t. Now, with the higher price of cloth the consumers would move up the demand curve from point *K* to point *S*. With this the quantity demanded of cloth falls to OQ ($= P_tS$). The consumers are harmed as their consumer surplus is reduced by the area P_wKSP_t which is equal to $a + b + c + d$ areas. Domestic producers will benefit from imposition of quota as at higher price P_t they will increase their output from *ON* to *OM* and as a result their producer surplus increases by the area P_wGRP_t which is marked as '*a*'. The crucial difference between the effects of tariff and quota is the area marked as '*c*'. In case of a tariff the area '*c*' is the amount of revenue received by the government *whereas in case of quota, the amount represented by the area* '*c*' *goes to foreign producers* which will lead to the increase in their profits. As in case of a tariff, the shaded areas marked as '*b*' and '*d*' are the ***dead-weight losses*** as while consumers lose, no one benefits in case of these shaded areas. They are therefore *efficiency loss for the society*. It is important to note that net welfare loss to the society is greater under a quota compared to a tariff. This is because in case of a tariff the revenue equal to the area '*c*' received by the government can be spent by it on the production of public goods which would increase the welfare of the people of the country, whereas in case of quota the area '*c* ' is lost to the foreign producers.

Two methods have been suggested so as to enable the government or society to capture the area '*c* ' under a quota rather than losing it to the foreign producers. First method suggested is to auction quotas to foreign producers in a free market. The advantage of acuctioning quota is that the foreign producers who are very keen to export cloth to India would pay the highest price. As a result, the Indian government will gain the area '*c* ' which was accruing to foreign producers.

The other method suggested for the government to gain area '*c* ' is to convert quota into equivalent tariff. The conversion of quota into equivalent tariff has the advantage that whereas under the new foreign trade regine under WTO (World Trade Organisation) while quota is illegal, tariff of a resonable magnitude is permitted under it.

QUESTIONS FOR REVIEW

1. Explain the various gaims from free trade. In view of these gaims from free trade, explain why the various countries put up various barriers to trade.
2. Critically examine the case for free trade in respect of developing countries like India.
3. What arguments are given in favour of protection? Do you think protection is right trade strategy for India in the emerging international economic environment ?
4. Critically examine the following arguments in support of protection.
   1. Infant industry argument
   2. Employment argument
   3. Balance of payment argument
   4. Promotion of *Swadeshi*
5. Explain the economic effects of tariffs. Compare them with effects of quota.
6. Tariffs cause misallocation of resources. Discuss.
7. The purpose and effect of protective tariffs is to reduce world trade. Discuss.
8. Explain the effects of quota. The fixation of quota for imports of goods reduces trade and raise prices. Examine critically.
9. What is tariff ? Why is it imposed by the government ? Show diagramatically its dead weight loss of welfare to the consumers.
10. Show how the imposition of quota by the government causes loss of economic efficiency in terms of consumer surplus ?
11. What is quota on imports. How does it affect consumers, producers and social welfare
12. Show that imposition of quota on imports causes dead weight loss of welfare. Does it affect government revenue.

Chapter 37

FOREIGN INVESTMENTS AND ROLE OF MULTINATIONAL CORPORATIONS

Introduction

After the Second World War, the developing countries are making concerted efforts to achieve rapid economic growth so as to alleviate problems of poverty and unemployment. Besides, there has been rapid growth of international trade. The developing countries like India have been facing the problem of shortage of capital. To meet this shortage, capital flows from the developed countries to the developing countries in the last two decades have substantially increased.

Immediately after the Second World War, capital flows were largely in the form of *foreign aid* from the Governments of the developed countries and international institutions such as IMF and World Bank to the Governments of the developing countries. In the last two decades the capital flows in the form of foreign aid on Government to Government basis have dried up, whereas aid from IMF and World Bank is based on certain conditionalities of undertaking some structural economic reforms which the developing countries often find difficult to implement to the entire satisfaction of these institutions. Besides, the capital needs of the developing countries are so large that IMF and World Bank alone cannot meet them. Therefore, there is great need for *capital flows on private account* on a large scale. Fortunately, capital flows on private account have substantially increased in the last two decades. These capital flows on private account are of the following two types:

1. Foreign Portfolio Investment
2. Foreign Direct Investment

We discuss below the roles of the above two types of capital flows briefly.

Foreign Portfolio Investment

This is an important type of capital flow under which foreign institutions such as banks, insurance companies, companies managing mutual funds and pension funds purchase stocks and bonds of companies of other countries in the secondary markets (*i.e.*, stock markets). They get returns in the form of capital gains and yearly payable dividends but do not exercise any direct control in running these companies. Pension funds, mutual funds and insurance companies have been very active in moving portfolio capital in the last two decades because restrictions on foreign equity investment by various countries have been reduced or removed in recent years allowing pension and mutual funds and insurance companies to diversify their portfolio in order to reduce risk. Besides, the growth of portfolio foreign capital in the last two decades has also been due to the policies of liberalisation followed by the developing countries. In India following the adoption of policy of liberalisation the flow of portfolio capital was permitted in 1991. Consequently, foreign portfolio capital flows have come to India in large amounts in the last ten years (1991-2001). However, Mexico has been the chief beneficiary of portfolio capital flows. Portfolio capital flows now account for one-third of net capital flows to the developing countries.

Benefits of Portfolio Capital Flows

Benefits and costs of portfolio capital flows have been a subject of severe controversy.

Private foreign portfolio investment in stocks, equities and bonds has been made by foreign investors in order to get higher return or higher interest on their investment and also to diversify their portfolio in order to reduce risk. Thus, Prof. Todaro rightly writes, "From the investor's point of view, investing in the stock markets of emerging countries permits them to increase their returns while diversifying their risks."[1]

In the early 1990s the return on the portfolio investment in the so called 'emerging' developing countries was quite high, for instance, it was 39 per cent during 1988-93 in Latin America which is main recipient of portfolio capital flows. But the high returns were marred by *high volatility* of stock markets of these countries. Many mutual funds based US companies which were lured by high returns had to suffer heavy losses when there was collapse of Mexican stock market following 35% devaluation of Mexican currency in Dec. 1994. The experience of these heavy losses resulted in drastic fall in portfolio capital flows in 1995 not only in Mexico but also in other developing countries. In fact, there was net capital outflows from the developing countries for some months.

From the perspective of developing countries private portfolio capital flows are generally considered as a welcome means for raising capital for private domestic companies. However, it has been argued whether it is prudent to rely on large and volatile private capital flows as they have a destabilising influence in both the financial markets and in the overall economy. It has been pointed out that when the developing countries run into trouble, the portfolio capital flows disappear in few days as happened in case of Mexico in 1995 and recently in South East Asian countries in 1997-98. However, it may be noted that a developing country should not cover up its structural weakness by obtaining private capital flows. If they want to ensure that their private portfolio capital flows continue occurring year after year on a sustained basis, they should manage the foreign exchange rates efficiently by controlling deficits on current account and also keeping adequate reserves of foreign exchange so that their currencies should not remain overvalued which might lead to severe long-term consequences.

From above it follows that private portfolio capital flows are welcome as they provide needed finance for developing countries and also bring in required foreign exchange. However, one should not consider them as substitutes for making structural changes in these economies such as maintaining exchange rate at the right level, controlling deficits in balance of payments, checking rate of inflation and adoption of proper monetary policy which should ensure relatively higher interest rates. We agree with Prof. Todaro who while commenting on portfolio capital flows writes, "Like multinational corporations, portfolio investors are not in the development business. If developed country interest rates rise or expected Third World profit rates decline, foreign speculators will withdraw their 'investments' as quickly as they brought them in. What LDCs need is true long-run economic investment (plants, equipment, physical and social infrastructure etc.) and not speculative equity capital."[2]

FOREIGN DIRECT INVESTMENT

The foreign direct investment (FDI) is the investment in the construction of physical capital such as building factories and infrastructure (*i.e.*, power, telecom, ports etc.) in the capital-importing country. It may be done in several ways. Companies or corporations may be specially set up for the purpose in the capital-exporting country to carry out trade and industry in an under--developed country. This is how the East India Company operated in India or railways were constructed in India. The head office is in the investing country and the operations are in the developing country. Another method is that an already existing corporation spreads out its business in another country by establishing branches. Thus, many foreign companies producing

1. Michael P. Todaro, *Economic Development*, 6th Edition, p. 544.
2. *Op. cit.*, p. 545.

cars, colour TVs have set up branches in India and are producing these items in India. There is another way open to the foreign entrepreneur, that is, to form companies and register them in the borrowing country without having any connection in the lending country. When India adopted protection, it became profitable, and quite fashionable too, for the foreign entrepreneurs to set up so called "India Limited," to jump over the tariff wall and to avail themselves of the various concessions which the government extended to national concerns. These were in fact disguised branches of the foreign firms. The Indian match industry, for instance, is dominated by Swedish concerns. There is still another possibility which is now becoming common in India, *viz.*, of joint ventures or joint participation. The foreign firms start industrial concerns in collaboration with Indian firms. This form of foreign investment has some special advantages. The standing and reputation of foreign firms inspire confidence in the domestic and foreign capital. The long experience and efficient techniques are placed at the disposal of the domestic companies. It avoids evils of absentee ownership; it provides full opportunities for developing local skills, and also a large proportion of profits is retained in the country. It is also advantageous to the foreign investor. He is able to ward off in this manner any danger of discriminatory treatment by the government. A developing country lacks capital for development and it will have to depend on foreign capital.

Merits of Foreign Direct Investment (FDI)

Several advantages can be claimed for foreign direct investment (FDI).

(1) Such investment does not burden the taxpayer since no interest at fixed rate is to be paid as in the case of foreign borrowing. The foreign investor is compensated by the profit he gets.

(2) In private investment, the investor is actuated by the profit motive; hence the business operations are subjected to careful calculations. This is a guarantee that the capital resources are most efficiently employed and are not frittered away in some reckless investment as may happen in the case of borrowing.

(3) Direct investment by foreign companies introduces, in the developing country, new technology, modern skills, innovations and new ideas. This is a great gain because the developing country is backward in technology and skills. The local entrepreneurs take a clue and start similar concerns. The Indian cotton textile industry was inspired by the Indian jute industry established by British entrepreneurs. Thus, direct foreign investment serves as an instrument for transferring modern technology to the developing countries.

(4) Another advantage is that a part of the profit is ploughed back into business and is not drained off from the country as it happens in the case of portfolio investment. The profits are invested either in modernisation and expansion of existing concerns or for establishing ancillary or subsidiary concerns in related fields. There is thus a continuing advantage for the developing country.

(5) Foreign direct investments are most likely to flow into export industries. By increasing exports and reducing imports, it will improve balance of payments of the developing country. It has a specially favourable effect on balance of payments position during recession because direct investment is serviced by dividends which are related to profits and not by fixed interest charges as in the case of loans. This flexibility of pressure on the balance of payments is of great advantage. It is worth mentioning that China in the last decades has industrially developed rapidly through foreign multinational firms to make investment in China to produce manufactured goods for exports. As a result, there has been unprecedented growth in industrial output and employment and as a result China has emerged as the fastest growing economy of the world. However, the latest trends show that in 2014-2015 and 2015-16, due to slowdown in world trade growth and protectionist policies of the developed countries, China's *overall* growth has fallen to 7% y-o-y as compared to over 9 per cent in the previous several years.

(6) Even otherwise flexible return on direct investment is a great advantage as compared with rigid interest and amortization requirements associated with public foreign loans.

(7) Direct foreign investment induces domestic investment also either in the form of joint participation or in local ancillary industries. Thus, foreign capital activates otherwise inert domestic capital. The domestic capital sheds off its shyness and enters into fields opened by direct foreign investment.

(8) The direct foreign investment makes a real addition to the productive capacity of the capital importing country. There is no question of foreign capital coming in this form being used for unproductive purposes. In the case of other types of foreign borrowings, there is nothing to prevent them from being utilised unproductively.

(9) Another important advantage of direct foreign investment is that it can be induced to be invested in infrastructure such as power, telecom, development of ports which is an obstacle to accelerating economic growth in the developing countries. Such direct foreign investment enables the developing countries to overcome supply-side bottlenecks which will spur domestic investment. It should be mentioned that in recent times the Indian Government has been wooing foreign investors to invest in the infrastructure sector. The foreign companies have the resources, technology and technical knowhow to start productive ventures in the infrastructure.

(10) The capital that comes through foreign direct investment has a distinct advantage over portfolio investment. While foreign institutional investors can sell their shares and take capital out of the developing countries in a very short time and thus destabilise these economies as recently happened during East Asia Crisis, it is not easy to close down foreign concerns established through direct investments. Foreign Direct Investment (FDI) enters the developing economies to build factories and these factories (*i.e.*, physical capital) stay even if the investors decide to sell out later to domestic buyers.

The foreign direct investment (FDI) is opposed on the ground that it seeks to establish *'financial imperialism'*. It leads to political domination and economic exploitation. That is why foreign capital was so unpopular in India. The strength of this objection really rests on the nature of political set-up. A free democratic country of the continental size such as India need not entertain such fears.

In order to promote private foreign direct investment, it is necessary both for the lending and borrowing countries to remove the impediments to free flow of capital and to grant the necessary facilities. The borrowing country should guarantee immunity from nationalisation and repatriation of profits. The crux of the problem is to assure higher returns and minimum risk. At present investment is deterred by political and social instability, uncertainty about the jurisdiction of courts, exchange controls and currency inconvertibility, controls on capital issues, fear of discriminatory legislation and fear of nationalisation, the practice of shutting out some industrial fields for foreign investors, employment of nationals in superior posts etc. Among the measures to minimise risks and allay fears may be mentioned investment treaties, government guarantees, tax incentives, joint ventures, relaxation of restrictions and granting of concessions. In short, the investment climate must be made most favourable by ensuring the following:

(*i*) Political stability and freedom from external aggression.

(*ii*) Security of life and property.

(*iii*) Availability of opportunities for earning profits.

(*iv*) Prompt payment for fair compensation and its remittance to the country of origin in the event of compulsory acquisition of a foreign enterprise.

(*v*) Facilities for the remittance of profits, dividends, interest, etc.

(*vi*) Facilities for the immigration and employment of foreign technical and administrative personnel.

(*vii*) A system of taxation that does not impose an excess burden on private foreign enterprise.

(*viii*) Freedom from double taxation.

(*ix*) A general spirit of friendliness for foreign investors.

The department of commerce of the United States mentions certain factors which stand in the way of private investment of American capital. They are uncertainty created by the present political situation, the policies and practices with respect to foreign investment, the relatively low level of economic infrastructure and the lack of trained labour, and the limited knowledge of the developing countries on the part of American businessmen. The particular impediments mentioned in the case of India are the nature of India's screening policy shutting out foreign investment in certain spheres and controls on imports, exports and foreign exchange and the absence of a double taxation agreement, obligation to employ and train Indian labour. India also offers some special inducements, *e.g.*, special tax exemptions, increased depreciation allowance and other benefits available to domestic industry, guarantee of exchange facilities for profit remittances, capital repatriation and import of essential requirements, government assistance in the acquisition of land, transport facilities, etc., the right to hold controlling interest. This compares quite favourably with many other countries. But there are still some countries with more attraction in the form of higher returns and better guarantees and they can absorb a substantial supply of foreign U.S. capital.

However, in the opinion of the present author direct foreign investment in the field of infrastructure in India may not be as much as expected because infrastructure projects relating to power, telecom, ports are of the nature of public utilities and therefore, the price of their end-product (for example, electricity charges) has to be controlled which may not yield sufficient profits to attract foreign investment.

NDA's 'Make in India' Policy

As part of structural economic reforms initiated in India from 1991 under the former Prime Minister, Dr. Manmohan Singh, there was *liberalisation of foreign investment policy*. To attract foreign investment, it was decided to grant automatic permission to private foreign investors to hold equity up to 51 per cent of total equity shares in high priority industries. The priority sector included power generation and petroleum refining. The government also gave a guarantee of 16 per cent return on foreign investment in priority sectors. As a result, foreign direct investment started coming into India and after 2003, it picked up but foreign direct investment into India was much less than that in China and some other developing countries. Now, after coming into power in 2014, NDA Government headed Mr. Modi has proposed a new scheme called '***Make in India***'. For this scheme, foreign investors have been invited to make large-scale investment in India for producing *manufacturing goods in India and mainly export them abroad*, just like China did three decades before. In this way, it is thought that industrial growth of output and employment will rise rapidly and overall growth of the economy will greatly pick up. The Modi Government intends to solve the problem of unemployment in this way.

To 'make in India' scheme a success, foreign investment norms in India have been greatly liberalised, land acquisition bill is being passed so that land for setting up industrial projects by foreigners be procured. In the budget for 2015-16, corporate income tax has been reduced from 30 per cent to 25 per cent in a phased manner in the five-year period. This will raise the return on foreign capital invested. No retrospective taxation has been promised to woo foreign investor. Minimum Alternative Tax (MAT) on the earnings of FIIs has been abolished with effect from 2015-16 so as to prevent them flowing out.

In the view of the present author the timing of '*Make for India*' is wrong as at present expansion in world trade has slowed down. The recovery in the US from 2008 crisis is fragile, Europe too is not yet developing and facing the problem of burden of external debt. It is due to these reasons that

China's industrial growth has slowed down. India's exports are also falling. How then in the present context under 'Make in India' by foreigners to export them would succeed. In our view, what is important is to launch a scheme of '*Make for India*', India's domestic product is quite large. What is needed is a well-designed development strategy which *uses labour-intensive technologies to produce mass consumer goods* and essential capital goods needed for them. Scarce resources should be discouraged to be used for manufacturing big luxury cars, colour TVs, Smart Mobiles etc. if growth with social justice in India is to be achieved.

India has the basic ingredients to attract much larger foreign direct investments. Its vast domestic market, a low-cost production base, along with the availability of skilled labour, adequate repatriation facilities and a good legal system have placed the country ahead of several others for foreign investors. If these favourable features are reinforced by a stable political environment, the country should be able to attract larger direct inflows of funds. China attracted over \$ 30 billion of direct foreign investment during four years (1991-95) accounting for over a third of the total funds earmarked for the Asia-Pacific region.

On the other hand, India has clearly emerged as a better market for foreign portfolio investment. Most fund managers rate India above China, the Philippines, South Korea and Indonesia among Asian markets. India is also rated higher than some of the Latin American markets like Colombia and Peru and European markets like Greece and Hungary.

ROLE OF MULTINATIONAL CORPORATIONS (MNCs)

Multinational corporations are those large firms which are incorporated in one country but which own, control or manage production and distribution facilities in several countries. Therefore, these multinational corporations are also known as *transnational corporations.* They transact business in a large number of countries and often operate in diversified business activities. The movements of private foreign capital take place through the medium of these multinational corporations. Thus multinational corporations are important source of foreign direct investment (FDI). Besides, it is through multinational corporations that modern high technology is transferred to the developing countries.

The important question about multinational corporations is why they exist. The multinational corporations exist because they are highly efficient. Their efficiencies in production and distribution of goods and services arise from *internalising* certain activities rather than contracting them out to other forms[3]. Managing a firm involves which production and distribution activities it will perform itself and which activities it will contract out to *other firms and individuals*. In addition to this basic issue, a big firm may decide to set up and operate business units in other countries to benefit from *advantages of location.* For examples, it has been found that giant American and European firms set up production units to explore and refine oil in Middle East Countries because oil is found there. Similarly, to take advantages of lower labour costs, and not strict environmental standards, multinational corporate firms set up production units in developing countries.

Alternative Methods of Foreign Investment by Multinational Companies

In order to increase their profitability many giant firms find it necessary to go in for horizontal and vertical integration. For this purpose they find it profitable to set up their production or distribution units outside their home country. The firms that sell abroad the products produced in the home country or the products produced abroad to sell in the home country must decide how to manage and control their assets in other countries. In this regard, there are three methods of foreign investment by multinational firms among which they have to choose which mode of control over their assets they adopt. There are three main modes of foreign investment:

3. See Ronald Coase, "The Nature of the Firm", *Economica*, Nov. 1937, pp.368-405.

1. Agreement with Local Firms for Sale of MNCs Products. A multinational firm can enter into an agreement with local firms for exporting the product produced by it in the home country to them for sale in their countries. In this case, a multinational firm allows the foreign firms to sell its product in the foreign markets and control all aspects of sale operations.

2. Setting up of Subsidiaries. The second mode for investment abroad by a multinational firm is to set up a *wholly owned subsidiary* to operate in the foreign country. In this case a multinational firm has complete control over its business operations ranging from the production of its product or service to its sale to the ultimate use or consumers. A subsidiary of a multinational corporation in a particular country is set up under the Companies Act of that country. Such subsidiary firm benefits from the managerial skills, financial resources, and international reputation of their parent company. However, it enjoys some independence from the parent company.

3. Branches of Multinational Corporation. Instead of establishing its subsidiaries, multinational corporations can set up their branches in other countries. Being branches they are not legally independent business unit but are linked with their parent company.

4. Foreign Collaboration or Joint Ventures. Thirdly, the multinational corporations set up joint ventures with foreign firms to either produce its product jointly with local companies of foreign countries for sale of the product in the foreign markets. A multinational firm may set up its business operation in collaboration with foreign local firms *to obtain raw materials* not available in the home country. More often, to reduce its overall production costs multinational companies set up joint ventures with local foreign firms to manufacture inputs or sub-components in foreign markets to produce the final product in the home country.

Some of the world's largest multinational corporations are given below.

Table 37.1. World's Some Important Non-Financial Multinational Corporations

S.No.	*International Corporation*	*Parent Country*	*Industry of Operation*
1.	General Electric	United States	Electronics
2.	Exxon Mobil Corporation	United States	Petroleum (exploring, refining and distributing)
3.	Royal Dutch Shell Group	Netherlands/UK	-do-
4.	General Motors	United States	Motor Vehicles
5.	Ford Motor Co.	United States	Motor Vehicles
6.	Toyota Motor Co.	Japan	Motor Vehicles
7.	IBM	United States	Computers
8.	BP	United Kingdom	Petroleum (Exp/Ref./Distt.)
9.	Nestle SA	Switzerland	Food/Beverages
10.	Nippon Oil Co.	Japan	Petroleum/Expl., Ref./Distt.
11.	Siemen AG	Germany	Electronics
12.	BMW AG	Germany	Motor Vehicles
13.	ABB	Switzerland	Electrical equipment
14.	Sony Corporation	Japan	Electronics
15.	Seagram	Canada	Food/Beverages
16.	Aventis	France	Pharmaceuticals/Chemicals
17.	Roche Group	Switzerland	Pharmaceuticals
18.	Honda Motor Co.	Japan	Motor Vehicles
19.	Phillips Electronics	Netherlands	Electronics
20.	Hewlett-Packard	United States	Electronics/Computers

Role of Multinational Corporations in the Indian Economy

Prior to 1991 multinational companies did not play much role in the Indian economy. In the pre-reform period the Indian economy was dominated by public enterprises. To prevent concentration of economic power Industrial Policy 1956 did not allow the private firms to grow in size beyond a point. By definition multinational companies are quite big and operate in several countries. While multinational companies played a significant role in the promotion of growth and trade in South-East Asian countries, they did not play much role in the Indian economy where import-substitution development strategy was followed. Since 1991with the adoption of industrial policy of liberalisation and privatisation role of private foreign capital has been recognised as important for rapid growth of the Indian economy. Since source of bulk of foreign capital and investment are multinational corporations, they have been allowed to operate in the Indian economy subject to some regulations. The following are the important reasons for this change in policy towards multinational companies in the post-reform period.

1. Promotion of Foreign Investment. In the recent years, external assistance to developing countries has been declining. This is because the donor developed countries have not been willing to part with a larger proportion of their GDP as assistance to developing countries. MNCs can bridge the gap between the requirements of foreign capital for increasing foreign investment in India. The liberalised foreign investment pursued since 1991 allows MNCs to make investment in India subject to different ceilings fixed for different industries or projects. However, in some industries 100 per cent export-oriented units (EOUs) can be set up. It may be noted, like domestic investment, foreign investment has also a multiplier effect on income and employment in a country. For example, the effect of Suzuki firm's investment in Maruti Udyog manufacturing cars is not confined to income and employment for the workers and employees of Maruti Udyog but goes beyond that. Many workers are employed in dealer firms who sell Maruti cars. Moreover, many intermediate goods are supplied by Indian suppliers to Maruti Udyog and for this many workers are employed by them to manufacture various parts and components used in Maruti cars. Thus their incomes also go up by investment by a Japanese multinational in Maruti Udyog Limited in India.

2. Non-debt Creating Capital Inflows. In pre-reform period in India when foreign direct investment by MNCs was discouraged, we relied heavily on *external commercial borrowing* (ECB) which was of debt-creating capital inflows. This raised the burden of external debt and debt service payments reached the alarming figure of 35 per cent of our current account receipts. This created doubts about our ability to fulfil our debt obligations and there was a flight of capital from India and this resulted in balance of payments crisis in 1991. As direct foreign investment by multinational corporations represents non-debt creating capital inflows we can avoid the liability of debt-servicing payments. Moreover, the advantage of investment by MNCs lies in the fact that servicing of non-debt capital begins only when the MNC firm reaches the stage of making profits to repatriate. Thus, MNCs can play an important role in reducing stress and strains on India's balance of payments (BOP).

3. Technology Transfer. Another important role of multinational corporations is that they transfer high sophisticated technology to developing countries which are essential for raising productivity of working class and enable them to start new productive ventures requiring high technology. Whenever multinational firms set up their subsidiary production units or joint-venture units, they not only import new equipment and machinery embodying new technology but also skills and technical know-how to use the new equipment and machinery. As a result, the Indian workers and engineers come to know of new superior technology and the way to use it. In India, the corporate sector spends only few resources on Research and Development (R&D). It is the giant multinational corporate firms (MNCs) which spend a lot on the development of new technologies which can greatly benefit the developing countries by transferring the new technology developed by them. Therefore, MNCs can play an important role in the technological upgradation of the Indian economy.

4. Promotion of Exports. With extensive links all over the world and producing products efficiently and therefore with lower costs multinationals can play a significant role in promoting exports of a country in which they invest. For example, the rapid expansion in China's exports in recent years is due to the large investment made by multinationals in various fields of Chinese industry. Historically in India, multinationals made large investment in planlations whose products they exported. In recent years, Japanese automobile company Suzuki made a large investment in Maruti Udyog with a joint collaboration with Government of India. Maruti cars are not only being sold in the Indian domestic market but are exported in a large number to the foreign countries.

As a matter of fact, until recently, when giving permission to a multinational firm for investment in India, Government granted the permission subject to the condition that the concerned multinational company would export the product so as to earn foreign exchange for India. However, in case of Pepsi, a famous cold-drink multinational company, while for getting a product licence in 1961 to produce Pepsi Cola in India it agreed to export a certain proportion of its product, but later it expressed its inability to do so. Instead, it ultimately agreed to export things other than what it produced such as tea.

5. Investment in Infrastructure. With a large command over financial resources and their superior ability to raise resources both globally and inside India it is said that multinational corporations could invest in infrastructure such as power projects, modernisation of airports and ports, telecommunication. The investment in infrastructure will give a boost to industrial growth and help in creating income and employment in the Indian economy. The external economies generated by investment in infrastructure by MNCs will therefore *crowd* in investment by the indigenous private sector and will therefore stimulate economic growth.

In view of above, even Common Minimum Programme of the UPA government provides that foreign direct investment (FDI) will be encouraged and actively sought, especially in areas of (*a*) infrastructure, (*b*) high technology, (*c*) exports, and (*d*) where domestic assets and employment are created on a significant scale.

A Critique of Multinational Corporations

In recent years foreign direct investment through multinational corporations has vastly increased in India and other developing countries. This vast increase in investment by multinational corporations in recent years is prompted by factors: (1) the liberalisation of industrial policy giving greater role to the private sector, (2) opening up of the economy and liberalisation of foreign trade and capital inflows. In this economic environment multinational corporations which are in search for global profits are induced to make investment in developing countries. As explained above, foreign direct investment by multinational firms bring many benefits to the recipient countries but there are many potential dangers and disadvantages from the viewpoint of economic growth and employment generation. Therefore, role of multinational corporations in India and other developing countries has been criticised on several grounds. We discuss below some of the criticisms levelled against multinational corporations.

1. Capturing Markets. First, it is alleged that multinational corporations invest their capital and locate their manufacturing units on their own or in collaboration with local firms in order *to sell their products and capture the domestic markets of the countries where they invest and operate.* With their vast resources and competitive strength, they can weed out their competitive firms. For example, in India if corporate multinational firms are allowed to sell or produce the products presently produced by small and medium enterprises, the latter would not be able to compete and therefore would be thrown out of business. This will lead to reduction in employment opportunities in the country.

2. Use of Capital-intensive Techniques. It has been seen that increasing capital intensity in modern manufacturing sector is responsible for slow growth of employment opportunities in India's industrial sector. These capital-intensive techniques may be imported by large domestic firms but presently they are being increasingly used by multinational corporations which bring their technology when they invest in India. Emphasising this factor, Thirlwall rightly writes, "*In this case the technology may be inappropriate not because there is not a spectrum of technology or inappropriate selection is made but because the technology available is circumscribed by the global profit maximising motives of multinational companies investing in the less-developed country concerned.*"[4]

3. Encouragement to Inessential Consumption. The investment by multinational companies leads to overall increase in investment in India but it is alleged that they encourage conspicuous consumption in the economy. These companies cater to the wants of the already well-to-do people. For example, in India very expensive cars (such as City Honda, Hyundai's Accent, Mercedes, Opal Astra, etc.), air conditioners, costly laptops, washing machines, expensive fridges, 29" and Plasma TVs are being produced/sold by multinational companies. Such goods are quite inappropriate for a poor country like India. Besides, their consumption has a demonstration effect on the consumption of others. This tends to raise the propensity to consume and adversely affects the increase in savings of the country.

4. Import of Obsolete Technology. Another criticism of MNCs is based on the ground that they import obsolete machines and technology. As mentioned above, some of the imported technologies are inappropriate to the conditions of Indian economy. It is alleged that India has been made a dumping ground for obsolete technology. Moreover, the multinational corporations do not undertake Research and Development (R&D) in India to promote local technologies suited to the Indian factor-endowment conditions. Instead, they concentrate R & D activity at their headquarters.

5. Setting up Environment-polluting Industries. It has been found that investment by multinational corporations in developing countries such as India is usually made for capturing domestic markets rather than for export promotion. Moreover, in order to evade strict environment control measures in their home countries they set up polluting industrial units in India. A classic example of this is a highly polluting chemical plant set up in Bhopal resulting in gas tragedy when thousands of people were either killed or made handicapped due to severe ailments. "With the tighening of environmental measures in such countries, there is a tendency among the MNCs to locate the polluting industries in the poor countries, where environmental legislation is non-existent or is not properly implemented, as exemplified in the Bhopal gas tragedy."[5]

6. Volatility in Exchange Rate. Another major consequence of liberalised foreign investment by multinational corporations is its impact on the foreign exchange rate of the host country. Foreign capital inflows affect the foreign exchange rate of the Indian rupee. A large capital inflows through foreign investment brings about increase in the supply of foreign exchange say of US dollars. With demand for foreign exchange being given, increase in supply of foreign exchange will lead to the appreciation of exchange rate of rupee. This appreciation of the Indian rupee will discourage exports and encourage imports causing deficit in balance of trade. For example, in India in the fiscal years 2004-05 and 2005-06, there was large capital inflows by FII (giant financial multinationals) in the Indian economy to take advantage of higher interest rates here and also booming of the Indian capital market.

---

4. A.P. Thirlwall, *Growth and Development,* Palgrave, Macmillan, Seventh Edition, 2003, p. 435.
5. Biplab Dasgupta, *Globalisation: India's Adjustment Experience,* Sage Publications, New Delhi, 2005, p.81.

On the other hand, when interest rates rise in the parent countries of these multinationals or rates of return from capital markets go up or when there is loss of confidence in the host country about its capacity to make payments of its debt as happened in case of South-East Asia in the late nineties, there is large outflow of capital by multinational companies resulting in the crisis and huge depreciation of their exchange rate. Thus, capital inflows and outflows by multinationals have been responsible for large volatility of exchange rate.

Then there is the question of *repatriation of profits* by the multinationals. Though a part of profit is reinvested by the multinational companies in the host country, a large amount of profits is remitted to their own parent countries. This has a potential disadvantage for the developing countries, especially when they are facing foreign exchange problem. Commenting on this Thirwall writes, "FDI has the potential disadvantage even when compared with loan finance, that there may be outflow of profits that lasts much longer."[6]

Transfer Pricing and Evasion of Local Taxes. Multinational corporations are usually vertically integrated. The production of a commodity by multinational firm comprises various phases in its production. The components used in the production of a final commodity may be produced in its parent country or in its affiliates in other countries. *Transfer pricing refers to the prices a vertically integrated multinational firm charges for its components or parts used for the production of the final commodity, say in India.* These prices of components or parts are not real prices as determined by demand for and supply of them. They are arbitrarely fixed by the companies so that they have to pay less taxes in India. They artificially inflate the transfer prices for intermediate products (*i.e.*, components) produced in their parent country or their overseas affiliates so as to show lower profits earned in India. As a result, they succeed in evading corporate income tax.

Conclusion. We have seen above that foreign investment by multinational companies has both advantages and disadvantages. Therefore, they need regulation and should be permitted in selected sectors and also subject to a cap on their investment in particular fields. If objective of economic growth with stability and social justice is to be achieved, there should not be complete open door policy for them.

It is true that multinational corporations take risk in making investment in India, they bring capital and foreign exchange which are non-debt creating, they generally promote technology and can help in raising exports. But they must be regulated so that they serve these goals. They should be allowed to invest in infrastructure, high-technology areas, and in industries whose products they can export and if they help in generating net employment opportunities. We agree with Colman and Nixon who write: "*Transnational corporations cannot be directly blamed for lack of development (or the direction development is taking) within less developed countries. Their prime objective is global profit maximisation and their actions are aimed at achieving that objective, not developing the host less developed country. If the technology and products that they introduce are inappropriate, if their actions exacerbate regional and social inequalities, if they weaken the balance of payments position, in the last resort it is up to the government of less developed country to pursue policies which will eliminate the causes of these problems.*"[7]

---

6. A.P. Thirlwall, *op.cit.*

7. D. Colman, D and F Nixon, *Economics of Change in Less Developed Countries,* Oxford University Press, 1978.

APPENDIX TO CHAPTER 37

FOREIGN DIRECT INVESTMENT (FDI) POLICY IN INDIA

Introduction

Foreign direct investment (FDI) is at present very necessary for India not only for accelerating economic growth but also for meeting current account deficit which in recent years has significantly increased and is posing a serious challenge for India's economy. In Oct. 2012, the Parliament approved 51% FDI in multi-brand retail business. But due to stringent conditions imposed not a single company has come forward to open a multi-brand retail store. Besides, in other fields or sectors where FDI has been permitted long ago, the caps or limits imposed are large enough to bring FDI in adequate quantity. Therefore, in 2013 government decided to raise these caps so as to attract more foreign direct investment. There is a need for building consensus on the role of foreign direct investment in India. In China's success growth story, foreign direct investment has played a significant role as China has not imposed much conditions on the entry of MNCs. Rate of return on investment in both India and China are much larger compared to the home countries of MNCs and, therefore, the right policy towards FDI will go a long way in attracting FDI to finance the projects needed for economic growth and to fill the gap in our current account deficit. To woo foreign direct investment we must give it a fair deal and ensure the security of their investment.

Our treatment of FDI has not been good in the past. In the 1970s India chucked out IBM and Coca-Cola in a fit of nationalist spirit. In the late 1990s we sent back a power producing company 'Enron' which lost billions of dollars in setting up a power plant in Maharashtra due the differences in pricing of power generated. Recently, Sweden's IKEA faced problems in selling Nordic Comfort food and furniture. Recent acquisitions by a multinational firm Vodafone, British mobile telecoms firm, and Dai- chi Sankyo a Japanese drug firm have recently been facing problems.

Despite the problems faced by MNCs in India, as per data furnished by the Reserve Bank of India the stock of FDI in India is now quite large equal to about $ 220 billion or 12% of our GDP. This includes all things from research centers in Bangalore to cement plants. The balance of payment data reveals that overall return on equity capital for foreign investors is probably below 10 percent and slumped in the year 2011-12. However, balance of payments data does not fully show the rate of return earned by foreign investors. A recent survey by the RBI estimated the return on equity capital by 745 firms with a significant foreign shareholders. According to it, the overall return on equity of these 745 firms with significant foreign shareholders was 13% in the year 2010-11.

Technology firms, car makers and chemical makers did well, transport and telecoms firms less so. *Unlike FDI in China, which has been directed at building factories for export, foreign direct investment into India is aimed at the domestic market—only 12% of the firms's sales were foreign.*

Successful Foreign Firms in India

It is worth mentioning that a number of foreign firms has been highly successful and they have invested a large amount of capital and have earned a lot. These high profile firms are listed in Indian stock market while being controlled by foreign firms such as Suzuki, Nestle, Unilever, Siemens (Suzuki, Glaxo-Smith kline and British American Tobacco Company). These are well established businesses with deep roots in India and high profitability.

Foreign-controlled firms among the top 100 companies listed in India made a 24% return on equity in the year 2010-11 better than domestic firms' returns. Their market value doubled over the five preceding years. That in turn has presented a different set of problems. Some foreign firms had the foresight to convert minority stakes in good Indian firms in to controlling ones. Suzuki took control of Maruti, India's biggest car manufacturer by volume, in 2002. Others have failed. After a botched effort to take control almost decades ago, British American Tabacco still owns only 31% of ITC, that has grown so fast it is now a global player in its own right. Faced with no prospect of control, some foreigners have pulled out. Honda sold its stake in Hero, an Indian motorbike-maker, in 2010 and is now competing directly against it. Those firms that have control of fast-growing subsidiaries have been keen to boost their stakes still further. Siemens raised its stake in its unit from 55% to 75 % in 2011. But often the valuations of Indian subsidiaries are quite high. Nestle India is valued at 50 times its profits—more than double the ratio of its Swiss parent.

Faced with lower stakes than they would like, foreigners have found another, shabby way, to extract more value from their Indian subsidiaries: charging them "royalty fees". They argue that these reflect the brand and technology benefits of being part of a global group. On 22nd January Hindustan Unilever became the latest firm to do so—announcing that the fee paid to its foreign parent would rise from 1.4% to 3.15% of sales. The minority investors in those companies complain that they are not being given fair deal. Still, for India the trend towards higher royalty payments is a backhanded compliment, It shows that some foreign-run firms in India make heavy profits. That is why their parent companies want to grab more of them. As a signal about outsiders' appetite for exposure to India, that may be rather important.

Liberalisation of Foreign Direct Investment (FDI)

Faced with a harsh economic environment of sluggish economic growth and widening current deficit (CAD) and rapidly depreciating Indian rupee, after 2010 UPA government in its significant latest reform opened the doors to greater foreign direct investment by *raising 'caps'* on foreign direct investment in about a dozen sectors including telecom and highly sensitive sector of defence production (for state of the art technology') Not only have the '*caps*' (that is, upper limits) have been raised in several sectors but the norms of their entry into India has been eased by allowing much foreign directs investment to come through **automatic route,** that is, without the approval of **Foreign investment promotion Board** (**FIPs**) which is a nodel agency to vet FDI applications in India.

Under UPA government in the Telecom Sector, FDI cap was raised to 100% from 74% (upto 49% through automatic route and 51% through approval of FIPB. Simitarly for *single brand retail*, FDI limit was 100% (49% through automatic route and the remaining with approval of FIPB. In the Insurance sector FDI limit was raised from 26 per cent to 49 per cent (Parliamentary approval for this was finally obtained under NDA government in 2014). In case of petroleum and Natural Gas Refineries, commodity exchanges stock exchanges and power exchanges FDI limit was fixed at 49% through automatic route.

It is evident from above that intended objective of liberalisation of foreign direct investment policy is to send right signals to the existing and potential foreign investors to revive Indian economic growth which until recently was an engine for global growth. Besides spurring economic growth, the new FDI policy has been guided to meet the current account deficit and to stem the slide of the Indian rupee. To ensure faster economic growth NDA government adopted a policy of 'Make in India' (discussed in the previous chapter. The aim of this policy is to attract foreign investors to set up or invest in the industries in India and export thier products so as boost India's export growth.

Big Push to Foreign Investment by NDA Government

In a major push to was foreign investment the NDA government in November 2015 relaxed FDI norms for 15 sectors including defence, construction and broadcasting. Fasing of FDI norms has been to ensure to the foreign investors that India is unstoppable on the path of economic reforms to boost economic growth. We explain below the various measures announced to boost foreign investment in various sectors of the Indian economy.

Defence. In its new initative the government has allowed foreign portfolio investors (FPIs) to increase stakes in local defence units to 49 per cent from the current 24 per cent through the automatic route. This move is expected to open up further finding for the defence sector. The government has also decided to foreign defence player to own up to 49 per cent the Indian joint ventures (IJVs) through the automatic route, meaning no prior approval of government will be required.

Plantations. Apart from tea, 100 per cent FDI has been allowed in Coffee, rubber, Cardamom, Palm oil and olive oil plantations. This is expected to give a big push to their production.

Construction and Realty Projects. As a part of the move to further open up the Indian economy the government has made FDI in real estate and construction projects more transparent and offered greater flexibility to foreign investors to invest in India's reality projects. It removed the minimum restrictions of 20,000 sq. meters for FDI and also the minimum investment of $ 5 million in real estate. New rules announced require that a foreign investor will be permitted to *exit* and repatriate of foreign investments before the completion of projects under the automatic route. As a result of this, white previously only large projects were getting foreign investment, the smaller projects can now get foreign investment. Besides, it has been decided that there will now no lock-in period for FDI investments in hotels and resorts, hospitals, SEZs, educational institutionals, old age homes and NRI investments.

Aviation. In the new policy measures 49 per cent FDI has been allowed under atuomatic route for regional our transport services.

Media and Broadcasting. The new policy has increased the role of FDI in the media and broadcasting sector. White the FDI investment cap for the news and current affairs channels and FM radio services has been increased to 49 per cent from 26 per cent, the ceiling on broadcasting carriage services like teleport, direct to home (DTH) cable services and head-end in the sky (HITS) has been enhanced to 100 per cent.

Retail Business. In the new FDI policy the government has allowed *single brand retailers* to directly enter on line retail segments including e-commerce without specific government approval. Earlier, the FDI policy for single brand retailing did not allow the companies to sell on line on their own. AS a result most of them adopted a franchisee model and did business with partnership with online retailers such as Flipkarts and Snap deals. Besides, companies such as Apple have been exempted from sourcing norms.

Banking. To make it easier for private sector banks to raise foreign capital, the new policy has permitted full fungability of foreign investment. There will be no sub-limits for foreign portfolio investors (FPI).

We have explained above changes in foreign investment policy in a crucial sectors which are expected to accelerate economic growth of the economy. These changes in FDI policy will give a big push to foreign investment in India.

CHAPTER 37A
GLOBALISATION, COMMERCIAL POLICY AND WTO

Meaning of Globalisation

In this chapter we will discuss the integration of the Indian economy with global economy. Globalisation is one of the most frequently used word which means different things to different people. In true economic meaning, globalisation refers to the increased openness of an economy to the international trade, capital flows (both portfolio and foreign direct investment, FDI), transfer of technology and free movement of labor or people. Thus **globalisation means the integration of economies of the world resulting from free flows of trade, capital, labor (or people) and technology.** The term globalisation signifies new international economic order which envisages the following:

(*a*) free flow of trade of goods and services between different countries of the world.

(*b*) free flow of capital among countries.

(*c*) free flow of technology among different countries of the world.

(*d*) free movement of labor or people internationally.

The growing interaction between different countries of the world has made the idea of globalisation increasingly popular in recent years. In the past there has been a movement to form regional groupings such as European Union (EU) formed by European countries and NAFTA (North American Free Trade Association) set up by North American countries to promote their regional interests by removing trade barriers among them. However, these regional groupings (also called *custom unions*) were found to be inadequate as they did not enable the countries forming these unions to produce on a global scale and enjoy all economies of scale and take full advantages from the modern technologies. Therefore, the developed countries pleaded for globalization to get benefits from widening of the size of the market for their products, locating and spreading optimally capital investment across different countries.

For developing countries like India globalisation carries benefits and opportunities as well as costs and risks. Both India and China have been recently using globalisation as an opportunity to accelerate rate of economic growth so as to catch up with the developed countries.

CASE FOR GLOBALISATION OF INDIAN ECONOMY

Several arguments have been given for integrating Indian economy with the global economy since 1991 when structural reforms were initiated in India. Tariffs and non-tariff barriers have been removed to promote free trade, the Indian economy has been opened up to foreign investment and import of technology from the developed countries. The following benefits of liberalisation and globalisation of Indian economy have been emphasised.

1. **Shift from Import-substitution to Export-led Growth Strategy.** The failure of import substitution strategy of industrial growth to achieve sustained growth forced India and other developing countries to pursue export-led growth strategy (which is also called outward looking strategy of development). It has been argued that by expanding exports to the other countries and getting required imports from them based on their respective *comparative costs*, developing countries will be

able to achieve faster rate of economic growth. An important argument for trade liberalisation from the viewpoint of the developing countries is that they will gain from it as they have a comparative advantage in abundant, low-cost unskilled labor. If they specialise in the production of those goods which are labour-intensive, greater integration into global markets would increase their exports and production. This will help in generating more employment opportunities for the poor.

The strategy of development focussed on export promotion requires that other countries, especially developed countries should not prevent the imports to their countries through imposition of tariffs and non-tariff barriers. This is possible in the framework of a global economy with free movements of trade, capital and technology among countries. The setting up of WTO on Jan 1, 1995 was a step towards that direction. Accordingly, the entry of India in WTO in 1995 was a step to further globalise its economy.

2. **Foreign Capital Inflows**. The globalisation or integration of the Indian economy with the world economy is also beneficial because it would give a boost to foreign capital inflows in the form of portfolio investment and foreign direct investment (FDI). Portfolio investment will bring valuable foreign exchange currencies in India and free us of balance of payments difficulties. With sufficient foreign exchange reserves, balance of payments constraint on accelerating the growth process will be removed.

In the eighties due to shrinkage of foreign assistance, India had to resort to external commercial borrowings (ECB) which carried relatively higher rates of interest and increased the burden of external public debt. It may be noted that small foreign exchange reserves led to the economic crisis of 1991. As shall be explained later, the role of foreign direct investment is more important than portfolio foreign investmenet as it raises the rate of real investment in the economy and helps us to achieve a faster rate of economic growth. Foreign direct investment (FDI), like domestic investmenet, has a multiplier effect on output and employment.

3. **Globalisation and Transfer of Technology**. Another benefit flowing from globalisation of the Indian economy is that it acts as a mechanism for the transfer of technology from the developed countries. Due to financial constraints, Indian companies are in a position to invest only a small amount of funds on R & D. Therefore, it is through globalisation of its economy that we will be able to get advanced technology from the developed countries. The technological upgradation of the Indian industries will lead to higher productivity and help us to achieve a higher rate of industrial growth. It is worth noting that it is the multinational corporations (MNCs) that are carriers of technolgy to the developing countries through technological and financial collaboration with doemstic enterprises.

Globalisation makes faster diffusion of new ideas and advanced technologies in the world. This will make possible for the developing countries like India to catch up the developed countries more quickly.

4. **Increased Market Access**. An important benefit of globalisation is increased market access. Long ago Adam Smith wrote in 1776 that division of labor is limited by the size of market. Free trade accompanying globalisation widens the markets for products of industries. The larger the market in which products can be sold, the greater the benefit that will accrue as a result of economies of scale and specialisation. This will lower unit cost of production and increase the competitiveness of manufactured products. Thus globalisation will ensure greater gain from trade. In addition, the wider market increases the incentives for investing in new innovations as the potential return on investment in them will increase.

5. **Faster Economic Growth and Poverty Reduction**. Above all, it has been argued by some prominent economists such as Jagdish Bhagwati, T.N. Srinivasan, Arvind Panagariya, that globalisation will help in faster rate of reduction in poverty through acceleration of economic growth. To quote professor Arvind Panagariya, "Countries that have achieved significant poverty reduction are generally those that have grown rapidly and have, in turn, been open to trade. The most obvious example

are the Newly Industrialized Economies (NIEs) including Hong Kong, Singapore, Republic of Korea and Taiwan that have entirely eliminated poverty according to the *dollar-a-day poverty line*. On the other hand, countries such as India that remained autarkic[1] and grew at less than 1.5% in per capita terms until late seventies experienced little reduction in the trend poverty ratio. Both India and China achieved poverty reduction after they began to dismantle *autarkic policies* and began to grow rapidly".[2]

6. **Employment Arguement**. An important arguement for libaration of trade and capital flows is that it will generate more employment oportunities. Employment increases on two counts. First, the growth in exports based on comparative cost advantage will lead to the creation of more employment opportunities. Secondly, employement opportunities also increase following the removal of restrictions on capital flows. The greater capital inflows, especially in the form of foreign direct investment (FDI), create not only more direct employment for labour but they have also a multiplier effect on the growth of employment and output.

MEASURES ADOPTED IN INDIA TO PROMOTE GLOBALISATION

The first Iraq war in 1991 involving gulf countries resulted in sharp rise in oil prices which landed India in acute balance of payments. To overcome this crisis and restore economic health of Indian economy many far-reaching changes in economic policy were made to promote globalisation of the Indian economy. The following measures are worth mentioning in this regard.

1. Import Liberalisation. For liberalising foreign trade, *import controls through licesing was* abolished. With this almost all items of capital goods, raw materials, intermediate goods can be freely imported subject only to payment of customs duties. For some time quantitative quantitative restrictions on consumer goods remained but with effect from 1995 all quantitative restrctions, even on imports of consumer goods, have been lifted.

To liberalise imports peak customs duties which in some cases were as high as over 300 per cent were lowered in stages to 150 per cent in July 1991, to 85 per cent in Feb. 1993, 50 per cent in 2002. The average import duty was further reduced to 31 per cent in 2003 and to 20 percent in Jan. 2004, to 15 per cent in 2005 and further to 12.5% in 2006. Import duties on capital goods have been reduced even below 20 per cent for certain categories. This phased reduction in exceptionally high custms duties and a phased removal of quantitative restrictions on imports has substantially reduced anti-export bias in the earlier trade and balance of payment policies.

2. Imports of Gold and Silver have been considerably liberalised. This reduced the incentive for smuggling. In Jan. 2004 imports of gold were made free from any commission charged for it.

3. Market-determined Exchange Rate. An important measure in external sector was to devalue the rupee in July 1991 and after about 2 years in 1993 exchange rate was changed from basket based pegged exchange rate system to *market-determined exchange rate*. With this the exchange rate of the rupee today is determined by demand and supply conditions in the foreign exchange markets.

4. Convertibility of Rupee. Another important reform for globalising the Indian economy was the convertibility of rupee on balance of payments on current account. This implies the importers can get their required quantity of foreign exchange by converting their rupee resources into dollars from the foreign exchange market. The exporters do not have to surrender their foreign exchange (US dollar or EU Euro) earned abroad to RBI but can now sell them in the foreign exhcnage markets.

5. Liberalisation of Foreign Investment. The new economic policy adopted since 1991 considerably liberalised the scope of foreign investment, both direct and portfolio. Earlier investmenet by foreign companies required prior approval of the government and was restricted to 40 per cent

---

1. Autarkic means the closed nature of the economy having not much trade transactions with other country and which aims at achieving self-sufficiency.
2. Arvind Panagariya, Open up trade, get rich, *The Economic Times*, April 25, 2003.

equity participation and was also subjected to the conditions of technology transfer to India. Besides, foreign investmenet was permitted in priority areas only. Foreign portfolio investmenet was allowed mainly into a limited number of public sectors bond issues. The new economic policy of 1991 plrovided for automatic approval of foreign direct investment, that is, no prior permission from the government is required upto 51 per cent of the total equity capital of the firms in 34 priority industries. In 1996, the government raised this ceiling limit for auotmatic approval from 51 per cent to 74 per cent for foreign equity participation.

Criteria for approval of foreign investment greater than 51 per cent equity participation prior to 1996 (and greater than 74 per cent after 1996) in priority industries and foreign equity participation in non priority industries which was approved on case to case basis was quite liberalised to attract more foreign investmenet. Besides, it is worth mentioning that elimination of industrial licensing restrictions, openning up to the private sector of a number of industries previously reserved for the public sector which increased the growht of domestic corporate sector also served to promote foreign private investment.

Portfolio Foreign Investment. An important feature of new economic policy adopted since 1991 is that it has laid stress on *non-debt creating capital inflows*, such as direct foreign investment and foreign portfolio investment for reducing reliance on *debt flows* as the chief source of external resources. Accordingly, in addition to foreign direct investment, foreign portfolio investment has also been liberalised. Portfolio foreign investment refers to the foreign firms, foreign institutional investors and NRI who invest in the equity bonds and securities of Indian firms. In case of foreign portfolio investment foreign investors get returns in the form of yearly payable dividents/interest and capital gains accrued but do not excercise any direct control in running these companies. Foreign private portfolio capital inflows are welcome as they provide needed finance to the Indian firms and also bring the much needed foreign exchange. But it is important to mention that they are prone to greater volatility. The currency crisis of East Asian countries in 1997-78 was mainly precipitated by the sudden flight of portfolio foreign capital. Therefore, sufficient foreign exchange reserves must be maintained to save the domestic currency from such sudden capital flight. In the new economic policy of 1991, foreign portfolio investment was also liberalised. In Feb. 1992 the Indian firms of good standing were allowed to raise funds through equity and convertible bond issues in Euromarkets and in September 1992 registered Foreign Institutional Investors (FII) were allowed to purchase both equity and debt securities directly in the Indian markets.

The Foreign Exchange Regulation Act (FERA) has been replaced by Foreign Exchange Managment Act (FEMA) to remove a number of constraints earlier applicable to firms with foreign equity operating in India. Besides, FEMA makes it easier for Indian companies to operate abroad. Further, procedure for Indian companies to invest abroad and also raise funds abroad through ECB (*External Commercial Borrowing*), that is, through ADR (American Depository Receipts) and GDR (General Depository Receipts have been simplified.

Foreign institutional investors (FII) found Indian stocks particularly attractive for the purpose of risk diversification. Both 'pull' factors and 'push' factors worked to attract foreign portfolio investment into India. The pull factors include the improvement in Indian macroeconomic environment following the abolition of licensing system and banking and financial sector reforms. On the other hand, 'push' factors included the fall in interest rates and yields and slowdown in economic growth in USA and other developed countries. These prompted foreign investors to invest in India and other developing countries, especially China, to get higher yields on their investment.

DANGERS AND RISKS OF GLOBALISATION

Globalisation is not without dangers and risks. Stakes of poor countries like India are much higher for pursuing pro-globalisation policies. The historical experience of the colonial period shows that free trade and capital inflows helped imperialist countries more than the poor underdeveloped countries. With help of free trade and capital investment in some particular spheres, the developed countries exploited the poor countries which happened to be their colonies and drew away resources from them. With achievement of independence by the developing countries, though the situation today is vastly different but the developed countries has not shed their behaviour of promoting their own interests at the cost of developing countries. In fact, as has been stressed by Joseph Stiglitz, a Nobel Laurate in economics, that while US and European countries preach free trade and globalisation, they actually practice protectionist polices to safeguard their domestic industries. Imposition of tariffs on imports of steel and grant of subsidies on farm products and cotton textiles by the developed countries such as the USA to protect their domestic industries which caused widespread protests around the world is a shining example of such double standards.[3] To quote Stiglitz, "*I have always been struck by the divergence between the policies that America pushes on developing countries and those practiced in the US itself*".[4] Therefore, he advises that the developing countries should look carefully not what America says but at what it did in the years when America emerged as the industrial power, and what it does today.

Even Dr. Manmohan singh who has been the votary of free trade and free capital flows between countries in his address to the United Nations General Assembly in September 2011 said that recent events had called into question that globalisation would always do social good. He lamented the tendency on the part of developed countries to adopt protectionist policies to protect their own interest at the cost of the developing countries. To quite him, "*Till a few years ago, the world had taken for granted the benefits of globalization and global interdependence. Today we are being called upon to cope with the negative dimensions of those very phenomena.*"

An important disadvantage of globalisation is the danger posed by free flows of capital, especially portfolio capital, which are highly volatile and are a source of great macroeconomic instability. When there are large capital inflows into a country, the currency of a currency appreciates which makes its exports costlier and therefore cause reduction in them which not only adversely affects its current account balance of payments but also adversely affects its levels of GPP and employment. Besides, excess capital inflows in an economy lead to the expansion on money supply and if not sterlised by the RBI will add to the inflationary pressures in the economy.

On the other hand, when there are large capital outflows from an economy, they cause depreciation of the national currency which though tend to increase its exports but carry some risks. First, depreciation makes the imports more expensive and therefore leads to the rise in prices of commodities and raw materials imported from abroad and causes cost-push inflation in the economy. Further, depreciation of a national currency, say the Indian rupee, will raise the burden of external debt as more rupees are required to pay for a US dollar. Further, countries like India badly need foreign capital flows to finance its current deficit. In the absence of adequate capital flows, some US dollars will have to withdrawn from foreign exchange reserves resulting in the decline in our foreign exchange reserves. As foreign exchange reserves are available in limited quantity, in the absence of adequate capital flows to India year after year deficit in current account cannot be easily financed.

Therefore, in view of the present author, though the case for globalisation is theoretically quite sound , but it carries with it many risks and dangers. If promoted in its true spirit by all countries, it would lead to rapid economic growth in both the developed and developing countries. But, the developed countries are interested in promoting their own interests even at the cost of the poor developing countries on whom they impose policies of liberalisation and globalisation through their greater say in World Bank and IMF. Therefore, globalisation and reduction of tariffs and removal of quantitative restrictions on imports from US, Japan and European countries, carry a good deal of risk for them. This is quite evident from the recent failure of talks held under the auspices of WTO at

3. Joseph Stiglitz, Protectionism : US Style, *Economic Times*, May 8, 2002.
4. Joseph Stiglitz, "Do what we did, not what we say" *The Economic Times*, Nov. 3, 2003.

Doha, Seatle, Singapore and Cancun. These talks to liberalise trade between the developed and developed countries failed because US and EU (European Union) were not willing to reduce huge subsidies they are providing to their farmers on agricultural products and also subsidies on exports of these products. This support by the developed countries to their farmers prevent agricultural exports from the developing countries. In this way exports from developing countries are prevented even when they enjoy comparative advantage in their production. That is why the people of the developing countries of Asia, Africa and Latin America are protesting against them during the multilateral talks being held under the auspices of WTO.

Protectionist Policies of US and European Countries

An important example of protectionist's policy of US is its imposition of the higher tariffs on imports of steel and subsidies on cotton testiles which prevent exports of these commodities from the developing countries, especially African countries and India. USA tried to use some provisions of WTO agreements regarding use of "safeguards for protecting employment" to defend these import tariffs and subsidies on cottons textiles. Recently, in Nov. 2003, WTO dispute settlement body has given its verdict against the imposition of higher tariffs on steel by the USA and has asked it to scrap them. The reluctance of USA to remove tariffs on imports of steel and subsidies on cotton textiles is against the spirit of globalisation.

Recent Protectionist Measures by the US: Through business process outsourcing (BPO), American firms and businesses are outsourcing various types of services such as medical transcription, insurance connected work, various types of office jobs done through telecommunication and computer internet in call-centres from India. This BPOs have created good deal of employment opportunities for Indian educated youth. But recently there is anti-outsourcing campaign in the USA and UK. The new US President Obama has withdrawn tax-breaks (i.e. tax concession) from those corporate enterprises which outsource their jobs, that is, get some services done from abroad (e.g. India and China). Besides, he has imposed visa restrictions for foreign skilled workers (e.g., software professionals) to protect jobs in the US. These protectionist measures will harm the Indians. Some US states are even trying to enact laws to prvent outsourcing and thereby use of foreign (Indian) cheap skilled labour. This is against the spirit of free trade as India has a comparative advantage in these business processing services because its educated skilled labor is available at comparatively much lower wages. Some Americans have argued "*our resources should not be used to create jobs overseas when we have skilled workers and unemployment at home*". But business processing services for American business firms are like exports from India based on comparative costs advantage and therefore to prevent them to protect employment of American labor is an argument in favour of protection.

Volatility in Exchange Rate and Economic Instability

Another adverse consequence of liberalisation and globalisation is the creation of volatility in foreign exchange rate of rupee which has adversely affected the Indian economy. Under the policy of economic reforms we allowed free movements of capital flows into and out of India and switched over to the market-determined exchange rate and made the rupee convertible. From 2003-04 to 2007-08 there were large capital inflows into the Indian economy, mainly through portfolio investment by FIIs which caused appreciation of Indian rupee whose value rose from above ₹ 46 to a US dollar in 2002-03 to ₹ 44.27 at the end of March 2006 and further to ₹ 39.4 to a US dollar in mid-Nov. 2007. This appreciation of rupee adversely affected our exports. Besides, large inflows of US dollars got converted into rupees which caued rapid increase in money supply generating inflationary pressures in the Indian economy. Further, large capital inflows, especially through portfolio investment by FIIs in the Indian stock market, artificially caused high rise in price of shares of Indian companies.

However, in 2008-09 the opposite happened when due to liquidity crunch in the banking system in the US caused by large-scale defaults on sub-prime housing loans and mortgage failures, FIIs started *net capital outflows* from India by selling shares of Indian companies there was heavy depreciation of rupee. The rupee which was around ₹ 39.4 to a US dollar in Nov. 2007 fell to around ₹ 50 to a dollar in the second week of Nov. 2008. After remaining stable in the range of ₹ 48 – 49 to a dollar for some months it again fell to around ₹ 52 to a US dollar in the first week of March 2009. Though this depreciation of Indian rupee will promote exports which at present (March 2009) have been slowing down for the last six months in succession, the imports will become costlier. Besides, weakening of rupee will further induce capital outflows which in turn will cause more depreciation of rupee. This increased pressure on rupee will compell RBI to intervene and supply US dollars from its reserves (which at present are of sizable size of $ 250 billion) to prevent further fall in the exchange rate of rupee against US dollar.

It may be noted that it is highly important for RBI to check further depreciation of rupee as the value of a currency is a reflection of the underlying strength of the economy whose rating outlook is downgraded when value of its currency falls drastically. Therefore, RBI must intervene in the economy to prevent sharp fall in the foreign exchange rate of rupee and to ensure exchange rate stability.

Besides, the large capital outflows from India by FIIs caused liquidity crunch in the Indian banking system of the second half of 2008-09 which adversely affected investment in the Indian economy whose growth rate is estimated to fall to less than 7 per cent in 2008-09 as compared to 9 per cent growth rate achieved in the previous four years (2004-08). It follows from above that globalisation of Indian economy with liberalisation of capital flows and market-determined exchange rate and convertibility of rupee has caused a lot of volatility in exchange rate of rupee and economic instability and is therefore not without risks and dangers.

GLOBAL COMMERCIAL POLICY

Commercial policy rates to policies regarding exports and imports of a country. Since trade between countries now plays a vital role in promoting growth and well-being of the people of a country, commercial policy has acquired great importance today. Before 1947, the countries were free to impose tariffs on the imports. But when one country raised tariffs to protect domestic industries and to promote employment at home, other countries retaliated and raised their tariffs too. This tariff wars resulted in reduction in world trade and output and lowered efficiency in production. It was therefore realised that to improve productivity and national income of all countries free trade between countries should be promoted and for that purpose tariffs imposed by different countries must be reduced.

Types of Regional Agreements

The first step in liberalisation of trade among countries has been the regional agreements between a small number of countries. A smaller number of countries agree to reduce or eliminate tariffs on each other's imports and thus liberalise trade among them. North American Free Trade Agreement (NAFTA) was the trade agreement reached between some countries of North America which agreed to reduce and ultimately eliminate tariffs on the imports of member countries. Another important regional agreement has been the agreement among a number of European countries, European Economic Community (EEC) which is now called European Union (EU). In order to obtain gain from trade, they removed tariffs on the imports of member countries. European Union is a highly successful common market. Through it, member European countries have achieved a high degree of economic integration. They have not only eliminated tariffs on imports of member countries but have also allowed free movement of capital and labour among the member countries. The European Union (EU) is a powerful economic block whose chief objective has been to compete with Japan and the USA. They pursue a common agricultural policy which subsidise their agricultural products and guarantee farm prices to protect the interest of their farmers. Since January 1999, through an agreement, a common currency, called '*Euro*' has been introduced.

It will be useful to describe the standard forms of regional trade agreements aimed at liberalising trade among them. They are (1) Free Trade Area (FTA), (2) Custom Union (CU) (3) Common market, and (4) *Economic Union* we explain them below briefly.

Free Trade Area (FTA). Under it, member countries reach an agreement to remove tariffs on the imports from member countries. But member countries are free to impose tariffs and other restrictions on imports from the other countries under it.

Customs Union. A customs union is a free trade area *plus* an agreement about *common tariff and non-tariff restrictions* on the imports of non-member countries. Because they impose a common tariffs on the imports of the rest of the world they do not keep any custom controls on the goods moving among the member countries. An important example of customs union has been European Union (EU), though in recent years it has moved towards a greater degree of integration, namely, common market as explained below. A customs union differs from free trade agreement in two ways. First important difference is that a customs union has a common external tariff. A common external tariff means that each member country of a customs union replaces its own tariff schedule by the same external tariff schedule agreed to by the members of a customs union. For example an Indian firm exporting to UK, West Germany, Portugal (who are the members of EU) faces same tariff rates on its exports to whichever of these countries it exports goods. The second important difference between free trade area (FTA) and customs union is that free trade area does not include trade in agricultural products, services and financial flows, while most of the customs unions cover a wide range of international trade that includes trade in agricultural products, services and capital flows.

Common Market. A common market type of regional trade agreement represents a great degree of economic integration among a number of countries. A common market is a customs union that not only impose a common tariff barriers on the imports from the rest of the world but also allow free movement of labour and capital among its members. The European Union (EU) in its current form comes near to the idea of a common market.

It is worth noting that free mobility of capital in common market ensures more efficient allocation of capital within the member countries of a common market which has a favourable effect on their output of goods and services. Besides, in a common market, labour is free to move within the member countries of a common market and this allows for more efficient allocation of human capital which also adds to output in the member countries of a common market. But it is worth noting that free mobility of labour between member countries of a common market is not without problems. This is because large wage differenticals among the member countries may result in a large migration of labour from one country to another which will affect their national wage levels.

Economic Union. Another form of regional agreement is economic union. Economic union is an agreement between countries that involve not only free trade of goods and services, free mobility of capital and labour within member countries and common external tariffs but also some degree of unification is monetary and other government policies.

An essential requirement of the formation of economic union is the creation of common currency which facilitates the transactions between member countries of economic union and adoption of uniform monetary policy. Thus an economic union represents a deeper form of economic integration. In order to move towards the ultimate goal of formation of an economic union, European Union adopted a common currency called *Euro* in 1995. Euro is thus currency of several European countries. However, the U.K has still its own currency, 'Pound Sterling'. Besides, European countries which constitute European Union (EU) has also adopted a uniform policies of providing agricultural subsidies to their farmers to protect them from cheaper imports from developing countries and the United States.

However the world economy has no full-fledged economic union containing all its features. However, European union is strongly moving in the direction of formation of fully integrated economic union.

Some Trade Agreements: NAFTA, European Union and ASEAN

Let us explain in brief some important free trade agreements in the world today. It is through these agreements that there is greater economic integration of various countries.

NAFTA (North American Free Trade Area). In 1988 a comprehensive agreement was signed between the US and Canada covering elimination of tariffs on all goods and most non-government services to promote trade between them. In 1993, Mexico joined this free trade agreement and this agreement between these three countries, namely, US Canada and Mexico has been called NAFTA which envisaged removal of all trade barriers between them after a 15 years phasing in period.

European Union (UE). The world's largest and most successful customs union is the European union which has increasingly moved in the direction of common market. European Union is the association of European countries that have agreed to free trade area by eliminating tariffs between them and also impose a common external tariff. Presently, European Union (EU) includes 15 countries, prominent among them are France, Belgium, West Germany, Luxembourg, Italy, the United Kingdom, Denmark, Ireland, Greece, Spain, Portugal, Austria, Sweden joined the European Union in 1986.

European started its growth in 1951 when the European *Coal and Steel Community* (ECSC) was formed to eliminate trade barriers (i.e. tariffs and quotas) to promote free trade in coal and steel between them. This association was later merged into European Union. Recently, member countries of European Union (EU) have agreed to issue a common currency called Euro.

ASEAN (Association of South East Asian Countries). A group of countries of South East Asia, namely, Indonesia, Malaysia, Philippines, Singapore, Thailand have formed a free trade group within which member countries have removed tariffs and also fixed a common external tariff for other countries. This has significantly promoted trade between them and also because common external tariff in low, it has no large trade diversion effect.

EFFECTS OF REGIONAL TRADE AGREEMENT: TRADE CREATION AND TRADE DIVERSION

Liberalisation of trade by eliminating tariffs within a free trade area or customs union has both benefits and costs as it affects resource reallocation within countries. The effects of economic integration through regional trade agreements have been divided into two categories : trade creation and trade diversion. In what follows we explain these effects.

Trade Creation

When through free trade agreements, the two countries mutually eliminate tariffs, trade between the two countries. This favourable effect on world trade is called *trade creation.* This trade creation occurs because producers in one member country to a free trade agreement are able to sell some products in another member country at cheaper prices than those by the domestic producers as the latter lose tariff protection as a result of free trade agreements. For example when North American Free Trade Agreement (NAFTA) eliminating tariffs between the member countries came into effect with both USA and Mexico as members some Mexican firms found that they could sell their some products in the USA cheaper than the domestic American firms, while some American firms found that they could sell some of their other products in Mexico at cheapter prices than the Mexican firms. This led to more trade between Mexico and the United States. This happens because elimination of previous differential tariffs between the member countries of free trade area will cause prices of goods to change in the member countries after the trade agreement eliminating tariffs. Suppose before NAFTA, Mexico was imposing 20 per cent tariff on imports of cars. Now, with the elimination of tariffs American cars became cheaper in Mexico than the prices of domestic producers of cars in Mexico. This led to more exports of American cars in Mexico after the implementation of NAFTA. Similarly, some Mexican firms found that when agreement came into effect they could sell some of their products in the USA at cheaper rates than the domestic American

Competitors. Thus NAFTA led to more trade between the member countries with beneficial effects on resource allocation. This trade creation leads to the gain in efficiency in the use and allocation of resources as more efficient member countries of free trade area displaced less efficient countries for production of various goods.

Trade Diversion

Apart from trade creation, regional free trade esuch as FTA, customs union has also bad effect of trade diversion. When as a result of elmination of tariffs trade between member countries of a free trade area (for example, between the US and Mexico) increases, it causes export losses to the non-member countries. These export losses to other (*i.e.* non-member) countries are known as trade diversion.Trade diversion causes loss of efficiency in resource use because less efficient member countries of a free trade area displace more efficient non-member countries. Thus, "*from global perspective trade diversion represents an inefficient use of resource.*"

Whether or not regional trade agreement between some countries increases world trade depends on as to whether trade creation effect is larger than trade diversion. However, from the narrow national point of view of members of free trade agreement, it causes some gain as well as some loss. In so far as there is common desire to increae domestic production in the member countries of free trade area, it brings mutual benefits to the member countries. It gives producers of the member countries of free trade agreement an advantage over producers of the rest of the world. This causes increase in production of goods and services in the member countries of free trade area or customs union and also trade between them. However, as explained above, it leads to reduction of trade of members of a union with non-member countries.

GATT (GENERAL AGREEMENT ON TARIFFS AND TRADE)

During the period of Great Depression (1929-33) and the Second World War various countries imposed high tariffs on the imports to protect domestic industries. They followed what is called '*Beggar Thy Neighbour*' policy. The high tariffs resulted in large shrinking of world trade bringing about loss of efficiency in resource use in the world economy. As a reaction against the devastation caused by high tariffs during the thirties and Second World War period, the General Agreement on Tariffs and Trade (GATT) was established in 1947 after the Second World War. GATT aimed at liberalisation of trade between countries through ***multilateral trading system*** and sought to promote trade between various countries by reducing tariffs significantly. GATT organised various rounds of negotiations between member countries that led to reduction of tariffs. As a result not only trade between countries increased but world GNP also significantly rose after 1947.

In one of the negotiations carried out by GATT, Kenedy Round which began in early 1960s member countries agreed to make large reduction in tariffs starting from 1967s. Another Round Tokyo Round, resulted in further cuts in tariffs in 1979.

In the latest round under GATT, **Uruguay Round** negotiations started in 1986 and concluded in 1994. After several years of negotiations comprehensive agreement was reached on a set of proposals which is popularly called Dunkel Draft as it was prepared by Arthur Dunkel who was then director General of GATT. In this agreement 40 per cent reduction in tariffs was reached. This sought to promote world trade significantly. Besides, Uruguay Round of talks established a world body called WTO (World Trade organisation) to achieve further liberalisation of free trade among countries. WTO came into existence in January 1995. WTO replaced GATT . The objective of WTO is to ensure efficiency in use of world resources and specialisation of various countries in accordance with their comparative advantage. We explain below WTO at same length.

WORLD TRADE ORGANISATION (WTO)

An important step in global commercial policy in setting up of World Trade Organisation (WTO) is a very important trade organisation to promote free trade among different countries. Protection has harmful effects on economic growth and consumer welfare. Bretton Woods

Conference held in 1944 which recommended the setting up of International Monetary Fund (IMF) and World Bank (WB) also recommended the establishment of an International trade organisation to promote free trade among nations. However, while IMF and World Bank were established in 1946, agreement to set up International Trade Organisation could not be finally reached. Instead, a less formal organisation called GATT (General Agreement on Tariffs and Trade) was set up to guide the international trading system based on rules and principles agreed between the member countries. However, it was found to be weak and ineffective although it succeeded in achieving its objective of promoting world trade to some extent by bringing down the level of tariffs on manufactured products imposed by the developed countries. The process of liberalisation of trade, however, suffered a set back since 1974. Although the process of reducing tariff barriers continued, Non-Tariff Barriers (NTB's) were substantially increased by various countries including the developed countries. Strangly enough, it was found that while the developed countries were adopting Non-Tariff Barriers such as quantitative restrictions to protect their industries, the developing countries were liberalising their trade by reducing tariff on the imports from the developed countries.

Unlike GATT, the WTO is a permanent organisation created by international treaty ratified by the governments and legislatures of member states. As the principal international body concerned with solving trade problems between countries and providing a forum for multilateral trade negotiations, it has global status similar to that of the International Monetary Fund and the World Bank.

Object of WTO

The objective of WTO agreements is *to establish **a multilateral trading system*** in order to promote free and fair trade among nations. Dunkel Act on which WTO has been founded is wedded to liberalisation of trade based on comparative costs. Accordingly, this provides for a multilateral framework for trade not only in industrial and agricultural goods but also in services. It also seeks to protect ***trade related intellectual property rights*** **(TRIPS).** By committing the member countries to give Most Favoured Nation (MFN) treatment to all trading partners, it has sought to discourage bilateral trading so as to encourage multilateral trading system.

At present there are 148 member countries of WTO and includes both developed and developing countries of the world.

WTO: Status, Administration and Functions

WTO is officially defined as the legal and institutional foundation of the *multilateral trade system.* Unlike GATT, the WTO is a permanent organisation created by international treaty ratified by the governments and legislatures of member states. As the principal international body concerned with solving trade problems between countries and providing a forum for multilateral trade negotiations, it has global status similar to that of the International Monetary Fund and the World Bank.

But unlike them, it is not a United Nations agency although it has a 'cooperative relationship' with the United Nations. Its underlying documents are a General Agreement consisting of a 38-article code aimed at ensuring open and fair trade in goods, services, agricultural produce and textiles and 500 pages of specific agreements reached at the Uruguay Round.

Administrative Structure of WTO. WTO is headed by a director-general (currently Mr. Pascal Lamy, former Italian trade minister) who has four deputies from different member states. The WTO's ruling body is the General Council, comprising each member country's permanent envoys. It sits in Geneva on an average once a month. Its supreme authority is the *ministerial conference,* to be held every two years. The General Council appoints the director-general for a four-year term after consultations among member countries.

WTO started with 125 countries. But seven more states including China, Russia were admitted later on as members. Members range from the top four world trade powers—the United States, the European Union, Japan and Canada—to the increasingly influential emerging developing economies of Asia such as India and China to some of the world's poorest countries, like Bangladesh, Guinea. such as a India and China.

Functions of WTO

WTO has the following five functions to perform :

(1) The WTO provides the framework for implementation, administration and operation of multilateral trade agreements reached at Uruguay Round.

(2) The WTO provides the forum for further negotiations among its member states concerning their multilateral trade relations with regard to the matters included in the agreements reached at Uruguay Round.

(3) The WTO undertakes the task of *settlement of disputes* among the member states, which arise from their different understandings of the rules and procedure agreed upon.

(4) The WTO administers the '*Trade Review Mechanism.*'

(5) In order to evolve a coherrent global economic policy to promote free and fair trade among the different countries, WTO cooperates in an appropriate manner with the IMF; World Bank and its affiliated agencies.

WTO Bodies for Administration

The two important bodies in WTO have been set up to perform its various functions.

1. Dispute Settlement Body (DSB). The DSB, on which all member countries can sit, usually meets twice a month to hear complaints of violations of WTO rules and agreements. It sets up expert panels to study disputes and decide if the rules are being broken. The DSB's final decisions, unlike those of a similar but less powerful body in the old GATT, cannot be challenged.

2. Trade Policy Review Body (TPRB). The TPRB is a forum for the entire membership to review the trade policies of all WTO member countries. Major trading policies are reviewed every two years, others every four years.

Other major bodies are the Council for Trade in Goods, the Council for Trade in Services and the Council for *Trade-Related Aspects of Intellectual Property Rights* (TRIPS).

THE GUIDING FUNDAMENTAL PRINCIPLES AND FEATURES OF WTO

The following fundamental principles have been provided for the functioning of WTO as a multiple trading system. These principles represent the salient features of WTO.

(*a*) **Non-discrimination.** This is the most important principle on which WTO has been founded. The principle of non-discrimination means two things. (1) All trading partners will be granted the *most favoured nation* (*MFN*) *status*, that is, each member state of WTO will treat every other member state equally as the most favoured nation doing trade. No discrimination will be done by a member of state between different trading states who are also members of WTO. However, some exceptions have been provided in this regard, for example, in case regional trade agreements exist. (1) Foreign goods, services, trade marks, patents and copyrights shall be given the same treatment as is given to nationals of a country.

(*b*) **Free Trade.** The objective of WTO, as in case of GATT, is to promote free trade among nations through negotiations. For this purpose WTO has to work for progressive liberalisation of trade through reduction in tariffs and removal of quantitative restrictions on imports by member countries.

(*c*) **Stability in the Trading System.** Under WTO agreements member states are committed not to raise tariff and non-tariff trade barriers arbitrarily. This provides stability and predictability to the trading system.

(*d*) **Promotion of Fair Competition.** WTO system of multilateral trading system provides for transparent, fair and undistorted competition among the various countries. Rules such as Most Favoured Nation (MFN) treatment to all trading parties, equal treatment to foreign goods, patents and copyrights as with nationals ensure fair competition among trading countries. Besides, WTO agreement provides for discouraging unfair competitive practices such as export subsidies and dumping (that is, selling products abroad below domestic prices to gain market access).

(*e*) **Special Concern for Developing Countries.** WTO has shown special concern for the developing countries as it has given them more time to adjust to agreements under it and also some special privileges. An important feature of WTO is that it would deal with not only the disputes in the area of trade in goods but a whole range of issues such services and intellectual property rights.

(*f*) **Market Access Commitment.** WTO agreements which seek to establish multilateral trading system require the member countries to undertake market access commitment on reciprocity basis. In fact, market access is ensured by abolsihing non-tariff barriers as well as by reducing tariffs. The understanding on market access requires that member countries will cut tariffs on industrial goods and agricultural products by about 37 per cent. In order to provide market access for the products of developing countries to the USA, USA agreed to cut down farm subsidies. The developing countries are also required to reduce agricultural subsidies to the level of 10 per cent of the value of agricultural produce.

In the area of trade in services, market access has been ensured by giving foreign service suppliers the same treatment as demestic service suppliers.

(*g*) **Decision at the Ministerial Level Meeting**. Another feature of WTO agreement is that it has upgraded decision-making at the ministeral level. Important decisions regarding trade related matters are to be taken at the Ministerial level meetings. Ministerial level meetings have now been incorporated in the legal structure of WTO.

(*h*) **Wider Range of Issues.** Another important feature of WTO is that it wil deal with not only issues and disputes relating to trade in goods but also the whole range of issues concerning *trade in services and intellectual property rights*.

(*i*) **Multilateral Trading System.** The most important features of WTO is that it seeks to establish just and fair multlateral system of international trade wherein the developed countries, the developing countries, and the least developing countries all have equal opportunities for market access of their products in foreign countries and wherin discriminatory trade barriers and unjust Government support to exports by different countries have to be eliminated.

WTO AGREEMENTS AND LIBERALISATION

We discuss below various WTO agreements which seek to promote liberalisation of trade.

(1) Liberalisation of trade in manufactures
(2) Liberalisation of agricultural trade
(3) Liberalisation of trade in services (GATS)
(4) Trade Related Intellectual Property Rights (TRIPS)
(5) Trade Related Investment Measures (TRIMS).

Liberalisation of Trade in Manufactures: Agreement on Tariff Bindings and Tariff Reducton

The main agreements with regard to trade liberalisation in manufactures are:

1. Firstly, the provision for *expansion of tariff bindings* which prevent the developed countries from increasing tariff rates in future beyond a particular level. Tariff bindings cover 99 per cent of imports. Secondly, the agreement provides for *reduction in tariff rates* by the developed countries by 40 per cent from 6.2 to 3.7 per cent. Thirdly, the agreement provides for *expansion of duty free access* from 20 to 43 per cent of their imports by the developed countries.

It is however important to note that gain to the developing countries from tariff reduction by the developed countries is not very significant. The average reduction of tariffs on their imports to the developed countries is estimated at 30 per cent and in case of labour-intensive industrial products such as textiles, clothing, leather goods and certain processed primary products such as fish products which are regarded as sensitive, the reduction in tariffs is less than even the average reduction of 30 per cent.

On the other hand, the offers of tariff cuts on imports of manufactures by the developing countries are estimated at about one-third of the average tariff reduction by the whole world. The expansion of tariff bindings which prevent the developing countries from increasing tariffs in future represent significant gains for the developed countries.

Agreement on Agriculture

To achieve liberalisation of agricultural trade, agreement on agriculture was also reached during multinational trade negotiations resulting in the establishment of WTO in 1995. Agreement on agriculture was highly significant as agriculture was a highly protected sectors in the developed countries. This distorted world prices of agricultural products prevented the developing countries to realise the benefits of their comparative advantage. As a result, exports of developing countries could not increase significantly. Agreement on agriculture includes the following conditions :

(*a*) **Tariffication.** It was agreed that *existing non-tariff barriers imposed by the member countries would by replaced by suitable tariffs* which would provide almost the same level of protection. This is called tariffication.

(*b*) **Tariff Bindings.** The agreement on agriculture also required that developed countries would reduce their tariff bindings by an average of 36 per cent within six years from 1995. The developing countries were required to reduce tariffs by an average 24 percent over a period of 10 years, while the least developed countries are not required to make any commitment regarding reduction of tariffs on their farm products.

The developing countries were permitted to indicate only their maximum bindings. For example, India declared maximum bindings of 100 per cent on farm products and 300 per cent on edible oils.

(*c*) **Subsidies and Domestic Support Policies.** Reduction in Agricultural subsidies and domestic support prices is another important aspect of agreement on agriculture. The developed countires such as USA and the countries of East European Union (EU) maintain high level of agricultural subsidies and other domestic support programmes to protect their farmers and prevent the imports from the developing countries. As a result, exports of farm products of the developing countries suffer a lot. WTO agreement on agriculture requires that subsidies on farm products should not exceed 10 per cent of the value of agricultural production.

Besides, it was agreed that product specific subsidies, non-product specific subsidies on fertilizers, irrigation, power and seeds etc were also not to exceed 10 per cent of the value of agricultural output. Agreement on agriculture also provided that countries with closed agricultural markets will have to import agricultural commodities to the extent of 3 per cent of their domestic consumption, going up to 5 per cent over six years. However, developing countries which were facing balance of payments problems were exempted from this rule of compulsory imports.

Agrement on agriculture also provides that countries are free either to give patents on agricultural products or to evolve an effective *sui genris* (*i.e.* special) system of protection for plant breeder rights. It implies freedom of multiplication and exchange of indigeneously produced seeds or plants in the country. In other words, the farmers are free to retain seeds from their own harvests for their uses and exchange with each other. The only exception is seeds evolved genetically through the tools of high-tech biotechnology. The patent fee would be payble if the seed or plant varieties bought are genetically superior to and more productive than the local varieties. Thus, the farmers will have to pay a patent fee on branded seeds or plant varieties imported from developed countries. The commerrcial sale of such seeds would be affected by patents.

GATS (General Agreement on Trade in Services)

The General Agreement on Trade in Services (GATS) has been one of the major achievements of Uruguay Round of negotiations which now forms part of WTO legal framework GATS covers all service sectors including financial services, telecommunications, transport, tourism, audio visual and professional services. GATS requires some basic obligations to be fulfilled by all member countries of WTO with regard to international trade in services. GATS applies to services provided by service suppliers of one country and sold to consumers of another country (for example tourism). It also applies to services provided by the commercial presence of a supplier of one country in the territory of another (for example bank of a country providing banking service in another country). It is equally applicable to services supplied by companies of one country in the territory of another. However, GATS shall not cover services provided by the government which are not supplied on a commercial basis.

GATS requires countries to provide most-favoured nation treatment (MFN) to all member countries. It implies that equally favourable treatment is to be accorded to service suppliers of all member foreign countries. GATS also requires that member countries should have transparency in their trade in services. This implies that each country to promptly publish all its relevant law and regulations pertaining to services including international agreements pertaining to trade in services to which the country is a signatory. Besides, each member country shall provide all information sought by any other member country relating to any service covered by GATS. Further, the countries are required to accord foreign service suppliers the same treatment as domestic service suppliers.

The countries may be exempted from GATS if they face balance of payments difficulties. They may also be exempted from GATS for national security reason as well as for purposes of protecting public order. The GATS requires that the countries would provide market access to foreign service suppliers. Hence, they would not impose restrictions on the number of service suppliers, total value of service transactions, the total number of service operations, joint ventures through which service may be supplied and the participation of foreign capital. The GATS has excluded labour movement from its purview. It allows countries to apply immigration laws to regulate the entry of persons into their territories. Countries are specifically allowed to apply *visa* requirements selectively to some countries and not to others.

TRIPS (Trade Related Intellectual Property Rights) Agreement

One of the most controversial agreement of Urugay Round of negotiations relates to Trade Related Intellectual Property Rights or briefly known as TRIPS. The agreement regarding TRIPS requires member countries to provide patent protection to all products or processes in all fields of technology. This protection is granted subject to the following three conditions:

(1) The product or the process is a new one
(2) It contains an inventive step
(3) It is capable of industrial application for 20 years from the grant of the patent.

TRIPS agreement corers the following seven intellectual properties.

(*i*) Patents
(*ii*) Copyrights and other related rights
(*iii*) Geographical indications
(*iv*) Industrial designs
(*v*) Trade marks
(*vi*) Layout designs of integrated circuits
(*vii*) Undisclosed information including trade secrets.

Patents shall be available without discrimination as to the place of invention, the field of technology and whether products are imported or locally produced. The patent-holder would enjoy exclusive marketing rights for a particular period. Any body else seeking to manufacture and sell the product would have to establish that there has been no violation of the patent-holder's rights. The term of patent under the new regime would be 20 years from the date of the filling of the application.

Since the patent would be available for products or processes, it would be possible in chemical-base products like drugs and pharmaceuticals, agro-chemicals, alloys and food products to take patents for new products for 20 years and process patents thereafter for another 20 years. The countries which do not have product patent in certain areas (in India, agriculture and horticulutre, technologies relating to atomic energy and chemical based products like chemicals, alloys, drugs and pharmaceuticals, agro-chemicals, food products and so on are exempt from product patents) are given 10 years transition period for the introduction of product patents.

The countries are free to give patents or adopt an *effective sui genris* system of protection for plant breeder rights enshrined in the International Union for Protection of new Plant Varieties UPOV) convention.

A period of five years was given to implement the provisions of the TRIPS agreement. In respect of Copyrights and related rights India is a signatory to the Berne convention. Likewise, in respect layout designs and integrated circuits India is signatory to Washington Treaty whose main obligations have also been incorporated in the TRIPS.

TRIMS (Trade Related Investment Measures) Agreement

TRIMS agreement refers to conditions or restrictions imposed on *foreign investors*. This agreement requires that investment regulations by member countries have to give same treatment to domestic products and imports. The TRIMS agreement specifically forbids imposing restrictions on operations of an enterprise which result in protecting domestic products and making imports disadvantageous. The foreign trade related restrictions on investment by foreign enterprises were generally imposed by developing countries including India. The following conditions which favour domestic production were prohibited under TRIMS agreement.

1. **Local content requirement.** That is, foreign enterprises must use a certain amount of locally produced inputs in production of products.

2. **Trade balancing requirement.** That is, imports by a foreign enterprise shall not exceed a certain proportion of exports by it.

3. Trade and foreign exchange balancing requirements.

4. **Domestic sales requirement.** This requires an enterprise to sell a certain proportion of its output locally.

However, it was agreed that subsidies applicable solely to domestic enterprises and government procurement policy in favour of domestic producers will not violate the TRIMS agreement. Thus, under TRIMS Agreement, investment regulations have to accord same treatment to domestic products and imports. *The TRIMS agreement requires removal of quantitative restrictions on imports and exports*. However, exemptions are allowed if a country is suffering from balance of payments problems.

The industrialised countries were required to eliminate conditins covered under TRIMS by July 1, 1997. Developing countries were required to do so by 2000 and least developed countries were required to eliminate them by 2002.

India notified the TRIMS required by it before 2000. It notified two TRIMS conditions (1) relating to local content requirements in the production of certain pharmaceutical products and (2) dividend balancing requirements in the case of investment in 22 categories of consumer items. These were to be eliminated by 1-1-2000. The developing countries, including India, requested for extension of transition period for the elimination of the notified TRIMS. In view of the failures of the Seatle Ministerial Conference and Cancun conference no final decision was taken on this request of the developing counties.

INDIA'S ROLE IN WTO TALKS

India has consistently taken the stand that the launch of any new round of talks depends on a full convergence of views amongst the entire WTO membership on the scope and framework for such negotiations. Our more urgent task is to resolve the concerns of developing countries on implementation of the Uruguay Round agreements. We are against calls for new commitments from the developing world for achieving symmetry and equity in the existing agreements. It is in favour of '*non-trade' issues* be permanently kept off the negotiating table.

Incorporating Livelihood Clause. Ensuring food and livelihood security is critical, particularly for a large agrarian economy like India. India's proposal in ongoing negotiations includes suggestions like allowing developing countries to maintain appropriate level of tariff bindings, commensurate with their developmental needs and the prevailing distortions in international markets. We are also seeking a separate safeguard mechanism including provision for imposition of quantitative restrictions under specified circumstances, particularly in case of a surge in imports or decline in prices; exemptions for developing countries from obligations to provide minimum market access; exemptions of all measures taken by developing countries for poverty alleviation, rural development and rural employment.

Our immediate priority is that the agreements reached earlier with the developing countries should be implemented so as to correct inherent imbalances in some of the Uruguay Round agreements. Sincere and meaningful implementation of commitments undertaken by developed countries and operationalisation of all special and differential treatment clauses for developing countries in the various agreements be made.

We also strongly favour extension of higher levels of protection to the geographical indications for products like Basmati rice, Darjeeling tea, and Alphonso mangoes at par with that provided to wines and spirits under the Trade-related Aspects of Intellectual Property Rights (TRIPS) agreement. In the TRIMS (Agreement on Trade-Related Investment Measures) review we want flexibility for developing countries in adopting appreopriate domestic policy while permitting foreign investment.

Developed Countries Demand and Singapore Issues. Developed countries are pushing for a comprehensive agenda like *rules on investment, environment, competition policy, trade facilitation, transparency in government procurement*, labour standards etc. They are pressing for incorporating non-trade issues of environment and labour standards. Using as an excuse that production of products in developing countries are not being done under proper environment and labour standards they can ban the imports of their products or impose other non-tariff restrictions. The developing countries are opposed to these non-trade issues

WTO and Indian Agriculture and Farmers

Some critics of WTO have expressed the fears that Indian farmers are threatened by the WTO. There is however no adverse impact. India has bound its tariff to the extent of 100 per cent for primary agricultural products, 150 per cent for processed agricultural products and 300 per cent for edible oils. A few agricultural products had been bound historically at low levels but these bindings have been raised following the Article XXVIII negotiations held in this regard.

It has also been possible to maintain without hindrance the domestic policy instruments for promotion of agriculture or for targeted supply of food grains. Domestic policy measures like the operation of minimum support price, public distribution system as well as provision of input subsidies to agriculture have not in any way been constrained by the WTO agreement. Certain provisions in the Agreement on Agriculture (AoA) also give us flexibility to provide support for research, pest and disease control, marketing and promotion services, infrastructure development, payments for relief from natural disasters, payments under the regional assistance programmes for disadvantaged regions and payments under environmental programmes. Indian farmers now need to take advantage of the opportunity provided by the AoA, by addressing productivity issues and making their products more competitive globally.

WTO and Indian Industries

Indian industry has had to face greater competition in the wake of globalisation. But it has successfully competed, as can be inferred from the fact that there has been no particular surge in imports. In fact, as per the provisional data for 2000-01 our non-oil imports declined by 14 per cent while our exports rose by over 20 per cent in the same period. A close watch is also being kept to ensure that Indian industry does not have to face unfair competition from dumped or subsidised imports of other countries.

As for drug prices, safeguards are provided like compulsory licensing, price controls, and parallel imports which should help address this concern. It must also be recognised that the prices of medicines are influenced by several factors including the level of competition, size of market, purchasing capacity etc. The issue of affordable access to treatment for AIDS, which has gathered international attention in recent months, is hopefully a pointer in the right direction. The TRIPS agreement should not be allowed to hinder the efforts of developing countries to provide affordable access to medicines.

The apex Indian organisations representing various industries are sincerely working towards ensuring a gainful transition with least disadvantage into the global economy. The government also has to strive to improve infrastructure and provide a facilitating environment for inducing acceleration in trade. Developed countries have been putting pressures on inclusion of non-trade issues such as labour standards, environment protection, human rights, rules on investment, competition policy in the WTO agreements. This is because by asserting that particular developing countries are not observing and implementing the rules in regard to the non-trade issues so that the developed countries can ban the imports of some goods in their countries, as the USA has been trying to do so from time to time. We are against any inclusion of non-trade issues that are directed in the long run at enforcing protectionist measures, particularly against developing countries.

INDIA'S GAIN FROM WTO

Now almost a decade has passed when India joined WTO from the very beginning in Jan. 1,1995. India has benefitted from joining WTO, despite the failures of two ministerial level conferences at Seatle and Cancun. We explain below the gains India has achieved by joining WTO.

Exports and Imports. According to the recent estimates. India's exports have almost doubled in less than a decade. With exports going up from $26.33 billion in 1994-1995 when India joined WTO to $51.7 billion in 2002-03. Besides, India's share in total world exports of goods and commercial services increased from 0.6 in 1995 to 0.86 in 2001 whereas its total world imports of goods and commercial services increased from 0.78 in 1995 to 0.99 per cent during the same period.

Exports of Textiles and Clothing. According to a WTO agreement known as Multi-Fibre Agreement (MFA) entire quotas in textile and clothing trade will come to end from January 1, 2005. Till now WTO agreement has required the member countries to phase out their existing quotas by the Dec. 31, 2004. It has further restrained them from expanding the size of quotas annually. These measures have helped India to increase its market access for its textile and clothing products. With effect from January 1, 2005, the entire textiles and clothing trade would get integrated into the multilateral trade framework of WTO. According to recent report (Feb. 2004), after China, India is largest gainer from the end of quotas and the consequent free trade in textiles and clothing. It is estimated that export market of $500 billion in garments alone with employment potential of 30 million jobs will be up for grabs from which India can get a good share.

Five years ago, Indian economists feared the country's fabrics sector would be steam-rollered by China. But recent reforms in the industry have given hope. India could see apparel exports rise between 40-100 per cent ($6.7 billion in 2001). A target of $50 billion in textile and clothing exports by 2010 is no longer a pipe dream.

Gain in Exports of Software Services. Further, thanks to WTO agreement on free trade in services India has become a world leader in software services which are contributing a lot to foreign exchange earnings and employment generation for Indians.

A Good Deal of Earnings from BPO. BPO (Business Processing Outsourcing Services) from USA and UK are coming to India which have enabled us to earn not only foreign exchange but also to generate a large number of employment opportunities for educated Indian youth. Recently, there is resentment in the USA against BPO to India and other countries because it is causing loss of jobs in the USA and UK. But BPO to India and other developing countries are in accordance with the comparative advantage principle and raises profits of American companies which on being used for further investment will generate more employment opportunities in the USA

India's Demand from the Industrialised Countries about Agricultural Subsidies and Movement of Labour. In its negotiating proposal, India has demanded substantial reduction in tariffs, elimination of trade-distorting domestic support and export subsidies in developed countries. The movement of professionals (*i.e.*, labour) from developing countries is constrained by a number of factors such as lack of specific sectorial commitments, lack of mutual recognition of qualifications, lack of transparency in administration of *visa* regimes, discriminatory practices in use of Economic Needs Test and social security contributions. India has, therefore, sought liberalisation of movement of professionals through removal of these constraints and submitted a paper for discussion at the ongoing negotiations.

Failure of WTO Talks at Cancun

From Sept. 10 to 14,2003 trade ministers of developed and developing countries met for the next stage of talks on multilateral trading system. In an earlier meeting at Doha is Nov. 2001 ministers recognised the inequities of Uruguay Round of negotiations. It was expected that the developing countries would get a just and fair global trading system at the conclusion of Cancun meeting. But these hopes were not fulfilled and Cancun meeting ended in a failure. At Cancun, a group of 21 developing countries led by India, Brazil and China frustrated the attempts of US-European Union Cambine to pressurise the developing countries to accept the agenda that suited them most and was against the interests of the developing countries. Draft declaration prepared by WTO allowed the developed countries to maintain or keep domestic subsidies and support for their agriculture and farmers and avoided eliminating export subsidies and credit concessions on their exports, especially on farm products and textiles. On the other hand, draft declaration sought even stiffer tariff cuts by developing countries. In this way, developing countries were badly treated with regard to market access of their products to the developed countries. All these were not acceptable to the developing countries including India. African countries who are badly hit by the subsidies and domestic support on the farm products by the developed countries walked out of the meeting. As a result, Cancun meeting ended without adopting any declaration. The lesson from the failure of Cancun meeting is that unless the developed countries look beyond their own interests and try to evolve multilateral trading system which is just and fair to the developing countries talks cannot succeed.

WTO'S HONG KONG DECLARATION

After failures at Seatle and Cancun, the ministerial meeting in Hong Kong in December 2005 negotiated trade liberalisation to conclude Doha Round on Development. In Nov. 2001 in the Ministerial meeting at Doha the developed countries (the US and EU) and developing countries, it was agreed to start new round of talks to *promote development through trade liberalisation*. Hence the name '*Doha Round on Development*'. The two successive ministerial meetings at Seatle and Cancun failed to arrive at an agreement acceptable to the LDCs and developing countries. It may be noted that Uruguay Round of 1994 resulted in setting up of WTO in Jan. 1995. It may be further noted that the agenda of Uruguay Round as well as its outcome discriminated against developing countries. The trade negotiations of Doha Round are meant to remove this discrimination against the developing countries[1]. During the negotiations developed countries have been raising irrelevant issues so as to protect their agriculture and textiles industry through tariffs and non-tariff barriers (NTB) so as to prevent free and fair trade from the viewpoint of developing countries. On the other hand, at the conclusion of Uruguay Round, the developed countries succeeded in extracting an agreement regarding the *inclusion of services and intellectual property rights* – the two issues which concern them most. In return for this agreement on services and intellectual property rights – they agreed to *eliminate* over a period of 10 years high tariff and non-tariff barriers on textiles, agricultural products, sugar – the products which are of special interest for developing countries However, the developed countries did not fulfil promise on one protext or the other. Commenting on this duplicity on the part of developed countries Stiglitz writes, "When the Uruguay Round began there was a grand bargain to expand the trade agenda to include services and intellectual property rights—the two issues which are of particular concern to developed countries. In return *developed countries were to make major concessions on agriculture*—the livelihood of the vast majority of people in developing countries — and textile quota the only trade area (besides sugur) in which quantitative restrictions persist. In the end developed countries got what they wanted, and developing countries were told to be patient : eventually the developed countries would fulfil their part of the deal. Even as the rich countries urged *developing countries to make quick adjustments*, they claimed that *they needed a decade to make the transition to a quota-free-textile regime.* In truth they were buying time; *they did nothing for a decade* and when the quota finally ended last January (*i.e.*, Jan. 2005) they pleaded they were still not prepared and thus concluded a three year extension with China.

Singapore Conference was meant to negotiate trade issues relating to agriculture, non-agriculture services and intellectual property rights so as achieve trade liberalisation for promoting economic development. The problem with regard to agriculture is that the developed countries, the US and EU provide export subsidies and also give financial support on a large scale to their agriculture resulting in cheapening of their agricultural products. This distorts trade between the countries as it prevents market access in developed countries for products of developing countries which have comparative advantage in their production. Developing countries discovered that their gain from trade with developed countries were far less than they were made out to be and LDCs (Least Developed Countries) felt than they were actually worse off by agreeing in Uruguay Round of talks resulting in establishment WTO. Thus developed countries lost their credibility.

With the above in background the Ministerial Conference at Hong Kong adopted a declaration which called for conclusion in 2006 of trade negotiations launched at Doha in Nov. 2001. This Hong Kong declaration established timeframe and targets in specific areas of trade liberalisation. The following agreement which addresses some of the concerns of developing countries related to agriculture was reached.

1. First it was resolved to complete Doha Round on Development in 2006 and conclude negotiations for liberalisation of trade between members of WTO.

2. **Elimination of Agricultural Export Subsidies.** The agreement was reached in Hong Kong that export subsidies given by the developed countries will be eliminated by the year 2013 in a phased manner. While the earlier proposal was to phase these export subsidies by 2010, the EU the largest provider of export subsidies sought a longer phase-out period. Thus, the agreed end-date of 2013 gives the EU a level of comfort of phasing out its export subsidies as a part of its next phase of reforms of the common agricultural policy (CAP). This is an achievement of the developing countries as it was contended earlier by its the earlier Trade Commissioner of WTO Pascal Lamy that Doha declaration had not mandated elimination of export subsidies in the developed countries. It may be noted further that under the agreement developing countries like India will continue to have the right to provide marketing and transport subsidies on agricultural exports for 5 years beyond the date for elimination of all forms of subsidies.

Agricultural Tariffs. To increase market access in the developed countries for the agricultural products of developing countries, the Hong Kong declaration has agreed to banding approach suggested by the group G-20 of developing Countries, According to this approach, agricultural tariffs have been put in *four bands* ranging from the lower to the high level, with the provision that tariff in the higher band will be subject to deeper cuts. The extent of cut in agricultural tariffs in each band is yet to be agreed upon in future negotiations. The higher cuts in tariffs in the higher bands will greatly bring down the agricultural tariffs of the developed countries and will therefore improve market access for agricultural exports of the developing countries in the developed industrialised economies. However, it will also mean substantial reduction in agricultural tariffs of developing countries such as India who have higher tariff bindings. Indias tariff bindings for agricultural products are in the range 100 to 150 per cent and will therefore require to reduction about 35 per cent which is quite large.

However, in the Hong Kong declaration, the developing countries have been provided some special safeguards in connection with the cuts in agricultural tariffs. Firstly, the developing countries have been given the flexibility to self-designate an appropriate number of tariff lines as *special products* (SPs). However, the self-designation of a number of tariff lines as special products will require the satisfaction of some criteria such as needs of food security, livelihood security and rural development.

Secondly, under Hong Kong agreement on agriculture (AoA) the developing countries are also entitled to adopt *Special Safeguard Mechanism* (SSM) based on import quantities and price triggers. This means under special safeguard mechanism, developing countries can raise their tariff on imports of agricultural products in the event of a sudden increase in imports or fall in prices of imports. However, they are not allowed to reintroduce quantitative restrictions (QRs) which both the developed and developing countries have already eliminated

Domestic Support for Agriculture. In Hong Kong developed countries agreed to the proposal of G-20 (*i.e.*, the group of developing countries including India and Brazil) for reduction of trade-distorting agricultural subsides with higher linear cuts in higher bands. However, the extent of reduction in these subsidies were left for future negotiations. It may be noted that Blue Box and Green Box agricultural subsidies have been excluded from the scope of reduction. But there is a provision for review of the criteria of the Green Box subsidies in which most of the EU domestic subsidies are included.

In Doha Round on development it was agreed to reduce the minimum level of support (de minimis) allowed to be exempted from reduction for both developed and developing countries. Developing countries are entitled to a de minimis of 10 per cent of the value of total production. In Hong Kong it was agreed that those developing countries whose domestic support to agriculture was below 10 per cent of the value of production and therefore no commitment to reduce them will be exempt from reductions in de minimis as well as overall trade-distorting agricultural subsidies.

Under the Hong Kong agreement export subsidies on cotton will be eliminated by developed countries in 2006 and trade distorting domestic subsidies on agricultural products will be reduced drastically over a storter period of time.

Non Agricultural Market Access (NAMA). The Hong Kong declaration provides for the reduction or elimination of tariffs including peak levels of tariffs, high levels of tarifs and tariff escalation on non-agricultural goods, particularly on the products of exports which are of great interest for the developing countries. In this context of market access for non-agricultural products it has been agreed to apply *Swis formula.* According to this formula, deeper cuts in tariff rates on a line-by-line basis have to be made by countries levying higher tariffs. How larger will be the cut in tariff levels will depend on the coefficients to be agreed.

However, in Hong Kong declaration the provision for *special and differential treatment* (S&DT) and *less than full reciprocity* for developing countries have been included. According to this provision, a certain proportion of tariff lines of the developing countries may be subjected to tariff cuts lower than that permitted by applying Swis formula and a certain percentage of the tariff lines of these countries may be kept unbound (that is, without any binding for tariff reduction). Thus flexibilities granted under S&DT include the longer implementation period as well as applying less than Swis formula cuts or no cuts for a specified list of tariff lines as unbound. These have been declared as an integral part of the modalities for negotiations under NAMA. On the sectoral initiatives, India and most of the developing countries has emphasized that formula approach should be the main modality for negotiations and sectoral initiative can only be a voluntary exercise.

Services. In the Hong Kong declaration it has been decided to achieve a higher level of liberalisation of trade in services and towards that negotiations will be intensified. New date lines were fixed for submission of revised offer and draft schedules of commitments by the member countries. These negotiations on trade in services will be conducted in accordance with principle of GATS and the Guidelines and Procedures for the Negotiations on Trade in Services. According to GATS and these guidelines, the developing countries are required to liberalise the service sectors and modes of supply of their own choice and determine the extent of liberalisation.

India is particularly concerned with liberalisation of trade in services as it wants to increase its access to the markets of developed countries for the services such as ITES in which it has a comparative advantage. In services, India submitted its Revised offer in August 2005. Eleven service sectors and 94 sub-sectors are covered in the Revised offer.

It is thus clear that with regard the services much depends on revised services offer by the member countries and the future negotiations.

Market Access for LDCs. An important feature of Hong Kong Declaration is that it was decided to provide *duty-free* and *quota free* market access for all LDCs (Least Developed Countries) products to *all developed countries.* The developing countries which are in a position to provide such access to LDCs are expected to do so.

Agreement on TRIPS. On December 6, 2005, the General Council of WTO adopted the amendment to the TRIPS to address public health concerns of developing countries which was reaffirmed by Hong Kong Declaration. This amendment enables manufacturing and export of pharmaceutical products under compulsory license to countries with limited or no manufacturing capacities in the pharmaceutical sectors. On Trips, CBD (Convention of Bio-Diversity) on relationship and protection of traditional knowledge, India along with a number of other developing countries which are rich in bio-diversity proposed that the Trips Agreement of WTO should be amended to provide for : (*i*) disclosure of source of the traditional knowledge used in the invention; (*ii*) disclosure of evidence of prior informed consent under the relevant national regime; and (*iii*) disclosure of evidence of benefit sharing under the relevant national regime. With this amendment the use of India's natural herbs such as Neem, Arjuna, Ashavgandha and others for preparing medicines and also certain plant varieties such as Basmati rice will get protection from the use by multinational companies.

In Hong Kong India made some modest gains in the field of agriculture, industrial products and TRIPS. A good number of issues concerning liberalisation of global trade are yet to be resolved. However, much depends on future negotiations during which several issues particularly regarding market access for industrial goods and services are to be resolved. Whether in these future negotiations, developed countries will adopt an attitude which ensure fair and free trade from the viewpoint of LDCs and developing countries. Already the developed countries propose to link lower farm subsidies by them to greater market access for non-agricultural goods in the developing countries. This is against the letter and spirit of Hong Kong Declaration. And it is these kinds of tactics of the developed countries that Juseph Stiglitz predicted and thought that due to this Doha Round of Development will collapse.

Trade Facilitation Agreement at Bali

Twenty years have passed when WTO was formed. The importance of WTO has been the establishment of *orderly multilateral trading system*. In 2001 Doha Round Development negotiations were started to ensure trade liberalistion and economic development especially of the developing countries. Under Doha Round, developed countries have to reduce their subsidies on their agricultural products, the developing countries have to cut import duties on agricultural products. It may be noted the subsidies on their form products by the developed countries which leads to their greater imports into the developing countries and thus harm their poor farmers.

In 2013, at Bali amid stalemates on major issues trade facilitation agreement (TFA) was reached under which the member countries of WTO agreed for easier clearances of goods at ports and airports. India blocked this trade facilitation agreement but later agreed to it after the *peace clause* tenure was made permanent as demanded by India. According to this peace clause no dispute complaint will be entertained by WTO even if there is a breach of the limit on procurements of foodgrains for public distribution system (PDS).

However, trade facilitation agreement at Bali was the first multilateral trade agreement since WTO was formed. A sense of the importance of trade facilitation agreement (TFA) is evident from WTO claim that is can lead to a cut in trade cost by about 14 per cent.

Having secured indefinite protection for tis PDS programme India is been to negotiate the other agenda of Doha Round. But their is a good deal of concern in India to find that developed countries, especially US and BU are busy making regional trade agreements and are no longer interested in pursing further multilateral trade agenda of Doha Round. This poses a challenge to India in the economy ministerial conference to be held at Nairobi at 15.17, December 2015.

India wants WTO members to allow developing countries to use special safeguards measures (SM) to raise traffic in case surge in imports in developing countries. Besides, India wants higher import duties on dairy and poultry products and apples as it fears a surge in their imports into India which will harm that poor Indian farmers. Besides, at Nairobi India is expected make a strong case for permanent solution of its food security problem. In this regard it will try that the formula of procurement of foodgrains be reworked to permit a larger level of permissible domestic procurement as the current one may restrict its ability to procure sufficient foodgrains from the farmers to ensure food security to poor people. The present commerce minister, Nirmla Sitaraman has said that The government is committed to the cause of the farmers and has taken a consistent stand at WTO to protect their interests.

QUESTIONS FOR REVIEW

1. Why has WTO been established? Explain its role in promoting world trade.
2. Explain how India has gained from becoming a member of WTO.
3. How agreements reached at WTO will affect (a) Indian agriculture and farmers (b) Indian industry.
4. Discuss the objects and functions of WTO (World Trade Organisation).
5. What are the main features of WTO? What are its functions?
6. What is meant by globalisation? How does WTO promote globalisation?
7. What is liberalisation of trade? What steps have been taken by WTO to promote trade liberalisation?
8. What is multilateral trading system? How is it better than bilateral trade agreement?
9. Explain briefly why ministerial level talks under the aegis of WTO at Cancun failed.
10. What is 'Doha Round of Development'? What progress has been made in this regard?
11. Explain the main features of WTO's Hong Kong Declaration. Critically examine how India and other developing countries will gain from it.
12. Explain the TRIPS agreement. What progress has been made in implementing this agreement?

PART XIV

Economics of Growth and Development

- Economic Growth and Development
- Underdevelopment and Common Characteristics of Developing Countries
- Obstacles to Economic Development
- Capital Formation: Determinants and Acceleration
- Technological Progress and Economic Development
- Human Resource Development: Education and Health
- Population Growth and Economic Development
- Harrod-Domar Model of Growth
- Neoclassical Theory of Growth
- New Theory of Economic Growth (Endogenous Growth Model)
- Mahalanobis Growth Model and Heavy-industry Strategy of Development
- Wage-goods Model and Strategy of Development
- Strategies of Balanced and Unbalanced Growth
- Lewis Model of Development with Unlimited Supply of Labour
- Choice of Techniques in Developing Countries
- Employment and Economic Growth in Developing Countries
- Poverty and Economic Growth
- Income Inequalities and Economic Growth
- Economic Reforms, Structural Adjustment and Growth
- Environment and Economic Development
- Sustainable Development and Climate Change

CHAPTER 38

ECONOMIC GROWTH AND DEVELOPMENT

Before explaining the factors which determine economic growth and development and the common characteristics of the developing countries, it is useful to describe how the terms economic growth and development are interpreted by the economists. In the fifties and sixties when the economics of growth and development were evolved, economic growth and development were considered to have the same meaning and therefore these two terms were used interchangeably. Today, however, an important distinction is drawn between economic growth and economic development. In what follows we first explain the meaning of these terms and the distinction between them.

Meaning of Economic Growth

Economic growth has been defined in two ways. In the first place, *economic growth is defined as a sustained annual increase in an economy's real national income over a long period of time.* In other words, economic growth means *rising trend of net national product at constant prices*. This definition has been criticised by some economists as inadequate and unsatisfactory. They argue that total national income may be increasing and yet the standard of living of the people may be falling. This can happen when the population is increasing at a faster rate than total national income. For instance, if national income is rising by 1% per year and population is increasing at 2% per year, the standard of living of the people will tend to fall. This is so because when population is increasing more rapidly than national income, per capita income will go on falling. Per capita income will rise when the national income increases faster than population. Therefore, the second and better way of defining economic growth is to do so in terms of per capita income. According to the second view, *"economic growth means the annual increase in real per capita income of a country over a long period.*" Thus, Professor Arthur Lewis says that "*economic growth means the growth of output per head of population*". Since the main aim of economic growth is to raise the standards of living of the people, therefore the second way of defining economic growth which runs in terms of per capita income or output is better.

Another point which is worth mentioning in regard to the definition of economic growth is that the increase in national income or more correctly increase in per capita income or output, must be a 'sustained increase' if it is to be called economic growth. By sustained increase in per capita income we mean the upward or rising trend in per capita income over a long period of time. A mere short-period rise in per capita income, such as that occurs over a business cycle, cannot be validly called economic growth.

Now, almost universally, rates of economic growth are measured both in terms of increase in overall Gross National Product (GNP) or Net National Product (NNP) and increase in per capita income. While Gross National Product (GNP) measures the total output of goods and services which an economy is capable of producing, per capita income measures how much of real goods and services which an average person of the community will have for consumption and investment, that is, average level of living of a citizen of a country.

Thus, world organisations such as World Bank and IMF have been employing both these measures of economic growth in their annual World Development Reports for comparing growth and levels of living of the developed and the developing countries. In India also our Planning Commission, Central Statistical Organisation (CSO), and Reserve Bank of India have been measuring economic growth on the basis of both overall GNP or NNP and per capita income. The recent estimates of annual growth in per capita income and GNP are given in Table 38.1. This table reveals an interesting feature *that economic growth achieved in recent years is higher in the developing countries than in the developed countries*. However, it should be noted that in the past several decades the present-day developed countries recorded much higher growth rates than the developing countries which remained static for a long period. As a result, per capita income and levels of living of the people of the developed countries are now much higher as compared to those of the developing countries. The problem of the developing countries is to catch up with the developed countries through attaining rapid economic growth so as to enjoy higher levels of living.

For the sake of comparison we have given in Table 38.1 the data of population, Gross National Income (GNI) per capita and growth rate of Gross Domestic Product (GDP) for three periods 1980-90, 1990-2000 and 2000-2012 of some developed and developing countries. It will be seen from this Table 38.1 that in the last two decades India has become the second fastest growing economy of the world.

Table 38.1. GNP Per Capita, Growth Rate and Population in Some Developed and Developing Countries

Country	Population	GNI per capita in 2011 (PPP $)	Average Annual Growth Rate of GDP (%)		
Developed Countries					
	(million) 2012	2014	1980-90	1990-2000	2000-2012
USA	314	52,947	3.0	3.6	1.7
Australia	23	42,261	3.5	3.9	3.1
UK	63	39,767	3.2	2.6	1.6
Canada	35		3.4	3.1	1.9
Japan	128	–	4.1	1.0	0.7
Developing Countries					
India	1,237	5,497	5.8	6.0	7.8
Brazil	199	15,175	–	2.7	3.6
China	1,351	12,541	10.2	10.6	10.6
Pakistan	179	4,866	6.3	3.8	4.4
Bangladesh	155	3,191	4.3	4.8	6.0
Sri Lanka	20	8,779	4.2	5.3	5.5

Source: World Bank, *Human Development Report,* 2015 and World Development Indicators, 2014.

Meaning of Economic Development: Traditional View

As mentioned above, no distinction was drawn between economic growth and development in the beginning of the evolution of economics of development. However, since the seventies it has been thought necessary to distinguish between economic growth and economic development. There are two views even about the concept of economic development. The traditional view has been to interpret it in terms of planned changes in the structure of national product and the occupational pattern of labour force and also the institutional and technological changes that bring about such changes or accompany such changes. It may be noted that Kuznets in his study of *Modern Economic Growth* interpreted the process of modern economic growth which involves these structural changes[1]. In this view during the process of economic growth share of agriculture in both national product and employment of labour force declines and that of industries and services increases. Various strategies of development which were suggested until 'seventies' generally focussed on rapid industrialisation so that structural transformation could be achieved. For this purpose appropriate institutional and technological changes were recommended to bring about such structural changes. Thus C.P. Kindleberger writes, "*Economic growth means more output and economic development*

1 Simon Kuznets, *Modern Economic Growth.*

implies both more output and changes in the technical and institutional arrangements by which it is produced."[2]

Thus, according to traditional view, economic development implies growth plus structural change. Structural change refers to changes in technological and institutional factors which cause shift of labour from agriculture to modern manufacturing and services sectors and also generate self-sustaining growth of output. An aspect of structural change which is of special mention is that during the process of economic development there occurs a shift of working population from low productivity employment in agriculture to the modern industrial and services sectors having higher levels of productivity of labour. That is, during the process of economic development percentage share of working population in agriculture sharply falls whereas percentage shares of working population employed in modern industrial and services sectors substantially increase. Along with this change in sectoral distribution of labour force there occurs a change in sectoral composition of national income in which percentage contribution of agriculture to national income declines and percentage contributions to national income of industrial and services sectors increase. This occurs due to the change in pattern of consumption of the people as economy grows and people's income increases as well as due to the changes in levels of productivity in the different sectors of the economy.

It is worth mentioning that in this view casual references were made to the role of some social factors such as growth of literacy, education and good health in economic development but they were considered to be of secondary importance. On the whole, in this view of economic development which generally prevailed till seventies, development was considered to be an economic phenomenon in which benefits from growth in overall GNP or per capita GNP and the structural changes accompanying it would trickle down to the poor and unemployed. No separate or special attention was paid to eliminate mass poverty and unemployment and to reduce inequalities in income distribution.[3]

Concept of Economic Development: Modern View

The experience of the developing countries during the sixties and seventies showed that whereas target rates of economic growth were in fact achieved, trickle-down effect in the form of creation of more employment opportunities, rise in wages and improvement in income distribution did not operate. The problems of poverty, unemployment and income inequality further worsened instead of getting reduced during the process of growth in the nineteen fifties and sixties in the developing countries. For instance, in India, Dandekar and Rath found that 40 per cent of rural population in India lived below the poverty line in 1968-69. Using somewhat different approach, B.S. Minhas estimated that 37 per cent of rural population in India lived below the poverty line in 1967-68. Similarly, the magnitude of poverty and unemployment and the extent of income inequalities also increased in many other developing countries.[4]

Thus, due to the failure of traditional strategies of development in solving the problems of poverty, unemployment and inequality, it was realised in the seventies that the concept of development should be broadened so that it should signify that well-being of the people has increased. This led to the view that economic development should not be judged on the basis of growth in GNP alone. Therefore, when we regard the well-being of the masses as the ultimate objective of development, we have to see whether poverty and unemployment are decreasing and how the increases in gross national product or national income are being distributed among the population. Economic development will take place in true terms only if the poor people are raised above the poverty line. Late Prof. Sukhamoy Chakravarty rightly writes, "*The rate of growth strategy is by itself an inadequate device to deal with the problems of generating employment opportunities and for reducing economic*

2 C.P. Kindleberger, *Economic Development,* McGraw-Hill, Second Edition, 1965, p. 3.

3 Dandekar and Rath, *Poverty in India*, 1971.

4 B.S. Minhas, "Poverty, Land Redistribution and Development Strategy", *Indian Economic Review*, 1971.

disparities. Much depends on the composition of the growth process and how growth is financed and how benefits from growth process are distributed."[5]

It is worth mentioning that there is no guarantee that when there is increase in GNP, employment will also increase. It can happen that with the use of more capital-intensive technique while production may be increasing at a rapid rate, employment may be falling instead of rising. According to the modern perception of economic development, rapid increase in GNP secured through displacing labour by machines and thus causing rise in unemployment and underemployment cannot be called true economic development. Professor Dudley Seers makes the meaning of economic development according to the new perception in the following words:

"*The questions to ask about a country's development are therefore: What has been happening to poverty? What has been happening to unemployment? What has been happening to inequality? If all three of these have declined from high levels, then beyond doubt this has been a period of development for the country concerned. If one or two of these central problems have been growing worse, especially if all three have, it would be strange to call the result development even if per capita income doubled.*"[6]

Recently, the concept of economic development has been further widened so that it now involves not only reduction in poverty, inequality and unemployment but also requires improvement in *quality of life which includes cleaner environment, better education, good health and nutrition*. Thus World Development Report 1991, published by the World Bank, asserts: "*The challenge of development is to improve the quality of life, especially in the world's poor countries, a better quality of life generally calls for higher incomes but it involves much more. It encompasses as ends in themselves better education, higher standards of health and nutrition, less poverty, a cleaner environment, more equality of opportunity.*" [7] Thus the concept of economic development has been greatly broadened. Today economic development is interpreted as not only in more growth in Gross Domestic Product (GDP) but also in terms of good quality of life which, according to Prof. Amartya Sen, '*consists in enlargement of opportunities for people and freedom of human choices*'. This new concept of development includes achievement of freedom from servitude to ignorance and illiteracy. It also includes enjoyment of human rights. Thus United Nations '*Human Development Report*' of 1994 in the writing of which Prof. Amartya Sen made a significant contribution, asserts, "Human beings are born with certain potential capabilities. *The purpose of development* is *to create an environment in which all people can expand their capabilities, and opportunities can be enlarged for both present and future generations*..... .Wealth is important for human life. but to concentrate on it exclusively is wrong for two reasons. First, accumulating wealth is not necessary for the fulfilment of some important human choices .... Second, human choices extend far beyond economic well-being." In his recent book, Amartya Sen writes, "Economic growth cannot be sensibly treated as an end in itself. Development has to be more concerned with enhancing the *lives we lead and the freedoms we enjoy*."[8]

On the basis of various ingredients of good quality of life and other criteria such as enlargement of human choices and freedom a human development index is prepared by United Nations Development Programme (UNDP). This human development index is considered as a better indicator of economic development index.

---

5 Sukhamoy Chakravarty, "Some Observations on Growth Process," *Seminar*, 1971.

6 Dudley Seers, "The Meaning of Development", a paper presented at the Eleventh World Conference of the Society for International Development (New Delhi), 1969.

7 World Bank, *World Development Report*, 1991.

8 Amartya Sen, *Development as Freedom,* Oxford University Press, 2001, p.175.

SEN'S 'CAPABILITIES' APPROACH TO DEVELOPMENT

Meaning of Development

Prof. Amartya Sen has put forward a new view of development based on his concepts of freedom and capabilities. According to Prof. Sen, "*Development can be seen as a process of expanding the real freedom that people enjoy.*"[9] Development consists of the *removal of various types of unfreedoms* that leave people with little choice and little opportunity of exercising their reasoned faculty. He further elaborates: "*Development requires the removal of major sources of unfreedoms*: *poverty as well as tyranny, poor economic opportunities as well as systematic social deprivation, neglect of public facilities as well as intolerance or overactivity of repressive states*". This Sen's view of development is in sharp contrast to traditional view of development in terms of growth of GNP or rise in personal incomes or industrialisation or technological advancement are important *as means* to expanding freedoms enjoyed by the people but they are not sufficient to remove unfreedoms suffered by a vast number of people in a society. In Sen's conception freedom has a very wide connotation. Thus, according to Sen, development means freedom from hunger and undernourishment, freedom from illiteracy, freedom from illness, freedom from unemployment, freedom from social deprivation, freedom from political intolerance and so forth. In his view, freedom also depends on factors such as social and economic arrangements (for example, facilities for education and health) as well as civil and political rights.[10]

It is worth noting that in Sen's approach expansion of freedom is viewed as both a primary as well as the principal objective of development. To quote him again, "Viewing development in terms of expanding substantive freedom *directs attention to the ends* that make development important rather than merely some of the means that, inter alia, play a prominent part in the process."[11]

Despite substantial achievement in terms of GDP and industrialisation, according to Sen, the contemporary society denies elementary freedoms to a vast number of people. Thus, despite social growth in GDP in several countries of the world (including India) there still prevails *abysmal poverty which deprives people of freedom to satisfy hunger, to get sufficient nutrition or get treatment for various diseases they suffer from and the opportunity to get adequate clothing, clean water and sanitation facilities*. He also mentions the lack of freedom or what he calls unfreedoms of the people relating to lack of public facilities for healthcare and educational facilities and above all lack of effective public institutions for maintenance of local law and order. Though certain political and social freedoms such as freedom of political participation and dissent or opportunities to get basic education and healthcare are 'conducive to development' but their significance lies in the fact that they are 'constituents of development'.

Some critics, especially Prof. Jagdish Bhagwati, have alleged that Sen is against market mechanism. This is however not correct. Like Adam Smith, Sen views market mechanism as '*freedom of exchange and transactions*'. He acknowledges that market mechanism is significant as a means of contributing to high economic growth and overall economic progress but emphasises the role of market *as a part of the process of development* in the sense of "freedom of exchange and transactions" of the people, not an end itself.

Freedoms and Capabilities

Sen's view of development as freedom forms the basis of his capabilities approach to development. *Capabilities refer to what a person can or cannot do, can or cannot be*. The relevant capabilities which people can have are of various types. Being free from starvation, from hunger, from undernourishment,

---

9 *Ibid.*

10 *Ibid.*

11 Amartya Sen, *op.cit.*, p. 3.

freedom to participate in communal life, being adequately clothed and sheltered, being free to travel etc. In this context, it is worthwhile to explain *Sen's concept of functionings* which refers to what a person does or can do with the commodities of given characteristics that he comes to possess or control. Explaining the meaning of functionalities Sen writes, "The concept of functionings reflects the various things a person may value doing or being. The value functions may vary from elementary ones, such as being adequately nourished and being free from avoidable diseases to very complex activities or personal states, such as being able to take part in the life of the community and having self-respect."[12]

Sen draws an important distinction between *realised functionings* and *capability* set. While the *realised* functionings refer to what a person is actually able to do, the capabilities is a set of alternatives open to him, that is, has freedom to choose from the opportunities available. Thus while realised functionings inform us about the things a person possesses, the "capabilities" tell us about the things a person is substantially free to do. Capability is thus a kind of freedom to achieve alternative combinations of functionings. These capabilities can be enhanced by public policy.

It follows from above that free from shackles of unfreedoms and the expansion of capabilities of persons are the central aspects of approach to development. It is because of Sen's emphasis on freedom and capabilities that in development economics today economists are laying stress on education and health, food security, nutrition, inclusive growth and empowerment of women. According to Sen, poverty must be seen as deprivation of basic capabilities rather than merely as *lowness of incomes* which is the standard criterion of identification of poverty.[13] He recognizes the major causes of poverty, but there are other factors that influence deprivation of capability that leads to poverty. Therefore, his view of development as freedom, and poverty as capability deprivation that he emphasizes that the objective of development must not be just maximisation of income or wealth as they are only one factor in the development process. Thus, according to him, "for a fuller understanding of the development process, it is simply not adequate to take as our basic objective just the maximisation of income or wealth . . . . economic growth cannot be treated as an end in itself. Development has to be more concerned with enhancing the lives we lead and the freedoms we enjoy. Expanding the freedoms that we reason to value not only make our lives richer and more unfettered but also allows us to be fuller social persons, exercising our own volitions and interacting with and influencing–the world in which we live".[14]

Prof. Amartya Sen on Growth, Poverty and Distribution in India

It will be noted from the above that Prof. Amartya Sen has been an advocate of the approach that considers building of capabilities of the people through more education and healthcare of the poor. Besides, he advocates for enlargement of job opportunities for the poor and freedom of human choice as a means of eliminating poverty and promoting well-being of the people. Unlike Jagdish Bhagwati, his chief critic, he does not consider growth first and then redistribution to share the problems of poverty and unemployment. In his latest book coauthered with Jean Dieze '*An Uncertain Glory: India and its Contradictions*, he points out that rapid growth in India has co-existed with grossly inadequate social services (education and health) and physical services (safe water, electricity, roads, drainage and sanitation).

Despite significant economic growth achieved in the last two decades (1991-2012), according to him, there still prevails widespread poverty in India. The poor and underprivileged people suffer from undernourishment, poor health, illiteracy, lack of education and skills to get gainful employment. Freedom from hunger and undernourishment, freedom from illness and freedom from illiteracy are

12 Amartya Sen, *op.cit.*, p. 75.

13 *op.cit.*, p. 87.

14 *op.cit.*, pp. 14-15.

essential requirements of economic development. In his new book he argues for integrating growth and development. He thus writes, "While growth is an important tool for enhancing living conditions, its *reach and impact depend greatly on what is being done with the fruits of growth*. The relation between economic growth and advancement of living standards depends on money factors, including economic and social inequality in general, and *no less importantly on what* the government does with the public revenue that is generated by economic growth. The importance of economic growth can be adequately understood only in this broader context"[15]. He thus writes, "It is necessary to recognise the role of growth in facilitating development in the form of enhancing human lives and freedoms, but it is also necessary in this context to appreciate how the possibilities of a country depends in turn on the advancement of human capabilities (through education, healthcare and other facilities) in which the State can play a useful role."

Thus, it is the building of capabilities of the poor and gain freedom from illness and illiteracy that Sen views as development. To quote him again, "*Economic growth cannot be sensibly treated as an end in itself. Development has to be more concerned with enhancing the lives we lead and freedoms*, we enjoy. Furthermore, the challenge of development is to improve the quality of life, especially in developing countries such as ours. A better quality of life generally calls for high standards of health and nutrition, less poverty, more equality of opportunity."[16] In his latest book, *An Uncertain Glory: India and its Contradictions*, he cites successful countries[17] like Korea and China to argue that fast economic growth is not sustainable without strong development of human capital and essential infrastructure. India lags behind many countries at a similar level of development and even behind poor Bangladesh. In an interview with *Times of India* at the time of publication of its above mentioned book in July 2013 he says, "Despite great success in many ways over the last two decades the fact is that India is in a dreadful state in many ways. *Inequality is the primary aspect of it.* About economic growth I think there are two things about it. First, the way the Indian dialogue has developed there seems to be basic involvement about how growth happens. *No country in the world with such a fast sustained rate of economic growth has done so little towards having an educated and healthy labour force.*" The way the debate has come it is now just '*we now have growth so we have to make it inclusive'*. Now, according to him, these are not two distinct things. If growth happens it depends on how it is happening. *If it happens because you have an educated and healthy labour force then a lot of people are better placed to earn more income*".

It is evident from the above that, according to Sen, the State should begin with investment in education and healthcare of the people right away without waiting for first to become rich and then think of redistribution for achieving inclusive growth. He gives the example of Singapore, Taiwan, Korea, China which began with investment in education and healthcare of the people without waiting for first becoming rich. In their case, according to him, growth happened along with the poor benefiting from it by making earnings from the education and health acquired by them in the process of growth and in this way become a productive part of the society. He further emphasises, as mentioned above, that growth will generate more resources and if they are ploughed back for further investment in building human capabilities, it will still generate rapid economic growth. He bemoans the fact that even after 60 years of development India has not achieved universal education. In response

15 Jean Dreze and Amartya Sen, *An Uncertain Glory: India and Its Contradictions*, Penguin Books, 1914, p. 38.

16 Amartya Sen, *Development as Freedom*, Oxford University Press, 2001, p. 14.

17 *Ibid*, p. 39.

to Bhagwati's criticism he says that he has not been against liberalisation and globalisation and did not support the licence, permit system. But he emphasises that while economic reforms initiated by Dr. Manmohan Singh in 1991 *attempted to correct the overactivity of government in some fields* (*i.e.*, removing licence-permit system by undertaking liberalisation and privatization), *the need to correct government underactivity in other areas has not really been addressed.* As back as in 1998, he wrote, "the success of liberalisation and closer integration with the world economy may be severely impaired by India's backwardness in basic education, elementary healthcare, gender inequality and other limitations".[18]

It is because of absence of the constructive activities of the State, namely, expansion in government spending on healthcare and nutrition of the people that he has criticised "Gujarat Model of Development".

Sen and his co-author Jean Dreze have supported the starting of *MGNREGA Scheme* because it helps the poor to obtain gainful employment giving them income which they can use to end their undernourishment which contribute to the improvement of their health. Similarly, they have supported the passage of Food Security Bill which would ensure the poor people adequate food to remove their hunger and undernourishment and therefore contribute to their capability to participate in the process of development.

Thus, Sen believes that the *State must directly deliver food, employment, education and health-care rather than relying on trickle-down effect of economic growth achieved through the dominant role of private sector in a predominantly free-market economy*. Besides, Sen is of the view that rapid economic growth achieved through liberalisation, privatisation and globalisation of the Indian economy has resulted in glaring income inequalities which besides being non-functional lead to conspicuous consumption of the few rich people. This has adversely affected resources available for growth and social spending by the government on food security, education, employment and healthcare of the poor people. Furthermore, according to him, rapid growth would have been brought about by involving in India's most abundant resource, namely, investing in education and health of its people and thereby building up their capabilities for productive work, the present glaring income inequalities would not have been there and inclusive growth in real terms would have been achieved.

Three Core Values of Development

We have explained above that, in the new view, development is used to describe the process of economic and social change in a society whereby there is not only growth in GDP but also there is wider distribution of benefits or gains of growth of GDP and expansion in employment opportunities for the people. As a result both poverty and unemployment fall. Prof. Amartya Sen, as seen above, further widened the meaning of development by viewing it as enhancing of freedoms, entitlements and capabilities of the people which are not necessarily generally accompanied with growth in GDP. Related to Sen's view of development is Goulet's[19] view of *core values* or *basic components* of development which reflect the wider meaning of real development and depends on the achievement of major economic and social objectives. These three core values are : (1) Life-sustenance; (2) Self-esteem; and (3) Freedom. We explain below these core values which constitute the three components of development.

---

18 Amartya Sen, "Theory and Practice of Development" in Isher Judge Ahluwalia and J.M.D. Little (eds), *India's Economic Reforms and Development*, Essays in Honour of Dr. Manmohan Singh, Oxford University Press, 1998, p. 82.

19 D. Goulet. *The Cruel Choice: A New Concept in the Theory of Development* (New York; Atheneum), 1971 and *Development Ethics at Work* (London: Routledge), 2006.

1. **Life-sustenance:** All human beings have certain *basic needs* which must be met if they have to live. These basic needs include food, clothing, shelter, and healthcare. If any of these essential things is not available or in critically short supply, it will not be possible for people to sustain their lives. A basic function of an economy is to provide people with food, clothing, shelter and healthcare to save them from miserable life. No doubt, the fulfilment of these basic needs of human beings requires economic growth with right type of product composition and equitable distribution of basic goods being produced. This essential objective of development will be realised if absolute poverty is eliminated, enough employment opportunities are created and present income inequalities are reduced.
2. **Self-esteem:** A second component of good life for the people is *to live with respect, honour or dignity*. All people and societies want to live with self-respect and dignity because only then they would have a sense of their own worth. The nature and form of self-esteem may vary from society to society and from culture to culture. A country will have true development if it gains self-esteem, being free from dominance by others. This is not only true of individuals but also of the society as a whole. As Goulet put it, "*Development is legitimized as a goal because it is an important, perhaps an indispensable, way of gaining esteem.*"[20] It may be noted for a nation as a whole to gain self-esteem and having a profound sense of its worth in the comity of nations it requires to grow economically and become prosperous. This is because, as Todaro and Smith write, "national prosperity has become an almost universal measure of wealth due to the significance attached to material values in developed nations, worthiness and esteem are nowadays increasingly conferred only on countries that possess economic wealth and technological power – those that have developed.[21]
3. **Freedom:** The third core value is the freedom from servitude. Freedom involves expanded range of choice for the individuals and society at large. As noted above, Amartya Sen views development as freedom. The term freedom is here used not only in the sense of political or social freedoms such as the freedom of political participation and dissent, the rule of law, freedom of expression and equality of opportunity but also the freedom from poverty, ignorance and squalor so that people can determine their own destiny.

For example, women in the Indian society do not enjoy freedom of choice as they are dominated by men. No person can be free if he has no freedom of choice, if he is imprisoned by living in abysmal poverty with no education and healthcare. There must be expansion in the range of human choice open to individuals and societies at large, if true development has to take place. Long back W. Arthur Lewis, a development economist, stressed the importance of freedom saying that "*the advantage of economic growth is not that wealth increases happiness but that it increases the range of human choice*."[22] The increase in wealth of individuals enhances freedom as it enables people to gain greater control over nature and physical environment through, for example, production of food, clothing, shelter, etc. and to provide education and healthcare. Such freedom cannot be enjoyed by the people if they remain poor. Using some of the above criteria UNDP in its annual Human Development Reports prepares human freedom index for the various countries covered by it. However, some countries which have achieved high rate of economic growth or high incomes, such as China, Singapore, Malaysia, Saudi Arabia, have lower human freedom indexes as they do not meet some human freedom criteria adopted by UNDP.

---

20 Denis Goulet, *The Cruel Choice, A New Concept in the Theory of Development* (New York: Atheneum, 1971, p. 90.

21. Tadaro and Smith, *Economic Development*, 9th Edition, p. 31.

22. W. Arthur Lewis, "Is Economic Growth Desirable" in *The Theory of Economic Growth* (London: Allen and Unwin, 1963, p. 420.

THE OBJECTIVES OF DEVELOPMENT

From our study of Sen's capability approach to development and Goulet's core values of development we are in a position to explain the objectives of development. We may mention here the basic objective of development is broadbased improvement in the economic and social conditions of our people so as to achieve better quality of life for them. We explain important objectives of development.

1. To Achieve a Higher Rate of GDP Growth. The first and foremost objective of development is to achieve a higher rate of GDP growth so as to raise the living standards of our people. Rapid growth of total GDP or per capita income is considered necessary because it ensures an expansion in the productive capacity of the economy without which broadbased improvement in living standards of the people is not possible. However, it should be recognised that faster economic growth, though necessary, is not a sufficient condition for raising the living standards of our teeming millions. This is because one can easily imagine a growth process which may not be sufficiently inclusive to ensure a spread of benefits to the mass of our population.

There are three reasons why GDP growth is necessary for raising the living standards of the population. First, rapid growth of GDP ensures a higher expansion in total income and production which, if growth process is sufficiently inclusive, will make available a larger output of goods and services to be consumed by the people and thus raise their living standards. Secondly, rapid economic growth generates more employment opportunities and income enhancing activities of the people, provided labour-saving technologies are not used for production of goods both in the industrial and agricultural sectors. Third, higher GDP growth is important because it generates higher revenues for government which helps to finance anti-poverty programmes started by the government.

2. To Eradicate Poverty and Unemployment. The second important objective of development is to eradicate poverty. As explained above in Amartya Sen's approach to development, poverty should be viewed as deprivation of basic capabilities rather than merely as low income. The existence of poverty or deprivation of basic capabilities is reflected in hunger, significant undernourishment, especially of children, premature mortality, permanent morbidity, widespread illness, and lack of basic education and other failures. Though economic growth is necessary for elimination of poverty but is not a sufficient condition for it because it is related to income distribution in a society as well.

Related to issue of poverty is the question of unemployment which exists on a large scale in developing countries, especially in labour-surplus countries such as ours. Chronic and long-term unemployment exists in the developing economies because due to higher population growth relative to capital formation it has not been possible to absorb the increasing number of workers in productive employment resulting in large-scale unemployment in developing countries. Gainful employment is the means of livelihood for the masses of the population. The growth of employment opportunities needs to be accelerated both in manufacturing and services sectors to provide employment to increasingly educated population which has high expectations and aspirations. Unemployment leads to the feelings of worthlessness and frustration among the youth leading to the increase in incidence of crime in the society.

3. To Expand the Freedoms Enjoyed by the People. The other important objective of development, as has been discussed in Amartya Sen's approach to development and Goulet's core values, is to expand the freedoms that people of a society enjoy. Growth of GDP or of individual incomes or industrialisation or technological progress are merely *means* to expanding human freedoms but freedoms depend on other factors as well. As Sen stresses viewing development as expansion of substantive freedoms directs attention to end or objective of development rather than means such as GDP growth or industrialisation which play an important role in the process of development.

If the objective of expansion of freedoms is to be achieved, the various sources of unfreedoms such as lack of public facilities of education and healthcare, denial of political liberty and basic civil rights (such as liberty to participate in public discussion) and denial of equal rights to women in the society have to be removed. It is worth mentioning here that some have supported the denial of political liberty and basic civil rights to the people on the ground that they promote economic growth. However, Amartya Sen has rightly pointed out that there is little evidence that authoritarian politics actually helps economic growth. Indeed, empirical evidence very strongly suggests that economic growth is more a matter of friendly economic climate than of a harsher political system.

HUMAN DEVELOPMENT INDEX (HDI)

Growth in national income (GNP) or per capita income considered above as indicator of economic growth does not represent a true and adequate measure of welfare and economic development of the people of a country. Therefore, a need was felt to put forward an index which should truly and correctly reflect the level of economic welfare and economic development of a nation. Thus, Human Development Report (1997) states, "Income is clearly only one option that people would like to have, though an important one. But it is not the sum total of their lives. Income is also a means, with human development, the end". *Human development index is a composite measure of three aspects of good life and well-being.* Three questions raised relate to what people want in order to have a good and cherished life. Answers provided are:

1. people want to lead a long and healthy life;
2. they want to be educated and to acquire knowledge; and
3. they want to have access to resources so that they enjoy a decent standard of living. In other words, they want to have a per capita income which is high enough to ensure them a decent standard of living.

The concept of human development has been developed by a number of economists prominent among whom are Dr. Mehbub-ul-Haq and Nobel Laureate Dr. Amartya Sen. Explaining the concept of *human development*, Prof. Amartya Sen writes, the process of *widening people's choices and the level of well-being they achieve are* the core of the notion of human development. But regardless of the level of development, the three essential choicess for people are to lead a long and healthy life, to acquire knowledge and to have access to the resources needed for a decent standard of living. Human development does not end there, however, other choices highly valued by many people range from *political, economic and social freedoms to opportunities for being creative and productive and enjoying self-respect and guaranteed human rights*."

It is clear from this that in the concept of human development Prof. Amartya Sen would like to include the social, economic and political freedoms as well as human rights they enjoy in a country along with the three indicators or criteria mentioned below. According to HDR 2015, India with HDI of 0.609 in 2014 had improved its ranking to a few number with its overall global ranking at 130 (out of 188 countries) as against 136 in 2011. India's HDI ranking is the lowest among the BRICS countries, with Russia at 50, Brazil at 75, South Africa at 116 and China at 90. Significantly, China improved its ranking by ten places between 2008 and 2014 mainly because of its higher growth in per capita income. Obviously, India has to improve its position regarding three indicators of good life to bridge the gap.

Three Indicators of Good Life. The three criteria or indicators which represent different aspects of good life or the three goals of human development are:

1. **Longevity**, which is measured by life expectancy at birth. Life expectancy at birth means how many years a newly born infant can hope to live in this world.
2. **Knowledge,** which is measured by the weighted average of adult literacy and mean years of schooling. For this 2/3rd weight is given to adult literacy and 1/3 weight is given to the mean years of schooling.
3. **Standard of living** which is measured by real per capita income of a country adjusted for purchasing power of currencies of different countries.

It is important to note that in 188 countries of the World, India's rank with regard to HDI was 130 in 2014. However India has improved its ranking from 135 in 2013. The existing gap in health and education indicators in India as compared to developed countries and also many of the developing countries highlights the need for much faster and wider spread of basic health and education. Life expectancy at birth in India was 68 years in 2014, compared to 81.8 years in Norway, 79.1 years in the United States, 74.5 years in Brazil, 74.9 years in Sri Lanka, 75.8 years in China, and the global average of 71 years. However the Indian life expectancy figure is significantly higher than that of South Africa (57.4 years) which has higher HDI rank and even higher per capita income within the same category. The Indian performance in mean years of schooling (5.4 years) is not only much below that of countries like China, Brazil, Sri Lanka, and Egypt which have higher per capita incomes but also below that of Bangladesh and Pakistan which have lower per capita incomes. It is also much lower than the global average of 7.7 years.

Table 38.2. Trends and India's Position in Global HDI, 2014

Country	HDI, 2014		GNI per capita in 2011 PPP$ 2014	Life Expectancy at birth (years)	Mean years of schooling 2013	Income inequality (Gini coefficient)	GII* 2013	
	Value	Rank					Value	Rank
Norway	0.944	1	64,992	81.8	12.6	25.8	0.068	9
Australia	0.935	2	42,261	82.4	13.0	–	–	–
Germany	0.918	6	43,919	80.9	13.1	28.3	0.046	3
United States	0.915	8	52,947	79.1	12.2	40.8	0.262	47
U.K	0.907	14	39,767	80.7	13.1	36.0	0.193	35
Russian Federation	0.798	50	22,362	70.1	12.0	40.1	0.314	32
Sri Lanka	0.757	73	8,779	74.9	10.8	36.4	0.383	75
Brazil	0.755	75	15,175	74.5	7.7	54.7	0.441	85
China	0.727	90	12,541	75.8	7.5	42.1	0.202	37
South Africa	0.686	116	12,122	57.4	9.9	63.1	0.461	94
India	**0.609**	**130**	**5,497**	**68.0**	**5.4**	**33.9**	**0.563**	**127**
Bangladesh	0.570	142	3,191	71.8	5.1	32.1	0.529	115
Pakistan	0.538	147	4,866	69.2	4.7	30.0	0.563	127

*GII is Gender Inequality Index. The lower the GII, the better it is.

Source: *Human Development Report*, 2015

Two things must be noted in regard to human development index (HDI). First, it measures relative and not absolute level of human development. Second, the focus of human development is on the goals or ends of development which are (1) long and healthy life, (2) the level of knowledge and (3) a decent standard of living as measured by real per capita income and not on real per capita income alone. In fact, human development index is a composite measure of well-being based on life expectancy, literacy level, and real per capita income. It is important to note that human development index attempts to make ranking of different countries on a scale of 0 (lowest human development) to 1 (the highest human development).

India's HDI value is also below the average of countries in both the *medium human development* group (0.614) and South Asia (0.588). Between 1980 and 2014, India's life expectancy at birth (LEB) increased by 11 years, mean years of schooling increased by 5.3 years while gross national income (GNI) per capita increased by about 306 per cent. As compared to BRICS nations and some neighbouring countries, India has the least mean years of schooling. China which recorded a slightly higher HDI than India in 1980 has widened the margin over India in 2014.

It follows from above that existing gap in health and education between India and developed countries and also many developing countries highlights the need for much faster and wide spread of basic health and education as has happened in China and Sri Lanka.

In terms of gender equality the HDR 2014 ranks India 127 out of 152 countries with a **gender inequality index** (GII) of 0.563 (see Table 38.2). GII for countries shows the extent to which gender inequality lowers national achievements in reproductive health, empowerment of women and labour force participation. The relatively lower Gender Inequality Index shows India in poor light on gender equality issues. Under the HDI, a higher GII value indicates poor performance.

Shortcomings and Critique of HDI

Some shortcomings of HDI as a true indicator of well-being for purposes of estimating development disparities among nations may be noted. First, for estimating literacy rate, expected years of schooling by children at the entrance age is used which overstates the literacy rate as in many countries (including India) many children who join primary school later drop out at some stage. Secondly, in preparing HDI equal weight of 1/3 is given to each of the three variables, namely, life expectancy, literacy rate and GNI per capita. This involves some value judgement and seems to be quite arbitrary. Besides, since these three components of HDI are measured in different units, to give equal weight to each component of HDI does not make much sense.

Thirdly, in constructing human development index, role of quality has been ignored. For example, there is a big difference between extra year of life for a healthy and well-educated person and extra one year life for a person who is bed-ridden and has limited capability to do work. Similarly, in constructing HDI only the *number of years of schooling* is taken into account while the quality of education also matters a lot for good living. Due to lack of adequate data about quality of health and education, it is not incorporated in the construction of HDI. But without the quality of health and education being considered, HDI does not represent the true index of human development.

But the important drawback of human development index (HDI) is that it is of composite character which makes it an imperfect indicator of development or well-being of the people. If these three components are highly correlated to each other, then a single one will serve the purpose of comparing the levels of development and well-being of the people of different countries. However, in our view it is not proper to put the three variables reflecting development into a single composite index. It is better and much more revealing if performance of different countries is compared in respect of *various indicators separately* because putting them together obscures the performance of development of various countries in respect of some crucial areas. Thus, in our view, *human development index hides more than it reveals*. Besides, in constructing human development index (HDI) there is a problem of what to be included and what to be excluded and what weights be assigned to the various variables included for construction of the index. In fact until recently human development index was prepared taking into account three variables, namely, life expectancy, literacy and per capita income and inequality in income distribution which also determines well-being of the people was ignored. Realising the importance of inequalities in income distribution, recent *Human Development Report*, 2013 has also given '*inequality-adjusted HDI*' in addition to the HDI without such adjustment.

Besides, human development index still does not take into account social, economic and political freedoms as well as human rights on which Amartya Sen laid a great stress in his book "*Development as Freedom*". Further, human development index as it is being presently constructed is not comprehensive as it does not include the two important indicators such as poverty and unemployment as reduction in them are important indicators of development. However, UNDP which constructs human development index separately calculates '*human poverty in`dex*' (HPI) which has now been replaced by '*multi-dimensioned poverty index*'. But the existence of unemployment which is an important aspect of human development still remains excluded. Thus in our view, the chief drawback of human development index (HDI) is that it obscures many dimensions of the concept of development. The concept of development is so much wider, deeper and richer that single composite measure like HDI cannot adequately measure it. Therefore, it is better to judge and assess the development performance of different countries by a number of indicators that reflect different aspects of development rather to judge it by a single composite index of HDI.

Factors Affecting Economic Growth

The process of economic growth is a highly complex phenomenon and is influenced by numerous and varied factors such as economic, political, social and cultural factors. It is believed by some economists that the capital is the main requirement for growth and therefore the greatest emphasis be laid on capital formation to bring about economic development. But this is wrong. As Professor Nurkse rightly remarks, "*Economic development has much to do with human endowments, social attitudes, political conditions and historical accidents. Capital is a necessary but not a sufficient condition of progress.*"[23]

Growth in gross domestic product (GDP) depends on the increase in the supply of various factors such as capital, labour, natural resources and the progress in technology. The economists often use the production function approach to explain the contributions of various factors that determine economic growth. The following aggregate production function is often used to explain the factors that contribute to growth, that is, increase in GDP.

$$Y = AF(L, K, N)$$

where Y is Gross domestic product

A is total factor productivity

L is the quantity of labour input

K is the size of capital stock

N is the quantity of natural resources

Now, in the above aggregate production function, the increase in A means more output can be produced with *the given amounts of labour, capital and natural resources* (such as land). Thus, A represents total factor productivity or technology, improvement in which affects the productivity of other factors of labour, capital and natural resources. The total factor productivity is often defined as the ratio of output produced to the quantity of inputs used in production of goods and services. In fact *technological progress in the economy is measured by the annual increase in total factor productivity.*

In the study of factors that determine economic growth natural resources (N) are taken as given and constant and they affect growth through *more efficient use* of them and improvement in technology. Besides, in developing countries where labour is abundant, the increase in their supply for bringing about economic growth poses no problem. It is the accumulation of capital along with the progress in technology is the crucial factor in determining economic growth. In the study of factors that affect economic growth, capital is now further differentiated as physical capital such as machines, tools and equipment, *human capital* which represents expansion and improvement in *education and health.* In fact, Prof. Amartya Sen, in his recent works, gives more importance to education and health as factors that determine economic growth.

We will study in detail in later chapters the contribution of capital formation, technological progress and education and health in determining economic growth. In popular Harrod Domar model of economic growth, economic growth rate depends on *rate of saving or investment* (*i.e.*, rate of capital formation) and *capital-output ratio*. The capital-output ratio or its inverse output-capital ratio depends on productivity of capital which depends to a great extent on technology used in the production process. Harrod-Domar growth equation can be written as

$$G = \frac{s}{v}$$

where G is growth rate of GDP, s is rate of saving which determines rate of investment and v is capital-output ratio which depends on productivity of capital. In a later chapter concerning capital

23. Ragnar Nurkse, *Problems of Capital Formation in Underdeveloped Countries*, Basil Blackwell, 1957.

formation we will discuss the factors which determine rate of saving and investment. We explain below capital-output ratio and the factors which determine it.

CAPITAL-OUTPUT RATIO

Capital-output ratio is the amount of capital required to produce output worth ₹ 1. If Y stands for output or income and K for the stock of capital used to produce that output, then $\frac{K}{Y}$ represents capital-output ratio. It is useful to distinguish between incremental or marginal capital-output ratio and average capital-output ratio. Whereas average capital-output ratio describes the ratio of total capital to total output or income of the economy, incremental capital-output ratIo is ratio of *increment* in the stock of capital to the *increment* in output. Therefore, marginal capital-output ratio (*ICOR*) can be written as $\frac{\Delta K}{\Delta Y}$ where ΔK is the increment in capital and ΔY is the resultant increase in income or output.

The rate of economic growth of a country depends upon the rate of capital formation and capital-output ratio. Capital-output ratio determines the rate at which output grows as a result of a given volume of capital investment. For example, a capital-output ratio of 4 would mean, in Indian rupees, that a capital investment of ₹ 4 results in the addition of output worth ₹ 1. Hence, in order to produce a given level of output, smaller capital investment would be needed if the capital-output ratio is lower than when it is higher.

Factors Determining Capital-output Ratio

It is difficult to estimate the capital-output ratio of an economy. The productivity of capital depends upon many factors such as the degree of technological improvement associated with capital investment, the efficiency of handling new types of equipment, the quality of managerial and organisational skill, the existence and the extent of the utilisation of economic overheads and the pattern and rate of investment. For instance, the higher the proportion of investment devoted to the production of light consumer goods, the lower the capital-output ratio; and higher the proportion of investment devoted to public utilities, *i.e.*, economic and social overheads, the higher will be the capital-output ratio, and *vice versa*. Likewise, the greater the investment devoted to basic heavy industries, the higher will be the capital-output ratio, and *vice versa*. The superior organising ability and the use of better technology will lower the capital-output ratio. The capital-output ratio also varies with the prices of inputs. .

It is agreed that capital-output ratio in developing countries is generally higher, *i.e.*, the capital is less productive in them than in developed countries. This is so because there is a relative inefficiency of the industries in developing countries which produce capital goods. There is the greater wastage of capital in the process of production due to low level of technical knowledge. Besides, owing to indivisibilities, certain kinds of investment are bound to be initially underutilised. Further, as development proceeds in the developing countries the pattern of demand shifts towards the more capital-intensive industries which tends to raise capital-output ratio.

In order to gain the most from capital formation, a country must also undergo technological and organizational progress, so that the capital-output ratio be reduced. Thus, the growth of the rate of output depends not only on the amount of capital accumulated but also on how much capital is required per unit increase in output (*i.e.*, incremental capital-output ratio). The lower the incremental capital-output ratio, the more accelerated is economic growth. A low capital-output ratio is, . thus, as significant as large capital accumulation. But it must also be pointed out that a lower incremental capital-output ratio (ICOR) requires technological and organisational progress so that capital becomes more productive.

Limitations of Capital-output Ratio

It may, however, be pointed out that the concept of capital-output ratio suffers from certain limitations. Its precise calculation presents some formidable difficulties. Hence, the quantitative relationship between capital investment and output, which the capital-output ratio suggests, may prove to be misleading. It would, therefore, be hazardous to base the estimates of capital requirements of an industry or economy on such ratios. Neither can the capital stock be assumed with any exactitude; nor is the other side of the ratio, *i.e.*, output capable of any precise measurement. Besides the index number problems, a clear distinction cannot be often made between capital goods and non-capital goods. Returns to social overheads, in particular, cannot be calculated accurately. Further, capital-output ratio is influenced by several variables, *e.g.*, technological improvements, better utilisation of equipment, organisational improvements, labour efficiency, and these factors elude quantitative measurement.

Hence, the concept of capital-output ratio has only a limited practical significance, because it cannot indicate the actual contribution of capital alone in a given scheme of investment. Great caution is, therefore, necessary in making use of a particular capital-output ratio in the adoption of actual investment policy.

Chapter 39

UNDERDEVELOPMENT AND COMMON CHARACTERISTICS OF DEVELOPING COUNTRIES

MEANING OF UNDERDEVELOPMENT

All economies of the world are not equally developed. Some are more developed and advanced than others. Therefore, the economies of the world have been classified into developed economies and underdeveloped economies. Previously, the underdeveloped economies were called backward economies and the developed economies were known as advanced or rich economies. But the terms backward and advanced are now considered unsatisfactory and misleading. Therefore, the terms developed and underdeveloped began to be used in place of advanced and backward. But it is not easy to give a precise definition of an underdeveloped economy. According to the United Nations experts, "*an underdeveloped country is one in which per capita real income is low when compared with the per capita real income of the United States of America, Canada, Australia and Western Europe. In this sense, an adequate synonym would be poor countries*." This definition brings out an important feature of underdevelopment, namely, poverty. But poverty is not always an adequate criterion to judge whether a country is developed or underdeveloped. For a country to be called underdeveloped, there must be unutilised or underutilised natural resources which may be used for production to raise per capita income and standards of living of the people. If the resources of a country are meagre and scanty, then even after having fully developed them, it will remain poor.

Professor Ragnar Nurkse also gave a definition of an underdeveloped country. According to him, "*underdeveloped countries are those which compared with the advanced countries are under-equipped with capital in relation to their population and natural resources*."[1] But this is also not a wholly correct and adequate definition of underdeveloped countries. This is because lack of capital is not the only cause of underdevelopment and poverty. There are other important factors which are responsible for low income and poverty of the people of the developing countries. Growth and development do not merely depend upon capital accumulation. Nurkse himself has pointed out that, "Economic development has much to do with human endowments, social attitudes, political conditions and historical accidents. Capital is a necessary but not a sufficient condition of progress."[2]

Professor Jacob Viner has provided a more satisfactory and adequate definition of an underdeveloped economy. According to Professor Viner, an underdeveloped country is one "*which has good potential prospect for using more capital or more labour or more available natural resources, or all of these to support its present population on a higher level of living or if its per capita income level is already fairly high to support a larger population on a not lower level of living*."[3] Thus, Professor Viner has emphasized that for a country to be called underdeveloped there must be possibilities and potentialities for its development. In other words, in an underdeveloped country resources must be

1. Ragnar Nurkse, *Problems of Capital Formation in Underdeveloped Countries*, 1953, p.1.
2. *Ibid.*.
3. Jacob Viner, "The Economics of Development", printed in A.N. Agarwala, and S.P. Singh (ed.), *The Economics of Underdevelopment*, Oxford University Press, 1958, p.12.

lying unutilised which can be used for increasing the rate of production and thereby raising the level of living of the people. Thus, according to him, the basic criterion of underdevelopment is whether there exist potential prospects for development and raising standards of living of the people. Similar to Professor Viner, Eugene Stanley has also provided a definition. According to him, an underdeveloped country is one which is "characterized (i) by mass poverty which is chronic and not the result of some temporary misfortune and (*ii*) by obsolete methods of production and social organization which means that poverty is not entirely due to poor natural resources and hence could presumably be lessened by methods already proved in other countries."[4] Thus Eugene Stanley lays emphasis on the existence of poverty as well as the potentialities for development and raising standards of living. We conclude that underdeveloped country is one which is poor but which has the future possibility and prospect of removing poverty and raising the levels of living of its people by utilising the idle and underutilised resources for production.

Since there exist potential prospects for development in almost all poor countries, therefore all poor countries are generally described as underdeveloped. These days the underdeveloped countries are usually called developing or less developed countries (LDCs). The developing countries are also called the Third World countries. The First World consists of the Western developed countries whereas the erstwhile Communist countries led by Soviet Russia were referred to as the Second World.

World Bank Classification of Low Income, Middle Income and High Income Countries

World Bank uses Gross National Income (GNI) per capita to classify different countries of the world into different countries. World Bank classifies the economies of the world into three groups:[5] (1) low income economies. (2) middle income economies, and (3) high income economies. Middle income economies are further divided into (a) lower middle income economies and (b) upper middle income economies. The low income economies are generally described as developing economies, while high income economies are called developed economies.

Table 39.1. Population and GNI Per Capita in 2011 PPP $ for the Year 2012

	Countries	Population (millions)	Gross National Income Per Capita, PPP $ (average)
1.	Low Income	846	1,387
2.	Middle Income (a) Lower Middle Income (b) Upper Middle Income	4,898 2,507 2,391	7,242 3,912 10,741
3.	Low and Middle Income Countries	5,744	6,376
4.	High Income Countries	1302	37,760
	World	7,046	12,129

Source: *World Development Report*, 2014. (Note that in this we have given the data for the year 2012 while in Table 38.1 we have given the data for the year 2014

Table 39.1 reveals that a large *development* gap exists between low income countries and high income developed countries (with $ 12,615 PPP or above) in 2012. In low income countries containing 846 million population average per capita income of $ 1,387 PPP as against average per capita income of $ 37,760 PPP $ of high income countries having population of 1,302 million in 2012. This is indeed a very large development gap between the low income countries and high income countries. Developing countries as generally defined contain both low-income and middle-income countries. It will be interesting to compare the gap between the developing countries as a whole on the one side and high income developed countries on the other. It will be seen from Table 39.1

4. Eugene Stanley, *The Future of Underdeveloped Countries*, New York, 1956, p. 18.
5. *World Development Report* 2014, Oxford University Press.

that developing countries (*i.e.*, low and middle-income countries) containing 5.744 billion population have average per capita income equal to $ 6,376 PPP $ in 2012 as against average per capita income of $ 37,760 PPP $ of high-income developed countries having total population of 1.30 billion in 2012. This is a very large income gap which has to be bridged by developing countries by achieving a much higher rate of GDP growth than that of developed countries, if they desire to catch up with them. We will explain this in detail later and see how East Asian countries joined the rank of high-income developed countries by registering a very high rate of GDP growth after 1965 till 1990. China and India, the *two fastest growing countries* of the world, are expected to catch up with the developed countries by 2050.

It may be noted that all high per capita income countries are not developed countries. For example, Kuwait, Saudi Arabia and some other oil-producing and exporting countries have a high per capita income but their richness is due to their being oil-abundant countries and higher prices of exports of oil since 1970s rather than the developed nature of their economies.

COMMON CHARACTERISTICS OF DEVELOPING (UNDERDEVELOPED) ECONOMIES

An idea of the characteristics of an underdeveloped or a developing economy must have been gathered from the above analysis of the definitions of an underdeveloped economy. Various developing countries differ a good deal from each other but even then there are some common features which are present in almost all the developing economies. We explain below some of the basic and important characteristics which are common to all developing economies.

Low per Capita Income or Poverty

The first important feature of the underdeveloped countries is their low per capita income. According to the World Bank Report 2014 in the year 2012 average per capita income *in terms of 2011* PPP $ of low income countries was 1,387 as compared to 7,243 of middle income countries and 37,760 of high income countries. This shows a *large development* gap between developing countries (*i.e.*, low income and middle income countries) and the rich developed countries (*i.e.*, high income countries). In the year 2014, per capita income in terms of 2011 PPP $ of was 5497 of India 12,541 of China, 15175 of Brazil, 8779 of Sri Lanka as against 52,941 of USA, 39,767 of UK and 64,982 of Norway. It may be noted that Norway is the richest country of the world in terms of per capita income.

It may however be noted that the extent of poverty prevailing in the developing countries is not fully reflected in the per capita income which is only an average income and also includes the incomes of the rich also. Large inequalities in income distribution prevailing in these economies have made the lives of the people more miserable. A large bulk of population of these countries live below the poverty line. The recent estimates as per Tendulkar Committee norms reveal that about 21.9 per cent of India's population (*i.e.*, about 270 million people) live below the poverty line in 2011-12, that is, they are unable to get even sufficient calories of food needed for minimum subsistence, not to speak of minimum clothing and housing facilities. The situation in other developing countries is no better.

The poverty in the underdeveloped countries is due to stagnation or lack of economic growth in the past and their technological backwardness, despite the vast resources existing there. By utilising their natural resources and making progress in technology they can increase their production and income and break the vicious circle of poverty operating in them.

It may however be noted that after the Second World War and with getting political freedom from colonial rule, in a good number of the underdeveloped countries the process of growth has been started and their gross domestic product (GDP) and per capita income are increasing. That is why they are now called the developing countries or less developed countries (LDCs). But it will take a long time before they are able to catch up with the present-day developed countries.

Excessive Dependence on Agriculture

An underdeveloped country is generally predominantly agricultural. About 60 to 70% of their population depends on agriculture and its allied activities for its livelihood. Further, about 30 to 50 per cent of national income of these countries is obtained from agriculture alone. This excessive dependence on agriculture is the result of low productivity and backwardness of their agriculture and lack of modern industrial growth. In the present-day developed countries, the modern industrial growth brought about structural transformation with the proportion of working population engaged in agriculture falling drastically and that employed in the modern industrial and services sectors rising enormously. This occurred due to the rapid growth of the modern sector on the one hand and tremendous rise in productivity in agriculture on the other.

In developing countries today, despite their modern industrial growth in the last four decades not much progress has been achieved towards structural transformation in the occupational structure of their economies. Due to the use of highly capital-intensive techniques in their industries very few employment opportunities have been created in their industrial sector. When increasing population cannot obtain employment in the modern non-agricultural occupations, such as industry, transport and other services, then the people remain on land and agriculture and do some work which they are able to get. This has resulted in excessive dependence on agriculture. During the last some decades because of population explosion the pressure of manpower on land in the developing countries has increased very much. Many bad results have followed from this. With the increase in man-land ratio the land has been divided into small holdings. According to the estimates of ILO in 2010 on an average 61 per cent of workforce of low income developing countries was employed in agriculture whereas only 19 per cent in industry and 20 per cent in services. On the contrary, in high-income developed countries, only 4 per cent of workforce was employed in agriculture as against 26 per cent in industry and 70 per cent in services.

Lack of Capital and Low Rate of Capital Formation

The insufficient amount of physical and human capital is so characteristic a feature in all undeveloped economies that they are often called simply `capital-poor' economies. One indication of the capital deficiency is the low amount of capital per head of population. Not only is the capital stock extremely small, but the current rate of capital formation is also very low. In early nineteen fifties most underdeveloped countries investment was only 5% to 8% of the national income, whereas in the United States, Canada[7] and Western Europe, it was generally from 15 per cent to 30 per cent. Since then there has been substantial increase in rate of saving and investment in the developing countries. However, the quantity of capital per head is still very low in them and therefore productivity remains low. For example, in India rate of investment has now risen to about 35 per cent in 2012-13 but still remains a poor country with low level of productivity. This is because of rapid population growth that capital per head is still very small.

The low level of capital formation in an underdeveloped country is due both to the weakness of the inducement to invest and to the low propensity and capacity to save. The saving in developing countries is low primarily because of the low level of national income. In such an economy, the low level of per capita income limits the size of the market demand for manufacturing output which weakens the inducement to invest. The low level of investment also arises as a result of the lack of dynamic entrepreneurship which was regarded by Schumpeter as the focal point in the process of economic development.

At the root of capital deficiency is the shortage of savings. The level of per capita income being quite low, most of it is spent on satisfying the bare necessities of life, leaving a very little margin of income for capital accumulation. Even with an increase in the level of individual incomes in an underdeveloped economy, there does not usually follow a higher rate of accumulation because of the tendency to copy the higher levels of consumption prevailing in the advanced countries. Nurkse has called this as "demonstration effect". It is usually caused through media like films, television or through foreign visits.

Generally, there exist large inequalities in the distribution of incomes in underdeveloped countries. This should have resulted in a greater volume of savings available for capital formation. But most often the sector in which the greatest concentration of incomes lies is the one which derives its income primarily from non-entrepreneurial sources such as unearned incomes of rents, interests and monopoly profits. The attitudes and social values of this sector are often such that it is prone to use its income for 'conspicuous consumption', investment in land and real estate, speculative transactions, inventory accumulation and hoarding of gold and jewellery. If these surpluses are channelled into productive investment, they would tend to increase substantially the level of capital formation.

Rapid Population Growth and Disguised Unemployment

The diversity among underdeveloped economies is perhaps nowhere to be seen so much in evidence as in respect of the facts of their population in respect of its size, density and growth. While we have examples of India and Pakistan with their teeming millions and galloping rates of population growth, there are the Latin American countries which are very sparsely populated and whose total population in some cases numbers less than a single metropolitan city in India and China. In several newly emerging countries of Africa too and in some of the Middle Eastern countries the size of their population cannot be regarded as excessive, considering their large expanse. The South-East and Eastern Asia, on the other hand, have large populations.

However, there appears to be a common measure, namely, a rapid rate of population increase. This rate has been rising still more in recent years, thanks to the advances in medical sciences which have greatly reduced the death rate due to epidemics and diseases. While the death rate has fallen sharply, birth rate does not yet show any significant decline so that the natural survival rate has become much larger. In countries like India, Pakistan, Burma, a veritable population explosion is feared. The great threat of this important trend consists in this is that it sets at nought all attempts at development inasmuch as the increased output is swallowed up by the increased population.

According to World Development Indicators Report 2011, between 1990-2009, annual population growth of low income countries was on an average 2.3 per cent as against 1.3 per cent of middle income countries and only 0.7 per cent of high-income developed countries.

One important consequence of this rapid rate of population growth is that it throws more and more people on land and into informal sector to eke out their living from agriculture, since alternative occupations do not simultaneously develop and thus are not there to absorb the increasing numbers seeking gainful employment. The resultant pressure of population on land and in informal sector thus gives rise to what has been called "disguised unemployment". Disguised unemployment means that there are more persons engaged in agriculture than are actually needed so that the addition of such persons does not add to agricultural output, or putting it alternatively, given the technology and organisation even if some of the persons are withdrawn from land, no fall in production will follow from such withdrawal. As a result, marginal productivity of a wide range of labour employed in agriculture is zero.

Underutilisation of Natural Resources. The natural resources in an underdeveloped countries are either unutilised or underutilised. Generally speaking, underdeveloped countries are not deficient in land, water, mineral, forest or power resources, though they may be untapped. In other words, they constitute only potential resources. The main problem in their case is that such resources have not been fully and properly utilised due to various difficulties such as shortage of capital, primitive technology and the small size of the market.

Economic Backwardness of the People. The people in underdeveloped countries are economically backward, that is, the quality of the people as productive agents is low. Instead of acquiring the greatest possible control over their physical environment, the people have struck a balance with nature at an elementary level. They have been relatively unsuccessful in solving the economic problem of man's conquest of his material environment. Particular manifestations of this are low labour efficiency, factor immobility, limited specialization in occupations and in trades, and a lack of entrepreneurship, illiteracy, ignorance, and conservative social values that minimise the incentives for economic change.

Human Development and Demographic Advantage

India with a large and young population has a great *demographic advantage*. The proportion of working-age population is likely to increase from approximately 58 per cent in 2001 to more than 64 per cent by 2021. While this provides opportunities, it also poses challenges. Policy makers have to design and execute development strategies that target this large young population. Demographic advantage is unlikely to last indefinitely. Therefore, timely action to make people healthy, educated, and adequately skilled is of paramount importance.

According to the United Nations Human Development Report (HDR) 2013, India with a human development index (HDI) of 0.554 in 2012 slipped down the global ranking to 136 from 134 as per HDR in 2012. India is in the medium human development category with countries including China, Egypt, Indonesia, South Africa, and Vietnam having better overall HDI ranking within the same category. The existing gap in health and education indicators in India as compared to developed countries and many developing countries calls for much faster spread of basic health and education. Life expectancy at birth in India was 65.8 years in 2012, compared to 75.1 years in Sri Lanka and 73.7 years in China. The expenditure on social service by the government as a percentage of GDP has increased from 6.8 per cent in 2008-09 to 7.2 per cent in 2013-14 (BE) with expenditure on education increasing from 2.9 per cent to 3.3 per cent and on health from 1.3 per cent to 1.4 per cent.

On the basis of poverty line (₹ 27 per capita for rural areas and ₹ 33 per capita for urban areas. The poverty ratio declined from 37.2 per cent in 2004-05 to 21.9 per cent in 2011-12. In absolute terms, according to Suresh Tendulkar Committee, the number of poor declined from 407.1 million in 2004-05 to 269.3 million in 2011-12. During 2004-05 to 2011-12, employment growth (CAGR) was only 0.5 per cent, compared to 2.8 per cent during 1999-2000 to 2004-05 as per usual status. However, the unemployment rate as per usual status (usual plus subsidiary) continued to have around 2 per cent.

Dualistic Structure of the Underdeveloped Economies

An important feature of developing economies, especially those which are marked by surplus labour, is that they have a dualistic structure. This dualistic character of these economies has been held to be the cause of unemployment and underemployment existing in them. Keeping in view this dualistic structure of less developed economies, important models of income and employment have been propounded. Famous Lewis model of economic development with unlimited supplies of labour and Fei-Ranis model of "*Development in a Labour Surplus Economy*" explain how in dualistic economies, the unemployed and underemployed labour in the traditional sector is drawn into a modern high productivity sector. The concept of dualism was first of all introduced into the development analysis by Dr. J.H. Boeke[6] but he emphasised the social dualism, according to which there is sharp contrast between the social systems characterising the two broad sectors of the economy, one in which the original social system with its subsistence or percapitalist nature, limited wants, non-economic behaviour and low level of economic and social welfare prevails, and the other where imported capitalist system with its modern system of industrial organisation, wage employment, unlimited wants and positive behaviour to economic incentives exists.

6. J.H. Boeke, *Economies and Economic Policy of Dual Societies*, New York, 1953.

However, it is technological dualism rather than Boeke's social dualism which has an important bearing on the problem of economic growth and surplus labour in the developing countries. According to the concept of technological dualism, the important difference between the traditional and the modern sectors lies in the difference between the production techniques or technologies used. In the small modern sector consisting of large-scale manufacturing and mining which provides wage employment, highly capital-intensive techniques imported from the developed countries are used. On the other hand, in the large traditional sector covering agriculture, handicrafts and allied activities, in which there exist extended family system and self-employment, labour-intensive technology is generally used. As a result of the difference in technologies used, the labour productivity and levels of earnings in the modern sector are much higher than those in the traditional sector. Moreover, since the technology used in the modern sector is highly capital-intensive, the growth of this sector has not absorbed adequate amount of labour in high productivity and high wage employment. With the explosive rate of growth of population and labour force and the limited creation of employment opportunities in the modern sectors because of the highly capital-intensive technology, surplus labour has emerged in agriculture and services. It has been possible for agriculture to contain the surplus labour because of the prevalence of extended family system in which both work and income are shared by the family members.

We thus see that the problem of unemployment and underemployment in less developed economies has been intensified by the technological dualism caused by the use, in the modern manufacturing and mining, of capital-intensive technology imported from abroad which is wholly unsuitable to the factor endowments of these less developed economies with abundant labour and small capital. The unemployment and underemployment in these less developed economies are not only due to the slow growth of capital or low rate of investment, it is also due to the highly capital-intensive techniques used in the modern sector. This technological dualism with the fact that modern sector has limited labour-absorptive capacity contains important implications for development strategy to be framed for less developed countries like India with surplus labour.

CHAPTER 40
OBSTACLES TO ECONOMIC DEVELOPMENT

In the previous two chapters we have explained the main common characteristics of developing countries and determinants of economic growth. Now the important question is why these countries have remained underdeveloped or, in other words, what have been the obstacles to their development and *constraints on their economic growth*. There is not one but several explanations, both economic and non-economic, for their underdevelopment. In the present chapter we are concerned with the study of some explanations of underdevelopment. A suitable strategy of development to accelerate the rate of economic growth can be framed if the correct explanation of their underdevelopment can be known.

IMBALANCE BETWEEN CAPITAL AND POPULATION

The first and foremost economic explanation of underdevelopment or the existence of mass unemployment and poverty in today's developing countries is that in them there has come about a serious imbalance between population and productive resources, especially the stock capital. This imbalance between resources and population has manifested itself not only in low per capita income and poverty but also in the existence of huge open unemployment, disguised unemployment in the underdeveloped countries. The present stock of capital (including land) is quite insufficient to employ all at a reasonable level of real per capita income.

In the present-day world man by himself can hardly produce anything. Even the primitive man needed some elementary tools like bow and arrow to engage in hunting for the earning of his livelihood. With the growth of technology and specialisation, he needs much more capital with which to be engaged in productive activity. If he is an agriculturist, he needs a piece of land and also a plough, a pair of oxen, seeds and some foodgrains and other necessities of life to sustain himself during the period of the sowing to the reaping of the harvest. In the industrial sector, he needs factories to work in and machines to work with. All these instruments of production belong to the nation's stock of capital. A nations's stock of capital can be enlarged by increased investment which in the absence of any utilised resources requires additional savings on the part of the community. The rate of capital formation should be kept sufficiently high so that employment opportunities are successively enlarged to absorb the increase in the labour force. Now, if the population grows faster than the stock of capital of a country, the entire addition to labour force cannot be absorbed in productive employment because not enough instruments of production will be there to employ them.

It is important to note that the existing balance between capital and land resources on the one hand and population on the other is a legacy of the past. Thus, the present state of underdevelopment must be viewed in the long-run setting of these countries. For several decades in the past the rate of investment and therefore the growth of the capital stock has not been keeping pace with the growth of population in underdeveloped countries like India. This implies that opportunities for productive employment have not been increasing commensurate with the growth of population and labour force. This has resulted in mass poverty and unemployment.

It may be noted that population growth is an exogenous factor. There is nothing in the system that should keep the rate of population growth equal to the rate of investment. Nor is there any mechanism that should automatically raise the rate of investment commensurate with any given rate of population growth. Thus, imbalance between resources especially capital stock and population emerges when population grows faster than investment.

Now, the question is when addition to labour force is unable to get employment in productive activities due to lack of capital, how exactly is it absorbed in the economy? Unable to get employment in the organised sector on a wage basis additional workers remain in their families and share work with other members of the family on the farm and other traditional occupations *i.e.*, whatever work their families have to do. Since the same amount of work is done by more workers, the additional workers do not add to total output. Land, technology and capital resources remaining the same, sharing of work by more workers lowers the hours and intensity of work put in by each worker engaged in agriculture and traditional occupations, resulting in mass disguised unemployment and lowering the standards of living of the people. Agricultural family farms and traditional occupations such as cottage industries, where disguised unemployment emerges, are run on family basis, do not represent modern form of economic organisation; they do not follow the principle of employing labour productivity by equating marginal product of labour with wage rate. They are, in fact, traditional forms of economic organisation. It follows therefore that "*it is through inferior forms of organisation and a reduction in the ratio of surplus to employed workers and/or through a general reduction in the levels of standards of living that the surplus population is absorbed.*"[1]

The lack of real capital per head of population is so characteristic a feature of the developing economies that they are often called "*capital-poor economies*". Low productivity of population and therefore their low income in the developing countries is mainly due to the small amount of capital per head of population. Thus the *most important economic explanation for underdevelopment and poverty is the imbalance between capital and population; capital is too small in relation to the huge amount of population.*

Not only was the existing stock of capital very small, but the rate of capital formation in the fifties when most of the developing countries started the process of development was also very low. In most underdeveloped countries rate of investment at that time was only 5% to 8%; it generally varied from 15% to 30% of the national income. The low rate of capital formation in under-developed countries is due to the following reasons: (1) The supply of domestic savings has been very small. (2) There is a lack of daring, honest and dynamic entrepreneurs who should perform the task of making investments and bearing risk. (3) Inducement to invest is very weak, that is, demand for making investment has been very small.

The basic solution to the problem of the above sort is the faster rate of capital formation so as to increase employment opportunities. For this purpose, every possible encouragement should be given to savings and their productive utilisation for increasing the rate of investment. In the developing countries investment incentives are very low and the State can assist in the process of capital formation directly as well as indirectly. Through a fiscal policy which encourages saving and investment and a sound monetary policy it can do much to encourage investors. The State itself can participate in the process of capital formation by investing in infrastructure and basic heavy industries in which the private entrepreneurs do not find it profitable to undertake investment. The developing countries suffer from a notorious shyness on the rate of investment. The other line of attack has got to be on the rate of population growth. Malthusian theory may be wrong but it remains true that if population grows at a rapid rate, then to maintain the people even at their existing levels of living large amounts of capital are needed which could otherwise had been used to raise the amount of capital available per man and hence raise the living standards at a faster rate.

1. Vakil and Brahmananda, *Planning for an Expanding Economy,* Vora & Co., Bombay, 1956.

The important point to be borne in mind is that even if at present the rate of investment is raised to keep pace with the rate of population growth, the already existing huge disguised unemployment and mass poverty which has come to exist as a result of the disproportion between investment and population will remain untouched. It follows, therefore, that for the solution of existing disguised unemployment and poverty, mere raising the rate of investment equal to the current rate of population growth is not enough. "*The central problem of economic growth in the context of India and other underdeveloped countries is that of the creation of those conditions under which the rate of investment can be stepped up above that of population growth.*"[2] It is heartening to note that as under the development policies pursued in the Five Year Plans in India, rate of domestic saving rose to 25.1 per cent in 1995-96 and came down to 24 per cent of national income in 2001-02. Apparently this seems to be high rate of saving for a developing country like India with a very low per capita income. Remember also that during the last 52 years of planned development, our population increased from 36 crore in 1951 to about 103 crore in 2002 and our population is currently increasing at the rate of around 2 per cent per annum. For achieving adequate growth rate in per capita income, in the Eleventh Five Year Plan (2007-12), target of growth rate of 8% per annum had been fixed to be achieved, for which it had been estimated that rate of saving had to be increased to 27% of GDP and rate of investment of 28% of GDP by 2006-2007.

VICIOUS CIRCLE OF POVERTY

Most widely propagated explanation why poor underdeveloped countries have failed to grow economically is that they are trapped in vicious circles of poverty. These vicious circles of poverty operate both on the supply and demand sides of capital formation. Supply side of capital formation refers to saving required to accelerate capital formation to raise productivity and per capita income. On the other hand, demand side of capital formation refers to the inducement to invest, that is, investment demand which depends on the size of market (*i.e.*, level of demand) which is small in the underdeveloped countries. Ragnar Nurkse in his now well known work[3] on underdeveloped countries attributed the persistence of poverty in them to these vicious circles of poverty and suggested policies to break these vicious circles. Both these vicious circles take poverty as their starting point. We explain below both these types of vicious circles of poverty.

Vicious Circle of Poverty (Supply Side)

In underdeveloped countries the rate of savings has been very low. In other words, the countries save a very small proportion of their current national income. In India the level of gross domestic saving has increased to about 30% of the national income in recent years because of the stimulus provided by the planned development under the Five Year Plans. At the time of independence the rate of saving in India was only 5 per cent of national income. The level of savings in underdeveloped countries is very small mainly because their level of national income and per capita income is very low. As a result, much of the income is consumed and little is left for investment purposes. Thus under-developed countries are caught up in the vicious circle of poverty. Because of the low level of income the capacity to save is very small. Since capacity to save is small, rate of saving and investment must be low. Investment being low leads to a lack of capital per head. Because there is small amount of capital per head the productivity of the people or real income per head will be low. Thus, the vicious circle is complete. Vicious circle of poverty on the supply side of capital formation is illustrated in Fig. 40.1.

In a slightly different way Professor Hicks describes the working of vicious circle in an under-developed country thus: "Productive power can be used either for the production of consumption

2. *Ibid.*, p. 240.
3. Ragnar Nurkse, *Problems of Capital Formation in Underdeveloped Countries*.

goods or of investment goods. Now when the average productivity of a community is low, it will have the greatest difficulty in producing enough consumption goods to satisfy the basic necessities of life; so it will have little productive power to spare for the production of investment goods. Countries which are in this position are involved in a vicious circle. A larger supply of capital equipment would enable them to escape from the toils of over-population but they are too deeply caught in those toils to be able to produce that equipment for themselves. Thus, they cannot escape without assistance from outside."[4]

Apart from the low level of per capita income, the low *relative* level of real income in underdeveloped countries as compared with the advanced and rich countries also makes their capacity to save very small. The great and growing inequalities between the income levels and, therefore, living standards of different countries, combined with increasing awareness of these inequalities have pushed up the general propensity to consume of the underdeveloped countries and have, therefore, reduced their capacity to save. The tendency of the people of underdeveloped countries to copy the higher levels of consumption prevailing in the advanced countries has been called "International demonstration effect" by Nurkse.[5] This also explains why the rate of savings and capital formation in developing countries has been low which is the main cause of their underdevelopment.

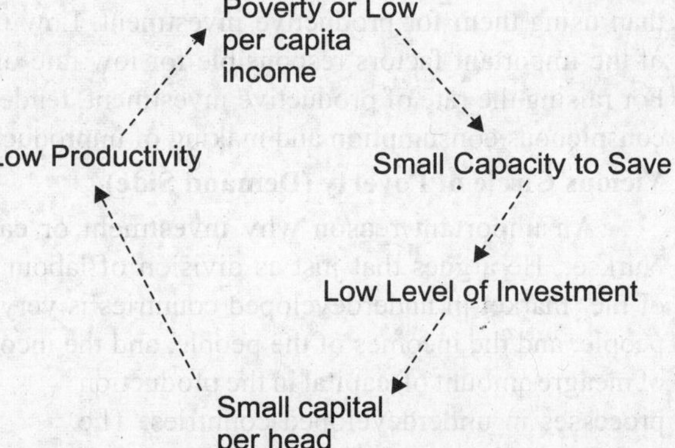

Fig. 40.1. *Vicious Circle of Poverty on the Supply-side of Capital Formation*

It may be noted that Nurkse used the idea of vicious circle of poverty to advocate that with adequate foreign aid or foreign direct investment the underdeveloped could succeed in breaking the viciou circle of poverty. He argued that with injection of foreign capital, productivity of labour would rise which would lead to higher incomes and eventually will generate higher savings. With higher savings made possible by injection of foreign capital momentum of capital formation will be built up which will help to break the vicious circle. Furthermore, higher incomes generated by greater investment with foreign capital would also increase aggregate demand in the economy which will stimulate capital formation by creating profitable investment opportunities.

A Critique

However, the above thesis of Nurkse is questionable. If this thesis were valid, then it would be difficult to explain the development and capital formation in the present-day developed countries which also started with low per capita incomes and many of them grew economically without much foreign aid. Of course, foreign aid and foreign investment can make an important contribution to economic development of the underdeveloped countries. What is being stressed here is that higher rate of investment and economic growth can be achieved without much inflow of foreign capital.

Though the majority of the people in poor developing countries are poor and are not in a position to save much but due to glaring inequalities in the distribution of income, there is minority of the people which is quite rich and wealthy and are in a position to save. *Savings, it is said, do not depend on per capita income but how income and wealth are distributed in a society*. For example, a poor Indian peasant may be unable to save because he has to spend all of whatever small income he earns to survive but there are many rich people in India who can save a lot. It is worth mentioning that some

4. J.R. Hicks, *Social Framework*, p. 35.
5. Ragnar Nurkse, *op. cit*.

poor countries have managed to achieve a high rate of savings and investment. Even Japan when it began developing at the end of 19th century had an average per capita income only slightly more than India in the 1950s and still it was able to raise its rate of saving and investment. Despite its poverty and low rate of economic growth (on average 3.5% per annum till the end of 1970s), even India was able to raise its rate of saving to 17 per cent in 1978-79 which far exceeded Arthur Lewis' prescription of raising rate of saving to 12 to 15% of national income to achieve self-sustained high growth rate.

But it is worth mentioning that in a country like India the people who get large incomes generally use much of their incomes for conspicuous consumption investment in land and real estate, speculative transactions, inventory accumulation and hoarding of gold and jewellery rather than using them for productive investment. Low use of savings for productive investment is one of the important factors responsible for low rate of economic growth in the developing countries. For raising the rate of productive investment, tendency for spending a good part of their income on conspicuous consumption and making of unproductive investment be discouraged.

Vicious Circle of Poverty (Demand Side)

An important reason why investment or capital formation is low has been advanced by Nurkse,. He argues that just as division of labour is limited by the size of the market. The size of the market in underdeveloped countries is very small due to the low per capita incomes of the people, and the incomes of the people, and the incomes of the people are low because there is a use of meagre amount of capital in the production processes in underdeveloped countries. The use of capital equipment in the production of goods and services for the domestic market is discouraged by the small size of the market. Thus a vicious circle operates on the demand side of capital formation. To quote Nurkse, "*The inducement to invest may be low because of the small buying power of the people, which is due to their small real income, which again is due to low productivity. The low level of productivity, however., is a result of the small amount of capital used in production, which in its turn may be caused at least partly by the small inducement to invest.*"[6] The vicious circle of poverty operating on the demand side is illustrated in Fig. 40.2.

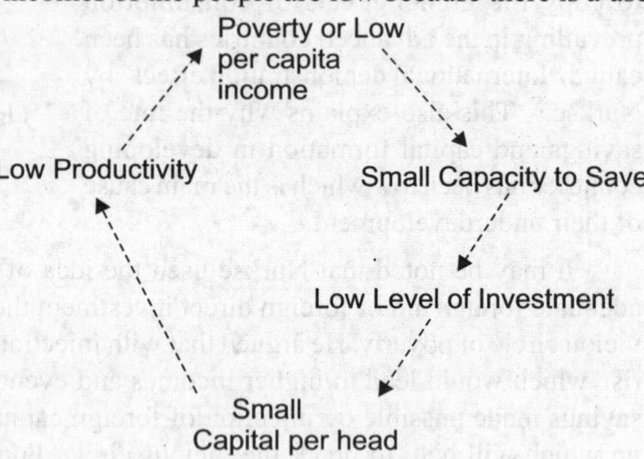

Fig. 40.2. *Vicious Circle of Poverty on the Demand-Side of Capital Formation*

It follows from above that underdeveloped countries have remained poor and economically backward because of low rate of capital accumulation as compared with the growth of population. As a result, capital per worker is very small and therefore his productivity is low. This, as seen above, gives rise to the vicious circle of poverty which furnishes a cause for persistent underdevelopment. If the underdeveloped countries want to accelerate the rate of capital accumulation so as to achieve rapid economic growth, then the vicious circles of poverty operating on the supply and demand sides of capital formation have to be broken. Once the vicious circles are broken and a country starts developing, economic growth becomes cumulative and these circles become beneficient.

A Critique

However, there are also flaws in the argument for the operation of vicious circle of poverty on the demand side of capital formation. In this regard, the argument of Nurkse that limited size of the

6. Ragnar Nurkse, *Problems of Capital Formation in Underdeveloped Countries*, p.5.

market is responsible for low rate of investment in poor underdeveloped countries is not entirely correct. The relationship between the limited size of the market and lower investment is based on the argument that economies of scale in industries are a key feature of their growth and for big industrial enterprises to produce on a large scale and to be economically profitable, the size of market, that is level of demand for a product, must be large.

However, Nurkse and others ignore the fact that economies of scale are important only for a few industries such as basic heavy industries of machine making steel, chemicals, fertilizers etc. but it is often feasible to produce quite efficiently in small enterprises in sectors such as textiles, shoes and leather products, food processing industries, jute manufactures, etc.

Besides, while per capita income in developing countries is small but due to inequalities in income distribution prevailing in the poor developing countries, there are quite a large number of people whose incomes are high enough to demand or buy the products of large-scale industrial enterprises which make them quite profitable to be set up in developing countries.

In view of the above it is possible to argue that limited size of the market is a problem for some industries but only in rare cases, it is a major cause of persistence of underdevelopment of the poor backward countries.

Breaking the Vicious Circle of Poverty

In order to break the vicious circle of poverty and bring about economic growth, an important thing to be noted is that in underdeveloped countries rate of saving and investment can be increased without the reduction in consumption. As is well known, rate of saving in underdeveloped countries is low. With a given rate of saving and investment, some rate of economic growth and as a result some increment in income must be occurring in these underdeveloped countries. If out of this increment in income proportionately greater part is saved, the rate of saving of the economy will rise. For instance, suppose as a result of 15% rate of saving and investment, an increment in income equal to ₹ 100 lakhs takes place. If out of these ₹ 100 lakhs, a greater part is saved, say 25%, and 75% is consumed, then the rate of saving will rise above 15%. It should be noted that in this process consumption will not decline, in fact it will increase by ₹ 75 lakhs, and along with this the rate of saving will also increase. If the increment in income of ₹ 100 lakhs is wholly consumed away, the rate of saving will decline.

The saving of ₹ 25 lakhs out of the increment of ₹ 100 lakhs means that the marginal rate of saving is 25%, which is higher than the previous average rate of saving, *i.e.*, 15%. Marginal rate of saving will go up. The sum and substance of the argument is that in *underdeveloped countries, the rate of investment and capital formation can be stepped up even without reduction in the consumption of the people, provided the marginal rate of saving is kept higher than the average rate of saving even though the current average rate of saving is low.* In this way, the rate of investment can be stepped up and standards of living of the people can be raised in future without the reduction in present consumption. This is how the vicious circle of poverty operating on the supply side of the capital formation can be overcome.

As explained before, vicious circle of poverty also operates on the demand side of capital formation. *To break the vicious circle on the demand side, Nurkse suggested the strategy of balanced growth.* According to Nurkse, if investment is made in an individual industry, it is likely to fail owing to low income and low purchasing power of the people. This is because investment in an individual industry does not raise the incomes of the people outside that industry. If the incomes of the people outside that industry do not rise, the demand for the product of that industry will not increase adequately to buy the larger production of the industry made possible by larger investment in it. Thus more investment in one individual industry makes possible large increase in production. But since the incomes of the people are low, the demand for the product of that industry in which more investment has been made will not rise and therefore productive capacity of the industry will not be fully utilised. As a result of this, larger investment in it may not be profitable. This would have an adverse effect on the inducement to invest in the industry.

But, according to Nurkse, *if investment is made in several industries simultaneously*, then the workers employed in various industries will become consumers of each other's products, and will thus create demand for one another. The simultaneous investment in a large number of industries has been called balanced growth. The balanced growth, *i.e.*, simultaneous investment in a large number of industries creates its own demand. Thus, according to Nurkse, through the strategy of *balanced growth* vicious circle of poverty operating on the demand side of capital formation can be broken.

Comments. In our view, Nurkse has exaggerated the vicious circle on the demand side of capital formation. Actually, there is enough demand for several products in underdeveloped countries like India and investment in the industries producing them is quite profitable. The reason for this is that in underdeveloped countries, owing to the large inequalities of income there are several people who have enough purchasing power to buy certain goods, despite the low per capita income. For instance, who can deny that in India investment in agriculture, sugar industry, refined oil, vanaspati ghee, etc., cannot be increased because of lack of demand for them though deficiency of aggregate demand due to low incomes is one of the factors. It is thus clear that production and capital formation in these countries are low not only due to the lack of demand. There are other factors which are responsible for low production and capital formation in these countries.

Another important thing to note in this connection is that several commodities are imported in large quantities by developing countries, which shows that demand for them exists in these countries. Substitution of these imports by producing them at home will not face any problem of lack of demand. Therefore, for the imported products, there is no problem of lack of demand or inadequate size of the market. We thus see that vicious circle of poverty can be broken in underdeveloped countries by making larger investment in a large number of import-substituting-industries. Besides, investment and industrial production can be substantially increased by producing for exports. East Asian countries of South Korea, Taiwan, Singapore, Hong Kong accelerated their rate of capital formation and economic growth during the seventies and eighties through this route. In this way economic growth can be accelerated to raise the standards of living of the people.

DEFICIENCY OF DYNAMIC ENTREPRENEURSHIP AND WEAK PROPENSITY TO ACCUMULATE

According to the classical economists, an entrepreneur or an organiser acts merely as an agency for bringing together the various agents of production and undertake to remunerate them for the work done. But the modern economists recognise the dynamic role that an entrepreneur plays in promoting economic development of the country. This was specially underlined by Schumpeter who thought that entrepreneur played a key role in economic development. According to him, the entrepreneur tries to maximise his profits by making innovations, that is, introducing new production techniques, bringing out a new product, tapping a new market, tapping new sources of raw materials and by adopting an optimum or most economical combination of the factors of production. In all these ways, while he succeeds in making higher profits for himself, he is making a significant contribution to the increase in the national income.

But there is a lack of dynamic entrepreneurs who can invest the savings in productive capital and carry out innovations. This explains why rate of economic development has been relatively low in them and underdevelopment persists. Whatever entrepreneurs are there in under-developed countries, they are *interested in quick returns* and are not daring enough to bear risks involved in making productive investment and in developing basic and heavy industries from whom profits accrue after a long time. The entrepreneurs in under developed countries try to avoid risk and are attracted by commerce and trade which provide them quick returns on the capital invested.

Thus, before independence private enterprise in India did not take India any far on the road of economic development. It left untouched and undeveloped some vital sectors such as basic heavy industries like steel, fertilizers, machine making and chemical industries. They were attracted more by commerce, trade and speculative activities than by industry. That is why after independence in the Five Year Plans, it became necessary for the Government to intervene and provide right type of entrepreneurship to bring about economic development.

Weak Propensity to Accumulate. Another characteristic of the entrepreneurs in developing countries is that they exhibit weak propensity to accumulate which is a great obstacle in the way of economic development. The entrepreneurs undertake investment and production of goods if there is a strong motivation to accumulate wealth. If instead of using wealth of income for a further investment a good part of it is consumed by the entrepreneurs, the motivation or propensity to accumulate will not be present in sufficient degree. The function of the capitalist (entrepreneur), according to Marx, is 'to accumulate' and this propensity to accumulate played a crucial role in the development of Western economies through promoting investment for more and more production. Marx has described capitalist (entrepreneur) as personified capital and accumulation as passion with him. To quote him, "Frantically bent upon the expansion of value, he (the capitalist) relentlessly drives human beings to production for production's sake, thus bringing about a development of social productivity and the creation of those material conditions of production which can alone form the real basis of a higher type of society .... only as the personification of capital is the capitalist respectable."[7] Thus the entrepreneurs (the capitalists) played an important role in the growth of Western economies as they did not spend their acquired wealth on their personal enjoyment and consumption. Instead, they used a major part of their income for further investment and production in order to gain power. For this they relatively saved more and consumed less. This greater propensity to invest for further production as compared to the propensity to consume helped to enlarge stock of capital and productive capacity in today's Western developed countries.

But the entrepreneurs in the developing countries of today have weak propensity to accumulate, and instead, indulge conspicuously in luxury consumption. That is why rate of capital accumulation and therefore of economic growth has been low in them and till today they remain underdeveloped. Commenting on this Late Prof. A.K. Dass Gupta writes, "*It is a tragedy that the underdeveloped countries have missed the early phase of capitalism when luxury goods were yet to arrive in the market and when the urge for power spent itself solely in accumulation. The conditions which stimulated accumulation in the early phase of capitalism in the West do not exist now. We have tasted the forbidden fruit and have fallen from paradise. The capitalist's urge for accumulation is now restrained everywhere by his love for display, and commodities making for display are multiplying*".[8]

INADEQUATE INFRASTRUCTURE

Economic growth in the developing countries has also been impeded by inadequate availability of infrastructure. Infrastructure includes power, transport and communication. It may also include credit facilities available from banks and other financial institutions and also the facilities for the education and training of labour. The availability of these infrastructures facilitates production in industry, agriculture and other productive sectors of the economy. These infrastructures give rise to external economies and thereby cause reduction in costs, facilitate production and increase efficiency in all productive sectors of the economy. Thus power (electricity) is used in the production process these days both in industry and agriculture. It is now too well known in India that non-availability of adequate amount of power lowers industrial and agricultural development. Likewise, the existence of the means of transport is essential for transportation of raw materials to the place of production and to sell the goods produced to far off places. In fact, the availability of transport widens the market for goods and thereby encourages their production.

7. Karl Marx, *Capital*, Vol. 1, (Everman's edition)
8. A.K. Dass Gupta, *The Economics of Austerity*, Oxford University Press.

One of the most important hurdles in the growth of entrepreneurship in developing countries is the *acute lack of external economies* which, as has been stated above, are provided by basic infrastructures such as transport, communication and power. In a developed country the economic system being highly diversified there are tremendous amounts of external economies which a new enterprise can draw upon. The various types of infrastructure in developed countries were generally built up by private enterprise although with the liberal help of the Government in the form of grants and loans. In contrast, the contemporary developing countries lack an adequate system of transport, communication and power. For instance, the inadequacy of railway network in most of the newly independent countries of Africa and Latin America is a major bottleneck in the expansion of. national market and growth of industries. Since development of power, transport, communication involves lumpy investment, has long gestation period and the returns accrue mainly in the form of external economies, the private enterprise is not attracted to build this infrastructure. The Government should therefore undertake the task of building up adequate infrastructure.

Likewise, there is a lack of adequate credit facilities or funds. For an entrepreneur who wishes to undertake business or set up some factory he should have sufficient funds to finance it. Credit facilities are also badly needed by farmers for agriculture. The growth of agriculture suffers if adequate credit facilities are not available. Therefore, the Government in the developing countries should give high priority to develop the facilities of providing adequate credit and finance for the development of industry and agriculture.

MYRDAL'S BACKWASH EFFECT AND ECONOMIC GROWTH

Gunnar Myrdal, in one of his seminal works[9] provided an important explanation of under-development. He argued that the very process of economic development in one part of the world created effect of impoverishing other parts or at least made it difficult for them to grow. Prior to Myrdal's thesis, there was a general consensus among the economists that economic growth in one part of the world would create beneficial effects, generally described as '*spread effects*', for the rest of the world. It was argued by these economists that development in one part would lead to the increase in the demand for exports of the poor countries. Secondly, it was thought that economic development of underdeveloped countries would also be promoted through the supply of foreign capital investment by the growing country. In this way they argued that development would create spread effects which would help in the process of economic development of the poor countries. Besides, it was pointed out that international trade between the growing rich countries and the poor countries based on the comparative cost principle would ensure that the benefits of growth in some countries would spread to the poor and underdeveloped countries.

Myrdal denied the existence of these spread effects to any significant degree. He also expressed doubts that free trade between the rich developed and the poor underdeveloped countries would tend to eliminate the large inequalities between the rich and poor countries. Myrdal further argued that in the absence of any counteracting measures trade between countries would not lead to the equalisation of incomes of the rich and poor countries but, on the contrary, would tend to increase the disparities between them. To quote Myrdal about the effects of international trade between the rich and underdeveloped countries, "this strengthens the rich and progressive countries whose manufacturing countries have the lead and are already fortified by the surrounding external economies while the underdeveloped countries are in continuous danger of seeing even what *they have of industry and, in particular, small scale industry and handicrafts priced out by cheap imports from industrial countries.*"[10]

9. Gunnar Myrdal, *Economic Theory and Underdeveloped Regions*, 1963.
10. Myrdal, *op. cit.* p. 51.

Comments

The above argument of Myrdal has some elements of truth. The decline of handicrafts in India during the colonial rule of the British was the result of cheap imports from British manufacturing industries. Further, the recent debates in WTO (World Trade Organisation) meetings reveal that the rich and developed world such as the US and countries of European Union (EU) are quite unwilling to reduce heavy subsidies they provide to their agriculture. The reduction of these agricultural subsidies would lead to the expansion in exports of agricultural products of the developing countries to the developed countries. On the contrary, the USA and European Union want the developing countries like India to make drastic cuts in tariffs so that the exports of their industrial products to the developing countries could rise. Given this attitude of the developed countries, the arguments of Myrdal cited above that trade between the rich developed and the poor developing countries cannot be fair to the latter and would work to widen the income inequalities rather than reducing them seem to be valid.

It is here noteworthy that in the past the main positive effect of international trade on the poor developing countries has been to promote the production of primary products such as products of plantations and mineral products which were exported to meet the needs of the expanding industries of the rich countries. Since primary products often meet inelastic demand in the export markets, they cannot be a source of much future economic development of the developing countries.

In the end, however, it may be noted that if the rich developed countries give up protectionist policies and also reduce agricultural subsidies, the trade between the developed and the developing countries can make important contributions to economic growth of the latter. Therefore, Myrdal's thesis of backwash effect, though containing some elements of truth as far as international trade between the two in the past is concerned, but in the present days, if the rich developed countries give up selfish and protectionist policies, international trade can go a long way in stimulating economic growth in developing countries. It is noteworthy here that East Asian countries such as South Korea, Taiwan, Hong Kong and Singapore which are called Asian Tigers have achieved unprecedented economic growth not only through appreciably raising the rate of domestic saving and investment but also adopting *outward-looking development strategies* (*i.e.*, pursuing export-led growth model). In India too under the new economic reforms started since 1991, export growth has been given an important role in the development strategy adopted.

DEMONSTRATION EFFECT AND ECONOMIC GROWTH

In raising rate of capital formation, the developing countries have to contend against one problem which arises from demonstration effect on consumption. Demonstration leads to imitation and imitation of superior consumption standards stimulates consumption among the poor which increases their propensity to consume and consequently reduce their capacity to save.

Nurkse laid great stress on this new theory of consumption and saving. We ordinarily think that a man's consumption depends on his income. But that is not quite correct. A person's consumption does not merely depend on his own income but also on the incomes and therefore the consumption behaviour of his friends and relations with whom he maintains social contacts. A man finds some of his friends using colour televisions, luxury cars, transistors, refrigerators air-conditioners, electric hot plates, electric washing machines and so on and experiences a sort of restlessness and a craving is generated in his mind to enjoy these amenities some immediately and others some later day. These desires for superior conspicuous consumption generally outrun the consumer's means. Thus consumption behaviour depends not on absolute real income but relative levels of real incomes. It does not depend on what we can afford but what the others afford and enjoy. This is what Duesenberry calls '*demonstration effect*'. The intensity of this demonstration effect on saving seriously impairs a person's capacity to save. He may save less even if his income has gone up; the ability to save may have increased but the willingness to save is decreased. It has been estimated that 75 per cent of the

Americans do not save. The reason is not that they are too poor to save but that they are adopting better ways of living seen among the upper classes. It is thus the interdependence of consumer's preferences which determines the choice between consumption and saving. This is Duesenberry's relative theory of consumption and saving.

What is true of individual applies also to nations. The disparities in the level of real incomes of nations have a profound effect on the economic development of nations. Whereas the rich nations can help the poor nations develop economically and break the vicious circle of poverty, they also pass on to them their consumption patterns. The poor countries may not find it easy to increase productivity and raise income but it is easy to imitate costly consumption. The demonstration effect aggravates propensity to consume and reduces capacity to save which is a serious impediment to economic development. The relative aspect of the problem need again to be emphasised. It is not the absolute level of income which determines consumption pattern but the relative levels of national income of connected countries which affect consumption of a less prosperous country. It follows therefore that even if the national income of a country goes up, it may not be able to save more than before simply because the income gap between this country and the other countries with which it has commercial or political intercourse has increased. The ability to save may have gone up but the willingness to save would have gone down on account of a keener desire to consume commodities which are consumed in the richer countries.

Thus the inter-personal and international consumption functions are inter-related and not independent. The consumption and saving habits of the economically backward countries are greatly influenced by those of the prosperous countries. As Nurkse puts it, "When people come into contact with superior goods or superior patterns of consumption, with new articles or new ways of meeting old wants, they are apt to feel after a while a certain restlessness and dissatisfaction. Their knowledge is extended, their imagination stimulated, new desires are aroused, the propensity to consume is shifted upward."[11]

It is not merely the superior consumption prevalent in rich countries which exercises influence on the consumption in the poor countries but also the knowledge of it. First we know and then we imitate. Knowledge opens eyes to the future possibilities. "It widens the horizons of imagination and desires." New products or new patterns or designs of old products are constantly being advertised and brought irresistibly to the notice of the consumer at home and abroad. These new goods enter into the standard of living of the poorer communities. "The presence or the mere knowledge of new goods and new methods of consumption tends to raise the general propensity to consume."[12] The movies, the radio and television, spread of education and modern travel facilities are the powerful media through which new goods or new and better ways of living are communicated far and wide. In the present state of the world attraction of advanced living standards have a fairly wide influence on the consumption of the people of low-income countries.

Propensity to save is directly influenced by propensity to consume. When the poor countries borrow living standards from the rich, they must pay the price. The price is that their capacity to save must diminish. This no doubt adds to the difficulties of bringing about economic development. This hits the saving potential associated with mass disguised unemployment prevailing in backward countries. We have already said that this saving potential can be mobilised only if consumption is kept at the same level as before. But we have seen that when such a country comes in contact with a rich country, its propensity to consume is likely to go up. The potential domestic sources of capital are seriously impaired by the impatience and dissatisfaction which the demonstration effect tends to generate. This impatience to raise living standards curtails the capacity to save. The conflict between the need to save and the desire to consume is intensified by the demonstration effect. Thus the small

11. Ragnar Nurkse, *Problems of Capital Formation in Underdeveloped Countries*, pp. 58-59.
12. Ragnar Nurkse, *op. cit.*

rate of saving in the underdeveloped countries may not be altogether due to low level of real income. It may also be due to the allurement of superior levels of living. This throws new light on international economic relations. Prosperity in one country may breed prosperity in another but it may also create difficulties by provoking consumption that it cannot afford and which it should avoid for some time at any rate. The high consumption of a rich country in this manner hinders capital formation in an underdeveloped economy.

We may conclude in Nurkse's words: "The great and growing gaps between the income levels and therefore living standards of different countries, combined with increasing awareness of these gaps, may tend to push up the general propensity to save." The entrepreneurial class in India is, in fact, aping western modes of living and squandering away profits which they should plough back into investment. The country is poorer for this senseless consumption behaviour.

How to Overcome Demonstration Effect

Curtailing the Consumption of Luxury Goods. The working of demonstration effect in developing economies implies that wants should be regulated or curtailed. If wants of the people go on multiplying under the influence of demonstration effect, the demand for luxuries or non-basic consumption goods will increase. This will raise the propensity to consume of the economy resulting in a low ratio of surplus to income. Further, the increased wants for luxuries and their consumption will divert the basic resources such as steel, cement, machines etc., towards their production. As a result, the production of basic consumption goods will be starved of essential resources. The use of scarce national resources to satisfy the non-basic wants of the people when millions who live below poverty line are unable to meet their basic wants for subsistence is purely a wastage of resources.

Thus, according to Prof. A. K. Dass Gupta, "*The implication of luxury consumption for an underdeveloped country is more obvious and direct. The underdeveloped countries will need a high rate of growth for quite some time to come. Their level of per capita income is too low. In India, for example, the per capita income is hardly adequate to afford a man his minimal subsistence. Even with a stationary population, the country will thus require years of accumulation for it to attain a civilized standard. Luxury consumption slows down the rate of accumulation and thus hampers growth.*"[13]

Limitation of Wants. Mahatma Gandhi had warned long ago against multiplication of wants. He thus stated, "I do not believe that multiplication of wants ... is taking the world a single step nearer its goal. I whole-heartedly detest this and desire to increase animal appetites and go to the ends of the earth to satisfy them. If modem civilization stands for all this and I have understood it does, I call it satanic."[14]

It may be noted that Mahatma Gandhi was not against the satisfaction of wants for consumption goods which are absolutely needed for maintenance of health and promotion of productivity. It is the multiplication of wants for non-basic consumption goods, that is, luxuries against which Mahatma wrote. According to him, there is no end to the multiplication of wants and efforts ought to be made to reduce them. To quote him again, "The mind is a restless bird, the more it gets the more it wants and still remains unsatisfied. Therefore, the ideal of creating an unlimited number of wants and satisfying them seems to be delusion and snare. Civilization in the real sense of the term consists not in multiplication but in the deliberate and voluntary reduction of wants."[15]

One may not entirely agree with Mahatma Gandhi regarding the limitation of wants, but in the context of poor underdeveloped economies in which glaring inequalities of income prevail, the restraint on the multiplication of wants of the richer sections of the society is essential if

13. A.K. Dass Gupta,. *The Economics of Austerity*, Oxford University Press, 1975, p. 12.
14. M.K. Gandhi, *Young India*, March 19, 1927.
15. M.K. Gandhi, *Hind Swaraj*, pp. 44–45.

economic surplus is to be generated and used for productive investment. Consumption of luxury and semi-luxury goods should therefore be controlled if rate of capital accumulation is to be stepped up. Russia and Japan were able to raise the level of their investment to 30 per cent of their national income and achieve a high rate of economic growth by cutting consumption to the minimum. Now, the question is how to regulate wants and curtail consumption of luxury goods. One way is the adoption of fiscal measures under which heavy taxes should be imposed on luxurious consumption expenditure by individuals and business companies. Heavy excise duties and sales taxes should also be imposed on luxury goods so as to curtail their consumption by the rich people. Physical controls can be imposed on the production and distribution of luxury goods such as cars, air conditioners. These goods should be distributed to the individuals on functional basis rather than on the basis of money power.

Restricting the Production of Luxury Goods. But the most effective method to curtail consumption of the affluent is to suitably restrict the production of luxury goods and a much high priority should be accorded to the production of basic consumption goods needed by the masses. This will raise the rate of capital accumulation on the one hand by releasing resources from the production of luxury goods and reduce inequalities between the rich and the poor on the other by raising the living standards of the latter. But the restriction on the production of luxury goods is easier said than done. This is because luxury consumer goods are produced by the private sector in mixed economies of underdeveloped countries. Guided as they are by the market or effective demand, the private capitalists prefer to produce luxury goods since there is ample market for them by the richer sections. Prof. A.K. Dass Gupta rightly states: "The Industrial Policy Resolution of 1956 gives private capitalists complete sway over the production of consumer goods. Restrictions on imports protect them from international competition. A sheltered market is created in the country for luxury goods. These industries thus receive high priority from India's capitalists."

APPENDIX TO CHAPTER 40
ROSTOW'S STAGES OF GROWTH

Prof. W.W. Rostow,[1] an eminent economic historian, has described the historical process of transition from underdevelopment to development in terms of a series of five stages of growth through which all countries must pass to reach the ultimate destination of 'a developed country'. These five stages of growth are: (1) the traditional society, (2) preconditions for take-off, (3) the take-off period, (4) the drive to maturity, and (5) stage of mass consumption.

It is important to note that these stages are not merely a way of generalising certain actual historical experiences about the sequence of development of today's developed countries but they have an inner logic and continuity. In fact, they constitute both a general theory of economic development and of social modem history as a whole. Rostow emphasises that gradual economic changes result in drastic changes in economic, social and political structure and institutions. Rostow argued that today's developed nations had all passed the stage of 'take-off into self-sustaining growth, while the underdeveloped countries were still in either the 'traditional society"or the "preconditions for take-off stage" and they had only to follow the path defined by the next stages of growth to reach the destination of a developed nation.

We explain below briefly Rostow's five stages of growth.

1. Traditional Society. This initial stage of traditional society signifies a primitive society having no access to modern science and technology. In other words, it is a society based on primitive technology and primitive attitude towards the physical world. Thus, Rostow defines a traditional society "as one whose structure is developed within the limited production function based on pre-Newtonian science and technology and as pre-Newtonian attitudes towards the physical world."[2]

However, Rostow does not view this traditional society as being completely static. In this stage of a society output could be increasing through the expansion of land area under cultivation or through the discovery and spread of a new crop. But the critical fact about this type of society is that there is limit to attainable output per head. This limit arises due to the absence of access to modern science and technology. This type of a society allocates a large proportion of its resources to agriculture and is characterised by a hierarchical social structure in which there is little possibility for vertical mobility. The value system that prevails in such a society is what Rostow calls a long-run fatalism. People of these societies think that not much economic progress is possible for them and for their future generations.

2. Preconditions or the Preparatory Stage. This covers a long period of a century or more during which the preconditions for take-off are established. These conditions mainly comprise fundamental changes in the social, political and economic fields; for example (*a*) a change in society's attitudes towards science, risk-taking and profit-earning; (*b*) the adaptability of the labour force; (*c*) political sovereignty; (*d*) development of a centralised tax system and financial institutions; and (*e*) the construction of certain economic and social infrastructure like railways, ports, power generation and educational institutions. India did some of these things in the First Five Year Plan period (1951-56).

1. W.W. Rostow, *The Stages of Economic Growth: A Non-Communist Manifesto*; Cambridge University Press, 1960.
2. Rostow, *op. cit.*

It is evident from above that in this second stage of growth foundations for economic transformation are laid. The people start using modern science and technology for increasing productivity in both agriculture and ·industry. Further, there is a change in the attitude of the people who start viewing the world where there are possibilities of future growth. A new class of entrepreneurs emerges in the society who mobilise savings and undertake investment in new enterprises and bear risks and uncertainties. In the sphere of political organisation, it is during this stage that an effective centralised nation-state starts emerging.

Thus in the stage of preconditions for take-off Rostow views agriculture as performing three roles. First, agriculture must produce sufficient food grains to meet the demand of growing population and of the workers who get employment in agriculture. Secondly, increase in agricultural incomes would lead to the demand for industrial products and stimulate industrial investment. Thirdly, expanding agriculture must provide much of the savings needed for the expansion of the industrial sector.

3. The "Take-off" Stage. This is the crucial stage which covers a relatively brief period of two to three decades in which the economy transforms itself in such a way that economic growth subsequently takes place more or less automatically. "*The take-off" is defined as "the interval during which the rate of investment increases in such a way that real output per capita rises and this initial increase carries with it radical changes in the techniques of production and the disposition of income flows which perpetuate the new scale of investment and prepetuate thereby the rising trend in per capita output*."

Thus, the term "take-off" implies three things. First, the proportion of investment to national income must rise from 5% to 10% and more so as to outstrip the likely population growth; secondly, the period must be relatively short so that it should show the characteristics of an economic revolution; and thirdly, it must culminate in self-sustaining and self-generating economic growth. Thus, during the take-off stage, the desire to achieve economic growth to raise the living standards dominates the society. Revolutionary changes occur in both agriculture and industry and productivity levels sharply increase. There is greater urbanisation and urban labour force increases. In a relatively short period of a decade or two, both the basic structure of the economy and social and political structure is changed so that a self-sustaining growth rate can be maintained.

It is worth noting that in the opinion of Rostow, the rise of a new elite (*i.e.*, new entrepreneurial class) and establishment of a nation-state are crucial for economic development.

4. Drive to Maturity: Period of Self-sustained Growth. This stage of economic growth occurs when the economy becomes mature and is capable of generating self-sustained growth. The rates of saving and investment are of such a magnitude that economic development becomes automatic. Overall capital per head increases as the economy matures. The structure of the economy changes increasingly. The initial key industries which sparked the take-off decelerate as diminishing returns set in. But the average rate of growth is maintained by a succession of new rapidly-growing sectors with a new set of leading sectors. The proportion of the population engaged in agriculture and other rural pursuits declines, and the structure of the country's foreign trade undergoes a radical change.

It is with both the problems and the cyclical movements of national income in such mature growing economies in this fourth stage that the bulk of modern theoretical economics is concerned. The students of contemporary developing countries and also of economic history are more likely to be concerned with the economics of the previous two stages, that is, the economics of the preparatory and the 'take-off' stages. If we are to have a useful and adequate theory of economic growth, it must obviously be comprehensive enough to embrace these two stages as well, especially the economics of the "take-off into self-sustaining growth."

5. Stage of Mass Consumption. In this stage of development per capita income of country rises to such a high level that consumption basket of the people increases beyond food, clothing and shelters to articles of comforts and luxuries on a mass scale. Further, with progressive industrialisation and urbanisation of the economy values of people change in favour of more consumption of luxuries and high styles of living. New types of industries producing durable consumer goods come into existence which satisfy the wants for more consumption. These new industries producing durable consumer goods become the new leading sectors of economic growth.

A Critique of Rostow's Stages of Growth

Rostow's stages of growth theory has come in for severe criticism. Gunnar Mydral has argued that there cannot an inevitable sequence of events described as successive stages of growth. According to him, economic growth is the result of certain economic policies adopted and not the other way around. Likewise, Meier argues that stages in the history of economic growth cannot be generalised from the development experience of some European countries as Rostow has done. To quote Meier, "Stage-making approaches are misleading when they consider a linear conception of history and imply that all economies tend to pass through the same series of stages. Although a particular sequence may correspond broadly to the historical experience of some economies, no single sequence fits the history of all countries. To maintain that every economy always follows the same course of development with a common past and the same future is to overschematize the complex forces of development, and to give the sequence of stages a generality that is unwarranted."[3]

3 Meier, *Leading Issues in Economic Development*, Oxford University Press.

CHAPTER 41
CAPITAL FORMATION: DETERMINANTS AND ACCELERATION

There are two important determinants of economic growth: (1) the rate of investment and (2) capital-output ratio. These factors were brought out by the growth models of Harrod and Domar. Therefore, if it is desired to step up the rate of economic growth, the rate of investment or capital formation should be raised and capital-output ratio should be lowered. An important factor which determines capital-output ratio in the economy is the technology that is being used for production in agriculture, industry and other activities. If the economy makes progress in technology, capital-output ratio will be lowered. It follows that rate of economic growth depends upon rate of investment or capital accumulation and technological progress. In the present chapter we shall discuss determinants of capital formation and measures to accelerate it. In the next chapter we will discuss technological change and its effect on productivity and growth.

Meaning of Capital Formation

Capital formation means the increase in the stock of real capital in a country. In other words, capital formation involves making of more capital goods as machines, tools, factories, transport equipment, materials, electricity, etc., which are all used for further production of goods. Thus, Prof. Gill defines capital accumulation in the following words: "*Capital accumulation is the process of adding to our stock of machinery, tools, buildings, and so on, over time. If our stock of capital at the end of the year is larger than it was at the beginning, the difference represents the amount of capital we have accumulated during the year. Another name for this is investment. Annual real investment is the addition to our capital stock over the course of a year.*"[1] Emphasising that for making additions to the stock of capital, saving and investment are essential, Ragnar Nurkse has defined capital formation as follows:

"The meaning of 'Capital Formation' is that society does not apply the whole of its current productive activity to the needs and desires of immediate consumption, but directs a part of it to the making of capital goods; tools and instruments, machines, and transport facilities, plant and equipment—all the various forms of real capital that can so greatly increase the efficacy of productive effort.... The essence of the process, then, is the diversion of a part of society's currently available resources to the purpose of increasing the stock of capital goods so as to make possible an expansion of consumable output in the future."[2]

Likewise, Professor J.R. Hicks remarks, "The way by which capital equipment is increased is the use of some part of the community's productive power during a period for the construction of new equipment (investment)."[3] It is thus evident that in order to accumulate capital goods some current consumption has to be sacrificed. The greater the extent that people are willing to abstain from present consumption, the greater the extent that society will devote resources to new capital formation. If society consumes all what it produces and saves nothing, future productive capacity of the economy will fall as the present capital equipment wears out. In other words, if whole of the

1. Richard T. Gill, *Economic Development: Past and Present*, 3rd edition, p. 14.
2. Ragnar Nurkse, *Problems of Capital Formation in Underdeveloped Countries*, p. 2.
3. J.R. Hicks, *The Social Framework*, p. 55.

current productive capacity is used to produce consumer goods and no new capital goods are made, production of consumer goods in the future will greatly decline. To cut down some of the consumption and to wait for more consumption in the future requires foresightedness on the part of the people. There is an old Chinese proverb: "He who cannot see beyond the dawn will have much good wine to drink at noon, much green wine to cure his headache at dusk, and only rain water to drink for the rest of days."

Widening of Capital and Deepening of Capital. A community could just go on building more transport facilities, more sources of power, more factories of the *existing type.* This process of duplicating capital of the existing technique is called "*widening of capital*", in contrast with "*deepening of capital*" which implies use of more capital-intensive techniques. In fact, capital accumulation and technological progress go hand in hand. Technological improvement is virtually impossible without capital accumulation. This is because new capital has to be built up according to the new superior techniques of production and since superior techniques are of higher capital intensity, larger investment has to be made. Thus, no nation which is not willing either to save or to get foreign aid for greater investment will enjoy the fruits of the advanced techniques and capital accumulation.

Human Capital

We have explained above the concept of *accumulation of physical capital.* For a long time in economics, it was thought that it was the physical capital which played a crucial role in expanding production. In recent years, a new concept of '*human capital*' has been evolved and emphasized. *By human capital is meant the stock of people equipped with education, skills, good health*, etc. It has now been found that the rate of growth achieved in the developed countries cannot be wholly explained by the increases in physical capital and advances in technology. A good part of economic growth has occurred due to the accumulation of human capital. It has now been realised that human capital formation is as important in increasing production and productivity as the physical capital formation. An educated, trained and skilled man is much more productive than an uneducated, untrained and unskilled man. Likewise, a person with good health contributes to production to a greater degree than the person with frail and poor health. Since the investment in education, skill and health adds greatly to the productivity of men, investment in human capital has also been called *investment in men,* or *investment in human beings*.

ROLE OF CAPITAL FORMATION IN ECONOMIC GROWTH

Capital plays a vital role in the modern productive system. Production without capital is hard for us even to imagine. Nature cannot furnish goods and materials to man unless he has the tools and machinery for mining, farming, forestry, fishing, etc. If man had to work with his hands on barren soil, productivity would be very low indeed. Even in the primitive stage, man used some tools and implements to assist him in the work of production. Primitive man made use of elementary tools like bow and arrow for hunting and fishing net for catching fishes. With the growth of technology and specialisation, capital has become more complex and is of superior and advanced type. More goods can be produced with the aid of capital. In fact, greater productivity of the developed economies like that of USA is mainly due to the extensive use of capital, *i.e.* machinery, tools or implements in the productive process. Capital adds greatly to the productivity of worker and hence of the economy as a whole.

Much economic development is not possible without making and using of industrial machinery, making of agricultural tools and implements, building of dams, bridges, factories, roads, railways, airports, ships, ports, harbours, etc., which are all capital. All these capital goods are man-made instruments of production and *increase the productive capacity* of the economy. Therefore, accumulation of capital goods every year greatly increases the national product or income. Capital accumulation is necessary to provide people with tools and implements of production. If the population goes on increasing and no net capital accumulation takes place, then the growing population would not be able to get necessary tools, instruments, machines and other means of production with the result that their capacity to produce would be seriously affected.

Besides this, *capital, accumulation makes possible the use of indirect or roundabout methods of production* which greatly increase the productivity of the workers. Under these indirect or roundabout methods of production, workers instead of working with bare hands, work with the aid of more productive tools, instruments and machinery. Under these indirect or roundabout methods some workers and other productive resources are first employed in producing capital goods and then with the help of these capital goods workers produce consumer goods. The greater the extent to which the methods of production would be indirect or roundabout, the greater their productivity and efficiency. But, as we have seen above, for the use of indirect or roundabout methods of production capital has to be accumulated. Therefore, we see that capital accumulation makes the use of indirect or roundabout methods of production possible and thereby greatly increases the national product and is helpful in bringing about rapid economic growth.

Moreover, *productivity of the workers depends upon the amount of capital per worker.* The greater the quantity of capital per worker, the greater the productivity of the workers. It is not capital accumulation alone that increases the amount of capital per worker. Capital per worker rises when the rate of capital accumulation is greater than the rate of population growth. With the increase in capital per worker, productivity per worker will increase with the result that national product and income will increase. Therefore, capital accumulation, by increasing the productivity of the workers, plays an important role in the growth of the economy.

From the viewpoint of economic growth capital formation is important also because *it makes large-scale production and greater degree of specialisation possible.* Thus, with capital accumulation the *advantages of large-scale production and specialisation* are obtained. The advantage of large-scale production and specialisation is that they greatly increase output and productivity and thereby bring down the *cost of production per unit.* Without adequate capital accumulation neither the scale of production can be increased nor greater specialisation and division of labour in the production process is possible. Hence, capital accumulation by enlarging the scale of production and specialisation increases the production and productivity in the economy and thereby promotes economic growth.

Another way in which capital accumulation contributes to growth is that *it makes the technological progress of the economy* possible. Different technologies need different types of capital goods. Therefore, when new, superior and better technology is discovered, its use can be made for production only if that technology is embodied in capital goods, that is, if capital goods according to that technology are made. Therefore, without capital accumulation, no technical progress can be made. If there is no capital accumulation, then the various new inventions or discoveries will remain unused for production. If is, therefore, clear that capital accumulation promotes technical progress in the country and through this accelerates the economic growth of the country.

Another important economic role of capital formation is the *creation of employment opportunities* in the country. Capital formation creates employment at two stages. First, when the capital is produced, some workers have to be employed to make capital like machinery, factories, dams, irrigation works, etc. Secondly, more men have to be employed when capital has to be used for producing further goods. In other words, many workers have to be engaged to produce goods with the help of machines, factories, etc. Thus, one sees that employment will increase as capital formation is stepped up in the economy. Now, if the population grows faster than the increase in the stock of capital, the entire addition to the labour force cannot be absorbed in productive employment because not enough instruments of production will be there to employ them. This results in unemployment. The rate of capital formation must be kept sufficiently high so that employment opportunities are enlarged to absorb the additions to the working force of the country as a result of population growth. In India the stock of capital has not been growing at a fast enough rate so as to keep pace with the growth of population. That is why there is huge unemployment and underemployment in both the urban and rural areas. The fundamental solution to this problem of unemployment and underemployment is to speed up the rate of capital formation so as to enlarge employment opportunities.

Role of Capital: Diagrammatic Illustration

A diagrammatic illustration will make it clear as to how a greater rate of capital accumulation steps up the growth rate and also what it costs to the society. A given amount of resources has to be distributed between the production of consumer goods and capital goods. But, the greater the amount of resources that are invested in production of capital goods, the smaller quantity of resources will be left for the production of consumer goods. Thus, greater accumulation and therefore greater rate of economic growth comes at the cost of present consumption. Of course, with greater rate of growth, the productive capacity and consumption in the future years will increase, but greater capital accumulation means less consumption in the present. This is what is generally described as "more jam tomorrow means less jam today". But, as said above, if a poor country wants to raise the standards of living of its people, it must step up its rate of economic growth through greater investment of resources in the production of capital goods. Consider Fig. 41.1 in which the *X*-axis measures consumer goods and the *Y*-axis measures capital goods. With a given amount of resources we have drawn a production possibility curve *PP′*. Suppose the allocation of resources between the consumer goods and capital goods is such that the economy is operating at point *D* on the production possibility curve so that it is producing *OC* amount of consumer goods and *OK* amount of capital goods. With the capital accumulation the productive capacity of the economy will increase and as a result the production possibility curve will shift outwards. If the same *proportion* of resources continues to be allocated to the production of capital goods, the economy will be growing at a certain rate, say 2% as is indicated by the ray *OA*.

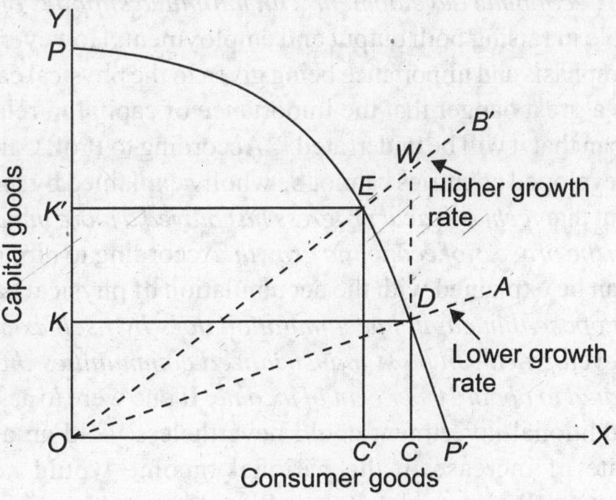

Fig. 41.1. *Capital Formation and Economic Growth*

Now, if the economy wants to step up the rate of economic growth, say to 3%, it will have to allocate a greater proportion of its resources towards the production of capital goods than before. Thus when the economy is operating at point *D* on production possibility curve *PP′*, and wants to build up more capital, it will have to allocate more resources to the production of capital goods and less to the production of consumer goods. Suppose the economy reallocates its resources so that it produces *OK′* capital goods and *OC′* consumer goods on the production possibility curve *PP′*. It is clear from the figure that with greater capital accumulation, the production of consumer goods has fallen from *OC* to *OC′*. Therefore, consumption will have to be cut down for the sake of more capital. But greater capital accumulation will mean greater growth of productive capacity with the result that production possibility curve will shift outwards more rapidly and if the same higher proportion of resources continues to be allocated, the economy will be having a higher rate of economic growth, say 3%, and will move along the ray *OB*.

Cost of Capital Formation. It is thus clear from above that the process of capital accumulation and economic growth is not a painless job. The price for it has to be paid and this price is paid in terms of the reduction in present consumption. But it should be remembered that greater capital formation will more than compensate this loss of present consumption. Thus, in our figure, when the economy has chosen the growth path of ray *OB* by having greater capital accumulation, then after some years it will reach point *W at which the consumption is OC*. Thus, at point *W*, the previous level of consumption has been restored. As the economy keeps up its higher rate of capital accumulation and moves along the growth path *OB* beyond *W*, it will be having higher consumption than along the growth path *OA*.

Role of Capital: A Dissenting View

It is generally agreed among economists that capital accumulation rate and progress of economic growth are closely correlated. In Harrod-Domar models of economic growth as well as Lewis' model of "*economic development with unlimited supplies of labour*"[4], capital accumulation plays a crucial role in raising both output and employment. However, some economists have objected to such a great emphasis and importance being given to the physical capital. Thus, Professor Cairncross remarks, "there is a great danger that the importance of capital in relation to economic progress will be exaggerated than that it will be underrated."[5] According to Prof. Cairncross, the rate of economic growth achieved in developed countries cannot be wholly explained by increases in labour and physical capital. He points out that *technological progress has played a more important role than accumulation of physical capital in the process of economic growth.* According to him, only one quarter of the rate of economic growth can be explained with the accumulation of physical capital. To quote him, *"There seems no reason to suppose that capital accumulation does by itself exercise so predominant an influence on economic development. In most industrialized communities the rate of capital accumulation out of savings is equal to about 10 per cent of income.* If one were to assume that innovation came to a standstill and that additional investment could nevertheless yield an average return of 5 per cent, the consequential rate of increase in the national income would normally be not more than ½ per cent per annum. We are told that the national income has in fact been rising in such communities at a rate of 2.3 per cent per annum. On this showing, capital accumulation could account for, at most, one-quarter of the recorded rate of economic progress. Nor were things very different in the nineteenth century."[6]

A.K. Cairncross thinks that capital accumulation need not necessarily take place along with technological progress. Technological progress can occur independently of any net capital accumulation. What is needed is that the funds kept for depreciation may be used for building up new assets and capital equipment, embodying new technology. He points out that new capital equipment embodying new advanced technology may even cost less and therefore technological progress is possible even with decline in physical capital. Moreover, Prof. Cairncross thinks that besides technological progress, improvements in social and economic organisation, trained management, new attitudes play an important role in raising production and promoting economic growth as the accumulation of physical capital. The process of economic development, according to him, "is a complex situation and it may exist in some underdeveloped countries. But it is by no means obvious that additional capital, whether borrowed from abroad or accumulated through the exertions

4. Arthur Lewis, "Economic Development with Unlimited Supplies of Labour", *The Manchester School*, 1954.
5. A.K. Cairncross, "The Place of Capital in Economic Progress", printed in G. Meier, *Leading Issues in Economic Development,* 2nd edition, 1971.
6. *Ibid.*

of surplus labour in the countryside, would by itself suffice to start off a cycle of industrialization. The problem is often one of organisation quite as much as of capital creation: of training management and men, of creating new attitudes towards industrial employment, of taking advantage of innovations that need little capital and using the resulting gains to finance investment elsewhere."

We agree with Prof. Cairncross that technological progress, human capital, *improvements in economic organisation, trained management.* etc. are also important factors in economic growth. However, the crucial importance of physical capital also cannot be denied. In our view, both physical and various forms of human capital are important in promoting economic growth.[7]

DETERMINANTS OF CAPITAL FORMATION

From above it is clear that saving is essential for capital formation. But, in a monetary economy, savings do not directly and automatically result in the production of capital goods. Savings must be invested in order to have capital goods. In a modern economy where savings and investment are done mainly by two different classes of people, there must be certain means or mechanism whereby savings of the people are obtained and mobilised in order to give them to businessmen or entrepreneurs to invest in capital goods. Therefore, in the modern free-market economy, the process of capital formation consists of the following three stages:

(*i*) An increase in the rate of real savings so that resources that would have been devoted to the production of consumption goods should be released for the purpose of capital formation.

(*ii*) Existence of a good financial system so that the available resources are obtained by private investors for capital formation.

(*iii*) The act of investment itself so that resources are actually used for the production of capital goods.

Rate of Savings

Savings in India are classified into three groups:

1. Household Savings;
2. Corporate Savings; and
3. Government Savings.

We discuss below these categories of savings and the factors determining them.

Household Savings. Household savings comprise savings of all households and individuals, both rural and urban, and non-corporate business undertakings, partnerships and non-profitable institutions like charitable trusts. Households save when they do not spend all their incomes on consumer goods. When individuals or households save, they release resources from the production of consumer goods. Resources (workers, capital, natural resources, materials, etc.) thus released are made available for the production of capital goods. The level of savings in a country depends upon *the power to save and the will to save*. The savings of the households are either in the form of *financial assets* like bank deposits, shares, bonds, provident fund, life insurance policies, national saving certificates etc, or in the form of *physical assets* such as various capital goods, houses, etc.

Level of per capita Income and Distribution of Income. The power to save or saving capacity of an economy mainly depends upon the level of per capita income and *the distribution of national income*. The greater the level of per capita income, the greater will be the amount of savings. The

7. In fact in new growth theory human capital is clubbed with physical capital to make up a single factor called '*Capital*'.

countries having higher levels of per capita income can save more than those with low levels of income. That is why the rate of savings in U.S.A. and Western European countries is much higher than that in developing and poor countries like India. Further, the greater the inequalities of income, the greater will be the amount of savings in the economy. If the national income is equally distributed, much of the income will be consumed and a little would be saved. On the contrary, if national income is unequally distributed, a larger amount of savings will be done by richer people.

Apart from the power to save, the total amount of savings depends upon the will to save. Various personal, family, or national considerations induce the people to save. People save in order to provide against old age and unforeseen contingencies. Some people desire to save a large sum to start their own business or to expand the existing business. Moreover, people want to make provision for education, marriages and a good start in business for their children.

Rate of Inflation. Another important factor which determines the household saving in the country is the price situation. In the economy when there is high inflation rate or prices are continuously rising, value of money declines. The rise in prices or the fall in the value of money has an adverse effect on the savings in the economy. When the value of money is declining every year due to the rise in prices, it is not profitable to save money and, therefore, propensity to save diminishes. Besides, inflation or rise in prices reduces the capacity to save of the people because a greater part of income has to be spent on essential consumption due to the rise in prices.

Growth Rate of per capita Income. In addition to the current level of per capita income, *growth rate of per capita income* as emphasised by life cycle theory of consumption is also an important determinant of household saving. If income per capita is not growing, the task of raising the marginal rate of saving is likely to be difficult because more saving in this case has to be achieved by curtailing present consumption. If per capita income is rising due to increase in productivity, it becomes possible to save relatively more out of increment in income and thereby rate of saving can be increased while increasing absolute consumption at the same time. It is noteworthy that the growth in per capita income can occur if investment in physical capital and human capital or its productivity increases. A higher rate of saving ensures higher rate of investment and thereby generates higher growth in per capita income. But the reverse is also true; a higher growth rate in per capita makes it possible to raise the rate of saving by keeping marginal rate of saving higher than average rate of saving.

Policies to Restrain Growth in Consumption. To increase the rate of saving deliberate policies to restrain the growth of consumption have also been adopted. For example in India prior to mid-1980's industrial licensing was actively used to ensure that there was no more investment in luxury durable consumer goods like automobiles, colour TVs, air conditioners, refrigerators except in small scale sectors and existing factories. Heavy excise duties were levied on consumer goods, especially durable ones, to discourage their production. Besides, to restrain the consumption of durable consumer goods heavy tariffs were levied on imports of such goods or imports of such goods were prohibited. It may however be noted that policies of using physical controls and detailed industrial licensing obstructed the rate of industrial growth. Therefore, in India in the beginning of the Sixth Plan (1980–85) output projections for the plan periods were based on consumption requirements of the people and physical controls on the production and consumption of goods were liberalised to step up economic growth.

Interest Rate Policy. To promote savings, policy of following positive real interest rate has been suggested by several economists. Positive real rate of interest has been suggested for raising household savings. Real rate of interest rate is nominal rate of interest rate minus rate of inflation. Advocates of the policy of real rate of interest argue that negative real interest rate will divert savings to informal financial markets, unproductive investment (such as gold, jewellery, luxurious houses) and will cause flight of capital from the country. Therefore, for promoting savings nominal rate of interest should be kept above rate of inflation. Positive real rate of interest also ensures efficient utilisation of resources, particularly scarce capital. Though rise in positive real rate of interest causes the *financial savings* to increase, there is no firm evidence about its positive effect on the increase in *overall savings* of the economy.

Table 41.1. Savings and Investment
(As per cent of GDP at current market prices)

	GDCF	GDS	Public Savings	Private Savings	Household Savings	Private Corporate Sector Savings
Average 1985-90	22.7	20.4	2.4	18.0	16.0	2.0
1990-91	26.3	23.1	1.1	22.0	19.3	2.7
1991-92	22.6	22.0	2.0	20.1	17.0	3.1
1992-93	23.6	21.8	1.6	20.2	17.5	2.7
1993-94	23.1	22.5	0.6	21.9	18.4	3.5
1994-95	26.0	24.8	1.7	23.2	19.7	3.5
1995-96	26.9	25.1	2.0	23.1	18.2	4.9
1996-97	24.5	23.2	1.7	21.5	17.0	4.5
1997-98	24.6	23.1	1.3	21.8	17.6	4.2
1998-99	22.7	21.5	–1.0	22.5	18.8	3.7
1999-2000	25.4	24.9	–0.9	25.7	21.3	4.5
2000-2001	23.9	23.5	–1.8	25.3	21.2	4.1
2001-2002	22.2	23.5	–2.0	25.5	22.1	3.4
2002-2003	25.0	26.4	–0.6	27.0	23.2	3.9
2003-2004	27.7	29.8	1.1	28.7	24.4	4.4
2004-2005	32.8	32.4	2.3	30.1	23.6	6.6
2005-06	33.7	33.5	2.4	31.0	23.5	7.5
2006-07	35.7	34.6	3.6	31.9	23.2	7.9
2007-08	38.1	36.8	5.0	31.0	22.5	9.4
2008-09	34.3	32.0	1.0	31.0	23.6	7.4
2009-10	36.3	33.7	0.2	33.6	25.7	8.4
2010-11	36.5	33.7	2.6	31.1	23.1	8.0
2011-12	35.5	31.3	1.2	30.1	22.8	7.3
2012-13	34.8	30.1	1.2	29.0	21.9	7.1

Public Savings

Public Savings are another source of savings. In Government savings are included the surplus of revenue obtained through taxes and surpluses from public undertakings over and above the current expenditure of the Government. Given the current consumption expenditure of the Government, the greater the revenue collected from taxes and the profits earned from public enterprises, the greater the Government or public savings. It is worth noting that all revenue earned from taxes, etc. does not constitute the public savings because a good part of the tax revenue is spent on current consumption expenditure of the Government such as civil administration, defence and other unproductive services. That part of income or revenue of the Government which is over and above the expenditure on civil administration, defence and other current consumption is called the public savings. The Government savings are used for public investment spending on building of factories, roads, irrigation facilities, and other various types of industries started by the Government.

In India public savings of both Centre and State governments was around 2.9 per cent of GDP at market price in 1970-71, it reached its peak of 4.9 per cent of GDP in 1976-77 and started declining thereafter and in 1989-90, it fell to 1.6 per cent of GDP. The decline in *budgetary savings* of government as distinguished from those of public enterprises was relatively much higher. It is noteworthy that trends in fiscal deficit determine the performance in public savings. Table 41.1 shows that public savings increased to 2.0 per cent of GDP in 1995-96, the year in which revenue deficit was at the lowest mark in the nineties. Subsequently, with the increase in revenue deficit, public savings declined and even became negative. Had the household savings not increased during this period, this would have led to the decline in gross domestic saving (GDS) and would have adversely affected investment and economic growth in the economy.

However, after 2003-04 there was appreciable increase in public saving as will be seen from Table 41.1. Rate of public saving as per cent of GDP which was –0.6% in 2002-03 became positive after it. In 2003-04 it was 1.1 per cent which went on increasing and in 2007-08 it was at peak of 5 per cent of GDP. This accounts for large increase in overall gross domestic saving (GDS) and gross domestic capital formation (GDCF). On account of this large increase in gross domestic saving and capital formation rate of growth of GDP rose above 9 per cent in three successive years, 2005-08, and the Indian Economy became the second highest growing economy of the world. It is worth noting that to prevent global financial crisis (2008-09) from causing drastic decline in economic growth the Indian Government reduced its several taxes and increased its expenditure to provide fiscal stimulus to the Indian economy. As a result, revenue deficit increased and public saving drastically declined to 1.0 per cent of GDP in 2008-09, 0.2 per cent in 2009-10, rose to 2.6% in 2010-11, but again fell to 1.2% of GDP as per 2004-05 series.

The huge combined fiscal deficit of the Centre and the States which presently constitutes about 10 per cent of Gross Domestic Product is adversely affecting saving and investment in the economy and therefore the rate of economic growth of the Indian economy. To quote Shankar Acharya, a former economic advisor to the Ministry of Finance, "It is quite interesting how the deterioration of about 3 per cent points of GDP in the consolidated revenue deficit between 1995-96, and 1998-99 is reflected strongly in the worsening of aggregate savings and investment ratios over the period. It would be hard to find more telling evidence of the adverse impact of fiscal deficit on savings and investment. Another way to look at this is that if the fiscal deficit deterioration since1995-96 had not occurred, saving and investment might well have been higher by around 3 per cent points of GDP in recent years."[8]

The above analysis underlines the importance of measures to reduce fiscal deficit, particularly revenue deficit, if rate of saving and investment in the Indian economy is to be raised to accelerate the rate of economic growth. For this purpose base of taxation must be widened by withdrawing the several exemptions in income and wealth taxes as suggested by various committees.[9] Besides, the subsidies must be well targeted so that only the poor should take advantage of them and the remaining subsidies must be phased out. Further, to reduce revenue deficit, non-plan non-essential government expenditure must be reduced along with the reforms in the taxation system. However, care should be taken to avoid any cut in Government expenditure on education and health which promote accumulation of human capital and have beneficial effect on economic growth.

It may be pointed out that savings may be either voluntary or forced. Voluntary savings are those which people make of their own free will. As explained above, voluntary savings depend upon the power to save and will to save of the people. On the other hand, taxes by the Government represent forced savings.

Private Corporate Sector Savings. Furthermore, savings may be done not only by households and Government but also by corporate business enterprises Corporate business enterprises save when they do not distribute whole of their profits but retain a part of them in the form of undistributed profits. They then use these undistributed profits for investment in real capital. In advanced countries like U.S.A. and England, these corporate savings form the largest part of the total savings. In U.S.A. and England, corporate savings, *i.e.*, undistributed profits constitute about 50% to 60% of the total annual domestic savings. Note that *savings of unincorporated*

8. Shankar Acharya, "Macroeconomic Management in the Nineties", *Economic and Political Weekly*, April 20, 2000, p.1525.
9. Recently Kelkar Committee has recommended reforms in taxation system on these lines.

business enterprises are included in household savings. Business savings of private corporate sector in India are shown separately in Table 41.1. It will be seen from this table that the private corporate sector, which operates in the organised sector, though contributes significantly to Gross National Product (GDP) their saving makes only a small contribution to gross domestic savings. The corporate sector savings which contributed on an annual average of 2.0 per cent of GDP to Gross Domestic Saving in 1990-91 reached its peak in 1995-96 when their savings constituted 4.9 per cent of GDP. Thereafter, it declined to 3.4 per cent of GDP in 2001-02 and to 3.9 per cent in 2002-03.

The lower corporate saving was attributed to high rental cost of capital (including high real rate of interest), high corporate tax rate, high rates of excise duties, tax on dividend distribution, inefficiency in use of investment funds and underutilisation of production capacity. Liberalisation regime initiated in India since 1991 providing for reduction in tariff rates on imported raw materials, reduction in corporate tax, and excise duties resulted in higher corporate saving in 1995-96 and 1997-98. Lowering of lending interest rates as a result of soft interest rate policy also boosted the profits of corporate sector and from 2002-2003 corporate saving has tended to rise (See Table 41.1). Corporate sector saving rose from 4.4% of GDP in 2003-04 to 6.6% in 2004-05, to 7.5% in 2005-06 and 7.9% in 2006-07 and further rose to 9.4% in 2007-08. The global financial crisis in 2008-09 affected the saving of the corporate sector which declined to 7.4 per cent in 2008-09 but again rose to 8.4% per cent in 2009-10 and 80% in 2010-11 with slowdown in economic growth of the Indian economy, corporate sector saving declined to 7.3% of GDP.

Financial System

Next step in the process of capital formation is that the savings of households must be mobilised and transferred to businessmen or entrepreneurs who require them for investment. In the capital market, funds are supplied by the individual investors (who may buy securities or shares issued by the companies). Banks, investment trusts, life insurance companies, finance corporations, are other financial institutions which mobilise savings for investment purposes. In India there is also unorganised capital market constituting indigenous bankers and money-lenders. One reason responsible for the low rate of capital formation in India has been the absence of well-developed capital market. Indian capital is said to be very shy. If the rate of capital formation in the private sector is to be stepped up, the development of financial system is very necessary. Well-developed financial market will ensure that the savings of the society will be mobilised and transferred to the entrepreneurs or businessmen who require them for investment.

Investment

For savings to result in capital formation and economic growth they must be invested. In order that investment of savings should take place, there must be a good number of honest and dynamic entrepreneurs in a country who are willing to take risks and bear uncertainty of production. Given that a country has got enough good and venturesome entrepreneurs, investment will be made by them only if there is sufficient *inducement to invest*. Inducement to invest depends on *marginal efficiency of capital*, *i.e.*, rate of return or profit on the one hand, and real rate of interest on the other. If prospects about profit are quite bright, that is, when entrepreneurs expect large profits from investment, investment will be made on a large scale. If profits are expected to be low, investment made will also be small. It is generally believed that at a lower real rate of interest, investment is more and at a higher real rate of interest, investment is less. In other words, it means that when credit is cheaper, businessmen will borrow more funds for investment purposes.

Of the two determinants of inducement to investment, namely, marginal efficiency of capital and the rate of interest, it is the former which is of greater importance. Marginal efficiency of capital depends upon the *cost or supply price of capital* as well as the *expectations of profits*. Fluctuations in investment are mainly due to the changes in expectations regarding profits. But it is *the size of market* (*i.e.*, demand for goods) which determines the scope for profitable investment. Thus primary factor which determines the level of investment or capital formation in an economy is the *size of market* for goods. Demand for capital equipment like machinery depends upon the demand for goods it helps to produce. The inducement to invest is limited by the size of the market, *i.e.,* the demand for the goods. The greater the size of the market, the greater will be the inducement to invest. The small size of the market will discourage the entrepreneurs to make investment. All poor countries have a small market in everything and this is a great hurdle in the way of capital formation in underdeveloped countries.

Animal Spirits and Investment. J.M. Keynes who emphasised that marginal efficiency of capital played a crucial role in determining investment also laid stress on animal spirits of entrepreneurs for causing fluctuations in investment. By animal spirits Keynes means the waves of optimism and pessimism which quite often grip business class and greatly affect their investment behaviour. Keynes attributed the drastic fall in private investment in 1929-33 to the collapse of marginal efficiency of capital which was caused by wave of pessimism among the business class which resulted in stock market crash in England. Even in India today lack of confidence among business class resulting from their animal spirits is responsible for slowdown in private investment since 1997.

Public Investment in Infrastructure. Private investment is also affected by the availability of infrastructure such as power, good roads and means of transport, good means of communication, good ports, etc. Slowdown in the Indian economy since 1996-97 is also partly due to lack of good infrastructure. It has therefore been suggested that *public investment in infrastructure not only generates demand for the industrial products but also leads* to the expansion of infrastructural facilities such as power, ports, roads etc. Thus, in the Indian context public investment helps to solve both the supply-side and demand-side problems of private sector investment. Therefore, in India the greater public investment *crowds in* private investment rather than *crowding it out*.

Taxation Policies. Taxation policies of the Government also affect investment behaviour of the corporate sector. High corporate taxes discourage private investment. For example, high corporation income tax, tax on dividend distribution, high excise duties and tariffs adversely affect corporate sector investment by lowering the profitability of investment. Therefore, to increase corporate investment, investment-friendly taxation policies need to be adopted.

Availability of Credit. Apart from internal savings, the corporate sector gets investment funds from the capital market and banks. To meet their needs for working capital they rely on bank credit. Therefore, to ensure greater private investment adequate credit facilities must be made available.

Improvement of Stock Market. Besides, corporate sector raises its equity capital from the stock market. Therefore, the working of stock markets must be improved and made fair and transparent, so that public should have confidence in fair dealings at the stock market. For instance, stock market crash in 1992 engineered by Harshad Mehta and recently by Ketan Parikh and scams involving UTI have badly shaken the confidence of small investors who have withdrawn from the market. This has created difficulties for private corporate sector for raising investment funds from the stock market through issue of equity capital.

Size of Market or Level of Demand

Ragnar Nurkse in his now well-known work, '*Problems of Capital Formation in Underdeveloped Countries*', emphasized that *size of market* (that is, *level of demand*) for a commodity is an important determinant of capital formation in developing countries. Furthermore, according to him, size of market for a commodity is determined by levels of incomes of the people. Since the production of industrial commodities in the modern times requires installation of machinery and other capital equipment which have large productive capacity, therefore, for making investment in them to be profitable there should be sufficient size of market or demand for that industrial commodity. According to Nurkse, since because of poverty in underdeveloped countries per capita income of the people is low, there will not be enough demand for an industrial product and as a result productive capacity created will remain underutilised. Therefore, it will not be profitable to make investment in the machinery and other capital equipment for the production of the commodity. Thus limited size of the market in underdeveloped countries lowers inducement to invest in them to overcome this problem of limited size of the market for commodities. Nurkse suggested '*balanced growth strategy*' under which *simultaneous investment* in a large number of industries be made so that the workers employed in various industries create demand for each other's product. We will discuss the issue of linked size of market adversely affecting inducement to invest and Nurkse's balanced growth strategy in detail in a later chapter.

CAPITAL FORMATION AND ECONOMIC GROWTH IN INDIA

In India also capital formation or rate of investment has played a significant role in generating higher economic growth rate. The higher annual growth rate of 5.8 per cent per annum during the eighties was due to higher rate of investment or capital formation as compared to the earlier decades of planned development which witnessed only 3.5 per cent average growth rate per annum.

The nineties represent a very important period in economic growth in India because far reaching economic reforms of liberalisation and privatisation were carried out during this period. As will be seen from Table 41.2, leaving the first two years as they represent the transition period of crisis and adjustment, in the post-reform period, the growth rate in 1992-93 to 1996-97 has been quite higher (6.6 per cent average annual growth rate) as campared to 5.8 per cent in the pre-reform period. This

Table 41.2. GDS, Capital Formation and Economic Growth (GDP Growth) 1990–2011

Year	Growth of GDP	(As per cent of GDP at current market prices) GDCF	GDS
1990–91	5.6	26.3	23.1
1991–92	1.3	22.6	22.0
1992–93	5.1	23.1	21.8
1993–94	5.9	26.0	22.5
1994–95	7.3	26.9	25.1
1995–96	7.3	24.5	23.2
1996–97	7.8	24.5	23.1
1997–98	4.8	24.6	21.5
1998–99	6.5	22.7	24.1
1999–2000	6.1	25.4	24.9
2000–2001	4.4	23.9	23.5
2001–2002	5.6	22.1	23.5
2002-2003	3.8	25.0	26.4
2003-2004	8.5	27.7	29.8
2004-2005	7.5	32.7	32.2
2005-2006	9.5	34.7	33.3
2006-2007	9.6	35.7	34.5
2007-2008	9.3	38.1	36.9
2008-2009	6.7	34.3	32.0
2009-2010	8.4	36.6	33.8
2010-2011	8.4	35.1	32.3
2011-2012	6.5	34.6	32.5
2012-2013	5.1	36.6	31.8
2013-14	6.8	32.3	30.6
2014-15	7.3	N.A	N.A

Note. Figures of GDP growth sets, capital formation and saving rates for the three years 2012-13, 2013-14 and 2014-15 are not strictly comparable with the earlier ones as they are based on new series (with 2011-12 as the base). Besides in this new series, GDP is measured at market prices as compared with at factor cost in the earlier series.

was due to pick up in corporate saving and investment following the abolition of industrial licensing and reduction in rental cost of capital due to reduction in tariffs on imports of raw materials, and reduction in corporate taxes which increased the profitability of private investment.

However, after 1996, higher growth rate could not be maintained and in fact Indian economy went into recession despite the continuation of economic reforms. Two factors are responsible for it. First, following the structural adjustment reforms and the need to contain fiscal deficit public investment declined. This adversely affected the overall growth in GDP as well as the industrial growth through adverse both supply-side and demand-side factors. It will be seen from Table 41.2 that during the whole of nineties (1991-2000) higher average growth rate of 6.2 per cent per annum was achieved mainly due to higher rate of gross domestic capital formation (GDCF). In the late nineties and in 2000-01 and 2001-02 when investment or GDCF was less especially due to decline in public investment, growth rate of GDP also fell.

After 2003 when both public investment and private investment had picked up, growth rate of GDP rose to over 9 per cent per annum during the three years (2005-08). In the Eleventh Five Year period there was average 8 per cent growth in GDP per annum. This was feasible as gross domestic saving rate (GDS) had gone up to 36.9 per cent per annum in 2007-08. However, the impact of global financial crisis (2008-09) brought down the growth rate of GDP to 6.7% in 2008-09 and to 8.4% in both 2009-10 and 2010-11. However, for the reasons explained below, for 2011-12, GDP growth rate was lower at 6.5 per cent, and 5.1 per cent in 2012-13.

MEASURES TO ACCELERATE CAPITAL FORMATION

As explained above, rate of capital formation is an important determinant of economic growth. Therefore, all efforts should be made to raise the rate of capital formation in the country, if the twin problems of mass poverty and unemployment are to be solved. We have also discussed the various problems, namely, the operation of vicious circles of poverty, the working of demonstration effect and the growth of wants, which come in the way of our efforts to accelerate capital formation in developing economies. These problems have to be overcome if the rate of capital formation is to be raised.

In order to raise capital formation in the economy we have first to raise *ratio of surplus to capital* or rate of surplus to national income and then measures have to be adopted to mobilize this surplus for purposes of productive investment.

Raising the Ratio of Surplus to Capital

In this connection the distinction between potential economic surplus and actual economic surplus should be borne in mind. Potential economic surplus is the difference between the national income of a country and the *essential or basic* consumption (which includes also capital replacement), while the actual economic surplus is the surplus actually achieved after spending on essential and non-essential consumption and also unproductive investment.

Now, if the ratio of actual surplus to capital is to be raised, unproductive uses of potential surplus, namely, non-essential consumption and unproductive investment have to be prevented. As has been mentioned before, in developing countries wasteful use of potential economic surplus in the form of non-essential consumption and unproductive investment is quite enormous both among private individuals and the Government. The unproductive consumption is indulged in by property owners such as big landlords and rich farmers, rich businessmen and industrialists when they spend a part of their income on luxuries such as purchase of expensive motor cars, air conditioners, construction of palatial houses, gold and jewellery, etc. The high wage earners and self-employed also indulge in luxurious consumption when they spend a good part of their income on relatively luxurious items such as costly sofa sets and other furniture, expensive clothes, wine, unnecessary entertainments and lavish expenditure on births, marriages and death ceremonies, etc. Even the Government spends a lot on unnecessary or unproductive consumption expenditure when it spends excessively on police and other administrative services, defence, large size of the ministries, liberal travelling allowances and other facilities to the officers and ministers, etc.

Productive surplus is also reduced by the various forms of unproductive investment made both by the private sector and the Government. The unproductive investment by the private sector takes place when the rich businessmen and landlords invest their incomes in real estate (*i.e.,* land), palatial houses, gold and jewellery. Likewise, even Government makes unproductive type of investment when it spends money on the construction and maintenance of luxurious hotels, office buildings, houses, parks, etc. It, therefore, follows that if the actual and productive surplus is to be raised the magnitudes of non-essential and unproductive investments are to be considerably reduced.

It may be noted that the propensity of the people of underdeveloped countries to use the economic surplus for conspicuous consumption has become stronger in recent years because of the *international demonstration effect.* The concept of international demonstration effect is that people of poor developing countries try to copy the high consumption standards of the people of affluent developed countries. As a result, unproductive or non-essential consumption of these countries increases and as a result surplus available for productive investment falls. This international demonstration effect has become quite strong these days because of the greater awareness on the part of the people in underdeveloped countries regarding the living standards of the people in rich nations. This greater awareness has emerged because of wide travel by the people of the poor countries in Western countries, propaganda and advertisement in radio, television and newspapers. Therefore, if the ratio of surplus to capital is to be raised significantly, the operation of the demonstration effect should be overcome by educating the people and also by providing various incentives for saving.

Incentives for Saving

An important measure to encourage savings and raise ratio of surplus to capital is that savings be exempted wholly or substantially from income taxes. When people know that they will not have to pay income taxes on the saved portion of their income, they will be greatly tempted not to consume a good part of their income and thereby save it. Secondly, in order to encourage savings, rates of interest on savings and time deposits with the banks be raised. In order to be able to earn more income because of higher interest rates, people will be induced to save more and make deposits in the banks. Thirdly, a great incentive for saving can be provided if the prices are prevented from rising. When the prices are continuously rising and as a result value of money is rapidly falling, it will not be worthwhile to save, for the money saved will greatly depreciate with the rise in the prices of goods.

We thus see that if proper fiscal, monetary and price policies are adopted incentives to save can be greatly increased and in this way ratio of surplus to capital can be raised. But contrary to what is required to raise the rate of saving, in India in recent years liberal loans for non-essential commodities such as purchase of cars, air conditioners, refrigerators and construction of luxury houses are being given by banks at cheap rates of interest. This tends to reduce the ratio of saving to national income.

Some development economists, especially Professor Arthur Lewis, say that the saving rate depends on the share of profit and wages. They have suggested that if the national income is distributed more in favour of profits, that is, if the share of profits in national income is greater, the saving ratio will be high and *vice versa*. This is because, according to them, recipients of profits, *i.e.*, capitalists or entrepreneurs save a major part of their income, their propensity to consume being very little.[10]

10. In fact, Lewis in his famous model of "*Economic Development with Unlimited Supply of Labour*" has assumed that all profits are saved and automatically invested and all wages are consumed.

On the other hand, if share of wages in national income is relatively high, the saving ratio will be low as the propensity to consume of the workers is very high. Thus, according to growth models of Kaldor, Robinson and Arthur Lewis, the saving or surplus ratio of a community depends on the relative shares of profits and wages. In order to raise the savings or surplus ratio they have proposed that wages in the modern industrial sector should be kept at a certain minimum level so that with the development of the sector, profits share goes on rising and as a result growth of investment or capital accumulation becomes cumulative. In this way the growth and employment objectives can be achieved in the shortest possible time.

Though the above mentioned relation between profits share and savings ratio is generally true, it may, however, be pointed out that a good part of the profits made by the capitalists in underdeveloped countries like India are spent on luxurious consumption and unproductive investment. Therefore, if rate of capital formation is to be raised through raising share of profits in national income, the tendency on the part of the capitalists to spend profits on luxurious consumption and unproductive investment has to be effectively curbed through taking suitable monetary and fiscal measures.

Mobilisation of resources for capital formation by the Government is of crucial importance in those underdeveloped countries where public sector plays an important role in the economic development. The Government can raise the resources for capital formation in the following ways: (*a*) by imposing taxes on the people, (*b*) by borrowing from the people, (*c*) by ploughing back the profits by the Government of public undertakings and (*d*) by creation of new money or borrowing from the central bank of the country which is also known as deficit financing. We explain below these various ways of mobilising surplus or resources by the Government to finance its investment expenditure.

Taxation

Taxation is an important method of raising resources by the Government. Taxation increases the amount of collective savings by restricting private consumption. Thus, as a result of taxation, the people in general or private sector are forced to save for accelerating capital formation. It should, however, be noted that the total tax revenue collected by the Government does not represent savings because a good deal of tax money is spent by the Government to meet its current expenditure such as defence, police and civil administration which is similar to the private consumption expenditure of the people. Therefore, it is only the surplus of tax revenue over and above such current expenditure of the Government that represents the net savings of the Government sector. Further, even the extra government savings achieved through taxation may not constitute net addition to the aggregate savings of the country. Net addition to aggregate savings will occur only to the extent the tax payments are made by sacrificing consumption. On the contrary, if taxes are paid by the individuals and firms by reducing private savings, then the increase in public savings may not result in any net increase in overall savings.

On the other hand, it has been asserted by some economists that the marginal propensity to save of the Government out of its tax revenue is lower than marginal propensity to save of the private individuals and firms. This is because the government uses a greater part of its income from taxation to meet its ever increasing current expenditure. If this is so, then the taxation instead of raising overall saving will actually reduce it. If the Government is to mobilise resources for development, it has to check the growth of its non-productive expenditure.

Both direct and indirect taxes play an important part in augmenting the resources of the Government for developmental purposes. It is best to levy taxes on non-entrepreneurial incomes and goods of luxury consumption. Such taxes will curtail unnecessary consumption and will have no disincentive effect on business enterprise. Hence, in order to raise adequate amount of resources, it becomes necessary to bring increasing number of '*services*' in the tax net. In addition, taxation of agricultural incomes has to play an important part in the mobilisation of resources for the public sector in a developing economy.

Taxation plays a double role in a developing economy. On the one hand, it mobilises resources for the public sector schemes and projects. On the other hand, if the taxation structure is well designed, it can direct the flow of the resources of the economy to productive channels of investment in the private sector. By discouraging speculative activity and other types of unproductive investment and also useless and luxury consumption, taxation promotes investment to accelerate economic growth.

The level of voluntary savings in underdeveloped countries is very low and the upper classes which are in a position to save prefer to invest these savings in real estate or jewellery or they simply hoard them. The result is that such savings are not available for productive investment. In these circumstances, compulsory saving enforced by taxation performs an essential function. By releasing resources from non-functional consumption and non-essential or unproductive investment, taxation makes these resources available for the public sector. In order to raise adequate resources for development, it may have to infringe on essential consumption too.

The role of taxation is not to be merely static. It has also to play a dynamic role in raising resources for capital formation. As economic development proceeds national output increases and new incomes are generated. The tax policy must make sure that these new incomes are not permitted to flow into unproductive channels. They must be directed towards productive investments. The tax structure should have *built-in flexibility* to perform this function automatically. As incomes grow, the tax base must also be deepened and broadened. For this purpose, the marginal rate of taxation will have to be kept high and Prof. Lewis suggests that the marginal rate of taxation should be considerably greater than the average rate[11]. Such a tax structure presupposes a progressive income tax and also high rates of indirect taxes on commodities having a large income elasticity of demand. In this way an increasingly larger proportion of increment in output would be converted into public savings. Considering that there is considerable non-monetised sector in backward economies and there are several institutional barriers, the tax structure will have to be considerably broadened, deepened and diversified. Otherwise there is no guarantee that adequate resources will be mobilised for developmental purposes.

Drawbacks. But taxation as a method of mobilisation of surplus presents some difficulties. In developing countries, only a narrow range of population is affected by taxation to any appreciable extent. The reason is that only a small number of people have their income reaching the level of direct taxes; and the indirect taxes on luxuries, which are the main source of income in advanced countries, fall on a small number of people. Hence the yield from taxes is small owing to their limited coverage. And the per capita revenue collected in the country is very low.

While taxation increases involuntary savings, it decreases voluntary savings to enable the people to maintain the former standard of consumption. The result is that the resources available for the private sector are thereby diminished. Another major drawback of taxation is that it has disincentive effect on business enterprises and discourages savings. The taxes on wage-earners diminish their incentive to work harder; taxes on profits of higher income groups reduce their incentive to save and to make investment in new enterprise; taxes on output or incomes of farmers diminish their incentives to improve agricultural techniques. Hence taxation militates against the efforts of those who can make an important contribution to economic growth.

Thus saving extracted from taxation is not an unmixed advantage. It is, therefore, very necessary to devise a system of taxation which will not weaken the incentives to work, save and invest; nor should it violate the accepted notion of equity. Taxation should not impair the productivity of the economic system. A high marginal rate of taxation may have a considerable disincentive effect on private enterprise.

11. W.A. Lewis. *The Theory of Economic Growth*.

Role of Indirect Taxes. It has been found that indirect taxes are better suited to the conditions obtaining in developing countries for reducing current consumption and mobilising resources for development because in such countries quite a large proportion of national income tends to be diverted to current consumption instead of being productively invested. The average propensity to consume in such countries is much higher than is the case in advanced countries. Therefore, in developing countries, indirect taxes must play a more important part. They will raise the rate of savings which are so essential for economic growth. It may be noted that with the adoption of new economic policy of liberalisation since 1991 rates of customs duties and excise duties in India have been reduced in order to reduce the costs of industries and thereby to increase their competitiveness However, this has reduced the tax-revenue of the Government and has also lowered the tax-GDP ratio.

But, "High rates of taxes on commodities with a high income elasticity of demand are quite effective in siphoning a substantial proportion of increase in output into the resources of the public sector needed for development financing and a higher rate of commodity taxes on luxury articles tends to introduce an element of progressiveness in an otherwise predominantly regressive tax structure in developing countries".[12] But in order to make sure that the resources raised through commodity taxes are adequate, it will be necessary to widen their base by withdrawing various exemptions from tariff duties and excise duties[13]. In the poor countries, it is not possible to exempt entirely goods of general consumption, because they are the only goods that provide a base broad enough to assure an adequate amount of resources.

Role of Taxation in Promoting Private Investment. It is generally considered that taxation is meant to mobilise resources for the public sector. But a little thought will show that even in promoting investment in the private sector taxation plays an important role. It plays an allocative role by directing the resources to desirable and useful channels. A developing economy can adopt taxation measures which will check conspicuous and non-functional consumption, especially the consumption of luxuries. Funds so saved will then be available for investment in the private sector. This role is played by import duties and excise duties on luxury or semi-luxury articles. A steady progressive tax on the upper income brackets also promotes saving. Diversion of savings into a real estate, Jewellery or other prestigious but unremunerative investment can be checked by taxes and so also investment for speculative purposes. Taxes for these purposes can be imposed on land and buildings, capital gains or increments in land values.

Thus taxation can provide incentives for desirable investments and can check undesirable and unproductive investment. For instance, there is a tax holiday for new industrial undertakings. Liberal depreciation allowances can be given on new buildings, plant and machinery. Development rebates may be given for reinvesting profits in business, income tax may be imposed on profits declared as dividends. Substantial carry-overs of business losses may be allowed.

Government Borrowing

Borrowing by the government is another method by which the savings of the community may be mobilised for economic development. In developing economies, the Governments resort to public borrowing in order to finance schemes of economic development. Public borrowing becomes necessary because taxation alone does not provide sufficient funds for economic development. Besides, too heavy taxation of income and wealth be avoided as it has an adverse effect on saving and investment.

Public borrowing takes two forms: (*a*) *market loans* and (*b*) *small savings*. In the case of market borrowing, the Government sells to the public negotiable Government securities of varying terms and duration and treasury bills. For financing capital projects long-term Government

12. Tripathy, R.N. *Public Finance in Underdeveloped Countries*, 1966, p. 200.
13. See Kelakar Committee Report, 2002.

bonds are floated in the capital market. This form of public borrowing is more important for mobilising resources for development. The treasury bills which represent short-term borrowing are intended to meet only the current Government expenditure. New bonds may be issued for meeting old maturing bonds. The small savings represent public borrowing which is not negotiable and are not bought and sold in the capital market. For mobilising small savings, various types of savings certificates are issued, *e.g.*, National Savings Certificates, National Development Certificates, Rural Development Bonds, Postal Certificates and Postal Accounts, Compulsory Deposits, etc. A widespread campaign is necessary to attract small savings.

Borrowing is the quickest mode of raising funds. On the other hand, tax finance is not so expeditious because passing of tax laws, assessment of taxes based on those laws and their collection involves considerable delay. Besides, public debt does not involve any burden if it is devoted to productive works. The subscribers to Government loans are able to find remunerative investment whereas the Government can pay the principal and interest out of the income yielded by investment financed from loans. Thus public borrowing is not only necessary but also desirable.

There is another advantage in public borrowing. *Public borrowing in anti-inflationary*. The underdeveloped countries are victims of inflation since they have to resort to deficit financing for finding funds for economic development. Since deficit financing is inflationary, public borrowing is preferable to deficit financing. Public borrowing mops up the surplus purchasing power with the people. It thus checks consumption and so a rise in prices. At the same time idle balances are absorbed in productive activity.

Public borrowing generates additional productive capacity. The funds raised by public borrowing can be *utilised for building up the infrastructure* for the economy through schemes for the development of roads, ports, highways, irrigation, transport and power. They help in building up of the agricultural and industrial base of the economy. All such investments facilitate further economic growth. There is no doubt, therefore, that the public borrowing plays a very useful and important role in accelerating economic development of developing economies. Public debt promotes saving and investment, the two most crucial determinants of economic growth.

But there exist a number of obstacles which hinder the success of borrowing policy in an underdeveloped economy. In many such countries, there are no organized money and capital markets and in those where such markets exist, they constitute a very small proportion of total money market of the country. There may not, besides, exist any organic relationship between the organized and unorganized part of the money market. Moreover, the resources of the organized capital market may be too inadequate to fulfil the needs both of the private and of the public sector. Further, in the capital market the competition for funds between the government and the private sector will raise rate of interest and this will have a highly disincentive effect on the expansion of investment in the private sectors. Besides, for the mobilisation of savings it is necessary to check and regulate the diversion of savings into unproductive investment such as real estate, gold and jewellery and inventory accumulation.

It may be noted here that there is now ample evidence in India that if resources through borrowing by the Government are spent for public investment in infrastructure such as power, communication, railways, roads, highways, *it crowds in* private investment rather than *crowding out* private investment. This is because private sector investment is impeded by the non-availability of good infrastructure. Further, public sector investment creates demand for products of private sector such as cement, steel and others in addition to the expansion of aggregate demand in general.

Suitable techniques of borrowing must also be devised. For example, bonds issued by the Government should be adjusted to the preferences of the general public; bonds of large denomination and long maturity may be offered to the institutional investors, whereas those of small denomination and short maturity may be reserved for the non-institutional public. If properly devised and conducted small savings campaign can mobilise a sizable amount of resources. Further, the mobilisation of the hoardings of gold and jewellery through government programmes constitutes a highly desirable source of resource mobilisation. Of course, suitable techniques of public borrowing for the mobilisation of these resources have to be evolved.

Profits of Public Undertakings

In developing countries with a mixed economy, the public sector has been expanded in the past especially when the objective was the establishment of a socialistic pattern of society in India. The result is that a large number of projects of irrigation, power and industrial projects in heavy and basic industries are being executed by the Government. Commercial enterprises like steel plants, oil and petroleum enterprises, fertilizer plants, machine tool factories and State trading organisations were set up by the Government. It was expected that these public undertakings would in course of time start yielding profits which would be available for reinvestment in the schemes of economic development. This is how public undertakings can make their contribution to self-generating growth.

The public undertakings are not to be run on no-profit no-loss basis known as the 'break-even' principle. But in the interest of a continuing process of economic growth they must be made to yield a reasonable return on investment. The price policy of State trading corporations must also be adjusted to mobilise resources in the form of trading profits. Like the private enterprises, they must pursue the principle of profitability implying efficiency and cost reduction to enlarge the profits accruing from such enterprises. Their working, therefore, has constantly to be reviewed and ordinary principles of accounting and auditing must be applied thereto. Thus, contrary to the usual conception, the public sector enterprises should not only pay their way but also make legitimate profits.

But it cannot be denied that the surpluses from public enterprises should constitute a reasonable proportion of the resources for development. In developing countries, where there is a paucity of investible resources, the process of accelerated economic development requires that these enterprises should make reasonable return on their investment and make their maximum contribution to capital formation. It is a very convenient and desirable method of resource mobilisation. The profit is concealed in the price charged. Thus it is administratively convenient and politically feasible.

The public enterprise is not now confined to providing environmental facilities or social and economic overheads but also extended to purely industrial and commercial business. They have, therefore, to pursue price-output policies which should generate enough resources for capital formation. Dr. V.K.R.V. Rao has rightly advocated that "*price policies in the public sector should be such as to yield a planned profit which will include not only depreciation expenditure but also an element of surplus for capital formation*".[14]

Raising Resources through Public Sector Disinvestment. It may be noted that with the adoption of new economic policy of liberalisation and privatisation since 1991 role of public sector in Indian economy has been greatly diluted and Government has been making disinvestment in public sector enterprises in which it holds a large share of equity capital and thereby raising resources for its various future investment projects. Accordingly, Government has sold some of its share capital in many public enterprises such as BALCO, VSNL, CMC, Modern Foods, some oil companies, Maruti to private sector firms. Recently, Government sold a good part of its share capital in Maruti Car Company through open public offer in the market and raised ₹ 900 crores. In the Tenth Plan period (2002-07), it was planned to raise resources of the order of ₹ 10,000 crores annually through public sector investment. Thus, disinvestment of public sector enterprises is yielding good revenue to the Government. Resources raised through public sector disinvestment if invested in infrastructure and social services such as education and health will accelerate economic growth in India.

14. Ooty Seminar Papers, 1959, p. 33.

CHAPTER 42
TECHNOLOGICAL PROGRESS AND ECONOMIC DEVELOPMENT

We have explained earlier that capital formation and technological change are two important factors that determine rate of economic growth. Having discussed capital formation in the preceding chapter, we now turn to the explanation of technological change and its role in expanding output and employment. In the beginning it may be noted that technological progress plays a more important role than capital formation. In the absence of technological progress capital formation can lead to increase in output per head only so long as all the workers in the given labour force have not been fully equipped with the capital equipment embodying the latest available techniques. Once all the workers in the labour force have been so equipped with capital, further economic growth, that is, increase in output or income per head comes to an end. *It is technological progress which can bring about continued increase in output per head of the population.* Technological progress is thus the prime-mover of economic growth.

Meaning of Technological Change

Technological change or progress refers to the discovery of the new and improved methods of producing goods. Sometimes technological advances result in the increase in available supplies of natural resources. But more generally technological changes result in increasing the productivity of labour, capital and other resources. The productivity of combined inputs of all factors is called *total factor productivity*. Thus technological progress means increase in total factor productivity. As a result of technological advance, it becomes possible to produce more output with same resources or the same amount of product with less resources.

But the question arises as to how does the technological progress take place. The technological progress takes place through *inventions and innovations*. The word invention is used for the new scientific discoveries, whereas the innovations are said to take place only when the new scientific discoveries are commercially used for actual production of goods. Some inventions may not be economically profitable to be used for actual production.

It follows from above that technological change brings about an increase in output per head. Thus technological change, or more precisely technological progress, is *the change in the production process which results in an increased output per unit of labour*. Technological change causes a shift in the production function embodying all known techniques. Technological change must be distinguished from a *change in technique*. While by technological change we mean advance in knowledge resulting in improved methods of production, change in technique refers to the use of a different but already known method of production.

TECHNOLOGICAL CHANGE AND ECONOMIC GROWTH

The process of economic growth involves the increase in the production of goods and services. Increase in production can be achieved either through the use of more resources and/or through the realisation of higher productivity by means of using the resources of labour, capital and land more efficiently. Technological change helps to promote growth in both these ways. It can help in the discovery of new natural resources in the country and thereby enhances the productive potential of the country. Technological change also increases the productivity of available resources. For instance, it can find out the productive uses of land that hitherto has been regarded as infertile or it can discover new economic use of a raw material that had previously been considered as useless. But, as stated above, *technological change more generally results in higher productivity of resources*. Technological change raises the productivity of worker through the provision of better machines, better methods and superior skills. In Table 42.1 we give the percentage increase in labour productivity in a number countries during the period 1970-1989. By bringing about increase in productivity of resources the progress in technology makes it possible to produce more output with the same resources or the same amount of output with less resources. Progress in technology causes improvement in technology through the provision of better machines, better methods and enhanced skills.

It is technology which underlies the process of producing new things with the existing resources or using the existing resources in new ways. This is what Schumpeter means when he says that, "the slow and continuous increase in the national supply of productive means and savings is obviously an important factor in explaining the course of economic history through the centuries, *but it is completely overshadowed by the fact that development consists primarily in employing existing resources in a different way, in doing new things with them, irrespective of whether those resources increase or not*."[1] It is important to emphasise that newly discovered techniques lead to the increase in output per worker.

Productivity of workers depends upon the quantity and quality of capital tools with which they work. For higher productivity the instruments of production have to be technologically efficient and superior. The technological options open to an economy determine the *input mix of production*. A commodity can be produced by various technologies. The quantity and quality of capital, skills and other factors required for production is directly dependent on the efficiency of the technique of production being used. Also, the managerial and organisational expertise has to be in line with the technological requirements of production. Viewed thus, technology in the present stage of economic development is an indispensable factor of production.

This is the age of technology. The developing countries are obsessed by the desire to make rapid progress in technology so as to catch up with the present-day developed countries. Frantic efforts are being made to install improved technology in agriculture, industries, health, sanitation and education; in fact in all walks of human life. Indeed, the newly emerging nations have come to regard technology as a bastion of national autonomy and as a status symbol in the international community.

Table 42.1. Percentage Increase in Labour Productivity, 1970-1989

Country	Increase %
United States	2.9
Canada	2.5
Denmark	3.3
France	4.1
Italy	4.9
Japan	6.0
Netherlands	4.9
Sweden	2.9
United Kingdom	3.8
Germany	3.2

Source: H. Craig Petersen and W. Cris Lewis, *Managerial Economics*, 3rd edition, 1995, p. 538.

It will be noticed that during 1970-89 rate of increase in labour productivity for the United States is less than other countries except Canada and Sweden. Slow growth in productivity in the United States during the period (1970-89) had been a major concern and was the subject of debate among

1. J.A. Schumpeter, *The Theory of Economic Development*, Cambridge, Harvard University Press, 1949, p. 68.

the economists. It will be further observed that during this period Japan had the highest growth rate in labour productivity of 6% followed by Italy and Netherlands.

Denison in his empirical study has estimated the sources of increase in real GNP between 1929 and 1982 of the United States. He found that technological change contributed 28% to the average annual growth in real GNP which was 2.9 per cent over this period. The estimates of contribution of various factors such as labour, education and capital are presented in Table 42.2.

Table 42.2. Sources of Growth in Real GNP in the United States (1929-82)

Source of Growth	Percentage Contribution to GNP growth due to
Growth in labour input	32
Growth in Labour Productivity	
Education per worker	14
Capital formation	19
Technological change	28
Economies of scale	9
Other factors*	–2

* These include worker safety, pollution abatement and fewer labour disputes.

Source: E. Denison, *Trends in American Economic Growth*, Washington, Brookings Institutions, 1985, p. 30.

From the table it will be seen that increase in the *quantity* of labour contributed to 32 per cent to 2.9 per cent annual growth in the United States. The other factors contribute to the annual growth by raising productivity of labour.

Technical progress manifests itself in the change in production function, So a simple measure of the technical progress would be the comparison of the position of production function at two points of time. The technological change may operate upon the production function through improvements of various sorts such as a superior equipment, an improved material, and superior organisational efficiency. Also, the technological progress may express itself in making available new products. The availability of a new product in many cases could be construed as a change in the production function as it can be regarded equivalent to a more efficient way of meeting an old want.

Technological Change and Capital Formation

The process of technological progress is inseparably linked with the process of capital formation. In fact, both go hand in hand. Technological progress is virtually impossible without prior capital formation. It is because the introduction of superior or more efficient techniques requires the building up of new capital equipment which incorporates new technology. In other words, new and superior technology can contribute to national product and its growth if it is first embodied in the new capital equipment. The capital accumulation has, therefore, been called the vehicle for the steady introduction of new technology into the economy:

It may be noted that Adam Smith viewed technological progress as a rise in productivity of workers as a result of increase in division of labour and specialization. This leads to the growth in national income. But it was J.A. Schumpeter who laid great stress on the role of technological innovations in bringing about economic growth. He visualised technical innovations in bringing about economic progress. It is the entrepreneur who carries out the innovations and organises the production structure more efficiently. As, according to Schumpter, the innovations occur in spurts rather than in a smooth flow, economic progress is not an uninterrupted process. The pace of economic progress is punctuated by the pace of innovations. Since innovations are governed by a host of factors, therefore, it is difficult to perdict about the economic progress resulting from technological changes.

Prof. Rostow has proposed four stages in the development of an economy. These stages are: (*i*) traditional society; (*ii*) preconditions for take-off; (*iii*) drive to maturity and (*iv*) stages of high mass consumption. It may be noted that the economic transformation of the society from one stage to another involves, along with other things, a change in the level and character of technology. In the present age of greater specialisation it is the technology factor that underlines all major aspects of the modern productive apparatus such as decision-making, production programming, skill requirements and market strategy.

The role of technological progress in economic development has not only been large but also increasing over time. It is beyond any doubt that the inventions and changes in production techniques played more important role in the late eighteenth and later centuries than in the earlier ones. The technical innovations provided the motive force for the industrial revolution in Western Europe. While the innovations mainly arose because of input shortages, they tended to overcome the impediments to mass production and the transport bottlenecks in the movement of industrial raw materials and labour. There were also improvements in managerial organisation with a view to reduce costs of production. The adjustments in the product designs also took place to meet the variations in the wants of the people.

Technical progress is the *prime mover of economic growth*. A continuing technical progress ensures a continuing growth. Where there is no technical progress, capital accumulation alone may fail to bring growth beyond a point. For capital accumulation without technical progress can bring about growth in income per head only so long as the available labour force has not been fully equipped with the latest techniques available, and only so long as all the wants of the consumers have not been supplied to the full extent. Once that point is reached where all labour has been equipped with the latest techniques and all wants of consumers have been met to the full extent, growth would come to an end unless technical progress occurs.

Technological Progress to Overcome Stationary State

It is now widely accepted that technological change raises productivity and that a continuous occurrence of technological change will enable the economy to escape from being driven to the stationary state or economic stagnation. Classical economists like David Ricardo and J.S. Mill expressed fear that the increase in the stock of capital will sooner or later land the economy into stationary state beyond which economic growth will come to an end. *Classical economists remained occupied with the idea of a stationary state because they did not take into account technological progress that could postpone the occurrence of a stationary state and ensure continued economic growth*. Indeed, if technological progress continuously takes place, demon of stationary state can be put off indefinitely.[2]

We can diagrammatically show how the technological change will raise productivity and thereby suspend the occurrence of stationary state with the help of Hicks's representation of classical model. Let us first represent the classical model of stationary state. Consider Fig. 42.1 where along the *X*-axis stock of capital is measured and up the *Y*-axis rate of return on capital is measured. This rate of return on capital is in effect the marginal productivity of capital. It may be noted that *rate of return* is here taken to be quite akin to the '*rate of interest*' which was regarded by the classical economists as the mechanism which determines investment to be made in any period of time.

Now, let us suppose that the economy has, at a particular moment, a stock of capital equal to K_1 in Fig. 42.1. Given this stock of capital, the rate of return which the businessmen would be able to earn from a *new investment* would be somewhat less than r_3. However, if the economy is to be kept growing through capital accumulation while maintaining full employment, the rate of interest must be less than r_3; let us say it is at the level r_2. In this situation, it would be profitable for the businessmen, in the given period, say, one year, to undertake investment (through borrowing or with their own funds) to an extent that results in a net addition of $K_1 K_2$ to the capital stock. Therefore, at the end of the year, the economy would come to have K_2 amount of capital stock. It is possible now to draw a curve *CC* in Fig. 42.1 which indicates the different rates of return obtainable from the different levels of capital stock.

2. J.R. Hicks, "Thoughts on the Theory of Capital," *Oxford Economic Papers* (New Series), 1960, p. 123.

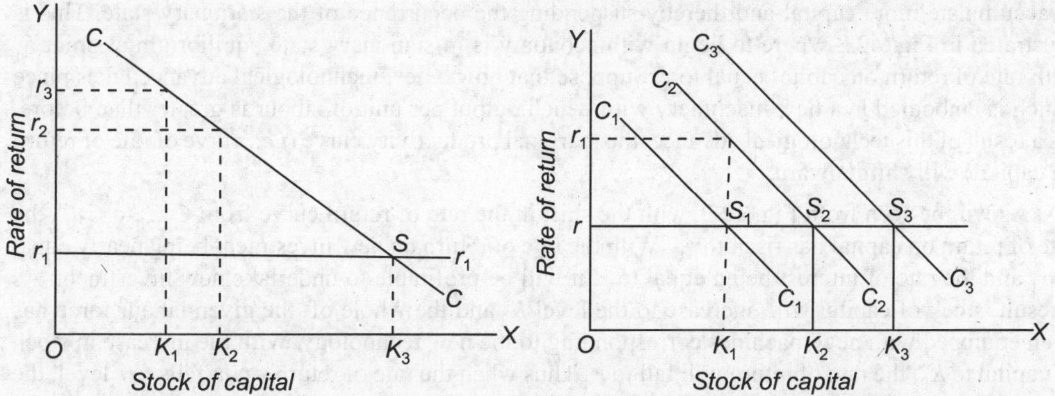

Fig. 42.1. *Classical Model of Stationary State* **Fig. 42.2.** *Stationary State and Technological Progress*

Assuming that the supply of labour is constant, in the absence of technological progress and consequently rate of return curve *CC* remaining the same, the expansion in the stock of capital from K_1 to K_2 will bring down the rate of return on capital from r_3 to r_2. As a result, the rate of return on new investment will now be slightly below r_2 after K_1K_2 investment has been undertaken in the year. Now, if there are still net *positive savings* in the community which is likely the case when capital stock has not yet sufficiently increased, then the rate of interest will fall below the level of r_2. With the fall in the rate of interest below r_2 and the rate of return on new investment being nearly equal to r_2 (*i.e.*, only slightly less than r_2), it will be worthwhile to undertake new investment in the next year and in this way to maintain full employment in the next year.

According to the classical economists, rate of interest will continue falling gradually over time till it reaches a level at which both net saving and net investment have fallen to zero. In Fig. 42.1 this will take place when the capital stock has increased to K_3 and rate of interest has fallen to r_1. Since there are no more net savings and investment, the further expansion in the stock of capital and therefore the process of economic growth will cease to occur and the economy will reach what classical economists called *stationary state*. However, according to the classical economists, the stationary state is a distant end which may not be reached at all. This view of 'classicals', according to Hicks, is based on the idea that the curve *CC* is highly elastic, so that the distances such as $K_1 K_2$ are, indeed, very small. Thus, capital accumulation takes place by very small steps and can go on for a very long time before the rate of return finally falls to r_1.

Stationary State can be Postponed through Technological Progress

It may be noted that the businessmen would continue undertaking investments so long as the capital stock is less than K_3. For, until K_3 level of capital stock is reached, the rate of return is high enough to induce businessmen to use the net savings for undertaking new investments. However, once the K_3 level of capital stock is reached, net savings would have fallen to the zero level and consequently both net investment and capital accumulation will cease to occur. While in this situation, net investment would necessarily have come to an end, gross investment may continue. But the gross investment does not add to capital stock; it is made just for the purposes of replacements of worn-out capital. When the stationary state has been reached, net investment would be zero because net savings are zero. Also, the population, income and standards of living would stabilise at a constant level.

Now, the importance of technological progress is that it can suspend the occurrence of stationary state equilibrium. Classical economists underestimated the role of technological progress in preventing the occurrence of stationary state equilibrium. When technical progress occurs it will raise the productivity of capital and labour. As a result of the rise in productivity, the rate of return curve *CC* (*i.e.*, marginal productivity curve of capital) will shift upward. This implies that it will be profitable to accumulate more capital and thereby suspending the occurrence of the stationary state. This is illustrated in Fig. 42.2 where to begin with economy is in stationary state equilibrium at point S_1 with rate of return on capital equal to r. Suppose that now a new technological advance takes place which is embodied in a new machinery with which output per unit of labour is greater than before. As a result of this technological advance, the marginal productivity curve (*i.e.*, curve of rate of return on capital) will shift upward.

As will be seen from Fig. 42.2, with the shift in the rate of return curve from C_1C_1 to C_2C_2 the rate of return on capital has risen to r_1. With the rate of return on new investment being nearly equal to r_1 and the rate of interest being equal to r, it will be profitable to undertake new investment. As a result, stock of capital will increase to the level K_2 and the whole of the given labour force has been equipped with new machines corresponding to the new technology. With the increase in stock of capital to K_2, the rate of return will fall to r. Thus when the rate of return again falls to r level, the net saving and the net investment will be reduced to zero and a new stationary state equilibrium will be reached at S_2. With the stock of capital equal to K_2 suppose there occurs a new burst of technical progress which causes the CC curve to shift further upward to C_3C_3. This will cause the capital accumulation to take place further to K_3. Thus, as a result of technological progress and more capital accumulation, output, income and the standards of living of the people will rise.

It follows from above that *technological progress holds the key to economic growth. As technological progress takes place, capital accumulation will proceed further and economic growth will take place.* Without technical progress, capital accumulation can raise the output and income to a limited extent. With constant labour force and the given technology, capital accumulation can lead to the increase in output so long as all the workers of the labour force are not equipped with the latest capital equipment embodying the latest technology. After this only technological progress can lead to the growth of output and income. Thus, Professors Stonier and Hague rightly state, "*Technical progress, therefore, turns out to be even more essential to economic growth than capital accumulation. Where there is no technical progress, capital accumulation can lead to growth in income per head only so long as labour force has not been fully equipped in accordance with the latest techniques. Once that happens, growth comes to an end. It is technical progress alone which ensures that the stationary state is not merely very far off but will never arrive so long as technical progress continues.*"[3]

Technological Change: Embodied and Disembodied

What factors cause sustained growth of national income and per capita income is an important issue that has been the concern of economists. Capital accumulation, growth in labour force and technological progress contribute to the growth of an economy. American economist R.M. Solow[4] used the aggregate production function that relates the level of output to the levels of various inputs such as growth of capital, labour and technological progress over time. Solow used the following production function:

$$Y = A(t)F(K, L) \qquad \ldots (1)$$

where Y stands for output, K for capital and L for labour. The term $A(t)$ represents technological change which is taken to depend merely on time, that is, as time passes, $A(t)$ increases showing that over time technological progress enables more to be produced with the given quantities of capital and labour. With such specification of production function, technological progress does not cause any change in the relative marginal productivities of capital and labour, that is, such technological progress brings about equal increases in the productivities of the two factors and is said to be *Hicks' neutral* in the sense that it favours neither capital nor labour.

3. Alfred W. Stonier and Douglas C. Hague, *A Textbook of Economic Theory,* 4th edition, 1972, p. 626.
4. R.M. Solow, "Technical Change and Aggregate Production Function", *Review of Economics and Statistics*, Aug. 1957.

But the important thing to note in the production function given above is that it represents *technological progress which is exogenous and disembodied.* By exogenous technological progress we mean it comes from outside the model and, therefore, the term $A(t)$ is shown outside the production function. Disembodied technological change raises the productivity of *both* old and new machines (*i.e.*, capital goods) and does not depend on the rate of new investment. *The disembedied technological change is purely organisational in nature which permits* more output to be produced from the given inputs without any new investment. That is, disembodied technological change causes a shift in the production function without disturbing capital-labour ratio. It is through organisational changes that disembedied technological change enables more output to be produced with existing inputs being used more effectively.

Disembodied technological change is represented in Fig. 42.3 where on the horizontal axis capital-labour ratio (K/L), that is, capital per worker is measured and along the vertical axis output per worker (Y/L) is measured. In the beginning, production function curve $Y = f_1(K/L)$ is given and,

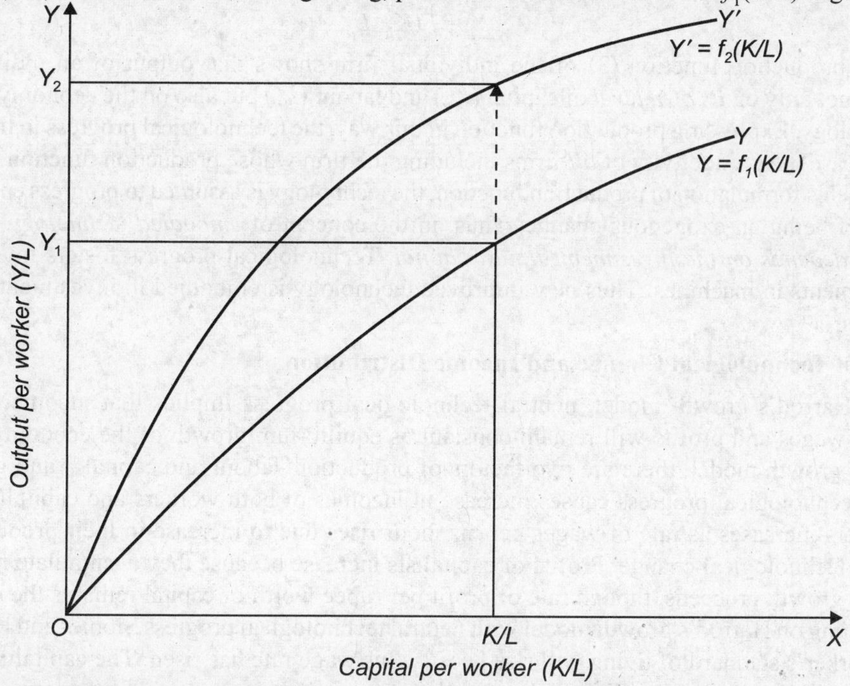

Fig. 42.3. *Shift in the Production Function due to Disembodied Technological Change*

according to this, with a capital per worker equal to K/L, level of output Y_1 is being produced. Suppose now that technological progress takes place and production function curve shifts upward to OY', that is, to $Y' = f_2(K/L)$. According to the new production function curve $Y' = f_2(K/L)$, with the given capital-labour ratio, K/L more output Y_2 is produced. Similarly, with any other capital-labour ratio on the new production function curve OY', *i.e.*, $Y' = f_2(K/L)$, more output will be produced as compared to that on the production function curve OY.

Embodied Technological Change. In the embodied technological change, the increase in investment or capital accumulation is regarded as the vehicle of technological progress. Technological progress increases the productivity of new machines built in any period compared with old machines built in the previous periods. This implies that embodied technological change does not cause increase in the productivity of machine already in existence. About embodied technological change, Hahn and Matthews in their review of the theory of economic growth write, "*Technological change is embodied in new machines. Machines unalterably embody the technology of their date of construction.*

Machines built in different dates ... are therefore qualitatively dissimilar and cannot in the general case be aggregated into a single measure of capital. A separate production function is needed for each vintage. Total output is the sum of output of all vintages in use."[5]

Production function of embodied technology can be written as

$$Y_t = F(K_t, L_t, A_t) \qquad \ldots (2)$$

where output Y_t depends on the amount of capital (K_t), labour (L_t) and the level of technology (A_t). Note that in the production function (2), the term A_t for technology appears *inside* the production function as one of the endogenous inputs.

It is important to note that in the production function (2), the relationship between output and technology differs from the relationship between output and other inputs, capital and labour. This can be better understood by considering this production function in case of an individual firm which can be stated as under:

$$Y_{it} = F(K_{it}, L_{it}, At) \qquad \ldots (3)$$

The production function (3) of the individual firm shows that output of an individual firm depends not only on *its own level* of capital (K_{it}) and labour (L_{it}) but also on the economy-wide level of technology. Expressing production function in this way, the technological progress in the economy will increase the productivity of *all* firms including the firm whose production function (3) is given above. In this formulation of production function, the technology is assumed to progress endogenously instead of being an exogenous change. Thus, in the concept of *embodied technology, production function depends on the investment in new capital*. Technological progress fosters inventions and improvements in machines. Thus new improved technology is embodied in investment in the new machines.

Nature of Technological Change and Income Distribution

In Harrod's growth model, neutral technological progress implies that income distribution between wages and profits will remain constant as equilibrium growth of the economy occurs. In Harrod's growth model, there are two factors of production–labour and capital – and growth with neutral technological progress causes increase in incomes of both workers and capitalists. Income of workers increases as rate of wages per manhour rises due to increase in their productivity as a result of technological change. Profits of capitalists increase because they accumulate more capital assets as growth proceeds, though rate of profit per rupee worth of capital remains the same. Thus, commenting on Harrod's growth model with neutral technological progress Stonier and Hague write, "The worker's standard of living will rise because his wage rate has risen. The capitalist's standard of living will rise too but it will do so because he gets much the same rate of return on each pound's worth of a bigger stock of capital."[6]

However, if in Harrod's growth theory *capital-saving technological progress* occurs, the share of profits in national income will fall and that of wages will rise. On the other hand, if *labour-saving technological progress* occurs the share of profits in national income will rise and share of wages will decline during the process of growth.

5. F.H. Hahn and R.C. Matthews, "Theory of Economic Growth", *A Survey of Economic Growth*, Vol. 74, 1964.
6. Stonier and Hague, *A Textbook of Economic Theory*, 5th edition, 1980, p. 654.

CHAPTER 43
HUMAN RESOURCE DEVELOPMENT: EDUCATION AND HEALTH

The Concept of Human Capital

Capital embodied in human beings in the form of education and health which make them more productive is called human capital. As explained in an earlier chapter, like physical capital human resources or human capital plays a significant role in determining economic development. More education makes the human beings more productive through enhancing their skills, abilities, knowledge and better health enables them not only to participate in the production process but increase their capabilities to produce more. Thus education and health can add to the value of production in the economy and also to the incomes of the persons who have been educated and made healthy.

The human capital is embodied in human beings. Human capital comprises the education, knowledge, the skills, better health and the capacities of all people in the society to undertake production. On the other hand, the physical capital consists of produced means of production such as machines that are used in producing other goods. Economic development calls for both forms of capital accumulation. And both forms of capital call for investment. Investment in human capital has therefore been called '*investment in man*'. Education plays a crucial role in the ability of a country to absorb modern technology and to generate the capacity for self-sustained growth and development. Health is also essential for increase in productivity of labour and, therefore, not only raises the private incomes but also contributes to the growth of GDP. Thus, Todaro and Smith write, "Both health and education can also be seen as vital *components of growth and development* as inputs to the aggregate production function. Their role as both inputs and output gives health and education their central importance in economic development."[1] In this chapter we explain the role of education in economic development and in the next chapter we will examine the role of health in economic development of the country.

Till recently economists have been considering physical capital as the most important factor determining economic growth and have been recommending that rate of physical capital formation in developing countries must be increased to accelerate the process of economic growth and raise the living standards of the people. But in the last three decades economic research has revealed the importance of education as a crucial factor in economic development. Education refers to the development of human skills and knowledge of the people or labour force. It is not only the quantitative expansion of educational opportunities but also the qualitative improvement of the type of education which is imparted to the labour force that holds the key to economic development. Because of its significant contribution to economic development, education has been called as human capital and expenditure on education of the people as investment in man or human capital. Speaking of the importance of education or human capital. Prof. Harbison writes: "*Human resources constitute the ultimate basis of production, human beings are the active agents who accumulate capital, exploit natural resources, build social, economic and political organisations, and carry forward national development. Clearly, a country which is unable to develop the skills and knowledge of its people and to utilise them effectively in the national economy will be unable to develop anything else.*"[2]

---

1. Michael P. Todaro and Stephen C. Smith, *Economic Development*, Tenth edition, 2012, Pearson Education, pp. 369-70.
2. Frederick Harbison, *Human Resources as the Wealth of Nations*, Oxford University Press, 1973, p. 3.

Role of Education or Human Capital in Economic Development

Gross domestic product of a country depends not only on the amount of labour (*i.e.*, work-hours) used for producing goods and services but also on its productivity. *One of the important factors that determines productivity of worker is human capital, that is, education.* As seen above, in the modern economics the concept of capital is not confined to physical capital such as machines, tractors, capital equipment, etc. that raises productivity of labour but also includes what is called human capital. By human capital we mean the skills and knowledge that workers acquire through education and training. This human capital, that is, knowledge and skills, are accumulated by human beings through education during the time they spend in primary and secondary schools and up to graduation and post-graduation in colleges or universities. Besides, specialised professional education such as engineering, computer training, management education and others raises greatly the productivity of workers. Therefore, *accumulation of human capital is generally called investment in people.*

Just as a firm considers whether to invest in physical capital, individuals decide whether to invest in their own human capital. When a firm purchases a machinery and other capital equipment to produce more output and increases its future profits, individuals invest in education or acquiring skills to raise their productivity and their future earnings. *Investment in education or acquiring new skills is called investment in human capital.* While usual models of labour supply assume that wage rate of individuals is fixed, however through investment in human capital the individuals can raise their wages or earnings as investment in human capital raises their productivity and it has been found that in the United States rate of return on investment in secondary education is 10 to 13 per cent higher.

According to Prof. Amartya Sen, the improvement in the availability and quality of education results in higher level of *functionings and capability* of labour. He has shown in his works[3] that for a country a low level of education lowers the growth of GDP due to shortages of labour with appropriate skills. Besides, the empirical evidence available suggests that countries which have invested more in education as measured by average years of schooling tend to experience, other things remaining constant, higher rate of growth. Furthermore, higher technical education helps to discover new inventions and innovations and thereby promotes technological progress.

Human capital, though less tangible than physical capital, is similar to it in many ways. First, like physical capital, accumulation of human capital increases the ability of a country to produce more goods and services. Secondly, like physical capital, human capital is a produced factor of production. Producing human capital requires investment in inputs such as student time, teachers, books and libraries, college buildings. Thirdly, like the physical capital, the accumulation of human capital increases the productivity of workers and therefore causes their wages to rise. Fourthly, *like the accumulation of physical capital, the accumulation of human capital requires the sacrifice of some present consumption so as to have more consumption in future.*

Cost of Human Capital or of Acquiring Skills

Human capital is built through acquiring more education or skills by spending more time in school, college or university. Thus, in spending more time in acquiring more education, one not only delays one's entry into labour force but also sacrifices income or wages which he could have earned by working during the time he spends for acquiring education. *This is the opportunity cost of acquiring more education, skills and grades (i.e., accumulating more human capital*). But they spend time in acquiring more human capital with the expectation that they would be able to earn higher income or wages in the future as more education and higher skills raise the productivity of the workers and therefore their wages. Thus, *the students, like the people who save, face a trade-off between less consumption today for more consumption in the future.*

Thus, "*spending more on education today (reducing consumption) raises future income but each*

3. See his works, (1) *Development as Freedom* and (2) *An Uncertain Glory: India and its Contradictions*.

additional investment in education provides a smaller and smaller return"[4]. It follows from above that, like physical capital, expenditure on education also represents investment in capital which raises productivity in the future. Besides, investment in education is tied to specific human being and therefore it is called human capital. Thus, according to Mankiw, "*Like all forms of capital, education represents an expenditure of resources at one point in time to raise productivity in the future. But unlike an investment in other forms of capital, investment in education is tied to a specific person and this linkage is what makes it human capital.*"[5] It is important to note that workers endowed with higher education (*i.e.*, more human capital) earn more income or wages than those having less education. The differences in wages between those with more education and those with less education are quite large and have been increasing. For example "*College graduates in the United States earn about twice as much as those workers who end their education with a high school diploma*"[6] *It may be noted that this difference in earnings between workers with more human capital and those with less human capital on an average tends to be even larger in developing countries where educated workers are in scarce supply.*"[7]

That investment in education or human capital has a cost can be easily illustrated. Economists are generally concerned with the *opportunity cost of time* spent in acquiring education or human capital. If you are attending a class acquiring education you cannot be working simultaneously at a job yielding you some income. Therefore, by attending computer classes you forgo some wages or earnings and hence some consumption which represents the cost of acquiring human capital.

Return on Investment in Education

The studies conducted by the World Bank show that economic returns on investment in education appears to be higher than in alternative kinds of investment, especially in physical capital. Further it has been found that rate of return to education of all the three levels in developing countries is higher than in developed countries. An important study based on available statistical data of both developing and developed countries made by George Psacharopoulos found that in the early 1990s private returns on primary, secondary and higher education were higher in developing countries of Sub-Saharan Africa, Asia and Latin America than the developed countries (see Table 43.1). For instance, private return to primary education was 41% in Sub-Saharan Africa, 39% in Asia, 26% in Latin America as compared to 22% in developed OECD countries.

Table 43.1. Private Rate of Return (%) of Different Levels of Education

Country Type and Region	Primary Education	Secondary Education	Higher Education
Developing			
Sub-Saharan Africa	41	27	28
Asia	39	19	20
Latin America	26	17	20
Developed	22	12	12

Source: George Psacharopoulos, "Returns to Education: A Global Update", *World Development* 22 (1994), pp. 1325-1343

Another fact worthwhile to note is that private return to education tends to fall as level of development (measured by the level of per capita income) increases.

Besides, there are *social returns* to investment in education which take into account the *externalities* which do not enter into the calculations of private individuals in their assessment of private return on education. We will explain these social returns on education later in this chapter.

---

4. Joseph E. Stiglitz, *Principles of Economics*, W.W. Noton & Co., New York, 1997, p. 170.
5. N. Gregory Mankiw, *Principles of Economics*, Thomson, 2nd edition, 2001, p. 419.
6. Mankiw, *op. cit.*, p. 419.
7. Stiglitz. *op. cit.*, p. 171.

Education and Sources of Economic Growth

The supply of human capital plays a crucial role in promoting economic growth. In an important study, Edward Denison[8] has measured the relative importance of various sources of economic growth in the United States for the 50 years period, 1929-69.

Table 43.2. Sources of US Economic Growth

Source	Per cent Contribution to Growth Rate	
	1929-1969	1948-1969
Advances in knowledge	31.1	34.1
More work done	28.7	23.9
Capital accumulation	15.8	21.6
Increased education	14.1	11.9
Others	10.3	8.5
Total	100	100

Rate of economic growth in the US economy was on an average 3.3 per cent during the period 1929-69 of the study. It will be seen from Table 43.2 that the most important source of economic growth (31.1 to 34 per cent) was advances in knowledge (which includes improvement in technological knowledge) that takes place as a result of increase in education and in research and development activity. The next important source of economic growth was increase in the quantity of labour force (*i.e.*, more work done) as measured by increase in hours worked by labour which contributed about 28.7 per cent during the period of study. Contrary to expectations of economists, accumulation of physical capital contributed only 15.8 per cent during the entire period (1929-69) and 21.6 per cent during 1948-69 to the growth in GDP in the United States. For the entire period of study (1929-69) increase in education (*i.e.*, number of years spent in schools, colleges and universities) contributed almost the same as accumulation of physical capital (14.1 per cent as against 15.8 per cent of physical capital). However, if we club the contribution of increased education and advances in knowledge, their contribution is as high as about 45 per cent to growth.

Other sources include the efficient use of available resources, economies of scale, a shift of labour force from low-productivity agricultural sector to higher-productivity industrial sector which contributed about 10 per cent to growth in GDP. It may however be noted that Denison could not measure directly the contribution of advances in knowledge. He measured the contributions of other four sources of growth, namely, more labour work, accumulation of physical capital, increased education, and others (as mentioned above) and assumed that all economic growth that could not be explained by other sources was the contribution of advances in knowledge.

As it pays to the individual to invest in human capital (education or skill formation), similarly, the community as a whole can raise its consumption or living standards by investing in human capital.

MEASURING CONTRIBUTION OF EDUCATION TO ECONOMIC GROWTH

Several empirical studies made in developed countries, especially the U.S.A., regarding the *sources of growth* or, in other words, contributions made by various factors such as physical capital, manhours, (*i.e.*, physical labour), education etc. have shown that education or the development of human capital is a significant source of economic growth.

It may be pointed out that economists generally hold that while some investments in education are '*economic*' in the sense that they promote growth directly, other expenditures on education and development of the human resources are basically of the form of '*social investment*' which, therefore, should be determined residually. It is not possible to separate out the consumption and investment parts of the expenditure on education so that it is very difficult to estimate the rate of financial return

8. Edward, Denison, *Accounting for US Economic Growth, 1929-1969,* Washington D.C., Brookings Institution, 1974.

on education in the same manner as in the case of a factory or a dam. At best, one can only recognise the probable importance of the expenditure on education in the process of economic growth.

The estimation of return on investments in education or human capital is beset with many difficulties. The fact is that human resource development encompasses a wide field. It cannot be solely examined in economic terms. For instance, it is not very realistic to estimate the return on education purely in terms of the increases in individual incomes or the income of the economy taken as a whole. Also, the increases in productivity do not by themselves constitute the effectiveness of human capital. Notwithstanding these considerations, economists have attempted to measure the contribution of education to economic growth solely on the basis of economic criteria.

We may mention here, in broad outline, the basis on which return and, therefore, role of investment in human resources, particularly education, has been sought to be incorporated in the mainstream of economic analysis. The following are the main approaches that have been developed to gauge the productivity of investment in education.

The Residual Approach or Production Function Approach

There have been some attempts to estimate the proportion of the measured increase in gross national product attributable to education. This is sought to be done by first determining the increase in gross national product on account of the measurable inputs of labour and capital. Then this figure is subtracted from the figure of GNP, to get a 'residue' which represents the increase in GDP due to the improvements in the quality of labour as a consequence of education.

Professor Solow[9], who was one of the first economists to measure the contribution of human capital to economic growth, estimated that for United States between 1909 and 1949, *57.5 per cent of the growth in output per manhour could be attributed to the residual factor which represents the effect of the technological change and of the improvement in the quality of labour mainly as a consequence of education*. He estimated this residual factor determining the increase in the total output on account of the measurable inputs of capital and labour (manhours). He then subtracted this figure from the total output to get the contribution of residual factor which represented the effect of education and technological change, the physically immeasurable factors.

As explained above, Denison[10] another American economist, made further refinement in estimating the contribution to economic growth of various factors. Denison tried to separate and measure the contributions of various elements of 'residual factor'. According to the estimates of Denison, over the period 1929-82 in the USA during which total national output grew at the rate of 2.9 per cent per annum. increase in labour input accounted for 32 per cent, the remaining 68 per cent was due to the *increase in productivity per worker*. He then measured the contributions of education per worker, capital formation, technological change and economies of scale. Denison found that 28 per cent of contribution to growth in output due to growth in labour productivity was due to technological change, 19 per cent to capital formation and 14 per cent due to education per worker and 9 per cent due to economies of scale. It is thus clear that education and technological progress together made 42 per cent (14 + 28) contribution to growth in national product.

Limitations. However, the above estimates are not fully reliable. The basic reason for this is that the methods used in calculating the contribution of education, are not free from flaws. Moreover, the data which have been used in the calculations applies only to formal education. *It completely ignores on-the-job training. Also, only private return on education is measured.* But education makes a considerable social contribution also in the form of increased mobility, adaptability and the growth

9. Robert M. Solow, "Technical Change and Aggregate Production Function" *Review of Economics and Statistics*, Aug. 1957.
10. Edward E. Denison, *Trends in American Economic Growth (1929-82)*, The Brookings Institution, 1983, p. 30.

of applied technology. In this sense, these calculations make underestimation of the contribution of education.

There being interrelation between capital formation, technology and growth of knowledge, the component in the residual ascribed to increased knowledge may, in fact, include some capital assets. But, in these methods no distinction has been made between formal and informal education. Again, the differences in the quality or content of education have been ignored.

Notwithstanding these limitations, these estimates highlight the significance of the improvements in the quality of human resources through education in the process of economic growth.

Rate of Return Approach: Further Elaboration

The contribution of education to economic growth has also been *measured through the rate of return approach*. In this approach rate of return is calculated from expenditure made by individuals on education and the measurement of the flow of an individual's future earnings expected to result from education. The present value of these is then calculated by using appropriate discount rate. For calculating the present value of future earnings with education the appropriate present value formula is used. This method has been used by Gary S. Backer who measured income differential arising from the cost or expenditure incurred on acquiring a college education in the United States. His estimates show that the rates of return on education in the U.S.A. for urban white population were 12.5 per cent in 1940 and 10 per cent in 1950.[11]

Another study on similar lines was made by E.F. Reneshaw. He used Schultz's earlier estimates, for USA of total earnings forgone and the cost of education in high schools, colleges and university. Doing this, he found that the average return on education ranged between 5 and 10 per cent for the period 1900 to 1950.[12]

It is worth noting that *estimates of rate of return on investment in education are based upon private rates of returns to individuals receiving education*. However, by assuming that differences in earnings in a market economy reflect differences in productivity, the rate of return on investment in education is taken to be the effect of education on the output of the country.

Limitations. However, these estimates of the return on education are valid only to the extent to which the underlying assumptions are valid. The expenditure on education has two components: the future earnings and the future consumption. While the educational investment results in increase in the future earnings, it also entails the consumption element in the sense of the direct satisfaction derived from education so obtained. The latter, *i.e.*, the consumption component of investment in education wherein resides the source of future utilities, nowhere enters into the measured national income. To the extent these estimates ignore the future utilities and other external economies of acquiring knowledge, they underestimate the returns on education.

Moreover, the earning capacity of individuals with varying educational levels is not entirely the function of formal education. Other factors such as on-the-job training, experience, family income, social status and natural abilities also influence the earning capacity. But the estimates of return on education take no account of these other factors.

Furthermore, these estimates concentrate only on the measurement of the private rates of return on educational investment. At best, they take into account merely the indirect effects of education on the output of the country on the assumption that the earnings differentials in a market economy reflect the differentials in productivity. But the various groups such as engineers, doctors, teachers and manual workers through collective efforts or trade unionism may distort the relative earnings in the economy. What is more, it is difficult to even estimate the private returns on education where

11. Gary S. Backer, "Underinvestment in College Education", *American Economic Review* (Papers and Proceedings), May 1960, p. 347.
12. E.F. Reneshaw, "Estimating the Returns to Education", *Review of Economics and Statistics*, Aug. 1970.

the costs of running an educational institution are negligible. For instance, in India there are a large number of single teacher schools where no fee is charged from the pupils.

Again, as has been argued by Eckaus, the cost (or price) of educated labour which is used in evaluating the rate of return ought to reflect the relative scarcities of the factors involved. However, when the primary burden of educational investment is borne by the government, the prices of educated labour fail to reflect the scarcities of factor inputs that are so used. Also, these estimates of rate of return on education undervalue the importance of education as a stepping stone to further education. The fact is that one level of education leads to another. As such, a comparison of individuals with primary education with those who lack it would simply underestimate the significance of primary education as the basic prerequisite to acquire further education. Further, these estimates of return on education provide no information as to the quantity and quality of education most conducive to economic development.

Schultz Approach: Comparing Expenditure on Education with Income Earned

Another approach to measure the contribution of education is based upon the analysis of the relationship between expenditure on education and income. Using this approach Schultz studied the relationship between expenditure on education and consumer's income and also the relationship between expenditure on education and physical capital formation for the United States during the period 1900 to 1956. He found that when measured in constant dollars, "the resources allocated to education rose about three and a half times (*a*) relative to consumer income in dollars, (*b*) relative to the gross formation of physical capital in dollars"[13].This implies that the "income elasticity" of the demand .for education was about 3.5 times over the period or in other words, *education considered as an investment could be regarded as 3.5 times* more attractive than *investment in physical capital*. It may however be noted that these estimates of Schultz only indirectly reflect the contribution of education to economic growth.[14]

In our above analysis we have explained that education is regarded as investment and like investment in physical capital, it raises productivity of the labour and thus contributes to growth of national income. The increased earnings or higher wages made by more educated workers have been considered as benefits not only to the private individuals, but also to the society as a whole. This is because higher earnings presumably reflect higher productivity, increased output in real as well as monetary terms.

Consumption Benefits of Education

We have explained above the investment benefits of education and therefore its effects on productivity and national output. But investment benefits are the only benefits flowing from education. Education also yields consumption benefits for the individual as he may "enjoy" more education, derive increased satisfaction from his present and future personal life. If the welfare of society depends on the welfare of its individual members, then the society as a whole also gains in welfare as a result of the increased consumption benefits of individuals from more education. Economic theory also helps us in quantifying the consumption benefits derived from education. In economic theory, to measure the marginal value of a product or service to a consumer we consider how much he has paid for it. An individual would not have purchased a product or service if it were not worth its price to him. Besides, an individual would have bought more units of a product if he thought that the marginal utility he was getting was more than the price he was paying. Thus *relative prices of various products reflect*

---

13. Schultz, "Education and Economic Growth", in Nelson B. Henry (ed.), *The Sixtieth Year Book of the National Society for the Study of Education*, University of Chicago Press. 1961.
14. R.S. Eckaus, "Economic Criterion for Education and Training", *Review of Economics and Statistics*, May, 1964.

the marginal values of different products and the amount consumed of various products multiplied by their prices would, therefore, indicate the consumption benefits derived by the individuals.

It may, however, be pointed out that the prices in a free economy are influenced by a given income distribution and the presence of monopolies and imperfections in the market structure and therefore they do not reflect the true marginal social values of different goods. However, an objective measure of consumption benefits of education is difficult and has yet to be found out. It may also be noted that, *according to the new view, economic development is not merely concerned with the growth of output but also with the increase in consumption and well-being of the society*. Thus Prof. Amartya Sen writes, "Education can add to the value of production in economy and also to the income of the person who has been educated. But even with the same level of income, *a person may benefit from education in reading, communicating, arguing, in being able to choose in a more informed way, in being taken more seriously by others and so on*."[15] Therefore, consumption benefits of education may also be regarded as developmental benefits.

External Benefits of Education

We have explained above the investment benefits and consumption benefits flowing from more education both for the individual and the society. The analysis of benefits has been based on the assumption that private interests of individuals are consistent with the social good. However, private and social benefits do not always coincide, for instance, social benefits may exceed private benefits. This is the case with the education of an individual which not only benefits individual privately but also others. First, education makes people better neighbours and citizens and makes social and political life more healthy and meaningful. Secondly, the most important external benefit of more education is its effect on technological change in the economy. More education, especially higher education, stimulates research and thereby raises productivity which undoubtedly benefits the society.

The individual inventor may not receive earnings equal to his contribution to the research. Denison's study of contribution of education to growth whose main findings have been mentioned above clearly shows the external benefits of education. After estimating the contribution of labour (including educated labour) and physical capital to economic growth he obtained an average residual of about 32 percentage to annual growth of 2.9 per cent in the U.S. during the period 1929-60. Denison attributed this to the *increase in knowledge* which is the direct result of research and indirectly of higher education. This is in addition to 14 per cent contribution to the annual growth made directly by increase in education. Therefore, Harris and James conclude, "*If the entire residual indeed stemmed ultimately from education, as some human capital enthusiasts have implied, this would mean that education, directly or indirectly, contributed over 40 per cent of total output growth and 80 per cent of increased productivity from 1929 to 1957*."[16] If Denison's residual is regarded as mainly due to research stimulated by additional education, then this is indeed a major external benefit of education.

Education, Inequality and Poverty

An evaluation of the role of education in economic development must not be confined to judging its impact on growth in output but should also include its impact on *structure and pattern of economic development as well as on the distribution of income and removal of past poverty*. In the 1950s, the most important objective of development has been the maximization of rate of economic growth, *i.e.*, growth of material output and in conformity with this the economics of education also focussed on estimating the contribution of education to the growth of national output. But nowadays policy of economic development has been increasingly concerned with the distribution of income, *i.e.*, how gains of economic growth are distributed and whether poverty is being reduced.

15. Amartya Sen, *Development as Freedom,* Oxford University Press, 2000.
16. S.E. Harris, and F. James, "Economics of Health and Education" in Brown, Neubergar and Palmatier (ed.), *Perspectives* in *Economics*, McGraw Hill, 1971, pp.165-166.

But recent studies have revealed that education, *given the present education system, has tended to increase the inequalities in income distribution rather than reducing them.*[17] The adverse effect of formal education on income distribution has been explained through establishing a positive correlation between level of education received by an individual and the level of his life-term earnings. It has been shown that those who are able to complete their secondary and university education earn *as high* as 300 to 800 percentage more income in their lifetime than those who complete a part or whole of their primary education. "*Since levels of earned income are so clearly dependent on years of completed schooling, it follows that large income inequalities will be reinforced and the magnitude of poverty perpetuated if students from middle and upper income brackets are represented disproportionately in secondary and university enrolments. If for financial and/or other reasons the poor are effectively denied access to secondary and higher education opportunities, then the educational system can actually perpetuate and even increase inequality in Third World nations.*"[18]

There are two important economic reasons why in the present education system, children and boys belonging to the poor families cannot complete their education up to the secondary level and in many cases even up to the primary level. First, the private costs, especially 'the opportunity costs' of primary education for the children belonging to the poor families are higher than for students belonging to the rich families. Children of the poor families are needed to do work on their family farms or in other family occupations, that is, the opportunity cost of studying in school is family work sacrificed. On the other hand, benefits of education to the poor students are also lower as compared to those to the rich students. This is because it is difficult for the poor students to be selected for the jobs because of poor contacts and influences as compared to rich students, even though they may possess the same level of education. Even in agriculture where it can be said that more education can benefit all equally because it raises the labour productivity, the *more benefits of education and consequently of higher productivity in agriculture are likely to be obtained by those who own land and have adequate resources to modernise their agriculture.* The benefits of more education and consequently higher productivity of landless labour may go to the landlords for whom they work.

It follows from above that as ***a result of higher private costs and lower expected benefits from education of the poor students, the poor family's rate of return from investment in education of a child is much lower.*** As a result of this, the children of poor families are likely to 'drop out' during the course of primary education. The fact that children and boys of poor family are unable to complete their secondary education coupled with the fact that there are large income or wage differentials between different persons of different levels of education explain that education in developing economies tends to increase income inequalities and perpetuates poverty rather than helps to reduce them.

Education and Rural Development

If the objective of raising the standards of living of the people in general and removal of mass poverty is to be attained in less developed countries like India, then rural development must get the highest priority. In the 1950's and 1960's in most of developing countries, the modernisation and development of the urban sector was given the highest priority in the development plans and more resources were allocated to this sector. But in recent years the thinking among economists all over the world has undergone a significant change, since the development of the large-scale industries and the urban sector has failed to solve the twin problems of poverty and unemployment. It has now been increasingly realised that it is through the emphasis on agricultural and rural development in the strategy of development that the problems of poverty and unemployment can be solved. Since over 65 per cent of the population of less developed countries directly or indirectly depends upon agriculture, rural area needs to be given the highest priority.

---

17. Jagdish Bhagwati, "Education, Class Structure and Income Inequalities", *World Development*, Vol. 1, No.5. 1973.
18. Michael P. Todaro, *Economic Development*,. Pearson Education (Singapore), 8th edition, 2003, p. 388.

Now, education can play an important role in agricultural and rural development provided it is suitably modified and given a rural bias. The present system of education has a strong urban bias so that it is ill-suited to the requirements of agricultural and rural development. Moreover, the emphasis in the present education system is on *general* education rather than on *vocational* education.

A relevant and meaningful education can raise the productivity of the rural labour in agricultural work. It can create new employment opportunities if during the schooling students are educated and trained in some useful skills or vocations. Moreover, education to the poor people will induce in them a desire to have fewer children with the result that not only their private level of living will rise, but it will also help the general economic development by checking the growth of population. Above all, education will bring about improvement in their health and nutrition. Phillip H. Coomb, an eminent educationist and economist, has classified different types of education that should be provided to the rural people so as to promote rapid rural and agricultural development.

HEALTH

Until recently improvement in health conditions like education was not given a priority by most of the developing countries though it was considered desirable for increasing welfare of the people. The improvement in conditions of health and nutrition was also not given much importance as a factor affecting economic growth in the studies of economic growth and development. Recently, more attention is being given to the health and nutrition of the people both at the national and international levels not only as important factors determining well-being of the people but also as factors determining economic growth and development. Health like education is now regarded as human capital which is as important as physical capital as determining productivity of workers. It is worth noting that it is poverty in the developing countries that is responsible for their poor health and nutrition. Mushkin rightly writes, "Improved health and education in developing countries help families escape vicious circle of poverty in which they are trapped".[19]

It may be noted that in the beginning, during the 1970s attention was given to the importance of health in ***basic needs approach to development.*** The rationale of basic needs approach has been that direct provision of such goods as health services along with elementary education, housing, access to sanitation facilities, access to safe drinking water supply was likely to alleviate absolute poverty more immediately than the alternative strategy that relies exclusively on accelerating economic growth. According to basic needs approach, benefits of rapid economic growth may trickle down to the poor. Besides, it was pointed out by the basic needs approach that it may take a very long time for economic growth to raise the incomes of the poor to enable them to spend adequately for fulfilment of their basic needs. Besides, it was pointed out that certain basic health facilities such as access to sanitation facilities, safe drinking water, can only be provided by the State.

Amartya Sen's Approach to Health and Nutrition

Amartya Sen in his now well-known work '*Development as Freedom*' and his new book '*An Uncertain Glory: India and its Contradictions*' co-authored with Jean Dreze has presented a new approach to improvement of health and nutrition. Sen's approach is based on his '*human capability approach*' to development. Improvement in health and nutrition, according to Sen, is needed to build up human capabilities for accelerating economic development. Further, according to him, the State can play a constructive role for improvement of healthcare services, as in case of promoting elementary school education. To quote Amartya Sen, "It is necessary to recognise the role of growth and development in the form of enhancing human lives and *freedoms but it is also necessary in this context to appreciate how the growth possibilities of a country depend on the advancement of human capabilities (through education, healthcare and other services in which State can play a constructive role*."[20]

---

19. Mushkin, Selma, "Health as Investment", *Journal of Political Economy*, 70, 1962, pp. 129-157.
20. Amartya Sen and Jean Dreze, *An Uncertain Glory: India and its Contradictions*, Penguin Books, 2014, pp. 38-39.

The main thrust of Amartya Sen's approach is that improvement in healthcare and nutrition is required not only as a welfare measure but is an important factor in determining economic growth by enhancing human capabilities. With their expanded human capabilities, the people will work more efficiently and productively. Amartya Sen laments that India has not gained much achievement in respect of various health indicators such as life expectancy, infant mortality rate, maternal mortality rate, access to sanitation facilities and safe drinking water facilities as is seen from Table 43.3. It will seen from this table that various health indicators such as life expectancy at birth, infant mortality rate, maternal mortality rate improved as level of national income increases. For example, life expectancy at birth in 2009 was 57 years in low income countries, 68 years in lower middle income countries, 72 years in upper middle income countries and 80 years in high income countries. Similarly, infant mortality rate and maternal mortality rate falls as we go from low income countries to high income countries Indeed, infant mortality rate is more than 10 times higher in low-income countries as compared to the high-income developed countries.

Another important fact revealed by Table 43.3 is that as economic development takes place, there is improvement in health conditions of the countries. Of course, this improvement in health conditions varies in different countries with economic development. This improvement in health conditions can be judged by comparing the numbers of health indicators for the year 2009 with the year 1990. For example, the infant mortality rate per 1000 live births in low income countries fell from 108 in 1990 to 76 in 2009. Similarly, in case of lower middle income countries maternal mortality rate (per 100,000 live births) fell from 400 in 1990 to 230 in 2009. This improvement in health conditions occurs mainly due to increase in access to improved water sources and improved sanitation facilities whose data are given in the last two columns of Table 43.3.

Table 43.3. Levels of Income and Health Measures

Level of Income	Life Expectancy at birth		Infant Mortality rate (per 1000 live births)		Under-Five Mortality rate (per 1000)		Maternal Mortality rate (per 100,000 live births)		Access to Improved water sources (% of population)		Access to Improved Sanitation Facilities (% of population)	
	1990	2009	1990	2009	1990	2009	1990	2008	1990	2008	1990	2008
Low Income	52	57	108	76	171	118	850	580	55	64	23	35
Lower Middle Income	63	68	66	43	93	57	400	230	70	86	37	50
Upper Middle Income	68	72	41	19	51	22	120	82	89	95	78	84
High Income	75	80	10	6	12	7	15	15	99	100	100	99
Euro Area	76	81	8	3	9	4	11	7	100	100	100	100
East Asia & Pacific	67	72	41	21	55	26	200	89	69	88	42	59
Europe & Central Asia	68	70	43	19	52	21	69	32	91	95	87	89
Latin America & Caribbean	68	74	42	19	52	23	140	86	85	93	69	79
Middle East & N. Africa	64	71	57	27	76	33	210	88	87	87	73	84
South Asia	58	64	89	55	125	71	610	290	74	87	22	36
Sub-Saharan Africa	50	53	109	81	181	130	870	650	49	60	27	31
World	52	57	108	76	171	118	400	260	77	87	52	61

Source: World Bank: *World Development Indicators,* 2011.

Importance of Healthcare as a Factor in Economic Development

It is evident from above that in economics today expenditure on health, like on education, is regarded as a form of investment in human capital. This is because better health increases the productivity of man and therefore expenditure on health and nutrition increases the productivity of man and therefore contributes to economic development of the country. It is however important to note that better health is a basic human need and increases welfare of the people. Therefore, to promote health and nutrition of the people is an important goal in its own right apart from its productivity enhancing effect. Since better health improves the quality of life, everyone gains from the fulfilment of his basic needs. With better health the young can contribute to national output for a longer period for they live longer and work and earn for a longer duration. Further, increases in health raises the return on education by extending the working life of the young.

As health, like education, is regarded as investment in human capital, it is an important factor in economic development. The higher expenditure on health improves the quality of human resources both in the present and future. Better health increases the strength, stamina and ability to concentrate on whatever work a person is doing. This raises the productivity of workers which will lead not only to higher wages of workers but will also increase their contribution to economic growth.

Return on Investment in Health

The return on investment in health is difficult to measure. Though investment in health increases the working life of the people but the problem faced in this regard is what value is assigned to it. In private-cost analysis discounted value of additional earnings made from extended life span as a result of improved health is measured as a return from it. However, from the viewpoint of social cost and benefits, it is harder to measure the return to society of investment in health improvement of the people. In a developing economy with surplus labour where marginal productivity of labour is small or even zero, increased benefits in the form of contribution to growth of output of extended working life resulting from investment in health will be obviously small. Besides, even when workers with better health are employed, increase in their productivity that is exclusively attributable to improvement in health is hard to measure.

However, in our view the quantitative effect on output or productivity of investment in health may be difficult to measure, but it cannot be denied that improvement in health and nutrition represents investment in human capital and makes significant contribution to economic growth. Before Amartya Sen, expenditure on health was considered as a mere welfare measure or fulfilment of a basic human need. It goes to the credit of Amartya Sen that expenditure on improvement of health is regarded as a means of enhancing human capability and as such is a factor in economic growth.

Two-way Relationship. It is worth noting that there is a two-way relationship between better health and economic growth. On the one hand, better health contributes to increase in rate of economic growth and, on the other hand, economic growth *not only generates* more individual incomes but also generates more resources to the government which can be used to spend more on further investment in improvement of health and nutrition of the people.

An important benefit of improved health is that *it raises the return on education in several ways*. First, the healthy children tend to have higher school attendance which helps them to learn more efficiently. Secondly, healthy children being more successful in school education are likely to get higher education which raises the chances of their getting higher paid jobs. Thirdly, by causing higher life span better health permits workers to work for a longer period resulting in higher return on education received by them. Finally, healthier persons are able to make more productive use of education at any point in life. Thus, Todaro and Smith rightly write, "Health status, once attained, also affects school performance . . . . . better health and nutrition leads to earlier and longer school enrolment, better school attendance and more effective learning."[21]

Our analysis of the importance of health and education for economic development and their poor conditions in developing countries shows that while economic growth is an important investment for improving the living conditions of the people of a country, economic growth alone is not enough for *much depends on how the fruits of economic growth are distributed.* If economic growth helps to improve health of the poor people and meet their nutritional requirements, it will not only *directly*

21. Todaro and Smith, *Economic Development*, Tenth Edition, 2012, p. 374.

increase their well-being but will also give a boost to further economic growth by raising their productivity and capability. If economic growth only makes the already well-off richer, it will not promote real development.

SELECTED HEALTH INDICATORS: A COMPARATIVE STUDY

As has been explained earlier, India's economic growth, especially since 1991, has been quite impressive. In the last two decades (1991-92-2013-14) India has achieved an average GDP growth rate of 7.2 per cent. As a result, it has become the second fastest growing economy of the world, next only to China, another developing country. However, even today India does not appear to fare well in respect of health condition as compared to the other poor developing countries of the world. We have given in Table 43.4 a comparative picture of selected health standards of some poor developing countries and rich developed countries along with their per capita GDP for the years 1990 and 2011.

As regards life expectancy at birth in the year 2011, it was 65 years in India as compared to 69 in Bangladesh and Nepal, 75 in Sri Lanka, 73 in China and 70 in Brazil. It will also be noticed from. Table 43.4 that with economic growth in the last two decades (1991-2011) India succeeded in raising life expectancy by 7 years from 58 in 1990 to 65 years in 2011. This has been possible by lowering maternal mortality rate, infant mortality rate, under –5 children mortality and preventing the spread of diseases such as malaria, tuberculosis, cholera which were some of the main causes of deaths in the country. Some other developing countries have done better than us, but as a highly populated country, India's achievement in this regard in our view is quite good.

Table 43.4. Health Indicators and Per Capita GDP of Some Developing and Developed Countries

Country	Per Capita GDP in 2005 PPP$		Life Expec-tancy at birth		Infant Mortal-ity rate		Under-5 Mortality rate		Maternal Mortality rate		Exp. on Health as % of GDP 2010
	1990	2011	1990	2011	1990	2011	1990	2011	1990	2011	2010
India	1,193	3,203	58	65	81	48	114	61	600	200	1.2
Bangladesh	741	1,569	59	69	97	37	139	46	800	240	1.2
Bhutan	678	5,088	53	67	96	42	138	54	1000	180	4.5
Nepal	716	1,102	54	69	94	39	135	48	770	170	1.8
Pakistan	1,624	2,424	61	65	98	59	122	72	490	260	0.8
Sri Lanka	2,017	4,529	70	75	24	11	29	12	85	35	1.3
China	1,121	7,418	69	73	37	13	49	15	120	39	2.7
Brazil	–	10,274	66	73*	46	14	56	16	120	55*	4.2
Russia	–	14,808	69	63*	23	10	27	12	74	39*	3.2
United States	–	42,488	75	79*	9	7	11	8	12	24*	9.5
Japan	–	30,680	79	83*	5	2	6	3	12	6	7.8
Korea, Rep. of	–	27,541	71	80*	8	5	9	5	18	18*	4.1
Singapore	–	53,581	74	81*	6	2	8	3	6	9*	1.4
France	–	28,809	77	81*	7	3	9	4	13	8*	9.3
U.K.	–	32,474	76	80*	8	5	10	5	10	12	8.1

Note: 1. Data marked star (*) belong to the year, 2009. 2. Infant and under -5 mortality rates are expressed as per 1000 live births 3. Maternal mortality rates are expressed as per 100,000 live births.

Source: *World Development Indicators,* 2011 *and* 2013, and *Human Development Report*, 2013.

The same is the situation with regard to infant mortality rate and maternal mortality rate. In 2011 infant mortality rate (per 1000 live births) was 48 in India, which was higher than even Bangladesh (37), Bhutan (42), Nepal (39), Brazil (14). Infant mortality rate was much higher than that in the rich developed countries such as 7 of the United States and 2 of Japan. The same is true of under-five mortality rate and maternal mortality rate (see Table 43.4). However, in the two decade period

(1990-2011) India succeeded in lowering under –5 mortality rate (per 1,000 live births) from 114 in 1990 to 61 in 2011 and maternal mortality rate (per 100,000 live births) from 600 in 1990 to 200 in 2011. Although this achievement by India appears to be lower as compared to those of some other developing countries, its achievement, according to our view, is commendable considering that India is a highly populated country with 125,000,000 population.

Making a comparative study of the above health indicators, Prof. Amartya Sen[22] and Jean Dreze in their jointly written book, *An Uncertain Glory: India and its Contradictions* (2014), have presented a highly dismal picture of India's achievement in respect of its improvement in health conditions. They thus write, "What is disturbing . . . . is how badly India does in terms of *non-income* features of living standards even with this group of poorest non-African countries." Further, they write, "Aside from GDP per capita India's rank among these 16 poor countries is 10th or worse in most cases. Not only are India's figures worse than the average for the other 15 countries for all social indicators presented here (except for the total fertility rate and male literacy rate), its rank in this group is inglorious 10th for child mortality, 11th for female literacy, 13th for access to improved sanitation and DPT immunisation and absolutely worse (along with Yemen) in terms of the proportion of underweight children." From their comparative study they conclude, "The basic concern remains that, for whatever reasons, India is not doing well at all in many respects even in comparisons with some of the poorest countries of the world. This is not to disparage India's achievements, but to put them in perspective and to focus on the deficiencies that plague India most and which have to be overcome."[23]

Evaluation of Views of Amartya Sen and Jean Dreze

Although India's position with regard to living conditions as measured by health indicators does not appear good in a comparative study, it is not correct to draw a dismal picture of conditions in India, as has been done by Amartya Sen and Jean Dreze. This is because it is not correct to make comparisons on the average rates or ratios such as GDP per capita, infant and under –5 mortality rate, maternal mortality rate and such other indicators because these average rates or ratios are calculated with a *large population as the base.* As India is a highly populated country with a population of 125,000,000 people, even if it makes a significant progress in absolute terms in certain fields, in terms of rate or ratio of this huge population as the base, this progress becomes too small. This is what has happened in India not only in case of per capita GDP which is a *ratio of total GDP to population* but in case of health and education indicators which are expressed as average rates. Although these ratios or rates are quite low as compared to other developing countries, it is not correct to draw dismal conclusions about India's progress in terms of these rates or ratios. As regards the other two highly populated countries, namely, China and Bangladesh, Jagdish Bhagwati and Arvind Panagariya have challenged the conclusions of Amartya Sen and Jean Dreze.[24]

However, we agree with the policy measures suggested by Sen and Jean Drez about making further progress in respect of health conditions in India. They rightly warn against "*to rely on income growth alone to bring about a transformation of living conditions*" of the people. They recommend that the State has to play a constructive role. Thus they point out "the overwhelming objective of ensuring access to health services and other requirements of good health to all members of the community irrespective of their ability to pay is intrinsically a public responsibility. Further, given the limitations of market arrangements and of private insurance in the field of healthcare, *public provision of health services* has an important foundational role to play in the realisation of universal health coverage".[25]

The need for government intervention is more strong for preventing rather than curing diseases. For this the government should take appropriate measures of immunization, improvement of sanitation, access to safe drinking water, management of waste disposal and adopt special measures to check the spread of malaria, tuberculosis etc. In India's new health policy 2014, the decision to raise health

22. Amartya Sen and Jean Dreze, *op.cit.*, p. 48.
23. *Ibid.* p. 50.
24. See Jagdish Bhagwati and Arvind Panagariya, *India's Tryst with Destiny,* Harper Collins, 2014, pp. 73-77.
25. Amartya Sen and Jean Dreze, *op.cit.*, p. 179.

expenditure from present 1.2 per cent of GDP to 2.5 per cent of GDP is a welcome measure. With this spurt in health expenditure it is hoped that our health conditions will improve significantly. As has been stressed earlier, it is level of poverty prevailing in India that is responsible for bad health conditions. If, in addition to increase in national expenditure on health, effective steps are taken to reduce poverty in the country, our health parameters will definitely improve.

UNDERNUTRITION AND DISEASE PREVENTION

An important gap between the rich and poor relates to the nutrition. It is the level of nutrition that determines health, energy intake and productivity of the people. Besides, the malnourishment and undernourishment make people, especially children, more prone to diseases and determine life expectancy at birth, maternal mortality rate and infant mortality rate. It has been estimated in World Bank studies that around 2 billion people of poor countries suffer from lack of one or more micro-nutrients.

Energy Intake and Level of Productivity

Let us first explain the relationship between energy intake and level of productivity. Energy intake depends upon the calories consumed by the human beings. It is intake of calories that determines the energy level which affects the health, work effort and productivity. Prof. Partha Dasgupta[26] in his important study on well-being of the poor in developing countries defines *undernourishment as "a state in which the physical functioning of a person is impaired to the point where he cannot maintain an adequate level of performance at physical work or at resisting or recovering from the effects of a variety of diseases*". As mentioned above, it is the daily calorie intake that determines the energy level. There is a certain minimum requirement (γ) of energy intake (*i.e.*, calorie consumption) which a person requires to perform his minimum activities of eating and maintaining essential personal hygiene. Beyond this minimum level (γ), calorie or energy intake affects his work efficiency and

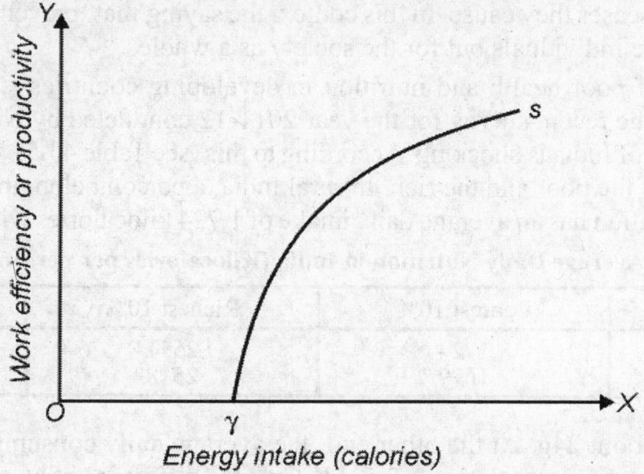

Fig. 43.1. *Relation between Energy Intake and Productivity*

productivity. This relation between energy intake and productivity is depicted in Fig. 43.1 where the curve γ_S traces the effect of increase in calorie intake on productivity beyond the minimum essential requirement (γ).

It will be seen from Fig. 43.1 that as energy intake increases beyond the minimum level γ, productivity or work efficiency increases but at a diminishing rate. However, the recent studies show that in the initial stages this productivity curve is linear rather than having a diminishing slope. Whatever be the case, the important fact revealed by γ curve is that *calorie intake determines the productivity level of human beings*. On the number of calories consumed depends the strength, fatigue and stamina of the people which affect their work efficiency and incomes earned by them. Prof. Partha

26. Partha Dasgupta, *An Enquiry into Well-Being and Destitution*, 1993.

Dasgupta points out the two-way relationship between food intake and income of the people. *The low income or poverty of the people is responsible for the undernourishment of the people which in turn is the cause of their poverty or low income as undernourishment or low food intake adversely affects their productivity and income earned.*

The root cause of many deaths in developing countries are the diseases which are attributed to malnutrition and micronutrient deficiencies. Diarrhoea, malaria, tuberculosis, acute respiratory infections, Hepatitis B, Dengu are some important diseases from which the people of a country suffer. AIDS is the latest addition to the cause of deaths in poor underdeveloped countries, especially sub-Saharan Africa. Diseases are caused by lack of food energy in terms of nutritional requirements of calorie intake and consumption of protein and deficiency of specific nutrients in diets of the people. The deficiency of calorie intake causes the loss of body weight, fatigue and mental faculties. Therefore, the persons with deficiency of calorie intake are less productive. Lack of protein in diet causes lower physical and mental development. The recent studies show that deficiency of protein can be made up by the extra calorie intake in diet. Thus calories are now regarded as a crucial factor in nutrition of the people and therefore its deficiency should be given higher priority for maintaining health of the people.

Besides, deficiencies of certain vitamins and minerals also causes several diseases and therefore to ensure adequate nutrition of the people steps should be taken to remove these deficiencies. Deficiency of vitamin A causes blindness and thus makes the children and adults suffering from it quite crippled and unable to do any productive work. Deficiency of iron causes anemia which badly impairs the health of the people. Therefore, in order to ensure adequate nutrition proper measures should be taken and government expenditure on health must be increased. Similarly, deficiency of iodine causes hypothyroidism (goitre) which leads to Cretinism and deaf mutism.

It is worth noting that cost of prevention of the above nutritional deficiencies is much less than the treatment of the diseases they cause. In this context the saying that 'prevention is better than cure' not only applies for the individuals but for the society as a whole.

The root cause of poor health and nutrition in developing countries is poverty itself. In this regard the finding of the recent survey for the year 2011-12 conducted by National Sample Survey Organisation (NSSO) of India is shocking. According to this (see Table 43.5), there is huge difference in nutritional intake of the poor and the rich. In rural India, a person belonging to the poorest 10 per cent of the rural population has an average daily intake of 1,724 kilocalories with protein consumption

Table 43.5. Average Daily Nutrition in India (Kilocalories per person) in 2011-12

	Poorest 10%	Richest 10%	Gap (%)
Rural	1724	2531	46%
Urban	1679	2518	50%

Source: NSSO, 2011-12

at about 45g and fat about 24g. At the other end, the average daily consumption per person of the richest 10 per cent of the rural population is 2,531 K.Cal., the difference of 46 per cent over the poorest 10 per cent. The similar difference is found in case of consumption of protein and fats.

As will be seen from Table 43.5, the nutritional gap between the rich and the poor in the urban areas is worse. In the urban areas in India, the average daily kilocalories intake per person of the poorest urban 10 per cent is 1679, whereas it is 2,518 of the richest 10 per cent which is 50 per cent more than that of the poorest urban 10 per cent. Another shocking result of 2011-12 National Sample Survey is that whereas average calorific intake per person and average daily protein consumption (gram per person) in both the rural and urban areas have declined marginally, the average fat consumption (gram per person) has increased in 2011-12 as compared to 1993-94. Two things are worth noting in this regard, first, as against the recommendation of National Institute of Nutrition requirement of 2320 Kcal a day for a man aged 18-29 years weighing 60 kg in a sedentary job both in the rural and urban areas, average calorific intake per person is much less than this and for the poorest 10% in both the rural and urban areas situation in this regard is worse (see Table 43.5).

CHAPTER 44

POPULATION GROWTH AND ECONOMIC DEVELOPMENT

Introduction

Having discussed the role of capital formation, technological progress and human resources (*i.e.*, education and health) in determining economic development we will now discuss how population growth affects economic development.

It is important to note here that in the present-day industrially developed countries, in spite of Mathus' view to the contrary population growth was beneficial for economic growth rather than retarding it. It has been argued by some that population growth leads to increase in labour force *which is an essential productive resource*. By increasing a productive resource population growth will help in producing more output. As someone has remarked, population growth brings in more hands to work for production and therefore contributes to economic growth. Secondly, it has been pointed out that the increase in population leads to the increase in demand for goods. Thus growing population means the growing market for goods is enlarged, they can be produced on a large scale and thus economies of large-scale production can be reaped. The economic history of USA and European countries shows that population growth in them contributed greatly to the increase in their national output.

But wheat has been true of USA and European countries may not be true in case of other countries. Whether or not the growth of population contributes to economic growth depends on the existing size of population, the available supplies of natural and capital resources, and the prevailing technology. In the United States, where supplies of natural and capital resources are comparatively abundant, growth in labour force caused by increasing population raises national output. In India where supplies of other economic resources, especially capital equipment, are relatively scarce, increase in population or labour force does not lead to the employment of all due to scarcity of capital resources. Unemployed people do not add to national output. As for the argument that population growth leads to increases in demand or market for goods, it may be noted that the demand or market for goods increases if the real purchasing power in the hands of the people increases. The mere growth of unemployed or paupers cannot lead to greater demand for goods or expansion in their markets.

Having ruled out the beneficial effects of population growth in the context of the Indian economy we discuss below how population growth in India retards economic development.

When population grows faster than GNP, the standard of living of the people does not improve. In fact rapid population growth has been obstructing economic growth in developing countries like India where since 1951 population has been growing at a relatively high rate. In Table 44.1 we present population growth in India. It will be seen that since 1951, population has been growing at about 2 per cent or more. In other developing countries such as Pakistan Bangladesh, rate of population has been greater than that of India. However, it may be noted that overpopulation and the rapid population rate is especially an Asian problem. The developing countries of Latin America and Africa do not face this population explosion problem. But for developing countries like India, population growth has to be checked if gains of development are not to be nullified by it. It is worthwhile to note here that it is only China which has brought population growth rate quite well under control by adopting *one child*

Table 44.1. Population Growth in India Since 1951

Year	*Population (in million)*	*Annual Average Growth Rate (%) during decade*
1951	361	–
1961	439	1.96
1971	548	2.20
1981	683	2.22
1991	843	2.14
2001	1028	1.93
2011	1210	1.6

per family norm. A democratic country like India cannot adopt coercive methods to control population. But, to show how rapid population growth retards economic development, it is necessary to mention that by economic development we mean not only increase in national income (GNP) or per capita income, but also reduction in unemployment as a result of the growth of employment opportunities and reduction in poverty and inequalities of income. Since economic growth depends on rate of saving and investment and productivity of labour, we will discuss the impact of population growth on these factors.

HOW POPULATION GROWTH ADVERSELY AFFECTS ECONOMIC DEVELOPMENT

1. Population Growth and Rate of Saving and Investment. Economic growth requires increasing supplies of capital goods. A higher rate of economic growth can be achieved by accelerating the rate of capital formation. Increasing supplies of capital goods become possible only with higher rate of investment. And a higher rate of investment, in turn, is possible if the rate of savings is high. Now, increase in population by adding to the number of people whose requirements of "feeding and clothing" have to be met which tends to raise consumption and, therefore, lowers both saving and investment. Coale and Hoover, in their famous work explained that saving rate was reduced by population growth because of *increase in burden of dependency.* He argued that with high fertility rate among the younger persons and declining mortality (death) rate among the old-age people, in the growing population the *proportion of non-working age groups* which depend on the working or earning members of their families increases. Since all must consume, in the absence of increase to productivity, saving per person must fall. Thus rapid growth of population by causing lower rate of savings and investment *tends to hold down* the rate of capital formation and therefore the rate of economic growth in developing countries like India. Under conditions like those in India population growth therefore actually impedes economic development rather than facilitates it.

2. Investible Resources and Raising Per Capita Income. While on the one hand rapid growth in population reduces investible resources for accelerating capital formation, *it raises the requirements for investment to achieve a given target increase in per capita income.* Suppose population of a country A is increasing at 1 per cent per annum and that of a country *B* at 3 per cent per annum. Given that capital-output ratio is 4 : 1, then country *A* would have to invest 4 per cent of its current income to maintain its per capita income, while country *B* would have to invest 12 per cent of its current income even to maintain its per capita output. Thus, when the population is increasing at a rapid rate, comparatively larger investments are needed to maintain the current level of income. Thus, given the scarcity of investible resources adequate resources are not left to *raise per capita income* significantly.

3. Lowers Growth of Per Capita Income. Like a thief in the night, population growth robs us of most of the gains in national income made from higher investment. *Rapid population growth nullifies our investment efforts to raise the living standards of our people.* In other words, a high rate of increase in population swallows up a large part of the increase in national income so that per capita income or living standards of the people do not rise much. This is precisely what has happened during the planning era in India. Thus, while the aggregate national income of India went up by 3.6% per annum in the First Plan period and 4.1% per annum in the Second Plan period, per capita income rose by only 1.8 per cent and 2 per cent per annum respectively. Average annual growth in national income and per capita income in various Five Year Plan periods is given in Table 44.2. It will be seen from this table that the annual growth in per capita income has been much less than the annual growth rate in the national income. It is the population growing at 2 per cent per annum or more during the planning period that has caused per capita income to rise much less than the increase achieved in national income.

Table 44.2. Annual Average Growth Rate (at 2004-05 prices)

Plan period	*National Income per cent*	*Per Capita Income per cent*
First Plan (1951-56)	4.2	2.4
Second Plan (1956-61)	4.2	2.2
Third Plan (1961-66)	2.6	0.3
Fourth Plan (1969-74)	3.2	0.9
Fifth Plan (1974-79)	4.9	2.6
Sixth Plan (1980-85)	5.4	3.1
Seventh Plan (1985-90)	5.5	3.3
Eighth Plan (1992-97)	6.7	4.6
Ninth Plan (1997-2002)	5.5	3.5
Tenth Plan (2002-07)	7.5	5.9
Eleventh Plan (2007-12)	7.7	6.3

Source : *Economic Survey*, 2002-03, 2007-08.

However, since 1991 population growth rate has been less than 2 per cent, it was 1.93 per cent between 1991 and 2001 and 1.6 per cent between 2001-2011, on the one hand and growth rate of national income was much higher on the other (see Table 44.2). Therefore, the growth rate of per capita income has been relatively higher. Per capita income (at 2004-05 prices) grew at the rate of 4.6 per cent in the Eighth Plan period (1992-97), 3.5 per cent in the 9th plan period (1997-2002), 5.9 per cent in the 10th Plan period and 6.3 per cent in the 11th Plan period. This higher per capita income growth rate since 1991 has tended to raise the standard of living of the people higher than before.

That the population growth prevents the rapid rise in per capita Income and therefore rise in living standards of the people can be expressed by the following growth formula:

$$g = I\alpha - r$$

where g stands for the rate of growth of per capita income, I represents rate of investment, α stands for output-capital ratio (or productivity of capital) and r represents rate of population growth.

Since rate of growth in national income is given by the rate of investment multiplied by the output-capital ratio, $I\alpha$ will signify the rate of growth of national income. Now, it will be seen that rate of population growth r appears as a negative factor and will therefore lower the rate of growth of per capita income g. It therefore follows that if rate of growth of per capita income g, and the rate of rise in living standards with a given rate of investment is to be raised, the rate of growth of population should be lowered.

4. Population Growth and Marketed Surplus of Foodgrains. Another way in which growth in population is impeding economic development is its effect on *marketed surplus of foodgrains*. The marketed surplus of foodgrains is a prerequisite for expansion in non-agricultural employment and output. When a country grows and accelerates its pace of industrialization, it requires foodgrains to feed the workers who are employed in industries. If enough surpluses of foodgrains are not forthcoming this acts as an important constraint on the industrial development. This prevents the living standards of the people to rise rapidly. Now, marketed surplus of foodgrains is the difference between the output of foodgrains by the agricultural population and their consumption of them. Thus,

$$\text{Marketed surplus of foodgrains} = (0 - C_s).$$

where 0 stands for output of foodgrains, and C_s for consumption of foodgrains by the farmers themselves. As about 65% of the population is engaged in agriculture, most of the increase in population also takes place there. This increase in population in the agriculture raises the consumption of foodgrains, *i.e.,* C_s in the above equation and therefore reduces the marketable surplus, if output remains the same. Even if output is rising, the extra consumption by the increase in population tends to lower the growth in marketed surplus for foodgrains. We thus see that the growth in population has an adverse effect on the marketed surplus of foodgrains and this acts as a drag on the growth of output and employment in industries. In India, in several years, increase in agricultural output has not been enough and further that the rapid growth in population has tended to reduce the growth of marketable surplus. This had an adverse effect on industrial development in India.

Rapid growth in population in an already overpopulated country also raises the problem of food security in the country. The cause of food problem in India is the rapid growth in population since 1951. In order to overcome the shortage of foodgrains and to prevent the occurrence of famines in the country, India was forced to import foodgrains and spend a good amount of valuable foreign exchange on them. This worsened the balance of payments problem of the country. As a result sufficient amount of foreign exchange to import materials. machines and equipment for our industries could not be made and this obstructed the growth of industrial output. This also shows how rapid growth in population by causing food shortage inhibits the rate of industrial development.

5. Population Growth and Unproductive Investment. In his study of population growth and economic development in India, Coale and Hoover,[1] focussed on the adverse effect of population growth on the resources available for productive investment. According to them, rapid population growth forces the country to make non-productive investment, that is, to invest in duplicating certain social welfare facilities such as the construction of parks, houses, social buildings, sanitation works. To the extent the Government has to increase its expenses on duplicating these social welfare facilities, investment resources for productive type of capital such as machines for industries, irrigation and fertilisers for agriculture, crucial basic goods such as steel, coal, electricity generation etc would be reduced. Thus, rapid population growth obstructs economic development by reducing the growth of productive capital.

6. Population Growth and Unemployment. Economic development requires that employment should increase adequately so that unemployment should decrease. Explosive growth in population has caused serious unemployment and underemployment problem in India. Due to explosive growth in population in India labour force has been increasing rapidly since 1951. In recent years labour force which was estimated at 309 million in 1983, went up to 333 million in 1988, to 382 million in 1994 and to 406[2] million in 1999-2000. As a result of this explosive increase in labour force demographic pressure on the economy has increased resulting in increase in backlog of unemployment and under-employment at the beginning of each successive Five Year Plan. In view of this much of our investment efforts are directed at absorbing the growing labour force in productive employment, and our ability to *raise productivity of labour* is severely constrained.

Since production processes in modem organised industrial sector is highly capital-intensive, much of the growing labour force cannot be employed there. As a result, demographic pressure on land and agriculture increases resulting in severe drop in the net sown area per capita. In agriculture, self-employment is predominant and the joint family system prevails under which both household's income and work are shared among the family members. Therefore, in the absence of employment

1. Coale and Hoover, *op. cit.*
2. Note that smaller labour force estimated in 1999-2000 is the result of estimated fall in labour force participation rate in NSS survey 1999-2000 as a result of certain change in methodology of data collection.

opportunities outside agriculture, much of the additional labour force is forced to remain in agriculture and allied activities. Agriculture performs the role of residual absorber. They share work in agriculture with other family members no matter how low the productivity per person becomes. Thus, with the fall in net sown area per person and increased population pressure, disguised unemployment emerges in agriculture. Disguised unemployment means more workers seem to be employed in it but quite a large number of additional workers do not add to agricultural output, that is, marginal productivity of workers in agriculture is zero or nearly zero. Since population growth reduces savings and investible resources, it is very difficult to withdraw any significant number of workers from agriculture so as to equal them with the required capital to provide them productive employment outside agriculture. To a certain extent lack of capital may be made up by harder work by workers in a country like India. But such a method of adjustment is not easy to achieve in India. This is because in the modern times man can produce little with bare hands. To provide them productive employment workers need to be equipped with enough capital goods. Even employment generation in agriculture apart from high yielding inputs such as fertilisers, HYV seeds, pesticides as requires irrigation works. an important capital needed for extension of double cropping which is highly employment generating way in agriculture. Due to lack of investible resources caused partly by population growth, it has not been possible to extend irrigation facilities to the currently known irrigation potential.

It follows from above that labour force consequence of population growth is to a good extent responsible for huge unemployment and underemployment prevailing in India.

7. **Population Growth and Poverty.** Last but not the least, the important consequence of rapid population growth is that it has made very difficult to make a significant dent into the problem of mass poverty prevailing in' the country. This is clear from the fact that population in India increased from 1028 million in 2001 to 1210 million in 2011. This gives rise to a huge problem of properly feeding and clothing them. Further, as has been explained in detail in the above sections, such large increase in population and consequently huge increment in labour force lowers our capacity to make productive investment and thereby to increase productivity of labour to ensure eradication of poverty Prof. K. Sundaram rightly writes, "The size of increments to population is itself of some consequence. Thus is because the resource requirements of feeding and clothing even at the current low levels are such that the incremental population itself constraints the ability of the economy to raise the living standards of the existing population."[3] A vicious circle of poverty operates in this regard. Rapid population growth leads to lower productivity which causes poverty, poverty causes high infant mortality rate which in turn causes high population growth. There is no wonder then even after over 50 years of planned economic development, 317 million people lived below the poverty line in 1993-94. The decline in number of poor people to 260 million in 1999-2000 is doubtful because of the change in methodology made in NSS of 1999-2000.

Demographic Changes in India and Population Dividend

India is passing through a phase of unprecedented demographic changes. These demographic changes are likely to contribute to a substantially increased labour force in the country. The census projection report shows that the proportion of working age population between 15 and 59 years is likely to increase from approximately 58 per cent in 2001 to more than 64 per cent by 2021. *In absolute numbers there will be approximately 63.5 million new entrants to the working age group between 2011 and 2016*. Further, it is important to note that the *bulk of this increase is likely to take place in the relatively younger age group of 20-35 years*. Such a trend would make India one of the youngest nations in the world. In 2020 average Indian will be only 29 years old. The comparable figures for China and US are 37 and for West Europe 45 and for Japan 48 years. *This higher proportion of young labour force in India has a great production potential* and has therefore been called '*demographic*

3. K. Sundaram, *Demography and Development*, 1951-2001, A Paper presented at the Institute of Economic Growth, April 27, 1984.

dividend'. This demographic dividend provides India large productive opportunities for rapid economic growth. However, this also poses a great challenge; *its benefit will be realised only if our population is healthy, educated and appropriately skilled.* Therefore, greater focus on human and inclusive development in India is necessary to best utilise the demographic dividend.

POPULATION CONTROL POLICY

Over-populated developing countries are currently facing the problem of population explosion. We have explained above in this chapter how population explosion is hampering economic growth in these countries. The population growth is swallowing up a large part of the gains in national income brought about by planned economic development. If we want that future generations should have at least as much good prospects of living as the present generation we must control population growth.

India's population which is presently about 1027 millions is increasing at a rate of about 2 per cent per annum. If current trends continue, India may overtake China in 2045 to become the most populous country in the world. At present about 50 per cent of India's population is below 20 years of age. Hence, a large number of our present population will live to see by the middle of the 21st century the disastrous consequences such as acute poverty conditions, widespread unemployment. A high degree of socio-economic tensions if current explosive growth in population is not checked.

The check on the growth of population is therefore absolutely essential if we want to solve problems of mass poverty and widespread unemployment. Further, if the economy is to be prevented from reaching the stationary state, controlling population growth is essential to achieve this aim. Stressing the importance of controlling population growth in India characterised by economic growth, the existence of mass poverty and unemployment, Prof. P.R. Brahmananda writes, "A stationary economy with an open ended population expansion angle will be the greatest permanent disaster for the country. All hopes of improvement in living standard, even at the meagre levels, of vast masses will have been forever dashed to pieces." He further adds, "Such an atmosphere cannot be conducive for economic progress. The prospect of more bread being ruled out, the flow of freedom may not flutter for long."

Now, an important question is what policies can be adopted to control population. Population growth depends on the fertility rate (birth rate) and death rate. Thus, policies that have to be adopted to check population growth should aim to reduce the birth rate. The increase in birth rate is ruled out for obvious reasons. Population growth can be reduced by reducing the birth rate. That is, each family should have smaller number of children. It may be noted that reducing fertility rate in under-developed countries is quite a formidable task. This is because, as has been pointed out by Kuznets, *"in underdeveloped countries people have an intense tendency to have more children because under the economic and social conditions large proportions of the population see their economic and social interests in more children as a supply 'family labour, as a pool for a genetic lottery, and as a matter of economic and social security a non-protecting society."*

The various policies that may be adopted to control the growth of population are: *(a)* Family Planning Programme, *(b)* Sterilisation, *(c)* Promotion of education, (*d*) Social and economic development, especially of the poor sections of the society. We shall now discuss each of them briefly.

Family Planning Programme

This is an important policy measure that has been adopted by several developing countries. India is the first country in the world to adopt family planning as a state policy. Under family planning programme various contraceptive devices and health services are provided to encourage the couples to adopt *small family norm,* that is, to have fewer children. Through various communication media a wide publicity is given that couples should go in for only two or three children and no more. An appeal is made to them that their life would be happy and prosperous if they have a small family.

Recently, policy of providing *economic incentives and discentives* has been adopted in some countries to discourage the people from having more children beyond a certain number which is generally fixed at two or three. Some of the incentives and disincentives adopted to promote family planning are: (1) Elimination or reduction of maternity leave and benefits beyond a certain number. This will discourage the employed women to bear more children. (2) The establishment of old-age social security provisions, such as old-age pension. This greatly helps in motivating the people to have smaller number of children. As is well known, in poor countries like India, people have a tendency to bear more children so that in their old age security be ensured. (3) Money payments to those who opt for small families and voluntarily go in for sterilisation operation, (4) Allotment of scarce public houses, housing plots, flats etc. on the basis of preference to those who have a small family. Besides these several other inentives and disincentives have been devised to encourage the small family norm.

For the success of the family planning programme a variety of *medically tested and safe contraceptives* such as IUD pills, condoms, foam tablets, jellies and others should be made freely available. It may be noted that a great boost was given to the family planning programme by the discovery of IUD (intrauterine device) popularly called as loop. But after some time an impress, ion was created among the female population that it was not very safe. The need of the hour is to strengthen the research in contraceptive technology and such devices should be discovered which should be effective but minimise the health risk. In India, after the experience of several years, there has been some reconsideration of the various aspects of the family planning programme. Family planning programme would now be an integral part of comprehensive family welfare policy covering education, health and child care, family planning and women's rights and nutrition. The accent in new policy is on integrating contraception with public health, medical care, maternal and child health and nutrition. In fact, the name of the ministry has been changed from Ministry for Family Planning to Ministry of Family Welfare. In so far as it goes, the new orientation may prove to be good but it needs to be stressed, as shall be brought out later, that the success of family planning programme depends heavily on the process of social and economic development.

Compulsory and Forced Sterilisation

As an effective policy measure to check the growth of population, compulsory sterilisation has been suggested by many. According to this, it should be made compulsory for the couples to undergo sterilisation beyond a certain number of children, say two or three. Those who violate this either penalties may be imposed or they are forced to undergo the operation. Admittedly, this is a drastic measure, but is considered very effective in the context of the failure of voluntary family planning programme.

Of all the measures involving several incentives and disincentives proposed the most controversial is the proposal of *compulsory sterilisation*. According to this proposal, compulsory sterilisation of every couple who has more than two children should be introduced. This suggestion of introducing an element of compulsion in the sterilisation programme implies that one is not hopeful of various incentives and disincentives succeeding in achieving the objective of stationaries in population size. It may be noted that people of India has already rejected the policy of compulsory sterilization adopted during emergency period 1975-77.

In India during the dark period of emergency the policy of forced sterilisation was in fact adopted, though officially it was described as voluntary. In the name of motivation campaign, family planning camps were set up where people were forcibly driven to undergo operation. Ambitious targets were laid down which were successively revised upwards. In fact *coercion* came to be used on a large scale, especially in the States in the North, so that the programme was voluntary only in name. The State Chief Ministers vied with each other to surpass their targets. and excesses were committed in the process. In 1976 while the target of sterilisation originally fixed was 43 lakhs, actually 75 Iakh sterilisations were effected. Little wonder, the official overenthusiasm and excesses made the whole programme highly unpopular and this was one of the main causes of Congress Party's utter rout in the Northern States in Lok Sabha Elections of March, 1977.

In the opinion of the present author, the policy of forced sterilisation is quite inhuman; it is treating human beings as animals. Public opinion is strongly opposed to it. Moreover, when there are several contraceptives available recourse to compulsory sterilisation is not needed.

Promotion of Education and Population Growth

Empirical studies conducted in several countries reveal that there is *inverse relation between the fertility rate and education, especially of women and girls.* Therefore, an important measure to reduce the birth rate is the promotion of education, especially among female population of the country. The educated men and women accept small family norm and more readily take family planning measures such as use of contraceptives. Moreover, when educated women are employed, their tendency to bear and rear more children falls. Dr. Ashish Bose, an eminent demographer, has pointed out that India's family planning campaign has been severely handicapped because of the slow rise in the *literacy level.* To quote him, *"The sad fact remains that in large parts of India, especially in U.P., Bihar, Madhya Pradesh and Rajasthan, women literacy rate in the rural areas* is *less than 10 per cent. And these are states where the progress of family planning is slow. These four States account for* 39 *per cent of India's population. So their impact (or lack of it) on the national birth rate is bound to be significant."*[4] It, therefore, follows that efforts to educate the, people especially women, both formally and informally, be strengthened to achieve the target of reduction in birth rate.

Economic Development and Population Growth

It has been said that *"development is the best contraceptive."* This implies that with social and economic development, the birth rate will go down, as people with higher levels of living prefer to have fewer children. First, people with a higher level of living do not need children to supplement the family's meagre income. Secondly, they come to prefer "quality" of children rather than to "quantity." Thirdly, their desire to further improve their level of living increases and this induces them to practise family planning measures. In the opinion of the present author if the population growth in India has remained virtually unchecked in spite of a nationwide family planning programme, the fault does not lie solely with the programme but to a much greater extent with the character of social and economic development in India which has failed to remove mass poverty and improve the levels of living of the poor and weaker sections of the society. Accordingly, if population is to be effectively checked, not only the tempo of economic development should be speeded up but strategy of development should be such as will promote employment opportunities for the poor especially landless labour, marginal and small farmers. It should be ensured that fruits of economic development should reach these poor people. For this land reform measures and income redistribution policies should also be effectively implemented. If the standards of living of the poor go up as a result of economic development, they will readily adopt small family norm.zzz

4. Ashish Bose, "New Directions in Family Planning Programme", *Asian Economist*, Vol. No. 889, May 23, 1978.

Chapter 45
HARROD-DOMAR MODEL OF GROWTH

Introduction

Keynes in his General Theory was concerned with the determination of income and employment in the short run. He explained that since in the short-run situation of developed capitalist economies aggregate demand was deficient in relation to the aggregate supply of output, the equilibrium will be established at less than full-employment level. Since the propensity to consume (and therefore saving propensity) is given and remains constant in the short run, if the amount of investment as determined by expected rate of profit and the market rate of interest is not equal to the amount of saving at the full-employment level of income, the economy will be in equilibrium at less than full capacity level (*i.e.* less than employment level) of output. He did not go into the question of the long-run growth of the economy. In fact, he overlooked the effect of investment in a given period on the increase in productive capacity. However, *investment has a dual effect.* Firstly, investment increases aggregate demand and income of the people through the multiplier process, and secondly, it raises the productive capacity of the economy through the addition it makes to the stock of capital. Indeed, investment by very definition means the addition to the stock of capital. While Keynes took into account the demand effect of investment, he ignored the capacity effect of investment.

Harrod[1] and Domar[2] extended the Keynesian analysis of income and employment to long-run problem of economic growth and therefore considered both the income and capacity effects of investment. Harrod and Domar models of economic growth explained at what rate investment should increase so that steady growth is possible in an advanced capitalist economy. In the growth models of Harrod and Domar, the rate of capital accumulation plays a crucial role in the determination of economic growth. The problem of present-day mature economies in the late forties and early fifties lay in averting both secular stagnation and secular inflation. It were the pioneer works of Harrod and Domar that set the ball rolling in regard to this issue of conditions required for the maintenance of steady economic growth in advanced industrialised countries. The *Harrod and Domar models seek to determine that unique rate at which investment and income must grow so that full-employment level is maintained over a long period of time, i.e., equilibrium growth is achieved.*

Harrod and Domar developed their models of steady growth quite independently, though Harrod published his theory earlier than Domar. Although their models of steady growth differ in details, yet the underlying basic idea is essentially the same. Both of them assigned to capital accumulation a crucial role in the development process. But they emphasised the double role of the investment process, namely, generating income (increasing demand) and adding to the productive capacity of the economy. The classical economists confined their attention to the capacity-side only, whereas the earlier Keynesian economists studied the problem of demand only whereas Harrod and Domar consider both sides.

They start with full employment equilibrium level of income. According to them, to maintain full employment equilibrium, aggregate demand (total spending) generated by investment must be sufficient to be the additional output caused by this investment. To ensure steady growth with full

---

1. R.F. Harrod, *Towards a Dynamic Economics*, Macmillan & Co. London, 1930.
2. E. Domar, "Capital Expansion, Rate of Growth and Employment", *Economica,* Vol.. 14, 1946.

employment the absolute amount of net investment must keep increasing and there must also be continuous growth of real national income. Because if demand and income did not increase while annual investment went on occurring, the additions made to the capital stock would remain unutilised and also employment could not be provided to the growing labour force which would result in unemployment of these two major resources. Obviously, such a situation is not conducive to steady economic growth.

DOMAR'S GROWTH MODEL

Fundamental Growth Equation

Let us first consider the supply side, that is, the capacity effect of investment. Increase in national output or national income of an economy during a period depends upon the increase in the stock of capital (which is represented by ΔK) during a period and the output-capital ratio or the productivity of capital. Assuming that both national income and the capital stock are measured in money, output-capital ratio can be written as $\Delta Y/\Delta K$, where ΔY stands for the increase in the national income and ΔK for the increase in the stock of capital. Thus if ₹4 worth of capital goods is required to produce one rupee worth of real output, marginal output-capital ratio $\left(\frac{\Delta Y}{\Delta K}\right)$ is equal to 1/4 or 0.25. Thus, the absolute increase in national income during a period (ΔY) can be obtained from increment in stock of capital ΔK multiplied by the output produced by a unit of capital $\left(i.e. \frac{\Delta Y}{\Delta K}\right)$. In symbolic terms we may express this as follows:

$$\Delta Y = \Delta K . \frac{\Delta Y}{\Delta K} \qquad \text{...}(i)$$

Now, change in capital stock (ΔK) is nothing else but investment. Therefore, following Domar in place of ΔK we can write I. The marginal output-capital ratio $\Delta Y/\Delta K$ which is assumed to be constant as well as assumed to be equal to average output-capital ratio $\left(\frac{Y}{K}\right)$ by Domar and Harrod, can be denoted by σ. Thus, as Domar puts it, growth in capacity-output can be written as under:

$$\Delta Y = I\sigma \qquad \text{...}(ii)$$

It may be noted that *output-capital ratio (σ) is reciprocal of capital-output ratio i.e.* $\left(\frac{\Delta K}{\Delta Y} \text{ or } \frac{K}{Y}\right)$. Let us give an example. If ₹ 500 crore is invested in a year and capital-output ratio is 4 (*i.e.*, output-capital ratio will be 1/4), then growth in output in a year will be

$$\Delta Y = 500 \times \tfrac{1}{4}$$
$$= 125 \text{ crore}$$

Demand or Income Effect of Investment

Now, according to Domar, growth in capacity-output will be realised only if aggregate demand or income of the people increases by a sufficient amount. The increase in aggregate demand or income is explained by the Keynesian theory of multiplier. Domar has based his analysis of demand or income effect of investment on the Keynesian theory of multiplier and income determination. According to this, *increase in income (or aggregate demand) is given by the increase in investment (ΔI) and the size of multiplier*, *i.e.*, *1/s* where s is the marginal propensity to save (assumed by Domar to be equal to the average propensity to save). Thus, according to income effect of investment,

$$\Delta Y = \frac{1}{s}\Delta I \qquad \text{...}(iii)$$

Note that $\frac{1}{s}$ represents the size of investment multiplier.

The growth in income (ΔY) must be large enough to generate demand equal to capacity growth in output as explained above.

Domar's Growth Equation in Terms of Rates of Growth

It is greatly helpful to express the above growth equation in terms of *rates of growth of income and capital*. That is, growth in income and capital should be expressed as ratios of total income. For doing so we divide both sides of equation (*i*) above by Y and obtain:

$$\frac{\Delta Y}{Y} = \frac{\Delta K}{Y} \cdot \frac{\Delta Y}{\Delta K} \qquad \text{. . . (iv)}$$

$\frac{\Delta Y}{Y}$ represents the rate of growth of income and is therefore written as simply G_y. Besides, ΔK stands for increment in capital during a given time period and is nothing else but investment. Therefore, for ΔK in the equation (*ii*) we can write I which represents investment. With these changes, we get the following equation:

$$G_y = \frac{I}{Y} \cdot \frac{\Delta Y}{\Delta K}$$

With the further assumption that output-capital ratio remains constant, then the marginal output-capital ratio $\left(\frac{\Delta Y}{\Delta K}\right)$ will be equal to average output-capital ratio $\left(\frac{Y}{K}\right)$. With this assumption and also expressing output-capital ratio by σ we can write the above equation as follows:

$$G_y = \frac{I}{Y} . \sigma \qquad \text{. . .(v)}$$

where G_y = Growth rate of output or income

$\frac{I}{Y}$ = Rate of investment as ratio of national income

σ = Output-capital ratio

From the growth equation (*iv*) above, it is clear that, given the output-capital ratio, rate of growth of output depends upon the rate of investment; the greater the rate of investment, the greater the rate of growth of output or income. If the economy is initially in full-employment equilibrium as investment is equal to saving at full capacity output, then *to maintain full-employment equilibrium as the economy grows at a steady rate, savings (S) must remain equal to investment* (*I*). Therefore, in equation (*iv*) we can write $\frac{S}{Y}$ for $\frac{I}{Y}$ and rewriting equation (*iv*) we have

$$G_Y = \frac{S}{Y} . \sigma$$

Since $\frac{S}{Y}$ represents ratio of savings to national income (which is also simply called saving ratio), we can write it as s. With rewriting the above equation we have,

$$G_Y = s.\sigma \qquad \text{. . . (vi)}$$

The above equation (*vi*) represents the productive capacity effect of investment and saving and therefore represents the supply side of the growth problem.

The Condition for Equilibrium Growth

To achieve and maintain equilibrium or balanced growth, aggregate demand (*i.e.*, productive aggregate expenditure) must increase at the rate which is large enough to use the increase in productive capacity which capital accumulation or investment generates. In his analysis Domar found that to remain on equlibrium growth path investment must increase at a rate large enough to absorb all savings which rising incomes make available.

Now, according to Domar, growth in capacity output will be realised only if aggregate demand or income of the people increases by a sufficient amount. The increase in aggregate demand or income is explained by the Keynesian theory of multiplier. Domar has based his analysis of demand or income effect of investment on the Keynesian theory of multiplier and income determination. According to this, *increase in income (or aggregate demand) is given by the increase in investment (ΔI) and the size of multiplier, i.e., 1/s* where s is the marginal propensity to save (assumed by Domar to be equal to the average propensity to save). Thus, according to income effect of investment,

$$\Delta Y = \frac{1}{s} \cdot \Delta I \qquad \ldots (vii)$$

Thus aggregate demand or income increases at the rate $\frac{1}{s} \cdot \Delta I$ where s is propensity to save and ΔI is the absolute increase in investment. On the other hand, as shown by equation(*ii*) above, the increase in capacity output occurs at the rate of $I\sigma$ where I is the absolute rate of investment and σ is the output-capital ratio. Thus, *steady equilibrium growth rate will be achieved only if rate of growth of aggregate expenditure (demand or income) equals the rate of growth in capacity output.*

Thus, it follows that to *maintain full-employment equilibrium growth of output* the following condition must hold :

$$\frac{1}{s} \cdot \Delta I = I\sigma$$

or

$$\frac{\Delta I}{I} = s\sigma$$

where $\frac{\Delta I}{I}$ represents the rate of growth of investment.

As seen above in equation (*v*), rate of growth of income $\left(\frac{\Delta Y}{Y} \text{ or } G_Y\right)$ is also equal to $s\sigma$,

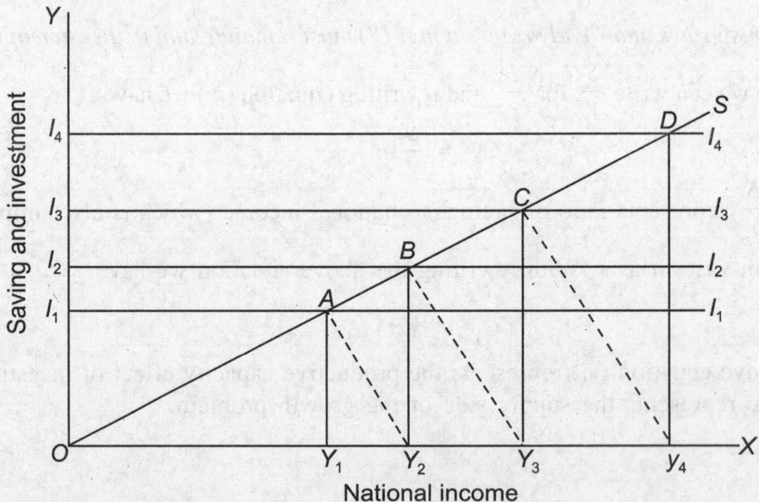

Fig. 45.1. *Harrod-Domar Model of Equilibrium Growth*

it follows that for equilibrium growth $G_y = \frac{\Delta Y}{Y} = \frac{\Delta I}{I} = s\sigma$

Thus, the essential condition for maintaining a continuous equilibrium state of full employment is that both investment and real income must grow at a constant annual rate. This rate should be equal to the propensity to save (s) multiplied by output-capital ratio, (σ) *i.e.*, $s\sigma$.

We can explain the model in geometrical terms with the help of Fig. 45.1 given above. Here real income is measured along the horizontal axis, while saving and investment (in real terms) are measured along the vertical axis. The saving function is represented by the line *OS* starting from the origin. Its slope is given by the marginal propensity to save (s) which is assumed to remain constant for a considerable period of time. The initial investment demand is represented by curve I_1I_1. This intersects the saving function *OS* at the point *A* so that the corresponding equilibrium level of income is Y_1. We assume that it corresponds to the full-employment level of national income.

Now, the new capital so created (represented by OI_1) will lead to the increase in productive capacity as determined by output-capital ratio. Given the output-capital ratio, investment OI_1 leads to Y_1Y_2 increase in output or income. As a result, national income will increase to Y_2. The ratio between the increase in income (ΔY or Y_1Y_2) and the increase in investment (OI_1) is given by the 'output-capital' ratio σ. But the new equilibrium level of income Y_2 will be realised or maintained only if the investment demand function shifts upward to I_2I_2 and intersects the saving function *OS* at the point *B*, which is vertically above Y_2.

However, as soon as the new capital equipment represented by the higher level OI_2 starts producing goods, capacity output or income will rise to Y_3 (indicating an increase by an amount σ times OI_2 over the previous level of income (Y_2). But the new level of income Y_3 will be maintained only if investment increases so much that the new investment demand curve I_3I_3 intersects the saving function *OS* at *C*. In this way, the process will continue so long as the investment increases by the appropriate amount in each period. The income would successively go on rising by an amount σ times the previous period's investment. And the investment in each period would rise by σ times the output-capital ratio. Thus, income would continue to grow at the steady rate of $s\sigma$.

It is evident from the basic equation $\frac{\Delta I}{I} = s\sigma$ that greater the saving rate(s), the greater will be the growth of investment needed to maintain the steady growth with full employment. Similarly, greater the value of σ (*i.e.*, the output-capital ratio), the greater should be the increase in income to avoid emergence of excess capacity. But greater income is possibly only through greater investment. Hence, if income is to grow at a steady rate, investment must also grow at the annual steady rate given by $s\sigma$.

If $\frac{\Delta I}{I} < s\sigma$, that is, if sufficient growth in investment does not take place, steady growth with full employment cannot be achieved. On the other hand, if today's investment is sufficient to achieve equilibrium growth with full employment, investment will have to be much more in the next period for generating sufficient increase in demand so as to utilise fully the expanded production capacity and to avoid the underutilisation of capital stock which will result in a fall in investment, and so cause depression. In other words, "*the economy must, so to speak, run faster and faster to stay in the same place, otherwise, it will slip downwards.*"[3]

HARROD'S GROWTH MODEL

Though Harrod's model of growth is similar to that of Domar, but it differs from the latter in details. Therefore, we will explain below the essential features of Harrod's growth theory separately.

3. Meier and Baldwin, *Economic Development*, 1966, p. 107.

In his essay "*Towards a Dynamic Economics*", Harrod put forward a theory which can be considered as truly dynamic. Explanation of secular trends is his main theme. He seeks to explain the secular causes of unemployment and inflation and the factors which determine the equilibrium and the actual rate of capital accumulation.

The classical economists considered economic development as a race between technological progress and capital accumulation on the one hand, and growing population and diminishing returns from land on the other. Harrod drops diminishing returns, regards technological progress and population growth as independent factors. In Harrod's analysis of economic growth there are three basic elements: (a) population growth, (b) output per head as determined by level of technique or inventions and (c) capital accumulation. Inventions may be neutral, *i.e.*, leave the capital coefficient unchanged, or capital-saving, *i.e.*, reducing the capital coefficient, or 'labour-saving' which will increase the capital-output ratio. It may be noted that capital-output ratio is the reciprocal of output-capital ratio (σ), the concept used by Domar. It is important to mention that Harrod uses the concept of incremental capital-output ratio, which is the reciprocal of marginal output-capital ratio (σ) of Domar's model.

While arriving at the behaviour of income as a response to entrepreneurial decisions regarding investment, Harrod makes two assumptions: (*i*) Saving in any period of time is a constant proportion s of the income received during that period, and (*ii*) the investment is proportional to the rate of increase of income. The second assumption *is in fact the acceleration principle which states that the increase in output or income that occurs induces an increase in the stock of capital.*

Harrod begins his analysis of growth by marrying together acceleration principle and the theory of investment multiplier. As in Domar's model, Harrod explains that growth rate $\left(G_y \text{ or } \frac{\Delta Y}{Y}\right)$ depends on the rate of capital formation (or investment) and capital-output ratio which he defines as "the value of the capital goods required for the production of a unit increment of output". He put forward three growth equations. He takes saving as a fixed proportion of national output or income. Presenting a more elaborative analysis of growth than Domar, Harrod advanced three growth equations.

Harrod writes his first growth equation as follows:

$$G_y = \frac{s}{v} \qquad \ldots (i)$$

where G_y is the rate of growth in a period $\left(\frac{\Delta Y}{Y}\right)$, s is the rate of saving (*i.e.* proportion of saving to national income $\left(\frac{S}{Y}\right)$, and v is the capital-output ratio. It is important to note that capital-output ratio v in Harrod's growth equation (*i*) above is the one that is *actually obtained* from the extra capital accumulation (ΔK) and increase in production of goods and services in a year (ΔY). Harrod derives this growth equation as follows.

Following the Keynesian framework, Harrod takes that actual saving must be equal to actual investment. Further, since Harrod takes saving (S) as a constant proportion of national income (Y) in a period we have

$$S = sY_t$$

where s is propensity to save.

Investment (I or ΔK) in a period t depends on the rate of increase in production or income), that is, ΔY (*or* $Y_t - Y_{t-1}$) and the actual *capital-output ratio* (v). Thus we have

$$\Delta K \text{ or } I = v(Y_t - Y_{t-1})$$

Since in a period *actual* saving must be equal to *actual* investment, we have

$$v\,(Y_t - Y_{t-1}) = sY_t$$

Dividing both sides by Y we have

$$\frac{v(Y_t - Y_{t-1})}{Y_t} = s$$

or

$$\frac{(Y_t - Y_{t-1})}{Y_t} = \frac{s}{v}$$

Since $\frac{(Y_t - Y_{t-1})}{Y_t}$ represents actual growth of output or income we can denote it by G_y. Thus

$$G_y = \frac{s}{v}$$

G_y is the growth of output or income which *actually* occurs in a period. The above growth equation is in fact a truism as it is always true by definition depending as it is on the accounting identity that actual investment equals actual savings of a period.

Warranted Rate of Growth

Harrod proposes a second growth equation which he calls a fundamental growth equation to describe the *equilibrium growth at a steady rate.* The warranted rate of growth is taken to be that rate of growth *which if it occurs will keep the entrepreneurs satisfied that they have produced neither more nor less than the right amount.* Being satisfied with the achievement of this growth rate, the entrepreneurs will maintain or perpetuate the same rate of growth. Warranted rate of growth is thus equilibrium rate of growth in the sense that producers, if they achieve it, will be induced to maintain it. The condition for warranted rate of growth is stated as under :

$$G_w = \frac{s}{v_r} \qquad \ldots (ii)$$

Harrod denotes capital-output ratio by the letter *C*, but following the modern practice we are using v for it.

G_w= "*Warranted rate of growth*", which is that rate of income growth of output or income. ($\Delta Y/Y$), which will keep entrepreneurs satisfied with the amount of investment they have actually made, *i.e.*, it is in fact the full capacity rate of growth.

v_r = required incremental capital-output ratio to sustain the warranted rate of growth and is determined by the state of technology and the nature of goods constituting the increment in output.

s = average propensity to save, that is, $\frac{S}{Y}$.

The sort of entrepreneurial behaviour envisaged by Harrod means that *to maintain full employment, the desired (ex-ante) saving out of the full-employment income must be offset by an equal amount of desired investment.* But to induce this much investment, income must be growing.

In both the above equations *(i) and (ii) above s* is the same because Harrod assumes that the saving intentions are always realised so that *ex-ante* savings are always equal to *ex-post* savings. Thus, Harrod is able to show that for dynamic equilibrium $G_w = \frac{s}{v_r}$.

It is important to note that v_r in growth equation (*ii*) is different from v of the growth equation (*i*), As noted above, v in Harrod's growth equation (*i*) shows the increment in the amount of new capital installed during a period divided by the *increase in output actually obtained* from it during that period. It shows what has been *actually* produced with the addition to the capital stock in a period and *not whether producers are satisfied with the increase in output actually realised.* For example, if there are boom conditions in the economy and as a result the increase in capital installed during the period is fully utilised, actual capital-output ratio (v) will be lower. On the other hand, if there is a demand recession in the economy, good amount of extra capital installed will not be utilised for production and consequently incremental capital-output ratio (v) will be higher.

But what determines the size of the required incremental capital-output ratio (v_r) which keeps entrepreneurs to perpetuate the growth. The size of the v_r is determined by technological conditions and the nature of goods comprising the increment of output. *This warranted rate of growth will be achieved if sufficient increase in income takes place during the growth process.*

The proportion of investment to income being fixed, an increase in income would mean that in the next period both income and investment must be higher. In such a situation the producers would like to perpetuate the rate of growth which they have already realised. Under such circumstances the producers invest in the hope that they will be able to sell what they have planned to produce. In other words, the producers will desire to invest an amount required by $G_w v_r$, which will be equal in magnitude to s, *i.e.*, the given proportional investment rate.

Condition for Equilibrium Growth Rate

Now, what is the condition for equilibrium growth rate? *In Harrod's model, if incremental capital-output (v) actually realised happens to be equal to required capital-output ratio (v_r), warranted by technological and other conditions, then the actual rate of growth, G_Y is equal to the warranted rate of growth (G_w), the rate which the circumstances of the economy warrant, the economy will be growing at the equilibrium rate* ($G_y = G_w$). It may be noted that actual rate of growth will be equal to warranted rate of growth when investment is increasing at the rate high enough to generate adequate demand to ensure capacity growth rate (G_w).

Harrod lays down the equilibrium condition for steady growth by saying that the actual rate of growth must be equal to the warranted rate of growth, i.e., the rate of increase in output or income should be just so much as to keep the entrepreneurs satisfied with the actual investment they have made.

Thus, so long $v_r = v$, the producers would like to perpetuate a growth rate that is equal to the actual or realised rate. In other words, G_y (the actual growth rate) will be the same as that which the producers want to perpetuate, *i.e.*, G_w. But we have seen above that G_w stands for the rate of growth which when realised leaves the entrepreneurs in a state of mind that they shall be prepared to undertake a similar advance in the future. Furthermore, if the income increases at this rate it will continue to increase at this rate. This is how a steady growth rate is assured. Income will have to grow faster and faster, if the entrepreneurs are to be convinced that the higher investment was desirable. In this way, both income and investment will go on increasing from one period to the next. There is thus a cumulative equilibrium growth of income and investment.

Natural Rate of Growth

Expansion, however, cannot go on indefinitely. The availability of labour and natural resources would put the limit. In other words, it is not necessary that the warranted rate of growth G_w (which is also equal to the actual rate of growth G_y) is the maximum attainable rate of growth. With this view in mind, Harrod introduces yet another rate of growth called the '*natural rate of growth*', G_n, which is the maximum rate of growth allowed by the increases of macro variables like population growth, technological improvements, and growth in natural resources. In fact, G_n *is the highest attainable growth rate which would bring about the fullest possible employment of the resources existing in the economy*. This may be considered as the ceiling rate of growth. Joan Robiuson calls it the maximum feasible rate of growth. If l stands for growth rate of population (or labour force) and t for technological progress (*i.e.*, rate of increase of productivity), then natural rate of growth can be written as

$$G_n = l + t$$

Hence for the equilibrium growth rate at full employment of all existing resources, the following condition must be satisfied:

$$G_n = G_w = G_y$$

Any deviation from this path would bring about instability in the economy.

The Golden Age

The equality of three growth rates ($G_y = G_w = G_n$) ensures that economy is in moving or dynamic equilibrium. This is also called *balanced growth equilirbrium. Joan Robinson describes the equality of these three growth rates as a golden age as it represents a very satisfactory and happy situation*. This is a happy situation because the equality of these three growth rates ($G_y = G_w = G_n$) will ensure steady equilibrium growth rate along with full employment of labour and without creating excess productive capacity. However, Joan Robinson has emphasised that the golden age, namely, the equality of three growth rates, "*represents a mythical state of affairs not likely to obtain in any actual economy*"[4]. This is because the four key variables, namely, propensity to save (*s*), required capital-output ratio (v_r) of the warranted rate of growth, the rate of growth of population (*l*) and rate of technological change (*t*) are determined quite independently of each other. While the warranted rate of growth (G_w) is determined by the value of *s* and v_r, the natural rate of growth is determined by rate of population growth (*l*) and the rate of technological progress (*t*). The golden age or balanced growth equilibrium of $G_y = G_w = G_n$ will occur only when the four variables, *s, v, l* and *t* have appropriate values. But this seems to be very unlikely to occur. It is only by chance that these four variables will have right or appropriate values to guarantee the golden age equilibrium.

Relevance of Harrod-Domar Growth Model for Developing Countries

Despite the fact that Harrod-Domar model was not conceived with developing countries in mind, it has nevertheless considerable influence in their development planning. An important aspect of the model is that it relates *rate of economic growth to investment which is often considered as a major limiting factor of economic growth*. Besides, it considers the level of savings which restricts investment in developing countries. The model has been used to answer the question as what rate of saving and therefore investment is required to achieve a certain target growth rate, given the assumed capital-output ratio. It has often been found that domestic resources or savings are inadequate to achieve target growth rate set for development plans of developi..g countries. Thus, there has been a saving gap in developing countries to achieve target growth rate. Therefore, the case of *foreign aid* or *foreign investment* has been made to supplement the domestic savings to achieve the planned growth rate.

However, there are some limitations of applying Harrod-Domar models to the conditions of developing countries. First, to assume away the role of government is to disown the realities completely. In fact, due to the vast structural changes to be effected government in these economies must step in a big way to initiate and accelerate economic development as an efficient manager of the whole economy, lest we may slip down the slope. Besides, the assumption of an initial full-employment level of income is not valid for the developing countries; disguised unemployment pervades these underdeveloped economies, especially the labour-surplus economies. It is a case of structural disequilibrium arising basically from the imbalance between labour and capital. Even if we make allowance for the entire savings to be invested, the growth of capital stock fails to match with the growth of labour force. Being based on fixed capital-output and capital-labour ratio, the

4. Joan Robinson, *The Accumulation of Capital*, London, 1956.

Harrod and Domar models have only limited applicability for the developing world. Their peculiar problems demand a different solution from the one suggested by these models. *To absorb the surplus labour force, there is a need to bring down both the capital-output and capital-labour ratios through reducing capital intensity*. The Harrod-Domar models by assuming a constant capital coefficients rule out such a possibility.

Secondly, the usefulness of models based on the concept of capital-output ratio is of little operational significance in developing economies. Depending on the nature and degree of various shortages, bottlenecks and market imperfections, the productivity of invested capital is amenable to considerable fluctuations. It is indeed very difficult to have an accurate and valid estimate of a concept like capital-output ratio under such fluid conditions. Commenting in this regard, Prof. Hirschman[5] remarks that the predictive and operational significance of a model based on the concept of *capital-output* ratio is far less for a developing economy than for advanced economies. Models such as these, therefore, cannot explain the mechanism through which economic growth can get under way and could be carried forward in the present-day developing economies.

Thirdly, the Harrod-Domar growth variables are aggregative in nature and, therefore, fail to show the sectorial interrelationship. The processes of development of the developing economies are, as is being increasingly acknowledged, fundamentally linked with structural and institutional changes. Their highly aggregative nature, comments Prof S. Chakravarty[6], '*prevents them from being used as a tool in detailed quantitative policy making and conceals many structural aspects of the problem of a steady rate of growth.*

Further, Harrod excludes autonomous investment as an explicit variable in his formulation of '*warranted*' saving-investment equality. But the exclusion of autonomous investment as an important factor in determining growth in developing countries by Harrod in his growth model renders Harrod's concept of 'warranted' growth rate analytically inadequate for the purpose of developing countries. He also ignored the role of public investment to which Keynes assigned a crucial role. But, autonomous investments, whether public or private, are of pivotal importance to the developing countries. Besides, Harrod-Domar growth models assume that propensity to save and the capital-output ratio are constant. But actually they are likely to change over the long run. Further, if the proportion of factors can be changed as labour may be substituted for capital, then adjustment within the economy can be easily made and steady growth made possible without any rigid conditions.

In spite of the fact that these models are of limited applicability to the developing countries and fail to highlight the crucial issues involved in the development process of these economies, they nevertheless *are useful in fixing the overall targets of income, investment and savings* and in checking the consistency of such targets. Prof. Kurihara states that "Harrod and Domar have made the essential nature of the growth mechanism operationally significant, for they stress saving ratio and the capital-output ratio (or its reciprocal) as measurable strategic variables to investigate and possibly to manipulate for a desired rate of growth. Because of the universal character of these strategic variables, the growth mechanism discussed by Harrod and Domar is applicable to all economic systems, albeit with the modification".[7]

The use of Harrod-Domar model has actually been made in some countries. For instance, in the First Five Year Plan of India, the rate of saving was planned to be raised through keeping the marginal rate of saving above the average rate of saving. And the current rate of capital formation and therefore growth of the economy was sought to be maximised through raising the marginal rate of saving. Thus, these models served to guide the planners in determining the growth rate of the Indian economy. Commenting on these models, Prof. S. Chakravarty remarks that "*The great*

---

5. Hirschman, *Strategy of Economic Development,* 1957.
6. Sukhamoy Chakravarty, *The Logic of Investment Planning,* p.43.
7. K.K. Kurihara, *The Keynesian Theory of Development*, 1959. Ch. 9, pp.153-154

service that these models perform is to indicate very roughly the dimensions of the problem involved in raising the per capita income level in an underdeveloped country".[8] Even India's Twelfth Five Year Plan points out that acceleration of economic growth requires increase in rate of investment and raising productivity of capital (*i.e.*, output-capital ratio).

As noted above, Harrod-Domar model brings out the crucial role for the continuous growth of investment to ensure sustained growth at a steady rate. If investment is not growing sufficiently, the problem of deficient demand will emerge which will bring about recesionary condition even in a developing country. The demand recession will result in rise in capital-output ratio due to the underutilisation of productive capacity. The Indian growth experience clearly brings out this fact. From the mid-sixties to the late seventies the Indian economy witnessed the problem of demand deficiency due to the fall in public investment resulting in lower industrial grwoth and increase in capital-output ratio. Again during 1997-2003 low industrial growth rate was achieved due to deficiency in demand resulting from stagnation in investment.

8. S. Chakravarty, *Op. cit.*

CHAPTER 46
NEOCLASSICAL THEORY OF GROWTH

Introduction

Having discussed Harrod-Domar growth model in the last chapter we will discuss neoclassical growth theory in the present chapter. The neoclassical growth theory was developed in the late 1950s and 1960s of the twentieth century as a result of intensive research in the field of growth economics. American economist Robert Solow[1], who won a Nobel Prize in Economics and British economist, J.E. Meade[2] are the two well known contributors to the neoclassical theory of growth. This neoclassical growth theory lays stress on capital accumulation and its related decision of saving as an important determinant of economic growth. Neoclassical growth model considered two-factor production function with capital and labour as determinants of output. Besides, it added *exogenously determined factor, technology*, to the production function. Thus neoclassical growth model uses the following production function:

$$Y = AF(K, L) \qquad \ldots (1)$$

where Y is Gross Domestic Product (GDP), K is the stock of capital, L is the amount of unskilled labour and A is *exogenously* determined level of technology. Note that change in this exogenous variable, technology, will cause a shift in the production function.

There are two ways in which technology parameter A is incorporated in the production function. One popular way of incorporating the technology parameter in the production function is to assume that technology is *labour augmenting* and accordingly the production function is written as

$$Y = F(K, AL) \qquad \ldots (2)$$

Note that labour-augmenting technological change implies that it increases productivity of labour.

The second important way of incorporating the technology factor in the production function is to assume that technological progress *augments all factors* (both capital and labour in our production function) and not just augmenting labour. It is in this way that we have written the production function equation (1) above. To repeat, in this neoclassical approach production function is written as

$$Y = AF(K, L)$$

Considering in this way *A represents total factor productivity* (that is, productivity of both factor inputs). When we empirically estimate production function specified in this way, then contribution of A to the growth in total output is called *Solow residual* which means that total factor productivity really measures the increase in output *which is not accounted for by changes in factors, capital and labour.*

Unlike the fixed proportion production function of Harrod-Domar model of economic growth, neoclassical growth model uses *variable proportion production function,* that is, it considers unlimited *possibilities of substitution between capital and labour* in the production process. That is why it is called neoclassical growth model as the earlier neoclassicals considered such a variable

1. Robert Solow, "A Contribution to the Theory of Economic Growth", *Quarterly Journal of Economics,* February 1956.

2. J. E. Meade, *Theory of Economic Growth.*

proportion production function. The second important departure made by neoclassical growth theory from Harrod-Domar growth model is that it assumes that *planned investment and saving are always equal* because of immediate adjustments in prices (including interest). With these assumptions, neoclassical growth theory focuses its attention on supply-side factors such as capital and technology for determining rate of economic growth of a country. Therefore, unlike Harrod-Domar growth model, it does not consider aggregate demand for goods limiting economic growth. Therefore, it is called '*classical*' along with '*neo*'. The growth of output in this model is achieved at least in the short run through higher rate of saving and therefore higher rate of capital formation, However, diminishing returns to capital limit economic growth in this model, though the neoclassical growth model assumes *constant returns to scale* which exhibits *diminishing returns to capital and labour separately.*

We explain below how neoclassical growth model explains economic growth through capital accumulation (*i.e.*, saving and investment) and how this growth process ends in *steady state equilibrium. By steady state equilibrium for the economy we mean that growth rate of output equals growth rate of labour force and growth rate of capital* $\left(i.e. \frac{\Delta Y}{Y} = \frac{\Delta L}{L} = \frac{\Delta K}{K}\right)$ *so that per capita income and per capita capital are no longer changing*. Note that for income per capita and capital per worker to remain constant in this steady state equilibrium when labour force is growing implies that *income and capital must be growing at the same rate as labour force.* Since growth in labour force (or population) is generally denoted by letter '*n*' in this steady state equilibrium, therefore, $\frac{\Delta Y}{Y} = \frac{\Delta K}{K} = \frac{\Delta N}{N} = n.$ Neoclassical growth theory explains the process of growth from any initial position to this steady state equilibrium.

Neoclassical Growth Theory: Production Function and Saving

As stated above, neoclassical growth theory uses following production function:

$$Y = AF(K, L) \qquad \ldots (1)$$

However, the neoclassical theory explains the growth process using the above production function in its intensive form, that is, in per capita terms. To obtain the above production function in per capita terms we divide both sides of the given production function by *L*, the number of labour force. Thus

$$\frac{Y}{L} = AF\left(\frac{K}{L}, \frac{L}{L}\right)$$

$$= AF\left(\frac{K}{L}; 1\right) = AF\left(\frac{K}{L}\right) \qquad \ldots (2)$$

To begin with we assume that there is no technological progress. With this assumption then equation (2) is reduced to

$$\frac{Y}{L} = F\left(\frac{K}{L}\right) \qquad \ldots (3)$$

The equation (3) states that output per head $\left(\frac{Y}{L}\right)$ is a function of capital per head $\frac{K}{L}$. Writing y for $\frac{Y}{L}$ and k for $\frac{K}{L}$, equation (3) can be written as

$$y = f(k) \qquad \ldots (4)$$

Now, in Fig. 46.1 we represent the production function (4) in per capita terms. It will be noticed from Fig. 46.1 that as capital per capita (k) increases output per head increases, that is, marginal product of labour is positive. But, as will be seen from Fig. 46.1, the slope of the production function curve decreases as capital per head increases. This implies that marginal product of capital diminishes. That is, the increase in capital per head causes output per head to increase but at a diminishing rate.

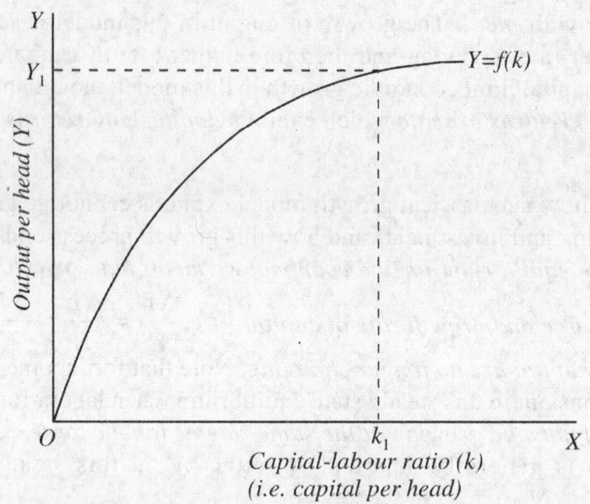

Fig. 46.1. *Production Function Relating Output per head to Capital per head*

It will be seen from the Fig. 46.1 that at capital-labour ratio (*i. e.* capital per worker) equal to k_1, output per head is y_1. Similarly we can read from the production function curve: $y = f(k)$ the output per head corresponding to any other capital per head.

Neoclassical Growth Theory: Fundamental Growth Equation

According to neoclassical theory, rate of saving plays an important role in the growth process of an economy. Like the Harrod-Domar model, neoclassical theory considers saving as a *constant fraction of income*. Thus,

$$S = sY \qquad \ldots (5)$$

where S = saving

Y = income

s = propensity to save

Since s is a constant fraction of income, average propensity to save is equal to marginal propensity to save. Further, since national income equals national product, we can also write equation (5) as

$$sY = sF(K, L)$$

As in neoclassical theory planned investment is always equal to planned saving, *net addition to the stock of capital* is ΔK, which is the same thing as investment (I), can be obtained by deducting *depreciation* of capital stock during a period from the planned saving. Thus,

$$\Delta K = I = sY - D \qquad \ldots (6)$$

where ΔK = net addition to the stock of capital, I stands for investment and D for depreciation. *Depreciation occurs at a certain percentage of the existing capital stock*. The total depreciation (D) can be written as

$$D = dK$$

Substituting dK for D in equation (6) we have

$$\Delta K = sY - dK$$

or
$$sY = \Delta K + dK \qquad \ldots (7)$$

Now dividing and multiplying the first term of the right side of equation (7) by K we have

$$sY = K \cdot \frac{\Delta K}{K} + dK \qquad \ldots (8)$$

We have seen above that *for the steady state equilibrium,* growth of capital $\left(\frac{\Delta K}{K}\right)$ must be equal to growth of labour force $\left(\frac{\Delta L}{L}\right)$, so that capital per worker and therefore income per head remains constant. If we denote growth rate of labour force $\left(\frac{\Delta L}{L}\right)$ by n, then in steady state $\frac{\Delta K}{K} = n$.

Substituting n for $\frac{\Delta K}{K}$ in equation (8) we have

$$sY = K.\, n + dK$$

or
$$sY = (n + d)\, K \qquad \ldots (9)$$

The above equation (9) is a fundamental growth equation of the neoclassical growth model and states the condition for the steady state equilibrium growth rate when capital per worker and therefore income per capita remains constant even though population or labour force is growing. Thus, for steady state growth equilibrium capital must be increasing equal to $(n + d) K$. Therefore $(n + d) K$ represents the *required investment* (or change in capital stock) which ensures steady state when capital and income must be growing at the same rate as labour force (or population)

The Growth Process

From the growth equation (9) it is evident that if planned *saving sY is greater than the required investment (i.e.* $(n + d) K$*) to keep per capita income constant, capital for worker will increase*. This increase in capital per worker will cause increase in productivity of worker. As a result, the economy will grow at higher rate than the steady-state equilibrium growth rate. However, this higher growth rate will not occur endlessly because diminishing returns to capital will bring it down to the steady rate of growth, though at a higher levels of per capita income and capital per worker.

In order to graphically show the growth process the growth equation is conventionally used in *intensive form*, that is, in per capita terms. In order to do so we divide both sides of equation (9) by L and have

$$\frac{sY}{L} = (n + d)\frac{K}{L}$$

where $\frac{Y}{L}$ represents income per capita and $\frac{K}{L}$ represents capital per worker (*i.e.* capital-labour ratio)

Writing y for $\frac{Y}{L}$ and k for $\frac{K}{L}$ we have

$$sy = (n + d)\, k \qquad \ldots (10)$$

The equation (10) represents fundamental neoclassical growth equation in per capita terms.

Growth Process and Steady Growth Rate. Fig. 46.2 shows the growth process that moves the economy over time from an initial position to the steady state equilibrium growth rate. In this Fig. 46.2 along with per capita production function ($y = f(k)$) we have also drawn per capita saving function curve sy. Besides, we have drawn $(n + d)k$ curve which depicts required investment per worker to keep constant the level of capital per capita when population or labour force is growing at a given rate n.

In Fig. 46.2 $y = f(k)$ is per capita production function curve as in Fig. 46.1. Since per capita saving is a constant fraction of per capita output (*i.e.*, income), the curve *sy* depicting per capita saving function is drawn below the per capita output function curve ($y = f(k)$) with the

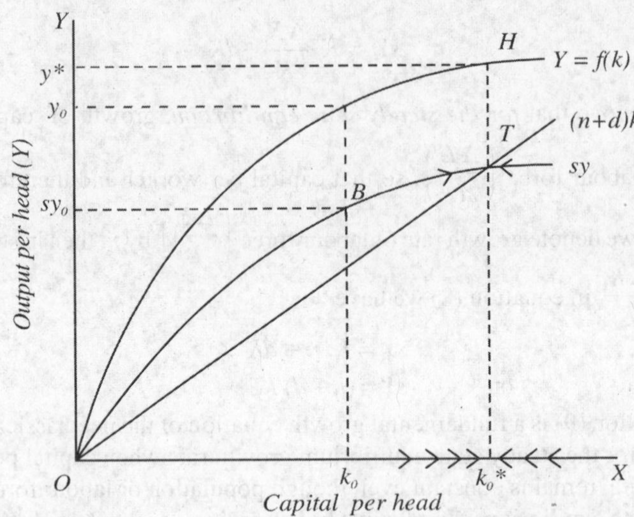

Fig. 46.2. *Neoclassical Model: Growth Process and Steady State Equilibrium*

same shape. Another straight line curve labelled as $(n + d)k$, is drawn which depicts the *required investment* to keep capital per head (*i.e.*, capital-labour ratio) constant at various levels of capital per head.

Now, let us assume the current capital per head is k_o at which per capita income (or output) is y_0 and per capita saving is sy_0. It will be seen from Fig. 46.2 that at capital per head k_0, per capita saving *sy* exceeds investment required to maintain capital per head equal to k_0 ($sy_0 > (n + d)k$). As a result, capital per head (k) will rise (as indicated by horizontal arrows) which will lead to increase in per capita income and the economy moves to the right. This adjustment process will continue so long as $sy > (n + d)\ k$. It will be seen when the economy reaches at capital per head equal to k^* and per capita income equal to y^* corresponding to which saving curve *sy* interseets the $(n + d)\ k$ curve at point *T*.

It will be noticed from Fig. 46.2 that the adjustment process comes to rest at capital per head equal to k^* because saving and investment corresponding to this state is equal to the investment required to maintain capital per head at k^*. Thus point *T* and its associated capital per head equal to k^* and income or output per head equal to y^* represent the steady state equilibrium. It is worth noting that whether the economy is initially at the left or right of k^*, the adjustment process leads to the steady state at point *T*. It may however be noted that in steady-state equilibrium, the economy is growing at the same rate as labour force (that is, equal to n or $\Delta L/L$).

It will be seen from Fig. 46.2 that although growth of economy comes down to the steady growth rate, its levels of per capita capital and per capita income at point *T* are greater as compared to the initial state at point *B*.

An important economic implication of the above growth process visualised in neoclassical growth model is that ***different countries having same saving rate and population growth rate and access to the same technology will ultimately converge to same per capita income although this convergence process*** **may take different time in different countries.**

Impact of Increase in the Saving Rate

As has been explained above that in the steady state, both capital per head (k) and income per head (y) remain constant when economy is growing at the rate of growth of population or labour force (*i.e. n*). In other words, in steady state equilibrium $\Delta k = 0$ and $\Delta Y = 0$. It follows from this that steady state growth rate or long-run growth rate which is equal to population or labour force growth rate n is not affected by changes in the saving rate. Changes in the saving rate affect only the *short-run growth rate* of the economy. This is an important implication of neoclassical growth model.

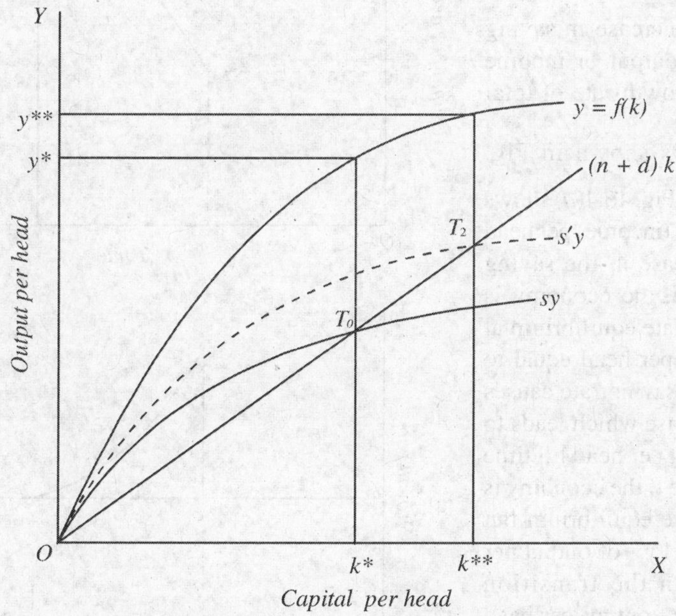

Fig. 46.3. *Impact of Increase in Saving Rate*

Now an important question is why do we get this apparently incredible result from the neoclassical growth theory. Impact of increase in the saving rate is illustrated in Fig. 46.3. It will be seen from this figure that initially with the saving curve sy, the economy is in steady state at point T_0 where the saving curve sy intersects required investment curve $(n + d)\, k$ with k^* as capital per head and y^* as income (output) per capita. Now suppose that saving rate increases, that is, individuals in the society decide to save a higher fraction of their income. As a result, saving curve shifts to the new higher position s′y (dotted). This higher saving curve $s'y$ intersects the $(n + d)k$ curve at point T_1 which therefore represents the new steady state. We thus see that increase in saving rate moves the steady state equilibrium to the right and causes both capital per head and income per head to rise to k^{**} and y^{**} respectively Note that in the new steady state the economy grows at the same rate as the growth rate of labour force (or population) which is denoted by n. It therefore follows that long-run growth rate of the economy remains unaffected by the increase in the saving rate though the steady state position has moved to the right.

Two points are worth noting here. First, though long-run growth rate of the economy remains the same as a result of increase in the saving rate, capital per head (k) and income per capita (y) have risen with the upward shift in the saving curve to $s'y$ and consequently the change in steady state from T_0 to T_1, capital per head has increased from k^* to k^{**} and income per head has risen from y^* to y^{**}.

However, it is important to note that *in the transition period* or *in the short run when the adjustment process is taking place from an initial steady state to a new steady state, a higher growth rate in per capita income is achieved.* Thus, in Fig. 46.3 when with the initial steady state point T_0, saving rate increases and saving curve shifts upward from *sy* to *s′y*, at the initial point T_0, planned saving or investment exceeds $(n + d)\,k$ which causes capital per head to rise in the short run resulting in a higher growth in per capita income than the growth rate in labour force (n) till the new steady state is reached.

The effect of increase in saving rate on growth in output or income per head (y) and growth rate of total output $\left(i.e. \frac{\Delta Y}{Y}\right)$ is shown in Fig. 46.4(*a*) and 46.4(*b*). Fig. 46.4(*a*) shows the growth in output (income) per head as a result of increase in the saving rate. To begin with, the economy is initially in steady state equilibrium at time t_0 with output per head equal to y^*. The increase in saving rate causes capital per head to rise which leads to the growth in output per head till time t_1 is reached. At time t_1 the economy is again in steady state equilibrium but now at a higher level y^{**} of output per head. Note that in the transition persued from t_0 to t_1, output per head increases but at a diminishing rate

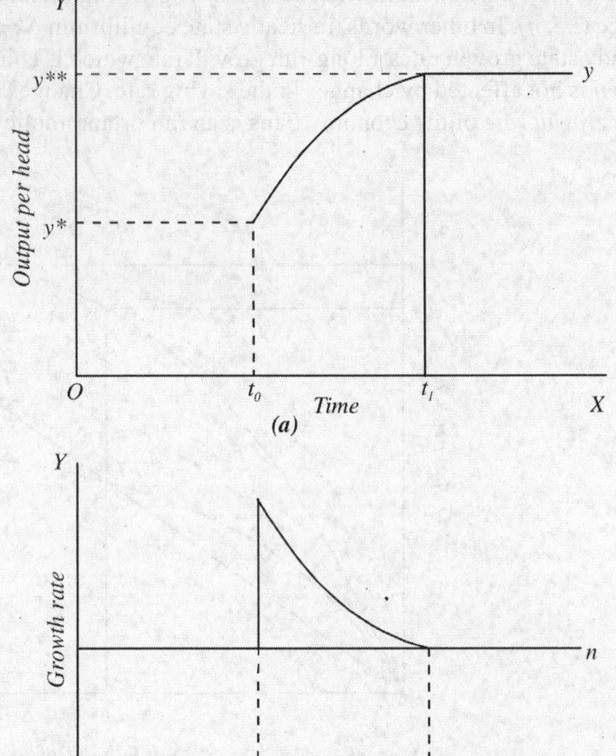

Fig. 46.4. *Adjustment Process*

Fig. 46.4 (*b*) illustrates the adjustment in growth rate of total output $\left(i.e. \frac{\Delta Y}{Y}\right)$. It will be seen from Fig. 46.4 (*b*) that starting from initial steady state at time t_0 the increase in saving rate and capital formation leads to growth rate in total output higher than the steady growth rate n in the period from t_0 to t_1 but in period t_1 it returns to the steady growth rate path n. It is thus evident that the higher saving rate leads to a higher growth rate in the short run only, while *long-run growth rate* in output remains unaffected. *The increase in the saving rate raises the growth rate of output in the short run due to faster growth in capital and therefore in output.* As more capital is accumulated, the growth rate decreases due to the diminishing returns to capital and eventually falls back to the population or labour force growth rate (n).

Effect of Increase in Population Growth Rate

For developing countries like India it is important to discuss the effect of increase in population growth rate on steady levels of capital per head (k) and output per head (y) and also on the steady-state rate of growth of aggregate output. Fig. 46.5 illustrates the effect of increase in population growth rate. An increase in population growth rate causes an upward shift in $(n + d)\,k$ line. Thus in Fig. 46.5, the increase in population growth rate from n to n' causes upward shifts of $(n + d)\,k$ to $(n' + d)\,k$ curve (dotted). It will be seen from Fig. 46.5 that the new $(n' + d)\,k$ curve cuts the given saving curve *sy* at point *T′* at which capital per head has decreased from k^*_1 to k^*_2 and output per capita has fallen from y^*_1 to y^*_2. This can be easily explained. Due to higher growth rate of population

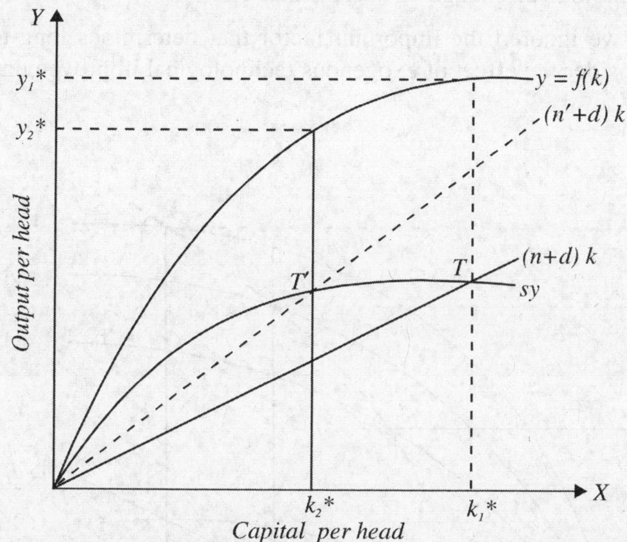

Fig. 46.5 *Effect of Population Growth Rate on Capital Per Head, Output Per Head, and Steady Growth Rate*

a given stock of capital is spread thinly over labour force which results in lower capital per head (*i.e.* capital-labour ratio). Decrease in capital per head causes decline in per capita output. *This is an important result of neoclassical growth theory which shows that population growth in developing countries like India impedes growth in per capita income and therefore multiplies our efforts to raise living standards of the people.*

Fig. 46.5 also shows that higher growth rate of population raises the steady-state growth rate. It will be seen from this figure that increase in population growth rate from *n* to *n´* causes $(n + d)\, k$ curve to shift upward to the new position $(n´ + d)\, k$ (dotted) which intersects the saving curve at new steady-state equilibrium point *T´*. The steady state growth rate has therefore risen to *n´*, that is, equal to the new growth rate of population. It may however be noted that *higher steady rate of growth* is not a desirable thing. As a matter of fact, a *higher steady growth means that to maintain a certain given capital-labour ratio and per capita income the economy has to save and invest more*. This implies that a higher rate of population acts as an obstacle to raise per capita income and therefore living standards of the people. Thus, this result provides a significant lesson for the developing countries like India, that is, if they want to achieve higher living standards for its people they should make efforts to control population growth rate.

Long-run Growth and Technological Change

Let us now analyse the effect of technological change on long-run growth of an economy. It is important to note that *neoclassical growth theory considers technological change as an exogenous variable*. By exogenous technological change we mean it is determined outside the model, that is, it is independent of the values of other factors, capital and labour. That is why neoclassical production function is written as

$$Y = AF\,(K, L)$$

where *A* represents exogenous technological change and appears outside the bracket.

In the foregoing analysis of neoclassical growth theory for the sake of simplification we have assumed that the technological change is absent, that is, $\frac{\Delta A}{A} = 0$. However, by assuming zero technological change we ignored the important factor that determines long-term growth of the economy. We now consider the effect of exogenous technological improvement over time, that is, when $\frac{\Delta A}{A} > 0$ over time.

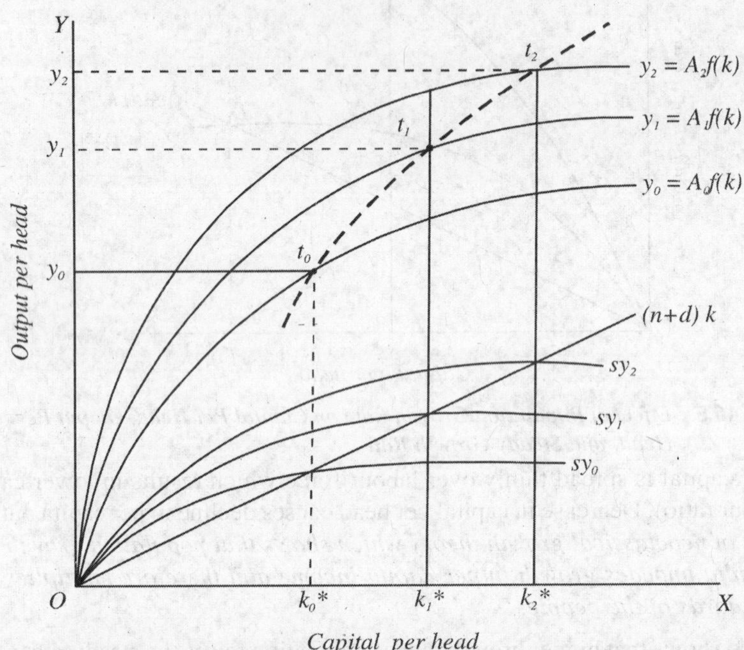

Fig. 46.6. *Growth in Capital Per Head and Output Per Head with Exogenous Technological Change*

The production function (in per capita terms), namely, $y = Af(k)$ considered so far can be taken as a snapshot in a year in which A is treated to be equal to 1. Viewed in this way, if technology improves at the rate of 1 per cent per year a snapshot taken in a year later will be $y = 1.01\, f(k)$, 2 years later, $y = (1.01)^2 f(k)$ and so forth. As a result of this technological change production function will shift upward. In general, if technological improvement $\frac{\Delta A}{A}$ per year is taken to be equal to g per cent per year, then production function shifts upward at g per cent per year as shown in Fig. 46.6 where to begin with production function curve in period t_0 is $y_0 = A_0 f(k)$ corresponding to which saving curve is sy_0. With this, in steady state equilibrium, capital per head is equal to k^*_0 and output (income) per head is y_0. With g per cent rate of technological progress in period t_1, production function shifts to $y_1 = A_1 f(k)$ and correspondingly saving curve shifts upward to sy_1. As a result in period t_1, in new steady state equilibrium capital per head rises to k^*_1 and per capita output to y_1. With a further g per cent rate of technological progress in period t_2, production function curve shifts to a higher level, $y_2 = A_2 f(k)$ and associated saving curve shifts to sy_2. As a result, capital per head rises to k^*_2 and per capita output to y_2 in period t_2. We thus see that progress in technology over time causes growth of per capita output (income). With this aggregate output will also increase over time as a result of technological progress.

The neoclassical growth theory has been successfully used to explain increase in per capita output and standard of living in the long term as a result of technological progress and capital accumulation.

Conclusion: Key Results of Solow's Neoclassical Model

Let us sum up the various key results of Solow's neoclassical growth model.

1. Neoclassical growth theory explains that output is a function of growth in factor inputs, especially capital and labour, and technological progress.
2. Contribution of increase in labour to the growth in output is the most important.
3. Growth rate of output in steady-state equilibrium is equal to the growth rate of population or labour force and *is exogenous of the saving rate, that is, it does not depend upon the rate of saving*.
4. Although saving rate does not determine the steady-state growth rate in output, it does cause *an increase in steady-state level of per capita income* (and therefore also total income) through raising capital per head.
5. Steady state rate of growth of per capita income, that is, *long-run growth rate is determined by progress in technology*.
6. *If there is no technical progress, then output per capita will ultimately converge to steady state level*.
7. A significant conclusion of neoclassical growth theory is that if two countries have the same rate of saving and same rate of population growth rate and have access to the same technology (*i.e.* production function), their levels of per capita income *will eventually converge*, that is, they will ultimately become equal. In this context it is worthwhile to quote Dornbusch, Fischer and Startz, "*The poor countries are poor because they have a less capital but if they save at the same rate as rich countries, and have access to the same technology, they will eventually catch up.*"[3]

SOURCES OF ECONOMIC GROWTH: PRODUCTION FUNCTION APPROACH

An important issue in growth economics is what contributions different factors, namely, capital, labour and technology make to economic growth. In other words, what is relative importance of these different factors as sources of economic growth. Robert Solow and Denison have attempted to study the relative importance of the various sources of economic growth by using the concept of production function. The rate of economic growth in an economy and differences in income levels of different countries and also their growth performance during a period can be explained in terms of the increase in these sources of economic growth.

It wll be recalled that the production function describes that the amount of total output produced depends on the amount of different factors used and the state of technology.

The following production function has been used to measure the various sources of economic growth:

$$Y = AF(K, L) \quad \ldots (1)$$

where Y = total national product
K = the quantity of physical capital used
L = the quantity of labour used
A = the state of technology

The production function equation (1) shows that increase in capital and labour and improvement in technology will lead to growth in national output.

Note that improvement in technology causes output increases with the given factor supplies. In other words, advancement in technology leads to the increase in productivity of factors used.

3. Dornbusch, Fischer, Startz, *op. cit.*, p. 61

Therefore, improvement in technology is generally measured by growth in **total factor productivity (TFP).**

It will also be noticed from the production function equation (1) that technology (*A*) has been taken to be a multiplicative factor. This implies that progress in technology increases the marginal productivity of both capital and labour uniformly. Such technological change is generally referred to as **neutral technological change**. Besides, we measure the sources of economic growth with the above production function by assuming *constant returns to scale.* Constant returns to scale implies that increase in inputs, that is, labour and capital, by a given percentage will lead to the same percentage increase in output. Further, the increase in improvement in technology (*A*) or what is also referred to as increase in total factor productivity causes a shift in the production function.

With the above assumptions it can be proved that the following factors represent the sources of economic growth:

or $$\frac{\Delta Y}{Y} = \left(\theta \times \frac{\Delta K}{K}\right) + \left(1-\theta) \times \frac{\Delta L}{L}\right) + \frac{\Delta A}{A} \qquad \ldots (2)$$

or $$\text{Growth of Output} = \begin{pmatrix} \text{Share} & & \text{Growth} \\ \text{of} & \times & \text{in} \\ \text{Capital} & & \text{Capital} \end{pmatrix} + \begin{pmatrix} \text{Share} & & \text{Growth} \\ \text{of} & \times & \text{in} \\ \text{Labour} & & \text{Labour} \end{pmatrix} + \text{Technical Progress}$$

where θ denotes share of capital in national product, $1 - \theta$ denotes share of labour in national product.

The above equation, which is generally referred to as *growth accounting equation,* shows the various sources of growth which are summarised below :

1. The contribution of increase in capital to the growth in output (G or $\frac{\Delta Y}{Y}$) is given by increase in $\left(\frac{\Delta K}{K}\right)$ capital multiplied by the share (θ) of capital in national product;
2. The increase in labour force contributes to rate of economic growth equal to the labour share ($1 - \theta$) in national product multiplied by the growth in labour force $\left(\frac{\Delta L}{L}\right)$; and
3. The technological improvement $\frac{\Delta A}{A}$ which is measured by the increase in total factor productivity also makes an important contribution to economic growth. As mentioned above, technological progress leads to the increase in total factor productivity (*TFP*) which implies that with the given resources (*i.e.* capital and labour) more output can be produced.

Proof

We can formally prove the growth accounting equation mentioned above. In the production function equation (1) the change in output (ΔY) depends on changes in various inputs or factors — capital and labour ΔK and ΔL and change in technology. This can be written as under :

$$\Delta Y = F(KL)\,\Delta A + MP_k \times \Delta K + MP_L \times \Delta L \qquad \ldots (3)$$

where MP_k and MP_L represent marginal products of labour and capital respectively.

Dividing both sides of equation (3) by Y we have

$$\frac{\Delta Y}{Y} = \frac{F(K,L)\,\Delta A}{Y} + \frac{MP_K \cdot \Delta K}{Y} + \frac{MP_L \cdot \Delta L}{Y}$$

Since $Y = AF(K, L)$ we have

$$\frac{\Delta Y}{Y} = \frac{F(K,L)\Delta A}{AF(K,L)} + \frac{MP_K \cdot \Delta K}{Y} + \frac{MP_L \cdot \Delta L}{Y}$$

or

$$\frac{\Delta Y}{Y} = \frac{\Delta A}{A} + \frac{MP_K}{Y} \cdot \Delta K + \frac{MP_L}{Y} \cdot \Delta L \quad \ldots (4)$$

Now multiplying and dividing the second term of the right-hand side of equation (4) by K and also multiplying and dividing the third term of right-hand side of the equation by L we have

$$\frac{\Delta Y}{Y} = \frac{\Delta A}{A} + \frac{K.MP_K}{Y} \cdot \frac{\Delta K}{K} + \frac{L.MP_L}{Y} \cdot \frac{\Delta L}{L} \quad \ldots (5)$$

Now, if rewards of factors of production are determined by marginal products of factors as actually is the case under perfect competition in neoclassical theory, then $\frac{K.MP_K}{Y}$ represents the share of capital in national product which we denote by θ and $\frac{L.MP_L}{Y}$ represents the share of labour in national product (Y) which we denote by $1 - \theta$, then substituting these in equation (5) we have:

$$\frac{\Delta Y}{Y} = \frac{\Delta A}{A} + \theta \frac{\Delta K}{K} + (1-\theta)\frac{\Delta L}{L}$$

The above is the same as growth accounting equation (2) which indicates the sources of growth of output.

Table 46.1. Sources of Economic Growth

	Growth of Output(%)	Growth of Capital (%)	Growth of Labour (%)	Increase in Total Factor Productivity (%)
United States				
1960–73	4.0	4.0	1.7	1.6
1973–90	2.5	3.4	2.1	0.0
Japan				
1960–73	10.0	10.6	1.3	5.9
1973–90	4.0	5.80	1.0	1.8
Europe				
1960–73	4.9	5.7	(–)10.1	3.2
1973–90	2.3	3.1	0.1	1.3

Source: Kumiharu, Shigehra, "*Causes of Declining Growth in Industrialised Countries in Policies for Long-run Economic Growth*", 1992.

In Table 46.1 we present the contributions made by capital, labour and total factor productivity (*i.e.*, technical improvement) in growth of output in the United States, Japan and the major countries of Europe in the two periods 1960-73 and 1973-90. It will be seen from the table that growth of capital and improvement in total factor productivity (*i.e.*, technological progress) have been the important sources of economic growth, especially in case of economic growth in Japan and European countries. Table 46.1 further reveals that it is decline in total factor productivity (*i.e.*, technological improvement) and in growth of capital that is responsible for slowdown of economic growth in the USA, Japan and European countries during the period 1973-90.

LIMITATIONS OF NEOCLASSICAL GROWTH MODEL

Knowledge or Education is the Missing Factor. In the above growth accounting equation one factor, namely, knowledge or education is missing which has been stressed among others by Nobel Laurette Prof. Amartya Sen as an important factor contributing to economic growth. It may be noted that increase in knowledge or education increases the productivity of workers by improving their productive skills and abilities. Besides, increased knowledge raises the productivity of capital and raises the return to investment in capital goods. Since investment in promotion of knowledge or education makes workers and machine more productive, the workforce equipped with knowledge and education is often called *human capital* which is regarded by modern economists as an important source of economic growth. Thus human capital or knowledge and education is the important missing factor in the growth equation of neoclassical economists, Solow and Denison. On including human capital as a separate factor which contributes to growth of output, the production function can be written as under:

$$Y = A\,F\,(K, L, H)$$

where H represents human capital which was omitted by Robert Solow in his growth accounting equation.

Ignores Economies of Scale as a factor Contributing to Economic Growth. Robert Solow in his study of sources of growth in real income did not consider economies of scale as a factor contributing to growth. Solow assumed constant returns to scale which implies that if each factor in the production function increases by one per cent, output also increases by one per cent. However, some economists such as Denison and those associated with World Bank emphasise *economies of scale* or what is also called increasing returns to scale as a separate factor determining the rate of economic growth. In case of the United States Denison estimated that of 2.92 per cent annual growth in national income recorded during the period 1929–1982, 0.26 per cent was due to economies of scale. However, whether there are increasing returns to scale or constant returns to scale is an empirical matter for investigation.

CHAPTER 47
NEW THEORY OF ECONOMIC GROWTH (ENDOGENOUS GROWTH MODEL)

Introduction

Neoclassical growth theory dominated the economic thinking for three decades (1955–1985), though it could not explain important facts about economic growth in various countries. By the late 1980s, the dissatisfaction with neoclassical growth theory to explain the real growth phenomenon increased so much that a need was greatly felt to propound a new theory of growth. *There are three basic difficulties with the neoclassical growth model.* First, neoclassical growth theory implies that increase in saving rate has only a temporary or short-run effect on rate of growth and it *does not affect the long-run rate of economic growth.* Secondly, neoclassical growth theory implies **convergence of growth rates.** That is, growth rates of countries whether having high or low saving rates will converge over time, though this convergence may be slow. This convergence implies that different countries with high and low saving rates but having same rate of population growth rate would over time achieve the same rate of growth in their national income in the long run $\left(i.e., \frac{\Delta Y}{Y} = \frac{\Delta N}{N} \right)$ while *growth rates in per capita income* $\left(\frac{\Delta y}{y} \right)$ *of these countries will tend to be equal to zero in long-run steady-state equilibrium.*

More importantly, neoclassical growth theory explained that long-run rate of economic growth depended on ***technological change which was considered as exogenous.*** By treating technological change as exogenous, neoclassical growth theory could not focus on the fundamental forces which determine long-run growth of nations. Furthermore, if there is free flow of technological knowledge between nations we should expect the rate of technological progress to converge and therefore cause long-run growth rates of different countries to converge. However, it was found in the real world that such convergence of long-run growth rates was not taking place or was occurring in case of only limited number of countries. Thus, ***neoclassical theory could not explain differences in growth rates of different countries nor it could explain greater proportion of economic growth observed in different countries.***

A basic concern of development economics is to explain why per capita incomes and growth rates differ among countries. Thus, the main motive of new growth theory is to explain differences in growth rates among countries and what are contributions of different factors to rates of growth observed in them.

ENDOGENOUS GROWTH MODELS

The new growth theory extends the neoclassical theory by making the *rate of technological progress as endogenous factor*. Therefore, new growth theory is also called *Endogenous Growth Theory* or *Endogenous Growth Model*. The new growth theory goes more deeply into ultimate sources of growth. Three different approaches have been adopted to make technological change as endogenous factor in determining economic growth.

Therefore, the main feature of endogenous growth theory is to explain the growth of productivity and thereby the growth rate of output *endogenously*, that is, within the model itself and does not depend on *exogenous technological change* as is done by neoclassical Solow model. As we will see that the *important implication of endogenous growth theory is that, contrary to Solow's neoclassical theory, the long-run rate of economic growth depends on the rate of saving and investment* (It may be recalled that in Solow's neoclassical theory the long-run economic growth is determined by *exogenous* technological change which causes increase in productivity by shifting upward the production function). There are a number of endogenous growth models. We start explaining the simplest endogenous growth model generally called *AK Model*.

AK Model

In AK Model it is assumed that the population or number of workers remains the same which implies that the growth rate of output per worker will be the growth rate of output of the economy. AK Model is based on the following aggregate production function:

$$Y = AK \qquad \ldots (1)$$

where Y is aggregate output, K is total capital stock and A is a positive constant. It is noteworthy that in the capital stock K, both physical and human capital are included.

The above AK production function shows that an additional unit increase in capital K causes increase in output by A units irrespective of how much capital is used in production. Thus A measures marginal product of capital that does not depend on the rise of capital stock in the production function. Therefore, in the production function $y = AK$ implies that *marginal product of capital does not diminish* as more units of capital are added to increase production and instead remains constant. This is contrary to neoclassical growth theory of Solow in which marginal productivity of capital diminishes.

Now, an important question arises: What accounts for constant marginal productivity of capital for the economy as a whole when there is increase in capital stock ? A number of explanations have been given for this. The important explanation lays stress on the role of *human capital* which is included in the K term of above aggregate production function. By human capital economists mean the knowledge, skills and training of the people, especially the workers who are engaged in the production process of goods. Imparting knowledge, skills to the people is described as *investment in people* as it raises their productivity. Increases in physical capital alone, keeping human capital constant, will cause diminishing marginal productivity of capital. According to endogenous growth theory, as accumulation of physical capital takes place in the economy, its human capital also increases in the same proportion through the process what is called *learning by investment* so that each additional unit of physical capital works with the same quantity of human capital which ensures that marginal productivity of capital does not diminish. Thus, accumulation of capital brings about continuous growth of output.

In this AK model of endogenous growth theory, a second explanation of marginal productivity of capital remaining the same as capital accumulation takes place in the economy is based on the fact that in a *growing economy the firms have incentives to undertake investment in research and development (R&D) activities*. These R&D activities lead to the discovery of new production techniques and new products. With the use of new production techniques and increase in technical know-how as capital accumulation takes place in the economy marginal productivity of capital does not diminish that makes continuous economic growth possible. Having explained how production function, $Y = AK$ ensures constant marginal product of capital when we incorporate into the analysis the role of human capital and research and development activities, we are in a position to bring out the implications of AK model of endogenous growth theory. Let us assume that saving, S, is a constant proportion, s of aggregate output Y, which is equal to AK in the model. Thus $S = sY = sAK$. Since a closed economy investment must equal saving, we have

$$I = sAK \qquad \ldots (2)$$

Now, investment is *net addition to the stock of capital* (ΔK) *plus depreciation* (dK), *that is* $I = \Delta K + dK$ (Note that K is the stock of capital and d is the rate of depreciation). Substituting $I = \Delta K + dK$ in equation (2) above, we have

$$\Delta K + dK = sAK \qquad \ldots (3)$$

Dividing both sides of equation (3) by K, we have

$$\frac{\Delta K}{K} + d = sA$$

Now, taking d on the right side of the above we have

$$\frac{\Delta K}{K} = sA - d \qquad \ldots (4)$$

$\frac{\Delta K}{K}$ is the rate of growth of capital stock. Since output is assumed to be proportional to the capital stock, rate of growth of output $\left(\frac{\Delta Y}{Y}\right)$ will equal the rate of growth of capital stock $\left(\frac{\Delta K}{K}\right)$. Thus, the above equation (4) can be rewritten as

$$\frac{\Delta Y}{Y} = sA - d \qquad \ldots (5)$$

The above equation (5) of the *AK* model of endogenous growth theory shows that growth of output $\left(\frac{\Delta Y}{Y}\right)$ depends on the saving rate (s). The higher the saving rate, the higher the rate of long-run rate of growth of output. As it has been assumed that the number of workers remains the same over time, the rate of growth of output per worker will be equal to the rate of growth of output as given by the equation (5) above, that is, equal to $sA–d$. The rate of growth of output per worker also depends on the saving rate. The finding that long-run rate of growth of output depends on the saving rate is in sharp contrast to Solow's neoclassical growth model in which rate of saving affects only the short-run growth rate and does not determine the long-run rate of growth of output. Commenting on the conclusion of endogenous growth model, A.B. Abel, B.S. Bernanke and D. Croushore write, "*Saving affects long-run growth in the endogenous growth framework because, in that framework, higher rates of saving and capital formation stimulates greater investment in human capital and R&D. The resulting increases in productivity help to spur long-run growth. In summary, in comparison to the Solow model, the endogenous growth model places greater emphasis on saving, human capital formation and R&D as sources of long-run growth.*"[1]

An important implication of endogenous growth model is that the Government can influence the long-run growth rate by taking suitable policies to raise saving rate of the economy.

Romer's Model of Endogenous Growth

In Romer's model, to incorporate endogenous technological change, the production function is modified as under:

$$Y_t = F(K_t, N_t, A_t) \qquad \ldots (1)$$

It will be seen from equation (1) that the level of aggregate output depends on the quantities of capital (K_t), and labour (N_t) used in the production as well as on technology (A_t) *which is treated as endogenous factor and therefore appears inside the production function as an endogenous input*. However, the relationship between output and technology is not the same as between output and other inputs, capital and labour. This is because in Romer's model, the output of an individual firm not only depends on its *own* level of capital and labour but also on the *technology used by other*

1. Andrew B. Abel, Ben S. Bernanke, and Dean Croushore, *Macroeconomics*, Pearson Education, 2011, p. 229.

firms whose benefits also accrue to it. Looking at the relationship between output and technology in this way we can write production function of an individual firm denoted by the subscript *i* as under:

$$Y_{it} = F(K_{it}, N_{it}, A_t) \qquad \ldots (2)$$

Note that in the production function in (2) above, the technology input subscript *t* appears along with subscript *i* because this technology may not be exclusive input of a firm, but may be copied or imitated from others. The technology is not assumed to progress exogenously *but to grow endogenously when more investment by firms takes place or may be the result of the efforts of labour*. In some models of endogenous technological change, more capital accumulation or investment by a firm promotes new inventions and improvements in machines (*i.e.*, capital goods) constituting the stock of capital through a process called *learning by doing*. In some other models of endogenous technological change, increase in labour inputs by firms also leads to the stock of technological knowledge as it is pointed out that increase in the use of labour input leads to the scope of learning and acquisition of new skills and discovering new better ways of doing things.

The *contribution of Paul Romer lies in explaining that the benefits of the new technology discovered and used by a firm will spill over to other firms in the economy.* Paul Romer,[2] one of the pioneers of endogenous growth theory, has put forward the view that *investment is a source of technological progress*. He distinguishes between *private returns* to capital and *social returns* to capital. According to Romer, an individual firm does not capture all the benefits of increase in its stock of capital, as it also creates benefits which are *external to the firm*. Capital accumulation by a firm, that is, investment by a firm causes not only increase in new *machines* but also *new ways of doing things*.[3] These new better methods of doing things result either from deliberate investment in research and development and sometimes as accidental by-products of investment activity by a firm. These new methods and ideas of doing things can be easily copied by others who also get benefits from these new methods.

We can consider investment as a source of technical progress for various reasons. If investment simply consists of duplication of existing capital, then that investment does not lead to innovation or technological change. However, investment may lead to innovation as firms attempt to solve their problem, especially to raise their productivity. They will try to discover new methods or designs of machines which can raise productivity and for this purpose they will make investment. Such investment will lead to innovation or technical progress. For example, investment by car manufacturers in India has led to new designs and models of cars which are fuel-efficient and pollution-free. Thus their investment in research and development has led to new innovations or technical progress.

Furthermore, if investment by a firm is successful, the other firms will also try to adopt them with some modifications which suit their needs. Thus expansion in investment may represent a sequence of innovations with each innovation improving upon the previous innovation. Such a type of technological progress has been called **learning by watching**. Beneficial external effects resulting from learning by watching are key determinants of economic growth. Therefore, policies to increase investment should be adopted. It is due to the externalities that *social returns* to investment exceed private returns. The rapid and quicker increase in overall growth can be achieved through Government adopting proper policies to promote investment in research and development.

This endogenous growth theory considers that whereas production function of a firm exhibits constant returns to scale *(i.e.,* constant returns to all factors) but there occur *external increasing returns to scale.* These external increasing returns are due to the endogenous technological improvements which result from (1) rate of investment, (2) size of the capital stock, and (3) *the stock of human capital.*

2. Paul Romer, *Increasing Returns and Long-Run Growth.*
3. For investment as a source of technical progress, see M. Scott, "A New Theory of Endogenous Economic Growth", *Oxford Review of Economic Policy,* Winter 1992 and M. King and M Robson, Investment and Technical Progress, *Oxford Review of Economic Policy*, 1992.

In the neoclassical model the increase in capital and labour will bring about increase in output proportionately, that is, constant return to labour and capital will occur. However, *in new endogenous growth theory increase in capital and labour will foster economy-wide technological progress*, that is, term A_t in the production function rises. This technological progress brings about extra expansion in output. Thus, *in the new growth theory, increase in capital and labour with endogenous technological advancement brings about increasing returns to scale*. "With increasing returns the changes in saving rate and therefore in the rate of capital formation can permanently affect the long-run equilibrium growth rate. This result contrasts the neoclassical growth theory in which the effect of changes in saving rate on the growth rate is temporary".[4]

Policy Implications. Endogenous growth model has important policy implications for the role of Government for accelerating economic growth. As just pointed out above, with endogenous technological change and therefore increasing returns to scale, the government policies that raise the saving rate and thereby the rate of capital formation can raise the long-run equilibrium growth rate of the economy. This shows Government can play an important role in accelerating economic growth.

Investment in Human Capital and Learning by Doing

In the above approach to endogenous technology the progress in technology is assumed to depend on growth of capital, that is, investment. New investment fosters innovations and improvements in machines and tools that also create external benefits and lead to increasing returns for the economy as a whole. In another important model of endogenous growth theory, *investment in human capital* (that is, labour input) is also assumed to be the source of technological progress. Arrow introduced the concept of *learning by doing*. According to this, the greater the level of labour input, the greater is the scope for learning and acquiring of new skills. An endogenous growth model which lays stress on the accumulation of human capital has been developed by Lucas.[5] According to Lucas' model, labour can be devoted either to production goods or to the accumulation of human capital, that is, acquiring of new skills and knowledge. Acquiring of new skills and knowledge will not only make a worker more productive but also increase the productivity of capital and other workers in the economy. Each new knowledge or skill makes the next idea possible and so the knowledge can grow indefinitely. Let e be the proportion of labour input directly devoted to production and H be the human capital, then the Cobb-Douglas type production function can be stated as under:

$$Y = AK^{a}(H.eL)^{1-a}$$

where the technology coefficient $A = H^b$ represents the external effect of human capital (H) on productivity of capital (K) and labour (eL). Since human capital (H) is a function of labour input, output per worker can be obtained as a function of capital and labour according to this approach, as with the earlier one of investment in physical capital. Though each firm faces constant returns to scale, there are increasing returns to the economy as a whole.

Policy Implications of New Growth Theory

New growth theory with its emphasis on endogenous technological change has important policy implications which in some important respects differ from the neoclassical growth theory. First. whereas neoclassical theory predicts that growth rates of different countries with same rates of saving and population growth and with access to the same technology will converge, *in the endogenous growth theory there is no force leading to the convergence of growth rates of different countries with closed economies*.[6] This is because of *the possibility of sustained long-term growth resulting from increasing returns to scale* which was not accounted for in the neoclassical theory. Growth rates of

4. Richard T. Froyan, *Macroeconomics, Theory and Policy,* Pearson Education, 8th edition, 2005, p. 108.
5. R. Lucas, "On the Mechanics of Economic Development"; *Journal of Monetary Economics*, 22 (1988). pp. 3–42.
6. An important empirical study by Robert Barro, "*Determinants of Economic Growth : A Cross-Country Empirical Study*", finds no evidence of convergence of growth rates.

different countries in new endogenous growth theory will remain constant or differ depending on their saving rates and technological levels. This is because of large differences in investment in human capital, research and development and increasing returns or external benefits accruing from them.

Furthermore, new growth theory emphasises that higher saving rate will ensure a higher long-run economic growth but there is no any mechanism working which ensures that per capita income in poor countries will catch up with those of the rich countries with similar saving rates.[7] The external increasing returns to the accumulation of capital, both physical and human, implies that rate of return on capital in capital-abundant rich countries will not fall in the developed rich countries relative to the poor less-developed countries. As a matter of fact, it is possible that due to endogenous technological improvement, rate of return to capital in capital-abundant rich countries exceeds that of capital-scarce countries and as a result capital may not flow from rich to poor countries. It is noteworthy that "potentially high rates of return on investment offered by developing countries with low capital ratios are greatly eroded by lower levels of complementary investments in human capital (education), infrastructure or research and development (R & D).[8]

Endogenous Growth Theory and Developing Countries

In sharp contrast to the policy implications derived by some from neoclassical theory, important implication of endogenous growth theory is that because individual firms do not capture full benefits from the positive externalities created by their own investments, the free market will lead to the accumulation of less than optimal level of complementary capital[9]. This will cause lower rate of economic growth in them. Lastly, because of the externalities created by investment, the new endogenous theory of growth envisages greater role of Government in improving the efficiency of resource allocation and promoting investment to raise rate of economic growth in the developing countries. The Government can directly adopt policies to raise the rate of saving and also make more investment in human capital (*i.e.,* education and healthcare) which promote private investment and generate increasing returns to scale. Though in many respects endogenous growth theory is a mere extension of neoclassical theory of growth, it however makes a departure from the neoclassical policy of free market and passive role of Government, To quote Todaro, "*In contrast to neoclassical counter revolution theories, models of endogenous growth suggest an active role for public policy in promoting economic development through direct and indirect investment in human capital formation and the encouragement of foreign private investment in knowledge-intensive industries such as computer software and telecommunications.*"[10]

Lastly, it may be noted that endogenous growth theory, like the neoclassical theory, focuses on supply-side factors determining economic growth and *totally neglects role of adequate growth in effective demand* for sustained long-run growth of the economy. As has been recently seen that in both the developed and the developing countries lack of effective demand due to sluggish investment, decline in exports and, in countries like India, due to monsoon-dependent decline in agricultural growth rate may come about. As a result of lack of effective demand recession occurs in the economy which slows down the short-run economic growth which may ultimately affect long-term growth rate. For instance, for the first time after the Second World War, the Japanese economy which achieved very high growth rates in the past has experienced recessionary conditions and a very slow growth rate for more than a decade now. This growth experience of the Japanese economy cannot be explained by new endogenous growth theory.

---

7. Michael Todaro, *Economic Development,* Addison-Wesley, 6th Edition, 1997, pp 92–93.
8. *Ibid*, p. 93.
9. Todaro, *op. cit.* p. 93.
10. Todaro, *Economic Development*, Pearson Education (Singapore), 8th Edition, p. 148.

Further, it has been pointed out that endogenous growth theory assumes that there is either a single sector of production or that all sectors are symmetrical. This ignores the importance of growth-generating reallocation of labour and capital among the sectors for bringing about structural change in developing countries.[11] Besides, economic growth in developing countries often faces obstructions arising from lack of adequate infrastructure, poor institutional framework, imperfect financial and money markets. To remove these obstructions is important for accelerating economic growth but are overlooked by the new endogenous growth theory and therefore it has limited applications for the developing countries.

Furthermore, there may be *lack of incentives* due to uncertainty of profit prospects to make investment which may be responsible for slowdown in economic growth in developing countries. This important factor has not been taken into account by the new endogenous growth theory. For example, India is presently facing slowdown in economic growth because of lack of incentives on the part of private sector to make more investment as they find bleak profit outlook.

11. See Syed Nawab Haider Naqvi, "The Significance of Development Economics", *World Development*, 24 (1996). p. 977

CHAPTER 48

MAHALANOBIS GROWTH MODEL AND HEAVY-INDUSTRY STRATEGY OF DEVELOPMENT

Having discussed Harrod-Domar model of growth and neoclassical and new theories of economic growth in the previous chapters, we proceed to discuss two alternative strategies of development which have been the subjects of a good deal of controversy in India. One is the strategy of development which accords top priority to basic and heavy industries. This basic heavy industry model of development was put forward by a renowned statistician late Professor P.C. Mahalanobis, who was friend and adviser to late Prime Minister Jawaharlal Nehru and who was one time member of the Planning commission. At the time of the formulation of the Second Five Year Plan, Prof. Mahalanobis prepared a growth model with which he showed that to achieve a rapid long-term rate of growth it would be essential to devote a major part of the investment outlay to building of basic heavy industries. Mahalanobis strategy of development emphasising basic heavy industries which was adopted first of all in the Second Plan also continued to hold the stage in Indian planning right up to the Fifth Plan which was terminated by the Janata Government in March 1978, a year before its full term of five years.

In critique of the Mahalanobis heavy industry development strategy, Professors Vakil and Brahmanand of Bombay University put forward a *wage-goods model of development* and suggested a development strategy which accorded a top priority to agriculture and other wage-goods industries in sharp contrast to the Mahalanobis heavy industry biased strategy of development. In the present chapter we shall critically examine the heavy industry biased strategy of development and in the next chapter we shall discuss wage-goods model of development.

Mahalanobis Model of Growth

It will be useful to explain first Mahalanobis model of growth which provided a rationale for the heavy industry biased development strategy. An important point to note is that Mahalanobis identifies the rate of growth of investment in the economy not with rate of growth of savings as is usually considered by the economists but with rate of growth of output in the capital goods sector within the economy. The growth of capital goods sector in turn depends upon the proportions of total investment allocated to the capital goods sector and output-capital ratio in the capital goods sector. Given the output-capital ratio in capital goods sector (*i.e.* heavy industries), he proves that if the proportion of total investment allocated to the capital goods is relatively greater, the rate of growth of output of capital goods will be greater and hence, given the Mahalanobis assumption, the future rate of growth of investment in the economy will be greater. Now, the greater the rate of investment, the greater will be the long-term rate of growth. We thus see that with the rate of growth of output of capital goods industries. Mahalanobis shows that the proportion of total investment resources allocated to the capital goods industries for each year is the most important factor determining the long-term rate of growth of national income. Let us represent his two-sector model in the mathematical form.

In his basic two-sector model Mahalanobis divides the economy into two sectors—the sector *C* produces consumer goods and sector K produces capital goods.

Let

t_0 = Initial rate of investment.

λ_k and λ_c = Proportions of total investment allocated to capital goods and consumer goods sectors respectively.

Therefore $\lambda_k + \lambda_c = 1$

β_k and β_c = Marginal output-capital ratio in the capital goods and consumer goods sectors respectively. In other words, they represent ratio of increment of income to investment in the sector *K* and sector *C* respectively.

Y_0, C_0, I_0 = The national income, consumption and investment in the base period.

Y_t, C_t, I_t = The national income, consumption and investment respectively in period *t*.

In Mahalanbis model, the net investment in any period can be divided into two components; the one $\lambda_k I_t$, going to capital goods sector *K* and $\lambda_c I_t$ going to consumer goods sector *C*. Therefore it follows that,

$$I_t = \lambda_k I_t + \lambda_c I_t$$

Let ΔI_t stands for increase in investment (*i.e.* addition to the stock of capital goods) and ΔC_t for increase in consumer goods in any period *t* depend on the net investment in the previous period *t*–1. Now, given the output-capital ratios, β_k and β_c of the captial goods and consumer goods sectors respectively, the relationship between investment and the resultant increment in output in capital goods can be worked out as follows:

$$\Delta I_t = \beta_k \lambda_k I_{t-1} \quad \text{Or} \quad I_t - I_{t-1} = \beta_k \lambda_k I_{t-1} \qquad \ldots (i)$$

This implies that the increase in investment in period *t* is equal to the increment in output of capital goods. The increase in output of capital goods ($\beta_K \lambda_K I_{t-1}$) in period *t* is given by investment in period I_{t-1} multiplied by the proportion of it going to capital goods sector (λ_k) and the output-capital ratio (B_k) in the capital goods sector.

It is clear from above that Mahalanobis takes into account only the physical aspect of investment and makes it dependent on the proportion of investment allocated to capital goods sector λ_k and output-capital ratio β_k in the capital goods sector.

Rationale of Mahalanobis Growth Model

It is worth noting that Mahalanobis recognises that output-capital ratio β_c in the consumer goods sector is greater than the output-capital ratio β_k in the captial goods sector. If this is the case, then it apparently implies that growth of output or income will be greater if more investment is made in the consumer goods sector. But in this case the higher rate of growth of income will be only in the short run. Mahalanobis showed that after a critical range of time, the larger the investment allocated to capital goods industries, (λ_k) *i.e.*, the higher will be growth in output or income. Elaborating this point Prof. Raj states, "The *logic here is the same as the more common proposition that a higher rate of investment* (*i.e., a larger proportion of the productive factors used for accumulation) would result in a smaller volume of output being available for consumption in the short run but that over a longer period, it would result in higher rate of growth of consumption; the difference is that the choice is here stated as between investment in capital goods and investment in consumer goods industries*."[1]

The rationale of Mahalanobis growth model and development strategy can be expressed in simple words without mathematical language. According to Mahalanobis, rate of economic growth depends upon the capital formation or real investment. The greater the rate of capital formation, the greater the rate of economic growth. The rate of capital formation in an economy, according to Mahalanobis, depends upon the capacity of the economy to produce capital goods. Thus, according to him, given

1. K.N. Raj, "Growth Models in Indian Planning", *The Indian Economic Review*, Feb. 1961. Reprinted in C.D. Wadhwa, "*Some Problems of India's Economic Policy*, Tata Publishing Co. Ltd., Bombay.

a closed economy, the rate of real capital formation depends not upon the savings of the economy but on the capacity to produce capital goods. Even if the rate of savings was substantially raised in order to accelerate the rate of capital formation, it would be futile, for required capital goods would not be there if there is a lack of capacity to produce capital goods. Of course, this is based on closed *economy* assumption.

Thus, according to him, if large investment is not made in the basic heavy industries producing capital goods, the country will for ever remain dependent on foreign countries for the imports of steel and capital goods like machinery for real capital formation. Since it is not possible for India to earn sufficient foreign exchange by increasing exports, the capital goods cannot be imported in sufficient quantities owing to foreign exchange constraint. The result will be that the rate of real capital formation and the rate of economic growth in the country will remain low. Thus, Mahalanobis was of the opinion that without adequate investment in basic heavy industries, it would not be possible to achieve rapid self-reliant economic growth. Therefore according to him, to achieve rapid economic growth and self-reliance, it would be necessary to give the highest priority to basic capital goods industries in the development strategy of a plan.

Employment Generation in Mahalanobis Model. It is necessary in this connection to mention Prof. Mahalanobis' views on increasing employment opportunities and to achieve a stage of full employment. According to him, productive employment can be increased only by increasing the production of capital goods like steel, electricity, machinery, fertilizers, etc. Whether it is increase in employment in the industrial sector or in the agricultural sector it cannot be achieved without increasing the output of capital goods. To quote him, "*The only way of eliminating unemployment in India is to build up a sufficiently large stock of capital which will enable all unemployed persons being absorbed into productive capacity. Increasing the rate of investment is, therefore, the only fundamental remedy for unemployment in India*."[2] Thus, in Prof. Mahalanobis' opinion, not only to achieve the objective of rapid economic growth but also to achieve the goal of full employment, it is necessary to accord high priority to capital goods industries in the development strategy.

Import-substituting Industrialisation. Mahalanobis' emphasis on basic heavy industries was also due to his objective of meeting the requirements of higher rate of capital accumulation from within the economy and therefore enabling the economy to stop imports of foreign capital equipment and machines. To quote him, "The proper strategy would be to bring about a rapid development of the industries producing investment goods in the beginning by increasing appreciably the proportion of investment in the basic heavy industries. As the capacity to manufacture both heavy and light machinery and other capital goods increases, the capacity to invest by using domestically produced capital goods would also increase steadily and India would become more and more independent of the imports of foreign machinery and capital."[3] In fact, Mahalanobis growth model advocates import-substitution type of industrial development strategy.

It is important to note that Mahalanobis assumed though implicitly that export earnings of India cannot be sufficiently increased. If this assumption is not valid, as has been pointed out by several critics, then he could not justifiably identify rate of investment in the economy with the domestic output of capital goods. If exports of a country can be adequately raised, the various capital goods can be imported in exchange for exports. In that case rate of investment or rate of capital accumulation in the economy can be stepped up without giving high priority to the basic heavy industries provided the exports can be adequately increased. Thus the assumption of stagnant exports is crucial in the Mahalanobis growth model for providing the rationale for a general shift in the investment pattern to the domestic production of capital goods.

Mahalanobis Growth Model and Development Strategy in India's Five Year Plans

2. P.C. Mahalanobis, "The Approach of Operational Research to Planning in India." *Sankya*, Vol. 16. December 1953.
3. *Ibid*, p.22.

As pointed out above, Mahalanobis heavy industry first strategy of development was put into actual practice in India's Five Year Plans beginning from the Second Plan. India started its planned development of its economy in 1951 when First Five-Year Plan was started. However, the Five Year Plan did not propose any explicit strategy of development; it took over several projects which had been worked out earlier and some of them were already in the process of being carried out. It laid emphasis on stepping up the rate of saving and therefore investment and growth by maintaining the marginal rate of saving at a substantially higher level than the average rate of saving. Although it did not present any explicit formulation of development strategy regarding the pattern of investment its emphasis was on agriculture, irrigation, power and transport aimed at creating the base for more rapid industrialisation of the economy in the future.

Second Five Year Plan, based as it was on Mahalanobis growth model, proposed an explicit strategy of development which gave top priority to basic heavy industries. Not only the objectives of rapid rate of economic growth and employment generation but also the aim of self-reliant and self-generating economy were sought to be achieved by *"the building up of economic and social overheads, exploration and development of minerals and the promotion of basic industries like steel, machine building, coal and heavy chemicals.*"[4] Identifying underdevelopment with dependence on agriculture and thinking industrial growth especially the development of heavy industries as the core of development formed the basis of the approach and strategy of the Second Five Year Plan. To quote from Second Plan again. "Low or static standards of living, underemployment and unemployment, and to a certain extent even the gap between the average incomes and the highest incomes are all manifestations of the basic underdevelopment which characterises an economy depending mainly on agriculture. Rapid industrialisation and diversification of the economy is thus the core of development. But *if industrialisation is to be rapid enough, the country must aim at developing basic industries and industries which make machines to make the machines needed for further development.*"[5]

It is clear from above that in the Second Plan there was clear shift of priorities from agriculture to industries and within industries to basic heavy industries. As mentioned above, the logic of Mahalanobis in emphasizing heavy industries was that the growth of basic heavy industries will enable the economy to accelerate the rate of capital formation and therefore economic growth. In fact, he identified the rate of growth of investment in the economy with the rate of growth of output in the capital goods (sector) industries within the economy.

Mahalanobis' Four-Sector Model. Mahalanobis realised that the basic heavy industries being capital-intensive will not ensure rapid expansion of employment opportunities and to bring the employment aspect into sharp focus he put forward a four-sector growth model in which he kept heavy industry sector (*i.e.* K-sector) intact but divided the C-sector (*i.e.* consumption goods sector) into three sub-sectors : C_1, C_2, and C_3 (sector C_1 represented factory enterprises using mechanised techniques and producing consumer goods; Sector C_2 represented the household and small-scale enterprises also producing consumer goods and Sector C_3 represented provision of services). It was sector C_2 representing households and small-scale industries which in Mahalanobis' four-sector model was visualised to ensure the increased supply of consumer goods to meet their rising demand for them and also to ensure, being labour-intensive, expansion of employment opportunities. In keeping with this approach, Second Five Year Plan put restrictions on the growth of capacity in factory enterprises engaged in commodity production. However, since for these household or cottage enterprises, adequate resources were not provided, nor any effort was made to improve their productivity, they could neither fulfil their target of production of consumer goods nor of generating enough employment opportunities.[6]

---

4. Government of India, Planning Commission, *Second Five Year Plan*, 1956, p. 24.
5. *Ibid*, p. 25.
6. See Amlan Datta, *Strategy of Economic Development*, 1973. In the spinning of cotton, main emphasis was on Ambar Charkha. Ambar Charkha experiment failed because productivity and earning capacity per person engaged in it was found to be too low.

In the Third Plan (1961–66) also the strategy of Second Plan was continued as is clear from the following: "In the Third Plan, as in the Second, the development of basic industries such as steel, fuel and power and machine building and chemical industries is fundamental to rapid economic growth. These industries largely determine the pace at which the economy can become self-reliant and self-generating."[7] Though, "to achieve self-sufficiency in foodgrains and increase agricultural production to meet the requirements of industry and exports" was stated to be one on the objectives of the Third Plan, actual allocation of resources between agriculture and other sectors did not exhibit any difference from that of the Second Plan. Therefore, the concern for food and agriculture in the Third Plan appears to be mere verbal and was not built into the strategy of development.

Critical Evaluation of Mahalanobis Heavy-industry Strategy of Development

Mahalanobis 'heavy industry first' strategy of development has come in for severe criticism right from the time of formulation of the Second Plan. First, **a serious mistake in Mahalanobis model is that he identified investment with saving**. That indicates lack of knowledge of economics. Economists have been emphasizing especially in the context of developing countries that increase in the rate of investment is governed by the increase in the rate of savings. If planned investment is not matched by savings, then inflationary gap will emerge which will cause prices to rise. It is bad economics to identify investment with savings. Savings in an economy are determined by behavioural characteristics of the decision-making units such as households, the corporate sector and the Government. Saving by the households depends upon the propensity to consume which in turn depends upon various subjective and objective factors. Savings by the corporate sector depend upon the policies regarding depreciation, distribution of dividends and undistributed profits. Savings of the Government are governed by its policies regarding taxation and consumption expenditure, efficiency and profitability of public enterprises.

Thus, when the planning authority allocates relatively more resources to heavy capital goods industries, as is envisaged in Mahalanobis growth model, it will lead to more physical investment or growth of capital stock. But there is no guarantee that savings, governed as they are by various behavioural characteristics of decision-making units, will rise to the level of planned investment. Therefore, by envisaging higher rate of physical investment without considering how savings of the community could be raised to the planned investment, Mahalanobis model contained built-in inflationary potential. Therefore, being unmatched by savings to fulfil the investment targets of the plan, recourse in actual practice had to be made to deficit financing. No wonder that due to the existence of inflationary gap in India's Five Year Plan prices began to rise from the very beginning of the Second Plan so much so that during the Fourth Plan period (1969–74) rate of inflation assumed serious proportion reaching as high as about 30% during 1973–74, the last year of the Fourth Plan.

A crucial weakness of Mahalanobis heavy industry strategy of development was pointed out by Professors Vakil and Brahmananda. They criticised heavy industry strategy in their now well known joint work, "*Planning for an Expanding Economy*." According to them, growth of national income and employment is determined by the supply of wage goods. They pointed out that when disguised unemployed are employed in productive work on the wage basis, the demand for wage goods will increase and unless the adequate amount of wage goods are not available, the disguised unemployed and openly unemployed cannot be put to work on the wage basis. According to them, there exists a *wage-goods gap* in underdeveloped countries like India and the basic cause of unemployment and disguised unemployment in them is the existence of this wage-goods gap. Wage goods act as a constraint on the growth of industrialisation and non-agricultural growth. Thus, in the determination of growth of income and employment, while Mahalanobis emphasised '*fixed capital*', Brahmananda and Vakil laid stress on the wage goods which are also called '*circulating capital*'. As mentioned earlier, Vakil and Brahmananda put forward a growth model in which wage goods played a key role.

7. Government of India, Planning Commission, *Third Five Year Plan.* 1961, p. 50.

Accordingly, they suggested a developed strategy in which top priority was given to the agriculture which produces most important form of wage goods *i.e.*, foodgrains. We shall discuss in detail the views of Professors Vakil and Brahmananda regarding the growth of income and employment in the next chapter. We shall also critically examine their wage-goods model of development in the next chapter.

There is no doubt that **Mahalanobis "heavy industry" strategy overlooked the importance of agriculture or wage goods in the growth process of the economy**. This manifested itself in the rapid rise of prices from the very beginning of the Second Plan. While a relatively large investment in the basic industries resulted in the creation of money incomes and consequently in a large increase in demand for wage goods, the supply of wage goods did not adequately increase due to the continued neglect of agriculture in India's planning strategy. This caused imbalance between demand for and supply of wage goods which has been responsible for inflationary situation in the Indian economy. The continued neglect of wage goods in the development strategy of Indian Plans, right up to the Fifth Plan, is quite surprising since, as pointed out above, this weakness of the strategy was pointed out by Professors Vakil and Brahmananda at the time of the formulation of the Second Plan.

The larger allocation of resources to the basic heavy industries deprived agriculture (including irrigation) and rural industries of enough resources required for their growth. It is now well-known as brought out by Professor Lipton, B.S. Minhas that India's Five Year Plans beginning from the Second one, based as they were on Mahalanobis growth model with its emphasis on heavy industries, suffered from '*relative neglect of agriculture*'. The view that there has been relative neglect of agriculture in Indian Plans has been forcefully and effectively expressed and provided by Professor Michael Lipton. He writes : "We have seen that neither allocations of public money, nor incentives to the movement of persons and other resources, have favoured agricultural development; that 70 per cent of the workers get less than 35 per cent of investment finance and a far smaller share of human skills. Several types of pressures on opinion and policy have combined to bias the allocation of cash, effort, personnel and research away from rural needs."[8]

Like Professor Lipton, B.S. Minhas also holds similar view about relative neglect of agriculture in Indian Planning. He thus states, "In practical terms the most unfortunate consequences of our adherence to this philosophy of development has been the appalling neglect of agriculture and rural development."[9] Before Lipton and Minhas, Professors C.N. Vakil and P.R. Brahmananda, as mentioned above, have been consistently expressing the view about the low priority assigned to agriculture since the beginning of Second Plan. Recently, C.H. Hanumantha Rao has also deplored the neglect of agriculture in the strategy of Indian Plans. He writes, "*The failure of the present strategy to step up public investment for agricultural development has accentuated the problem of unemployment in rural areas*."

A serious failure of Mahalanobis development strategy with its emphasis on basic heavy industries is that **it failed to generate adequate amount of employment opportunities.** Pattern of industrialization has been based on capital-intensive technology imported from abroad and has been oriented towards urban large-scale industries. If the emphasis were on rural-oriented industrialization promoting cottage and small-scale industries, using intermediate technologies, the problem of labour-surplus and unemployment would not have been as acute as it is now. Besides this, attention was not paid to absorb labour in the agricultural sector, and the agricultural development strategy that could absorb more labour in productive employment was not adopted. Necessary technological and institutional changes for designing such an agricultural strategy required to create more employment were not in actual practice made. Many land reform measures, apart from abolition of big Zamindars and Jagirdars, remained mostly on paper. Legislations regarding tenancy reforms and ceiling on land

8. M. Lipton, "Strategy for Agriculture: Urban Bias and Rural Planning", printed in P. Streeman and M. Lipton (ed.) *The Crisis of Indian Planning*, Oxford University Press, 1960, p. 130.
9. B.S. Minhas, *Self-reliance and India's Planning Strategy*, Indian Statistical Institute (Discussion Paper No. 138), March 1976, p.14.

holdings were not actually enforced and implemented. Being highly capital-intensive the basic heavy industries have themselves created very little employment opportunities.

It is no doubt true that certain products of heavy industries such as fertilizers, steel, cement, etc. are certainly required for the growth of agriculture, especially in the context of new agricultural technology and should, therefore, be produced, but Mahalanobis strategy envisaged achievement of self-sufficiency all along the line, involving production of highly sophisticated machines to make machines. This *autarchic* nature of Mahalanobis development strategy has been rightly criticised. B.S. Minhas rightly writes, "*The logic of the Heavy Industry First Strategy in India sprouted from a number of philosophical riddles. One important conundrum runs as follows: should you import food, or import fertilisers to produce food at home? Rather than import fertilisers, should you not import fertiliser materials and machinery in order to make fertilisers at home? But why import fertiliser machinery? Should you not import machines to make fertiliser machinery at home? On and on goes the riddle.*[10]

In view of the limited resources available for investment, the inclusion of highly capital-intensive industries producing machines and mother machines, exhaust a great deal of resources, leaving relatively little for agriculture and rural industries which contain large employment potential. Therefore, in our view, in initial stages of development some intermediate products, machines, and capital goods should be imported if they are urgently needed for some industries, even if in their case, we possess comparative advantage in the dynamic context.

However, it needs to be emphasised again that the form of capital is crucial for employment generation. What is required is the small and simple types of machines and tools which are required for the growth of agriculture and small-scale rural industries. Further, the production of those types of capital goods such as fertilizers, pesticides, cement which will help in raising agricultural production and employment need to be expanded. Some large-scale industries, if their large size is necessary in view of technological considerations, should play a supporting role to the development of agriculture and rural industries. Thus, *the strategy of growth in India should be reoriented so that agriculture and rural industries constitute the main thrust of the development effort.*

From the foregoing analysis, it follows that if the mounting unemployment problem is to be tackled, the strategy of development adopted need to be revised and modified. In the new strategy, agriculture has to play a key role in generating enough employment opportunities for a long time to come. The fact that 68 per cent of the Indian labour force and majority of the unemployed and underemployed reside in the rural areas and further that the employment potential of the industries of large-scale type is very little, the promotion of agricultural and rural development must be made the springboard and the leading sector for generating productive employment for millions. Therefore, *the strategy for agricultural development should be such as will absorb productively the largest possible number of workers.*

Another weakness of the "heavy industry first" strategy and development was that it chose pattern of investment with *higher capital-output ratio*. In basic heavy industries capital-output ratio was admittedly high and in agriculture and related activities, capital output ratio was low. Thus, the highest priority to basic heavy industries and low priority to agriculture in the pattern of investment meant the choice of high macro-value of incremental capital-output ratio (ICOR). As has been made clear in earlier chapters that if a higher growth rate is desired, given the initial value of a rate of saving and annual increments therein over the projected period, then the criterion should be to choose a pattern of investment with a lower macro-value of capital-output ratio. In the case of India, this would have implied a high priority to be given to agricultural and related sectors over a long period of time. Therefore, the choice of the pattern of investment containing a high priority to basic heavy industries and low priority to agriculture meant a low rate of economic growth. As has been pointed

10. B.S. Minhas, *op.cit.* p.12.

out above, Mahalanobis and others supporting his strategy contended that their strategy would ensure high rate of growth in the long run. But in actual practice even after 25 years (1956-80) Mahalanobis development strategy of 1955–56 failed to ensure more than 3.5 per cent average annual rate of growth. In the meantime people suffered a lot due to the inflation generated by the strategy and the non-availability of requisite amount of essential consumer goods.

Mahalanobis heavy industry strategy suffers from another weakness in that it heavily depends upon foreign exchange requirements. Though the strategy assumed that exports from the Indian economy could not be sufficiently increased it required a large amount of foreign exchange resources to establish a network of heavy capital goods industries which required the imports of capital equipment and machinery on a large scale from other countries. For a self-contained growth model framed in the context of a closed economy or stagnant exports, this was an inner contradiction. This also had two evil effects. Because of the low priority to agriculture and consumer goods industries which had export potentials, export could not rise much, and secondly, due to the highest priority to heavy capital goods industries large imports of capital equipment and materials had to be made. Such was the large requirement of foreign exchange to import capital equipment that despite the liberal foreign aid received from countries, especially the U.S.A., the country had to face a serious foreign exchange crisis.

It was said in the defence of the heavy industry strategy that it would ultimately help in substituting imports of heavy industrial products such as steel, cement, various kinds of machines, fertilizers etc. This implies that import substitution when accomplished would enable the economy to dispense with foreign aid. However, it was said that in the meantime aid was required to import equipment and materials to set up heavy capital goods industries. Thus it was claimed that the heavy industry strategy envisaged getting "*aid to end aid*".

However, the fact remains that far from moving the economy towards self-reliance, the heavy industry strategy increased the dependence on foreign aid. Because of low priority to agriculture in the strategy, production of foodgrains did not increase adequately and as a result the imbalance between population and foodgrains further increased during the planning period. In order to correct this imbalance foodgrains had to be imported on a large scale from other countries. Whereas the imports of foodgrains into India during the First Plan period were of the order of 12 million tonnes, it rose to 17 million tonnes in the Second Plan period. The situation worsened in the Third Plan period as the implementation of heavy industry strategy further proceeded and 26 million tonnes of foodgrains were imported during the Third Plan period. During the three years of plan-holiday period (1966–69), the imports of foodgrains jumped to 25 million tonnes in *three* years period. In the Fourth Plan period in which hangover of the heavy industry strategy persisted, foodgrains imports continued unabated with the exception of the year 1972. During the six years period 1970–76, 26.4 million tonnes of foodgrains were imported.[11]

It is thus clear that the imbalance between foodgrains and population was the direct result of the neglect of agriculture in India's heavy industry first strategy. Thus the most distressing effect of the strategy was that the country came to be dependent on even foodgrains. But for the imports of foodgrains under PL 480 and liberal foreign aid which was made available, India could not have been able to implement the Mahalanobis strategy for its inner contradictions and its inappropriateness to the Indian reality would have come to surface much earlier. The situation in the beginning of the Fifth Plan, *i.e.,* in 1974 was that India's large foreign exchange resources were being used to meet the needs of current consumption (*i.e.*, to import foodgrains) rather than to import capital equipment and materials to sustain *new investment* or capital accumulation. Thus for stepping up the rate of capital accumulation and therefore economic growth we came to be dependent on all the three types of commodities (*i*) capital equipment, (*ii*) raw materials, and (*iii*) foodgrains. Thus, despite the objective

11. Government of India, *Economic Survey*, 1977–78, p. 69.

of achieving self-reliance, the strategy adopted drove the economy away from it. The philosophy of "*aid to end aid*" proved to be myth.

Heavy industry strategy has also been criticised for its *ignoring the relevance of the principle of comparative advantage* and the gains to be obtained from specialisation in some selected fields. As mentioned before, Mahalanobis formulated his growth model on the assumption that exports could not be expanded sufficiently. With the help of his model he showed that it is through the adoption of the strategy of substituting imports by domestic production of heavy industry products such as steel, fertilizers, various kinds of machines and machines to make machines, heavy chemicals that high long-term rate of growth could be achieved. Thus Mahalanobis strategy involved industrialisation of import-substitution type. The question whether the new heavy industries would have a comparative advantage or not was not considered as a relevant consideration.

The reckless pursuit of import substitution all along the line and accordingly setting up of all basic heavy industries such as steel, fertilizers, heavy chemicals, machine building and also industries which make machines to make machines (*i.e.*, industries producing mother machines), and so on without considering whether we had comparative advantage in their production. If we had comparative advantage in *some* basic heavy industries we should have concentrated our resources and enjoyed advantages of quick specialisation in selected lines rather than *thinly* spreading our resources on all types of basic heavy industries irrespective of whether we have comparative advantage or not in them caused misuse of scarce investment resources. As seen above, as a result of this strategy of all-round import substitution to build basic heavy industries we came to be dependent on imports from the rest of the world than we ever were or needed to be.

It may also be pointed out that in the Mahalanobis heavy industry strategy alternative policy of development of industries which had large export potential was not at all given any consideration. That supplies of some capital goods could be procured from abroad by development of export industries was ruled out. In this connection it was pointed out by the followers of Mahalanobis that the expansion of India's exports would mean the decline in its terms of trade. In our view, the contention that the increase in India's exports would necessarily mean decline in terms of trade was not justified. In fact, up to 1972 (*i.e.* before the oil crisis) the terms of trade had been in India's favour. It may be pointed out that declining terms of trade argument was advanced by supporters of Mahalanobis strategy only after the Mahalanobis model had been accepted and put into actual practice.

Finally, it may be noted that for achieving high long-term rate of economic growth Mahalanobis and later Raj and Sen[12] and Gautam Mathur[13] laid the greatest stress upon high magnitude of the ratio of incremental plough back of heavy industry products to the expansion of heavy industries sector itself as compared to all other sectors. This pattern of industrialisation, according to them, would automatically raise the saving-income ratio and generate a high rate of economic growth in the long run. However this argument is full of deficiencies and infirmities. The all-round development of heavy industry sector and the highest ratio of plough back of heavy industry products to the heavy industry sector involves, given the scarcity of resources, not only the appalling neglect of agriculture and rural development but also the neglect of the development of export industries and thereby enjoying the benefits of specialisation.

---

12. K.N. Raj and A.K. Sen, "Alternative Patterns of Growth under Conditions of Stagnant Export Earnings", *Oxford Economic Papers,* Feb. 1961.
13. Gautam Mathur, *Planning for Steady Growth*, Oxford University Press, 1965.

CHAPTER 49

WAGE-GOODS MODEL AND STRATEGY OF DEVELOPMENT

As stated in the previous chapter, an alternative model of development was developed by Vakil and Brahmananda. Professors Vakil and Brahmananda adapted and modified classical theory of growth of income and employment in the context of today's developing countries characterised by disguised unemployment. In sharp contrast to Mahalanobis model which emphasised the role of fixed capital Vakil and Brahmananda laid stress on wage goods or what they called liquid capital in determining the growth of employment and income. In conformity with their emphasis on wage goods or liquid capital in their model they propounded a strategy of development which accorded the highest priority to wage goods industries, especially agriculture, in allocation of investment resources. In the present chapter we shall explain their wage goods model and development strategy only briefly.

Wage-goods Model

As in their old work *'Planning for an Expanding Economy'* written in 1956, Vakil and Brahmananda and attribute the prevailing poverty and unemployment to the existence of wage-goods gap. They, however, later defined their wage-goods gap with reference to the magnitude of poverty rather than with reference to magnitude of disguised unemployment as they did in their old work. Admittedly, poverty and unemployment are the two most pressing problems facing the underdeveloped countries today. Poverty is defined as living by households below minimum level of subsistence or real income.

According to Vakil and Brahmananda, to remove this poverty it is essential to increase the aggregate supply of wage-goods. Unless and until the wage-goods gap is bridged, poverty will not be eliminated. In an article published in 1976 Dr. Brahmananda explained the causes of poverty in the following terms. "Poverty in 1975-76 could have been eliminated if the aggregate supply position in regard to basic wage-goods was higher, let us say, about 25% to 30%, and poverty would not raise its ugly head, if from 1975-76 onwards, each year our aggregate supplies of these goods would be growing at the same rate as our population would be growing. Thus, *there is a basic wage-goods gap in the economy which explains poverty*."[1]

Components of Wage Goods

It is useful to note what are the components of wage goods whose output expansion is stressed in the wage goods strategy. According to Brahmananda, they are: (1) Foodgrains; Cereals, Pulses, (2) Milk and Milk Products (3) Edible Oils, (4) Fish, Eggs and Meat, (5) Sugar and Sugar Products, (6) Fruits and Vegetables, (7) Spices. (8) Tea, (9) Coffee, (10) Cloth, (11) Matches, (12) Soap (13) Salt, (14) Kerosene.

In addition to the above wage goods which are of the nature of private consumption goods, there are goods or services which fulfil minimum needs of the people and are generally provided at the collective level. These are "drugs for common use, medical and hospital facilities, minimum educational and library facilities, minimum utility services like water, electricity, roads etc., recreational services and facilities". Besides these, there may be other wage-goods in different regions depending upon the conventional and physiological-cum-nutritional requirements of workers.

1. P.R. Brahmananda, "How Well Have We Planned", *The Illustrated Weekly of India*, October 24, 1976.

Now, the question arises as to why there is a deficiency of wage goods in the economy. This is due to the fact that the capital stock or productive capacity designed and directed to produce wage goods is deficient. To eliminate poverty capital stock designed for the production of wage-goods needs to be expanded. It may be noted that overall capital stock may be adequate but the capital stock in the specific form of productive capacity to produce basic consumer goods may be deficient. This is because all types of capital stocks cannot be used for the production of wage goods. For instance, steel plants and machine making factories cannot be utilized for producing wage goods such as foodgrains, kerosene oil, cloth, sugar, etc. In this connection economists generally draw distinction between capital in "putty" form and "clay' form. While *putty capital* is generalised form of capital which is capable of being used to manufacture all types of goods, that is, *capital goods, luxury goods and basic wage goods, capital in a clay form is suited to the production of specific goods for which it is designed.* Thus, according to Dr. Brahmananda, "*So long as capital is in a putty, all purpose suited form, it can be directed towards creation of any sort of specific capacity but when once putty capital becomes capital goods in the form of factories, projects like mills, machinery, plants, mining and power stations and transport networks, it partakes the nature of clay, fit for a particular purpose.*"[2]

It is only the stock of wage-goods and the raw materials and accessories used in the production of these wage-goods that represents putty capital or generalised capital because "*if stocks of wage goods exist, labour power can be purchased and deployed for whatever specific capacities are needed in the country.*"[3] We can represent the argument of Brahmananda in the following type of production process:

Wage goods → Labour Employment → Capital → Wage goods.

This means that if wage goods are somehow made available, they can be used to employ labour for producing capital. If the new capital is designed to manufacture wage goods, the growth process can be started which will become cumulative and self-sustaining. The general argument of the most modern economists is that for providing productive employment to the workers and to raise their productive efficiency and thereby accelerate the rate of economic growth, it is the deficiency of cooperating factors such as capital equipment, machines, cement, etc. which stand in the way. Vakil and Brahmananda hold this generally held view as quite fallacious since, according to them, this is based upon the assumption that adequate quantities of wage goods are already available. Thus Vakil and Brahmananda lay greater stress on wage goods (*i.e.*, liquid or circulating capital) than fixed capital in the growth process of the economy and for eradication of poverty and unemployment. "If only sufficient quantities of real liquid capital were available, the labour force can be diverted towards the manufacture of the tools and other equipment."[4]

It follows from above that, according to Vakil and Brahmananda, for removal of poverty, promotion of economic growth or capital accumulation alone is not enough. "The way out of poverty is, therefore, to pay immediate attention to making good the capital gap in respect of wage-goods capacity. This, in its turn, would imply drastic alteration in the pattern of investment in the economy."[5]

Disguised Unemployment and Wage Goods. Not only does the removal of poverty but also the eradication of disguised unemployment requires the need for the expansion of basic wage goods. We will explain in a later chapter how Vakil and Brahmananda visualise saving potential in the form of wage goods in the disguised unemployment. They take it granted that a huge amount of disguised unemployment exists in Indian agriculture and to eradicate it they propose a strategy of withdrawing the disguised unemployment from agriculture and using the wage goods surplus left behind by them. They recognise that there will be leakage of wage goods in the process of withdrawal of disguised

2. P.R. Brahmananda, *op. cit.*, p. 31.
3. *Ibid*, p.33.
4. Vakil and Brahmananda, *Planning for an Expanding Economy*, *op. cit.*, p. 250.
5. P.R Brahmananda, *op.cit.*, 1976 p, 35.

unemployed. They propounded in this connection a consumption multiplier or wage goods multiplier. In their scheme the disguised unemployed can be productively employed in capital construction works provided the additional supplies of wage goods are made available. They will produce those kinds of capital goods which will be used for manufacturing wage goods. In this way productive capacity designed for the production of wage goods will increase which will set in motion a cumulative process of growth in employment and output.

As a matter of fact, according to them, poverty and disguised unemployment are not two different problems; they are the two aspects of the same problem. Further, the solution of the problem of poverty as well as of disguised unemployment lies in raising the supply of wage goods. To quote Brahmananda, "Granted the existence of disguised unemployment, its eradication becomes conditional upon additional supplies of wage goods; for their employment in full-bodied productive capacity would create additional demand for such goods, unless the system is able to procure these wage goods additional employment will not be stabilised. Additional supply of wage goods leading to an improvement in per capita real income in terms of wage goods and/or in the real wage rate is the mechanism by means of which unemployment gets eradicated. Poverty in terms of inadequate supplies of wage goods is the barrier to full employment or fuller employment. In fact, there are no two problems like poverty and unemployment; there is only one problem–inadequate supply of wage goods."[6]

In the wage goods model of Vakil and Brahmananda "*the rate of growth of real income is determined by the proportion of savings to income and the portion of such savings devoted to the expansion of the wage goods sector*." It follows, therefore, that in order to accelerate the increase in the capital stock (or productive capacity) designed and directed to produce wage goods the proportion of saving to national income must be raised. Given the capital-output ratio, the greater the rate of saving, the greater will be the rate of investment or capital accumulation. To raise the proportion of saving to national income, Vakil and Brahmananda[7] recommended the increase in the rate of interest especially on bank deposits, fiscal incentives to those who are in a position to save and check on inflation or rising prices as they reduce the value of money and adversely affect the propensity to save. But, according to wage goods model, to achieve faster rate of growth of employment and real income increase in the rate of saving is not enough. The pattern of investment must also be oriented towards building of capital stock designed for the production of wage goods rather than building capital stock in the form of heavy industries.

Wage Goods Strategy Further Developed and Modified

In recent years Brahmananda has further developed and modified his wage goods model and development strategy thereof. He now calls his strategy as the "Extended Wage Goods Strategy" for he now includes in his priority sector what he calls '*integrated wage goods complex*' not only the wage goods but also those capital goods which are required for the production of wage goods. To quote him, "Since not all capital goods (machinery and plant, tools and raw materials) are directly and/or indirectly but necessarily involved in the production process of wage goods, priority in development of capital goods has to be assigned to those capital goods whose supply (and/or production) leads to the production of wage goods".

It is therefore clear that Prof. Brahmananda later realised the importance of capital goods and modified his model in the light of the criticism levelled against his earlier model that it was based on simple type of production process which considered that only labour is required for production of wage goods or capital goods. Thus, the modification made by him is in the right direction and makes his model more consistent and realistic. In order that not only the growing population but also the existing unemployed and disguised unemployed are provided with employment, the extended

6. Brahmananda, *op.cit.*, 1976 p, 35.
7. C.N. Vakil, *Growth With Stability*, Dr. Gadgil Memorial Lectures, Oct. 9, 1977.

wage goods strategy suggests that the integrated wage goods complex (that is, basic wage goods and capital goods required for their production) should grow at a much higher rate than population. Accordingly, in the development strategy a very high priority must be accorded to the investment in wage goods complex.

Real-income Multiplier. Besides in the wage goods model, real income multiplier plays a significant role in process of economic growth. It may, however, be noted that Vakil-Brahmananda's concept of consumption multiplier has undergone some change in recent years. While their old concept of consumption multiplier expounded in their earlier work "*Planning for an Expanding Economy*" was defined with particular reference to the removal of disguised unemployed from agriculture they have now put forward a concept of multiplier which relates the increase in the supply of wage goods to the increase in real income. In a later econometric study, Dr. Brahmananda found that ₹ 100 *worth of increase in the supply of wage goods leads to an increase in real domestic product (national income) equivalent to more than* ₹ 300. *There is a stable multiplier.*"[8] This has, therefore, been called consumption-real income multiplier and is an important factor in the determination of the magnitude of real income in the economy. This multiplier works on account of the fact that net productivity or output per unit of labour employed far exceeds the amount of wage goods consumed per unit of labour. The surplus of wage goods so obtained can be further used to provide more labour employment to yield more output. "There is thus a multiplier process in regard to supply of output, the value of the multiplier being dependent upon the fractional excess of the net output efficiency of labour over and above the portion contributing to meeting the wage goods requirements of performance of labour."[9]

Let us briefly state the wage-goods model of development.

1. The poverty and unemployment exist because there is a wage-goods gap.
2. It follows from first proposition that for solving the problem of poverty and unemployment, expansion in the supply of wage goods is required.
3. To secure a rapid increase in the supply of wage goods, capital stock (or productive capacity) designed for the production of wage goods must be expanded. In other words, agriculture, other wage goods industries and capital goods needed for production in them must be given topmost priority in the investment pattern.
4. To enlarge the capacity to produce wage goods, proportion of saving to national income should be raised and a relatively high proportion of national saving should be invested in wage goods complex.
5. The importance of increase in the capacity to produce wage goods is enhanced because of the working of consumption-real income multiplier.

Arguments in Favour of Wage-goods Strategy

Many arguments have been given in defence of wage-goods model as compared to Mahalanobis Heavy Industries model of growth. Firstly, given the investment resources, rate of economic growth will be faster if a high priority is given to the wage goods sector in the pattern of investment. This is because capital-output ratio is much lower in wage goods industries, especially agriculture, than in basic heavy industries sector. It has been contended by Dr. Brahmananda that within the framework of heavy industries model we cannot achieve a growth rate higher than 4% given the saving and investment rate of about 15% because of the high capital-output ratio in the heavy industries. But, according to him, if we make a shift to wage-goods model, the achievement of a higher growth rate becomes possible because of lower capital-output ratio in wage goods sector.

Secondly, the implementation of the wage goods model will not involve much foreign exchange requirements both for building the wage goods capacity and for current purposes of using the capacity

8. P.R. Brahmananda, "*Determinants of Real National Income and Prices Level in India*, University of Bombay.
9. Vakil and Brahmananda, *op,cit.*, 1976 p. 215.

to produce wage-goods. As a result, as compared to Mahalanobis heavy industry strategy, the balance of payments gap in 'wage-goods bias' development strategy would be quite negligible.

Further, another merit of wage-goods model of development which was pointed out was that the wage-goods industries such as agriculture, cotton textile, sugar had a large export potential. Their products tend to have a large foreign demand. Thus, with the expansion of exports on the one hand and low import requirements on the other, the wage-goods strategy will result in net earning of adequate amount of foreign exchange which can be utilised for importing capital equipment required for building capital goods base needed for producing capital goods and materials for the expansion of industrial and agricultural sectors. In this way the adoption of wage-goods model will ensure the attainment of self-reliance in a short time.

Thirdly, the most important advantage of wage goods model is that the high rate of growth of the agriculture and the related sectors under it would provide potentially increasing amount of wage goods or circulating capital. The expansion in supplies of wage-goods or circulating capital will enable us to provide employment to the open and disguised unemployed and will thus help in solving the problems of poverty and unemployment. Fourthly, as the majority of the new workers getting employment following the implementation of the wage-goods strategy will be drawn from the poverty-stricken households, the wage-goods strategy will have a good income redistribution effect. It was thus claimed by Dr. Brahmananda that the wage goods strategy put forth by him would promote economic growth with social justice.

Lastly, the expansion of wage-goods sector, that is, agriculture will generate a large demand for the products of manufacturing industries. This will ensure rapid industrial development with proper linkage with agriculture. The manufacturing industries would obtain foodgrains and raw materials from agriculture and in turn would feed latter with inputs and industrial consumer goods. It is here worth mentioning that Vakil and Brahmananda, like Ragnar Nurkse, believed that disguised unemployment in labour-surplus developing countries contained saving potential in the form of wage goods. By withdrawing the disguised unemployed people from agriculture, some amount of wage goods can be released for providing productive employment in the creation of capital assets (that is, in the investment sector). But Vakil and Brahmananda thought that the surplus of wage goods so released would not be sufficient to provide employment to all the potentially available labour force. Therefore, they also emphasised increase in the production of wage-goods, especially foodgrains, so as to fill in the wage-goods gap and thus to achieve full-employment conditions.

The most important kind of wage-goods are foodgrains which are the product of agriculture and for which income elasticity of demand in the less developed countries is very high. Therefore, the approach to unemployment which visualises wage-goods gap as the cause of unemployment assigns a paramount role to agricultural development in the strategy of growth so that food surpluses are made available to the newly employed workers outside agriculture. That the growth in natural employment through creating capital assets under rural works programme or through the development of manufacturing industries (including the capital goods industries) depends upon the marketable food surpluses has long been recognised in development economics. Development experience of Western capitalist countries as well as of socialist Russia shows that for growth of industrial employment, availability of food surpluses is essential prerequisite, though the methods of mobilising food surpluses have been different in the capitalist and socialist worlds. Whereas in today's Western developed countries such as U.K., U.S.A. and France. required food surpluses were obtained through agricultural growth by adopting capitalist mode of production in Soviet Russia collectivisation of agriculture helped a good deal in extracting food surplus for growth in industrial employment.[10]

10. See M.M. Dobb, *Soviet Economic Development since* 1917. See also his *Economic Growth and Underdeveloped Countries*, 1963, p. 49.

The Critical Evaluation of Wage-goods Strategy

That the growth of non-agricultural sector depends on the availability of foodgrains or wage goods is quite obvious. When people are employed in rural public works and in industries producing capital goods and consumer goods outside agriculture, there will be need for food to feed them. If food is not made available to them directly or through market mechanism, their employment cannot be sustained. It may be asked even when unemployed they must be consuming food to make their living possible and therefore no extra food would be required when they are given employment. That is, whether employed or unemployed people must consume food in order to live. However, when employed and paid wages for their work, they would definitely increase their demand for food, income elasticity of demand for food being quite high in underdeveloped countries. If more food is not forthcoming pressure of demand would push up prices of foodgrains which will result in the redistribution of foodgrains. It is thus clear that the supply of foodgrains is an important bottleneck in the way of expansion of employment outside agriculture. Thus, Prof. Vakil and Brahmananda's stress on wage-goods as an important factor in determining employment outside agriculture is quite proper.

If b stands for food output per person employed in agriculture, w for consumption of food per person in agriculture, then food surplus per man employed in agriculture will be $(b-w)$ and if E_1 is the total number of persons employed in agriculture, then the total food surplus will be $(b-w)E_1$. Now, the magnitude of this $(b-w)\,E_1$ will determine the generation of employment opportunities outside agriculture. If we designate employment outside agriculture by E_2 then we can express employment outside agriculture in the following functional form:

$$E_2 = f(b-w)E_1$$

In other words, viewed in this way, employment outside agriculture depends upon the food surplus in the agricultural sector. And the food surplus in agriculture depends upon the productivity per head (b) in agriculture and the consumption per head of the persons E_1 engaged in it. If productivity per head (b) in agriculture is increased substantially, opportunities of wage-employment outside agriculture can be expanded rapidly.

However, the assertion by Vakil and Brahmananda that the growth in employment solely and exclusively depends upon the supply of wage goods and that the capital goods (that is, heavy industry products) play no important role in the creation of employment opportunities is open to question. Under adequate organisation and entrepreneurial ability as well as optimum-institutional forms, two important bottlenecks, namely capital goods bottleneck and wage-goods bottleneck, are faced in creating opportunities for productive employment. For expansion in productive employment both these bottlenecks have to be overcome and to emphasise the breaking of one bottleneck, while making no attempt to tackle the other, is an unrealistic approach to the employment problem. Wage-goods are required to meet the demands of the newly employed people but if some people are to be employed in productive activities in the very first instance, they are required to be equipped with some capital goods. Further, for the production of wage-goods themselves, capital goods are required. That is, labour is to be equipped with capital goods to produce wage-goods.[11] Professor Dantwala rightly points out that, "*it is a mistake to consider capital goods and wage goods as exclusive and unrelated categories. Anyone who studies the composition of inputs needed for transformation of traditional agriculture should be able to appreciate the inter-connection between the two.*"[12]

11. This does not mean that capital goods industries should get high priority in the pattern of investment. What has been emphasised is that to make the employment productive, labour is required to be equipped with capital. Besides, for generation of employment opportunities, form of capital is as important as its amount. Whether capital goods needed for growth of industries or for the growth of agriculture to be given higher priority, depends upon whether industry or agriculture constitutes the main thrust of the development strategy.
12. M.L. Dantwala, "Approaches to Growth and Employment", *Economic and Political Weekly*, December 1972, p. 2456.

Thus, development strategy for rapid expansion of employment has to be such as will envisage the increase in production of both wage-goods and capital goods. Thus, to talk of wage-goods versus capital goods is a false issue and an unrealistic approach to the problem of employment since both are needed to generate employment; capital goods are required to make the employment productive and wage-goods are needed to feed the workers provided with new employment, that is, to sustain their employment. To say this, however, does not imply that all capital goods industries should be necessarily developed at home and given higher priority in the allocation of resources. Which specific capital goods should be produced at home and which be imported from abroad should be decided on the basis of resource endowments and comparative advantage conceived in the dynamic context.

Further, it needs to be emphasised again that crucial factor in the generation of employment opportunities is the production and use of appropriate types of capital which aid labour in production rather than displace it. Thus, *the use of appropriate technology for production is as important as the accumulation of capital if employment opportunities are to be increased and poverty eradicated.* In other words, form of capital is as important as its amount. Moreover, in view of the limited employment potential of large-scale industries using capital-intensive technology, if a development strategy is to generate sufficient employment opportunities, agriculture and rural industries must play a leading role and be given higher priority in the pattern of investment, for it is in the agriculture and rural industries that vast employment potential exists. Industries should play supporting role to the development of agriculture, *i.e.*, industry should produce such capital goods and other inputs such as fertilizers, cement, pesticides that are required for the growth of agriculture. Thus, whereas in the development designs of Western countries as well as of Soviet Russia, agriculture played a supporting role to urban industrialisation, in India with surplus labour industrialisation must serve the needs of agricultural growth. Thus, if for generation of employment opportunities agriculture is to constitute the main thrust, it is the basic and capital goods needed for the expansion of agriculture such as irrigation facilities, fertilisers, HYV improvement and development of land that need to be given a high priority.

However, the wage-goods approach to employment, as enunciated by Vakil and Brahmananda, has not considered the creation of employment opportunities in wage-goods industries themselves, especially in agriculture. Emphasis on agricultural development in wage-goods approach has been laid for increasing the supply of wage-goods so that they can be used for employment generation in non-wage goods sector, especially the rural public works or accumulation of capital. Professers Vakil and Brahmananda in their proposed growth strategy had visualised withdrawing disguised unemployed from agriculture rather than creating employment opportunities in agriculture itself. That the problem of unemployment and underemployment or disguised unemployment in agriculture can be solved through agricultural growth itself, if proper strategy for agricultural development is adopted, was ignored by them. In our view, in overpopulated and agriculturally backward developing countries like India, agriculture can still absorb more labour if appropriate technological and institutional changes are brought about. Thus, agriculture is to be given high priority not only because it will yield wage goods which are badly needed for the expansion of non-agricultural employment but also because it is agriculture which, if proper strategy is adopted, contains immense potential for the absorption of labour in productive employment. Therefore, a great weakness of the strategy proposed by Vakil and Brahmananda is that it ignores the need for bringing about appropriate technological and institutional changes to generate enough employment opportunities in the agricultural sector itself.

Further, Vakil and Brahmananda's strategy of employing labour in public works or in constructing other forms of capital is based upon a simple production process which assumes that labour alone can produce capital goods. This is far from the prevailing situation in the real world. Labour with bare hands can do little. Extra capital equipment and materials are required to employ labour in capital creating works. This necessitates a more elaborate approach to the growth of employment in developing economies, involving proper balance between wage-goods, labour employment, capital goods and materials.

Chapter 50
STRATEGIES OF BALANCED AND UNBALANCED GROWTH

There have been, among the development specialists, two major schools of thought regarding the strategy of economic development that should be adopted in developing countries. On the one side, there are economists like Ragnar Nurkse and Rosenstein-Rodan who are of the view that the strategy of investment should be so designed as to ensure a balanced development of the various sectors of the economy. They, therefore, advocate simultaneous investment in a number of industries so that there is a balanced growth of different industries. Economists, like H.W. Singer and A.O. Hirschman, on the other side, believe that for rapid economic growth there should be concentration of investment in certain strategic industries rather than an even distribution of investment among the various industries. In other words, in the view of these latter economists, unbalanced growth is more conducive to economic development than a balanced one. We may now consider both these views at some length.

STRATEGY OF BALANCED GROWTH

In an earlier chapter, we explained how the underdeveloped countries are caught up in a vicious circle of poverty. We also pointed out how difficult it was to break this vicious circle. We explained there how the vicious circle of poverty operates both on supply and demand sides of capital formation. Nurkse put forward the doctrine of balanced growth in order to break the vicious circle of poverty on the demand side of capital formation. It will be useful to have again a cursory look at this vicious circle.

In an underdeveloped country, the level of per capita income is low which means that the people's purchasing power is low. Owing to small incomes and low purchasing power their demand for consumer goods is low. As a result of low demand for goods, the inducement for investment is less and capital equipment per capita (*i.e.*, per worker) is small. Since the amount of capital per capita is small, productivity per worker is low. Low per capita productivity means low per capita income, *i.e.*, poverty. This completes the vicious circle of poverty. In a poor country, the size of the market for goods is small so that sufficient opportunities for profitable investment in industries are lacking. This is the main reason for lack of inducement to invest which we discuss presently.

Size of Market and Inducement to Invest

Investment means the expenditure on the making and installation of capital goods, e.g., construction of factories and the making of machines and their installation, execution of irrigation power projects, the construction of roads, railway, etc. Obviously, an entrepreneur will be induced to invest in factories, machinery, etc., if he expects sufficient return on his investment. Businessmen will have incentive to invest only from a motive of earning a profit. It is the expectation of profit which is a fundamental factor influencing the amount of investment in a country at a given time. In a poor country, the low level of investment is due to low expectations of profit because of less demand for goods or a small size of the market.

Let us understand clearly why there is .less inducement to invest in a poor country. It is easily understandable that in underdeveloped countries there is a great need for capital for economic development. People are too poor even to have two square meals a day or a reasonable housing accommodation or clothing to cover their bodies. Hence, there is an urgent need for large-scale production of consumer goods, but it cannot be done without the production and use of capital goods in large quantities. Agricultural improvements, the establishment and expansion of industries, the optimum use of the natural resources and harnessing the natural resources into the service of the people, all require capital. The need for capital can be great but the inducement to invest can be weak. The level of investment depends not on the need of capital but on the inducement to invest in the form of attraction to earn profit from the capital invested. Without reasonable expectation of profit, much capital will not flow into investment

The quantity of profitable investment in a country depends on the size of the market. Adam Smith said, "*Division of labour is limited by the size of the market.*" We can say in the same manner that inducement to invest depends on the size of the market, that is, on the level of demand. The small size of the market or the low level of demand for the products concerned discourages the entrepreneurs from investment in industries.

This will be clear from an illustration. In a modern dairy, milking, filling up bottles and their loading all these operations are done with the aid of automatic machinery. Will the installation of such a machinery in every Indian town be profitable for individual entrepreneurs. Obviously, it will not be profitable. Per capita income being low in India, the demand for milk in each town will be too small to make the full use of such automatic machinery. Such costly plant and machinery will remain mostly idle and there will be work for such a machinery only for a few hours during a week. This means a great waste of valuable capital asset. Which entrepreneur dare start such a business? As an inducement to invest, the entrepreneurs should be sure that the capital equipment will be profitably employed. This will be possible only if the machinery can be kept in continuous use, and this cannot be done unless there is sufficient demand for the product made by this machinery.

Take another example. Suppose a cloth of a special design is very attractive and it can fetch a very high price. But it will not be economical to install a big machine to make a cloth of a special design, because owing to its high price and low incomes of the people, there will not be sufficient demand for this type of cloth, *i.e.*, the market will be too small. In America, the cars are cheap but they are very expensive in India. What is the reason? One reason is that the demand for cars in India as compared with that in America is so small that manufacturers of cars cannot be induced to make them in large quantities which would have made them cheap on account of the economies of scale. Examples can be multiplied. The conclusion is clear that inducement to invest depends on the size of the market or the purchasing power of the people.

It may be clearly understood that in the underdeveloped countries, *demand for consumer goods cannot be increased merely by the expansion of money supply in the country. The real demand will increase only if there is increase in the productivity per worker and as a result thereof there is increase in the real per capita income.* But mere expansion of money supply and thus putting more money into people's pockets demand can increase only in the form of money which will result in inflation or higher prices, but not increase in the real aggregate demand

Similarly, the demand for goods or the size of the market cannot be large merely because size of a country is big or its population is large. If the purchasing power of the people is low because of their extreme poverty, the demand for goods in that country will be small or the size of the market will be small even though the country is big in size or its population is large. Further, in poor countries, where the people's purchasing power is low on account of low per capita income, the demand for goods, and hence the size of the market, cannot be increased by high pressure salesmanship and vigorous advertising campaigns. There should be enough people to buy them. Thus, it is clear that, in the underdeveloped countries, the demand for goods, or the size of the market, cannot be increased by increasing money supply or by increase in population or by salesmanship and advertisement or the large size of the country. The size of the market can be increased only by increasing productivity. As Nurkse puts, it, "The crucial determinant of the size of the market is productivity."[1] Increase in productivity will increase people's income and hence their purchasing power.

1. Ragnar Nurkse, *Problems of Capital Formation in Underdeveloped Countries,* p. 8.

The level of people's income in any country can be raised and consequently their purchasing power can be increased by increasing productivity and aggregate output or, in other words, by increasing productive employment. A situation of higher productivity, the greater employment and incomes and high purchasing power of the people will provide a profitable field for investment. It may be said that the size of the market can be enlarged by lowering the price of the products. But this is no solution of the problem. The real solution of the problem is only an increase in productivity of the people by raising productive employment. Only as a result of increase in productivity, there is increase in income and increase in purchasing power which will increase demand and enlarge the size of the market.

Say's law propounded by classical economists, which we have studied earlier, tells us that production or supply creates its own demand. But this law cannot be accepted in the sense that the production of cloth creates its own demand because the workers engaged in the making of cloth will not spend their entire earnings on the purchase of cloth. In the same way, production of shoes cannot create its own demand. The reason lies in the variety of man's demands. However, Say's law can be applied to some extent to the developing countries. If, in the developing countries, investment is made simultaneously in a large number of industries, incomes of a large number of workers engaged in these industries will increase. This will create demand for goods produced by one another. In other words, if investment is made simultaneously in a number of industries and production is increased, the supply will create its own demand. Say's law will hold good in such a situation.

Thus, we see that investment in a particular industry and the resultant production or supply cannot create its own demand but simultaneous investment in a number of industries can. As Nurkse puts it, *"An increase in production over a wide range of consumables so proportioned as to correspond with the pattern of consumers' preference, does create its own demand."*[2]

Nurksian Strategy of Balanced Growth

We have explained above how, in the underdeveloped countries, the small size of the market or the limited demand for goods acts as a hindrance in the way of their economic growth or capital formation. When an entrepreneur wants to set up a factory or install plant and machinery, he makes sure whether there is enough demand for the goods he proposes to manufacture and whether the investment will be profitable. We have seen that owing to low demand for industrial goods investment is discouraged because of low profitability. That is why the vicious circle of poverty operates on the demand side of capital formation. The people in the underdeveloped countries are poor and their per capita income is low. This keeps the demand limited and size of the market small. Since the market is small, the entrepreneurs are discouraged from investment in plant and machinery in which only large-scale production is possible and economical.

The result is that capital formation in the country is discouraged. Owing to lack of capital, productivity is low and since productivity per worker is low, the per capita income is low which means there is poverty. This is how the vicious circle of poverty operates in the underdeveloped countries.

According to Nurkse, it is the vicious circle operating in the underdeveloped countries, which stands in the way of their economic development, and accordingly, if this vicious circle can be broken, economic development will follow. The operation of the vicious circle can also be described thus: Inducement to invest depends ultimately upon demand, *i.e.*, the size of the market. And the size of the market in turn depends upon productivity, because the capacity to buy is ultimately based on the capacity to produce. Productivity, in its turn, largely depends on the use of capital. But, for an entrepreneur, the small size of the market will limit the use of capital so that productivity will remain low, thus keeping the size of the market small. The vicious circle will then repeat itself. This *vicious*

2. Nurkse, *op cit.*, p. 12.

circle of poverty, according to Nurkse, can be broken by a simultaneous investment in a large number of industries, i.e., by a balanced economic growth.

We have explained above how Say's law cannot be helpful in underdeveloped countries, if investment is made only in one industry. The output of any single industry newly set up with capital equipment cannot create its own demand. Human wants being diverse, the people engaged in the new industry will not wish to spend all their income on their own products. Suppose a shoe manufacturing industry is set up. If in the rest of the economy, nothing happens to increase productivity and employment, and hence the buying power of the people, the market for the additional output of shoes is likely to be deficient. People outside the shoe industry will not give up the consumption of essential food, clothing, etc., to create a sufficient demand for shoes every year. The supply of shoes is likely to outrun demand, and if investment is confined only to one particular industry, it cannot prove fruitful.

But if investment is made simultaneously in a large number of industries, it will provide work for a large number of people producing diverse commodities. It will increase their income and they will be in a position to buy the consumer goods made by one another. This is how supply can create its own demand (as Say's Law asserts) through the strategy of balanced growth. The workers employed in different industries become customers of one another's goods and demand is increased or the size of the market is enlarged. The expansion of one industry helps in the expansion of others and there is all-round growth. This is how the difficulty arising from small size of the market is overcome and the obstacle in the way of economic growth cleared. In Nurkse's words, "The difficulty caused by the small size of the market relates to individual investment incentives in any single line of production taken by itself. At least in principle, the difficulty vanishes in the case of different industries. Here is an escape from the deadlock; here the result is an overall enlargement of the market. *People, working with more and better tools in a number of complementary projects become each other's customers. Most industries catering for mass consumption are complementary in the sense that they provide a market for, and thus support, each other. This basic complementarity stems in the last analysis from the diversity of human wants. The case for balanced growth rests on the need for a balanced diet."*[3] Taken separately, a number of industries may be unprofitable so that the private profit motive would not suffice to induce investment in these industries. However, undertaken together in a synchronized manner, a balanced increase in production would enlarge the size of the market for each firm or industry so that "synchronized undertaking" would become profitable. *This wave of capital investment in a number of different industries is called by Nurkse as "balanced growth."*

In this way, as we have already said, the hindrance to economic growth owing to the small size of the market is removed, The aggregate demand is increased owing to simultaneous investment in a large number of industries, because the incomes go up and productivity levels of persons employed in different industries go up. Hence, the underdevelopment equilibrium trap and the vicious circle of poverty can be broken by balanced growth. If once this circle is broken then, since there is circular connection, this circle will turn from poverty to balanced growth and to all-round development of the economy. In this way, the circle can be given beneficial form.

Now the question arises Which industries should be selected for investment? The answer is to be found in the above solution offered by Nurkse. Investment should be made simultaneously in such industries the manufactured products of which are in accordance with the demand or the preferences of the consumers or on which the persons engaged in different industries would spend their incomes. There should be investment in a large number of complementary industries in the sense that persons employed in them become each other's customers. Only by a simultaneous investment in such industries, production or supply will create its own demand.

3 Nurkse, *op. cit.*, p. 11.

Then the question is: How is it to be made sure that simultaneous investment in a large number of industries is actually made? Nurkse answers that, if in the country there are dynamic and constructive entrepreneurs and industrialists, they can be induced to make investment simultaneously in different industries. If there is lack of such entrepreneurs, then the government can take the work of balanced growth in its own hands. That is, the government can itself make simultaneous investment in several industries and can thus increase people's incomes and productivity. As a result of investment in several industries, it will be possible to increase the use of capital goods in large quantities which will raise the level of productivity and there will be large increase in the aggregate output of consumer goods and services. As a result of this, the level of national income will rise which will help to raise the standards of living of the people. In this way, the poverty of the people will be eliminated. What is needed to remove the poverty of the people is to launch an attack on the various sectors of the economy simultaneously. This will remove the obstacle arising from limited demand or narrowness of the market and the inducement to investment will increase.

External Economies and Balanced Growth. It seems to be proper to refer in this connection to external economies. When one industry creates demand for another, it will be profitable to the other industry. When one industry benefits from the growth of another industry, then we say that external economies are available from one industry to another. We have seen above that it proves profitable to make investment in complementary industries, because people engaged in such industries become one another's customers or create demand for one another. It is clear, *therefore, that the doctrine of balanced growth is based on the concept of external economies.*

It is to be noted that here we do not use the term 'external economies' in the sense in which Marshall used it. By 'external economies' Marshall meant those economies which arise from the localisation of a certain industry in a particular place and these economies are enjoyed by each firm in the industry by the establishment of numerous firms there. But in development economics, *by external economies we mean those benefits which accrue to other industries by the establishment of new industries or the expansion of the existing industries.* We have seen above how, according to Nurkse's doctrine of balanced growth, these benefits accrue to the other industries by the establishment of new industries or the expansion of old industries through simultaneous investment in such industries in the form of increased demand or extension of the market. In fact, the increasing returns which arise from the process of economic growth, are mainly due to the creation of external economies in the form of extension of the market or increase in demand and not due to the external economies mentioned by Marshall such as technical information from the journals, improvement in the technical skill of labour, development in the means of communication and transport, etc. which arise from the location of an industry in a particular place.

It is worthwhile knowing whether, in the balanced growth, investment will be made in agriculture or not. Nurkse has not discussed this point in his book *'Problems of Capital Formation in Underdeveloped Countries.'* But later on he made it clear that in his balanced growth strategy, appropriate investment will be made in agriculture. Thus, he has not ignored agricultural development in his doctrine of balanced growth. In fact, investment in agriculture is implied in his book referred to above, because he has said that investment would be made in such industries simultaneously as produce goods conforming to consumer's demand or preferences. Since when, in the underdeveloped countries, people will get employment in the various industries, they mostly spend their incomes on the foodgrains, investment in agriculture will be necessary to meet their demand and to promote balanced growth.

Nurkse has also not said in his doctrine of balanced growth whether investment should be made in capital goods industries and social overhead capital like transport and communications to promote balanced growth. Actually, Nurkse has suggested investment in consumer goods industries. But how will the machinery and capital equipment required in these industries be obtained? If they are not to be imported from abroad, they will have to be produced in the country and for that purpose investment will have to be made in their production.

Thus, we see if the doctrine of balanced growth is to be fully implemented, then investment will have to be made in consumer goods industries, agriculture, capital goods industries and social overhead capital. But when investment is to be made in all such sectors and industries, then, in order to bring about balanced economic growth, large quantities of resources will be required. It is doubtful if the underdeveloped countries have the means to mobilise resources in such large quantities.

A Critique of Balanced Growth Doctrine

Prof. Hans Singer and Albert Hirschman, eminent American economists, have criticized Nurkse's doctrine of balanced growth. They contend that what is needed is not balanced growth, but a strategy of judiciously-planned unbalanced growth. According to Singer, balanced growth cannot solve the problem of the underdeveloped countries, nor do they have sufficient resources to achieve balanced growth. Singer[4] maintains that balanced growth doctrine might be better expressed as follows: "As hundred flowers may grow whereas a single flower would wither away for lack of nourishment." But where are the resources to grow hundred flowers? Singer argues that the slogan "stop thinking piecemeal and start thinking big" is a sound advice for underdeveloped countries but he also feels that there are "several areas of doubt" about the balanced growth theory in its Nurksian form.

First, if the balanced growth doctrine is interpreted to advise the underdeveloped countries to embark on a large and varied package of industrial investment with no attention to agricultural productivity, it can lead to trouble. At the initial stages of development, as the income grows with new industrial investment and employment, the relatively greater demand would be created for food and other agricultural goods. In order to sustain industrial investment, the agricultural productivity would have to be greatly raised. Thus, the big push in industry must be accompanied by a big push in agriculture as well, if the country is not to run short of foodstuffs and agricultural raw materials during the transition to an industrialised society.

But when we start talking about varied investment package for industry and "major additional blocks of investment in agriculture" at the same time, we run into serious doubts about the capacity of underdeveloped countries to follow the balanced growth path. According to Marcus Fleming, "Whereas the balanced growth doctrine assumes that the relationship between industries is for the most part complementary, the limitation of factor supply assures that the relationship is for the most part competitive."[5] Singer adds: "*The resources required for carrying out the policy of balanced growth ... are of such an order of magnitude that a country disposing of such resources would in fact not be underdeveloped.*"[6]

Investment may be of whatever type, it necessarily induces some additional investment and some other productive activities. According to Singer, *the expansion of social overhead capital and the growth of consumer goods industries and improvement of production techniques in them to raise productivity cannot take place simultaneously, because the underdeveloped countries have only limited capabilities of making use of their resources.* In the under developed countries, not only are the resources and the capabilities to bring about balanced growth lacking but, according to Hirschman, balanced growth is not even desirable. His view is that if economic growth is to be accelerated, it will have to be brought about by unbalanced growth. If we promote growth by creating imbalances in the economy, the growth will be accelerated, because it will produce such incentives and pressures which will encourage development in the private sector. "The balanced growth doctrine is premature rather than wrong." Singer concludes, *It is applicable to a subsequent stage of self-sustained growth rather than to the breaking of a deadlock.* For launching growth "it may well be a better development strategy to concentrate available resources on those types of investment which help to make the economic system more elastic, more capable of expansion under the stimulus of expanded markets

4. H.W. Singer, *Economic Progress in Underdeveloped Countries*
5. Marcus Fleming, "External Economies and Balanced Growth", printed in Agarwala and Singh, *Economics of Underdevelopment,* Oxford University Press.
6. *Op. cit.*

and expanding demand". He mentions investments in social overhead capital and removal of special bottlenecks as examples of such "strategic" investment.

The fundamental trouble with the balanced growth doctrine, according to Singer, is its failure to come to grips with the true problem of underdeveloped countries, the shortage of resources. "Think Big" is a sound advice to underdeveloped countries, but "Act Big" is unwise counsel, if it spurs them to bite more than they can possibly chew. Moreover, the balanced growth doctrine assumes that an underdeveloped country starts from a scratch. In reality, every underdeveloped country starts from a position that reflects previous investment and previous development. Thus, at any point of time, there are some highly desirable investment programmes which are not in themselves balanced investment packages but which represent unbalanced investment to complement existing imbalances.

HIRSCHMAN'S STRATEGY OF UNBALANCED GROWTH

Professor Albert Hirschman in his book, *"Strategy of Economic Development,"* carried Singer's idea further ard contended that deliberate unbalancing of an economy, in accordance with a predetermined strategy, was the best way of achieving economic growth. Like Singer, he argues that balanced growth theory requires huge amounts of precisely those abilities which have been identified as likely to be very limited in supply in the underdeveloped countries. He characterises the balanced growth doctrine as *"the application to underdevelopment of a therapy originally devised for an underemployment situation" by J.M. Keynes. In an advanced country, during depression, "industries, machines, managers, and workers as well as the consumption habits" are present, while in under developed countries this is obviously not so.*

As an underdeveloped country is incapable of financing and managing simultaneously a balanced "investment package" in industry and the needed investment in agriculture, in order to give a big push to lift an underdeveloped economy from a position of stagnation, Hirschman prescribes big push in strategic selected industries or sectors of the economy. After all, he points out, the industrialised countries did not get to where they are now through "balanced growth." True, if you compared the economy of the United States in 1950 with the situation in 1850, you will find that many things have grown, but not everything grew at the same rate throughout the whole century. *Development has proceeded "with growth being communicated from the leading sectors of the economy to the followers, from one industry to another; from one firm to another."*[7]

According to Professor Hirschman, the real scarcity in underdeveloped countries is not the resources themselves "but the ability to bring them into play." He divides the initial investment into two related activities: (*a*) directly productive activities (DPA) and (*b*) social overhead capital (SOC). An underdeveloped country may follow the method of unbalanced growth by undertaking initial investment either in social overhead capital or the directly productive activities. Whichever the type of investments, it will yield an 'extra dividend' of induced decisions resulting in additional investment and output. He contends that social overhead capital, and directly productive activities cannot be expanded simultaneously because of the limited ability to utilise resources. Thus, the planning problem is to determine the sequence of expansion that will maximize induced decision-making. Balanced growth (of social overhead capital and directly productive activities) is not only unattainable in most underdeveloped countries, it may also not be desirable. The rate of growth is likely to be faster with crucial imbalances precisely because of "the incentives and pressures" it sets up.

Hirschman's Illustration of Balanced and Unbalanced Growth Path

Now, should we choose 'development via excess capacity of SOC' or development via shortage of SOC? According to Hirschman, the sequence of development which is vigorously self-propelling should be adopted. We may explain the rationale behind this contention with the help of Hirschman's diagram as shown in Fig. 50.1. In this diagram, the units of investment in *SOC* are measured along

7. A Hirschman, *Strategy of Economic Development*, pp. 63-64 (italics added).

the vertical axis, while the units of investment in DPA are measured along the horizontal axis. The curves *I, II, III* are the isoquants, reflecting the different combinations of *SOC* and *DPA* that result in the same gross national products at a given time. As we move successively from curve *I* to *II* to Ill, we reach a higher level of gross national product. For the sake of analytical simplicity, the curves have been drawn in such a way that their optimal points lie on the 45° line. In fact, this line gives the locus of balanced growth of *DPA* and *SOC.* Assuming that balanced growth of *SOC* and *DPA* is not possible because of the inherent limited ability of the under-developed countries to utilise resources, we have to determine that sequence of development which maximises induced decision-making.

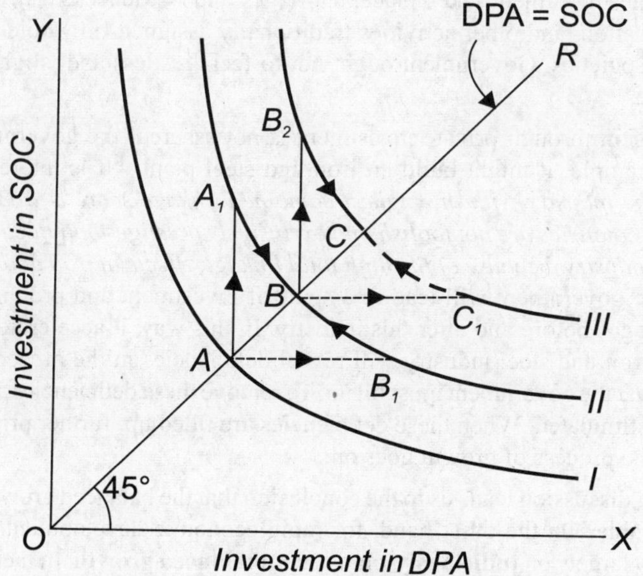

Fig. 50.1. *Balanced and Unbalanced Growth Paths*

Let us first consider the *sequence of development via excess capacity of SOC*. The path of development assumed by the economy would then be given by the heavy line $A \to A_1 \to B \to B_2 \to C$. Starting from *A*, the increase in *SOC* to A_1 invites increase in *DPA* till the balance is achieved at *B*. With the increased gross national product, the government may undertake further investment in *SOC* to B_2. This in turn will induce the *DPA* to increase to the point *C*.

If, on the other hand, the economy adopts the sequence of development via shortage of *SOC*, the course followed by the economy would be the one shown by the dotted line AB_1BC_1C. In this case to start with, we increase *DPA* from *A* to B_1. To restore balance, this will be followed by the increase in *SOC* from B_1 to *B*. If there is a further increase in *DPA* to C_1, *SOC* will have to follow suit until balance is restored at *C*.

It needs to be noted that unbalanced growth via both the paths yields an "extra dividend" of "induced easy-to-take, or compelled decisions resulting in additional investment and output". However, the *sequence of development via excess capacity of SOC is what Hirschman calls "self-propelling" in the sense that it is more continuous and smooth.* The second path *i.e.*, 'development via shortage of *SOC* lacks this attribute since it may take some time for the political pressure to be generated so that the adjustment in *SOC* is delayed. And thus the *DPA* cost of producing an amount of output is pushed up. In Hirschman's terminology, the 'development via excess *SOC* is basically a permissive sequence, while the development via shortage of *SOC* is essentially a *compulsive sequence.*

Having demonstrated the virtues of strategic imbalances, we are left with the problem of discovering what kind of imbalance is likely to be most effective. Any particular investment project may have both *"forward linkage"* (that is, it may encourage investment in subsequent stages of production) and *"backward linkage"* (that is, it may encourage investment in earlier stages of production). The task is to find the *projects with greatest "total linkage."* The projects, with the greatest total linkage, will vary from country to country and from time to time and can be discovered only by empirical studies of the "input-output tables."

In determining the sequence of projects, planning authorities should also give attention to the alteration of *"pressure-creating" and "pressure-relieving"* investments. In countries with vigorously expanding private enterprise sectors, the government's function can be largely limited to "pressure-relieving." As private investment takes place, shortages and bottlenecks will appear in transport, public utilities, education, and other activities traditionally assigned (in whole or in part) to public enterprises in such societies. Government ought not to feel "restless and slighted" when confined to this "induced role."

Where expansion through private investment is not assured, the government's role must be more active. For example, it might build an iron and steel plant. "It is interesting to note," says Hirschman, *"that the industry with the highest combined linkage score is iron and steel. Perhaps the underdeveloped countries are not foolish and exclusively prestige-motivated in attributing prime importance to this industry, because of the high total linkage effects of iron and steel industry."* The building of it by the government will lead to a spurt of investment and production in a variety of fields both in the stages before and after this industry. In this way, it accelerates economic growth. The investment in iron and steel industry will reveal deficiencies in the preceding and succeeding sectors of industry that the government must fill up. To remove these deficiencies and obstacles, further investment will be stimulated. When these deficiencies are filled up, further private investment will take place, and so the process of growth goes on.

The foregoing discussion leads us to the conclusion that the balanced growth doctrine is neither attainable nor desirable. On the other hand, for rapid economic development the underdeveloped countries should rely largely on **judiciously-planned unbalanced growth.** In fact, under Mahalanobis strategy of development, India has been following this course.

A Critique of Unbalanced Growth Strategy

The strategy of unbalanced growth has come in for severe criticism. First, it has been pointed out that unbalanced growth strategy is based on wrong assumption that only factor constraining economic growth is the scarcity of decision making ability in respect of investment. According to it all that is needed for accelerating growth in less developed countries is to provide inducements and incentives to private enterprise to undertake investment projects. Once this is done, supply of financial resources will adequately flow into investment projects. This is not a realistic assumption to make in the context of the developing economies. In the developing countries supply of financial resources is scarce due to low rate of saving and this hampers economic growth. Hirschman paid little attention to overcome this bottleneck to accelerate growth. Thus, not only the supply of physical resources are limited but also the availability of financial resources for funding the developmental projects are scarce.

Hirschman's unbalanced growth strategy has also been criticised on the ground that it will generate inflationary pressures in the economy. Whether more investment is undertaken in social overhead capital (*SOC*) or directly productive activities (*DPA*) incomes of the people will rise which will lead to the increased demand for consumer goods, especially foodgrains. If sufficient investment in agriculture and other consumer goods is not made, it will cause rise in prices as was actually witnessed in India during the second and third five year plans.

Thirdly, it has been pointed out that in case response from private enterprises to the inducements and pressures created by unbalanced growth strategy is not adequate, imbalances will be created in the economy without causing expansion in the other linked sectors *resulting in excess capacity in some industries or sectors*. This excess capacity represents waste of resources.

Lastly, it has been pointed out by Paul Streeten that unbalanced growth strategy *neglects the possibility of resistances* for adjustment to imbalances created by the unbalanced growth strategy. These resistances to growth may occur in a variety of forms. There may come to existence monopolies which have vested interests in restricting expansion in output. In the background of imbalances and shortages private enterprises which are interested in making quick profits will be more willing to raise prices of products rather than expanding their quantities. As Paul Streeten emphasises *"The theory of unbalanced growth concentrates on stimuli to expansion and tends to neglect or minimise resistances caused by unbalanced growth."*[8]

We however conclude that despite some shortcomings in the unbalanced growth strategy, laying stress on the decision-making ability for accelerating economic growth and on the need for building up social overhead capital, Hirschman has made a valuable contribution to development economics.

8. Paul Streeten, "Balanced and Unbalanced Growth", printed in G. Meier, *Leading Issues in Economic Development*, Oxford University Press, 1995.

CHAPTER 51

LEWIS MODEL OF DEVELOPMENT WITH UNLIMITED SUPPLY OF LABOUR

An eminent development economist, Arthur Lewis, put forward his model of "*Economic Development with Unlimited Supplies of Labour*"[1] which envisages the capital accumulation in the modern industrial sector so as to draw labour from the subsistence agricultural sector. Lewis model has been somewhat modified and extended by Fei and Ranis[2] but the essence of the two models is the same. Both the models (that is, one by Lewis and the other modified one by Fei-Ranis) assume the existence of surplus labour in the economy, the main component of which is the enormous disguised unemployment in agriculture. Further, they visualise '*dual economic structure*' with manufacturing, mines and plantations representing the modern sector, the salient features of which are the use of reproducible capital, production for market and for the profit, employing labour on wage-payment basis and modern methods of industrial organisation. On the other hand, agriculture represents the subsistence or traditional sector using non-reproducible land on self-employment basis and producing mainly for self-consumption with inferior techniques of production and containing surplus labour in the form of disguised unemployment. As a result, the productivity or output per head in the modern sector is much higher than that in agriculture. Though the marginal productivity in agriculture over a wide range is taken to be zero, the average productivity is assumed to be positive and equal to the bare subsistence level.

Lewis Model of Development with Surplus Labour

In the labour-surplus models of Lewis and Fei-Ranis, the wage rate in the modern industrial sector is determined by the average productivity in the agriculture. To this average productivity is added a margin (Lewis fixes this margin at 30%) which is required for furnishing an incentive for labourers to transfer themselves from the countryside to the urban industries as well as for meeting the higher cost of urban living. In this setting, the model shows how the expansion in the industrial investment and production or, in other words, capital accumulation outside agriculture will generate sufficient employment opportunities so as to absorb all the surplus labour from agriculture and elsewhere.

The process of expansion and capital accumulation in the modern sector and the absorption of labour by it is explained by Fig. 51.1. *OS* represents the real wages which a worker would be getting in the subsistence sector, that is, *OS* is the average product per worker in the subsistence sector. *OW* is the wage rate fixed in the modern sector which is greater than *OS* (*i.e.*, average product in agriculture) by 30%. So long as surplus labour exists in the economy any amount of labour will be available to the modern sector at the given wage rate *OW*, which will remain constant. With a given initial amount of industrial capital, the demand for labour is given by the marginal productivity curve MP_1. On the basis of the principle of profit maximisation, at the wage rate *OW*, the modern sector will employ OL_1 labour at which marginal product of labour equals the given wage rate *OW*. With this the total share of labour, *i.e.*, wage in the modern sector, will be OWQ_1L_1 and WQ_1D will be the capitalists' surplus. Now, Lewis assumes that all wages are consumed and all profits saved and invested.

1. Arthur Lewis. "Economic Development with Unlimited Supplies of Labour," *Manchester School*, 22, May 1954.
2. C.H. Fei and Gustav Ranis, *Development of the Labour Surplus Economy*, Irwin, 1964.

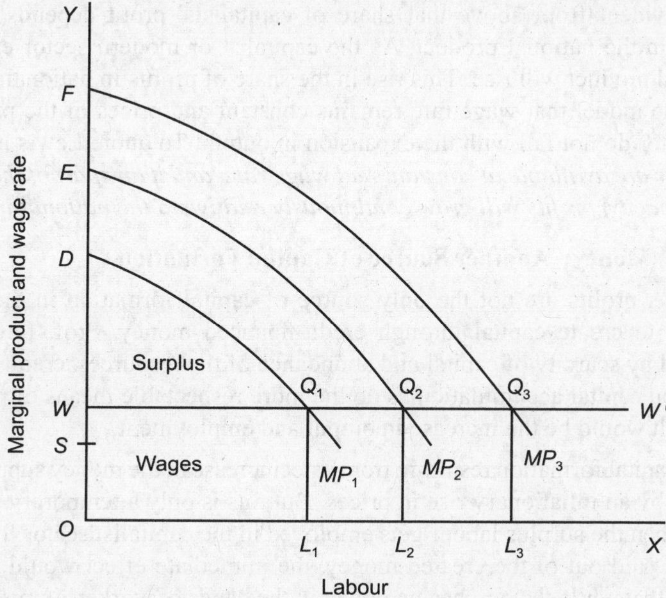

Fig. 51.1. *Capital expansion and growth in labour employment*

Reinvestment of Capitalist Surplus (Profits)

When the capitalists will reinvest their profits for setting up new factories or expanding the old ones, the stock of capital assets in the modern sector will increase. As a result of the increase in the stock of industrial capital, the demand for labour or magrinal productivity curve of labour will shift outward, for instance, from MP_1 to MP_2 in our diagram. With MP_2 as the new demand curve for labour and the wage rate remaining constant at OW, amount of labour OL_2 will be employed in the modern sector. In this new equilibrium situation profit or surplus accruing to the capitalist class will rise to WQ_2E which is larger than the previous WQ_1D. The new surplus or profits of WQ_2E will be further invested with the result that capital stock will increase and the demand or marginal productivity curve for labour will further shift upward, say to MP_3 position. When the demand curve for labour is MP_3 employment of labour will rise to OL_3. In this way, the profits earned will go on being reinvested and the expansion of the modern sector will go on absorbing surplus labour from the subsistence sector until all the labour surplus is fully absorbed in productive employment.

It is worth mentioning that in Lewis model, the rate of accumulation of industrial capital and, therefore, the absorption of surplus labour depends upon the distribution of income. *With the aid of classical assumption that all wages are consumed and all profits saved, Lewis shows that the share of profits and therefore rate of saving and investment will rise continuously in the modern sector and capital will continue to be expanded until all the surplus labour has been absorbed.* Rising share of profits serves as an incentive to reinvest them in building industrial *capacity as well as a source of savings to finance it.*

Profit as the Main Source of Capital Formation

It is clear from the above analysis of Lewis model with unlimited supply of labour that profits constitute the main source of capital formation. The greater the share of profits in national income, the greater the rate of savings and capital accumulation. Thus with the expansion of the modern or capitalist sector, the rate of saving and investment as percentage of national income will continuously rise. As a result, rate of capital accumulation will also increase relative to national income. It is of course assumed that all profits or a greater part of the profits is saved and automatically invested.

It is also evident from above that share of capitalists' profit depends on the share of the capitalist sector in the national product. As the capitalist or modern sector expands, the share of profits in national product will rise. This rise in the share of profits in national product is due to the assumption of the model that wage rate remains constant and prices of the products produced by the capitalist sector do not fall with the expansion in output. To quote Lewis himself, "*If unlimited supplies of labour are available at constant real wage rate, and if any part of the profits is reinvested in productive capacity, profits will grow continuously relative to the national income.*"[3]

Credit-financed Money: Another Source of Capital Formation

Furthermore, profits are not the only source of capital formation in the Lewis model. It is equally possible to create capital through credit-financed money. Prof. Lewis holds that in an economy marked by scarcity of capital and abundance of the resources, credit creation would have the same effect on capital accumulation as do the more respectable means of profits. In both cases, the ultimate result would be the increase in output and employment.

However, capital formation resulting from a net increase in the money supply would necessarily be accompanied by an inflationary rise in prices. But this is only a temporary phase – a short-lived phenomenon. When the surplus labour gets employed in the capitalist sector for purposes of capital formation and is paid out of the created money, the immediate effect would be the rise in prices. What happens is that while the purchasing power in the hands of workers immediately increases, the output of the consumer goods remains constant for a time. However, the moment the newly formed capital created by credit money is put to use, the output of consumption goods would also rise. Therefore, after a time-lag, the output of consumer goods catches up with the increased purchasing power, and thus the prices start taking the downturn.

Besides, the inflationary process would also be made to die away through another mechanism that operates simultaneously. With the monetary expansion, not only the output and employment rise, but also the profits. As the profits increase relative to national income, so will do the savings. Therefore, the volume of investment being financed out of created money would go on diminishing with the increase in voluntary savings. In the ultimate analysis, the seeds of inflation would be completely killed when the voluntary savings catch up with the 'inflated level of investment'. Once the equilibrium is restored, new investments, made thereafter, could be financed without having to fall back upon bank credit. With the savings growing into equilibrium with investment, the rise in prices is bound eventually to peter out.

Introducing Technical Progress. What about the role of technical progress in the type of economic expansion proposed in the Lewis model? Prof. Lewis contends that for the purposes of his analysis, the growth of technical knowledge and the growth of productive capital can be taken to work in the same direction – *i.e.*, to increase profits and wage employment. He argues that to be able to apply new technical knowledge, we ought to have new investments. If the new technical knowledge is capital-saving, it could be regarded as equivalent to an increment in capital. And if it happens to be labour-saving, it could be regarded as equivalent to an increment in the marginal productivity of labour. In effect, therefore, both cases would produce the same result—an outward shift in the marginal productivity schedule. As such, *Prof. Lewis in his model takes the growth of knowledge and the growth of productive capital to be essentially a 'single phenomenon'*.

However, the growth of technical knowledge in the subsistence sector is going to be a different kettle of fish. Its effect would be to raise the level of wages. As such, the capitalists' surplus would be reduced. It is, thus, that the capitalists have a profound and direct interest in holding down the productivity of labour in the subsistence sector.

Tapering off of the Expansionary Process

Now, if there is no hitch in the process of expansion and things go on smoothly, the capitalist

3. Arthur Lewis, *op. cit.*

sector will continue to expand till it has completely absorbed the surplus labour. Then comes the stage where capital accumulation has matched the excess supply of labour. Thereafter, the supply of labour ceases to be perfectly elastic. As a result, *real wages, instead of remaining constant, start rising. Also, the share of profits in the national income ceases to show any further rise*. And, thus, investment rather than growing relative to the national income comes almost to a standstill.

However, the expansion process may come to a halt much before the surplus labour has completely been absorbed. Prof. Lewis suggests three ways whereby the expansionary process might be halted before its natural conclusion. These are:

(*a*) The expansion of the capitalist sector may be rapid enough, so that the absolute population in the subsistence sector is greatly reduced, without, of course, affecting the total product. As a result, the average productivity of labour in the subsistence sector will rise. Consequently, the level of subsistence earnings (*S*) and the capitalist wage (*W*) are raised thereby reducing the volume of capitalists' surplus. However, *this shall not be if the growth of population happens to be greater than the rate of growth of the capitalist sector*.

(*b*) Technical progress may occur in the subsistence sector and may, therefore, raise the productivity there. This will, in turn, be reflected in a rise in the level of subsistence earnings (*S*) and the capitalist wage (*W*). The result, again, would be the squeezing of the capitalist surplus.

(*c*) *The terms of trade may turn against the capitalist sector due to a rise in the prices of raw materials and food*. This is very likely to happen, especially where the modern capitalist sector produces no food products. Now, as the capitalist sector expands, the demand for food will increase. As such, the food prices must rise in terms of the prices of the products of capitalist sector. In other words, the terms of trade would become unfavourable to the capitalist sector. But if the real income of the worker is to be kept constant, the capitalist wage (*W*) must rise (and so also the subsistence earnings (*S*)). The effect is that the capitalists' surplus is reduced.

However, if none of these methods operate or do so only mildly, the capital accumulation will continue to take place. And the modern capitalist sector will continue expanding till the whole surplus labour is absorbed. In any case, the process of expansion cannot proceed indefinitely as wage rate in the modern industrial sector will rise, profits (or capitalist surplus) will decline and as a result capital accumulation will decline and ultimately bringing the growth process to a halt.

A Critical Appraisal of Lewis Model

The validity and usefulness of the labour-surplus model of Lewis for developing countries like India depend of course on the extent to which their underlying assumptions are valid for the economies in question. We are here not interested in validity of all the assumptions, explicitly or implicitly, made in this model. In our view the basic premise of this model is wrong and that makes it unrealistic and irrelevant for framing a suitable development strategy to solve the problem of surplus labour and unemployment. The basic premise of the model is that industrial growth can generate adequate employment opportunities so as to draw away all the surplus labour from agriculture in an over populated developing country like India where population is currently increasing at the annual rate of around 1.6 per cent. This premise has been proved to be a myth in the light of generation of little employment opportunities in the organised industrial sector during over sixty years of economic development in India, Latin American and African countries.

For instance, in the 30 years (1951-81) of industrial development in India during which fairly good rates of industrial production had been achieved, the organised industrial employment increased by only 3 million which was too meagre to make any significant impact on the urban unemployment situation, far from providing a solution to the labour-surplus problem in agriculture.

Thus, the generation of adequate employment opportunities and as a result the absorption of surplus labour from agriculture in the expanding industrial sector has not proceeded as predicted by the Lewis model.

It may be pointed out here that migration of some workers from the rural to the urban areas in India has occurred as shown by the slight increase in the degree of urbanisation noticed in the various censuses but these immigrants to the urban areas have not been absorbed into the modern high-productivity employment, as envisaged by Lewis and Fei-Ranis. This is evident from the statistical data about meagre increase in employment in the organised sector. These immigrants to the urban areas have been mainly employed in petty trade, domestic service and casual work in which the disguised unemployment and poverty exist as acutely as in agriculture. Thus, as things stand, the traditional sector of the economy is simply moving from the countryside into the cities in apparent contrast to the Lewis model.

Lewis Model Neglects the Importance of Labour Absorption in Agriculture. A grave weakness of the models of Lewis and Fei-Ranis is that they have ignored the generation of productive employment in agriculture. No doubt, Lewis in his later writings and Fei-Ranis in their modified and extended version of Lewis model have envisaged an important role for agricultural development so as to sustain industrial growth and capital accumulation. But they visualise such an agricultural development strategy that will *release labour force from agriculture* rather than absorbing them in agriculture. Thus to quote Fei and Ranis: "*In such a dualistic setting the heart of the development problem lies in the gradual shifting of the economy's centre of gravity from the agricultural to the industrial sector through labour reallocation.* In this process each sector is called upon to perform a special role: productivity in the agricultural sector must rise sufficiently so that *smaller fraction of the total population can support the entire economy* with food and raw materials, thus enabling agricultural workers to be released; simultaneously, the industrial sector must expand sufficiently to provide employment opportunities for the released workers ... *labour reallocation must be rapid enough to swamp massive population increases if the economy's centre of gravity is to be shifted over time.*"[4]

We have shown above that employment potential of organised industrial sector is so little that labour reallocation between agriculture and industry and "smaller fraction of the total population being employed in agriculture" is just not possible in labour-surplus developing countries like India. Indeed, a good amount of employment opportunities can be generated in agriculture itself by capital accumulation in agriculture, adopting proper agricultural technologies and making appropriate institutional reforms in the pattern of land ownership.

Even about the African countries most of which do not suffer from the Malthusian problem of overpopulation but are currently faced with acute urban unemployment (especially of what is known as "*Unemployment of School Leavers*" majority of which have migrated from the villages to the urban areas) the expert opinion has veered round to the view of seeking solution of labour-surplus problem within agriculture. Thus Sara S. Berry remarks about the African experience: "*Most students of the problem of rising African urban unemployment agree that the solution to the problem lies in raising incomes and employment opportunities in agriculture so as to ensure a new market equilibrium with more people productively employed in agriculture.*"[5]

Assumption of Adequate Labour-absorptive Capacity of the Modern Industrial Sector. Another related shortcoming of development models of Lewis, Fei and Ranis is their assumption that the growth of industrial employment (in absolute amount) will be greater than the growth in labour

4. Fei and Ranis, *op. cit.*
5. Sara S. Berry, "Economic Development with Surplus Labour: Further Complications Suggested by Contemporary African Experience," *Oxford Economic Papers*, June 1970.

force (which in India at present is of the order of about 12 million people per year). Because only then the organised industrial sector can absorb surplus labour from agriculture. The employment potential of industrial sector is so little that far from withdrawing labour currently employed in agriculture, it does not seem to be possible for the organised industries and services, on the basis of existing capital-intensive technologies, even to absorb the new entrants to the labour force.

An important drawback of the Lewis model is that it has neglected the importance of agricultural growth in sustaining capital formation in the modern industrial sector. When as a result of the expansion of capitalist modern sector, transfer of labour from agriculture to industry takes place, the demand for foodgrains will rise. If the output of foodgrains does not increase through agricultural development to meet the additional demand for foodgrains, their prices will rise. With the rise in prices of foodgrains wages of industrial labour will increase. Rise in wages will lower the share of profits in the industrial product which in turn will slow down or even choke off the process of capital accumulation and economic development. Thus, if no allowance is made for agricultural growth, the expansion of modern sector and capital accumulation is bound to be halted. Thus, neglect of agriculture in the development strategy pursued in India since the Second Plan virtually resulted in stagnation in the industrial sector, during the period 1966-1979.

The Assumption of Constant Real Wage Rate in the Modern Sector. The assumption of *constant* real wages to be paid by the urban industrial sector until the entire labour surplus in agriculture has been drawn away by the expanding industrial sector is quite unrealistic. The actual experience has revealed a striking feature that in the urban labour markets where trade unions play a crucial role in wage determination, there has been a tendency for the urban wages to rise substantially over time, both in absolute terms and relative to average real wages even in the presence of rising levels of urban open unemployment. The rise in wages, as explained above, seriously impairs the development process of the modern sector.

It Neglects the Labour-saving Nature of Technological Progress. A serious lacuna of the Lewis model from the viewpoint of employment creation is its neglect of the labour-saving nature of technological progress. It is assumed in the model, though implicitly, that rate of employment creation and therefore of labour transfer from agriculture to the modern urban sector will not be proportional to the rate of capital accumulation in the industrial sector. Accordingly, the greater the rate of growth of capital formation in the modern sector, the greater the creation of employment opportunities in it. *But if capital accumulation is accomplished by labour-saving technological change, that is, if the profit made by the capitalists are reinvested in more mechanised labour-saving capital equipment rather than in existing types of capital, then employment in the industrial sector may not increase at all.*

The Lewis model has been reproduced in Fig. 51.2. with a modification that profits made are reinvested in labour-saving capital equipment due to the technological change that has taken place. As a result of this, marginal productivity curve does not shift uniformly outward but crosses the original marginal productivity curve from above. It is evident from Fig. 51.2 that with the constant wage rate *OW*, the employment of labour does not increase even though marginal productivity curve has shifted to the right.

It will be observed from Fig. 51.2 that though employment of labour and total wage (*OWQL*) have remained the same, the total output has increased substantially, the area *OEQL* is much greater than the area *ODQL*. This illustration points to the fact that while the industrial output and profits of the capitalist class can increase, the employment and incomes of labour class remain unchanged. Although GNP has increased, labouring class has not received any benefit from it. It is not just theoretical illustration but has been actually borne out by the experience of industrial development of several developing countries. This experience shows that while industrial output has significantly increased, employment has lagged far behind.

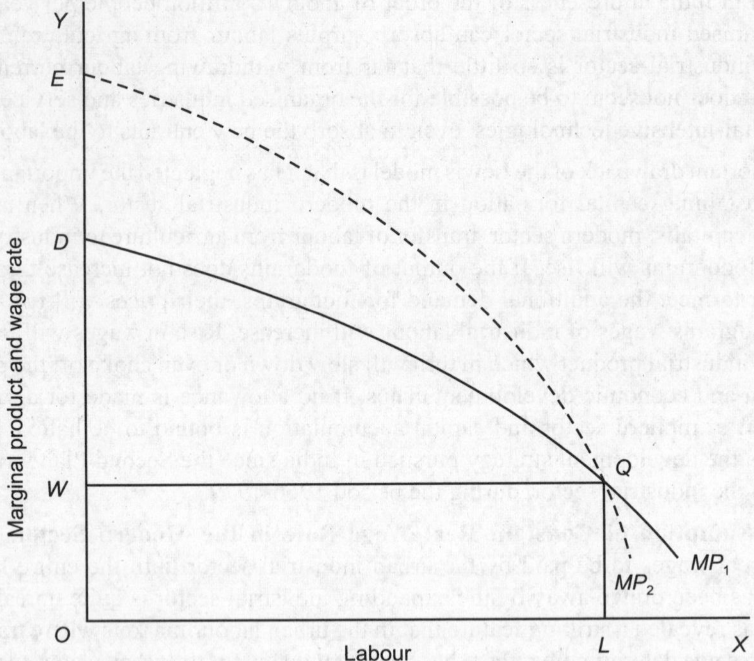

Fig. 51.2. *Capital expansion with labour-saving technological change*

Lewis Model Ignores the Problem of Aggregate Demand. A serious factor which can slow down or even halt the expansionary process in the Lewis model is the problem of deficiency of aggregate demand. Lewis assumes, though implicitly, that no matter how much is produced by the capitalist or modern sector, it will find a market. Either the whole increment in output will be demanded by the people in the modern sector itself or it will be exported. But to think that entire expansion in output will be disposed of in this manner is not valid. This is because a good part of the demand for industrial products comes from the agricultural sector. If agricultural productivity and therefore incomes of the farming population do not increase, the problem of shortage of aggregate demand will emerge which will choke off the growth process in the capitalist industrial sector. However, once an allowance is made for the increase in agricultural productivity through a priority to agricultural development, the basic foundations of the Lewis model crumble down. This is because a rise in agricultural productivity in Lewis model will mean a rise in wage rate in the modern capitalist sector. The rise in the wage rate will reduce the captalists' profits which in turn will bring about a premature halting of the expansionary process.

Conclusion

Despite several limitations and drawbacks, the Lewis model retains a high degree of analytical value. It clearly points out the role of capital accumulation in raising the level of output and employment in labour-surplus developing countries. The model makes a systematic and penetrating analysis of the growth problem of dual economies and brings out some of crucial importance of such factors as profits and wages rates in the modern sector for determinating the rate of capital accumulation and economic growth. It underlines the importance of inter-sectoral relationship (*i.e.*, the relationship between agriculture and the modern industrial sector) in the growth process of a dual economy.

CHAPTER 52

CHOICE OF TECHNIQUES IN DEVELOPING COUNTRIES

A choice between alternative techniques of production is a major problem in the planning for developing countries. This is because a particular choice of techniques of production affects not only the magnitude of employment but also the rate of economic growth. Several alternative techniques of production are available to produce a commodity and these differ with regard to the amount of capital being used with a unit of labour for production. In other words, the various techniques differ with regard to capital intensity which is generally measured by the magnitude of capital-labour (*K/L*) ratio. Thus, the higher the capital intensity, the more quantity of capital as compared to labour will be used to produce a given level of output. On the other hand the lower the capital intensity, more employment for labour will be created. Thus, the lower capital intensity implies the higher labour intensity. Therefore, in labour-surplus developing countries it is generally believed that for rapid growth of employment labour-intensive techniques (*i.e.*, less capital-intensive techniques) should be preferred.

Factor-price Ratio and Choice of Technique

In economic theory based on the perfect competition model where the factor prices are given and constant for a firm and factor proportions are variable, the choice of a technique or capital-labour combination is easily made by a firm which aims to minimise cost for a given level of output through equating relative prices of factors to their relative marginal products. Consider Fig. 52.1, where curve *q* is an isoquant which represents a given level of output.

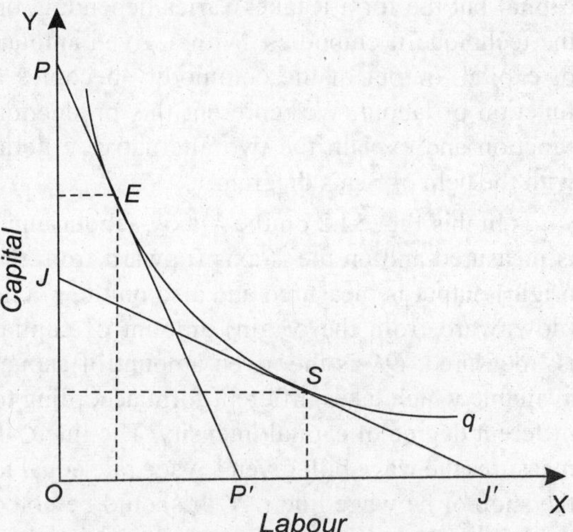

Fig. 52.1. *Maximum Reinvestible Surplus and Choice of Techniques*

The slope of line *PP'* measures the relative prices of capital and labour. The firm will be minimizing cost by choosing capital intensity represented by the point *E* at isoquant *q*. At *E* the price line *PP'* is tangent to the isoquant *q*. The slope of the isoquant *q* at point *E* measures the ratio of marginal products of capital and labour which is equal to the slope of the price line representing factor price ratio.

However, if a country has abundant labour and shortage of capital, in the absence of any price distortions, labour will be relatively cheaper than capital and price line representing this will be less steep, say it is *JJ '*. As will be seen from Fig. 52.1, the new price line *JJ '* is tangent to the isoquant *q* at point *S*. Accordingly, the capital intensity chosen by a cost minimising firm will be given by the point *S*. This technique of production is more labour-intensive than the technique corresponding to point *E*. However, it is worth mentioning that in a labour-surplus economy, labour-intensive technique may be chosen by a cost-minimising firm because of the existence of distortions of factor prices from their true scarcity values.

Now, the question arises why in labour-surplus developing countries factor prices are distorted? The pressure of trade unions of labour for higher wages, minimum wage legislation fixing wage at higher level make labour costlier on the one hand and lower interest rates fixed either by the government if interest rates are administered by it or by cheap credit policy followed by the Central Bank make capital relatively cheaper. Further, liberal investment allowance (*i.e.,* tax break) on capital investment also lowers the cost of capital and encourages the use of more capital. Besides, *overvalued exchange rate* makes the import of capital equipment cheaper.

Therefore, in order to ensure the use of labour-intensive or less capital-intensive techniques in order to maximise employment of labour it has been suggested that *factor prices must be set right*, that is, labour should be made cheaper and capital relatively more expensive.

Choice of Techniques: Maximum Reinvestible Surplus Criterion

In the planning for developing countries, the choice of techniques is not to be decided from the point of view of private profit maximization of private cost minimisation. In it the choice of capital intensity has to be decided keeping in view the problem of mass unemployment and the need for rapid economic growth to raise levels of living of the people. The problem is made difficult because the achievements of the twin objectives of reducing unemployment and promoting rapid economic growth in the choice of techniques clash with each other at least in the short run. For the optimum choice of technique of production or capital intensity two alternative criteria have been put forward. They are the *maximum output and maximum reinvestible surplus criteria.*

To explain these criteria let us take a single product model in which two factors, capital and labour, are used to produce a commodity. We further assume that there is a given amount of capital but the form it takes varies depending on the technique it embodies. With a given amount of capital, output of the commodity becomes a function of labour. We represent this production function and explain the two alternative criteria with the help of Sen's diagram.[1]

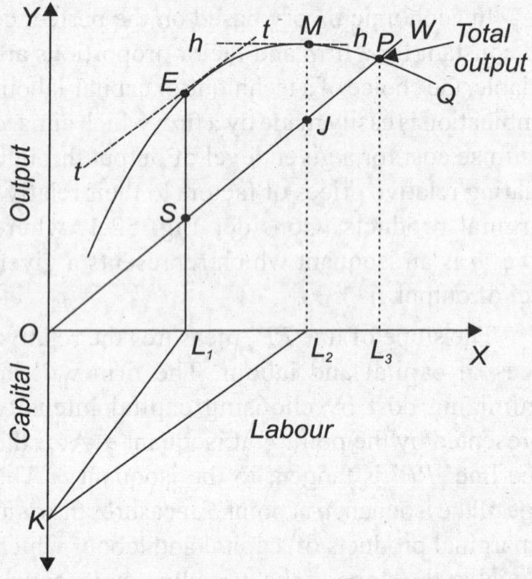

Fig. 52.2. *Maximum Reinvestible Surplus and Choice of Technique*

In this Fig. 52.2 on the *X*-axis, labour input is measured and on the *Y*-axis (upward from the origin) output is measured and also on the *Y*-axis (downward from the origin) amount of capital is measured. *OK* is the given amount of capital available which takes different form according to different degree of capital intensity. The line *OW* measures the wage bill, given a wage rate equal to the shop of ḥe wage line *OW*. It should be noted that with increase in labour employed, given the stock of capital *OK*, capital-labour ratio falls (or labour-capital ratio increases). Thus, as we use more labour input, capital intensity will vary along the line *OX*. It should be further noted that, given the wage rate, as more labour is used total wage bill will be increasing. Thus with OL_1 labour employed L_1S is the wage bill and with OL_2 labour employed L_2J is the total wage bill. With capital stock equal to *OK*, output is a function of labour which is given by the production function curve *Q*. In drawing this production function we have

1. Amartya Sen, *Choice of I Techniques, Oxford University Press, 1962.*

assumed that as more labour is used with a given capital stock there occurs diminishing returns to labour and ultimately with increasing labour intensity, total output declines, so that a certain labour-capital ratio corresponds to maximum output.

We are now in a position to explain the choice of technique on the basis of maximum output and maximum reinvestible surplus criteria. If *the planner wants to choose capital intensity (i.e., technique of production) to maximise output,* then he will choose point L_2 where output is maximum (at OL_2 marginal product of labour is equal to zero). With the given capital stock *OK*, capital intensity chosen will be equal to the slope of the line L_2K, *i.e.,* OK/OL_2. With this choice of capital intensity, OL_2 labour is employed. If maximisation of employment in the present is desired , then, obviously, capital-intensity OK/OL_2 is the optimum choice.

However, *maximisation of present employment may not yield a satisfactory rate of growth.* Maximum reinvestible criterion was put forward by Galenson and Leibenstein. According to them, the surplus of output over total wage bill at OL_2 level of employment is *MJ* which is not the largest. If the maximum of surplus over wage bill is required, then the capital intensity (or, in other words, employment of labour with the given capital stock) at which the wage rate equals the marginal product of labour should be chosen.[2] From Fig. 52.2 it will be noticed that at OL_1 use of labour input or capital-intensity represented by L_1, the marginal product of labour (measured by the slope of the production function curve *Q* at point *E*) equals the wage rate (as measured by the wage line *OW*). The surplus of output over the wage bill at capital intensity at L_1 (which is equal to OK/OL_1) is *ES* which is the largest under the given circumstances. At L_1, the capital intensity is higher but employment smaller than with capital intensity at L_2 Thus, the largest surplus *ES* is obtained with a higher capital intensity and lower labour employment at the present.

If it is assumed, as is done by the exponents of maximum reinvestible surplus criterion, that the whole surplus is reinvested and the whole wages are consumed, then this larger surplus on being reinvested would yield a higher rate of economic growth. On the other hand, with lower capital intensity at L_2 though the level of present employment is larger, surplus *MJ* is smaller which when reinvested would yield a lower rate of growth. *With a higher capital intensity and higher rate of growth, the rate of growth of employment will be higher, though the level of present employment will be less. On the contrary, with a lower capital intensity the surplus is smaller and consequently rate of growth of output and employment will be smaller, though the present level of employment will be large.*

Thus the choice of capital intensity implies the choice between the higher levels of present employment and output on the one hand and the higher rates of growth of employment and output on the other. Thus, it is argued *if you are interested in maximising the current level of employment (and production), choose a lower capital-intensive technique such as the one represented by L_2. On the other hand, if you want a higher rate of growth of employment and output, choose a higher capital-intensive technique such as the one represented by L_1. Thus we find that there is a conflict between maximising present employment (or consumption) and maximising employment or consumption at some future date.*

Galenson and Leibenstein's maximum reinvestible surplus criterion has been criticised for its assumption that all profits are saved and reinvested and all wages are consumed. It has been pointed out that capitalists consume a lot of profits earned by them as they indulge in conspicuous consumption. Further, they point out that all wages may not be consumed and workers do save from their wages. In case of State enterprises it has been pointed out that due to their inefficiency, they may not generate much surplus and if they have some surplus they may be spent by the government for their current revenue expenditure. Furthermore, in case of multinational corporate companies (MNCs), operating in developing countries, they are likely to repatriate their profits (*i.e.,* surplus) to their home countries rather than reinvesting them in the developing countries. A more important

2. W. Galenson and H. Leibenstein, "Investment Criteria, Productivity and Economic Development," *Quarterly Journal of Economics,* Vol. 69, 1955.

critique of maximum reinvestment criterion is that it is best to raise necessary savings or surplus through the use of appropriate fiscal and monetary measures rather than using choice of technique for the purpose. Further, it is unrealistic on the part of Galenson and Leibenstein to assume that the wage rate is constant whatever technique of production is chosen. In this connection it may be noted that MNCs using capital-intensive techniques in their production process are likely to pay higher wages to their workers which will reduce the reinvestible surplus and promote higher consumption on the part of their workers.

Lastly, Galenson and Leibenstein ignore the role of other factors such as the availability of raw materials, skilled labour and foreign exchange to import capital goods required by the investment plan.

Amartya Sen's Time Series Criterion and Choice of Technique

To resolve the above conflict or dilemma, Prof. Sen in his famous work '*Choice of Techniques*' has proposed what is called *Time Series Criterion*. According to it, choice of technique depends upon the time horizon of output or employment generation, the time preference and social welfare function. To make an optimum choice of technique, alternative time series of employment and output following the adoption of different techniques have to be obtained. This is illustrated in Fig. 52.3. Let the vertical axis measure labour-employment and output of consumer goods and the horizontal axis measure time. The more steep curve *AK* represents growth of output or employment over time with using capital-intensive technique *K* whereas the less steep curve *BL* shows the growth of output and employment over time with adopting labour-intensive' technique of production *L*.

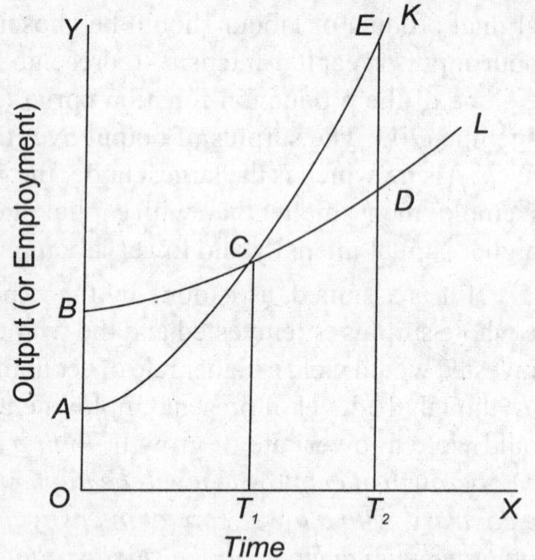

Fig. 52.3. *Amartya Sen's Criterion of Choice of Technique*

Till time T_1, labour-intensive technique generates more output and employment as compared with the capital-intensive technique. But, since the growth of output or employment is greater in case of capital-intensive technique due to its yielding larger surplus and reinvestment, beyond time T_1 the output or employment is greater in case of capital-intensive technique *K*. On account of higher growth of output and employment the capital-intensive technique fully compensates for the initial loss of output (or employment) by the time T_2. Note that the area *CED* equals the area *ABC* in Fig. 52.3). It is evident from Fig. 52.3 that after time T_2, the society would have higher levels of output and employment by choosing capital-intensive technique. Thus though the choice of labour-intensive technique yields higher levels of output and employment *in the present* it yields lower rate of growth due to smaller reinvestible surplus. On the basis of this time series criterion, Amartya Sen argues that if social welfare function of a society is such that it is prepared to wait up to time T_2, (say thirty years) output and employment sacrificed at present could be fully compensated and after T_2, the society would enjoy higher levels of consumption and employment by choosing capital-intensive technique. However, if social welfare function is such that society prefers present output (consumption) and employment to future consumption and employment, then it may choose labour-intensive technique.

Many of the criticisms levelled against Galenson and Leibenstein's maximum reinvestible criterion also apply to Sen's time-criterion for choice of technique as it is also based on maximisation of surplus. As mentioned above, this criterion depends on a particular choice of technique to generate more saving or surplus for investment to bring about higher rate of growth of output and employment. As has been stated above, in his analysis Sen assumes that a higher proportion of profits is saved than that from wages (in the extreme case it is assumed that all profits are saved and all wages are consumed). The critics have pointed out that this is not true. In fact a lot of profits are consumed instead of being saved and further that employees do save a part of their wages or salaries.

Furthermore, it is assumed by Sen that wage rates do not depend on techniques. In fact, as with capital-intensive technique productivity of workers is higher, their wages will also be higher and this will lower the surplus made on using capital-intensive technique. Sen has also been criticised that saving or surplus for investment can be increased by the choice of capital-intensive technique alone. Rather than depending on choice of technique, in a planned economy saving can be increased by the government through adopting appropriate fiscal and monetary policies. In a later article Sen concedes, *"The total amount of income to be saved can be determined by the planner in any way he likes ... if this is true then the link snaps between the choice of technique and the proportion of income saved. Then technical choice can be made with the main purpose of maximising output."*[3]

Furthermore, as will be seen from Fig. 52.2 that if wage rate is lower, then the wage line will lie below the wage line *OW* and its slope will be less and as a result surplus maximising technique will lie to the right of L_1 and will be relatively labour-intensive and nearer to the maximum output point. Given the fact that in a labour.-surplus economy wage rate is lower, it follows that surplus-maximising technique will be relatively less capital-intensive *i.e.*, relatively labour-intensive. Even Sen in his later work writes, "The proportion that the techniques maximising output would tend to be relatively less capital-intensive in the underdeveloped economies compared with those in the advanced economies seems to be valid."[4]

Lastly, Sen has been criticised that in his analysis of choice of technique his emphasis is on raising saving and assumes that consumption makes no contribution to future, growth. It is now well recognised that level of consumption demand or growth in demand plays an important role for sustained growth of the economy.

CHOICE BETWEEN MAXIMISING EMPLOYMENT AND MAXIMISING PRODUCTION

When new projects for investment are planned, there is likely to be greater scope for varying the amount of labour employment. For example, a given amount of resources may be invested in either handlooms or automatic looms which employ quite different amounts of labour. It may, however, be noted that a given amount of new investment in labour-intensive and capital-intensive techniques may result in different levels of output; labour-intensive techniques may yield less output as compared with capital-intensive techniques. Thus, while the labour-intensive techniques are characterised by *higher labour-output ratios,* it is not necessary that all such techniques have *lower capital-output ratio*. For instance, it is worthwhile to note that Amartya Sen has shown that the labour-intensive technique of Ambar Charkha has higher capital-output ratios than the factory methods. Dhar and Lydall have also found that some labour-intensive small-scale industries have higher capital-output ratios. Hence, a *conflict between output and employment arises*.

However, the decision whether output should be sacrificed or employment will depend upon the social welfare function. Given the extent of unemployment and underemployment and the glaring disparities of income we feel that some output is worth sacrificing for more employment. Provision of employment is by far the most important way of raising the people above the poverty line and of ensuring widespread sharing in the fruits of economic development. In addition, employment gives individual a sense of participation in socially gainful activity and prevents the feeling of being not wanted which has a great demoralising effect. Indeed, as has been remarked by Barbra Ward, *"of all the evils worklessness is the worst."*

---

3. Amartya Sen, "Choice of Technology: A Critical Survey of a Class of Debates,", in G. M. Meier, *Leading Issues in Economic Development*, 1975.
4. Amartya Sen, *Ibid.*

It should however be noted that all *labour-intensive techniques do not have higher capital-output ratio* and therefore in their case conflict does not arise between maximizing employment and maximizing production. In fact, many labour-intensive techniques and small-scale industries, capital-output ratio is lower than the corresponding large-scale industries. Hence a case for encouraging such industries in developing countries like India.

Conflict between Maximising Current Employment and Maximising Rate of Growth of Production and Employment

Conflict which generally arises relates to the *maximisation of current employment and maximisation of rate of growth* of production and employment. As seen above, from Sen's analysis of choice of techniques, maximisation of output and employment in the current period will not lead to the higher rate of economic growth of both. output and employment. This question is of particular importance in respect of (*a*) allocation of resources for investment between capital goods and consumer goods and (*b*) the choice between labour-intensive and capital-intensive techniques. Maximising rate of growth of production now at the cost of present employment makes it possible to use extra output produced for generating more employment in the future. It is worth noting that more output (*i.e.*, more reinvestible surplus) is useful not only for its own sake but enables the planners to generate more employment opportunities in the future. Thus, a conflict (and therefore inter-temporal trade-off) arises between *more employment now and more employment tomorrow.*

Let us explain how conflict arises in the two cases mentioned above. Allocation of more resources to the production of investment goods by sacrificing some employment in the present (that is, if extra output is in the form of capital goods such as machines etc.) will enable us to provide employment to more men in the future. Similarly, if extra production is in the form of more wage goods, liquid capital as they are sometimes called, enables planners to create more jobs in the future as the availability of wage goods limits the opportunities for creation of employment opportunities.

On the contrary, raising employment now means sacrificing not only some production in the current period but also lowering the rate of growth of employment. Thus with the allocation of resources to the consumer goods industries ensuring more employment now, level of employment at some future date would be lower than would have been possible if more allocation of resources to capital goods industries in the present was preferred.

Turning to the conflict between current employment and the rate of growth of employment arising out of the choice of techniques of production, it has been shown by Sen and others that the choice of labour-intensive techniques, though maximizes employment in the current period, will reduce the share of profits (or investible surplus) as compared to wages. And the reduction in profits will adversely affect the rate of savings and investment and therefore the growth of output and employment. On the contrary, if capital-intensive techniques are chosen, they may yield less employment (and output) in the present but they will yield more surplus or profits as compared to wages.[5] Therefore, the choice of capital-intensive techniques will ensure higher rate of growth and therefore more employment in the future. Here, the conflict or trade-off involved does not arise between current output and current employment. Instead, there is inter-temporal trade-off, that is, more employment *now* or more employment *tomorrow*. It means that some more unemployment in the present may be tolerated for the sake of generating more investible surpluses (profits) so that higher growth rate is achieved and more employment opportunities are provided in the future.

5. Generally models concerning this aspect of development assume that all profits are saved and all wages are consumed. On the contrary, an ILO Report on Columbia found that "*notwithstanding the extreme degree of economic inequality that has prevailed in Columbia, total personal savings appear to be quite low*".

APPROPRIATE OR INTERMEDIATE TECHNOLOGY

As has been stated above, the growth of organised or large-scale industries with modem capital-intensive technology yields only meagre employment opportunities. Therefore, in order to create adequate employment opportunities through industrial growth, there is a need for developing and adopting an '*intermediate technology*' which requires less capital per workplace without loss of efficiency. Besides, to prevent excessive urbanisation and concentration and to arrest the tendency for mass migration into the urban areas, there is need for adopting a strategy f*or rural industrialisation*, based on small-scale manufacturing using intermediate technology dispersed throughout the countryside to increase rural employment and income. Dr. E.F. Schumacher, the advocate of the adoption of intermediate technology, rightly says, *"It is surely an astonishing error to assume that the technology developed in the West is necessarily appropriate to the developing countries. Granted that their technological backwardness is an important reason for their poverty: granted too, that their traditional methods of production, in their present condition of decay, lack essential viability: it by no means follows that the technology of the richest countries is necessarily suitable for the advancement of the poor.* It must never be forgotten that modern technology is the product of the countries which are 'long' in capital and 'short' in labour, and that its main purpose, abundantly demonstrated by the trend towards automation, is to substitute machines for men. How could this technology fit the conditions of countries which suffer from a surplus of labour and a shortage of machines?"[6] He further remarks *"that technology devised primarily for the purpose of saving labour should be inappropriate in a country troubled with a vast labour surplus could hardly be called surprising."*[7]

It is worth noting here that research in India, after making proper cost and price calculations, has already discovered intermediate technologies for producing some thirty types of such goods as agricultural implements, processed foods and consumer goods which are actually cheaper and more efficient than those produced by the advanced machinery.[8] It should be further noted that the evolving of intermediate technologies for various industries does not imply discovery of altogether new principles of science and engineering. What is required is the application of the basic principles of modern science and engineering to evolve the production techniques which conform to the factor endowments of the labour-surplus economies. These appropriate or intermediate technologies may be obtained by scaling down the advanced techniques by adapting them so as to make them more labour-intensive, or by scaling up handicrafts techniques with the introduction of new tools and simple machines and thus improving economic efficiency of these techniques while maintaining their labour intensity. In this process of adaptation product design itself will also have to be modified and standardised. Further, entirely new intermediate technologies have also to be evolved from the applications of the basic principles of modern science and engineering so as to suit the labour-surplus conditions of developing countries. For that, a good deal of research and development is required to be undertaken.

It may be noted again that Japan's remarkable adaptability has been reflected not only in agricultural techniques but also in the extent to which it has organised its new industries efficiently on a cottage industry basis. Here, as in agriculture, Japan's production methods are especially well-suited to the conditions of labour-abundant economies. Lastly, it may be noted that recently, four East Asian countries, namely, South Korea, Taiwan, Hong-Kong and Singapore which have achieved remarkably higher growth rates and are therefore called *Asian Tigers followed labour-intensive path of economic growth*. That is, they used relatively much more labour in their industrial development than India.

---

6. "Industrialisation through Intermediate Technology", a paper presented by Dr. Schumacher to the Cambridge Conference on Industrialisation in Developing Countries in September 1964. Published in Ronals Robbinon, *Developing the Third World: The Experience of the Nineteen-Sixties* (Cambridge University Press,19741), pp. 88-89.
7. *lbid.*, 97.
8. *Appropriate Technologies for India Industries,* Small Industry Extension Training Institute (Hyderabad, India,) *pp.* 11-12.

CHAPTER 53
EMPLOYMENT AND ECONOMIC GROWTH IN DEVELOPING COUNTRIES

Introduction

We have explained in detail in a previous chapter that most of the unemployment in underdeveloped countries is of a different nature from that in advanced and developed countries. A major part of unemployment in developed countries is of cyclical nature which is due to *deficiency of aggregate effective demand.* But most of the unemployment in underdeveloped countries is not cyclical. Instead, it is a long-term problem. The major cause of unemployment and underemployment in underdeveloped countries like India is *the deficiency of the stock of capital* in relation to the needs of the growing labour force. Therefore, increase in employment opportunities requires rapid capital formation.

A nation's stock of capital can be enlarged by increased investment which, in the absence of any unutilised resources, requires additional savings on the part of the community. The concern of the classical economists was to ensure that the rate of capital formation was kept sufficiently high so that employment opportunities were successively enlarged to absorb the additions to the working force of a country as a result of population growth. This is also the problem that developing countries like India are facing today. In recent times, the labour force in India has been growing at an annual rate of 2.2 per cent yet our *rate of investment expressed as a percentage of our stock of capital* has not been growing at a fast enough rate so as to keep pace with the growth of population. The country's ability to offer productive employment to the new entrants to the labour market has been severely limited. This manifests itself in two things—first, the prevalence of large-scale open unemployment in the urban areas as evidenced by the statistics of employment exchanges; second, it manifests itself in the form of open unemployment as well as disguised unemployment in agriculture.

Since employment opportunities in the non-agricultural sector have not been growing rapidly, the new entrants to the workforce are compelled to remain in agriculture and perpetuate the phenomenon of *disguised unemployment* which means that people are engaged in occupations where their marginal productivity is very low (if not zero or negative) and that a shift to alternative occupations will improve their marginal productivity and add to national income of the country.

The basic solution to the problem of this sort is the faster rate of capital formation so as to enlarge employment opportunities. For this purpose, every possible encouragement should be given to the growth of domestic savings and their productive utilisation in increasing the rate of investment. In developing countries investment incentives for the private sector are very low and the State can assist in the process of capital formation directly as well as indirectly. Through a fiscal policy which encourages savings and investment and a sound monetary policy it can do much to encourage investors. The State itself can participate in the process of capital formation by undertaking such developmental activities as building infrastructure which do not attract the private investors. Therefore, the State has got to assume a special role in speeding up the rate of economic development. The other line of attack has got to be on the rate of population growth. If population grows at a rapid rate, then to maintain the people even at their existing levels large amounts of capital are needed which could otherwise have been used to raise the amount of capital available per man and hence to raise the living standards at a faster rate. In a later section we critically examine at length the various strategies of employment generation to solve the unemployment problem in the developing countries.

FACTORS DETERMINING GROWTH OF EMPLOYMENT

Capital Accumulation and GDP Growth Rate. It is important to explain what factors determine growth of employment and therefore the emergence of unemployment. As explained above, unemployment in labour-surplus developing countries is due to shortage of cooperating factors such as capital stock (including land), relatively to labour force. In developing countries capital stock has not been increasing fast enough to keep pace with the growth of labour force. It is through rapid increase in investment or faster capital accumulation that the GDP growth can be accelerated.

According to ***Harrod-Domar model,*** the rate of growth of output depends upon the ratio of total investment to national income divided by the capital-output ratio, *i.e.*,

$$G = \frac{I}{v}$$

where

G = rate of growth of output,
I = rate of investment (as ratio of national income),
v = capital-output ratio.

As Harrod-Domar model assumes capital-output and capital-labour ratios to remain constant, increase in capital stock would imply increase in both output and employment. Therefore, Harrod-Domar model suggests that the rate of growth of both output and employment is determined by the growth of capital stock. Thus, according to this model, solution to the problem of unemployment lies in sufficient increase in the rate of investment or capital accumulation. Assuming a fixed capital-output ratio, Harrod-Domar model lays stress on capital formation or investment to achieve a higher growth of output and employment. It is worth mentioning that this model does not distinguish between capital formation in industry and agriculture. Now, given the capital-output and capital-labour ratio, faster growth of capital formation and GDP growth will lead to more employment being generated. Further, according to the model, investment or capital accumulation depends on rate of saving in the economy.

The basic solution to the problem of unemployment is therefore the faster rate of capital formation so as to enlarge employment opportunities. For this purpose, every possible encouragement should be given to the growth of domestic savings and their productive utilisation in increasing the rate of investment. In developing countries investment incentives for the private sector are very low and the State can assist in the process of capital formation directly as well as indirectly. Through a fiscal policy which encourages savings and investment and a sound monetary policy it can do much to encourage investors. The State itself can participate in the process of capital formation by undertaking such developmental activities as building infrastructure which do not attract the private investors. Therefore, the State has got to assume a special role in speeding up the rate of economic development. The other line of attack has got to be on the rate of population growth. If population grows at a rapid rate, then to maintain the people even at their existing levels, large amounts of capital are needed which could otherwise have been used to raise the amount of capital available per man and hence to raise the living standards at a faster rate.

Employment Elasticity of Output

To what extent GDP growth will generate employment depends on *employment elasticity of output*. By employment elasticity of output we mean percentage growth in employment caused by one per cent growth in output. For the 1970s this author in his Ph.D dissertation[1] had calculated employment elasticity of industrial growth in India being equal to 0.5. Recent estimates show that it has declined substantially. It was thought once that industrial growth would solve the unemployment problem in

1 *Development Strategy in Labour-Surplus Developing Economies*, 1982, Unpublished Ph.D. Dissertation, Delhi School of Economics, Delhi University.

developing countries. Thus, India's first three Five Year Plans were based on sufficiently absorbing labour through industrial growth, so much that it would be able to withdraw surplus labour in the long term from agriculture and thus bring about structural change in the Indian economy. However, actual employment generation by industrial growth has been very disappointing. This is because of lower employment elasticity of growth of industrial output. As a result, employment growth has lagged behind. Let us explain employment growth with the following formula:

$$g(E_i) = e_{iY}g(y_i)S_i$$

where $g(E_i)$ is employment growth in the *organised* industry as a percentage of total employment, e_{iY}= elasticity of employment with respect to growth of industrial output (*i.e.*, gross value added by the industry) and S_i is the *initial* share of industrial employment as a fraction of total employment. In India if initially the employment in the organised industrial sector as a fraction of labour force was 0.10, employment elasticity of growth of industrial output (*i.e.*, gross value added) as estimated by many being equal to 0.5 and if industrial growth rate of output $g(Y_i)$, was 6 per cent, then growth of industrial employment is given as under:

$$g(E_i) = 0.5 \times 6\% \times 0.10 = 0.3$$

Given the industrial growth equal to 6%, it means only 0.3 per cent of the labour force will get additional employment each year by the growth of the organised industrial sector while the labour force was growing at the rate of 2.2 per cent of labour force. This will actually be the case if *more capital-bias and labour-saving technologies* are not adopted by the industries. However, the trend in India has been to adopt highly capital-intensive or labour-saving technologies by importing from the industrialised developed countries with the result that in the organised sector, employment growth has been very little.

Table 53.1. Employment Growth, GDP Growth and Employment Elasticity

	CENSUS		NSS		Labour Bureau
	1991 to 2001	2001 to 2011	1993-94 to 1999-2000	1999-2000 to 2011-12	2011-12 to 2013-14
Change in Employment (Millions)	88.4	79.2	25.5	73.4	9.15
Employment Growth (% per annum)	2.5	1.8	1.1	1.4	1.0
GDP Growth (% per annum)	5.7	7.7	6.8	7.3	4.6
Employment Elasticity	0.44	0.29	0.16	0.19	0.22

Source: *Economic Survey*, 2014-15 Vol. I, p. 11.

An important factor responsible for slow growth of employment in India is the decline in employment elasticity of growth of output. *India's Economic Survey,* 2014-15 has clearly brought out this by using the data of employment growth in the economy as a whole and overall GDP growth of India. We give the data regarding measures of elasticity of employment from Census, NSS and Labour Bureau in Table 53.1. The measures of elasticity of employment of growth of output from Census figures for the period 1991 to 2001 was 0.44 which fell to 0.29 in the period 2001 to 2011. According to the Census 2011, the rate of growth of employment per annum during the period 2001-11 was 1.8 per cent per annum despite a high GDP growth of 7.7% per annum achieved during this period. Thus, even according to census estimates employment elasticity declined. Similarly, according to NSS data, during the period, 1993-94 to 1999-2000 employment elasticity of growth of output was still less at 0.16 and in the period 1999-2000 – 2011-12, it was found to be 0.19 even though GDP growth picked up to 7.3 per cent per annum. Labour Bureau has estimated employment elasticity of growth of output equal to 0.22 between 2011-12 and 2013-14.

It is evident from above that employment elasticity of growth of output has declined since 2000. This reduction in employment elasticity has been mainly caused by the use of more capital-intensive technologies in production. Accordingly, the *relation between GDP growth and employment generation has greatly weakened.* It is worth quoting Economic Survey, 2014-15 which states that "Estimates of employment growth and its elasticity relative to economic growth vary widely. However tentatively one might say that employment growth and elasticity have declined in the 2000s compared to the 1990s. Since labour force is in excess of employment growth, labour absorption will be a challenge[2].

Factor Price Distortions

The other important cause of poor growth of employment opportunities and the existence of unemployment in labour-abundant developing countries is the prevailing factor price distortions. Ideally, the prices of capital and labour in an economy should reflect their *scarcity or shadow prices.* In labour-abundant developing countries since capital is scarce, its price should be relatively higher and as labour is abundant, its price (*i.e.,* wages) should be lower. However, actually, in developing countries factor prices are distorted in the ways that capital is made artificially cheaper relative to labour, or to put in other words, price of labour (that is, wages) is made relatively higher and capital cheaper. As a result of these factor price distortions, the rational firms use capital-intensive methods to minimise their costs for a given level of output and thereby maximise profits. This makes employment of labour to grow slowly as industrial expansion takes place.

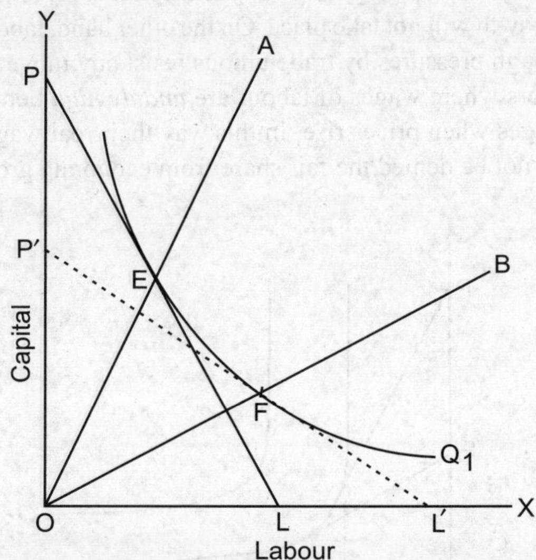

Fig. 53.1. *Factor Price Distortions and Choice of a Factor Combination (i.e.Technique)*

Now the question arises why wages of labour are relatively higher. Pressures of trade unions of labour and fixation of minimum wages by the Government make wages higher for those fortunate enough to get jobs in the modern sector. It is often pointed out that because of political pressures of trade unions the Government fixes wages at a relatively higher levels.

On the other hand, capital is artificially made cheaper by keeping interest rates lower which make capital equipment cheaper to buy by borrowing from the banks or other financial institutions. Besides, there is often *overvaluation* of local currency in terms of foreign exchange which also makes the import of capital equipment cheaper. More important, in their fiscal policy Government provides liberal depreciation allowances, investment breaks, long tax holiday to induce capital investment. As a result of capital being made artificially cheaper and labour costlier, with steeper iso-cost line *PL*, the producer firms will choose a relatively capital-intensive method or technique of production

2. GOI, *Economic Survey* 2014-15, Vol. 1, p. 10.

as shown in Fig. 53.1 by point *E* on isoquant Q_1. In Fig. 53.1 capital is measured along *Y*-axis and labour is measured along *X*-axis. An isoquant Q_1 indicating a certain level of output has been drawn. When due to factor price distortions, capital is relatively cheaper and labour more expensive, we have iso-cost line *PL* which is tangent to isoquant Q_1 at point *E*. Thus, the producer will choose production technique (*i.e.*, factor combination) which is highly capital-intensive. Given these distorted factor prices, the producer firms will expand along the capital-intensive path shown by the line *OA* which will generate fewer employment opportunities.

Now, if the distortions in factor prices are removed and factor prices are made equal to their scarcity values, we will get the iso-cost line *P′L′* with which to produce output level Q_1 the producer will choose factor combination *F* which is relatively more labour-intensive. Given this labour-intensive technique, the firms will expand along the expansion path *OB* which indicates large increases in labour employment as output grows.

If the Government is serious about adequate increase in employment opportunities to solve the unemployment problem, it must correct the factor price distortions and *withdraw concessions given on the investment of capital* and try to fix wages at the *fair level* and not at unduly higher levels through labour reforms. However, the Governments are often reluctant to remove these factor price distortions as often there are political pressures especially by the capitalist class which through their lobbies insist that without adequate incentives in the form of liberal depreciation allowances, tax breaks on investment, longer tax holidays and other exemptions, investment will not increase and as a result rapid economic growth will not take place. On the other hand, labour in the organised industries and services sector through pressures by trade unions resist any move to lower their money wages. In our view, in the sectors where wages of labour are *unduly high* better way to reduce them is not to raise their money wages when prices rise. In this way their real wages will fall. However, in our view, the labour should not be denied the fair share from economic growth in developing countries

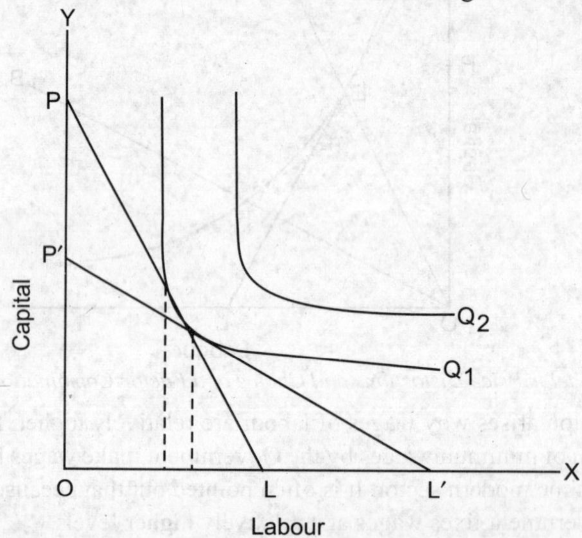

Fig. 53.2. *Relatively Fixed Factor Proportion Production Function with Low Elasticity of Factor Substitution*

as the objective in developing countries is now not merely rapid economic growth but also inclusive growth. Profits of capitalist class must be adequately taxed and several exemptions in taxation which have been provided as incentives to them to make more capital investment must be withdrawn to make capital expensive. This will ensure choice of labour-intensive techniques.

Now, to what extent, the removal of factor price distortions on the use of labour will occur depends on *elasticity of factor substitution*. The elasticity of factor substitution is the percentage change in capital-labour ratio, $\frac{d(K/L)}{K/L}$, in response to the change in the ratio of price of labour to the price of capital $\left(\frac{d(w/r)}{(w/r)}\right)$. The expression w/r represents what is called *wage-rental ratio* where rental is the price of the use of capital. The greater the elasticity of factor substitution, the greater will be the effect of removal of factor-price distortions on the use of labour.

Some economists have pointed out that elasticity of substitution between factors is very low *as relatively fixed factor proportions* are actually found in the production processes in the real world as shown by the isoqants in Fig. 53.2. It will be seen from this figure that isoquants are almost right-angled and removal of factor price distortions, that is, raising the price of capital and lowering the wage rate will have a *very small effect* on the choice of technique or labour use. However, that is a very pessimistic view about the elasticity of factor substitution. Many empirical studies have found the elasticity of factor substitution ranges from 0.4 to 0.5. Even though in some industries the technology used is such that in the *core production process* factor proportions are relatively rigid, but in *subsidiary operations* such as material handling, packaging etc. there exists good scope for greater use of labour.

Various Development Strategies for Employment Generation in Developing Countries

In what follows we shall critically examine some well-known strategies of development, especially from the point of view of whether they provide an adequate answer to the problem of surplus labour in developing countries with special reference to India. Their critical evaluation will enable us to bring out the various policy measures which need to be adopted for designing an appropriate development strategy to ameliorate the problem of unemployment and underemployment in India. It is important to note that a development strategy for promoting the growth of employment is based on a particular diagnosis and explanation of the problem of surplus labour. As an explanation of unemployment and from the viewpoint of generation of new employment opportunities, the various employment strategies and aproaches have been proposed from time to time by different economists. The first important approach for generation of employment is what can be called **industrialisation-led strategy of development**. Some economists, prominent among whom are Nurkse, Lewis, Eckaus, Robinson, Mahalanobis, Hirschman, Fei-Ranis and Jorgenson, attribute unemployment in the developing countries to the lack of industrialisation and therefore suggest stepping up the rate of investment or capital formation for the expansion of modern industrial sector in which to absorb surplus labour.

On the other hand, Vakil and Brahmananda laid stress on wage-goods gap as a cause of unemployment and put forward a development strategy which assigned a high priority to wage-goods industries (especially agriculture) in the allocation of investment resources so as to increase the supply of wage goods which would enable us to generate more employment opportunities. Another group of economists led by Schumacher[3], Myrdal [4]and Singer hold the use of capital-intensive technology in the large-scale industries responsible for the prevailing unemployment in developing countries. This approach leads to the adoption of labour-intensive techniques as an essential part of the development strategy to create more employment for labour. Finally, some economists, more prominently Dandekar and Rath, J.P. Lewis, B.S. Minhas, have proposed undertaking of rural public works on a large scale as a solution to the problem of unemployment and poverty. According to them, adequate employment opportunities cannot be generated, at least in the short run, through any

3 E.F. Schumacher, *Small is Beautiful*, Board Briggs, London.

4 Gunnar Myrdal, *Asian Drama: An Enquiry into the Poverty of Nations*, Penguin Books, New York, 1968.

emphasis on particular technological aspects or the pattern of investment within the agricultural and industrial sectors. Hence the argument for undertaking rural public works to generate employment opportunities. In what follows, we shall examine these various strategies of development for productive absorption of surplus labour.

INDUSTRIALISATION-LED STRATEGY OF DEVELOPMENT

Influenced by the growth experience of the western developed countries, the several developing countries adopted industrialisation-led strategy of development for absorbing surplus labour in productive employment in the modern industrial sector. An essential feature of this approach is that ***it seeks to generate more employment opportunities by achieving a higher rate of industrial growth through accelerating investment or capital formation in it.***

Besides, this approach regards labour and capital as complementary inputs. In the developing countries the stock of capital and modern industries have not been growing at a rate fast enough to keep pace with the growth of population. As a result, the whole of growing labour force cannot be absorbed in productive employment because not enough capital is available to employ them. Thus this approach attributes the existence of unemployment and underemployment in developing countries to the lack of capital relative to the magnitude of labour force. A variety of growth models and development strategies come under this approach, all of these, though differing in some other aspects, lay stress on the importance of capital accumulation for generation of employment opportunities. We shall explain below some of the models of this industrial growth-led approach towards employment generation that fall in this category.

Industrialisation-Led Strategy of Development: Lewis Model

In their growth models for a dualistic economy with surplus labour, Lewis[5] and Fei-Ranis[6] regard capital as the crucial factor for the expansion of employment in the modern industrial sector. While their models suggest that rate of growth of employment depends upon the rate of capital accumulation and economic growth, it differs from Harrod-Domar model in the sense that they draw distinction between the two sectors of a less developed economy: (1) the subsistence agricultural sector which is characterised by surplus labour, low productivity and self-employment; and (2) the modern industrial sector which is characterised by wage employment and high productivity.

These growth models emphasise transfer of surplus labour from the subsistence agricultural sector to the modern industrial sector. And this *transfer of labour and expansion of employment opportunities in the modern industrial sector is determined by the rate of capital accumulation and industrial growth.* Therefore, the higher rate of capital accumulation and economic growth in the modern industrial sector over a number of years would enable us withdraw all the disguisedly unemployed labour from agriculture.

It was assumed in this Lewis model of industrialisation that agriculture would adjust itself endogenously to the requirements of industrial growth. The assumption of unlimited supply of labour at *a given wage rate* implied that agriculture would adjust endogenously and release labour for industrial growth without lowering of agricultural production. Lewis realised that if on withdrawal of labour from agricultural productivity of labour falls, it would cause prices of agricultural output, especially food, to rise. The rise in food prices will lead to the rise in wage rates of labour in the industrial sector. The rise in wages would lower profits in the industrial sector and bring about stoppage of further industrial growth.

Heavy-industry Strategy and Mahalanobis Growth Model

This growth model which formed the basis of development strategy adopted in India's Second

5 Arthur Lewis, "Economic Development with Unlimited Supplies of Labour", *Manchester School*, Vol. 22, 1954.

6 J.C.H. Fei, G. Ranis, *Development of the Labour-Surplus Economy, Theory and Policy*, Irwin, 1964.

and Third Plans also laid a great stress on the rate of capital accumulation for determining the rate of growth of employment. According to Mahalanobis, unemployment of labour is caused by the lack of capital stock. To employ a person in productive work, he needs to be equipped with adequate capital. Therefore, in order to generate more opportunities of productive employment more capital has to be accumulated. *Given a closed economy, the rate of capital formation, according to him, depends on the capacity of the economy to produce capital goods*. With this premise he showed that in order to achieve a high rate of economic growth and employment it *is necessary to accord a high priority to basic heavy industries producing capital goods* in the development strategy of a plan. It is worthwhile to mention that in Mahalanobis' opinion it is necessary to allocate relatively greater resources to basic heavy industries to achieve not only the objective of rapid economic growth but also of expanding opportunities for productive employment. With this, the stock of basic goods like steel, machinery, fertilizers, should be sufficiently increased to eliminate unemployment.[7]

It may be noted that Mahalanobis later realised that basic heavy industries, though important for determining the *long-term growth of employment*, will themselves not create adequate employment opportunities. In response to the need to create enough employment opportunities *in the short run*, he extended his two-sector model to four sectors wherein he kept the basic heavy industries sector intact but divided the consumption goods sector into three sub-sectors, namely: (1) factory enterprises sector, (2) household and small-scale sector, and (3) services sector.

It was the household and small-scale sector in his four-sector model which being labour-intensive was expected to create a good amount of employment opportunities. However, his four-sector model did not represent any essential change over his two-sector model in as much as it maintained the emphasis on basic heavy industries. Moreover, since adequate resources were not provided for the household and small-scale industries, nor was any effort made to improve their productivity so that they could successfully compete with the large-scale industries, they recorded a low growth rate in output and employment.

A Critical Appraisal of Industrialisation-led Strategy of Development and Employment

Industrialisation-led strategy of development rightly lays stress on the importance of capital goods for generating employment. Without some capital goods such as tools, implements, machinery and equipment, it is hard to generate productive employment in both the large-scale and small-scale industries. Likewise, capital inputs such as irrigation facilities, fertilizers, etc., play a crucial role for the productive absorption of labour in agriculture. Above all, shortage of infrastructure such as steel, power, cement and transport facilities serves as an important bottleneck for the expansion of output and employment in both industry and agriculture. Therefore, expansion in their supply would promote the growth of output and employment in the economy.

However, *industrialisation-led strategy of development did not sufficiently recognize the importance of wage-goods constraint for generation of employment opportunities*. Thus, at a time when fixed capital goods alone were thought to be necessary, Vakil and Brahmananda[8] made a valuable contribution by highlighting the importance of wage goods for expansion of employment opportunities. But the assertion that the increase in the supply of wage goods alone is both a necessary and sufficient condition for creation of employment sounds quite unrealistic.[9] It is not a right approach to think of capital goods versus wage goods. Indeed, increase in the supply of both is needed to solve

7 P.C. Mahalanobis, "The Approach of Operational Research to Planning in India", *Sankhya*, Vol. 16, 1955.

8 C.N. Vakil and P.R. Brahmananda, *Planning in an Expanding Economy*, Vohra & Co., Bombay, 1966.

9 It may be mentioned that Brahmananda later modified his wage woods strategy by incorporating in what he calls "*Integrated Wage-goods Complex*" which includes not only the wage-goods but also the capital goods required for the production of wage-goods. But the role of intermediate capital goods such as steel, cement, coal and power required for the production of wage-goods and also for investment sector had not been recognised by him even in his '*Extended Wage-goods Strategy*'.

the problem of unemployment. The availability of wage goods serves as a constraint for generating wage-employment as without them the demand of the newly employed persons for wage goods would not be met. But if opportunities for productive employment are to be generated in the very first instance either in industry or in agriculture, labour needs to be provided with some capital goods.

It is worthwhile to note that growth of the modern industrial sector was expected to absorb not only the openly unemployed but also the disguised unemployed persons from agriculture. However, the actual experience belies the hopes of generating sufficient employment opportunities through industrial growth. For instance, an annual average rate of growth of around 6 per cent achieved by the industrial sector in India during the four decades (1956-96) and a very high increase in the rate of net capital formation (from 5.5 per cent in 1950-51 to 25 per cent of national income in 1995-96) had not been able to solve the problem of unemployment and underemployment in India. It is thus clear that not much employment opportunities can be generated *directly* through growth of modern industrial sector even with high rate of increase in its output.[10]

Industrialisation-led strategy based on Mahalanobis and Lewis models underplayed the importance of growth of agricultural output. Without the growth of marketable surplus of agricultural output, especially food, the sustained growth in the industrial sector could not be achieved. In the case of erstwhile Soviet Russia, the required food surplus was procured *through forcible collection* from the farmers. In democratic countries like India forcible collection was ruled out. The neglect of agriculture in India in the Second Five Year Plan led to the food shortage in the early sixties. This caused not only inflation in the economy but we had to import foodgrains from the USA to feed our growing population.

A serious drawback of strategies of employment generation through industrial growth is that *they have ignored the possibilities of absorbing labour productively in agriculture*. These strategies regarded agriculture merely as the source of labour supply to the expanding industries. Even in growth models of Lewis and Fei-Ranis of a labour-surplus dual economy, and also of *Mahalanobis heavy-industry strategy of development* both surplus labour and dualism have been sought to be eliminated by way of shifting labour from agriculture to the modern industrial sector.[11] But there is ample scope for absorbing labour in the Indian agriculture provided proper agricultural policies are adopted. India employs much less labour per hectare than other countries like Japan.[12] A good amount of labour can be productively absorbed in the Indian agriculture through rapid growth of its output by increasing cropping intensity and irrigation facilities and through the adoption of new high yielding technology. Reckless mechanisation, that is, use of labour-displacing agricultural machines such as tractors, combine harvestors is to be avoided if more employment opportunities are to be created in agriculture.

Finally, most of the growth strategies focussed on industrial growth for employment generation and *ignored the importance of institutional reforms, especially in agriculture, for promoting the expansion of employment opportunities*. Land reforms in agriculture such as redistribution of

---

10 Note that the growth of modern industrial sector generates few employment opportunities was also corroborated by the experience of Latin American countries such as Brazil, Argentina, Mexico, Columbia, Peru, Venezuela which have achieved annual rate of industrial growth ranging from 7.6 per cent to 10 per cent during the decades of nineteen fifties and sixties and still they faced the mounting problem of surplus labour.

11 It is important to note that Fei-Ranis have emphasised the importance of agricultural development in the growth process of a labour-surplus economy, but they have envisaged such a pattern of agricultural development which ensures more and more agricultural output for less and less use of labour input. Besides, the Second Five Year Plan based on Mahalanobis development strategy visualised reducing labour force in agriculture and employing a larger proportion of labour force in the modern industrial sector in the next 20 years.

12 In the mid-1950s, labour input per hectare of net sown area was 530 working man days in Japan, as against only 137 in West Bengal in India.

land and changes in the tenancy system can play an important role in increasing employment opportunities for landless labourers, small and marginal farmers and share-croppers. There exist large inequalities in land-ownership in India with the result that whereas a large number of rural households have either insufficient or no land to provide them adequate employment and income, a few families, on the other hand, own such a large amount of land that they cannot manage it efficiently. Consequently, there is underutilization of land and irrigation potential. Further, it has been found from 'Farm Management Studies' in India that employment per hectare and cropping intensity are higher on small farms as compared to those on large farms. This may be taken to imply that redistribution of land through imposition and effective implementation of ceilings on land holdings will bring about increase in employment opportunities in agriculture.

To sum up, agriculture contains immense employment potential provided proper strategy for its development involving technological and institutional changes is adopted. This has, however, been ignored by the mainstream of the industrialisation-led strategies of development.

WAGE-GOODS STRATEGY OF DEVELOPMENT AND EMPLOYMENT IN THE DEVELOPING COUNTRIES

The industrialisation-led strategy of development explained above neglects the constraint of wage goods availability on employment generation. In their criticism of the development strategy of Second and Third Five Year Plans, Vakil and Brahmananda advanced what is known as wage-goods strategy of development. According to them, unemployment and disguised unemployment in less developed countries are due to the deficiency of the supply of wage goods. They have argued that the magnitude of employment in less developed countries is determined by the available amount of wage goods. This is because whenever employment for new persons is created, say, in rural public works, it cannot be sustained unless adequate amount of wage goods, especially foodgrains, is available. Of course, when people are openly or disguisedly unemployed, they would be consuming some foodgrains. But in that situation their consumption would be very low because their budget would not allow them to consume more. Therefore, when they are given wage employment, their effective demand for wage goods will tend to increase. Hence, given the reasonable real wage rate (in terms of wage goods) we can calculate the total amount of wage goods required to obtain full employment. They call the difference between the required magnitude of wage goods and the actually available supply of wage goods as the '*wage goods gap.*' Open and disguised unemployment in less developed countries like India, according to them, is due to the existence of this wage goods gap. The level of full employment can be attained only when this wage-goods gap is bridged by expanding the capacity to produce wage-goods.

We thus see that whereas in Mahalanobis' approach to growth and employment "the entire emphasis was on the role of fixed capital, Vakil and Brahmananda's entire emphasis was on the role of wage goods as capital." [13] It may be noted that in his later works Brahmananda modified his strategy in that he later assigned a high priority to what he calls '*Integrated Wage Goods Complex*' which includes not only wage goods but also capital goods which are used for the production of wage goods. Since they consider the supply of wage-goods as the most important determinant of employment in less developed economies Vakil and Brahmananda proposed a development strategy which assigned the top priority to wage goods industries, especially agriculture, in the pattern of investment. That is why they severely criticized the development strategy adopted in India's Second and Third Plans which gave high priority to basic heavy industries producing fixed capital goods.

It may be noted that, like Nurkse, Vakil and Brahmananda believe that disguised unemployment in developing countries contains saving potential (in the form of wage goods) for capital accumulation. By withdrawing the disguised unemployed from agriculture, some amount of wage-goods would be

13 Jagdish Bhagwati and S. Chakravarty, *Indian Economic Analysis*, (1971), p. 14.

released for providing wage employment in the investment sector. But, unlike Nurkse, they think that the surplus of wage goods so released will not be sufficient to provide employment to all the potentially available labour force. Therefore, they emphasised the expansion of the capacity to produce wage goods and for that purpose advanced a development strategy which envisaged greater allocation of investible resources to agriculture and other wage-goods industries.

A Critical Evaluation of Wage-goods Strategy

The assertion by Vakil and Brahmananda that the growth in employment *solely and exclusively* depends upon the supply of wage goods and that the capital goods (that is, heavy-industry products) play no important role in the creation of employment opportunities is open to question. Under adequate organisation and entrepreneurial ability as well as optimum institutional forms, two important bottlenecks, namely, capital goods bottleneck and wage goods bottleneck, are faced in creating opportunities for productive employment. For expansion in productive employment both these bottlenecks have to be overcome and to emphasise the breaking of one bottleneck, while making no attempt to tackle the other, is an unrealistic approach to the employment problem. Wage goods are required to meet the demands of the newly employed people, but if some people are to be employed in productive activities in the very first instance, they are required to be equipped with some capital goods. Further, for the production of wage goods themselves, capital goods are required. That is, labour is to be equipped with capital goods to produce wage goods. Professor Dantwala rightly points our that, *"It is a mistake to consider capital goods and wage goods as exclusive and unrelated categories. Anyone who studies the composition of inputs needed for transformation of traditional agriculture should be able to appreciate the inter-connection between the two."*

Thus, development strategy for rapid expansion of employment has to be such as will envisage the increase in production of both wage goods and capital goods. Thus, to talk of wage goods versus capital goods is a false issue and an unrealistic approach to the problem of employment since both are needed to generate employment; capital goods are required to make the employment productive and wage goods are needed to feed the workers provided with new employment, that is, to sustain their employment. To say this, however, does not imply that all capital goods industries should be necessarily developed at home and given higher priority in the allocation of resources. Which specific capital goods should be produced at home and which imported from abroad should be decided on the basis of resource endowments and comparative advantage conceived in the dynamic context.

DEVELOPMENT STRATEGY OF USING LABOUR-INTENSIVE TECHNOLOGY

An important school of thought led by Schumacher, Singer and Myrdal attributes the mounting problem of unemployment and underemployment in the developing countries to the use of capital-intensive technology. It has been observed that while industrial output in developing countries has increased at a reasonably good rate, growth of employment has lagged far behind. And this is said to be due to the use of capital-intensive technology. Thus, according to this school of thought, the creation of meagre amount of employment opportunities due to the use of capital-intensive technology on the one hand and the growth of population at an alarming rate on the other has resulted in the huge magnitude of surplus labour.

It is worthwhile to note here that the modern industries using capital-intensive technology not only create few new opportunities of employment but also destroy employment in the traditional household industries. This unfavourable impact of modern sector on the traditional employment is what Myrdal calls *backwash effect*. According to this school of thought, modern manufacturing enterprises using capital-intensive technology produce goods on a mass scale at lower cost per unit which drive out the products of traditional household industries from the market. As a result, a good number of people engaged in these household industries are displaced. They add to either open or disguised unemployment in agriculture.

In this context, **labour migration model of Todaro** has often been invoked, especially by Singer, to explain the mounting problem of urban unemployment in developing countries. According to Todaro,[14] decision of rural workers to migrate to urban areas rests on the higher level of wage rate and the expectations of getting employment in the modern urban sector. A few employment opportunities at relatively higher wage levels created by capital-intensive technology in the modern urban sector attract many more rural workers than it is possible for it to absorb. This results in enormous increase in urban unemployment and underemployment.

Of course, Galenson-Leibenstein, M. Dobb, Amartya Sen have pointed out that *the use of capital-intensive techniques would generate more surplus resources which would be ploughed back into further investment and it would result in greater growth of employment than is possible through labour-intensive techniques*. But this has not been borne out by empirical evidence. For instance, estimates of saving and investment in India reveal that contributions of the public and the private corporate sectors to the domestic savings have not only been too small but have also recorded slow growth in the last six decades in spite of the fact that the capital-intensive technology has been used in them. On the contrary, it is the savings of the household sector (consisting of households and incorporated enterprises) which often use labour-intensive technology that have greatly increased. This raises doubts about the argument that capital-intensive techniques necessarily generate more surplus and therefore greater growth of employment.

There is another important aspect of the use of capital-intensive technology which is worth mentioning. The *use of capital-intensive techniques creates high profit-wage ratio and thereby accentuates inequalities of income*. These income inequalities create demand pattern oriented towards luxuries which are again produced by the use of capital-intensive techniques. This limits the size of market for labour-intensive goods consumed by the working classes. Thus, the use of capital-intensive techniques gives rise to a vicious circle which causes slow growth not only of direct employment but also of indirect employment through its effects on the pattern of demand.

Why in Developing Countries Capital-intensive Technology is Used? Now, a pertinent question is why have the developing countries been using capital-intensive technology despite the fact that surplus labour prevails in them? There are two views about this. First, the alternative technology which though efficient but labour-intensive is generally not available. The available technology is the capital-intensive technology that was developed in the Western countries to suit their factor endowments. The Western technology is quite inappropriate for the developing countries which suffer from surplus of labour and shortage of capital. Now, given the technology imported from the Western countries, there is not much possibility of substitution of capital for labour. In view of this, Schumacher and Singer advocate for the development of what has been called *intermediate or appropriate technology*. Appropriate technology for the developing countries is that which though efficient should be labour-intensive so that it conforms to their factor endowments.

According to Schumacher, the development of appropriate technologies does not imply the discovery of altogether new principles of science and technology. What is required is the application of basic principles of modern science and technology to evolve the appropriate production techniques. These appropriate production techniques may be obtained by scaling down the advanced techniques by adapting them so as to make them more labour-intensive, or by scaling up handicrafts technique with the introduction of new tools and simple machines and thus improving economic efficiency of these techniques while maintaining their labour intensity.

On the other hand, several other scholars such as Ranis, Blaug, Layard and Woodhall, Little, Scott and Scitovsky are of the view that the slow growth of employment opportunities in the industrial sector is not due so much to the lack of flexibility in the production function but to the **distortions**

14. Michael, P. Todaro, "Income Expectations, Rural-Urban Migration and Employment in Africa", *International Labour Review*, Nov. 1971.

in factor prices.[15] Due to the various concessions and subsidies provided in taxation structure by the Government such as liberal investment or development rebate, policy of low interest rate and over-valued foreign exchange rate have caused the price of capital to be very low relative to its scarcity. On the other hand, the prevailing wage rates are too high for a labour-surplus economy due to the strong bargaining power of the trade unions in the organised sector. According to them, these distortions in factor prices induce the adoption of capital-intensive techniques in industries with the consequence that less labour is employed. In order to promote the use of more labour-intensive techniques, it is therefore imperative that all distortions in factor prices be corrected.

By emphasising the use of labour-intensive technology in developing countries, this strategy has brought into sharp focus the question of appropriate technological choices. In a labour-surplus economy the importance of the use of appropriate technology which is labour-absorptive and at the same time quite efficient can hardly be denied. However, appropriate technologies for various manufacturing industries just do not exist and a good deal of more research and development (R & D) would have to be undertaken before appropriate techniques are evolved for several industries. Of course, to encourage the development and use of appropriate techniques the various factor price distortions ought to be removed. Whereas, in view of the prevailing price situation, wage rates of industrial workers do not seem to be high, the effective price of capital is low. Therefore, the various concessions and rebates in taxes on various types of capital investment be drastically reduced and the lending rates of interest charged by banks and other financial institutions be raised so as to curb the tendency to substitute capital for labour.

As for technological choices in agriculture it may be noted that new high-yielding technology represented by a package of high-yielding varieties, fertilizers and pesticides is quite appropriate for a labour-surplus economy such as ours, for it not only raises yield per hectare substantially but is also labour-absorptive.

STRATEGY OF RURAL PUBLIC WORKS FOR EMPLOYMENT GENERATION

This approach seeks solution to the unemployment problem not through emphasising any particular pattern of resource allocation or technological choice but through special employment schemes, especially rural public works. This approach regards the regular development process as being incapable of alleviating the problem of unemployment and underemployment in the short run. The launching of rural works programme on a large scale had been especially advanced by Dandekar and Rath[16] to solve the problem of mass poverty which, they maintain, is mainly caused by unemployment and underemployment. Rural public works also constituted the main thrust of the recommendations of Bhagwati Committee on unemployment in India to tackle the problem of mass rural poverty and unemployment. Further, the scheme of rural public works as a major source of employment generation in rural areas has also been strongly recommended by J.P. Lewis, K.N. Raj, M.L. Dantwala, and Raj Krishna. Besides, B.S. Minhas[17] also suggested massive rural works programme in his "*integrated programme of compulsory consolidation and land development*."[18]

Employment generation through Rural Public Works was undertaken in 1977-78 on a regular basis under the '*Food for Work Programme*' during the regime of Janata Government and was later

---

15. Eckaus in his famous work '*Factor-Proportions Problem in Underdeveloped Countries*' showed that either factor-price disequilibrium or limited possibilities of substituting labour for capital can result in unemployment in less developed economies. Blaug, Layard and Woodhall in their study on educated unemployment have held high wage rates responsible for educated unemployment in India.
16. Dandekar, V.M. and Rath, N., *Poverty in India*, Indian School of Political Economy, Poona, 1971.
17. Minhas, B.S., "Rural Poverty, Land Redistribution and Development Strategy", *Indian Economic Review*, Vol. 5, 1970.
18. In the strategy proposed by Minhas the scheme of rural public works was to be used for developing land and water resources after consolidation of land has been accomplished.

adopted as a part of the Sixth Five Year Plan (1978-83). Under this scheme a major part of wages is paid in terms of foodgrains. The constraint of wage goods (*i.e.*, foodgrains) which stood in the way of implementation of earlier schemes of rural public works did not then exist. In the Sixth and Seventh Plans rural public works programme continued under the scheme called **'National Rural Employment Programme' (NREP).** This programme intended to provide supplementary employment opportunities to the rural poor seeking work during the lean periods. Projects relating to soil and water conservation, irrigation, flood protection and drainage, field channels in irrigation command areas, construction and improvement of village tanks and ponds, school and dispensary buildings, improvements of village environment, hygiene and sanitation had been undertaken under this programme.

In the Eleventh Five Year Plan period a new scheme called **Mahatma Gandhi National Rural Employment Guarantee Scheme** (MGNREGS) with focus on employment generation through rural public works was started under which one member of poor households has been guaranteed employment for hundred days in a year at a specified wage rate. Though it has been reported that a large number of poor persons have been given employment under the scheme, the scheme suffers from various lacunae such as lack of awareness about rights and entitlements, lack of durable assets, faulty job cards and muster rolls. Therefore, there has been demand for recasting the scheme so as to create durable productive assets under the scheme and to stop the leakage of funds.

In recent years the scheme has come in for a lot of criticism. A major criticism is that MGNREGS funds are used for works of a non-permanent nature and therefore divert Government spending from the more productive sectors of the economy. To prevent wastes of scarce financial resources there is need to employ persons for making more durable and productive assets. Without the creation of permanent productive assets the money would evaporate. In view of the non-productive nature of assets created by employing poor persons merely to fulfil the targets by district administration, it has been asserted that *this employment guarantee scheme is just a cash transfer to the poor* rather than serving as a means of productive employment for the manpower.

Besides, in the scheme there has been a lot of leakages due to which the poor neither get full wages nor entire 100 days of employment are actually provided to a person of each poor rural household. This means a good deal of funds are siphoned off by the middlemen and do not reach the poor households. As no useful productive work is being done under the scheme, a former Rural Development Minister Jairam Ramesh had said that MGNREGA scheme had become a plan only for *'digging ditches.'* and the budget expenditure of ₹ 36,000 crore being annually spent on the scheme was just being wasted.

A MGNREGA survey conducted in Andhra Pradesh, Madhya Pradesh and Rajasthan and its report submitted recently to the Government has revealed that the average number of days worked per household in the projects under the scheme in 2010-11 was 45, 37, and 52 days for these States respectively as against 100 days per household guaranteed under the scheme. This shows hardly 50 per cent of the days that are being claimed by the poor, that is, the poor are not getting full employment opportunities guaranteed under the MGNREGS. It is therefore evident that MGNREGS suffers from implementation problems. In the recent years not only employment provided to the number of people has decreased but also *average days* of employment provided per household was 46 days in 2012-13 and 2013-14 but only 34 days in 2014-15 as against 100 days of employment per household in a year. Besides, another big problem being faced by the scheme is *large delay* in wage payments to the persons who worked under the scheme. Therefore, NDA government has decided to make *direct transfer of money* to the bank accounts of the households.

One of the persistent complaints about the scheme is that *full wage* is not being received by the poor under the scheme due to various leakages from the funds allocated for the scheme. The findings of the survey conducted for AP, MP and Rajasthan also reveal that average wage received per day by the workers employed under the scheme was ₹ 94, 98 and 79 in Andhra Pradesh; Madhya Pradesh and Rajasthan respectively which are far less than the wages fixed by the government. It is therefore evident that there is a lot of leakages in the scheme at the local level. That is, a good chunk of benefits under the scheme are grabbed by the middlemen who arrange employment work under the scheme. It has often been found that district administration is hard put to spend sanctioned amounts and often resort to sanctioning unnecessary works and local functionaries hoarding job cards are able to grab a good deal of funds sanctioned for the works. This is a great lacuna in the scheme. Besides, due to lack of useful or productive projects in individual villages the poor are employed in unproductive and useless works. This prompted the former Union Minister for Rural Development to say that "how many ditches will you dig in a village, how much labour work you can do in building community assets such as schools, rural roads in a village or nearby year after year, how many trees you can plant in a village or nearby year after year. There is a limit to such works or projects in individual villages".

In our view for the scheme to have utility in terms of creating productive assets *there is need for integrating the scheme with the rural development projects included in the five year plant.* This would require to employ people demanding work under the scheme far off from their villages. But this can be done by making available transport facilities to bring them to project places and back to their villages.

Further, the permissible works need to be expanded; especially *farm work* should be included in this guaranteed employment scheme. This will help in raising agricultural output apart from generating employment opportunities. This will also meet the complaint often made against the scheme that diversion of workers to the guaranteed employment scheme is affecting farm output. There is already demand that the MGNREGS be taken off during the peak season in agriculture when workers are badly needed for various agricultural operations. For engaging workers under the guaranteed scheme in farm work of private landlonds some funds can be collected from them to pay wages to the workers supplied to them.

Besides, the Government should include new types of rural works such as soil conservation measures, products of composts and bio-pesticides, other works related to sowing and harvesting activities such as threshing would help in raising farm output. Some of the funds can be collected from the landlords on whose land such farming related activities are undertaken under MGNREGA scheme. All these require to recast the employment guarantee scheme. However, for the success of such extended employment guarantee scheme there is need for **micro-level planning** which is not yet being done.

There can be hardly any two opinions that rural public works can be helpful in providing employment to the rural poor, especially the landless labourers and marginal farmers who remain without gainful work for a large number of days in a year. At the same time the programme can be used to build *durable community assets* which would promote rural development. Further, by providing alternative avenues of wage employment the rural works programme can ensure reasonable wages to the agricultural workers. But *the view that these rural public works should serve as a substitute for employment generation through a proper development strategy as was conceived by Dandekar and Rath and some others is open to question.* For, being construction activities, rural public works are, at best, an interim or short-run solution of the problem of unemployment and leave the question unanswered as to how the growing number of workforce will be productively employed in the long run. The lasting and long-term solution to the problem of unemployment and underemployment lies in framing a development strategy that involves not only acceleration of rate of capital formation but also technological and institutional changes which are appropriate for labour-surplus conditions of the Indian economy. However, the rural public works have an important role to play in an employment-oriented strategy. As stated above, with the scheme of rural public works we can build up durable assets or infrastructural facilities. Now, these assets or facilities will help in creating more employment opportunities in agriculture and rural industries in future. To fulfil this purpose, rural public works need to be integrated with the overall strategy of rural development rather than being formulated and implemented in isolation of it.

CHAPTER 54
POVERTY AND ECONOMIC GROWTH

POVERTY AND ECONOMIC GROWTH

Despite significant economic growth achieved during the last half a century, there still prevails widespread poverty in developing countries. The poor people suffer from undernourishment, poor health, illiteracy. Freedom from undernourishment, freedom from illness and freedom from illiteracy are the essential requirements of economic development. *It is the building up capabilities of the poor to gain freedoms in these respects which Amartya Sen calls development.*[1] Now, an important question is whether faster economic growth will necessarily solve the problems of poverty and development. In the fifties and sixties it was widely held that benefits of economic growth would trickle down to the poor. This is generally referred to as '*trickle down theory*'. It was then maintained that if you take care of economic growth, poverty will take care of itself.

But in the seventies there was a drastic change in the views of prominent development economists. They pointed that economic growth of GNP, though necessary, was not sufficient for eliminating poverty. This has been generally referred to as '*Dethronement of GNP*'. During the 1970s in India, it was realised that when the economy was growing at the rate of 4 to 5 per cent and with population growth of around 2 per cent annually at that time one could not rely on the 'trickle down' effect to reduce the problem of mass poverty. Therefore, to reduce poverty measures for making '*direct attack on poverty*' were suggested along with taking measures to accelerate economic growth. For this purpose development economists suggested special anti-poverty schemes such as starting of employment schemes, especially construction of rural public works, and strengthening of public distribution system for providing to the poor foodgrains and some other essential commodities at subsidised prices. However, in the late eighties and early nineties some World Bank economists with the officials of U.S. Government who met in Washington again reverted to the old view that economic growth benefits poor substantially even if no special pro-poor measures are adopted[2]. This provided a rationale for what is popularly called **Washington Consensus** which envisaged fiscal and structural adjustment reforms to accelerate economic growth in developing countries, first imposed on the Latin American countries and then on the developing countries of Malaysia, Indonesia and India.

In India there was an economic crisis in 1990-91 resulting from serious balance of payments problem and reduction of foreign exchange reserves to a very low level and flight of capital abroad. There was an urgent need for foreign aid from IMF and World Bank to tide over the crisis. But IMF and World Bank provided aid only if certain conditionalities were fulfilled. These conditionalities were nothing but implementation of fiscal and structural adjustment programme (SAP) as contained in Washington Consensus. These economic reforms of fiscal consolidation and structural adjustment involved reduction in fiscal deficit and carrying out the policies of economic liberalisation, privatisation and globalisation. In 1991-93 most of these policies were implemented in India to accelerate economic growth and hoping that it will help to reduce poverty.

It may be further noted that the data collected by World Bank reveals that economic growth has led to the reduction in absolute poverty where it has been accompanied by *more equal distribution*

1 See Amartya Sen, *Development as Freedom*, Alfred Knoph, New York, 1999, p.43.

2 Author of the various provisions of Washington Consensus John Williamson writes, "The evidence says pretty clearly that growth benefits the poor, even if nothing is consciously done to make it pro-poor growth. *Benefits do trickle down*", in 'Washington Consensus and Beyond', *EPW*, April 12, 2003.

of assets, especially land. Stressing this point Michael Bruno and Lyn Squire, two development economists of World Bank, write: "*Using the new data, we found that developing countries with a more equal distribution of assets—specifically land – grew more rapidly than countries with a less equal distribution of assets. Why is a more equal distribution of assets good for growth? One likely explanation involves credit. We know that investment is crucial for growth. The poor often cannot invest because they lack capital and they lack the collateral to borrow. In countries with a very unequal distribution of assets, many people find it difficult or impossible to invest, even in their own health and education.*"[3] They further write, "Economies in East Asia that had comparatively equal land distribution and aggressively pursued economic growth have achieved dramatic reductions in poverty. In Latin America, land distribution has generally been less equal, growth has been slower and less consistent, and poverty has remained stubbornly high."[4]

To reduce absolute poverty and to accelerate progress in human development, economic growth is of course necessary but not enough. World Development Report 2004 by World Bank rightly stressed that broad improvements in human welfare would not occur unless poor people receive wider access to affordable, better quality services in health, education, water, sanitation, and electricity. Without such improvements in services, freedom from illness and freedom from illiteracy — two of the most important ways poor people can escape poverty – will remain elusive to many.[5]

As we shall see in the next chapter, poverty in India has declined since 2004-05 but to attribute this decline in poverty to *economic growth alone* is not correct. In fact various special employment schemes, promotion of literacy, and strengthening of public distribution system to provide foodgrains at subsidised prices have also made important contribution to reduce poverty. Besides, as we shall see below to what extent poverty has declined is also a matter of great controversy because of changes in methodology used for collecting household consumption expenditure data in NSS survey for 1999-2000, 2004-05 and 2009-10.

In our above analysis we have only given introduction to the issue of how economic growth affects poverty; we will discuss in detail in the next chapter the relation between economic growth and poverty.

In what follows we shall explain the concept of absolute and relative poverty and in the next chapter we will discuss the magnitude and trends of poverty in India. We shall also explain who are the poor and what are the causes of poverty in India. Besides, we will explain in the next chapter the different strategies adopted in different States of India to tackle the problem of poverty.

CONCEPTS OF POVERTY

It is not easy to measure poverty. Poverty is generally defined in absolute sense, though the concept of relative poverty is also important. We first explain the concept of absolute poverty. It may be noted that the concept of absolute poverty is measured with reference to income of the people which is required to meet the minimum basic needs. Through a survey method, consumption expenditure of the households is calculated to measure *income poverty*. Thus consumption expenditure is considered as a proxy for income. Both the concepts of poverty, namely *Head Count Ratio* (HCR) and *Poverty Gap* (PG) are measures of income poverty.

Absolute Poverty

In this concept of poverty a person is poor if his income falls below a certain minimum level which is required to satisfy *minimum basic needs*. Now, the problem faced in measuring poverty is to specify the minimum basic needs or what is also called *minimum living standard* (MLS) that should be defined as *poverty line*. Then a person having an income or consumption expenditure below that poverty line will be described as poor in the absolute sense. Thus *norms* have to be fixed for minimum *calorie intake*

3 Michael Bruno and Lyn Squire, "Why Worry about Inequality?" *Economic Times*, Oct. 8, 1996.
4 *Ibid*.
5 World Bank, *World Development Report*, 2004.

of food, minimum clothing, minimum shelter, minimum healthcare which must be met for fulfilment of minimum basic needs or minimum standard of living. Then these norms have to be converted into *minimum consumption expenditure* or *income per capita* which will meet minimum basic needs or minimum living standard. Persons having income or consumption expenditure less than the minimum consumption expenditure defined as poverty line are called *poor in absolute sense*, that is, *living below the poverty line*. These poor will be living in condition of absolute human misery.

It is evident from above that to specify the poverty line in terms of consumption expenditure per capita, one needs the data about prices of goods that enter into basket of minimum basic needs. The two concepts of measuring absolute income poverty have been suggested: (1) *Head Count Ratio* (HCR) which is most widely used to estimate the incidence of poverty in countries. This refers to the number of people in a society living below the poverty line fixed on the basis of some minimum consumption level. (2) The *Poverty Gap* which refers to the extent the income consumption of an individual falls short of the poverty line, that is, minimum income or consumption level. We explain below these two concepts in some detail.

Head Count Ratio. Thus, absolute poverty is measured by counting the number of persons or heads living below the specified poverty line in terms of minimum consumption expenditure or income per capita. Since this approach to measure absolute poverty is concerned with counting the number of persons or heads, this is generally referred to as *Head Count Ratio* approach to measure absolute poverty. If N stands for the number of population, H for the number of persons having consumption expenditure or income below that, then head count ratio of absolute poverty can be written as

$$\text{Head Count Ratio (HCR)} = \frac{H}{N}$$

It may be noted that to measure absolute poverty at the international level and to compare the incidence of extreme poverty in various countries, World Bank has specified \$ 1.25 a day in PPP (purchasing power parity) US dollar as poverty line norm. Recently \$2 a day in PPP is also being used by World Bank to estimate the incidence of poverty prevailing in various countries of the world. A shortcoming of head count ratio as a measure of poverty is that it *ignores the extent to which a poor person's income falls* below the poverty line. It is noteworthy that changes in the incidence of poverty over time in a country also poses a number of problems because it raises the question as to which price index number ought to be used to find the real income or consumption expenditure required to meet the minimum basic needs.

Poverty Gap. However the head count ratio as a measure of absolute poverty suffers from a major drawback. The head count ratio does not indicate the *severity or depth* of a poor person. For example, if poverty line for rural India is fixed at monthly per capita consumption expenditure at ₹ 150, then all those who have a per capita monthly income or consumption expenditure less than this would be counted as poor, no matter whether a person's monthly consumption expenditure is ₹ 75 or ₹ 125. Obviously, poverty is more severe in case of persons who have monthly consumption expenditure of ₹ 75 than of those who have monthly consumption expenditure or income of ₹ 125. To overcome this drawback absolute poverty is measured by poverty gap approach. Poverty gap concept seeks to measure severity or depth of poverty of a person. If Y_p stands for minimum income or consumption expenditure representing poverty line, Y_i stands for income or consumption expenditure of any individual, then $Y_p - Y_i$ will indicate the *severity or depth of this poverty*. In accordance with this, the total poverty gap (TPG) in an economy or a sector can be written as

$$TPG = \sum_{i=1}^{H} (Y_p - Y_i)$$

where H is the number of persons having per capita income or consumption expenditure less than the minimum living standard or consumption expenditure specified as poverty line.

Total poverty gap is considered as a better index of the extent of poverty in a country as it takes into account the severity or depth of the poverty of persons comprising a society.

Absolute poverty is bad because no civilised society can feel satisfied if its people exist in conditions of absolute human misery. Absolute poverty not only creates miserable living conditions for individuals, it also imperils social and political stability in a society. It is now a well recognised fact that the existence of absolute poverty raises the incidence of crime and creates law and order problems in a society. Therefore, all modern societies have started taking steps to alleviate absolute poverty.

Relative Poverty

As is well known, per capita income is only an average income per person and therefore does not represent a true index of social welfare. Social welfare in a society also depends on how the national income is distributed among the population. Poverty is in fact a relative concept. Relative poverty refers to *the people having income less than the average or mean per capita income of a country*. While the level of economic development of a country determines the mean per capita income of a country, inequalities in the distribution of income determines the number of poor people who have less than mean or average per capita income of a country.

Income of an individual in relation to the incomes of others in a country or region is an important determinant of welfare. Thus, greater the inequalities in income distribution, the lower the level of social welfare. It has been realised that greater income inequalities affect the capabilities of those who have relatively lower income through adversely affecting their access to credit from banks and other financial institutions and their capability to provide education to their children. Relative poverty reflects income inequalities in a country and like absolute poverty negatively affects social welfare. Income inequalities in a society is generally measured through estimating the value of *Gini coefficient*.

Social welfare function can be written as

$$W = (Y, I, P)$$

where W = Social welfare
Y = Per capita income
I = Inequality index
P = Absolute poverty

Level of per capita income has a positive effect on welfare, whereas inequality (I) and absolute poverty (P) negatively affect welfare. That is, given the per capita income, the higher income inequality and absolute poverty the lower will be social welfare.

Why is Relative Poverty Bad? But what is so bad about relative poverty, that is, income inequality? Long ago Simon Kuznets using historical data from the first half of the 20th century from three countries — England, Germany and the U.S. pointed out that as a country industrialises and achieves economic growth, the gap between the rich and poor first widens (that is, income inequality increases) and then gradually narrows (that is, income inequality decreases). This is generally known as *Kuznets inverted U hypothesis*. To support their views, some advocates of free market economy argued that increased inequality was a necessary condition for accelerating economic growth. This view is not supported by the data collected by World Bank in case of countries.[6] In fact, it has now been found that income inequality adversely affects economic growth. This is due to the following factors.

First, existence of income inequalities lowers the access to credit of those who are relatively poor. It is noteworthy that when low-income individuals cannot have adequate access to credit, they cannot provide good education to their children or start and expand business. Further, the recent findings indicate that with high inequality saving rate in the economy tends to be lower[7]. This is

6 See Michael Bruno and Lyn Squire, *op. cit.* and Todaro and Smith, *Economic Development*, 8th edition, 2003, pp. 209-210.

7 See Joseph Zeira, "Income Distribution and Macroeconomics", *Review of Economic Studies*, 60 (January 1993), and Gary Fields, *Distribution and Development: A New Look at the Developing World*, Cambridge, MIT Press, 2000, Ch. 10.

because empirical findings reveal that highest rate of marginal saving is found among the middle classes, whereas the rich spend a relatively larger fraction of their incomes on luxury goods, gold and jewellery, luxurious houses or seek safe havens abroad for depositing their savings in foreign banks and this causes capital flight. In short, the rich save and invest domestically a relatively smaller proportion of their income. Commenting on worldwide data Bruno and Squire write, "*Research shows that developing countries with a more equal distribution of assets grew more rapidly than countries with a less equal distribution of assets.*"[8]

REGIONAL POVERTY ESTIMATES OF THE WORLD

World Bank has set a poverty line at $1.25 a day in terms of PPP 2005 US dollars. In Table 54.1, we have given the incidence of the poverty, given the poverty line at $1.25 a day PPP. It will be seen from this table that in the world as a whole (*i.e.*, total) 16.99 per cent of world population, which in absolute number amounts to over 1 billion people (1011.37 millions to be exact), lived below the poverty line in 2011. We will see later that both the percentage of world population and the number of world population living below this poverty line have been continuously declining.

Table 54.1. Regional Poverty Estimates of 2011

Region	Population below $1.25 PPP a day (Head Count Ratio)%	Poverty Gap at $ 1.25 PPP a day (%)	Number of Poor (Million)	Population (Million)
East Asia & Pacific	7.43	1.56	160.7	2027.27
Europe & Central Asia	0.49	0.14	2.35	479.13
Latin America & the Caribbean	4.63	2.17	27.63	596.77
Middle East and North Africa	1.69	0.35	5.64	333.78
South Asia	24.50	5.16	398.95	1628.35
Sub-Saharan Africa	46.85	19.20	415.7	883.43
Total World	16.99	5.05	1,011.37	5,952

Source: *World Development Indicators,* 2015.

From the region-wise data it will be seen from Table 54.1 that in the year 2011 in Sub-Saharan Africa, poverty ratio was the highest at 46.85 per cent. This was followed by South Asia (which includes India) where the percentage of population living below the poverty line was estimated 24.50 per cent in 2011. The number of people living below the poverty line ($1.25 a day PPP) was 415.7 million in Sub-Saharan Africa and 398.95 million in South Asia in 2011. These two regions together accounted for 80 per cent of the poor population of the world. In the East Asia and Pacific region 7.43% of the population was living below the poverty line which in absolute number was 160 million. In developing countries of Latin America and the Caribbean, there prevails less poverty as head count ratio of poverty with international poverty line of $1.25 a day was estimated to be equal to 4.63 per cent.

In Europe and Central Asia where mostly developed countries exist, the incidence of poverty at $ 1.25 a day is very low with only 0.49 per cent of population below the poverty line (head count ratio). The region of Middle East and North Africa covers oil-rich countries and earn large income from export of oil. Therefore, the incidence of poverty is very low; only 1.69% of their population was below the poverty line of $ 1.25 a day.

8 Bruno and Squire, "Why Worry about Inequality?" *op.cit.*

As in case of Head Count Ratio (HCR), measure of poverty in terms of poverty gap at $ 1.25 a day varies in the different regions (See Column III in Table 54.2). Again in Sub-Saharan Africa and South Asia, poverty gap was larger with 19.20 and 5.0 per cent respectively.

The magnitude of poverty in various developing countries is quite uneven. In Table 54.2 we have given both head count ratio of poverty and poverty gap at the international poverty line $ 1.25 a day and $ 2 a day. In this connection, the incidence of poverty in China and India are worth mentioning. China which achieved around 10 per cent GDP growth in the last over three decades has succeeded in substantial reduction in poverty. In 2010, percentage of population living below the International poverty line of $ 1.25 a day was only 6.3 per cent in China compared to 23.6 per cent in India. Similarly, poverty on the basis of poverty gap at $1.25 a day in China was quite low at 1.3 per cent compared to 4.8 per cent of India. It will also be seen that East Asian countries of Indonesia, Malaysia, Thailand which through their export-led growth strategy achieved quite a high GDP growth from the mid-1960s to 2000, poverty (both head count ratio and poverty gap) has substantially declined. In African countries, of Zambia, Guinea, Nigeria, Kenya, Central African Republic the incidence of poverty is very high. On the other hand, in Latin American Countries of Mexico and Philippines the poverty both in respect of Head count Ratio (HCR) and Poverty Gap at $1.25 a day is quite low.

Table 54.2. Percentage of Population below International Poverty Line of $ 1.25 day

Country	Survey year	Population below $ 1.25 a day (%)	Poverty gap at $1.25 a day (%)
Bangladesh	2010	43.3	11.2
Benin	2011	51.6	18.8
Bhutan	2012	2.4	0.5
Botswana	2009	13.4	4.0
Brazil	2012	3.8	2.1
Central African Republic	2008	62.8	31.3
China	**2010**	**6.3**	**1.3**
Guinea	2012	40.9	12.7
India	**2011**	**23.6**	**4.8**
Indonesia	2011	16.2	2.7
Jordan	2010	2.0	0.5
Kenya	2005	43.4	16.9
Malaysia	2009	2.0	0.5
Mexico	2012	2.82	0.5
Philippines	2012	19.0	4.0
Pakistan	2010	12.7	1.9
South Africa	2011	9.4	1.2
Sri Lanka	2009	4.1	9.7
Thailand	2010	2.0	0.5
Nigeria	2010	62.6	27.5
Nepal	2010	23.7	5.2
Zambia	2010	74.3	41.8

Source: *World Development Indicators*, 2015.

Declining Poverty Rates at International Poverty Line

An important fact revealed by the poverty rates at international poverty line fixed at $1.25 a day of the world as a whole and in different regions is that they have been falling since 1980s (see Table 54.3). The percentage of population living below the international poverty line of $1.25 a day of the whole world fell from 51.9 per cent in 1981 to 41.7 per cent in 1990—a 10 percentage points fall in the decade. The percentage of total world population below the poverty line ($1.25 a day) further declined to 33.7 per cent in 1999 to 25.2 per cent in 2005 and to 16.99 per cent in 2011.

As will be seen from Table 54.3, the largest decline in poverty ratio with $1.25 a day as the poverty line took place in East Asia and Pacific where the poverty ratio declined from around 78 per cent in 1981 to 17 per cent in 2005 and the **number** of persons living below $1.25 a day fell more than 750 million in this region. It is worth mentioning that a largest part of decline in poverty ratio in this region was in China where incidence of poverty decreased from 84 per cent of population in 1981 to 16 per cent in 2005. It is in this period that China made a transition from socialist regime to privatisation and liberalisation of the economy with foreign investment playing an important role in its industrial development and succeeded in achieving around 10 per cent GDP growth rate.

Table 54.3. Falling Poverty Ratio (HCR) at International Poverty Line of $1.25 a day (per cent)

Region	1981	1990	1999	2002	2005	2011
East Asia and Pacific	77.7	54.7	35.5	27.6	16.8	7.43
China	84.0	60.2	35.6	28.4	15.9	6.30
Europe and Central Asia	1.7	2.0	5.1	4.6	3.7	0.49
Latin America and Caribbean	12.9	11.3	10.9	10.7	8.2	4.63
Middle East & North Africa	7.9	4.3	4.2	3.6	3.6	1.69
South Asia	59.4	51.7	44.1	43.8	40.3	24.5
India	59.8	51.3	44.8	43.9	41.6	23.6
Sub-Saharan Africa	53.4	57.6	58.4	55.0	50.9	46.85
Total	51.9	41.7	33.7	30.5	25.2	16.99

Source: *World Development Indicators*, 2011 and 2015.

In South Asia where India is situated the decline in poverty ratio was not so dramatic; poverty ratio in South Asia dropped from 59 per cent in 1981 be 40 per cent in 2005 and further to 24.5 per cent in 2011. It will be observed from Table 54.3 that in India percentage of population living below $ 1.25 a day fell from 59.8 in 1981 to 41.6 per cent in 2005 and further to 23.6 per cent in 2011. In sharp contrast to other regions, in Sub-Saharan Africa the poverty ratio declined only slightly during this period, it fell from 57.6 per cent in 1990 to 50 per cent in 2005 (However, the decline in poverty ratio in Sub-Saharan Africa is not consistent as it rose in some years, for example, it increased between 1981 and 1990) (See Table 54.3).

It may be noted that the target of Millennium Development Goal was to reduce the world poverty ratio to half by 2015 compared to what prevailed in 1990. As will be seen from Table 54.3 this target was achieved by 2011 as in 2011, world poverty ratio declined to 16.99 per cent compared to 41.7 per cent in 1990. It may be further noted that to reduce poverty ratio, the faster and sustained GDP growth rate is required. But it is now well recognised that GDP growth is necessary but not a sufficient condition for reduction in poverty. What is needed is that the character of GDP growth should be more inclusive with generation of adequate opportunities of gainful employment for the people.

The New Index of Poverty: Multidimensional Poverty Index (MPI) by the United Nations

Human Development Report (HDR) 2010 measures poverty in terms of a new parameter namely "*Multidimensional Poverty Index* (MPI) which has replaced the *human Poverty Index* (HPI) since 1997. *The multidimensional poverty index* (MPI) *indicates the share of the population that is multi dimensionally poor adjusted by the intensity of deprivation in terms of living standards, health and education.* According to this parameter, as will be seen from Table 54.4, India with a multidimensional poverty index of 0.296 and poverty ratio of 41.6 per cent in terms of purchasing power parity (PPP) of $ 1.25 a day and 28.6 per cent (as per national poverty line) in 2004-05 is not favourably placed when compared with countries like China and Sri Lanka. In fact, the difference in population below the poverty line (BPL) as per multidimensional poverty index (MPI) and national poverty indicator widens substantially in case of India while for other countries there is less of difference and in some cases even a fall (see Table 54.4). This multidimensional poverty index has been prepared by the '*Oxford Poverty and Human Development Initiative*' for United Nations Development Programme (UNDP).

India's appalling record of inability to ensure basic levels of nutrition and other basic needs to its people is responsible for its huge poverty level as measured by the new international Multidimensional Poverty Index (MPI). About 645 million people, that is, *55% of India's population is poor as measured by this composite index made up of 10 markers of health, education and standard of living achievement levels. This MPI attempts to capture more than just income poverty at the household level. It is composed of ten indicators: years of schooling and child enrolment (education),*

Table 54.4. Multidimensional Poverty Index

Country	Multidimensional Poverty Index*	Population below Income Poverty Line	
		PPP $ 1.25 a day 2000-2008**	National Poverty Line 2000-2008**
Poland	(41)	Less than 2	14.8
Malaysia	(57)	Less than 2	12.8
Russia	0.005 (65)	Less than 2	19.6
Brazil	0.039 (73)	5.2	21.5
Turkey	0.039 (83)	2.6	27
China	0.056 (89)	15.9	2.8
Sri Lanka	0.021 (91)	14	22.7
Thailand	0.006 (92)	Less than 2	–
Philippines	0.067 (97)	22.6	–
Egypt	0.026 (101)	Less than 2	16.7
Indonesia	0.095 (108)	29.4	16.7
South Africa	0.014 (110)	26.2	22.0
Vietnam	0.075 (113)	21.5	28.9
India	0.296 (119)	41.6	28.6
Pakistan	0.275(125)	22.6	–
Kenya	0.302(128)	19.7	46.6
Bangladesh	0.291(129)	49.6	40.0

Source: HRD 2010.

Note: * Not all indicators were available for all countries; Caution should thus be used in making cross-country comparisons

** Data refer to the most recent year available during the period specified. Data regarding national poverty refers to the year 2004-05

Figures in parenthesis in column 2 *give ranking among 169 countries.*

child mortality and nutrition (health), electricity, flooring, drinking water, sanitation, cooking fuel and assets. Standard of living indicators of education and health has a 1/6 weight each living indicator has a 1/8 weight.

The data for the year 2008 also shows that even in relatively rich Indian States such as Haryana, Gujarat and Karnataka more than 40 per cent of population is poor by the new multidimensional poverty index, while Kerala is the only State in which the poor are less than 20 per cent by this composite measure of poverty.

It is worth noting that **MPI measures both the incidence of poverty as well as its intensity.** In this multidimensional approach, a person is defined as poor if he or she is deprived of at least three of the ten indicators. By this definition 55% of India's poor is close to double of much criticised figure of official national poverty ratio of around 29 per cent. Nutritional deprivation is overwhelming which is the largest factor responsible for overall poverty in India as half of all children in India are undernourished according to the National Family Health Survey III (2005-06). Close to 40% of those who are defined as poor are also nutritionally deprived. In fact, the contribution of nutrition to the overall multidimensional poverty index (MPI) is greater in urban than in rural India.

Poverty by Caste Categories as per MPI. In Hindu population of India multidimensional poverty is highest among Scheduled Tribes with 81.4 per cent of them being poor followed by Scheduled Castes with 65.8 persons of poor, other backward classes (OBC) with 58.3% of them being poor and finally general Hindu population with 33 per cent counted as poor as per this composite measure. There is significant variation between the poverty incidence in various States as per the MPI and as per the Indian Planning Commission official figures based on head-count ratio. Bihar has by far the most poor of any State in the country with 81.4 per cent of its population defined as poor which is close to 12 per cent more than the next worst State of Uttar Pradesh. As per the Planning Commission figures 41.4 per cent of Bihar and 32.8 per cent of UP is poor. In a possible indication of inadequate access to health and education facilities which do not show up in income poverty, almost 60 per cent of North-East India and nearly 50 per cent of population of Jammu and Kashmir are poor as per the MPI while the Planning Commission figures of income poverty of these are around 16 per cent and 5 per cent respectively.

It is evident from above that India's method of poverty estimates by Planning Commission with focus primarily on subsistence expenditure or income are quite inadequate to capture the extent of poverty or deprivation in the country.

WORLD BANK'S NOTION OF SHARED PROSPERITY

Until recently World Bank has been concerned with reducing poverty in the developing countries and set a target of reducing poverty in 2015 half of what prevailed in 1990. Although poverty rates have fallen rapidly over the last three decades but as we have seen above more than 1 billion people of the world still live below the extreme poverty line of $ 1.25 a day. The existence of this extreme poverty is morally unacceptable given the world's resources and technology available. Now, the World Bank has set a target of reducing the extreme poverty of the people living below $ 1.25 a day to 3 per cent globally by 2030. This requires rapid and sustained growth in the developing countries and the measures to be taken by the government to ensure that benefits of growth reach the poor people.

Besides the goal of almost eliminating extreme poverty within a generation a special World Bank Group (WBG) has set a more ambitious goal of promoting what has been called *'Shared Prosperity'* By shared prosperity the World Bank Group means that a *sustainable growth in developing countries must ensure increase in the well-being of the less well-off sections of the society*. More specifically, under the goal of shared prosperity the World Bank will *strive to foster income growth of the bottom 40 per cent of the population in every country*. This goal of achieving shared prosperity is proposed to be well-aligned with the overall objectives of the Millennium Development Goals in the post-2015 Agenda. It has been pointed out by the World Bank Group that in the world there has been rising inequality in income distribution and social exclusion that has accompanied GDP growth in many countries poses a threat to social stability. While in the past the focus of the World Bank has been to achieve faster GDP growth and it was assumed that the benefits of economic growth would trickle down to the poor. Now, under the leadership of its new head Dr. Kaushik Basu, it has been recognised by the World Bank Group that the benefits of GDP growth will not automatically trickle down to the poor. Therefore, under the goal of share prosperity, according to it, *conscious efforts by the state have to be made so that the gains of economic growth must be adequately shared by the bottom 40 per cent of the population*. Thus, according to World Bank Group, "Our shared prosperity implies a *direct focus on the income of the less well-off* (that is, bottom 40 per cent of population) as opposed to the common practice of focusing only on growth of GDP and implicitly relying on the "trickle down" impact of growth on the bottom of the distribution[9]". It may be noted that 'shared prosperity' notion of World Bank seems to be the same as the concept of *inclusive growth* laid down in the *Twelfth Five Year Plan* (2012–17) of India.

The three elements considered to be central to the notion of shared prosperity are:

(1) Sustained economic growth.
(2) Reduction in income inequalities and achieving equity in income distribution.
(3) Differences in opportunities for various sections of the society.

We explain below the above three elements of shared prosperity.

Shared Prosperity and Economic Growth

Economic growth is thought to be necessary because it ensures an expansion in productive capacity of the economy which when utilised causes increase in the amount of goods and services required for increase in consumption of the people. Therefore, without economic growth, the living standards of the less well-off are unlikely to rise in the medium to the long run.

For example, growth of agriculture will increase the opportunities for a large part of the population (including small farmers and landless labourers). Economic growth is also important because it generates more employment for the people. Like agricultural growth, growth of small and medium enterprises will create higher incomes and larger employment opportunities for the poor people.

But for growth to be inclusive, the priority ought be given in the policy framework that ensures equitable distribution of growth in GDP or national income. The World Bank document on shared prosperity emphasises that for achievement of shared prosperity the growth in incomes should be accompanied by *decline in income inequality* as has been in the case of many Latin American countries during the past decade (2000-2010) and further that "sustained progress in shared prosperity is incompatible with a long-term increase in inequality.'[10]

It is evident from above that there is now distinct shift in the approach to development of the World Bank which along with IMF previously emphasised GDP growth alone in its structural adjustment reforms to be undertaken by the developing countries. Now, it refers to the two empirical research studies at the International Monetary Fund which find that the inequality of incomes is an obstacle to economic growth. Thus one such study finds that '*high inequality is associated with shorter growth episodes*'.[11] Similarly, another research study by W. Easterly suggests that '*structural inequality is a significant barrier to prosperity and growth in per capita income.*'[12]

9 World Bank, "*End Extreme Poverty and Promote Shared Prosperity*", 2013, p. 20.

10 World Bank, *op.cit.*, p. 22.

11 A Berg, J. Osty and J. Zettelmeyer. 'What Causes Growth Sustained' *Journal of Development Economics*, 2012, 98 (2).

12 W. Easterly, 'Inequality Does Cause Underdevelopment', *Journal of Development Economics*, 2007, 84 (2).

More importantly, Ferreira and Ravallion[13] have found that sustained economic growth that raises the welfare of those at the bottom is not compatible with long run increase in income inequality. They point to the fact that in general *no country has grown beyond middle income level while maintaining a high level of inequality.* This is evident from Fig. 54.1 where data of income inequality and GDP per capita of various countries is shown by various dots representing them. A trend curve in the data is drawn which clearly shows that in countries with higher GDP per capita income inequality has declined. However, there is large variation in inequality among countries representing low and middle income countries. It further shows that where *income inequality is higher promoting shared prosperity will need that income of the bottom 40 per cent grow faster than average income of the population.*

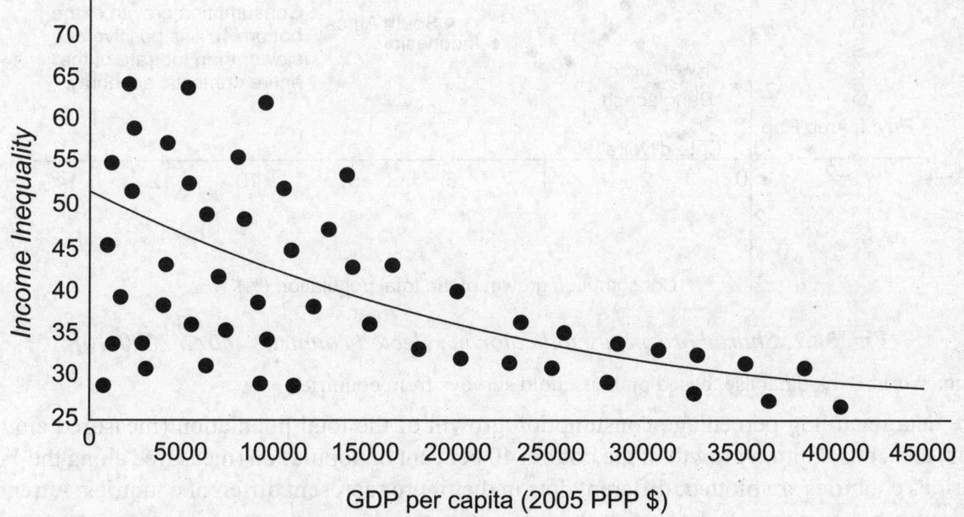

Fig. 54.1. *Inequality and GDP per capita in low, middle and high income countries* (*circa* 2010)

It may be noted that for achievement of shared prosperity, income of the bottom 40 per cent of population must rise faster than the growth of the average income of the total population. *Thus though growth in GDP is necessary for higher performance in achievement of shared prosperity but it is not a sufficient condition for it.* This is because economic growth does not necessarily lead to the gains in incomes of the bottom 40 per cent of population. This is evident from the experience of the United States where per capita income of the bottom 40 per cent of population rose by just 7.5 per cent between 1975 and 2012 while mean income of the total households income increased by 40.5 per cent in the same period.[14]

The higher initial income inequality is a significant barrier to shared prosperity as well as GDP growth rate. It has been found that in 20 out of 55 developing countries in the decade of 2000-2010, real per capita consumption (or income) of the bottom 40 per cent of the population increased at the average rate of more than 5 per cent annually which is quite satisfactory. It is worth mentioning the empirical findings about the relationship between overall consumption (or income) growth and the consumption (or income) of the bottom 40 per cent of the population. As will be seen from Fig. 54.2

---

13. F. Ferreira and M. Ravallion (2009), "Poverty and Inequality: The Global Context in W Salverda et al" *The Oxford Handbook of Economic Inequality,* Oxford University Press.
14. See *Shared Prosperity, Links to Growth, Inequality and Inequality of Opportunity*, World Bank's Policy Research Working Paper 6649 by Ambar Narayan, Jaime Saavedra Chanduvi and Sailesh Tiwari, 2013, p.12.

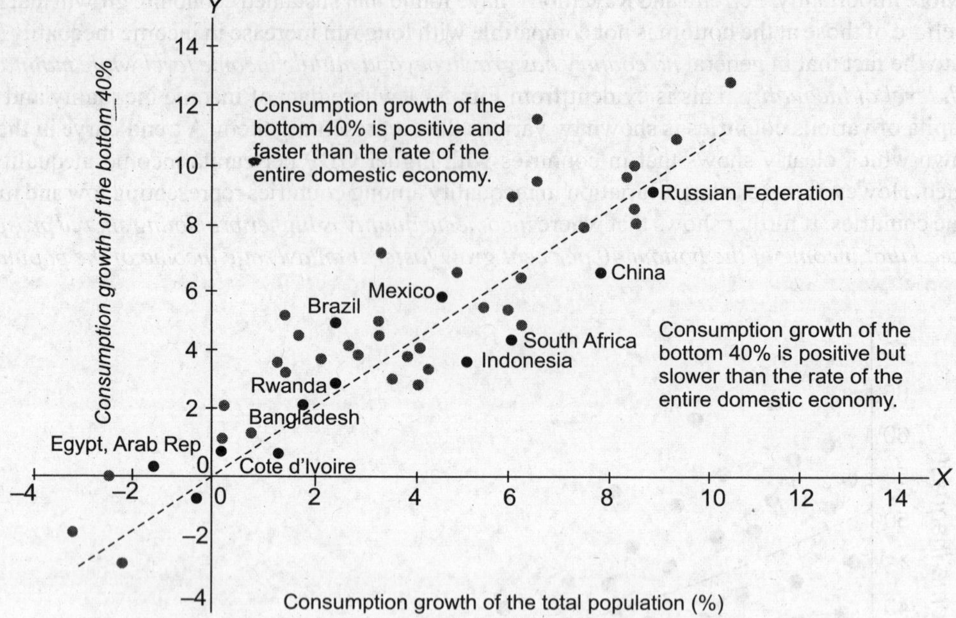

Fig. 54.2. *Shared prosperity indicator in selected countries (circa 2000-10)*

Source: World Bank database, based on household surveys from countries.

where data regarding percentage consumption growth of the total population (measured along the *X*-axis) and consumption growth of the bottom 40 per cent of population (measured along the *Y*-axis) of various countries are plotted; different dots in the figure represent different countries. A trend line has also been drawn which shows that there is a strong and positive relation between consumption of the bottom 40 per cent of population and consumption growth of the total population, This means *for achievement of shared prosperity growth in national income or consumption is necessary*.

The second important fact revealed by Fig. 54.2 is that in the decade 2000-2010, 60 per cent of developing countries which include large Latin American countries of Brazil and Mexico income (or consumption) of the bottom 40 per cent of population increased at a higher rate than the average growth of consumption (or income) of the total population.[15] This could *occur only if there was a decline* in income inequality. This clearly shows that *fall in inequality in income distribution is essential for promoting shared prosperity*. It will be seen from Fig. 54.2 that the countries which lies above the trend line such as Brazil, Mexico, though their percentage consumption growth of the total population is low, the percentage consumption growth of the bottom 40 per cent of population is relatively higher. Thus the *notion of shared prosperity that focusses on growth of bottom 40 per cent of population rather than average rate of GDP growth ensures improvement in living standards or welfare of the less well-off sections of population in a country*. It will also be observed from Fig. 54.2 that position of countries such as China, Indonesia, South Africa which lie below the trend line have relatively higher consumption growth of total population but greater income inequality, the consumption growth of bottom 40 per cent of population is lower.

Equality of Opportunity and Shared Prosperity

To foster shared prosperity requires ensuring equality of opportunity to all people, both children and male adults as well as women. It is only by providing equality of opportunity that production potential of every individual is maximised. Equality of opportunity in childhood and youth requires

15 World Bank, *End Extreme Poverty and Promote Shared Prosperity,* 2013, p. 24.

adequate nutrition, healthcare and education so as to build human capital and raise productivity in the production process. During adulthood, ensuring equality of opportunity implies imparting appropriate skills and education, taking care of their health and creation of employment opportunities and providing credit facilities to the young entrepreneurs so that they can set up their micro and small enterprises if not employed on wage basis in the organised sector. Thus, a World Bank research paper on shared prosperity[16] states that "*Efforts to improve and equalise opportunities would lead to immediate rise in welfare along multiple (including non-monetary) dimensions particularly among the poor and the excluded. Over long term, it would also promote social mobility, reduce income inequality and enhance economic dynamism and prosperity for the economy as a whole*". While the definitions of equality of opportunity by different economists and social thinkers vary significantly but the minimal view is that all children and adults must have access to basic goods and services such as safe water, adequate nutrition and healthcare, primary education so as to maximise human capability. In India, for example, there exists a gender gap in providing basic goods and services so that the girls and women are denied access to education, nutrition, healthcare and employment opportunities. To achieve shared prosperity women must be ensured equal opportunity with men.

Pathways to Shared Prosperity

There are many pathways for achievement of shared prosperity. Economic growth can insure improvement in living standards of bottom 40 per cent of population if it is broadbased and generates a lot of opportunities for better quality jobs or gainful employment for all sections of the population. The state has to play a key role in promoting climate for investment in labour-intensive sectors. For generation of employment opportunities the government should link investment allowances to the magnitude of employment created by the industrial enterprises rather than to the amount of capital investment. Through fiscal incentives, the government should ensure that growth is diversified across multiple sectors. Besides, the Government should ensure "*ease of doing business*", improve competitiveness by preventing the emergence of monopolies and oligopolies in the production of goods and services. Besides, for the creation of adequate jobs, the government should encourage innovations or technological changes that are labour-absorptive rather than labour-displacing.

The second pathway to promote shared prosperity is to ensure that *growth is inclusive* through a specific structure of taxation, government expenditure and social programmes initiated by the government, especially those that make *investment in people* which build their capabilities and increase productivity and thereby promote economic growth with equity. For fostering shared prosperity, the tax structure should be such that gives incentives to the private sector but at the same time it should ensure equitable distribution of income in a society.

The third path to promote shared prosperity is to make investment by the government in social institutions so as to ensure equal *opportunities for all* including the marginalised sections of the society. For this purpose, the World Bank Group emphasizes that *rather than redistributing resources from one section of the society to another, the investment should be made for improving the capabilities of the people that increases their productivity and therefore promote growth and equity over time and across generations. With this the people will be able to better their lives on their own.*[17]

The evidence furnished by the World Bank Group gives many examples relating to developing countries where inequality of opportunity is still very high and despite growth made by them the circumstances at the time of birth of the individuals, the background and status of their parents, their gender and ethnicity still play a large part in determining his opportunity. To foster shared prosperity the World Bank document states, "The social contract for promoting equity and growth should aim to create an opportunity society where the human and productive potential of every individual regardless of gender, region, ethnicity or parental background is maximised. In childhood and youth, this would

16. See Ambar Narayan, Jaime Saavedra Chanduvi and Sailesh Tiwari, *op. cit.*
17. See World Bank, *"End Extreme Poverty and Promote Shared Prosperity"*, 2013, p. 28.

imply providing universal access to early childhood development, health, nutrition, education and basic infrastructure to build human capital and increase long-term productivity. This will enhance mobility on the economic and social ladder within and across generations."[18]

Further, a significant part of social contract aimed at creating conditions to promote shared prosperity is to ensure *gender equality.* Having equal opportunity with men and same access to facilities of education, healthcare, nutrition, sanitation would enable women to contribute to the best of their abilities in the development process. To quote the said World Bank Document, "Gender equality is not just a core development objective; it also enhances productivity, improves development outcomes for the next generation, and makes institutions more productive."[19]

LATEST ESTIMATES OF GLOBAL POVERTY AT UPDATED POVERTY LINE OF 1.90 PPP $ A DAY BY WORLD BANK

The World Bank. On October 6, 2015, released its new estimates of global poverty in its report titled "*Ending Extreme Poverty and Sharing Prosperity: Progress and Policies*". In this report World Bank has used a new updated international poverty line of $1.90 PPP a day per person in 2011 PPP prices to measure extreme poverty in the world. *Thus in these latest estimates, the extreme poverty is defined as living at less than $ 1.90 a day*. It is important to note that this new poverty line has maintained the purchasing power of previous international poverty line of $ 1.25 (in 2005 PPP prices) living below which represented the extreme poverty earlier. Besides, the new extreme poverty line of $ 1.90 a day incorporates the new information on *differences in cost of living across countries* and the new PPP exchange rates. The estimates of global poverty based on the new international poverty line of $ 1.90 a day are given in Table 54.5.

Table 54.5. Estimates of Extreme Poverty based on the New International Poverty Line ($ 1.90 a day in 2011 PPP prices)

Year	Number of poor (millions)	Poverty Ratio
1990	1,958	37.1
1999	1,747	29.0
2011	987	14.2
2012	902	12.8
2015*	702	9.6

* Projected

Source: World Bank Report, "*Ending Extreme Poverty and Sharing Prosperity: Progress and Policies*, 2015.

As will be seen from Table 54.6 that the important fact revealed from the poverty estimates based on the new international poverty line of $ 1.90 a day is that the ratio of the poor people (*i.e.*, the people living below the new poverty line) to the total world population has declined from 37.1 per cent in 1990 to 29 per cent in 1999 and further to 12.8 per cent in 2012 and is projected to fall to the single digit of 9.6% in 2015. Similarly, in absolute number, the persons living below the new poverty line of $ 1.90 a day declined from 1,958 million in 1990 to 1,747 million in 1999 to 902 million in 2012 and is estimated to fall further to 702 million in 2015. This shows a significant achievement of the world community in reducing extreme poverty in the last 25 years, especially since 1999. About 1 billion people of the world have been lifted out of extreme poverty since 2000 (1747–702 = 1045 million).

Most developing countries have experienced significant income growth of the bottom 40 per cent of their population in their income distribution. These poverty estimates based on new international poverty line and the new country level data on living standards shows that a quarter century long (1990-2015) reduction in extreme poverty in the developing countries (especially China and India) is moving the world closer to the historic goal of the UN of eradicating extreme poverty by the end of 2030 under its new sustainable development goals.

18 World Bank, *op. cit.*

19 *Ibid.*

The new estimates of poverty are graphically shown in Fig. 54.3.

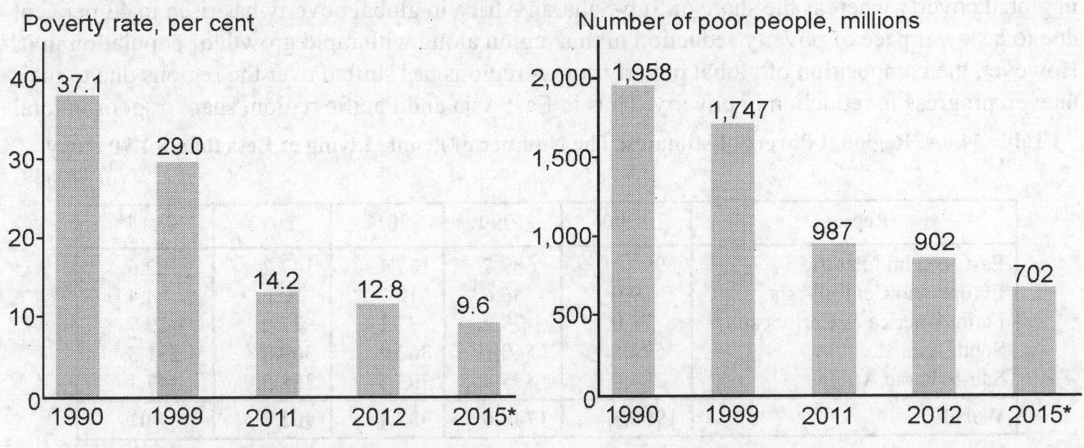

(*a*) *the global poverty rate has declined significantly over the last 25 years*

(*b*) *The most rapid decline occurred during the 2000s*

Fig. 54.3. *Estimates of global poverty as per the new international poverty line of $ 1.90 a day*

The estimate of 1 billion people of the world being lifted out of poverty since 2000 and that in 2015, the global poverty would fall to single digit of 9.6 % are good news. Jim Yonk Kim, the president of World Bank group which prepared the report has said that, "This is the best story in the world today . . ., these projections show us that we are in the first generation in human history that can end extreme poverty." He however has cautioned that with slowing global economic growth and many of the world's remaining poor people live in fragile and conflict affected states, the goal to end extreme poverty by 2030 is a highly ambitious target.

It may be further noted that the decline in poverty to 9.6 per cent of world population in 2015 is quite impressive but the number of people living in extreme poverty conditions, namely, 702 million persons in 2015, still remains unacceptably high.

Regionwise New Poverty Estimates

We have seen above that global poverty has declined significantly over the last three decades. However, the progress in poverty reduction in different regions of the world has been quite uneven. As will be seen from Tables 54.6 and 54.6A that extreme poverty is now more concentrated in Sub-Saharan Africa and South Asia. For the past several decades, the same three regions, namely, East

Table 54.6. Regional Poverty Estimates with updated International Poverty Line ($ 1.90 a day)

Share of Population living below $ 1.90 a day (%)

Region	1990	1999	2011	2012	2015*
East Asia and Pacific	60.8	37.5	8.5	7.2	4.1
Europe and Central Asia	1.9	7.8	2.7	2.5	1.7
Latin America & Caribbean	17.7	14.1	6.5	6.2	5.6
South Asia	50.6	41.2	22.2	18.8	13.5
Sub-Saharan Africa	56	58.1	44.3	42.6	35.2
Developing World	44.3	34.2	16.6	15.0	11.9
World	**37.1**	**29.0**	**14.2**	**12.8**	**9.6**

***Projected**

Asia and Pacific (which includes China), South Asia (which includes India) and Sub-Saharan Africa account for 95 per cent of global poverty. The latest 2012 poverty estimates confirm this high degree of concentration (see Tables 54.6 and 54.6A). Yet the composition of global poverty across these three regions has shifted overtime. There is large reduction in relative share of East Asia and Pacific in global poverty whereas the share of Sub-Saharan Africa in global poverty has risen to 43 per cent due to a slower pace of poverty reduction in this region along with rapid growth of population in it. However, the composition of global poverty across regions has shifted over the regions due to their uneven progress in reduction in poverty. Thus in East Asia and Pacific region, share of poor in total

Table 54.6A. Regional Poverty Estimates: The Number of People Living at Less than $ 1.90 a day

(*in Million*)

Region	1990	1999	2011	2012	2015*
East Asia and Pacific	999.3	689.7	177.1	147.2	82.6
Europe and Central Asia	9.0	36.6	12.7	12.6	4.4
Latin America & Caribbean	78.0	77.2	37.2	37.1	29.7
South Asia	574.5	560.1	362.3	309.2	231.3
Sub-Saharan Africa	284.0	375.4	393.5	388.5	347.1
World	**1958.5**	**1746.6**	**987.4**	**902.0**	**702**

***Projected**

Source: World Bank Report '*Ending Extreme Poverty and Sharing Prosperity*': Progress and Policies, Oct. 2015

population which was 60.8 per cent in 1990, fell to 37.5 per cent in 1999 and to 7.2 per cent in 2012 and is projected to fall further to 4.1 per cent in 2015. This has been due to rapid progress made in China in poverty reduction due to very high GDP growth rate. Similarly in South Asia, there has been good progress in poverty reduction due *mainly to substantial reduction in poverty in India in the last over two decades*. Thus, in South Asia poverty ratio which was 50.6 per cent in 1990, fell to 41.2 per cent in 1999 and further to 18.8 per cent in 2012 and is projected to fall further to 13.5 per cent in 2015.

The share of Sub-Saharan Africa in global poverty has risen to 43 per cent due to a slower pace of poverty reduction in the region amidst rapid population growth. As will be seen from Table 54.6, in Sub-Saharan Africa the poverty ratio fell from 56 to 42.6 per cent between 1990 and 2012. Though South Asia achieved more rapid poverty reduction over the past 30 years, yet it is still home to about a third of the world's poor (see Table 54.6A). It will be seen from Table 54.6A that in 2012, as per new international poverty line of $ 1.90 a day in 2011 PPP prices, out of 902 million total world poor, 309 million people lived in South Asia and 388 million in Sub-Saharan Africa. Thus, *it is evident that extreme poverty is now concentrated in South Asia and Sub-Saharan Africa*. More focus is therefore needed to reduce extreme poverty in Sub-Saharan Africa and South Asia where most of the poor live.

In case of India, the World Bank Report 2015 says that, according to Uniform Recall Period (URP) methodology, the per cent of population living below the poverty line (as per national estimates made by Tendulkar Committee) for 2011-12 is 21.2 per cent. According to it, though India has succeeded in significantly reducing poverty ratio but "*India was home to the largest number of poor in 2012* but its poverty rate is one of the lowest among those countries with the largest number of poor."

According to the report, strong economic growth and greater investment in physical and human capital in emerging economies, particularly in India and China, have led to the significant decline in poverty in the developing world.

PROBLEM OF POVERTY IN INDIA

To eliminate poverty has been an important objective of planned economic development in India. Even before independence it was thought that without economic freedom and eradication of poverty, political freedom would be meaningless. In designing our strategy of development, the earlier Five Year Plans assumed that a fast rate of economic growth would by itself create adequate employment opportunities and produce higher living standards for the poor. In addition, in the earlier Plans, land reforms especially redistribution of land by imposing ceiling on land holdings, tenancy reforms aimed at protecting the tenants from eviction and charging high rents by landlords were expected to improve the living conditions of the poor.

But economic growth in the first two decades (1951–71) resulted in increase in national income at an average rate of 3.7 per cent per annum. Yet the achievement of this growth rate made little impact on the problem of poverty in India. Land reform legislations were not effectively implemented and remained on paper only.

Dandekar-Rath Measure of Poverty

In the early seventies Dandekar and Rath[20] in their important study on poverty in India estimated that in rural areas *per capita monthly consumption expenditure* of at least ₹ 14.2 per person at 1960-61 prices was required to obtain a minimum subsistence diet of 2250 calories of food. On the basis of this norm of the poverty line, *they estimated that, in 1960-61, 33 per cent of the Indian rural population lived below the poverty line*, that is, they did not get even the above-mentioned minimum subsistence diet. For the *urban areas* they fixed ₹ 22.6 per person at the 1960-61 prices as the monthly minimum consumption expenditure required for minimum subsistence and estimated that in *1960-61, 48.6 per cent of urban population did not obtain this subsistence, that is, they lived below the poverty line. Further, by making adjustments for changes in prices, they estimated that in the year 1968-69, 40 per cent of the rural population and 50 per cent of the urban population lived below the poverty line*. Thus, according to the estimates of Dandekar and Rath, during 1961-69 the extent of rural poverty substantially increased, while urban poverty increased only marginally. However, contrary to popular perception, according to Dandekar and Rath study, the incidence of urban poverty was much greater than that of rural poverty. According to it half of urban people lived below the poverty line.

Dandekar and Rath study on poverty prompted other scholars such as Dr. B.S. Minhas[21] Pranab Bhardan[22] to make their own studies for estimating poverty in India, though there are some important differences in their estimates of poverty. A fact brought to light by these studies on poverty was that the *first two decades of planned development did not make much impact on the problem of poverty*. This gave rise to rethinking on the part of the Government and Planning Commission about revising strategy of development to solve the problem of poverty. Thus, in its Approach to Fifth Five Year Plan (June 1972), Planning Commission stated, "Economic development in the last two decades has resulted in an all-round increase in per capita income. The proportion of the poor, defined as those living below a basic minimum standard of consumption, has slightly come down. Yet the *absolute number of people below the poverty line today is just as large as it was two decades ago* and these people living in object poverty constitute between 40 and 50 per cent of all citizens."[23] Further it was realised in the approach paper that *to rely on economic growth alone* will involve too long a period for the solution of the problem of poverty. It was therefore suggested that *a direct attack on the problem of poverty and unemployment be made if the objective of assuring minimum consumption levels were to be achieved within a foreseeable future*. Direct attack on poverty implies that *special employment schemes* be started so as to assure more and fuller employment to the poor.

The study by Dandekar and Rath on poverty also gave rise to a severe controversy about the proper estimates of poverty and to what extent it had declined. This controversy still exists today with different scholars making different estimates of poverty. In what follows we shall study the

20. V.M. Dandekar and N. Rath, Poverty in India, *Economic and Political Weekly*, Jan. 2 and Jan. 9, 1971.
21. B.S. Minhas, *Planning and the Poor*, S. Chand and Co. 1973
22. Pranav K. Bhardan, On the Incidence of Poverty in Rural India, *Economic and Political Weekly*, Annual Number Feb. 1973.
23. Planning Commission, *Approach to the Fifth Five Year Plan*, 1972, p. 3.

various issues involved in estimating poverty and the extent to which it has declined during the last five decades of development (1951–2000). We shall also study the causes of prevailing poverty and the measures or strategy that should be adopted to tackle this problem.

Issues in Estimating Poverty

An important issue for estimating *poverty line* is to determine the minimum nutritional requirement of a person in the rural and urban areas. In this regard widely accepted norm is the consumption of 2400 calories of food per day per person in the rural areas and 2200 calories per day per person of food consumption in the urban areas. The next step is to convert the minimum consumption into money terms, that is, into minimum consumption expenditure. For this, appropriate prices have to be used.

To estimate the variation in poverty over a period of time, poverty line has to be adjusted on the basis of change in prices over a period of time. For this appropriate price deflator has to be used and in regard to this several difficulties are faced. Thus, the following three elements affect the fixation of poverty line which determines the people below and above the poverty line.

1. Norm regarding minimum nutritional requirements of a person.
2. Use of what prices be made to convert the minimum nutritional requirement into money terms.
3. Choice of a price deflator to update the poverty line.
4. To adjust the number of households in different expenditure classes to determine the number of households below and above poverty line.

Various scholars have made different assumptions and different methods have been used to determine the above elements. This accounts for a wide difference in their estimates of the incidence of poverty.

Trends in Poverty in India Since 1973-74

The estimates of poverty made by the Planning Commission from 1973-74 to 2004-05 are given in Table 54.8 for the national level. It may however be noted that the estimates of poverty for the year 1999-2000 are not quite comparable to those of the earlier years, as the *questionaire* used for the survey was changed. Therefore, the decline in poverty ratio in 1999-2000 was highly controversial. However, poverty estimates for 2004-05 are based on uniform recall period (URP) as the earlier ones and therefore they are comparable with 1993-94 estimates and of the previous years.

Table 54.7. Trends in Poverty in India

Years	*Poverty Ratio (%)*			*Poor (million)*		
	Rural	*Urban*	*(All India Combined)*	*Urban*	*Rural*	*Total*
1973–74	56.4	49.0	54.9	261	60	321
1977–78	53.1	45.2	51.3	264	65	329
1983	45.7	40.8	44.5	252	71	323
1987–88	39.1	38.2	38.9	232	75	307
1993–94	37.3	32.4	36.0	244	76	320
1999–2000 (MRP)	27.1	23.6	26.1	193	67	260
2004-2005 (URP)	28.3	25.7	27.8	221	80	301

Source: Planning Commission.

Three important changes had taken place in regard to the extent of poverty during this period. First, as is evident from Table 54.7 there has been a *continuous reduction in the incidence of poverty since 1973-74*. The percentage of households below the poverty line for all India (that is, rural and urban India combined) which was 54.9 per cent in 1973-74 declined to 51.3 per cent in 1977-78 and further to 44.5 per cent in 1983. For all India, poverty ratio further declined to around 38.9 per cent in 1987-88 and to 36 per cent in 1993-94. The poverty ratio for all India declined to 26 per cent in 1999-2000, which was based on *mixed recall* period (MRP) and is not comparable with the 1993-94 poverty data which was based on *uniform recall period* (URP). According to uniform recall period (URP) poverty data of *National Sample Survey 2004-05,* poverty ratio in case of all India declined to 27.8 per cent.

A noteworthy fact is that rate *of decline in poverty ratio during this period was much lower than the rate of economic growth*. Thus while annual growth in GDP in real terms during 1993-94 to 2004-05 was 6.3%, the rate of reduction in poverty was much slower, 0.74 per cent per annum. Secondly, the rate of reduction in poverty ratio slowed down in the period 1993-94 to 2004-05, as compared to the pre-reform period. Whereas decline in the poverty ratio was about 1.24 per cent per annum in the pre-reform period 1977-78 to 1987-88, it was only 0.74 per cent per annum during the post-reform period (1993-94 to 2004-05).

Second, the incidence of both rural poverty and urban poverty separately had also declined between 1973-74 and 2004-05. Thus, rural poverty ratio, which was 56.4 per cent in 1973-74, fell to 45.7 per cent in 1983 and further to 37.3 per cent in 1993-94. In 2004-05, it was estimated to have fallen to 28.3 per cent (URP). Thus there had been a marked decline in the *ratio of poor rural households* when comparing 2004-05 estimates with those of 1973-74. Likewise, *poverty ratio in urban India* declined from 49 per cent in 1973-74 to 40.8 per cent in 1983 and to 32.4 per cent in 1993-94. In 2004-05, poverty ratio (URP) in urban India further fell to 25.7 per cent. ***An important fact revealed from these rural and urban poverty ratios is that contrary to the estimates of Dandekar and Rath the incidence of poverty in rural areas was greater than that in urban areas.***

Third, the data regarding the incidence of poverty in various States of India during this period (1973-2004-05) reveal that the *secular decline in rural poverty ratio occurred in every individual state of the country,* although the extent of decline varied. However, it is important to note that in 2004-05 *rural poverty ratio* was still relatively high in Orissa, Bihar, Uttar Pradesh, Madhya Pradesh, Uttarakhand, Jharkhand and West Bengal as compared to 28.3 per cent at the all India level.

It may be noted that the NSS data about incidence of poverty reveal that there has also been a *secular decline in urban poverty ratio* in every state of the country, although magnitude of this decline varies a good deal in different states . *It should be further noted that in 2004-05, in Orissa, Madhya Pradesh, Bihar and Uttar Pradesh, urban poverty ratio was relatively much higher than the poverty ratio at the urban national level of 25.7. It was 44.3% in Orissa, 34.6% in Bihar, 42% in Madhya Pradesh, 30.6 % in Uttar Pradesh, and 32.6% in Karnataka.*

Latest Estimates of Poverty in India: 68th Round of NSSO

In June 2013 the Planning Commission released the poverty estimates based on special 68th round of NSS made by an expert committee headed by Late Prof. Suresh Tendulkar. These estimates gave rise to a lot of controversy about the true extent of poverty prevailing in India. Adjusted for price change, the *poverty line* based on Tendulkar Committee methodology was fixed at a *daily expenditure of ₹ 27.20 per capita for rural areas and ₹ 33.3 per capita for urban areas for the year 2011-12*. Based on this poverty line Planning Commission calculated the poverty ratio and the absolute number of people living below the poverty line based on National Sample Survey on Employment and Unemployment for the year 2011-12. It may be noted that for the sake of comparison, household expenditure data for the year 2004-05 have been used to make estimates of poverty for 2004-05 according to Tendulkar Committee methodology and poverty norms. As will be seen from Table 54.8, *during the period 2004-05 and 2011-12 there has been massive decline in poverty ratio. According to these estimates poverty ratio for all India declined from 37.2 per cent in 2004-05 to* 21.9 per cent in 2011-12, that is, 15.3 per cent fall in poverty in just seven years which amounts to

Table 54.8. Estimates of Poverty in 2004-05 and 2011-12
Based on Tendulkar Committee Methodology

	Rural	Urban	Total
2004-05	41.8	25.7	37.2
2011-12	25.7	13.7	21.9

2.2 per cent per year which is quite a large decline. Further, it will be seen from Table 54.8 that there has been much larger decline in poverty ratio in rural areas, from 41.8 per cent in 2004-05 to 25.7 per cent in 2011-12, that is, 16.1 per cent decline in seven years as compared to 12 per cent in urban areas in these seven years. Based on the poverty line cut-off, there were 270 million poor in India in 2011-12 against 407 million in 2004-05, that is, around 137 million persons were lifted out of below poverty line. This means *poverty declined at faster rate of 2.2 per year every year as against* the reduction of 0.74 per cent per year during 1993-94 and 2004-05.

For calculating the above poverty estimates the Planning Commission used *Dr. Suresh Tendulkar Committee's methodology which also factors in spending on education and health besides calorie intake to arrive at a poverty line in urban and rural areas*. Accordingly, those whose daily consumption expenditure per capita on goods and services exceed ₹ 33.33 in urban India and ₹ 27.20 per capita in rural India at 2011-12 prices are considered not poor.

Better Performance in Poverty Reduction in Odisha and Bihar

An interesting feature of NSSO data about poverty reveals that poverty line in Odisha and Bihar have recorded the sharpest decline in poverty levels between 2004-05 and 2011-12, although the proportion of the poor in these states remains well above the national average. Latest data released by Planning Commission. reveals that in Odisha, the proportion of people below the poverty line (BPL) in the total population came down from 57.2% in 2004-05 to 32.6% in 2011-12, a decline of 24.6 percentage points.

In Bihar, which registered the fastest growth rate in the 11th Five-Year Plan (2007-12), the share of BPL in total population was estimated at 33.7% in 2011-12, compared to 54.4% in 2004-05, a reduction by 20.7 percentage points in seven years.

At the all-India level, BPL population was 21.9% – almost 270 million. This means every fifth Indian lives below the poverty line. The Planning Commission has set the poverty line low, defining as BPL anyone earning ₹ 27.20 a day or less in villages and up to ₹33.30 in urban areas at 2011-12 prices.

Table 54.9. BPL as % of Total Population

State	2004-05	2011-12	% change
Worst Performing States			
Chhattisgarh	49.4	39.93	–9.75
Jharkhand	45.3	36.96	–8.36
Manipur	37.9	36.89	–1.01
Arunachal Pradesh	31.4	34.67	+ 3.27
Puducherry	14.2	9.69	–2.51
Best Performing States			
Odisha	57.2	32.6	–24.6
Bihar	54.4	33.74	–20.66
Goa	24.9	5.09	–19.81
Kerala	19.6	7.05	–12.58
Himachal Pradesh	22.9	8.06	–14.84
Punjab	20.9	8.26	–12.64

The poverty estimates however, vary from state to state. Although things seem to be looking up in the poor states, especially Bimaru States, which in absolute numbers amounts to still remain home to the maximum number of poor people in the country. While Uttar Pradesh has just under 30% of its population in the BPL group, the number adds up to almost 60 million. Bihar, despite the improvement, still has 33.7% which in absolute numbers amounts to 35.8 million poor, and ranks second, followed by Madhya Pradesh where 23.4 million or 31.6% of the population is below poverty line (BPL).

Among the Bimaru states, only Rajasthan has managed to do better than the national average (21.9%) with the share of BPL in total population estimated to fall to 14.7% in 2011-12, compared to 34.4% in 2004-05. In fact, Rajasthan now is a better performer than Gujarat, famed for its rapid growth and good infrastructure. Gujarat ruled by Narendra Modi had 16.6% people below the poverty line in 2011-12.

The other important trend coming from the latest poverty estimates, which have traditionally created controversy, is the fact *that rural India has seen faster improvement in poverty reduction than urban centres. The decline in poverty was steeper in rural areas as BPL population came down to 25.7% (2011-12)* from 42% (2004-05), around 17 percentage points, as against around 12 percentage points in urban areas.

On an all-India basis, there were 217 million poor in rural areas and 53 million poor in urban areas in 2011-12 as against 326 million and 81 million respectively, in 2004-05. That is, in absolute numbers, the number of poor people has declined from 407 million in 2004-05 to 270 million in 2011-12.

Planning Commission estimates based on Tendulkar Committee's methodolgoy were greatly criticized. It was pointed out that ₹ 27.7 of daily expenditure per capita in rural areas and ₹ 33.3 of daily per capita expenditure in urban area fixed as poverty line were too low and persons could hardly survive on such low expenditure. Many leaders said, it was *joke on the poor.* In the year 2011, even Supreme Court criticised these crieria of poor in the rural and urban India. This led the Government to appoint a committee headed by Dr. C. Rangarajan to redefine the poverty line and prepare estimates of poverty ratio on that basis.

The poverty numbers for 2004-05 had been arrived at by the Planning Commission after taking into account per capita expenditure numbers from the 66th round of the National Sample Survery (NSS) as recommended by the Suresh Tendulkar expert committee on poverty estimation. ***Prof. Tendulkar's panel had suggested moving away from taking the calorie intake as the defining parameter for poverty line to a more comprehensive method of using per capita expenditure data on basic needs such as food, clothing, housing and services like health and education.***

The Planning Commission explained this sharp fall in poverty ratio, especially in the rural areas when India saw one of the worst droughts in 2009-10, with high growth of average 8.5 per cent achieved in India during 2004-05 to 2011-12.

It was asserted by the Planning Commission Chairman, Dr. Montek Singh Ahluwalia, that *benefits of high economic growth achieved during the period (2004-05 to 2011-12) trickled down to the poor and this resulted in the drastic fall in poverty.* The higher average of about 8.5 per cent growth in GDP was achieved during this period which, as claimed by the Planning Commission, brought down the poverty levels. This is despite the fact that India saw one of the worst droughts in 2009-10 and slowdown in economic growth in 2008-09 to 6.7% due to global financial crisis.

However, in our view a significant factor that accounted for decline in poverty during this period was *larger spending by the government on social welfare programmes* such as Mahatma Gandhi Rural Employment Guarantee Scheme (MGREGS) which was started in 2006. This scheme entitles 100 days of employment for one person per household at minimum of ₹ 100 per day to all rural households. This scheme not only provided employment to the poor households but also helped *pushing up wages in rural areas* and thereby further ensuring reduction in rural poverty. Since no such programme as MGREGS had been started in the urban areas there had been, as seen above, lower decline in poverty ratio in urban India during this period.

Critique of Poverty Estimates

A severe controversy broke out when the Planning Commission released the estimates of poverty for 2011-12 based on the methodology recommended by an Expert Committee headed by Late Prof. Suresh Tendulkar. The adjusted for inflation figure of poverty line as per Prof. Tendulkar Committee was fixed at ₹ 33.3 in urban areas and ₹ 27.2 in rural areas at 2011-12 prices. The critics pointed out *whether anybody could survive* for a day on ₹ 33.3 in the urban areas and ₹ 27.2 in rural areas. Besides, it was reported by the Planning Commission that all India poverty ratio fell from 37.2 per cent in 2004-05 to 21.9 per cent in 2011-12, a sharp fall of 15.3 percentage points in the poverty level in just 7 years, that is, at the rate of 2.2 per cent per year as against 0.74 per cent per annum in the period 1993-94 to 2004-05. This was quite unacceptable and critics asserted that Planning Commission indulged in statistical jugglery to show grand success of the UPA Government in reducing poverty significantly and achieving inclusive growth.

Besides, while the earlier poverty estimates were based largely on food intake of 2400 calories per person per day in rural areas and 2100 calories per person per day in urban areas, it was regarded as bare subsistence and the estimates of poor below the poverty line (BPL) on this benchmark was in no way linked with the social benefits, especially the food and other items distributed through Public Distribution System. When the approach of BPL (Below Poverty Line) *targeting* was started under which social benefits provided by the Government was linked to the people below the poverty line (BPL), the use of poverty index for this purpose was quite unwarranted. Estimates of people below poverty line was a statistical exercise to measure the extent of poverty in India and to judge the performance of planned development with regard to the reduction of overall poverty in India. It was not good and fair to be used as benchmark for distribution of social benefits provided by the government.

For distribution of social benefits we need to *identify actual persons* who are poor and need help by the Government. Prof Jean Dreze, an eminent Professor at the Delhi School of Economics, writes, "The whole idea of *'BPL targeting'* that is, of restricting various social benefits (in particular the public distribution system) to households below the poverty line, quickly transforms poverty line from a statistical benchmark into a real life social division. The division was all the more artificial as the identification of BPL households was highly unreliable. Indeed, the Planning Commission uses one method to count the poor and the Ministry of Rural Development uses another method to identify them. This cannot work",[24] Criticizing the methodology he writes that the Tendulkar Committee "further complicated the matter by claiming for the first time that poverty line ensures *adequacy of actual private expenditure on food, education and health*. The ₹ 33 per person per day (the Tendulkar poverty line for the urban areas at today's prices) is wholly insufficient for the purpose is self-evident from a common sense point of view. This is all the more glaring when one looks at the break down of ₹ 32 packet which includes, for instance, just about *one daily rupee for health expenditure*. Yet the experts managed to rustle up technical arguments to substantiate their startling claims".[25]

Even as the critics accused the Government and the Planning Commission of tampering with poverty figures, the Planning Commission, especially its Vice Chairman, stuck to its stand that poverty declined by 15.3 per cent between 2004-05 and 2011-12, a period when Congress and UPA were in power. However, due to strong criticisim of poverty estimates, changing its stand Deputy Chairman

24. Jean Dreze, "The Poverty Trap", *The Hindustan Times*, July 2012.
25. ₹ 32 per person per day for the urban areas was calculated by the Planning Commission as per the poverty line drawn by Tendulkar committee in June 2011 prices and submitted to the Supreme Court.

of Planning Commission Montek Singh Ahluwalia conceded that *poverty line was meant to reflect conditions of absolute poverty*. This poverty line is not the poverty line we think a *"person can comfortably survive* — the poverty line is a rock bottom bare subsistence kind of line."[26]

Further, due to severe criticism levelled against the Government for a planning to link the social benefits of Government social welfare schemes to these poverty estimates, Planning Commission clarified that food subsidies and other entitlements would not be linked to the new poverty line. Thus, it said that "entitlements for food subsidy benefits will not be linked to the number of persons estimated to be below the official poverty line". Further, *it clarified that poverty estimates will be used only to gauge the performance* of the government in combating poverty.

In our view, the fall in poverty ratio by 15.3 per cent in 2011-12 compared to 2004-05 as estimated by the Planning Commission appears to be too impressive. Besides, it cannot be claimed that poverty ratio declined due to the trickle-down effect of high economic growth rate achieved during 2004-05 and 2011-12. The fall in poverty, if any, is largely because of enhanced government expenditure on social welfare schemes such as Mahatma Gandhi National Rural Employment Guarantee Scheme (MGNREGS) which has a lot of leakages and have *not created much durable productive assets* and therefore merely represents *cash transfers.* This way of poverty reduction is not a durable solution of the poverty problem. Besides, increase in wages in rural areas in general during the period (2004-05 to 2011-12) following the introduction of MGNREGS also caused reduction in poverty in rural areas. Besides, sharp increase in procurement prices of wheat and rice also resulted in increase in incomes of the farmers. However, it cannot be denied that higher economic growth would have some effect on poverty reduction as under the development programmes, a lot of construction activity has been involved in which *casual labour* has been employed on daily wages.

In our view *reduction in poverty as a result of economic growth depends on the character of economic growth.* Economic growth can lead to reduction in poverty if it generates adequate opportunities of productive employment at fair wages and if gross inequalities of income decline. On the contrary, Economic Survey 2012-13 of Government of India itself admitted that economic growth in India had not generated much employment in the organised sector and further it had led to a large increase in income inequalities. Further, still in 2011 it was estimated that over 40 per cent of children in India were undernourished and malnourished despite an 8% plus average GDP growth rate achieved. Under these circumstances, how can it be said that the poverty declined due to higher growth rate alone. This clearly shows that poverty estimates for 2011-12 released by the Planning Commission in June 2013 claimed to be based on Tendulkar Committee recommendations are not correct.

The criticism of these poverty estimates inside Parliament and outside was so severe that the Government appointed another committee headed by Dr. Rangarajan to propose proper norms for poverty line and prepare new estimates of poverty level in India.

Rangarajan Committee Report on Poverty-July, 2014

The findings of expert committee headed by Late Prof. Suresh Tendulkar was criticised on the ground that poverty line fixed at ₹ 27 for rural areas and ₹ 33 for urban areas was too low. According to the critics, no one can even subsist at these extremely low poverty norms of Tendulkar Committee. Besides, Tendulkar Committee estimated that poverty ratio has declined from 37.2 per cent of population in 2004-05 to 21.9 per cent in 2011-12. In absolute terms 269 million was estimated by the committee to be living below the poverty line in 2011-12. It was claimed by UPA government that this showed that it was higher growth rate of over 8 per cent during the period of its rule which had contributed to the significant decline in poverty ratio and the number of people living below the poverty line.

---

26. Statement published in *The Times of India,* March 21, 2012,

The Rangarajan Committee report which was submitted in July 2014 on poverty estimates estimated that India's poor to be much more than the previous estimates by Tendulkar Committee. The Rangarajan Committee estimated that the poor accounted for 29.6 per cent of population instead of 21.9 per cent of Tendulkar Committee's estimate. In absolute numbers in 2011-12, according to Rangarajan Committee, 363 million Indians of the country's 1.2 billion people were poor, that is, lived below the poverty line as against 269 million poor estimated by Tendulkar Committee, that is, Rangarajan Committee estimated *94 million more poor people than Tendulkar Committee*.

Dr. Rangarajan emphasizes that their numbers measure poverty and in no way can be considered as a criterion of entitlement to receive benefits from any government programme meant to help the people. According to him, these estimates of poverty are a mere measure of understanding how the economy is moving, that is, how economic growth is affecting the poor of the economy. According to the present general perception, it matters little to the standard of living of those who are counted and classified whether they are placed by one method, propounded by one eminent economist above the poverty line or below that as per another method propounded by another eminent economist. The fact, according to the committee, remains that for most Indians their quality of life as measured by how much they consume has been improving steadily. The task of the government is to accelerate this process. However, this is to belittle the importance of measuring poverty, however defined. The purpose of counting and identifying the poor is to *provide benefits* such as employment, food security, education by the Government to lift them out of poverty. Thus, estimating and identifying the poor serves a useful purpose and not a mere academic exercise as stated by Dr. Rangarajan. In fact, Dr. Rangarajan's definition of poverty is based on "***basic standard***" of living including spending on food, education, transport and house rent.

According to the Rangarajan panel, on statewise poverty ratios *Bihar witnessed the maximum dip in poverty ratios with the fall from 63.9% in 2009-10 to 41.3% in 2011-12 – a dip of about 22 percentage points*. In absolute terms, it translates into 21 million lesser poor people in the State. The fact resulted in Bihar losing the dubious tag of having the highest poverty ratio in India.

Bihar has been replaced by BJP-ruled Chhattisgarh, where the panel has categorised 47.9% of the people as being poor. In 2009-10, 53.8% of Chhattisgarh's population was categorised as poor as per Rangarajan's definition of poverty based on "*basic standard of well-being*" including spending on food, education, transport and house rent.

The panel's estimation shows that poverty is still a huge concern in Maoist affected States even though the UPA government had spent over ₹ 4,600 crore in 83 districts over four years starting 2010-11 under its ambitious Integrated Action Plan (IAP) for bridging development deficiency in Left Wing extremism hit areas. Apart from Maoist-hit Chhattisgarh, two other affected States Odisha and Jharkhand, find themselves high on the poverty ratio scale.

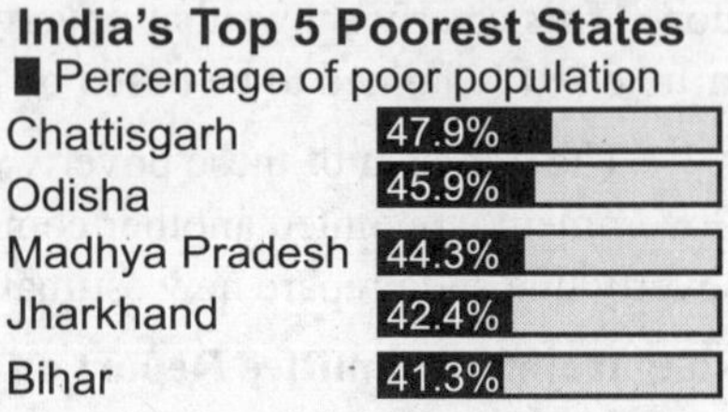

India's Top 5 Poorest States

■ Percentage of poor population

Chattisgarh	47.9%
Odisha	45.9%
Madhya Pradesh	44.3%
Jharkhand	42.4%
Bihar	41.3%

Fig. 54.4. *Rangarajan Committee Estimates*

According to Rangarajan Committee, around 45.9% of people in Odisha were poor in 2011-12 as compared to 48.3% in 2009-10. In Jharkhand, 42.4% were poor in 2011-12 as compared to 52.1% in 2009-10. A member of the panel said the estimation clearly showed that *effective implementation of poverty alleviation programmes can result in dip in numbers*. "Bihar is a case for study on how government programmes can reduce poverty. But in contrast the same programmes have not helped in substantially reducing poverty in other Maoist affected States where impact of these programmes is not much", he aded. Eight districts of Bihar were also covered under IAP.

Who are the Poor in India?

The knowledge about this is necessary for devising effective anti-poverty policies. Different sections of the poor face different structural and institutional obstacles which may require different strategies for reduction of poverty.

The distribution of total poor across different socio-economic groups reveals that around 45 per cent of poor rural households in 1993-94 belonged to *agricultural labour.* These agricultural labour are landless or own very small farming land. Next major group of poor is "*self-employed in agriculture*" which accounted for around 28 per cent of total rural poor. These self-employed in agriculture are small families who suffer from acute underemployment. "*Self-employed in non-agriculture*" which accounted for around 12 per cent of the rural poor. However, in various states the relative proportions of 'agricultural labour' and self-employed in agriculture in the total poor vary depending on the available employment opportunities on the one hand and land holding system on the other. It therefore follows *that access to productive assets such as land is an important determinant of poverty in a country*. Therefore, as a measure for alleviation of poverty in the earlier Five Year Plans in India emphasis was laid on the implementation of land reforms which sought to ensure some land to the landless labourers and to ensure security of tenure to the tenants. However, in the last 20 years with the exception of W. Bengal no importance has been given by the States to the implementation of land reforms.

CAUSES OF POVERTY IN INDIA

As seen above, estimates of poverty as made by NSS 2011–2012 based on Tendulkar Committee recommendations show that 270 million persons or 21.9 per cent of population live below the poverty line, that is, do not have sufficient consumption expenditure to meet minimum food intake in terms of calories. 70 per cent of the poor live in the rural areas. What are the causes of this mass poverty? We explain below some important causes of poverty in India.

1. Lack of Inclusive Growth. The first important reason for mass poverty prevailing in India is lack of inclusive economic growth in India. Besides, because of prevailing large income inequalities per capita consumption expenditure of the poor did not rise much.

In the next two decades of development (1981-2000) national income increased on an average by 5.5 per cent per annum. With about 2 per cent per annum growth in population, per capita income rose by 3.5 per cent which was quite high and, as seen above, did result in some decline in poverty ratio both in the rural and urban areas. But this could not sufficiently reduce poverty because land reforms could not be implemented in agriculture and industrial growth did not generate much employment opportunities, Besides, strategy of growth benefited the rich more than it helped the poor. Capital-intensive and labour-displacing technology was adopted in the growing industries. As a result, unemployment and underemployment increased. Besides, due to the increase in income inequalities during this period, rise in average per capita income could not bring about significant rise in per capita income of the weaker sections of the society. Further, trickle down effect of overall economic growth was operating only to a small extent. This is manifest from the fact that employment opportunities had been increasing at a very slow pace during this period. No wonder that even in 2011-12, 270 million persons or 21.9 per cent of population lived below the poverty line.

2. Sluggish Agricultural Performance and Poverty. Many economists have pointed out that in the year of good agricultural output, poverty ratio declines. Good performance in agriculture leads to more employment opportunities and fall in prices of food grains. More employment opportunities and lower food price cause poverty ratio to decline. The experience of Punjab and Haryana shows that with agricultural growth through use of new high yielding technology (popularly called green revolution), poverty ratio can be significantly reduced. However, in various States of the country such as Orissa, Bihar, Madhya Pradesh, Assam, East Uttar Pradesh, where poverty ratio is still very high new high-yielding technology has not been adopted on a significant scale and as a result agricultural performance has not been good. As a result, poverty prevails to a larger extent in them.

Further, *Indian policy makers have neglected public sector investment in agriculture, particularly irrigation,* for quite a long period since 1980-81. As a result, irrigation facilities whose availability ensures adoption of new high-yielding technology and leads to higher productivity, income and employment, are available in not more than 35 per cent of cultivable land. As a result, many parts of the country remain semi-arid and rain-fed areas where agricultural productivity, income and employment cannot be sufficient to ensure significant reduction in poverty.

3. Non-implementation of Land Reforms. Equitable access to land is an important measure of poverty reduction. Access to adequate land, a productive asset, is necessary for fuller employment of members of an agricultural household. Most of the rural poor are agricultural labourers (who are generally landless) and self-employed small farmers owning less than 2 acres of land. They are unable to find employment throughout the year. As a result, they remain unemployed and underemployed for a large number of days in a year. No wonder that these landless agricultural labourers and self-employed small farmers remain poor. Land-reforms were intended to provide them equitable access to land.

An important type of land reform was to redistribute land through implementation of ceilings on land holdings so that poor landless labourers and small farmers should have access to adequate land for having fuller employment and enough income to meet their basic needs.

Another important land reform measure related to tenancy reforms, which were intended to protect the poor tenants from eviction of land cultivated by them and fixation of fair rents to be charged by the landlords. Further, ultimately ownership rights over land are to be conferred on the tenants cultivating the land. With security of tenure and fair rents they would have incentives to cultivate land and adopt the new green revolution technology. In this way with the implementation of land reforms, poor agricultural labourers, small farmers and tenants were expected to rise above the poverty line. But with the exception of West Bengal, and to some extent Kerala, land reforms remain unimplemented with the result the rural poverty did not decline adequately. On the other hand, in West Bengal implementation of land reforms has led to a faster decline in rural poverty.

4. Rapid Population Growth. Rapid population growth since 1951 is another important factor responsible for persisting poverty in India. Population in India has increased from 36 crore in 1951 to 121 crore in 2011, that is, 85 crore people have been added to the Indian population in the last about 50 years since independence. Rapid population growth causes excessive sub-division and fragmentation of holdings. As a result, per person available land has greatly declined so that households do not have access to sufficient land to produce enough output and income for them. Further, rapid growth of population increases the dependency ratio, that is, an earning member has more persons to support. This causes lower per capita consumption expenditure which is not enough even to meet the basic needs.

Rapid growth in population in India since 1951 has *caused lower growth in per capita income* causing lower living standards of the people. Besides, explosive growth in population has worked to reduce rates of saving and capital formation which are crucial factors for determining growth in income and employment. Though the rate of gross saving and investment rose from about 9 per cent of GNP to about 21 per cent in 1978-79, the rate of saving and real investment (*i.e.,* capital formation) remained almost stagnant at around 22 to 23 per cent per annum during the eighties and nineties. This has especially caused low rate of real investment in agriculture. This has adversely affected the growth in irrigation facilities which are important factors in determining income and employment in the rural economy.

We thus see that rapid population growth has adversely affected growth in per capita income, saving and capital formation which are so essential for reduction in poverty.

5. Unemployment and Underemployment. The existence of unemployment and underemployment in the Indian economy is another cause of poverty in India. Unemployment prevails more among casual labour whose proportion in labour force has been increasing and in their case unemployment and poverty go together. The unemployment has been caused by rapid growth of population and labour force on the one hand and relatively low rate of capital formation and economic growth on the other. Besides, the generation of employment opportunities by the organised sector has been quite insignificant. As a result, the demographic pressure on land has been increasing resulting in unemployment and disguised unemployment in agriculture and the informal sector. This has caused low productivity, low incomes and poverty.

According to the NSS survey 1999-2000 7.2 per cent of rural males and 7.0 per cent of rural females in the labour force were unemployed on the basis of daily status approach. In the urban areas, 7.3 per cent males and 9.4 per cent females were unemployed on the daily status basis. It is worth pointing out that even daily status unemployment rate does not reflect accurately the extent of unemployment in India. In a country like India where social security system does not exist, poor cannot afford to remain unemployed for long. He takes up whatever work is available no matter how low the remuneration is. In fact, the poor work very hard and even for long hours and still are unable to earn enough income to meet their basic needs.

6. Slow Growth of Employment Opportunities in the Organised Sector. In the beginning of the planning era it was thought the organised sector over a period of 20 years will generate sufficient employment opportunities for the unemployed poor. The actual experience has belied these hopes. Take, for instance, the experience of the decade (1990–2000) of economic development. It is estimated that employment in the *total organised* sector (including both public sector and private sector) increased from about 26.4 million in 1990 to about 28 million in 2000, that is, in 10 years period, the organised sector which includes industrial and services sectors as well as organised agriculture could generate employment for only 1.6 million people whereas labour force increased from about 340 million in 1990 to 406 million in 2001, that is, 66 million increase. The reason for this is the *use of capital-intensive techniques* in the production precesses. These techniques are intended to save labour rather than employ it. Further, relative to the size of the total Indian economy, the organised sector is of very small size so that it is wishful thinking to expect that the growth of organised sector, that is, factory type industries and services using capital-intensive technologies, generally imported from the Western countries, would be able to generate enough employment opportunities.

7. Inflation and Food Prices. Rate of inflation and level of food prices are an important factor that causes poverty. Inflation, especially rise in food prices, raises the cost of minimum consumption expenditure required to neet the basic needs. Thus, inflation, especially rise in food prices, pushes down many households below the poverty line. That is why, Public Distribution System has been designed to provide foodgrains and other essential items such as kerosene oil, standard cloth, pulses at subsidised prices, that is, prices which are lower than the free market prices. However, most of Public Distribution ration shops are located in the urban areas. Therefore, the rural poor who constitute 70 per cent of the total poor cannot get much benefit from the food subsidies provided by the Government.

POVERTY AND ECONOMIC GROWTH

Whether economic growth necessarily reduces poverty has been a controversial issue. As pointed out above, it was first generally believed that benefits of economic growth would trickle down to the poor and thus alleviate poverty in the economy. This trickle-down effect of economic growth was thought to operate through an increase in employment opportunities and rise in real wages as a result of increase in productivity of workers. The favourable employment effect of economic growth depends on employment elasticity of output growth. The rise in real wages was thought to occur as a result of rise in wages of workers employed in agriculture and non-agricultural sector, both organised and informal.[27]

27. GOI, *Economic Survey,* 2000–2001 p. 240 and S 50.

The agricultural growth was visualised to cause reduction in food prices and greater availability of foodgrains along with employment growth in agriculture.[28] This would ensure reduction of poverty as it would enable the poor to meet their minimum consumption requirements Besides, in this approach it was thought that the poor will also be benefited by the multiplier effect of economic growth. According to this, benefits of growth are reaped first by the rich and subsequently by the poor when the rich spend their incomes leading to the rise in income and employment for the poor.[29] Further, as explained by T. N. Srinivsan[30] the multiplier effect also operates through economic growth raising the revenue of government which if invested in *public goods and services* which are relatively consumed more by the poor would reduce non-income facets of poverty such as reduction in *infant mortality rate* and rise in *life expectancy* of the poor.

Further, some Indian economists, especially Prof. Jagdish Bhagwati and Arvind Panagariya of Columbia University (USA), have argued that economic growth leads to the increase in revenue of the government which can be spent on special anti-poverty schemes. This will help in reducing poverty. This direct approach to poverty alleviation has in fact been adopted by the Indian government by starting various special employment schemes such as National Rural Employment Gurantee Scheme (MGNREGS), Prime Minister's Sarak Yojana, mid-day meals in government schools and provision of food subsidy through Public Distribution Scheme.

The reduction in poverty in the last 20 years (*i.e.,* since 1980) has been mainly due to the increase in government expenditure on special anti-poverty schemes. However, it may be noted that increase in government expenditure might be the result of higher fiscal deficit incurred rather than due to the large revenue generated through taxation as a result of higher rate of economic growth. As Paul Streeten has observed that *trickle down effect does not happen where assets and power are concentrated.* The government did not act like Platonic guardian but reinforced the accrual of wealth to the rich and powerful. And it was not realised that investing in the developing countries' most abundant resource, its masses of poor people could be a paying proposition. This raises two issues. First, whether economic growth will cause reduction in poverty depends on the initial conditions of asset distribution. Since main asset in developing countries is land, if it is grossly unequally distributed, the benefits of agricultural growth will mainly go to the landlords with only small benefits to the landless labour, small and marginal farmers.

In India the economic reforms initiated in 1991 involved the policy of liberalisation, privatisation and globalisation that has not paid any attention to the question of asset distribution (including the distribution of surplus land held by big landlords), and to the tenancy reforms. This explains why agricultural growth in India, though low it has been since 1991, has not led to any significant reduction in rural poverty, despite claims to the contrary by the government and the Planning Commission.

Since the strategy of growth pursued by government failed to reduce poverty significantly, it has undertaken *special employment and anti-poverty schemes* to alleviate poverty in the country. Besides, benefits of growth have not reached the poor that government has been providing subsidies on food, sugar, kerosene oil and distributing through public distribution system to help them. The quotation from Paul Streeten[31] given above indicates that ***"solution to the problem of poverty lies in developing countries undertaking greater investment in the masses of poor people.*** This can happen if government spends a significant amount of its investment expenditure on human development, that is, on promotion of education and health of the poor. *Labour is essentially the only source of income for the poor people.* Therefore, increasing the productivity of labour through providing education

28. See Montek Singh Ahluwalia, "Rural Poverty and Agricultural Performance in India", *Journal of Development Studies*, 1978, 14:13, pp. 298-323.
29. N. Kakwani and E.M. Pernia, "What is Pro-Poor Growth?" *Asian Development Review*, (2000) Vol. 18, Number 1.
30. T.N. Srinivsan, "Growth and Poverty Alleviation, Lessons from Development Experience", (2000), Paper presented at the Symposium, Asian Development Bank Institute, Tokyo.
31. Paul Streeten, *Thinking About Development*, Cambridge University Press, 1995.

and healthcare and technical skills to them provides a sustainable means of poverty reduction. Dreze and Amartya Sen have laid great stress on instrumental role of education in enabling people to seize economic opportunities created by the growth process.[32]

The best example of this is Kerala where a very high rate of literacy of about 90 per cent has provided most successful performance in rural poverty reduction. In contrast, Bihar, the weakest of the large Indian States in terms of poverty reduction, is *characterised by an abysmally low literacy rate*. In an important study Datt and Ravallion have also found that *initial levels of education affect the poverty reducing effect of economic growth.*[33]

Moreover, the impact of economic growth on poverty reduction also depends on to what extent it is accompanied by inflation. Inflation determines the purchasing power of money wages and incomes earned by the poor people. A higher rate of inflation erodes to a large extent the purchasing power of wage earners as it affects the ability of the people to meet the specified level of consumption. Given the money wages, a higher rate of inflation will send more people below the poverty line. In the unorganised labour market wages of workers are not indexed to the cost of living. Thus rate of inflation accompanying economic growth has an adverse effect on poverty through its effect on real wages.

Further, for economic growth to have a significant impact on poverty depends not much on its trickle down effect but whether economic growth is pro-poor or what has been recently described as '*inclusive growth*'. However, our Planning Commission considers inclusive growth merely in terms of starting some special employment or anti-poverty schemes in which not only there are a lot of leakages but also do not result in the creation of durable productive assets. If economic growth has to be pro-poor or inclusive, the strategy of growth should be such that generates a lot of employment opportunities for the poor people in agricultural, industrial and services sectors at reasonable wages that enable them to meet their basic minimum needs. Such a pro-poor or inclusive growth strategy will, as mentioned above, involves adequate public expenditure on human resource development in the fields of education and healthcare which benefit the poor and enable them to participate in the activities that bring about expansion in output. Therefore, Kakwani and Pernia rightly define pro-poor growth as "*growth that enables the poor to actively participate and significantly benefit from economic activity*".[34]

However, as pointed out by Gaurav Nayyar, "*Owing to the inherent advantages enjoyed by the rich in terms of material and human capital, the growth process generally tends to benefit the rich proportionately more than the poor*". Therefore, pro-poor and inclusive growth strategy requires building up human development such as increasing public expenditure on education and healthcare of the poor people. Besides, pro-poor growth strategy also involves institutional reforms such as land reforms in agriculture aimed at distributing surplus land of big landlords among the landless agricultural workers and marginal and small farmers and also taking effective measures to protect the tenants and sharecroppers against exploitation by landlords.

Besides, inclusive growth strategy requires the adoption of labour-intensive technologies wherever feasible so as to generate adequate employment opportunities for the poor and unemployed. For this purpose reckless mechanisation of agriculture should be discouraged and appropriate fiscal measures be taken so that modern industries and service sector labour-displacing technologies imported from abroad and unsuited to labour-surplus conditions of the Indian economy are not used in production of goods and services. Moreover, to provide gainful employment opportunities to the poor greater emphasis should be given to **investment in physical infrastructure**. It has been empirically found that *development of physical infrastructure* such as roads, irregation facilities, and power

32. See Dreze and Amartya Sen, *India's Development and Participation,* Oxford University Press, 2002.
33. See G. Datt, and M. Ravallion, *When is Growth Pro-Poor: Evidence from the Diverse Experiences of Indian States"* Policy Research Paper, World Bank, Washington DC.
34. Kakwani, N. and E.M. Pernia, "What is Pro-Poor Growth", *Asian Development Review*, Vol. 18, Number 11, 2000.

(electricity) apart from creating direct employment has a large multiplier effect on both output and employment in agricultural and non-agricultural sectors.

Conclusion. To sum up, government in India has relied on two approaches to alleviate poverty in India. The first approach is based on the thinking that benefit of economic growth will trickle down to the poor in the long run and if higher rate of economic growth is achieved it will result in removal of poverty. Therefore, in this approach, emphasis was laid exclusively on attaining a higher economic growth to eradicate poverty. After 1980, it was realised that it will take a long time for economic growth, even the higher one, that will cause reduction in poverty.

Therefore, according to the second approach, in the short run until economic growth alleviates poverty significantly special anti-poverty schemes such as National Rural Employment Scheme (MGNRES) providing food subsidies to the poor through public distribution system. Evidence gathered so far indicates that both approaches have only limited success in reduction of poverty because of a lot of leakages in the money spent on the projects, a large chunk of benefits are bagged by the middle men. In 2009-10, according to 66th the round of NSS survey 29.6 per cent or in absolute numbers 360 million people were still living below the poverty line. Even this estimate was considered to be a gross underestimate of poverty in India. Therefore, as a result of severe criticism of these estimates by members of Parliament and many academicians and experts the Indian Government appointed another expert committee to go into the question again of proper methodology of estimating poverty in India and count the extent of poverty prevailing in India. Even the '*poverty line norm*' recommended by Suresh Tendulkar Committee has been challenged. It is only after the expert committee submits its report, we will be able to tell to what extent economic growth and special anti-poverty schemes have been able to reduce poverty in the country.

On the basis of 2011-2012 National Sample, survey, the claim of the Planning Commission, that the higher rate of economic growth of average 8.5 per cent between 2004-05 and 2011-12 resulted in the decline in poverty at the rate of 2.2 per cent per annum during this period was not trustworthy as the growth strategy pursued during this period has been pro-rich rather than pro-poor. The policy of economic reforms of liberalisation, privatisation and globalisation has no doubt generated higher economic growth but its impact on the poverty has been very little as it has benefited the rich more. *Achievement of higher economic growth is not enough, much depends on the character of economic growth regarding whether growth strategy is pro-poor or pro-rich.* The higher rate of economic growth can help in reducing poverty significantly if it is employment oriented and involves human resource development of the poor, lowers inflation, especially food inflation, institutional changes to redistribute productive assets such as implementation of land reforms in agriculture, the greater priority to building up physical infrastructure which has a large multiplier effect. Such a pro-poor growth strategy has not been pursued in India.

STRATEGY TO REDUCE POVERTY

We explain below various measures as a part of a strategy which should be taken to reduce poverty in India.

1. Accelerating Labour-Intensive Industrial Growth. In the fifties and sixties it was generally thought that poverty in India can be significantly reduced by accelerating economic growth. According to this view, benefits of economic growth will trickle down to the poor in the form of more employment opportunities, greater productivity and higher wages. With this it was expected that the poor will be raised above the poverty line.[35]

Various growth models put forward in the fifties and sixties such as Harrod-Domar growth model, Mahalanobis' growth model, Lewis' model of economic development with unlimited supplies of labour suggested rapid growth of the modern industrial sector to tackle the problem of poverty

35. See Jean Dreze, the "The Poverty Trap" in *The Hindustan Times*, July 2011.

in the long run. For this purpose they suggested for increasing the rate of capital formation so as to generate more employment opportunities and increase productivity of labour. Though this is correct that higher rate of capital formation is necessary for accelerating economic growth and thereby for solving the problem of poverty, but this *will not generate sufficient employment opportunities if labour-saving capital-intensive techniques of production are used in the process of growth*. This has been clearly brought out by the actual experience in India in the last three decades of development (1985-2015). Whereas in the two decades of development, rate of growth in GDP achieved is in the range of 7 per cent to 8 per cent per annum, there has been only little increase in employment opportunities, especially in the organised industrial sector. Therefore, while efforts should be made to accelerate economic growth but if it has to make a significant dent on the problem of poverty the use of capital-intensive technologies imported from the Western countries should be avoided. In fact, we should pursure labour-intensive path of economic growth. Such monetary and fiscal policies should be adopted that provide incentives for using labour-intensive techniques.

2. Agricultural Growth and Poverty Alleviation. Agricultural growth has been recognized as an important factor that contributes to marked reduction in poverty. A study made by Montek Singh Ahluwalia, former member of Planning Commission, brought out clearly that agricultural growth and poverty are inversely related; the higher agricultural growth leads to lower poverty ratio. The experience of Punjab and Haryana in the late sixties and in the seventies confirmed this inverse relation between agricultural growth and poverty. The growth in agricultural output in these States propelled by the adoption of new-high-yielding technology caused a marked reduction in poverty. Rural poverty ratio in Punjab and Haryana was 6.4 and 8.3 per cent respectively in 1999-2000.

Therefore, other States have been urged to follow the path of Punjab and Haryana for reduction of rural poverty. Thus, Late Prof. S. Chakravarty states, "*The solution to the problem of rural poverty requires that small farmers must also be given access to land-augmenting innovations*".[36] By land augmenting innovations he means the new high-yielding technology represented by green revolution that occurred first in Punjab and Haryana.

However, *in the recent years relationship between agricultural growth and poverty reduction seems to have weakened*. For example, at all India level, employment elasticity of growth in agricultural output has been found to be zero during 1993-94–1999-2000 whereas it was 0.45 during 1977-78–1983.[37] It appears that at the all India level employment generated by new green revolution technology has been cancelled out by increasing mechanisation of agricultural operations in various parts of the country. Thus, even in the light of the finding of zero employment elasticity of agricultural output at the all India level, positive impact of agricultural growth on the incomes of small farmers and, more particularly on the wage income of agricultural labourers, cannot be denied.

To ensure marked decline in rural poverty through agricultural growth, rate of agricultural growth should be accelerated by increasing public investment in irrigation and other infrastructure. In recent years since 1980, rate of capital formation in agriculture has been declining. This trend has to be reversed by increasing public investment in agriculture, especially irrigation. Besides, higher agricultural growth can be achieved in semi-arid and rain-fed areas by increasing public investment in infrastructure and ensuring adequate access to credit to the small farmers.

3. Speedy Development of Infrastructure. An important measure to generate employment opportunities for the poor and to raise their productivity is the speedy development of infrastructure. Since private sector is not attracted to make adequate investment in infrastructure, public investment needs to be stepped up for its development. Infrastructure development consists of building of roads, highways, ports, telecommunication, power and irrigation. They involve mainly construction work which is highly labour-intensive.

36. Sukhamoy Chakravarty. *Development Planning. The Indian Experencie* (1989) OUP, Delhi, p. 88.
37. Planning Commission (2001), *Report of the Task Force on Employment.*

Besides, the availability of infrastructure such as power and irrigation greatly raises productivity of labour. C.H. Hanumantha Rao in his study of East and South East Asian countries finds that reduction in rural poverty in them achieved through economic growth occurred due to the fact that physical infrastructure in them were already highly developed. To quote him, "*The experience of East and South East Asian countries shows that the impact of development on rural poverty reduction has been greatest in situations where land reforms have been implemented effectively and a high priority has been accorded to infrastructure development, agriculture and human resource development* within a liberalised economic policy framework. These facts suggest, in the first place, that in a country like India where the physical and social infrastructure is inherited from the pre-liberalisation period is not strong and redistribution of land on a significant scale is not feasible, public investment needs to be stepped up for expanding physical infrastructure in the less developed areas."[38]

4. Accelerating Human Resource Development. Besides physical infrastructure development, poverty can also be reduced through human resource development. Human resource development requires greater investment in educational facilities such as schools to promote literacy, technical training institutes and vocational colleges to impart skills to the people. Further, human resource development requires healthcare by public investment in Primary Health Centres, dispensaries and hospitals.

The human resource development not only generates a good deal of employment opportunities but also raises productivity and income of the poor. Further, people equipped with skills, education and good health can easily get wage employment or self-employment with higher productivity. In this way human resource development helps in reducing poverty. The experience of East and South-East Asian countries, referred to above, and Kerala in our country shows that poverty can be significantly reduced through investment in human resource development. However, private sector, guided as it is by profit motive, will not adequately invest in the human resource development.

5. Growth of Non-farm Rural Employment. For reduction of poverty growth of non-farm employment in the rural areas is of special importance. Non-farm employment is created in marketing (*i.e.,* petty trade), transportation, handicrafts, dairying, forestry, processing of food and other agricultural products, repair workshops. A study of poverty alleviation in Haryana[39] brings out that significant reduction in rural poverty in Haryana in spite of a reduction in employment opportunities in agriculture was due to the remarkable increase in non-farm employment. Similarly, a study of Andhra Pradesh[40] also shows that poverty declined more rapidly in the districts adjoining Hyderabad city due to large increase in non-farm employment.

6. Access to Assets. Rapid growth of population after independence has led to greater subdivision and fragmentation of agricultural holdings and lack of employment opportunities in industries and other non-farm sectors has worsened the conditions of agricultural labour and self-employed small farmers. With no land or little land they cannot engage themselves in self-employment activities for earning adequate income to meet their basic needs. Redistribution of land through effective redistribution, implementation of tenancy reforms so as to ensure security of tenure and fixation of fair rent would be an important measure of reducing rural poverty. Except in case of West Bengal and Kerala land reforms have not been implemented to reduce rural poverty.

---

38. C.H. Hanumantha Rao, *Economic Reforms, Agricultural Growth and Rural Poverty,* Sukhamoy Chakravarty Memorial Lecture delivered at the 17th Annual Conference of the Indian Economic Association, Chandigarh, Dec. 28-30, 1995.
39. Sheila Bhalla, "Development, Poverty and Policy. The Haryana Experience" *Economic and Political Weekly*, Oct. 14-21, 1995.
40. G. Parthasarthy, "Public Intervention and Rural Poverty, Case of Non-Sustainable Reduction in Poverty in Andhra Pradesh, *Economic and Political Weekly*, Oct. 14-21, 1995.

However, abolition of Zamindari system is the only land reform measure which has been faithfully implemented in several parts of the country. Other land reform measures, namely, ceilings on land holdings, protection of tenants against eviction, and regulation of fair rent by the Government have remained unimplemented except in case of West Bengal and Kerala. But without land reforms, a significant dent on the problem of rural poverty cannot be made. On small farms labour employment per hectare, output per hectare and double cropping are greater as compared to those of large farmers. Tenants who enjoy no security of tenure do not invest adequately in yield-increasing inputs to raise their productivity. The effective implementation of land reforms to ensure access to land and security of tenure are essential for the reduction of poverty.

But, unfortunately, the present Neeti Aayog and the Government do not attach much importance to these much awaited land reforms. Citing the experience of East and South East Asia in which both rate of growth and poverty reduction have been remarkable, Prof. C.H. Hanumantha Rao states, "their experience shows that the impact of development on rural poverty reduction has been greatest in situation where land reforms have been implemented effectively.[36]

7. Access to Credit. Availability of credit to the poor on easy terms can create the conditions for small farmers gaining access to productive resources such as HYV seeds, fertilizers, construction of minor irrigation such as wells and tubewells. This will enable the small farmers to adopt high-yielding technology to raise their productivity. The new technology is size-neutral, that is, it can be adopted equally well by small farmers. But the adoption of new technology requires financial resources which are lacking with the small farmers. Besides, the non-farmer poor need credit for marketing, food processing, dairying, forestry, development of handicrafts which can provide them gainful employment. Important changes have been introduced in the credit delivery system in India. Expansion of network of rural branches of commercial banks after nationalisation and fixation of limits for compulsory lending to the priority sectors (which include agriculture, small-scale industries) and fixation of lower interest rates to be charged from the poor farmers and artisans have led to some progress in this regard.

However, banks and other financial institutions have not shown adequate response to provide adequate credit to them as they *consider the poor to be non-creditworthy.* Unless the banks and other financial institutions change their attitude toward supplying credit to the poor much success cannot be achieved to ensure adequate credit to the poor farmers and artisans. What they do need is short-term loan for working capital such as for purchasing raw materials, fertilizers, pesticides.

An important step in credit delivery system for the poor has been setting up of regional rural banks (RRB). Regional rural banks are primarily meant to meet the credit needs of the poor. The Government should take effective steps to improve the functioning of these financial institutions so as to ensure availability of adequate credit to the poor.

8. Public Distribution System (PDS). Poor households spend nearly 80 per cent of their income on food. Therefore, an effective way of raising rural incomes and ensuring food security to the poor households is an assured supply of adequate quantity of foodgrains and other essential commodities at subsidised prices, that is, at prices which are lower than the market prices. A properly functioning public distribution system which is targeted to the poor households is an important element of the strategy for poverty reduction. The Central Government organisation 'Food Corporation of India' procures the foodgrains from the farmers at the minimum support prices (MSP) and stores them in their warehouses located throughout the country. The foodgrains so procured are allotted to the State governments to be sold through the public distribution systems (*i.e.,* ration shops) at subsidised prices which are below market prices. The difference between the two prices is paid by the Central Government as subsidy.

The expenditure on food subsidy has greatly increased in recent years. The ratio of release of foodgrains through PDS to total foodgrains is around 10 to 13 per cent. Some economists have suggested lowering of subsidies. What is needed is to ensure subsidised foodgrain supply through PDS be made only to the targeted group of households living below the poverty line and not to all households. In this way the expenditure on food subsidy can be reduced significantly and only the poor will get the benefits of the subsidy. It is worth mentioning that Andhra Pradesh is a shining example of using PDS to help the poor under ₹ 2 per kilogram of rice supplied through ration shops.

9. Direct Attack on Poverty: Special Employment Schemes for the Poor. It was realised in the early seventies that it would take a very long time for economic growth to generate enough employment opportunities to provide productive employment to all the unemployed and poor in the country. Therefore, a strategy of providing employment to the poor in the short run, special schemes of employing poor on rural public works was proposed by Dandekar and Rath in their now famous work "*Poverty in India*".

The special employment scheme of rural public works which was launched by the Government in 5th Five Year Plan constitutes a direct attack on poverty as it does not depend on the trickle down effect of economic growth on the poor. There are mainly two types of such special anti-poverty schemes launched by the Government from time to time. First, there are several special schemes of providing *wage employment to the poor.* The first special employment scheme is MGNREGA (Mahatma Gandhi National Rural Employment Guarantee Act). It is a centrally-sponsored scheme implemented by Gram Panchayats to generate wage employment for the rural poor. The approach of this scheme is to employ the poor on building *durable and productive community assets* such as roads, small irrigation facilities, school buildings, rural electrification. These durable productive assets after completion would create employment opportunities on sustained basis.

Similarly, a second such special employment scheme is TRYSEM (*Training of Rural Youth for Self-Employment*) under which rural youth are given training and equipped with skills so that they can start some self-employment activity. Bank credit is arranged to financially support them.

The third special scheme called *Swarn Jayanti Gram Swarozgar Yojana* (SGSY) which aims at promoting micro-enterprises by helping the poor to form *self-help groups* with the assistance from the Centre and States. If implemented in a right spirit, these schemes can make useful contribution to poverty reduction in the rural areas. However, the actual performance of these schemes has not been satisfactory. While a large amount of money is spent by the Government on payment of wages, durable productive assets to be used for future use are not generally created. Similarly, Swarn Jayanti Gram Swarozgar Yojana (SGSY) for promotion of self-employment is merely used to get loans from the banks and financial assistance from the States. This is not generally used for investment for viable schemes of productive self-employment. This highlights the need for making reforms in these schemes so that they lead to the expansion of productive types of wage-employment and self-employment.

CHAPTER 55
INCOME INEQUALITIES AND ECONOMIC GROWTH

This chapter attempts to explain economic inequality, particularly income inequality, and how it changes as economic development proceeds in a developing economy. Its discussion is divided into three parts. In the first part we explain the meaning and concept of economic inequality and the methods of its measurement. In the second part we critically examine the rationale for social acceptance of some degree of income inequality. In this part we will critically examine how much inequality is justified on the basis of incentives to work more, to acquire skills and education and to promote more saving for increasing the rate of economic growth. In the third part we discuss Kuznets' famous *inverted U-hypothesis* about the impact of economic growth on the distribution of income. In the next part four we examine the relationship between economic growth and income inequality in the context of the Indian economy in the last six decades of planned economic development (1951-2011).

In the modern economics as we saw in Chapter 38, the distinction is now commonly made between the concepts of economic growth and economic development. Economic development means economic growth plus more desirable changes that must occur to raise the levels of living of the people at large. These other desirable things for economic development to occur are (1) reduction in economic inequalities, both of income and wealth, (2) reduction of absolute poverty which is the extreme form of inequality and (3) reduction in unemployment and underemployment.

The economic development requires a faster and more inclusive growth in GDP. The basic issue in economic development is not only how to achieve faster growth in GDP but also who benefits from growth.The principal objective of developing countries, especially of India, is to raise the levels of living of the people of a country. Therefore, economic growth will be of little value if it benefits only a few rich, keeping a good number of the population in the state of poverty and unemployment. This question of extreme inequality, that is, absolute poverty, and its relationship with economic growth will be examined in the next chapter.

It has now been found that many developing countries have experienced relatively higher rates of economic growth but much of the benefits of growth have not reached to the bottom of around 33 to 40 per cent of population and as a result they remained poor.

It may be noted that prime cause of inequality is not so much due to the differences in earnings from work *i.e.,* labour services but because of the *highly skewed distribution of property or assets* and also of the *differential access of the people* to such facilities as education, health, housing and employment. Therefore, if the objective of economic equality is to be realised we have not only to improve the access of the people in general to the facilities of education, health, housing but also to redistribute assets (for example, land reforms). Besides, to ensure relative equality, adequate employment opportunities are to be generated to ensure earnings from work to labourers.

Inequality: Meaning and Measurement

Economists are primarily concerned with these inter-related types of inequalities, namely, inequality in *size distribution* of (a) income, (b) earnings from work (*i.e.,* wages and salaries), and

(c) wealth. Inequalities in earnings from work or labour services arise either from paid employment (*i.e.,* wages and salaries) or earnings of individuals from self-employment in various types of economic activities.

Inequalities in incomes are described with reference to households or families and refers to the amounts received by all members of a household from all sources and shared by all its members, both earners and non-earners (including dependants and children). In the discussion of income inequalities what matters is how much a person or household receives regardless of whether income received is solely from work, from wage employment or from other sources such as rent from land, interest on financial capital, profits from business or dividends from shares of corporate companies or gifts or from inherited wealth. That is, in the concept of income inequalities we are not concerned with *functional distribution* but with the *size distribution* of income. Of course, functional distribution of income affects the size distribution of income. Incomes accrue to the households from the following three sources:

1. Income from work or earnings from labour services of the members of the family who are employed.
2. Income from wealth or assets possessed by a household or its individual members. The wealth consists of land and other non-renewable natural resources, tangible capital assets such as machines, equipment, buildings and financial assets belonging to households. Thus, rent from land, interest on financial capital and bank deposits and profits from enterprise and investment undertaken by the households or its members. Obviously, the distribution of income from wealth depends on distribution of wealth or assets among households in a society. The greater the inequality in the distribution of wealth or assets, the greater the inequality in income distribution.
3. The third source of income by households is the private or public transfer of income to the households as a result of unilateral income flows and cash transfers such as unemployment allowances, sickness benefits, old age pension, private transfers such as gifts, and remittances received from abroad by the households.

The income transfers from government also include free or subsidised supply of public services such as health and education to those who are incapable of funding them privately. It may be noted that transfer of income from government to individuals or households is essential in a welfare state and widely accepted as the function of the modern government in a democratic country.

MEASUREMENT OF INCOME INEQUALITY

To measure income inequalities economists arrange all individuals or households in ascending order of personal incomes. Then total population of individuals or households is divided into *quintiles* (*i.e.,* fifths) or *deciles* (*i.e.,* tenths) from the bottom to the top income groups, that is, in ascending order of income levels. It is then determined what percentage or proportion of total national income accrues to different quintiles or deciles of households. To give a concrete illustration we give in Table 55.1 the distribution of national income of India according to the ascending order of household disposable income of India for the years 1999-2000, 2005 and 2011.

It will be observed from Table 55.1 that in the year 1999-2000 the top 20 per cent of households in India had 43.3 per cent share of consumption expenditure, whereas bottom 20 per cent of population had only 8.9 per cent share of consumption expenditure in India. This shows the gross inequalities in consumption expenditure and therefore income distribution in the year 1999-2000. In the year 2005, household survey of consumption expenditure revealed that the highest (*i.e.,* top) 20 per cent of the people had 45.3 per cent share of total consumption whereas the share of the lowest 20 per cent of population in consumption expenditure was only 8.1 per cent. This shows that between 1999-2000 and 2005 the inequality in consumption expenditure increased. This is also revealed by Gini Coefficient which was 36.8 in 2005 compared to 32.5 in 1999-2000.

Table 55.1. Distribution of Income/Consumption among Population in India: Percentage Share of Income/Consumption Expenditure

Percentage of Population	1999-2000	2005	2011
1. Lowest 10 %	3.9	3.6	3.7
2. Lowest 20%	8.9	8.1	8.5
3. Second 20%	12.3	11.3	12.1
4. Third 20%	16.0	14.9	15.7
5. Fourth 20%	21.2	20.4	20.8
6. Highest 20%	43.3	45.3	42.8
7. Highest 10%	28.5	31.3	28.8
Gini Coefficient	32.5	36.8	33.6

Source: World Bank, *World Development Indicators* for the years 2006 and 2015 and are based on '*Household Expenditure Survey*' data for the years.

Kuznets' Ratio

A popular method of measuring degree of income inequality is Kuznet' ratio after the name of Simon Kuznets who has been a pioneer in the study of income inequality. *Kuznets' ratio is given by the income received by the top 20 per cent divided by the income of the bottom 40 per cent of the households or persons.* In India, as is seen from Table 55.1 in 1999-2000 top 20 per cent of households had 43.3 per cent share of national consumption expenditure and the bottom 40 per cent households 21.2 per cent of national consumption expenditure. This gives us Kuznets' ratio to be equal to 43.3/21.2 = 2.04. Thus, the degree of consumption or income inequality in India in the year 1999-2000 as measured by Kuznets' ratio was 2.04 which is quite high. In 2005, Kuznets' ratio is 45.3/19.4 = 2.3 which is higher than in 1999-2000.

It is worth mentioning here that Kuznets' ratio measures the degree of income inequality between the two extremes of very poor and very rich in the country. How is income distributed within the remaining range of income distribution is not considered by Kuznets' ratio.

Kuznets' Curve

A very popular method of measuring income inequality is Kuznets' Curve which graphically measures inequality in income distribution. In order to draw Lorenz Curve we have first to find out *cumulative* percentage of households corresponding to cumulative percentage of income accruing to them. This is shown in Fig. 55.1 where cumulative percentage of households (*i.e.,* recipients of income) arranged from the poorest to the richest (in terms of size of income) are represented on the horizontal axis and cumulative percentages of income of the households starting from the poorest to the richest are represented on the horizontal axis. Straight diagonal line *OD* shows equal distribution of income according to which every household receives the same income and therefore Lorenz Curve representing perfect or absolute equality of income will be a diagonal line. When the distribution of income is less than perfect equality, Lorenz Curve will be below it, as shown by *OD* curve in Fig. 55.1. The line of perfect equality serves as a point of reference to which we compare actual or observed distribution of income in a society. The shaded area between the line of perfect equality and the actually obtained Lorenz Curve such as *OD* represents the observed distribution of income, that is, inequality in a society. This shaded area is therefore generally described as '*the region of inequality*'. The larger the deviation of Lorenz Curve depicting actual income distribution from the line of perfect equality the greater the inequality of actual distribution of income.

Thus, in Fig. 55.2, the Lorenz Curve *H* (dotted) will indicate greater inequality of income distribution than the Lorenz Curve *L*. The maximum deviation of actual distribution from the line of perfect equality will be represented by the area of a triangle *OAD* in Fig. 55.2. According to

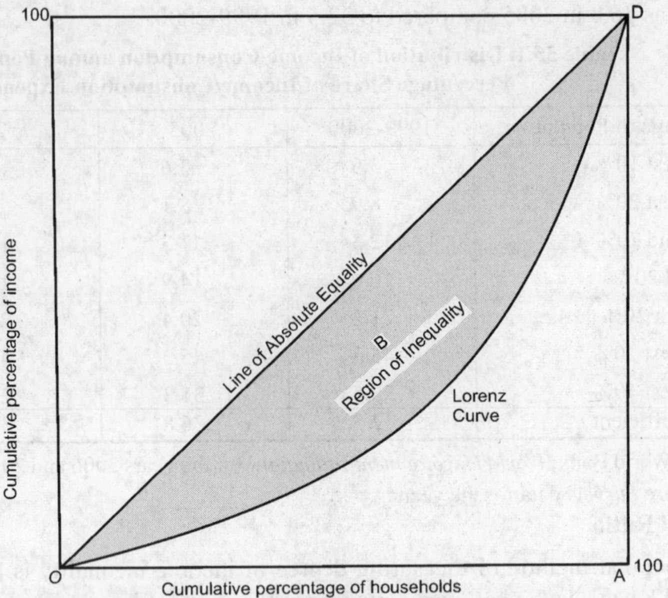

Fig. 55.1. *Lorenz curve showing income distribution*

Lorenz Curve criterion of income distribution, the farther the Lorenz curve to the line of absolute equality, the greater the inequality in the distribution of income. Thus in Fig. 55.2, the Lorenz curve *H* represents greater inequality in income distribution than Lorenz curve *L*.

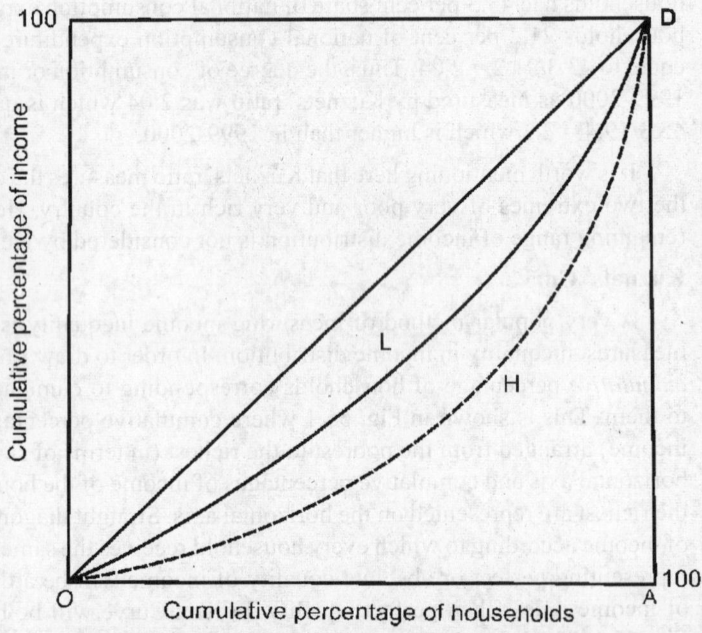

Fig. 55.2. *The Lorenz curves showing less and more inequality in income distribution*

Gini Coefficient as a Measure of Inequality

A conventional summary measure of inequality based on Lorenz curve is the Gini Coefficient named after the Italian economist who first suggested it in 1912. Gini Coefficient in Fig. 55.1 is given by the ratio of shaded area (*representing the actual deviation of observed income distribution from the line of perfect equality*) *to the area of triangle OAD*. Thus in Fig. 55.1, Gini Coefficient is given by the ratio of shaded area *B* to the total area of triangle *OAD*. Thus

$$\text{Gini Coefficient} = \frac{\text{The Shaded Area } B}{\text{The Total Area of Triangle } OAD}$$

It is worth mentioning the following four properties of Gini Coefficient:

1. First, the value of Gini Coefficient will be between zero (in case of perfect equality) and unity (in case of perfect inequality).
2. Second, since the Gini Coefficient is based on ratio of shares represented by a Lorenz curve, it is a relative measure of inequality in income distribution. This implies that the same Lorenz curve may refer to the distribution around different levels of income. This must be taken into account while comparing different income distributions.
3. In the third place, Gini Coefficient attempts to capture the extent of dispersion underlying the size distribution of income or consumption. Therefore, any statistical measure of distribution can be used to measure inequality in a given distribution of income.
4. Fourthly, the same Gini Coefficient may result from different Lorenz curves representing different income distribution patterns and therefore a *unique relationship* between the two does not exist. Let us illustrate this point with the help of Lorenz curve. Consider two Lorenz curves *A* and *B* with the same levels of mean income which cross each other as shown in Fig. 55.3. This situation of intersecting Lorenz curves represents conflict situation. In this case of income distribution share of income *accruing to certain sections* of the population is lower and in others higher in one situation than in the other. In Fig. 55.3, according to Lorenz

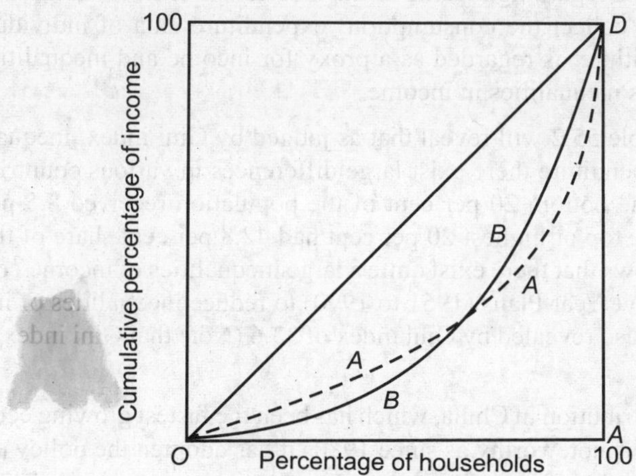

Fig. 55.3. *Two Lorenz curves representing conflict situation*

curve *A* in the households at the **bottom** of income distribution there is **more equality** and in households with higher income levels the income distribution is **more unequal**. The opposite is case *in Lorenz curve B in Fig. 55.3 where in the households at the bottom level there is more inequality*, while at the higher income level, income inequality is less. With the *level of mean income* being the same, the two Lorenz curves show different patterns of income distribution. The choice between such income distribution patterns represented by two Lorenz curves intersecting each other requires the *resolution of conflict situations* over which agreement regarding extent of inequality of income distribution may not be possible.

Coefficient of Variation (*CV*)

The coefficient of variation is a statistical measure of dispersion of income distribution. This measure gives more weight to larger deviation of households' income from mean and therefore represents better measure of income inequality than the '*mean absolute deviation*'. Coefficient of variation is the most common measure of dispersion of income that is generally used by statisticians and is based on standard deviation (σ) which squares all deviations from mean. Since square of a number rises more than proportionately to the number itself, this implies greater weight being assigned to the larger deviations from mean. *The coefficient of variation is just the standard deviation divided by the mean* and, therefore, it reflects only relative incomes. Let there be n individuals and m is the number of distinct incomes and j denotes an income class and nj denotes the number of individuals earning income in each income class. The coefficient of variation of an income distribution is written as

$$CV = \frac{1}{\bar{Y}} \sqrt{\sum_{J=1}^{m} \frac{nj}{n} \left(Y_j - \bar{Y}\right)^2}$$

In the above formula Y_j is income of a household, $\bar{Y}$ is the mean income, then $(Y_j - \bar{Y})^2$ is the square of deviation of income of a household from the mean. With summing up of such squared deviations of incomes of all households and then taking a square root of it we get standard deviation. On dividing this standard deviation by the mean $(\bar{Y})$ we get the value of coefficient of variation.

Income Inequality: A Global Comparison

In Table 55.2, we have given the distributions of income or consumption of both developed and developing countries. It may be noted that, the method of collecting data, degree of coverage of income or consumption data differ from country to country. However, some adjustments in income or consumption data have been made by the World Bank to make them comparable. It is worth mentioning that data regarding income of the individuals or households is difficult to collect. Therefore, most surveys collect the consumption expenditure data of individuals or households and consumption expendtiure is regarded as a proxy for income and inequalities in consumption expenditure are treated as inequalities in income.

Now, a glance at Table 55.2 will reveal that as judged by Gini index, inequality in distribution income/consumption expenditure there exist large differences in various countries of the world. In India in the year 2011 the bottom 20 per cent of the population received 8.5 per cent of income/ consumption whereas the top or highest 20 per cent had 42.8 per cent share of the national income or consumption. This shows that there exist quite a large inequalities of income/consumption despite the aim of the first six Five Year Plans (1951 to 1990) to reduce inequalities of income. The degree of inequality in India is also revealed by Gini index of 33.6 (Note that Gini index is Gini Coefficient multiplied by 100).

Now, the income distribution in China, which has been the fastest growing economy of the world in the last three decades, is noteworthy as since 1980s it has adopted the policy of privatisation and liberalisation with greater role to foreign investment to accelerate economic growth . It will be observed from Table 55.2 that in China in 2011, the top or the highest 20 per cent of population received 47.1 per cent of income/consumption while the bottom 20 per cent got only 4.7 per cent share of national income or consumption . It is the policy of privatisation and liberalisation of the economy that is responsible for the current large inequalities of income in China. Gini Coefficient in China in 2011 was 37.0 which is also quite high.

In Table 55.2, the worst developing country from the viewpoint of inequality is Brazil where Gini Coefficient in 2012 is 52.7 which is very high. Besides, in Brazil the top 20 per cent of population had 57.2 per cent share in national income/consumption, whereas the bottom 20 per cent received only 3.4 per cent share of it.

In the developed country of the US, inequalities in income distribution are quite large. In 2010, its Gini index was 41.0 which is quite high and in it while the top 20 per cent of population received 46 per cent of income, the bottom 20 per cent had only 4.7 per cent share in national income. In Japan, there exist less inequalities in income distribution where the highest 20 per cent of population had a share of 39.7 per cent in national income/consumption, the bottom 20 per cent received 7.4 per cent of it.

Table 55.2. Distribution of Income Consumption

Country	Year of Survey	Gini Index	Percentage Share of Income /Consumption of Subgroups						
			Lowest 10%	Lowest 20%	Second 20%	Third 20%	Fourth 20%	Highest 20%	Highest 10%
India	2011	33.6	3.7	8.5	12.1	15.7	21.8	42.8	28.8
Pakistan	2010	29.6	4.2	9.6	13.1	16.5	21.3	39.5	25.6
Bangladesh	2010	32.1	4.0	8.9	12.4	16.1	21.3	41.4	27.0
Sri Lanka	2009	36.4	3.4	7.7	11.4	15.3	21.0	44.6	30.0
China	2011	37.0	1.7	4.7	9.7	15.3	23.2	47.1	30.0
Brazil	2012	52.7	1.0	3.4	7.7	12.4	19.3	57.2	41.7
Kenya	2005	47.7	2.0	4.8	8.7	13.2	20.0	53.2	38.0
Indonesia	2011	38.1	3.4	7.6	11.3	15.6	21.8	43.7	28.2
Malaysia	2009	46.2	1.8	4.5	8.7	13.7	21.6	51.5	34.7
South Africa	2011	65.0	1.1	2.5	4.3	7.7	15.7	69.9	53.8
Thailand	2010	39.4	2.8	6.8	11.5	14.6	21.5	46.7	31.0
Philippines	2012	43.0	2.5	5.9	9.4	13.9	21.2	49.7	33.5
U.K.	2010	38.6	1.9	5.8	11.4	16.2	22.6	44.1	28.6
Canada	2010	33.7	2.7	7.1	12.4	16.8	22.7	41.0	25.7
U.S.	2010	41.0	1.4	4.7	10.4	15.8	23.1	46.0	29.6
Japan	2008	32.1	2.7	7.4	12.9	17.3	22.7	39.7	24.8
Sweden	2005	26.1	3.8	9.5	14.3	17.8	22.7	35.7	21.4
Norway	2000	25.8	3.9	9.6	14.0	17.2	22.0	37.2	23.4

Source: World Bank, *World Development Indicators,* 2015.

Of special mention are the Scandinavian countries of Sweden and Norway which have developed and become rich while maintaining relative equality in income distribution. It will be seen from Table 55.2 that in Sweden and Norway Gini Index is 26.1 and 25.8 respectively which show lesser income inequalities in the distribution of income. In Sweden, the top 20 per cent of population had a share of 35.7 per cent in income/consumption whereas the lowest 20 per cent received 9.5 per cent of income. Gini index of income distribution of Sweden is equal to 26.1 which is quite small. The almost same situation regarding income distribution exists in Norway. The readers can themselves calculate Kuznets' ratio from income distribution data of various countries given in Table 55.2.

ECONOMIC GROWTH AND INEQUALITY: KUZNETS' INVERTED-U HYPOTHESIS

In his presidential address to the American Economic Association in 1955, Simon Kuznets put forward the hypothesis that relationship between per capita national income and the degree of inequality in income distribution may be of the form of inverted-U. Due to limitations of data he used an inequality measure of *"the ratio of income share of the richest 20 per cent of the population to the bottom 60 per cent of the population"*[1] known, as Kuznets' ratio. According to Kuznets' Inverted U-hypothesis, as per capita national income of a country increases, in the initial stages of

1. Simon Kuznets, "Economic Growth and Inequality", *American Economic Review,* 1955.

growth, inequality in income distribtion rises and after reaching the highest *degree in the intermediate* level the income inequality falls. This is shown in Fig. 55.4 where as a country develops and its per capita income rises, the degree of income inequality initially rises and after reaching the maximum level, it falls as GDP per capita increases further. As time series data of the transition of the poor underdeveloped countries from underdeveloped stage to the developed stage was not available, he used the data of cross-section of countries including both developed and developing countries. In his 1955 study he calculated Kuznets' ratios and found that *the developing countries tend to have a higher degree of inequality whereas the rich developed countries tend to have a lower degree of inequality.*

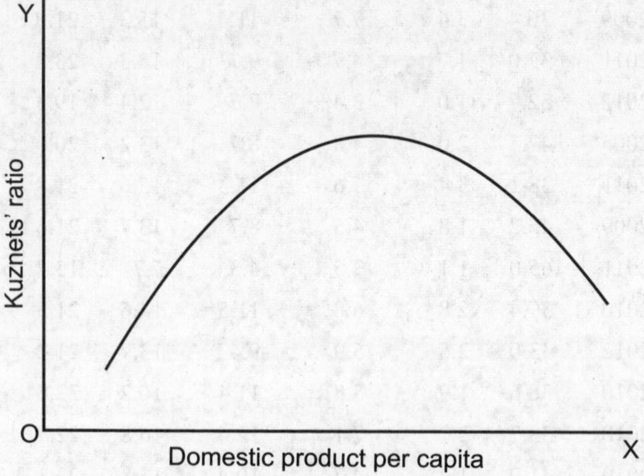

Fig. 55.4. *Kuznets' Inverted-U curve*

Later in his1963 study Kuznets provided further evidence of his inverted-U hypothesis regarding the relationship between inequality and economic growth. In this study he included eighteen countries, as in the earlier study, he included in this sample both the developed and developing countries. *From this later study he concluded that the share of upper income groups in the rich developed countries was signifcantly lower than their counterparts in the developing countries.*[2] This means income inequalities were higher in developing countries compared to those in the developed countries.

Other economists have also carried out studies to test Kuznets' inverted-U hypothesis. Due to the non-availability of income distribution data of an individual country over time as it grows over time from an underdeveloped stage, like Kuznets, others have also generally used cross-section data of countries with a mixture of developed and developing countries to test Kuznets' inverted U-hypothesis regarding the relationship between changes in income inequality and economic growth. One such cross-section study with data of forty-six countries classified into different income categories according to the per capita GDP in 1965 in US dollars was made by Paukert[3] using Gini Coefficient as a measure of inequality. Paukert's analysis of cross-section of countries also confirmed the inverted U-hypothesis of Kuznets and his findings are given in Table 55.3. As will be seen from Table 55.3, in less than $100 per capita GDP category countries, Gini Coefficient is 0.419 and as we go to the next category of countries with per capita GDP between $101 and $200, Gini Coefficient rises to 0.468 and in still higher categories of per capita GDP between $201 and $ 300 inequality as measured by

2. Simon Kuznets, "Quantitative Aspects of Economic Growth, Distribution of Income by Size", *Economic Development and Cultural Change*, Vol. 12, 1963.
3. Paukert, F. (1973), "Income Distribution at Different Levels of Development: A Survey of Evidence," *International Labour Review*, 108, pp. 97-125.

Table 55.3. Gini Coefficient in Cross-section of Countries

Income Category (1965 US $)	Average Gini Coefficient in Various Income Categories of Countries
Less than $100	0.419
$101 to $200	0.468
$201 to $300	0.499
$301 to $500	0.494
$501 to $1000	0.438
$1001 to $2000	0.401
$2001 and higher	0.365

Source: Paukert, F (1973), *op. cit.*

Gini Coefficient rises to 0.499. However, beyond this in still high income categories of countries, the value of Gini Coefficient goes on falling and in the highest income category of countries with per capita GDP $2001 and above, Gini Coefficient falls to 0.365. This is in accordance with Kuznets' inverted U-hypothesis regarding changes in income inequality as economic growth occurs.

Kuznets' inverted U-hypothesis seems to hold well in later years, at least up to the year 1970. Montek Singh Ahluwalia used income distribution data of cross-section of countries and made estimates for the countries in near about the year 1970. Results of his study are given in Table 55.4.

Table 55.4. Inequalities of Income in GDP Per Capita

GNP Per Capita (in 1970 in $)	Number of Countries Included	Percentage of Income Recorded		Gini Coefficient	GNP Per Capita Mean ($)
		Lowest 40%	Highest 40%		
Under $ 150	10	15.6	51.3	0.402	101
$150 – $500	19	11.2	57.9	0.479	301
$ 500 – $1500	12	12.2	56.3	0.461	754
Developed Market Economies	13	16.1	43.5	0.358	2849
Socialist Countries	6	24.1	34.6	0.238	913

Source: Montek S. Ahluwalia, "Inequality, Poverty and Development", *Journal of Development Economics* 3, (1978), pp. 340-41.

From the above table inequality can be judged by any three measures, namely, share of bottom 40% of population in GNP, share of top 40% of population in GNP and Gini Coefficients in different income categories countries. It is worth mentioning that column 6th of Table 55.4 GNP per capita indicates the level of development of the economy. Using any of the three inequality measures it is found that the inequality first rises, then falls as per capita GNP increases as Kuznets' inverted U-hypothesis suggested. Changes in Gini ratio reveals that as average GNP per capita of countries increased from $101 to $301 Gini Coefficient increases from 0.402 to 0.479 and in countries with per capita GNP of $ 754 Gini Coefficient falls to 0.461 and then at mean GNP per capita of $ 2849, Gini Coefficient falls to 0.358.

Similarly, the share of **bottom 40%** of population in GNP indicates that it first falls and then rises again showing that inequality first rises and then falls. In accordance with this the share of highest 40 per cent in GNP first rises and then falls.

An interesting fact is revealed by the last row of Table 55.4 which gives the data of 6 socialist countries around the year 1970. This reveals that degree of income inequality as per all the three inequality measures in them was much less compared to market capitalist countries. This is because in erstwhile socialist countries private ownership of tangible physical assets was generally abolished and therefore inequalities of income that arise mainly due to highly skewed distribution of assets and property did not exist in these socialist countries at that time. Even wage differentials in these countries were found to be less.

East Asian Countries and Kuznets' Inverted-U Hypothesis

It is woth mentioning that development experience in East Asian countries does not conform to the inverted-U. hypothesis of Kuznets. In East Asian countries such as Japan, South Korea, Taiwan, Thailand, Indonesia and Malaysia, contrary to Kuznets' inverted U-hypothesis, in the initial stages growth was not associated with increase in inequality. Instead, the increase in national income was widely shared among its population and millions were lifted out of poverty. For example, in Malaysia and Thailand the incidence of poverty declined from about 50 per cent in 1960s to less than 20 per cent by the end of the 20th century[4]. Though the policies pursued by the various East Asian countries differed a lot, *but the common features of these countries were "high rates of investment in physical and human capital, rapid growth of agricultural productivity and declining fertility"*.[5] All these were conducive to economic growth with decline inequality in income distribution.

Explanation of Kuznets' Inverted-U Hypothesis

Leaving aside the case of East Asian countries where Kuznets' Inverted-U hypothesis regarding changes in inequality as a result of economic growth does not apply to a number of other countries, both the developed and developing countries, Kuznets' inverted-U hypothesis seems to apply well. What explanations have been offered for this inverted U-hypothesis. It is Arthur Lewis's famous model of '*Economic Development with Unlimited Supply of Labour*' that was put forward to explain the development process in a dual developing economy has also found to be highly useful to explain Kuznets' inverted-U hypothesis regarding the relation between economic growth and changes in income distribution over time. It may be noted that Lewis's development model like those of Ricardo and Marx was based on the assumption that the unlimited amount of labour was available for employment in modern industrial sector at a fixed real wage rate which was unlike the neoclassical economists who assumed that labour was a scarce factor of production to be bid away from alternative uses and therefore was not available in unlimited quantity for being employed for the growth of modern industrial sector.

Lewis's labour-surplus model suggests that as economic growth takes place with withdrawal of surplus labour from low-productivity agriculture to the high-productivity modern industrial sector, income inequality will first increase and then after a point tends to decrease. This is therefore consistent with Kuznets' inverted-U hypotheses. In Lewis's labour-surplus model there are two reasons for increase in income inequality in the initial phase of economic development. First, the share of *profits or surplus of the capitalists* (who in the Lewis's version of the model may be either private capitalists or government running State-owned enterprises) increases as the modern industrial sector expands. The second reason for increase in income inequality in the early stages of economic growth is that in the beginning of urban industrialisation, wages in the modern industrial sector, according to Lewis, are 30 per cent higher in real terms compared to the subsistence wage level in the traditional agricultural sector. This tends to increase the overall income inequality in the economy.

However, according to Lewis's model, *this tendency toward increasing inequalities is finally reversed as growth proceeds further and when all surplus labour has been fully withdrawn* from subsistence sector and absorbed in the modern sector. This is because when all surplus labour is productively absorbed in the modern sector, labour becomes scarce and as demand for labour by the modern sector further increases, the increment in wages are needed to bid away workers from the marginal uses. Thus, it is eventually rise in general wage level in Lewis's model that causes decline

4 See Joseph E. Stiglitz, *Fair Trade for All* (New York: Oxford University Press, 2005), pp. 14-15.

5 *Ibid.*

in income inequality and also leads to the elimination of poverty. Thus commenting on changes in income inequality as economic growth takes place in Lewis's model, Debraj Ray writes, "*It should be quite clear that economic development viewed in this way cannot be evenly spread across the entire population at any one point in time. Initially only a few people get access to the progressive or modern sector. This view also suggests that the developed countries that have completed the transition from 'old' to the 'new' sectors should exhibit less inequality than the developing countries that are in the middle of transition process where individuals are in both sectors. The argument suggests that change is first uneven and then compensatory*."[6]

The second factor that causes increase in income inequalities as a result of economic growth is that technological progress initially benefits relatively few industries that employ a relatively small number of workers. Thus technological progress in the developing countries initially can benefit only a fraction of the industrial economy benefiting only a small number of workers employed in the industrial sector. The modern capital-intensive farming is less applicable to the agriculture of the developing countries *that have a large number of workers in the rural sector* who are poor. The small and marginal farmers do not have access to bank credit to buy and use modern inputs such as fertilizers, high-yielding varieties (HYV) of seeds, pesticides, modern tools which are important inputs for adoption of green revolution technology. That is why, in the initial stages of the green revolution in India only rich farmers and landlords were mainly able to get benefits.

There is another way in which the technological progress can lead to the increase in inequality in developing countries. This is because *superior technology is biased against unskilled labour* and can be used only by the trained skilled manpower. However, in developing countries initially only unskilled labour is in surplus and needs to be productively employed and their larger supply tends to lower wage rate and this tends to increase inequality in income distribution in the initial stages of growth . To quote Debraj again, "The skill differentials are ultimately compensated by the growing educational status of the labour force but this is a slower process. When it does occur, inequality tends to decline."[7]

Lastly, without the bias of technological progress and without the dual economy structure industrialisation itself also leads to the large profits accruing to the capitalist class which owns financial resources and entrepreneurial ability and seize the opportunities of making profits from industrial growth that causes increase in income inequalities. However, it has been a general view among the economists that benefits of growth will *ultimately trickle down to the poor working class in the form of more employment and higher wages*, especially when the labour is organised into trade unions. But, as emphasized by Prof. Debraj Ray, in many developing countries labour is initially in surplus so that wages do not rise right away and laws that protect labour are either absent or difficult to implement. However, it has been a general belief among economists that with the passage of time, rise in wages will ultimately eat into profits and lower the share of profits. This will therefore eventually bring about greater equality in income distribution. It was therefore claimed that inequality in the present-day developed countries was less as compared to developing countries.

Conclusion

We have examined above Kuznets' inverted-U hypothesis that suggests that in the growth process inequality first rises and then decline. The various factors and arguments have been advanced in favour of inverted U-hypothesis. However, as pointed out above, in case of East Asian countries such as Japan, South Korea, Taiwan, Thailand, Singapore where, contrary to Kuznets' inverted U-hypothesis, economic growth has resulted in the reduction in income inequality. This is *because the effect of economic growth on income distribution* has been influenced by economic policies pursued in these countries. For example, in countries of South Korea, Japan, Taiwan and other East Asian countries there

6 Debraj Ray, *Development Economics* (New Delhi: Oxford University Press, 1998), p. 210.

7 Debraj Ray, *op. cit.*

had been redistribution of land and also other interventions by government in influencing economic activities that growth process worked to lower inequality in income distribution. *In our view, there is no single path of growth which first increases inequality and then decreases it and much depends on the character of growth and policies followed by the governments of countries in the growth process*. Many factors and policies influence growth and income distribution and *the view that each country* must travel through the inverted-U hypothesis is quite unwarranted.[8]

INCOME INEQUALITY, ECONOMIC EFFICIENCY AND ECONOMIC GROWTH

Inequality, Incentives and Efficiency

While movement towards reduction in economic inequality whether of income or wealth is generally regarded as socially desirable as it leads to increase in welfare of the society as a whole, but attempts have been made by some, especially those who are supporters of capitalism or a free market economy to defend income inequalities on grounds of adverse effect on economic efficiency and economic growth of reducing income inequalities. It is argued that a certain income distribution cannot be settled once and for all through redistribution because income distribution is a continuous process as production of goods and services goes on year after year and rewards for those who contribute to the production of goods and services are distributed every year. Therefore, the question arises if inequality in distribution of income or rewards for the services is reduced, *it may adversely affect the incentives to work hard and invest and as a result the production of goods and services may decline, that is, GDP or economic growth will be hurt*.

It is thus argued that attempts to bring about equality, that is, to reduce income inequalities may reduce the size of pie which is to be distributed. The lower size of pie will reduce the well-being of the people. On the other hand, the attempts to reduce income inequalities will lower the efficiency or productivity as the people with higher abilities or productivities will not utilize their abilities or capabilities fully or put in best of their efforts. As a result, productive efficiency of the economy will decline which will lower economic growth of the country, which it is argued is a necessary condition for improving the levels of living of the population. According to Late Prof. Suresh Tendulkar, "The work-incentive effect stands on a firmer ground. For one thing, interpersonal differences in the capacity to work and productivity of such work are established facts, although there is no agreement on the causes or measures of these differences. Again, while it may be possible to force a person to do a given job it is admitted that the efficiency with which he works will depend on the effort put in. This in turn is determined by the inherently individualised motivation factors which are influenced largely by the incentives provided in the shape of economic rewards."[9]

Related to the incentives required for full use of their capabilities is the question of incentives needed for the individuals to acquire education or higher skills or for the entrepreneurs to innovate. How can individuals or workers be induced to acquire higher skills and education to become more productive or entrepreneurs to innovate if they are to receive no more than the ordinary unskilled workers? It is thus argued that differential incomes or rewards are essential to achieve higher levels of productivity and growth.

The above incentive argument for inequalities of income has, however, a major loophole. Above argument would have been perfectly valid if perfect equality of income was intended to be established. As a matter of fact, nobody, not even the socialists, seeks to establish perfect or absolute

8 For similar view, see Debraj Ray, *op.cit.,* p. 211 and Todaro and Smith, *op.cit*, Tenth Edition, 2012, p. 233.

9 Suresh D. Tendulkar, "Economic Inequality in the Indian Context" in Andre Beteille (ed.), *Equality and Inequality,* Oxford University Press, Delhi, 1983.

equality of incomes. *What is wanted is that reasonable inequalities of incomes should be established. The reasonable inequalities of incomes are those that are just sufficient to furnish or maintain the incentives*. Inequalities of income beyond the point necessary to furnish incentives are not justified on grounds of higher productivity and growth. *Inequalities of income that prevail today are much more than what are required to furnish incentives to work hard and innovate or use their capabilities fully*. Prevailing inequalities of incomes are so glaring and non-functional that even a substantial reduction in them will not adversely affect the incentives and therefore the fear that the national production or growth will fall as a result of equalitarian policies is unfounded.

Besides, the present large income inequalities found in the free market economies are quite unrelated to the differential capabilities of the individuals and are mainly based on extraneous factors such as ownership of property or assets, the existence of certain economic and social institutions. Furthermore, the important question is how to find reasonable or functional income inequalities which should therefore be socially acceptable. This is a difficult question which cannot be answered unambiguously as this depends on value judgements of a society which can differ from one individual to another. Based on the value judgements of a society, preferably arrived at through democratic ways, we find *socially acceptable degree of inequality* which is of functional nature and sufficient to furnish incentives for individuals to put in their best efforts and use their capabilities fully and to acquire higher skills or education.

It is worth noting that there may be some trade-off between efficiency and equity, that is , some loss of efficiency may be sacrificed to have an equitable distribution of income, that is, a certain socially acceptable degree of income inequality.

It may be further noted that argument put forward for income inequality based on work-incentives does not apply to the people below the poverty line defined by minimum normative level of income just sufficient for subsistence. This is because *lower than minimum level of income and consumption prevent them to be effective participant in economic activity.* Due to their extremely low income they suffer from *undernourishment*, *have inadequate clothing, shelter* and healthcare. As a matter of fact, *the eradication of poverty is a precondition for building up their capabilities to work and improve their efficiency*.

Economists since Adam Smith have acknowledged the need for economic incentives for promoting economic growth and consequent inequality in economic rewards as the necessary price for furnishing economic incentives. An eminent economist, A.P. Lerner, drew distinction between *functional and dysfunctional* inequalities. The functional income inequalities are those that promote economic growth and dysfunctional are those that do not do so. Nevertheless, there is general abhorrence for growing income inequality and any reduction in it is universally taken as advancing equity. It is, therefore, generally believed that government should intervene to bring about growth with equitable distribution of income especially in developing countries.

Therefore, in India, in the latest Five Year Plans (Eleventh and Twelfth Plans) the objective has been to achieve *inclusion growth*. It may be recalled that the concept of economic efficiency or what is called Pareto optimality takes the *distribution of income in a society as given* and explains how social optimum resource allocation is achieved. *It does not go into the question of optimum distribution of income in a society*. With a different income distribution, there will be different Pareto optimum. As a matter of fact, most government policies affect the distribution of income. This issue was even raised by classical economists Adam Smith and Ricardo who advocated free trade which benefits some and harms other members of the society. N. Kaldor and J.R. Hicks provided a welfare criterion according to which if as a result of government policy those who gain can compensate the loss and still remain better off, the new situation would represent an improvement over the existing one. However, the serious drawback of this criterion is that the *compensation visualised is potential and not real*. Therefore, for increase in actual welfare when any growth project is undertaken those who are displaced must adequately be compensated and also alternative gainful employment opportunities are provided to them.

Furthermore, the present inequalities of income reflect the *differences in opportunities* faced by various individuals. A man is the product of numerous forces acting upon him. Some forces work before his birth. The result of these forces is that an individual possesses certain inborn qualities. Other influences upon man come from environment in which he grows up. This environment consists of conditions of home, family, friends, teachers, and general surroundings. Both these hereditary and environmental forces come from outside man himself. An individual who is born with the silver spoon in his mouth and is therefore brought up in superior home influences and receives good education and training is not likely to acquire higher skills and education. He will be able to earn large income.

On the other hand, a person who is born in a poor family and is brought up in the environment of dust, disease and despair and who cannot obtain sufficient education and training because of the poverty of his parents is not likely to acquire higher skills and education. He will be able to earn only a meagre income. Under these circumstances, it cannot be said that the former individual *deserved success* and large income and the latter failure and meagre income. It is mere accident of birth and social conditions which have determined the success of the former individual and the failure of the latter. Each individual is the product of society. The wealth and richness of a man is mainly made possible by good heredity and favourable social environments. The poverty of the person who fails in life is due to bad heredity and unfavourable social environments. No one can claim with justification that his success and wealth are purely due to his personal efforts and that society has not contributed anything towards it.

Inequality, Saving and Capital Formation

From the viewpoint of economic growth, a forceful argument advanced to justify income inequalities is that inequality in the distribution of income promotes savings and capital formation. Because of their higher incomes rich people are in a position to make more savings, and invest them. As a result, capital formation takes place in the society. The greater the inequality in the distribution of income, the greater will be the rate of saving and investment. This increases the rate of capital formation. Capital formation is key to economic growth. A higher rate of capital formation brings about rapid economic growth and greatly increases the level of national production. It is argued that if the national income is equally distributed, that is, if income equality is brought about, average rate of saving will fall. With the fall in saving the rate of investment and capital formation will decline. As a result, level of national production and rate of GDP growth will go down making everybody poorer than before.

It is thus concluded that large incomes which accrue to the rich people are really beneficial since a good part of them is saved and invested. This promotes capital formation and economic growth which in turn raise the levels of living of the people. In this way, according to this view, *the rich are public benefactors*.

The relation between inequality and capital formation established above is doubtful and uncertain. It may be true that in a capitalist system large savings are made possible by the huge incomes accruing to the few rich people. But it is quite wrong to hold that the rate of savings and capital formation will necessarily decline in any other economic system having relative equality of incomes. Indeed it may be said that the large incomes accruing to the rich people are not adequately utilised for the purposes of capital formation. Quite a good amount of these high incomes is squandered away in *conspicuous consumption*. It is more conducive to capital formation if the surplus incomes of the rich come in the hands of the Government which can fully utilise them for investment in desired directions. Professor

J.K. Galbraith, an eminent economist, apply remarks, "*To give individuals large incomes to encourage savings has elements of illogic. The rich man saves because he is able to satisfy all his wants and then have some left over. Such a saving, in other words, is residual after luxurious consumption. This obviously is not an especially efficient way to promote capital formation.*"[10]

Even empirical evidence does not support the view that large inequality in the distribution of income is helpful to capital formation or that the drastic reduction in inequality of incomes will discourage saving and capital formation. For instance, Norway, which had adopted effective policies to promote equality of incomes, has achieved since the Second World War one of the highest rates of capital formation and economic growth in the non-communist countries. In sharp contrast, Middle East countries, where inequality of income is perhaps the greatest, are among those with lowest rates of capital formation.

Further, large inequalities of income *may create economic instability* which may adversely affect investment and growth. Thus, Nobel Laureate Joseph E. Stiglitz in his recent book, "*The Price of Inequality*", points out that there is clear *association between inequality and economic instability*. According to him, in the US the people in top income-groups save a lot and relatively spend little, whereas the people in low and middle income groups save very little and spend all or most part of their incomes. This is the result of highly skewed income distribution and this reduces an aggregate consumption demand in a society which tends to reduce the level of GDP and thus causes economic instability.

However, in case of India when the *problem is more on supply side*, though large savings by the rich are good as they make large investment and growth possible but to rely on greater income inequalities to generate higher saving and investment is not the right approach. *Many East Asian countries such as Singapore, Malaysia, Taiwan, South Korea which achieved a higher rate of GDP growth from the mid-sixties to the mid-nineties did not experience the increase in income inequalities*. They were able to achieve high rates of economic growth with active participation of Government in their economies. Writing about growth miracle of East Asian countries, Joseph Sitglitz writes, "*Their economic model included a strong role for the public sector. They clearly did not believe in free and unfettered markets but government played a critical role. Governments acted as catalyst which helped markets by providing the requisite physical and institutional infrastructure, by remedying market failures and by promoting saving and technology.*"[11]

Further, Stiglitz writes, contrary to what has happened in India, "East Asian countries proved that the initial stages of development did not have to be associated with an increase in inequality. Instead the new prosperity was widely shared among the population and millions were lifted out of poverty. For example, in Malaysia and Thailand, the incidence of poverty declined from almost 50 per cent in the 1960s to less than 20 per cent by the end of the century."

Inequality is bad partly due to the fact that for any given average per capita income, the higher the inequality a larger number of people *lack the collateral to get loans from banks* or any other source because they are not creditworthy. With low incomes, the people cannot educate their children and cannot spend adequately on their healthcare. As a result, their children remain uneducated and undernourished. Besides, in case of India due to large income inequalities, as in other developing countries, the people in high income-brackets such as big capitalists, landlords, high-salaried persons spend lavishly on imported goods, luxury cars, palatial houses, gold and jewellery etc. which have *demonstration effect* on the middle-income group people and tend to lower overall saving rate of the society by increasing their propensity to consume. It is thus evident that income inequalities are bad for achieving higher rate of saving and economic growth.

---

10 Galbraith, J.K., *The Affluent Society*, p. 63.

11 Stiglitz, *op. cit.*

High income inequalities also promote *rent seeking* by the rich people who through excessive lobbying, providing large donations to the political parties, bribing the ministers and bureaucrats succeed in getting licences and permits as happened in India in case of allotment of 2G spectrum and coal blocks. When large financial resources are used for such rent seeking activities, they are diverted from their productive use which could have generated rapid growth.

We have seen earlier that, according to inverted-U hypothesis, as developing countries start growing in the initial stage inequalities of income will increase. On the basis of this hypothesis, the supporters of free market economy argued that increased inequality is a necessary condition for economic growth. However, relevant data collected by World Bank show that income inequality adversely affects economic growth.[12] Expressing a similar view Todaro and Smith write: "*A growth strategy based on sizable and gross income inequalities may in reality be nothing more than an opportunistic myth designed to perpetuate the vested interests and maintain status quo of the economic and political elites of developing nations, often at the expense of the great majority of general population.*"[13]

Society's Tolerance of Inequalities

Recently, a view has been expressed about the conditions or circumstances in which income inequalities can be socially tolerated. Late Prof. Suresh Tendulkar provided an economic rationale for rising income inequalities during periods of rapid economic growth because they cause upward *income mobility* for the people. Rapid economic growth throws up opportunities for the people to move from low income to higher income levels despite the fact that income inequalities as measured by Gini Coefficient remain quite large. Thus, rapid economic growth generates hope for the people to move to higher income level category and makes the existence of large income inequalities socially acceptable. However, according to Tendulkar, equity would be achieved in the *normal course of market functioning without* government intervention for reducing inequalities *if technological changes reduce the relative price of one or more necessities of life or by enhancing the supply and consumption of necessities like food.*[14] Two things must be noted about this view of Late Prof. Tendulkar. First, if a free market economy experiences rapid economic growth, it will promote equity without intervention by the government to reduce income inequalities. As it ensures upward income mobility, it will make the existing income inequalities socially acceptable Second, Tendulkar considers that if food and other necessities of life become cheaper and become more available for consumption as a result of rapid economic growth, income inequalities will become socially acceptable. In other words, *if functioning of free market economy brings about rapid growth which causes absolute poverty to fall, the existing income inequalities will be socially acceptable*. In our view, this is an indirect argument for a free market economy established by the policy of liberalisation and privatisation adopted under the structural adjustment reforms.

Like Tendulkar, long ago Hirschman had also argued that in a rapidly growing economy the improvement in the well-being of others may signal a benign external environment and hope for the rest. This hope or 'expectations calculus' produces gratification that overcomes envy and raises society's tolerance for inequalities. A noted welfare economist. Tibor Scitovsky, was one of the earlier economists who suggested three criteria for social acceptability of inequalities. First, those inequalities that are based on merit or people's contribution to social product that can be considered as 'functional inequalities'. Second, those inequalities that arise even when there exist conditions of equality of opportunity which allow everyone to aspire to move up the income ladder. Third, those inequalities that prevail along with the improvement in the well-being of those at the bottom of the income ladder. This third criterion advanced by Scitovsky implies that if economic growth benefits the poor, the prevailing income inequalities will be socially tolerable.

12. See Michael Bruno and Lyn Squire, "Why Worry About Inequality", *Economic Times*, Oct. 8, 1996.
13. Todaro and Smith, *Economic Development*, 8th edition, 2003, p. 210.
14. Suresh Tendulkar, "Inequality and Equity during Rapid Growth Process" printed in *India's Economic Performance and Challenges*, Essays in Honour of Montek Singh Ahluwalia.

Chapter 56
ECONOMIC REFORMS, STRUCTURAL ADJUSTMENT AND GROWTH

Rationale for Economic Reforms

An important issue in the development policy has been whether the economic growth should be promoted by adopting planning and controls or through free market system in which private sector and foreign investors are permitted to play an important role. In the fifties and sixties, this issue was resolved in favour of planning and regulation of the private sector. It was felt that intervention by the Government was essential to initiate the process of development and thereby to break the vicious circle of poverty. However, in the eighties it was felt that excessive regulation and control of the private sector by the Government and high importance given to the public sector in the strategies of development of the developing countries were obstructing economic growth and *promoting inefficiency*. As a result, there was a slowdown in growth of these countries and further they experienced serious problems of large fiscal deficit resulting in a high rate of inflation and deficits in the balance of payments resulting in heavy indebtedness of the developing countries.

To overcome these problems some economists such as I.M.D. Little, Jagdish Bhagwati, Bela Balassa who had been advisors to the World Bank and IMF argued for the adoption of the policy of economic liberalisation by the developing countries to promote growth and check inflation and solve the problems of balance of payments. They advocated that free markets and greater role of private sector (including foreign investors) would ensure efficiency by encouraging competition. Under these reforms direct controls had to be given up, the role of planning and public sector to be greatly diluted. Further, liberalisation of imports was suggested by drastically cutting down tariffs. Foreign private investment had been given greater importance and incentives had to be provided to attract them.

Economic Crisis in India in 1991

We explain below these structural adjustment reforms, with special reference to India. After independence, India avoided the extremes of both capitalism and socialism and opted for a mixed economy where planning and the public sector played an important role in the economic development of the country. The experiment of a mixed economy as evolved in India seemed to be a success in the beginning but during the seventies and eighties several shortcomings in its working had been noticed. First, the public sector which was expected to generate adequate resources for the growth of the economy not only failed to do so but in fact was incurring huge losses which raised the expenditure of the Government. The losses of the public sector were said to be due to the *inefficiency in the working of public enterprises*. Secondly, the problem of *macroeconomic imbalances*, both in the internal and external sectors, emerged and assumed serious proportions in 1990-91. The huge deficit in the successive budgets of the government due to internal imbalance of Government expenditure and revenue led to the creation of more money supply which created the serious problem of rising prices, so much so that rate of inflation went up to about 17 per cent per annum during August 1991. This compelled the newly elected Congress government to reduce Government expenditure and fiscal deficit for stabilising prices.

Like many other developing countries, India in early 1991 faced a serious economic crisis. Fiscal deficit of the Centre and States combined went up to more than 12 per cent of GDP with the result that inflation rate shot up to around 13 y-o-y. In the external sector, *the huge deficits in the balance of payments due to the rising imports and sluggish exports persisted for a long time and, to meet this deficit, the government resorted to commercial borrowing from abroad at higher market rates of interest*. As a result of persistent deficits on the current account, the foreign exchange reserves declined to a very small sum in June 1991 which could last for payments for necessary imports only for about few weeks. This compelled the Government to approach IMF (International Monetary Fund) for loans to make necessary payments and enhance the foreign exchange reserves.

Though the annual growth of national income *during the eighties* went up to over 5 per cent and of the industrial sector to 8 per cent, which was considered to be a good performance, but due to macroeconomic imbalances which emerged in the economy they could no longer be sustained any more. In the external sector, the situation became so worse in 1991 that India came to the brink of default. To avoid the default and to win over the confidence of international business community, the government had to sell a part of its gold reserves and a certain amount of gold was even pledged with Bank of England to obtain the foreign exchange reserves to meet its obligations. Further, it had to approach the IMF for loans to tide over the foreign exchange crisis.

Excessive Government Controls over Private Sector. Besides, excessive government regulation and control over the private sector proved to be counter-productive. The following drawbacks of excessive regulation of the private sector were emphasised and case for liberalisation of private sector was built up, especially by Jagdish Bhagwati and T.N. Srinivasan, the two economists associated with World Bank and International Monetary Fund. The following arguments were given in favour of deregulation of the private sector and assigning to it a greater role in economic growth of the Indian economy.

The industrial licensing and controls had degenerated into arbitrary bureaucratic controls on the allocation of resources in competing industries. The licences were issued to make investment in those industries which bureaucrats chose in tune with the policies they administered. This licensing system replaced the efficient market system of resource allocation by personal system of resource allocation. Thus excessive regulation of the private sector, according to the advocates of economic reforms, had resulted in the sub-optimal resource allocation in industries. Hence, there was a need for liberalisation. Under the regime of licensing, the industrial units which tried to increase output by making full utilisation of *installed capacity* were penalised on the ground that their production exceeded the *licensed capacity* even if goods produced were much needed and were in short supply. Hence, *industrial liberalisation* became essential from the point of view of promoting industrial growth for creation of productive capacity and its optimal utilisation. Furthermore, industrial policy had failed to achieve its objectives of balanced regional industrial development and prevention of economic power. In fact, during licensing regime, the regional industrial imbalances and concentration of economic power had got intensified. Hence there was a strong case for scrapping licensing system and switching over to the market system.

WASHINGTON CONSENSUS, DEVELOPMENT POLICY AND STRUCTURAL ADJUSTMENT REFORMS

Before explaining the economic reforms initiated in 1991 under the pressure of IMF and World Bank, it may be mentioned that these reforms mainly constituted what is now labelled as *Washington Consensus*. This consensus was formulated by John Williamson, with the other officials of World Bank, IMF and key US government agencies. This consensus contains 10 elements of development policy which were suggested to promote economic growth in the developing countries and also help them solve their problems of indebtedness, balance of payments difficulties and high rate of inflation. *It is noteworthy that this Washington Consensus followed a free market approach to development and, contrary to the views of many development economists, it did not emphasise elimination of poverty and reducing inequalities to achieve real development, as it is now widely understood.* The Washington Consensus and fiscal and structural reforms it envisages views that poverty will be automatically removed if rapid economic growth is achieved. Its message is 'take care of growth, poverty will take care of itself'. Economic policy reforms by IMF-World Bank are in fact a 'standard blueprint', a sort of 'same size fit all' regardless of their special problems.

Let us list below the various elements of Washington Consensus.

1. **Fiscal Adjustment.** This means that developing countries should take steps to reduce fiscal deficit through curtailing government expenditure by drastically reducing explicit and implicit subsidies provided by the government of developing countries.

2. **Tax Reforms.** It was suggested that tax rates should be cut substantially to promote saving and private investment. This will promote investment and ensure a higher rate of economic growth. This was in line with supply-side economics. In fact, Jagdish Bhagwati and T.N. Srinivasan invoked *Laffer Curve* concept of supply-side economics to argue that reduction in rates of taxes would cause not only greater private saving and investment but will lead to greater revenue for the government.

Besides, it was suggested to broaden the tax base by withdrawing exemptions and plugging the loopholes in taxes.

3. **Deregulation.** This was most significant policy measure under which it was recommended that industrial licensing controls be abolished but also such measures as Monopolistic and Restrictive Trade Practices Commission (MRTPC) and FERA (Foreign Exchange Regulation Act) be done away with so that private sector should grow without any obstructions.

4. **Privatisation.** This is another significant measure of development policy under which it was proposed that there should be disinvestment of public sector enterprises and accordingly either they should be sold outrightly to the private sector or government stake should be reduced and its shares sold or transferred to private enterprises.

5. **Trade Liberalisation.** Under this, it was suggested that tariffs on imports should be drastically reduced *to promote free trade*. Besides, all quantitative restrictions on imports were also to be eliminated to permit free trade.

6. **Competitive Exchange Rate.** Under this, it was recommended that developing countries like India should devalue their over-valued currencies and ultimately adopt flexible exchange rate system. Besides, it was suggested to make the currency convertible so that obstacles to growth of free trade and capital mobility be removed.

7. **Removal of Barriers to Foreign Investment.** It was stressed that economic growth in developing countries could be accelerated through larger foreign direct investment (FDI). Therefore, all barriers put up by developing countries should be eliminated to attract foreign investment in their countries. To facilitate foreign investment it was also suggested to adopt *convertibility of currency on current account as well as on capital account*.

8. **Financial Reforms.** They involve reforms in the banking and insurance system and also in capital market.

9. **Protection of Property Rights.** According to this, suitable legislative measures should be taken to protect property rights. Besides, labour law should be amended so that it becomes easy for *private enterprise to enter and exit* the industries. In this, *freedom to exit* was emphasised and it was recommended that labour laws be made flexible so that it should be easy to retrench workers.

10. **Redirection of Public Sector Investment.** Lastly, it was emphasised that public sector investment should be redirected towards education, health and infrastructure only and also leave these and other fields open to private sector operation.

STRUCTURAL ADJUSTMENT REFORMS IN INDIA

However, to evolve a long-term solution to promote price stability, to bring about equilibrium in balance of payments and, above all, to ensure higher rate of economic growth, the elected Congress government with Dr. Manmohan Singh as Finance Minister adopted in 1991 what is popularly called *New Economic Policy* on the advice of IMF and World Bank. According to this policy, the government was to prune down its expenditure so as to *reduce budget deficits for achieving price stability. But more important element of new economic policy is the adoption of what are called Structural Adjustment Reforms* which sought to change the nature of Indian economic system by ultimately establishing *a predominantly free market economy*. Under this structural adjustment programme, the role of public sector in economic development had been diluted and that of private sector enhanced and expanded.

It may be noted that the policy of free-market economy also got strong stimulus due to the collapse of socialist system in Soviet Russia, China and other centrally planned socialist economies in Eastern Europe. Accordingly, under the new economic policy, the idea of the predominant role of the public sector was abandoned and the private sector was given a greater role in promoting economic growth through competition and efficiency. Instead of being directed by the planning and licensing, the private sector was now to be governed by the market forces, that is, by the free play of demand and supply of commodities and factors. We describe below the new economic reforms undertaken in India since 1991 under the *New Economic Policy* or *Structural Adjustment Programme*.

However, it is worth noting that it is generally believed that Dr. Manmohan Singh formulated these reforms. This is quite wrong. As has been seen above that these reforms were proposed by IMF and World Bank and they used them as '*conditionalities*' for giving aid. They asked 'devalue the rupee', he devalued it; they asked cut tariffs for trade liberalisation, he did so and so on.

Dereservation of Industries of the Public Sector

The new industrial policy 1991 was adopted under which far-reaching structural reforms have been initiated to lift excess direct controls and regulations on industries and to ensure a free market-oriented economic system. The list of industries reserved for the public sector has been pruned. Now, only 6 industries remain reserved for the public sector. Among the industries reserved earlier for the public sector were many core industries like iron and steel, electricity, air transport, ship building, heavy machinery industries such as heavy chemical plants and telecommunication, cables and instruments. The new industrial policy of 1991 threw open all these industries for the private sector for investment and growth. Thus the new policy indicates the government's intention to invite a greater degree of participation by the private sector in important areas of the economy. The six industries which continue to be reserved for the public sector are in areas where security and strategic considerations predominate. According to the new policy, the resources in the public sector will now be used for the development of strategic, high-technology industries and essential infrastructure areas and for social sectors such as education, public health and poverty alleviation programmes.

Liberalisation: Abolition of Industrial Licensing System

Until 1991 the Indian industrial sector had functioned under a system of tight controls and regulations represented by industrial licensing which meant to allocate the scarce resources towards building the industrial base of the economy. By the year 1991, the Indian industrial economy had a quite wide and diversified base, the new policy abolished all industrial licensing irrespective of the level of investment except for 15 industries for which licence was still required. These industries are those which are essential for security and safety purposes and for protection of environment. For granting of licence for even these 15 industries procedure has been greatly simplified. For getting licences only certain locational guidelines are to be fulfilled so that polluting industries should not cluster around the major urban centres. Besides, restrictions by Monopolistic and Restrictive Trade Practices Act (MRTPA) for expansion of big firms were also relaxed.

Privatisation of Public Sector Enterprises

An important aspect of new industrial policy of the government is that it should not operate commercial enterprises. With that end in view the government decided to disinvest the public enterprises. The government can sell its enterprises completely to the private sector or disinvest a part of its equity capital held by it to the private sector companies or in the open market. Therefore *this policy of public sector disinvestment has also been called the policy of privatisation.* Through disinvestment or privatisation, the government can mop up a good amount of resources which can be used for various purposes. The released resources can be used to restructure and strengthen the public sector enterprises which are potentially viable. These resources can also be used to pay back a part of public debt. These resources can also be used to finance budget deficits.

What are the reasons for the policy of privatisation or public sector disinvestment? First, resources available with the government are scarce. The government needs resources to reduce its budget deficit. The government urgently requires resources to make investment in infrastructure, education, public health and for poverty alleviation programmes. Resources released through disinvestment can be used for investment in these crucial sectors. Second, a good number of existing public enterprises are working inefficiently and incurring huge losses.

Accordingly, as part of the economic reforms policy, the government started reforms in public sector enterprises. The main elements of government policy towards Public Sector Enterprises (PSU) are:

1. Disinvestment of government equity in all *non-strategic* public sector undertakings (PSU) to 26 per cent or lower if necessary.

2. Those public sector enterprises which are potentially viable have to be restructured and revived.

3. Those public sector undertakings (PSU) which cannot be revived would be closed down.

Table 56.1. Disinvestment of Public Sector Undertakings in India Since 1991

Year	*Target (₹ Crore)*	*Actual Receipts (₹ Crore)*
1991-92	2,500	3,038
1992-93	2,500	1,913
1993-94	3,500	NIL
1994-95	4,000	4,843
1995-96	7,000	168
1996-97	5,000	380
1997-98	4,800	910
1998-99	5,000	5,371
1999-2000	10,000	1,860
2000-01	10,000	1,871
2001-02	12,000	5,632
2002-03	12,000	3,348
2003-04	14,500	15,547
2004-05	4,000	2765
2005-06	no target	1568
2007-08	no target	2367
2008-09	no target	546
2009-10	no target	1120
2010-11	40,000	13,854
2012-13	30,000	NA

Source: GOI, *Economic Surveys*, 2003-04, 2005-06, 2006-07, 2007-08, 2008-09, 2009-10, 2010-11 and 2011-12.

The government has now evolved a fair, transparent and equitable procedure for disinvestment in selected public sector enterprises. The achievement made with regard to disinvestment of public sector undertakings which started in 1991-92 with the sale of minority stakes in some public sector undertakings are given in Table 56.1.

Globalisation

The following measures for globalisation of the Indian economy were taken.

Welcoming Private Foreign Investment and Foreign Technology. Till 1991, foreign investment and import of foreign technology was regulated tightly in India. In the case of foreign technology agreements sought by Indian firms as well as foreign investment, it was necessary to obtain specific prior approval from the government for each project. This involved delays and hampered business decision-making in the import of technology by Indian firms. Under the new economic policy for a selected list of high technology and high investment priority industries, firms receive *automatic approval* to make foreign technology agreements within certain guidelines. This policy measure is expected to remove the delay and uncertainty which earlier clouded the relationships between Indian and foreign firms.

Further, the government has *liberalised its policy regarding foreign investment.* To attract private foreign investment, it has been decided to grant *automatic permission to private foreign investors to invest up to 51 per cent of the total equity shares in 34 high priority industries if certain norms are fulfilled.* This facility will be available to those firms which are able to finance their capital equipment imports through their foreign equity. This is a major departure from the previous policies which required case-by-case approval of foreign investment normally limited to 40 per cent equity participation. The priority sector includes power generation and petroleum refining. Further, the government has given a guarantee of 16 per cent return on foreign investment in priority sectors. There is also proposal to allow automatic clearance for foreign *financial* and technical collaboration with Indian firms.

Trade Liberalisation. In a bid to open up the economy, under the new economic policy *quantitative restrictions*, that is, import licensing for imports of goods have been removed. Further, to promote competitiveness, efficiency and globalisation of the Indian economy import duties have been reduced to 20 per cent (excluding agricultural and dairy products) with effect from January 9, 2004. The new thinking is that with the cheapening of imported goods, the Indian firms will try to increase their efficiency and reduce costs so that their products can compete with the foreign products. In this way, they will be able to expand their exports.

Currency Convertibility and Floating of Indian Rupee. Another major step taken towards liberalisation of the Indian economy has been to make the Indian rupee fully convertible on current account since March 1993. This implies that for the purpose of foreign trade and travel you can convert the rupees into dollars and dollars into rupees in the foreign exchange market at the market determined exchange rate. That is, the foreign exchange rate of rupee is now not fixed but liable to change as demand for and supply of foreign exchange change. In July 1991, the Indian rupee was devalued by about 20 per cent. But now rupee has been made flexible whose price is determined by demand and supply. The Indian rupee can depreciate or appreciate depending upon demand and supply conditions.

STRUCTURAL ECONOMIC REFORMS: A CRITICAL EVALUATION

It is evident from the reforms introduced in the Indian economy that from a planned economy it has moved towards a free-market economy. Though we still have mixed economy with both the public and private sectors coexisting but the role of private sector which is governed by market forces has been greatly increased and that of the public sector greatly diluted. So we now have a mixed economy with greater orientation towards a free market and private sector.

Now, 25 years have passed when we initiated economic reforms in 1991 laying greater emphasis on liberalisation and privatisation. Now, the pertinent question is whether these economic reforms have ensured higher and sustained economic growth which the supporters of these reforms argued they would, whether these reforms have led to a higher growth rate and to the greater reduction of poverty and unemployment which was expected of them. The answer to these questions is a mixed one. Foreign exchange reserves which had gone too low in 1991 due to profligacy of the late 1980s and temporary oil shock of first Iraq (Kuwait) war in 1990, have substantially increased in the post-reform period.

Foreign Exchange Reserves and Balance of Payments. Foreign exchange reserves are now, in August 2013, at around 285 billion US dollars. In 2007-08, our growth in export was 29.0 per cent and in the later half of 2008-09 due to global financial crisis our export started falling and for the whole year 2008-09 there was 13.6 per cent growth rate in exports and due to global slowdown in 2009-10 there was negative growth in exports to the extent of 3.5 per cent. But in 2010-11 and 2011-12 exports registered substantial growth of 40.5 and 21.3 per cent respectively but again there is negative growth of exports (– 4.9 per cent) in 2012-13.

Following the adoption of export-oriented strategy in place of import-substitution strategy, exports on an average have grown at a higher rate in the post-reforms period. An important achievement in this regard has been that India has become a leading country in *software exports*. India's software exports rose from near zero in 1991 to 17 billion US dollars in 2004-05. Foreign investment in India which was barely US $ 1 million in 1990-91 rose to $ 33 billion in 2011-12. But a serious problem which has risen recently is of *large current account deficit* (CAD) due to import liberalisation of economic reforms. Current account deficit (CAD) was within comfortable zone until 2010-11, it was 1.5% of GDP in 2007-08, 2.3% in 2008-09, –2.8 per cent in 2009-10 and 2010-11 but worsened thereafter. In 2011-12 it was estimated at –4.2 per cent of GDP and in 2012-13 CAD rose to 4.8% of GDP. This had been due to imports of large quantity of **gold** and our dependence on foreign supply of crude oil. Therefore, many persons said we were back to the state of crisis of 1991. However, Dr. Manmohan Singh refuted this by saying that in 1991 our foreign reserves were sufficient for only 15 days of imports and we had to pledge gold with central banks of England and Japan to obtain US dollars and UK pound sterlings to escape from defaulting on paying our foreign debt. According to him, in August 2013 we had $ 285 billion foreign exchange reserves which were sufficient to cover 7 months of imports.

But whatever Dr. Manmohan Singh says, the high magnitude of our current account deficit poses a serious challenge. There are two ways to solve this problem. First, we must expand our exports and restrict imports by raising customs duty on imports, especially which are of luxury types. We have already raised the duty on gold to 10% which will help to bring down the imports of gold. Second, we must attract *foreign direct investment* (FDI) which is a stable source of foreign funds and reduce our dependence on portfolio investment by foreign investment institutions (FIls). In July 2013 to solve our current account deficit and to prevent the depreciation of the rupee Government had raised the **caps** of foreign direct investment in various sections such as telecom, power, oil refineries etc and made their entry easier by permitting a proportion of their investment through '*automatic route*'. The present author is of the view that these steps may take a long time and therefore we may approach IMF again to bail out.

Control of Inflation. Another achievement of economic reforms *in the initial stages* was that inflation rate had been controlled. The inflation rate as measured by change in Wholesale Price Index (WPI) which reached a peak of over 15 per cent in August 1991 was brought down to 4.5 per cent in 1997-98. The inflation rate as measured by WPI (1999-2000 = 100) varied somewhat since then but it had been small on an average till 2008-09. Annual rate of inflation as measured by WPI (1999-2000 = 100) was 7.1 per cent in 2000-01, 3.6 % in 2003-04, 6.5% in 2004-05 and 5.4% in 2006-07. It was about 4% in 2007-08. However, inflation rate flared up in 2008-09 when year-on-year WPI inflation rate started rising from March 2008 and crossed 9% mark in June 2008 and 11% in August 2008 and reached a peak level of 12.63% in mid-Sept. 2008 and after it started falling and was estimated at 8.1 per cent during 2008-09. It further fell to 3.8 per cent during 2009-10.

However, WPI inflation flared up again during 2010-11 and was estimated at 9.6% in 2010-11 and 9.4 per cent in 2011-12 and in Aug. 2012 it was at around 8 per cent despite tightening of monetary policy by RBI in the last three years. The inflation regarding consumer price index

(CPI-IW) was estimated at 9.1% in 2008-09, 12.4 per cent in 2009-10, 10.4 per cent in 2010-11 and 9.1 per cent in 2011-12. It crossed 11 per cent in August 2012. Thus, inflation at present is a matter of serious concern. However, it may be noted that current inflation rate is not due to economic reforms undertaken since 1991. It is the result of both demand and supply-side pressures working in the Indian economy. Government's inability to reduce large fiscal deficit and to increase the supply of essential goods (especially food items) are responsible for current inflation.

WPI inflation has been brought down to zero per cent in March 2015 but CPI (IW) inflation was still at 5.5 per cent y-o-y basis in March 2015.

Economic Growth. A significant achievement of economic reforms has been that rate of economic growth on an average rose to 6.1 per cent per annum during 1992-2002 as compared to the previous decade of 1980's when GDP grew at 5.6 per cent per annum (See Table 56.2). After 2003-04 rate of economic growth rose further to 8% to 9% with the exception of year 2008-09 when there was 6.7% growth rate due to global financial crisis.

It will be seen from Table 56.2 that growth rate of industry where major reforms have been undertaken was estimated to be lower at 6.4 per cent per annum in the immediate post-reforms period during 1992-93-2000-01 (even omitting the year 1991-92) as compared to 7.1 per cent achieved in the pre-reforms decade 1981-82-1990-91. The industrial growth rate was still lower at 4.6 per cent per annum during the five years period of Ninth Plan, 1997-98-2000-02. The overall growth rate in GDP during the post-reforms period has been 7.2 per cent which is higher than that of pre-reforms period.

Table 56.2. Average Annual Growth Rate

	1981-82 to 1990-91	1992-93 to 2000-01	1997-98 9th Plan	2002-07 10th Plan	2007-08	2008-09	2009-10	2010-2011	2011-2012	2012 2013
Agriculture	3.6	3.2	2.0	2.5	2.5	0.1	1.0	7.0	3.6	1.9
Industry	7.1	6.4	4.6	8.9	9.0	4.5	8.3	7.9	2.7	1.2
Services	6.7	7.8	8.1	9.3	10.2	10.9	10.0	9.4	7.9	6.8
GDP (factor cost)	5.6	6.1	5.5	7.8	9.3	6.7	8.6	9.3	6.2	5

Source: 1. Dr. Shankar Acharya, "Growth Has Slowed", *Economic Times*, 22 May, 2001.
2. GOI, *Economic Surveys* 2007-08, 2008-09, 2009-10, 2010-11, 2011-12, and 2012-13

However, industrial growth rate picked up to 8.9% in the 10th Plan period 2002-07. It is important to note that overall higher growth rate of 7.2% per annum in 1992-93 to 2011-12 was achieved due to higher growth rate (7.8%) achieved in services. As has been pointed out by Dr. Shankar Acharya[1], the higher growth rate in both industry and services was due to large increase in government administration services where as a result of 5th Pay Commission large increases in salaries to the Government employees (including army) were given which pulled up the overall growth rate. However, during the 9th Plan period (1992-98 to 2002) and 10th Plan period (2002-07) growth rate in services has been quite high, 8.1 % and 9.3% respectively. *In fact, services have been the important driving force of economic growth in India in the last 20 years.*[2]

From 2003 to 2011-12 onward, industry also achieved appreciable growth rate of 8.9 per cent in the 10th Plan period (2002-07) and 9 per cent in 2007-08. Leaving the year 2008-09 when industrial growth rate fell to 4.5 per cent it picked up to 8.3 per cent in 2009-10 and 7.9 per cent in 2010-11. However, during 2011-12, it fell to 2.8% and during 2012-13, industrial growth further fell to 1.2 per cent only due to adverse global conditions created by slow growth in the US and stagnation in European countries due to Eurozone crisis. Even overall GDP growth in 2011-12 has fallen to 6.2 per cent as against the original target of 9 per cent for this year. Even industrial growth during 2012-13 as stated above has fallen to 1.2 per cent due to the above mentioned factors which are still operating to lower economic growth rate of India. However, it is agriculture which has registered

1 Shankar Acharya, "Macroeconomic Management in the Nineties", *EPW*, April 2002.
2 Shankar Acharya, "*India Rising*, 2004-05", *Outlook*, April 4, 2005.

a low growth in the last decade; agriculture grew at the rate of only 2 per cent per annum in the 9th Plan period (1997-98 to 2002) and 2.5 per cent per annum in the 10th Plan period (2002-07).This is mainly because in the post-reforms period investment in agriculture has been very low.

A significant achievement of economic reforms has been a higher overall rate of economic growth in the post-reforms period. *A high overall growth rate of 6.2 per cent was achieved during 1991-2003, and about 7.8 per cent growth in GDP in the five years (2002-07) of 10th Plan. India has become one of the fastest growing economies of the world.* In fact in the three years period (2005-08) annual average growth had been over 9 per cent which is quite a high rate of growth. However, the benefits of this high growth rate had not trickled down much to the poor. Therefore, the Eleventh Plan laid stress on **inclusive growth** along with the target growth of 9 per cent during the plan period. However, in the 11th Plan period (2007-12) India succeeded in achieving average 8 per cent growth rate despite global financial crisis of 2007-09 which affected India also.

As stated above, quite a *dismal performance* in the post-reforms period has been witnessed in respect of agricultural growth. Agricultural growth has greatly decelerated after mid-1990s. Agriculture which had grown at 3.2 per cent per annum from 1980 to 1996 deteriorated to 2.0 per cent during the 9th Plan (1997-2002). In the five years (2002-07) of the 10th Plan agricultural growth was estimated at 2.5 per cent per annum[3] against the target of 4 per cent per annum. This shows whatever the objectives, in actual investment effort and implementation of programmes, agriculture, especially irrigation, has been grossly neglected.

It may be noted that **for the two years (2011-12** and **2012-13) Indian economy had been in crisis,** its GDP growth rate had fallen drastically. In 2011-12 its GDP growth was 6.2 per cent and in 2012-13, it fell to 5 per cent – the lowest in a decade. *Lack of infrastructure is a key constraint* which as a result of **policy paralysis** on the part of government was holding down growth rate. Growth of exports which is now an important part of our GDP had been sluggish partly because of policy paralysis and partly due to recession in the US and European countries. Sentiments of investors have been quite subdued because of policy uncertainty of the government and dear monetary policy of the RBI which had to be adopted to fight inflation and widening current account deficit (CAD) which was 4.2 per cent in 2011-12 and 4.8 per cent of GDP in 2012-13. On the one hand, lack of infrastructure and the delays in environment clearances of projects have held up various investment projects and adversely caused uncertainty in the minds of investors. Further, due to lavish unproductive expenditure aimed at getting votes in the future parliamentary elections has resulted in macroeconomic imbalances in the economy which has resulted in large fiscal deficit of the central government which was 5.7 per cent in 2011-12 and 4.8% in 2012-13 creating the problem of inflation and widening of current account deficit (CAD) which has caused among other factors fast depreciation of rupee. Our dependence on volatile capital flows brought in by FIIs to meet current account deficit and spur economic growth is now creating problems. The large outflows of capital by FIIs caused sharp depreciation of rupee.

However, the growth rate of the Indian economy picked up to 6.9% in 2013-14, to 7.3% in 2014-15 and is now estimated at 7.5 per cent in 2015-16. With this India has now become the fastest growing economy of the world.

Employment and Unemployment: Jobless Growth

But a more dismal performance of the Indian economy has been in respect of growth of employment opportunities. The growth of employment in the organised sector (consisting of industrial and services sectors, both public and private sectors) of the Indian economy has been negative in the last 15 years since 1991. In fact in April 2010 about 40 million jobless persons were registered with the employment exchanges. In the post-reforms period 1994-2008 while rate of growth of employment in the public sector was negative (–0.65% per annum), in the organised private sector employment increased at the rate of 1.75 per cent per annum. The increase in employment in the total organised sector was only 0.05 per cent per annum in 1994-2008 whereas in the pre-reforms period (1983-1994) the growth of employment in total organised sector was 1.20 per cent per annum. Thus, *Economic Survey* 2014-15 states that a cause for concern is the deceleration in the compound *annual* growth of employment during 2004-05 to 2011-12 to only

3. Planning Commission, *Tenth Plan: Mid-Term Appraisal Report*, 2005.

0.5 per cent as against 2.9 per cent annual growth in labour force during the same period.[4] This has been due to the *decline in employment elasticity of growth of output*. Thus, the policy of economic reforms, namely, *liberalisation, privatisation and globalisation*, failed to generate higher growth of employment despite the achievement of higher average GDP growth of around 8 per cent per annum in the period 2004-05 to 2011-12.

The generation of fewer employment opportunities knocks out the argument advanced by pro-reforms economists that higher economic growth by itself will generate enough employment opportunities. The dismal performance in respect of employment growth is due to three factors. First, capital-intensive technologies have been used in the organised sector which have tended to displace labour rather than absorbing it in productive employment. Secondly, no attention has been paid to the implementation of land reforms and improvement of the credit delivery system in agriculture. Thirdly, investment in agriculture, especially in irrigation, has been neglected.

The contention of the Indian Planning Commission and many economists that economic growth generates sufficient employment opportunities has further proved to be wrong by special employment and unemployment survey carried out by NSSO in 2011-12. This survey shows dramatic drip in employment generation. According to it, on the basis of *usual principal and subsidiary status* the proportion of population gainfully employed declined in 2011-12 compared to 2009-10. In 2009-10, 36.5 per cent of population was gainfully employed for the better part of the year. By 2011-12, the proportion of population employed declined to 35.4 per cent. Accordingly, *unemployment rate as per usual principal and subsidiary status went up from 2.5 per cent to 2.7 per cent*. Further, the survey reveals that the share of employed men remained roughly constant between 2009 and 2011, women's employment dropped from 18 per cent to 16 per cent. This may appear a small decline but translated into real numbers, the crisis in employment is starkly revealed. In rural areas about 90 lakh women lost their jobs in the two years period. This would have been catastrophic but for the fact that in urban areas about 35 lakh new women got jobs in these two years.

However, unemployment rate on the daily status basis fell from 6.6 per cent in 2009-10 to 5.6 per cent in 2011-12. This is due to a large number of mandays of employment created under Mahatma Gandhi National Rural Unemployment Guarantee Scheme (MGNREGS) which was implemented on a large scale in these two years. It is now a common knowledge that employment created in this scheme is of very low quality and does not create any durable assets. Thus, fall in daily status unemployment rate does not show creation of productive employment. It is important to note that dismal performance of employment generation in these years 2009-10, 2010-11 and 2011-12 has been despite a relatively higher GDP growth; 8.6 per cent in 2009-10, 9.3 per cent in 2010-11 and 6.2 per cent in 2011-12.

It is worth mentioning that this higher growth in employment has been registered in the unorganised and informal sector where productivity of labour is quite low. The growth of employment in the organised sector has been negligible (0.05 per cent) per annum in the post-reforms period (1994-2008).

The low growth of employment was expected. As mentioned above, the chief drawback of New Economic Policy has been that it involves the use of labour-saving modern technology, the competition from the imported products to be faced by the domestic industries, and public sector disinvestment on a large scale. The use of capital-intensive technology not only displaces labour already employed but also lowers the growth of employment from the new investment and capital accumulation. Besides, due to the greater role assigned to the private sector, big business houses, foreign capital (MNCs) the already existing inequalities will further increase. This is against the objective of social justice which has been the cherished goal of the Indian planning.

Poverty Reduction and Economic Reforms

The impact of economic reforms on poverty in India is a highly controversial issue. The critics point out that the economic reforms aimed at liberalisation, privatisation and globalisation of the Indian economy has slowed down the reduction in poverty ratio (*i.e.*, the percentage of population living below the poverty line) despite a higher rate of economic growth achieved in the post-reforms period. On the basis of uniform recall period (that is, 30 days recall period for all items included in

4 GOI, *Economic Survey* 2014-15, p 135.

the consumption basket for estimating poverty ratio), the poverty at all India level fell from 36 per cent in 1993-94 to 27.8 per cent in 2004-05, that is, about 0.79 per cent per annum fall in the poverty ratio in this period. This is much lower than the fall in poverty ratio (on URP basis) which declined from 51.3 per cent in 1977-78 to 38.9 per cent in 1987-88, that is, about 1.24 per cent annual fall in poverty ratio in these 10 years of pre-reforms period.

According to critics several factors are responsible for the existence of poverty in India despite over 60 years of planned development. First, there had been slow growth of productive employment opportunities relative to population growth resulting in increase in unemployment which on daily status basis, as mentioned above, rose from 6.1 per cent in 1993-94 to 7.3 per cent in 1999-2000 and further to 8.2 per cent in 2004-05. Secondly, due to liberalisation of cheap imports of capital goods abroad, the small and medium enterprises (SME) could not compete with them resulting in their closure and decrease in employment opportunities. Thirdly, in the post-reforms period the rate of investment in agriculture has been significantly reduced. As a result, there has been a low rate of growth of employment in agriculture in the post-reforms period. This caused a relative reduction in the supply of wage goods, especially foodgrains, which increased the problem of hunger and malnutrition in various vulnerable sections of population in various parts of the country. It was expected that higher rate of growth will ensure greater reduction in poverty. However, this did not happen as benefits of economic growth have not trickled down to poor.

The debate on poverty flared up when the Planning Commission released the data on poverty estimates for the 66th round of NSS for the year 2009-10. These poverty estimates were based on the *new poverty line* norm suggested by an expert committee headed by late Prof. Suresh Tendulkar which submitted its report in 2009. The committee deviated from the calorie intake as norm for poverty estimates and included in their poverty line poverty indices for health and education. Thus contrary to popular belief, the poverty line was higher in Tendulkar committee approach. Therefore, to compare the estimates for 2009-10, poverty estimates were prepared afresh on the basis of new poverty line suggested by Tendulkar committee approach. The poverty line in 2009-10 was fixed at ₹ 4,298 per month for a family in urban areas and ₹ 3,364 per month for a family in rural areas. *According to estimates on the basis of new poverty line, the people living below the poverty line fell to 29.8% of its population in 2009-10 from 37.2% in 2004-05.* That is, poverty declined by 1.5 percentage points per annum between 2004-05 and 2009-10. It is the fastest decline in poverty compared to the earlier periods. The several economists and social scientists challenged the new poverty norm and the significant fall in the poverty ratio keeping in view that 2009-10 was one of the worst drought years. However, the Planning Commission attributed it to high rate of economic growth and government intervention by way of starting special anti-poverty schemes such as MGNREGA on a large scale during 2009-10. The critics also raised the question regarding whether a person can live or survive with the meagre amount of money expenditure fixed as poverty line norm by the Tendulkar committee. The criticism of poverty estimates was so severe that government appointed a new expert committee headed by C. Rangarajan to suggest appropriate methodology for fixing poverty line norm and prepare estimates of poverty in India on the basis of new norm.

However, to estimate poverty new special survey for the year 2011-12 was ordered on the basis of Tendulkar committee methodology because 2009-10 as the drought year poverty estimates for it did not present true picture about poverty. The result of this survey was released by the Planning Commission in June 2013. Adjusted for price change, *on the basis of Tendulkar committee methodology, poverty line for 2011-12 was fixed at ₹ 27.20 for rural areas and ₹ 33.40 for urban areas. On the basis of these poverty norms, Planning Commission estimated that poverty ratio for India declined from 37.2 per cent in 2004-05 to 21.9 per cent of population in 2011-12, a decline of 15.3% in just seven years*. This means poverty declined at the rate of 2.2 per cent a year between 2004-05 and 2011-12 compared to 0.74 per cent a year between 1993-94 and 2004-05. According to this the number of population declined from 400 million in 2004-05 to 270 million in 2011-12, that is, 137 million persons were lifted out of poverty. This was described as a big achievement of economic reforms carried out by UPA government and as a result higher economic growth of around 8 per cent was achieved during these years.

After both Congress and opposition leaders debunked expenditure of ₹ 27 per person per day for rural areas and ₹ 33 per person per day for urban areas as a benchmark, the government said the line had not been cast in stone. "*The poverty line is being reassessed. The Rangarajan committee is doing just that. In any case, the line is not linked to pro-poor programmes like rural job guarantee or food security*".

The problem for the government lies in the perception that ₹ 27 and ₹ 33 seem embarrassingly low consumption limits, particularly in the context of persistent high inflation rates. The politically incorrect figures mean the government is unable to admit that incomes have risen and absolute or extreme poverty has declined. Wary of being seen as callous in an election year the government is set to revise the poverty line but in the process it will dent its claim that 137 million people were brought above the below poverty line since the UPA government took over. Upward revision of the poverty line will send millions of people back into poverty in terms of head-count ratio. This is not a comfortable prospect of any government.

Adverse Effects of Globalisation

Economic reforms undertaken since 1991 brought about globalisation of the Indian economy as under them the Indian economy was opened up. Trade liberalisation occurred following removal of quantitative restrictions on imports and large reduction in tariffs. Reforms also led to the free flow of foreign capital. This benefited India as its exports increased at a higher rate in the initial years and balance of payments on current account improved. Further foreign investment in India contributed to a higher growth rate achieved in the post-reforms period. However, globalisation is not without dangers. These benefits of globalisation were proved to be only temporary.

Foreign Exchange Rate Volatility. The adverse effect of globalisation has been clearly seen in the two years, 2007 and 2008, and recently since May 2013. First, the free flow of portfolio capital has created a lot of volatility in exchange rate of rupee. Due to large capital inflows the value of rupee which was around ₹ 46 to a US dollar in end-March 2004, appreciated to ₹ 44.27 to a US dollar in end-March 2006 and further to ₹ 39.4 to a dollar in January 2008. However, most often due to capital outflows there has been depreciation of rupee. Recently, after May 2013, there has been a sharp depreciation of rupee which has been triggered by the announcement of governor of US Reserve System that it would start relaxing its ***Quantitative Easing*** (*QE*) Programme from September 2013 onwards due to the signs of revival of the US economy. As a result of this it was feared that current account deficit (CAD) would further widen and India's paying capacity will be adversely affected. This also contributed to the capital outflows by FIIs. First, they pulled out from debit market and then from equity market. As a result of these, rupee depreciated sharply, the value of rupee which was around ₹ 56 to a US dollar in the early June 2013 first declined to ₹ 60 to a US dollar in the end of June 2003, and then to ₹ 61.21 to a US dollar on 8th July 2013 and then to ₹ 68.8 to a dollar on 28th August 2011.

Global Financial Crisis. However, when financial crisis in the US which was triggered by defaults of sub-prime housing loans given by American banks and other financial institutions in the US deepened in 2008, creating liquidity crunch in the American economy, capital outflows started occurring from India. The flight of capital from India led to depreciation of Indian rupee. To meet liquidity requirements in their parent country, foreign financial institutions (FII) started selling shares of Indian corporate companies and got rupee receipts converted into US dollars which they repatriated to their parent country. These outflows of dollars led to the increase in demand for US dollars in the foreign exchange market which caused the depreciation of Indian rupee. The depreciation of rupee will make our imports, especially crude oil, costlier and can adversely affect our balance of payments. The higher prices of imported products, especially capital goods and metals, will tend to raise inflation rate in India.

From August 2011 there was again outflow of capital due to uncertainty caused by worsening of sovereign debt crises of European countries and slowdown of the US economy. The capital outflows led to a large depreciation of rupee from September 2011. The Indian rupee which was around ₹ 44 to a US dollar in August 2011 fell to nearly ₹ 52 to a US dollar on November 2011 and further to ₹ 53 to a US dollar in mid-December 2011. From January to March 2012 value of the Indian rupee remained within range of ₹ 51 to 52 to a US dollar.

The sharp depreciation of rupee is harmful for the Indian economy. To prevent excessive depreciation of the Indian rupee, the RBI uses its foreign exchange reserves to sell US dollars in the foreign exchange market. As a result, there is decline in foreign exchange reserves which are needed to cover imports of goods and services needed for industrial growth and to pay interest on the foreign debt every year. Further, lack of adequate foreign exchange reserves with the RBI leads to the loss of confidence of the foreign investors about the ability of the country to pay back the loans. Besides, weakening of the rupee is likely to cause losses to the foreign investors who repatriate their profits to their home countries. This discourages inflows of foreign direct investment needed for economic growth of the country.

It is thus evident that capital outflows and consequent depreciation of rupee causes a lot of macroeconomic instability in the economy. Therefore, RBI intervenes in the foreign exchange market to check the sharp depreciation of rupee through sale of US dollars in the foreign exchange market. Besides, it takes other measures to encourage capital flows into the Indian economy. It relaxed the norms for external commercial borrowing (ECB) so that the Indian companies directly borrow foreign money from abroad rather than through foreign exchange market. RBI also raised the interest rate cap on NRI deposits of foreign exchange in our banks. Further, it eased the conditions for foreign direct investment (FDI) to encourage more capital inflows through this route. These measures help in increasing capital inflows and prevent excessive depreciation of the rupee.

It is evident from above that globalisation with its free capital flows causes a lot of volatility in exchange rate which has adversely affected the Indian economy. Besides volatility in exchange rate, financial crisis in the US affected our stock market and banking system. The financial crisis in the second half of 2007 following the defaults in sub-prime housing loans given by the American banks and other financial institutions resulted in the collapse of the banking system and crash in American stock markets. This caused liquidity crunch in America and affected credit flow in the economy. The effects of this financial crisis were not only confined to America but also spilled over to other countries not only of Europe but also of Asia (including India) as they were integrated with America. As a result of this world-wide financial crisis global meltdown occurred. Under pressure of redemption of securities of investment banks and other financial institutions, foreign institutional investors started selling shares in the Indian stock market and repatriating dollars to their parent countries. These outflows of dollars also led to the liquidity crunch in the Indian banking system which affected credit flow for investment in the Indian economy. The recessionary conditions in the United States and global meltdown caused by the financial crisis also adversely affected the growth of our exports. The lower exports and costlier imports affected our current account balance of payments.

Besides, the aggressive selling of shares by FIIs which held substantial stock in the Indian corporate companies resulted in crash in Dalal Street, Sensex of Bombay Stock Exchange which reached the peak of 20,827 on January 11,2008 fell to 15,976 in May 2008, to 13,531 on September 15, 2008 and went below 10,000 mark on 17-10-2008. It reached a three year low of 8,000 in the last week of October 2008. This caused huge losses to the Indian investors and washed away the vast amount of their wealth. Even the share prices of banks (both of public and private sectors) registered a sharp decline. The negative wealth effect of the fall in the value of shares has affected the investment spending and demand of the investor class which adversely affect our industrial growth. Again since May 2013, the decision to unwind the *Quantitative Easing* (QE) by the US Federal Reserve System has triggered capital outflows from India resulting in sharp depreciation of rupee.

We have seen above that due to the globalisation of the Indian economy the bad effects of financial crisis originated in the United States have spilled over to the Indian economy.

Impact of Globalisation on Economic Growth

In the year 2008-09 global financial crisis affected the Indian economy as it reduced the growth of our exports and caused capital outflows from the Indian economy as due to uncertain global conditions capital flows went to the safe destination of government bonds of the US which is considered as quite safe investment. As a result of this, investment or gross domestic capital formation (GDCF) in India fell to 34.3 per cent of GDP bringing down the growth rate to 6.7 per cent in 2008-09. As a result of various fiscal and monetary stimulus measures by the government investment rate as per cent of GDP rose to 36.6 per cent in 2009-10 and 35.1 per cent in 2010-11 which caused higher GDP growth rate of 8.4 per cent in both these years. But in 2011-12, the global environment again changed. The deepening financial crisis in European countries, especially Greece and Spain, slow growth in the US and Japanese nuclear disaster resulted in a sharp global slowdown during 2011-12. As this happened quickly after the global recession of 2008-09 when the economies of the US and Europe had not yet fully recovered from the previous crisis, it was difficult to deal with Eurozone crisis. Much of the fiscal and monetary policies to revive and stimulate the earlier recession of 2008-09 had not yet been fully withdrawn that threat of new global meltdown had to be faced. Thanks to India's rapid growth over the last two decades and its increasing integration with the world it could no longer be impervious to global developments. Not surprisingly the Indian economy was adversely affected and its GDP growth fell to 6.5 per cent in 2011-12, the lowest in nine years since 2003. Further, even for the year 2012-13, RBI has forecast 6.5 per cent growth rate of GDP.

Besides adverse global environment, domestic factors also contributed to slowdown in economic growth in 2011-12 and 2012-13. The persistent inflation in the last three years required to control inflation by tightening monetary policy (*i.e.,* raising interest rate and checking excessive growth in liquidity in the banking system by the RBI) raised the borrowing costs of the private corporate sector which forced them to defer various investment projects. This resulted in lower *investment in fixed capital assets* and as a result as per cent of GDP it fell to 32.3 per cent in 2010-11 and further to 30 per cent in 2011-12. Further, policy paralysis caused by various scams and corruption charges on Ministers and top bureaucrats and hurdles put by various coalition partners of the government stalled the process of economic reforms.

Market Failures and Role of Government

The supporters of economic reforms who argued for diluting the role of public sector or government as it does not ensure efficiency in production laid great stress on '*Government failures*' in promoting economic growth. They however turned a blind eye to the *market failures*. To quote an eminent American economist, Dr. Joseph Stiglitz, who won Nobel Prize in economics in recent years, "*Market failures are a fact of life, as are government failures. New liberal ideologies assume perfect markets, perfect information, and a host of other things which even the best performing market economies cannot satisfy.*" He therefore forcefully argues for a *balanced approach to government* and *markets recognising* that both are important and complementary. This means both public sector and private sector should play complementary roles in bringing about economic growth. Due to market failures we cannot rely on private sector alone to bring about sustained economic growth and adequate expansion in employment opportunities.

In our view reduction in fiscal deficit should not be elevated to a *dogma* as is generally considered by the advocates of economic reforms. A reasonable amount of fiscal deficit is actually needed for an economy which is growing at over 7 per cent per annum. What is relevant is that government *should spend its borrowing for investment purposes*, especially infrastructure, and on development of human capital, education and healthcare. What is needed is that government should not borrow for consumption purposes. Prof. Amartya Sen rightly writes that *the success of liberalisation and closer integration with the world economy* may be severely impaired by India's backwardness in basic education, *elementary healthcare, gender inequality and limitations of land reforms.*[5] While Manmohan Singh did initiate the correction of governmental *over-activity* in some fields, the need to correct the governmental *under-activity* in other areas has not really been addressed. Further, referring to the higher growth rate in some East Asian countries which have received high growth rate with considerable equality, he writes, "These countries all share some conditions that are particularly favourable to widespread participation of the population in economic change. *The relevant features include high rates of literacy, a fair degree of female empowerment, and quite radical land reforms*. Can we expect in India results similar to those that the more socially egalitrian countries have achieved, given that half the Indian population (and two-thirds of the women) are still illiterate, that female empowerment is very little achieved in most parts of India, that credit is very hard to secure by the rural poor, and that land reforms remain only partially and unevenly expected? Should we not have expected more in the direction of breaking the terrible neglect of these concerns, on which the basic capabilities of the Indian masses depend?"[6]

---

5 Amartya Sen, "Theory and Practice of Development", in *India's Economic Reforms and Development, Essays for Manmohan Singh*, ed. by Isher Judge Ahluwalia and I.M.D. Little, Oxford University Press, 1998, p. 83.

6 *Ibid*, p. 84.

CHAPTER 57

ENVIRONMENT AND ECONOMIC DEVELOPMENT

Growth and Environment

In recent years there has been growing concern about degradation and pollution of environment and climate change as they impact on future development of both the developing and developed countries. In 1992, representatives of over 150 countries met at Rio in Brazil to discuss the environmental issues and their implications for future development of the world. This meeting at Rio is called the '**Earth Summit**' or the **United Nations Conference on Environment and Development** (UNCED). This conference clearly spelled out the linkages between natural environment and development and put forward the concept of **"sustainable development"**. This has produced greater awareness about environmental issues and facilitated the cooperation between different countries to reduce environmental degradation, particularly to reduce emission of greenhouse gases (GHG) such as carbon dioxide to prevent adverse climate change, that is, global warning in future which if not prevented will have disastrous consequences for the welfare and development of the population, both in developing and developed countries.

The *conflict between economic growth and environment is sharper today than ever before, particularly in developing countries like India with fast growing population and mass poverty.* The developing countries are making strenuous efforts to balance their need for rapid economic growth with the environmental concerns for keeping their natural resource base intact. In India, as in other developing countries, the adoption of development strategy based primarily on large-scale industrialisation, energy-intensive technologies and biochemical-based agricultural technology, ignoring indigenous development paradigm based on locally self-sufficient technologies, has led to environmental degradation. In this chapter, we shall explain various environmental issues and the adverse impact of climate change and explain the need for sustainable development.

In a recent conference State environment ministers held in August 2009, the former Prime Minister Manmohan Singh said that the ecological crisis that confronts the country has led to '**alarming situation**'. He urged upon the State governments to control environment pollution, clean rivers and fight climate change. This statement highlights the fact that India's growth is increasingly taking place at the cost of its environment. This shows that we have not been taking adequate measures to ensure sustainable development. To stop further decline in our natural resources and protect environment from pollution stringent regulations and incentives are needed. But, due to rampant corruption prevailing in the bureaucracy the rules regarding the use of our natural resources and protection of environment have been flouted with impunity. There are numerous instances when the Ministry of Environment and Forests has flouted all norms to give clearance to dubious projects at the cost of environment and communities that depend on it.

Population and Environment

Concern over environment issues arises because it is thought that the population may increase to a level whose needs cannot be adequately met by earth's natural resources resulting in fall in living standards of the people. This is because increasing population causes environmental degradation. Increases in population cause over-exploitation of natural resources such as forests, water, fisheries and minerals at a rate far greater than their capacity to regenerate. Besides, population pressure on land compels us to cultivate arable land more intensively by using chemical inputs such as fertilizers, pesticides which cause soil degradation. Further, increase in population through its effect on deforestation by the rural and urban population for timber and fuel leads to the increase in carbon dioxide in the atmosphere and therefore causes air pollution. Thus the population impacts on environment primarily through the use and depletion of natural resources and is associated with environmental problems such as air and water pollution and loss of biodiversity and increased pressure on arable land. As an example of estimate of loss, Tata Energy Research Institute (TERI) estimates that economic losses due to soil degradation, diseases caused by pollution and forest degradation is between ₹ 100,000 to ₹ 450,000 crore every year and this does not include a series of other losses due to ecological destruction.[1]

Population increases cause over-exploitation of land and water resources and loss of biodiversity and forests and will therefore endanger sustainability of agriculture and food security in the country. According to Prof. M.S. Swaminathan, a noted agriculture scientist, " the capacity to support even the existing human and animal population has been exceeded in many parts of the world"[2] and if population growth is not checked, it will endanger the attainment of food security. To quote him again, "In the recent years there has been considerable concern about human capacity to produce adequate food to meet the needs of the growing population and the increase in purchasing power resulting in higher consumption of animal products."[3]

It is manifest from above that rapidly increasing population in developing countries will lead to the over-exploitation and degradation of land and depletion of fisheries which will threaten the achievement of food security in the developing countries.

Besides, the growth of urban population in the absence of adequate infrastructure facilities has caused the **lack of clean water to drink** and given rise to **slums with poor sanitation** which has increased the vulnerability of the people to several diseases. It follows from above that to meet the expanding needs of the developing countries, environment degradation must be halted. For that purpose, reduction in population growth will greatly help in easing the intensification of many environmental problems.

Poverty and Environment

Poverty in developing countries is also said to be responsible for environmental degradation. Poor people rely on natural resources more than the rich. For survival the rural poor are forced to cut forests for timber and fuel as well as graze animals on pasture lands more than the reproductive capacity of these natural resources. Besides, when the cultivable land becomes short relative to population, the poor are forced to make their subsistence by cultivating fragile land on hills and mountains resulting in soil erosion on a large scale. It is in such environment that poverty becomes a vicious circle. Poverty leads to land degradation and land degradation accelerates the process of impoverishment because the poor people depend directly on exploitation of natural resources on which property rights are not properly assigned. Moreover, poverty results in malnutrition and reduces poor people's capacity to do work.

With growing population the poor encroach upon large remote areas of forests and grasslands to make their livelihood. If there are strict regulations to prevent such encroachment, it is opposed with stiff resistance by the poor, especially in the tribal areas.

The solution to the poverty problem of these teeming millions lies in land reforms, generation of more employment opportunities and improvement in productivity of arable land already in use, for example, shifting the poor from poor resource-based to modern science-based agriculture. This is what has been sought to be achieved through green revolution technology.

However, the green revolution has been criticised for environmental reasons, especially the use of fertilizers and pesticides that increase soil salinity. Besides, ample use of irrigation in green revolution technology without adequate drainage facilities results in soil degradation through salinity and waterlogging. Adequate and appropriate steps should be taken to make the green revolution

1. Quoted by Ashish Kothari, in "Unsustainable Development, Unfulfilled Promises" in the **Survey of Environment.** *The Hindu*, 2002.
2. M.S. Swaminathan, "Agriculture: Route to an Evergreen Revolution " in **the Survey of Environment**, *2002, The Hindu*, p. 34.
3. *Ibid*, p. 33.

technology clean and environment-friendly. There can be no two opinions that major efforts must be made to overcome these defects through development of **less poisonous chemicals**, pest and insect control with reduced chemical application and improved drainage facilities. However, if the efforts to develop modern technology were abandoned because of these defects, employment and income-earning opportunities for marginal farmers and agricultural labourer would continue to be reduced under population pressure. As a result, many would be forced to push cultivation frontiers into ecologically fragile lands resulting in increased incidence of food and soil erosion."[4]

It is important to note that the environment representing forests, mines, sources of water, land that provide employment and livelihood to the poor people, especially those living in the tribal regions and it is they who in India are opposing the various development projects, even those cleared by the government Sunita Narain, a noted environmentalist, rightly writes, "*Let us be clear, it is not the minister or his ministry, which is opposing development projects - from Vedanta to Posco or Jindal's steel or thermal plant or Nirma's cement plant. It is people, often the poorest, saying these projects will devastate their environment, their forests, which is their source of water, land and livelihood. They are saying, we are poor, but your development will make us poorer. The environmental movement of the country is being led from the bottom today. It is not in the hands of middle-class environmentalists like me*."[5]

SOURCES OF POLLUTION AND ENVIRONMENTAL DEGRADATION

We explain below the various sources of pollution and environmental degradation. They must be checked if the objective of sustainable development has to be achieved.

Industrialisation, Urbanisation and Environment

Industrialisation of the economy and urban development go together. In the post-independence period, especially since the beginning of the Second Five Year Plan (1956 – 61) high priority has been accorded to industrialisation to achieve the goals of rapid growth of output and employment. New industrial factories and plants have been set up in the towns and cities or near to them. This industrialisation and urban development may bring about rise in material levels of living but they also cause environmental degradation, especially the pollution of air and water and creation of hazardous wastes. Across the world in both the developed and developing countries air and water pollution are being considered as a matter of serious concern as they not only cause harm to the health of the present population but endanger future generation by causing global warming with disastrous consequences.

It is worth noting that we need various projects, which relate to the manufacture of steel, cars cement, nuclear plants for industrial development of the country. But, as mentioned above, they are opposed by the people, especially the poor ones, because they will be displaced without being sure of the alternative avenues of their livelihood. Though we need these projects for our growth but we cannot build them against the will of the people. While approving the industrial projects we must adopt such strategy that ensures that the people who live on the rich lands with minerals, water and forests and who are displaced by the development projects must benefit from them. In this regard we need urgent reforms in policies for land acquisition, forest management and mineral development. Besides, we must approve the development project subject to the conditions that while they take local resources—mineral water or land, the project developers or the government must provide jobs to the people who are displaced so as to provide livelihood to them. It is because of the failure to provide such guarantee that people are opposing and fighting for the setting up of projects.

It is worth mentioning that urbanisation, that is, the growth of the urban population has been increasing at a faster rate than industrial growth causing not only air and water pollution but also putting increased stress on the existing water supplies and sanitation facilities which create health hazards for the urban population. The governments of developing countries are generally unable to

4. Yujiro Hayami and Yoshihisa Godo, *Development Economics* Oxford University Press, 3rd ed. 2005. p. 227.
5. Sunita Narain, "Fix Growth, Not Environment" in *Times of India*.

cope with the vastly increased needs for clean water and sanitation facilities resulting from rapid growth of urban population. Urban population has been increasing not only at the rate of natural population growth but also through rural-urban migration induced by industrialisation. Thus the lack of clean water and sanitation has led to several diseases from which urban people suffer resulting in their poor health, Besides, due to poor housing facilities the growth of urban population has be led to the **growth of slums** without clean water and hygienic conditions.

Let us understand the Indian urban scene as unfolded by the census 2001. Between 1991 and 2001, the number of towns and cities increased from 3,696 to 4,368. The total urban population in India rose to 285 million in 2001. Of this 108 million, that is, nearly 40 per cent lived in 35 cities with a population of one million or more each. Such a large increase in urban population has led to the growth of slums with acute scarcity of clean drinking water and poor sanitation facilities. Apart from growth of slums, industrialisation and urbanisation has led to the problems of air and water pollution and piles of hazardous wastes which we discuss below in some detail. Though these problems exist both in the developed and developing countries but they are often more serious and acute in developing countries. It is worth mentioning that it is consumption of energy from fossil fuels such as petroleum, coal and gas in various urban activities such as industrial production, vehicular traffic that is the basic source of pollution. In this early stage of industrialisation, as we find today in the case of developing countries, as the share of industry in domestic product increases, it results in energy consumption faster than the growth in national product. If a developing country gives a higher priority to basic heavy and chemical industries, as was the case in India in the second, third and fourth five year plans, consumption of energy takes place at a higher rate resulting in higher degree of pollution.

Air Pollution and Urban Development

Two main urban activities, namely, industrial production and vehicular traffic are the principal man-made source of air pollution as energy is consumed. Therefore, as more industrial growth takes place in developing countries, it results in more air pollution. Industries emit smoke containing carbon dioxide, carbon monoxide, methane gas in the atmosphere causing air pollution. This air pollution is quite harmful for human beings as it causes tuberculosis, asthma, and other respiratory diseases and, according to World Development Report 1992, causes 3 to 7 lakh premature deaths annually. Besides, respiratory diseases adversely affect health of the people and reduce their productivity and working capacity. The people have to spend a lot of money for treatment of the ill-effects of toxic emissions by the factories which represents social cost which is ignored by the private entrepreneurs in their cost-profit accounting.

Since anti-pollution regulations are quite weak in developing countries, the emission of toxic gases and pollution of water have been increasing enormously resulting from increased energy consumption in the industrial sector.

Vehicular Traffic and Environment

Besides industrial production, toxic gases are also emitted from vehicles plying in the urban areas. Motor vehicles cause both air and noise pollutants. Motor vehicles such as cars, buses, trucks, scooters and bikes emit pollutants such as carbon monoxide, hydrocarbon and nitrogen oxide which cause immense damage to the health of the human beings. At present (2009) more than 1.5 mullion registered motor vehicles are there in Delhi alone and about 150,000 are being added every year. These emit around 1000 tonnes of toxic gases into Delhi's atmosphere every day. In other metropolitan cities, Mumbai, Kolkata, Chennai, Bangalore air pollution due to vehicular transport is also enormous.

In recent years under pressure of High Courts and Supreme Court which have delivered judgements for improvement of environment some success has been achieved towards the use of lead-free fuels, better emission standards for cars and buses plying on the city roads. As a result, new models of cars with more fuel efficiency and lower emission levels have come in the market. In Delhi DTC buses and three-wheelers have been using less toxic CNG gas instead of diesel and petrol. Similarly, government has taken some measures to reduce emission of **lead** by vehicular traffic. It has been found that dangerous level of lead in the blood of young children exposes them to the various health problems including neurological deficiencies. Elevated levels of lead in the atmosphere expose even adults to higher blood pressure, higher risk of heart attack and stroke.

Water Pollution and Urban Development

Water pollution is another major environment problem in urban areas. Water is contaminated mostly from wastes of expanding industries in the urban areas. In the towns and cities where industrial growth is taking place, rivers become polluted with toxic chemicals and heavy metals such as lead and mercury thrown into them by expanding industries and mining activities. The rivers like Ganga and Yamuna have become highly polluted over time. It is very difficult to remove pollutants from drinking water with standard purification facilities. Therefore, special measures and instruments are being used to clear rivers from pollutants. As a result, not only there is scarcity of safe drinking water for human beings, even the capacity of rivers to support aquatic life has decreased.

Not only surface water near towns and cities has been contaminated over the years but also ground water has been polluted through seepage of chemical wastes, heavy metals and other hazardous wastes created by industries and thrown into rivers and drains. Water pollution is a very serious problem in the urban areas, especially is cities like Delhi, Mumbai and Kolkata, especially for the urban poor who live in slums with no proper facilities for clean drinking water and sanitation. It is mainly these urban poor who bear the direct impact of water-borne diseases which are responsible for a large number of deaths, especially of children. According to **World Development Report 1992,** in the year 1990 one billion people in developing countries had no access to clean water and 1.7 billion lacked access to proper sanitation. After two decades of development now in 2009 the number of population without access to clean water and proper sanitation must have increased substantially despite good rate of economic growth achieved during this period.

As mentioned above, the various development projects are being vehemently opposed by the people who are affected and displaced by them. Sunita Narain, a noted environmentalist, has given the right suggestion that for making environmental assessment of the project the people affected must be heard and their interests protected. If proper inclusive industrial development has to occur, the words of Sunita Narain, "we must strengthen institutions of regulation and the procedures to listen to people. This will build credibility of the process of governance and it will also mitigate risks for investors. Public hearings held during the environmental impact assessment provide the platform for people to voice their concerns. In most cases today, the effort is to rig and undermine the hearing. The environmental impact assessment is seen by industry as a document, which must clear the project, not seriously assess it. Then even if a project is cleared with environmental conditions (like Posco or Navi Mumbai airport), there is no capacity to monitor compliance."[6]

RURAL AND AGRICULTURAL DEVELOPMENT AND ENVIRONMENT

About 75 per cent of India's population live in rural areas. Agriculture is the main occupation of rural people. Population in India has been increasing at the rate of 1.90 per cent per annum. Adequate food production is needed to feed the teeming millions. In recent years there has been considerable debate about the need to ensure food security for the rising population. Food security, as is now generally understood, refers not only to the adequate food production but also sufficient increase in purchasing power of the people for having access to the consumption of required food and animal products. An eminent agricultural scientist, Dr M.S. Swaminathan, writes, "*The future of food security depends on the conservation and care of arable land through attention on soil health and replenishment of fertility and the conservation and careful management of all water sources so that more crops can be produced. Improvement of production in perpetuity without associated social or ecological harm is vital to safeguard food security.*"[7]

Land degradation is the chief environment problem in rural areas. **Land degradation occurs through soil erosion, salinisation of soil, and waterlogging.** The land degradation lowers agricultural production and adversely affects the fertility or productivity of the soil. Reduction in agricultural production and productivity threatens the livelihood of the rural people. To quote Dr Swaminathan again, "Land degradation, diversion of arable land and the depletion and pollution of water resources

6. Sunita Narain, op. cit.
7. M.S. Swaminathan, "Agriculture Route to an Evergreen Revolution", in **Hindu Survey of Environment,** 2002, p. 34.

result in the loss of rural livelihood. This triggers unplanned migration of the rural poor to towns and cities resulting in the proliferation of urban slums."[8]

It is worth mentioning that degraded land belongs either to the poor farm households or it is common property which has been overused and overgrazed. To ensure food security and livelihood to the rural population land and water resources ought to be protected from degradation and pollution. Land degradation of various forms in the regions of both the irrigated agriculture and rain-fed agriculture is causing serious threats to sustainability of agricultural growth. A serous problem that affects the sustainability of rain-fed agriculture is the **top-soil erosion** which occurs as a result of rain water draining away the top soil of the land and thereby lowers its fertility. It has been estimated that 6 billion tons of soil in India are eroded from Indian crop land each year.[9] The soil erosion adversely affects agricultural growth and achievement of food security.

Though land degradation due to soil erosion is taking place gradually farmers do not take effective measures to prevent or halt the process of soil erosion, Kerr and Sanghi[10] observe that the most important factor responsible for farmers not adopting the recommended practices to control soil erosion is **due to their perception** about soil erosion and other forms of land degradation. It is therefore important that government should integrate the use of technologies for preventing land degradation with its programme and strategy of agricultural growth.

Deforestation: Another important factor causing land degradation is deforestation, that is, cutting of trees for fuel and timber. A good deal of deforestation has also been due to commercial logging. Forests are highly useful for protecting environment. Trees of forests during the process of photosynthesis absorb carbon dioxide from atmosphere and give out oxygen. Thus, deforestation reduces CO_2 in the atmosphere that prevents the increase in temperature of the earth surface. Therefore, deforestation (*i.e.*, cutting of trees for fuel and timber and for extending cultivation for agricultural crops) leads to the increase in CO_2 in the atmosphere which is responsible for global warming. Besides, deforestation causes floods, waterlogging and sail erosion in the local areas and thereby leads to land degradation.

Overgrazing on Common Land. The other factor causing land degradation is **overgrazing on common land** and intensification of agriculture through use of excess irrigation water and chemicals such as fertilizers and pesticides. According to a global survey conducted by the UN Environmental Programme, land degradation during five decades (1945-95) was of the order of 2 billion hectares of total cultivable land area of the world. Of this 80 per cent of degraded area was located in Asia, Africa and Latin America. According to this global survey, 30 per cent of land degradation was caused by deforestation ,7 per cent by over-exploitation for getting fodder and fuel, 35 per cent by overgrazing, 28 per cent by agricultural production activities and only 1 per cent by industrialisation. Though agricultural growth with the use of modern agricultural technology such as the use of excess irrigation water, chemical inputs has played a part in causing land degradation, the main factor responsible for land degradation is over-exploitation of natural resources by poverty-driven population.

Water Scarcity. Another important environmental problem facing India is the **decline in underground water resources**. A recently released **'State of the Environment Report'** says that the amount of water being used in India from underground sources is estimated to be 45 per cent more than what natural system can replenish. A National Aeronautics and Space Administration study released in August 2009 found that ground water levels in northern India–Punjab, Haryana and Rajasthan–are declining at an alarming rate as much as one foot per year over the last decade. Its reason is that water is being pumped and consumed by human activities, mainly for irrigation, faster than aquifiers can be replenished by natural processes. The problem of water shortage therefore poses a serious threat to the sustained agricultural growth. Therefore, it is necessary to ensure scientific delivery and usage of water for irrigation besides encouraging the development and use of drought-resistent seeds and change in cropping pattern where necessary. The use of **drip irrigation** should be encouraged as in

8. *Op. Cit*, p.35.
9. See V.V. Naryana and Ram Babu, "Estimation of Soil Erosion in India" in *Journal of Irrigation and Drainage Engineering* ASCE, 109(4) pp 419-33.
10. Kerr T.M. and V. Sanghi, "Indigenous Soil and Water Conservation in India's Semi-Arid Tropics", *Gatekeepers,* Series 34,1992, International Institute of Environment and Development.

this water is delivered in measured dozes through slender pipes close to the plant, thereby avoiding waste from spillovers and is being used successfully in some countries where water is scarce. Besides, water-intensive crops like sugarcane and rice ought to be confined to the areas where these crops are grown traditionally because of availability of enough water.

India depends on ground water reserves for most of its agricultural and domestic water requirement. Hence there is increased stress on ground water reservoirs. Urban and rural **rain water harvesting** therefore should be explored fully. Check dams can be thought of as an alternative to mega dam projects to provide local water support. In our view the way to stop environmental degradation in the rural areas is not to curb the use of modern green revolution technology for fear of its defects but efforts should be made to overcome its defects through undertaking new scientific research.

Lack of clean water and sanitation facilities. Another important environmental problem faced by rural population, especially the rural poor, arises due to lack of clean drinking water and sanitation. These greatly contribute to many water-borne diseases such as typhoid, cholera, amoebic infections, bacillary dysentery and diarrhea which account for 80 per cent of all diseases that prevail in developing countries and account for a large number of child deaths each year. Apart from responsible for deaths in rural areas these diseases lower the **health** of the people which in modern economic theory is considered as **human capital** and its deterioration in the rural areas lowers productivity and economic growth.

Similarly, air pollution in the rural areas which arises due to greater dependence on **biomass fuels** such as wood, straw, manure that is closely related to the poverty of the rural population impacts their health. About the harmful effects of air pollution due to greater dependence on biomass fuel, Todaro and Smith[11] write, "The burning of biomass fuels for cooking and boiling of water create dangerously high levels of indoor pollution to which 400 million to 700 million people, mostly women and children, are exposed each year. Smoke and fumes from indoor stoves are believed to contribute significantly to up to 4.3 million childhood deaths each year from respiratory diseases and to an ever-larger number of chronic respiratory illnesses."

In the context of rising consumption standards both in the developing and developed countries along with rapid growth of population in developing countries achievement of the goal of sustainable development is a major challenge facing the world today. It is evident that meeting the consumption needs of future generations when the world population is estimated to increase by an additional 3 billion by 2050 require substantial changes in consumption and production pattern.

Rio conference held in 1992 where 150 countries participated was an important milestone in defining the concept of "sustainable development". This was spelled out in considerable detail in agenda 21, which set out general principles for sustainable development and framed blueprint for conservation and use of forests and suggested important steps that are required to be taken to generate an environmentally stable and sustainable planet.

NEGATIVE EXTERNALITIES AND ENVIRONMENT RESOURCES

Externalities is the most important case of market failure and one most directly relevant to the use of environmental resources. Externalities refer to the beneficial and detrimental effects of the economic activity of an agent (a firm or a consumer or industry) on others. To be more precise, when an economic agent through his activity benefits others for which he does not receive any payment, there exist beneficial or positive externalities. On the other hand, detrimental or negative *externalities occur* when the activity of an economic agent harms others for which it is not required to pay. It is important to note that when there are individual property rights over goods, services and factors of production, the markets function efficiently (that is, allocate resources optimally). This is because when there is a complete set of individual property rights, the market exercises control over their activities to ensure optimum consumption and production of goods or services or use of a factor as they require payment to be made to the owner. However, when externalities occur and there are no individual property rights on goods, services or factors, market will not ensure optimum or socially desirable output due to the non-payment for the benefits or harms created for others. In case of positive externalities less will be produced and in case of negative (*i.e.*, detrimental) externalities more of a good is produced than is socially desirable.

11. Michael P.Tadoro and Stephen C.Smith, *Economic Development,* 8th edition, 2003, Pearson Education (Singapore): p.471.

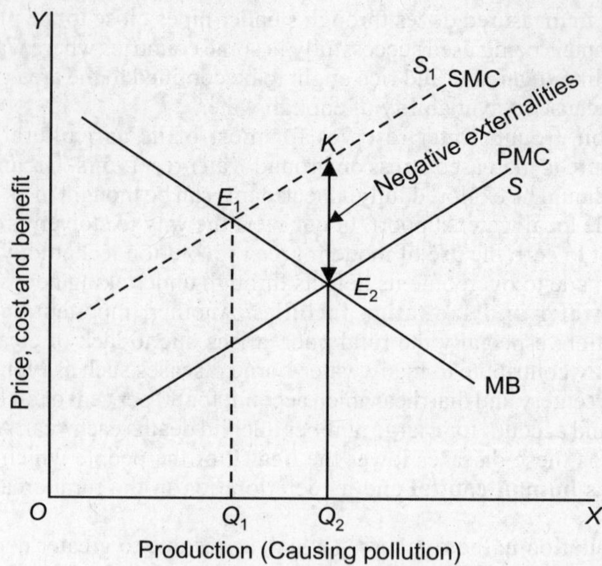

Fig. 57.1. *In the presence of negative externalities more output is produced than is socially desirable*

Let us give an example of how negative externalities lead to more than socially desirable output. The chief example the negative externalities occurs in case of electricity producing firm which pollutes air which harms others but they are not paid for this by the firm. Note that *no one has individual property rights over the air* which is provided by the environment. Therefore, the firm can extend its production to the level which maximises its profits or benefits for the industry. In the case of a firm, which pollutes air and does not pay those who suffer as a result of pollution, *marginal private cost (MPC)* will be smaller than the *social marginal cost* (SMC) as the latter will also include negative externalities of pollution. How, in this case, more of output of good will be produced than is socially desirable? This is illustrated in Fig. 57.1 where MB is marginal benefit curve of the firm whose production pollutes the air and is sloping downward. This is because this marginal benefit curve depends on the demand curve of the product of the firm. As the firm produces and sells more of its product, *its price falls.* This price is the benefit for the firm. Therefore, as it produces more, its marginal benefit (MB) declines. On the other hand, as more is produced by the firm its private marginal cost (PMC) will rise in accordance with the principle of diminishing returns of traditional microeconomic theory, and rising private marginal cost (PMC) of the firm will cause the private marginal cost curve of the firm to slope upward as shown in Fig. 57.1 by PMC. As explained above, *the firm does not take into account the harms done to others* especially those living in the surrounding areas of air pollution generated by it. The question is how it is allowed to pollute the air to extend its production to the level beyond what is socially desirable? This is because, as mentioned above, no one has the ownership rights over the air which is supplied by the environment (*i.e.*, air is environmental good). It will be seen from Fig. 57.1 that considering only its private marginal cost the firm will equate private marginal cost (PMC) with its marginal benefit (MB) and produce output OQ_2 to maximize its profits.

Now, if negative externalities represented by creation of air pollution, which inflicts costs on others, are into the account, we will add to private marginal cost, these negative externalities to get *social marginal cost* (SMC) curve which lies above the *private marginal cost* (PMC) curve. The socially desirable output which represents optimum allocation of resources is equal to OQ_1 at which social marginal cost (SMC) equals marginal benefits (MB). Thus, by producing OQ_2 output, the firm in case of the existence of negative externalities produces more than what is socially optimum output level OQ_1.

Another important example of negative external economies is found in case of a firm discharging its toxic effluent into the water of a lake which is used by the fishermen to catch fishes to make their living. The toxic effluent in the lake kills many fishes which reduces the number of fishes captured by the fishermen and thus harms them. The harms done to the fishermen by the firm dumping its toxic product into lake do not enter into the calculations of the polluting firm. As no one has ownership right over water of lake, the firm produces more product than is socially desirable. In this case also *social marginal cost* (SMC) curve will be above the *private marginal cost* (PMC).

COMMON PROPERTY RESOURCES

Common property resources are the resources to which everyone has *free access*, that is, anyone can use the property without making any payment. Parks, rivers, ocean, fisheries, public roads and highways are examples of common property resources. For example, anyone can have a free entry into the India Gate parks in Delhi and enjoy there. However, *each person's use of common property reduces its value to others and thus creates **negative externality***. We explain this below.

Since people do not have to pay for the use of common property resources, they tend to be overused or overexploited. For example, parks with free entry often become overcrowded which results in lowering of enjoyment of parks by everybody. Further, the use of rivers, Ganga and Yamuna, in India is quite free and have become so much polluted because factories dump their waste products into them without paying any cost. Even people in general throw all sorts of their wastes (even dead bodies, and their remains) into these rivers. Similarly, costless use of public common land for grazing of cattle, or hunting results in overuse of these common property resources. **This *overuse* of common property resources where there is free access to them is often referred to as the tragedy of commons.** This is also called the *common problem.*

An important common property resource is fisheries in which no fee is charged for catching fishes from them. Each fisherman wants to catch fish before others do the fishing. *This results in overfishing which may ultimately* lead to the destruction of stock of fish in the fisheries for use in future years.

*It is worth noting that **lack of property rights over the common resources** leads to their overuse. Fishermen when they have costless access to the use of common property resources have incentive to catch more fish than they would if the fisheries were private property, whose owner would charge fee or price for catching fishes by fishermen.* Suppose now each fishery has been given exclusive private property right to a lake where fishes are caught by fishermen. Now, since there exist well-defined property rights, there are no externalities. Therefore, each owner of lake would not allow overfishing in any year so as to maintain the stock (*i.e.*, the number) of fish for the future years.

On the other hand, most *ocean fisheries* are common property to which fishermen have free access. As in case of private manufacturers who create external diseconomy by polluting air or water, fishermen operating in ocean fisheries take into account only their *private costs*. Thus in calculating their costs they include in them the cost of boats, other complementary equipment, a crew etc. *They do not take into account the costs they impose on future generation as their fishing in the present leads to the decrease in the stock of fish in the ocean in the future years*. Thus Professor Perloff writes, "The fewer fish there are, the harder it is to catch any, so reducing the population of fish today raises the cost of catching fish in the future. As a result, fishers do not forego fishing now to leave fish for the future. *The social cost is the private cost plus the externality cost from reduced future population of fish*."[12] It may be emphasised again that it is the lack of clearly defined property rights of fisheries that leads to overfishing. Fishermen have an incentive to catch more fish than they would if the fishery was private property.

Roads and Highways. Let us take another case of common property resources: roads and highways. Everyone has access to them for driving their cars and other vehicles. Since owner of a car has no exclusive right to use the highway on which he drives, he cannot exclude others from driving on the same highway and must share it with others. This leads to the too many drivers using the highway for driving their vehicles which leads to congestion and often jams (*i.e.*, a negative externality). This slows down driving on the highway. Thus this is another example of overuse of the common property resource.

Diagrammatic Illustration of Overuse of Common Property Resources

Let us illustrate diagrammatically *how in case of common property resources, there is overuse of them which is often referred to as 'tragedy of commons'*. Let us take the case of a lake which is the common property of nearby town and this lake is used for fishing. Every fisherman has free (*i.e.*, costless) access to catch fishes from the lake and then sell it in the market. But by catching fish from the lake, fishermen reduce the population of fish in the lake and

12. Jeffery M. Perloff, Microeconomics, Addison-Wesley, 1998, p.679.

thereby raise the cost of catching fishes for others. That is, by catching fishes from the lake, fishermen create external diseconomy for other fishermen.

In Fig. 57.2 we have drawn a demand curve *DD* for fish (in pounds) in the market which reflect consumer's marginal benefits (MB) from fish and *SS* is the market supply curve of fish based on marginal private cost (MPC) of fishermen. If there is free access to lake, the number of fish caught and supplied in the market is *OQ* pounds. Since fishermen do not take into account the negative externality they create, the fishermen's marginal private costs (therefore the supply curve *SS* based on them) will be less than the marginal social cost (MSC). Therefore, the supply curve *S′S′* representing the marginal social costs intersects the market demand curve (*i.e.*, marginal benefit curve) at point *R* and thus determine the *OL* quantity of fish caught and sold. (Note that the marginal social cost (MSC) equals marginal private cost (MPC) plus the negative externality created). Thus *OL* quantity of fish (in pounds) is the *socially optimum* output. But as fishermen do not take into account the external diseconomy or social marginal cost and operate on the basis of their marginal private cost (MPC), they catch more fish equal to *OQ* which thus *leads to the overuse of the fishery*. This is therefore called **tragedy of commons.**

Assigning Property Rights for Common Property Resources: Coase Theorem

We have explained above that common property resources tend to be overused and therefore there is economically inefficient use of them. R. Coase,[13] a Nobel Prize winner of 1991, has proposed a solution for this problem to ensure economically efficient use of common property resources. According to Coase, if costs of negotiating between the parties is quite small, the parties that create negative externality and the victims of their activity negotiate, they can arrive at a price which properly takes into account the negative external effect and thus ensures economic efficiency in the use of common property resources. For example, if *downstream water users* are assigned the property right *to obtain water of a particular quality*, a firm that wants to pollute the stream will be prepared to give compensation to them (*i.e.*, to downstream water users who have been given rights to obtain water of certain specified quality). *Since the firm will now have to give compensation to the downstream water users, it will not find it worthwhile to pollute the stream beyond the economically efficient point.* This is the gist of Coase theorem and its importance lies in its suggestion that assignment of well-defined property rights might help to promote economic efficiency.

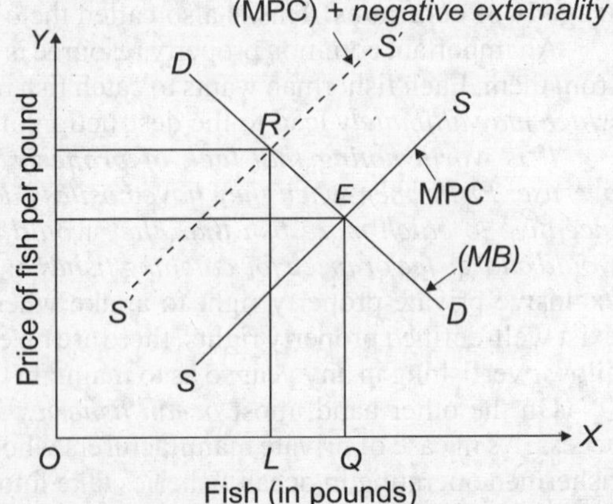

Fig. 57.2. *Overuse of Common Property Resources*

It may be emphasised that Coase theorem assumes that transaction costs, that is, costs of negotiating and contracting by the interested parties are very small. For instance, in our above example, the downstream water users join together and then effectively negotiate with the polluting firm without much cost about the price to be charged from the latter for polluting the stream. Besides, even if the transaction costs are small, negotiations with the polluting firm may not be practical because *it is difficult to measure precisely the external cost* (*i.e.*, negative externality) of pollution imposed by the polluting firm on the downstream water users. Further, if the number of interested parties in negotiating a deal is large, it may not be possible to get unanimity required to make the negotiation effective. To conclude in the words of Edwin Mansfield, Coase theorem suggests that "the assignment of well-defined property rights might help to promote economic efficiency. *For example, to get around the difficulties caused by external diseconomies arising from waste disposal, society might find it useful to try to establish more unambiguous property rights for individuals and firms with respect to environment quality. Then, assuming that the relevant negotiations are feasible, the interested parties in a particular area might try to negotiate to determine how much pollution will occur.*"[14]

13. R. Coase, *The Problem of Social Choice*.
14. Edwin Mansfield, *Microeconomics*, Seventh Edition, 1991, p. 472.

CHAPTER 58
SUSTAINABLE DEVELOPMENT AND CLIMATE CHANGE

SUSTAINABLE DEVELOPMENT: MEANING AND CONDITIONS

Economists have used the term **sustainable development** in an attempt to clarify the balance between economic growth on the one hand and conservation and protection of environment on the other. Sustainable development refers to **"meeting the needs of the present generation without compromising the needs of future generation"**[1] Thus economic growth will be sustainable if the stock of capital assets including land remains constant or increases over time. Though the stock of natural resources are a part of capital assets of a society, they are only limited substitutes of other types of capital assets. It may however be noted that future economic development and quality of life crucially depends on the natural resource base and quality of the environment, *i.e.*, the quality of land, water and air. To destroy and over-exploit the natural resources indiscriminately and pollute the environment will though raise the short-term growth rate and living standards of the people will have adverse effect on the long-term future growth and the quality of life of the future generations as the latter will have smaller natural resource base and poor quality of environment.

It is therefore necessary that while deciding about growth environmental issues should be factored in. For instance, loss or preservation of important environmental resources should be taken into account while preparing estimates of growth and wellbeing of the people. Alternatively, economic policy makers may opt for framing a growth strategy of **no net loss of environmental assets** so as to ensure sustainable development. In this latter case if environmental resources are destroyed or depleted in one area, an equal or greater amount of environmental resources are replenished or regenerated elsewhere so that future economic growth is not adversely affected.

Pearce and Warford[2] have put forward a view which takes into account preservation of environment for sustainability of growth. In the overall capital stock they include not only the man-made physical capital (such as machines, factories, roads) but also human capital (education, skills, healthcare) and environmental capital (such as forests, water resources, climate, arable land, soil quality). According to Pearson and Warford, sustainable development implies that overall capital stock, as defined above, does not decrease in the growth process. In this regard, they point out that the sustainable measure of net national product (NNP*) can be obtained as the amount that is consumed during the growth process without any decrease in the overall capital stock during a year. Thus according to Pearce and Warford the condition for sustainable development can be written as:

$$\text{NNI}^* = \text{GNI} - D_m - D_n$$

where NNI* is sustainable level of national income, GNI is gross national income, D_m is depreciation of manufactured capital assets, and D_n is the depreciation of environmental capital resource. It is only when the above condition is fulfilled, NNI* will remain constant as the growth of GNI will

1. Michael P. Todaro and Stephen C. Smith, *Economic Development,* 8th edition, 2003, Pearson Education (Singapore), p.471.
2. David W. Pearce and Jeremy J. Warford, *World without End: Economic Environment and Sustainable Development: A Summary* (Washington DC: World Bank, 1993).

make up the loss due to depreciation of manufactured capital assets, and use of environmental resources.

In the context of rising consumption standards both in the developing and developed countries along with rapid growth of population in developing countries achievement of the goal of sustainable development is a major challenge facing the world today. It is evident that meeting the consumption needs of future generations when the world population is estimated to increase by an additional 2 billion by 2050 require substantial changes in consumption and production pattern.

Rio Conference held in 1992 where 150 countries participated was an important milestone in defining the concept of "sustainable development". This was spelled out in considerable detail in agenda 21, which set out general principles for sustainable development and framed blueprint for conservation and use of forests and suggested important steps that are required to be taken to generate an environmentally stable and sustainable planet.

Over the last two decades there has been increased awareness about the need for preventing environmental degradation, checking pollution and averting global warming as a result of climate change but not significant practical steps have been taken in this regard and more world conferences on environment and climate change have taken place since 1992. These conferences were expected to resolve the issues eoncering environmental pollution and climate change between the developing and developed countries. The difficult problem is that the development and technological path pursued by the western developed countries is highly energy-intensive and exploits huge quantities of natural resources and leaves a highly degraded environment with emission of greenhouse gases (GHG) in the atmosphere. It is this path that developing countries are also following for their economic and social development which is quite unsustainable and leads to a large-scale pollution and ecological degradation.

To ensure sustainable development in the world today both the developed and developing countries have to take appropriate environment-friendly steps, especially by the developed ones as they have in the past through reckless industrialisation and use of energy-intensive technology, emitted greenhouse gases to a larger extent. A noted Indian environmentalist, Sunita Narain, rightly writes, "*The process of ecological globalisation is driven by the fact that levels of production and consumption have reached a stage where what one does in one's country can have major impact on neighbouring countries or even on the rest of the world. Even simple things like using a refrigerator or an air-conditioner can today destroy the world's ozone layer, running an automobile or cutting a tree without planting another one can destabilize the world's climate. And, using a persistent organic compound like DDT can mean life threatening pollution for human beings and other life forms in the remote polar regions of the world as these compounds are slowly but steadily carried to these regions by the world's oceanic currents and air streams. Never before have human beings needed to learn to live in 'one world' as now.*"[3]

The Size of World's Greenhouse Gas Emission Problem

The Intergovernmental Panel on Climate Change (IPCC) which reviews and assesses the worl wide information regarding climate change in its recent report published in 2014 has observed that there has been increasing trend in the omissions of greenhouse gases (GHG) since the advent of the industrial revolution with about half of CO_2 emissions during this period occurring in the last forty years. The period 1983-2012 is likely to have been the warmest thirty-year period of the last 1400 years. CO_2 emissions from fossil fuel combustion and industrial processes have contributed a major portion of total GHG emissions during the period 1970-2010.

3. Sunita Narain, "Southern Challenges: Need for One Version," in *The Hindu*, "Survey of Environment," 2002, p. 13.

It is important to note that the change in the climate system is likely to have adverse impact on the economy, livelihood, cropping pattern, and food security. According to the various projections by the IPCC, extreme heat event are likely to be longer and more intense in addition to changes in precipitation pattern. The change in climate could affect the production of wheat, rice and maize in the tropical and temperate regions, have negative impact on health by worsening health problems that already exist, especially in developing countries, and adversely affect productive activities like growing food and working outdoors.

The ecological overshoot problem being larger than the bio-capacity of the population is an important issue in global climate negotiations that are taking place in the last over two decades. The ecological overshoot problem can also be understood in terms of carbon budget. The risk of climate change is largely a function of total cumulative emission of GHG in the atmosphere. IPCC has estimated that for temperature increase to remain below 2°C of pre-industrial levels the world can emit only about 2900 Guga tonnes (Gt) of CO_2 from all sources from the industrial revolution till 2100. Till 2011, the world 1900 Gt of CO_2, thus already consuming around two-thirds of this carbon budget of 2900 Gt, only 2000 Gt remains to be used between now and 2100. The World Resources Institute estimates that if emissions continue unabated it will last only next 30 more years.

Allocation of Sparse Carbon Budget

The important issue therefore at present is how to allocate the remaining sparse carbon budget between countries in a manner which is both fair and achievable. This certainly should involve an assessment of historic responsibility based on how much various countries have contributed to cumulative emissions of global CO_2 (1850-2011) so far. In Fig. 58.1 we have shown the contribution by different countries in cumulative global CO_2 emissions so far (1850-2011). It will be seen from Fig. 58.1 that India's contribution to cumulative global CO_2 (1850-2011) was a meagre 2.8 per cent as against 21.2% by the USA, 18% by the EU and 10.7% by China. The allocation of the remaining carbon space needs to be made with social justice and equity in this regard. For developing countries, their future commitment will also be determined by what kind and level of financial capacity-building and other support is provided by developed countries that have contributed most to cumulative global GHGs emissions so far.

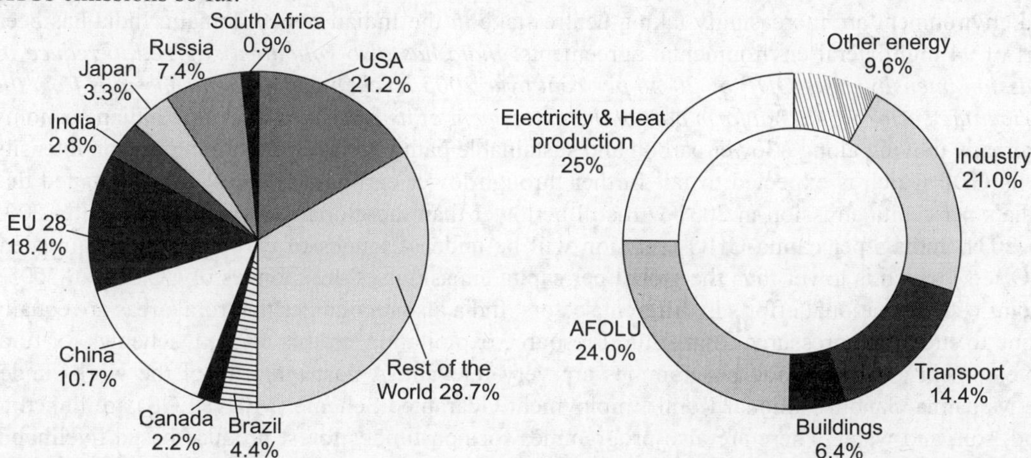

Fig. 58.1. *Contribution by Different Countries to Cumulative Global CO_2*

Fig. 58.2. *Global GHG Emissions from Different Sectors* in 2010.

Source: Centre for Science and Environment and IPCC AR5.

Global GHS Emissions from Different Sectors

We have shown in Fig. 58.2 Global GHG Emission from Major Sectors Countries since 2000. GHG emissions have been increasing in all sectors *except* agriculture, forestry and other land use

(AFOLU). Of the (494.5) Gt CO_2 eq. (CO_2 equivalent) emissions in 2010, 35 per cent were *released in the energy supply* sector, 24 per cent in AFOLU, 21 per cent in industry, 14 per cent in transport, and 6.4 per in buildings (see Fig. 58.2).

There are substantial variations in total and per capita emissions of GHG of different countries. As per report of IPCC, per capita GHG emissions in 2010 were highly unequal. With median per capita emissions for the group of low-income countries (1.4 CO_2 eq/cap) being 9 times lower than median per capita emissions of high-income countries (13t CO_2 eq/cap). In terms of absolute CO_2 emission from fossil fuels use cement production in 2013. China, the USA and EU hold the first three positions respectively with India having a distant 4th position. However in terms of per capita CO_2 emissions from the same sector in 2013, countries like India, Brazil and South Africa fall in the bottom 100 among 196 countries.

Sustainable Development and Climate Change in the Indian Context

The key environmental challenges in India have been sharper in the past two decades. The State of the Environment Report by the MoEF clubs the issues under five key challenges faced by India, which are climate change, food security, water security, energy security, and managing urbanization. Climate change is impacting the natural ecosystems and is expected to have substantial adverse effects in India, mainly on agriculture on which 50 per cent of the population still depends for livelihood, water storage in the Himalayan glaciers which are the source of major rivers and groundwater recharge, sea-level rise, and threats to a long coastline and habitations. Climate change will also cause increased frequency of extreme events such as floods and droughts. These in turn will impact India's food security problems and water security. As per the Second National Communication submitted by India to the UNFCCC, it is projected that the annual mean surface air temperature rise by the end of the century ranges from 3.5°C to 4.3°C whereas the sea level along the Indian coast has been rising at the rate of about 1.3 mm/year on an average. These climate change projections are likely to impact human health, agriculture, water resources, natural ecosystems, and biodiversity.

Wary of the threats imposed by climate change and pressures on natural resources, sustainability and environment are increasingly taking centre stage in the Indian policy domain. India has been part of 94 multilateral environmental agreements. *India has also voluntarily agreed to reduce its emission intensity of its GDP by 20-26 per cent over 2005 levels by 2020, and emissions from the agriculture sector would not form part of the assessment of its emissions intensity*. Indian economy is already moving along a lower carbon and sustainable path in terms of declining carbon intensity of its GDP which is expected to fall further through lower carbon strategies. It is estimated that India's per capita emission in 2031 will still be lower than the global per capita emission in 2005 (in 2031, India's per capita GHG emission will be under 4 tonnes of carbon dioxide equivalent (CO_2 eq.) which is lower than the global per capita emissions of 4.22 tonnes of CO_2 eq. in 2005). Along with the national efforts in different sectors, India also recognizes that rural areas are equally prone to stress and pressures from natural resource exploitation. In this context, schemes for rural development and livelihood programmes are very relevant. A vast majority of the works under the Mahatma Gandhi National Rural Employment Guarantee Scheme (MGNREGS) are linked to land, soil, and water. There are also programmes for non-timber forest produce-based livelihood, promotion of organic and low-chemical agriculture, and increased soil health and fertility to sustain agriculture-based livelihoods. These schemes help mobilize and develop capacities of community institutions to utilize natural resources in a sustainable manner and their potential can be further developed.

Together with efforts to incorporate sustainability in the rural development process, India is increasingly making efforts to *integrate the three pillars of sustainable development into its national policy space*. In fact, environment protection is enshrined in our Constitution (Articles 48A and

51A[g]). Various policy measures are being implemented across the domains of forestry, pollution control, water management, clean energy, and marine and coastal environment. Some of these are policies like Joint Forest Management, Green Rating for Integrate Habitat Assessment, Coastal Zone Regulation Zone, eco labelling and energy efficiency labelling, fuel efficiency standards etc. Over a period of time, a stable organizational structure has been developed for environment protection. *The country has been making fast progress in increasing its renewable energy capacity and has displayed the fastest expansion rate of investment of any large renewables market in the world in 2011, with a 62 per cent increase to $12 billion (Frankfurt School of Finance and Management 'Global Trends in Renewable Energy Investment 2012'*). The Twelfth Five Year Plan with a prominent focus on sustainability makes provision and provides for many more opportunities like these.

Economic development without environmental considerations can cause serious environmental damage, in turn impairing the quality of life of present and future generations. Such environmental degradation imposes a cost on the society and needs to be explicitly factored into economic planning, with necessary remedial measures incorporated. The challenge of sustainable development thus requires integration of the country's quest for economic development with its environmental concerns. Environment management in India has, over the years, recognized these sustainable development concerns. The National Environment Policy 2006 has attempted to mainstream environmental concerns in all our developmental activities. It underlines that '*while conservation of environmental resources is necessary to secure livelihoods and well-being of all, the most secure basis for conservation is to ensure that people dependent on particular resources obtain better livelihoods from the fact of conservation, than from degradation of the resource*'. The Government of India, through its various policies, has been factoring ecological concerns into the development process so that economic development can be achieved without critically damaging the environment. The strong sustainable development agenda followed by India incorporates rigorous environmental safeguards for infrastructure projects, strengthening of the environmental governance system, revitalizing of regulatory institutions, focusing on river conservation, and efforts for improvements in air and water quality, on a continuous basis.

Our environmental standards are set through Government policies aimed at a development process that is environmentally sustainable and ensures well-being of the people. The broad objectives of our environmental policies and programmes are:

1. Conservation of flora, fauna, forests, and wildlife;
2. Prevention and control of pollution;
3. Afforestation and regeneration of degraded areas; and
4. Protection of the environment.

As a country, India has been in the forefront of *preserving biodiversity, sustainable management of forests, reducing emissions intensity of the economy*, and following sustainable consumption and production patterns. Specifically, India has been following a development path that takes into consideration the needs of the present generation without compromising the ability of future generations to meet their needs. Suitable attention has been given to protecting and conserving critical ecological systems and resources and invaluable natural and man-made heritage, which are essential for life-support, livelihoods, economic growth, and a broad conception of human well-being. Moreover, the effort has been to ensure equitable access to environmental resources and quality for all sections of society, in particular to ensure that poor communities which are most dependent on environmental resources for their livelihoods are assured secure access to these resources.

Sustainable Development and Poverty Eradication

Working on the *social and economic pillars of sustainable development policies, programmes and targeted schemes have been introduced to eradicate poverty*. This is done either through a direct focus on economic indicators like employment generation, youth mobilization, and building up assets of the poor, or indirectly through social indicators of human development with emphasis on health, education, and women's empowerment. Many parameters on this front have shown improvement. The poverty head-count ratio declined by 15.3 percentage points from 37.2% in 2004-05 to 21.9% in 2011-12, maternal mortality rate (MMR) dropped from 301 per 100,000 live births in 2001-03 to 212 in 2007-09; literacy rates have been constantly rising and are estimated to be 82.14 per cent for men and 66.46 per cent for women as per the 2011 Census of India. However, India is still not on target to meet some key Millennium Development Goals (MDGs) by 2015.

Over the years arguments in favour of looking beyond the conventional measure of GDP and taking into account the environmental damage caused by production of goods and services have received attention. An expert group under the chairmanship of Prof Sir Partha Dasgupta has been set up to develop a framework for '*Green National Accounts*' for India. In fact, the Central Statistics Organisation (CSO) under the Ministry of Statistics & Programme Implementation (MOSPI) has been publishing comprehensive environment statistics since 1997. The process of putting in place a system for *natural resources accounting was initiated by* MOSPI way back in 2002.

Despite all these efforts, the challenges that confront us on the environmental front continue to be harsh and complex. Increasing population, urbanization, and growing demand for water and land resources have severely impacted the quality and availability of water and soil resources. Rising energy needs is another area of concern. Besides, rapid growth will require corresponding growth in energy supply. Presently a large share of our energy demand is met through coal and oil and this trend will continue, given the unprecedented surge in energy demand and resource constraints. Energy issues have become more complex with existing energy shortage and rise in energy prices. There is considerable scope for increasing efficiency in the use of energy and water in India together with other development indicators like infant mortality rate, maternal mortality rate (MMR), sanitation facilities, and public health services. Economic instruments, regulatory measures, and market mechanisms can play an important role in helping to achieve development and growth in a sustainable manner.

THE UNITED NATIONS CONFERENCE ON SUSTAINABLE DEVELOPMENT (UNCSD), RIO +20, 2012

Recognising the well-founded concerns on the need to redress environmental problems, there have been global calls for cooperation, action, and innovation. World leaders in 2012 continued to engage and deliberate in international forums dedicated to climate and environment and also in forums like the G-20 where sustainable development and climate change were an integral part of the discussions. Ambition or goal setting to reach targets, provision of finance and technology for developing countries, and institutions and mechanisms for capacity building were the common threads of negotiations running through all these forums. One such event in 2012 was the United Nations Conference on Sustainable Development (UNCSD) which was held at Rio in Brazil in June 2012. This is also known as Rio + 20 Conference since the first United Nations Conference on Environment and Development was held at Rio in Brazil in 1992. Therefore, the name Rio + 20. It was attended by the heads of states of the member countries.

The objective of the Rio+20 Conference was to secure renewed political commitment for sustainable development, review progress made and identify implementation gaps, and assess new and emerging challenges since the UNCSD held 20 years ago in Rio in 1992. Towards this end, the Conference had two themes, viz, (a) *green economy* in the context of sustainable development and poverty eradication, and (b) *institutional framework for sustainable development*. The most significant outcomes of the Rio Summit have been the ***restoration of the principles of equity and of common but differentiated responsibilities (CBDR) in the global environmental discourse***

and placing poverty eradication at the centre of the global development agenda. The outcome also ensures the required domestic policy space to countries on a green economy **and launched four processes or mechanisms,** *i.e.* developing sustainable development goals (SDGs), financing strategy, technology transfer, and defining the format and organizational aspects of the proposed high-level political forum to follow up on the implementation of sustainable development.

Equity or 'Fairness' as an issue received attention. It is a matter of satisfaction and achievement for India that the Rio outcome document reaffirms equity and the principle of common but differential responsibility (CBDR) among other Rio principles. India together with other developing countries played an instrumental role in this. CBDR is especially important for developing countries, as it *implies that while all countries should take sustainable development actions*, *the developed countries have to take the leading role in environmental protection*, as they have contributed the most to environmental problems. Also they should support developing countries with finance and technology in their sustainable development efforts. *India has always held that the eradication of poverty should be the over-arching goal of sustainable development*. This was given due recognition in the deliberations at the Rio Summit and in the outcome document.

On the issue of Green economy, the outcome document affirms that there are different approaches, visions, models, and tools available to each country, in accordance with its national circumstances and priorities, for achieving sustainable development. *It identifies green economy in the context of sustainable development and poverty eradications as one of the important tools for achieving sustainable development* but specifies that while it could provide options for policy-making, it should not be a rigid set of rules. The outcome document clearly states what green economy policies should result in and what they should not. It is a matter of satisfaction that the document firmly rejects prescriptive policies, unilateral measures, and trade barriers as well as unwarranted conditionality of official developmental assistance in this context.

The Rio+20 Conference will also be remembered for kick-starting the process on developing sustainable development goals (SDG). The SDGs would address and incorporate in a balanced way, all the three dimensions of sustainable development and their inter-linkages. The SDGs would be universal, global, and voluntary. Since the SDGs are expected to become a part of the post-2015 UN development agenda, they would hopefully guide the international community towards inclusive sustainable development.

From India's point of view, SDGs need to bring together development and environment into a single set of targets. The fault line, as ever in global conferences, is the inappropriate balance between environment and development. Developing countries do not want any bindings on their efforts towards poverty eradication or any agreement that comes with such a price tag. Therefore, we could also view the SDGs and the post-2015 agenda as an opportunity for revisiting and fine-tuning the *millennium development goals* (MDGs) framework and sustainable regaining focus on developmental issues. India and many developing countries are slow or offtrack in achieving targets under some of the MDGs, which have concrete areas of overlap between environment and development. This is another reason why these MDGs should continue to be a part of the post-2015 global policy architecture.

The Rio Summit did not lead to any specific commitments on the finance and technology front. The developed countries, having obligations and responsibilities, need to commit to provision of adequate public funds to provide assistance including transfer of technology and capacity building to developing countries.

CLIMATE CHANGE: ISSUES, CAUSES AND EFFECTS

Climate Change: The Issue

The climate change has become one of the most important environmental issues that confronts the human beings both in the developed and developing countries. The issue is that the various economic activities by the people, especially the burning of fossil fuels such as coal, petroleum, natural gas in various productive activities, is likely to disturb the planet's heat and radiation balance. It will lead to an increase in earth's temperature (*i.e.,* warmer climate) in the next 50 years which would have disatrous consequences not only for the quality of life (*i.e.,* clean environment) of the people but also for economic development. Climate change has therefore become a global issue and negotiations among the countries, both developed and developing, are going on to mitigate global warming through reduction in emission of greenhouse gases such as carbaon dioxide (CO_2) and thereby save the humanity.

What is Climate Change?

Now, what is climate change? 'Climate change' means a change of climate which is attributed directly or indirectly to human activity that alters the composition of the global atmosphere and which is in addition to natural climate variability observed over comparable time periods. Increasing levels of fossil fuel burning and land use changes have emitted, and are continuing to emit, greenhouse gases (mainly carbon dioxide (CO_2), methane, and nitrous oxide) into the earth's atmosphere. This increasing level of emissions of greenhouse gases has caused a rise in the amount of heat from the sun trapped in the earth's atmosphere, heat that would normally be radiated back into space. This has led to the **greenhouse effect**, resulting in climate change. The major characteristics of climate change are rise in average global temperature, ice cap melting, changes in precipitation, and increase in ocean temperature. The efforts needed to *address the climate change problem include greenhouse gas emissions on the one hand and building of capacities to cope with the adverse impacts of climate change on various sectors of the society and economy on the other*. Scientists who have been working on the subject for a long time point out that greenhouse gases (GHG) such as carbon dioxide in the atmosphere trap the incoming heat from the sun and thereby leads to the increase in earth's temperature which is often referred to as **global warming.** Besides carbon dioxide, other greenhouse gases are methane, nitrous oxide and water vapour. This climate change is likey to cause adverse change in weather patterns and extreme events such as cyclones, hurricanes, floods, rise in sea level. How to **reduce the emission of greenhouse gases** and also to **adapt to the changes in climate** are the issues that confront the humanity.

Why does carbon dioxide concentration in the atmosphere increase? Though human beings also emit carbon dioxide while they breathe, the most important source of carbon dioxide emission is **burning of fossil fuels** such as coal, oil, gas during the process of industrial production. As most of the industrual growth has taken place in the developed countries such as the USA, Europe and Japan over the last more than a century, it is they who have contributed most to the increase in carbon dioxide concentration in the atmosphere. It is the unsustainable consumption pattern of the rich developed countries that is responsible for much of the emission of carbon dioxide and other greenhouse gases by burning fossil fuels. While 25 per cent of world population live in these rich industrialised countries, they emit more than 70 per cent of total global carbon dioxide emission.[4] In per capita terms disparities in emission of carbon are also very large. Thus, while a citizen of USA on an average emits more than 5.5 tonnes carbon per year, an Indian citizen emits only 0.25 tonnes of carbon per year.[5]

It may be noted that though the developed countries emit more carbon dioxide and other greenhouse gases (GHG) into the atmosphere, it is not only their local atmosphere that gets polluted, but the whole **global atmosphere** is adversely affected. The air and winds know no national

4. Jyoti K. Parikh and Kirti Parikh, "Climate Change: India's Perceptions, Position, Policies and Possibilities", in *Climate Change and Development*, OECD, 2002, p.5.
5. Ibid, p-5.

boundaries; the emission of toxic gases into the air by one nation easily travels to the whole global atmosphere. Thus, there is climate change for India and other developing countries though it is the rich industrialised nations of the USA, European countries and Japan which have contributed most to the concentration of greenhouse gases (GHG) in the atmosphere. Similarly, the flow of sea water knows no national boundaries and therefore rise in sea level as a result of climate change will affect the people of coastal areas of many countries. Therefore, global cooperation and efforts are required to mitigate climate change.

India is a country which will be severely impacted by climate change. This puts additional hurdles in its developmental path in addition to the challenges of poverty eradication and growing population. The projected impacts of climate change cut across water resources, forests, agriculture, and health. With a large agrarian population, India is vulnerable to changes in weather parameters. Further, rainfall, variability and melting of glaciers will impact replenishment of rivers, thereby affecting availability of water in river basins and watersheds. In India, most of the rivers flowing in the northern regions are dependent on snow and glacial melt, thus climate change threatens the perennial nature of these rivers. This has huge implications for agriculture and allied activities and resultant livelihoods. This is a serious concern for an economy that is tied to its natural resource base along its developmental path.

Why should India be Concerned about Climate Change?

India is deeply concerned about climate change because it will have substantial adverse effects on the Indian people. As will be explained later in detail, climate change or global warming will (1) lower agriculture production, (2) raise sea level leading to submergence of coastal areas, and (3) increase the frequencies of extreme events such as cyclones, hurricanes, floods. These are long-term issues. Negotiations at the global level are taking place to reduce global warming. India must play an active role in negotiations so that agreement about the extent of global level reduction in emission of greenhouse gases is reached and just and fair distribution of individual countries' contribution to it.

Since the developed countries have made substantial contribution to the concentration of CO_2 and other greenhouse gases (GHG), they have to accept greater responsibility in reducing emission of these gases in future. Besides, they have to give financial and technical support to the developing countries for reduction of emission of GHG.

It is worth mentioning here that **India's main energy source is coal which is highly polluting.** With the threat of climate change India is being urged by developed countries to change its energy strategy based on coal and instead use other energy sources such as oil, natural gas, nuclear energy which are more expensive. Thus immediate important issue facing India and other developing countries is to reach an agreement with the developed countries so that we have more freedom to decide which type of energy source we use, how we generate power, how to reduce methane emissions by agricultural practices or forestry and so on. Negotiations are important for us as a means to reduce or postpone future vulnerability by getting the developed countries to reduce their emissions."[6]

Adverse Effects of Climate Change

The climate change or global warming poses a threat to the sustainable economic development and therefore to the future generations. The global warming has the following effects:

1. Risk of Lower Agricultural Growth. Climate change has a significant adverse impact on the agricultural growth of developing countries. As regards India, it is a low latitude tropical country highly vulnerable to the consequences of increase in atmospheric temperature (*i.e.,* global warming). The global weather variability is a major constraint on agricultural growth and welfare in

6. Jyoti Parikh and Kirti Parikh, *op cit*; p. 6.

India. Increase in temperature in the Tibetan plateau, the Hindu Kush and Himalayas will affect the volume and timing of river flows with adverse effect on **crop production.** Kumar and Parikh[7] have shown that even with farm level adaptations, the impact of climate change on Indian agriculture would be significant. According to their estimates a temperature change of + 2°C and accompanying precipitation change of +7 per cent, farm level total revenue would fall by 9 per cent whereas with a temperature change of +3.5°C and precipitation change of +15 per cent the fall in farm level total revenue product would decline by nearly 25 per cent. This is a very large change in agricultural production which can cause much hardship for the Indian people, especially the poor. These estimates reveal that even 2°C increase in temperature will be quite substantial for India.

2. Rise in Sea Level. There is a great risk of rise in sea level as a result of global warming. It is expected that a large coastal area will be submerged with the rise in sea level. This will cause large-scale emigration from the coastal settlements. These climate migrants, often called climate refugees, will create problems for India for their settlement and cause much human hardship. The Ganges-Brahmaputra delta of India has been perceived to be exceptionally vulnerable area as water level in this area will rise submerging most of the land. Early signs of the potential impact can already be seen in the Sunderbans and Bangladesh. In view of this India faces a serious prospect of climate migrants.

According to the estimates of Asian Development Bank (ADB) made in 1995, if there is one metre rise in sea level today, it would displace 7 million people in India. With one metre rise in sea level 35 per cent of land in Bangladesh would be submerged. The cost of building a wall along the coastal line to prevent submerging is quite enormous.

3. Risk of Extreme Events. As a result of climate change extreme events like cyclones, hurricanes will more frequently occur causing a lot of human misery in terms of both death and loss of livelihood of the people. For example, in Andhra Pradesh in India the cyclone of 1996 and again of 2009 caused deaths of more than 1000 people. Further, it caused loss of large property owned by the people, especially the poor.

4. Climate Change and Economic Growth. Climate change is intimately linked to the process of economic growth and development. Since greenhouse gases (GHG) are generated by burning fossil fuels for energy use for industrial production and vehicular transport, industrial factories, power plants, automobiles generate CO_2 and other greenhouse gases. The reduction in their emission will mean lower GDP and economic growth and consequently rise in poverty. Therefore, in developing countries where absolute poverty still prevails on a large scale and the need for sustained economic growth is therefore very necessary, it is very difficult for them to reduce emission of GHG to achieve the global target. Thus, in case of developing countries the larger reduction in GHG will come at the expense of economic growth.

On the other hand, the rich developed countries which contributed most to CO_2 and other GHG emissions in the past and where absolute poverty has been largely eliminated are in a better position to reduce emission of these gases. Besides, they have both the finance and technology to mitigate climate change.

Murthy, Panda and Parikh[8] have shown that reduction in CO_2 emission in developing countries imposes costs in terms of lower GDP and higher degree of poverty. Using a multisectoral and intertemporal model they have estimated the effect of reduction of CO_2 emission on GDP and incidence of poverty in developing countries. Their estimates show that in India reduction of emission of $C0_2$ over the period of 30 years (taking the year 2000 as baseline) and using annual emission target will lead to a fall in GDP by 4 per cent and increase in the number of poor by 17.5 per cent. Therefore, in view of this an important problem faced in this regard in democratic countries

7. Kumar, K.S. Kavi and Jyoti Parikh, "Socio-Economic Impacts of Climate Change on Indian Agriculture," *International Review of Environment Strategies,* 2(2), 2001.
8. Murthy, N.S. Manoj Panda and Jyoti Parikh, "Economic Growth, Energy Demand and Carbon Dioxide Emission in India, 1999-2020" in *Environment and Development Economics*, Vol 2, 2009.

like India is to get public support for this reduction in CO_2 emission. While the cost will be borne by the present generation in the immediate period, the benefits of climate change will accrue to the future generations. Anand Patwardhan rightly poses this problem when he writes: "Any serious attempt to cut emission will have clear and immediate costs but the benefits may not appear for a long time. To the extent that the benefits may be disasters that did not happen, they may never be obvious that the costs will be. As the debate develops **much of it is being cast in terms of the restraint that** the present generation owes to future generations."[9]

Mitigating Climate Change to Ensure Sustainable Development

As seen above, climate change or global warming has disastrous consequences. The global negotiations are going on to reach an agreement regarding emission of greenhouse gases by the developed and developing countries as per certain time framework. However, realising that it is in our own interest, India has been for quite some time pursuing environment-friendly policies and taken steps to reduce emission of GHG through reduction in energy consumption. It is worth noting that India is committed to eradication of poverty. This requires stepping up rate of economic growth (*i.e.*, increase in GDP). Therefore, emphasis on high economic growth to reduce poverty results in higher energy consumption which is likely to adversely affect climate change. Given this constraint the following steps have been taken in India to mitigate climate change:

1. Promoting energy conservation and efficiency
2. Promotion of renewable energy sources
3. Reduction of air pollution
4. Afforestation and wasteland development
5. Fuel substitution policies

We discuss these policies below in some detail.

Energy Conservation and Efficiency. In India energy conservation and greater efficiency in energy use play an important role in national policy to reduce GHG emissions. It has been estimated by Jyoti Parikh and Kirti Parikh that power sector is the highest direct CO_2 emission in India (48%) followed by transport sector, *i.e.*, road, railways, air transport (10%), iron and steel (10%). Increase in the use of coal and electricity is estimated to result in significant reduction in CO_2 emission. Nag and Jyoti Parikh[10] have found that *greater efficiency in energy* use has been actually achieved in India as a result of government efforts.

Production of Non-conventional Energy Sources. Apart from energy conservation and increase in efficiency in energy use, steps have also been taken in India by Ministry of Non-Conventional Energy Sources to develop and exploit non-conventional energy sources which are clean and renewable. Various renewable sources, namely, wind farms, micro-hydroelectric plants, biomas and cogeneration power plants, biomass gasfiers system and solar photovoltaic system of aggregate installed capacity of 2302 MW was planned to set up by the end of the Eighth Plan in 1997 and this was planned to rise to 6500 MW by the end of Ninth Plan (1997-2002).

According to data furnished by Jyoti Parikh and Kirti Parikh, about 2.6 million family size biogass plants and 25 million improved cooking gas stoves that use less fuelwood and 43,000 solar cookers had been cumulatively installed by the end of Eighth Plan. The expected energy savings from these non-commercial energy sources amounts to 12 to 15 per cent in total wood consumption which is quite significant.

Reduction of Air Pollution. Measures taken by the Government to increase efficiency in use of fossil fuels has led to dual benefits. First, they have succeeded to a great extent in reducing local

9. Anand Patwardhan, "Global Warming and India," *Seminar,* 486, Feb. 2000, pp. 63-69.
10. Nag, Barnali and Jyoti Parikh, "Indicators of Carbon Emission Intensity from Commercial Energy Use in India", *Energy Economics,* Vol. 22, pp. 441-461.

area pollution. Secondly, they have achieved abatement in GHG emission. Pollution levels have been prescribed for vehicles in cities like Delhi to reduce urban pollution. Thus by providing for more energy efficiency in vehicles attempt has been made to reduce air pollution. This regulation of pollution has also resulted in higher fuel quality. It is estimated that pollution control measures for vehicular traffic will cause reduction of GHG in the range of 10 to 15 per cent by 2015 by which time the current stock of vehicles would be largely replaced.

Afforestation and Wasteland Development. Afforestation and wasteland development can contribute significantly to reduction in emission of CO_2 if proper policy is adopted and enforced. Biomass is widely used in India for various purposes such as fuel, timber and feedstock. This causes degradation of forests and as a result degradation of soil. Recently, Government has initiated programme for afforestation and wasteland development. This programmne is meant not only to halt deforestation but also plant more trees to increase green cover. Under this programme various agencies spend about ₹ 50 billion every year on wasteland development. If we succeed in our efforts to achieve the targets of afforestation and wasteland development there will be significant decline in India's emission of CO_2. Afforestation and regreening of wasteland is also desirable to arrest land degradation, to improve soil fertility, to provide renewable source of fuel and timber and non-timber forest products which provide livelihood to the millions of poor people.

Coal Substitution. In India till recently coal was the main source of commercial energy. After the policy of liberalisation initiated in 1991, the use of oil and natural gas has been increasingly made in sectors where its use was not permitted earlier. That is, the use of oil and gas has increasingly substituted coal since 1991. For example, power plants now use natural gas instead of coal. Similarly, fertilizer plants no longer use coal for its production and instead use other fuels. The use of coal in railways has been completely stopped.

Increases in Energy Efficiency

Due to import liberalisation strong competition from imports has induced the private firms including joint ventures producing consumer goods such as cars, televisions, refrigerators, air conditioners, consumer appliances to become more fuel efficient as price of electricity has been raised by removing subsidy on it.

Impacts of Climate Change in India

Climate change has enormous implications for the natural resources and livelihoods of the people. It will have wide-ranging effects on the environmental and socio-economic and related sectors. Various studies indicate that the key sectors in India such as agriculture, water, natural ecosystem, biodiversity, and health are vulnerable to climate change. This is happening precisely at a time when it is confronted with huge development imperatives. The Indian Network for Climate Change Assessment (INCCA) released a report in November 2010 on assessment of the impact of climate change on key sectors and regions of India in the 2030s. The assessment covers four key sectors of the Indian economy, namely, agriculture, water, natural ecosystems and biodiversity, and health in four climate sensitive regions, namely, the Himalayan region, the Western Ghats, the coastal area, and the North-east region. The report warns of impacts such as sea-level rise, increase in cyclonic intensity, reduced crop yield in rainfed crops, stress on livestock, reduction in milk productivity, increased flooding, and spread of malaria. This calls for urgency of action in reducing vulnerability to adverse impacts of climate change and enhancing adaptive capacity through sector-specific interventions and efforts.

On the request of Government of India World Bank carried out a *study of cost of environment damage in India* whose report was brought out in July 2013. According to this study outdoor air pollution in urban areas by itself claims 1.09 lakh lives of adults and 7,513 lives of children below

5 years, annually. The study revealed that the annual cost of environmental degradation in India is about ₹ 3.75 lakh crore, which is equivalent to 5.7% of the country's GDP in 2009. However, the report says, India can make green growth a reality by putting in place strategies to reduce the damage at a minimal cost of 0.02% to 0.04% of its average annual GDP growth rate. The World Bank has brought out how urban growth centres are choking and claims that outdoor and indoor air pollution have the maximum share of this annual burden on India's economy. According to the estimates of the Bank, outdoor air pollution accounts for 29% of this, followed by indoor air pollution (23%), cropland degradation (19%), water supply and sanitation (14%), pasture (11%) and forest degradation (4%). This is shown in the following Table 58.1.

Table 58.1. In 2009, India's Environmental Degradation Cost is Pegged ₹ 3.75 Trillion Annually Cost of Degradation

Environmental Categories	₹ Billion	% age of Total Damage
Outdoor Air Pollution	1,100	29
Indoor Air Pollution	870	23
Croplands Degradation	703	19
Water Supply, Sanitation & Hygiene	540	14
Pastures Degradation	405	11
Forest Degradation	133	4
Total Environmental Damage	**3.75 lakh crore**	

Besides, according to World Bank study environmental degradation and diseases are responsible for 22.8% of *child mortality* in India. As will be seen from Table 58.2, 0.4% of child mortality was due to outdoor air pollution, 9.8 per cent was due to indoor air pollution and 12.6% of children's death was caused by polluted water supply, poor sanitation and hygiene

Table 58.2. Estimated Share of Child Mortality because of Pollution

Cause	% share of Child Mortality due to Pollution
Outdoor Air Pollution	0.4
Indoor Air Pollution	9.8
Inadequate Water Supply, Sanitation and Hygiene	12.6
Total	22.8

India's Policies to Tackle Climate Change

India's total CO_2 emissions are about 4 per cent of total global CO_2 emissions and the energy intensity of India's output has been falling with improvements in energy efficiency, autonomous technological changes, and economical use of energy. India's climate modelling studies show that even with 8-9 per cent gross domestic product (GDP) growth every year for the next decade or two, its per capita emissions will be around 3-3.5 tonnes of CO_2 eq. by 2030, as compared to the present 1-1.2 tonnes. These are well below developed country averages by any estimation. India's determination in addressing climate change is evident from the fact that an indicative target of increasing energy efficiency by 20 per cent by 2016-17 was included in the Eleventh Five Year Plan. This has now been supplemented with the domestic mitigation goal of reducing emissions intensity of the GDP by 20-25 per cent of the 2005 level by 2020 through proactive policies.

The resources for the measures required to achieve this objective will need to be mobilized from various sources including the national planning process. Studies in respect of a low carbon strategy for development aimed at ensuring inclusive growth are being conducted with the aim of including this as one of the key pillars in the Twelfth Five Year Plan. Second, India is taking conscious steps to *diversify the energy fuel mix such as setting up of 20,000 MW of solar power-*

generating capacity by 2022, doubling the present share of 3 per cent of nuclear power in the energy mix over the next decade, putting in place a major market-based programme to stimulate energy efficiency, imposing clean energy cess on coal for funding research and development (R&D) of clean energy technologies even though coal will continue to play a key role in our future energy strategy, and aggressively expanding the use of natural gas in power production. Third, India has been pursuing aggressive strategies for forestry and coastal management to increase the quality and quantity of forest cover and has launched a major new programme on coastal zone management to address the adaptation challenges facing over 300 million people in our country who live in vulnerable areas near our coast.

As part of its international obligations under the United Nations Framework Convention on Climate Change (UNFCCC) India periodically prepares the National Communication (NATCOM) that gives an inventory of the GHG emissions in India, assesses the vulnerability and impacts, and makes appropriate recommendations regarding social, economic and technological measures for addressing climate change. The First NATCOM was presented to the UNFCCC in 2004. The Government's NATCOM II was presented to the UNFCCC in 2011. Preparation of NATCOM is an exercise of high scientific rigour based on an extensive network of research and scientific institutions in India and draws upon expertise and excellence from different institutions. India has set up an elaborate Indian Network for Comprehensive Climate Change Assessment of some 250 scientists and 120 research institutions to assist in this work. This Network has already published India's GHG inventory for the year 2007 and a 4 × 4 assessment of climate change impacts on four key sectors and four key regions of the country for the 2030s, a time frame for which decisive interventions can be made now. This network is expected to put in place a programme for measuring, monitoring, and modelling the impact of black carbon which could have climate change and public health impacts.

India's strategy for *enhancing its adaptive capacity to climate* variability is reflected in many of its social and economic development programmes. For developing countries like India, adaptation ultimately boils down to assisting the vulnerable population during exigencies and empowering them to build their lives and cope with uncertainties in the long run. Several of India's social-sector schemes, with their emphasis on livelihood security and welfare of the weaker sections, aim to do just that. India implements a series of Central sector and centrally sponsored schemes under different Ministries/Departments aimed at achieving social and economic development. Many of these schemes contain elements (objectives and targets) that are decidedly geared to adaptation. In other words, there is substantial adaptation orientation in many of the sectoral schemes currently under operation. An exercise has been carried out to measure the expenditure on adaptation-related programmes with critical adaptation components: (*a*) crop improvement and research, (*b*) poverty alleviation and livelihood preservation, (*c*) drought proofing and flood control, (*d*) risk financing, (*e*) forest conservation, (*f*) health, and (*g*) rural education and infrastructure. It has been found that India's expenditure on these *adaptation oriented schemes* has increased from 1.45 per cent of GDP in the year 2000-01 to 2.84 per cent during 2009-10. This is a fairly impressive level of spending and is an obvious reflection of the multiplicity of economic and social welfare.

India's Progress on Climate Change

India was one of the early adopters of a national climate change plan. Launched in the year 2008, the National Action Plan on Climate Change (NAPCC) outlines policies directed at mitigation and adaptation to reduce climate change. India is also working on the voluntary goal of reducing the emissions intensity of GDP (excluding emission from agriculture) by 20-25 per cent by 2020 as compared to the base year of 2005. It has been recognised by a United Nations body that India is one of the countries that is making efforts to achieve its voluntary pledges regarding emission of GHG. *India is also laking proactive steps in enhancing energy efficiency and expanding renewable energy resources to combat climate change*. At the same time adaptation measures in agriculture, water resources and urban areas remain its key priorities. Prime Minister Narendra Modi has suggested that he wants 100 GW of solar power and 60 GW of *wind power* by 2020. This is quite a high target but if achieved it can save fuel energy which is highly polluting.

The NAPCC is coordinated by the Ministry of Environment and Forests and implemented through the nodal Ministries and is aimed at advancing relevant actions in specific sectors/areas. Eight national missions in the areas of solar energy, enhanced energy efficiency, sustainable agriculture, sustainable habitat, water, Himalayan ecosystem, increasing the forest cover, and strategic knowledge for climate change form the core of NAPCC. State Governments are also preparing, under the advice of the Central Government, State Action Plans aimed at creating institutional and programme-oriented capacities to address climate change. These, together with the National Missions, will enhance climate change-related actions in the public and private domains.

CLIMATE CHANGE NEGOTIATIONS

Negotiations on climate change began under the auspices of the United Nations to reach an agreement. These negotiations resulted in the **United Nations Framework Convention on Climate Change** (UNFCCC) at the Earth Summit at Rio in 1992. This convention required some binding rules to be pursued by the participating countries regarding GHG emissions. The convention required the developed countries to formulate and adopt policies that aim at stabilising greenhouse gases emission at 1990 level. Besides, this convention adopted the notion of **common but differentiated responsibility** to achieve the goal of stabilising GHG emission. This followed from the recognition that global climate was a common resource and responsibility. It was however emphasized that there were asymmetries between the developed and developing countries regarding the past and present contribution of GHG emissions and regarding financial resources required to reduce future emissions. The developed countries had contributed most to GHG concentration in the atmosphere during the process of industrial growth in the past. Therefore, they should bear substantial part of the cost to be borne by the developing countries to achieve abatement of climate change. Moreover, the developed countries had the necessary technology to reduce GHG emissions.

Berlin Meeting and Kyoto Protocol

After Rio Summit of May 1992, the advocates of more specific and legally binding targets and appropriate timeframe for reducing GHG emission succeeded in 1995 at a meeting of the parties to the Convention on Climate Change held at **Berlin** in Germany which concluded that commitments made in 1992 at Rio to reduce greenhouse gas emission were inadequate to meet the objective of climate change abatement. As a result, parties meeting at Berlin agreed to set forth a process of analysis and assessment of what next steps might be taken to reduce GHG emissions. This process resulted in negotiation of a protocol and to adopt it. A meeting of the parties to the UN Framework Convention on Climate Change was held at **Kyoto** in Japan in Dec. 1997. **This Kyoto protocol committed rich industrialized countries to specific, legally binding emission targets for six greenhouse gases, namely, carbon dioxide, methane, nitrous oxide, hydrofluorocarbon and sulphuric hexafluoride.** However the USA which is the largest emitter of GHG refused to sign it pointing to its needs for higher industrial growth which results in higher emission of GHG. As a result, negotiations about fixing a limit of GHGs emission has remained in a limbo since then. Under UN auspices negotiators have been trying to hammer out a deal to cut global emission of GHG. Now, a meeting of parties to UN Framework Convention on climate change had been called to be held in Copenhagen in December 2009 arrive at an agreement on mitigating climate change.

The adoption of the Kyoto Protocol contains a legally binding quantitative time-bound target for developed countries. The Kyoto Protocol set a target for developed countries (individually or jointly) to reduce overall emissions by at least 5 per cent below 1990 levels in the first commitment period, 2008 to 2012. Recognizing that relying on domestic measures alone to meet the target could be onerous, the Kyoto Protocol offers considerable flexibility through three mechanisms: **Clean Development Mechanisms** (CDM), **Joint Implementation** (JI), and **Emissions Trading** (ET). Through the CDM, industrial countries can finance mitigation projects in developing countries contributing to their sustainable development. *Credits received from such projects can be used to meet commitments under the Kyoto Protocol.* Through it, industrialized countries acquire emissions credit by financially supporting projects in other industrialized countries. ET allows countries that expect their emissions to be above target to buy unused quotas from other countries. All major countries except the USA have ratified the Kyoto Protocol.

Bali Action Plan and Copenhagen Accord

The Conference of Parties (CoP), which is the supreme body of the Convention, meets annually. During the 13th CoP held at Bali, Indonesia, in December 2007, a comprehensive process called the Bali Action Plan to enable the full, effective, and sustained implementation of the Convention through Long Term Cooperative Action, now, up to, and beyond 2012 was launched. Currently, international actions for addressing climate change are being pursued under the Bali Action Plan and the mandate of the Kyoto Protocol. The 15th CoP held at Copenhagen in December 2009 made some advance in the form of the 'Copenhagen Accord', which reflects the political understanding reached by a select group of countries. However, this was *noted* and *not adopted by* the parties to the Convention.

Copenhagen Accord. In December 2009 climate talks were held at Copenhagen to arrive at new policy framework on climate change and review the progress under Kyoto Protocol. It may be recalled that Kyoto Protocol requires the developed countries to bring down the emission of greenhouse gases by at least 5 per cent below the 1990 levels. In subsequent negotiations the developed countries took upon themselves the commitment of emission reduction targets of 20 to 25 per cent below 1990 levels to achieve the objective of limiting the rise of global temperature to 2° Celsius. The developing countries had made no such commitment as they had not contributed much to the global warming and their emission levels were far below those of the rich developed nations due to lack of industrialisation in them.

At Copenhagen not only the developed countries went back on their emission reduction commitment under the Kyoto Protocol, they also succeeded in securing the new framework which is an alternative to Kyoto Protocol. This para 4 of Copenhagen Accord requires the developed countries to commit themselves to emission reduction targets for 2020 which were to be communicated to the UN Secretariat by Jan 31, 2010. This was to be kept as INF document which essentially means 'for information' only. Thus in Copenhagen Accord, the legally binding targets for reduction in emissions by developed countries were given up. Thus the legal obligations of the rich developed nations under Kyoto Protocol were done way with in Copenhagen accord.

With the present attitude of developed countries in climate talks, in all likelihood, these talks are bound to fail. Thus while the efforts to check further global warming are halting, the greenhouse gases are building up in the atmosphere with the disastrous consequences, the possibility of meeting even agreed upon target of linking global warming to 2°C seems to be quite remote.

Cancun Agreement. The climate negotiations held at Cancun during 29 November 2010 to 11 December 2010 that resulted in a set of decisions that cover various areas of action, for example mitigation, adaptation, technology and finance as outlined in the Bali Action Plan, while agreeing to work towards an ambitious target of emission reduction under the Kyoto Protocol. The Cancun Agreements include decisions under both the Convention and Kyoto Protocol negotiating tracks. They were widely perceived as a modest, small step forward and a reaffirmation of faith in the multilateral process. As *per the Cancun Agreements, all parties to the Convention (including the developed and developing countries) have agreed to report their voluntary mitigation goals for implementation.* These will be subject to measurement and verification or international consultation as appropriate, in accordance with agreed international guidelines. Decisions were taken at Cancun to set up a *Green Climate Fund*, *a Technology Mechanism*, and an *Adaptation Committee* at global level to support developing country actions for adaptation and mitigation. These decisions are significant because they reflect, to a large degree, the political understanding that was reached by a select group of countries in the form of the Copenhagen Accord in December 2009.

Efforts to Back Out by Developed Countries

In what could turn into a deal-breaker at the climate talks, US and other indutrialized countries in the Major Economies Forum meet at Washington in the second week of Sept. 2009 told India that they wanted to do away with the Kyoto Protocol, a move opposed by India. The industrialized countries proposed that the developing world should agree to the Australian proposal, which demands all countries, regardless of development levels, take on commitments and fix timetable by when they start reducing emissions. It demands that all key countries, including India, curb their emissions by 2020. India's stand is that the official negotiations at UN were already grappling with the question of strong mid-term targets for rich nations under the second phase of Kyoto Protocol and to do away with the protocol was not a part of the mandate of negotiations. US and some other industrialized countries, European Union (EU) have for long opposed strong mid-term targets and demanded China and India to get on board on a long-term agreement.

Besides, both the **USA and EU are trying to link foreign trade with mitigating climate change**. The provisions in the US domestic bill on climate change to impose trade tariffs on goods imported from countries that don't agree to emission reductions were like 'sanctions' against 'renegade' countries that did not sign up the international agreements. India has opposed the provisions, which the Indian Government perceives as protectionism hidden in the garb of climate change concerns. It's rather ironic for us to say so after the US backed out of Kyoto Protocol and now finds it politically improbable to pass even an extremely weak domestic legislation on climate change. The attempts by US to throw up completely unacceptable proposals on the table at the last moment with less than three months left of the final round of negotiations in December as well as becoming aggressive on trade barriers is being seen as a ploy to deflect attention from its failure to move ahead on its domestic climate change bill.

Durban Conference in December 2011

A significant achievement of Durban agreement reached at conference at Durban in South Africa in December 2011 was the second commitment period (2013-2017) secured for Kyoto Protocol which was to expire in 2012. Though no fresh emission cuts were proposed for the second commitment period, the developed countries which were legally bound to certain emission cuts by 2012 under Kyoto Protocol would now have to be fulfilled till 2017. Thus at Durban the Kyoto Protocol got a new lease of life for a few more years. The Kyoto Protocol sets binding targets for 37 industrialised nations and the Europeon Union (EU) to slash carbon emissions to 5% below the 1990 levels by 2012. Until now, China and India have been exempt from any constraints because they are developing countries. Besides, the United States which was not the signatory to Kyoto Protocol came on board in Durban agreement and accepted the transparency of its climate action and to emission cuts that will be decided under the agreement to be operational beyond 2020.

Green Climate Fund (GCF) and Technology Transfer at Durban Conference

A significant achievement at Durban for India, China and other developing countries was that they were able to get commitment about Green Climate Fund (GCF) of $ 100 billion (already committed but not implemented) from rich nations. Under this India will get long -term finance to adapt to climate change, transfer and development of clean technology.

Besides, the Green Climate Fund to finance the adaptation to and mitigation of climate change,

technological transfer talk gained momentum at Durban. India's proposal to break the logjam on transferring green technologies gained centre stage when other intractable issues appeared to be resolved. Under the Cancun Agreements, countries had decided to set up centres that would help transfer technologies from rich countries and develop new ones in collaboration. But the talks on operationalising these centres had got stuck since countries were unable to come to a consensus over several key elements. The key point of dissension had been how the UN convention would control the flow of technologies and ensure they were meeting the need of the developing world that would put forth its requirements .The talks on technology had got stuck since developed countries were not keen to have a review process to assess if they were providing the kind of technological help required and demanded by developing countries.

It was decided to select a panel that would oversee the entire process of technology transfer and guide why it has run into rough weather because developing countries are keen to have clean technologies. The developing countries want a transparent process of selection of this panel.

Intellectual Property Rights. Besides equity, India raised the issue of intellectual property rights for clean technology transfer. India wants intellectual property rights should not be enforced in case of clean technology transfer. However, for this. India was told to take up the issue at '*World Intellectual Property Organisation*'. In this regard India has China's support but not from other. BRICS nations–Brazil and South Africa.

Unilateral Trade Barriers. Another issue that India raised at Durban was that unilateral trade barriers imposed by the developed countries should be done away with. Most countries wanted India to take up the issue at World Trade Organisation (WTO). On the other hand, in this connection a paper on carbon tax on aviation and maritime transport was approved by a UN Committee despite India's protest.

Conclusion. It is evident from above that at Durban Conference India failed to secure commitment regarding equity in emission cuts. Instead it accepted its clubbing with developed countries which has historically contributed most to the pollution of environment. Besides, its other demands regarding intellectual property rights and unilateral trade barriers were not accepted.

It may be noted that new climate treaty with binding emission cuts after 2020 has important implications for India's economic growth and energy strategy. For both industrial and agricultural growth we require the use of energy which in India are mostly supplied by thermal power plants which use coal for generation of power. If we accept the larger carbon emission cuts, we will have to restrict the use of coal-based energy generation and make large investment in development of other sources of alternative clean technologies. To conclude in the words of one environmentalist. "At Durban Conference the differentiation between the developed and developing countries - which has been the bedrock of the global climate regime - has already been written out, even before work on the Durban Plan on Enhanced Action could begin. While there are specific timelines for actions or requirements that developing countries have to meet, there is no timeline for support, neither technology nor finance, from developed countries."[11]

Doha Climate Change Conference 2012

The 18th session of the Committee of Parties (COP) to the UNFCCC, that started on 26 November concluded on 8 December 2012 in Doha, Qatar, has resulted in a set of decisions (clubbed together as '*Doha Climate Gateway*') aimed at advancing the implementation of the UNFCCC and its Kyoto Protocol (KP). The key issues for the Doha conference were; amending the KP to implement the second commitment period under the Protocol; successfully concluding the worked of the Bali Action Plan (BAP) within which there was urgent need for a clear path to climate finance; and planning the work under the *Durban Platform* (DP) for enhanced action. The Conference addressed all three issues and came out with a package which balanced the interests and obligations of various countries.

At the Doha Conference, the three issues of equity, technology-related IPRs, and unilateral

11. Urmi A. Goswami, "Only Noble Talks, No. Bid to Cut Emission," The Economic Times, Dec. 16, 2012.

measures raised by India resounded in the decisions. These outstanding or unresolved issues under the BAP are now part of the planned or continuing work of various bodies of the Convention. At Doha, India also ensured that no hasty decision is taken on aspects related to mitigation in agricultural sector at global level as agriculture is a sensitive sector for developing countries. The Conference has explicitly recognized that the action of Parties will be based on equity and CBDR including the need for equitable access to sustainable development. The Conference also recognized that issues relating to global peaking that could place a cap on emissions of developing countries and restrict their development space were controversial and best avoided at this stage of development. At the same time, in an effort to cater to the interest of all countries and come up with a balanced package, some elements of the package required compromise or deferral. In many cases, ambitious and strong demands were collectively made by developing countries, but in the act of balancing, countries were made to accept the mellowed down and subtle versions of their demands. Among the key concerns which the Conference could not address were those relating to financing commitments of developed countries and sectorial actions. No specific targets for mid-term financing (2013-2020) were adopted. While the Conference stopped short of giving a mandate to the International Civil Aviation Organization (ICAO) or International Maritine Organization (IMO) to initiate steps for curtailing emissions in their respective sectors, the absence of a decision on sectoral framework for such actions has left open the possibility of such actions being initiated in such sectors by the respective international organizations. Considering the fact that some of the leading members of ICAO prefer a global market based mechanism to be the vehicle of such actions, the framework and the principle on the basis of which such actions will be taken are likely to be a bone of contention for quite sometime. Also, despite vociferous demand from vulnerable countries, there could be no satisfactory agreement on a compensation mechanism for loss and damage resulting from climate change.

On the positive ride, the Doha Conference succeeded in carrying out amendments to the KP to ensure a second commitment period. The second commitment period will last for a period of eight years as of 1st January 2013.

This decision has ensured that there will be no gap between the first commitment period under the Kyoto Protocol (KP) ending on 31 December 2012 and the second one commencing on 1 January 2013. With the exception of Russia, New Zealand, Japan, and Canada, all other countries that were part of the first commitment period entered into the second round, with some new countries joining as well. It has been agreed that the KP Parties will revisit their targets in 2014 with a view to increasing their ambition. The emission reduction obligations undertaken by the Kyoto Protocol Parties are not as ambitious as required by science; however, they provide a relative degree of certainty to the carbon markets. The EU will reduce its emissions by 20 per cent by 2020 compared to 1990. Governments also agreed to speedily work under the Durban Platform (DP) to evolve a new set of arrangements for mitigation commitments and actions applicable to all countries from 2020, and to adopt it by 2015. In a significant and positive advancement, it has been agreed that the work of the DP will be based on the principles of the convention.

Key Outcomes of Doha Conference 2012 on Climate Change

1. It has been agreed that the KP, as the only existing and binding agreement under which developed countries commit to cutting emissions of GHG, will enter a second commitment period that will run for eight years.
2. Governments have agreed to speedily work toward a climate change agreement under DP applicable to all countries from 2020, to be adopted by 2015. Further governments have decided to find ways to scale up efforts before 2020 to meet the gap in global ambition for emissions reduction.
3. Governments have launched a robust process to review the long-term temperature goal. This will start in 2013 and conclude by 2015 and is a reality check on the advance to the climate change threat and the possible need to mobilize further action.

4. The Work Programme on Long Term Finance launched last year has been extended for another year to contribute to the ongoing efforts to scale up mobilization of climate finance. Developed countries have reiterated their commitment to deliver on promises of mobilizing US $100 billion both for adaptation and mitigation by 2020.
5. Decision also encourages developed countries to increase efforts for providing finance between 2013 and 2015, and at least to the average annual level provided during the 2010-2012 fast-start finance period.
6. Finance pledges of about $6 billion for period up to 2015 announced by Germany, the UK, France, Denmark, Sweden and the EU Commission.
7. The selection of the Republic of Korea by the Board of the Green Climate Fund (GCF) to host the GCF has been endorsed.
8. The unresolved issues of technology-related Intellectual Property Rights (IPR) and the Unilateral Measures under the BAP are now part of the planned or continuing work of various bodies of the Convention. Based on the decisions, the Technology Executive Committee (TEC) will initiate exploration of issues relating to enabling environments and barriers, including IPRs in its future work-plan. The TEC has already identified IPRs as one of the key messages on which further work is necessary. Similarly, a decision has been taken for facilitating discussion on the issue of unilateral measures under the existing forum on implementation of response measures.

PARIS TREATY ON CLIMATE CHANGE, 2015

In December 2015, the representatives of 196 countries of the world met at Paris in France to decide about the future plan for mitigating climate change. They agreed on a treaty which has been described as historic, 'A New Era of Opportunity', 'Message of Hope', a 'Beautiful Document' and so on. In the Paris deal, which will take effect in 2020, it is the first time that all member countries of the United Nations have joined the fight against climate change. Ban Ki-moon, the Secretary General of UN has said that, "The Paris agreement is a triumph for people, the environment and for multilateralism. It is a health insurance policy for the planet. For the first time every country in the world has pledged to curb their emissions, strengthen resilience to climate change and act internationally and domestically to address to climate change."[12]

We describe below the major components of Paris treaty on climate change.

1. Limit on Rise in Temperature. The agreement aims to keep the rise in global temperature well below 2°C from the pre-industrial times to the end of the 21st century and strive to limit even further to 1.5°C. This is especially important for the nations of Africa, Small Island developing countries and least developed countries. The treaty also recognises that in order to meet the target set in the treaty, the countries should peak greenhouse gas emissions as soon as possible.

2. Submission of Nationally Determined Contributions to Reduction in GHG Emissions. All the countries have agreed to submit their 'nationally determined contributions' to global GHG emissions to a United Nations body which show what they are prepared to do to fight climate change. 168 countries have already submitted their 'Intended Nationally Determined Contributions" and the remaining ones are expected to do so soon. It may be noted that presently these nationally determined targets leave us with dangerous 3°C rise in global temperature. Therefore, under the Paris treaty, they have agreed to review their national climate mitigation plans every five years, beginning in 2018. This will allow them to increase ambitions in line with what science and technology require.

From above it is significant to note that unlike Kyoto Protocol to which U.S. did not agree to sign, Paris deal does not make any reduction in greenhouse gas emissions *legally binding,* despite the efforts made by India and other developing countries to make the reduction in GHG emissions obligatory on the part of the developed countries who are responsible for making large contributions

12. Ban Ki-moon, "A New Era of Opportunity", in *Times of India,* December 19, 2015.

to GHG emissions in the past as a result of their rapid industrial growth based on the large use of energy from fossil fuel. This implies that, the notion of '*historical responsibility*' has been bypassed by the Paris treaty.

3. Support to Developing Countries. The Paris pact also ensures support to balanced adaptation and climate mitigation plans of developing countries, especially the poorest and the most vulnerable among them, by transfer of cheap and clean technology by the developed countries. This is expected to scale up global efforts to *address and minimise loss and damage from the climate change.*

4. Addressing Climate Change according to their Capabilities. In the Paris treaty, the governments of the countries have agreed to binding, robust and transparent rules to ensure that all countries do what they have said and communicated to the UN regarding climate mitigation and adaptation. Besides, developed countries have agreed to lead in mobilising finance amounting to $ 100 billion and to scale up technology support and capacity building. Further, the developed countries have assumed increasing responsibility to *address climate change in line with their capabilities*.

Critical Evaluation

Paris treaty has come under heavy attack by experts on the subject of climate change and sustainable development. Our Prime Minister, Narender Modi has said in agreeing to the treaty '*no one is gainer, no one is loser*'. But there were great expectations from the developed countries that they would realise their historic responsibility for their greater GHG emissions in the past and would agree to do more in making contributions to mitigation of climate change and also support in providing adequate finance and transfer of cheap and clean technology to the developing countries. The Paris deal does not make it obligatory for developed countries to not only make relatively more emission cuts but also provide adequate financial and technological support to developing countries to compensate them for their greater emissions in the past. The Paris agreement has failed miserably to do so and the notion of historic responsibility has been erased from it. There is nothing in the agreement to enforce the carbon emission cuts required to get near to 2°C rise in temperature by the end of the country. What the Paris agreement does, then, is to urge and prod countries to make emission cuts more deeply over time. About the Paris agreement Sunita Narain, the director general of CSE, rightly says that, 'Many details were left out for the future of developing countries. India will have to fight for this over the next 10 years. Not much enhancement in ambitions will happen in the next 10 years *and the developed* countries have not committed any significant finance or emission cuts before 2020'.[13] Thus, India and its allied countries including Brazil, South Africa and China of BASIC block failed to get from developed countries the legally binding commitments with regard to relatively more emission cuts and to providing finance and transfer of technology to developing countries to fight climate change. Though the Paris deal document talks about support for loss and damage suffered by poor developing countries in the past but says that this would not be considered as liability of the developed countries for compensation to the poor developing countries. This further weakens the obligation of developed countries for ensuring 'climate justice' in the fight for climate change.

Another damaging part of the Paris agreement is its failure to consider ***fair allocation of carbon budget*** and as a result developed countries can continue to disproportionaly appropriate more carbon space in the future as they have done in the past. In this context it is worth mentioning that developed countries have got a *carbon market* through which they can offset their emissions and therefore they will do even less *domestically* than what they have promised in their '*Intended Nationally Determined Contributions* (*INDC*). Therefore, Sunita Narain rightly says that "Rich nations will continue to operate in business in their usual manners, making the developing world reduce emissions for them."[14]

India's Minister for Environment, Mr. Prakash Javadekar thinks that India has not to do much before 2030 but the experts from Centre for Science and Environment (CSE) believe that India will

13. Reported in Times of India, December 14, 2015
14. *Ibid.*

be under constant pressure to take more burden for mitigating climate change which will affect its industrial growth. Furthermore, Sunita Narain rightly says that 'on the whole the Paris agreement continues to be weak and unambitious, as it does not reduce the emissions of the developed countries. It notes that climate injustice is a concern of some and it maintains that agreement will be under the UN convention. But as *it does not operationalise equity and the term carbon budget did not even find mention in the text*, this will end up furthering climate apartheid."[15]

But the crucial question is what India has got from treaty. It has got right words like equity, common but differentiated responsibilities and climate justice mentioned in the document. But they are not incorporated in the operational parts of the agreement. So beyond the right words, India and other developing countries got nothing from the so called historic treaty on climate change.

---

15. Reported in *Times of India*, December 14, 2015.